MW00396576

DESIGN

By ELWYN E. SEELYE

DATA BOOK for
CIVIL ENGINEERS

VOLUME ONE—DESIGN
Third edition

VOLUME TWO—SPECIFICATIONS
AND COSTS—Third edition

VOLUME THREE—FIELD PRACTICE
Second edition

•

FOUNDATIONS—Design and Practice

DATA BOOK FOR CIVIL ENGINEERS

DESIGN

THIRD EDITION

ELWYN E. SEELYE

With special acknowledgment to
MORTON C. SIMMONS

JOHN WILEY AND SONS, Inc.
New York · Chichester · Brisbane · Toronto · Singapore

25 24 23

ISBN 0 471 77286 0

Library of Congress Catalog Card Number: 58-5932

PRINTED IN THE UNITED STATES OF AMERICA

Preface to the Third Edition

The continued demand for this book after it has been on the market for nearly 15 years proves that it is doing a useful job among engineers.

Because of the progress the science of civil engineering has made over the past few years, it has become necessary to bring the book up to date.

The major new items added are:

A. Structural:

Concrete column tables for axial and eccentric loads (both uncracked and cracked sections) have been expanded to provide a pushbutton solution for a wide range of sizes and concrete strengths and are all in accordance with ACI rules.

Prestressed Concrete.

Ultimate Strength Design of Concrete.

Plastic Design of Steel.

Composite Beam Design.

Aluminum Design.

Lightweight Aggregates in Concrete.

Unit Stress for Structural Glued Lumber.

Unit Stress for Reinforced Masonry.

Wind pressures on Chimneys, Tanks, Towers, and Roofs, from American Standards Association.

B. Civil Engineering:

Pavements—Complete design data from Portland Cement Association, Asphalt Institute, Civil Aeronautics Administration, and Corps of Engineers.

Highways—A great many necessary highway design criteria which is needed is only found in AASHO Handbooks. A number of these have been reproduced with the kind permission of AASHO enabling the designer to have as many design data as possible in one book without having to go to other reference material. Included now are the following:

Complete geometrics for highways.

Highway capacity.

Parking data.

Airports

Complete airport design data on capacity, configuration, navigational aids etc., in accordance with the Federal Aviation Agency.

Heliport design data.

Terminal buildings.

Bridges

Typical designs and details for composite beams, reinforced concrete, simple span, and continuous span bridges.

Typical bearing and expansion joint details.

Cathodic protection data.

Waterfront structures.

Wave action, shore protection, waterfront structures, small piers etc.

C. Sanitary and Hydraulics:

Drainage

Three edge bearing strengths for rigid and flexible pipe with various bedding conditions and heights of fill.

Manhole covers.

Sewage

Latest screening equipment.

Sludge heating by external heat exchangers.
Sludge digestion.
Sludge conditioning by thickening and elutriation.
Typical electrical installation for a sewage pumping station.
Chlorination.
Small systems.

Water supply, purification and distribution has been brought up to date in accordance with the latest standards.
Petroleum products handling.

In addition to this all material has been revised according to the latest codes and data available.

The headings above in italics represent subchapters in new fields in engineering.

The theory of the approach behind these subjects has been worked out logically with a view to being readily understood by the designers and also available for teaching.

ELWYN E. SEELYE

October, 1959

Acknowledgments, Third Edition

The author wishes to acknowledge the contributions made to this edition by the following persons:

Mr. R. C. Kasser of ALCOA for his assistance on *structural aluminum.*

Mr. William E. Cullinan, Jr., District Airport Engineer, District 1, Regional, FAA for his assistance on *Airports.*

Others who have given advice or criticism are:

Mr. C. F. Blanchard, Deputy Chief Engineer, State of New York, Department of Public Works.
Mr. Vernon J. Burns, Associate Civil Engineer, State of New York, Department of Public Works.
Col. Leonard C. Urquhart of Porter, Urquhart, McCreary and O'Brien.
Mr. Louis Cutler of District 1D, New York State of Public Works.
Mr. Irving Sheinbart of District 1D, New York State of Public Works.
Mr. Herbert H. Howell, Chief, Washington International Airport Division, FAA.
Mr. P. A. Hahn, Chief, Engineering Division, Office of Airports, FAA.
Mr. Joseph H. Tippets, Director, Office of Air Navigation Facilities, FAA.
Mr. R. M. Brown, Chief, Air Navigation Facilities Division, FAA.
Mr. Edward A. Robinson of Concrete Plank Co., Inc.
Mr. T. R. Higgins of American Institute of Steel Construction.
Mr. R. Estes, Jr. of American Institute of Steel Construction.
Mr. Alexander P. Beakley of New York Central Railroad.
Mr. Ralph H. Mann of American Wood Preservers' Association.

The author wishes to make special mention of the assistance received from the following members of his organization:

Mr. F. P. Davis for his work on *Prestressed Concrete.*

Mr. W. A. Faiella for his advice on conforming to the New York City Building Code and the American Concrete Institute Code.

Mr. J. D. Goldreich for his work on *Composite Beam Design.*

Messrs. F. J. Kircher, W. Siskos, and P. C. Wang for their work on the development of charts for design of *Concrete Columns.*

Mr. P. Mason for his work on *Ultimate Strength Design of Concrete and Plastic Design of Steel.*

And for their work on the *structural chapters:*

Messrs. G. D. Fish
H. S. Woodward
I. Hooper
P. P. Page, Jr.
S. Czark
C. L. Dayton
M. A. Shubs
H. Thropp

Messrs. N. Blumenthal
M. P. Capurso
L. Goldman
F. J. Mayer
F. Plastini
B. Rifkin
L. Rushefsky
G. Wallen

Messrs. M. Ciappa and G. Shearer for their work on *Pavements.*
Mr. A. De Konkoly for his work on tables for *Loads on Pipes.*
Mr. R. A. Freeman for his work on *Airfields.*
Mr. G. Getter for his work on *Highways and Bridges.*
Mr. J. Keck for his work on *Highway and Street Lighting.*
Mr. P. A. Malafronte for his work on *Airfield Lighting.*
Mr. W. E. Phelps for his work on *Railroads Chapters.*

Dr. F. C. Sanderson for his work on *Cathodic Protection.*

Mr. H. Singer for his work on *Parking and Heliports.*

Messrs. J. L. Staunton and E. LeClercq for their work on *Drainage, Sewerage, Sewage Treatment, Water Supply, Water Purification,* and *Water Distribution.*

Mr. A. Vollmar for his work on *Waterfront Structures* and *Petroleum Products Handling.*

Mr. F. Zeigler for his work on *Chlorination.*

Messrs. R. E. Dougherty, S. D. Teetor, J. Michelson Jr., R. Shumaker, R. Gunn, R. Thackaberry, G. E. Landgren, F. Bielefeld, and S. Rosanoff for their aid on the chapters on *Civil, Hydraulics* and *Sanitary Engineering.*

Other members of this firm who have contributed to this edition are (Index): Mr. J. Gragnano and Mr. N. McGrath (Drafting): Mr. M. Dzinisewich, Miss A. Arvai, and Mrs. M. Voisinet (Secretarial): Miss J. M. Luckett, Miss D. Donnelly, and Miss M. Zippo.

Preface to the Second Edition

The prime purpose of this data book was to furnish an engineer with sufficient data so that he could design any civil engineering work without other reference books. And in so doing, the greatest care has been taken to make the data simple and usable, and to guide the user in their use.

Judging from the reception since publication this objective was attained. After five years, however, it was necessary to bring the book up to date and to expand its scope. This has been done in this new edition, where the major additions include:

1. A wider range of "push button" design data, such as an expansion of spread footing tables.
2. Simple, non-technical explanation of soil behavior and the application of the principles of mechanics to soils, designed to enable the engineer to utilize soil mechanics without being a specialist in the field.
3. Up-to-date airfield design data, including the latest airport layouts.
4. Standards for swimming pools.
5. Rigid frames have been extended to help the engineer obtain his solutions as simply as possible by:
   - (*a*) "Push button" solutions of a large number of frames.
   - (*b*) Numerical examples of common rigid frame problems, including variable sections.
   - (*c*) Solution of rigid frame bridge bents by influence line data.
   - (*d*) Design data of a large number of existing rigid frame bridges.
6. Data on industrial waste treatment in the Sanitary Section.
7. Under Dams will be found details of equipment such as gates, flash boards, fish ladders.
8. To Highways have been added further data such as sight distances, maximum curvature, and grades. Also the New York Thruway and New Jersey Turnpike standards for intersections have been added.
9. An interesting addition to hydrology is a problem in computing runoff by the unit hydrograph method.

Elwyn E. Seelye

September, 1951

Acknowledgments, Second Edition

The author wishes to acknowledge the contributions made to this edition by the following persons:

Mr. William E. Cullinan, Jr., Chief, Airports Division, Region One, and Mr. George Milnes of the Airports Division, Civil Aeronautics Administration, for their assistance on preparation of airfield data.

Mr. E. T. Gawkins, Deputy Chief Engineer, State of New York Department of Public Works, for his advice on highway standards.

Mr. Howard O. Burgess, of Nepsco Services, Inc., for his advice on the details of dams.

Mr. Elliot A. Haller for his advice on soil mechanics.

Mr. Eudell G. Whitten for his advice on fish ladders for dams.

The author wishes to make special mention of the assistance received from the following members of his organization:

Mr. Jack L. Staunton for his work on industrial waste treatment, drainage, and swimming pools.

Mr. Harold Hayward for his contribution to industrial waste treatment.

Mr. S. D. Teetor for his work on soil mechanics.

Other members of this firm who have contributed to this edition are Mr. Paul Kluger, Mr. Alexander Kolenda, Mr. Rolf Sahlberg, Mr. M. A. Shubs, Mr. C. Mortimer Throop, and Miss Jean M. Luckett.

Foreword to the First Edition

The author of this volume has made an important contribution to his profession by making available design data developed in the course of his experience, covering a period of 35 years, supplemented by similar data drawn from the latest publications of outstanding authorities.

The compilation is of an inclusive character and will be useful to all workers in the field of civil engineering, not only as a reference manual but also as offering basic data concerning current practice which could otherwise be obtained only after extensive research. This volume, which will be supplemented by others covering specifications, costs, and field practice, will fill a long-felt want.

The book contains many original plates of unusual interest, such as those showing methods for rapidly figuring combined flexure and axial stress in columns, those showing the development of highway transition curves, and those showing safe loads on manhole and catch basin covers and on sewer pipes. Explanations and derivations have been largely eliminated, as the notes on the plates give the necessary explanation and guidance. The volume may be described as unique in engineering literature, for it combines maximum data with a minimum explanation. The author is to be congratulated on the clarity of presentation, which he has achieved by full use of illustrative details and examples.

Arthur S. Tuttle

Preface to the First Edition

Most civil engineers collect data which are particularly pertinent to their work. Indeed, the engineer's efficiency depends largely on the quality of the data with which he equips himself.

The growth of technical codes of practice promulgated by technical societies, trade associations, and government bureaus, while advancing the science of civil engineering, has tended to render the practice of engineering more dependent on data than formerly. This has created an unhealthy degree of specialization. For instance, a structural engineer is likely to balk at designing a drain because he does not have the data on runoff and flow in pipes readily available.

The purposes of this volume are:

To make readily available in one place effective data in each main field of civil engineering—structures, sanitation, water supply, drainage, roads, airfields, dams, docks, bridges, and soils.

To make this book complete in itself so that a civil engineer isolated from other reference books and technical data could function.

To furnish data which will eliminate tedious office computing and thus minimize errors made in such computations.

To aid the engineer in attaining results in accordance with the best engineering practice by making available data based on such practice.

To expand the field of activity of the specialized engineer by providing him with handy data in fields allied to his specialty.

In general, desirable engineering information embraces model specifications, and data on design, cost, and field practice. This volume embraces the design data and has been arranged to be as nearly self-indexed as possible. A field manual and a book containing cost data and model specifications will be issued as subsequent volumes.

This book has been written with the view that the practicing engineer, who must produce working designs, needs data carried to final conclusions, even though much technical writing dodges these conclusions on account of the common inhibition among engineers against expressing themselves definitely and without reservations. The directness of the data in this book, which have been presented without these usual reservations, renders it incumbent upon the writer to warn that data cannot be considered a substitute for intelligent judgment based on practice and consideration of all factors and phases of the specific problem at hand. For this reason, collateral reading is encouraged by abundant references, but in general no collateral reading is required to arrive at a direct solution of an engineering problem.

Elwyn E. Seelye

January 30, 1945

Acknowledgments, First Edition

The author wishes to acknowledge the contributions made to this edition by the following persons:

Dr. Arthur S. Tuttle, for his help and advice and for making available to the author his wide contacts for the collection of information.

Dean S. C. Hollister, of the College of Engineering, Cornell University, who has given the author advice and encouragement, particularly in regard to the general scope of the book.

Mr. Elson T. Killam, who has given extensive advice and assistance in preparing chapters on sanitation and water supply.

Mr. Clinton L. Bogert, who has also assisted on sanitation and water supply.

Mr. Gilbert D. Fish, who has aided in the preparation of data on arc-welding.

Col. F. W. Scheidenhelm, who has given the author general advice on dams.

Col. Marcel Garsaud, who has given us constructive criticism on the dock section.

Mr. Shortridge Hardesty, who has given the author advice in preparing the data on bridges.

Mr. Bedrich Fruhauf, who has assisted on soil mechanics.

Professor Hamilton Gray, of New York University, who has given the author advice on soils.

Mr. C. Joseph Klueh, of the New York Central Railroad, who has given advice on railroads.

Mr. Lazarus White, who has given advice on dams and foundations.

Mr. Harry T. Immerman, of Spencer, White & Prentis, for advice on foundations.

Mr. E. P. Albright, of Underpinning and Foundation Co., for his advice on foundations.

Mr. F. Ellwood Allen, for his advice on athletic fields.

Mr. George Seybolt, of Electric Bond and Share Co., for his advice on transmission towers.

Mr. Robert C. Dennett, Assistant Chief Engineer of the National Board of Fire Underwriters, for his advice on fire flow tests.

Mr. Albert E. Cummings, of Raymond Concrete Pile Co., for advice on piles.

Mr. Isaac J. Stander, for advice on docks.

Mr. Walter D. Binger, Commissioner of Borough Works, for making available New York City standards.

Mr. Robert E. Horton, for advice on drainage.

Mr. Richard M. Hodges, for advice on highway bridges.

Mr. Fred J. Biele, for advice on incinerators.

Others who have given advice or criticism are:

Col. Carlton S. Proctor.

Mr. Prevost Hubbard and Mr. Bernard E. Gray, of The Asphalt Institute.

Mr. Allan S. Platt, of Structural Clay Products Institute.

Mr. James C. Harding, Commissioner of Public Works, Westchester County, New York.

Mr. T. W. Pinard, Engineer of Bridges and Buildings, Pennsylvania Railroad.

Mr. William S. Elliott, Chief of the Materials Section, Department of Public Works, City of New York.

Mr. George Pistor, Vice-President of Bethlehem Steel Co.

Mr. Wilton Bentley, Vice-President of K. W. Battery Company.

Mr. S. Neatwait, Chief Engineer, Republic Fireproofing Co.

Mr. J. F. Dealy, Vice President, MacArthur Concrete Pile Corporation.

Mr. E. M. Fleming, District Manager, Portland Cement Association.

Mr. Bertrand H. Wait, President, The Wait Associates, Inc.

Mr. L. J. Markwadt, of Forest Products Laboratory.

Mr. Edward A. Robinson, of Colonial Sand and Gravel Co.

Professor Henry N. Ogden, of Cornell University.

Mr. Ralph H. Mann, of American Wood-Preservers' Association.

Mr. Frank H. Alcott, of National Lumber Manufacturers Association.

Mr. T. R. Higgins, Director of Engineering, American Institute of Steel Construction, Inc.
Mr. Orvil S. Tuttle, Chief Engineer, U. S. Plywood Corporation.
Mr. Elliott Haller, of Haller Engineering Associates, Inc.
Major Rolf Eliassen.
Mr. W. A. Darby and Mr. R. S. Rankin, of the Dorr Co., Inc.
The late Mr. A. M. Brosius.

The author wishes to make special mention of the self-sacrificing support received from the following members of his organization:

Mr. A. L. Stevenson, who has provided the initiative for a large part of the structural section.
Messrs. V. J. Pacello and C. Mortimer Throop, for the structural section.
Mr. J. U. Wiesendanger, for general editing and for the sections on airports and railroads.
Mr. E. G. Whitten, for sections on roads, airports, and soils.
Ensign Harold Hayward and Lt. Jack L. Staunton, for sections on sewage treatment and water supply.
Mr. A. F. Stine, for the section on drainage.
Col. Burnside R. Value, Mr. Rolf Sahlberg, Mr. Laurence Vought, Mr. Julius G. Lenart, Mr. Karl Bloch, Mr. E. R. Kroyer, Mr. C. H. Hamilton, Mr. A. H. Jorgensen, and Mr. Kenneth Whitney, for general assistance.

Contents

10. PAVEMENTS

11. EARTHWORK

12. HIGHWAYS, ROADS AND STREETS

13. BRIDGES

14. RAILROADS

15. AIRPORTS

16. DAMS AND CATHODIC PROTECTION

17. WATERFRONT STRUCTURES

18. DRAINAGE AND SEWERAGE

19. SEWAGE TREATMENT

20. WATER SUPPLY

21. WATER PURIFICATION

22. WATER DISTRIBUTION

23. PETROLEUM PRODUCTS HANDLING

24. ATHLETIC FIELDS

25. MISCELLANEOUS

INDEX

STRUCTURAL-GENERAL-UNIT STRESSES-1

TABLE A—ALLOWABLE WORKING STRESSES IN CONCRETE BASED ON ULTIMATE STRENGTH = f′c

Description	Allowable Working Stresses		
	N.Y.C. Controlled	A.C.I., 1957	Maximum Value, A.C.I.
FLEXURE			
Extreme fiber stress in compression.	0.40 f′c	0.45 f′c	
Extreme fiber stress in compression adjacent to supports of continuous or fixed beams, or slabs, or of rigid frames.	0.45 f′c		
Tension in plain concrete footings.	0.03 f′c*	0.03 f′c	
SHEAR			
Beams with no web reinforcement and without special anchorage of longitudinal steel.	0.02 f′c	0.03 f′c	90
Beams with no web reinforcement, but with special anchorage of longitudinal steel, see p. 2-02.	0.03 f′c		
Beams with properly designed web reinforcement, but without special anchorage of longitudinal steel, see p. 2-14.	0.06 f′c	0.08 f′c	240
Beams with properly designed web reinforcement and with special anchorage of longitudinal steel, see pp. 2-02, 2-03 and 2-04.	0.09 f′c	0.12 f′c	360
Flat slabs at distance *d* from edge of column cap or drop panel.			
(a) When at least 50% of total neg. reinf. passes over the column cap.	0.03 f′c	0.03 f′c	100
(b) When 25% or less of total neg. reinf. passes over the column cap.	0.025 f′c	0.025 f′c	85
(c) For intermediate percentages, use intermediate values.			
Footings where longitudinal bars are without special anchorage.	0.02 f′c	0.03 f′c	75
Footings where longitudinal bars have special anchorage.	0.03 f′c		
BOND Top bars with 12 in. or more concrete below: (a) Plain.		0.03 f′c	105
(b) Deformed.		0.07 f′c	245
Beams, slabs & one-way footings: (a) Plain bars or structural shapes.	0.04 f′c	0.045 f′c	158
(b) Deformed bars.	0.05 f′c	0.10 f′c	350
In two-way footings: (a) Plain bars or structural shapes.	0.03 f′c	0.036 f′c	126
(b) Deformed bars.	0.0375 f′c	0.08 f′c	280
(N.Y.C. only) Where special anchorage is provided, one and one-half times these values in bond may be used.			
(A.C.I. only) Bars not conforming to A.S.T.M. designation A-305 shall be treated as plain bars. Plain bars must have hooks.			
BEARING			
Direct bearing on full area.	0.25 f′c	0.25 f′c	
Direct bearing on one-third full area or less.	0.375 f′c	0.375 f′c†	
The allowable bearing stress on an area greater than one-third, but less than the full area shall be interpolated.			
AXIAL COMPRESSION			
In columns with lateral ties.	0.25 f′c		
In pedestals.	0.25 f′c	0.25 f′c	
For E see p. 1-08. n =	$\frac{30,000}{f'_c}$	$\frac{30,000}{f'_c}$	

TABLE B—REINFORCEMENT (Values in p.s.i.)

Structural-grade billet steel.	Tension = 18,000	Web reinforcement:	Tension N.Y.C.	= 16,000
Intermediate-grade billet steel.	Tension = 20,000	(See Table A, p. 2-14.)	Tension A.C.I.	= 20,000
Rail steel, straight or machine-bent.	Tension = 20,000	Cold-drawn steel wire.	Tension	= 20,000

*For N.Y. City Code requirements add 4 in. to depth of footing (see p. 7-03.)

†This increase shall be permitted only when the least distance between the edges of the loaded and unloaded areas is a minimum of one-fourth of the parallel side dimension of the loaded area. The allowable bearing stress on a reasonably concentric area greater than one-third but less than the full area shall be interpolated between the values given.

STRUCTURAL – GENERAL – UNIT STRESSES – 2

TABLE A – ALLOWABLE WORKING STRESSES IN STEEL (in p.s.i.)

	A.I.S.C., 1957	N.Y.C., 1957
TENSION		
Structutal steel section.	20,000	20,000
Rivets or high-strength bolts – on area based on nominal diameter – tension only.	20,000	20,000
Rivets or high-strength bolts – tension combined with shear.		15,000
Bolts and other threaded parts – on area at root of thread.	20,000	15,000
COMPRESSION*		
Structural rolled steel on short lengths – lateral deflection prevented.		20,000
Columns – gross section axially loaded: main member $\frac{L}{r} < 120$.	$17{,}000 - 0.485\frac{L^2}{r^2}$	Same as A.I.S.C.
secondary member $\frac{L}{r}$ 120 to 200.	$\dfrac{18{,}000}{1+\dfrac{L^2}{18{,}000r^2}}$	
Plate girder stiffeners – gross section.	20,000	
Webs of rolled section at toe of fillet.	24,000	
BENDING (FLEXURE)†		
Rolled sections and built-up members – extreme fiber. Tension on net section (except no deduction for rivet holes) < 15% flange area.	20,000	20,000
Compression‡ (L for cantilever unstayed at outer end = two times length of compression flange.) $\frac{Ld}{bt} < 600$.	20,000	Same as A.I.S.C.
$\frac{Ld}{bt} > 600$.	$\dfrac{12{,}000{,}000}{\dfrac{Ld}{bt}}$	
Rolled beam encased in stone concrete (Not more than 1:5½ mix) as follows:		22,000

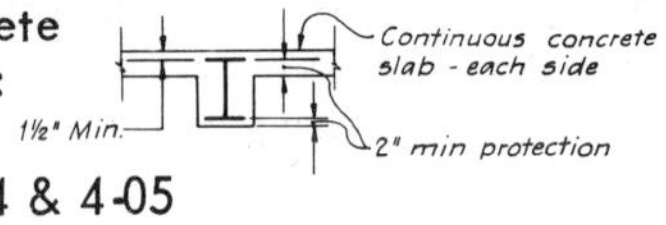
Continuous concrete slab - each side
1½" Min.
2" min protection

	A.I.S.C., 1957	N.Y.C., 1957
Pins – extreme fiber.	30,000	30,000
COMBINED STRESSES see pp. 4-04 & 4-05		
SHEARING		
Rivets – power-driven, high strength bolts, pins and turned bolts in reamed or drilled holes.	15,000	15,000
Unfinished bolts.	10,000	10,000
Webs of beams and plate girders, gross section.	13,000	13,000
BEARING	Double Shear / Single Shear	Double Shear / Single Shear
Rivets, high-strength bolts and turned bolts in reamed or drilled holes.	40,000 / 32,000	40,000 / 32,000
Unfinished bolts.	25,000 / 20,000	25,000 / 20,000
Pins.	32,000	32,000
Contact area-milled stiffeners and other milled surfaces.	30,000	30,000
fitted stiffeners.	27,000	27,000
Expansion rollers and rockers (lbs. per linear inch).	600 diam.	600 diam.
CAST STEEL		
Tension. Note: For A.I.S.C.: compression and bearing same as for structural steel. Other unit stresses 75% of those for structural steel.	See note at left.	16,000
Compression.		16,000
CAST IRON		
Tension, shear.		3,000
Bending – extreme fiber, compression side.		16,000
tension side.		3,000
compression on columns – maximum $\frac{L}{r} = 70$.		$9{,}000 - 40\frac{L}{r}$
WELDED JOINTS		
Compression on section through throat.	20,000§	20,000§
Tension on section through throat.	20,000§	20,000§
Shear on section through throat of fillet weld.	13,600	13,500
Shear on section through throat of butt weld.	13,000	13,500

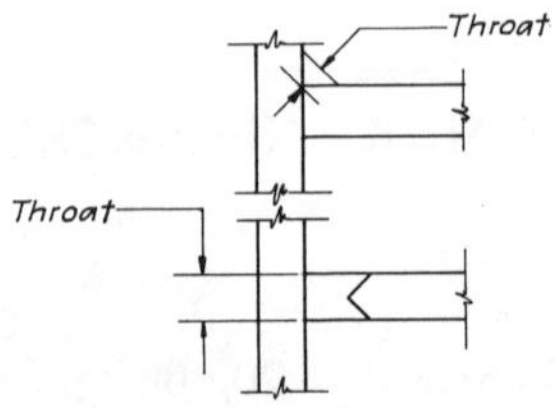
Throat
Throat

Bending – fiber stresses shall not exceed values given above for compression and tension, respectively.

Notation in Table A:
- L = unbraced length in inches
- r = radius of gyration in inches
- b = width of compression flange in inches
- t = thickness of compression flange in inches
- d = depth in inches

*A.I.S.C. allows main members with $\frac{L}{r}$ from 120 to 200 if not subject to shock or vibrating loads – allowable stress $= \dfrac{18{,}000}{1+\dfrac{L^2}{18{,}000r^2}} \times \left(1.6 - \dfrac{\frac{L}{r}}{200}\right)$.

For main compression members 120. For bracing and other secondary members in compression 200. For main tension members 240. For bracing and other secondary members in tension 300.

†A.I.S.C. only: Use 20% higher stress for negative moments at supports in continuous beams provided that section is not less than that required by maximum positive moments and compression flange is figured as unsupported from point of contraflexure to support. ‡See also p. 4-32. §Butt-weld only.

STRUCTURAL–GENERAL–UNIT STRESSES–3

LIMITATIONS OF TABLE A, pages 1-03, 1-04, and 1-06*

The allowable unit stresses in Table A below apply to sawn lumber under continuously dry conditions. The Allowable Unit Stresses in Table A, p. 1-06, apply to structural glued laminated lumber.

Reduce allowable unit stresses in Table A below when the moisture content is at or above the fiber saturation point, as follows:

In compression parallel to grain 10%
In compression perpendicular to grain 33⅓%
Modulus of elasticity $\frac{1}{11}$

When member is fully stressed to maximum allowable stress for many years (either continuously or cumulatively), use 90% of values in Table A, pp. 1-03, 1-04 and 1-06.

Increase allowable unit stresses in Table A, pp. 1-03, 1-04, 1-06 when the duration of the full maximum load does not exceed the period indicated below:

15% for 2 months' duration (as for snow)
25% for 7 days' duration
33⅓% for wind or earthquake
100% for impact

When stress induced by impact does not exceed the allowable unit stress for normal loading, disregard impact. The resulting structural members shall not be smaller than required for a longer duration of loading.

TABLE A – ALLOWABLE UNIT STRESSES FOR STRESS-GRADE LUMBER, 1957

Species	Use	Grade	Allowable Unit Stresses (in p.s.i.)				
			f & t Ⓐ	H	c⊥	c Ⓑ	E
Cypress, southern coast type (Tidewater red)	Joist and planks }	1700f	1700	145	360	1425	1,320,000
	Beams and stringers }	1300f	1300	120	360	1125	1,320,000
	Posts and columns	1450c	–	–	360	1450	1,320,000
		1200c	–	–	360	1200	1,320,000
Cypress, southern inland type (swamp)	Joist and planks }	1700f	1700	145	360	1425	1,320,000
	Beams and stringers }	1300f	1300	120	360	1125	1,320,000
	Posts and columns	1450c	–	–	360	1450	1,320,000
		1200c	–	–	360	1200	1,320,000
Douglas fir, coast region	Light framing	Dense select structural Ⓒ	2050	120 Ⓓ	455	1500	1,760,000
		Select structural	1900	120 Ⓓ	415	1400	1,760,000
		1500f Industrial	1500	120	390	1200	1,760,000
		1200f Industrial	1200	95	390	1000	1,760,000
	Joist and planks	Dense select structural Ⓒ	2050	120 Ⓓ	455	1650	1,760,000
		Select structural	1900	120 Ⓓ	415	1500	1,760,000
		Dense construction Ⓒ	1750	120 Ⓔ	455	1400	1,760,000
		Construction	1500	120 Ⓔ	390	1200	1,760,000
		Standard	1200	95 Ⓔ	390	1000	1,760,000
	Beams and stringers	Dense select structural Ⓒ	2050	120 Ⓕ	455	1500	1,760,000
		Select structural	1900	120 Ⓕ	415	1400	1,760,000
		Dense construction Ⓒ	1750	120 Ⓕ	455	1200	1,760,000
		Construction	1500	120 Ⓕ	390	1000	1,760,000
	Posts and columns	Dense select structural Ⓒ	1900	120 Ⓕ	455	1650	1,760,000
		Select structural	1750	120 Ⓕ	415	1500	1,760,000
		Dense construction Ⓒ	1500	120 Ⓕ	455	1400	1,760,000
		Construction	1200	120 Ⓕ	390	1200	1,760,000
Douglas fir, inland region	Joist and planks Ⓖ	Dense select structural Ⓒ	2050	120 Ⓓ	455	1650	1,760,000
		Select structural	1900	120 Ⓓ	415	1500	1,760,000
		Dense structural Ⓒ	1750	120 Ⓔ	455	1400	1,760,000
		Structural	1500	120 Ⓔ	390	1200	1,760,000
		Standard structural	1200	95 Ⓔ	390	1000	1,760,000
	Posts and columns	Dense select structural Ⓒ	1900	120 Ⓕ	455	1650	1,760,000
		Select structural	1750	120 Ⓕ	415	1500	1,760,000
		Dense structural Ⓒ	1500	120 Ⓕ	455	1400	1,760,000
		Structural	1200	120 Ⓕ	390	1200	1,760,000
Hemlock, eastern	Joist and planks Ⓗ	Common structural	1100	60	360	650	1,210,000
		Utility structural	950	60	360	600	1,210,000
	Beams and stringers Ⓖ	Select structural	1300	85	360	850	1,210,000
	Posts and columns	Select structural	–	–	360	850	1,210,000

STRUCTURAL– GENERAL– UNIT STRESSES – 4

TABLE A – ALLOWABLE UNIT STRESSES FOR STRESS-GRADE LUMBER, 1957 (Continued)*

Species	Use	Grade	Allowable Unit Stresses (in p.s.i.)				
			f & t (A)	H	c⊥	c (B)	E
Hemlock, west coast	Light framing	1500f Industrial	1500	100	365	1000	1,540,000
		1200f Industrial	1200	80	365	900	1,540,000
	Joist and planks	Construction	1500	100 (I)	365	1100	1,540,000
		Standard	1200	80 (I)	365	1000	1,540,000
	Beams and stringers	Construction	1500	100 (J)	365	1000	1,540,000
	Posts and columns	Construction	1200	100 (J)	365	1100	1,540,000
Oak, red and white	Joist and planks	1700f	1700	145	600	1200	1,650,000
		1450f	1450	120	600	1050	1,650,000
	Beams and stringers	1450f	1450	120	600	1050	1,650,000
		1300f	1300	120	600	950	1,650,000
	Posts and columns	1200c	–	–	600	1200	1,650,000
		1075c	–	–	600	1075	1,650,000
Pine, Norway	Joist and planks (H)	Common, structural	1100	75	360	775	1,320,000
		Utility structural	950	75	360	650	1,320,000
Pine, southern (K)	2" thick	No. 1 Dense KD (L)	2050	135	455	1750	1,760,000
		No. 1 KD (M)	1750	135	390	1500	1,760,000
		No. 2 Dense KD (L)	1750	120	455	1300	1,760,000
		No. 2 KD (M)	1500	120	390	1100	1,760,000
		No. 1 Dense (N)	1750	120	455	1550	1,760,000
		No. 1	1500	120	390	1350	1,760,000
		No. 2 Dense (N)	1400	105	455	1050	1,760,000
		No. 2	1200	105	390	900	1,760,000
	3" and 4" thick	No. 1 Dense SR (N)	1750	120	455	1750	1,760,000
		No. 1 SR	1500	120	390	1500	1,760,000
		No. 2 Dense SR (N)	1400	105	455	1050	1,760,000
		No. 2 SR	1200	105	390	900	1,760,000
	5" thick and up	No. 1 Dense SR (N)	1600 (O)	120	455	1500	1,760,000
		No. 1 SR	1400 (O)	120	390	1300	1,760,000
		No. 2 Dense SR (N)	1400 (O)	105	455	1050	1,760,000
		No. 2 SR	1200 (O)	105	390	900	1,760,000
	1", 1¼" and 1½" thick	Industrial 58 KD (M)	1750	120	390	1400	1,760,000
		Industrial 50 KD (M)	1500	120	390	1100	1,760,000
		Industrial 58	1500	105	390	1250	1,760,000
		Industrial 50	1200	105	390	900	1,760,000
Redwood	Joist and planks, Beams and stringers (G)	Dense structural (C)	1700	110	320	1450	1,320,000
		Heart structural	1300	95	320	1100	1,320,000
	Posts and columns	Dense structural (C)	–	–	320	1450	1,320,000
		Heart structural	–	–	320	1100	1,320,000
Spruce, eastern	Joist and planks (G)	1450f Structural	1450	110	300	1050	1,320,000
		1300f Structural	1300	95	300	975	1,320,000
		1200f Structural	1200	95	300	900	1,320,000

f = extreme fiber in bending t = tension parallel to grain H = horizontal shear c⊥ = compression perpendicular to grain
c = compression parallel to grain E = modulus of elasticity

*From "*National Design Specification for Stress-Grade Lumber and its Fastenings*," by National Lumber Manufacturers Association, 1957 Edition.

STRUCTURAL-GENERAL-UNIT STRESSES-5

*FOOTNOTES FOR TABLE A, 1-03 and 1-04**

A. In tension members, the slope of grain limitations applicable to the middle portion of the length of the joist and plank and beam and stringer grades used shall apply throughout the length of the piece.

B. Value for c shown is maximum unit stress. To determine the allowable unit stress (which cannot exceed value c), use the following formula:

$$\frac{P}{A} = \frac{0.3\,E}{(l/d)}\,2$$

Where P/A = load per unit of cross-sectional area
E = modulus of elasticity
l = unsupported over-all length in inches
d = dimension of least side in inches

C. These grades meet the requirements for density.

D. Value applies to pieces used as planks. Value applies to 2"-thick pieces of Select Structural grade used as joists. For 3"-thick pieces of Select Structural, Construction, and Standard grades used as joists:
H = 120 when length of split is approximately 2-1/4".
H = 80 when length of split is approximately 4-1/2".
For 4"-thick pieces of Select Structural, Construction, and Standard grades used as joists:
H = 120 when length of split is approximately 3".
H = 80 when length of split is approximately 6".

E. Value applies to pieces used as planks. For 2"-thick pieces of Construction and Standard grades used as joists:
H = 120 when length of split is approximately equal to 1/2 the width of piece.
H = 100 when length of split is approximately equal to the width of piece.
H = 70 when length of split is approximately equal to 1-1/2 times width of piece.
For 3"-thick pieces of Select Structural, Construction, and Standard grades used as joists:
H = 120 when length of split is approximately 2-1/4".
H = 80 when length of split is approximately 4-1/2".
For 4"-thick pieces of Select Structural, Construction, and Standard grades used as joists:
H = 120 when length of split is approximately 3".
H = 80 when length of split is approximately 6".

F. For beams and stringers and for posts and timbers:
H = 120 when length of split is equal to 1/2 the nominal narrow face dimension.
H = 100 when length of split is equal to the nominal narrow face dimension.
H = 80 when the length of split is equal to 1-1/2 times the nominal narrow face dimension.

G. The allowable unit stresses for tension parallel to grain t and for compression parallel to grain c given for these joist and plank and beam and stringer grades are applicable when the following additional provisions are applied to the grades:
The sum of the sizes of all knots in any 6 in. of the length of the piece shall not exceed twice the maximum permissible size of knot. Two knots of maximum permissible size shall not be within the same 6 in. of length of any face.

H. These grades applicable to 2" thickness only. The allowable unit stresses for tension parallel to grain t and for compression parallel to grain c given for these joist and plank and beam and stringer grades are applicable when the following additional provisions are applied to the grades:
The sum of the sizes of all knots in any 6 in. of the length of the piece shall not exceed twice the maximum permissible size of knot. Two knots of maximum permissible size shall not be within the same 6 in. of length of any face.

I. Value applies to pieces used as planks. For 2"-thick pieces of Construction and Standard grades used as joists:
H = 100 when length of split is approximately equal to 1/2 the width of piece.
H = 80 when length of split is approximately equal to the width of the piece.
H = 60 when length of split is approximately equal to 1-1/2 times the width of piece.
For 3"-thick pieces of Select Structural, Construction, and standard grades used as joists:
H = 100 when length of split is approximately 2-1/4".
H = 70 when length of split is approximately 4-1/2".
For 4"-thick pieces of Select Structural, Construction, and Standard Grade used as joists:
H = 100 when length of split is approximately 3".
H = 70 when length of split is approximately 6".

J. For beams and stringers and for posts and timbers:
H = 100 when length of split is equal to 3/4 nominal narrow face dimension.
H = 90 when length of split is equal to the nominal narrow face dimension.
H = 70 when length of split is equal to 1-1/2 times the nominal narrow face dimension.

K. All stress grades under the 1956 Grading Rules are all-purpose grades and apply to all sizes. Pieces so graded may be cut to shorter lengths without impairment of the stress rating of the shorter pieces.

Grade restrictions provided by the 1956 Grading Rules apply to the entire length of the piece, and each piece is suitable for use in continuous spans, over double spans, or under concentrated loads, without regrading for special shear of other special stress requirements.

The following apply to lumber in service under wet conditions or where the moisture content is at or above fiber saturation point, as when continuously submerged: (a) The allowable unit stresses in bending, tension parallel to grain, and horizontal shear shall be limited to all thicknesses to the stresses indicated for thickness of 5" and up. (b) The allowable unit stresses for compression parallel to grain shall be limited to the stresses indicated for thicknesses of 5" and up reduced by 10%. (c) The allowable unit stresses for compression perpendicular to grain shall be reduced one third. (d) The values for modulus of elasticity shall be reduced one eleventh.

L. These grades meet the requirements for density. KD= Kiln dried in accordance with the provisions of the 1956 Grading Rules. Longleaf may be specified by substituting "Longleaf" for "Dense" in the grade name, and, when so specified, the same allowable stresses shall apply.

M. KD= Kiln dried in accordance with the provisions of the 1956 Grading Rules.

N. These grades meet the requirements for density. Longleaf may be specified by substituting "Longleaf" for "Dense" in the grade name, and, when so specified, the same allowable stresses shall apply. SR = Stress rated.

O. These stresses apply for loading either on narrow face or on wide face.

*From "*National Design Specification for Stress-Grade Lumber and Its Fastenings*," by National Lumber Manufacturers Association, 1957 Edition.

STRUCTURAL—GENERAL—UNIT STRESSES—6

TABLE A — ALLOWABLE UNIT STRESSES FOR STRUCTURAL GLUED LAMINATED LUMBER*

Combination Number	Outer Laminations		Inner Laminations	Extreme Fiber in Bending f[1,2]		Tension Parallel to Grain t[1]		Compression Parallel to Grain c[1]		Horizontal Shear H	Compression Perpendicular to Grain $c\perp$
	Grade	Number, Each Side	Grade	From 4 to 14 Laminations	15 or More Laminations	From 4 to 14 Laminations	15 or More Laminations	From 4 to 14 Laminations	15 or More Laminations		
PART I — DRY CONDITIONS OF USE[3]											
DOUGLAS-FIR, COAST REGION[4]											
1	Clear (Dense)[5]	One	Dense Select Structural	3000	3000	3000	3000	2400	2500	165	450
2	Clear (Dense)[5]	One	Dense Construction	3000	3000	2600	3000	2200	2300	165	450
3	Dense Select Structural	All	Dense Select Structural	2800	3000	3000	3000	2400	2500	165	450
4	Clear (Close-Grain)[5]	One	Select Structural	2800	2800	2800	2800	2200	2200	165	415
5	Select Structural	All	Select Structural	2600	2800	2800	2800	2200	2200	165	415
6	Select Structural	Two	Construction	2600	2600	2600	2600	2000	2000	165	415
7	Clear (Medium Grain)[5]	One	Construction	2600	2600	2200	2400	1900	2000	165	385
8	Dense Construction	All	Dense Construction	2400	2600	2600	3000	2200	2300	165	450
9	Dense Construction	1/14 of total	Construction	2400	2600	2200	2400	1900	2000	165	450
10	Select Structural	One	Construction	2400	2600	2200	2400	1900	2000	165	415
11	Select Structural	Two	Standard[9]	2600	2600	2400	2400	2000	2000	165	415
12	Clear (Medium Grain[5]	One	Standard[9]	2200	2200	2000	2400	1800	1900	165	385
13	Select Structural	One	Standard[9]	2200	2200	2000	2400	1800	1900	165	415
14	Construction	All	Construction	2000	2200	2200	2400	1900	2000	165	385
15	Construction	One	Standard[9]	2000	2200	2000	2400	1800	1900	165	385
16	Standard	All	Standard[9]	1600	2000	2000	2400	1800	1900	165	385
PINE, SOUTHERN[6,7]											
1-1	No. 1 Dense	All	No. 1 Dense	3000	3000	3000	3000	2400	2500	200	450
1-2	B & B Dense	One	No. 1	3000	3000	2600	2600	2100	2100	200	450
1-3	No. 1 Dense	1/14 of total	No. 1	3000	3000	2600	2600	2100	2100	200	450
1-4	B & B Dense	One	No. 2 Dense	2800	2800	3000	3000	2400	2400	200	450
1-5	No. 1 Dense	1/5 of total	No. 2 Dense	2800	3000	2800	3000	2300	2400	200	450
1-6	No. 1	All	No. 1	2600	2600	2600	2600	2100	2100	200	385
1-7	B & B Dense	1/14 of total	No. 2	2400	2800	2600	2600	2000	2000	200	450
1-8	B & B	One	No. 2	2400	2400	2600	2600	2000	2000	200	385
1-9	No. 1	1/5 of total	No. 2	2400	2600	2400	2600	2000	2000	200	385
1-10	No. 2 Dense	All	No. 2 Dense	2000	2600	2600	3000	2200	2300	200	450
1-11	No. 2 Dense	1/14 of total	No. 2	2000	2600	2200	2600	1900	2000	200	450
1-12	No. 2	All	No. 2	1800	2200	2200	2600	1900	2000	200	385
1-13	No. 1	1/5 of total	No. 3 †	1800 ‡	2400	2400 ‡	2500	1900 ‡	2000	200	385
1-14	No. 2	1/5 of total	No. 3 †	1800 ‡	2200	2200 ‡	2300	1900 ‡	1900	200	385
PART II — WET CONDITIONS OF USE[8]											
DOUGLAS-FIR, COAST REGION[4]											
1	Clear (Dense)[5]	One	Dense Select Structural	2400	2400	2400	2400	1700	1800	145	305
2	Clear (Dense)[5]	One	Dense Construction	2400	2400	2000	2400	1600	1700	145	305
3	Dense Select Structural	All	Dense Select Structural	2200	2400	2400	2400	1700	1800	145	305
4	Clear (Close-Grain)[5]	One	Select Structural	2200	2200	2200	2200	1600	1600	145	275
5	Select Structural	All	Select Structural	2000	2200	2200	2200	1600	1600	145	275
6	Select Structural	Two	Construction	2000	2200	2000	2000	1500	1500	145	275
7	Clear (Medium Grain)[5]	One	Construction	2000	2000	1800	2000	1400	1400	145	260
8	Dense Construction	All	Dense Construction	2000	2200	2000	2400	1600	1700	145	305
9	Dense Construction	1/14 of total	Construction	2000	2200	1800	2000	1400	1400	145	305
10	Select Structural	One	Construction	2000	2000	1800	2000	1400	1400	145	275
11	Select Structural	Two	Standard[9]	2000	2000	2000	2000	1400	1400	145	275
12	Clear (Medium Grain)[5]	One	Standard[9]	1800	1800	1600	1800	1300	1400	145	260
13	Select Structural	One	Standard[9]	1800	1800	1600	1800	1300	1400	145	275
14	Construction	All	Construction	1600	1800	1800	2000	1400	1400	145	260
15	Construction	One	Standard[9]	1600	1800	1600	1800	1300	1400	145	260
16	Standard	All	Standard[9]	1200	1600	1600	1800	1300	1400	145	260
PINE, SOUTHERN[6,7]											
2-1	No. 1 Dense	All	No. 1 Dense	2400	2400	2400	2400	1800	1800	175	300
2-2	B & B Dense	One	No. 1	2400	2400	2000	2000	1500	1500	175	300
2-3	No. 1 Dense	1/14 of total	No. 1	2400	2400	2000	2000	1500	1500	175	300
2-4	B & B Dense	One	No. 2 Dense	2200	2200	2400	2400	1700	1700	175	300
2-5	No. 1 Dense	1/5 of total	No. 2 Dense	2200	2400	2200	2400	1700	1700	175	300
2-6	No. 1	All	No. 1	2000	2000	2000	2000	1500	1500	175	260
2-7	B & B Dense	1/14 of total	No. 2	1800	2200	2000	2000	1500	1500	175	300
2-8	B & B	One	No. 2	1800	2000	2000	2000	1500	1500	175	260
2-9	No. 1	1/5 of total	No. 2	2000	2000	2000	2000	1400	1500	175	260
2-10	No. 2 Dense	All	No. 2 Dense	1600	2000	2000	2400	1600	1700	175	300
2-11	No. 2 Dense	1/14 of total	No. 2	1600	2000	1800	2000	1400	1400	175	300
2-12	No. 2	All	No. 2	1400	1800	1800	2000	1400	1400	175	260
2-13	No. 1	1/5 of total	No. 3 †	1400 ‡	1900	1900 ‡	2000	1400‡	1400	175	160
2-14	No. 2	1/5 of total	No. 3 †	1400 ‡	1700	1800 ‡	1900	1400‡	1400	175	260

1. For special slope of grain requirements see applicable specifications listed in Notes 4 and 6.
2. The allowable unit stresses in bending apply only when the wide faces of the laminations are placed normal to the direction of the load. For allowable stresses in bending when the loading is applied parallel to the planes of the laminations, see the applicable specification indicated in Notes 4 and 6.
3. The Modulus of Elasticity E is 1,800,000 p.s.i. for dry conditions of use.
4. *"Standard Specifications for Structural Glued Laminated Douglas-Fir (Coast Region) Lumber"*, by West Coast Lumbermen's Association, applies.
5. The rate of growth and density requirements of inner laminations shall apply to clear outer laminations.
6. *"Standard Specifications for Structural Glued Laminated Southern Pine "Lumber"*, by Southern Pine Inspection Bureau, applies.
7. In grade combinations 1-1, 1-6, 2-1, and 2-6 no provision has been made for use of B & B grade in outer laminations because higher stress rating would not be justified. If, in these combinations, B & B quality is desired for one or both faces of a member to improve appearance, it should be particularly specified, keeping in mind that B & B Dense is required when inner laminations are dense.
8. The Modulus of Elasticity E is 1,600,000 p.s.i. for wet conditions of use.
9. When the grade of standard is used, the lumber must be of medium grain.

* Adapted from *"National Design Specification for Stress-Grade Lumber and its Fastenings" 1957 Edition*, by National Lumber Manufacturers Association.

† Pitch, pitch pockets, pitch streaks, redheart, wane, shakes and decay in No. 3 southern pine boards and dimension used for structural gluing shall not exceed that permitted in No. 2 dimension.

‡ 6–14 laminations.

NOTE: Members with intermediate working stresses will satisfy most design requirements. Members with the highest working stresses are available when special strenght requirements are needed. Where larger sizes are desired, those combinations with lower stresses should be used.

STRUCTURAL–GENERAL–UNIT STRESSES–7

TABLE A – WORKING STRESSES IN STONE MASONRY

Kind of Masonry		Compression, Gross Area of Cross Section			
		New York City 1957		Dept of Commerce 1955	
		Cement[1] Lime Mortar	Cement[2] Mortar	Cement[1] Lime Mortar	Cement[2] Mortar
Granite		640	800	640	800
Gneiss		600	750	–	–
Limestone		400	500	400	300
Marble		400	500	400	500
Sandstone		250	300	320	400
Bluestone		300	400	–	–
Natural Stone	Cut	110	140	–	–
	Uncut	110	140	–	–

TABLE B – WORKING STRESSES IN REINFORCED SOLID AND HOLLOW UNIT MASONRY*

Type of Stress	
Compression, axial	0.20 f′m†
Compression, flexural	0.33 f′m†
Shear, no shear reinforcement	0.02 f′m†
Shear, shear reinforcement Taking 2/3 of entire shear	0.04 f′m†
Bearing	0.25 f′m†
Modulus of elasticity	1000 f′m†
Bond	
Plain bars	30 p.s.i.‡
Deformed bars	90 p.s.i.§

TABLE C – WORKING STRESSES IN UNREINFORCED UNIT MASONRY*

Material: Grade of Unit	Stresses (in p.s.i. of Gross Area)			
	Type-A Mortar[3]		Type-B Mortar[4]	
	Compression	Tension in Flexure or Shear ‡	Compression	Tension in Flexure or Shear ‡
Plain solid-brick masonry:				
f′m = 4500 p.s.i., plus	250	10	200	7.5
f′m = 2500 – 4500 p.s.i.	175	10	140	7.5
f′m = 1500 – 2500 p.s.i.	125	10	100	7.5
Hollow unit masonry	85	6‖	70	5‖
Cavity wall masonry:				
Solid units, 2500 p.s.i., plus	140‖	6‖	110‖	5‖
Solid units, 1500 – 2500 p.s.i.	100‖	6‖	80‖	5‖
Hollow units	70‖	6‖	50‖	5‖
Stone masonry:				
Cast stone	400	4	320	4
Natural stone	140	4	100	4
Solid-concrete brick units:				
Grade A	175	6	125	6
Grade B	125	6	100	6

NOTE:
1. Cement lime mortar composed of 1 part Portland cement, 1 part hydrated lime, to not more than 6 parts sand, proportioned by volume.
2. Cement mortar: a mortar composed of 1 part Portland cement to not more than 3 parts of sand, proportioned by volume, with a allowable addition of hydrated lime, or lime putty, not to exceed 15% of the cement by volume.
3. Type A: by volume, 1 part Portland cement, 4½ parts sand maximum, and ½ part hydrated lime maximum.
4. Type B: by volume, 1 part Portland cement, 6 parts sand maximum, and 1 part hydrated lime maximum.

* Tables B and C from Uniform Building Code, 1955.
† f′m equals compressive strength of masonary at 28 days.
‡ These stresses may be doubled with continuous inspection.
§ With continuous inspection, 130 p.s.i. may be used.
‖ Net area, all others gross cross-sectional area.

STRUCTURAL-GENERAL-PROPERTIES OF MATERIALS

TABLE A – PROPERTIES OF STRUCTURAL MATERIALS

Substance	Ultimate Stress, p.s.i.	Yield Point, p.s.i.	Modulus of Elasticity, p.s.i.	Coefficient of Linear Expansion for 1° F.
Aluminum, Structural alloy 2014-T6	65,000t	32,000tc	10,600,000	0.00000128
Structural alloy 6061-T6	42,000t	35,000tc	10,000,000	0.00000130
Iron, Cast, gray	18,000 - 24,000t 25,000 - 33,000B			
Wrought	48,000		28,000,000	0.0000067
Steel, Structural A.S.T.M. A7*	60,000 - 72,000t	33,000t	29,000,000	0.0000063
Structural A.S.T.M. A8 (3% to 4% nickel)	90,000 - 115,000t	55,000t	29,000,000	0.0000063
Structural A.S.T.M. A94 (0.2% silicon)	80,000 - 95,000t	45,000t	29,000,000	0.0000063
Structural A.S.T.M. A242 (low alloy)	70,000t	50,000t	29,000,000	0.0000063
Structural A.S.T.M. A373 (for welding)	58,000 - 75,000t	32,000t	29,000,000	0.0000077
Structural U.S. steel (T-1)	105,000 - 135,000t	90,000t	29,000,000	
Clay, Brick	1000c, 200T, 600B		2,000,000	0.0000031
Masonry, Granite	420c,† 600B†			
Limestone & bluestone	350c,† 500B†			
Sandstone	280c,† 400B†			
Rubble	140c,† 250B†			0.0000035
Stone, Bluestone	12,000c, 1200t, 2500B		7,000,000	0.00000
Granite/gneiss	12,000c, 1200t, 1600B		7,000,000	0.0000047
Limestone/marble	8,000c, 800t, 1500B		7,000,000	0.0000044/0.0000056
Sandstone	5,000c, 150t, 1200B		3,000,000	0.0000061
Slate	10,000c, 3000t, 5000B		14,000,000	0.0000058
Wood, Douglas fir			1,760,000	0.0000021 parallel 0.000032 perpendicular
Southern pine			1,760,000	0.000003 parallel 0.000019 perpendicular
Concrete			1000 f′c	0.0000079

t = ultimate stress tension.
c = ultimate stress compression.
B = ultimate stress bending.
* = for welding A.S.T.M. A-373 preferred.
† = safe working stress.

STRUCTURAL–GENERAL–UNIT WEIGHTS

TABLE A

Type of Substance	p.s.f.
ROOF AND WALL COVERINGS	
Wood shingles	3
Asphalt shingles	2
Cement asbestos shingles	4
Cement tile	16
Clay tile (for mortar add 10 lb.)	
2" book tile	12
3" book tile	20
Roman	12
Spanish	19
Ludowici	10
Sheet-metal roofing	2
Corrugated roofing (no. 20)	2
Corrugated asbestos	3-4
Copper or tin	1
Corrugated iron	2
Gypsum sheathing 1/2"	2
Wood sheathing	3
Slag roofing	5
Composition roofing	
3-ply ready roofing	1
3-ply felt tar and gravel roofing	4
4-ply felt tar and gravel roofing	5½
5-ply felt tar and gravel roofing	6
Slate roofing laid in place with 3" double lap, 3/16" thickness	7
Slate roofing laid in place with 3" double lap, 1/4" thickness	10
Slate roofing laid in place with 3" double lap, 3/8" thickness	14½
Slate roofing laid in place with 3" double lap, 1/2" thickness	19½
Skylight, metal frame, 3/8" wire glass	8
Plate glass per in. of thickness	14
Fiberboard 1/2"	¾
CEILINGS	
Plaster on tile or concrete	5
Suspended metal lath and gypsum plaster	10
Suspended metal lath and cement plaster	15
Plaster on wood lath	8

Type of Substance	p.s.f.	p.c.f.
MASONARY WALLS AND PARTITIONS—DEAD LOAD WITHOUT PLASTER		
Solid brickwork: Clay brick		120
Sand lime		105
Concrete (heavy aggregate)		130
Concrete (light aggregate)		98
Concrete, stone		144
Granite, blue stone, marble		165
Limestone		156
Sandstone		144
4" brick + 4" load-bearing structural clay tile backing.	60	
4" brick + 6" hollow block backing	75	
4" brick + 8" hollow block backing		80
Concrete block: Heavy aggregate		85
Light aggregate		55
2" furring tile	12	
2" T.C. or cinder block (non-bearing)	16	
3" T.C. or cinder block (non-bearing)	17	
4" T.C. or cinder block (non-bearing)	18	
6" T.C. or cinder block (non-bearing)	28	
8" T.C. or cinder block (non-bearing)	34	
4" load-bearing T.C. or cinder block	22	Multiply by 1.40 for concrete block (gravel or stone aggregate)
6" load-bearing T.C. or cinder block	32	
8" load-bearing T.C. or cinder block	39	
12" load-bearing T.C. or cinder block	56	
2" solid gypsum	10	
3" solid gypsum	13	
3" hollow gypsum	12	
4" hollow gypsum	13	
6" hollow gypsum	19	
2" solid plaster	20	
Wood studs 2 + 4 (unplastered)	4	
Plaster add each side	8	
4" glass block masonry	18	

Adapted from *A.C.I. Handbook, C.R.S.I. Handbook,* and American Standards Association.

RIB SLABS

Depth in Inches (Rib Depth + Slab Thickness)	Width of Rib 4	5	6	7	8	Add for Tapered Ends
12-inch clay-tile fillers						
4 + 2	49	51	52	54	–	–
6 + 2	60	63	65	67	–	–
8 + 2½	79	82	85	87	–	–
10 + 3	96	100	103	106	–	–
12 + 3	108	112	116	120	–	–
20-inch metal fillers						
6 + 2	41	43	45	47	–	4
8 + 2½	51	54	57	60	–	5
10 + 3	63	67	70	74	–	5
12 + 3	69	74	78	82	86	5
14 + 3	75	81	82	87	91	5
30-inch metal fillers						
6 + 2½	41	43	45	47	–	3
8 + 2½	45	48	50	54	–	4
10 + 3	56	59	61	65	–	4
12 + 3	–	63	67	70	73	4
14 + 3	–	69	72	76	80	4
2-way clay tile fillers (12 × 12)						
4 + 2	61	62	64	–	–	–
6 + 2	87	87	90	–	–	–
8 + 2½	100	103	107	–	–	–
10 + 3	121	126	131	–	–	–
12 + 3	136	141	146	–	–	–
2-way metal fillers (16 × 16)						
4 + 2	44	47	50	–	–	–
6 + 2	55	60	63	–	–	–
8 + 2½	72	78	83	–	–	–
10 + 3	91	96	103	–	–	–
12 + 3	103	111	118	–	–	–
14 + 3	116	125	133	–	–	–
2-way metal fillers (20 + 20)						
4 + 2	42	44	46	–	–	–
6 + 2	50	54	58	–	–	–
8 + 2½	66	71	76	–	–	–
10 + 3	83	88	94	–	–	–
12 + 3	93	100	107	–	–	–
14 + 3	105	113	120	–	–	–

Type of Substance	p.s.f.
FLOORS—FLOOR FILL AND FINISHES AND FLOOR SLABS AND WATERPROOFING	
1" terrazzo + 2" stone concrete	38
1½" terrazzo	18
Tile and setting bed	15-23
Marble and setting bed	25-30
1½" asphalt mastic flooring	18
2" asphalt block + ½" mortar	30
3" wood block + ½" mortar	16
Linoleum or asphalt finish	2
Hardwood flooring, 7/8" thick	4
Softwood subflooring, 7/8" thick	3
Oak and yellow pine	48*
Spruce, fir, hemlock, white pine	30*
Concrete, reinforced stone, per inch of thickness	12½
Concrete, reinforced, cinder, per inch of thickness	9¼
Concrete, reinforced, lightweight, per inch of thickness	9
Concrete, plain, stone, per inch of thickness	12
Concrete, plain, cinder, per inch of thickness	9
Concrete, plain, lightweight, per inch of thickness	7
WATERPROOFING	
Five-ply membrane, 1/2" thick	5
Five-ply membrane, mortar, stone concrete, 5" thick	55
2" split tile + 3" stone concrete, 5" thick	45
PATENTED STRUCTURAL SYSTEMS	
2" Gypsteel plank	12
2½" Sheetrock – Pyrofil	12
2½" Lightweight nailing concrete	19
2¾" lightweight channel slabs	10
2½" Porete roof slabs (with 1/2" nailing fin)	15
3½" Porete channel slabs	12
2" Cinder Plank	15
Aerocrete lightweight concrete	50-80*
Nalecode	75*

*lb. per cu. ft.

STRUCTURAL – GENERAL – LIVE LOADS – 1

TABLE A – LIVE LOADS IN POUNDS PER SQUARE FOOT

Occupancy	American Standards Association, 1955	Nat. Bd. of Fire Underwriters, 1955	Pacific Coast Bd. Officials Conference,† 1955	Southern Std. Bldg. Code,* 1954	New York, 1957	Chicago,* 1956	Philadelphia*	Dallas, 1951	Detroit*
Dwellings, apartment and tenement houses, hotels, clubhouses, hospitals, and places of detention:									
Dwellings, private rooms, and apartments	40[1]	40[11]	40	40[25]	40[27]	40	40	40[43]	40
Public corridors, lobbies, and dining rooms	100[2]	100[2]	100[17]	100	100	100	100	80	100
School buildings:									
Classromms and rooms of similar use	40	40[3]	40	40	60[28]	40	50[38]	50[44]	50[38]
Corridors and public parts of building	100[3]	100	100[18]	100	100	100	100	80	100
Theaters, assembly halls, and other places of assemblage:									
Auditoriums with fixed seats	60	60	50	50	75[29]	60	60[39]	50	60
Lobbies, passageways, gymnasiums, grandstands and auditoriums of places of assemblage without fixed seats	100[4]	100[4]	100[4]	100[4]	100	100	100	100[45]	100
Office Buildings:									
Office space	80[5]	80	50[12,19]	50	50[27]	50[12,19]	60	50	50[51]
Corridors and other public places	100	100	100	100	100[30]	100	100	80	100
Workshops, factories, and mercantile establishments									
Manufacturing – light	125	125[12]	75	100	120	100[36]	120[40]	75	125[52]
Manufacturing – heavy	125		125[20]	150	120[31]	100[36]	200[40]	100-150[46]	125
Storage – light	125	125[12]	125	125	120	100[36]	120-150[40]	125	125[53]
Storage – heavy	250	250[12]	250[21]	250	120[31]	100[36]	200[40]	125[47]	250
Stores – retail	100[6]	75[13]	75	75	75[32]	100[36]	100[40]	100	100[52]
Stores – wholesale	125	125[12]	100	100	120	100[36]	100[40]	100	125
Garages:									
All types of vehicles		[14]	100[22]	120[22]	175[33]	100[22]	100[8]	100[48]	175[54]
Passenger cars only	100[7]	100[12]	50	75	75[5]	50[13]	75	80[48]	80[55]
All stairs and fire escapes, except in private residences	100	100[15]	100	100	100	100	100	100	100[56]
Roofs (flat)	20	20	20[23]	20	40	25	30	30[49]	30
Sidewalks	250[8,9]	250[8]	250	200[8]	300[34]		150[41]	300[8]	250
Wind	–[10]	15-40[16]	15-20[24]	10-40[26]	20[35]	20[37]	15-25[42]	20[50]	20[57]

* Taken from C.R.S.I., 1957; others from local codes. †Or Uniform Building Code, 1955.

FOOTNOTES FOR TABLE A

1. 30 for second floor dwellings and habitable attics; 20 for uninhabitable attics; 60 for operating rooms.
2. 60 for corridors in apartment and tenement houses and public corridors in hotels; 40 for private corridors in hotels.
3. 60 for library reading rooms; 150 for stack rooms.
4. 150 for armories and stage floors.
5. Or 2000 concentrated.
6. 75 for upper floors.
7. Floors shall be designed to carry 150% of the maximum wheel load anywhere on the floor.
8. Or 8000 concentrated.
9. 100 for yards and terraces for pedestrians only.
10. See p. 1-14.
11. On first floor, 40 p.s.f.; upper floors for dwellings 30 p.s.f.
12. Or 2000 on any space 2½ ft. square.
13. 100 on first floor and alternate of 2000 lb. on area 2½ feet square.
14. 100 p.s.f. for passenger cars. Trucks: 150 p.s.f. (3 to 10 tons including load), 200 p.s.f. (over 10 tons including load). Concentrated load; 150% maximum wheel load for passenger cars; 125% maximum axle load for trucks.
15. 300 lb. on 2½ ft. square at any location.

STRUCTURAL – GENERAL – LIVE LOADS – 2

FOOTNOTES FOR TABLE A, p. 1-10 (Continued)

16. 15 p.s.f. for building height less than 30 ft.; increase to 40 p.s.f. for building height equal to or greater than 1200 feet.
17. 40 for private corridors in hotels; 50 for rest rooms.
18. 60 for library reading rooms; 125 for stack rooms.
19. Where partitions are subject to change 20 p.s.f. to all other loads.
20. 100 for composing and linotype rooms; 150 for press rooms.
21. 250 minimum. Load to be determined by use.
22. Or concentrated rear wheel of loaded truck in any position.
23. 20 p.s.f. for under 200 sq. ft.; 16 p.s.f. for 201 to 600 sq. ft.; 12 p.s.f. for over 600 sq. ft. loaded area on a member.
24. 15 for portions below 60 ft.; 20 for portions above 60 ft. 30 lb. for roof tanks, signs, and exposed roof structures.
25. 30 p.s.f. for one-story, one- and two-family dwellings.
26. 10 p.s.f. for portions below 30 ft.; 40 p.s.f. for portions above 400 ft. in Southern Inland Regions. Varies from 25 to 50 p.s.f. for Southern Coastal Regions.
27. Including corridors.
28. For rooms with fixed seats or classrooms not exceeding 900 sq. ft. with movable seats; 120 for library stack rooms.
29. 60 for churches.
30. Including entire first floor.
31. The minimum for storage or manufacturing is 120 p.s.f., but floors must be designed for any heavier loads contemplated and for any concentrations.
32. 100 for entire first floor.
33. Or 6000 concentrated. Trucking space and driveways 12,000 concentrated. (For beams, columns, and girders 120 p.s.f. live load.)
34. Or 12,000 concentrated for driveways over sidewalks.
35. 20 p.s.f. for structures over 100 ft. high. 30 p.s.f. for tank towers, stacks, and isolated chimneys.
36. The minimum for storage or manufacturing is 100 p.s.f., but floors must be designed for any heavier loads contemplated and for any concentrations.
37. 20 for buildings less than 300 ft. high; add 0.025 p.s.f. per ft. above 300 ft.
38. Only school classrooms with fixed seats. (Removable seats 80 p.s.f.)
39. Churches only.
40. Every floor beam 4000 concentrated.
41. Interior courts, sidewalks, etc., not accessible to a driveway.
42. 15 p.s.f. up to 50 ft. high, 20 p.s.f. from 50 to 200 ft., 25 p.s.f. over 200 ft. high. Roofs over 30 degrees, 20 p.s.f. on windward side, 10 p.s.f. on leeward.
43. 100 for public rooms; 40 for guest corridors in hotels; 50 for rest rooms.
44. 60 for laboratories, library reading rooms; 20 plus actual weight of stack and contents.
45. 150 for armories and theater stages; 120 for drill rooms.
46. 100 for composing and linotype rooms; 150 for press rooms; for all others, load is determined by use.
47. Load is determined by use.
48. For unloaded vehicles.
49. 25 for flat roofs without parapet walls.
50. 20 for structures less than 300 ft. high. Above 300 ft., $p = 20 + (H - 300) \times 0.025$ for tanks, signs, and exposed structures 30 lb.
51. Above first floor including corridors.
52. 125 for first floor.
53. 150 for first floor.
54. Or 2500 lb. concentrated on area 6 in. square with such concentrations spaced alternately 2 ft. 4 in. and 4 ft. 8 in. in one direction and 5 ft. and 10 ft. in the other direction.
55. Only structures with clear headroom of 8 ft. 6 in. or less. Or 1500 lb. concentrated, spaced as in note 54.
56. 50 for dwellings and apartments under 3 stories.
57. For buildings less than 500 ft. high.

TABLE A – LIVE LOAD REDUCTIONS – NEW YORK CITY BUILDING CODE, 1956

(a) In structures for storage purposes all columns, piers, walls, and foundations may be designed for 85% of live load.

(b) In structures intended for other purposes, live load reductions for columns, piers, walls, and foundations are as follows: 100% L.L. on roof, 85% top floor, 80% next floor, and 5% reduction for each successive lower floor, provided that in all cases at least 50% of live load is assumed.

(c) Girder members, except in roofs and as specified below, carrying floor loads the equivalent of 200 sq. ft. or more may be designed for 85% of live load.

(d) For trusses and girders supporting columns and for determining area of footings the full dead load and live load may be taken with the reductions as permitted above.

STRUCTURAL-GENERAL-LIVE LOADS-3

TABLE A - MISCELLANEOUS LIVE LOADS

GRANDSTAND LOADING

LIVE LOAD 100 p.s.f. of gross horizontal projection. 120 lb./lin. ft. for designing seats and footboards.

HORIZONTAL FORCES A horizontal swaying force parallel to the seats of 24 lb./ft. and perpendicular to the seats of 10 lb./ft.

WIND 30 p.s.f. on vertical projection. (See also p. 1 – 14.)

WIND STRESS REDUCTIONS

CONCRETE STRUCTURES Increase allowable stresses one third where wind loads are added to live and dead loads.†

STEEL STRUCTURES For members subject only to wind, increase allowable stresses one third. For members subject to wind and other forces, increase one third, but section to be not less than that required for dead, live, and impact loads.‡

WOOD STRUCTURES Increase allowable stresses one third, but section to be not less than that required for dead and live loads.§

IMPACT‡

For structures carrying live loads which induce impact or vibration, the assumed live load shall be increased sufficiently to provide for same. If not otherwise specified, the increase shall be:

For supports of elevators	100%
For traveling crane support girders and their connections	25
For supports of light machinery, shaft – or motor-driven, not less than	20
For supports of reciprocating machinery or power units, not less than	50
For threaded hanger rods supporting floors and balconies	$33\frac{1}{3}$

CRANE RUNWAY HORIZONTAL FORCES‡

The lateral force on crane runways to provide for the effect of moving crane trolleys shall, if not otherwise specified, be 20% of the sum of the weights of the lifted load and of the crane trolley (but exclusive of other parts of the crane), applied at the top of rail one half on each side of runway; and shall be considered as acting in either direction normal to the runway rail.

The longitudinal force shall, if not otherwise specified, be taken as 10% of the maximum wheel loads of the crane applied at the top of rail.

UPLIFT ON ROOFS

Design roof for factories, hangers, armories, etc., which have large open interiors for an uplift of 25 p.s.f.

FOOTNOTES

- * American Standards Association, Places of Outdoor Assembly.
- † American Concrete Institute, 1956.
- ‡ American Institute of Steel Construction, 1954.
- § National Lumber Manufacturers Association, National Design Specifications for Stress-Grade Lumber & its Fastenings, 1955.

STRUCTURAL-GENERAL-LIVE LOADS-4

TABLE A - SNOW LOAD*

Where no values for snow loads appear on the map, they have been omitted because of irregular distribution associated with rugged terrain. Consult local Weather Bureau or Building Code.

The value of 20 lb. on the projected area has been selected as a minimum because it has been considered necessary to provide for occasional loading due to workmen and materials during repair operations. Where load exceeds 20 p.s.f., the excess may be reduced from its full value at a 20-degree slope to zero at a 60-degree slope.

For big flat roofs and high parapets special attention is required.

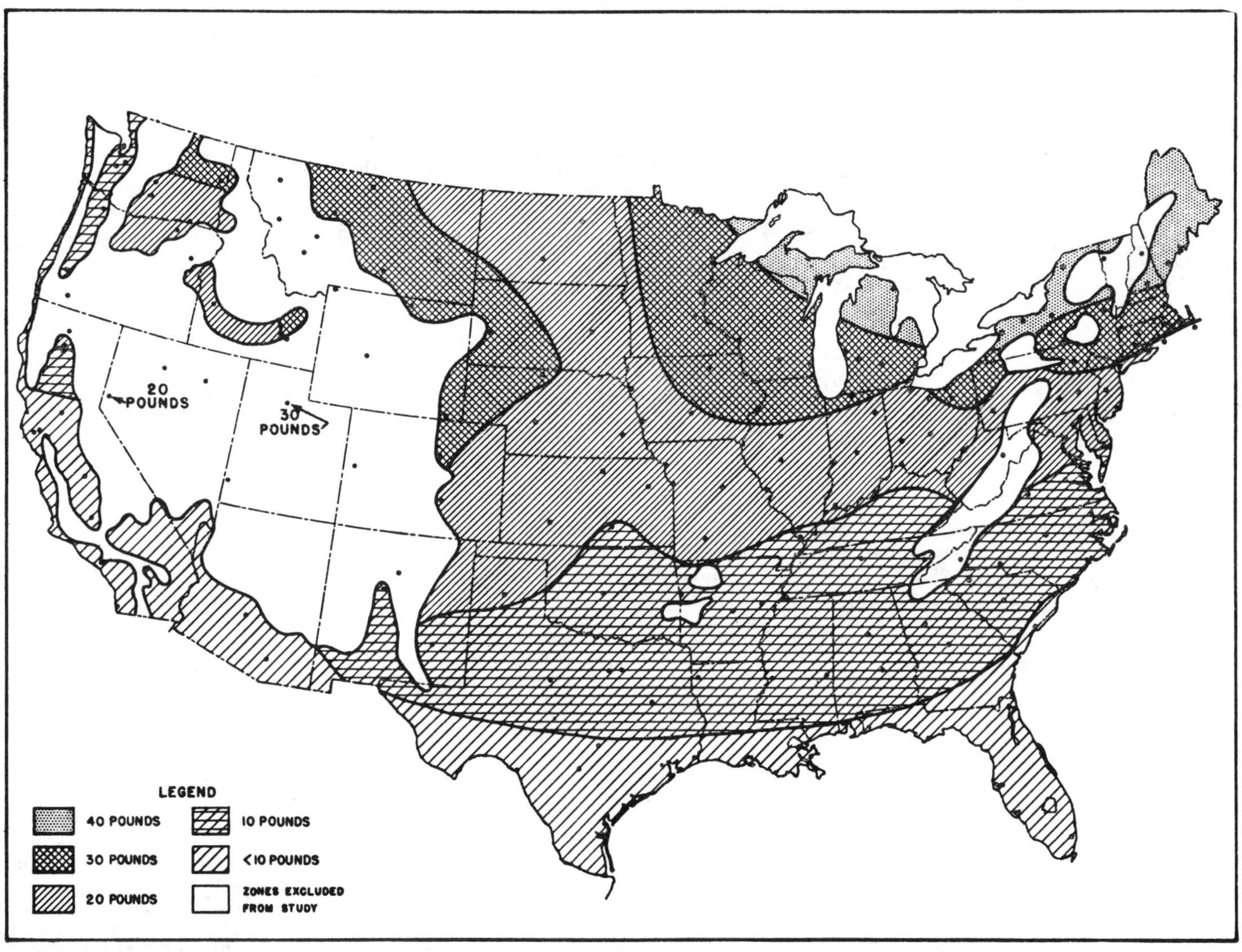
20 POUNDS
30 POUNDS
LEGEND
40 POUNDS
30 POUNDS
20 POUNDS
10 POUNDS
<10 POUNDS
ZONES EXCLUDED FROM STUDY

FIG. B. ESTIMATED WEIGHT OF SEASONAL SNOWPACK EQUALED OR EXCEEDED ONE YEAR IN TEN (p.s.f.)†

* Based on American Standard Building Code Requirements for Minimum Design Loads in Buildings and Other Structures, A58. 1-1955, by American Standards Association.

† This material is reproduced from the American Standards Building Code Requirements for Minimum Design Loads in Building and other Structures, A58. 1—1955, copyrighted by American Standards Association.

STRUCTURAL-GENERAL-LIVE LOADS-5

MINIMUM DESIGN PRESSURE

Buildings, chimneys, tanks, and solid towers shall be designed and constructed to withstand the applicable horizontal pressures shown in Tables B, C, or D, allowing for wind in any direction. The height is to be measured above the average level of the ground adjacent to the building, chimney, tank, or solid tower. Figures do not provide for extreme wind conditions of short duration such as tornadoes (gusts are included). The factors in Table B or C already take into consideration the 1.3 shape factor used in determining the wind pressures noted in Fig. A. The design pressure selected from Table B should be multiplied by the correct factor from Tables C or D.

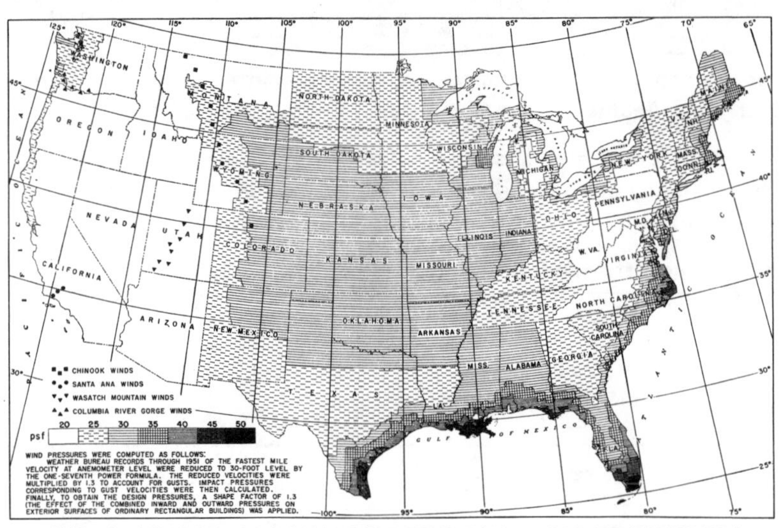

WASHINGTON
OREGON
IDAHO
MONTANA
NORTH DAKOTA
SOUTH DAKOTA
MINNESOTA
WISCONSIN
MICHIGAN
NEW YORK
PENNSYLVANIA
NEVADA
UTAH
WYOMING
NEBRASKA
IOWA
ILLINOIS
INDIANA
OHIO
CALIFORNIA
COLORADO
KANSAS
MISSOURI
KENTUCKY
W. VA.
VIRGINIA
ARIZONA
NEW MEXICO
OKLAHOMA
ARKANSAS
TENNESSEE
NORTH CAROLINA
SOUTH CAROLINA
MISS.
ALABAMA
GEORGIA
TEXAS
LA.
GULF OF MEXICO
CHINOOK WINDS
SANTA ANA WINDS
WASATCH MOUNTAIN WINDS
COLUMBIA RIVER GORGE WINDS
psf 20 25 30 35 40 45 50
WIND PRESSURES WERE COMPUTED AS FOLLOWS:
WEATHER BUREAU RECORDS THROUGH 1951 OF THE FASTEST MILE VELOCITY AT ANEMOMETER LEVEL WERE REDUCED TO 30-FOOT LEVEL BY THE ONE-SEVENTH POWER FORMULA. THE REDUCED VELOCITIES WERE MULTIPLIED BY 1.3 TO ACCOUNT FOR GUSTS. IMPACT PRESSURES CORRESPONDING TO GUST VELOCITIES WERE THEN CALCULATED. FINALLY, TO OBTAIN THE DESIGN PRESSURES, A SHAPE FACTOR OF 1.3 (THE EFFECT OF THE COMBINED INWARD AND OUTWARD PRESSURES ON EXTERIOR SURFACES OF ORDINARY RECTANGULAR BUILDINGS) WAS APPLIED.

FIG. A. MINIMUM ALLOWABLE RESULTANT WIND PRESSURES *

TABLE B – WIND PRESSURES FOR VARIOUS HEIGHT ZONES ABOVE GROUND*

Height of Zone, ft.	Wind-pressure-map Areas, p.s.f.						
	20	25	30	35	40	45	50
Less than 30	15	20	25	25	30	35	40
30 to 49	20	25	30	35	40	45	50
50 to 99	25	30	40	45	50	55	60
100 to 499	30	40	45	55	60	70	70†
500 to 1199	35	45	55	60	70	70†	70†
1200 and over	40	50	60	70	70†	70†	70†

NOTE: Recommended application:

Windward face: $\frac{1.00}{1.30} = 77\%$

Leeward face: $\frac{.30}{1.30} = 23\%$

† 70-lb. upper limit recommendations of Dean Sherlock at American Society of Civil Engineers Convention, October 1957.

TABLE C – FACTORS FOR TRUSSED TOWERS*

Wind pressure on tower = p × A × factor shown below.
p = pressure from Fig. A, modified with Table B for tower height.
A = total normal projected area of all the elements of one face of the tower.

Type	Factor ‡	Illustrations
Wind normal to one face of tower, 4-cornered, flat or angular sections, steel or wood.	2.20	
3-cornered, flat or angular sections, steel or wood	2.00	
Wind on corner, 4-cornered tower, flat or angular sections.	2.40	
Wind parallel to one face of 3-cornered tower, flat or angular sections.	1.50	
Wind on individual members.		
Cylindrical 2 in. or less in diameter.	1.00	
Cylindrical over 2 in. in diameter.	0.80	
Flat or angular sections	1.30	

NOTE: Factors for towers with cylindrical elements are approximately two thirds of those for similar towers with flat or angular sections.

TABLE D – FACTORS FOR CHIMNEYS, TANKS, AND TOWERS*

Horizontal Cross Section	Factor
Square or rectangular	1.00
Hexagonal or octagonal	0.80
Round or elliptical	0.60

EXAMPLE:

Given: Radio tower located in eastern New York State. Tower base 50' above average elevation of surrounding terrain.

Find: Wind pressure Pt.

Procedure: From Fig. A, eastern New York State is in the 25 m.p.h. zone. From Table B, select pressures for various height zones as in Fig. E. From Table C, select the appropriate factor. Assume a 4-cornered tower with wind normal to one face and members angular. (Factor = 2.20.) Therefore, total pressures due to wind at various elevations are:

Elev. 50 to 99 Pt = 2.2 × 30 = 66 p.s.f. × A
Elev. 100 to 200 Pt = 2.2 × 40 = 88 p.s.f. × A

200' 40 p.s.f.
Tower
150'
100' 30 p.s.f.
50' Tower base
50'
Average terrain elevation.

FIG. E

TABLE F – REQUIREMENTS FOR COMBINED LOADINGS ON ROOFS *

Slope	Exterior load coefficients (factors to be applied to wind pressures specified in 5.1) Windward Slope	Leeward Slope
20° or less	−0.60	−0.45
Between 20° and 30°	0.06-1.8	−0.45
Between 30° and 60°	0.015A-0.45	−0.45
Between 60° and 90°	0.45	−0.45

Positive values indicate an inward load; negative values, an outward load. A is the roof slope in degrees.

*This material is reproduced from the American Standards Building Code Requirements for Minimum Design Loads in Building and other Structures, A58.1–1955, copyrighted by American Standards Association

‡ Although the factors vary with the solidity ratio, these factors in Table C may be considered as average values.

STRUCTURAL – GENERAL – WIND DESIGN

APPROXIMATE WIND DESIGN METHODS FOR MULTI-STORIED FRAME BUILDINGS

NOTE:
See p. 4-36 for Riveted Wind Connection Moments & Details.
See p. 4-59 for Welded Wind Connection Moments & Details.

I – PORTAL METHOD

Assumptions:

1. Each bay an independent frame.
2. The point of contraflexure of each column is at midheight.
3. The point of contraflexure of each girder is at its midlength.
4. The horizontal shear on any plane is divided equally among the number of aisles. An outer column takes 1/2 the shear of an interior column.
5. The wind load is resisted by the frame.
6. All direct stress in outside columns.

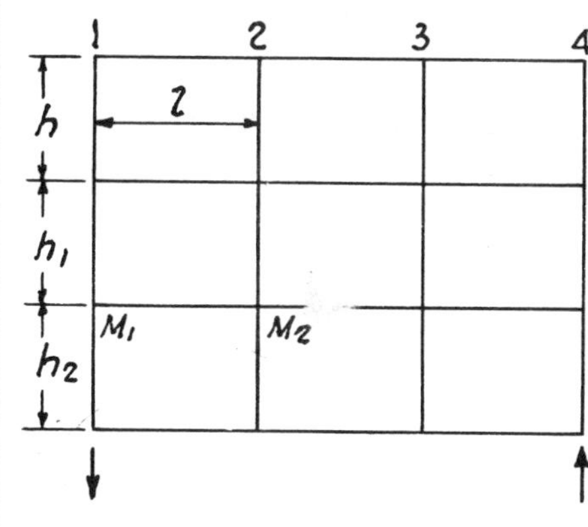
1
2
3
4
h
l
h1
h2
M1
M2
Wind

Let S = total shear in bay above floor line.
S_1, S_2, S_3, S_4 = proportional shears taken on ea. col.
Then: $S_1 + S_2 + S_3 + S_4 = S$ *

$$S_1 = S_4 = \frac{S}{6}\ ;\ S_2 = S_3 = \frac{S}{3}$$

$$\text{Moment per girder connection} = \frac{S \times h_{average}}{\text{No. of girder Connects.}}$$

Given $S = 20$ Kips; $\dfrac{h_1 + h_2}{2} = 12'$

$$M_1 = M_2 = \frac{20 \times 12}{6} = 40' \text{Kips}$$

* Shear may be proportioned to the moment of inertia of the column.

II – SHEAR DISTRIBUTION BY STIFFNESS OF CELLS OF RESISTANCE

AFTER E.E. SEELYE, E.N.R., MAR. 20, 1930.

1. Total horizontal Shear $= S = S_1 + S_2 + S_3 + S_4$

2. Total Stiffness Coef. $= F = F_1 + F_2 + F_3 + F_4$

3.
$$F_1 = \frac{K_1(0 + K_5)}{0 + K_1 + K_5 + K_8}$$
$$F_2 = \frac{K_2(K_5 + K_6)}{K_5 + K_2 + K_6 + K_9}$$
$$F_3 = \frac{K_3(K_6 + K_7)}{K_6 + K_3 + K_7 + K_{10}}$$
$$F_4 = \frac{K_4(K_7 + 0)}{K_7 + K_4 + 0 + K_{11}}$$

4.
$$S_1 = \frac{F_1}{F} \times S$$
$$S_2 = \frac{F_2}{F} \times S$$
$$S_3 = \frac{F_3}{F} \times S$$
$$S_4 = \frac{F_4}{F} \times S$$

$$K = \frac{I}{\ell} \text{ or } \frac{I}{h}$$

DIRECT STRESSES from wind in columns may be neglected, except in unusual cases where they exceed allowable wind overstress or require anchorage.

STRUCTURAL–GENERAL–WALL THICKNESS

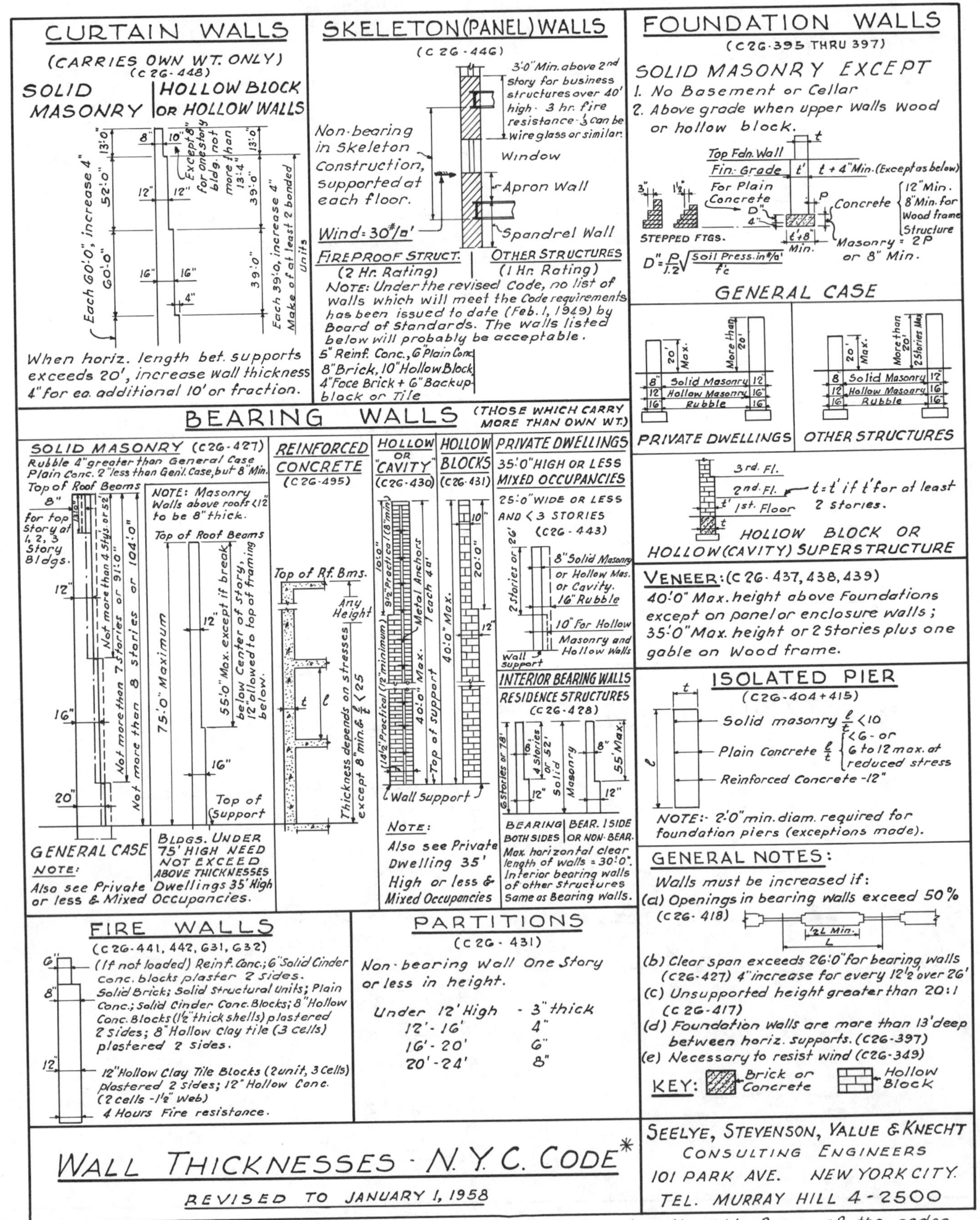

CURTAIN WALLS
(CARRIES OWN WT. ONLY)
(C 26-448)
SOLID MASONRY
HOLLOW BLOCK OR HOLLOW WALLS
Each 60'-0", increase 4"
Except 8" for one story bldg. not more than 13'-4"
Each 39'-0, increase 4". Make of at least 2 bonded units
When horiz. length bet. supports exceeds 20', increase wall thickness 4" for ea. additional 10' or fraction.
SKELETON (PANEL) WALLS
(C 26-446)
3'-0" Min. above 2nd story for business structures over 40' high - 3 hr. fire resistance - 1/3 can be wire glass or similar.
Non-bearing in Skeleton Construction, supported at each floor.
Window
Apron Wall
Spandrel Wall
Wind = 30#/□'
FIREPROOF STRUCT. (2 Hr. Rating)
OTHER STRUCTURES (1 Hr. Rating)
NOTE: Under the revised Code, no list of walls which will meet the Code requirements has been issued to date (Feb. 1, 1949) by Board of Standards. The walls listed below will probably be acceptable.
5" Reinf. Conc., 6" Plain Conc.
8" Brick, 10" Hollow Block
4" Face Brick + 6" Backup block or Tile
FOUNDATION WALLS
(C 26-395 THRU 397)
SOLID MASONRY EXCEPT
1. No Basement or Cellar
2. Above grade when upper walls wood or hollow block.
Top Fdn. Wall
Fin. Grade
t + 4" Min. (Except as below)
For Plain Concrete
Concrete {12" Min. 8" Min. for Wood frame Structure
STEPPED FTGS.
Masonry = 2P or 8" Min.
$D'' = \frac{P}{1.2}\sqrt{\frac{\text{Soil Press. in \#/□'}}{f'_c}}$
GENERAL CASE
PRIVATE DWELLINGS
OTHER STRUCTURES
Solid Masonry 12"
Hollow Masonry 16"
Rubble 16"
BEARING WALLS (THOSE WHICH CARRY MORE THAN OWN WT.)
SOLID MASONRY (C 26-427)
Rubble 4" greater than General Case
Plain Conc. 2" less than Genl. Case, but 8" Min.
Top of Roof Beams
8" for top story at 1, 2, 3 story bldgs.
NOTE: Masonry walls above roofs < 12' to be 8" thick.
Not more than 4 stys. or 52'
Not more than 7 stories or 91'-0"
Not more than 8 stories or 104'-0"
75'-0" Maximum
55'-0" Max. except if break below Center of story, 12' allowed to top of framing below.
Top of Support
GENERAL CASE
NOTE: Also see Private Dwellings 35' High or less & Mixed Occupancies.
BLDGS. UNDER 75' HIGH NEED NOT EXCEED ABOVE THICKNESSES
REINFORCED CONCRETE
(C 26-495)
Top of Rf. Bms.
Any Height
Thickness depends on stresses except 8" min. & l/t < 25
HOLLOW OR "CAVITY"
(C 26-430)
HOLLOW BLOCKS
(C 26-431)
10'-0" 9½" Practical (8" min.)
(14½" Practical (12" minimum))
Metal Anchors 1 each 4□'
40'-0" Max.
Top of Support
Wall Support
NOTE: Also see Private Dwelling 35' High or less & Mixed Occupancies
PRIVATE DWELLINGS
35'-0" HIGH OR LESS
MIXED OCCUPANCIES
25'-0" WIDE OR LESS AND < 3 STORIES
(C 26-443)
2 Stories or 26'
8" Solid Masonry or Hollow Mas. or Cavity. 16" Rubble
10" For Hollow Masonry and Hollow Walls
Wall support
INTERIOR BEARING WALLS
RESIDENCE STRUCTURES
(C 26-428)
6 Stories or 78'
4 Stories or 52'
Solid Masonry
55' Max.
BEARING BOTH SIDES
BEAR. 1 SIDE OR NON-BEAR.
Max. horizontal clear length of walls = 30'-0".
Interior bearing walls of other structures same as Bearing Walls.
3rd. Fl.
2nd. Fl.
1st. Floor
t = t' if t' for at least 2 stories.
HOLLOW BLOCK OR HOLLOW (CAVITY) SUPERSTRUCTURE
VENEER: (C 26-437, 438, 439)
40'-0" Max. height above Foundations except on panel or enclosure walls; 35'-0" Max. height or 2 Stories plus one gable on Wood frame.
ISOLATED PIER
(C 26-404 + 415)
Solid masonry l/t < 10
Plain Concrete l/t {< 6 - or 6 to 12 max. at reduced stress
Reinforced Concrete - 12"
NOTE: - 2'-0" min. diam. required for foundation piers (exceptions made).
GENERAL NOTES:
Walls must be increased if:
(a) Openings in bearing walls exceed 50% (C 26-418)
½L Min.
(b) Clear span exceeds 26'-0" for bearing walls (C 26-427) 4" increase for every 12½' over 26'
(c) Unsupported height greater than 20:1 (C 26-417)
(d) Foundation walls are more than 13' deep between horiz. supports. (C 26-397)
(e) Necessary to resist wind (C 26-349)
KEY: Brick or Concrete; Hollow Block
FIRE WALLS
(C 26-441, 442, 631, 632)
(If not loaded) Reinf. Conc.; 6" Solid Cinder Conc. blocks plaster 2 sides.
Solid Brick; Solid Structural Units; Plain Conc.; Solid Cinder Conc. Blocks; 8" Hollow Conc. Blocks (1½" thick shells) plastered 2 sides; 8" Hollow clay tile (3 cells) plastered 2 sides.
12" Hollow Clay Tile Blocks (2 unit, 3 Cells) plastered 2 sides; 12" Hollow Conc. (2 cells - 1½" web)
4 Hours Fire resistance.
PARTITIONS
(C 26-431)
Non-bearing Wall One Story or less in height.
Under 12' High - 3" thick
12'-16' 4"
16'-20' 6"
20'-24' 8"
WALL THICKNESSES - N.Y.C. CODE*
REVISED TO JANUARY 1, 1958
SEELYE, STEVENSON, VALUE & KNECHT
CONSULTING ENGINEERS
101 PARK AVE. NEW YORK CITY
TEL. MURRAY HILL 4-2500

*This code is incorporated as typical of modern engineering thought of any of the codes.
For walls subject to earth pressure – see page 7-31

STRUCTURAL-GENERAL-FIREPROOFING REQUIREMENTS

TABLE A – SUGGESTED FIREPROOFING REQUIREMENTS*

Protection of Structural Parts

Member to Be Protected	Fireproofing Material	Minimum Thickness for Hour Ratings: 4	3	2	1
Steel columns equal or greater than 6 x 6	Group 1 concrete†	2½	2	1½	1
	Group 2 concrete†	3	2	1½	1
	Brick, clay (with brick fill)	3¾	3¾	3¾	2¼
	Concrete block hollow (mortar and broken block fill)	3	3	3	3
	Clay tile (two 2" layers)	4	2‡	2	2
	Solid gypsum block with ties	2 pl	2 pl	2	2
	Hollow gypsum block with ties	3 pl	3 pl	3	3
	Poured gypsum mesh	2	1½	1	1
	Gypsum – Vermiculite on self-furring metal	1¾	1⅜	1⅜	1⅜
	Gypsum – Vermiculite on gypsum lath reinforcement	1½	1½	1½	1½
	Gypsum – Perlite on self-furring metal	1¾	1⅜	1	1
	Gypsum – Perlite on gypsum lath reinforcement	1½	1½	1	1
	Sprayed fiber – directly to steel	–	2	1½	1½
Steel beams, girders, and truss members (individually protected)	Group 1 concrete	2	2	1½	1
	Group 2 concrete	2½	2½	2	1½
	Brick of clay, concrete, or sand lime	3¾	3¾	2¼	2¼
	Gypsum blocks, solid	2 pl	2 pl	2	2
	Gypsum blocks, hollow	3 pl	3	2	2
	Gypsum, poured	2	2	1½	1
	Gypsum – Vermiculite on self-furring metal lath cage □	–	–	⅞	⅞
	Gypsum – Perlite on self-furring metal lath cage	1½	1½	1	1
	Sprayed fiber – directly to adhesive coated steel	3⅛	2	2	2
Structural members♦ with floor and ceiling protection	Floor = 2½" concrete slab on metal lath and gypsum vermiculite ceiling	1	¾	¾	¾
	Floor = 2" precast gypsum slab + ½ cover of mortar and gypsum vermiculite ceiling	1	¾	¾	¾
	Floor = 2" concrete slab on metal lath and gypsum perlite ceiling	1	½	½	½
	Floor = 2" concrete slab on metal lath and gypsum vermiculite ceiling	–	1	1	1
	Floor = 2¾" concrete plank and gypsum vermiculite ceiling	–	1	1	1
	Floor = 2" cinder concrete slab on steel cellular units and gypsum vermiculite ceiling	⅞	⅞	⅞	⅞
	Floor = 2" perlite concrete on steel cellular units and gypsum perlite ceiling	1	1	1	1
	Floor = 2" reinforced concrete on steel cellular units and gypsum perlite ceiling	⅞	⅞	⅞	⅞
	Floor = 2" concrete on steel cellular units and sprayed fiber ceiling	1⅛	1⅛	1⅛	1⅛
	Floor = 1½" concrete + ½ cement mortar finish on steel cellular units and gypsum vermiculite ceiling	1	1	1	1
	Floor = 2" concrete on steel cellular units and gypsum perlite or gypsum vermiculite ceiling	–	⅞	⅞	⅞
Reinforced-concrete beams, girders and trusses – reinforcing bars	Coarse aggregate air-cooled slag, expanded slag, crushed limestone, calcareous gravel, siliceous gravel, or traprock	1½	1½	1½	1½
Reinforced-concrete slabs – reinforcing bars	Group 1 concrete	1	1	¾⊕	¾
Reinforced-concrete columns – reinforcing bars	Group 1 concrete (12" or larger round or square columns). Not including cinder aggregate	1½	1½	1½	1½
	Group 2 concrete (16" or larger round or square columns). Not including cinder aggregate	2½	1½	1½	1½

Protection of Walls and Partitions

Material	Construction		Bearing§ Minimum Thickness for Hour Ratings: 4	3	2	1	Non-Bearing Minimum Thickness for Hour Ratings: 4	3	2	1
Brick of clay, shale, concrete, or sand lime	Solid unplastered		8	8	4	–	–	–	–	4
	Solid plastered one side		8	8	8	4	–	–	4	4
	Solid plastered both sides		8	8	4	4	–	–	4	4
	Hollow cavity		10	10	10	10	–	–	–	–
Hollow structural tile, clay, or shale	Unplastered	Note: For number of units and cells see N.B.F.U. Fire Resistance Ratings.	12	12	12	8	–	–	6	4
	Plastered one side		12	12	8	8	–	–	–	–
	Plastered both sides		12	8	8	8	–	6	4	4
Hollow structural tile, clay, or shale bonded to 4" brick facing	Unplastered		8	4	–	–	–	–	–	–
	Plastered tile side		4	–	–	–	–	–	–	–
Hollow structural facing tile, clay	Unplastered		–	–	–	–	–	6	–	–
	Plastered one side		–	–	–	–	–	6	4	4
Gypsum block	Solid unplastered		–	–	–	–	5	3	3	2
	Hollow Unplastered		–	–	–	–	–	–	–	3
	Hollow plastered		–	–	–	–	4	3	3	3
Hollow concrete masonry units	Bonded to 4" brick facing		4 pl	–	–	–	4 pl	–	–	–
	Aggregate expanded burnt clay or shale, crushed lime stone, unexpanded slag, or cinders	Unplastered	10	8	8	–	–	–	6	4
		Plastered	8	8	–	–	6	6	4	3
	Expanded slag or pumice aggregate	Unplastered	8	6	–	–	–	6	6	3
		Plastered	6	–	–	–	6	6	3	–
	Calcareous sand and gravel	Unplastered	–	–	–	10	–	–	–	–
		Plastered	10	–	–	–	–	–	–	4
	Siliceous sand and gravel	Unplastered	12	–	–	–	–	–	–	–
		Plastered	12	–	–	–	–	–	–	4
Plain concrete	Solid monolithic walls		7½	6½	5½	–	–	–	–	4
Reinforced concrete (monolithic)	Solid walls unplastered	Group 1	6½	6	5	3½	–	–	–	–
		Group 2	7½	6½	5½	–	–	–	–	4
	Solid walls plastered both sides	Group 1	5	4	–	–	–	–	3	3
		Group 2	6	5	4	–	–	–	–	3
Stone masonry	Solid walls		12	12	12	8	–	–	–	–

NOTE: Group 1 concrete aggregates are blast-furnace slag, limestone, calcareous gravel, traprock, burnt clay or shale, cinders containing not more than 25% of combustible material and not more than 5% of volatile material, and other materials meeting the requirements of these specifications and containing not more than 30% quartz, chert, flint, and similar materials.

Group 2 concrete aggregates are granite, quartzite, siliceous gravel, sandstone, gneiss, cinders containing more than 25%, but not more than 40% of combustible material and not more than 5% of volatile material and other materials meeting requirements of these specifications, and containing more than 30% of quartz, chert, flint, and similar materials.

* Based on author's interpretation of National Board of Fire Underwriters (N.B.F.U.) publication, *Fire Resistance Ratings*, 1957. For Further details on construction and materials, see N.B.F.U. publication, *Fire Resistance Ratings*.

† Not more than 10% unburned coals and not over 5% ash.

‡ With fill and ties.

|| Steel cellular units or other steel structural units.

¶ 1" for calcareous aggregate.

§ None or noncombustible members framing into wall.

REINFORCED-CONCRETE FLOORS

4-Hr. Rating: 4½" slab, expanded slag aggregate (¾" protection to reinforcement).
6" slab, air-cooled slag, aggregate (1" protection to reinforcement).
5" slab, limestone aggregate and 1" gypsum vermiculite plaster ceiling, metal lath (electrical raceways and junction boxes in slab.
3" slab, limestone aggregate and 1" gypsum vermiculite plaster ceiling, metal lath.

3-Hr. Rating: 6" slab, traprock, or crushed limestone, or calcareous gravel, or siliceous gravel aggregate (1" protection to reinforcement).
4" slab, limestone aggregate hung ceiling of ¾" gypsum vermiculite plaster metal lath (electrical raceways and junction boxes in slab).
2" slab, limestone aggregate hung ceiling of ¾" gypsum vermiculite plaster, metal lath.

2-Hr. Rating: 4¾" slab, traprock, or siliceous gravel aggregate (¾" protection to reinforcement).
4¾" slab, calcareous gravel, or crushed limestone aggregate (1" protection to reinforcement).

1-Hr. Rating: 4" slab, siliceous gravel aggregate (¾" protection to reinforcement).

STRUCTURAL–GENERAL–GAGES

WIRE AND SHEET METAL GAGES
IN DECIMALS OF AN INCH

Name of Gage	United States Standard Gage*		The United States Steel Wire Gage	American or Brown & Sharpe Wire Gage	New Birmingham Standard Sheet & Hoop Gage	British Imperial or English Legal Standard Wire Gage	Birmingham or Stubs Iron Wire Gage	Name of Gage
Principal Use	Uncoated Steel Sheets and Light Plates		Steel Wire except Music Wire	Non-Ferrous Sheets and Wire	Iron and Steel Sheets and Hoops	Wire	Strips, Bands, Hoops and Wire	Principal Use
Gage No.	Weight Oz. per Sq. Ft.	Approx. Thickness Inches	Thickness, Inches					Gage No.
7 0's			.4900		.6666	.500		7/0's
6/0's			.4615	.5800	.625	.464		6/0's
5 0's			.4305	.5165	.5883	.432	.500	5/0's
4 0's			.3938	.4600	.5416	.400	.454	4/0's
3/0's			.3625	.4096	500	.372	.425	3/0's
2/0's			.3310	.3648	.4452	.348	.380	2/0's
0			.3065	.3249	.3964	.324	.340	0
1			.2830	.2893	.3532	.300	.300	1
2			.2625	.2576	.3147	.276	.284	2
3	160	.2391	.2437	.2294	.2804	.252	.259	3
4	150	.2242	.2253	.2043	.250	.232	.238	4
5	140	.2092	.2070	.1819	.2225	.212	.220	5
6	130	.1943	.1920	.1620	.1981	.192	.203	6
7	120	.1793	.1770	.1443	.1764	.176	.180	7
8	110	.1644	.1620	.1285	.1570	.160	.165	8
9	100	.1495	.1483	.1144	.1398	.144	.148	9
10	90	.1345	.1350	.1019	.1250	128	.134	10
11	80	.1196	.1205	.0907	.1113	.116	.120	11
12	70	.1046	.1055	.0808	.0991	.104	.109	12
13	60	.0897	.0915	.0720	.0882	.092	.095	13
14	50	.0747	.0800	.0641	.0785	.080	.083	14
15	45	.0673	.0720	.0571	.0699	.072	.072	15
16	40	0598	.0625	.0508	.0625	.064	.065	16
17	36	.0538	.0540	.0453	.0556	.056	.058	17
18	32	.0478	.0475	.0403	.0495	.048	.049	18
19	28	.0418	.0410	.0359	.0440	.040	.042	19
20	24	.0359	.0348	.0320	.0392	.036	.035	20
21	22	0329	.0318	.0285	.0349	.032	.032	21
22	20	.0299	.0286	.0253	.0313	.028	.028	22
23	18	.0269	.0258	.0226	.0278	.024	.025	23
24	16	.0239	.0230	.0201	.0248	.022	.022	24
25	14	.0209	.0204	.0179	.0220	.020	.020	25
26	12	0179	.0181	.0159	.0196	.018	.018	26
27	11	.0164	.0173	.0142	.0175	.0164	.016	27
28	10	.0149	.0162	.0126	.0156	.0148	.014	28
29	9	.0135	.0150	.0113	.0139	.0136	.013	29
30	8	.0120	.0140	.0100	.0123	.0124	.012	30
31	7	.0105	.0132	.0089	.0110	.0116	.010	31
32	6.5	.0097	.0128	.0080	.0098	.0108	.009	32
33	6	.0090	.0118	.0071	.0087	.0100	.008	33
34	5.5	.0082	.0104	.0063	.0077	.0092	.007	34
35	5	.0075	.0095	.0056	.0069	.0084	.005	35
36	4.5	.0067	.0090	.0050	.0061	.0076	.004	36
37	4.25	.0064	.0085	.0045	.0054	.0068		37
38	4	.0060	.0080	.0040	.0048	.0060		38
39			.0075	.0035	.0043	.0052		39
40			.0070	.0031	.0039	.0048		40

* U. S. Standard Gage is officially a weight gage, in oz. per sq. ft. as tabulated. The Approx. Thickness shown is the "Manufacturers' Standard" of the American Iron and Steel Institute, based on steel as weighing 501.81 lbs. per cu. ft. (489.6 true weight plus 2.5 percent for average over-run in area and thickness). The A.I.S.I. standard nomenclature for flat rolled carbon steel is as follows:

Widths, Inches	Thicknesses, Inch							
	0.2500 and thicker	0.2499 to 0.2031	0.2030 to 0.1875	0.1874 to 0.0568	0.0567 to 0.0344	0.0343 to 0.0255	0.0254 to 0.0142	0.0141 and thinner
To 3½ incl.	Bar	Bar	Strip	Strip	Strip	Strip	Sheet	Sheet
Over 3½ to 6 incl.	Bar	Bar	Strip	Strip	Strip	Sheet	Sheet	Sheet
" 6 to 12 "	Plate	Strip	Strip	Strip	Sheet	Sheet	Sheet	Sheet
" 12 to 32 "	Plate	Sheet	Sheet	Sheet	Sheet	Sheet	Sheet	Black Plate
" 32 to 48 "	Plate	Sheet	Sheet	Sheet	Sheet	Sheet	Sheet	Sheet
" 48	Plate	Plate	Plate	Sheet	Sheet	Sheet	Sheet	—

TABLE A

Adapted from American Institute of Steel Construction, 1947.

STRUCTURAL-GENERAL-CHAINS & WIRE ROPES

TABLE A – SAFE WORKING LOADS OF CHAINS (in kips) A.S.T.M. A56-39

Nominal Size of Chain Bar, inches	1/4	5/16	3/8	7/16	1/2	9/16	5/8	3/4	7/8	1	1 1/8	1 1/4	1 3/8	1 1/2	1 5/8	1 3/4	1 7/8	2
Crane Chain	1.0	1.6	2.3	3.2	4.2	5.3	6.6	9.5	12.9	16.9	20.0	24.7	29.9	35.6	41.8	48.4	55.3	63.3
Proof Coil	.85	1.3	1.9	2.6	3.4	4.3	5.3	7.6	10.4	13.6								

TABLE B – NOMINAL WEIGHTS AND DIMENSIONS OF CHAINS A.S.T.M. A56-39

Normal Size of Chain Bar, inches	Actual Size of Material, inches	Nominal Length of 100 Links, in.		Nominal Weight per 100 ft., lb.		Nominal Dimensions of Links, inches							
						Crane Chain				Proof Coil			
						Outside		Inside		Outside		Inside	
		Crane Chain	Proof Coil	Crane Chain	Proof Coil	Length	Width	Length	Width	Length	Width	Length	Width
1/4	9/32	86	100	78	70	1 27/64	1	55/64	7/16	1 9/16	1 1/16	1	1/2
5/16	11/32	100	111	115	105	1 11/16	1 3/16	1	1/2	1 51/64	1 3/16	1 7/64	1/2
3/8	13/32	109	123	166	158	1 29/32	1 7/16	1 3/32	5/8	2 3/64	1 7/16	1 15/64	5/8
7/16	15/32	122	138	220	210	2 5/32	1 5/8	1 7/32	11/16	2 5/16	1 11/16	1 3/8	3/4
1/2	17/32	134	150	275	265	2 13/32	1 13/16	1 11/32	3/4	2 9/16	1 7/8	1 1/2	13/16
9/16	19/32	156	175	350	335	2 3/4	1 15/16	1 9/16	3/4	2 15/16	2 1/16	1 3/4	7/8
5/8	21/32	169	188	430	410	3	2 3/16	1 11/16	7/8	3 3/16	2 5/16	1 7/8	1
3/4	25/32	188	213	615	580	3 7/16	2 9/16	1 7/8	1	3 11/16	2 11/16	2 1/8	1 1/8
7/8	29/32	225	250	820	780	4 1/16	3 1/16	2 1/4	1 1/4	4 5/16	3 3/16	2 1/2	1 3/8
1	1 1/32	256	275	1045	1000	4 5/8	3 7/16	2 9/16	1 3/8	4 13/16	3 9/16	2 3/4	1 1/2
1 1/8	1 5/32	288		1310		5 3/16	3 15/16	2 7/8	1 5/8				
1 1/4	1 9/32	306		1600		5 5/8	4 5/16	3 1/16	1 3/4				
1 3/8	1 13/32	363		1930		6 7/16	4 11/16	3 5/8	1 7/8				
1 1/2	1 17/16	387		2335		6 15/16	5 1/16	3 7/8	2				
1 5/8	1 21/32	425		2740		7 9/16	5 7/16	4 1/4	2 1/8				
1 3/4	1 25/32	475		3180		8 5/16	5 13/16	4 3/4	2 1/4				
1 7/8	1 29/32	525		3650		9 1/16	6 5/16	5 1/4	2 1/2				
2	2 1/32	575		4100		9 13/16	6 11/16	5 3/4	2 5/8				

TABLE C – WIRE ROPE 6 x 19 STANDARD HOISTING PLOW STEEL*

Diameter, in.	2 3/4	2 1/4	2	1 7/8	1 3/4	1 5/8	1 1/2	1 3/8	1 1/4	1 1/8	1	7/8	3/4	5/8	9/16	1/2	7/16	3/8
Breaking strength,† tons of 2000 lb.	254.0	174.0	139.0	123.0	108.0	93.4	80.0	67.5	56.2	45.7	36.4	28.0	20.7	14.5	11.8	9.35	7.19	5.31

*From John A. Roebling's Sons Co. †Use factor of safety from 5 to 8.

STRUCTURAL-GENERAL-PROPERTIES OF SECTIONS-1

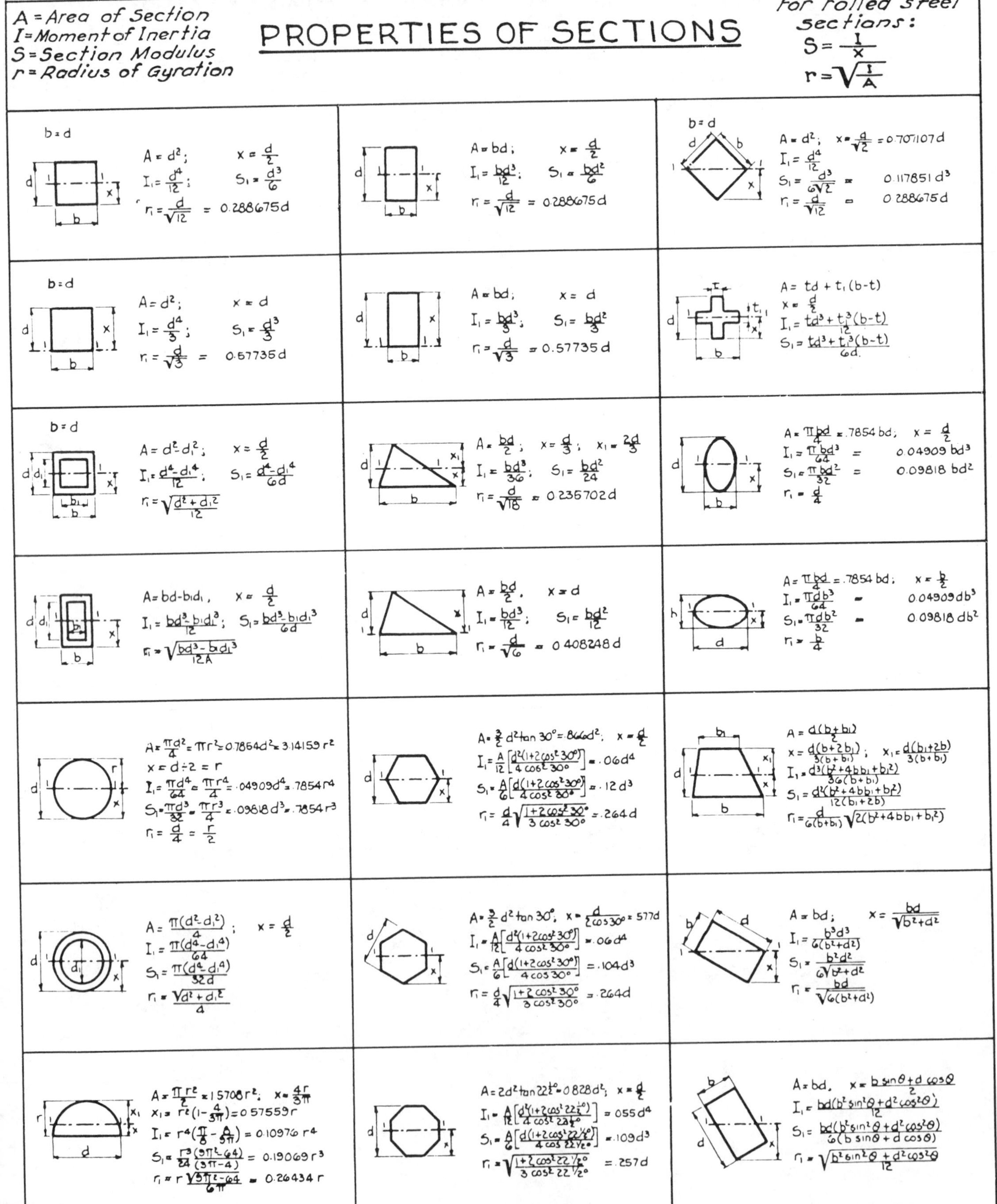
A = Area of Section
I = Moment of Inertia
S = Section Modulus
r = Radius of Gyration
PROPERTIES OF SECTIONS
For rolled steel sections:
$S = \frac{I}{x}$
$r = \sqrt{\frac{I}{A}}$
b = d
$A = d^2$; $x = \frac{d}{2}$
$I_1 = \frac{d^4}{12}$; $S_1 = \frac{d^3}{6}$
$r_1 = \frac{d}{\sqrt{12}} = 0.288675d$
$A = bd$; $x = \frac{d}{2}$
$I_1 = \frac{bd^3}{12}$; $S_1 = \frac{bd^2}{6}$
$r_1 = \frac{d}{\sqrt{12}} = 0.288675d$
b = d
$A = d^2$; $x = \frac{d}{\sqrt{2}} = 0.707107d$
$I_1 = \frac{d^4}{12}$
$S_1 = \frac{d^3}{6\sqrt{2}} = 0.117851d^3$
$r_1 = \frac{d}{\sqrt{12}} = 0.288675d$
b = d
$A = d^2$; $x = d$
$I_1 = \frac{d^4}{3}$; $S_1 = \frac{d^3}{3}$
$r_1 = \frac{d}{\sqrt{3}} = 0.57735d$
$A = bd$; $x = d$
$I_1 = \frac{bd^3}{3}$; $S_1 = \frac{bd^2}{3}$
$r_1 = \frac{d}{\sqrt{3}} = 0.57735d$
$A = td + t_1(b-t)$
$x = \frac{d}{2}$
$I_1 = \frac{td^3 + t_1^3(b-t)}{12}$
$S_1 = \frac{td^3 + t_1^3(b-t)}{6d}$
b = d
$A = d^2 - d_1^2$; $x = \frac{d}{2}$
$I_1 = \frac{d^4 - d_1^4}{12}$; $S_1 = \frac{d^4 - d_1^4}{6d}$
$r_1 = \sqrt{\frac{d^2 + d_1^2}{12}}$
$A = \frac{bd}{2}$; $x = \frac{d}{3}$; $x_1 = \frac{2d}{3}$
$I_1 = \frac{bd^3}{36}$; $S_1 = \frac{bd^2}{24}$
$r_1 = \frac{d}{\sqrt{18}} = 0.235702d$
$A = \frac{\pi bd}{4} = .7854bd$; $x = \frac{d}{2}$
$I_1 = \frac{\pi bd^3}{64} = 0.04909bd^3$
$S_1 = \frac{\pi bd^2}{32} = 0.09818bd^2$
$r_1 = \frac{d}{4}$
$A = bd - b_1d_1$, $x = \frac{d}{2}$
$I_1 = \frac{bd^3 - b_1d_1^3}{12}$; $S_1 = \frac{bd^3 - b_1d_1^3}{6d}$
$r_1 = \sqrt{\frac{bd^3 - b_1d_1^3}{12A}}$
$A = \frac{bd}{2}$, $x = d$
$I_1 = \frac{bd^3}{12}$; $S_1 = \frac{bd^2}{12}$
$r_1 = \frac{d}{\sqrt{6}} = 0.408248d$
$A = \frac{\pi bd}{4} = .7854bd$; $x = \frac{b}{2}$
$I_1 = \frac{\pi db^3}{64} = 0.04909db^3$
$S_1 = \frac{\pi db^2}{32} = 0.09818db^2$
$r_1 = \frac{b}{4}$
$A = \frac{\pi d^2}{4} = \pi r^2 = 0.7854d^2 = 3.14159r^2$
$x = d \div 2 = r$
$I_1 = \frac{\pi d^4}{64} = \frac{\pi r^4}{4} = .04909d^4 = .7854r^4$
$S_1 = \frac{\pi d^3}{32} = \frac{\pi r^3}{4} = .09818d^3 = .7854r^3$
$r_1 = \frac{d}{4} = \frac{r}{2}$
$A = \frac{3}{2}d^2 \tan 30° = .866d^2$; $x = \frac{d}{2}$
$I_1 = \frac{A}{12}\left[\frac{d^2(1+2\cos^2 30°)}{4\cos^2 30°}\right] = .06d^4$
$S_1 = \frac{A}{6}\left[\frac{d(1+2\cos^2 30°)}{4\cos^2 30°}\right] = .12d^3$
$r_1 = \frac{d}{4}\sqrt{\frac{1+2\cos^2 30°}{3\cos^2 30°}} = .264d$
$A = \frac{d(b+b_1)}{2}$
$x = \frac{d(b+2b_1)}{3(b+b_1)}$; $x_1 = \frac{d(b_1+2b)}{3(b+b_1)}$
$I_1 = \frac{d^3(b^2+4bb_1+b_1^2)}{36(b+b_1)}$
$S_1 = \frac{d^2(b^2+4bb_1+b_1^2)}{12(b_1+2b)}$
$r_1 = \frac{d}{6(b+b_1)}\sqrt{2(b^2+4bb_1+b_1^2)}$
$A = \frac{\pi(d^2-d_1^2)}{4}$; $x = \frac{d}{2}$
$I_1 = \frac{\pi(d^4-d_1^4)}{64}$
$S_1 = \frac{\pi(d^4-d_1^4)}{32d}$
$r_1 = \frac{\sqrt{d^2+d_1^2}}{4}$
$A = \frac{3}{2}d^2 \tan 30°$; $x = \frac{d}{2\cos 30°} = .577d$
$I_1 = \frac{A}{12}\left[\frac{d^2(1+2\cos^2 30°)}{4\cos^2 30°}\right] = .06d^4$
$S_1 = \frac{A}{6}\left[\frac{d(1+2\cos^2 30°)}{4\cos 30°}\right] = .104d^3$
$r_1 = \frac{d}{4}\sqrt{\frac{1+2\cos^2 30°}{3\cos^2 30°}} = .264d$
$A = bd$; $x = \frac{bd}{\sqrt{b^2+d^2}}$
$I_1 = \frac{b^3d^3}{6(b^2+d^2)}$
$S_1 = \frac{b^2d^2}{6\sqrt{b^2+d^2}}$
$r_1 = \frac{bd}{\sqrt{6(b^2+d^2)}}$
$A = \frac{\pi r^2}{2} = 1.5708r^2$; $x = \frac{4r}{3\pi}$
$x_1 = r(1-\frac{4}{3\pi}) = 0.57559r$
$I_1 = r^4(\frac{\pi}{8} - \frac{8}{9\pi}) = 0.10976r^4$
$S_1 = \frac{r^3}{24}\frac{(9\pi^2-64)}{(3\pi-4)} = 0.19069r^3$
$r_1 = r\frac{\sqrt{9\pi^2-64}}{6\pi} = 0.26434r$
$A = 2d^2\tan 22\frac{1}{2}° = 0.828d^2$; $x = \frac{d}{2}$
$I_1 = \frac{A}{12}\left[\frac{d^2(1+2\cos^2 22\frac{1}{2}°)}{4\cos^2 22\frac{1}{2}°}\right] = .055d^4$
$S_1 = \frac{A}{6}\left[\frac{d(1+2\cos^2 22\frac{1}{2}°)}{4\cos^2 22\frac{1}{2}°}\right] = .109d^3$
$r_1 = \sqrt{\frac{1+2\cos^2 22\frac{1}{2}°}{3\cos^2 22\frac{1}{2}°}} = .257d$
$A = bd$, $x = \frac{b\sin\theta + d\cos\theta}{2}$
$I_1 = \frac{bd(b^2\sin^2\theta + d^2\cos^2\theta)}{12}$
$S_1 = \frac{bd(b^2\sin^2\theta + d^2\cos^2\theta)}{6(b\sin\theta + d\cos\theta)}$
$r_1 = \sqrt{\frac{b^2\sin^2\theta + d^2\cos^2\theta}{12}}$

Adapted from Singleton, Manual of Structural Design, H.M. Ives & Sons

NATALE S. NAPPI, P.E.
14104 PARKVALE RD.
ROCKVILLE, MD 20853

STRUCTURAL-GENERAL-PROPERTIES OF SECTIONS-2

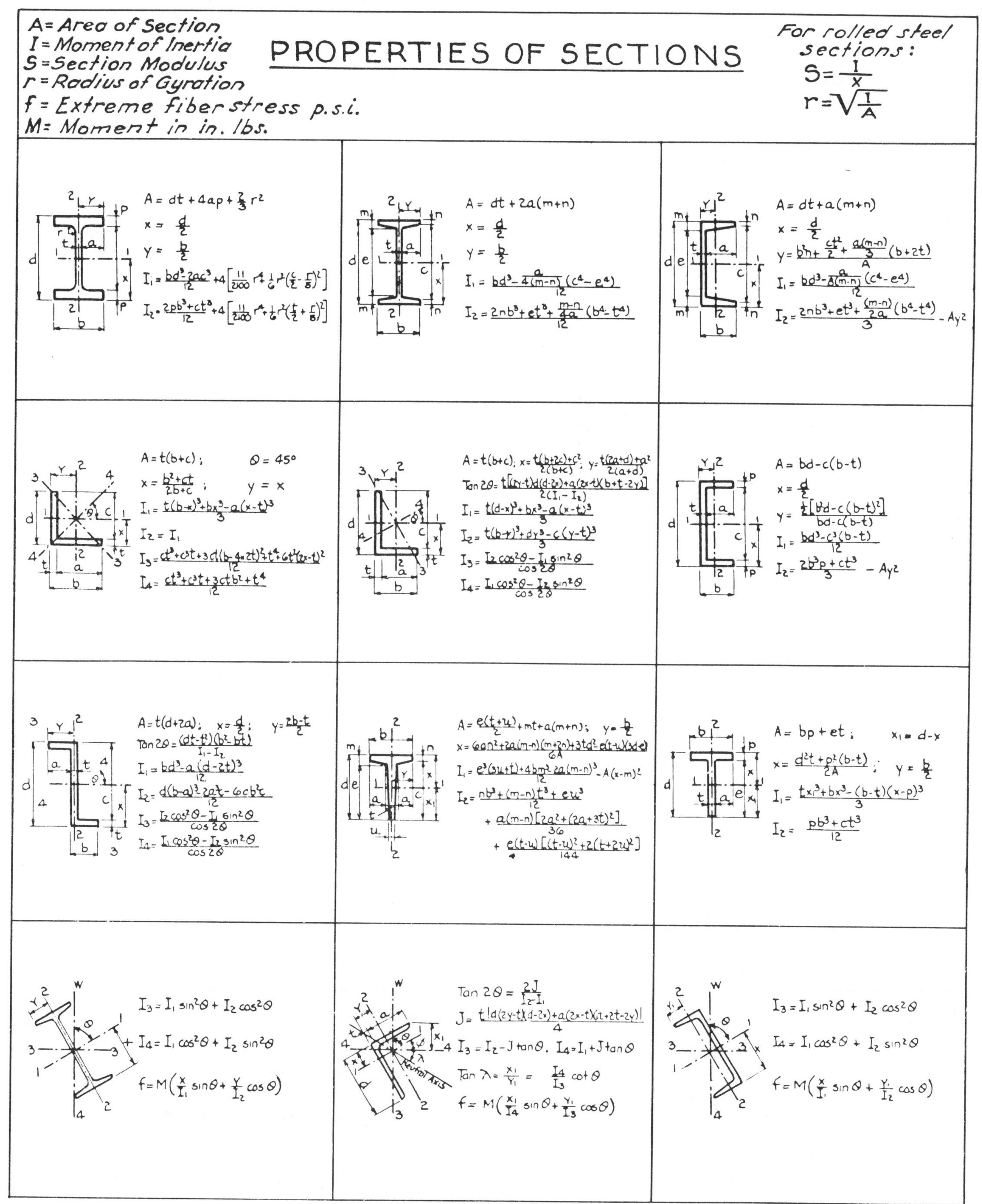
A= Area of Section
I= Moment of Inertia
S= Section Modulus
r= Radius of Gyration
f= Extreme fiber stress p.s.i.
M= Moment in in. lbs.
PROPERTIES OF SECTIONS
For rolled steel sections:
S= I/x
r= √(I/A)
A = dt + 4ap + 2/3 r²
x = d/2
y = b/2
A = dt + 2a(m+n)
x = d/2
y = b/2
I₂ = 2nb³ + et³ + (m-n)/4a (b⁴-t⁴) / 12
A = dt + a(m+n)
x = d/2
A = t(b+c);
θ = 45°
y = x
I₂ = I₁
A = bd - c(b-t)
x = d/2
I₁ = bd³ - c³(b-t) / 12
I₂ = 2b³p + ct³ / 3 - Ay²
A = t(d+2a);
x = d/2;
I₁ = bd³ - a(d-2t)³ / 12
I₃ = I₂ cos²θ - I₁ sin²θ / cos 2θ
I₄ = I₁ cos²θ - I₂ sin²θ / cos 2θ
A = bp + et;
x₁ = d-x
y = b/2
I₂ = pb³ + ct³ / 12
I₃ = I₁ sin²θ + I₂ cos²θ
I₄ = I₁ cos²θ + I₂ sin²θ
f = M(x/I₁ sinθ + y/I₂ cosθ)
Tan 2θ = 2J / I₂-I₁
I₃ = I₂ - J tanθ. I₄ = I₁ + J tanθ
Tan λ = x₁/y₁ = I₄/I₃ cot θ
f = M(x₁/I₄ sinθ + y₁/I₃ cosθ)
Neutral Axis

Adapted from Singleton, Manual of Structural Design, H. M. Ives & Sons.

STRUCTURAL-GENERAL-APPROXIMATE SOLUTION

TABLE A – CONTINUOUS BEAMS AND SLABS, A.C.I.. 1956

l = Clear span for positive moment and shear, and the average of the two adjacent clear spans for negative moment
w = Uniformly distributed load per unit of length of beam or per unit area of slab

Applicable if (1) Longer of 2 adjacent spans does not exceed shorter by more than 20%.
(2) Live load is not more than 3 times dead load.

Positive moment	
End spans	
If discontinuous end is unrestrained	$\frac{1}{11} wl'^2$
If discontinuous end is integral with the support	$\frac{1}{14} wl'^2$
Interior spans	$\frac{1}{16} wl'^2$
Negative moment at exterior face of first interior support	
Two spans	$\frac{1}{9} wl'^2$
More than two spans	$\frac{1}{10} wl'^2$
Negative moment at other faces of interior supports	$\frac{1}{11} wl'^2$
Negative moment at face of all supports for (a) slabs with spans not exceeding 10 ft., and (b) beams and girders where ratio of sum of column stiffnesses to beam stiffness exceeds 8 at each end of the span	$\frac{1}{12} wl'^2$
Negative moment at interior faces of exterior supports for members built integrally with their supports	
Where the support is a spandrel beam or girder	$\frac{1}{24} wl'^2$
Where the support is a column	$\frac{1}{16} wl'^2$
Shear in end members at first interior support	$1.15 \frac{wl'}{2}$
Shear at all other supports	$\frac{wl'}{2}$

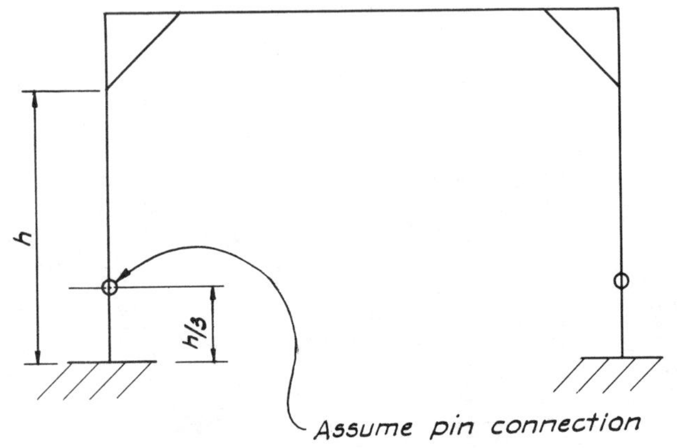
h
h/3
Assume pin connection

RIGID FRAMES – FIXED AT BASE

STEEL BEAMS

Approximate section modulus of steel WF = $\frac{d}{11} \times$

STEEL COLUMNS

Safe load in kips on WF column, normal story height 4 to 5 times weight per foot. Use 4 for light sections, 5 for heavy sections.

DEFLECTION

Steel Beams

Uniform load – see pp. 4-02 and 4-03.

Other loads – use formulas on pp. 8-01 through 8-04.

Concrete Beams

Use formulas on pp. 8-01 through 8-04.

E = 3,000,000 for short time loads to 1,000,000 for permanent loads

I = I of transformed section

CONCRETE BEAMS

(3000 lb. concrete; 12 in. width, d in ft.)

Allowable moment foot kips = $34 \times d^2$

Reaction – No stirrups = 10d
Reaction – Maximum = 40d

CONCRETE TIED COLUMNS

(3000 lb. concrete)

Maximum concentric load in kips = 150 × area in sq. ft.

STRUCTURAL - GENERAL - EARTHQUAKES

LATERAL BRACING OF STRUCTURES

PACIFIC COAST UNIFORM BUILDING CODE - 1952

SUPERSTRUCTURE:

Design structure as a whole and every portion of same to resist lateral forces applied at each floor or roof level above foundations. Horizontal Force, "F" = CW in lbs. Where: C = Force factor; factor from table A. W = total D.L. at and above point under consideration except for warehouses and tanks where: W = Total D.L. + Total L.L.

FOUNDATIONS:

All foundations on piles or soil with less than 2000#/□' bearing value shall have footings interconnected with ties in two directions at 90° to each other. Tie shall transmit 10% of vertical load of heavier footing connected. Concrete slab min. 6" or 1/48 span, and not more than 12" above top of footing may be used in place of ties. Min. reinf. 3/8"φ at 12" O.C. each way.

GENERAL DESIGN DATA:

Stresses shall not be more than 33 1/3% above allowable working stresses in code except that shear in concrete walls 6" or thicker shall not exceed .05 f'c. Over turning moment shall not exceed 2/3 of moment of stability both calculated using the same loads. Calculation of shears and moments shall be in accordance with the wind design.

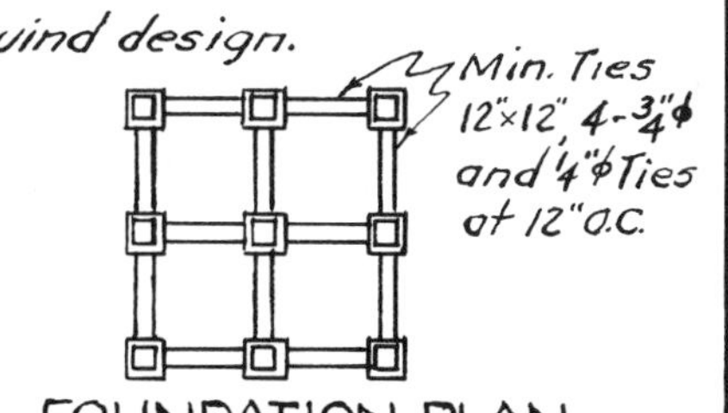

F
F1
F2
F3
Min. Ties 12"×12", 4-3/4"φ and 1/4"φ Ties at 12" O.C.

TYPICAL BENT — FOUNDATION PLAN

TABLE A - HORIZONTAL FORCE FACTORS

PART OR PORTION	VALUE OF "C"*	DIRECTION OF FORCE
Floors, roofs columns and bracing in any story of a bldg. or the structure as a whole **	$\frac{.15}{N+4.5}$	Any direction horizontally
Bearing walls, non-bearing walls, partitions, free standing masonry walls over 6' in height	.05 with min. of 5#/□'	Normal to surface of wall
Cantilever parapet and other cantilever walls, except retaining walls.	.25	Normal to surface of wall
Exterior and Interior ornamentations & appendages	.25	Any direction horizontally
When connected to or a part of a building: towers, tanks, towers and tanks plus contents, chimneys, smokestacks & penthouses.	.05	Any direction horizontally
Elevated water tanks and other tower supported structures not supported by a building	.03	Any direction horizontally

* See B for zones. The values given "C" are minimum and should be adopted in locations not subject to frequent seismic disturbances as shown in zone 1. For location in zone 2, "C" shall be doubled. For locations in zone 3, "C" shall be multiplied by 4.

** Where wind load would produce higher stresses this load shall be used in lieu of factor shown. N is number of stories above the story under consideration, provided that for floors or horizontal bracing, N shall be only the number of stories contributing loads.

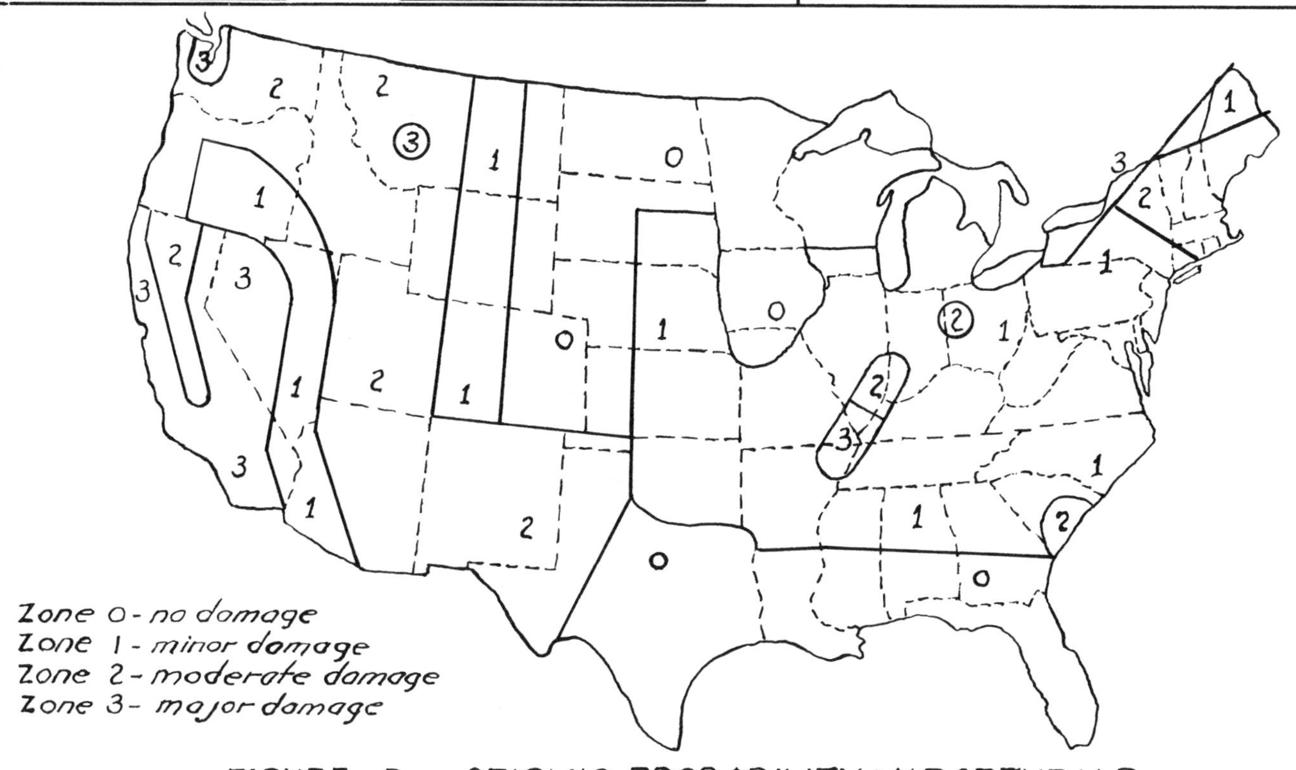

Zone 0 - no damage
Zone 1 - minor damage
Zone 2 - moderate damage
Zone 3 - major damage

FIGURE B - SEISMIC PROBABILITY MAP OF THE U.S.

STRUCTURAL-GENERAL-TORSION & SKIN STRESSES

TORSION

θ = *Angle of twist (radians)*
T = *Twisting moment (in. lb.)*
L = *Length (in.)*
S_s = *Unit Shear Stress #/□" due to torsion moment.*
G = *Modulus of rigidity #/□" = 11,000,000 approximately for structural steel and .45E (E = Modulus of Elasticity) for concrete.*
K *(in.^{4}) and Q (in.^{3}) are functions of the cross section.*
Safe working stress in torsional shear = 0.03 f'_c (concrete)
Safe working stress in torsional shear = 12,000 #/□" (steel)
r = *Fillet radius*
D = *Diameter largest inscribed circle.*

CROSS SECTION: FOR SHEAR	CROSS SECTION: FOR θ	MAXIMUM SHEARING UNIT - STRESS *	FORMULA FOR K IN $\theta = \frac{TL}{KG}$ †
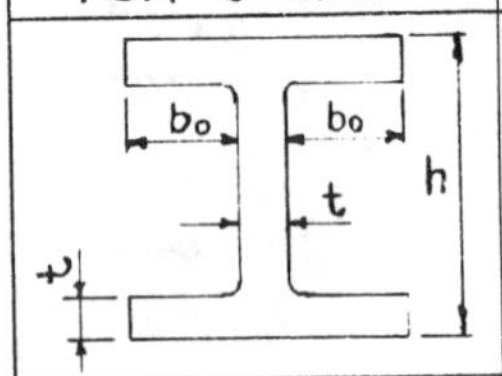	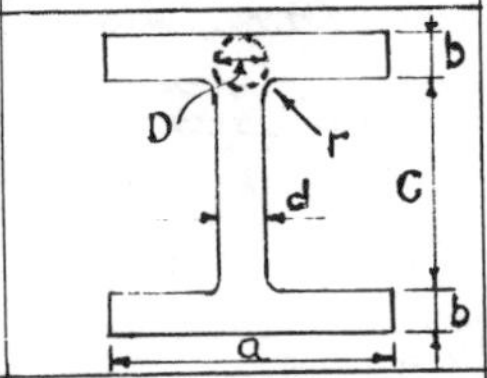	$S_s = \frac{9}{2}\ \frac{T}{t^2\ h + 4b_o}$	$K = 2K_1 + K_2 + 2\alpha D^4$, *where* $K_1 = ab^3\left[\frac{1}{3} - 0.21\frac{b}{a}\left(1 - \frac{b^4}{12a^4}\right)\right]$ $K_2 = \frac{1}{3}cd^3 \quad \alpha = \frac{t}{t_1}\left(0.15 + 0.1\frac{r}{b}\right)$, *where* t = (b) or (d) *whichever the thinner.* t_1 = (b) or (d) *whichever the thicker.*
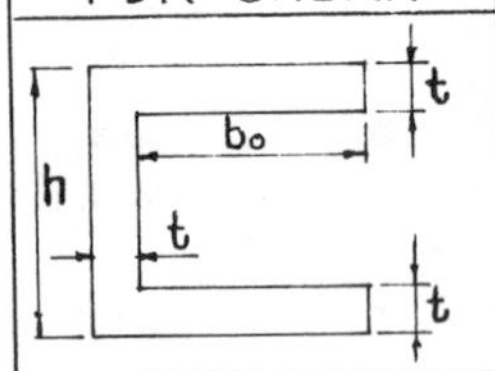	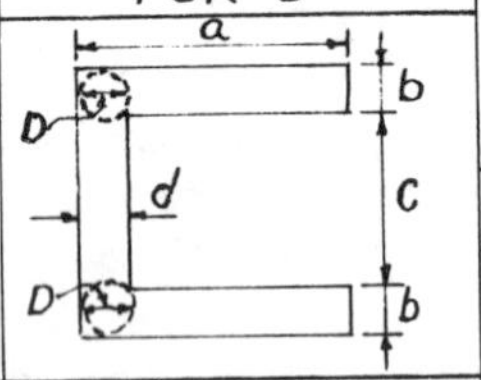	$S_s = \frac{9T}{2t^2h}$ $S_s = \frac{9}{2}\ \frac{T}{t^2(h + 2b_o)}$	$K = K_1 + K_2 + K_3 + \alpha D^4$ $K_1 = ab^3\left[\frac{1}{3} - 0.21\frac{b}{a}\left(1 - \frac{b^4}{12a^4}\right)\right]$ $K_3 = K_1$ $K_2 = cd^3\left[\frac{1}{3} - 0.105\frac{d}{c}\left(1 - \frac{d^4}{192c^4}\right)\right]$ $\alpha = \frac{b}{d}\left(0.07 + 0.076\frac{r}{b}\right)$
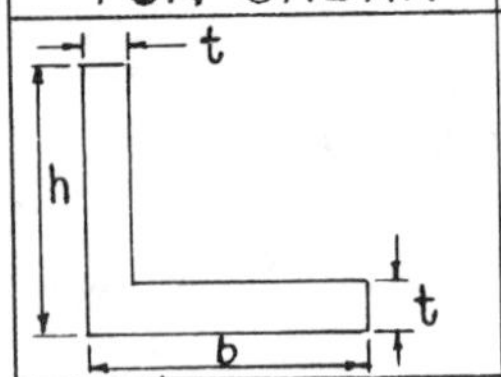	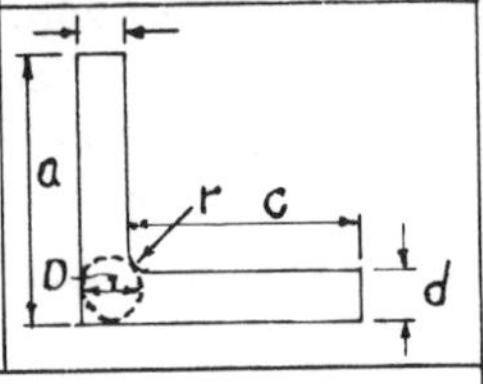	$S_s = \frac{9}{2}\ \frac{T}{t^2(h + b - t)}$	$K = K_1 + K_2 + \alpha D^4$ *where* $K_1 = ab^3\left[\frac{1}{3} - 0.21\frac{b}{a}\left(1 - \frac{b^4}{12a^4}\right)\right]$ $K_2 = cd^3\left[\frac{1}{3} - 0.105\frac{d}{c}\left(1 - \frac{d^4}{192c^4}\right)\right]$ $\alpha = \frac{b}{d}\left(0.07 + 0.076\frac{r}{b}\right)$
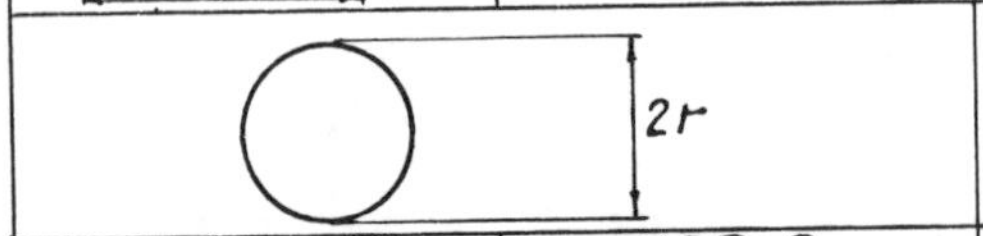		$S_s = \frac{2T}{\pi r^3}$	$K = \frac{1}{2}\pi r^4$
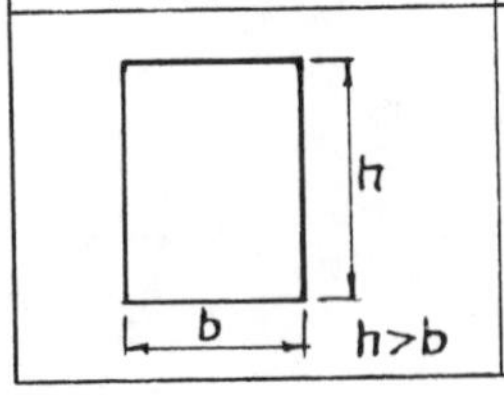	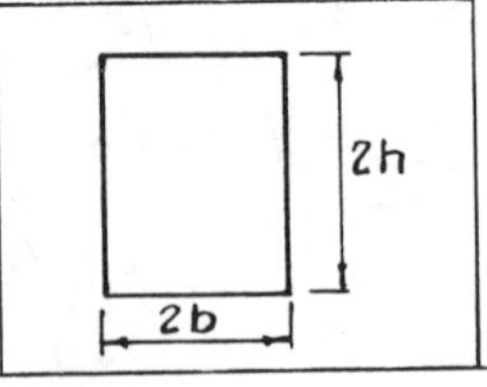	$S_s = \frac{9T}{2b^2h}$	$K = ab^3\left[\frac{16}{3} - 3.36\frac{b}{a}\left(1 - \frac{b^4}{12a^4}\right)\right]$

SKIN STRESSES †

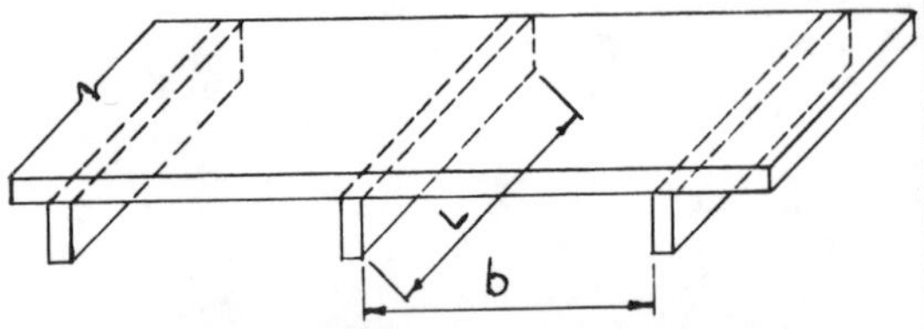
L
b

BEAMS WITH WIDE FLANGES
Effective flange width = C×L
Values of C in table opposite

TYPE OF BEAM	$\frac{b}{\ell} = 0.1$	$\frac{b}{\ell} = 0.2$	$\frac{b}{\ell} = 0.4$	$\frac{b}{\ell} = \infty$
Continuous, uniform load				
at support	0.066	0.085	0.105	0.120
at mid-span	0.090	0.150	0.190	0.220
Continuous, center loads				
at support	0.076	0.105	0.130	0.150
at mid-span	0.076	0.105	0.130	0.150
Simple beam, uniform load		0.175	0.290	0.360
Simple beam, center load		0.150	0.210	0.310

† *Adapted from "Formulas for Stress and Strain" by R. J. Roark - McGraw-Hill 1943.*
✱ *Adapted from "Advanced Mechanics of Materials" by F. B. Seely - John Wiley 1947.*

STRUCTURAL-GENERAL-TRUSSES

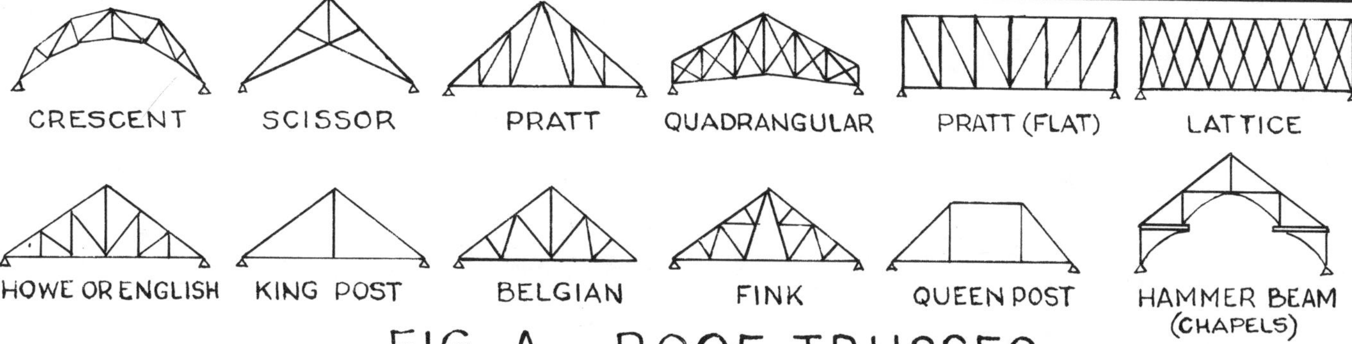
CRESCENT
SCISSOR
PRATT
QUADRANGULAR
PRATT (FLAT)
LATTICE
HOWE OR ENGLISH
KING POST
BELGIAN
FINK
QUEEN POST
HAMMER BEAM (CHAPELS)

FIG. A ROOF TRUSSES

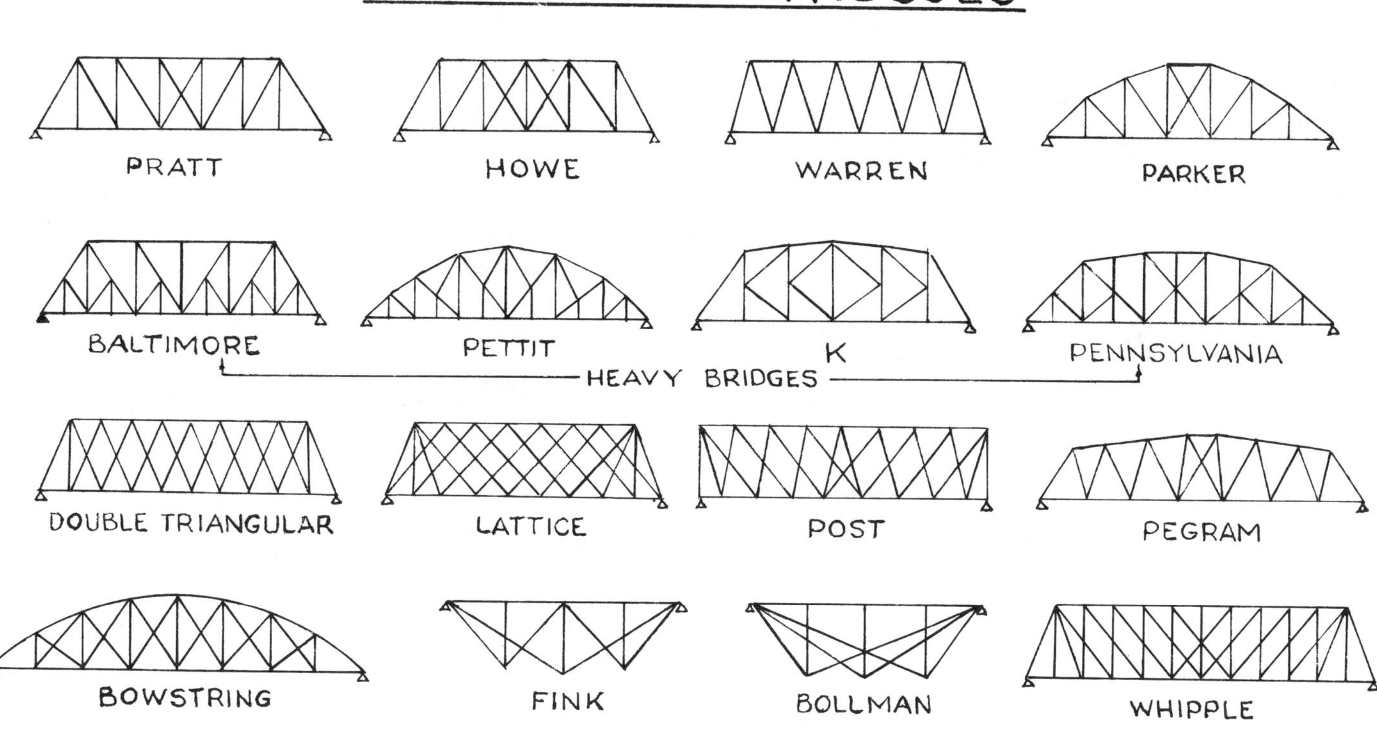
PRATT
HOWE
WARREN
PARKER
BALTIMORE
PETTIT
K
PENNSYLVANIA
HEAVY BRIDGES
DOUBLE TRIANGULAR
LATTICE
POST
PEGRAM
BOWSTRING
FINK
BOLLMAN
WHIPPLE

FIG. B BRIDGE TRUSSES

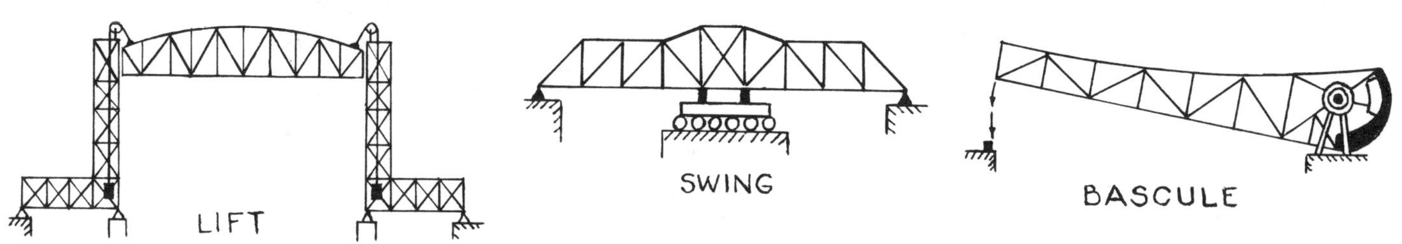
LIFT
SWING
BASCULE

FIG. C MOVABLE BRIDGES

TABLE D - TRUSS WEIGHTS *

ROOFS	$W = \frac{3pL}{450 + 15L + ps}$	W = Wt. in lbs/sq.ft. of horiz. covered surface; L = Truss span-ft. S = Truss spacing - ft.; and p = load, in lbs./sq.ft. of horiz. covered area, carried by truss. (40 lbs/sq.ft. min.)
R.R. BRIDGES (Single Track)	$W = 2(2L + 5E)$	W = Wt. in lbs./ft. of span of one truss, ½ of lateral bracing, & ½ of the floor system.; L = span length - ft.; E = the number of the class of Cooper's loading governing the design.
HIGHWAY BRIDGES	$W = \frac{w_1}{9} + L$	W = wt. per ft. of one truss & ½ of the bracing; w_1 = total superimposed load per ft. brought to each truss: - wt. of floor plus live load + impact; L = span in ft.

*Adopted from Civil Engineering Handbook by L. C. Urquhart - McGraw-Hill.

STRUCTURAL-GENERAL-ERRORS & REMEDIES

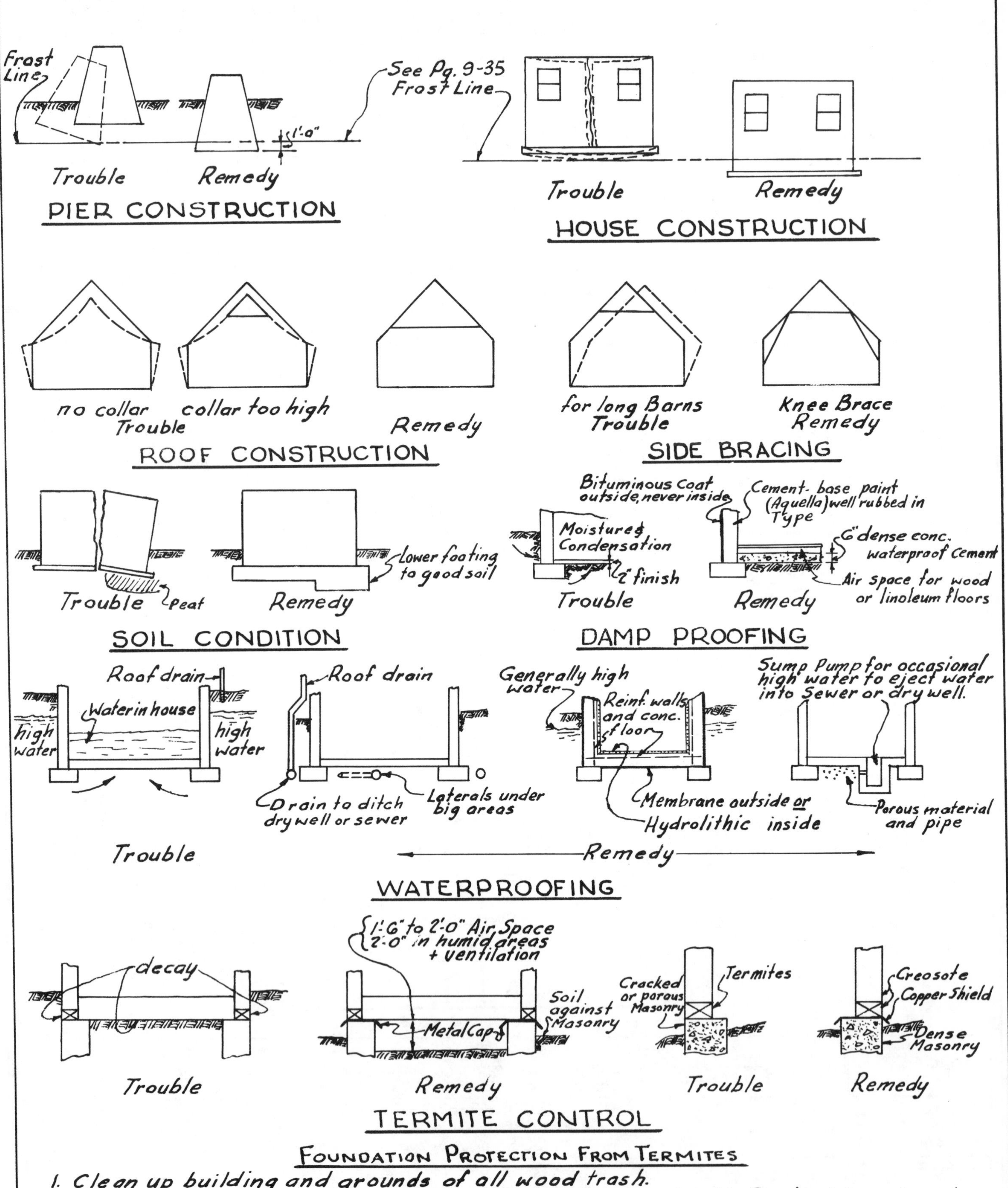
Frost Line
See Pg. 9-35 Frost Line
1'-0"
Trouble
Remedy
PIER CONSTRUCTION
Trouble
Remedy
HOUSE CONSTRUCTION
no collar
collar too high
Trouble
Remedy
ROOF CONSTRUCTION
for long Barns
Trouble
Knee Brace
Remedy
SIDE BRACING
Lower footing to good soil
Trouble
Peat
Remedy
SOIL CONDITION
Bituminous Coat outside, never inside
Moisture & Condensation
2" finish
Cement-base paint (Aquella) well rubbed in Type
6" dense conc. waterproof Cement
Air space for wood or linoleum floors
Trouble
Remedy
DAMP PROOFING
Roof drain
Water in house
high water
high water
Roof drain
Drain to ditch dry well or sewer
Laterals under big areas
Generally high water
Reinf. walls and conc. floor
Membrane outside or Hydrolithic inside
Sump Pump for occasional high water to eject water into Sewer or dry well.
Porous material and pipe
Trouble
Remedy
WATERPROOFING
decay
1'-6" to 2'-0" Air Space
2'-0" in humid areas + ventilation
Metal Cap
Soil against Masonry
Cracked or porous Masonry
Termites
Creosote
Copper Shield
Dense Masonry
Trouble
Remedy
Trouble
Remedy
TERMITE CONTROL

FOUNDATION PROTECTION FROM TERMITES

1. Clean up building and grounds of all wood trash.
2. Break all wood to earth contacts and treat ground with Pentachlorophenol.

STRUCTURAL - GENERAL - BRICK MASONRY

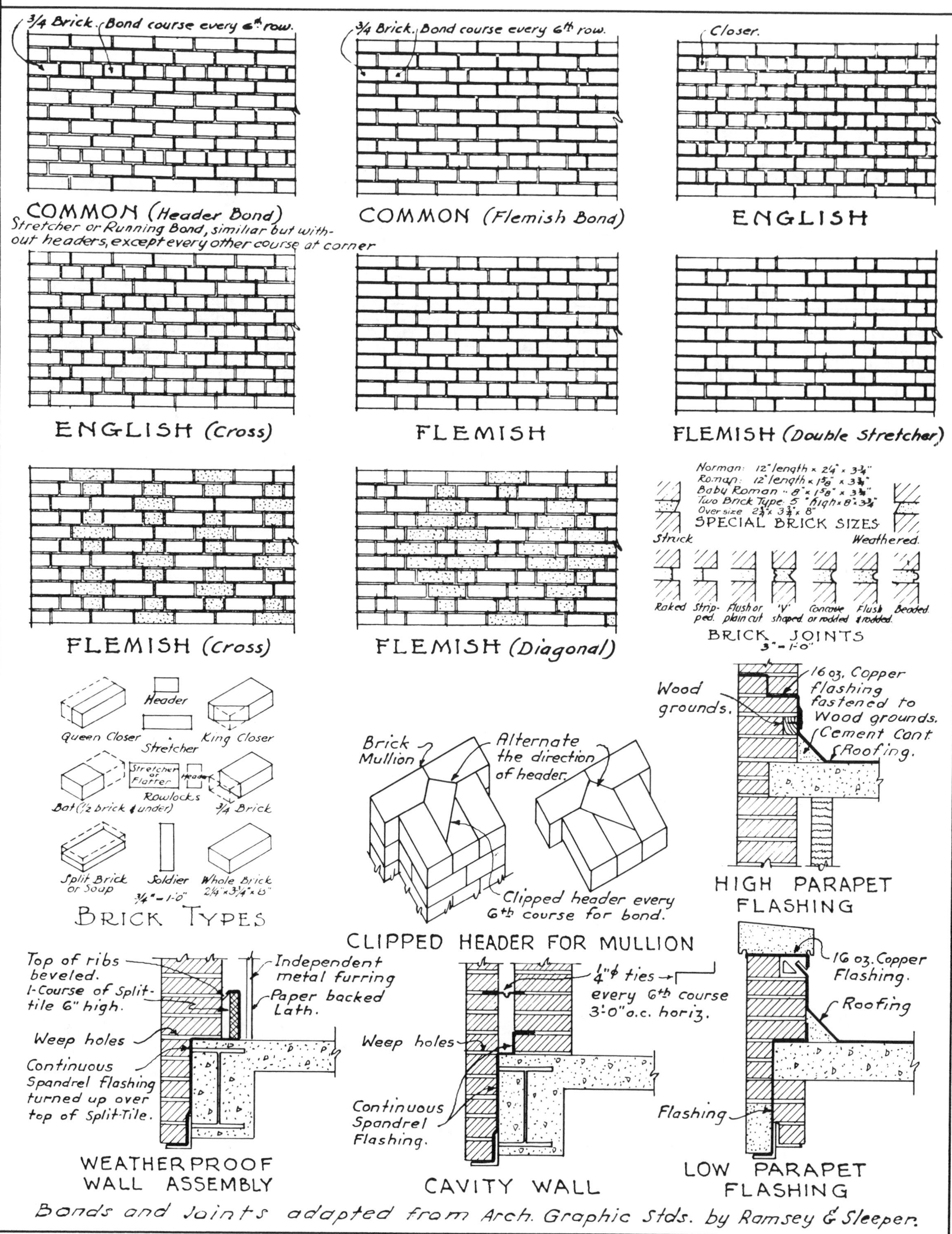
3/4 Brick.
Bond course every 6th row.
COMMON (Header Bond)
Stretcher or Running Bond, similiar but without headers, except every other course at corner
3/4 Brick.
Bond course every 6th row.
COMMON (Flemish Bond)
Closer.
ENGLISH
ENGLISH (Cross)
FLEMISH
FLEMISH (Double Stretcher)
FLEMISH (Cross)
FLEMISH (Diagonal)
Norman: 12" length × 2¼" × 3¾"
Roman: 12" length × 1⅝" × 3¾"
Baby Roman " 8" × 1⅝" × 3¾"
Two Brick Type: 5" high × 8" × 3¾"
Oversize 2¾" × 3¾" × 8"
SPECIAL BRICK SIZES
Struck
Weathered.
Raked
Stripped.
Flush or plain cut
'V' shaped.
Concave or rodded
Flush & rodded.
Beaded.
BRICK JOINTS
3" = 1'-0"
Header
Queen Closer
Stretcher
King Closer
Stretcher or Flatter
Header
Rowlocks
Bat (½ brick & under)
3/4 Brick
Split Brick or Soap
Soldier
Whole Brick 2¼" × 3¾" × 8"
3/4" = 1'-0"
BRICK TYPES
Brick Mullion
Alternate the direction of header.
Clipped header every 6th course for bond.
CLIPPED HEADER FOR MULLION
Wood grounds.
16 oz. Copper flashing fastened to Wood grounds.
Cement Cant.
Roofing.
HIGH PARAPET FLASHING
Top of ribs beveled.
1-Course of Split-tile 6" high.
Independent metal furring
Paper backed Lath.
Weep holes
Continuous Spandrel flashing turned up over top of Split-Tile.
WEATHERPROOF WALL ASSEMBLY
¼"φ ties
every 6th course
3'-0" o.c. horiz.
Weep holes
Continuous Spandrel Flashing.
CAVITY WALL
16 oz. Copper Flashing.
Roofing
Flashing
LOW PARAPET FLASHING
Bonds and Joints adapted from Arch. Graphic Stds. by Ramsey & Sleeper.

STRUCTURAL-GENERAL-STONE MASONRY

-STONEWORK-

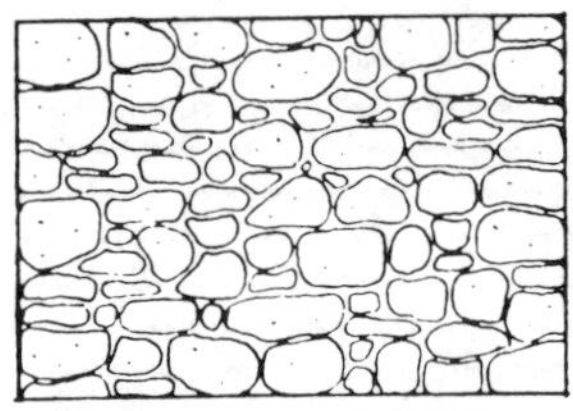

UNCOURSED FIELDSTONE ROUGH OR ORDINARY.

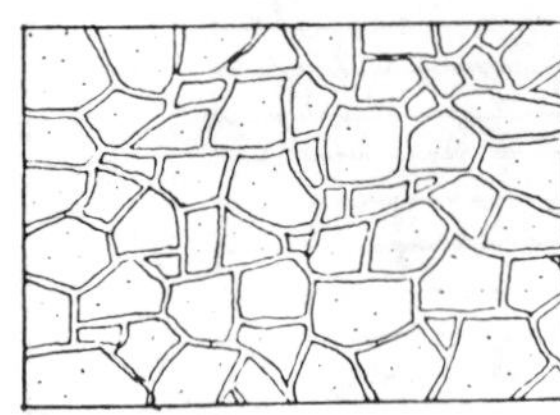

POLYGONAL, MOSAIC OR RANDOM.

COURSED

TYPES OF RUBBLE MASONRY

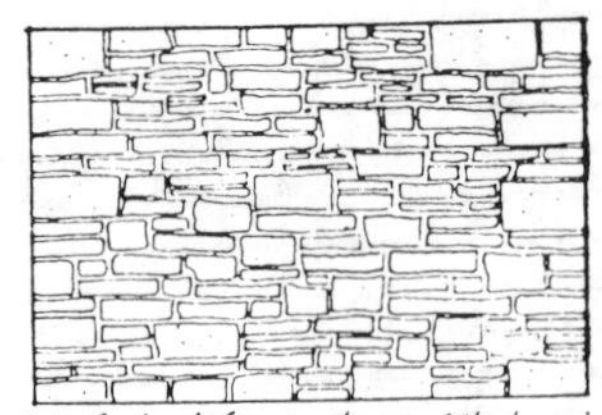

Laid of stratified stone fitted on job. It is between rubble & ashlar. Finish is quarry face, seam face or split. Called rubble ashlar in granite.

SQUARED-STONE MASONRY.

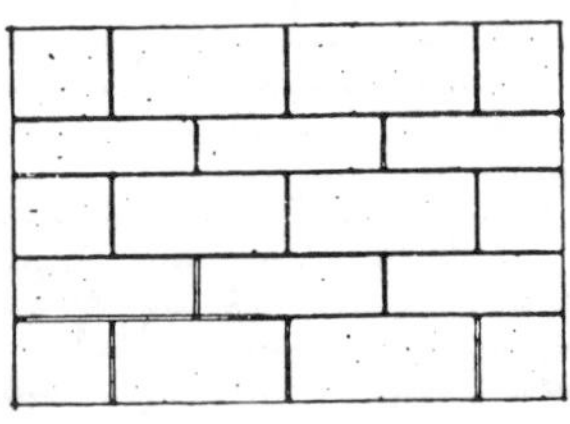

RANGE.
Coursed

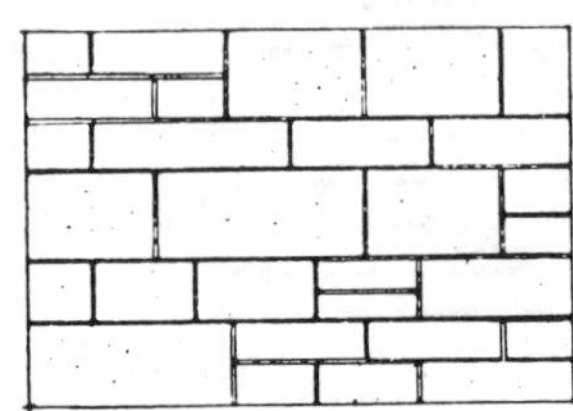

BROKEN RANGE.

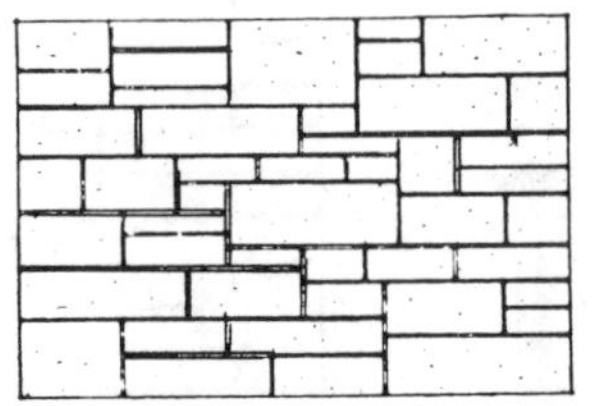

RANDOM
Interrupted coursed

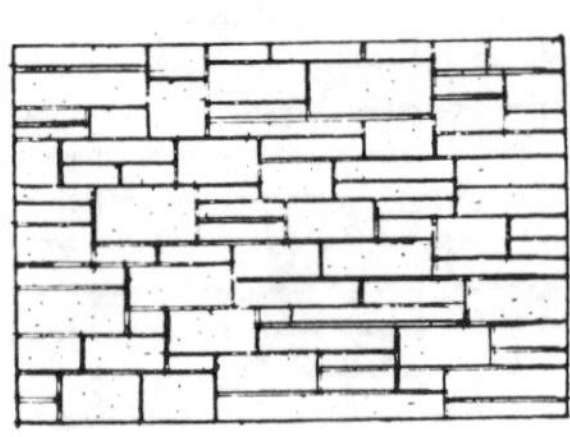

RANGE.
Coursed (Long stones)

TYPES OF ASHLAR MASONRY
This is stone that is sawed, dressed, squared or Quarry faced.

ELEVATIONS SHOWING FACE JOINTING FOR STONE.

Draft line

For both hard and soft stones.
Rock or Pitch Face.

Smooth but saw mark visible. All stones.
Sawed Finish (Gang).

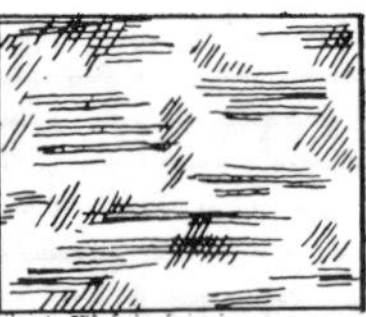

More marked than sawed. Soft stones.
Shot Sawed (Rough).

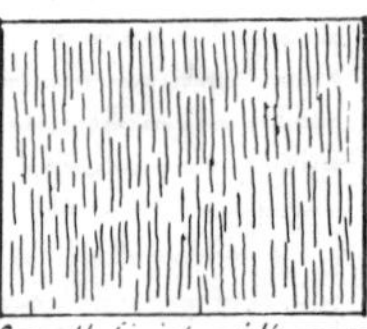

Smooth finish with some texture. Soft stones.
Machine Finish (Planer).

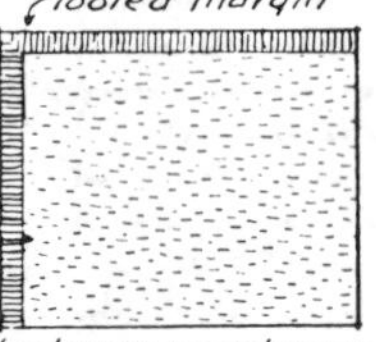
Tooled margin

May be coarse, medium or fine. Usually on hard stones.
Pointed Finish.

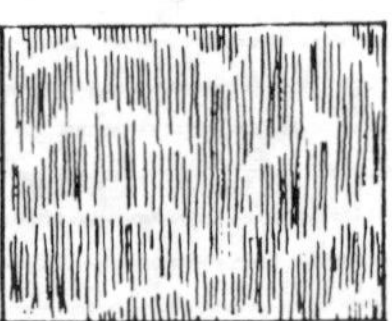

After pointing on hard stones.
Pean Hammered.

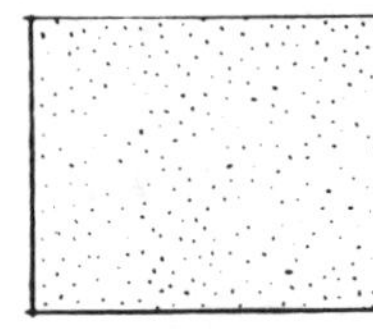

For soft stones
Bush-hammered.

All stones. Used much on granite 4 to 8 cut in 7/8".
Patent Bush-hammer.

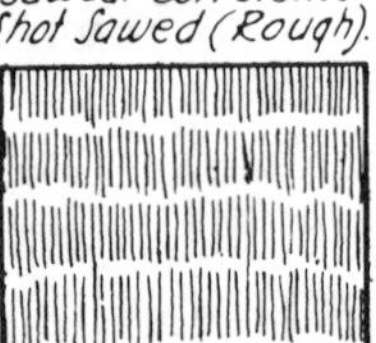

For soft stones.
Drove or Boasted.

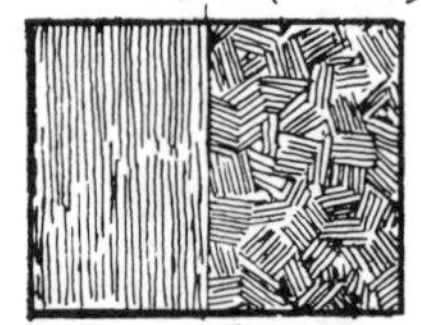

Random
For soft stones.
Hand Tooled.

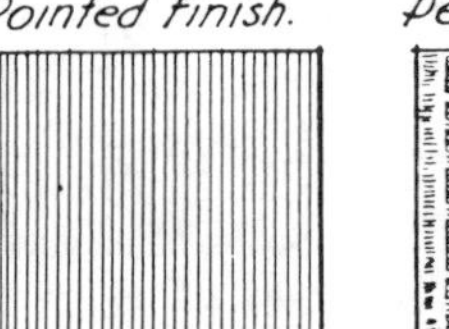

Tool marks may be 2 to 10 per inch.
Machine Tooled.

For soft stones.
Tooth-chisel.

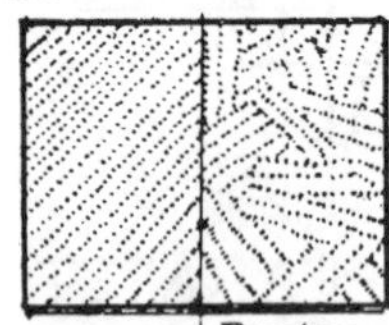

Random
For soft stones.
Crandalled.

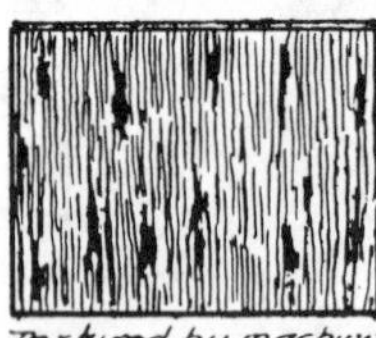

Textured by machine
For Limestone
Plucker Finish.

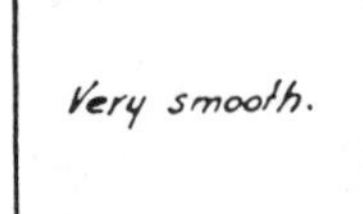
Very smooth.

For Limestone
Done by machine
Carborundum Finish.

Smooth

All stones. May use sand or carborundum.
Rubbed (Wet).

Very Smooth

Marble, granite. For interior work. Soft stones.
Honed (rubbed first).

Very smooth
Has high gloss.

Marble and Granite.
Polished (honed first).

STONE FINISHES.
Seam face and split face (or Quarry face) not shown as they are not worked finishes.

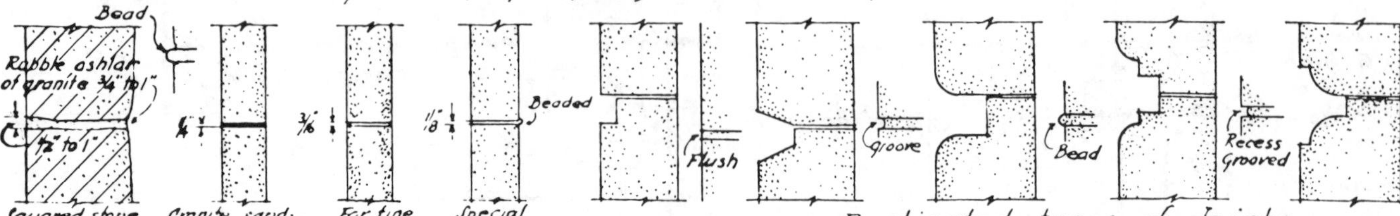
Bead
Rubble ashlar of granite 3/4" to 1"
1/2" to 1"
1/4"
3/16"
1/8"
Beaded
Flush
Groove
Bead
Recess Grooved

Squared stone masonry

Granite, sandstone & limestone ashlar general use.

For fine work Limestone

Special interiors

Rusticated types of Joints.

STONE JOINTS

TYPES, FINISH AND JOINTING OF STONE MASONRY.

A perch is nominally 16'6" long, 1'0" high & 1'6" thick = 24 3/4 cu.ft. In some localities 16 1/2 & 22 cu.ft. are used.

Adapted from Arch. Graphic Stds. by Ramsey & Sleeper.

STRUCTURAL-GENERAL-NOTES

CONCRETE

DESIGN DRAWINGS

1. Concrete shall develop a strength of --p.s.i. at 28 days. Reinforcement shall be intermediate-grade deformed bars conforming to A.S.T.M. A-15 and A-305.
2. In two-way slabs, place short-span bars in bottom layer.
3. Concrete beams to have 8-in. bearing on walls unless otherwise noted.
4. Concrete joists to have not less than 6-in. bearing on walls.
5. Solid slabs to have 4-in. bearing on walls.
6. Fireproofing thickness shall be 3/4-in. for slabs and joists, 1-1/2 in. for beams, girders, and columns.
7. See specifications for instructions to detailers.
8. Add notes as may be required by local codes.

WORKING DRAWINGS (include inspecifications.)

9. Provide adequate ties for all steel bars and stirrups in slabs and beams. All reinforcing steel in slabs and beams to be held at correct distance from forms by adequate concrete blocks, steel chairs, or ties in accordance with latest A.C.I. detailing practices.
10. Follow A.C.I. rules as to stirrups, column ties, and anchorage.
11. Where concrete beams frame into steel, provide two 3/4-in. round anchor bolts at top of concrete beam with double nuts through web or hooks over flanges except as otherwise noted.
12. Provide--header bars in bottom of slab in front of all openings and cases in bearing walls unless otherwise shown.
13. Metal tile slabs to be anchored to walls with 5/8-in. round anchors 4 ft. long with 6-in. hook at wall end, placed in top of alternate joists.
14. Metal tile joists to be 25-in. on center except as shown. Steel shown in slab schedules is per joist.
15. Topping of metal tile joist slabs to be 2 in. thicker than typical each side of concrete beams parallel to joists.
16. Where joists run parallel to wall, provide 1/2-in. round anchors 3'-6" long with 6-in. hooks at right angles to joists 4 ft. on center in topping.
17. Working drawings to show T flanges on beams of sufficient width and thickness to develop full strength of steel.
18. Provide no. 2 stirrups 12 in. on center for beams and joists with top steel for sections where no stirrups are called for.
19. Provide two no. 3 tie rods in top of all concrete beams to fasten stirrups where no top steel occurs.
20. Topping of metal tile joist slabs to be reinforced with 66-66 steel mesh.
21. Provide longitudinal or temperature reinforcement in solid slabs in accordance with A.C.I. code. (See p. 2-04.)
22. Provide 100% continuity over supports for all continuous slabs, beams, and joists unless otherwise noted.
23. In all solid slabs, bend up alternate bars, and extend into adjacent spans where continuous, unless otherwise noted on plans.
24. Where main slab reinforcement is parallel to beam, place no. 3 rods, spacing equal or less than 18 in. or 5 times the thickness of the slab by 6 ft. long, except where otherwise shown over beam.
25. All negative-moment rods to extend to quarter points of spans unless otherwise shown.
26. (a) At ends of non-continuous beams, provide anchor bars in top of beam. Anchors to be equivalent to one third of main reinforcement, but not less than two no. 6 4 ft. long, unless noted otherwise. (See p. 2-02.)

    (b) At ends of non-continuous slabs, provide anchor bars in top of slab. Anchors to be equivalent to one-half main reinforcement but not less than no. 3 bars spaced 12 in. on center 3'-0" long. (See p. 2-03.)

STEEL

DESIGN DRAWINGS

1. All lintels shall have 6-in. minimum bearing each side unless otherwise noted.
2. Steel joists to have not less than 4-in. bearing on walls.
3. Where open-web steel joists span more than 14 ft., provide safety headers of approved type.
4. All rivets to be --diameter unless otherwise noted. (Include on shop drawings.)
5. Provide 8" × 5/8" × 0'-8" bearing plates and government anchors for all steel beams resting on masonry, except as otherwise shown.
6. Double steel beams and girders to be provided with pipe separators.
7. Provide soffit clips for bottom flanges of all steel beams to be fireproofed, and mesh for all steel columns to be fireproofed.
8. Fireproofing thickness shall be as follows: Beams--, columns--.
9. Steel beams carrying chimneys or other masonry to be fireproofed with stone concrete.

WOOD

DESIGN DRAWINGS

1. All wood shall be (give species and stress grade).
2. Provide 1-1/4" × 2" cross-bridging not over 6 ft. on center for all wood joists.
3. Where wood joists frame into or rest on top of steel beams, provide 1-1/2" × 3/16" steel strap anchors spiked to wood joists and hooked over flange of beam at every third joist on each side of beam.
4. Where wood joists rest on masonry, provide metal anchors at maximum intervals of 4 ft. on center having a minimum cross section 1/4" × 1-1/4" by minimum length of 16 in. securely fastened to joist and built into masonry with split or upset ends. Where joists are parallel to walls, provide similar anchors at maximum spacing of 6 ft., engaging 3 joists.
5. Bolts of ring connectors shall be retightened periodically during the process of seasoning.
6. Provide standard beam hangers for the beams framed into girders except as otherwise shown.
7. Joists to be double under partitions parallel to joists except as otherwise shown.

MASONRY

(Include in Specification)

1. Mortar to be (Engineer to fill in).
2. All masonry work to be bonded as shown or as approved by engineer.
3. Brick pilasters in tile backup walls to be bonded into adjoining masonry.

FOUNDATIONS

DESIGN DRAWINGS

1. Foundations designed for--tons per sq. ft. (check soil in field.) Client or Engineer will check the assumption.
2. Concrete shall develop a strength of --p.s.i. at 28 days.
3. No backfilling against foundation walls to be done until after superstructure is in place.

*4. Where reinforced mat slabs are erected and ground-water pressure is likely to occur, relief holes should be left to relieve water pressure until approved by Engineer.

5. Extreme high water level has been assumed at elevation--. This assumption must be checked in the field. Client or Engineer will check the assumption.
6. See specifications for instructions to detailers.

WORKING DRAWINGS (Include in Specifications).

7. Wall footings to be stepped where elevation changes 1 vertical to 2 horizontal, except where otherwise shown. Maximum vertical slip not to exceed 2'-0".
8. Provide 6 no. 4 rods continuous in top and 2 no. 6 continuous in bottom of all exterior walls, except as otherwise shown. Lap 40 diameters, and bend around corners.
9. Provide 2 no. 4 rods continuous in top and bottom of all interior concrete walls, except as otherwise shown.
10. All wall, pier, and stack footings shall be 12 in. thick and project 6 in. beyond all faces of walls, piers, and stacks, except as otherwise noted.
11. Provide 2 no. 6 rods, all sides, for all openings in concrete walls unless otherwise noted. Extend 2 ft. beyond openings or hook ends.
12. Provide pockets in walls for all concrete beams and slabs at first floor.
13. Construction joints in foundation walls, interior and exterior, shall be placed not more than 30 ft. apart and shall be V-chamfered unless otherwise shown. Location of joints shall be as shown on the drawings or approved by the Engineer. Sections of walls shall be poured alternately.

*This note to be placed on working drawings.

STRUCTURAL CONCRETE – MIXES

APPROXIMATE DATA ON CONCRETE MIXES

TABLE A – WATER CEMENT RATIO (W/C) FOR VARIOUS STRENGTHS

Water Content, gallons per sack of Cement	Compressive Strength of Concrete at 28 days, p.s.i.
5 max.	5000
6 max.	4000
7 max.	3200
8 max.	2500

NOTE: – W/C ratio should be determined by trial mixes when practicable. To allow for field conditions, the values shown in table should be reduced by about 20%.

TABLE B – RECOMMEND CONSISTENCY OR SLUMP OF CONCRETE

Type of Structure	Slump in inches (approx.)
Reinforced foundation walls and footings	4
Plain footings and substructure walls	4
Slabs, beams, columns, and reinforced walls	4
Pavement and mass concrete	4

TABLE C – EXPOSED CONCRETE – MAXIMUM WATER CONTENT IN GALLONS PER SACK

Type or Location of Concrete	Mild Climate
At water line (intermittent saturation) – Sea Water	5½
At water line (intermittent saturation) – Fresh water	6
Not at water line but frequent wetting – sea water	6½
Not at water line but frequent wetting – fresh water	7
Ordinary exposed structures	7
Completely submerged – sea water	6½
Completely submerged – fresh water	7
Concrete deposited through water	5½
Pavement slabs on ground wearing slab	6
Pavement slabs on ground base slab	7

TABLE D – RECOMMENDED PER CENT OF SAND TO TOTAL AGGREGATE BY WEIGHT

Crushed stone	Max. 1½ in. size	38 – 42%
Crushed stone	Max. ¾ in. size	43 – 49%
Gravel	Max. 1½ in. size	36 – 40%
Gravel	Max. ¾ in. size	39 – 44%

Sand-aggregate ratio, or percentage by weight or volume of sand to total aggregate in mix, should be from 35% to 45%, with extreme limits of 28% and 49%. The most economical mix will be that with the lowest-aggregate ratio which produces the desired plasticity, workability, and consistency.

TABLE E – PHYSICAL PROPERTIES OF LIGHTWEIGHT CONCRETE

Type of Aggregate	Use	28-day Compressive Strength, p.s.i.	Oven-Dry Weight, p.c.f.
Expanded clay, slate, shale: e.g., Lelite, Haydite; Galite, Idealite	Structural and some insulation	1000 to 5500, some to 7000	70 - 118
Expanded slag e.g. Waylite	Structural and insulation	750 to 4000	75 - 110
Expanded perlite (Volcanic glass)	Mostly insulation, some structural	125 to 1500	21 - 50
Vermiculite (expanded mica)	Mostly insulation, little structural	100 to 700	23 - 50
Pumice (volcanic glass)	Insulation, some structural	1000 to 2000	60 - 85
Scoria (volcanic lava)	Insulation, some structural	200 to 2500	75 - 95
Grit crete	Short-span arch and fill	1000 min.	90 - 110 p.s.i.
Aerocrete	Insulation and fill	1000	70 - 80 p.s.i.

OBSERVATION

Structural lightweight must be designed and investigated on an individual basis. There are many lightweight aggregates in existence necessitating from 0 to 100% additional cement for a required strength. This is also true of lightweight aggregates that are similar in appearance and are produced by similar processes.

STRUCT. CONCR.-TYP. BEAM DETAILS & ANCHORAGE REQUIREMENTS

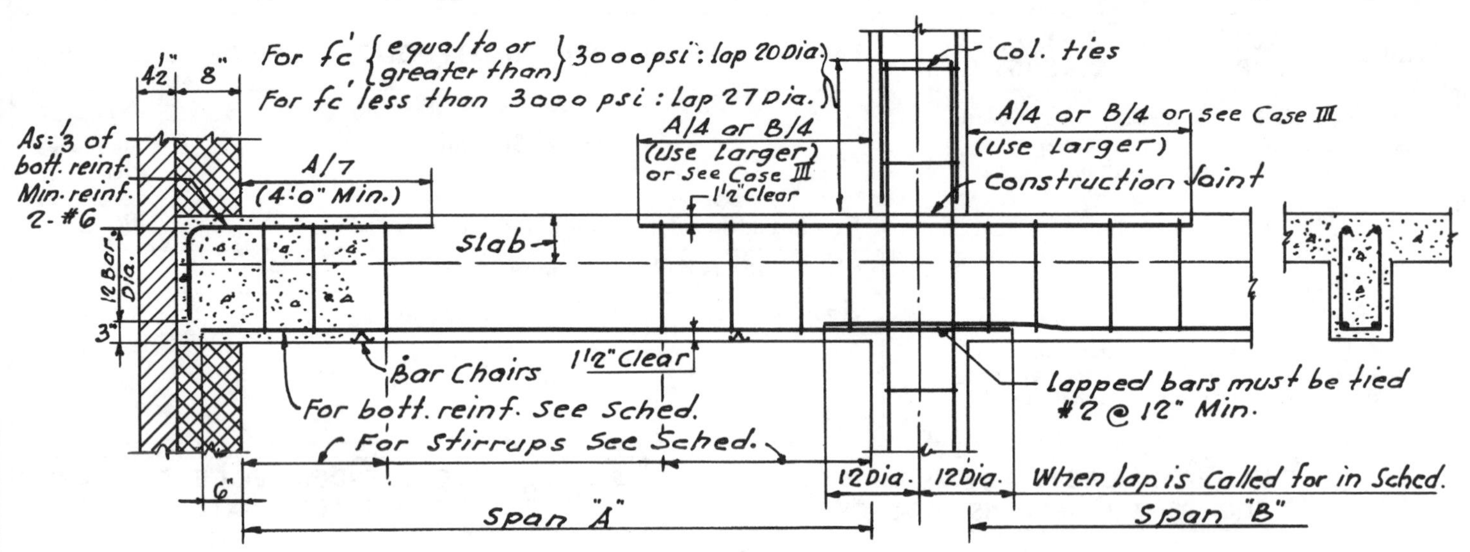

For fc' {equal to or greater than} 3000 psi : lap 20 Dia.
For fc' less than 3000 psi : lap 27 Dia.
4½" 8"
As: ⅓ of bott. reinf. Min. reinf. 2-#6
A/7 (4'0" Min.)
Col. ties
A/4 or B/4 (Use Larger) or See Case III
1½" Clear
A/4 or B/4 or see Case III (Use Larger)
Construction Joint
Slab
12 Bar Dia.
3"
Bar Chairs
1½" Clear
For bott. reinf. See Sched.
For Stirrups See Sched.
Lapped bars must be tied #2 @ 12" Min.
6"
12 Dia. 12 Dia.
When lap is Called for in Sched.
Span "A"
Span "B"

TYPICAL BEAM AND COLUMN DETAIL

FIG. A

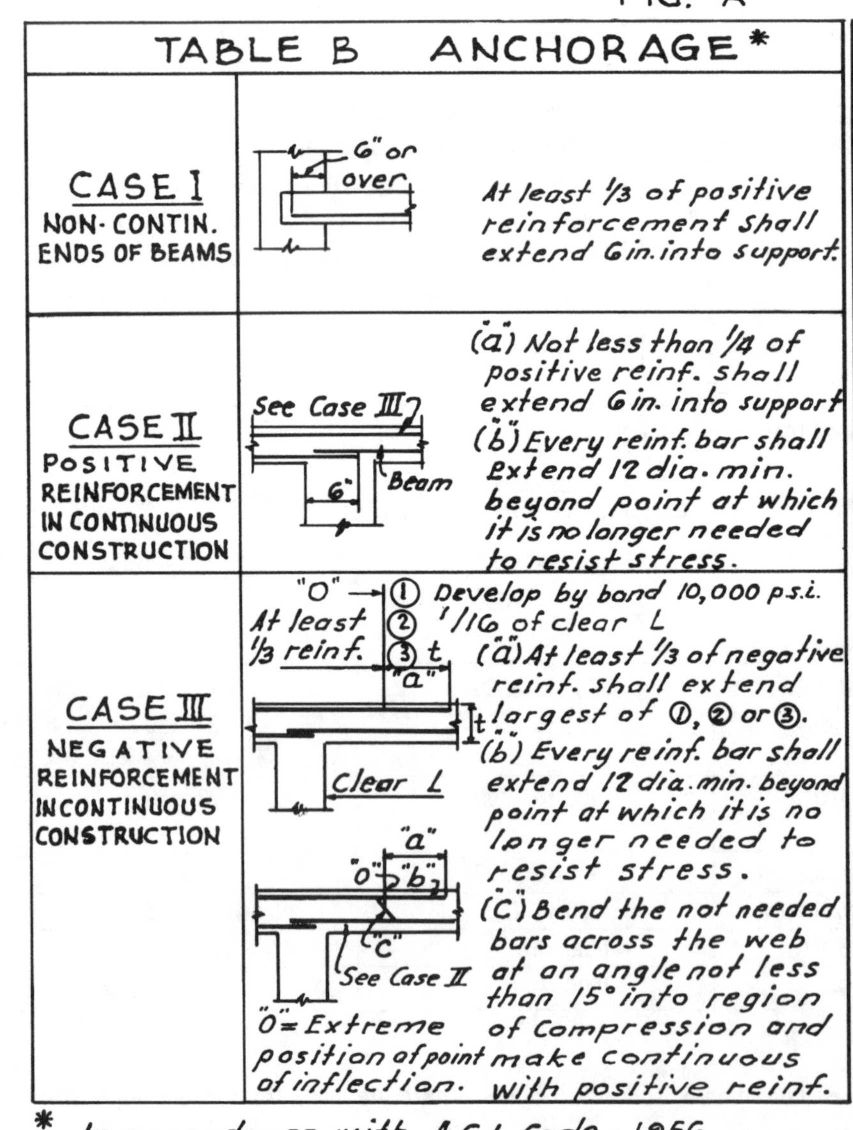

TABLE B ANCHORAGE*

CASE I NON-CONTIN. ENDS OF BEAMS	6" or over	At least ⅓ of positive reinforcement shall extend 6 in. into support.
CASE II POSITIVE REINFORCEMENT IN CONTINUOUS CONSTRUCTION	See Case III; 6"; Beam	("a") Not less than ¼ of positive reinf. shall extend 6 in. into support ("b") Every reinf. bar shall extend 12 dia. min. beyond point at which it is no longer needed to resist stress.
CASE III NEGATIVE REINFORCEMENT IN CONTINUOUS CONSTRUCTION	"O" → ① Develop by bond 10,000 p.s.i. At least ⅓ reinf. ② 1/16 of clear L ③ t "a"; t; Clear L "a"; "O"; "b"; "c"; See Case II "O" = Extreme position of point of inflection.	("a") At least ⅓ of negative reinf. shall extend largest of ①, ② or ③. ("b") Every reinf. bar shall extend 12 dia. min. beyond point at which it is no longer needed to resist stress. ("C") Bend the not needed bars across the web at an angle not less than 15° into region of compression and make continuous with positive reinf.

* In accordance with A.C.I. Code - 1956
For bars conforming to A-305 deformations.

PIPES EMBEDDED IN CONCRETE

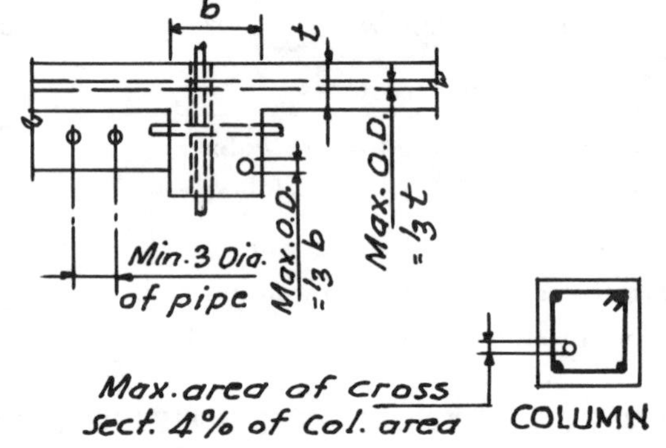

b
t
Min. 3 Dia. of pipe
Max. O.D. = ⅓ b
Max. O.D. = ⅓ t
Max. area of cross Sect. 4% of Col. area
COLUMN

NOTE:
Pipes Containing Liquid, Gas or Vapor may be embedded in Structural Concrete if tested prior to Concreting at 1½ times design pressure or minimum pressure of 150 p.s.i. for 4 hours. Maximum pressure to be 200 p.s.i.; Max. temperature 150° F.
No liquid, Gas or Vapor to be placed in the pipes until the Concrete has thoroughly set.
Piping Shall be placed between top & bottom steel.
Concrete covering of pipes to be min. 1".
Reinforcement with a min. area of 0.2 percent of concrete section Area to be provided normal to piping.
No liquid, Gas or Vapor which may be injurious or detrimental to pipes Shall be placed in them.
Uncoated pipes or sleeves of 2 in or less in diameter and not less in thickness than standard steel or wrought iron pipe may be considered to replace the displaced concrete.

STRUCTURAL CONCRETE – TYPICAL SLAB DETAILS

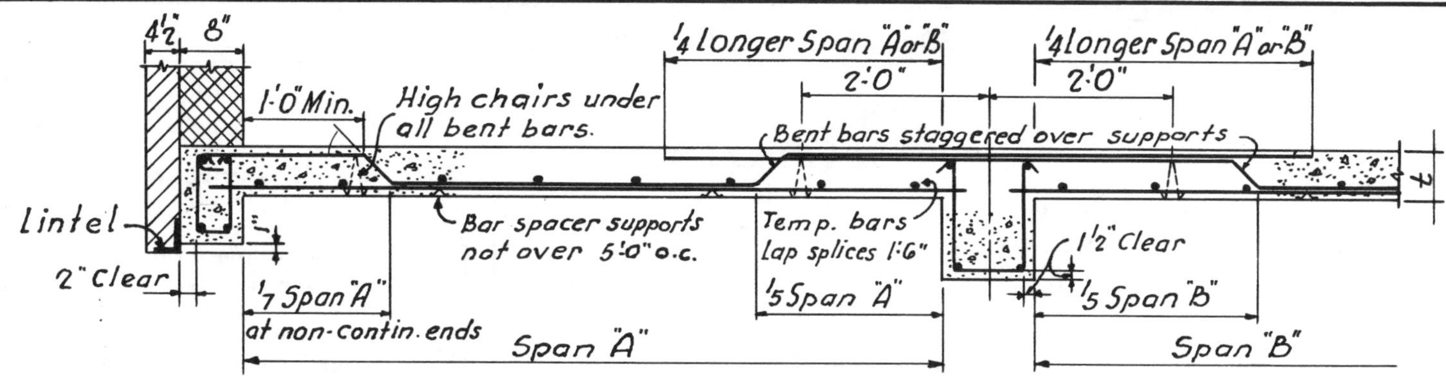

4½" 8"
1'-0" Min.
High chairs under all bent bars.
¼ Longer Span "A" or "B"
2'-0"
¼ Longer Span "A" or "B"
2'-0"
Bent bars staggered over supports
Lintel
2" Clear
Bar spacer supports not over 5'-0" o.c.
Temp. bars lap splices 1'-6"
1½" Clear
⅐ Span "A" at non-contin. ends
⅕ Span "A"
⅕ Span "B"
Span "A"
Span "B"
4"

NOTE: Main reinforcement: alternate bars to be bent up over support and staggered with bars from adjacent span.
At non-continuous ends of slabs, bent bars shall extend 3'-0" beyond ℄ of support.
Temperature bars shall project 4" into walls and beams at end conditions.

Provide #3 @ 12" x 6'-0" in top of slabs over all supports where main reinforcement runs Parallel to supports on both sides; and #3 @ 12" x 3'-0" in top of slabs over supports which are adjacent to openings where slabs frame parallel to supports.

FIG. A

DETAIL OF TYPICAL SLAB USING BENT BARS

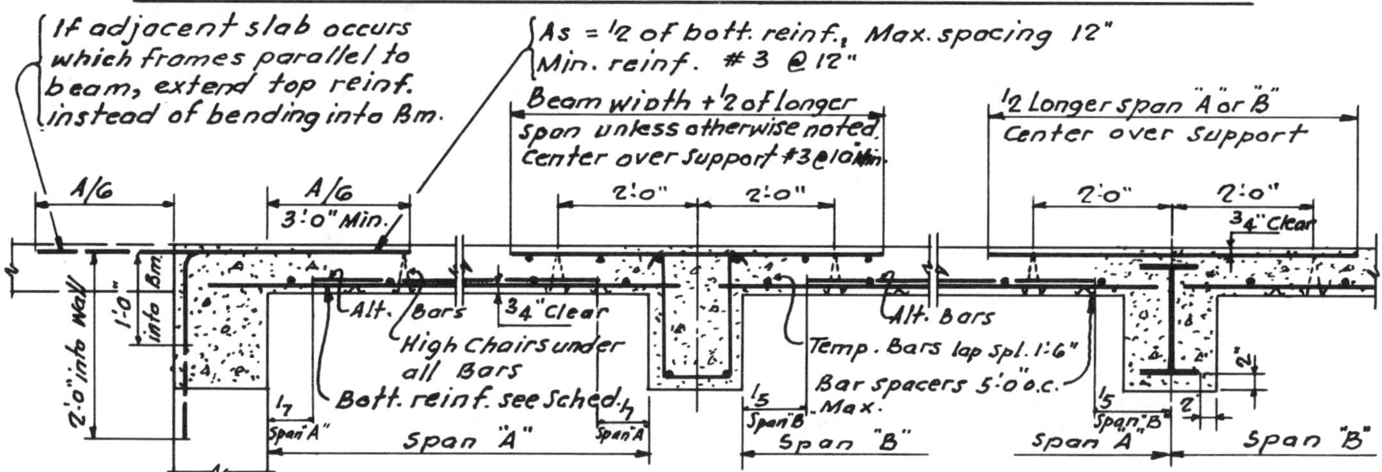

If adjacent slab occurs which frames parallel to beam, extend top reinf. instead of bending into Bm.
As = ½ of bott. reinf., Max. spacing 12"
Min. reinf. #3 @ 12"
Beam width + ½ of longer span unless otherwise noted. Center over Support #3 @ 10" Min.
½ Longer span "A" or "B" Center over Support
A/6
A/6
3'-0" Min.
2'-0"
2'-0"
2'-0"
2'-0"
¾" Clear
2'-0" into Wall
1'-0" into Bm.
Alt. Bars
¾" Clear
High Chairs under all Bars
Bott. reinf. see Sched.
Alt. Bars
Temp. Bars lap Spl. 1'-6"
Bar spacers 5'-0" o.c. Max.
⅐ Span "A"
⅐ Span "A"
Span "A"
⅕ Span "B"
Span "B"
⅕ Span "B"
Span "A"
Span "B"
2"
2"

CASE I – DIRECTION OF SLAB FRAMING PERPENDICULAR TO SUPPORT

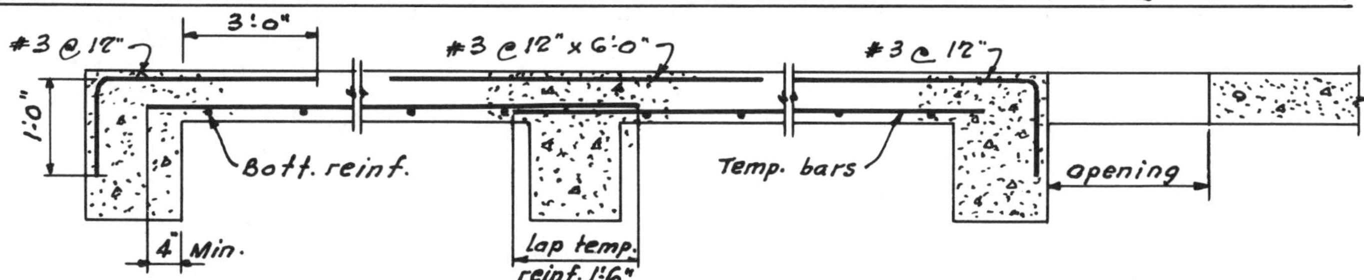

#3 @ 12"
3'-0"
#3 @ 12" x 6'-0"
#3 @ 12"
1'-0"
Bott. reinf.
Temp. bars
opening
4" Min.
lap temp. reinf. 1'-6"

CASE II – DIRECTION OF SLAB FRAMING PARALLEL TO SUPPORT

SLAB THICKNESS		4"	4½"	5"	5½"	6"	6½"	7"	7½"	8"
TEMP. REINF.	PLAIN BARS	#3 @ 11"	#4 @ 17"	#4 @ 16"	#4 @ 14"	#4 @ 13"	#4 @ 12"	#4 @ 11½"	#4 @ 10½"	#4 @ 10"
	DEFORMED BARS	#3 @ 14"	#3 @ 12"	#3 @ 11"	#3 @ 10"	#4 @ 16"	#4 @ 15"	#4 @ 14"	#4 @ 13"	#4 @ 12"

TYPICAL CONCRETE SLAB DETAILS USING STRAIGHT BARS

FIG. B

STRUCT. CONCR.–MISCELLANEOUS REQUIREMENTS OF ACI CODE

MISCELLANEOUS REQUIREMENTS OF A.C.I. CODE, 1956

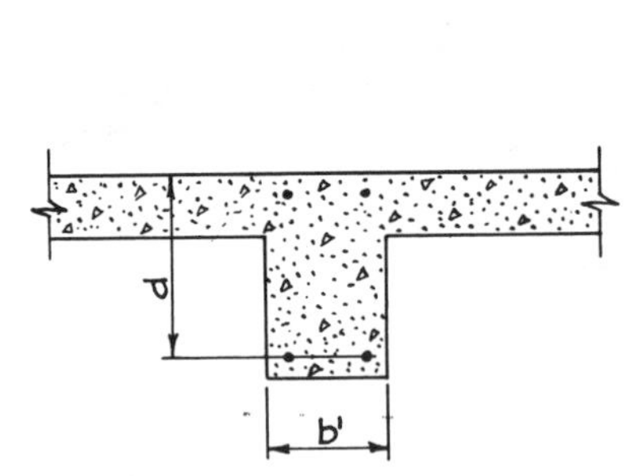
d
b'

Minimum positive reinforcement > 0.005 db'.
Minimum negative reinforcement > 0.005 db' at outer end of members built integrally with supports.

BEAMS AND GIRDERS

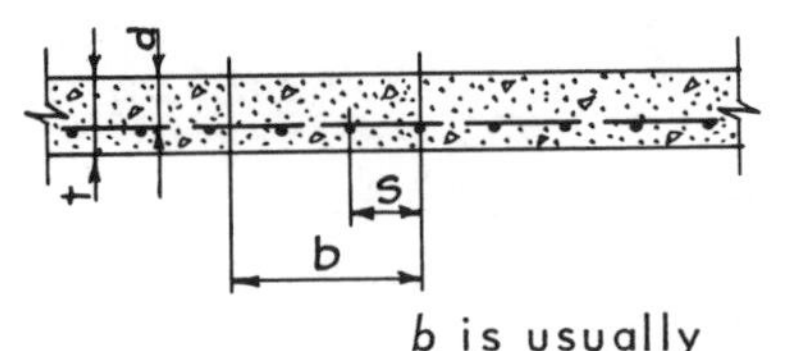
d
t
s
b

b is usually assumed 12" wide.

Main reinforcement:
Minimum area = 0.0025 db.
Maximum spacing of bars (s) = 3*t*.

Temperature and shrinkage reinforcement.
Minimum area = 0.002 bt for floors and roofs for deformed bars.
Maximum spacing (*s*) = 5*t*, but not over 18".

ONE-WAY SLAB, UNIFORM THICKNESS

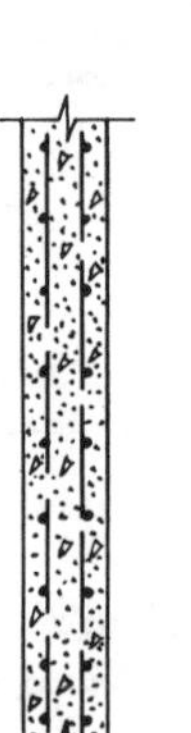
t

Minimum area horizontal reinforcement shall be 0.0025 times cross-sectional area of wall. Vertical reinforcement 0.0015 times cross-sectional area of wall and not less than three fourths as much if of welded wire fabric.
Minimum size of reinforcement: Deformed bars: #3-18" on center. Wire mesh: #10 AS & W gage. At openings, 2 #5 minimum diagonal reinforcement at corners or 24" minimum beyond corners.
Walls over 10" thick to have two layers of reinforcement: 1/2 to 2/3 of total reinforcement in exterior face, balance in interior face.

WALLS*

FIG. A. MINIMUM REINFORCEMENT

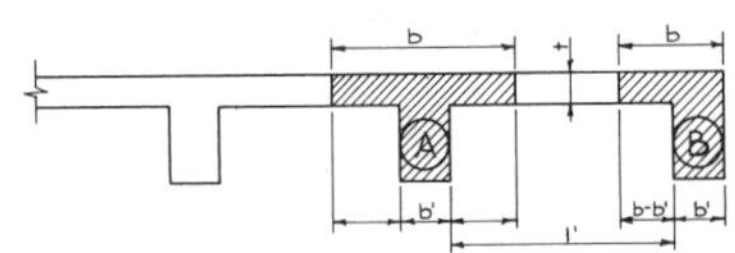
b
t
b
A
B
b'
b-b'
b'
l'

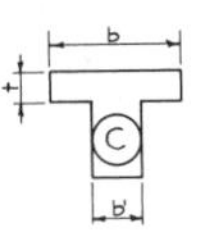
b
t
C
b'

(1) SYMMETRICAL T BEAM
b < 1/4 span length of beam.
$\frac{b - b'}{2} < 8t$ or $1/2\ l'$.

(2) ONE-SIDED T BEAM
b − b' < 1/12 span length of beam.
b − b' < 6*t* or 1/2 *l*'.

(3) ISOLATED T BEAM
t > 1/2 b'
b < 4 b'

FIG. B. T BEAMS

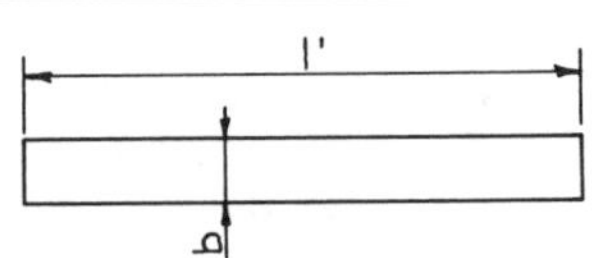
l'
b

PLAN

l' < 32b
l' = clear span.
b = least width of compression flange.

FIG. C. LATERAL SUPPORT OF BEAMS

K – STIFFNESS FACTORS FOR ANALYSIS OF CONTINUOUS FRAMES

For columns: $K = \frac{EI}{h}$; for slabs, beams, girders: $K = \frac{EI}{l}$.

Where E = modulus of elasticity, I = moment of inertia,
h = unsupported length of column, *l* = length of span.

I for a rectangular section $= \frac{bd^3}{12}$ in.^{4}

I for a square section $= \frac{d^4}{12}$ in.^{4}

I for a circular section $= \frac{\pi r^4}{4}$ in.^{4}

For I of a T section see below.

Example: I of a T section.
Given: b = 60", b' = 12", d = 24", and *t* = 8".
Required: To find moment of inertia.
Solution: $\frac{b}{b'} = \frac{60}{12} = 5$. Enter chart at $\frac{b}{b'} = 5$, go vertically to the curve, read c = 1.9.

$$I = 1.9 \times \frac{12 \times 24^3}{12} = 26{,}265 \text{ in.}^4$$

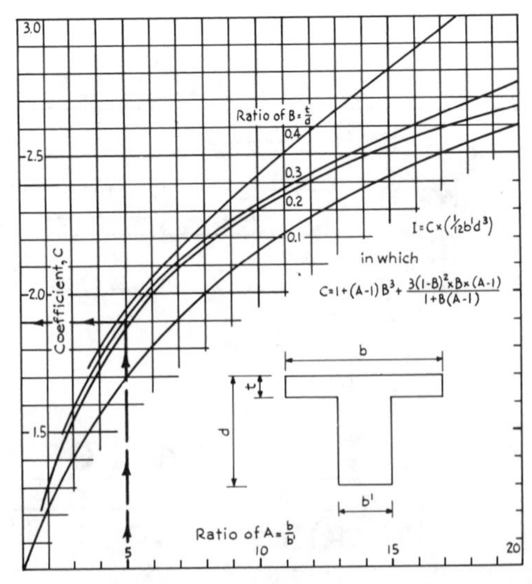
3.0
2.5
2.0
1.5
Coefficient, C
Ratio of B = t/d
0.4
0.3
0.2
0.1
I = C × (1/12 b'd³)
in which
C = 1 + (A−1)B³ + 3(1−B)²B×(A−1) / 1+B(A−1)
b
t
d
b'
Ratio of A = b/b'
5
10
15
20

FIG. D. EFFECT OF FLANGE ON MOMENT OF INERTIA OF T BEAMS.

REQUIREMENTS FOR PIPES: No deduction in strength need be made for iron or steel pipes or sleeves under 2" in diameter and spaced 3 diameters on centers. Embedded pipes shall not be larger than one-third the thickness of slab, beam, or wall.

*One layer of reinforcement at least 3" from exposed face recommended by author.

STRUCTURAL CONCRETE – FLAT SLAB – I

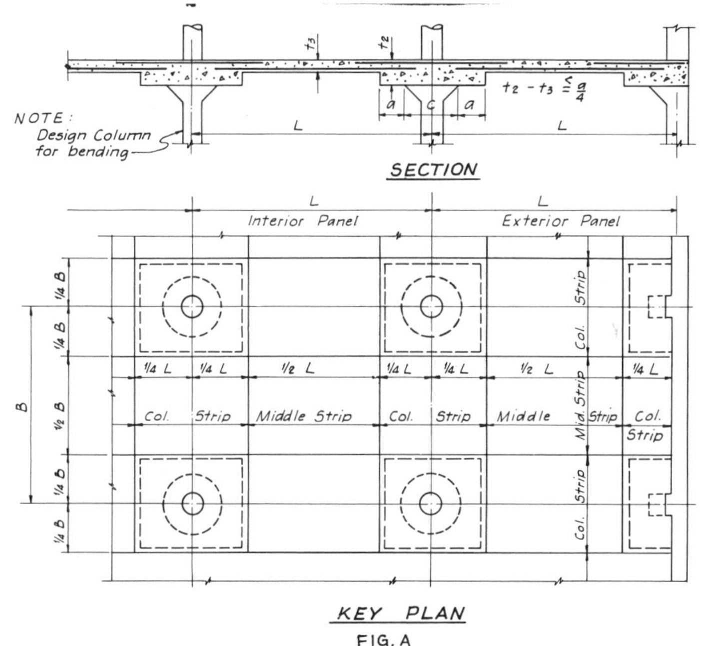

t_3
t_2
$t_2 - t_3 \leq \frac{a}{4}$
a
c
a
L
L
NOTE: Design Column for bending
SECTION
L
L
Interior Panel
Exterior Panel
¼ B
¼ B
½ B
¼ B
¼ B
B
¼ L
¼ L
½ L
¼ L
¼ L
½ L
¼ L
Col. Strip
Middle Strip
Col. Strip
Middle
Mid Strip
Strip
Col. Strip
Col. Strip
Col. Strip
KEY PLAN

FIG. A

$$M_o = 0.09WLF\left(1 - \frac{2C}{3L}\right)^2$$

$$M_o = 0.09WBF\left(1 - \frac{2C}{3B}\right)^2$$

$$F = 1.15 - \frac{C}{L} \text{ but not less than } 1$$

$$F = 1.15 - \frac{C}{B} \text{ but not less than } 1$$

TABLE B – MOMENTS IN FLAT SLAB PANELS IN PERCENTAGES OF M_o†

Strip	Column Head	Side Support Type	End Support Type	Exterior Panel: Exterior Negative Moment	Exterior Panel: Positive Moment	Exterior Panel: Interior Negative Moment	Interior Panel: Positive Moment	Interior Panel: Negative Moment
Column Strip	With Drop		A	44	24	56	20	50
			B	36				
			C	6	36	72		
	Without Drop		A	40	28	50	22	46
			B	32				
			C	6	40	66		
Middle Strip	With Drop		A	10	20	17*	15	15*
			B	20				
			C	6	26	22*		
	Without Drop		A	10	20	18*	16	16*
			B	20				
			C	6	28	24*		
Half Column Strip Adjacent to Marginal Beam or Wall	With Drop	1	A	22	12	28	10	25
			B	18				
			C	3	18	36		
		2	A	17	9	21	8	19
			B	14				
			C	3	14	27		
		3	A	11	6	14	5	13
			B	9				
			C	3	9	18		
	Without Drop	1	A	20	14	25	11	23
			B	16				
			C	3	20	33		
		2	A	15	11	19	9	18
			B	12				
			C	3	15	25		
		3	A	10	7	13	6	12
			B	8				
			C	3	10	17		

* See p. 2 – 06.

† A.C.I. Building Code (318-56).

STRUCTURAL CONCRETE — FLAT SLAB —2

TABLE A†

Percentage of Panel Load to Be Carried by Marginal Beam or Wall in Addition to Loads Directly Superimposed Thereon	Type of Support as Listed on p. 2-05		
	Side Support Parallel to Strip	Side or End Edge Condition of Slabs at Depth t	End Support at Right Angles to Strip
0	1	Columns with no beams	A
20	2	Columns with beams of total Depth $1\frac{1}{4}\ t$	
40	3	Columns with beams of total depth 3 t or more	B
		Reinforced-concrete bearing walls integral with slab	
		Masonry or other walls providing negligible restraint	C

*Increase negative moments 30% of tabulated values when middle strip is continuous across support of type B or C. No other values need be increased.

†A.C.I. Building Code (318-56).

NOTE: For intermediate proportions of total beam depth to slab thickness, values for loads and moments may be obtained by interpolation,

LIMITATIONS FOR USE OF TABLE

1. L/B = Not more than 1.33.
   Slab continuous over 3 or more panels in each direction.
3. The successive span lengths in each direction differ by not more than 20% of the longer span. Columns may be offset a maximum of 10% of the span in the direction of the offset from either axis between center line of successive columns.

COMPRESSION DUE TO BENDING

¾ of the width of the strip or drop panel shall be taken as the width of the section in computing compression due to bending. (For positive and negative moments, tension reinforcement to be distributed over entire strip.) Account shall be taken of any recesses which reduce the compressive area.

THICKNESS OF SLABS

1. L/36 Without drop panel, but not less than 5" nor t_1.
   L/40 With drop panel, but not less than 4" nor t_2.

*2. $t_1 = 0.028L\left(1-\frac{2c}{3L}\right)\frac{W'}{\sqrt{f'_c/2000}} + 1\frac{1}{2}"$

*3. $t_2 = 0.024L\left(1-\frac{2c}{3L}\right)\frac{W'}{\sqrt{f'_c/2000}} + 1"$

4. Where the exterior supports provide only negligible restraint to the slab, the values of t_1 and t_2 for the exterior panel shall be increased by at least 15%.
5. The maximum total thickness at the drop panel used in computing negative steel area shall be $1.5t_2$. The side or diameter of the drop panel shall be at least 0.33 times the span in the parallel direction.
6. The minimum thickness of slabs where drop panels at wall columns are omitted shall equal $(t_1 + t_2)/2$ provided the value of c used in the computations complies with General Notes No.3.

*t_1 and t_2 in inches, L and c in feet.
W' = uniformly distributed unit dead and live load.

SHEAR

Shearing unit stress V on a vertical section which follows a periphery b at distance d beyond the edge of the column or column capital and parallel or concentric with it, shall not exceed the following values when computing $v = \frac{V}{bjd}$.

(a) $0.03f'_c$ but not more than 100 p.s.i. when at least 50% of the total negative reinforcement in the column strip passes through the periphery.

(b) $0.025f'_c$ but not more than 85 p.s.i. when 25%, which is the least value permitted, of the total negative reinforcement in the column strip passes through the periphery.

(c) Proportionate values of the shearing unit stress for intermediate percentages of reinforcement.

(d) Where drop panels are used, the shearing unit stress on vertical sections, which lie at distance d beyond the edges of the drop panel and parallel with them, shall not exceed $0.03f'_c$ nor 100 p.s.i. At least 50% of the total negative reinforcement in the column strip shall be within the width of the strip directly above the drop panel.

REINFORCEMENT

The ratio of reinforcement in any strip shall not be less than 0.0025bd. Spacing of bars shall not exceed 2 times the slab thickness. Length of splice — 36 diameter.

GENERAL NOTES

1. The coefficients of the table may be varied by no more than 10% provided the numerical sum of the + and − moments remains unchanged.
2. For columns without a capital the distance c shall be taken as the dimension of the column in the direction considered.
3. For columns with brackets take c equal to twice the distance from center line of column to the point where the thickness of the bracket is 1½".
4. Panels supported by marginal beams on opposite sides shall be designed as one or two-way slabs.

OPENINGS IN FLAT SLAB

1. Openings of any size may be provided in a flat slab in the area common to two intersecting middle strips provided the total positive and negative steel areas are maintained.
2. In the area common to two column strips, not more than ⅛ of the width of strip in any span shall be interrupted by openings. The equivalent of all bars interrupted shall be provided by extra steel on all sides of the openings.
3. In any area common to one column strip and one middle strip openings may interrupt ¼ of the bars in either strip. The equivalent of the interrupted bars shall be provided on all sides of the openings.
4. Any opening larger than described above shall be analyzed by accepted engineering principles and shall be completely framed as required to carry the loads to the columns.

STRUCTURAL CONCRETE–TWO WAY SLAB–I

TABLE A–MOMENT COEFFICIENTS FOR TWO WAY SLABS–A.C.I. CODE 1956

L/S		A	B	C	D	E	F	G	H	K
1.00	S	.500	.508	.381	.400	.285	.305	.389	.318	.250
	L	.500	.381	.508	.400	.388	.318	.285	.305	.250
1.05	S	.546	.549	.421	.437	.318	.337	.416	.344	.274
	L	.501	.378	.515	.400	.398	.322	.278	.303	.251
1.10	S	.590	.588	.460	.472	.348	.368	.442	.367	.296
	L	.503	.372	.522	.403	.410	.326	.273	.299	.252
1.15	S	.635	.624	.500	.507	.384	.400	.464	.390	.318
	L	.502	.369	.528	.402	.417	.330	.266	.295	.252
1.20	S	.680	.665	.540	.544	.420	.432	.495	.416	.340
	L	.501	.363	.530	.400	.424	.332	.261	.290	.251
1.25	S	.720	.701	.580	.576	.456	.465	.521	.438	.361
	L	.496	.355	.532	.397	.427	.333	.255	.285	.248
1.30	S	.764	.736	.618	.610	.491	.495	.548	.460	.382
	L	.489	.353	.535	.391	.432	.334	.249	.281	.245
1.35	S	.805	.773	.655	.645	.525	.524	.579	.484	.403
	L	.483	.345	.532	.386	.433	.333	.242	.276	.242
1.40	S	.847	.810	.693	.678	.560	.555	.600	.507	.424
	L	.480	.336	.530	.384	.436	.332	.236	.269	.240
1.45	S	.885	.846	.730	.709	.594	.585	.627	.530	.443
	L	.471	.330	.524	.377	.436	.328	.230	.264	.237
1.50	S	.927	.885	.765	.742	.628	.612	.657	.553	.465
	L	.462	.323	.518	.370	.436	.324	.224	.258	.232

L/S		A	B	C	D	E	F	G	H	K
1.55	S	.965	.920	.804	.772	.662	.644		.576	.483
	L	.453	.315	.514	.362	.436	.321		.252	.227
1.60	S	1.01	.958	.835	.803	.692	.668		.600	.502
	L	.450	.308	.506	.360	.436	.318		.246	.225
1.65	S	1.05	.995	.873	.836	.728	.698		.629	.524
	L	.440	.300	.500	.353	.427	.313		.240	.221
1.70	S	1.08	1.04	.907	.866	.758	.725		.648	.542
	L	.433	.293	.493	.347	.427	.308		.234	.217
1.75	S	1.12	1.07	.942	.900	.790	.754		.673	.561
	L	.421	.288	.490	.337	.423	.307		.230	.211
1.80	S	1.16		.978	.932	.820	.782			.583
	L	.414		.480	.331	.417	.300			.208
1.85	S	1.20		1.01	.965	.850	.810			.603
	L	.405		.472	.324	.414	.295			.203
1.90	S	1.24		1.05	.995	.880	.836			.629
	L	.399		.463	.319	.410	.290			.199
1.95	S	1.29		1.08	1.03	.914	.864			.645
	L	.391		.456	.313	.403	.286			.196
2.00	S	1.34		1.12	1.07	.940	.892			.667
	L	.378		.446	.302	.400	.280			.189

Number of Short Sides of Panel Restrained	Number of Long Sides of Panel Restrained: 0	1	2
0	A	B	G
1	C	D	H
2	E	F	K

CONDITION OF RESTRAINT.

Formula:-

$$t = \frac{Perimeter}{180}$$

Where:-
t = Slab thickness - 4" minimum

L = Long span; S = Short span in In.
M_1, M_2 & M_3 = Maximum Positive Moments in direction considered.

M_1
1.11x M_1
1.11x M_2
M_2

Span direction
Short or long Sides

M_1
1.0x M_1
1.6x M_2
M_2
1.45x M_2
1.45x M_3
M_3
1.45 M_3

SKETCHES FOR NEGATIVE MOMENTS AT SUPPORTS.

The positive moment M for a slab strip one foot wide is expressed by the formula $M = WS^2c$, where M is in in.lbs.; W = total live and dead load in lbs.per sq. ft.; S = short span in feet and c = a coefficient given in table depending on condition of restraint at edges of panel and is determined as follows: Select proper case denoted by letters A to K inclusive from "Condition of Restraint." Determine ratio between long and short panel sides = L/S. The value of "c" is given in 9 at intersection of proper case and ratio L/S.

Negative moments at continuous supports are determined from the positive moments, as noted adjacent to supports in "Sketches for Negative Moments at Supports." Where the moments as determined from each of two adjacent panels differ, use the average value. Coefficients in this table apply where the larger of two adjacent values of 1/1 or 1/5 does not exceed the smaller by more than 20%. For other cases, refer to A.C.I. Code for procedure.

At non-continuous supports provide hooked top bars of sectional area equal to 50% of bottom steel and extend 1/5 of clear short span into panel. Bottom steel may be reduced 25% for a distance equal to 1/4 of clear short span adjacent to continuous edges only. Data from Republic Fireproofing Company, Inc., New York City.

STRUCTURAL CONCRETE–TWO WAY SLAB–2

TABLE A – EQUIVALENT LOAD & MOMENT COEFFICIENTS FOR BEAMS FROM TWO-WAY SLABS-A.C.I. CODE- 1956

L/S		CASE A		CASE B		CASE C		CASE D		CASE E	
		a	b	a	b	a	b	a	b	a	b
1.00	S	.250	1.33	.199	1.45	.302	1.24	.153	1.58	.348	1.17
	L	.250	1.33	.302	1.24	.199	1.45	.348	1.17	.153	1.58
1.05	S	.243	1.37	.190	1.50	.298	1.27	.145	1.62	.348	1.19
	L	.269	1.30	.319	1.21	.217	1.41	.362	1.15	.169	1.53
1.10	S	.236	1.41	.183	1.54	.293	1.30	.136	1.66	.352	1.21
	L	.286	1.27	.334	1.19	.233	1.37	.376	1.13	.180	1.49
1.15	S	.228	1.45	.174	1.58	.287	1.33	.129	1.68	.344	1.24
	L	.302	1.24	.349	1.17	.250	1.33	.388	1.12	.201	1.45
1.20	S	.220	1.49	.165	1.62	.280	1.37	.122	1.68	.341	1.27
	L	.317	1.21	.362	1.15	.267	1.30	.399	1.10	.216	1.41
1.25	S	.211	1.53	.157	1.65	.272	1.40	.115	1.66	.336	1.29
	L	.331	1.19	.374	1.13	.283	1.27	.408	1.09	.231	1.38
1.30	S	.203	1.57	.150	1.67	.266	1.44	.108	1.63	.330	1.32
	L	.344	1.17	.385	1.12	.295	1.25	.417	1.08	.246	1.34
1.35	S	.195	1.60	.143	1.68	.258	1.47	.102	1.56	.325	1.35
	L	.356	1.16	.395	1.11	.309	1.23	.424	1.08	.260	1.31
1.40	S	.187	1.63	.136	1.68	.249	1.51	.097	1.46	.318	1.38
	L	.367	1.14	.403	1.10	.323	1.21	.431	1.07	.273	1.29
1.45	S	.179	1.66	.129	1.66	.241	1.54	.091	1.30	.310	1.42
	L	.377	1.13	.411	1.09	.333	1.19	.437	1.06	.286	1.26
1.50	S	.171	1.68	.123	1.62	.233	1.58	.086	1.10	.302	1.45
	L	.386	1.12	.418	1.08	.345	1.17	.443	1.05	.299	1.24
1.55	S	.164	1.68	.116	1.55	.224	1.60			.293	1.48
	L	.394	1.11	.425	1.07	.356	1.16			.311	1.22
1.60	S	.157	1.68	.111	1.46	.217	1.63			.285	1.51
	L	.402	1.10	.431	1.07	.365	1.15			.322	1.21
1.65	S	.150	1.66	.105	1.34	.209	1.65			.276	1.54
	L	.409	1.09	.436	1.06	.374	1.14			.333	1.19
1.70	S	.144	1.64	.101	1.16	.200	1.68			.269	1.56
	L	.416	1.08	.441	1.06	.382	1.12			.341	1.18
1.75	S	.138	1.60	.096	1.00	.194	1.68			.261	1.59
	L	.421	1.08	.445	1.05	.389	1.11			.351	1.16
1.80	S	.132	1.52			.186	1.68			.253	1.62
	L	.427	1.07			.397	1.10			.360	1.15
1.85	S	.126	1.44			.179	1.67			.245	1.64
	L	.432	1.07			.403	1.10			.368	1.14
1.90	S	.121	1.33			.172	1.67			.237	1.65
	L	.437	1.06			.410	1.09			.376	1.13
1.95	S	.116	1.18			.165	1.65			.229	1.67
	L	.441	1.06			.416	1.08			.383	1.12
2.00	S	.111	1.00			.160	1.61			.222	1.68
	L	.445	1.05			.420	1.08			.389	1.11

Number Of Short Sides Of Panel Restrained.	Number Of Long Sides Of Panels Restrained. 0	1	2
0	A	B	D
1	C	A	B
2	E	C	A

CONDITION OF RESTRAINT.

NOTE:
Use coefficient "a" for determining end reactions.
Use coefficient "a×b" for determining bending moments.

The load per linear foot of beam from a two-way panel, if uniformly distributed, would equal W×S×a, where W= total live and dead load per sq. ft., S= short span in feet and a= the coefficient given in Table 10. Coefficients given in S lines apply to short span beams and coefficients in L lines apply to long span beams.

Since the load on the beams is not considered uniform but varying from a Max. at center and diminishing towards each support, the equivalent uniformly distributed load for moment = W×S×a×b where coefficient "b" is also given in Table 10.

Coefficients "a" and "b" are obtained as follows: Find proper case denoted A to E incl. from "Conditions of Restraint." Determine ratio between long and short panel sides L/S. Coefficients "a" and "b" are then found at intersection of proper case and ratio L/S. Data from Republic Fireproofing Company, Inc. New York City.

STRUCTURAL CONCRETE–BEAM TABLES–1

EXAMPLES FOR USE OF BEAM TABLES – pp. 2-10, 11, 12, 13

Assumptions for All Examples: f_s = 20,000 p.s.i., f'_c = 3000, n = 10
f_c = 0.45 × 3000 = 1350. Design for a minimum & balanced reinforcement.

Example No. 1.

Given: Rectangular beam, M = 90,000 ft.-lb., b = 12 in.

Required: Effective depth d steel "A_s".

Solution: From Table A, p. 2-12, the minimum effective depth d for moment of 90,000 ft.-lb. required is 20 for 12 in. width good for 94,000 ft.-lb. requiring A_s = 3.26 in.^{2} ∴ Total depth of beam = 20" + 2" = 22" (if stirrups required add ½") and required.

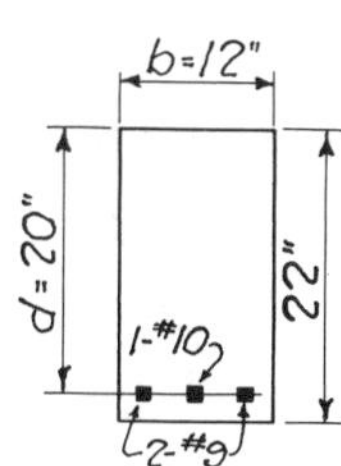
b=12"
d=20"
22"
1-#10
2-#9

$$A_s = 3.26 \times \frac{90,000}{94,200} = 3.11 \text{ in.}^2$$

Use 2 #9 and 1 #10.

Referring to Table A, p. 2-17 these 3 bars may be placed in 1 row for 12"–wide beam.

Check shear and bond stresses.

Example No. 2. Given: Rectangular beam with compression steel, M = 130 ft.-lb., d = 19.5 in. b = 14 in.

Required: Tension steel A and compression steel A'_s.

Solution: From Table A, p. 2-12, a beam 12 in. wide and of depth 19.5 in. has a resisting moment of approximately 90 ft.-kips working with 3.2 in.^{2} of bottom steel For 14 in. width:

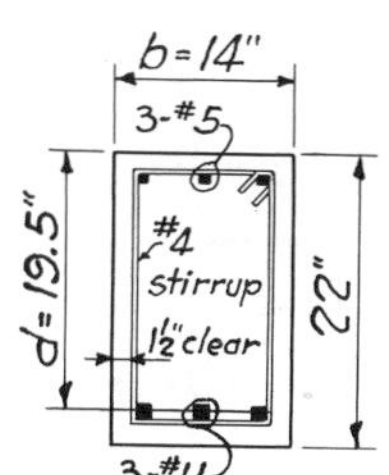
b=14"
3-#5
#4
stirrup
1½"clear
d=19.5"
22"
3-#11

$$M = 90 \times \frac{14}{12} = 105 \text{ ft.-kips}$$

$$A_s = 3.2 \times \frac{105}{90} = 3.7 \text{ in.}^2$$

Compression steel must take 130 – 105 = 25 ft.-kips from Table A, p. 2-13, 1 in.^{2} of compression steel, gives a moment of M = 27.7 ft.-kips (A.C.I.). $A'_s = \frac{25}{27.7} \times 1$ = 0.9 in.^{2} furnished by 3 #5 = 0.93 in.^{2}. From Table A, tension steel required to balance compression steel = 0.95 $\times \frac{0.9}{1}$ = 0.9 in.^{2}. Total A_s required = 3.7 + 0.9 = 4.6 in.^{2}. Furnished by 3 #11 = 4.68 in.^{2}, ϵ_o = 13.3 in. Check Table A, p 2-17 for placing one row in width of 14 in.. Check for shear and bond requirements.

Example No. 3. Given: T-Beam, M = 105,000 ft.-lb., slab thickness 4 in. width, b = 2 ft. – 0 in., width of stem, b' = 12".

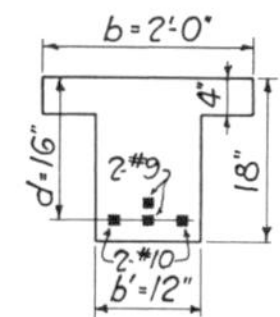
b=2'-0"
4"
d=16"
18"
2-#9
2-#10
b'=12"

Required: Depth of beam d, steel A_s.

Solution: From Table A, p. 2-12, valve of moment to be carried by 12 in. of width $M = \frac{105,000}{2}$ = 52,000 ft.-lb. in same table, in column for 4 in. slabs, the minimum effective depth "d" for moment of 52,500 ft.-lb. is 16 in. for 12 in. width and is good for 53,300 ft.-lb., requiring A_s = 2.23 in.^{2} ∴ Total depth of beam = 16" + 2 = 18, and required $A_s = \frac{105,000}{53,000} \times 2.23$ = 4.4 in.^{2}.∴ Use 2 #9 and 2 #10. Referring to Table A, p. 2-17 place 2 #10 in bottom layer, 1 #9 in bottom layer, and 1 #9 in second layer.

Example No. 4. Given: T-Beam with compression steel. Total moment M = 132,000 ft.-lb. slab thickness 4", b = 2' – 0", d = 16".

Required: Tension and compression reinforcement.

(a) Concrete Compression in Stem Neglected.

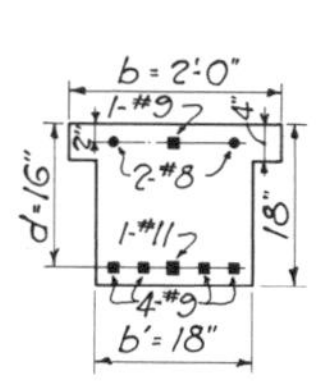
b=2'-0"
1-#9
2-#8
d=16"
18"
1-#11
4-#9
b'=18"

Solution: From Table A, p. 2-12, in column for 4" slab and d = 16", find M = 53,300 ft.-lb. and A_s = 2.23 in.^{2} ∴ The difference in moments M_c = 132,000 – 2 × 53,200 = 25,400 ft.-lb. which must be taken up by compression steel. From Table A, p. 2-12, in column for #9 compression steel, M = 9800 ft.-lb. for d = 16". Total compression steel required $A'_s = \frac{25,400}{9,800}$ = 2.6 in.^{2} ∴ Use 2 #8 and 1 #9*. This compression steel must be balanced by additional tension steel. Referring to last column in Table A, p. 2-12: A_r = 0.42 in.^{2} per #9 of compression steel. Total tension steel, A_s = 2.23 × 2 + 0.42 × 2.6 = 5.55 in.^{2}. ∴ Use 4 #9 and 1 #11. Referring to Table A, p. 2-17 minimum width of stem for 4 #9 and 1 #11, b' = 17½", use 18", and bars may be placed in 1 row for 18"-wide beam. Check shear and bond stresses.

(b) Concrete Compression in Stem Considered.

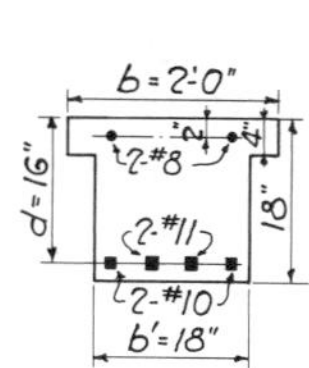
b=2'-0"
2-#8
d=16"
18"
2-#11
2-#10
b'=18"

Solution: From Table A, p. 2-12, in column for 4" slab and d = 16", find M = 53,300 ft.-lb. and A_s = 2.23 in.^{2} A_s = 2.23 × 6/12 = 1.12 in.^{2}*. From Table A, p. 2-12, in rectangular beam column and d = 16", find M = 60,300 ft.-lb. requiring steel A_s = 2.61 in.^{2}. ∴ For 18"-wide beam and 16" effective depth, M = 60,300 × $^{18}/_{12}$ = 90,500 ft.-lbs., A_s = 2.61 × $^{18}/_{12}$ = 3.92 in.^{2} The difference in moments M_c = 132,000 – 26,700 – 90,500 = 14,800 ft.-lb.. From Table A, p. 2-12, in column for #9 compression steel, M = 9800 ft.-lbs. for d = 16". Total compression steel required, $A_s = \frac{14,800}{9,800}$ = 1.5 in.^{2}. This compression steel must be balanced by additional tension steel. Referring to Table A, p. 2-12 A_T = 0.42 in.^{2} ∴ use 2 #8. Total tension steel required, A_s = 1.12 + 3.92 + 0.42 × 1.5 = 5.67 in.^{2}. ∴ Use 2 #10 and 2 #11, and bars may be placed on 1 row for 18"-wide beam. Check Table A, p 2-17. Check shear and bond stresses.

* Compression steel designed by N.Y.C. Code. See pp. 2-10 through 2-13 for A.C.I. Code.

STRUCTURAL CONCRETE – BEAM TABLES – 2

TABLE A – RESISTING MOMENTS OF CONCRETE BEAMS.*

Slab; d; d′

$f_s = 20{,}000$ #/□″ $f_c = 900$

$f'_c = 2{,}000$ $n = 15$ $R = 157.0$

M = Moment of resistance of beam one foot wide in 1000 foot lbs.
A_s = Tensile steel area in sq. inches for moment M.
A_T = Sq. in. additional tensile steel for each 1□″ compression steel.

d	TEE BEAMS FOR SLAB THICKNESSES SHOWN												RECTANGULAR BEAM		MOM. 1″□ COMP. STEEL			
	2″ SLAB		2½″ SLAB		3″ SLAB		4″ SLAB		5″ SLAB		6″ SLAB				A.C.I.		N.Y.C.	
	M	A_s	M	A_s	M	A_s	M	A_s	M	A_s	M	A_s	M	A_s	M′	A_T	M′	A_T
6″	5.5	.63											5.7	.65	1.5	.23	.73	.11
6½	6.4	.67											6.6	.71	2.3	.31	1.1	.15
7	7.2	.70	7.6	.75									7.7	.76	3.2	.38	1.5	.18
7½	8.0	.72	8.6	.79									8.8	.82	4.1	.44	2.0	.21
8	8.9	.74	9.7	.83	9.9	.87							10.0	.87	5.0	.50	2.4	.24
8½	9.7	.76	10.2	.86	11.2	.91							11.3	.92	5.9	.55	2.9	.26
9	10.6	.78	11.8	.88	12.4	.95							12.7	.98	6.8	.59	3.3	.28
9½	11.5	.80	12.8	.91	13.7	.99							14.2	1.03	7.8	.63	3.8	.30
10	12.3	.81	13.9	.93	14.9	1.02	15.7	1.09					15.7	1.09	8.8	.66	4.2	.32
10½	13.2	.83	15.0	.95	16.2	1.05	17.3	1.14					17.3	1.14	9.8	.69	4.7	.34
11	14.1	.84	16.0	.97	17.4	1.07	18.9	1.19					19.0	1.20	10.8	.72	5.2	.35
11½	15.0	.85	17.1	.99	18.7	1.10	20.5	1.23					20.8	1.25	11.8	.75	5.7	.36
12	15.8	.86	18.2	1.00	20.0	1.12	22.1	1.27					22.6	1.31	12.8	.77	6.2	.37
12½	16.7	.86	19.3	1.02	21.3	1.14	23.8	1.31					24.5	1.36	13.8	.79	6.7	.38
13	17.6	.87	20.4	1.03	22.6	1.16	25.4	1.34	26.5	1.41			26.5	1.41	14.8	.81	7.1	.39
13½	18.5	.88	21.4	1.04	23.9	1.18	27.1	1.37	28.5	1.46			28.7	1.47	15.8	.83	7.6	.40
14	19.4	.89	22.5	1.05	25.1	1.19	28.7	1.39	30.5	1.50			30.8	1.52	16.9	.84	8.1	.41
15	21.1	.90	24.7	1.07	27.7	1.22	32.1	1.45	34.6	1.58	35.3	1.63	35.3	1.63	19.0	.87	9.1	.42
16	22.9	.91	26.9	1.09	30.4	1.24	35.5	1.49	38.7	1.65	40.0	1.73	40.2	1.74	21.1	.90	10.1	.43
17	24.7	.92	29.2	1.11	33.0	1.27	39.0	1.53	42.8	1.72	44.8	1.82	45.4	1.85	23.1	.93	11.2	.45
18	26.5	.93	31.4	1.12	35.6	1.29	42.4	1.56	47.0	1.77	49.8	1.90	50.9	1.96	25.2	.95	12.2	.46
19	28.3	.94	33.6	1.13	38.3	1.30	45.9	1.60	51.3	1.82	54.7	1.97	56.7	2.07	27.3	.97	13.2	.47
20	30.0	.95	35.8	1.14	40.9	1.32	49.3	1.62	55.5	1.86	59.7	2.04	62.8	2.18	29.0	.97	14.2	.47
21	31.8	.95	38.0	1.15	43.6	1.33	52.8	1.65	59.8	1.90	64.7	2.09	69.2	2.28	30.6	.97	15.2	.48
22	33.6	.96	40.3	1.16	46.2	1.35	56.3	1.67	64.0	1.94	69.7	2.14	76.0	2.39	32.2	.97	16.3	.49
23	35.4	.96	42.5	1.17	48.9	1.36	59.8	1.69	68.3	1.97	74.8	2.19	83.0	2.50	33.8	.97	17.3	.49
24	37.2	.97	44.7	1.18	51.5	1.37	63.3	1.71	72.7	2.00	79.9	2.23	90.4	2.61	35.4	.96	18.3	.50
25	38.8	.97	46.9	1.18	54.2	1.38	66.9	1.72	76.9	2.02	85.0	2.27	98.1	2.73	37.0	.96	19.3	.51
26			49.2	1.19	56.9	1.39	70.4	1.75	81.4	2.06	90.2	2.31	106.1	2.83	38.5	.96	20.4	.51
27			51.4	1.20	59.5	1.40	74.0	1.77	86.0	2.08	94.8	2.35	114.5	2.95	40.2	.96	21.5	.51
28			53.7	1.20	62.2	1.40	77.5	1.78	90.2	2.10	100.6	2.38	123.1	3.05	41.8	.96	22.5	.52
29			55.5	1.20	64.8	1.41	80.7	1.79	94.2	2.12	105.0	2.41	132.0	3.17	43.3	.96	23.5	.52
30			58.2	1.21	67.6	1.42	84.6	1.80	99.0	2.14	111.0	2.44	141.3	3.26	44.8	.96	24.5	.53
32					73.0	1.43	91.6	1.82	107.8	2.18	121.3	2.48	160.8	3.48	48.1	.96	26.6	.53
34					78.3	1.44	98.7	1.84	116.6	2.21	131.9	2.53	181.5	3.70	51.3	.96	28.7	.54
36					83.6	1.45	105.8	1.86	125.4	2.23	142.5	2.57	203.5	3.92	54.4	.96	30.8	.54
38							113.0	1.88	134.2	2.26	152.9	2.60	226.7	4.13	57.6	.96	32.9	.55
40							120.1	1.89	143.1	2.28	163.5	2.64	251.2	4.35	60.8	.96	34.9	.55
44									160.9	2.32	184.9	2.69	303.9	4.79	67.3	.96	39.1	.56
48									178.7	2.35	206.0	2.74	361.7	5.22	73.5	.96	43.3	.56

*Adapted from Singleton, Manual of Structural Design, H.M. Ives & Sons.

See Pg. 2-09 for Examples for use of Beam Tables.

STRUCTURAL CONCRETE-BEAM TABLES-3

TABLE A-RESISTING MOMENTS OF CONCRETE BEAMS.*

Slab d d' d d'

$f_s = 20{,}000 \,\#/\square''$ $f_c = 1125$

$f'_c = 2{,}500$ $n = 12$ $R = 196.2$

M = Moment of resistance of beam one foot wide in 1000 foot lbs.

A_s = Tensile steel area in sq. inches for moment M.

A_T = Sq. in. additional tensile steel for each $1\square''$ compression steel.

d	TEE BEAMS FOR SLAB THICKNESSES SHOWN												RECTANGULAR BEAM		MOM. $1\square''$ COMP. STEEL			
	2" SLAB		2½" SLAB		3" SLAB		4" SLAB		5" SLAB		6" SLAB				A.C.I.		N.Y.C.	
	M	A_s	M	A_s	M	A_s	M	A_s	M	A_s	M	A_s	M	A_s	M'	A_T	M'	A_T
6"	6.9	.79											7.1	.82	1.5	.23	.7	.11
6½	7.9	.84											8.3	.88	2.3	.31	1.1	.15
7	9.0	.87	9.5	.94									9.6	.95	3.1	.38	1.5	.18
7½	10.1	.90	10.8	.99									11.0	1.02	4.0	.44	2.0	.21
8	11.1	.93	12.1	1.04	12.3	1.09							12.6	1.09	4.9	.49	2.4	.24
8½	12.2	.95	12.8	1.08	13.9	1.14							14.2	1.16	5.8	.54	2.8	.26
9	13.2	.98	14.7	1.10	15.5	1.19							15.9	1.22	6.8	.58	3.2	.28
9½	14.3	1.00	16.0	1.13	17.1	1.23							17.7	1.29	7.8	.62	3.7	.30
10	15.4	1.02	17.3	1.16	18.7	1.27	19.6	1.36					19.6	1.36	8.7	.65	4.2	.31
10½	16.5	1.04	18.7	1.19	20.2	1.31	21.6	1.43					21.6	1.43	9.7	.68	4.7	.33
11	17.6	1.05	20.0	1.21	21.8	1.34	23.6	1.49					23.7	1.50	10.7	.71	5.1	.34
11½	18.7	1.06	21.4	1.23	23.4	1.37	25.6	1.54					26.0	1.56	11.7	.74	5.6	.35
12	19.8	1.07	22.7	1.25	25.0	1.40	27.6	1.59					28.3	1.63	12.7	.76	6.0	.36
12½	20.9	1.08	24.1	1.27	26.6	1.43	29.7	1.64					30.7	1.70	13.7	.78	6.5	.37
13	22.0	1.09	25.4	1.29	28.2	1.45	31.8	1.68	33.1	1.76			33.2	1.77	14.7	.80	7.0	.38
13½	23.1	1.10	26.8	1.31	29.8	1.47	33.8	1.71	35.6	1.82			35.9	1.84	15.7	.82	7.5	.39
14	24.2	1.11	28.2	1.32	31.4	1.49	35.9	1.74	38.1	1.87			38.5	1.90	16.7	.84	8.0	.40
15	26.4	1.13	30.9	1.34	34.7	1.53	40.1	1.80	43.2	1.97	44.2	2.04	44.2	2.04	18.8	.87	9.0	.41
16	28.6	1.14	33.7	1.36	37.9	1.56	44.4	1.86	48.3	2.07	50.0	2.16	50.2	2.18	20.9	.89	10.0	.43
17	30.9	1.15	36.5	1.38	41.3	1.59	48.7	1.91	53.5	2.15	56.1	2.27	56.7	2.31	22.9	.92	11.0	.44
18	33.1	1.17	39.2	1.40	44.5	1.61	53.0	1.95	58.7	2.21	62.2	2.36	63.6	2.45	25.0	.94	12.0	.45
19	35.3	1.18	42.0	1.42	47.8	1.63	57.3	1.99	64.1	2.27	68.3	2.45	70.8	2.58	27.1	.96	13.0	.46
20	37.5	1.19	44.8	1.43	51.1	1.65	61.7	2.02	69.4	2.32	74.6	2.54	78.5	2.72	28.8	.96	14.0	.47
21	39.8	1.19	47.5	1.44	54.5	1.66	66.1	2.06	74.7	2.37	80.8	2.61	86.5	2.86	30.3	.96	15.0	.47
22	42.0	1.20	50.4	1.46	57.8	1.68	70.4	2.09	80.0	2.42	87.1	2.67	95.0	2.99	31.9	.96	16.0	.48
23	44.2	1.21	53.0	1.47	61.0	1.70	74.8	2.12	85.4	2.46	93.5	2.73	103.8	3.13	33.5	.96	17.0	.49
24	46.5	1.21	55.9	1.48	64.4	1.71	79.2	2.14	90.9	2.50	99.9	2.79	113.0	3.26	35.1	.96	18.0	.49
25	48.8	1.22	58.8	1.48	67.5	1.72	83.8	2.16	96.3	2.54	106.3	2.84	122.5	3.39	36.6	.96	19.0	.50
26			61.6	1.49	71.0	1.73	87.9	2.18	101.8	2.57	112.7	2.88	132.7	3.54	38.2	.95	20.0	.50
27			64.2	1.50	74.4	1.74	92.4	2.20	107.4	2.60	118.7	2.93	142.9	3.67	39.8	.95	21.0	.51
28			67.1	1.50	77.8	1.75	96.8	2.21	112.8	2.62	125.7	2.97	153.8	3.81	41.4	.95	22.1	.51
29			69.5	1.50	81.1	1.76	101.8	2.23	118.5	2.65	132.0	3.00	164.9	3.95	42.9	.95	23.1	.51
30			72.8	1.51	84.3	1.77	105.7	2.24	123.7	2.67	138.8	3.04	176.6	4.08	44.4	.95	24.1	.52
32					91.1	1.79	114.5	2.27	134.7	2.72	151.7	3.10	200.9	4.35	47.6	.95	26.1	.52
34					97.8	1.80	123.4	2.30	145.7	2.76	164.9	3.15	226.8	4.62	50.8	.95	28.2	.53
36					104.5	1.81	132.3	2.32	156.7	2.79	178.1	3.20	254.3	4.90	54.0	.95	30.2	.53
38							141.3	2.34	167.7	2.81	191.2	3.25	283.4	5.17	57.2	.95	32.3	.54
40							150.1	2.36	178.9	2.85	204.4	3.30	314.0	5.44	60.3	.95	34.3	.54
44									201.1	2.90	231.2	3.37	379.9	5.98	66.5	.95	38.4	.55
48									212.2	2.92	257.2	3.42	452.1	6.53	72.8	.95	42.5	.55

* Adapted from Singleton, Manual of Structural Design, H.M. Ives & Sons.

See Pg. 2-09 for Examples for use of Beam Tables.

STRUCTURAL CONCRETE—BEAM TABLES—4

TABLE A– RESISTING MOMENTS OF CONCRETE BEAMS.*

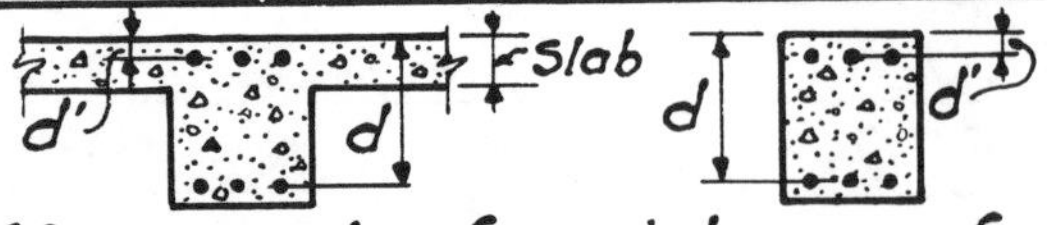

$f_s = 20{,}000$ #/□" $f_c = 1350$

$f_c' = 3{,}000$ $n = 10$ $R = 235.6$

M = Moment of resistance of beam one foot wide in 1000 foot lbs.

A_s = Tensile steel area in sq. inches for moment M.

A_T = Sq. in. additional tensile steel for each 1□" compression steel.

d	TEE BEAMS FOR SLAB THICKNESSES SHOWN												RECTANGULAR BEAM		MOM. 1□" COMP. STEEL			
	2" SLAB		2½" SLAB		3" SLAB		4" SLAB		5" SLAB		6" SLAB				A.C.I.		N.Y.C.	
	M	A_s	M	A_s	M	A_s	M	A_s	M	A_s	M	A_s	M	A_s	M'	A_T	M'	A_T
6"	8.3	.94											8.5	.98	1.5	.22	.7	.11
6½	9.5	1.01											9.9	1.06	2.3	.30	1.1	.15
7	10.8	1.05	11.4	1.12									11.5	1.14	3.1	.38	1.5	.18
7½	12.1	1.08	12.9	1.18									13.2	1.22	4.0	.44	1.9	.21
8	13.3	1.11	14.5	1.24	14.8	1.30							15.1	1.31	4.9	.49	2.3	.23
8½	14.6	1.14	15.3	1.29	16.7	1.36							17.0	1.39	5.8	.53	2.8	.25
9	15.9	1.17	17.6	1.33	18.6	1.42							19.1	1.47	6.7	.58	3.2	.27
9½	17.2	1.20	19.2	1.37	20.5	1.47							21.3	1.55	7.7	.62	3.7	.29
10	18.5	1.22	20.8	1.40	22.4	1.52	23.6	1.63					23.5	1.63	8.6	.65	4.1	.31
10½	19.8	1.24	22.4	1.42	24.2	1.57	25.9	1.71					26.0	1.71	9.6	.68	4.6	.32
11	21.1	1.26	24.0	1.45	26.1	1.61	28.3	1.78					28.5	1.80	10.6	.71	5.0	.33
11½	22.4	1.27	25.7	1.48	28.0	1.65	30.7	1.84					31.1	1.88	11.6	.73	5.5	.35
12	23.7	1.28	27.2	1.50	30.0	1.68	33.2	1.90					33.9	1.96	12.5	.75	5.9	.36
12½	25.1	1.29	28.9	1.53	31.9	1.71	35.6	1.96					36.8	2.04	13.5	.78	6.4	.37
13	26.4	1.30	30.5	1.55	33.8	1.74	38.1	2.01	39.8	2.12			39.8	2.12	14.5	.80	6.9	.38
13½	27.7	1.31	32.2	1.57	35.7	1.77	40.6	2.05	42.7	2.19			43.0	2.20	15.6	.82	7.4	.39
14	29.0	1.32	33.8	1.58	37.7	1.80	43.1	2.09	45.7	2.25			46.2	2.28	16.6	.83	7.8	.39
15	31.7	1.34	37.1	1.61	41.6	1.84	48.2	2.17	51.8	2.37	53.0	2.45	53.0	2.45	18.6	.86	8.8	.41
16	34.4	1.36	40.4	1.64	45.5	1.88	53.3	2.23	58.0	2.49	60.0	2.59	60.3	2.61	20.6	.89	9.8	.42
17	37.1	1.38	43.7	1.66	49.5	1.90	58.4	2.30	64.2	2.58	67.3	2.73	68.1	2.77	22.7	.91	10.8	.43
18	39.7	1.40	47.0	1.68	53.5	1.93	63.5	2.35	70.5	2.65	74.6	2.85	76.3	2.94	24.8	.93	11.7	.44
19	42.4	1.42	50.3	1.70	57.4	1.96	68.8	2.39	76.9	2.72	82.0	2.96	85.0	3.10	26.9	.95	12.7	.45
20	45.1	1.43	53.7	1.71	61.3	1.98	74.0	2.43	83.2	2.79	89.5	3.06	94.2	3.26	28.5	.95	13.7	.46
21	47.7	1.44	57.0	1.73	65.4	2.00	79.3	2.47	89.7	2.85	97.0	3.14	103.8	3.43	30.1	.95	14.7	.46
22	50.5	1.45	60.5	1.75	69.4	2.02	84.5	2.51	96.0	2.90	104.6	3.21	114.0	3.59	31.6	.95	15.7	.47
23	53.0	1.46	63.7	1.76	73.3	2.04	89.7	2.54	102.5	2.95	112.2	3.28	124.6	3.75	33.2	.95	16.7	.48
24	55.8	1.47	67.2	1.77	77.2	2.06	95.0	2.57	109.0	3.00	119.9	3.34	135.6	3.92	34.7	.95	17.7	.48
25	58.8	1.47	70.6	1.78	81.3	2.07	100.0	2.59	115.6	3.04	127.5	3.38	147.5	4.08	36.3	.95	18.7	.49
26			73.9	1.78	85.3	2.08	105.5	2.62	122.1	3.08	135.3	3.46	159.2	4.24	37.8	.94	19.7	.49
27			77.2	1.79	89.3	2.09	110.8	2.64	128.3	3.12	142.2	3.50	172.0	4.41	39.4	.94	20.7	.50
28			80.5	1.80	93.4	2.10	116.2	2.66	135.3	3.15	150.8	3.55	184.6	4.57	40.9	.94	21.7	.50
29			84.1	1.81	97.6	2.11	121.9	2.68	141.3	3.18	158.1	3.60	198.5	4.74	42.5	.94	22.7	.50
30			87.4	1.82	101.3	2.12	126.9	2.70	148.4	3.21	166.5	3.64	211.9	4.90	44.1	.94	23.7	.51
32					109.4	2.14	137.4	2.73	161.7	3.27	182.0	3.72	241.1	5.22	47.2	.94	25.7	.51
34					117.4	2.16	148.1	2.76	174.9	3.32	197.8	3.80	272.2	5.55	50.3	.94	27.7	.52
36					125.4	2.17	158.8	2.79	188.0	3.36	213.7	3.86	305.2	5.88	53.4	.94	29.7	.52
38							169.6	2.82	201.2	3.39	229.4	3.91	340.0	6.20	56.5	.94	31.7	.53
40							180.1	2.84	214.6	3.42	245.3	3.96	376.8	6.53	59.6	.94	33.7	.53
44									241.3	3.48	277.4	4.04	455.9	7.18	65.8	.94	37.7	.54
48									268.1	3.53	309.1	4.11	542.5	7.83	72.0	.94	41.8	.55

*Adapted from Singleton, Manual of Structural Design, H.M. Ives & Sons.

See Pg. 2-09 for Examples for use of Beam Tables.

STRUCTURAL CONCRETE—BEAM TABLES—5

TABLE A— RESISTING MOMENTS OF CONCRETE BEAMS.*

$f_s = 20{,}000\ \#/\square''$ $f_c = 1700$

$f_c' = 3{,}750$ $n = 8$ $R = 298.0$

M = Moment of resistance of beam one foot wide in 1000 foot lbs.

A_S = Tensile steel area in sq. inches for moment M.

A_T = Sq. in. additional tensile steel for each 1□" compression steel.

d	TEE BEAMS FOR SLAB THICKNESSES SHOWN												RECTANGULAR BEAM		MOM. 1□" COMP. STEEL			
	2" SLAB		2½" SLAB		3" SLAB		4" SLAB		5" SLAB		6" SLAB				A.C.I.		N.Y.C.	
	M	A_s	M	A_s	M	A_s	M	A_s	M	A_s	M	A_s	M	A_s	M'	A_T	M'	A_T
6"	10.5	1.22											10.7	1.24	1.5	.22	.7	.11
6½	12.0	1.26											12.6	1.34	2.3	.31	1.1	.15
7	13.5	1.30	14.4	1.43									14.6	1.44	3.1	.37	1.5	.18
7½	15.2	1.35	16.4	1.50									16.7	1.55	4.0	.44	1.9	.21
8	16.7	1.39	18.3	1.57	19.0	1.66							19.1	1.66	4.9	.49	2.3	.23
8½	18.4	1.43	20.2	1.64	21.2	1.73							21.5	1.76	5.8	.53	2.7	.25
9	19.8	1.46	22.2	1.70	23.5	1.80							24.1	1.87	6.7	.57	3.1	.27
9½	21.5	1.49	24.2	1.75	25.8	1.87							26.8	1.96	7.7	.61	3.6	.29
10	23.3	1.52	26.2	1.79	28.2	1.93							29.8	2.07	8.6	.65	4.0	.30
10½	25.0	1.55	28.3	1.82	30.6	1.99							32.8	2.17	9.6	.68	4.4	.32
11	26.8	1.58	30.4	1.85	33.0	2.05	35.9	2.28					36.0	2.27	10.5	.70	4.9	.33
11½	28.3	1.60	32.4	1.88	35.4	2.08	38.7	2.34					39.3	2.38	11.5	.73	5.4	.34
12	30.0	1.62	34.5	1.90	37.8	2.12	41.9	2.41					42.9	2.48	12.5	.75	5.8	.35
12½	31.5	1.63	36.5	1.92	40.2	2.15	44.9	2.48					46.6	2.59	13.5	.77	6.3	.36
13	33.2	1.65	38.6	1.94	42.5	2.18	48.2	2.55					50.4	2.69	14.5	.79	6.8	.37
13½	35.0	1.67	40.8	1.96	44.8	2.22	51.2	2.62					54.2	2.79	15.5	.81	7.2	.38
14	36.6	1.69	42.9	1.99	47.5	2.27	54.3	2.69	57.5	2.88			58.4	2.90	16.5	.83	7.7	.39
15	39.9	1.71	47.0	2.01	52.4	2.31	60.7	2.76	65.5	3.01			67.2	3.11	18.5	.86	8.7	.40
16	43.4	1.73	50.8	2.04	57.4	2.34	67.0	2.82	73.0	3.13	76.2	3.37	76.3	3.31	20.5	.88	9.6	.41
17	46.8	1.75	55.0	2.07	62.4	2.37	73.4	2.88	81.0	3.23	85.2	3.49	86.0	3.51	22.6	.91	10.6	.42
18	50.1	1.76	59.3	2.10	67.4	2.40	79.5	2.94	89.0	3.32	94.2	3.61	96.5	3.72	24.6	.93	11.5	.43
19	53.0	1.77	63.5	2.12	72.0	2.43	86.3	3.00	96.4	3.40	103.5	3.72	107.5	3.93	26.6	.94	12.5	.44
20	56.9	1.78	67.8	2.14	77.2	2.46	93.6	3.05	104.8	3.48	112.5	3.83	119.2	4.14	28.1	.94	13.4	.45
21	60.5	1.79	71.4	2.16	82.4	2.49	99.6	3.09	112.6	3.56	124.0	3.93	131.2	4.34	29.6	.93	14.4	.46
22	63.5	1.80	76.0	2.18	87.3	2.51	106.5	3.13	120.8	3.63	131.5	4.02	144.0	4.54	31.1	.93	15.4	.46
23	66.6	1.81	79.3	2.20	92.0	2.53	113.0	3.17	129.0	3.70	141.0	4.11	157.5	4.75	32.7	.93	16.4	.47
24	70.2	1.82	84.0	2.22	97.4	2.55	119.8	3.21	137.0	3.77	151.0	4.20	171.8	4.96	34.2	.93	17.3	.47
25			88.7	2.23	102.5	2.57	126.1	3.25	145.5	3.83	160.6	4.28	186.3	5.18	35.7	.93	18.3	.48
26			93.3	2.24	106.4	2.59	132.5	3.28	153.5	3.88	169.2	4.35	201.5	5.38	37.2	.93	19.3	.48
27			97.7	2.25	112.0	2.61	139.0	3.31	162.2	3.92	178.8	4.42	217.3	5.59	38.8	.93	20.3	.49
28			101.0	2.26	117.3	2.63	146.5	3.33	170.8	3.96	188.8	4.48	233.3	5.78	40.3	.93	21.3	.49
29			105.8	2.27	122.5	2.65	152.8	3.35	178.3	4.00	199.0	4.54	250.5	5.99	41.8	.93	22.2	.49
30			110.4	2.28	127.8	2.67	159.3	3.37	187.2	4.03	209.7	4.59	268.2	6.22	43.3	.93	23.2	.50
32					137.5	2.70	173.0	3.41	203.0	4.09	230.0	4.67	304.0	6.60	46.4	.93	25.1	.50
34					146.8	2.72	187.5	3.45	220.0	4.15	250.0	4.75	345.5	7.06	49.5	.93	27.1	.51
36					158.0	2.74	200.0	3.49	237.5	4.21	270.0	4.82	387.0	7.47	52.5	.93	29.1	.51
38							212.0	3.53	252.5	4.27	288.8	4.89	429.0	7.85	55.6	.93	31.0	.52
40							227.2	3.57	270.4	4.32	308.8	4.96	476.8	8.26	58.6	.93	33.0	.52
44							254.0	3.62	304.2	4.38	349.0	5.06	577.0	9.15	64.6	.93	37.0	.53
48							283.0	3.67	337.0	4.43	390.0	5.16	685.0	9.92	70.7	.93	41.0	.54

*Adapted from Singelton Manual of Structural Design, H.M. Ives & Sons.

See Pg. 2-09 for use of Beam Tables.

STRUCTURAL CONCRETE–STIRRUP DATA

STIRRUP DATA
WEB REINFORCEMENT, A. C. I.

TABLE A – MINIMUM DEPTHS FOR STIRRUP EMBEDMENT*

Deformed Bars: $d = [f_v - 10{,}000]\,\frac{5c}{f_c} + 7c + 2$ Plain Bars: $d = [f_v - 10{,}000]\,\frac{11.1c}{f_c} + 7c + 2$

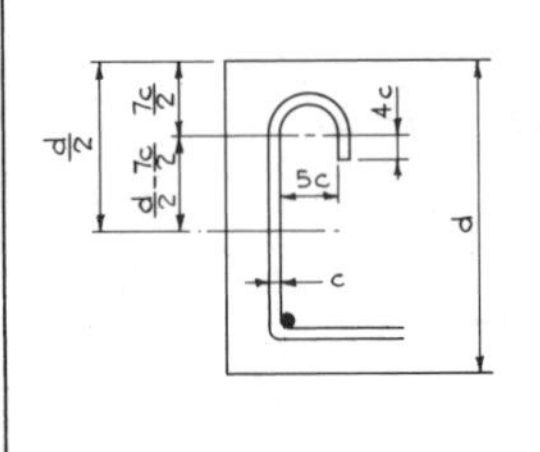

5c
4c
c
d

f_c	Size: fv = 20,000 p.s.i. Plain #2	Deformed #3	Deformed #4	Deformed #5	Size: fv = 18,000 p.s.i. Plain #2	Deformed #3	Deformed #4	Deformed #5	Size: fv = 16,000 p.s.i. Plain #2	Deformed #3	Deformed #4	Deformed #5
2000	17.6	14.0	18.0	22.0	14.9	12.1	15.5	18.9	12.1	10.3	13.0	15.8
2500	14.9	12.1	15.5	18.9	12.6	10.6	13.5	16.4	10.4	9.1	11.5	13.9
3000	13.0	10.9	13.8	16.8	11.2	9.6	12.2	14.7	9.3	8.4	10.5	12.6
3750	11.2	9.6	12.2	14.7	9.7	8.6	10.8	13.0	8.2	7.6	9.5	11.4
Stirrup Value (in pounds)	2000	4400	8000	12,000	1800	3960	7200	11,160	1600	3520	6400	9920

TABLE B – FORMULAS FOR *N* FOR UNIFORMLY LOADED BEAMS (f_v = 20,000 p.s.i.)

Size of Stirrups	#2	#3	#4	#5
N,* Theoretical Number of Stirrups in One End of Beam	$\frac{(V-V_c)\cdot a}{3.5\,d}$	$\frac{(V-V_c)\cdot a}{7.7\,d}$	$\frac{(V-V_c)\cdot a}{4.0\,d}$	$\frac{(V-V_c)\cdot a}{21.7\,d}$

*Increase proportionately for fv less than 20,000 p.s.i.

Where: V = total shear in kips.
V_c = shear carried by concrete in kips.
a = distance in inches from end of beam to point where shear : 1/c
d = effective depth of beam in inches.

TABLE C – SPACING OF U SHAPED STIRRUPS FOR UNIFORMLY LOADED BEAMS

N Number of Stirrups in One End of Beam	Distance from 1st Stirrup to Face of Support	Spacing, Center to Center of Stirrups in Terms of a — 1st Group No.	Spacing	2nd Group No.	Spacing	3rd Group No.	Spacing	4th Group No.	Spacing	5th Group No.	Spacing
20	0.013 a	8	0.03 a	7	0.04 a	2	0.06 a	1	0.08 a	1	0.11a
19	0.013 a	7	0.03 a	6	0.04 a	3	0.06 a	1	0.08 a	1	0.12 a
18	0.014 a	6	0.03 a	5	0.04 a	4	0.06 a	1	0.08 a	1	0.12 a
17	0.015 a	5	0.03 a	5	0.04 a	4	0.06 a	1	0.09 a	1	0.13 a
16	0.016 a	3	0.03 a	5	0.04 a	5	0.06 a	1	0.09 a	1	0.13 a
15	0.017 a	2	0.03 a	5	0.04 a	4	0.06 a	2	0.08 a	1	0.14 a
14	0.018 a	5	0.04 a	4	0.05 a	2	0.08 a	1	0.09 a	1	0.14 a
13	0.019 a	4	0.04 a	3	0.05 a	3	0.08 a	1	0.09 a	1	0.14 a
12	0.021 a	6	0.05 a	3	0.07 a	1	0.12 a	1	0.15 a		
11	0.023 a	5	0.05 a	3	0.08 a	1	0.12 a	1	0.15 a		
10	0.025 a	3	0.05 a	4	0.08 a	1	0.12 a	1	0.16 a		
9	0.028 a	3	0.06 a	3	0.09 a	1	0.12 a	1	0.17 a		
8	0.032 a	2	0.07 a	3	0.09 a	1	0.13 a	1	0.18 a		
7	0.036 a	3	0.08 a	2	0.13 a	1	0.20 a				
6	0.04 a	3	0.10 a	1	0.15 a	1	0.22 a				
5	0.05 a	2	0.12 a	1	0.16 a	1	0.23 a				
4	0.07 a	2	0.16 a	1	0.26 a						
3	0.09 a	1	0.21 a	1	0.30 a						
2	0.13 a	1	0.37 a								
1	0.29 a										

NOTE: Stirrups are to be carried on for a distance a.

Carried by stirrups
Carried by concrete
$V-V_c$
V
V_c
d
a
Half of Span

EXAMPLE:

Given: Beam with b = 12", d = 20", span 1 = 24'-0", and load = 3000 lb. per lin. ft.

Required: Stirrup spacing using f'_c = 3000 p.s.i.

Solution: $V = 3000 \times 12 = 36.0$ kips, $V_c = 12 \times 20 \times \frac{7}{8} \times (0.03\,f'_c) = 18.9$ kips $V - V_c = 17.1$ kips to be carried by stirrups. Assuming ⅜" φ U stirrups, from Table A, with f_v = 20,000 p.s.i. and f'_c = 3000 p.s.i. #3 U stirrups are satisfactory for d = 20". Solve for a (inches) $= \frac{17.1}{36.0} \times 12 \times 12 = 68.4$". Solve for *N*, using formula in Table B: $N = \frac{17.1 \times 68.4}{7.7 \times 20}$ 7.6. Enter Table C at N = 8, and find stirrup spacing: 1 @ 2", 2 @ 4", 3 @ 6", 1@ 8", and 1 @ 12". Reduce spacing of 1@ 12" to 1 @ 10", as the A.C.I. Code requires that the maximum spacing of stirrups when required shall not be greater than 0.5 d. Also, when maximum unit shear is greater than 0.06 f'_c, this spacing should be reduced to 0.25 d. If number of stirrups from Table C does not extend to full *a* distance plus *d*, then add sufficient stirrups @ 0.5 *d* spacing.

*A.C.I. Code, 1956.

STRUCTURAL CONCRETE-FLOOR SLABS

TABLE A - RESISTING MOMENTS OF CONCRETE SLABS

Temp. reinforcing
Main reinforcing
d"

M = Moment of resistance of slabs one foot wide in 1000 foot lbs.
A_s = Tensile steel area in square inches for moment M.

f_s	f'_c	f_c	n	R	p	Effective Depth d in inches: 2		2½		3		3½		4		4½		5		5½		6	
						M	A_s	M	A_s	M	A_s	M	A_s	M	A_s	M	A_s	M	A_s	M	A_s	M	A_s
20,000	2000	900	15	157.0	.0091	.63	.22	.98	.27	1.41	.33	1.92	.38	2.51	.44	3.2	.49	3.9	.54	4.7	.60	5.7	.65
	2500	1125	12	196.2	.0113	.78	.27	1.23	.34	1.76	.41	2.40	.48	3.14	.54	4.0	.61	4.9	.68	5.9	.75	7.1	.82
	3000	1350	10	235.6	.0136	.94	.33	1.48	.41	2.12	.49	2.89	.57	3.78	.65	4.9	.73	5.9	.82	7.1	.90	8.3	.98
	3750	1700	8	298.0	.0172	1.19	.41	1.86	.52	2.68	.62	3.65	.72	4.77	.82	6.0	.93	7.5	1.03	9.0	1.13	10.7	1.23

CINDER CONCRETE FLOOR SLABS

NEW YORK CITY BLDG. CODE 1948 - EMPIRICAL FORMULAS.

Non-continuous ends anchored.
1.43 As Cinder / 1.53 As Stone } non-continuous spans.
1" ; 1" ; As ; As ; t = inches ; L feet ; L feet ; 2" ; 2"
{10'-0" Max. for W < 200#/sq. ft.
{8'-0" " " W > 200#/sq. ft.

$$t = \frac{L}{2} + \frac{W - 75}{200}$$

$$\dagger A_s = \frac{WL^2}{3C}$$

†Min. A_s = .018(t-1)

t = 4" if L = 8'-0" or less and W = 200 or less.
W = Total uniform load in #/□' (Liveload + Dead load.)
* C = 26,000 for steel fabric of ultimate strength of 71,500#/□".

TABLE B - CROSS SECTIONAL AREA OF WELDED FABRIC PER FT. WIDTH OF SLAB = A_s IN □".

L	3'-0"	4'-0"	5'-0"	6'-0"	7'-0"	8'-0"	9'-0"	10'-0"
W \ t	4" SLAB						5" SLAB	5½" SLAB
80	.054	.054	.054	.054	.054	.066	.083	.103
90	.054	.054	.054	.054	.057	.074	.094	.115
100	.054	.054	.054	.054	.063	.082	.104	.128
110	.054	.054	.054	.054	.069	.090	.114	.141
120	.054	.054	.054	.056	.076	.098	.125	.154
130	.054	.054	.054	.060	.082	.107	.135	.167
140	.054	.054	.054	.065	.088	.115	.146	.180
150	.054	.054	.054	.069	.095	.123	.156	.192
160	.054	.054	.054	.074	.101	.131	.166	.205
170	.054	.054	.054	.079	.107	.140	.177	.218
							5½" SLAB	6" SLAB
180	.054	.054	.058	.083	.113	.148	.187	.230
190	.054	.054	.061	.088	.120	.156	.197	.243
200	.054	.054	.064	.092	.126	.164	.208	.256
W \ t	4½" SLAB					5" SLAB		
210	.063	.063	.067	.097	.132	.173		
220	.063	.063	.071	.102	.138	.181		
230	.063	.063	.074	.106	.145	.189		
240	.063	.063	.077	.111	.151	.197		
250	.063	.063	.080	.115	.157	.206		
260	.063	.063	.083	.120	.163	.214		
270	.063	.063	.087	.125	.170	.222		

MIX
1-Part cement
2-Parts sand
5-Parts cinders
(Clean hard burned anthracite.)

For sizes and spacing of steel mesh see 18 on page 2-16.
* C = 29,900 for stone concrete floor slabs (1:2:5 Mix.) Multiply values in table by .87 for area of reinf. for stone concrete.

STRUCTURAL CONCRETE–REINFORCEMENT–I

REINFORCEMENT

TABLE A – WELDED WIRE MESH BY AMERICAN STEEL & WIRE CO.

Spacing of Wires, in.		Size of Wires, A.S. & W. Gage		Sect. Area, sq. in. per ft.		Weight, lb. per 100 sq. ft.	Spacing of Wires, in.		Size of Wires, A.S. & W. Gage		Sect. Area, sq. in. per ft.		Weight, lb. per 100 sq. ft.
Longit.	Trans.	Longit.	Trans.	Longit.	Trans.		Longit.	Trans.	Longit.	Trans.	Longit.	Trans.	
2	2	10	10	.086	.086	60*	4	12	5	7	.101	.025	45
2	2	12	12	.052	.052	37*†	4	12	5	10	.101	.014	42
2	2	14	14	.030	.030	21*†	4	12	6	10	.087	.014	36
2	12	0	6	.443	.029	166	4	12	7	11	.074	.011	31
2	16	0	6	.443	.022	163	4	12	8	12	.062	.009	26
2	16	1	7	.377	.018	140	4	12	9	12	.052	.009	22
2	16	2	8	.325	.015	119	4	8	7	11	.074	.017	33
2	16	3	8	.280	.015	104	4	8	8	12	.062	.013	27
2	16	4	9	.239	.013	89	4	8	9	12	.052	.013	23
2	16	5	10	.202	.011	75	4	8	10	12	.043	.013	20
2	16	6	10	.174	.011	65	6	6	0	0	.148	.148	107*
2	16	7	11	.148	.009	55	6	6	1	1	.126	.126	91*
3	3	8	8	.082	.082	58*	6	6	2	2	.108	.108	78*
3	3	10	10	.057	.057	41*	6	6	3	3	.093	.093	68*
3	3	12	12	.035	.035	25*†	6	6	4	4	.080	.080	58*
3	3	14	14	.020	.020	14*†	6	6	4	6	.080	.058	50*
3	16	2	8	.216	.015	83	6	6	5	5	.067	.067	49*
3	16	3	8	.187	.015	72	6	6	6	6	.058	.058	42*
3	16	4	9	.159	.013	61	6	6	7	7	.049	.049	36*
4	4	4	4	.120	.120	85*	6	6	8	8	.041	.041	30*
4	4	6	6	.087	.087	62*	6	6	9	9	.035	.035	25*
4	4	8	8	.062	.062	44*	6	6	10	10	.029	.029	21*
4	4	10	10	.043	.043	31*	6	12	00	4	.172	.040	78
4	4	12	12	.026	.026	19*†	6	12	0	0	.148	.074	81
4	4	13	13	.020	.020	14*†	6	12	0	3	.148	.047	72
4	16	3	8	.140	.015	56	6	12	1	1	.126	.063	69
4	16	4	9	.120	.013	48	6	12	1	4	.126	.040	61
4	16	5	10	.101	.011	40	6	12	2	2	.108	.054	59
4	16	6	10	.087	.011	35	6	12	2	5	.108	.034	52
4	16	7	11	.074	.009	30	6	12	3	3	.093	.047	51
4	16	8	12	.062	.007	25	6	12	4	4	.080	.040	44
4	16	9	12	.052	.007	21	6	12	6	6	.058	.029	32
4	12	4	9	.120	.017	49							

*Two-way type. †Usually furnished only in Galvanized wire.

TABLE B – AREA OF STEEL PER FOOT OF WIDTH.

Bar No.	Bar Spacing, inches																			
	4	$4\frac{1}{2}$	5	$5\frac{1}{2}$	6	$6\frac{1}{2}$	7	$7\frac{1}{2}$	8	$8\frac{1}{2}$	9	$9\frac{1}{2}$	10	$10\frac{1}{2}$	11	$11\frac{1}{2}$	12	13	14	15	16	17	18
#2	0.15	0.13	0.12	0.11	0.10	0.09	0.09	0.08	0.08	0.07	0.07	0.06	0.06	0.06	0.05	0.05	0.05	–	–	–	–	–	–
#3	0.33	0.29	0.26	0.24	0.22	0.20	0.19	0.18	0.17	0.16	0.15	0.14	0.13	0.13	0.12	0.11	0.11	0.10	0.09	0.09	–	–	–
#4	0.60	0.53	0.48	0.44	0.40	0.37	0.34	0.32	0.30	0.28	0.27	0.25	0.24	0.23	0.22	0.21	0.20	0.18	0.17	0.16	0.15	0.14	0.13
#5	0.93	0.83	0.74	0.68	0.62	0.57	0.53	0.50	0.47	0.44	0.41	0.39	0.37	0.35	0.34	0.32	0.31	0.29	0.27	0.25	0.23	0.22	0.21
#6	1.32	1.17	1.06	0.96	0.88	0.81	0.75	0.70	0.66	0.62	0.59	0.56	0.53	0.50	0.48	0.46	0.44	0.41	0.38	0.35	0.33	0.31	0.29
#7	1.80	1.60	1.44	1.31	1.20	1.11	1.03	0.96	0.90	0.85	0.80	0.76	0.72	0.69	0.65	0.63	0.60	0.55	0.51	0.48	0.45	0.42	0.40
#8	2.37	2.11	1.90	1.72	1.58	1.46	1.35	1.26	1.19	1.12	1.05	1.00	0.95	0.90	0.86	0.82	0.79	0.73	0.68	0.63	0.59	0.56	0.53
#9	3.00	2.67	2.40	2.18	2.00	1.85	1.71	1.60	1.50	1.41	1.33	1.26	1.20	1.14	1.09	1.04	1.00	0.92	0.86	0.80	0.75	0.71	0.67
#10	3.81	3.39	3.05	2.77	2.54	2.35	2.18	2.03	1.91	1.79	1.69	1.60	1.52	1.45	1.39	1.33	1.27	1.17	1.09	1.02	0.95	0.90	0.85
#11	4.68	4.16	3.74	3.40	3.12	2.88	2.67	2.50	2.34	2.20	2.08	1.97	1.87	1.78	1.70	1.63	1.56	1.44	1.34	1.25	1.17	1.10	1.04

STRUCTURAL CONCRETE - REINFORCEMENT - 2

REINFORCEMENT

TABLE A – MINIMUM BEAM WIDTHS, A. C. I. CODE (1) *†								
Size of Bars	Number of Bars in Single Layer of Reinforcing							Add for Each Added Bar
	2	3	4	5	6	7	8	
#4	5¾	7¼	8¾	10¼	11¾	13¼	14¾	1½
#5	6	7¾	9¼	11	12½	14¼	15¾	1⅝
#6	6¼	8	9¾	11½	13¼	15	16¾	1¾
#7	6½	8½	10¼	12¼	14	16	17¾	1⅞
#8	6¾	8¾	10¾	12¾	14¾	16¾	18¾	2
#9	7¼	9½	11¾	14	16¼	18½	20¾	2¼
#10	7¾	10¼	12¾	15¼	17¾	20¼	23	2⅝
#11	8	11	13¾	16½	19½	22¼	25	2⅞

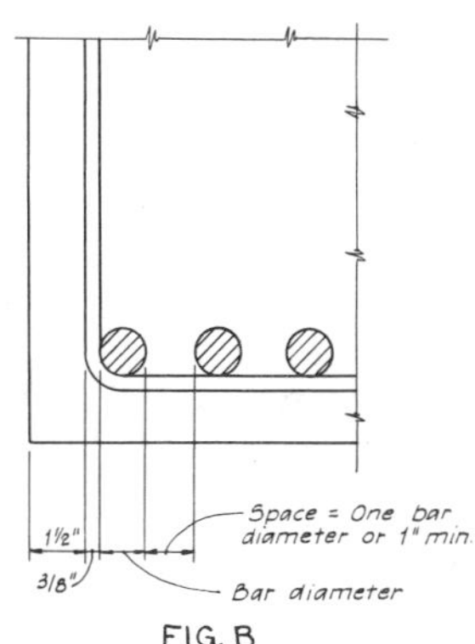
Space = One bar diameter or 1" min.
1½"
3/8"
Bar diameter
FIG. B

(1) Table includes #3 stirrup. (If no stirrups, deduct ¾" from figures shown.) The clear distance between parallel bars shall not be less than the nominal diameter of the bars, 1⅓ times the maximum-size coarse aggregate, nor 1".

TABLE B – MINIMUM BEAM WIDTHS, A. A. S. H. O. SPECIFICATIONS (2) *†								
Size of Bars	Number of Bars in Single Layer of Reinforcing							Add for Each Added Bar
	2	3	4	5	6	7	8	
#4	7	8½	10					1½
#5	7	9	10½	12	13½			1⅝
#6	7½	9½	11½	13	15	17		1⅞
#7	8	10	12½	14½	17	19	21	2 3/16
#8	8½	11	13½	16	18½	21	23½	2½
#9	9	12	14½	17½	20	23	26	2 13/16
#10	9½	12½	16	19	22	25½	28½	3 3/16
#11	10	13½	17	20½	24	27½	31	3½

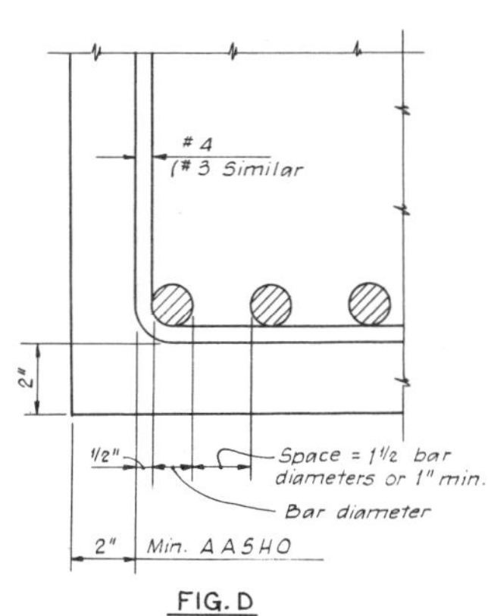
#4
(#3 Similar
2"
1/2"
Space = 1½ bar diameters or 1" min.
Bar diameter
2"
Min. AASHO
FIG. D

(2) Table includes #3 and #4 stirrups. (If no stirrups, deduct 1" from figures shown.) The clear distance between parallel bars shall not be less than 1½ times the nominal diameter of the bars, nor 1".

*For bars of different sizes, select the beam width for smaller bars, and add last column figure for each larger bar used.

TABLE C – AREAS AND PERIMETERS OF VARIOUS NUMBERS OF BARS												
Bar No.		Number of Bars										
		1	2	3	4	5	6	7	8	9	10	12
#2	As	0.05										
	εo	0.8										
#3	As	0.11										
	εo	1.2										
#4	As	0.20	0.40	0.60	0.80	1.00	1.20	1.40	1.60	1.80	2.00	2.40
	εo	1.6	3.1	4.7	6.3	7.9	9.4	11.0	12.6	14.1	15.7	18.8
#5	As	0.31	0.62	0.93	1.24	1.55	1.86	2.17	2.48	2.79	3.10	3.72
	εo	2.0	3.9	5.9	7.9	9.8	11.8	13.7	15.7	17.7	19.6	23.5
#6	As	0.44	0.88	1.32	1.76	2.20	2.64	3.08	3.52	3.96	4.40	5.28
	εo	2.4	4.7	7.1	9.4	11.8	14.1	16.5	18.8	21.2	23.6	28.3
#7	As	0.60	1.20	1.80	2.40	3.00	3.60	4.20	4.80	5.40	6.00	7.20
	εo	2.7	5.5	8.2	11.0	13.7	16.5	19.2	22.0	24.7	27.5	33.0
#8	As	0.79	1.58	2.37	3.16	3.95	4.74	5.53	6.32	7.11	7.90	9.48
	εo	3.1	6.3	9.4	12.6	15.7	18.9	22.0	25.1	28.3	31.4	37.7
#9	As	1.00	2.00	3.00	4.00	5.00	6.00	7.00	8.00	9.00	10.00	12.00
	εo	3.5	7.1	10.6	14.2	17.7	21.3	24.8	28.4	31.9	35.5	42.6
#10	As	1.27	2.54	3.81	5.08	6.35	7.62	8.89	10.16	11.43	12.70	15.24
	εo	4.0	8.0	12.0	16.0	20.0	24.0	27.9	31.9	35.9	39.9	47.9
#11	As	1.56	3.12	4.68	6.24	7.80	9.36	10.92	12.48	14.04	15.60	18.72
	εo	4.4	8.9	13.3	17.7	22.2	26.6	31.0	35.4	39.9	44.3	53.2

As = Area steel; spacing of bars in slab (in inches) = $\frac{12\ As}{A's}$ A's = Area steel required; εo = perimeter

†From *Manual of Standard Practice for Detailing Reinforced Concrete Structures* (ACI 315-57).

STRUCTURAL CONCRETE-REINFORCEMENT-3

REINFORCEMENT*

RECOMMENDED A. C. I. REQUIREMENTS FOR HOOKS

TABLE A – 180° HOOKS

Bar Size d	Hook A or G	J	Approx. H
#2	4	2	3½
#3	5	3	4
#4	6	4	4½
#5	7	5	5
#6	8	6	6
#7	10	7	7
#8	1-1	10	9
#9	1-3	11¼	10¼
#10	1-5	1-0½	11¼
#11	1-7	1-2	1-0¾

TABLE B – 90° HOOKS

Bar Size d	Hook A or G	J
#2	3½	4
#3	5½	6
#4	7½	8¼
#5	9	10¼
#6	10½	1-0½
#7	1-0½	1-2½
#8	1-2½	1-5
#9	1-4½	1-7
#10	1-6½	1-9½
#11	1-8½	2-0

LAP SPLICES – NUMBER OF BAR DIAMETERS REQUIRED FOR LAP

TABLE C – TENSION SPLICES (NOT INCLUDING TOP BARS)

Concrete, Strength, f′c	Steel Stress, fs					
	16,000	18,000	20,000	24,000	27,000	30,000
2000	24	24	25	30	34	38
2500	24	24	24	24	27	30
3000	24	24	24	24	24	25
3500 or more	24	24	24	24	24	24

TABLE E – TENSION SPLICES (TOP BARS)†

Concrete, Strength, f′c	Steel Stress, fs					
	16,000	18,000	20,000	24,000	27,000	30,000
2000	29	32	36	43	48	54
2500	24	26	29	34	39	43
3000	24	24	24	29	32	36
3500 or more	24	24	24	24	28	31

TABLE D – INCHES OF LAP CORRESPONDING TO NUMBER OF BAR DIAMETERS

No. of Diam.	Size of Bar								
	#3	#4	#5	#6	#7	#8	#9	#10	#11
20	–	–	13	15	18	20	23	26	29
21	–	–	14	16	19	21	24	27	30
22	–	–	14	17	20	22	25	28	31
23	–	12	15	18	21	23	26	30	33
24	–	12	15	18	21	24	28	31	34
25	–	13	16	19	22	25	29	32	36
26	–	13	17	20	23	26	30	33	37
27	–	14	17	21	24	27	31	35	39
28	–	14	18	21	25	28	32	36	40
29	–	15	19	22	26	29	33	37	41
30	12	15	19	23	27	30	34	39	43
32	12	16	20	24	28	32	36	41	45
34	13	17	22	26	30	34	39	44	48
36	14	18	23	27	32	36	41	46	51
38	15	19	24	29	34	38	43	49	54
40	15	20	25	30	35	40	46	51	57

*From *Manual of Standard Practice for Detailing Reinforced Concrete Structures* (ACI 315-57).

†Top bars are those near top of beams, girders and walls.

STRUCTURAL CONCRETE —TYPICAL TIED COLUMN DETAILS

TYPICAL TIED COLUMN DETAILS*

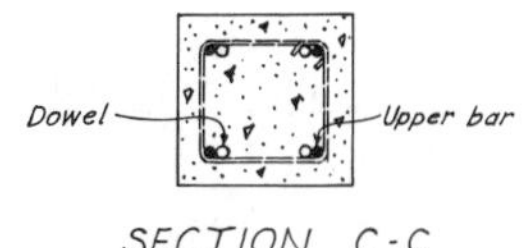
Dowel
Upper bar

SECTION C-C

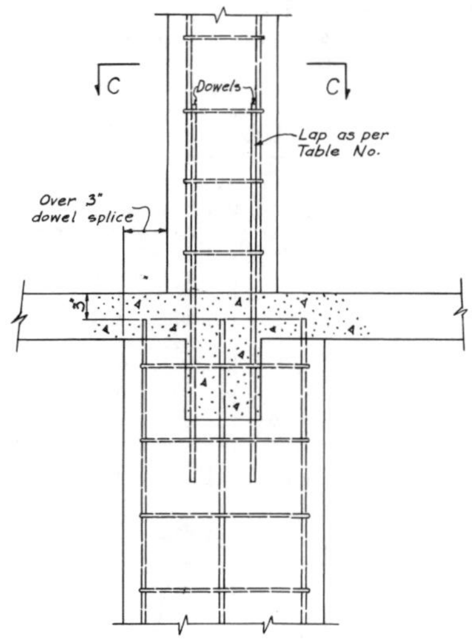
C
Dowels
C
Lap as per Table No.
Over 3" dowel splice
3"

DETAIL SHOWING RECTANGULAR TIED COLUMN

FIG. A—DETAIL SHOWING RECTANGULAR COLUMN

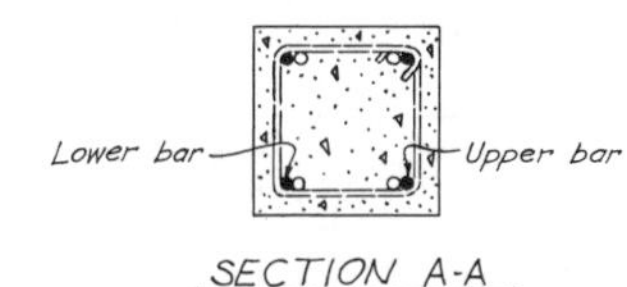
Lower bar
Upper bar

SECTION A-A

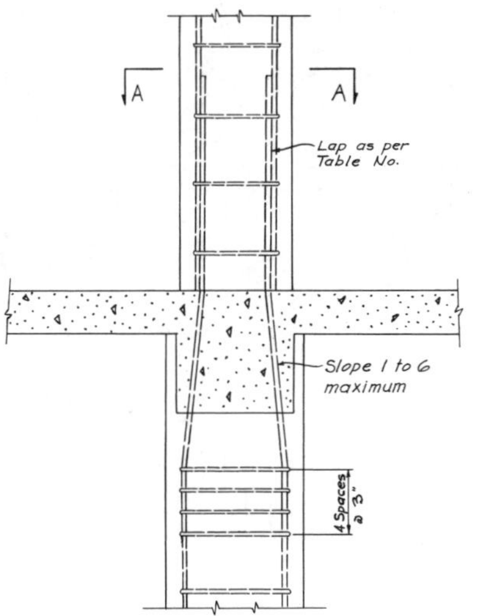
A
A
Lap as per Table No.
Slope 1 to 6 maximum
4 Spaces @ 3"

DETAIL SHOWING TYPICAL TIED COLUMN

FIG. B—DETAIL SHOWING TYPICAL TIED COLUMN

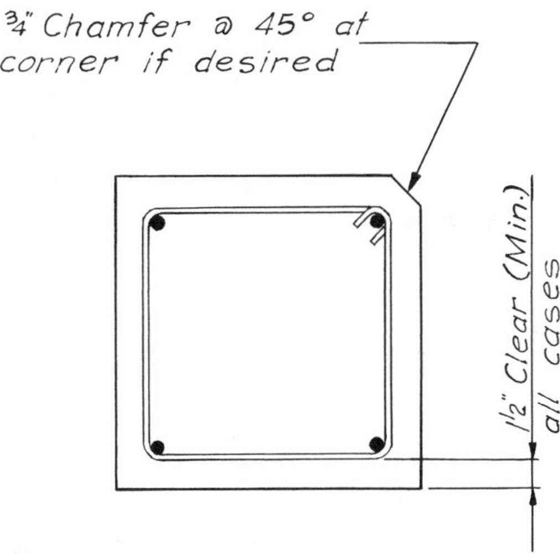
¾" Chamfer @ 45° at corner if desired
1½" Clear (Min.) all cases

4 - BARS
1 TIE PER SET

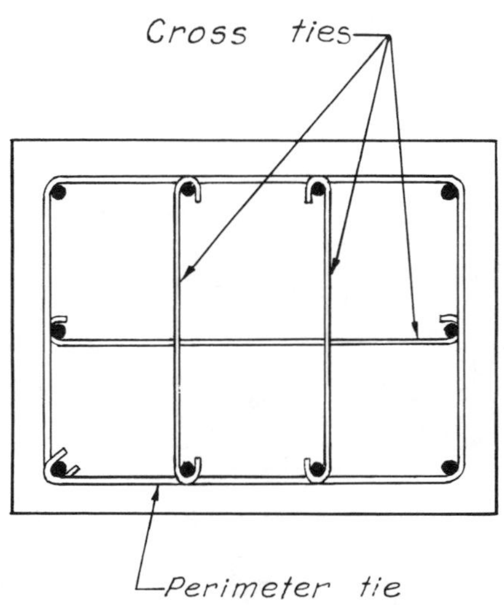
Cross ties
Perimeter tie

6 - OR MORE BARS
1 PERIMETER TIE PLUS 1 CROSS TIE FOR EVERY 2 ADDITIONAL VERTICAL BARS OVER 4

FIG. C—LATERAL TIE DETAILS

*From *Manual of Standard Practice for Detailing Reinforced Concrete Structure* (ACI 315-57).

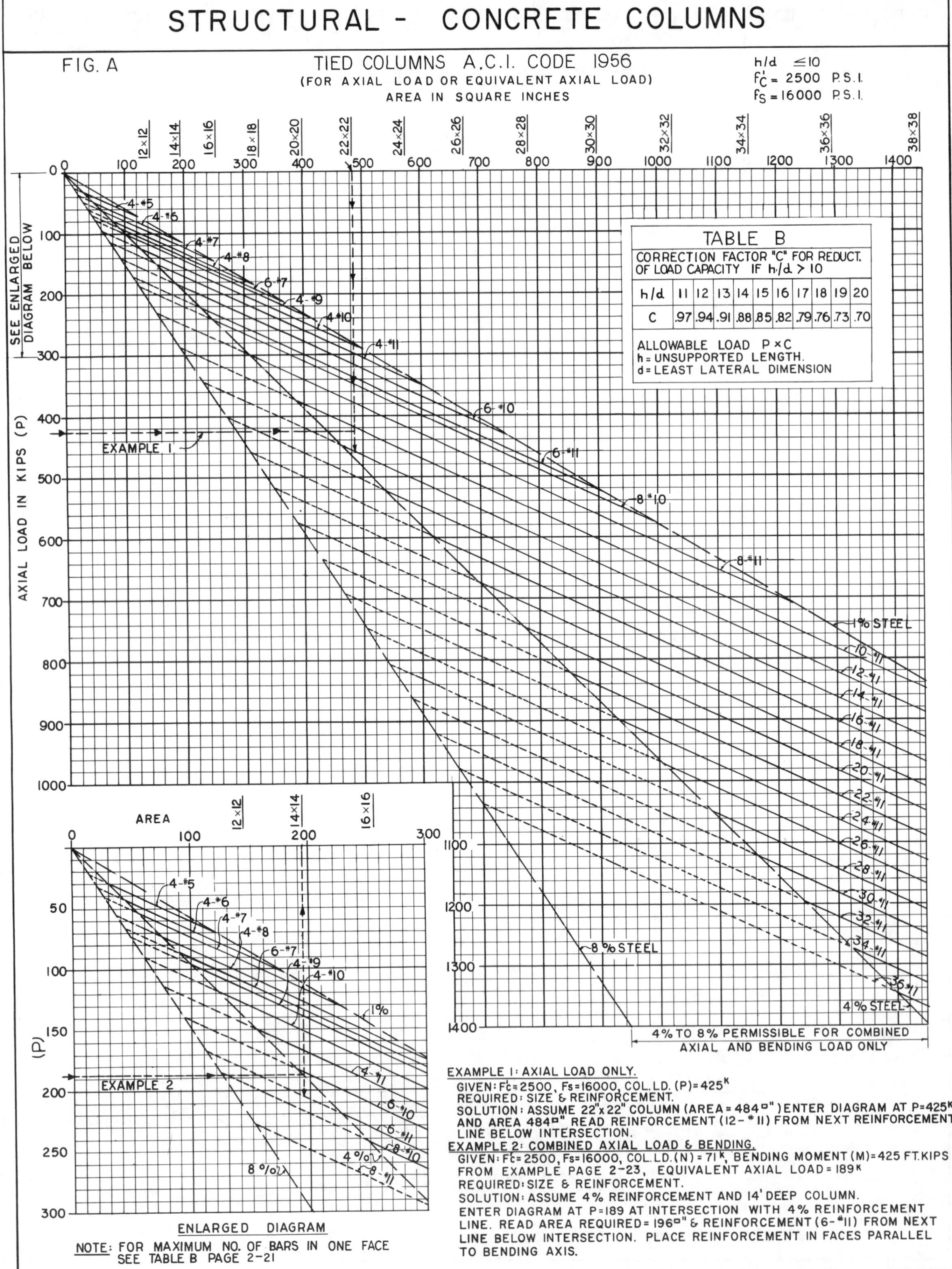
STRUCTURAL - CONCRETE COLUMNS
FIG. A
TIED COLUMNS A.C.I. CODE 1956
(FOR AXIAL LOAD OR EQUIVALENT AXIAL LOAD)
AREA IN SQUARE INCHES
h/d ≤10
f'c = 2500 P.S.I.
fs = 16000 P.S.I.
12×12
14×14
16×16
18×18
20×20
22×22
24×24
26×26
28×28
30×30
32×32
34×34
36×36
38×38
0
100
200
300
400
500
600
700
800
900
1000
1100
1200
1300
1400
SEE ENLARGED DIAGRAM BELOW
AXIAL LOAD IN KIPS (P)
4-#5
4-#6
4-#7
4-#8
6-#7
4-#9
4-#10
4-#11
6-#10
6-#11
8-#10
8-#11
1% STEEL
10-#11
12-#11
14-#11
16-#11
18-#11
20-#11
22-#11
24-#11
26-#11
28-#11
30-#11
32-#11
34-#11
36-#11
4% STEEL
8% STEEL
EXAMPLE 1
TABLE B
CORRECTION FACTOR "C" FOR REDUCT. OF LOAD CAPACITY IF h/d > 10
h/d | 11 | 12 | 13 | 14 | 15 | 16 | 17 | 18 | 19 | 20
C | .97 | .94 | .91 | .88 | .85 | .82 | .79 | .76 | .73 | .70
ALLOWABLE LOAD P × C
h = UNSUPPORTED LENGTH.
d = LEAST LATERAL DIMENSION
4% TO 8% PERMISSIBLE FOR COMBINED AXIAL AND BENDING LOAD ONLY
AREA
0
100
200
300
50
150
250
(P)
4-#5
4-#6
4-#7
4-#8
6-#7
4-#9
4-#10
1%
4-#11
6-#10
6-#11
8-#10
4%
8-#11
8%
EXAMPLE 2
ENLARGED DIAGRAM
NOTE: FOR MAXIMUM NO. OF BARS IN ONE FACE SEE TABLE B PAGE 2-21
EXAMPLE 1: AXIAL LOAD ONLY.
GIVEN: Fc=2500, Fs=16000, COL. LD. (P)=425K
REQUIRED: SIZE & REINFORCEMENT.
SOLUTION: ASSUME 22"x22" COLUMN (AREA = 484□") ENTER DIAGRAM AT P=425K AND AREA 484□" READ REINFORCEMENT (12-#11) FROM NEXT REINFORCEMENT LINE BELOW INTERSECTION.
EXAMPLE 2: COMBINED AXIAL LOAD & BENDING.
GIVEN: Fc=2500, Fs=16000, COL. LD. (N)=71K, BENDING MOMENT (M)=425 FT. KIPS FROM EXAMPLE PAGE 2-23, EQUIVALENT AXIAL LOAD = 189K
REQUIRED: SIZE & REINFORCEMENT.
SOLUTION: ASSUME 4% REINFORCEMENT AND 14' DEEP COLUMN. ENTER DIAGRAM AT P=189 AT INTERSECTION WITH 4% REINFORCEMENT LINE. READ AREA REQUIRED = 196□" & REINFORCEMENT (6-#11) FROM NEXT LINE BELOW INTERSECTION. PLACE REINFORCEMENT IN FACES PARALLEL TO BENDING AXIS.

STRUCTURAL - CONCRETE COLUMNS

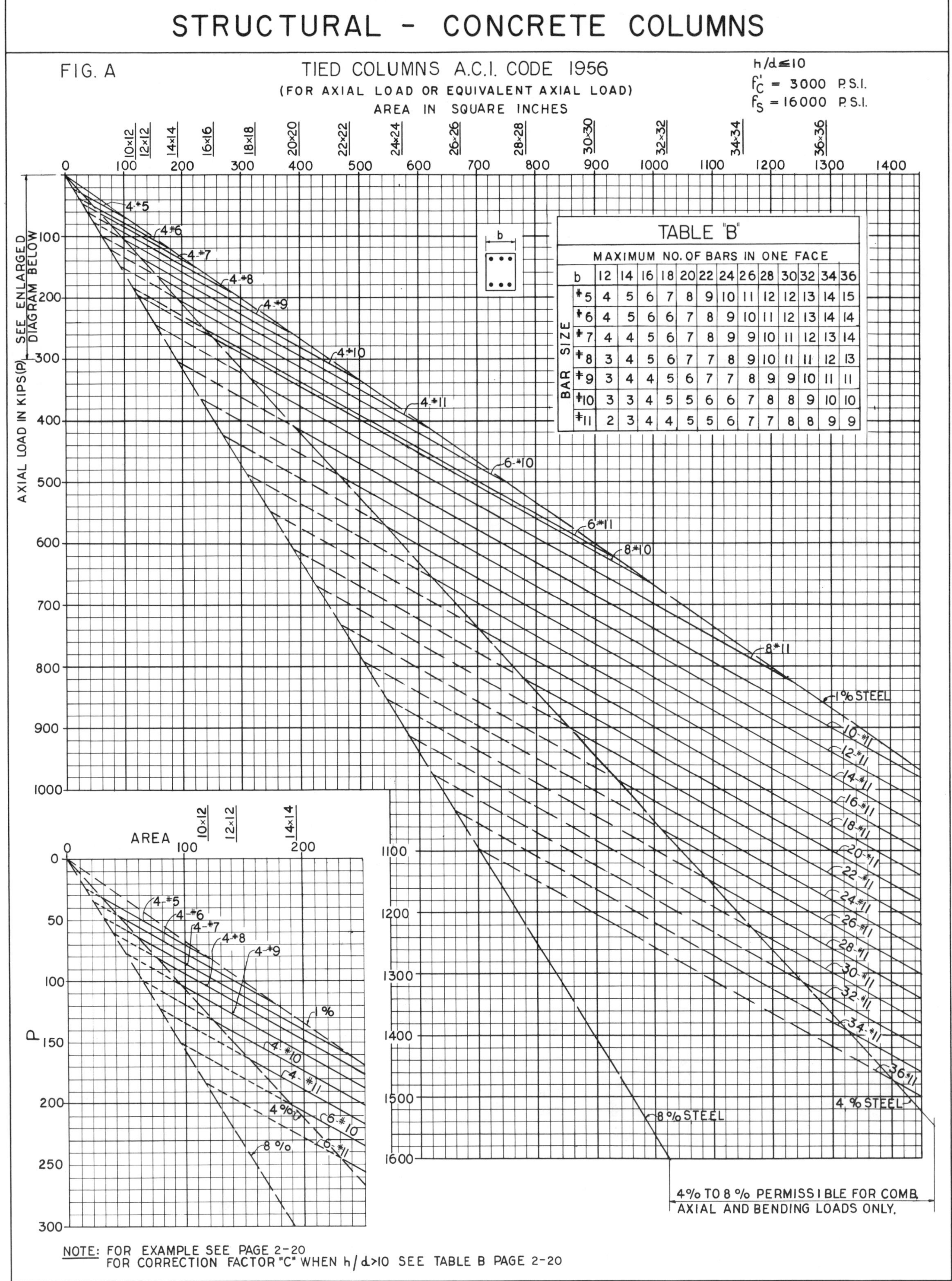
FIG. A
TIED COLUMNS A.C.I. CODE 1956
(FOR AXIAL LOAD OR EQUIVALENT AXIAL LOAD)
AREA IN SQUARE INCHES
h/d ≤ 10
f'c = 3000 P.S.I.
fs = 16000 P.S.I.
10×12 12×12 14×14 16×16 18×18 20×20 22×22 24×24 26×26 28×28 30×30 32×32 34×34 36×36
AXIAL LOAD IN KIPS (P)
SEE ENLARGED DIAGRAM BELOW
4-#5
4-#6
4-#7
4-#8
4-#9
4-#10
4-#11
6-#10
6-#11
8-#10
8-#11
1% STEEL
10-#11
12-#11
14-#11
16-#11
18-#11
20-#11
22-#11
24-#11
26-#11
28-#11
30-#11
32-#11
34-#11
36-#11
4 % STEEL
8 % STEEL
4% TO 8 % PERMISSIBLE FOR COMB. AXIAL AND BENDING LOADS ONLY.
AREA
P
1%
4-#10
4-#11
4%
6-#10
6-#11
8 %

TABLE "B"

MAXIMUM NO. OF BARS IN ONE FACE

BAR SIZE \ b	12	14	16	18	20	22	24	26	28	30	32	34	36
#5	4	5	6	7	8	9	10	11	12	12	13	14	15
#6	4	5	6	6	7	8	9	10	11	12	13	14	14
#7	4	4	5	6	7	8	9	9	10	11	12	13	14
#8	3	4	5	6	7	7	8	9	10	11	11	12	13
#9	3	4	4	5	6	7	7	8	9	9	10	11	11
#10	3	3	4	5	5	6	6	7	8	8	9	10	10
#11	2	3	4	4	5	5	6	7	7	8	8	9	9

NOTE: FOR EXAMPLE SEE PAGE 2-20
FOR CORRECTION FACTOR "C" WHEN h/d>10 SEE TABLE B PAGE 2-20

STRUCTURAL - CONCRETE COLUMNS

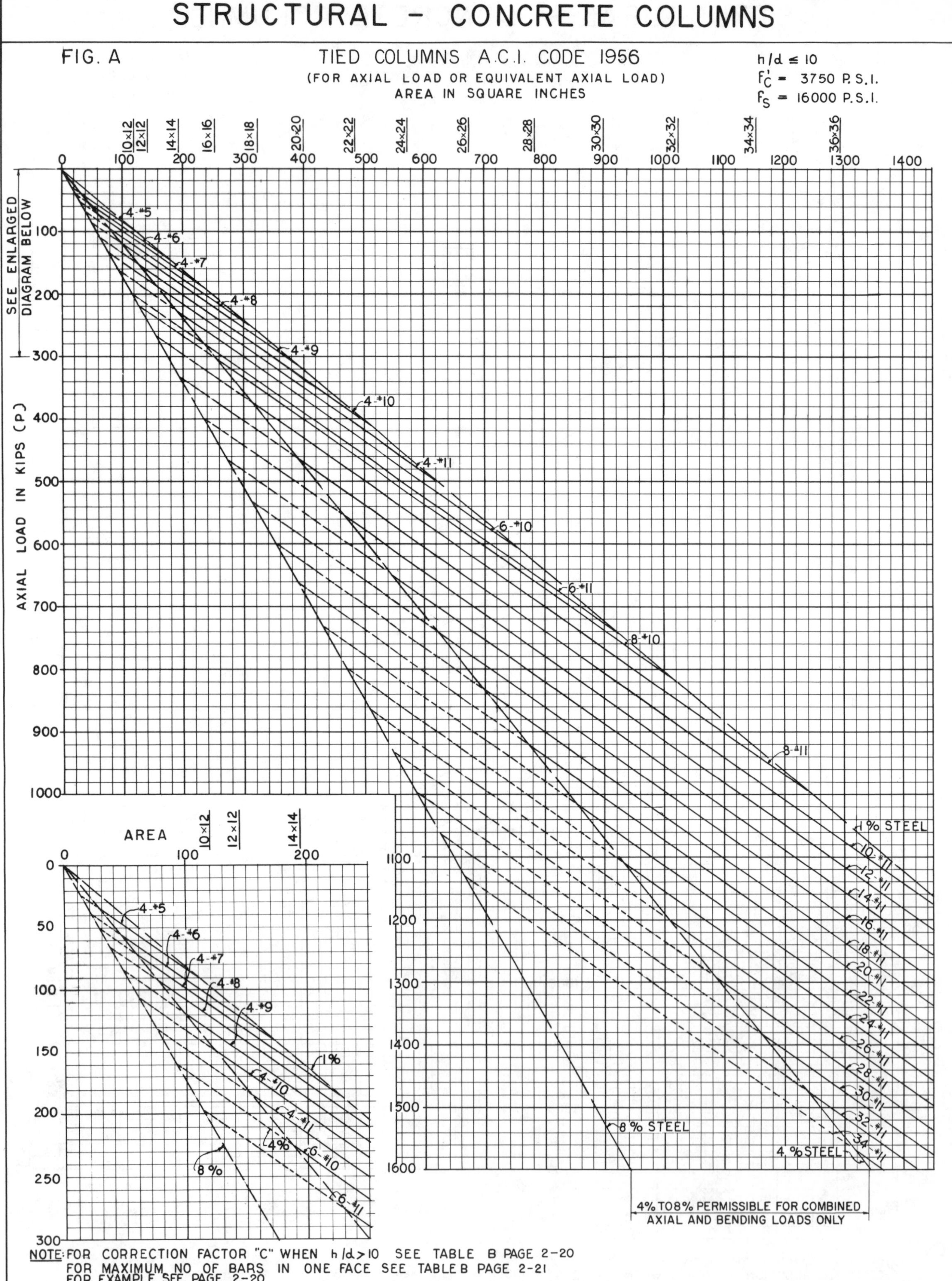
FIG. A
TIED COLUMNS A.C.I. CODE 1956
(FOR AXIAL LOAD OR EQUIVALENT AXIAL LOAD)
AREA IN SQUARE INCHES
h/d ≤ 10
f'c = 3750 P.S.I.
fs = 16000 P.S.I.
10×12
12×12
14×14
16×16
18×18
20×20
22×22
24×24
26×26
28×28
30×30
32×32
34×34
36×36
0
100
200
300
400
500
600
700
800
900
1000
1100
1200
1300
1400
SEE ENLARGED DIAGRAM BELOW
AXIAL LOAD IN KIPS (P)
1500
1600
4-#5
4-#6
4-#7
4-#8
4-#9
4-#10
4-#11
6-#10
6-#11
8-#10
8-#11
1% STEEL
10-#11
12-#11
14-#11
16-#11
18-#11
20-#11
22-#11
24-#11
26-#11
28-#11
30-#11
32-#11
34-#11
8% STEEL
4%STEEL
4% TO 8% PERMISSIBLE FOR COMBINED AXIAL AND BENDING LOADS ONLY
AREA
50
150
250
1%
4%
8%

NOTE: FOR CORRECTION FACTOR "C" WHEN h/d>10 SEE TABLE B PAGE 2-20
FOR MAXIMUM NO OF BARS IN ONE FACE SEE TABLE B PAGE 2-21
FOR EXAMPLE SEE PAGE 2-20

STRUCTURAL - CONCRETE COLUMNS

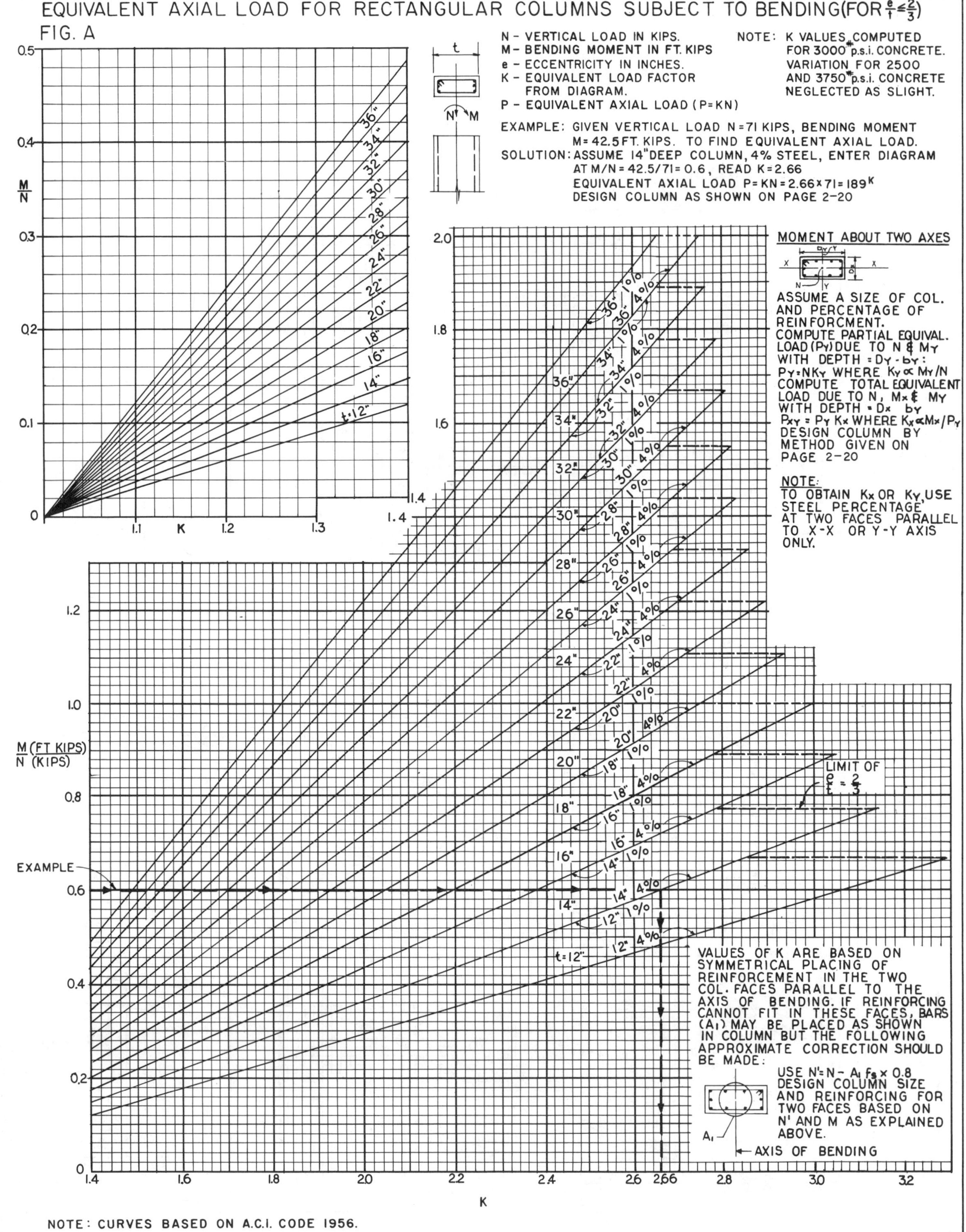

EQUIVALENT AXIAL LOAD FOR RECTANGULAR COLUMNS SUBJECT TO BENDING (FOR $\frac{e}{t} \leq \frac{2}{3}$)
FIG. A
N - VERTICAL LOAD IN KIPS.
M - BENDING MOMENT IN FT. KIPS
e - ECCENTRICITY IN INCHES.
K - EQUIVALENT LOAD FACTOR FROM DIAGRAM.
P - EQUIVALENT AXIAL LOAD (P=KN)
NOTE: K VALUES COMPUTED FOR 3000# p.s.i. CONCRETE. VARIATION FOR 2500 AND 3750# p.s.i. CONCRETE NEGLECTED AS SLIGHT.
EXAMPLE: GIVEN VERTICAL LOAD N=71 KIPS, BENDING MOMENT M=42.5 FT. KIPS. TO FIND EQUIVALENT AXIAL LOAD.
SOLUTION: ASSUME 14" DEEP COLUMN, 4% STEEL, ENTER DIAGRAM AT M/N=42.5/71=0.6, READ K=2.66
EQUIVALENT AXIAL LOAD P=KN=2.66×71=189K
DESIGN COLUMN AS SHOWN ON PAGE 2-20
MOMENT ABOUT TWO AXES
ASSUME A SIZE OF COL. AND PERCENTAGE OF REINFORCMENT.
COMPUTE PARTIAL EQUIVAL. LOAD (P_Y) DUE TO N & M_Y WITH DEPTH = D_Y - b_Y:
$P_Y = NK_Y$ WHERE $K_Y \propto M_Y/N$
COMPUTE TOTAL EQUIVALENT LOAD DUE TO N, M_X & M_Y WITH DEPTH = D_X b_Y
$P_{XY} = P_Y K_X$ WHERE $K_X \propto M_X/P_Y$
DESIGN COLUMN BY METHOD GIVEN ON PAGE 2-20
NOTE: TO OBTAIN K_X OR K_Y USE STEEL PERCENTAGE AT TWO FACES PARALLEL TO X-X OR Y-Y AXIS ONLY.
$\frac{M}{N}$
K
t=12" 14" 16" 18" 20" 22" 24" 26" 28" 30" 32" 34" 36"
$\frac{M\ (FT\ KIPS)}{N\ (KIPS)}$
EXAMPLE
LIMIT OF $\frac{e}{t} = \frac{2}{3}$
36" 1% 36" 4% 34" 1% 34" 4% 32" 1% 32" 4% 30" 1% 30" 4% 28" 1% 28" 4% 26" 1% 26" 4% 24" 1% 24" 4% 22" 1% 22" 4% 20" 1% 20" 4% 18" 1% 18" 4% 16" 1% 16" 4% 14" 1% 14" 4% 12" 1% 12" 4%
VALUES OF K ARE BASED ON SYMMETRICAL PLACING OF REINFORCEMENT IN THE TWO COL. FACES PARALLEL TO THE AXIS OF BENDING. IF REINFORCING CANNOT FIT IN THESE FACES, BARS (A_1) MAY BE PLACED AS SHOWN IN COLUMN BUT THE FOLLOWING APPROXIMATE CORRECTION SHOULD BE MADE:
USE $N' = N - A_1 f_s \times 0.8$
DESIGN COLUMN SIZE AND REINFORCING FOR TWO FACES BASED ON N' AND M AS EXPLAINED ABOVE.
AXIS OF BENDING
NOTE: CURVES BASED ON A.C.I. CODE 1956.

STRUCTURAL—

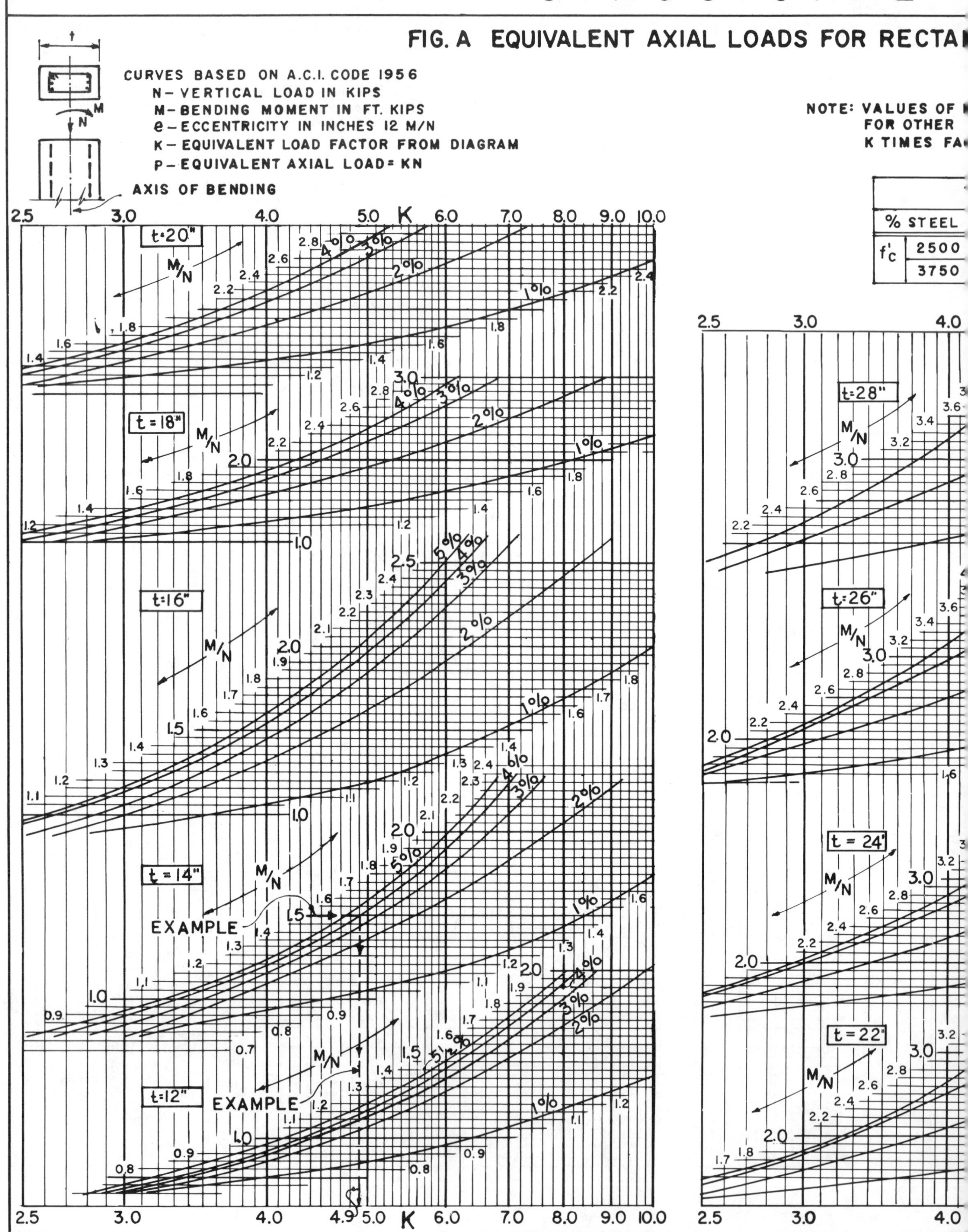

FIG. A EQUIVALENT AXIAL LOADS FOR RECTA
CURVES BASED ON A.C.I. CODE 1956
N – VERTICAL LOAD IN KIPS
M – BENDING MOMENT IN FT. KIPS
e – ECCENTRICITY IN INCHES 12 M/N
K – EQUIVALENT LOAD FACTOR FROM DIAGRAM
P – EQUIVALENT AXIAL LOAD = KN
AXIS OF BENDING
NOTE: VALUES OF
FOR OTHER
K TIMES FA
% STEEL
f'c
2500
3750
t=20"
t=18"
t=16"
t=14"
t=12"
t=28"
t=26"
t=24"
t=22"
M/N
EXAMPLE
EXAMPLE
K
2.5
3.0
4.0
4.9
5.0
6.0
7.0
8.0
9.0
10.0

NCRETE COLUMNS

R COLUMNS SUBJECT TO BENDING (FOR $\frac{e}{t} > \frac{2}{3}$)

ED FOR f'_c = 3000 p.s.i.
E STRENGTHS USE
WN BELOW.

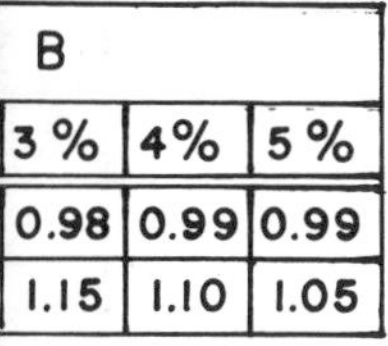

B		
3 %	4 %	5 %
0.98	0.99	0.99
1.15	1.10	1.05

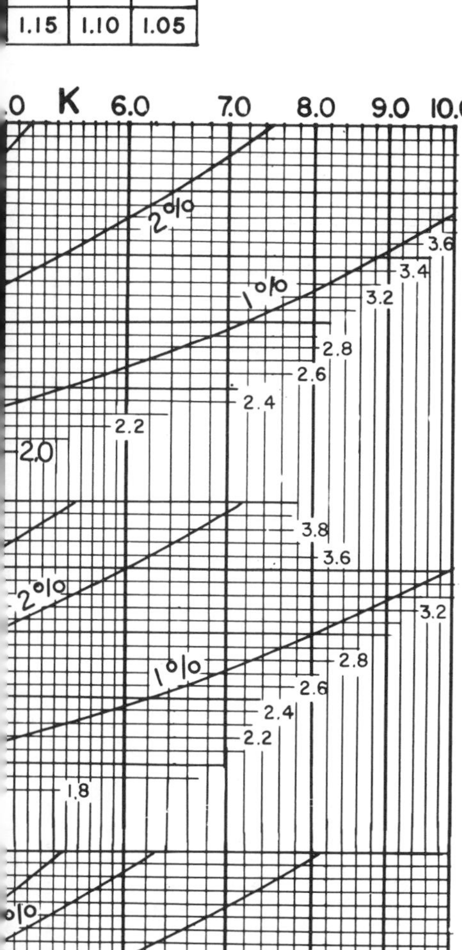

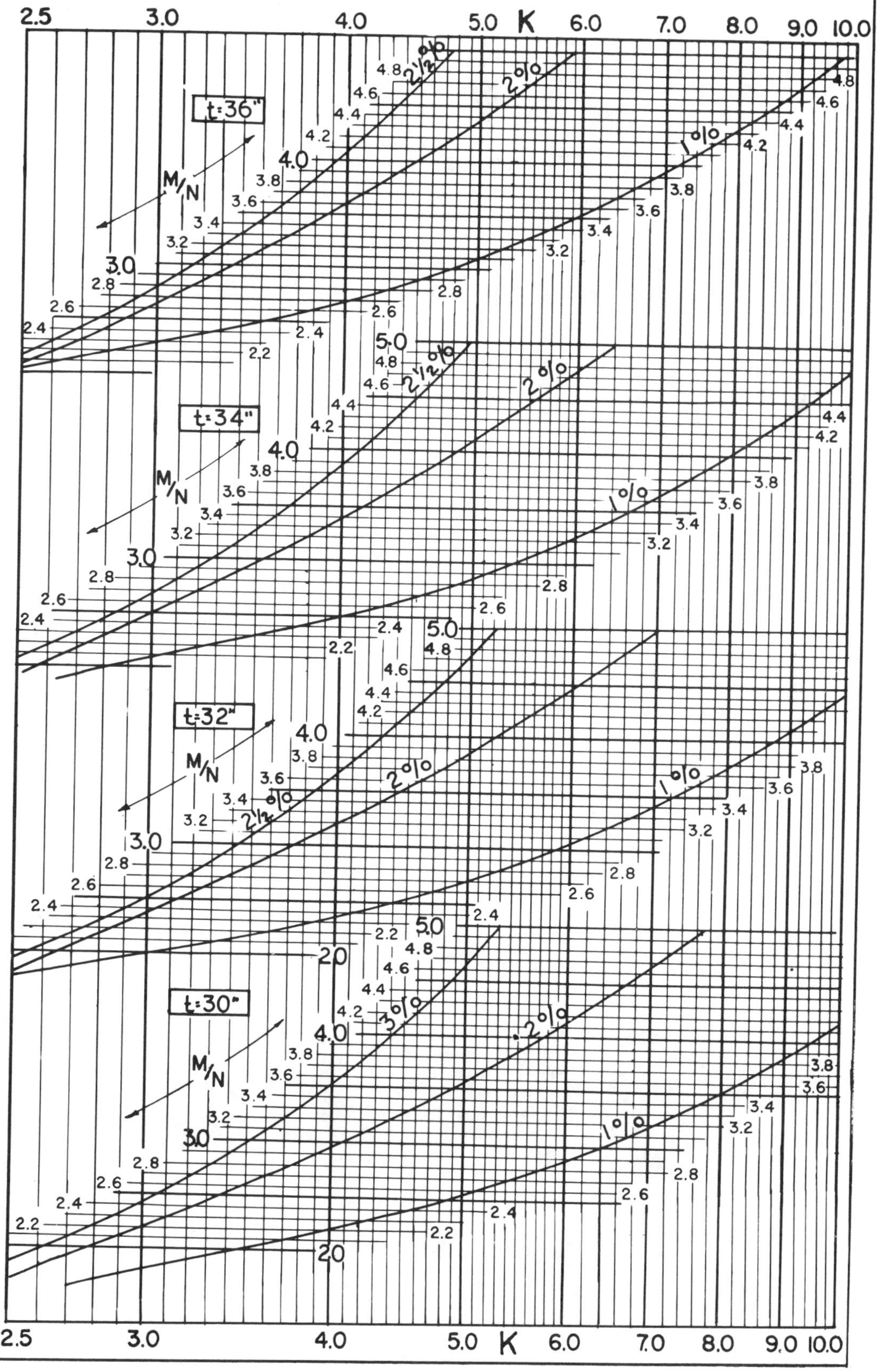

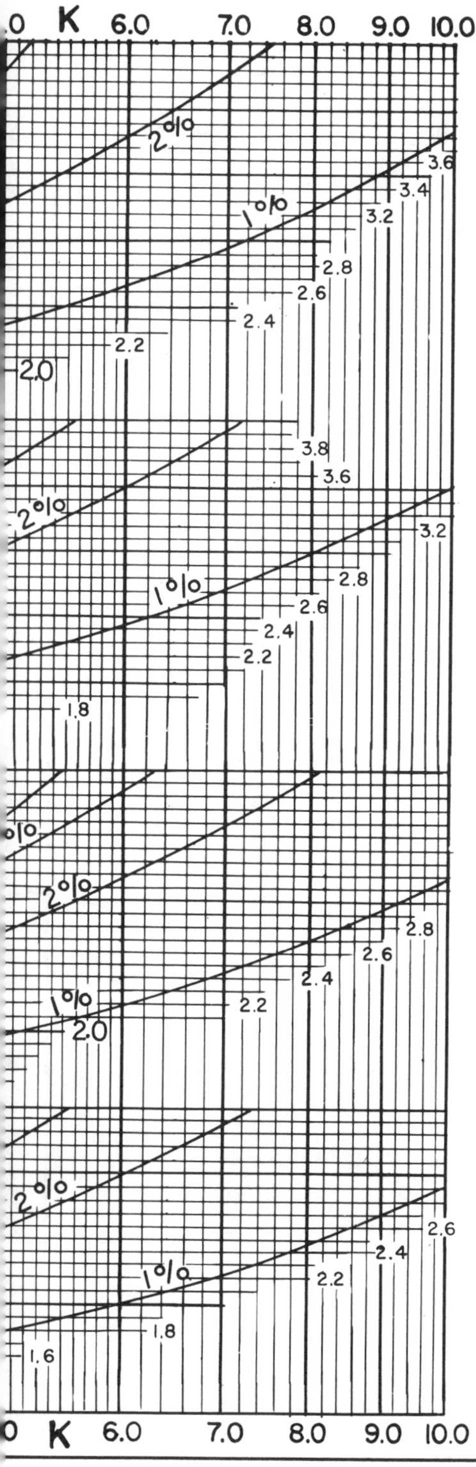

K
6.0 7.0 8.0 9.0 10.0

EXAMPLE: GIVEN VERTICAL LOAD N = 39^k, BENDING MOMENT M = 59^k f'_c = 2500,
TO FIND EQUIVALENT AXIAL LOAD,
SOLUTION: ASSUME t = 14" COL. WITH 4% STEEL, ENTER FIG. A AT M/N 59/39 = 1.5, READ TO 4% STEEL, READ DOWN K = 4.9 FOR f'_c = 2500 p.s.i., SELECT FACTOR FROM TABLE B (FOR 4% STEEL = 0.99)
EQUIVALENT AXIAL LOAD P = 39 X 4.9 X .99 = 189^k
FOR DESIGN OF COLUMN SEE METHOD GIVEN ON PAGE 2-20

2.5 3.0 4.0 5.0 K 6.0 7.0 8.0 9.0 10.0

t = 36" t = 34" t = 32" t = 30"

STRUCTURAL CONCRETE – TYPICAL SPIRAL COLUMN DETAILS

SPIRAL COLUMN DETAILS*

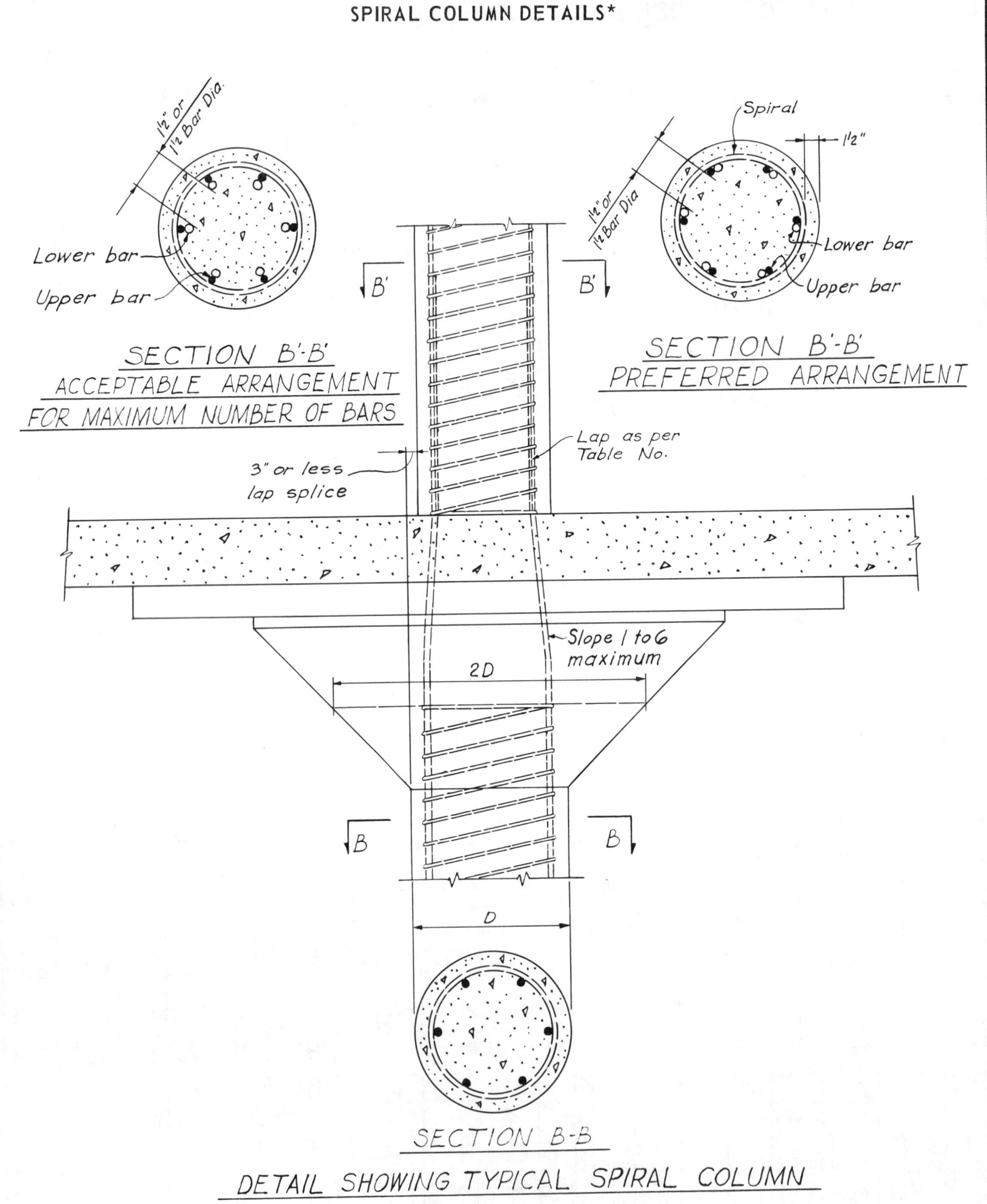
1½" or
1½ Bar Dia.
Lower bar
Upper bar
SECTION B'-B'
ACCEPTABLE ARRANGEMENT
FOR MAXIMUM NUMBER OF BARS
Spiral
1½"
1½" or
1½ Bar Dia
Lower bar
Upper bar
SECTION B'-B'
PREFERRED ARRANGEMENT
B'
B'
Lap as per
Table No.
3" or less
lap splice
Slope 1 to 6
maximum
2D
B
B
D
SECTION B-B
DETAIL SHOWING TYPICAL SPIRAL COLUMN

FIG. A

*From *Manual of Standard Practice for Detailing Reinforced Concrete Structure* (ACI 315-57).

STRUCTURAL - CONCRETE COLUMNS

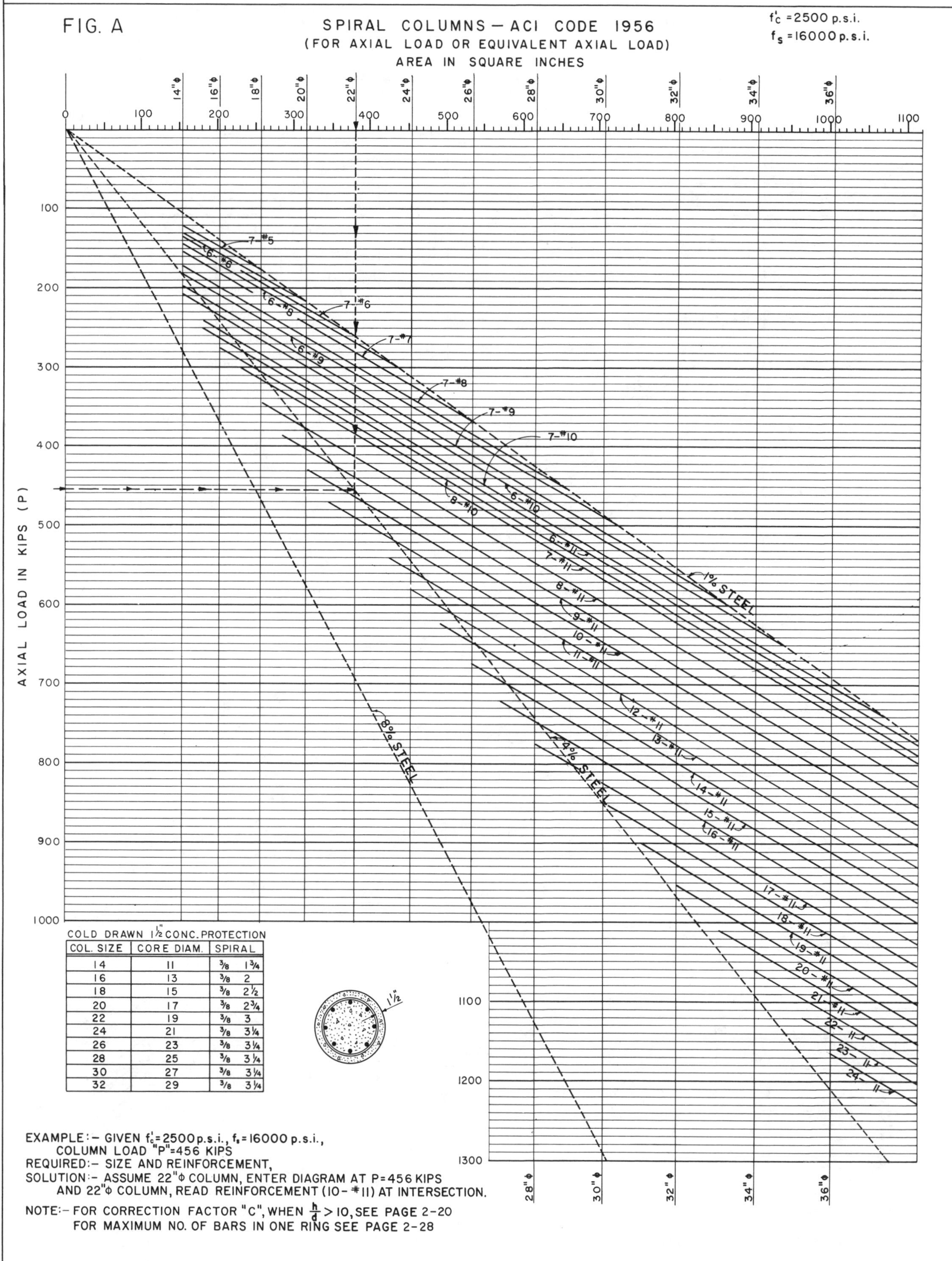

FIG. A
SPIRAL COLUMNS — ACI CODE 1956
(FOR AXIAL LOAD OR EQUIVALENT AXIAL LOAD)
AREA IN SQUARE INCHES
f'_c = 2500 p.s.i.
f_s = 16000 p.s.i.
14"φ 16"φ 18"φ 20"φ 22"φ 24"φ 26"φ 28"φ 30"φ 32"φ 34"φ 36"φ
0 100 200 300 400 500 600 700 800 900 1000 1100
AXIAL LOAD IN KIPS (P)
100 200 300 400 500 600 700 800 900 1000 1100 1200 1300
7-#5
6-#6
7-#6
6-#8
7-#7
6-#9
7-#8
7-#9
7-#10
8-#10
6-#10
6-#11
7-#11
8-#11
9-#11
10-#11
11-#11
12-#11
13-#11
14-#11
15-#11
16-#11
17-#11
18-#11
19-#11
20-#11
21-#11
22-11
23-11
24-11
1% STEEL
4% STEEL
8% STEEL
1½"
EXAMPLE:– GIVEN f'_c=2500 p.s.i., f_s=16000 p.s.i.,
COLUMN LOAD "P"=456 KIPS
REQUIRED:– SIZE AND REINFORCEMENT,
SOLUTION:– ASSUME 22"φ COLUMN, ENTER DIAGRAM AT P=456 KIPS
AND 22"φ COLUMN, READ REINFORCEMENT (10-#11) AT INTERSECTION.
NOTE:– FOR CORRECTION FACTOR "C", WHEN $\frac{h}{d}$ > 10, SEE PAGE 2-20
FOR MAXIMUM NO. OF BARS IN ONE RING SEE PAGE 2-28

COLD DRAWN 1½" CONC. PROTECTION

COL. SIZE	CORE DIAM.	SPIRAL
14	11	3/8 1¾
16	13	3/8 2
18	15	3/8 2½
20	17	3/8 2¾
22	19	3/8 3
24	21	3/8 3¼
26	23	3/8 3¼
28	25	3/8 3¼
30	27	3/8 3¼
32	29	3/8 3¼

STRUCTURAL - CONCRETE COLUMNS

FIG. A

SPIRAL COLUMNS A.C.I. CODE 1956
(FOR AXIAL LOAD OR EQUIVALENT AXIAL LOAD)

$f'_c = 3000$ p.s.i
$f_s = 16000$ p.s.i

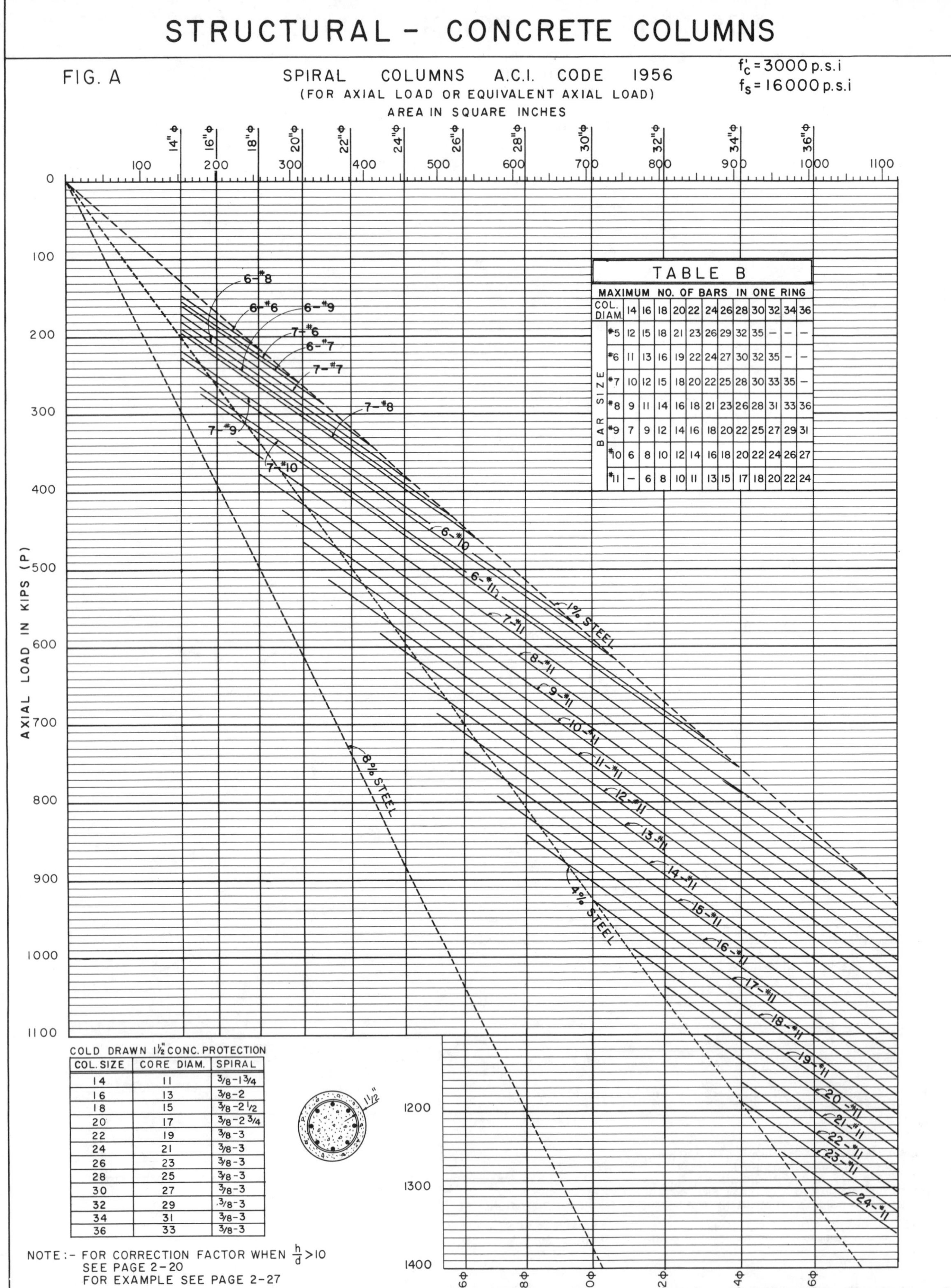

AREA IN SQUARE INCHES
14"Φ
16"Φ
18"Φ
20"Φ
22"Φ
24"Φ
26"Φ
28"Φ
30"Φ
32"Φ
34"Φ
36"Φ
100
200
300
400
500
600
700
800
900
1000
1100
AXIAL LOAD IN KIPS (P)
0
100
200
300
400
500
600
700
800
900
1000
1100
1200
1300
1400
6-#8
6-#6
6-#9
7-#6
6-#7
7-#7
7-#8
7-#9
7-#10
6-#10
6-#11
7-#11
8-#11
9-#11
10-#11
11-#11
12-#11
13-#11
14-#11
15-#11
16-#11
17-#11
18-#11
19-#11
20-#11
21-#11
22-#11
23-#11
24-#11
1% STEEL
4% STEEL
8% STEEL
1½"
26Φ
28Φ
30Φ
32Φ
34Φ
36Φ

TABLE B

MAXIMUM NO. OF BARS IN ONE RING

	COL. DIAM.	14	16	18	20	22	24	26	28	30	32	34	36
BAR SIZE	#5	12	15	18	21	23	26	29	32	35	–	–	–
	#6	11	13	16	19	22	24	27	30	32	35	–	–
	#7	10	12	15	18	20	22	25	28	30	33	35	–
	#8	9	11	14	16	18	21	23	26	28	31	33	36
	#9	7	9	12	14	16	18	20	22	25	27	29	31
	#10	6	8	10	12	14	16	18	20	22	24	26	27
	#11	–	6	8	10	11	13	15	17	18	20	22	24

COLD DRAWN 1½" CONC. PROTECTION

COL. SIZE	CORE DIAM.	SPIRAL
14	11	3/8-1 3/4
16	13	3/8-2
18	15	3/8-2 1/2
20	17	3/8-2 3/4
22	19	3/8-3
24	21	3/8-3
26	23	3/8-3
28	25	3/8-3
30	27	3/8-3
32	29	3/8-3
34	31	3/8-3
36	33	3/8-3

NOTE:- FOR CORRECTION FACTOR WHEN $\frac{h}{d} > 10$
SEE PAGE 2-20
FOR EXAMPLE SEE PAGE 2-27

STRUCTURAL — CONCRETE COLUMNS

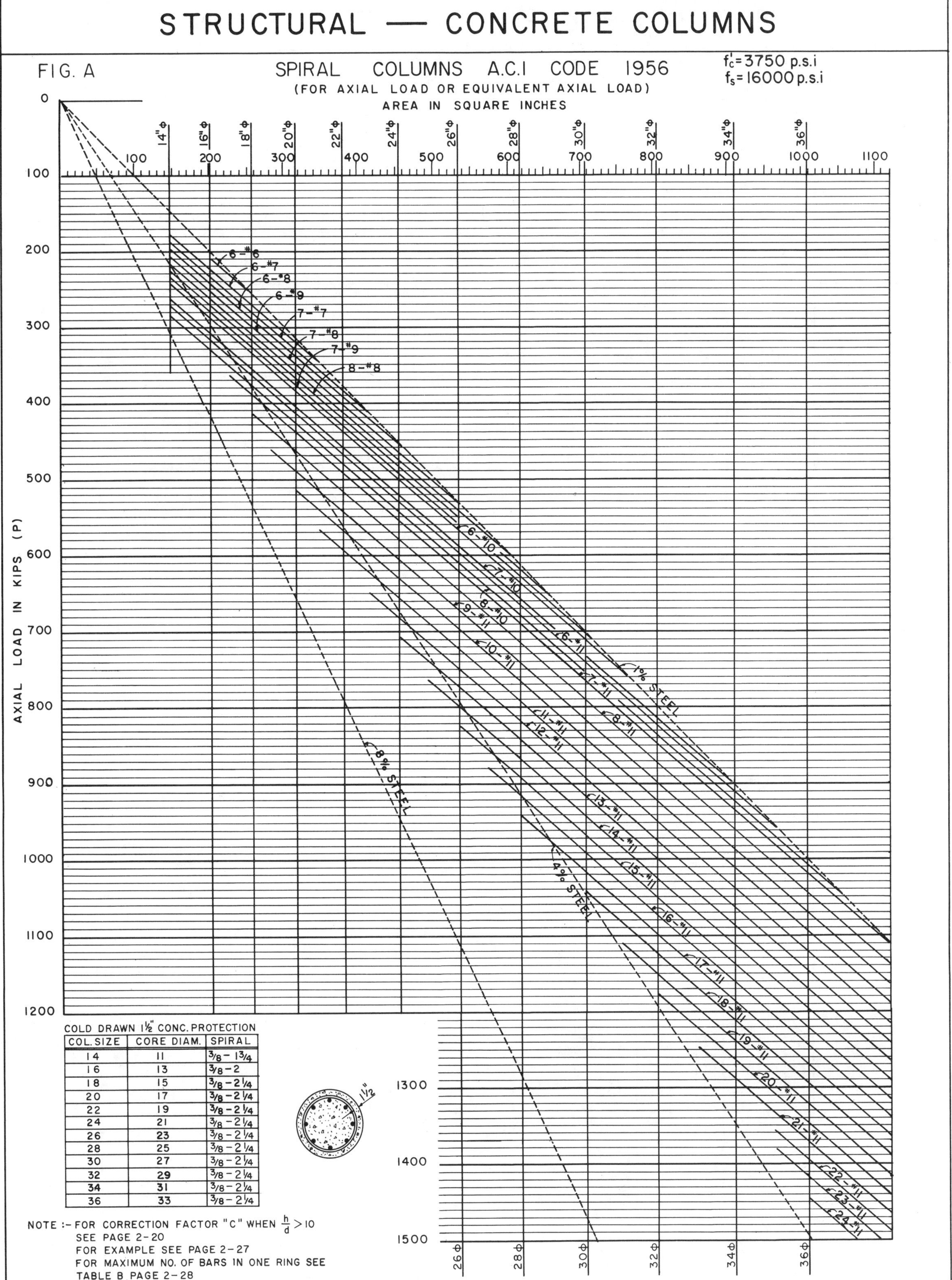
FIG. A
SPIRAL COLUMNS A.C.I CODE 1956
(FOR AXIAL LOAD OR EQUIVALENT AXIAL LOAD)
f'_c = 3750 p.s.i
f_s = 16000 p.s.i
AREA IN SQUARE INCHES
AXIAL LOAD IN KIPS (P)
14"ϕ 16"ϕ 18"ϕ 20"ϕ 22"ϕ 24"ϕ 26"ϕ 28"ϕ 30"ϕ 32"ϕ 34"ϕ 36"ϕ
0 100 200 300 400 500 600 700 800 900 1000 1100
100 200 300 400 500 600 700 800 900 1000 1100 1200 1300 1400 1500
6-#6
6-#7
6-#8
6-#9
7-#7
7-#8
7-#9
8-#8
6-#10
7-#10
8-#10
9-#11
10-#11
6-#11
7-#11
8-#11
11-#11
12-#11
13-#11
14-#11
15-#11
16-#11
17-#11
18-#11
19-#11
20-#11
21-#11
22-#11
23-#11
24-#11
1% STEEL
4% STEEL
8% STEEL
1½"
26ϕ 28ϕ 30ϕ 32ϕ 34ϕ 36ϕ

COLD DRAWN 1½" CONC. PROTECTION

COL. SIZE	CORE DIAM.	SPIRAL
14	11	3/8 - 1 3/4
16	13	3/8 - 2
18	15	3/8 - 2 1/4
20	17	3/8 - 2 1/4
22	19	3/8 - 2 1/4
24	21	3/8 - 2 1/4
26	23	3/8 - 2 1/4
28	25	3/8 - 2 1/4
30	27	3/8 - 2 1/4
32	29	3/8 - 2 1/4
34	31	3/8 - 2 1/4
36	33	3/8 - 2 1/4

NOTE :- FOR CORRECTION FACTOR "C" WHEN $\frac{h}{d} > 10$ SEE PAGE 2-20
FOR EXAMPLE SEE PAGE 2-27
FOR MAXIMUM NO. OF BARS IN ONE RING SEE TABLE B PAGE 2-28

STRUCTURAL – CONCRETE COLUMNS

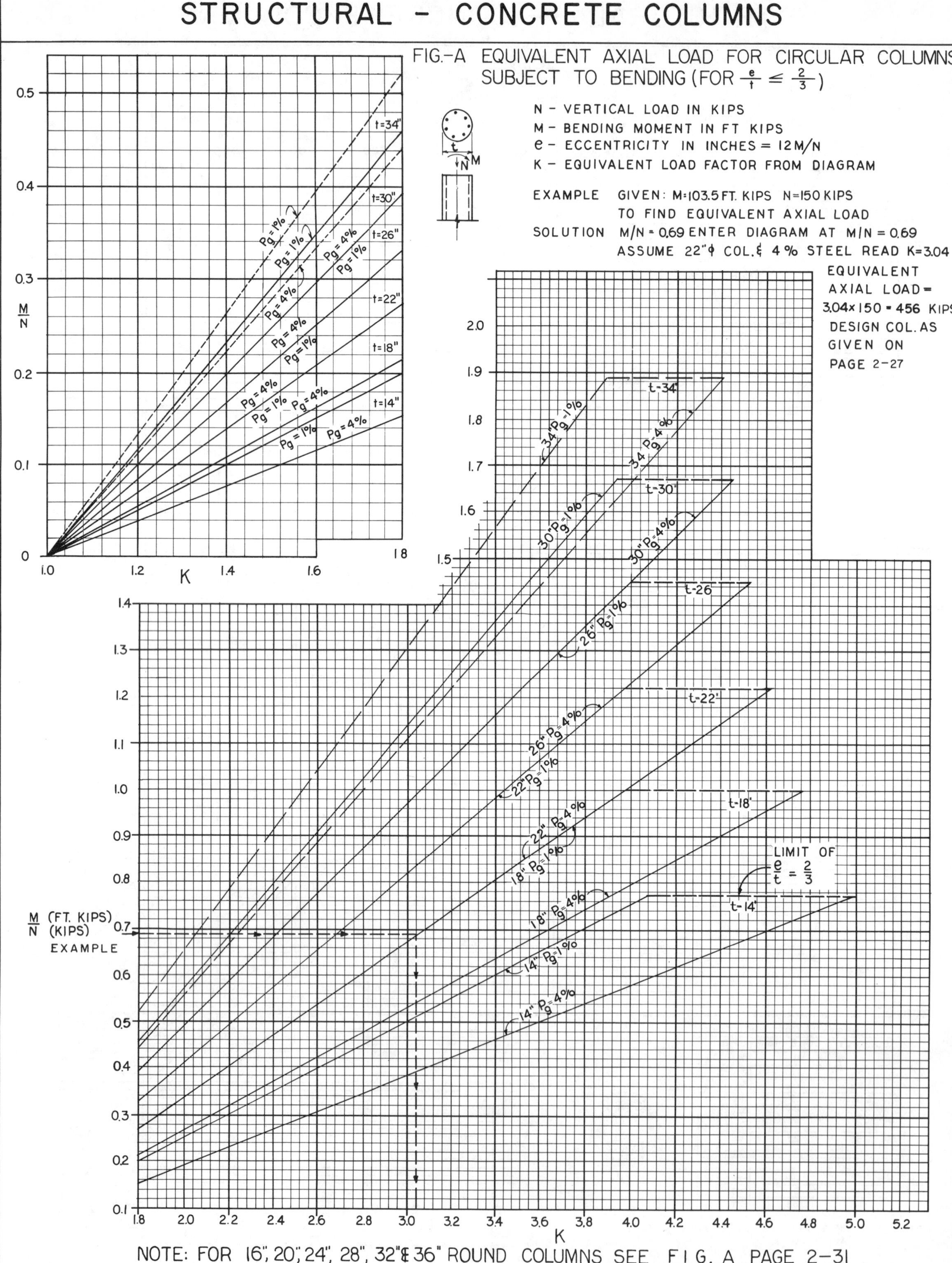

FIG.-A EQUIVALENT AXIAL LOAD FOR CIRCULAR COLUMNS SUBJECT TO BENDING (FOR $\frac{e}{t} \leq \frac{2}{3}$)
N - VERTICAL LOAD IN KIPS
M - BENDING MOMENT IN FT KIPS
e - ECCENTRICITY IN INCHES = 12M/N
K - EQUIVALENT LOAD FACTOR FROM DIAGRAM
EXAMPLE GIVEN: M=103.5 FT. KIPS N=150 KIPS
TO FIND EQUIVALENT AXIAL LOAD
SOLUTION M/N = 0.69 ENTER DIAGRAM AT M/N = 0.69
ASSUME 22"φ COL. & 4% STEEL READ K=3.04
EQUIVALENT AXIAL LOAD = 3.04×150 = 456 KIPS
DESIGN COL. AS GIVEN ON PAGE 2-27
t=34"
t=30"
t=26"
t=22"
t=18"
t=14"
Pg=1%
Pg=4%
M/N
K
t-34"
t-30"
t-26
t-22"
t-18"
t-14"
34" Pg=1%
34" Pg=4%
30" Pg=1%
30" Pg=4%
26" Pg=1%
26" Pg=4%
22" Pg=1%
22" Pg=4%
18" Pg=1%
18" Pg=4%
14" Pg=1%
14" Pg=4%
LIMIT OF e/t = 2/3
M/N (FT. KIPS)/(KIPS) EXAMPLE

NOTE: FOR 16", 20", 24", 28", 32" & 36" ROUND COLUMNS SEE FIG. A PAGE 2–31

STRUCTURAL - CONCRETE COLUMNS

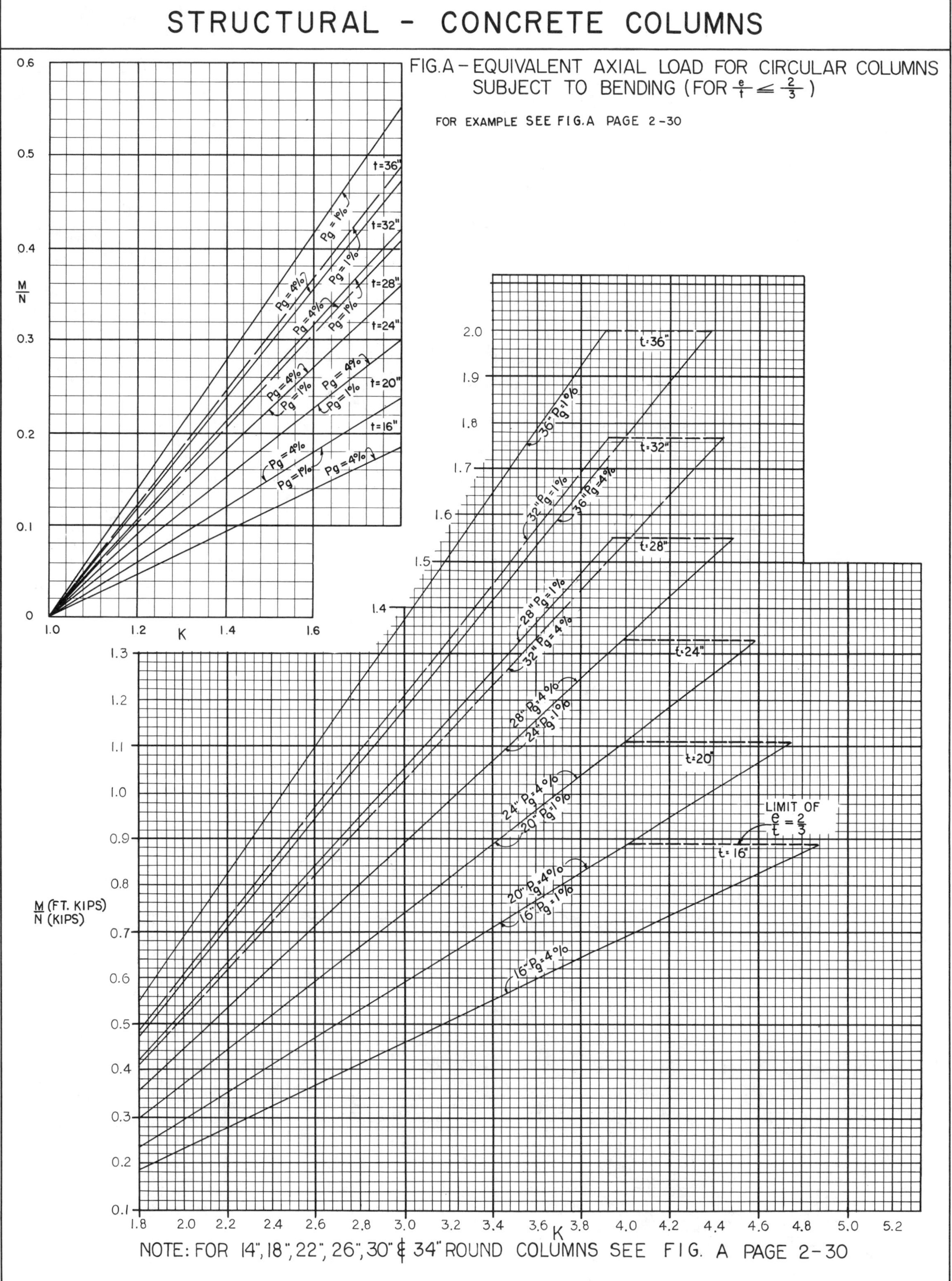
FIG. A – EQUIVALENT AXIAL LOAD FOR CIRCULAR COLUMNS SUBJECT TO BENDING (FOR $\frac{e}{t} \leq \frac{2}{3}$)
FOR EXAMPLE SEE FIG. A PAGE 2-30
M/N
K
t=36"
t=32"
t=28"
t=24"
t=20"
t=16"
Pg=1%
Pg=4%
M (FT. KIPS) / N (KIPS)
36" Pg=1%
36" Pg=4%
32" Pg=1%
32" Pg=4%
28" Pg=1%
28" Pg=4%
24" Pg=1%
24" Pg=4%
20" Pg=1%
20" Pg=4%
16" Pg=1%
16" Pg=4%
LIMIT OF $\frac{e}{t}=\frac{2}{3}$
NOTE: FOR 14", 18", 22", 26", 30" & 34" ROUND COLUMNS SEE FIG. A PAGE 2-30

STRUCTURAL—

FIG. A EQUIVALENT AXIAL LOADS FOR CIRCU

FOR EXAMPLE. SEE RECT

N – VERTICAL LOAD IN KIPS

M – BENDING MOMENT IN FT. KIPS

e – ECCENTRICITY IN INCHES 12 M/N

K – EQUIVALENT LOAD FACTOR FROM DIAGRAM

P – EQUIVALENT AXIAL LOAD = KN

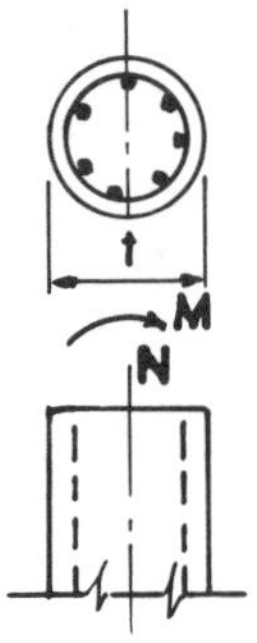

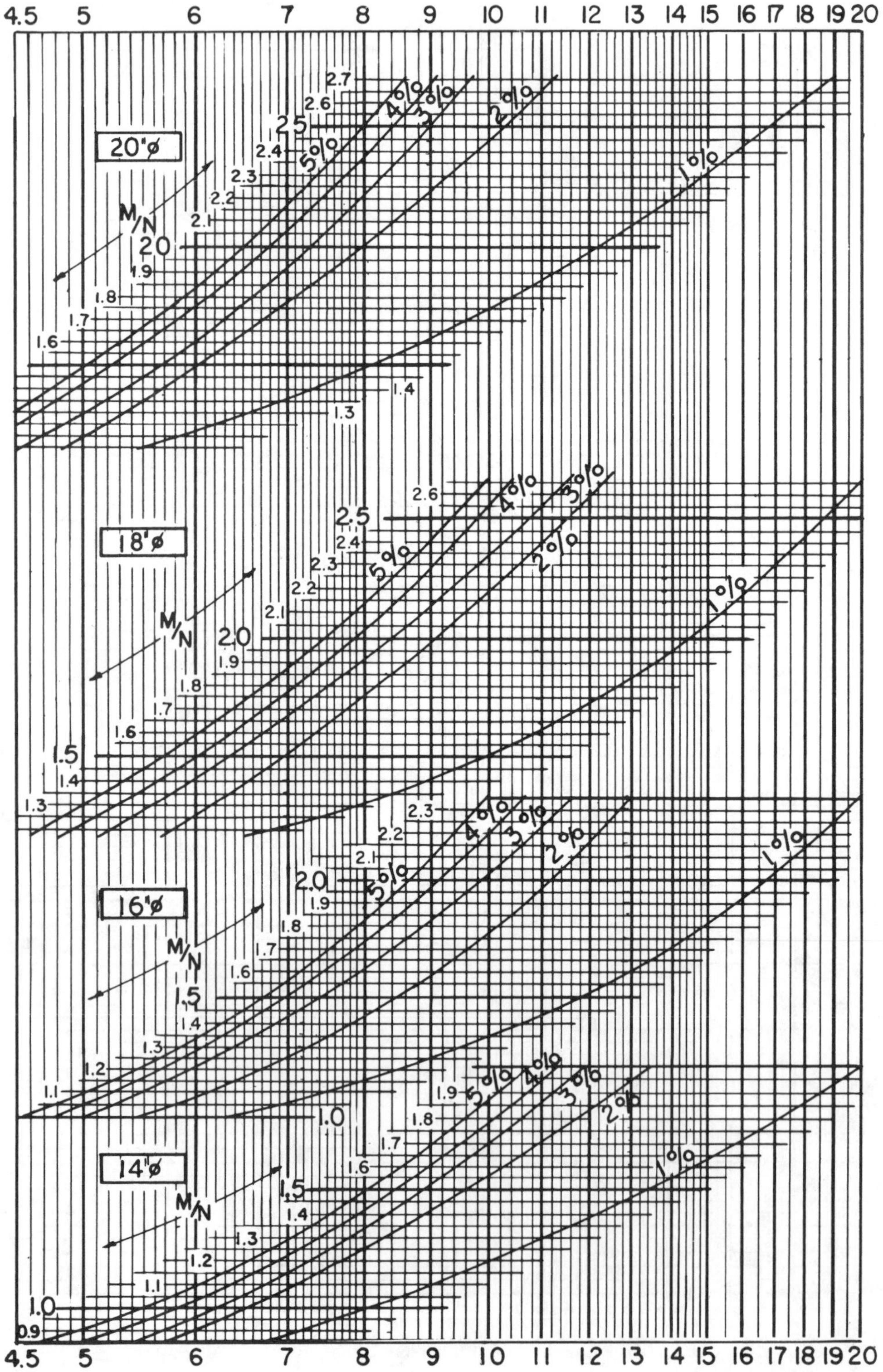

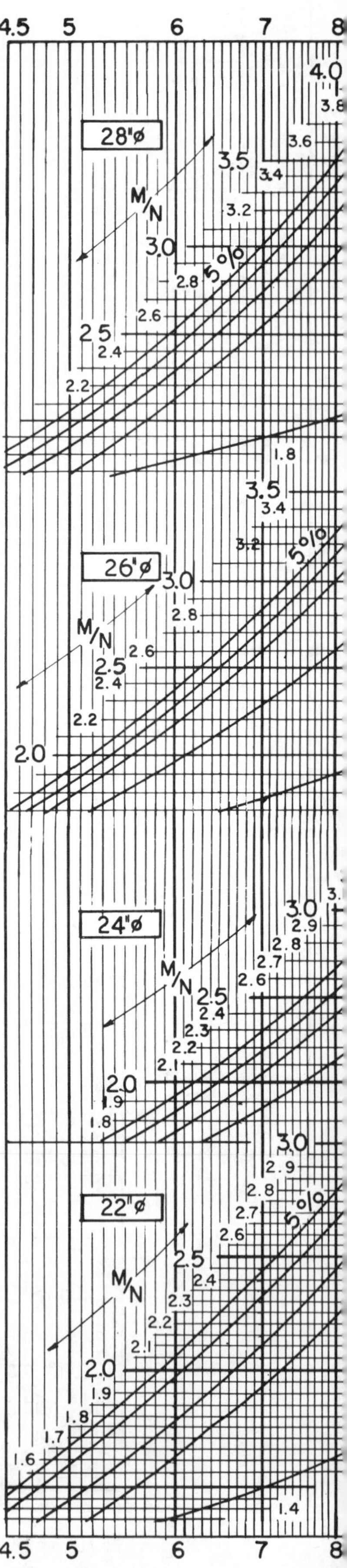

4.5
5
6
7
28"ø
26"ø
24"ø
22"ø
M/N
5%

NCRETE COLUMNS

OLUMNS SUBJECT TO BENDING (FOR $\frac{e}{t} > \frac{2}{3}$)

R COL. ON PAGE 2-24

NOTE: VALUES OF K COMPUTED FOR f'_c = 3000 p.s.i.
FOR OTHER CONCRETE STRENGTHS USE
K TIMES FACTOR SHOWN BELOW

TABLE B						
% STEEL		1%	2%	3%	4%	5%
f'_c	2500	0.92	0.96	0.98	0.99	0.99
	3750	1.20	1.18	1.15	1.10	1.05

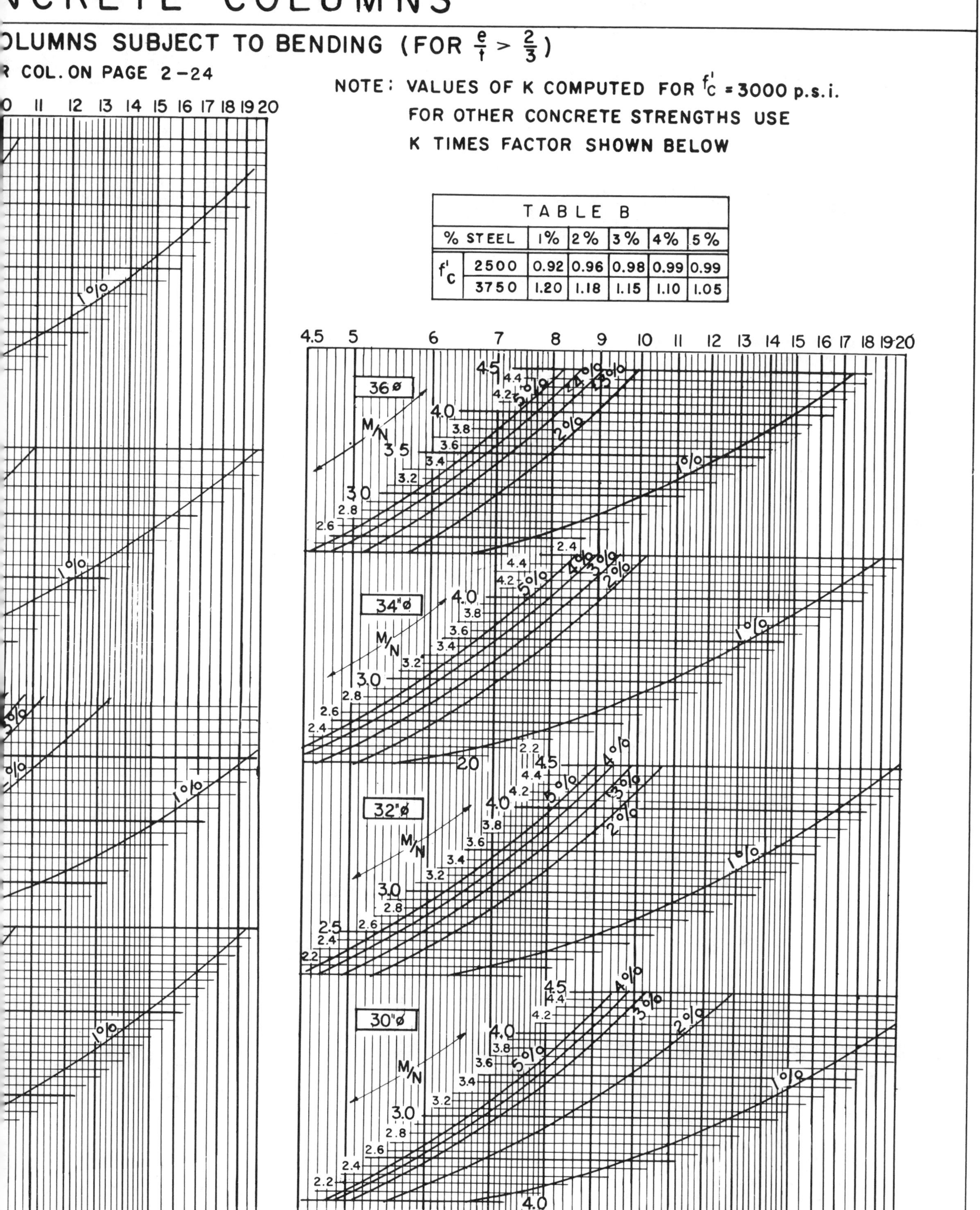

36"ø
34"ø
32"ø
30"ø
M/N
1%
2%
3%
4%
5%
4.5 5 6 7 8 9 10 11 12 13 14 15 16 17 18 19 20

STRUCTURAL – PRESTRESSED CONCRETE – I

GENERAL USAGE Prestressed concrete can be used for long-span construction mainly in bridges. Also for circular structures and prestressed pavement.

Methods	Illustration	Usage
POSTTENSIONING (prestressing steel tensioned after concrete has set, cured and attained required strength, steel then grouted in; end anchorage. Sheating is used to provide a slot in which to place the prestressing steel.)	PRESTRESSING (HIGH STRENGTH) STEEL INSERTED AND USUALLY GROUTED IN AFTER PRESTRESSING. FIXED ANCHORAGE JACK TO PRESTRESS PARABOLIC CURVE IS IDEAL CONDITION. CONCRETE BEAM PRECAST WITH PREFORMED SLOT TO RECEIVE PRESTRESSING STEEL. STEEL ANCHORED AFTER STRESSING. *Fig. A**	Used primarily at site for long-span and non-repetitious members, and special structures: i.e., beams and girders more than 75 ft. long., continuous girders, lift slabs, frames, arches, trusses, piles.
PRETENSIONING (prestressing steel tensioned before concrete is poured and released after concrete has set, cured and attained required strength; anchorage by bond.)	FIXED ANCHORAGE JACK FIXED ABUTMENTS CASTING BED PRESTRESSING STEEL CUT AT ANCHORAGES AFTER CONCRETE HAS SET. *Fig. B*	Used primarily in central plants, where mass production is possible and transportation to site is not difficult: i.e., slabs and short-span beams and girders, piles, poles, railway ties.
SPECIAL METHOD FOR CIRCULAR STRUCTURES (basically, an application of post-tensioning in which specialized self-propelled equipment is used to wind stressed wires continuously around a cylindrical structure.)	GUNITE COVER FOR PRESTRESSING STEEL CONCRETE WALL PLACED FIRST VERTICAL PRESTRESSING STEEL IN VERTICAL CHASE OUTSIDE INSIDE CIRCUMFERENTIAL PRESTRESSING STEEL. ANCHORAGE VERTICAL SECTION CIRCULAR WALL OR STRUCTURE *Fig. C*	All types of circular structures: i.e., tanks, silos, large-diameter pipes, etc.

TYPICAL POST-TENSIONED ANCHORAGES

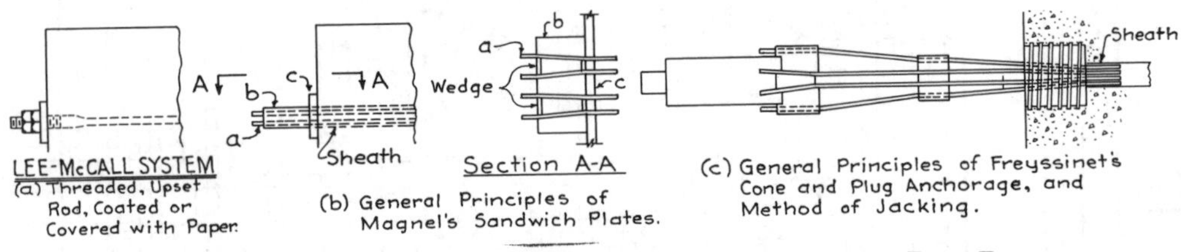

A
b
c
A
a
Wedge
a
b
c
Sheath
Sheath
LEE-McCALL SYSTEM
(a) Threaded, Upset Rod, Coated or Covered with Paper.
Section A-A
(b) General Principles of Magnel's Sandwich Plates.
(c) General Principles of Freyssinet's Cone and Plug Anchorage, and Method of Jacking.

Fig. D *Fig. E*

TABLE F – PROPERTIES OF PRESTRESSED CONCRETE REINFORCEMENT

	Nominal Diameter, inches	Area, square inches	Ultimate Strength, pounds	× %‡	Initial Tensioning Force, pounds	× %‡	Final Design Force, pounds
Uncoated Stress-Relieved Strand for Pretensioned Bonded Design,† A.S.T.M.-4416-57T	1/4	.0356	9,000	70	6,300	80	5,040
	5/16	.0578	14,500	70	10,150	80	8,120
	3/8	.0799	20,000	70	14,000	80	11,200
	7/16	.1089	27,000	70	18,900	80	15,120
	1/2	.1438	36,000	70	25,200	80	20,160
Galvanized Strand for Post-tensioned†	0.600	.215	46,000	68	31,200	83.5	26,000
	0.835	.409	86,000	68	58,500	83.5	49,000
	1.0	.577	122,000	68	82,600	83.5	69,000
	1-1/8	.751	156,000	69	108,000	83.5	90,000
	1-1/4	0.931	192,000	69.5	134,000	83.5	112,000
	1-3/8	1.12	232,000	69.5	160,000	83.5	134,000
	1-1/2	1.36	276,000	71	196,000	83.5	163,000
	1-9/16	1.48	300,000	71	212,500	83.5	177,000
	1-5/8	1.60	324,000	71	230,000	83.5	192,000
	1-11/16	1.73	352,000	71	250,000	83.5	208,000

	Nominal Diameter, inches	Area, square inches	Ultimate Strength, pounds	× %‡	Initial Tensioning Force, pounds	× %‡	Final Design Force, pounds
Uncoated Stress-Relieved Wire†	.192	.02895	7,300	70	5,100	80	4,100
	.196	.03017	7,500	70	5,250	80	4,200
	.250	.04909	11,800	70	8,250	80	6,600
	.276	.05963	14,200	70	9,950	80	7,950
Stress Steel Bars§	1/2	.196	28,400	69	19,600	85	16,680
	5/8	.306	44,400	69	30,600	85	26,050
	3/4	.442	64,100	69	44,200	85	37,600
	7/8	.601	87,200	69	60,100	85	51,100
	1	.785	113,900	69	78,500	85	66,800
	1-1/8	.994	114,000	69	99,400	85	84,500

*Materials used are flexible rubber or metal tubing or a bond braking paper wrapping on the prestressing steel. Methods apply only to posttensioning.
†Adapted from John A. Roebling's Sons Corp. ‡Percentage factor for reducing ultimate strength (or initial force). §Adapted from Stressteel Corp.

NOTE: See also p. 1-08 for high-strength structural steel. E.g.: Ultimate strength = **9000 lb.** Therefore 9000 × 70 = 6300 lb. = initial tensioning force. Final design force = 6300 × 0.80 = **5040 lb.**

STRUCTURAL–PRESTRESSED CONCRETE–2

TABLE A – ALLOWABLE STRESS IN PRESTRESSED CONCRETE*

Description		Allowable Working Stresses	
		Temporary†	Permanent‡
FLEXURE			
Extreme fiber stress in compression	Pre-tensioned	0.60 f'_{ci}	0.40 f'_c
	Post-tensioned	0.55 f'_{ci}	0.40 f'_c
Tension	(Without conventional reinforcement to resist tension)	0.05 f'_{ci}	0
	(With conventional reinforcement to resist tension)	0.05 f'_{ci}	0.08 f'_c
SHEAR	At distance X from support (X being distance equivalent to 1½ × depth of beam) without stirrups		0.03 f'_c
BEARING	(between prestressing anchorage and concrete) Use $f_b = 0.4 f'_c \sqrt[3]{\frac{A_c}{A_p}}$ if $< f'_c$ (see Fig. E, p. 2-52)	1.00 f'_c	1.00 f'_c

TABLE B – PRESTRESSING STEEL REINFORCEMENT*

Description	Allowable Working Stresses	
	Temporary	Permanent
TENSION	0.80 f'_{sy}	0.60 f'_s¶

TABLE C – ULTIMATE STRENGTH*

The ultimate strength must be able to withstand the greater of *a* or *b*.

a = Dead load plus 3 times live plus impact loads
b = Twice the sum of dead, live, and impact loads

TABLE D – LOSSES IN PRESTRESS*

Decrease in prestress (p.s.i.) due to creep, shrinkage, and deformation	
Pre-tension	6000 + 16 fcs + 0.04 fsi.
Post-tension	3000 + 11 fcs + 0.04 fsi.

NOTE: Loss in prestress may be taken as 0.15 f'_s as an approximation for preliminary design

fsi = initial prestress, p.s.i.
fcs = concentrated stress at c.g. of prestressing steel, p.s.i.
(average concentrated stress between support)

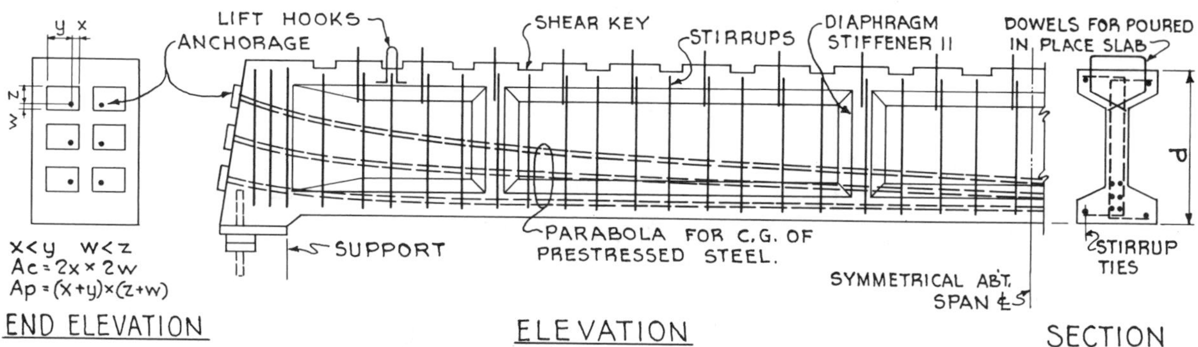

y x
LIFT HOOKS
ANCHORAGE
SHEAR KEY
STIRRUPS
DIAPHRAGM STIFFENER §
DOWELS FOR POURED IN PLACE SLAB
z
w
d
x<y w<z
Ac = 2x × 2w
Ap = (x+y)×(z+w)
SUPPORT
PARABOLA FOR C.G. OF PRESTRESSED STEEL.
SYMMETRICAL ABT. SPAN ℄
STIRRUP TIES
END ELEVATION
ELEVATION
SECTION

FIG. E DETAIL OF TYPICAL PRESTRESSED CONCRETE MEMBER

PRESTRESS STEEL REQUIREMENT*

(1) For beams where tension in the prestressing steel is maintained by bond:
  (a) When single wires are used, maximum size 0.2 in.
  (b) When wires are used in seven-wire strands, the maximum strand size is ⅜ in.
  (c) Minimum spacing vertically and horizontally for prestressed steel shall be 3 times the diameter of wire or strand used measured from center to center. The clear distance shall not be less than 1½ times the maximum size of coarse aggregate.
  (d) The minimum cover shall be 1½" or one diameter of bar strand or duct, whichever is greater.

(2) For beams where tension in the prestressing steel is maintained by adequate end anchorage, the above limitations are not applicable, except that a clear spacing horizontally of 1½ times the maximum size of coarse aggregate shall be maintained.

STIRRUPS*

Stirrups are recommended, whether or not computations show they are needed. Maximum spacing no greater than three fourths of depth of beam. The sum of the cross-sectional area of the stirrups should not be less than .08% of the cross-sectional area.

*Based on *Criteria for Prestressed Concrete Bridges*, U. S. Dept. of Commerce Bureau of Public Roads, 1954. For nomenclature see Table A, p. 2-52. †Before creep and shrinkage. ‡After creep and shrinkage; due to dead, live, impact loads, and any combination thereof. ¶Use 0.80 f'_{sy} if 0.6 f'_s. (f'_{sy} = stress at 0.2% of plastic set.) §Portland Cement Association recommends diaphragms not less than 6" wide and spaced 20'-0" center to center maximum.

STRUCTURAL–PRESTRESSED CONCRETE-3

TABLE A – BASIC FORMULAS AND NOMENCLATURE

FLEXURE

Stress due to dead, live, and impact loads

$= \frac{My}{I}$

Stress due to prestressing

$= \frac{P}{A} \pm \frac{Pey}{I}$ (initial)

$= \frac{NP}{A} \pm \frac{NPey}{I}$ (final)

SHEAR

Principal tensile stress

$= \sqrt{v^2 + \frac{c^2}{4}} - \frac{c}{2}$

ULTIMATE STRENGTH

$Pb = 0.23 \times \frac{0.8 f'c}{f's}$

For $p \leq Pb$, $Mu = 0.9$ as $f's$ d

For $p > Pb$, $Mu = 0.9 \sqrt{as\ abs}\ f's d$

A = cross-sectional area
Ac = effective area of concrete under bearing plate
Ap = area of bearing plate
a_{bs} = steel area for a balanced section, sq. in.
as = steel area, sq. in.
c = unit compression
d = depth of section from compression face to centroid of steel
e = eccentricity of prestressing group from neutral axis, in.
fc = allowable stress in concrete
f'c = ultimate 28-day strength of concrete
f'ci = ultimate strength of concrete at time of prestress
fs = allowable stress in prestressing steel
f's = ultimate stress of prestressing steel
I = moment of inertia
M = bending moment
Mu = ultimate moment
N = proportion of P remaining permanently*
NP = permanent prestress force
P = initial prestress force
p = as/wd
Pb = value of p for a balanced section abs/wd
v = unit shear
w = average width of the compression area, in.
y = distance from neutral axis to point at which stress is to be computed

SUBSCRIPTS FOR y

b = bottom fibers
t = top fibers

CHECK OF PRESTRESSED CONCRETE BEAM

Given: Prestressed beam in Fig. A

f'c = 5000 p.s.i. f's = 145,000 p.s.i.

Moment exclusive of beam weight = 164' kips dead load; 200' kips live load; 364' kips

Moment due to beam weight = 102' kips
Initial prestress force, (p) = 284' kips
Losses due to shrinkage and creep = 15%
Final prestress force (after 15% loss) = 242 kips

Allowable Stresses

Concrete: Compression 0.40 × 5000 = 2000 p.s.i.
Tension = 0
Steel: Initially = 100,000 p.s.i.
After losses = 85,000 p.s.i.

Required: To check beam for bending and diagonal tension.

A of shaded half above neutral axis = 130 □"
Q = 12.9×130 = 1675
C.G. of shaded area
12.9
Neutral Axis
4"
As = 3 sq. in.
39"
yt = 21.1"
yb = 17.9"
e = 12.9"
A conc. = 270 sq. in.
I = 49,500 in.4
Reaction = 1.28×27 = 34.6K
LOAD = 1K/FT. (D.L.+L.L.)
BEAM WT. = 0.28K/FT.
1.28 K/FT.
℄ SPAN
1.5×3.25 = 4.87'
Ⓐ
L = 54 FT.
C.G. of prestressing steel (Parabolic curve.)

SECTION AT ℄ SPAN ELEVATION

FIGURE B

Solution: Step 1: Compute top and bottom fiber stresses due to beam weight, applied moment, and prestressing force, before and after losses, using basic formulas. Results are shown and tabulated in Table A, p. 2-53.

FLEXURE FORMULAS

Stress due to dead, live, and impact loads $\frac{M \times y(t \text{ or } b)}{I}$

Stress due to prestressing force $= \frac{P}{A} + \frac{Pe \times y(t \text{ or } b)}{I}$

Compression = positive (+) ; tension = negative (–)

*The loss in the prestress force *P* is due to the shrinkage in the concrete, plastic flow of the concrete due to stress, and the creep in the concrete and steel. This loss (percentagewise) = 1 – N.

STRUCTURAL–PRESTRESSED CONCRETE–4

TABLE A – CHECK OF PRESTRESSED CONCRETE BEAM*

	Fiber Stresses, p.s.i.							
Stage	Initial (Immediately after Prestressing)				Final (All Loads – after Losses)			
Loading \ Fiber	Top		Bottom		Top		Bottom	
Beam wt. only	$\frac{102\times 12\times 1000\times 21.1}{49,500}$	+522	$\frac{102\times 12\times 1000\times 17.9}{49,500}$	−442	$\frac{102\times 12\times 1000\times 21.1}{49,500}$	+522	$\frac{102\times 12\times 1000\times 17.9}{49,500}$	−442
Dead + live loads (exclusive of beam wt.)	–	–	–	–	$\frac{364\times 12\times 1000\times 21.1}{49,500}$	+1860	$\frac{364\times 12\times 1000\times 17.9}{49,500}$	−1580
Prestress force	$\frac{284,000}{270}-\frac{284,000\times 12.9\times 21.1}{49,500}$	−510	$\frac{284,000}{270}+\frac{284,000\times 12.9\times 17.9}{49,500}$	+2372	$\frac{242,000}{270}-\frac{242,000\times 12.9\times 21.1}{49,500}$	−433	$\frac{242,000}{270}+\frac{242,000\times 12.9+17.9}{49,500}$	+2026
Total Stress (sum of values above)	–	+12	–	+1930	–	+1949	–	+4

Stresses are found to be less than allowable values as given above – O.K.

Check stresses in prestressing steel before losses = $\frac{284,000}{3}$ = 94,670 p.s.i.; after losses = $\frac{242,000}{3}$ = 80,667 p.s.i. – O.K. (less than allowable values as "given".)

Step 2: Check diagonal tension at point *A*, 1.5 times depth of beam, from support (B.P.R. Code) = 1.5 × 3.25 = 4.87. At point *A* compute vertical shear due to beam weight and applied loads = (34.6 − 1.28 × 4.87) = 28.4 K. Vertical component of prestressing force for parabolic curve = $\frac{4Ph}{[L - 2(1.5d)]}$ where *h* = rise in curve from ℄ span; $h = \frac{(27 - 4.87)^2}{27^2} \times 12.9 = 8.62$"; vertical component = $\frac{4 \times 242 \times 8.62}{(54 - 2 \times 4.87)12}$ = 15.7 K (upward)

Total shear at section *V* = 28.4 − 15.7 = 12.7 kips.

Principal tensile stress at neutral axis of section; unit vertical shearing stress = $v = \frac{Vq}{It} = \frac{12,700 \times 1675}{49,500 \times 4}$ = 107 p.s.i.; unit compression stress at neutral axis = $c = \frac{P}{A} = \frac{242,000}{270}$ = 896 p.s.i.

Principal tensile stress = $\sqrt{v^2 + \frac{c^2}{4}} - \frac{c}{2} = \sqrt{(107)^2 + \frac{(896)^2}{4}} - \frac{896}{2}$ = 13 p.s.i.

Allowable stress = 0.03 f'_c = 150 p.s.i. – O.K.

Principal tension stress should be investigated at other critical points, these being the points of change in section.

Step 3: Check Ultimate strength of member. According to the B.P.R. specifications, the ultimate strength of the beam must exceed the greater of

(a) D.L. + 3 (L.L. + Impact) = 164 + 102 + 3(200) = 866 ft.-kips

(b) 2(D.L. + L.L. + Impact) = 2(164 + 102 + 200) = 932 ft.-kips (this value governs).

The percentage of steel for balanced design (pb) = $0.23 \times \frac{0.8 f'_c}{f'_s} = \frac{.23 \times 0.8 \times 5000}{145,000}$ = 0.635%. The actual percentage of steel (*p* actual) = $\frac{A_s}{Wd}$. The depth of the compressive area ≈ 0.23d* = 0.23 × 34 = 7.82, and the section area corresponding to this depth = 74.9 in.^{2}

Therefore, the average compression width (w) = $\frac{74.9}{7.82}$ = 9.58. ∴ *p* actual = $\frac{3 \text{ in.}^2}{34 \times 9.58}$ = .915%, ∴ a_{bs} = pb × wd = 0.00635 × 9.58 × 34 = 2.07 in.^{2}

Since *p* actual > pb, (0.915 > 0.635).

The ultimate moment (Mu) = $0.9 \sqrt{a_s\, a_{bs}}\; f'_s\; d = 0.9 \left(\sqrt{3 \times 2.07}\times\right) 145,000 \times 34$ = 11,200,000 in.-lb.

$$Mu = \frac{11,200,000 \text{ in.-lb.}}{12 \times 1000} = 933 \text{ ft.-kips.}$$

933 ft.-kips > 932 ft.-kips ∴ O.K.

NOTE: Before beam is considered satisfactory, check the following: curvature of prestressing elements so that stresses at all sections are within allowable values; requirements of ultimate strength in diagonal tension, end block stresses, and anchorage.

* Based on *Criteria for Prestressed Concrete Bridges*, U. S. Dept. of Commerce, Bureau of Public Roads, 1954.

STRUCTURAL CONCRETE – ULTIMATE STRENGTH–I

ULTIMATE STRENGTH DESIGN is based on a plastic theory which assumes, at ultimate stress, parabolic distribution of stresses in a beam (see Fig. A-1) rather than straight line (see Fig. A-2).

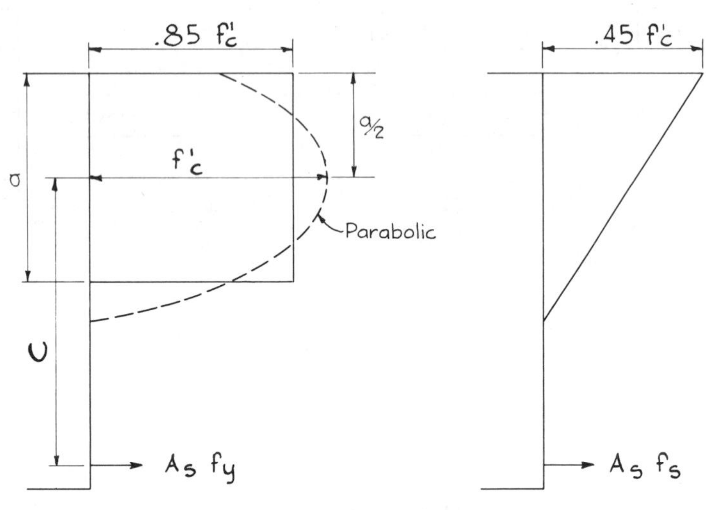

.85 f'c
a/2
f'c
a
Parabolic
c
As fy
.45 f'c
As fs

FIG. A

METHOD:

Determine moments in the member by conventional methods.

As the stress distribution in Fig. A-1 or C is reached at yield-point stress in the steel, we design the steel area for yield-point stress, using the moment *M* multiplied by a load or safety factor U.

Determine Mu from Table B.

Mu = moment to be used in design. Yield-point steel stress $= f_{ly} = \dfrac{Mu}{cAs}$ or $As = \dfrac{Mu}{cf_y}$

$$c = d - a/2$$

TABLE B LOAD FACTORS (SAFETY FACTORS)*

Where Wind May Be Neglected	Where Wind Must Be Included
I. $U = 1.2B + 2.4L$ II. $U = K(B + L)$ } Use maximum value.	Ia. $U = 1.2B + 2.4L + 0.6W$ Ib. $U = 1.2B + 0.6L + 2.4W$ IIa. $U = K(B + L + \frac{W}{2})$ IIb. $U = K(B + \frac{L}{2} + W)$ } Use maximum value

When earthquake must be considered, replace *W* by *E* in above formulas.
K = 1.8 for members subject to bending only.
k = 2 for members subject to combined bending and axial load.
U = ultimate strength of section = maximum combination of loads that the member can sustain before failure.
B = effect of dead load
L = effect of live load plus impact
W = effect of wind load
E = effect of earthquake forces

DERIVATION OF BASIC FORMULAS.
1. RECTANGULAR BEAMS WITH TENSILE REINFORCEMENT ONLY

Replace parabolic stress distribution with rectangular stress distribution equal to 0.85 f'_c (A.C.I.)

From $\Sigma H = 0 \quad a = \dfrac{As\,f_y}{0.85bf'_c}$

or $Mu = bd^2 f'_c q\,(1 - 0.59q)$

$q = P\dfrac{f_y}{f'_c} \quad P = \dfrac{As}{bd}$

From $\Sigma M = 0 \quad Mu = As\,f_y\,(d - \frac{a}{2})$

where Mu = ultimate moment capacity
f'_c = 28-day strength of concrete
f_y = yield stress of steel
b = beam width

LIMITATIONS (A.C.I.)

If $q > 0.18$, check for deflection.

If $q > 0.40$, compressive steel is required.

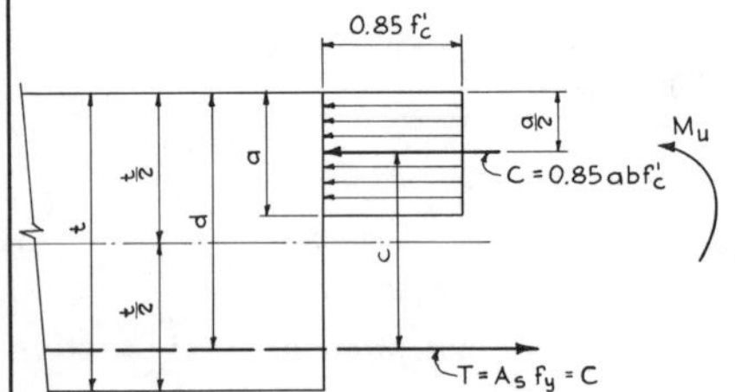

0.85 f'c
Mu
C = 0.85abf'c
T = As fy = C

FIG. C. FORCES ACTING ON BEAM

*From A.C.I. Code, paragraph A604, 1956.

STRUCTURAL CONCRETE - ULTIMATE STRENGTH-2

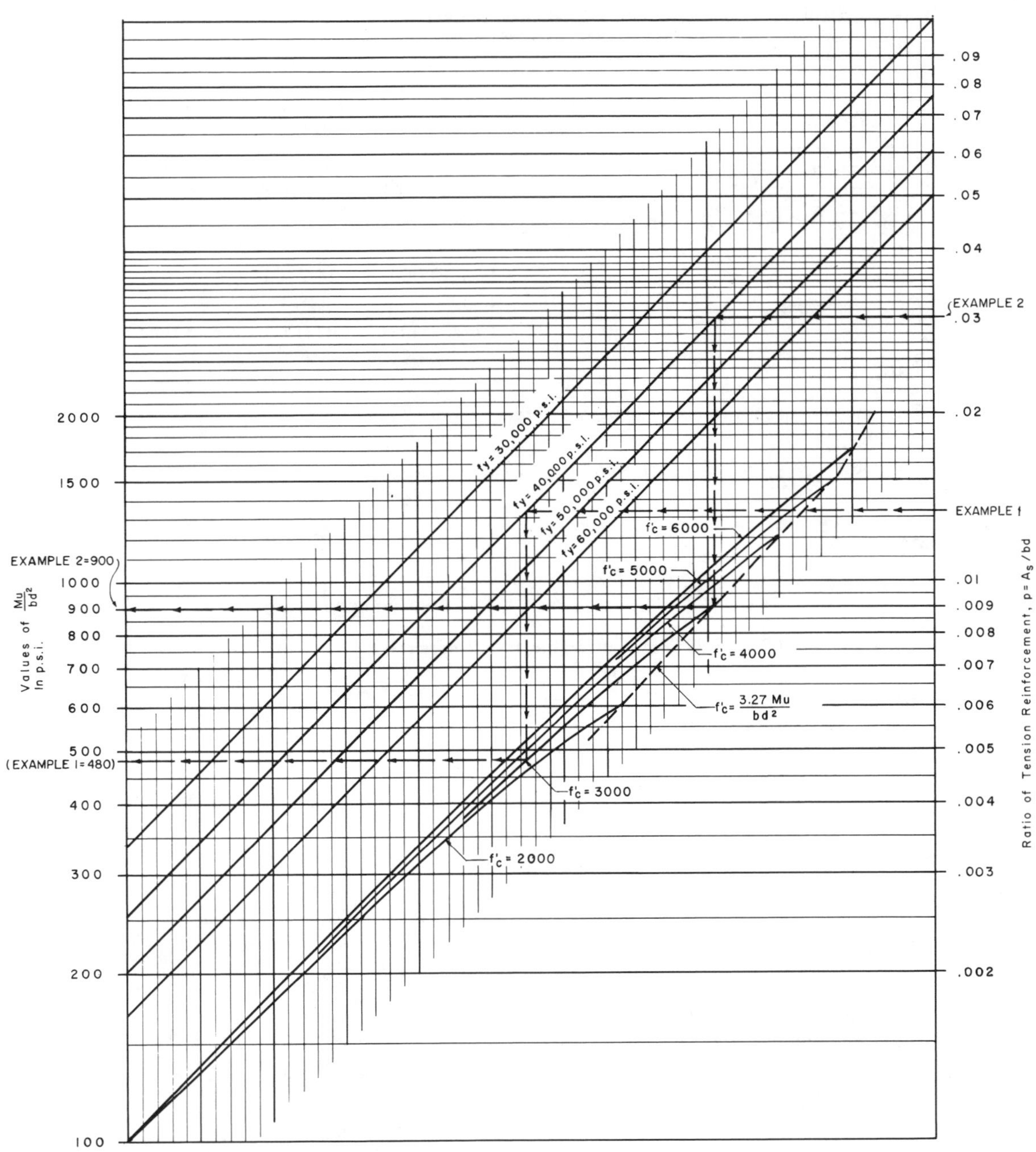

fy = 30,000 p.s.i.
fy = 40,000 p.s.i.
fy = 50,000 p.s.i.
fy = 60,000 p.s.i.
f'c = 6000
f'c = 5000
f'c = 4000
f'c = 3.27 Mu / bd²
f'c = 3000
f'c = 2000
EXAMPLE 2 = 900
EXAMPLE 2
EXAMPLE 1
(EXAMPLE 1 = 480)
Values of Mu/bd² In p.s.i.
Ratio of Tension Reinforcement, p = As/bd
100
200
300
400
500
600
700
800
900
1000
1500
2000
.002
.003
.004
.005
.006
.007
.008
.009
.01
.02
.03
.04
.05
.06
.07
.08
.09

FIG. A. MOMENT CAPACITY OF RECTANGULAR SECTIONS WITHOUT COMPRESSION REINFORCEMENT †

EXAMPLE 1: *Given* Mu* = 204 kip – ft. = 2,450,000 in. lb., b = 10 in., fy = 40,000, f'_c = 3,000.

Find d and As.

For smallest section requiring no deflection analysis, q = 0.18. Compute $P = q\frac{f'_c}{fy} = \frac{0.18 \times 3000}{40,000} = 0.0135$. Enter chart at right with P = 0.0135, and read $\frac{Mu}{bd^2} = 480$. Compute d. $d^2 = \frac{2,450,000}{480 \times 10} = 510$; d = 22.6; say 23. Compute As = pbd = 0.0135 (10)(23) = 3.10 in².

**Computation of Mu (ultimate moment)*

Given: e = 20'. Live load = 1.2 k./ft. Dead load = 1 k./ft. (no wind).

Equation I: U = 1.28 + 2.4L = 1.2 + 2.4(1.2) = 4.08. (See Table B, p. 2-76.)

Equation II: U = K (B + L) = 1.8 (1 + 1.2) = 3.96 (See Table B, p. 2-76.)

Use largest value $Mu = \frac{WL^2}{8} = \frac{(4.08)(400)}{8} = 204$ kip – ft.

† "Guide for Ultimate Strength Design of Reinforced Concrete," by Charles S. Whitney and Edward Cohen, *A.C.I. Journal* (Nov. 1956, p. 476).

STRUCTURAL CONCRETE – ULTIMATE STRENGTH-3

2. RECTANGULAR BEAM WITH COMPRESSION REINFORCEMENT

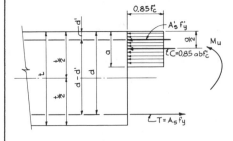

Consider $M_u = M_1 + M_2$

where M_1 = resisting moment of rectangular beam with no compressive reinforcement.

M_2 = resisting moment of additional steel.

EXAMPLE 2: *Given:* M_u = 1,440,000 in lb., b = 8", D = 14", f'_c = 3,000, f_y = 40,000, d = 14-2 = 12", d' = 2".

Find: Area of steel required in tension and compression.

FIG. A

Let q = 0.40 (A.C.I. limit for beam without compression steel; see p. 2.76).

Compute $p = q\frac{f'_c}{f_y} = 0.40 \times \frac{3,000}{40,000} = 0.03$. Enter chart (p. 2-77) with p = 0.03. Read $\frac{M_1}{bd^2} = 900$.

Compute $M_1 = \frac{900 \times 8 \times 144}{12 \times 1000} = 86.5$ kip – ft.

Compute $M_2 = M_u - M_1 = 120 - 86.5 = 33.5$

Compute $A_{s1} = pbd = 0.03 \times 8 \times 12 = 2.88$ in.^{2}

$A_{s2} = \frac{33.5 \times 12 \times 1,000}{40,000\,(12-2)} = 1.005$ additional tensile steel.

Total Tensile Steel = 2.88 + 1.005 = 3.89 in.^{2} Correcting the compressive stress in the steel for the loss of concrete area:

Compute $A'_s = \frac{33.5 \times 12 \times 1,000}{(40,000 - .85 \times 3,000)(12-2)} = 1.075$ in.^{2}

Compressive steel = 1.075 in.^{2}

3. T-BEAMS

3. T-BEAMS In general, it will not be practical to place enough tensile steel in the stem to develop the full compressive strength of the flange. Therefore, use formula 1 Rectangular Beam with Tensile Reinforcement, only with b equal to width of T-flange.

4. COLUMNS

4. COLUMNS Design columns using charts on following pages. The A.C.I. code requires that all members subjected to axial loads shall be designed for a minimum eccentricity for spiral reinforcement columns equal to 0.050 and for tied columns .10*t*

EXAMPLE 3:

Given: P_u = 400 kips, f'_c = 3,000, f_y = 40,000. Assume p_t (total steel) = 0.02 Find column size, b, t, d, and As.

Assume $d = 0.8t$, $m = \frac{f_y}{0.85 f'_c}$.

Compute $p_t m = \frac{0.02 \times 40,000}{0.85 \times 3,000} = 0.314$.

Enter chart (Fig. B) at ptm = 0.314 and $\frac{e'}{t} = 0.10$.

Read $\frac{P_u}{f'_c bt} = 0.81$.

Compute $bt = \frac{P_u}{0.81 f'_c} = \frac{400,000}{0.81 \times 3,000} = 165$.

Let b = 12, t = 14.

Recompute $\frac{P_u}{f'_c bt} = \frac{400,000}{3,000 \times 12 \times 14} = 0.794$, and enter chart (Fig. B) at K = 0.794 and, $\frac{e'}{t} = 0.10$.

Read $p_t m = 0.30$.

Compute $p_t = p_t m \times \frac{1}{m} = 0.30 \times \frac{0.85 \times 3,000}{40,000} = 0.0191$.

$A_s = 0.0191(12)(14) = 3.21$ in.^{2}

Answer: b = 12 in., t = 14 in., d = 11 in., As = 3.21 in.^{2}

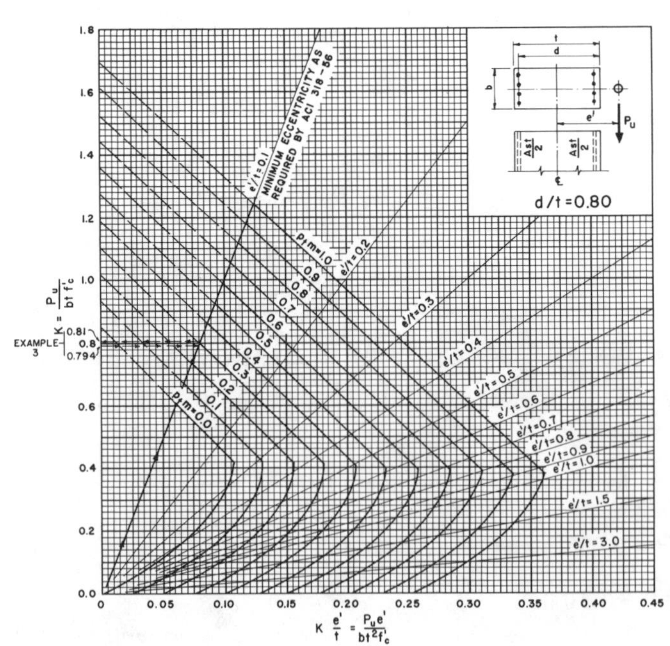

MINIMUM ECCENTRICITY AS REQUIRED BY ACI 318-56
d/t = 0.80
EXAMPLE 3
$K = \frac{P_u}{bt f'_c}$
$K \frac{e'}{t} = \frac{P_u e'}{bt^2 f'_c}$

FIG. B. BENDING & AXIAL LOAD $\frac{d}{t} = 0.80$ †

(Rectangular Section with Symmetrical Reinforcement)

† "Guide for Ultimate Strength Design of Reinforced Concrete," by Charles S. Whitnen and Edward Cohen, A.C.I. *Journal* (Nov. 1956, p. 482).

STRUCTURAL CONCRETE – ULTIMATE STRENGTH-4

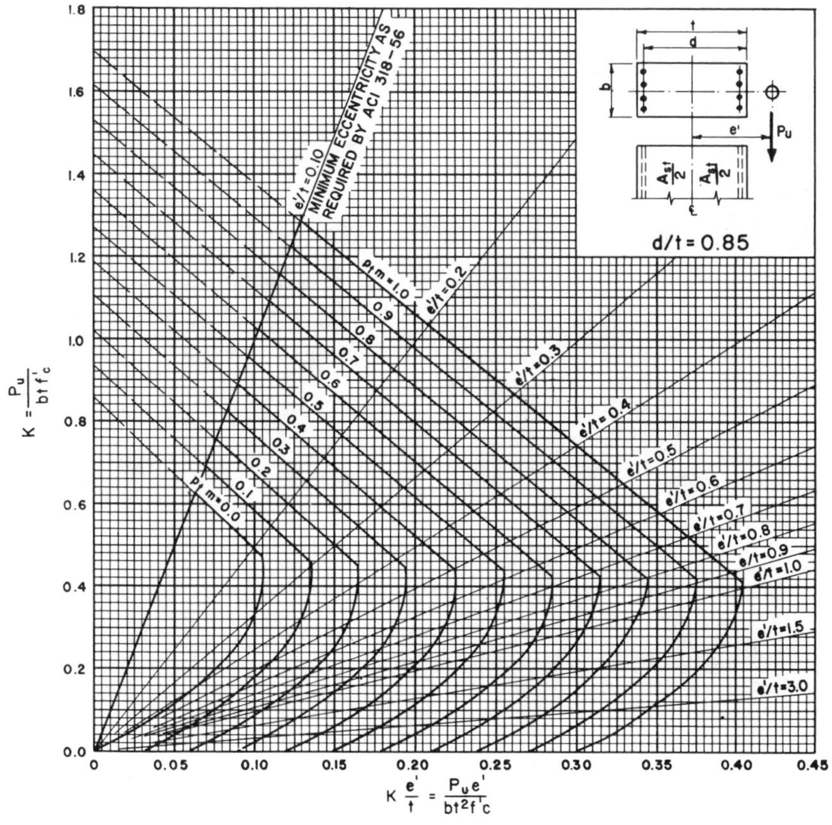

FIG. A. BENDING AND AXIAL LOAD†
d/t = 0.85 rectangular section with symmetrical reinforcement.

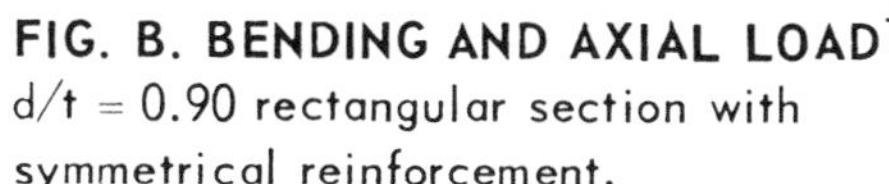

FIG. B. BENDING AND AXIAL LOAD†
d/t = 0.90 rectangular section with symmetrical reinforcement.

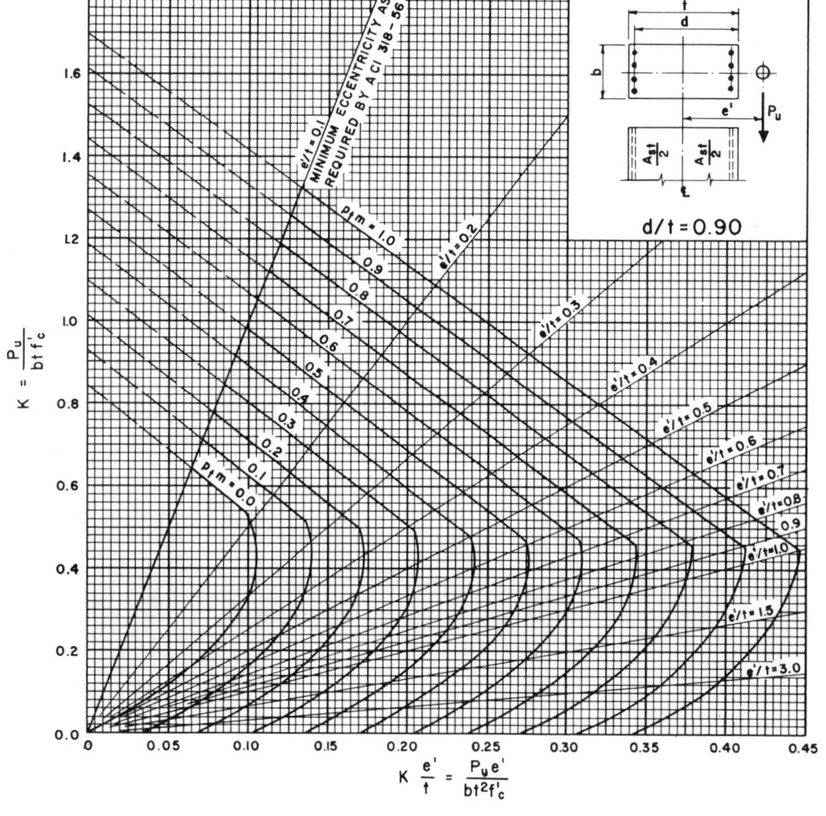

† "Guide for Ultimate Strength Design of Reinforced Concrete," by Charles S. Whitney and Edward Cohen, *A.C.I. Journal* (Nov. 1956. pp. 480 and 481).

STRUCTURAL CONCRETE – ULTIMATE STRENGTH-5

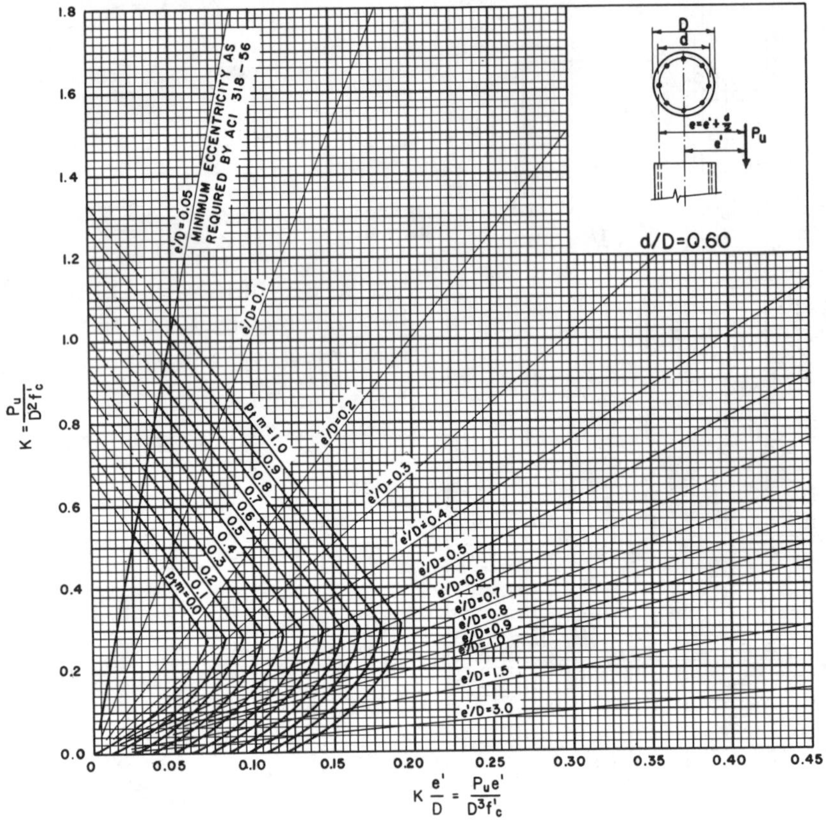

e'/D=0.05 MINIMUM ECCENTRICITY AS REQUIRED BY ACI 318-56
d/D=0.60
$K = \frac{P_u}{D^2 f'_c}$
$K \frac{e'}{D} = \frac{P_u e'}{D^3 f'_c}$

FIG. A. BENDING AND AXIAL LOAD†
d/D = 0.60 circular sections with spiral reinforcement

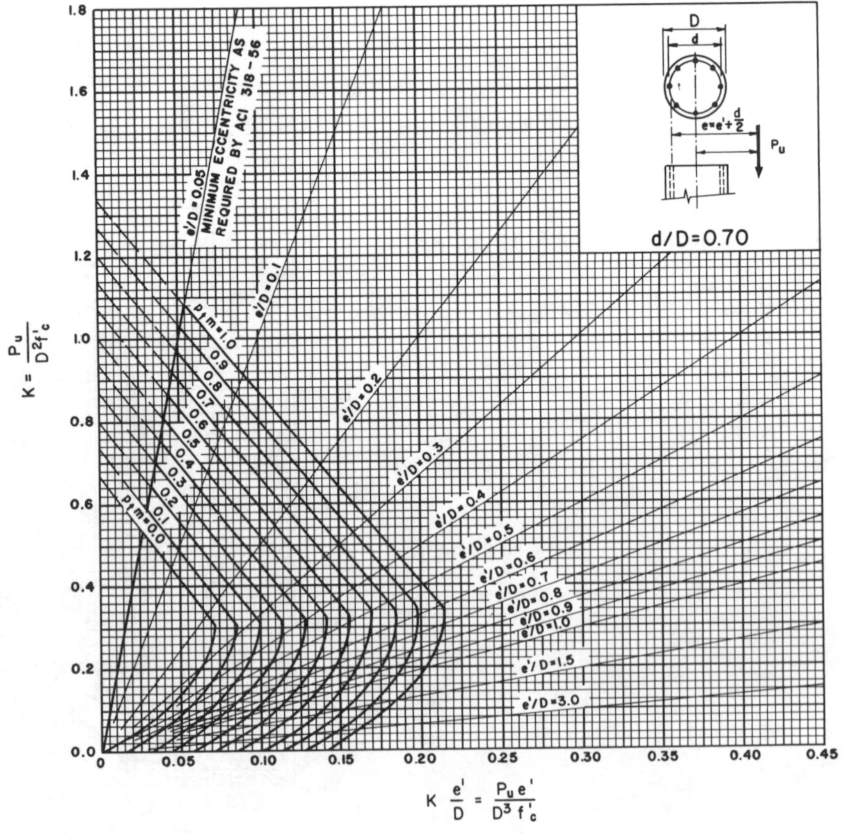

e'/D=0.05 MINIMUM ECCENTRICITY AS REQUIRED BY ACI 318-56
d/D=0.70
$K = \frac{P_u}{D^2 f'_c}$
$K \frac{e'}{D} = \frac{P_u e'}{D^3 f'_c}$

FIG. B. BENDING AND AXIAL LOAD†
d/D = 0.70 circular sections with spiral reinforcement

† "Guide for Ultimate Strength Design of Reinforced Concrete," by Charles S. Whitney and Edward Cohen, *A.C.I. Journal* (Nov. 1956, pp. 489-490).

STRUCTURAL CONCRETE – ULTIMATE STRENGTH-6

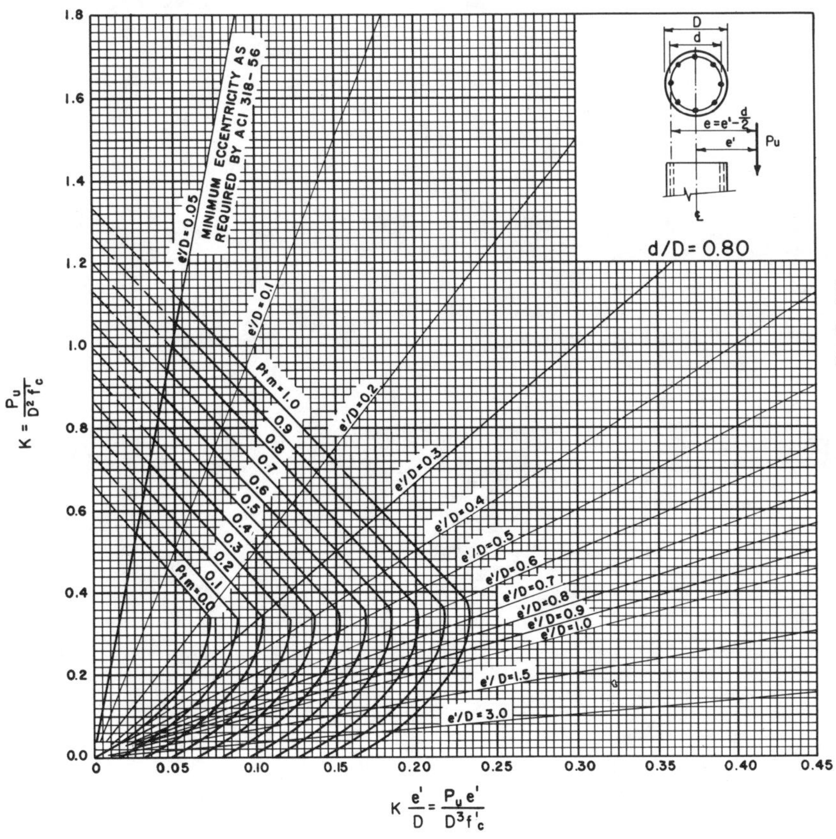

d/D= 0.80
MINIMUM ECCENTRICITY AS REQUIRED BY ACI 318-56
$K = \frac{P_u}{D^2 f'_c}$
$K \frac{e'}{D} = \frac{P_u e'}{D^3 f'_c}$

FIG. A. BENDING AND AXIAL LOAD†
d/D = 0.80 circular sections with spiral reinforcement

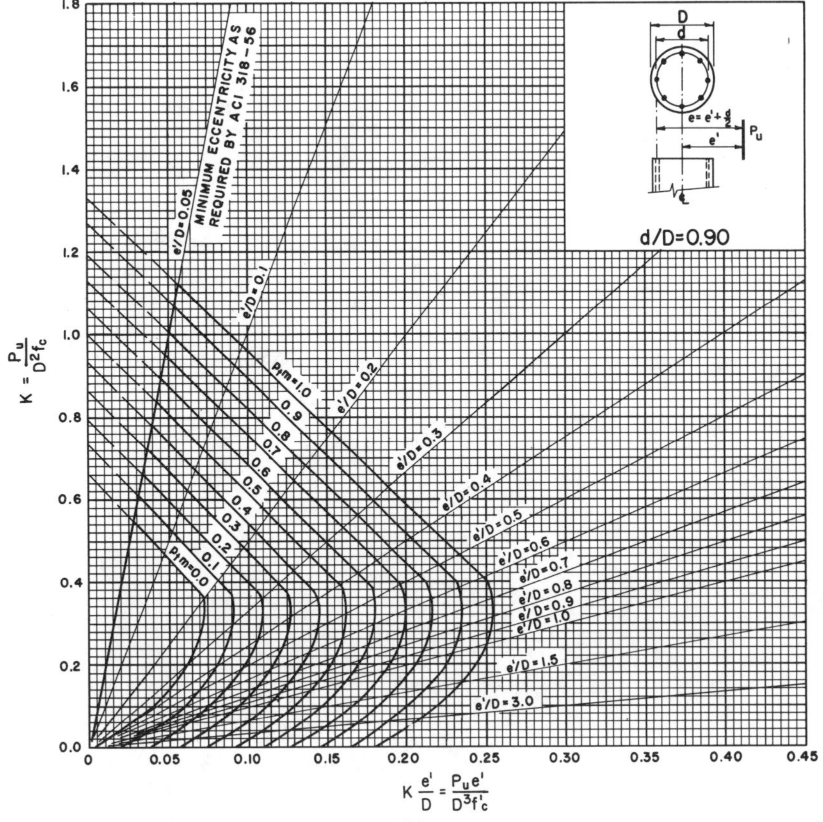

d/D=0.90
MINIMUM ECCENTRICITY AS REQUIRED BY ACI 318-56
$K = \frac{P_u}{D^2 f_c}$
$K \frac{e'}{D} = \frac{P_u e'}{D^3 f'_c}$

FIG. B. BENDING AND AXIAL LOAD†
d/D = 0.90 circular sections with spiral reinforcement

† "Guide for Ultimate Strength Design of Reinforced Concrete," by Charles S. Whitney and Edward Cohen, *A.C.I. Journal* (Nov. 1956, pp. 487 and 488).

STRUCTURAL-RIGID FRAMES-FIXED END MOMENTS

TABLE A - PARTIAL UNIFORM LOAD *

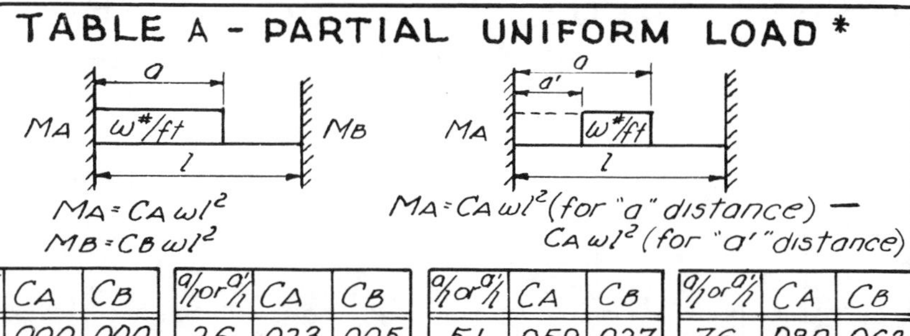

a
MA
w#/ft
MB
l
a'
MA = CA wl²
MB = CB wl²
MA = CA wl² (for "a" distance) — CA wl² (for "a'" distance)

a/l or a'/l	CA	CB	a/l or a'/l	CA	CB	a/l or a'/l	CA	CB	a/l or a'/l	CA	CB
.01	.000	.000	.26	.023	.005	.51	.059	.027	.76	.080	.063
.02	.000	.000	.27	.025	.005	.52	.060	.029	.77	.080	.064
.03	.000	.000	.28	.026	.006	.53	.061	.030	.78	.080	.066
.04	.001	.000	.29	.028	.006	.54	.062	.031	.79	.081	.067
.05	.001	.000	.30	.029	.007	.55	.063	.033	.80	.081	.068
.06	.002	.000	.31	.031	.008	.56	.064	.034	.81	.081	.070
.07	.002	.000	.32	.032	.008	.57	.065	.035	.82	.082	.071
.08	.003	.000	.33	.033	.009	.58	.066	.037	.83	.082	.072
.09	.004	.000	.34	.035	.010	.59	.067	.038	.84	.082	.073
.10	.004	.000	.35	.036	.011	.60	.068	.040	.85	.082	.074
.11	.005	.000	.36	.038	.011	.61	.069	.041	.86	.083	.075
.12	.006	.001	.37	.039	.012	.62	.070	.043	.87	.083	.076
.13	.007	.001	.38	.041	.013	.63	.071	.044	.88	.083	.077
.14	.008	.001	.39	.042	.014	.64	.072	.045	.89	.083	.078
.15	.009	.001	.40	.044	.015	.65	.073	.047	.90	.083	.079
.16	.010	.001	.41	.045	.016	.66	.074	.048	.91	.083	.080
.17	.011	.001	.42	.047	.017	.67	.074	.050	.92	.083	.080
.18	.013	.002	.43	.048	.018	.68	.075	.051	.93	.083	.081
.19	.014	.002	.44	.049	.019	.69	.076	.053	.94	.083	.082
.20	.015	.002	.45	.051	.020	.70	.076	.054	.95	.083	.082
.21	.016	.003	.46	.052	.021	.71	.077	.056	.96	.083	.083
.22	.018	.003	.47	.053	.022	.72	.078	.057	.97	.083	.083
.23	.019	.003	.48	.055	.024	.73	.078	.059	.98	.083	.083
.24	.020	.004	.49	.056	.025	.74	.079	.060	.99	.083	.083
.25	.022	.004	.50	.057	.026	.75	.079	.062	1.00	.083	.083

TABLE B CONCENTRATED LOADING *

Mc, a, P, l

$M_C = C'_C \times P \times l$

a/l	Cc	a/l	Cc	a/l	Cc	a/l	Cc
.01	.010	.26	.142	.51	.123	.76	.044
.02	.019	.27	.144	.52	.120	.77	.041
.03	.028	.28	.145	.53	.117	.78	.038
.04	.037	.29	.146	.54	.114	.79	.035
.05	.045	.30	.147	.55	.111	.80	.032
.06	.053	.31	.148	.56	.108	.81	.029
.07	.061	.32	.148	.57	.105	.82	.027
.08	.068	.33	.148	.58	.102	.83	.024
.09	.075	.34	.148	.59	.099	.84	.022
.10	.081	.35	.148	.60	.096	.85	.019
.11	.087	.36	.148	.61	.093	.86	.017
.12	.093	.37	.147	.62	.090	.87	.015
.13	.098	.38	.146	.63	.086	.88	.013
.14	.104	.39	.145	.64	.083	.89	.011
.15	.108	.40	.144	.65	.080	.90	.009
.16	.113	.41	.143	.66	.076	.91	.007
.17	.117	.42	.141	.67	.073	.92	.006
.18	.121	.43	.140	.68	.070	.93	.005
.19	.125	.44	.138	.69	.066	.94	.003
.20	.128	.45	.136	.70	.063	.95	.002
.21	.131	.46	.134	.71	.060	.96	.002
.22	.134	.47	.132	.72	.056	.97	.001
.23	.136	.48	.130	.73	.053	.98	.000
.24	.139	.49	.127	.74	.050	.99	.000
.25	.141	.50	.125	.75	.047	1.00	.000

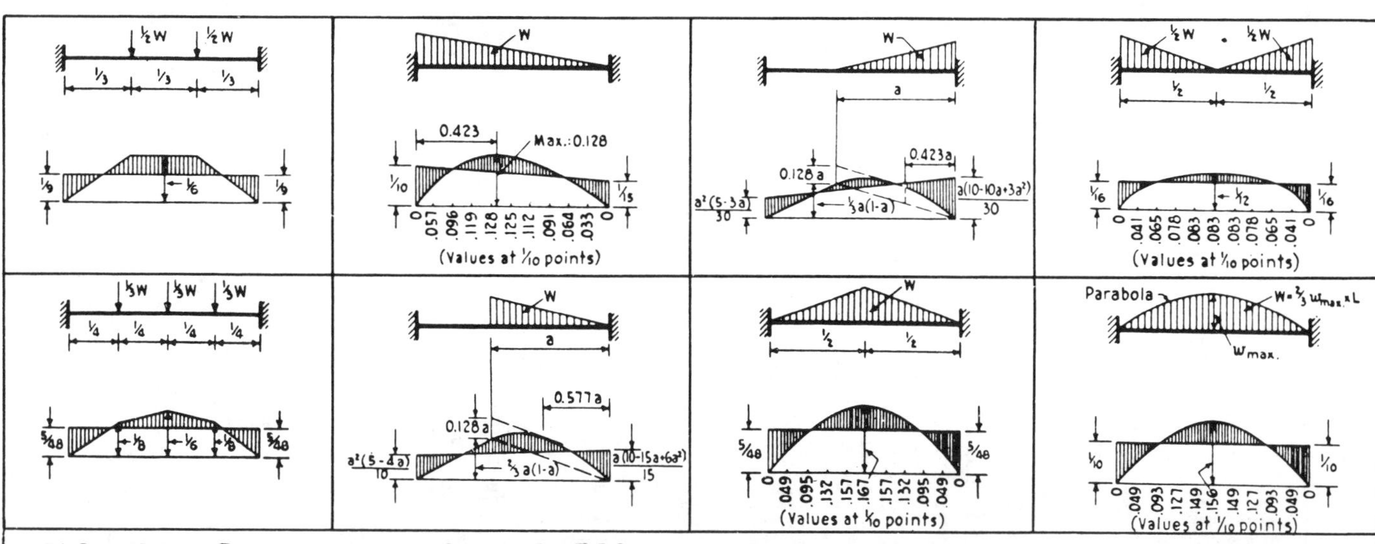

½W ½W
⅓ ⅓ ⅓
1/9 1/6 1/9
W
0.423
Max.: 0.128
1/10 1/15
0 .057 .096 .119 .128 .125 .112 .091 .064 .033 0
(Values at 1/10 points)
W
a
0.128a
0.423a
a²(5-3a)/30
a(10-10a+3a²)/30
⅓a(1-a)
½W ½W
½ ½
1/16 1/12 1/16
0 .041 .065 .078 .083 .083 .083 .078 .065 .041 0
(Values at 1/10 points)
⅓W ⅓W ⅓W
¼ ¼ ¼ ¼
5/48 1/8 1/6 1/8 5/48
W
a
0.577a
0.128a
a²(5-4a)/10
a(10-15a+6a²)/15
⅔a(1-a)
W
½ ½
5/48 5/48
0 .049 .095 .132 .157 .167 .157 .132 .095 .049 0
(Values at 1/10 points)
Parabola
W = ⅔ wmax × L
wmax.
1/10 1/10
0 .049 .093 .127 .149 .156 .149 .127 .093 .049 0
(Values at 1/10 points)

FIGURE C – COEFFICIENT FOR MOMENTS IN BEAMS OF CONSTANT SECTION AND WITH FIXED ENDS **

$M = c \times W \times L$

c = coefficient taken from diagram
W = total load on beam
L = length of beam
a = length in terms of "L"

* Data from Eng News Record, June 18, 1942, Article by M.W. Rostenstein.
** From Portland Cement Association.

STRUCTURAL - RIGID FRAMES - VALUE OF K FOR CONC. BEAMS & COLS.

TABLE A - STIFFNESS OF BEAMS - VALUES OF K FOR TEE-BEAMS *

$+K = \frac{2I}{10L}$ $\quad I = \frac{b'd^3}{12}$ $\quad b'$ = width of stem. $\quad d$ = depth of beam.

c = Factor for Tee-Beams when $\frac{\text{flange width}}{\text{stem width}} = 6.0$

See also page 2-04

d	b'	I	L: Length of beam (feet) 8	10	12	14	16	20	24	30
8	6	256	6	5	4	4	3	3	2	2
	8	341	9	7	6	5	4	3	3	2
	10	427	11	9	7	6	5	4	4	3
	11½	491	12	10	8	7	6	5	4	3
	13	555	14	11	9	8	7	6	5	4
	15	640	16	13	11	9	8	6	5	4
	17	725	18	15	12	10	9	7	6	5
	19	811	20	16	14	12	10	8	7	5
10	6	500	13	10	8	7	6	5	4	3
	8	667	17	13	11	10	8	7	6	4
	10	833	21	17	14	12	10	8	7	6
	11½	958	24	19	16	14	12	10	8	6
	13	1083	27	22	18	15	14	11	9	7
	15	1250	31	25	21	18	16	13	10	8
	17	1417	35	28	24	20	18	14	12	9
	19	1583	40	32	26	23	20	16	13	11
12	6	864	22	17	14	12	11	9	7	6
	8	1152	29	23	19	16	14	12	10	8
	10	1440	36	29	24	21	18	14	12	10
	11½	1656	41	33	28	24	21	17	14	11
	13	1872	47	37	31	27	23	19	16	12
	15	2160	54	43	36	31	27	22	18	14
	17	2448	61	49	41	35	31	25	20	16
	19	2736	68	55	46	39	34	27	23	18
14	6	1372	34	27	23	20	17	14	11	9
	8	1829	46	37	30	26	23	18	15	12
	10	2287	57	46	38	33	29	23	19	15
	11½	2630	66	53	44	38	33	26	22	18
	13	2973	74	59	50	42	37	30	25	20
	15	3430	86	69	57	49	43	34	29	23
	17	3887	97	78	65	56	49	39	32	26
	19	4345	109	87	72	62	54	43	36	29
16	6	2048	51	41	34	29	26	20	17	14
	8	2731	68	55	46	39	34	27	23	18
	10	3413	85	68	57	49	43	34	28	23
	11½	3925	98	79	65	56	49	39	33	26
	13	4437	111	89	74	63	55	44	37	30
	15	5120	128	102	85	73	64	51	43	34
	17	5803	145	116	97	83	73	58	48	39
	19	6485	162	130	108	93	81	65	54	43
18	6	2916	73	58	49	42	36	29	24	19
	8	3888	97	78	65	56	49	39	32	26
	10	4860	122	97	81	69	61	49	41	32
	11½	5589	140	112	93	80	70	56	47	37
	13	6318	158	126	105	90	79	63	53	42
	15	7290	182	146	122	104	91	73	61	49
	17	8262	207	165	138	118	103	83	69	55
	19	9234	231	185	154	132	115	92	77	62
20	6	4000	100	80	67	57	50	40	33	27
	8	5333	133	107	89	76	67	53	44	36
	10	6667	167	133	111	95	83	67	56	44
	11½	7667	192	153	128	110	96	77	64	51
	13	8667	217	173	144	124	108	87	72	58
	15	10000	250	200	167	143	125	100	83	67
	17	11333	283	227	189	162	142	113	94	76
	19	12667	317	253	211	181	158	127	106	84
22	6	5324	133	106	89	76	67	53	44	36
	8	7099	177	142	118	101	89	71	59	47
	10	8873	222	177	148	127	111	89	74	59
	11½	10204	255	204	170	146	128	102	85	68
	13	11535	288	231	192	165	144	115	96	77
	15	13310	333	266	222	190	166	133	111	89
	17	15085	377	302	251	215	189	151	126	101
	19	16859	421	337	281	241	211	169	141	112
24	8	9216	230	185	155	130	115	90	75	60
	10	11520	290	230	190	165	145	115	95	75
	11½	13248	330	265	220	190	165	130	110	90
	13	14976	375	300	250	215	185	150	125	100
	15	17280	430	345	290	245	215	175	145	115
	17	19584	490	390	325	280	245	195	165	130
	19	21888	545	440	365	315	275	220	180	145
	21	24192	605	485	405	345	300	240	200	160
26	8	11717	295	235	195	165	145	115	100	80
	10	14647	365	295	245	210	185	145	120	100
	11½	16844	420	335	280	240	210	170	140	110
	13	19041	475	380	315	270	240	190	160	125
	15	21970	550	440	365	315	275	220	185	145
	17	24899	620	500	415	355	310	250	205	165
	19	27829	695	555	465	400	350	280	230	185
	21	30758	770	615	515	440	385	310	255	205
28	8	14635	365	295	245	210	185	145	120	100
	10	18293	455	365	305	260	230	185	150	120
	11½	21037	525	420	350	300	265	210	175	140
	13	23781	595	475	395	340	295	240	200	160
	15	27440	685	550	455	390	345	275	230	185
	17	31099	775	620	520	445	390	310	260	205
	19	34757	870	695	580	495	435	350	290	230
	21	38416	960	770	640	550	480	385	320	255
30	8	18000	450	360	300	255	225	180	150	120
	10	22500	565	450	375	320	280	225	190	150
	11½	25875	645	520	430	370	325	260	215	175
	13	29250	730	585	490	420	365	295	245	195
	15	33750	845	675	565	480	420	340	280	225
	17	38250	955	765	640	545	480	385	320	255
	19	42750	1070	855	715	610	535	430	355	285
	21	47250	1180	945	790	675	590	475	395	315
36	8	31104	780	620	520	445	390	310	260	205
	10	38880	970	780	650	555	485	390	325	260
	11½	44712	1120	895	745	640	560	445	375	300
	13	50544	1260	1010	840	720	630	505	420	335
	15	58320	1460	1170	970	835	730	585	485	390
	17	66096	1650	1320	1100	945	825	660	550	440
	19	73872	1850	1480	1230	1060	925	740	615	490
	21	81648	2040	1630	1360	1170	1020	815	680	545
42	8	49392	1230	990	825	705	615	495	410	330
	10	61740	1540	1230	1030	880	770	615	515	410
	11½	71001	1780	1420	1180	1010	890	710	590	475
	13	80262	2010	1610	1340	1150	1000	805	670	535
	15	92610	2320	1850	1540	1320	1160	925	770	615
	17	104958	2620	2100	1750	1500	1310	1050	875	700
	19	117306	2930	2350	1950	1680	1470	1170	975	780
	21	129654	3240	2590	2160	1850	1620	1300	1080	865
48	8	73728	1840	1470	1230	1050	920	735	615	490
	10	92160	2300	1840	1540	1320	1150	920	770	615
	11½	105984	2650	2120	1770	1510	1320	1060	885	705
	13	119808	3000	2400	2000	1710	1500	1200	1000	800
	15	138240	3460	2760	2300	1970	1730	1380	1150	920
	17	156672	3920	3130	2610	2240	1960	1570	1310	1040
	19	175104	4380	3500	2920	2500	2190	1750	1460	1170
	21	193536	4840	3870	3230	2760	2420	1940	1610	1290

TABLE B - STIFFNESS OF COLUMNS - VALUES OF K FOR COLUMNS *

$+K = \frac{I}{10h}$ $\quad I = \frac{bd^3}{12}$ $\quad b$ = width $\quad d$ = depth

d	b	I	h: Height of column (feet) 8	9	10	11	12	14	16	20
8	10	427	5	5	4	4	4	3	3	2
	12	512	6	6	5	5	4	4	3	3
	14	597	7	7	6	5	5	4	4	3
	18	768	10	9	8	7	6	5	5	4
	22	939	12	10	9	9	8	7	6	5
	26	1109	14	12	11	10	9	8	7	6
	30	1280	16	14	13	12	11	9	8	6
	36	1536	19	17	15	14	13	11	10	8
10	10	833	10	9	8	8	7	6	5	4
	12	1000	13	11	10	9	8	7	6	5
	14	1167	15	13	12	11	10	8	7	6
	18	1500	19	17	15	14	13	11	9	8
	22	1833	23	20	18	17	15	13	11	9
	26	2167	27	24	22	20	18	16	14	11
	30	2500	31	28	25	23	21	18	16	13
	36	3000	38	33	30	27	25	21	19	15
12	10	1440	18	16	14	13	12	10	9	7
	12	1728	22	19	17	16	14	12	11	9
	14	2016	25	22	20	18	17	14	13	10
	18	2592	32	29	26	24	22	19	16	13
	22	3168	40	35	32	29	26	23	20	16
	26	3744	47	42	37	34	31	27	23	19
	30	4320	54	48	43	39	36	31	27	22
	36	5184	65	58	52	47	43	37	32	26
14	10	2287	29	25	23	21	19	16	14	11
	12	2744	34	30	27	25	23	20	17	14
	14	3201	40	36	32	29	27	23	20	16
	18	4116	51	46	41	37	34	29	26	21
	22	5031	63	56	50	46	42	36	31	25
	26	5945	74	66	59	54	50	42	37	30
	30	6860	86	76	69	62	57	49	43	34
	36	8232	103	91	82	75	69	59	51	41
16	10	3413	43	38	34	31	28	24	21	17
	12	4096	51	46	41	37	34	29	26	20
	14	4779	60	53	48	43	40	34	30	24
	18	6144	77	68	61	56	51	44	38	31
	22	7509	94	83	75	68	63	54	47	38
	26	8875	111	99	89	81	74	63	55	44
	30	10240	128	114	102	93	85	73	64	51
	36	12288	154	137	123	112	102	88	77	61
18	10	4860	61	54	49	44	41	35	30	24
	12	5832	73	65	58	53	49	42	36	29
	14	6804	85	76	68	62	57	49	43	34
	18	8748	109	97	87	80	73	62	55	44
	22	10692	134	119	107	97	89	76	67	53
	26	12636	158	140	126	115	105	90	79	63
	30	14580	182	162	146	133	122	104	91	73
	36	17496	219	194	175	159	146	125	109	87
20	10	6667	83	74	67	61	56	48	42	33
	12	8000	100	89	80	73	67	57	50	40
	14	9333	117	104	93	85	78	67	58	47
	18	12000	150	133	120	109	100	86	75	60
	22	14667	183	163	147	133	122	105	92	73
	26	17333	217	193	173	158	144	124	108	87
	30	20000	250	222	200	182	167	143	125	100
	36	24000	300	267	240	218	200	171	150	120
22	10	8873	111	99	89	81	74	63	55	44
	12	10648	133	118	106	97	89	76	67	53
	14	12422	155	138	124	113	104	89	78	62
	18	15972	200	177	160	145	133	114	100	80
	22	19521	244	217	195	177	163	139	122	98
	26	23071	288	256	231	210	192	165	144	115
	30	26620	333	296	266	242	222	190	166	133
	36	31944	399	355	319	290	266	228	200	160
24	12	13824	175	155	140	125	115	100	85	70
	14	16128	200	180	160	145	135	115	100	80
	18	20738	260	230	205	190	175	150	130	105
	22	25344	315	280	255	230	210	180	160	125
	26	29952	375	335	300	270	250	215	185	150
	30	34560	430	385	345	315	290	245	215	175
	36	41472	520	460	415	375	345	295	260	205
	42	48384	605	540	485	440	405	345	300	240
26	12	17576	220	195	175	160	145	125	110	90
	14	20505	255	230	205	185	170	145	130	105
	18	26364	330	295	265	240	220	190	165	130
	22	32223	405	360	320	295	270	230	200	160
	26	38081	475	425	380	345	315	270	240	190
	30	43940	550	490	440	400	365	315	275	220
	36	52728	660	585	525	480	440	375	330	265
	42	61516	770	685	615	560	515	440	385	310
28	12	21952	275	245	220	200	185	155	135	110
	14	25611	320	285	255	235	215	185	160	130
	18	32928	410	365	330	300	275	235	205	165
	22	40245	505	445	400	365	335	285	250	200
	26	47563	595	530	475	430	395	340	295	240
	30	54880	685	610	550	500	455	390	345	275
	36	65856	825	730	660	600	550	470	410	330
	42	76832	960	855	770	700	640	550	480	385
30	12	27000	340	300	270	245	225	195	170	135
	14	31500	395	350	315	285	265	225	195	160
	18	40500	505	450	405	370	340	290	255	205
	22	49500	620	550	495	450	415	355	310	250
	26	58500	730	650	585	530	490	420	365	295
	30	67500	845	750	675	615	565	480	420	340
	36	81000	1010	900	810	735	675	580	505	405
	42	94500	1180	1050	945	860	790	675	590	475
32	12	32768	410	365	330	300	275	235	205	165
	14	38229	480	425	380	350	320	275	240	190
	18	49152	615	545	490	445	410	350	305	245
	22	60075	750	670	600	545	500	430	375	300
	26	70997	885	790	710	645	590	505	445	355
	30	81920	1020	910	820	745	685	585	510	410
	36	98304	1230	1090	985	895	820	700	615	490
	42	114688	1430	1270	1150	1040	955	820	715	575
34	12	39304	490	435	395	355	330	280	245	195
	14	45855	575	510	460	415	380	330	285	230
	18	58956	735	655	590	535	490	420	370	295
	22	72057	900	800	720	655	600	515	450	360
	26	85159	1060	945	850	775	710	610	530	425
	30	98260	1230	1090	985	895	820	700	615	490
	36	117912	1470	1310	1180	1070	980	840	735	590
	42	137564	1720	1530	1380	1250	1150	985	860	690
36	12	46656	585	520	465	425	390	335	290	235
	14	54432	680	605	545	495	455	390	340	270
	18	69984	875	780	700	635	585	500	435	350
	22	85536	1070	950	855	780	715	610	535	430
	26	101088	1260	1120	1010	920	840	720	630	505
	30	116640	1460	1300	1170	1060	970	835	730	585
	36	139968	1750	1560	1400	1270	1170	1000	875	700
	42	163296	2040	1810	1630	1480	1360	1170	1020	815

+ Values of K in table divided by 10 for convenience. * From Portland Cement Assoc.

STRUCTURAL-RIGID FRAMES-HARDY CROSS METHOD-1

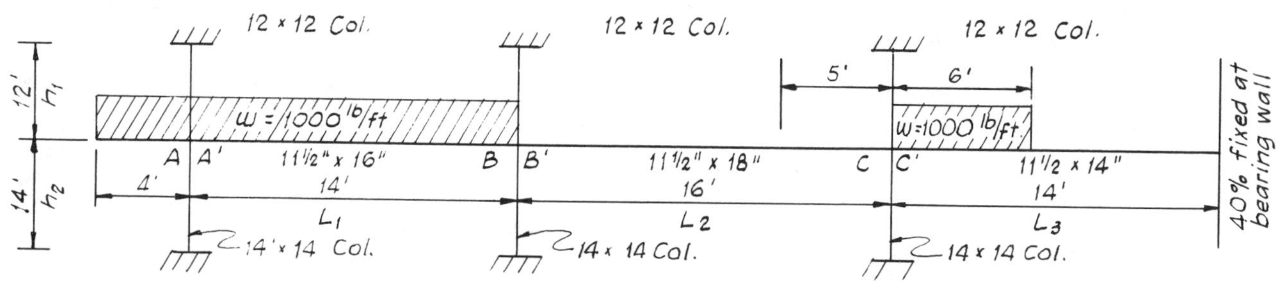

12 × 12 Col.
12 × 12 Col.
12 × 12 Col.
12'
h_1
14'
h_2
w = 1000 lb/ft
w = 1000 lb/ft
5'
6'
A A'
11½" x 16"
B B'
11½" x 18"
C C'
11½ x 14"
4'
14'
16'
14'
L_1
L_2
L_3
14' x 14 Col.
14 x 14 Col.
14 x 14 Col.
40% fixed at bearing wall

Given : Reinforced concrete rigid frame L_1, L_2, L_3, h_1, h_2 loads as shown and $\frac{\text{Flange Width}}{\text{Stem Width}} = 6.0$

To Find : The moments and reactions in frame. Answers shown 7.4 below.

Solution : Find K* for beams and columns. See pg. 3-02

Member	K
12 × 12 Column	14
14 × 14 Column	23
Beam A'B	56
Beam B'C	70
Beam C'D	38

Note : See Fig. 2 pg. 2-04 for additional data on "K"

Draw diagram and find stiffness ratios as indicated below.
Find fixed end and moments (FEM) as noted. Balance moments at least 3 times and carry over.
Compute negative moments at supports, moments in columns, reactions and positive moments as indicated

1.00 for simple support; K from table above

		(14) A'	(56) B	(14) B'	(70) C	(14) C'	(38) D	
Stiffness Ratio always for cantilever	0.0	.60	.34	.43	.48	.26	.60	.40 (given)
FEM	-8.0	-16.4	-16.4	-10.6	-23.6	-9.4	-3.5	
Balance Moments	0	+5.0	+2.0	-2.5	+6.8	-3.7	+2.1	
Carryover	0	1.0	-2.5	-3.4	+1.3	-1.1	+1.9	
Balance	0	+.6	.3	+.4	-1.2	+.6	-1.1	
Carryover	0	+.2	-.3	+.6	-.2	+.6	-.3	
Balance		-1	+.3	-.4	+.4	-.2	+.2	
Neg. Moments	-8.0	-11.7	-17.2	-15.9	-16.5	-13.2	-7	
Unbalanced M		-3.7 = (11.7 - 8.0)	-1.3	= (17.2 - 15.9)	3.3	= (16.5 - 13.2)		
Moment In Cols. upper tier		1.4 = 3.7 × 14/(14+23)	.5	= 1.3 × 14/(14+23)	1.3	= 3.3 × 14/(14+23)		
lower tier		2.3 = 3.7 - 1.4	.8	= 1.3 - .5	2.0	= 3.3 - 1.3		
Beam Reactions		40.66			6.9	5.6	.4	

Stiffness ratios: $.60 = \frac{56}{14+23+56+0}$; $.34 = \frac{56}{14+23+56+70}$; $.60 = (1.00 - .40)$

FEM: $1000 \times \frac{14^2}{12} = 16.4$; -10.6, -23.6 and -9.4, -3.5 From pg. 1-50

$(-8.0 + 16.4) \times 0 = 0$; $+5.0 = (-16.4 + 8.0) \times -6.0$; $+2.0 = (-16.4 + 10.6) \times -34$; $+6.8 = (-10.6 + 16.4) \times -43$; $+2.1 = (-3.5 + 0) -6$

Carryover: $(+5.0 \times -\frac{1}{2})$; $x - \frac{1}{2}$ =; $= (+6.8 \times -\frac{1}{2})$; $1.0 = (-1.0 + 0.0) \times -0.6)$; $.3 = (-2.5 + 3.4) \times 0.34$

Reaction = Reaction as simple beam − [Moment at reaction end − Moment at other end of beam] / Span

$R_B = 7.0 \frac{[-17.2 - (11.7)]}{14} = 7.4K$

$R_{A'} = 14 - 7.4 = 6.6K$

Positive Moments:

$\frac{6.6}{1} = 6.6$ $+M = 6.6 \times 6.6 = +43.6$; $1 \times 6.6 \times \frac{6.6}{2} = -21.8$; $+21.8$; -11.7; $+10.1$

$+M = 3.1 \times 11 = +34.1$; $-M = -15.9$; $+18.2$

$5.6 = 5.6$ $+M = 5.6 \times 5.6 = +31.4$; $1 \times 5.6 \times \frac{5.6}{2} = -15.7$; $+15.7$; $-M = -13.2$; $+2.5$

*Solution for steel frame similar. $K = \frac{EI}{L}$ where E = modulus of elasticity, I = moment of inertia and L = length of member E cancels except if structural steel and concrete is used in same frame.

STRUCTURAL-RIGID FRAMES-HARDY CROSS METHOD-2

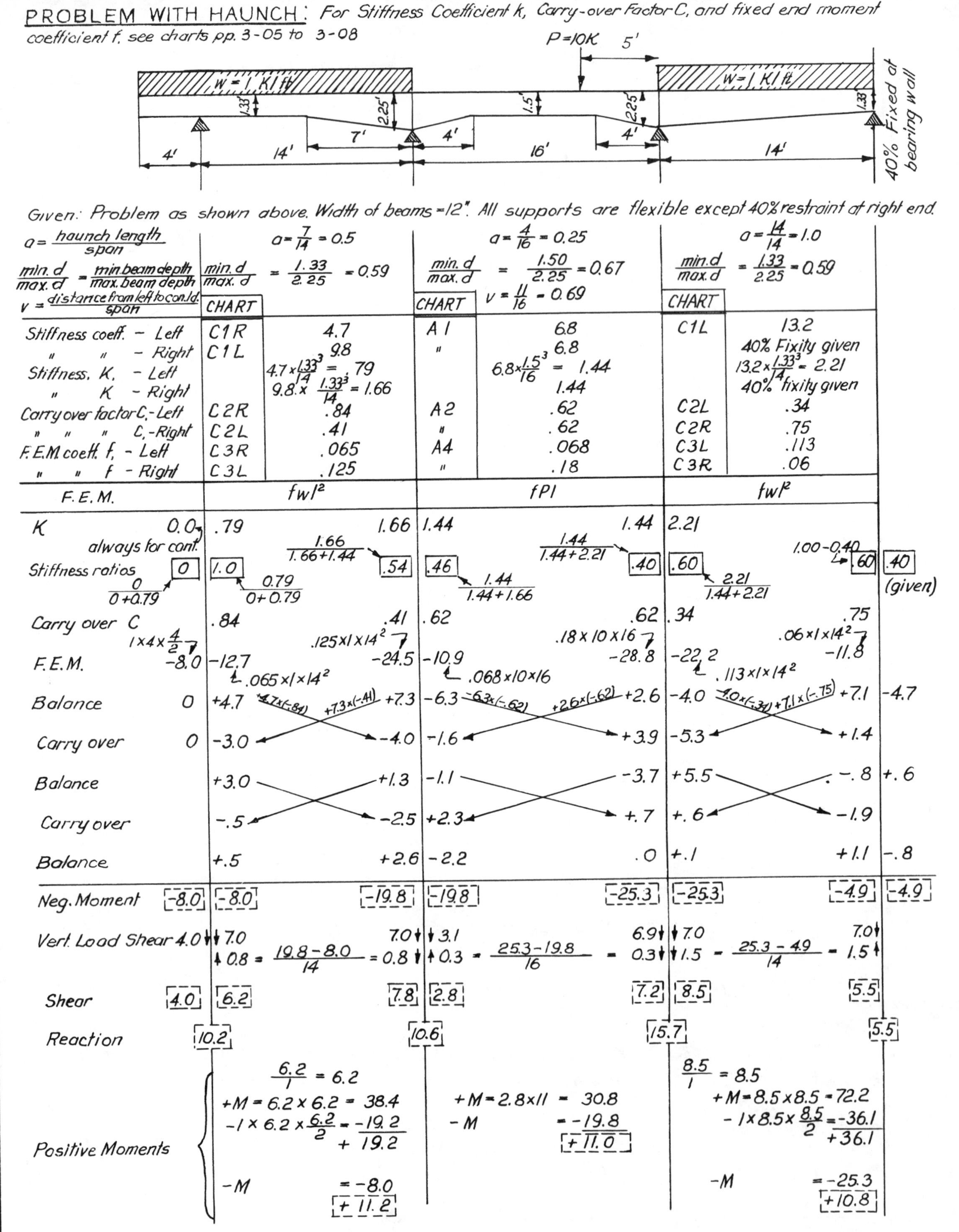

PROBLEM WITH HAUNCH: For Stiffness Coefficient k, Carry-over Factor C, and fixed end moment coefficient f, see charts pp. 3-05 to 3-08

P=10K
5'
W=1 K/ft
W=1 K/ft
1.33'
2.25'
1.5'
2.25'
1.33'
7'
4'
4'
4'
14'
16'
14'
40% Fixed at bearing wall

Given: Problem as shown above. Width of beams = 12". All supports are flexible except 40% restraint at right end.

$a = \frac{\text{haunch length}}{\text{span}}$ | $a = \frac{7}{14} = 0.5$ | $a = \frac{4}{16} = 0.25$ | $a = \frac{14}{14} = 1.0$

$\frac{\text{min. d}}{\text{max. d}} = \frac{\text{min. beam depth}}{\text{max. beam depth}}$ | $\frac{\text{min. d}}{\text{max. d}} = \frac{1.33}{2.25} = 0.59$ | $\frac{\text{min. d}}{\text{max. d}} = \frac{1.50}{2.25} = 0.67$ | $\frac{\text{min. d}}{\text{max. d}} = \frac{1.33}{2.25} = 0.59$

$v = \frac{\text{distance from left to con. ld.}}{\text{span}}$ | | $v = \frac{11}{16} = 0.69$ |

	CHART	Span 1	CHART	Span 2	CHART	Span 3
Stiffness coeff. – Left	C1R	4.7	A1	6.8	C1L	13.2
" " – Right	C1L	9.8	"	6.8		40% Fixity given
Stiffness, K, – Left		$4.7 \times \frac{1.33^3}{14} = .79$		$6.8 \times \frac{1.5^3}{16} = 1.44$		$13.2 \times \frac{1.33^3}{14} = 2.21$
" K – Right		$9.8 \times \frac{1.33^3}{14} = 1.66$		1.44		40% fixity given
Carry over factor C, – Left	C2R	.84	A2	.62	C2L	.34
" " " C, – Right	C2L	.41	"	.62	C2R	.75
F.E.M. coeff. f, – Left	C3R	.065	A4	.068	C3L	.113
" " f – Right	C3L	.125	"	.18	C3R	.06
F.E.M.		fwl^2		fPl		fwl^2

	Cantilever	Span 1 Left	Span 1 Right	Span 2 Left	Span 2 Right	Span 3 Left	Span 3 Right	Wall
K	0.0 always for cont.	.79	1.66	1.44	1.44	2.21		
Stiffness ratios	0 ($\frac{0}{0+0.79}$)	1.0 ($\frac{0.79}{0+0.79}$)	.54 ($\frac{1.66}{1.66+1.44}$)	.46 ($\frac{1.44}{1.44+1.66}$)	.40 ($\frac{1.44}{1.44+2.21}$)	.60 ($\frac{2.21}{1.44+2.21}$)	.60 (1.00−0.40)	.40 (given)
Carry over C		.84	.41	.62	.62	.34	.75	
F.E.M.	−8.0 ($1 \times 4 \times \frac{4}{2}$)	−12.7 ($.065 \times 1 \times 14^2$)	−24.5 ($.125 \times 1 \times 14^2$)	−10.9 ($.068 \times 10 \times 16$)	−28.8 ($.18 \times 10 \times 16$)	−22.2 ($.113 \times 1 \times 14^2$)	−11.8 ($.06 \times 1 \times 14^2$)	
Balance	0	+4.7	+7.3	−6.3	+2.6	−4.0	+7.1	−4.7
Carry over	0	−3.0 ($4.7 \times (-.84)$)	−4.0 ($+7.3 \times (-.41)$)	−1.6 ($2.6 \times (-.62)$)	+3.9 ($-6.3 \times (-.62)$)	−5.3 ($+7.1 \times (-.75)$)	+1.4 ($4.0 \times (-.34)$)	
Balance		+3.0	+1.3	−1.1	−3.7	+5.5	−.8	+.6
Carry over		−.5	−2.5	+2.3	+.7	+.6	−1.9	
Balance		+.5	+2.6	−2.2	.0	+.1	+1.1	−.8
Neg. Moment	−8.0	−8.0	−19.8	−19.8	−25.3	−25.3	−4.9	−4.9
Vert. Load Shear	4.0↓	↓7.0, ↑0.8	7.0↓, 0.8↓	↓3.1, ↑0.3	6.9↓, 0.3↓	↓7.0, ↓1.5	7.0↓, 1.5↑	
Shear	4.0	6.2	7.8	2.8	7.2	8.5	5.5	
Reaction		10.2		10.6		15.7		5.5

Vert. load shear: $0.8 = \frac{19.8 - 8.0}{14} = 0.8$; $0.3 = \frac{25.3 - 19.8}{16} = 0.3$; $1.5 = \frac{25.3 - 4.9}{14} = 1.5$

Positive Moments

Span 1: $\frac{6.2}{1} = 6.2$; $+M = 6.2 \times 6.2 = 38.4$; $-1 \times 6.2 \times \frac{6.2}{2} = -19.2$; $+19.2$; $-M = -8.0$; $+11.2$

Span 2: $+M = 2.8 \times 11 = 30.8$; $-M = -19.8$; $+11.0$

Span 3: $\frac{8.5}{1} = 8.5$; $+M = 8.5 \times 8.5 = 72.2$; $-1 \times 8.5 \times \frac{8.5}{2} = -36.1$; $+36.1$; $-M = -25.3$; $+10.8$

STRUCTURAL–RIGID FRAMES–HAUNCHED BEAM PROPERTIES-I

MEMBERS WITH STRAIGHT HAUNCHES

Fig. A–Stiffness Coefficient **A1**

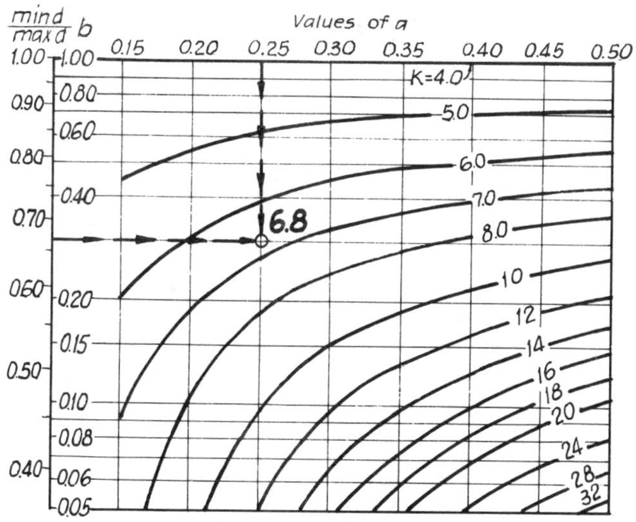

$\frac{min\,d}{max\,d}$ b — Values of a

0.15 0.20 0.25 0.30 0.35 0.40 0.45 0.50

K=4.0, 5.0, 6.0, 7.0, 8.0, 10, 12, 14, 16, 18, 20, 24, 28, 32; 6.8

Fig. C–Carry-over, Factor C **A2**

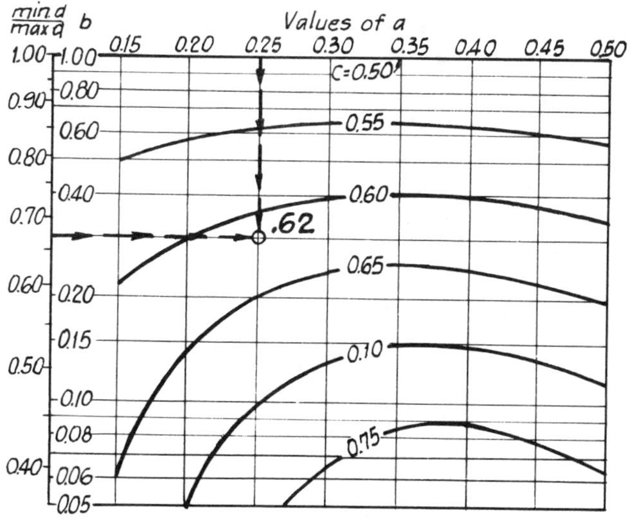

$\frac{min\,d}{max\,d}$ b — Values of a

0.15 0.20 0.25 0.30 0.35 0.40 0.45 0.50

C=0.50, 0.55, 0.60, 0.65, 0.70, 0.75; .62

Fig. D–Uniform Load f.e.m. Coefficient, f **A3**

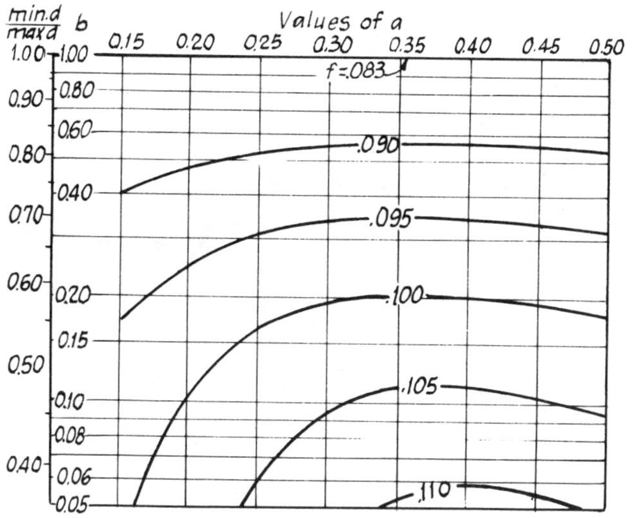

$\frac{min\,d}{max\,d}$ b — Values of a

0.15 0.20 0.25 0.30 0.35 0.40 0.45 0.50

f=.083, .090, .095, .100, .105, .110

Fig. B–Concentrated Load F.E.M. Coefficient **A4**

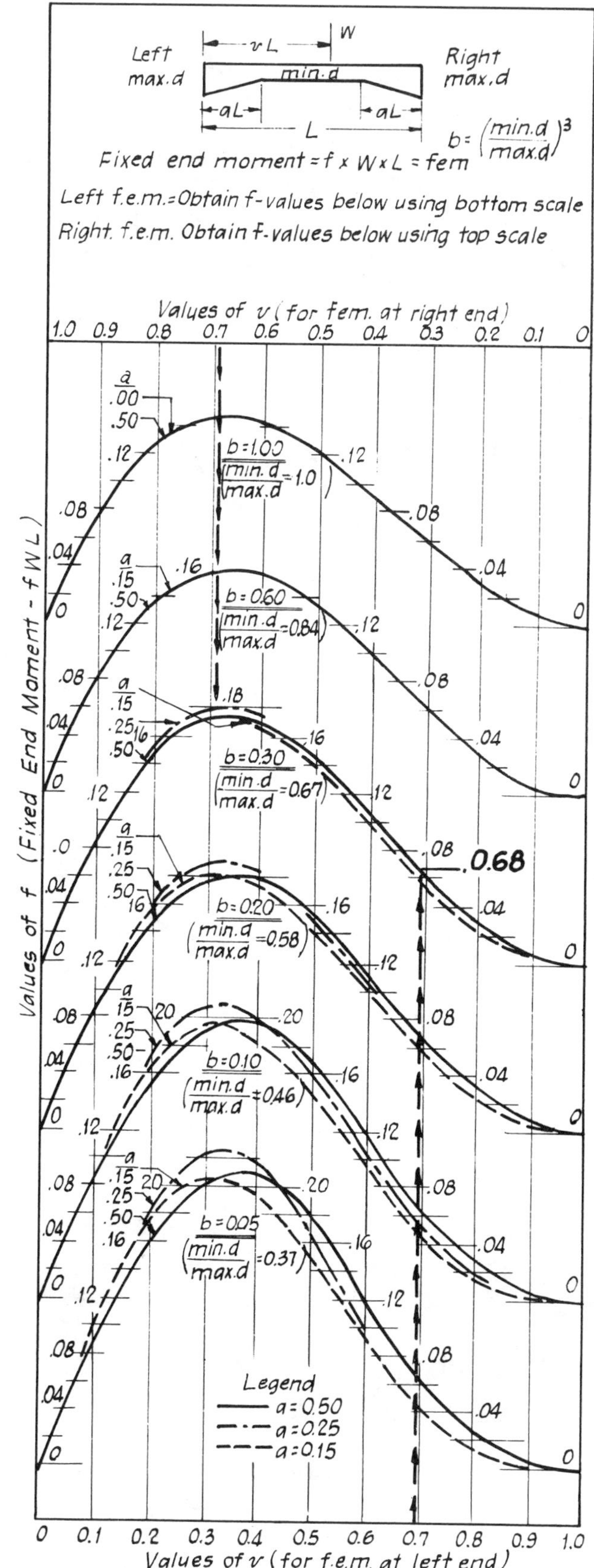

Left max.d
Right max.d
min.d
vL
W
aL
L
$b = \left(\frac{min.d}{max.d}\right)^3$
Fixed end moment = f × W × L = fem
Left f.e.m.=Obtain f-values below using bottom scale
Right f.e.m. Obtain f-values below using top scale
Values of v (for fem. at right end)
Values of f (Fixed End Moment - fWL)
b=1.00 (min.d/max.d = 1.0)
b=0.60 (min.d/max.d = 0.84)
b=0.30 (min.d/max.d = 0.67)
b=0.20 (min.d/max.d = 0.58)
b=0.10 (min.d/max.d = 0.46)
b=0.05 (min.d/max.d = 0.37)
.68
Legend
a=0.50
a=0.25
a=0.15
Values of v (for f.e.m. at left end)

* From Portland Cement Association

STRUCTURAL–RIGID FRAMES–HAUNCHED BEAM PROPERTIES-2

SYMMETRICAL MEMBERS WITH PARABOLIC HAUNCHES

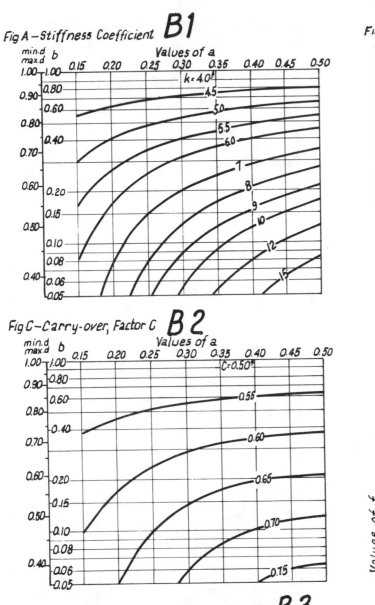

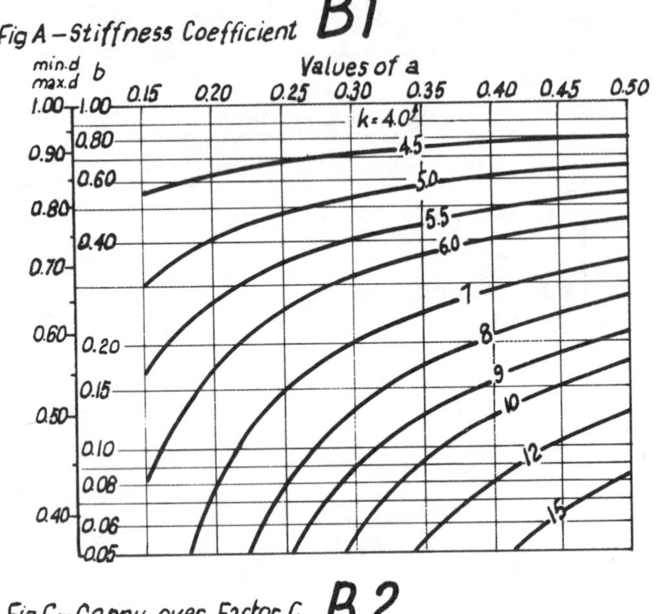

B1
Fig A – Stiffness Coefficient
Values of a
min.d/max.d
b
k = 4.0*

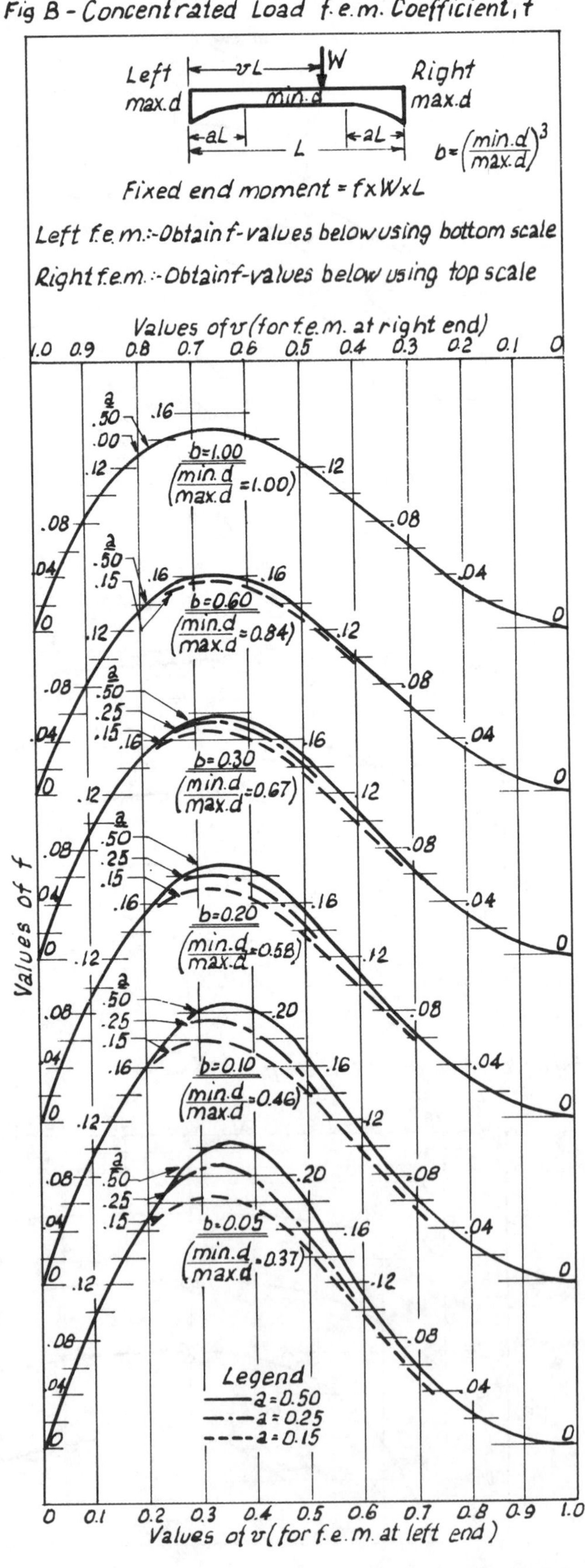

B4
Fig B – Concentrated Load f.e.m. Coefficient, f
Fixed end moment = f x W x L
Left f.e.m.:- Obtain f-values below using bottom scale
Right f.e.m.:- Obtain f-values below using top scale
Values of v (for f.e.m. at right end)
Values of v (for f.e.m. at left end)
Values of f
Legend
a = 0.50
a = 0.25
a = 0.15

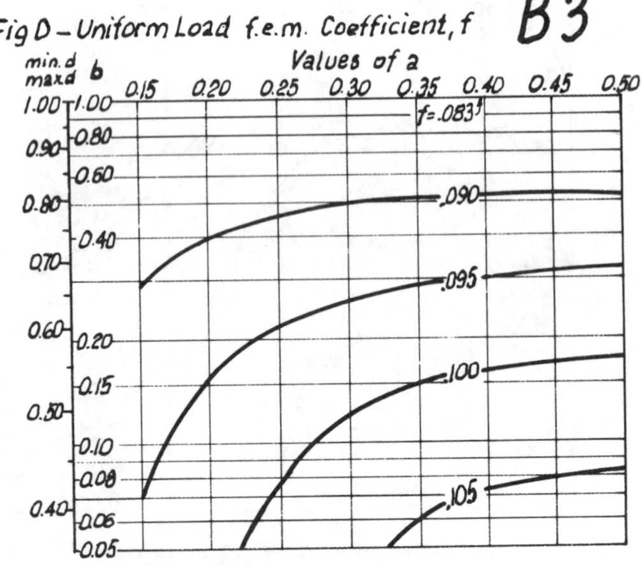

B2
Fig C – Carry-over, Factor C
B3
Fig D – Uniform Load f.e.m. Coefficient, f
Values of a
C = 0.50*
f = .083*

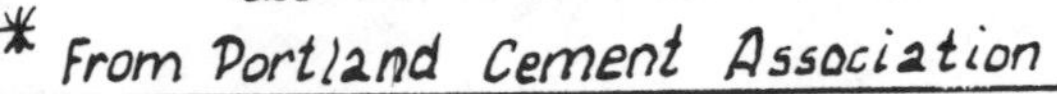

* From Portland Cement Association

STRUCTURAL-RIGID FRAMES-HAUNCHED BEAM PROPERTIES-3

UNSYMMETRICAL MEMBERS WITH STRAIGHT HAUNCH AT ONE END

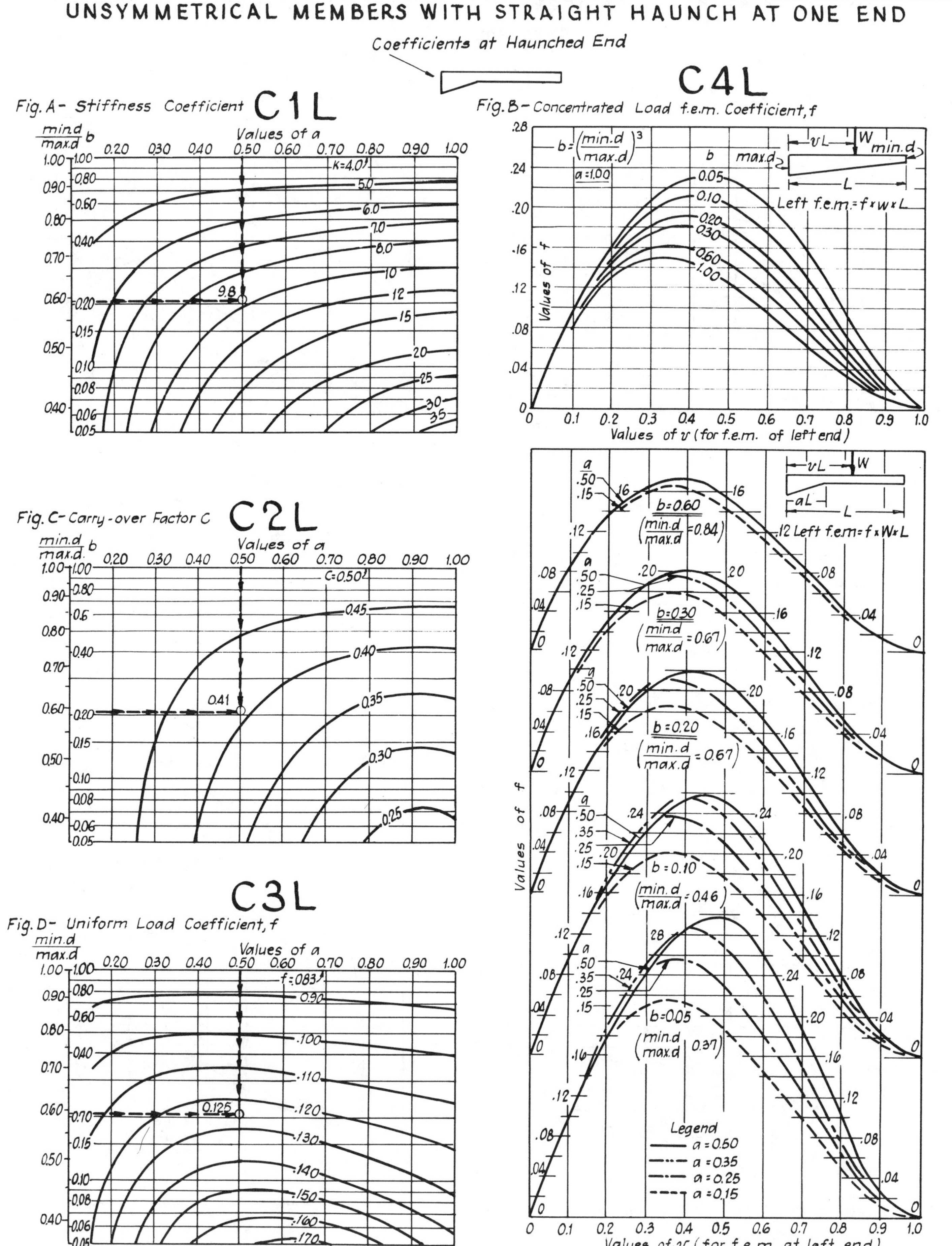

Coefficients at Haunched End
C1L
Fig. A- Stiffness Coefficient
C4L
Fig. B- Concentrated Load f.e.m. Coefficient, f
Left f.e.m. = f × W × L
Values of v (for f.e.m. of left end)
C2L
Fig. C- Carry-over Factor C
C3L
Fig. D- Uniform Load Coefficient, f
Values of a
Values of f
Legend
a = 0.50
a = 0.35
a = 0.25
a = 0.15
Values of v (for f.e.m. at left end)

✱ From Portland Cement Association

STRUCTURAL–RIGID FRAMES–HAUNCHED BEAM PROPERTIES–4

UNSYMMETRICAL MEMBERS WITH STRAIGHT HAUNCH AT ONE END

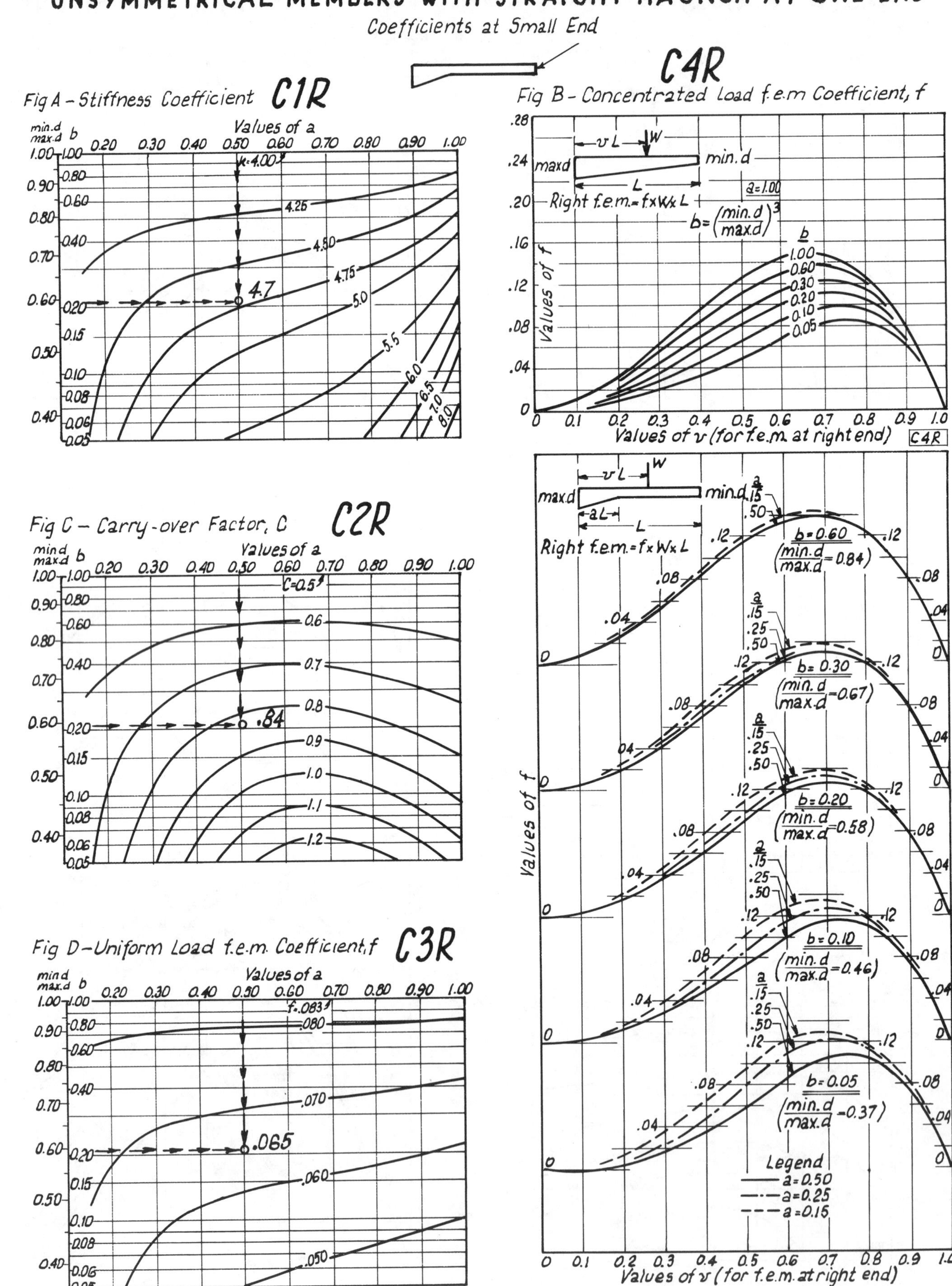

Coefficients at Small End
C1R
Fig A – Stiffness Coefficient
min.d / max.d
b
Values of a
0.20 0.30 0.40 0.50 0.60 0.70 0.80 0.90 1.00
k=4.00
4.25
4.50
4.75
5.0
5.5
6.0
6.5
7.0
8.0
4.7
C4R
Fig B – Concentrated Load f.e.m Coefficient, f
vL
W
max d
min. d
L
Right f.e.m.=f×W×L
a=1.00
$b=\left(\frac{min.d}{max.d}\right)^3$
b
1.00
0.60
0.30
0.20
0.10
0.05
Values of f
Values of v (for f.e.m. at right end)
C4R
C2R
Fig C – Carry-over Factor, C
Values of a
C=0.5
0.6
0.7
0.8
.84
0.9
1.0
1.1
1.2
aL
Right f.e.m.=f×W×L
b=0.60 (min.d/max.d=0.84)
b=0.30 (min.d/max.d=0.67)
b=0.20 (min.d/max.d=0.58)
b=0.10 (min.d/max.d=0.46)
b=0.05 (min.d/max.d=0.37)
a
.15
.25
.50
C3R
Fig D–Uniform Load f.e.m. Coefficient, f
Values of a
f=.083
.080
.070
.065
.060
.050
Legend
a=0.50
a=0.25
a=0.15
Values of f
Values of v (for f.e.m. at right end)

* From Portland Cement Association

STRUCTURAL-RIGID FRAMES-VARIABLE SECTION-I

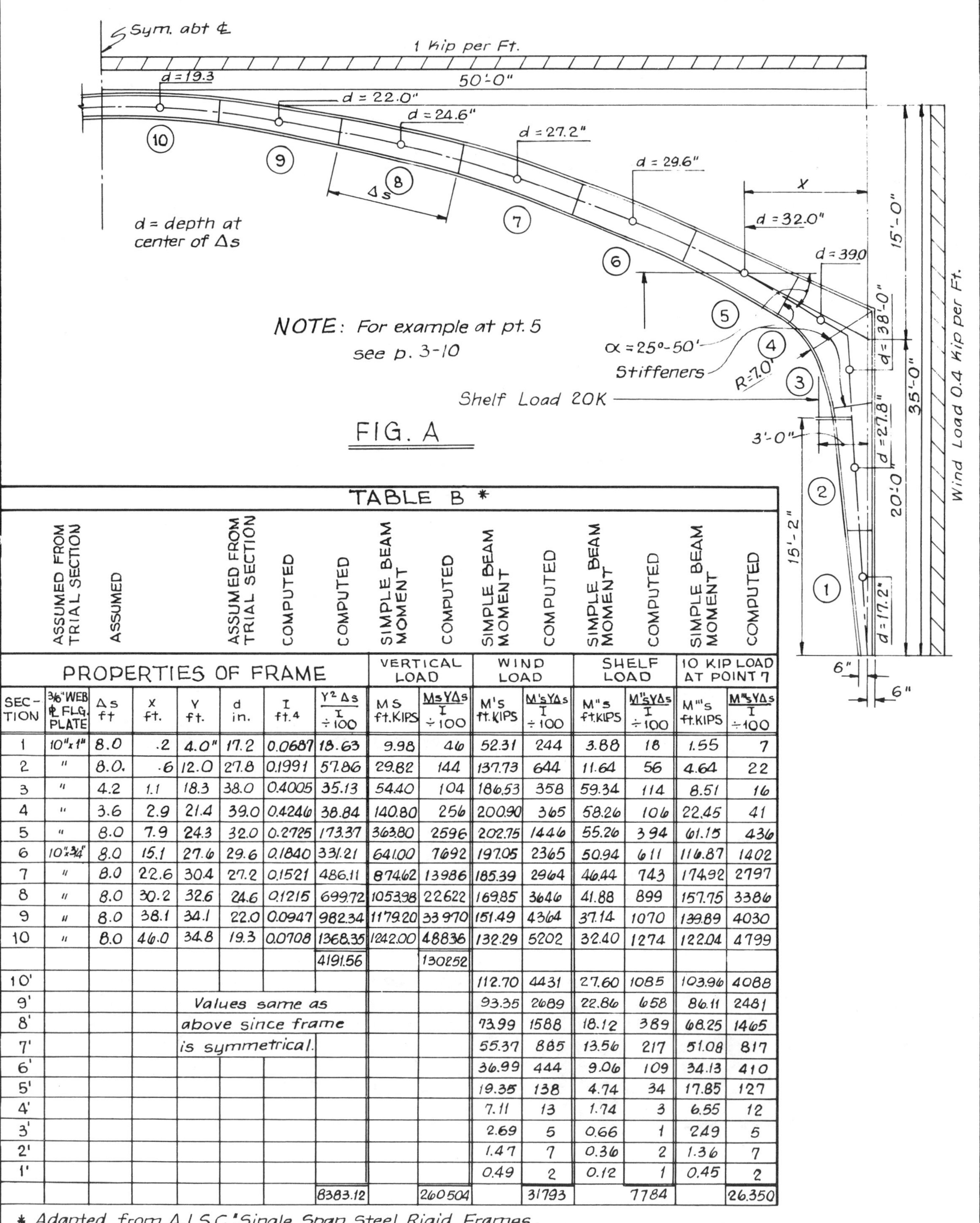

Sym. abt ℄
1 Kip per Ft.
50'-0"
d = 19.3
d = 22.0"
d = 24.6"
d = 27.2"
d = 29.6"
d = 32.0"
d = 39.0
X
Δs
d = depth at center of Δs
NOTE: For example at pt. 5 see p. 3-10
α = 25°-50'
Stiffeners
R=7.0'
Shelf Load 20K
3'-0"
d = 38'-0"
d = 27.8"
d = 17.2"
15'-0"
35'-0"
20'-0"
15'-2"
6"
6"
Wind Load 0.4 Kip per Ft.
1
2
3
4
5
6
7
8
9
10

FIG. A

TABLE B *

	ASSUMED FROM TRIAL SECTION	ASSUMED			ASSUMED FROM TRIAL SECTION	COMPUTED	COMPUTED	SIMPLE BEAM MOMENT	COMPUTED	SIMPLE BEAM MOMENT	COMPUTED	SIMPLE BEAM MOMENT	COMPUTED	SIMPLE BEAM MOMENT	COMPUTED
PROPERTIES OF FRAME								VERTICAL LOAD		WIND LOAD		SHELF LOAD		10 KIP LOAD AT POINT 7	
SECTION	3/16" WEB ℄ FLG. PLATE	Δs ft	x ft.	Y ft.	d in.	I ft.^{4}	$\frac{Y^2 \Delta s}{I} \div 100$	MS ft.KIPS	$\frac{M_s Y \Delta s}{I} \div 100$	M's ft.KIPS	$\frac{M'_s Y \Delta s}{I} \div 100$	M"s ft.KIPS	$\frac{M''_s Y \Delta s}{I} \div 100$	M'''s ft.KIPS	$\frac{M'''_s Y \Delta s}{I} \div 100$
1	10"x1"	8.0	.2	4.0"	17.2	0.0687	18.63	9.98	46	52.31	244	3.88	18	1.55	7
2	"	8.0.	.6	12.0	27.8	0.1991	57.86	29.82	144	137.73	644	11.64	56	4.64	22
3	"	4.2	1.1	18.3	38.0	0.4005	35.13	54.40	104	186.53	358	59.34	114	8.51	16
4	"	3.6	2.9	21.4	39.0	0.4246	38.84	140.80	256	200.90	365	58.26	106	22.45	41
5	"	8.0	7.9	24.3	32.0	0.2725	173.37	363.80	2596	202.75	1446	55.26	394	61.15	436
6	10"x3/4"	8.0	15.1	27.6	29.6	0.1840	331.21	641.00	7692	197.05	2365	50.94	611	116.87	1402
7	"	8.0	22.6	30.4	27.2	0.1521	486.11	874.62	13986	185.39	2964	46.44	743	174.92	2797
8	"	8.0	30.2	32.6	24.6	0.1215	699.72	1053.98	22622	169.85	3646	41.88	899	157.75	3386
9	"	8.0	38.1	34.1	22.0	0.0947	982.34	1179.20	33970	151.49	4364	37.14	1070	139.89	4030
10	"	8.0	46.0	34.8	19.3	0.0708	1368.35	1242.00	48836	132.29	5202	32.40	1274	122.04	4799
							4191.56		130252						
10'										112.70	4431	27.60	1085	103.96	4088
9'				Values same as						93.35	2689	22.86	658	86.11	2481
8'				above since frame						73.99	1588	18.12	389	68.25	1465
7'				is symmetrical.						55.37	885	13.56	217	51.08	817
6'										36.99	444	9.06	109	34.13	410
5'										19.35	138	4.74	34	17.85	127
4'										7.11	13	1.74	3	6.55	12
3'										2.69	5	0.66	1	2.49	5
2'										1.47	7	0.36	2	1.36	7
1'										0.49	2	0.12	1	0.45	2
							8383.12		260504		31793		7784		26.350

* Adapted from A.I.S.C.,"Single Span Steel Rigid Frames.

STRUCTURAL-RIGID FRAMES-VARIABLE SECTION-2

FORCE DIAGRAMS FOR SIMPLE BEAM VARIOUS LOADING

I	II	III	IV
UNIFORM VERTICAL LOAD	WIND LOAD	SHELF LOAD	10KIP LOAD AT POINT 7

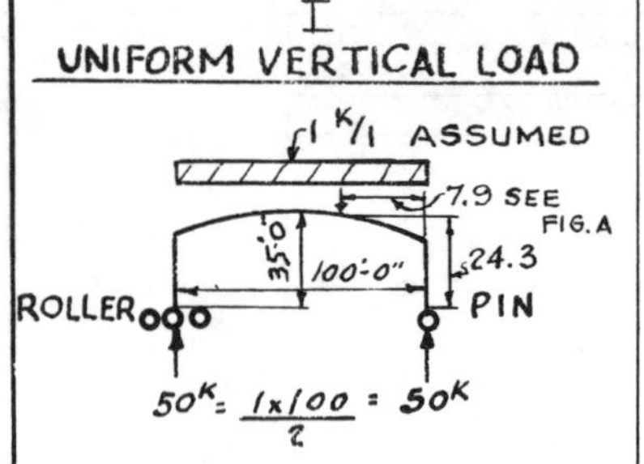
1 K/1 ASSUMED
7.9 SEE FIG. A
35'-0
100'-0"
24.3
ROLLER
PIN
50K = 1x100/2 = 50K

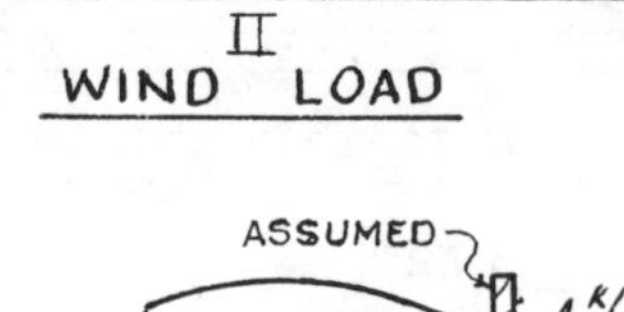
ASSUMED
.4 K/1

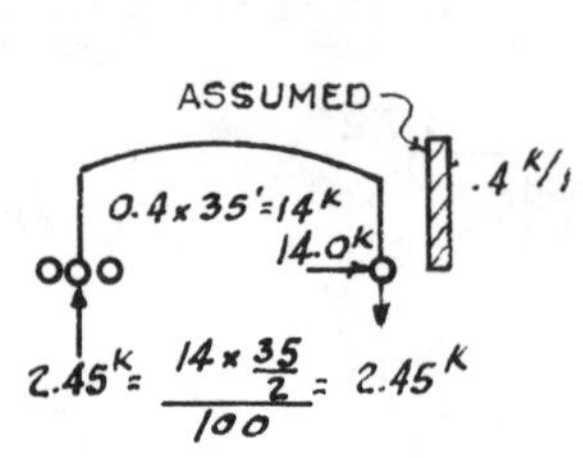
0.4 x 35' = 14K
14.0K
2.45K = (14 x 35/2)/100 = 2.45K

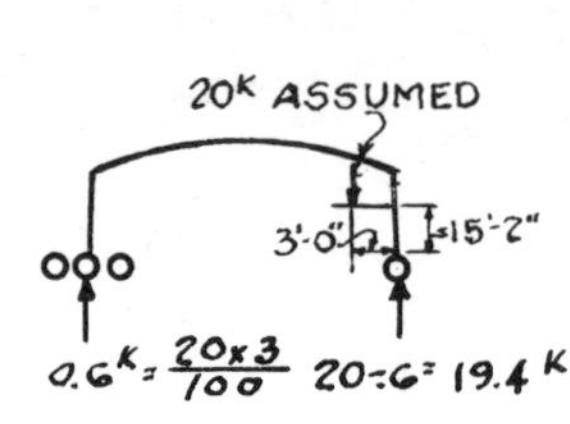
20K ASSUMED
3'-0"
15'-2"
0.6K = 20x3/100
20-.6 = 19.4K

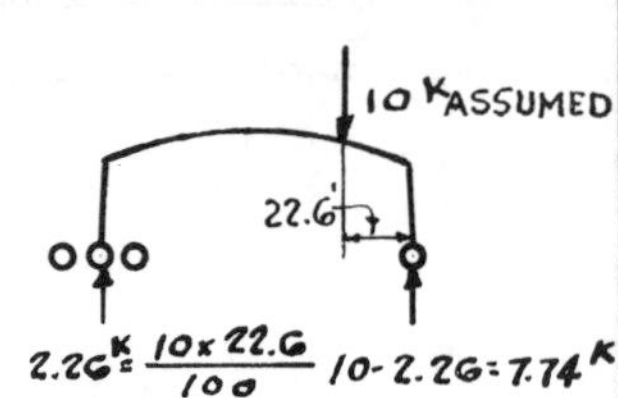
10K ASSUMED
22.6'
2.26K = 10x22.6/100
10-2.26 = 7.74K

COMPUTATION OF M_S, M'_S ETC. AT PT. 5 FOR TABLE B FOR PT 5 X=7.9, Y=24.3

I	II	III	IV
$M_S = 50 \times 7.9 = +395.0$	$M'_S = 14 \times 24.3 = +340.2$	$M''_S = 0.6 \times 92.1 = +55.26$	$M'''_S = 7.74 \times 7.9 = 61.15$
$1 \times 7.9 \times \frac{7.9}{2} = -31.2$	$0.4 \times 24.3 \times \frac{24.3}{2} = -118.1$		
$+363.8$	$2.45 \times 7.9 = -19.35$		
	$+202.75$		

FINAL FORCE DIAGRAMS

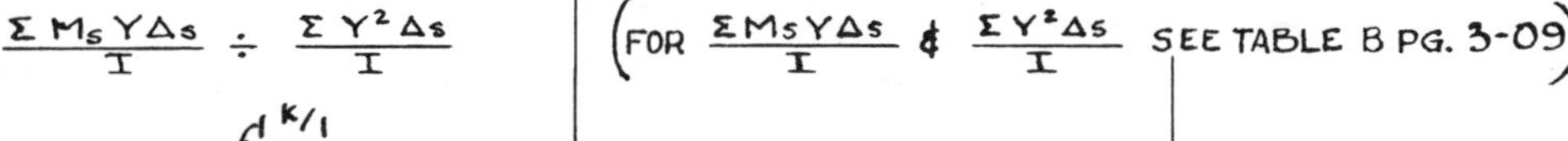

$H = \frac{\Sigma M_S Y \Delta s}{I} \div \frac{\Sigma Y^2 \Delta s}{I}$ (FOR $\frac{\Sigma M_S Y \Delta s}{I}$ & $\frac{\Sigma Y^2 \Delta s}{I}$ SEE TABLE B PG. 3-09)

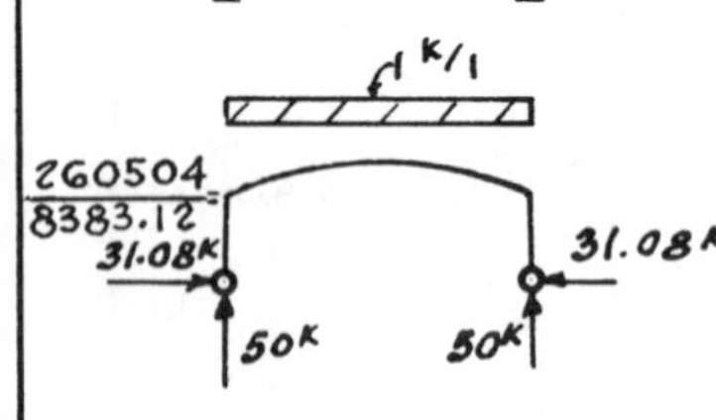
1 K/1
260504/8383.12 =
31.08K
31.08K
50K
50K

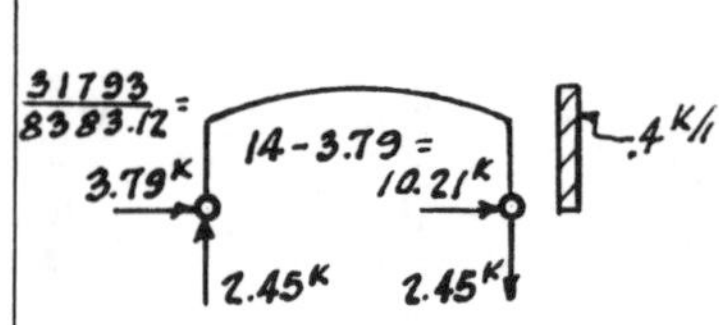
31793/8383.12 =
3.79K
14-3.79 = 10.21K
.4K/1
2.45K
2.45K

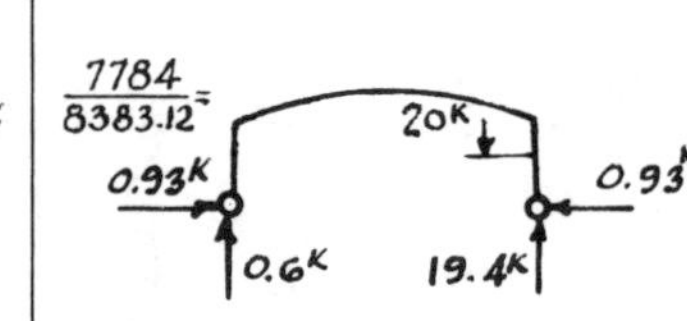
7784/8383.12 =
0.93K
20K
0.93K
0.6K
19.4K

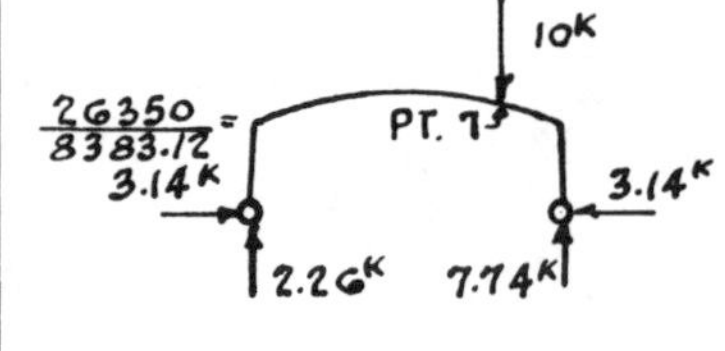
10K
26350/8383.12 =
3.14K
PT. 7
3.14K
2.26K
7.74K

FINAL MOMENTS AT PT. 5

I	II	III	IV
$50 \times 7.9 = +395$	$10.21 \times 24.3 = +248$	$0.6 \times 92.1 = +55$	$7.74 \times 7.9 = +61$
$1 \times 7.9 \times \frac{7.9}{2} = -31$	$.4 \times 24.3 \times \frac{24.3}{2} = -118$	$0.93 \times 24.3 = -23$	$3.14 \times 24.3 = -76$
$31.05 \times 24.3 = -755$	$2.45 \times 7.9 = -19$	$+32^{1K}$	-15
-391^{1K}	$+111^{1K}$		

THRUST $\alpha = 26° - 50'$ GIVEN

I	II	III	IV
$(50 - 7.9 \times 1) \sin\alpha + 31.08 \cos\alpha = 46.7^K$	$2.45 \sin\alpha + (10.21 - .4 \times 24.3)\cos\alpha = 1.5^K$	$0.6 \sin\alpha + 0.93 \cos\alpha = 1.2^K$	$7.74 \sin\alpha + 3.14 \cos\alpha = 6.3^K$

WIND FROM RIGHT AND CRANE LOAD DO NOT GOVERN MOMENTS AT PT. 5 NOTE HORIZONTAL FORCE OF CRANE IS NOT INCLUDED.

$M = -391^{1K}$, -15^{1K}, total -406^{1K} $P = 46.7^K + 6.3^K = 53.0^K$

UNBRACED LENGTH = 15' $\therefore \frac{ld}{bt} = \frac{15 \times 12 \times 32}{10 \times 1.0} = 576 < 600$ $F_B = 20{,}000\ \#/\square''$

NOTE: Bottom flange in comp. at pt. 5. 15' is distance between longitudinal bott. flange braces. $\frac{l}{r} = \frac{180}{2.3} = 78.3$ $F_A = 14{,}000\ \#/\square''$

MAXIMUM STRESS @ PT. 5 $= \frac{P}{A} + \frac{M}{S}$

$$\frac{53.0}{31.25} + \frac{406 \times 12}{353} = 15.5\ ^K/\square''$$

CHECK COMBINED STRESS $\frac{f_a}{F_A} + \frac{f_b}{F_B}$

$$\frac{1.7}{14} + \frac{13.8}{20} = 0.81 < 1 \text{ OK}$$

HORIZONTAL TIE FORCE = 31.08 + .93 + 3.14 = 35.15^K

WIND FORCE, *Take directly on foundations* = 10.21^K *each side.*

STRUCTURAL-RIGID FRAMES-SIDE-SWAY

SIDE-SWAY PROBLEM FOR CASES OF ECCENTRIC LOAD, HORIZONTAL LOAD AND UNSYMMETRICAL COLUMNS

1- The overturning moment, O.T.M., is 1000# x 30' = 30,000 ft. lbs. The legs not being of equal stiffness, this O.T.M. is initially apportioned $\frac{3}{4}$ to left leg and $\frac{1}{4}$ to right leg; $\frac{1}{2}$ of the said portion to top and $\frac{1}{2}$ to bottom, $30{,}000 \times \frac{3}{4} \times \frac{1}{2} = 11{,}250$, $30{,}000 \times \frac{1}{4} \times \frac{1}{2} = 3750$. The relation of rigidity of beam to columns is as shown in the loading diagram below.

2- BJ₁ The joints are then balanced according to the rigidity of members. At B joint: $\frac{11250}{2} = 5625$.

3- C.O. Carry over one half, changing signs.

4- BT₁ -5625 +937

+2813 / 8438   -3691   +3691   +1230   -1230   -468 / 1405

8438
1405
$9843 \times \frac{3}{4} \times \frac{1}{2} = 3691$
$9843 \times \frac{1}{4} \times \frac{1}{2} = 1230$

The above is the correction for side-sway to eliminate O.T.M. and arrive at true rotation of joint. Correct every cycle by turning back as indicated by dotted lines.

5- BJ₂   +1406   +3691   +1142   -1142   +2285

These values are counter to each other 3691 − 1406 = 2285 Net. The joint is then balanced according to rigidity. At left joint: $2285/_2 = 1142$. This completes one cycle. Continue cycles until values are negligible.

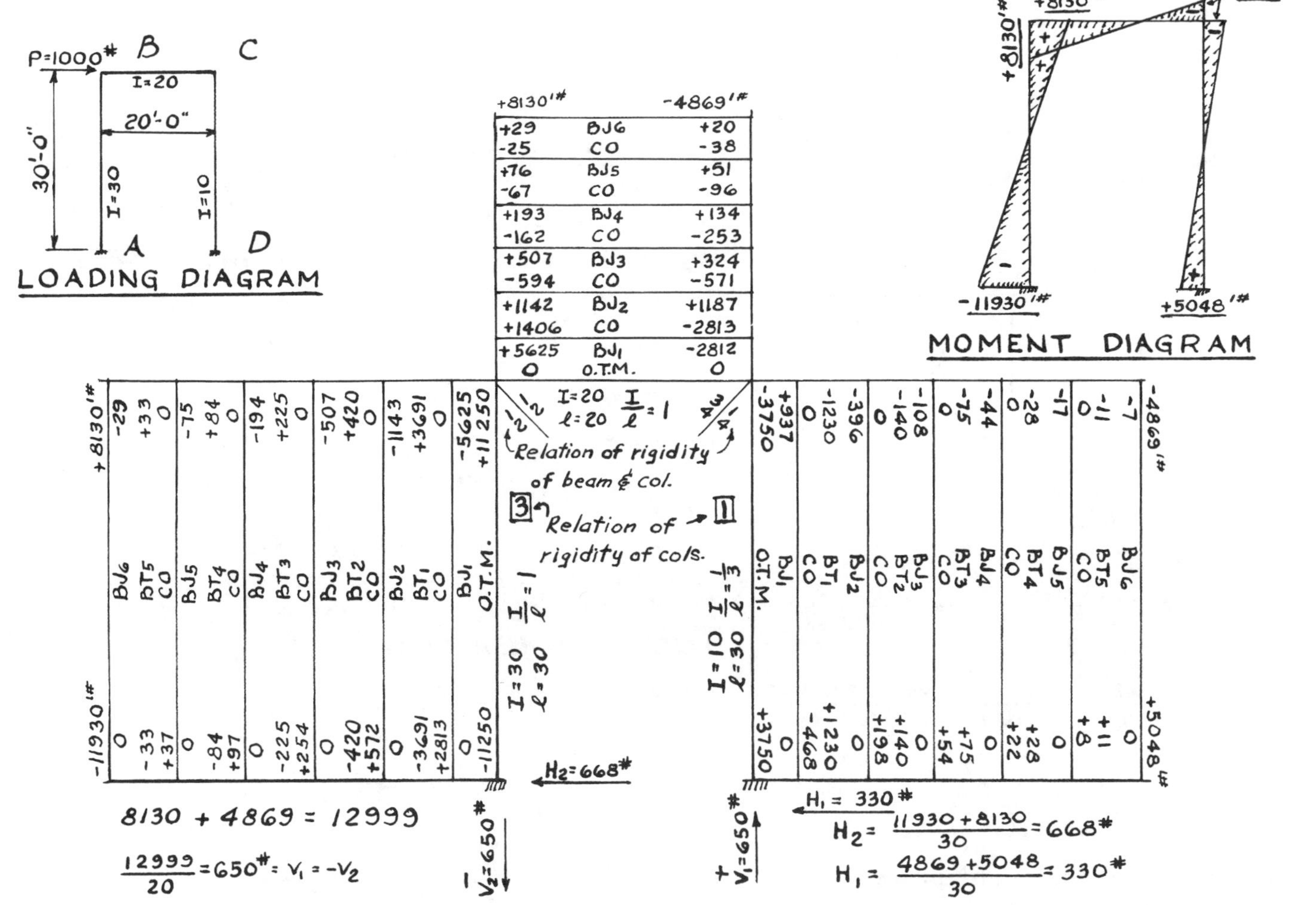

P=1000#
B
C
I=20
20'-0"
30'-0"
I=30
I=10
A
D
LOADING DIAGRAM
+8130'#
-4869'#
+29 BJ6 +20
-25 CO -38
+76 BJ5 +51
-67 CO -96
+193 BJ4 +134
-162 CO -253
+507 BJ3 +324
-594 CO -571
+1142 BJ2 +1187
+1406 CO -2813
+5625 BJ1 -2812
0 O.T.M. 0
I=20 ℓ=20 I/ℓ=1
Relation of rigidity of beam & col.
3 Relation of rigidity of cols. 1
I=30 ℓ=30 I/ℓ=1
I=10 ℓ=30 I/ℓ=1/3
+8130'# -29 +33 0 -75 +84 0 -194 +225 0 -507 +420 0 -1143 +3691 0 -5625 +11250
BJ6 BT5 CO BJ5 BT4 CO BJ4 BT3 CO BJ3 BT2 CO BJ2 BT1 CO BJ1 O.T.M.
-11930'# 0 -33 +37 0 -84 +97 0 -225 +254 0 -420 +572 0 -3691 +2813 0 -11250
+937 -3750 -1230 0 -396 0 -108 -140 0 -75 0 -44 0 -28 0 -17 -11 0 -7 -4869'#
BJ1 O.T.M. BJ1 CO BT1 BJ2 CO BJ3 BT2 CO BT3 BJ4 CO BT4 BJ5 CO BT5 BJ6
+3750 0 +1230 -468 0 +140 +198 0 +140 +54 0 +75 0 +28 +22 0 +11 +8 0 +5048'#
H2=668#
H1=330#
8130 + 4869 = 12999
12999/20 = 650# = V1 = -V2
V2=650#
V1=650#
H2 = (11930+8130)/30 = 668#
H1 = (4869+5048)/30 = 330#
+8130'#
-4869'#
-11930'#
+5048'#
MOMENT DIAGRAM

STRUCTURAL–RIGID FRAMES–FORMULAS–I

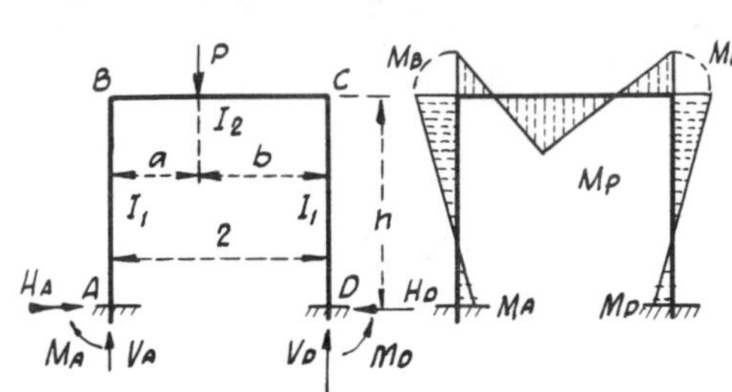
B
P
C
M_B
M_C
I_2
a
b
M_P
I_1
I_1
h
ℓ
H_A
A
D
H_D
M_A
M_D
M_A
V_A
V_D
M_D

$$V_A=\frac{Pb}{\ell}\cdot\frac{1+s-2s^2+6k}{1+6k}\qquad k=\frac{I_2}{I_1}\cdot\frac{h}{\ell}$$

$$V_D=\frac{Pa}{\ell}\cdot\frac{3s-2s^2+6k}{1+6k}\qquad s=\frac{a}{\ell}$$

$$H_A=H_D=\frac{3\,Pab}{2h\ell(2+k)}\qquad M_A=\frac{Pab}{2\ell}\cdot\frac{5k-1+2s(2+k)}{(2+k)(1+6k)}$$

$$M_D=\frac{Pab}{2\ell}\cdot\frac{3+7k-2s(2+k)}{(2+k)(1+6k)}\qquad M_B=M_A+H_Ah$$

$a=b$

$$V_A=V_D=\frac{P}{2}\qquad k=\frac{I_2}{I_1}\cdot\frac{h}{\ell}$$

$$H_A=H_D=\frac{3P\ell}{8h(2+k)}$$

$$M_A=M_D=\frac{P\ell}{8(2+k)}=H_A\cdot\frac{h}{3}$$

$$M_B=M_C=-\frac{P\ell}{4(2+k)}=H_A\cdot\frac{2h}{3}$$

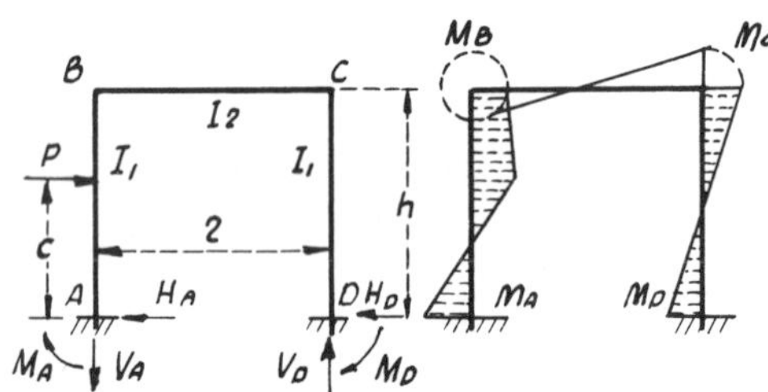
B
C
M_B
M_C
I_2
P
I_1
I_1
h
ℓ
c
A
H_A
D
H_D
M_A
M_D
M_A
V_A
V_D
M_D

$$-V_A=V_D=\frac{3Pks}{2(1+6k)}\qquad k=\frac{I_2}{I_1}\cdot\frac{h}{\ell}\qquad s=\frac{c}{h}$$

$$H_A=P-H_D\qquad H_D=\frac{Ps^2}{2(2+k)}\left[3(1+k)-s(1+2k)\right]$$

$$M_A=\frac{-Pcs}{2}\left[\frac{2}{s}-\frac{3+2k-s(1+k)}{2+k}-\frac{3k}{1+6k}\right]$$

$$M_A=\frac{Pcs}{2}\left[\frac{2}{s}-\frac{3+2k-s(1+k)}{2+k}-\frac{3k}{1+6k}\right]$$

$$M_B=-M_A-H_Dh+Pc$$

$$M_C=M_D-H_Dh\qquad -M_P=M_A+H_Ac$$

$c=h$

$$-V_A=V_D=\frac{3Phk}{\ell(1+6k)}\qquad k=\frac{I_2}{I_1}\cdot\frac{h}{\ell}$$

$$H_A=H_D=\frac{P}{2}$$

$$M_A=\frac{-Ph}{2}\cdot\frac{1+3k}{1+6k}\qquad M_D=-\frac{Ph}{2}\cdot\frac{1+3k}{1+6k}$$

$$M_B=\frac{Ph}{2}\cdot\frac{3k}{1+6k}\qquad M_C=\frac{-Ph}{2}\cdot\frac{3k}{1+6k}$$

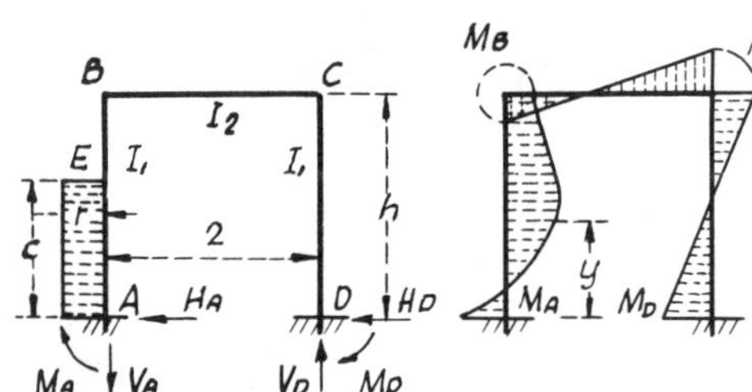
B
C
M_B
M_C
I_2
E
I_1
I_1
h
r
c
ℓ
A
H_A
D
H_D
y
M_A
M_D
M_A
V_A
V_D
M_D

$$-V_A=V_D=\frac{rc^2ks}{2(1+6k)}\qquad k=\frac{I_2}{I_1}\cdot\frac{h}{\ell}\qquad s=\frac{c}{h}$$

$$H_D=\frac{-rcs^2}{8}\cdot\frac{4(1+k)-s(1+2k)}{2+k}\qquad H_A=rc-H_D$$

$$M_A=\frac{-rc^2s}{24}\left[\frac{12}{s}-\frac{4(3+2k)-3s(1+k)}{2+k}-\frac{12k}{1+6k}\right]$$

$$M_D=\frac{-rc^2s}{24}\left[\frac{4(3+2k)-3s(1+k)}{2+k}-\frac{12k}{1+6k}\right]$$

$$-M_y=M_A+H_Ay-\frac{py^2}{6c}(3c-y)$$

$$-M_B=M_A+H_Dh-\frac{rc^2}{2}\qquad M_C=-M_D-H_Dh$$

$c=h$

$$-V_A=V_D=\frac{rh^2k}{2(1+6k)}\qquad k=\frac{I_2}{I}\cdot\frac{h}{\ell}$$

$$H_D=\frac{rh}{8}\cdot\frac{3+2k}{2+k}\qquad H_A=rh-H_D$$

$$M_A=\frac{-rh^2}{24}\left(12-\frac{9+5k}{2+k}-\frac{12k}{1+6k}\right)$$

$$M_D=\frac{-rh^2}{24}\left(\frac{9+5k}{2+k}-\frac{12k}{1+6k}\right)$$

$$-M_y=M_A+H_Ay-\frac{ry^2}{2}$$

$$-M_B=M_A+H_Ah-\frac{rh^2}{2}\qquad M_C=M_D-H_Dh$$

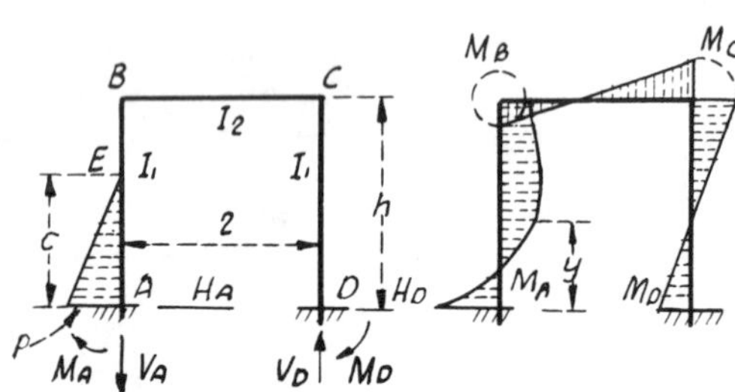
B
C
M_B
M_C
I_2
E
I_1
I_1
h
c
ℓ
A
H_A
D
H_D
y
P
M_A
M_D
M_A
V_A
V_D
M_D

$$-V_A=V_D=\frac{pkc^2s}{4\ell(1+6k)}\qquad k=\frac{I_2}{I_1}\cdot\frac{h}{\ell}\qquad s=\frac{c}{h}$$

$$H_D=\frac{pcs^2}{40}\cdot\frac{5(1+k)-s(1+2k)}{2+k}\qquad H_A=\frac{pc^2}{2}-H_D$$

$$M_A=\frac{-pc^2s}{120}\left[\frac{20}{s}-\frac{5(3+2k)-3s(1+k)}{2+k}-\frac{15k}{1+6k}\right]$$

$$M_D=\frac{-pc^2s}{120}\left[\frac{5(3+2k)-3s(1+k)}{2+k}-\frac{15k}{1+6k}\right]$$

$$-M_y=M_A+H_Ay-\frac{py^2}{6c}(3c-y)$$

$$-M_E=M_A+H_Ac-\frac{pc^2}{3}\qquad M_B=M_A-H_Dh+\frac{pc^2}{6}$$

$$M_C=-M_D-H_Dh$$

$c=h$

$$-V_A=V_D=\frac{pkh^2}{4\ell(1+6k)}\qquad k=\frac{I_2}{I_1}\cdot\frac{h}{\ell}$$

$$-H_D=\frac{ph}{40}\cdot\frac{4+3k}{2+k}\qquad H_A=\frac{ph}{2}-H_D$$

$$M_A=\frac{ph^2}{120}\left(20-\frac{12+7k}{2+k}-\frac{15k}{1+6k}\right)$$

$$M_D=\frac{ph^2}{120}\left(\frac{12+7k}{2+k}-\frac{15k}{1+6k}\right)$$

$$-M_y=M_A+H_Ay-\frac{py^2}{6h}(3h-y)$$

$$M=M_A-M_Dh+\frac{ph^2}{6}\qquad M_C=M_D-H_Dh$$

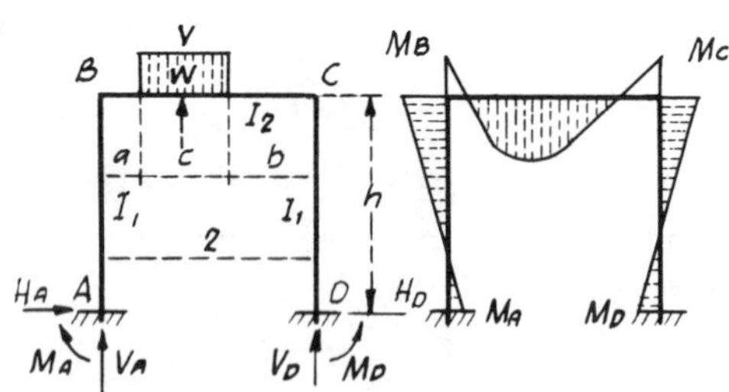
W
B
C
M_B
M_C
I_2
a
c
b
I_1
I_1
h
ℓ
H_A
A
D
H_D
M_A
M_D
M_A
V_A
V_D
M_D

$$V_A=\frac{wc}{\ell^3(1+6k)}\left[3abc+b^2(3\ell-2b)+c^2\left(\ell+b-\frac{c}{2}\right)+3\ell^2(2b+c)k\right]$$

$$V_D=wc-V_A\qquad k=\frac{I_2}{I_1}\cdot\frac{h}{\ell}$$

$$H_A=H_D=\frac{wc}{4h\ell(2+k)}(6ab+3c\ell-2c^2)$$

$$M_A=\frac{wc}{4\ell^2(2+k)(1+6k)}\left[6abc\ell+4a^2c+8ab^2c+\frac{10}{3}\cdot c^3\ell-c^2\ell^2-2c^4+(14abc\ell+2a^2c-4ab^2c-2c^3\ell+5c^2\ell^2-c^4)k\right]$$

$$M_D=\frac{w}{4\ell^2(2+k)(1+6k)}\left[6abc\ell+4a^2c^2+8ab^2c+\frac{10}{3}\cdot c^3\ell-c^2\ell^2-2c^4+(14abc\ell+2a^2c-4ab^2c-2c^3\ell+5c^2\ell^2-c^4)k\right]$$

$$M_B=M_A-H_Ah\qquad M_C=-M_D-H_Dh$$

$a=0,\ c=b$

$$V_A=\frac{w\ell}{32}\cdot\frac{13+72k}{1+k}=\frac{w\ell}{2}-V_D$$

$$V_D=\frac{3w\ell}{32}\cdot\frac{1+8k}{1+6k}\qquad k=\frac{I_2}{I_1}\cdot\frac{h}{\ell}$$

$$H_A=H_D=\frac{w\ell^2}{8h(2+k)}$$

$$M_A=\frac{w\ell^2}{192}\cdot\frac{2+45k}{(2+k)(1+6k)}\qquad M_D=\frac{w\ell^2}{192}\cdot\frac{14+51k}{(2+k)(1+6k)}$$

$$M_B=M_A-H_Ah\qquad M_C=M_D-H_Dh$$

$$X_0=\frac{13+72k}{1+6k}\cdot\frac{\ell}{32}=\frac{V_A}{w}$$

Top = Full uniform load

$$V_A=V_D=\frac{w\ell}{2}\qquad k=\frac{I_2}{I_1}\cdot\frac{h}{\ell}$$

$$H_A=H_D=\frac{w\ell^2}{4h(2+k)}$$

$$M_A=M_D=\frac{+w\ell^2}{12(2+k)}\qquad M_B=M_C=\frac{-w\ell^2}{6(2+k)}$$

$$M_x=\frac{w\ell}{2}\cdot x-\frac{wx^2}{2}-\frac{w\ell^2}{6(2+k)}$$

$$M_{max}=\frac{w\ell^2}{8}-\frac{w\ell^2}{6(2+k)}=\frac{w\ell^2}{24}\cdot\frac{2+3k}{2+k}$$

* From Ketchum's Structural Engineer's Handbook, McGraw-Hill.

STRUCTURAL-RIGID FRAMES-FORMULAS-2

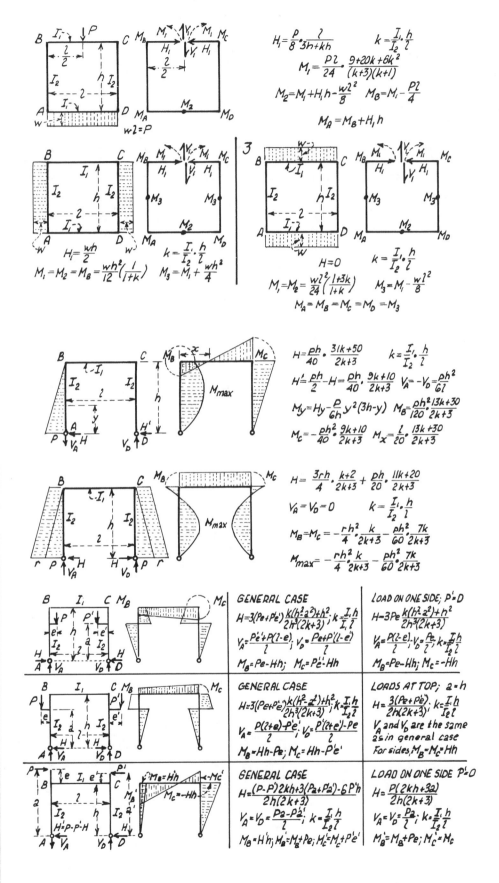

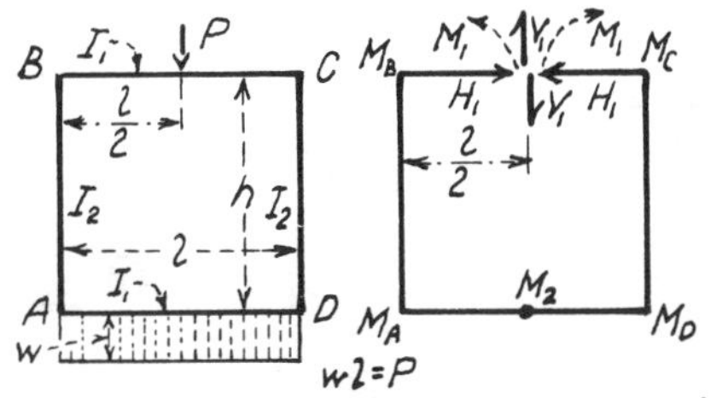

$$H_1=\frac{P}{8}\cdot\frac{l}{3h+kh}\qquad k=\frac{I_1}{I_2}\cdot\frac{h}{l}$$

$$M_1=\frac{Pl}{24}\cdot\frac{9+20k+6k^2}{(k+3)(k+1)}$$

$$M_2=M_1+H_1h-\frac{wl^2}{8}\qquad M_B=M_1-\frac{Pl}{4}$$

$$M_A=M_B+H_1h$$

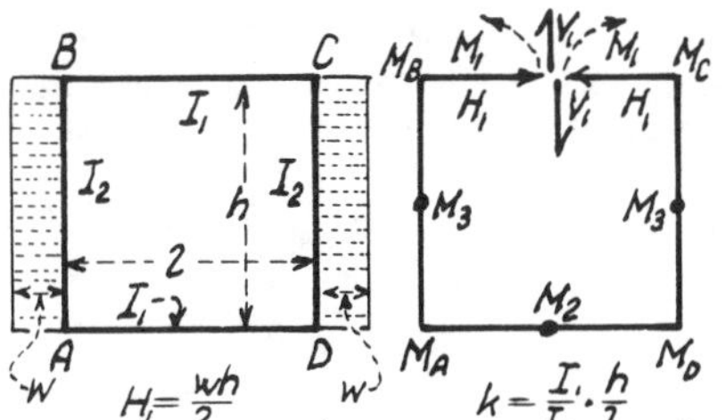

$wl=P$

$$H_1=\frac{wh}{2}\qquad k=\frac{I_1}{I_2}\cdot\frac{h}{l}$$

$$M_1=M_2=M_B=\frac{wh^2}{12}\left(\frac{1}{1+k}\right)\qquad M_3=M_1+\frac{wh^2}{4}$$

3

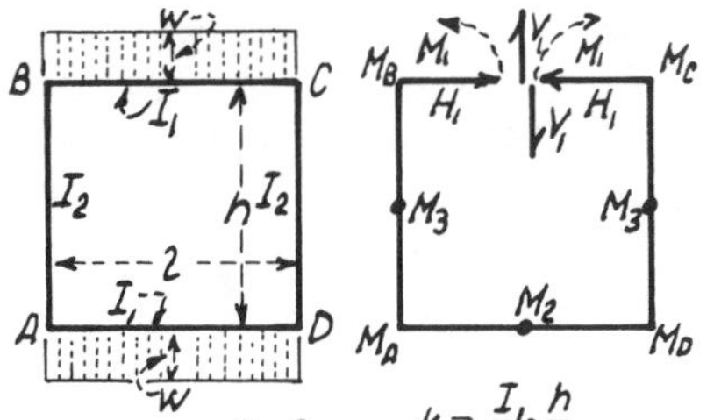

$$H=0\qquad k=\frac{I_1}{I_2}\cdot\frac{h}{l}$$

$$M_1=M_2=\frac{wl^2}{24}\left(\frac{1+3k}{1+k}\right)\qquad M_3=M_1-\frac{wl^2}{8}$$

$$M_A=M_B=M_C=M_D=M_3$$

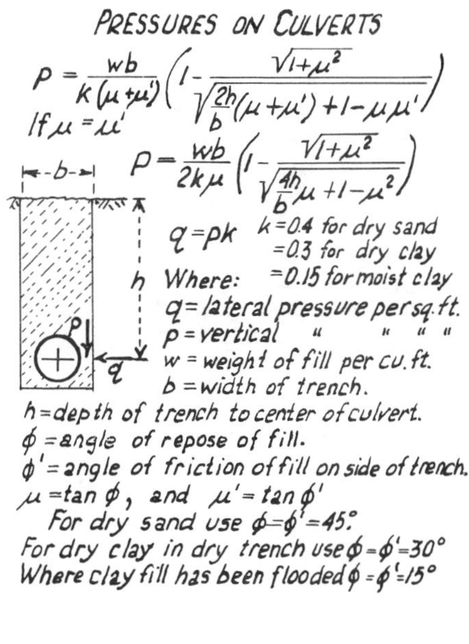

PRESSURES ON CULVERTS

$$p=\frac{wb}{k(\mu+\mu')}\left(1-\frac{\sqrt{1+\mu^2}}{\sqrt{\frac{2h}{b}(\mu+\mu')+1-\mu\mu'}}\right)$$

If $\mu=\mu'$

$$p=\frac{wb}{2k\mu}\left(1-\frac{\sqrt{1+\mu^2}}{\sqrt{\frac{4h}{b}\mu+1-\mu^2}}\right)$$

$q=pk$ — $k=0.4$ for dry sand; $=0.3$ for dry clay; $=0.15$ for moist clay

Where: q = lateral pressure per sq. ft.
p = vertical " " "
w = weight of fill per cu. ft.
b = width of trench.
h = depth of trench to center of culvert.
ϕ = angle of repose of fill.
ϕ' = angle of friction of fill on side of trench.
$\mu=\tan\phi$, and $\mu'=\tan\phi'$

For dry sand use $\phi=\phi'=45°$.
For dry clay in dry trench use $\phi=\phi'=30°$
Where clay fill has been flooded $\phi=\phi'=15°$

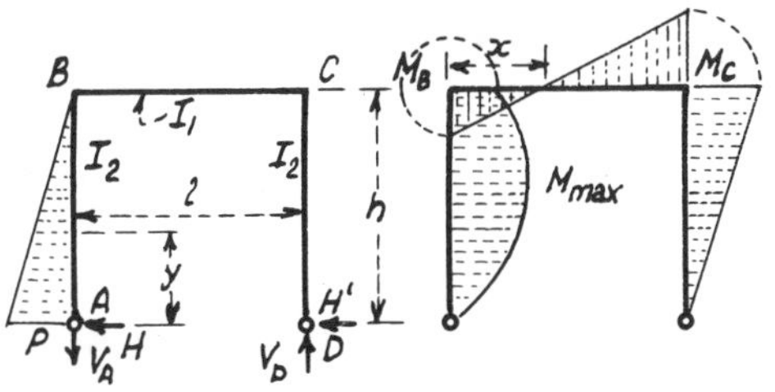

$$H=\frac{ph}{40}\cdot\frac{31k+50}{2k+3}\qquad k=\frac{I_1}{I_2}\cdot\frac{h}{l}$$

$$H'=\frac{ph}{2}-H=\frac{ph}{40}\cdot\frac{9k+10}{2k+3}\qquad V_A=-V_D=\frac{ph^2}{6l}$$

$$M_y=Hy-\frac{p}{6h}y^2(3h-y)\qquad M_B=\frac{ph^2}{120}\cdot\frac{13k+30}{2k+3}$$

$$M_C=-\frac{ph^2}{40}\cdot\frac{9k+10}{2k+3}\qquad M_x=\frac{l}{20}\cdot\frac{13k+30}{2k+3}$$

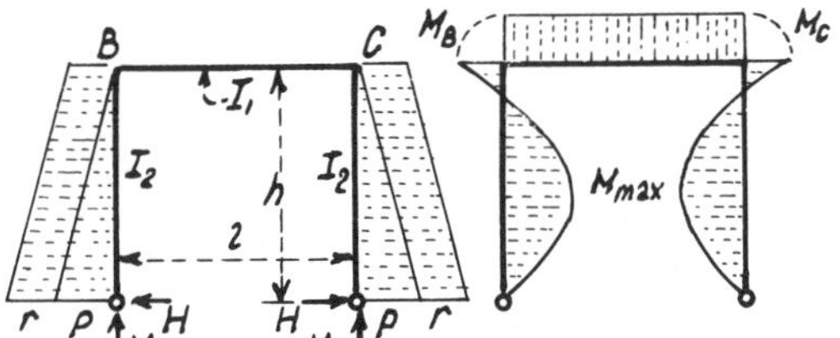

$$H=\frac{3rh}{4}\cdot\frac{k+2}{2k+3}+\frac{ph}{20}\cdot\frac{11k+20}{2k+3}$$

$$V_A=V_D=0\qquad k=\frac{I_1}{I_2}\cdot\frac{h}{l}$$

$$M_B=M_C=-\frac{rh^2}{4}\cdot\frac{k}{2k+3}-\frac{ph^2}{60}\cdot\frac{7k}{2k+3}$$

$$M_{max}=-\frac{rh^2}{4}\cdot\frac{k}{2k+3}-\frac{ph^2}{60}\cdot\frac{7k}{2k+3}$$

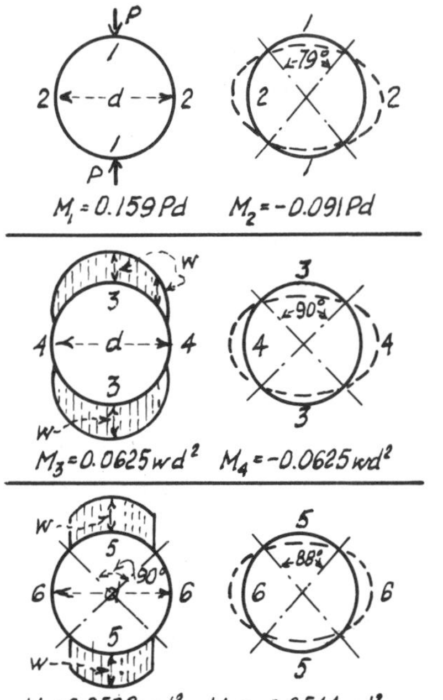

$M_1=0.159Pd$
$M_2=-0.091Pd$
$M_3=0.0625wd^2$
$M_4=-0.0625wd^2$
$M_5=0.0598wd^2$
$M_6=-0.0544wd^2$

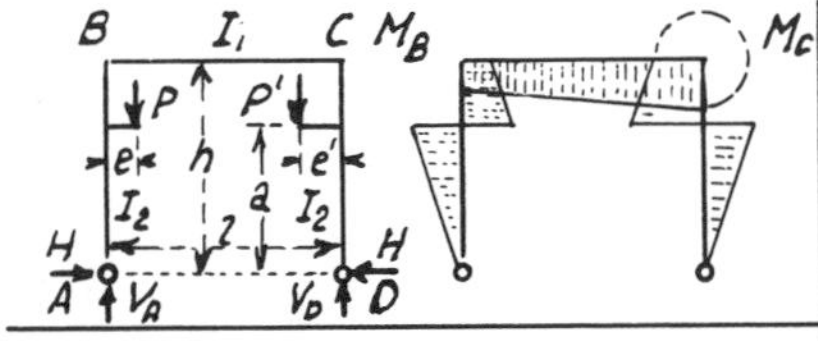

	GENERAL CASE	
	$H=3(Pe+P'e')\frac{k(h^2-a^2)+h^2}{2h^3(2k+3)}$; $k=\frac{I_1h}{I_2l}$ $V_A=\frac{P'e'+P(l-e)}{l}$; $V_D=\frac{Pe+P'(l-e')}{l}$ $M_B=Pe-Hh$; $M_C=P'e'-Hh$	LOAD ON ONE SIDE; $P'=0$ $H=3Pe\frac{k(h^2-a^2)+h^2}{2h^3(2k+3)}$ $V_A=\frac{P(l-e)}{l}$; $V_D=\frac{Pe}{l}$; $k=\frac{I_1h}{I_2l}$ $M_B=Pe-Hh$; $M_C=-Hh$
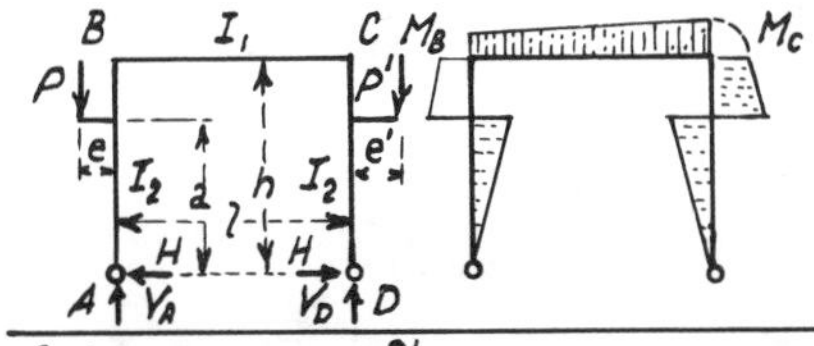	GENERAL CASE $H=3(Pe+P'e')\frac{k(h^2-a^2)+h^2}{2h^3(2k+3)}$; $k=\frac{I_1h}{I_2l}$ $V_A=\frac{P(l+e)-P'e'}{l}$; $V_D=\frac{P'(l+e')-Pe}{l}$ $M_B=Hh-Pe$; $M_C=Hh-P'e'$	LOADS AT TOP; $a=h$ $H=\frac{3(Pe+P'e')}{2h(2k+3)}$; $k=\frac{I_1h}{I_2l}$ V_A and V_D are the same as in general case For sides, $M_B=M_C=Hh$
$M_B=H'h$; $M_C=-Hh$; $H'=P-P'-H$	GENERAL CASE $H=\frac{(P-P')2kh+3(Pa+P'a')-6P'h}{2h(2k+3)}$ $V_A=V_D=\frac{Pa-P'a'}{l}$; $k=\frac{I_1h}{I_2l}$ $M_B=H'h$; $M_B'=M_B+Pe$; $M_C'=M_C+P'e'$	LOAD ON ONE SIDE $P'=0$ $H=\frac{P(2kh+3a)}{2h(2k+3)}$ $V_A=V_D=\frac{Pa}{l}$; $k=\frac{I_1h}{I_2l}$ $M_B'=M_B+Pe$; $M_C'=M_C$

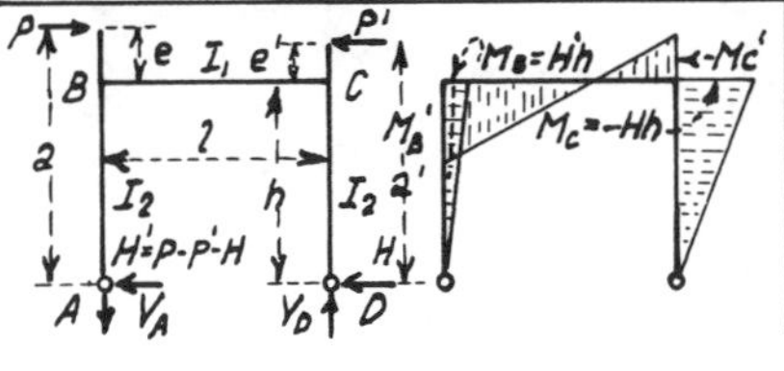

From Ketchum's Structural Engineer's Handbook, McGraw-Hill

STRUCTURAL–RIGID FRAMES–FORMULAS–3

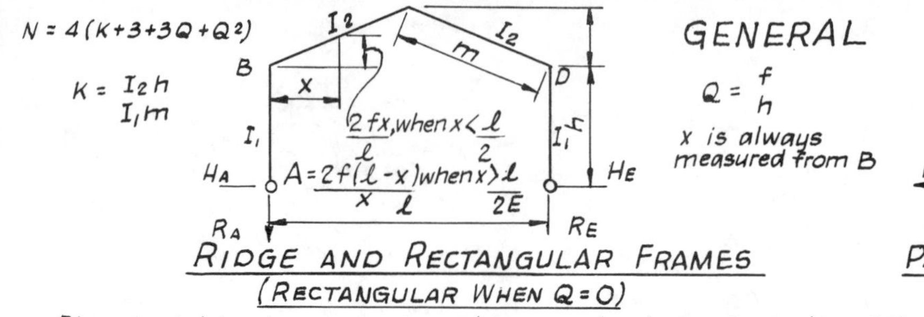
N = 4(K+3+3Q+Q²)
K = I₂h / I₁m
I₂
m
x
2fx/l, when x < l/2
A = 2f(l−x)/l when x > l/2
I₁
h
D
f
HA
HE
RA
RE
l
GENERAL
Q = f/h
x is always measured from B

RIDGE AND RECTANGULAR FRAMES
(RECTANGULAR WHEN Q = 0)

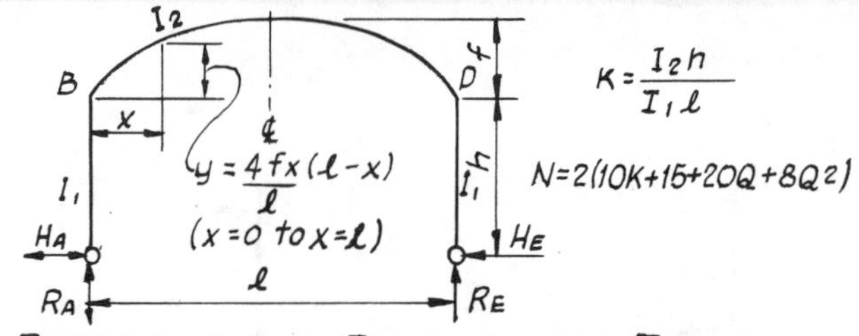
I₂
B
D
f
x
℄
y = 4fx(l−x)/l
(x = 0 to x = l)
I₁
h
HA
HE
RA
RE
l
K = I₂h / I₁l
N = 2(10K+15+20Q+8Q²)

PARABOLIC AND RECTANGULAR FRAMES
(RECTANGULAR WHEN Q = 0)

Plus sign (+) denotes moments which cause tension on the inside of the frame when the vertical load acts downward and the horizontal loads, applied to the left side of the frame, act towards the right. The direction of the reaction, and the signs of all terms in the moment formulas are shown correctly for this condition. When the direction of the loads (but not their position) is reversed the direction of the reactions may, and the signs for all moments will be reversed.

UNIFORMLY DISTRIBUTED VERTICAL LOAD ENTIRE SPAN

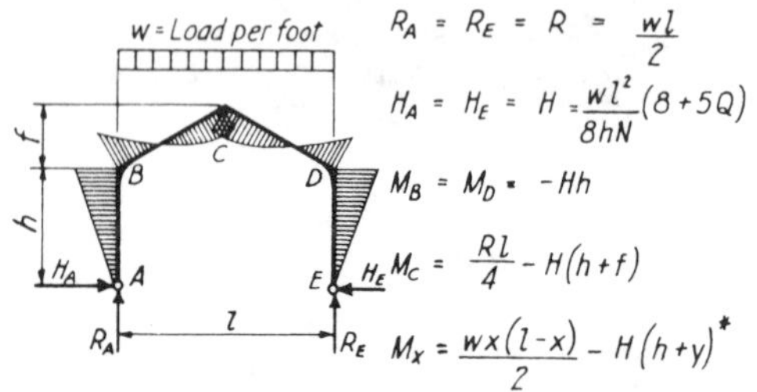
w = Load per foot

$$R_A = R_E = R = \frac{wl}{2}$$
$$H_A = H_E = H = \frac{wl^2}{8hN}(8+5Q)$$
$$M_B = M_D = -Hh$$
$$M_C = \frac{Rl}{4} - H(h+f)$$
$$M_x = \frac{wx(l-x)}{2} - H(h+y)\,*$$

UNIFORMLY DISTRIBUTED VERTICAL LOAD HALF SPAN (LEFT)

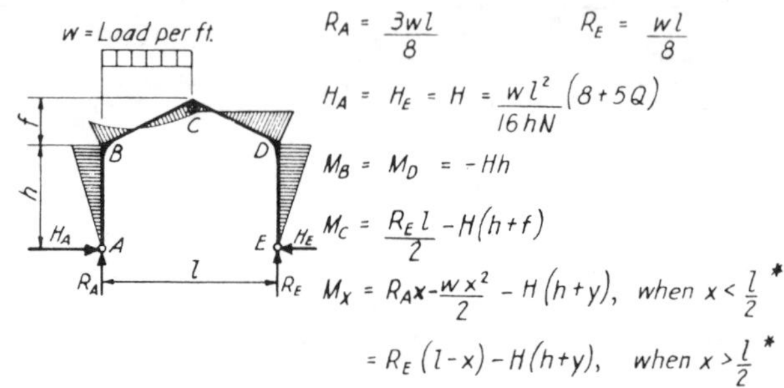
w = Load per ft.

$$R_A = \frac{3wl}{8} \qquad R_E = \frac{wl}{8}$$
$$H_A = H_E = H = \frac{wl^2}{16hN}(8+5Q)$$
$$M_B = M_D = -Hh$$
$$M_C = \frac{R_E l}{2} - H(h+f)$$
$$M_x = R_A x - \frac{wx^2}{2} - H(h+y), \text{ when } x < \frac{l}{2}\,*$$
$$= R_E(l-x) - H(h+y), \text{ when } x > \frac{l}{2}\,*$$

ONE CONCENTRATED ROOF LOAD AT ANY POSITION

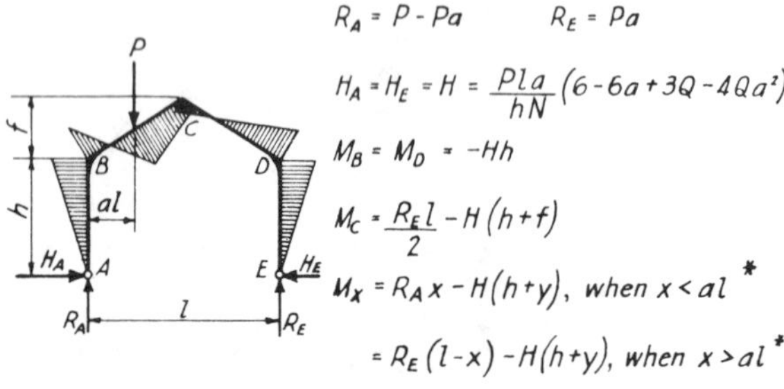
P
al

$$R_A = P - Pa \qquad R_E = Pa$$
$$H_A = H_E = H = \frac{Pla}{hN}(6 - 6a + 3Q - 4Qa^2)$$
$$M_B = M_D = -Hh$$
$$M_C = \frac{R_E l}{2} - H(h+f)$$
$$M_x = R_A x - H(h+y), \text{ when } x < al\,*$$
$$= R_E(l-x) - H(h+y), \text{ when } x > al\,*$$

BRACKET LOAD ON ONE COLUMN (LEFT COLUMN)

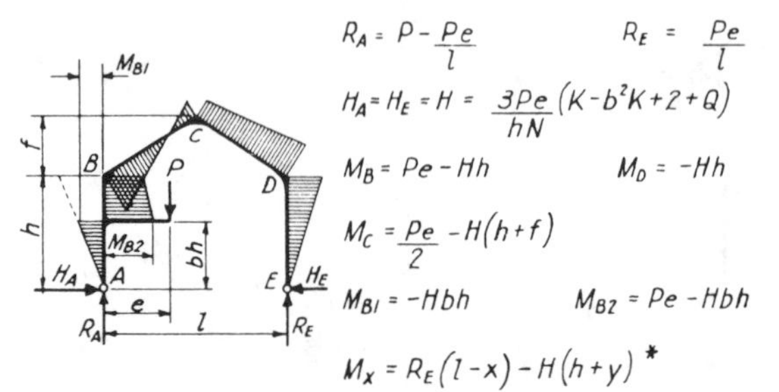
MB1
MB2
P
bh
e

$$R_A = P - \frac{Pe}{l} \qquad R_E = \frac{Pe}{l}$$
$$H_A = H_E = H = \frac{3Pe}{hN}(K - b^2K + 2 + Q)$$
$$M_B = Pe - Hh \qquad M_D = -Hh$$
$$M_C = \frac{Pe}{2} - H(h+f)$$
$$M_{B1} = -Hbh \qquad M_{B2} = Pe - Hbh$$
$$M_x = R_E(l-x) - H(h+y)\,*$$

UNIFORM HORIZONTAL LOAD ON BOTH INCLINED AND VERTICAL SURFACES

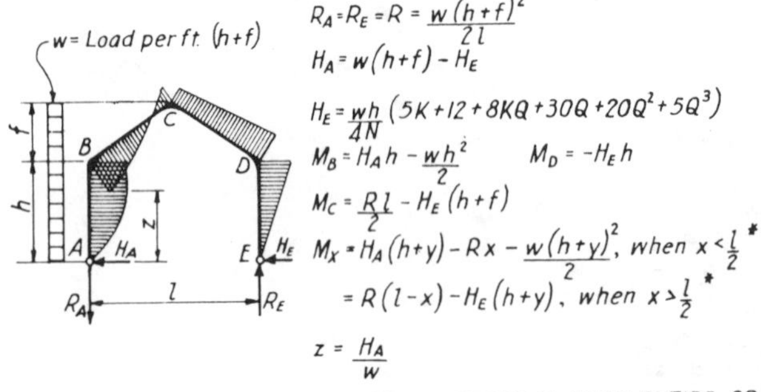
w = Load per ft. (h+f)
z

$$R_A = R_E = R = \frac{w(h+f)^2}{2l}$$
$$H_A = w(h+f) - H_E$$
$$H_E = \frac{wh}{4N}(5K + 12 + 8KQ + 30Q + 20Q^2 + 5Q^3)$$
$$M_B = H_A h - \frac{wh^2}{2} \qquad M_D = -H_E h$$
$$M_C = \frac{Rl}{2} - H_E(h+f)$$
$$M_x = H_A(h+y) - Rx - \frac{w(h+y)^2}{2}, \text{ when } x < \frac{l}{2}\,*$$
$$= R(l-x) - H_E(h+y), \text{ when } x > \frac{l}{2}\,*$$
$$z = \frac{H_A}{w}$$

ONE HORIZONTAL CONCENTRATED LOAD AT ANY POSITION ON COLUMN ($b \leq 1.0$)

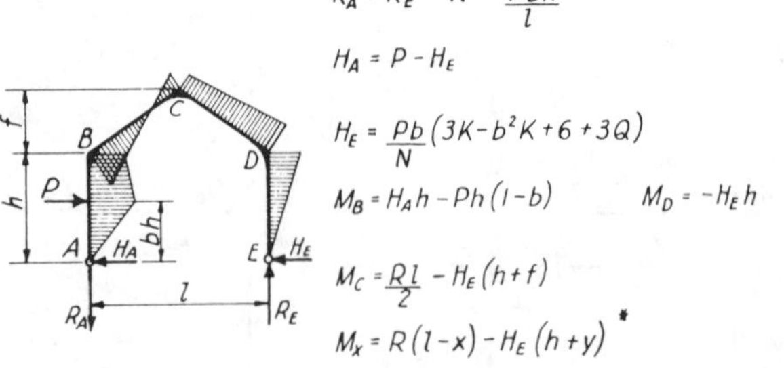
P
bh

$$R_A = R_E = R = \frac{Pbh}{l}$$
$$H_A = P - H_E$$
$$H_E = \frac{Pb}{N}(3K - b^2K + 6 + 3Q)$$
$$M_B = H_A h - Ph(1-b) \qquad M_D = -H_E h$$
$$M_C = \frac{Rl}{2} - H_E(h+f)$$
$$M_x = R(l-x) - H_E(h+y)\,*$$

UNIFORMLY DISTRIBUTED VERTICAL LOAD ENTIRE SPAN

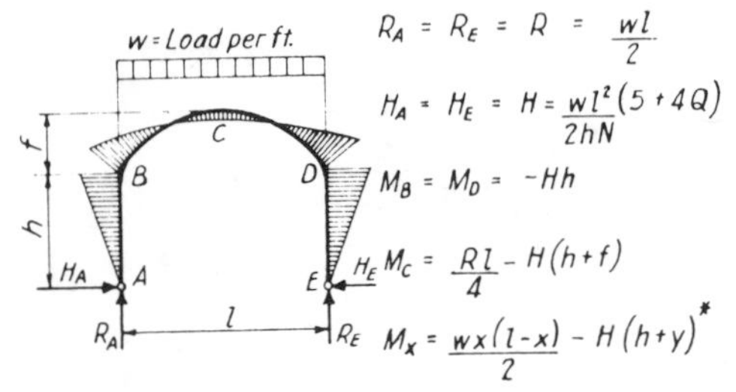
w = Load per ft.

$$R_A = R_E = R = \frac{wl}{2}$$
$$H_A = H_E = H = \frac{wl^2}{2hN}(5+4Q)$$
$$M_B = M_D = -Hh$$
$$M_C = \frac{Rl}{4} - H(h+f)$$
$$M_x = \frac{wx(l-x)}{2} - H(h+y)\,*$$

UNIFORM HORIZONTAL LOAD ON BOTH INCLINED AND VERTICAL SURFACES

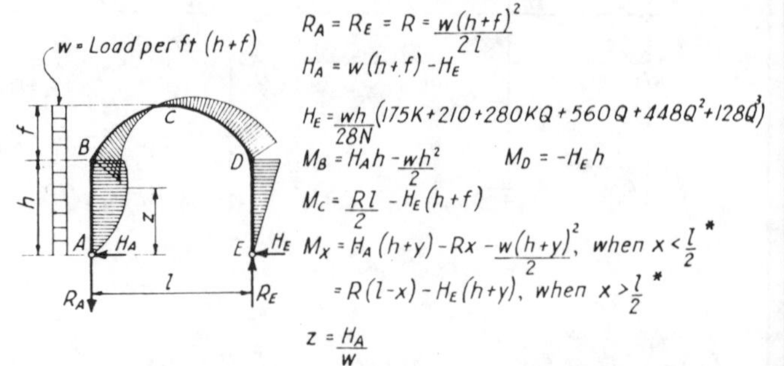
w = Load per ft (h+f)
z

$$R_A = R_E = R = \frac{w(h+f)^2}{2l}$$
$$H_A = w(h+f) - H_E$$
$$H_E = \frac{wh}{28N}(175K + 210 + 280KQ + 560Q + 448Q^2 + 128Q^3)$$
$$M_B = H_A h - \frac{wh^2}{2} \qquad M_D = -H_E h$$
$$M_C = \frac{Rl}{2} - H_E(h+f)$$
$$M_x = H_A(h+y) - Rx - \frac{w(h+y)^2}{2}, \text{ when } x < \frac{l}{2}\,*$$
$$= R(l-x) - H_E(h+y), \text{ when } x > \frac{l}{2}\,*$$
$$z = \frac{H_A}{w}$$

* From American Institute of Steel Construction.

STRUCTURAL STEEL

RED LIGHTS IN STRUCTURAL STEEL DESIGN

BEAMS

Compute section modulus, and select size. See pp. 4-02 and 4-03. Reaction must be less than allowable web shear V. See pp. 4-02 and 4-03. Brace top flange laterally: $\frac{bd}{bt} = 600$ maximum. For design, see p. 4-32. Strength of brace is to be not less than 2% of flange stress. Consider relative deflection also. Limit of $\frac{1}{360}$ of span for live load is commonly used to prevent plaster cracks. Horizontal shear is limited by $S = \frac{QV^*}{It}$. S = maximum at neutral axis. Check web crippling at bearing concentrations. Use R and G, p. 4-02, or maximum end reaction = 24,000t (atk); maximum interior load = 24,000t (a_1 + 2k). Resolve loads on beams not set vertically in components parallel and perpendicular to web. Check extreme fiber stress.** Eliminate beams subject to torsion wherever possible.

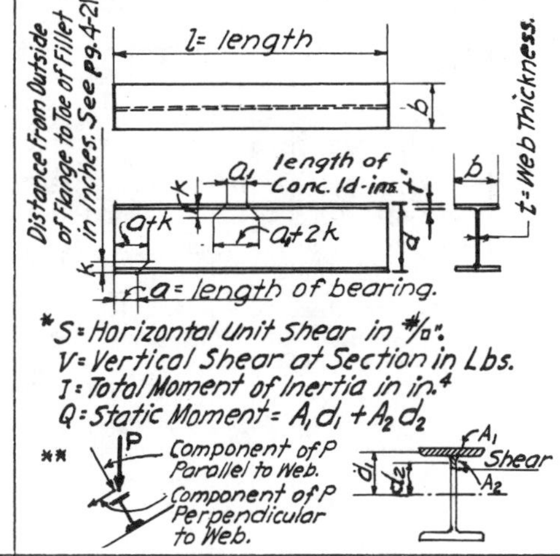
Distance from Outside of Flange to Toe of Fillet in Inches. See p. 4-21
l = length
t = Web Thickness.
length of Conc. ld-ins.
a+k
a₁+2k
a = length of bearing.
*S = Horizontal Unit Shear in #/□".
V = Vertical Shear at Section in Lbs.
I = Total Moment of Inertia in in.⁴
Q = Static Moment = A₁d₁ + A₂d₂
** Component of P Parallel to Web.
Component of P Perpendicular to Web.
Shear

PLATE GIRDERS

Design girder by moment of inertia of gross section using tables of *I*. Correct for rivet hole deductions of more than 15%. Check web shear. Allowable stress = 13,000 p.s.i. for $\frac{h}{t} < 70$. Maximum $\frac{h}{t} = 170$. Reduce web stress between $\frac{h}{t} = 70$ and 170 to $v = (\frac{8000}{h/t})^2$, or add stiffeners. Stiffeners required at points in girder where $\frac{h}{t} \geq \frac{8000}{\sqrt{v}}$ spaced. $\frac{11,000t}{\sqrt{v}}$ or 84" maximum. $P' = 8''$ diameter maximum. Provide bearing stiffeners for all bearing concentrations. Determine length of cover plates from bending-moment diagram. Determine rivet pitch, $p = \frac{RH}{V}$ (approx.) or $\frac{RI}{VQ}$ (exact), 6 in. maximum. For lateral bracing and deflection, see *BEAMS*.

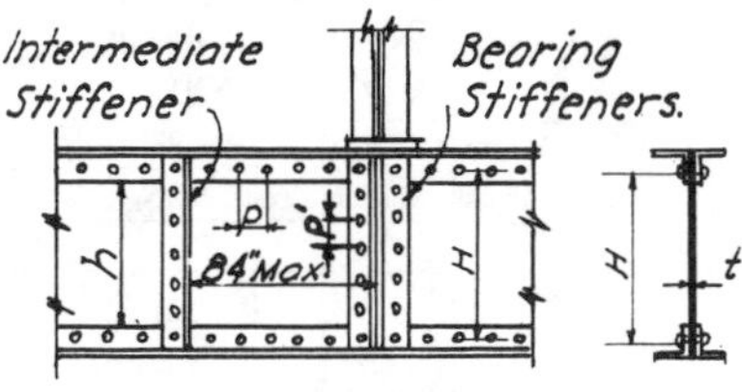
Intermediate Stiffener
Bearing Stiffeners.
84" Max.

R = Resistance of one rivet.
h = Unsupported Height of Web in Ins.
t = Web Thickness in Ins.
p = Rivet Pitch in Ins.
v = Unit Shear #/□"
H = Distance Between Rivet Lines in Ins.
For V, Q & I see Beams.

COLUMNS

With direct load only, select column from tables; see pp. 4-04 and 4-05. With direct load and bending, combined stress should satisfy criterion $\frac{f_a}{F_A} + \frac{f_b}{F_B} \leq$ unity, where f_a = direct stress, f_b = bending stress and F_A and F_B are respective allowable unit stresses. Brace all columns in two directions at splices. Minimum = 2% of load. Eliminate eccentricity by using stiff connections, or increase column section. Distribute column moment above and below floor according to column stiffness.

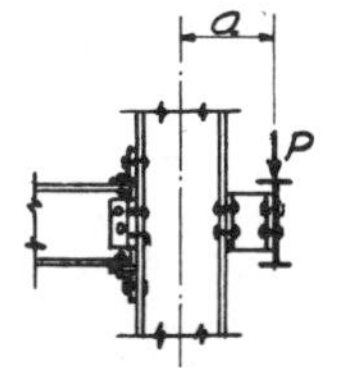
a
P

Moment in Column = Pa unless balanced by connection on opposite face.

Rivets for Shear.
Rivets for Shear and Moment.

Moment taken in beam to limit eccentricity in column.

TRUSSES

Determine stresses by graphic or analytical method.
For size of members, see angle strut tables, pp. 4-07 to 4-10 and angle area tables, p. 4-06.
Check size and thickness of important gussets for shear and bending.
Check top chord for lateral bracing.
Chord members with direct load and bending should be designed for combined stresses.
Design columns for moment due to eccentricity e.

Check Shear
Check Gusset for Shear & Moment.
e
3/8" Min. gusset
Design Col. for Eccentric Load.
END GUSSET.

BENT BEAMS

Cut bottom flange, notch web, bend. Butt-weld to develop full flange and web. Add stiffener plates to each side as shown.

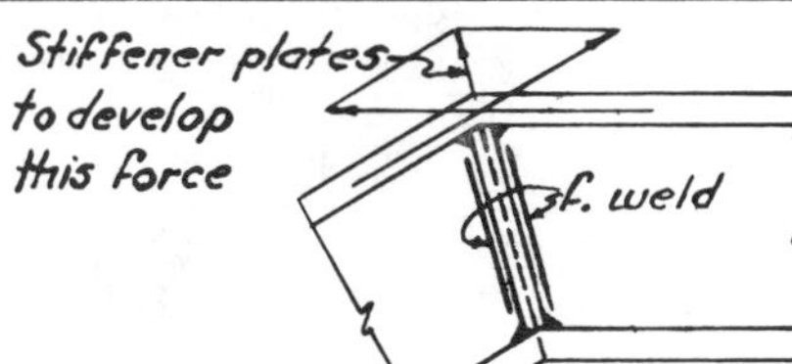
Stiffener plates to develop this force
F. weld
Butt weld

STRUCTURAL STEEL

BETHLEHEM AND CARNEGIE STRUCTURAL SEC

D	Wt	S	t	L_u	V	R	G	d	b	t'	r'	Wh
36 WF	300	1105.1	.94	38.0	451	143	22.7	36 3/4	16 5/8	1.68	3.73	
	280	1031.2	.88	35.5	420	132	21.2	36 1/2	16 5/8	1.57	3.70	790C
	260	951.1	.84	32.5	398	123	20.3	36 1/4	16 1/2	1.44	3.65	950S
	245	892.5	.80	30.5	376	115	19.2	36	16 1/2	1.35	3.62	
	230	835.5	.76	28.5	357	108	18.4	35 7/8	16 1/2	1.26	3.59	
	194	663.6	.77	20.5	365	104	18.5	36 1/2	12 1/8	1.26	2.49	
	182	621.2	.72	19.5	342	97	17.4	36 3/8	12 1/8	1.18	2.47	590C
	170	579.1	.68	18.0	320	89	16.3	36 1/8	12	1.10	2.45	710S
	160	541.0	.65	17.0	306	84	15.7	36	12	1.02	2.42	
	150	502.9	.62	15.5	291	80	15.0	35 7/8	12	.94	2.38	
33 WF	240	811.1	.83	33.0	362	118	19.9	33 1/2	15 7/8	1.40	3.52	700C
	220	740.6	.77	30.0	335	108	18.6	33 1/4	15 3/4	1.27	3.48	840S
	200	669.6	.71	27.0	307	98	17.2	33	15 3/4	1.15	3.43	
	152	486.4	.63	18.0	277	82	15.2	33 1/2	11 5/8	1.05	2.39	
	141	446.8	.60	16.5	262	76	14.5	33 1/4	11 1/2	.96	2.35	510C
	130	404.8	.58	15.0	250	72	13.9	33 1/8	11 1/2	.85	2.29	620S
30 WF	210	649.9	.77	32.5	306	108	18.6	30 3/8	15 1/8	1.31	3.38	
	190	586.1	.71	29.5	278	97	17.0	30 1/8	15	1.18	3.34	600C
	172	528.2	.65	26.5	254	87	15.7	29 7/8	15	1.06	3.30	730S
	132	379.7	.61	17.5	242	77	14.8	30 1/4	10 1/2	1.00	2.18	
	124	354.6	.58	16.0	229	72	14.0	30 1/8	10 1/2	.93	2.16	430C
	116	327.9	.56	15.0	220	69	13.5	30	10 1/2	.85	2.12	530S
	108	299.2	.54	13.0	212	66	13.2	29 7/8	10 1/2	.76	2.06	
27 WF	177	492.8	.72	30.5	257	98	17.4	27 1/4	14 1/8	1.19	3.16	510C
	160	444.5	.66	27.5	232	88	15.8	27 1/8	14	1.07	3.12	620S
	145	402.9	.60	25.0	210	78	14.4	26 7/8	14	.97	3.09	
	114	299.2	.57	17.0	202	70	13.7	27 1/4	10 1/8	.93	2.11	
	102	266.3	.52	15.0	182	63	12.4	27 1/8	10	.83	2.08	370C
	94	242.8	.49	13.5	171	59	11.8	26 7/8	10	.75	2.04	460S

D	Wt	S	t	L_u	V	R	G	d	b	t'	r'	Wh
24 WF	160	413.5	.66	32.0	211	87	15.7	24 3/4	14 1/8	1.13	3.23	470C
	145	372.5	.61	29.0	194	79	14.6	24 1/2	14	1.02	3.19	570S
	130	330.7	.56	26.0	178	71	13.6	24 1/4	14	.90	3.13	
	120	299.1	.56	23.0	176	69	13.3	24 1/4	12 1/8	.93	2.68	
	110	274.4	.51	21.0	160	63	12.2	24 1/8	12	.85	2.66	390C
	100	248.9	.47	19.0	146	57	11.2	24	12	.77	2.63	470S
	94	220.9	.52	16.0	163	61	12.4	24 1/4	9	.87	1.92	
	84	196.3	.47	14.5	147	55	11.3	24 1/8	9	.77	1.89	310C
	76	175.4	.44	12.5	137	50	10.6	23 7/8	9	.68	1.85	380S
21 WF	142	317.2	.66	33.5	184	85	15.8	21 1/2	13 1/8	1.09	3.04	390C
	127	284.1	.59	30.0	162	74	14.1	21 1/4	13	.98	3.01	470S
	112	249.6	.53	27.0	144	65	12.7	21	13	.86	2.96	
	96	197.6	.57	20.0	158	70	13.8	21 1/8	9	.93	1.97	290C
	82	168.0	.50	17.0	135	59	12.0	20 7/8	9	.79	1.93	350S
	73	150.7	.45	14.5	126	53	10.9	21 1/4	8 1/4	.74	1.76	
	68	139.9	.43	13.5	118	49	10.3	21 1/8	8 1/4	.68	1.74	250C
	62	126.4	.40	12.0	109	45	9.6	21	8 1/4	.61	1.71	310S
18 WF	114	220.1	.59	31.5	143	74	14.3	18 1/2	11 7/8	.99	2.76	330C
	105	202.2	.55	29.0	132	68	13.3	18 3/8	11 3/4	.91	2.73	390S
	96	184.4	.51	27.0	121	62	12.3	18 1/8	11 3/4	.83	2.71	
	85	156.1	.53	22.0	125	63	12.6	18 3/8	8 7/8	.91	2.00	
	77	141.7	.47	20.0	112	56	11.4	18 1/8	8 3/4	.83	1.98	240C
	70	128.2	.44	18.0	103	51	10.5	18	8 3/4	.75	1.95	290S
	64	117.0	.40	16.5	94	46	9.7	17 7/8	8 3/4	.69	1.93	
	60	107.8	.42	14.5	99	47	10.0	18 1/4	7 1/2	.69	1.63	200C
	55	98.2	.39	13.0	92	43	9.4	18 1/8	7 1/2	.63	1.61	240S
	50	89.0	.36	12.0	84	39	8.6	18	7 1/2	.57	1.59	
16 WF	96	166.1	.53	31.0	114	66	12.8	16 3/8	11 1/2	.87	2.71	300C
	88	151.3	.50	28.0	106	61	12.1	16 1/8	11 1/2	.79	2.67	360S
	78	127.8	.53	23.0	112	64	12.7	16 3/8	8 5/8	.87	1.95	
	71	115.9	.49	21.0	102	57	11.7	16 1/8	8 1/2	.79	1.93	210C
	64	104.2	.44	19.0	92	51	10.6	16	8 1/2	.71	1.91	260S
	58	94.1	.41	17.0	84	46	9.8	15 7/8	8 1/2	.64	1.88	
	50	80.7	.38	13.5	80	42	9.1	16 1/4	7 1/8	.63	1.54	
	45	72.4	.35	12.0	73	38	8.3	16 1/8	7	.56	1.52	170C
	40	64.4	.31	11.0	64	33	7.4	16	7	.50	1.50	210S
	36	56.3	.30	9.5	62	32	7.2	15 7/8	7	.43	1.45	

D	Wt	S	t	L_u	V	R	G	d	b	t'	r'	Wh
14 WF	426	707.4	1.87	134.0	455	321	45.0	18 3/4	16 3/4	3.03	4.34	670C
	398	656.9	1.77	129.0	422	295	42.5	18 1/4	16 5/8	2.84	4.31	760S
	370	608.1	1.65	122.0	386	268	39.7	18	16 1/2	2.66	4.27	
	342	559.4	1.54	115.0	353	244	37.1	17 1/2	16 3/8	2.47	4.24	590C
	314	511.9	1.41	108.0	316	217	34.0	17 1/4	16 1/4	2.28	4.20	670S
	287	465.5	1.31	99.7	286	194	31.4	16 3/4	16 1/8	2.09	4.17	570C
	264	427.4	1.20	94.1	258	175	28.9	16 1/2	16	1.94	4.14	590S
	320	492.8	1.89	104.0	413	281	45.4	16 3/4	16 3/4	2.09	4.17	570C 650S
14 WF	246	397.4	1.12	89.0	238	160	27.0	16 1/4	16	1.81	4.12	
	237	382.2	1.09	86.4	228	154	26.2	16 1/8	15 7/8	1.75	4.11	
	228	367.8	1.04	83.7	217	146	25.1	16	15 7/8	1.69	4.10	460C
	219	352.6	1.00	80.8	207	139	24.1	15 7/8	15 7/8	1.62	4.08	530S
	211	339.2	.98	78.3	201	134	23.5	15 3/4	15 3/4	1.56	4.07	
	202	324.9	.93	75.6	189	125	22.3	15 5/8	15 3/4	1.50	4.06	
	193	310.0	.89	73.1	179	119	21.4	15 1/2	15 3/4	1.44	4.05	
	184	295.8	.84	70.4	168	111	20.2	15 3/8	15 5/8	1.38	4.04	
	176	281.9	.82	67.4	163	107	19.7	15 1/4	15 5/8	1.31	4.02	400C
	167	267.3	.78	64.5	154	100	18.7	15 1/8	15 5/8	1.25	4.01	460S
	158	253.4	.73	61.7	142	93	17.5	15	15 1/2	1.19	4.00	
	150	240.2	.69	59.0	134	88	16.7	14 7/8	15 1/2	1.13	3.99	
	142	226.7	.68	55.6	130	85	16.3	14 3/4	15 1/2	1.06	3.97	
14 WF	136	216.0	.66	53.0	127	82	15.8	14 3/4	14 3/4	1.06	3.77	
	127	202.0	.61	50.2	116	75	14.6	14 5/8	14 3/4	1.00	3.76	
	119	189.4	.57	47.0	107	69	13.7	14 1/2	14 5/8	.94	3.75	310C
	111	176.3	.54	44.0	101	65	13.0	14 3/8	14 5/8	.87	3.73	370S
	103	163.6	.49	41.0	92	59	11.9	14 1/4	14 5/8	.81	3.72	
	95	150.6	.46	38.0	85	55	11.2	14 1/8	14 1/2	.75	3.71	
	87	138.1	.42	35.5	76	49	10.1	14	14 1/2	.69	3.70	
14 WF	84	130.9	.45	33.0	83	53	10.8	14 1/8	12	.78	3.02	
	78	121.1	.43	30.5	78	49	10.3	14	12	.72	3.00	
	74	112.3	.45	28.0	83	53	10.8	14 1/4	10 1/8	.78	2.48	220C
	68	103.0	.42	25.5	76	48	10.0	14	10	.72	2.46	260S
	61	92.2	.38	23.0	68	43	9.1	13 7/8	10	.64	2.45	
	53	77.8	.37	19.0	67	42	8.9	14	8	.66	1.92	
	48	70.2	.34	17.0	61	38	8.1	13 3/4	8	.59	1.91	160C
	43	62.7	.31	15.5	55	34	7.4	13 5/8	8	.53	1.89	200S
	38	54.6	.31	12.5	58	34	7.5	14 1/8	6 3/4	.51	1.49	
	34	48.5	.29	11.0	52	31	6.9	14	6 3/4	.45	1.46	
	30	41.8	.27	9.5	49	28	6.5	13 7/8	6 3/4	.38	1.41	

D	Wt	S	t
12 WF	190	263.2	1.0
	161	222.2	.9
	133	182.5	.7
	120	163.4	.7
	106	144.5	.6
	99	134.7	.5
	92	125.0	.5
	85	115.7	.4
	79	107.1	.4
	72	97.5	.4
	65	88.0	.3
12 WF	58	78.1	.3
	53	70.7	.3
	50	64.7	.3
	45	58.2	.3
	40	51.9	.2
	36	45.9	.3
	31	39.4	.2
	27	34.1	.2
	22	25.3	.2
	19	21.4	.2
	16.5	17.5	.2
	14	14.8	.2
10 WF	112	126.3	.7
	100	112.4	.6
	89	99.7	.6
	77	86.1	.5
	72	80.1	.5
	66	73.7	.4
	60	67.1	.4
	54	60.4	.3
	49	54.6	.3
10 WF	45	49.1	.3
	39	42.2	.3
	33	35.0	.2
	29	30.8	.2
	25	26.4	.2
	21	21.5	.2
	19	18.8	.2
	17	16.2	.2
	15	13.8	.2
	11.5	10.5	.1

COLUMN WORKING STRESSES - KIPS/SQ. IN.
A.I.S.C. MAIN MEMBERS

l/r	f	l/r	f	l/r	f	l/r	f	l/r	f	l/r	f	l/r	f	l/r	f
15	16.9	61	15.2	71	14.6	81	13.8	91	13.0	101	12.1	111	11.0	125	9.4
20	16.8	62	15.1	72	14.5	82	13.7	92	12.9	102	12.0	112	10.9	130	8.8
25	16.7	63	15.1	73	14.4	83	13.7	93	12.8	103	11.9	113	10.8	135	8.3
30	16.6	64	15.0	74	14.3	84	13.6	94	12.7	104	11.8	114	10.7	140	7.8
35	16.4	65	15.0	75	14.3	85	13.5	95	12.6	105	11.7	115	10.6	150	6.8
40	16.2	66	14.9	76	14.2	86	13.4	96	12.5	106	11.6	116	10.5	160	5.9
45	16.0	67	14.8	77	14.1	87	13.3	97	12.4	107	11.5	117	10.4	170	5.2
50	15.8	68	14.8	78	14.1	88	13.2	98	12.3	108	11.3	118	10.3	180	4.5
55	15.5	69	14.7	79	14.0	89	13.2	99	12.3	109	11.2	119	10.1	190	3.9
60	15.3	70	14.6	80	13.9	90	13.1	100	12.2	110	11.1	120	10.0	200	3.4

DEFLECTION COEFFICIENT - UNIFORM LOADS 20,000 #/□"

$$\text{DEFL. INCHES} = \frac{\text{DEFL. COEF.}}{d} = \frac{.02069 L^2}{d}$$

L	DEFL. COEF.	L	DEFL. COEF.	L	DEFL. COEF.	L	DEFL. COEF.	L	DEFL. COEF.
21	9.124	31	19.883	41	34.780	51	53.815	62	79.532
22	10.014	32	21.187	42	36.497	52	55.946	64	84.746
23	10.945	33	22.531	43	38.256	53	58.118	66	90.126
24	11.917	34	23.918	44	40.056	54	60.332	68	95.671
25	12.931	35	25.345	45	41.897	55	62.587	70	101.381
26	13.986	36	26.814	46	43.780	56	64.884	72	107.257
27	15.083	37	28.325	47	45.704	57	67.222	74	113.298
28	16.221	38	29.876	48	47.670	58	69.601	76	119.505
29	17.400	39	31.469	49	49.677	59	72.022	78	125.878
30	18.621	40	33.104	50	51.725	60	74.484	80	132.416

SHEAR @ 15,000 — VALUE FOR POWER DRIVEN RIVETS AND TURNED BOLTS IN REAMED HOLES — BEARING SINGLE = 32,000, DOUBLE = 40,000

RIVET DIAM.	5/8"		3/4"		7/8"		1"		1 1/8"	
SINGLE SHEAR	4600		6630		9020		11780		14910	
DOUBLE SHEAR	9200		13250		18040		23560		29820	
PLATE THICKNESS	BEARING (IN KIPS)		BEARING (IN KIPS)		BEARING (IN KIPS)		BEARING (IN KIPS)		BEARING (IN KIPS)	
	32.0	40.0	32.0	40.0	32.0	40.0	32.0	40.0	32.0	40.0
1/4"	—	6.25	6.00	7.50	7.00	8.75	8.00	10.0	—	—
5/16"	—	7.81	—	9.38	8.75	10.9	10.0	12.5	11.3	14.1
3/8"	—	9.38	—	11.3	—	13.1	12.0	15.0	13.5	16.9
7/16"	—	—	—	13.1	—	15.3	—	17.5	—	19.7
1/2"	—	—	—	—	—	17.5	—	20.0	—	22.5
9/16"	—	—	—	—	—	—	—	22.5	—	25.3
5/8"	—	—	—	—	—	—	—	—	—	28.1

TENSION IN RIVETS 20,000 #/□"

DIA.	TENSION IN KIPS
1/2"	3.9
5/8"	6.1
3/4"	8.8
7/8"	12.0
1"	15.7
1 1/8"	19.8
1 1/4"	24.4

SHEAR @ 10,000 — UN

BOLT DIAM.	5/8"	
SINGLE SHEAR	3070	
DOUBLE SHEAR	6140	
PLATE THICKNESS	BEARING (IN KIPS)	
	20.0	25.0
1/4"	3.13	3.91
5/16"	—	4.88
3/8"	—	5.86
7/16"	—	—
1/2"	—	—
9/16"	—	—
5/8"	—	—

Sections revised according to Bulletin R-216-46 U.S. Dept. of Commerce. Dated Feb. 15, 1946.

BEAM SECTIONS

d	b	t'	r'	Wh
3/8	12 5/8	1.74	3.25	
7/8	12 1/2	1.49	3.20	340C
3/8	12 3/8	1.24	3.16	390S
1/8	12 3/8	1.11	3.13	
7/8	12 1/4	.99	3.11	
3/4	12 1/4	.92	3.09	
5/8	12 1/8	.86	3.08	
1/2	12 1/8	.80	3.07	230C
3/8	12 1/8	.74	3.05	280S
1/4	12	.67	3.04	
1/8	12	.61	3.02	
1/4	10	.64	2.51	180C
	10	.58	2.48	220S
1/4	8 1/8	.64	1.96	
	8	.58	1.94	150C
	8	.52	1.94	180S
1/4	6 5/8	.54	1.50	
1/8	6 1/2	.46	147	120C
	6 1/2	.40	1.44	150S
1/4	4	.42	.84	
1/8	4	.35	.81	
	4	.27	.76	80C
7/8	4	.22	.74	100S
3/8	10 3/8	1.25	2.67	230C
1/8	10 3/8	1.12	2.65	260S
7/8	10 1/4	1.00	2.63	
5/8	10 1/4	.87	2.60	
1/2	10 1/8	.81	2.59	
3/8	10 1/8	.75	2.58	160C
1/4	10 1/8	.68	2.57	190S
1/8	10	.62	2.56	
	10	.56	2.54	
1/8	8	.62	2.00	120C
	8	.53	1.98	140S
9 3/4	8	.43	1.94	
1/4	5 3/4	.50	1.34	
1/8	5 3/4	.43	1.31	90C
9 7/8	5 3/4	.34	1.25	110S
1/4	4	.39	.86	
1/8	4	.33	.83	70C
	4	.27	.80	80S
9 7/8	4	.20	.77	

D	Wt	S	t	L_u	V	R	G	d	b	t'	r'	Wh
8 WF	67	60.4	.57	42.9	67	67	13.8	9	8 1/4	.93	2.12	
	58	52.0	.51	37.9	58	57	12.2	8 3/4	8 1/4	.81	2.10	130C
	48	43.2	.40	32.6	45	44	9.7	8 1/2	8 1/8	.68	2.08	150S
	40	35.5	.36	27.4	39	39	8.8	8 1/4	8 1/8	.56	2.04	
	35	31.1	.31	24.0	33	33	7.6	8 1/8	8	.49	2.03	100C
	31	27.4	.29	21.5	30	30	6.9	8	8	.43	2.01	110S
	28	24.3	.29	18.5	30	30	6.8	8	6 1/2	.46	1.62	80C
	24	20.8	.24	16.0	25	25	5.9	7 7/8	6 1/2	.40	1.61	90S
	20	17.0	.25	12.0	26	25	6.0	8 1/8	5 1/4	.38	1.20	70C
	17	14.1	.23	10.0	24	23	5.5	8	5 1/4	.31	1.16	80S
	15	11.8	.24	7.5	26	24	5.9	8 1/8	4	.31	.86	
	13	9.9	.23	6.0	24	22	5.5	8	4	.25	.83	50C
	10	7.8	.17	5.0	18	16	4.1	7 7/8	4	.20	.82	60S
8C	34.3	28.9	.37	21.9	39	39	9.0	8	8	.44	1.87	100C 110S
6 WF	25	16.8	.32	21.8	27	33	7.7	6 3/8	6	.46	1.52	
	20	13.4	.26	17.8	21	26	6.2	6 1/4	6	.37	1.50	60C
	15.5	10.1	.24	13.5	19	24	5.8	6	6	.27	1.45	70S
	16	10.1	.26	13.0	21	26	6.2	6 1/4	4	.40	.96	50C
	12	7.24	.23	9.0	18	22	5.5	6	4	.28	.90	60S
	8.5	5.07	.17	6.5	13	16	4.1	5 7/8	4	.19	.87	
6C	25	15.7	.31	25.0	24	32	7.5	6	6	.50	1.43	60C
	20	12.9	.25	18.8	20	26	6.0	6	6	.37	1.39	70S
5 WF	18.9	9.5	.31	21.9	20	32	7.5	5	5	.44	1.20	40C
	18.5	9.9	.26	20.6	18	27	6.4	5 1/8	5	.42	1.28	50S
	16	8.5	.24	18.0	16	24	5.8	5	5	.36	1.26	
4B	13	5.45	.28	16.8	15	27	6.7	4 1/8	4	.34	.99	30C 40S
4C	13	5.2	.25	18.8	13	25	6.0	4	4	.37	.94	
4 WF	10	4.1	.22	13.2	11	22	5.3	4	4	.26	.97	

Junior Beams
(By Jones & Laughlin Steel Corp. Only)

D	Wt	S	t	L_u	V	R	G	d	b	t'	r'
12	11.8	12.0	.175	3.2	27.0	16.8	4.2	12	3	1/4	.53
11	10.3	9.6	.165	2.4	24.0	15.8	4.0	11	2 7/8	3/16	.50
10	9.0	7.8	.155	2.5	20.0	14.7	3.7	10	2 3/4	3/16	.48
9	7.5	5.8	.145	2.5	17.0	13.7	3.3	9	2 3/8	3/16	.42
8	6.5	4.7	.135	2.7	14.0	12.6	1.4	8	2 1/4	3/16	.42
7	5.5	3.5	.126	2.8	11.5	11.5	—	7	2 1/8	3/16	.39
6	4.4	2.4	.114	2.9	8.9	8.9	—	6	1 7/8	3/16	.36

Junior Channels
(By Jones & Laughlin Steel Corp. Only)

D	Wt	S	t	L_u	V	R	G	d	b	t'	r'
12	10.6	9.3	.19	2.0	29.6	18.6	4.5	12	1 1/2	5/16	.35
10	8.4	6.5	.17	1.9	22.1	16.0	4.0	10	1 1/2	1/4	.37
10	6.5	4.4	.15	2.3	19.5	13.9	3.6	10	1 1/8	13/32	.25

Amer. Std. Channel Sect.

D	Wt	S	t	L_u	V	R	G	d	b	t'	r'	Wh
18 [	58	74.5	.70	7.3	164	81	16.8	18	4 1/4	.62	1.04	
	51.9	69.1	.60	7.1	140	69	14.4	18	4 1/8	.62	1.06	140C
	45.8	63.7	.50	6.9	117	58	12.0	18	4	.62	1.09	170S
	42.7	61.0	.45	6.9	105	52	10.8	18	4	.62	1.10	
15 [	50	53.6	.72	8.0	140	83	17.2	15	3 3/4	.65	.87	120C
	40	46.2	.52	7.6	101	60	12.5	15	3 1/2	.65	.89	140S
	33.9	41.7	.40	7.4	78	46	9.6	15	3 3/8	.65	.91	
13 [	50	48.1	.79	10.3	133	91	19.0	13	4 3/8	.61	1.07	
	40	41.7	.56	9.8	94	65	13.4	13	4 1/8	.61	1.09	100C
	35	38.6	.45	9.5	76	52	10.8	13	4 1/8	.61	1.10	120S
	31.8	36.5	.38	9.4	64	43	9.1	13	4	.61	1.11	
12 [	30	26.9	.51	6.6	80	56	12.2	12	3 1/8	.50	.77	80C
	25	23.9	.39	6.4	60	42	9.3	12	3	.50	.79	100S
	20.7	21.4	.28	6.1	44	31	6.7	12	3	.50	.81	
10 [	30	20.6	.67	6.6	88	72	16.2	10	3	.44	.67	
	25	18.1	.53	6.3	68	56	12.6	10	2 7/8	.44	.68	70C
	20	15.7	.38	6.0	49	40	9.1	10	2 3/4	.44	.70	90S
	15.3	13.4	.24	5.7	31	26	5.8	10	2 5/8	.44	.72	
9 [	20	13.5	.45	6.1	52	47	10.8	9	2 5/8	.41	.65	60C
	15	11.3	.28	5.7	33	30	6.8	9	2 1/2	.41	.67	70S
	13.4	10.5	.23	5.6	27	24	5.5	9	2 3/8	.41	.67	
8 [	18.75	10.9	.49	6.2	51	50	11.7	8	2 1/2	.39	.60	50C
	13.75	9.0	.30	5.7	32	31	7.3	8	2 3/8	.39	.62	60S
	11.5	8.1	.22	5.5	23	23	5.3	8	2 1/4	.39	.63	
7 [	14.75	7.7	.42	6.0	38	43	10.1	7	2 1/4	.37	.57	40C
	12.25	6.9	.31	5.7	29	33	7.5	7	2 1/4	.37	.58	50S
	9.8	6.0	.21	5.5	19	22	5.0	7	2 1/8	.37	.59	
6 [	13.0	5.8	.44	6.2	34	45	10.5	6	2 1/8	.34	.53	30C
	10.5	5.0	.31	5.8	25	32	7.5	6	2	.34	.53	40S
	8.2	4.3	.20	5.5	16	20	4.8	6	1 7/8	.34	.54	
5 [	9.0	3.5	.32	6.0	21	33	7.8	5	1 7/8	.32	.49	25C
	6.7	3.0	.19	5.6	12	19	4.6	5	1 3/4	.32	.50	35S
4 [	7.25	2.3	.32	6.4	17	32	7.7	4	1 3/4	.30	.46	20C
	5.4	1.9	.18	5.9	9	18	4.3	4	1 5/8	.30	.45	30S
3 [	6.0	1.4	.36	7.2	14	35	8.5	3	1 5/8	.27	.42	
	5.0	1.2	.26	6.8	10	26	6.2	3	1 1/2	.27	.41	10C
	4.1	1.1	.17	6.4	7	17	4.1	3	1 3/8	.27	.41	10S

Amer. Std. Beam Sect.

D	Wt	S	t	L_u	V	R	G	d	b	t'	r'	Wh
24 I	120	250.9	.80	18.5	249	104	19.2	24	8	1.10	1.56	320C
	105.9	234.3	.62	18.0	195	82	15.0	24	7 7/8	1.10	1.60	380S
	100	197.6	.75	13.0	233	92	17.9	24	7 1/4	.87	1.29	
	90	185.8	.62	13.0	195	77	15.0	24	7 1/8	.87	1.32	280C 330S
	79.9	173.9	.50	12.5	156	62	12.0	24	7	.87	1.36	
20 I	95	160.0	.80	16.5	208	101	19.2	20	7 1/4	.92	1.35	
	85	150.2	.65	16.0	170	82	15.7	20	7	.92	1.38	240C
	75	126.3	.64	12.5	167	78	15.4	20	6 3/8	.79	1.17	280S
	65.4	116.9	.50	12.0	130	61	12.0	20	6 1/4	.79	1.21	
18 I	70	101.9	.71	12.0	166	83	17.1	18	6 1/4	.69	1.09	190C
	54.7	88.4	.46	11.5	108	54	11.0	18	6	.69	1.15	230S
15 I	50	64.2	.55	11.5	107	63	13.2	15	5 5/8	.62	1.05	160C
	42.9	58.9	.41	11.0	80	47	9.8	15	5 1/2	.62	1.08	190S
12 I	50	50.3	.69	15.0	107	79	16.5	12	5 1/2	.66	1.05	
	40.8	44.8	.46	14.5	72	53	11.0	12	5 1/4	.66	1.08	120C
	35	37.8	.43	11.5	67	48	10.3	12	5 1/8	.54	.99	140S
	31.8	36.0	.35	11.5	55	39	8.4	12	5	.54	1.01	
10 I	35	29.2	.59	12.0	77	64	14.3	10	5	.49	.91	90C
	25.4	24.4	.31	11.5	40	34	7.4	10	4 5/8	.49	.97	110S
8 I	23.0	16.0	.44	11.0	46	46	10.6	8	4 1/8	.42	.81	70C
	18.4	14.2	.27	10.5	28	28	6.5	8	4	.42	.84	80S
7 I	20.0	12.0	.45	11.0	41	47	10.8	7	3 7/8	.39	.74	50C
	15.3	10.4	.25	10.0	23	26	6.0	7	3 5/8	.39	.78	60S
6 I	17.25	8.7	.47	10.5	36	47	11.0	6	3 5/8	.36	.68	30C
	12.5	7.3	.23	10.0	18	24	5.5	6	3 3/8	.36	.72	40S
5 I	14.75	6.0	.49	10.5	32	50	12.0	5	3 1/4	.33	.63	30C
	10.0	4.8	.21	10.0	14	21	5.0	5	3	.33	.65	40S
4 I	9.5	3.3	.33	10.0	17	32	8.0	4	2 3/4	.29	.58	30C
	7.7	3.0	.19	9.5	10	19	4.6	4	2 5/8	.29	.59	40S
3 I	7.5	1.9	.35	11.0	14	34	8.4	3	2 1/2	.26	.53	20C
	5.7	1.7	.17	10.0	7	17	4.1	3	2 3/8	.26	.53	30S

Bearing Single=20,000 Double=25,000		Tension in Bolts 20,000#/□"	
1"		Dia	Tension in Kips
7850		1/2"	2.5
15710		5/8"	4.0
Bearing (in Kips)		3/4"	6.0
20.0	25.0	7/8"	8.3
5.00	6.25	1"	11.0
6.25	7.81	1 1/8"	13.8
7.50	9.38	1 1/4"	17.8
—	10.9	1 3/8"	21.0
—	12.5	1 1/2"	25.8
—	14.1	1 3/4"	34.8
—	15.6	2"	46.0

Equal Leg Angles

Size	S_{1-1}	Area	r_{1-1}	r_{3-3}	Size	S_{1-1}	Area	r_{1-1}	r_{3-3}
2 1/2×2 1/2×1/4	.39	1.19	.77	.49	5×5×3/8	2.4	3.61	1.56	.99
2 1/2×2 1/2×3/8	.57	1.73	.75	.49	5×5×1/2	3.2	4.75	1.54	.98
3×3×1/4	.58	1.44	.93	.59	6×6×3/8	3.5	4.36	1.88	1.19
3×3×5/16	.71	1.78	.92	.59	6×6×7/16	4.1	5.06	1.87	1.19
3 1/2×3 1/2×1/4	.79	1.69	1.09	.69	6×6×1/2	4.6	5.75	1.86	1.18
3 1/2×3 1/2×3/8	1.2	2.48	1.07	.69	6×6×3/4	6.7	8.44	1.83	1.17
3 1/2×3 1/2×1/2	1.5	3.25	1.06	.68	6×6×1	8.6	11.00	1.80	1.17
4×4×1/4	1.1	1.94	1.25	.80	8×8×1/2	8.4	7.75	2.50	1.59
4×4×3/8	1.5	2.86	1.23	.79	8×8×3/4	12.2	11.44	2.47	1.57
4×4×1/2	2.0	3.75	1.22	.78	8×8×1	15.8	15.00	2.44	1.56
4×4×3/4	2.8	5.44	1.19	.78	8×8×1 1/8	17.5	16.73	2.42	1.56

Unequal Leg Angles

Size	S_{1-1}	S_{2-2}	Area	r_{1-1}	r_{2-2}	r_{3-3}	Size	S_{1-1}	S_{2-2}	Area	r_{1-1}	r_{2-2}	r_{3-3}
2 1/2×2×1/4	.38	.25	1.06	.78	.59	.42	7×4×1/2	5.8	2.1	5.25	2.25	1.11	.87
3×2 1/2×1/4	.56	.40	1.31	.95	.75	.53	7×4×3/4	8.4	3.0	7.69	2.22	1.09	.86
3 1/2×3×1/4	.78	.59	1.56	1.11	.91	.63	8×4×1/2	7.5	2.2	5.75	2.59	1.08	.86
4×3×1/4	1.0	.60	1.69	1.28	.90	.65	8×4×3/4	10.9	3.1	8.44	2.55	1.05	.85
4×3×5/16	1.2	.73	2.09	1.27	.89	.65	8×4×1	14.1	3.9	11.00	2.52	1.03	.85
5×3 1/2×5/16	1.9	1.0	2.56	1.61	1.03	.76	8×6×1/2	8.0	4.8	6.75	2.56	1.79	1.30
5×3 1/2×3/8	2.3	1.2	3.05	1.60	1.02	.76	8×6×3/4	11.7	6.9	9.94	2.53	1.76	1.29
5×3 1/2×1/2	3.0	1.6	4.00	1.58	1.01	.75	8×6×1	15.1	8.9	13.00	2.49	1.73	1.28
6×4×3/8	3.3	1.6	3.61	1.93	1.17	.88	9×4×1/2	9.3	2.2	6.25	2.92	1.05	.85
6×4×1/2	4.3	2.1	4.75	1.91	1.15	.87	9×4×3/4	13.6	3.1	9.19	2.88	1.02	.84
6×4×3/4	6.3	3.0	6.94	1.88	1.12	.86	9×4×1	17.6	4.0	12.00	2.84	1.00	.83

Nomenclature

D = NOMINAL DEPTH IN INCHES.
Wt = WEIGHT PER FOOT IN LBS.
S = SECTION MODULUS.
t = WEB THICKNESS IN INCHES.
L_u = MAXIMUM LENGTH @ 20,000#/□" WITH UNBRACED COMP. FLANGE.
V = MAXIMUM WEB SHEAR IN KIPS FOR 13,000#/□"
R = ALLOWABLE END REACTION IN KIPS FOR 3 1/2" BEARING (A.I.S.C.)
G = INCREASE IN R IN KIPS FOR 1" ADDITIONAL BEARING (A.I.S.C.)
d = ACTUAL DEPTH IN INCHES.
b = FLANGE WIDTH IN INCHES.
t' = FLANGE THICKNESS IN INCHES.
Wh = WEIGHT OF HAUNCH & SECTION IN #/'. S = STONE CONC. C = CINDER CONC.
B = BETHLEHEM STEEL CO. SECTION.
C = U.S. STEEL CORP. SECTION.
I = AMERICAN STANDARD SECTION.
l = LENGTH OF MEMBER IN INCHES.
r = RADIUS OF GYRATION.
r' = LEAST RADIUS OF GYRATION.
f = MAXIMUM ALLOWABLE STRESS INTENSITY PER SQ. INCH.

SEELYE, STEVENSON, VALUE & KNECHT.
CONSULTING ENGINEERS
101 PARK AVE., NEW YORK CITY.

Diagonal lines in boxes indicate web shear governs.

STRUCTURAL STEEL

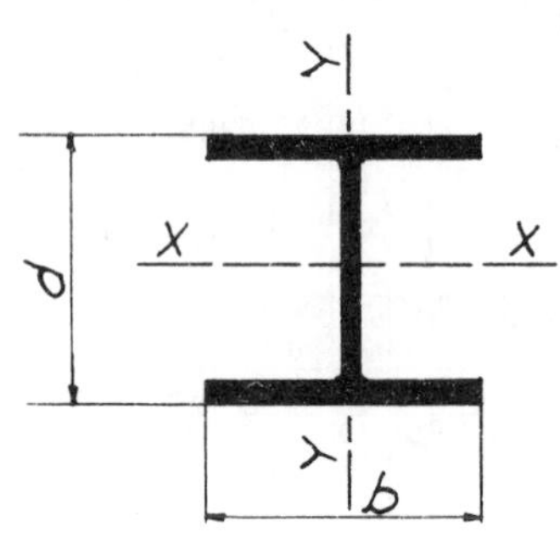
Y
X
X
d
Y
b

FORMULA

$F_A = 17{,}000 - 0.485 \frac{l^2}{r^2}$ for $\frac{l}{r} < 120$

$F_A = \frac{18{,}000}{1 + l^2/18{,}000r^2} \times (1.6 - \frac{l/r}{200})$ for l/r = 120 to 200

For columns subject to axial load and bending, combined stresses $\frac{f_a}{F_A} + \frac{f_b}{F_B}$ shall be less than unity.

NOTE: Values to right of heavy line are for l/r ratios between 120 and 200–main members.

The 14 WF 320 section is used as column core.

For secondary members where l/r = 120 to 200, $F_A = \frac{18{,}000}{1 + l^2/r^2 \times 18{,}000}$.

NOMENCLATURE

l = unbraced length of column in inches

F_A = allowable axial stress, p.s.i.

F_B = allowable bending stress, p.s.i.

f_a = actual axial stress, p.s.i.

f_b = actual bending stress, p.s.i.

A = area of column, sq. in.

S_x, S_y = section moduli, cu. in.

B_x, B_y = bending factors

$I_x = S_x \cdot d/2$ = moment of inertia, in.^{4}

$I_x = S_y \cdot d/2$ = moment of inertia, in.^{4}

$\frac{r_x}{r_y}$ = ratio of radii of gyration

r = least radius of gyration

NOTE: l for X axis must not be greater than r_x/r_y times l for Y axis.

Example: Design of a column with eccentric loading.
Given: Axial load P = 400 kips; moment about X axis, M = 1900 in. – kips, l = 18'-0"
Required: To find size of column.
Solution: Assume 14" column, B_x = 0.18. Equivalent direct load from bending = $M \cdot B_x$ = 1900 × 0.18 = 342 kips. Approximate column load = 400 + 342 = 742 kips. From table for load and length, approximate column size = 14 WF 167. Check column size using criterion $f_a/f_A + f_b/f_B = 1$. $F_A = \frac{\text{load in table}}{A}$; F_B = 20,000 p.s.i.

Using 14 WF 167; F_A = 15,600 p.s.i.
F_B = 20,000 #/□, $f_a = \frac{P}{A}$ = 8140 p.s.i.
$f_b = \frac{M}{S}$ = 7120 #/□. Criterion gives 0.876 – column column too strong.

Using 14 WF 150; F_A = 15.600 p.s.i.
F_B = 20,000 #/□; $f_a = \frac{P}{A}$ = 9060 p.s.i.
$f_b = \frac{M}{S}$ = 7900 #/□. Criterion gives 0.98 – correct size.

COLUMNS

Table A—ALLOWABLE CONCENTRIC LOADS IN KIPS FOR WIDE FLANGE COLUMNS—A.I.S.C., 1947

Nominal Size	Weight, lb. ft.	Area, sq. in.	Dimension d	Dimension b	Axis X–X S_x	Axis X–X B_x	Axis Y–Y S_y	Axis Y–Y B_y	Radius of Gyration r	Radius of Gyration r_x/r_y	10	12	14	16	18	20	22	24	26	28	30	32	34	36	40
											Unbraced Length in feet for Least Radius of Gyration														
6	15.5	4.62	6	6	10.1	0.46	3.2	1.44	1.45	1.77	63	56	48	39	31	25	20	15							
	20	5.90	6-1/4	6	13.4	0.44	4.4	1.34	1.50	1.77	82	74	64	53	43	35	28	22							
	25	7.37	6-3/8	6	16.8	0.44	5.6	1.32	1.52	1.77	103	93	82	68	55	45	36	28							
8	24	7.06	7-7/8	6-1/2	20.8	0.34	5.6	1.26	1.61	2.12	101	93	83	71	59	49	40	32	26						
	28	8.23	8	6-1/2	24.3	0.34	6.6	1.24	1.62	2.13	118	108	96	83	69	58	47	39	31						
	31	9.12	8	8	27.4	0.33	9.2	0.99	2.01	1.73	139	132	124	115	104	92	79	68	57	49	42	35			
	35	10.30	8-1/8	8	31.1	0.33	10.6	0.97	2.03	1.72	158	150	141	131	119	105	91	78	67	57	48	41			
	40	11.76	8-1/4	8-1/8	35.5	0.33	12.1	0.97	2.04	1.73	180	172	161	149	136	121	105	89	77	65	56	47	40		
	48	14.11	8-1/2	8-1/8	43.2	0.33	15.0	0.94	2.08	1.74	217	207	195	182	166	149	129	112	96	82	70	60	50		
	58	17.06	8-3/4	8-1/4	52.0	0.33	18.2	0.94	2.10	1.74	263	251	237	221	202	182	159	137	118	102	87	73	62		
	67	19.70	9	8-1/4	60.4	0.33	21.4	0.92	2.12	1.75	304	291	275	256	236	212	186	161	139	120	102	87	74		
10	33	9.71	9-3/4	8	35.0	0.28	9.2	1.06	1.94	2.16	147	139	130	119	107	92	79	68	57	48	40	34			
	39	11.47	10	8	42.2	0.27	11.2	1.02	1.98	2.16	175	166	155	143	129	113	97	83	71	60	50	42			
	45	13.24	10-1/8	8	49.1	0.27	13.3	1.00	2.00	2.17	202	192	180	166	150	133	114	98	84	71	60	50			
	49	14.40	10	10	54.6	0.26	18.6	0.77	2.54	1.71	229	222	214	205	194	182	169	155	139	123	109	96	85	74	57
	54	15.88	10-1/8	10	60.4	0.26	20.7	0.77	2.56	1.71	253	246	237	227	215	202	188	172	156	138	122	108	96	84	64
	60	17.66	10-1/4	10-1/8	67.1	0.26	23.1	0.77	2.57	1.72	282	274	264	252	240	226	210	193	174	154	137	121	106	94	72
	66	19.41	10-3/8	10-1/8	73.7	0.26	25.5	0.76	2.58	1.72	309	301	290	278	264	249	231	213	192	171	152	134	118	105	80
	72	21.18	10-1/2	10-1/8	80.1	0.26	27.9	0.76	2.59	1.72	338	328	317	303	289	272	253	233	211	187	167	148	130	115	88
	77	22.67	10-5/8	10-1/4	86.1	0.26	30.1	0.75	2.60	1.73	362	352	340	326	310	292	272	250	227	202	180	159	140	124	95
	89	26.19	10-7/8	10-1/4	99.7	0.26	35.2	0.74	2.63	1.73	419	407	394	378	360	340	317	293	267	237	212	188	167	147	113
	100	29.43	11-1/8	10-3/8	112.4	0.26	39.9	0.74	2.65	1.74	471	458	443	425	406	383	359	332	302	270	241	214	190	168	131
	112	32.92	11-3/8	10-3/8	126.3	0.26	45.2	0.73	2.67	1.75	528	513	496	477	455	431	403	374	341	306	273	243	216	191	149
12	40	11.77	12	8	51.9	0.23	11.0	1.07	1.94	2.64	178	169	157	144	129	113	96	81	69	58	49	41			
	45	13.24	12	8	58.2	0.23	12.4	1.07	1.94	2.65	200	190	177	162	145	127	108	92	78	65	55	46			
	50	14.71	12-1/4	8-1/8	64.7	0.23	14.0	1.05	1.96	2.64	223	212	198	182	163	142	122	104	88	75	63	53			
	53	15.59	12	10	70.7	0.22	19.2	0.81	2.48	2.11	248	240	230	220	208	194	179	163	145	128	113	99	87	76	58
	58	17.06	12-1/4	10	78.1	0.22	21.4	0.80	2.51	2.10	271	263	253	241	229	214	198	181	161	143	126	111	98	86	65
	65	19.11	12-1/8	12	88.0	0.22	29.1	0.66	3.02	1.75	310	304	296	288	278	266	254	241	226	210	193	175	158	142	115
	72	21.16	12-1/4	12	97.5	0.22	32.4	0.65	3.04	1.75	344	337	328	319	308	296	282	268	252	234	216	196	176	160	130
	79	23.22	12-3/8	12-1/8	107.1	0.22	35.8	0.65	3.05	1.75	378	370	361	350	338	325	310	294	277	258	238	215	196	177	143
	85	24.98	12-1/2	12-1/8	115.7	0.22	38.9	0.64	3.07	1.75	406	398	388	377	365	351	335	318	300	280	258	235	212	192	156
	92	27.06	12-5/8	12-1/8	125.0	0.22	42.2	0.64	3.08	1.75	440	431	421	409	395	380	364	346	326	304	281	255	232	209	170
	99	29.09	12-3/4	12-1/4	134.7	0.22	45.7	0.64	3.09	1.76	473	464	453	440	426	410	392	372	351	328	303	276	250	226	184
	106	31.19	12-7/8	12-1/4	144.5	0.22	49.2	0.63	3.11	1.76	508	498	486	473	457	440	421	400	378	354	327	299	271	245	200
	120	35.31	13-1/8	12-3/8	163.4	0.22	56.0	0.63	3.13	1.76	575	564	551	536	519	500	478	455	430	403	374	341	310	281	230
	133	39.11	13-3/8	12-3/8	182.5	0.21	63.1	0.62	3.16	1.77	638	626	611	595	576	555	532	507	480	451	419	384	348	316	260
	161	47.38	13-7/8	12-1/2	222.2	0.21	77.7	0.61	3.20	1.78	773	759	742	723	701	676	649	619	587	552	515	475	431	392	322
	190	55.86	14-3/8	12-5/8	263.2	0.21	93.1	0.60	3.25	1.79	913	896	877	855	830	802	771	737	700	660	617	571	521	474	392
14	43	12.65	13-5/8	8	62.7	0.20	11.3	1.12	1.89	3.08	190	179	166	152	135	116	98	83	70	59	49				
	48	14.11	13-7/8	8	70.2	0.20	12.8	1.10	1.91	3.07	213	201	187	171	152	131	112	95	80	67	56				
	53	15.59	14	8	77.8	0.20	14.3	1.09	1.92	3.07	235	222	207	190	169	146	125	106	90	75	63	52			
	61	17.94	13-7/8	10	92.2	0.19	21.5	0.83	2.45	2.44	284	275	264	252	238	222	204	185	164	144	127	112	97	85	64
	68	20.00	14	10	103.0	0.19	24.1	0.83	2.46	2.45	317	307	295	281	265	248	228	207	184	162	142	125	109	96	72
	74	21.76	14-1/4	10-1/8	112.3	0.19	26.5	0.82	2.48	2.44	345	334	321	307	290	271	250	228	202	179	157	139	121	106	80
	78	22.94	14	12	121.1	0.19	34.5	0.67	3.00	2.03	372	364	355	344	332	319	304	288	270	250	230	207	188	169	136
	84	24.71	14-1/8	12	130.9	0.19	37.5	0.66	3.02	2.03	401	393	383	372	359	345	329	311	292	272	250	226	204	184	149
	87	25.56	14	14-1/2	138.1	0.19	48.2	0.53	3.70	1.66	422	416	409	401	392	382	372	360	346	332	317	301	284	265	226
	95	27.94	14-1/8	14-1/2	150.6	0.19	52.8	0.53	3.71	1.66	461	455	447	439	429	418	406	393	379	364	347	330	311	291	249
	103	30.26	14-1/4	14-5/8	163.6	0.19	57.6	0.53	3.72	1.67	499	493	485	475	465	453	441	427	411	395	377	358	338	317	270
	111	32.65	14-3/8	14-5/8	176.3	0.19	62.2	0.53	3.73	1.67	539	531	523	513	502	490	476	461	444	426	408	387	366	343	293
	119	34.99	14-1/2	14-5/8	189.4	0.19	67.1	0.52	3.75	1.67	578	570	561	551	539	525	511	495	477	459	438	417	394	370	317
	127	37.33	14-5/8	14-3/4	202.0	0.19	71.8	0.52	3.76	1.67	616	608	598	587	575	561	545	528	510	490	469	446	421	396	339
	136	39.98	14-3/4	14-3/4	216.0	0.19	77.0	0.52	3.77	1.67	660	651	641	629	616	601	585	567	547	526	503	478	453	425	365
	142	41.85	14-3/4	15-1/2	226.7	0.19	85.2	0.49	3.97	1.59	693	685	675	664	652	638	622	605	586	566	544	522	497	471	414
	150	44.08	14-7/8	15-1/2	240.2	0.18	90.6	0.49	3.99	1.60	730	722	712	700	687	672	656	638	619	598	575	552	526	499	439
	158	46.47	15	15-1/2	253.4	0.18	95.8	0.49	4.00	1.60	770	761	750	738	724	709	692	673	653	631	607	582	556	527	465
	167	49.09	15-1/8	15-5/8	267.3	0.18	101.3	0.49	4.01	1.60	813	804	793	780	765	749	731	712	690	667	643	616	588	558	493
	176	51.73	15-1/4	15-5/8	281.9	0.18	107.1	0.48	4.02	1.60	857	847	836	822	807	790	771	751	728	704	678	651	621	589	522
	184	54.07	15-3/8	15-5/8	295.8	0.18	112.7	0.48	4.04	1.61	896	886	874	860	844	827	807	786	763	738	711	683	652	619	549
	193	56.73	15-1/2	15-3/4	310.0	0.18	118.4	0.48	4.05	1.61	940	930	917	903	886	868	848	825	801	775	747	717	685	651	578
	202	59.39	15-5/8	15-3/4	324.9	0.18	124.4	0.48	4.06	1.61	984	973	960	945	928	909	888	865	840	812	783	752	719	684	607
	211	62.07	15-3/4	15-3/4	339.2	0.18	130.2	0.48	4.07	1.61	1029	1018	1004	988	971	951	928	904	878	850	819	787	753	716	637
	219	64.36	15-7/8	15-7/8	352.6	0.18	135.6	0.48	4.08	1.62	1067	1055	1041	1025	1007	986	963	938	911	882	851	818	782	744	662
	228	67.06	16	15-7/8	367.8	0.18	141.8	0.47	4.10	1.61	1112	1100	1086	1069	1050	1029	1005	980	952	927	889	854	818	779	694
	237	69.69	16-1/8	15-7/8	382.2	0.18	147.7	0.47	4.11	1.62	1156	1143	1128	1111	1091	1070	1045	1019	990	959	925	890	851	811	724
	246	72.33	16-1/4	16	397.4	0.18	153.9	0.47	4.12	1.62	1200	1187	1171	1154	1133	1110	1085	1058	1029	996	962	925	886	844	753
	264	77.63	16-1/2	16	427.4	0.18	166.1	0.47	4.14	1.63	1288	1274	1258	1239	1217	1193	1166	1137	1106	1071	1035	995	954	910	814
	287	84.37	16-3/4	16-1/8	465.5	0.18	181.8	0.46	4.17	1.63	1400	1386	1368	1348	1325	1299	1270	1239	1205	1168	1130	1087	1043	995	892
	314	92.30	17-1/4	16-1/4	511.9	0.18	201.0	0.46	4.20	1.64	1533	1516	1497	1476	1450	1423	1392	1359	1322	1283	1240	1195	1147	1095	984
	320	94.12	16-3/4	16-3/4	492.8	0.19	195.7	0.48	4.17	1.59	1562	1546	1526	1503	1478	1449	1417	1382	1345	1303	1260	1213	1163	1110	995
	342	100.59	17-1/2	16-3/8	559.4	0.18	220.8	0.46	4.24	1.65	1671	1654	1633	1610	1584	1554	1521	1485	1446	1404	1358	1310	1259	1203	1085
	370	108.78	18	16-1/2	608.1	0.18	241.1	0.45	4.27	1.66	1808	1789	1768	1742	1714	1683	1648	1610	1567	1522	1474	1423	1367	1309	1183
	398	116.98	18-1/4	16-5/8	656.9	0.18	261.6	0.45	4.31	1.66	1945	1925	1902	1876	1846	1813	1775	1736	1691	1643	1593	1538	1480	1419	1285
	426	125.25	18-3/4	16-3/4	707.4	0.18	282.7	0.44	4.34	1.67	2083	2062	2038	2011	1979	1944	1905	1862	1815	1765	1711	1653	1593	1528	1386

STRUCTURAL STEEL-EFFECTIVE NET AREAS

TABLE A- EFFECTIVE NET AREAS OF DOUBLE ANGLES IN TENSION

One Hole Out. Two Holes Out.

(Size of holes 1/8" larger than rivet size.)

SIZE OF ANGLES	THICKNESS (inches)	GROSS AREA OF ANGLES (sq. in.)	NET AREA IN SQUARE INCHES					
			ONE HOLE OUT			TWO HOLES OUT		
			SIZE OF HOLE			SIZE OF HOLE		
			7/8"	1"	1 1/8"	7/8"	1"	1 1/8"
8 x 8	1 1/8	33.46	31.50	31.21	30.92	29.54	28.96	28.38
	1	30.00	28.25	28.00	27.75	26.50	26.00	25.50
	7/8	26.46	24.92	24.71	24.50	23.38	22.96	22.54
	3/4	22.88	21.56	21.38	21.20	20.24	19.88	19.52
	5/8	19.22	18.12	17.97	17.82	17.02	16.72	16.42
	1/2	15.50	14.62	14.50	14.38	13.74	13.50	13.26
8 x 6	1	26.00	24.25	24.00	23.75	22.50	22.00	21.50
	7/8	22.96	21.42	21.21	21.00	19.88	19.46	19.04
	3/4	19.88	18.56	18.38	18.20	17.24	16.88	16.52
	5/8	16.72	15.62	15.47	15.32	14.52	14.22	13.92
	1/2	13.50	12.62	12.50	12.38	11.74	11.50	11.26
6 x 6	1	22.00	20.25	20.00	19.75	18.50	18.00	17.50
	7/8	19.46	17.92	17.71	17.50	16.38	15.96	15.54
	3/4	16.88	15.56	15.38	15.20	14.24	13.88	13.52
	5/8	14.22	13.12	12.97	12.82	12.02	11.72	11.42
	1/2	11.50	10.62	10.50	10.38	9.74	9.50	9.26
	3/8	8.72	8.06	7.97	7.88	7.40	7.22	7.04
5 x 5	1	18.00	16.25	16.00	15.75	14.50	14.00	13.50
	7/8	15.96	14.42	14.21	14.00	12.88	12.46	12.04
	3/4	13.88	12.56	12.38	12.20	11.24	10.88	10.52
	5/8	11.72	10.62	10.47	10.32	9.52	9.22	8.92
	1/2	9.50	8.62	8.50	8.38	7.74	7.50	7.26
	3/8	7.22	6.56	6.47	6.38	5.90	5.72	5.54
8 x 4	1	22.00	20.25	20.00	19.75	18.50	18.00	17.50
	7/8	19.46	17.92	17.71	17.50	16.38	15.96	15.54
	3/4	16.88	15.56	15.38	15.20	14.24	13.88	13.52
	5/8	14.22	13.12	12.97	12.82	12.02	11.72	11.42
	1/2	11.50	10.62	10.50	10.38	9.74	9.50	9.26
7 x 4	1	20.00	18.25	18.00	17.75	16.50	16.00	15.50
	7/8	17.72	16.18	15.97	15.76	14.64	14.22	13.80
	3/4	15.38	14.06	13.88	13.70	12.74	12.38	12.02
	5/8	12.98	11.88	11.73	11.58	10.78	10.48	10.18
	1/2	10.50	9.62	9.50	9.38	8.74	8.50	8.26
	3/8	7.98	7.32	7.23	7.14	6.66	6.48	6.30
6 x 4	7/8	15.96	14.42	14.21	14.00	12.88	12.46	12.04
	3/4	13.88	12.56	12.38	12.20	11.24	10.88	10.52
	5/8	11.72	10.62	10.47	10.32	9.52	9.22	8.92
	1/2	9.50	8.62	8.50	8.38	7.74	7.50	7.26
	3/8	7.22	6.56	6.47	6.38	5.90	5.72	5.54

SIZE OF ANGLES	THICKNESS (inches)	GROSS AREA OF ANGLES (sq. in.)	NET AREA IN SQUARE INCHES					
			ONE HOLE OUT			TWO HOLES OUT		
			SIZE OF HOLE			SIZE OF HOLE		
			7/8"	1"	1 1/8"	7/8"	1"	1 1/8"
5 x 3 1/2	3/4	11.62	10.30	10.12	9.94	8.98	8.62	8.26
	5/8	9.84	8.74	8.59	8.44	7.64	7.34	7.04
	1/2	8.00	7.12	7.00	6.88	6.24	6.00	5.76
	3/8	6.10	5.44	5.35	5.26	4.78	4.60	4.42
4 x 4	3/4	10.88	9.56	9.38	9.20	8.24	7.88	7.52
	5/8	9.22	8.12	7.97	7.82	7.02	6.72	6.42
	1/2	7.50	6.62	6.50	6.38	5.74	5.50	5.26
	3/8	5.72	5.06	4.97	4.88	4.40	4.22	4.04
	1/4	3.88	3.44	3.38	3.32	3.00	2.88	2.76
4 x 3 1/2	3/4	10.12	8.80	8.62	8.44	7.48	7.12	6.76
	5/8	8.60	7.50	7.35	7.20	6.40	6.10	5.80
	1/2	7.00	6.12	6.00	5.88	5.24	5.00	4.76
	3/8	5.34	4.68	4.59	4.50	4.02	3.84	3.66
4 x 3	5/8	7.96	6.86	6.71		5.76	5.46	
	1/2	6.50	5.62	5.50		4.74	4.50	
	3/8	4.96	4.30	4.21		3.64	3.46	
	1/4	3.38	2.94	2.88		2.50	2.38	
3 1/2 x 3 1/2	3/4	9.38	8.06	7.88		6.74	6.38	
	5/8	7.96	6.86	6.71		5.76	5.46	
	1/2	6.50	5.62	5.50		4.74	4.50	
	3/8	4.96	4.30	4.21		3.64	3.46	
	1/4	3.38	2.94	2.88		2.50	2.38	
3 1/2 x 3	5/8	7.34	6.24	6.09		5.14	4.84	
	1/2	6.00	5.12	5.00		4.24	4.00	
	3/8	4.60	3.94	3.85		3.28	3.10	
	1/4	3.12	2.68	2.62		2.24	2.12	
3 x 3	5/8	6.72	5.62	5.47		4.52	4.22	
	1/2	5.50	4.62	4.50		3.74	3.50	
	3/8	4.22	3.56	3.47		2.90	2.72	
	1/4	2.88	2.44	2.38		2.00	1.88	
3 x 2 1/2	3/8	3.84	3.18			2.52		
	1/4	2.62	2.18			1.74		
2 1/2 x 2 1/2	1/2	4.50	3.62			2.74		
	3/8	3.46	2.80			2.14		
	1/4	2.38	1.94			1.50		
2 1/2 x 2	3/8	3.10	2.44			1.78		
	1/4	2.12	1.68			1.24		
2 x 2	3/8	2.72	2.06			1.40		
	1/4	1.88	1.44			1.00		

STRUCTURAL STEEL—STRUTS—1

TABLE A - ALLOWABLE CONCENTRIC LOADS ON STRUTS OF ONE ANGLE (IN KIPS).

A.I.S.C. CODE 1947 MAXIMUM $l/r = 120$

Size L	t	Unbraced Length 6	7	8	9	10	11	12	13	14	15	Max. L
8×8	1⅛	267	261	254	246	237	227	215	203	190	175	15.6
	1	239	234	228	220	212	203	193	182	170	158	15.6
	⅞	212	207	200	195	187	179	171	162	151	141	15.7
	¾	183	178	173	168	162	155	148	140	131	122	15.7
	⅝	154	150	146	142	136	131	125	118	111	103	15.8
	½	124	121	118	114	110	106	101	95	90	83	15.9
8×6	1	201	194	185	176	166	154	141	130			12.8
	⅞	177	172	164	156	146	136	125	115			12.8
	¾	154	149	142	135	127	118	109	99			12.9
	⅝	129	125	119	114	107	100	92	84			12.9
	½	105	101	97	92	87	81	74	67			13.0

Size L	t	Unbraced Length 3	4	5	6	7	8	9	10	11	12	Max. L
6×6	1	182	178	173	167	159	151	141	131	119	110	11.7
	⅞	161	158	153	147	141	134	125	116	105	97	11.7
	¾	140	137	133	128	122	116	108	100	91	84	11.7
	⅝	118	115	112	108	103	98	92	85	78	71	11.8
	½	95	93	91	87	84	79	74	69	63	57	11.8
	⅜	72	71	69	66	63	60	57	53	48	44	11.9
5×5	1	147	142	136	129	120	110	99	90			9.7
	⅞	130	126	121	114	107	98	88	80			9.7
	¾	113	110	105	99	93	85	76	69			9.7
	⅝	96	93	89	84	79	72	65	59			9.8
	½	78	75	72	68	64	58	53	47			9.8
	⅜	59	57	55	52	48	44	40	36			9.8
8×4	1	177	170	160	149	135	119	110				8.5
	⅞	157	150	141	132	120	105	97				8.5
	¾	136	130	122	114	104	91	84				8.5
	⅝	115	110	104	97	88	78	71				8.6
	½	93	89	84	78	71	63	57				8.6
7×4	1	161	154	145	136	123	108	100				8.5
	⅞	143	137	130	121	110	97	89				8.6
	¾	124	119	113	105	95	84	77				8.6
	⅝	105	100	95	88	80	71	65				8.6
	½	85	82	77	72	66	58	52				8.7
	⅜	64	62	59	55	50	45	40				8.7
6×4	⅞	129	124	117	109	99	88	80				8.6
	¾	112	107	102	95	86	76	69				8.6
	⅝	95	91	86	80	73	64	59				8.6
	½	77	74	70	65	59	53	47				8.7
	⅜	58	56	53	50	45	41	36				8.8

Size L	t	Unbraced Length 2	3	4	5	6	7	8	Max. L
5×3½	¾	96	92	87	81	73	64	58	7.5
	⅝	81	78	74	68	62	54	49	7.5
	½	66	63	60	56	50	44	40	7.5
	⅜	50	49	46	43	39	34	30	7.6
4×4	¾	90	87	82	77	70	62	54	7.8
	⅝	76	74	70	65	59	53	46	7.8
	½	62	60	57	53	48	43	37	7.8
	⅜	47	46	44	41	37	33	29	7.9
	¼	32	31	30	28	25	23	19	8.0
4×3½	¾	83	80	75	69	61	53	51	7.2
	⅝	71	68	64	59	52	45	43	7.2
	½	58	55	52	48	43	36	35	7.2
	⅜	44	42	40	37	33	28	27	7.3
4×3	⅝	65	61	57	51	43	40		6.4
	½	53	50	46	41	35	32		6.4
	⅜	40	38	35	32	27	25		6.4
	¼	28	26	24	22	19	17		6.5
3½×3½	¾	77	73	68	62	54	47		6.8
	⅝	65	62	58	52	46	40		6.8
	½	53	51	47	43	38	32		6.8
	⅜	41	39	36	33	29	25		6.9
	¼	28	26	25	23	20	17		6.9
3½×3	⅝	60	56	52	46	38	37		6.2
	½	49	46	42	37	31	30		6.2
	⅜	37	35	32	29	24	23		6.2
	¼	25	24	22	20	17	16		6.3
3×3	⅝	54	51	46	40	34			5.8
	½	44	42	37	32	27			5.8
	⅜	34	32	29	25	21			5.8
	¼	23	22	20	17	14			5.9
3×2½	⅜	31	28	25	20	19			5.2
	¼	21	19	17	15	13			5.3
2½×2½	½	36	32	28	22				4.9
	⅜	27	25	21	17				4.9
	¼	19	17	15	12				4.9
2½×2	⅜	24	21	16	15				4.2
	¼	16	14	12	11				4.2
2×2	⅜	20	17	14					3.9
	¼	14	12	9					3.9

In using this table allowance must be made for eccentric loading.
Numbers at right of heavy lines give loads for maximum length "L".

STRUCTURAL STEEL – STRUTS–2

TABLE A–ALLOWABLE CONCENTRIC LOADS ON STRUTS FORMED BY TWO EQUAL ANGLES IN KIPS.

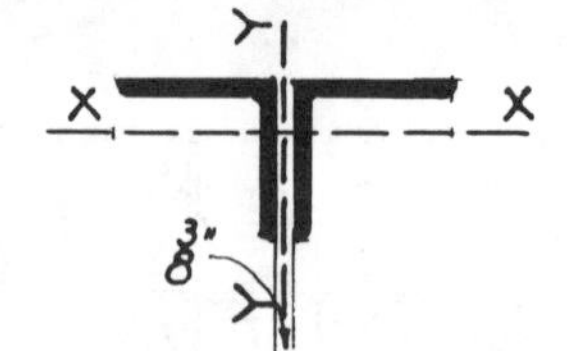

Y
X
X
3/8"
Y

A.I.S.C. CODE - 1947 MAXIMUM l/r = 120

Size	t	Area	r X-X Axis	r Y-Y Axis	Unbraced Length in Feet 3/8" Back to Back — Axis X-X: 4	5	6	8	10	12	14	16	18	20	* Max. L	Axis Y-Y: 6	7	8	10	12	14	16	18	20	22	24	* Max. L
8 x 8	1⅛	33.46	2.42	3.55	562	559	554	543	529	512	490	467	440	412	24.2	563	560	557	550	542	533	521	510	495	479	463	35.5
	1	30.00	2.44	3.53	505	501	497	488	475	459	440	420	396	369	24.4	504	502	499	493	486	477	467	455	443	428	413	35.3
	⅞	26.46	2.45	3.51	445	443	438	430	420	407	390	372	350	327	24.5	445	443	440	434	428	421	413	402	390	378	364	35.1
	¾	22.88	2.47	3.49	385	382	380	373	363	352	338	322	304	284	24.7	384	382	380	376	370	363	355	346	337	325	313	34.9
	⅝	19.22	2.49	3.47	323	321	319	312	305	296	284	272	256	240	24.9	323	322	320	316	310	304	298	291	282	271	261	34.7
	½	15.50	2.50	3.45	261	258	257	252	247	239	230	219	207	194	25.0	260	259	258	254	251	246	240	234	227	219	211	34.5
6 x 6	1	22.00	1.80	2.72	366	362	357	344	327	306	281	252	221	221	18.0	367	364	361	353	344	333	320	307	291	274	254	27.2
	⅞	19.46	1.81	2.70	324	320	316	304	289	271	250	225	197	195	18.1	324	322	319	312	304	295	283	271	256	240	223	27.0
	¾	16.88	1.83	2.68	281	278	274	265	252	236	219	197	172	169	18.3	282	279	276	271	263	255	245	234	222	208	193	26.8
	⅝	14.22	1.84	2.66	237	234	231	223	213	200	185	167	147	142	18.4	237	235	233	228	221	214	206	197	186	174	161	26.6
	½	11.50	1.86	2.64	192	190	187	181	172	162	151	136	120	115	18.6	192	190	188	184	179	173	166	158	149	140	129	26.4
	⅜	8.72	1.88	2.62	146	143	141	137	131	123	114	104	92	87	18.8	145	144	142	139	135	131	125	119	113	105	97	26.2
5 x 5	1	18.00	1.48	2.33	297	292	286	270	249	224	194	180			14.8	297	295	291	283	273	261	247	231	213	194	180	23.3
	⅞	15.96	1.49	2.31	263	259	254	240	221	200	173	160			14.9	264	261	258	250	241	230	218	204	188	170	160	23.1
	¾	13.88	1.51	2.28	230	226	221	219	194	175	153	139			15.1	230	227	224	218	210	199	188	176	162	146	139	22.8
	⅝	11.72	1.52	2.26	194	191	187	177	164	148	130	117			15.2	194	192	189	183	176	168	158	147	135	122	117	22.6
	½	9.50	1.54	2.23	157	154	151	144	133	121	107	95			15.4	156	155	153	148	142	135	127	118	108	97	95	22.3
	⅜	7.22	1.56	2.22	119	117	115	109	102	93	82	72			15.6	119	117	116	112	107	102	96	90	82	73	72	22.2
4 x 4	¾	10.88	1.19	1.88	176	171	166	151	131	110	109				11.9	178	175	171	164	154	143	130	115	109			18.8
	⅝	9.22	1.20	1.86	150	145	140	128	112	93	92				12.0	150	147	144	138	129	120	109	96	92			18.6
	½	7.50	1.22	1.83	122	119	115	105	92	77	75				12.2	122	120	118	112	105	97	87	77	75			18.3
	⅜	5.72	1.23	1.81	93	91	88	80	71	59	57				12.3	93	91	89	85	80	73	66	58	57			18.1
	¼	3.88	1.25	1.79	63	62	60	55	49	41	39				12.5	63	62	61	58	54	50	44	39				17.9
3½ x 3½	¾	9.38	1.03	1.69	149	144	137	120	97	94					10.3	151	148	144	137	126	114	101	94				16.9
	⅝	7.96	1.04	1.66	127	122	117	102	84	80					10.4	128	125	122	115	106	96	84	80				16.6
	½	6.50	1.06	1.64	104	101	96	85	70	65					10.6	104	102	99	94	86	78	67	65				16.4
	⅜	4.96	1.07	1.61	79	77	74	65	54	50					10.7	80	78	76	71	65	58	50	50				16.1
	¼	3.38	1.09	1.59	54	52	50	45	37	34					10.9	54	53	51	48	44	39	34					15.9
3 x 3	⅝	6.72	0.88	1.46	104	99	92	75	67						8.8	106	103	100	92	82	71	67					14.6
	½	5.50	0.90	1.43	86	82	77	63	55						9.0	87	85	82	75	66	57	55					14.3
	⅜	4.22	0.91	1.41	66	63	59	49	42						9.1	66	64	62	57	50	43	42					14.1
	¼	2.88	0.93	1.38	45	43	41	34	29						9.3	45	44	42	38	34	29						13.8
2½ x 2½	½	4.50	0.74	1.24	67	62	56	45							7.4	69	66	63	56	47	45						12.4
	⅜	3.46	0.75	1.21	52	48	43	35							7.5	53	51	48	42	35	35						12.1
	¼	2.38	0.77	1.19	36	33	30	24							7.7	36	35	33	29	24							11.9

*Values to right of heavy line indicate Max. Loads for l/r = 120.

By courtesy of the American Institute of Steel Construction.

STRUCTURAL STEEL- STRUTS-3

TABLE A - ALLOWABLE CONCENTRIC LOADS ON STRUTS FORMED BY TWO UNEQUAL ANGLES - LONG LEGS BACK TO BACK.

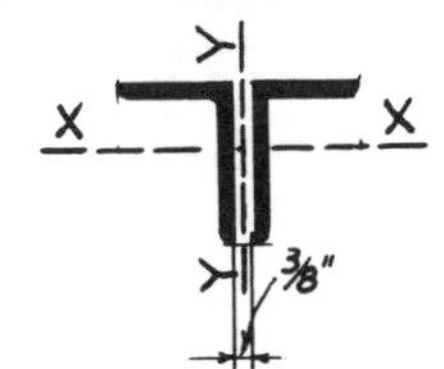

X
X
3/8"

LOADS IN KIPS

A.I.S.C. CODE-1947

MAXIMUM $l/r = 120$

Size	t	Area	r		Unbraced Length in Feet - 3/8" Back to Back.																						
			X-X Axis	Y-Y Axis	Axis X-X										Max.	Axis Y-Y											Max.
					4	6	8	10	12	14	16	18	20	24	L	4	6	8	10	12	14	16	18	20	22	24	L
8x6	1	26.00	2.49	2.52	437	432	423	413	400	384	368	347	325	274	24.9	437	432	424	414	401	387	370	350	328	305	274	25.2
	7/8	22.96	2.51	2.50	386	381	374	366	354	340	325	308	288	244	25.1	386	382	374	365	354	340	324	307	289	267	243	25.0
	3/4	19.88	2.53	2.48	335	330	325	317	307	295	283	264	251	214	25.3	335	330	324	316	306	294	280	265	248	229	208	24.8
	5/8	16.72	2.54	2.46	282	278	273	267	259	249	238	226	212	181	25.4	282	278	272	265	257	247	235	222	207	192	173	24.6
	1/2	13.50	2.56	2.44	227	224	220	215	209	202	193	183	172	147	25.6	227	224	220	214	207	199	189	178	166	153	138	24.4
8x4	1	22.00	2.52	1.61	370	365	359	350	342	327	312	296	277	235	25.2	364	353	336	315	289	258	223	220				16.1
	7/8	19.46	2.53	1.58	327	323	317	310	302	289	277	262	246	208	25.3	322	311	296	276	252	225	195					15.8
	3/4	16.88	2.55	1.55	284	280	276	269	262	252	241	229	215	183	25.5	279	269	256	238	216	191	169					15.5
	5/8	14.22	2.57	1.53	239	236	233	227	222	213	204	193	182	155	25.7	235	227	215	200	181	159	142					15.3
	1/2	11.50	2.59	1.51	194	191	188	184	179	172	165	157	148	127	25.9	190	183	173	160	145	126	115					15.1
7x4	1	20.00	2.18	1.67	336	329	321	311	298	282	265	245	223	200	21.8	332	322	308	290	268	242	212	200				16.7
	7/8	17.72	2.20	1.64	297	292	285	276	265	251	236	219	199	177	22.0	294	285	272	256	235	211	184	177				16.4
	3/4	15.38	2.22	1.62	258	254	248	240	230	219	205	193	174	154	22.2	255	247	236	220	203	181	157	154				16.2
	5/8	12.96	2.24	1.59	218	214	209	203	195	185	174	162	148	130	22.4	215	208	198	185	169	150	130					15.9
	1/2	10.50	2.25	1.57	176	173	170	164	158	150	141	132	121	105	22.5	174	168	160	149	136	120	105					15.7
	3/8	7.96	2.27	1.55	134	131	128	124	119	114	107	100	92	80	22.7	132	127	120	112	102	90	80					15.5
6x4	7/8	15.96	1.86	1.71	267	260	251	239	224	208	189	167	160		18.6	266	258	247	233	216	197	174	160				17.1
	3/4	13.88	1.88	1.69	232	226	218	209	197	183	166	147	139		18.8	231	224	214	202	187	170	149	139				16.9
	5/8	11.72	1.90	1.66	196	191	185	177	167	155	141	126	117		19.0	195	189	181	170	156	141	123	117				16.6
	1/2	9.50	1.91	1.65	159	155	149	143	135	125	114	103	95		19.1	158	152	146	137	126	113	99	95				16.5
	3/8	7.22	1.93	1.62	120	118	114	109	103	96	88	79	72		19.3	1.19	116	110	103	95	85	74	72				16.2
5x3½	3/4	11.62	1.55	1.54	192	185	176	164	149	132	116				15.5	192	185	176	163	148	131	116					15.4
	5/8	9.84	1.56	1.51	162	157	149	139	126	111	98				15.6	162	156	148	137	123	107	98					15.1
	1/2	8.00	1.58	1.49	132	128	121	113	103	92	80				15.8	131	126	119	110	99	87	80					14.9
	3/8	6.10	1.60	1.46	101	98	93	87	80	71	61				16.0	100	97	91	84	75	65	61					14.6
4x3½	3/4	10.12	1.20	1.63	164	155	141	123	101	101					12.0	168	163	155	145	134	120	103	101				16.3
	5/8	8.60	1.22	1.60	140	132	120	106	88	86					12.2	142	138	131	123	117	100	86	86				16.0
	1/2	7.00	1.23	1.58	114	107	98	88	73	70					12.3	116	112	106	100	91	81	70					15.8
	3/8	5.34	1.25	1.56	87	82	75	67	57	53					12.5	89	85	81	76	69	61	53					15.6
4x3	5/8	7.96	1.23	1.36	129	122	112	100	83	80					12.3	131	124	116	105	92	80	NOTE:					13.6
	1/2	6.50	1.25	1.33	106	100	92	82	69	65					12.5	106	101	94	85	74	65	Values to right of					13.3
	3/8	4.96	1.26	1.31	81	76	70	63	53	50					12.6	81	77	71	64	55	50	heavy lines indicate					13.1
	1/4	3.38	1.28	1.29	55	52	48	43	37	34					12.8	55	52	48	43	37	34	Max. Load for $l/r = 120$					12.9
3½x3	5/8	7.34	1.06	1.41	117	108	96	79	73						10.6	120	115	108	99	88	75	73					14.1
	1/2	6.00	1.07	1.38	96	89	79	65	60						10.7	99	94	88	80	72	60						13.8
	3/8	4.60	1.09	1.36	74	68	61	51	46						10.9	76	72	67	61	53	46						13.6
	1/4	3.12	1.11	1.34	50	47	42	35	31						11.1	51	49	45	41	36	31						13.4
3x2½	3/8	3.84	0.93	1.16	60	54	46	38							9.3	62	58	53	45	38							11.6
	1/4	2.62	0.95	1.13	41	37	32	26							9.5	42	40	35	30	26							11.3
2½x2	3/8	3.10	0.77	0.96	47	40	31								7.7	49	44	38	31								9.6
	1/4	2.12	0.78	0.94	32	27	21								7.8	33	30	26	21								9.4

By courtesy of the American Institute of Steel Construction.

STRUCTURAL STEEL – STRUTS–4

TABLE A – ALLOWABLE CONCENTRIC LOADS ON STRUTS FORMED BY TWO UNEQUAL ANGLES – SHORT LEGS BACK TO BACK.

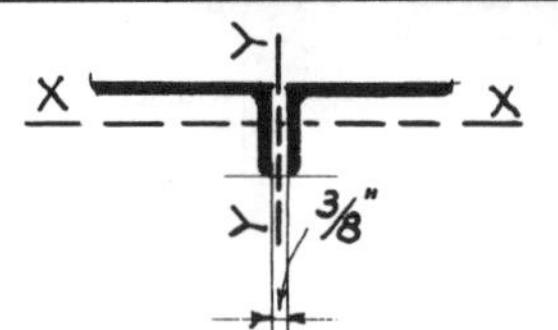

X
X
Y
Y
3/8"

LOADS IN KIPS

A.I.S.C. CODE–1947

MAXIMUM l/r = 120

Size	t	Area	r X-X Axis	r Y-Y Axis	Axis X-X: 4	5	6	8	10	12	14	16	Max. L	Axis Y-Y: 6	8	10	12	14	16	18	20	22	24	28	32	36	Max. L
8×6	1	26.00	1.73	3.78	432	428	420	404	382	355	323	286	17.3	437	434	429	424	417	410	402	392	381	369	343	312	278	37.8
	7/8	22.96	1.74	3.76	382	378	372	357	335	314	286	253	17.4	386	383	379	374	369	362	354	346	336	325	302	275	243	37.6
	3/4	19.88	1.76	3.73	331	328	322	310	293	274	250	224	17.6	335	332	329	324	319	313	306	298	290	281	260	236	209	37.3
	5/8	16.72	1.77	3.72	278	275	271	261	247	231	211	189	17.7	282	279	276	272	268	263	257	251	244	236	218	198	175	37.2
	1/2	13.50	1.79	3.69	225	222	219	211	200	187	172	154	17.9	227	225	223	220	216	212	207	202	196	190	176	158	140	36.9
8×4	1	22.00	1.03	4.10	351	338	322	282	230	220			10.3	371	369	365	361	356	351	345	338	330	322	302	281	256	41.0
	7/8	19.46	1.04	4.07	310	299	286	251	206	195			10.4	328	326	323	319	315	310	304	298	291	283	267	247	225	40.7
	3/4	16.88	1.05	4.04	270	260	249	218	180	169			10.5	285	283	280	277	273	269	264	258	252	245	230	213	193	40.4
	5/8	14.22	1.07	4.02	228	220	211	187	155	142			10.7	240	238	236	233	230	226	222	217	212	207	194	178	163	40.2
	1/2	11.50	1.08	4.00	185	179	171	151	126	115			10.8	194	192	190	188	186	183	179	175	171	167	156	144	130	40.0
7×4	1	20.00	1.06	3.54	320	300	295	260	216	200			10.6	336	333	329	324	318	312	304	295	286	275	252	226	200	35.4
	7/8	17.72	1.07	3.51	284	274	263	233	194	177			10.7	298	295	292	287	282	276	269	261	253	244	222	199	177	35.1
	3/4	15.38	1.09	3.49	247	239	230	204	171	154			10.9	259	256	253	249	244	239	233	226	219	211	193	171	154	34.9
	5/8	12.96	1.10	3.47	208	202	194	173	146	130			11.0	218	216	213	210	206	202	197	191	184	177	161	143	130	34.7
	1/2	10.50	1.11	3.45	169	164	157	141	118	105			11.1	176	175	173	170	167	163	159	154	149	143	131	115	105	34.5
	3/8	7.96	1.13	3.42	128	124	120	107	92	80			11.3	134	132	130	128	126	123	120	116	112	108	98	87	80	34.2
6×4	7/8	15.96	1.11	2.97	257	248	238	214	181	160			11.1	267	264	260	254	246	238	230	220	210	199	173	160		29.7
	3/4	13.88	1.12	2.95	224	216	208	187	159	139			11.2	232	229	225	220	214	208	200	192	182	172	148	139		29.5
	5/8	11.72	1.13	2.92	189	183	177	158	136	117			11.3	196	193	190	186	181	175	168	161	153	144	124	117		29.2
	1/2	9.50	1.15	2.90	153	149	143	129	110	95			11.5	158	156	153	150	146	141	136	130	123	116	99	95		29.0
	3/8	7.22	1.17	2.87	117	113	109	99	86	72			11.7	121	118	116	114	111	107	103	98	93	88	75	72		28.7
5×3½	3/4	11.62	0.98	2.48	184	177	167	143	116				9.8	193	189	184	179	172	164	155	145	133	122	116			24.8
	5/8	9.84	0.99	2.45	155	149	142	122	98				9.9	163	160	155	150	144	138	130	121	112	101	98			24.5
	1/2	8.00	1.01	2.43	127	122	116	101	82	80			10.1	132	130	126	122	117	112	105	98	91	82	80			24.3
	3/8	6.10	1.02	2.40	97	94	89	78	63	61			10.2	101	99	96	93	89	85	80	74	68	61	61			24.0
4×3½	3/4	10.12	1.01	1.94	161	155	147	127	104	101			10.1	165	160	153	145	135	124	111	101						19.4
	5/8	8.60	1.03	1.91	137	132	126	110	90	86			10.3	140	135	130	122	114	104	93	86						19.1
	1/2	7.00	1.04	1.89	112	108	103	90	74	70			10.4	114	110	105	99	92	84	75	70						18.9
	3/8	5.34	1.06	1.88	86	83	79	70	58	53			10.6	87	84	80	75	70	64	57	53						18.8
4×3	5/8	7.96	0.85	1.99	123	116	108	86	80				8.5	130	126	121	115	108	100	90	80						19.9
	1/2	6.50	0.86	1.96	101	95	89	71	65				8.6	106	103	99	94	88	81	72	65						19.6
	3/8	4.96	0.88	1.94	77	73	67	56	50				8.8	81	78	75	71	66	61	55	50						19.4
	1/4	3.38	0.90	1.92	53	50	47	39	34				9.0	55	53	51	48	45	41	37	34						19.2
3½×3	5/8	7.34	0.87	1.72	114	108	100	82	73				8.7	118	113	107	100	91	80	73							17.2
	1/2	6.00	0.88	1.70	94	89	83	67	60				8.8	97	93	88	81	74	65	60							17.0
	3/8	4.60	0.90	1.67	72	68	64	53	46				9.0	74	71	67	62	56	49	46							16.7
	1/4	3.12	0.91	1.65	49	47	44	36	31				9.1	50	48	45	41	37	33	31							16.5
3×2½	3/8	3.84	0.74	1.48	57	53	48	38					7.4	61	57	53	48	41	38								14.8
	1/4	2.62	0.75	1.45	40	37	33	26					7.5	41	39	36	32	28	26								14.5
2½×2	3/8	3.10	0.58	1.27	42	37	31						5.8	48	44	40	34	31									12.7
	1/4	2.12	0.59	1.25	29	25	21						5.9	33	30	27	22	21									12.5

NOTE: Values to right of heavy lines indicate Max. Load for l/r = 120

By courtesy of American Institute of Steel Construction.

STRUCTURAL STEEL-MOMENT OF INERTIA-I

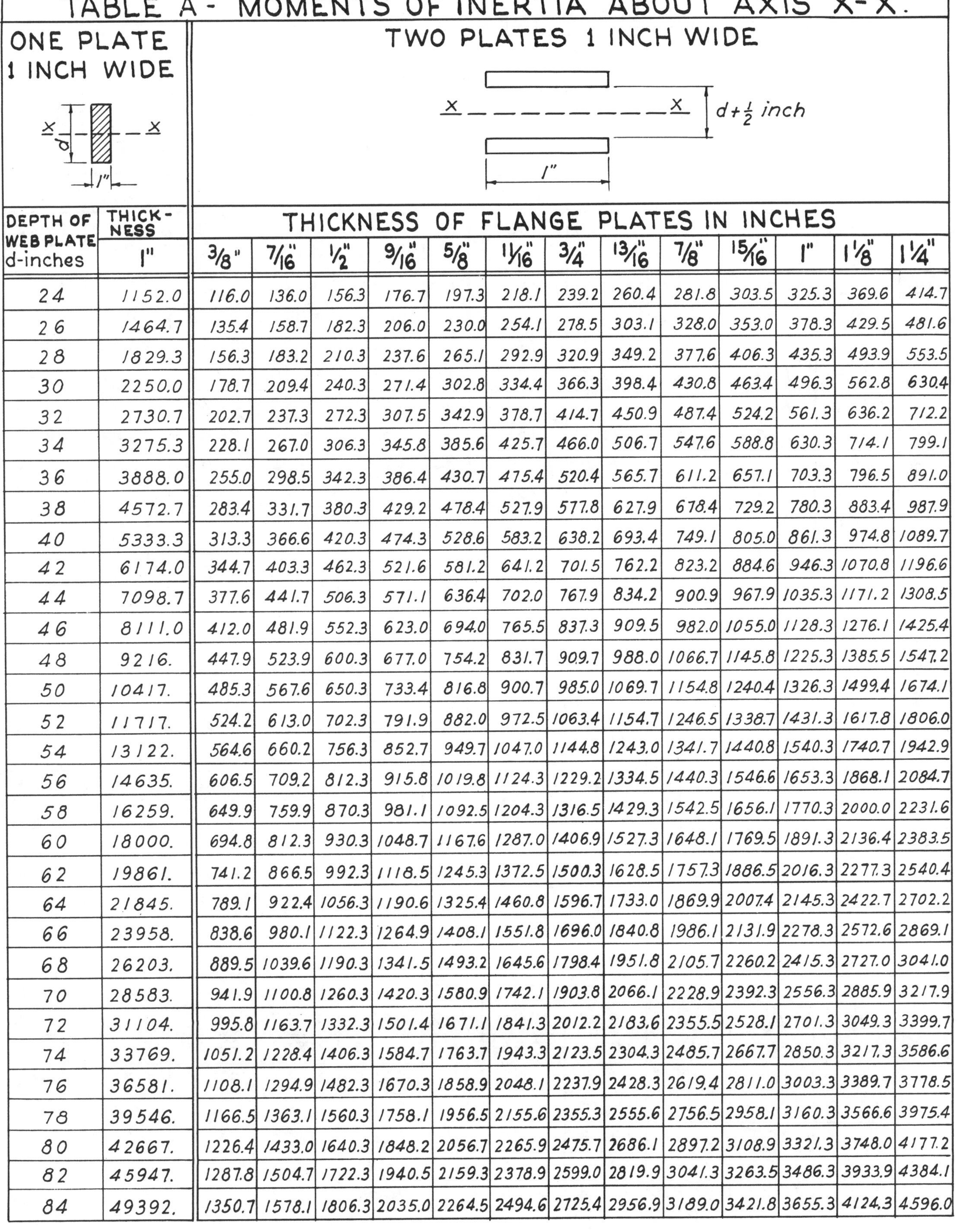
ONE PLATE 1 INCH WIDE
x
x
d
1"
TWO PLATES 1 INCH WIDE
x
x
d + 1/2 inch
1"

TABLE A- MOMENTS OF INERTIA ABOUT AXIS X-X.

ONE PLATE 1 INCH WIDE		TWO PLATES 1 INCH WIDE												
DEPTH OF WEB PLATE d-inches	THICKNESS 1"	THICKNESS OF FLANGE PLATES IN INCHES												
		3/8"	7/16"	1/2"	9/16"	5/8"	11/16"	3/4"	13/16"	7/8"	15/16"	1"	1 1/8"	1 1/4"
24	1152.0	116.0	136.0	156.3	176.7	197.3	218.1	239.2	260.4	281.8	303.5	325.3	369.6	414.7
26	1464.7	135.4	158.7	182.3	206.0	230.0	254.1	278.5	303.1	328.0	353.0	378.3	429.5	481.6
28	1829.3	156.3	183.2	210.3	237.6	265.1	292.9	320.9	349.2	377.6	406.3	435.3	493.9	553.5
30	2250.0	178.7	209.4	240.3	271.4	302.8	334.4	366.3	398.4	430.8	463.4	496.3	562.8	630.4
32	2730.7	202.7	237.3	272.3	307.5	342.9	378.7	414.7	450.9	487.4	524.2	561.3	636.2	712.2
34	3275.3	228.1	267.0	306.3	345.8	385.6	425.7	466.0	506.7	547.6	588.8	630.3	714.1	799.1
36	3888.0	255.0	298.5	342.3	386.4	430.7	475.4	520.4	565.7	611.2	657.1	703.3	796.5	891.0
38	4572.7	283.4	331.7	380.3	429.2	478.4	527.9	577.8	627.9	678.4	729.2	780.3	883.4	987.9
40	5333.3	313.3	366.6	420.3	474.3	528.6	583.2	638.2	693.4	749.1	805.0	861.3	974.8	1089.7
42	6174.0	344.7	403.3	462.3	521.6	581.2	641.2	701.5	762.2	823.2	884.6	946.3	1070.8	1196.6
44	7098.7	377.6	441.7	506.3	571.1	636.4	702.0	767.9	834.2	900.9	967.9	1035.3	1171.2	1308.5
46	8111.0	412.0	481.9	552.3	623.0	694.0	765.5	837.3	909.5	982.0	1055.0	1128.3	1276.1	1425.4
48	9216.	447.9	523.9	600.3	677.0	754.2	831.7	909.7	988.0	1066.7	1145.8	1225.3	1385.5	1547.2
50	10417.	485.3	567.6	650.3	733.4	816.8	900.7	985.0	1069.7	1154.8	1240.4	1326.3	1499.4	1674.1
52	11717.	524.2	613.0	702.3	791.9	882.0	972.5	1063.4	1154.7	1246.5	1338.7	1431.3	1617.8	1806.0
54	13122.	564.6	660.2	756.3	852.7	949.7	1047.0	1144.8	1243.0	1341.7	1440.8	1540.3	1740.7	1942.9
56	14635.	606.5	709.2	812.3	915.8	1019.8	1124.3	1229.2	1334.5	1440.3	1546.6	1653.3	1868.1	2084.7
58	16259.	649.9	759.9	870.3	981.1	1092.5	1204.3	1316.5	1429.3	1542.5	1656.1	1770.3	2000.0	2231.6
60	18000.	694.8	812.3	930.3	1048.7	1167.6	1287.0	1406.9	1527.3	1648.1	1769.5	1891.3	2136.4	2383.5
62	19861.	741.2	866.5	992.3	1118.5	1245.3	1372.5	1500.3	1628.5	1757.3	1886.5	2016.3	2277.3	2540.4
64	21845.	789.1	922.4	1056.3	1190.6	1325.4	1460.8	1596.7	1733.0	1869.9	2007.4	2145.3	2422.7	2702.2
66	23958.	838.6	980.1	1122.3	1264.9	1408.1	1551.8	1696.0	1840.8	1986.1	2131.9	2278.3	2572.6	2869.1
68	26203.	889.5	1039.6	1190.3	1341.5	1493.2	1645.6	1798.4	1951.8	2105.7	2260.2	2415.3	2727.0	3041.0
70	28583.	941.9	1100.8	1260.3	1420.3	1580.9	1742.1	1903.8	2066.1	2228.9	2392.3	2556.3	2885.9	3217.9
72	31104.	995.8	1163.7	1332.3	1501.4	1671.1	1841.3	2012.2	2183.6	2355.5	2528.1	2701.3	3049.3	3399.7
74	33769.	1051.2	1228.4	1406.3	1584.7	1763.7	1943.3	2123.5	2304.3	2485.7	2667.7	2850.3	3217.3	3586.6
76	36581.	1108.1	1294.9	1482.3	1670.3	1858.9	2048.1	2237.9	2428.3	2619.4	2811.0	3003.3	3389.7	3778.5
78	39546.	1166.5	1363.1	1560.3	1758.1	1956.5	2155.6	2355.3	2555.6	2756.5	2958.1	3160.3	3566.6	3975.4
80	42667.	1226.4	1433.0	1640.3	1848.2	2056.7	2265.9	2475.7	2686.1	2897.2	3108.9	3321.3	3748.0	4177.2
82	45947.	1287.8	1504.7	1722.3	1940.5	2159.3	2378.9	2599.0	2819.9	3041.3	3263.5	3486.3	3933.9	4384.1
84	49392.	1350.7	1578.1	1806.3	2035.0	2264.5	2494.6	2725.4	2956.9	3189.0	3421.8	3655.3	4124.3	4596.0

STRUCTURAL STEEL–MOMENT OF INERTIA–2

TABLE A

MOMENT OF INERTIA OF A PAIR OF UNIT AREAS ABOUT AXIS X-X

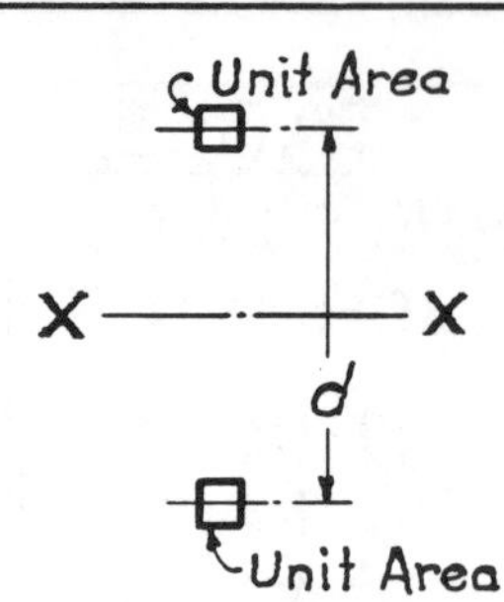

Unit Area
X
X
d
Unit Area

d	.0	.2	.4	.6	.8
10	50	52	54	56	58
11	61	63	65	67	70
12	72	74	77	79	82
13	85	87	90	92	95
14	98	101	104	107	110
15	113	116	119	122	125
16	128	131	134	138	141
17	145	148	151	155	158
18	162	166	169	173	177
19	181	184	188	192	196
20	200	204	208	212	216
21	221	225	229	233	238
22	242	246	251	255	260
23	265	269	274	278	283
24	288	293	298	303	308
25	313	318	323	328	333
26	338	343	348	354	359
27	365	370	375	381	386
28	392	398	403	409	415
29	421	426	432	438	444
30	450	456	462	468	474
32	512	518	525	531	538
34	578	585	592	598	606
36	648	655	662	670	677
38	722	730	737	745	753
40	800	808	816	824	832
42	882	890	899	907	916
44	968	977	986	995	1004
46	1058	1067	1076	1086	1095
48	1152	1162	1171	1181	1191
50	1250	1260	1270	1280	1290
52	1352	1362	1373	1383	1394
54	1458	1469	1480	1491	1502
56	1568	1579	1590	1602	1613

d	.0	.2	.4	.6	.8
58	1682	1694	1705	1717	1729
60	1800	1812	1824	1836	1848
62	1922	1934	1947	1959	1972
64	2048	2061	2074	2087	2100
66	2178	2191	2204	2218	2231
68	2312	2326	2339	2353	2367
70	2450	2464	2478	2492	2506
72	2592	2606	2621	2635	2650
74	2738	2753	2768	2783	2798
76	2888	2903	2918	2934	2949
78	3042	3058	3073	3089	3105
80	3200	3216	3232	3248	3264
82	3362	3378	3395	3411	3428
84	3528	3545	3562	3579	3596
86	3698	3715	3732	3750	3767
88	3872	3890	3907	3925	3943
90	4050	4068	4086	4104	4122
92	4232	4250	4269	4287	4306
94	4418	4437	4456	4475	4494
96	4608	4627	4646	4666	4685
98	4802	4822	4841	4861	4881
100	5000	5020	5040	5060	5080
102	5202	5222	5243	5263	5284
104	5408	5429	5450	5471	5492
106	5618	5639	5660	5682	5703
108	5832	5854	5875	5897	5919
110	6050	6072	6094	6116	6138
112	6272	6294	6317	6339	6362
114	6498	6521	6544	6567	6590
116	6728	6751	6774	6798	6821
118	6962	6986	7009	7033	7057
120	7200	7224	7248	7272	7296
122	7442	7466	7491	7515	7540
124	7688	7713	7738	7763	7788

STRUCTURAL STEEL-MOMENT OF INERTIA-3

TABLE A - MOMENTS OF INERTIA OF FOUR ANGLES.

8 X 8 ANGLES

THICKNESS	1/2"	9/16"	5/8"	11/16"	3/4"	13/16"	7/8"	15/16"	1"	1 1/16"	1 1/8"
AREA 4 Ls	31.00	34.72	38.44	42.12	45.76	49.36	52.92	56.48	60.00	63.48	66.92
d IN INCHES	MOMENTS OF INERTIA ABOUT AXIS X-X IN INCHES⁴ FOR VARIOUS DISTANCES BACK TO BACK OF ANGLES										
24	3332	3716	4097	4471	4828	5186	5536	5884	6213	6646	6871
26	3987	4448	4906	5355	5786	6217	6640	7060	7458	7861	8255
28	4703	5249	5792	6324	6835	7348	7850	8349	8824	9303	9773
30	5482	6120	6754	7377	7977	8577	9166	9751	10310	10872	11425
32	6323	7060	7794	8514	9209	9904	10587	11266	11915	12569	13210
34	7225	8070	8910	9736	10534	11331	12114	12893	13641	14392	15129
36	8190	9149	10103	11041	11950	12856	13748	14634	15486	16342	17183
38	9217	10298	11373	12431	13457	14480	15487	16488	17452	18419	19369
40	10306	11516	12720	13905	15056	16203	17331	18454	19538	20623	21690
42	11456	12803	14144	15464	16746	18024	19282	20534	21743	22954	24145
44	12669	14160	15645	17107	18528	19944	21338	22726	24069	25412	26733
46	13944	15586	17222	18833	20401	21963	23501	25032	26514	27997	29456
48	15280	17082	18877	20645	22366	24081	25769	27450	29080	30709	32312
50	16679	18647	20608	22540	24423	26291	28143	29982	31766	33548	35302
52	18140	20282	22416	24520	26571	28612	30623	32626	34571	36513	38426
54	19663	21986	24301	26584	28810	31026	33208	35384	37497	39606	41684
56	21247	23759	26263	28732	31141	33538	35900	38254	40542	42826	45075
58	22894	25602	28302	30964	33564	36149	38697	41237	43708	46172	48601
60	24603	27515	30418	33281	36078	38859	41600	44333	46994	49646	52260
62	26376	29497	32610	35681	38683	41668	44609	47543	50399	53247	56053
64	28206	31548	34880	38167	41381	44575	47724	50865	53925	56974	59980
66	30101	33669	37226	40736	44169	47581	50945	54300	57570	60829	64040
68	32058	35859	39650	43389	47049	50686	54272	57848	61336	64810	68235
70	34076	38118	42150	46127	50021	53889	57704	61509	65222	68919	72563
72	36157	40447	44727	48949	53084	57191	61242	65283	69227	73154	77025
74	38300	42846	47381	51856	56239	60592	64886	69170	73353	77516	81621
76	40505	45314	50111	54846	59485	64092	68636	73170	77598	82006	86351
78	42771	47851	52919	57921	62823	67690	72492	77283	81964	86622	91215
80	45100	50458	55804	61080	66252	71387	76453	81509	86450	91365	96213
82	47491	53134	58765	64323	69773	75183	80521	85847	91055	96235	101344
84	49943	55880	61803	67651	73385	79077	84694	90299	95781	101233	106609

STRUCTURAL STEEL - MOMENT OF INERTIA - 4

TABLE A - MOMENTS OF INERTIA OF FOUR ANGLES.

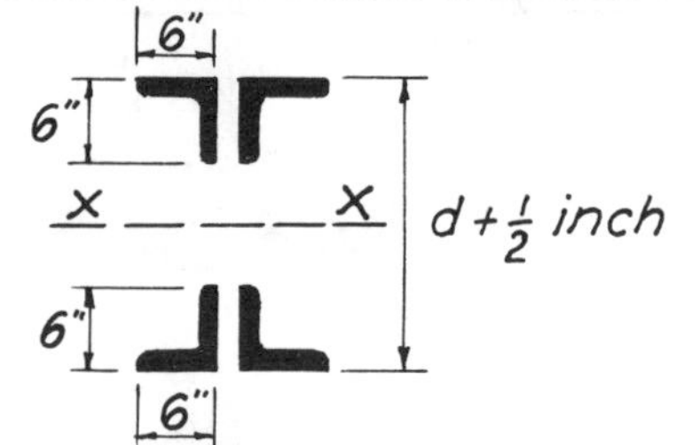
6"
6"
X
X
d + ½ inch
6"
6"

6 x 6 ANGLES

THICKNESS	3/8	7/16	1/2	9/16	5/8	11/16	3/4	13/16	7/8	15/16	1"
AREA 4 Ls	17.4	20.2	23.0	25.7	28.4	31.1	33.8	36.4	38.9	41.5	44.0
d IN INCHES	MOMENTS OF INERTIA ABOUT AXIS X-X IN INCHES^4 FOR VARIOUS DISTANCES BACK TO BACK OF ANGLES										
24	2025	2341	2649	2946	3244	3536	3813	4091	4362	4630	4892
26	2412	2790	3159	3513	3871	4220	4554	4887	5212	5535	5850
28	2835	3279	3714	4133	4555	4967	5362	5756	6141	6523	6896
30	3292	3809	4315	4804	5295	5776	6238	6698	7147	7594	8031
32	3784	4379	4962	5526	6093	6648	7181	7712	8232	8748	9253
34	4311	4990	5655	6299	6947	7581	8192	8799	9394	9985	10563
36	4873	5641	6395	7125	7858	8577	9270	9959	10634	11305	11962
38	5470	6333	7180	8001	8826	9635	10416	11192	11952	12708	13448
40	6102	7065	8011	8929	9851	10756	11629	12497	13347	14194	15022
42	6768	7838	8888	9909	10933	11938	12910	13875	14821	15762	16685
44	7470	8651	9812	10939	12072	13183	14259	15326	16372	17414	18435
46	8206	9505	10781	12022	13268	14490	15675	16850	18001	19149	20273
48	8977	10399	11796	13155	14520	15859	17158	18446	19709	20966	22200
50	9783	11334	12857	14341	15829	17291	18709	20115	21493	22867	24214
52	10624	12309	13964	15577	17196	18785	20327	21856	23356	24850	26316
54	11500	13325	15118	16865	18619	20341	22013	23671	25297	26917	28507
56	12411	14381	16317	18205	20099	21959	23767	25558	27315	29066	30785
58	13356	15478	17562	19596	21636	23639	25588	27518	29411	31299	33151
60	14337	16615	18853	21038	23230	25382	27476	29550	31585	33614	35605
62	15352	17792	20191	22532	24880	27187	29432	31655	33837	36013	38148
64	16402	19010	21574	24077	26588	29054	31456	33833	36167	38494	40778
66	17488	20269	23003	25673	28352	30984	33547	36084	38575	41058	43496
68	18608	21568	24478	27322	31073	32975	35706	38407	41060	43706	46303
70	19762	22907	25999	29022	32052	35029	37932	40803	43623	46436	49197
72	20952	24287	27567	30773	33987	37145	40225	43272	46264	49249	52179
74	22177	25708	29180	32575	35979	39324	42587	45814	48983	52145	55250
76	23436	27169	30839	34429	38027	41564	45015	48428	51780	55124	58408
78	24731	28670	32544	36334	40133	43867	47512	51115	54655	58186	61654
80	26060	30212	34295	38291	42296	46232	50075	53875	57607	61331	64989
82	27424	31794	36093	40299	44515	48660	52707	56707	60638	64559	68411
84	28823	33417	37936	42359	46792	51149	55405	59612	63746	67870	71921

STRUCTURAL STEEL - MOMENT OF INERTIA - 5

TABLE A - MOMENTS OF INERTIA OF FOUR ANGLES.

5 X 5 ANGLES

5" 5" 5" 5"
X — — — X
$d + \frac{1}{2}$ inch

THICKNESS	3/8	7/16	1/2	9/16	5/8	11/16	3/4	13/16	7/8	15/16	1"
AREA 4 Ls	14.44	16.72	19.00	21.24	23.44	25.60	27.76	29.88	31.92	34.00	36.00
d IN INCHES	MOMENTS OF INERTIA ABOUT AXIS X-X IN INCHES4 FOR VARIOUS DISTANCES BACK TO BACK OF ANGLES										
24	1738	2002	2269	2518	2769	3017	3263	3490	3721	3939	4154
26	2066	2381	2700	2997	3296	3593	3888	4160	4437	4697	4957
28	2423	2793	3168	3520	3870	4221	4568	4890	5216	5524	5830
30	2809	3239	3674	4081	4491	4899	5304	5679	6060	6424	6776
32	3224	3718	4218	4687	5159	5624	6095	6528	6967	7382	7794
34	3667	4230	4800	5336	5873	6409	6942	7437	7939	8413	8884
36	4140	4776	5420	6026	6635	7241	7844	8406	8974	9512	10046
38	4641	5355	6078	6759	7443	8125	8802	9434	10074	10679	11280
40	5171	5968	6775	7535	8298	9059	9816	10523	11237	11914	12586
42	5730	6614	7509	8353	9200	10045	10885	11671	12465	13217	13964
44	6318	7293	8281	9213	10154	11081	12010	12879	13756	14587	15414
46	6935	8006	9101	10116	11144	12169	13190	14194	15112	16056	16936
48	7583	8752	9939	11061	12186	13309	14426	15474	16541	17533	18531
50	8256	9531	10825	12048	13275	14499	15717	16862	18015	19108	20197
52	8959	10344	11750	13078	14401	15737	17065	18309	19562	20751	21931
54	9692	11191	12712	14151	15594	17033	18468	19816	21174	22462	23745
56	10453	12070	13712	15275	16823	18377	19926	21382	22850	24241	25627
58	11243	12984	14750	16422	18099	19773	21440	23017	24589	26088	27581
60	12062	13930	15826	17622	19422	21219	23009	24695	26392	28002	29607
62	12910	14910	16940	18864	20792	22714	24635	26441	28260	29985	31705
64	13787	15923	18093	20148	22199	24265	26315	28247	30191	32036	33875
66	14692	16970	19283	21475	23672	25865	28052	30113	32187	34155	36107
68	15627	18050	20511	22844	25183	27517	29844	32039	34246	36343	38431
70	16590	19163	21777	24275	26740	29219	31691	34024	36370	38597	40817
72	17583	20311	23081	25709	28343	30973	33595	36069	38557	40920	43275
74	18604	21491	24423	27206	29994	32777	35553	38174	40809	43311	47806
76	19654	22705	25804	28744	31692	34633	37568	40339	43125	45769	48408
78	20733	23952	27222	30325	33436	36541	39638	42564	45504	48296	51082
80	21841	25233	28678	31949	35227	38499	41763	44848	47948	50891	53828
82	22978	26547	30172	33615	37065	40509	43945	47192	50455	53554	56646
84	24143	27894	31704	35323	38949	42569	46181	49596	53027	56285	59536

STRUCTURAL STEEL – MOMENT OF INERTIA – 6

TABLE A – MOMENTS OF INERTIA OF FOUR ANGLES.

8 x 6 ANGLES

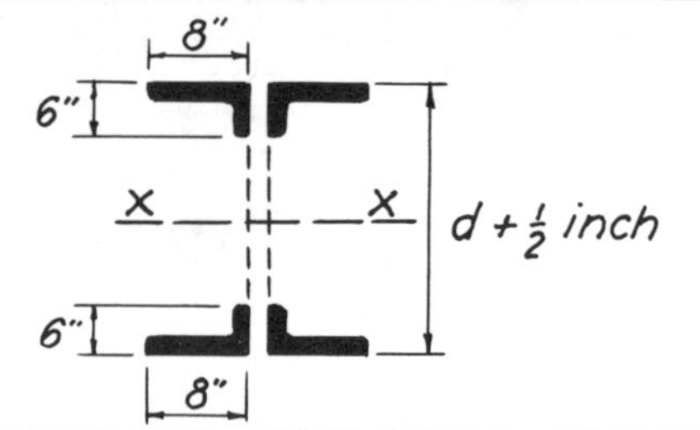

8"
6"
x
x
d + ½ inch
6"
8"

THICKNESS	7/16	1/2	9/16	5/8	11/16	3/4	13/16	7/8	15/16	1"
AREA 4 Ls	23.72	27.00	30.24	33.44	36.60	39.76	42.88	45.92	49.00	52.00
d IN INCHES	MOMENTS OF INERTIA ABOUT AXIS X-X IN INCHES^4 FOR VARIOUS DISTANCES BACK TO BACK OF ANGLES									
24	2844	3224	3591	3955	4312	4667	5004	5338	5674	5998
26	3380	3833	4271	4706	5133	5556	5961	6361	6764	7152
28	3963	4497	5012	5524	6027	6526	7004	7476	7951	8411
30	4594	5214	5813	6409	6994	7575	8133	8683	9237	9773
32	5273	5985	6675	7361	8034	8703	9347	9982	10621	11239
34	5999	6810	7598	8379	9147	9911	10647	11372	12103	12810
36	6772	7689	8580	9465	10334	11198	12033	12854	13682	14484
38	7593	8622	9624	10617	11594	12565	13505	14428	15360	16263
40	8461	9609	10727	11836	12927	14012	15062	16094	17136	18145
42	9376	10650	11892	13123	14333	15538	16705	17852	19010	20131
44	10339	11746	13116	14476	15812	17143	18434	19702	20981	22222
46	11350	12895	14402	15895	17365	18828	20249	21643	23051	24416
48	12408	14098	15747	17382	18990	20593	22149	23677	25219	26715
50	13513	15355	17153	18936	20689	22437	24135	25082	27485	29117
52	14666	16666	18620	20556	22462	24360	26207	28019	29848	31623
54	15866	18031	20147	22244	24307	26364	28365	30328	32310	34234
56	17114	19450	21735	23998	26226	28446	30608	32728	34870	36948
58	18409	20923	23383	25819	28217	30608	32938	35221	37528	39767
60	19751	22450	25091	27707	30282	32850	35353	37805	40284	42689
62	21141	24032	26860	29662	32420	35171	37853	40482	43137	45715
64	22579	25667	28690	31684	34632	37572	40440	43250	46089	48846
66	24064	27356	30580	33772	36916	40052	43112	46110	49139	52080
68	25596	29099	32530	35928	39274	42612	45870	49061	52287	55419
70	27176	30896	34541	38150	41705	45251	48714	52105	55532	58861
72	28803	32747	36613	40440	44209	47970	51644	55240	58876	62407
74	30478	34652	38745	42796	46787	50768	54659	58468	62318	66058
76	32200	36611	40937	45219	49437	53646	57760	61787	65858	69812
78	33969	38625	43190	47709	52161	56603	60947	65198	69495	73671
80	35786	40692	45503	50266	54958	59640	64220	68700	73231	77633
82	37651	42813	47877	52889	57828	62757	67578	72295	77065	81699
84	39562	44988	50312	55800	60771	65953	71022	75981	80997	85870

STRUCTURAL STEEL-MOMENT OF INERTIA-7

TABLE A - MOMENTS OF INERTIA OF FOUR ANGLES.

6 x 4 ANGLES

THICKNESS	3/8	7/16	1/2	9/16	5/8	11/16	3/4	13/16	7/8
AREA 4 Ls	14.44	16.72	19.00	21.24	23.44	25.60	27.76	29.88	31.92
d IN INCHES	MOMENTS OF INERTIA ABOUT AXIS X-X IN INCHES4 FOR VARIOUS DISTANCES BACK TO BACK OF ANGLES								
24	1867	2154	2434	2711	2981	3238	3498	3752	3993
26	2208	2548	2881	3210	3530	3837	4146	4448	4736
28	2578	2976	3366	3751	4127	4486	4850	5204	5542
30	2977	3437	3889	4335	4770	5187	5609	6020	6412
32	3404	3931	4450	4961	5460	5939	6423	6895	7346
34	3861	4459	5048	5629	6197	6743	7293	7830	8344
36	4346	5021	5685	6341	6981	7597	8219	8825	9406
38	4861	5616	6360	7094	7811	8503	9200	9880	10531
40	5404	6244	7073	7890	8689	9460	10236	10995	11720
42	5976	6906	7824	8729	9613	10468	11328	12169	12974
44	6577	7601	8613	9610	10585	11527	12476	13403	14291
46	7207	8330	9440	10533	11603	12638	13679	14697	15671
48	7866	9092	10305	11499	12668	13800	14938	16050	17116
50	8553	9887	11207	12508	13780	15012	16252	17464	18625
52	9270	10716	12148	13559	14939	16277	17622	18937	20197
54	10015	11579	13127	14652	16145	17592	19047	20470	21833
56	10789	12475	14144	15788	17397	18958	20527	22062	23533
58	11593	13404	15199	16967	18697	20376	22064	23715	25297
60	12425	14367	16292	18187	20043	21845	23655	25427	27125
62	13286	15363	17423	19451	21437	23365	25303	27199	29017
64	14175	16392	18592	20757	22877	24937	27006	29030	30972
66	15094	17455	19799	22105	24364	26559	28764	30922	32991
68	16042	18552	21043	23496	25898	28233	30578	32873	35074
70	17018	19682	22326	24929	27478	29958	32447	34885	37221
72	18023	20845	23647	26405	29106	31734	34372	36955	39432
74	19057	22042	25006	27923	30781	33561	36352	39086	41707
76	20121	23272	26403	29484	32502	35440	38388	41276	44045
78	21212	24536	27838	31087	34270	37370	40480	43526	46447
80	22333	25833	29311	32733	36086	39350	42627	45836	48914
82	23483	27164	30822	34421	37948	41383	44829	48205	51444
84	24662	28528	32370	36151	39857	43466	47087	50634	54037

STRUCTURAL STEEL – MOMENT OF INERTIA – 8

TABLE A - MOMENTS OF INERTIA OF FOUR ANGLES.

5 x 3½ ANGLES
LONG LEGS OUT

5″
3½″
X X $d + \frac{1}{2}$ inch.
3½″
5″

THICKNESS	5/16	3/8	7/16	1/2	9/16	5/8	11/16	3/4
AREA 4 Ls	10.24	12.20	14.12	16.00	17.88	19.68	21.48	23.24
d IN INCHES	MOMENTS OF INERTIA ABOUT AXIS X-X IN INCHES⁴ FOR VARIOUS DISTANCES BACK TO BACK OF ANGLES							
7	98	115	131	145	160	174	187	198
8	130	153	175	195	215	233	252	268
9	167	197	226	252	279	303	328	349
10	210	248	284	318	351	382	414	442
12	311	367	422	472	524	571	620	663
14	432	511	587	659	732	800	868	930
16	573	679	781	878	976	1067	1159	1244
18	735	872	1004	1129	1256	1374	1493	1604
20	918	1088	1254	1412	1571	1721	1871	2011
22	1121	1330	1533	1727	1922	2107	2291	2464
24	1344	1595	1840	2074	2309	2532	2754	2964
26	1588	1886	2175	2453	2732	2997	3260	3510
28	1852	2200	2539	2864	3190	3501	3809	4102
30	2137	2539	2930	3306	3684	4044	4401	4741
32	2443	2902	3350	3781	4214	4626	5036	5427
34	2768	3290	3798	4288	4780	5248	5714	6159
36	3115	3702	4275	4827	5381	5909	6435	6938
38	3482	4139	4779	5398	6019	6610	7199	7763
40	3869	4600	5312	6001	6692	7350	8005	8634
42	4277	5085	5873	6636	7400	8129	8855	9552
44	4705	5595	6463	7303	8145	8948	9748	10517
46	5153	6129	7080	8001	8925	9806	10683	11527
48	5623	6687	7726	8732	9741	10703	11662	12585
50	6112	7270	8400	9495	10593	11640	12684	13688
52	6623	7878	9103	10290	11481	12616	13748	14839
54	7153	8509	9833	11117	12404	13632	14856	16035
56	7704	9165	10592	11976	13363	14687	16006	17279
58	8276	9846	11379	12867	14358	15781	17199	18569
60	8868	10551	12194	13790	15389	16914	18436	19905
62	9481	11280	13038	14744	16455	18087	19715	21288

STRUCTURAL STEEL-MOMENT OF INERTIA -9

TABLE A - MOMENTS OF INERTIA OF FOUR ANGLES.

6 x 4 ANGLES

THICKNESS	3/8	7/16	1/2	9/16	5/8	11/16	3/4	13/16	7/8
AREA 4 Ls	14.44	16.72	19.00	21.24	23.44	25.60	27.76	29.88	31.92
d IN INCHES	MOMENTS OF INERTIA ABOUT AXIS X-X IN INCHES^4 FOR VARIOUS DISTANCES BACK TO BACK OF ANGLES								
24	1589	1832	2070	2304	2533	2749	2969	3183	3387
26	1901	2193	2479	2760	3035	3297	3562	3819	4066
28	2242	2587	2925	3259	3585	3895	4210	4516	4808
30	2612	3015	3410	3800	4181	4545	4913	5272	5614
32	3011	3476	3933	4384	4824	5246	5672	6087	6484
34	3439	3971	4494	5010	5514	5998	6486	6963	7418
36	3895	4499	5093	5679	6251	6801	7356	7898	8416
38	4381	5060	5730	6390	7035	7656	8282	8893	9478
40	4895	5655	6405	7143	7866	8562	9263	9948	10603
42	5438	6283	7118	7940	8743	9519	10300	11062	11793
44	6010	6945	7868	8778	9668	10527	11392	12237	13046
46	6611	7640	8657	9659	10529	11586	12539	13471	14363
48	7241	8369	9484	10583	11658	12697	13742	14764	15744
50	7900	9131	10349	11549	12722	13858	15001	16118	17189
52	8588	9927	11252	12557	13834	15071	16315	17531	18697
54	9304	10756	12193	13608	14993	16335	17685	19004	20269
56	10049	11618	13172	14701	16199	17651	19110	20537	21906
58	10824	12514	14189	15837	17452	19016	20591	22130	23606
60	11627	13443	15244	17016	18751	20434	22127	23782	25370
62	12459	14406	16336	18237	20097	21903	23719	25494	27197
64	13320	15420	17467	19500	21491	23423	25366	27266	29089
66	14210	16432	18636	20806	22931	24994	27069	29098	31045
68	15128	17495	19843	22154	24418	26617	28827	30989	33064
70	16076	18591	21088	23545	25952	28291	30641	32940	35147
72	17052	19721	22371	24978	27533	30016	32510	34951	37294
74	18058	20885	23692	26454	29160	31792	34435	37022	39505
76	19092	22082	25051	27972	30835	33619	36416	39152	41780
78	20155	23312	26447	29533	32556	35498	38452	41343	44118
80	21247	24576	27882	31136	34325	37427	40543	43593	46520
82	22368	25873	29355	32782	36140	39408	42690	45902	48986
84	23517	27203	30866	34470	38002	41440	44892	48272	51516

STRUCTURAL STEEL - MOMENT OF INERTIA-10

TABLE A- MOMENTS OF INERTIA OF FOUR ANGLES.

5 x 3½ ANGLES
SHORT LEGS OUT

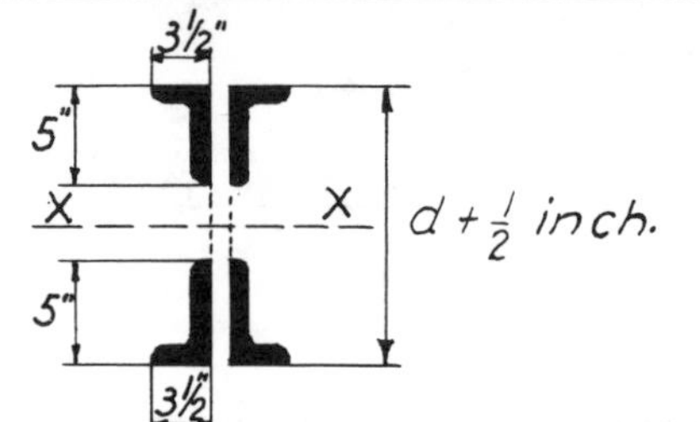
3½"
5"
X
X
d + ½ inch.
5"
3½"

THICKNESS	5/16	3/8	7/16	1/2	9/16	5/8	11/16	3/4
AREA 4 Ls	10.24	12.20	14.12	16.00	17.88	19.68	21.48	23.24
d IN INCHES	MOMENTS OF INERTIA ABOUT AXIS X-X IN INCHES^4 FOR VARIOUS DISTANCES BACK TO BACK OF ANGLES							
16	481	569	654	735	817	892	968	1038
18	627	743	856	962	1070	1170	1270	1363
20	794	942	1085	1221	1357	1487	1615	1735
22	982	1165	1342	1511	1682	1843	2003	2153
24	1190	1412	1628	1834	2042	2239	2434	2618
26	1419	1684	1942	2189	2438	2674	2908	3129
28	1668	1980	2284	2576	2869	3148	3424	3687
30	1937	2301	2655	2995	3337	3661	3984	4291
32	2227	2646	3054	3446	3840	4214	4587	4942
34	2538	3015	3481	3929	4379	4807	5233	5639
36	2869	3409	3936	4444	4953	5439	5921	6383
38	3220	3827	4419	4990	5564	6110	6653	7173
40	3592	4270	4931	5569	6210	6820	7427	8010
42	3984	4737	5471	6180	6892	7570	8245	8893
44	4397	5228	6039	6823	7610	8359	9105	9822
46	4831	5744	6636	7498	8363	9188	10009	10798
48	5284	6285	7260	8205	9152	10055	10955	11821
50	5759	6849	7913	8944	9977	10963	11945	12890
52	6254	7438	8594	9715	10838	11909	12977	14005
54	6769	8052	9304	10518	11734	12895	14052	15167
56	7305	8689	10041	11352	12667	13921	15170	16376
58	7861	9352	10807	12219	13635	14985	16332	17631
60	8437	10038	11601	13118	14639	16089	17536	18932
62	9035	10749	12424	14049	15678	17233	18783	20280
64	9652	11485	13274	15012	16753	18416	20073	21675
66	10291	12245	14153	16007	17864	19638	21406	23116
68	10949	13029	15060	17034	19011	20899	22782	24603
70	11628	13837	15996	18093	20194	22200	24201	26137
72	12328	14670	16959	19183	21412	23540	25663	27717
74	13048	15528	17951	20306	22666	24920	27168	29344
76	13789	16409	18971	21461	23956	26339	28716	31017

STRUCTURAL STEEL—BEARING & BASE PLATES

BEARING PLATES FOR STEEL BEAMS

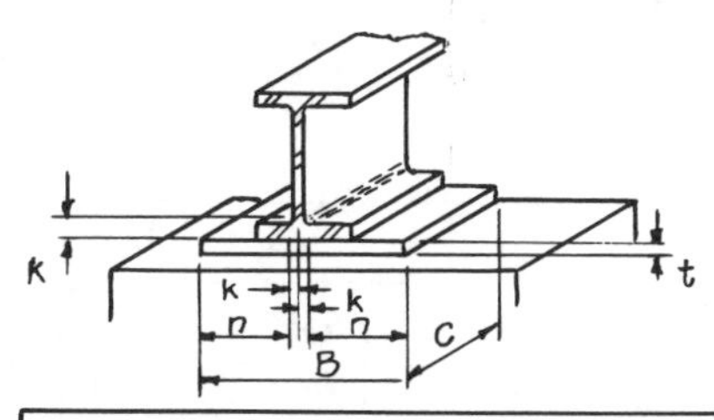

k
k
k
n
n
B
C
t

EXAMPLE: Given- 18WF 85; Load 40,000#; safe bearing pressure: 200#/□"
C = 10"; k = $1\frac{1}{2}$" (from table B)
To find size and thickness of plate.
$B = \frac{40{,}000}{10 \times 200} = 20"$ $n = \frac{B}{2} - k = \frac{20}{2} - 1\frac{1}{2} = 8\frac{1}{2}$
∴ From table → $t = 1\frac{1}{2}"$
Bearing Plate = 10" × 20" × $1\frac{1}{2}$"

Unit steel stress = 20,000#/□" (A.I.S.C)
p = Safe pressure on masonry #/□"
t = Thickness of plate in inches as computed by

$$t = n\sqrt{\frac{.15p}{1000}}$$

TABLE A - VALUES OF "n"

t \ p	70	100	110	125	150	200	250	300	325	350	400	450	500	550	600	625	750	875
3/8	3.7	3.1	2.9	2.7	2.5	2.2	1.94	1.77	1.70	1.64	1.54	1.45	1.37	1.31	1.26	1.23	1.12	1.04
1/2	4.9	4.1	3.9	3.7	3.3	2.9	2.6	2.36	2.27	2.19	2.05	1.92	1.83	1.74	1.67	1.64	1.49	1.38
5/8	6.1	5.1	4.9	4.6	4.2	3.6	3.2	2.9	2.84	2.73	2.55	2.40	2.29	2.17	2.08	2.04	1.87	1.73
3/4	7.3	6.1	5.9	5.5	5.0	4.3	3.9	3.5	3.4	3.28	3.06	2.89	2.74	2.62	2.50	2.45	2.24	2.07
7/8	8.5	7.2	6.8	6.4	5.8	5.1	4.5	4.1	3.97	3.83	3.58	3.37	3.20	3.05	2.92	2.87	2.61	2.41
1	9.8	8.2	7.8	7.3	6.7	5.8	5.2	4.7	4.53	4.37	4.08	3.86	3.66	3.49	3.34	3.27	2.98	2.76
1¼	12.2	10.2	9.8	9.1	8.4	7.2	6.5	5.9	5.68	5.46	5.14	4.82	4.56	4.38	4.17	4.08	3.74	3.45
1½	14.7	12.2	11.7	11.1	10.0	8.7	7.8	7.1	6.81	6.57	6.13	5.80	5.48	5.23	5.00	4.91	4.48	4.14
2	19.6	16.3	15.6	14.6	13.4	11.6	10.4	9.4	9.06	8.75	8.18	7.70	7.33	6.98	6.68	6.55	5.99	5.53
2½	24.4	20.5	19.5	18.3	16.7	14.5	12.9	11.8	11.4	10.9	10.2	9.65	9.15	8.72	8.35	8.18	7.45	6.92
3	29.3	24.5	23.4	21.9	20.0	17.3	15.5	14.2	13.6	13.1	12.3	11.6	11.0	10.5	10.0	9.8	9.0	8.3
3½	34.2	28.6	27.3	25.7	23.4	20.2	18.1	16.5	15.9	15.3	14.3	13.5	12.8	12.2	11.7	11.5	10.5	9.7
4	39.1	32.7	31.2	29.3	26.7	23.2	20.7	18.9	18.1	17.5	16.4	15.4	14.6	13.9	13.4	13.1	12.0	11.1

TABLE B - VALUES OF "k" FOR BEARING PLATE DESIGN

SIZE	k	SIZE	k	SIZE	k	SIZE	k	SIZE	k	SIZE	k	SIZE	k	SIZE	k
36WF300	2 13/16	30WF132	1 11/16	24WF 76	1¼	18WF 60	1 3/16	14WF 74	1⅜	12WF 22	¾	8WF 15	⅝	18I 70	1⅜
36WF230	2⅜	30WF108	1½	21WF142	1⅞	18WF 50	1 1/16	14WF 61	1¼	10WF 66	1¼	6WF 25	¾	15I 50	1¼
36WF194	2⅛	27WF177	2⅛	21WF112	1⅝	16WF 96	1⅝	14WF 53	1¼	10WF 49	1 1/16	6WF 16	11/16	12I 50	1 5/16
36WF150	1 13/16	27WF145	1 15/16	21WF 96	1 9/16	16WF 78	1½	14WF 38	1	10WF 39	1 1/16	5WF18.5	¾	12I 35	1⅛
33WF240	2 7/16	27WF114	1⅝	21WF 73	1 5/16	16WF 58	1¼	14WF 30	⅞	10WF 29	⅞	4WF 13	⅝	10I 35	1
33WF200	2 3/16	27WF 94	1 7/16	21WF 62	1 3/16	16WF 50	1⅛	12WF 85	1⅜	10WF 21	11/16			8I 23	⅞
33WF152	1⅞	24WF160	2	18WF114	1 11/16	16WF 36	15/16	12WF 58	1¼	10WF19	11/16	24I 120	1 15/16	7I 20	13/16
33WF130	1 11/16	24WF130	1¾	18WF 96	1½	14WF 136	1 11/16	12WF 50	1¼	8WF 35	⅞	24I 100	1⅝	6I 17.25	¾
30WF210	2 5/16	24WF100	1 9/16	18WF 85	1½	14WF 87	1 5/16	12WF 36	15/16	8WF 28	13/16	20I 95	1¾	5I 14.75	11/16
30WF172	2 1/16	24WF 94	1 7/16	18WF64	1¼	14WF 84	1⅜	12WF 27	13/16	8WF 20	11/16	20I 75	1 9/16	4I 9.5	⅝

COLUMN BASE PLATES

Moment figured about these lines

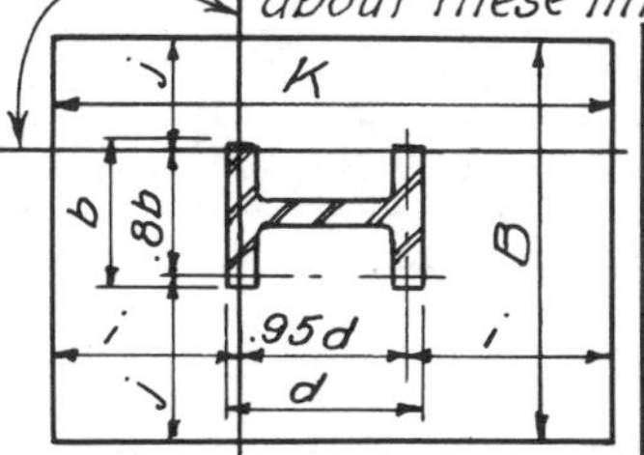

j
K
b
.8b
B
i
.95d
i
j
d

EXAMPLE:- Given: Load = 400,000#; safe bearing pressure p = 875#/□"; B = 20"; b = 12"; d = 12". To find size and thickness of plate. Area required = $\frac{400{,}000}{875} = 457$□"
$K = \frac{457}{20} = 23"$
$2j = B - .8b \therefore j = \frac{20 - 9.6}{2} = 5.2$ Use larger
$2i = K - .95d \therefore i = \frac{23 - 11.4}{2} = 5.8$ one i = 5.8
∴ From table $t = 2\frac{1}{4}"$ — B.P. = 20" × 23" × $2\frac{1}{4}$"

f = Unit steel stress 20,000#/□" (A.I.S.C.)
p = Safe pressure on masonry #/□"
t = Thickness of plate in inches as computed by $t = i \text{ or } j\sqrt{\frac{p}{6666}}$
(Use i or j whichever is greater)

TABLE C - VALUES OF i OR j FOR VARIOUS THICKNESSES OF PLATES

t \ p	110	200	250	500	625	750	875	t \ p	110	200	250	500	625	750	875
1	7.78	5.77	5.17	3.65	3.26	2.98	2.76	3½	27.2	20.2	18.1	12.8	11.4	10.4	9.65
1¼	9.73	7.22	6.45	4.57	4.07	3.73	3.45	4	31.1	23.1	20.6	14.6	13.0	11.9	11.0
1½	11.6	8.66	7.75	5.48	4.88	4.47	4.14	4½	35.0	25.9	23.2	16.4	14.7	13.4	12.4
1¾	13.6	10.1	9.04	6.38	5.71	5.22	4.83	5	38.9	28.8	25.8	18.2	16.3	14.9	13.8
2	15.5	11.5	10.3	7.30	6.52	5.96	5.52	5½	42.7	31.7	28.4	20.1	17.9	16.4	15.2
2¼	17.5	13.0	11.6	8.22	7.33	6.71	6.27	6	46.6	34.6	31.0	21.9	19.6	17.9	16.5
2½	19.4	14.4	12.9	9.13	8.15	7.45	6.90	6½	50.6	37.5	33.6	23.7	21.2	19.4	17.9
2¾	21.4	15.9	14.2	10.0	8.97	8.20	7.58	7	54.4	40.3	36.2	25.6	22.8	20.8	19.3
3	23.3	17.3	15.5	11.0	9.77	8.94	8.28	8	62.2	46.2	41.3	29.2	26.1	23.8	22.0

STRUCTURAL STEEL-ECCENTRIC CONNECTIONS

TABLE A - SAFE LOADS ON ECCENTRIC CONNECTIONS.

For $\frac{7}{8}$"ø Rivets-Max. Stress on Extreme Rivet = 9020# - Single Shear, A.I.S.C.-1947.
L = Moment Arm in Inches. N = Total No. of Rivets in group. P = Load in kips.

b = 3"

N \ L	1	2	3	4½	6	9	12	15	18
2	15.3	10.8	8.0	5.8	4.3	3.0	2.2	1.8	1.5
3	24.3	18.9	15.3	11.1	8.6	5.9	4.4	3.6	3.0
4	33.4	28.0	23.4	17.4	13.5	9.9	7.4	6.0	5.0
5	42.3	37.9	31.6	25.0	19.8	14.4	10.8	8.8	7.4
6	52.2	47.0	41.3	32.6	27.0	19.8	15.2	12.6	9.9
7	61.3	56.8	50.5	41.8	35.0	25.2	19.8	16.2	13.5
8	70.3	65.8	60.4	51.2	43.3	32.4	25.2	20.7	17.1
9	79.3	75.8	69.5	60.4	52.5	39.7	31.6	25.2	21.6
10	88.3	84.8	79.4	69.2	61.3	46.9	37.8	30.7	26.2
11	98.4	94.8	88.3	79.4	70.2	55.0	44.2	37.0	31.6
12	107.0	103.8	98.4	88.6	79.3	63.1	51.4	43.3	37.0

b = 3"

EXAMPLE #1: Given: L = 3", N = 10, $\frac{7}{8}$"ø Rivets, Single Shear. Required: P
From table P = 63.2.

N \ L	1	2	3	4½	6	9	12	15	18
4	30.6	21.6	16.0	11.6	8.6	6.0	4.4	3.6	3.0
6	48.6	37.8	30.6	22.2	17.2	11.8	8.8	7.2	6.0
8	66.8	56.0	46.8	34.8	27.0	19.8	14.6	12.0	10.0
10	84.6	75.8	63.2	50.0	39.6	28.8	21.6	17.6	14.8
12	104.4	94.0	82.6	65.2	54.0	39.6	30.4	25.2	19.8
14	122.6	113.6	101.0	83.6	70.0	50.4	39.6	32.4	27.0
16	140.6	131.6	120.8	102.4	86.6	64.8	50.4	41.4	34.2
18	158.6	151.6	139.0	120.8	105.0	79.4	63.2	50.4	43.2
20	176.6	169.6	158.8	138.4	121.6	93.8	75.6	61.4	52.4
22	196.8	189.6	176.6	158.8	140.4	110.0	88.4	74.0	63.2
24	214.0	207.6	196.8	177.2	158.2	126.2	102.8	86.6	74.0

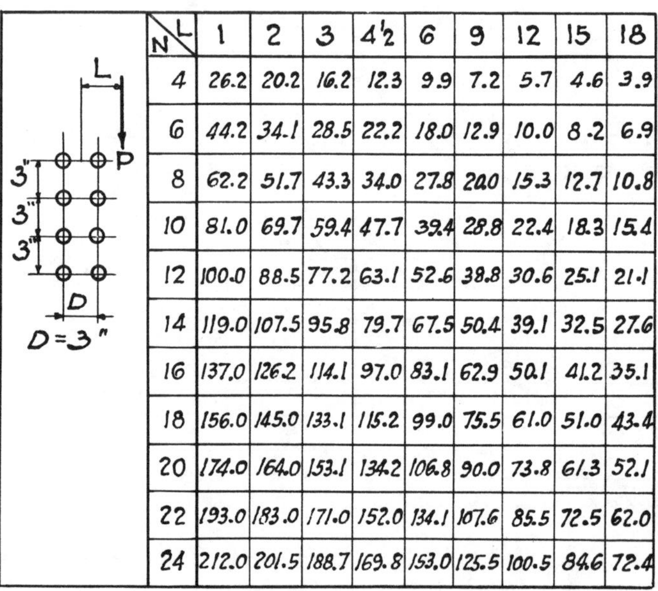

D = 3"

N \ L	1	2	3	4½	6	9	12	15	18
4	26.2	20.2	16.2	12.3	9.9	7.2	5.7	4.6	3.9
6	44.2	34.1	28.5	22.2	18.0	12.9	10.0	8.2	6.9
8	62.2	51.7	43.3	34.0	27.8	20.0	15.3	12.7	10.8
10	81.0	69.7	59.4	47.7	39.4	28.8	22.4	18.3	15.4
12	100.0	88.5	77.2	63.1	52.6	38.8	30.6	25.1	21.1
14	119.0	107.5	95.8	79.7	67.5	50.4	39.1	32.5	27.6
16	137.0	126.2	114.1	97.0	83.1	62.9	50.1	41.2	35.1
18	156.0	145.0	133.1	115.2	99.0	75.5	61.0	51.0	43.4
20	174.0	164.0	153.1	134.2	106.8	90.0	73.8	61.3	52.1
22	193.0	183.0	171.0	152.0	134.1	107.6	85.5	72.5	62.0
24	212.0	201.5	188.7	169.8	153.0	125.5	100.5	84.6	72.4

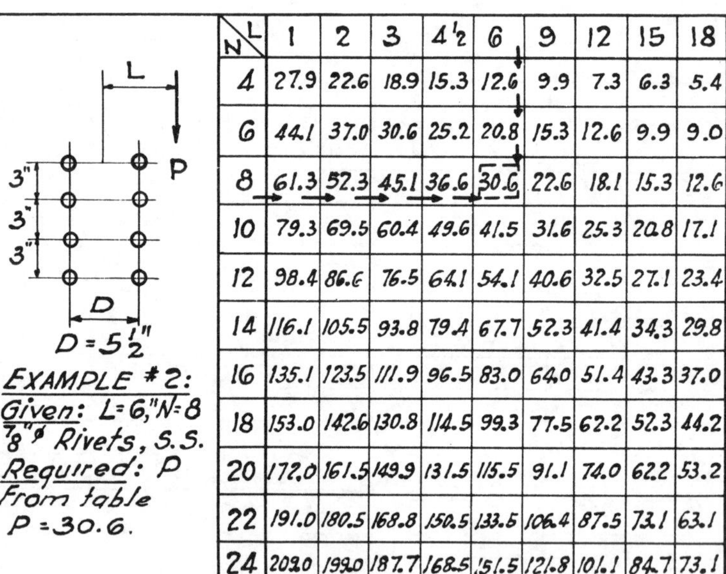

D = 5½"

EXAMPLE #2: Given: L = 6", N = 8, $\frac{7}{8}$"ø Rivets, S.S. Required: P
From table P = 30.6.

N \ L	1	2	3	4½	6	9	12	15	18
4	27.9	22.6	18.9	15.3	12.6	9.9	7.3	6.3	5.4
6	44.1	37.0	30.6	25.2	20.8	15.3	12.6	9.9	9.0
8	61.3	52.3	45.1	36.6	30.6	22.6	18.1	15.3	12.6
10	79.3	69.5	60.4	49.6	41.5	31.6	25.3	20.8	17.1
12	98.4	86.6	76.5	64.1	54.1	40.6	32.5	27.1	23.4
14	116.1	105.5	93.8	79.4	67.7	52.3	41.4	34.3	29.8
16	135.1	123.5	111.9	96.5	83.0	64.0	51.4	43.3	37.0
18	153.0	142.6	130.8	114.5	99.3	77.5	62.2	52.3	44.2
20	172.0	161.5	149.9	131.5	115.5	91.1	74.0	62.2	53.2
22	191.0	180.5	168.8	150.5	133.5	106.4	87.5	73.1	63.1
24	202.0	199.0	187.7	168.5	151.5	121.8	101.1	84.7	73.1

TABLE B - COEFFICIENT "K".

K × P in Table 21 gives loads for other conditions.
S.S. = Single Shear, t = Web or Plate thickness, D.S. = Double Shear.

RIVETS

	SINGLE BEAR.			S.S.	DOUBLE BEARING						D.S.
t	1/4	5/16	3/8		1/4	5/16	3/8	7/16	1/2	9/16	
1"ø	.88	1.11	1.33	1.30	1.10	1.39	1.66	1.94	2.21	2.49	2.61
7/8"ø	.77	.97	-	1.00	.97	1.21	1.45	1.70	1.94	-	2.00
3/4"ø	.66	-	-	.73	.83	1.04	1.25	1.45	-	-	1.47

BOLTS

	SINGLE BEAR.				S.S.	DOUBLE BEARING							D.S.
t	1/4	5/16	3/8	7/16		1/4	5/16	3/8	7/16	1/2	9/16	5/8	
1"ø	.55	.69	.83	-	.87	.69	.87	1.03	1.21	1.38	1.56	1.73	1.74
7/8"ø	.48	.60	-	-	.67	.60	.75	.91	1.05	1.21	-	-	1.33
3/4"ø	.41	-	-	-	.49	.51	.65	.77	.90	-	-	-	.98

EXAMPLE: In Example #2, if $\frac{7}{8}$"ø Rivets are in Double Bearing, A.I.S.C. Code, on ½" Plate find in table above K = 1.94, then K × P = 1.94 × 30.6 = 59.4 Kips.

Note: Where b = 2½ for Single line and b = 2½ and D = 2½ for Double line, use 85% of "P" for approximate values.

STRUCTURAL STEEL– LALLY COLUMNS-I

TABLE A – ROUND LALLY COLUMNS* – Page 1 of 2

SAFE LOADS IN 1000 LB.

Outside Diameter of Column, in inches		Standard Heavyweight														
		3-1/2	4	4-1/2	5	5-1/2	6-5/8	7-5/8	8-5/8	9-5/8	10-3/4	12-3/4	14	16	18	20
Weight, lb. per ft.		15	20	24	29	36	49	64	81	100	123	169	193	245	304	370
Section Modulus		1.9	2.7	3.7	4.9	6.4	10.2	14.6	20.8	29.2	38.0	57.5	72.3	111.7	130.2	165.5
Area of Steel, sq. in.		2.2	2.7	3.2	3.7	4.3	5.6	6.9	8.4	10.0	11.9	14.6	16.1	18.5	20.8	23.1
Area of Concrete, sq. in.		7.4	9.9	12.7	16.0	20.0	28.9	38.7	50.0	62.8	78.9	113.1	137.9	182.7	233.7	281.4
Radius of Gyration (Steel Only)		1.2	1.3	1.5	1.7	1.9	2.2	2.6	2.9	3.3	3.7	4.4	4.8	5.1	6.2	6.9
Unbraced Length of Column, feet	6	38	49	62	76	92	128	166	211	259	319	422	488	604	724	850
	7	35	46	59	72	88	124	161	206	254	313	415	482	595	717	840
	8	32	43	55	69	85	120	157	201	249	307	409	475	588	710	833
	9	29	40	52	65	81	116	152	196	243	301	402	468	580	704	826
	10	27	37	49	62	77	112	148	191	237	295	396	460	574	697	820
	11	24	34	46	58	73	108	143	186	232	289	389	454	565	685	810
	12		31	42	55	70	103	139	181	227	284	383	448	560	678	800
	13		28	39	51	66	99	134	176	221	278	376	440	552	670	789
	14			36	48	62	95	130	171	216	272	370	434	544	663	780
	15			33	44	58	91	125	166	210	266	363	427	537	655	773
	16				41	55	87	121	161	205	260	357	422	530	644	764
	17				37	51	83	116	156	199	254	350	415	524	637	756
	18					47	78	111	151	194	248	344	408	516	630	749
	19					43	74	107	146	189	242	337	400	508	623	740
	20						70	102	141	183	236	331	394	500	616	732
	30								90	105	146	266	325	427	540	652
	40										122	202	258	356	464	574
	50												189	282	385	490
	60														308	410
	70															330
Equivalent Direct Load to be added	10,000 in.-lb. Bending Moment†	14.8	12.9	11.5	10.3	9.4	7.9	7.0	6.1	5.2	4.8	4.2	3.7	3.1	3.0	2.9
	Standard Bracket‡	5.2	5.2	4.8	4.6	4.4	4.2	4.0	3.8	3.7	3.6	3.5	3.4	3.4	3.4	3.4
	Thru Plate§	1.5	1.5	1.5	1.5	1.5	1.5	1.6	1.6	1.6	1.6	1.6	1.6	1.6	1.7	1.7

Column formula: $P = (A_c + 12 A_s)(1600 - 24\, l/d)$, in which P = safe carrying capacity in pounds, A_c = area of concrete in square inches, A_s = area of steel in square inches, l = length of column inches, d = diameter of column in inches.

*Adapted from the Lally Column Co.

†For each 10,000 in.-lb. unbalanced moment on the column, add the number of kips shown to the sum of all the vertical loads; or

‡Multiply the average factor of 4 by unbalanced vertical load as a trial; then use correct factor from table; or

§Multiply by the factor 1½ for trial when thru-plate connections are used. See examples on p. 4-26. Based on a moment connection of the beam or girder for thru-plate.

STRUCTURAL STEEL-LALLY COLUMNS-2

TABLE A – ROUND LALLY COLUMNS* – Page 2 of 2

SAFE LOADS IN 1000 LB.

Outside Diameter of Column, inches		Extra Heavyweight							Double Extra Heavyweight				
		4	4-1/2	5-1/2	6-5/8	8-5/8	10-3/4	12-3/4	4	4-1/2	5-1/2	6-5/8	8-5/8
Weight, lb. per ft.		21	27	39	56	91	133	178	29	35	52	72	110
Section Modulus		3.4	4.7	8.3	13.6	27.3	46.4	68.9	5.0	7.0	12.7	20.8	39.7
Area of Steel, sq. in.		3.7	4.4	6.1	8.4	12.8	16.1	19.2	6.7	8.1	11.3	15.6	21.3
Area of Concrete, sq. in.		8.9	11.5	18.2	26.1	45.7	74.7	108.4	5.9	7.8	12.9	18.9	37.1
Radius of Gyration (Steel Only)		1.3	1.5	1.8	2.2	2.8	3.6	4.3	1.2	1.4	1.7	2.1	2.8
Unbraced Length of Column, feet	6	63	78	117	169	279	385	495	101	127	191	276	410
	7	59	74	113	164	272	378	490	95	121	184	268	400
	8	55	70	108	159	265	370	484	89	114	176	259	390
	9	51	66	104	153	258	364	476	82	107	169	250	380
	10	47	62	99	148	250	356	468	76	100	161	241	370
	11	43	57	94	143	244	349	460	70	94	152	232	360
	12	39	52	89	137	239	342	452	64	87	145	223	350
	13	35	47	84	132	232	335	444	58	81	137	214	340
	14		43	79	126	225	328	436		74	128	204	330
	15		39	74	121	219	321	428		67	120	195	320
	16			70	115	212	314	420			113	186	310
	17			65	110	206	306	412			105	177	300
	18			60	104	199	298	405			97	168	290
	19			55	99	192	292	397			90	159	280
	20			51	93	185	285	389			83	150	270
	24					157	255	360					230
	28					125	225	328					
	32						197	296					
	36						169	264					
	40							232					
Equivalent Direct Load to be added	10,000 in.-lb. Bending Moment†	12.9	11.4	9.2	7.8	6.0	4.7	4.1	14.3	12.4	9.8	8.4	6.2
	Standard Bracket‡	5.4	4.9	4.3	4.1	3.7	3.5	3.4	5.7	5.3	4.7	4.4	3.3
	Thru Plate§	1.5	1.5	1.5	1.6	1.6	1.6	1.6	1.7	1.6	1.6	1.6	1.6

*Adapted from the Lally Column Co.

†‡§ See p. 4-23. For examples see p. 4-26.

STRUCTURAL STEEL- LALLY COLUMNS-3

TABLE A – RECTANGULAR LALLY COLUMNS*

SAFE LOADS IN 1000 LB.

Exterior Dimensions, inches			Rectangular Column									
			$3\times3\times\frac{3}{16}$	$3\times3\frac{1}{4}$	$3\frac{1}{2}\times3\frac{1}{2}\times\frac{3}{16}$	$3\frac{1}{2}\times3\frac{1}{2}\times\frac{1}{4}$	$4\times4\times\frac{3}{16}$	$4\times4\times\frac{1}{4}$	$4\times6\times\frac{1}{4}$	$4\times8\times\frac{1}{4}$	$5\times5\times\frac{1}{4}$	$6\times6\times\frac{1}{4}$
Weight, lb. per ft.			14	16	18	20	23	26	36	46	37	50
Section Modulus			2.1	2.5	3.0	3.6	4.1	4.9	9.2	14.5	8.3	12.7
Area of Steel, sq. in.			2.1	2.8	2.5	3.3	2.9	3.8	4.8	5.8	4.8	5.8
Area of Concrete, sq. in.			6.9	6.3	9.8	9.0	13.1	12.3	19.3	26.3	20.3	30.3
Least Radius of Gyration (Steel Only)			1.2	1.1	1.4	1.3	1.6	1.5	1.6	1.7	1.9	2.3
Unbraced Length of Column, feet		6	37	43	47	56	59	71	94	118	102	135
		7	34	40	45	53	56	67	90	112	98	131
		8	31	37	43	50	53	64	85	106	94	127
		9	28	33	40	47	51	60	80	100	90	123
		10	25	30	37	43	48	57	76	94	86	119
		11	23	27	34	40	45	53	71	88	83	115
		12	20	24	31	37	42	50	66	83	80	111
		13			28	33	39	46	62	77	76	107
		14			26	30	37	43	57	71	72	103
		15					34	40	53	65	68	100
		16					31	36	48	60	64	96
		17									61	92
		18									57	88
		19									53	84
		20									50	80
		21										76
		22										72
		23										68
		24										64
Equivalent Direct Load to Be Added	Axis A–A	10,000 in.-lb. Bending Moment†	12.8	13.0	11.0	11.0	9.7	9.5	9.1	8.8	7.8	6.5
		Standard Bracket‡	4.5	4.6	5.0	4.2	3.9	3.7	3.6	3.4	3.7	3.2
		Thru Plate§	1.2	1.2	1.2	1.2	1.2	1.2	1.2	1.2	1.2	1.2
	Axis B–B	10,000 in.-lb. Bending Moment†	12.8	13.0	11.0	11.0	9.7	9.5	6.9	5.5	7.8	6.5
		Standard Bracket‡	4.5	4.6	5.0	4.2	3.9	3.7	3.5	3.3	3.7	3.2
		Thru Plate§	1.2	1.2	1.2	1.2	1.2	1.2	1.3	1.3	1.2	1.2

Section modulus given on stronger axis.

*Adapted from the Lally Column Co.

†‡§ See p. 4-23. For example see p. 4-26.

STRUCTURAL STEEL– LALLY COLUMNS–4

EXAMPLES OF DESIGN OF LALLY COLUMNS WITH ECCENTRIC LOADS*

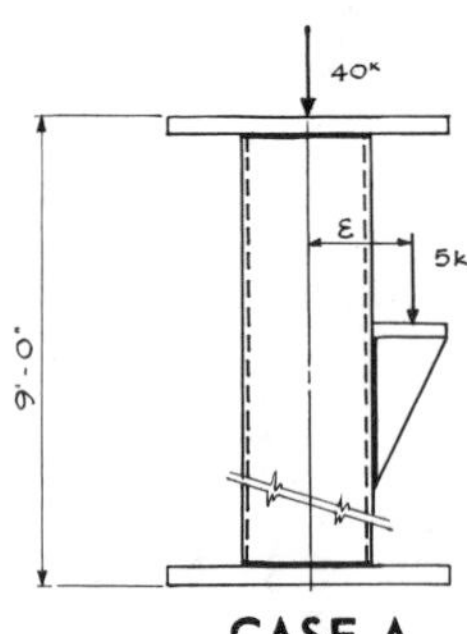

40k
ε
5k
9'-0"

CASE A
STANDARD BRACKET, SINGLE ECCENTRIC LOAD

Direct load = 40 kips + 5 kips = 45 kips
Eccentric load = 5 kips
Equivalent direct load = 5 kips × 4 = 20 kips
Equivalent total load = 45 kips + 20 kips = 65 kips
Requires 5" ϕ H. W. Lally column.
Actual factor for 5"ϕ column is 4.6
Equivalent to 5 kips × 4.6 = 23.0 kips
Total load = 68 kips

NOTE: $\epsilon = \frac{1}{2}d + 2$; $\epsilon' = 0.3d$.

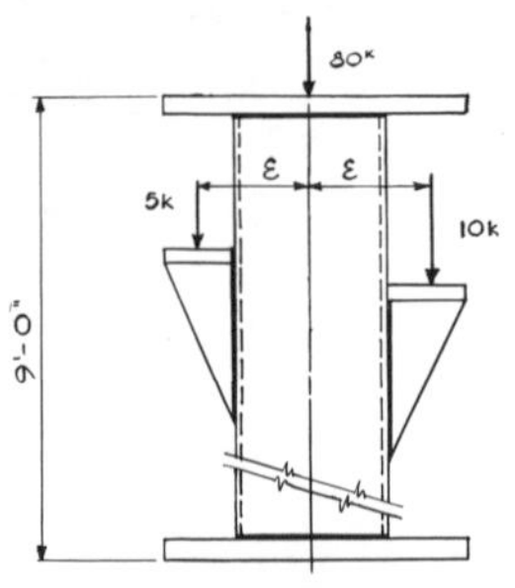

80k
ε
ε
5k
10k
9'-0"

CASE B
STANDARD BRACKET, 2 ECCENTRIC LOADS

Direct load = 80 kips + 10 kips + 5 kips = 95 kips
Resultant eccentric load = 10 kips – 5 kips = 5 kips
Equivalent direct load = 5 kips × 4 = 20 kips
Equivalent total load = 95 kips + 20 kips = 115 kips
Requires 6-5/8" ϕ H. W. Lally column.
Actual factor for 6-5/8" ϕ column is 4.2,
Equivalent to 5 kips × 4.2 = 21.0 kips
Total load = 116 kips

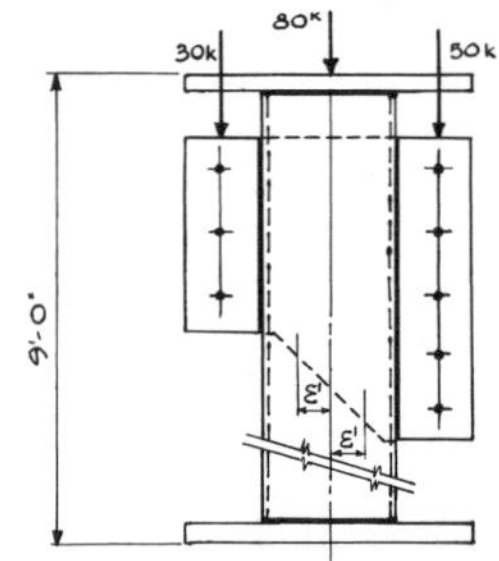

80k
30k
50k
9'-0"

CASE C
THRU PLATE, 2 ECCENTRIC LOADS

Direct load = 80 kips + 50 kips + 30 kips = 160 kips
Resultant eccentric load = 50 kips – 30 kips = 20 kips
Equivalent direct load = 20 kips × 1.5 = 30 kips
Equivalent total load = 160 kips + 30 kips = 190 kips
Requires 8-5/8" ϕ H. W. Lally column.
Actual factor for 8-5/8" ϕ column is 1.6,
Equivalent to 20 kips × 1.6 = 32 kips
Total load = 192 kips

TABLE A – STANDARD BASES AND CAPS FOR LALLY COLUMNS

STANDARD BASES

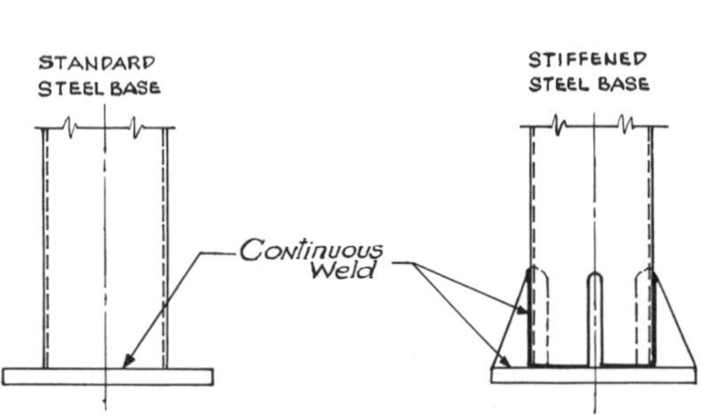

STANDARD STEEL BASE
STIFFENED STEEL BASE
Continuous Weld

STANDARD CAPS

STANDARD CAP

STIFFENED CAP

"G"

"D" "D"

Continuous Weld

Column Diameter, in.	Size of Base Plate, in.	Safe Load, kips	Thickness of Base Plate, in.	
			Standard Base	Stiffened Base
3-1/2	8 x 8	32.0	5/8	1/2
4	9 x 9	40.5	3/4	1/2
4-1/2	10 x 10	50.0	7/8	1/2
5	12 x 12	72.0	1	1/2
5-1/2	14 x 14	98.0	1-1/4	3/4
6-5/8	16 x 16	128.0	1-1/4	3/4
7-5/8	18 x 18	162.0	1-1/2	3/4
8-5/8	20 x 20	200.0	1-3/4	3/4
9-5/8	22 x 22	242.0	1-3/4	3/4
10-3/4	24 x 24	288.0	2	7/8
12-3/4	28 x 28	392.0	2-1/4	7/8

Thickness of Cap Plate, in.	Distance D, in.	
	Standard Cap	Stiffened Cap
1/2	3-1/4	4-1/4
1/2	3-1/2	4-1/2
1/2	3-3/4	4-3/4
5/8	4	5
5/8	4-1/4	5-1/4
3/4	4-3/4	6-1/4
3/4	5	6-3/4
3/4	5-3/4	7-1/4
3/4	6-1/4	7-3/4
3/4	7	8-1/4
3/4	8	9

NOTES: (1) Distance G equals the standard gage of the supported beam. (2) The width of plate is determined by the beam flange. (3) Distance D is constant for 1-, 2-, 3-and 4-way caps. (4) Caps may be used for steel, wood, or concrete construction. Safe loads are based on 500 p.s.i. bearing on concrete.
*Adapted from the Lally Column Co.

STRUCTURAL STEEL- PIPE COLUMNS

PIPE COLUMNS*
SAFE LOADS (in 1000 lb.)

TABLE A — STANDARD

Unbraced Length, ft.	Nominal Diameter – Weight per ft.											
	12		10			8		6	5	4	3½	3
	49.56	43.77	40.48	34.24	31.20	28.55	24.70	18.97	14.62	10.79	9.11	7.58
6	246	217	200	169	154	140	121	92	70	50	42	33
8	244	216	199	168	153	138	120	90	68	47	38	30
10	243	214	196	166	151	136	118	86	64	44	35	26
12	240	212	194	164	149	133	115	82	61	40	30	
14	237	210	190	161	147	129	112	79	56	34		
16	234	207	187	158	144	125	109	74	51			
18	231	204	182	154	141	121	105	69	45			
20	227	200	178	151	137	115	100	63				
22	222	196	172	146	133	109	95	56				

TABLE B — EXTRA STRONG

Unbraced Length, ft.	Nominal Diameter – Weight per ft.							
	12	10	8	6	5	4	3½	3
	65.42	54.74	43.39	28.57	20.78	14.98	12.51	10.25
6	325	271	213	139	99	70	58	45
8	323	268	210	135	96	65	53	40
10	320	265	206	131	91	60	47	35
12	317	261	201	125	85	54	40	
14	313	257	196	119	79	47		
16	309	252	189	112	71			
18	304	246	182	103	63			
20	299	239	173	94				
22	293	232	164	84				
24	286	224	155					

TABLE C — DOUBLE EXTRA STRONG

Unbraced Length, ft.	Nominal Diameter – Weight per ft.					
	8	6	5	4	3½	3
	72.42	53.16	38.55	27.54	22.85	18.58
6	355	257	183	130	103	80
8	350	249	176	118	93	70
10	343	240	165	108	82	59
12	334	228	154	94	68	
14	324	213	140			
16	312	200	125			
18	299	182				
20	284	163				
22	269					
24	250					
26	230					

Note: Loads are for maximum height in feet for $\frac{l}{r}$ = 120 or less.

*By Courtesy of the American Institute of Steel Construction.

STRUCTURAL STEEL-OPEN WEB JOISTS-I*

TABLE A – ALLOWABLE TOTAL SAFE LOADS FOR SHORT SPAN JOISTS IN POUNDS PER LINEAR FOOT

Joist Designation	8S2	10S2	10S3	10S4	12S3	12S4	12S5	12S6	14S4	14S5	14S6	14S7	16S5	16S6	16S7	16S8	18S6	18S7	18S8	20S6	20S7	20S8	22S7	22S8	24S8
Depth, in.†	8	10	10	10	12	12	12	12	14	14	14	14	16	16	16	16	18	18	18	20	20	20	22	22	24
Resisting Moment, inch-kips	52.5	65.0	83.0	103.0	100.0	125.0	149.0	182.0	147.0	175.0	214.0	251.0	202.0	242.0	290.0	336.0	272.0	328.0	380.0	295.0	358.0	420.0	385.0	456.0	500.0
Approximate Weight,‡ lb. per linear ft.	4.0	4.5	5.0	6.0	5.0	6.0	6.5	8.0	6.5	7.5	8.0	9.5	7.5	8.5	10.0	11.5	9.0	10.0	11.5	9.5	10.5	12.0	11.5	12.5	13.0
Maximum End Reaction, lb.	1900	1900	2000	2200	2200	2300	2500	2800	2600	2900	3100	3400	3000	3300	3600	3900	3600	3800	4100	3700	3900	4200	4000	4300	4500
Span, ft.																									
4	950	950																							
5	760	760	800	880	880																				
6	633	633	667	735	735																				
7	543	543	572	630	630																				
8	475	475	500	550	550																				
9	422	422	445	488	488																				
10	350	380	400	440	440																				
11	289	345	364	400	400	418	455	508																	
12	243	301	333	367	367	383	417	467																	
13	207	256	308	338	338	354	385	431																	
14	179	221	282	314	314	329	357	400	371	414	443	486	428	472	515	556									
15	156	193	246	293	293	307	333	373	347	387	413	453	400	440	480	520									
16	137	169	216	268	260	288	313	350	325	363	388	425	375	413	450	488									
17		150	191	238	231	271	294	329	306	341	365	400	353	388	424	459									
18		134	171	212	206	256	278	311	289	322	344	378	333	367	400	433	400	422	456						
19		120	153	190	185	231	263	295	271	305	326	358	316	347	379	411	379	400	432						
20		108	138	172	167	208	248	280	245	290	310	340	300	330	360	390	360	380	410	370	390	420			
21					151	189	225	267	222	265	295	324	286	314	343	371	343	362	390	352	371	400			
22					138	172	205	251	202	241	282	309	273	300	327	355	327	345	373	336	355	382	364	391	410
23					126	158	188	229	185	221	270	296	255	287	313	339	313	330	357	322	339	365	348	374	392
24					116	145	172	211	170	203	248	283	234	275	300	325	300	317	342	308	325	350	333	358	375
25									157	187	228	268	215	258	288	312	288	304	328	296	312	336	320	344	360
26									145	173	211	248	199	239	277	300	268	292	315	285	300	323	308	331	346
27									134	160	196	230	185	221	265	289	249	281	304	270	289	311	296	319	333
28									125	149	182	213	172	206	247	279	231	271	293	251	279	300	286	307	321
29													160	192	230	266	216	260	283	234	269	290	276	297	310
30													150	179	215	249	201	243	273	219	260	280	267	287	300
31													140	168	201	233	189	228	264	205	248	271	258	277	290
32													132	158	189	219	177	214	247	192	233	263	250	269	281
33																	167	201	233	181	219	255	236	261	273
34																	157	189	219	170	206	242	222	253	265
35																	148	179	207	161	195	229	210	246	257
36																	140	169	195	152	184	216	198	235	250
37																				144	174	205	187	222	243
38																				136	165	194	178	211	231
39																				129	157	184	169	200	219
40																				123	149	175	160	190	208
41																							153	181	198
42																							146	172	189
43																							139	164	180
44																							133	157	172
45																									165
46																									158
47																									151
48																									145

TABLE B – BRIDGING

Clear Span	Number of Lines of Bridging
Up to 14'	One row near center.
14' to 21'	Two rows placed at approximately 1/3 points of span.
21' to 32'	Three rows placed at approximately 1/4 points of span.
32' to 40'	Four rows placed at approximately 1/5 points of span.
40' to 48'	Five rows placed at approximately 1/6 points of span.

TABLE C – TOTAL ALLOWABLE UNIFORM LOAD IN POUNDS PER LINEAR FOOT OF EXTENDED END

Unsupported Length of Extended End	Type of Standard Extended Ends 2½" Depth			
	No. 1	No. 2	No. 3	No. 4
2' - 6"	250	250	250	250
3' - 0"	231	250	250	250
3' - 6"	170	207	243	250
4' - 0"	131	159	187	234
4' - 6"	104	126	148	186
5' - 0"		102	120	150
5' - 6"			100	123

NOTE: Loads above heavy zigzag line are governed by shear.
Spacing for above joists assumed at 12 in. o.c. In floors, it is recommended that the maximum spacing be 24 in., but in no case greater than the safe span of the floor slab or deck. In roofs, the spacing shall not exceed the safe span of the roof deck.

* Adapted by the Steel Joist Institute, August 20, 1929, and revised to June 26, 1958. Effective January 1, 1959.
† Indicates nominal depth of steel joists only.
‡ Approximate weights per linear foot of joists includes accessories, but does not include wood nailer strip.

STRUCTURAL STEEL-OPEN WEB JOISTS-2

TABLE A – LONGSPAN STEEL JOISTS* ALLOWABLE TOTAL SAFE LOADS (in lb. per linear ft. of Span)

Joist Designation	Approximate Weight, lb. per linear ft.	Depth, in.	Maximum End Reaction	Clear Opening or Net Span, ft.															
				25	26	27	28	29	30	31	32	33	34	35	36				
18L02	13	18	3,632	283	267	251	237	224	211	200	190	180	171	163	155				
18L03	14	18	4,094	319	300	283	267	253	239	227	215	204	194	185	176				
18L04	16	18	4,941	385	361	339	319	301	284	268	254	241	229	217	207				
18L05	17	18	5,364	418	394	372	351	331	313	298	282	268	254	242	231				
18L06	19	18	6,417	500	469	440	414	391	369	349	330	313	297	282	268				
18L07	21	18	6,880	536	516	486	458	432	408	386	365	346	329	313	296				
18L08	23	18	7,482	583	561	541	522	491	463	437	414	392	371	352	335				
18L09	25	18	7,697	600	577	556	537	519	502	474	449	425	403	383	364				
18L10	27	18	8,265	644	620	597	577	557	539	522	493	466	442	419	398				
18L11	29	18	8,753	682	656	633	611	590	571	553	536	520	493	469	445				
18L12	31	18	9,166	714	687	663	639	618	598	579	561	544	529	514	488				
				25	26	27	28	29	30	31	32	33	34	35	36	37	38	39	40
20L03	14	20	4,235	330	312	296	280	266	252	240	228	217	207	197	188	180	172	164	157
20L04	16	20	5,185	404	381	360	340	320	304	288	273	259	247	235	224	213	204	194	186
20L05	17	20	5,557	433	409	387	367	348	331	314	299	285	271	259	247	236	226	216	207
20L06	19	20	6,763	527	496	467	441	417	395	374	355	337	320	305	290	277	264	252	241
20L07	21	20	7,110	554	533	514	486	459	435	412	391	372	354	337	321	306	292	279	267
20L08	23	20	7,832	610	587	566	546	528	499	472	447	425	403	383	365	348	332	317	303
20L09	25	20	8,107	632	608	586	566	547	529	512	485	460	437	416	396	377	360	344	329
20L10	27	20	8,568	668	643	619	598	578	559	541	525	509	483	459	436	415	396	378	361
20L11	29	20	9,095	709	682	657	634	613	593	574	557	540	525	510	485	462	441	421	403
20L12	31	20	9,605	748	720	694	670	647	626	607	588	571	554	539	524	510	486	463	442
20L13	36	20	10,533	821	790	761	735	710	687	665	645	626	608	591	575	559	545	531	518
				33	34	35	36	37	38	39	40	41	42	43	44	45	46	47	48
24L04	16	24	4,798	285	272	260	249	238	228	219	210	201	193	186	179	172	166	160	154
24L05	17	24	5,117	304	292	279	268	257	247	237	228	219	211	203	196	189	182	175	169
24L06	19	24	6,245	371	354	339	324	310	297	284	273	262	251	242	232	224	215	207	200
24L07	21	24	6,868	408	390	373	357	342	328	314	301	289	278	267	257	248	238	230	222
24L08	23	24	7,996	475	453	432	412	394	377	361	346	332	318	306	294	283	272	262	252
24L09	25	24	8,652	514	490	468	447	427	409	391	375	360	345	331	319	306	295	284	274
24L10	27	24	9,345	555	539	524	500	477	456	436	417	400	383	368	353	339	326	314	302
24L11	29	24	9,686	575	559	543	528	514	501	480	460	441	424	407	391	376	362	349	336
24L12	31	24	10,431	619	601	585	569	554	539	526	513	491	471	452	434	417	401	386	371
24L13	36	24	11,479	682	662	644	626	610	594	579	565	551	538	526	514	494	475	457	440
24L14	38	24	12,087	718	697	678	659	642	625	609	594	580	567	554	541	529	518	496	476
				41	42	43	44	45	46	47	48	49	50	51	52	53	54	55	56
28L06	19	28	5,875	282	272	262	253	244	235	227	220	212	205	199	192	186	180	175	170
28L07	21	28	6,479	311	300	289	279	269	260	251	243	235	227	220	213	206	200	194	188
28L08	23	28	7,542	362	348	335	323	312	300	290	280	270	261	252	244	236	229	221	215
28L09	25	28	8,167	392	377	363	350	337	325	314	303	293	283	274	265	256	248	240	233
28L10	27	28	9,208	442	425	408	393	378	365	351	339	327	316	305	295	285	276	267	259
28L11	29	28	10,000	480	463	445	429	414	399	385	372	359	347	336	325	314	304	295	286
28L12	31	28	10,960	526	514	502	483	465	448	432	417	402	388	375	363	351	339	328	318
28L13	36	28	12,202	586	572	559	546	534	523	512	494	477	460	445	430	415	402	389	377
28L14	38	28	12,793	614	600	586	573	561	549	537	526	515	505	488	471	455	440	426	412
28L15	43	28	13,443	645	630	616	602	589	576	564	552	541	531	520	510	501	482	465	449
				49	50	51	52	53	54	55	56	57	58	59	60	61	62	63	64
32L07	21	32	6,159	248	240	233	226	220	213	207	201	196	190	185	180	175	171	166	162
32L08	23	32	7,177	289	280	271	263	256	248	241	234	227	220	214	208	202	197	191	186
32L09	25	32	7,798	314	304	295	285	277	269	260	253	246	239	232	225	219	213	207	202
32L10	27	32	8,791	354	343	332	321	311	302	292	283	275	267	259	252	245	238	231	225
32L11	29	32	9,586	386	374	362	351	340	330	321	311	302	294	285	277	270	262	255	249
32L12	31	32	10,827	436	422	409	396	383	371	360	349	339	329	319	310	301	293	285	277
32L13	36	32	12,667	510	500	485	469	453	440	427	414	401	390	378	367	357	347	338	328
32L14	38	32	13,470	543	532	522	512	502	486	471	457	443	429	417	404	393	382	371	360
32L15	43	32	14,445	582	570	559	549	538	528	519	510	501	484	468	452	439	424	411	398
32L16	48	32	15,729	633	621	609	597	586	575	565	555	546	536	527	519	510	502	487	472

STRUCTURAL STEEL-OPEN WEB JOISTS-3

TABLE A (Continued) – LONGSPAN STEEL JOISTS*
ALLOWABLE TOTAL SAFE LOADS
(in lb. per linear ft. of Span)

Joist Designation	Approximate Weight, lb. per linear ft.	Depth, in.	Maximum End Reaction	Clear Opening or Net Span, ft.															
				57	58	59	60	61	62	63	64	65	66	67	68	69	70	71	72
36L08	23	36	6,920	240	234	227	221	216	210	205	199	194	189	185	180	176	172	167	164
36L09	25	36	7,497	260	253	246	240	233	227	221	216	210	205	200	195	191	186	182	177
36L10	27	36	8,506	295	287	279	271	264	257	250	243	237	231	225	219	214	209	204	199
36L11	29	36	9,198	319	310	302	294	286	279	272	265	258	252	246	240	234	228	223	218
36L12	31	36	10,467	363	352	343	333	324	316	307	299	291	284	277	270	263	257	250	244
36L13	36	36	12,398	430	418	406	395	384	374	364	355	346	337	328	320	312	304	297	290
36L14	38	36	13,782	478	464	451	438	426	414	403	392	382	372	362	353	344	336	327	319
36L15	43	36	15,275	530	521	512	497	484	471	458	446	434	423	412	400	389	378	368	357
36L16	48	36	16,482	572	562	552	543	535	526	518	510	502	489	476	464	453	442	431	420
36L17	54	36	17,765	616	606	595	586	576	567	558	549	541	533	525	517	510	497	485	473
				65	66	67	68	69	70	71	72	73	74	75	76	77	78	79	80
40L09	25	40	7,223	220	215	210	205	200	196	191	187	183	179	175	171	168	164	161	157
40L10	27	40	8,208	250	244	238	233	227	222	217	212	207	202	198	193	189	185	181	177
40L11	29	40	8,865	270	264	258	252	246	241	235	230	225	220	215	211	206	202	198	193
40L12	31	40	10,113	308	301	294	287	280	273	267	261	255	249	243	238	233	228	223	218
40L13	36	40	12,017	366	357	348	340	332	324	316	309	302	295	289	282	276	270	264	259
40L14	38	40	13,396	408	397	387	378	369	360	351	343	335	327	320	312	305	299	292	286
40L15	43	40	15,136	461	450	439	428	418	408	399	389	380	372	363	355	347	341	332	324
40L16	48	40	17,187	523	516	508	495	483	472	461	450	440	430	420	410	401	392	384	376
40L17	54	40	18,421	561	553	545	537	529	521	514	507	495	484	473	463	452	442	433	423
40L18	61	40	19,981	609	599	591	582	574	566	558	550	542	535	528	521	515	508	496	485
				73	74	75	76	77	78	79	80	81	82	83	84	85	86	87	88
44L10	27	44	7,993	217	212	208	203	199	195	191	187	183	179	176	172	169	165	162	159
44L11	29	44	8,582	233	228	224	219	215	210	206	202	198	194	191	187	183	180	177	173
44L12	31	44	9,835	267	261	256	250	245	240	235	230	225	221	216	212	208	204	200	196
44L13	36	44	11,639	316	310	303	297	290	284	278	273	267	262	257	251	246	242	237	232
44L14	38	44	13,039	354	346	338	331	324	317	310	304	297	291	285	279	274	268	263	258
44L15	43	44	14,733	400	392	383	375	367	359	352	344	337	330	324	317	311	305	299	293
44L16	48	44	17,054	463	453	443	434	424	415	407	398	390	382	374	367	360	352	345	339
44L17	54	44	19,040	517	510	499	489	478	468	458	449	439	430	422	413	405	397	389	382
44L18	61	44	20,743	563	556	548	541	534	527	521	514	508	497	487	477	467	457	448	439
44L19	68	44	22,311	606	598	590	582	575	567	560	553	546	540	533	527	521	515	509	498
				81	82	83	84	85	86	87	88	89	90	91	92	93	94	95	96
48L11	29	48	8,330	204	200	197	193	189	186	183	179	176	173	170	167	164	162	159	156
48L12	31	48	9,596	235	230	226	221	217	213	209	205	201	198	194	191	187	184	181	178
48L13	36	48	11,352	278	273	268	262	257	253	248	243	239	234	230	226	222	218	214	211
48L14	38	48	12,740	312	305	299	294	288	282	277	272	266	261	257	252	247	243	238	234
48L15	43	48	14,373	352	345	338	332	326	319	313	308	302	296	291	285	280	275	270	266
48L16	48	48	16,660	408	400	392	384	377	370	363	356	349	343	337	331	325	319	313	308
48L17	54	48	18,743	459	450	441	433	425	416	409	401	394	386	379	372	366	359	353	346
48L18	61	48	21,336	523	516	510	504	494	485	475	466	457	448	440	432	424	416	408	401
48L19	68	48	23,029	564	557	550	544	538	531	525	519	514	508	498	489	480	471	462	454

This load table applies to "longspan" joists with either parallel chords or standard pitched top chords.

The carrying capacities of "longspan" joist with top chords pitched is determined by the nominal depth of the "longspan" joists at the center of the span.

Standard pitch is 1/8" per foot. If pitch exceeds this standard, the load table does not apply.

Loads below heavy lines are governed by maximum end reaction.

The weight of dead loads, including the weight of "longspans," must in all cases be deducted to determine the live load-carrying capacities which must be reduced for concentrated loads (accessories included).

Figures to the right of heavy vertical line to be used for roof construction only.

When holes are required in top or bottom chords, the above carrying capacities must be reduced in proportion to reduction of chord areas.

The top chords are considered as being stayed laterally by floor slab or roof deck.

Spacing of "longspan" steel joists shall not exceed the safe span of the floor slab or roof deck.

* Adopted by the Steel Joist Institute, April 28, 1953. Effective April 28, 1953.

STRUCTURAL STEEL-OPEN WEB JOISTS-4

STEEL JOIST DETAILS *

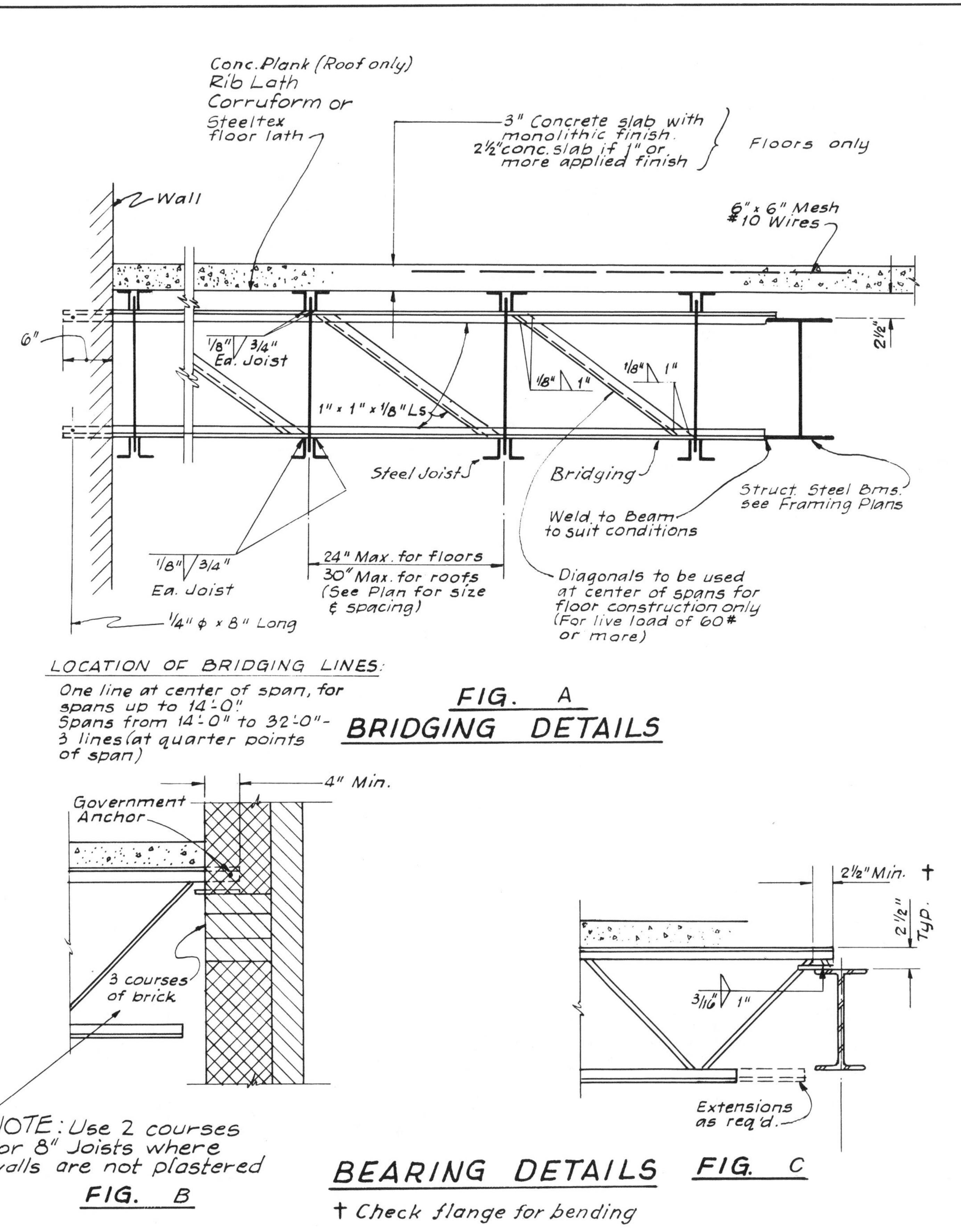

Conc. Plank (Roof only)
Rib Lath
Corruform or
Steeltex
floor lath
3" Concrete slab with monolithic finish.
2½" conc. slab if 1" or more applied finish
Floors only
Wall
6" x 6" Mesh
#10 Wires
6"
2½"
1/8" 3/4"
Ea. Joist
1/8" 1"
1/8" 1"
1" x 1" x 1/8" Ls
Steel Joist
Bridging
Struct. Steel Bms. see Framing Plans
Weld to Beam to suit conditions
1/8" 3/4"
Ea. Joist
24" Max. for floors
30" Max. for roofs
(See Plan for size & spacing)
Diagonals to be used at center of spans for floor construction only (For live load of 60# or more)
1/4" φ x 8" Long
LOCATION OF BRIDGING LINES:
One line at center of span, for spans up to 14'-0"
Spans from 14'-0" to 32'-0" - 3 lines (at quarter points of span)
FIG. A
BRIDGING DETAILS
4" Min.
Government Anchor
3 courses of brick
2½" Min. †
2½" Typ.
3/16" 1"
Extensions as req'd.
NOTE: Use 2 courses for 8" Joists where walls are not plastered
FIG. B
BEARING DETAILS
FIG. C
† Check flange for bending

* For Short Span Joists only. Long Span Joist Details similar.

STRUCTURAL STEEL-ECONOMICAL SELECTION OF BEAMS

TABLE FOR ECONOMICAL SELECTION OF STEEL BEAMS WITH UNBRACED FLANGES*-AISC 1947

Example: Given - M = Moment = 89 ft. kip ; L = Unbraced length = 14 ft.
Steel Working Stress = 20,000 p.s.i.
To Find - Economical Wide-Flange Beams

$$\text{Solution - } S = \text{Section Modulus} = \frac{89{,}000 \times 12}{20{,}000} = 53.4$$

$$S \times L = 53.4 \times 14 = 747$$

Enter table at first figure of SL_u greater than 747, which is 759. Select first beam with section modulus greater than required, 53.4. Beam is 16 WF 45. Check for deflection.

SL.	S	Section	L.
WIDE-FLANGE BEAMS			
142	14.1	8 WF 17	10.1
208	17.0	8 WF 20	12.2
213	21.5	10 WF 21	9.9
324	26.4	10 WF 25	12.3
339	20.8	8 WF 24	16.3
372	34.1	12 WF 27	10.9
389	41.8	14 WF 30	9.3
437	30.8	10 WF 29	14.2
457	24.3	8 WF 28	18.8
493	39.4	12 WF 31	12.5
529	48.5	14 WF 34	10.9
529	56.3	16 WF 36	9.4
592	27.4	8 WF 31	21.6
616	35.0	10 WF 33	17.6
665	45.9	12 WF 36	14.5
671	54.6	14 WF 38	12.3
708	64.4	16 WF 40	11.0
759	31.1	8 WF 35	24.4
890	72.4	16 WF 45	12.3
895	42.2	10 WF 39	21.2
898	51.9	12 WF 40	17.3
966	62.7	14 WF 43	15.4
1060	89.0	18 WF 50	11.8
1100	80.7	16 WF 50	13.6
1120	58.2	12 WF 45	19.2
1200	49.1	10 WF 45	24.5
1210	70.2	14 WF 48	17.2
1290	98.2	18 WF 55	13.1
1370	64.7	12 WF 50	21.2
1480	77.8	14 WF 53	19.0
1520	54.6	10 WF 49	27.9
1530	126.4	21 WF 62	12.1
1550	107.8	18 WF 60	14.4
1620	94.1	16 WF 58	17.2
1690	70.7	12 WF 53	23.9
1850	60.4	10 WF 54	30.6
1880	139.9	21 WF 68	13.4
1950	117.0	18 WF 64	16.7
1980	104.2	16 WF 64	19.0
2050	78.1	12 WF 58	26.3
2130	92.2	14 WF 61	23.1
2170	150.7	21 WF 73	14.4
2250	175.4	24 WF 76	12.8
2260	67.1	10 WF 60	33.6
2330	128.2	18 WF 70	18.2
2430	115.9	16 WF 71	21.0
2630	88.0	12 WF 65	29.9
2640	103.0	14 WF 68	25.6
2690	73.7	10 WF 66	36.5
2830	196.3	24 WF 84	14.4
2850	141.7	18 WF 77	20.1
2870	168.0	21 WF 82	17.1
2940	127.8	16 WF 78	23.0
3120	112.3	14 WF 74	27.8
3210	97.5	12 WF 72	32.9
3380	242.8	27 WF 94	13.9
3420	156.1	18 WF 85	21.9
3600	220.9	24 WF 94	16.3
3720	121.1	14 WF 78	30.7
3860	107.1	12 WF 79	36.0
3950	197.6	21 WF 96	20.0
4010	299.2	30 WF 108	13.4
4070	266.3	27 WF 102	15.3
4270	151.3	16 WF 88	28.2
4310	130.9	14 WF 84	32.9
4490	115.7	12 WF 85	38.8
4830	248.9	24 WF 100	19.4
4930	138.1	14 WF 87	35.7
4890	327.9	30 WF 116	14.9
4960	184.4	18 WF 96	26.9
5130	166.1	16 WF 96	30.9
5150	299.2	27 WF 114	17.2
5740	354.6	30 WF 124	16.2
5800	150.6	14 WF 95	38.5
5870	274.4	24 WF 110	21.4
5900	202.2	18 WF 105	29.2
6030	404.8	33 WF 130	14.9
6610	379.7	30 WF 132	17.4
6660	249.6	21 WF 112	26.7
6910	299.1	24 WF 120	23.1
6960	220.1	18 WF 114	31.6
7420	446.8	33 WF 141	16.6
7900	502.9	36 WF 150	15.7
8610	284.1	21 WF 127	30.3
8570	330.7	24 WF 130	25.9
8850	486.4	33 WF 152	18.2
9200	541.0	36 WF 160	17.0
10230	402.9	27 WF 145	25.4
10600	579.1	36 WF 170	18.3
10660	317.2	21 WF 142	33.6
10880	372.5	24 WF 145	29.2
12180	621.2	36 WF 182	19.6
12360	444.5	27 WF 160	27.8
13360	413.5	24 WF 160	32.3
13870	663.6	36 WF 194	20.9
14100	528.2	30 WF 172	26.7
15110	492.8	27 WF 177	30.7
17340	586.1	30 WF 190	29.6
18410	669.6	33 WF 200	27.5
21250	649.9	30 WF 210	32.7
22440	740.6	33 WF 220	30.3
24150	835.5	36 WF 230	28.9
26850	811.1	33 WF 240	33.1
27540	892.5	36 WF 245	30.9
31290	951.1	36 WF 260	32.9
36810	1031.2	36 WF 280	35.7
42180	1105.1	36 WF 300	38.2
LIGHT BEAMS			
62	9.9	8 LB 13	6.3
67	7.2	6 LB 12	9.3
74	13.8	10 LB 15	5.4
79	17.5	12 LB 16-1/2	4.5
91	11.8	8 LB 15	7.7
105	16.2	10 LB 17	6.5
122	21.4	12 LB 19	5.7
131	10.1	6 LB 16	13.1
145	18.8	10 LB 19	7.7
175	25.3	12 LB 22	6.9
JOISTS			
33	5.1	6x4 J 8-1/2	6.5
40	7.8	8x4 J 10	5.1
43	10.5	10x4 J 11-1/2	4.1
55	14.8	12x4 J 14	3.7
STANDARD MILL BEAMS			
144	14.0	8 M 17	10.3
159	15.2	8 M 20	10.5
236	21.7	10 M 21	10.9
262	23.6	10 M 25	11.1
320	21.0	8 M 24	15.2
350	22.5	8 M 28	15.6
JUNIOR BEAMS			
7	2.4	6 JR 4.4	2.9
10	3.5	7 JR 5.5	2.8
13	4.7	8 JR 6.5	2.7
14	5.8	9 JR 7.5	2.5
20	7.8	10 JR 9.0	2.5
23	9.6	11 JR 10.3	2.4
38	12.0	12 JR 11.8	3.2
JUNIOR CHANNELS*			
10	4.4	10 [6.5	2.3
12	6.5	10 [8.4	1.9
18	9.3	12 [10.6	2.0
STANDARD I-BEAMS			
17	1.7	3 I 5.7	10.1
21	1.9	3 I 7.5	10.8
29	3.0	4 I 7.7	9.7
34	3.3	4 I 9.5	10.2
47	4.8	5 I 10.0	9.8
64	6.0	5 I 14.75	10.7
73	7.3	6 I 12.5	10.0
93	8.7	6 I 17.25	10.7
106	10.4	7 I 15.3	10.2
130	12.0	7 I 20.0	10.8
151	14.2	8 I 18.4	10.6
178	16.0	8 I 23.0	11.1
278	24.4	10 I 25.4	11.4
353	29.2	10 I 35.0	12.1
407	36.0	12 I 31.8	11.3
435	37.8	12 I 35.0	11.5
645	44.8	12 I 40.8	14.4
671	58.9	15 I 42.9	11.4
751	64.2	15 I 50.0	11.7
760	50.3	12 I 50.0	15.1
1020	88.4	18 I 54.7	11.5
1220	101.9	18 I 70.0	12.0
1440	116.9	20 I 65.4	12.3
1590	126.3	20 I 75.0	12.6
2210	173.9	24 I 79.9	12.7
2400	185.8	24 I 90.0	12.9
2430	150.2	20 I 85.0	16.2
2590	197.6	24 I 100.0	13.1
2640	160.0	20 I 95.0	16.5
4240	234.3	24 I 105.9	18.1
4640	250.9	24 I 120.0	18.5
STANDARD CHANNELS*			
7	1.1	3 [4.1	6.4
8	1.2	3 [5.0	6.8
10	1.4	3 [6.0	7.2
11	1.9	4 [5.4	5.9
15	2.3	4 [7.25	6.4
17	3.0	5 [6.7	5.6
21	3.5	5 [9.0	6.0
24	4.3	6 [8.2	5.5
29	5.0	6 [10.5	5.8
33	6.0	7 [9.8	5.5
36	5.8	6 [13.0	6.2
39	6.9	7 [12.25	5.7
45	8.1	8 [11.5	5.5
46	7.7	7 [14.75	6.0
51	9.0	8 [13.75	5.7
59	10.5	9 [13.4	5.6
64	11.3	9 [15.0	5.7
67	10.9	8 [18.75	6.2
76	13.4	10 [15.3	5.7
82	13.5	9 [20.0	6.1
94	15.7	10 [20.0	6.0
114	18.1	10 [25.0	6.3
131	21.4	12 [20.7	6.1
136	20.6	10 [30.0	6.6
153	23.9	12 [25.0	6.4
178	26.9	12 [30.0	6.6
309	41.7	15 [33.9	7.4
351	46.2	15 [40.0	7.6
421	61.0	18 [42.7	6.9
429	53.6	15 [50.0	8.0
440	63.7	18 [45.8	6.9
491	69.1	18 [51.9	7.1
544	74.5	18 [58.0	7.3

*These sections should also be investigated for torsion.

*Data from Eng. News Record, March 18, 1948. Article by William P. Stewart

STRUCTURAL STEEL—RIVET DATA

TABLE A—BEAM SEATS WITHOUT STIFFENERS—ALLOWABLE LOAD IN KIPS[1]

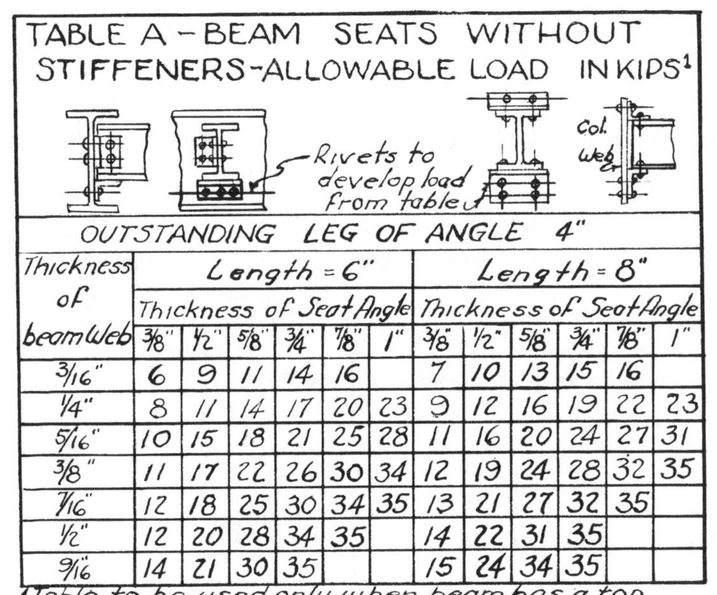
Rivets to develop load from table
Col. Web

OUTSTANDING LEG OF ANGLE 4"

Thickness of beam Web	Length = 6"						Length = 8"					
	Thickness of Seat Angle						Thickness of Seat Angle					
	3/8"	1/2"	5/8"	3/4"	7/8"	1"	3/8"	1/2"	5/8"	3/4"	7/8"	1"
3/16"	6	9	11	14	16		7	10	13	15	16	
1/4"	8	11	14	17	20	23	9	12	16	19	22	23
5/16"	10	15	18	21	25	28	11	16	20	24	27	31
3/8"	11	17	22	26	30	34	12	19	24	28	32	35
7/16"	12	18	25	30	34	35	13	21	27	32	35	
1/2"	12	20	28	34	35		14	22	31	35		
9/16"	14	21	30	35			15	24	34	35		

1 Table to be used only when beam has a top angle or side lug. Over 35 kips use stiffened seats. Table based on effective bearing beginning 1/2" from back of seat angle. Interpolate for other than 6" or 8" lengths.

TABLE B—DRIVING CLEARANCE

Dia. Rivet, inches		1/2	5/8	3/4	7/8	1	1 1/8	1 1/4	1 3/8	1 1/2
Driving Clearance Inches	E (Min.)	3/4	7/8	1	1 1/8	1 1/4	1 3/8	1 1/2	1 5/8	1 3/4
	E (Pref.)	1	1 1/8	1 1/4	1 3/8	1 1/2	1 5/8	1 3/4	1 7/8	2

TABLE C—MINIMUM EDGE DISTANCE TO ¢ OF PUNCHED HOLES

Rivet Dia. inches	In Sheared Edge	In Rolled Edge of Plates	In Rolled Edge of Structural Shapes †
1/2	1	7/8	3/4
5/8	1 1/8	1	7/8
3/4	1 1/4	1 1/8	1
7/8	1 1/2	1 1/4	1 1/8
1	1 3/4	1 1/2	1 1/4
1 1/8	2	1 3/4	1 1/2
1 1/4	2 1/4	2	1 3/4

† May be decreased 1/8" when holes are near end of beam

TABLE-D USUAL GAGES FOR ANGLES, INCHES

Leg	8	7	6	5	4	3 1/2	3	2 1/2	2	1 3/4	1 1/2	1 3/8	1 1/4	1
g	4 1/2	4	3 1/2	3	2 1/2	2	1 3/4	1 3/8	1 1/8	1	7/8	7/8	3/4	5/8
g_1	3	2 1/2	2 1/4	2										
g_2	3	3	2 1/2	1 3/4										

CRIMPS

$b = t + 1\frac{1}{2}"$ Min. = 2"

SPACING OF RIVETS

MINIMUM PITCH:—preferably not < 3 rivet diameters

MAXIMUM PITCH:—in lines of stress of compression members of plates and shapes, not > 16 × thickness of thinnest outside plate, nor > 20 × thickness of thinnest enclosed plate, max. 12". At right angles to direction of stress, distance between lines of rivets not > 32 × thickness of thinnest plate if more than 1 ply. For angles in built-up sections with 2 gage lines, rivets staggered, max. pitch in line of stress in each gage line not > 24 × thickness of thinnest plate with max. of 18 inches.

END PITCH IN COMPRESSION MEMBERS:—not > 4 rivet diam. for a length = 1 1/2 × max. width of member.

TWO-ANGLE MEMBERS:—pitch of 3'-6" in tension and 2'-0" in compression, but 1/r for each angle between rivets not > 3/4 of that for whole member.

MINIMUM EDGE DISTANCE IN LINE OF STRESS:—¢ rivet to edge not < the shearing area of the rivet shank (single or double shear) divided by the plate thickness; minimum edge distance may be decreased in ratio of $\frac{\text{actual stress}}{\text{allowable stress}}$ and requirement may be disregarded if rivet in question is 1 of 3 or more in line parallel to direction of stress.

MAXIMUM EDGE DISTANCE:—= 12 × thickness of plate, but not > 6".

GRIP—Max. rivet grip 5 diam. Increase number of rivets 1% for each additional 1/16" of rivet grip.

ECCENTRIC LOADS—For rivets with eccentric loads—see pg. 4-22.

TABLE E—BOLT HEADS AND NUTS

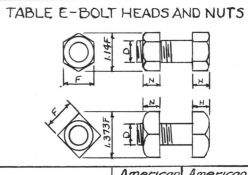

HEADS AND NUTS		American Standard Regular	American Standard Heavy
HEAD	Height, H	2/3 D	3/4 D + 1/16"
	Short dia, F	1 1/2 D	1 1/2 D + 1/8"
NUT	Height, N	7/8 D	D
	Short dia., F	1 1/2 D + 1/16 (D = 5/8 or less); 1 1/2 D (D > 5/8")	1 1/2 D + 1/8"

* Data adopted from A.I.S.C. Manual 1947

STRUCTURAL STEEL–GRATING & FLOOR PLATES

TABLE A –GRATING– SAFE LOAD AND DEFLECTION *

SIZE BRG. BAR	WT. lbs/ft²		SPAN IN FEET 2-0	2-6	3-0	3-6	4-0	4-6	5-0	5-6	6-0	6-6	7-0	8-0	9-0
3/4 × 1/8	4.0	W	371	250	161										
		D	.085	.134	.192										
		P	371	298	242										
		D	.068	.108	.154										
3/4 × 3/16	5.7	W	563	360	244										
		D	.085	.134	.192										
		P	563	450	366										
		D	.068	.108	.154										
1 × 1/8	5.1	W	675	432	300	212	169								
		D	.064	.099	.143	.195	.256								
		P	675	540	450	371	338								
		D	.051	.080	.115	.156	.205								
1 × 3/16	7.7	W	1013	653	450	322	253								
		D	.064	.099	.143	.195	.256								
		P	1013	816	675	563	506								
		D	.051	.080	.115	.156	.205								
1 1/4 × 1/8	6.2	W	1069	675	473	341	261	207	164	135					
		D	.051	.081	.115	.157	.205	.259	.321	.389					
		P	1069	844	709	596	523	467	411	371					
		D	.041	.064	.092	.125	.163	.207	.256	.310					
1 1/4 × 3/16	9.3	W	1603	1013	712	514	394	313	248	205					
		D	.051	.081	.115	.157	.205	.259	.321	389					
		P	1603	1266	1069	900	788	703	619	563					
		D	.041	.064	.092	.125	.163	.207	.256	.310					
1 1/2 × 1/8	7.3	W	1536	990	689	501	383	299	248	205	169	144	124		
		D	.043	.067	.094	.131	.166	.216	.267	.324	.385	.440	.522		
		P	1536	1238	1029	883	765	675	619	563	506	467	433		
		D	.034	.053	.077	.104	.137	.173	.214	.259	.308	.361	.418		
1 1/2 × 3/16	11.0	W	2306	1485	1032	756	576	450	371	307	253	216	185		
		D	.043	.067	.094	.131	.166	.216	.267	.324	.385	.440	522		
		P	2306	1856	1547	1322	1153	1013	928	844	759	703	647		
		D	.034	.053	.077	.104	.137	.173	.214	.259	.308	.361	.418		
1 3/4 × 3/16	12.7	W	3150	2003	1384	1029	788	612	495	410	347	295	257	197	150
		D	.038	.057	.082	.112	.147	.185	.229	.276	.330	.387	.450	.580	.737
		P	3150	2503	2093	1800	1575	1378	1238	1125	1041	956	900	788	675
		D	.029	.046	.066	.090	.117	.148	.183	.221	.264	.308	.358	.468	.593
2 × 3/16	14.6	W	4106	2633	1820	1350	1026	813	653	542	450	389	330	253	200
		D	.032	.050	.072	.099	.128	.163	.201	.243	.289	.341	.397	.516	.651
		P	4106	3291	2728	2363	2053	1828	1631	1491	1350	1266	1153	1013	900
		D	.026	.040	.057	.078	.102	.129	.160	.193	.230	.269	.314	.409	.518
2 1/4 × 3/16	16.5	W	5231	3330	2323	2094	1294	1026	833	684	581	493	426	324	.257
		D	.027	.044	.064	.087	.113	.148	.177	.214	255	.305	.349	.455	.574
		P	5231	4163	3488	2981	2588	2106	2081	1884	1744	1603	1491	1194	1153
		D	.023	.035	.051	.070	.091	.115	.142	.172	.204	.240	.279	.364	.460

W = Safe uniform load in lbs. per sq. ft.
P = Safe concentrated load in lbs. per ft. of width
D = Deflection in inches

Spans to right of heavy line not recommended.

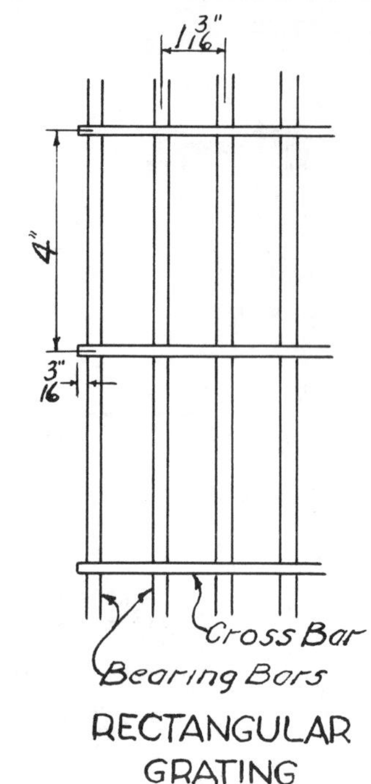

1 3/16"
4"
3/16"
Cross Bar
Bearing Bars

RECTANGULAR GRATING

*This table is for Reliance type IR4; Irving type AA; and Blaw-Knox Standard type grating.
Maximum allowable fiber stress 18,000 #/□".
The weights are approx.; They vary with the different manufacturers.

TABLE B – FLOOR PLATES – SAFE UNIFORM LOAD IN POUNDS PER SQUARE FOOT

GAGE	WT. lbs/ft²	SPAN IN FEET 1-6	2-0	2-6	3-0	3-6	4-0	4-6	5-0	6-0
3/16	8.70	333	188	120	84	61	47			
1/4	11.25	593	333	213	148	109	83	66	53	
5/16	13.80	925	520	333	232	170	130	103	83	58
3/8	16.35	1335	750	480	333	245	188	148	120	84
7/16	18.90	1810	1020	655	453	333	255	204	164	113
1/2	21.45	2370	1330	852	592	435	333	264	213	148
9/16	24.00	3000	1690	1080	750	550	423	333	270	187
5/8	26.55	3700	2080	1330	925	680	520	411	333	232
3/4	31.65	5340	3000	1920	1330	980	750	593	480	333
Deflection Coefficient		.035	.066	.095	.149	.250	.265	.338	.415	.595

Notes:

The thickness of plate is thru body, does not include projections. Loads include weight of plates.
Deflection above zig-zag line will exceed 1/100 of the span. Deflection in inches with maximum safe uniform load = Deflection coefficient divided by Thickness of plate in inches. Deflection in inches with any uniform load within the elastic limit = Deflection coefficient times actual load per sq. ft., all divided by maximum safe load per sq. ft.
Maximum allowable fiber stress 16,000 lbs. per sq. in.

STRUCTURAL STEEL– HANGERS

FIG. A. TYPE A (RIVETED)

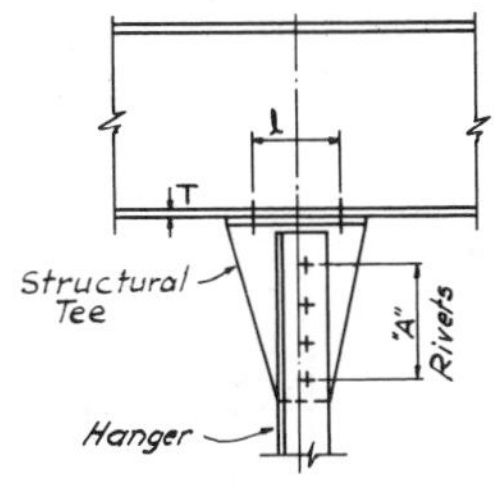

T
Structural Tee
"A" Rivets
Hanger

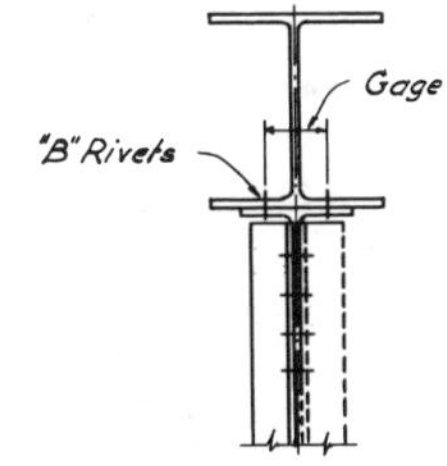

Gage
"B" Rivets

RIVETS – 3/4"

FIG. B. TYPE C

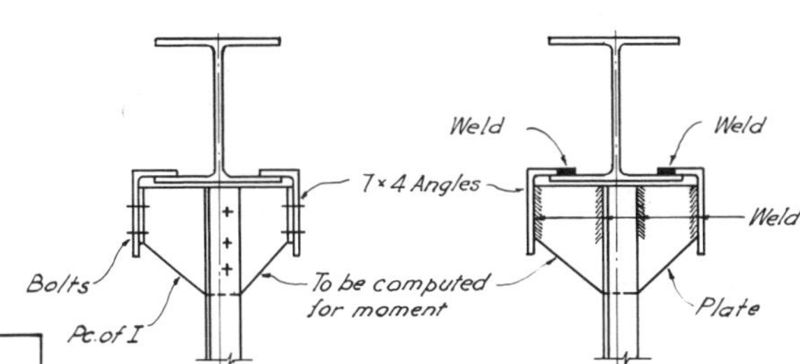

Weld
Weld
7×4 Angles
Weld
Bolts
To be computed for moment
Plate
Pc. of I

BOLTED WELDED

Connection to Existing Beam Where Inadvisable to Drill or Weld to Bottom Flange

TABLE C–RIVETED CONNECTIONS

Load, kips	Gage, in.	No. of A Rivets, Double Shear	No. of B Rivets	Rivets Staggered, One Hole Out: Size of Structural Tee	Rivets Staggered, One Hole Out: l, in.	Rivets in Line, Two Holes Out: Size of Structural Tee	Rivets in Line, Two Holes Out: l, in.
13	3-1/2	2	2	16 WF 45	3-1/4	–	–
	5-1/2	2	2	16 WF 45	3-1/4	–	–
26	5-1/2	2	4	16 WF 64 or 18 WF 70	9-3/4	18 WF 70	4
39	5-1/2	3	6	18 WF 70	16-1/4	18 WF 70	8
53	5-1/2	4	8	18 WF 70	22-3/4	18 WF 77	9
66	5-1/2	5	10	18 WF 70	29-1/4	18 WF 77	12

FIG. D. TYPE D

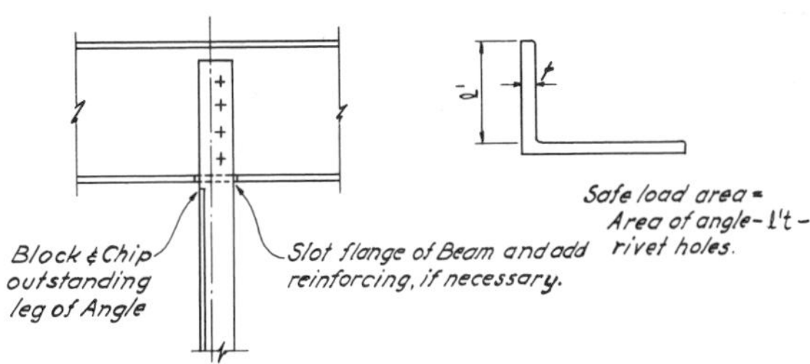

l'
t
Safe load area = Area of angle – 1¼t – rivet holes.
Block & Chip outstanding leg of Angle
Slot flange of Beam and add reinforcing, if necessary.

SUGGESTED DETAIL FOR CONNECTION TO BEAM WEB

FIG. E. TYPE B (WELDED)

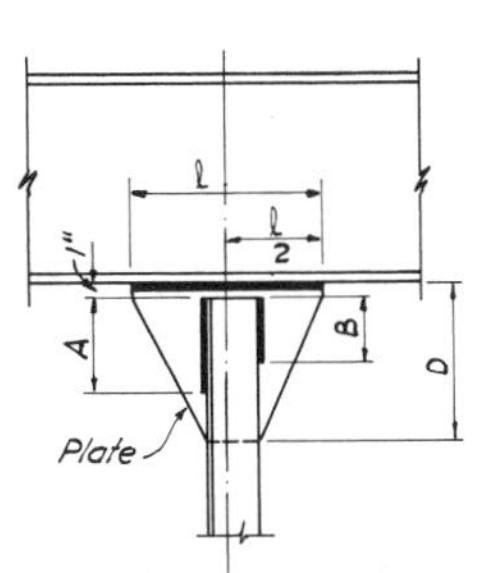

l
l/2
A
B
D
Plate

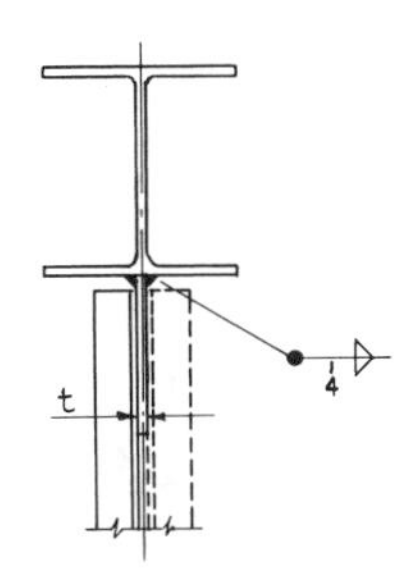

t
1/4

TABLE F–WELDED CONNECTIONS

Load, kips	ℙ t, in.	ℙ D, in.	ℙ l, in.	Weld ℙ to ∠ Fillet Weld Size, in.	Weld ℙ to ∠ A, in.	Weld ℙ to ∠ B, in.	Hanger Angle Size, in.
10	5/16	5	6	1/4	2-1/2*	1-1/2*	1∠ 3 x 3 x 5/16*
20	5/16	7	6	1/4	5*	2*	1∠ 3 x 3 x 5/16*
30	5/16	6	7	5/16	4	1-1/2	2∠s 3 x 3 x 3/8
40	5/16	7	9	5/16	5	2	2∠s 3 x 3 x 3/8
50	5/16	9	11	5/16	6	2-1/2	2∠s 3 x 3 x 3/8
60	5/16	10	13	5/16	7	3	2∠s 3 x 3 x 3/8

*Where 2 ∠s are used, use 1/2 of A and B values with minimum of 1½".

GENERAL NOTE: Where t is less than 5/8", check flange for bending over length 2 × ∠.

STRUCTURAL STEEL—WIND BRACING

WIND BRACING COLUMN CONNECTIONS - AISC 1949 - RESISTING MOMENT (R.M.) FOOT KIPS*

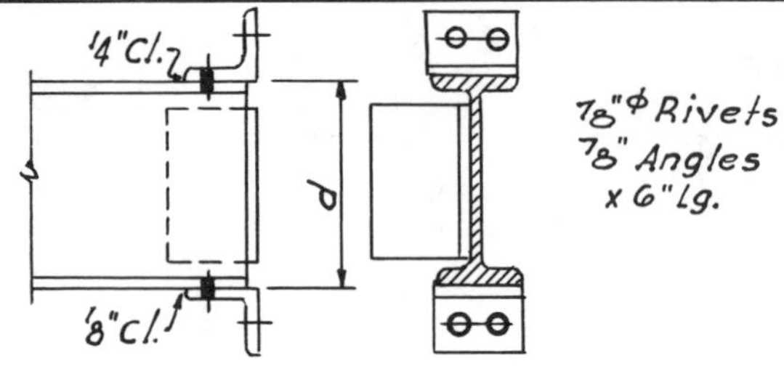

CONNECTION "a"

¼" Cl. ⅛" Cl. d

⅞"φ Rivets
⅞" Angles
x 6" Lg.

Normal 2-rivet connection to web or flange of columns.

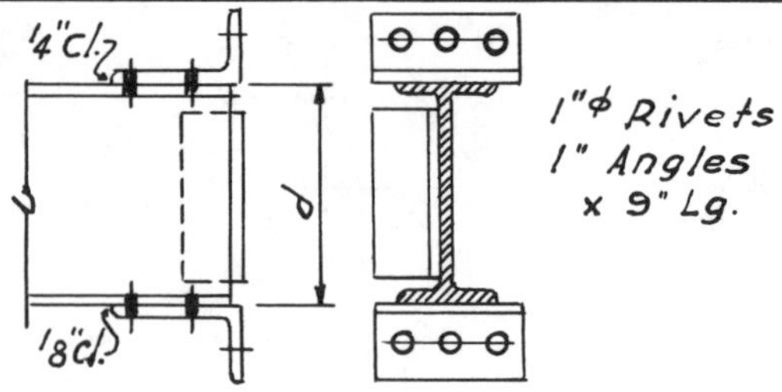

d	R.M.
12	24
14	28
15	30
16	32
18	36
20	40
21	42
24	48
27	54
30	60
33	66
36	72

CONNECTION "b"

¼" Cl. ⅛" Cl. d

1"φ Rivets
1" Angles
x 9" Lg.

Normal 3 rivet connection to web of columns.

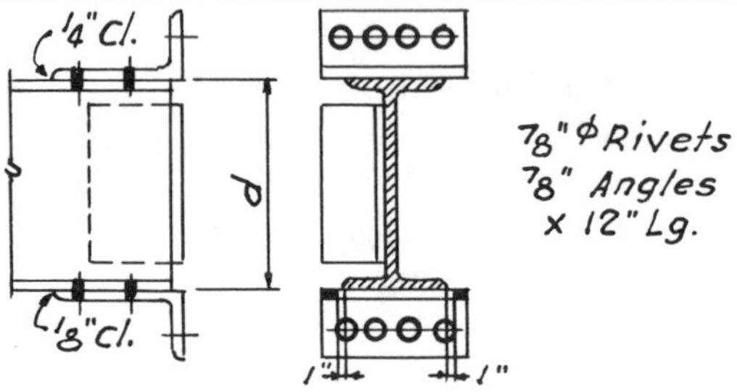

d	R.M.
12	48
14	56
15	60
16	63
18	69
20	75
21	78
24	87
27	96
30	105
33	114
36	123

CONNECTION "c"

¼" Cl. ⅛" Cl. d 1" 1"

⅞"φ Rivets
⅞" Angles
x 12" Lg.

Special 4 rivet connection for flanges of Col's. over ⅝" thick and 12" wide, min.

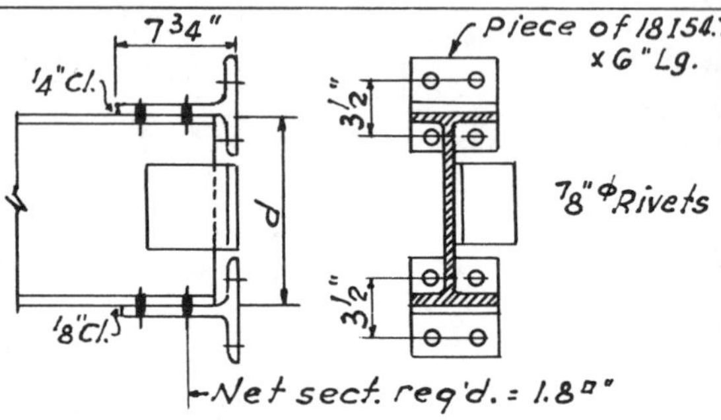

d	R.M.
12	48
14	56
15	60
16	64
18	72
20	80
21	84
24	96
27	108
30	120
33	132
36	144

CONNECTION "d"

7¾" ¼" Cl. ⅛" Cl. d 3½" 3½"

Piece of 18 I 54.7 x 6" Lg.

⅞"φ Rivets

Net sect. req'd. = 1.8□"

Normal 4 rivet connection for webs, thin flanges, narrow flanges.

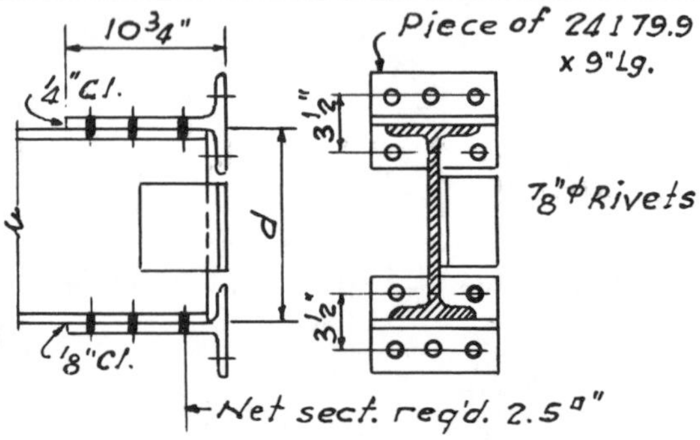

d	R.M.
12	48
14	56
15	60
16	64
18	72
20	80
21	84
24	96
27	108
30	120
33	132
36	144

CONNECTION "e"

10¾" ¼" Cl. ⅛" Cl. d 3½" 3½"

Piece of 24 I 79.9 x 9" Lg.

⅞"φ Rivets

Net sect. req'd. 2.5□"

Normal 5 rivet connection for web of Col's. only.

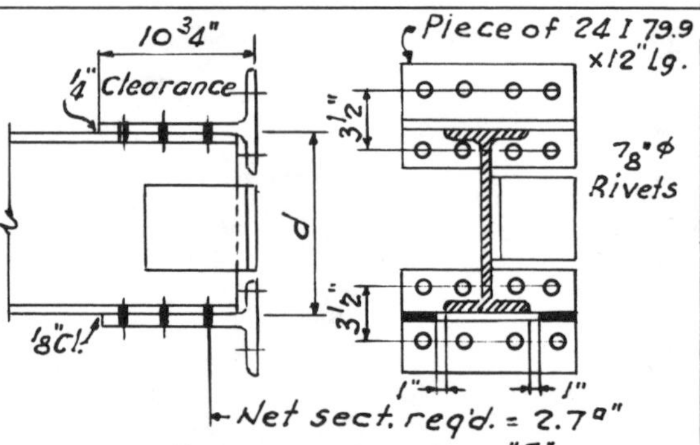

d	R.M.
12	66
14	76
15	81
16	86
18	96
20	106
21	111
24	126
27	141
30	156
33	171
36	186

CONNECTION "f"

10¾" ¼" Clearance ⅛" Cl. d 3½" 3½" 1" 1"

Piece of 24 I 79.9 x 12" Lg.

⅞"φ Rivets

Net sect. req'd. = 2.7□"

Normal 6 rivet connection for flange of Col's. only.

d	R.M.
12	72
14	84
15	90
16	96
18	108
20	120
21	126
24	144
27	162
30	180
33	198
36	216

CONNECTION "g"

13½" ¼" Cl. ⅛" Cl. d 3½" 3½" 1" 1"

Piece of 27 WF 102 x 12" Lg.

⅞"φ Rivets

Net sect. req'd. 3.6□"

Normal 8 rivet connection for flange of Col's. only.

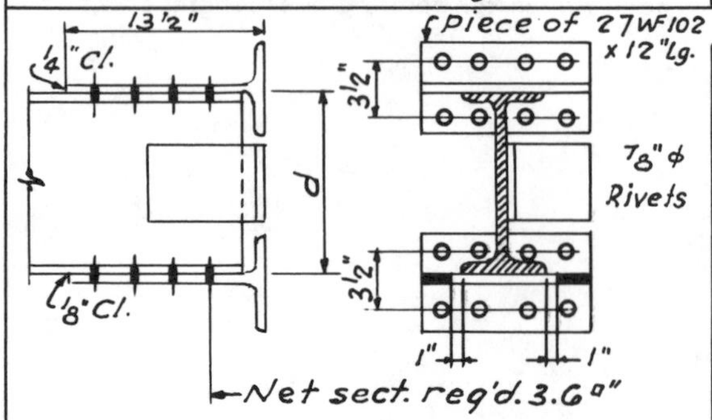

d	R.M.
12	96
14	112
15	120
16	128
18	144
20	160
21	168
24	192
27	216
30	240
33	264
36	288

NOTES:

1. Shear and Pull on head:
   a.- ⅞" φ rivet = 9020# x 1.33 = 12030#
   b.- 1" φ rivet = 11780# x 1.33 = 15710#
2. Above values apply to N.Y.C. Code 1949 except as follows:
   a. Connection "a" - d = 36" - Use 1" angles x 6" lg.
   b. Connections "d" thru "g" inclusive:- Net sections required are ⅓ greater than shown above for AISC Spec.
   c. Connection "d" - Use 18 I 65 in place of 18 I 54.7.

For Welded Connections, See p. 4-59.
For Wind Design, See p. 1-15.

*Courtesy of Austin M. Rice

STRUCTURAL-WELDING

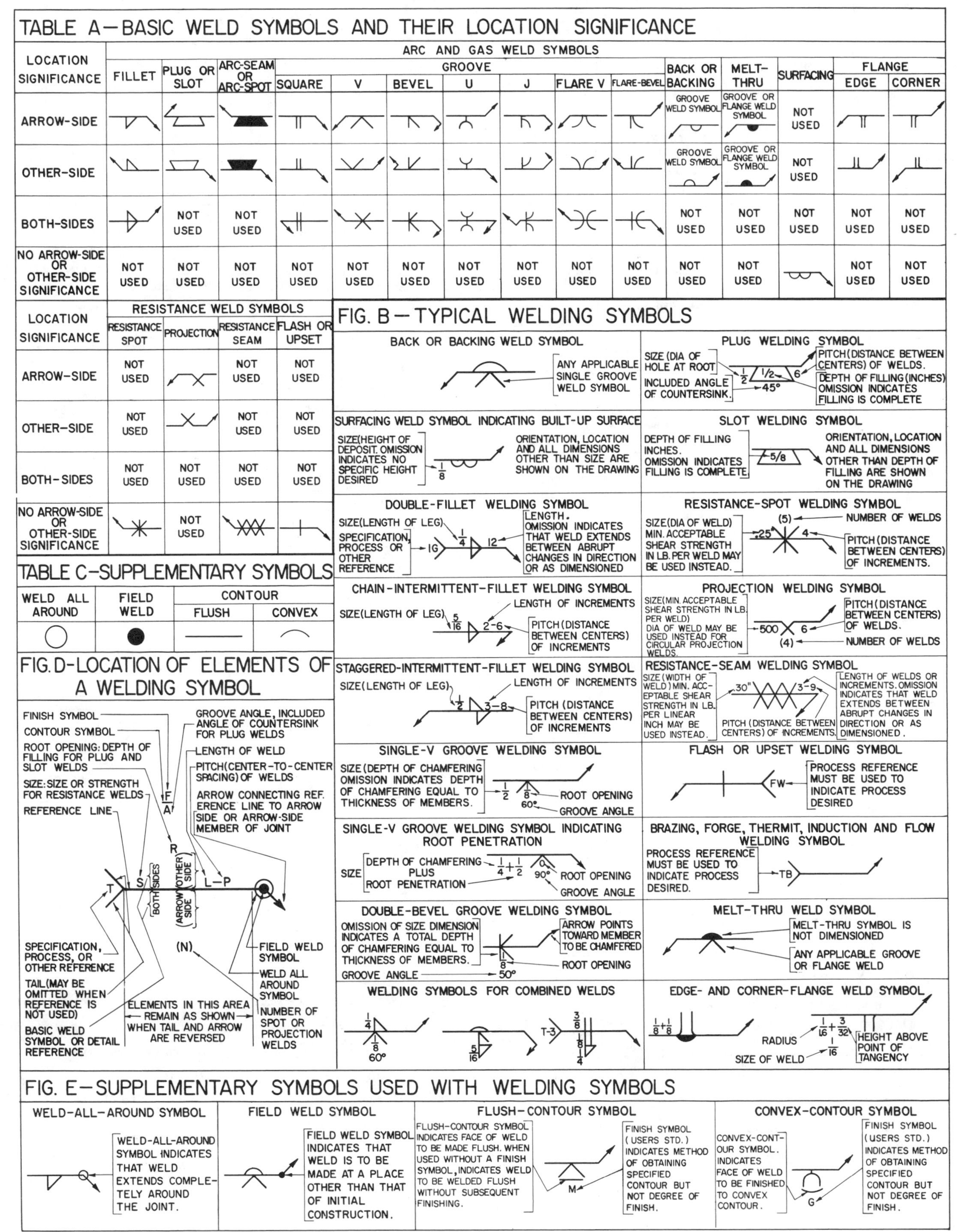

TABLE A—BASIC WELD SYMBOLS AND THEIR LOCATION SIGNIFICANCE
LOCATION SIGNIFICANCE
ARC AND GAS WELD SYMBOLS
FILLET
PLUG OR SLOT
ARC-SEAM OR ARC-SPOT
GROOVE
SQUARE
V
BEVEL
U
J
FLARE V
FLARE-BEVEL
BACK OR BACKING
MELT-THRU
SURFACING
FLANGE
EDGE
CORNER
ARROW-SIDE
OTHER-SIDE
BOTH-SIDES
NO ARROW-SIDE OR OTHER-SIDE SIGNIFICANCE
GROOVE WELD SYMBOL
GROOVE OR FLANGE WELD SYMBOL
NOT USED
RESISTANCE WELD SYMBOLS
RESISTANCE SPOT
PROJECTION
RESISTANCE SEAM
FLASH OR UPSET
TABLE C—SUPPLEMENTARY SYMBOLS
WELD ALL AROUND
FIELD WELD
CONTOUR
FLUSH
CONVEX
FIG. D-LOCATION OF ELEMENTS OF A WELDING SYMBOL
FINISH SYMBOL
CONTOUR SYMBOL
ROOT OPENING: DEPTH OF FILLING FOR PLUG AND SLOT WELDS
SIZE: SIZE OR STRENGTH FOR RESISTANCE WELDS
REFERENCE LINE
GROOVE ANGLE, INCLUDED ANGLE OF COUNTERSINK FOR PLUG WELDS
LENGTH OF WELD
PITCH (CENTER-TO-CENTER SPACING) OF WELDS
ARROW CONNECTING REFERENCE LINE TO ARROW SIDE OR ARROW-SIDE MEMBER OF JOINT
F
A
R
T
S
L—P
(BOTH SIDES)
(ARROW SIDE)
(OTHER SIDE)
(N)
SPECIFICATION, PROCESS, OR OTHER REFERENCE
TAIL (MAY BE OMITTED WHEN REFERENCE IS NOT USED)
BASIC WELD SYMBOL OR DETAIL REFERENCE
ELEMENTS IN THIS AREA REMAIN AS SHOWN WHEN TAIL AND ARROW ARE REVERSED
FIELD WELD SYMBOL
WELD ALL AROUND SYMBOL
NUMBER OF SPOT OR PROJECTION WELDS
FIG. B—TYPICAL WELDING SYMBOLS
BACK OR BACKING WELD SYMBOL
ANY APPLICABLE SINGLE GROOVE WELD SYMBOL
PLUG WELDING SYMBOL
SIZE (DIA OF HOLE AT ROOT)
INCLUDED ANGLE OF COUNTERSINK
PITCH (DISTANCE BETWEEN CENTERS) OF WELDS.
DEPTH OF FILLING (INCHES) OMISSION INDICATES FILLING IS COMPLETE
1/2 1/2 6 45°
SURFACING WELD SYMBOL INDICATING BUILT-UP SURFACE
SIZE (HEIGHT OF DEPOSIT. OMISSION INDICATES NO SPECIFIC HEIGHT DESIRED
ORIENTATION, LOCATION AND ALL DIMENSIONS OTHER THAN SIZE ARE SHOWN ON THE DRAWING
1/8
SLOT WELDING SYMBOL
DEPTH OF FILLING INCHES. OMISSION INDICATES FILLING IS COMPLETE.
5/8
ORIENTATION, LOCATION AND ALL DIMENSIONS OTHER THAN DEPTH OF FILLING ARE SHOWN ON THE DRAWING
DOUBLE-FILLET WELDING SYMBOL
SIZE (LENGTH OF LEG)
SPECIFICATION, PROCESS OR OTHER REFERENCE
LENGTH. OMISSION INDICATES THAT WELD EXTENDS BETWEEN ABRUPT CHANGES IN DIRECTION OR AS DIMENSIONED
IG 1/4 12
RESISTANCE-SPOT WELDING SYMBOL
SIZE (DIA OF WELD) MIN. ACCEPTABLE SHEAR STRENGTH IN LB. PER WELD MAY BE USED INSTEAD.
(5) NUMBER OF WELDS
.25 4
PITCH (DISTANCE BETWEEN CENTERS) OF INCREMENTS.
CHAIN-INTERMITTENT-FILLET WELDING SYMBOL
SIZE (LENGTH OF LEG)
LENGTH OF INCREMENTS
5/16 2-6
PITCH (DISTANCE BETWEEN CENTERS) OF INCREMENTS
PROJECTION WELDING SYMBOL
SIZE (MIN. ACCEPTABLE SHEAR STRENGTH IN LB. PER WELD) DIA OF WELD MAY BE USED INSTEAD FOR CIRCULAR PROJECTION WELDS.
500 6 (4)
PITCH (DISTANCE BETWEEN CENTERS) OF WELDS.
NUMBER OF WELDS
STAGGERED-INTERMITTENT-FILLET WELDING SYMBOL
SIZE (LENGTH OF LEG)
LENGTH OF INCREMENTS
1/2 3-8
PITCH (DISTANCE BETWEEN CENTERS) OF INCREMENTS
RESISTANCE-SEAM WELDING SYMBOL
SIZE (WIDTH OF WELD) MIN. ACCEPTABLE SHEAR STRENGTH IN LB. PER LINEAR INCH MAY BE USED INSTEAD.
.30" 3-9
PITCH (DISTANCE BETWEEN CENTERS) OF INCREMENTS.
LENGTH OF WELDS OR INCREMENTS. OMISSION INDICATES THAT WELD EXTENDS BETWEEN ABRUPT CHANGES IN DIRECTION OR AS DIMENSIONED.
SINGLE-V GROOVE WELDING SYMBOL
SIZE (DEPTH OF CHAMFERING OMISSION INDICATES DEPTH OF CHAMFERING EQUAL TO THICKNESS OF MEMBERS.
1/2 1/8 60°
ROOT OPENING
GROOVE ANGLE
FLASH OR UPSET WELDING SYMBOL
FW
PROCESS REFERENCE MUST BE USED TO INDICATE PROCESS DESIRED
SINGLE-V GROOVE WELDING SYMBOL INDICATING ROOT PENETRATION
SIZE
DEPTH OF CHAMFERING PLUS ROOT PENETRATION
1/4 + 1/2 0 90°
ROOT OPENING
GROOVE ANGLE
BRAZING, FORGE, THERMIT, INDUCTION AND FLOW WELDING SYMBOL
PROCESS REFERENCE MUST BE USED TO INDICATE PROCESS DESIRED.
TB
DOUBLE-BEVEL GROOVE WELDING SYMBOL
OMISSION OF SIZE DIMENSION INDICATES A TOTAL DEPTH OF CHAMFERING EQUAL TO THICKNESS OF MEMBERS.
GROOVE ANGLE
ARROW POINTS TOWARD MEMBER TO BE CHAMFERED
ROOT OPENING
1/8 50°
MELT-THRU WELD SYMBOL
MELT-THRU SYMBOL IS NOT DIMENSIONED
ANY APPLICABLE GROOVE OR FLANGE WELD
WELDING SYMBOLS FOR COMBINED WELDS
1/4 1/8 60° 5/16 T-3 3/8 1/8 1/4
EDGE- AND CORNER-FLANGE WELD SYMBOL
1/8 + 1/8
RADIUS
SIZE OF WELD
1/16 + 3/32 1/16
HEIGHT ABOVE POINT OF TANGENCY
FIG. E—SUPPLEMENTARY SYMBOLS USED WITH WELDING SYMBOLS
WELD-ALL-AROUND SYMBOL
WELD-ALL-AROUND SYMBOL INDICATES THAT WELD EXTENDS COMPLETELY AROUND THE JOINT.
FIELD WELD SYMBOL
FIELD WELD SYMBOL INDICATES THAT WELD IS TO BE MADE AT A PLACE OTHER THAN THAT OF INITIAL CONSTRUCTION.
FLUSH-CONTOUR SYMBOL
FLUSH-CONTOUR SYMBOL INDICATES FACE OF WELD TO BE MADE FLUSH. WHEN USED WITHOUT A FINISH SYMBOL, INDICATES WELD TO BE WELDED FLUSH WITHOUT SUBSEQUENT FINISHING.
M
FINISH SYMBOL (USERS STD.) INDICATES METHOD OF OBTAINING SPECIFIED CONTOUR BUT NOT DEGREE OF FINISH.
CONVEX-CONTOUR SYMBOL
CONVEX-CONTOUR SYMBOL. INDICATES FACE OF WELD TO BE FINISHED TO CONVEX CONTOUR.
G
FINISH SYMBOL (USERS STD.) INDICATES METHOD OF OBTAINING SPECIFIED CONTOUR BUT NOT DEGREE OF FINISH.

STRUCTURAL—WELDING

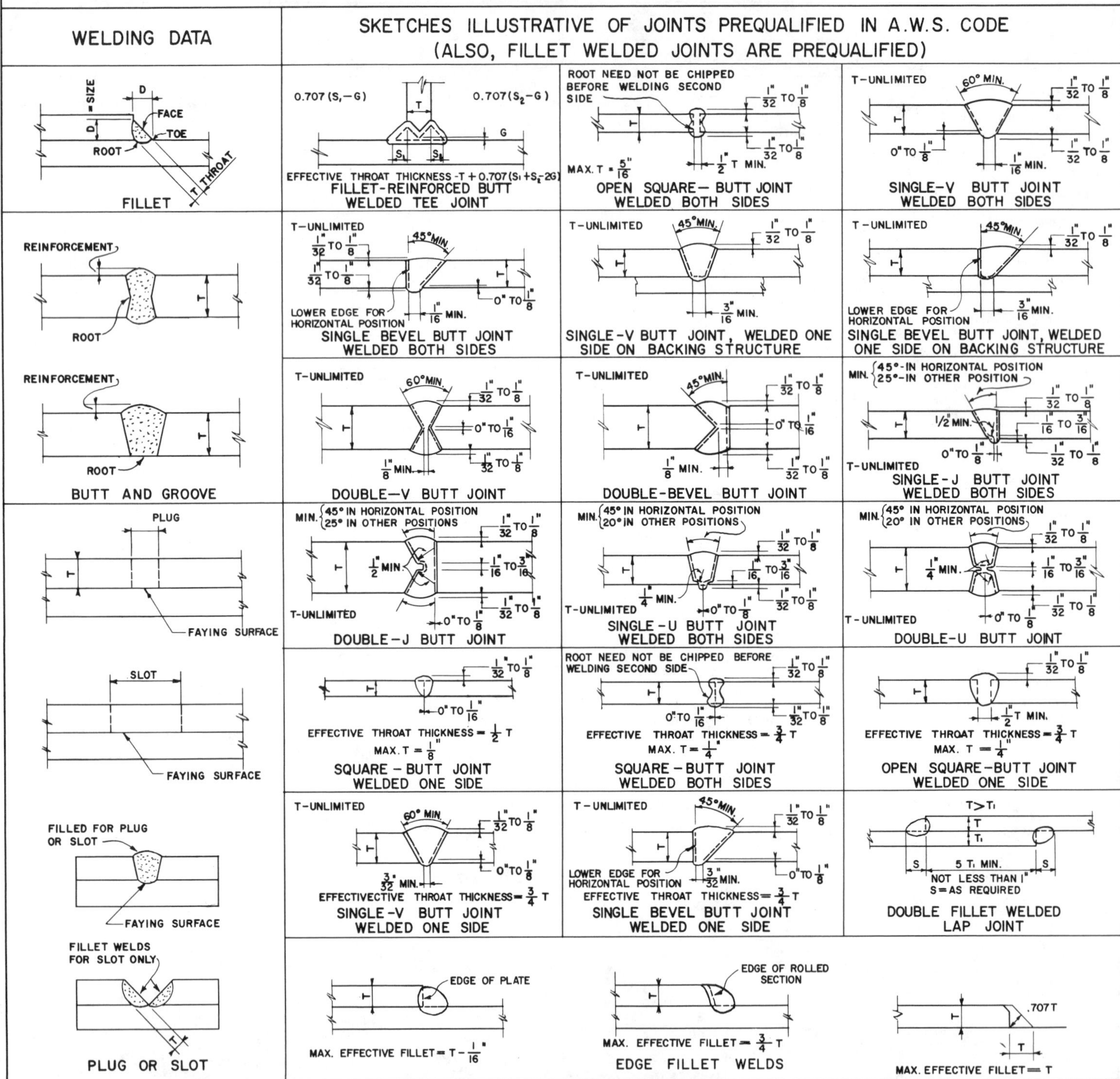

WELDING DATA
SKETCHES ILLUSTRATIVE OF JOINTS PREQUALIFIED IN A.W.S. CODE
(ALSO, FILLET WELDED JOINTS ARE PREQUALIFIED)
D = SIZE
FACE
TOE
ROOT
THROAT
FILLET
0.707 (S₁−G)
0.707 (S₂−G)
EFFECTIVE THROAT THICKNESS -T + 0.707 (S₁+S₂-2G)
FILLET-REINFORCED BUTT WELDED TEE JOINT
ROOT NEED NOT BE CHIPPED BEFORE WELDING SECOND SIDE
MAX. T = 5/16"
OPEN SQUARE— BUTT JOINT WELDED BOTH SIDES
T-UNLIMITED
60° MIN.
SINGLE-V BUTT JOINT WELDED BOTH SIDES
REINFORCEMENT
ROOT
45° MIN.
LOWER EDGE FOR HORIZONTAL POSITION
SINGLE BEVEL BUTT JOINT WELDED BOTH SIDES
SINGLE-V BUTT JOINT, WELDED ONE SIDE ON BACKING STRUCTURE
SINGLE BEVEL BUTT JOINT, WELDED ONE SIDE ON BACKING STRUCTURE
BUTT AND GROOVE
DOUBLE—V BUTT JOINT
DOUBLE-BEVEL BUTT JOINT
MIN. 45°-IN HORIZONTAL POSITION
25°-IN OTHER POSITION
SINGLE-J BUTT JOINT WELDED BOTH SIDES
PLUG
FAYING SURFACE
MIN. 45° IN HORIZONTAL POSITION
25° IN OTHER POSITIONS
DOUBLE-J BUTT JOINT
MIN. 45° IN HORIZONTAL POSITION
20° IN OTHER POSITIONS
SINGLE-U BUTT JOINT WELDED BOTH SIDES
DOUBLE-U BUTT JOINT
SLOT
EFFECTIVE THROAT THICKNESS = 1/2 T
MAX. T = 1/8"
SQUARE - BUTT JOINT WELDED ONE SIDE
ROOT NEED NOT BE CHIPPED BEFORE WELDING SECOND SIDE
EFFECTIVE THROAT THICKNESS = 3/4 T
MAX. T = 1/4"
SQUARE - BUTT JOINT WELDED BOTH SIDES
OPEN SQUARE-BUTT JOINT WELDED ONE SIDE
FILLED FOR PLUG OR SLOT
EFFECTIVE THROAT THICKNESS = 3/4 T
SINGLE-V BUTT JOINT WELDED ONE SIDE
SINGLE BEVEL BUTT JOINT WELDED ONE SIDE
T>T₁
5 T₁ MIN.
NOT LESS THAN 1"
S = AS REQUIRED
DOUBLE FILLET WELDED LAP JOINT
FILLET WELDS FOR SLOT ONLY
PLUG OR SLOT
EDGE OF PLATE
MAX. EFFECTIVE FILLET = T − 1/16"
EDGE OF ROLLED SECTION
MAX. EFFECTIVE FILLET = 3/4 T
EDGE FILLET WELDS
.707T
MAX. EFFECTIVE FILLET = T

TABLE A – SIZES OF FILLET WELD		
MINIMUM SIZE OF FILLET WELD-INCHES	MINIMUM THICKNESS OF PART IN INCHES	MAXIMUM SIZE
3/16"	1/2"	SIZE ASSUMED IN DESIGN OF CONNECTION SHALL BE SUCH THAT ALLOWABLE STRESSES IN ADJACENT BASE MATERIAL SHALL NOT BE EXCEEDED.
1/4"	3/4"	
5/16"	1 1/4"	
3/8"	2"	
1/2"	6"	
5/8"	OVER 6"	

NOTE:

1. LENGTH OF WELDS

THE EFFECTIVE LENGTH OF FILLET WELDS SHALL BE THE OVERALL LENGTH OF FULL SIZE FILLET INCLUDING RETURNS AROUND CORNERS.

THE EFFECTIVE LENGTH OF BUTT WELD SHALL BE THE WIDTH OF PART JOINED.

THE EFFECTIVE LENGTH OF FILLET WELDED PLUG OR SLOT WELDS SHALL BE THE LENGTH OF THE CENTER LINE THRU THE CENTER PLANE THRU THROAT.

2. EFFECTIVE AREAS OF WELD METAL

BUTT AND FILLET WELD = EFFECTIVE LENGTH TIMES THROAT THICKNESS.

FILLED PLUG OR SLOT WELD = NOMINAL CROSS-SECTION OF HOLE OR SLOT IN PLANE OF FAYING SURFACE.

FILLET WELDED PLUG OR SLOT WELDS = SAME AS FILLET WELDS.

ADAPTED FROM WELDING HANDBOOK, AMERICAN WELDING SOCIETY.

STRUCTURAL STEEL-WELDING-DESIGN

DESIGN OF WELDS

P = LOAD IN KIPS. D = WELD SIZE IN INCHES. T = THROAT IN INCHES. L = LENGTH IN INCHES.

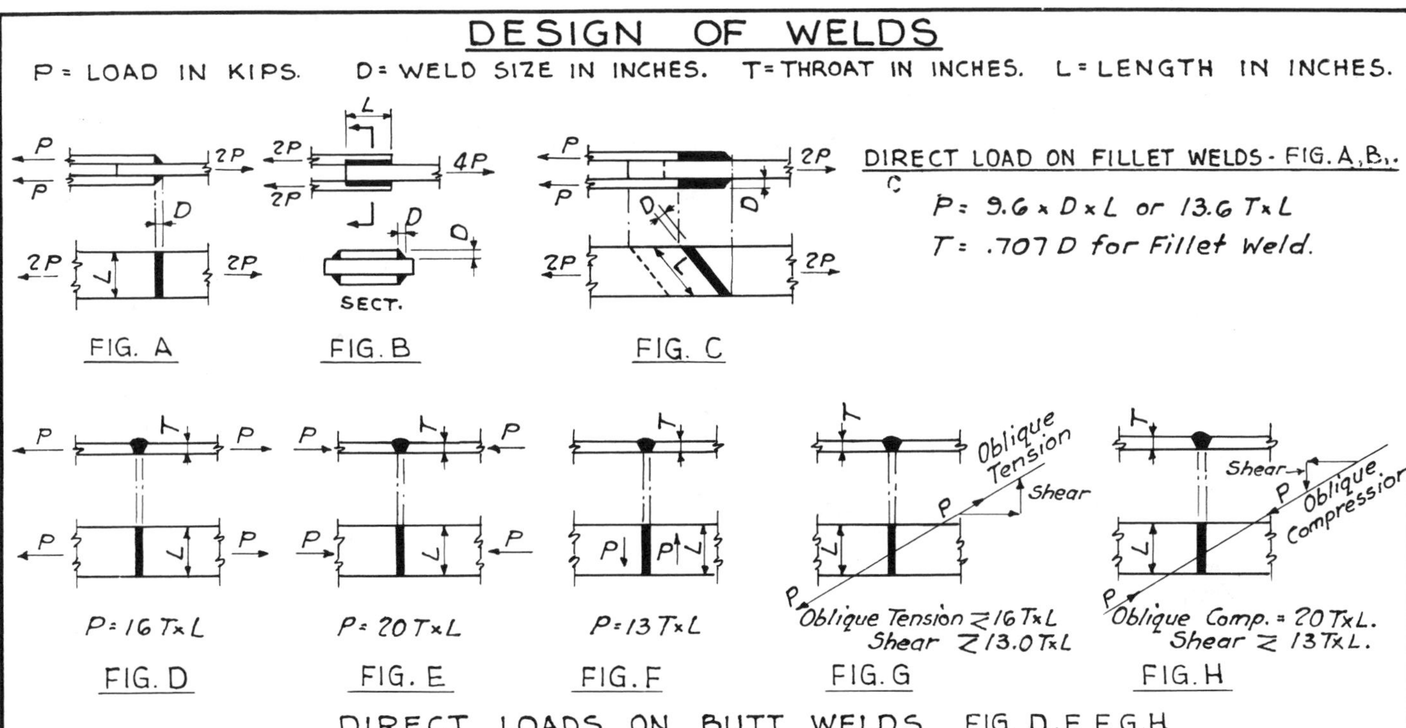

P
2P
4P
L
D
SECT.
FIG. A
FIG. B
FIG. C
Oblique Tension
Shear
Oblique Compression
FIG. D
FIG. E
FIG. F
FIG. G
FIG. H

DIRECT LOAD ON FILLET WELDS - FIG. A, B, C

$P = 9.6 \times D \times L$ or $13.6\ T \times L$

$T = .707\ D$ for Fillet Weld.

FIG. D: $P = 16\ T \times L$

FIG. E: $P = 20\ T \times L$

FIG. F: $P = 13\ T \times L$

FIG. G: Oblique Tension $\geqq 16\ T \times L$; Shear $\geqq 13.0\ T \times L$

FIG. H: Oblique Comp. $= 20\ T \times L$; Shear $\geqq 13\ T \times L$

DIRECT LOADS ON BUTT WELDS FIG. D, E, F, G, H

ECCENTRIC LOADING

MANY WELDED CONNECTIONS INVOLVE COMBINED STRESSES WHICH MAY BE SOLVED AS FOLLOWS:-

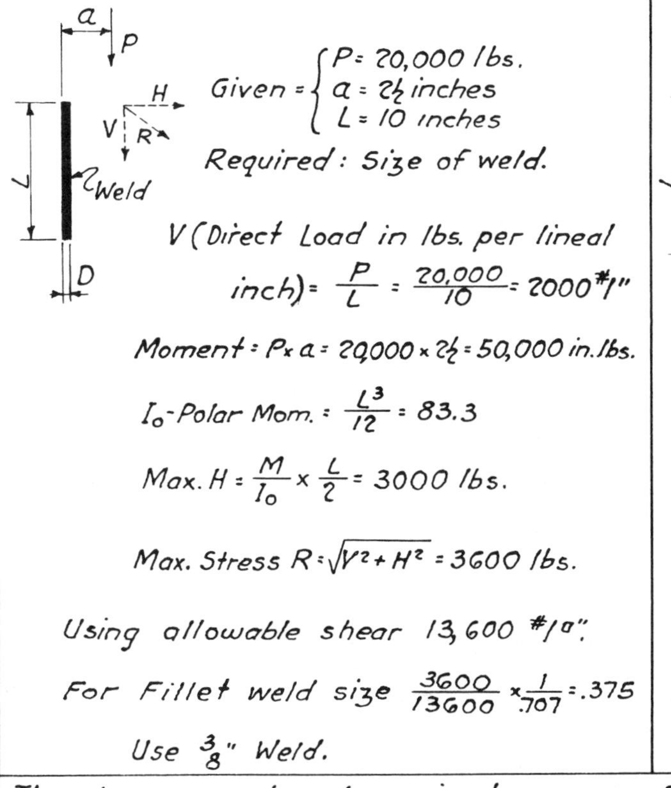

a
P
H
V
R
L
Weld
D

Given = { P = 20,000 lbs., a = 2½ inches, L = 10 inches }

Required: Size of weld.

V (Direct Load in lbs. per lineal inch) $= \frac{P}{L} = \frac{20,000}{10} = 2000\#/''$

Moment $= P \times a = 20,000 \times 2\frac{1}{2} = 50,000$ in. lbs.

I_o-Polar Mom. $= \frac{L^3}{12} = 83.3$

Max. $H = \frac{M}{I_o} \times \frac{L}{2} = 3000$ lbs.

Max. Stress $R = \sqrt{V^2 + H^2} = 3600$ lbs.

Using allowable shear 13,600 #/□".

For Fillet weld size $\frac{3600}{13600} \times \frac{1}{.707} = .375$

Use $\frac{3}{8}$" Weld.

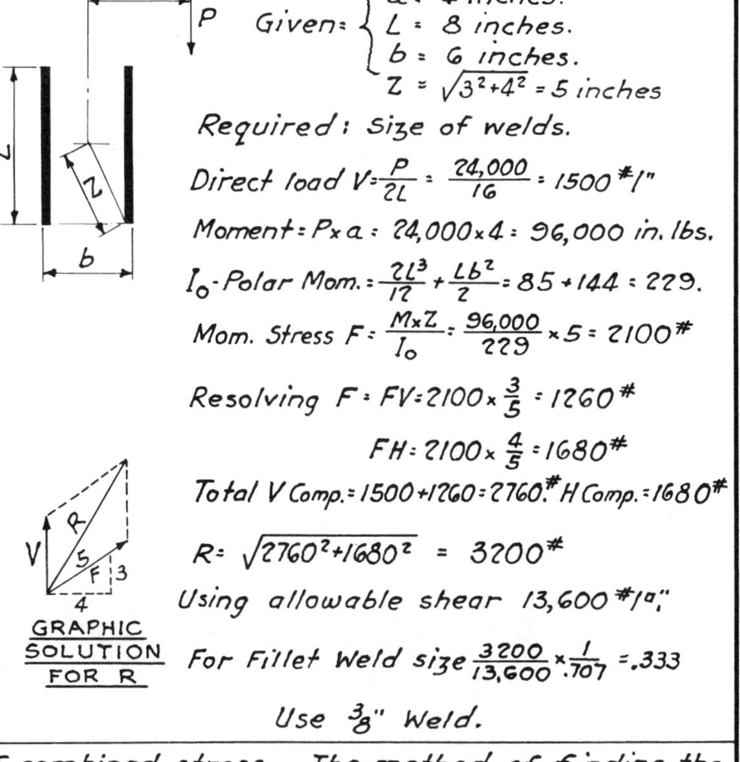

a
P
L
Z
b
V
R
5
F
3
4
GRAPHIC SOLUTION FOR R

Given: { P = 24,000 lbs., a = 4 inches, L = 8 inches, b = 6 inches, $Z = \sqrt{3^2+4^2} = 5$ inches }

Required: Size of welds.

Direct load $V = \frac{P}{2L} = \frac{24,000}{16} = 1500\#/''$

Moment $= P \times a = 24,000 \times 4 = 96,000$ in. lbs.

I_o-Polar Mom. $= \frac{2L^3}{12} + \frac{Lb^2}{2} = 85 + 144 = 229.$

Mom. Stress $F = \frac{M \times Z}{I_o} = \frac{96,000}{229} \times 5 = 2100\#$

Resolving $F = FV = 2100 \times \frac{3}{5} = 1260\#$

$FH = 2100 \times \frac{4}{5} = 1680\#$

Total V Comp. = 1500 + 1260 = 2760.# H Comp. = 1680#

$R = \sqrt{2760^2 + 1680^2} = 3200\#$

Using allowable shear 13,600 #/□".

For Fillet Weld size $\frac{3200}{13,600} \times \frac{1}{.707} = .333$

Use $\frac{3}{8}$" Weld.

The above examples show simple cases of combined stress. The method of finding the stress due to moment by determining the Polar Moment of Inertia of weld group, solving for moment stress and combining with direct stress may be used for any weld pattern. Tables showing certain useful groups and formulas are shown on pp. 4-54 to 4-57.

The above data are based on allowable unit stresses of A.I.S.C. Code 1951, as shown on Page 1-02.

STRUCTURAL STEEL-ALLOWABLE ECCENTRIC LOADINGS-I

ECCENTRIC LOADS ON WELD GROUPS

TABLE A

a	K															
	0	0.1	0.2	0.3	0.4	0.5	0.6	0.7	0.8	0.9	1.0	1.2	1.4	1.6	1.8	2.0
0.2	.545	.553	.576	.611	.652	.698	.743	.787	.827	.866	.899	.956	1.00	1.04	1.06	1.08
0.3	.429	.436	.457	.489	.529	.573	.620	.666	.710	.752	.790	.858	.913	.958	.995	1.02
0.4	.353	.359	.378	.407	.444	.486	.530	.575	.619	.662	.702	.775	.836	.888	.931	.968
0.5	.300	.306	.323	.349	.383	.422	.463	.506	.548	.590	.630	.704	.771	.825	.872	.913
0.6	.261	.266	.281	.306	.336	.372	.411	.451	.491	.532	.571	.644	.710	.768	.818	.873
0.7	.231	.236	.250	.272	.300	.333	.369	.407	.445	.483	.521	.593	.658	.717	.769	.815
0.8	.207	.211	.224	.244	.270	.301	.335	.373	.406	.443	.479	.548	.613	.672	.724	.771
0.9	.187	.189	.203	.222	.246	.275	.306	.339	.374	.408	.443	.510	.573	.631	.683	.731
1.0	.171	.175	.186	.203	.226	.253	.282	.314	.346	.379	.412	.476	.537	.594	.646	.694
1.2	.146	.150	.159	.174	.194	.218	.244	.272	.301	.331	.360	.420	.477	.531	.582	.629
1.4	.128	.131	.139	.153	.170	.191	.215	.240	.266	.293	.320	.375	.428	.479	.528	.573
1.6	.113	.116	.123	.136	.151	.170	.192	.215	.239	.263	.288	.339	.388	.436	.482	.526
1.8	.102	.104	.111	.122	.136	.154	.173	.194	.216	.239	.262	.309	.355	.405	.444	.485
2.0	.092	.094	.101	.111	.124	.140	.158	.177	.197	.219	.240	.283	.327	.369	.410	.450
2.2	.084	.086	.092	.102	.114	.129	.145	.163	.182	.201	.221	.262	.303	.343	.382	.419
2.4	.078	.080	.085	.094	.105	.119	.134	.151	.168	.187	.205	.243	.282	.319	.356	.392
2.6	.072	.074	.079	.087	.098	.110	.125	.140	.157	.174	.192	.227	.263	.299	.334	.369
2.8	.067	.069	.074	.081	.091	.103	.117	.131	.147	.163	.180	.213	.247	.281	.315	.328
3.0	.063	.065	.069	.076	.086	.097	.109	.123	.138	.153	.169	.201	.233	.265	.297	.328

P = permissible eccentric load in kips
l = length of each weld in inches
D = number of sixteenths of an inch in fillet-weld size
C = coefficients tabulated below

$$P = CDl$$

Required minimum

$$C = \frac{P}{Dl} \qquad D = \frac{P}{Cl} \qquad l = \frac{P}{CD}$$

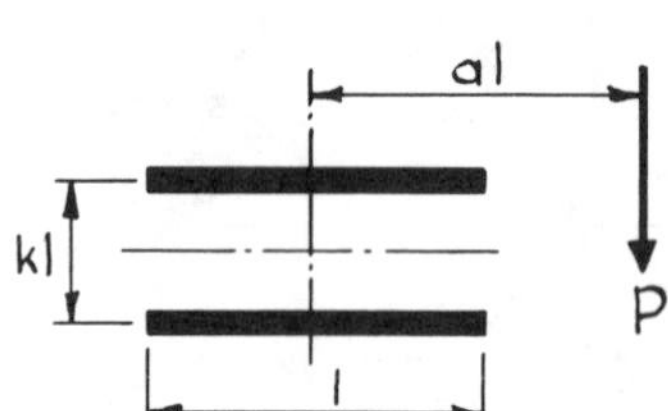
al
kl
P
l

TABLE B

a	K															
	0	0.1	0.2	0.3	0.4	0.5	0.6	0.7	0.8	0.9	1.0	1.2	1.4	1.6	1.8	2.0
0.2	.768	.744	.741	.753	.773	.796	.819	.842	.862	.882	.899	.930	.955	.976	.994	1.01
0.3	.582	.570	.577	.597	.625	.655	.686	.716	.742	.768	.790	.830	.863	.890	.914	.934
0.4	.462	.455	.466	.489	.519	.552	.586	.618	.648	.677	.702	.747	.785	.817	.845	.869
0.5	.380	.377	.389	.412	.441	.475	.509	.542	.573	.603	.630	.678	.719	.754	.785	.811
0.6	.322	.320	.332	.355	.383	.415	.449	.481	.513	.543	.571	.620	.663	.700	.732	.760
0.7	.278	.278	.290	.311	.338	.369	.401	.433	.464	.493	.521	.571	.615	.653	.686	.716
0.8	.244	.246	.257	.276	.302	.331	.362	.392	.423	.452	.479	.529	.572	.611	.645	.675
0.9	.218	.220	.230	.249	.273	.300	.329	.359	.388	.416	.443	.492	.536	.574	.608	.639
1.0	.198	.199	.209	.226	.249	.274	.302	.330	.358	.386	.412	.460	.503	.541	.576	.607
1.2	.166	.167	.176	.191	.211	.234	.259	.285	.311	.336	.360	.406	.448	.486	.520	.551
1.4	.142	.144	.152	.165	.183	.204	.227	.250	.274	.298	.320	.364	.404	.440	.473	.504
1.6	.124	.126	.133	.146	.162	.181	.201	.223	.245	.266	.288	.329	.367	.402	.435	.464
1.8	.110	.112	.119	.130	.145	.162	.181	.201	.221	.242	.262	.301	.337	.375	.402	.430
2.0	.100	.101	.107	.118	.131	.147	.165	.183	.202	.221	.240	.277	.311	.343	.374	.401
2.2	.090	.092	.098	.107	.120	.134	.151	.168	.186	.204	.221	.256	.289	.320	.348	.375
2.4	.084	.085	.090	.099	.110	.124	.139	.155	.172	.189	.205	.238	.269	.299	.327	.353
2.6	.076	.078	.083	.091	.102	.115	.129	.144	.160	.176	.192	.223	.253	.284	.308	.333
2.8	.072	.073	.077	.085	.095	.107	.120	.135	.149	.164	.180	.209	.238	.265	.291	.315
3.0	.066	.068	.072	.079	.089	.100	.113	.126	.140	.155	.169	.197	.224	.251	.275	.300

P = Permissible eccentric load in kips
l = length of each weld in inches
D = number of sixteenths of an inch in fillet-weld size
C = coefficients tabulated below

$$P = CDl$$

Required minimum

$$C = \frac{P}{Dl} \qquad D = \frac{P}{Cl} \qquad l = \frac{P}{CD}$$

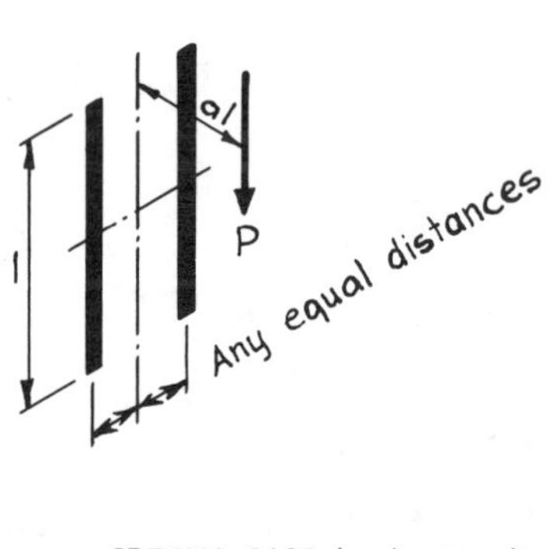
al
P
l
kl
Any equal distances

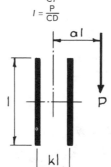

SPECIAL CASE: Load not in plane of weld group— use C values given in column headed *K* = 0.

TABLE C

a	K															
	0.0	0.1	0.2	0.3	0.4	0.5	0.6	0.7	0.8	0.9	1.0	1.2	1.4	1.6	1.8	2.0
0.2	.384	.516	.575	.654	.731	.810	.889	.969	1.05	1.14	1.22	1.40	1.58	1.78	1.97	2.17
0.3	.291	.422	.472	.543	.612	.680	.749	.820	.893	.966	1.04	1.20	1.37	1.54	1.72	1.91
0.4	.231	.351	.395	.460	.522	.583	.644	.708	.772	.839	.908	1.05	1.20	1.36	1.52	1.69
0.5	.190	.298	.338	.397	.453	.508	.564	.621	.680	.740	.802	.932	1.07	1.21	1.36	1.52
0.6	.161	.258	.294	.348	.399	.450	.500	.552	.605	.661	.718	.836	.963	1.09	1.24	1.38
0.7	.139	.227	.260	.310	.357	.403	.449	.497	.546	.596	.649	.759	.875	.998	1.13	1.27
0.8	.122	.202	.233	.278	.322	.364	.407	.451	.497	.543	.592	.694	.802	.917	1.04	1.17
0.9	.109	.182	.211	.253	.293	.333	.372	.414	.455	.499	.544	.639	.741	.849	.963	1.09
1.0	.099	.166	.192	.231	.269	.306	.343	.381	.420	.461	.503	.592	.687	.788	.897	1.01
1.2	.083	.140	.163	.198	.231	.263	.296	.330	.364	.400	.437	.516	.601	.691	.788	.892
1.4	.071	.121	.142	.172	.202	.231	.260	.290	.321	.353	.387	.457	.533	.615	.703	.796
1.6	.062	.107	.125	.153	.179	.205	.232	.259	.287	.316	.346	.410	.480	.554	.634	.719
1.8	.055	.096	.112	.137	.161	.185	.209	.234	.259	.286	.314	.372	.436	.504	.578	.656
2.0	.050	.086	.102	.124	.146	.168	.190	.213	.237	.261	.287	.341	.399	.462	.531	.603
2.2	.045	.079	.093	.114	.134	.154	.175	.196	.218	.240	.264	.314	.368	.427	.490	.558
2.4	.042	.072	.085	.105	.124	.142	.162	.181	.201	.221	.244	.291	.342	.397	.456	.519
2.6	.038	.067	.079	.097	.115	.132	.150	.168	.187	.207	.228	.271	.319	.370	.420	.485
2.8	.036	.062	.074	.091	.107	.124	.140	.157	.175	.194	.213	.254	.299	.347	.399	.456
3.0	.033	.058	.069	.085	.100	.116	.132	.143	.164	.182	.200	.239	.281	.327	.376	.429
X	.0	.008	.029	.056	.089	.125	.164	.204	.246	.289	.333	.424	.516	.601	.704	.800

P = permissible eccentric load in kips
l = length of weld parallel to load *P* in inches
D = number of sixteenths of an inch in fillet-weld size
C = coefficients tabulated below
xl = Distance from vertical weld to center of gravity of weld group

$$P = CDl$$

Required minimum

$$C = \frac{P}{Dl} \qquad D = \frac{P}{Cl} \qquad l = \frac{P}{CD}$$

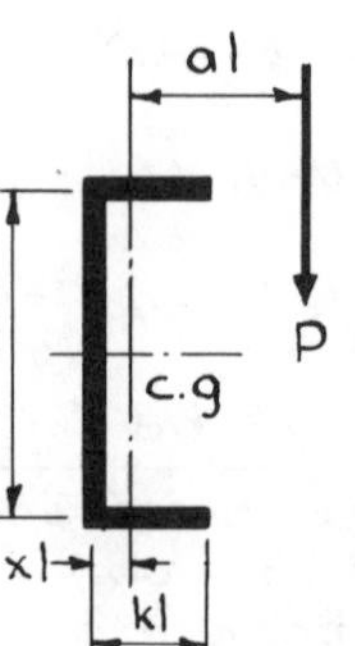
al
l
P
c.g
xl
kl

STRUCTURAL STEEL–ALLOWABLE ECCENTRIC LOADINGS–2

ECCENTRIC LOADS ON WELD GROUPS

TABLE A

a	K															
	0	0.1	0.2	0.3	0.4	0.5	0.6	0.7	.08	0.9	1.0	1.2	1.4	1.6	1.8	2.0
0.2	.384	.522	.655	.765	.868	.965	1.06	1.15	1.25	1.34	1.43	1.61	1.80	1.99	2.19	2.38
0.3	.291	.428	.544	.650	.748	.842	.930	1.02	1.10	1.19	1.27	1.44	1.61	1.79	1.97	2.17
0.4	.231	.357	.455	.554	.647	.735	.820	.902	.932	1.06	1.14	1.30	1.46	1.62	1.79	1.96
0.5	.190	.303	.387	.478	.564	.647	.727	.804	.881	.955	1.03	1.18	1.33	1.48	1.64	1.80
0.6	.161	.263	.335	.418	.498	.575	.650	.724	.795	.866	.936	1.08	1.22	1.36	1.51	1.66
0.7	.139	.231	.294	.369	.443	.516	.587	.655	.723	.790	.856	.988	1.12	1.26	1.40	1.54
0.8	.122	.206	.262	.331	.399	.467	.533	.598	.662	.726	.787	.913	1.04	1.17	1.30	1.44
0.9	.109	.186	.236	.299	.362	.425	.487	.549	.610	.670	.729	.847	.968	1.08	1.22	1.35
1.0	.099	.169	.214	.272	.331	.390	.449	.507	.564	.622	.678	.791	.905	1.02	1.14	1.26
1.2	.083	.143	.181	.231	.283	.335	.337	.439	.491	.542	.594	.697	.801	.906	1.02	1.13
1.4	.071	.124	.156	.200	.246	.293	.340	.387	.434	.481	.528	.622	.717	.815	.915	1.02
1.6	.062	.109	.137	.177	.217	.257	.300	.345	.388	.432	.474	.561	.649	.739	.832	.926
1.8	.056	.097	.122	.156	.191	.227	.263	.302	.344	.387	.431	.511	.593	.676	.762	.850
2.0	.050	.088	.111	.140	.171	.202	.235	.269	.305	.343	.380	.469	.545	.623	.703	.785
2.2	.045	.080	.100	.127	.154	.182	.211	.242	.274	.308	.344	.423	.505	.578	.653	.730
2.4	.042	.074	.091	.116	.141	.166	.192	.220	.249	.276	.312	.382	.462	.539	.609	.682
2.6	.038	.068	.084	.107	.129	.152	.176	.201	.228	.255	.285	.349	.421	.501	.571	.640
2.8	.036	.063	.077	.099	.120	.141	.163	.186	.210	.235	.262	.321	.382	.460	.537	.602
3.0	.033	.059	.073	.092	.113	.131	.151	.173	.195	.218	.243	.297	.358	.425	.500	.569
X	.0	.008	.029	.056	.089	.125	.164	.204	.246	.289	.333	.424	.516	.601	.704	.800

P = permissible eccentric load in kips
l = length of weld parallel to load P in inches
D = number of sixteenths of an inch in fillet-weld size
C = coefficients tabulated below.
xl = distance from vertical weld to center of gravity of weld group

$$P = CDl$$

Required Minimum

$$C = \frac{P}{Dl}$$

$$D = \frac{P}{Cl}$$

$$l = \frac{P}{CD}$$

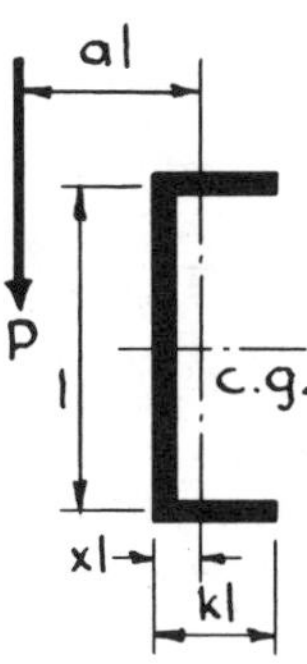

al
P
l
c.g.
xl
kl

TABLE B

a	K										
	0	0.1	0.2	0.3	0.4	0.5	.06	0.7	0.8	0.9	1.0
0.2	.768	.892	1.01	1.11	1.21	1.31	1.41	1.50	1.60	1.70	1.80
0.3	.582	.703	.814	.920	1.02	1.12	1.21	1.30	1.40	1.49	1.58
0.4	.462	.570	.675	.775	.871	.963	1.05	1.14	1.23	1.32	1.41
0.5	.380	.474	.571	.665	.756	.844	.929	1.01	1.10	1.18	1.26
0.6	.322	.407	.494	.580	.665	.747	.828	.908	.985	1.06	1.14
0.7	.278	.355	.434	.514	.592	.670	.745	.821	.894	.968	1.04
0.8	.244	.314	.386	.460	.533	.606	.677	.748	.819	.888	.958
0.9	.218	.281	.348	.416	.585	.553	.620	.687	.754	.820	.885
1.0	.198	.255	.316	.380	.444	.508	.571	.635	.698	.760	.823
1.2	.166	.214	.267	.323	.380	.437	.493	.550	.607	.664	.721
1.4	.142	.185	.231	.281	.331	.382	.434	.485	.537	.589	.641
1.6	.124	.162	.204	.248	.294	.340	.387	.434	.481	.529	.577
1.8	.110	.145	.182	.222	.264	.306	.348	.392	.436	.479	.523
2.0	.100	.131	.165	.201	.239	.278	.318	.358	.398	.439	.480
2.2	.090	.119	.150	.184	.219	.255	.292	.329	.367	.404	.443
2.4	.084	.109	.138	.169	.202	.235	.269	.304	.339	.375	.411
2.6	.076	.101	.128	.156	.187	.218	.250	.283	.316	.349	.383
2.8	.072	.094	.119	.146	.174	.204	.234	.264	.296	.327	.359
3.0	.066	.088	.111	.136	.163	.191	.219	.248	.278	.308	.338

P = permissible eccentric load in kips
l = length of longer welds in inches
D = number of sixteenths of an inch in fillet-weld size
C = coefficients tabulated below
NOTE: When load P is perpendicular to longer side l, use Table C.

$$P = CDl$$

Required minimum

$$C = \frac{P}{Dl}$$

$$D = \frac{P}{Cl}$$

$$l = \frac{P}{CD}$$

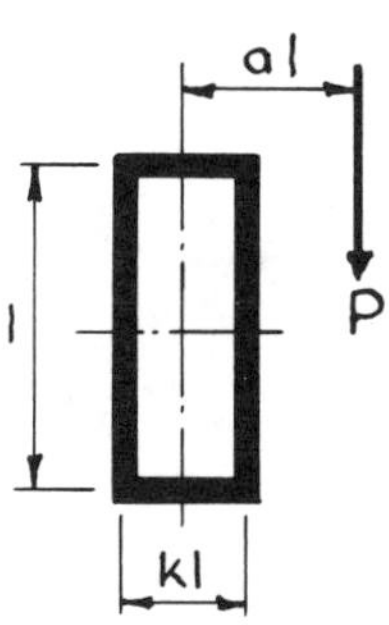

al
l
P
kl

TABLE C

a	K										
	0	0.1	0.2	0.3	0.4	0.5	0.6	0.7	0.8	0.9	1.0
0.2	.545	.662	.782	.905	1.03	1.16	1.28	1.41	1.54	1.67	1.80
0.3	.429	.530	.636	.746	.860	.976	1.10	1.22	1.34	1.46	1.58
0.4	.353	.442	.536	.635	.738	.844	.952	1.06	1.17	1.29	1.41
0.5	.300	.379	.463	.552	.645	.742	.842	.943	1.05	1.15	1.26
0.6	.261	.331	.407	.488	.573	.662	.753	.846	.943	1.04	1.14
0.7	.231	.294	.364	.438	.516	.597	.681	.768	.857	.948	1.04
0.8	.207	.265	.328	.397	.469	.544	.622	.703	.785	.871	.958
0.9	.187	.241	.299	.363	.429	.499	.574	.647	.725	.804	.885
1.0	.171	.221	.275	.334	.396	.462	.529	.599	.672	.746	.823
1.2	.146	.189	.237	.288	.343	.402	.460	.523	.587	.653	.721
1.4	.128	.166	.208	.253	.302	.354	.407	.463	.521	.580	.641
1.6	.113	.147	.185	.226	.270	.317	.365	.416	.468	.521	.577
1.8	.102	.132	.167	.204	.244	.287	.330	.377	.425	.474	.523
2.0	.092	.120	.162	.186	.223	.262	.303	.345	.389	.434	.480
2.2	.084	.110	.139	.171	.205	.241	.279	.318	.358	.400	.443
2.4	.078	.102	.129	.158	.190	.223	.258	.295	.332	.371	.411
2.6	.072	.095	.120	.147	.177	.208	.240	.275	.310	.346	.383
2.8	.067	.088	.112	.138	.165	.195	.227	.257	.290	.324	.359
3.0	.063	.083	.105	.129	.155	.183	.212	.242	.273	.305	.338

P = permissible eccentric load in kips
l = length of longer welds in inches
D = number of sixteenths of an inch in fillet-weld size
C = coefficients tabulated below
NOTE: When load P is parallel to longer side l, use Table B.

$$P = CDl$$

Required minimum

$$C = \frac{P}{Dl}$$

$$D = \frac{P}{Cl}$$

$$l = \frac{P}{CD}$$

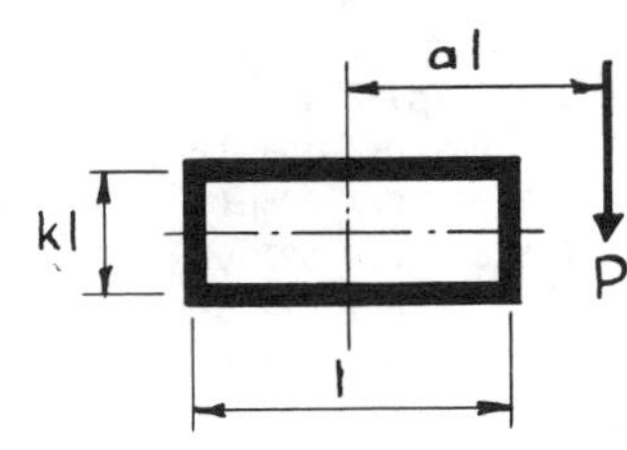

al
kl
P
l

STRUCTURAL STEEL-ALLOWABLE ECCENTRIC LOADINGS-3

ECCENTRIC LOADS ON WELD GROUPS

TABLE A

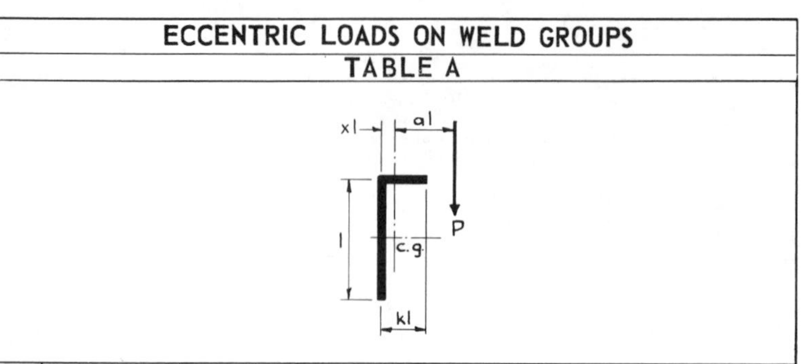

xl
al
P
l
c.g.
kl

a	K 0	0.1	0.2	0.3	0.4	0.5	0.6	0.7	0.8	0.9	1.0	1.2	1.4	1.6	1.8	2.0
0.2	.384	.434	.462	.484	.505	.527	.552	.582	.615	.651	.691	.778	.875	.977	1.09	1.20
0.3	.291	.337	.375	.394	.412	.431	.451	.476	.503	.534	.569	.645	.731	.823	.920	1.02
0.4	.231	.268	.301	.331	.347	.363	.381	.402	.426	.453	.483	.550	.627	.711	.800	.894
0.5	.190	.221	.248	.276	.299	.313	.329	.347	.368	.392	.419	.480	.549	.625	.707	.794
0.6	.161	.188	.211	.234	.263	.275	.290	.306	.325	.346	.370	.425	.489	.558	.634	.713
0.7	.139	.162	.182	.203	.225	.246	.258	.273	.290	.310	.332	.382	.440	.504	.574	.648
0.8	.122	.143	.161	.178	.198	.221	.233	.247	.262	.280	.300	.346	.400	.459	.525	.593
0.9	.109	.128	.143	.159	.176	.197	.212	.225	.239	.256	.274	.317	.367	.422	.483	.548
1.0	.099	.115	.129	.144	.159	.178	.195	.207	.220	.235	.252	.292	.338	.390	.447	.508
1.2	.083	.097	.108	.120	.133	.148	.167	.178	.189	.203	.218	.253	.293	.340	.390	.444
1.4	.071	.083	.093	.103	.114	.127	.143	.156	.166	.178	.191	.224	.259	.300	.346	.395
1.6	.062	.073	.081	.090	.100	.111	.125	.139	.148	.159	.171	.199	.232	.269	.310	.355
1.8	.055	.065	.073	.080	.089	.099	.111	.125	.134	.143	.154	.180	.210	.244	.282	.323
2.0	.050	.058	.065	.072	.080	.089	.100	.112	.122	.130	.140	.164	.191	.223	.258	.296
2.2	.045	.053	.059	.066	.073	.081	.091	.102	.112	.120	.129	.151	.176	.205	.238	.273
2.4	.042	.049	.055	.060	.066	.074	.083	.093	.103	.111	.119	.139	.163	.190	.220	.253
2.6	.038	.044	.050	.056	.062	.068	.077	.086	.096	.103	.111	.130	.152	.177	.205	.236
2.8	.036	.041	.047	.052	.057	.063	.071	.080	.090	.096	.105	.121	.142	.166	.192	.222
3.0	.033	.038	.044	.050	.054	.060	.066	.075	.084	.090	.097	.114	.133	.156	.181	.209
X	.0	.005	.017	.035	.057	.083	.113	.144	.178	.213	.250	.327	.408	.492	.579	.667

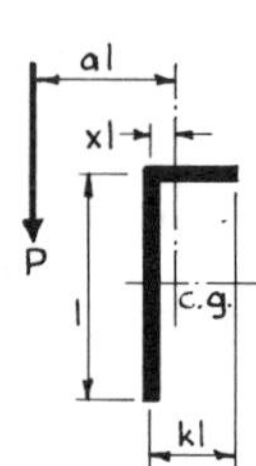

al
xl
P
l
c.g.
kl

TABLE C

a	K 0	0.1	0.2	0.3	0.4	0.5	0.6	0.7	0.8	0.9	1.0	1.2	1.4	1.6	1.8	2.0
0.2	.384	.436	.478	.516	.554	.595	.638	.684	.733	.784	.837	.945	1.06	1.17	1.28	1.39
0.3	.291	.335	.369	.400	.432	.467	.505	.547	.593	.641	.691	.796	.904	1.01	1.12	1.24
0.4	.231	.267	.295	.321	.348	.379	.412	.450	.490	.534	.581	.680	.783	.889	.996	1.10
0.5	.190	.220	.244	.266	.290	.316	.346	.379	.415	.456	.498	.590	.688	.788	.890	.993
0.6	.161	.187	.207	.227	.247	.270	.296	.326	.359	.395	.435	.520	.611	.706	.803	.902
0.7	.139	.162	.178	.197	.215	.236	.259	.286	.316	.349	.385	.463	.548	.638	.730	.824
0.8	.122	.143	.159	.174	.190	.209	.230	.254	.281	.312	.344	.417	.497	.581	.669	.758
0.9	.109	.127	.142	.156	.170	.187	.207	.229	.253	.281	.312	.379	.454	.533	.616	.702
1.0	.099	.115	.128	.141	.154	.170	.187	.208	.231	.256	.284	.347	.417	.493	.571	.652
1.2	.083	.096	.107	.118	.129	.143	.158	.175	.195	.217	.242	.297	.359	.427	.498	.572
1.4	.071	.083	.092	.102	.112	.123	.136	.151	.169	.189	.210	.259	.315	.376	.440	.508
1.6	.062	.073	.081	.089	.098	.108	.120	.133	.149	.171	.186	.233	.280	.336	.395	.457
1.8	.055	.065	.072	.079	.087	.096	.107	.119	.133	.149	.166	.207	.253	.303	.358	.421
2.0	.050	.058	.065	.072	.079	.087	.096	.108	.120	.135	.151	.187	.230	.276	.327	.380
2.2	.045	.053	.059	.065	.072	.079	.088	.098	.110	.123	.138	.171	.210	.254	.301	.351
2.4	.042	.049	.054	.060	.066	.073	.081	.090	.100	.113	.127	.158	.194	.235	.279	.325
2.6	.038	.045	.050	.055	.061	.067	.075	.083	.093	.105	.117	.147	.180	.218	.259	.303
2.8	.036	.042	.047	.051	.056	.062	.069	.077	.087	.097	.109	.137	.168	.204	.243	.284
3.0	.033	.039	.043	.048	.053	.058	.065	.072	.081	.092	.102	.128	.158	.191	.228	.267
X	.0	.005	.017	.035	.057	.083	.113	.144	.178	.213	.250	.327	.408	.492	.579	.667

TABLE B – STIFFENED WELDED BEAM SEATS

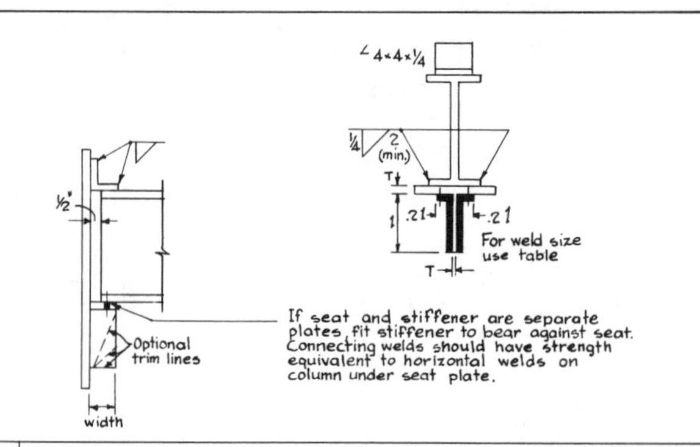

∠4×4×¼
¼
2 (min)
T
½"
.21
.21
For weld size use table
T
Optional trim lines
width
If seat and stiffener are separate plates, fit stiffener to bear against seat. Connecting welds should have strength equivalent to horizontal welds on column under seat plate.

	Width of Seat 4 in. Size of Welds 1/4"	4 in. 5/16"	4 in. 3/8"	4 in. 7/16"	5 in. Size of Welds 5/16"	5 in. 3/8"	5 in. 7/16"	5 in. 1/2"	6 in. Size of Welds 5/16"	6 in. 3/8"	6 in. 7/16"	6 in. 1/2"
6	15	18	22	26	15	18	21	24	–	–	–	21
7	19	24	29	34	21	25	29	33	–	21	25	28
8	24	31	37	43	26	31	36	41	22	27	31	35
9	30	37	45	52	32	38	44	51	28	33	39	44
10	35	44	53	62	38	46	53	61	33	40	47	53
11	41	51	62	72	45	54	63	72	39	47	55	63
12	47	59	71	83	52	62	73	83	46	55	64	73
13	–	67	80	93	59	71	83	94	52	63	73	84
14	–	74	89	104	66	80	93	106	59	71	83	95
15	–	82	99	115	74	89	103	118	66	80	93	106
16	–	90	108	126	81	98	114	130	74	89	103	118
17	–	98	117	137	89	107	125	143	81	97	114	130
18	–	106	127	148	97	119	138	158	89	106	124	142
19	–	113	136	–	105	126	146	167	96	115	135	154
20	–	121	145	–	112	135	157	–	104	125	145	166
21	–	128	–	–	120	144	168	–	112	134	156	178
22	–	137	–	–	128	154	–	–	119	143	167	191
23	–	145	–	–	136	163	–	–	127	153	178	–
24	–	–	–	–	144	173	–	–	135	162	189	–
25	–	–	–	–	152	–	–	–	143	171	–	–
26	–	–	–	–	159	–	–	–	151	181	–	–
27	–	–	–	–	161	–	–	–	158	190	–	–

	Width of Seat 7 in. Size of Welds 5/16"	7 in. 3/8"	7 in. 7/16"	7 in. 1/2"	8 in. Size of Welds 5/16"	8 in. 3/8"	8 in. 1/2"	8 in. 5/8"	9 in. Size of Welds 5/16"	9 in. 3/8"	9 in. 1/2"	9 in. 5/8"
11	35	42	49	56	–	–	–	63	–	–	–	–
12	41	49	57	65	–	–	59	73	–	–	–	–
13	47	56	66	75	–	–	68	84	–	–	–	77
14	53	64	75	85	–	58	77	97	–	–	–	88
15	60	72	84	96	55	65	87	109	–	–	80	100
16	67	80	94	107	61	73	98	122	–	–	90	112
17	74	89	104	118	68	81	108	135	–	75	100	124
18	81	97	114	130	75	89	119	149	–	82	110	137
19	88	106	124	142	82	98	130	163	75	90	121	151
20	96	115	134	153	89	106	142	177	82	99	131	164
21	103	124	145	165	96	115	153	192	89	107	143	178
22	111	133	155	178	103	124	165	206	96	115	154	192
23	119	142	166	190	111	133	177	221	103	124	165	207
24	126	152	177	202	118	142	189	236	111	133	177	221
25	134	161	188	214	126	151	201	–	118	142	189	236
26	142	170	199	–	133	160	213	–	125	151	201	251
27	150	180	209	–	141	169	226	–	133	160	212	266
28	157	189	220	–	149	179	238	–	141	169	225	–
29	165	198	–	–	157	188	–	–	148	178	237	–
30	173	208	–	–	164	197	–	–	156	187	249	–
31	181	217	–	–	172	207	–	–	164	196	262	–
32	189	–	–	–	180	216	–	–	171	206	–	–

STRUCTURAL STEEL—ALLOWABLE ECCENTRIC LOADINGS—4

TABLE A—WELDED BEAM WEB CONNECTIONS

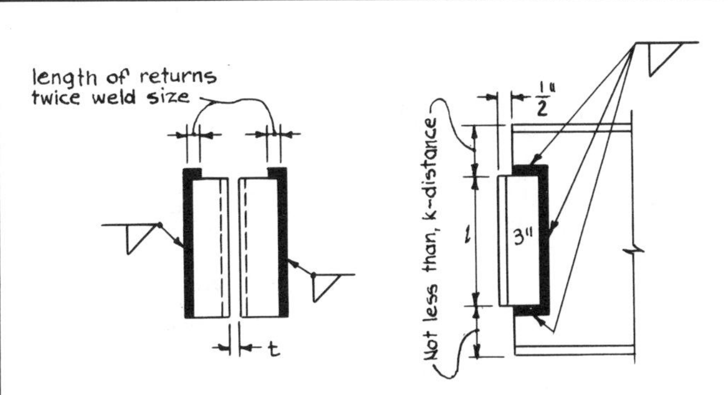

length of returns
twice weld size
t
1"/2
Not less than, k-distance
l
3"

l, in.	Size of Angles, in.	Weld Size Shop	Weld Size Field	R Value Shop Welds	R Value Field Welds	Minimum Beam Web Thickness
32	4x3x7/16	5/16	3/8	192	210	Web thickness of all 27, 30, 33, and 36 WF adequate to develop tabulated R Values.
31	4x3x7/16	5/16	3/8	186	202	
30	4x3x7/16	5/16	3/8	180	195	
29	4x3x7/16	5/16	3/8	173	187	
28	4x3x7/16	5/16	3/8	167	179	
27	4x3x7/16	5/16	3/8	160	172	
26	4x3x7/16	5/16	3/8	154	164	
25	4x3x7/16	5/16	3/8	147	156	
24	4x3x7/16	5/16	3/8	141	148	
23	4x3x7/16	5/16	3/8	139	140	
22	4x3x7/16	5/16	3/8	128	133	
21	4x3x7/16	5/16	3/8	121	125	
20	4x3x7/16	5/16	3/8	115	117	
19	4x3x7/16	5/16	3/8	108	109	.47
18	4x3x7/16	5/16	3/8	102	101	.47
18	3x3x3/8	5/16	5/16	102	93	.43
17	3x3x3/8	5/16	5/16	95	86	.42
16	3x3x3/8	5/16	5/16	89	80	.42
15	3x3x3/8	5/16	5/16	82	73	.42
14	3x3x7/16	5/16	3/8	76	80	.47
14	3x3x3/8	5/16	5/16	76	66	.41
13	3x3x7/16	5/16	3/8	69	72	.47
13	3x3x3/8	5/16	5/16	69	60	.40
12	3x3x7/16	5/16	3/8	64	64	.47
12	3x3x3/8	5/16	5/16	64	54	.40
11	3x3x7/16	5/16	3/8	57	57	.47
11	3x3x3/8	1/4	5/16	46	47	.38
10	3x3x7/16	5/16	3/8	51	49	.45
10	3x3x3/8	1/4	5/16	41	41	.38
9	3x3x7/16	5/16	3/8	45	41	.43
9	3x3x3/8	1/4	5/16	36	35	.38
8	3x3x7/16	5/16	3/8	39	34	.41
8	3x3x3/8	1/4	5/16	31	29	.35
7	3x3x7/16	5/16	3/8	33	27	.38
7	3x3x3/8	1/4	5/16	27	23	.32
6	3x3x7/16	1/4	3/8	22	21	.35
6	3x3x3/8	1/4	5/16	22	17.5	.29
5	3x3x7/16	1/4	3/8	18	15	.31
5	3x3x3/8	3/16	5/16	13.5	12.5	.26
4	3x3x7/16	3/16	3/8	10.5	10	.27
4	3x3x3/8	3/16	5/16	10.5	8.5	.23

TABLE B—UNSTIFFENED WELDED SEAT ANGLES

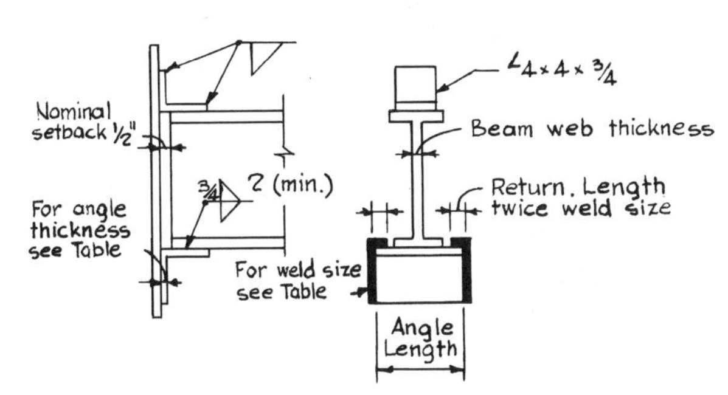

Nominal setback 1/2"
For angle thickness see Table
3/4
2 (min.)
For weld size see Table
∠4×4×3/4
Beam web thickness
Return, Length twice weld size
Angle Length

Thickness of Beam Web, in.	Thickness of Seat Angle, in. 3/8	1/2	5/8	3/4	7/8	1
	Length = 6"					
3/16	6	9	11	14	16	16
1/4	8	11	14	17	20	23
5/16	10	15	18	21	25	28
3/8	11	17	22	26	30	34
7/16	12	18	25	30	34	39
1/2	12	20	28	34	39	44
9/16	14	21	30	39	44	50
	Length = 8"					
3/16	7	10	13	15	16	16
1/4	9	12	16	19	22	23
5/16	11	16	20	24	27	31
3/8	12	19	24	28	32	36
7/16	13	21	27	32	37	42
1/2	14	22	31	37	42	47
9/16	15	24	34	42	48	50

Weld Size Required, in.	Load on Seat and Size of Seat Angle 4x3½	5x3½	6x4	7x4	8x4	9x4
1/4	7	11	14	18	23	27
5/16	9	13	17	23	28	34
3/8	11	16	21	27	34	41
7/16	13	19	24	32	40	48
1/2	14	22	28	36	46	50
5/8	–	–	35	46	50	–

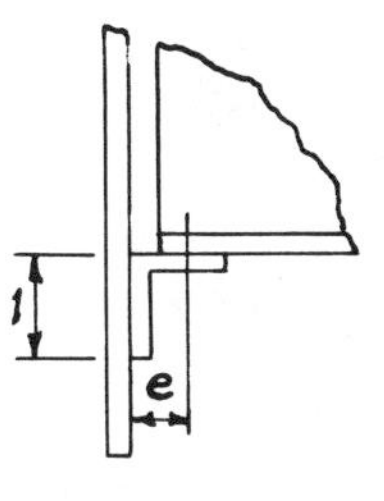

l
e

STRUCTURAL STEEL-WELDING-DETAILS

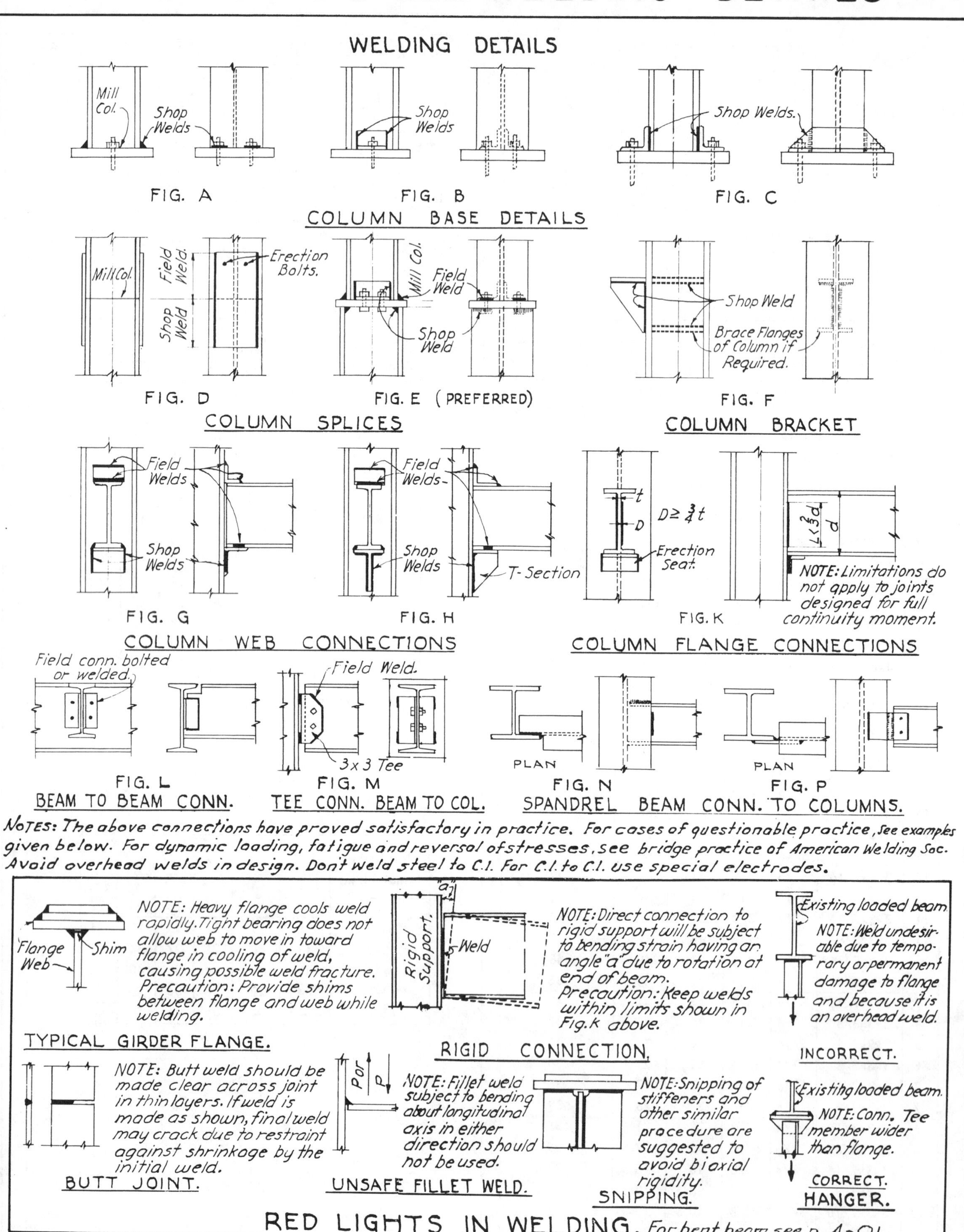

WELDING DETAILS
Mill Col.
Shop Welds
Shop Welds
Shop Welds.
FIG. A
FIG. B
FIG. C
COLUMN BASE DETAILS
Mill Col.
Field Weld.
Shop Weld
Erection Bolts.
Mill Col.
Field Weld
Shop Weld
Shop Weld
Brace Flanges of Column if Required.
FIG. D
FIG. E (PREFERRED)
FIG. F
COLUMN SPLICES
COLUMN BRACKET
Field Welds
Shop Welds
Field Welds
Shop Welds
T-Section
t
D
D ≥ 3/4 t
Erection Seat.
L < 2/3 d
d
NOTE: Limitations do not apply to joints designed for full continuity moment.
FIG. G
FIG. H
FIG. K
COLUMN WEB CONNECTIONS
COLUMN FLANGE CONNECTIONS
Field conn. bolted or welded.
Field Weld.
3 x 3 Tee
PLAN
PLAN
FIG. L
BEAM TO BEAM CONN.
FIG. M
TEE CONN. BEAM TO COL.
FIG. N
FIG. P
SPANDREL BEAM CONN. TO COLUMNS.

NOTES: The above connections have proved satisfactory in practice. For cases of questionable practice, see examples given below. For dynamic loading, fatigue and reversal of stresses, see bridge practice of American Welding Soc. Avoid overhead welds in design. Don't weld steel to C.I. For C.I. to C.I. use special electrodes.

Flange
Web
Shim
NOTE: Heavy flange cools weld rapidly. Tight bearing does not allow web to move in toward flange in cooling of weld, causing possible weld fracture. Precaution: Provide shims between flange and web while welding.
TYPICAL GIRDER FLANGE.
"a"
Rigid Support.
Weld
NOTE: Direct connection to rigid support will be subject to bending strain having an angle "a" due to rotation at end of beam. Precaution: Keep welds within limits shown in Fig. K above.
RIGID CONNECTION.
Existing loaded beam.
NOTE: Weld undesirable due to temporary or permanent damage to flange and because it is an overhead weld.
INCORRECT.
NOTE: Butt weld should be made clear across joint in thin layers. If weld is made as shown, final weld may crack due to restraint against shrinkage by the initial weld.
BUTT JOINT.
P or P
NOTE: Fillet weld subject to bending about longitudinal axis in either direction should not be used.
UNSAFE FILLET WELD.
NOTE: Snipping of stiffeners and other similar procedure are suggested to avoid biaxial rigidity.
SNIPPING.
Existing loaded beam.
NOTE: Conn. Tee member wider than flange.
CORRECT.
HANGER.

RED LIGHTS IN WELDING. For bent beam see p. 4-01

STRUCTURAL STEEL-WELDING CONNECTIONS

COLUMN CONNECTIONS-1951 AISC STRESSES*

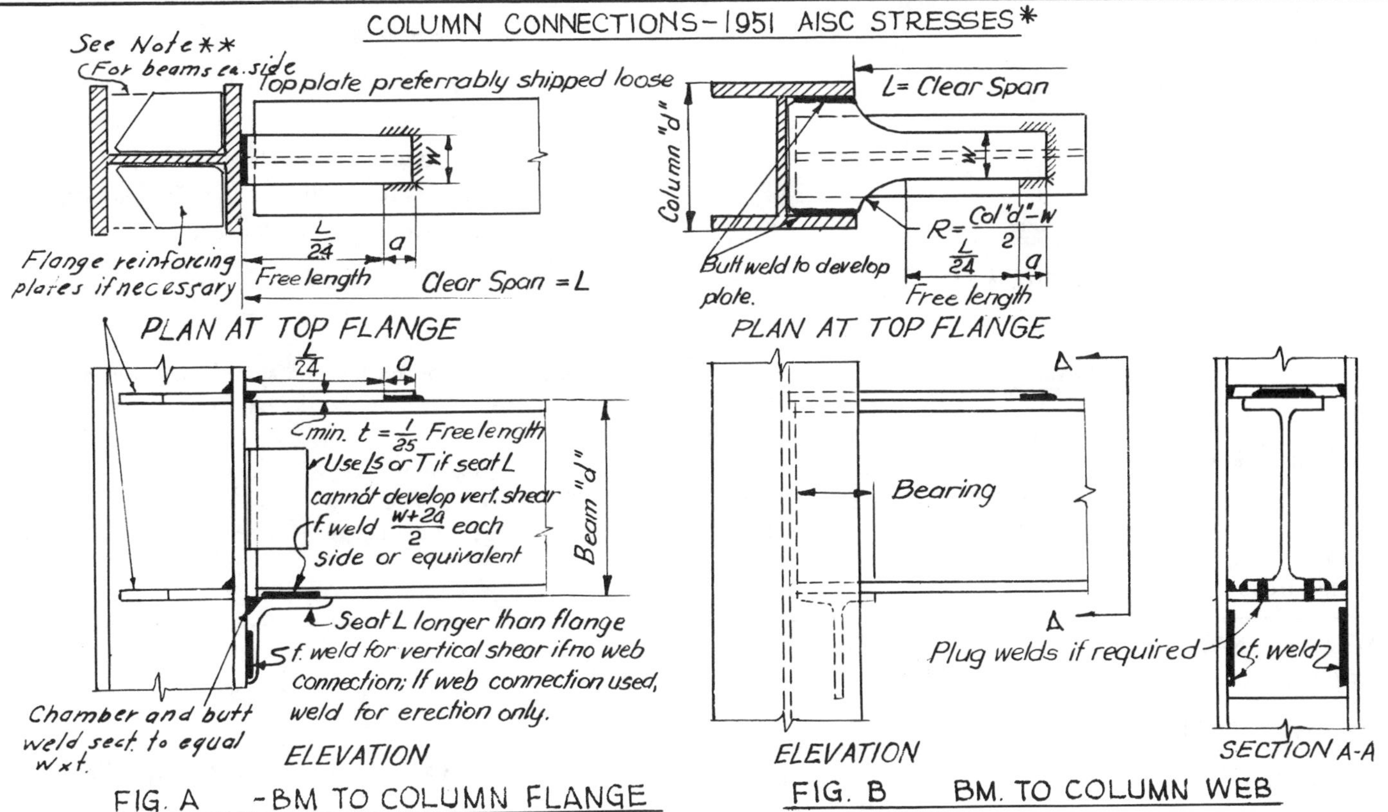

See Note**
(For beams ea. side
Top plate preferrably shipped loose
W
L/24
a
Flange reinforcing plates if necessary
Free length
Clear Span = L
PLAN AT TOP FLANGE
L/24
a
min. t = 1/25 Free length
Use Ls or T if seat L cannot develop vert. shear
f. weld (W+2a)/2 each side or equivalent
Beam "d"
Seat L longer than flange
f. weld for vertical shear if no web connection; If web connection used, weld for erection only.
Chamber and butt weld sect. to equal W×t.
ELEVATION
L= Clear Span
Column "d"
W
R= (Col "d" − W)/2
Butt weld to develop plate.
L/24
a
Free length
PLAN AT TOP FLANGE
A
Bearing
A
Plug welds if required
f. weld
ELEVATION
SECTION A-A

FIG. A -BM TO COLUMN FLANGE

FIG. B BM. TO COLUMN WEB

TABLE 48 - RESISTING MOMENTS- FT. KIPS & WELD LENGTHS-INCHES

BM "d" Inch. \ PL DIM W×t	$2\times\frac{3}{8}$	$3\times\frac{3}{8}$	$3\times\frac{1}{2}$	$4\times\frac{1}{2}$	$5\times\frac{1}{2}$	$6\times\frac{1}{2}$	$6\times\frac{5}{8}$	$6\times\frac{3}{4}$	$6\times\frac{7}{8}$
12	20	30	40	53	67	80	100	120	140
14	23	35	46	62	77	93	116	140	163
15	25	37	50	66	83	100	125	150	175
16	26	40	53	71	89	106	133	160	186
18	30	45	60	80	100	120	150	180	210
20	33	50	66	88	111	133	166	200	233
21	35	52	70	93	116	140	175	210	245
24	40	60	80	106	133	160	200	240	280
27	45	67	90	120	150	180	225	270	315
30	50	75	100	133	166	200	250	300	350
33	55	82	110	146	183	220	275	330	384
36	60	90	120	160	200	240	300	360	420
WELD SIZE	WELD LENGTH = (W + 2a) - inches.								
$\frac{1}{4}$	7	10	13	17	21	25	–	–	–
$\frac{5}{16}$	5	8	10	14	17	20	25	–	–
$\frac{3}{8}$	5	7	9	12	14	17	21	25	–
$\frac{1}{2}$	–	–	–	–	–	–	16	19	22

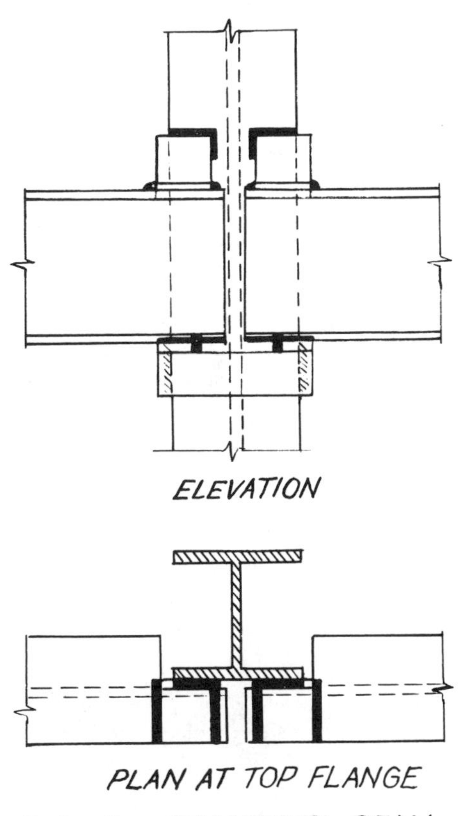

ELEVATION
PLAN AT TOP FLANGE

FIG. C SPANDREL BEAM ON OUTER FACE OF COLUMN

** A.W.S. and A.I.S.C. codes call for backing strip under butt weld made from one side unless stresses are reduced 25 percent.

* Based on AISC, modified by recommendation of Gilbert D. Fish.

STRUCTURAL STEEL-PLASTIC DESIGN-I

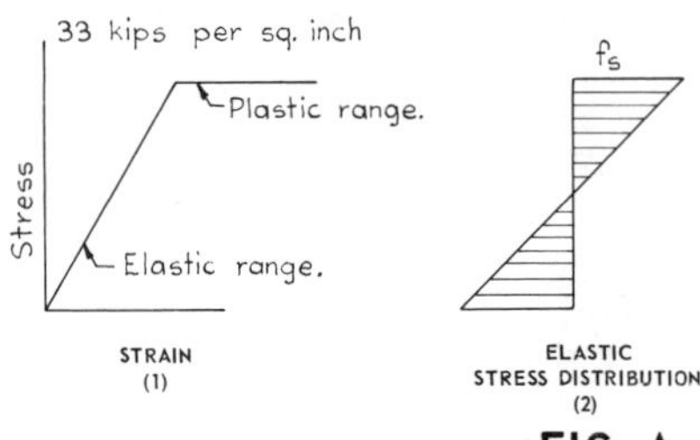

33 kips per sq. inch
Plastic range.
Stress
Elastic range.
STRAIN
(1)
f_s
ELASTIC
STRESS DISTRIBUTION
(2)
f_y
FULL PLASTIC
STRESS DISTRIBUTION
(3)

FIG. A

PLASTIC DESIGN in steel is based on the theory that plastic hinges will form at points of maximum moment in the structure when the stress distribution is as shown in Fig. A (3). When enough plastic hinges form, the structure becomes unstable and collapses.

ELASTIC

$$F_s = \frac{M}{S}$$

PLASTIC

$$f_y = \frac{M_p}{Z}$$

S = section modulus (elastic range)

Z = plastic modulus = approximately 1.12 S (I section)

Mp = plastic moment

fy = yield stress

FACTOR OF SAFETY is based on the additional load that a simple beam can carry by plastic analysis compared to elastic analysis.

For a working stress of 20 kips per sq. in., the factor of safety against yield is 33/20 h 1.65. The factor of safety against developing the full plastic stress distribution is 1.65 × 1.12 = 1.85.

METHOD OF ANALYSIS

EXAMPLE 1

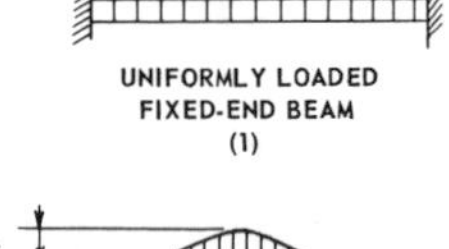

UNIFORMLY LOADED
FIXED-END BEAM
(1)

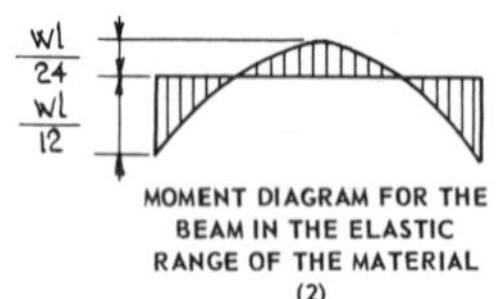

wl/24
wl/12
MOMENT DIAGRAM FOR THE
BEAM IN THE ELASTIC
RANGE OF THE MATERIAL
(2)

FIG. B

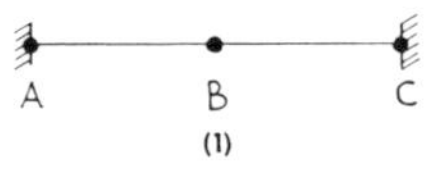

A
B
C
(1)

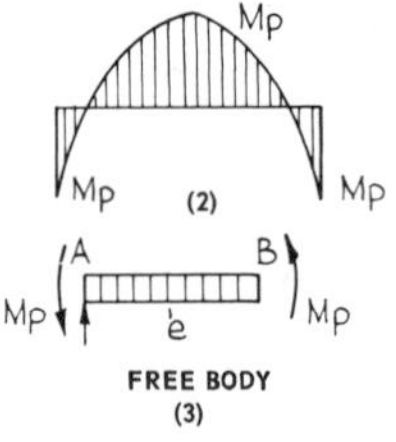

Mp
Mp
(2)
Mp
A
B
Mp
Mp
FREE BODY
(3)

FIG. C. PLASTIC MOMENT DIAGRAMS

With an increase in load the beam yields first at A and C, developing *plastic hinges*. When this occurs, points A and C can provide no further increase in moment resistance, and, with a greater increase in load, a redistribution of moment occurs, until a *plastic hinge* forms at B. The plastic moments at A, B, and C are all equal.

To compute Mp

$$\Sigma M_B = 0 = -M_p + \frac{wl}{2}\cdot\frac{l}{2} - wl\cdot\frac{l}{4} - M_p$$

$$M_p = \frac{wl^2}{16}$$

EXAMPLE 2: Design a beam to carry the load shown.

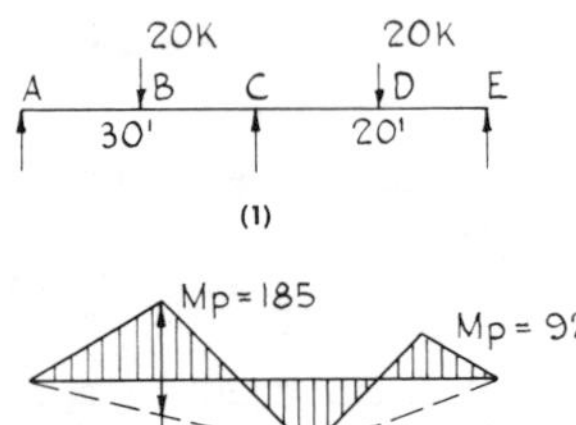

20K
20K
A
B
C
D
E
30'
20'
(1)
Mp=185
Mp=92.5
Mp=185
Pl/4
(2)

FIG. D.

RED LIGHT: In no case may the moment exceed Mp.

SOLUTION:

Plastic hinges will form at points of maximum moment B and C to create a mechanism.

From Fig. D (1):

$$\frac{Pl}{4} - \frac{M_p}{2} = M_p \quad \text{or} \quad M_p = \frac{Pl}{6}$$

Factor of safety = 1.85

Pu = 1.85 × 20 = 37.0 kips

$$M_p = \frac{(37)(30)}{6} = 185 \text{ ft.-kips} = 2220 \text{ in.-kips}$$

$$Z = \frac{M_p}{f_y} = \frac{2220}{33} = 67.2, \text{ from Table B p. 4-71.}$$

Select most economical section; use 16 WF − 40 (Z = 72.7).

NOTE: For moment at point D:

$$M_D = \frac{(37)(20)}{4} - \frac{185}{2} = 92.5$$

STRUCTURAL STEEL–PLASTIC DESIGN–2

TABLE A – ADDITIONAL CONSIDERATIONS

The following criteria should be met in order to insure the development of the plastic hinges.

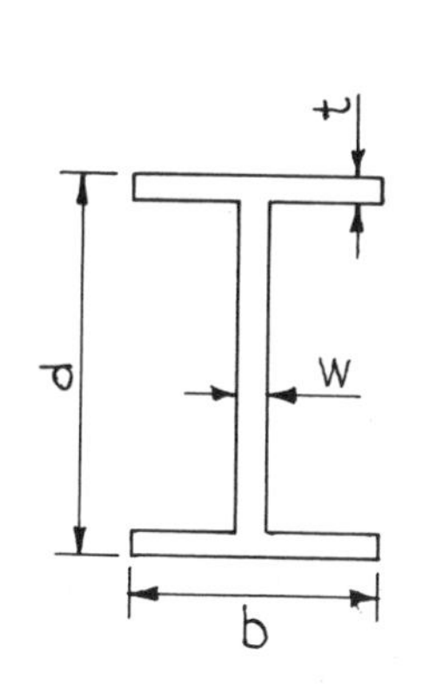

t
d
W
b

Criteria	Guide and Formula
Moment plus compression	$Z \text{ required} = \dfrac{0.85Z}{1 - \dfrac{P}{Py}}$ $Py = fyA$ Z required = section required to carry moment and loads Neglect if $P < 0.15Py$.
Maximum Allowable Shear	V maximum = 18,000Wd
Local buckling	$\dfrac{b}{t} < 17$ (beams and columns) $\dfrac{d}{w} \leq 70 - 100\dfrac{P}{Py}$ but need not be less than 40
Lateral buckling	$l = \left(60 - \dfrac{40M}{Mp}\right) r_y$, but need not be less than $35r_y$. l = distance between bracing points r = radius of gyration NOTE: Provide lateral bracing at all plastic hinge locations.
Column limits	$\dfrac{P}{Py} < 0.60$ $\dfrac{l}{r} < 120$ in plane of bending

TABLE B – ECONOMICAL SELECTION OF BEAMS BY PLASTIC SECTION MODULUS*

Z †	Section	Z †	Section	Z †	Section	Z †	Section
1255.0	**36 WF 300**	247.9	18 WF 114	117.9	16 WF 64	38.8	10 WF 33
1167.0	**36 WF 280**	242.7	14 WF 136	114.8	14 WF 68	**38.0**	**12 WF 27**
1076.0	**36 WF 260**	238.8	24 I 100	114.4	10 WF 89	35.2	10 I 35
1007.0	**36 WF 245**	226.5	18 WF 105	**111.6**	**18 WF 55**	34.7	10 WF 29
941.8	**36 WF 230**	226.3	21 WF 96	106.2	16 WF 58	34.7	8 WF 35
916.4	33 WF 240	225.9	14 WF 127	**103.5**	**18 I 54.7**	32.8	8 M 34.3
836.4	**33 WF 220**	**224.0**	**24 WF 84**	102.4	14 WF 61	30.4	8 WF 31
767.3	**36 WF 194**	220.5	24 I 90	**100.8**	**18 WF 50**	**29.5**	**10 WF 25**
756.4	33 WF 200	210.9	14 WF 119	97.3	10 WF 77	**29.4**	**12 B 22**
734.5	30 WF 210	209.7	12 WF 133	**92.7**	**16 WF 50**	28.0	10 I 25.4
717.1	**36 WF 182**	206.0	18 WF 96	90.7	10 WF 72	27.1	8 WF 28
666.5	**36 WF 170**	**203.0**	**24 I 79.9**	87.1	14 WF 53	**24.8**	**12 B 19**
659.6	30 WF 190	**200.1**	**24 WF 76**	86.5	12 WF 58	**24.1**	**10 WF 21**
623.3	**36 WF 160**	196.0	14 WF 111	82.8	10 WF 66	23.4	8 M 24
593.0	30 WF 172	192.0	20 I 95	**82.0**	**16 WF 45**	23.1	8 WF 24
579.8	**36 WF 150**	191.6	21 WF 82	78.5	14 WF 48	**21.6**	**10 B 19**
558.2	33 WF 152	186.4	12 WF 120	78.2	12 WF 53	**20.6**	**12 B 16.5**
557.1	27 WF 177	186.0	16 WF 96	76.5	15 I 50	19.2	8 I 23
513.1	**33 WF 141**	177.6	18 WF 85	75.1	10 WF 60	19.1	8 WF 20
504.3	27 WF 160	177.3	20 I 85	**72.7**	**16 WF 40**	19.0	6 WF 25
466.2	**33 WF 130**	**172.1**	**21 WF 73**	72.6	12 WF 50	18.6	10 B 17
463.7	24 WF 160	169.0	16 WF 88	70.1	8 WF 67	17.9	6 M 25
452.0	27 WF 145	163.4	12 WF 106	69.6	14 WF 43	17.5	8 M 20
436.7	30 WF 132	160.5	18 WF 77	68.6	15 I 42.9	**17.4**	**12 B 14**
416.0	24 WF 145	**159.8**	**21 WF 68**	67.0	10 WF 54	16.3	8 I 18.4
407.4	**30 WF 124**	151.8	12 WF 99	64.9	12 WF 45	16.0	10 B 15
377.6	**30 WF 116**	151.5	20 I 75	**63.9**	**16 WF 36**	15.8	8 WF 17
369.2	24 WF 130	147.5	10 WF 112	61.5	14 WF 38	15.7	8 B 17
357.0	21 WF 142	145.5	16 WF 78	60.6	12 I 50	15.0	6 WF 20
345.5	**30 WF 108**	145.4	14 WF 84	60.3	10 WF 49	14.6	6 M 20
342.8	27 WF 114	144.7	18 WF 70	59.9	8 WF 58	14.4	7 I 20
336.6	24 WF 120	**144.1**	**21 WF 62**	57.6	12 WF 40	**14.2**	**12 B 11.8**
317.8	21 WF 127	140.2	12 WF 92	55.0	10 WF 45	13.6	8 B 15
311.5	12 WF 190	137.3	20 I 65.4	**54.5**	**14 WF 34**	**12.2**	**10 B 11.5**
307.7	24 WF 110	134.0	14 WF 78	52.4	12 I 40.8	11.9	7 I 15.3
304.4	**27 WF 102**	131.8	18 WF 64	51.4	12 WF 36	11.6	6 B 16
298.2	24 I 120	131.6	16 WF 71	49.0	8 WF 48	11.4	8 B 13
278.3	**24 WF 100**	130.7	10 WF 100	**47.1**	**14 WF 30**	11.3	6 M 15.5
278.0	21 WF 112	129.1	12 WF 85	47.0	10 WF 39	10.5	6 I 17.25
277.7	**27 WF 94**	125.6	14 WF 74	44.4	12 I 35	8.4	6 I 12.5
273.0	24 I 105.9	123.8	18 I 70	44.0	12 WF 31	8.2	6 B 12
259.2	12 WF 161	**122.6**	**18 WF 60**	41.7	12 I 31.8		
253.0	**24 WF 94**	119.3	12 WF 79	39.9	8 WF 40		

NOTE: Boldface type indicates most economical section.

*See Fig. A (3), p. 4-70. † Plastic section modulus.

STRUCTURAL STEEL-PLASTIC DESIGN-3

$Q = \dfrac{f}{h}$

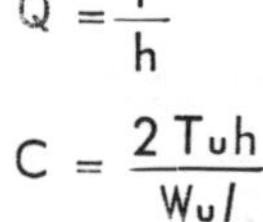

$C = \dfrac{2\,T_u h}{W_u l}$

$W_u = wl$

$T_u h$ = moment of total horizontal loading about point A

Z = plastic modulus (assumed constant throughout frame)

MOMENTS AND REACTIONS AT ULTIMATE LOAD IN ACCORDANCE WITH PLASTIC THEORY

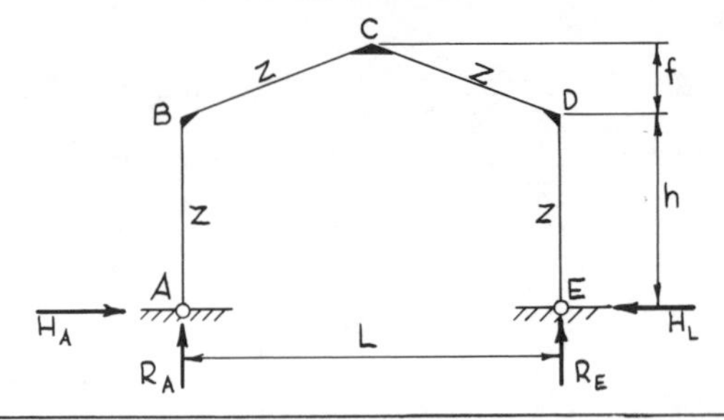

C
Z
Z
B
D
f
Z
Z
h
A
E
H_A
L
H_L
R_A
R_E

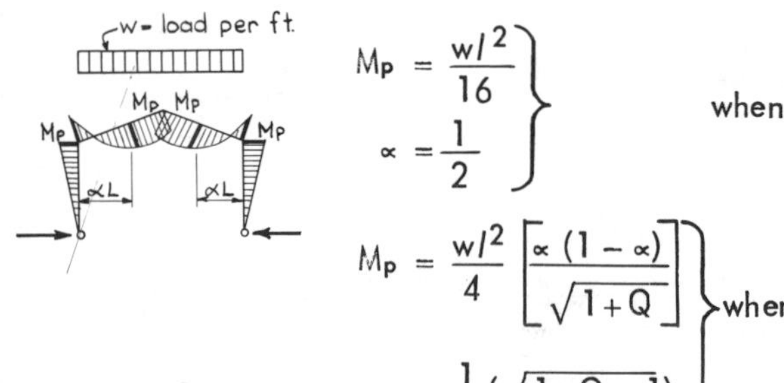

w- load per ft.
Mp Mp
Mp
Mp
∝L
∝L

$R_A = R_E = \dfrac{wl}{2}$

$H_A = H_E = \dfrac{M_p}{h}$

$\left.\begin{aligned} M_p &= \dfrac{wl^2}{16} \\ \alpha &= \dfrac{1}{2} \end{aligned}\right\}$ when $Q = 0$

$\left.\begin{aligned} M_p &= \dfrac{wl^2}{4}\left[\dfrac{\alpha(1-\alpha)}{\sqrt{1+Q}}\right] \\ \alpha &= \dfrac{1}{Q}\left(\sqrt{1+Q}-1\right) \end{aligned}\right\}$ when $Q > 0$

FIG. A. PINNED COLUMN BASES

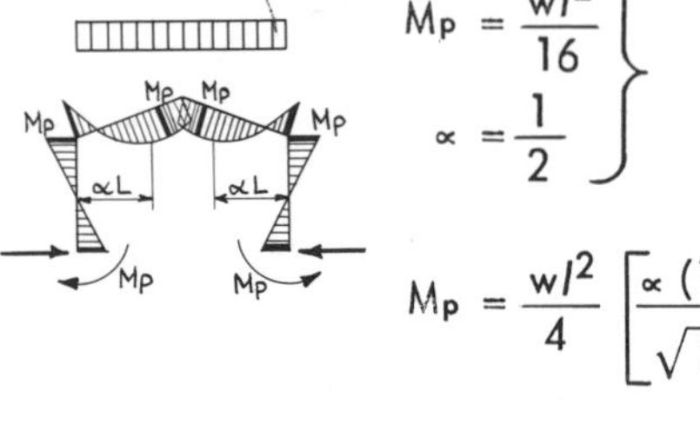

Mp Mp
Mp
Mp
∝L
∝L
Mp
Mp

$R_A = R_E = \dfrac{wl}{2}$

$H_A = H_E \leq \dfrac{2\,M_p}{h}$

$\left.\begin{aligned} M_p &= \dfrac{wl^2}{16} \\ \alpha &= \dfrac{1}{2} \end{aligned}\right\}$ when $Q = 0$

$\left.\begin{aligned} M_p &= \dfrac{wl^2}{4}\left[\dfrac{\alpha(1-\alpha)}{\sqrt{1+2Q}}\right] \\ \alpha &= \dfrac{1}{2Q}\left(\sqrt{1+2Q}-1\right) \end{aligned}\right\}$ when $Q > 0$

FIG. B. FIXED COLUMN BASES

UNIFORMLY DISTRIBUTED VERTICAL LOAD ALONE

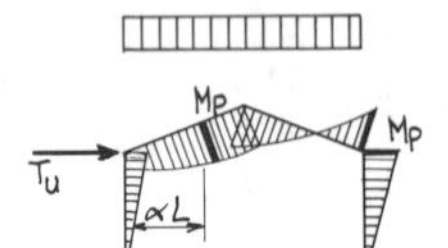

Mp
Mp
Tu
∝L

$R_A = \dfrac{wl}{2} - \dfrac{T_u h}{l}$

$R_E = \dfrac{wl}{2} + \dfrac{T_u h}{l}$

$H_A = T_u - H_E$

$H_E = \dfrac{M_p}{h}$

When $C > \dfrac{1}{1+Q}$ (possible condition):

$\left.\begin{aligned} M_p &= \dfrac{wl^2}{4}\cdot C \\ \alpha &= 0 \end{aligned}\right\}$ Plastic hinges at B and D only

When $C < \dfrac{1}{1+Q}$ (usual condition):

$\left.\begin{aligned} M_p &= \dfrac{wl^2}{16}(1+C)^2 \\ \alpha &= \dfrac{1-C}{2} \end{aligned}\right\}$ when $Q - 0$

$\left.\begin{aligned} M_p &= \dfrac{wl^2}{4}\left[\dfrac{(1-\alpha)(C+\alpha)}{\sqrt{(1+Q)(1-QC)}}\right] \\ \alpha &= \dfrac{1}{Q}\left(\sqrt{(1+Q)(1-QC)}-1\right) \end{aligned}\right\}$ when $Q > 0$

FIG. C. PINNED COLUMN BASES

$R_A = \dfrac{wl}{2} - \dfrac{T_u h}{l} + \dfrac{2M_p}{l}$

$R_E = \dfrac{wl}{2} + \dfrac{T_u h}{l} - \dfrac{2M_p}{l}$

$H_A = T_u = H_E$

$H_E = \dfrac{2\,M_p}{h}$

When $C > \dfrac{2}{1+2Q}$ (possible condition):

$\left.\begin{aligned} M_p &= \dfrac{wl^2}{8}\cdot C \\ \alpha &= 0 \end{aligned}\right\}$ Plastic hinges at A, B, D, and E only

When $C < \dfrac{2}{1+2Q}$ (usual condition):

$\left.\begin{aligned} M_p &= \dfrac{wl^2}{4}\left(3+C-2\sqrt{2+C}\right)^* \\ \alpha &= 2-\sqrt{2+C}\,^* \end{aligned}\right\}$ when $Q = 0$

$\left.\begin{aligned} M_p &= \dfrac{wl^2}{4}\left[\dfrac{(1-\alpha)(C+\alpha)}{\sqrt{2+C-4CQ^2+4Q}}\right]^* \\ \alpha &= \dfrac{1}{1-2Q}\left(2-\sqrt{2+C-4CQ^2+4Q}\right)^* \end{aligned}\right\}$ when $Q > 0$

FIG. D. FIXED COLUMN BASES

UNIFORMLY DISTRIBUTED VERTICAL LOAD COMBINED WITH HORIZONTAL LOADING

*When $C < 0.25$, using these formulas, corresponding value for Mp may be less than that computed for vertical load alone, and latter will govern.

From *Plastic Design in Steel*, American Institute of Steel Construction, Inc.

STRUCTURAL STEEL-COMPOSITE BEAM DESIGN-I

USAGE

Composite construction is used to take advantage of the extra strength of an I beam integrated with a concrete slab. It has been used on highways very extensively, and, less frequently but with success on buildings where the integration is obtained by shear connectors welded to the flange of the beam. The saving in structural steel might run from 25% to 40%. The actual economy depends on an economic study but is substantial, even after taking into account the cost of the integration.

DESIGN CRITERIA FOR COMPOSITE CONSTRUCTION FOR BUILDINGS	DESIGN STEPS IN COMPOSITE CONSTRUCTION FOR BUILDINGS
1. Width of slab effective as T-flange is taken as the least of the following values: (a) 1/4 span length of beam. (b) The distance center to center of beams. (c) 16 times the least thickness of the slab. For fully encased beams, the effective width may be taken as 16 times the least slab thickness plus the stern width. 2. Full shoring is supplied, so that the composite section carries both dead and live loads. In actual practice, if the shores are omitted, the safety factor on ultimate strength will not be changed. 3. The allowable load on the shear connectors is based on their useful capacity with a safety factor of 2.4.	1. Choose a trial section composed of concrete slab thickness, steel beam, and steel cover plate. 2. Compute neutral axis of steel section. 3. Compute moment of inertia of steel section about its own neutral axis. 4. Compute neutral axis of total steel section plus transformed concrete area. 5. Compute moment of inertia of total steel section plus transformed concrete area about its neutral axis. 6. Compute section moduli for steel and concrete in composite section. 7. Compute extreme fiber stresses in steel and concrete. 8. Compute length of cover plate. Steps 4,5, and 6 must be repeated, omitting the cover plate from the calculations. 9. Design the shear connectors.

NOMENCLATURE

$\overline{Y}$ = distance from top of slab to neutral axis of combined steel beam and plate
Y = distance from top of slab to center line of steel area listed
Ay = moment of area about top of slab
I_o = moment of inertia of steel about its own axis
Ss = section modulus of steel
Ss = section modulus of concrete
fs = extreme fiber stress of steel
fc = extreme fiber stress of concrete
h = horizontal shear
V = total shear (end reaction)
Q = statical moment of steel beam about that portion of cross section of composite section lying above steel beam

PROBLEM: Given: Beam moment = 192 foot-kips
Design: Composite beam.

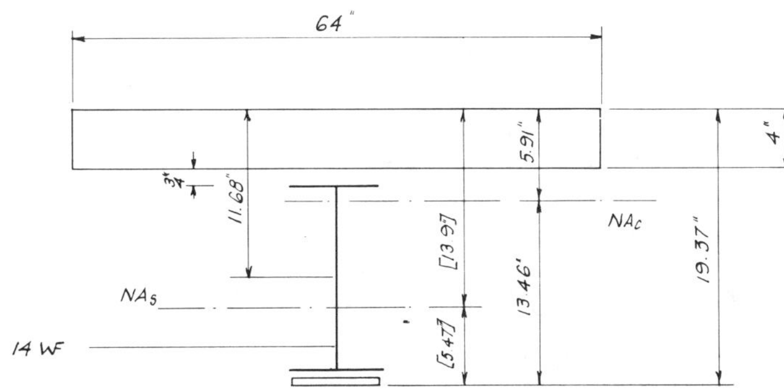

64"
5.91"
4"
3/4
11.68"
[13.9]
NAc
19.37"
13.46'
NAs
[5.47]
14 WF

Dimensions shown thus [ ] are computed values.

Assume 4" concrete slab
$n = Es/Ec = 10$

STEP 1

14 WF 30 with 5 × 3/4 bottom cover plate.
3/4" haunch between top of beam and bottom of concrete slab.

STEP 2

Member	Area	Y	Ay
Plate 5 × 3/4	3.75	19.0	71.2
14 WF 30	8.81	11.68	103.0
	12.56		174.2

$$y = \frac{174.2}{12.56} = 13.9''$$

STEP 3

$Byl_0 + \Sigma Ay^2$

I_0 (14 WF 30)		= 289.6
Ay^2 (14 WF)	$8.81 \times \overline{2.19}^2$	42.3
Ay^2 (plate)	$3.75 \times \overline{5.09}^2$	97.0
		428.9

STEP 4

Take moments about top of slab.

Transformed area of slab $= \frac{64 \times 4}{10}$

$$kd = \frac{12.56 \times 13.9 + \frac{64 \times 4}{10} \times \frac{4}{2}}{12.56 + \frac{64 \times 4}{10}} = \frac{174.5 + 51.2}{12.56 + 25.6} = 5.91''$$

STRUCTURAL STEEL–COMPOSITE BEAM DESIGN–2

STEP 5

I of transformed slab area about $NA_c (I = \frac{bd^3}{3})$

+ Ay^2 (steel beam + plate)
+ I_0 (steel beam + plate)

$$I \text{ composite} = \frac{64}{3 \times 10}(5.91^3 - 1.91^3) + 12.56 \times 7.96^2 + 429$$
$$= 425 + 796 + 430$$
$$= 1651$$

STEP 6

$$S_s = \frac{1658}{13.46} = 122.7$$

$$S_c = 10 \times \frac{1651}{5.91} = 2790$$

STEP 7

$$f_s = \frac{192000 \times 12}{123} = 18700 \text{ psi}$$

$$f_c = \frac{192000 \times 12}{2800} = 823 \text{ psi}$$

STEP 8

The theoretical length of cover plate is determined from the parabolic moment curve by the given equation where

l = span length

S_s = section modulus of composite section plus plate

S_0 = section modulus of composite section without plate

l' = theoretical length of cover plate

The actual length of the cover plate should extend 1'-0" beyond the theoretical points of cut-off.

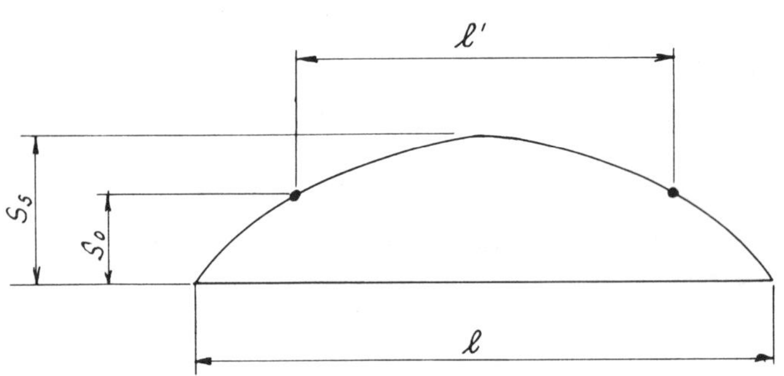

ℓ'
S_s
S_o
ℓ

Repeat STEP 4 – Without Plate

Total depth of section = 4.75 + 13.87 = 18.62"
Distance from top of slab to ℄ of steel beam
= 4.75 + 6.93 = 11.68"

$$kd = \frac{8.81 \times 11.68 + \frac{64 \times 4}{10} \times \frac{4}{2}}{8.81 + \frac{64 \times 4}{10}} = 4.48$$

Repeat STEP 5

$$I = \frac{64}{30}(4.48^3 - 0.48^3) + 8.81 \times 7.20^2 + 289$$
$$= 938$$

Repeat STEP 6

$$S_0 = \frac{938}{(18.62 - 4.48)} = 66.3$$

Length of cover plate for a 30 ft. beam $= 30\sqrt{1 - \frac{67}{123}} = 20.4$

Use plate length = 22' – 3"

STEP 9

The horizontal shear per unit length of beam $= H = \frac{VQ}{I}$.

The spacing of shear connectors equals the total working capacity of the connectors divided by the horizontal shear per unit length of beam. Thus $S = \frac{UC}{FS} \times \frac{I}{VQ}$

Given: Beam length = 30'-0"
End reaction = 25.5 kips
Use: 2-3/4 ϕ stud shear connectors.
Working load for 2 studs = 2 × 4.25 = 8.5 kips.
(See following table for values.)

No Plate $Q = \frac{64 \times 4}{10}(4.48 - 2) = 58.5$ *With Plate* $Q = 25.6(5.84 - 2) = 98.4$

Maximum V at end of beam = 25.5 kips $S = \frac{8.5 \times 938}{25.5 \times 58.5} = 5.3"$

V at 4'-0" = 18.75 kips $S = \frac{8.5 \times 1658}{18.75 \times 98.4} = 7.64"$

V at 7'-0" = 13.70 kips $S = 7.64 \times \frac{18.75}{13.70} = 10.5"$

V at 10'-0" = 8.6 kips $S = 7.64 \times \frac{18.75}{8.6} = 16.7"$

Use the following pitch of connectors: 4'-0" at 5", 4'-6" at 7½", balance at 12".

STRUCTURAL STEEL–COMPOSITE BEAM DESIGN – 3

TABLE A – SECTION MODULUS S_s AND S_c FOR GIVEN BEAMS

NOTE: For computing length of Cover plates and shear connectors, see problem, pp. 4-89 and 4-90. These tables have been figured for 3000 p.s.i. concrete, n = 10.

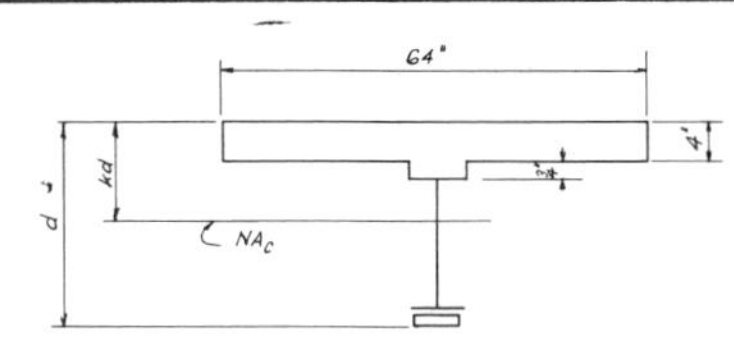

Section	Steel Weight, lb./ft.	I	Kd	d – Kd	S_s	S_c
14WF30	30	938	4.48	14.13	66.3	2095
14WF30 6 x 1 PL	54.4	2030	6.64	12.97	156.7	3060
14WF43	43	1241	5.17	13.26	93.6	2400
14WF43 6 x 1 PL	63.4	2220	7.04	12.39	179	3155
14WF61	61	1669	6.0	12.66	132	2780
14WF61 6 x 1 PL	81.4	2585	7.58	12.08	214	3415
14WF78	78	2042	6.61	12.20	168	3095
14WF78 6 x 1 PL	98.4	2900	8.02	11.79	246	3620
16WF36	36	1333	5.11	15.49	86	2610
16WF36 9 x 1 PL	66.6	3178	8.29	13.31	238	3620
16WF40	40	1481	5.38	15.37	96	2755
16WF40 9 x 1 PL	70.6	3311	8.46	13.29	249	3910
16WF50	50	1792	5.96	15.04	119	3010
16WF50 9 x 1 PL	80.6	3568	8.80	13.20	270	4050
16WF64	64	2122	6.55	14.20	149.5	3240
16WF64 6 x 1 PL	84.4	3263	8.28	13.47	242	3940

Section	Steel Weight, lb./ft.	I	Kd	d – Kd	S_s	S_c
18WF50	50	2126	6.29	15.46	129	3380
18WF50 9 x 1 PL	80.6	4242	9.38	14.37	296	4520
18WF60	60	2493	6.84	16.16	154	3640
18WF60 9 x 1 PL	90.6	4559	9.70	14.3	319	4700
18WF64	64	2558	6.93	15.69	163	3690
18WF64 8 x 1 PL	91.2	4340	9.4	14.22	305	4610
21WF62	62	3228	7.52	18.22	177	4300
21WF62 10 x 1 PL	96	6087	10.97	15.77	386	5540
21WF82	82	3941	8.4	17.21	229	4700
21WF82 8 x 1 PL	109.2	6106	10.85	15.76	388	5620
24WF76	76	4709	8.85	19.81	238	5315
24WF76 8 x 1 PL	103.2	7540	11.76	17.9	421	6410
27WF94	94	6790	10.4	2126	320	6530
27WF94 8 x 1 PL	121.2	10086	13.25	19.41	518	7600

Intermediate cover-plate sizes may be obtained by interpolation.

EXAMPLE: Find cover-plate size for a 14 WF 30 that will provide a steel section modulus of 130 in.³

14 WF 30 } 6 × 1 PL } A = 14.81 ; 14 WF 30 A = 8.81 } Area of PL = 6.0

S_s = 156.7 ; S_s = 66.3 } S_s of PL = 90.4

$$\begin{array}{r} 130 \\ -66.3 \\ \hline 63.7 \end{array} = \text{Req'd } S_S \qquad 6.0 \times \frac{63.7}{90.4} = 4.23 \text{ Area Req'd}$$

Since this ratio is dependent on plates of equal thickness, use 4½" × 1" PL.

TABLE B – WORKING LOAD OF ONE STUD CONNECTOR, LB. FOR $\frac{h}{d} > 4.2$

Stud Diameter, in.	Concrete Strength, p.s.i. 2500	3000	3500	4000	Detail
½	2680	2940	3180	3400	WELD
⅝	3870	4250	4580	4910	
¾	5290	5790	6250	6700	Studs

TABLE C – SPIRAL CONNECTORS

Spiral Bar Diameter, in.	Working Load Per Weld – Concrete Strength, p.s.i. 2500	3000	3500	4000	Effective Length of Weld, in. per Pitch – Weld Size, in.	Concrete Strength, p.s.i. 2500	3000	3500	4000
½	5650	5920	6150	6360	¼	1.05	1.10	1.14	1.18
⅝	7070	7400	7690	7950	¼ 5/16	1.18 1.01	1.23 1.06	1.28 1.10	1.33 1.14
¾	8480	8880	9220	9540	5/16 ⅜	1.12 1.01	1.17 1.06	1.21 1.10	1.25 1.13

Details

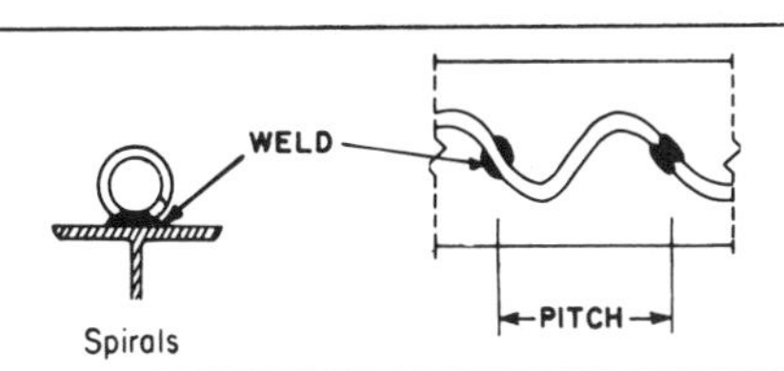

WELD
PITCH
Spirals

TABLE D – CHANNEL CONNECTORS

Channel Type Size	Working Load for 1 in. of Width – Concrete Strength, p.s.i. 2500	3000	3500	4000	Required Length of Weld for 1 in. of Width – Weld Size, in.	Concrete Strength, p.s.i. 2500	3000	3500	4000
Am. Std. 3"⊔ 4.1#	1730	1900	2050	2190	⅛	.98	1.08	1.16	1.24
5.0#	1910	2090	2260	2420	⅛	1.08	1.18	1.28	1.37
6.0#	2080	2280	2460	2630	⅛	1.18	1.29	1.39	1.49
4"⊔ 5.4#	1890	2080	2240	2390	⅛	1.07	1.18	1.27	1.35
7.25#	2150	2360	2540	2720	⅛	1.22	1.34	1.44	1.54
Car and Ship 3"⊔ 7.1#	2050	2250	2430	2600	⅛ 3/16 ¼	1.16 .78 .58	1.27 .85 .64	1.37 .92 .69	1.47 .99 .74
9.0#	2400	2630	2840	3040	⅛ 3/16 ¼	1.36 .91 .67	1.49 1.00 .75	1.61 1.08 .80	1.74 1.15 .86
4"⊔ 13.8#	2930	3210	3470	3710	⅛ 3/16 ¼ 5/16 ⅜	1.66 1.11 .83 .67 .56	1.82 1.22 .91 .73 .61	1.96 1.35 .98 .79 .66	2.10 1.41 1.05 .84 .70

Details

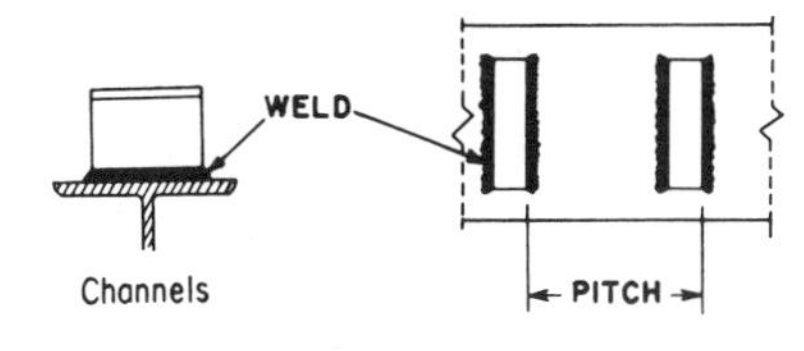

WELD
PITCH
Channels

Formula from which Tables B, C, and D are derived.

Formulas: Use formulas 1 and 2 for Table B, formulas 1 and 3 for Table C, and formulas 1 and 4 for Table D.

1. $WL = \frac{U.C.}{F.S.\,(2.4)}$
   - WL = working load
   - UC = useful capacity
   - FS = factor of safety (2.4)
2. $UC = 330\,d^2\sqrt{f'_c}$ $(h/d > 4.2)$
   $= 80hd\,\sqrt{f'_c}$ $(h/d < 4.2)$
   - h = height of stud
   - d = stud diameter
3. $UC = 3840 d_{sp} \sqrt[4]{f'_c}$
   - dsp = diameter of spiral bar
4. $UC = 180(h + 0.5t)\, w\sqrt{f'_c}$
   - h = maximum thick- of channel flange
   - t = thickness of channel web
   - w = width of channel connector, inches

STRUCTURAL ALUMINUM – ALLOWABLE WORKING STRESSES-1

SELECTION OF ALLOY:
Alloy 6061: General purpose. Moderate-strength alloy of high resistance to corrosion. Good weldability.
Alloy 2014: Heavy-duty high-strength alloy with moderate resistance to corrosion. Welding not recommended.
Temper T6 is always used for both alloys, as it gives greater strength.

TABLE A ALLOWABLE WORKING STRESSES FOR ALUMINUM

Type of Stress	Aluminum Alloy	Stress, Kips/in.^{2}
Axial tension net section	6061-T6 2014-T6	15 22
Tension in extreme fibers of shapes girders and built-up members subject to bending, net section	6061-T6 2014-T6	15 22
Stress in extreme fibers of pins	6061-T6 2014-T6	22 34
Shear in plates and webs. See Figs. A and B, p. 5-03	6061-T6 2014-T6	
Shear in rivets cold driven	6061-T6 2117-T3	10 10
Shear in rivets cold driven at temperature from 990°F. to 1050°F.	6061-T43	8
Axial compression and compression in extreme fibers of shapes girders and built-up members subject to bending	6061-T6 2014-T6	(see below)
Shear in turned bolts in reamed holes	2024-T4 2014-T4	12 12
Shear in pins	6061-T6 2014-T6	10 16
Bearing on pins	6061-T6 2014-T6	22 30
Bearing on hot driven or cold driven rivets, milled stiffeners, turned bolts in reamed holes, and other parts in fixed contact	6061-T6 2014-T6	27 36
Welding in tension	6061-T6	8
Longitudinal shear in fillet welds	6061-T6	4
Transverse shear in fillet welds	6061-T6	5.5

BEAM AND COLUMN – ALLOWABLE COMPRESSIVE STRESSES

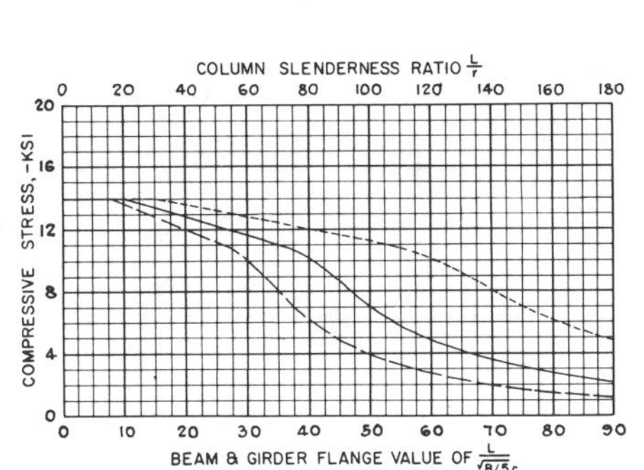

COLUMN SLENDERNESS RATIO L/r
COMPRESSIVE STRESS, -KSI
BEAM & GIRDER FLANGE VALUE OF L/√B/Sc

Beam and Girder Flange Values of $\frac{L}{\sqrt{B/S_c}}$

Fig. B. Alloy 6061-T6

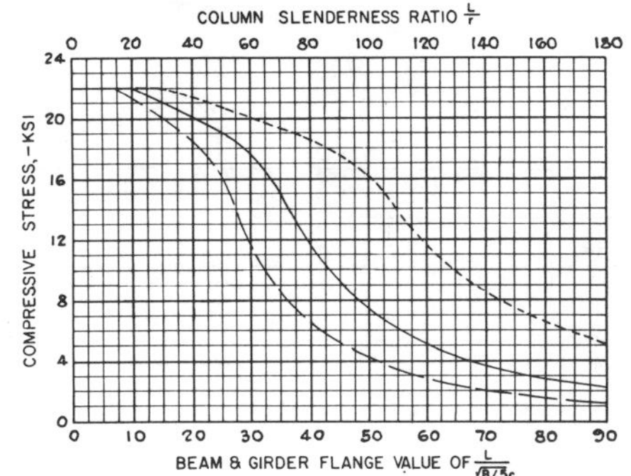

COLUMN SLENDERNESS RATIO L/r
COMPRESSIVE STRESS, -KSI
BEAM & GIRDER FLANGE VALUE OF L/√B/Sc

Beam and Girder Flange Values of $\frac{L}{\sqrt{B/S_c}}$

Fig. C. Alloy 2014-T6

KEY

– – – – Curve for fixed-end condition: K = 0.5

— — — Curve for pin ends: K = 1.0 and values of $\frac{L}{\sqrt{B/S_c}}$

———— Partial restraint at ends: K = 0.75

DEFINITION OF TERMS FOR USE WITH FIGS. B AND C

BEAM AND GIRDER FLANGE COMPRESSIVE STRESSES

L = Clear distance between supports, providing lateral support (for cantilever use 4/3 length).
r = Least radius of gyration.
S_c = Section modulus for beam about axis normal to web (compression side) in inches cubed.

$$B = Iyyd\sqrt{11.7 + \frac{J}{Iyy}\left(\frac{L}{d}\right)^2}$$

Iyy = Moment of interia about y–y axis in inches to the fourth power.
J = Torsion factor in inches to the fourth power.
d = Depth of beam in inches.

COLUMNS – AXIALLY LOADED

L = Column length.
r = Minimum radius of gyration.
When L/r > 120 special attention should be given to any possible lateral loads which might not otherwise be considered such as; wind on a strut, workmen and their equipment and dead load of the member.
The allowable compressive stress should be computed by the method given under columns – combined bending and compression.

COLUMNS – COMBINED BENDING AND COMPRESSION

$$f_b = f_B\left(1 - \frac{P/A}{f_c}\right)\left(1 - \frac{P/A}{f_{ce}}\right)$$

f_b = Maximum bending stress (in compression) that may be permitted at or near the center of the unsupported length in addition to uniform compression. P/A.

f_B = Allowable compressive working stress for the member considered as a beam. See Fig. B or C.
f_c = Allowable working stress for member considered as an axially loaded column. See Fig. B or C.

$$fce = \frac{70{,}000}{\left(\frac{L}{r}\right)^2} \text{ (for aluminum alloy 6061-T6)}$$

$$fce = \frac{74{,}000}{\left(\frac{L}{r}\right)^2} \text{ (for aluminum alloy 2014-T6)}$$

COLUMNS – TRANSVERSE SHEAR

$$V = P\,\frac{4.5r^2\,(fB - P/A)}{f_c\,L_c} + V_t$$

Note: V min. = $0.02P + V_t$

V = Maximum shear on any transverse section of a column in the outer eighth of the length at each end, in the direction of assumed bending, in kips.

c = Distance from the centroidal axis to the extreme fiber in inches.

V_t = Shear due to any transverse loads on column in kips.

Adapted from ALCOA *Structural Handbook*, 1956.

STRUCTURAL ALUMINUM - ALLOWABLE WORKING STRESSES-2

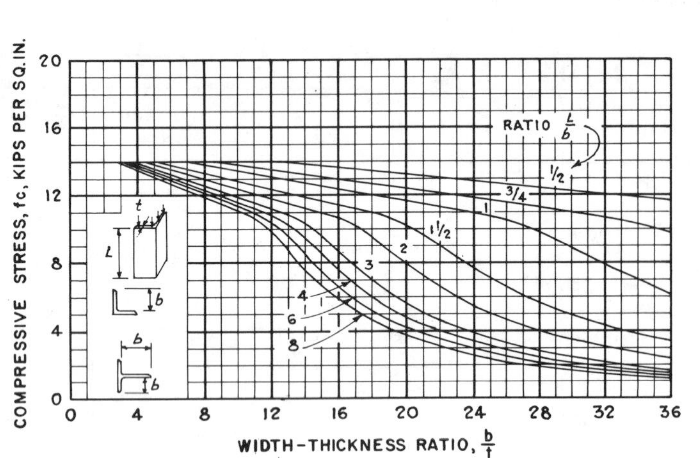
COMPRESSIVE STRESS, fc, KIPS PER SQ.IN.
WIDTH-THICKNESS RATIO, b/t
RATIO L/b
1/2
3/4
1
1 1/2
2
3
4
6
8

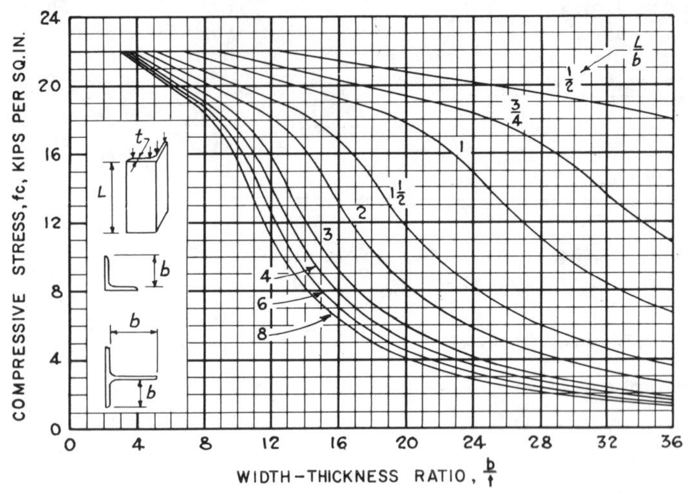
COMPRESSIVE STRESS, fc, KIPS PER SQ.IN.
WIDTH-THICKNESS RATIO, b/t
L/b
1/2
3/4
1
1 1/2
2
3
4
6
8

FIG. A. ALLOY 6061-T6 **FIG. B. ALLOY 2014-T6**

Allowable Compressive Stresses in Outstanding Legs of Single Angle and T-Section Struts (Gross Section)

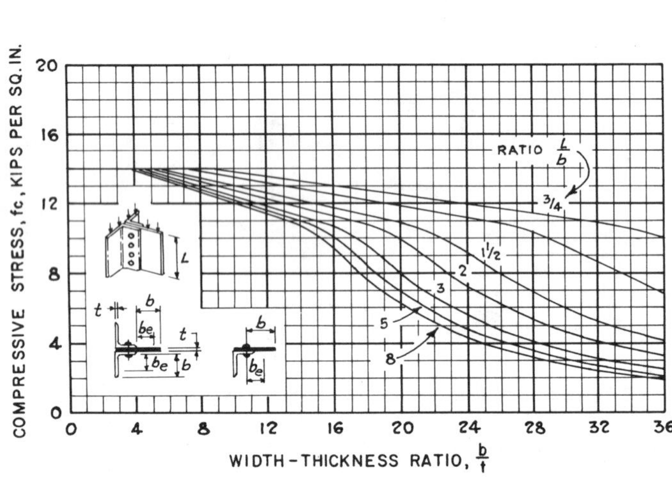
COMPRESSIVE STRESS, fc, KIPS PER SQ.IN.
WIDTH-THICKNESS RATIO, b/t
RATIO L/b
3/4
1 1/2
2
3
5
8

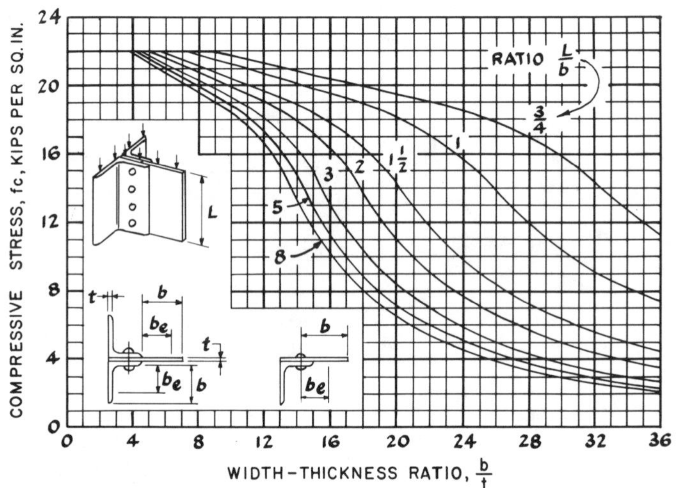
COMPRESSIVE STRESS, fc, KIPS PER SQ.IN.
WIDTH-THICKNESS RATIO, b/t
RATIO L/b
3/4
1
1 1/2
2
3
5
8

FIG. C. ALLOY 6061-T6 **FIG. D. ALLOY 2014-T6**

Chart for Determining Effective Width for Outstanding Legs of Angles Built into Other Parts and for Plates Built in along One Edge

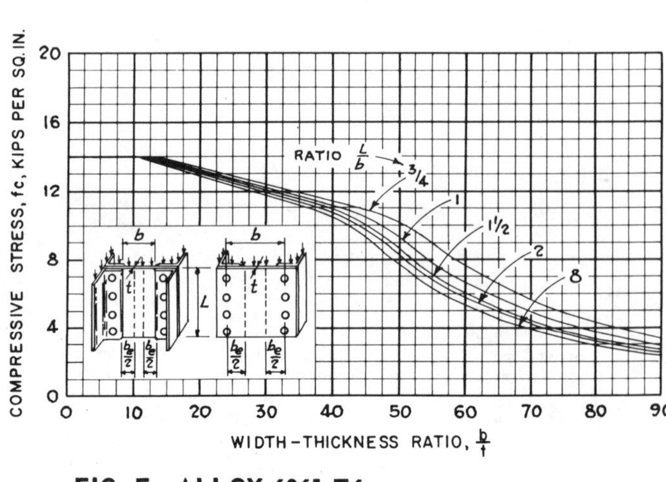
COMPRESSIVE STRESS, fc, KIPS PER SQ.IN.
WIDTH-THICKNESS RATIO, b/t
RATIO L/b
3/4
1
1 1/2
2
8

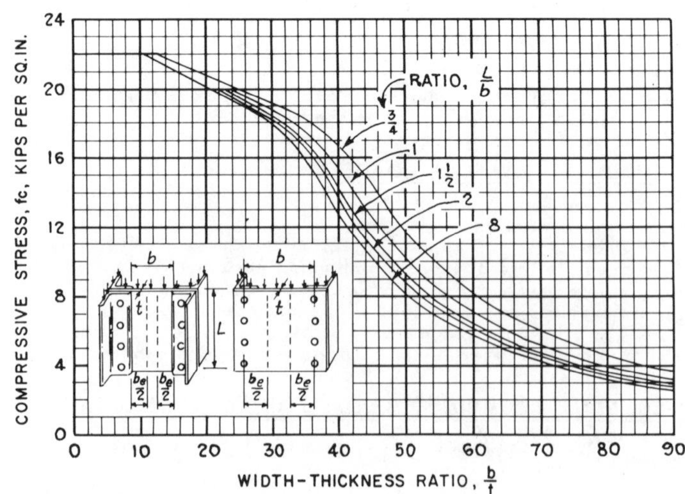
COMPRESSIVE STRESS, fc, KIPS PER SQ.IN.
WIDTH-THICKNESS RATIO, b/t
RATIO, L/b
3/4
1
1 1/2
2
8

FIG. E. ALLOY 6061-T6 **FIG. F. ALLOY 2014-T6**

Chart for Determining Effective Width for Flat Plates Built in along Two Edges

STRUCTURAL ALUMINUM – ALLOWABLE WORKING STRESSES-3

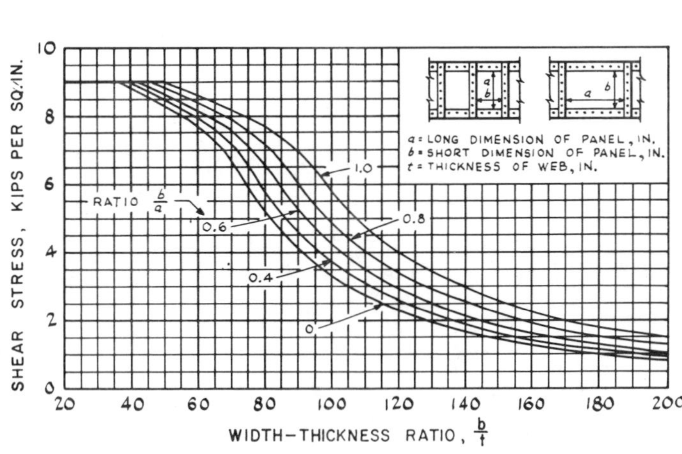
SHEAR STRESS, KIPS PER SQ. IN.
WIDTH–THICKNESS RATIO, b/t
RATIO b/a
a = LONG DIMENSION OF PANEL, IN.
b = SHORT DIMENSION OF PANEL, IN.
t = THICKNESS OF WEB, IN.

FIG. A. ALLOY 6061-T6

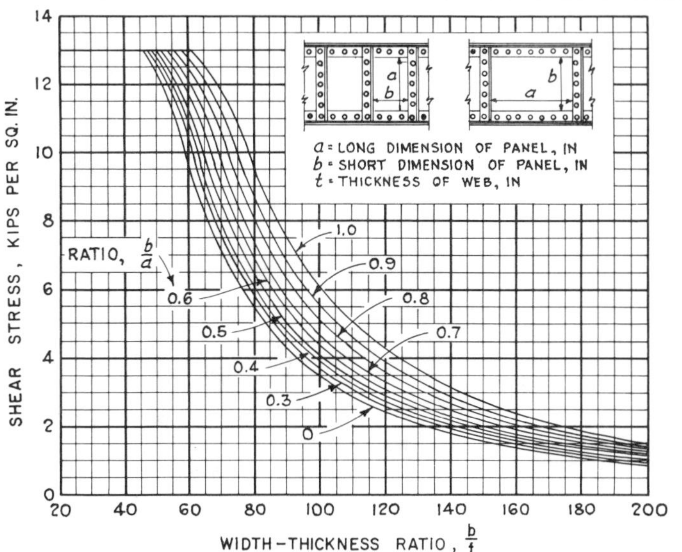
SHEAR STRESS, KIPS PER SQ. IN.
WIDTH–THICKNESS RATIO, b/t
RATIO, b/a
a = LONG DIMENSION OF PANEL, IN
b = SHORT DIMENSION OF PANEL, IN
t = THICKNESS OF WEB, IN

FIG. B. ALLOY 2014-T6

Allowable Shear Stresses on Webs; Partial Restraint Assumed at Edges of Rectangular Panels (Gross Section)

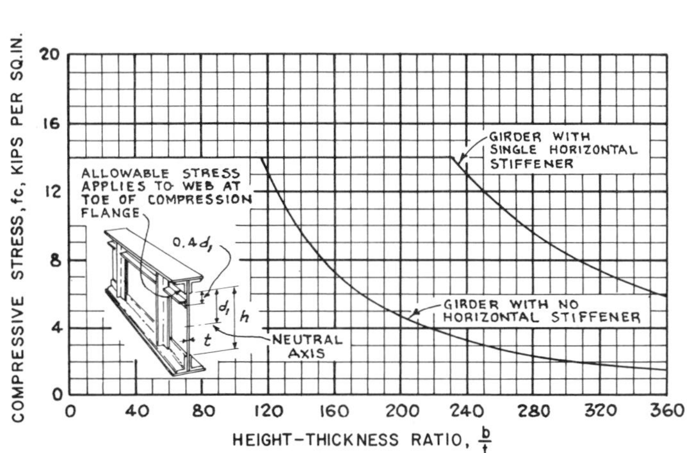
COMPRESSIVE STRESS, fc, KIPS PER SQ.IN.
HEIGHT–THICKNESS RATIO, b/t
ALLOWABLE STRESS APPLIES TO WEB AT TOE OF COMPRESSION FLANGE
GIRDER WITH SINGLE HORIZONTAL STIFFENER
GIRDER WITH NO HORIZONTAL STIFFENER
NEUTRAL AXIS

FIG. C. ALLOY 6061-T6

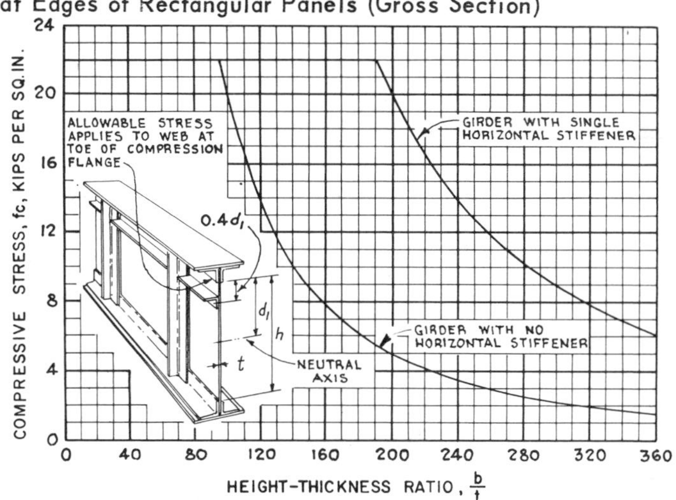
COMPRESSIVE STRESS, fc, KIPS PER SQ.IN.
HEIGHT–THICKNESS RATIO, b/t
ALLOWABLE STRESS APPLIES TO WEB AT TOE OF COMPRESSION FLANGE
GIRDER WITH SINGLE HORIZONTAL STIFFENER
GIRDER WITH NO HORIZONTAL STIFFENER
NEUTRAL AXIS

FIG. D. ALLOY 2014-T6

Allowable Longitudinal Compressive Stresses for Webs of Girders

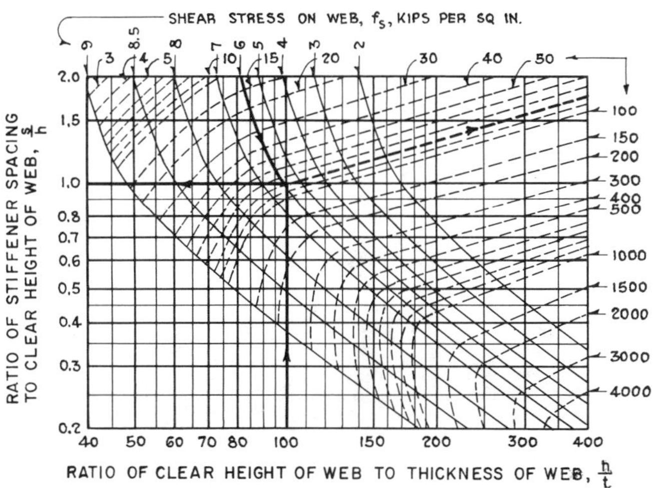
SHEAR STRESS ON WEB, f_s, KIPS PER SQ IN.
RATIO OF STIFFENER SPACING TO CLEAR HEIGHT OF WEB, s/h
RATIO OF CLEAR HEIGHT OF WEB TO THICKNESS OF WEB, h/t

FIG. E. ALLOY 6061-T6

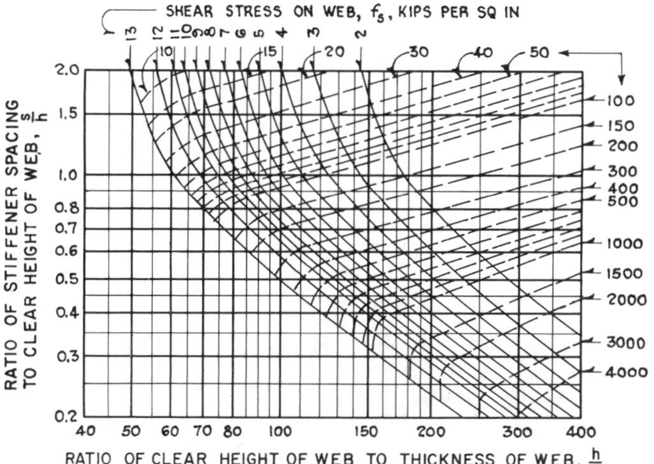
SHEAR STRESS ON WEB, f_s, KIPS PER SQ IN
RATIO OF STIFFENER SPACING TO CLEAR HEIGHT OF WEB, s/h
RATIO OF CLEAR HEIGHT OF WEB TO THICKNESS OF WEB, h/t

FIG. F. ALLOY 2014-T6

Spacing and Moment of Inertia of Vertical Stiffeners to Resist Shear Buckling on Webs of Plate Girders

Given: Shear Stress = 6 kips/sq. in., ratio h/t = 100
Find: Stiffener spacing = s, Moment of inertia for stiffener = Is
Solution: Enter chart E with shear stress. Follow curve until intersection with ratio h/t reading to the left give ratio s/h. (Multiply this by h, as previously determined, to get spacing.) Reading to the right along the dotted line gives the ratio I = /t^4. (Multiply this by t^4, as previously determined, to get Is.)

STRUCTURAL ALUMINUM

TABLE A - PROPERTIES OF ROUND TUBE SECTIONS.

D	t	WT.	A	I	S	r
1/2	1/32	.054	.046	.001	.005	.166
	1/16	.101	.086	.002	.008	.156
	3/32	.141	.120	.003	.010	.147
	1/8	.173	.147	.003	.012	.140
	5/32	.199	.169	.003	.012	.134
3/4	1/32	.083	.071	.005	.012	.254
	1/16	.159	.135	.008	.021	.244
	3/32	.227	.193	.011	.028	.234
	1/8	.288	.245	.012	.033	.225
	3/16	.389	.331	.015	.039	.210
1	1/32	.112	.095	.011	.022	.342
	1/16	.216	.184	.020	.041	.332
	1/8	.405	.344	.034	.067	.313
	3/16	.563	.479	.042	.083	.295
	1/4	.693	.589	.046	.092	.280
1 1/4	1/32	.141	.120	.022	.036	.431
	1/16	.274	.233	.041	.066	.420
	1/8	.520	.442	.071	.113	.400
	3/16	.736	.626	.091	.146	.381
	1/4	.923	.785	.104	.167	.364
1 1/2	1/32	.169	.144	.039	.052	.519
	1/16	.332	.282	.073	.097	.509
	1/8	.635	.540	.129	.172	.488
	3/16	.909	.773	.170	.227	.469
	1/4	1.155	.982	.199	.266	.451
1 3/4	1/32	.199	.169	.062	.071	.608
	1/16	.389	.331	.118	.135	.597
	1/8	.750	.638	.212	.242	.576
	3/16	1.082	.920	.285	.326	.556
	1/4	1.385	1.178	.341	.389	.538
2	1/16	.447	.380	.179	.179	.685
	1/8	.865	.736	.325	.325	.664
	3/16	1.256	1.068	.443	.443	.644
	1/4	1.616	1.374	.537	.537	.625
	5/16	1.949	1.657	.610	.610	.607
2 1/4	1/16	.506	.430	.257	.229	.774
	1/8	.981	.834	.473	.420	.753
	3/16	1.429	1.215	.651	.579	.732
	1/4	1.847	1.571	.798	.709	.713
	5/16	2.237	1.902	.916	.814	.694
2 1/2	1/16	.563	.479	.356	.285	.862
	1/8	1.097	.933	.660	.528	.841
	3/16	1.602	1.362	.917	.733	.820
	1/4	2.078	1.767	1.132	.906	.800
	5/16	2.526	2.148	1.311	1.049	.781
2 3/4	1/16	.621	.528	.477	.347	.950
	1/8	1.212	1.031	.890	.647	.929
	3/16	1.776	1.510	1.246	.906	.908
	1/4	2.310	1.964	1.549	1.127	.888
	5/16	2.814	2.393	1.807	1.314	.869
3	1/8	1.517	1.129	1.169	.779	1.017
	3/16	1.949	1.657	1.646	1.097	.997
	1/4	2.540	2.160	2.059	1.373	.976
	5/16	3.103	2.639	2.414	1.610	.957
	3/8	3.637	3.093	2.718	1.812	.938
3 1/4	1/8	1.443	1.227	1.501	.923	1.106
	3/16	2.122	1.804	2.123	1.306	1.085
	1/4	2.771	2.356	2.669	1.643	1.064
	5/16	3.392	2.884	3.146	1.936	1.044
	3/8	3.983	3.387	3.559	2.190	1.025
3 1/2	1/8	1.558	1.325	1.890	1.080	1.194
	3/16	2.294	1.951	2.685	1.534	1.173
	1/4	3.002	2.553	3.390	1.937	1.153
	5/16	3.680	3.129	4.013	2.293	1.132
	3/8	4.330	3.682	4.559	2.605	1.113
3 3/4	1/8	1.675	1.424	2.341	1.249	1.282
	3/16	2.468	2.099	3.339	1.781	1.261
	1/4	3.233	2.749	4.231	2.256	1.241
	5/16	3.969	3.375	5.026	2.681	1.220
	3/8	4.676	3.976	5.732	3.057	1.201
4	1/8	1.790	1.522	2.859	1.429	1.371
	3/16	2.641	2.246	4.090	2.045	1.350
	1/4	3.463	2.945	5.200	2.600	1.329
	5/16	4.257	3.620	6.198	3.098	1.308
	3/8	5.023	4.271	7.090	3.544	1.289
4 1/4	1/8	1.905	1.620	3.449	1.623	1.459
	3/16	2.814	2.393	4.948	2.328	1.438
	1/4	3.695	3.142	6.308	2.968	1.417
	5/16	4.546	3.866	7.539	3.548	1.397
	3/8	5.368	4.565	8.649	4.070	1.376

D	t	WT.	A	I	S	r
4 1/2	1/8	2.02	1.718	4.114	1.829	1.547
	3/16	2.987	2.54	5.917	2.63	1.526
	1/4	3.925	3.338	7.563	3.62	1.505
	5/16	4.835	4.111	9.062	4.028	1.485
	3/8	5.715	4.86	10.42	4.633	1.464
4 3/4	1/8	2.136	1.816	4.86	2.047	1.636
	3/16	3.161	2.688	7.006	2.95	1.615
	1/4	4.156	3.534	8.975	3.779	1.594
	5/16	5.124	4.357	10.78	4.538	1.573
	3/8	6.061	5.154	12.42	5.231	1.553
5	3/16	3.334	2.835	8.22	3.288	1.703
	1/4	4.388	3.731	10.55	4.220	1.682
	5/16	5.412	4.602	12.70	5.078	1.661
	3/8	6.408	5.449	14.67	5.866	1.641
	7/16	7.375	6.271	16.47	6.587	1.621
5 1/4	3/16	3.507	2.982	9.568	3.645	1.791
	1/4	4.618	3.927	12.30	4.687	1.770
	5/16	5.700	4.847	14.83	5.650	1.749
	3/8	6.754	5.743	17.16	6.538	1.729
	7/16	7.779	6.615	19.31	7.356	1.709
5 1/2	3/16	3.680	3.129	11.05	4.020	1.879
	1/4	4.849	4.123	14.24	5.178	1.858
	5/16	5.989	5.093	17.19	6.252	1.837
	3/8	7.101	6.038	19.93	7.247	1.817
	7/16	8.183	6.958	22.46	8.166	1.797
6	3/16	4.027	3.424	14.47	4.825	2.056
	1/4	5.311	4.516	18.70	6.232	2.035
	5/16	6.567	5.584	22.65	7.547	2.014
	3/8	7.793	6.627	26.33	8.774	1.993
	1/2	10.16	8.639	32.94	10.98	1.953
6 1/2	3/16	4.372	3.718	18.54	5.703	2.233
	1/4	5.773	4.909	24.01	7.385	2.212
	5/16	7.144	6.075	29.15	8.966	2.190
	3/8	8.486	7.216	33.97	10.45	2.170
	1/2	11.08	9.425	42.71	13.14	2.129
7	3/16	4.719	4.013	23.30	6.656	2.410
	1/4	6.235	5.302	30.23	8.638	2.388
	5/16	7.722	6.566	36.78	10.51	2.367
	3/8	9.179	7.805	42.96	12.27	2.346
	1/2	12.01	10.21	54.24	15.50	2.305
7 1/2	3/16	5.065	4.307	28.81	7.683	2.586
	1/4	6.696	5.694	37.46	9.989	2.565
	5/16	8.298	7.056	45.66	12.18	2.544
	3/8	9.871	8.394	53.42	14.24	2.523
	1/2	12.94	11.00	67.70	18.05	2.481
8	3/16	5.412	4.602	35.13	8.78	2.763
	1/4	7.158	6.087	45.75	11.43	2.742
	5/16	8.875	7.547	55.84	13.96	2.720
	3/8	10.56	8.983	65.45	16.36	2.699
	1/2	13.85	11.78	83.21	20.80	2.658
8 1/2	3/16	5.758	4.896	42.32	9.956	2.940
	1/4	7.620	6.480	55.18	12.98	2.918
	5/16	9.453	8.038	67.45	15.87	2.897
	3/8	11.26	9.572	79.16	18.63	2.876
	1/2	14.78	12.57	100.90	23.75	2.834
9	3/16	6.105	5.191	50.42	11.20	3.116
	1/4	8.081	6.872	65.82	14.63	3.095
	5/16	10.03	8.529	80.57	17.91	3.074
	3/8	11.95	10.16	94.67	21.04	3.052
	1/2	15.70	13.35	121.00	26.89	3.010
9 1/2	3/16	6.450	5.485	59.49	12.52	3.293
	1/4	8.544	7.265	77.76	16.36	3.272
	5/16	10.61	9.020	95.28	20.05	3.250
	3/8	12.64	10.75	112.1	23.59	3.229
	1/2	16.63	14.14	143.6	30.22	3.187
10	3/16	6.797	5.780	69.59	13.92	3.470
	1/4	9.006	7.658	91.06	18.21	3.448
	5/16	11.18	9.510	111.7	22.34	3.427
	3/8	13.34	11.34	131.5	26.30	3.406
	1/2	17.55	14.92	168.8	33.76	3.363
10 1/2	3/16	7.143	6.074	80.78	15.39	3.647
	1/4	9.467	8.050	105.8	20.15	3.625
	5/16	11.76	10.00	129.9	24.74	3.604
	3/8	14.03	11.93	153.1	29.15	3.582
	1/2	18.47	15.71	196.8	37.49	3.540
11	3/16	7.490	6.369	93.1	16.93	3.823
	1/4	9.929	8.443	122.0	22.19	3.802
	5/16	12.34	10.49	149.9	27.26	3.780
	3/8	14.72	12.52	176.9	32.16	3.759
	1/2	19.39	16.49	227.8	41.42	3.717

TABLE B - PROPERTIES OF WIDE FLANGE STANDARD I-BEAMS AND STANDARD CHANNELS.

D	WT.	t	A	d	b	I	S	r′	J	C_s	RIVET DATA DIA.	RIVET DATA g	RIVET DATA u
4 WF	4.71	.313	4.00	4.00	4.00	10.72	5.36	.94	.22	11.63	5/8	1 1/8	
5 WF	6.45	.313	5.48	5.00	5.00	23.82	9.53	1.19	.34	40.85	3/4	1 3/8	
6 WF	4.16	.230	3.54	6.00	4.00	21.75	7.25	.92	.082	24.38	7/8	1 3/4	
	5.40	.240	4.59	6.00	6.00	30.17	10.06	1.45	.11	79.57	7/8	1 3/4	
	7.81	.250	6.64	6.00	5.94	44.06	14.69	1.46	.45	108.74	7/8	1 3/4	
	8.26	.313	7.02	6.00	6.00	45.19	15.06	1.44	.50	112.30	7/8	1 3/4	
	9.14	.438	7.77	6.00	6.12	47.44	15.81	1.42	.62	119.86	7/8	1 3/4	
8 WF	5.90	.230	5.02	8.00	5.25	56.73	14.18	1.22	.13	110.05	7/8	1 3/8	
	8.32	.245	7.08	8.00	6.50	84.15	21.04	1.61	.31	263.38	7/8	1 3/4	
	10.72	.288	9.12	8.00	8.00	109.66	27.41	2.01	.50	529.22	7/8	2 3/4	
	11.19	.313	9.52	8.00	7.94	112.94	28.23	1.89	.68	483.94	7/8	2 3/4	
	11.77	.375	10.01	8.00	8.00	115.58	28.90	1.87	.75	496.00	7/8	2 3/4	
	12.95	.500	11.01	8.00	8.12	120.92	30.23	1.83	.96	520.95	7/8	2 3/4	
10 WF	7.30	.240	6.21	9.90	5.75	106.74	21.56	1.32	.20	246.08	3/4	1 3/8	
3 I	1.96	.170	1.67	3.00	2.33	2.52	1.68	.52	.045	.85	3/8	3/4	5/16
	2.25	.251	1.91	3.00	2.41	2.71	1.80	.52	.061	.94	3/8	3/4	5/16
	2.59	.349	2.21	3.00	2.51	2.93	1.95	.52	.093	1.08	3/8	3/4	5/16
4 I	2.64	.190	2.25	4.00	2.66	6.06	3.03	.58	.074	2.58	1/2	3/4	5/16
	3.28	.326	2.79	4.00	2.80	6.79	3.39	.57	.12	3.04	1/2	3/4	5/16
5 I	3.43	.210	2.92	5.00	3.00	12.26	4.90	.64	.12	6.54	1/2	7/8	3/8
	4.23	.347	3.60	5.00	3.14	13.69	5.48	.63	.19	7.60	1/2	7/8	3/8
	5.10	.494	4.34	5.00	3.28	15.22	6.09	.62	.33	8.92	1/2	7/8	3/8
6 I	4.30	.230	3.66	6.00	3.33	22.08	7.36	.71	.17	14.34	5/8	1	3/8
	5.10	.343	4.34	6.00	3.44	24.11	8.04	.69	.24	16.04	5/8	1	3/8
	5.96	.465	5.07	6.00	3.56	26.31	8.77	.68	.38	18.12	5/8	1	3/8
7 I	5.27	.250	4.48	7.00	3.66	36.69	10.48	.77	.25	28.45	5/8	1 1/8	3/8
	6.05	.345	5.15	7.00	3.76	39.40	11.26	.75	.32	31.11	5/8	1 1/8	3/8
	6.92	.450	5.88	7.00	3.86	42.40	12.12	.73	.46	34.19	5/8	1 1/8	3/8
8 I	6.35	.270	5.40	8.00	4.00	57.55	14.39	.83	.34	53.04	3/4	1 1/8	7/16
	7.96	.441	6.77	8.00	4.17	64.85	16.21	.80	.56	61.16	3/4	1 1/8	7/16
	8.81	.532	7.49	8.00	4.26	68.73	17.18	.79	.75	66.04	3/4	1 1/8	1/2
9 I	7.51	.290	6.38	9.00	4.33	85.90	19.09	.89	.46	92.09	3/4	1 1/4	1/2
10 I	8.76	.310	7.45	10.00	4.66	173.39	24.68	.95	.62	152.04	3/4	1 3/8	1/2
	10.37	.447	8.82	10.00	4.80	134.81	26.96	.92	.86	167.94	3/4	1 3/8	1/2
	12.10	.594	10.29	10.00	4.94	147.06	29.41	.90	1.31	186.92	3/4	1 3/8	1/2
12 I	10.99	.350	9.35	12.00	5.00	218.13	36.35	1.00	.92	304.64	3/4	1 1/2	9/16
	12.09	.428	10.28	12.00	5.08	229.36	38.23	.98	1.10	321.25	3/4	1 1/2	9/16
	14.08	.460	11.97	12.00	5.25	272.15	45.36	1.06	1.78	432.08	3/4	1 1/2	3/4
	15.56	.565	13.23	12.00	5.36	287.27	47.88	1.05	2.19	462.30	3/4	1 1/2	3/4
	17.28	.687	14.70	12.00	5.48	304.84	50.81	1.03	2.85	500.35	3/4	1 3/4	3/4
3 [	1.42	.17	1.21	3.00	1.41	1.66	1.10	.40	.031	.27	1/2	7/8	1/4
	1.48	.19	1.26	3.00	1.43	1.69	1.13	.41	.033	.29	1/2	7/8	1/4
	1.73	.26	1.47	3.00	1.50	1.85	1.24	.41	.047	.34	1/2	7/8	1/4
	1.95	.32	1.66	3.00	1.56	1.99	1.33	.41		.38	1/2	7/8	1/4
	2.07	.36	1.76	3.00	1.60	2.07	1.38	.42	.08	.42	1/2	7/8	1/4
4 [	1.85	.18	1.57	4.00	1.58	3.83	1.92	.45	.045	.80	1/2	1	5/16
	2.16	.25	1.84	4.00	1.65	4.19	2.10	.45	.062	.93	1/2	1	5/16
	2.50	.32	2.13	4.00	1.72	4.58	2.29	.45	.090	1.09	1/2	1	5/16
5 [	2.32	.19	1.97	5.00	1.75	7.49	3.00	.49	.064	1.92	1/2	1 1/8	5/16
	3.11	.32	2.64	5.00	1.	8.90	3.56	.49	.12	2.52	1/2	1 1/8	5/16
	3.97	.47	3.38	5.00	2.03	10.43	4.17	.49	.25	3.34	1/2	1 1/8	5/16
6 [	2.83	.20	2.40	6.00	1.92	13.12	4.37	.54		4.06	5/8	1 1/8	5/16
	3.00	.22	2.55	6.00	1.94	13.57	4.52	.54	.097	4.27	5/8	1 1/8	5/16
	3.63	.31	3.09	6.00	2.03	15.18	5.06	.53	.14	5.14	5/8	1 1/8	3/8
	4.48	.44	3.82	6.00	2.16	17.39	5.80	.52	.26	6.34	5/8	1 3/8	3/8
7 [	3.38	.21	2.87	7.00	2.	21.27	6.08	.58	.12	7.77	5/8	1 1/4	3/8
	3.54	.23	3.01	7.00	2.11	21.84	6.24	.58	.13	8.03	5/8	1 1/4	7/16
	4.23	.31	3.60	7.00	2.19	24.24	6.93	.57	.18	9.58	5/8	1 1/4	3/8
	5.10	.42	4.33	7.00	2.30	27.24	7.78	.56	.29	11.42	5/8	1 1/4	7/16
	5.96	.52	5.07	7.00	2.40	30.25	8.64	.56	.47	13.44	5/8	1 1/2	7/16
8 [	4.25	.25	3.62	8.00	2.29	33.85	8.46	.62	.17	14.83	3/4	1 3/8	3/8
	4.75	.30	4.04	8.00	2.34	36.11	9.03	.61	.21	16.28	3/4	1 3/8	3/8
	5.62	.40	4.78	8.00	2.44	40.04	10.01	.61	.32	19.15	3/4	1 1/2	7/16
	6.48	.49	5.51	8.00	2.53	43.96	10.99	.60	.47	21.57	3/4	1 1/2	7/16
9 [	4.60	.23	3.91	9.00	2.43	47.68	10.60	.67	.20	23.60	3/4	1 3/8	7/16
	5.19	.28	4.41	9.00	2.48	51.02	11.34	.66	.24	25.87	3/4	1 3/8	7/16
	6.91	.45	5.88	9.00	2.65	60.92	13.54	.64	.47	33.70	3/4	1 1/2	1/2
	8.65	.61	7.35	9.00	2.81	70.89	15.75	.63	.92	42.36	3/4	1 1/2	1/2
10 [	5.28	.24	4.49	10.00	2.60	67.37	13.47	.71	.25	37.97	3/4	1 1/2	7/16
	6.91	.38	5.88	10.00	2.74	78.95	15.79	.69	.41	47.29	3/4	1 1/2	7/16
	8.64	.53	7.35	10.00	2.89	91.20	18.24	.68	.75	58.59	3/4	1 3/4	1/2
	10.37	.67	8.82	10.00	3.03	103.45	20.69	.67	1.32	70.94	3/4	1 3/4	1/2
12 [	7.41	.30	6.30	12.00	2.96	131.84	21.97	.80	.46	96.84	7/8	1 3/4	1/2
	8.64	.39	7.35	12.00	3.05	144.37	24.06	.78	.61	109.72	7/8	1 3/4	1/2
	10.37	.51	8.82	12.00	3.17	162.08	27.01	.76	.95	129.02	7/8	1 3/4	1/2
	12.10	.63	10.29	12.00	3.29	179.65	29.94	.75	1.48	148.64	7/8	2	5/8
15 [	11.71	40		15.00	3.40	314.76	41.97	.90	1.17	361.54	1	2	5/8
	17.28	.72	14.70	15.00	3.72	403.64	53.87	.92	2.89	511.46	1	2	5/8

RIVET DATA IN PREPARATION BY ALCOA. DATA SHOWN IS FOR SIMILAR STEEL SHAPES.

TABL

LEGS	t	S 1-1	S 2-2
2 1/2 × 2	1/8	.19	.13
	3/16	.29	.19
	1/4	.38	.25
	5/16	.46	.30
	3/8	.54	.36
2 1/2 × 2 1/2	1/8	.20	—
	3/16	.30	—
	1/4	.39	—
	5/16	.48	—
	3/8	.56	—
	7/16	.64	—
	1/2	.72	—
3 × 1 1/2	1/4	.51	.14
3 × 2	3/16	.40	.19
	1/4	.52	.25
	5/16	.65	.30
	3/8	.76	.36
	7/16	.88	.41
	1/2	.99	.46
3 × 2 1/2	1/4	.53	.38
	5/16	.66	.47
	3/8	.78	.55
	7/16	.90	.64
	1/2	1.01	.72
3 × 3	3/16	.42	—
	1/4	.54	—
	5/16	.67	—
	3/8	.80	—
	7/16	.92	—
	1/2	1.04	—
3 1/2 × 2 1/2	1/4	.72	.38
	5/16	.89	.48
	3/8	1.06	.57
	7/16	1.22	.65
	1/2	1.37	.73
3 1/2 × 3	1/4	.74	.57
	5/16	.92	.69
	3/8	1.09	.82
	7/16	1.26	.94
	1/2	1.42	1.06
3 1/2 × 3 1/2	1/4	.76	—
	5/16	.94	—
	3/8	1.11	—
	7/16	1.28	—
	1/2	1.45	—
4 × 3	1/4	.96	.56
	5/16	1.19	.70
	3/8	1.42	.83
	7/16	1.63	.96
	1/2	1.85	1.08
	5/8	2.25	1.32

TABLE G - PERCE
OF ALUMINUM AL
USE IN THIN PL

RATIO[a] D/t	LOSS IN DOUBLE SHEAR[b]	RATIO D/t
1.5	0	2.2
1.6	1.3	2.3
1.7	2.6	2.4
1.8	3.9	2.5
1.9	5.2	2.6
2.0	6.5	2.7
2.1	7.8	2.8

D = *Diameter*
t = *Thickness*
leg of ang
WT. = *Weight in*
A = *Area in*
I = *Moment*
S = *Section*
r = *Radius*
d = *Actual*
b = *Width in*
r′ = *Least r*
J = *Torsion*
C_s = *Torsion*
g = *Usual r*
center of
u = *Nominal*

PROPERTIES OF SECTIONS

OF ANGLES.

EGS	t	S_{1-1}	S_{2-2}	A	r_{1-1}	r_{2-2}	r_{3-3}	J
	5/16	1.19	.93	2.23	1.23	1.04	.72	.076
×3½	3/8	1.43	1.11	2.66	1.23	1.04	.72	.132
	7/16	1.65	1.29	3.08	1.22	1.03	.71	.209
	1/2	1.87	1.46	3.49	1.22	1.03	.71	.313
	1/4	1.00	—	1.94	1.23	—	.79	.042
	5/16	1.24	—	2.41	1.23	—	.78	.081
	3/8	1.48	—	2.86	1.23	—	.78	.141
	7/16	1.71	—	3.31	1.21	—	.78	.223
×4	1/2	1.93	—	3.75	1.21	—	.78	.333
	9/16	2.15	—	4.19	1.20	—	.77	.475
	5/8	2.36	—	4.61	1.19	—	.77	.651
	11/16	2.57	—	5.03	1.19	—	.77	.867
	3/4	2.77	—	5.44	1.18	—	.77	1.125
×2½	1/2	2.77	.77	3.50	1.58	.64	.52	.313
×3	3/8	2.15	.84	2.85	1.59	.82	.64	.141
	1/2	2.83	1.10	3.74	1.57	.81	.64	.333
	5/16	1.85	.96	2.56	1.58	1.00	.75	.09
	3/8	2.21	1.15	3.05	1.58	1.00	.75	.15
×3½	7/16	2.56	1.33	3.53	1.57	.99	.75	.24
	1/2	2.90	1.50	4.00	1.56	.99	.74	.35
	5/8	3.56	1.84	4.92	1.55	.98	.74	.69
	3/4	4.16	2.14	5.81	1.53	.96	.74	1.20
	3/8	2.30	—	3.60	1.52	—	.98	.18
	7/16	2.67	—	4.18	1.52	—	.97	.28
×5	1/2	3.03	—	4.74	1.52	—	.97	.42
	5/8	3.73	—	5.85	1.50	—	.97	.81
	3/4	4.41	—	6.93	1.49	—	.96	1.41
	5/16	2.64	.98	2.88	1.92	.97	.76	.10
×3½	3/8	3.15	1.17	3.43	1.92	.96	.75	.17
	1/2	4.14	1.53	4.51	1.90	.95	.75	.40
	5/8	5.09	1.88	5.56	1.89	.94	.74	.77
	3/8	3.17	1.50	3.60	1.90	1.13	.86	.18
	7/16	3.69	1.74	4.18	1.90	1.13	.86	.28
×4	1/2	4.19	1.98	4.74	1.89	1.13	.86	.42
	9/16	4.69	2.21	5.30	1.88	1.12	.85	.59
	5/8	5.17	2.44	5.85	1.88	1.11	.85	.81
	3/4	6.11	2.87	6.93	1.86	1.10	.85	1.41
	3/8	3.38	—	4.35	1.85	—	1.18	.21
	7/16	3.93	—	5.05	1.84	—	1.18	.34
	1/2	4.46	—	5.74	1.84	—	1.17	.50
6×6	9/16	4.99	—	6.43	1.83	—	1.17	.71
	5/8	5.51	—	7.10	1.82	—	1.17	.98
	11/16	6.02	—	7.77	1.82	—	1.17	1.30
	3/4	6.52	—	8.43	1.81	—	1.16	1.69
	5/8	9.74	5.77	8.37	2.53	1.76	1.29	1.14
8×6	11/16	10.58	6.25	9.15	2.52	1.75	1.28	1.52
	3/4	11.47	6.77	9.93	2.51	1.74	1.28	1.97
	1/2	8.16	—	7.77	2.48	—	1.58	0.67
8×8	3/4	11.99	—	11.46	2.45	—	1.57	2.25
	1	15.60	—	15.02	2.42	—	1.56	5.33

R STRENGTH
ROM THEIR

RATIO[a] D/t	LOSS IN: SINGLE SHEAR	LOSS IN: DOUBLE SHEAR
3.5	2.0	26.0
3.6	2.4	27.3
3.7	2.8	28.6
3.8	3.2	29.9
3.9	3.6	31.2
4.0	4.0	32.5
—	—	—

[a] Ratio of the rivet diameter, D, to the plate thickness, t. The thickness used is that of the thinnest plate in a single shear joint or of the middle plate in a double shear joint. [b] The percentage loss of strength in single shear is zero for D/t less than 3.0.

es.
[,

inches.

inches.[6]

TABLE D –ALLOWABLE DESIGN LOADS, IN KIPS PER RIVET, FOR HOT-DRIVEN 6061-T43 RIVETS IN 6061-T6 STRUCTURES. (RIVETS DRIVEN AT 990°F TO 1050°F; SHEAR, 8 KIPS PER SQ. INCH; AND BEARING, 27 KIPS PER SQ. INCH.[a])

DIMENSIONS IN INCHES																
RIVET DIAMETER	3/8		7/16		1/2		9/16		5/8		3/4		7/8		1	
HOLE DIAMETER	0.397		0.469		0.531		0.594		0.656		0.781		0.922		1.063	
DRILL SIZE	X		15/32		17/32		19/32		21/32		25/32		59/64		1 1/16	
THICKNESS OF PLATE OR SHAPE, IN INCHES	RIVET IN SINGLE SHEAR (ss) OR IN DOUBLE SHEAR (ds)															
	ss	ds	ss	ds	ss	ds	ss	ds	ss	ds	ss	ds	ss	ds	ss	ds
1/8	0.99	1.34[b]	1.35[c]	1.58[b]	1.70[c]	1.79[b]	2.00[c]	2.00[b]	2.21[b]	2.21[b]	—	—	—	—	—	—
3/16	0.99	1.85[c]	1.38	2.37[b]	1.77	2.69[b]	2.22	3.01[b]	2.66[c]	3.32[b]	3.68[c]	3.95[b]	4.67[b]	4.67[b]	—	—
1/4	0.99	1.98	1.38	2.68[c]	1.77	3.31[c]	2.22	4.00[c]	2.70	4.43[b]	3.83	5.27[b]	5.23[c]	6.22[b]	6.82[c]	7.18[b]
5/16	0.99	1.98	1.38	2.76	1.77	3.50[c]	2.22	4.26[c]	2.70	5.06[c]	3.83	6.59[b]	5.34	7.78[b]	7.04[c]	8.97[b]
3/8	0.99	1.98	1.38	2.76	1.77	3.54	2.22	4.43	2.70	5.29[c]	3.83	7.17[c]	5.34	9.34[b]	7.10	10.76[b]
7/16	—	—	1.38	2.76	1.77	3.54	2.22	4.43	2.70	5.41	3.83	7.46[c]	5.34	9.99[c]	7.10	12.56[b]
1/2	—	—	—	—	1.77	3.54	2.22	4.43	2.70	5.41	3.83	7.67	5.34	10.34[c]	7.10	13.28[c]
9/16	—	—	—	—	—	—	2.22	4.43	2.70	5.41	3.83	7.67	5.34	10.61[c]	7.10	13.69[c]
5/8	—	—	—	—	—	—	—	—	2.70	5.41	3.83	7.67	5.34	10.68	7.10	14.02[c]
3/4	—	—	—	—	—	—	—	—	—	—	3.83	7.67	5.34	10.68	7.10	14.20
7/8	—	—	—	—	—	—	—	—	—	—	—	—	5.34	10.68	7.10	14.20
1	—	—	—	—	—	—	—	—	—	—	—	—	—	—	7.10	14.20

TABLE E –ALLOWABLE DESIGN LOADS, IN KIPS PER RIVET, FOR HOT-DRIVEN 6061-T43 RIVETS IN 2014-T6 STRUCTURES. (RIVETS DRIVEN AT 990°F TO 1050°F; SHEAR, 8 KIPS PER SQ. INCH; AND BEARING, 36 KIPS PER SQ. INCH.[a])

DIMENSIONS IN INCHES																
RIVET DIAMETER	3/8		7/16		1/2		9/16		5/8		3/4		7/8		1	
HOLE DIAMETER	0.397		0.469		0.531		0.594		0.656		0.781		0.922		1.063	
DRILL SIZE	X		15/32		17/32		19/32		21/32		25/32		59/64		1 1/16	
THICKNESS OF PLATE OR SHAPE, IN INCHES	RIVET IN SINGLE SHEAR (ss) OR IN DOUBLE SHEAR (ds)															
	ss	ds	ss	ds	ss	ds	ss	ds	ss	ds	ss	ds	ss	ds	ss	ds
1/8	0.99	1.60[c]	1.35[c]	2.05[c]	1.70[c]	2.39[b]	2.08[c]	2.67[b]	2.49[c]	2.95[b]	—	—	—	—	—	—
3/16	0.99	1.85[c]	1.38	2.47[c]	1.77	3.01[c]	2.22	3.57[c]	2.66[c]	4.12[c]	3.68[c]	5.17[c]	4.98[c]	6.22[b]	—	—
1/4	0.99	1.98	1.38	2.68[c]	1.77	3.31[c]	2.22	4.00[c]	2.70	4.71[c]	3.83	6.17[c]	5.23[c]	7.91[c]	6.82[c]	9.59[c]
5/16	0.99	1.98	1.38	2.76	1.77	3.50[c]	2.22	4.26[c]	2.70	5.06[c]	3.83	6.77[c]	5.34	8.88[c]	7.04[c]	10.88[c]
3/8	0.99	1.98	1.38	2.76	1.77	3.54	2.22	4.43	2.70	5.29[c]	3.83	7.17[c]	5.34	9.53[c]	7.10	12.04[c]
7/16	—	—	1.38	2.76	1.77	3.54	2.22	4.43	2.70	5.41	3.83	7.46[c]	5.34	9.99[c]	7.10	12.77[c]
1/2	—	—	—	—	1.77	3.54	2.22	4.43	2.70	5.41	3.83	7.67	5.34	10.34[c]	7.10	13.28[c]
9/16	—	—	—	—	—	—	2.22	4.43	2.70	5.41	3.83	7.67	5.34	10.61[c]	7.10	13.69[c]
5/8	—	—	—	—	—	—	—	—	2.70	5.41	3.83	7.67	5.34	10.68	7.10	14.02[c]
3/4	—	—	—	—	—	—	—	—	—	—	3.83	7.67	5.34	10.68	7.10	14.20
7/8	—	—	—	—	—	—	—	—	—	—	—	—	5.34	10.68	7.10	14.20
1	—	—	—	—	—	—	—	—	—	—	—	—	—	—	7.10	14.20

TABLE F –ALLOWABLE DESIGN LOAD, IN KIPS PER RIVET FOR COLD-DRIVEN 6061-T6 RIVETS IN 6061-T6 STRUCTURES (SHEAR, 10 KIPS PER SQ. INCH; AND BEARING, 27 KIPS PER SQ. INCH.[a])

DIMENSIONS IN INCHES																
RIVET DIAMETER	3/8		7/16		1/2		9/16		5/8		3/4		7/8		1	
HOLE DIAMETER	0.386		0.453		0.516		0.578		0.641		0.766		0.891		1.016	
DRILL SIZE	W		29/64		33/64		37/64		41/64		49/64		57/64		1 1/64	
THICKNESS OF PLATE OR SHAPE, IN INCHES	RIVET IN SINGLE SHEAR (ss) OR IN DOUBLE SHEAR (ds)															
	ss	ds	ss	ds	ss	ds	ss	ds	ss	ds	ss	ds	ss	ds	ss	ds
1/8	1.17	1.30[b]	1.53[b]	1.53[b]	1.74[b]	1.74[b]	1.95[b]	1.95[b]	2.16[b]	2.16[b]	—	—	—	—	—	—
3/16	1.17	1.95[b]	1.61	2.29[b]	2.09	2.61[b]	2.62	2.93[b]	3.18[c]	3.25[b]	3.88[b]	3.88[b]	4.51[b]	4.51[b]	—	—
1/4	1.17	2.34	1.61	3.06[b]	2.09	3.48[b]	2.62	3.90[b]	3.23	4.33[b]	4.61	5.17[b]	6.02[b]	6.02[b]	6.86[b]	6.86[b]
5/16	1.17	2.34	1.61	3.22	2.09	4.13[c]	2.62	4.88[b]	3.23	5.41[b]	4.61	6.46[b]	6.24	7.52[b]	8.04[c]	8.57[b]
3/8	1.17	2.34	1.61	3.22	2.09	4.18	2.62	5.25	3.23	6.31[c]	4.61	7.76[b]	6.24	9.02[b]	8.11	10.29[b]
7/16	—	—	1.61	3.22	2.09	4.18	2.62	5.25	3.23	6.45	4.61	8.97[c]	6.24	10.52[b]	8.11	12.00[b]
1/2	—	—	—	—	2.09	4.18	2.62	5.25	3.23	6.45	4.61	9.22	6.24	12.03[b]	8.11	13.72[b]
9/16	—	—	—	—	—	—	2.62	5.25	3.23	6.45	4.61	9.22	6.24	12.39[c]	8.11	15.43[b]
5/8	—	—	—	—	—	—	—	—	3.23	6.45	4.61	9.22	6.24	12.47	8.11	16.00[c]
3/4	—	—	—	—	—	—	—	—	—	—	4.61	9.22	6.24	12.47	8.11	16.22
7/8	—	—	—	—	—	—	—	—	—	—	—	—	6.24	12.47	8.11	16.22
1	—	—	—	—	—	—	—	—	—	—	—	—	—	—	8.11	16.22

TABLE H– ALLOWABLE DESIGN LOAD, IN KIPS PER RIVET, FOR COLD-DRIVEN 2117-T3 RIVETS IN 2014-T6 STRUCTURES (SHEAR, 10 KIPS PER SQ. INCH AND BEARING, 36 KIPS PER SQ. INCH.[a])

DIMENSIONS IN INCHES																
RIVET DIAMETER	3/8		7/16		1/2		9/16		5/8		3/4		7/8		1	
HOLE DIAMETER	0.386		0.453		0.516		0.578		0.641		0.766		0.891		1.016	
DRILL SIZE	W		29/64		33/64		37/64		41/64		49/64		57/64		1 1/64	
THICKNESS OF PLATE OR SHAPE, IN INCHES	RIVET IN SINGLE SHEAR (ss) OR IN DOUBLE SHEAR (ds)															
	ss	ds	ss	ds	ss	ds	ss	ds	ss	ds	ss	ds	ss	ds	ss	ds
1/8	1.17	1.74[b]	1.58[c]	2.04[b]	2.01[c]	2.32[b]	2.52[c]	2.60[b]	2.88[b]	2.88[b]	—	—	—	—	—	—
3/16	1.17	2.19[c]	1.61	2.88[c]	2.09	3.48[b]	2.62	3.90[b]	3.18[c]	4.33[b]	4.42[c]	5.17[b]	5.82[c]	6.02[b]	—	—
1/4	1.17	2.34	1.61	3.12[c]	2.09	3.91[c]	2.62	4.74[c]	3.23	5.62[c]	4.61	6.89[b]	6.11[c]	8.02[b]	7.78[c]	9.14[b]
5/16	1.17	2.34	1.61	3.22	2.09	4.13[c]	2.62	5.04[c]	3.23	6.03[c]	4.61	8.14[c]	6.24	10.02[b]	8.04[c]	11.43[b]
3/8	1.17	2.34	1.61	3.22	2.09	4.18	2.62	5.25	3.23	6.31[c]	4.61	8.62[c]	6.24	11.12[c]	8.11	13.72[b]
7/16	—	—	1.61	3.22	2.09	4.18	2.62	5.25	3.23	6.45	4.61	8.97[c]	6.24	11.66[c]	8.11	14.58[c]
1/2	—	—	—	—	2.09	4.18	2.62	5.25	3.23	6.45	4.61	9.22	6.24	12.07[c]	8.11	15.16[c]
9/16	—	—	—	—	—		2.62	5.25	3.23	6.45	4.61	9.22	6.24	12.39[c]	8.11	15.64[c]
5/8	—	—	—	—	—	—	—	—	3.23	6.45	4.61	9.22	6.24	12.47	8.11	16.00[c]
3/4	—	—	—	—	—	—	—	—	—	—	4.61	9.22	6.24	12.47	8.11	16.22
7/8	—	—	—	—	—	—	—	—	—	—	—	—	6.24	12.47	8.11	16.22
1	—	—	—	—	—	—	—	—	—	—	—	—	—	—	8.11	16.22

[a] Assuming distance from center of rivet to edge of member toward which the pressure of the rivet is directed is not less than twice the nominal rivet diameter. [b] These values are governed by bearing. [c] These values are governed by reduced shear strengths as indicated in Table . All other values are governed by basic allowable shear stress.

Adapted from ALCOA STRUCTURAL HANDBOOK, 1956

STRUCTURAL WOOD – PROPERTIES OF LUMBER & COLUMNS

TABLE A – PROPERTIES OF STRUCTURAL LUMBER

Nominal Size, in.	American Standard Dressed Size, in.	Area of Section, in.^{2}	Weight per foot, lb.	Moment of Inertia, in.^{4}	Section Modulus, in.^{3}
2 x 4	1⅝ x 3⅝	5.89	1.64	6.45	3.56
6	5⅝	9.14	2.54	24.1	8.57
8	7½	12.2	3.39	57.1	15.3
10	9½	15.4	4.29	116	24.4
12	11½	18.7	5.19	206	35.8
14	13½	21.9	6.09	333	49.4
16	15½	25.2	6.99	504	65.1
18	17½	28.4	7.90	726	82.9
3 x 4	2⅝ x 3⅝	9.52	2.64	10.4	5.75
6	5⅝	14.8	4.10	38.9	13.8
8	7½	19.7	5.47	92.3	24.6
10	9½	24.9	6.93	188	39.5
12	11½	30.2	8.39	333	57.9
14	13½	35.4	9.84	538	79.7
16	15½	40.7	11.3	815	105
18	17½	45.9	12.8	1172	134
4 x 4	3⅝ x 3⅝	13.1	3.65	14.4	7.94
6	5⅝	20.4	5.66	53.8	19.1
8	7½	27.2	7.55	127	34.0
10	9½	34.4	9.57	259	54.5
12	11½	41.7	11.6	459	79.9
14	13½	48.9	13.6	743	110
16	15½	56.2	15.6	1125	145
18	17½	63.4	17.6	1619	185
6 x 6	5½ x 5½	30.3	8.40	76.3	27.7
8	7½	41.3	11.4	193	51.6
10	9½	52.3	14.5	393	82.7
12	11½	63.3	17.5	697	121
14	13½	74.3	20.6	1128	167
16	15½	85.3	23.6	1707	220
18	17½	96.3	26.7	2456	281
20	19½	107.3	29.8	3398	349
8 x 8	7½ x 7½	56.3	15.6	264	70.3
10	9½	71.3	19.8	536	113
12	11½	86.3	23.9	951	165
14	13½	101.3	28.0	1538	228
16	15½	116.3	32.0	2327	300
18	17½	131.3	36.4	3350	383
20	19½	146.3	40.6	4634	475
22	21½	161.3	44.8	6211	578
10 x 10	9½ x 9½	90.3	25.0	679	143
12	11½	109	30.3	1,204	209
14	13½	128	35.6	1,948	289
16	15½	147	40.9	2,948	380
18	17½	166	46.1	4,243	485
20	19½	185	51.4	5,870	602
22	21½	204	56.7	7,868	732
24	23½	223	62.0	10,274	874
12 x 12	11½ x 11½	132	36.7	1,458	253
14	13½	155	43.1	2,358	349
16	15½	178	49.5	3,569	460
18	17½	201	55.9	5,136	587
20	19½	224	62.3	7,106	729
22	21½	247	68.7	9,524	886
24	23½	270	75.0	12,437	1058
14 x 14	13½ x 13½	182	50.6	2,768	410
16	15½	209	58.1	4,189	541
18	17½	236	65.6	6,029	689
20	19½	263	73.1	8,342	856
22	21½	290	80.6	11,181	1040
24	23½	317	88.1	14,600	1243
16 x 16	15½ x 15½	240	66.7	4,810	621
18	17½	271	75.3	6,923	791
20	19½	302	83.9	9,578	982
22	21½	333	92.5	12,837	1194
24	23½	364	101	16,763	1427
18 x 18	17½ x 17½	306	85.0	7,816	893
20	19½	341	94.8	10,813	1109
22	21½	376	105	14,493	1348
24	23½	411	114	18,926	1611
26	25½	446	124	24,181	1897
20 x 20	19½ x 19½	380	106	12,049	1236
22	21½	419	116	16,150	1502
24	23½	458	127	21,089	1795
26	25½	497	138	26,945	2113
28	27½	536	149	33,795	2458
24 x 24	23½ x 23½	552	153	25,415	2163
26	25½	599	166	32,472	2547
28	27½	646	180	40,727	2962
30	29½	693	193	50,275	3408

TABLE B – SAFE AXIAL LOAD (KIPS) DRESSED COLUMNS

Nominal Size, in.	Unsupported Length						
	7' - 0"	8' - 0"	9' - 0"	10' - 0"	12' - 0"	14' - 0"	16' - 0"
4 x 4	12.09	9.26	7.32	5.94	4.13	3.04	
4 x 6	18.76	14.38	11.36	9.22	6.40	4.71	
4 x 8	25.02	19.17	15.15	12.29	8.54	6.28	
6 x 6	30.25	30.25	30.25	30.25	21.81	16.09	12.28
6 x 8	41.25	41.25	41.25	41.25	29.74	21.95	16.75
6 x 10	52.25	52.25	52.25	52.25	37.67	27.80	21.21
6 x 12	63.25	63.25	63.25	63.25	45.60	33.65	25.68
8 x 8	56.25	56.25	56.25	56.25	56.25	55.46	42.47
8 x 10	71.25	71.25	71.25	71.25	71.25	70.25	53.79
8 x 12	86.25	86.25	86.25	86.25	86.25	85.04	65.12
8 x 14	101.25	101.25	101.25	101.25	101.25	99.83	76.44
10 x 10	90.25	90.25	90.25	90.25	90.25	90.25	90.25
10 x 12	109.25	109.25	109.25	109.25	109.25	109.25	109.25
10 x 14	128.25	128.25	128.25	128.25	128.25	128.25	128.25
12 x 12	132.25	132.25	132.25	132.25	132.25	132.25	132.25

Simple solid wood Columns – Douglas fir, oak, southern pine. E = 1,650,000 p.s.i. C = compression parallel to grain = 1000 p.s.i.
For woods having other moduli of elasticity, multiply above loads by the ratio $\frac{E}{1,650,000}$.

Figures beneath and to left of heavy line are based on compression parallel to grain.
Adapted from Wood Structural Design Data by National Lumber Manufacturers Association.

STRUCTURAL WOOD — JOISTS & STUD WALLS

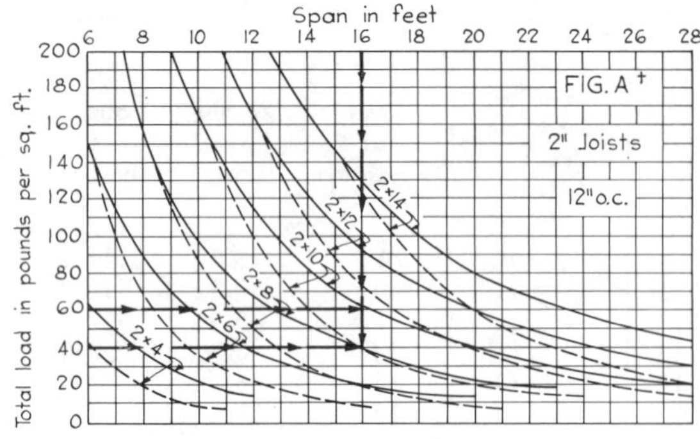

Span in feet
6 8 10 12 14 16 18 20 22 24 26 28
Total load in pounds per sq. ft.
200 180 160 140 120 100 80 60 40 20 0
FIG. A †
2" Joists
12"o.c.
2×4
2×6
2×8
2×10
2×12
2×14

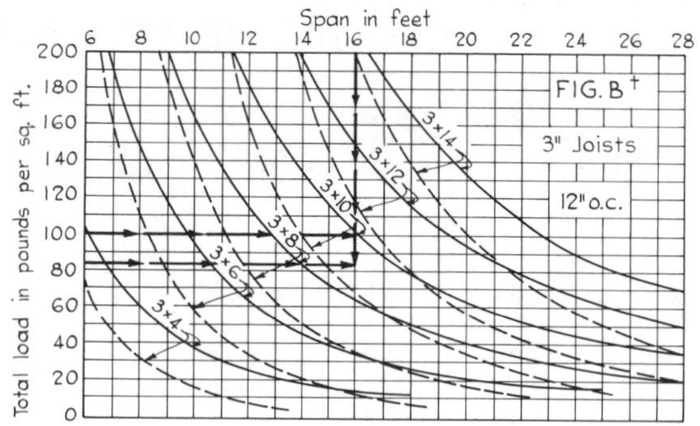

Span in feet
6 8 10 12 14 16 18 20 22 24 26 28
Total load in pounds per sq. ft.
200 180 160 140 120 100 80 60 40 20 0
FIG. B †
3" Joists
12"o.c.
3×4
3×6
3×8
3×10
3×12
3×14

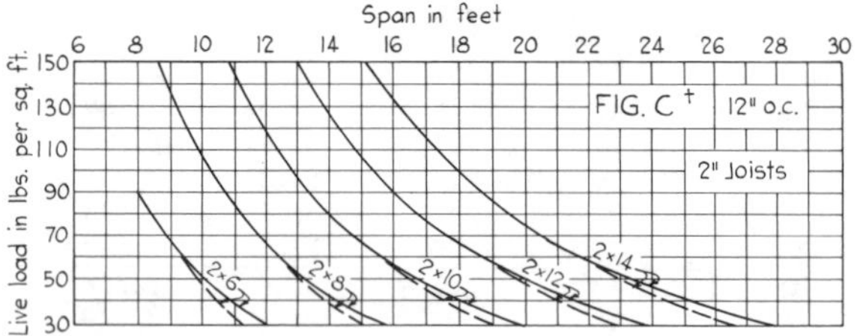

Span in feet
6 8 10 12 14 16 18 20 22 24 26 28 30
Live load in lbs. per sq. ft.
150 130 110 90 70 50 30
FIG. C †
12" o.c.
2" Joists
2×6
2×8
2×10
2×12
2×14

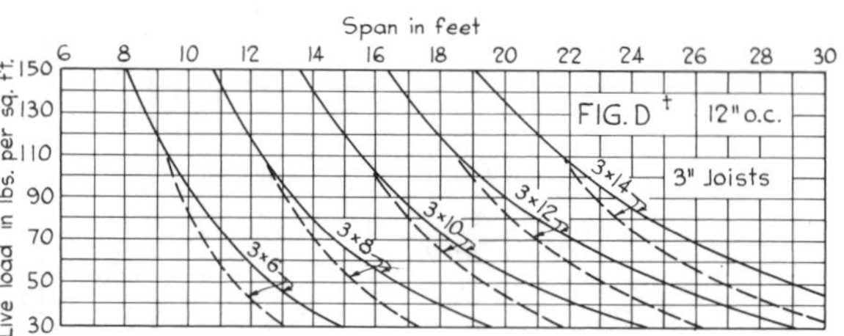

Span in feet
6 8 10 12 14 16 18 20 22 24 26 28 30
Live load in lbs. per sq. ft.
150 130 110 90 70 50 30
FIG. D †
12" o.c.
3" Joists
3×6
3×8
3×10
3×12
3×14

FIGS. A and B Solid curves show span and load of which bending stress in joists is 1000 p.s.i. Dashed curves show span and load which will cause a deflection of 1/360 of the span based on modulus of elasticity of E = **1,000,000 p.s.i.**
FIGS. C and D Solid curves show span and load of which bending stress in joists is 1200 p.s.i. Dashed curves show span and load which will cause a deflection of 1/360 of the span based on modulus of elasticity of E = 1,600,000 **p.s.i.**
Bending stress governs the size of joists unless plastered ceilings or specifications call for limited deflection. All joists with loads greater than those shown in Figs. A and B must be checked for horizontal shear.

EXAMPLE 1
Given: Live load (L.L.) = 40 p.s.f., dead load (D.L.) = 20 p.s.f., span (L) = 16' - 0", stress (f) = 1000 p.s.i., E = **1,000,000 p.s.i.**
Required: Size and spacing of 2" joists.
Solution: Enter Fig. A at 16' span, and 60 lb. total load, and find intersection below solid curve of 2 x 10 - 12" O.C. 2 x 10 - 12" O.C. are good for bending.
Check deflection for live load of 40 lb. Enter Fig. A at 16' span and 40 lb. load, and find intersection below dashed curve of 2" x 10" - 12" O.C. 2 x 10 - 12" O.C. are good for deflection. If spacing of 16" O.C. is desired, enter Fig. A with load of $60 \times \frac{16}{12} = 80$ lb. for bending, and load of $40 \times \frac{16}{12} = 53$ lb. for deflection. Proceed the same as above.

EXAMPLE 2
Given: L.L. = 100 p.s.f., D.L. = 20 p.s.f., span = 16' - 0", working stress = 1200 p.s.i. and E = **1,200,000 p.s.i.**
Required: Size and spacing of 3" joists.
Solution: Decrease total load in proportion to stresses. $W = 120 \times \frac{1000}{1200} =$ 100 p.s.f. Enter Fig. B at 16' span and 100 lbs. load, and find intersection below solid curve for 3 x 10 - 12" O.C. 3 x 10 - 12" O.C. are good for bending.
Check deflection. Decrease deflection for live load in proportion to E, or $100 \times \frac{1,000,000}{1,200,000} = 83$ lb. Enter Fig. B at 16' span and 83 lb. Find intersection above dashed curve of 3 x 10. Use 3 x 12 if deflection has to be less than 1/360 and if 12" O.C. desired. If 16" O.C. spacing is desired, decrease load as for Example 1, and proceed as in Example 1.

TABLE E — STUD PARTITIONS — SAFE LOAD (kips per linear ft.)*

Height in feet \ No. of Rows of Bridging	SIZE OF STUDS — 16" O.C.																		
	2 x 4		2 x 6			2 x 8				3 x 4		3 x 6			3 x 8			4 x 4
	1	2†	1	2	3†	1	2	3	4†	0	1†	0	1	2†	0	1	2†	0†
7' - 0"	2.84	3.25	4.40	5.80	5.98	5.88	7.75	8.07	8.15	3.06	5.26	4.75	9.60	9.66	6.34	12.79	13.18	7.25
8' - 0"	2.20	2.67	3.41	5.56	5.84	4.56	7.44	7.97	8.09	2.32	4.44	3.61	9.30	9.43	4.80	12.40	13.08	5.95
9' - 0"	1.74	2.18	2.70	5.23	5.64	3.60	7.00	7.84	8.01	1.36	3.53	2.89	8.86	9.12	3.85	11.81	12.95	4.86
10' - 0"	1.41	1.77	2.18	4.74	5.37	2.92	6.33	7.63	7.89	1.48	2.86	2.30	8.27	8.68	3.06	11.02	12.74	3.94
11' - 0"	1.18	1.46	1.83	4.13	4.99	2.44	5.45	7.35	7.74	1.25	2.36	1.94	7.46	8.07	2.59	9.95	12.50	3.26
12' - 0"	0.98	1.23	1.52	3.42	4.51	2.08	4.57	6.96	7.53		1.99		6.45	7.29		8.60	12.17	2.74
13' - 0"	0.84	1.05	1.30	2.93	3.92	1.74	3.92	6.54	7.25		1.70		5.53	6.17		7.37	11.71	2.34
14' - 0"		0.84		2.53	3.36		3.36	5.92	6.93		1.36		4.75	5.43		6.33	11.20	2.01
15' - 0"		0.79		2.20	2.93		2.92	5.34	6.52		1.28		4.12	4.73		5.50	10.55	1.76
16' - 0"				1.93	2.57		2.58	4.60	6.00				3.74	4.15		4.98	9.70	

Based on Timber with E = 1,600,000 P.S.I. and compression stress = 900 P.S.I.
*For allowable working stresses of various lumber, see pp. 1-03, 1-04, and 1-05.
Adapted from *Wood Structural Design Data* by National Lumber Manufacturers Association.
†Use values in this line either for given rows of bridging or for no rows of bridging where studs are braced laterally in weak direction by sheathing and plaster, or similar stiff materials. For spacing other than 16" O.C., multiply table values by 16/spacing in inches. Values given are based on column formulas recommended by *Forest Products Laboratory, Forest Service, Department of Agriculture.* Studs must be checked for wind and any other special conditions.

EXAMPLE 3
Given: Height of partition = 10' - 0", braced laterally. Load per foot = 2500 lb.
Required: Size and spacing of studs.
Solution: From Table E:

Using 2 x 4 studs: Spacing $= \frac{1770}{2500} \times 16 = 11.3"$ maximum.

Using 2 x 6 studs: Spacing $= \frac{5370}{2500} \times 16 = 34.4"$ maximum.

Using 3 x 4 studs: Spacing $= \frac{2860}{2500} \times 16 = 18.3"$ maximum.

EXAMPLE 4
Given: Height of partition = 10' - 0", braced laterally. with 1 row of bridging; Load per foot = 2500 lb.
Required: Size and spacing of studs.
Solution: From Table A:

Using 2 x 4 studs: Spacing $= \frac{1410}{2500} \times 16 = 9"$ maximum.

Using 2 x 6 studs: Spacing $= \frac{2180}{2500} \times 16 = 14"$ maximum.

Using 2 x 8 studs: Spacing $= \frac{2920}{2500} \times 16 = 18.7"$ maximum.

Using 3 x 4 studs: Spacing $= \frac{2860}{2500} \times 16 = 18.3"$ maximum.

STRUCTURAL WOOD — NAIL DATA

TABLE A – LATERAL RESISTANCE OF WIRE NAILS (P) POUNDS PER NAIL †

Nails in pennyweight	SOFTWOODS			HARDWOODS
	Hemlock: eastern Spruce: eastern	Hemlock, western Pine, Norway Red Cypress, southern Redwood Douglas-fir, inland type	Douglas-fir, coast type Larch, western Pine, southern yellow	Oak: red white
	K = 900	K = 1125	K = 1375	K = 1700
2	17	22	27	33
4	28	35	42	52
6	34	43	52	65
8	43	53	65	81
10	51	64	78	97
12	51	64	78	97
16	59	73	90	111
20	76	95	116	143
30	85	106	130	160
40	96	120	147	182
50	108	136	166	205
60	121	151	185	228

† This table is based on seasoned wood and a safety factor of about 5.

NOTE: Increase values 25% where metal side plates are used. Decrease values 33% for hardwoods; 40% for softwoods; when nails are driven in end grain. Decrease values 25% for side grain of unseasoned wood which will remain wet, or will be loaded before seasoning.

Resistance $P = KD^{3/2}$, K = constant, D = diameter of nail in inches.

TABLE B – RESISTANCE OF WIRE NAILS TO WITHDRAWAL FROM SIDE GRAIN (P), POUNDS PER INCH OF PENETRATION

Nails in pennyweight	2	4	6	8	10	12	16	20	30	40	50	60
Douglas-fir, coast	16	21	25	29	32	32	35	42	45	49	53	57
Douglas-fir, inland	12	16	18	21	24	24	26	31	33	36	39	42
Hemlock, eastern	10	14	16	18	20	20	22	27	29	31	34	36
Hemlock, western	11	15	17	20	22	22	24	29	31	34	37	39
Oak, red	29	39	45	53	60	60	65	77	83	91	98	105
Oak, white	35	47	55	63	71	71	78	93	100	109	118	126
Pine, southern	22	30	35	41	46	46	50	60	64	70	76	81
Redwood	9	12	14	17	19	19	21	24	26	29	31	33
Spruce	9	12	14	17	19	19	21	24	26	29	31	33

Resistance $P = 1150\,G^{5/2}\,D$ where G = specific gravity.
When driven in unseasoned wood which will season subsequently under load, use 25% of values.
Do not design for withdrawal from end grain.

TABLE C – WIRE NAIL DATA

Nails in pennyweight	2	4	6	8	10	12	16	20	30	40	50	60
Length, in.	1	1½	2	2½	3	3¼	3½	4	4¼	5	5½	6
Diameter, in.	0.072	0.098	0.113	0.131	0.148	0.148	0.162	0.192	0.207	0.225	0.244	0.262
Minimum Penetration, inches												
For Dense woods – Z 10 D	¾	1	1¼	1⅜	1½	1½	1⅝	2	2⅛	2¼	2½	2¾
For Lightweight woods – Z 14 D	1⅛	1½	1⅝	1⅞	2⅛	2⅛	2⅜	2¾	3	3¼	3½	3¾

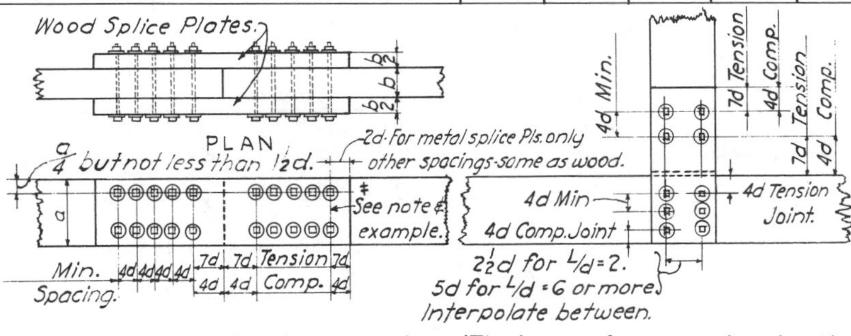

Wood Splice Plates.
PLAN
a/4 but not less than ½d.
2d-For metal splice Pls. only other spacings-same as wood.
‡ See note & example.
Min. Spacing.
4d 4d 4d 4d
7d 7d Tension 7d
4d 4d Comp. 4d
4d Min.
4d Comp. Joint
4d Tension Joint.
7d Tension
4d Comp.
7d Tension
4d Comp.
2½d for L/d = 2.
5d for L/d = 6 or more.
Interpolate between.

L = length of bolt in bearing inches. (Thickness of main member, L = B).
D = diameter of bolt, inches.

ELEVATIONS

FIG. D–BASIC DATE FOR BOLTED WOOD CONNECTIONS

‡NOTE: The net tension area of the critical section shall be ≥ 80% of bolt-bearing area for softwoods and 100% of bolt-bearing area for hardwoods.

EXAMPLE:

Given: Main member Douglas-fir (softwood) 6 × 10 (area = 52.25 sq. in.) 10¾" ϕ bolts.

Required: Net tension area of critical section.
Net area = 52.25 − 2(.75 × 5.5) = 44 sq. in.
Check that this area is ≥ 80% of bearing area under all bolts = 0.8 × 10(.75 × 5.5) = 33 sq. in.

See pp. 6-04, 6-05, and 6-06.

*Adapted from *Wood Handbook* of Forest Products Laboratory, U. S. Dept. of Agriculture.

STRUCTURAL WOOD — CONNECTORS-I

TABLE A – UNIT BEARING STRESSES FOR CALCULATING SAFE LOADS IN POUNDS PER SQUARE INCH FOR BOLTED JOINTS IN DOUBLE SHEAR

Species	L/d*	Parallel to Grain = P		Perpendicular to Grain = Q					
		Wood Splice Plates	Metal Splice Plates	Diameter of Bolt					
				5/8"	3/4"	7/8"	1"	1-1/4"	1-1/2"
SOFTWOODS									
Hemlock, eastern	2	760	950	456	423	399	381	357	342
	4	756	945	456	423	399	381	357	342
	6	651	814	456	423	399	381	357	342
	8	488	610	402	373	352	336	314	302
	10	390	488	307	284	268	256	240	229
	12	326	407	237	220	207	198	186	178
Pine, Norway Red	2	840	1050	334	310	292	279	262	251
	4	818	1022	334	310	292	279	262	251
	6	636	795	334	310	292	279	262	251
	8	477	597	321	298	281	269	252	242
	10	382	478	355	236	222	212	199	191
	12	319	398	204	189	178	170	160	153
Spruce, eastern	2	840	1050	380	352	332	318	297	285
	4	818	1022	380	352	332	318	297	285
	6	636	795	380	352	332	318	297	285
	8	477	597	365	339	320	305	286	274
	10	382	478	290	268	253	242	226	217
	12	319	398	232	215	203	194	181	174
Douglas-fir, inland	2	840	1050	425	395	372	356	333	319
	4	818	1022	425	395	372	356	333	319
	6	636	795	425	395	372	356	333	319
	8	477	597	409	380	358	342	320	307
	10	382	478	324	301	284	271	254	243
	12	319	398	260	241	227	217	203	195
Hemlock, western	2	960	1200	456	423	399	381	357	342
	4	935	1170	456	423	399	381	357	342
	6	729	910	456	423	399	381	357	342
	8	546	683	402	373	352	336	314	302
	10	437	546	307	284	268	256	240	229
	12	364	455	237	220	207	198	186	178
Redwood	2	1080	1350	380	352	332	318	297	285
	4	999	1249	380	352	332	318	297	285
	6	726	906	380	352	332	318	297	285
	8	544	680	365	339	320	305	286	274
	10	435	544	290	268	253	242	226	217
	12	363	454	232	215	203	194	181	174
Cypress, southern	2	1160	1450	456	423	399	381	357	342
	4	1071	1340	456	423	399	381	357	342
	6	779	974	456	423	399	381	357	342
	8	585	731	402	373	352	336	314	302
	10	467	584	307	284	268	256	240	229
	12	390	487	237	220	207	198	186	178
Douglas-fir, coast Pine, southern yellow	2	1160	1450	486	451	425	406	381	365
	4	1071	1340	486	451	425	406	381	365
	6	779	974	486	451	425	406	381	365
	8	585	731	428	398	375	359	336	322
	10	467	584	327	303	286	274	256	245
	12	390	487	253	235	221	212	198	190
HARDWOODS									
Oak, red and white	2	1080	1350	760	705	665	635	595	570
	4	999	1249	760	705	665	635	595	570
	6	725	906	732	679	640	612	573	549
	8	544	680	570	529	499	476	446	428
	10	435	544	421	391	368	352	329	316
	12	363	454	323	300	282	270	253	242

ALLOWABLE BOLT VALUE *N* FOR BOLT LOAD AT ANY ANGLE TO GRAIN IN POUNDS

Hankinson Formula: $N = \dfrac{PQA}{P \sin^2\theta + Q \cos^2\theta}$

For *P* and *Q*, see table A:

A = projected atea (diameter × length in main member).

Q = angle between direction of load and direction of grain.

N = *PA* when angle of load to grain = 0° (or parallel to grain).

N = *QA* when angle of wood to grain = 90° (or perpendicular to grain).

TABLE B – FUNCTIONS OF ANGLE OF LOAD (θ) BETWEEN 0° AND 90°

Function \ Angle	0°	20°	30°	45°	60°	90°
Sine² θ	0	0.12	0.25	0.50	0.75	1.0
Cosine² θ	1.0	0.88	0.75	0.50	0.25	0

EXAMPLE 1:
Given: 7/8" ϕ bolts, 2" wood splice plates, 4" main member, load parallel to grain using Norway pine (softwoods).
Required: To find bolt value *N*, using formula.
From Table A, interpolate L/d = 4.57, value of *P* = 766.
$N = 766 \times 4 \times 7/8 = 2680$ lb.

EXAMPLE 2:
Given: Same conditions as Example 1, except when angle of load to grain (θ) = 30°.
Required: To find *N*, using Hankinson formula.
From Table A, *P* = 766 lb, *Q* = 292 lb. From Table B, $\sin^2\theta = 0.25$, $\cos^2\theta = 0.75$ for $\theta = 30°$.

$$N = \frac{766 \times 292 \times 4 \times 7/8}{766 \times .25 + 292 \times .75} = 1230 \text{ lb.}$$

NOTE: Values in Table A are based on seasoned material in dry locations. For green material which may season after installation, use one third of the tabulated bolt loads. For material occasionally wet, use three fourths of the tabulated bolt loads. For material usually wet, use two thirds of the tabulated bolt loads.

*For spacing of bolts, see p. 6-03.

Adopted from *Wood Handbook* of Forest Products Laboratory, U. S. Dept. of Agriculture.

STRUCTURAL WOOD — CONNECTORS-2

SPLIT RING

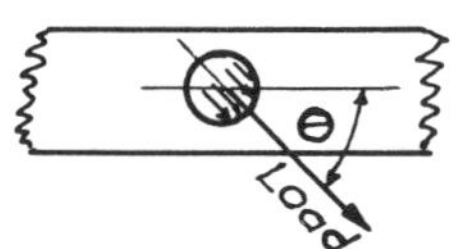
θ
Load

ANGLE OF LOAD TO GRAIN θ

FIGURE 1

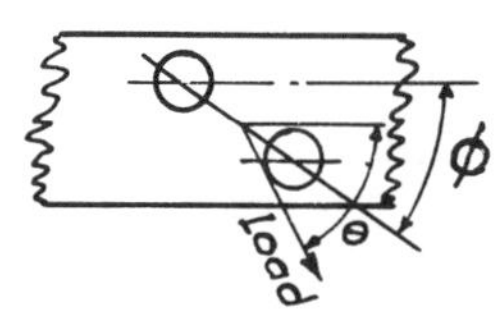
φ
θ
Load

ANGLE OF CONNECTOR AXIS TO GRAIN φ

FIGURE 2

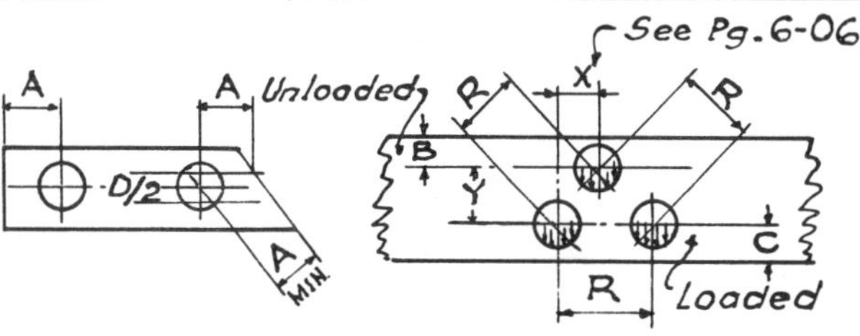
See Pg. 6-06.
A
A
Unloaded
D/2
A
MIN
R
X
R
B
Y
C
R
Loaded

END DISTANCE "A"

FIGURE 3

EDGE DISTANCES "B & C" AND SPACING R

FIGURE 4

FIG. A – SPLIT-RING TIMBER CONNECTORS

TABLE B – CONNECTOR LOAD GROUPING OF SPECIES WHEN STRUCTURALLY GRADED		% of Values, Table C
Group 4	Oak (red and white	116%
Group 3	Douglas-fir* (coast), southern pine*	100%
Group 2	Cypress, hemlock (western), Norway pine, redwood, spruce (eastern)	84%
Group 1	Hemlock (eastern)	72%

* When Dense Grades are used (for stresses, see pp. 1-03, 1-04, and 1-05), use 116%.

TABLE C – STANDARD DESIGN LOADS FOR ONE SPLIT RING AND BOLT IN SINGLE SHEAR GROUP *B*

SPLIT RING CONNECTOR			Net Lumber Thickness for Connector Used		Allowable Loads in pounds per Connector and Bolt at Angle of Load to Grain θ						
Inside Diameter, in.	Washer, in.	Bolt Diameter, in.			0°	15°	30°	45°	60°	75°	90°
2½	2⅛ φ x 3/32 or 2 x 2 x ⅛	½	1" minimum	1⅝" minimum	2480	2410	2240	2088	1900	1790	1750
			1⅝" & thicker	2" & thicker	2975	2900	2700	2470	2280	2150	2105
4	3 φ x 5/32 or 3 x 3 x 3/16	¾	1" minimum		3800	3710	3460	3120	2900	2700	2665
				1⅝" minimum	4025	3910	3620	3300	3030	2860	2800
				2" minimum	4630	4500	4180	3800	3500	3290	3200
				2⅝" minimum	5635	5500	5100	4620	4260	4010	3925
			1⅝" & thicker	3" & thicker	5735	5580	5190	4710	4340	4100	3995

NOTES:
1. Table C is for seasoned lumber. Connectors act as keys, not as pins; therefore their bolts must remain tight, by continuous retightening, until lumber is thoroughly seasoned.
2. For angles to grain θ other than shown. Values may be interpolated.
3. Lumber thicknesses less than those shown are not recommended.
4. Minimum width of lumber is 3⅝" for 2½" ring, and 5½" for 4" ring for θ = 0.

TABLE D – ADJUSTMENT TO ALLOWABLE LOADS FROM TABLE C

PART I. LOADING CONDITIONS

Type of Loading	%
2 months' loading (as for snow)	115
7 days	125
Wind or earthquake	133⅓
Impact	200
Permanent	90

PART II. LUMBER CONDITIONS

Condition	%
Seasoned when fabricated, seasoned when used	100
Unseasoned when fabricated, seasoned when used	80
Unseasoned when fabricated, unseasoned or wet when used	67

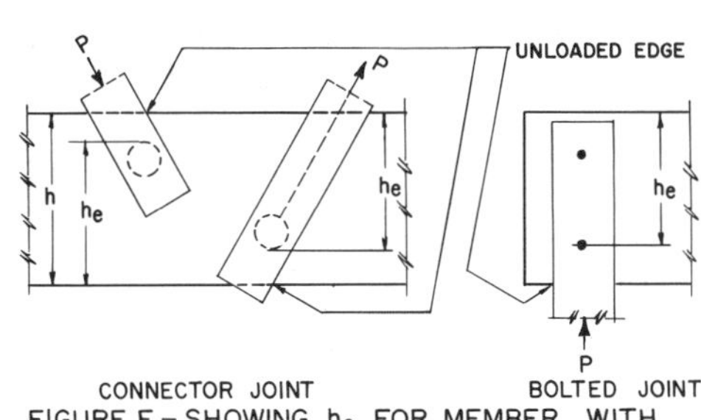
P
P
UNLOADED EDGE
h
he
he
he
P
CONNECTOR JOINT
BOLTED JOINT

FIGURE E – SHOWING h_e FOR MEMBER WITH VARIOUS FASTENINGS.

DESIGN OF ECCENTRIC JOINTS AND OF BEAMS SUPPORTED BY FASTENINGS

Eccentric connector and bolted joints and beams supported by connectors or bolts shall be designed so that *H* in the following formula does not exceed the allowable unit stresses in horizontal shear.

$$H = \frac{3v}{2bh_e}$$

in which

he (with connectors) = the depth of the member less the distance from the unloaded edge of the member to the nearest edge of the nearest connector.

he (with bolts only) = the depth of the member less the distance from the unloaded edge of the member to the center of the nearest bolt.

Adopted from *Teco Design Manual* No. 1, Timber Engineering Co.

STRUCTURAL WOOD — CONNECTORS-3

SPLIT RING TIMBER CONNECTORS.

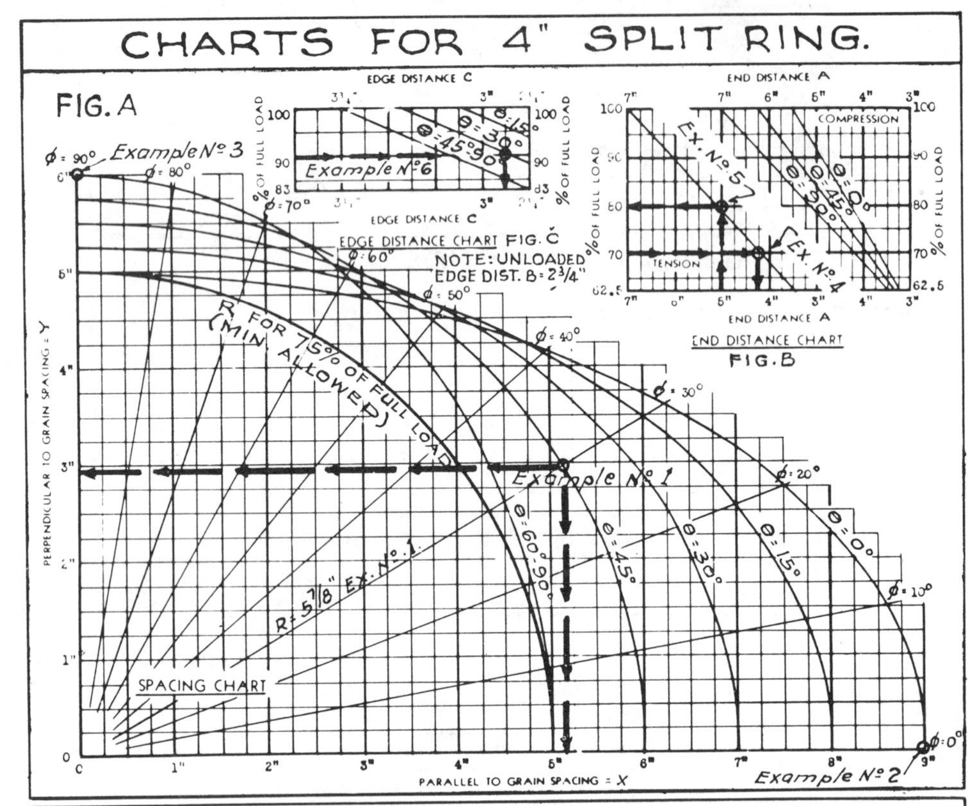

CHARTS FOR 4" SPLIT RING.
FIG. A
EDGE DISTANCE C
Example Nº 6
EDGE DISTANCE CHART FIG. C
NOTE: UNLOADED EDGE DIST. B = 2 3/4"
END DISTANCE A
COMPRESSION
TENSION
EX. Nº 5
EX. Nº 4
END DISTANCE CHART
FIG. B
% OF FULL LOAD
Example Nº 3
R FOR 75% OF FULL LOAD (MIN. ALLOWED)
Example Nº 1
R = 5 7/8" EX. Nº 1
SPACING CHART
PERPENDICULAR TO GRAIN SPACING = Y
PARALLEL TO GRAIN SPACING = X
Example Nº 2

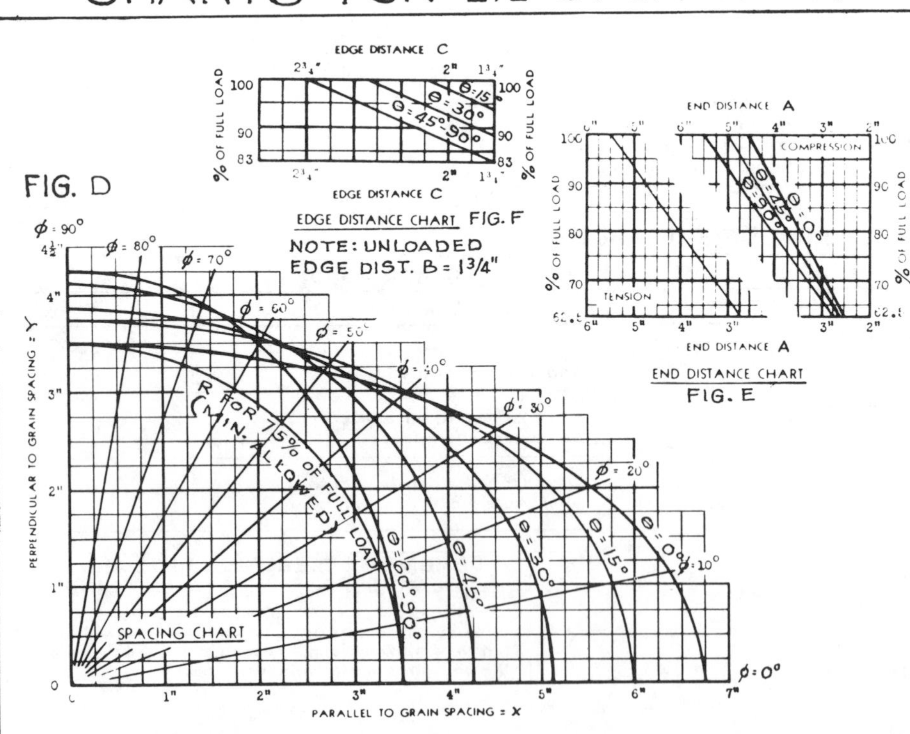

CHARTS FOR 2½" SPLIT RING.
EDGE DISTANCE C
EDGE DISTANCE CHART FIG. F
NOTE: UNLOADED EDGE DIST. B = 1 3/4"
FIG. D
END DISTANCE A
COMPRESSION
TENSION
END DISTANCE CHART
FIG. E
% OF FULL LOAD
R FOR 75% OF FULL LOAD (MIN. ALLOWED)
SPACING CHART
PERPENDICULAR TO GRAIN SPACING = Y
PARALLEL TO GRAIN SPACING = X

EXPLANATION OF CHARTS USING 4" CONNECTORS. FOR DETAILS, SEE FIGURES 1, 2, 3, & 4, Pg. 6-05.

SPACING CHART, FIG. A.

Example Nº 1. Given: $\theta = 45°$ & $\phi = 30°$
Required: R
Solution: Enter Chart at intersection of $\theta = 45°$ and $\phi = 30°$, scale R = 5 7/8" or read components Y = 2 15/16"; X = 5 1/8" for Full Load.
If in Example Nº 1, R = 5", use 75% of Full Load which is Min. spacing for 4" Rings, See 75% Curve. For values of R between 5" & 5 7/8" interpolate between 75% & 100%.

Example Nº 2. Given: $\theta = 0°$, $\phi = 0°$
Required: R. Solution: Enter Chart for $\theta = 0°$ & $\phi = 0°$
Read R or X = 9, Y = 0

Example Nº 3. Given: $\theta = 90°$, $\phi = 90°$.
Required: R. Solution: Enter Chart for $\theta = 90°$, $\phi = 90°$
Read R = 6, X = 0, Y = 6.

END DISTANCE CHART, FIG. B.

Example Nº 4. Given: Tension end with 70% of Full Load.
Required: A.
Solution: Enter Chart for 70%, carry over to line and Read A = 4 1/4".

Example Nº 5. Given: Tension end with A = 5".
Required: % of Full Load.
Solution: Enter Chart for A = 5", carry up to line & read % = 80%. For Compression ends use same procedure using line for proper θ.

EDGE DISTANCE CHART, FIG. C.

Example Nº 6. Given: Loaded edge with $\theta = 30°$ & 90% of Full Load.
Required: C.
Solution: Enter Chart for 90%, carry over to $\theta = 30°$ & read 2 7/8".
For Unloaded edge distances see charts.

Use Charts for 2½" Connector in similar manner.

NOTE: When values of Connectors are reduced for decreases in spacing, end distances or edge distances, reductions are not cumulative.

Adapted from Teco Design Manual Nº 1, Timber Engineering Co.

STRUCTURAL WOOD – PROPERTIES OF LAMINATED LUMBER

TABLE A – PROPERTIES OF STRUCTURAL GLUED LAMINATED LUMBER

Nominal Width in inches b	Number of 1-5/8" Laminations*	Net Finished Size* in inches b x h	Area of Section A = bh sq. in.	Section Modulus $S = \frac{bh^2}{6}$
3	4	2-1/4 x 6-1/2	14.6	15.8
4	4	3-1/4 x 6-1/2	21.1	22.9
3	5	2-1/4 x 8-1/8	18.3	24.8
5	4	4-1/4 x 6-1/2	27.6	29.9
6	4	5 x 6-1/2	32.5	35.2
3	6	2-1/4 x 9-3/4	21.9	35.7
4	5	3-1/4 x 8-1/8	26.4	35.8
6	4	5-1/4 x 6-1/2	34.1	37.0
5	5	4-1/4 x 8-1/8	34.5	46.8
3	7	2-1/4 x 11-3/8	25.6	48.5
4	6	3-1/4 x 9-3/4	31.7	51.5
6	5	5 x 8-1/8	40.6	55.0
6	5	5-1/4 x 8-1/8	42.7	57.8
5	6	4-1/4 x 9-3/4	41.4	67.3
4	7	3-1/4 x 11-3/8	37.0	70.1
8	5	7 x 8-1/8	56.9	77.0
6	6	5 x 9-3/4	48.8	79.2
6	6	5-1/4 x 9-3/4	51.2	83.2
4	8	3-1/4 x 13	42.3	91.6
5	7	4-1/4 x 11-3/8	48.3	91.7
6	7	5 x 11-3/8	56.9	108.
8	6	7 x 9-3/4	68.3	111.
6	7	5-1/4 x 11-3/8	59.7	113.
4	9	3-1/4 x 14-5/8	47.5	116.
5	8	4-1/4 x 13	55.3	120.
6	8	5 x 13	65.0	141.
4	10	3-1/4 x 16-1/4	52.8	143.
10	6	9 x 9-3/4	87.8	143.
6	8	5-1/4 x 13	68.3	148.
8	7	7 x 11-3/8	79.6	151.
5	9	4-1/4 x 14-5/8	62.2	152.
6	9	5 x 14-5/8	73.1	178.
5	10	4-1/4 x 16-1/4	69.1	187.
6	9	5-1/4 x 14-5/8	76.8	187.
10	7	9 x 11-3/8	102.0	194.
8	8	7 x 13	91.0	197.
6	10	5 x 16-1/4	81.3	220.
5	11	4-1/4 x 17-7/8	76.0	226.
6	10	5-1/4 x 16-1/4	85.3	231.
12	7	11 x 11-3/8	125.0	237.
8	9	7 x 14-5/8	102.0	250.
10	8	9 x 13	117.0	254.
6	11	5 x 17-7/8	89.4	266.
5	12	4-1/4 x 19-1/2	82.9	269.
6	11	5-1/4 x 17-7/8	93.8	280.
8	10	7 x 16-1/4	114.0	308.
12	8	11 x 13	143.0	310.
5	13	4-1/4 x 21-1/8	89.8	316.
6	12	5 x 19-1/2	97.5	317.
10	9	9 x 14-5/8	132.	321
6	12	5-1/4 x 19-1/2	102.	333.
14	8	12-1/2 x 13	163.	352.
6	13	5 x 21-1/8	106.	372.
8	11	7 x 17-7/8	125.	373.
6	13	5-1/4 x 21-1/8	111.	390.
12	9	11 x 14-5/8	161.	392.
10	10	9 x 16-1/4	146.	396.
6	14	5 x 22-3/4	114.	431.
8	12	7 x 19-1/2	137.	444.
14	9	12-1/2 x 14-5/8	183.	446.
6	14	5-1/4 x 22-3/4	119.	453.
10	11	9 x 17-7/8	161.	479.
12	10	11 x 16-1/4	179.	484.
6	15	5 x 24-3/8	122.	495.
16	9	14-1/2 x 14-5/8	212.	517.
6	15	5-1/4 x 24-3/8	128.	520.
8	13	7 x 21-1/8	148.	521.
14	10	12-1/2 x 16-1/4	203.	550.
6	16	5 x 26	130.	563.
10	12	9 x 19-1/2	176.	570.
12	11	11 x 17-7/8	197.	586.
6	16	5-1/4 x 26	137.	592.
8	14	7 x 22-3/4	159.	604.
16	10	14-1/2 x 16-1/4	236.	638.

Nominal Width in inches b	Number of 1-5/8" Laminations*	Net Finished Size* in inches b x h	Area of Section A = bh sq. in.	Section Modulus $S = \frac{bh^2}{6}$
14	11	12-1/2 x 17-7/8	223.	666.
10	13	9 x 21-1/8	190.	669.
8	15	7 x 24-3/8	171.	693.
12	12	11 x 19-1/2	215.	697.
16	11	14-1/2 x 17-7/8	259.	772.
10	14	9 x 22-3/4	205.	776.
8	16	7 x 26	182.	789.
14	12	12-1/2 x 19-1/2	244.	792.
12	13	11 x 21-1/8	232.	818.
8	17	7 x 27-5/8	193.	890.
10	15	9 x 24-3/8	219.	891.
16	12	14-1/2 x 19-1/2	283.	919.
14	13	12-1/2 x 21-1/8	264.	930.
12	14	11 x 22-3/4	250.	949.
8	18	7 x 29-1/4	205.	998.
10	16	9 x 26	234.	1014.
14	14	12-1/2 x 22-3/4	284.	1078.
16	13	14-1/2 x 21-1/8	306.	1078
12	15	11 x 24-3/8	268.	1089.
8	19	7 x 30-7/8	216.	1112.
10	17	9 x 27-5/8	249.	1145.
8	20	7 x 32-1/2	228.	1232.
14	15	12-1/2 x 24-3/8	305.	1238.
12	16	11 x 26	286.	1239.
16	14	14-1/2 x 22-3/4	330.	1251.
10	18	9 x 29-1/4	263.	1283.
8	21	7 x 34-1/8	239.	1359.
12	17	11 x 27-5/8	304.	1399.
14	16	12-1/2 x 26	325.	1408.
10	19	9 x 30-7/8	278.	1430.
16	15	14-1/2 x 24-3/8	353.	1436.
8	22	7 x 35-3/4	250.	1491.
12	18	11 x 29-1/4	322.	1569.
10	20	9 x 32-1/2	293.	1584.
14	17	12-1/2 x 27-5/8	345.	1590.
16	16	14-1/2 x 26	377.	1634.
10	21	9 x 34-1/8	307.	1747.
12	19	11 x 30-7/8	340.	1748.
14	18	12-1/2 x 29-1/4	366.	1782.
16	17	14-1/2 x 27-5/8	401.	1844.
10	22	9 x 35-3/4	322.	1917.
12	20	11 x 32-1/2	358.	1936.
14	19	12-1/2 x 30-7/8	386.	1986.
16	18	14-1/2 x 29-1/4	424.	2068.
10	23	9 x 37-3/8	336.	2095.
12	21	11 x 34-1/8	375.	2135.
14	20	12-1/2 x 32-1/2	406.	2201.
10	24	9 x 39	351.	2282.
16	19	14-1/2 x 30-7/8	448.	2304.
12	22	11 x 35-3/4	393.	2343.
14	21	12-1/2 x 34-1/8	427.	2426.
10	25	9 x 40-5/8	366.	2476.
16	20	14-1/2 x 32-1/2	471.	2553.
12	23	11 x 37-3/8	411.	2561.
14	22	12-1/2 x 35-3/4	447.	2663.
10	26	9 x 42-1/4	380.	2678.
12	24	11 x 39	429.	2789.
16	21	14-1/2 x 34-1/8	495.	2814.
10	27	9 x 43-7/8	395.	2888.
14	23	12-1/2 x 37-3/8	467.	2910.
12	25	11 x 40-5/8	447.	3026.
16	22	14-1/2 x 35-3/4	518.	3089.
10	28	9 x 45-1/2	410.	3105.
14	24	12-1/2 x 39	488.	3169.
12	26	11 x 42-1/4	465.	3273.
16	23	14-1/2 x 37-3/8	542.	3376.
14	25	12-1/2 x 40-5/8	508.	3438.
12	27	11 x 43-7/8	483.	3529.
16	24	14-1/2 x 39	566.	3676.
14	26	12-1/2 x 42-1/4	528.	3719.
12	28	11 x 45-1/2	501.	3795.
16	25	14-1/2 x 40-5/8	589.	3988.
14	27	12-1/2 x 43-7/8	548.	4010.
12	29	11 x 47-1/8	518.	4071.
14	28	12-1/2 x 45-1/2	569.	4313.
16	26	14-1/2 x 42-1/4	613.	4314.
12	30	11 x 48-3/4	536.	4357.
14	29	12-1/2 x 47-1/8	589.	4627.
12	31	11 x 50-3/8	554.	4652.
16	27	14-1/2 x 43-7/8	636.	4652.
14	30	12-1/2 x 48-3/4	609.	4951.
16	28	14-1/2 x 45-1/2	660.	5003.
14	31	12-1/2 x 50-3/8	630.	5287.
16	29	14-1/2 x 47-1/8	683.	5367.
16	30	14-1/2 x 48-3/4	707.	5743.
16	31	14-1/2 x 50-3/8	730.	6133.

*With glued laminated structural lumber, many additional sizes may be obtained. Greatest economy will result from using standard widths and depths that are multiples of standard board and dimension lumber thicknesses.

For working stresses see p. 1-06.

Adapted from *Wood Structural Design Data* by National Lumber Manufacturers Association.

†To conform to National Lumber Manufacturers Association and Forest Products Laboratory specification.

STRUCTURAL PLYWOOD – WORKING STRESSES

TABLE A – DOUGLAS FIR PLYWOOD WORKING STRESSES DAMP OR WET LOCATION – TO BE APPLIED TO STRESS RESISTING PLIES ONLY.† SEE Pg. 6-27.

STRESS \ GRADE	(These two exterior grades available only on Special Mill Order) Clear 100% G-2-S	95% G-1-S	87½% So-2-S	80% Plywall or So-1-S	75% Sheathing
Extreme Fibre in Bending	2000	1900	1750	1600	1500
Compression perpendicular to grain	325	325	325	325	325
Compression parallel to grain—3-ply only; for 5-ply and thicker Use next lower grade except no change for 75% grade.	1466	1390	1285	1170	1100
Maximum horizontal shear	120	114	105	96	90
Modulus of Elasticity	1,600,000	1,600,000	1,600,000	1,600,000	1,600,000

NOTES: For basic working stresses of lumber other than Douglas Fir, See pp. 1-03, 1-04 & 1-05

Where moisture content of plywood under load does not exceed 16%, above working stresses may be increased 25%.

G-2-S = good 2 sides; G-1-S = good 1 side; So-2-S = Sound 2 sides; So-1-S = Sound 1 side.

TABLE B – MOMENTS OF INERTIA, SECTION MODULI and VENEER AREAS for SELECTED PLYWOOD CONSTRUCTIONS.†

(12" WIDTHS)

Plywood Thickness (net)	No. of Plies	Veneer Thickness (Nominal) in Inches: Faces*	Centers	Crossband	Parallel[3] Plies Only: Area (Sq. Ins.)	Moment of Inertia I (Inches⁴)	Section Modulus S (Inches³)	Perpendicular[3] Plies Only: Area (Square Inches)	Moment of Inertia I (Inches⁴)	Section Modulus S (Inches³)	Weight, lbs. per 1000 sq. ft. (Approx.) (As shipped from Mill)
1/8"—R(1)	3	1/24	1/24		1.00	.0019	.030	0.50	.0001	.0034	490
1/8"—S(2)	3	1/16	1/16		0.75	.0017	.027	0.75	.0002	.0077	490
3/16"—R	3	1/16	1/16		1.50	.0064	.068	0.75	.0002	.0077	640
3/16"—S	3	1/12	1/12		1.25	.0060	.064	1.00	.0006	.0139	640
1/4"—R	3	1/12	1/12		2.00	.0150	.120	1.00	.0006	.0139	790
1/4"—S	3	1/9	1/9		1.67	.0143	.114	1.33	.0014	.0247	790
5/16"—R	3	1/10+	1/10+		2.50	.0294	.188	1.25	.0011	.0215	950
5/16"—S	3	1/8	1/8		2.25	.0286	.183	1.50	.0020	.0312	950
3/8"—R	3	1/8	1/8		3.00	.0509	.271	1.50	.0020	.0312	1125
3/8"—S	3	1/8	3/16		2.25	.0461	.246	2.25	.0066	.0704	1125
3/8"—S	5	1/10	1/12	2@1/12	2.50	.0377	.201	2.00	.0150	.120	1125
7/16"—R	3	1/8	3/16		3.00	.0772	.353	2.25	.0066	.0704	1300
7/16"—R	5	1/10	1/12	2@1/12	3.25	.0688	.314	2.00	.0150	.120	1125
7/16"—S	5	1/10	1/10	2@1/10	2.85	.0575	.263	2.40	.0260	.1735	1300
1/2"—R	5	1/10	1/10	2@1/10	3.60	.0990	.396	2.40	.0260	.1735	1525
1/2"—S	5	1/8	1/8	2@1/10	3.60	.0926	.370	2.40	.0324	.1995	1525
9/16"—R	5	1/8	1/8	2@1/10	4.35	.1457	.517	2.40	.0324	.1995	1675
9/16"—S	5	1/8	1/8	2@1/8	3.75	.1273	.452	3.00	.0507	.271	1675
5/8"—R	5	1/8	1/8	2@1/8	4.50	.1934	.619	3.00	.0507	.271	1825
5/8"—S	5	1/8	3/16	2@1/8	4.50	.1670	.534	3.00	.0771	.352	1825
11/16"—R	5	1/8	3/16	2@1/8	5.25	.2478	.720	3.00	.0771	.352	2000
11/16"—S	5	1/8	1/8	2@3/16	3.75	.202	.588	4.50	.123	.492	2000
3/4"—R	5	1/8	1/8	2@3/16	4.50	.299	.798	4.50	.123	.492	2225
3/4"—S	5	1/8	3/16	2@3/16	4.50	.251	.670	4.50	.171	.608	2225
3/4"—S	7	1/8	2@1/12	3@1/8	4.50	.286	.763	4.50	.136	.503	2225
13/16"—R	5	1/8	3/16	2@3/16	5.25	.365	.898	4.50	.171	.608	2375
13/16"—R	7	1/8	2@1/12	3@1/8	5.25	.401	.988	4.50	.136	.503	2225
13/16"—S	7	1/8	2@1/8	3@1/8	5.25	.343	.845	4.50	.193	.617	2375
7/8"—R	7	1/8	2@1/8	3@1/8	6.00	.477	1.090	4.50	.193	.617	2600
7/8"—S	7	1/8	2@5/32	3@1/8	6.00	.427	.976	4.50	.243	.707	2600
15/16"—R	7	1/8	2@5/32	3@1/8	6.75	.581	1.241	4.50	.243	.707	2800
15/16"—S	7	1/8	2@3/16	3@1/8	6.75	.525	1.120	4.50	.299	.797	2800
1" —R	7	1/8	2@3/16	3@1/8	7.50	.701	1.402	4.50	.299	.797	3000
1" —S	7	1/8	2@1/8	3@3/16	5.25	.540	1.080	6.75	.460	1.131	3000
1-1/16"—R	7	1/8	2@1/8	3@3/16	6.00	.740	1.393	6.75	.460	1.131	3175
1-1/16"—S	7	1/8	2@1/6	3@3/16	6.00	.615	1.157	6.75	.585	1.305	3175
1-1/8"—R	7	1/8	2@1/6	3@3/16	6.75	.839	1.490	6.75	.585	1.305	3350
1-1/8"—S	7	1/8	2@3/16	3@3/16	6.75	.771	1.371	6.75	.653	1.395	3350
1-3/16"—R	7	1/8	2@3/16	3@3/16	7.50	1.022	1.725	6.75	.653	1.395	3525
1-3/16"—S	7	1/8	2@7/32	3@3/16	7.50	.912	1.538	6.75	.763	1.526	3525

[1]Rough; [2]Sanded; [3]Refers to direction of face grain; *For sanded panels, thickness is before sanding.

† Adapted From Perkins And Countryman, "Technical Data on Plywood", Douglas Fir Plywood, Assoc.

STRUCTURAL PLYWOOD — DESIGN

DESIGN METHOD AND ALLOWABLE STRESSES FOR CALCULATING THE STRENGTH AND STIFFNESS OF PLYWOOD*

Property, and Direction of Stress with Respect to Direction of Face Grain	Area to Be Considered	Unit Stress to Be Used
Tension: Parallel or perpendicular	Parallel plies† only	Unit stress for extreme fiber in bending
±45°	Full cross-sectional area	⅙ unit stress for extreme fiber in bending
Compression: Parallel or perpendicular	Parallel plies† only	Unit stress in compression parallel to grain
±45°	Full cross-sectional area	⅓ unit stress in compression parallel to grain
Bearing at right angles to plane of plywood.	Loaded area	Unit stress in compression perpendicular to grain
Load in bending, parallel or perpendicular	Bending moment $M = KSI/c$ where S = unit stress for extreme fiber in bending, I = moment of inertia computed on basis of parallel plies only, c = distance from neutral axis to outer fiber of outermost ply having its grain in the direction of the span, $K = 1.50$ for 3-ply plywood having the grain of the outer plies perpendicular to the span, $K = 0.85$ for all other plywood.	Unit stress for extreme fiber in bending
Deflection in bending, parallel or perpendicular	Deflection may be calculated by the usual formulas, taking as the moment of inertia that of the parallel plies plus $\frac{1}{20}$ that of the perpendicular plies. (When face plies are parallel, the calculation may be simplified, with but little error, by taking the moment of inertia as that of the parallel plies only.)	Unit value for modulus of elasticity
Deformation in tension or compression, parallel or perpendicular	Parallel plies† only	Unit value for modulus of elasticity
Shear through thickness: Parallel or perpendicular	Full cross-sectional area	Double unit stress in horizontal shear‡
±45°	Full cross-sectional area	4 times unit stress in horizontal shear
Shear in plane of plies: Parallel or perpendicular	Full shear area: Plywood beams. Horizontal shear	¾ unit stress in horizontal shear
	Area of contact between plywood and flange or framing member: I- or box-beams with plywood webs	⅜ unit stress in horizontal shear
	Shear between plies of web or between web and flange:	
	Panels having plywood covers stressed in compression or tension, or both. Shear between plies or between cover and framing members when depth of member exceeds twice its width and end headers are used or when depth is not more than twice the width and no headers are used	
	Interior framing members	¾ unit stress in horizontal shear
	Framing members at edge of panel	⅜ unit stress in horizontal shear
±45°	Area of contact between plywood and flange or framing member: I- or box-beams with plywood webs	½ unit stress in horizontal shear
	Shear between plies of web or between web and flange:	
	Panels having plywood covers stressed in compression, or tension, or both. Shear between plies or between cover and framing members when depth of member exceeds twice its width and end headers are used or when depth is not more than twice the width and no headers are used.	
	Interior framing members	Unit stress in horizontal shear
	Framing members at edge of panel	½ unit stress in horizontal shear

* The suggested simplified methods of calculation apply reasonably well with usual plywood types under ordinary conditions of service. It is recognized, however, that they are not entirely valid for all types of plywood and plywood constructions, or for all spans and span-depth ratios. Also the methods given are not applicable to structures so proportioned that the plywood is in the buckling range, in which event the results will be too high.

† By "parallel plies" is meant those plies whose grain direction is parallel to the direction of principal stress.

‡ This value should be reduced if gaps occur at edge joints (see p. 283 of Handbook 72.

Source: Handbook 72, U.S. Department of Agriculture, Forest Products Laboratory.

STRUCTURAL PLYWOOD-SKIN STRESSED PANELS

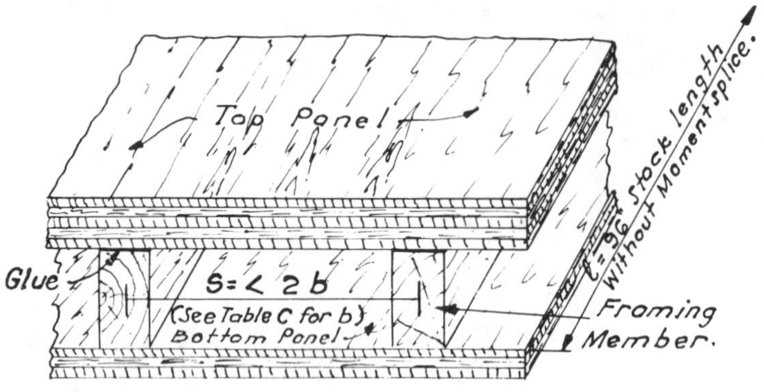
Top Panel
ℓ = 96" stock length without Moment splice.
Glue
S = < 2 b
(See Table C for b) Bottom Panel
Framing Member.

PROBLEM *is similar to Design of Box Girder with span ℓ and width S.*

SOLUTION:

1. Select top & bottom panels to span "S" distance with required loading by flexure formulas using Table B - Pg. 6-26 & "Load in bending," Pg. 6-27. Check max. horizontal shear. See "Shear in Plane of Plies", Pg. 6-27
2. Consider an I section. Find Neutral Axis and Moment of Inertia (I) about Neutral Axis.

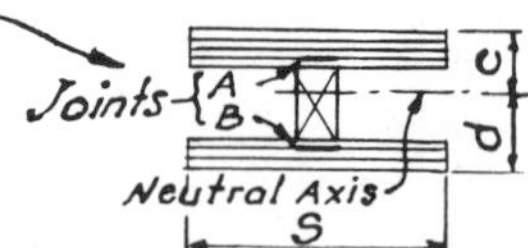
Joints
A
B
c
d
Neutral Axis
S

3. See Fig. B and Table C below for reduction of Compression stresses in top panel and tension stresses in bottom panel.
4. Resisting $M = \frac{sI}{c}$ or $\frac{s'I}{d}$; s = Compression Stresses. s' = Tension Stresses. See "Compression" & "Tension", Pg. 6-27.
5. Check rolling shear at $A = \frac{\text{Reaction} \times Q}{I \times \text{thickness of framing member}}$; Check Joint B similarly.

I = Moment of Inertia component section.
Q = Static M above A = $t \times S \times (c - t_1)$ where t = thickness of parallel plies above Joint A and $t_1 = \frac{1}{2}$ y distance shown.
Rolling shear to be less than allowable "Shear in plane of plies", Pg. 6-27

y — Parallel plies. — Joint A

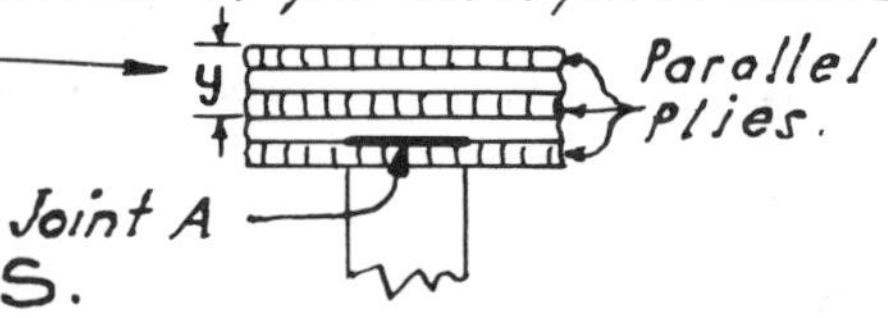

FIG. A - DESIGN OF SKIN STRESSED PANELS.

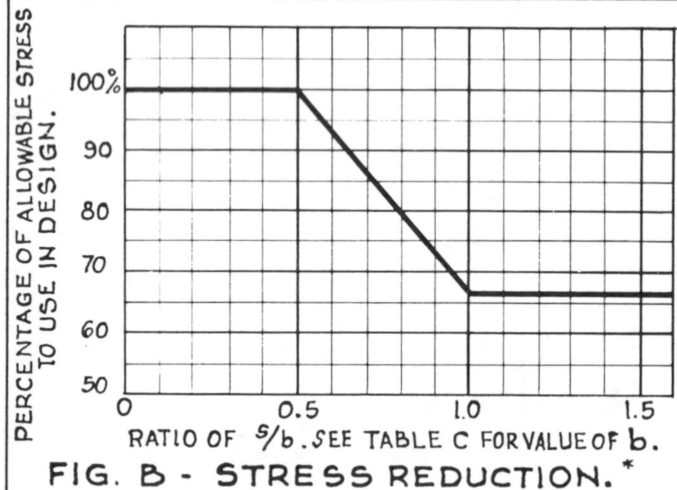
PERCENTAGE OF ALLOWABLE STRESS TO USE IN DESIGN.
100%
90
80
70
60
50
0
0.5
1.0
1.5
RATIO OF S/b. SEE TABLE C FOR VALUE OF b.

FIG. B - STRESS REDUCTION.*

TABLE C - VALUES OF "b" FOR USE IN FIG. B.*

Plywood Thickness	Basic Spacing "b_1," inches — Face grain parallel to longitudinal members	Face grain perpendicular to longitudinal members
1/4" Sanded	10.35"	11.61"
5/16" Rough	11.87	16.80
3/8" Rough (3-ply)	14.25	20.13
3/8" Sanded (3-ply)	16.43	16.43
3/8" Sanded (5-ply)	18.10	20.25
1/2" Rough and Sanded	23.25	28.5
5/8" Rough and Sanded	29.1	35.6
3/4" Sanded (5-ply and 7-ply)	38.2	38.2
7/8" Rough and Sanded	41.6	48.1
1" Rough	45.5	58.9
1" Sanded	54.5	47.9

$M = F_{P.L.} \times \frac{SI}{C}$:

Check rolling shear by "Shear in Plane of Plies, Pg. 6-27 - Par. 2B.

$F_{P.L.}$ = Form Factor of Proportional Limit. Find from Chart at right.

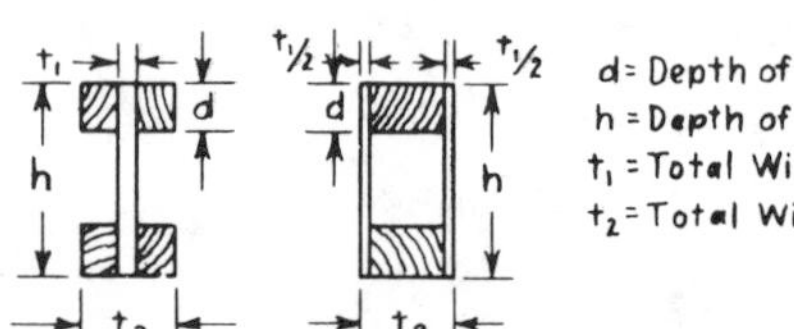
t1
t1/2
t1/2
d
h
t2

d = Depth of Compression Flange
h = Depth of Beam
t_1 = Total Width of Webs
t_2 = Total Width of Beam

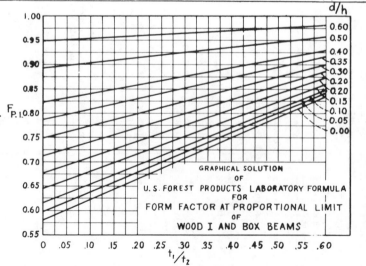

GRAPHICAL SOLUTION OF U.S. FOREST PRODUCTS LABORATORY FORMULA FOR FORM FACTOR AT PROPORTIONAL LIMIT OF WOOD I AND BOX BEAMS

($F_{P.L.}$ from 0.55 to 1.00 vs. t_1/t_2 from 0 to .60; curves for d/h = 0.60, 0.50, 0.40, 0.35, 0.30, 0.25, 0.20, 0.15, 0.10, 0.05, 0.00)

FIG. D - DESIGN OF I AND BOX BEAMS.*

*Adapted From Perkins And Countryman, "Technical Data on Plywood," Douglas Fir Plywood. Assoc.

STRUCTURAL PLYWOOD — DATA

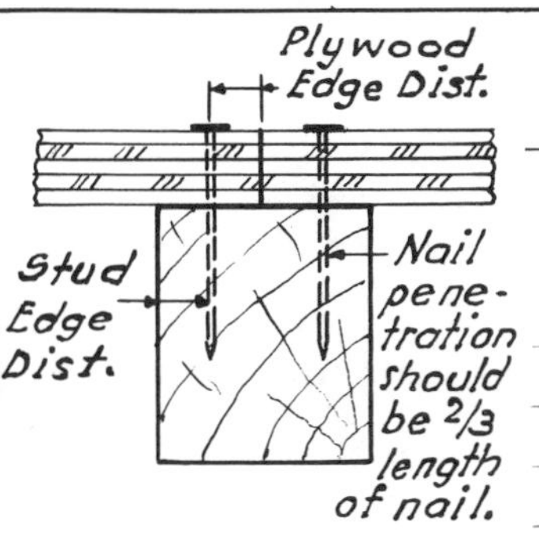
Plywood Edge Dist.
Stud Edge Dist.
Nail penetration should be 2/3 length of nail.

TABLE A - ULTIMATE STRENGTH IN SHEAR OF NAILS IN PLYWOOD. USE 1/6 OF VALUES FOR DESIGN.*

PLYWOOD THICKNESSES	NAIL SIZE: 6d COM. LOAD	6d COM. PLYWOOD EDGE-DISTANCE	8d COM. LOAD	8d COM. PLYWOOD EDGE-DISTANCE	10d COM. LOAD	10d COM. PLYWOOD EDGE-DISTANCE	16d COM. LOAD	16d COM. PLYWOOD EDGE-DISTANCE
5/16"	335#	1/2"	420#	5/8"				
3/8"	345#	1/2"	470#	5/8"				
1/2"			470#	5/8"	500#	1/2"		
5/8"			470#	5/8"	575#	5/8"	620#	5/8"
3/4"					575#	3/4"	675#	3/4"

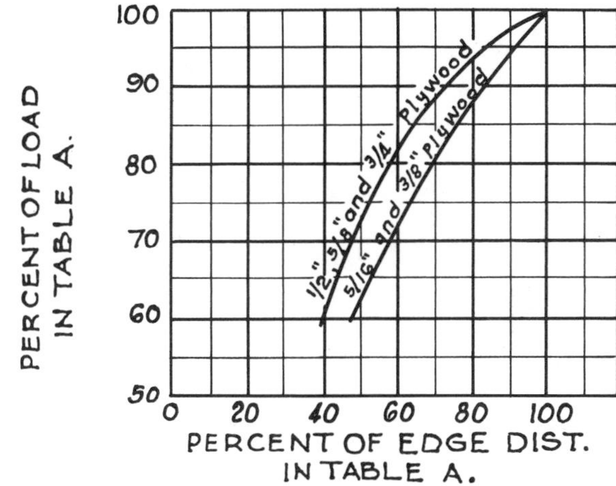
PERCENT OF LOAD IN TABLE A.
100
90
80
70
60
50
0
20
40
60
80
100
1/2", 5/8" and 3/4" Plywood
5/16" and 3/8" Plywood
PERCENT OF EDGE DIST. IN TABLE A.

FIG. B - REDUCTION OF VALUES IN TABLE A FOR PLYWOOD EDGE DISTANCES LESS THAN THOSE SHOWN IN TABLE A.*

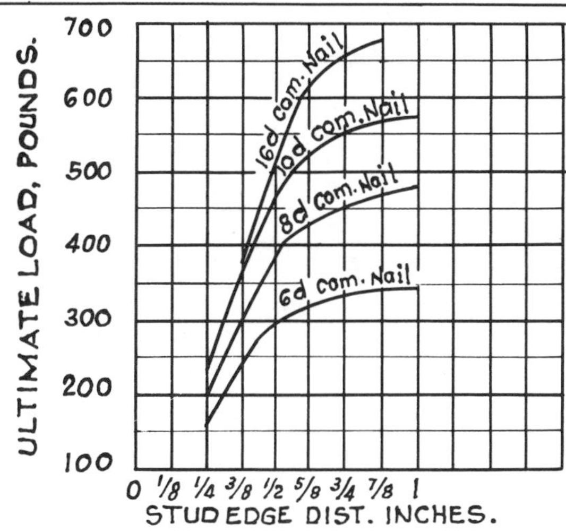
ULTIMATE LOAD, POUNDS.
700
600
500
400
300
200
100
16d com. Nail
10d com. Nail
8d com. Nail
6d com. Nail
0 1/8 1/4 3/8 1/2 5/8 3/4 7/8 1
STUD EDGE DIST. INCHES.

FIG. C - ULTIMATE STRENGTH IN SHEAR OF NAILS IN STUDS. USE 1/6 FOR DESIGN.*

EXTERIOR TYPE

Item	Width (Inches)	Length (Inches)	Thickness[1] (Inches) (after sanding)
Standard Panels (G2S-Ext.) (G1S-Ext.) (SO2S-Ext.) (SO1S-Ext.)	12 26 14 28 16 30 18 36 20 42 22 48 24	48 60 72 84 96	3/16 (3 ply) 3/4 (5 ply) 1/4 (3 ply) 13/16 (5 ply) 5/16 (3 ply) 7/8 (7 ply) 3/8 (3 ply) 15/16 (7 ply) 7/16 (5 ply) 1 (7 ply) 1/2 (5 ply) 1 1/16 (7 ply) 9/16 (5 ply) 1 1/8 (7 ply) 5/8 (5 ply) 1 3/16 (7 ply) 11/16 (5 ply)
Sheathing Exterior	48	96	5/16 (5 ply unsanded) 3/8 (3 ply unsanded) 1/2 (3 ply unsanded) 5/8 (3 ply unsanded)
Industrial Exterior	As Ordered	As Ordered	1/4 (3 ply unsanded) 5/16 (3 ply unsanded) 3/8 (3 ply unsanded) 7/16 (3 ply unsanded) 1/2 (5 ply unsanded) 9/16 (5 ply unsanded) 5/8 (5 ply unsanded) 11/16 (5 ply unsanded) 3/4 (5 ply unsanded) 7/8 (5 ply unsanded)
Concrete Form Panels Exterior	Same as Standard Panels	Same as Standard Panels	5/8 (3 ply sanded 2 sides) 3/4 (5 ply sanded 2 sides)

MOISTURE-RESISTANT TYPE

Item	Width (Inches)	Length (Inches)	Thickness (Inches) (after sanding)
Standard Panels (SO2S) (SO1S)	24 30 36 48	60 72 84 96	1/8 (3 ply) 3/16 (3 ply) 1/4 (3 ply) 3/8 (3 ply) 1/2 (5 ply) 5/8 (5 ply) 3/4 (5 ply)
Wallboard	48	60 72 84 96	1/4 (3 ply sanded 2 sides) 3/8 (3 ply sanded 2 sides) 1/2 (5 ply sanded 2 sides)
Sheathing	36 48	96	5/16 (3 ply unsanded) 3/8 (3 ply unsanded) 1/2 (3 or 5 ply unsanded) 5/8 (3 or 5 ply unsanded)
Automobile and Industrial	As Ordered up to 48	As Ordered up to 96	1/4 (3 ply unsanded) 5/16 (3 ply unsanded) 3/8 (3 ply unsanded) 1/2 (5 ply unsanded) 9/16 (5 ply unsanded) 5/8 (5 ply unsanded) 11/16 (5 ply unsanded) 3/4 (5 ply unsanded) 7/8 (5 ply unsanded) 7/8 (7 ply unsanded)
Concrete Form Panels	36 48	60 72 84 96	1/4 (3 ply sanded 2 sides) 1/2 (5 ply sanded 2 sides) 9/16 (5 ply sanded 2 sides) 5/8 (5 ply sanded 2 sides) 3/4 (5 ply sanded 2 sides)

FIG. D - STANDARD DOUGLAS FIR PLYWOOD SIZES.*

*Adapted From Perkins And Countryman, "Technical Data on Plywood," Douglas Fir Plywood Assoc.

STRUCTURAL FOUNDATIONS – LOAD TEST – I

INSTRUCTIONS TO INSPECTORS – FOUNDATIONS

METHOD OF CONDUCTING SOIL LOAD TEST – N.Y. CITY CODE – 1956

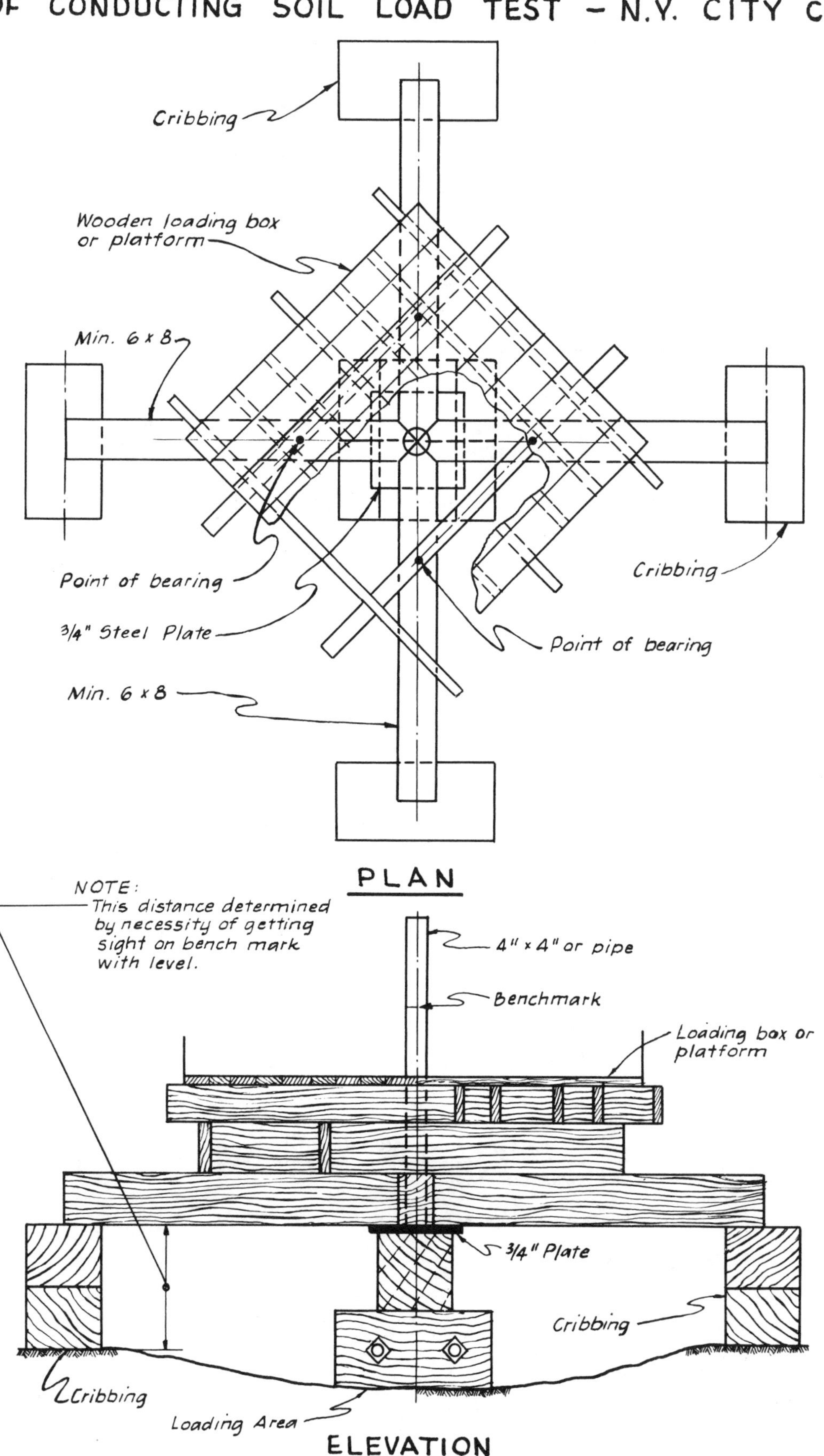
Cribbing
Wooden loading box or platform
Min. 6 x 8
Point of bearing
3/4" Steel Plate
Min. 6 x 8
Cribbing
Point of bearing
PLAN
NOTE:
This distance determined by necessity of getting sight on bench mark with level.
4" x 4" or pipe
Benchmark
Loading box or platform
3/4" Plate
Cribbing
Cribbing
Loading Area
ELEVATION

STRUCTURAL FOUNDATIONS – LOAD TEST – 2

PROCEDURE

For bearing materials 1 to 4, inclusive, the loaded area shall be 1 sq. ft., and for other classes at least 4 sq. ft. For materials of classes 4 to 12, inclusive, the loaded area shall be full size of pit and shall be at such depth that the ratio of width of loaded area to its depth below adjacent ground surface is the same as or greater than the larger of the following:

(a) Ratio of width of any footing to its depth below adjacent ground surface.

(b) Ratio of width of entire foundation or group of footings to its depth below average surrounding ground surface.

Apply sufficient load uniformly on platform to produce a center load equal to proposed design load. Read settlement every 24 hours, until no settlement occurs in 24 hours.

Add a 50% excess load in increments of 25% design load.

For each increment, read settlement every hour during first 6 hours then once every 12 hours thereafter, until no settlement occurs in 24 hours.

The gross settlement under proposed load upon materials of classes 1 to 10, inclusive, shall not exceed 1/2-in., and total gross settlement after the 50% excess load shall not exceed 1 in.

For materials below class 10, a report is required by a licensed professional engineer experienced in soil testing and analysis, and based on laboratory tests of undisturbed samples. The report will contain a foundation analysis indicating probable total magnitude, distribution, and time rate of settlement to be expected for the proposed structure, and that these will not be excessive.

NEW YORK CITY CODE

PRESUMPTIVE BEARING CAPACITY OF SOILS

For definition of terms see p. 9-30.

Class	Material	Tons per sq. ft.
1	Hard sound rock	60
2	Medium hard rock	40
3	Hard pan overlying rock	12
4	Compact gravel & boulder-gravel formations; very compact sandy gravel	10
5	Soft Rock	8
6	Loose gravel & sandy gravel; compact sand & gravelly sand; very compact sand - inorganic silt soils	6
7	Hard dry consolidated clay	5
8	Loose coarse to medium sand; medium compact fine sand	4
9	Compact sand-clay soils	3
10	Loose fine sand; medium compact sand - inorganic silt soils	2
11	Firm or stiff clay	1.5
12	Loose saturated sand clay soils, medium soft clay	1

STRUCTURAL FOUNDATIONS - PLAIN CONCRETE WALL FOOTINGS

PLAIN CONCRETE WALL FOOTINGS.

2000 LB. CONCRETE

Thickness "t" (in inches.)

Soil Pressures (in lbs. per sq. ft.)

20,000 16,000 12,000 8,000 6,000 5,000 4,000 3,000 2,000

Projection "a" (in inches.)

3000 LB. CONCRETE

Thickness "t" (in inches.)

Soil Pressures (in lbs. per sq. ft.)

20,000 16,000 12,000 8,000 6,000 5,000 4,000 3,000 2,000

Projection "a" (in inches.)

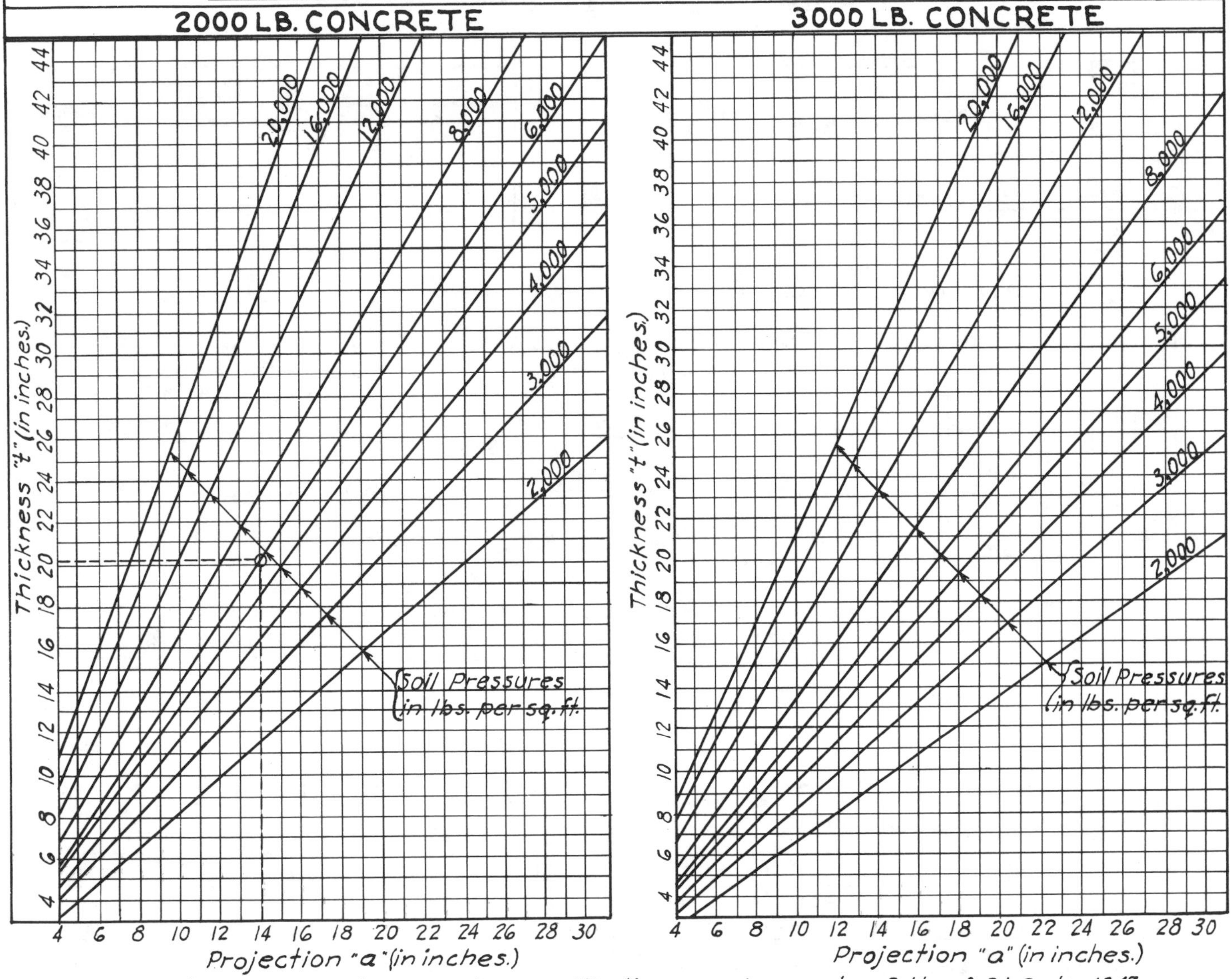

NOTE: The diagrams are in accordance with the requirements of the A.C.I. Code - 1947.
For Joint Committee requirements add 2 inches to the thickness "t" from the diagrams above.
For New York City Code requirements add 4 inches to the thickness "t" from the diagrams above.

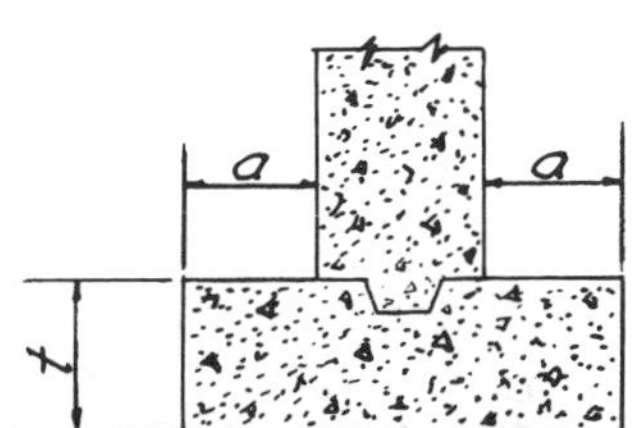
a
a
t

Formula for plain concrete wall footings:-

$$t = a\sqrt{\frac{S}{48 f_c}}$$

In Which: t = Total depth of footing in inches.
a = Projection in inches.
S = Soil pressure in lb. per sq. ft.
$f_c = 0.03 f_c'$. e.g. 0.03 x 3,000 = 90 lb. per sq. in. for 3,000 lb. concrete.

EXAMPLE:- Given:- Wall thickness = 12 inches ; Soil pressure = 6,000 lb. per sq. ft. ; and Wall load = 20,000 lb. per linear ft. of wall; 2,000 lb. concrete.

Solution:- $2a = \frac{20,000}{6,000} - 1.0 = 2.33$ ft. = 28 in. $a = 14$ in.

From diagram above for 2,000 lb. concrete, with a = 14 in. and 6,000 lb. soil
t = 20.2 in. or 21 in. for A.C.I. requirements.
t = 23" for Joint Committee requirements.
t = 25" for New York City Code requirements.

STRUCTURAL FOUNDATIONS-SQUARE COLUMN FOOTINGS-I

SQUARE COLUMN FOOTINGS.*

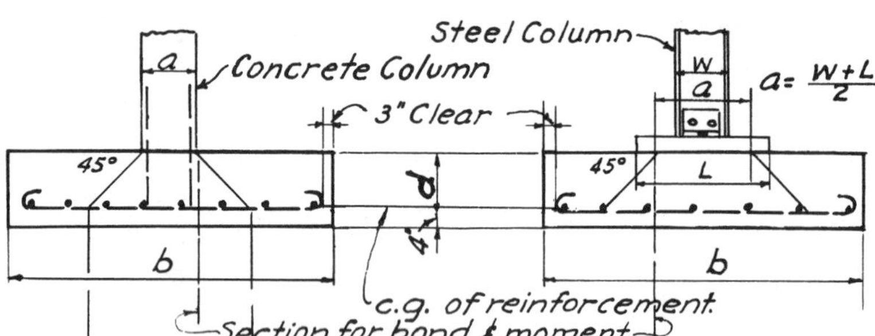
Steel Column
Concrete Column
a
W
a
$a = \frac{W+L}{2}$
3" Clear
45°
45°
L
d
4"
b
b
c.g. of reinforcement.
Section for bond & moment.

DESIGN DATA.		
f'_c = 2,000 #/□"	f'_c = 3,000 #/□"	f'_c = 3,750 #/□"
f_c = 900 #/□"	f_c = 1,350 #/□"	f_c = 1,688 #/□"
f_s = 20,000 #/□"	f_s = 20,000 #/□"	f_s = 20,000 #/□"
v = 60 #/□"	v = 75 #/□"	v = 75 #/□"
u = 112.5 #/□"	u = 168 #/□"	u = 200 #/□"

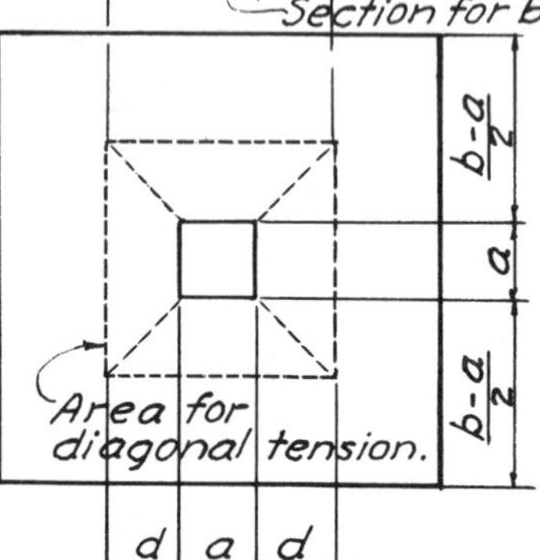
b-a/2
a
b-a/2
Area for diagonal tension.
d
a
d

1. Depth of footing is determined by diagonal tension.

$$v = \frac{(W-g)[b^2-(a+2d)^2]}{0.875 \times 4(a+2d)d}$$ Where: W = Soil pressure in #/□'. g = Weight of footing in #/□'.

2. $\Sigma_o = \frac{(W-g)\,b(b-a)}{2} \times \frac{0.85}{u \times 0.875d}$ Sum of perimeter of bars in inches.

3. Moment $= \frac{(W-g)\,b(b-a)^2}{8}$ (in ft. lbs.) (Check for f_c.)

4. $A_s = \frac{M \times 12}{20{,}000 \times 0.875 \times d} \times 0.85$ (in sq. inches.)

5. Reinforcement must satisfy both Σ_o and A_s.
6. Number of bars will vary directly with increased depth.
7. If size of bars is changed:-Provide equivalent perimeter when larger bars are used; and equivalent area when smaller bars are used.

TABLE A-SOIL BEARING VALUE. 2,000 LB. PER SQ. FT.

COLUMN LOAD IN 1000 LB.	b	d (IN)	f'_c=2,000#/□" MIN. a	f'_c=2,000#/□" REINF. EACH WAY	f'_c=3,000#/□" MIN. a	f'_c=3,000#/□" REINF. EACH WAY	f'_c=3,750#/□" MIN. a	f'_c=3,750#/□" REINF. EACH WAY
29	4'-0"	12	9"	6-1/2"φ	9"	5-1/2"φ	9"	5-1/2"φ
45	5'-0"	12	9"	9-1/2"φ	9"	6-1/2"φ	9"	5-1/2"φ
65	6'-0"	12	9"	13-1/2"φ	9"	9-1/2"φ	9"	6-5/8"φ
89	7'-0"	12	10"	17-1/2"φ	9"	10-5/8"φ	9"	10-5/8"φ
114	8'-0"	13	11"	22-1/2"φ	10"	10-3/4"φ	9"	10-3/4"φ
143	9'-0"	15	11"	19-5/8"φ	10"	12-3/4"φ	10"	12-3/4"φ
174	10'-0"	17	12"	21-5/8"φ	11"	15-3/4"φ	10"	15-3/4"φ
208	11'-0"	18	13"	20-3/4"φ	11"	19-3/4"φ	11"	19-3/4"φ
245	12'-0"	20	14"	21-3/4"φ	13"	21-3/4"φ	12"	21-3/4"φ
283	13'-0"	22	14"	23-3/4"φ	13"	18-7/8"φ	13"	18-7/8"φ
324	14'-0"	24	14"	20-7/8"φ	14"	20-7/8"φ	13"	20-7/8"φ
369	15'-0"	25	16"	23-7/8"φ	15"	23-7/8"φ	14"	23-7/8"φ
416	16'-0"	27	16"	20-1"φ	15"	20-1"φ	14"	20-1"φ

TABLE B-SOIL BEARING VALUE. 3,000 LB. PER SQ. FT.

COLUMN LOAD IN 1000 LB.	b	d (IN.)	f'_c=2,000#/□" MIN. a	f'_c=2,000#/□" REINF. EACH WAY	f'_c=3,000#/□" MIN. a	f'_c=3,000#/□" REINF. EACH WAY	f'_c=3,750#/□" MIN. a	f'_c=3,750#/□" REINF. EACH WAY
45	4'-0"	12"	9"	9-1/2"φ	9"	6-1/2"φ	9"	5-1/2"φ
70	5'-0"	12"	9"	14-1/2"φ	9"	9-1/2"φ	9"	8-1/2"φ
101	6'-0"	12"	10"	20-1/2"φ	9"	14-1/2"φ	9"	10-5/8"φ
136	7'-0"	14"	11"	24-1/2"φ	10"	13-5/8"φ	9"	9-3/4"φ
176	8'-0"	16"	12"	27-1/2"φ	11"	12-3/4"φ	10"	12-3/4"φ
221	9'-0"	18"	13"	24-5/8"φ	11"	15-3/4"φ	11"	11-7/8"φ
270	10'-0"	20"	14"	27-5/8"φ	12"	18-3/4"φ	11"	14-7/8"φ
324	11'-0"	22"	16"	30-5/8"φ	14"	16-7/8"φ	13"	16-7/8"φ
382	12'-0"	24"	17"	26-3/4"φ	15"	19-7/8"φ	13"	15-1"φ
444	13'-0"	26"	18"	28-3/4"φ	15"	17-1"φ	14"	18-1"φ
510	14'-0"	28"	18"	26-7/8"φ	16"	20-1"φ	15"	20-1"φ
580	15'-0"	30"	20"	28-7/8"φ	18"	22-1"φ	17"	22-1"φ
654	16'-0"	32"	21"	31-7/8"φ	19"	25-1"φ	17"	20-1"□

*Based on A.C.I. Code - 1947. Special bond bars (A.S.T.M. A-305) not required. Hooks required.

STRUCTURAL FOUNDATIONS - SQUARE COLUMN FOOTINGS-2

SQUARE COLUMN FOOTINGS.*

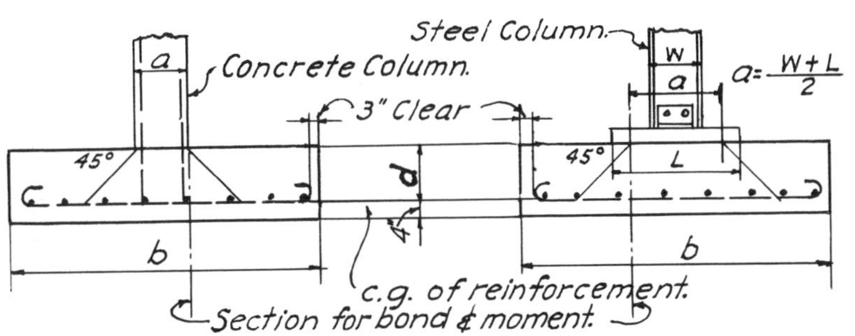
Steel Column.
Concrete Column.
3" Clear
a
W
a = (W+L)/2
45°
L
d
b
c.g. of reinforcement.
Section for bond & moment.

DESIGN DATA.

f'_c = 2,000#/□"	f'_c = 3,000#/□"	f'_c = 3,750#/□"
f_c = 900#/□"	f_c = 1,350#/□"	f_c = 1,688#/□"
f_s = 20,000#/□"	f_s = 20,000#/□"	f_s = 20,000#/□"
v = 60#/□"	v = 75#/□"	v = 75#/□"
u = 112.5#/□"	u = 168#/□"	u = 200#/□"

TABLE A - SOIL BEARING VALUE. 4,000 LB. PER SQ. FT.

COLUMN LOAD IN 1000 LB.	b	d (IN.)	f'_c=2,000#/□" MIN. a	f'_c=2,000#/□" REINF. EACH WAY	f'_c=3,000#/□" MIN. a	f'_c=3,000#/□" REINF. EACH WAY	f'_c=3,750#/□" MIN a	f'_c=3,750#/□" REINF. EACH WAY
34.2	3'-0"	12	9"	6-1/2"φ	9"	5-1/2"φ	9"	5-1/2"φ
46.5	3'-6"	12	9"	9-1/2"φ	9"	6-1/2"φ	9"	5-1/2"φ
60.8	4'-0"	12	9"	12-1/2"φ	9"	8-1/2"φ	9"	7-1/2"φ
77	4-6"	12	10"	12-5/8"φ	9"	10-1/2"φ	9"	9-1/2"φ
95	5'-0"	12	10"	15-5/8"φ	9"	13-1/2"φ	9"	11-1/2"φ
115	5'-6"	12	11"	18-5/8"φ	10"	15-1/2"φ	9"	11-5/8"φ
136	6'-0"	13	11"	20-5/8"φ	10"	17-1/2"φ	9"	12-5/8"φ
159	6'-6"	14	12"	22-5/8"φ	10"	15-5/8"φ	10"	14-5/8"φ
184	7'-0"	16	12"	23-5/8"φ	11"	15-5/8"φ	10"	11-3/4"φ
211	7'-6"	17	12"	24-5/8"φ	12"	13-3/4"φ	11"	12-3/4"φ
240	8'-0"	18	14"	26-5/8"φ	13"	15-3/4"φ	12"	14-3/4"φ
268	8'-6"	19	14"	27-5/8"φ	13"	16-3/4"φ	12"	16-3/4"φ
300	9'-0"	20	14"	29-5/8"φ	14"	17-3/4"φ	13"	13-7/8"φ
335	9'-6"	22	16"	29-5/8"φ	14	14-7/8"φ	13"	14-7/8"φ
369	10'-0"	23	16"	31-5/8"φ	15"	15-7/8"φ	14"	12-1"φ
405	10'-6"	24	16"	33-5/8"φ	15"	17-7/8"φ	14"	13-1"φ
443	11'-0"	25	18"	28-3/4"φ	15"	14-1"φ	14"	15-1"φ
484	11'-6"	27	18"	29-3/4"φ	16"	15-1"φ	15"	16-1"φ
524	12'-0"	28	18"	30-3/4"φ	16"	17-1"φ	15"	17-1"φ
610	13'-0"	30	20"	33-3/4"φ	17"	20-1"φ	16"	16-1"□
700	14'-0"	32	22"	38-3/4"φ	19"	18-1"□	17"	19-1"□
795	15'-0"	34	23"	33-7/8"φ	20"	21-1"□	18"	21-1"□
900	16'-0"	36	25"	30-1"φ	22"	23-1"□	20	24-1"□

TABLE B - SOIL BEARING VALUE. 5,000 LB. PER SQ. FT.

COLUMN LOAD IN 1000 LB.	b	d (IN.)	f'_c=2,000#/□" MIN. a	f'_c=2,000#/□" REINF. EACH WAY	f'_c=3,000#/□" MIN. a	f'_c=3,000#/□" REINF. EACH WAY	f'_c=3,750#/□" MIN. a	f'_c=3,750#/□" REINF. EACH WAY
43.5	3'-0"	12	9"	8-1/2"φ	9"	5-1/2"φ	9"	5-1/2"φ
59	3'-6"	12	9"	11-1/2"φ	9"	8-1/2"φ	9"	6-1/2"φ
77	4'-0"	12	10"	12-5/8"φ	9"	10-1/2"φ	9"	9-1/2"φ
97	4'-6"	12	10"	15-5/8"φ	9"	13-1/2"φ	9"	11-1/2"φ
120	5'-0"	12	11"	18-5/8"φ	10"	16-1/2"φ	9"	14-1/2"φ
145	5'-6"	13	11"	17-3/4"φ	10	18-1/2"φ	10"	12-5/8"φ
172	6'-0"	14"	12"	19-3/4"φ	11"	20-1/2"φ	10"	13-5/8"φ
202	6'-6"	16	13"	20-3/4"φ	11"	16-5/8"φ	10"	15-5/8"φ
232	7'-0"	17	13"	21-3/4"φ	12"	17-5/8"φ	11"	13-3/4"φ
266	7'-6"	19	14"	22-3/4"φ	12"	19-5/8"φ	12"	14-3/4"φ
301	8'-0"	20	15"	24-3/4"φ	14"	21-5/8"φ	13"	15-3/4"φ
340	8'-6"	21	16"	25-3/4"φ	14"	18-3/4"φ	13"	13-7/8"φ
380	9'-0"	22	17"	27-3/4"φ	15"	20-3/4"φ	14"	15-7/8"φ
420	9'-6"	24"	17"	28-3/4"φ	15"	16-7/8"φ	14"	12-1"φ
464	10'-0"	25	18"	29-3/4"φ	16"	18-7/8"φ	14"	14-1"φ
520	10'-6"	26	19"	31-3/4"φ	16"	20-7/8"φ	15"	15-1"φ
560	11'-0"	27	19"	33-3/4"φ	17"	21-7/8"φ	16"	17-1"φ
610	11'-6"	28	21"	34-3/4"φ	18"	18-1"φ	17"	15-1"□
661	12'-0"	29	21"	36-3/4"φ	19"	20-1"φ	17"	16-1"□
770	13'-0"	31	22"	40-3/4"φ	20"	18-1"□	18"	19-1"□
890	14'-0"	33	26"	37-7/8"φ	22"	21-1"□	20"	22-1"□
1015	15'-0"	35	27"	39-7/8"φ	23"	25-1"□	21"	26-1"□
1150	16'-0"	37	28"	45-7/8"φ	24"	28-1"□	22"	29-1"□

* Based on A.C.I. Code - 1947. Special bond bars (A.S.T.M. A-305) not required. Hooks required.

STRUCTURAL FOUNDATIONS - SQUARE COLUMN FOOTINGS-3

SQUARE COLUMN FOOTINGS.*

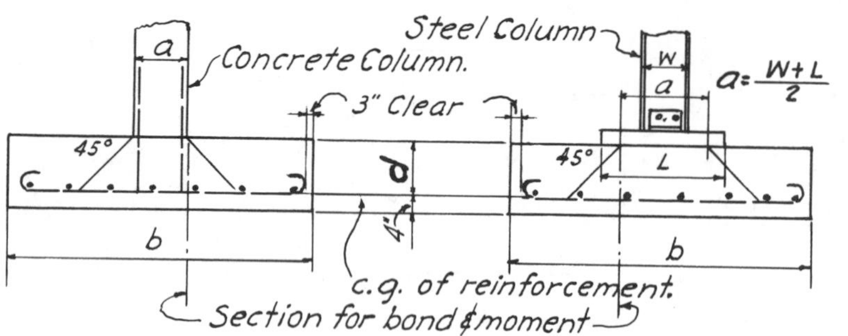

a
Concrete Column.
Steel Column
W
a
$a = \frac{W+L}{2}$
3" Clear
45°
45°
d
L
4"
b
b
c.g. of reinforcement.
Section for bond & moment

DESIGN DATA.		
f'_c = 2,000#/□"	f'_c = 3,000#/□"	f'_c = 3,750#/□"
f_c = 900#/□"	f_c = 1,350#/□"	f_c = 1,688#/□"
f_s = 20,000#/□"	f_s = 20,000#/□"	f_s = 20,000#/□"
v = 60#/□"	v = 75#/□"	v = 75#/□"
u = 112.5#/□"	u = 168#/□"	u = 200#/□"

TABLE A - SOIL BEARING VALUE. 6,000 LB. PER SQ. FT.

COLUMN LOAD IN 1000 LB.	b	d (IN.)	f'_c = 2,000#/□"		f'_c = 3,000#/□"		f'_c = 3,750#/□"	
			MIN. a	REINF. EACH WAY	MIN. a	REINF. EACH WAY	MIN. a	REINF. EACH WAY
52	3'-0"	12'	9"	9 - 1/2"φ	9"	6 - 1/2"φ	9"	6 - 1/2"φ
71	3'-6"	12	9"	10 - 5/8"φ	9"	9 - 1/2"φ	9"	8 - 1/2"φ
93	4'-0"	12	10"	14 - 5/8"φ	9"	12 - 1/2"φ	9"	10 - 1/2"φ
118	4'-6"	12	11"	17 - 5/8"φ	10"	15 - 1/2"φ	9"	13 - 1/2"φ
145	5'-0"	13	11"	17 - 3/4"φ	10"	18 - 1/2"φ	10"	15 - 1/2"φ
175	5'-6"	14	12"	19 - 3/4"φ	11"	19 - 1/2"φ	10"	13 - 5/8"φ
208	6'-0"	15	13"	21 - 3/4"φ	11"	17 - 5/8"φ	11"	15 - 5/8"φ
243	6'-6"	17	14"	22 - 3/4"φ	12"	18 - 5/8"φ	11"	17 - 5/8"φ
281	7'-0"	18	15"	24 - 3/4"φ	13"	20 - 5/8"φ	13"	14 - 3/4"φ
322	7'-6"	20	16"	24 - 3/4"φ	14"	21 - 5/8"φ	13"	15 - 3/4"φ
365	8'-0"	21	16"	27 - 3/4"φ	14"	18 - 3/4"φ	14"	13 - 7/8"φ
410	8'-6"	22	17"	29 - 3/4"φ	15"	20 - 3/4"φ	14"	15 - 7/8"φ
460	9'-0"	23	18"	31 - 3/4"φ	16"	22 - 3/4"φ	15"	16 - 7/8"φ
510	9'-6"	25	18"	32 - 3/4"φ	16"	19 - 7/8"φ	15"	11 - 1"□
563	10'-0"	26	20"	34 - 3/4"φ	17"	20 - 7/8"φ	15"	12 - 1"□
620	10'-6"	27	21"	35 - 3/4"φ	18"	14 - 1"□	17"	14 - 1"□
680	11'-0"	28	21"	38 - 3/4"φ	19"	15 - 1"□	17"	15 - 1"□
740	11'-6"	29	22"	40 - 3/4"φ	19"	16 - 1"□	18"	17 - 1"□
805	12'-0"	30	23"	42 - 3/4"φ	20"	18 - 1"□	18"	18 - 1"□
940	13'-0"	33	25"	38 - 7/8"φ	22"	21 - 1"□	20"	21 - 1"□
1080	14'-0"	36	26"	40 - 7/8"φ	23"	23 - 1"□	21"	24 - 1"□
1230	15'-0"	38	29"	43 - 7/8"φ	25"	27 - 1"□	23"	27 - 1"□
1400	16'-0"	40	30"	49 - 7/8"φ	26"	31 - 1"□	24"	32 - 1"□

TABLE B - SOIL BEARING VALUE. 8,000 LB. PER SQ. FT.

COLUMN LOAD IN 1000 LB.	b	d (IN.)	f'_c = 2,000#/□"		f'_c = 3,000#/□"		f'_c = 3,750#/□"	
			MIN. a	REINF. EACH WAY	MIN. a	REINF. EACH WAY	MIN. a	REINF. EACH WAY
70	3'-0"	12	9"	8 - 3/4"φ	9"	8 - 1/2"φ	9"	7 - 1/2"φ
96	3'-6"	12	10"	11 - 3/4"φ	9"	10 - 5/8"φ	9"	10 - 1/2"φ
125	4'-0"	12	11"	13 - 7/8"φ	10"	12 - 5/8"φ	9"	13 - 1/2"φ
158	4'-6"	12	12"	16 - 7/8"φ	10"	16 - 5/8"φ	10"	14 - 5/8"φ
194	5'-0"	14	12"	18 - 7/8"φ	11"	17 - 5/8"φ	10"	15 - 5/8"φ
235	5'-6"	15	13"	20 - 7/8"φ	12"	19 - 5/8"φ	11"	16 - 5/8"φ
280	6'-0"	17"	14"	21 - 7/8"φ	12"	20 - 5/8"φ	11"	18 - 5/8"φ
327	6'-6"	18	16"	23 - 7/8"φ	14"	22 - 5/8"φ	13"	19 - 5/8"φ
378	7'-0"	19	17"	25 - 7/8"φ	15"	24 - 5/8"φ	13"	18 - 3/4"φ
432	7'-6"	21	17"	27 - 7/8"φ	15"	26 - 5/8"φ	14"	19 - 3/4"φ
492	8'-0"	22	18"	29 - 7/8"φ	16"	23 - 3/4"φ	15"	21 - 3/4"φ
553	8'-6"	24	19"	30 - 7/8"φ	17"	24 - 3/4"φ	15"	18 - 7/8"φ
620	9'-0"	25	21"	32 - 7/8"φ	18"	26 - 3/4"φ	17"	20 - 7/8"φ
690	9'-6"	26	22"	34 - 7/8"φ	19"	29 - 3/4"φ	18"	14 - 1"□
760	10-0"	28	22"	35 - 7/8"φ	19"	24 - 7/8"φ	18"	15 - 1"□
835	10'-6"	29	23"	38 - 7/8"φ	20"	26 - 7/8"φ	19"	16 - 1"□
915	11'-0"	30	26"	39 - 7/8"φ	23"	28 - 7/8"φ	21"	18 - 1"□
1000	11'-6"	32	27"	40 - 7/8"φ	23"	19 - 1"□	22"	19 - 1"□
1090	12'-0"	33	28"	42 - 7/8"φ	24"	20 - 1"□	22"	21 - 1"□
1270	13'-0"	36	29"	45 - 7/8"φ	25"	24 - 1"□	23"	25 - 1"□
1460	14'-0"	39	30"	49 - 7/8"φ	26"	28 - 1"□	24"	29 - 1"□
1670	15-0"	42	32"	51 - 7/8"φ	28"	32 - 1"□	26"	32 - 1"□
1900	16'-0"	44	34"	56 - 7/8"φ	29"	37 - 1"□	27"	37 - 1"□

*Based on A.C.I. Code - 1947. Special bond bars (A.S.T.M. A-305) not required. Hooks required.

STRUCTURAL FOUNDATIONS-SQUARE COLUMN FOOTINGS-4

SQUARE COLUMN FOOTINGS.*

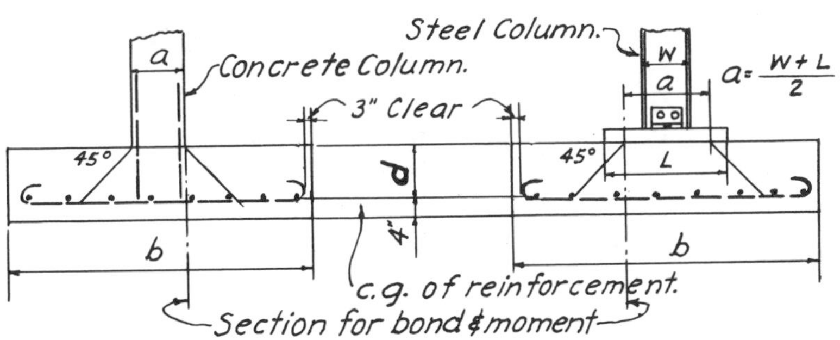

Steel Column.
a
Concrete Column.
W
a
a = (W+L)/2
3" Clear
45°
45°
d
L
4"
b
b
c.g. of reinforcement.
Section for bond & moment

DESIGN DATA.		
f'_c = 2,000#/□"	f'_c = 3,000#/□"	f'_c = 3,750#/□"
f_c = 900#/□"	f_c = 1,350#/□"	f_c = 1,688#/□"
f_s = 20,000#/□"	f_s = 20,000#/□"	f_s = 20,000#/□"
v = 60#/□"	v = 75#/□"	v = 75#/□"
u = 112.5#/□"	u = 168#/□"	u = 200#/□"

TABLE A-SOIL BEARING VALUE. 12,000 LB. PER SQ. FT.

COLUMN LOAD IN 1000 LB.	b	d (IN.)	f'_c = 2,000#/□" MIN. a	f'_c = 2,000#/□" REINF. EACH WAY	f'_c = 3,000#/□" MIN. a	f'_c = 3,000#/□" REINF. EACH WAY	f'_c = 3,750#/□" MIN. a	f'_c = 3,750#/□" REINF. EACH WAY
106	3'-0"	12	10"	7-1"□	9"	8-3/4"φ	9"	9-5/8"φ
145	3'-6"	12	11"	10-1"□	10"	11-3/4"φ	10"	10-3/4"φ
188	4'-0"	12	12"	13-1"□	10"	13-7/8"φ	10"	13-3/4"φ
240	4'-6"	14	13"	14-1"□	11"	14-7/8"φ	11"	14-3/4"φ
295	5'-0"	15	14"	16-1"□	13"	16-7/8"φ	13"	16-3/4"φ
356	5'-6"	17	16"	17-1"□	14"	19-7/8"φ	13"	19-3/4"φ
423	6'-0"	18	17"	20-1"□	15"	20-7/8"φ	14"	20-3/4"φ
495	6'-6"	20	18"	21-1"□	16"	21-7/8"φ	15"	21-3/4"φ
573	7'-0"	21	19"	23-1"□	17"	23-7/8"φ	16"	23-3/4"φ
657	7'-6"	23	21"	24-1"□	18"	24-7/8"φ	17"	24-3/4"φ
747	8'-0"	24	22"	26-1"□	19"	26-7/8"φ	18"	28-3/4"φ
840	8'-6"	26	23"	27-1"□	20"	28-7/8"φ	19"	23-7/8"φ
940	9'-0"	27	26"	29-1"□	23"	29-7/8"φ	21"	25-7/8"φ
1050	9'-6"	28	27"	31-1"□	24"	32-7/8"φ	22"	29-7/8"φ
1158	10'-0"	30	28"	32-1"□	24"	33-7/8"φ	23"	19-1"□
1275	10'-6"	32	29"	33-1"□	25"	34-7/8"φ	23"	21-1"□
1390	11'-0"	33	30"	36-1"□	26"	37-7/8"φ	24"	23-1"□
1525	11'-6"	35	31"	37-1"□	27"	40-7/8"φ	25"	25-1"□
1657	12'-0"	36	32"	39-1"□	28"	27-1"□	26"	28-1"□
1940	13'-0"	40	33"	42-1"□	29"	32-1"□	27"	32-1"□
2240	14'-0"	42	37"	45-1"□	33"	36-1"□	30	37-1"□
2560	15'-0"	46	39"	47-1"□	34"	41-1"□	32	43-1"□
2900	16'-0"	48	41"	51-1"□	36"	47-1"□	33	49-1"□

TABLE B-SOIL BEARING VALUE. 16,000 LB. PER SQ. FT.

COLUMN LOAD IN 1000 LB.	b	d (IN.)	f'_c = 2,000#/□" MIN. a	f'_c = 2,000#/□" REINF. EACH WAY	f'_c = 3,000#/□" MIN. a	f'_c = 3,000#/□" REINF. EACH WAY	f'_c = 3,750#/□" MIN. a	f'_c = 3,750#/□" REINF. EACH WAY
142	3'-0"	12	11"	9-1"□	10"	9-7/8"φ	9"	8-7/8"φ
194	3'-6"	14	12"	11-1"□	11"	11-7/8"φ	10"	10-7/8"φ
252	4'-0"	15	14"	13-1"□	12"	9-1"□	11"	12-7/8"φ
319	4'-6"	16	16"	15-1"□	14"	11-1"□	13"	14-7/8"φ
393	5'-0"	18	17"	17-1"□	15"	12-1"□	14"	15-7/8"φ
475	5'-6"	20	18"	19-1"□	16"	13-1"□	15"	16-7/8"φ
565	6'-0"	21	20"	21-1"□	18"	15-1"□	16"	18-7/8"φ
662	6'-6"	23	21"	23-1"□	19"	16-1"□	17"	19-7/8"φ
765	7'-0"	25	23"	25-1"□	19"	17-1"□	19"	20-7/8"φ
880	7'-6"	26	25"	27-1"□	20"	19-1"□	20"	23-7/8"φ
1000	8'-0"	27	27"	29-1"□	22"	20-1"□	21"	26-7/8"φ
1127	8-6"	28	28"	32-1"□	24"	22-1"□	22"	28-7/8"φ
1260	9'-0"	29	29"	35-1"□	25"	25-1"□	23"	21-1"□
1405	9'-6"	30	30"	38-1"□	26"	26-1"□	24"	23-1"□
1555	10'-0"	32	31"	40-1"□	27"	28-1"□	25"	24-1"□
1712	10'-6"	33	32"	42-1"□	28"	29-1"□	26"	26-1"□
1880	11'-0"	35	33"	44-1"□	29"	30-1"□	27"	28-1"□
2045	11'-6"	37	34"	45-1"□	30"	31-1"□	28"	30-1"□
2230	12'-0"	38	38"	47-1"□	33"	33-1"□	30"	33-1"□
2615	13'-0"	42	40"	50-1"□	34"	36-1"□	32"	38-1"□
3020	14'-0"	45	42"	54-1"□	36"	43-1"□	34"	44-1"□
3460	15'-0"	48	45"	59-1"□	39"	49-1"□	35"	52-1"□
3930	16'-0"	51	48"	63-1"□	42"	56-1"□	37"	59-1"□

*Based on A.C.I. Code - 1947. Special bond bars (A.S.T.M. A-305) not required. Hooks required

STRUCTURAL FOUNDATIONS–SQUARE COLUMN FOOTINGS-1

SQUARE COLUMN FOOTINGS.* PAGE 1 OF 4

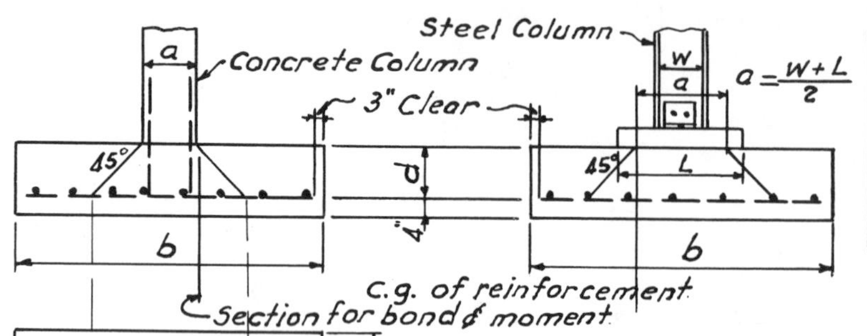
Concrete Column
Steel Column
3" Clear
45°
d
4"
b
W
a
L
$a = \frac{W+L}{2}$
C.g. of reinforcement
Section for bond & moment

DESIGN DATA			
f'_c = 2,000#/□"	f'_c = 2,500#/□"	f'_c = 3,000#/□"	f'_c = 3,750#/□"
f_c = 900#/□"	f_c = 1,125#/□"	f_c = 1,350#/□"	f_c = 1,688#/□"
f_s = 20,000#/□"	f_s = 20,000#/□"	f_s = 20,000#/□"	f_s = 20,000#/□"
V = 60#/□"	V = 75#/□"	V = 75#/□"	V = 75#/□"
u = 160#/□"	u = 200#/□"	u = 240#/□"	u = 280#/□"

Area for diagonal tension

$\frac{b-a}{2}$ a $\frac{b-a}{2}$

d   a   d

1. Depth of footing is determined by diagonal tension.
   $$V = \frac{(W-g)[b^2-(a+2d)^2]}{0.866 \times 4(a+2d)d}$$ Where: W = Soil pressure in #/□'. g = Weight of footing in #/□'.
2. $$\Sigma_o = \frac{(W-g)\,b\,(b-a)}{2} \times \frac{0.85}{u \times 0.866\,d}$$ Sum of perimeter of bars in inches.
3. $$\text{Moment} = \frac{(W-g)\,b\,(b-a)^2}{8}$$ (in ft. lbs.) (Check for f_c)
4. $$A_s = \frac{M \times 12}{20{,}000 \times 0.866 \times d} \times 0.85$$ (in sq. inches)
5. Reinforcement must satisfy both Σ_o and A_s.
6. Number of bars varies inversely with increased depth.
7. If size of bars is changed:- Provide equivalent perimeter when larger bars are used; and equivalent area when smaller bars are used.

TABLE A - SOIL BEARING VALUE. 2,000 LB. PER SQ. FT.

COLUMN LOAD IN 1000 LB	b	d (IN.)	f'_c=2,000#/□"		f'_c=2,500#/□"		f'_c=3,000#/□"		f'_c=3,750#/□"	
			MIN. a (IN.)	REINF. EACH WAY	MIN. a (IN.)	REINF. EACH WAY	MIN. a (IN.)	REINF. EACH WAY	MIN. a (IN.)	REINF. EACH WAY
29	4'-0"	12	9	5-#3**	9	5-#3	9	5-#3	9	5-#3
45	5'-0"	12	9	9-#3	9	6-#4	9	6-#4	9	6-#4
65	6'-0"	12	9	10-#4	9	7-#5	9	7-#5	9	7-#5
89	7'-0"	12	10	10-#5	9	10-#5	9	10-#5	9	10-#5
114	8'-0"	13	11	10-#6	10	10-#6	10	10-#6	9	10-#6
143	9'-0"	15	11	12-#6	10	12-#6	10	9-#7	10	9-#7
174	10'-0"	17	12	10-#7	11	8-#8	11	8-#8	10	8-#8
208	11'-0"	18	13	10-#8	12	8-#9	11	8-#9	11	8-#9
245	12'-0"	20	14	15-#7	13	9-#9	13	9-#9	12	9-#9
283	13'-0"	22	14	11-#9	13	11-#9	13	13-#8	13	13-#8
324	14'-0"	24	14	12-#9	14	12-#9	14	12-#9	13	12-#9
369	15'-0"	25	16	11-#10	15	11-#10	15	11-#10	14	11-#10
416	16'-0"	27	16	12-#10	15	12-#10	15	12-#10	14	12-#10

TABLE B - SOIL BEARING VALUE. 3,000 LB. PER SQ. FT.

COLUMN LOAD IN 1000 LB	b	d (IN.)	f'_c=2,000#/□"		f'_c=2,500#/□"		f'_c=3,000#/□"		f'_c=3,750#/□"	
			MIN. a (IN.)	REINF. EACH WAY	MIN. a (IN.)	REINF. EACH WAY	MIN. a (IN.)	REINF. EACH WAY	MIN. a (IN.)	REINF. EACH WAY
45	4'-0"	12	9	8-#3	9	5-#4	9	5-#4	9	5-#4
70	5'-0"	12	9	10-#4	9	8-#4	9	8-#4	9	8-#4
101	6'-0"	12	10	14-#4	9	10-#5	9	7-#6	9	7-#6
136	7'-0"	14	11	13-#5	10	9-#6	10	7-#7	9	7-#7
176	8'-0"	16	12	16-#5	11	9-#7	11	7-#8	10	7-#8
221	9'-0"	18	13	14-#6	12	11-#7	11	7-#9	11	7-#9
270	10'-0"	20	14	13-#7	13	10-#8	12	8-#9	11	8-#9
324	11'-0"	22	16	15-#7	15	12-#8	14	8-#10	13	8-#10
382	12'-0"	24	17	14-#8	16	11-#9	15	9-#10	13	9-#10
444	13'-0"	26	18	13-#9	16	13-#9	15	11-#10	14	11-#10
510	14'-0"	28	18	12-#10	17	10-#11	16	10-#11	15	10-#11
580	15'-0"	30	20	13-#10	19	11-#11	18	11-#11	17	11-#11
654	16'-0"	32	21	15-#10	20	12-#11	19	15-#10	17	13-#11

*Based on A.C.I. Code - 1951

Hooks not required. ** #1–#8 = 1/8"φ – 8/8"φ ; #9, 10, 11 ≡ 1"□, 1 1/8"□, 1 1/4"□

STRUCTURAL FOUNDATIONS – SQUARE COLUMN FOOTINGS-2

SQUARE COLUMN FOOTINGS*

For Design Data see pg. 7-08.

TABLE A - SOIL BEARING VALUE 4,000 LB. PER SQ. FT.

COLUMN LOAD IN 1000 LB.	b	d (IN.)	f'c=2,000#/□" MIN. a (IN.)	f'c=2,000#/□" REINF. EACH WAY	f'c=2,500#/□" MIN. a (IN.)	f'c=2,500#/□" REINF. EACH WAY	f'c=3,000#/□" MIN. a (IN.)	f'c=3,000#/□" REINF. EACH WAY	f'c=3,750#/□" MIN. a (IN.)	f'c=3,750#/□" REINF. EACH WAY
34.2	3'0"	12	9	6-#3**	9	5-#3	9	4-#3	9	4-#3
46.5	3'6"	12	9	8-#3	9	5-#4	9	4-#4	9	4-#4
60.8	4'0"	12	9	8-#4	9	9-#3	9	6-#4	9	5-#4
77	4'6"	12	10	10-#4	9	9-#4	9	7-#4	9	5-#5
95	5'0"	12	10	13-#4	9	11-#4	9	7-#5	9	7-#5
115	5'6"	12	11	16-#4	10	10-#5	10	7-#6	9	7-#6
136	6'0"	13	11	18-#4	10	12-#5	10	8-#6	9	7-#7
159	6'6"	14	12	15-#5	11	10-#6	10	8-#7	10	7-#7
184	7'0"	16	12	15-#5	11	11-#6	11	8-#7	10	6-#8
211	7'6"	17	12	17-#5	12	12-#6	12	7-#8	11	7-#8
240	8'0"	18	14	19-#5	13	10-#7	13	8-#8	12	6-#9
268	8'6"	19	14	16-#6	13	11-#7	13	7-#9	12	7-#9
300	9'0"	20	14	17-#6	14	10-#8	14	8-#9	13	8-#9
335	9'6"	22	16	18-#6	15	10-#8	14	8-#9	13	7-#10
369	10'0"	23	16	15-#7	15	12-#8	15	9-#9	14	9-#9
405	10'6"	24	16	14-#8	15	10-#9	15	8-#10	14	8-#10
443	11'0"	25	18	18-#7	16	11-#9	15	9-#10	14	9-#10
484	11'6"	27	18	15-#8	17	12-#9	16	9-#10	15	12-#9
524	12'0"	28	18	16-#8	17	10-#10	16	10-#10	15	9-#11
610	13'0"	30	20	15-#9	18	10-#11	17	10-#11	16	10-#11
700	14'0"	32	22	17-#9	20	11-#11	19	14-#10	17	12-#11
795	15'0"	34	23	16-#10	21	13-#11	20	13-#11	18	14-#11
900	16'0"	36	25	15-#11	23	15-#11	22	15-#11	20	15-#11

TABLE B - SOIL BEARING VALUE 5,000 LB. PER SQ. FT.

COLUMN LOAD IN 1000 LB.	b	d (IN.)	f'c=2,000#/□" MIN. a (IN.)	f'c=2,000#/□" REINF. EACH WAY	f'c=2,500#/□" MIN. a (IN.)	f'c=2,500#/□" REINF. EACH WAY	f'c=3,000#/□" MIN. a (IN.)	f'c=3,000#/□" REINF. EACH WAY	f'c=3,750#/□" MIN. a (IN.)	f'c=3,750#/□" REINF. EACH WAY
43.5	3'0"	12	9	7-#3	9	6-#3	9	5-#3	9	4-#3
59	3'6"	12	9	10-#3	9	8-#3	9	5-#4	9	5-#4
77	4'0"	12	10	13-#3	9	8-#4	9	7-#4	9	5-#5
97	4'6"	12	10	17-#3	9	11-#4	9	7-#5	9	6-#5
120	5'0"	12	11	16-#4	10	13-#4	10	7-#6	9	7-#6
145	5'6"	13	11	18-#4	10	12-#5	10	8-#6	10	6-#7
172	6'0"	14	12	20-#4	11	13-#5	11	9-#6	10	7-#7
202	6'6"	16	13	21-#4	12	14-#5	11	8-#7	10	6-#8
232	7'0"	17	13	18-#5	12	12-#6	12	9-#7	11	7-#8
266	7'6"	19	14	19-#5	13	13-#6	12	10-#7	12	6-#9
301	8'0"	20	15	21-#5	14	12-#7	14	11-#7	13	7-#9
340	8'6"	21	16	23-#5	15	12-#7	14	8-#9	13	8-#9
380	9'0"	22	17	19-#6	16	14-#7	15	9-#9	14	7-#10
420	9'6"	24	17	20-#6	16	15-#7	15	9-#9	14	8-#10
464	10'0"	25	18	22-#6	17	13-#8	16	10-#9	14	11-#9
520	10'6"	26	19	19-#7	17	14-#8	16	9-#10	15	8-#11
560	11'0"	27	19	21-#7	18	13-#9	17	10-#10	16	10-#10
610	11'6"	28	21	22-#7	19	14-#9	18	9-#11	17	9-#11
661	12'0"	29	21	19-#8	20	12-#10	19	10-#11	17	10-#11
770	13'0"	31	22	22-#8	21	14-#10	20	14-#10	18	12-#11
890	14'0"	33	26	20-#9	24	13-#11	22	14-#11	20	14-#11
1015	15'0"	35	27	23-#9	25	15-#11	23	16-#11	21	16-#11
1150	16'0"	37	28	21-#10	26	18-#11	24	18-#11	22	18-#11

*Based on A.C.I. Code - 1951

Hooks not required. ** #1–#8 = $\frac{1}{8}$"φ–$\frac{8}{8}$"φ ; #9, 10, 11 = 1"□, 1$\frac{1}{8}$"□, 1$\frac{1}{4}$"□

STRUCTURAL FOUNDATIONS–SQUARE COLUMN FOOTINGS-3

SQUARE COLUMN FOOTINGS*

For Design Data see pg. 7-08.

TABLE A - SOIL BEARING VALUE. 6,000 LB. PER SQ. FT.

COLUMN LOAD IN 1000LB.	b	d (IN.)	fc'=2,000#/□" MIN. a (IN.)	fc'=2,000#/□" REINF. EACH WAY	fc'=2,500#/□" MIN. a (IN.)	fc'=2,500#/□" REINF. EACH WAY	fc'=3,000#/□" MIN. a (IN.)	fc'=3,000#/□" REINF. EACH WAY	fc'=3,750#/□" MIN. a (IN.)	fc'=3,750#/□" REINF. EACH WAY
52	3'-0"	12	9	9-#3**	9	5-#4	9	6-#3	9	5-#3
71	3'-6"	12	9	12-#3	9	10-#3	9	6-#4	9	6-#4
93	4'-0"	12	10	12-#4	9	10-#4	9	8-#4	9	8-#4
118	4'-6"	12	11	15-#4	10	13-#4	10	8-#5	9	6-#6
145	5'-0"	13	11	18-#4	10	15-#4	10	10-#5	10	7-#6
175	5'-6"	14	12	20-#4	11	13-#5	11	11-#5	10	7-#7
208	6'-0	15	13	18-#5	12	15-#5	11	10-#6	11	8-#7
245	6'-6"	17	14	23-#4	13	16-#5	12	11-#6	11	7-#8
281	7'-0"	18	15	20-#5	14	14-#6	13	10-#7	13	10-#7
322	7'-6"	20	16	20-#5	15	14-#6	14	11-#7	13	7-#9
365	8'-0"	21	16	24-#5	15	13-#7	14	10-#8	14	8-#9
410	8'-6"	22	17	20-#6	16	14-#7	15	9-#9	14	7-#10
460	9'-0"	23	18	21-#6	17	13-#8	16	12-#8	15	8-#10
510	9'-6"	25	18	23-#6	17	14#8	16	13-#8	15	7-#11
563	10'-0	26	20	25-#6	18	15-#8	17	12-#9	15	8-#11
620	10'-6"	27	21	21-#7	19	16-#8	18	10-#10	17	9-#11
680	11'-0"	28	21	23-#7	20	14-#9	19	10-#11	17	10-#11
740	11'-6"	29	22	25-#7	20	16-#9	19	10-#11	18	11-#11
805	12'-0"	30	23	22-#8	21	17-#9	20	11-#11	18	12-#11
940	13'-0"	33	23	24-#8	23	16-#10	22	13-#11	20	13-#11
1080	14'-0"	36	26	22-#9	24	15-#11	23	15-#11	21	15-#11
1230	15'-0"	38	29	25-#9	27	20-#10	25	17-#11	23	17-#11
1400	16'-0"	40	30	23-#10	28	19-#11	26	20-#11	24	20-#11

TABLE B - SOIL BEARING VALUE. 8,000 LB. PER SQ. FT.

COLUMN LOAD IN 1000 LB.	b	d (IN.)	fc'=2,000#/□" MIN. a (IN.)	fc'=2,000#/□" REINF. EACH WAY	fc'=2,500#/□" MIN. a (IN.)	fc'=2,500#/□" REINF. EACH WAY	fc'=3,000#/□" MIN. a (IN.)	fc'=3,000#/□" REINF. EACH WAY	fc'=3,750#/□" MIN. a (IN.)	fc'=3,750#/□" REINF. EACH WAY
70	3'-0"	12	9	9-#4	9	9-#3	9	6-#4	9	5-#4
96	3'-6"	12	10	12-#4	9	13-#3	9	8-#4	9	7-#4
125	4'-0"	12	11	16-#4	10	13-#4	10	11-#4	9	10-#4
158	4'-6"	12	12	20-#4	11	16-#4	10	11-#5	10	10-#5
194	5'-0"	14	12	22-#4	11	18-#4	11	12-#5	10	9-#6
235	5'-6"	15	13	20-#5	12	16-#5	12	14-#5	11	8-#7
280	6'-0"	17	14	21-#5	13	17-#5	12	12-#6	11	9-#7
327	6'-6"	18	16	23-#5	15	18-#5	14	13-#6	13	10-#7
378	7'-0"	19	17	25-#5	16	17-#6	15	12-#7	13	10-#8
432	7'-6"	21	17	26-#5	16	18-#6	15	13-#7	14	9-#9
492	8'-0"	22	18	28-#5	17	16-#7	16	12-#8	15	12-#8
553	8'-6"	24	19	31-#5	18	17-#7	17	13-#8	15	11-#9
620	9'-0"	25	21	26-#6	19	18-#7	18	12-#9	17	9-#10
690	9'-6"	26	22	28-#6	20	21-#7	19	13-#9	18	13-#9
760	10'-0	28	22	24-#7	20	18-#8	19	14-#9	18	10-#11
835	10'-6"	29	23	26-#7	21	20-#8	20	16-#9	19	10-#11
915	11'-0"	30	26	27-#7	24	17-#9	23	13-#10	21	11-#11
1000	11'-6"	32	27	28-#7	25	18-#9	23	12-#11	22	12-#11
1090	12'-0"	33	28	31-#7	26	19-#9	24	13-#11	22	13-#11
1270	13'-0"	36	29	28-#8	27	18-#10	25	15-#11	23	16-#11
1460	14'-0"	39	30	27-#9	28	17-#11	26	18-#11	24	18-#11
1670	15'-0"	42	32	30-#9	30	20-#11	28	20-#11	26	21-#11
1900	16'-0"	44	34	27-#10	31	23-#11	29	23-#11	27	24-#11

*Based on A.C.I. Code - 1951

Hooks not required. ** #1–#8 = 1/8"φ – 8/8"φ; #9, 10, 11 ≡ 1"□, 1 1/8"□, 1 1/4"□

STRUCTURAL FOUNDATIONS - SQUARE COLUMN FOOTINGS-4

SQUARE COLUMN FOOTINGS* PAGE 4 OF 4

For Design Data see pg. 7-08.

TABLE A - SOIL BEARING VALUE. 12,000 LB. PER SQ. FT.

COLUMN LOAD IN 1000 LB.	b	d (IN.)	f'_c=2,000#/□"		f'_c=2,500#/□"		f'_c=3,000#/□"		f'_c=3,750#/□"	
			MIN. a (IN.)	REINF. EACH WAY	MIN. a (IN.)	REINF. EACH WAY	MIN. a (IN.)	REINF. EACH WAY	MIN. a (IN.)	REINF. EACH WAY
106	3'-0"	12	10	13-#4**	9	11-#4	9	12-#3	9	10-#3
145	3'-6"	12	11	14-#5	10	14-#4	10	12-#4	10	10-#4
188	4'-0"	12	12	18-#5	11	15-#5	10	16-#4	10	11-#5
240	4'-6"	14	13	20-#5	12	17-#5	11	18-#4	11	12-#5
295	5'-0"	15	14	23-#5	13	19-#5	13	16-#5	13	14-#5
356	5'-6"	17	16	25-#5	15	20-#5	14	18-#5	13	13-#6
423	6'-0"	18	17	23-#6	16	23-#5	15	16-#6	14	12-#7
495	6'-6"	20	18	25-#6	17	24-#5	16	17-#6	15	13-#7
573	7'-0"	21	19	27-#6	18	22-#6	17	16-#7	16	12-#8
657	7'-6"	23	21	28-#6	19	23-#6	18	17-#7	17	13-#8
747	8'-0"	24	22	31-#6	20	26-#6	19	16-#8	18	13-#9
840	8'-6"	26	23	32-#6	21	23-#7	20	17-#8	19	14-#9
940	9'-0"	27	26	34-#6	24	24-#7	23	19-#8	21	12-#10
1050	9'-6"	28	27	37-#6	25	26-#7	24	17-#9	22	14-#10
1158	10'-0"	30	28	38-#6	26	29-#7	24	18-#9	23	12-#11
1275	10'-6"	32	29	39-#6	27	24-#8	25	16-#10	23	13-#11
1390	11'-0"	33	30	36-#7	28	27-#8	26	22-#9	24	15-#11
1525	11'-6"	35	31	37-#7	29	29-#8	27	16-#11	25	16-#11
1657	12'-0"	36	32	34-#8	30	25-#9	28	17-#11	26	18-#11
1940	13'-0"	40	33	37-#8	31	24-#10	29	20-#11	27	20-#11
2240	14'-0"	42	37	42-#8	35	27-#10	33	23-#11	30	24-#11
2560	15'-0"	46	39	37-#9	36	25-#11	34	26-#11	32	27-#11
2900	16'-0"	48	41	44-#9	38	29-#11	36	30-#11	33	31-#11

TABLE B - SOIL BEARING VALUE. 16,000 LB. PER SQ. FT.

COLUMN LOAD IN 1000 LB.	b	d (IN.)	f'_c=2,000#/□"		f'_c=2,500#/□"		f'_c=3,000#/□"		f'_c=3,750#/□"	
			MIN. a (IN.)	REINF. EACH WAY	MIN. a (IN.)	REINF. EACH WAY	MIN. a (IN.)	REINF. EACH WAY	MIN. a (IN.)	REINF. EACH WAY
142	3'-0"	12	11	13-#5	10	13-#4	10	11-#4	9	10-#4
194	3'-6"	14	12	15-#5	11	13-#5	11	13-#4	10	10-#5
252	4'-0"	15	14	16-#6	13	15-#5	12	16-#4	11	15-#4
319	4'-6"	16	16	18-#6	15	18-#5	14	19-#4	13	14-#5
393	5'-0"	18	17	20-#6	16	20-#5	15	17-#5	14	16-#5
475	5'-6"	20	18	22-#6	17	22-#5	16	19-#5	15	14-#6
565	6'-0"	21	20	25-#6	19	25-#5	18	21-#5	16	16-#6
662	6'-6"	23	21	27-#6	20	27-#5	19	19-#6	17	14-#7
765	7'-0"	25	23	25-#7	21	29-#5	19	21-#6	19	13-#8
880	7'-6"	26	25	27-#7	22	26-#6	20	19-#7	20	15-#8
1000	8'-0"	27	27	29-#7	24	29-#6	22	21-#7	21	14-#9
1127	8'-6"	28	28	32-#7	26	32-#6	24	20-#8	22	19-#8
1260	9'-0"	29	29	35-#7	27	29-#7	25	22-#8	23	18-#9
1405	9'-6"	30	30	33-#8	28	31-#7	26	25-#8	24	16-#10
1555	10'-0"	32	31	35-#8	29	34-#7	27	22-#9	25	15-#11
1712	10'-6"	33	32	37-#8	30	31-#8	28	24-#9	26	16-#11
1880	11'-0"	35	33	39-#8	31	32-#8	29	21-#10	27	18-#11
2045	11'-6"	37	34	40-#8	32	35-#8	30	23-#10	28	19-#11
2230	12'-0"	38	38	37-#9	35	30-#9	33	24-#10	30	21-#11
2615	13'-0"	42	40	39-#9	37	34-#9	34	23-#11	32	24-#11
3020	14'-0"	45	42	38-#10	39	32-#10	36	27-#11	34	28-#11
3460	15'-0"	48	45	41-#10	42	30-#11	39	31-#11	35	33-#11
3930	16'-0"	51	48	39-#11	45	33-#11	42	35-#11	37	37-#11

*Based on A.C.I. Code - 1951.

Hooks not required. ** #1–#8 = 1/8"Ø – 8/8"Ø; #9,10,11 ≡ 1"□, 1 1/8"□, 1 1/4"□

STRUCTURAL FOUNDATIONS–RECTANGULAR FOOTINGS

RECTANGULAR COLUMN FOOTINGS - USING TABLES FOR SQUARE COLUMN FOOTINGS.

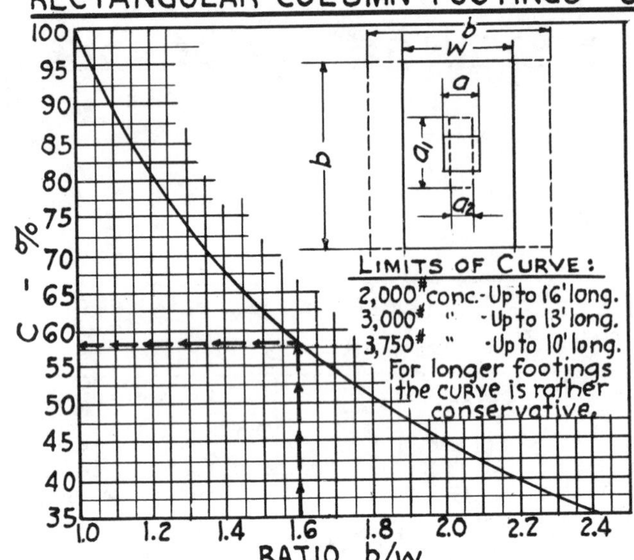

100
95
90
85
80
75
70
65
60
55
50
45
40
35
C - %
1.0
1.2
1.4
1.6
1.8
2.0
2.2
2.4
RATIO b/W
b
W
a
a_1
a_2
b
LIMITS OF CURVE:
2,000# conc.-Up to 16' long.
3,000# " -Up to 13' long.
3,750# " -Up to 10' long.
For longer footings the curve is rather conservative.

FIG. A- VALUE OF "C" FOR TRANSVERSE REINFORCEMENT.

GIVEN: Footing 7'-6" x 12'; 3,000 lb. concrete; soil pressure = 12,000 lb./sq.ft.
REQUIRED: Depth of footing and reinforcement.
SOLUTION: From Table A, page 7-11 for 12' square footing depth = 36 in.
Longitudinal reinforcement = $\frac{7.5}{12}$ x 17-#11 bars = 11-#11 bars.
Transverse reinforcement: From Fig. A at left, C = 58% for b/W = 1.6. 17 x 58% = 9.9. Use 10-#11 bars.

EFFECT OF RECTANGULAR COLUMNS ON SQUARE OR RECTANGULAR FOOTINGS.

Let "a" = One side of square column given in Square Column Footing Tables; a_1 = long side of rectangular Col. a_2 = short side of rectangular Col.
Select reinforcement parallel to a_1 from square footing whose size is $= b - a_1 + a$.
Select reinforcement parallel to a_2 from square footing whose size is $= b + a - a_2$.

RECTANGULAR COMBINED FOOTINGS.†

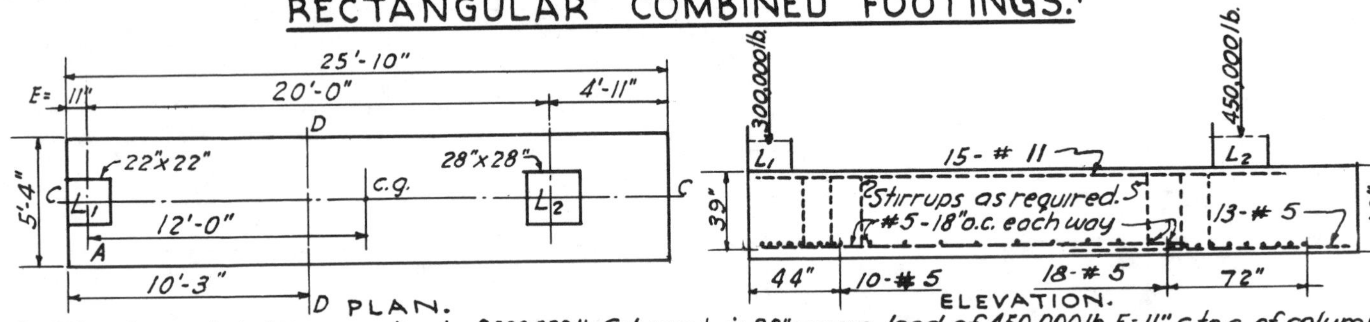

25'-10"
E = 11"
20'-0"
4'-11"
D
22" x 22"
28" x 28"
5'-4"
C
L_1
c.g.
L_2
12'-0"
A
10'-3"
D PLAN.
300,000 lb.
450,000 lb.
L_1
L_2
15-#11
Stirrups as required.
#5-18" o.c. each way
13-#5
39"
43"
44"
10-#5
18-#5
72"
ELEVATION.

GIVEN: Column L_1 is 22" square, load of 300,000 lb. Column L_2 is 28" square, load of 450,000 lb. E = 11"; c to c of columns is 20'-0". Allowable soil pressure is 6,000 lb./sq.ft. 2,000 lb. concrete with intermediate grade steel @ 20,000 p.s.i.
REQUIRED: Size of footing and necessary reinforcement.
Center of gravity of loads from A = $\frac{450{,}000 \times 20}{300{,}000 + 450{,}000} = 12.0'$. This should also be the center of the footing area. Then the required length of footing = $2 \times 12 + \frac{11}{12} \times 2 = 25'\text{-}10''$. Assume the footing weight is 500 lb./sq.ft.
The net bearing value = 6,000 - 500 = 5,500 lb./sq.ft. Required area = $\frac{750{,}000}{5{,}500} = 136.5$ sq.ft. Required width = $\frac{137}{25.83} = 5.3'$
Maximum moment at line of zero shear, D-D, from left end obtained by equation is $300{,}000 = 5{,}500 \times 5.3X$, whence $X = 10.3'$. $M = 300{,}000 \times 9.4 - 5{,}500 \times 5.3 \times \frac{10.3^2}{2} = 1{,}275{,}000$ ft.lb. ∴ For rectangular beam $d = \sqrt{\frac{1{,}275{,}000 \times 12}{157 \times 5.3 \times 12}} = 39''$ or a total depth of 43". The assumed weight of footing is sufficiently exact.
Then from the formula $A_s = \frac{M}{f_s jd} = \frac{1{,}275{,}000 \times 12}{20{,}000 \times .866 \times 39} = 22.7$ sq.in. = 15-#11 rods
Next investigate bond at columns L_1 and L_2. Shear at column $L_1 = 300{,}000 - 5{,}500 \times 5.3 \times 1.83 = 247{,}000$ lb.
*Unit shear = $\frac{247{,}000}{64 \times .866 \times 39} = 114$ lb./sq.in. requiring stirrups. Then no. of rods for bond, $u = \frac{V}{\Sigma_o jd} = \frac{247{,}000}{4.43 \times 140 \times .866 \times 39} = 12$ rods.
Shear at column $L_2 = 450{,}000 - 5{,}500 \times 5.3 \times 6.08 = 273{,}000$ lb. *Unit shear = $\frac{273{,}000}{64 \times .866 \times 39} = 126$ lb./sq.in. requiring stirrups.
No. of #11 rods for bond = $\frac{273{,}000}{4.43 \times 140 \times .866 \times 39} = 13$. Therefore steel for moment governs.
Between column L_2 and the right end of footing the moment is positive and the bottom bars are determined by the moment at the right face of the column. $M = 5.3 \times 5{,}500 \times 3.75 \times \frac{3.75}{2} = 206{,}000$ ft.lb.
$A_s = \frac{206{,}000 \times 12}{20{,}000 \times .866 \times 39} = 3.7$ sq. in. Use 13-#5 rods. Shear = 5.3 x 5,500 x 3.75 = 109,500 lb. No. of #5 rods for bond = $\frac{109{,}500}{1.96 \times 160 \times .866 \times 39} = 11$ rods. ∴ Use 13-#5 rods.
Transverse reinforcement is calculated in a similar manner. With the depth of slab used, the transverse bars are hardly needed as the projection beyond the column is only about ½ the depth of the slab. The theoretical steel area along the line C-C for column L_2 is as follows: Moment = $\frac{450{,}000}{5.3} \times \frac{L.5^2}{2} = 96{,}000$ ft.lb.
$A_s = \frac{96{,}000 \times 12}{20{,}000 \times .866 \times 39} = 1.7$ sq.in. $\Sigma_o = \frac{190{,}000}{160 \times .866 \times 39} = 35$ sq.in. Therefore use 18-#5 rods. Similarly steel may be figured for column L_1 and found to be 10-#5 bars.
It is customary to place a few rods in the bottom of the footing as shown in the sketch (#5-18" o.c. each way) to provide for any possible defects in the bed of the foundation.

* Some authorities recommend figuring shear at a distance from face of column equal to the effective depth of footing.

† Adapted from Principles of Reinforced Conc. Construction by Turneaure & Maurer.

STRUCTURAL FOUNDATIONS—TRAPEZOIDAL FOOTINGS

TRAPEZOIDAL COMBINED FOOTING

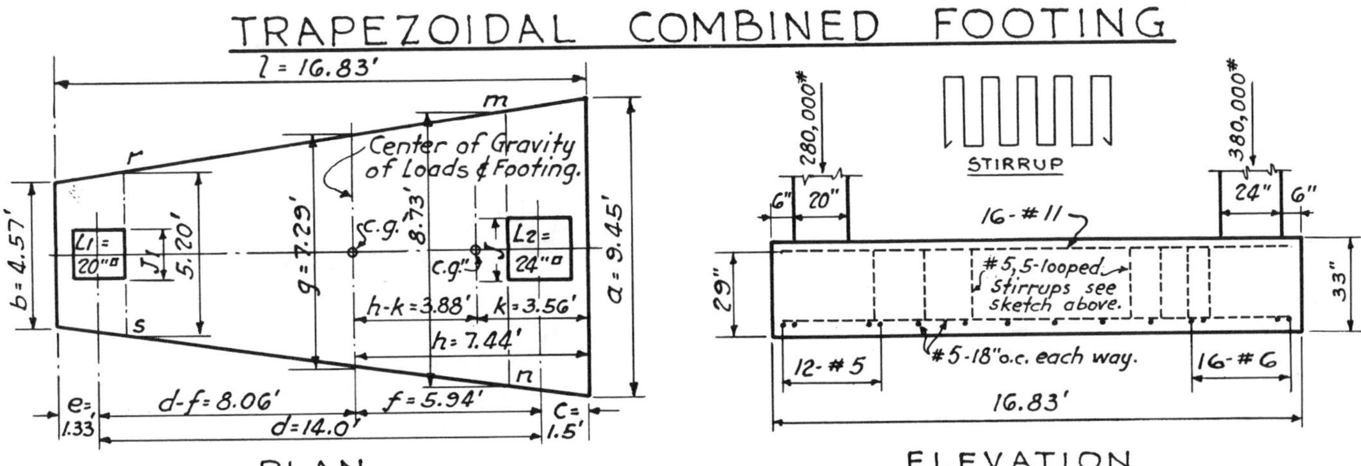

l = 16.83'
m
r
Center of Gravity of Loads & Footing.
b = 4.57'
L1 = 20"□
J1
5.20'
g = 7.29'
c.g.
8.73'
c.g."
j
L2 = 24"□
a = 9.45'
h-k = 3.88'
k = 3.56'
h = 7.44'
s
n
e = 1.33'
d-f = 8.06'
f = 5.94'
C = 1.5'
d = 14.0'
PLAN
280,000#
380,000#
STIRRUP
6"
20"
24"
6"
16-#11
#5,5-looped Stirrups see sketch above.
29"
33"
12-#5
#5-18"o.c. each way.
16-#6
16.83'
ELEVATION

GIVEN:- Column L_1 is 20" square, load of 280,000#; Column L_2 is 24" square, load of 380,000#; $e = 1.33'$; $c = 1.5'$; c.to c. of columns is 14'-0"; allowable soil pressure is 6,000#/□'; 2,000# concrete with intermediate grade steel @ 20,000#/□".

REQUIRED:- Size of footing and necessary reinforcement.

Let A' = Area of entire footing; P' = allowable soil pressure minus weight of footing; A'' = Area of portion of footing from center of gravity to wider end.

(1)- $A' = \frac{L_1 + L_2}{P'} = \frac{280{,}000 + 380{,}000}{5600} = 118$ □'. (2)- Center of gravity (c.g.') of loads which must coincide with center of gravity of footing for uniform soil pressure $= f = \frac{L_1 d}{L_2 + L_1} = \frac{280{,}000 \times 14}{660{,}000} = 5.94$ feet.

$\therefore h = f + c = 5.94' + 1.5' = 7.44$ feet.

(3)- From formula $a = \frac{A'(4l - 6h)}{l^2} = \frac{118(4 \times 16.83 - 6 \times 7.44)}{16.83^2} = 9.45$ feet.

(4)- From formula for area of trapezoid $A' = \frac{a+b}{2}(l)$, then $b = \frac{2A'}{l} - a = \frac{118 \times 2}{16.83} - 9.45 = 4.57$ feet.

(5)- $g = a - (a-b)\frac{h}{l} = 9.45 - (9.45 - 4.57)\frac{7.44}{16.83} = 7.29$ feet. (6)- c.g." or $k = \frac{h}{3}\left(\frac{a+2g}{a+g}\right) = \frac{7.44}{3}\left(\frac{9.45 + 2 \times 7.29}{9.45 + 7.29}\right) = 3.56$ feet.

(7)- $A'' = \frac{g+a}{2} \times h = \frac{7.29 + 9.45}{2} \times 7.44 = 62.3$ square feet.

(8) Maximum Moment $= L_2 \times f - A''(h-k)P' = 380{,}000 \times 5.94 - 62.3 \times 3.88 \times 5600 = 905{,}000$ ft. lbs.

From formula for rectangular beams $d = \sqrt{\frac{M}{kg}} = \sqrt{\frac{905{,}000 \times 12}{157 \times 12 \times 7.29}} = 28.1$ use 29" or total depth of 33".

Then from formula $A_s = \frac{M}{f_s jd} = \frac{905{,}000 \times 12}{20{,}000 \times .866 \times 29} = 21.7$ sq. in. or 14-#11

Next investigate bond at Cols. L_1 and L_2.

Shear at Col. $L_2 = 380{,}000 - \frac{8.73 + 9.45}{2} \times 2.5 \times 5600 = 253{,}000$ lbs. Bond $u = \frac{V}{\Sigma_o jd}$ or No. of rods $= \frac{253{,}000}{4.43 \times .866 \times 29 \times 140} = 16.$

$\therefore$ Use 16-#11 rods

Shear at Col. $L_1 = 280{,}000 - \frac{4.57 + 5.20}{2} \times 2.17 \times 5{,}600 = 221{,}000$ lbs..

Transverse reinforcement is provided to prevent bending of the projection of the footing. Considering Col. L_2, and assuming the width of the distributing beam as 3'-6", the load $= \frac{L_2}{2} \times \frac{a-j}{a} = \frac{380{,}000}{2} \times \frac{9.45 - 2.0}{9.45} =$ 149,500 lbs. Then Mom = load $\times \frac{a-j}{4} = 149{,}500 \times \frac{9.45 - 2.0}{4} = 278{,}000$ ft. lbs. Then $d = \sqrt{\frac{278{,}000 \times 12}{157 \times 3.5 \times 12}} = 22.5$, which is smaller than depth of footing; and $A_s = \frac{278{,}000 \times 12}{20{,}000 \times .866 \times 29} = 6.7$□" or 16-#6

For bond $u = \frac{V}{\Sigma_o jd}$ or No. of rods $= \frac{149{,}500}{2.36 \times .866 \times 160 \times 29} = 16.$ $\therefore$ Use 16-#6

In the same manner the distributing steel for column L_1 is determined. Load $= \frac{280{,}000}{2} \times \frac{4.57 - 1.67}{4.65} = 88{,}700$ lbs.

Mom. $= 88{,}700 \times \frac{4.57 - 1.67}{4} = 64{,}300$ ft. lbs. and $A_s = \frac{64{,}300 \times 12}{20{,}000 \times 29 \times .866} = 1.53$ or 6-#5

For bond the no. of rods $= \frac{88{,}700}{1.96 \times .866 \times 160 \times 29} = 12$ $\therefore$ use 12-#5

Diagonal tension reinforcement is not needed in the distribution beams since each distributing beam and column has a similar load to that of a single footing. For such footings the intensity of shearing stress as a measure of diagonal tension is computed on a section at a distance from the face of the column equal to the depth of the footing to the steel. Compute the shear, in the transverse direction, on a section at the supports as mn and rs. The spacing and size of the stirrups are determined as in a single beam. Place rods in the bottom of footing as shown in sketch (#5 - 18"o.c. each way) to provide for possible defects in the foundation bed.

STRUCTURAL FOUNDATIONS—CANTILEVER FOOTINGS

CANTILEVER FOOTING (COMMONLY KNOWN AS PUMP HANDLE FOOTING)†

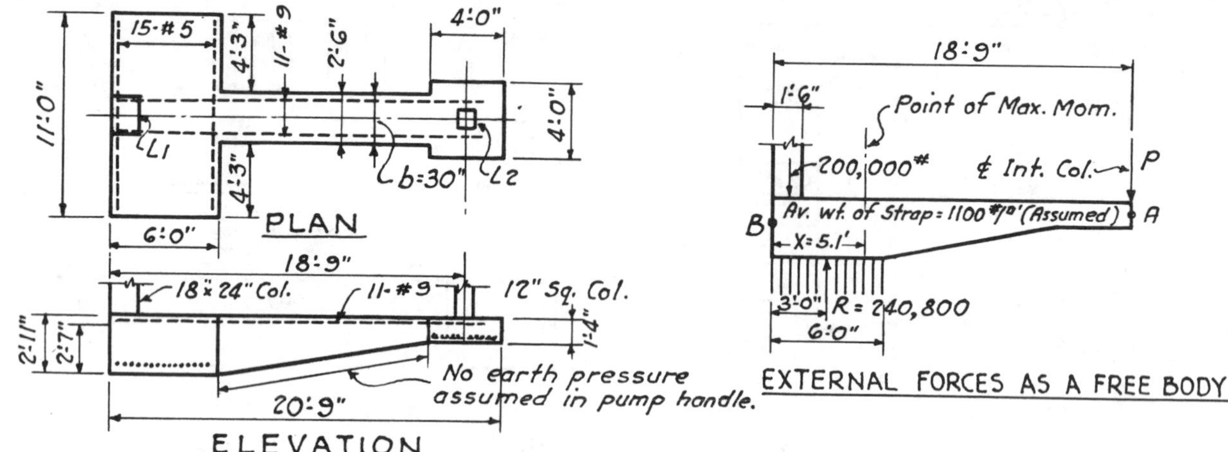

15-#5
4'-3"
11-#9
2'-6"
4'-0"
11'-0"
L1
4'-3"
b=30"
L2
4'-0"
6'-0"
PLAN
18'-9"
18"x24" Col.
11-#9
12" Sq. Col.
2'-1"
2'-7"
1'-4"
No earth pressure assumed in pump handle.
20'-9"
ELEVATION
18'-9"
1'-6"
Point of Max. Mom.
200,000#
℄ Int. Col.
P
B
Av. wt. of Strap = 1100#/' (Assumed)
A
X=5.1'
3'-0"
R = 240,800
6'-0"
EXTERNAL FORCES AS A FREE BODY

GIVEN: Column L_1 = 18"×24", load of 200,000#; Column L_2 = 12" square, load of 60,000# of which 30,000# is Dead Load; 18'-9" from ℄ of interior col. L_2 to the outside face of exterior column L_1; allowable soil pressure of 4,000 lbs. per sq. ft.; 2,000 lb. concrete with intermediate grade reinforcement.

REQUIRED:- Size and reinforcement of the eccentrically loaded footing and the connecting beam or strap.

The interior column footing is designed in the usual manner and it is assumed this design has been made.

The area required for the exterior column load, allowing about 30 per cent for the weight of the footing and strap. =

260,000/4,000 = 65.0 sq. ft. A base 6'-0" × 11'-0" is selected.

Considering the strap as a free body with the external forces as illustrated above, the reaction R is determined by taking moments about A, as follows:- $-200,000 \times 18 - 1100 \times 18.75 \times 18.75/2 + R \times 15.75 = 0$, from which $R = 240,800$.

The downward force P is determined by taking moments about B, as follows:- $200,000 \times .75 + 1100 \times 18.75 \times 18.75/2 + P \times 18.75 = 240,800 \times 3$, from which $P = 20,200$. Since P is less than the dead load (30,000#) of Col. L_2, there will be no uplift at column L_2.

The maximum moment in the strap occurs at the point of zero shear at a distance X from B as follows:- $-200,000 + 240,800/6\,X - 1100X = 0$, from which $X = 5.1'$.

The maximum moment at this section is then, $M = -200,000(5.1-0.75) - 1100 \times \frac{5.1^2}{2} + 240,800 \times \frac{5.1^2}{6 \cdot 2} = -363,300'^{\#}$ or $-4,360,000"^{\#}$ with $b = 30"$, $d = \sqrt{\frac{4,360,000}{157 \times 30}} = 31"$ or total depth = 35"

The area of steel for moment $= A_s = \frac{4,360,000}{20,000 \times .866 \times 31} = 8.1$ □" No. of rods for bond $= \frac{141,500}{3.54 \times 140 \times .866 \times 31} = 10.6$ rods.

Therefore use 11-#9 rods in the top of the strap.

The maximum shear occurs at the inside face of the exterior column $V = -200,000 - 1100 \times 1.5 + 240,800/6 \times 1.5 = 141,500$#. The critical width for shear in the plane of the longitudinal bars will include not only the width of the strap but also the extra width* of the exterior footing at this plane. With the proposed arrangement of the exterior footing, the shearing width in the plane of the longitudinal reinforcement is approximately $30 + 2 \times \frac{1}{8} \times 51 = 43"$ ∴ shear $= \frac{141,500}{43 \times 31 \times .866} = 122$#/□" requiring stirrups which are figured in the same manner as for an ordinary beam. The unit shear at the inner edge of the exterior footing is considerably less than 60#/□".

The average soil pressure under the exterior footing inclusive of the weight of the strap is $\frac{240,800 + 21,000 \text{ (wt. of footing)}}{11.0 \times 6.0} = 3970$#/□' which is less than 4,000#/□' allowable.

Next consider the exterior footing design. The net upward pressure on each of the cantilever portions of the footing is $\frac{240,800}{11 \times 6} = 3650$ lbs. per sq. ft.

The maximum moment in the cantilever at the edge of the strap is =

$3650 \times 6 \times 4.25 \times \frac{4.25}{2} \times 12 = 2,370,000"^{\#}$ Then $A_s = \frac{2,370,000}{20,000 \times .866 \times 31} = 4.42$ □" = 15-#5

For bond - no. of #5 bars $= \frac{3650 \times 4.25 \times 6}{1.96 \times 160 \times 31 \times .866} = 11$ ∴ use 15-#5

* $\frac{1}{8}$ of overhang.

† Adapted from Urquhart & O'Rourke, Design of Conc. Structures, McGraw-Hill.

STRUCTURAL FOUNDATIONS—PILE TYPES-2

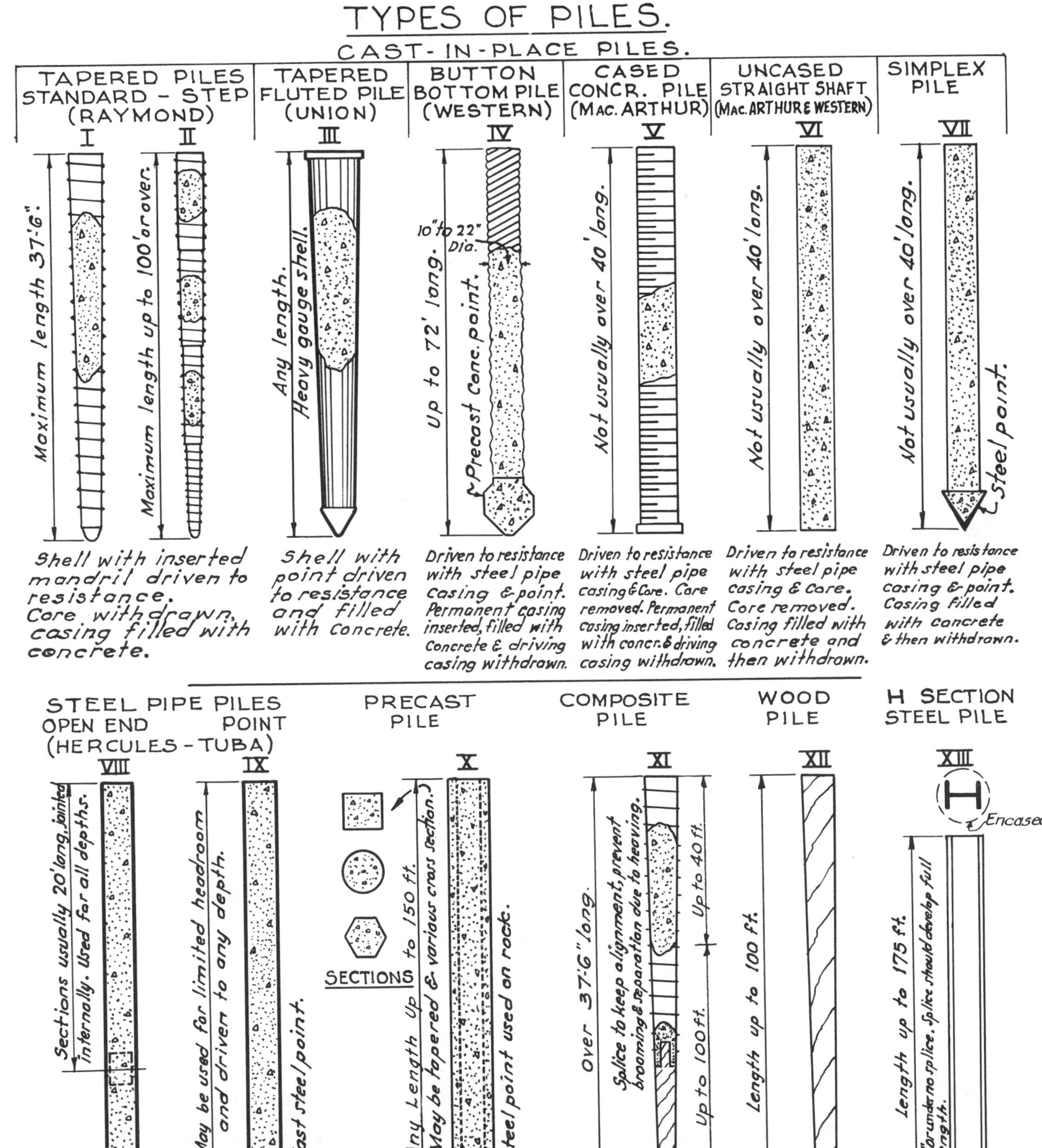

TYPES OF PILES.
CAST-IN-PLACE PILES.
TAPERED PILES STANDARD - STEP (RAYMOND)
TAPERED FLUTED PILE (UNION)
BUTTON BOTTOM PILE (WESTERN)
CASED CONCR. PILE (Mac. ARTHUR)
UNCASED STRAIGHT SHAFT (Mac. ARTHUR & WESTERN)
SIMPLEX PILE
I
II
III
IV
V
VI
VII
Maximum length 37'-6".
Maximum length up to 100' or over.
Any length. Heavy gauge shell.
Up to 72' long.
10" to 22" Dia.
Precast Conc. point.
Not usually over 40' long.
Not usually over 40' long.
Not usually over 40' long.
Steel point.
Shell with inserted mandril driven to resistance. Core withdrawn, casing filled with concrete.
Shell with point driven to resistance and filled with Concrete.
Driven to resistance with steel pipe casing & point. Permanent casing inserted, filled with Concrete & driving casing withdrawn.
Driven to resistance with steel pipe casing & Core. Core removed. Permanent casing inserted, filled with concr. & driving casing withdrawn.
Driven to resistance with steel pipe casing & core. Core removed. Casing filled with concrete and then withdrawn.
Driven to resistance with steel pipe casing & point. Casing filled with concrete & then withdrawn.
STEEL PIPE PILES OPEN END POINT (HERCULES-TUBA)
PRECAST PILE
COMPOSITE PILE
WOOD PILE
H SECTION STEEL PILE
VIII
IX
X
XI
XII
XIII
Sections usually 20' long, jointed internally. Used for all depths.
May be used for limited headroom and driven to any depth.
Cast steel point.
SECTIONS
Any Length Up to 150 ft.
May be tapered & various cross section.
Steel point used on rock.
Over 37'-6" long.
Splice to keep alignment, prevent brooming & separation due to heaving.
Up to 40 ft.
Up to 100 ft.
Length up to 100 ft.
Encased
Length up to 175 ft.
40'-0" or under no splice. Splice should develop full strength.

Earth blown out with air jet as driven to refusal and loaded as column.

Driven to refusal or to resistance.

Driven to refusal or to resistance.

Wood pile driven below water level & to resistance. Upper section spliced to wood pile & filled with concrete.

Driven to resistance.

Driven to refusal or resistance. Used where penetration is in hard material.
Encased if H is placed in 18 ga. shell. Concrete is poured. H is driven on rock to refusal.

For general notes see Pg. 7-16.

STRUCTURAL FOUNDATIONS–PILE DATA –1

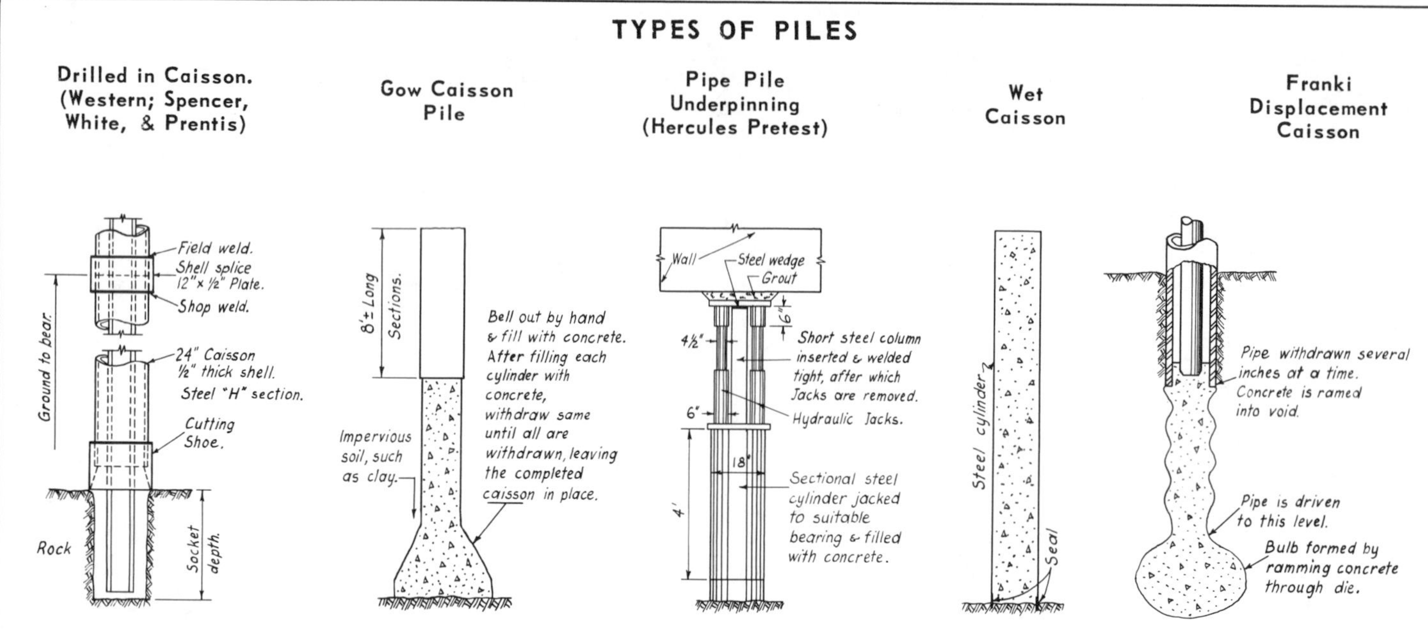
TYPES OF PILES
Drilled in Caisson. (Western; Spencer, White, & Prentis)
Gow Caisson Pile
Pipe Pile Underpinning (Hercules Pretest)
Wet Caisson
Franki Displacement Caisson
Field weld.
Shell splice 12" x ½" Plate.
Shop weld.
Ground to bear.
24" Caisson ½" thick shell.
Steel "H" section.
Cutting Shoe.
Rock
Socket depth.
8'± Long Sections.
Bell out by hand & fill with concrete. After filling each cylinder with concrete, withdraw same until all are withdrawn, leaving the completed caisson in place.
Impervious soil, such as clay.
Wall
Steel wedge
Grout
4½"
6¼"
Short steel column inserted & welded tight, after which Jacks are removed.
6"
Hydraulic Jacks.
18"
4'
Sectional steel cylinder jacked to suitable bearing & filled with concrete.
Steel cylinder
Seal
Pipe withdrawn several inches at a time. Concrete is ramed into void.
Pipe is driven to this level.
Bulb formed by ramming concrete through die.

(a) Shell driven to rock, cleaned out. Rock socket drilled, core inserted, and shell filled with concrete.

(b) Shallow pit excavated by hand, and top cylinder placed in pit. Second cylinder placed inside first, and process repeated until caisson reaches its full depth.

(c) Cylinder tested with jacks to an overload capacity.

(d) Steel cylinder sunk to rock as earth is removed. Bottom sealed, water removed, and cylinder filled with concrete.

(e) Steel pipe driven to desired depth.

Pile No.*	Notes
I, II, III, V, XI	Precautions are required to prevent collapsing of shell when driving adjoining piles.
VI, VII	With uncased piles, precautions should be taken to prevent damage due to driving adjoining piles because pile has no sheet casing around it.
IV, V	When shell is inserted inside driving casing and casing withdrawn, soil must be relied on to grip pile as firmly as if it had been driven without casing.
VIII	Concrete steel pipe piles sometimes driven open-ended, to predetermined depth, and filled with concrete. After concrete has set, pile is driven to required resistance. This is done so as not to disturb adjoining wall and foundation and also when driving cast-in-place piles to prevent heaving.
X	Precast piles are used for marine structure, require heavy handling equipment.
XI, XII	Cut off untreated wood pile below permanent water level. Pressure-creosoted wood piles used with cut-off above permanent water table. Precautions against overdriving should be taken.
I to IX	These piles have less give under hammer than piles of more flexible material such as wood or concrete, and consequently, if driven to the same resistance, have a greater safety factor.
XIII	Steel H sections should not be used through cinders, ash fill, or normally active rust-producing material without adequate protection.
Caissons	These are generally used when sinking foundation to considerable depths with heavy loads. This is done by wood sheeting, steel sheeting, and steel cylinders. In case of water condition, operations are carried on under compressed air.

* For sketch see p. 7-15.

STRUCTURAL FOUNDATIONS—PILE TYPES—I

PILE DATA

TABLE A—PILE SPACING AND LOADS [†]

TYPE OF PILE	FRICTION OR BEARING	LOADS [1]: LIMITING LOADS	LOADS [1]: REMARKS	SPACING: MIN. C. TO C.	SPACING: REMARKS
Wood	Friction	20 Tons 6" Point 25 Tons 8" Point or >	15 & 20 Ton piles in many codes	2'-6"	1. Piles bearing on rock or penetrating into rock shall have a minimum spacing center to center of twice the average diameter or 1.75 times the diagonal of the pile, but not less than twenty four inches. 2. All other piles shall have a minimum spacing center to center of twice the average diameter or 1.75 times the diagonal of the pile but not less than thirty inches, except that all piles located in groups or abutting groups that receive their principal support in materials such as clay, sand clay or uncompacted sand shall have their spacing increased above the minimum values by 10% for each interior pile up to a maximum of 40%.
Composite	Friction	Load of weaker section	—	2'-6"	
Cast in place conc.	Friction	Max. 60 Tons	Maximum axial load $= f_c \times A_c$	2'-6"	
Pre-Cast Concrete	Friction	Max. 60 Tons	Maximum axial load $= f_c \times (A_c + nA_s)$, $p = 2\%$ min. 4% max.	2'-6"	
Concrete filled steel pipe	Friction—driven with shoe	Max. 60 Tons	$A_s \times 9{,}000 + A_c \times f_c$	2'-6"	
	Bearing—driven open-ended to rock and cleaned out.	Max. 200 Tons or ½ jacking pressure	See p. 7-18 for sizes and loads	2'-0"	
Steel H piles	Friction	Max. 60 Tons	$A_s \times 9{,}000$	2'-6"	
	Bearing—hard rock	120 Tons maximum on rock. 80 Tons on hardpan or gravel boulders overlying rock. (over 40 Tons requires load test)	See p. 7-18 for sizes & loads	2'-0"	
	Bearing-encased in concrete and steel shell	200 T. max. on hard rock (over 100 Tons requires load test.)	See p. 7-18 for sizes & loads	2'-0"	
Drilled in Caissons	Bearing	—	Generally from 200 to 2500 tons.	—	Generally one under ea. col.

1 Load test required for friction piles over 30 tons. Use Engineering News formula. If piles are driven into soft clays, silts or mud without reaching hard stratum, formula should be checked by load test.

TABLE B—PROBABLE PENETRATION EXPECTANCY FOR FRICTION PILES.*

MATERIAL	PENETRATION	REMARKS
Clean compact sand.	Slight	Usually jetted.
Other sands.	20 ft.	
Sandy clay.	30 ft.	
Pure clay.	35 ft.	
Clay and silt.	45 ft.	
Silt and mud.	50 to 100 ft.	
Glacial till.	Slight	Piles can not be driven.

*Penetration can be best determined by test piles or test rods. This table is to be used only for a rough indication from borings.

SAFE UPLIFT STRENGTH OF PILES.

Friction Piles—in sand, clay or gravel. Use one-half the safe bearing load.

Point Bearing Piles (Piles driven to hard stratum). Compute by contact surface area x shearing strength of material penetrated. See p. 9-08 for soil shearing values—Note: for sand 250 lb. per sq. ft. suggested.

Full size tests are desirable on account of the difficulty of establishing data for computation.

Lateral Force at top of piles = 1 kip allowable without test. Also 1 kip allowed on batter piles in addition to horizontal component.

† Based on New York City Building Code, 1950.

STRUCTURAL FOUNDATIONS-PILE DATA-2

STEEL PIPE AND STRUCTURAL STEEL PILES ON ROCK

TABLE-A STEEL PIPE PILES ON ROCK*

DIAMETER	WALL THICKNESS	LOAD IN TONS
8 3/4	5/16	40
10 3/4	5/16	53
	3/8	59
	7/16	66
12 3/4	5/16	67
	3/8	75
	7/16	83
	1/2	91
14	3/8	85
	7/16	94
	1/2	103
	5/8	120
15	3/8	93
	7/16	103
	1/2	113
	5/8	131
16	3/8	103
	7/16	113
	1/2	123
	5/8	143
18	3/8	122
	7/16	133
	1/2	144
	5/8	167
20	3/8	142
	7/16	154
	1/2	167
	5/8	192

TABLE-B STRUCTURAL STEEL PILES ON ROCK*

SIZE	NOT ENCAS'D	ENCASED	
	LOAD IN TONS	LOAD IN TONS	DIA. OF ENCAS.
8BP36	47	86	15 1/2
10BP42	55	107	18
10BP57	75	129	18 1/4
12BP53	70	140	21
12BP74	98	171	21 1/4
14BP73	96	189	24
14BP89	118	212	24 1/4
14BP102	135	232	24 1/2
14BP117	155	254	24 3/4
8WF31	41	78	15 1/4
8WF58	76	118	16
8WF67	88	133	16 1/2
10WF49	64	118	18 1/4
10WF60	79	133	18 1/4
10WF77	102	156	18 1/4
12WF85	112	187	21 1/2
14WF84	111	195	22 3/4

TABLE-C PILE DIMENSIONS (USE WITH FIG-D BELOW)

CENTER TO CENTER OF PILES "D" INCHES	3-PILE PIERS a INCHES	3-PILE PIERS b INCHES	5 & 13 PILE PIERS a INCHES	7, 8, 10, 11 & 14 PILE PIERS a INCHES
24	14	7	17	21
25	15	7 1/2	18	22
26	15	7 1/2	18 1/2	22 1/2
27	16	8	19	23 1/2
28	16	8	20	24 1/2
29	17	8 1/2	20 1/2	25
30	17	8 1/2	21 1/2	26
31	18	9	22	27
32	19	9 1/2	22 1/2	28
33	19	9 1/2	23 1/2	28 1/2
34	20	10	24	29 1/2
35	21	10 1/2	25	30 1/2
36	21	10 1/2	25 1/2	31
37	22	11	26 1/2	32
38	22	11	27	33
40	23	11 1/2	28 1/2	35
42	24	12	30	36 1/2
44	26	13	31 1/2	38 1/2
46	27	13 1/2	32 1/2	40
48	28	14	34	41 1/2
50	29	14 1/2	35 1/2	43 1/2
52	30	15	37	45
54	31	15 1/2	38 1/2	46 1/2
56	33	16 1/2	40	48 1/2

FIG.-D PILE GROUP SPACING FOR TABLE C

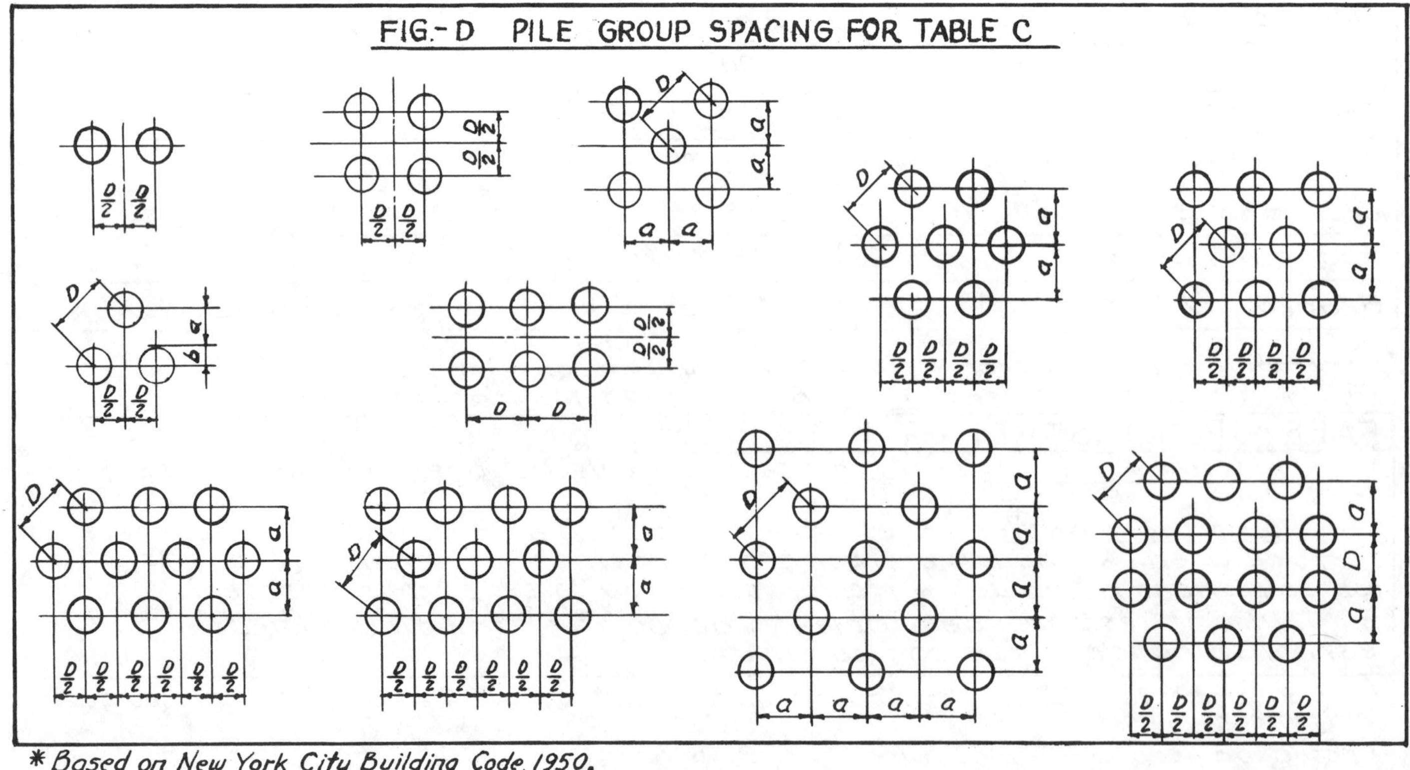
D/2
D
a
b

* *Based on New York City Building Code, 1950.*

STRUCTURAL FOUNDATIONS–PILE CAPS–I

PILE FOOTINGS

Designed according to ACI 1947. Pg. 7-19 to 7-23 are designed for ordinary bond bars, hooks required. Pg. 7-24 to 7-29 designed for special bond bars (A.S.T.M. A-305); hooks not required.

WARNING – *Check overloading of strata on which piles rest.*

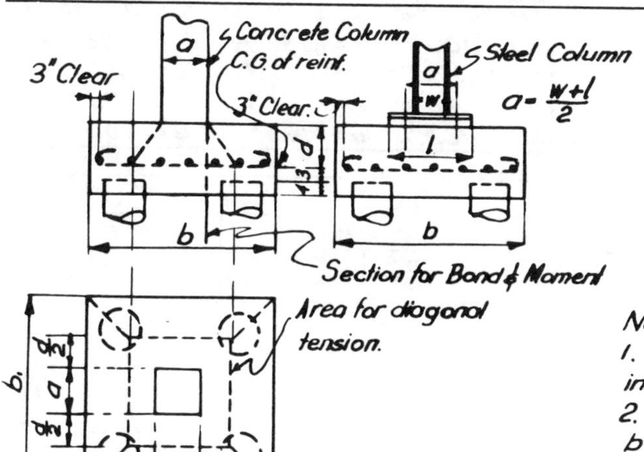

Concrete Column
C.G. of reinf.
3" Clear
3" Clear
Steel Column
$a = \frac{w+l}{2}$
a
w
l
d
b
b
Section for Bond & Moment
Area for diagonal tension.
b₁
d/2
a
d/2
a
a/2
a/2

DESIGN DATA		
$f_c' = 2{,}000$ #/□"	$f_c' = 3{,}000$ #/□"	$f_c' = 3{,}750$ #/□"
$f_c = 900$ #/□"	$f_c = 1.350$ #/□"	$f_c = 1.688$ #/□"
$f_s = 20{,}000$ #/□"	$f_s = 20{,}000$ #/□"	$f_s = 20{,}000$ #/□"
$v = 60$ #/□"	$v = 75$ #/□"	$v = 75$ #/□"
$u = 112.5$ #/□"	$u = 168$ #/□"	$u = 200$ #/□"

Notes:

1. *If depth of pile caps is increased steel perimeter may be decreased in proportion to depth.*
2. *If size of bars is changed:– Provide equivalent perimeter when larger bars are used, and equivalent area when smaller bars are used.*

Number of Piles	Plan	For All Footings: Pile Value (in kips)	For All Footings: Column Load in Kips	d (in.)	$f_c' = 2{,}000$ #/□": a (in.)	Reinforcement: Long Way	Reinforcement: Short Way	$f_c' = 3{,}000$ #/□": a (in.)	Reinforcement: Long Way	Reinforcement: Short Way	$f_c' = 3{,}750$ #/□": a (in.)	Reinforcement: Long Way	Reinforcement: Short Way
1	2'-6" × 2'-6"; 1'-3" 1'-3"	20	18.5	12	–	3-⅝"φ	3-⅝"φ	–	3-⅝"φ	3-⅝"φ	–	3-⅝"φ	3-⅝"φ
		30	28.5	12	–	3-⅝"φ	3-⅝"φ	–	3-⅝"φ	3-⅝"φ	–	3-⅝"φ	3-⅝"φ
		40	38.5	12	–	3-⅝"φ	3-⅝"φ	–	3-⅝"φ	3-⅝"φ	–	3-⅝"φ	3-⅝"φ
		50	48.5	12	–	3-⅝"φ	3-⅝"φ	–	3-⅝"φ	3-⅝"φ	–	3-⅝"φ	3-⅝"φ
		60	58.5	12	–	3-⅝"φ	3-⅝"φ	–	3-⅝"φ	3-⅝"φ	–	3-⅝"φ	3-⅝"φ
2	5'-0" × 2'-6"; 1'-3" 1'-3" 1'-3" 1'-3"	20	37	12	7	6-⅝"φ	4-⅝"φ	7	4-⅝"φ	4-⅝"φ	7	4-⅝"φ	4-⅝"φ
		30	57	15	7	8-⅝"φ	4-⅝"φ	7	5-⅝"φ	4-⅝"φ	7	5-⅝"φ	4-⅝"φ
		40	76	17	9	9-⅝"φ	4-⅝"φ	9	6-⅝"φ	4-⅝"φ	9	5-⅝"φ	4-⅝"φ
		50	96	19	9	8-¾"φ	4-⅝"φ	9	7-⅝"φ	4-⅝"φ	9	6-⅝"φ	4-⅝"φ
		60	116	20	10	8-⅞"φ	4-⅝"φ	9	8-⅝"φ	4-⅝"φ	9	7-⅝"φ	4-⅝"φ
3	5'-5" × 4'-8"; 1'-5½" 1'-3" 1'-3" 1'-5½"	20	54	15	8	3 Bands of 4-⅝"φ		7	3 Bands of 3-⅝"φ		7	3 Bands of 3-⅝"φ	
		30	83	17	9	" " " 5-⅝"φ		9	" " " 4-⅝"φ		9	" " " 3-⅝"φ	
		40	113	19	10	" " " 6-⅝"φ		9	" " " 4-⅝"φ		9	" " " 4-⅝"φ	
		50	143	20	11	" " " 8-⅝"φ		10	" " " 5-⅝"φ		9	" " " 4-⅝"φ	
		60	172	21	12	" " " 7-¾"φ		10	" " " 6-⅝"φ		10	" " " 5-⅝"φ	
4	5'-0" × 5'-0"; 1'-3" 1'-3" 1'-3" 1'-3"	20	72	14	8	14-½"φ	14-½"φ	8	10-½"φ	10-½"φ	8	8-½"φ	8-½"φ
		30	112	15	10	17-⅝"φ	17-⅝"φ	9	14-½"φ	14-½"φ	9	12-½"φ	12-½"φ
		40	152	16	12	15-⅞"φ	15-⅞"φ	10	14-⅝"φ	14-⅝"φ	9	15-½"φ	15-½"φ
		50	192	17	12	16-1"φ	16-1"φ	11	17-⅝"φ	17-⅝"φ	10	14-⅝"φ	14-⅝"φ
		60	232	18	13	14-1"□	14-1"□	12	16-¾"φ	16-¾"φ	12	16-⅝"φ	16-⅝"φ

STRUCTURAL FOUNDATIONS-PILE CAPS-2

PILE FOOTINGS.*

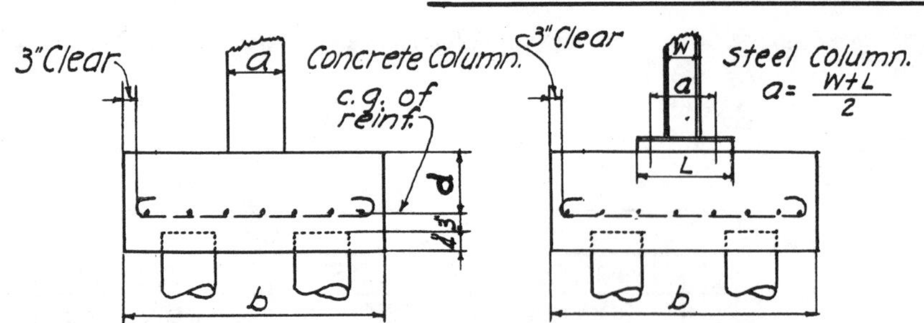
3"Clear
a
Concrete Column.
c.g. of reinf.
d
4.5"
b
3"Clear
W
a
Steel Column.
a = (W+L)/2
L
b

DESIGN DATA.		
f'_c = 2,000#/□"	f'_c = 3,000#/□"	f'_c = 3,750#/□"
f_c = 900#/□"	f_c = 1,350#/□"	f_c = 1,688#/□"
f_s = 20,000#/□"	f_s = 20,000#/□"	f_s = 20,000#/□"
v = 60#/□"	v = 75#/□"	v = 75#/□"
u = 112.5#/□"	u = 168#/□"	u = 200#/□"

Number of Piles	Plan	For All Footings: Pile Value in Kips	For All Footings: Column Load in Kips	d (in.)	f'_c = 2,000#/□": a (in.)	f'_c = 2,000#/□": Reinforcement Long Way	f'_c = 2,000#/□": Reinforcement Short Way	f'_c = 3,000#/□": a (in.)	f'_c = 3,000#/□": Reinforcement Long Way	f'_c = 3,000#/□": Reinforcement Short Way	f'_c = 3,750#/□": a (in.)	f'_c = 3,750#/□": Reinforcement Long Way	f'_c = 3,750#/□": Reinforcement Short Way
5	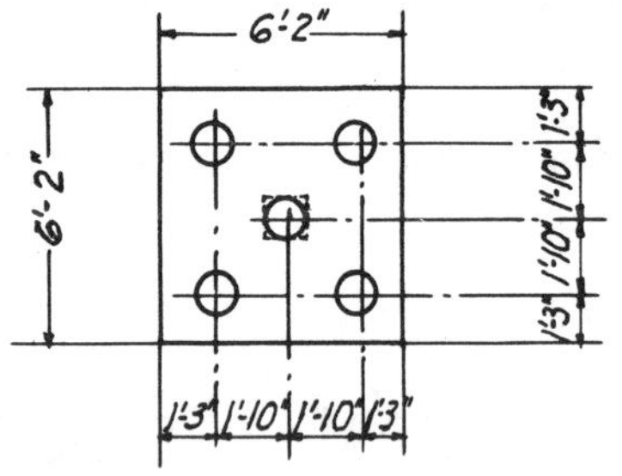	20	90	15	10	13-1/2"φ	13-1/2"φ	9	7-5/8"φ	7-5/8"φ	9	8-1/2"φ	8-1/2"φ
		30	137	18	11	17-1/2"φ	17-1/2"φ	10	12-1/2"φ	12-1/2"φ	10	12-1/2"φ	12-1/2"φ
		40	186	21	13	20-1/2"φ	20-1/2"φ	11	14-1/2"φ	14-1/2"φ	10	14-1/2"φ	14-1/2"φ
		50	235	22	13	19-5/8"φ	19-5/8"φ	12	13-5/8"φ	13-5/8"φ	11	11-5/8"φ	11-5/8"φ
		60	284	24	14	18-3/4"φ	18-3/4"φ	14	14-5/8"φ	14-5/8"φ	12	12-5/8"φ	12-5/8"φ
6	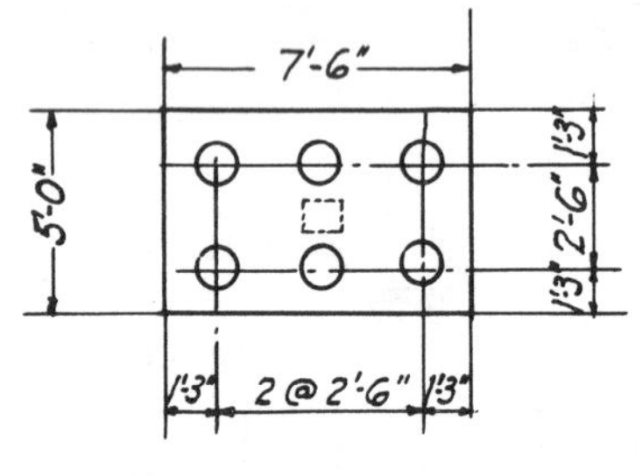	20	106	22	10	10-1/2"φ	14-1/2"φ	9	7-5/8"φ	9-1/2"φ	8	7-5/8"φ	8-1/2"φ
		30	164	27	12	12-1/2"φ	17-1/2"φ	10	8-5/8"φ	11-1/2"φ	10	6-3/4"φ	10-1/2"φ
		40	222	32	14	13-1/2"φ	19-1/2"φ	12	9-5/8"φ	13-1/2"φ	11	7-3/4"φ	11-1/2"φ
		50	280	35	14	15-1/2"φ	22-1/2"φ	13	7-3/4"φ	15-1/2"φ	13	7-3/4"φ	13-1/2"φ
		60	338	36	16	17-1/2"φ	26-1/2"φ	15	8-3/4"φ	18-1/2"φ	14	8-3/4"φ	15-1/2"φ
7	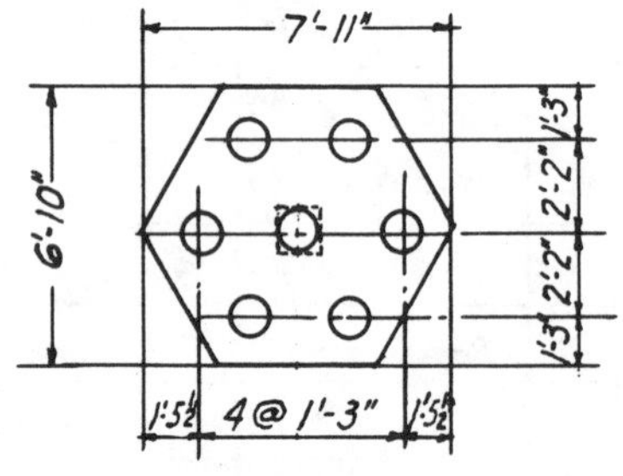	20	122	21	11	14-1/2"φ	9-1/2"φ	10	9-1/2"φ	9-1/2"φ	9	10-1/2"φ	9-1/2"φ
		30	188	27	13	17-1/2"φ	11-1/2"φ	12	11-1/2"φ	10-1/2"φ	11	11-1/2"φ	10-1/2"φ
		40	255	29	14	17-5/8"φ	11-5/8"φ	14	14-1/2"φ	12-1/2"φ	13	13-1/2"φ	12-1/2"φ
		50	325	31	15	20-5/8"φ	13-5/8"φ	15	17-1/2"φ	14-1/2"φ	14	15-1/2"φ	14-1/2"φ
		60	393	32	16	23-5/8"φ	16-5/8"φ	16	20-1/2"φ	15-1/2"φ	15	16-1/2"φ	16-1/2"φ
8	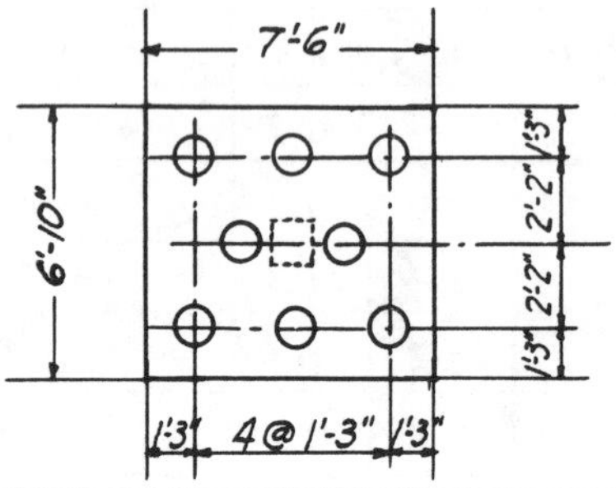	20	142	21	11	13-1/2"φ	13-1/2"φ	10	9-5/8"φ	9-5/8"φ	9	6-3/4"φ	6-3/4"φ
		30	220	24	13	20-1/2"φ	20-1/2"φ	12	11-5/8"φ	11-5/8"φ	11	8-3/4"φ	8-3/4"φ
		40	298	27	14	23-1/2"φ	23-1/2"φ	13	13-5/8"φ	13-5/8"φ	12	10-3/4"	10-3/4"φ
		50	375	30	16	21-5/8"φ	21-5/8"φ	15	14-5/8"φ	14-5/8"φ	14	10-3/4"φ	10-3/4"φ
		60	454	32	18	24-5/8"φ	24-5/8"φ	17	16-5/8"φ	16-5/8"φ	16	11-3/4"φ	11-3/4"φ.

* Based on A.C.I. Code-1947. Special bond bars (A.S.T.M. A-305) not required. Hooks required.

STRUCTURAL FOUNDATIONS – PILE CAPS – 3

PILE FOOTINGS.*

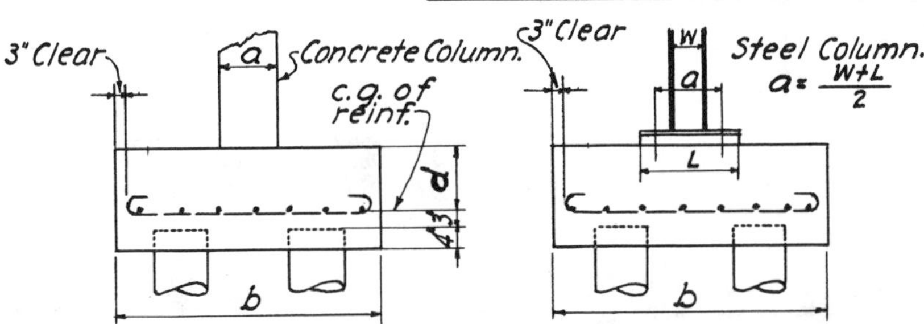
3" Clear
a
Concrete Column.
c.g. of reinf.
d
4
b
3" Clear
W
a
Steel Column.
$a = \frac{W+L}{2}$
L
b

DESIGN DATA.		
f'_c = 2,000#/□"	f'_c = 3,000#/□"	f'_c = 3,750#/□"
f_c = 900#/□"	f_c = 1,350#/□"	f_c = 1,688#/□"
f_s = 20,000#/□"	f_s = 20,000#/□"	f_s = 20,000#/□"
v = 60#/□"	v = 75#/□"	v = 75#/□"
u = 112.5#/□"	u = 168#/□"	u = 200#/□"

Number of Piles	Plan	For All Footings: Pile Value in Kips	For All Footings: Column Load in Kips	d (in.)	f'_c = 2,000#/□": a (in.)	f'_c = 2,000#/□": Reinforcement Long Way	f'_c = 2,000#/□": Reinforcement Short Way	f'_c = 3,000#/□": a (in.)	f'_c = 3,000#/□": Reinforcement Long Way	f'_c = 3,000#/□": Reinforcement Short Way	f'_c = 3,750#/□": a (in.)	f'_c = 3,750#/□": Reinforcement Long Way	f'_c = 3,750#/□": Reinforcement Short Way
9	7'-6" × 7'-6"; 1'-3", 2'-6", 2'-6", 1'-3"	20	160	21	12	15-1/2"φ	15-1/2"φ	10	8-3/4"φ	8-3/4"φ	9	8-3/4"φ	8-3/4"φ
		30	246	27	14	17-1/2"φ	17-1/2"φ	12	8-3/4"φ	8-3/4"φ	11	9-3/4"φ	9-3/4"φ
		40	334	31	16	20-1/2"φ	20-1/2"φ	14	10-3/4"φ	10-3/4"φ	13	10-3/4"φ	10-3/4"φ
		50	421	34	18	23-1/2"φ	23-1/2"φ	16	11-3/4"φ	11-3/4"φ	15	11-3/4"φ	11-3/4"φ
		60	509	36	20	26-1/2"φ	26-1/2"φ	18	12-3/4"φ	12-3/4"φ	17	12-3/4"φ	12-3/4"φ
10	10'-5" × 6'-10"; 1'-3", 2'-2", 2'-2", 1'-3"; 1'-5½", 6 @ 1'-3", 1'-5½"	20	174	25	12	17-1/2"φ	12-1/2"φ	12	11-5/8"φ	7-5/8"φ	11	11-5/8"φ	7-5/8"φ
		30	268	31	14	19-1/2"φ	14-1/2"φ	14	13-5/8"φ	8-5/8"φ	13	13-5/8"φ	8-5/8"φ
		40	366	34	16	24-1/2"φ	18-1/2"φ	15	16-5/8"φ	10-5/8"φ	14	16-5/8"φ	10-5/8"φ
		50	461	37	18	22-5/8"φ	16-5/8"φ	17	17-5/8"φ	11-5/8"φ	16	17-5/8"φ	11-5/8"φ
		60	559	40	20	24-5/8"φ	19-5/8"φ	19	18-5/8"φ	13-5/8"φ	18	18-5/8"φ	13-5/8"φ
11	10'-0" × 6'-10"; 1'-3", 2'-2", 2'-2", 1'-3"; 1'-3", 6 @ 1'-3", 1'-3"	20	191	27	12	14-5/8"φ	14-1/2"φ	11	9-3/4"φ	9-5/8"φ	10	9-3/4"φ	9-5/8"φ
		30	296	33	14	18-5/8"φ	18-1/2"φ	13	11-3/4"φ	10-5/8"φ	12	11-3/4"φ	11-5/8"φ
		40	402	37	16	22-5/8"φ	22-1/2"φ	15	13-3/4"φ	12-5/8"φ	14	10-7/8"φ	12-5/8"φ
		50	509	40	18	25-5/8"φ	26-1/2"φ	17	15-3/4"φ	14-5/8"φ	16	11-7/8"φ	13-5/8"φ
		60	617	43	20	23-3/4"φ	23-5/8"φ	19	16-3/4"φ	16-5/8"φ	18	12-7/8"φ	14-5/8"φ
12	10'-0" × 7'-6"; 1'-3", 2'-6", 2'-6", 1'-3"; 1'-3", 3 @ 2'-6", 1'-3"	20	209	26	14	18-5/8"φ	16-1/2"φ	12	11-3/4"φ	10-5/8"φ	11	11-3/4"φ	10-5/8"φ
		30	322	33	15	21-5/8"φ	18-1/2"φ	14	13-3/4	12-5/8"φ	13	13-3/4"φ	12-5/8"φ
		40	436	39	17	25-5/8"φ	21-1/2"φ	16	14-3/4"φ	13-5/8"φ	15	14-3/4"φ	13-5/8"φ
		50	552	43	19	28-5/8"φ	24-1/2"φ	18	16-3/4"φ	15-5/8"φ	17	16-3/4"φ	15-5/8"φ
		60	669	47	21	25-3/4"φ	26-1/2"φ	20	17-3/4"φ	16-5/8"φ	19	16-3/4"φ	16-5/8"φ

* *Based on A.C.I. Code - 1947. Special bond bars (A.S.T.M. A-305) not required. Hooks required.*

STRUCTURAL FOUNDATIONS–PILE CAPS–4

PILE FOOTINGS.*

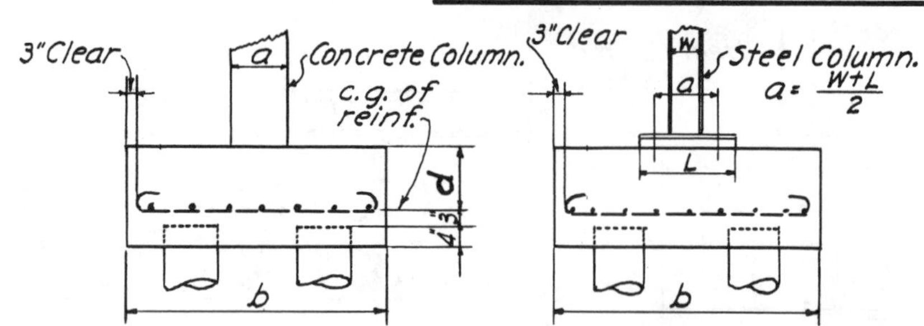
3"Clear
a
Concrete Column.
c.g. of reinf.
d
b
3"Clear
W
a
Steel Column.
a = (W+L)/2
L
b

DESIGN DATA.		
f'_c = 2,000#/□"	f'_c = 3,000#/□"	f'_c = 3,750#/□"
f_c = 900#/□"	f_c = 1,350#/□"	f_c = 1,688#/□"
f_s = 20,000#/□"	f_s = 20,000#/□"	f_s = 20,000#/□"
v = 60#/□"	v = 75#/□"	v = 75#/□"
u = 112.5#/□"	u = 168#/□"	u = 200#/□"

NUMBER OF PILES	PLAN	FOR ALL FOOTINGS: PILE VALUE IN KIPS	FOR ALL FOOTINGS: COLUMN LOAD IN KIPS	d (IN.)	f'_c = 2,000#/□": a (IN.)	f'_c = 2,000#/□": REINFORCEMENT LONG WAY	f'_c = 2,000#/□": REINFORCEMENT SHORT WAY	f'_c = 3,000#/□": a (IN.)	f'_c = 3,000#/□": REINFORCEMENT LONG WAY	f'_c = 3,000#/□": REINFORCEMENT SHORT WAY	f'_c = 3,750#/□": a (IN.)	f'_c = 3,750#/□": REINFORCEMENT LONG WAY	f'_c = 3,750#/□": REINFORCEMENT SHORT WAY
13	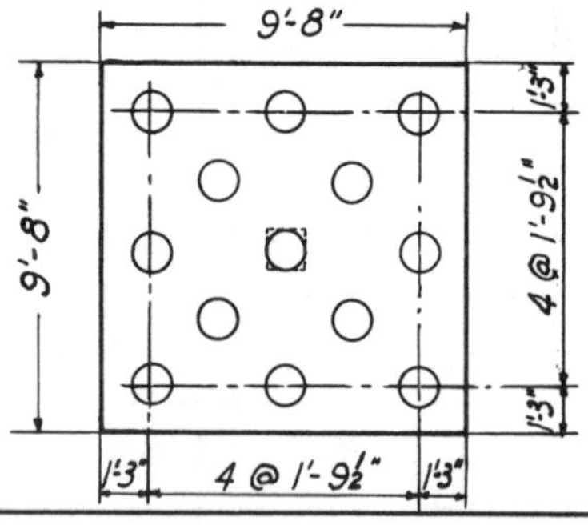	20	221	24	14	16-5/8"φ	16-5/8"φ	12	11-3/4"φ	11-3/4"φ	11	9-7/8"φ	9-7/8"φ
		30	348	29	16	20-5/8"φ	20-5/8"φ	15	14-3/4"φ	14-3/4"φ	14	11-7/8"φ	11-7/8"φ
		40	475	32	18	25-5/8"φ	25-5/8"φ	17	17-3/4"φ	17-3/4"φ	16	13-7/8"φ	13-7/8"φ
		50	602	34	20	30-5/8"φ	30-5/8"φ	19	19-3/4"φ	19-3/4"φ	18	14-7/8"φ	14-7/8"φ
		60	729	37	22	33-5/8"φ	33-5/8"φ	21	20-3/4"φ	20-3/4"φ	19	16-7/8"φ	16-7/8"φ
14	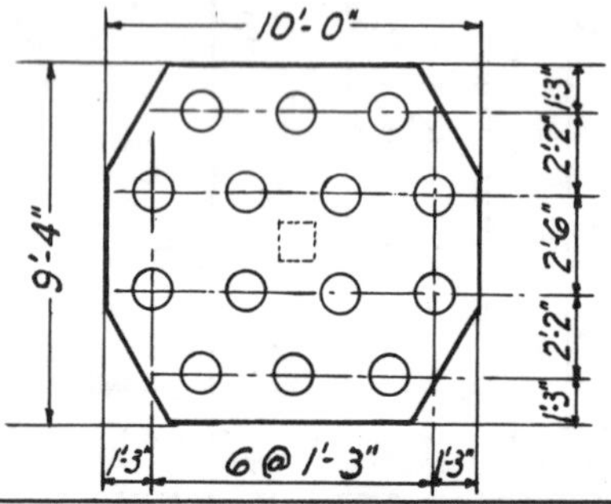	20	242	26	14	18-5/8"φ	20-5/8"φ	13	15-5/8"φ	15-5/8"φ	12	11-3/4"φ	11-3/4"φ
		30	375	32	16	22-5/8"φ	26-5/8"φ	16	18-5/8"φ	18-5/8"φ	14	13-3/4"φ	13-3/4"φ
		40	508	38	18	25-5/8"φ	29-5/8"φ	18	19-5/8"φ	20-5/8"φ	17	14-3/4"φ	14-3/4"φ
		50	643	42	22	27-5/8"φ	30-5/8"φ	20	21-5/8"φ	22-5/8"φ	19	15-3/4"φ	16-3/4"φ
		60	779	46	24	29-5/8"φ	32-5/8"φ	22	22-5/8"φ	23-5/8"φ	20	16-3/4"φ	17-3/4"φ
15	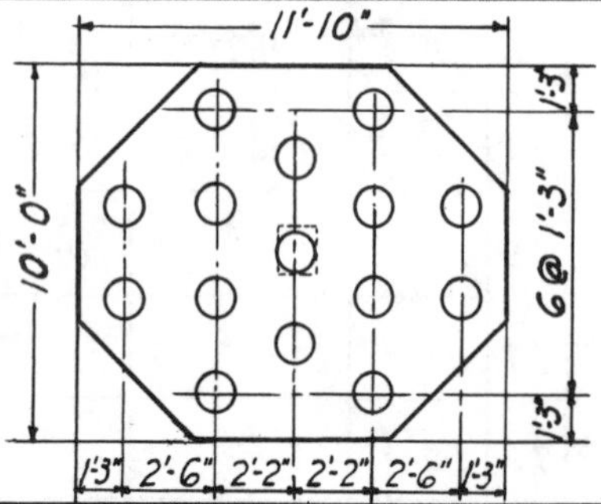	20	254	30	14	16-5/8"φ	22-1/2"φ	13	12-3/4"φ	13-5/8"φ	12	12-3/4"φ	13-5/8"φ
		30	400	34	16	21-5/8"φ	30-1/2"φ	16	11-7/8"φ	16-5/8"φ	15	11-7/8"φ	16-5/8"φ
		40	546	36	20	27-5/8"φ	37-1/2"φ	18	14-7/8"φ	21-5/8"φ	17	11-1"φ	19-5/8"φ
		50	693	38	22	32-5/8"φ	34-5/8"φ	20	16-7/8"φ	24-5/8"φ	19	12-1"φ	22-5/8"φ
		60	841	40	24	36-5/8"φ	37-5/8"φ	22	18-7/8"φ	26-5/8"φ	20	11-1"□	25-5/8"φ
16	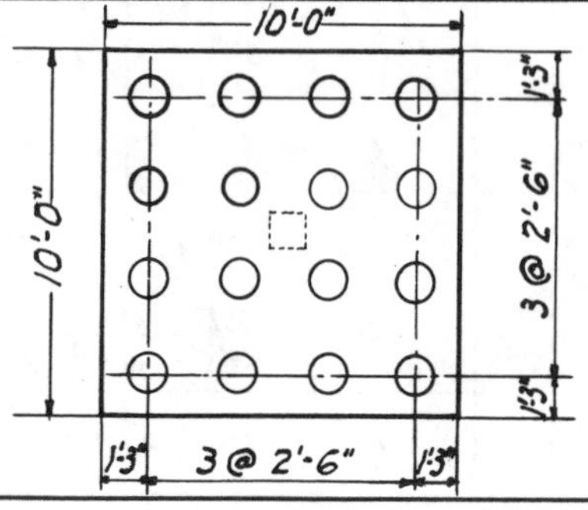	20	279	26	14	24-5/8"φ	24-5/8"φ	14	14-3/4"φ	14-3/4"φ	13	10-7/8"φ	10-7/8"φ
		30	431	32	18	30-5/8"φ	30-5/8"φ	16	17-3/4"φ	17-3/4"φ	15	12-7/8"φ	12-7/8"φ
		40	585	37	20	34-5/8"φ	34-5/8"φ	18	20-3/4"φ	20-3/4"φ	17	14-7/8"φ	14-7/8"φ
		50	739	42	22	36-5/8"φ	36-5/8"φ	20	21-3/4"φ	21-3/4"φ	19	15-7/8"φ	15-7/8"φ
		60	894	46	25	37-5/8"φ	37-5/8"φ	22	22-3/4"φ	22-3/4"φ	21	16-7/8"φ	16-7/8"φ

* Based on A.C.I. Code - 1947. Special bond bars (A.S.T.M. A-305) not required. Hooks required.

STRUCTURAL FOUNDATIONS – PILE CAPS – 5

PILE FOOTINGS.*

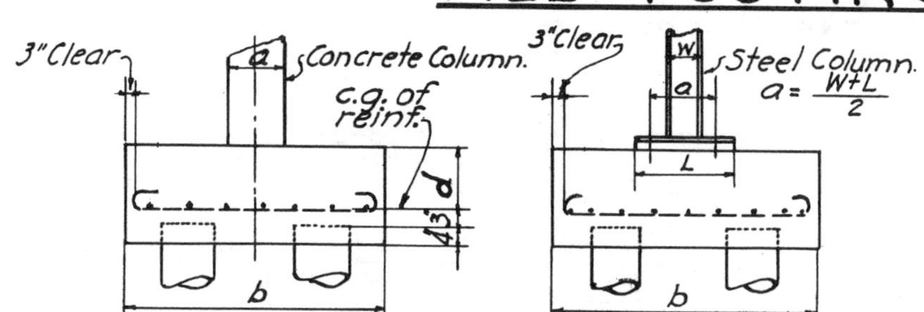
3" Clear
a
Concrete Column.
c.g. of reinf.
d
4 1/2"
b
3" Clear
W
a
Steel Column.
a = (W+L)/2
L
b

DESIGN DATA.

f'_c = 2,000 #/□"	f'_c = 3,000 #/□"	f'_c = 3,750 #/□"
f_c = 900 #/□"	f_c = 1,350 #/□"	f_c = 1,688 #/□"
f_s = 20,000 #/□"	f_s = 20,000 #/□"	f_s = 20,000 #/□"
v = 60 #/□"	v = 75 #/□"	v = 75 #/□"
u = 112.5 #/□"	u = 168 #/□"	u = 200 #/□"

Number of Piles.	Plan	For All Footings: Pile Value in Kips	For All Footings: Column Load in Kips	d (in.)	f'_c = 2,000 #/□": a (in.)	f'_c = 2,000: Reinforcement Long Way	f'_c = 2,000: Reinforcement Short Way	f'_c = 3,000 #/□": a (in.)	f'_c = 3,000: Reinforcement Long Way	f'_c = 3,000: Reinforcement Short Way	f'_c = 3,750 #/□": a (in.)	f'_c = 3,750: Reinforcement Long Way	f'_c = 3,750: Reinforcement Short Way
17	11'-2"; 10'-0"; 8 @ 1'-3"; 1'-3", 4 @ 2'-2", 1'-3"	20	290	32	14	18-5/8"φ	17-5/8"φ	14	9-7/8"φ	10-3/4"φ	13	9-7/8"φ	10-3/4"φ
		30	454	36	18	23-5/8"φ	23-5/8"φ	17	12-7/8"φ	13-3/4"φ	16	12-7/8"φ	9-7/8"φ
		40	620	40	20	28-5/8"φ	28-5/8"φ	19	15-7/8"φ	16-3/4"φ	18	11-1"φ	12-7/8"φ
		50	786	42	24	33-5/8"φ	32-5/8"φ	21	17-7/8"φ	18-3/4"φ	20	13-1"φ	13-7/8"φ
		60	952	45	26	38-5/8"φ	35-5/8"φ	23	19-7/8"φ	20-3/4"φ	21	14-1"φ	15-7/8"φ
18	11'-10"; 10'-0"; 8 @ 1'-3"; 1'-3", 2'-2", 2 @ 2'-6", 2'-2", 1'-3"	20	307	34	14	19-5/8"φ	18-5/8"φ	14	8-1"φ	11-3/4"φ	13	8-1"φ	8-7/8"φ
		30	481	39	18	25-5/8"φ	24-5/8"φ	18	10-1"φ	13-3/4"φ	17	10-1"φ	10-7/8"φ
		40	656	43	22	29-5/8"φ	28-5/8"φ	20	12-1"φ	16-3/4"φ	19	12-1"φ	12-7/8"φ
		50	832	46	24	33-5/8"φ	32-5/8"φ	22	13-1"φ	19-3/4"φ	20	13-1"φ	14-7/8"φ
		60	1009	48	27	36-5/8"φ	35-5/8"φ	24	15-1"φ	21-3/4"φ	22	15-1"φ	15-7/8"φ
19	12'-11"; 6'-5"; 1'-3", 4 @ 2'-2", 1'-3"; 11'-2"	20	327	32	16	19-5/8"φ	19-5/8"φ	14	13-3/4"φ	13-3/4"φ	13	10-7/8"φ	10-7/8"φ
		30	512	37	18	26-5/8"φ	23-5/8"φ	18	16-3/4"φ	16-3/4"φ	17	12-7/8"φ	12-7/8"φ
		40	698	41	22	31-5/8"φ	28-5/8"φ	20	19-3/4"φ	19-3/4"φ	19	14-7/8"φ	14-7/8"φ
		50	886	43	26	36-5/8"φ	34-5/8"φ	22	22-3/4"φ	22-3/4"φ	21	16-7/8"φ	16-7/8"φ
		60	1071	45	27	40-5/8"φ	40-5/8"φ	24	25-3/4"φ	25-3/4"φ	23	18-7/8"φ	18-7/8"φ
20	12'-6"; 10'-0"; 1'-3", 3 @ 2'-6", 1'-3"; 1'-3", 4 @ 2'-6", 1'-3"	20	332	36	16	16-3/4"φ	21-5/8"φ	15	12-7/8"φ	17-5/8"φ	14	9-1"φ	12-3/4"φ
		30	523	42	18	20-3/4"φ	28-5/8"φ	18	15-7/8"φ	21-5/8"φ	17	12-1"φ	15-3/4"φ
		40	719	45	22	25-3/4"φ	32-5/8"φ	20	19-7/8"φ	26-5/8"φ	19	14-1"φ	18-3/4"φ
		50	914	48	26	28-3/4"φ	35-5/8"φ	22	21-7/8"φ	22-3/4"φ	21	16-1"φ	20-3/4"φ
		60	1104	52	28	31-3/4"φ	37-5/8"φ	24	23-7/8"φ	23-3/4"φ	22	18-1"φ	22-3/4"φ

* Based on A.C.I. Code-1947. Special bond bars (A.S.T.M. A-305) not required. Hooks required.

STRUCTURAL FOUNDATIONS - PILE CAPS - I

PILE FOOTINGS PAGE 1 OF 4

Designed according to ACI 1947. Pg. 7-14 to 7-23 are designed for ordinary bond bars, hooks required. Pg. 7-24 to 7-29 designed for special bond bars (A.S.T.M. A-305); hooks not required. See also note bottom p. 7-25.

WARNING - *Check overloading of strata on which piles rest.*

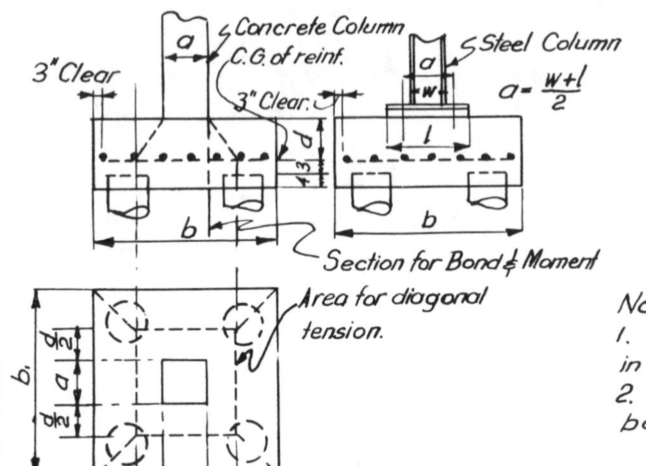

Concrete Column
C.G. of reinf.
3" Clear
3" Clear
Steel Column
a = (w+l)/2
d
1/3 d
b
b
Section for Bond & Moment
Area for diagonal tension.
d/2
a
d/2
b

DESIGN DATA			
f'c = 2000#/□"	f'c = 2500#/□"	f'c = 3000#/□"	f'c = 3750#/□"
fc = 900#/□"	fc = 1125#/□"	fc = 1350#/□"	fc = 1688#/□"
fs = 20,000#/□"	fs = 20,000#/□"	fs = 20,000#/□"	fs = 20,000#/□"
v = 60#/□"	v = 75#/□"	v = 75#/□"	v = 75#/□"
u = 160#/□"	u = 200#/□"	u = 240#/□"	u = 280#/□"

Notes:

1. If depth of pile caps is increased steel perimeter may be decreased in proportion to depth.
2. If size of bars is changed:- Provide equivalent perimeter when larger bars are used, and equivalent area when smaller bars are used.

NUMBER OF PILES	PLAN	FOR ALL FOOTINGS: PILE VALUE (Kips)	COLUMN LOAD (Kips)	d (IN)	f'c = 2000#/□": a (IN)	REINFORCEMENT LONG WAY	REINFORCEMENT SHORT WAY	f'c = 2500#/□": a (IN)	REINFORCEMENT LONG WAY	REINFORCEMENT SHORT WAY	f'c = 3000#/□": a (IN)	REINFORCEMENT LONG WAY	REINFORCEMENT SHORT WAY	f'c = 3750#/□": a (IN)	REINFORCEMENT LONG WAY	REINFORCEMENT SHORT WAY
1		20	18.5	12	—	3-#5*	3-#5	—	3-#5	3-#5	—	3-#5	3-#5	—	3-#5	3-#5
		30	28.5	12	—	3-#5	3-#5	—	3-#5	3-#5	—	3-#5	3-#5	—	3-#5	3-#5
		40	38.5	12	—	3-#5	3-#5	—	3-#5	3-#5	—	3-#5	3-#5	—	3-#5	3-#5
		50	48.5	12	—	3-#5	3-#5	—	3-#5	3-#5	—	3-#5	3-#5	—	3-#5	3-#5
		60	58.5	12	—	3-#5	3-#5	—	3-#5	3-#5	—	3-#5	3-#5	—	3-#5	3-#5
		80	78.5	12	8	3-#5	3-#5	8	3-#5	3-#5	7	3-#5	3-#5	7	3-#5	3-#5
		100	98.5	12	—	3-#5	3-#5	8	3-#5	3-#5	—	3-#5	3-#5	7	3-#5	3-#5
		120	118.5	12	10	3-#5	3-#5	9	3-#5	3-#6	9	3-#5	3-#5	8	3-#5	3-#5
2		20	37	12	7	6-#4	7-#4	7	3-#6	7-#4	7	3-#6	7-#4	7	3-#6	7-#4
		30	57	15	7	7-#4	7-#4	7	5-#5	7-#4	7	3-#6	7-#4	7	3-#6	7-#4
		40	76	17	9	8-#4	7-#4	9	7-#4	7-#4	9	3-#7	7-#4	9	3-#7	7-#4
		50	96	19	9	9-#4	4-#5	9	8-#4	4-#5	9	5-#5	4-#5	9	3-#7	4-#5
		60	116	20	10	10-#4	4-#5	10	9-#4	4-#5	9	6-#5	4-#5	9	4-#6	4-#5
		80	156	21	11	8-#7	4-#5	11	9-#5	4-#5	10	10-#4	4-#5	10	10-#4	4-#5
		100	195	21	12	9-#8	4-#5	11	10-#6	4-#5	11	10-#5	4-#5	10	7-#6	4-#5
		120	235	22	14	9-#9	4-#5	13	10-#7	4-#5	12	9-#6	4-#5	12	10-#5	4-#5
3		20	54	15	8	3 Bands of 4-#4		8	3 Bands of 3-#4		7	3 Bands of 2-#5		7	3 Bands of 2-#5	
		30	83	17	9	" " " 5-#4		9	" " " 4-#4		9	" " " 2-#6		9	" " " 2-#6	
		40	113	19	10	" " " 6-#4		10	" " " 5-#4		9	" " " 3-#5		9	" " " 2-#6	
		50	143	20	11	" " " 7-#4		11	" " " 6-#4		10	" " " 3-#6		9	" " " 2-#7	
		60	172	21	12	" " " 6-#5		11	" " " 6-#4		10	" " " 4-#5		10	" " " 3-#6	
		80	234	19	13	" " " 9-#6		12	" " " 9-#5		11	" " " 8-#5		11	" " " 9-#4	
		100	293	22	14	" " " 9-#6		13	" " " 9-#5		12	" " " 8-#5		11	" " " 9-#4	
		120	353	23	16	" " " 9-#6		15	" " " 9-#5		14	" " " 8-#5		13	" " " 9-#4	
4		20	72	14	8	10-#4	10-#4	8	9-#4	9-#4	8	7-#4	7-#4	8	5-#5	5-#5
		30	112	15	10	15-#4	15-#4	10	12-#4	12-#4	9	10-#4	10-#4	9	7-#5	7-#5
		40	152	16	12	19-#4	19-#4	11	15-#4	15-#4	10	13-#4	13-#4	9	7-#6	7-#6
		50	192	17	12	18-#5	18-#5	12	18-#4	18-#4	11	15-#4	15-#4	10	10-#5	10-#5
		60	232	18	13	16-#6	16-#6	13	17-#5	17-#5	12	17-#4	17-#4	12	12-#5	12-#5
		80	312	18	15	19-#7	19-#7	15	21-#5	21-#5	14	22-#4	22-#4	13	19-#4	19-#4
		100	392	19	17	18-#9	18-#9	15	18-#7	18-#7	15	21-#5	21-#5	14	18-#5	18-#5
		120	472	20	18	16-#11	16-#11	16	18-#8	18-#8	15	17-#7	17-#7	14	21-#5	21-#5

* #1–#8 = 1/8"φ – 8/8"φ; #9, 10, 11 ≡ 1"□, 1 1/8"□, 1 1/4"□

STRUCTURAL FOUNDATIONS – PILE CAPS – 2

PILE FOOTINGS*

PAGE 2 OF 4

For basis of design and design data see P. 7-24.

NUMBER OF PILES	PLAN	FOR ALL FOOTINGS			f'c = 2000#/□"			f'c = 2500#/□"			f'c = 3000#/□"			f'c = 3750#/□"		
		PILE VALUE kips	COLUMN LOAD kips	d In.	a In.	REINFORCEMENT LONG WAY	REINFORCEMENT SHORT WAY	a In.	REINFORCEMENT LONG WAY	REINFORCEMENT SHORT WAY	a In	REINFORCEMENT LONG WAY	REINFORCEMENT SHORT WAY	a In	REINFORCEMENT LONG WAY	REINFORCEMENT SHORT WAY
5	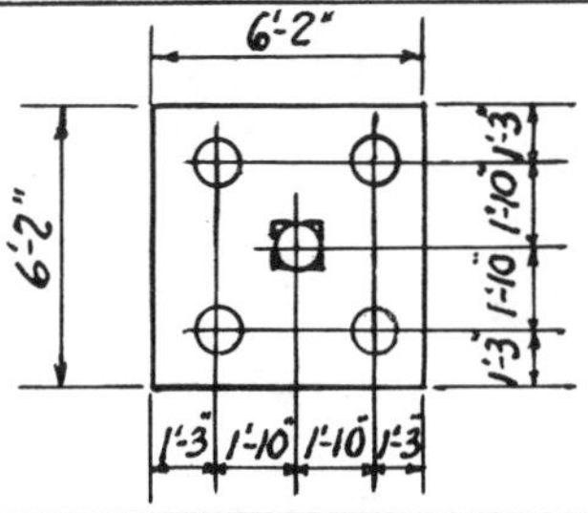 6'-2" × 6'-2"; 1'-3, 1'-10, 1'-10, 1'-3	20	90	15	10	10-#4**	10-#4	10	7-#5	7-#5	9	7-#5	7-#5	9	7-#5	7-#5
		30	137	18	11	13-#4	13-#4	11	8-#5	8-#5	10	6-#6	6-#6	10	6-#6	6-#6
		40	186	21	13	14-#4	14-#4	12	9-#5	9-#5	11	7-#6	7-#6	10	5-#7	5-#7
		50	235	22	13	13-#5	13-#5	13	11-#5	11-#5	12	8-#6	8-#6	11	6-#7	6-#7
		60	284	24	14	15-#5	15-#5	14	12-#5	12-#5	14	7-#7	7-#7	12	6-#7	6-#7
		80	385	24	17	25-#4	25-#4	16	16-#5	16-#5	14	11-#6	11-#6	14	8-#7	8-#7
		100	485	25	18	20-#6	20-#6	17	19-#5	19-#5	16	14-#6	14-#6	14	10-#7	10-#7
		120	584	26	19	23-#6	23-#6	18	22-#5	22-#5	17	19-#5	19-#5	15	12-#7	12-#7
6	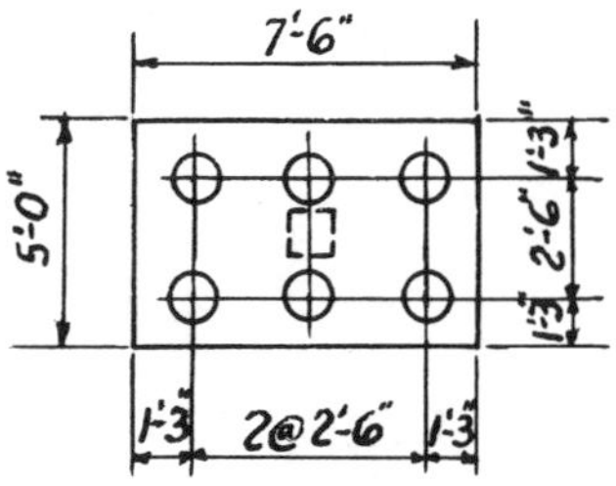 7'-6" × 5'-0"; 1'-3, 2@2'-6", 1'-3; 1'-3, 2'-6", 1'-3	20	106	22	10	5-#6	10-#4	10	5-#6	8-#4	9	5-#6	7-#4	8	5-#6	6-#4
		30	164	27	12	6-#6	12-#4	11	4-#7	10-#4	10	4-#7	8-#4	10	4-#7	7-#4
		40	222	32	14	6-#6	14-#4	13	6-#6	11-#4	12	5-#7	9-#4	11	5-#7	8-#4
		50	280	35	14	7-#6	16-#4	14	5-#7	13-#4	13	5-#7	11-#4	13	4-#8	9-#4
		60	338	36	16	8-#6	15-#5	16	6-#7	15-#4	15	6-#7	13-#4	14	4-#9	11-#4
		80	459	38	18	11-#6	23-#4	16	7-#7	19-#4	15	6-#8	16-#4	14	4-#10	14-#4
		100	578	40	19	12-#6	26-#4	18	11-#6	22-#4	17	7-#8	19-#4	15	4-#11	16-#4
		120	698	39	22	19-#5	29-#4	20	13-#6	25-#4	19	10-#7	23-#4	17	7-#9	20-#4
7	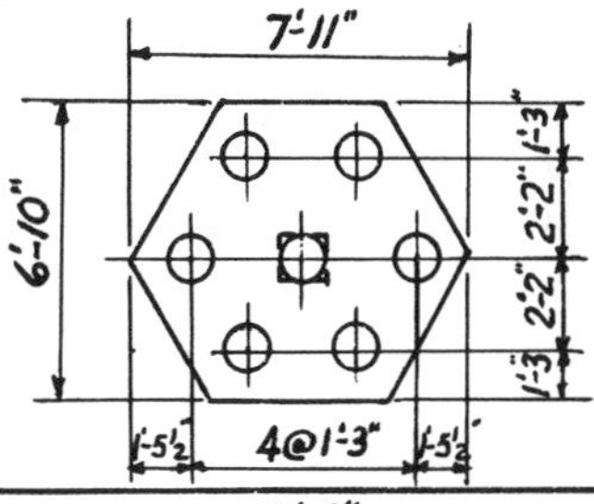 7'-11" × 6'-10"; 1'-5½", 4@1'-3", 1'-5½"; 1'-3, 2'-2", 2'-2", 1'-3	20	122	21	11	10-#4	6-#5	11	9-#4	6-#5	10	6-#5	6-#5	9	5-#6	6-#5
		30	188	27	13	12-#4	7-#5	13	10-#4	7-#5	12	7-#5	7-#5	11	5-#6	7-#5
		40	255	29	14	15-#4	8-#5	14	12-#4	8-#5	14	8-#5	6-#6	13	6-#6	6-#6
		50	325	31	15	18-#4	10-#5	15	15-#4	9-#5	15	10-#5	6-#6	14	7-#6	7-#6
		60	393	32	16	16-#5	11-#5	16	17-#4	10-#5	16	11-#5	7-#6	15	8-#6	7-#6
		80	534	32	19	22-#5	20-#4	17	22-#4	10-#6	16	15-#5	7-#7	15	10-#6	7-#6
		100	674	32	21	18-#7	23-#4	20	18-#5	16-#5	19	19-#5	11-#6	17	17-#5	7-#8
		120	812	34	23	17-#8	26-#4	21	19-#6	17-#5	20	20-#5	12-#6	18	18-#5	10-#7
8	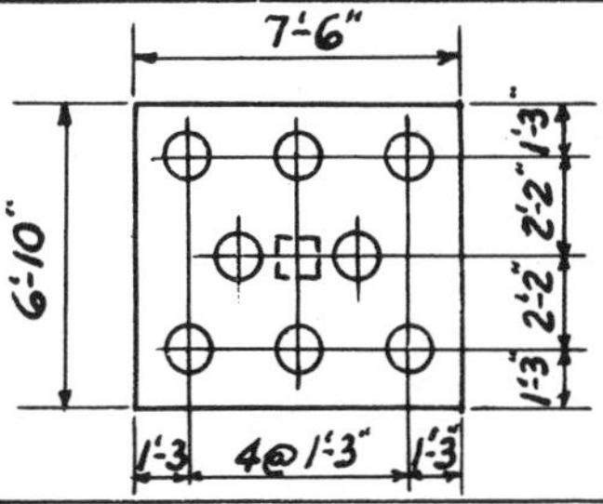 7'-6" × 6'-10"; 1'-3, 4@1'-3", 1'-3; 1'-3, 2'-2", 2'-2", 1'-3	20	142	21	11	8-#5	8-#5	11	6-#6	6-#6	10	6-#6	6-#6	9	6-#6	6-#6
		30	220	24	13	11-#5	11-#5	13	8-#6	8-#6	12	6-#7	6-#7	11	6-#7	6-#7
		40	298	27	14	13-#5	13-#5	14	9-#6	9-#6	13	7-#7	7-#7	12	7-#7	7-#7
		50	375	30	16	15-#5	15-#5	16	10-#6	10-#6	15	7-#7	7-#7	14	6-#8	6-#8
		60	455	32	18	17-#5	17-#5	18	15-#5	15-#5	17	8-#7	8-#7	16	6-#8	6-#8
		80	615	32	20	27-#4	29-#4	18	18-#5	19-#5	17	13-#6	11-#7	16	11-#7	11-#7
		100	774	33	22	26-#5	34-#4	21	21-#5	22-#5	20	15-#6	16-#6	18	12-#7	10-#8
		120	934	33	26	29-#5	33-#5	24	24-#5	26-#5	22	17-#6	18-#6	21	13-#7	14-#7
9	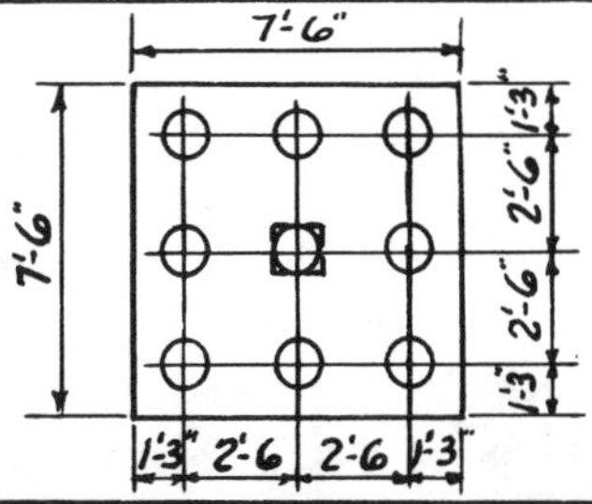 7'-6" × 7'-6"; 1'-3", 2'-6, 2'-6, 1'-3"; 1'-3", 2'-6", 2'-6", 1'-3"	20	160	21	12	7-#6	7-#6	11	7-#6	7-#6	10	6-#7	6-#7	9	6-#7	6-#7
		30	246	27	14	8-#6	8-#6	13	6-#7	6-#7	12	6-#7	6-#7	11	6-#7	6-#7
		40	333	31	16	8-#7	8-#7	15	7-#7	7-#7	14	7-#7	7-#7	13	7-#7	7-#7
		50	421	34	18	14-#5	14-#5	17	8-#7	8-#7	16	6-#8	6-#8	15	6-#8	6-#8
		60	509	36	20	15-#5	15-#5	19	11-#6	11-#6	18	8-#7	8-#7	17	7-#8	7-#8
		80	690	36	22	20-#5	20-#5	20	11-#7	11-#7	19	11-#7	11-#7	17	7-#9	7-#9
		100	868	37	23	24-#5	24-#5	21	16-#6	16-#6	20	10-#8	10-#8	19	8-#9	8-#9
		120	1049	37	27	29-#5	29-#5	25	19-#6	19-#6	23	14-#7	14-#7	22	11-#8	11-#8
10	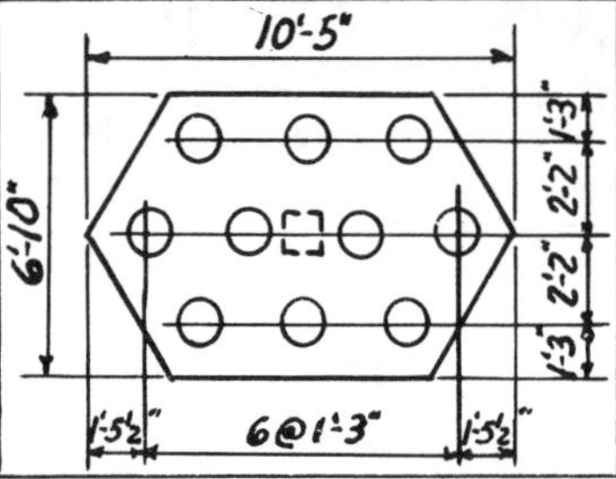 10'-5" × 6'-10"; 1'-5½", 6@1'-3", 1'-5½"; 1'-3, 2'-2", 2'-2", 1'-3	20	174	25	12	8-#6	7-#5	12	6-#7	7-#5	12	6-#7	7-#5	11	6-#7	7-#5
		30	268	31	14	9-#6	8-#5	14	7-#7	6-#6	14	7-#7	8-#5	13	7-#7	8-#5
		40	366	34	16	11-#6	10-#5	16	8-#7	7-#6	15	6-#8	7-#6	14	8-#7	7-#6
		50	461	37	18	13-#6	12-#5	18	12-#6	8-#6	17	7-#8	8-#6	16	7-#8	8-#6
		60	559	40	20	14-#6	17-#4	20	10-#7	11-#5	19	7-#8	8-#6	18	6-#9	8-#6
		80	769	35	22	26-#5	25-#4	21	15-#7	16-#5	19	11-#8	12-#6	18	9-#9	10-#6
		100	970	35	26	31-#5	32-#4	24	22-#6	20-#5	23	14-#8	14-#6	21	11-#9	11-#7
		120	1169	36	28	26-#7	37-#4	26	24-#6	31-#4	24	18-#7	17-#6	23	12-#9	12-#7

*Based on A.C.I. Code-1947 with bond stresses as recommended by A.C.I. Committee 208 for deformed bars complying with A.S.T.M. A-305, 49. Hooks not required. **#1–#8 = 1/8"φ–8/8"φ; #9, 10, 11 ≡ 1"□, 1 1/8"□, 1 1/4"□

STRUCTURAL FOUNDATIONS – PILE CAPS – 3

PILE FOOTINGS* PAGE 3 OF 4

For basis of design and design data see P. 7-24.

NUMBER OF PILES	PLAN	FOR ALL FOOTINGS			f'_c = 2000#/□"			f'_c = 2500#/□"			f'_c = 3000#/□"			f'_c = 3750#/□"		
		PILE VALUE (Kips)	COLUMN LOAD (Kips)	d (IN)	a (IN)	REINFORCEMENT LONG WAY	REINFORCEMENT SHORT WAY	a (IN)	REINFORCEMENT LONG WAY	REINFORCEMENT SHORT WAY	a (IN)	REINFORCEMENT LONG WAY	REINFORCEMENT SHORT WAY	a (IN)	REINFORCEMENT LONG WAY	REINFORCEMENT SHORT WAY
11	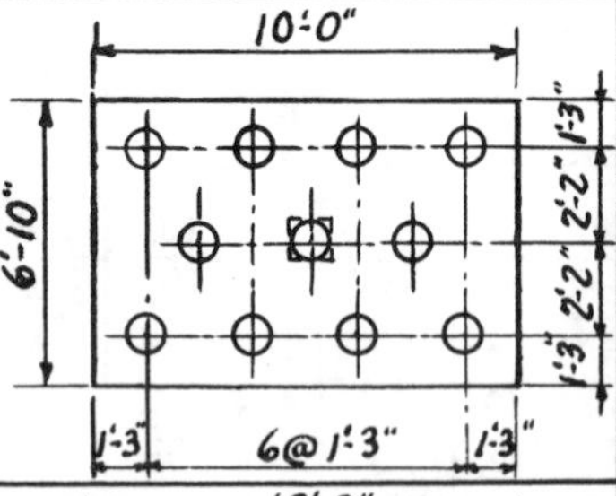 10'-0" × 6'-10"; 6@1'-3"; 1'-3", 2'-2", 2'-2", 1'-3"	20	191	27	12	7-#7**	9-#5	12	7-#7	9-#5	11	7-#7	9-#5	10	7-#7	9-#5
		30	296	33	14	11-#6	10-#5	14	8-#7	10-#5	13	8-#7	10-#5	12	8-#7	10-#5
		40	402	37	16	13-#6	12-#5	16	9-#7	9-#6	15	7-#8	8-#6	14	7-#8	9-#6
		50	509	40	18	19-#5	20-#4	18	11-#7	10-#6	17	10-#7	9-#6	16	6-#9	9-#6
		60	617	43	20	20-#5	21-#4	20	12-#7	14-#5	19	11-#7	8-#7	18	7-#9	8-#7
		80	836	45	23	25-#5	26-#4	21	14-#7	17-#5	20	11-#8	12-#6	18	9-#9	9-#7
		100	1054	47	27	27-#5	31-#4	25	16-#7	20-#5	23	12-#8	14-#6	22	8-#10	11-#7
		120	1270	51	29	29-#5	35-#4	27	21-#6	28-#4	25	13-#8	20-#5	23	10-#9	12-#7
12	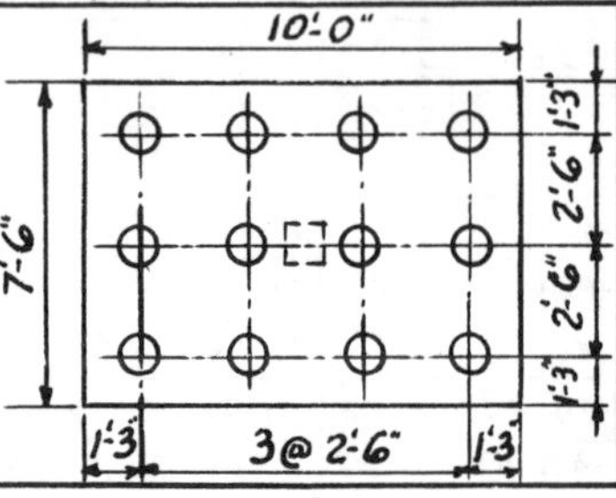 10'-0" × 7'-6"; 3@2'-6"; 1'-3", 2'-6", 2'-6", 1'-3"	20	209	26	14	11-#6	10-#5	13	8-#7	10-#5	12	6-#8	10-#5	11	8-#7	8-#6
		30	322	33	15	13-#6	9-#6	15	9-#7	9-#6	14	7-#8	9-#6	13	6-#9	9-#6
		40	436	39	17	19-#5	10-#6	17	11-#7	9-#6	16	10-#7	10-#6	15	7-#9	10-#6
		50	552	43	19	16-#6	14-#5	19	12-#7	8-#7	18	11-#7	8-#7	17	7-#9	8-#7
		60	669	47	21	22-#5	15-#5	21	16-#6	11-#6	20	11-#7	8-#7	19	7-#9	8-#7
		80	902	55	24	23-#5	17-#5	22	13-#7	10-#7	21	10-#8	9-#7	19	8-#9	7-#8
		100	1141	56	27	26-#5	27-#4	25	15-#7	14-#6	24	12-#8	10-#7	22	8-#10	8-#8
		120	1379	58	30	28-#5	30-#4	27	17-#7	20-#5	26	13-#8	12-#7	24	11-#9	9-#8
13	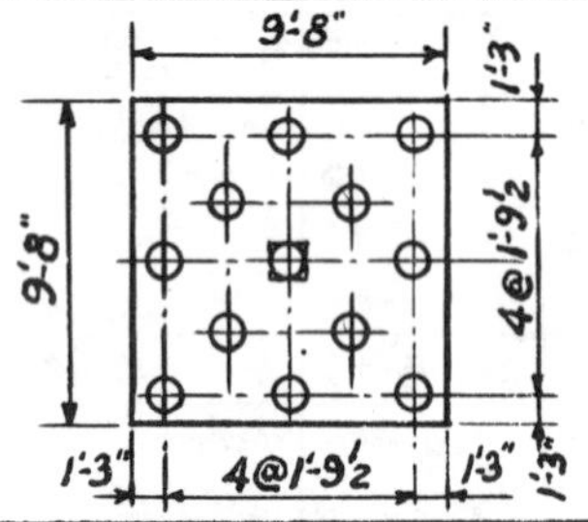 9'-8" × 9'-8"; 4@1'-9½"; 1'-3"	20	221	24	14	8-#7	8-#7	13	8-#7	8-#7	12	8-#7	8-#7	11	11-#6	11-#6
		30	348	29	16	11-#7	11-#7	16	10-#7	10-#7	15	10-#7	10-#7	14	10-#7	10-#7
		40	475	32	18	16-#6	16-#6	18	12-#7	12-#7	17	9-#8	9-#8	16	8-#9	8-#9
		50	602	34	20	18-#6	18-#6	20	11-#8	11-#8	19	9-#9	9-#9	18	9-#9	9-#9
		60	729	37	22	20-#6	20-#6	22	15-#7	15-#7	21	9-#9	9-#9	19	9-#9	9-#9
		80	983	42	26	28-#5	28-#5	24	16-#7	16-#7	23	13-#8	13-#8	21	7-#11	7-#11
		100	1240	44	29	34-#5	34-#5	26	19-#7	19-#7	25	15-#8	15-#8	23	10-#10	10-#10
		120	1496	48	30	37-#5	37-#5	28	21-#7	21-#7	26	16-#8	16-#8	24	13-#9	13-#9
14	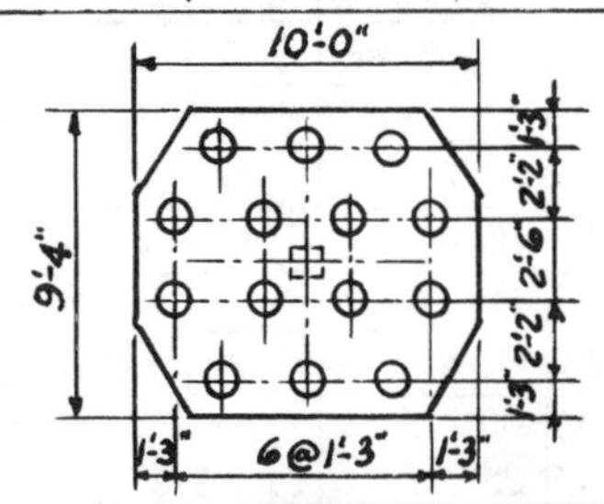 10'-0" × 9'-4"; 6@1'-3"; 1'-3", 2'-2", 2'-6", 2'-2", 1'-3"	20	242	26	14	10-#6	15-#5	14	8-#7	10-#6	13	8-#7	8-#7	12	8-#7	8-#7
		30	375	32	16	13-#6	18-#5	16	12-#6	13-#6	16	7-#8	9-#7	14	10-#7	10-#7
		40	508	38	18	19-#5	21-#5	18	13-#6	18-#5	18	10-#7	10-#7	17	10-#7	10-#7
		50	643	42	22	20-#5	22-#5	21	14-#6	19-#5	20	11-#7	11-#7	19	7-#9	9-#8
		60	779	46	24	21-#5	23-#5	23	15-#6	20-#5	22	11-#7	12-#7	20	7-#9	11-#7
		80	1056	48	27	26-#5	34-#4	25	18-#6	24-#5	23	12-#8	18-#6	22	10-#8	11-#8
		100	1335	49	29	30-#5	38-#4	27	22-#6	28-#5	26	16-#7	20-#6	24	11-#9	13-#8
		120	1613	50	31	35-#5	34-#5	29	24-#6	30-#5	27	19-#7	22-#6	25	12-#9	18-#7
15	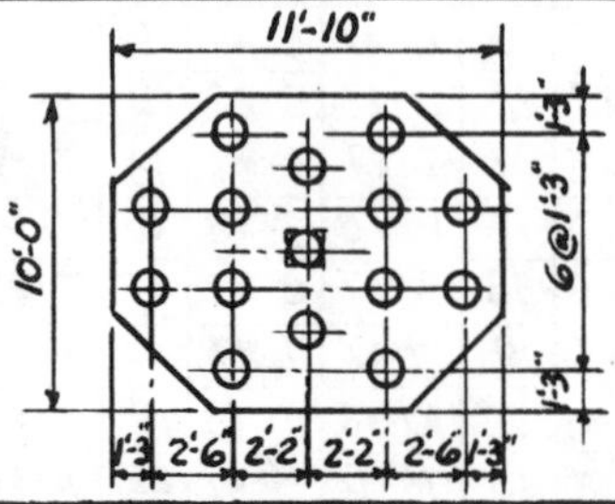 11'-10" × 10'-0"; 6@1'-3"; 1'-3", 2'-6", 2'-2", 2'-2", 2'-6", 1'-3"	20	254	30	14	8-#7	13-#5	14	11-#6	9-#6	13	9-#7	9-#6	12	9-#7	9-#6
		30	400	34	16	11-#7	17-#5	16	11-#7	12-#6	16	11-#7	11-#6	15	11-#7	9-#7
		40	546	36	20	14-#7	21-#5	19	10-#8	18-#5	18	8-#9	13-#6	17	8-#9	10-#7
		50	693	38	22	16-#7	24-#5	21	12-#8	20-#5	20	8-#10	12-#7	19	8-#10	12-#7
		60	841	40	24	22-#6	26-#5	23	17-#7	22-#5	22	10-#9	16-#6	20	11-#9	13-#7
		80	1139	44	27	26-#6	36-#4	25	19-#7	25-#5	24	12-#9	18-#6	22	8-#11	14-#7
		100	1439	44	30	33-#6	34-#5	28	24-#7	29-#5	26	15-#9	21-#6	24	10-#11	17-#7
		120	1732	49	32	36-#6	34-#5	30	25-#7	30-#5	28	16-#9	22-#6	26	11-#11	17-#7
16	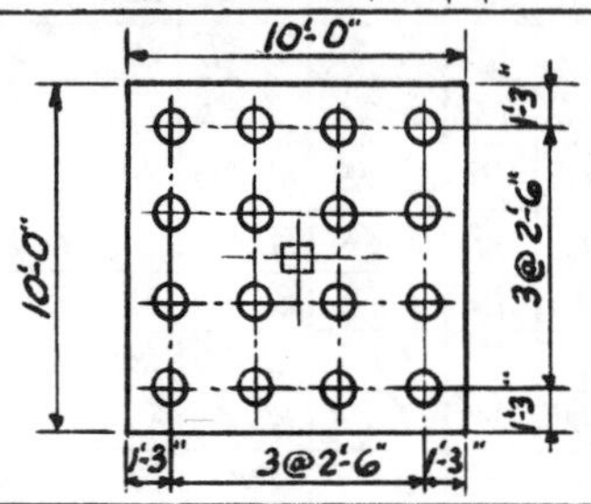 10'-0" × 10'-0"; 3@2'-6"; 1'-3"	20	279	26	14	14-#6	14-#6	14	10-#7	10-#7	14	10-#7	10-#7	13	8-#8	8-#8
		30	431	32	18	18-#6	18-#6	17	16-#6	16-#6	16	12-#7	12-#7	15	8-#9	8-#9
		40	585	37	20	25-#5	25-#5	19	14-#7	14-#7	18	10-#9	10-#9	17	9-#9	9-#9
		50	739	42	22	21-#6	21-#6	21	15-#7	15-#7	20	11-#8	11-#8	19	9-#9	9-#9
		60	894	46	25	21-#6	21-#6	24	19-#6	19-#6	22	15-#7	15-#7	21	8-#10	8-#10
		80	1205	53	28	29-#5	29-#5	26	17-#7	17-#7	25	13-#8	13-#8	23	13-#8	13-#8
		100	1524	54	31	33-#5	33-#5	28	20-#7	20-#7	27	15-#8	15-#8	25	12-#9	12-#9
		120	1842	55	33	29-#6	29-#6	31	22-#7	22-#7	29	17-#8	17-#8	27	14-#9	14-#9

*Based on A.C.I. Code-1947 with bond stresses as recommended by A.C.I. Committee 208 for deformed bars complying with A.S.T.M. A-305, 49. Hooks not required. ** #1-#8 = ⅛"φ-⅞"φ; #9,10,11 ≡ 1"□, 1⅛"□, 1¼"□.

STRUCTURAL FOUNDATIONS – PILE CAPS – 4

PILE FOOTINGS* PAGE 4 OF 4

For basis of design and design data see P. 7-24.

NUMBER OF PILES	PLAN	FOR ALL FOOTINGS			f'c = 2000#/□"			f'c = 2500#/□"			f'c = 3000#/□"			f'c = 3750#/□"		
		PILE VALUE (Kips)	COLUMN LOAD (Kips)	d (IN)	a (IN)	REINFORCEMENT LONG WAY	REINFORCEMENT SHORT WAY	a (IN)	REINFORCEMENT LONG WAY	REINFORCEMENT SHORT WAY	a (IN)	REINFORCEMENT LONG WAY	REINFORCEMENT SHORT WAY	a (IN)	REINFORCEMENT LONG WAY	REINFORCEMENT SHORT WAY
17	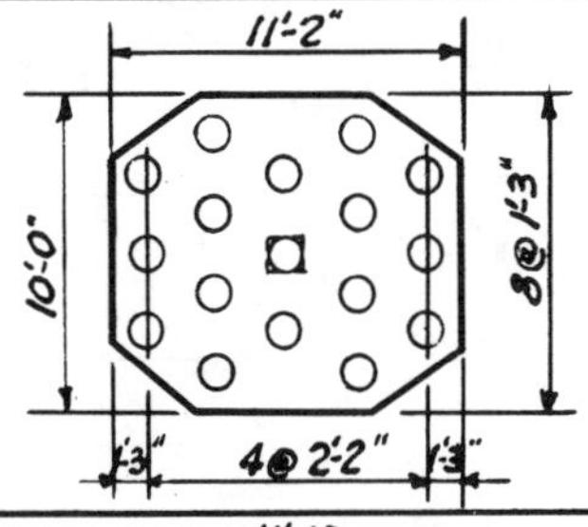	20	290	32	14	9-#7**	10-#6	14	9-#7	9-#6	14	9-#7	9-#6	13	10-#7	8-#7
		30	454	36	18	12-#7	14-#6	18	9-#8	10-#7	17	9-#8	9-#7	16	8-#9	10-#7
		40	620	40	20	14-#7	20-#5	20	9-#9	14-#6	19	9-#9	11-#7	18	9-#9	11-#7
		50	786	42	24	17-#7	22-#5	23	16-#7	16-#6	21	8-#10	12-#7	20	8-#10	8-#9
		60	952	45	26	23-#6	24-#5	25	14-#8	17-#6	23	11-#9	13-#7	21	9-#10	8-#9
		80	1291	48	29	28-#6	30-#5	27	17-#8	20-#6	25	13-#9	16-#7	23	9-#11	13-#8
		100	1628	51	32	34-#6	34-#5	29	24-#7	23-#6	27	15-#9	18-#7	25	11-#11	14-#8
		120	1965	53	34	39-#6	38-#5	31	27-#7	26-#6	29	21-#8	20-#7	27	14-#10	16-#8
18	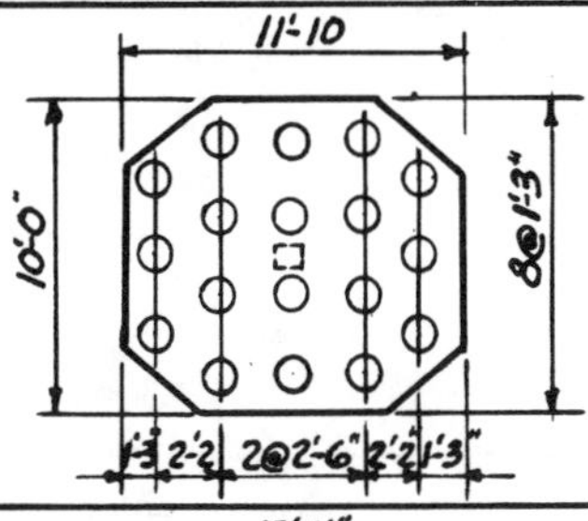	20	307	34	14	10-#7	9-#7	15	9-#7	9-#7	14	10-#7	10-#6	13	10-#7	11-#6
		30	481	39	18	16-#6	12-#7	18	12-#7	9-#8	18	8-#9	10-#7	17	8-#9	10-#7
		40	652	43	22	18-#6	14-#7	21	14-#7	11-#8	20	9-#9	11-#7	19	8-#10	9-#8
		50	828	46	26	21-#6	20-#6	24	15-#7	16-#7	22	13-#8	10-#8	20	11-#9	10-#8
		60	1009	48	27	23-#6	23-#6	26	17-#7	14-#8	24	9-#10	11-#8	22	8-#11	9-#9
		80	1360	52	30	22-#7	29-#5	27	14-#9	21-#6	26	11-#10	13-#8	24	10-#11	13-#8
		100	1719	53	32	28-#7	34-#5	30	17-#9	24-#6	28	13-#10	19-#7	26	11-#11	15-#8
		120	2075	56	35	37-#6	38-#5	32	23-#8	27-#6	30	15-#10	20-#7	27	13-#11	14-#9
19	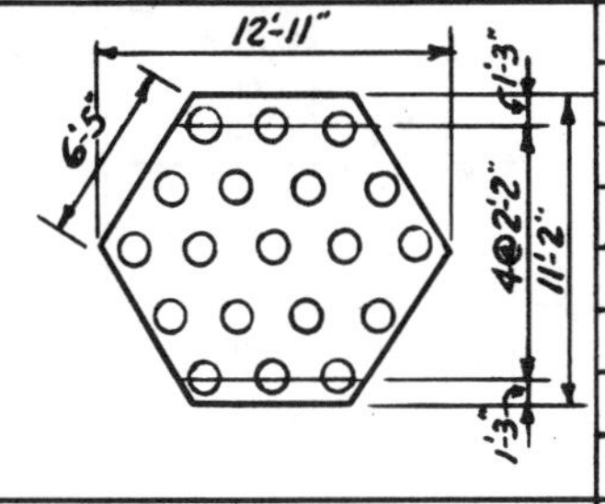	20	327	32	16	10-#7	9-#7	15	9-#7	9-#7	14	10-#7	10-#7	13	10-#7	10-#7
		30	512	37	18	16-#6	12-#7	18	12-#7	9-#8	18	9-#8	9-#8	17	9-#8	9-#8
		40	694	41	22	18-#6	14-#7	21	14-#7	11-#8	20	9-#9	9-#9	19	9-#9	9-#9
		50	883	43	26	21-#6	20-#6	24	15-#7	16-#7	22	12-#8	12-#8	21	10-#9	10-#9
		60	1071	45	27	23-#6	23-#6	26	17-#7	14-#8	24	11-#9	11-#9	23	11-#9	11-#9
		80	1441	51	30	27-#6	27-#6	28	19-#7	20-#7	26	15-#8	20-#7	24	10-#10	10-#11
		100	1823	52	33	40-#5	33-#6	30	23-#7	21-#8	29	12-#10	15-#9	26	12-#10	10-#11
		120	2194	56	36	34-#6	37-#6	33	25-#7	26-#7	31	19#8	19-#8	29	13-#10	13-#10
20	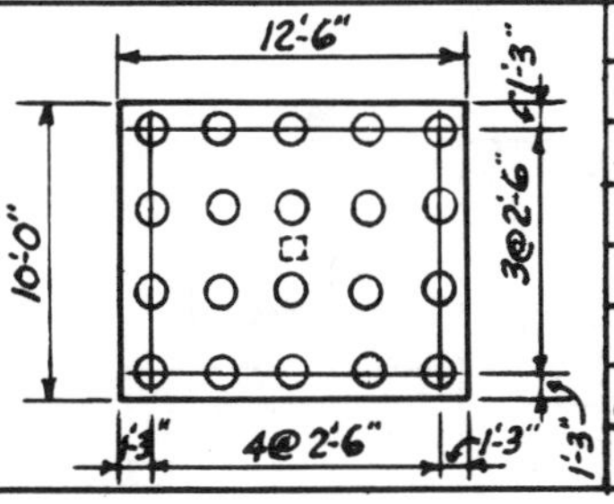	20	332	36	16	11-#7	12-#6	16	9-#8	9-#7	15	9-#8	9-#7	14	9-#8	9-#7
		30	523	42	18	9-#9	16-#6	18	9-#9	14-#6	18	9-#9	11-#7	17	9-#9	11-#7
		40	719	45	22	11-#9	24-#5	21	11-#9	17-#6	20	11-#9	10-#8	19	11-#9	10-#8
		50	914	48	26	15-#8	26-#5	24	10-#10	19-#6	22	8-#11	15-#7	21	10-#10	9-#9
		60	1104	52	28	21-#7	27-#5	26	11-#10	16-#7	24	9-#11	12-#8	22	10-#11	10-#9
		80	1506	53	31	22-#8	27-#6	28	14-#10	20-#7	26	11-#11	15-#8	25	14-#10	12-#9
		100	1902	55	34	31-#7	38-#5	31	20-#9	22-#7	29	13-#11	17-#8	27	14-#11	14-#9
		120	2297	59	37	34-#7	30-#6	34	21-#9	23-#7	32	15-#11	18-#8	29	15-#11	15-#9

*Based on A.C.I. Code 1947 with bond stresses as recommended by A.C.I. Committee 208 for deformed bars complying with A.S.T.M. A.-305,49. Hooks not required. ** #1–#8 = $\frac{1}{8}$"ϕ – $\frac{8}{8}$"ϕ; #9,10,11 = 1"□, 1$\frac{1}{8}$"□, 1$\frac{1}{4}$"□

STRUCTURAL FOUNDATIONS-PILE CAPS-I

PILE FOOTINGS PAGE 1 OF 2

Designed according to A.C.I. Code, 1956. Pile spacing 3'-0" on center. (Preferred for Friction Bearing Piles.) For 2'-6" spacing, see pp. 7-24 through 7-27. Special bond bars (A.S.T.M. A-305) used; hooks not required.

DESIGN DATA
f'c = 3000 p.s.i.
fc = 1350 p.s.i.
v = 75 p.s.i.
u = 240 p.s.i.

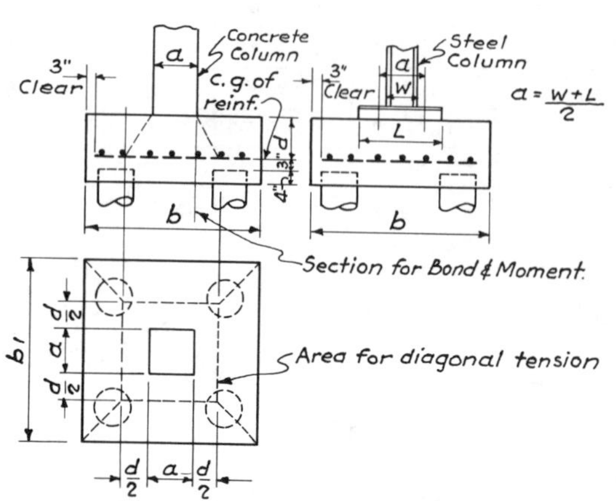

Concrete Column
3" Clear
c.g. of reinf.
Steel Column
a = (W+L)/2
Section for Bond & Moment
Area for diagonal tension

No. of Piles	Plan	Pile Value, kips	Column Load, kips	d, in.	a, in.	Reinforcement (f'c = 3000 p.s.i.) Long Way	Short Way
1		40	38.5	12	–	3-#5	3-#5
		50	48.5	12	–	3-#5	3-#5
		60	58.5	12	–	3-#5	3-#5
		80	78.5	12	7	3-#5	3-#5
2		40	76	18	9	6-#5	4-#5
		50	95	20	9	6-#5	4-#5
		60	115	21	10	6-#6	4-#5
		80	154	26	7	6-#7	6-#5
3		40	112	21	10	3 bands 6-#5	
		50	142	23	10	3 bands 6-#5	
		60	171	24	11	3 bands 7-#5	
		80	233	26	11	3 bands 6-#6	
4		40	150	20	10	9-#5	9-#5
		50	190	20	11	11-#5	11-#5
		60	229	21	12	12-#5	12-#5
		80	308	24	12	10-#7	10-#7

No. of Piles	Plan	Pile Value, kips	Column Load, kips	d, in.	a, in.	Reinforcement (f'c = 3000 p.s.i.) Long Way	Short Way
5		40	183	21	11	9-#6	9-#6
		50	233	23	12	14-#5	14-#5
		60	282	25	14	10-#6	10-#6
		80	380	28	14	16-#5	16-#5
6		40	218	30	12	9-#6	8-#5
		50	276	34	13	10-#6	9-#5
		60	335	36	15	11-#6	10-#5
		80	454	38	15	9-#8	12-#5
7		40	257	29	14	8-#6	7-#6
		50	325	32	15	9-#6	8-#6
		60	393	35	16	9-#6	9-#6
		80	533	35	16	17-#5	11-#6
8		40	293	26	14	12-#6	12-#6
		50	370	29	15	13-#6	10-#7
		60	449	31	17	14-#6	15-#6
		80	606	35	17	13-#7	13-#7

STRUCTURAL FOUNDATIONS–PILE CAPS–2

PILE FOOTINGS* PAGE 2 OF 2

No. of Piles	Plan	Pile Value, kips	Column Load, kips	d, in.	a, in.	$f'_c = 3000$ p.s.i. Reinforcement Long Way	Short Way
9	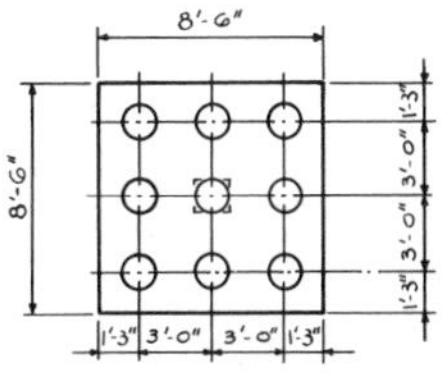	40	327	29	14	13-#6	13-#6
		50	414	33	16	14-#6	14-#6
		60	502	35	18	11-#7	11-#7
		80	677	41	19	13-#7	13-#7
10	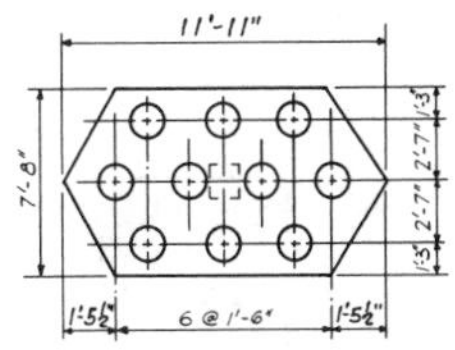	40	360	36	15	10-#7	9-#6
		50	457	39	17	11-#7	10-#6
		60	554	42	19	12-#7	10-#6
		80	751	46	19	9-#9	9-#7
11	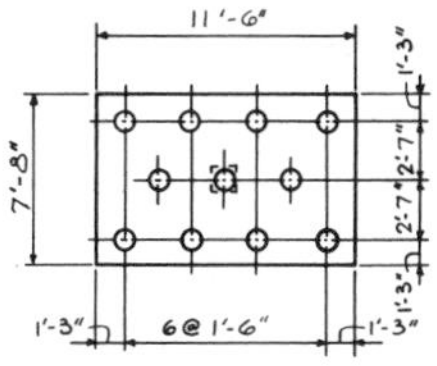	40	393	36	15	10-#8	8-#7
		50	497	41	17	13-#7	12-#6
		60	604	44	19	14-#7	9-#7
		80	819	48	20	11-#9	9-#8
12	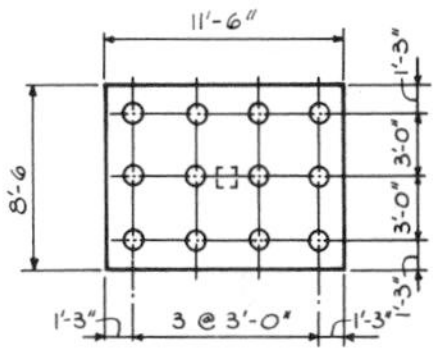	40	429	35	16	9-#9	10-#7
		50	543	40	18	12-#8	8-#8
		60	657	45	20	10-#9	11-#7
		80	890	50	21	15-#8	13-#7
13	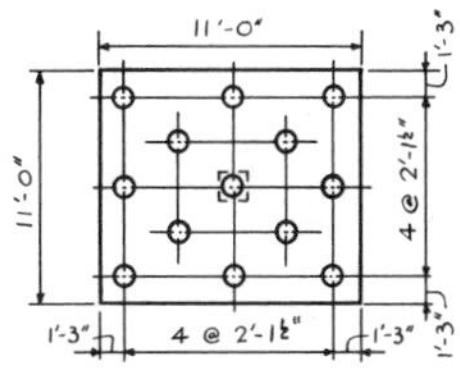	40	461	32	17	12-#8	12-#8
		50	586	35	19	18-#7	18-#7
		60	713	37	21	15-#8	15-#8
		80	969	40	23	14-#9	14-#9
14	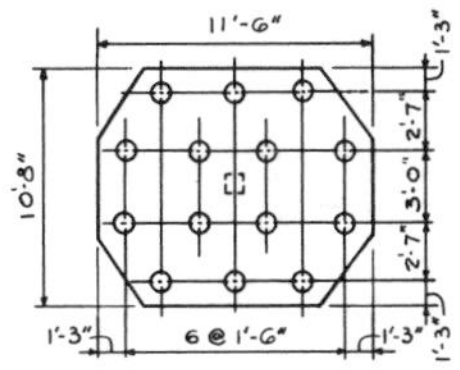	40	503	35	18	11-#8	11-#8
		50	636	40	20	12-#8	11-#9
		60	772	43	22	12-#8	13-#8
		80	1044	49	23	15-#8	18-#7

No. of Piles	Plan	Pile Value, kips	Column Load, kips	d, in.	a, in.	$f'_c = 3000$ p.s.i. Reinforcement Long Way	Short Way
15	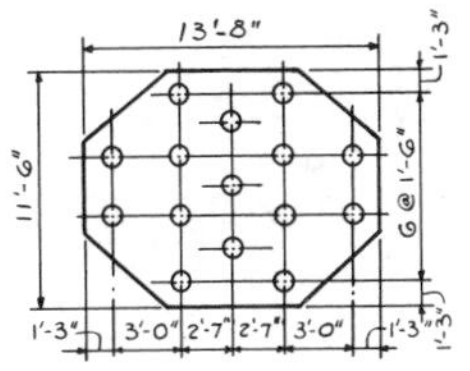	40	529	37	18	11-#8	10-#8
		50	674	40	20	15-#8	11-#8
		60	821	42	22	13-#9	16-#7
		80	1114	46	24	20-#8	18-#7
16	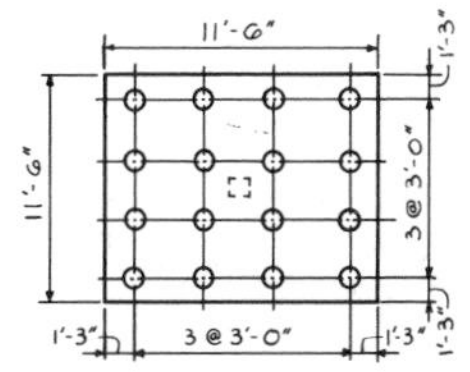	40	570	35	18	15-#8	15-#8
		50	722	40	20	14-#7	14-#7
		60	877	43	22	13-#9	13-#9
		80	1186	50	25	18-#8	18-#8
17	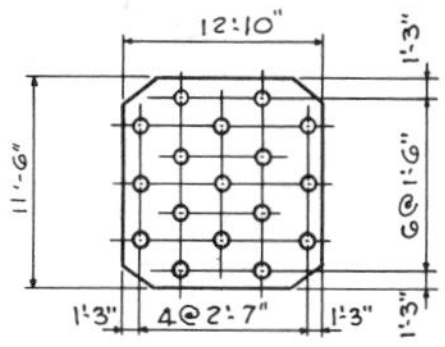	40	599	41	19	11-#9	8-#9
		50	764	44	21	16-#8	12-#8
		60	931	46	23	14-#9	13-#8
		80	1264	50	25	17-#9	16-#8
18	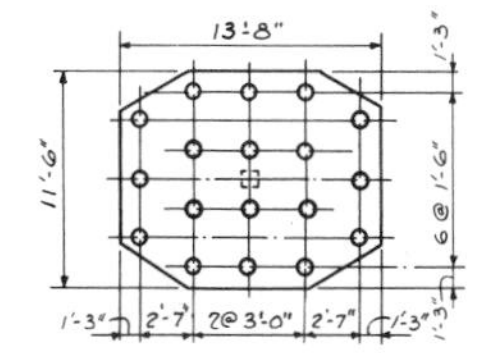	40	630	43	20	14-#9	9-#9
		50	803	47	22	13-#9	10-#9
		60	980	49	24	15-#9	14-#9
		80	1332	53	26	18-#9	17-#9
19	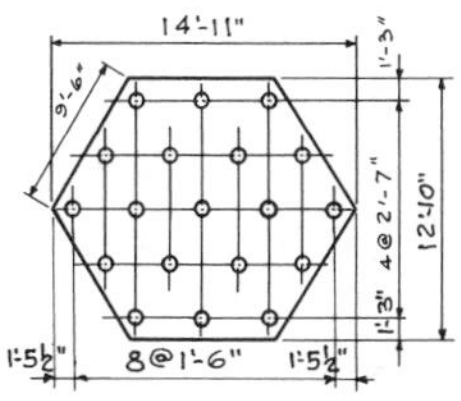	40	673	41	20	11-#9	11-#9
		50	858	44	22	16-#8	16-#8
		60	1044	46	24	14-#9	14-#9
		80	1415	51	26	21-#8	21-#8
20	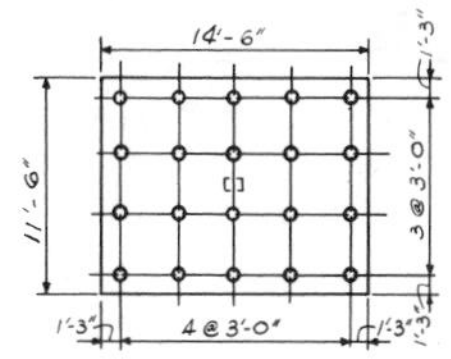	40	687	47	20	13-#9	10-#9
		50	879	51	22	15-#9	11-#9
		60	1075	53	24	18-#9	16-#8
		80	1465	58	26	17-#10	19-#8

* Based on A.C.I. Code, 1956.

STRUCTURAL FOUNDATIONS—RETAINING WALLS-I

CANTILEVER AND GRAVITY TYPE RETAINING WALLS

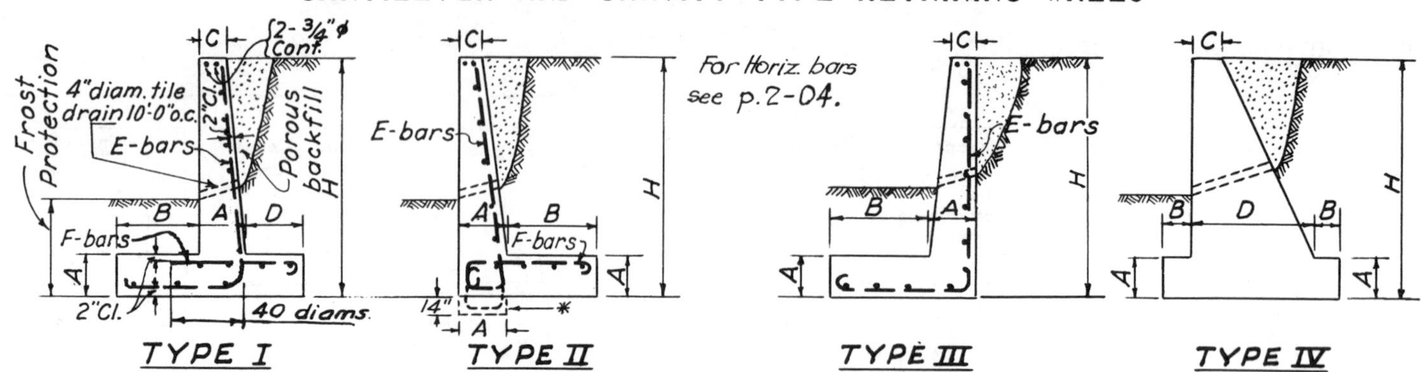

C
2-3/4"φ Cont.
Frost Protection
4"diam. tile drain 10'-0" o.c.
2"Cl.
E-bars
Porous backfill
H
B
A
D
F-bars
A
2"Cl.
40 diams.
TYPE I
E-bars
F-bars
14"
A
*
TYPE II
For Horiz. bars see p.2-04.
E-bars
TYPE III
TYPE IV

TYPE I							
H (ft.-in.)	A (in.)	B (ft.-in.)	D (ft.-in.)	C (in.)	Toe Pressure p.s.i.	E-Bars (in.)	F-Bars (in.)
5-0	8	1-2	0-6	8	765	#3 – 12 o.c.	#3 – 18 o.c.
6-0	8	1-5	0-8	8	865	#3 – 12 o.c.	#3 – 18 o.c.
7-0	8	1-8	0-10	8	930	#3 – 9 o.c.	#3 – 18 o.c.
8-0	12	1-11	0-9	8	1125	#3 – 11½ o.c.	#3 – 18 o.c.
9-0	12	2-2	1-0	8	1230	#3 – 7½ o.c.	#3 – 18 o.c.
10-0	12	2-5	1-2	8	1315	#4 – 10. o.c.	#3 – 12 o.c.
11-0	12	2-8	1-5	8	1420	#5 – 10½ o.c.	#3 – 12 o.c.
12-0	12	2-11	1-8	8	1515	#6 – 12 o.c.	#4 – 14 o.c.
13-0	12	3-2	1-11	10	1630	#7 – 12 o.c.	#4 – 12 o.c.
14-0	12	3-5	2-2	10	1735	#8 – 12½ o.c.	#4 – 10 o.c.
15-0	14	3-8	2-3	12	1895	#8 – 12½ o.c.	#4 – 12 o.c.
16-0	15	3-11	2-4	12	2010	#8 – 11½ o.c.	#4 – 10 o.c.
17-0	16	4-2	2-6	12	2130	#9 – 13 o.c.	#4 – 9½ o.c.
18-0	17	4-4	2-7	12	2260	#9 – 11½ o.c.	#4 – 9½ o.c.
20-0	19	4-10	2-11	12	2510	#9 – 10 o.c.	#5 – 10½ o.c.
22-0	21	5-4	3-3	12	2750	#9 – 8 o.c.	#6 – 11½ o.c.
24-0	24	5-10	3-5	12	3020	#9 – 7½ o.c.	#6 – 11½ o.c.
26-0	26	6-4	3-8	12	3240	#9 – 6 o.c.	#6 – 11 o.c.
28-0	28	6-10	4-0	12	3500	#9 – 5½ o.c.	#6 – 9 o.c.
30-0	31	7-3	4-2	12	3780	#9 – 5 o.c.	#6 – 9 o.c.

TYPE II						
H (ft.-in.)	A (in.)	B (ft.-in.)	C (in.)	Toe Pressure p.s.i.	E-Bars (in.)	F-Bars (in.)
5-0	8	2-1	8	1170	#3 – 12 o.c.	#3 – 12 o.c.
6-0	8	2-7	8	1380	#3 – 12 o.c.	#3 – 12 o.c.
7-0	8	3-2	8	1610	#3 – 9 o.c.	#3 – 8 o.c.
8-0	12	3-7	8	1825	#3 – 11½ o.c.	#3 – 8 o.c.
9-0	12	4-0	8	2030	#4 – 14 o.c.	#4 – 10½ o.c.
10-0	12	4-7	8	2230	#4 – 10 o.c.	#5 – 11 o.c.
*11-0	12	5-2	8	2440	#5 – 10½ o.c.	#6 – 12 o.c.
*12-0	12	5-9	8	2640	#6 – 12 o.c.	#7 – 12 o.c.
*13-0	12	6-4	10	2850	#7 – 12 o.c.	#7 – 9½ o.c.
*14-0	12	6-11	10	3050	#7 – 9½ o.c.	#8 – 10 o.c.
*15-0	14	7-3	12	3290	#8 – 12½ o.c.	#8 – 10½ o.c.
*16-0	15	7-8	12	3510	#8 – 11½ o.c.	#9 – 12 o.c.
*17-0	16	8-3	12	3730	#9 – 13 o.c.	#9 – 10 o.c.
*18-0	18	8-6	12	3950	#9 – 12½ o.c.	#9 – 10 o.c.
20-0	20	9-6	12	4390	#9 – 10 o.c.	#9 – 8 o.c.
22-0	22	10-5	12	4820	#9 – 8½ o.c.	#9 – 7 o.c.
24-0	25	11-3	12	5270	#9 – 7½ o.c.	#9 – 6 o.c.
26-0	28	12-2	12	5690	#9 – 7 o.c.	#9 – 5½ o.c.
28-0	30	13-1	12	6160	#9 – 6 o.c.	#9 – 4½ o.c.
30-0	33	13-10	12	6580	#9 – 5½ o.c.	#9 – 4 o.c.

TYPE III					
H (ft.-in.)	A (in.)	B (ft.-in.)	C (in.)	Toe Pressure p.s.i.	E-Bars (in.)
5-0	8	1-8	8	572	#3 – 12 o.c.
6-0	8	2-5	8	545	#3 – 12 o.c.
7-0	8	3-3	8	525	#4 – 14 o.c.
8-0	12	3-2	8	720	#3 – 10 o.c.
9-0	12	4-1	8	694	#4 – 12 o.c.
10-0	12	5-0	8	675	#5 – 12½ o.c.
11-0	12	6-0	8	657	#6 – 13 o.c.
12-0	12	7-1	8	640	#6 – 10 o.c.
13-0	12	7-9	10	677	#7 – 11 o.c.
14-0	12	9-0	10	657	#8 – 11 o.c.
15-0	14	8-10	12	800	#8 – 11 o.c.
16-0	15	9-6	12	838	#9 – 12 o.c.
17-0	16	10-3	12	875	#9 – 11 o.c.
18-0	18	10-9	12	955	#9 – 10½ o.c.
20-0	20	12-2	12	1030	#9 – 9 o.c.
22-0	22	13-8	12	1104	#9 – 7½ o.c.
24-0	25	14-10	12	1225	#9 – 6½ o.c.
26-0	28	16-2	12	1328	#9 – 6 o.c.
28-0	30	17-5	12	1425	#9 – 5 o.c.
30-0	33	18-7	12	1545	#9 – 4½ o.c.

TYPE IV					
H (ft.-in.)	A (ft.-in.)	B (in.)	D (ft.-in.)	C (in.)	Toe Pressure p.s.i.
5-0	1-0	6	1-4	8	1046
6-0	1-0	6	1-8	8	1360
7-0	1-0	6	2-4	8	1540
8-0	1-0	6	2-10	8	1760
9-0	1-0	6	3-4	8	2040
10-0	1-0	6	3-10	12	2300
11-0	1-0	6	4-4	12	2560
12-0	1-0	6	4-10	12	2800
13-0	1-0	6	5-5	12	3040
14-0	1-0	6	5-11	12	3140
15-0	1-0	6	6-6	12	3500
16-0	1-0	6	7-0	16	3780
17-0	1-0	6	7-6	16	4040
18-0	1-0	6	8-0	16	4340
20-0	1-0	6	9-4	16	4750

Design based on weight of earth 100 p.c.f., and angle of repose assumed 33° and no surcharge. Designed for 2000-lb. controlled c rete. A low water-cement ratio recommended for permanency. Will pass for New York City Class B concrete.

The resultant pressure on walls above is at the outer edge of middle third.

Alternate vertical E-bars in types I, II, and III walls may be cut at $\frac{1}{2}$ H.

Expansion joints in walls should not be over 75'-0" on center; construction joints 30'0" on center.

For additional data on retaining walls, se p. 9-32.

STRUCTURAL FOUNDATIONS-RETAINING WALLS-2

BASEMENT RETAINING WALLS

fc′ = 2,000 p.s.i., fs = 20,000 p.s.i.

TABLE A – REINFORCEMENT FOR EARTH PRESSURE (DRY)

Angle of repose ϕ = 33°.
Earth W = 100 p.c.f.
⅜" ϕ 18" o.c. horizontal
See A.C.I. requirements, p. 2-04 for exposed walls.
Vertical steel in table.

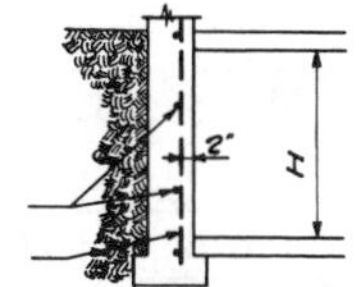

2'
H

H ft.-in.	12" Wall in.	16" Wall in.	20" Wall in.
8-0	#3 – 12 o.c.	#3 – 12 o.c.	#3 – 12 o.c.
9-0	#3 – 12 o.c.	#3 – 12 o.c.	#3 – 12 o.c.
10-0	#3 – 10 o.c.	#3 – 12 o.c.	#3 – 12 o.c.
11-0	#3 – 7½ o.c.	#3 – 11 o.c.	#3 – 12 o.c.
12-0	#4 – 10½ o.c.	#3 – 8 o.c.	#3 – 10½ o.c.
13-0	#4 – 8 o.c.	#4 – 12 o.c.	#3 – 8½ o.c.
14-0	#5 – 10 o.c.	#4 – 9 o.c.	#4 – 12 o.c.
15-0	#5 – 8 o.c.	#5 – 11 o.c.	#4 – 9½ o.c.
16-0	#6 – 10 o.c.	#5 – 9½ o.c.	#4 – 7½ o.c.
17-0	#7 – 11 o.c.	#6 – 11½ o.c.	#5 – 10 o.c.
18-0	#7 – 9½ o.c.	#6 – 9½ o.c.	#5 – 8½ o.c.
19-0	#8 – 10½ o.c.	#7 – 11 o.c.	#6 – 10½ o.c.
20-0	#9 – 11½ o.c.	#7 – 9½ o.c.	#6 – 9 o.c.
21-0		#8 – 11 o.c.	#7 – 10½ o.c.

TABLE B – REINFORCEMENT FOR SATURATED EARTH PRESSURE

Saturated earth,
Assumed equivalent fluid weight w′ = 75 p.c.f.
Horizontal bars – same as table.
Vertical steel in table.

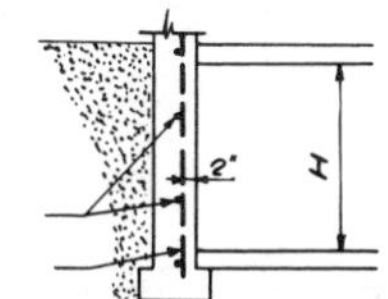

2'
H

H ft.-in.	12" Wall in.	16" Wall in.	20" Wall in.
8-0	#3 – 7½ o.c.	#3 – 11 o.c.	#3 – 12 o.c.
9-0	#4 – 10 o.c.	#3 – 7½ o.c.	#3 – 9½ o.c.
10-0	#5 – 11 o.c.	#4 – 10 o.c.	#3 – 7 o.c.
11-0	#6 – 12 o.c.	#5 – 11½ o.c.	#4 – 9½ o.c.
12-0	#7 – 12 o.c.	#5 – 9 o.c.	#5 – 11½ o.c.
13-0	#7 – 10 o.c.	#6 – 10 o.c.	#5 – 9 o.c.
14-0	#8 – 10½ o.c.	#7 – 11 o.c.	#6 – 10 o.c.
15-0		#8 – 12 o.c.	#7 – 11½ o.c.
16-0		#9 – 12 o.c.	#7 – 9½ o.c.
17-0		#9 – 10 o.c.	#8 – 10½ o.c.
18-0		#9 – 8½ o.c.	#9 – 11 o.c.
19-0			#9 – 9½ o.c.
20-0			#10 – 10 o.c.
21-0			#11 – 11 o.c.

Table A is computed without surcharge: Mom. = 0.128 hp. (See Table C.)

With surcharge, increase Mom. by ratio $\frac{P + SWH}{P}$

W = weight of soil, S = height of surcharge in feet.

TABLE C – EARTH PRESSURES

Angle of repose ϕ = 33°
W = 100 p.c.f.

H
P

$P = \frac{1}{2} WH^2 \frac{1 - \sin\phi}{1 + \sin\phi}$

per ft. length of wall (Rankine).

H, ft.	P, lb.	P/sq. ft. max.	H, ft.	P, lb.	P/sq. ft. max.
5	370	147	14	2,890	413
6	530	177	16	3,770	472
7	720	206	18	4,780	531
8	940	236	20	5,900	590
9	1200	266	22	7,130	649
10	1470	295	24	8,490	708
11	1780	324	26	9,960	766
12	2120	354	28	11,560	825
13	2490	383	30	13,260	885

TABLE D – SATURATED EARTH PRESSURE

W′ = 75 p.c.f.
$P = \frac{1}{2} W' H^2$
Per ft. length of wall.

H, ft.	P, lb.	P/sq. ft. max.	H, ft.	P, lb.	P/sq. ft. max.
1	38	75	12	5,400	900
2	150	150	13	6,330	975
3	338	225	14	7,350	1050
4	600	300	15	8,440	1125
5	940	375	16	9,600	1200
6	1350	450	17	10,800	1275
7	1840	525	18	12,150	1350
8	2400	600	19	13,500	1425
9	3040	675	20	15,000	1500
10	3750	750	21	16,600	1575
11	4530	825	22	18,100	1650

TABLE E – COAL PRESSURES

Total pressure for depth H for bituminous coal on vertical walls.
Weight per cu. ft. 50 lb. ϕ = 35°

H, ft.	P, lb.	P/sq. ft. max.	H, ft.	P, lb.	P/sq. ft. max.
5	170	68	11	820	149
6	240	81	12	980	163
7	330	95	14	1330	190
8	430	108	16	1730	217
9	550	122	18	2190	244
10	680	135	20	2710	271

NOTE: Use Table A for permanent dry earth only. Interpolate between Tables A and B for seasonal rains resulting in saturated earth against walls.

For angles of repose, see p. 9-08; for additional data on retaining walls, see p. 9-32.

STRUCTURAL FOUNDATIONS — BEARING PILES & SHEET PILING

TABLE A – H-BEARING PILES – DIMENSIONS AND PROPERTIES FOR DESIGNING

Section Number and Nominal Size, in.	Weight per foot, lb.	Area of Section A, in.²	Depth of Section d, in.	Flange Width b, in.	Flange Thickness t, in.	Web Thickness W, in.	Axis X-X I, in.⁴	Axis X-X S, in.³	Axis X-X r, in.	Axis Y-Y I′, in.⁴	Axis Y-Y S′, in.³	Axis Y-Y r′, in.
BP-14 14 × 14½	117	34.44	14.234	14.885	0.805	0.805	1228.5	172.6	5.97	443.1	59.5	3.59
	102	30.01	14.032	14.784	0.704	0.704	1055.1	150.4	5.93	379.6	51.3	3.56
	89	26.19	13.856	14.696	0.616	0.616	909.1	131.2	5.89	326.2	44.4	3.53
	73	21.46	13.636	14.586	0.506	0.506	733.1	107.5	5.85	261.9	35.9	3.49
BP-12 12 × 12	74	21.76	12.122	12.217	0.607	0.607	566.5	93.5	5.10	184.7	30.2	2.91
	53	15.58	11.780	12.046	0.436	0.436	394.8	67.0	5.03	127.3	21.2	2.86
BP-10 10 × 10	57	16.76	10.012	10.224	0.564	0.564	294.7	58.9	4.19	100.6	19.7	2.45
	42	12.35	9.720	10.078	0.418	0.418	210.8	43.4	4.13	71.4	14.2	2.40
BP-8 8 × 8	36	10.60	8.026	8.158	0.446	0.446	119.8	29.9	3.36	40.4	9.9	1.95

TABLE B – STEEL SHEET PILING

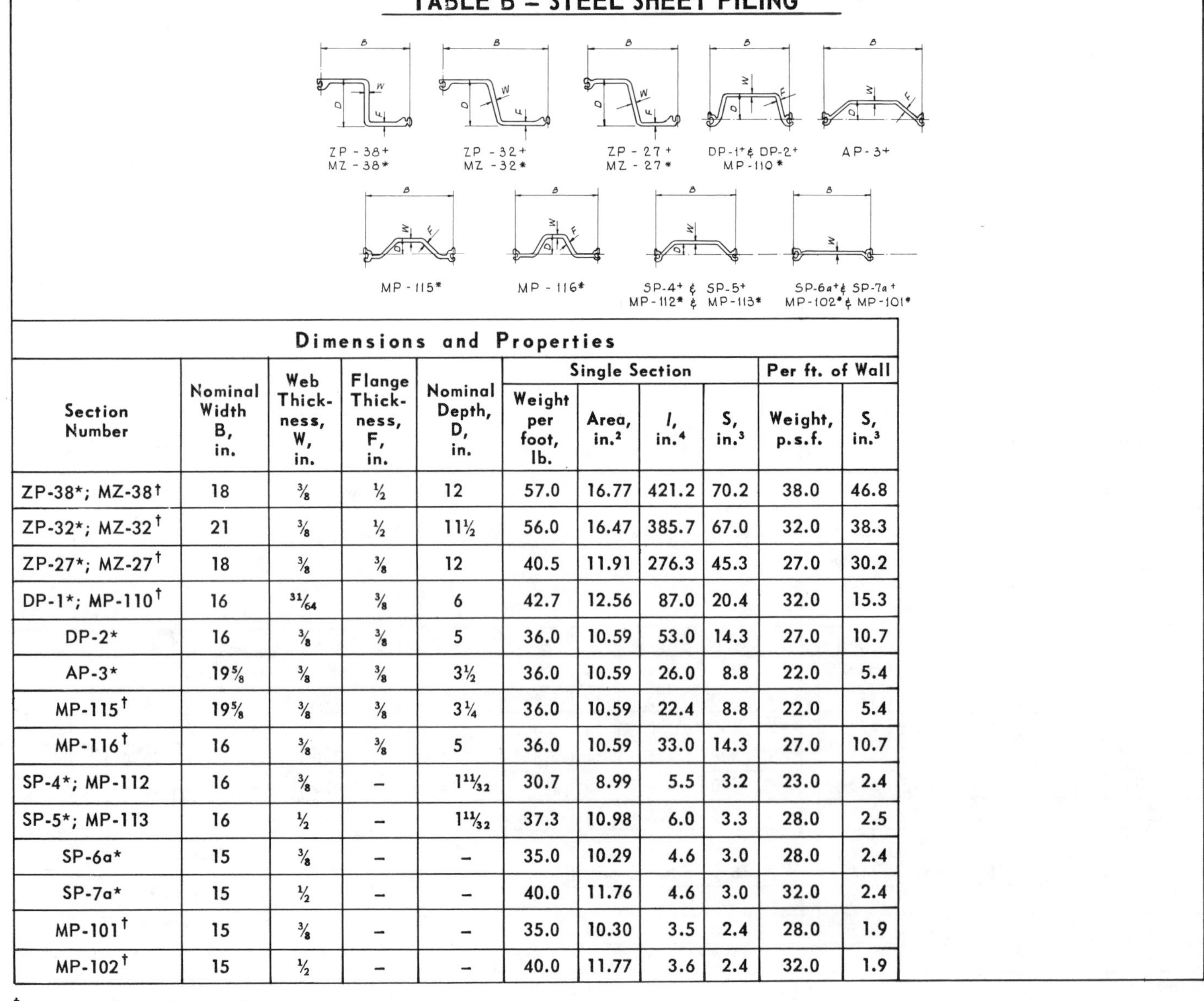
ZP - 38+
MZ - 38*
ZP - 32+
MZ - 32*
ZP - 27+
MZ - 27*
DP-1+ & DP-2+
MP-110*
AP-3+
MP-115*
MP-116*
SP-4+ & SP-5+
MP-112* & MP-113*
SP-6a+ & SP-7a+
MP-102* & MP-101*

Dimensions and Properties

Section Number	Nominal Width B, in.	Web Thickness, W, in.	Flange Thickness, F, in.	Nominal Depth, D, in.	Single Section: Weight per foot, lb.	Single Section: Area, in.²	Single Section: I, in.⁴	Single Section: S, in.³	Per ft. of Wall: Weight, p.s.f.	Per ft. of Wall: S, in.³
ZP-38*; MZ-38†	18	⅜	½	12	57.0	16.77	421.2	70.2	38.0	46.8
ZP-32*; MZ-32†	21	⅜	½	11½	56.0	16.47	385.7	67.0	32.0	38.3
ZP-27*; MZ-27†	18	⅜	⅜	12	40.5	11.91	276.3	45.3	27.0	30.2
DP-1*; MP-110†	16	31/64	⅜	6	42.7	12.56	87.0	20.4	32.0	15.3
DP-2*	16	⅜	⅜	5	36.0	10.59	53.0	14.3	27.0	10.7
AP-3*	19⅝	⅜	⅜	3½	36.0	10.59	26.0	8.8	22.0	5.4
MP-115†	19⅝	⅜	⅜	3¼	36.0	10.59	22.4	8.8	22.0	5.4
MP-116†	16	⅜	⅜	5	36.0	10.59	33.0	14.3	27.0	10.7
SP-4*; MP-112	16	⅜	–	1 11/32	30.7	8.99	5.5	3.2	23.0	2.4
SP-5*; MP-113	16	½	–	1 11/32	37.3	10.98	6.0	3.3	28.0	2.5
SP-6a*	15	⅜	–	–	35.0	10.29	4.6	3.0	28.0	2.4
SP-7a*	15	½	–	–	40.0	11.76	4.6	3.0	32.0	2.4
MP-101†	15	⅜	–	–	35.0	10.30	3.5	2.4	28.0	1.9
MP-102†	15	½	–	–	40.0	11.77	3.6	2.4	32.0	1.9

† Indicates United States Steel Corp. piles.

* Indicates Bethleham Steel Co. piles.

FOUNDATIONS - MANHOLES & INLETS

BASIS

DETAILS

LOADING:

75 tons on dual tandem wheels (18.75 T on each of 4 wheels).

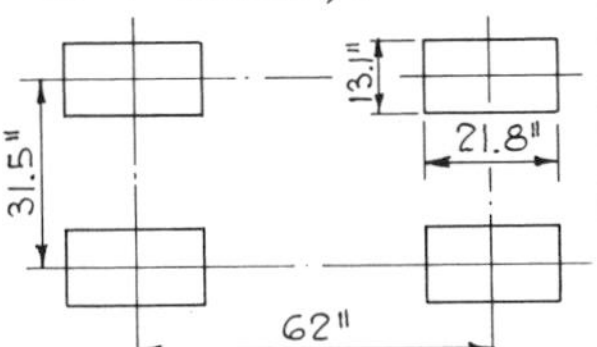

13.1"
21.8"
31.5"
62"

IMPRINT PATTERN

Size of manhole
4'-0" sq. x 6'-0" deep
2'-0" dia. opening as shown.

f'_c = 3000 p.s.i.
f_s = 20 000 p.s.i.
A.C.I. code 1956

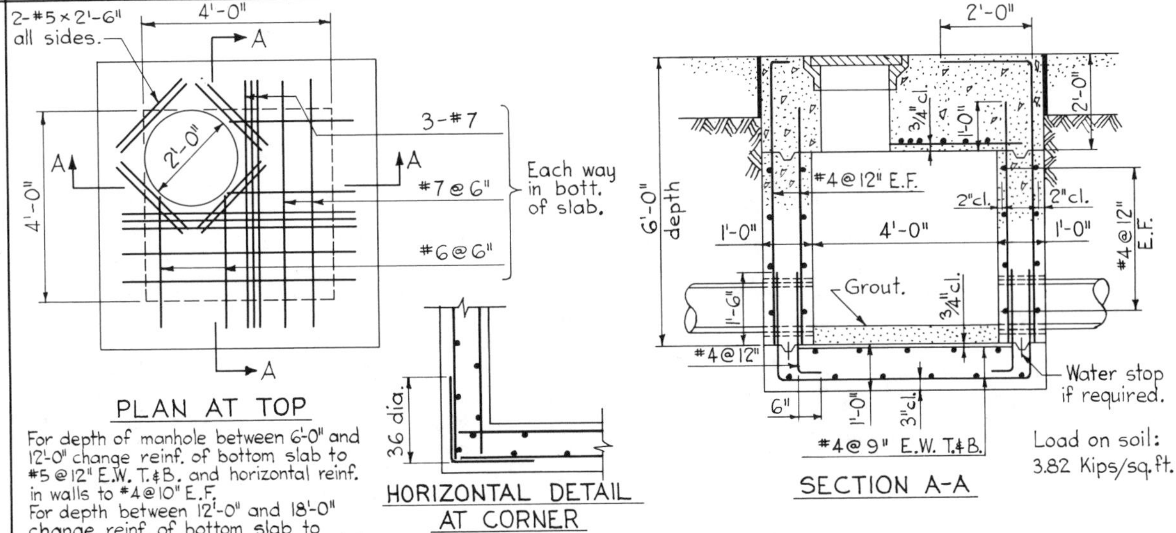

2-#5 x 2'-6" all sides.
4'-0"
A
2'-0"
4'-0"
3-#7
#7 @ 6"
#6 @ 6"
Each way in bott. of slab.
PLAN AT TOP
36 dia.
HORIZONTAL DETAIL AT CORNER
2'-0"
6'-0" depth
3/4"cl.
1'-0"
2'-0"
#4 @ 12" E.F.
2"cl.
1'-0"
4'-0"
Grout.
1'-6"
#4 @ 12"
6"
3"cl.
#4 @ 9" E.W. T.&B.
Water stop if required.
SECTION A-A
Load on soil: 3.82 Kips/sq.ft.

For depth of manhole between 6'-0" and 12'-0" change reinf. of bottom slab to #5 @ 12" E.W. T.&B. and horizontal reinf. in walls to #4 @ 10" E.F.
For depth between 12'-0" and 18'-0" change reinf. of bottom slab to #5 @ 11" E.W. T.&B. and horizontal reinf. in walls to #4 @ 10" E.F.

FIG. A

LOADING:

A.A.S.H.O. H20-44
(8 ton wheel load)

Other conditions same as Fig. A.

The design shown can be used to a depth of 15'-0". It can be used to a depth of 20'-0" if the reinforcement of the bottom slab is changed to #4 @ 10" E.W. T.&B.

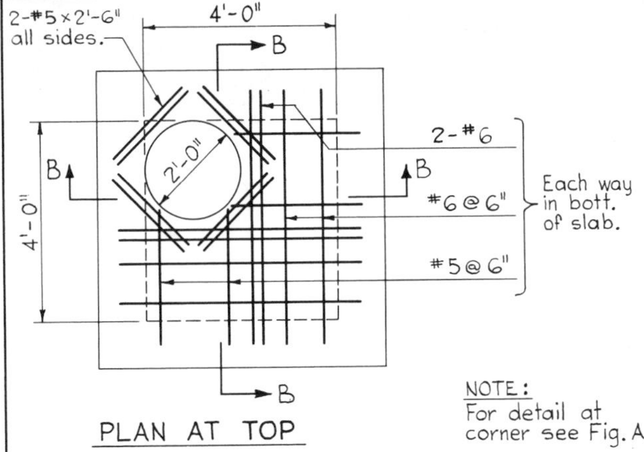

2-#5 x 2'-6" all sides.
4'-0"
B
2'-0"
4'-0"
2-#6
#6 @ 6"
#5 @ 6"
Each way in bott. of slab.

PLAN AT TOP

NOTE: For detail at corner see Fig. A

2'-0"
1'-3"
6'-0" depth
2"cl.
3/4"cl.
#4 @ 12" E.F.
1'-0"
4'-0"
Grout.
1'-6"
#4 @ 12"
6"
1'-0"
3"cl.
#4 @ 12" E.W. T.&B.
Water stop if required.

SECTION B-B

Load on soil: 1.62 Kips/sq.ft.

FIG. B

LOADING:

Light
(3½ ton wheel load)

Other conditions same as Fig. A.

The design shown can be used to a depth of 12'-0". For depths to 16'-0" change reinforcement in bottom slab to #3 @ 7" E.W. T.&B.
For depths to 20'-0" change reinforcement in bottom slab to #4 @ 10" E.W. T.&B. and horizontal reinforcement in lower part of wall to #4 @ 10" E.F.

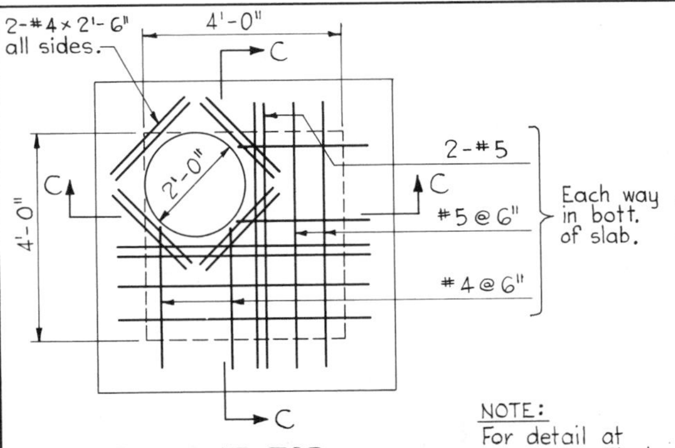

2-#4 x 2'-6" all sides.
4'-0"
C
2'-0"
4'-0"
2-#5
#5 @ 6"
#4 @ 6"
Each way in bott. of slab.

PLAN AT TOP

NOTE: For detail at corner see Fig. A

2'-0"
1'-0"
6'-0" depth
3/4"cl.
#3 @ 9" E.F.
2"cl.
10"
4'-0"
Grout.
1'-6"
#3 @ 9"
6"
3"cl.
#3 @ 9" E.W. T.&B.
Water stop if required.

SECTION C-C

Load on soil: 1.08 Kips/sq.ft.

FIG. C

RED LIGHTS:
Concrete dimensions and reinforcement should be checked and revised if necessary for the following:

1. Excessively large pipes in walls;
2. Other sizes and shapes of manholes;
3. Variations in loading;
4. Poor soil conditions.

1. In general, pipes of any reasonable size can be accomodated in the walls of the above manholes. Add bars above, below, and at sides of large openings.
2. Rungs should be provided.
3. Electrical manholes can be similar to the manholes detailed here if suitable adjustments are made. Provide "pulling irons" opposite each duct bank.

* Inlets generally classified structurally as "Manholes".

By C.L. Dayton

TRANSMISSION TOWERS-LIVE LOADS & UNIT STRESSES

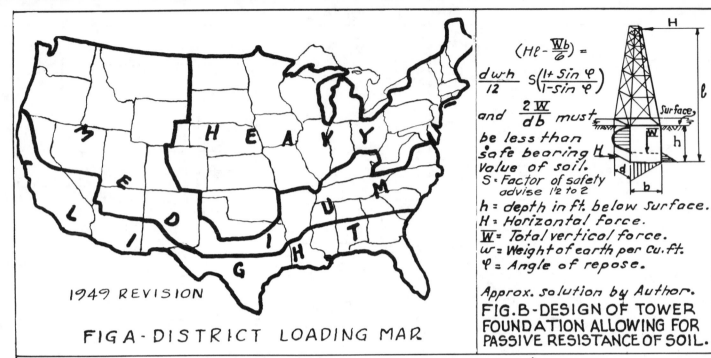

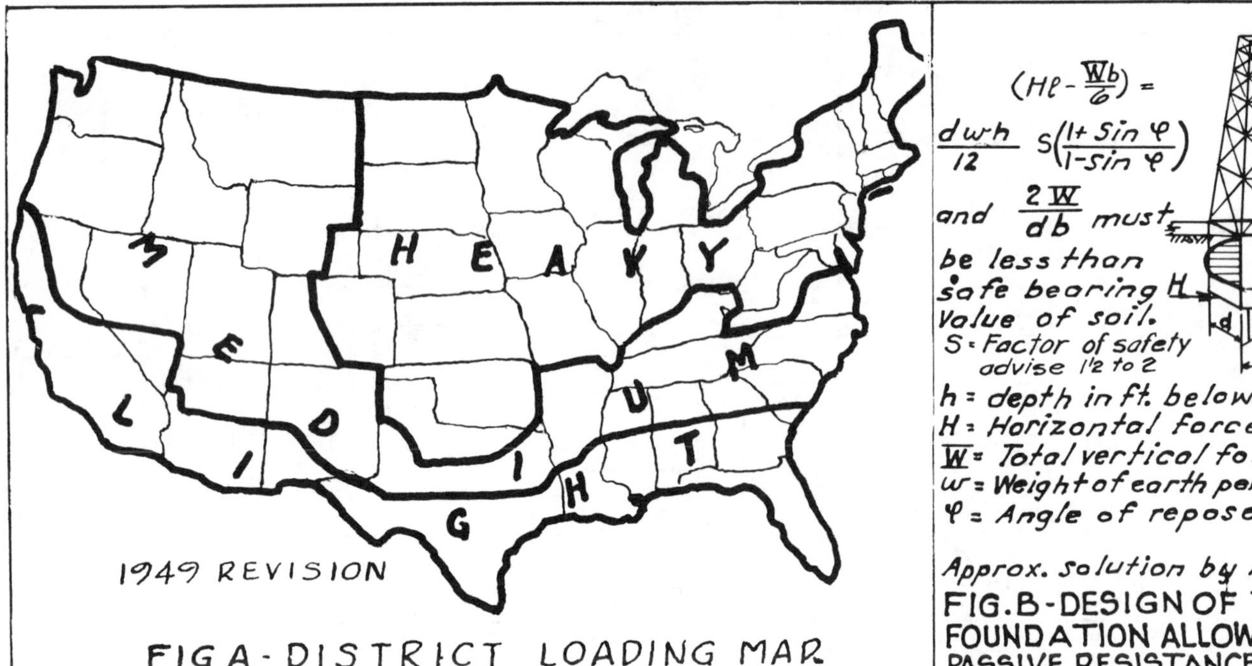
HEAVY
MEDIUM
LIGHT
1949 REVISION
H
l
Surface
W
h
H
d
b

FIG A - DISTRICT LOADING MAP

$$\left(Hl - \frac{Wb}{6}\right) = \frac{dwh}{12}\, S\left(\frac{1+\sin\varphi}{1-\sin\varphi}\right)$$

and $\frac{2W}{db}$ must be less than safe bearing value of soil.

S = Factor of safety advise 1½ to 2

h = depth in ft. below surface.
H = Horizontal force.
W = Total vertical force.
w = Weight of earth per cu. ft.
φ = Angle of repose.

Approx. solution by Author.

FIG. B - DESIGN OF TOWER FOUNDATION ALLOWING FOR PASSIVE RESISTANCE OF SOIL.

TABLE C - LOADS ON TOWER.*

LONGITUDINAL	VERTICAL	TRANSVERSE
1. Unbalanced loads due to change in grade of construction broken wires, jointly used poles, dead ends, and any other unbalanced conductors.	1. Weight of Tower & accessories. 2. Weight of wire & cables. 3. Radial thickness of ice on wires. Heavy District .50 in. ice. Medium " .25 " " Light " .00 " " 4. Any effect of difference in elevation of supports.	1. Horizontal wind on wires. H = 4# per sq. ft. on projected area with .50 in. Ice. M = 4# " " " " " " .25 " " L = 9# " " " " " " .00 " " 2. Horizontal wind on Tower with no ice. H = 13# per sq. ft. M = 13# " " " L = 19.2# " " " } on flat surface. With lattice Tower use 1½ times projected areas of members. 3. Resultant of loads due to change in direction of Conductors.

SIMULTANEOUS APPLICATION OF LOADS.*

1. When calculating transverse strength, the assumed transverse and vertical load should be taken as acting simultaneously.
2. In calculating longitudinal strength, the assumed longitudinal loads shall be taken without consideration of the vertical or transverse loads.

Table D —Allowable Unit Stresses in Steel for Transverse and Longitudinal Strengths.*

	Allowable stresses for transverse strength			Allowable stresses for longitudinal strength	
	Grade A	Grade B	Grade C	Grades A and B crossings	Grades A and B except at crossings
	Lbs. per sq. in.	Lbs. per sq. in.	Lbs. per sq. in.	Lbs. per sq. in.	Lbs. per sq. in.
Structural steel:					
Tension	20,000	26,000	30,000	30,000	33,000
Compression	20,000 −80 L/R	26,000 −90 L/R	30,000 −100 L/R	30,000 −100 L/R	33,000 −100 L/R
Bolts:					
Shear	20,000	24,000	35,000	35,000	40,000
Bearing	40,000	48,000	70,000	70,000	80,000
Rivets:					
Shear	18,000	22,000	30,000	30,000	33,000
Bearing	36,000	44,000	60,000	60,000	66,000

*Adapted from National Electrical Safety Code, U.S. Bureau of Standards.
The Author recommends these unit stresses be used only on large projects. For small projects, use A.I.S.C. stresses.

Table E —Thickness of Steel.*

Kind of member	Thickness of main members of cross arms and legs	Thickness of other members
Galvanized:	Inches	Inches
For localities where experience has shown deterioration of galvanized material is rapid	¼	3/16
For other localities	3/16	⅛
Painted	¼	•¼

• Painted bracing members having L/R not exceeding 125 may be 3/16 inch in thickness.

Table F —L/R for Compression Members

Kind of compression member	L/R
Leg members	150
Other members having figured stresses	200
Secondary members without figured stresses	250

STRUCTURAL — BEAM FORMULAS-I

NOMENCLATURE

R = reaction *M = moment*

V = shear *Δ = deflection*

1. SIMPLE BEAM—UNIFORMLY DISTRIBUTED LOAD

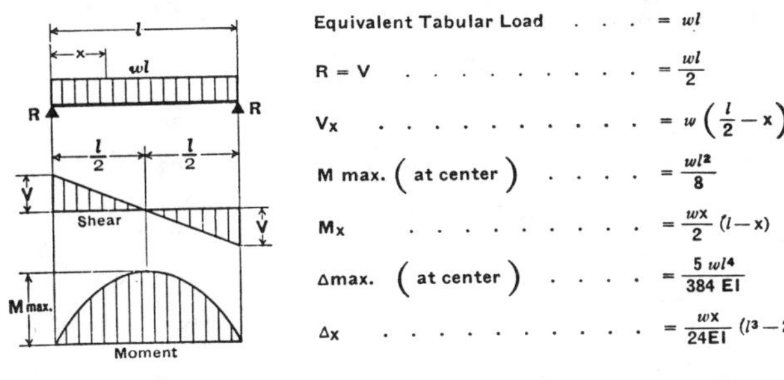

l
x
wl
R
R
l/2
l/2
V
Shear
V
M max.
Moment

Equivalent Tabular Load	$= wl$
R = V	$= \frac{wl}{2}$
V_x	$= w\left(\frac{l}{2}-x\right)$
M max. (at center)	$= \frac{wl^2}{8}$
M_x	$= \frac{wx}{2}(l-x)$
Δmax. (at center)	$= \frac{5wl^4}{384EI}$
Δ_x	$= \frac{wx}{24EI}(l^3-2lx^2+x^3)$

2. SIMPLE BEAM—LOAD INCREASING UNIFORMLY TO ONE END

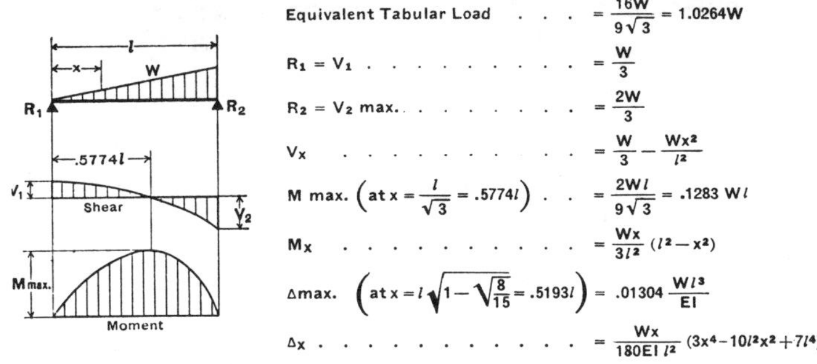

l
x
W
R1
R2
.5774l
V1
Shear
V2
M max.
Moment

Equivalent Tabular Load	$= \frac{16W}{9\sqrt{3}} = 1.0264W$
$R_1 = V_1$	$= \frac{W}{3}$
$R_2 = V_2$ max.	$= \frac{2W}{3}$
V_x	$= \frac{W}{3} - \frac{Wx^2}{l^2}$
M max. $\left(\text{at } x = \frac{l}{\sqrt{3}} = .5774l\right)$	$= \frac{2Wl}{9\sqrt{3}} = .1283\,Wl$
M_x	$= \frac{Wx}{3l^2}(l^2-x^2)$
Δmax. $\left(\text{at } x = l\sqrt{1-\sqrt{\frac{8}{15}}} = .5193l\right)$	$= .01304\,\frac{Wl^3}{EI}$
Δ_x	$= \frac{Wx}{180EI\,l^2}(3x^4-10l^2x^2+7l^4)$

3. SIMPLE BEAM—LOAD INCREASING UNIFORMLY TO CENTER

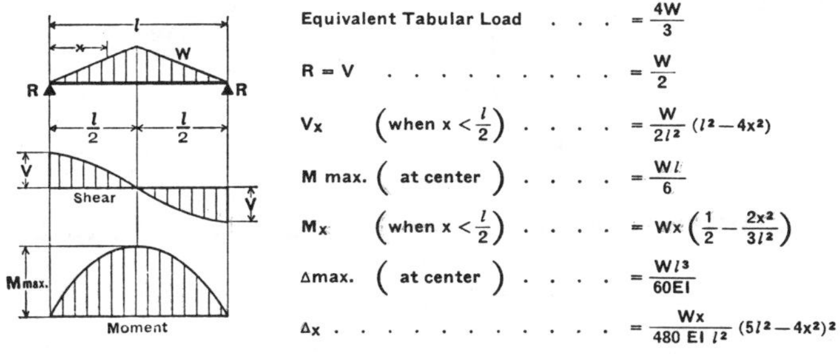

l
x
W
R
R
l/2
l/2
V
Shear
V
M max.
Moment

Equivalent Tabular Load	$= \frac{4W}{3}$
R = V	$= \frac{W}{2}$
V_x $\left(\text{when } x < \frac{l}{2}\right)$	$= \frac{W}{2l^2}(l^2-4x^2)$
M max. (at center)	$= \frac{Wl}{6}$
M_x $\left(\text{when } x < \frac{l}{2}\right)$	$= Wx\left(\frac{1}{2}-\frac{2x^2}{3l^2}\right)$
Δmax. (at center)	$= \frac{Wl^3}{60EI}$
Δ_x	$= \frac{Wx}{480\,EI\,l^2}(5l^2-4x^2)^2$

4. SIMPLE BEAM—UNIFORM LOAD PARTIALLY DISTRIBUTED

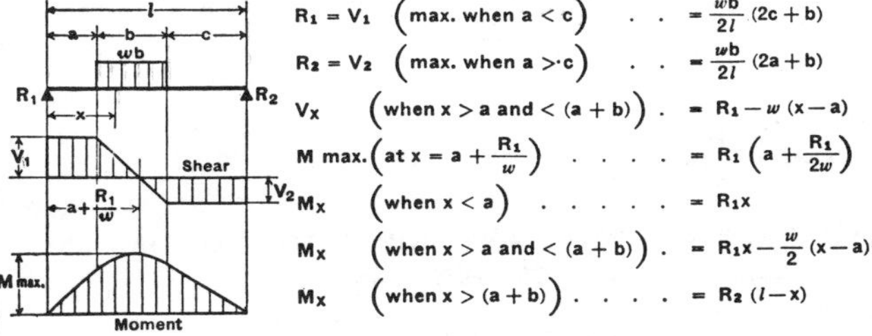

l
a
b
c
wb
R1
R2
x
V1
Shear
V2
a + R1/w
M max.
Moment

$R_1 = V_1$ (max. when a < c)	$= \frac{wb}{2l}(2c+b)$
$R_2 = V_2$ (max. when a > c)	$= \frac{wb}{2l}(2a+b)$
V_x (when x > a and < (a + b))	$= R_1 - w(x-a)$
M max. $\left(\text{at } x = a + \frac{R_1}{w}\right)$	$= R_1\left(a+\frac{R_1}{2w}\right)$
M_x (when x < a)	$= R_1x$
M_x (when x > a and < (a + b))	$= R_1x - \frac{w}{2}(x-a)^2$
M_x (when x > (a + b))	$= R_2(l-x)$

5. SIMPLE BEAM—UNIFORM LOAD PARTIALLY DISTRIBUTED AT ONE END

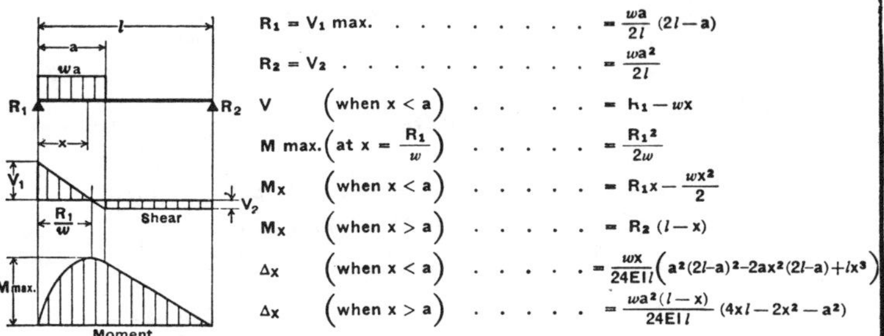

l
a
wa
R1
R2
x
V1
R1/w
Shear
V2
M max.
Moment

$R_1 = V_1$ max.	$= \frac{wa}{2l}(2l-a)$
$R_2 = V_2$	$= \frac{wa^2}{2l}$
V (when x < a)	$= h_1 - wx$
M max. $\left(\text{at } x = \frac{R_1}{w}\right)$	$= \frac{R_1^2}{2w}$
M_x (when x < a)	$= R_1x - \frac{wx^2}{2}$
M_x (when x > a)	$= R_2(l-x)$
Δ_x (when x < a)	$= \frac{wx}{24EIl}\left(a^2(2l-a)^2-2ax^2(2l-a)+lx^3\right)$
Δ_x (when x > a)	$= \frac{wa^2(l-x)}{24EIl}(4xl-2x^2-a^2)$

6. SIMPLE BEAM—UNIFORM LOAD PARTIALLY DISTRIBUTED AT EACH END

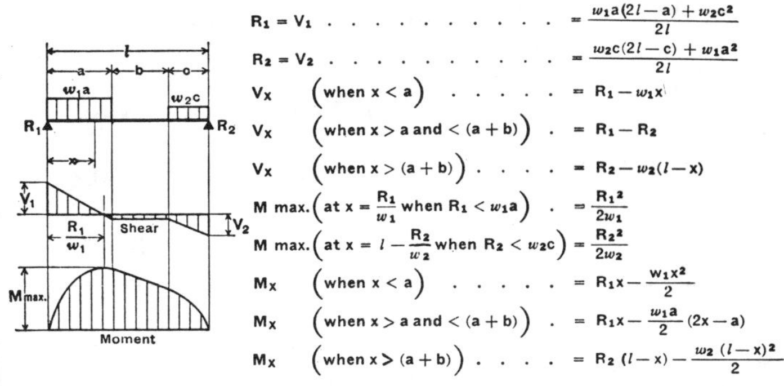

l
a
b
c
w1a
w2c
R1
R2
x
V1
R1/w1
Shear
V2
M max.
Moment

$R_1 = V_1$	$= \frac{w_1a(2l-a)+w_2c^2}{2l}$
$R_2 = V_2$	$= \frac{w_2c(2l-c)+w_1a^2}{2l}$
V_x (when x < a)	$= R_1 - w_1x$
V_x (when x > a and < (a + b))	$= R_1 - R_2$
V_x (when x > (a + b))	$= R_2 - w_2(l-x)$
M max. $\left(\text{at } x = \frac{R_1}{w_1} \text{ when } R_1 < w_1a\right)$	$= \frac{R_1^2}{2w_1}$
M max. $\left(\text{at } x = l - \frac{R_2}{w_2} \text{ when } R_2 < w_2c\right)$	$= \frac{R_2^2}{2w_2}$
M_x (when x < a)	$= R_1x - \frac{w_1x^2}{2}$
M_x (when x > a and < (a + b))	$= R_1x - \frac{w_1a}{2}(2x-a)$
M_x (when x > (a + b))	$= R_2(l-x) - \frac{w_2(l-x)^2}{2}$

7. SIMPLE BEAM—CONCENTRATED LOAD AT CENTER

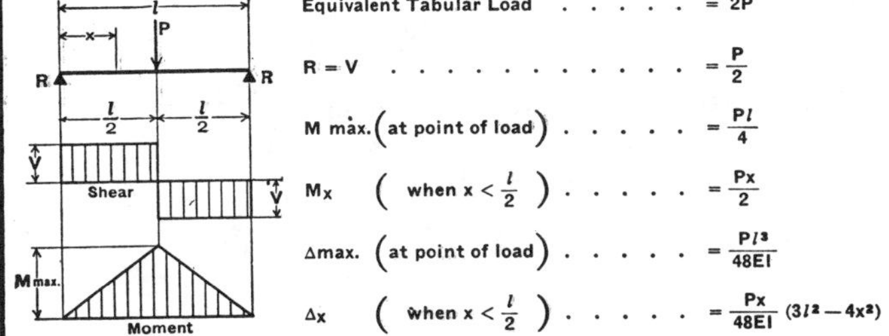

l
x
P
R
R
l/2
l/2
V
Shear
V
M max.
Moment

Equivalent Tabular Load	$= 2P$
R = V	$= \frac{P}{2}$
M max. (at point of load)	$= \frac{Pl}{4}$
M_x $\left(\text{when } x < \frac{l}{2}\right)$	$= \frac{Px}{2}$
Δmax. (at point of load)	$= \frac{Pl^3}{48EI}$
Δ_x $\left(\text{when } x < \frac{l}{2}\right)$	$= \frac{Px}{48EI}(3l^2-4x^2)$

8. SIMPLE BEAM—CONCENTRATED LOAD AT ANY POINT

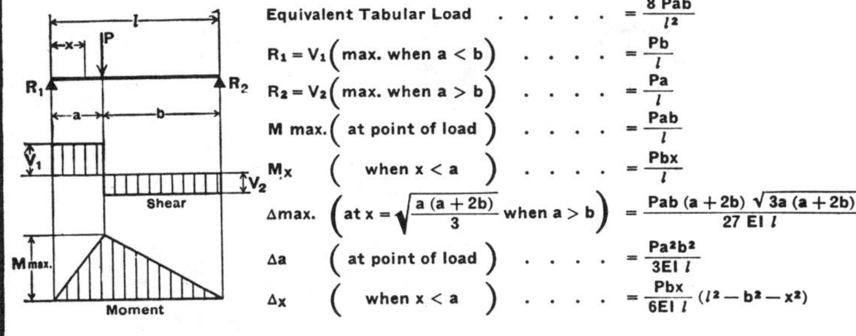

l
x
P
R1
R2
a
b
V1
Shear
V2
M max.
Moment

Equivalent Tabular Load	$= \frac{8Pab}{l^2}$
$R_1 = V_1$ (max. when a < b)	$= \frac{Pb}{l}$
$R_2 = V_2$ (max. when a > b)	$= \frac{Pa}{l}$
M max. (at point of load)	$= \frac{Pab}{l}$
M_x (when x < a)	$= \frac{Pbx}{l}$
Δmax. $\left(\text{at } x = \sqrt{\frac{a(a+2b)}{3}} \text{ when } a > b\right)$	$= \frac{Pab(a+2b)\sqrt{3a(a+2b)}}{27\,EI\,l}$
Δa (at point of load)	$= \frac{Pa^2b^2}{3EI\,l}$
Δ_x (when x < a)	$= \frac{Pbx}{6EI\,l}(l^2-b^2-x^2)$

From A.I.S.C. Steel Construction Manual- 1950.

STRUCTURAL — BEAM FORMULAS-2

For nomenclature see page 8-01

9. SIMPLE BEAM—TWO EQUAL CONCENTRATED LOADS SYMMETRICALLY PLACED

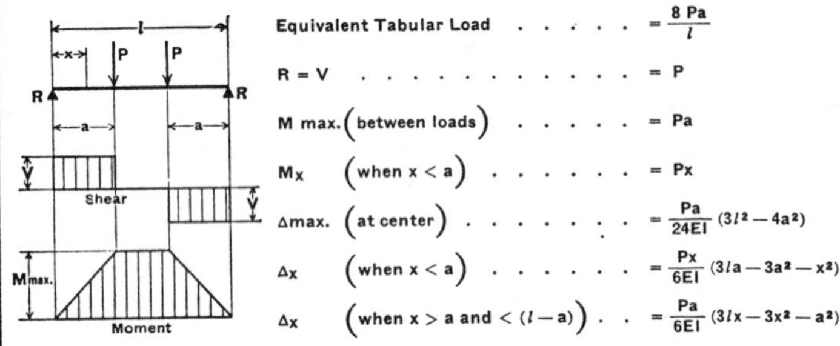

l
x
P
P
R
R
a
a
V
Shear
M max.
Moment

Equivalent Tabular Load . . . . . . $= \frac{8\,Pa}{l}$

R = V . . . . $= P$

M max. (between loads) . . . . . . $= Pa$

M_x (when $x < a$) . . . . $= Px$

Δmax. (at center) . . . . $= \frac{Pa}{24EI}(3l^2 - 4a^2)$

Δ_x (when $x < a$) . . . . $= \frac{Px}{6EI}(3la - 3a^2 - x^2)$

Δ_x (when $x > a$ and $< (l - a)$) . . $= \frac{Pa}{6EI}(3lx - 3x^2 - a^2)$

10. SIMPLE BEAM—TWO EQUAL CONCENTRATED LOADS UNSYMMETRICALLY PLACED

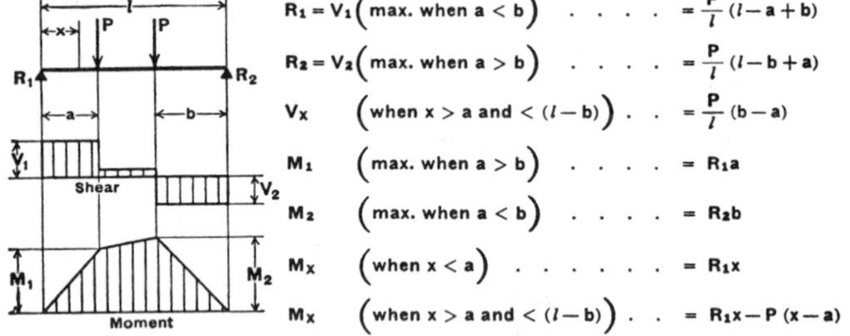

l
x
P
P
R1
R2
a
b
V1
V2
Shear
M1
M2
Moment

$R_1 = V_1$ (max. when $a < b$) . . . . . $= \frac{P}{l}(l - a + b)$

$R_2 = V_2$ (max. when $a > b$) . . . . . $= \frac{P}{l}(l - b + a)$

V_x (when $x > a$ and $< (l - b)$) . . $= \frac{P}{l}(b - a)$

M_1 (max. when $a > b$) . . . . . $= R_1a$

M_2 (max. when $a < b$) . . . . . $= R_2b$

M_x (when $x < a$) . . . . $= R_1x$

M_x (when $x > a$ and $< (l - b)$) . . $= R_1x - P(x - a)$

11. SIMPLE BEAM—TWO UNEQUAL CONCENTRATED LOADS UNSYMMETRICALLY PLACED

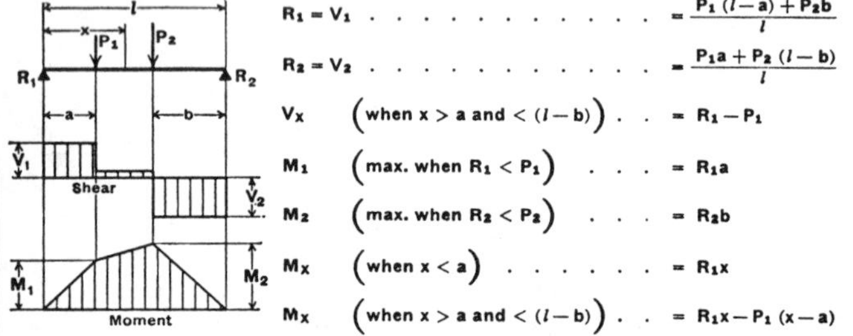

l
x
P1
P2
R1
R2
a
b
V1
V2
Shear
M1
M2
Moment

$R_1 = V_1$ . . . . $= \frac{P_1(l - a) + P_2b}{l}$

$R_2 = V_2$ . . . . $= \frac{P_1a + P_2(l - b)}{l}$

V_x (when $x > a$ and $< (l - b)$) . . $= R_1 - P_1$

M_1 (max. when $R_1 < P_1$) . . . . $= R_1a$

M_2 (max. when $R_2 < P_2$) . . . . $= R_2b$

M_x (when $x < a$) . . . . $= R_1x$

M_x (when $x > a$ and $< (l - b)$) . . $= R_1x - P_1(x - a)$

12. BEAM FIXED AT ONE END, SUPPORTED AT OTHER—UNIFORMLY DISTRIBUTED LOAD

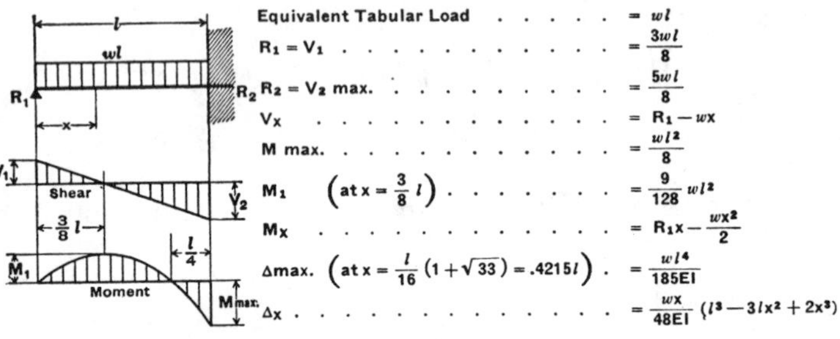

l
wl
R1
R2
x
V1
V2
Shear
3/8 l
l/4
M1
Moment
M max.

Equivalent Tabular Load . . . . . . $= wl$

$R_1 = V_1$ . . . . $= \frac{3wl}{8}$

$R_2 = V_2$ max. . . . . $= \frac{5wl}{8}$

V_x . . . . $= R_1 - wx$

M max. . . . . $= \frac{wl^2}{8}$

M_1 (at $x = \frac{3}{8}l$) . . . . $= \frac{9}{128}wl^2$

M_x . . . . $= R_1x - \frac{wx^2}{2}$

Δmax. (at $x = \frac{l}{16}(1 + \sqrt{33}) = .4215l$) . $= \frac{wl^4}{185EI}$

Δ_x . . . . $= \frac{wx}{48EI}(l^3 - 3lx^2 + 2x^3)$

13. BEAM FIXED AT ONE END, SUPPORTED AT OTHER—CONCENTRATED LOAD AT CENTER

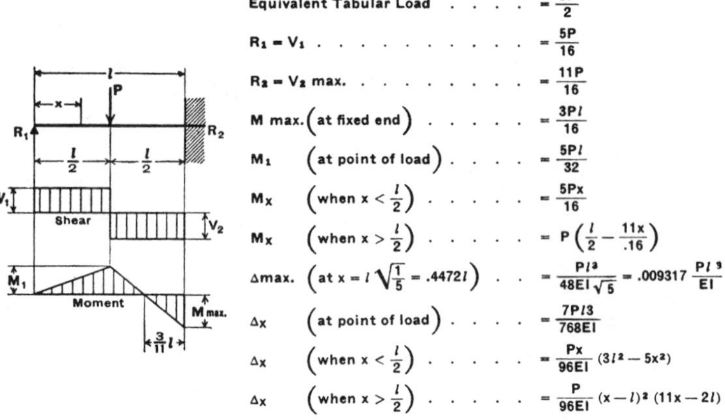

l
x
P
R1
R2
l/2
l/2
V1
V2
Shear
M1
Moment
M max.
3/11 l

Equivalent Tabular Load . . . . $= \frac{3P}{2}$

$R_1 = V_1$ . . . . $= \frac{5P}{16}$

$R_2 = V_2$ max. . . . . $= \frac{11P}{16}$

M max. (at fixed end) . . . . . . $= \frac{3Pl}{16}$

M_1 (at point of load) . . . . $= \frac{5Pl}{32}$

M_x (when $x < \frac{l}{2}$) . . . . . . $= \frac{5Px}{16}$

M_x (when $x > \frac{l}{2}$) . . . . . . $= P\left(\frac{l}{2} - \frac{11x}{16}\right)$

Δmax. (at $x = l\sqrt{\frac{1}{5}} = .4472l$) . . $= \frac{Pl^3}{48EI\sqrt{5}} = .009317\frac{Pl^3}{EI}$

Δ_x (at point of load) . . . . $= \frac{7Pl^3}{768EI}$

Δ_x (when $x < \frac{l}{2}$) . . . . . . $= \frac{Px}{96EI}(3l^2 - 5x^2)$

Δ_x (when $x > \frac{l}{2}$) . . . . . . $= \frac{P}{96EI}(x - l)^2(11x - 2l)$

14. BEAM FIXED AT ONE END, SUPPORTED AT OTHER—CONCENTRATED LOAD AT ANY POINT

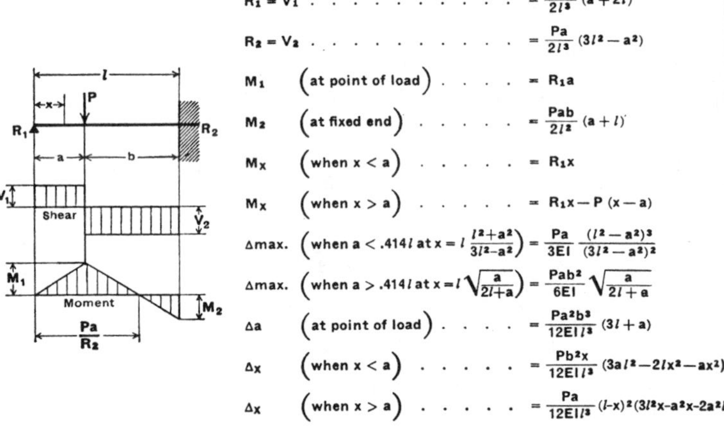

l
x
P
R1
R2
a
b
V1
V2
Shear
M1
M2
Moment
Pa/R2

$R_1 = V_1$ . . . . $= \frac{Pb^2}{2l^3}(a + 2l)$

$R_2 = V_2$ . . . . $= \frac{Pa}{2l^3}(3l^2 - a^2)$

M_1 (at point of load) . . . . . $= R_1a$

M_2 (at fixed end) . . . . . . $= \frac{Pab}{2l^2}(a + l)$

M_x (when $x < a$) . . . . . . $= R_1x$

M_x (when $x > a$) . . . . . . $= R_1x - P(x - a)$

Δmax. (when $a < .414l$ at $x = l\frac{l^2 + a^2}{3l^2 - a^2}$) $= \frac{Pa}{3EI}\frac{(l^2 - a^2)^3}{(3l^2 - a^2)^2}$

Δmax. (when $a > .414l$ at $x = l\sqrt{\frac{a}{2l + a}}$) $= \frac{Pab^2}{6EI}\sqrt{\frac{a}{2l + a}}$

Δ_a (at point of load) . . . . $= \frac{Pa^2b^3}{12EIl^3}(3l + a)$

Δ_x (when $x < a$) . . . . . . $= \frac{Pb^2x}{12EIl^3}(3al^2 - 2lx^2 - ax^2)$

Δ_x (when $x > a$) . . . . . . $= \frac{Pa}{12EIl^3}(l - x)^2(3l^2x - a^2x - 2a^2l)$

15. BEAM FIXED AT BOTH ENDS—UNIFORMLY DISTRIBUTED LOADS

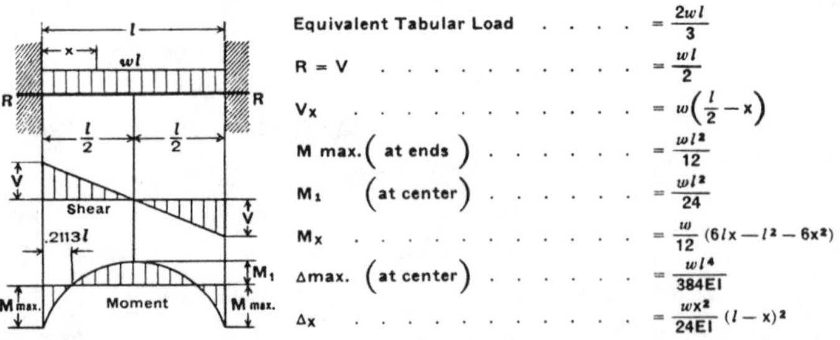

l
x
wl
R
R
l/2
l/2
V
V
Shear
.2113l
M1
M max.
Moment
M max.

Equivalent Tabular Load . . . . $= \frac{2wl}{3}$

R = V . . . . $= \frac{wl}{2}$

V_x . . . . $= w\left(\frac{l}{2} - x\right)$

M max. (at ends) . . . . $= \frac{wl^2}{12}$

M_1 (at center) . . . . $= \frac{wl^2}{24}$

M_x . . . . $= \frac{w}{12}(6lx - l^2 - 6x^2)$

Δmax. (at center) . . . . $= \frac{wl^4}{384EI}$

Δ_x . . . . $= \frac{wx^2}{24EI}(l - x)^2$

16. BEAM FIXED AT BOTH ENDS—CONCENTRATED LOAD AT CENTER

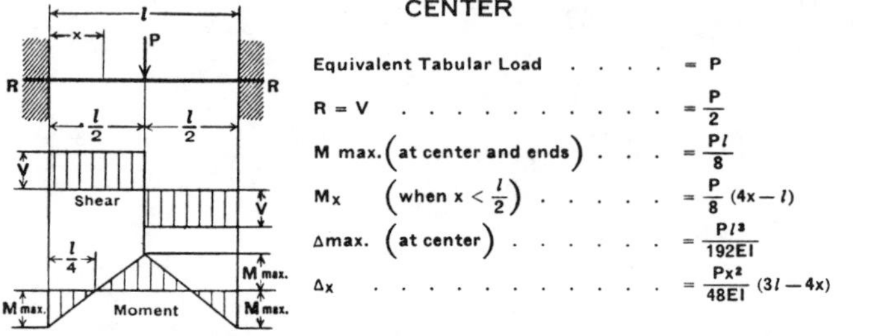

l
x
P
R
R
l/2
l/2
V
V
Shear
l/4
M max.
Moment
M max.
M max.

Equivalent Tabular Load . . . . $= P$

R = V . . . . $= \frac{P}{2}$

M max. (at center and ends) . . . $= \frac{Pl}{8}$

M_x (when $x < \frac{l}{2}$) . . . . . . $= \frac{P}{8}(4x - l)$

Δmax. (at center) . . . . $= \frac{Pl^3}{192EI}$

Δ_x . . . . $= \frac{Px^2}{48EI}(3l - 4x)$

From A.I.S.C. Steel Construction Manual - 1950.

STRUCTURAL — BEAM FORMULAS–3

For nomenclature see page 8-01

17. BEAM FIXED AT BOTH ENDS—CONCENTRATED LOAD AT ANY POINT

$R_1 = V_1$ (max. when $a < b$) . . . . $= \frac{Pb^2}{l^3}(3a + b)$

$R_2 = V_2$ (max. when $a > b$) . . . . $= \frac{Pa^2}{l^3}(a + 3b)$

M_1 (max. when $a < b$) . . . . $= \frac{Pab^2}{l^2}$

M_2 (max. when $a > b$) . . . . $= \frac{Pa^2b}{l^2}$

M_a (at point of load) . . . . $= \frac{2Pa^2b^2}{l^3}$

M_x (when $x < a$) . . . . . $= R_1x - \frac{Pab^2}{l^2}$

Δmax. (when $a > b$ at $x = \frac{2al}{3a+b}$) . $= \frac{2Pa^3b^2}{3EI\,(3a+b)^2}$

Δa (at point of load) . . . . $= \frac{Pa^3b^3}{3EIl^3}$

Δx (when $x < a$) . . . . . $= \frac{Pb^2x^2}{6EIl^3}(3al - 3ax - bx)$

18. CANTILEVER BEAM—LOAD INCREASING UNIFORMLY TO FIXED END

Equivalent Tabular Load . . . . . $= \frac{8}{3}W$

$R = V$ . . . . $= W$

V_x . . . . $= W\frac{x^2}{l^2}$

M max. (at fixed end) . . . . . $= \frac{Wl}{3}$

M_x . . . . $= \frac{Wx^3}{3l^2}$

Δmax. (at free end) . . . . . $= \frac{Wl^3}{15EI}$

Δx . . . . $= \frac{W}{60EIl^2}(x^5 - 5l^4x + 4l^5)$

19. CANTILEVER BEAM—UNIFORMLY DISTRIBUTED LOAD

Equivalent Tabular Load . . . . . $= 4wl$

$R = V$ . . . . $= wl$

V_x . . . . $= wx$

M max. (at fixed end) . . . . . $= \frac{wl^2}{2}$

M_x . . . . $= \frac{wx^2}{2}$

Δmax. (at free end) . . . . . $= \frac{wl^4}{8EI}$

Δx . . . . $= \frac{w}{24EI}(x^4 - 4l^3x + 3l^4)$

20. BEAM FIXED AT ONE END, FREE BUT GUIDED AT OTHER—UNIFORMLY DISTRIBUTED LOAD

The deflection at the guided end is assumed to be in a vertical plane.

Equivalent Tabular Load . . . . . $= \frac{8}{3}wl$

$R = V$ . . . . $= wl$

V_x . . . . $= wx$

M max. (at fixed end) . . . . . $= \frac{wl^2}{3}$

M_1 (at guided end) . . . . . $= \frac{wl^2}{6}$

M_x . . . . $= \frac{w}{6}(l^2 - 3x^2)$

Δmax. (at guided end) . . . . . $= \frac{wl^4}{24EI}$

Δx . . . . $= \frac{w\,(l^2 - x^2)^2}{24EI}$

21. CANTILEVER BEAM—CONCENTRATED LOAD AT ANY POINT

Equivalent Tabular Load . . . . . $= \frac{8Pb}{l}$

$R = V$ (when $x < a$) . . . . . $= P$

M max. (at fixed end) . . . . . $= Pb$

M_x (when $x > a$) . . . . . $= P\,(x - a)$

Δmax. (at free end) . . . . . $= \frac{Pb^2}{6EI}(3l - b)$

Δa (at point of load) . . . . . $= \frac{Pb^3}{3EI}$

Δx (when $x < a$) . . . . . $= \frac{Pb^2}{6EI}(3l - 3x - b)$

Δx (when $x > a$) . . . . . $= \frac{P\,(l - x)^2}{6EI}(3b - l + x)$

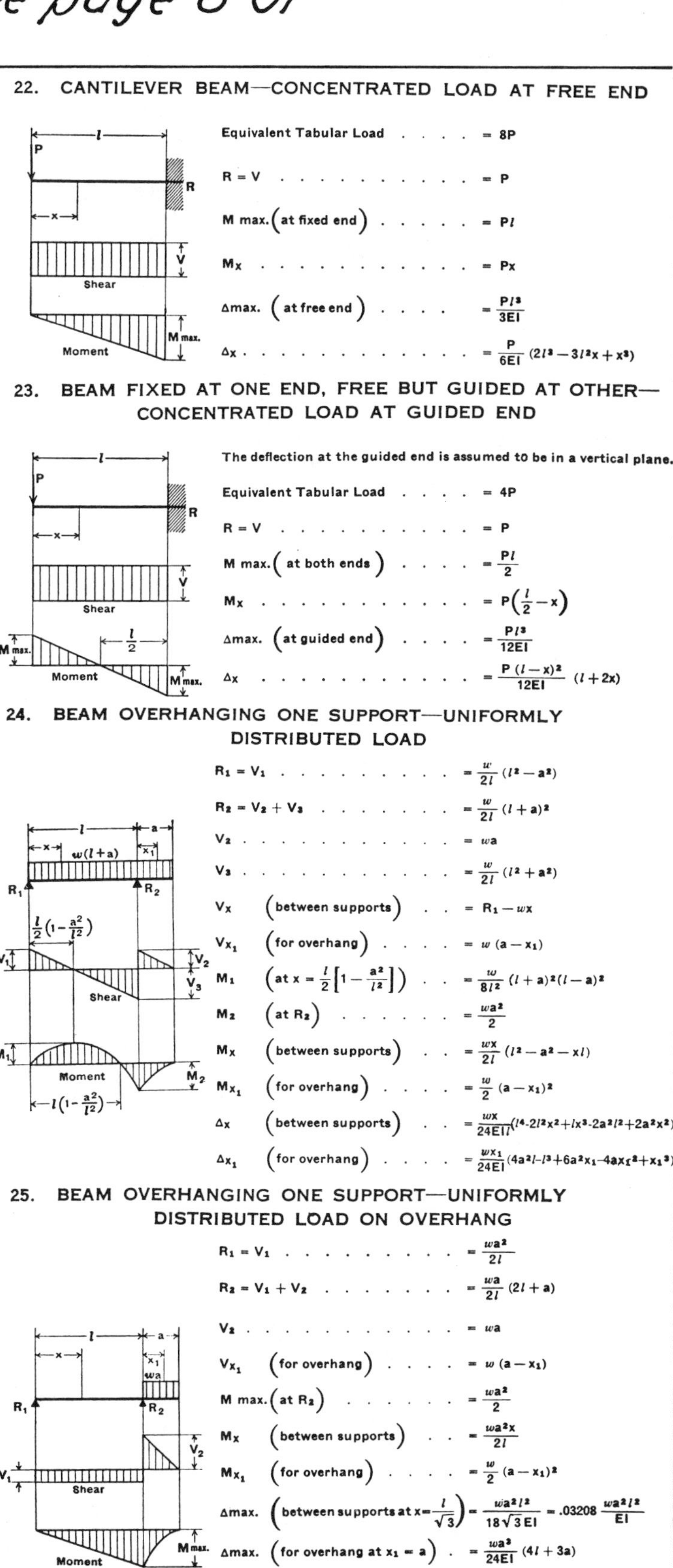

22. CANTILEVER BEAM—CONCENTRATED LOAD AT FREE END

Equivalent Tabular Load . . . . . $= 8P$

$R = V$ . . . . $= P$

M max. (at fixed end) . . . . . $= Pl$

M_x . . . . $= Px$

Δmax. (at free end) . . . . . $= \frac{Pl^3}{3EI}$

Δx . . . . $= \frac{P}{6EI}(2l^3 - 3l^2x + x^3)$

23. BEAM FIXED AT ONE END, FREE BUT GUIDED AT OTHER—CONCENTRATED LOAD AT GUIDED END

The deflection at the guided end is assumed to be in a vertical plane.

Equivalent Tabular Load . . . . . $= 4P$

$R = V$ . . . . $= P$

M max. (at both ends) . . . . . $= \frac{Pl}{2}$

M_x . . . . $= P\left(\frac{l}{2} - x\right)$

Δmax. (at guided end) . . . . . $= \frac{Pl^3}{12EI}$

Δx . . . . $= \frac{P\,(l - x)^2}{12EI}(l + 2x)$

24. BEAM OVERHANGING ONE SUPPORT—UNIFORMLY DISTRIBUTED LOAD

$R_1 = V_1$ . . . . $= \frac{w}{2l}(l^2 - a^2)$

$R_2 = V_2 + V_3$ . . . . $= \frac{w}{2l}(l + a)^2$

V_2 . . . . $= wa$

V_3 . . . . $= \frac{w}{2l}(l^2 + a^2)$

V_x (between supports) . . . $= R_1 - wx$

V_{x_1} (for overhang) . . . . . $= w\,(a - x_1)$

M_1 (at $x = \frac{l}{2}\left[1 - \frac{a^2}{l^2}\right]$) . . . $= \frac{w}{8l^2}(l + a)^2(l - a)^2$

M_2 (at R_2) . . . . . . $= \frac{wa^2}{2}$

M_x (between supports) . . . $= \frac{wx}{2l}(l^2 - a^2 - xl)$

M_{x_1} (for overhang) . . . . . $= \frac{w}{2}(a - x_1)^2$

Δx (between supports) . . . $= \frac{wx}{24EIl}(l^4 - 2l^2x^2 + lx^3 - 2a^2l^2 + 2a^2x^2)$

Δx_1 (for overhang) . . . . . $= \frac{wx_1}{24EI}(4a^2l - l^3 + 6a^2x_1 - 4ax_1^2 + x_1^3)$

25. BEAM OVERHANGING ONE SUPPORT—UNIFORMLY DISTRIBUTED LOAD ON OVERHANG

$R_1 = V_1$ . . . . $= \frac{wa^2}{2l}$

$R_2 = V_1 + V_2$ . . . . $= \frac{wa}{2l}(2l + a)$

V_2 . . . . $= wa$

V_{x_1} (for overhang) . . . . . $= w\,(a - x_1)$

M max. (at R_2) . . . . . . $= \frac{wa^2}{2}$

M_x (between supports) . . . $= \frac{wa^2x}{2l}$

M_{x_1} (for overhang) . . . . . $= \frac{w}{2}(a - x_1)^2$

Δmax. (between supports at $x = \frac{l}{\sqrt{3}}$) $= \frac{wa^2l^2}{18\sqrt{3}\,EI} = .03208\,\frac{wa^2l^2}{EI}$

Δmax. (for overhang at $x_1 = a$) . $= \frac{wa^3}{24EI}(4l + 3a)$

Δx (between supports) . . . $= \frac{wa^2x}{12EIl}(l^2 - x^2)$

Δx_1 (for overhang) . . . . . $= \frac{wx_1}{24EI}(4a^2l + 6a^2x_1 - 4ax_1^2 + x_1^3)$

From A.I.S.C. Steel Construction Manual - 1950.

STRUCTURAL — BEAM FORMULAS–4

For nomenclature see page 8-01

26. BEAM OVERHANGING ONE SUPPORT—CONCENTRATED LOAD AT END OF OVERHANG

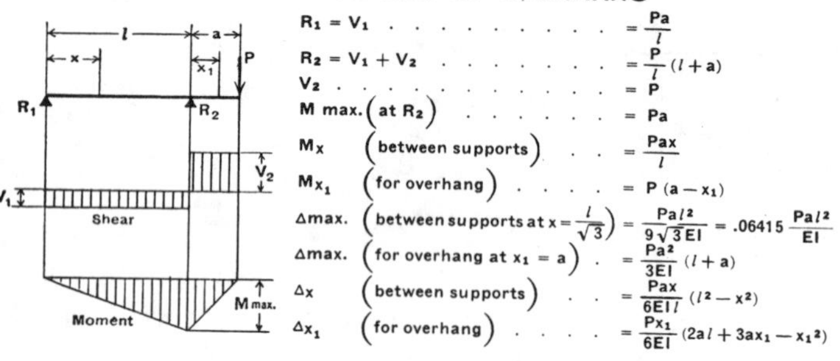

Shear
Moment

$R_1 = V_1 \dots\dots = \frac{Pa}{l}$

$R_2 = V_1 + V_2 \dots\dots = \frac{P}{l}(l + a)$

$V_2 \dots\dots = P$

M max. (at R_2) $\dots\dots = Pa$

M_x (between supports) $\dots\dots = \frac{Pax}{l}$

M_{x_1} (for overhang) $\dots\dots = P(a - x_1)$

Δmax. (between supports at $x = \frac{l}{\sqrt{3}}$) $= \frac{Pal^2}{9\sqrt{3}EI} = .06415\frac{Pal^2}{EI}$

Δmax. (for overhang at $x_1 = a$) $= \frac{Pa^2}{3EI}(l + a)$

Δx (between supports) $\dots\dots = \frac{Pax}{6EIl}(l^2 - x^2)$

Δx_1 (for overhang) $\dots\dots = \frac{Px_1}{6EI}(2al + 3ax_1 - x_1^2)$

27. BEAM OVERHANGING ONE SUPPORT—UNIFORMLY DISTRIBUTED LOAD BETWEEN SUPPORTS

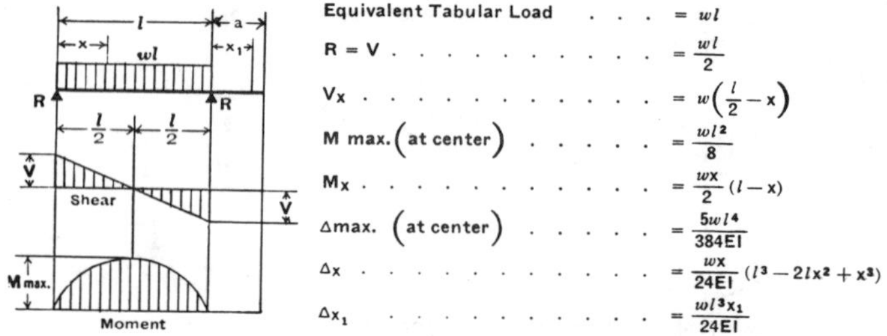

Shear
Moment

Equivalent Tabular Load $\dots\dots = wl$

$R = V \dots\dots = \frac{wl}{2}$

$V_x \dots\dots = w\left(\frac{l}{2} - x\right)$

M max. (at center) $\dots\dots = \frac{wl^2}{8}$

$M_x \dots\dots = \frac{wx}{2}(l - x)$

Δmax. (at center) $\dots\dots = \frac{5wl^4}{384EI}$

$\Delta x \dots\dots = \frac{wx}{24EI}(l^3 - 2lx^2 + x^3)$

$\Delta x_1 \dots\dots = \frac{wl^3x_1}{24EI}$

28. BEAM OVERHANGING ONE SUPPORT—CONCENTRATED LOAD AT ANY POINT BETWEEN SUPPORTS

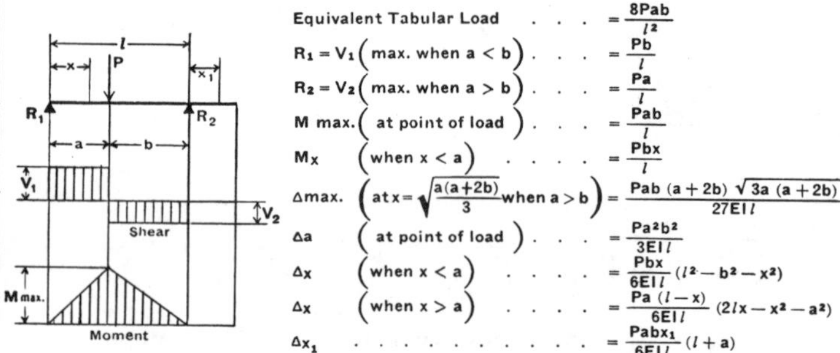

Shear
Moment

Equivalent Tabular Load $\dots\dots = \frac{8Pab}{l^2}$

$R_1 = V_1$ (max. when $a < b$) $\dots\dots = \frac{Pb}{l}$

$R_2 = V_2$ (max. when $a > b$) $\dots\dots = \frac{Pa}{l}$

M max. (at point of load) $\dots\dots = \frac{Pab}{l}$

M_x (when $x < a$) $\dots\dots = \frac{Pbx}{l}$

Δmax. (at $x = \sqrt{\frac{a(a+2b)}{3}}$ when $a > b$) $= \frac{Pab\,(a + 2b)\sqrt{3a\,(a + 2b)}}{27EIl}$

Δa (at point of load) $\dots\dots = \frac{Pa^2b^2}{3EIl}$

Δx (when $x < a$) $\dots\dots = \frac{Pbx}{6EIl}(l^2 - b^2 - x^2)$

Δx (when $x > a$) $\dots\dots = \frac{Pa\,(l - x)}{6EIl}(2lx - x^2 - a^2)$

$\Delta x_1 \dots\dots = \frac{Pabx_1}{6EIl}(l + a)$

29. CONTINUOUS BEAM—TWO EQUAL SPANS—UNIFORM LOAD ON ONE SPAN

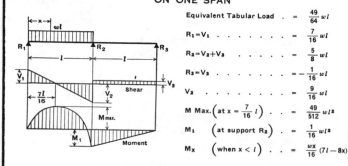

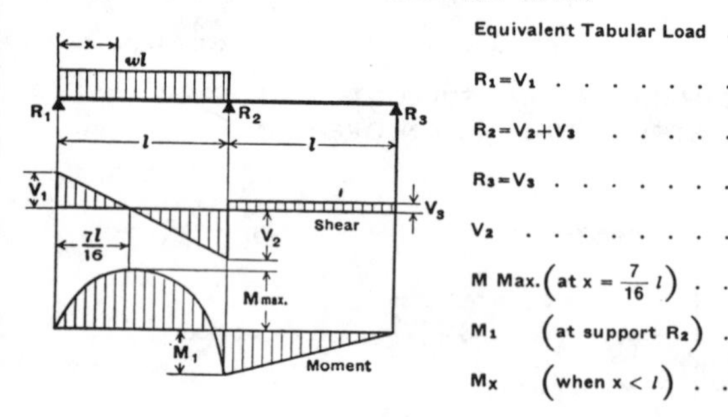

Shear
Moment

Equivalent Tabular Load $\dots = \frac{49}{64}wl$

$R_1 = V_1 \dots\dots = \frac{7}{16}wl$

$R_2 = V_2 + V_3 \dots\dots = \frac{5}{8}wl$

$R_3 = V_3 \dots\dots = -\frac{1}{16}wl$

$V_2 \dots\dots = \frac{9}{16}wl$

M Max. (at $x = \frac{7}{16}l$) $\dots = \frac{49}{512}wl^2$

M_1 (at support R_2) $\dots = \frac{1}{16}wl^2$

M_x (when $x < l$) $\dots = \frac{wx}{16}(7l - 8x)$

30. CONTINUOUS BEAM—TWO EQUAL SPANS—CONCENTRATED LOAD AT CENTER OF ONE SPAN

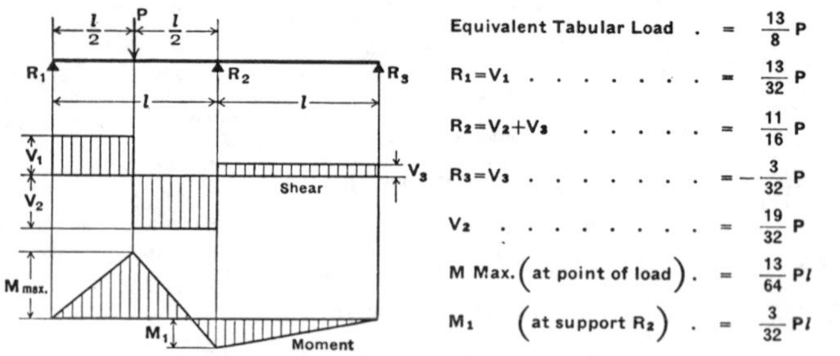

Shear
Moment

Equivalent Tabular Load $\dots = \frac{13}{8}P$

$R_1 = V_1 \dots\dots = \frac{13}{32}P$

$R_2 = V_2 + V_3 \dots\dots = \frac{11}{16}P$

$R_3 = V_3 \dots\dots = -\frac{3}{32}P$

$V_2 \dots\dots = \frac{19}{32}P$

M Max. (at point of load) $\dots = \frac{13}{64}Pl$

M_1 (at support R_2) $\dots = \frac{3}{32}Pl$

31. CONTINUOUS BEAM—TWO EQUAL SPANS—CONCENTRATED LOAD AT ANY POINT

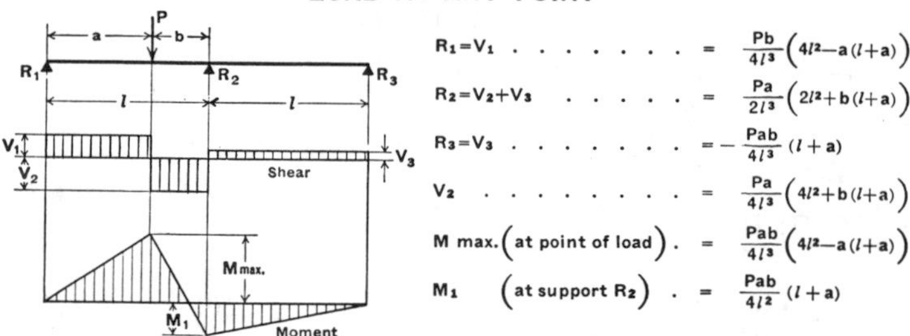

Shear
Moment

$R_1 = V_1 \dots\dots = \frac{Pb}{4l^3}\left(4l^2 - a(l + a)\right)$

$R_2 = V_2 + V_3 \dots\dots = \frac{Pa}{2l^3}\left(2l^2 + b(l + a)\right)$

$R_3 = V_3 \dots\dots = -\frac{Pab}{4l^3}(l + a)$

$V_2 \dots\dots = \frac{Pa}{4l^3}\left(4l^2 + b(l + a)\right)$

M max. (at point of load) $\dots = \frac{Pab}{4l^3}\left(4l^2 - a(l + a)\right)$

M_1 (at support R_2) $\dots = \frac{Pab}{4l^2}(l + a)$

32. SIMPLE BEAM—ONE CONCENTRATED MOVING LOAD

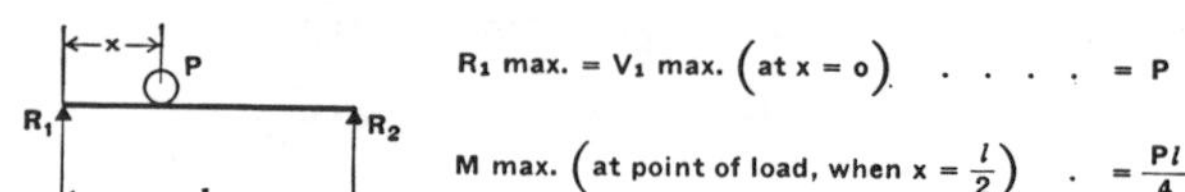

R_1 max. $= V_1$ max. (at $x = o$) $\dots\dots = P$

M max. (at point of load, when $x = \frac{l}{2}$) $\dots = \frac{Pl}{4}$

33. SIMPLE BEAM—TWO EQUAL CONCENTRATED MOVING LOADS

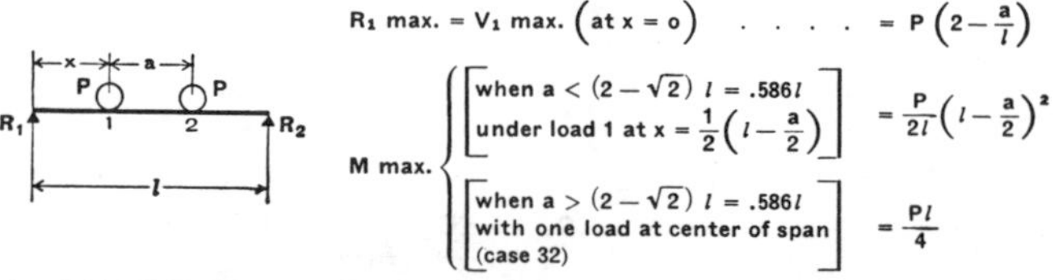

R_1 max. $= V_1$ max. (at $x = o$) $\dots\dots = P\left(2 - \frac{a}{l}\right)$

M max.
- [when $a < (2 - \sqrt{2})\,l = .586l$ under load 1 at $x = \frac{1}{2}\left(l - \frac{a}{2}\right)$] $= \frac{P}{2l}\left(l - \frac{a}{2}\right)^2$
- [when $a > (2 - \sqrt{2})\,l = .586l$ with one load at center of span (case 32)] $= \frac{Pl}{4}$

34. SIMPLE BEAM—TWO UNEQUAL CONCENTRATED MOVING LOADS

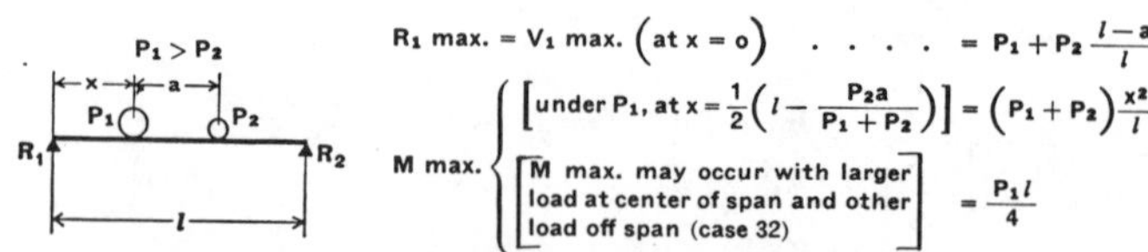

R_1 max. $= V_1$ max. (at $x = o$) $\dots\dots = P_1 + P_2\frac{l - a}{l}$

M max.
- [under P_1, at $x = \frac{1}{2}\left(l - \frac{P_2a}{P_1 + P_2}\right)$] $= (P_1 + P_2)\frac{x^2}{l}$
- [M max. may occur with larger load at center of span and other load off span (case 32)] $= \frac{P_1l}{4}$

GENERAL RULES FOR SIMPLE BEAMS CARRYING MOVING CONCENTRATED LOADS

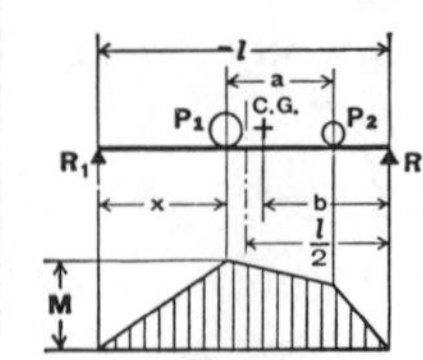

Moment

The maximum shear due to moving concentrated loads occurs at one support when one of the loads is at that support. With several moving loads, the location that will produce maximum shear must be determined by trial.

The maximum bending moment produced by moving concentrated loads occurs under one of the loads when that load is as far from one support as the center of gravity of all the moving loads on the beam is from the other support.

In the accompanying diagram, the maximum bending moment occurs under load P_1 when $x = b$. It should also be noted that this condition occurs when the center line of the span is midway between the center of gravity of loads and the nearest concentrated load.

From A.I.S.C. Steel Construction Manual - 1950.

SOIL BEHAVIOR - I

SOIL MECHANICS - SOIL BEHAVIOR

SHEARING RESISTANCE vs. STABILITY OF SOILS

All stability in soils is derived from shearing strength.

Corollaries

(a) Ability of soils to support spread footings is directly proportional to their shearing strength.

(b) The steepness of the safe slope of embankments depends on the shearing strength of the soil.

(c) The safety of the supporting soil under the embankment from mud wave action depends on the shearing strength of the soil.

SHEARING RESISTANCE - GRANULAR vs. COHESIVE SOILS

As granular soils have internal friction resistance, their shearing strength increases in relation to the normal pressure to which they are subjected.

Granular soils have no cohesion.

Clays are cohesive soils and have no angle of internal friction. They have a resistance to shearing due to their cohesive or molecular strength.

Hence their shearing strength is the same regardless of the normal pressure to which they are subjected.

Soil mixtures of clay and sand partake of both cohesive and internal friction resistance. Thus, most sands have some cohesion; most clays have some internal friction.

Corollaries

(a) The safe slope of a granular bank does not decrease as the height increases because its shearing strength increases to resist the increasing shearing stresses as the bank becomes higher.

(b) The safe slope of a clay bank becomes flatter as the bank becomes higher because its shearing strength does not increase to resist the corresponding increasing shearing stress.

FIELD TEST FOR SHEARING STRENGTH (See p. 9-26.)

(a) The so-called unconfined compression test is very simple and may be improvised in the field. (See p. 9-26.)

(b) For granular soils, the angle of repose is approximately equal to the angle of internal friction.

(c) Load tests are of little value except on cohesive soils.

JUDGING STRENGTH OF SOILS

(a) Classification is by grain size such as clay, silt, sand, and gravel. It is usually used in connection with empirical tables to give strength for building foundations.

(b) Refinement of this classification is the measurement of resistance to a standard driving spoon which is tied into the empirical strength table. See p. 9-12.

(c) Load tests on soils and piles, see p. 7-01.

(d) For determining the wheel bearing capacity of composite sections of flexible pavements and bases, there is the California Bearing Ratio, which is the resistance of a standard plunger to being forced into a layer of the base or subbase compared to its resistance to being forced into a standard layer of broken stone. Empirical curves are available. See pp. 10-58 and 10-59.

(e) For rigid pavements, a plate loaded according to a standard procedure gives a value known as the modulus of subgrade reaction, <u>K</u>, which is used in connection with established formulas to estimate the bearing capacity of concrete pavements. See p. 10-57.

(f) In general, the heavier the unit weight of soil, the greater the strength.

(g) In general, the less voids, the greater the strength.

WORKED-OVER CLAYS

Clays are subject to remolding; that is, an apparently stiff clay if worked will give off water and become soft. This is because a certain amount of water is fixed or attracted to the submicroscopic particles of clay, and the adhered water is broken off by working. If the clay is allowed to rest, the water will readhere and the clay will regain some of its stiffness.

Corollaries

(a) Piles driven into clay, if left overnight, will set up and be difficult to start.

(b) Freezing and thawing works the clay and weakens it, causing settlement or slides.

SETTLEMENT OF FOOTINGS ON CLAYS

Clay settlement varies directly with water content and inversely with the cohesive strength.

For a given unit pressure and a given clay, a definite amount of settlement may be expected.

For instance, clay sample No. 1 under pressure of 2 tons to the square foot might have an expected total settlement of 2 inches. After this settlement has taken place and the pressure is increased to 3 tons, there would be a certain additional settlement of, let us say, 1-1/2 inches.

SOIL BEHAVIOR - 2

There are also definite periods of time for these settlements to become complete.
The settlement is caused by the squeezing out of water.

Corollaries

(a) The larger the footing, the greater the settlement, because the area of the bulb of pressure will increase in proportion to the sizes of the footings. See Fig. 1.

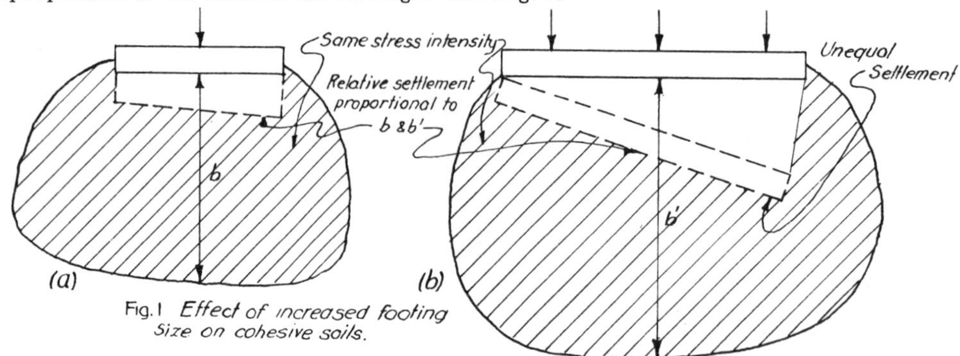
Same stress intensity
Relative settlement proportional to b & b'
Unequal Settlement
b
b'
(a)
(b)

Fig. 1 Effect of increased footing size on cohesive soils.

(b) While increase in size of footing increases the amount of settlement in the relation of b' to b, it does not increase the danger of failure because the intensity of shearing stresses in 1(a) are no greater than in 1(b).

(c) However, unequal settlement as shown in 1(b) might produce critical shearing stresses, or a vicious cycle of rotation.

SHRINKAGE

Clays tend to hold free water in addition to their adhered water. They do not drain nor do they dry out rapidly.
Granular soils do not hold water readily.

Corollaries

(a) Clays are subject to a large amount of shrinkage, but the loss of water that causes this shrinkage is slow.

(b) These shrinkages might amount to as much as 20% in volume.

(c) Granular soils do not shrink much when drying, and they shrink more rapidly.

SETTLEMENT OF FOOTINGS ON SANDS

Compressibility of sands and silts varies with density.

PORE WATER PRESSURE

When granular soils are saturated with water and the water is trapped, the footing may be supported on hydraulic pressure. This is called pore water pressure. Examples are boils or soft spots in a subgrade, or a ground with quakes under the passage of trucks.

Corollaries

(a) Soils under pore water pressure are without shearing strength.

(b) The seepage out of pore water under pressure will cause settlement.

(c) Drainage to relieve the water is indicated.

COMPACTION-OPTIMUM MOISTURE

For most soils there is a percentage of moisture at which the soil will compact to its greatest density.
This optimum moisture may run from 8% for sands, 15% for silts, to 15-20% for clays.

Corollaries

(a) Water content should be controlled in making fills.

(b) Reduction of water in clays is very difficult. Suggestions for doing this are as follows:
Spread out in thin layers to dry.
Choose a borrow dryer than optimum moisture.
Modern construction equipment supplemented by rollers is recommended for consolidation of fills.

(c) Deep trenches with clay backfill may require puddling, which will result in considerable shrinkage, or power tampers may be used to break up the lumps.

(d) Sand backfill may be inundated in compacting around a foundation wall, which will leave the sand loose and subject to settlement, but it may be of benefit as it will improve the drainage around the basement walls.

(e) For pavements and floor fills, optimum moisture and heavy rolling and tamping are the only answers.

(f) A well-graded gravel may be compacted with a bulldozer in 12-inch layers.

SOIL BEHAVIOR – 3

SOIL INVESTIGATIONS

(a) Borings giving the material and the resistance to driving with a standard spoon.

(b) Undisturbed samples which can be tested for density, shearing resistance, and settlement.

(c) Atterberg Limit Tests, which are tests of fluidity and sensitivity to vibration. Their use includes determination of resistance of soils to frost, soil classification, compressibility, and shrinkage. See p.9-21.

CAPILLARY ACTION

Soils possess capillary action similar to a dry cloth with one end immersed in water.

Corollaries

(a) Coarse gravel: no capillary action

(b) Coarse sand: up to 12 inches.

(c) Fine sands and silts: up to 3 feet, low in clay.

(d) Pure clay: low value.

FROST ACTION

Most damage from frost action occurs at the time of thawing.

Corollaries

(a) With granular soils and mixtures, frost aided by capillary action may suck up water and cause actual ice lenses. When this material thaws, the soil is over-irrigated and incapable of supporting a load. A pore water condition may result.

(b) With a clay soil, the heave may be less because the clay soil does not readily suck up water from below on account of its imperviousness, but when it thaws the frozen clay will lose its strength as when it is worked.

(c) Frost boils in pavement occur from similar causes.

CURE

All structural foundations to extend below frost. Subgrades of roads to have a gravel capillary cut-off.

QUICKSAND

Quicksand occurs where a current of water passing upward through a soil is of sufficient velocity to cause a flotation or boiling up of the particles.

Corollaries

(a) The bearing value of the material is destroyed.

(b) Material may be carried off from under a structure or adding material.

(c) Correct by lowering ground water level.

SAND FILTERS

Water passing out of a fine soil into a coarse aggregate may tend to carry material with it and clog the drain.

Corollary

(a) The grains around the drain should be graded so that there is no abrupt change in size. See p.9-07.

VIBRATION

Vibration may cause damage to structures by:

(a) Pile driving, which
  1. May endanger tender adjoining masonry.
  2. May start overloaded skin friction piles down.
  3. May cause cracks in plaster or masonry.

(b) Effect of resonance: The natural period of the earth crust may vary between 1000 and 2000 cycles per minute, and machinery, for instance, vibrating within these limits might cause trouble.

(c) Truck vibrations may act similarly to pile driving.

(d) Blasts are not continuous, and a large part of the effect of blast is an air wave. However, blasts are a distinct danger.

Precautions

(a) Shoring of weak masonry and foundations.

(b) Insulation of foundations of vibrating machinery.

(c) Open windows during blasting.

(d) Line hole rock supporting adjacent buildings.

(e) Shore exposed rock faces.

(f) Protect client by preconstruction survey of adjoining properties together with photographs.

SOIL BEHAVIOR – 4

EFFECT OF WATER ON SOILS

Water in soil is the all-round enemy of the engineer and constructor. Some of the troubles caused by water are:

(a) Frost failures.
(b) Settlement due to drying out of water in soil.
(c) Reduction of the bearing power of soil due to weakening of the shearing strength.
(d) Removal of soil under foundations due to pumping operations.
(e) Cavitation in piping in dams.
(f) Construction difficulties.

Water is fought by:

(a) Gravity drainage.
(b) Ordinary pumping.
(c) Pumping with well points in soils that are somewhat granular.
(d) Sealing of the casing of well points to produce better vacuum.

Advantages of Water:

(a) Protection of untreated timber below water line.
(b) Prevention of shrinkage in soils due to drying out.

FLOW THROUGH SOILS

Flow through soils is dependent on the hydraulic gradient and the fineness of the material, and can be estimated. See p. 9-34.

(a) The uplift on the base of a dam should be estimated.
(b) Piping which is a channelized outflow of soil carried away by this seepage must be controlled.

SOILS – APPLICATION OF MECHANICS – I

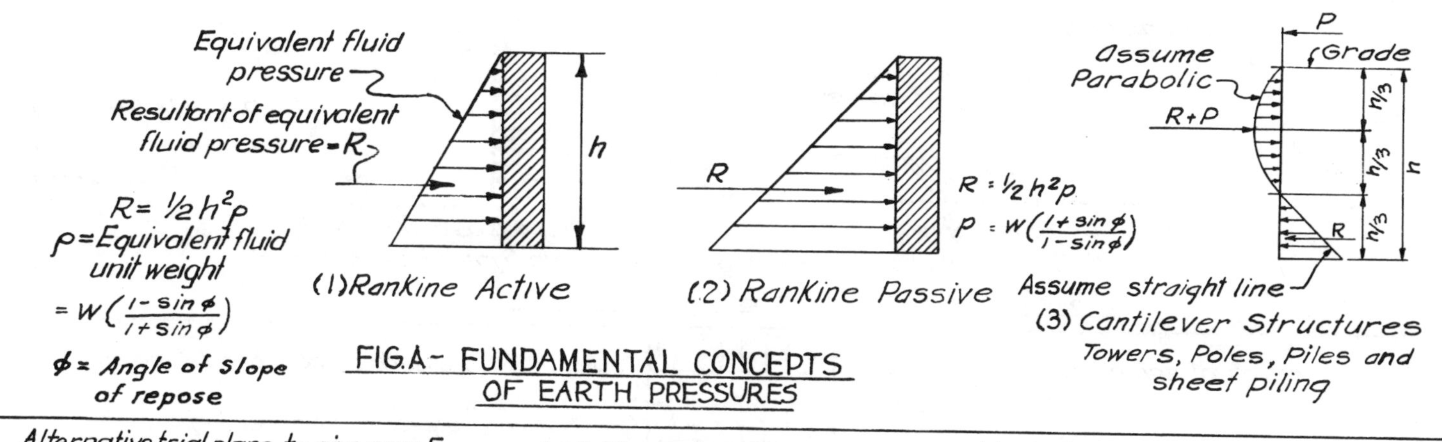

Equivalent fluid pressure
Resultant of equivalent fluid pressure = R
h
$R = \frac{1}{2} h^2 p$
p = Equivalent fluid unit weight
$= w\left(\frac{1-\sin\phi}{1+\sin\phi}\right)$
φ = Angle of slope of repose
(1) Rankine Active
R
$R = \frac{1}{2} h^2 p$
$p = w\left(\frac{1+\sin\phi}{1-\sin\phi}\right)$
(2) Rankine Passive
P
Grade
Assume Parabolic
R+P
h/3
h/3
h/3
h
R
Assume straight line
(3) Cantilever Structures
Towers, Poles, Piles and sheet piling

FIG. A – FUNDAMENTAL CONCEPTS OF EARTH PRESSURES

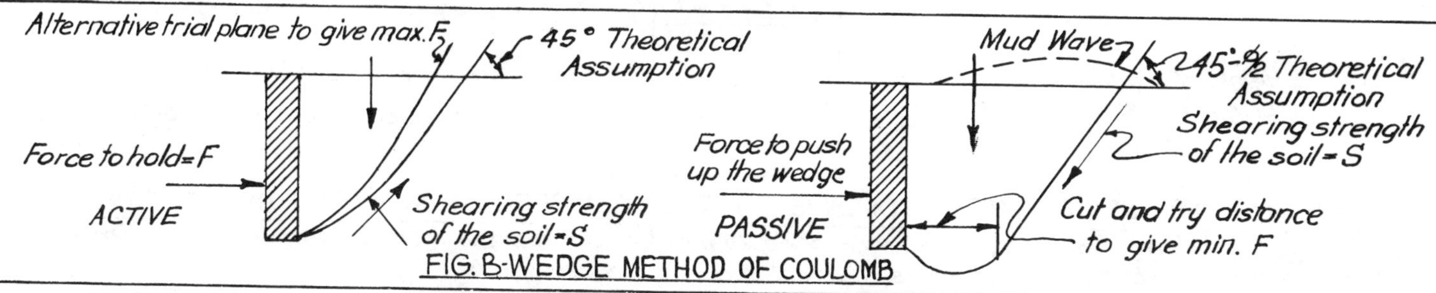

Alternative trial plane to give max. F
45° Theoretical Assumption
Force to hold = F
ACTIVE
Shearing strength of the soil = S
Mud Wave
45°-φ/2 Theoretical Assumption
Shearing strength of the soil = S
Force to push up the wedge
PASSIVE
Cut and try distance to give min. F

FIG. B – WEDGE METHOD OF COULOMB

SUBMERGED SATURATED, AND MOIST WEIGHTS.

Moist weight increases if sample is saturated and decreases if submerged on account of flotation. Most unfavorable condition should be assumed.

p = Equivalent fluid unit weight
w = Unit weight of soil behind wall
Let $k = p/w$
l = lbs. per sq. ft of surcharge
h = Equivalent fluid head without surcharge.
h' = Head with surcharge.

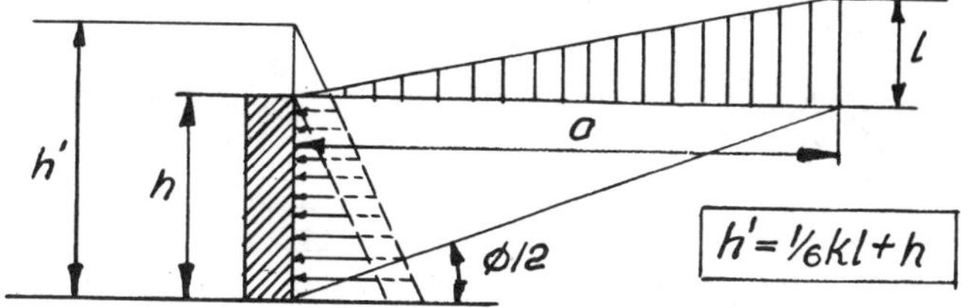

l
a
h'
h
φ/2
$h' = \frac{1}{6}kl + h$

Case I – Surcharge due to sloping bank

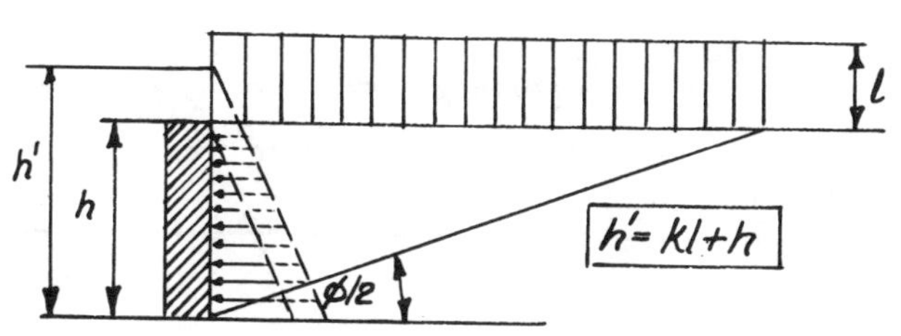

l
h'
h
$h' = kl + h$
φ/2

Case II – Uniform surcharge

Conception of equivalent fluid head behind a wall with surcharge.

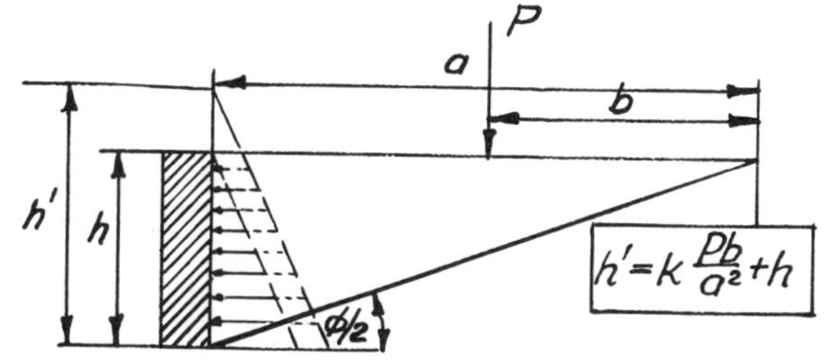

P
a
b
h'
h
$h' = k\frac{Pb}{a^2} + h$
φ/2

Case III – Concentrated load

FIG. C – RELATION OF SURCHARGE TO EQUIVALENT FLUID PRESSURE

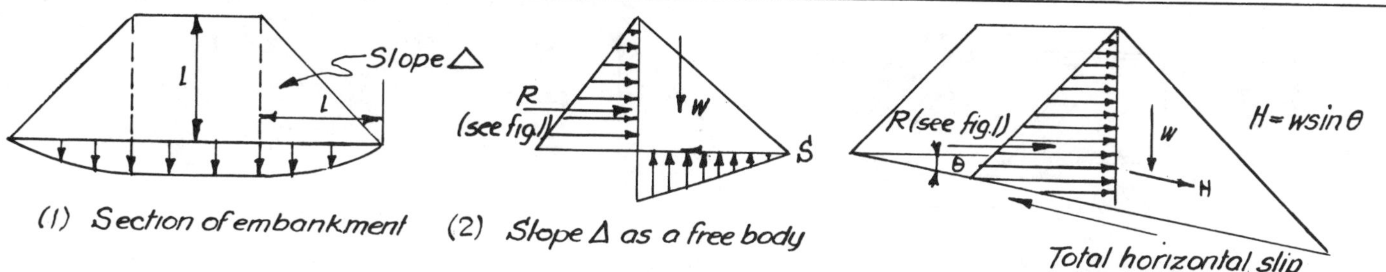

Slope △
l
l
R (see fig. 1)
W
S
R (see fig. 1)
W
θ
H
H = w sin θ

(1) Section of embankment (2) Slope △ as a free body

Check of a safe shear = S
S = Friction + cohesion

Total horizontal slip on base = R cos θ + H

(3) Embankment on a slope

FIG. D – EMBANKMENT STABILITY – "SLIDING BLOCK" METHOD

SOILS - APPLICATION OF MECHANICS - 2

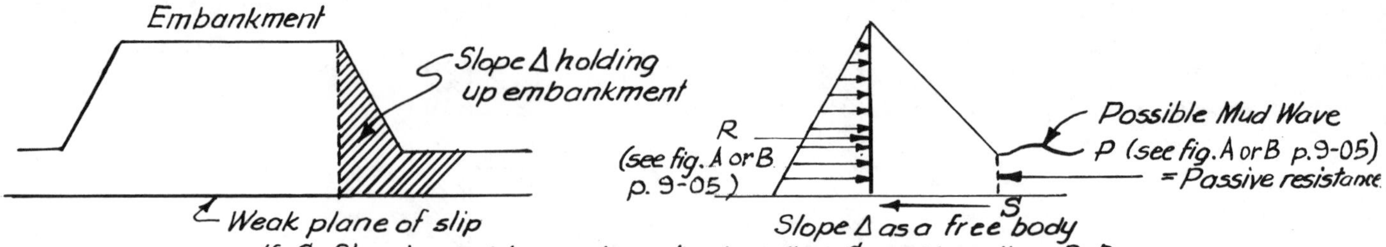

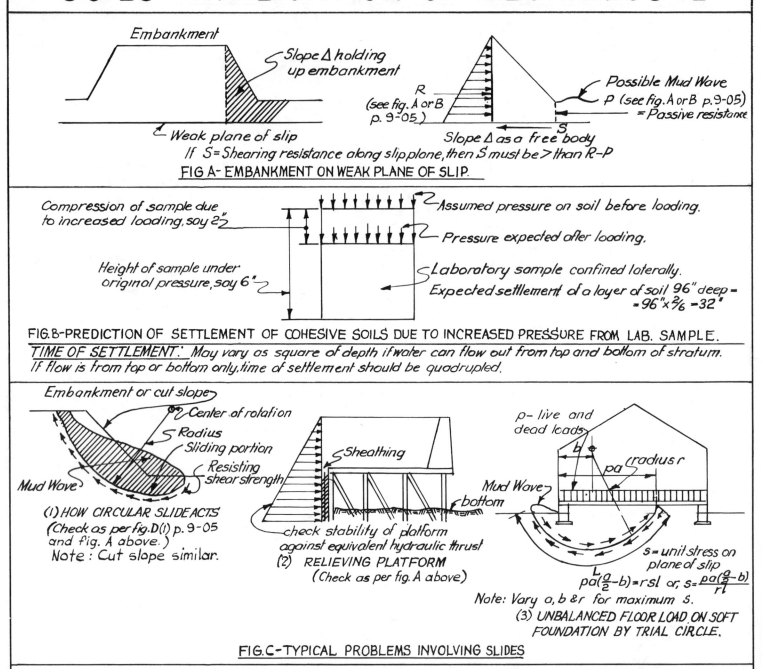

Embankment

Slope Δ holding up embankment

Weak plane of slip

R (see fig. A or B p. 9-05)

Possible Mud Wave

P (see fig. A or B p. 9-05) = Passive resistance

S

Slope Δ as a free body

If S = Shearing resistance along slip plane, then S must be > than R-P

FIG A- EMBANKMENT ON WEAK PLANE OF SLIP.

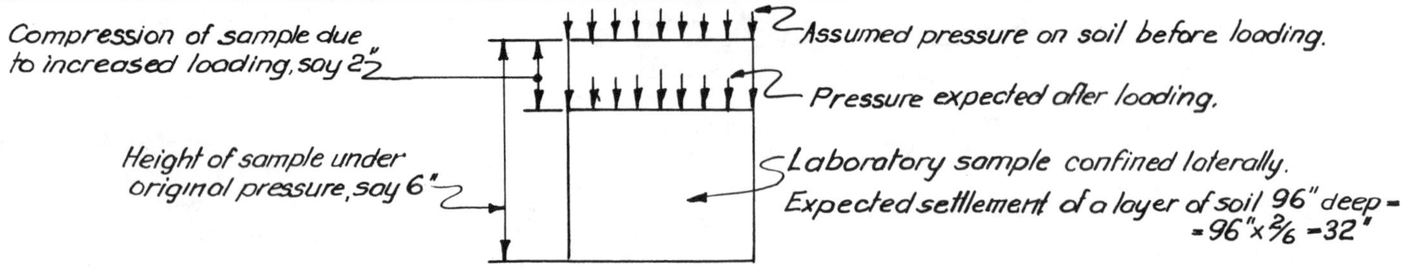

Compression of sample due to increased loading, say 2"

Assumed pressure on soil before loading.

Pressure expected after loading.

Height of sample under original pressure, say 6"

Laboratory sample confined laterally.

Expected settlement of a layer of soil 96" deep = $96'' \times 2/6 = 32''$

FIG. B-PREDICTION OF SETTLEMENT OF COHESIVE SOILS DUE TO INCREASED PRESSURE FROM LAB. SAMPLE.

TIME OF SETTLEMENT: May vary as square of depth if water can flow out from top and bottom of stratum. If flow is from top or bottom only, time of settlement should be quadrupled.

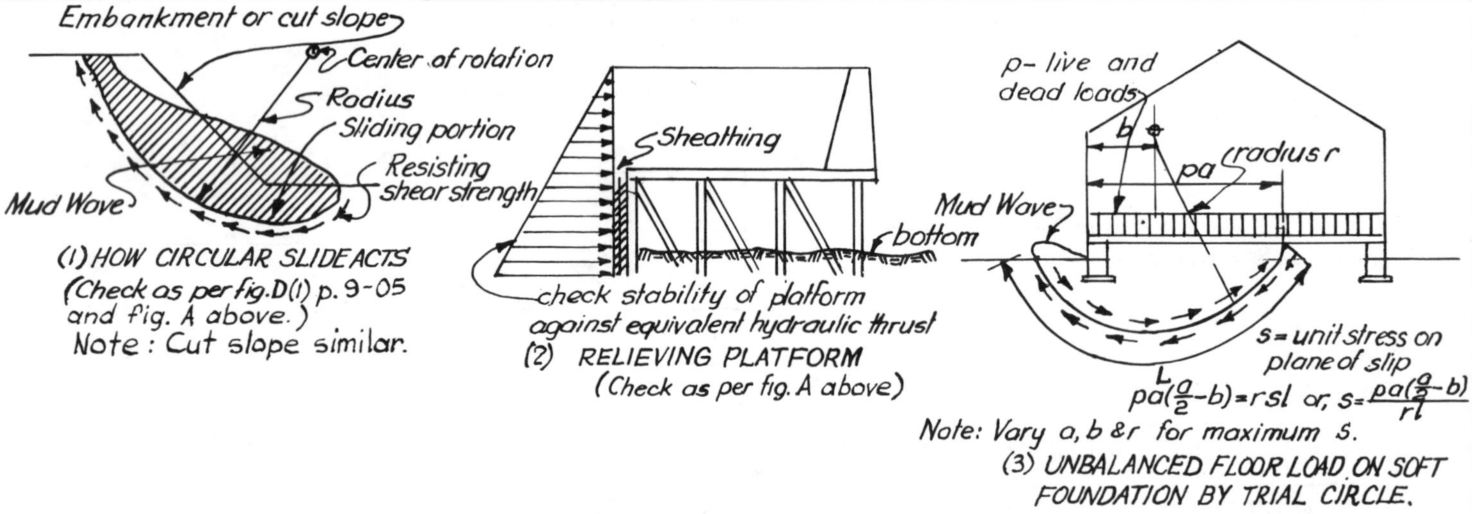

Embankment or cut slope

Center of rotation

Radius

Sliding portion

Resisting shear strength

Mud Wave

(1) HOW CIRCULAR SLIDE ACTS
(Check as per fig. D(1) p. 9-05 and fig. A above.)
Note: Cut slope similar.

Sheathing

bottom

check stability of platform against equivalent hydraulic thrust

(2) RELIEVING PLATFORM
(Check as per fig. A above)

p - live and dead loads

b

a

radius r

Mud Wave

L

s = unit stress on plane of slip

$pa(\frac{a}{2}-b) = rsl$ or, $s = \frac{pa(\frac{a}{2}-b)}{rl}$

Note: Vary a, b & r for maximum s.

(3) UNBALANCED FLOOR LOAD ON SOFT FOUNDATION BY TRIAL CIRCLE.

FIG. C-TYPICAL PROBLEMS INVOLVING SLIDES

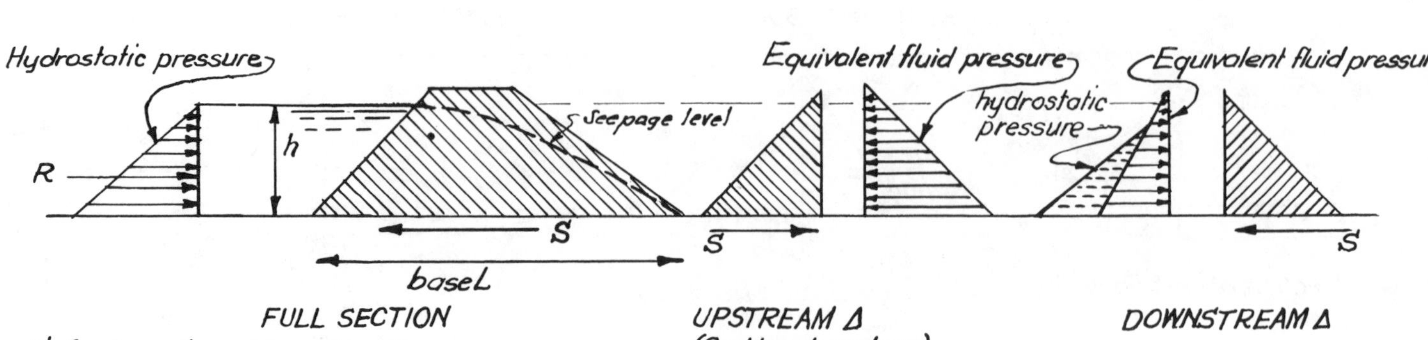
Hydrostatic pressure
R
h
Seepage level
S
base L
FULL SECTION
Safe shear, S must be > than R
Equivalent fluid pressure
S
UPSTREAM Δ
(Sudden drawdown)
Equivalent fluid pressure
hydrostatic pressure
S
DOWNSTREAM Δ

FIG. D-STRESS DIAGRAMS FOR EARTH DAMS

SOILS – APPLICATION OF MECHANICS – 3

I. ARBITRARY STRENGTH PER SQUARE FOOT

By Soil Classification, see pages 9-08 to 9-11 and 9-15 to 9-17.

II. BY DISPLACEMENT:

For cohesive soils such as clays or clay mixtures.

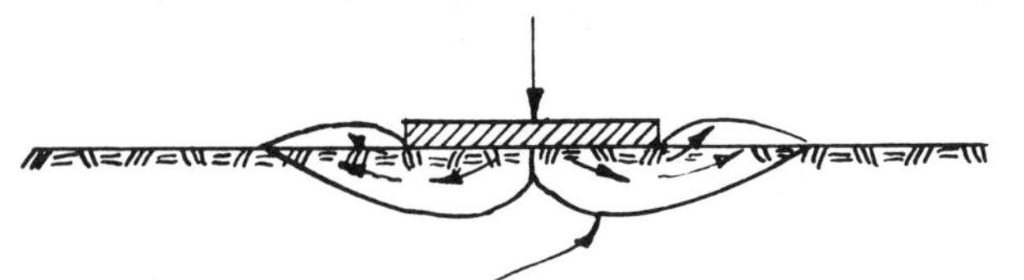

Load causes soil to shear on such plane as this.

III. ELASTIC THEORY: Assumes soil truly Elastic

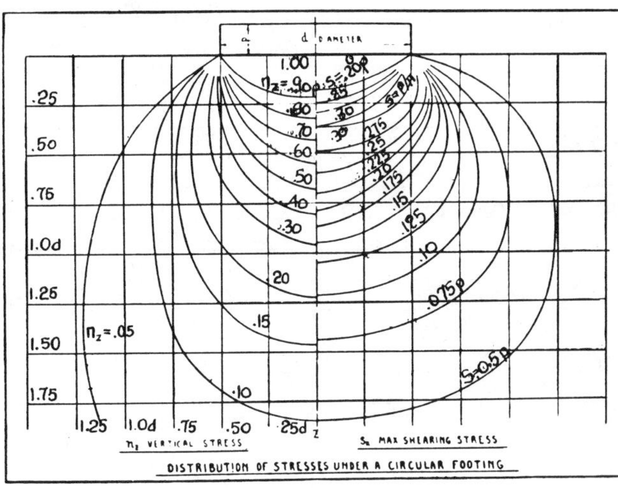

d DIAMETER
.25
.50
.75
1.0d
1.25
1.50
1.75
n_z = .05
n_z VERTICAL STRESS
S_z MAX SHEARING STRESS
DISTRIBUTION OF STRESSES UNDER A CIRCULAR FOOTING

Unit shear and direct stress given for a circular footing.

FIG. A DESIGN OF SPREAD FOOTINGS

D'ARCY'S LAW:

Seepage volume varies directly with area.

Volume of seepage $= kA\frac{h}{L}$

L = Average length of flow

W = Average width of flow path.

A = w x length of dam

k = coefficient of permeability See page 9-34

boundary

h

L

w

FIG. B – COMPUTATION OF SEEPAGE

Flow lines without filter

Flow lines with filter

Filters tend to bend seepage lines down.

Filter drain

Note: Filters are of sand and gravel and are graded from coarse to fine to prevent the fine from the soil from being carried into the filter and clogging it

FIG. C CONTROL OF DAM SEEPAGE BY FILTERS

Base strong

Sub-base medium strong

Sub grade medium weak

Sub-soil weak

Figure shows spreading of load in accordance with strength of underlying strata.

FIG. D BASIS OF FLEXIBLE PAVEMENT DESIGN

Design pavement for flexural strength

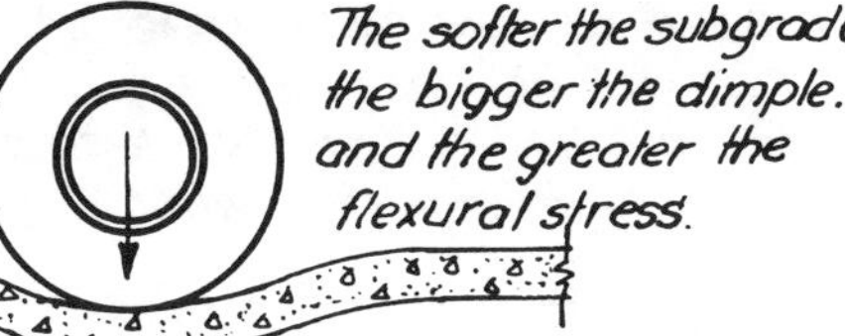

The softer the subgrade the bigger the dimple. and the greater the flexural stress.

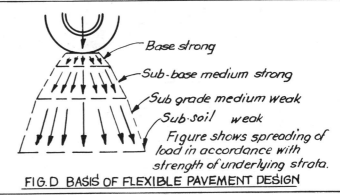

Intensity of dimple estimated from plate load test on soil, called subgrade reaction.

FIG. E BASIS OF RIGID PAVEMENT DESIGN.

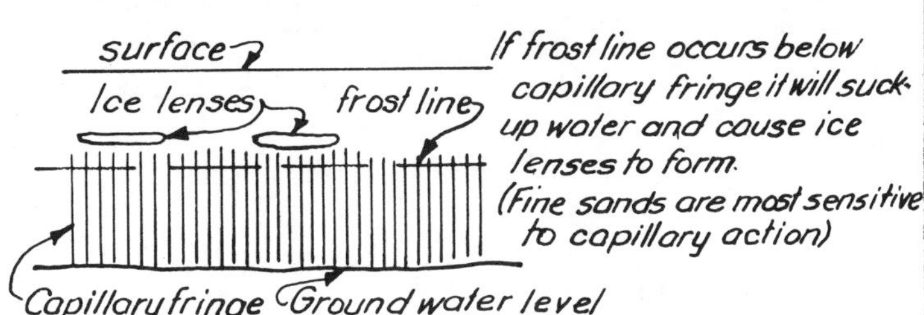

surface
Ice lenses
frost line
Capillary fringe
Ground water level

If frost line occurs below capillary fringe it will suck up water and cause ice lenses to form. (Fine sands are most sensitive to capillary action)

FIG. F RELATION OF CAPILLARY PROPERTIES TO FROST BOILS IN HIGHWAYS.

SOILS — FACTUAL DATA

TABLE A - SLOPES OF REPOSE

KIND OF EARTH	SLOPE OF REPOSE	
	NON-SUBMERGED	SUBMERGED
Sand, clean	1 on 1.5	1 on 2
Sand and clay	1 on 1.33	1 on 3
Clay, dry	1 on 1.75	
Clay		1 on 3.5
Clay, damp, plastic	1 on 3	
Gravel, Clean	1 on 1.33	1 on 2
Gravel and clay	1 on 1.33	1 on 3
Gravel, sand, and clay	1 on 1.5	1 on 3
Soft, rotten rock	1 on 1	1 on 1
Hard, rotten rock	1 on 1	
Hard rock, riprap		1 on 1
Bituminous cinders	1 on 1	
River mud		1 on 3 to 1 on 20
Anthracite Ashes	1 on 1.5 to 1 on 2	

Rule of thumb for submerged excavated slopes:
Sand- 1 on 2   Clay-1 on 1.5 to vertical
Stiff mud-1 on 1 to vertical   Sluiced Mud-1 on 10 to 1 on 20

TABLE D - SOIL SHEAR VALUES

KIND OF EARTH	COHESION, lbs/ft²	ANG. OF INT. FR.°
Clay - liquid†	100	0°
" very soft	200	2°
" soft	400-500	4°
" firm	1000	6°
" stiff	2000	12°
" very stiff	2000-4000	14
Sand wet	0	10°-15°
Sand dry or unmoved	0	34°
Silt	0	± 20°
Cemented Sand & Gravel - Wet	500	34°
" " " " - Dry	1000	34°

TABLE G - WEIGHT OF SOLIDS SUBMERGED IN SEA WATER

MATERIAL	POUND PER CUBIC FOOT		
	MAXIMUM	MINIMUM	AVERAGE
Gravel and Marl		42.0	62.9
Gravel and Sand	73.0	42.0	62.4
Sand	66.0	42.0	58.3
Gravel, Sand, and Clay	80.9	51.2	70.0
Stiff Clay	64.8	38.4	47.8
Stiff Clay and Gravel	70.3	44.8	52.6

* C.B.R. = California Bearing Ratio.

TABLE B - UNIT WEIGHT OF SOILS

KIND OF EARTH	UNIT WEIGHTS - lb/cu.ft.
Moist Soils	110
Medium or stiff clay	120
Saturated earth	110 + % voids × 62.5 = say 132
Submerged earth	132 − 62.5 = say 70
Soft clay or mud	100

TABLE C - UNIT WEIGHTS AND C.B.R. VALUES FOR COMPACTED SOILS

SOLIDS AT OPTIMUM COMPACTION	WELL GRADED		NOT GRADED	
	UNIT WT.	C.B.R.*	UNIT WT.	C.B.R.
Sand and Silt	120	–	105	8-30
Sand and Clay (Binder)	125	20-60	105	8-30
Sands	120	20-60	100	10-30
Gravel	130	>50	115	25-60
Silts inorganic	–	–	100	6-25
Organic Silts	–	–	90	3-8

TABLE E - ROUGH DATA FOR EQUIVALENT FLUID PRESSURES OF SOILS

MULTIPLY UNIT WEIGHT BY "K"

KIND OF EARTH	"K"
Granular Sand	0.33
Mixtures of Clay and Granular Soils	0.50
Soft Clays, Silts, Organic Soils	1.00
Stiff Clays	1.00

TABLE F - EQUIVALENT FLUID PRESSURES FOR SOILS SUBMERGED IN SEA WATER**

SLOPE OF REPOSE OF EARTH	WEIGHT 'W' OF SUBMERGED EARTH LB./CU.FT.							
	40	44	48	52	56	60	64	68
1 on 1/2	66.2	66.4	66.7	66.9	67.1	67.3	67.6	67.8
1 on 3/4	68.2	68.9	69.3	69.8	70.2	70.7	71.1	71.6
1 on 1	70.9	71.6	72.2	72.9	73.6	74.3	75.0	75.7
1 on 1 1/4	73.2	74.2	75.1	76.0	76.9	77.9	78.8	79.7
1 on 1 1/2	75.4	76.6	77.7	78.9	80.0	81.2	82.3	83.5
1 on 1 3/4	77.5	78.8	80.2	81.5	82.9	84.2	85.6	86.9
1 on 2	79.3	80.8	82.3	83.9	85.4	86.9	88.4	90.0
1 on 2 1/2	82.3	84.2	86.0	87.8	89.7	91.5	93.3	95.2
1 on 3	84.8	86.9	88.9	91.0	93.1	95.2	97.2	99.3
1 on 3 1/2	86.8	89.0	91.3	93.6	95.9	98.2	100.0	102.0
1 on 4	88.4	90.8	93.3	95.7	98.1	101.0	103.0	105.0
1 on 5	90.9	93.6	96.3	99.0	102.0	104.0	107.0	109.0
1 on 6	104.0	108.0	112.0	116.0	120.0	124.0	128.0	132.0

To obtain equivalent fluid pressure for a given slope of repose, say 1¾:1, get the weight of the submerged soil from table G, say 60; enter table F at column marked nearest to this value. Obtain equivalent weight, 84.2 from row of given slope of repose.

† For Classification see page 9-30 (N.Y.C. Code)

** Adapted from "American Civil Engineers' Handbook" by Merriman & Wiggin.

SOILS - RED LIGHTS

FOUNDATIONS:

1. Foundations of structures designed on the basis of empirical bearing values of soils should be considered adequate and safe only where such values have been thoroughly established by local practice and have proven sound by adequate experience in similar structures in the same localities. Otherwise a complete analysis of the geology of the site should be determined, securing the necessary data by borings, test pits, and other soil investigations. Complex geological formations or unusual foundation conditions may frequently require the advice and experience of specialists in foundation design and soil analysis.*

2. Investigation of all, even thin layers of material under a foundation should be made, as thin layers may be weak in shear and thus cause a settlement. Even at great depths weak layers can cause important settlement.

3. Silts are not necessarily a poor foundation if dense. They are, however, likely to be found in a loose state.

4. Foundations should not be constructed on fill, peat, or organic materials.

5. Foundations should go down to maximum frost penetration. See page 9-35 and check local building codes.

6. Clay bearing values vary with moisture content. Adjoining deeper excavation may dry out clays and cause settlement.

7. Settlement on clays should be studied. For important structures, time settlement relations should be studied. Interior footings on clay may settle more than wall footings if some unit pressures are used; this, however, is not always true for pre-compressed clay.**

8. Foundations should be insulated from vibrating machinery.

9. Pumping operations should not be permitted to carry away soil from under foundations. Sands and silts should be particularly watched.

RETAINING WALLS:

1. Resultant of retaining wall should be at or near middle of base for cohesive soils because unequal settlement at the toe may start a vicious cycle.

2. Too much emphasis cannot be placed on drainage of rear of retaining walls.

3. Watch out for slip on foundation.

EMBANKMENTS AND DAMS:

1. Watch out for weak planes of slip such as thin layers of silt or sand which may irrigate the soil and weaken its shearing resistance.

2. Water seeping through cut slopes or sudden draw downs such as a tidal drop produce unbalanced hydraulic pressure as well as weakening the shearing resistance.

3. The angle, ø, of internal friction is lower for saturated soils than for moist soils thus increasing active and decreasing passive resistance.

PILES:

1. Settlements of group piles may be greater than those for single piles.

2. Drive and inspect all shells in a group for a cast in place pile before filling with concrete.

3. Wood piles must be cut off below permanent water level.

4. Steel "H" piles must be cut off below an acid bearing back fill such as cinders.

5. Piles which permit inspection after driving have an advantage in soils containing boulders or other obstructions.

SAMPLING:

The use of remoulded samples instead of undisturbed samples is invalid if the soil has been subjected to a larger load in the past as in the case of a glacier or overlying strata since eroded. Undisturbed samples should, therefore, be used.

Shelby sampling tubes with a diameter less than 3" should never be used to obtain undisturbed samples. Samples obtained with smaller diameter tubes are subject to disturbance because of friction along walls of tube during sampling and extrusion.

* Col. Carlton S. Proctor of Moran, Proctor, Freeman & Mueser. ** See page 9-30.

SOILS-IDENTIFICATION OF PRINCIPAL TYPES

TABLE A - MAJOR DIVISIONS OF SOILS.

COARSE GRAINED (GRANULAR)		FINE GRAINED		ORGANIC	
GRAVEL	SAND	SILT	CLAY	MUCK	PEAT

TABLE B - IDENTIFICATION - VISUAL & BY TEXTURE.

GRAVEL

Rounded or waterworn pebbles or bulk rock grains. No cohesion. No plasticity. Gritty & granular. Crunchy under foot. As a soil, over 1/10" in size. As an aggregate, over 1/4" in size.

SAND

Granular, gritty, loose grains, passing № 10 and retained on № 270 sieve. Individual grains readily seen and felt. No plasticity or cohesion. When dry a cast formed in the hands will fall apart. When moist, a cast will crumble when touched. The coarse grains are rounded - the fine grains are visible and angular. As an aggregate for Construction Sand consists of mineral grains between 1/4" and 1/200".

SILT

Fine, barely visible grains, passing № 270 sieve and over .005 mm. in size. Little or no plasticity. No Cohesion. A dried cast is easily crushed in the hands. Permeable - movement of water through voids occurs easily and is visible. When mixed with water the grains will settle in from 30 min. to 1 hour. Feels gritty when bitten. Will not form a ribbon. Care must be used to distinguish fine sand from silt & fine silt from clay.

CLAY

Invisible particles under 0.005 mm. (or 0.002 mm. in M.I.T. scale) in size. Cohesive. Highly plastic when moist. When pinched between the fingers will form a long thin, flexible ribbon. Can be rolled into a thread to a pin point. When bitten with the teeth will not feel gritty. Will form hard lumps or clods when dry, difficult or impossible to crush in hands. Impermeable no movement of water apparent through voids. Will remain suspended in water 3 hrs. to indefinitely.

MUCK and ORGANIC SILT

Thoroughly decomposed organic material with considerable mineral soil material. Usually black in color with a few fibrous remains. Odorous when dried and burnt. Found as deposits in swamps, peat bogs and muskeg. Easily identified. May contain some sand or silt.

PEAT

Partly decayed plant material. Mostly organic. Highly fibrous with visible plant remains.

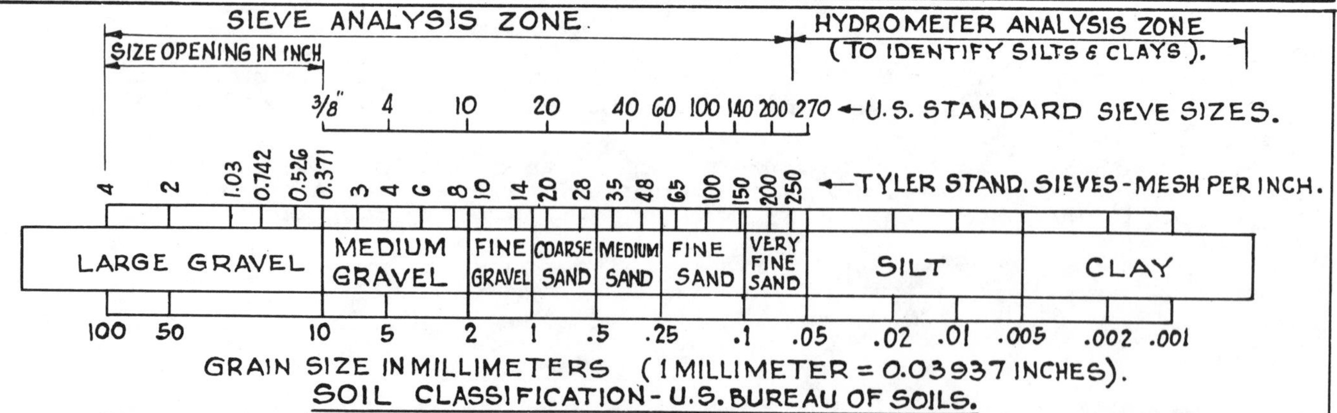
SIEVE ANALYSIS ZONE
HYDROMETER ANALYSIS ZONE
(TO IDENTIFY SILTS & CLAYS).
SIZE OPENING IN INCH
3/8" 4 10 20 40 60 100 140 200 270 ← U.S. STANDARD SIEVE SIZES.
4 2 1.03 0.742 0.526 0.371 3 4 6 8 10 14 20 28 35 48 65 100 150 200 250 ← TYLER STAND. SIEVES - MESH PER INCH.
LARGE GRAVEL
MEDIUM GRAVEL
FINE GRAVEL
COARSE SAND
MEDIUM SAND
FINE SAND
VERY FINE SAND
SILT
CLAY
100 50 10 5 2 1 .5 .25 .1 .05 .02 .01 .005 .002 .001
GRAIN SIZE IN MILLIMETERS (1 MILLIMETER = 0.03937 INCHES).

SOIL CLASSIFICATION - U.S. BUREAU OF SOILS.

FIG. C - IDENTIFICATION BY MECHANICAL GRAIN SIZE ANALYSIS.

NOTES: Mechanical analysis is necessary to identify soils into the various divisions, and into PRA & Casagrande systems. In general, the value of soils as a foundation for structures and as a material of Construction is determined by the grain sizes and the gradation of the soil mixture. Other widely used grain size classifications are International, Natl. Pk. Serv., A.S.T.M.

SOILS — CLASSIFICATION

TABLE A - CLASSIFICATION OF SOIL MIXTURES.*

CLASS	PER CENT SAND	SILT	CLAY
SAND	80-100	0-20	0-20
SANDY LOAM	50-80	0-50	0-20
LOAM	30-50	30-50	0-20
SILT LOAM	0-50	50-80	0-20
SANDY CLAY LOAM	50-80	0-30	20-30
CLAY LOAM	20-50	20-50	20-30
SILTY CLAY LOAM	0-30	50-80	20-30
SANDY CLAY	50 70	0-20	30-50
CLAY	0-50	0-50	30-100
SILTY CLAY	0-20	50-70	30-45

NOTE: Determine proportions of sand, silt & clay by sieve analysis or inspection. (Natural soils seldom exist separately as gravel, sand, silt, clay, but are found as mixtures.)

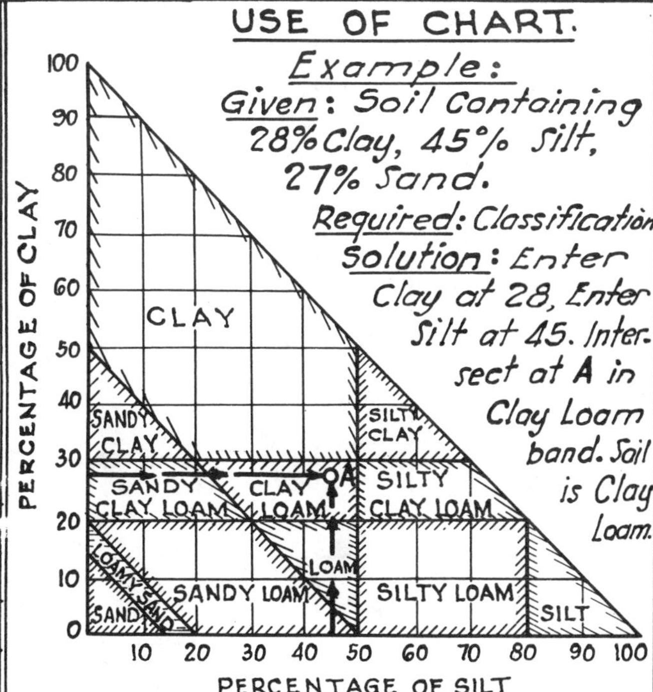
USE OF CHART.
Example:
Given: Soil Containing 28% Clay, 45% Silt, 27% Sand.
Required: Classification
Solution: Enter Clay at 28, Enter Silt at 45. Intersect at A in Clay Loam band. Soil is Clay Loam.
PERCENTAGE OF CLAY
100
90
80
70
60
50
40
30
20
10
0
CLAY
SANDY CLAY
SILTY CLAY
SANDY CLAY LOAM
CLAY LOAM
A
SILTY CLAY LOAM
LOAM
LOAMY SAND
SAND
SANDY LOAM
SILTY LOAM
SILT
10 20 30 40 50 60 70 80 90 100
PERCENTAGE OF SILT

FIG. B-RIGHT ANGLE SOIL CHART.*

Surface
A_0 A_1 A_2
B_1 B_2 B_3
C
D

Soil Profile: A vertical cross-section of the soil layers from the surface downwards through the parent material.

A HORIZON The upper layer, surface soil or topsoil. The upper part is designated A_0 and is humus or organic debris. Indices are used for sub division into transition zones as shown for A_1, A_2, etc. May range to 24" in depth.

B HORIZON The heavier textured under layer or subsoil. May range from 6" to 8' in depth. May be subdivided into transition zones B_1, B_2, etc., as shown. The products of the leaching or eluviation of the A horizon may be deposited in horizon B.

C HORIZON The unweathered or incompletely weathered parent material.

D HORIZON The underlying stratum such as hard rock, hardpan, sand or clay.

FIG. C-CLASSIFICATION OF SOILS BY HORIZONS.*

Widely used in soil stabilization.

NOTES: Structures or pavements are not usually placed on A horizon soils. Also the organic content of these soils may adversely affect stabilization. In cuts the C horizon soil does not usually have as good bearing value as the more weathered B horizon. Foundations for heavy structures are preferably founded on the D horizon where it is bedrock or unyielding.

TABLE D — CLASSIFICATION OF SOILS BY ORIGIN.

Residual:		Rock weathered in place - Wacke, laterite, podzols, residual sands, clays & gravels.
Cumulose		Organic accumulations - peat, muck, swamp soils, muskeg, humus, bog soils.
Transported	Glacial	Moraines, eskers, drumlins, kames - till, drift, boulder clay, glacial sands & gravels.
Transported	Alluvial	Flood planes, deltas, bars - Sedimentary clays & silts, Alluvial sands & gravels.
Transported	Aeolian	Wind-borne deposits - Blow sands, dune sands, loess, adobe.
Transported	Colluvial	Gravity deposits - Cliff debris, talus, avalanches, masses of rock waste.
Transported	Volcanic	Volcanic deposits - Dakota bentonite, Volclay, Volcanic ash, lava.
Transported	Fill	Man made deposits - may range from waste & rubbish to carefully built embankments.

NOTE: In general, residual or glacial deposits are preferable for heavy foundations.

Important in soil surveys & Eng. Reports.

*Adapted from Soil Cement Laboratory Handbook, Portland Cement Assoc.

References: Engineering Geology, by Ries & Watson.

SOILS

TABLE A - CASAGRANDE

Major Division		Soil Group Symbols	Soil Groups & Typical Names	General Identification: Dry Strength	General Identification: Other Pertinent Examinations	Observations and Tests Relating to Material in Place	Principal Classifica[tion] Tests o[n] Disturbed Sa[mples]
Coarse-Grained Soils.	Gravel & Gravelly Soils.	GW	Well-graded gravel and gravel-sand mixtures; little or no fines.	None.	Gradation, Grain shape.	Dry unit weight or void ratio; degree of compaction; cementation; durability of grains; stratification and drainage characteristics; ground-water conditions; traffic tests; large-scale load tests; or California bearing tests.	Sieve analy
		GC	Well-graded gravel-sand-clay mixtures; excellent binder.	Medium to High	Gradation, grain shape, binder examination, wet & dry.		Sieve anal. liquid and p limits on bin
		GP	Poorly graded gravel & gravel-sand mixtures; little or no fines.	None.	Gradation, grain shape.		Sieve analy
		GF	Gravel with fines, very silty gravel, clayey gravel poorly graded gravel-sand-clay mixtures.	Very slight to high.	Gradation, grain shape, binder examination, wet and dry.		Sieve anal liquid and p limits on bin if applicabl
	Sands & Sandy Soils.	SW	Well-graded sands & gravelly sands; little or no fines.	None.	Gradation, grain shape.		Sieve anal
		SC	Well-graded sand-clay mixtures; excellent binder.	Medium to high.	Gradation, grain shape, binder exam. wet & dry.		Sieve analysis, plastic limits c
		SP	Poorly graded sands; little or no fines.	None.	Gradation, grain shape.		Sieve analy.
		SF	Sand with fines, very silty sands, clayey sands, poorly graded sand-clay mixtures.	Very slight to high.	Gradation, grain shape, binder examination, wet and dry.		Sieve analys liquid and p limits on bin if applicable.
Fine-Grained Soils (containing little or no coarse-grained material).	Fine-grained soils having low to medium compressibility.	ML	Silts (inorganic) and very fine sands, Mo, rock flour, silty or clayey fine sands with slight plasticity.	Very slight to medium.	Examination wet (shaking test and plasticity).	Dry unit weight, water content, and void ratio. Consistency undisturbed and remoulded. Stratification. Root holes. Fissures, etc. Drainage and ground-water condition. Traffic tests, large-scale load tests, California bearing tests, or compression tests.	Sieve analy. liquid and p limits, if ap able.
		CL	Clays (inorganic) of low to medium plasticity, sandy clays, silty clays, lean clays.	Medium to high.	Examination in plastic range.		Liquid and limits.
		OL	Organic silts and organic silt-clays of low plasticity.	Slight to medium.	Examination in plastic range, odor.		Liquid & plastic from natural co and after oven
	Fine-grained soils having high compressibility.	MH	Micaceous or diatomaceous fine sandy and silty soils; elastic silts.	Very slight to medium.	Examination wet (shaking test and plasticity).		Sieve analysis & plastic limi applicable.
		CH	Clays (inorganic) of high plasticity; fat clays.	High.	Examination in plastic range.		Liquid and limits.
		OH	Organic clays of medium to high plasticity.	High.	Examination in plastic range, odor.		Liquid and plas from natural c and after ove
Fibrous organic soils with very high compressibility.		Pt	Peat and other highly organic swamp soils.	Readily identified.		Consistency, texture and natural water content.	

Legend for Soil Group Symbols.

C - Clay, plastic-inorganic soil.
F - Fines, material < 0.1 mm.
G - Gravel, gravelly soil.
H - High compressibility.
L - Relatively low to medium compressibility.
M - Mo, very fine sand, silt, rock flour.
O - Organic silt, silt clay or clay.
P - Poorly graded.
Pt - Peat, highly organic fib
S - Sand, sandy soil.
W - Well graded.

CASAGRANDE CLASSIFICATION

OIL CHART.

AS ATION T SUBJECT T ACTION	VALUE AS WEARING SURFACE.*† WITH DUST PALLIATIVE.	WITH BIT. SURF. TREAT.	POTENTIAL FROST ACTION†	SHRINKAGE EXPANSION ELASTICITY	DRAINAGE CHARACTERISTICS	COMPACTION CHARACTERISTICS & EQUIPMENT.	SOLIDS AT OPTIMUM COMPACTION. u, lb/cu.ft.** e, Void Ratio	CALIFORNIA BEARING RATIO FOR COMPACTED AND SOAKED SPECIMEN	COMPARABLE GROUP IN PUBLIC ROADS CLASS. (P.R.A.)
lent.	Fair to Poor.	Excellent.	None to very slight.	Almost none.	Excellent.	Excellent, Tractor.	u > 125 e < 0.35	> 50	A-3
lent.	Excellent.	Excellent.	Medium.	Very slight.	Practically impervious.	Excellent, Tamping Roller.	u > 130 e < 0.30	> 40	A-1
to lent.	Poor.	Poor to Fair	None to very slight.	Almost none.	Excellent.	Good, Tractor.	u > 115 e < 0.45	25-60	A-3
to lent.	Poor to Good	Fair to Good.	Slight to medium.	Almost none to slight.	Fair to practically impervious.	Good, Close Control Essential, Rubber Tired Roller, Tractor.	u > 120 e < 0.40	> 20	A-2
llent ood.	Poor.	Good.	None to very slight.	Almost none.	Excellent.	Excellent, Tractor.	u > 120 e < 0.40	20-60	A-3
llent ood	Excellent.	Excellent.	Medium.	Very slight.	Practically impervious.	Excellent, Tamping Roller.	u > 125 e < 0.35	20-60	A-1
to d.	Poor.	Poor.	None to very slight.	Almost none.	Excellent.	Good, Tractor	u > 100 e < 0.70	10-30	A-3
to d.	Poor to Good.	Poor to Good.	Slight to high.	Almost none to medium.	Fair to practically impervious.	Good, Close Control Essential, Rubber Tired Roller.	u > 105 e < 0.60	8-30	A-2
to :	Poor.		Medium to very high.	Slight to medium.	Fair to poor.	Good to Poor, Close Control Essential, Rubber Tired Roller.	u > 100 e < 0.70	6-25	A-4
to r.	Poor.		Medium to high.	Medium.	Practically impervious.	Fair to Good, Tamping Roller.	u > 100 e < 0.70	4-15	A-4 A-6 A-7
r.	Very Poor.		Medium to high.	Medium to high.	Poor.	Fair to Poor Tamping Roller.	u > 90 e < 0.90	3-8	A-4 A-7
or.	Very Poor.		Medium to very high.	High.	Fair to poor.	Poor to Very Poor.	u > 100 e < 0.70	< 7	A-5
to poor.	Very Poor.		Medium.	High.	Practically impervious.	Fair to Poor, Tamping Roller.	u > 90 e < 0.90	< 6	A-6 A-7
poor.	Useless.		Medium.	High.	Practically impervious.	Poor to Very Poor.	u > 100 e < 0.70	< 4	A-7 A-8
emely or.	Useless.		Slight.	Very high.	Fair to poor.	Compaction not Practical. Replace with Compactible Material.			A-8

Notes:

* Values are for subgrade and base courses, except for base courses directly under wearing surface.

† Values are for guidance only. Design should be based on test results.

** Unit weights apply only to soils with specific gravities ranging between 2.65 and 2.75.

SOILS — CONSTANTS FOR PAVEMENT DESIGN

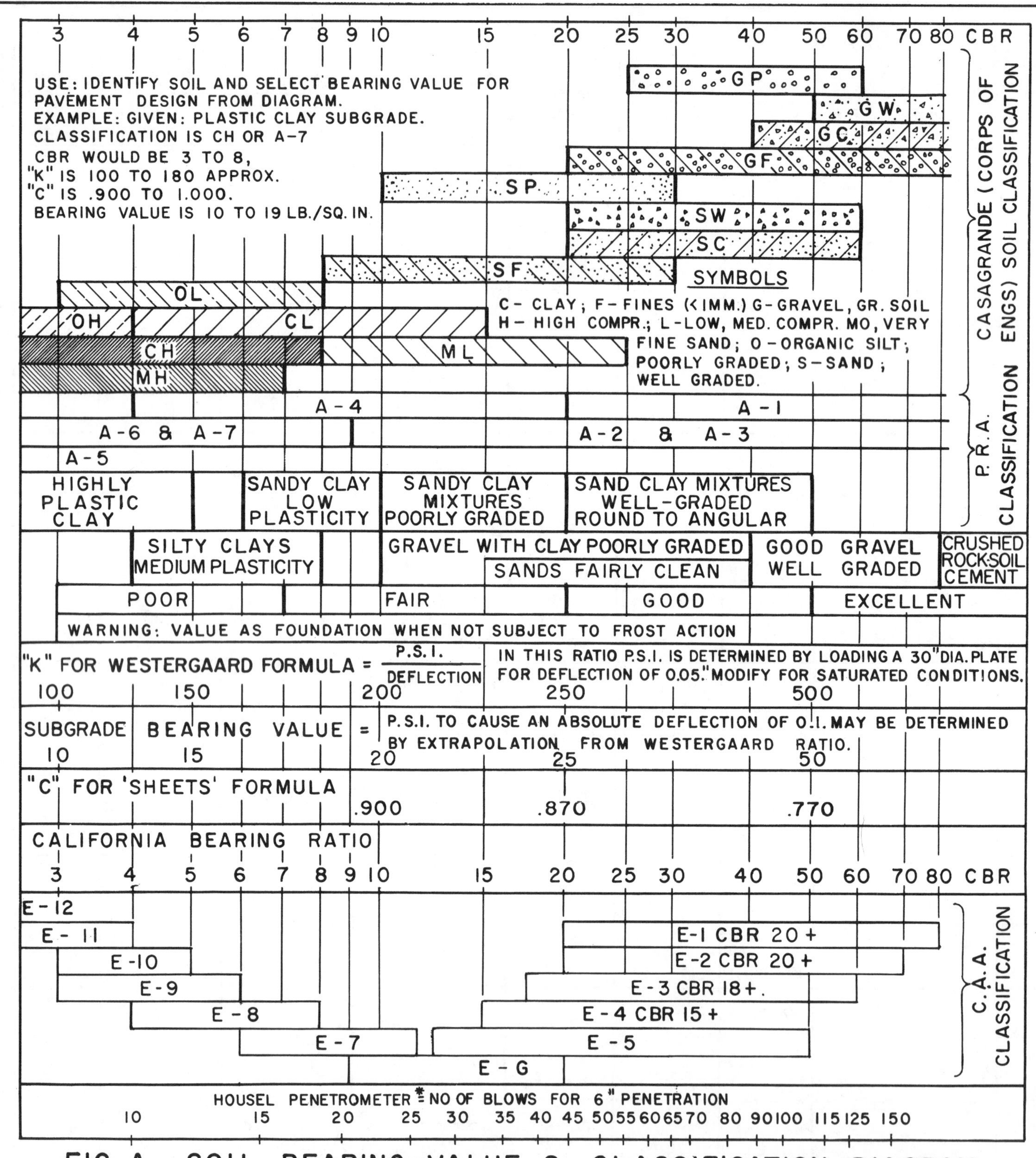
3 4 5 6 7 8 9 10 15 20 25 30 40 50 60 70 80 CBR
USE: IDENTIFY SOIL AND SELECT BEARING VALUE FOR PAVEMENT DESIGN FROM DIAGRAM.
EXAMPLE: GIVEN: PLASTIC CLAY SUBGRADE.
CLASSIFICATION IS CH OR A-7
CBR WOULD BE 3 TO 8,
"K" IS 100 TO 180 APPROX.
"C" IS .900 TO 1.000.
BEARING VALUE IS 10 TO 19 LB./SQ. IN.
GP
GW
GC
GF
SP
SW
SC
SF
SYMBOLS
OL
OH
CL
CH
MH
ML
C- CLAY; F- FINES (<1MM.) G-GRAVEL, GR. SOIL
H- HIGH COMPR.; L-LOW, MED. COMPR. MO, VERY FINE SAND; O-ORGANIC SILT; POORLY GRADED; S-SAND; WELL GRADED.
CASAGRANDE (CORPS OF ENGS) SOIL CLASSIFICATION
A-4
A-1
A-6 & A-7
A-2 & A-3
A-5
P.R.A. CLASSIFICATION
HIGHLY PLASTIC CLAY
SANDY CLAY LOW PLASTICITY
SANDY CLAY MIXTURES POORLY GRADED
SAND CLAY MIXTURES WELL-GRADED ROUND TO ANGULAR
SILTY CLAYS MEDIUM PLASTICITY
GRAVEL WITH CLAY POORLY GRADED
SANDS FAIRLY CLEAN
GOOD GRAVEL WELL GRADED
CRUSHED ROCK-SOIL CEMENT
POOR
FAIR
GOOD
EXCELLENT
WARNING: VALUE AS FOUNDATION WHEN NOT SUBJECT TO FROST ACTION
"K" FOR WESTERGAARD FORMULA = P.S.I. / DEFLECTION
IN THIS RATIO P.S.I. IS DETERMINED BY LOADING A 30" DIA. PLATE FOR DEFLECTION OF 0.05." MODIFY FOR SATURATED CONDITIONS.
100 150 200 250 500
SUBGRADE BEARING VALUE = P.S.I. TO CAUSE AN ABSOLUTE DEFLECTION OF 0.1. MAY BE DETERMINED BY EXTRAPOLATION FROM WESTERGAARD RATIO.
10 15 20 25 50
"C" FOR 'SHEETS' FORMULA
.900 .870 .770
CALIFORNIA BEARING RATIO
3 4 5 6 7 8 9 10 15 20 25 30 40 50 60 70 80 CBR
E-12
E-11
E-10
E-9
E-8
E-7
E-G
E-1 CBR 20+
E-2 CBR 20+
E-3 CBR 18+.
E-4 CBR 15+
E-5
C.A.A. CLASSIFICATION
HOUSEL PENETROMETER * NO OF BLOWS FOR 6" PENETRATION
10 15 20 25 30 35 40 45 50 55 60 65 70 80 90 100 115 125 150

FIG. A - SOIL BEARING VALUE & CLASSIFICATION DIAGRAM

NOTES:

BEARING VALUES GIVEN ARE APPROXIMATE AND SHOULD NOT BE USED AS A SUBSTITUTE FOR C.B.R. OR FIELD LOAD TESTS.

FOR DESIGN OF FLEXIBLE PAVEMENTS, SEE PP. 10-05 TO 10-10 FOR ROADS AND PP. 10-52/54/56/58/59 FOR AIRFIELDS.

FOR DESIGN OF RIGID PAVEMENTS, SEE PP. 10-11 TO 10-17 FOR ROADS & PP. 10-51/53/57/61 TO 65 FOR AIRFIELDS. THIS DIAGRAM IS COMPILED FROM MATERIAL IN FOLLOWING REFERENCES: SOIL TESTS FOR DESIGN OF RUNWAY PAVEMENTS BY MIDDLEBROOKS (U.S.E.D) & BERTRAM (A.A.F.). PROCEEDINGS HIGHWAY RESEARCH BOARD, DECEMBER 1942. ENGINEERING NEWS RECORD, VOLUME 130, No. 4, JAN. 28, 1943. DESIGN OF AIRPORT RUNWAYS, OFFICE, CHIEF OF ENGINEERS, U.S. ARMY. MANUAL No. 3 U.S. NAVY, BUREAU YARDS AND DOCKS. CHAPTER XX, ENG. MANUAL, WAR DEPT. CONCRETE ROAD DESIGN BY FRANK T. SHEETS & C.A.A. — * SEE PAGE 9–29

SOILS – USE & TREATMENT – P.R.A. GROUPS

TABLE A –	FOR ROADS AND AIRFIELDS*. See also Table A-Pg. 9-17
A-1 SOILS:	Well graded gravels & Sand-clays, as Florida sand-clay or Georgia Topsoil. Satisfactory treated surface. Good base with thin pavement. Excellent fill. Frost heave & break-up in North if plastic. Use sub-drainage to lower water table. Stabilize; mechanically, chlorides or Portland Cement.
A-2 SOILS:	Poorly graded sands & gravels, as S. Carolina Topsoil or Bank Run. Good base for moderate flexible or thin rigid pavement. Good fill. Frost heave, break-up if plastic. Softens when wet if plastic. Use base course when subgrade P.I. > 6. Sub-drainage effective. Stabilize; with bitumen, chlorides, Cement or Admixture soils.
A-3 SOILS:	Clean sands & Gravels, as Florida sand, glacial gravel, beach sand, wash gravel. Ideal base for moderate flexible or thin rigid pavement. Good fill. No frost heave or break up. Sub-drainage only thru impervious shoulders. Stabilize: with soil binder, bituminous, or chemical admixtures.
A-4 SOILS:	Silty soils as N.H. silt or Minn. silt. No good for surface. Poor base. Absorbs water. Unstable when wet. Bad frost heave & break-up. Use Sub-drainage and/or base and sub-base with flexible pavement. Use bituminous sub-grade prime. Use thick concrete pavement (7" to 10") with steel reinforcement and crack control.
A-5 SOILS:	Elastic silts as N. Carolina micaceous silt or Maryland micaceous Sandy loam. Use sub-drainage and/or granular base and sub-base with bitum. Sub-grade prime. Use thick conc. pavement, reinforced with crack control.
A-6 SOILS:	Clays, as Miss. Gumbo, Missouri collodial clay, sandy clays. Impermeable & stable when dry and undisturbed (hard clay). Plastic & absorbent if disturbed. Bad pumping into porous base, Macadam or pavement joints. Shrinks & cracks when dry. Use granular base & sub-base. Use sub-drainage only when made pervious by cracks, root holes & laminations. Frost heave slight when impermeable, bad when pervious. Use sub-grade prime. Use thick, strong, dense flexible pvmt. or reinf'd. crack controlled concrete.
A-7 SOILS:	Expansive, plastic clays, as Adobe, Missouri Clay, Illinois or Red River Gumbo. Excessive Volume change. Bad frost heave & break up. Sub-drainage not effective. Use thick, dense flexible pavement with base & sub-base over sub-grade prime or reinforced crack controlled concrete placed on impervious paper.
A-8 SOILS:	Muck & Peat. No good for construction purposes. Excavate to solid stratum & replace with selected fill. Displacement by superimposed fill is doubtful. Displacement by explosive under superimposed fill is sometimes effective.

USE AS FOUNDATIONS FOR STRUCTURES.

A-1 to A-3 Soils: BEST. A-6 & A-7 Soils: NEXT BEST, when hard, undisturbed & not plastic. A-5, A-8, Plastic A-6 & A-7 Soils: Require special treatment in each case.

USE IN EARTH DAM.†

HOMOGENEOUS TYPE LOW PERCOLATION.

Use: A-1, A-2 Plastic, A-4 Plastic, Better grades of A-6 & A-7.

Homogeneous

IMPERVIOUS CORE TYPE.

CORE.

Use: A-1, A-2 Plastic, A-4 Plastic, Better grades of A-6 & A-7.

POROUS FACES.

Non Plastic A-2 & A-4.

Faces
Core

Do not use: A-3, A-5, A-8 or Highly Plastic A-6 or A-7 in Dam Construction. A-4, A-6 & A-7 Soils: Use Controlled compaction to maximum density at optimum moisture content.

† Adapted from Hogentogler, Engineering Properties of Soil, McGraw-Hill. Refs.-Highway Subgrades by A.G. Bruce, Principles of Highway Const., Public Rds. Adm., Highway Research Board – Wartime Road Problems. *See also pp. 9-18 & 9-19.

SOILS — P.R.A. CLASSIFICATION

TABLE A - CHARACTERISTICS FOR IDENTIFYING P.R.A. SOIL GROUPS.*

Established by Public Roads Administration & Highway Research Board· Classification as shown is latest modification. Extensively used by engrs for highways, airfields & dams.

CHARACTERISTICS / GROUP		A-1 NON-PLASTIC	A-1 PLASTIC	A-2 NON-PLASTIC	A-2 PLASTIC	A-3	A-4 and A-4-7†	A-5 and A-5-7†	A-6	A-7	A-8
TEXTURAL CLASS		UNIFORMLY GRADED GRANULAR COARSE TO FINE		POORLY GRADED GRANULAR· COARSE AND FINE		CLEAN SAND OR GRAVEL	SILT OR SILT-LOAM	SILT OR SILT LOAM	PLASTIC CLAY	PLASTIC CLAY LOAM	MUCK & PEAT
Soil Constants.	Internal Friction	High	High	High	High	High	Variable	Variable	Low	Low	Low
	Cohesion	High	High	Low	High	None	"	Low	High	High	Low
	Shrinkage	Not detrimental		Not Significant	Detrimental if poorly graded	Not Significant	"	Variable	Detrimental	Detrimental	Detrimental
	Expansion	None		None	Some	Slight	"	High	High	"	"
	Capillarity	"		"	"		Detrimental	"	"	High	"
	Elasticity	"		"	"	None	Variable	Detrimental	None	"	"
Capillary Rise		Low	High	36" Max.	Over 36"	6" Max.	High	High	High	High	"
Atterberg Limits	Liquid Limit	25 Max.	35 Max.	35 Max.	40 Max.	Non Plastic	40 Max.	Over 40	35 Min.	35 Min.	35-400
	Plasticity Index	6 Max.	4-9	Non Plastic	15 Max.	"	0-15	0-60	18 Min.	12 Min.	0-60
	Shrinkage Limit	14 - 20		15-25	25 Max.	Not Essential.	20-30	30-120	6-14	10-30	30-120
Field Moisture Equivalent		Not Essential	Not Essential	Not Essential	Not Essential	Not Essential	30 Max.	30-120	50 Max.	30-100	30-400
Centrifuge Moisture Equivalent		15 Max.		12-25	25 Max.	12 Max.	Not Essential	Not Essential	Not Essential	Not Essential	Not Essential
Shrinkage Ratio		1.7 - 1.9		1.7 - 1.9	1.7 - 1.9	Not Essential	1.5 - 1.7	0.7 - 1.5	1.7 - 2.0	1.7 - 2.0	0.3 - 1.4
Volume Change		0 - 10		0 - 6	0 - 6	None	0 - 16	0 - 16	17 Min.	17 Min.	4-200
Lineal Shrinkage		0 - 3		0 - 2	0 - 4	"	0 - 4	0 - 4	5 Min.	5 Min.	1 - 30
Grading (Grain Size).	% Sand	70 - 85		55-80	55-80	75-100	55 Max.	55 Max.	55 Max.	55 Max.	55 Max.
	% Silt	10 - 20		0-45	0-45		High	Medium	Medium	Medium	Not
	% Clay	5 - 10		0-45	0-45		Low	Low	30 Min	30 Min.	Significant
	% Passing № 10	20-100	40-100								
	% Passing № 40	10 - 70	25-70								
	% Passing № 200	3 - 25	8 - 25	Less than 35	Less than 35	0 - 10					

† A-4 or A-5 soil with A-7 characteristics.

Subgrade or Surface.

CLAY SAND SILT

Rapid succession of soil types, pockets of different soil types. Changes in soil profile or changes in soil structure.
Non-uniform support.

GROUP B-1

Subgrade or Surface.

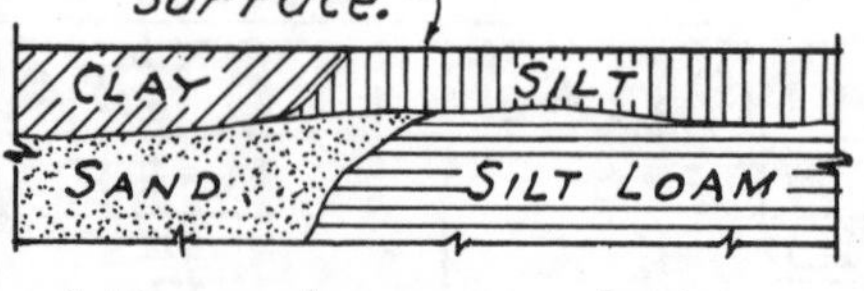
CLAY
SILT
SAND
SILT LOAM

Adjacent parts of fills constructed of widely differing soil types, as where hauling is from different sources.
Non-uniform support & settlement.

GROUP B-2

Orig. Surface
A Horizon
B Horizon
C Horizon
Subgrade.
WEAK ZONE
FILL
D Horizon

Cut through side hill creating sub grade on several types of natural soil and on fill within a short space.
Non-uniform support & settlement.

GROUP B-3

FIG. B - CLASSIFICATION OF NON-UNIFORM SUBGRADE SOILS.*

See Page 9-17 for treatment of these Subgrade Groups.

*Adapted from Public Roads Administration & Highway Research Board Publications.

SOILS-CHARACTERISTICS & PERFORMANCE-P.R.A.GROUPS

TABLE A

SOIL GROUP	A-1 Non Plastic	A-1 Plastic	A-2 Non Plastic	A-2 Plastic	A-3	A-4 and A-4-7†	A-5 and A-5-7†	A-6	A-7	A-8
Stability	High	High when dry	High	Good when dry	Ideal when Confined	Good When dry	Doubtful	Good, when properly compacted or undisturbed		None
Base	Good	Fair	Fair	Fair	Excellent	N.G.	N.G.	N.G.	N.G.	N.G.
Sub-base	Excellent	Good	Excellent	Good	"	"	"	"	"	"
Sub-grade	"	"	"	"	"	Poor	Poor	Poor	Bad	"
Fills under 50'	"	Excellent	"	"	Good	Good to Poor	Poor to Very Poor	Bad	Fair to Poor	"
Fills over 50'	Good	Good	Good to Fair	Good to Fair	Good to Fair	Fair to Poor	Very Poor	Bad	Very Poor	"
Frost Action	Slight	Subject to	Slight	Subject to	None	Bad	Bad	Slight to Bad	Slight to Bad	Slight
Dry Density	over 130 lb.	over 130 lb.	120-130 lb.	120-130 lb.	120-130 lb.	110-120 lb.	80-100 lb.	80-110 lb.	80-110 lb.	under 90 lb.
Optimum Moisture	9%	9%	9-12%	9-12%	9-12%	12-17%	22-30%	17-28%	17-28%	—
Required Compaction	90-95%	90-95%	90-95%	90-95%	90-95%	95%	100%	100%	100%	Waste
Compaction Methods	Rolling with smooth face, tamping or rubber tire roller.				Tractor, Disking, Vibration.	Tamping or Sheeps-foot roller.	Tamping or Rubber tire roller.	Heavy Sheeps foot or tamping roller		"
Compaction Abilities	Excellent	Excellent	Good with close control		Good	Poor to Good	Very Poor	Poor to Good	Poor to Fair	N.G.
Pumping Action	None	Slight	None	Slight	None			Bad	Bad	—
Bearing Value	Good to Excellent	Good to Excellent	Fair to Excellent			Poor to Fair	Poor	Poor	Poor	N.G.
Drainage	Drains Freely	Impervious	Fair to practically impervious		Drains Freely	Fair to Impervious	Fair to Impervious	Impervious	Poor	Poor
Flex. Pavement & Base Required	0"-6"	0"-6"	0"-6"	2"-8"	0"-6"	9"-18"	9"-24"	12"-24"	12"-24"	

† A-4 or A-5 soil with A-7 characteristics.

NOTES: <u>A-1 to A-3 Soils:</u> When used as base, Plasticity Index and Liquid Limit should not exceed 6 and 25 respectively. <u>A-1 to A-3 Soils</u>: Best for soil cement stabilizing, use 8 to 12% cement. <u>Non-plastic A-1 to A-3 Soils</u>: May require vibration and saturation for compaction. <u>A-4 to A-7 Soils</u>: Fills should be placed in dry season at not over optimum moisture content. <u>A-4 Silts</u>: Will settle rapidly in fills and are liable to erosion. <u>A-5 Soils</u>: Very difficult to compact because of expansion and rebound. <u>A-6 Soils</u> (clays): Will pump badly into porous bases, cracks and R.R. ballast. Fills will settle over long period of time. High banks in cuts and fills very liable to slide.

TREATMENT FOR NON-UNIFORM SUB-GRADES. (See Pg. 9-16.)

B-1: Loosen, pulverize, mix and recompact the soil to maximum density or remove to depth of 12" to 36" and substitute uniform base course material. Provide adequate sub-drainage for water trapped in perches or porous pockets.

B-2: Avoid this condition if possible. If unavoidable provide a uniform base of selected soil or thoroughly manipulate the soil for 12" to 24" depth and re-compact to maximum density.

B-3: Provide a uniform layer of base material 18" to 36" deep across the weak zone. Compact the fill to same density as adjacent bank. Provide outlet sub-drainage for seepage that is following porous soil strata or moving between strata.

References: Engineering Properties of Soil by C.A. Hogentogler.
" " Highway Subgrades by A.G. Bruce.
" " Public Roads, Public Roads Administration.
" " Wartime Road Problems, Highway Research Board.

SOILS - AIRPORTS - C.A.A. - 1

GENERAL: The C.A.A. method of soil classification is based on 3 tests; (1) mechanical analysis; (2) liquid limit, and (3) plastic limit. See p. 9-23 for determination of liquid limit and plasticity Index.

TABLE A - C.A.A. SOIL CLASSIFICATION						
Soil Group *	MECHANICAL ANALYSIS				Liquid Limit	Plasticity Index
	Retained on #10 Sieve %	MAT'L FINER THAN #10 SIEVE				
		Coarse Sand % Pass #10 Ret. #60	Fine Sand % Pass #60 Ret. #270	Combined Silt & Clay % Pass #270		
E-1	0-45	40+	60-	15-	25-	6-
E-2	0-45	15+	85-	25-	25-	6-
E-3	0-45			25-	25-	6-
E-4	0-45			35-	35-	10-
E-5	0-45			45-	40-	15-
E-6	0-55			45+	40-	10-
E-7	0-55			45+	50-	10-30
E-8	0-55			45+	60-	15-40
E-9	0-55			45+	40+	30-
E-10	0-55			45+	70-	20-50
E-11	0-55			45+	80-	30+
E-12	0-55			45+	80+	—
E-13	MUCK AND PEAT - FIELD EXAMINATION					

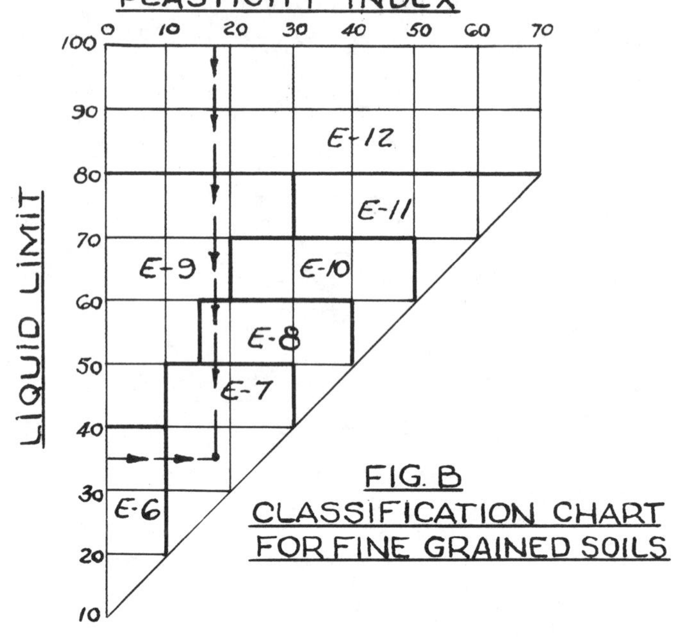
PLASTICITY INDEX
0
10
20
30
40
50
60
70
LIQUID LIMIT
100
90
80
70
60
50
40
30
20
10
E-12
E-11
E-10
E-9
E-8
E-7
E-6

FIG. B
CLASSIFICATION CHART FOR FINE GRAINED SOILS

Example: #1

Given: 1000 gm. sample, 300 gms. ret. #10 sieve 240 gms. ret. #60 sieve, 320 gms. ret. #270 sieve, 140 gms. pass #270 sieve. Liquid limit = 20, plasticity index = 4. Required: C.A.A. Soil Group of sample.

Solution: 1000 - 300 = 700 gms. passing #10 sieve (i.e. < 45% retained) 240/700 = 34% pass #10 ret #60. 320/700 = 46% pass #60 ret. #270. 140/700 = 20% pass #270. Referring to Table A sample is in E-2 group.

Example #2

Given: 1000 gm sample, 200 ret #10 sieve, 90 gms ret. #60 sieve, 240 gms. ret. #270 sieve, 470 gms. ret. #270 sieve, 470 gms. ret. #270 sieve. Liquid limit = 35, plasticity index = 18. Required: C.A.A. Soil Group of Sample.

Solution: 1000 - 200 = 800 gms pass #10 sieve (less than 45% retained) 90/800 = 11% pass #10 ret #60. 240/800 = 30% pass #60 ret #270. 470/800 = 59% pass #270. Referring to Table A sample E-7 or E-8. Entering Fig. B sample is in E-7 group.

* If percent retained on #10 sieve is over 45% for E1 to E5 soils and over 55% for others, and coarse fraction well graded and durable, upgrading of 1 or 2 groups is permitted.

CLASSIFICATION BY NAME

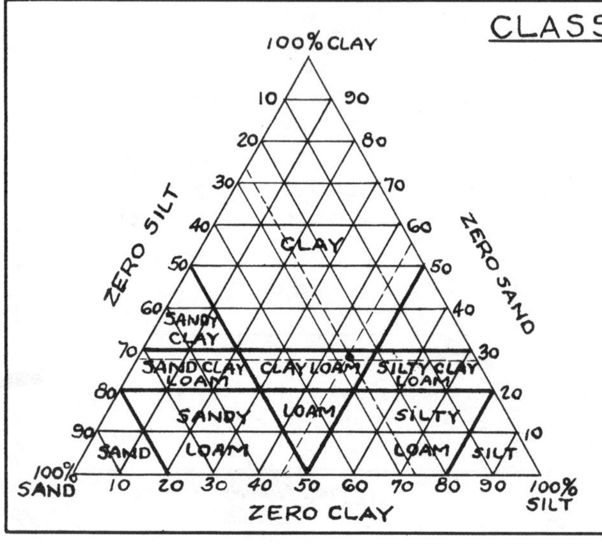
100% CLAY
ZERO SILT
ZERO SAND
ZERO CLAY
100% SAND
100% SILT
10
20
30
40
50
60
70
80
90
CLAY
SANDY CLAY
SAND CLAY LOAM
CLAY LOAM
SILTY CLAY LOAM
SANDY LOAM
LOAM
SILTY LOAM
SAND
SILT

FRACTION	SIEVE SIZE	GRAIN SIZE MM
Coarse Sand	#10 - #60	2.0-0.25
Fine Sand	#60 - #270	0.25-0.05
Silt	< #270	0.05-0.005
Clay	—	<0.005

Example:

Given: Soil with 28% clay, 45% silt and 27% sand

Required: Texture of soil

Solution: Enter textural percentages and find intersection of 3 lines. Soil is clay loam.

Adapted from "Airport Paving" by C.A.A. Oct. 1956

SOILS - AIRPORTS - C.A.A. - 2

TABLE A - C.A.A. SOIL CLASSIFICATION*	
Group E-1	Well graded coarse granular soils, stable under poor drainage condition and not subject to detrimental frost heave. These soils may conform to requirements for soil type base courses such as well graded sand clays with excellent binder.
Group E-2	Similar to group E-1 but less coarse sand and may contain greater percentages of silt and clay. Consequently, soils of this group may become unstable when poorly drained and be subject to frost heave to a limited extent.
Groups E-3 and E-4	Fine sandy soils of inferior grading. Fine cohesionless sand or sand clay types with a fair to good quality of binder. Less stable than group E-2 soils under adverse conditions of drainage and frost action.
Group E-5	All poorly graded granular soils having more than 35% and less 45% of silt and clay combined. All soils with less than 45% of silt and and plasticity indices greater than 10. A plasticity index greater than 15 even though the soil may have more than 55% sand, would cause it to be classified with fine grained soils.
Group E-6	Silts and silty loam soils having zero to low plasticity These soils have little cohesion, are quite stable when dry at low moisture contents, but lose stability and become very spongy when wet and are difficult to compact unless the moisture content is carefully controlled. Capillary rise in the soils is very rapid and the group is subject to detrimental frost heave.
Group E-7	Clay loams, silty clays, clays and some sandy clays. They range from friable to hard consistency when dry and are plastic when wet. These soils are stiff and dense when compacted at the proper moisture content. Variations in moisture are apt to produce detrimental volume change. Capillary forces acting in the soil are strong but the rate of capillary rise is relatively slow, and frost heave, while detrimental, is not as severe as in E-6 soils
Group E-8	Similar to E-7, but the higher liquid limits indicate a greater degree of compressibility, expansion and shrinkage and lower stability under adverse moisture conditions.
Group E-9	Silts and clays containing micaceous and diatomaceous materials. They are highly elastic and very difficult to compact. They have low stability in both the wet and dry state and and are subject to frost heave.
Group E-10	Silty clay and clay soils that form hard clods when dry and very plastic when wet. They are very compressible possess the properties of expansion, shrinkage and elasticity to a high degree and are subject to frost heave. More difficult to compact than those of the E-7 and E-8 groups and require careful control of moisture to produce a dense, stable fill.
Group E-11	Similar to E-10 but higher liquid limits. This group includes all soils with liquid limits between 70 and 80 and plasticity over 30.
Group E-12	Soils having liquid limits over 80 regardless of their plasticity indices. They may be highly plastic clays that are extremely unstable in the presence of moisture or they may be very elastic soils containing mica, diatoms, or organic matter in excessive amounts. Whatever the cause of their instability, they will require the maximum in corrective measures.
Group E-13	Organic swamp soils such as muck and peat which are recognized by examination in the field. Their range in test values is too great to be of any value in a system of identification and classification. They are characterized by very low stability, very low density in their natural state and very high moisture contents.

* See also pp. 9-15, 9-16 & 9-17 Adapted from "Airport Paving" by C.A.A. Oct. 1956

SOILS – SURVEYING & SAMPLING METHODS

SEE ALSO SOILS-SURVEYING & SAMPLING DEVICES-Pg. 9-14.

TABLE A – EXPLORATION & SAMPLING METHODS.*

METHOD	MATERIAL IN WHICH USED	PENETRATION METHOD	SAMPLING METHOD	TYPE OF SAMPLE	PURPOSE OR VALUE
Rod Sounding or Jet Probing.	All Soils except Hardpan or Boulders.	Driving 1" Steel Rod or 3/4" jet pipe with hand pump.	No Sample.		To obtain depth of Muck or soft strata. Location Ledge or Boulders. Otherwise Valueless.
Wash Borings.	All Soils except Hardpan or Boulders.	Washing inside 2½" driven casing with chopping bit on end of 1" Ex. heavy pipe.	Sample recovered from sediment in wash water.	Disturbed-Sedimentary. Coarse grains only.	Depth to Ledge or Boulders. Otherwise valueless. Results deceptive & dangerous.
Dry Sample Boring.	"	"	Open end pipe or split spoon sampler driven into soil.	Disturbed but not Separated.	Density Data from penetration of spoon. Fairly reliable and inexpensive.
Special Sampling Devices.	Cohesive Soils.	Driven Casing or Auger boring	By special sampling spoon or device.	Undisturbed.	To obtain samples for laboratory study.
Auger Boring.	Cohesive Soils. Cohesionless Soils above ground water.	Soil, Wood or Post Hole. Auger rotated by hand or machine & withdrawn.	Sample recovered from Soil brought up by Auger.	Disturbed but better than wash samples.	To locate soil strata & ground water. Roads, Airfields, Canals & Railroads. Samples for visual inspection and soil profile.
Well or Churn Drilling.	All soils Including Boulders Rock & Gravel.	Churn Drilling by Power.	Bailed sample of churned material or use of "Clay socket".	"Clay Socket" or "Dry" Sample.	Occasionally used for foundations. "Bailed" Samples worthless.
Rotary Drilling	"	Rotating Bit.	From Circulating liquid.	Fluid Sample.	Samples worthless.
Core Drill Borings	Large Boulders & Solid Rock.	Diamond, Shot or Sawtooth Cutters.	Cores cut & recovered.	Rock Cores 7/8" and over in Dia.	Best method to obtain type & condition of rock.
Test Pits & Caissons.	All soils. Below ground water use pneumatic Caisson or lower water table.	Excav. by hand or Power. Pit over 6' sheeted or lagged.	Bulk sample by hand. Undisturbed sample with spoon, tube or spec. device.	Disturbed or Undisturbed.	Most satisfactory Method. Should supplement others. To obtain undisturbed sample Cohesionless Soil. Soil Can be inspected in natural condition.
Geophysical Seismic, Elec. Resistance, Elec. Potential.	No samples. Continuous vibration or impulse from dynamite explosion. Device to register vibrations. Mostly patented methods.				Primary exploration. Will indicate earth, loose rock or solid rock. Interpretation uncertain.

TABLE B – SPACING & DEPTH OF BORINGS & TEST PITS OR TEST HOLES.

HIGHWAYS:** At 100' stations plus additional necessary at culverts, bridges, weak zones, wide cuts & fills, muck deposits, borrow pits and sources of base material. Depth not less than 3' below subgrade. Locate ground water table, seepage sources & direction of flow.

AIRFIELDS:*** At 100' to 1000' spacing on ℄, edge of pavement & edge of shoulders. Depth not less than 4' to 6' below subgrade in cut or ground surface in fill. Not less than twice diameter of tire contact area nor less than frost penetration. Locate ground water table & seepage data. Make Field Load Bearing Tests at time of survey (from 5 to 10 usual for each Airfield).

BRIDGES, DAMS & PIERS:† Borings spaced as needed to bed rock or well below foundation level. Make borings at least 20' into solid rock. Make one or more borings at each Pier 50' min. into solid rock. Use open pit exploration on land and in shallow water. Make soil bearing tests and Pile loading tests.

BUILDING FOUNDATIONS, TOWERS, CHIMNEYS, ETC.:† Borings spaced not over 50' c. to c. Depth 15' to 20' min. below foundation level. Initial borings to depth = 2 x width loaded area. Core borings into rock greater than minimum design depth of rock required. Supplement borings with test pits, load tests and test piles.

*Adapted from Low Dams by Nat. Resources Comm. based on Harvard Grad. Eng. School pub. #208 by H. A. Mohr.
References **A.S.T.M.-D.420, C.A.A. Specs., *** P.R.A., U.S.E.D., A.A.F. & C.A.A. † Man. Eng. Practice No 8, A.S.C.E.

SOILS – SURVEYING & SAMPLING DEVICES

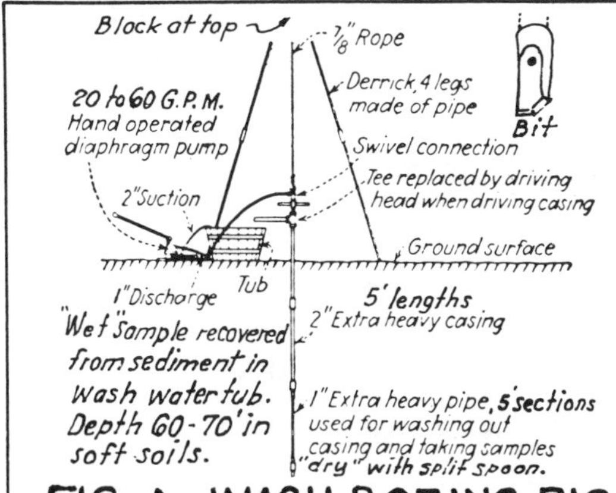

Block at top
7/8" Rope
Derrick, 4 legs made of pipe
Bit
20 to 60 G.P.M. Hand operated diaphragm pump
Swivel connection
Tee replaced by driving head when driving casing
2" Suction
Ground surface
Tub
1" Discharge
5' lengths
2" Extra heavy casing
"Wet" Sample recovered from sediment in wash water tub. Depth 60-70' in soft soils.
1" Extra heavy pipe, 5' sections used for washing out casing and taking samples "dry" with split spoon.

FIG. A - WASH BORING RIG.*

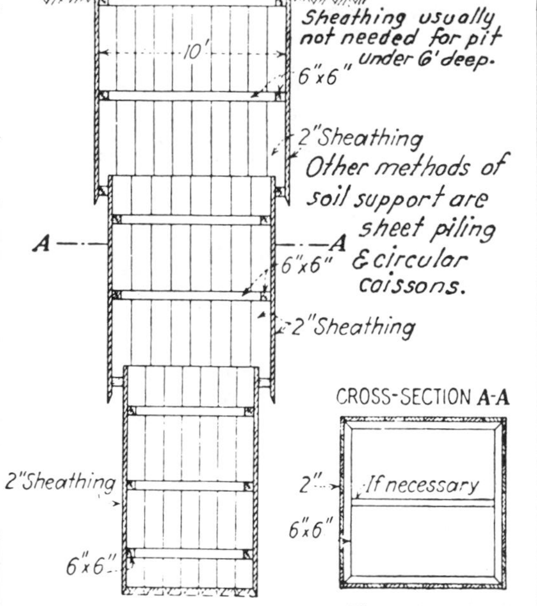

Sheathing usually not needed for pit under 6' deep.
10'
6"x6"
2" Sheathing
Other methods of soil support are sheet piling & circular caissons.
A — — A
6"x6"
2" Sheathing
CROSS-SECTION A-A
2"Sheathing
2"
If necessary
6"x6"
6"x6"

FIG. B - TEST PIT.** SHEATHED & BRACED.

Note: Auger borings may be carried to average depth of 20' by hand. Use cased borings for penetrating cohesionless soils below ground water table.

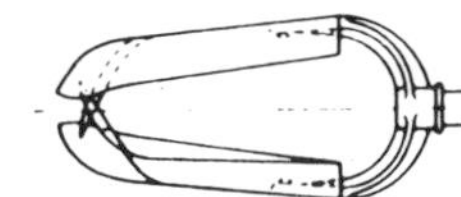

Iwan Pattern Earth Auger

Other auger types:

2" to 3" spiral auger for clay soils & muck.
Wood augers for hard soils, glacial till, etc.
10" to 20" power driven augers for gravel, etc.

FIG. C - SOIL AUGER***

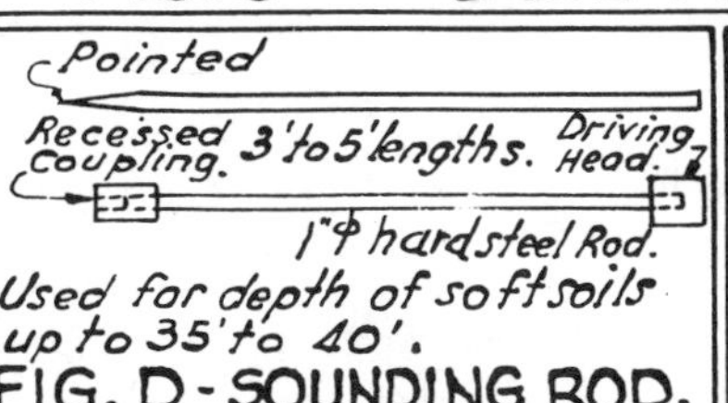

Pointed
Recessed Coupling.
3' to 5' lengths.
Driving Head.
1"φ hard steel Rod.

Used for depth of soft soils up to 35' to 40'.

FIG. D - SOUNDING ROD.

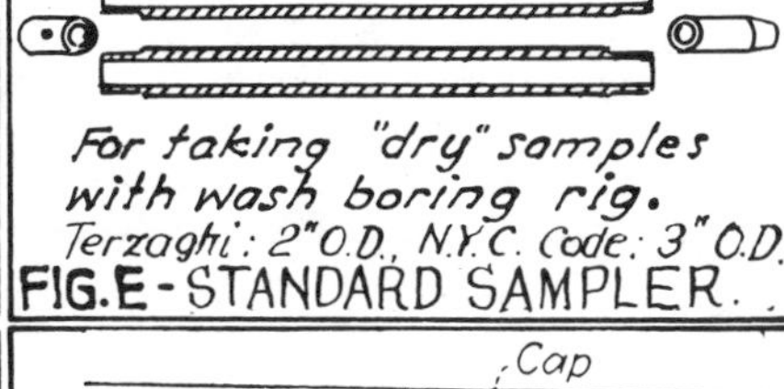

For taking "dry" samples with wash boring rig.
Terzaghi: 2" O.D., N.Y.C. Code: 3" O.D.

FIG. E - STANDARD SAMPLER.

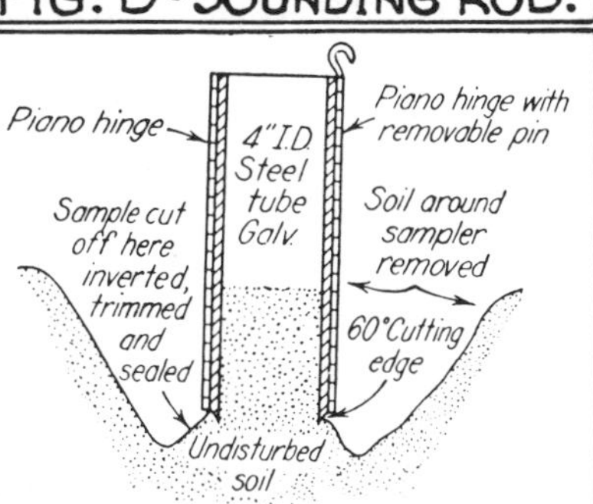

Piano hinge
4" I.D. Steel tube Galv.
Piano hinge with removable pin
Sample cut off here inverted, trimmed and sealed
Soil around sampler removed
60° Cutting edge
Undisturbed soil

FIG. F - SHALLOW SAMPLER FOR COHESIVE SOIL.****

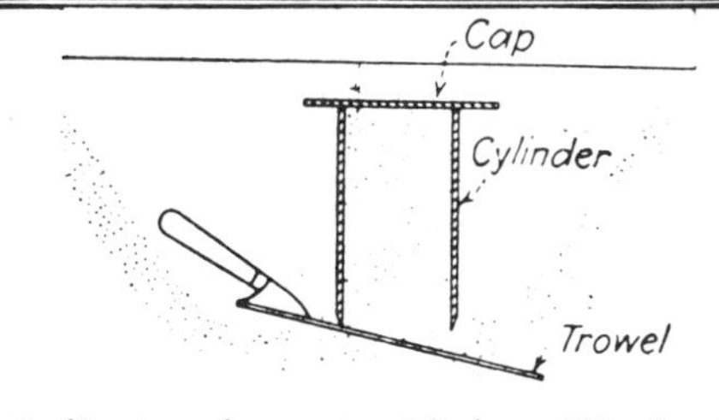

Cap
Cylinder
Trowel

Cylinder is worked into soil by hand. Sample is reversed, excess soil trimmed and sample sealed.

FIG. G - SHALLOW SAMPLING, COHESIONLESS SOIL (SAND)**

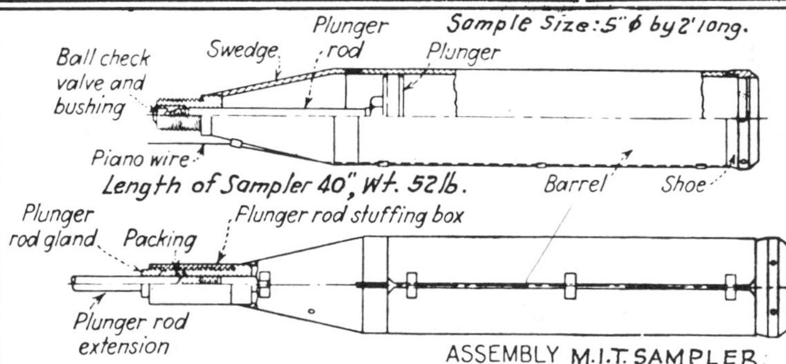

Plunger rod
Sample Size: 5"φ by 2' long.
Ball check valve and bushing
Swedge
Plunger
Piano wire
Length of Sampler 40", Wt. 52 lb.
Barrel
Shoe
Plunger rod gland
Packing
Plunger rod stuffing box
Plunger rod extension
ASSEMBLY M.I.T. SAMPLER

FIG. H - DEEP SAMPLER, COHESIVE SOILS.†

Advantages of plunger type samplers:

1. Plunger prevents soil from entering tube 'till right level is reached.
2. Sampler gives per-cent recovery.
3. Sample is prevented from falling out by vacuum created.

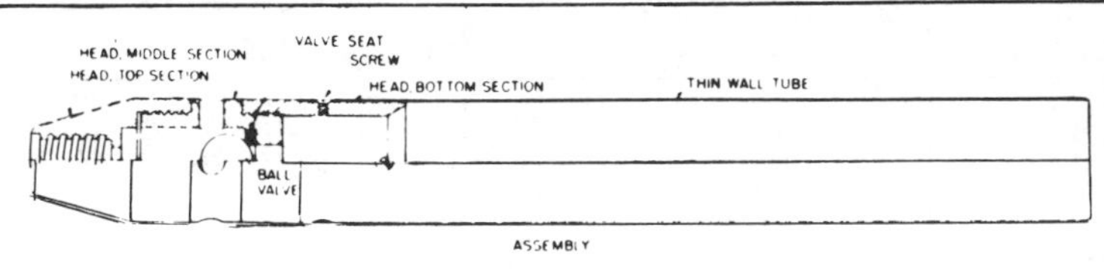

HEAD, MIDDLE SECTION
HEAD, TOP SECTION
VALVE SEAT SCREW
HEAD, BOTTOM SECTION
THIN WALL TUBE
BALL VALVE
ASSEMBLY

This sampler is used for obtaining undisturbed samples from cohesive soils only.

FIG. J "SHELBY-TUBE" SAMPLER***

TABLE K - SIZE OF SAMPLES.	
Visual inspection & record	1 Qt. Mason Jar.
California bearing ratio	125 lbs.
Soil stabilization	125 lbs.
Physical Constants & Mech. Analysis	5-15 lbs.
Aggregates for Construction (Concrete)	35 lbs.
Moisture-Density (Proctor Tests)	10-35 lbs.
Undisturbed Sample	12" to 2' long x 3" to 5" diam.
Rock Core	Usually 7/8" to 1½" diam.

Note: Seal undisturbed samples in tube with paraffin so structure and moisture content are not disturbed. Place bulk (disturbed) samples in bag or container tight enough so fines will not be lost.

*After Mohr. ** Krynine, Soil Mech., McGraw-Hill. *** Sprague and Henwood, Inc, Scranton, Pa. **** After Taylor.
† From Gilboy & Buchanan Proc. A.S.C.E. Vol. 59, 1933.

SOILS — PROFILES & BORING LOGS

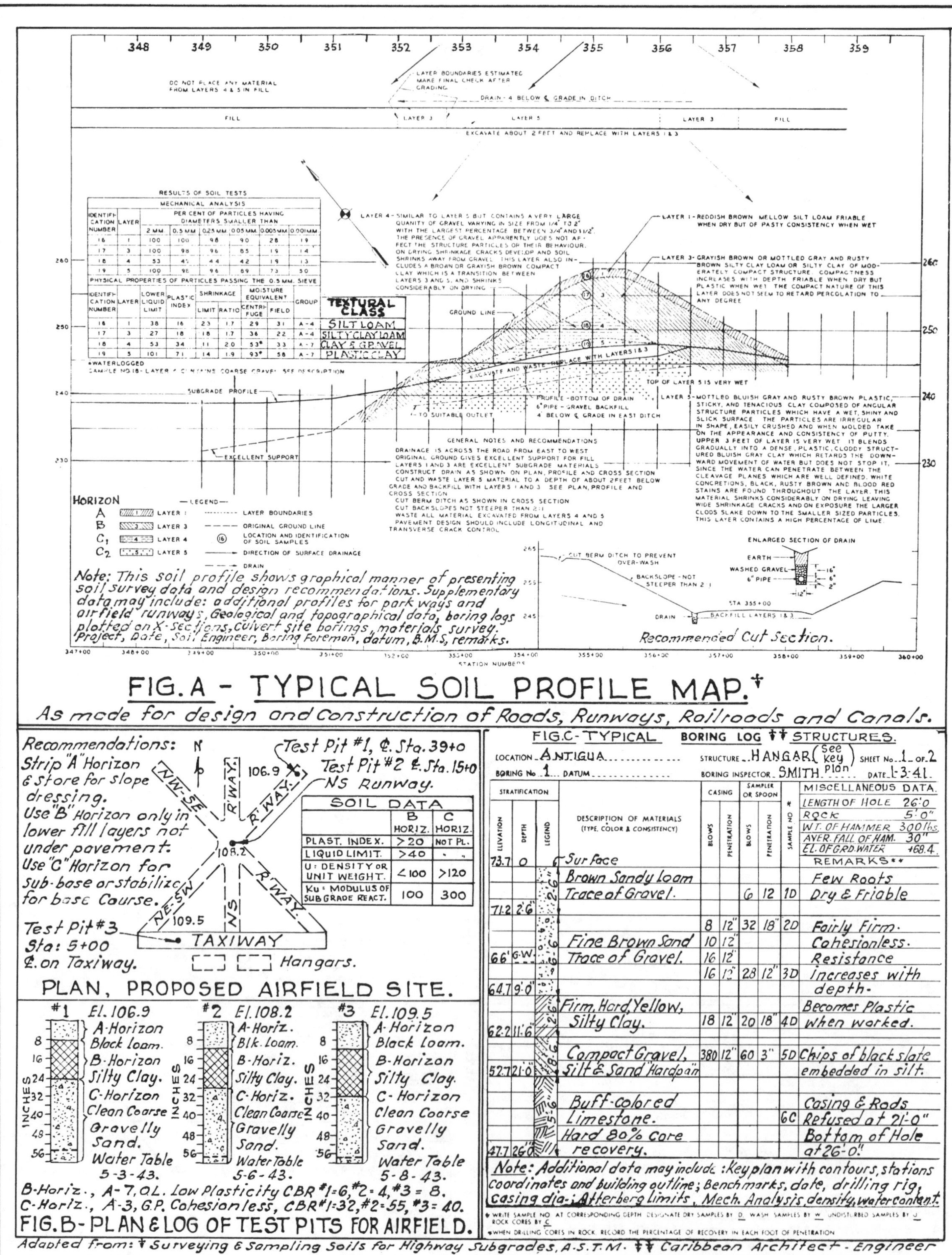

RESULTS OF SOIL TESTS

MECHANICAL ANALYSIS

IDENTIFICATION NUMBER	LAYER	PER CENT OF PARTICLES HAVING DIAMETERS SMALLER THAN 2 MM.	0.5 MM	0.25 MM	0.05 MM.	0.005 MM	0.001 MM
16	1	100	100	98	90	28	19
17	3	100	98	96	85	19	14
18	4	53	45	44	42	19	13
19	5	100	98	96	89	73	50

PHYSICAL PROPERTIES OF PARTICLES PASSING THE 0.5 MM. SIEVE

IDENTIFICATION NUMBER	LAYER	LOWER LIQUID LIMIT	PLASTIC INDEX	SHRINKAGE LIMIT	SHRINKAGE RATIO	MOISTURE EQUIVALENT CENTRIFUGE	MOISTURE EQUIVALENT FIELD	GROUP	TEXTURAL CLASS
16	1	38	16	23	1.7	29	31	A-4	SILT LOAM
17	3	27	18	18	1.7	36	22	A-4	SILTY CLAY LOAM
18	4	53	34	11	2.0	53*	33	A-7	CLAY & GRAVEL
19	5	101	71	14	1.9	93*	58	A-7	PLASTIC CLAY

*WATERLOGGED

DO NOT PLACE ANY MATERIAL FROM LAYERS 4 & 5 IN FILL
LAYER BOUNDARIES ESTIMATED MAKE FINAL CHECK AFTER GRADING
DRAIN 4' BELOW ℄ GRADE IN DITCH
FILL
LAYER 3
LAYER 5
LAYER 3
FILL
EXCAVATE ABOUT 2 FEET AND REPLACE WITH LAYERS 1 & 3
LAYER 4 - SIMILAR TO LAYER 5 BUT CONTAINS A VERY LARGE QUANITY OF GRAVEL VARYING IN SIZE FROM 1/4" TO 2" WITH THE LARGEST PERCENTAGE BETWEEN 3/4" AND 1 1/2". THE PRESENCE OF GRAVEL APPARENTLY DOES NOT AFFECT THE STRUCTURE PARTICLES OR THEIR BEHAVIOUR. ON DRYING SHRINKAGE CRACKS DEVELOP AND SOIL SHRINKS AWAY FROM GRAVEL. THIS LAYER ALSO INCLUDES A BROWN OR GRAYISH BROWN COMPACT CLAY WHICH IS A TRANSITION BETWEEN LAYERS 3 AND 5, AND SHRINKS CONSIDERABLY ON DRYING
LAYER 1 - REDDISH BROWN MELLOW SILT LOAM FRIABLE WHEN DRY BUT OF PASTY CONSISTENCY WHEN WET
LAYER 3 - GRAYISH BROWN OR MOTTLED GRAY AND RUSTY BROWN SILTY CLAY LOAM OR SILTY CLAY OF MODERATELY COMPACT STRUCTURE. COMPACTNESS INCREASES WITH DEPTH. FRIABLE WHEN DRY BUT PLASTIC WHEN WET. THE COMPACT NATURE OF THIS LAYER DOES NOT SEEM TO RETARD PERCOLATION TO ANY DEGREE
GROUND LINE
EXCAVATE AND WASTE REPLACE WITH LAYERS 1 & 3
TOP OF LAYER 5 IS VERY WET
SUBGRADE PROFILE
PROFILE - BOTTOM OF DRAIN 6" PIPE - GRAVEL BACKFILL 4' BELOW ℄ GRADE IN EAST DITCH
TO SUITABLE OUTLET
EXCELLENT SUPPORT
LAYER 5 - MOTTLED BLUISH GRAY AND RUSTY BROWN PLASTIC, STICKY, AND TENACIOUS CLAY COMPOSED OF ANGULAR STRUCTURE PARTICLES WHICH HAVE A WET, SHINY AND SLICK SURFACE. THE PARTICLES ARE IRREGULAR IN SHAPE, EASILY CRUSHED AND WHEN MOLDED TAKE ON THE APPEARANCE AND CONSISTENCY OF PUTTY. UPPER 3 FEET OF LAYER IS VERY WET. IT BLENDS GRADUALLY INTO A DENSE, PLASTIC, CLODDY STRUCTURED BLUISH GRAY CLAY WHICH RETARDS THE DOWNWARD MOVEMENT OF WATER BUT DOES NOT STOP IT, SINCE THE WATER CAN PENETRATE BETWEEN THE CLEAVAGE PLANES WHICH ARE WELL DEFINED. WHITE CONCRETIONS, BLACK, RUSTY BROWN AND BLOOD RED STAINS ARE FOUND THROUGHOUT THE LAYER. THIS MATERIAL SHRINKS CONSIDERABLY ON DRYING LEAVING WIDE SHRINKAGE CRACKS AND ON EXPOSURE THE LARGER CLODS SLAKE DOWN TO THE SMALLER SIZED PARTICLES. THIS LAYER CONTAINS A HIGH PERCENTAGE OF LIME.
GENERAL NOTES AND RECOMMENDATIONS
DRAINAGE IS ACROSS THE ROAD FROM EAST TO WEST. ORIGINAL GROUND GIVES EXCELLENT SUPPORT FOR FILL. LAYERS 1 AND 3 ARE EXCELLENT SUBGRADE MATERIALS. CONSTRUCT DRAIN AS SHOWN ON PLAN, PROFILE AND CROSS SECTION. CUT AND WASTE LAYER 5 MATERIAL TO A DEPTH OF ABOUT 2FEET BELOW GRADE AND BACKFILL WITH LAYERS 1 AND 3. SEE PLAN, PROFILE AND CROSS SECTION. CUT BERM DITCH AS SHOWN IN CROSS SECTION. CUT BACKSLOPES NOT STEEPER THAN 2:1. WASTE ALL MATERIAL EXCAVATED FROM LAYERS 4 AND 5. PAVEMENT DESIGN SHOULD INCLUDE LONGITUDINAL AND TRANSVERSE CRACK CONTROL.
HORIZON A LAYER 1; B LAYER 3; C1 LAYER 4; C2 LAYER 5
LEGEND: LAYER BOUNDARIES; ORIGINAL GROUND LINE; LOCATION AND IDENTIFICATION OF SOIL SAMPLES; DIRECTION OF SURFACE DRAINAGE; DRAIN
Note: This soil profile shows graphical manner of presenting soil survey data and design recommendations. Supplementary data may include: additional profiles for parkways and airfield runways, Geological and topographical data, boring logs plotted on X-sections, culvert site borings, materials survey. Project, Date, Soil Engineer, Boring Foreman, datum, B.M.S, remarks.
CUT BERM DITCH TO PREVENT OVER-WASH
BACKSLOPE - NOT STEEPER THAN 2:1
STA 355+00
DRAIN
BACKFILL LAYERS 1 & 3
ENLARGED SECTION OF DRAIN: EARTH; WASHED GRAVEL; 6" PIPE
Recommended Cut Section.
STATION NUMBERS
FIG. A - TYPICAL SOIL PROFILE MAP.†
As made for design and Construction of Roads, Runways, Railroads and Canals.
Recommendations: Strip "A" Horizon & store for slope dressing. Use "B" Horizon only in lower fill layers not under pavement. Use "C" Horizon for sub-base or stabilize for base Course.
Test Pit #1, ℄ Sta. 39+0
Test Pit #2 ℄ Sta. 15+0
NS Runway.
Test Pit #3 Sta: 5+00 ℄ on Taxiway.
TAXIWAY
Hangars.
SOIL DATA: B HORIZ. / C HORIZ.; PLAST. INDEX >20 / NOT PL.; LIQUID LIMIT >40; U: DENSITY OR UNIT WEIGHT <100 / >120; Ku: MODULUS OF SUBGRADE REACT. 100 / 300
PLAN, PROPOSED AIRFIELD SITE.
#1 El. 106.9 A-Horizon Black loam. B-Horizon Silty Clay. C-Horizon Clean Coarse Gravelly Sand. Water Table 5-3-43.
#2 El. 108.2 A-Horiz. Blk. loam. B-Horiz. Silty Clay. C-Horiz. Clean Coarse Gravelly Sand. Water Table 5-6-43.
#3 El. 109.5 A-Horizon Black loam. B-Horizon Silty Clay. C-Horizon Clean Coarse Gravelly Sand. Water Table 5-8-43.
INCHES
B-Horiz., A-7, O.L. Low Plasticity CBR #1=6, #2=4, #3=8.
C-Horiz., A-3, G.P. Cohesionless, CBR #1=32, #2=55, #3=40.
FIG. B - PLAN & LOG OF TEST PITS FOR AIRFIELD.
FIG. C - TYPICAL BORING LOG †† STRUCTURES.
LOCATION ANTIGUA
STRUCTURE HANGAR (See key plan) SHEET No. 1 OF 2
BORING No 1 DATUM
BORING INSPECTOR SMITH DATE 1-3-41
MISCELLANEOUS DATA. LENGTH OF HOLE 26'0; ROCK 5'0"; WT. OF HAMMER 300 lbs.; AVER. FALL OF HAM. 30"; EL. OF GRD. WATER +68.4

ELEVATION	DEPTH	DESCRIPTION OF MATERIALS (TYPE, COLOR & CONSISTENCY)	CASING BLOWS	CASING PENETRATION	SAMPLER OR SPOON BLOWS	SAMPLER OR SPOON PENETRATION	SAMPLE NO*	REMARKS**
73.7	0	Surface						
		Brown Sandy Loam						Few Roots
		Trace of Gravel.			6	12	1D	Dry & Friable
71.2	2'6"							
			8	12"	32	18"	2D	Fairly Firm.
		Fine Brown Sand	10	12"				Cohesionless.
66'	G.W.	Trace of Gravel.	16	12"				Resistance
			16	12"	28	12"	3D	increases with depth.
64.7	9'0"							
		Firm, Hard Yellow,						Becomes Plastic
		Silty Clay.	18	12"	20	18"	4D	when worked.
62.2	11'6"							
		Compact Gravel,	380	12"	60	3"	5D	Chips of black slate
		Silt & Sand "Hardpan"						embedded in silt.
52.7	21'0"							
		Buff-colored						Casing & Rods
		Limestone.					6C	Refused at 21'-0"
		Hard 80% Core						Bottom of Hole
47.7	26'0"	recovery.						at 26'-0".

Note: Additional data may include: Key plan with contours, stations, coordinates and building outline; Benchmarks, date, drilling rig, casing dia.; Atterberg Limits, Mech. Analysis, density, water content.

* WRITE SAMPLE NO. AT CORRESPONDING DEPTH. DESIGNATE DRY SAMPLES BY D, WASH SAMPLES BY W, UNDISTURBED SAMPLES BY U, ROCK CORES BY C.

** WHEN DRILLING CORES IN ROCK RECORD THE PERCENTAGE OF RECOVERY IN EACH FOOT OF PENETRATION.

Adapted from: † Surveying & Sampling Soils for Highway Subgrades, A.S.T.M. †† Caribbean Architect-Engineer

SOILS – ATTERBERG LIMIT TESTS

PURPOSE:

1. To classify soils into P.R.A., Casagrande, or C.A.A. groups.
2. To assign soils an approximate value as a foundation or construction material.
3. High values of liquid limit and plasticity index indicate high compressibility and low bearing capacity.
4. To determine soil suitability for road construction.

TABLE A- LIMITING VALUES	BASE COURSE	SUBGRADE	SUB-BASE	STAB. SURF.	SOIL CEMENT	CEM. TREATED BASE
	No Shrinkage L.L.=25; P.I.=6 max	Lineal Shrinkage 3% to 5%	L.L.=35; P.I.=15 max.	P.I.=4 to 9	L.L.=40; P.I.=18 max.	L.L.=25 P.I.=6 to 9

The water content or moisture content is expressed as a percentage of the oven dried weight of the soil sample. These soil constants are determined from the soil fraction passing the № 40 (420 micron) sieve.

The Liquid Limit (L.L.) of a soil is the water content at which the groove formed in a soil sample with a Std. grooving tool will just meet when the dish is held in one hand & tapped lightly 10 blows with the heel of the other hand. In the machine method the L.L. is the water content when the soil sample flows together for ½" along the groove with 25 shakes of the machine at 2 drops per sec. Diameter of brass cup or evaporating dish about 4½".

EXAMPLE OF FLOW CURVE*

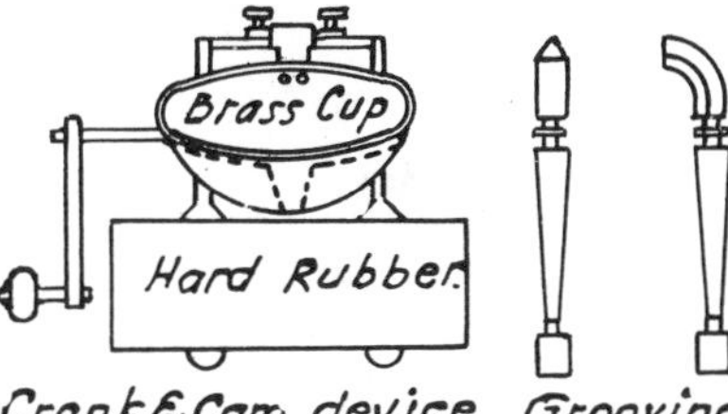
Brass Cup
Hard Rubber
Crank & Cam device to produce 1 centimeter drop of cup.
Grooving Tool.
CASAGRANDE LIQUID LIMIT MACHINE.

Size of sample: By hand 30 grams: By machine 100 grams. Several trials are made, the moisture content being gradually increased. Blows are plotted against water content and the Liquid Limit picked off from the curve as shown

or $L.L. = \frac{\text{Weight of water}}{\text{Weight of oven dried soil}} \times 100.$

Soil Cake After Test ‡

FIG. B - LIQUID LIMIT (L.L.) A.S.T.M. D 423, A.A.S.H.O., T-89.

The Plastic Limit (P.L.) is the lowest water content at which a thread of the soil can be just rolled to a diam. of ⅛" without cracking, crumbling or breaking into pieces. $P.L. = \frac{\text{Weight of water}}{\text{Wt. of oven dried soil}} \times 100$

Size of soil sample is 15 grams.
Soil which cannot be rolled into a thread is recorded as Non-Plastic (N.P.)

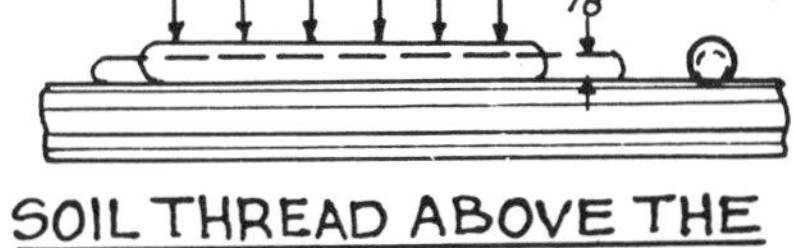
⅛"
SOIL THREAD ABOVE THE PLASTIC LIMIT.

CRUMBLING OF SOIL THREAD BELOW THE PLASTIC LIMIT.

FIG. C - PLASTIC LIMIT (P.L.) A.S.T.M. D 424, A.A.S.H.O., T-90.

PLASTICITY INDEX (P.I.): A.A.S.H.O., T-91. numerical difference between LIQUID LIMIT (L.L.) & PLASTIC LIMIT (P.L.) or P.I. = L.L. - P.L. Example: Given L.L. = 28, P.L. = 24, P.I. = 4. Cohesionless soils are reported as Non-Plastic (N.P.). When Plastic Limit is equal to or greater than Liquid Limit the P.I. is reported as 0. See Table A, Pg. 9-10.

SHRINKAGE RATIO (R): = bulk specific gravity of the dried soil pat used in obtaining Shrinkage Limit. $R = \frac{\text{Weight of oven-dried soil pat in grams}}{\text{Volume of oven dried soil pat in c.c.}}$ or $\frac{W_o}{V_o}$

SHRINKAGE LIMIT (S): A.S.T.M., A.A.S.H.O., T-92. Water content at which there is no further decrease in volume with additional drying of the soil but at which an increase in water content will cause an increase in volume.

$S = \left(\frac{1}{\text{Shrinkage Ratio}} - \frac{1}{\text{Spec. Gravity}}\right) \times 100.$ Size of sample 30 grams. 1¾" Dia. x ½" High Milkdish.

Wet Soil — BEFORE SHRINKAGE.
Dry Soil — AFTER SHRINKAGE.

LINEAL SHRINKAGE: is the decrease in one dimension of the soil mass when the water content is reduced to the Shrinkage Limit or the % change in length occurring when a moist sample has dried out.

FIELD TESTS. Shrinkage Soil; 1" x 1" x 10" Mold; Cigar Shaped; Pins; 5"; Soil

Adapted from *Krynine, Soil Mechanics, McGraw-Hill. ‡ Hogentogler, Engineering Properties of Soil, McGraw-Hill.

SOILS – MOISTURE DETERMINATION

PURPOSE: 1. To determine moisture content for optimum moisture and maximum density relations.

2. To determine the amount of water in aggregates for concrete, bituminous and other mixtures.

Gravelly Soils: Use Pycnometer Method - Fig. A or heat method described below.
Sandy Soils : Use Chapman Flask - Fig. B or heat method described below.
Silts and Clays: Use heat method described below.

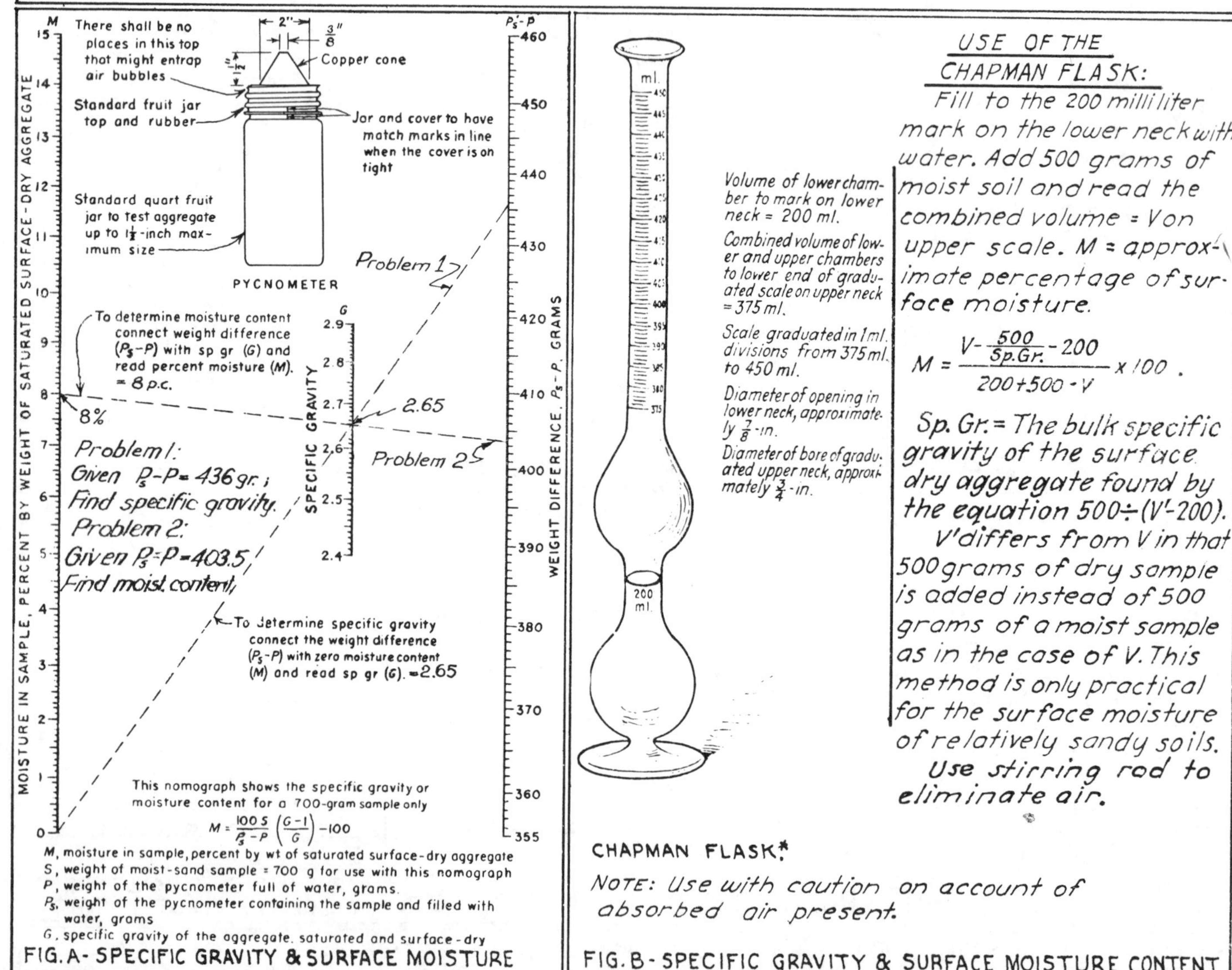
There shall be no places in this top that might entrap air bubbles
2"
3/8"
Copper cone
Standard fruit jar top and rubber
Jar and cover to have match marks in line when the cover is on tight
Standard quart fruit jar to test aggregate up to 1½-inch maximum size
PYCNOMETER
MOISTURE IN SAMPLE, PERCENT BY WEIGHT OF SATURATED SURFACE-DRY AGGREGATE
SPECIFIC GRAVITY
WEIGHT DIFFERENCE, $P_s'-P$, GRAMS
To determine moisture content connect weight difference (P_s-P) with sp gr (G) and read percent moisture (M). = 8 p.c.
8%
2.65
Problem 1
Problem 2
Problem 1: Given P_s-P = 436 gr.; Find specific gravity.
Problem 2: Given P_s-P = 403.5; Find moist. content.
To determine specific gravity connect the weight difference (P_s-P) with zero moisture content (M) and read sp gr (G). = 2.65
This nomograph shows the specific gravity or moisture content for a 700-gram sample only
$M = \frac{100S}{P_s-P}\left(\frac{G-1}{G}\right)-100$
M, moisture in sample, percent by wt of saturated surface-dry aggregate
S, weight of moist-sand sample = 700 g for use with this nomograph
P, weight of the pycnometer full of water, grams.
P_s, weight of the pycnometer containing the sample and filled with water, grams
G, specific gravity of the aggregate, saturated and surface-dry
Volume of lower chamber to mark on lower neck = 200 ml.
Combined volume of lower and upper chambers to lower end of graduated scale on upper neck = 375 ml.
Scale graduated in 1 ml. divisions from 375 ml. to 450 ml.
Diameter of opening in lower neck, approximately 7/8-in.
Diameter of bore of graduated upper neck, approximately 3/4-in.
200 ml.
USE OF THE CHAPMAN FLASK:
Fill to the 200 milliliter mark on the lower neck with water. Add 500 grams of moist soil and read the combined volume = V on upper scale. M = approximate percentage of surface moisture.
$M = \frac{V - \frac{500}{Sp.Gr.} - 200}{200 + 500 - V} \times 100$.
Sp. Gr. = The bulk specific gravity of the surface dry aggregate found by the equation 500÷(V'-200). V' differs from V in that 500 grams of dry sample is added instead of 500 grams of a moist sample as in the case of V. This method is only practical for the surface moisture of relatively sandy soils.
Use stirring rod to eliminate air.
CHAPMAN FLASK.*
NOTE: Use with caution on account of absorbed air present.

FIG. A - SPECIFIC GRAVITY & SURFACE MOISTURE CONTENT OF AGGREGATE - PYCNOMETER METHOD.

FIG. B - SPECIFIC GRAVITY & SURFACE MOISTURE CONTENT OF AGGREGATE - CHAPMAN FLASK METHOD.

HEAT METHOD: FOR TOTAL MOISTURE CONTENT OR SURFACE MOISTURE CONTENT.

1. Obtain a representative sample. If a metric scale is available the sample should not be smaller than 100 grams. If an avoirdupois scale graduated by ½ ounces is used, the sample should contain at least 50 ounces.
2. Weigh sample and record weight.
3. Place sample in pan and spread to permit uniform drying. Set pan in oven or on top of stove in a second pan to prevent burning of soil.
4. Dry to constant weight when total moisture is to be found; dry until surface moisture disappears when surface moisture content is desired. Temperature should not exceed 105°C (221°F). Stir constantly to prevent burning.
5. After the sample has been dried to constant weight, remove from oven and allow to cool sufficiently to permit absorption of hygroscopic moisture. Weigh dried sample and record weight.
6. Compute the moisture content as follows:

$$\text{Percent Moisture} = \frac{\text{weight of wet soil} - \text{weight of dry soil}}{\text{weight of dry soil}} \times 100.$$

*From A.S.T.M. Specifications.

SOILS- MAXIMUM DENSITY OPTIMUM MOISTURE & PROCTOR NEEDLE PLASTICITY TEST

PURPOSE: of Maximum Density - Optimum Moisture is to determine the % of moisture at which the Maximum Density can be obtained when soil is compacted in fill, earth dams, embankments, etc.
Purpose of the Proctor Needle test is to obtain a measure of the degree of compaction of a soil by measuring its resistance to penetration. Also a method of determining soil moisture. Cannot be used in soils with coarse particles. Used mostly in earth dam construction.

MAX. DENSITY - OPTIMUM MOISTURE.

as per A.S.T.M.-D.698 - A.A.S.H.O.-D:T.99.

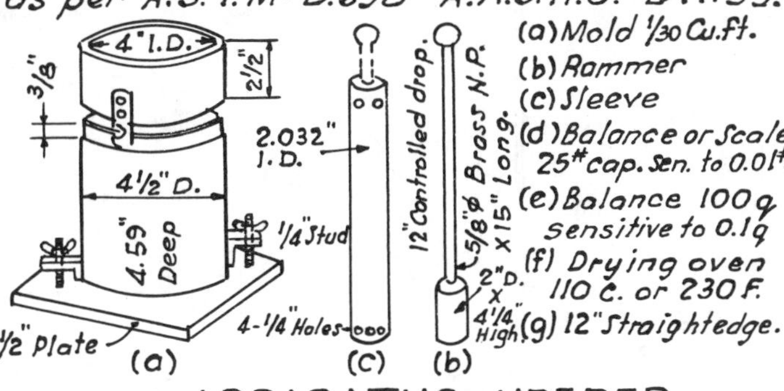
4" I.D.
3/8"
2½"
2.032" I.D.
4½" D.
4.59" Deep
¼" Stud
12" Controlled drop.
5/8" ⌀ Brass N.P. x 15" Long.
2" D. x 4¼" High
4-¼" Holes
½" Plate
(a)
(c)
(b)
(a) Mold 1/30 Cu. ft.
(b) Rammer
(c) Sleeve
(d) Balance or Scale 25# cap. Sen. to 0.01#
(e) Balance 100 g sensitive to 0.1 g
(f) Drying oven 110 C. or 230 F.
(g) 12" Straightedge.

APPARATUS NEEDED.

TESTING PROCEDURE:

6 lb. ± (3000 grams) of air dried soil slightly damp & passing the No. 4 sieve is mixed thoroughly, then compacted in the mold of 1/30 of cu. ft. capacity in 3 equal layers. Each layer receiving 25 blows from the rammer with a controlled drop of 1 ft. The collar is removed and the soil struck off level and weighed.

(Wt. of soil plus mold − Wt. of mold) x 30 = Wet weight per cubic foot or wet density.

A 100-g. sample from the center of the mass is weighed, dried at 230 F. and moisture content determined.

Pulverize original 6 lb., add about 1% water and repeat test. Repeat until soil becomes saturated (about 5 times). Plot Wet-Density Curve. See Fig. B. Compute Dry Density by formula and plot curve:

$$\text{Dry Density} = \frac{\text{Wet Wt., lb. per Cu. ft.}}{\%\ \text{moisture} + 100} \times 100$$

In Fig. B - Enter at top of Dry Density Curve & read optimum moisture & max. wt. of soil 20.2% & 103.5 lb.

MODIFIED A.A.S.H.O. METHOD.**

Same as above except:

1. Rammer to weigh 10 lb.
2. Rammer to have controlled drop of 18".
3. Soil compacted in mold in 5 equal layers. 25 blows to each layer.

The highest dry density is recorded as lab. unit wt.

NOTE: Modern Air-field compaction equipment can secure greater densities than can be obtained by the standard Proctor or A.A.S.H.O. Test. If field compaction or vibration will give greater densities on any job than the test, the higher density should be used to control compaction.

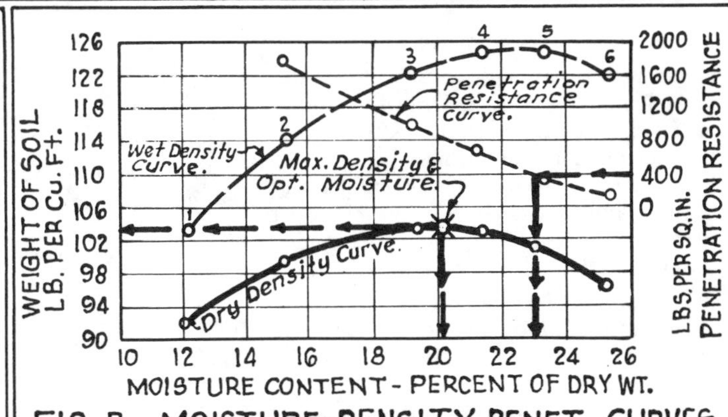
WEIGHT OF SOIL LB. PER Cu. Ft.
126
122
118
114
110
106
102
98
94
90
2000
1600
1200
800
400
0
LBS. PER SQ. IN. PENETRATION RESISTANCE
10 12 14 16 18 20 22 24 26
MOISTURE CONTENT - PERCENT OF DRY WT.
Wet Density Curve.
Max. Density & Opt. Moisture
Penetration Resistance Curve.
Dry Density Curve
1 2 3 4 5 6

FIG. B - MOISTURE-DENSITY PENET. CURVES.

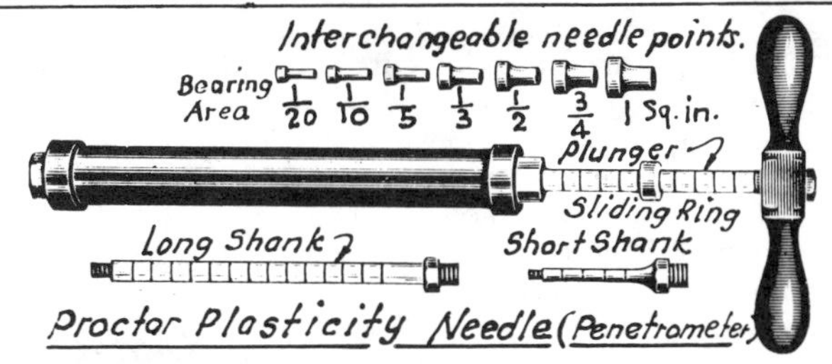
Interchangeable needle points.
Bearing Area 1/20 1/10 1/5 1/3 1/2 3/4 1 Sq. in.
Plunger
Sliding Ring
Long Shank
Short Shank
Proctor Plasticity Needle (Penetrometer)

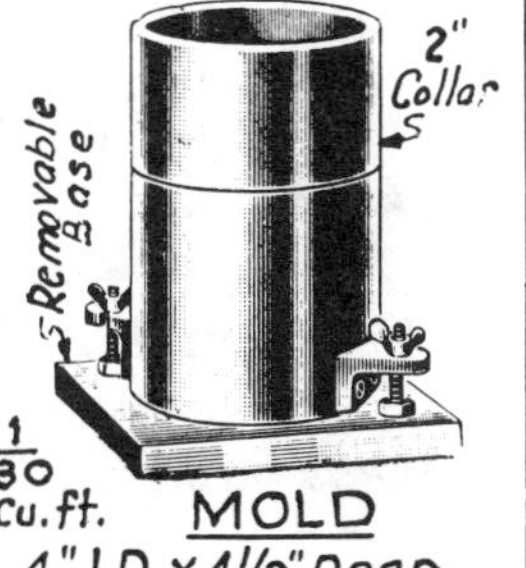
2" Collar
Removable Base
1/30 Cu. ft.
MOLD
4" I.D. x 4½" Deep.

Needle shanks are graduated at ½" intervals to indicate penetration. Plunger rod calibrated for every 10 lb. up to 110 lb. pressure. Ring indicates max. pressure.

5½ lb. Rammer
With end area of 3 Sq. Inches.

FIG. C - APPARATUS.*

PROCTOR TEST:† 5 lb. of dry soil passing a No. 10 sieve is mixed thoroughly with just enough water to make it slightly damp, then compacted in the mold in 3 layers. Each layer is given 25 blows with the rammer dropped 1 ft. The soil is then struck off level with the cylinder, weighed, and the stability determined with the plasticity needle by measuring the force required to press it into the soil at the rate of ½" per sec. A small portion of the soil is oven dried to determine the moisture content. This procedure is repeated 3 to 6 or more times, each time adding about 1% more water until the soil becomes very wet. The density & plasticity needle readings are plotted against moisture content; see Fig. B. Thus in Fig. B a needle reading of 400 gives a moisture content of 23%.

* Adapted from: Humboldt Mfg. Co., ** Engineering Manual, O.C.E., War Dept.,
† Engineering News Record, Aug. 31 to Sept. 28, 1933, R. R. Proctor.

SOILS–FIELD DENSITY TEST

PURPOSE: 1. To obtain the natural density of soil in place (a) as an indication of its stability or bearing value as foundation, (b) to compute the shrinkage or swell when the soil is removed and placed in embankment at a higher or lower density. 2. To determine the per cent of compaction being obtained to check against requirements of specifications.

METHOD OF DETERMINING WEIGHT PER CUBIC FT. OF SOIL IN PLACE.

CALIBRATED SAND METHOD

The density of a soil layer may be determined by finding the weight of a disturbed sample and measuring the volume of the space occupied by the sample prior to removal. This volume may be measured by filling the space with a weighed quantity of a medium of predetermined weight per unit volume. Sand, heavy lubricating oil or water in a thin rubber sack may be used.

1. Determine the weight per cubic foot of the dry sand by filling a measure of known volume. The height and diameter of the measure should be approximately equal and its volume should be not less than 0.1 cu. ft. The sand should be deposited in the measure by pouring through a funnel or from a measure with a funnel spout from a fixed height. The measure is filled until the sand overflows and the excess is struck off with a straightedge. The weight of the sand in the measure is determined and the weight per cubic foot computed and recorded.
2. Remove all loose soil from an area large enough to place a box similar to the one shown in Figure B and cut a plane surface for bedding the box firmly. A dish pan with a circular hole in the bottom may be used.
3. With a soil auger or other cutting tools bore a hole the full depth of the compacted lift.
4. Place in pans all soil removed, including any spillage caught in the box. Remove all loose particles from the hole with a small can or spoon. Extreme care should be taken not to lose any soil.
5. Weigh all soil taken from the hole and record weight.
6. Mix sample thoroughly and take sample for water determination.
7. Weigh a volume of sand in excess of that required to fill the test hole and record weight.
8. Deposit sand in test hole by means of a funnel or from a measure as illustrated in Figure B by exactly the same procedure as was used in determination of unit weight of sand until the hole is filled almost flush with original ground surface. Bring the sand to the level of the base course by adding the last increments with a small can or trowel and testing with a straightedge.
9. Weigh remaining sand and record weight.
10. Determine moisture content of soil samples in percentage of dry weight of sample.
11. Compute dry density from the following formulas:

$$\text{Vol. Soil} = \frac{\text{Wt. of sand to replace soil}}{\text{Wt. per Cu. ft. of sand.}}$$

$$\%\text{ moisture} = \frac{\text{Wt. moist soil} - \text{Wt. dry soil}}{\text{Wt. of dry soil}} \times 100$$

$$\text{Moist density} = \frac{\text{Weight of soil}}{\text{Volume of soil.}}$$

$$\text{Dry density} = \frac{\text{Moist density}}{1 + \frac{\%\text{ of moisture}}{100}}$$

$$\%\text{ Compaction} = \frac{\text{Dry density}}{\text{Maximum density}} \times 100$$

EXAMPLE: Given:

Wt. per Cubic ft. of sand = 100 Lb.
Wt. of moist soil from hole = 5.7 Lb.
Moisture content of soil = 15%
Wt. of sand to fill hole = 4.5 Lb.

Required: Density & % Compaction.

Solution: $\text{Vol. Soil} = \frac{4.5}{100} = 0.045$ Cu. ft.

$$\text{Moist density} = \frac{5.7}{0.045} = 126.7 \text{ Lbs.}$$

$$\text{Dry density} = \frac{126.7}{1 + 15/100} = 110.0 \text{ Lbs.}$$

Given Maximum density = 115 Lbs (from density test Pg. 9-23).

$$\%\text{ Compaction} = \frac{110}{115} \times 100 = 95.7\%$$

Note: In gravel soils material over 1/4" is screened out and correction made.

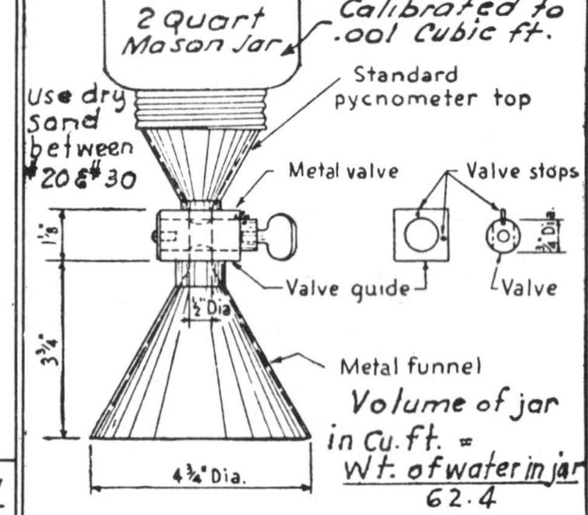

2 Quart Mason Jar
Calibrated to .001 Cubic ft.
Standard pycnometer top
Use dry sand between #20 & #30
Metal valve
Valve stops
Valve guide
Valve
1/2" Dia.
Metal funnel
4 3/4" Dia.
Volume of jar in Cu. ft. = Wt. of water in jar / 62.4

FIG. A - FIELD DENSITY DETERMINATION APPARATUS. DRY SAND METHOD.

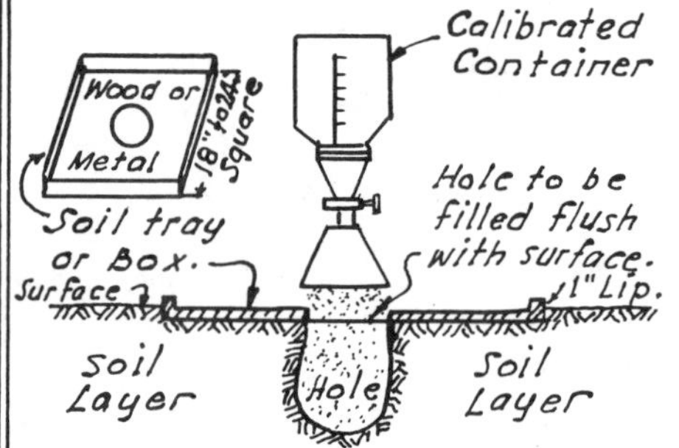

Calibrated Container
Wood or Metal
18" to 24" Square
Soil tray or Box.
Hole to be filled flush with surface.
Surface
1" Lip.
Soil Layer
Hole
Soil Layer

FIG. B - FIELD DENSITY TEST.

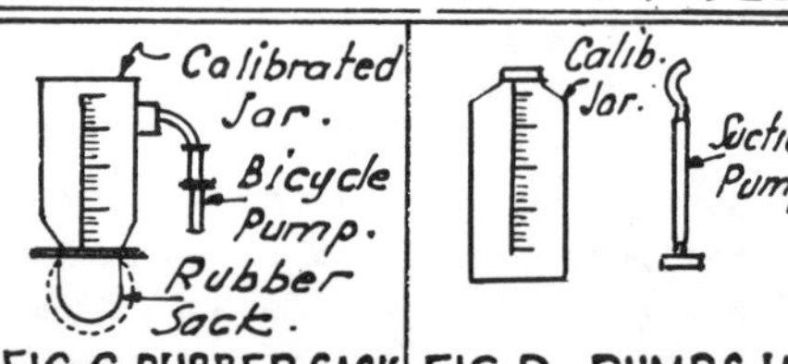

Calibrated Jar.
Bicycle Pump.
Rubber Sack.
Calib. Jar.
Suction Pump

FIG. C - RUBBER SACK inflated to fill hole with known Volume of Water.

FIG. D - PUMP & JAR to fill hole with known Volume of oil. S.A.E. #40.

TABLE E - REL. BEARING VALUES & % COMPACTION REQUIRED.

MAX. DRY DENSITY	SOIL RATING	RECOMMEND. COMPACTION
90 Lb. & less	N.G.	–
90 Lb - 100 Lb	Very Poor	95 - 100%
100 - 110 Lb.	Poor to Very Poor	95 - 100%
110 - 120 Lb.	Poor to Fair	90 - 95%
120 - 130 Lb.	Good	90 - 95%
130 Lb. & over	Excellent	90 - 95%

Note: Density or $\frac{\text{Wt.}}{\text{Vol.}}$ may be expressed as lb. per Cu. ft. or grams per Cu. Centimeter. Density in grams per C.C. = bulk specific gravity.

Chunk Sample Method: 1. Cut sample 4"-5" Dia. full depth of layer. 2. Determine % moisture. 3. Trim sample & weigh to 1/2 oz. 4. Immerse sample in hot paraffin, remove, cool & weigh again. 5. Compute vol. of paraffin using 55 lb. per Cu. ft. 6. Compute Vol. of sample by weighing in water (correcting for Vol. of paraffin). 7. Compute density data by formulas above.

Adopted from: Public Roads, Vol. 22, No. 12 by Harold Allen, Public Roads Administration.

SOILS-MECHANICAL ANALYSIS

PURPOSE: 1. To identify homogeneous soils in the major divisions. See Pg. 9-08.
2. To classify soil mixtures occurring in a natural state, Table A & Fig. C, Pg. 9-08.
3. To classify soil into the P.R.A. or Casagrande groups. Pages 9-10, 9-11, 9-15 & 9-17.
4. To design or control stabilized soil mixtures.
5. To determine frost heaving potentialities. See Pgs. 9-11, 9-15, 9-16 and 9-17
6. To determine Effective Size (D_{10}) & Uniformity Coefficient (C_u) for the design and control of filters and sub-drainage back fill.

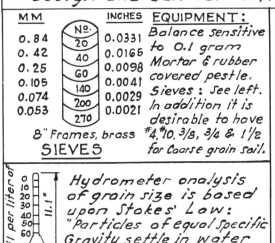

MM	No.	INCHES
0.84	20	0.0331
0.42	40	0.0165
0.25	60	0.0098
0.105	140	0.0041
0.074	200	0.0029
0.053	270	0.0021

8" Frames, brass

SIEVES

EQUIPMENT: Balance sensitive to 0.1 gram. Mortar & rubber covered pestle. Sieves: See left. In addition it is desirable to have #4, #10, 3/8, 3/4 & 1 1/2 for Coarse grain soil.

Grad. in grams of soil per liter of suspension. 0, 10, 20, 30, 40, 50, 60. 11.1". 1.20" Dia.

Hydrometer analysis of grain size is based upon Stokes' Law: "Particles of equal Specific Gravity settle in water at a rate which is in proportion to the size of the particle." NOTE: This test requires Laboratory Technique.

HYDROMETER TEST.

SIEVE ANALYSIS.

Size of sample to be 400 to 750 grams - the coarser the material the larger the sample required.

Take sample by quartering or with sample splitter.

Dry surface moisture by heating the quartered sample at less than 212°F., or boiling point of water at high altitudes, in open pan until surface water disappears & sample is apparently dry and will not lose more weight with additional heating.

Break up cakes with mortar & pestle.

Record dry weight of sample.

Proceed to pass material through screens by placing sample in a stack of sieves, largest size on top, & shake vigorously with horizontal rotating motion balancing on bumper or pad until no more material will pass through each screen.

Weigh amount retained on each sieve, compute per cent of total weight of sample and plot curve.

Washing is recommended for No. 200 sieves and smaller. Partly immerse the largest sieve in a pan of water and agitate. Take material and water from pan and repeat for next smaller size sieve. Agitate smallest size sieve in several water baths until water remains clear. Air-Dry portions retained in sieves, weigh & plot curve.

FIG. A - MECHANICAL ANALYSIS OF SOILS.

3 4 10 20 40 60 80 100 200 ← U.S. STANDARD SIEVE NUMBERS

4" 3" 2" 1 1/2" 1" 3/4" 1/2" 3/8" 1/4" ← SIZE OF SIEVE OPENINGS IN INCHES

PERCENT FINER BY WEIGHT — PERCENT RETAINED BY WEIGHT

SIEVE ANALYSIS → ← WET ANALYSIS ← HYDROMETER

GRAIN SIZE IN MILLIMETERS

GRAIN SIZE IN INCHES

A.S.T.M. & A.A.S.H.O — COARSE SAND | FINE SAND | SILT | CLAY | COLLOIDS

AV. ENG. — GRAVEL | COARSE SAND FINE | SILT | CLAY

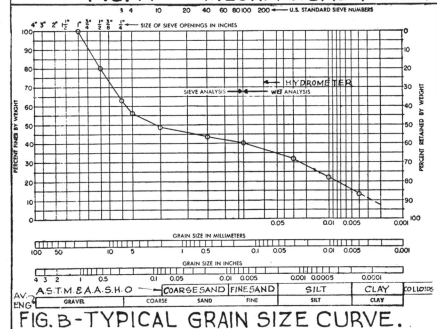

FIG. B - TYPICAL GRAIN SIZE CURVE.

EFFECTIVE SIZE (D_{10}): of a soil is the particle size that is coarser than 10% (by weight) of the soil, that is, 10% of the soil consists of particles smaller than the Effective Size (D_{10}) and 90% consists of larger particles. Example: in Fig. C Chart, Effective Size (D_{10}) is 0.02 mm.

UNIFORMITY COEFFICIENT (C_u): is computed by first determining the size that is coarser than 60% of the soil and dividing that size by the Effective Size (D_{10}), ie,

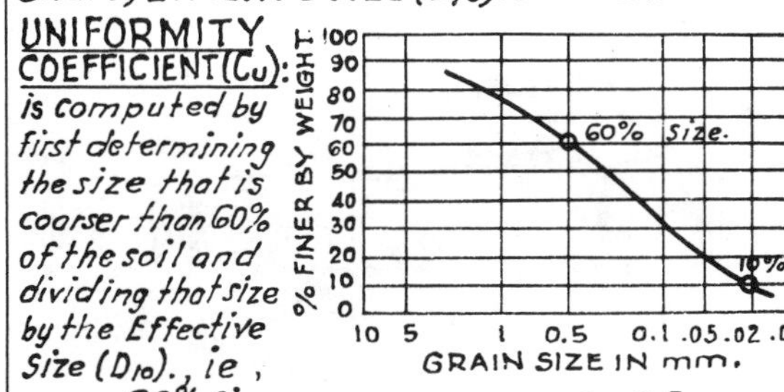

% FINER BY WEIGHT
100 90 80 70 60 50 40 30 20 10 0
60% Size.
10%
10 5 1 0.5 0.1 .05 .02 .01
GRAIN SIZE IN mm.

$C_u = \frac{60\%\ Size}{10\%\ Size}$. Example: in Chart, $C_u = \frac{0.5}{0.02} = 25$.

Note: The C_u of filter backfill should not be over 20. The D_{10} of non-frost heaving uniform soil is 0.02 mm. minimum.

FIG. C - EFFECTIVE SIZE (D_{10}) & UNIFORMITY COEFFICIENT (C_u).

SOILS-SHEAR TESTS

PURPOSE: *To determine maximum shearing stresses, cohesion, and angle of internal friction from soil samples.*

1. UNCONFINED SHEAR TEST:

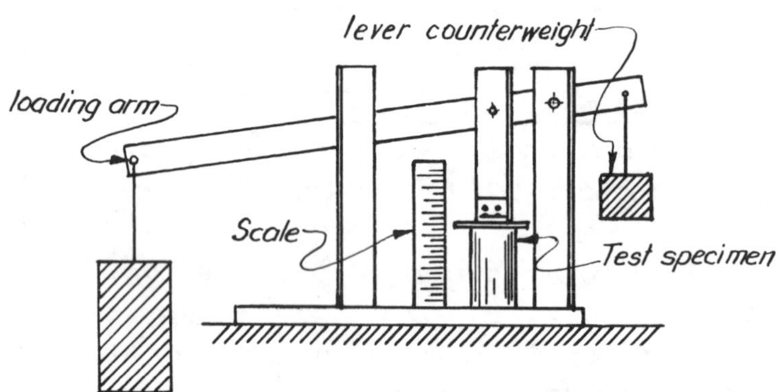

lever counterweight
loading arm
Scale
Test specimen

FIG. A– UNCONFINED SHEAR TEST APPARATUS.

When Used: Field test for undisturbed soils. For clays and clayey mixtures. Cannot be used with materials without a binder.

Method: Specimen loaded to failure by weights added to loading arm
Approximately: Shear = $\frac{1}{2}$ x p.s.i. on specimen.

Advantages: Test is easily performed; apparatus is portable and can be home-made. Test can be improvised. It could be made by adding water to a bucket resting on specimen. Unconfined shear test is more accurate for clays or mixtures of clay.

Note: Height of specimen = 2 x width

2. ANGLE OF REPOSE METHOD:

For sands and free draining deposits of sand. Determine angle of repose which equals lower limit of angle of friction.

3. SHEAR BOX METHOD:

Normal load
Ames gage to show vert. movemt.
Movable top frame
Ames gage to measure lat. movement
sample
lateral force
Fixed lower frame

FIG. B – SHEAR BOX

When Used: To simulate effect of material under pressure. Gives the maximum angle of internal friction. Use for free draining deposits, soil mixtures above ground water, or for clays.

Method: With a constant normal load a lateral force is applied until slip occurs. A curve as shown at left may then be plotted. For clays and mixtures a quick shear test may be desirable.

Advantages: Apparatus is portable and easy to set up.

Note: Sample should be at field density

Lateral Shear S (p.s.i.)
Test points
Angle of internal friction
ø
Cohesion = Intercept
Normal Pressure (p.s.i.) N

FIG. C – SHEAR TEST CURVE

4. TRIAXIAL SHEAR TEST:

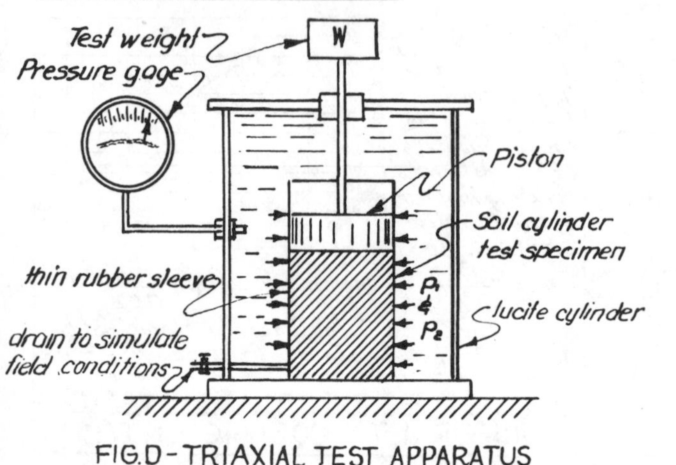

W
Test weight
Pressure gage
Piston
Soil cylinder test specimen
thin rubber sleeve
P_1 & P_2
lucite cylinder
drain to simulate field conditions

FIG. D – TRIAXIAL TEST APPARATUS

When Used: For silty soils generally. Can be used for all soils. Since apparatus is bulky, test is performed in a laboratory.

Method: Lateral hydrostatic pressure is maintained constant and a vertical force is applied until failure occurs. Test is repeated with increased lateral pressure and results are plotted as shown, based on Mohr's circle theorem.

Advantages: Test can be used for all soils, gives most accurate results as actual drainage conditions are simulated.

P_{1-2} = lat. press. for runs 1 & 2 (p.s.i.)
W_{1-2} = Vert. press. for runs 1 & 2 (p.s.i.)
ø = Angle of internal friction (determined)
c = cohesion (determined from diagram)
N = normal pressure on soil (p.s.i.)
S = total shear strength (p.s.i.)

$S = c + N \tan ø$

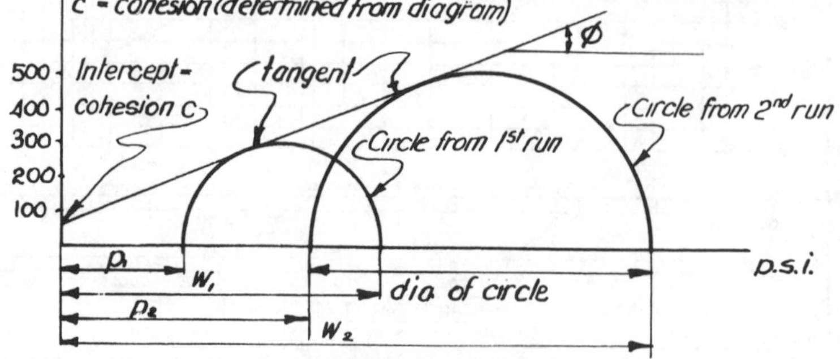

p.s.i.
ø
500
400
300
200
100
Intercept-cohesion c
tangent
Circle from 2nd run
Circle from 1st run
P_1
W_1
P_2
W_2
dia. of circle
p.s.i.

FIG. E – MOHR'S CIRCLE DIAGRAM

SOILS - MAXIMUM DENSITY FROM SIEVE ANALYSIS & HOUSEL PENETROMETER

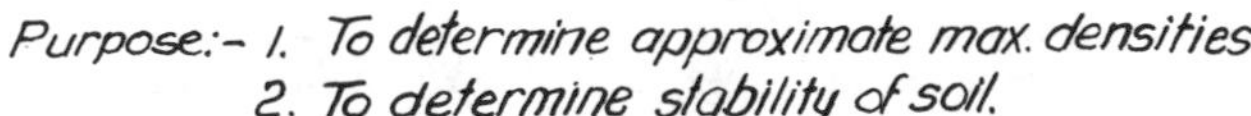

Purpose:- 1. To determine approximate max. densities.
2. To determine stability of soil.

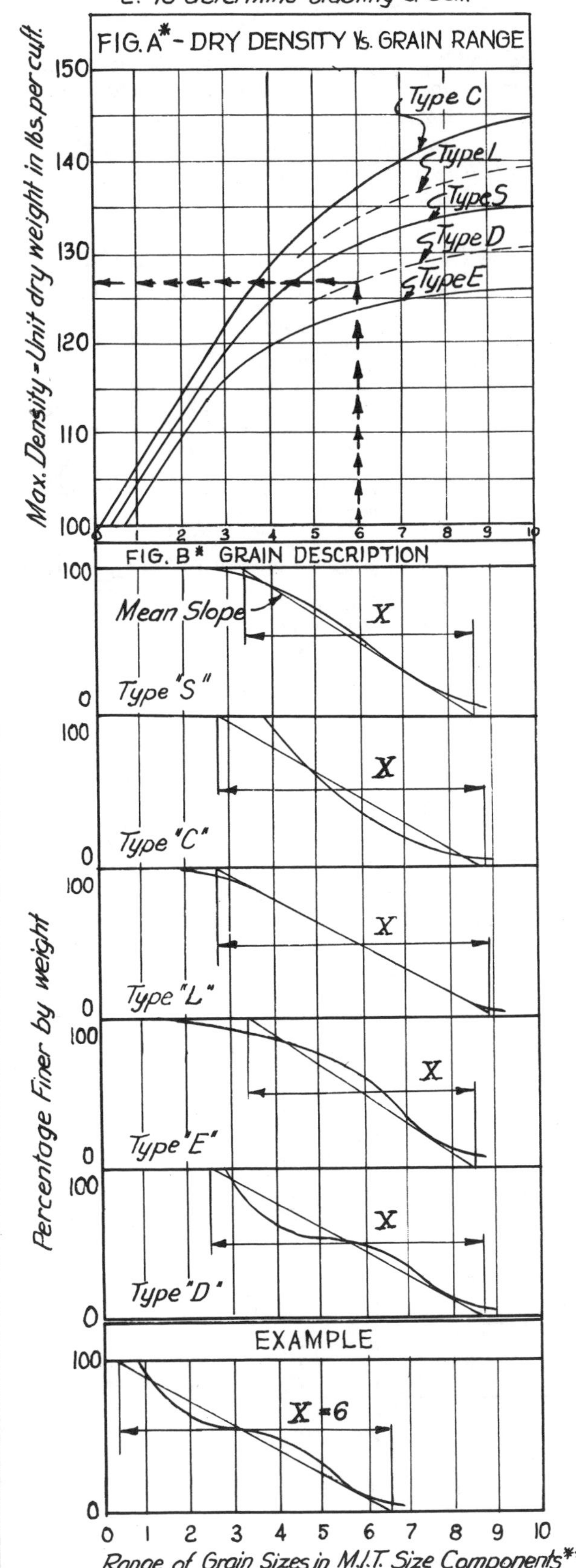

FIG. A* - DRY DENSITY Vs. GRAIN RANGE
Max. Density - Unit dry weight in lbs. per cu. ft.
150
140
130
120
110
100
Type C
Type L
Type S
Type D
Type E
FIG. B* GRAIN DESCRIPTION
Percentage Finer by weight
Mean Slope
X
Type "S"
Type "C"
Type "L"
Type "E"
Type "D"
EXAMPLE
X = 6
0 1 2 3 4 5 6 7 8 9 10
Range of Grain Sizes in M.I.T. Size Components**

Example: Plot a given sample's analysis on example form, Fig. B. Note value of X=6 in this case. Form of curve is similar to type D, fig. B. Enter fig. A at X=6 and obtain max. density = 127 p.c.f.

* Fig. A & B empirical by D.M. Burmister.

** See pg. 9-08.

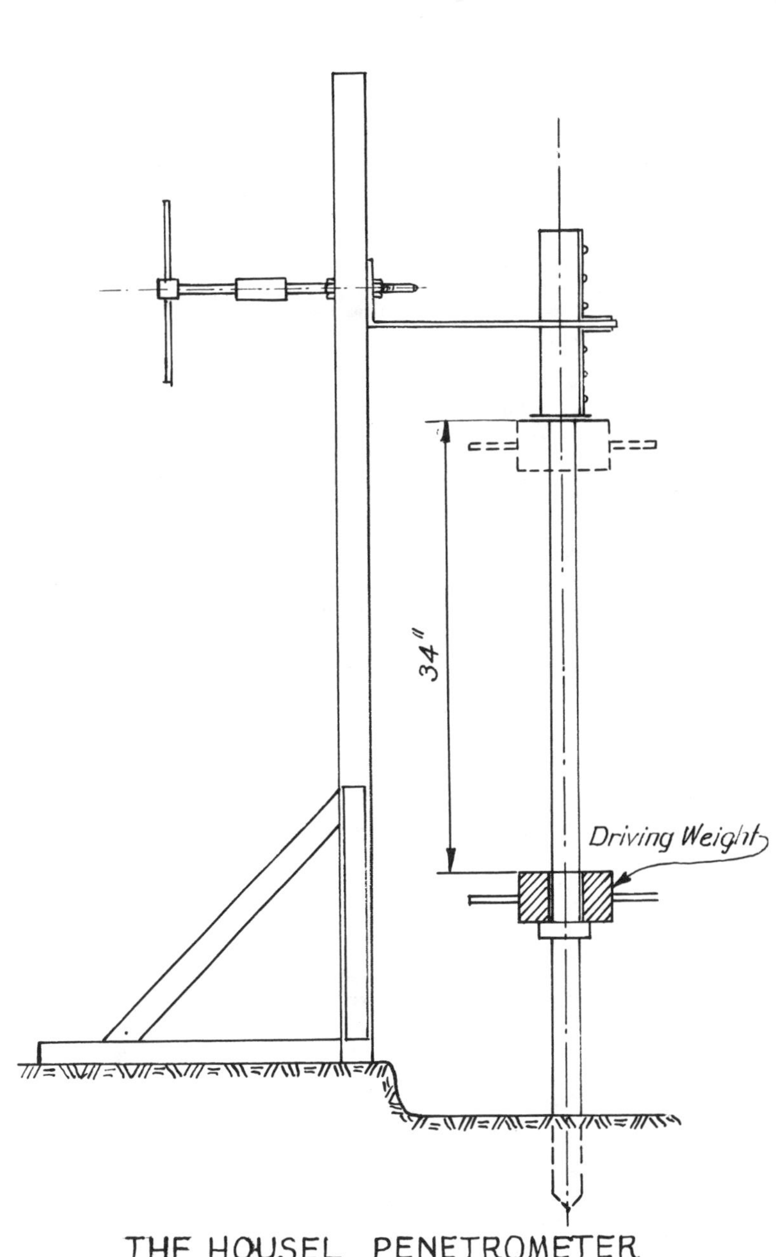

34"
Driving Weight

THE HOUSEL PENETROMETER
(NOT TO SCALE)

The penetrometer equipment consists of a sharpened 1.25-in. diameter standard pipe, which with accessories weighs exactly 20 lb. exclusive of the driving weight, which also weighs exactly 20 lb. Stops on the barrel of the 1.25 in. pipe, control the height of drop of the driving weight to exactly 34 in. The test consists of determining the number of blows of the 20 lb. driving weight falling exactly 34 in. required to drive the sharpened pipe 6 in. into the soil. A cardboard strip firmly attached to the barrel of the pipe is marked with a pencil at the beginning of the test and after the penetration of each blow.

SOILS-SPREAD FOOTINGS

TABLE-A CLASSIFICATION OF SUPPORTING SOILS*

Class	Material	Maximum allowable presumptive bearing values in tons per square foot
1	Hard sound rock	60
2	Medium hard rock	40
3	Hardpan overlaying rock	12
4	Compact gravel and boulder-gravel formations; very compact sandy gravel	10
5	Soft rock	8
6	Loose gravel and sandy gravel; compact sand and gravelly sand; very compact sand-inorganic silt soils	6
7	Hard dry consolidated clay	5
8	Loose coarse to medium sand; medium compact fine sand	4
9	Compact sand-clay soils	3
10	Loose fine sand; medium compact sand-inorganic silt soils	2
11	Firm or stiff clay	1.5
12	Loose saturated sand-clay soils; medium soft clay	1

Explanation of Terms

Compaction Related to Spoon Blows; Sand

Descriptive Term	Blows/Foot	Remarks
Loose	15 or less	These figures approximate for medium sand, 2½-inch spoon, 300-pound hammer, 18-inch fall. Coarser soil requires more blows, finer material, fewer blows.
Compact	16 to 50	
Very compact	50 or more	

Consistency Related to Spoon Blows; Mud, Clay, Etc.

Descriptive Term	Blows/Foot	Remarks
Very soft	push to 2	Molded with relatively slight finger pressure.
Soft	3 to 10	
Stiff	11 to 30	Molded with substantial finger pressure; might be removed by spading.
Hard	30 or more	Not molded by fingers, or with extreme difficulty; might require picking for removal.

Soil Sizes

Descriptive Term	Pass Sieve Number	Retained Sieve Number	Size Range
Clay	200	Hydrometer	.006 mm.
Silt	200	analysis	.006 to .074 mm.
Fine sand	65	200	.074 to .208 mm.
Medium sand	28	65	.208 to .589 mm.
Coarse sand	8	28	.589 to 2.362 mm.
Gravel	—	8	2.362 mm.
Pebble	—	—	2.362 mm. to 2½"
Cobble	—	—	2½" to 6"
Boulder	—	—	6"

TABLE-B PROPOSED BEARING VALUES FOR CLAY**

N = number of blows per foot in standard penetration test.[1]

q_u = unconfined compressive strength in tons per sq. ft.

q_d = ultimate bearing capacity of continuous footing in tons/□'

q_{ds} = ultimate bearing capacity of square footing in tons/□'

q_a = proposed normal allowable bearing value in tons/□' ($G_s = 3$)

q_a' = proposed maximum tolerable bearing value in tons/□' ($G_s = 2$)

G_s = factor of safety with respect to base failure.

[1] Standard Penetration Test = 140# weight, 30" drop, 2" o.d. sampler

Description of Clay	N	q_u	q_d	q_{ds}	q_a Square $1.2q_u$	q_a Cont. $0.9q_u$	q_a' Square $1.8q_u$	q_a' Cont. $1.3q_u$
Very Soft*	Less than 2	Less than 0.25	Less than 0.71	Less than 0.92	Less than 0.30	Less than 0.22	Less than 0.45	Less than 0.32
Soft*	2 to 4	0.25 to 0.50	0.71 to 1.42	0.92 to 1.85	0.30 to 0.60	0.22 to 0.45	0.45 to 0.90	0.32 to 0.65
Medium	4 to 8	0.50 to 1.00	1.42 to 2.85	1.85 to 3.70	0.60 to 1.20	0.45 to 0.90	0.90 to 1.80	0.65 to 1.30
Stiff	8 to 15	1.00 to 2.00	2.85 to 5.70	3.70 to 7.40	1.20 to 2.40	0.90 to 1.80	1.80 to 3.60	1.30 to 2.60
Very Stiff	15 to 30	2.00 to 4.00	5.70 to 11.40	7.40 to 14.80	2.40 to 4.80	1.80 to 3.60	3.60 to 7.20	2.60 to 5.20
Hard	Over 30	Over 4.00	Over 11.40	Over 14.80	Over 4.80	Over 3.60	Over 7.20	Over 5.20

* If clay is normally loaded settlement can be important even under smallest allowable soil pressure.

FIG. A SAND BEARING CURVES**

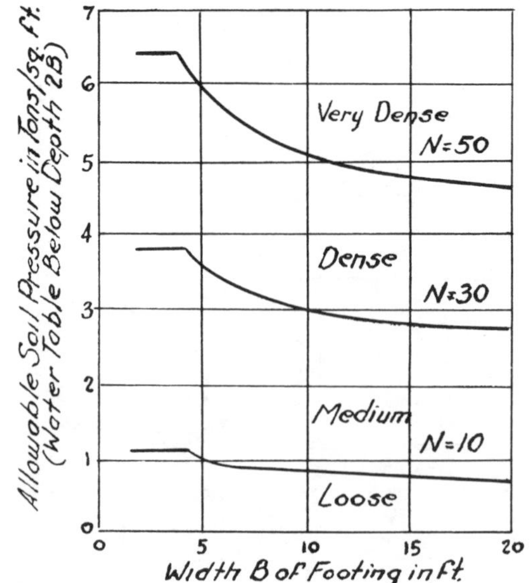

Allowable Soil Pressure in Tons/sq. ft. (Water Table Below Depth 2B)
0
1
2
3
4
5
6
7
Very Dense
N=50
Dense
N=30
Medium
N=10
Loose
0
5
10
15
20
Width B of Footing in Ft.

* Adapted from N.Y.C. Building Code, 1951

** From "Soil Mechanics in Engineering Practice" Terzaghi & Peck, John Wiley, 1948

SOILS—EMBANKMENT FOUNDATIONS

ELASTIC THEORY:- *Based on Boussinesq equation.*

CAUTION:- *In applying this to soils, it is to be remembered that it is based on a foundation of homogeneous, isotropic material of infinite depth, modulus of elasticity independent of depth, which are not the usual properties of soils.*

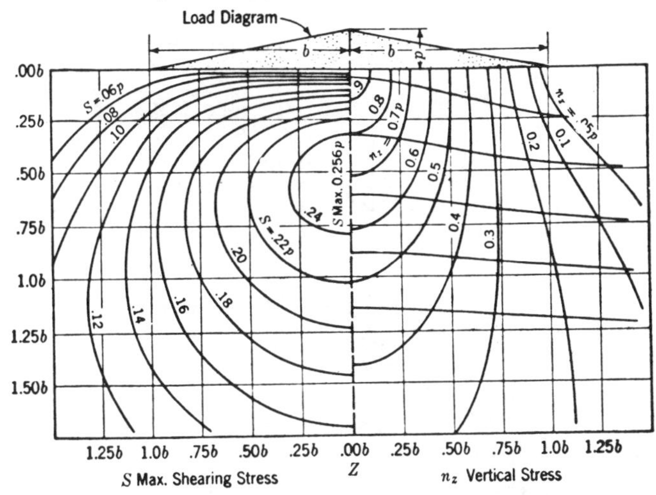
Load Diagram
S Max. Shearing Stress
Z
n_z Vertical Stress
S Max. 0.256p
S = .22p

FIG. A - TRIANGULAR LOADING.

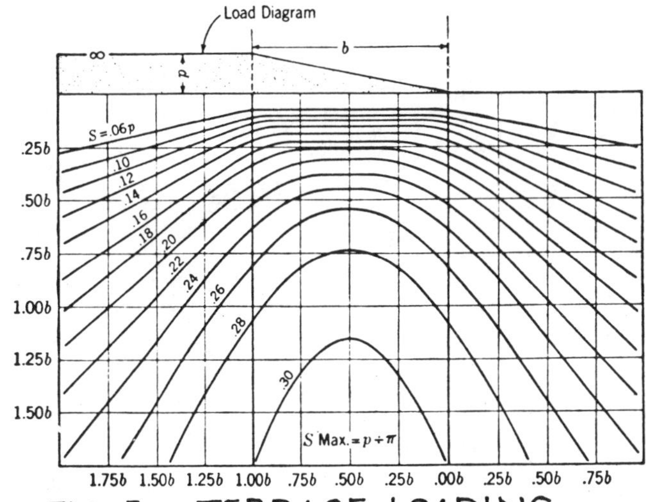
Load Diagram
S = .06p
S Max. = p ÷ π

FIG. B - TERRACE LOADING.

DISTRIBUTION OF SHEARING STRESSES UNDER EMBANKMENTS.*

S = Shear per sq. ft.; p = Load per sq. ft. on foundation at ℄ of dam.
Foundation of homogeneous isotropic material, infinite depth.

USE OF DIAGRAMS:* *Given:- Embankment 100 ft. high; foundation material = fine silt with tan ϕ = 0.4 and C = 0, weight = 120#/cu.ft., b = 200 ft.*

Required:- Factor of Safety at depth 0.5 b. Solution:- In Terrace loading - Fig. B diagram at 0.5b, max. unit shear = 0.25 p = $\frac{0.25 \times 100 \times 120}{2000}$ = 1.5 tons/sq.ft. Unit Shearing strength = W. tan ϕ + C. W = unit weight x total depth at point of max. unit shear (midpoint of slope = 100 + 50) ∴ Unit shearing strength = $\frac{150 \times 120}{2000} \times 0.4 + 0 = 3.6$ tons/sq.ft. factor of safety = $\frac{3.6}{1.5}$ = 2.4.

MIN. FACTOR OF SAFETY:-*Try solution at other depths, 0.25 b, 0.375 b, 0.75 b, etc.*

Notes: 1.- Justin, Hinds & Creager prefer using Terrace Loading rather than Triangular loading for dam embankments.
2.- Applicable only to homogeneous isotropic foundations of infinite depth.
3.- Applicable to fill such as on soft harbor foundations. Test as in Fig. D below.

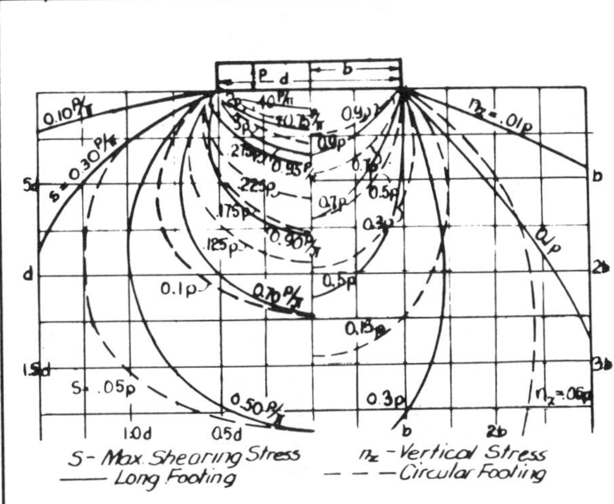
S- Max. Shearing Stress
—— Long Footing
n_z - Vertical Stress
- - - Circular Footing

P - Lbs. per sq. ft. distributed Load.

FIG. C - DISTRIBUTION OF SHEARING STRESSES UNDER A LONG FOOTING* (for circular ftgs. see p. 9-07)

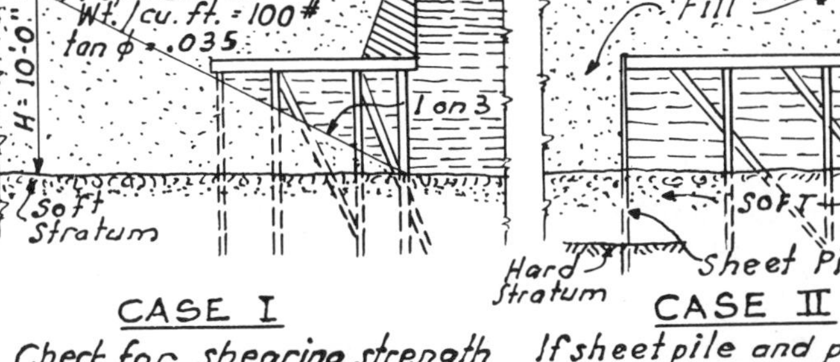
30 ft
Proposed fill - C = 200
Wt./cu.ft. = 100#
tan φ = .035
H = 10'-0"
1 on 3
Soft Stratum
Fill
Soft
Hard Stratum
Sheet Piling

CASE I

Check for shearing strength of soft foundation stratum as follows:

CASE II

If sheet pile and platform are strong enough to resist horizontal thrust of soft foundation and fill and adequately tied into hard stratum no further check of shearing stress is necessary.

EXAMPLE: *Test soft stratum for support of Embankment.*

*By Fig. B - Max. Unit Shear = 0.25p = 0.25 x 10 x 100 = **250**.*
*Unit Shearing Strength at 0.5 b = W. tan ϕ + C = 25 x 100 x .035 + 200 = **287.5***
[Where depth at 0.5 b = (10 + $\frac{30}{2}$) = 25]
Safety factor = $\frac{287.5}{250}$ = 1.1 since this is low, check by friction circle theory

FIG. D - CHECK FOUNDATION OF FILL BEHIND RELIEVING PLATFORM.**

* *From The Application of Theories of Elasticity and Plasticity to Foundation Problems by Leo Jürgenson, J. Boston Soc. Civil Engrs., July 1934, adapted by Justin, Hinds & Creager.* ** *Adapted from Plummer & Dore, Soil Mechanics & Foundations, Pitman Pub. Corp.*

SOILS-EMBANKMENTS & RETAINING WALLS

LIMITATIONS: *All slopes of cohesive material require flatter angles as the height is increased. This limiting height will vary with the degree of compaction, cohesive strength and angle of friction. It will also vary with the strength of the foundation on which it rests. Granular soils have no limiting heights.*

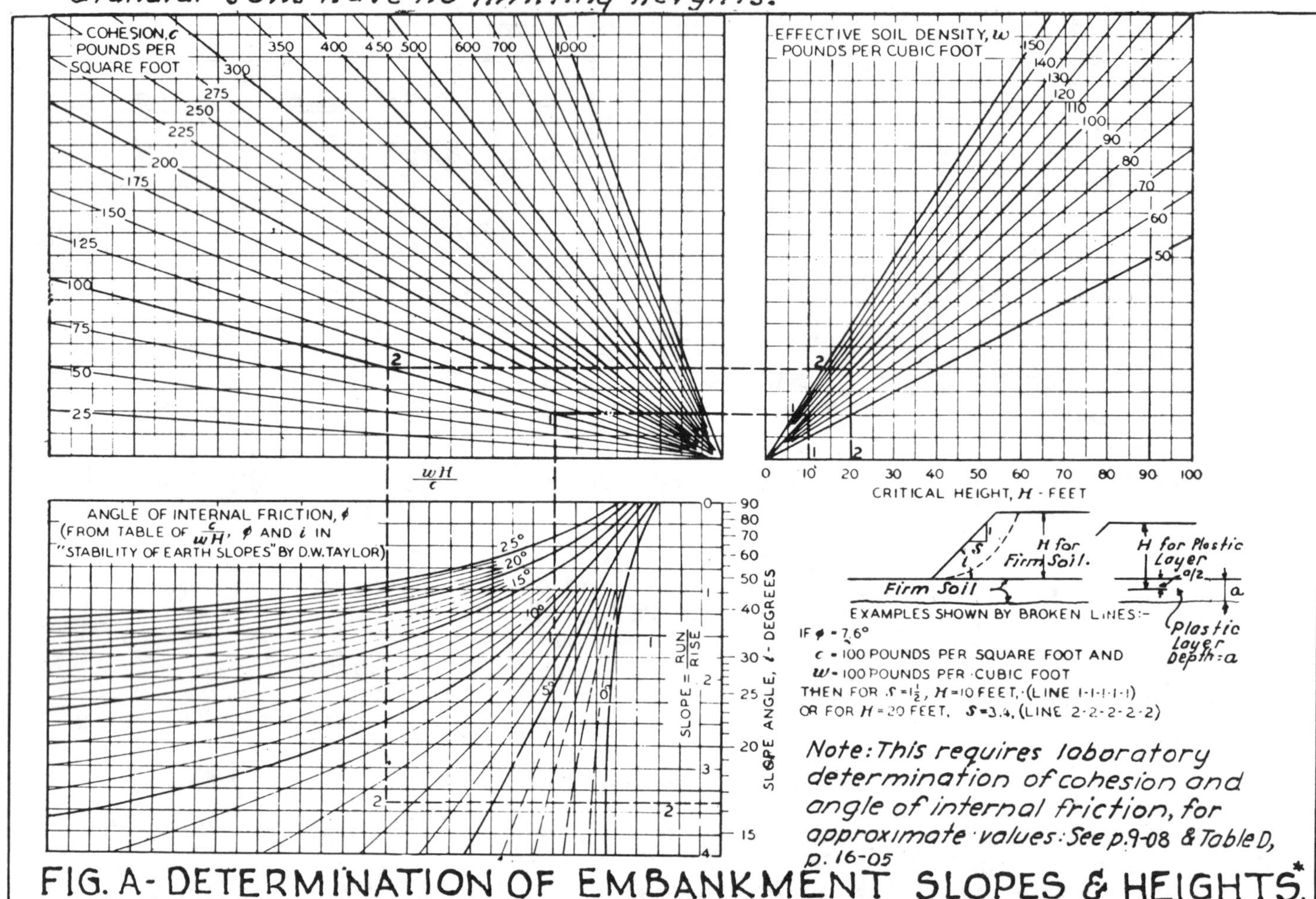
COHESION, c POUNDS PER SQUARE FOOT
EFFECTIVE SOIL DENSITY, w POUNDS PER CUBIC FOOT
CRITICAL HEIGHT, H - FEET
ANGLE OF INTERNAL FRICTION, φ (FROM TABLE OF c/wH, φ AND i IN "STABILITY OF EARTH SLOPES" BY D.W. TAYLOR)
SLOPE = RUN/RISE
SLOPE ANGLE, i - DEGREES
H for Firm Soil.
Firm Soil
H for Plastic Layer
Plastic Layer Depth: a
EXAMPLES SHOWN BY BROKEN LINES:-
IF φ = 7.6°
c = 100 POUNDS PER SQUARE FOOT AND
w = 100 POUNDS PER CUBIC FOOT
THEN FOR S = 1½, H = 10 FEET, (LINE 1-1-1-1-1)
OR FOR H = 20 FEET, S = 3.4, (LINE 2-2-2-2-2)
Note: This requires laboratory determination of cohesion and angle of internal friction, for approximate values: See p.9-08 & Table D, p. 16-05

FIG. A- DETERMINATION OF EMBANKMENT SLOPES & HEIGHTS.*

RETAINING WALLS† (After Dr. Karl Terzaghi).

TABLE B. DATA FOR ESTIMATING EARTH PRESSURE ON THE BASIS OF EQUIVALENT FLUID EARTH.

DESCRIPTION OF BACKFILL.	w lb/cu.ft.	w' lb/cu.ft.	Cohesion lb./sq.ft.	H_o ft.
1. Coarse grained backfill without admixture of fine soil particles, very permeable.	110	27	–	0
2. Coarse grained backfill with low permeability due to admixture of soil particles with silt size.	115	35	–	0
3. Residual soil with stones, loamy sand and other backfill materials with conspicuous clay content.	115	45	–	0
4. Plastic clay.	120	120	200	4.3
5. Soft clay or mud.	100	100	0	0

H_o Top of w'

w' = Equivalent Fluid weight
w = Weight of material lb./cu.ft.
c = Cohesion lb./sq.ft.

For clay: $w = w'$; $H_o = \frac{2.5c}{w}$.
For soils #1 to #3 in Table B, on compressible footing, increase w' by 50%.

RETAINING WALLS WITH SURCHARGE LOADS.

CASE I - UNIFORM SURCHARGE.

Uniform Surcharge – q lb/sq.ft.

increase pressure per unit area against wall by : $q\frac{w'}{w}$.

CASE II - SURCHARGE DUE TO AN ADJACENT VERTICAL LOAD.

L = Load per linear foot.

40° P

Horizontal load per l.f. on back of wall caused by $L = P = L\frac{w'}{w}$.

CASE III - SLOPING SURCHARGE.

Angle of Repose, H', H

Top of equivalent fluid.

$w'' = w'\left[2 - \left(\frac{H - H_1}{H}\right)^2\right]$

w'' = Equivalent unit fluid weight behind wall.
if $H' > H$, $w'' = 2w'$.

Check for Sliding on Clays: Shear must be less than ½ bearing capacity of soil.

* Adapted from Eng. News Record, Vol. 128 - No. 7, dated Feb. 12, 1942 article by P.R.A.
† See also Pg. 7-30, 7-31 for data on retaining walls.

SOILS-EMBANKMENT SHRINKAGE & PERMEABILITY

EMBANKMENT SHRINKAGE

CASE I - *Yield of borrow in finished embankment.*

$$\text{Volume} = \frac{\text{Weight}}{\text{Density}}$$: *Volume varies inversely as density.*

Assume { *borrow pit dry density = 110 lbs. per cu. ft.*
embankment dry density = 125 lbs. per cu. ft.

Then 1 cu. yd. of borrow will yield 110/125 cu. yd. embankment.

CASE II - *Shrinkage after embankment compaction.*
This will be zero if embankment is compacted to maximum density. In ordinary practice, i.e. power equipment wheel compaction - no rollers, no moisture control - it may be taken as:
Sand and gravel - 1 to 2 % — Clay or loam - 2 to 3 %**
Certain deposits such as expansive clays swell in embankments.

CASE III - *Loss by waste, of improper material such as top soil or muck, should be allowed for.*

TABLE A - VOLUME OF BORROW REQUIRED FOR 1 CU. YD. OF EMBANKMENT.

(Empirical rules for use where densities are not obtainable. Power equipment wheel compaction, not rolled.)

Material	*	**	Material	*	**
Gravel	1.09 *	1.12 **	Light sandy earth	1.14*	1.20 **
Gravel and sand	1.10 *	1.13 **	Vegetable surface soil	1.18*	1.24 **
Clay and clayey earth	1.11 *	1.15 **	Ledge rock		0.70 **

PERMEABILITY See also Pg. 9-34.

TABLE B - PERMEABILITY DATA.†

COEFFICIENT OF PERMEABILITY (k) in cm. per sec. (Log scale)

Turbulent flow (k = 10^2 to 1.0) — Perfect validity of Darcy's Law (k = 10^{-7} and below)

k =	10^2	10^1	1.0	10^{-1}	10^{-2}	10^{-3}	10^{-4}	10^{-5}	10^{-6}	10^{-7}	10^{-8}	10^{-9}
DRAINAGE PROPERTY		Good drainage					Poor drainage		Practically impervious			
APPLICATION IN EARTH DAMS AND DIKES		Pervious sections of dams and dikes					Impervious sections of earth dams and dikes					
TYPES OF SOIL	Clean Gravel		Clean sands; Clean sand & gravel Mixtures			Very fine sands; organic & inorganic silts; mixtures of sand, silt & clay; glacial till; stratified clay deposits; etc.				"Impervious" soils e.g. homogeneous clays below zone of weathering		
					"Impervious soils" which are modified by the effects of vegetation and weathering							
DIRECT DETERMINATION of coefficient of permeability	Direct testing of soil in its original position (e.g. well points) if properly conducted - reliable - considerable experience required											
	CONSTANT HEAD PERMEAMETER Little experience required											
	FALLING HEAD PERMEAMETER		Reliable little experience required			Unreliable Much experience necessary for correct interpretation			Fairly reliable Considerable experience necessary			
INDIRECT DETERMINATION of coefficient of permeability	COMPUTATION from the grain size distribution (e.g. Hazen's formula) only applicable to clean cohesionless sands & gravels											
				HORIZONTAL CAPILLARITY TEST Very little experience necessary. Especially useful for rapid testing of a large number of samples in the field without Lab. facilities						COMPUTATIONS from consolidation tests; expensive Lab. equip. & considerable experience required		

* From Am. C. E. Handbook by Merriman & Wiggin. ** In accordance with modern highway practice.
† Adapted from Publication No. 268, Harvard Grad. School of Eng.

SOILS-VOIDS RATIO & PERMEABILITY

DETERMINATION OF VOIDS RATIO = e

EXAMPLE: Weight of Soil = 120 pounds per cubic foot.
Specific Gravity of grains = 2.70
2.70 x 62.4 = 169 lb./cu. ft. - Theoretical weight of solid material.
120/169 = 0.71 cu. ft. - Volume of Solid material.
1.00 - 0.71 = 0.29 cu. ft. - Volume of voids.

$$\frac{0.29}{0.71} = 0.41 \text{: VOIDS RATIO.}$$

PERMEABILITY

Darcy's Law : $Q = kA\frac{H}{L}$ for flow of water through soil.

WHERE:
Q = Volume of Flow in cu. ft. per min.
k = Coefficient of permeability - for values see Table A.
A = Area in sq. ft. of cross-section under consideration.
$\frac{H}{L}$ = Hydraulic Gradient.

TABLE A - COEFFICIENTS OF PERMEABILITY (k) in ft./min. for 60° F.

EFFECTIVE SIZE	Porosity = $\frac{e}{1+e}$										
	30 percent	32 percent	34 percent	36 percent	38 percent	40 percent	42 percent	44 percent	46 percent	48 percent	50 percent
0.01	0.00003	0.00004	0.00005	0.000060	0.000072	0.000085	0.000101	0.000123	0.000149	0.000181	0.000215
0.02	.00013	.00016	.00020	.000239	.000286	.000339	.000405	.000492	.000597	.000724	.000861
0.03	.00030	.00036	.00045	.000538	.000645	.000763	.000911	.00111	.00134	.00163	.00194
0.04	.00053	.00065	.00079	.000958	.001145	.001355	.00162	.00197	.00239	.00290	.00344
0.05	.00082	.00101	.00124	.001495	.001790	.002120	.00253	.00307	.00373	.00452	.00538
0.06	.00118	.00146	.00178	.002150	.002580	.003050	.00364	.00442	.00537	.00652	.00775
0.07	.00161	.00198	.00243	.002930	.003510	.004155	.00496	.00602	.00732	.00887	.0105
0.08	.00211	.00259	.00218	.003825	.004585	.005425	.00647	.00787	.00956	.0116	.0138
0.09	.00266	.00328	.00402	.004845	.005800	.006860	.00820	.00995	.0121	.0147	.0174
0.10	.00328	.00405	.00496	.005980	.007170	.008480	.0101	.0123	.0149	.0181	.0215
0.12	.00473	.00583	.00713	.008620	.01032	.01220	.0146	.0177	.0215	.0261	.0310
0.14	.00643	.00794	.00972	.01172	.01404	.01662	.0198	.0241	.0293	.0355	.0422
0.15	.00739	.00912	.01115	.01345	.01611	.01910	.0228	.0277	.0336	.0407	.0484
0.16	.00841	.01036	.01268	.01531	.01835	.02170	.0259	.0315	.0382	.0463	.0551
0.18	.01064	.01311	.01605	.01940	.02320	.02745	.0328	.0398	.0484	.0586	.0697
0.20	.01315	.0162	.01983	.02390	.02865	.03390	.0405	.0492	.0597	.0724	.0861
0.25	.020	.0253	.03100	.03740	.04480	.05300	.0632	.0768	.0933	.113	.134
0.30	.0296	.0364	.04460	.05380	.06450	.07630	.0911	.111	.134	.163	.194
0.35	.0403	.0496	.0608	.07330	.08790	.1039	.124	.151	.183	.222	.264
0.40	.0527	.0648	.07940	.09575	.1145	.1355	.162	.197	.239	.290	.344
0.45	.0665	.0820	.1005	.1211	.1450	.1718	.205	.249	.302	.366	.436
0.50	.0822	.1012	.1240	.1495	.1780	.2120	.253	.307	.373	.452	.538
0.55	.0994	.1225	.1500	.1810	.2165	.2565	.306	.372	.452	.547	.651
0.60	.1182	.1458	.1784	.2150	.2580	.3050	.364	.442	.537	.652	.775
0.65	.1390	.1710	.2095	.2530	.3030	.3580	.428	.519	.631	.765	.909
0.70	.1610	.1983	.2430	.2930	.3510	.4155	.496	.602	.732	.887	1.05
0.75	.1850	.2278	.2785	.3365	.4030	.4770	.569	.691	.840	1.02	1.21
0.80	.2105	.2590	.3175	.3825	.4585	.5425	.648	.787	.956	1.16	1.38
0.85	.2375	.2925	.3580	.4325	.5175	.6125	.731	.888	1.08	1.31	1.55
0.90	.2660	.3280	.4018	.4845	.5800	.6860	.820	.995	1.21	1.47	1.74
0.95	.2965	.3650	.4470	.5400	.6460	.7650	.913	1.11	1.35	1.63	1.94
1.00	.3282	.4050	.4960	.5880	.7170	.8480	1.01	1.23	1.49	1.81	2.15
2.00	1.315	1.620	1.983	2.390	2.865	3.390	4.05	4.92	5.97	7.24	8.61
3.00	2.960	3.640	4.460	5.380	6.450	7.630	9.11	11.1	13.4	16.3	19.4
4.00	5.270	6.480	7.940	9.575	11.45	13.55	16.2	19.7	23.9	29.0	34.4
5.00	8.220	10.12	12.40	14.95	17.90	21.20	25.3	30.7	37.3	45.2	53.8

TABLE B - TEMPERATURE CORRECTIONS TO TABLE A.

See below for use of table.

TEMP. IN °F.	t_c
32	0.64
35	.67
40	.73
45	.80
50	.86
55	0.93
60	1.00
65	1.08
70	1.15
75	1.23
80	1.30
85	1.39
90	1.47
95	1.55
100	1.64

Use of Table B:-
If temperature is other than 60° F., multiply k value in Table A by value of t_c opposite applicable temperature in Table B.

* Adapted from Low Dams by National Resources Committee.

SOILS — FROST

FROST ACTION SOILS: *Soils subject to frost action are well-graded soils containing more than 3 per cent by dry weight of particles less than 0.02 mm. (0.0008 inch) in size, and uniformly graded soils containing more than 10 per cent of particles less than 0.02 mm. in size, See Fig. B.

GENERAL: To prevent frost damage, the combined thickness of pavement and non-frost action base should be equal to the average depth of frost penetration as shown in Fig. A except as limited by Table C below.

Plasticity Index should be less than 6, preferably non plastic and liquid limit less than 25 if soils are not subject to frost damage.

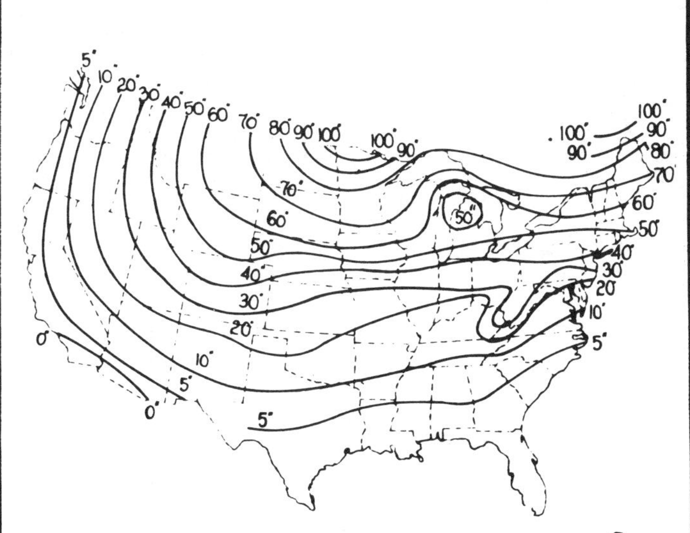

FIG A: MAXIMUM FROST PENETRATION IN INCHES[©] USE FOR FOUNDATIONS

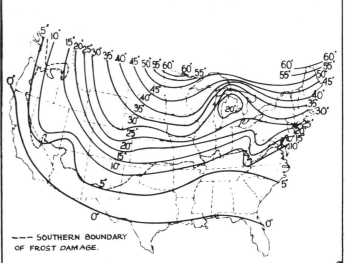

--- SOUTHERN BOUNDARY OF FROST DAMAGE.

FIG A′ - AVERAGE ANNUAL FROST PENET. & SO. LIMIT OF FROST DAMAGE[©] USE FOR HIGHWAY SUBGRADE

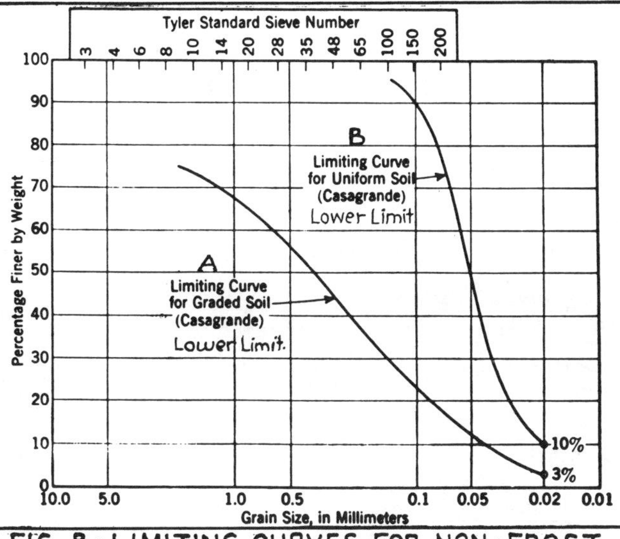
Tyler Standard Sieve Number
3 4 6 8 10 14 20 28 35 48 65 100 150 200
Percentage Finer by Weight
Grain Size, in Millimeters
10.0 5.0 1.0 0.5 0.1 0.05 0.02 0.01
B
Limiting Curve for Uniform Soil (Casagrande)
Lower Limit
A
Limiting Curve for Graded Soil (Casagrande)
Lower Limit
10%
3%

FIG. B - LIMITING CURVES FOR NON-FROST HEAVING SOILS (AFTER CASAGRANDE). †

See explanation for use of chart at right.

TABLE C - MAXIMUM REQUIRED THICKNESS IN INCHES OF FLEXIBLE PAVEMENT AND NON-FROST ACTION BASE[©©]

DESIGN WHEEL LOAD (POUNDS)	FINENESS OF SUBGRADE - % PASSING SIEVE NO. 200		
	OVER 25%	10 - 25%	UNDER 10%
15,000 and under	24 in.	20 in.	15 in.
40,000	38 "	26 "	26 "
60,000	46 "	38 "	32 "
150,000	72 "	60 "	48 "

USE OF FIG. B.

Plot curve for soil in question on Fig. B. If the curve is close to curve A, the 3% limit for dry weight of particles less than .02 mm. is to be used. If the curve is close to curve B, the 10% limit shall be used. The Highway Research Board recommends all soils containing more than 8% by weight of particles, finer than 200 mesh be considered as soils subject to frost action.

*Adapted from: Aviation Engineers Manual, 1944. ** C.A.A. Design Manual *** Highway Research Board 1943 Proceedings, D.J. Belcher. † Civil Engineering, Vol. 12. © U.S. Weather Bureau, 1951, ©© War Deptmt. C.E.

SOILS—SUBDRAIN FILTER

SEE SECTION ON DAMS FOR DESIGN OF DAM FILTERS.

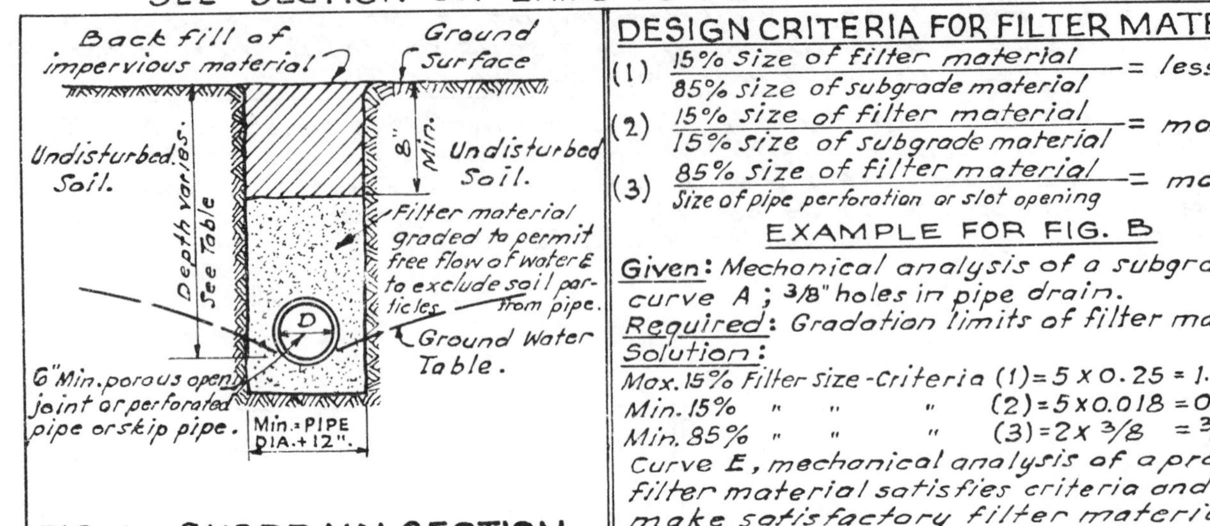
Back fill of impervious material
Ground Surface
8" Min.
Undisturbed Soil.
Depth varies. See Table
Undisturbed Soil.
Filter material graded to permit free flow of water & to exclude soil particles from pipe.
D
Ground Water Table.
6" Min. porous open joint or perforated pipe or skip pipe.
Min. = PIPE DIA. + 12".

FIG. A - SUBDRAIN SECTION.

DESIGN CRITERIA FOR FILTER MATERIAL.

(1) $\frac{\text{15\% size of filter material}}{\text{85\% size of subgrade material}}$ = less than 5

(2) $\frac{\text{15\% size of filter material}}{\text{15\% size of subgrade material}}$ = more than 5

(3) $\frac{\text{85\% size of filter material}}{\text{Size of pipe perforation or slot opening}}$ = more than 2

EXAMPLE FOR FIG. B

Given: Mechanical analysis of a subgrade, curve A; 3/8" holes in pipe drain.

Required: Gradation limits of filter material.

Solution:

Max. 15% Filter size-Criteria (1) = 5 x 0.25 = 1.25 (Pt. B).

Min. 15% " " " (2) = 5 x 0.018 = 0.09 (Pt. C).

Min. 85% " " " (3) = 2 x 3/8 = 3/4 (Pt. D).

Curve E, mechanical analysis of a proposed filter material satisfies criteria and should make satisfactory filter material for subgrade of this example.

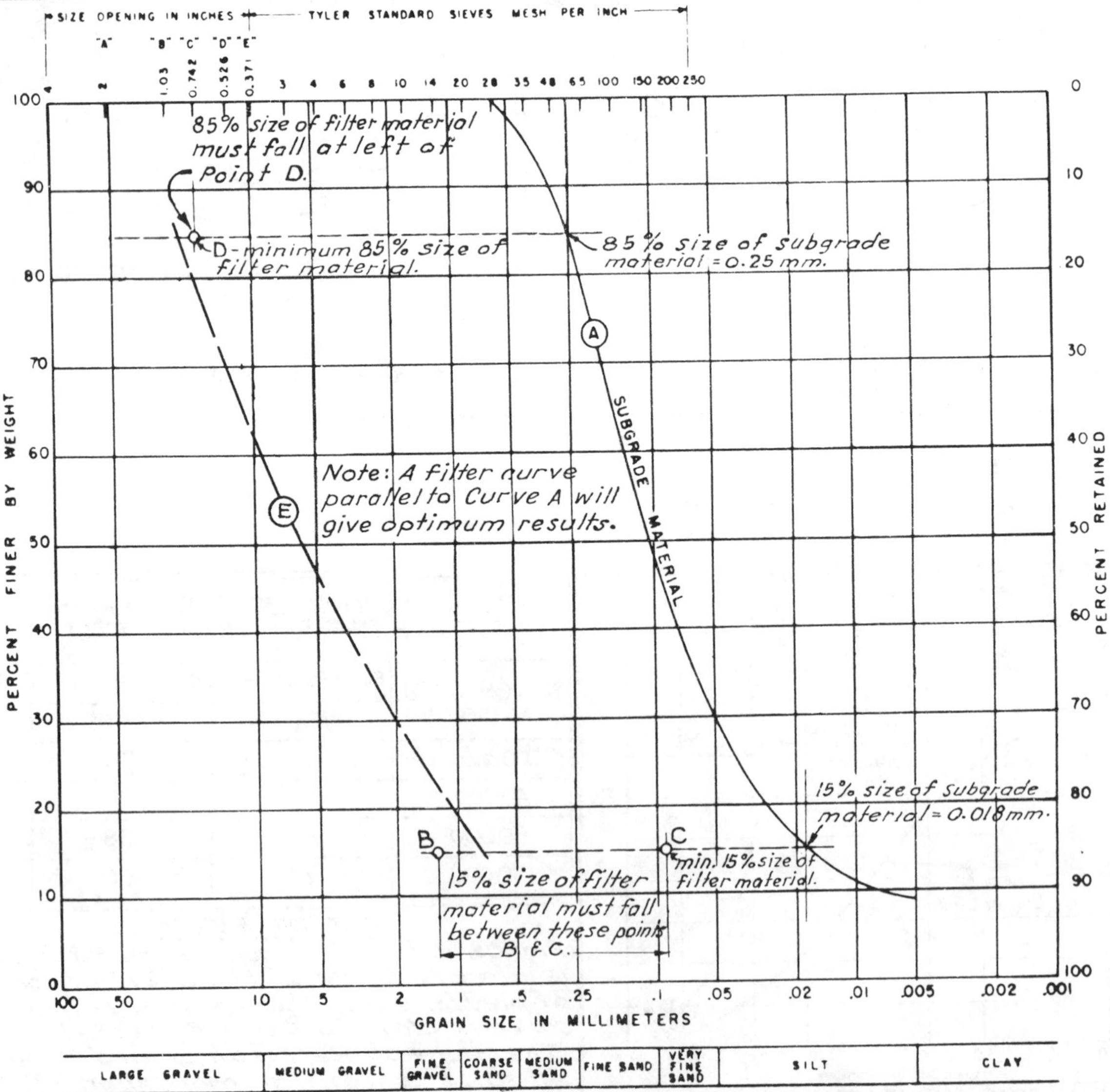

SIZE OPENING IN INCHES — TYLER STANDARD SIEVES MESH PER INCH

"A" "B" "C" "D" "E"

4, 2, 1.05, 0.742, 0.526, 0.371, 3, 4, 6, 8, 10, 14, 20, 28, 35, 48, 65, 100, 150, 200, 250

85% size of filter material must fall at left of Point D.

D - minimum 85% size of filter material.

85% size of subgrade material = 0.25 mm.

(A) SUBGRADE MATERIAL

Note: A filter curve parallel to Curve A will give optimum results.

(E)

15% size of subgrade material = 0.018 mm.

B, C

15% size of filter material must fall between these points B & C.

min. 15% size of filter material.

PERCENT FINER BY WEIGHT: 0, 10, 20, 30, 40, 50, 60, 70, 80, 90, 100

PERCENT RETAINED: 0, 10, 20, 30, 40, 50, 60, 70, 80, 90, 100

GRAIN SIZE IN MILLIMETERS: 100, 50, 10, 5, 2, 1, .5, .25, .1, .05, .02, .01, .005, .002, .001

LARGE GRAVEL	MEDIUM GRAVEL	FINE GRAVEL	COARSE SAND	MEDIUM SAND	FINE SAND	VERY FINE SAND	SILT	CLAY

FIG. B - CURVES OF SUBGRADE & FILTER MATERIAL.
See Design Criteria and Example above.

After Dr. K. Terzaghi (1932) subsequently modified by U.S. Army Engineers (Manual - Chapt. XX). Feb. 1943

SOILS–STABILIZATION

STABILIZED SOILS: *Should be sealed when used as a wearing course. Stabilization is mostly used to form base courses for thin bituminous or Portland cement concrete pavements.*

MECHANICAL STABILIZATION: *Plot curve of soil on chart in Fig. A. Curve should fall within shaded zone; otherwise coarse or fine material must be added to soil.*

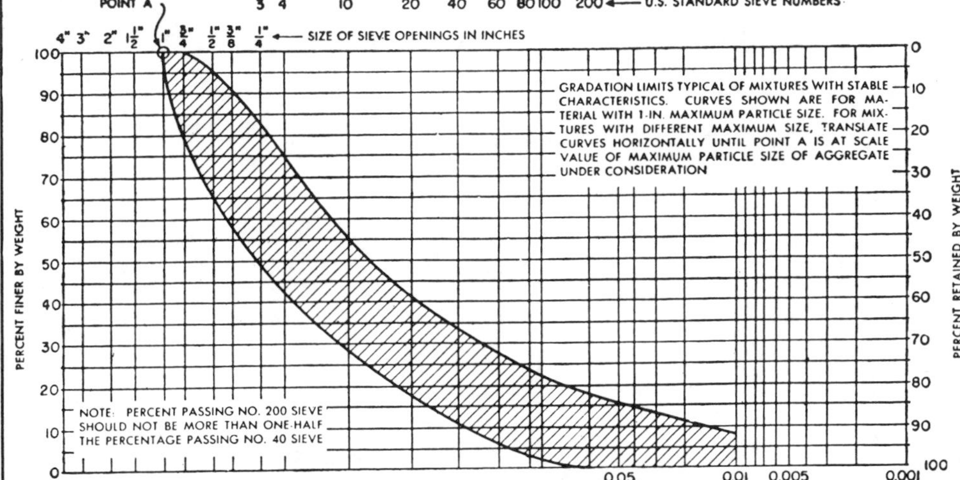

POINT A
3 4 10 20 40 60 80 100 200 ← U.S. STANDARD SIEVE NUMBERS
4" 3" 2" 1½" 1" ¾" ½" ⅜" ¼" ← SIZE OF SIEVE OPENINGS IN INCHES
GRADATION LIMITS TYPICAL OF MIXTURES WITH STABLE CHARACTERISTICS. CURVES SHOWN ARE FOR MATERIAL WITH 1-IN. MAXIMUM PARTICLE SIZE. FOR MIXTURES WITH DIFFERENT MAXIMUM SIZE, TRANSLATE CURVES HORIZONTALLY UNTIL POINT A IS AT SCALE VALUE OF MAXIMUM PARTICLE SIZE OF AGGREGATE UNDER CONSIDERATION
NOTE: PERCENT PASSING NO. 200 SIEVE SHOULD NOT BE MORE THAN ONE-HALF THE PERCENTAGE PASSING NO. 40 SIEVE
PERCENT FINER BY WEIGHT
PERCENT RETAINED BY WEIGHT
0.05 0.01 0.005 0.001

FIG. A - GRADATION LIMITS FOR STABLE MIXTURES.*

CRITERIA FOR MECH. STABILIZATION:

Plasticity Index (P.I.) of fraction of soil passing № 40 sieve should be from 3 to 9 but not over 6 if bituminous surface treatment is to be added.

The fraction passing № 10 sieve should show no appreciable shrinkage.

PORTLAND CEMENT STABILIZATION: *The amount of cement to be added should be determined by laboratory analysis, Fig. B to be used as a guide only.*

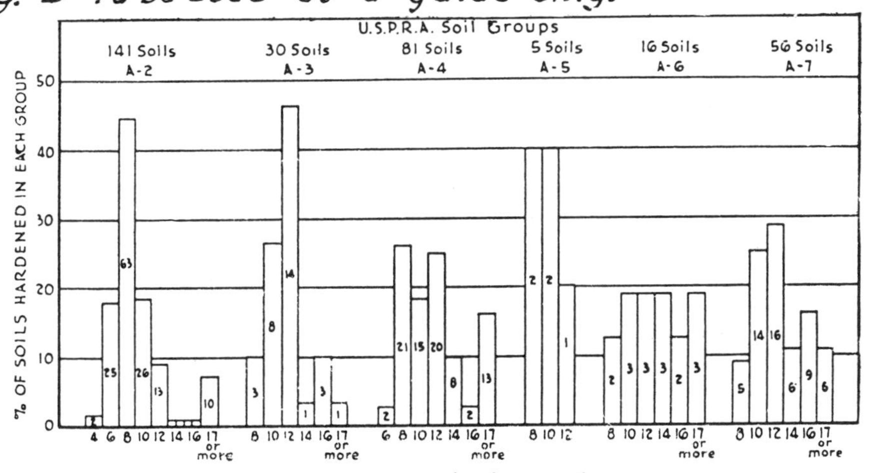

U.S.P.R.A. Soil Groups
141 Soils A-2
30 Soils A-3
81 Soils A-4
5 Soils A-5
16 Soils A-6
56 Soils A-7
% OF SOILS HARDENED IN EACH GROUP
PER CENTS BY VOLUME

FIG. B - RESULTS OF TESTS OF SOILS TO DETERMINE CEMENT REQUIRED FOR ADDITIONAL BINDING.**

CRITERIA FOR CEMENT STABILIZATION.

Liquid Limit (L.L.) should not be more than 40.

Plasticity Index (P.I.) not more than 18.

Maximum depth of construction = 6".

Note: L.L. and P.I. values as recommended by the Portland Cement Assoc.

For cement treated base suggest A-3 type soil. *See also Table A, Pg. 9-21*

BITUMINOUS STABILIZATION: *Soil should be selected or native granular materials A1 to A3 types. The usual depth of construction is 2½" to 4". Amount of bituminous material would vary according to size with maximum requirement for fine soils. An average would be ½ gal. per sq. yd. for each inch of stabilized depth.*
Bituminous materials usually used are cut-back asphalt MC2 or MC3, Road tars RT2 to RT5, Asphalt emulsion & slow curing liquid road oils. *See Tables Pg. 10-08 & 10-09.*

RESIN STABILIZATION: *About ¼ of 1% by weight is required for normal application. The powdered resins (Vinsol, Stabinol, Pextite) are added to the soil in the same manner as cement. Up to the present, 1949, resins have been used successfully only with acid soils.*

CALCIUM CHLORIDE STABILIZATION: *Current practice is to use ½ lb. per sq. yd. per inch of compacted depth with a maximum of 3 lb. per sq. yd. Valuable in reaching desired densities and in preventing frost action.*

LIST OF EQUIPMENT USED IN STABILIZATION: *Chisel tooth, spike tooth and offset disc harrows; disc plows; motor patrol graders, dozers & other spreading equipment; rubber tired tractors; wobbly wheel, rubber tired, sheepsfoot, tandem & 3 wheel rollers, special mixing equipment as premixing plants (Barber Greene type) & speed rotary tillers (Pulvi-mixers, Roto-Tillers).*

*Adapted from Aviation Eng. Manual, Apr. 1944 **Highway Research Bd. Wartime Road Problems № 7.

ROADS – WHEEL LOADS FOR DESIGN

TABLE A TYPE OF HIGHWAY – STREET OR ROAD

DESCRIPTION	P.C.A.				C. OF E.	
	HIGHWAY		STREET & ROAD		HWY., ST. & RD.	
	CLASS	WHEEL LOAD*	CLASS	WHEEL LOAD*	CLASS	WHEEL LOAD*
TRAFFIC ROUTES HAVING A LARGE VOLUME OF BOTH WHEELED AND TRACK-LAYING VEHICLES, SUCH AS HEAVY-DUTY TRUCK, TRACTORS, TANKS, SELF PROPELLED ARTILLERY VEHICLES, ETC., OR MATERIAL HANDLING EQUIPMENT EQUIPPED WITH EITHER HARD RUBBER OR STEEL TREADS.	—	—	—	—	AA	16,000 LBS. PLUS
HEAVY DUTY THOROUGHFARES. PRIMARY ROUTES IN OR NEAR A METROPOLITAN AREA. INCLUDES DOWNTOWN BUSINESS STREETS, STATE AND COUNTY TRUCKLINE ROUTES THROUGH CITIES, PRINCIPAL TRAFFIC ARTERIES SERVING LARGE MANUFACTURING DISTRICTS. FREIGHT TERMINALS, DOCKS AND WAREHOUSES.	I	13,000 LBS. TO 14,000 LBS.	I	10,000 LBS. TO 11,000 LBS.	A	12,000 LBS.
MEDIUM TRAFFIC ROUTES CONTAINING UNIFORM PASSENGER AND COMMERCIAL VEHICLES. INCLUDES ARTERIAL ROUTES NOT INCLUDED IN CLASS ABOVE. MAIN TRAFFIC ROUTES CARRYING INTRACITY TRAFFIC AND PRIMARY ROUTE IN RURAL AREA.	II	11,000 LBS. TO 12,000 LBS.	II	8,000 LBS. TO 9,000 LBS.	B	9,000 LBS.
LIGHTLY TRAVELLED PRIMARY ROUTE.	III	10,000 LBS. TO 11,000 LBS.	—	—	—	—
INCLUDES SECONDARY ARTERIAL STREETS OR FEEDER ROUTES USED FOR COLLECTING AND DISPERSING TRAFFIC TO AND FROM THE ARTERIAL STREET SYSTEM.	IV	9,000 LBS. TO 10,000 LBS.	III	6,000 LBS. TO 7,000 LBS.	—	—
STREETS OR ROADS CARRYING LITTLE TRAFFIC. LIMITED TO RESIDENTIAL AREAS WHERE HEAVIEST WHEEL LOADS EXPECTED ARE PASSENGER VEHICLES AND LIGHT TRUCKS.	—	—	IV	5,000 LBS. TO 6,000 LBS	C	6,000 LBS.

* - STATIC WHEEL LOADS

NOTES:

1. ABOVE WHEEL LOADS ARE PERMITTED WHERE INFREQUENT TRAFFIC USE BY VEHICLES WITH WHEEL LOADS ONE CLASS HIGHER THAN NOTED ARE ANTICIPATED.
2. WHEEL LOAD SELECTED SHOULD BE BASED ON LOCAL LAWS AND REGULATIONS AND, IF POSSIBLE, ON TRAFFIC STUDIES OF LOAD INTENSITY AND FREQUENCY. PAVEMENT DESIGN IS BASED ON THE WHEEL LOAD AND NOT ON THE GROSS WEIGHT OF VEHICLE. RIGID PAVEMENTS ARE DESIGNED WITH A SAFETY FACTOR (USUALLY 2) ALLOWING PRACTICALLY UNLIMITED STRESS REPETITIONS CAUSED BY THE DESIGN LOAD, THUS AN OCCASIONAL OVERLOAD UP TO AS HIGH AS TWICE THE DESIGN LOAD WILL NOT BE DESTRUCTIVE. FLEXIBLE PAVEMENTS, WHILE NOT ADAPTED TO SUCH EXACT ANALYSIS, WILL ALSO CARRY OCCASIONAL OVERLOADS IF CONSERVATIVELY DESIGNED. DUAL WHEELS ARE CONSIDERED AS ONE WHEEL LOAD AND ONE CONTACT AREA IF TIRES ARE WITHIN 3' CENTERS.

TABLE B IMPACT FACTORS

STATIC WHEEL LOAD	P.C.A.	C. OF E.
16,000 LBS.	1.20	1.24
12,000 LBS.	1.20	1.24
9,000 LBS.	1.20	1.33
6,000 LBS.	1.20	1.50

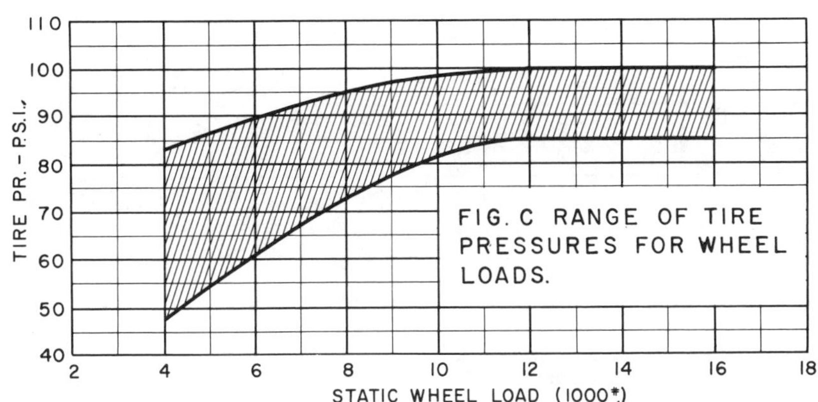
110
100
90
80
70
60
50
40
TIRE PR. – P.S.I.
2
4
6
8
10
12
14
16
18
STATIC WHEEL LOAD (1000*)

FIG. C RANGE OF TIRE PRESSURES FOR WHEEL LOADS.

References: Corps of Engineers Manual Part X Chapter 1; P.C.A. Concrete Pavement Design; Design of Concrete Pavements for City Streets.

ROADS & AIRFIELDS-GENERAL SOIL RATINGS

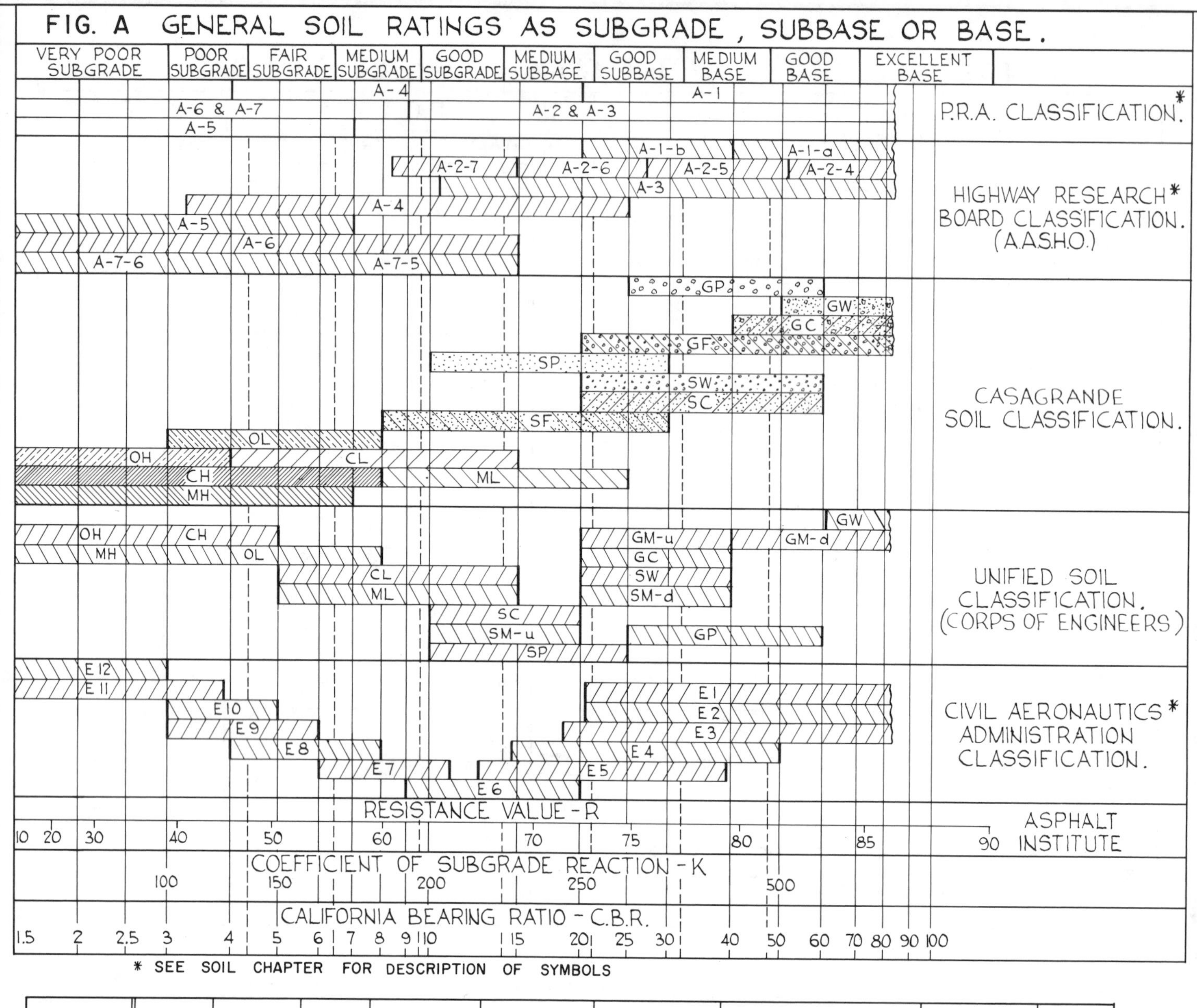
FIG. A GENERAL SOIL RATINGS AS SUBGRADE, SUBBASE OR BASE.
VERY POOR SUBGRADE
POOR SUBGRADE
FAIR SUBGRADE
MEDIUM SUBGRADE
GOOD SUBGRADE
MEDIUM SUBBASE
GOOD SUBBASE
MEDIUM BASE
GOOD BASE
EXCELLENT BASE
A-4
A-1
A-6 & A-7
A-2 & A-3
A-5
P.R.A. CLASSIFICATION.*
A-1-b
A-1-a
A-2-7
A-2-6
A-2-5
A-2-4
A-3
A-4
A-5
A-6
A-7-6
A-7-5
HIGHWAY RESEARCH* BOARD CLASSIFICATION. (A.A.S.H.O.)
GP
GW
GC
GF
SP
SW
SC
SF
OL
OH
CL
CH
ML
MH
CASAGRANDE SOIL CLASSIFICATION.
GW
OH
CH
GM-u
GM-d
MH
OL
GC
CL
SW
ML
SM-d
SC
SM-u
GP
SP
UNIFIED SOIL CLASSIFICATION. (CORPS OF ENGINEERS)
E 12
E 11
E1
E10
E2
E9
E3
E8
E4
E7
E5
E6
CIVIL AERONAUTICS* ADMINISTRATION CLASSIFICATION.
RESISTANCE VALUE - R
10 20 30 40 50 60 70 75 80 85 90
ASPHALT INSTITUTE
COEFFICIENT OF SUBGRADE REACTION - K
100 150 200 250 500
CALIFORNIA BEARING RATIO - C.B.R.
1.5 2 2.5 3 4 5 6 7 8 9 10 15 20 25 30 40 50 60 70 80 90 100
* SEE SOIL CHAPTER FOR DESCRIPTION OF SYMBOLS

AASHO	BOULDERS	COARSE GRAVEL	MEDIUM GRAVEL	FINE GRAVEL	COARSE SAND	FINE SAND	SILT	CLAY	COLLOIDS
CAA	GRAVEL				COARSE SAND	FINE SAND	SILT	CLAY	
UNIFIED (C of E)	COBBLES	COARSE GRAVEL	FINE GRAVEL	COARSE SAND	MEDIUM SAND	FINE SAND	FINES (SILT OR CLAY)		
SIEVE SIZES	3"	1-1/2"	3/4"	1/2"	#4	#10	#20	#40	#60
	#100	#200	#270						
PARTICLE SIZE mm	100.0 80.0 60.0 40.0 30.0 20.0 10.0 8.0 6.0 4.0 3.0 2.0 1.0 0.8 0.6 0.4 0.3 0.2 0.1 0.08 0.06 0.04 0.03 0.02 0.01 0.008 0.006 0.004 0.003 0.002 0.001								

FIG. B TEXTURAL CLASSIFICATION FOR SOILS

USE: IDENTIFY SOIL AND SELECT BEARING VALUE FOR PAVEMENT. DESIGN FROM DIAGRAM.

EXAMPLE GIVEN: HIGHLY COMPRESSIBLE CLAY SUBGRADE.
CLASSIFICATION IS CH, A-7 OR A-8
CBR = 4 TO 8
K = 125 TO 180
R = 45 TO 60

NOTE: BEARING VALUES ARE APPROXIMATE AND SHOULD NOT BE USED AS A SUBSTITUTE FOR ACTUAL SOIL TESTS.

LEGEND OF SYMBOLS

G—gravel
S—sand
M—mo, very fine sand, silt, rock flour
C—clay
F—fines (- 0.1mm)
O—organic
W—well graded
P—poorly graded
L—low to medium compressibility
H—high compressibility
u—undesirable
d—desirable

ROADS - SUBGRADE

DRAINAGE

Adequate subdrainage should be provided for roads to:
1. Stabilize the subgrade.
2. Reduce frost damage by removing source of water.
3. Prevent formation of slippage planes on slopes due to reduced cohesion caused by saturation.
4. Prevent accumulations of water and ice on pavement from side-hill seepage.
5. Prevent "pumping" of the subgrade (buildup of pore-water pressure).

Subdrains may be used to draw down the water level under the road. Intercepting, or slope drains may be used to prevent slides caused by excess ground water.

A base filter is used under pavements to drain the subgrade. It stabilizes the subgrade and removes the water to prevent frost action or pumping. See drainage section.

PREVENTION OF FROST DAMAGE

Soils subject to frost action are well-graded soils containing more than 3% by dry weight of particles less than 0.02 mm. (0.0008 inch) in size, and uniformly graded soils containing more than 10% of particles less than 0.02 mm in size. For preventive steps see Soils section and page 10-13.

SUBGRADE PUMPING UNDER PAVEMENT

Subgrade pumping of pavement under a moving load occurs in fine-grained soils which hold water and do not drain easily. Water enters through joints or cracks producing an emulsion with the soil below as it does so. The application of loads on pavement causes a buildup of pore-water pressure below. The passage of traffic forces this emulsion out under pressure, through the joints, leaving voids in the subgrade. The entire process is progressive, leading to destruction of the pavement. See Table A, page 9-17, and page 10-13.

WARPING OF PAVEMENT

Warping of concrete pavement is caused by differential swelling or shrinking of the subgrade soil due to excessive variations in moisture content. This condition is most likely to occur in clay belts of semi-arid regions. See Table A, page 9-13, and page 10-13.

RED LIGHTS

Do not figure the safe slope of high embankments of cohesive soils on the basis of observed slopes of low embankments.

The presence of ground water may affect the cohesion of soils. Heavy rains may reduce the cohesion.

In artesian condition, water trapped under a hydrostatic head by an overlying impervious layer will change the results of time consolidation calculations.

In consolidation by the sand-drain method, fill should not be placed so fast that the shearing strength of the soil is exceeded. This (the shearing strength) may be correlated with the pore pressure and measured in the field with cells or piezometer.

In cuts, water from the interception of sloping or water-bearing strata must be caught by intercepting drains and taken into account in design so as to prevent slides.

Avoid settlement due to squeezing out of water in deep layers of clay or plastic material caused by the excess weight of the structure imposed.

This is more likely to occur under large fills of considerable area than it is under buildings where the added weights at lower depth may be inconsiderable because of distribution.

The overloading of foundation by an embankment on soft material or peat or dense silt matter will cause a flow out and rise, popularly known as a mud wave.

RESIDUARY HYDRAULIC PRESSURE

In case of sudden low water such as tides or drawdowns in a reservoir, the exposed surface embankments will be loaded with unbalanced pore water, and the stability will be reduced by seepage pressure so that a slide may occur.

EVALUATING THE SUBGRADE

In designing the pavement the designer assumes a value for the subgrade and checks this value in the field. The soil is generally classified by one of the recognized systems (see page 10-02 for classification) and the C.B.R. or *K* value is read off the chart. This value is used in the pavement design which follows. The true *K* or C.B.R. value should be determined by field tests; see chapter on Soils.

ROADS - DESIGN OF PAVEMENTS

AS THE PAVEMENT SECTION IS PART OF THE FOUNDATION, THE DESIGN OF THE COMPLETE SECTION IS GIVEN BELOW. WEAK SUBGRADES REQUIRE THICKER PAVEMENTS THAN STRONG ONES.

THE THEORY OF LOAD DISTRIBUTION IS SHOWN BELOW:

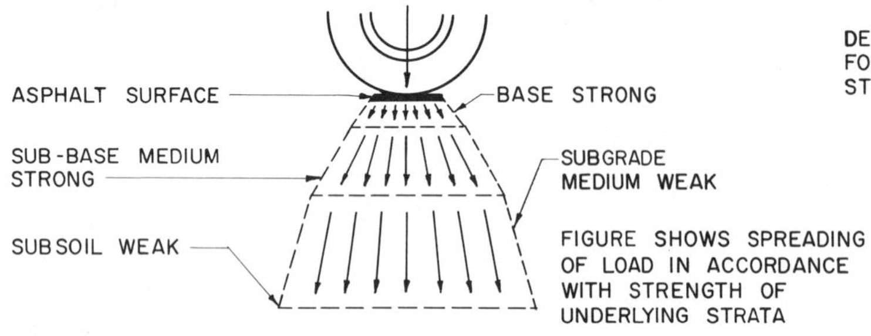
ASPHALT SURFACE
BASE STRONG
SUB-BASE MEDIUM STRONG
SUBGRADE MEDIUM WEAK
SUBSOIL WEAK
FIGURE SHOWS SPREADING OF LOAD IN ACCORDANCE WITH STRENGTH OF UNDERLYING STRATA

BASIS OF FLEXIBLE PAVEMENT DESIGN

DESIGN PAVEMENT FOR FLEXURAL STRENGTH

THE SOFTER THE SUBGRADE THE BIGGER THE DIMPLE AND THE GREATER THE FLEXURAL STRESS

INTENSITY OF DIMPLE ESTIMATED FROM PLATE LOAD TEST ON SOIL CALLED SUBGRADE REACTION.

BASIS OF RIGID PAVEMENT DESIGN

FIG. A THEORY OF LOAD DISTRIBUTION THROUGH PAVEMENTS

TABLE B FACTORS – SELECTION OF RIGID OR FLEXIBLE TYPE.	
RIGID (CONCRETE)	FLEXIBLE (BITUMINOUS)
LOW MAINTENANCE COST - LONG LIFE - HIGH SALVAGE VALUE AS BASE FOR FUTURE RESURFACING - HIGH VISIBILITY AND REFLECTION AT NIGHT - SPREADS LOAD OVER WIDE AREA - MAY BE PLACED DIRECTLY ON WEAK OR SANDY SUBGRADES- RESISTS TURNING STRESSES - UNDAMAGED BY OIL OR GAS DRIP - LOW TRACTIVE RESISTANCE - CAN BE DESIGNED AND BUILT CLOSE TO TOLERANCES.	ADAPTABLE TO STAGE CONSTRUCTION - LOW COST TYPES MAY BE BUILT WITH LOCAL LABOR, MATERIALS AND EQUIPMENT - EASILY OPENED AND PATCHED - LOW INITIAL COST (EXCEPT HIGH TYPES) - FROST HEAVE AND SETTLEMENT EASILY REPAIRED - RESISTS FORMATION OF ICE GLAZE - GREAT VARIETY OF TYPES TO FIT A WIDE RANGE OF CONDITIONS - HAS NO JOINTS, HAS RESILIENT RIDING QUALITIES.
NOTE: BITUM. SURFACE MAY BE PLACED ON RIGID BASE. CONCRETE MAY BE DARKENED & BITUM. MAY BE LIGHTENED IN COLOR.	

FIG. C YARDSTICK FOR PAVEMENT THICKNESS

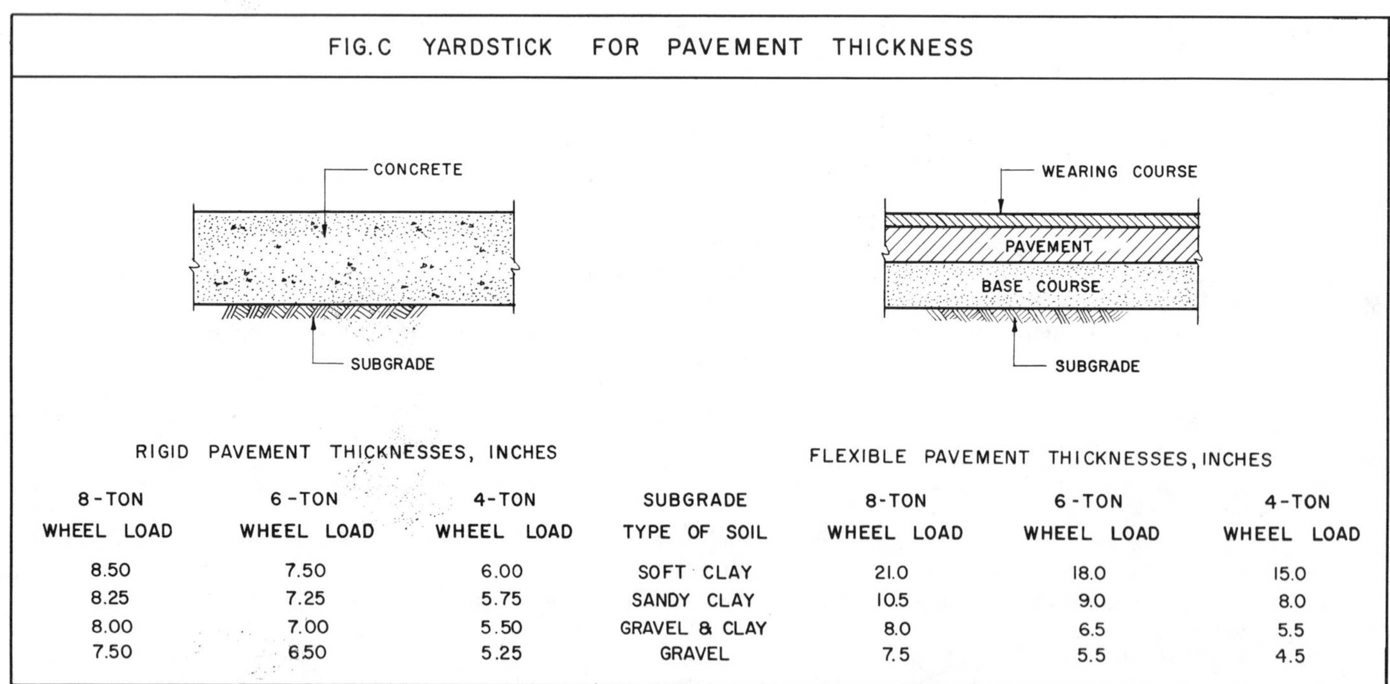
CONCRETE
SUBGRADE
WEARING COURSE
PAVEMENT
BASE COURSE
SUBGRADE

RIGID PAVEMENT THICKNESSES, INCHES				FLEXIBLE PAVEMENT THICKNESSES, INCHES		
8-TON WHEEL LOAD	6-TON WHEEL LOAD	4-TON WHEEL LOAD	SUBGRADE TYPE OF SOIL	8-TON WHEEL LOAD	6-TON WHEEL LOAD	4-TON WHEEL LOAD
8.50	7.50	6.00	SOFT CLAY	21.0	18.0	15.0
8.25	7.25	5.75	SANDY CLAY	10.5	9.0	8.0
8.00	7.00	5.50	GRAVEL & CLAY	8.0	6.5	5.5
7.50	6.50	5.25	GRAVEL	7.5	5.5	4.5

NOTE:
THE ABOVE TABLE FOR PAVEMENT THICKNESS AND ACCOMPANYING COST TABLE ARE NOT INTENDED TO SUPPLEMENT DESIGN, THEY ARE PRESENTED TO GIVE AN ENGINEER A PERSPECTIVE FOR DESIGN.

WHEEL LOADING	PAVEMENT COST PER YARD *	
TONS	FLEXIBLE	RIGID
8	$3.30—$2.25	$4.00—$3.50
6	2.90— 2.00	3.50— 3.10
4	2.25— 1.50	2.90— 2.50

* *From Perspective Data on Paving Design by E. E. Seelye, March 1958 "Consulting Engineer" ENR Construction Index 744*

ROADS - DESIGN OF FLEXIBLE PAVEMENT

PROCEDURE:

1. ON THE BASIS OF DATA FROM TRAFFIC STUDY, DETERMINE TRAFFIC CLASSIFICATION FOR DESIGN FROM TABLE A.
2. DETERMINE MAXIMUM SINGLE AXLE LOAD ANTICIPATED DURING LIFE OF PAVEMENT.
   a. FOR BUS STOP AREAS, THE PAVEMENT SHOULD BE DESIGNED FOR THE "HEAVY" TRAFFIC CLASSIFICATION WHERE THE NORMAL PAVEMENT DESIGN IS FOR "LIGHT" OR "MEDIUM" TRAFFIC AND FOR "VERY HEAVY" TRAFFIC WHERE THE NORMAL PAVEMENT DESIGN IS FOR "HEAVY" OR "VERY HEAVY" TRAFFIC.
   b. FOR PARKING LOTS, THE FLEXIBLE PAVEMENT SHOULD BE DESIGNED FOR "HEAVY" TRAFFIC EXCEPT THAT WHEN ONLY PASSENGER CARS AND LIGHT TRUCKS OF 6,000 LB. AXLE LOAD OR LESS ARE EXPECTED TO USE THE PARKING FACILITY, THE MINIMUM REQUIRED THICKNESS OF BASE AND PAVEMENT MAY BE REDUCED TO 6".
3. MINIMUM BASE COURSE IS EQUAL TO 4" WITH A CBR OF 80 PLUS.

TABLE A

CLASSIFICATION OF TRAFFIC	TRAFFIC DENSITY MAXIMUM, PER LANE, PER DAY	
TRAFFIC CLASSIFICATION	DAILY VOLUME PASSENGER CARS & LIGHT TRUCKS UNDER 6000 #	DAILY VOLUME COMMERCIAL TRUCKS AND BUSES OVER 6000 #
LIGHT	25	5
MEDIUM	500	25
HEAVY	UNLIMITED	250
VERY HEAVY	UNLIMITED	UNLIMITED

RESISTANCE VALUE - R

10 20 30 40 50 60 70 75 80 85

CALIFORNIA BEARING RATIO - C.B.R.

1·5 2 3 4 5 6 7 8 9 10 15 20 30 40 50 60 70 80 90

STEP 1

3,000 LB. SINGLE AXLE LOAD

6,000

9,000

12,000

15,000

18,000

24,000

30,000

36,000

42,000

TRAFFIC CLASSIFICATION

VERY HEAVY

HEAVY

MEDIUM

LIGHT

STEP 2

TOTAL THICKNESS OF FLEXIBLE PAVEMENT IN INCHES

0 5 10 15 20 25 30 35 40

THE MINIMUM TOTAL THICKNESS OF BASE AND PAVEMENT SHALL BE AS FOLLOWS:

LIGHT TRAFFIC ········ 5 IN.

MEDIUM TRAFFIC ······ 6 IN.

HEAVY TRAFFIC ······· 8 IN.

VERY HEAVY TRAFFIC··· 10 IN.

FIG.B DESIGN OF TOTAL PAVEMENT AND BASE THICKNESS

DESIGN OF PAVEMENT

PROBLEM:

GIVEN:

PLANT MIX
TRAFFIC CLASSIFICATION = MEDIUM
MAXIMUM SINGLE AXLE LOAD = 18,000 LBS
SUBGRADE CBR = 4

FIND:

REQUIRED PAVEMENT THICKNESS
SUBBASE COURSE CBR

SOLUTION:

STEP 1 — ENTER CHART (FIG.B) AT CBR = 4; READ DOWN TO INTERSECT SINGLE AXLE LOAD CURVE = 18,000 LBS; PROJECT HORIZONTLY TO LEFT EDGE OF CHART; THEN FROM INTERSECTION OF EDGE OF CHART THRU PIVOT POINT FOR MEDIUM TRAFFIC DRAW LINE TO GET TOTAL PAVEMENT THICKNESS OF 12 INCHES.

STEP 2 — FROM TABLE C BELOW FOR PLANT MIX
BITUMINOUS SURFACE = 3 IN.
MINIMUM BASE COURSE = 4 IN. (See note 3 above)
TOTAL = 7 IN.

ENTER CHART (FIG. B) WITH 7 IN. THICKNESS (AT RIGHT) AND DRAW LINE THRU PIVOT POINT FOR MEDIUM TRAFFIC TO LEFT EDGE; PROJECT TO THE RIGHT TO SINGLE AXLE LOAD CURVE = 18000 LBS, READ UP TO CBR = 10. THIS IS THE MIN. CBR FOR THE REMAINING SUBBASE OF 5 IN.

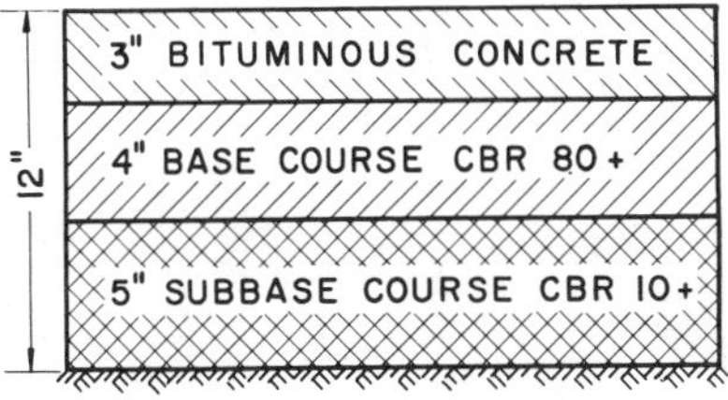

12"
3" BITUMINOUS CONCRETE
4" BASE COURSE CBR 80+
5" SUBBASE COURSE CBR 10+

SUBGRADE CBR 4

TABLE—C

SELECTION OF BITUMINOUS PAVEMENT TYPES FOR DIFFERENT DENSITIES OF TRAFFIC AS INFLUENCED BY SERVICE LIFE, QUALITY, AND COST FACTORS — BITUMINOUS PAVEMENT SURFACES SUGGESTED THICKNESSES

EXPECTED SERVICE	BITUMINOUS PAVEMENT TYPE	LIGHT TRAFFIC QUALITY CHOICE	LIGHT TRAFFIC COST CHOICE	MEDIUM TRAFFIC QUALITY CHOICE	MEDIUM TRAFFIC COST CHOICE	HEAVY TRAFFIC QUALITY CHOICE	HEAVY TRAFFIC COST CHOICE	VERY HEAVY TRAFFIC QUALITY CHOICE	VERY HEAVY TRAFFIC COST CHOICE	PAVEMENT SURFACE THICKNESS IN INCHES: LIGHT TRAFFIC	MEDIUM TRAFFIC	HEAVY TRAFFIC	VERY HEAVY TRAFFIC
SHORT LIFE	SINGLE SURFACE TREATMENT	6 th	1 st	6 th	1 st	—	—	—	—	1−	1−	—	—
MEDIUM LIFE	DOUBLE SURFACE TREATMENT	5 th	2 nd	5 th	2 nd	—	—	—	—	1	1	—	—
MEDIUM LIFE	ROAD MIX	4 th	3 rd	4 th	3 rd	4th	1 st	—	—	2	3	3	—
LONG LIFE	PLANT MIX	3 rd	4 th	3 rd	4 th	3 rd	2 nd	—	—	2	3	3	—
LONG LIFE	MACADAM *	2 nd	5 th	2 nd	5 th	2nd	3 rd	—	—	2	2-½	2-½	—
LONG LIFE	ASPHALTIC CONCRETE	1 st	6th	1 st	6 th	1 st	4 th	1 st	1 st	2	3	3	4

* COST CHOICE FOR THIS TYPE OF PAVEMENT IS VARIABLE, DEPENDING UPON LOCAL CONDITIONS AND THE AVAILABILITY OF CRUSHED AGGREGATES.

Adapted from Asphalt Institute - Handbook, and Thickness Design Flexible Pavements for Streets and Highways.

ROADS-TYPICAL SECTIONS FLEXIBLE PAVEMENT

BITUMINOUS SURFACE TREATMENT, 1-2 COATS, 3/8" TO 3/4"
6"-8" LIMEROCK, GRAVEL, CRUSHED STONE OR SOIL CEMENT STABILIZATION. (BASE PRIMED)

LOW TYPE - SURFACE TREATMENT

2"-3" AGGREGATE MIXED IN PLACE SURFACE COURSE
6" OR MORE GRAVEL BASE

INTERMEDIATE TYPE - ROAD MIX

2½" TO 3" STONE SURFACE COURSE PENETRATED WITH BITUMEN
4" TO 6" CRUSHED STONE BASE. (VOIDS FILLED WITH SAND OR SCREENINGS.)

INTERMEDIATE TYPE - PENETRATION

2" BITUMINOUS CONCRETE SURFACE
3" BITUMINOUS CONCRETE BASE OR PENETRATION MACADAM BASE
SUB-BASE AS REQUIRED

HIGH TYPE - BITUMINOUS CONCRETE

1½" BITUMINOUS CONCRETE SURFACE
1½" BITUMINOUS CONCRETE BINDER
6" TO 9" CRUSHED STONE BASE (VOIDS FILLED WITH SAND OR SCREENINGS)

HIGH TYPE - BITUMINOUS CONCRETE

1½" SHEET ASPHALT SURFACE
1½" BITUMINOUS BINDER COURSE
6" TO 9" PORTLAND CEMENT CONCRETE BASE

HIGH TYPE - SHEET ASPHALT

1½" SHEET ASPHALT SURFACE
3½" BITUMINOUS CONCRETE BINDER
0" TO 8" SUB-BASE AS REQUIRED

HIGH TYPE - SHEET ASPHALT

1" SHEET ASPHALT OR BITUMINOUS CONCRETE SURFACE
1½" BITUMINOUS CONCRETE BINDER
SAND OR BITUMINOUS LEVELING COURSE
EXISTING PAVEMENT FOR BASE

HIGH TYPE - RESURFACE

FIG. A TYPICAL SECTIONS – BITUMINOUS ROADS

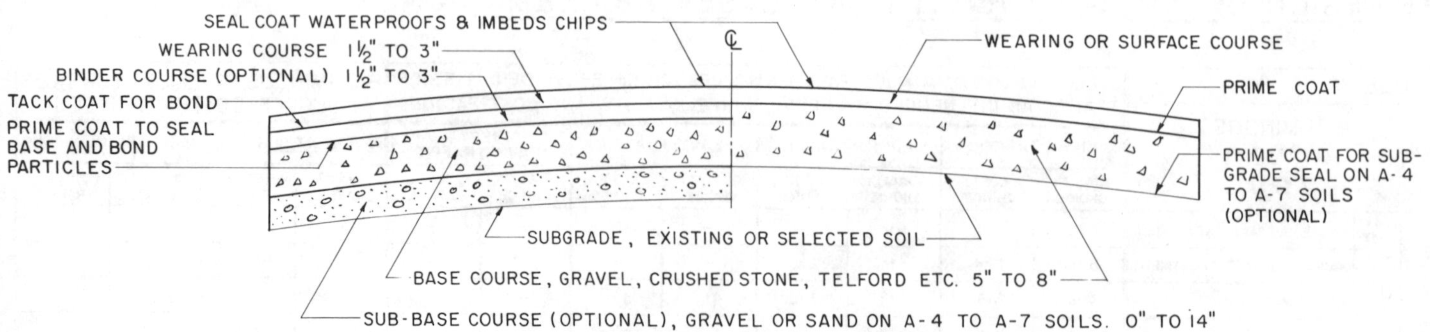
SEAL COAT WATERPROOFS & IMBEDS CHIPS
WEARING COURSE 1½" TO 3"
BINDER COURSE (OPTIONAL) 1½" TO 3"
TACK COAT FOR BOND
PRIME COAT TO SEAL BASE AND BOND PARTICLES
℄
WEARING OR SURFACE COURSE
PRIME COAT
PRIME COAT FOR SUB-GRADE SEAL ON A-4 TO A-7 SOILS (OPTIONAL)
SUBGRADE, EXISTING OR SELECTED SOIL
BASE COURSE, GRAVEL, CRUSHED STONE, TELFORD ETC. 5" TO 8"
SUB-BASE COURSE (OPTIONAL), GRAVEL OR SAND ON A-4 TO A-7 SOILS. 0" TO 14"

FIG. B ILLUSTRATION OF ELEMENTS OF BITUMINOUS ROADS

ROADS-FLEXIBLE PAVEMENTS RECOMMENDED THICKNESSES

TABLE A RECOMMENDED THICKNESSES FOR STREETS AND HIGHWAYS

Subgrade Classification		Fine-Grained Soils, Plastic						Fine-Grained Soils, Slight to Non-plastic						Gravelly or Sandy Soils, Well Drained					
Wheel Load, lb.		6000		9000		12,000		6000		9000		12,000		6000		9000		12,000	
Traffic Classification	Bituminous Surface, Inches	B	S	B	S	B	S	B	S	B	S	B	S	B	S	B	S	B	S
Light	2	4	–	5	–	4	4	4	–	4	–	4	–	*	–	*	–	*	–
Medium	3	4	4	4	4	4	7	4	–	4	–	5	–	4	–	4	–	4	–
Heavy	3	5	4	5	5	5	9	5	–	6	–	7	–	5	–	5	–	5	–
Very heavy	4	6	4	6	4	6	8	6	–	6	–	7	–	6	–	6	–	6	–

TABLE B RECOMMENDED THICKNESSES FOR PARKING LOTS

Subgrade Classification	Bituminous Pavement, Thickness, inches					
	Passenger Cars			Heavy Trucks		
	Bituminous Surface	Base Course†	Subbase Course	Bituminous Surface	Base Course†	Subbase Course
Gravelly or sandy soils, well drained	1 to 3	4	–	3	5	–
Fine-grained soils, slight to non-plastic	2 to 3	4	–	3	5	5
Fine-grained soils, plastic	2 to 3	4	4	3	5	11

B = base course (inches)

S = subbase course (inches)

Bituminous surface (inches) includes wearing and binder courses.

*Stabilization of subgrade recommended.

† Base Course shall consist of material classified as "Excellent Base", see page 10-02.

ROADS - USE OF BITUMINOUS MATERIALS -1

TABLE A PRINCIPAL USES OF BITUMINOUS MATERIALS FOR PAVING

MATERIAL	ROAD TAR (R.T.) CUTBACK ROAD TAR (R.T.C.B.)														RAPID CURING* (R.C)						MEDIUM CURING* (M.C.)						SLOW CURING (S.C.)						EMULSIFIED AS.						ASPHALT CEMENTS (A.C.)							
	R.T.												R.T.C.B.																				RS	MS	MS	MS	SS	SS								
COMMERCIAL GRADE & PENETRATION NO.	1	2	3	4	5	6	7	8	9	10	11	12	5	6	0	1	2	3	4	5	0	1	2	3	4	5	0	1	2	3	4	5	1	1	2	3	1	2	50-60	60-70	70-85	85-100	100-120	120-150	150-200	200-300
APPLICATION TEMPERATURE (F)	60-125	60-125	80-150	80-150	80-150	80-150	150-225	150-225	150-225	175-250	175-250	175-250	60-120	60-120	50-135	80-180	80-210	125-240	150-255	175-285	50-140	80-185	100-215	150-250	175-265	200-290	50-140	80-185	150-215	175-250	175-265	200-290	75-130	50-100	100-160	50-110	75-130	75-130	275-350	275-350	275-350	275-350	275-350	275-350	275-350	200-325
DUST PALLIATIVE	X																										X	X																		
STABILIZED BASE		X	X	X	X	X																																								
PRIMING																																														
TIGHTLY BONDED SURFACES		X																			X						X																			
LOOSELY BONDED FINE GRAINED SURFACES			X																			X						X																		
LOOSELY BONDED COARSE GRAINED SURFACES				X																			X						X																	
SURFACE TREATMENT & COLOR COATS																																														
WITH OR WITHOUT LIGHT SAND COVER															X																															
COARSE SAND COVER					X										X	X							X																							
CLEAN 1/4" AGGREGATE COVER					X	X										X							X																							
CLEAN 1/2" AGGREGATE COVER						X	X	X	X								X																X													
CLEAN 5/8" AGGREGATE COVER								X	X	X								X															X												X	X
CLEAN 3/4" AGGREGATE COVER									X	X	X								X	X						X																			X	X
GRADED GRAVEL AGGREGATE COVER									X	X													X	X						X																
GRAVEL MULCH					X	X	X	X															X						X																	
MIXED-IN-PLACE																																														
OPEN GRADED AGGREGATE																																														
SAND						X	X	X	X								X	X					X	X																						
MAX. DIA. 1", HIGH % PASS NO. 10 SIEVE						X	X	X	X															X	X									X												
MACADAM AGGREGATE					X	X	X	X	X								X	X							X	X																				
DENSE GRADED AGGREGATE																																														
HIGH % PASS NO. 200 SIEVE							X	X	X														X						X	X																
MAX. DIA. 1", MED. PASS NO. 200 SIEVE					X	X	X	X															X	X						X																
FINE SOIL			X	X	X	X																	X															X								
MODIFIED PENETRATION										X	X	X								X													X													
COLD PATCH																																														
OPEN GRADED AGGREGATE																X	X							X													X									
DENSE GRADED AGGREGATE																							X						X																	
TACK COAT (FOG SPRAY)				X	X	X	X									X	X	X															X													
PLANT MIX																																														
COLD LAID																																														
SAND																	X	X																												
OPEN GRADED, HIGH % PASS NO. 10 SIEVE								X	X	X								X					X														X									
MACADAM AGGREGATE					X	X	X	X	X									X	X						X	X																				
PRECOATING TO BE FOLLOWED WITH SOFT A.C.																					X																					X				
DENSE GRADED, HIGH % PASS NO. 200 SIEVE						X	X	X	X	X			X	X										X						X																
DENSE GRADED, LOW % PASS NO. 200 SIEVE						X	X	X	X	X			X	X				X						X						X																
HOT LAID																																														
SAND																																										X	X			
DENSE GRADED										X	X	X							X	X					X	X					X	X														X
ASPHALT MACADAM, PENETRATION METHOD																																										X	X	X		
HOT MIX, ASPHALTIC CONCRETE																																								X	X	X	X	X	X	
STONE FILLED SHEET ASPHALT																																							X	X	X	X				
SHEET ASPHALT, BINDER & SURFACE COURSES																																							X	X	X	X				
GROUT FILLER FOR STONE BLOCK																																							X	X	X					

* CUT-BACK MATERIAL

NOTE: FOR CUT-BACK ASPHALTS, THE HIGHER THE NUMBER THE MORE VISCOUS THE MATERIAL. FOR ASPHALT CEMENTS, THE HIGHER THE PENETRATION NUMBER THE SOFTER THE ASPHALT; THUS FOR HEAVY TRAFFIC IN HOT WEATHER A STIFF GRADE SUCH AS 50 TO 60 OR 60 TO 70 IS USED AND FOR COOL WEATHER OR LIGHT TRAFFIC SOFTER GRADES ARE USED. FOR ASPHALT EMULSIONS: SS= SLOW SETTING; MS= MEDIUM SETTING; RS= RAPID SETTING.

FOR BOTH TARS AND SLOW CURING ROAD OILS, THE HIGHER THE NUMBER THE MORE VISCOUS THE MATERIAL. THUS AN RT-1, RT-2, AND SC-0 ARE NON-VISCOUS LIQUIDS AT ORDINARY TEMPERATURES SUITED FOR SOAKING INTO A TIGHTLY BOUND BASE LIKE CLAY-GRAVEL. RT-4 TO RT-7 AND SC-2 TO SC-4 WILL REMAIN SEMI-LIQUID AND ARE SUITED FOR MIXING IN PLACE. RT-9 TO RT-12 AND SC-5 OR SC-6 BECOME SOLID AT AIR TEMPERATURES AND ARE SUITED FOR HOT PLANT MIXES, SEALING AND MACADAM.

TABLE B

BITUMINOUS PAVEMENT TYPES	SUGGESTED USES (MINIMUM RECOMMENDATIONS)
SURFACE TREATMENT	MAINTENANCE - ON ALL TYPES OF BITUMINOUS SURFACES, BADLY WORN CONCRETE, STONE & BRICK TYPES. LIGHT DUTY ROADS - (RESIDENTIAL) GENERALLY CONSIDERED AS AN ECONOMY PAVEMENT.
MIX IN PLACE (ROAD MIX) MODIFIED PENETRATION	LOCAL ROADS AND STREETS CARRYING MODERATE VOLUME OF MIXED TRAFFIC.
BITUMINOUS-MACADAM PENETRATION AND PLANT MIX.	SECONDARY HIGHWAYS, IMPORTANT LOCAL ROADS CARRYING MEDIUM TO HEAVY TRAFFIC.
ASPHALT CONCRETE	PRIMARY HIGHWAYS, CITY STREETS, THOROUGHFARES SUBJECT TO HEAVY AND VERY HEAVY TRAFFIC.

Adapted from Asphalt Institute Handbook, Koppers Company Tarmac Handbook, Barber-Green Bituminous Construction Handbook, and Highway Research Board Bulletin No. 105

ROADS- USE OF BITUMINOUS MATERIALS -2

TABLE A PRINCIPAL USES OF MINERAL AGGREGATES FOR TYPES OF PAVEMENT

IDENTIFICATION	SIMPLIFIED PRACTICE *												
	3	357	4	467	5	57	6	67	68	7	78	8	9
COVER FOR SURFACE TREATMENTS													
ALTERNATES							X	X		X	X	X	X
MIXED IN PLACE													
MACADAM AGGREGATE TYPE													
COARSE AGGREGATE					X								
SEAL COAT										X			
DENSE GRADED AGGREGATE TYPE													
MODIFIED PENETRATION													
WITH EMULSIFIED ASPHALT													
1½ INCHES THICK			X								X		X
1¼ INCHES THICK					X							X	X
1 INCH THICK							X					X	X
¾ INCH THICK								X					X
WITH CUT-BACK ASPHALT													
1½ TO 2 INCHES THICK			X								X		
1 INCH THICK					X						X		
COLD PATCH													
OPEN GRADED OVER 1½ INCHES THICK		X											
UNDER 1½ INCHES THICK								X					
DENSE GRADED STOCK PILE													
PLANT MIX													
COLD LAID													
PRECOATED MACADAM AGGREGATE TYPE					X				X				
MACADAM AGGREGATE TYPE													
COARSE AGGREGATE					X								
SEAL COAT										X			
DENSE GRADED TYPE WITH CUT-BACK													
DENSE GRADED WITH EMULSIFIED ASPHALT													
BASE OR BINDER													
BASE OR SURFACE COURSE													
COARSE SURFACE COURSE													
FINE SURFACE COURSE													
HOT LAID GRADED AGGREGATE													
COARSE MIX													
FINE MIX													
ASPHALT MACADAM PENETRATION METHOD	X							X			X		
HOT MIX ASPHALTIC CONCRETE													
GRADED AGGREGATE TYPE													
BASE COURSE, ALTERNATES													
SURFACE COURSE, ALTERNATES													
DENSE GRADED AGGREGATE TYPE													
COARSE AGGREGATE, ALTERNATES		X		X		X		X			X		
FINE AGGREGATE													
MINERAL FILLER													
BASE COURSE MIX, ALTERNATES													
SURFACE COURSE MIX, ALTERNATES													
STONE FILLED SHEET ASPHALT													
CONSTITUENTS												X	
MIX													
SHEET ASPHALT													
BINDER COURSE													
CONSTITUENTS									X				
MIX													
SURFACE COURSE													
CONSTITUENTS													
MIX													

IDENTIFICATION	MISCELLANEOUS * "TOTAL PASSING"														
	A	B	C	D	E	F	G	H	J	K	L	M	N	O	P
COVER FOR SURFACE TREATMENTS															
ALTERNATES										X				X	
MIXED IN PLACE															
MACADAM AGGREGATE TYPE															
COARSE AGGREGATE															
SEAL COAT															
DENSE GRADED AGGREGATE TYPE									X						
MODIFIED PENETRATION															
WITH EMULSIFIED ASPHALT															
1½ INCHES THICK															
1¼ INCHES THICK															
1 INCH THICK															
¾ INCH THICK															
WITH CUT-BACK ASPHALT															
1½ TO 2 INCHES THICK															
1 INCH THICK															
COLD PATCH															
OPEN GRADED OVER 1½ INCHES THICK															
UNDER 1½ INCHES THICK															
DENSE GRADED STOCK PILE							X								
PLANT MIX															
COLD LAID															
PRECOATED MACADAM AGGREGATE TYPE														X	
MACADAM AGGREGATE TYPE															
COARSE AGGREGATE															
SEAL COAT															
DENSE GRADED TYPE WITH CUT-BACK								X							
DENSE GRADED WITH EMULSIFIED ASPHALT				X											
BASE OR BINDER					X										
BASE OR SURFACE COURSE						X									
COARSE SURFACE COURSE													X		
FINE SURFACE COURSE														X	
HOT LAID GRADED AGGREGATE															
COARSE MIX								X							
FINE MIX											X				
ASPHALT MACADAM PENETRATION METHOD															
HOT MIX ASPHALTIC CONCRETE															
GRADED AGGREGATE TYPE															
BASE COURSE, ALTERNATES	X	X	X												
SURFACE COURSE, ALTERNATES			X				X					X			
DENSE GRADED AGGREGATE TYPE															
COARSE AGGREGATE, ALTERNATES															
FINE AGGREGATE															
MINERAL FILLER															X
BASE COURSE MIX, ALTERNATES															
SURFACE COURSE MIX, ALTERNATES															
STONE FILLED SHEET ASPHALT															
CONSTITUENTS															X
MIX															
SHEET ASPHALT															
BINDER COURSE															
CONSTITUENTS															
MIX															
SURFACE COURSE															
CONSTITUENTS															X
MIX															

IDENTIFICATION	MISCELLANEOUS * "PASSING AND RETAINED"										
	I	II	III	IV	V	VI	VII	VIII	IX	X	XI
COVER FOR SURFACE TREATMENTS											
ALTERNATES											
MIXED IN PLACE											
MACADAM AGGREGATE TYPE											
COARSE AGGREGATE											
SEAL COAT											
DENSE GRADED AGGREGATE TYPE											
MODIFIED PENETRATION											
WITH EMULSIFIED ASPHALT											
1½ INCHES THICK											
1¼ INCHES THICK											
1 INCH THICK											
¾ INCH THICK											
WITH CUT-BACK ASPHALT											
1½ TO 2 INCHES THICK											
1 INCH THICK											
COLD PATCH											
OPEN GRADED OVER 1½ INCHES THICK											
UNDER 1½ INCHES THICK											
DENSE GRADED STOCK PILE											
PLANT MIX											
COLD LAID											
PRECOATED MACADAM AGGREGATE TYPE											
MACADAM AGGREGATE TYPE											
COARSE AGGREGATE											
SEAL COAT											
DENSE GRADED TYPE WITH CUT-BACK											
DENSE GRADED WITH EMULSIFIED ASPHALT											
BASE OR BINDER											
BASE OR SURFACE COURSE											
COARSE SURFACE COURSE											
FINE SURFACE COURSE											
HOT LAID GRADED AGGREGATE											
COARSE MIX											
FINE MIX											
ASPHALT MACADAM PENETRATION METHOD											
HOT MIX ASPHALTIC CONCRETE											
GRADED AGGREGATE TYPE											
BASE COURSE, ALTERNATES											
SURFACE COURSE, ALTERNATES											
DENSE GRADED AGGREGATE TYPE											
COARSE AGGREGATE, ALTERNATES											
FINE AGGREGATE									X		
MINERAL FILLER											
BASE COURSE MIX, ALTERNATES	X	X	X								
SURFACE COURSE MIX, ALTERNATES			X		X	X					
STONE FILLED SHEET ASPHALT											
CONSTITUENTS									X		
MIX							X				
SHEET ASPHALT											
BINDER COURSE											
CONSTITUENTS								X			
MIX				X							
SURFACE COURSE											
CONSTITUENTS										X	
MIX											X

* SEE PAGE 10-10

TABLE B BITUMEN REQUIREMENTS FOR VARIOUS PAVEMENT TYPES

PAVEMENT TYPE	GAL./S.Y.	% BITUMEN BY WEIGHT
PRIMING		
TIGHT FINE GRAIN	03	—
LOOSE BONDED COARSE GRAIN	08	—
TACK COAT (FOG SPRAY)	0.10-0.25	—
SEAL COAT	02 - 0.4	—
SINGLE SURFACE TREATMENT		
SIMPLIFIED PRACTICE No. 9	01 - 0.15	—
No. 78	0.15-0.25	—
No. 7	0.25-0.35	—
No. 67	0.35-0.45	—
No. 6	0.45-0.55	—
LOOSELY BONDED SURFACE	0.25-0.40	—
EMULSIFIED SURFACE TREATMENT		
SINGLE	0.30-0.40	—
DOUBLE - FIRST APPLICATION	0.20-0.30	—
- SECOND APPLICATION	0.40-0.50	—

PAVEMENT TYPE	GAL./S.Y.	% BITUMEN BY WEIGHT
MIXED IN PLACE		
MACADAM TYPE BINDER	0.3 - 0.9	—
DENSE GRADED AGGREGATE BINDER	—	4-7
SAND BINDER	—	5-10
MODIFIED PENETRATION		
EMULSIFIED ASPHALT		
FIRST APPLICATION	0.30-0.60	—
SECOND APPLICATION	0.35-0.70	—
CUT-BACK BITUMEN BINDER	0.40-1.20	—
MACADAM PENETRATION		
SURFACE COURSE BINDER	2.0 (±0.1)	—
BASE COURSE		
FIRST APPLICATION	1.25 (±0.1)	—
SECOND APPLICATION	0.4 (±0.1)	—

Adapted from Asphalt Institute Handbook

PAVEMENT TYPE	GAL./S.Y.	% BITUMEN BY WEIGHT
PLANT MIX - COLD LAID		
PRECOATED MACADAM AGGREGATE	—	1.5-9.0
MACADAM AGGREGATE	—	3.5-5.5
DENSE GRADED AGGREGATE	—	4.0-6.0
EMULSIFIED ASPHALT		
BASE COURSE	—	6.5-14.0
SURFACE COURSE	—	8.0-16.0
PLANT MIX - HOT LAID		
ASPHALT CONC. - DENSE GRAD AGG.	—	4.5-8.0
- GRADED AGG. TYPE	—	4.0-8.5
STONE FILLED SHEET ASPHALT	—	7.0-9.5
SHEET ASPHALT - BINDER COURSE	—	4.0-6.0
- SURFACE COURSE	—	9.5-12.0
SAND ASPHALT - BASE COURSE	—	4.0-7.0
- SURFACE COURSE	—	5.0-10.0

ROADS - BITUMINOUS MIXTURES

To design a bituminous mix for highways, select type from page 10-09, i.e. plant mix, cold laid, precoated macadam aggregate type. Next select method, i.e. simplified practice. Table A on page 10-09 shows that simplified practice no. 5 or 68 may be used. From Table A below, select either one.

Various state specifications may supersede these mixes. For military work the Corps of Engineers have their own specifications, while for Airports the C.A.A. has specifications which should be conformed to when possible.

MINERAL AGGREGATE GRADATION REQUIREMENTS FOR BITUMINOUS MIXTURES

TABLE A SIMPLIFIED PRACTICE

Simp. Prac. No.	3	357	4	467	5	57	6	67	68	7	78	8	9
	Total Passing Sieve, Per Cent												
2½ in.	100	100	–	–	–	–	–	–	–	–	–	–	–
2 in	90-100	95-100	100	100	–	–	–	–	–	–	–	–	–
1½ in.	35-70	–	90-100	95-100	100	100	–	–	–	–	–	–	–
1 in.	0-15	35-70	20-35	–	90-100	90-100	100	100	100	–	–	–	–
¾ in.	–	–	0-15	35-70	40-75	–	90-100	90-100	90-100	100	100	–	–
½ in.	0-5	10-30	–	–	15-35	25-60	20-55	–	–	90-100	90-100	100	–
⅜ in.	–	–	0-5	10-30	0-15	–	0-15	20-55	30-65	40-70	40-75	85-100	100
No. 4	–	0-15	–	0-15	0-5	0-15	0-5	0-15	5-25	0-15	5-25	10-30	85-100
No. 8	–	–	–	–	–	0-5	–	0-5	0-5	0-5	0-5	0-10	10-40
No. 10	–	0.5	–	0.5	–	0.5	–	0.5	–	–	0-5	–	–
No. 16	–	–	–	–	–	–	–	–	–	–	–	–	0-10

TABLE B MISCELLANEOUS "TOTAL PASSING"

Identi-fication	A	B	C	D	E	F	G	H	J	K	L	M	N	O	P
	Total Passing Sieve, Per Cent														
2½ in.	100	–	–	–	–	–	–	–	–	–	–	–	–	–	–
2 in.	95-100	100	–	–	–	–	–	–	–	–	–	–	–	–	–
1½ in.	80-95	95-100	100	–	–	–	–	–	–	–	–	–	–	–	–
1¼ in.	–	–	–	85-100	100	100	–	–	–	–	–	–	–	–	–
1 in.	–	–	95-100	–	–	–	100	100	100	–	–	–	–	–	–
¾ in.	60-75	70-85	–	65-90	75-100	90-100	95-100	–	–	100	100	100	–	–	–
½ in.	–	–	70-85	50-80	60-80	80-100	75-90	75-90	–	90-100	95-100	95-100	100	–	–
⅜ in.	–	–	–	–	–	–	–	–	–	0-5	–	–	–	100	–
No. 4	30-45	35-50	40-55	35-60	40-60	60-85	45-60	50-70	50-70	0-4	45-70	60-80	75-100	95-100	–
No. 8	–	–	–	–	–	–	–	–	–	–	–	–	–	–	–
No. 10	20-35	25-37	30-42	20-45	25-40	40-60	35-47	35-50	35-60	–	35-50	40-55	45-70	–	–
No. 16	–	–	–	–	–	–	–	–	–	–	–	–	–	45-80	–
No. 30	–	–	–	–	–	–	–	–	–	–	–	–	–	–	100
No. 40	12-22	15-25	20-30	10-20	10-20	15-30	23-33	20-30	–	–	20-30	25-35	15-30	–	–
No. 50	–	–	–	–	–	–	–	–	–	–	–	–	–	5-30	–
No. 80	6-16	6-16	12-22	5-10	5-10	5-15	16-24	–	–	–	–	18-27	5-15	–	–
No. 100	–	–	–	–	–	–	–	–	–	–	–	–	–	0-8	85-100
No. 200	0-4	2-6	5-10	0-8	0-8	0-8	6-12	0-8	0-14	0-1	0-8	8-15	3-8	–	65-100

TABLE C MISCELLANEOUS "PASSING AND RETAINED"

Identification		I	II	III	IV	V	VI	VII	VIII	IX	X	XI
Passing Sieve	Retained in Sieve	Per Cent										
2 in.	1 in.	15-40	–	–	–	–	–	–	–	–	–	–
1½ in.	¾ in.	–	14-48	–	–	–	–	–	–	–	–	–
1 in.	½ in.	3-45	–	17-52	–	–	–	–	–	–	–	–
1 in.	¼ in.	–	–	–	52-72	–	–	–	–	–	–	–
¾ in.	⅜ in.	–	3-45	–	–	18-50	–	–	–	–	–	–
½ in.	⅜ in.	–	–	–	–	–	8-39	–	–	–	–	–
½ in.	No. 4	–	–	6-42	–	–	–	–	–	–	–	–
½ in.	No. 10	–	–	–	–	–	–	20-35	–	–	–	–
⅜ in.	No. 4	–	–	–	–	3-36	8-45	–	–	–	–	–
No. 4	–	–	–	–	–	–	–	–	98-100	98-100	–	–
No. 4	No. 10	5-15	5-15	5-15	8-20	9-22	9-27	–	8-25	0-15	–	–
Total	No. 10	60-80	55-75	55-70	–	50-65	50-65	–	–	–	–	–
No. 10	No. 40	3-19	3-21	4-20	–	5-22	5-22	7-30	15-50	15-50	15-50	10-40
No. 40	No. 80	5-22	6-25	8-25	–	9-27	9-27	11-40	22-63	30-60	30-60	20-45
No. 80	No. 200	3-15	3-16	4-16	–	5-18	5-18	10-30	7-40	15-40	15-40	12-32
No. 200	–	3-5	4-6	4-7	–	5-8	5-8	7-12	0-8	0-5	0-5	10-20
No. 10	Total	20-40	25-45	30-45	15-35	35-50	35-50	–	–	–	98-100	–

Adapted from Asphalt Institute Handbook.

ROADS - DESIGN OF RIGID PAVEMENTS - I

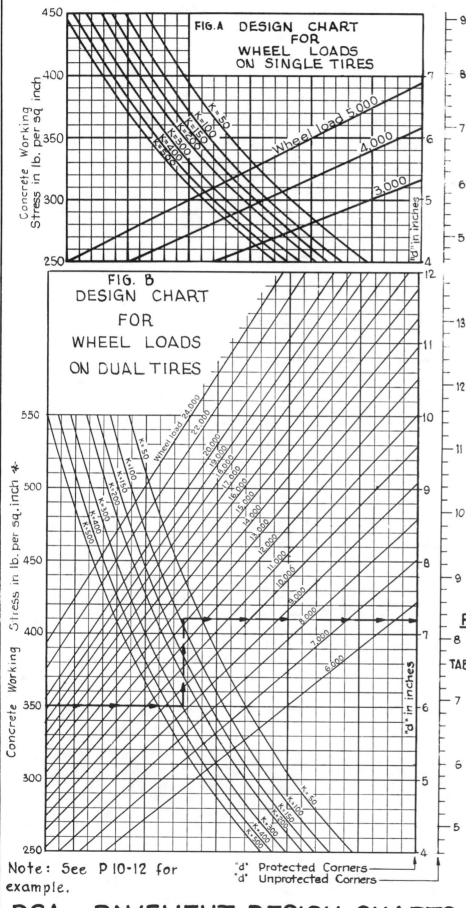

FIG. A DESIGN CHART FOR WHEEL LOADS ON SINGLE TIRES

Concrete Working Stress in lb. per sq inch: 250, 300, 350, 400, 450

K=50, K=100, K=150, K=200, K=300, K=400, K=500

Wheel load 5,000; 4,000; 3,000

"d" in inches: 4, 5, 6, 7 (Protected Corners); 5, 6, 7, 8, 9 (Unprotected Corners)

FIG. B DESIGN CHART FOR WHEEL LOADS ON DUAL TIRES

Concrete Working Stress in lb. per sq. inch *: 250, 300, 350, 400, 450, 500, 550

K=50, K=100, K=150, K=200, K=300, K=400, K=500

Wheel load 24,000; 22,000; 20,000; 19,000; 18,000; 17,000; 16,000; 15,000; 14,000; 13,000; 12,000; 11,000; 10,000; 9,000; 8,000; 7,000; 6,000

"d" in inches: 4, 5, 6, 7, 8, 9, 10, 11, 12 (Protected Corners); 5, 6, 7, 8, 9, 10, 11, 12, 13 (Unprotected Corners)

"d" Protected Corners
"d" Unprotected Corners

Note: See P 10-12 for example.

PROTECTED CORNERS ARE THOSE AT WHICH PROVISION IS MADE FOR TRANSFERRING AT LEAST 20% OF THE LOAD ACROSS THE INTERVENING JOINT THROUGH AGGREGATE INTERLOCK DOWELS OR KEYS.

UNPROTECTED CORNERS ARE THOSE AT WHICH THERE IS NO ADEQUATE PROVISION FOR LOAD TRANSFERENCE. THIS CONDITION EXISTS AT UNDOWELED EXPANSION JOINTS OR AT CRACKS OR UNDOWELED CONTRACTION JOINTS WHICH DO NOT MEET THE ABOVE REQUIREMENTS.

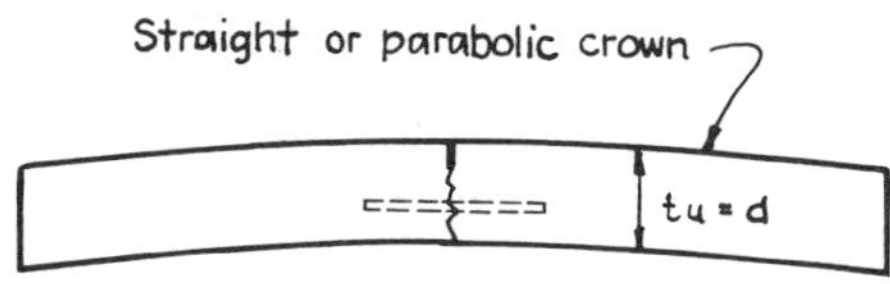

Straight or parabolic crown
$t_u = d$

UNIFORM (THICKNESS) SECTION

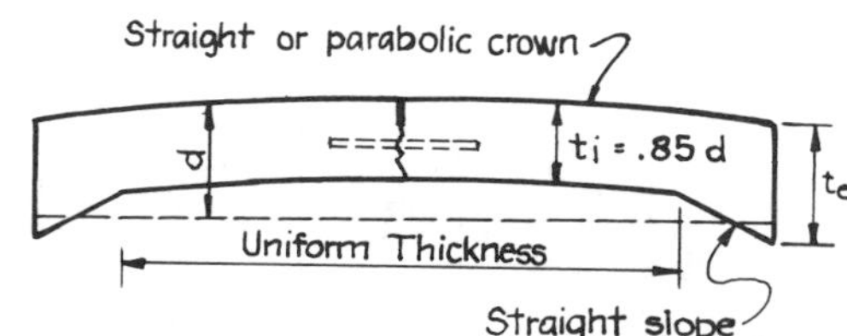

Straight or parabolic crown
d
$t_i = .85\,d$
t_e
Uniform Thickness
Straight slope

THICKENED EDGE SECTION
(NOT PRACTICAL FOR UNPROTECTED CORNER DESIGN)

PRINCIPAL TYPES OF X-SECTIONS

FIG. C

TABLE D INTERIOR THICKNESS $t_i = 0.85d$

SHAPE OF EDGE THICKENING	REQUIRED $\frac{t_e}{d}$	REQUIRED $\frac{t_e}{t_i}$
STRAIGHT SLOPE OF		
2 FT. 0 IN.	1.275	1.500
2 FT. 3 IN.	1.227	1.444
2 FT. 6 IN.	1.189	1.399
2 FT. 9 IN.	1.159	1.364
3 FT. 0 IN.	1.134	1.334
3 FT. 3 IN.	1.116	1.313
3 FT. 6 IN.	1.102	1.296
3 FT. 9 IN.	1.091	1.284
4 FT. 0 IN.	1.082	1.273

FOR FINAL PAVEMENT THICKNESS FRACTIONAL VALUES SHOULD BE ADJUSTED TO NEXT EVEN ¼".

PCA - PAVEMENT DESIGN CHARTS

* Concrete Working Stress = Modulus of Rupture ÷ Safety Factor

Adapted from P.C.A. Concrete Pavement Design

ROADS - DESIGN OF RIGID PAVEMENTS - 2

DESIGN EXAMPLE

Design concrete pavement with protected corners given: Modulus of rupture = 700 p.s.i.* Coefficient of subgrade reaction (K) = 300. Wheel load on dual tires = 11,000 lb.

$$\text{Concrete working stress} = \frac{\text{modulus of rupture} = 350 \text{ p.s.i.}}{2 \text{ (recommended safety factor)}}$$

Gross wheel load = 11,000 × 1.2 = 13,200 lb. (Includes 20% impact.)

By method illustrated by arrows on design chart (Fig. B, Page 10-11), obtain required pavement thickness (d) of 7.2 inches.

For uniform (thickness) section tu = d. Use tu = 7½ inches.

For thickened-edge section:

Interior thickness $ti = 0.85d$ (Fig. C, page 10-11) = 0.85(7.2) = 6.1 inches. Use 6½ inches.

Edge thickness (2'-0" slope) - from table D p. 10-11, $\frac{te}{d}$ = 1.275 therefore te = 1.275 d = 9.2 inches. Use 9½ inches.

TABLE A STREETS AND HIGHWAYS – TYPICAL SECTIONS

Class of		Controlling Wheel Load[†] (Static)	Concrete Pavement Thickness, Protected Corners fc = 350					
			Fine-Grained Soils, Plastic		Fine-Grained Soils, Slight to Non-Plastic		Gravelly or Sandy Soils, Well Drained	
Highway	Street		U[‡]	Thickened Edge	U[‡]	Thickened Edge	U[‡]	Thickened Edge
I		14,000	9	11½-7½-11½	8½	11 -7½-11	8½	11 -7½-11
II		12,000	8½	10½-7 -10½	8	10 -7 -10	8	10 -6½-10
III	I	11,000	8	10 -7 -10	7½	9½-6½- 9½	7½	9½-6½- 9½
IV		10,000	7½	9½-6½- 9½	7½	9 -6½- 9	7	9 -6 - 9
	II	9,000	7	9 -6 - 9	7	9 -6 - 9	7	8½-6 - 8½
	III	7,000	6½	8½-5½- 8½	6	8 -5½- 8	6	7½-5 - 7½
	IV	6,000	6	7½-5 - 7½	5½	7½-5 - 7½	5½	7½-5 - 7½

DESIGN CONSIDERATIONS Loaded concrete pavements impose very low unit pressures on the subgrade. Increase in "K" value by use of subbase course is uneconomical. However, subbase course is required if pavement damage may occur from frost action, swell, and shrinkage in high-volume-change soils or pumping of fine-grained soils.

COMPACTION REQUIREMENTS Subbase course and top 6 inches of subgrade should be compacted to at least 95% of maximum density (A.A.S.H.O. T99).

* 650 to 700 psi is usually specified for rigid pavements.

[†] Impact factor = 1.2.

[‡] U = Uniform.

ROADS-RIGID PAVEMENT SUBBASES — GRADATION LIMITS

Generally, concrete pavements constructed on subgrades containing large percentages of silt and clay will be subject to failures from warping, pumping, and frost damage (if in freezing zones). Granular subbases 4 to 6 inches that meet the grading and plasticity requirements shown below have prevented this type of failure under adverse conditions. See page 10-03.

When DENSELY GRADED materials are used, the subbase should be constructed about 2 feet wider than the pavement. No special provisions for draining them need be made, for, when properly compacted, the subbase is practically impervious. May be used to prevent frost damage* if ground water is not a problem.

When OPEN-GRADED (free-draining) materials are used, some positive means of preventing the accumulation of water in the subbase should be provided. This material is normally limited to the prevention of frost damage.* Construction should extend a minimum of 12 inches beyond the edges of the pavement.

All granular subbases should be compacted to 95% of materials maximum density.

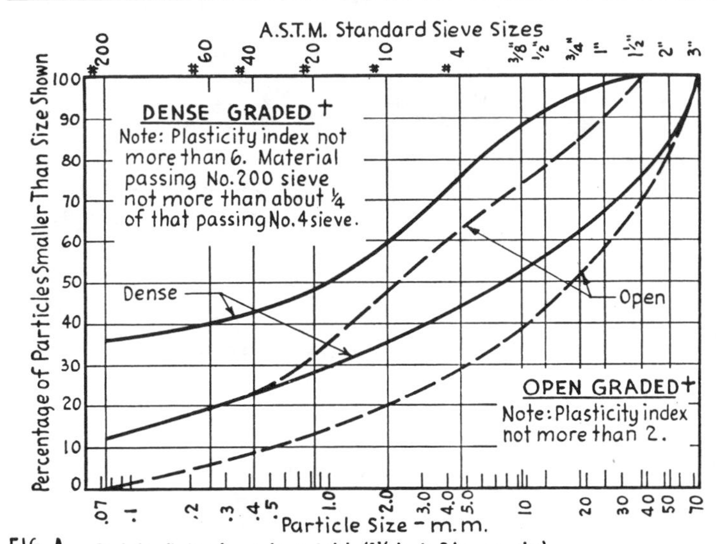
A.S.T.M. Standard Sieve Sizes
Percentage of Particles Smaller Than Size Shown
Particle Size - m.m.
DENSE GRADED†
Note: Plasticity index not more than 6. Material passing No.200 sieve not more than about ¼ of that passing No.4 sieve.
Dense
Open
OPEN GRADED†
Note: Plasticity index not more than 2.

FIG. A Gradation limits of granular materials (1½-in. to 3-in. max. size)

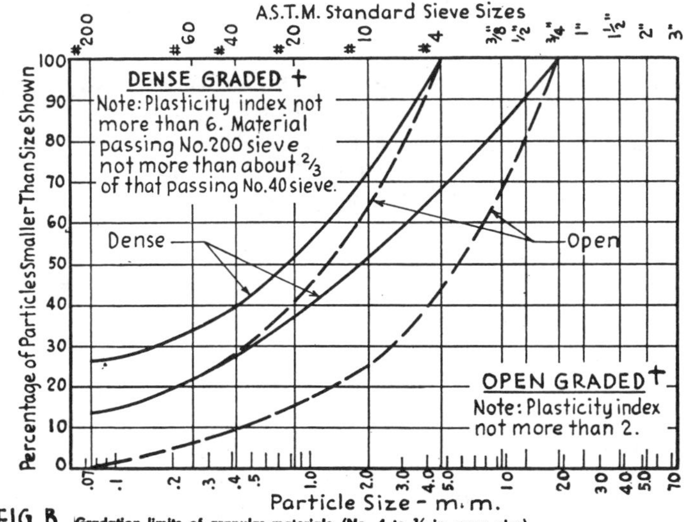
A.S.T.M. Standard Sieve Sizes
Percentage of Particles Smaller Than Size Shown
Particle Size - m.m.
DENSE GRADED †
Note: Plasticity index not more than 6. Material passing No.200 sieve not more than about ⅔ of that passing No.40 sieve.
Dense
Open
OPEN GRADED†
Note: Plasticity index not more than 2.

FIG. B Gradation limits of granular materials (No. 4 to ¾-in. max. size)

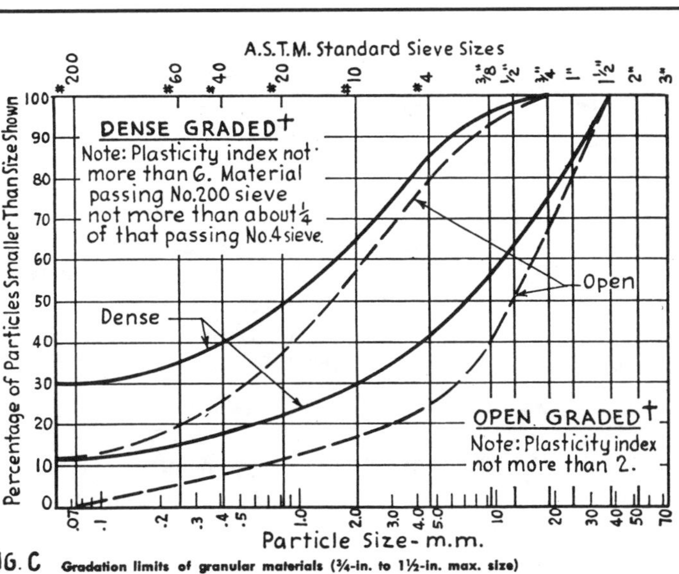
A.S.T.M. Standard Sieve Sizes
Percentage of Particles Smaller Than Size Shown
Particle Size - m.m.
DENSE GRADED†
Note: Plasticity index not more than 6. Material passing No.200 sieve not more than about ¼ of that passing No.4 sieve.
Dense
Open
OPEN GRADED†
Note: Plasticity index not more than 2.

FIG. C Gradation limits of granular materials (¾-in. to 1½-in. max. size)

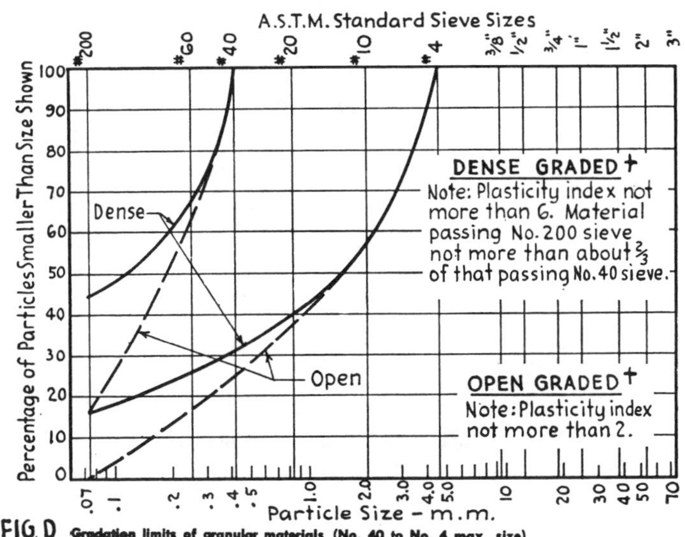
A.S.T.M. Standard Sieve Sizes
Percentage of Particles Smaller Than Size Shown
Particle Size - m.m.
DENSE GRADED †
Note: Plasticity index not more than 6. Material passing No. 200 sieve not more than about ⅔ of that passing No.40 sieve.
Dense
Open
OPEN GRADED†
Note: Plasticity index not more than 2.

FIG. D Gradation limits of granular materials (No. 40 to No. 4 max. size)

* Material when used for this purpose should contain less than 3% by weight of particles smaller than 0.02 mm.

† See pages 9-16 and 9-23 for Plasticity Index.

Adapted from P.C.A. Concrete Pavement Design.

ROADS - DOWELS - TIEBARS

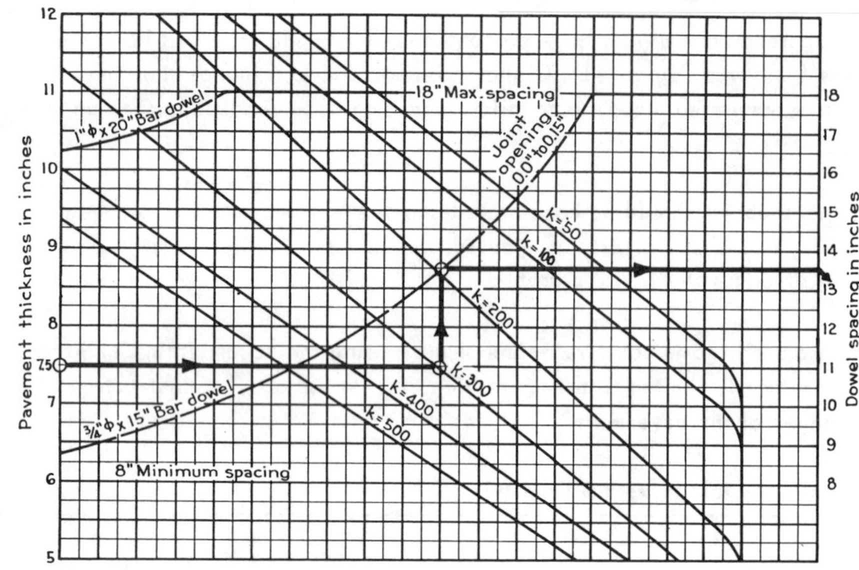

Pavement thickness in inches
Dowel spacing in inches
18" Max. spacing
1"ϕ x 20" Bar dowel
Joint opening 0.0" to 0.15"
k=50
k=100
k=200
k=300
k=400
k=500
3/4"ϕ x 15" Bar dowel
8" Minimum spacing

FIGURE A. DESIGN CHART FOR SMOOTH ROUND STEEL DOWELS USED ACROSS CONTRACTION JOINTS.

NOTE: CHART BASED ON CONTRACTION JOINT SPACE OF 0.15". IF JOINT SPACE EXCEEDS 0.15" USE FIGURE C FOR CONTRACTION JOINT DOWEL DESIGN.

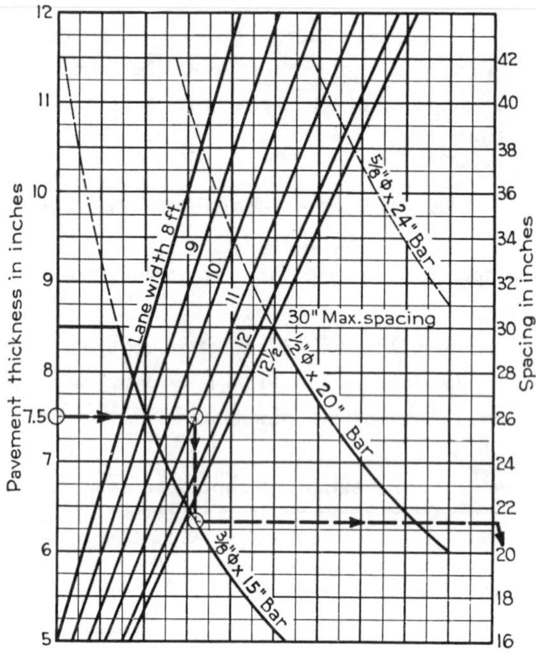

Pavement thickness in inches
Spacing in inches
Lane width 8 ft.
9
10
11
12
12½
30" Max. spacing
5/8"ϕ x 24" Bar
1/2"ϕ x 20" Bar
3/8"ϕ x 15" Bar

FIGURE B. DIAMETER, LENGTH AND SPACING OF TIEBARS.

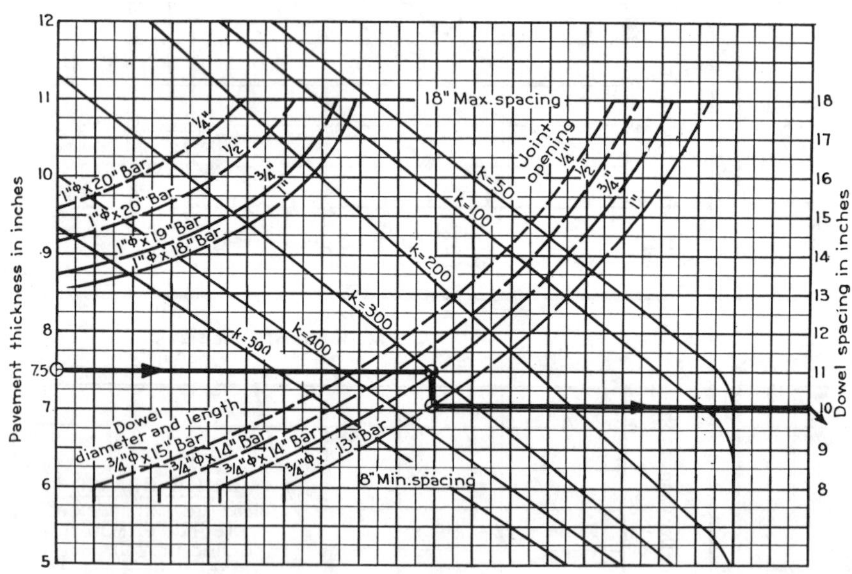

Pavement thickness in inches
Dowel spacing in inches
18" Max. spacing
Joint opening
1/4"
1/2"
3/4"
1"
1"ϕ x 20" Bar
1"ϕ x 20" Bar
1"ϕ x 19" Bar
1"ϕ x 18" Bar
k=50
k=100
k=200
k=300
k=400
k=500
Dowel diameter and length
3/4"ϕ x 15" Bar
3/4"ϕ x 14" Bar
3/4"ϕ x 14" Bar
3/4"ϕ x 13" Bar
8" Min. spacing

FIGURE C. DESIGN CHART FOR SMOOTH ROUND STEEL DOWELS USED ACROSS EXPANSION JOINTS.

TIEBAR DESIGN CHART BASED ON:

(a) COEFFICIENT OF FRICTION BETWEEN PAVEMENT AND SUB-GRADE = 1.5

(b) CONCRETE WEIGHT 150 LB. PER CUBIC FOOT.

(c) ALLOWABLE WORKING STEEL STRESS 25,000 P.S.I.

DOWEL DESIGN CHARTS BASED ON:

(a) TRANSFER CAPACITY THAT EXCEEDS MINIMUM REQUIREMENTS OF PROTECTED CORNERS DESIGN.

(b) INTERMEDIATE OR HARD BILLET, INTERMEDIATE OR HARD GRADE AXLE OR RAIL STEEL.

(c) f_c BEARING = 1800 P.S.I.

NOTE: 1) FOR THICKENED-EDGE DESIGN SECTION USE THE VALUE OF "d" RATHER THAN THE AVERAGE THICKNESS.

2) FOR LOCATION OF DOWELS AND TIEBARS SEE P 10-15.

USE OF DESIGN CHARTS

DESIGN EXAMPLE GIVEN: UNIFORM (THICKNESS) SECTION, t_u = 7½", K = 300 ; PROTECTED CORNERS.

FIND: SIZE, LENGTH, AND SPACING OF STEEL BY METHOD ILLUSTRATED BY ARROWS ON DESIGN CHARTS.

DOWELS (CONTRACTION JOINTS): FIGURE A - USE ¾" ϕ DOWELS X 15" LG. X 13" O.C.

DOWELS (EXPANSION JOINTS): FIGURE C - USE ¾" ϕ DOWELS X 13" LG. X 10" O.C.

TIEBARS: FIGURE B - LANE WIDTH = 11 FT., USE ⅜" ϕ TIEBARS X 15" LG. X 21" O.C.

Adapted from Portland Cement Association Concrete Pavement Design Manuals.

ROADS — JOINT DETAILS

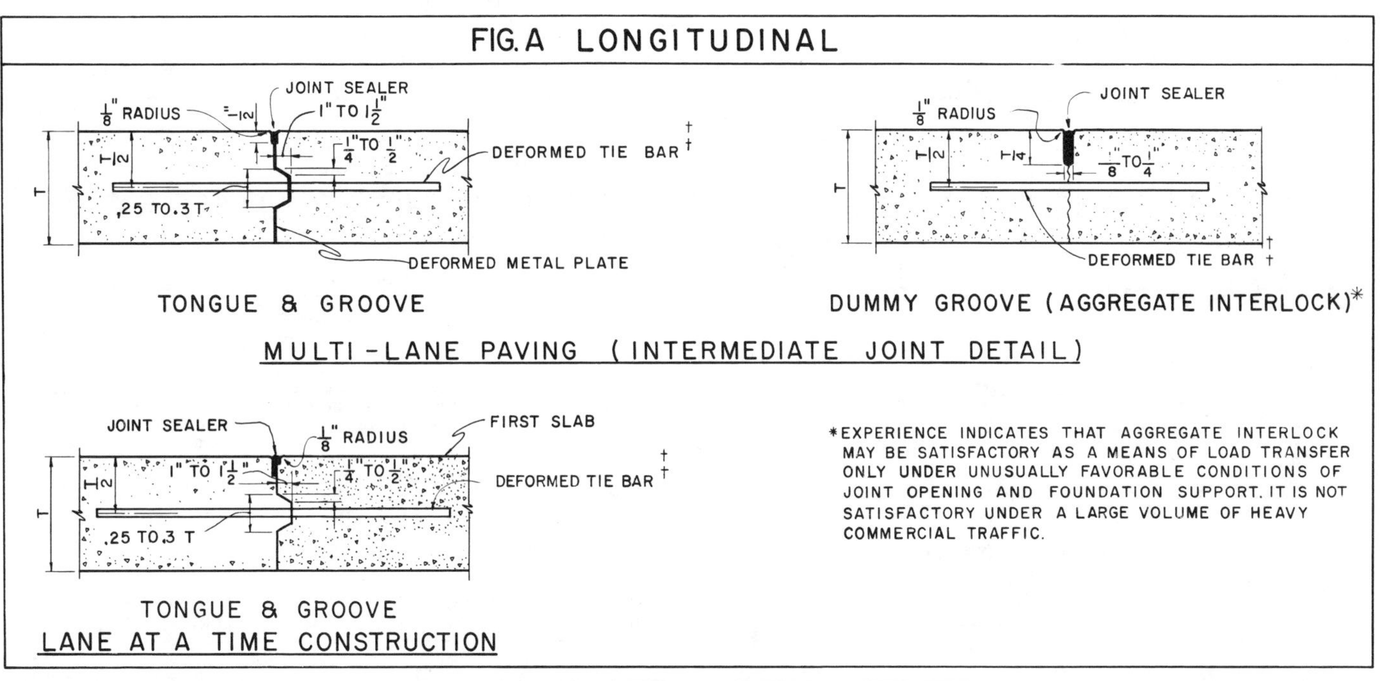

FIG. A LONGITUDINAL
JOINT SEALER
1/8" RADIUS
1" TO 1 1/2"
1/4" TO 1/2"
DEFORMED TIE BAR ††
.25 TO .3 T
DEFORMED METAL PLATE
TONGUE & GROOVE
1/8" TO 1/4"
DEFORMED TIE BAR ††
DUMMY GROOVE (AGGREGATE INTERLOCK)*
MULTI-LANE PAVING (INTERMEDIATE JOINT DETAIL)
FIRST SLAB
TONGUE & GROOVE
LANE AT A TIME CONSTRUCTION

*EXPERIENCE INDICATES THAT AGGREGATE INTERLOCK MAY BE SATISFACTORY AS A MEANS OF LOAD TRANSFER ONLY UNDER UNUSUALLY FAVORABLE CONDITIONS OF JOINT OPENING AND FOUNDATION SUPPORT. IT IS NOT SATISFACTORY UNDER A LARGE VOLUME OF HEAVY COMMERCIAL TRAFFIC.

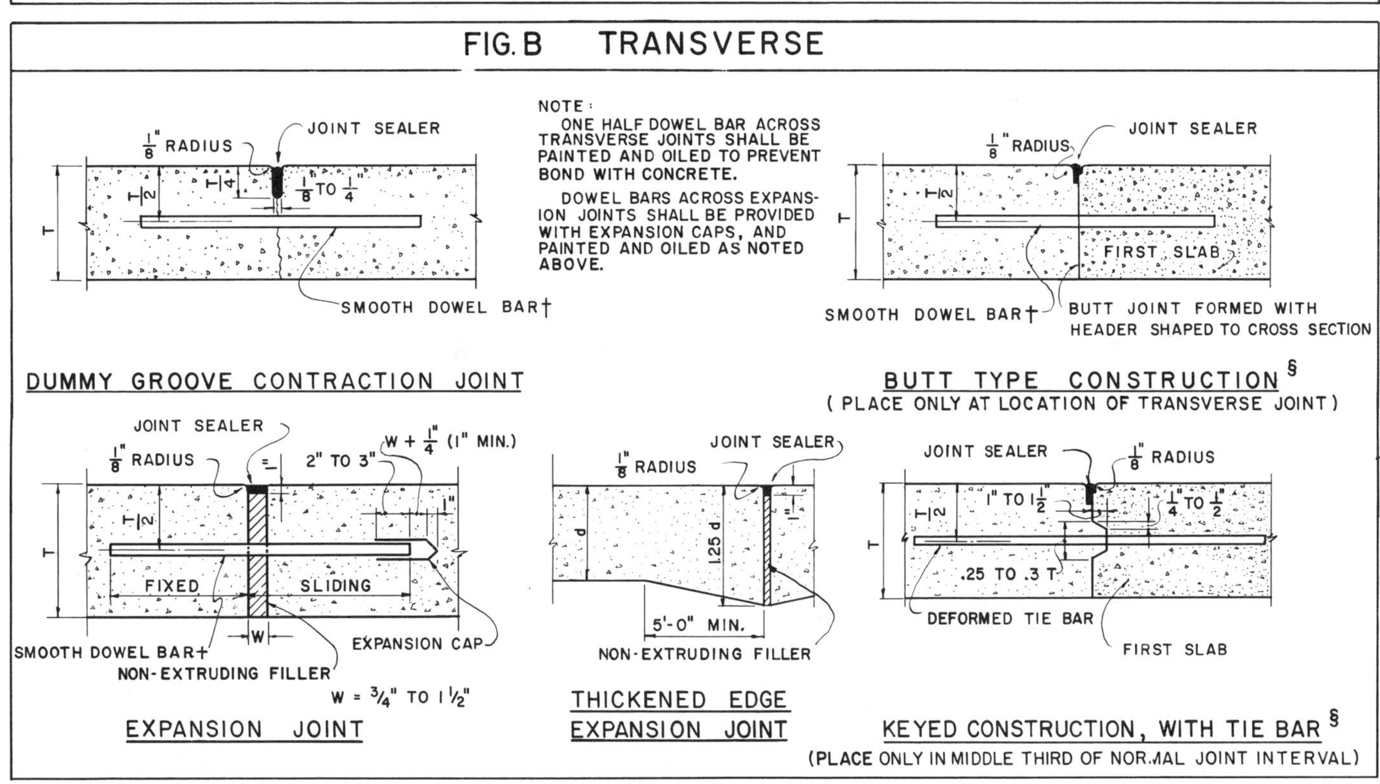

FIG. B TRANSVERSE
JOINT SEALER
1/8" RADIUS
1/8" TO 1/4"
SMOOTH DOWEL BAR†
DUMMY GROOVE CONTRACTION JOINT
FIRST SLAB
BUTT JOINT FORMED WITH HEADER SHAPED TO CROSS SECTION
BUTT TYPE CONSTRUCTION §
(PLACE ONLY AT LOCATION OF TRANSVERSE JOINT)
2" TO 3"
W + 1/4" (1" MIN.)
FIXED
SLIDING
SMOOTH DOWEL BAR†
NON-EXTRUDING FILLER
EXPANSION CAP
W = 3/4" TO 1 1/2"
EXPANSION JOINT
1.25 d
5'-0" MIN.
NON-EXTRUDING FILLER
THICKENED EDGE EXPANSION JOINT
1" TO 1 1/2"
1/4" TO 1/2"
.25 TO .3 T
DEFORMED TIE BAR
FIRST SLAB
KEYED CONSTRUCTION, WITH TIE BAR §
(PLACE ONLY IN MIDDLE THIRD OF NORMAL JOINT INTERVAL)

NOTE:
ONE HALF DOWEL BAR ACROSS TRANSVERSE JOINTS SHALL BE PAINTED AND OILED TO PREVENT BOND WITH CONCRETE.
DOWEL BARS ACROSS EXPANSION JOINTS SHALL BE PROVIDED WITH EXPANSION CAPS, AND PAINTED AND OILED AS NOTED ABOVE.

† DOWELS ARE NOT REQUIRED WHEN PAVEMENT IS DESIGNED WITH UNPROTECTED CORNERS. IN PAVEMENTS DESIGNED WITH PROTECTED CORNERS, DOWELS ARE REQUIRED ACROSS: (1) TRANSVERSE EXPANSION JOINTS, (2) PLAIN (BUTT) CONSTRUCTION JOINTS, (3) CONTRACTION JOINTS WHERE PANEL LENGTHS ARE GREATER THAN ABOUT 20 FEET OR WHERE EXPERIENCE INDICATES THAT AGGREGATE INTERLOCK WILL NOT PROVIDE ADEQUATE LOAD TRANSFERENCE, (4) FIRST 6 TO 10 CONTRACTION JOINTS EACH SIDE OF EXPANSION JOINT.

†† TIEBARS ARE REQUIRED ACROSS LONGITUDINAL JOINTS IN THE THIN INTERIORS OF THICKENED EDGE SLABS (30 INCH MAXIMUM SPACING). MAY BE USED ACROSS LONGITUDINAL JOINTS IN UNIFORM THICKNESS SLABS OR ACROSS THICKENED EDGES OF THICKENED-EDGE SLABS TO PREVENT SEPARATION OR DIFFERENTIAL SETTLEMENT. NOT MORE THAN 4 LANES SHOULD BE TIED TOGETHER.

GROOVES IN JOINTS MAY BE FORMED BY: (1) TEMPORARY EMBEDMENT OF A SUITABLE MANDREL, (2) INSTALLATION OF A THIN STRIP OF PREMOLDED JOINT FILLER MATERIAL, (3) SAWING THE PAVEMENT AFTER THE CONCRETE HAS HARDENED.

§ TRANSVERSE CONSTRUCTION JOINTS ARE REQUIRED AT END OF DAY'S PAVING OPERATION OR WHERE PLACING OF CONCRETE IS DISCONTINUED A SUFFICIENT TIME FOR CONCRETE TO SET.

Adapted from P.C.A. Concrete Pavement Design Manual.

ROADS – RIGID PAVEMENT DETAILS

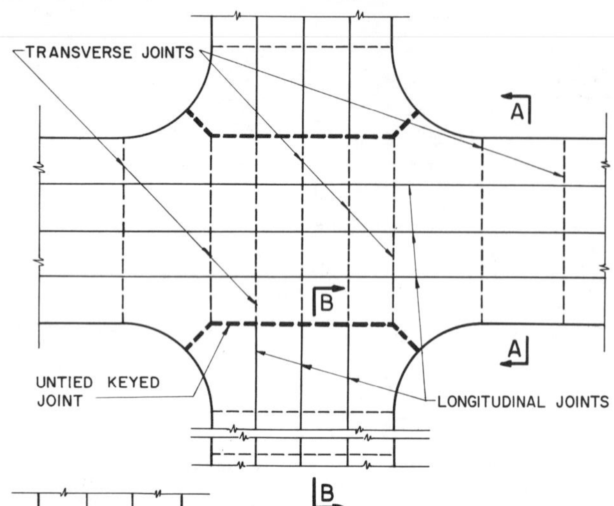
TRANSVERSE JOINTS
A
B
A
UNTIED KEYED JOINT
LONGITUDINAL JOINTS
B

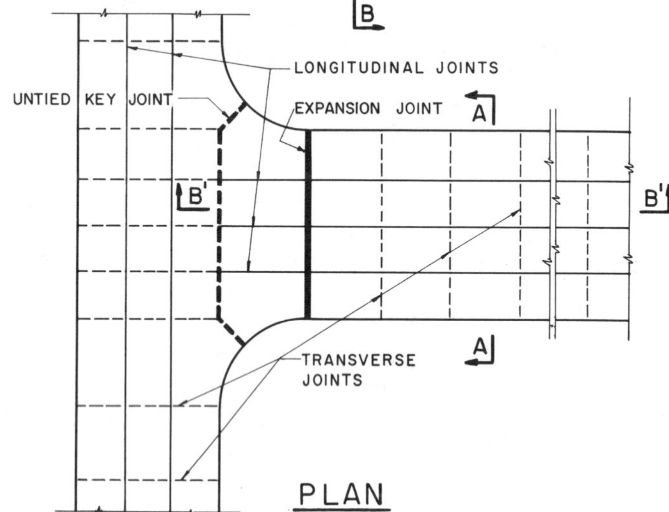
LONGITUDINAL JOINTS
UNTIED KEY JOINT
EXPANSION JOINT
A
B'
B'
A
TRANSVERSE JOINTS

PLAN

FIG. A DETAILS

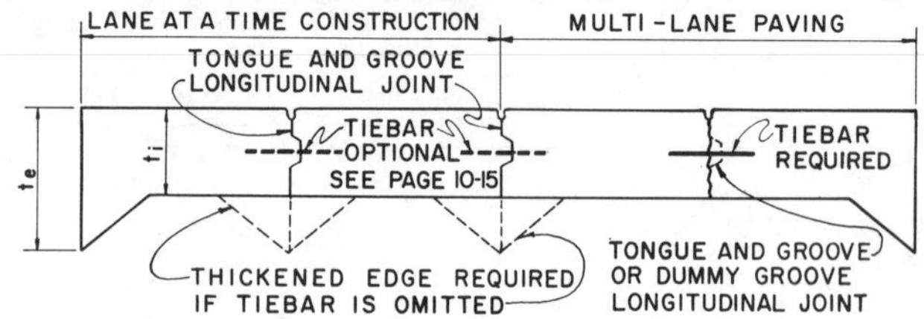
LANE AT A TIME CONSTRUCTION
MULTI-LANE PAVING
TONGUE AND GROOVE LONGITUDINAL JOINT
TIEBAR OPTIONAL SEE PAGE 10-15
TIEBAR REQUIRED
te
ti
THICKENED EDGE REQUIRED IF TIEBAR IS OMITTED
TONGUE AND GROOVE OR DUMMY GROOVE LONGITUDINAL JOINT

THICKENED – EDGE TRAVERSE SECTION

(PROTECTED CORNERS ONLY)*

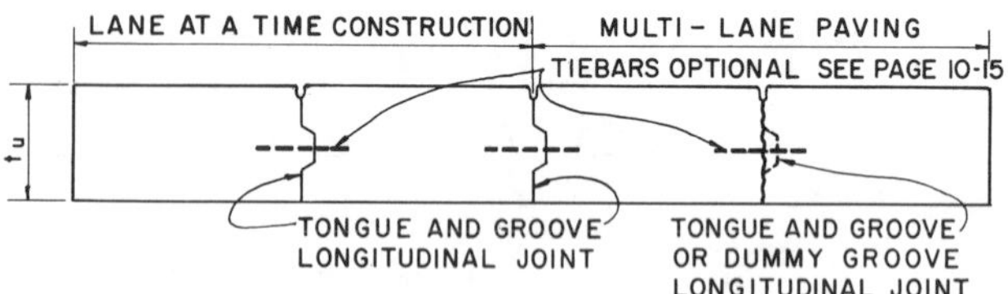
LANE AT A TIME CONSTRUCTION
MULTI - LANE PAVING
TIEBARS OPTIONAL SEE PAGE 10-15
tu
TONGUE AND GROOVE LONGITUDINAL JOINT
TONGUE AND GROOVE OR DUMMY GROOVE LONGITUDINAL JOINT

UNIFORM THICKNESS TRAVERSE SECTION

(PROTECTED AND UNPROTECTED CORNERS)

SECTIONS A-A

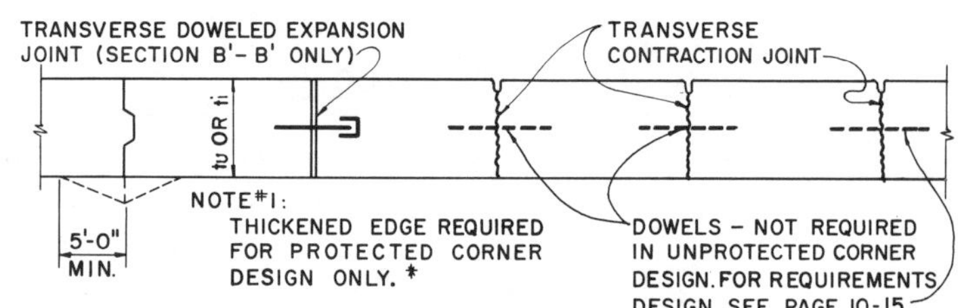
TRANSVERSE DOWELED EXPANSION JOINT (SECTION B'-B' ONLY)
TRANSVERSE CONTRACTION JOINT
tu OR ti
NOTE#1:
THICKENED EDGE REQUIRED FOR PROTECTED CORNER DESIGN ONLY. *
5'-0" MIN.
DOWELS – NOT REQUIRED IN UNPROTECTED CORNER DESIGN. FOR REQUIREMENTS DESIGN SEE PAGE 10-15

LONGITUDINAL SECTION

(THICKENED EDGE SECTION – PROTECTED CORNERS)*
(UNIFORM THICKNESS SECTION – PROTECTED AND UNPROTECTED CORNERS)

SECTION B-B AND B'-B'

* THICKENED EDGE PAVEMENT SECTIONS HAVE NO PRACTICAL APPLICATION FOR DESIGNS BASED ON UNPROTECTED CORNERS.

JOINT SPACING

TABLE B – TRANSVERSE JOINTS

KIND OF COARSE AGGREGATE (IN CONC.)	JOINT SPACING (FT.)
GRANITE	25–30
LIMESTONE	20–30
FLINTY LIMESTONE	20–25
GRAVEL – CALCAREOUS	20–25
SILICEOUS	15–20
LESS THAN 3/4 INCH SIZE	15
SLAG	15

LONGITUDINAL JOINTS

A LONGITUDINAL JOINT IS PLACED AT THE CENTERLINE OF TWO-LANE PAVEMENTS UP TO 25 FEET IN WIDTH. LONGITUDINAL JOINTS IN MULTI-LANE PAVEMENTS ARE SPACED SO THEY COINCIDE WITH THE LANE MARKINGS; THIS SPACES THEM AT 8 TO 12 1/2 FT. INTERVALS.

NOTE: TRANSVERSE & EXPANSION JOINT SPACES NOTED ARE FOR PLAIN CONCRETE PAVEMENTS ONLY. FOR REINFORCED CONCRETE JOINT SPACING SEE PAGE 10-17

EXPANSION JOINTS

UNDER USUAL CONDITIONS, WHEN THE PAVEMENT WILL BE CONSTRUCTED DURING NORMAL WEATHER USING NORMAL AGGREGATES, EXPANSION JOINTS ARE NOT REQUIRED EXCEPT AT STRUCTURES AND CERTAIN INTERSECTIONS SUCH AS 3 WAY INTERSECTIONS AND 4 WAY OFFSET (JAGGED) INTERSECTIONS. CONCRETE PAVEMENTS BUILT DURING COLD WEATHER OR CONSTRUCTED OF MATERIALS HAVING HIGH COEFFICIENTS OF EXPANSION SHOULD HAVE 3/4" TO 1 1/2" EXPANSION JOINTS SPACED AT 600 TO 800 FT. INTERVALS.

Adapted from P.C.A. Design of Concrete Pavements for City Streets.

ROADS – REINFORCED RIGID PAVEMENTS

DISTRIBUTED STEEL – MAY BE EITHER WELDED WIRE FABRIC OR BAR MATS. PRINCIPAL FUNCTION IS TO HOLD TOGETHER THE FRACTURED FACES OF SLABS IF CRACKS SHOULD FORM, ASSURING AGGREGATE INTERLOCK. IT DOES NOT SIGNIFICANTLY INCREASE THE FLEXURAL STRENGTH OF AN UNBROKEN SLAB. REQUIRED WHEN TRANSVERSE JOINT SPACINGS ARE IN EXCESS OF THOSE NOTED ON PAGE 10-16

NOTE – EXPANSION JOINTS SPACED 90'-0" TO 120'-0" CONTRACTION JOINTS 30'-0" TO 60'-0".
FOR TIEBAR AND DOWEL REQUIREMENTS SEE PAGE 10–15
GROOVE TYPE HINGE JOINTS ARE RECOMMENDED BETWEEN CONTRACTION JOINTS WHERE CONTRACTION JOINTS EXCEED SPACING NOTED ON PAGE 10-16

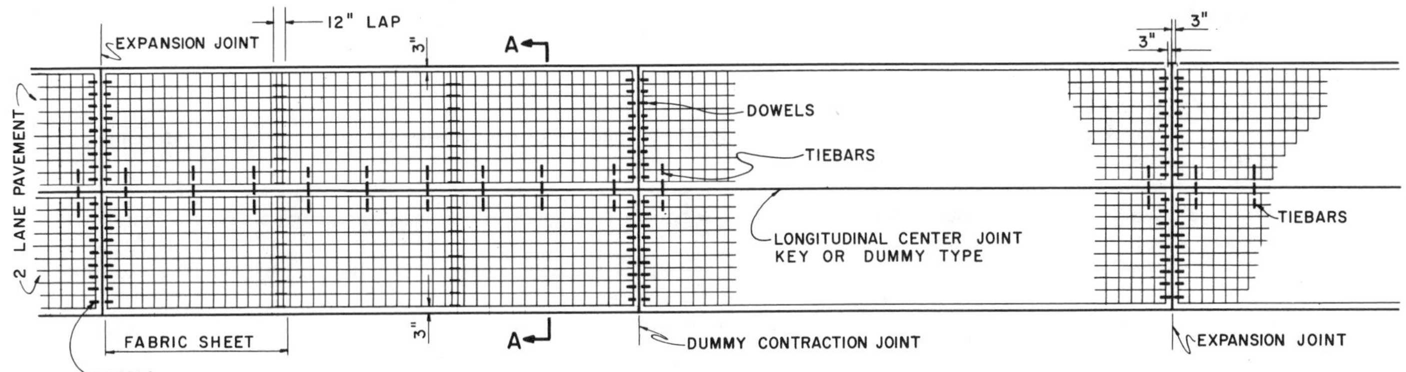
12" LAP
EXPANSION JOINT
3"
A
3"
3"
2 LANE PAVEMENT
DOWELS
TIEBARS
TIEBARS
LONGITUDINAL CENTER JOINT
KEY OR DUMMY TYPE
FABRIC SHEET
3"
A
DUMMY CONTRACTION JOINT
EXPANSION JOINT
DOWELS

FIG. A PLAN TYPICAL REINFORCED SLAB

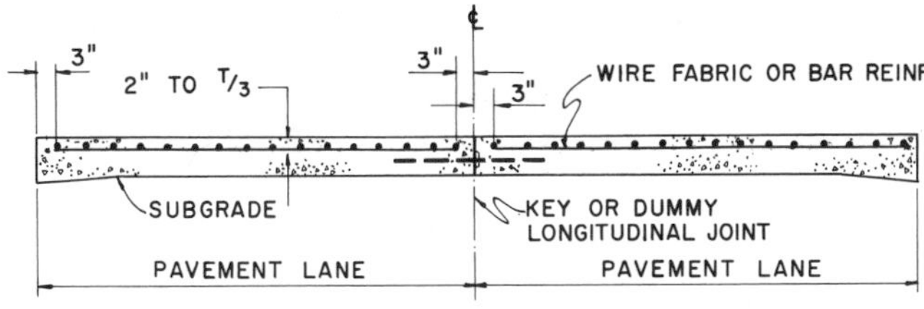
3"
2" TO T/3
3"
3"
WIRE FABRIC OR BAR REINF.
SUBGRADE
KEY OR DUMMY
LONGITUDINAL JOINT
PAVEMENT LANE
PAVEMENT LANE

FIG. B TRANSVERSE SECTION A-A

SHOWING LOCATION OF SINGLE BAR MAT OR WIRE FABRIC

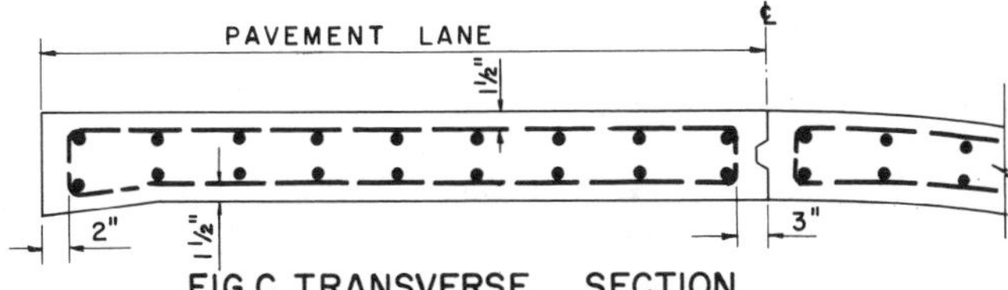
PAVEMENT LANE
1½"
2"
1½"
3"

FIG. C TRANSVERSE SECTION

SHOWING LOCATION OF DOUBLE BAR MAT.
USED FOR BRIDGE APPROACHES, EXTREME FROST CONDITIONS, BAD SUBGRADES AND OVER CULVERTS.

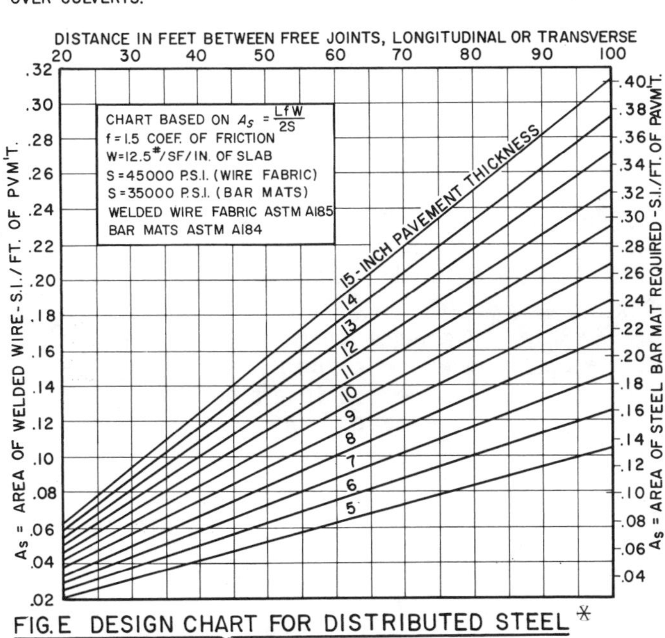
DISTANCE IN FEET BETWEEN FREE JOINTS, LONGITUDINAL OR TRANSVERSE
20 30 40 50 60 70 80 90 100
CHART BASED ON $A_S = \frac{LfW}{2S}$
f = 1.5 COEF. OF FRICTION
W = 12.5#/SF/IN. OF SLAB
S = 45000 P.S.I. (WIRE FABRIC)
S = 35000 P.S.I. (BAR MATS)
WELDED WIRE FABRIC ASTM A185
BAR MATS ASTM A184
15-INCH PAVEMENT THICKNESS
14
13
12
11
10
9
8
7
6
5
A_S = AREA OF WELDED WIRE - S.I./FT. OF PVM'T.
.32 .30 .28 .26 .24 .22 .20 .18 .16 .14 .12 .10 .08 .06 .04 .02
A_S = AREA OF STEEL BAR MAT REQUIRED - S.I./FT. OF PAVM'T.
.40 .38 .36 .34 .32 .30 .28 .26 .24 .22 .20 .18 .16 .14 .12 .10 .08 .06 .04

FIG. E DESIGN CHART FOR DISTRIBUTED STEEL *

* BASED ON PCA RECOMMENDED USE OF 80% OF YEILD POINT.

STYLE	SPACING OF WIRES IN.		STEEL WIRE GAUGE NO.		CROSS-SECTIONAL AREA SQ. IN./FT. OF WIDTH	
	LONG.	TRANS.	LONG	TRANS.	LONG.	TRANS.
316-610	3	16	6	10	.022	.011
412-67	4	12	6	7	.029	.025
412-4/04	4	12	4/0	4	.365	.040
412-5/03	4	12	5/0	3	.437	.047
412-6/02	4	12	6/0	2	.502	.054
412-7/01	4	12	7/0	1	.566	.063
66-88	6	6	8	8	.041	.041
66-77	6	6	7	7	.049	.049
66-66	6	6	6	6	.048	.048
66-55	6	6	5	5	.067	.067
66-46	6	6	4	6	.080	.048
66-44	6	6	4	4	.080	.080
66-33	6	6	3	3	.093	.093
66-22	6	6	2	2	.108	.108
66-11	6	6	1	1	.126	.126
66-00	6	6	0	0	.148	.148
68-11	6	8	1	1	.126	.094
612-77	6	12	7	7	.049	.025
612-66	6	12	6	6	.058	.029
612-55	6	12	5	5	.067	.034
612-46	6	12	4	6	.080	.029
612-44	6	12	4	4	.080	.040
612-43	6	12	4	3	.080	.047
612-36	6	12	3	6	.093	.029
612-34	6	12	3	4	.093	.040
612-33	6	12	3	3	.093	.047
612-25	6	12	2	5	.108	.034
612-24	6	12	2	4	.108	.040
612-22	6	12	2	2	.108	.054
612-17	6	12	1	7	.126	.025
612-11	6	12	1	1	.126	.063
612-06	6	12	0	6	.148	.029
612-00	6	12	0	0	.148	.074
612-2/04	6	12	2/0	4	.172	.040
612-2/03	6	12	2/0	3	.172	.047
612-2/0 1/4	6	12	2/0	1/4	.172	.049
612-21/64 3	6	12	21/64	3	.169	.047
612-21/64 1/4	6	12	21/64	1/4	.169	.049
612-3/04	6	12	3/0	4	.206	.040
612-4/04	6	12	4/0	4	.244	.040
612-5/03	6	12	5/0	3	.291	.047
612-6/02	6	12	6/0	2	.335	.054
612-7/01	6	12	7/0	1	.377	.063
88-33	8	8	3	3	.070	.070

TABLE D STYLES OF WELDED WIRE FABRIC

BAR SIZES		NOMINAL DIMENSIONS– ROUND SECTIONS	
OLD, DIA., IN.	NEW, NO.	DIAMETER, IN.	CROSS-SECTIONAL AREA, SQ. IN.
1/4	2	0.250	0.05
3/8	3	0.375	0.11
1/2	4	0.500	0.20
5/8	5	0.625	0.31
3/4	6	0.750	0.44
7/8	7	0.875	0.60
1	8	1.000	0.79

TABLE F BAR MAT STEEL

Adapted From P.C.A. Pamphlet. Distributed steel for Concrete Pavement.

AIRFIELD PAVEMENTS—CAA DESIGN METHOD

C.A.A. design* is based on assignment of the subgrade to arbitrary classifications and determination of pavement thickness from design curves.

The following tests are required to classify subgrade and correctly analyze site conditions. See pages 9-23 through 9-29.

1. Mechanical analysis.
2. Atterberg limits.
3. Maximum density and optimum moisture determination.
4. Additional tests, such as those for shrinkage, moisture equivalent, and C.B.R., should be made if required, to aid in estimating the performance of a soil.

DESIGN PROCEDURE – GENERAL

1. Subgrade Classification: Determine C.A.A. classification from pages 9-18 to 9-19.
2. Wheel Load: C.A.A. pavement-design charts are based on single-wheel loads. For multiple-wheel undercarriages, determine equivalent single-wheel load from pages 10-51 and 10-52.
3. Drainage: Determine drainage classification – good or poor – as follows:
   (*a*) Good drainage: Conditions where no accumulations of water will take place that would develop spongy areas in subgrade, low ground-water table, rapid surface-water runoff.
   (*b*) Poor drainage: Inadequate drainage due to character of soil profile, high ground-water table, poor surface runoff or any other cause that may result in instability or produce saturation of the subgrade.
4. Frost Action: Determine frost classification – no frost or severe frost – as follows:
   (*a*) Determine average depth of annual frost penetration from Fig. A, page 9-35.
   (*b*) Severe frost conditions are assumed to exist if depth of frost penetration is greater than anticipated total pavement thickness determined by trial design for "no frost" and drainage condition as defined above.
   (*c*) No frost: Condition where depth of frost penetration is less than anticipated total pavement thickness determined in 4*b*.
   (*d*) Rigid pavements have considerable insulation value. Depth of frost penetration may therefore be reduced by an amount equal to one-half the thickness of the concrete slab.
   (*e*) For rigid pavements, judgment of local climatic and soil characteristics should be used in determining the required pavement thickness to prevent frost heave and resulting pavement deterioration. Conditions may occur where the thickness of the subbase course must be increased over that obtained from design curves.
5. Thickened Areas: Taxiways, aprons, turnarounds, warmup pads, and runway ends.
6. Miscellaneous
   (*a*) Wearing course should be dense and well bonded to prevent loose aggregates from damaging propellers or jet engines, and should be of a texture to provide non-skid properties.
   (*b*) The surface of concrete pavements should have a granular texture such as is obtained by a burlap drag finish.
7. Arid Regions: The subbase beneath rigid pavements may be reduced considerably below the value given on the curves; such a reduction, however, must be predicated on the knowledge of the behavior of the particular soil as a subgrade pavement foundation. However, in no case should the subbase be reduced below 6 inches.
8. Pavement Surfaces: These are either (1) rigid type (Portland cement concrete) or (2) flexible type (bituminous materials). See pages 10-51 to 10-56 and 10-61 to 10-64 for design of pavements.
9. Bases: These can be (1) non-bituminous or dry or water-bound macadam (P-205 and P-206),† crushed stone (P-209), caliche (P-210), lime rock (P-211), shell (P-212), and lean-mix rolled concrete (P-302), (2) Bituminous central plant hot mix (P-201); and (3) bituminous coated aggregates, mixed in place (P-204), or emulsified asphalt, plant mix (P-215). See C.A.A. Standard Specifications for Construction of Airports.
10. Subbase: This is generally a well-graded granular material. May be bank-run gravel. See P-154 of C.A.A. Standard Specifications for Construction of Airports.

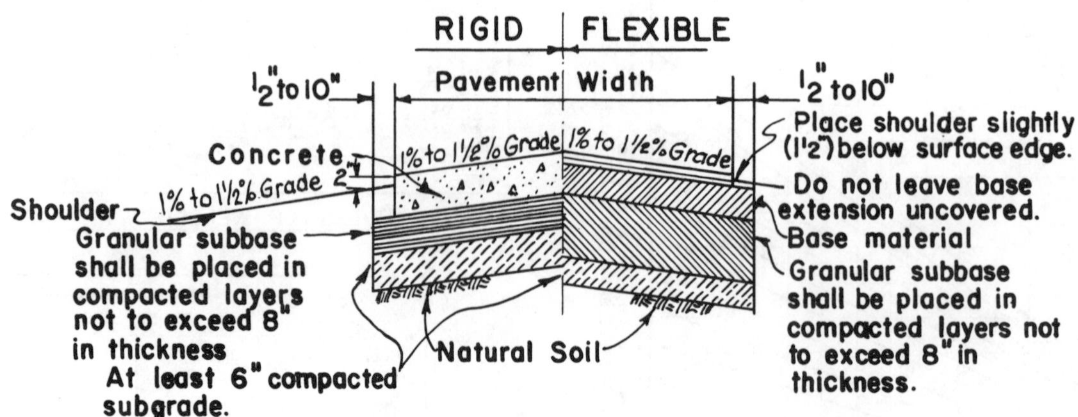
RIGID
FLEXIBLE
½" to 10"
Pavement Width
½" to 10"
Place shoulder slightly (1½") below surface edge.
Concrete
1% to 1½% Grade
1% to 1½% Grade
1% to 1½% Grade
2"
Shoulder
Do not leave base extension uncovered.
Granular subbase shall be placed in compacted layers not to exceed 8" in thickness
Base material
Granular subbase shall be placed in compacted layers not to exceed 8" in thickness.
Natural Soil
At least 6" compacted subgrade.

FIG. A – CROSS SECTION OF RUNWAY

* Adapted from "Airport Paving", C.A.A. October 1956.

† (P-205) etc., refer to C.A.A. Specification Items.

AIRFIELD PAVEMENTS -CAA EQUIVALENT SINGLE WHEEL-LOAD-I

C.A.A. METHOD FOR CONVERSION OF MULTIPLE-WHEEL ASSEMBLY LOADS TO EQUIVALENT SINGLE-WHEEL LOADS FOR RIGID-PAVEMENT DESIGN

TABLE A VALUES OF RADIUS OF RELATIVE STIFFNESS

Values of Radius of Relative Stiffness, l, in inches
$E = 4{,}000{,}000$ p.s.i. and $\mu = 0.15$

Radius of relative stiffness $= l = \sqrt[4]{\dfrac{Eh^3}{12(1-\mu^2)k}}$

Thickness in inches	k=50	k=100	k=150	k=200	k=250	k=300	k=350	k=400	k=500
6	34.84	29.30	26.47	24.63	23.30	22.26	21.42	20.72	19.59
7	39.11	32.89	29.72	27.65	26.15	24.99	24.04	23.25	21.99
8	43.23	36.35	32.85	30.57	28.91	27.62	26.58	25.70	24.31
9	47.22	39.71	35.88	33.39	31.58	30.17	29.03	28.08	26.55
10	51.10	42.97	38.83	36.14	34.17	32.65	31.42	30.39	28.74
11	54.89	46.16	41.71	38.81	36.71	35.07	33.75	32.64	30.87
12	58.59	49.27	44.52	41.43	39.18	37.44	36.02	34.84	32.95
13	62.22	52.32	47.27	43.99	41.61	39.75	38.25	36.99	34.99
14	65.77	55.31	49.98	46.51	43.98	42.02	40.44	39.11	36.99
15	69.27	58.25	52.63	48.98	46.32	44.26	42.58	41.19	38.95
16	72.70	61.13	55.24	51.41	48.62	46.45	44.70	43.23	40.88
17	76.08	63.98	57.81	53.80	50.88	48.61	46.77	45.24	42.78
18	79.41	66.78	60.35	56.16	53.11	50.74	48.82	47.22	44.66

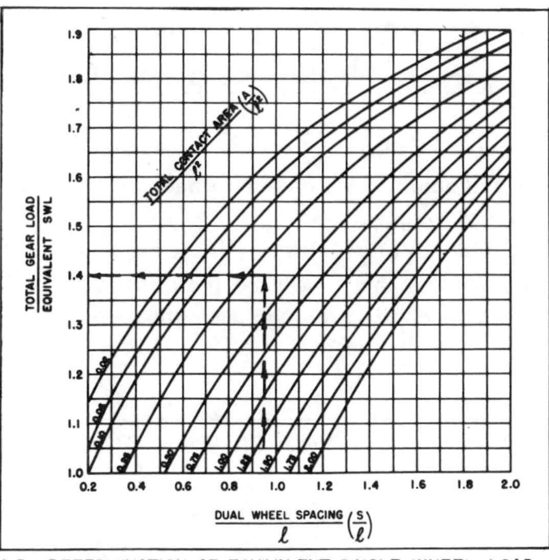

TOTAL GEAR LOAD / EQUIVALENT SWL
TOTAL CONTACT AREA (A/l²)
DUAL WHEEL SPACING (S/l)

FIG. B—DETERMINATION OF EQUIVALENT SINGLE WHEEL LOAD RIGID PAVEMENTS — DUAL WHEEL GEAR.

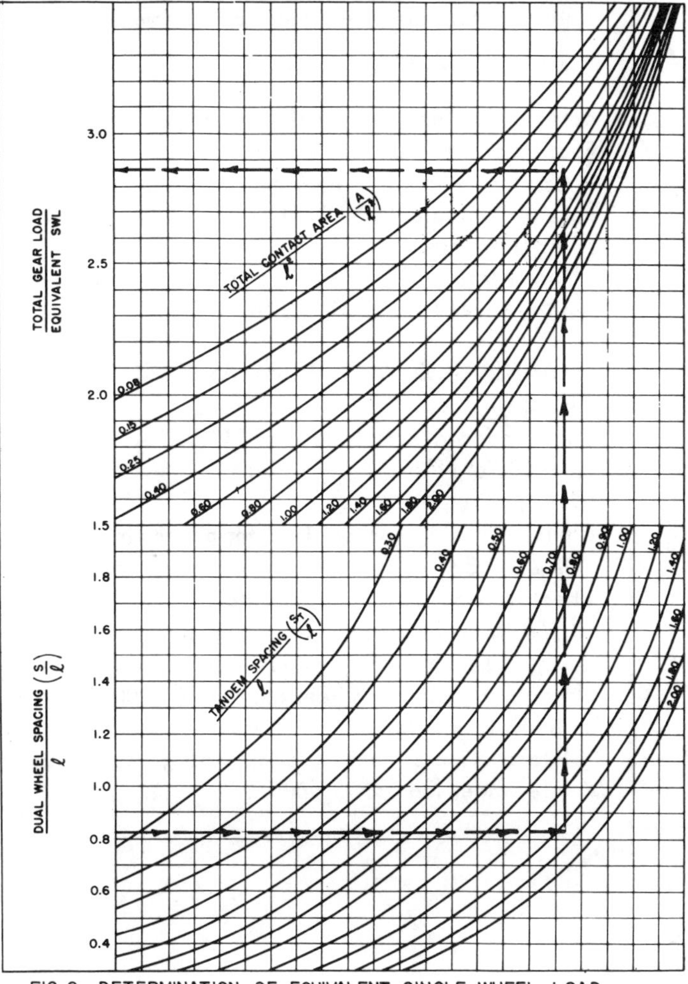

TOTAL GEAR LOAD / EQUIVALENT SWL
TOTAL CONTACT AREA (A/l²)
TANDEM SPACING (S_T/l)
DUAL WHEEL SPACING (S/l)

FIG. C—DETERMINATION OF EQUIVALENT SINGLE WHEEL LOAD RIGID PAVEMENTS—DUAL TANDEM GEAR.

EXAMPLE DUAL-WHEEL GEAR

Given: Aircraft gear assembly load = 48,400 lb. Spacing of wheels S = 31". Total five-print area A = 378 sq. in.

Find: Equivalent single-wheel load for concrete pavement depth h* = 10", k = 300.+

From Table A, $l = 32.65$". $S/l = 0.95$, $A/l^2 = 0.35$. Enter Fig. B with value for $S/l = 0.95$, and carry to intersection with curve $A/l^2 = 0.35$ (interpolation necessary). From this point, intercept vertical axis where value of the ratio of total gear load to equivalent single-wheel load = 1.4. Equivalent single-wheel load $= \dfrac{48{,}400}{1.4} = 34{,}600$ lb.

EXAMPLE DUAL-TANDEM ASSEMBLY

Given: Aircraft-gear assembly load = 161,000 lb. S = 31.25", S_T = 61.25", A = 824 sq. in.

Find: Equivalent single-wheel load for concrete pavement depth, h† = 12", k = 300.+

From Table A, $l = 37.44$", $S/l = 0.83$, $S_T/l = 1.63$, $A/l^2 = 0.59$.

Enter the vertical axis of Fig. C with the value $S/l = 0.83$, and intercept curve for $S_T/l = 1.63$. Extend value vertically to intersect curve for $A/l^2 = 0.59$, and intercept vertical axis at that point. Ratio of total gear load to equivalent single-wheel load = 2.87. Equivalent single-wheel load = $\dfrac{161{,}000}{2.87} = 56{,}000$ lb.

Given: Dual wheel landing gear 48,400 lb.

Find: Approximate "h"

Approximate single-wheel load $= \dfrac{48{,}400}{1.35} = 35{,}850$ lb. from chart on p. 10-53. Select "h", using 35,850 lb. h = 9", + say 10.

Approximate Conversion Factor
1.35 dual-wheel landing gear.
3.00 dual-tandem landing gear.

* Determination of "h" based on approximate conversion factor.

† Determination of "h" similar to above.

Data from C.A.A., *Airport Paving,* Oct. 1956.

C.A.A. METHOD FOR CONVERSION OF MULTIPLE-WHEEL ASSEMBLY LOADS TO EQUIVALENT SINGLE-WHEEL LOADS FOR FLEXIBLE-PAVEMENT DESIGN

LOAD DISTRIBUTION – DUAL WHEEL ASSEMBLY

d/2 is the depth to which each wheel of the assembly acts as an individual single wheel.

2S is the depth at which the effects produced in the subgrade are approximately the same as if the total load on the dual assembly were applied through a single wheel.

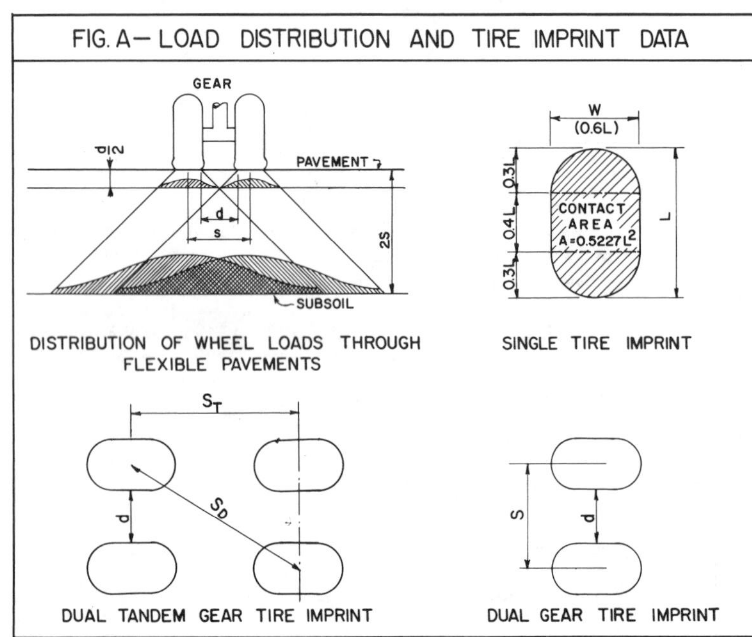
GEAR
d/2
PAVEMENT
d
s
2S
SUBSOIL
W (0.6L)
.3L
.4L
.3L
CONTACT AREA A=0.5227L²
L
S_T
S_D
d
S
d

FIG. A— LOAD DISTRIBUTION AND TIRE IMPRINT DATA

DISTRIBUTION OF WHEEL LOADS THROUGH FLEXIBLE PAVEMENTS

SINGLE TIRE IMPRINT

DUAL TANDEM GEAR TIRE IMPRINT

DUAL GEAR TIRE IMPRINT

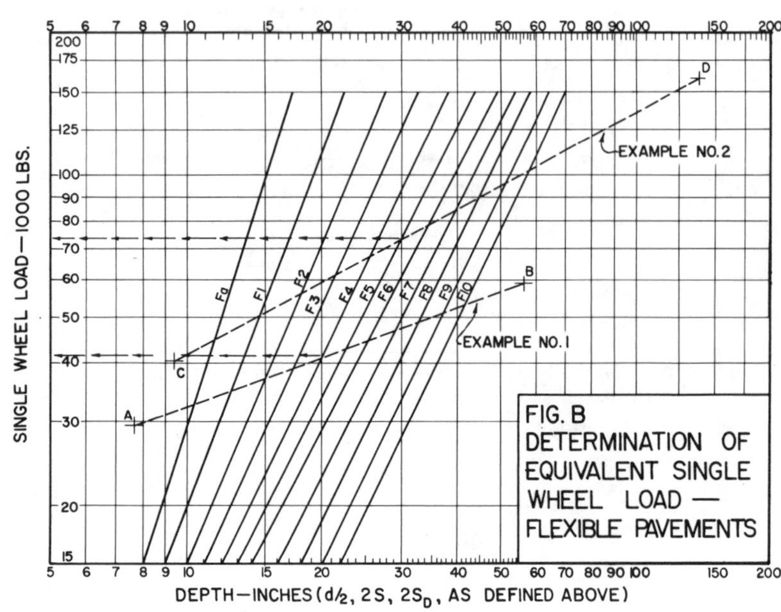
SINGLE WHEEL LOAD—1000 LBS.
DEPTH—INCHES ($d/2$, 2S, $2S_D$, AS DEFINED ABOVE)
EXAMPLE NO. 2
EXAMPLE NO. 1
A
B
C
D
F0 F1 F2 F3 F4 F5 F6 F7 F8 F9 F10

FIG. B DETERMINATION OF EQUIVALENT SINGLE WHEEL LOAD — FLEXIBLE PAVEMENTS

LOAD DISTRIBUTION – DUAL-TANDEM ASSEMBLY

d/2 is the depth to which each wheel of the assembly acts as an individual single wheel.

$2S_D$ is the depth at which the effects produced in the subgrade are approximately the same as if the total load on the dual-tandem assembly were applied through a single wheel.

EXAMPLE 1: DUAL-WHEEL ASSEMBLY

Given:

Dual-wheel assembly load of 59,000 lb., d = 15.4", S = 28.0", F-4 subgrade class.

Find:

Equivalent single-wheel load for pavement design.

$\frac{d}{2}$ = 7.7" = depth of pavement to which each wheel acts as an individual single-wheel load of 29,500 lb. Plot as point A on Fig. B.

2S = 56" = depth of pavement at which assembly acts as a single-wheel load of 59,000 lb. Plot as point B on Fig. B.

Intersection of Lines A-B and F-4 gives equivalent single-wheel load of 41,000 lb.

EXAMPLE 2: DUAL-TANDEM ASSEMBLY

Given:

Dual-tandem assembly load of 161,000 lb., d = 18.75", S_D = 68.76", F-5 subgrade class.

Find:

Equivalent single-wheel load for pavement design.

d/2 = 9.38" = depth of pavement to which each wheel acts as an individual single-wheel load of 40,250 lb. Plot as point C on Fig. B.

$2S_D$ = 137.5" = depth of pavement at which assembly acts as a single-wheel load of 161,000 lb. Plot as point D on Fig. B.

Intersection of Lines C-D and F-5 gives equivalent single-wheel load of 72,500 lb.

AIRFIELD PAVEMENTS — RIGID DESIGN CHARTS — C.A.A.

TABLE A — SUBGRADE CLASSES FOR RIGID PAVEMENTS.

SOIL GROUP	SUBGRADE CLASS			
	GOOD DRAINAGE		POOR DRAINAGE	
	NO FROST	SEVERE FROST	NO FROST	SEVERE FROST
E-1	Ra	Ra	Ra	Ra
E-2	Ra	Ra	Ra	Ra
E-3	Ra	Ra	Ra	Ra
E-4	Ra	Ra	Rb	Rb
E-5	Ra	Rb	Rb	Rb
E-6	Rb	Rb	Rb	Rc
E-7	Rb	Rb	Rb	Rc
E-8	Rb	Rc	Rc	Rd
E-9	Rc	Rc	Rc	Rd
E-10	Rc	Rc	Rc	Rd
E-11	Rd	Rd	Rd	Re
E-12	Rd	Re	Re	Re
E-13	Not suitable for subgrade.			

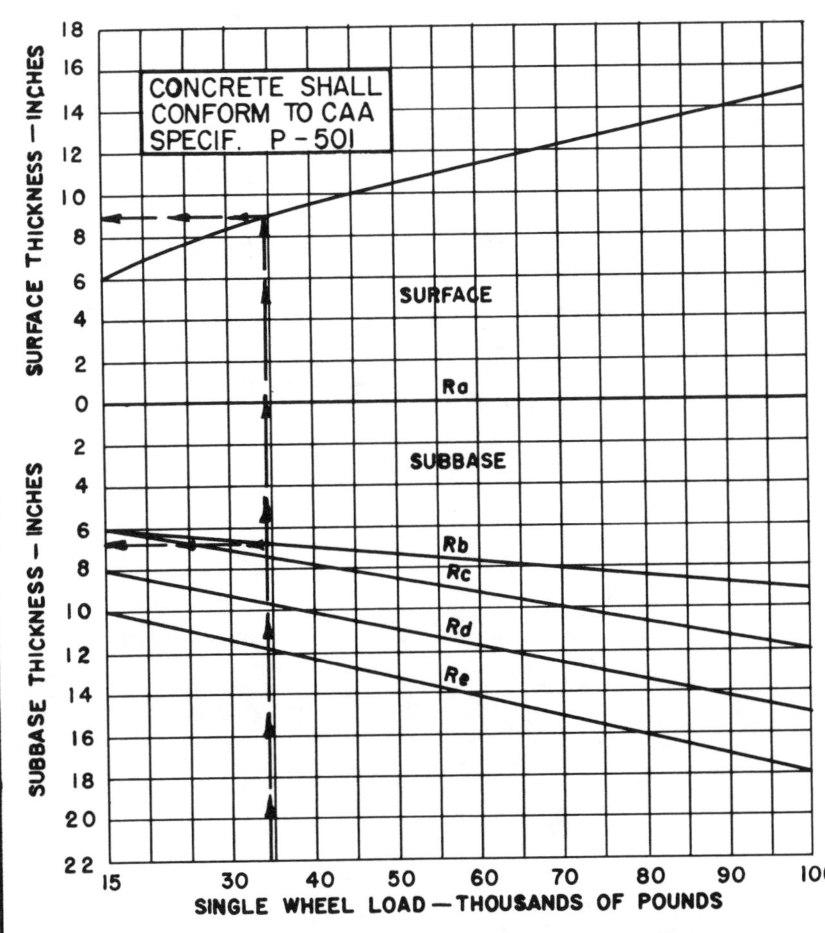
CONCRETE SHALL CONFORM TO CAA SPECIF. P-501
SURFACE THICKNESS — INCHES
SUBBASE THICKNESS — INCHES
SURFACE
SUBBASE
Ra
Rb
Rc
Rd
Re
SINGLE WHEEL LOAD — THOUSANDS OF POUNDS

FIG. B — DESIGN CURVES FOR RIGID PAVEMENTS — TAXIWAYS, APRONS, AND RUNWAY ENDS.

From C.A.A. "Airport Paving" Oct. 1956.

Surface Thickness curve in Fig. B is for critical pavement areas such as aprons, taxiways, turn-arounds, and runup areas at runway ends. To determine the surface thickness for non-critical pavement area (runway pavement between runway ends), use 80% of surface thickness as determined from curve in Fig. B. See page 10-55 for illustration of critical and non-critical runway-pavement areas.

Subbase Thickness remains the same for critical and non-critical areas of a given subgrade and wheel loading.

NOTE: When a fractional thickness of ½ inch or more is indicated, use the next full-inch thickness.

EXAMPLE:

Given:
Aircraft given in Example 1, page 10-51, having dual-wheel assembly load of 48,400 lb., E-8 subgrade, good drainage conditions, average depth of annual frost penetration of 18 inches.

Find:
Concrete pavement and subbase course thickness required.

Solution:

1. As determined in Example 1, page 10-51, equivalent single-wheel load for E-8 subgrade is 34,600 lb.
2. Make trial design, assuming "no frost", to determine if severe frost condition. . . . From Table A, subgrade class is Rb. From Fig. B, thickness of concrete pavement and subbase course required for taxiways, aprons, and runway ends are 9 and 7 inches respectively. For non-critical areas, concrete pavement required is $0.8 \times 9 = 7.2$ inches. Use 7½ inches.
3. Considering insulating value of concrete pavement, reduced depth of frost penetration (18 in. − ½ thickness of concrete pavement) is less than total thickness of concrete pavement and subbase course required in step 2. Therefore no frost condition exists, and pavement thicknesses obtained in step 2 are applicable.

AIRFIELD PAVEMENTS —FLEXIBLE DESIGN CHARTS—CAA

TABLE-A SUBGRADE CLASSES FOR FLEXIBLE PAVEMENTS				
SOIL GROUP	SUBGRADE CLASS			
	GOOD DRAINAGE		POOR DRAINAGE	
	No Frost	Severe Frost	No Frost	Severe Frost
E-1	Fa	Fa	Fa	Fa
E-2	Fa	Fa	Fl	F2
E-3	Fl	Fl	F2	F2
E-4	Fl	Fl	F2	F3
E-5	Fl	F2	F3	F4
E-6	F2	F3	F4	F5
E-7	F3	F4	F5	F6
E-8	F4	F5	F6	F7
E-9	F5	F6	F7	F8
E-10	F5	F6	F7	F8
E-11	F6	F7	F8	F9
E-12	F7	F8	F9	F10
E-13	NOT SUITABLE FOR SUBGRADE			

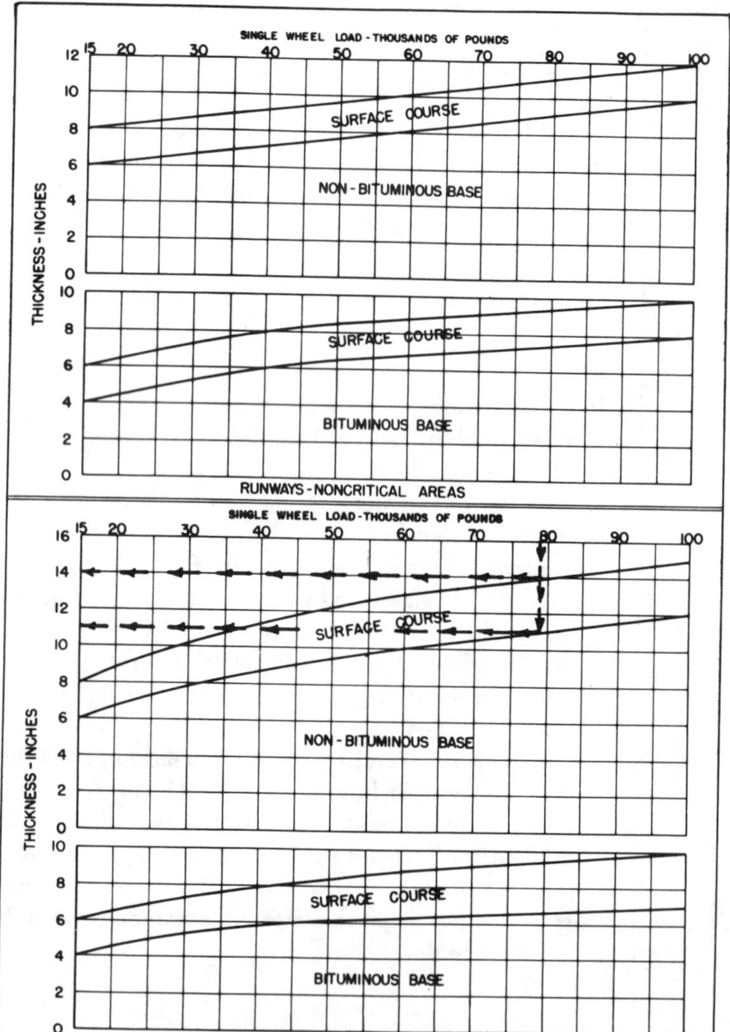
SINGLE WHEEL LOAD - THOUSANDS OF POUNDS
THICKNESS - INCHES
SURFACE COURSE
NON-BITUMINOUS BASE
BITUMINOUS BASE
RUNWAYS-NONCRITICAL AREAS
TAXIWAYS, APRONS, RUNWAY ENDS

FIG. D-FLEXIBLE PAVEMENTS—SURFACE AND BASE COURSE THICKNESS

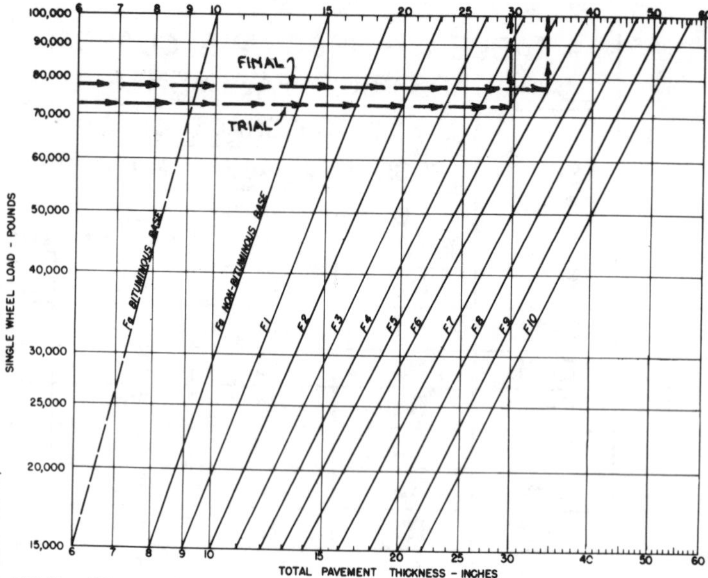
FINAL
TRIAL
SINGLE WHEEL LOAD - POUNDS
Fa BITUMINOUS BASE
Fa NON-BITUMINOUS BASE
F1
F2
F3
F4
F5
F6
F7
F8
F9
F10
TOTAL PAVEMENT THICKNESS - INCHES

FIG. B— DESIGN CURVES FOR TAXIWAYS, APRONS, AND RUNWAY ENDS

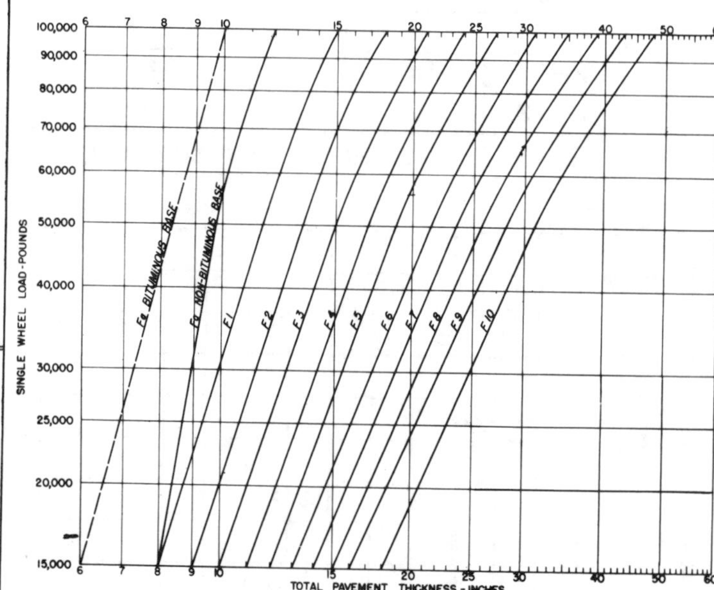
SINGLE WHEEL LOAD-POUNDS
Fa BITUMINOUS BASE
Fa NON-BITUMINOUS BASE
F1
F2
F3
F4
F5
F6
F7
F8
F9
F10
TOTAL PAVEMENT THICKNESS - INCHES

FIG. C—DESIGN CURVES FOR NONCRITICAL RUNWAY AREAS

Given: Aircraft with dual-tandem gear load of 161,000 lb. E-7 subgrade, poor drainage conditions, average depth of annual frost penetration of 36 inches.

Find: The pavement requirements for taxiways, aprons, and runway ends. A nonbituminous vase is to be used.

Solution:

1. Make trial design, assuming "no frost" and above drainage condition to determine if severe frost condition exists, as defined on page 10-50.
   (*a*) For subgrade conditions outlined above, from Table A, subgrade class is F-5.
   (*b*) The equivalent single-wheel load for this aircraft and F-5 subgrade class is 72,500 lb. (See page 10-52.)
   (*c*) From Fig. B, the total thickness required pavement is 30 inches.
   (*d*) Since the required pavement thickness of 30 inches is less than the depth of frost penetration, a severe frost condition exists.
2. Final design based on severe frost and poor drainage conditions:
   (*a*) For subgrade conditions outlined above, from Table A, subgrade class is F-6.
   (*b*) From Fig. B, page 10-52, the equivalent single-wheel load for this aircraft and F-6 subgrade class is 78,500 lb.
   (*c*) From Figs. B and D, the profile of the required pavement is:
   Surface .... 3"
   Non-bituminous base course .... 11"
   Subbase course .... 20"

NOTE:
1. *Fa* curves represent combined thickness of surface and base courses.
2. No subbase required on *Fa* subgrades.
3. Horizontal increment, at design wheel load, between a particular subgrade curve and *Fa* curve represents the subbase thickness required.

AIRFIELD PAVEMENTS - PAVEMENT LAYOUT & SECTIONS - CAA

TAXIWAY FILLET RADII
ANGLES UP TO 85° = 25'R ANGLES OF 85° TO 115° = 50'R ANGLES GREATER THAN 115° = 100'R

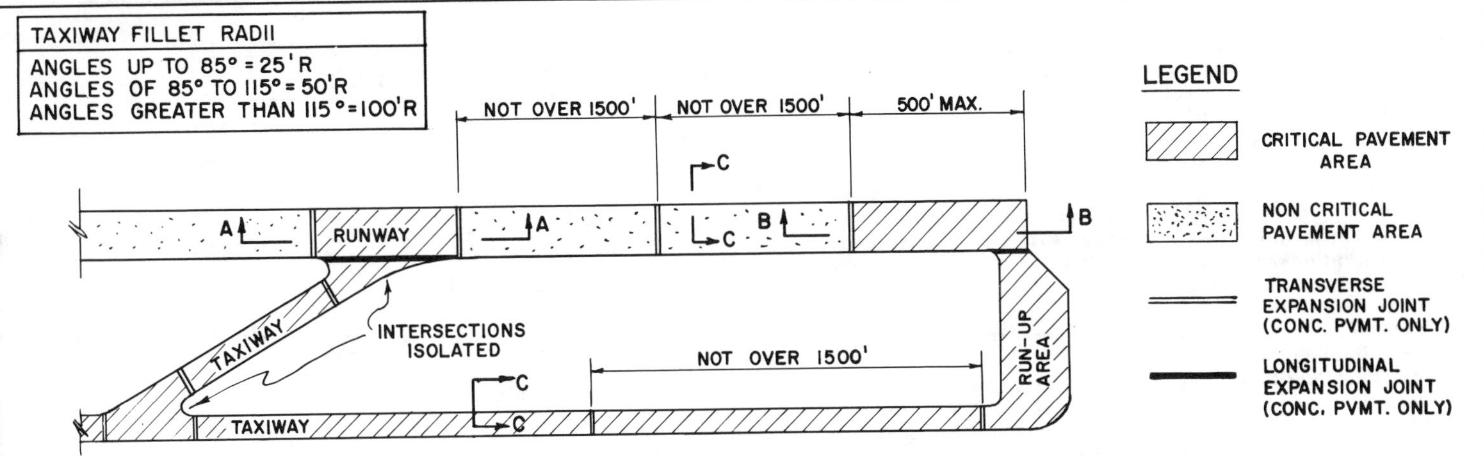

NOT OVER 1500'
NOT OVER 1500'
500' MAX.
RUNWAY
A
A
B
B
C
C
C
C
INTERSECTIONS ISOLATED
TAXIWAY
TAXIWAY
NOT OVER 1500'
RUN-UP AREA
LEGEND
CRITICAL PAVEMENT AREA
NON CRITICAL PAVEMENT AREA
TRANSVERSE EXPANSION JOINT (CONC. PVMT. ONLY)
LONGITUDINAL EXPANSION JOINT (CONC. PVMT. ONLY)

TYPICAL AIRFIELD LAYOUT SHOWING AREAS OF CRITICAL AND NONCRITICAL PAVEMENT

CONCRETE

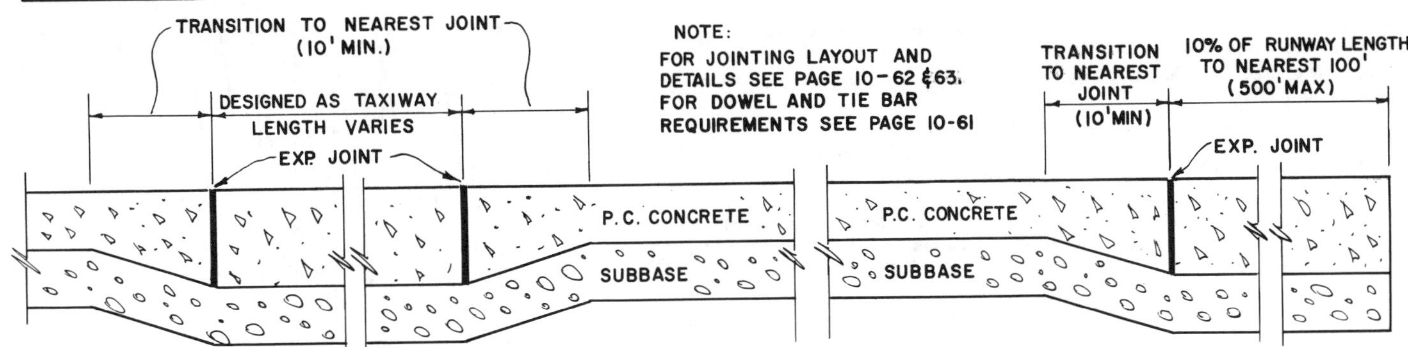

TRANSITION TO NEAREST JOINT (10' MIN.)
DESIGNED AS TAXIWAY
LENGTH VARIES
EXP. JOINT
P.C. CONCRETE
SUBBASE
NOTE:
FOR JOINTING LAYOUT AND DETAILS SEE PAGE 10-62 & 63.
FOR DOWEL AND TIE BAR REQUIREMENTS SEE PAGE 10-61
TRANSITION TO NEAREST JOINT (10' MIN)
10% OF RUNWAY LENGTH TO NEAREST 100' (500' MAX)
EXP. JOINT
P.C. CONCRETE
SUBBASE

SECTION A-A

TAXIWAY INTERSECTION SHOWING TRANSITION TO NON-CRITICAL PAVEMENT SECTION

SECTION B-B

RUNWAY END SHOWING TRANSITION TO THICKENED PAVEMENT SECTION

FLEXIBLE

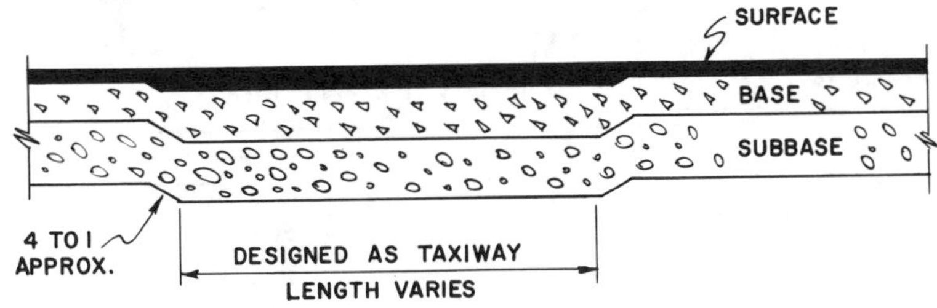

SURFACE
BASE
SUBBASE
4 TO 1 APPROX.
DESIGNED AS TAXIWAY
LENGTH VARIES

SECTION A-A

TAXIWAY INTERSECTION SHOWING TRANSITION TO NON-CRITICAL PAVEMENT SECTION

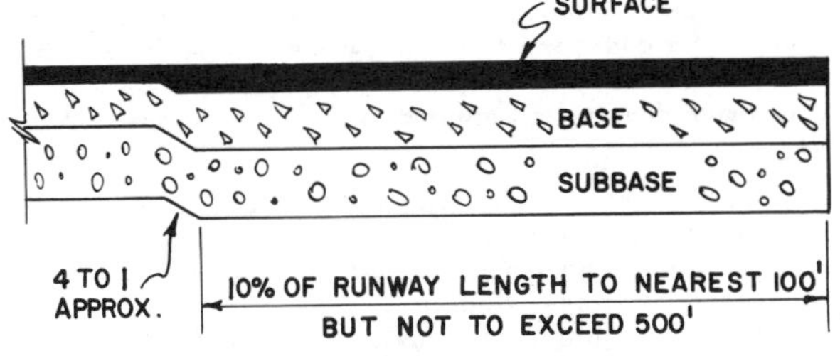

SURFACE
BASE
SUBBASE
4 TO 1 APPROX.
10% OF RUNWAY LENGTH TO NEAREST 100'
BUT NOT TO EXCEED 500'

SECTION B-B

RUNWAY END SHOWING TRANSITION TO THICKENED PAVEMENT SECTION

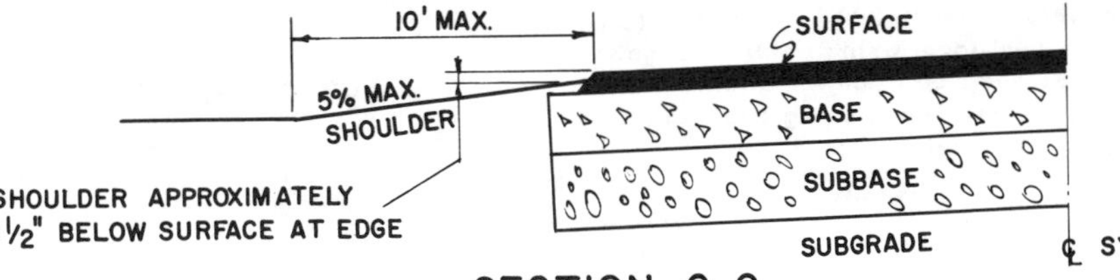

10' MAX.
SURFACE
5% MAX.
SHOULDER
BASE
SUBBASE
SHOULDER APPROXIMATELY 1½" BELOW SURFACE AT EDGE
SUBGRADE
℄ SYM. ABOUT ℄

SECTION C-C

TYPICAL CROSS-SECTION SHOWING DETAIL AT EDGE OF PAVEMENT

ADAPTED FROM CAA "AIRPORT PAVING" MANUAL DATED OCTOBER 1956

AIRFIELD PAVEMENTS — SECONDARY AIRPORTS - DESIGN - CAA

Secondary Airports provide landing facilities to accommodate personal aircraft and other aircraft normally engaged in non-scheduled flying activities. Secondary airports seldom will be required to handle aircraft of gross weights exceeding 15,000 lb.

Pavement generally will consist of:

1. Thin bituminous surface placed on prepared base course, or
2. Rigid surface of minimum allowable thickness (6 in. concrete). The design data on this page deal with flexible pavements only. No special design criteria for secondary-airport rigid pavement, since the 6 in. minimum-thickness pavement will serve all aircraft which normally operate from secondary airports.

Surface Courses:

1. Bituminous surface treatment for wheel-load designs to 5000 lb. C.A.A. specification P-609.
2. Keystone mat C.A.A. specification P-405 with bituminous surface treatment for wheel load design in excess of 5000 lb.

NOTE: When economically feasible, a more durable dense-graded plant-mix bituminous concrete should be used in lieu of that noted above.

Base and Subbase Courses: In addition to the base and subbase specified on page 10-50, satisfactory base courses can be provided for secondary airports by soil stabilization. Compaction control for secondary airports is based on the C.A.A. compaction control test T-611.

C.A.A. Recommended Methods of Soil Stabilization Soil stabilization is that procedure whereby the properties of a soil are improved to the extent that it will meet requirements for pavement bases or subbases. Stabilized soils are not intended to serve as surface courses.

1. Mechanical Stabilization – the mixing of aggregate and soil to produce interlocking of aggregate and soil particles and other desirable physical properties. See C.A.A. specifications P-208 and P-213 for gradation requirements.
2. Stabilization of improperly graded soils by means of binders such as Portland Cement, bituminous material, and lime, to produce a material that will provide adequate load support and will not soften in the presence of water.
   (a) *Bituminous Stabilization* This should be restricted to soils of a granular nature. Most commonly used grades of bituminous binders are RC-1, RC-4, MC-1, MC-4, SC-1, SC-4, RT-3, RT-7, and slow-curing emulsified asphalt. (See C.A.A. specification P-216.)
   (b) *Soil Cement Stabilization* All types of soils and materials such as shale, gravel, screenings, slag, and mine tailings can be stabilized by the addition of Portland cement in the correct quantity. The minimum allowable thickness shall be 6 in. See C.A.A. specification P-301.
   (c) *Lime-Soil Stabilization* Lime in small percentages (2 or 3%) will accomplish a marked improvement in stability of gravel, disintegrated granite, crusher-run stone and caliche for base-course use. Lime-stabilized plastic soils should be treated with hydrated lime in amounts ranging from 2 to 10%, recommended for subbase only. See C.A.A. specification P-301.

Aggregate Turf landing strip is that in which stability of soil has been increased by the addition of granular materials before establishment of turf. Economy design to serve 10,000 lb. gross weight aircraft. Compact – 70 to 90% maximum density to permit growth of grass. See C.A.A. specification P-217.

EXAMPLE: Given: 7000 lb. single-wheel load, E-8 subgrade, 6 inches frost penetration, good drainage.

Required: Flexible-pavement non-bituminous base for main landing strip, aggregate turf for cross-wind strip.

Solution: 1. Aggregate Turf: Provide 6 in. soil-stabilized aggregate. Gradation per specification P-217. Compact 70 to 90% maximum density. Loosen 1 1/2 in. of stabilized material and seed.

2. Flexible Pavement: From page 10-54, E8 soil = F4. From Fig. B stabilize 2 in. of subbase with lime. Stabilize 6 in. base with gravel. Provide 1 in. dense-graded plant mix.

NOTE: When designing for single wheels above 10,000 lb., design should be based on 15,000 lb. single-wheel load, using chart on page 10-54. No reduction in thickness shall be made for intermediate areas of runways on secondary airports when using Fig. B on this page. Adapted from C.A.A., *Airport Paving*, Oct. 1956.

TABLE A

Soil Group	Aggregate Turf Thickness, inches
E1 to E5	4 to 6
E6 to E7	6 minimum
E8 to E12	6 to 10*
E13	Not suitable

*Thickness dependent on climatic and drainage conditions.

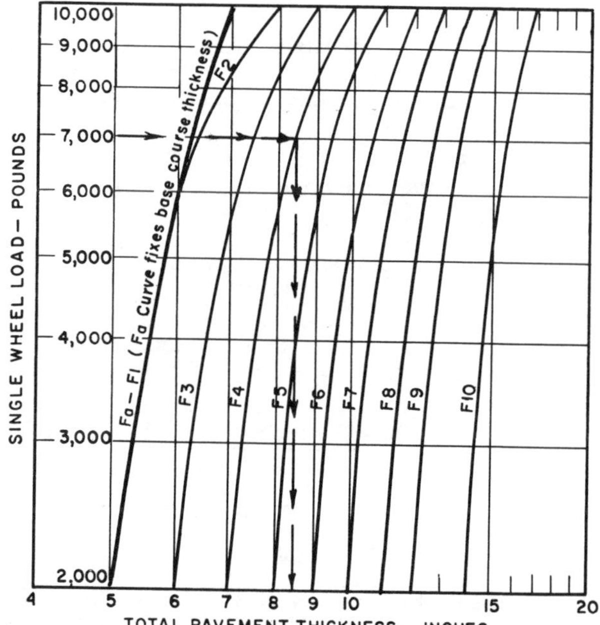

SINGLE WHEEL LOAD – POUNDS
10,000
9,000
8,000
7,000
6,000
5,000
4,000
3,000
2,000
Fa – F1 (Fa Curve fixes base course thickness)
F2
F3
F4
F5
F6
F7
F8
F9
F10
4
5
6
7
8
9
10
15
20
TOTAL PAVEMENT THICKNESS — INCHES

Figure B – Design Curves for Flexible Pavements – Secondary Airports

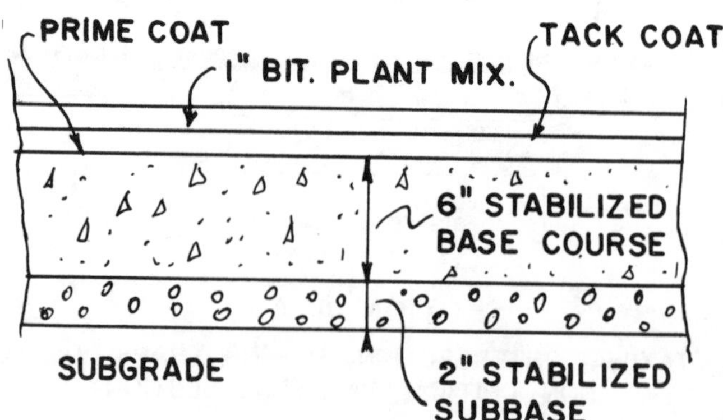

PRIME COAT
TACK COAT
1" BIT. PLANT MIX.
6" STABILIZED BASE COURSE
SUBGRADE
2" STABILIZED SUBBASE

FLEXIBLE PAVEMENT SECTION

AIRFIELD PAVEMENTS — RIGID PAVEMENT DESIGN — CORPS OF ENGINEERS

CORPS OF ENGINEERS DESIGN CHARTS given below are based on the theoretical analyses of Westergaard supplemented by empirical modifications based on accelerator traffic tests and observation of pavement behavior under service conditions.

MODULUS OF SUBGRADE REACTION The supporting value of the pavement subgrade is determined by the "Field Load-Bearing Test." This test consists of subjecting the subgrade to known pressures (up to 30 p.s.i.) at a predetermined rate, by means of a hydraulic jack working against a jacking frame and through a rigid 30-inch-diameter bearing plate. From the test data a load-deformation curve is plotted, and the uncorrected value of the "Modulus of Subgrade Reaction *(K)*" is determined by the formula Ku = 10 p.s.i. per average deflection under 10 p.s.i. load. Corrections must be made to Ku for bending of the plate, moisture, density, and type of material, to determine the design value of K that will be representative of ultimate field conditions. Approximate values of K are given on page 10-02.

WHEEL OR GEAR LOADS AND CONFIGURATIONS See pages 15-12 and 15-13.

CONCRETE FLEXURAL STRENGTH (Modulus of Rupture) Corps of Engineer design is based on using the 90-day flexural strength of concrete.

TRAFFIC AREAS For definition of and typical layout of traffic areas see page 10-60.

PAVEMENT THICKNESS Required pavement thickness t_B for type-B traffic areas is taken directly from design curves given below. Required pavement thickness t_A for type-A traffic areas is obtained from formula $t_A = 1.125t_B$. Required pavement thickness t_C for type-C traffic areas is obtained from formula $t_C = 0.9t_B$.

THICKENED EDGES are provided at longitudinal (undoweled) expansion joints, at juncture between two pavement features, at free edges in approaches to buildings and around pavement level structures, and long free edges of pavements less than 8 inches in thickness where pavement extension is feasible. Edge thickness (te) is generally designed as 1.25 of the thickness of main portion of pavement.

NOTE: When final pavement thicknesses indicate a fractional value greater than ¼ inch, the next full-inch thickness will be used.

PAVEMENT DETAILS See pages 10-62 and 10-63. FROST PROTECTION REQUIREMENTS See Soil Section.

DESIGN EXAMPLE: Given: Subgrade modulus (K) = 300. Twin wheels spaced 37½ inches center to center, 125,000-lb. gear load. Concrete flexural strength = 610 p.s.i. at 90 days.
Required: Pavement thickness for airfield runway, taxiway, and apron.
Solution: From Fig. C, by method illustrated by arrows on design chart, obtain required pavement thickness t_B of 16.7 inches. Use 17 inches.
From formulas noted above, $t_A = 1.125(16.7) = 18.8$ inches. Use 19 inches. $t_C = 0.9(16.7) = 15$ inches.

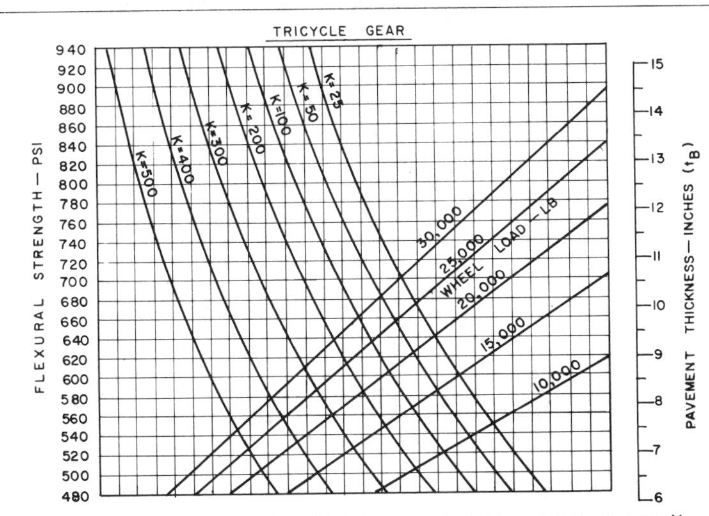
TRICYCLE GEAR
FLEXURAL STRENGTH — PSI
PAVEMENT THICKNESS — INCHES (t_B)
K=500
K=400
K=300
K=200
K=100
K=50
K=25
30,000
25,000
WHEEL LOAD — LB
20,000
15,000
10,000

FIG. A—SINGLE WHEEL—100 PSI INFLATION PRESSURE.*

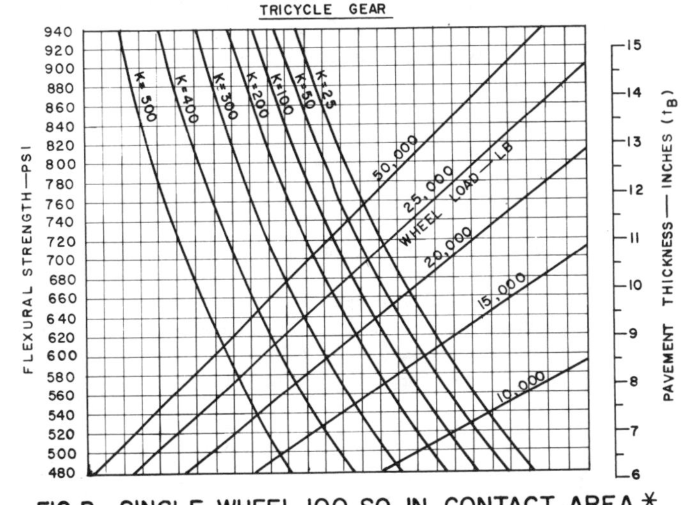
TRICYCLE GEAR
FLEXURAL STRENGTH —PSI
PAVEMENT THICKNESS — INCHES (t_B)
K=500
K=400
K=300
K=200
K=100
K=50
K=25
50,000
25,000
WHEEL LOAD — LB
20,000
15,000
10,000

FIG. B—SINGLE WHEEL 100 SQ. IN. CONTACT AREA.*

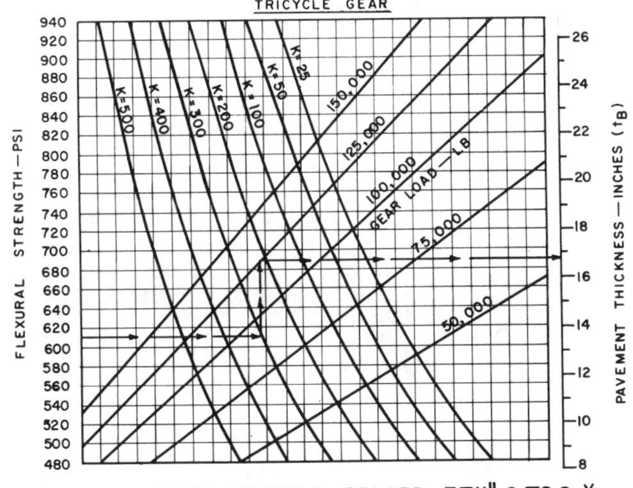
TRICYCLE GEAR
FLEXURAL STRENGTH —PSI
PAVEMENT THICKNESS — INCHES (t_B)
K=500
K=400
K=300
K=200
K=100
K=50
K=25
150,000
125,000
100,000
GEAR LOAD — LB
75,000
50,000

FIG. C—TWIN WHEELS SPACED 37½" C. TO C.*
(267 SQ. IN. CONTACT AREA EACH WHEEL)

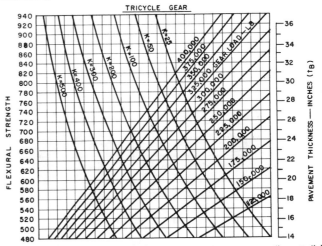

FIG. D—TWIN—TANDEM WHEELS SPACED 31¼"x 62¾".*
(267 SQ IN. CONTACT AREA EACH WHEEL)

CONCRETE PAVEMENT DESIGN CURVES.

* Adapted from Corps of Engineers Rigid Airfield Pavement Manual EM 1110-45-303 dated 3 Feb. '58.

AIRFIELD PAVEMENTS — FLEXIBLE PAVEMENT DESIGN — CORPS OF ENGINEERS

The Corps of Engineer flexible-pavement design method is adopted from the empirical method used by the California Division of Highways known as the C.B.R. method. The required thicknesses of subbase, base, and bituminous pavement can be obtained from the design charts when the California Bearing Ratio (C.B.R.) of the materials is known.

DETERMINATION OF C.B.R. consists of testing a compacted, 4-day-soaked sample of soil at specified construction density in a 6 in. diameter cylinder with a 3 sq.in.-area circular piston. Enough load is applied to penetrate the sample at a rate of 0.05 in. per minute to a total penetration of 0.50 in.; the load is measured and stated in ratio to that supported by crushed stone under the same conditions. The ratio obtained is the C.B.R. Approximate values of C.B.R. are given on p. 10-02.

WHEEL OR GEAR LOADS AND CONFIGURATIONS See pages 15-12 and 15-13.

TRAFFIC AREAS For definition and typical layout of traffic areas see page 10-60.

PROPORTIONING COMBINED THICKNESS INTO BITUMINOUS PAVEMENT, BASE, AND SUBBASE See figures on page 10-59.

The total pavement thickness required above a given layer of material is obtained by entering the design charts, as illustrated below, with the design C.B.R. value of the material.

The required pavement thickness t_B for type-B traffic areas is taken directly from Figs. A, B, C, and E.

The required pavement thickness t_A for type-A traffic areas is taken directly from Figs. D and F.

The required pavement thickness t_C for type-C traffic areas is obtained from the formula $t_c = 0.9 t_B$.

FROST PROTECTION REQUIREMENTS See Soil section.

COMPACTION REQUIREMENTS

Base Courses – Maximum that can be obtained, generally in excess of 100% of modified A.A.S.H.O. maximum and never less than 100%.

Subbase Courses – 100% of modified A.A.S.H.O. maximum, except where it is known that a higher density can be obtained practically, in which case the higher density should be required.

Select Materials and Subgrade – See Table D.

DESIGN EXAMPLE

Given: Twin assembly, tricycle gear, 100,000-lb. assembly load, spacing 37 inches, contact area 267 sq. in. each wheel.

C.B.R. 8 subgrade, C.B.R. 17 select material, C.B.R. 50 subbase, C.B.R. 100 base.

Required: Pavement thickness for airfield runway, taxiway, and apron.

Solution: From Figs. C and D, by method illustrated by arrows on design charts, obtain required thicknesses for subbase, and select material for type-B and A traffic areas. Required thickness of subbase and select material for type C traffic area are obtained from formula given above. Required minimum thickness of pavement and base is obtained from design charts.

Layer	Traffic Area A	Traffic Area B	Traffic Area C
Prime Coat			
Tack Coat			
Surface Course	1½"	1½"	1½"
Binder Course	2½"	1½"	1½"
Base Course C.B.R. 100	6"	6"	6"
Subbase Course C.B.R. 50	13"	12"	10"
Select Material C.B.R. 17	14"	12"	11"
Subgrade C.B.R. 8			

REQUIRED PAVEMENT SECTIONS FOR DESIGN EXAMPLE

TABLE A PROPORTIONING BITUMINOUS PAVEMENT INTO BINDER AND SURFACE COURSES

Pavement Thickness, in.	Binder-Course Thickness, in.	Surface-Course Thickness, in.
2	–	2
3	1½	1½
4	2½	1½
5	2, 1½	1½
6	2½, 2	1½

TABLE B TYPES OF BASE-COURSE MATERIALS (C. OF E.)

The Corps of Engineers have prepared guide specifications for the six types of base courses shown below. Numbers 1 through 5 may be used without limitation. Number 6 is limited to other than heavily and intensely loaded areas. It may not be used in type-A traffic areas designed for any loading or types-B and C traffic areas for heavy-load designs for bicycle-type landing-gear aircraft.

No.	Type	Design C.B.R.
1	Graded crushed aggregate	100
2	Water-found macadam	100
3	Dry-bound macadam	100
4	Bituminous base course, central plant, hot mix	100
5	Lime rock	80
6	Stabilized aggregate	80

TABLE C SELECTION OF DESIGN C.B.R. FOR SELECT MATERIAL AND SUBBASES (C. OF E.)

The Corps of Engineers arbitrarily classify materials having a design C.B.R. value below 20 as "select materials" and those with C.B.R. values of 20 or above as "subbases." Minimum thicknesses of pavement and base given on design charts eliminate need for subbase with design C.B.R. values above 50. In addition to requirements shown below, the materials must also have laboratory C.B.R. values equal to or higher than the design C.B.R.

Material	Maximum Design C.B.R.	Maximum Permissible Value				
			Gradation Requirements, % Passing		Atterberg Limits	
		Size, in.	No. 10	No. 200	LL	PI
Subbase	50	3*	50	15	25	5
Subbase	40	3	80	15	25	5
Subbase	30	3	100	15	25	5
Select Material	20	3	–	100*	35*	12*

*Suggested limits.

TABLE D COMPACTION REQUIREMENTS FOR SELECT MATERIALS AND SUBGRADE

Studies made by C. of E. indicate that aircraft traffic endures a high-compaction effort in the layers of the pavement structure. The compaction effort is higher near the surface, decreasing with depth, and varies with gear load, tire pressure, and repetitions of load. Because of the wide variations in the compaction requirements for select materials and subgrade, the requirements for only two aircraft are shown below. For requirements for other aircraft, see Corps of Engineers Manual EM 1110-45-302.

Required % Modified A.A.S.H.O. Compaction	Twin-Tandem Tricycle, Spacing 31-63 in. Contact Area 267 sq. in. Gear Load 170,000 lb.						Single-Wheel Contact Area 100 sq. in. Gear Load 30,000 lb.			
	Cohesive Material Depth, feet			Cohesionless Material Depth, feet			Cohesive Material Depth, feet		Cohesionless Material Depth, feet	
	Traffic Areas						Traffic Areas			
	A	B	C	A	B	C	B	C	B	C
100	2	2	1.5	3.5	3	3	1	1	1.5	1.5
95	3.5	3	3	6.5	5.5	5	1.5	1.5	2.5	2.5
90	5	4.5	4	9	8.5	7.5	2	2	4	3.5
85	6.5	6	5.5	12	11	10	3	2.5	4.5	4
80	8	7.5	7	–	–	–	3.5	3	–	–

Fill: As shown above except that in no case will cohesionless fill be placed at less than 95% nor cohesive fill at less than 90%.

Cut: Subgrade in cut must have natural densities equal to or greater than the values listed above.

Adapted from Flexible Airfield Pavement Manual EM 1110-45-302, dated Aug. 15, 1958

AIRFIELD PAVEMENTS — FLEXIBLE DESIGN CHARTS† — CORPS OF ENGINEERS

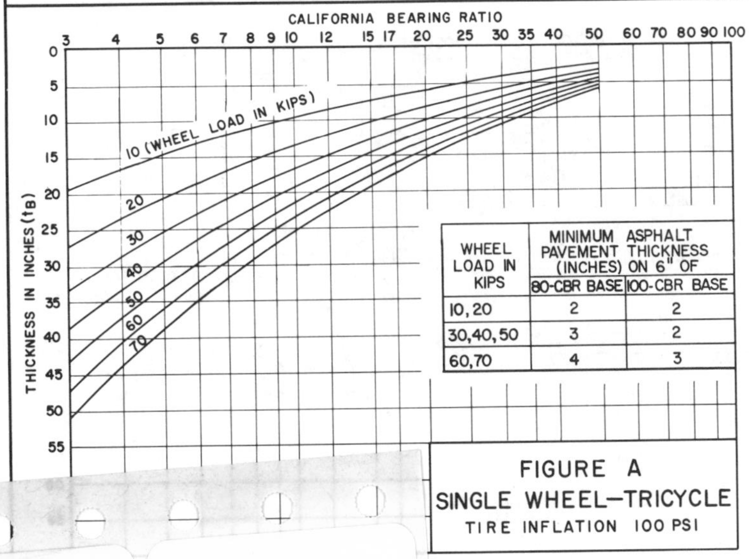

CALIFORNIA BEARING RATIO

THICKNESS IN INCHES (t_B)

10 (WHEEL LOAD IN KIPS), 20, 30, 40, 50, 60, 70

WHEEL LOAD IN KIPS	MINIMUM ASPHALT PAVEMENT THICKNESS (INCHES) ON 6" OF	
	80-CBR BASE	100-CBR BASE
10, 20	2	2
30, 40, 50	3	2
60, 70	4	3

FIGURE A
SINGLE WHEEL—TRICYCLE
TIRE INFLATION 100 PSI

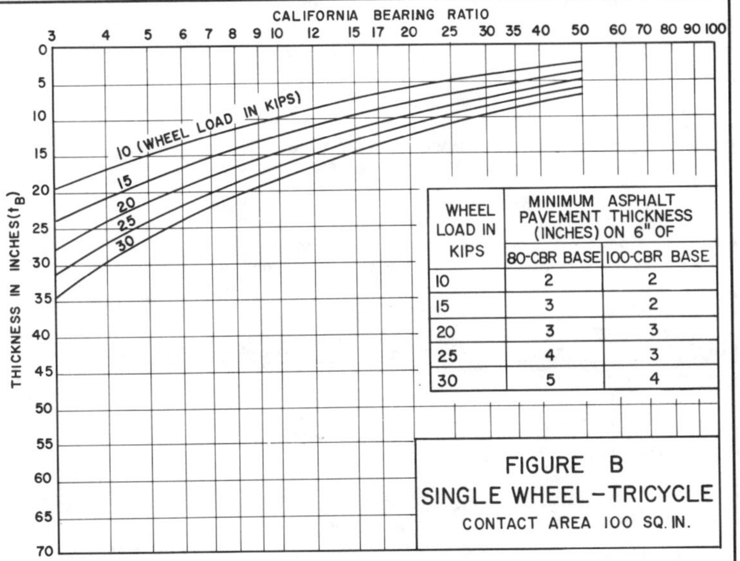

CALIFORNIA BEARING RATIO

THICKNESS IN INCHES (t_B)

10 (WHEEL LOAD IN KIPS), 15, 20, 25, 30

WHEEL LOAD IN KIPS	MINIMUM ASPHALT PAVEMENT THICKNESS (INCHES) ON 6" OF	
	80-CBR BASE	100-CBR BASE
10	2	2
15	3	2
20	3	3
25	4	3
30	5	4

FIGURE B
SINGLE WHEEL—TRICYCLE
CONTACT AREA 100 SQ. IN.

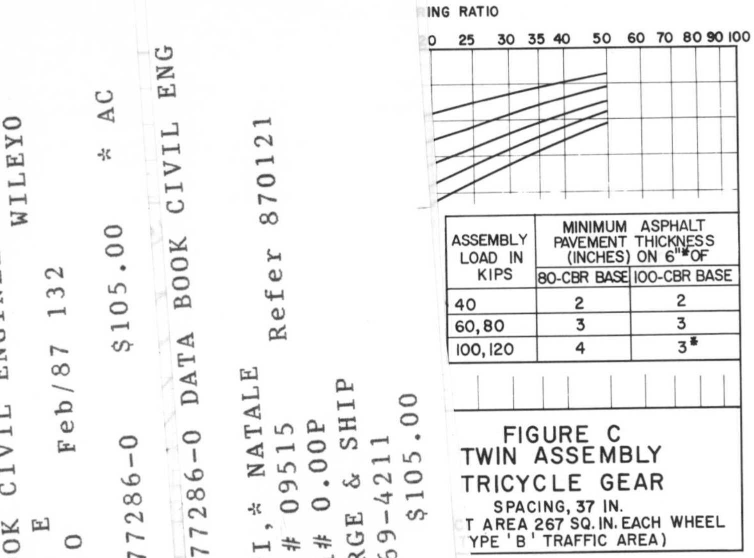

ING RATIO

ASSEMBLY LOAD IN KIPS	MINIMUM ASPHALT PAVEMENT THICKNESS (INCHES) ON 6"* OF	
	80-CBR BASE	100-CBR BASE
40	2	2
60, 80	3	3
100, 120	4	3*

FIGURE C
TWIN ASSEMBLY
TRICYCLE GEAR
SPACING, 37 IN.
T AREA 267 SQ. IN. EACH WHEEL
YPE 'B' TRAFFIC AREA)

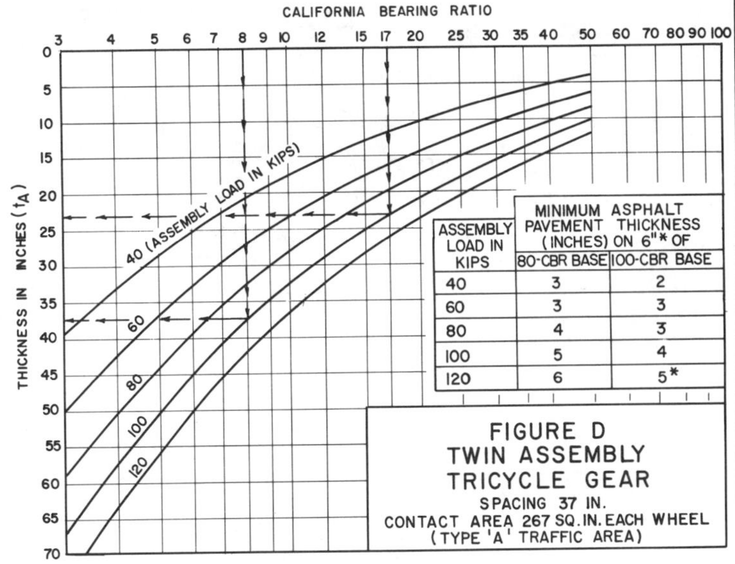

CALIFORNIA BEARING RATIO

THICKNESS IN INCHES (t_A)

40 (ASSEMBLY LOAD IN KIPS), 60, 80, 100, 120

ASSEMBLY LOAD IN KIPS	MINIMUM ASPHALT PAVEMENT THICKNESS (INCHES) ON 6"* OF	
	80-CBR BASE	100-CBR BASE
40	3	2
60	3	3
80	4	3
100	5	4
120	6	5*

FIGURE D
TWIN ASSEMBLY
TRICYCLE GEAR
SPACING 37 IN.
CONTACT AREA 267 SQ. IN. EACH WHEEL
(TYPE 'A' TRAFFIC AREA)

*7" MIN. 100 CBR BASE REQUIRED FOR 120 KIP ASSEMBLY ONLY

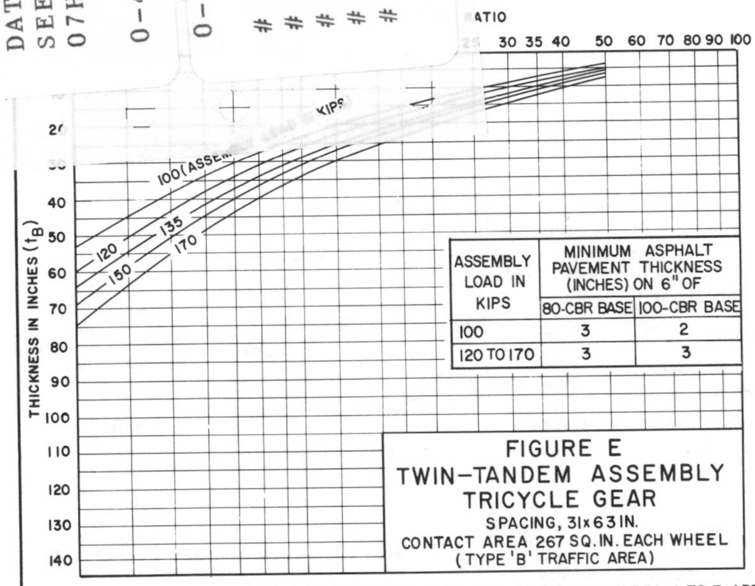

ATIO

THICKNESS IN INCHES (t_B)

100 (ASSEMBLY LOAD IN KIPS), 120, 135, 150, 170

ASSEMBLY LOAD IN KIPS	MINIMUM ASPHALT PAVEMENT THICKNESS (INCHES) ON 6" OF	
	80-CBR BASE	100-CBR BASE
100	3	2
120 TO 170	3	3

FIGURE E
TWIN-TANDEM ASSEMBLY
TRICYCLE GEAR
SPACING, 31 x 63 IN.
CONTACT AREA 267 SQ. IN. EACH WHEEL
(TYPE 'B' TRAFFIC AREA)

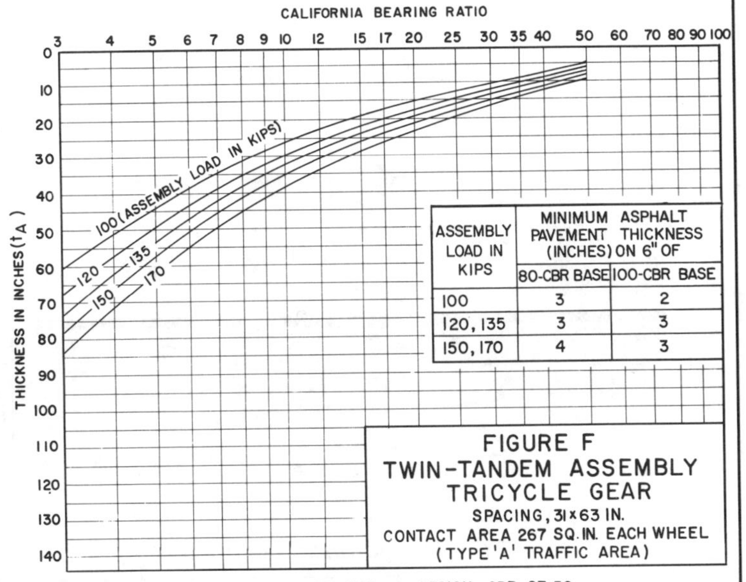

CALIFORNIA BEARING RATIO

THICKNESS IN INCHES (t_A)

100 (ASSEMBLY LOAD IN KIPS), 120, 135, 150, 170

ASSEMBLY LOAD IN KIPS	MINIMUM ASPHALT PAVEMENT THICKNESS (INCHES) ON 6" OF	
	80-CBR BASE	100-CBR BASE
100	3	2
120, 135	3	3
150, 170	4	3

FIGURE F
TWIN-TANDEM ASSEMBLY
TRICYCLE GEAR
SPACING, 31 x 63 IN.
CONTACT AREA 267 SQ. IN. EACH WHEEL
(TYPE 'A' TRAFFIC AREA)

NOTE:
1. MINIMUM THICKNESSES OF PAVEMENT AND BASE GIVEN IN FIGURES C TO F APPLY WHEN LAYER DIRECTLY UNDER BASE HAS A DESIGN CBR OF 50.
2. FLORIDA LIMEROCK OR STABILIZED AGGREGATE PERMITTED FOR 80 CBR BASE IN TYPE 'B' AND 'C' TRAFFIC AREAS.
3. 80 CBR BASE FOR TYPE 'A' TRAFFIC AREAS RESTRICTED TO FLORIDA LIMEROCK.

† Adapted from Corps of Engineers Flexible Airfield Pavement Manual EM 1110-45-302 dated 15 Aug. '58.

AIRFIELD PAVEMENTS – PAVEMENT LAYOUT AND SECTIONS§ – CORPS OF ENGINEERS

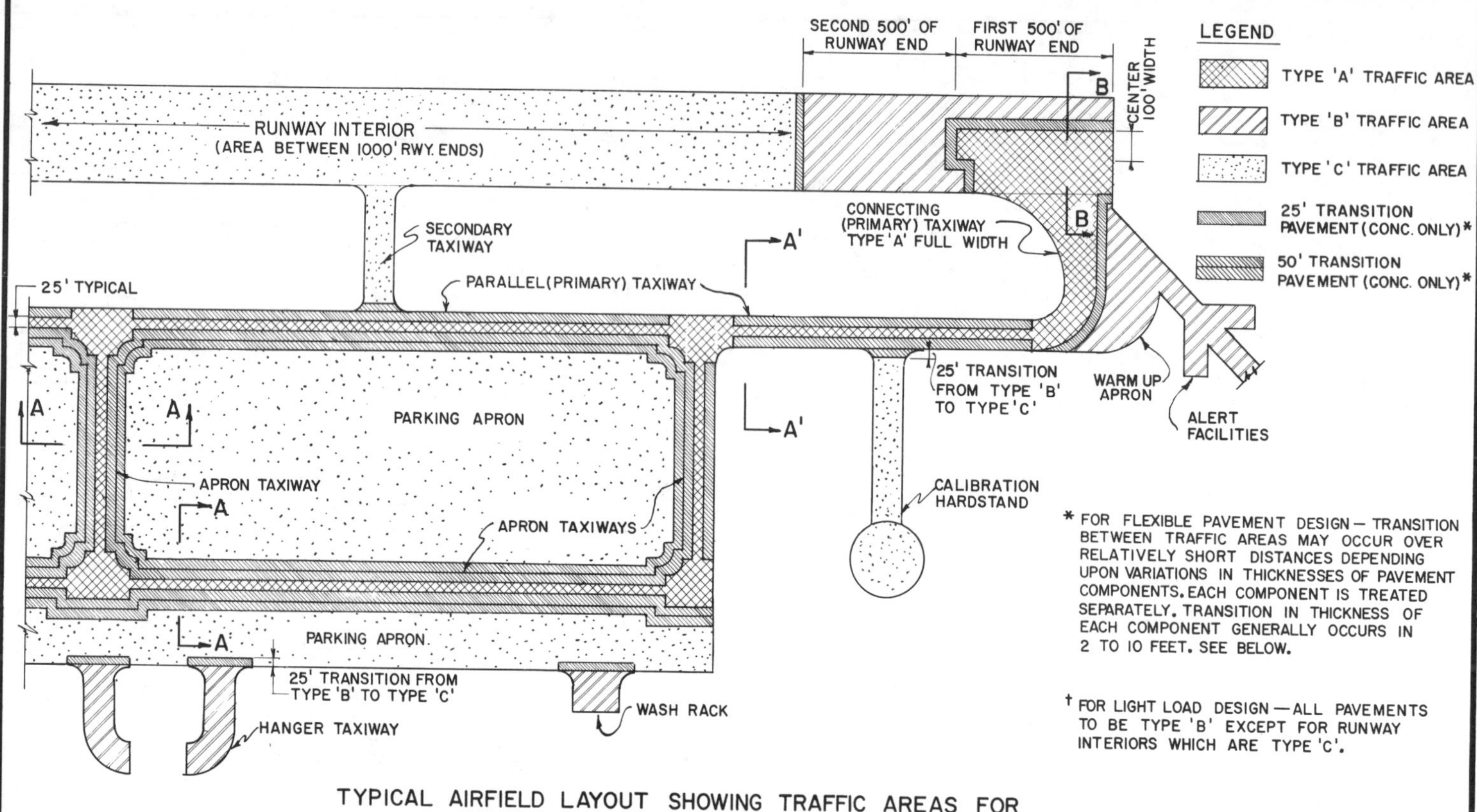

SECOND 500' OF RUNWAY END
FIRST 500' OF RUNWAY END
CENTER 100' WIDTH
B
RUNWAY INTERIOR
(AREA BETWEEN 1000' RWY. ENDS)
SECONDARY TAXIWAY
CONNECTING (PRIMARY) TAXIWAY TYPE 'A' FULL WIDTH
A'
25' TYPICAL
PARALLEL (PRIMARY) TAXIWAY
A
PARKING APRON
25' TRANSITION FROM TYPE 'B' TO TYPE 'C'
WARM UP APRON
ALERT FACILITIES
APRON TAXIWAY
APRON TAXIWAYS
CALIBRATION HARDSTAND
PARKING APRON
25' TRANSITION FROM TYPE 'B' TO TYPE 'C'
WASH RACK
HANGER TAXIWAY
LEGEND
TYPE 'A' TRAFFIC AREA
TYPE 'B' TRAFFIC AREA
TYPE 'C' TRAFFIC AREA
25' TRANSITION PAVEMENT (CONC. ONLY)*
50' TRANSITION PAVEMENT (CONC. ONLY)*

* FOR FLEXIBLE PAVEMENT DESIGN – TRANSITION BETWEEN TRAFFIC AREAS MAY OCCUR OVER RELATIVELY SHORT DISTANCES DEPENDING UPON VARIATIONS IN THICKNESSES OF PAVEMENT COMPONENTS. EACH COMPONENT IS TREATED SEPARATELY. TRANSITION IN THICKNESS OF EACH COMPONENT GENERALLY OCCURS IN 2 TO 10 FEET. SEE BELOW.

† FOR LIGHT LOAD DESIGN – ALL PAVEMENTS TO BE TYPE 'B' EXCEPT FOR RUNWAY INTERIORS WHICH ARE TYPE 'C'.

TYPICAL AIRFIELD LAYOUT SHOWING TRAFFIC AREAS FOR HEAVY LOAD† DESIGN PAVEMENTS
(NOT TO SCALE)

CONCRETE

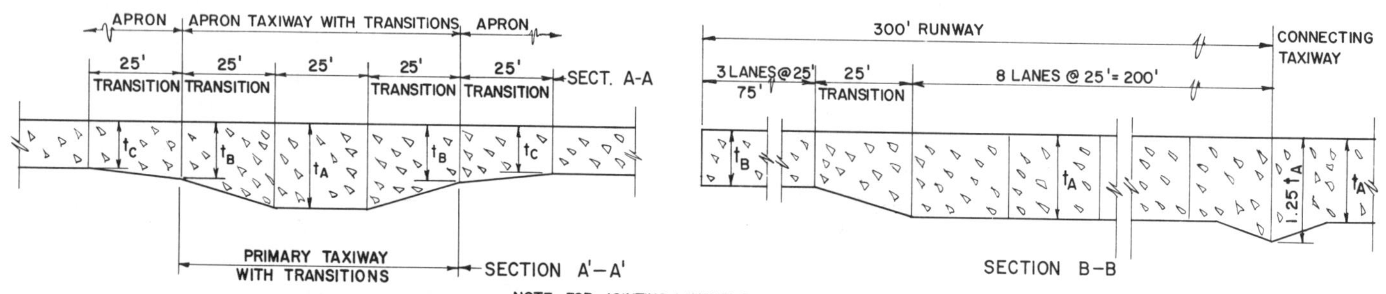

APRON
APRON TAXIWAY WITH TRANSITIONS
APRON
25' TRANSITION
25' TRANSITION
25'
25' TRANSITION
25' TRANSITION
SECT. A-A
t_C
t_B
t_A
t_B
t_C
PRIMARY TAXIWAY WITH TRANSITIONS
SECTION A'–A'
300' RUNWAY
CONNECTING TAXIWAY
3 LANES @ 25' 75'
25' TRANSITION
8 LANES @ 25' = 200'
t_B
t_A
$1.25\,t_A$
t_A
SECTION B–B

NOTE: FOR JOINTING LAYOUT & DETAILS SEE PAGES 10-62 & 10-63 FOR DOWEL AND TIE BAR REQUIREMENTS SEE PAGE 10-61

TYPICAL TAXIWAY SECTIONS

RUNWAY END WITH CONNECTING TAXIWAY
(FOR TYP. 300' WIDE RWY.)

FLEXIBLE

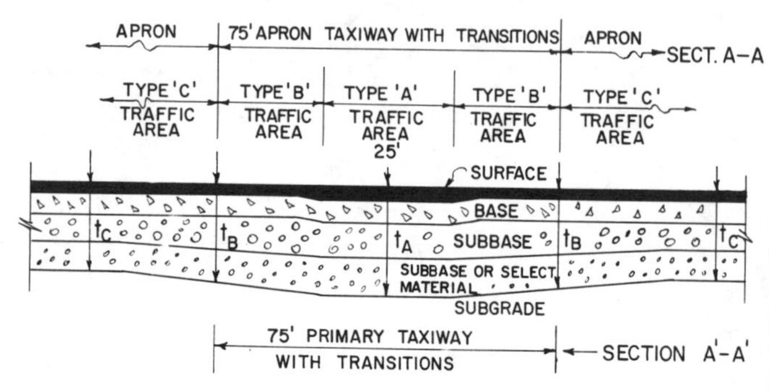

APRON
75' APRON TAXIWAY WITH TRANSITIONS
APRON
SECT. A–A
TYPE 'C' TRAFFIC AREA
TYPE 'B' TRAFFIC AREA
TYPE 'A' TRAFFIC AREA 25'
TYPE 'B' TRAFFIC AREA
TYPE 'C' TRAFFIC AREA
SURFACE
BASE
t_C
t_B
t_A
SUBBASE
t_B
t_C
SUBBASE OR SELECT MATERIAL
SUBGRADE
75' PRIMARY TAXIWAY WITH TRANSITIONS
SECTION A'-A'

TYPICAL TAXIWAY SECTIONS

§ Adapted from Corps. of Engineers Rigid Airfield Pavement Manual EM. 1110–45–303 dated 3. Feb. '58. and Flexible Airfield Pavement Manual EM. 1110–45–302 dated 15 Aug. '58.

DESIGN OF JUNCTION BETWEEN RIGID AND FLEXIBLE PAVEMENTS*

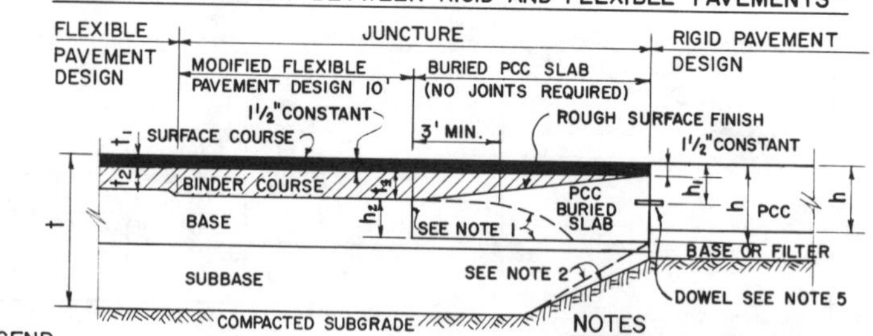

FLEXIBLE PAVEMENT DESIGN
JUNCTURE
RIGID PAVEMENT DESIGN
MODIFIED FLEXIBLE PAVEMENT DESIGN 10'
BURIED PCC SLAB (NO JOINTS REQUIRED)
1½" CONSTANT
3' MIN.
ROUGH SURFACE FINISH
1½" CONSTANT
SURFACE COURSE
BINDER COURSE
BASE
SEE NOTE 1
PCC BURIED SLAB
PCC
BASE OR FILTER
SUBBASE
SEE NOTE 2
DOWEL SEE NOTE 5
COMPACTED SUBGRADE

LEGEND

h = DESIGN THICKNESS OF PCC

$h_1 = \frac{h + 1\frac{1}{2}}{2}$

$h_2 = h - 4\frac{1}{2}$ BUT NOT LESS THAN 4" FOR LIGHT LOAD DESIGN AND NOT LESS THAN 6" FOR HEAVY LOAD DESIGN.

t = DESIGN THICKNESS OF FLEXIBLE PAVEMENT

t_1 = DESIGN THICKNESS OF A.C. SURFACE COURSE

t_2 = DESIGN THICKNESS OF A.C. BINDER COURSE

$t_3 = h - h_2$ BUT NOT LESS THAN t_2

NOTES

1. COMPACT FLEXIBLE PVMT. TO DOTTED LINE. CUT OUT TO SOLID LINE NOT DISTURBING THE MATERIALS OUTSIDE LIMITS OF BURIED SLAB.
2. EXCAVATE AND COMPACT SUBGRADE TO DOTTED LINE IF A BASE OR FILTER IS NOT USED BENEATH PCC. EXCAVATE AND COMPACT SUBGRADE TO SOLID LINE WHEN BASE OR FILTER COURSE IS USED BENEATH PCC.
3. PLACE PCC BURIED SLAB DIRECTLY AGAINST CUT BACK BASE COURSE. NO FORM WILL BE USED.
4. WHEN NEW RIGID PVMT. JOINS AN EXISTING FLEXIBLE PVMT., EXISTING PVMT. WILL BE CUT BACK TO REQUIRED DIMENSIONS OF "BURIED PCC SLAB ONLY. SECOND 10 FT. NOT APPLICABLE.
5. SEE PG. 10-61 FOR DOWEL SIZE AND SPACING, PAINT AND GREASE END INSTALLED IN RIGID PVMT. TO PERMIT REMOVAL DURING PREPARATION OF BASE AND SUBBASE.

*For critical traffic areas or areas where even slight deviations from design grade are objectionable. Specifically, all junctures in a transverse direction in runways, between runways and taxiways, and in Type 'A' traffic areas.

AIRFIELD PAVEMENTS —DOWELS AND TIEBARS—CAA— CORPS OF ENGINEERS

Corps of Engineers requires:

A. *Dowels* at the following locations:
1. Transverse expansion joints.
2. Transverse construction joints.
3. Longitudinal construction joints in pavement less than 8 inches in thickness except where the thickened-edge-type joint is used.
4. Longitudinal construction joints in type-A traffic areas except where the thickened-edge-type joint is used.
5. Longitudinal construction joints in rigid-overlay pavements.
6. Longitudinal dummy joints in rigid-overlay pavements in type-A traffic areas when the longitudinal dummy joint in the overlay coincides with a longitudinal expansion joint in the base pavement.
7. Last three transverse dummy joints back from the ends of all runways, and for the ends of other large paved areas where local experience indicates that they are needed.

B. *Tiebars* consisting of 5/8" ϕ deformed steel bars, 2'-6" in length and spaced 2'-6" apart only across longitudinal dummy groove joints next to the free edges of major paved areas.

TABLE A—ROUND DOWELS—C of E REQUIREMENTS

PAVEMENT THICKNESS	MINIMUM LENGTH	MAXIMUM SPACING	TYPE OF DOWEL*
LESS THAN 8"	16"	12"	3/4"ϕ BAR
8" TO 11"	16"	12"	1"ϕ BAR
12" TO 15"	20"	15"	1¼"ϕ BAR OR 1" EX. ST. PIPE†
16" TO 20"	20"	18"	1½"ϕ BAR OR 1½"ϕ EX. ST. PIPE†
21" TO 25"	24"	18"	2"ϕ BAR OR 2" EX. ST. PIPE†
OVER 25"	30"	18"	3"ϕ BAR OR 3" EX. ST. PIPE†

C.A.A. Requires:

A. *Dowels* at the following locations:
1. Transverse expansion joints.
2. Butt-type construction joints.
3. Across all transverse contraction joints in apron, taxiways, and thickened ends of runways.
4. The first two transverse contraction joints on each side of an expansion joint.

B. *Tiebars* consisting of 5/8" ϕ deformed steel bars, 2'-6" in length and spaced 2'-6" apart across longitudinal hinge-type groove joints and longitudinal keyed construction joints which occur within 25 ft. of the free edge of the pavement.

TABLE B—ROUND DOWELS—CAA REQUIREMENTS

PAVEMENT THICKNESS	LENGTH	SPACING	TYPE OF DOWEL*
6" TO 7"	16"	12"	3/4"ϕ BAR
8" TO 10"	16"	12"	1"ϕ BAR OR EX. ST. PIPE†
11" TO 15"	20"	15"	1¼"ϕ BAR OR EX. ST. PIPE†
16" TO 20"	24"	15"	1½"ϕ BAR OR EX. ST. PIPE†

---

*All dowels shall be straight, smooth, round bars free from burrs, painted and greased before installation.

†When extra-strength pipe is used for dowels, the pipe will be filled with a stiff mixture of sand-asphalt or cement mortar, or pipe will have plugged ends. If the ends of the pipe are plugged, the plugs must fit inside of the pipe and be cut off flush with the ends of the pipe, to permit free movement of dowel.

AIRFIELD PAVEMENTS — JOINT DETAILS — CAA — CORPS OF ENGINEERS

FIG. A EXPANSION JOINTS

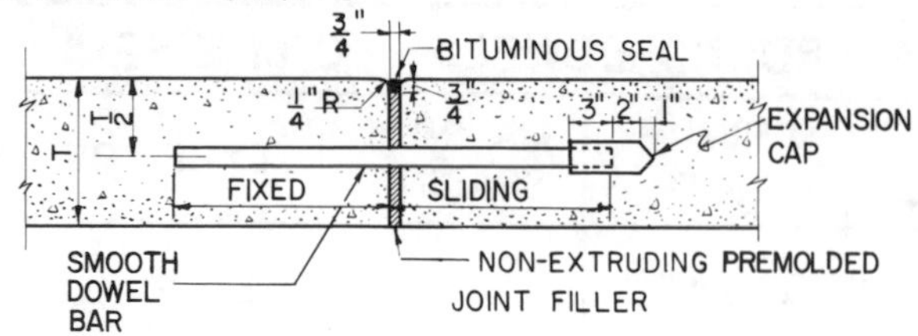

3"/4
BITUMINOUS SEAL
1/4"R
3"/4
3" 2" 1"
EXPANSION CAP
T/2
T
FIXED
SLIDING
SMOOTH DOWEL BAR
NON-EXTRUDING PREMOLDED JOINT FILLER

DOWELED TRANSVERSE EXPANSION JOINT

3"/4 BITUMINOUS SEAL
1/4"R
NON-EXTRUDING PREMOLDED JOINT FILLER
T
Te = 1.25 T
CAA — TO NEAREST JOINT NOT LESS THAN 10'
C OF E — 1/2 SLAB WIDTH MIN.
CAA RECOMMENDS Te TO NEAREST 1" BUT NOT LESS THAN T + 2".

LONGITUDINAL EXPANSION & THICKENED EDGE FREE JOINT

FIG. B CONSTRUCTION JOINTS

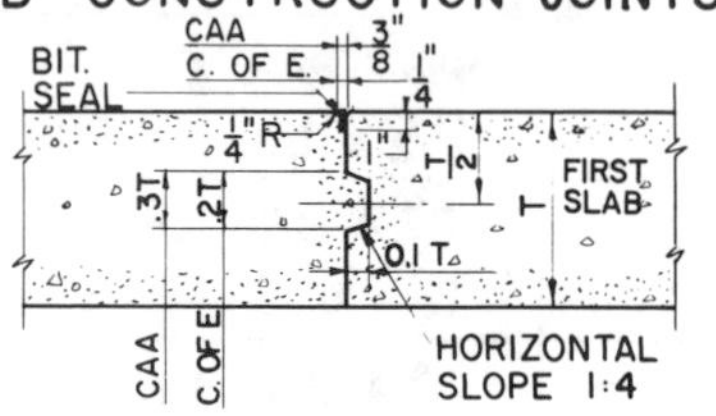

BIT. SEAL
CAA
C. OF E.
3"/8
1"/4
1/4"R
1"
T/2
FIRST SLAB
T
.3T
.2T
0.1 T
CAA
C. OF E.
HORIZONTAL SLOPE 1:4

LONGITUDINAL KEY TYPE JOINT [1]

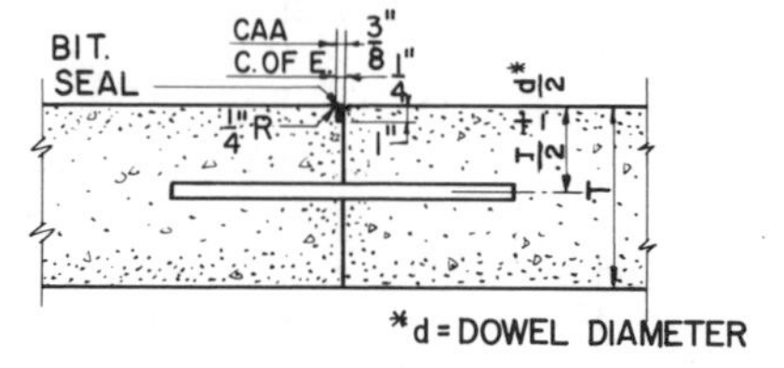

BIT. SEAL
CAA
C. OF E.
3"/8
1"/4
1/4"R
1"
T/2 + d*/2
T
*d = DOWEL DIAMETER

DOWELED LONGITUDINAL OR TRANSVERSE BUTT TYPE JOINT [2]

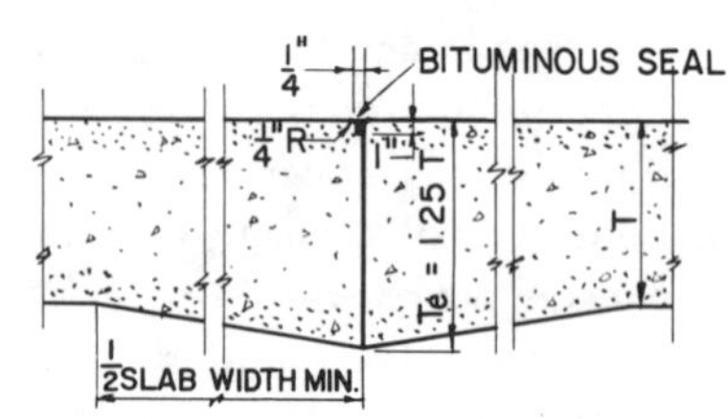

1"/4
BITUMINOUS SEAL
1/4"R
1"
T
Te = 1.25 T
1/2 SLAB WIDTH MIN.

LONGITUDINAL THICKENED EDGE BUTT TYPE JOINT

FIG. C DUMMY GROOVE JOINTS [3]

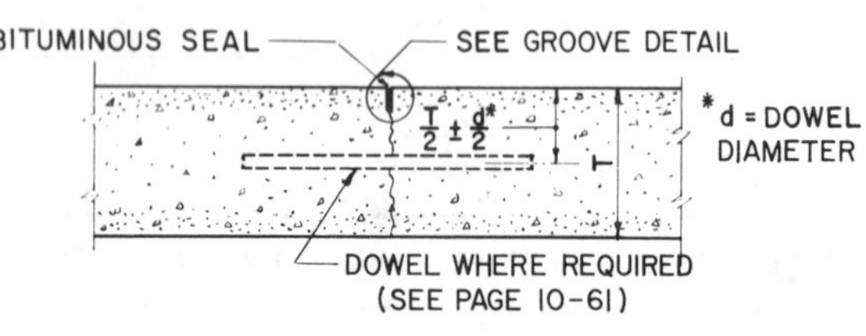

BITUMINOUS SEAL
SEE GROOVE DETAIL
T/2 ± d*/2
T
*d = DOWEL DIAMETER
DOWEL WHERE REQUIRED (SEE PAGE 10-61)

LONGITUDINAL OR TRANSVERSE CONTRACTION JOINT [4]

T/6 TO T/4
1/4" MIN.

SAWED GROOVE

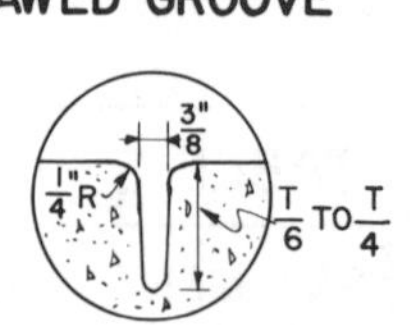

3"/8
1/4"R
T/6 TO T/4

FORMED GROOVE

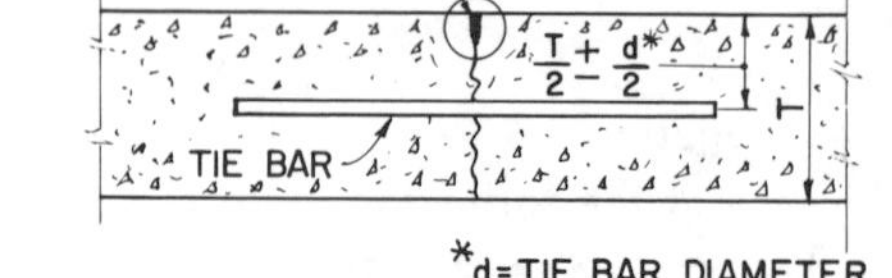

BITUMINOUS SEAL
SEE GROOVE DETAIL
T/2 − d*/2
T
TIE BAR
*d=TIE BAR DIAMETER

LONGITUDINAL HINGE TYPE JOINT

NOTE:
1. C. OF E. DOES NOT PERMIT KEYED JOINTS WHERE $T < 8$ INCHES.
2. TRANSVERSE CONSTRUCTION JOINTS ARE REQUIRED AT END OF DAYS PAVING OPERATION OR WHERE PLACING OF CONCRETE IS DISCONTINUED A SUFFICIENT TIME FOR CONCRETE TO SET. LOCATE AT REGULARLY SPACED TRANSVERSE JOINT IF PRACTICABLE OR MIDDLE THIRD OF SLAB.
3. C. OF E. REQUIRES DUMMY-JOINT GROOVE TO BE SAWED.
4. C. OF E. PERMITS INCREASE IN GROOVE WIDTH TO 3/8 OR 1/2 INCH AS NECESSARY, WHERE EXPERIENCE HAS SHOWN THAT 1/4 INCH WIDTH WILL NOT PROVIDE ENOUGH JOINT SEAL MATERIAL TO WITHSTAND LARGE MOVEMENTS AT THE JOINT.

JOINT SPACING — PLAIN CONCRETE

TRANSVERSE JOINTS

THE MAXIMUM SPACING BETWEEN JOINTS IS DEPENDENT UPON PAVEMENT THICKNESS, CLIMATIC CONDITIONS, SUBGRADE RESTRAINT, AND PROPERTIES OF THE AGGREGATE AND CONCRETE.
CAA RECOMMENDS THAT SPACING NORMALLY SHOULD NOT EXCEED 20 FEET. CORPS OF ENGINEERS JOINT SPACINGS GIVEN IN TABLE A, BELOW, IS TO BE USED AS A GUIDE SUBJECT TO MODIFICATION BASED ON AVAILABLE INFORMATION REGARDING LOCAL CONDITIONS.

TABLE A — CORPS OF ENGINEERS*

PAVEMENT THICKNESS	TRANSVERSE JOINT SPACING
LESS THAN 8 INCHES	12.5 TO 15 FEET
8 INCHES TO 10 INCHES	15 TO 20 FEET
MORE THAN 10 INCHES	20 TO 25 FEET

LONGITUDINAL JOINTS

TABLE B	CORPS OF ENGINEERS*	CAA
LONGITUDINAL CONSTRUCTION JOINT	GENERALLY SPACED 20 TO 25 FEET APART.	12 1/2 FT. MAXIMUM FOR PAVEMENTS 10 IN. OR LESS IN THICKNESS. 25 FT. MAXIMUM FOR PAVEMENTS GREATER THAN 10 IN. IN THICKNESS.
LONGITUDINAL DUMMY GROOVE JOINT	REQUIRED ALONG CENTERLINE OF PAVING LANES THAT HAVE A WIDTH GREATER THAN MAXIMUM ALLOWABLE TRANSVERSE JOINT SPACING.	REQUIRED ALONG CENTERLINE OF PAVING LANE FOR PAVEMENTS 10 INCHES OR LESS IN THICKNESS WHERE SLAB WIDTH EXCEEDS 12 1/2 FEET.

*It is desirable, insofar as practiable, to keep length-to-width dimension of slab as near equal as possible. In no case should transverse joint spacing be greater than 1.25 longitudinal joint spacing.

EXPANSION JOINTS

EXPANSION JOINTS ARE USED AT ALL INTERSECTIONS OF PAVEMENTS WITH STRUCTURES, AND WHERE REQUIRED WITHIN PAVEMENT FEATURES. LONGITUDINAL EXPANSION JOINTS AND EXPANSION JOINTS AROUND PAVEMENT LEVEL STRUCTURES WILL BE OF THE THICKENED-EDGE TYPE. TRANSVERSE EXPANSION JOINTS WILL BE DOWELLED TYPE EXCEPT WHERE PAVEMENT FEATURES DICTATE THICKENED-EDGE TYPE. FOR CAA JOINT LOCATION AND SPACING SEE PAGE 10-55. CORPS. OF ENGINEERS — RECOMMENDS THAT EXPANSION JOINTS BE OMITTED WITHIN PAVEMENTS 10 IN. OR MORE IN THICKNESS AND IN PAVEMENTS LESS THAN 10 IN. IN THICKNESS WHEN CONCRETE IS PLACED DURING WARM WEATHER. SEE PAGE 10-55.

Adapted from Corps of Engineers Flexible Airfield Pavement Manual EM 1110-45-302 dated 15 Aug. '58 and from CAA "Airport Paving" Manual dated Oct. '56.

AIRFIELD PAVEMENTS - RIGID - JOINT LAYOUT

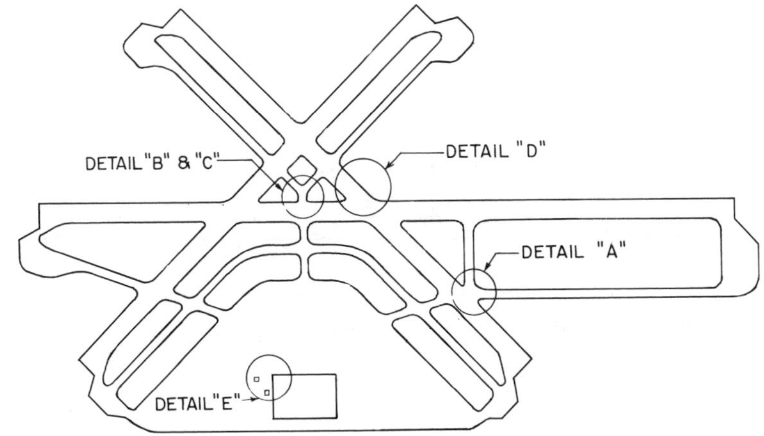
DETAIL "B" & "C"
DETAIL "D"
DETAIL "A"
DETAIL "E"

LEGEND

- LONGITUDINAL EXPANSION JOINT
- LONGITUDINAL CONSTRUCTION JOINT
- LONGITUDINAL DUMMY GROOVE JOINT
- TRANSVERSE EXPANSION JOINT (CAA ONLY)
- TRANSVERSE CONTRACTION JOINT

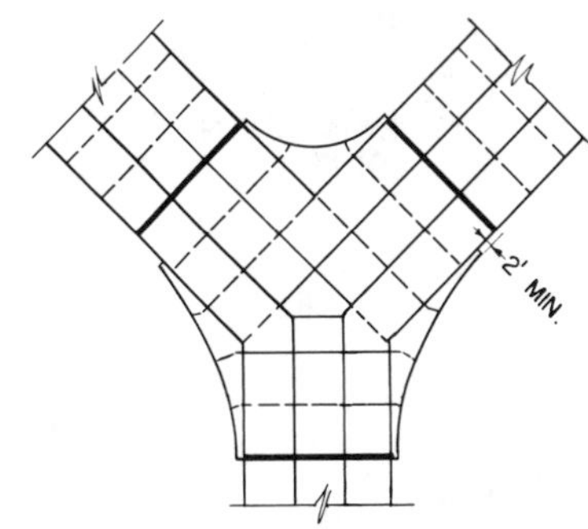
2' MIN.

DETAIL "B"—JOINTING PLAN—INTERSECTION OF THREE TAXIWAYS

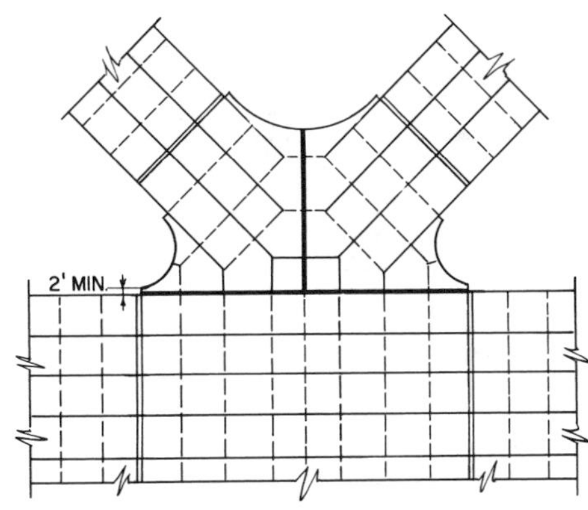
2' MIN.

DETAIL "A"— JOINTING PLAN— INTERSECTION OF TWO TAXYWAYS AND RUNWAY

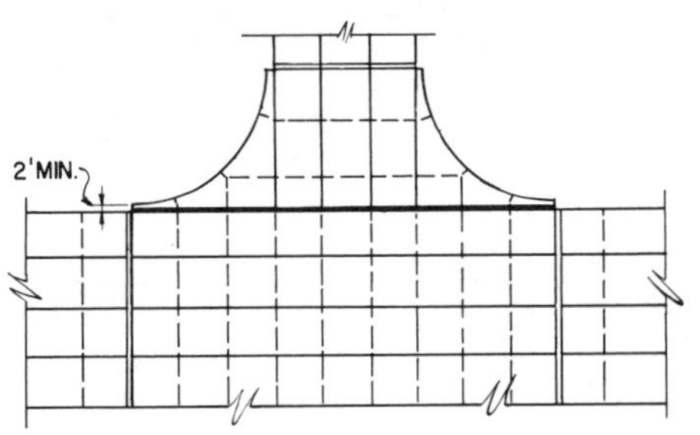
2' MIN.

DETAIL "C"—JOINTING PLAN - RIGHT ANGLE INTERSECTION OF TAXIWAY AND RUNWAY

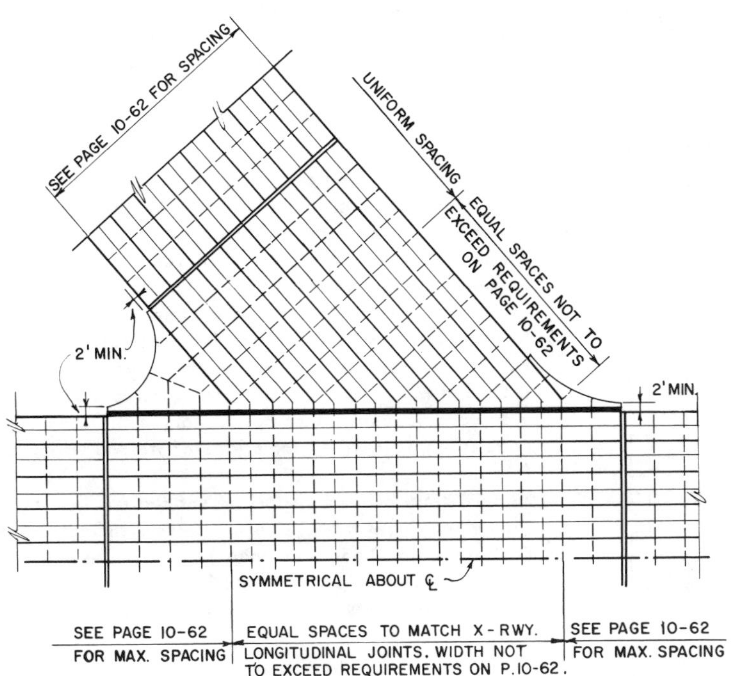
SEE PAGE 10-62 FOR SPACING
UNIFORM SPACING EQUAL SPACES NOT TO EXCEED REQUIREMENTS ON PAGE 10-62
2' MIN.
2' MIN.
SYMMETRICAL ABOUT ℄
SEE PAGE 10-62 FOR MAX. SPACING
EQUAL SPACES TO MATCH X-RWY. LONGITUDINAL JOINTS. WIDTH NOT TO EXCEED REQUIREMENTS ON P.10-62.
SEE PAGE 10-62 FOR MAX. SPACING

DETAIL "D"— JOINTING PLAN — ANGULAR INTERSECTION OF TWO RUNWAY

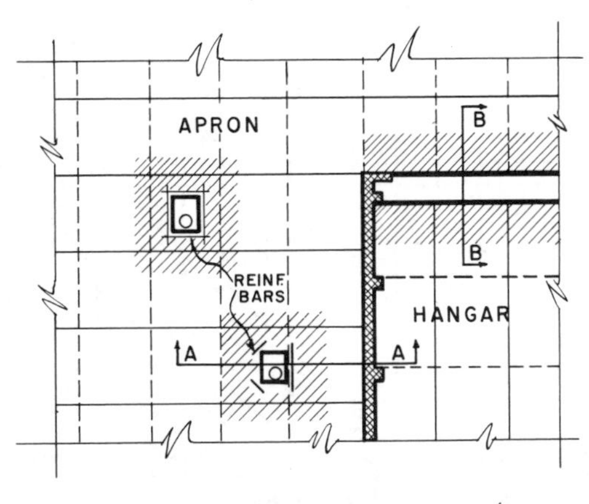
APRON
B
B
REINF. BARS
HANGAR
A
A

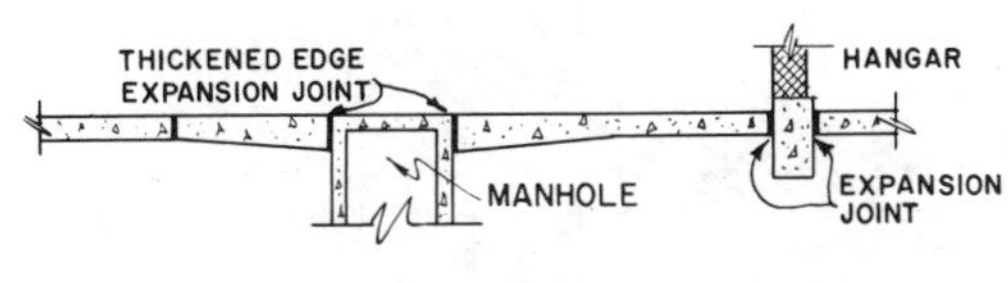
THICKENED EDGE EXPANSION JOINT
HANGAR
MANHOLE
EXPANSION JOINT

SECTION A-A

HANGAR DOOR PAD

THICKENED EDGE EXPANSION JOINT

SECTION B-B

DETAIL "E"— JOINTING PLAN — APRON — SHOWING DETAILS AT PAVEMENT LEVEL STRUCTURES AND HANGAR

AIRFIELD PAVEMENTS — REINFORCED RIGID-DESIGN-DETAILS — CAA

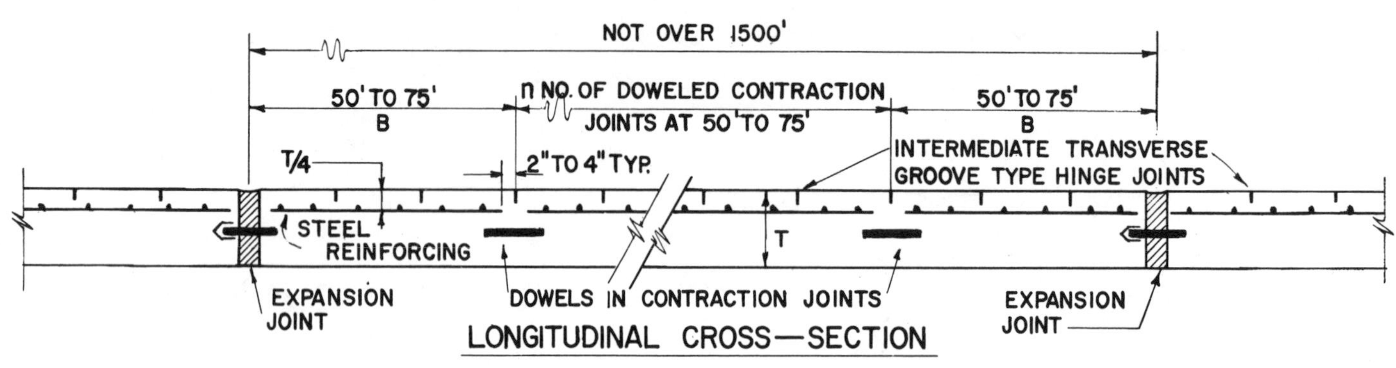
NOT OVER 1500'
50' TO 75'
B
n NO. OF DOWELED CONTRACTION
JOINTS AT 50' TO 75'
50' TO 75'
B
T/4
2" TO 4" TYP.
INTERMEDIATE TRANSVERSE
GROOVE TYPE HINGE JOINTS
STEEL
REINFORCING
T
EXPANSION
JOINT
DOWELS IN CONTRACTION JOINTS
EXPANSION
JOINT
LONGITUDINAL CROSS—SECTION

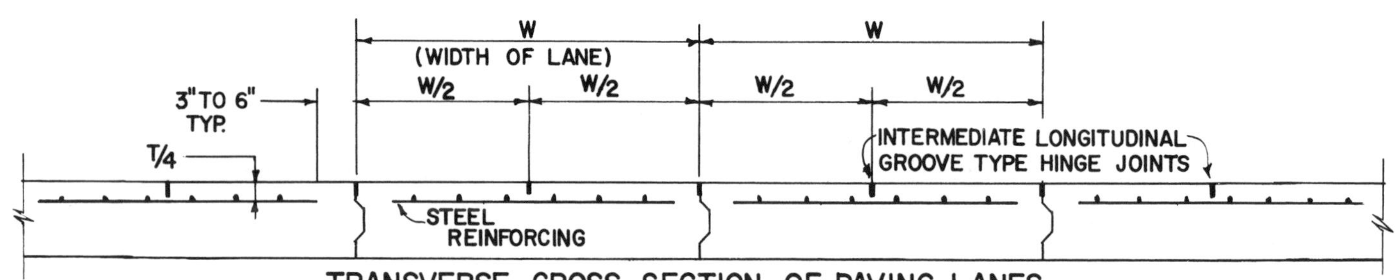
W
(WIDTH OF LANE)
W
W/2
W/2
W/2
W/2
3" TO 6"
TYP.
T/4
INTERMEDIATE LONGITUDINAL
GROOVE TYPE HINGE JOINTS
STEEL
REINFORCING
TRANSVERSE CROSS—SECTION OF PAVING LANES

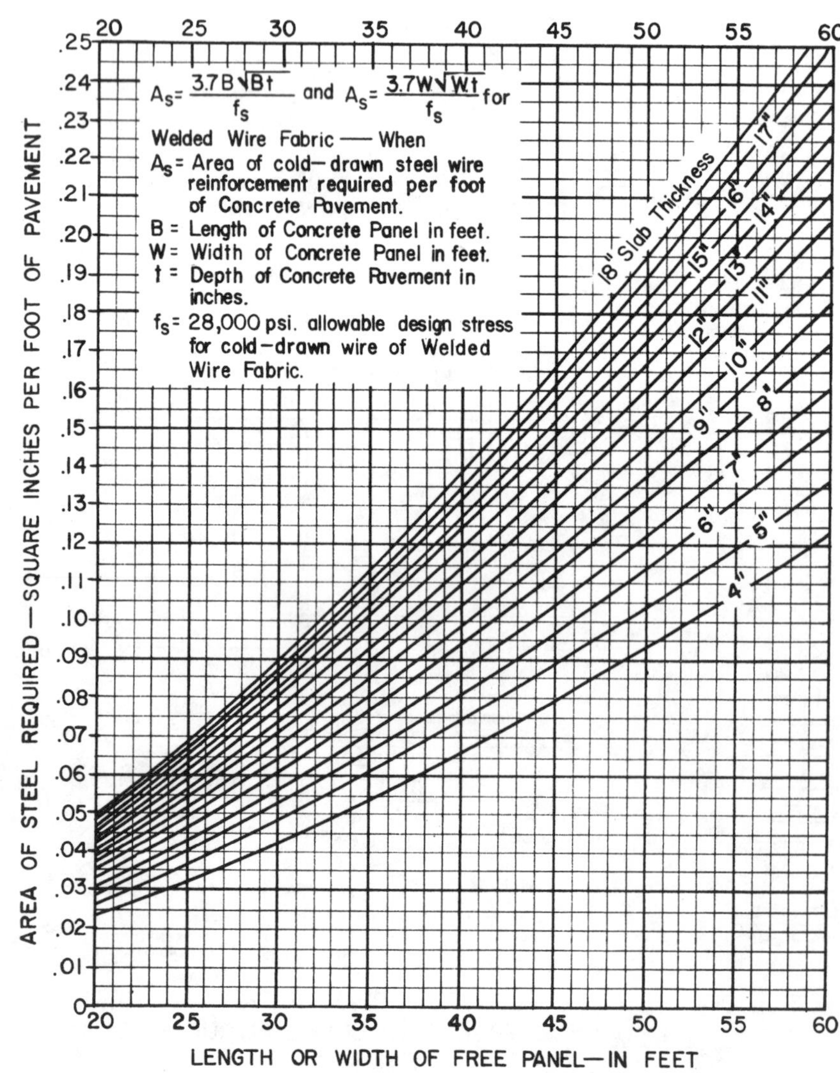
$A_s = \frac{3.7B\sqrt{Bt}}{f_s}$ and $A_s = \frac{3.7W\sqrt{Wt}}{f_s}$ for
Welded Wire Fabric — When
A_s = Area of cold-drawn steel wire reinforcement required per foot of Concrete Pavement.
B = Length of Concrete Panel in feet.
W = Width of Concrete Panel in feet.
t = Depth of Concrete Pavement in inches.
f_s = 28,000 psi. allowable design stress for cold-drawn wire of Welded Wire Fabric.
18" Slab Thickness
17"
16"
15"
14"
13"
12"
11"
10"
9"
8"
7"
6"
5"
4"
AREA OF STEEL REQUIRED — SQUARE INCHES PER FOOT OF PAVEMENT
LENGTH OR WIDTH OF FREE PANEL— IN FEET
20
25
30
35
40
45
50
55
60
0
.01
.02
.03
.04
.05
.06
.07
.08
.09
.10
.11
.12
.13
.14
.15
.16
.17
.18
.19
.20
.21
.22
.23
.24
.25

Distributed reinforcement presented herein serves as a means of crack control and provides load transference by means of aggregate interlock across slabs. It does not increase to any appreciable degree the load-carrying capacity of uncracked slabs. Any cracking that does form is held tightly together by the reinforcement so that the interlock of the irregular faces of the slab ends provide load transference.

GENERAL NOTES

1. Details and spacing of transverse and contraction joints are identical with plain concrete requirements, except for reinforcing.
2. Thickness requirements for reinforced-concrete pavements shall be the same as for plain concrete for a given condition. See page 10-53.
3. Reinforcing may be either welded wire fabric or bar mats, furnished in flat sheets or mats and installed with end and side laps so as to provide continuous reinforcing throughout the slab panel. See page 10-17 for styles of welded wire fabric.
4. Longitudinal members shall be spaced not less than 6" nor more than 12"; transverse members not less than 6" nor more than 18".
5. Grade of material, fabrication, lapping, and other requirements are to conform to standard specifications.

AIRFIELD PAVEMENTS — REINFORCED RIGID-DESIGN-DETAILS — CORPS. OF ENGINEERS

The Corps of Engineers design method utilizes the principle of allowing a reduction in the required non-reinforced rigid pavement thickness by the addition of varying amounts of steel reinforcement.

Figure A can be used to determine either (1) the required thickness of reinforced pavement if the per cent steel to be used is known, or (2) the per cent steel required if the desired thickness of reinforced pavement is known. The per cent steel in Fig. A refers to the total area of the steel in one direction. In all cases, equal percentages of steel shall be provided in both the transverse and longitudinal direction of the slab.

DETERMINATION OF MAXIMUM REINFORCED-PAVEMENT SLAB LENGTH

Condition 1. Reinforced rigid-slab overlay cast directly on a rigid slab permitting no slippage between new and old slab.

Design: Provide joints in the reinforced overlay to coincide with all joints in the base pavement. It is not necessary for joints to be over like joints: (i.e. contraction joint over construction joint, etc.)

Condition 2. Reinforced rigid slab placed on a material where the reinforced slab is not bonded to the underlying material, such as subgrade, base course, flexible pavement, or another rigid slab when a bond breaking course is placed between the two rigid pavements.

Design: Slab length (L) (Ft) $= 2.19 \sqrt[3]{S^2 h_r}\,(2/3 f_s)^2$

Where S = per cent steel used, and f_s = yield strength of reinforcement.

Width of reinforced rigid pavement slab will generally be controlled by the concrete paving equipment — normally 25 feet — unless smaller widths are necessary to meet dimensional requirements.

Reinforcement Steel may be either deformed bars or welded wire fabric. If deformed steel bars are used, the minimum requirements for the deformations will be in accordance with A.S.T.M. Designation A305.

REINFORCEMENT STEEL shall conform to one of the following:

Type of Steel	A.S.T.M. Designation
Billet steel bars	A15
Axle steel bars	A160
Rail steel bars	A16
Cold drawn steel wire	A82

FABRICATED STEEL BAR MATS OR WELDED WIRE FABRIC shall conform to one of the following:

Type	A.S.T.M. Designation
Fabricated steel bar or rod mat	A184
Welded steel wire fabric	A185

METHOD OF DESIGN

First Step: Determine the required plain (non-reinforced) pavement thickness h_d from appropriate design curves on page 10-57.

Second Step: From Fig. A determine the amount of reinforcement steel to use and the thickness of reinforcement pavement (h_r).

Third Step: Determine the maximum length of slabs for design conditions noted above.

NOTES:

1. Reinforcement steel will not carry through any type joint with the exception of the longitudinal dummy-type joint.
2. All longitudinal construction joints, transverse construction joints, and transverse contraction (dummy) joints will be of the doweled type and will be designed in accordance with page 10-61, using the thickness, h_r, of the reinforced pavement.
3. Expansion joints will be kept to a minimum and used only when absolutely necessary. The requirements for expansion joints in reinforced pavements will be the same as for non-reinforced pavements.
4. When transverse expansion joints are used, they will be of the doweled type.
5. Longitudinal expansion joints for overlays will be of the doweled type, while for single-slab construction the longitudinal expansion joint will be of the thickened-edge type.

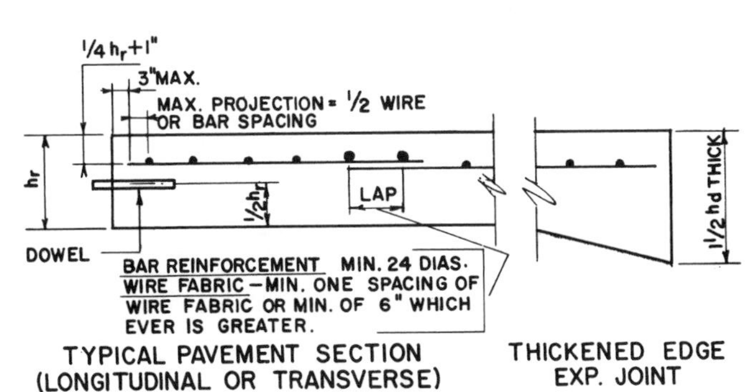
1/4 hr+1"
3" MAX.
MAX. PROJECTION = 1/2 WIRE OR BAR SPACING
hr
1/2hr
LAP
DOWEL
BAR REINFORCEMENT MIN. 24 DIAS.
WIRE FABRIC – MIN. ONE SPACING OF WIRE FABRIC OR MIN. OF 6" WHICH EVER IS GREATER.
1 1/2 hd THICK

TYPICAL PAVEMENT SECTION (LONGITUDINAL OR TRANSVERSE)

THICKENED EDGE EXP. JOINT

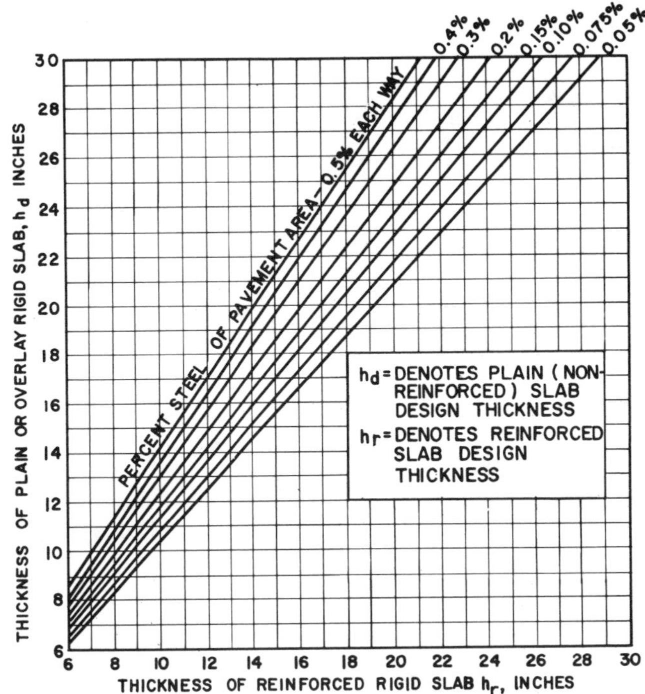
0.4%
0.3%
0.2%
0.15%
0.10%
0.075%
0.05%
PERCENT STEEL OF PAVEMENT AREA – 0.5% EACH WAY
hd = DENOTES PLAIN (NON-REINFORCED) SLAB DESIGN THICKNESS
hr = DENOTES REINFORCED SLAB DESIGN THICKNESS
THICKNESS OF PLAIN OR OVERLAY RIGID SLAB, hd INCHES
THICKNESS OF REINFORCED RIGID SLAB hr, INCHES
6
8
10
12
14
16
18
20
22
24
26
28
30

REINFORCED RIGID PAVEMENT DESIGN

LIMITATIONS to Corps of Engineers Reinforced Rigid-Pavement Design Method:

1. No reduction in the required thickness for non-reinforced rigid pavement shall be allowed for percentages of steel reinforcement less than 0.05%.
2. No further reduction in the required thickness for non-reinforced rigid pavement shall be allowed over that indicated for 0.5% steel reinforcement in Fig. A, regardless of the per cent steel used.
3. The maximum length of any reinforced slab will be 100 feet, regardless of the per cent steel used or slab thickness.
4. The minimum thickness of reinforced rigid pavement or reinforced rigid overlay shall be 6 inches.

EARTHWORK — COMPUTATIONS

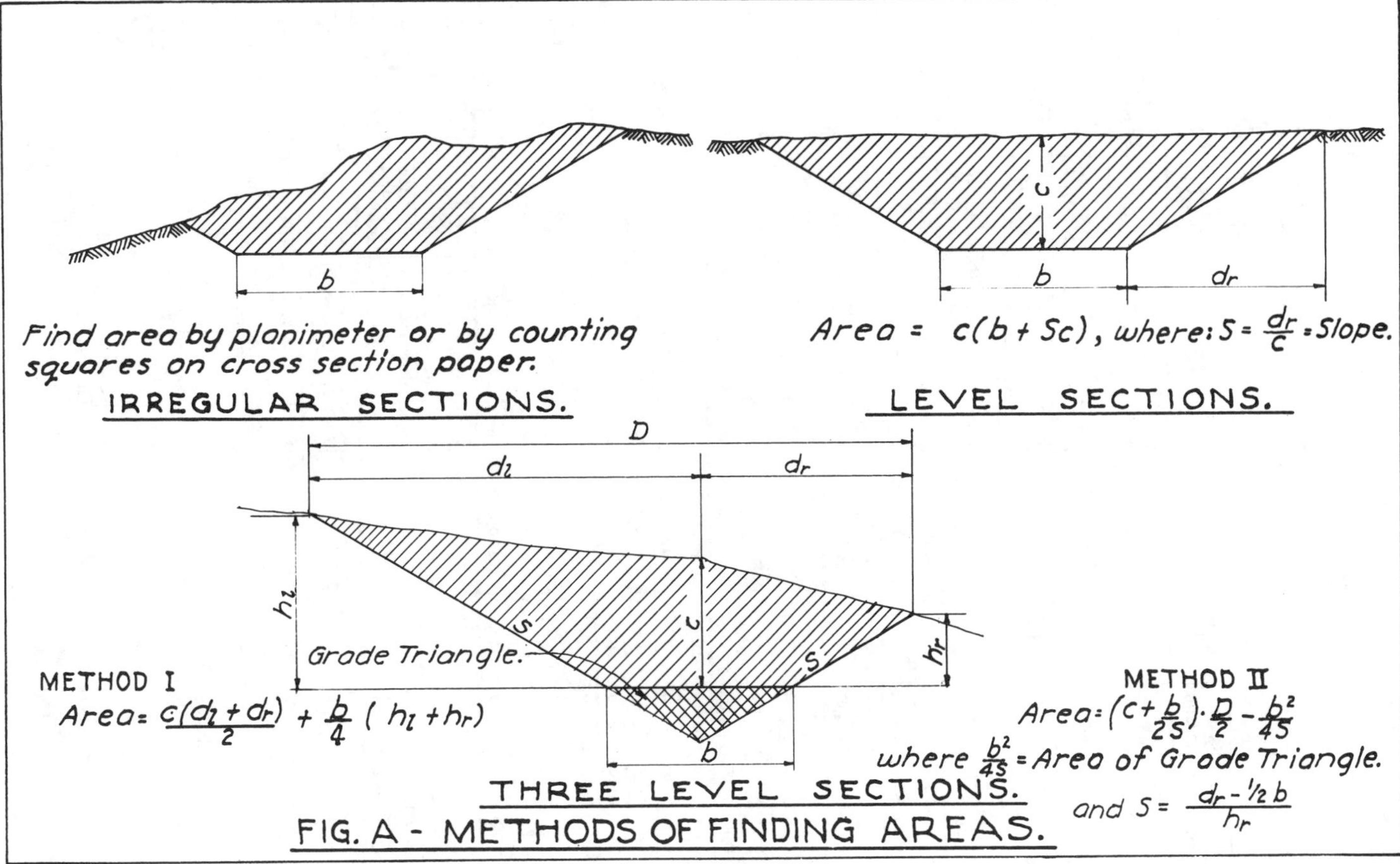

b
Find area by planimeter or by counting squares on cross section paper.
IRREGULAR SECTIONS.
c
b
d_r
Area = $c(b+Sc)$, where: $S = \frac{d_r}{c}$ = Slope.
LEVEL SECTIONS.
D
d_l
d_r
h_l
c
S
S
h_r
Grade Triangle.
METHOD I
Area = $\frac{c(d_l+d_r)}{2} + \frac{b}{4}(h_l+h_r)$
b
METHOD II
Area = $\left(c+\frac{b}{2S}\right)\cdot\frac{D}{2} - \frac{b^2}{4S}$
where $\frac{b^2}{4S}$ = Area of Grade Triangle.
and $S = \frac{d_r - \frac{1}{2}b}{h_r}$
THREE LEVEL SECTIONS.

FIG. A - METHODS OF FINDING AREAS.

1. By Average End Areas*: Volume in cubic yards = $\frac{A_0+A_1}{2}\cdot\frac{l}{27}$ where l = distance in feet between section A_0 and A_1. Compute end areas as indicated in Fig. A above. Use Table A, pg. 11-02 & 11-03 also see example page 11-02.
2. By Prismoidal Formula: Volume in cubic yards = $\frac{A_0+4M+A_1}{6}\cdot\frac{l}{27}$, where l = distance in feet between sections A_0 and A_1, M = area at section midway between section A_0 and A_1.
3. Using Prismoidal Corrections: Subtract volume in Table A page 11-04 from volume found using Average End Areas method.
4. To find volume of excavation on curves use average end area method with "l" between sections as indicated below. Fill volumes can be computed similarly.

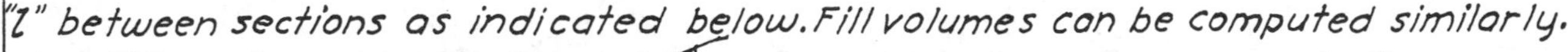

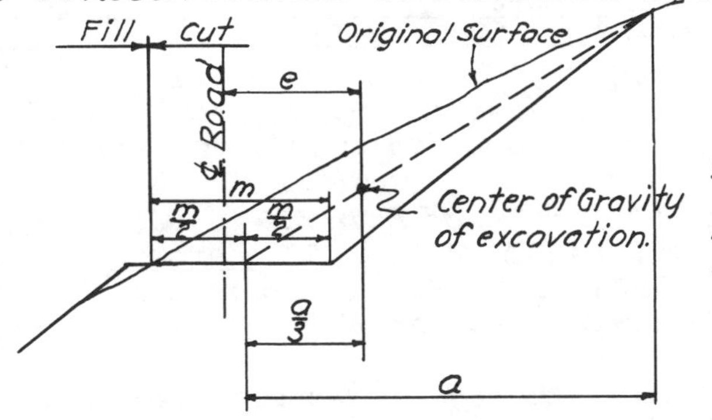

Fill
Cut
Original Surface
℄ Road
e
m
$\frac{m}{2}$
$\frac{m}{2}$
Center of Gravity of excavation.
$\frac{a}{3}$
a

l = Distance beteen centers of gravity of adjacent sections. Locate c.g. as shown on left - plot "e" on plan and scale "l" along curve as indicated at right.

e_1

℄ Road

l

$R - \frac{(e_0+e_1)}{2}$

e_0

FIG. B - METHODS OF FINDING VOLUMES.

* Used by most State Highway Departments and Public Roads Administration. Recommended for roads and airports.

EARTHWORK – DOUBLE END-AREA VOLUMES-1

TABLE A - CUBIC YARDS FOR SUM OF END AREAS FOR DISTANCE BETWEEN STATIONS OF 50 FEET.*

D.A. = Sum of end areas in square feet																				DISTANCE BETWEEN SECTIONS	CONSTANT
D.A.	C.Y.	D.A.	C.Y.	D.A.	C.Y.	D.A.	C.Y.	D.A.	C.Y.	D.A.	C.Y.	D.A.	C.Y.	D.A.	C.Y.	D.A.	C.Y.	D.A.	C.Y.		
0	0	50	46	100	93	150	139	200	185	250	231	300	278	350	324	400	370	450	417	0'	.0000
1	1	51	47	101	94	151	140	201	186	251	232	301	279	351	325	401	371	451	418	1'	.0185
2	2	52	48	102	94	152	141	202	187	252	233	302	280	352	326	402	372	452	419	2'	.0370
3	3	53	49	103	95	153	142	203	188	253	234	303	281	353	327	403	373	453	419	3'	.0556
4	4	54	50	104	96	154	143	204	189	254	235	304	281	354	328	404	374	454	420	4'	.0741
5	5	55	51	105	97	155	144	205	190	255	236	305	282	355	329	405	375	455	421	5'	.0926
6	6	56	52	106	98	156	144	206	191	256	237	306	283	356	330	406	376	456	422	6'	.1111
7	6	57	53	107	99	157	145	207	192	257	238	307	284	357	331	407	377	457	423	7'	.1296
8	7	58	54	108	100	158	146	208	193	258	239	308	285	358	331	408	378	458	424	8'	.1482
9	8	59	55	109	101	159	147	209	194	259	240	309	286	359	332	409	379	459	425	9'	.1667
10	9	60	56	110	102	160	148	210	194	260	241	310	287	360	333	410	380	460	426	10'	.1852
11	10	61	56	111	103	161	149	211	195	261	242	311	288	361	334	411	381	461	427	11'	.2037
12	11	62	57	112	104	162	150	212	196	262	243	312	289	362	335	412	381	462	428	12'	.2222
13	12	63	58	113	105	163	151	213	197	263	244	313	290	363	336	413	382	463	429	13'	.2407
14	13	64	59	114	106	164	152	214	198	264	244	314	291	364	337	414	383	464	430	14'	.2593
15	14	65	60	115	106	165	153	215	199	265	245	315	292	365	338	415	384	465	431	15'	.2778
16	15	66	61	116	107	166	154	216	200	266	246	316	293	366	339	416	385	466	431	16'	.2963
17	16	67	62	117	108	167	155	**217**	**201**	267	247	317	294	367	340	417	386	467	432	17'	.3148
18	17	68	63	118	109	168	156	218	202	268	248	318	294	368	341	418	387	468	433	18'	.3333
19	18	69	64	119	110	169	156	219	203	269	249	319	295	369	342	419	388	469	434	19'	.3519
20	19	70	65	120	111	170	157	220	204	270	250	320	296	370	343	420	389	470	435	20'	.3704
21	19	71	66	121	112	171	158	221	205	271	251	321	297	371	344	421	390	471	436	21'	.3889
22	20	72	67	122	113	172	159	222	206	272	252	322	298	372	344	422	391	472	437	22'	.4074
23	21	73	68	123	114	173	160	223	206	273	253	323	299	373	345	423	392	473	438	23'	.4259
24	22	74	69	124	115	174	161	224	207	274	254	324	300	374	346	424	393	474	439	24'	.4445
25	23	75	69	125	116	175	162	225	208	275	255	325	301	375	347	425	394	475	440	25'	.4630
26	24	76	70	126	117	176	163	226	209	276	256	326	302	376	348	426	394	476	441	26'	.4815
27	25	77	71	127	118	177	164	227	210	277	256	327	303	377	349	427	395	477	442	27'	.5000
28	26	78	72	128	119	178	165	228	211	278	257	328	304	378	350	428	396	478	443	28'	.5185
29	27	79	73	129	119	179	166	229	212	279	258	329	305	379	351	429	397	479	444	29'	.5370
30	28	80	74	130	120	180	167	230	213	280	259	330	306	380	352	430	398	480	444	30'	.5556
31	29	81	75	131	121	181	168	231	214	281	260	331	306	381	353	431	399	481	445	31'	.5741
32	30	82	76	132	122	182	169	232	215	282	261	332	307	382	354	432	400	482	446	32'	.5926
33	31	83	77	133	123	183	169	233	216	283	262	333	308	383	355	433	401	483	447	33'	.6111
34	31	84	78	134	124	184	170	234	217	284	263	334	309	384	356	434	402	484	448	34'	.6296
35	32	85	79	135	125	185	171	235	218	285	264	335	310	385	356	435	403	485	449	35'	.6482
36	33	86	80	136	126	186	172	236	219	286	265	336	311	386	357	436	404	486	450	36'	.6667
37	34	87	81	137	127	187	173	237	219	287	266	337	312	387	358	437	405	487	451	**37'**	**.6852**
38	35	88	81	138	128	188	174	238	220	288	267	338	313	388	359	438	406	488	452	38'	.7037
39	36	89	82	139	129	189	175	239	221	289	268	339	314	389	360	439	406	489	453	39'	.7222
40	37	90	83	140	130	190	176	240	222	290	269	340	315	390	361	440	407	490	454	40'	.7408
41	38	91	84	141	131	191	177	241	223	291	269	341	316	391	362	441	408	491	455	41'	.7593
42	39	92	85	142	131	192	178	242	224	292	270	342	317	392	363	442	409	492	456	42'	.7778
43	40	93	86	143	132	193	179	243	225	293	271	343	318	393	364	443	410	493	456	43'	.7963
44	41	94	87	144	133	194	180	244	226	294	272	344	319	394	365	444	411	494	457	44'	.8148
45	42	95	88	145	134	195	181	245	227	295	273	345	319	395	366	445	412	495	458	45'	.8333
46	43	96	89	146	135	196	181	246	228	296	274	346	320	396	367	446	413	496	459	46'	.8519
47	44	97	90	147	136	197	182	247	229	297	275	347	321	397	368	447	414	497	460	47'	.8704
48	44	98	91	148	137	198	183	248	230	298	276	348	322	398	369	448	415	498	461	48'	.8889
49	45	99	92	149	138	199	184	249	231	299	277	349	323	399	369	449	416	499	462	49'	.9074
50	46	100	93	150	139	200	185	250	231	300	278	350	324	400	370	450	417	**500**	**463**	50'	.9259
1000 = 926				2000 = 1852				3000 = 2778				4000 = 3704				5000 = 4630					

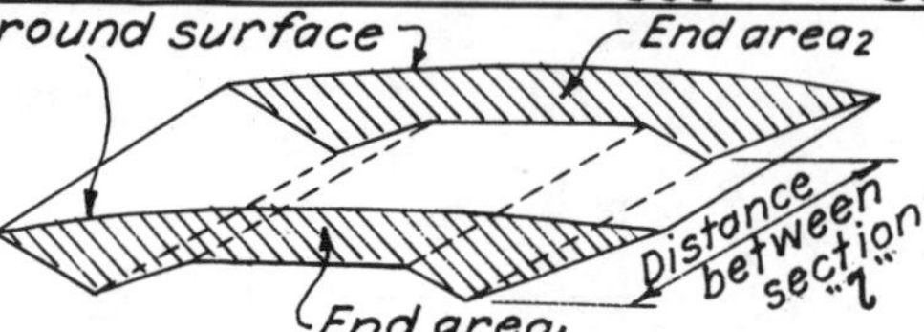

Ground surface
End area₂
Distance between sections "l"
End area₁

Example 1.- Given: End Area$_1$ = 97 sq.ft. End Area$_2$ = 120 sq.ft., l = 50'. Required: Cubic yards between sections. Solution: D.A. = 97 + 120 = 217 sq.ft. Enter D.A. column and to right of 217 find C.Y. = 201 in C.Y. column.

Use Table A - page 11-03 for D.A. of from 500 to 1000 cu.yds.

Example 2: Given - D.A. = 2751 sq.ft., l = 50'. Required: Cubic yards between stations. Solution - D.A. of 2000 = 1852 cu.yds. - Find at bottom of Table A p. 3-37 ; D.A. of 751 sq.ft = 695 cu.yd. Therefore cubic yards for D.A of 2751 sq.ft = 1852 + 695 = 2547 Cubic yards.

Example 3: When "l" is less than 50'. Given: D.A. = 217 sq.ft., l = 37'. Required: C.Y. between sections. Solution: Enter column "Distance between Sections" and to right of 37 find "Constant" .6852. Then .6852 x 217 = 149 C.Y.

* Based on average end area formula. Not as accurate as prismoidal formula, but as accurate as usual field measurements warrant. Specified for payment quantities by most State Highway Depts.

EARTHWORK – DOUBLE END-AREA VOLUMES-2

TABLE A - CUBIC YARDS FOR SUM OF END AREAS FOR DISTANCE BETWEEN STATIONS OF 50 FEET.*

D.A. = Sum of end areas in square feet. (Double End Area)

D.A.	C.Y.	D.A.	C.Y.	D.A.	C.Y.	D.A.	C.Y.	D.A.	C.Y.	D.A.	C.Y.	D.A.	C.Y.	D.A.	C.Y.	D.A.	C.Y.	D.A.	C.Y.	DISTANCE BETWEEN SECTIONS	CONSTANT
500	463	550	509	600	556	650	602	700	648	750	694	800	741	850	787	900	833	950	880	1	0.0185
501	464	551	510	601	556	651	603	701	649	751	695	801	742	851	788	901	834	951	881	2	0.0370
502	465	552	511	602	557	652	604	702	650	752	696	802	743	852	789	902	835	952	881	3	0.0556
503	466	553	512	603	558	653	605	703	651	753	697	803	744	853	790	903	836	953	882	4	0.0741
504	467	554	513	604	559	654	606	704	652	754	698	804	744	854	791	904	837	954	883	5	0.0926
505	468	555	514	605	560	655	606	705	653	755	699	805	745	855	792	905	838	955	884	6	0.1111
506	469	556	515	606	561	656	607	706	654	756	700	806	746	856	793	906	839	956	885	7	0.1296
507	469	557	516	607	562	657	608	707	655	757	701	807	747	857	794	907	840	957	886	8	0.1482
508	470	558	517	608	563	658	609	708	656	758	702	808	748	858	794	908	841	958	887	9	0.1667
509	471	559	518	609	564	659	610	709	656	759	703	809	749	859	795	909	842	959	888	10	0.1852
510	472	560	519	610	565	660	611	710	657	760	704	810	750	860	796	910	843	960	889	11	0.2037
511	473	561	519	611	566	661	612	711	658	761	705	811	751	861	797	911	844	961	890	12	0.2222
512	474	562	520	612	567	662	613	712	659	762	706	812	752	862	798	912	844	962	891	13	0.2407
513	475	563	521	613	568	663	614	713	660	763	706	813	753	863	799	913	845	963	892	14	0.2593
514	476	564	522	614	569	664	615	714	661	764	707	814	754	864	800	914	846	964	893	15	0.2778
515	477	565	523	615	569	665	616	715	662	765	708	815	755	865	801	915	847	965	894	16	0.2963
516	478	566	524	616	570	666	617	716	663	766	709	816	756	866	802	916	848	966	894	17	0.3148
517	479	567	525	617	571	667	618	717	664	767	710	817	756	867	803	917	849	967	895	18	0.3333
518	480	568	526	618	572	668	619	718	665	768	711	818	757	868	804	918	850	968	896	19	0.3519
519	481	569	527	619	573	669	619	719	666	769	712	819	758	869	805	919	851	969	897	20	0.3704
520	481	570	528	620	574	670	620	720	667	770	713	820	759	870	806	920	852	970	898	21	0.3889
521	482	571	529	621	575	671	621	721	668	771	714	821	760	871	806	921	853	971	899	22	0.4074
522	483	572	530	622	576	672	622	722	669	772	715	822	761	872	807	922	854	972	900	23	0.4259
523	484	573	531	623	577	673	623	723	669	773	716	823	762	873	808	923	855	973	901	24	0.4445
524	485	574	531	624	578	674	624	724	670	774	717	824	763	874	809	924	856	974	902	25	0.4630
525	486	575	532	625	579	675	625	725	671	775	718	825	764	875	810	925	856	975	903	26	0.4815
526	487	576	533	626	580	676	626	726	672	776	719	826	765	876	811	926	857	976	904	27	0.5000
527	488	577	534	627	581	677	627	727	673	777	719	827	766	877	812	927	858	977	905	28	0.5185
528	489	578	535	628	581	678	628	728	674	778	720	828	767	878	813	928	859	978	906	29	0.5370
529	490	579	536	629	582	679	629	729	675	779	721	829	768	879	814	929	860	979	906	30	0.5556
530	491	580	537	630	583	680	630	730	676	780	722	830	769	880	815	930	861	980	907	31	0.5741
531	492	581	538	631	584	681	631	731	677	781	723	831	769	881	816	931	862	981	908	32	0.5926
532	493	582	539	632	585	682	631	732	678	782	724	832	770	882	817	932	863	982	909	33	0.6111
533	494	583	540	633	586	683	632	733	679	783	725	833	771	883	818	933	864	983	910	34	0.6296
534	494	584	541	634	587	684	633	734	680	784	726	834	772	884	819	934	865	984	911	35	0.6482
535	495	585	542	635	588	685	634	735	681	785	727	835	773	885	819	935	866	985	912	36	0.6667
536	496	586	543	636	589	686	635	736	681	786	728	836	774	886	820	936	867	986	913	37	0.6852
537	497	587	544	637	590	687	636	737	682	787	729	837	775	887	821	937	868	987	914	38	0.7037
538	498	588	544	638	591	688	637	738	683	788	730	838	776	888	822	938	869	988	915	39	0.7222
539	499	589	545	639	592	689	638	739	684	789	731	839	777	889	823	939	869	989	916	40	0.7408
540	500	590	546	640	593	690	639	740	685	790	731	840	778	890	824	940	870	990	917	41	0.7593
541	501	591	547	641	593	691	640	741	686	791	732	841	779	891	825	941	871	991	918	42	0.7778
542	502	592	548	642	594	692	641	742	687	792	733	842	780	892	826	942	872	992	919	43	0.7963
543	503	593	549	643	595	693	642	743	688	793	734	843	781	893	827	943	873	993	919	44	0.8148
544	504	594	550	644	596	694	643	744	689	794	735	844	781	894	828	944	874	994	920	45	0.8333
545	505	595	551	645	597	695	644	745	690	795	736	845	782	895	829	945	875	995	921	46	0.8519
546	506	596	552	646	598	696	644	746	691	796	737	846	783	896	830	946	876	996	922	47	0.8704
547	506	597	553	647	599	697	645	747	692	797	738	847	784	897	831	947	877	997	923	48	0.8889
548	507	598	554	648	600	698	646	748	693	798	739	848	785	898	831	948	878	998	924	49	0.9074
549	508	599	555	649	601	699	647	749	694	799	740	849	786	899	832	949	879	999	925	50	0.9259

1000 = 926 **2000 = 1852** 3000 = 2778 4000 = 3704 5000 = 4630 6000 = 5556 7000 = 6481 8000 = 7407

* Based on average end area formula. Not as accurate as prismoidal, but as accurate as usual field measurements warrant. Specified for payment quantities by most State Hiway Depts. For examples illustrating use of table see page 11-02.

EARTHWORK – PRISMOIDAL CORRECTIONS

TABLE A – PRISMOIDAL CORRECTIONS FOR L = 100' STATIONS.*

$c_1 - c_2 =$ $D_1 - D_2$	1	2	3	4	5	6	7	8	9
5.1	1.57	3.15	4.72	6.30	7.87	9.44	11.02	12.59	14.17
5.2	1.60	3.21	4.81	6.42	8.02	9.63	11.23	12.84	14.44
5.3	1.64	3.27	4.91	6.54	8.18	9.81	11.45	13.09	14.72
5.4	1.67	3.33	5.00	6.67	8.33	10.00	11.67	13.33	15.00
5.5	1.70	3.40	5.09	6.79	8.49	10.19	11.88	13.58	15.28
5.6	1.73	3.46	5.19	6.91	8.64	10.37	12.10	13.83	15.56
5.7	1.76	3.52	5.28	7.04	8.80	10.56	12.31	14.07	15.83
5.8	1.79	3.58	5.37	7.16	8.95	10.74	12.53	14.32	16.11
5.9	1.82	3.64	5.46	7.28	9.10	10.93	12.75	14.57	16.39
6.0	1.85	3.70	5.56	7.41	9.26	11.11	12.96	14.81	16.67
6.1	1.88	3.77	5.65	7.53	9.41	11.30	13.18	15.06	16.94
6.2	1.91	3.83	5.74	7.65	9.57	11.48	13.40	15.31	17.22
6.3	1.94	3.89	5.83	7.78	9.72	11.67	13.61	15.56	17.50
6.4	1.98	3.95	5.93	7.90	9.88	11.85	13.83	15.80	17.78
6.5	2.01	4.01	6.02	8.02	10.03	12.04	14.04	16.05	18.06
6.6	2.04	4.07	6.11	8.15	10.19	12.22	14.26	16.30	18.33
6.7	2.07	4.14	6.20	8.27	10.34	12.41	14.48	16.54	18.61
6.8	2.10	4.20	6.30	8.40	10.49	12.59	14.69	16.79	18.89
6.9	2.13	4.26	6.39	8.52	10.65	12.78	14.91	17.04	19.17
7.0	2.16	4.32	6.48	8.64	10.80	12.96	15.12	17.28	19.44
7.1	2.19	4.38	6.57	8.77	10.96	13.15	15.34	17.53	19.72
7.2	2.22	4.44	6.67	8.89	11.11	13.33	15.56	17.78	20.00
7.3	2.25	4.51	6.76	9.01	11.27	13.52	15.77	18.02	20.28
7.4	2.28	4.57	6.85	9.14	11.42	13.70	15.99	18.27	20.56
7.5	2.31	4.63	6.94	9.26	11.57	13.89	16.20	18.52	20.83
7.6	2.35	4.69	7.04	9.38	11.73	14.07	16.42	18.77	21.11
7.7	2.38	4.75	7.13	9.51	11.88	14.26	16.64	19.01	21.39
7.8	2.41	4.81	7.22	9.63	12.04	14.44	16.85	19.26	21.67
7.9	2.44	4.88	7.31	9.75	12.19	14.63	17.07	19.51	21.94
8.0	2.47	4.94	7.41	9.88	12.35	14.81	17.28	19.75	22.22
8.1	2.50	5.00	7.50	10.00	12.50	15.00	17.50	20.00	22.50
8.2	2.53	5.06	7.59	10.12	12.65	15.19	17.72	20.25	22.78
8.3	2.56	5.12	7.69	10.25	12.81	15.37	17.93	20.49	23.06
8.4	2.59	5.19	7.78	10.37	12.96	15.56	18.15	20.74	23.33
8.5	2.62	5.25	7.87	10.49	13.12	15.74	18.36	20.99	23.61
8.6	2.65	5.31	7.96	10.62	13.27	15.93	18.58	21.23	23.89
8.7	2.69	5.37	8.06	10.74	13.43	16.11	18.80	21.48	24.17
8.8	2.72	5.43	8.15	10.86	13.58	16.30	19.01	21.73	24.44
8.9	2.75	5.49	8.24	10.99	13.73	16.48	19.23	21.97	24.72
9.0	2.78	5.56	8.33	11.11	13.89	16.67	19.44	22.22	25.00
9.1	2.81	5.62	8.43	11.23	14.04	16.85	19.66	22.47	25.28
9.2	2.84	5.68	8.52	11.36	14.20	17.04	19.88	22.72	25.56
9.3	2.87	5.74	8.61	11.48	14.35	17.22	20.09	22.96	25.83
9.4	2.90	5.80	8.70	11.60	14.51	17.41	20.31	23.21	26.11
9.5	2.93	5.86	8.80	11.73	14.66	17.59	20.52	23.46	26.39
9.6	2.96	5.93	8.89	11.85	14.81	17.78	20.74	23.70	26.67
9.7	2.99	5.99	8.98	11.98	14.97	17.96	20.96	23.95	26.94
9.8	3.02	6.05	9.07	12.10	15.12	18.15	21.17	24.20	27.22
9.9	3.06	6.11	9.17	12.22	15.28	18.33	21.39	24.44	27.50
10.0	3.09	6.17	9.26	12.35	15.43	18.52	21.60	24.69	27.78
$c_1 - c_2 =$	1	2	3	4	5	6	7	8	9

$c_1 - c_2 =$ $D_1 - D_2$	1	2	3	4	5	6	7	8	9
0.1	0.03	0.06	0.09	0.12	0.15	0.19	0.22	0.25	0.28
0.2	0.06	0.12	0.19	0.25	0.31	0.37	0.43	0.49	0.56
0.3	0.09	0.19	0.28	0.37	0.46	0.56	0.65	0.74	0.83
0.4	0.12	0.25	0.37	0.49	0.62	0.74	0.86	0.99	1.11
0.5	0.15	0.31	0.46	0.62	0.77	0.93	1.08	1.23	1.39
0.6	0.19	0.37	0.56	0.74	0.93	1.11	1.30	1.48	1.67
0.7	0.22	0.43	0.65	0.86	1.08	1.30	1.51	1.73	1.94
0.8	0.25	0.49	0.74	0.99	1.23	1.48	1.73	1.98	2.22
0.9	0.28	0.56	0.83	1.11	1.39	1.67	1.94	2.22	2.50
1.0	0.31	0.62	0.93	1.23	1.54	1.85	2.16	2.47	2.78
1.1	0.34	0.68	1.02	1.36	1.70	2.04	2.38	2.72	3.06
1.2	0.37	0.74	1.11	1.48	1.85	2.22	2.59	2.96	3.33
1.3	0.40	0.80	1.20	1.60	2.01	2.41	2.81	3.21	3.61
1.4	0.43	0.86	1.30	1.73	2.16	2.59	3.02	3.46	3.89
1.5	0.46	0.93	1.39	1.85	2.31	2.78	3.24	3.70	4.17
1.6	0.49	0.99	1.48	1.98	2.47	2.96	3.46	3.95	4.44
1.7	0.52	1.05	1.57	2.10	2.62	3.15	3.67	4.20	4.72
1.8	0.56	1.11	1.67	2.22	2.78	3.33	3.89	4.44	5.00
1.9	0.59	1.17	1.76	2.35	2.93	3.52	4.10	4.69	5.28
2.0	0.62	1.23	1.85	2.47	3.09	3.70	4.32	4.94	5.56
2.1	0.65	1.30	1.94	2.59	3.24	3.89	4.54	5.19	5.83
2.2	0.68	1.36	2.04	2.72	3.40	4.07	4.75	5.43	6.11
2.3	0.71	1.42	2.13	2.84	3.55	4.26	4.97	5.68	6.39
2.4	0.74	1.48	2.22	2.96	3.70	4.44	5.19	5.93	6.67
2.5	0.77	1.54	2.31	3.09	3.86	4.63	5.40	6.17	6.94
2.6	0.80	1.60	2.41	3.21	4.01	4.81	5.62	6.42	7.22
2.7	0.83	1.67	2.50	3.33	4.17	5.00	5.83	6.67	7.50
2.8	0.86	1.73	2.59	3.46	4.32	5.19	6.05	6.91	7.78
2.9	0.90	1.79	2.69	3.58	4.48	5.37	6.27	7.16	8.06
3.0	0.93	1.85	2.78	3.70	4.63	5.56	6.48	7.41	8.33
3.1	0.96	1.91	2.87	3.83	4.78	5.74	6.70	7.65	8.61
3.2	0.99	1.98	2.96	3.95	4.94	5.93	6.91	7.90	8.89
3.3	1.02	2.04	3.06	4.07	5.09	6.11	7.13	8.15	9.17
3.4	1.05	2.10	3.15	4.20	5.25	6.30	7.35	8.40	9.44
3.5	1.08	2.16	3.24	4.32	5.40	6.48	7.56	8.64	9.72
3.6	1.11	2.22	3.33	4.44	5.56	6.67	7.78	8.89	10.00
3.7	1.14	2.28	3.43	4.57	5.71	6.85	7.99	9.14	10.28
3.8	1.17	2.35	3.52	4.69	5.86	7.04	8.21	9.38	10.56
3.9	1.20	2.41	3.61	4.81	6.02	7.22	8.43	9.63	10.83
4.0	1.23	2.47	3.70	4.94	6.17	7.41	8.64	9.88	11.11
4.1	1.27	2.53	3.80	5.06	6.33	7.59	8.86	10.12	11.39
4.2	1.30	2.59	3.89	5.19	6.48	7.78	9.07	10.37	11.67
4.3	1.33	2.65	3.98	5.31	6.64	7.96	9.29	10.62	11.94
4.4	1.36	2.72	4.07	5.43	6.79	8.15	9.51	10.86	12.22
4.5	1.39	2.78	4.17	5.56	6.94	8.33	9.72	11.11	12.50
4.6	1.42	2.84	4.26	5.68	7.10	8.52	9.94	11.36	12.78
4.7	1.45	2.90	4.35	5.80	7.25	8.70	10.15	11.60	13.06
4.8	1.48	2.96	4.44	5.93	7.41	8.89	10.37	11.85	13.33
4.9	1.51	3.02	4.54	6.05	7.56	9.07	10.50	12.10	13.61
5.0	1.54	3.09	4.63	6.17	7.72	9.26	10.80	12.35	13.89
$c_1 - c_2 =$	1	2	3	4	5	6	7	8	9

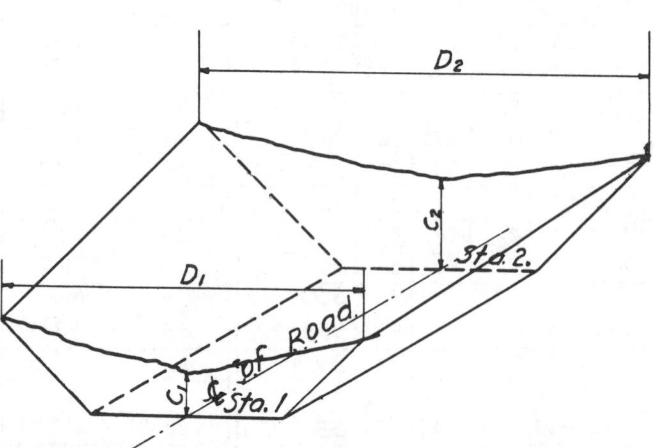

D2
C2
Sta. 2.
D1
℄ of Road
C1
℄ Sta. 1

$c_1, c_2, D_1, \& D_2$ are shown for a three level section. Volume by Average End Area ± Prismoidal Correction = Volume by Prismoidal Formula.

When $(c_2 - c_1)(D_2 - D_1)$ is +, subtract correction.

When $(c_2 - c_1)(D_2 - D_1)$ is −, add correction.

Irregular Sections are generally treated the same as three level sections.

Example: Given: $C_1 = 4'$, $D_1 = 130'$, $C_2 = 8'$, $D_2 = 138'$. Required: Prismoidal Correction Value. Solution: $C_1 - C_2 = 4$; $D_1 - D_2 = 8$. Enter table as indicated; read correction = 9.88 cu. yds. $(C_2 - C_1)(D_2 - D_1) = (8-4)(138-130) = +$. Subtract Correction from Volume by Aver. End Area Method. See Pg. 11-02

*Adapted from American Civil Engineers Handbook by Merriman & Wiggin.

EARTHWORK-MASS DIAGRAM

A MASS DIAGRAM is a graphical solution of movement of earth, or "haul," plotted on cross-section paper-above or below a profile of a road or railroad at the same horizontal scale, with stations on the profile projected on the diagram. Each station is assumed to be the center of gravity of a volume of earth extending 50' on either side, and the volumes for the stations are first computed, cut being plus and fill minus. Ordinates of the mass diagram are cumulative algebraic sums of station volumes from the point of beginning. At stations in fill it is customary to make an allowance for shrinkage of earth or swelling of ledge rock which is computed as a percentage of the material involved and added algebraically to the ordinate. Ordinates at station points are connected by a smooth curve, completing the mass diagram.

A MASS DIAGRAM is used –

(a) to determine volume, location and distances of haul and "overhaul" and thus estimate cost of earth movement.
(b) to plan earthwork economically with respect to disposition of fill, location of borrow and waste, limits of profitable haul and points where borrow and waste are economical.
(c) to determine the most efficient type of construction equipment.
(d) to balance cut and fill in design of vertical alignment. This balance should be secondary compared to a finished design with recommended grades, alignments, clearances, widths, etc.

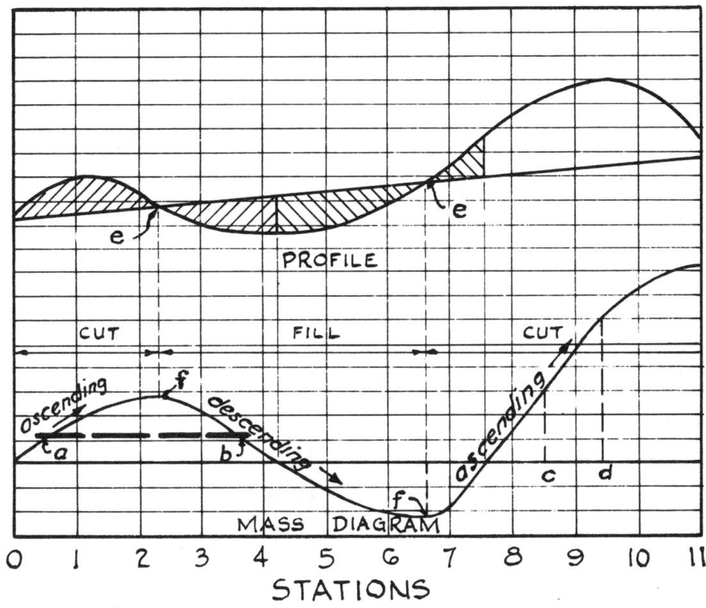
e
e
PROFILE
CUT
FILL
CUT
ascending
f
descending
ascending
a
b
c
d
f
MASS DIAGRAM
0 1 2 3 4 5 6 7 8 9 10 11
STATIONS

MASS DIAGRAM COMPUTATION

STATION	THEORETICAL VOLUME CUT	THEORETICAL VOLUME FILL	FILLS PLUS SHRINKAGE All. (15%)	MASS DIAGRAM ORDINATE
0				0
1	+182			+182
2	+ 78			+260
3		− 84	− 97	+163
4		−123	−141	+ 22
5		−107	−123	−101
6		− 92	−106	−207
7	+ 64			−143
8	+ 251			+108
9	+ 332			+440
10	+ 287			+727
11	+ 76			+803

FIG. A-EXAMPLE OF MASS DIAGRAM CONSTRUCTION.

Note: Shrinkage allowance of 15% used in above computation.

PROPERTIES OF MASS DIAGRAM

(1) Ascending curves denote excavation and descending curves embankment.
(2) Between any two stations such as "a" & "b" where the curve is intersected by a horizontal line, excavation equals embankment.
(3) The area cut off by this horizontal line and the curve is the measure of the haul between the two stations cut by that line, "a" & "b".
(4) The difference in length between any two vertical ordinates of the diagram is the volume between the stations at which the ordinates are erected, such as "c" & "d."
(5) Hill sections in the diagram represent haul forward on the profile, and valley sections haul backward.
(6) Grade points of the profile as at "e" correspond to maximum and minimum points of the mass diagram as at "f."

HIGHWAYS - DESIGN PROCEDURE - 1

I. Determine where the traffic wants to go, often called desire lines. Determine the current quantity and type of traffic along each desire line, including the peak loads and directions. Sufficient data must be obtained to account for daily, weekly, and seasonal variations.

TRAFFIC SURVEYS

1. *Traffic Count*
   a. *Manual Counters:* Used to determine percentages of trucks and other types of vehicles.
   Used at intersections to determine turning movements.
   b. *Automatic Counters:* Fixed or portable. Record continuous count with subtotals per unit of time. Fifteen-minute interval common.
2. *Origin – Destination*
   a. *Driver Interview:* Minimum sampling 25%.
   b. *Driver Postal Card:* 100% distribution desirable.
3. *Speed, Travel Time, and Route*
   a. *Test Runs:* Conducted under varied traffic conditions.
   b. *Moving Vehicle License Plate:* Suited to machine computer operation.
   c. *Vehicle Tag:* Simplified versions give quick results.

II. Project traffic data to future design year. Twenty-year period is generally considered maximum that can be reasonably subject to statistical analysis. Key factors include the following:

1. Regional population trends. Review census data; transportation studies available from railroads, bus companies and airlines; and utility expansion studies generally available from telephone, electric, gas, and water-supply companies.
2. Regional trends of motor-vehicle registration, motor-vehicle mileage, and public transit use. Relate to trend of regional commerce and industry.
3. Effect of regional master planning including new highways or transit scheduled for activation prior to design year.

NOTE: No fine precision can be expected, or is required, in projecting traffic figures many years into the future so that rounded figures and simplified assumptions should be used in the computation stages and in the presentation of results.

III. Select type and size of highway to accomodate the estimated traffic demand.

A basic design decision as to the degree of access control must be made, and the desirable design speed established.

Table A, p. 12-03, Tables A and B, p. 12-04, and Table A, p. 12-05 provide average values for judging highway capacity. These may be modified in accordance with pertinent local experience.

IV. Make preliminary plans and profiles based on field studies for all reasonable alternate routes which satisfy the desire lines. Analyze and compare alternate routes. Select preferred location by comparing cost estimates; road user benefits; capability of incorporating the alternates into regional and local master plans; types of vehicles versus feasible design and operating speeds, safety, and other pertinent controls and factors including the following:

1. Location of interchanges and intersections with respect to traffic needs and local topography.
2. Frequency of steep grades and need for "creeper" lanes or reduced speed for heavy trucking, thus reducing highway capacity and safety.
3. Right-of-way problems such as time delay required by difficult acquisition, disruption of existing economy, and presence of "untouchables" such as cemeteries and historical sites.
4. Effect of local climate; e.g., low-road alternate might hit fog pockets, or flood plain; high-road alternate might be more exposed to snow drift and icing.
5. Suitability to stage construction and, in some instances, ease of maintaining existing traffic during construction period.
6. Location with respect to regional transportation problem so that different forms of transportation may be coordinated to provide best over-all service to the public.
7. Pavement design and cost may vary greatly if the subgrade at two alternate routes varies greatly. See Section 10 for pavement design.
8. Maintenance difficulties. Although geology is considered in construction cost, it must also be considered in connection with subgrade stability, possibility of washouts, earth and rock slides, and similar items.
9. Financial feasibility of project, no matter how theoretically desirable.

V. Study selected route to determine detailed criteria.

Establish geometric criteria and design speed for each major segment of new highway in conjunction with preliminary alignment and grade.

Standards may be modified to suit serious local construction problems, to avoid monotony, or to provide greater safety where special traffic conditions exist.

Preliminary intersection design must include, in addition to geometrics, the layout of instruction signs and pavement markings.

The esthetics of landscaping, of views from the highway, and of structures, are also to be considered in the preliminary stage.

HIGHWAYS – DESIGN PROCEDURE-2

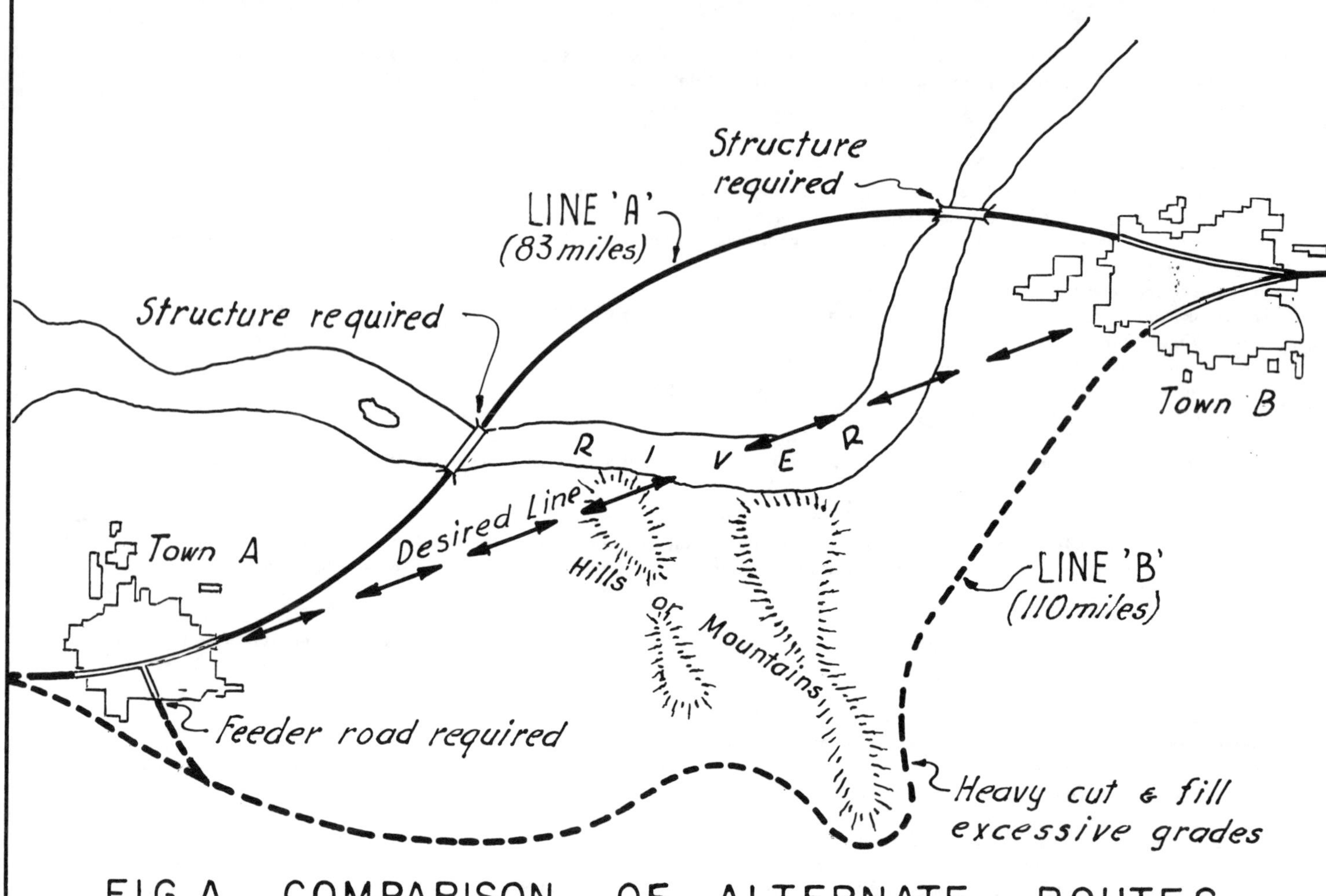

Structure required
LINE 'A' (83 miles)
Structure required
R I V E R
Desired Line
Town A
Town B
Hills or Mountains
LINE 'B' (110 miles)
Feeder road required
Heavy cut & fill excessive grades

FIG. A COMPARISON OF ALTERNATE ROUTES

ORIGIN AND DESTINATION *

FIELD SHEET
METHOD #1

STATION #______
INBOUND______
LOCATION______ WEATHER______ OUTBOUND______
TIME START______ A.M./P.M. TIME STOP______ A.M./P.M.

1	2	3	4	5
ORIGIN	DESTINATION	ROUTE USED	PARKING	OTHER
INDICATE BY BLOCK, STREET, ZONE, OR OTHER CITY.		INDICATE STREETS, ZONES OR HIGHWAYS	LOCATION AND TYPE	

A.C.S.C. No. 142 DATE______ RECORDER______

(STATE) | O | D | S

Trip is in Direction Traveling When Card Was Received
Where did this trip START? ....
If in (Locality). ....
(Write in Number and Name of Street or nearest Steet intersection or Plant or Well-known Building)
If outside (Locality) ....
(Write in Locality and State)

Trip is in Direction Traveling When Card Was Received
Where did this trip END? ....
If in (Locality). ....
(Write in Number and Name of Street or nearest Street intersection or Plant or Well known Building)
If outside (Locality) ....
(Write in Locality and State)

Routes followed. ....
(Write in Names of Major Streets or Route Numbers, in the order traveled in or near [Locality] on this trip)

Check One *(Only)* In Each Group

DIRECTION OF TRIP
1. ☐ From Home Garage
2. ☐ To Home Garage

1. ☐ Passenger Car
2. ☐ Light Truck
3. ☐ Heavy Truck

1. ☐ Work
2. ☐ Shopping
3. ☐ Others

1. ☐ Parked at Curb
2. ☐ Parking Lot
3. ☐ Storage Garage
4. ☐ Did Not Park

Parked for....... Hrs.

THANK YOU!

A. M. | P. M. | (STATE). ....

WHERE DID THIS TRIP START?
If in (Locality). ....
(Write in Number and Name of Street of nearest Street intersection or Plant or Well-known Building)
If outside (Locality) ....
(Write in Locality and State)

WHERE DID THIS TRIP END?
If in (Locality). ....
(Write in Number and Name of Street or nearest Street intersection or Plant or Well-known Building)
If outside (Locality) ....
(Write in Locality and State)

PLEASE CHECK ☑ INTERSECTION PASSED

1 ☐ Lake Shore Drive & Central
2 ☐ Lake Shore Drive & Roberts
3. ☐ Central & 3rd
4. ☐ Central & 4th
5. ☐ Central & 7th
6. ☐ 5th & Swan
7. ☐ 6th & Park
8. ☐ Main & 2nd
9. ☐ Main & 4th
10. ☐ Main & 7th

S
O
D

☑ Check One In Each Group

1. ☐ Passenger Car
2. ☐ Light Truck
3. ☐ Heavy Truck

1. ☐ Business-Work
2. ☐ Shopping
3. ☐ Other

1. ☐ Parked at Meter
2. ☐ Parking Lot
3. ☐ Garage

THANK YOU!

* From "Manual of Traffic Engineering Studies", Association of Casualty and Surety Companies.

HIGHWAYS – CAPACITY – I

Average and recommended design capacities for different types of highways that have been established by the various state highway departments, are listed in the *Highway Capacity Manual* of the Bureau of Public Roads, and other official sources.

It is well to remember that difference in driving speed between various drivers and various vehicles as often sets the capacity limit of one direction of travel of a multilane separated highway as does the geometrics of the highway. The conflict generated by faster drivers attempting to pass the slower drivers may set the capacity limit for the highway. Steep grades, which slow down trucking, thus increasing the difference in speed between them and passenger vehicles, are an important factor in reducing highway capacity and may create the need for an additional ascending lane. Because experience has indicated that the new generation of drivers (conditioned to more powerful automobiles, to superhighways, and to blowout-resistant tires) will set higher self-imposed limits than have previously been in use, the design speed of a modern highway often is made greater than is required for the currently intended speed limit, and the possible capacity may well be greater in the future. For example the 2000-car-per-hour-per-lane basic capacity for a multilane highway (listed in Table A below, long accepted as the theoretical maximum, is now regularly exceeded on some of the California freeways. The speed at which maximum capacity of a facility is achieved is still well below the highest regularly observed vehicle speeds on that facility.

Expressions used to define highway capacity in terms of the maximum number of passenger cars that can pass a given point on a lane or roadway during one hour are as follows:

BASIC – Maximum under ideal conditions.
POSSIBLE – Maximum under existing conditions.
PRACTICAL – Desirable maximum under existing conditions.

The traffic survey and projection data required to determine highway needs and capacity usually include the following, for which standard abbreviations* and generally representative values or factors are indicated:

ADT (Current) current Average Daily Traffic, year specified, that would use the improvement.

ADT (Future) future Average Daily Traffic, year specified, that would use the improvement, including traffic increases (normal traffic growth, generated traffic, and development traffic).

DHV Design Hour Volume, representative of the 30th highest hourly volume of the future year chosen for design, a two-way volume unless otherwise specified.

K Ratio of DHV to ADT (future), is normally 10 to 12% for major streets, 12 to 14% for freeways, and 12 to 18% for rural highways. The maximum hourly volume in urban areas averages 1.2 to 1.4 times the 30th highest hourly volume.

D Directional distribution of DHV, one-way volume in predominant direction of travel as a percentage of total. D varies widely, but representative values are: downtown areas, 55%, intermediate areas, 60%, outlying areas, 65 to 75%. Downward trend is likely as urban development becomes more concentrated.

T Trucks, exclusive of light delivery trucks as a percentage of DHV. T percentage of urban DHV is usually ½ to ¾ of T percentage of ADT. A value of 10% is not uncommon on principal routes through cities, particularly where local bus lines utilize the same route. On arterials, which carry traffic primarily between suburbs and central business districts, T may be 5% or less. (Special note must be made of those routes that have a high percentage of the larger, heavier trucks.)

TABLE A – ROADWAY CAPACITIES FOR UNINTERRUPTED FLOWS UNDER IDEAL TRAFFIC AND ROADWAY CONDITIONS†

	Two-Lane, Two-Way Highway: Total for Both Lanes‡	Three-Lane, Two-Way Highway: Total for All Lanes§	Multilane Highway: Average per Lane for Direction of Heavier Flow‖
Basic capacity¶	2000	4000	2000
Practical capacity** for urban conditions (35-40 m.p.h.)	1500	2000	1500
Practical capacity** for rural conditions (45-50 m.p.h.)	900	1500	1000

* Abbreviations and general values from *A Policy on Arterial Highways in Urban Areas*, A.A.S.H.O.
† *Highway Capacity Manual* 1950, U.S. Dept. of Commerce, Bureau of Public Roads.
‡ Distribution by direction is not a factor.
§ Although distribution by direction is not a factor, the use of three-lane, two-way highways is not recommended by the author except where the third lane is a passing lane only.
‖ During periods of peak flow, traffic in one direction may be much heavier than in the other direction.
¶ Same as possible capacity for ideal conditions.
** Generally equivalent to design capacity.

HIGHWAYS - CAPACITY-2

FACTORS THAT MAY REDUCE DESIGN CAPACITY

1. Narrow lanes. } See Table A.
2. Restricted lateral clearances. } See Table A.
3. Narrow shoulders. } See Table A.
4. Commercial vehicles. See Table B.
5. Imperfect alignment.
6. Excessive grades.
7. Signalized intersections at grade.

EFFECT OF FACTORS THAT REDUCE CAPACITIES*

TABLE A – COMBINED EFFECT OF LANE WIDTH AND EDGE CLEARANCES ON HIGHWAY CAPACITIES†

Clearance from Pavement Edge to Obstruction, ft.	Capacity Expressed as % of Capacity of two 12-ft. Lanes with No Restrictive Lateral Clearances							
	Obstruction on One Side				Obstruction on Both Sides‡			
	12-ft. Lanes	11-ft. Lanes	10-ft. Lanes	9-ft. Lanes	12-ft. Lanes	11-ft. Lanes	10-ft. Lanes	9-ft. Lanes
Possible Capacity of Two-Lane Highway								
6	100	88	81	76	100	88	81	76
4	97	85	79	74	94	83	76	71
2	93	81	75	70	85	75	69	65
0	88	77	71	67	76	67	62	58
Practical Capacity of Two-Lane Highway								
6	100	86	77	70	100	86	77	70
4	96	83	74	68	92	79	71	65
2	91	78	70	64	81	70	63	57
0	85	73	66	60	70	60	54	49
Possible and Practical Capacities of Two Lanes‡ For One Direction of Travel on Divided Highways								
6	100	97	91	81	100	97	91	81
4	99	96	90	80	98	95	89	79
2	97	94	88	79	94	91	86	76
0	90	87	82	73	81	79	74	66

†Effects of lane widths and lateral clearances on driver comfort, accident rates, etc., are not included in Table A.
‡When there are more than two lanes for one direction of travel, interior lanes may be assumed to have the same capacity as lanes with lateral clearances equivalent to the distance between the edge of the lane and a vehicle centered in the adjacent lane.

TABLE B – EFFECT OF PASSING SIGHT DISTANCE§ RESTRICTION ON PRACTICAL CAPACITIES OF TWO-LANE HIGHWAYS WHEN ADEQUATE STOPPING SIGHT DISTANCES ARE ALWAYS PRESENT*‖

% of Total Length of Highway on Which Sight Distance Is Restricted to Less than 1500 ft.	Practical Capacity, Passenger Cars per hr.	
	For Operating Speed¶ of 45-50 m.p.h.	For Operating Speed¶ of 50-55 m.p.h.
0	900	600
20	860	560
40	800	500
60	720	420
80	620	300
100	500	160

§ For discussion of sight distance, see p. 12-09.

‖ The data in this table apply to sections with 12-ft. traffic lanes, shoulder adequate for parking disabled vehicles clear of the traffic lanes, and a continuous stopping sight distance corresponding to the design speed. Also the sight distance on the restricted portions of the section must be uniformly distributed between the required stopping sight distance for the design speed and 1500 ft.

¶ Average speed for drivers trying to travel at maximum safe speed.

* *Highway Capacity Manual*, 1950, U. S. Department of Commerce, Bureau of Public Roads.

HIGHWAYS – CAPACITY–3

TABLE A – EFFECT OF COMMERCIAL VEHICLES ON PRACTICAL CAPACITIES ON MULTILANE FACILITIES* †

Commercial Vehicles, %	Capacity Expressed as % of Passenger-Car Capacity on Level Terrain	
	Level Terrain	Rolling Terrain
None	100	100
10	91	77
20	83	63

*On 2-lane highways, the effect is about 25% greater than on multilane expressways.

PROBLEMS

EXAMPLE 1:

Determine the design hourly practical capacity for a two-lane rural highway with a 22 ft. surface and with frequent obstructions within 4 ft. of both pavement edges. The highway is located in rolling terrain where 10% of the peak hour traffic is commercial vehicles, and the sight distance is restricted to less than 1500 ft. over 60% of its length.

Solution. For *ideal conditions*, from Table A, p. 12-03:
Design capacity = 900 passenger cars per hr. to provide operating speed of 45 to 50 m.p.h.

Adjustments

Surface width 22 ft. and lateral clearance 4 ft. both sides.
0.79% from Table A, p. 12-04.

Commercial vehicles

100% – 77% = 23% reduction — from Table A, p. 12-05.
Apply 25% factor for 2-lane highway: 23% × 1.25 = 29%
100% - 29% = 71% capacity

Alignment 720/900 — from Table B, p. 12-04.

Application of Adjustments

Design capacity = $900 \times 0.79 \times 0.71 \times \frac{720}{900}$ = 404 vehicles per hr., say 400 vehicles per hr.
Two-way volume.

EXAMPLE 2:

Determine the practical (design) capacity of two 11 ft. lanes for one direction of travel on a 4-lane divided rural highway with a wide median and 10 ft. shoulders. (Assume level terrain and no commercial vehicles.)

Solution. For *ideal conditions*, from Table A, p. 12-03:
Design capacity = 2000 passenger cars per hr. (1000 each lane) to provide operating speed of 45 to 50 m.p.h.

Adjustments

Surface width (two 11-ft. lanes) and lateral clearance — 0.97% from Table A, p. 12-04.

Applications of Adjustments

2000 × 0.97 = 1940 vehicles per hr., say 1950 vehicles per hr. one-way volume.

† *Highway Capacity Manual*, 1950, U.S. Dept. of Commerce, Bureau of Public Roads.

HIGHWAYS – DESIGN SPEED

DESIGN AND RUNNING SPEEDS*

A design speed should be selected and used for correlation of the physical features of a highway that influence vehicle operation. The assumed design speed should be a logical one with respect to the character of terrain and the type of highway. Some geometric features of highway design, such as superelevation rate, critical length of grade, intersections, curves, etc., require consideration of averagr running speeds.

Design speeds normally recommended are as follows:

Design speed, m.p.h.	30	40	50	60	70
Corresponding average running speed, m.p.h.	27	34	40	45	49

As high a design speed as practicable should be used, preferably a constant value for any one highway. Where there is variation in terrain and other physical controls, changes in design speed for some sections of highway may be necessary.

The following table contains design speeds used by various state highway agencies (For the year 1957).

DESIGN SPEEDS IN MILES PER HOUR

State	Terrain		
	Level or Flat	Rolling	Mountainous
New York	30 - 60 (urban) 50 - 70 (rural)		35 - 50
Montana	70	60	50
Illinois	40 - 70	30 - 60	
Washington	40 - 70	30 - 60	25 - 50
Arizona	50 - 80	40 - 70	30 - 60
New Mexico	50 min. (urban) 75 min. (rural)		55 min.
Virginia	40 - 70	30 - 70	20 - 60
Ohio	40 - 70	30 - 70	25 - 50

CRITICAL LENGTH OF GRADES

Where feasible, upgrade should not be of a length that causes loaded trucks to reduce speed unduly. The critical grade length is determined for a selected truck as that which will cause a 15-m.p.h. reduction in speed below the average running speed on the approach to the upgrade. On this basis, critical lengths of upgrades when approached by level or nearly level sections of road, are as follows:

Upgrade, %	3	4	5	6	7	8
Critical length of upgrade, ft.	1600	1100	800	650	550	500

Where critical lengths of upgrade cannot be avoided, consideration should be given to providing climbing lanes, particularly where truck volume is high.

* From *A Policy on Geometric Design of Rural Highways*, A.A.S.H.O.

HIGHWAYS - GRADES

Either speed or pay load must be drastically reduced on sustained grades of over 3%.

+3% Grade
3'
100'

COMMERCIAL VEHICLES.

Heavy trucks can maintain a speed of 20 miles per hour on sustained grades of 3%. Passing facilities (added lanes or long sight distance) should be provided on mixed traffic roads where grades can not be reduced to 3%.

Reduction of grades of less than 5% or 6% is not warranted for passenger traffic only.

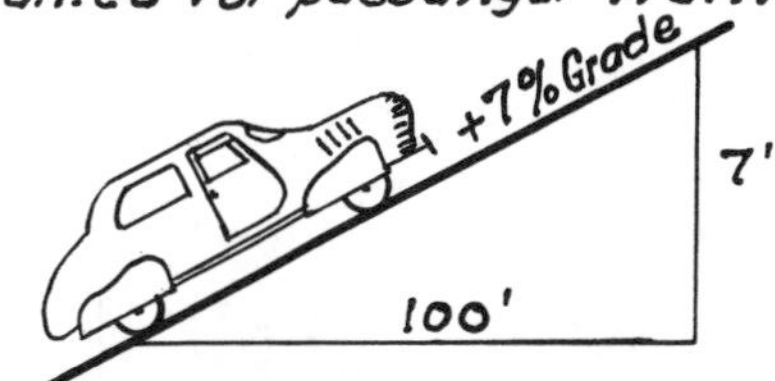
+7% Grade
7'
100'

PASSENGER VEHICLES.

Automobiles can normally operate in high gear on maximum sustained grades up to 7%. Reduction of sustained grades to 5% or 6% for automobile traffic is justified, however, by the need for safety.

6% is the maximum sustained grade for safe operation of trucks and automobiles. On mountain roads, in high altitudes and areas subject to frequent ice, snow, sleet and fog the maximum safe sustained grade is about 5% for all vehicles.

FIG. A - MAXIMUM SUSTAINED GRADES.*

℄ Road Grade - 0.50%

Parallel ditch or subdrain grade

The minimum grade for good ditch drainage is 0.50%. 1% is preferable and 0.25% is the absolute minimum.

ROADS

℄ Street Grade - 0.30%

Parallel gutter grade

The minimum grade for good gutter drainage is 0.30%. With great care in construction an absolute minimum of 0.10% may be used. (Not recommended).

STREETS

FIG. B - MINIMUM GRADES FOR DRAINAGE.**

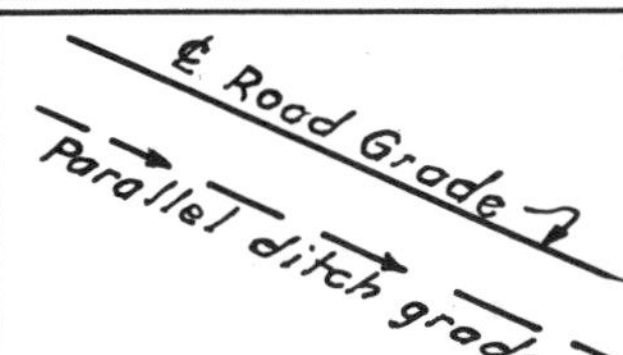
℄ Road Grade
Parallel ditch grade

1% to 3% - Sodded ditches.
2% to 5% - Ditch checks, sod or paved ditch of concrete, bitum. or rubble.

Silty soils will erode on grades over 1%
Most soils will erode on grades over 2%
See Table A-Pg. 5-09 for data on velocities.

FIG. C - MAXIMUM DITCH GRADES.

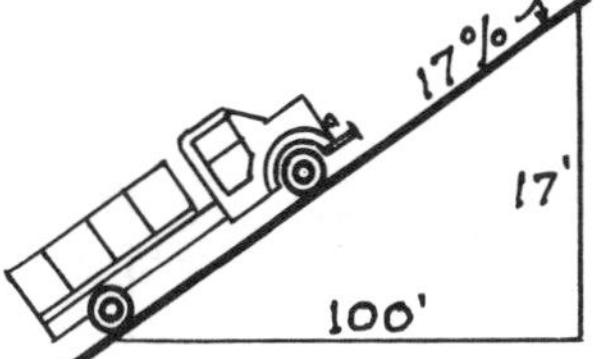
17%
17'
100'

The steepest grades on existing paved highways or streets in the U.S.A. are 9% to 12% for highways and 30% to 32% for urban streets.

The average commercial vehicle can ascend a continuous 17% grade in low gear; use only for ramps, access, drive ways.

FIG. D - ABSOLUTE MAXIMUM GRADE.

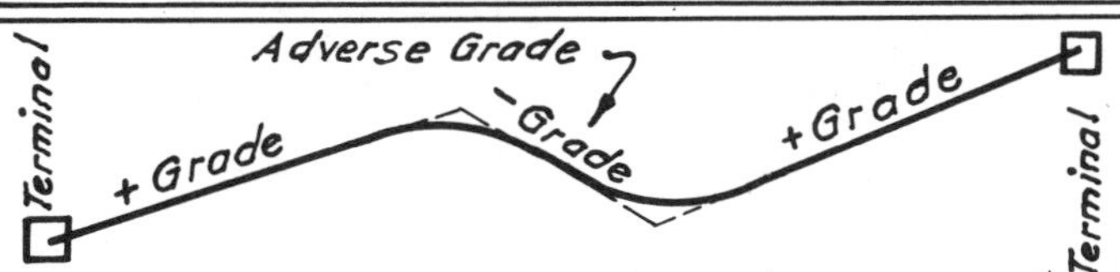
Adverse Grade
Terminal
+ Grade
-Grade
+Grade
Terminal

A grade which is contrary to the general rise or fall between terminals is an adverse grade and should be avoided when practicable because of wasted energy.

FIG. E - ADVERSE GRADES.

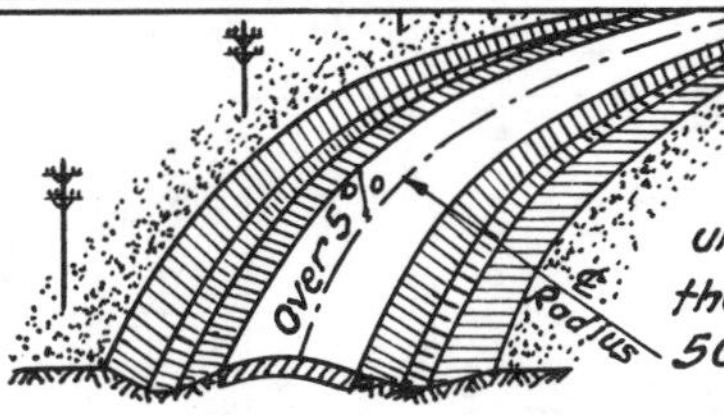
over 5%
℄ Radius

Compensate the grade if grade is over 5% on curves under 500' Rad. by reducing the grade 0.5% for each 50' of rad. less than 500'.

(Combined maximum curve and grade should be avoided if at all practicable)

FIG. F - GRADE COMPENSATION.

TABLE G - LIST OF PHYSICAL GRADE CONTROL POINTS.

1. Elevation of existing & proposed intersecting roads, streets, bridges, separations & railroad tracks.
2. High water at streams, swamps & lowlying areas.
3. Access to adjacent buildings & properties.
4. Maximum gradient for type of traffic expected.
5. Sight distance at sags, summits & intersections.
6. Elevation of ground water table.
7. Unstable soil strata & rock or ledge elevation.
8. Cut and fill balance; do not sacrifice design criteria to balance quantities.
9. Adequate cover over culverts.
10. Minimum & Max. grades for drainage & erosion.
11. Snow drifting areas & late melting location.

* Reference: Public Roads, Vol. 23, No. 3, and Bulletin 65, Iowa Experiment Station, Ames, Iowa.
** Reference: Concrete Pavement Manual, Portland Cement Assoc.

HIGHWAYS - EXAMPLE OF N.Y. STATE GEOMETRIC DESIGN

NEW YORK STATE GEOMETRIC HIGHWAY STANDARDS

Department Of Public Works — Bureau Of Highway Planning

Remarks — BRIDGE NOTES:

2-lane bridges over 500' long with 12' lanes use 88%, 11' lanes use 75%, 10' lanes use 67% of free flow capacity for highway of corresponding width (last column to right of table).

4-lane and over, 12' lanes use 98%.

Area	Terrain	Design Class	Forecast Traffic Counts: Design hour One-Way Capacity — Total	Express Service	Width Data: Pavement, Travel Width	Travel Lane	Parking Lane in Addition to Travel Lanes	Shoulder or Clearance Lanes	Maximum Grade %	% Mileage With Less Than 1500' Passing Sight Distance	Speed, m.p.h. — Design	Running Peak, hours	Trucks, %	1-Way Free Flow Capacity per hr., Ideal Conditions	Remarks
Urban	Level or Rolling	1	4300	3400 900	72÷ 24'≠24'	12' 12'	No parking 8'	10'	4	Not Critical	60 30	45 25	0-10 0-10	3600 2000	Expressway, full control of access, acceleration lanes, etc. No crossings at grade. 24' service roads, parking lane one side, expressway entrance at main streets.
		2	3300	2400 900	72÷ 24'≠24'	12' 12'	No parking 8'	10'	5	Not Critical	60 30	45 25	0-10 0-10	3600 2000	Expressway, partial control of access. Eliminations for cross traffic above 100 per hr. per lane. 24' service roads, parking lane one side, expressway entrance at all streets.
		3	2800	2200 600	48÷ 20'≠20'	12' 10'	No parking 8'	10'	4	Not Critical	60 30	45 25	0-10 0-10	2400 1800	Expressway, full control of access, acceleration lanes, etc. 20' service road, parking lane one side, expressway entrance at main streets.
		4	1500		48÷	12'	12'	8'	5	Not Critical	40	35	0-10	2100	Urban surface arterial, cross roads, right and left turn lanes and signals. Parking restricted at intersections.
		5	1100		48÷	12'	No parking	2'-2'	5	Not Critical	40	35	0-10	2000	Standard arterial street, no service roads, with cross roads and signals. Parking prohibited, 4' median, 2' clearance lanes at curb and median.
		6	800		48'	12'	8'		5	Not Critical	40	35	0-10	1800	Standard arterial street, parking lanes provided.
		7	450 or less		24'	12'	8'		5	30	40	35	0-10	1000	Standard arterial street, parking lanes provided.
Rural	Level or Rolling	1	2800		72÷	12'		10'	3	Not Critical	70	50	0-10	3000	Expressway, full control of access.
		2	1800		48÷	12'		10'	4	Not Critical	70	50	0-10	2000	Expressway, full control of access.
		3	1500		48÷	12'		10'	5	Not Critical	60	45	10-20	2000	4-lane divided, control of access not possible.
		4	1100		48'	12'		10'	5	Not Critical	60	45	10-20	1800	Minimum 4-lane highway. Median consistent with conditions.
		5	500		24'	12'		10'	5	20	60	45	10	600	
		6	300		22'	11'		10'	5	60	60	45	10	600	
		7	200		20'	10'		8'	5	70	60	45	10	600	
		8(S)	100		20'			6'	7		50				Use only on free access, secondary routes, not on the state system.
		9(S)	40		18'			6'	7		50				
		10(S)	10		18'			4'	8		50				
	Hilly or Mountainous	5(M)	350		24'	12'		8'	7'	70	50	40	10	600	Provide truck lane on critical grades.
		6(M)	250		22'	11'		8'	7'	80	50	40	10	600	Provide minimum of one passing opportunity per mile. (1500' sight distance or a passing lane).
		7(M)	150		20'	10'		8'	8'	90	40	35	10	600	Provide minimum of one passing opportunity per mile. (1500' sight distance or a passing lane).
		8(MS)	100		20'			5'	10'		40				Use only on free access, secondary routes, not on the state system.
		9(MS)	40		18'			5'	10'		40				
		10(MS)	10		18'			4'	12'		35				

GENERAL NOTES:

1. *Guide Rail:* Where guide rail is used, add 2' to shoulder width on 8' shoulders and 1'-6" on 10' shoulders, and set road face of posts 2'-0" in from new shoulder line.
2. *Deviation from Standard:* If the minimum standard cannot be obtained or if materially better than the minimum is economically obtainable, a written explanation and the computed capacity of the proposed design shall be submitted for approval.
3. *Medians:* Unless otherwise noted, divided medians shall have a minimum width of 20'.
4. *Show on Plans:* Each plan shall show *(a)* current design hour and annual average daily traffic volumes, *(b)* Highway Design Class, and *(c)* design criteria used.
5. *New Routes:* When enoute is on new alignment, and the forecast traffic exceeds the capacity of the highest-class two-lane facility, the appropriate CONTROLLED-ACCESS design class shall be used.

HIGHWAYS — SUPERELEVATION AND WIDENING - I

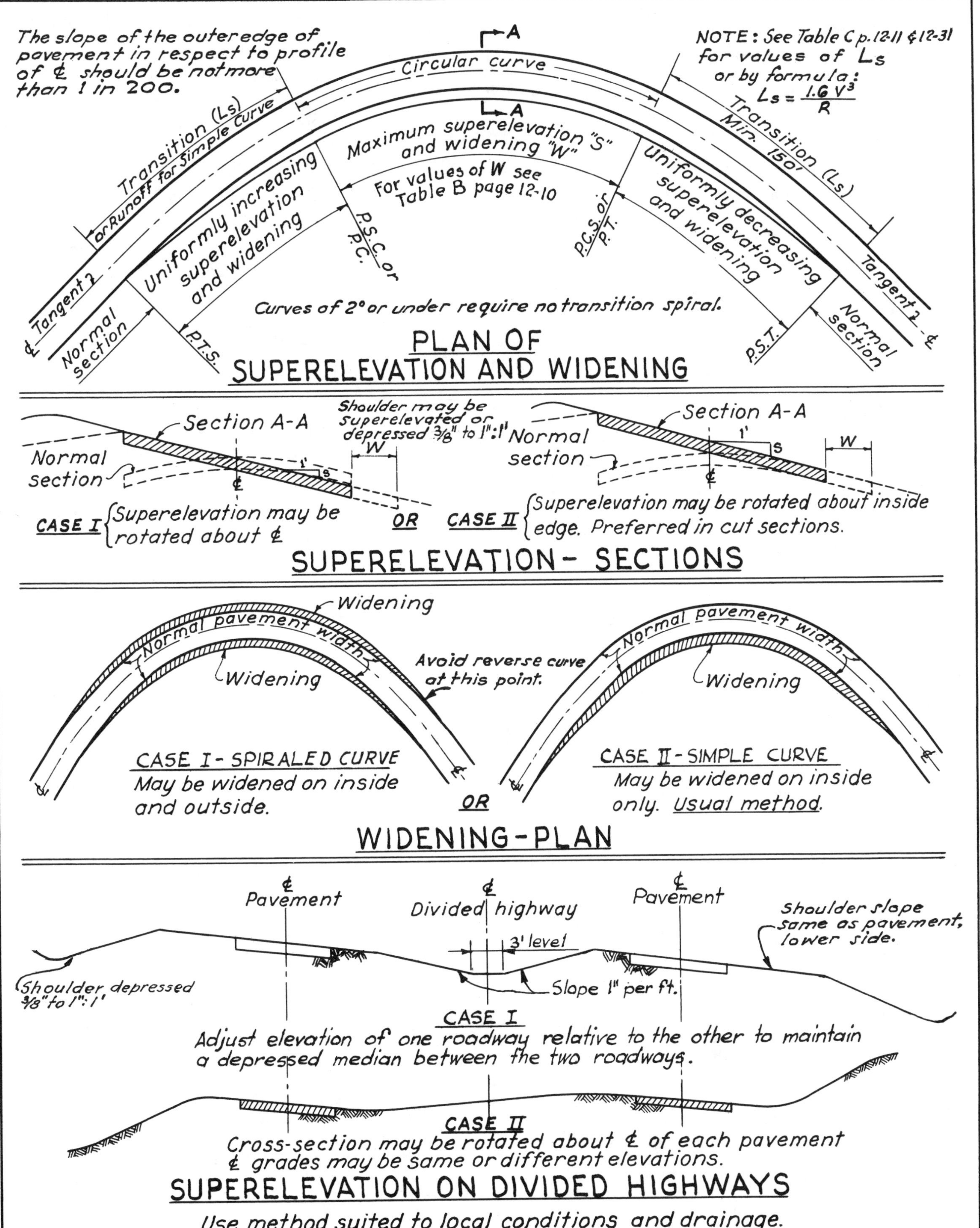

The slope of the outeredge of pavement in respect to profile of ℄ should be not more than 1 in 200.
A
Circular curve
A
NOTE: See Table C p. 12-11 & 12-31 for values of L_s or by formula: $L_s = \frac{1.6V^3}{R}$
Transition (L_s) or Runoff for Simple Curve
Maximum superelevation "S" and widening "W"
Transition (L_s) Min. 150'
For values of W see Table B page 12-10
Uniformly increasing superelevation and widening
Uniformly decreasing superelevation and widening
P.S.C. or P.C.
P.C.S. or P.T.
℄ Tangent ℄
Tangent ℄ ℄
Normal section
P.T.S.
P.S.T.
Normal section
Curves of 2° or under require no transition spiral.
PLAN OF SUPERELEVATION AND WIDENING
Section A-A
Shoulder may be superelevated or depressed 3/8" to 1":1'
Normal section
W
1'
S
℄
CASE I {Superelevation may be rotated about ℄
OR
Section A-A
Normal section
1'
S
W
℄
CASE II {Superelevation may be rotated about inside edge. Preferred in cut sections.
SUPERELEVATION - SECTIONS
Widening
Normal pavement width
Widening
Avoid reverse curve at this point.
CASE I - SPIRALED CURVE
May be widened on inside and outside.
OR
Normal pavement width
Widening
CASE II - SIMPLE CURVE
May be widened on inside only. Usual method.
WIDENING-PLAN
℄ Pavement
℄ Divided highway
℄ Pavement
Shoulder slope same as pavement, lower side.
3' level
Shoulder depressed 3/8" to 1":1'
Slope 1" per ft.
CASE I
Adjust elevation of one roadway relative to the other to maintain a depressed median between the two roadways.
CASE II
Cross-section may be rotated about ℄ of each pavement ℄ grades may be same or different elevations.
SUPERELEVATION ON DIVIDED HIGHWAYS
Use method suited to local conditions and drainage.

HIGHWAYS – SUPERELEVATION AND WIDENING–2

SUPERELEVATION ON CURVES

The basic formula for vehicle operation on a curve is:

$$e + f = \frac{0.067V^2}{R} = \frac{V^2}{15R}$$

where e = rate of superelevation, foot per foot

f = side friction factor.

V = vehicle speed, miles per hour.

R = radius of curve in feet.

TABLE A – MAXIMUM CURVATURE FOR NORMAL CROWN SECTION*

Design Speed, m.p.h.	Average Running Speed, m.p.h.	Maximum degree of Curve	Minimum Curve Radius, ft.
30	27	1° 30'	3,800
40	34	0° 45'	7,600
50	40	0° 30'	11,450
60	45	0° 20'	17,200
70	49	0° 15'	22,900

PAVEMENT WIDENING ON CURVES

Required because:

1. Vehicle occupies greater width in tracking.
2. Driver difficulty in holiday lane.

TABLE B – CALCULATED AND DESIGN VALUES FOR PAVEMENT WIDENING ON OPEN HIGHWAY CURVES*

2-Lane Pavements, One-Way or Two-Way

Degree of Curve	W (Widening). In feet, for 2-lane Pavements on Curves for Width of Pavement on tangent of:														
	20 feet					22 feet					24 feet				
	Design Speed, m.p.h.					Design Speed, m.p.h.					Design Speed, m.p.h.				
	30	40	50	60	70	30	40	50	60	70	30	40	50	60	70
1.5	0.5	0.5	1	1	1	0	0	0	0	0	0	0	0	0	0
2	0.5	1	1	1	1.5	0	0	0	0	0.5	0	0	0	0	0
3	1	1	1.5	1.5	2	0	0	0	0.5	1	0	0	0	0	0
4	1	1.5	1.5	2	2	0	0	0.5	1	1	0	0	0	0	0
5	1.5	1.5	2	2		0	0.5	1	1		0	0	0	0	
6	1.5	2	2			0.5	1	1			0	0	0		
7	1.5	2	2.5			0.5	1	1.5			0	0	0.5		
8	2	2	2.5			1	1.5	1.5			0	0	0.5		
9-10	2	2				1	1.5				0	0.5			
11-13	2.5	3				1.5	2				0.5	1			
14-17	3					2					1				
18-21	3.5					2.5					1.5				
22-24	4					3					2				

NOTE: Values in bold type are suggested for design.

3-lane highways: Use above values where volumes are well below capacity, where curves are operated as 3 lanes and where volumes are high, multiply above values by 1.5.

4-lane undivided highways: Multiply above values by 2.

Where semitrailers are significant, increase tabular values of widening by 0.5 for curves of 10 to 16 degrees, and by 1.0 for curves 17 degrees and sharper.

TABLE C – MOST COMMONLY USED VALUES OF e AND THEIR APPLICATION

e, ft./ft.	Application
0.06	Where traffic density and marginal development tend to reduce speeds. Important intersections.
0.08	Where snow and ice are factors (a desirable maximum value).
0.10	Where slow speeds are required on curves. Where snow and ice are factors (an absolute maximum value).
0.12	Where snow and ice problems do not exist. Maximum practical value.

TABLE D – VALUES OF *f* FOR VARIOUS DESIGN SPEEDS*

Design speed	30	40	50	60	70
f (safe value)	0.16	0.15	0.14	0.13	0.12

TABLE E – MAXIMUM DEGREE OF CURVE AND MINIMUM RADIUS DETERMINED FOR LIMITING VALUES OF e and *f*

Design Speed, m.p.h.	Maximum e	Maximum *f*	Total (e + *f*)	Minimum Radius, ft.	Maximum degree of Curve, degrees
30	0.06	0.16	0.22	273	21.0
40	0.06	0.15	0.21	508	11.3
50	0.06	0.14	0.20	833	6.9
60	0.06	0.13	0.19	1263	4.5
70	0.06	0.12	0.18	1815	3.2
30	0.08	0.16	0.24	250	22.9
40	0.08	0.15	0.23	464	12.4
50	0.08	0.14	0.22	758	7.6
60	0.08	0.13	0.21	1143	5.0
70	0.08	0.12	0.20	1633	3.5
30	0.10	0.16	0.26	231	24.8
40	0.10	0.15	0.25	427	13.4
50	0.10	0.14	0.24	694	8.3
60	0.10	0.13	0.23	1043	5.5
70	0.10	0.12	0.22	1485	3.9
30	0.12	0.16	0.28	214	26.7
40	0.12	0.15	0.27	395	14.5
50	0.12	0.14	0.26	641	8.9
60	0.12	0.13	0.25	960	6.0
70	0.12	0.12	0.24	1361	4.2

*From *A Policy on Geometric Design of Rural Highways,* A.A.S.H.O.

HIGHWAYS - SUPERELEVATION-

DESIGN VALUES FOR RATE OF SUPERELEVATION (e) AND MINIMUM LENGTH OF RUNOFF OR SPIRAL CURVE (l in feet)

e max. = 0.06

D	R	V = 30			V = 40			V = 50			V = 60			V = 70		
			L			L			L			L			L	
Deg.-Min.	Min.	e	2 Lane	4 Lane	e	2 Lane	4 Lane	e	2 Lane	4 Lane	e	2 Lane	4 Lane	e	2 Lane	4 Lane
0 – 15	22918	NC	0	0	NC	0	0	NC	0	0	NC	0	0	RC	200	200
0 – 30	11459	NC	0	0	NC	0	0	RC	150	150	RC	175	175	RC	200	200
0 – 45	7639	NC	0	0	RC	125	125	RC	150	150	.017	175	175	.021	200	200
1 – 00	5730	NC	0	0	RC	125	125	.018	150	150	.023	175	175	.028	200	200
1 – 30	3820	RC	100	100	.018	125	125	.025	150	150	.033	175	175	.040	200	200
2 – 00	2865	RC	100	100	.023	125	125	.031	150	150	.041	175	175	.050	200	220
2 – 30	2292	RC	100	100	.028	125	125	.037	150	150	.048	175	190	.056	200	250
3 – 00	1910	.022	100	100	.032	125	125	.042	150	150	.053	175	210	.060	200	270
3 – 30	1637	.025	100	100	.036	125	125	.046	150	170	.056	175	230	.060	200	270
4 – 00	1432	.027	100	100	.039	125	125	.050	150	180	.059	175	240	D max. = 3.3°		
5 – 00	1146	.032	100	100	.045	125	140	.056	150	200	.060	175	240			
6 – 00	955	.036	100	100	.049	125	150	.059	150	210	D max. = 4.6°					
7 – 00	819	.040	100	110	.053	125	170	.060	150	220						
8 – 00	716	.043	100	120	.056	125	180	D max. = 6.9°								
9 – 00	637	.045	100	120	.058	125	180									
10 – 00	573	.048	100	130	.059	125	190									
11 – 00	521	.050	100	130	.060	130	190									
12 – 00	477	.052	100	140	.060	130	190									
13 – 00	441	.053	100	140	D max. = 11.3°											
14 – 00	409	.055	100	150												
16 – 00	358	.057	100	150												
18 – 00	318	.059	110	160												
20 – 00	286	.060	110	160												
21 – 00	273	.060	110	160												
		D max. = 21.0°														

e max. = 0.10

D	R	V = 30			V = 40			V = 50			V = 60			V = 70		
			L			L			L			L			L	
Deg.-Min.	Min.	e	2 Lane	4 Lane	e	2 Lane	4 Lane	e	2 Lane	4 Lane	e	2 Lane	4 Lane	e	2 Lane	4 Lane
0 – 15	22918	NC	0	0	NC	0	0	NC	0	0	NC	0	0	RC	200	200
0 – 30	11459	NC	0	0	NC	0	0	RC	150	150	RC	175	175	RC	200	200
0 – 45	7639	NC	0	0	RC	125	125	RC	150	150	.018	175	175	.020	200	200
1 – 00	5730	NC	0	0	RC	125	125	.018	150	150	.022	175	175	.028	200	200
1 – 30	3820	RC	100	100	.020	125	125	.027	150	150	.034	175	175	.042	200	200
2 – 00	2865	RC	100	100	.027	125	125	.036	150	150	.046	175	190	.055	200	250
2 – 30	2292	.020	100	100	.033	125	125	.045	150	160	.059	175	240	.069	210	310
3 – 00	1910	.024	100	100	.038	125	125	.054	150	190	.070	190	280	.083	250	370
3 – 30	1637	.027	100	100	.045	125	140	.063	150	230	.081	220	330	.096	290	430
4 – 00	1432	.030	100	100	.050	125	160	.070	170	250	.090	240	360	.100	300	450
5 – 00	1146	.038	100	100	.060	130	190	.083	200	300	.099	270	400	D max. = 3.9°		
6 – 00	955	.044	100	120	.068	140	210	.093	220	330	.100	270	400			
7 – 00	819	.050	100	140	.076	160	240	.097	230	350	D max. = 5.5°					
8 – 00	716	.055	100	150	.084	180	260	.100	240	360						
9 – 00	637	.061	110	160	.089	190	280	.100	240	360						
10 – 00	573	.065	120	180	.093	200	290	D max. = 8.3°								
11 – 00	521	.070	130	190	.096	200	300									
12 – 00	477	.074	130	200	.098	210	310									
13 – 00	441	.078	140	210	.099	210	310									
14 – 00	409	.082	150	220	.100	210	320									
16 – 00	358	.087	160	240	D max. = 13.4°											
18 – 00	318	.093	170	250												
20 – 00	286	.096	170	260												
22 – 00	260	.099	180	270												
24 – 00	239	.100	180	270												
24.8°	231	.100	180	270												
		D max. = 24.8°														

e max. = 0.08

D	R	V = 30			V = 40			V = 50			V = 60			V = 70		
			L			L			L			L			L	
Deg.-Min.	Min.	e	2 Lane	4 Lane	e	2 Lane	7 Lane	e	2 Lane	4 Lane	e	2 Lane	4 Lane	e	2 Lane	4 Lane
0 – 15	22918	NC	0	0	NC	0	0	NC	0	0	NC	0	0	RC	200	200
0 – 30	11459	NC	0	0	NC	0	0	RC	150	150	RC	175	175	RC	200	200
0 – 45	7639	NC	0	0	RC	125	125	RC	150	150	.018	175	175	.021	200	200
1 – 00	5730	NC	0	0	RC	125	125	.018	150	150	.022	175	175	.028	200	200
1 – 30	3820	RC	0	0	.019	125	125	.027	150	150	.035	175	175	.042	200	200
2 – 00	2865	.016	100	100	.025	125	125	.035	150	150	.047	175	190	.056	200	250
2 – 30	2292	.020	100	100	.030	125	125	.043	150	150	.057	175	230	.069	210	310
3 – 00	1910	.023	100	100	.035	125	125	.050	150	180	.066	180	270	.077	230	350
3 – 30	1637	.026	100	100	.040	125	125	.056	150	200	.072	190	290	.080	240	360
4 – 00	1432	.029	100	100	.044	125	140	.062	150	220	.076	210	310	.080	240	360
5 – 00	1146	.035	100	100	.053	125	170	.070	170	250	.080	220	320	D max. = 3.5°		
6 – 00	955	.041	100	110	.060	125	190	.076	180	270	.080	220	320			
7 – 00	819	.045	100	120	.066	140	210	.079	190	280	D max. = 5.0°					
8 – 00	716	.050	100	140	.071	150	220	.080	190	290						
9 – 00	637	.054	100	150	.074	160	230	D max. = 7.6°								
10 – 00	573	.058	100	160	.077	160	240									
11 – 00	521	.061	110	160	.079	170	250									
12 – 00	477	.065	120	180	.080	170	250									
13 – 00	441	.067	120	180	.080	170	250									
14 – 00	409	.070	130	190	D max. = 12.4°											
16 – 00	358	.074	130	200												
18 – 00	318	.077	140	210												
20 – 00	286	.079	140	210												
22 – 00	260	.080	140	220												
22.9°	250	.080	140	220												
		D max. = 22.9°														

e max. = 0.12

D	R	V = 30			V = 40			V = 50			V = 60			V = 70		
			L			L			L			l			l	
Deg.-Min.	Min.	e	2 Lane	4 Lane	e	2 Lane	4 Lane	e	2 Lane	4 Lane	e	2 Lane	4 Lane	e	2 Lane	4 Lane
0 – 15	22918	NC	0	0	NC	0	0	NC	0	0	NC	0	0	RC	200	200
0 – 30	11459	NC	0	0	NC	0	0	RC	150	150	RC	175	175	RC	200	200
0 – 45	7639	NC	0	0	RC	125	125	RC	150	150	.018	175	175	.021	200	200
1 – 00	5730	NC	0	0	RC	125	125	.018	150	150	.023	175	175	.029	200	200
1 – 30	3820	RC	100	100	.020	125	125	.027	150	150	.035	175	175	.043	200	200
2 – 00	2865	RC	100	100	.026	125	125	.036	150	150	.046	175	190	.057	200	260
2 – 30	2292	.021	100	100	.033	125	125	.045	150	160	.058	175	230	.071	210	320
3 – 00	1910	.025	100	100	.040	125	125	.055	150	200	.070	190	280	.086	260	390
3 – 30	1637	.029	100	100	.046	125	140	.064	150	230	.081	220	330	.100	300	450
4 – 00	1432	.033	100	100	.052	125	160	.073	180	260	.094	250	380	.114	340	510
5 – 00	1146	.040	100	110	.064	130	200	.090	220	320	.110	300	450	D max. = 4.2°		
6 – 00	955	.047	100	130	.075	160	240	.104	250	370	.110	300	450			
7 – 00	819	.054	100	150	.085	180	270	.113	270	410	D max. = 5.7°					
8 – 00	716	.061	110	160	.094	200	300	.118	280	430						
9 – 00	637	.067	120	180	.102	210	320	.120	290	430						
10 – 00	573	.073	130	200	.109	230	340	D max. = 8.9°								
11 – 00	521	.078	140	210	.112	240	350									
12 – 00	477	.083	150	220	.116	240	370									
13 – 00	441	.088	160	240	.119	250	380									
14 – 00	409	.093	170	250	.120	250	380									
16 – 00	358	.101	180	270	.120	250	380									
18 – 00	318	.107	190	290	D max. = 14.5°											
20 – 00	286	.112	200	300												
22 – 00	260	.116	210	310												
24 – 00	239	.119	210	320												
26 – 00	220	.120	220	320												
26.7°	214	.120	220	320												
		D max. = 26.7°														

NC Normal crown section
RC Remove adverse crown superelevate at normal crown slope

Spirals desirable but not as essential above heavy line.
Lengths rounded in multiples of 25 or 50 feet permit simpler calculations.

HIGHWAYS - HORIZONTAL SIGHT DISTANCE

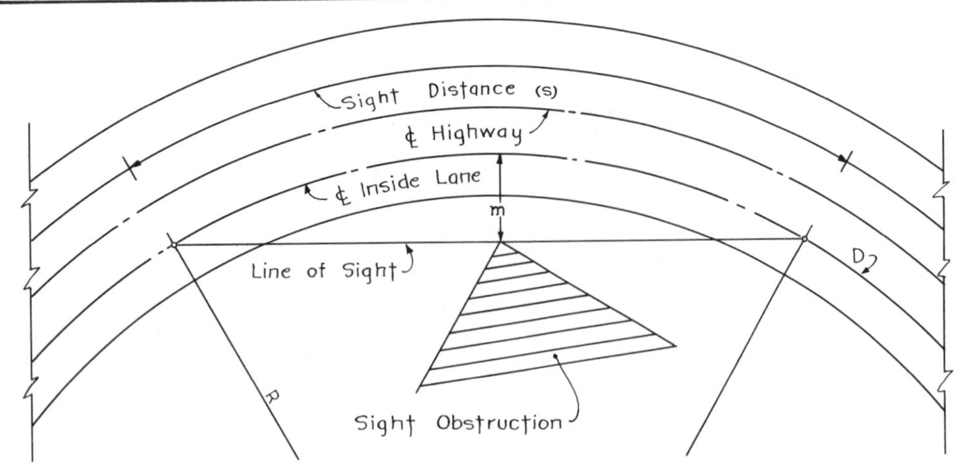
Sight Distance (s)
℄ Highway
℄ Inside Lane
m
Line of Sight
D
R
Sight Obstruction

NOTE:
m = middle ordinate.
= distance in feet from sight obstruction to center line of inside lane measured perpendicular to the line of sight.
D = degree of curve
R = radius of center-line inside lane.

FIG. A – SIGHT DISTANCE ON HORIZONTAL CURVES

$$m = \frac{5730}{D} \text{ vers } \frac{50}{200}$$

Also $m = R\left(\text{vers } \frac{28.65\ S}{R}\right)$

And $S = \frac{R}{28.65} \cos^{-1} \frac{R-M}{R}$

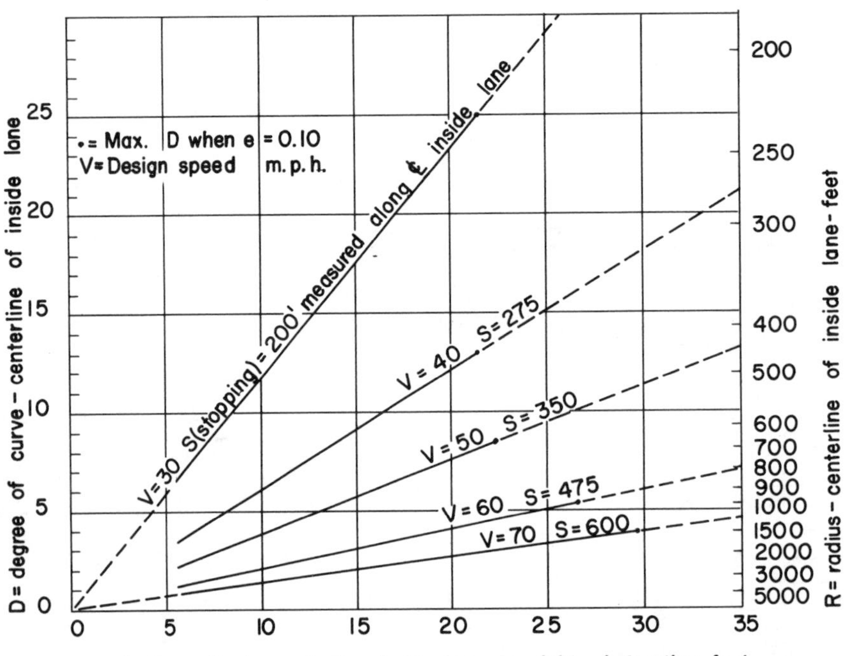
• = Max. D when e = 0.10
V = Design speed m.p.h.
V=30 S(stopping)=200' measured along ℄ inside lane
V=40 S=275
V=50 S=350
V=60 S=475
V=70 S=600
D = degree of curve - centerline of inside lane
R = radius - centerline of inside lane - feet
m = middle ordinate: centerline inside lane to sight obstruction - feet
0
5
10
15
20
25
30
35
200
250
300
400
500
600
700
800
900
1000
1500
2000
3000
5000

FIG. B – MIDDLE ORDINATE STOPPING SIGHT DISTANCE ON HORIZONTAL CURVES (OPEN-ROAD CONDITION)*

TABLE C – MIDDLE ORDINATES FOR PASSING SIGHT DISTANCE ON CURVES*

Degree of Curve		Two-Lane Highway: Required middle ordinate† in feet for design speed, m.p.h., of:					Three-Lane Highway: Required middle ordinate† in feet for design speed, m.p.h., of:				
deg.	min.	30	40	50	60	70	30	40	50	60	70
0	15	3	9	16	22	29	Not Applicable	Not Applicable			
0	30	7	18	32	44	58			16	21	28
0	45	10	28	47	65	86			24	32	42
1	00	14	37	63	87	115			31	43	56
1	15	17	46	79	109	144			39	53	70
1	30	21	55	94	130	172			47	64	83
2	00	28	73	125	173				63	85	111
3	00	42	110	186					93	127	165
4	00	55	145						124	168	
5	00	69	179						154		
6	00	83									
8	00	109									
10	00	134									
12	00	158									

*A policy on Geometric Design of Rural Highways, A.A.S.H.O.
†Measured from center line of inside lane.

HIGHWAYS–VERTICAL SIGHT DISTANCE–I

SIGHT DISTANCE ON VERTICAL CURVES Basic formulas for length of a parabolic curve in terms of algebraic difference in grade and sight distance are:

When S is less than L,

$$L = \frac{AS^2}{100(\sqrt{2h_1} + \sqrt{2h_2})^2}.$$

When S is greater than L,

$$L = 2S - \frac{200(\sqrt{h_1} + \sqrt{h_2})^2}{A}.$$

where L = length of vertical curve, ft.
S = sight distance, ft.
A = algebraic difference in grade, %
h_1 = height of eye above roadway surface, ft.
h_2 = height of object above roadway surface, ft.

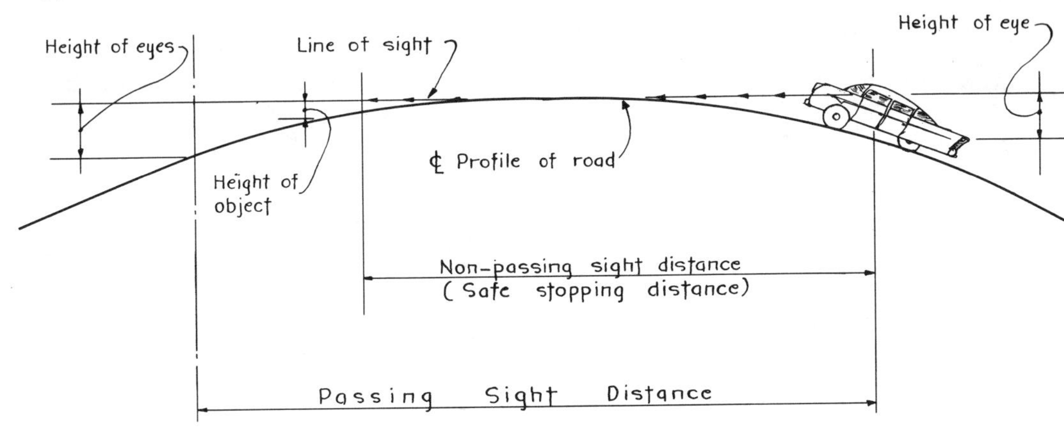
Height of eyes
Line of sight
Height of eye
Height of object
℄ Profile of road
Non-passing sight distance
(Safe stopping distance)
Passing Sight Distance

PROFILE

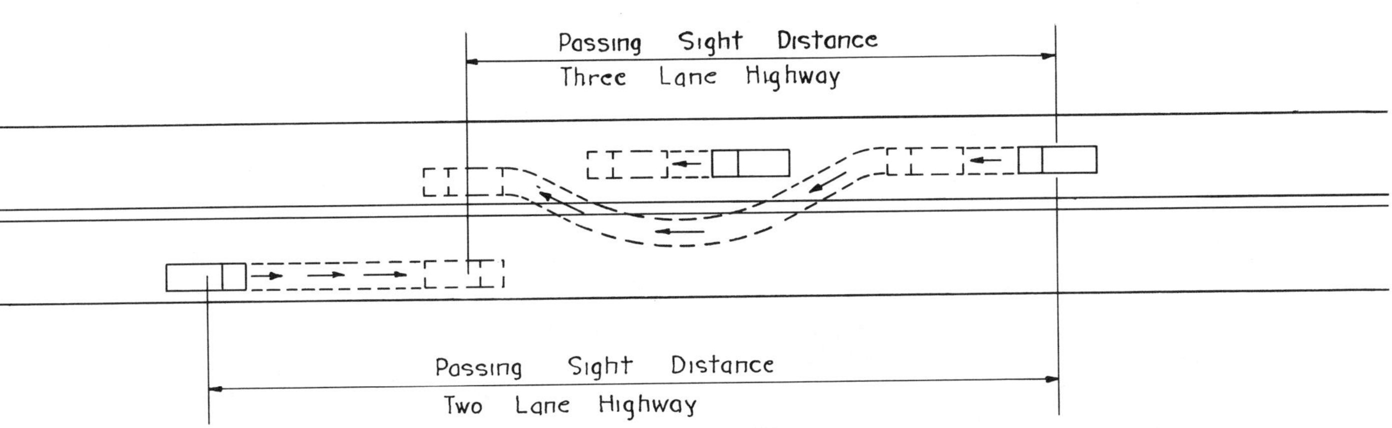
Passing Sight Distance
Three Lane Highway
Passing Sight Distance
Two Lane Highway

PLAN

VERTICAL SIGHT DISTANCE

On three-lane highway, opposing traffic should be restricted by pavement marking from using center lane in passing zones.

MINIMUM SIGHT DISTANCE*			
Design Speed, m.p.h.	Minimum Passing Sight Distance		Minimum Stopping Sight Distance, ft.†
	2-Lane Highway	3-Lane Highway	
30	800		200
40	1300		275
50	1700	1200	350
60	2000	1400	475
70	2300	1600	600

* *A Policy on Geometric Design of Rural Highways*, A.A.S.H.O.
† Measured from driver's eye to top of object.

HIGHWAYS—VERTICAL SIGHT DISTANCE-2

NON-PASSING SIGHT DISTANCE CHARTS

PASSING SIGHT DISTANCE CHARTS

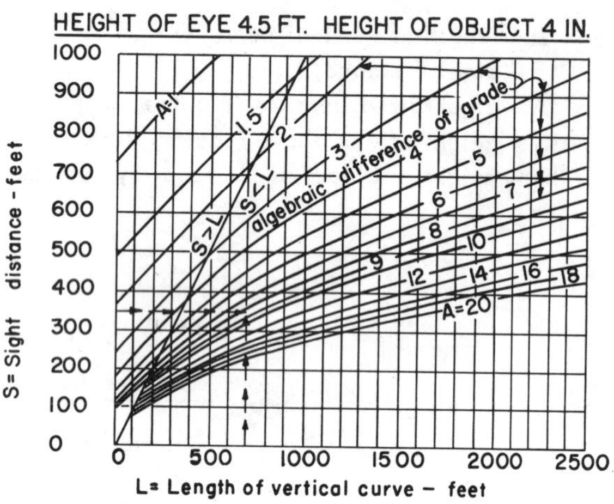

HEIGHT OF EYE 4.5 FT. HEIGHT OF OBJECT 4 IN.
S= Sight distance - feet
L= Length of vertical curve - feet
algebraic difference of grade
S>L
S<L
A=1
A=20

EXAMPLE:

Given:
Design speed 50 m.p.h.
A = 8.
S = 350'.

Required: L.

Solution from chart:
L = 700'.

A = algebraic difference of grades ÷ 100.
Formula:

When $S > L$, $S = \frac{7.28}{A} + \frac{L}{2}$. When $S < L$, $S = 3.82\sqrt{\frac{L}{A}}$

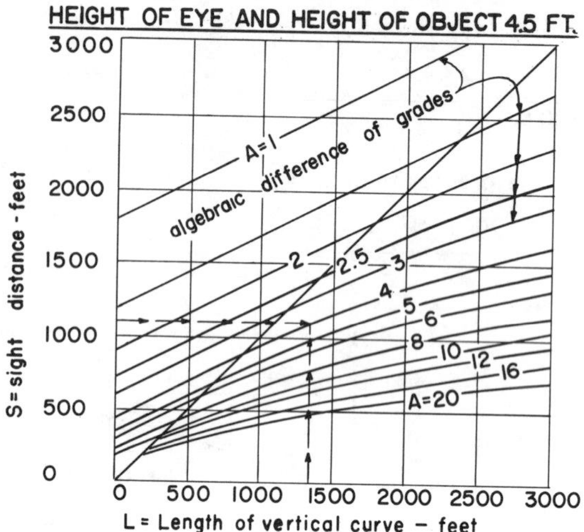

HEIGHT OF EYE AND HEIGHT OF OBJECT 4.5 FT.
S= sight distance - feet
L= Length of vertical curve - feet
algebraic difference of grades
A=1
A=20

EXAMPLE:

Given:
Design speed 40 m.p.h.
A = 4.
S = 1100'.

Required: L.

Solution from chart:
L = 1350'.

A = algebraic difference of grades ÷ 100.
Formula:

When $S > L$, $S = \frac{18}{A} + \frac{L}{2}$. When $S < L$, $S = 6\sqrt{\frac{L}{A}}$.

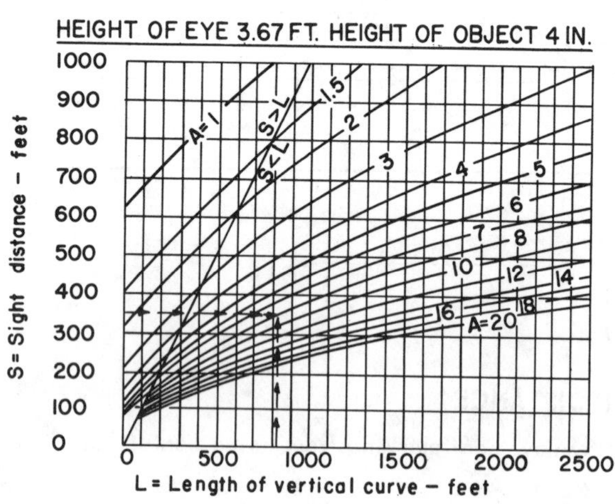

HEIGHT OF EYE 3.67 FT. HEIGHT OF OBJECT 4 IN.
S= Sight distance - feet
L= Length of vertical curve - feet
S>L
S<L
A=1
A=20

EXAMPLE:

Given:
Design speed 50 m.p.h.
A = 8.
S = 350'

Required: L.

Solution from chart:
L = 830'.

A = algebraic difference of grades ÷ 100.
Formula:

When $S > L$, $S = \frac{6.21}{A} + \frac{L}{2}$. When $S < L$, $S = 3.52\sqrt{\frac{L}{A}}$.

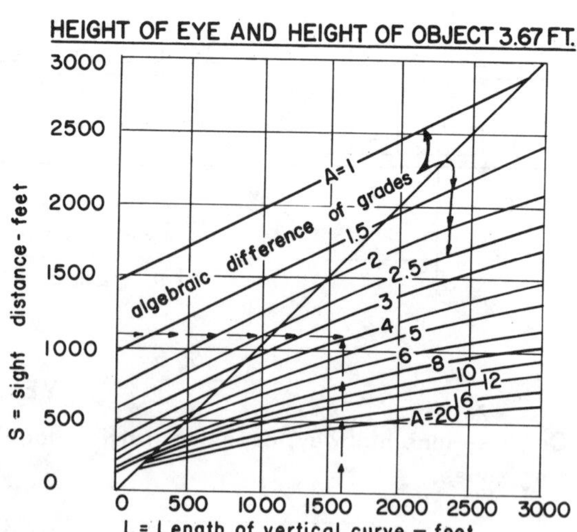

HEIGHT OF EYE AND HEIGHT OF OBJECT 3.67 FT.
S = sight distance - feet
L= Length of vertical curve - feet
algebraic difference of grades
A=1
A=20

EXAMPLE:

Given:
Design speed 40 m.p.h.
A = 4.
S = 1100'.

Required: L.

Solution from chart:
L = 1600'.

A = algebraic difference of grades ÷ 100.
Formula:

When $S > L$, $S = \frac{14.65}{A} + \frac{L}{2}$. When $S < L$, $S = 5.42\sqrt{\frac{L}{A}}$.

Adapted from AASHO

HIGHWAYS – VERTICAL SIGHT DISTANCE – 3

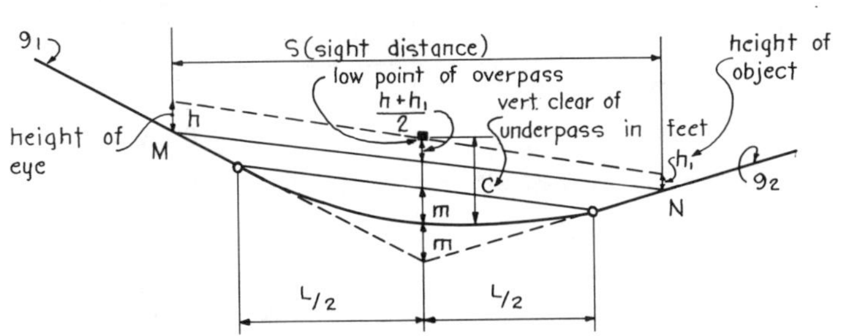

g_1
S (sight distance)
low point of overpass
$\frac{h+h_1}{2}$
vert. clear of underpass in feet
height of object
height of eye
M
h
h_1
C
m
m
N
g_2
L/2
L/2

CASE I – SIGHT DISTANCE GREATER THAN LENGTH OF VERTICAL CURVE, OR $S > L$

PASSING

Formula: (Based on C = 14'-6", h = 4'-6", h_1 = 4'-6")

$$L = 2S - \frac{80}{A} \text{ (Where } S \text{ is in 100' stations).}$$

EXAMPLE:
Given: S = 6, A (algeb. diff.) = 8.
Required: Length of vertical curve, *L*.
Solution: $L = 12 - \frac{80}{8} = 2$ (100' sta.).

NON-PASSING

Formula: (Based on C = 14'-6", h = 4'-6", h_1 = 6") $L = 2S - \frac{96}{A}$.

EXAMPLE:
Given: S = 8, A = 8.
Required: Length of vertical curve, *L*.
Solution: $L = 16 - \frac{96}{8} = 4$ (100' sta.).

PASSING

Formula: (Based on C = 14'-0", h = 3'-8", h_1 = 3'-8")

$$L = 2S - \frac{82.64}{A} \text{ (where } S \text{ is in 100' sta.)}$$

EXAMPLE:
Given: S = 6, A (algeb. diff.) = 8.
Required: Length of vertical curve, *L*.
Solution: $L = 12 - \frac{82.64}{8} = 1.67$ (100' sta.).

NON-PASSING

Formula: (Based on C = 14'-0", h = 3'-8", h_1 = 4") $L = 2S - \frac{96}{A}$.

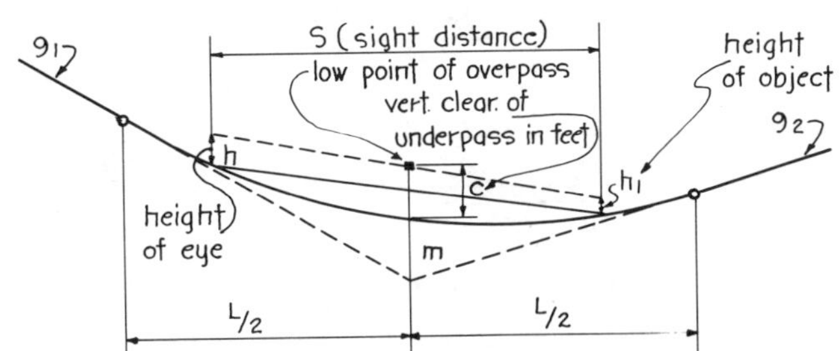

g_1
S (sight distance)
low point of overpass
vert. clear. of underpass in feet
height of object
g_2
h
C
h_1
height of eye
m
L/2
L/2

CASE II – SIGHT DISTANCE LESS THAN LENGTH OF VERTICAL CURVE, OR $S < L$

PASSING

Formula: (Based on C = 14'-6", h = 4'-6", h_1 = 4'-6")

$$L = \frac{S^2A}{80} \text{ (where } S \text{ is in 100' stations).}$$

EXAMPLE:
Given: S = 11, A (algeb. diff.) = 8.
Required: Length of vertical curve, *L*.
Solution: $L = \frac{11^2 \times 8}{80} = 12.1$ (100' sta.).

NON-PASSING

Formula: (Based on C = 14'-6", h = 4'-6", h_1 = 6") $L = \frac{S^2A}{96}$.

EXAMPLE:
Given: S = 13, A = 8.
Required: Length of vertical curve, *L*.
Solution: $L = \frac{13^2 \times 8}{96} = 14.08$ (100' sta.).

PASSING

Formula: (Based on C = 14'-0", h = 3'-8", h_1 = 3'-8")

$$L = \frac{S^2A}{82.64} \text{ (where } S \text{ is in 100' stations.)}$$

EXAMPLE:
Given: S = 11, A (algeb. diff.) = 8.
Required: Length of vertical curve, *L*.
Solution: $L = \frac{11^2 \times 8}{82.64} = 11.7$ (100' sta.).

NON-PASSING

Formula: (Based on C = 14'-0", h = 3'-8", h_1 = 4") $L = \frac{S^2A}{96}$.

SIGHT DISTANCE AT UNDERPASS (Minimum Stopping Sight Distance)*

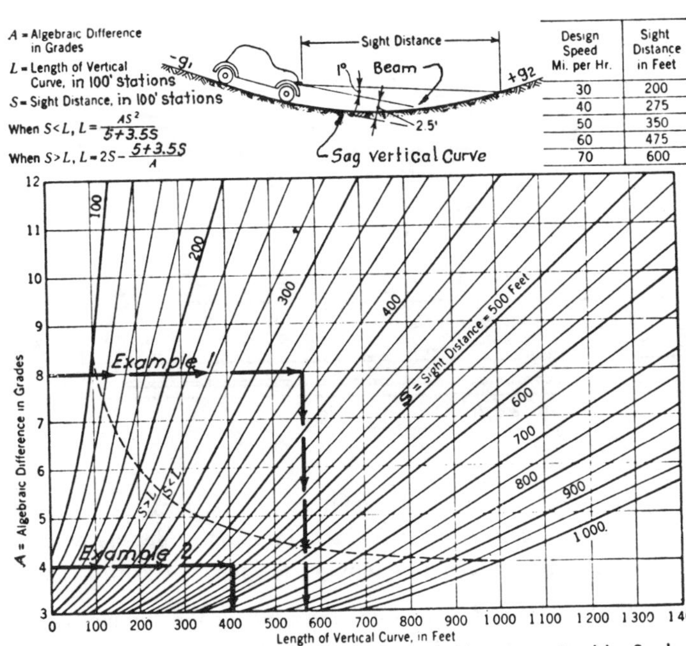

A = Algebraic Difference in Grades
L = Length of Vertical Curve, in 100' stations
S = Sight Distance, in 100' stations
When $S<L$, $L = \frac{AS^2}{5+3.5S}$
When $S>L$, $L = 2S - \frac{5+3.5S}{A}$
Sight Distance
$-g_1$
$+g_2$
1°
Beam
2.5'
Sag Vertical Curve
Example 1
Example 2
S = Sight Distance = 500 Feet
100
200
300
400
600
700
800
900
1000
S>L
S<L
A = Algebraic Difference in Grades
Length of Vertical Curve, in Feet

Design Speed Mi. per Hr.	Sight Distance in Feet
30	200
40	275
50	350
60	475
70	600

EXAMPLE 1. – $S < L$.
Given: Design speed – 50 m.p.h., and A = 8.
Required: Sight distance and length of vertical curve, *L*.
Solution: From table, sight distance = 350 ft., and, from chart, *L* = 568 ft.

EXAMPLE 2. – $S > L$.
Given: Design speed – 60 m.p.h., and A = 4.
Required: Sight distance and length of vertical curve, *L*.
Solution: From table, sight distance = 475 ft., and, from chart, *L* = 410 ft.

Based on headlight illumination for safe stopping distance.

HEADLIGHT SIGHT DISTANCE† CHART – SAG VERTICAL CURVES

* Adapted from *American Highway Practice*, by L.I. Hewes; revised by Seelye, Stevenson, Valne, & Knecht.
† Adapted from *Civil Engineering*, Jan. 1944, p. 22, Fig. 2, by D. Thompson.

HIGHWAYS–CROSS SECTION ELEMENTS–I

TABLE A – PAVEMENT CROSS SLOPES

Surface Type	Range in Rate of Cross Slope	
	Inch per foot	Foot per foot
High	1/8 – 1/4	0.01 – 0.02
Intermediate	3/16 – 3/8	0.015 – 0.03
Low	1/4 – 1/2	0.02 – 0.04

PAVEMENT LANE WIDTHS

Lane widths of 11 and 12 feet should be provided on modern main highways. Use of 13-foot lanes should be considered on high-speed, high-volume highways. Nine- and 10-foot lanes should be reserved only for low-volume highways, as dictated by economic considerations.

TABLE B – SHOULDER CROSS SLOPES

Types of Surface	Shoulder Cross Slope	
	Inch per foot	Foot per foot
No pavement edge curbs:		
Bituminous	3/8 – 1/2	0.03 – 0.04
Gravel or crushed stone	1/2 – 3/4	0.04 – 0.06
Turf	1	0.08
With shoulder curbs at pavement edge:		
Bituminous	1/4	0.02
Gravel or crushed stone	1/4 – 1/2	0.02 – 0.04
Turf	3/8 – 1/2	0.03 – 0.04

TABLE C – SHOULDER WIDTHS

Type of Highway	Width of Shoulder
Heavily traveled highway. High-speed highway.	10' minimum, 12' desirable
Low-type highway. Highway in difficult terrain	4' minimum, 6' or 8' desirable

REMARKS:
Full-width usable shoulders should be provided on all highways on which the D.H.V. is more than 100, when they can be obtained at reasonable cost.

NOTE:
In general, a width of 10 feet is a desirable minimum.

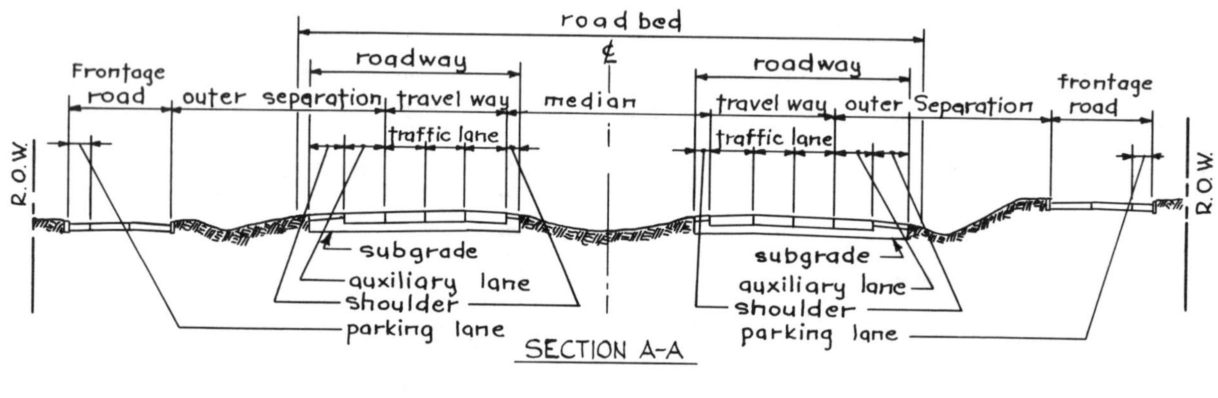
road bed
roadway
roadway
Frontage road
outer separation
travel way
median
travel way
outer Separation
frontage road
traffic lane
traffic lane
R.O.W.
R.O.W.
subgrade
subgrade
auxiliary lane
auxiliary lane
shoulder
shoulder
parking lane
parking lane

SECTION A–A

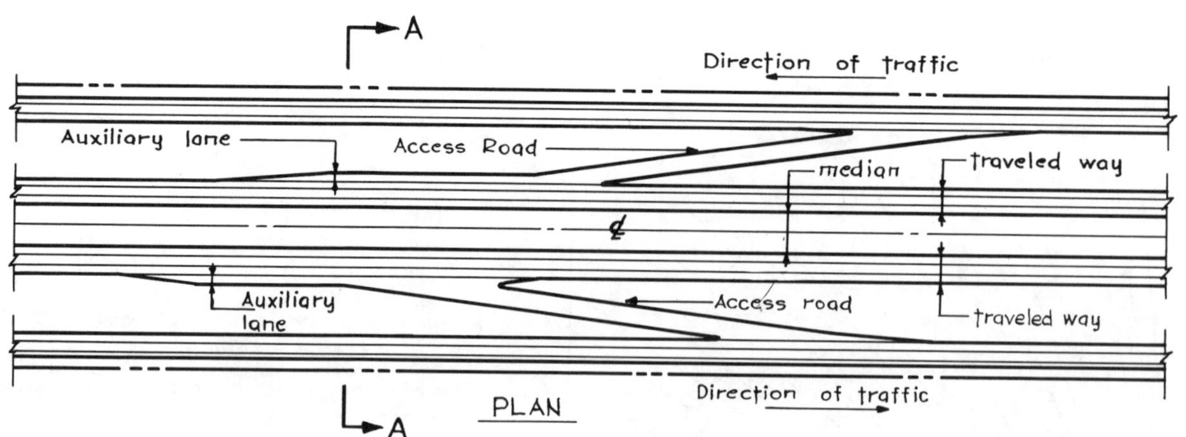
A
Direction of traffic
Auxiliary lane
Access Road
median
traveled way
Auxiliary lane
Access road
traveled way
Direction of traffic
A

PLAN

Roadway (General). The portion of a highway, including shoulders, for vehicular use. A divided highway has two or more roadways.

Parking lane An auxiliary lane primarily for the parking of vehicles.

Speed-change lane An auxiliary lane, including tapered areas, primarily for the acceleration or deceleration of vehicles entering or leaving the through-traffic lanes.

Median lane A speed-change lane within the median to accommodate left-turning vehicles.

From, "A Policy on Geometric Design of Rural Highways," AASHO.

HIGHWAYS - CROSS SECTION ELEMENTS - 2

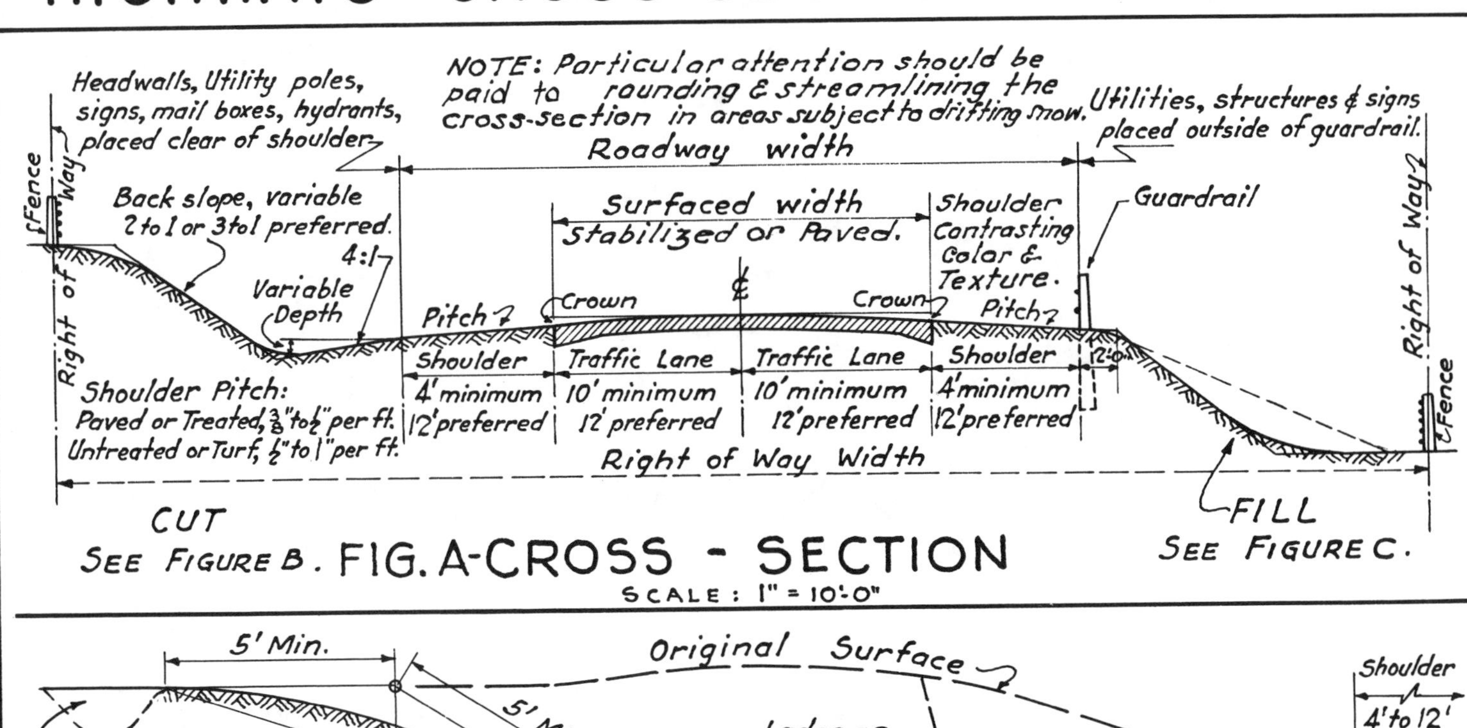
NOTE: Particular attention should be paid to rounding & streamlining the cross-section in areas subject to drifting snow.
Headwalls, Utility poles, signs, mail boxes, hydrants, placed clear of shoulder.
Utilities, structures & signs placed outside of guardrail.
Roadway width
Fence
Right of Way
Back slope, variable 2 to 1 or 3 to 1 preferred.
4:1
Variable Depth
Surfaced width Stabilized or Paved.
Shoulder Contrasting Color & Texture.
Guardrail
Pitch
Crown
Crown
Pitch
Shoulder
Traffic Lane
Traffic Lane
Shoulder
2'-0"
4' minimum 12' preferred
10' minimum 12' preferred
10' minimum 12' preferred
4' minimum 12' preferred
Shoulder Pitch:
Paved or Treated, 3/8" to 1/2" per ft.
Untreated or Turf, 1/2" to 1" per ft.
Right of Way Width
CUT
See Figure B.
FILL
See Figure C.

FIG. A - CROSS - SECTION
SCALE: 1" = 10'-0"

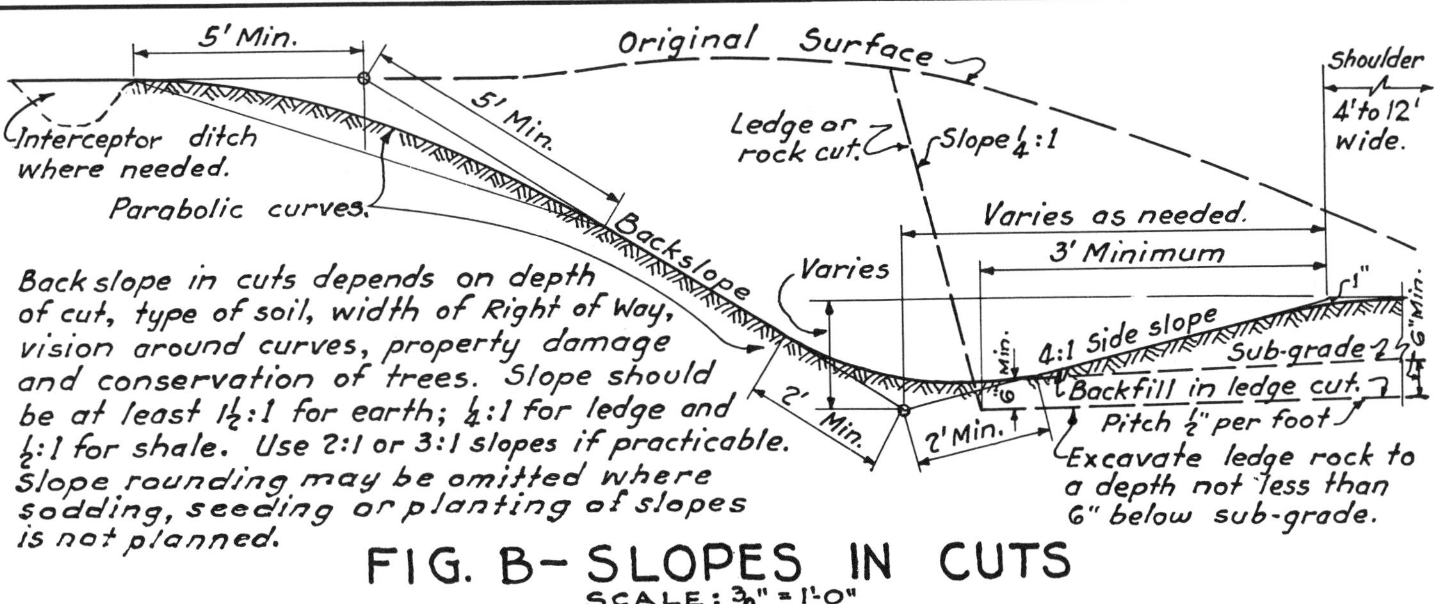
5' Min.
Original Surface
Shoulder 4' to 12' wide.
5' Min.
Interceptor ditch where needed.
Ledge or rock cut.
Slope 1/4:1
Parabolic curves.
Backslope
Varies as needed.
3' Minimum
Varies
1"
4:1 Side slope
Sub-grade
6" Min.
6" Min.
Backfill in ledge cut.
2' Min.
2' Min.
Pitch 1/2" per foot
Excavate ledge rock to a depth not less than 6" below sub-grade.
Back slope in cuts depends on depth of cut, type of soil, width of Right of Way, vision around curves, property damage and conservation of trees. Slope should be at least 1½:1 for earth; ¼:1 for ledge and ½:1 for shale. Use 2:1 or 3:1 slopes if practicable. Slope rounding may be omitted where sodding, seeding or planting of slopes is not planned.

FIG. B - SLOPES IN CUTS
SCALE: 3/8" = 1'-0"

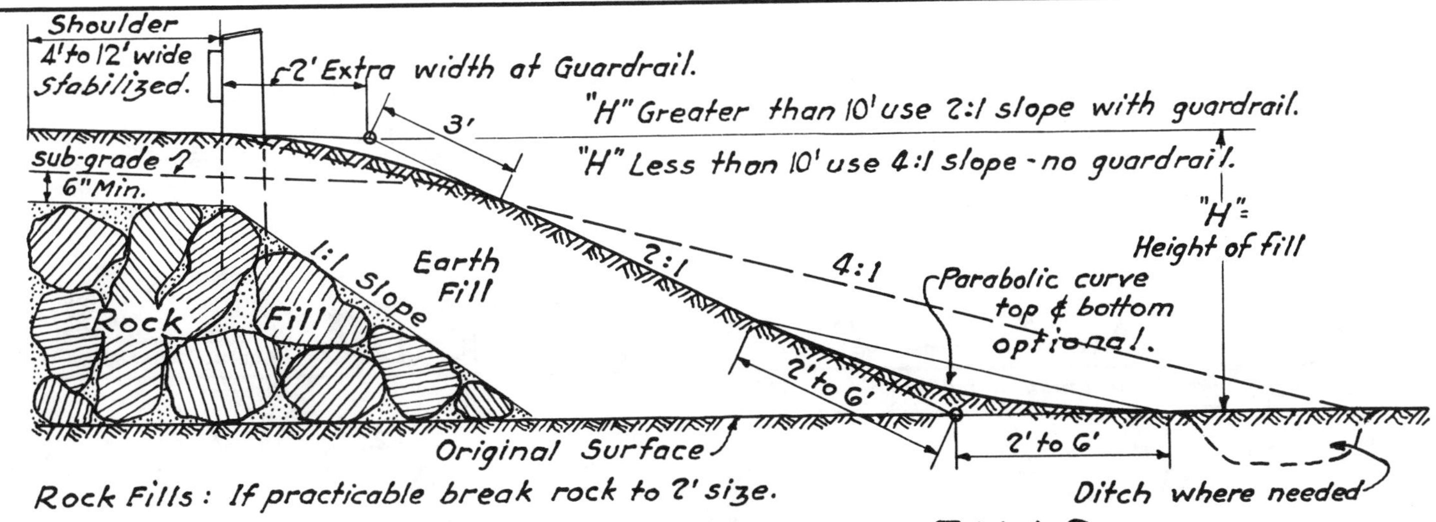
Shoulder 4' to 12' wide Stabilized.
2' Extra width at Guardrail.
3'
"H" Greater than 10' use 2:1 slope with guardrail.
"H" Less than 10' use 4:1 slope - no guardrail.
sub-grade
6" Min.
"H" = Height of fill
Earth Fill
1:1 Slope
Rock Fill
2:1
4:1
Parabolic curve top & bottom optional.
2' to 6'
2' to 6'
Original Surface
Rock Fills: If practicable break rock to 2' size.
Ditch where needed

FIG. C - SLOPES IN FILLS
SCALE: 3/8" = 1'-0"

Ref.: Highway Design Policies by A.A.S.H.O.

HIGHWAYS - BARRIERS

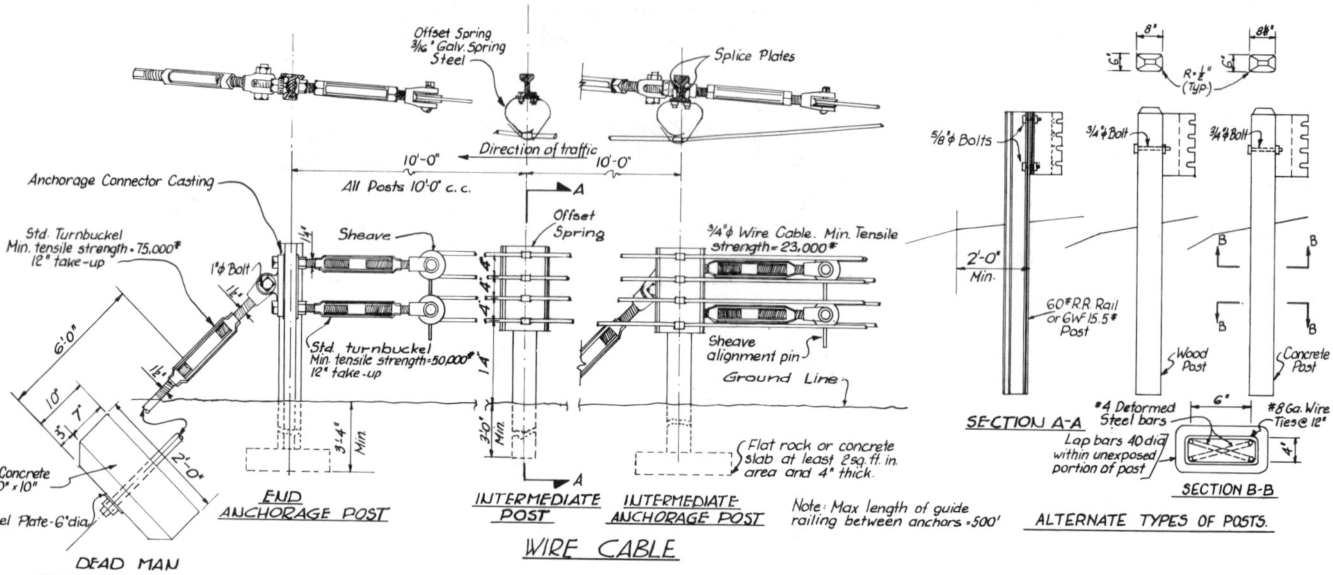

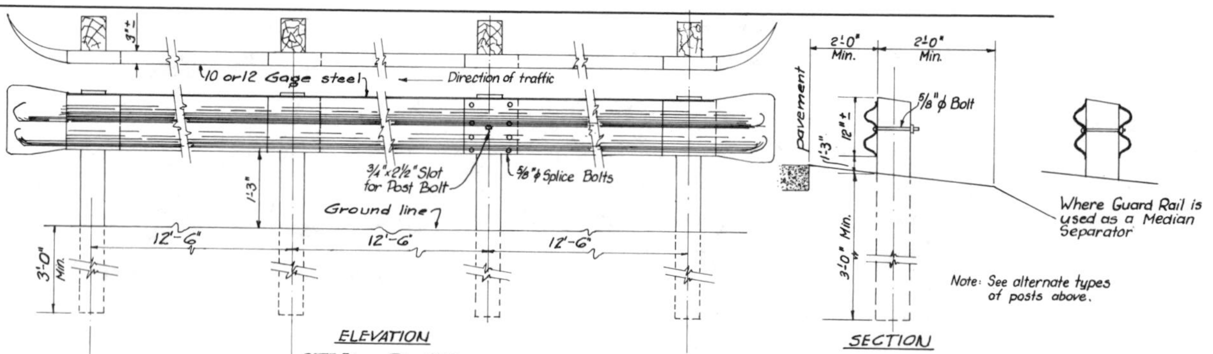

Offset Spring 3/16" Galv. Spring Steel

Splice Plates

Direction of traffic

10'-0" 10'-0"

All Posts 10'-0" c. c.

Anchorage Connector Casting

Std. Turnbuckel Min. tensile strength = 75,000# 12" take-up

1"ϕ Bolt

Sheave

Offset Spring

3/4"ϕ Wire Cable. Min. Tensile strength = 23,000#

Std. turnbuckel Min. tensile strength = 50,000# 12" take-up

Sheave alignment pin

Ground Line

6'-0"

10" 7" 3"

2'-0"

1×2×3½ Concrete 2'-0"×2'-0"×10"

½"ϕ Steel Plate-6" dia.

3'-4" Min.

3'-0" Min.

Flat rock or concrete slab at least 2 sq. ft. in. area and 4" thick.

DEAD MAN

END ANCHORAGE POST

INTERMEDIATE POST

INTERMEDIATE ANCHORAGE POST

Note: Max length of guide railing between anchors = 500'

WIRE CABLE

5/8"ϕ Bolts

2'-0" Min.

60# RR Rail or GW 15.5# Post

SECTION A-A

8" 8½" 6" R = ½" (Typ.)

3/4"ϕ Bolt

Wood Post

Concrete Post

\#4 Deformed Steel bars

Lap bars 40 dia. within unexposed portion of post

\#8 Ga. Wire Ties @ 12"

6" 4"

SECTION B-B

ALTERNATE TYPES OF POSTS.

3"±

10 or 12 Gage steel

Direction of traffic

3/4"×2½" Slot for Post Bolt

5/8"ϕ Splice Bolts

1'-3"

Ground line

12'-6" 12'-6" 12'-6"

3'-0" Min.

ELEVATION

STEEL PLATE

For heavy traffic

2'-0" Min. 2'-0" Min.

pavement

5/8"ϕ Bolt

12"± 1'-3"

3'-0" Min.

Where Guard Rail is used as a Median Separator

Note: See alternate types of posts above.

SECTION

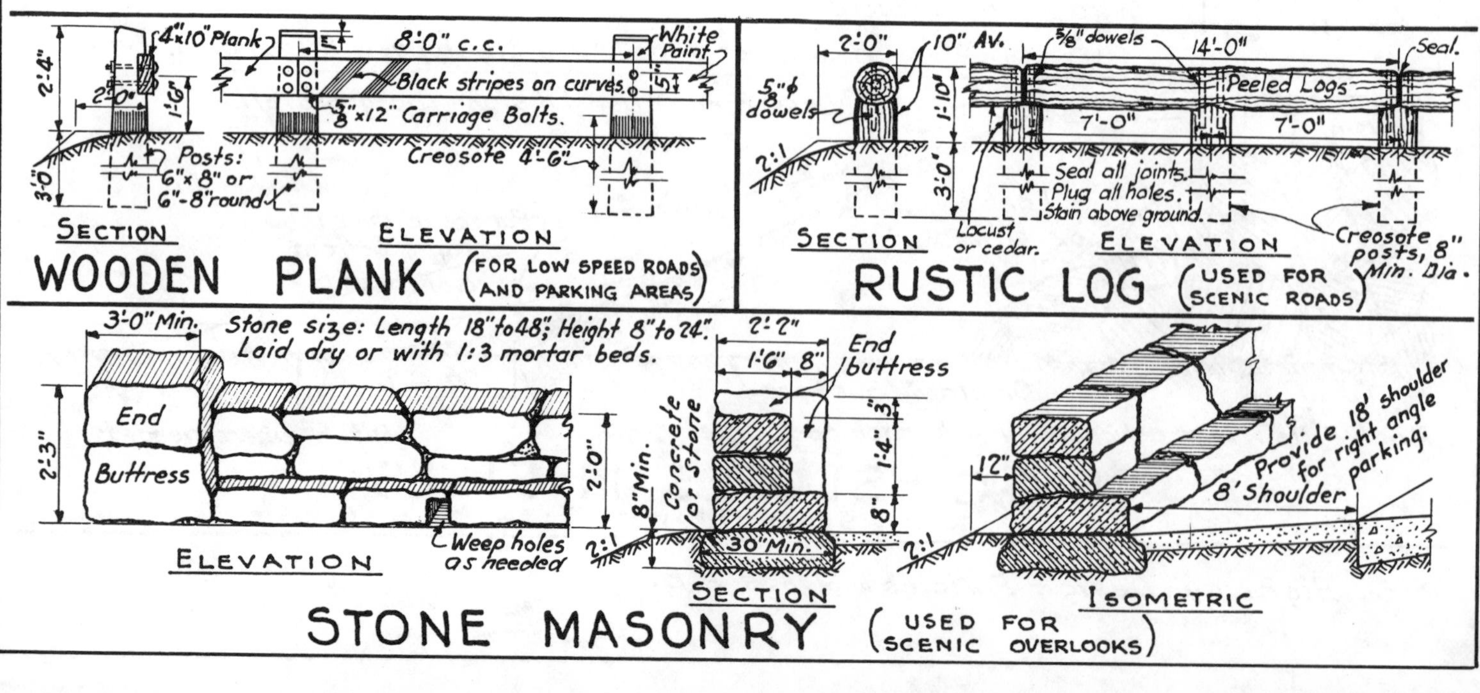
2'-4"
4"×10" Plank
8'-0" c.c.
White Paint
2'-0"
1'-6"
5"
Black stripes on curves.
5/8"×12" Carriage Bolts.
Creosote 4'-6"
3'-0"
Posts: 6"×8" or 6"-8" round
SECTION
ELEVATION
WOODEN PLANK (FOR LOW SPEED ROADS AND PARKING AREAS)
2'-0"
10" AV.
5/8" dowels
14'-0"
Seal.
5/8"ϕ dowels
1'-10"
Peeled Logs
7'-0"
7'-0"
2:1
3'-0"
Seal all joints.
Plug all holes.
Stain above ground.
Locust or cedar.
Creosote posts, 8" Min. Dia.
SECTION
ELEVATION
RUSTIC LOG (USED FOR SCENIC ROADS)
3'-0" Min.
Stone size: Length 18" to 48"; Height 8" to 24".
Laid dry or with 1:3 mortar beds.
End
Buttress
2'-3"
2'-0"
Weep holes as needed
ELEVATION
2'-2"
1'-6" 8"
End buttress
3"
1'-4"
8"
8" Min.
Concrete or Stone
2:1
30" Min.
SECTION
12"
Provide 18' shoulder for right angle parking.
8' Shoulder
2:1
ISOMETRIC
STONE MASONRY (USED FOR SCENIC OVERLOOKS)

HIGHWAYS – INTERCHANGES – I

ELEMENTS OF DESIGN (CONT.)

GENERAL TYPES OF INTERCHANGES

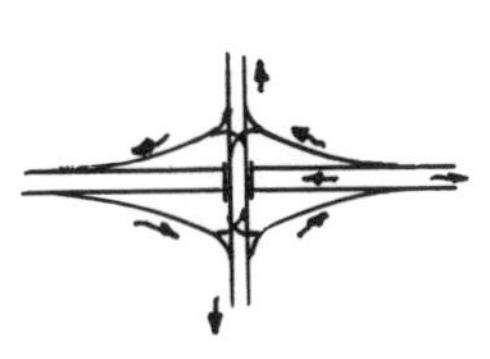

Conventional Diamond

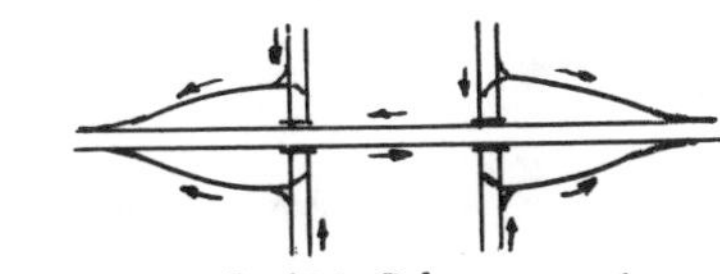

Split Diamond

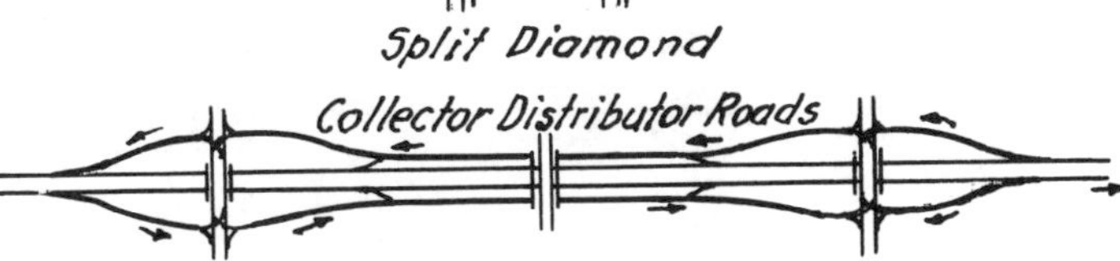
Collector Distributor Roads

Successive Diamond with C-D Road

Split Diamond One way

DIAMOND INTERCHANGES

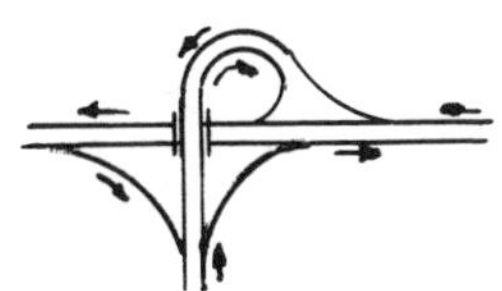

Trumpet

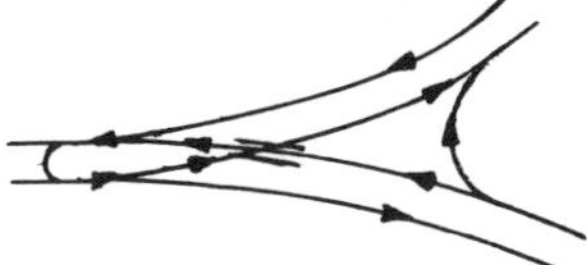

Single Structure-Y

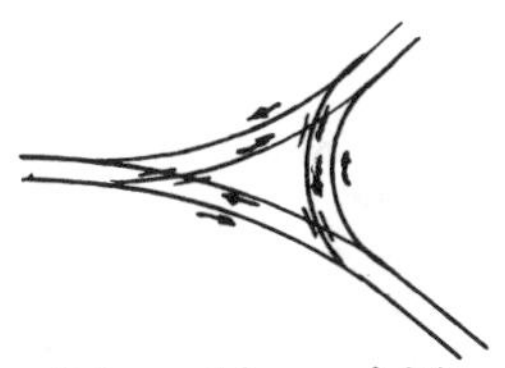

Directional-Y

T & Y INTERCHANGES

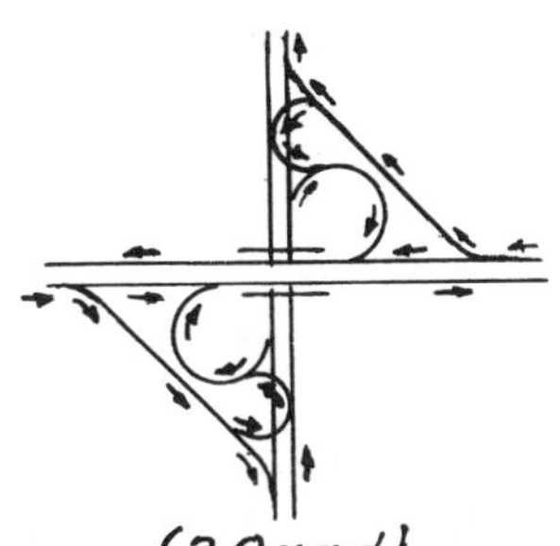

(2 Quad.)

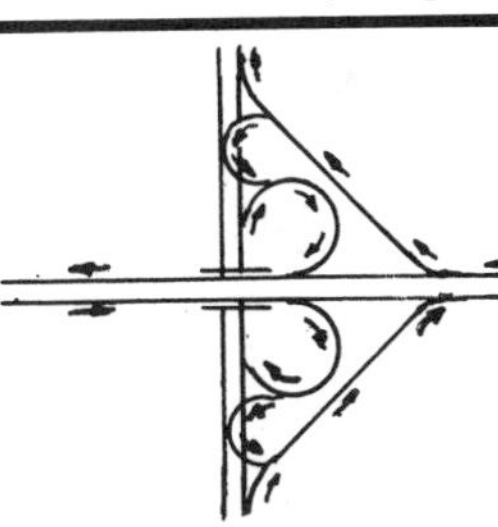

(2 Quad.)

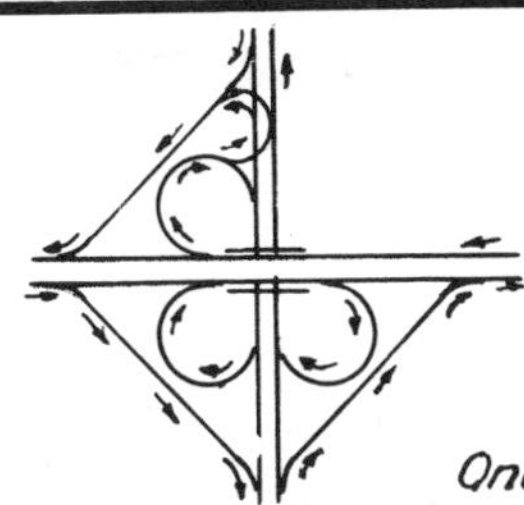

(3 Quad.)

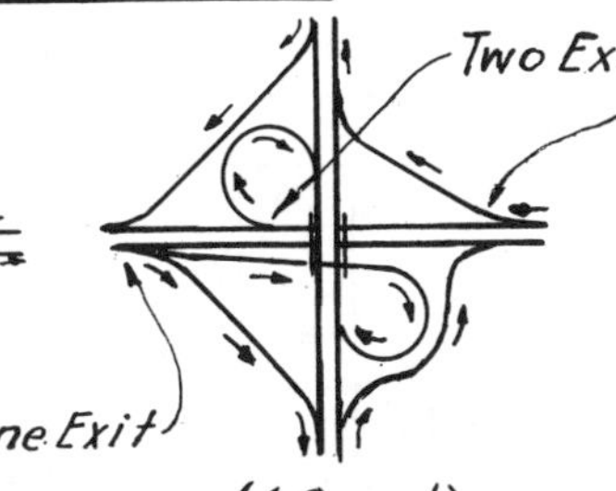
Two Exit
One Exit

(4 Quad.)

PARTIAL CLOVERLEAF INTERCHANGES

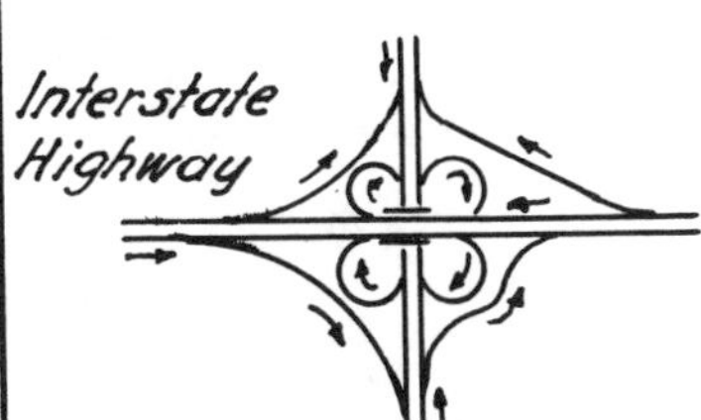
Interstate Highway

Cloverleaf

Cloverleaf with C-D Roads

CLOVERLEAF INTERCHANGES

Directional with Loops

All Directional

DIRECTIONAL INTERCHANGES

SELECTION OF INTERCHANGES

Characteristic of Interchange	Area	Adaptable Interchanges
Expressway with minor crossroad	Rural	Partial cloverleaf, conventional diamond
Expressway with minor crossroad	Urban	Conventional diamond; Split diamond; Partial cloverleaf, 2 quad
Expressway with primary highway	Rural Suburban	Cloverleaf; directional Partial cloverleaf
Interchanges with primary highway	Urban	Diamond with collector distributor roads
Expressway with expressway or interstate	Rural and urban	Directional and cloverleaf

*Adapted from American Association of State Highway Officials.

HIGHWAYS – INTERCHANGES-2

DESIGN OF TURNING ROADWAYS

Ramps All types, arrangements, and sizes of turning roadways that connect two or more legs at an interchange.

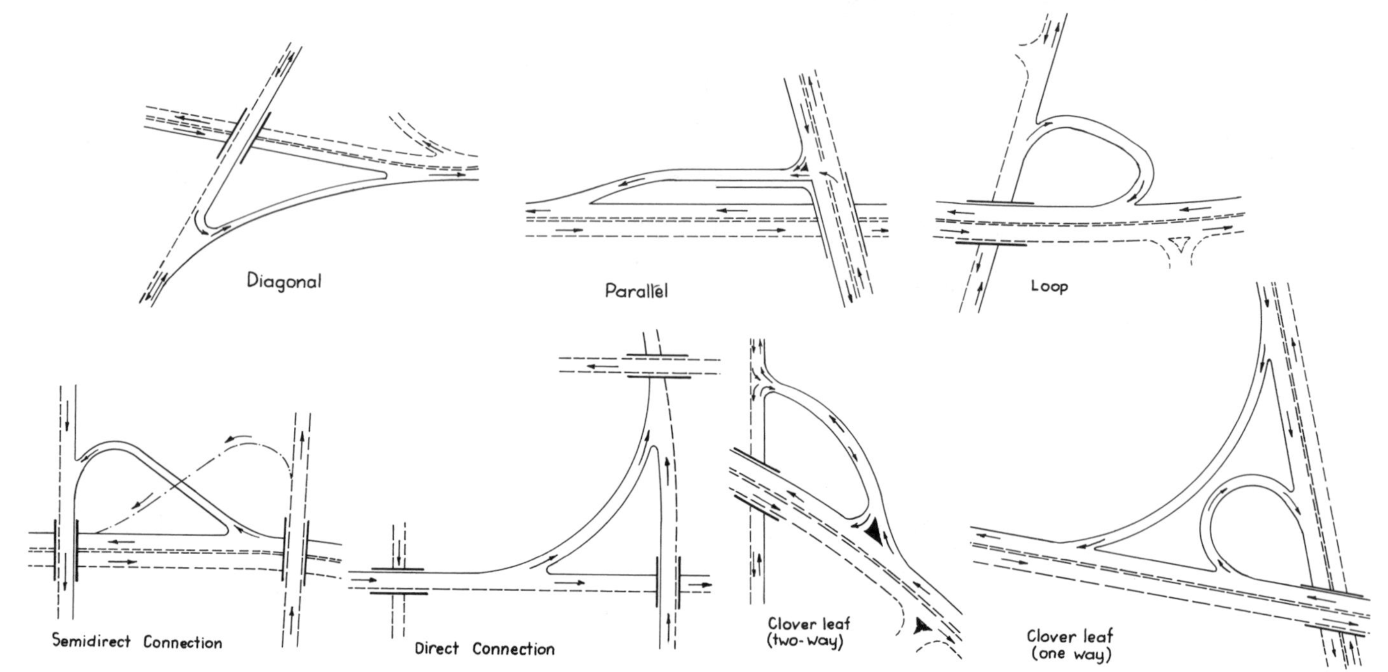
Diagonal
Parallel
Loop
Semidirect Connection
Direct Connection
Clover leaf
(two-way)
Clover leaf
(one way)

FIG. A – GENERAL TYPES OF RAMPS*

TABLE B – RAMP DESIGN SPEED*

Highway Design Speed, m.p.h.	30	40	50	60	70
Ramp design speed, m.p.h.					
Desirable	25	35	40	45	50
Minimum	15	20	25	30	30
Corresponding minimum radius, ft.					
Desirable	150	300	430	550	690
Minimum	50	90	150	230	230

TABLE D – MAXIMUM RAMP GRADIENTS*

%	Design Condition
4-6	High-standard highway
5	Area subject to snow or ice.
3-4	Significant volume of trucks and buses.
8-10	Only if necessary; avoid in snow areas and on upgrades.

TABLE E – SHOULDER WIDTH OR EQUIVALENT LATERAL CLEARANCE OUTSIDE THE PAVEMENT EDGES OF TURNING ROADWAYS*
(All dimensions Greater Where Necessary for Sight Distance)

Roadway Condition	Design	Shoulder Width or Lateral Clearance outside of Pavement Edge, ft.	
		At Left	At Right
At ground level, short length, usually within channelized intersection	Minimum	Posts or rails at least 2 ft. removed	
	Desirable	4	4
At ground level, length intermediate to long, or in cut or on fill	Minimum	4	6
	Desirable	6 to 10	8 to 12

*Adopted from American Association of State Highway Officials.

TABLE C – DESIGN WIDTHS OF PAVEMENTS FOR TURNING ROADWAYS*

R, Radius on Inner Edge of Pavement ft.	Pavement Width in feet for:								
	Case I 1-lane, 1-way Operation, no Provision for Passing			Case II 1-lane, 1-way Operation, with Provision for Passing a Stalled Vehicle			Case III 2-lane Operation, Either 1-way or 2-way		
	Design Traffic Condition†								
	A	B	C	A	B	C	A	B	C
50	16	17	20	21	24	27	30	33	37
75	15	16	18	20	22	25	28	31	34
100	14	16	17	19	21	24	27	30	33
150	13	15	16	18	20	23	26	29	31
200	13	15	16	18	20	22	26	28	29
300	12	15	15	17	19	21	25	27	28
400	12	14	15	17	19	21	25	27	28
500	12	14	15	17	19	21	25	27	27
Tangent	12	14	14	16	18	20	22	24	24
Width Modification regarding Edge of Pavement Treatment:									
No stabilized shoulder	None			None			None		
Mountable curb	None			None			None		
Barrier curb: One side Two sides	add 1' add 2'			None add 1'			add 1' add 2'		
Stabilized shoulder, one or both sides	None			Deduct shoulder width; minimum pavement width as under case I.			Deduct 2' where shoulder is 4' or wider.		

NOTE: † Design traffic condition:

Traffic condition A Predominantly P vehicles but some consideration for SU trucks.

Traffic condition B Sufficient SU vehicles to govern design, but some consideration for semitrailer vehicles.

Traffic condition C Sufficient semitrailer, C43 or C50 vehicles to govern design.

HIGHWAYS – INTERCHANGES – 3

DESIGN OF TURNING ROADWAYS

Speed-Change Lanes Acceleration and deceleration lanes are added pavements which enable drivers to maneuver and change speed between through-traffic pavements and turning roadways.

TABLE A
DESIGN LENGTHS OF SPEED-CHANGE LANES
Flat Grades – 2% or Less

Design Speed of Turning Roadway Curve, m.p.h. (Stop condition)		–	15	20	25	30	35	40	45	50
Minimum Curve Radius, ft.		–	50	90	150	230	310	430	550	690
Design Speed of Highway, m.p.h.	**Length of Taper, ft.**	**Total Length of Deceleration Lane, including Taper, ft. (All Main Highways)**								
40	175	300	250	250	200	175	†	–	–	–
50	200	400	350	350	300	250	250	200	†	–
60	225	450	400	400	350	350	300	250	225	†
70	250	500	450	450	400	400	350	350	300	250
Design Speed of Highway, m.p.h.	**Length of Taper ft.**	**Total Length of Acceleration Lane, including Taper, ft.**								
Case I – High Volume Highways										
40	175	–	450	400	350	250	*	–	–	–
50	200	–	700	650	600	500	400	250	*	–
60	225	–	1000	950	900	800	700	550	400	250
70	250	–	1300	1200	1100	1000	900	800	700	500
Case II – Other Main Highways										
40	175	–	300	250	200	†	–	–	–	–
50	200	–	500	450	350	300	†	–	–	–
60	225	–	750	700	600	500	400	250	†	–
70	250	–	950	900	800	750	600	500	350	†

†Less than length of taper; use compound curve or partial taper.

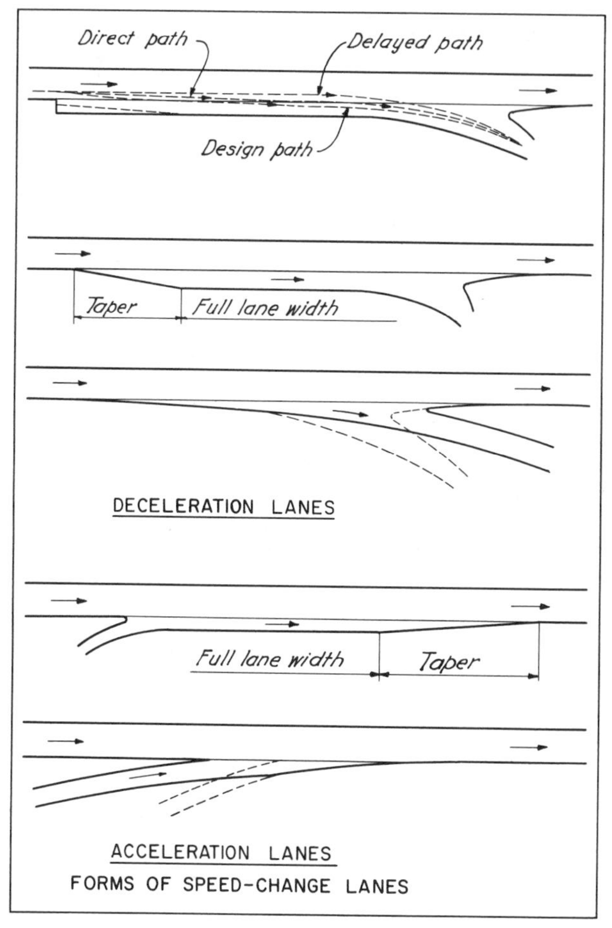

Direct path
Delayed path
Design path
Taper
Full lane width
DECELERATION LANES
Full lane width
Taper
ACCELERATION LANES

FORMS OF SPEED-CHANGE LANES

TABLE B – RATIO OF LENGTH OF SPEED-CHANGE LANES ON GRADE TO LENGTH ON LEVEL

Design Speed of Highway, m.p.h.	Deceleration Lanes: Ratio of Length on Grade to Length on Level‡ for:	
All	3 to 4% upgrade 0.9	3 to 4% downgrade 1.2
All	5 to 6% upgrade 0.8	5 to 6% downgrade 1.35

‡ Ratio from this table multiplied by length in Table A gives length of speed-change lane on grade.

Design Speed of Highway, m.p.h.	Acceleration Lanes: Ratio of Length on Grade to Length on Level‡ for: Design Speed of Turning Roadway Curve, m.p.h.				
	20	30	40	50	All Speeds
	3 to 4% upgrade				3 to 4% downgrade
40	1.3	1.3	–	–	0.7
50	1.3	1.4	1.4	–	0.65
60	1.4	1.5	1.5	1.6	0.6
70	1.5	1.6	1.7	1.8	0.6
	5 to 6% upgrade				5 to 6% downgrade
40	1.5	1.5	–	–	0.6
50	1.5	1.7	1.9	–	0.55
60	1.7	1.9	2.2	2.5	0.5
70	2.0	2.2	2.6	3.0	0.5

*A.A.S.H.O.

HIGHWAYS - TURNING - I

DESIGN OF TURNING ROADWAY

Design Vehicles - *A design vehicle is a selected motor vehicle of a designated type, the weight, dimensions and operating characteristics of which are used to establish highway design controls to accomodate vehicles of that type. The dimensions and minimum turning path of a design vehicle is a design control that affects principally the radius and width of pavement in intersection areas.*

DESIGN VEHICLES & MINIMUM TURNING PATHS

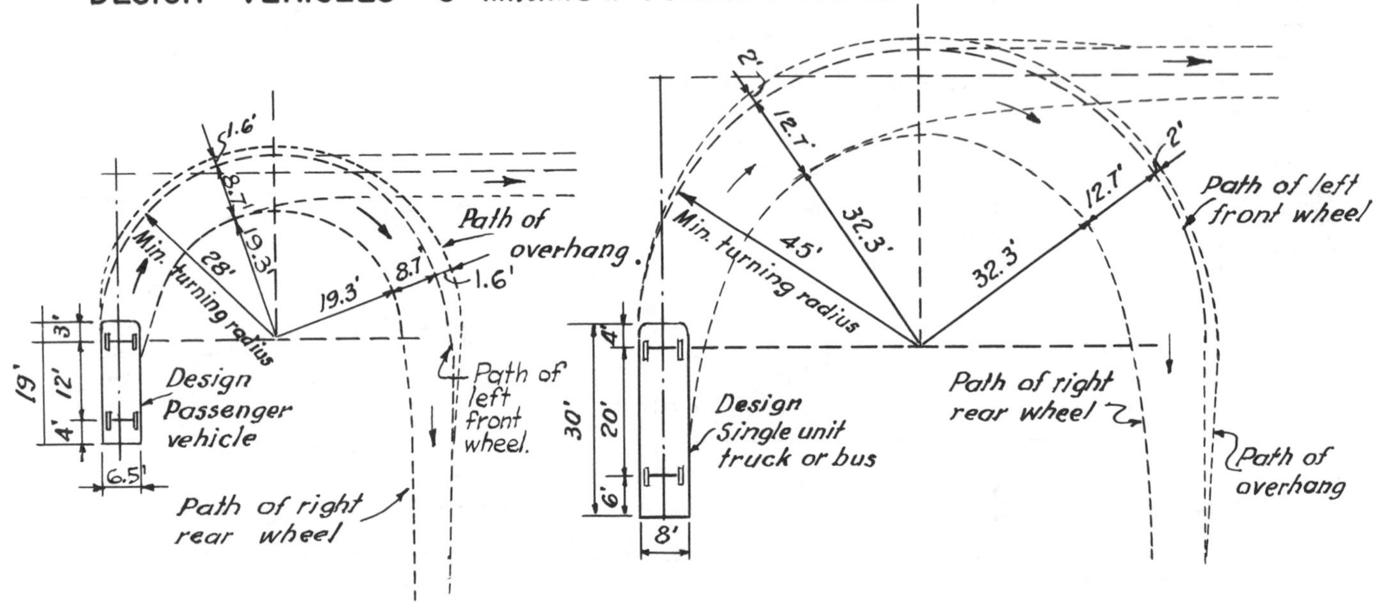
1.6'
8.7'
19.3'
28'
Min. turning radius
19.3'
8.7'
1.6'
Path of overhang
Path of left front wheel
Path of right rear wheel
Design Passenger vehicle
19'
3'
12'
4'
6.5'
2'
12.7'
32.3'
45'
Min. turning radius
32.3'
12.7'
2'
Path of left front wheel
Path of right rear wheel
Path of overhang
Design Single unit truck or bus
30'
4'
20'
6'
8'

Passenger Design Vehicle(1)

Single Unit Truck or Bus Design Vehicle.(2)

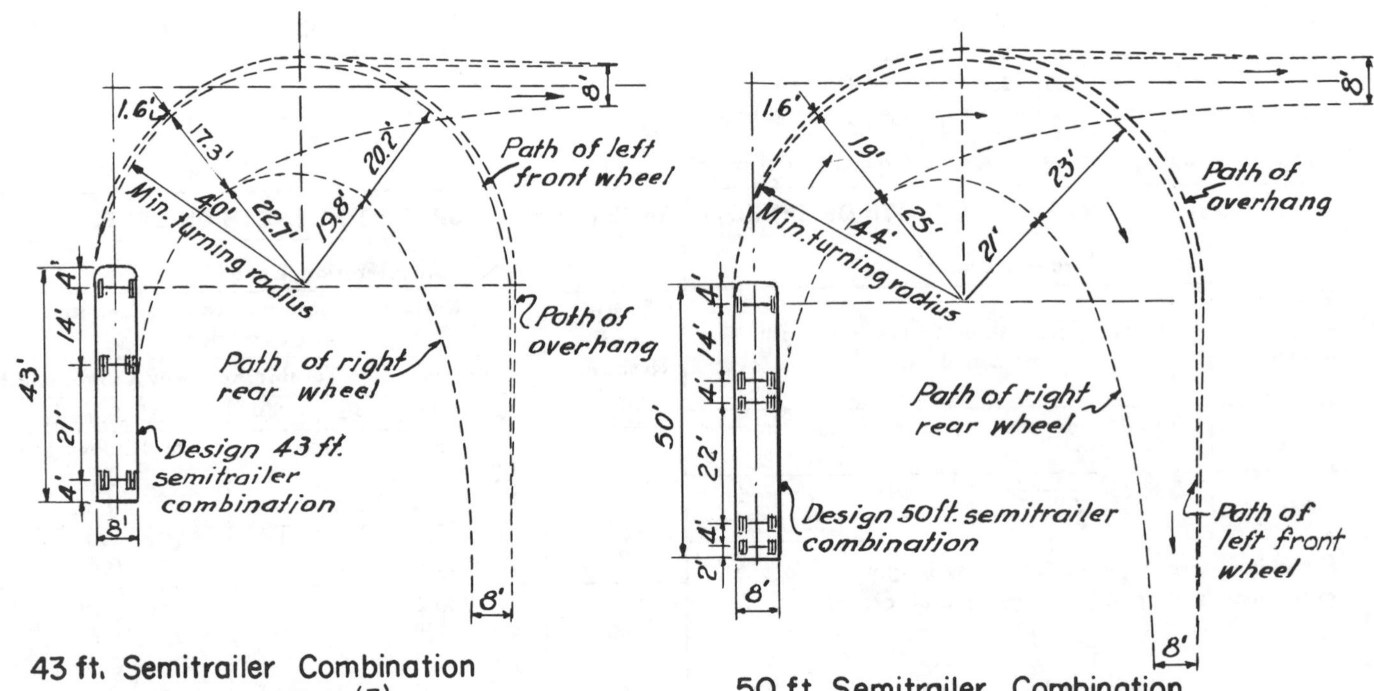
1.6'
17.3'
40'
22.7'
19.8'
20.2'
8'
Min. turning radius
Path of left front wheel
Path of overhang
Path of right rear wheel
Design 43 ft. semitrailer combination
43'
4'
14'
21'
4'
8'
8'
1.6'
19'
44'
25'
21'
23'
8'
Min. turning radius
Path of overhang
Path of right rear wheel
Design 50 ft. semitrailer combination
Path of left front wheel
50'
4'
14'
4'
22'
4'
2'
8'
8'

43 ft. Semitrailer Combination Design Vehicle.(3)

50 ft. Semitrailer Combination Design Vehicle.(4)

(1)- A.A.S.H.O. Designation = P
(2)- A.A.S.H.O. Designation = S.U.
(3)- A.A.S.H.O. Designation = C-43
(4)- A.A.S.H.O. Designation = C-50

HIGHWAYS - TURNING-2

DESIGN OF TURNING ROADWAYS

Minimum Designs for Turning Roadways Where the inner edges of pavement for right turns at intersections are designed to accommodate semitrailer combinations, or where the design permits passenger vehicles to turn at speeds of 15 m.p.h. or more, the pavement area of the intersection may become excessively large for proper control of traffic. To avoid this, a corner island should be provided to form a separate turning roadway. Corner islands are of minimum practical size, placed within unused areas. Islands should clear through pavements by 2 ft. and be edged with mountable curbs, except that large islands may not require curbs.

TABLE A

Angle of Turn, degrees	Design Classification*	3-Centered Compound Curve, ft. Radii	3-Centered Compound Curve, ft. Offset	Width of Lane, ft.	Approximate Island Size, sq. ft.
75	A	150-75-150	3.5	14	60
	B	150-75-150	5.0	16	50
	C	180-90-180	3.5	18	50
90	A	150-50-150	3.0	14	50
	B	150-50-150	5.0	16	110
	C	180-65-180	4.5	18	210
105	A	120-40-120	2.0	15	70
	B	120-40-120	4.5	20	60
	C	150-40-150	7.5	26	50
120	A	100-30-100	2.5	16	120
	B	100-30-100	5.0	24	70
	C	120-35-120	7.0	28	150
135	A	100-30-100	2.5	16	460
	B	100-30-100	5.0	26	370
	C	120-30-120	8.0	28	500
150	A	100-30-100	2.5	16	1400
	B	100-30-100	5.0	28	1250
	C	120-30-120	7.5	32	1500

A Primarily passenger vehicles; permits occasional design single-unit truck to turn with restricted clearances.

B Provides adequately for SU; permits occasional C50 to turn with slight encroachment on adjacent traffic lanes.

C Provides fully for C50.

TABLE B – SPEED-CURVATURE RELATIONS –

The minimum radii for design of intersection curves are as follows:

Design (turning) speed, m.p.h.	15	20	25	30	35	40
Suggested curvature for design:						
Radius (minimum), ft.	50	90	150	230	310	430
Degree of curve—(maximum)	—	64	38	25	18	13
Corresponding average running speed, m.p.h.	14	18	22	26	30	34

The above radii preferably are applied to the inner edge of the pavement and are based on reasonably attainable rates of superelevation which should be up to 0.08 where feasible and greater where snow and ice is not a factor.

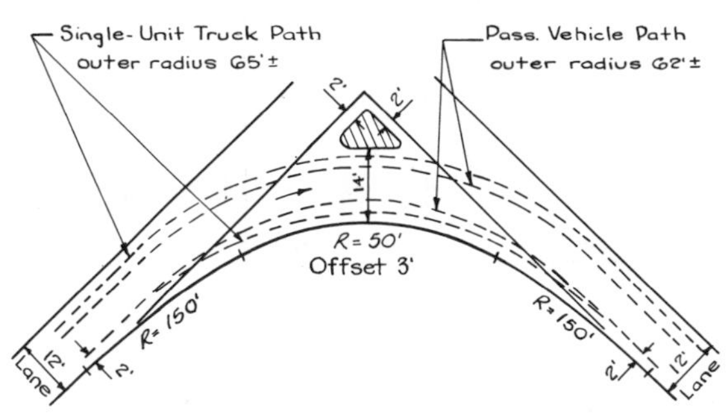
Single-Unit Truck Path
outer radius 65'±
Pass. Vehicle Path
outer radius 62'±
14'
R=50'
Offset 3'
R=150'
R=150'
12'
Lane
2'
12'
Lane
2'

3-Centered Curve: 150' - 50' - 150', Offset 3'
Equivalent Simple Curve Radius, 60'

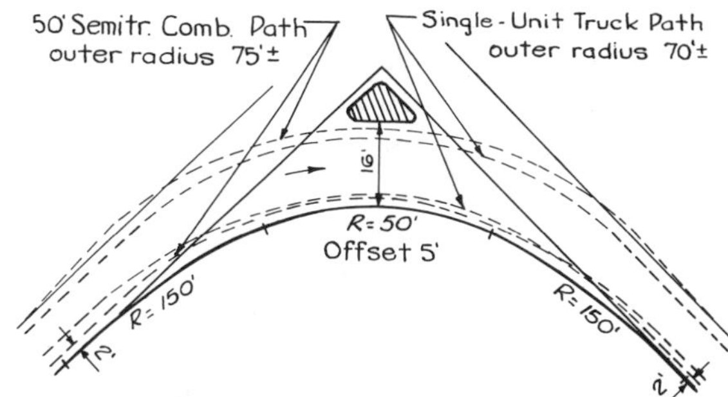
50' Semitr. Comb. Path
outer radius 75'±
Single-Unit Truck Path
outer radius 70'±
16'
R=50'
Offset 5'
R=150'
R=150'
2'
2'

3-Centered Curve: 150' - 50' - 150', Offset 5'
Equivalent Simple Curve Radius, 67'

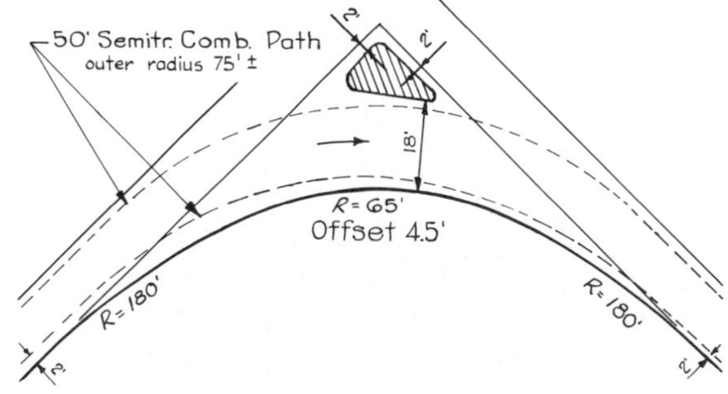
50' Semitr. Comb. Path
outer radius 75'±
18'
R=65'
Offset 4.5'
R=180'
R=180'
2'
2'

3-Centered Curve: 180'- 65' - 180', Offset 4.5'
Equivalent Simple Curve Radius, 80'

DESIGN FOR TURNING ROADWAYS WITH MINIMUM CORNER ISLAND, 90-DEGREE RIGHT TURN

HIGHWAYS – TURNING – 3

DESIGN OF TURNING ROADWAY

Intersection Curves. Minimum Design for Sharpest Turns Minimum edge of pavement designs for turns at intersections for the four design vehicles operating at less than 10 m.p.h. are given in Table A below. Other combinations of radii also may be satisfactory.

TABLE A

Design Vehicle	Angle of Turn (Δ), degrees	Simple Curve Radius, ft.	3-Centered Compound Curve, ft.	
			Radii	Offset
P	30	60		
SU		100		
C43		150		
C50		200		
P	45	50		
SU		75		
C43		120		
C50		150	200-100-200	3.0
P	60	40		
SU		60		
C43		100		
C50			200-75-200	3.5
P	75	35	100-25-100	2.0
SU		55	120-45-120	2.0
C43		75	120-45-120	4.0
C50			150-50-150	5.5
P	90	30	100-20-100	2.5
SU		50	120-40-120	2.0
C43			120-40-120	5.0
C50			150-50-150	5.0
P	105		100-20-100	2.5
SU			100-35-100	3.0
C43			100-35-100	5.0
C50			150-40-150	6.5
P	120		100-20-100	2.0
SU			100-30-100	5.0
C43			100-30-100	5.5
C50			120-35-120	7.0
P	135		100-20-100	1.5
SU			100-30-100	5.0
C43			100-30-100	5.0
C50			120-30-120	8.0
P	150		75-18-75	2.0
SU			100-30-100	5.0
C43			100-30-100	5.0
C50			120-30-120	7.5
P	180 U-turn		50-15-50	5.0
SU			100-30-100	5.0
C43			100-20-100	10.0
C50			120-25-120	11.0

Adapted from AASHO.

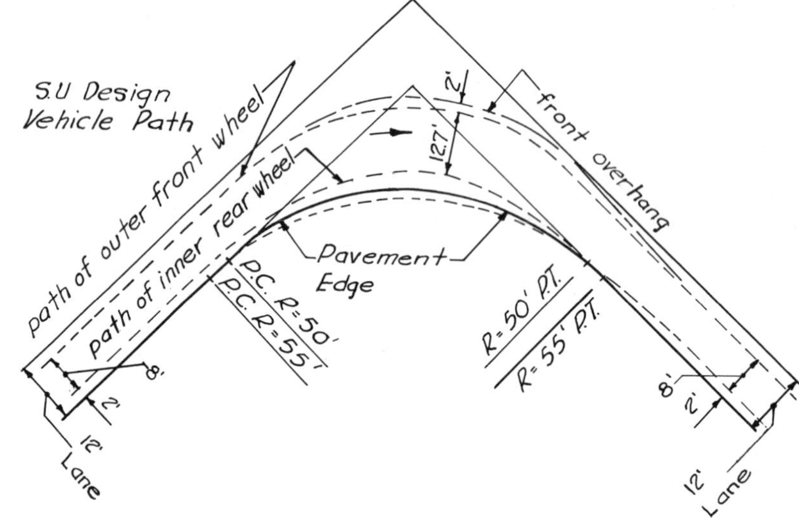
S.U Design Vehicle Path
path of outer front wheel
path of inner rear wheel
front overhang
12.7'
2'
Pavement Edge
P.C. R=50'
P.C. R=55'
R=50' P.T.
R=55' P.T.
8'
2'
12' Lane
12' Lane

Minimum Simple Curve, 50' or 55' Radius

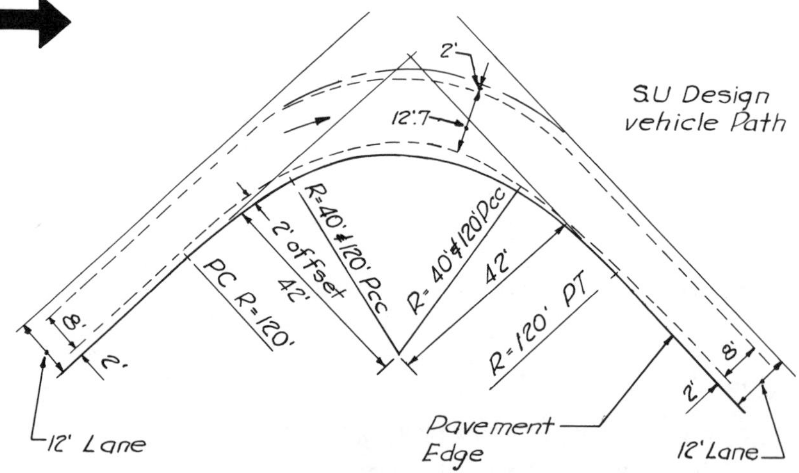
S.U Design vehicle Path
2'
12.7'
R=40'+120' Pcc
R=40'+120' Pcc
2' offset
42'
42'
PC R=120'
R=120' PT
8'
2'
12' Lane
Pavement Edge
12' Lane

3-Centered Compound Curve, 120' - 40' - 120' Radii, offset 2.0'

EDGE OF PAVEMENT CURVE FOR 90-DEGREE TURN

MINIMUM DESIGN FOR SINGLE-UNIT TRUCKS and BUSES

HIGHWAYS – CURVES-I

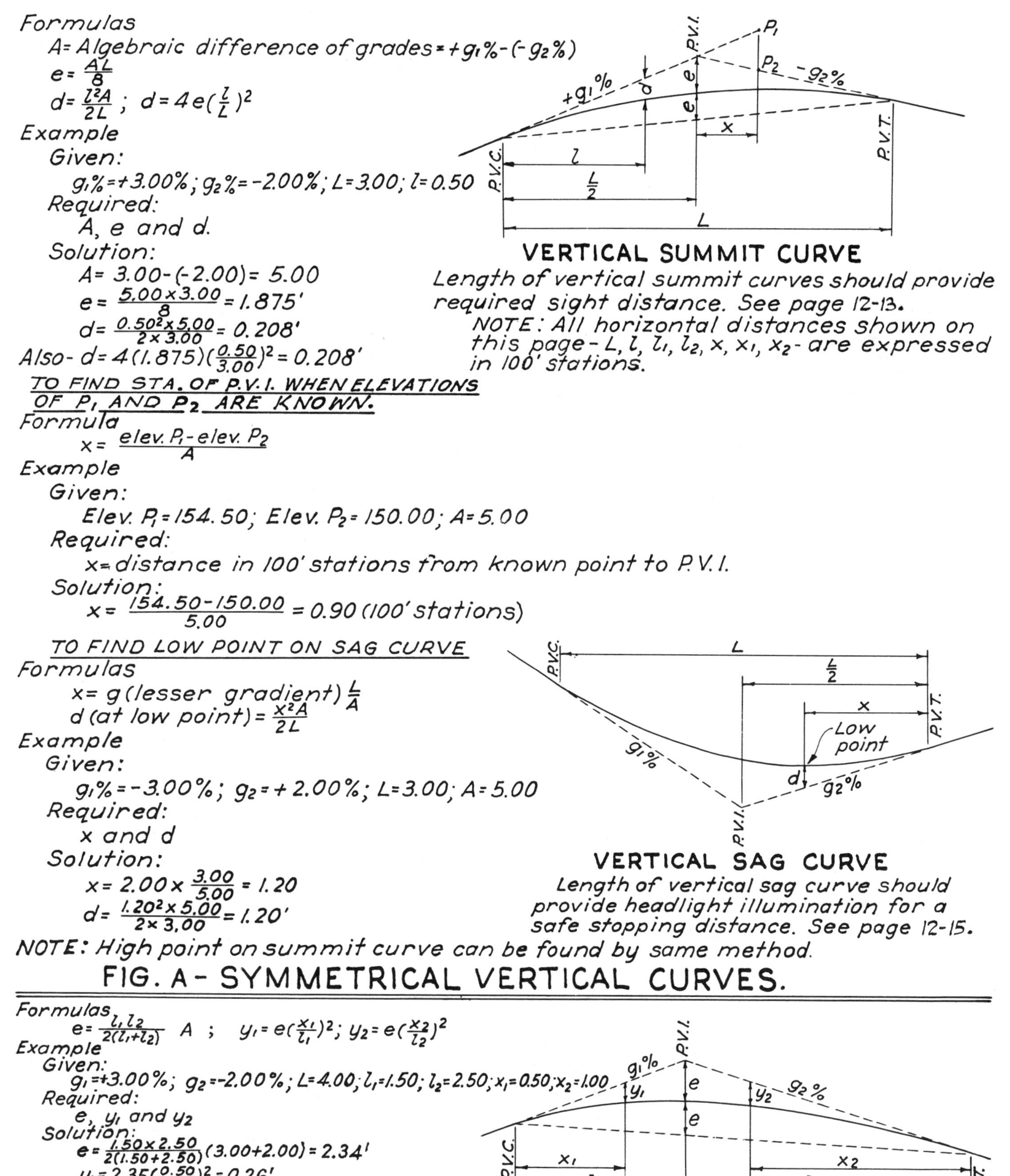

P.V.I.
P_1
P_2
$+g_1\%$
$-g_2\%$
d
e
e
x
l
L/2
L
P.V.C.
P.V.T.

Formulas

A = Algebraic difference of grades = $+g_1\% - (-g_2\%)$

$e = \frac{AL}{8}$

$d = \frac{l^2 A}{2L}$; $d = 4e\left(\frac{l}{L}\right)^2$

Example

Given:

$g_1\% = +3.00\%$; $g_2\% = -2.00\%$; $L = 3.00$; $l = 0.50$

Required:

A, e and d.

Solution:

$A = 3.00 - (-2.00) = 5.00$

$e = \frac{5.00 \times 3.00}{8} = 1.875'$

$d = \frac{0.50^2 \times 5.00}{2 \times 3.00} = 0.208'$

Also- $d = 4(1.875)\left(\frac{0.50}{3.00}\right)^2 = 0.208'$

VERTICAL SUMMIT CURVE

Length of vertical summit curves should provide required sight distance. See page 12-13.

NOTE: All horizontal distances shown on this page – L, l, l_1, l_2, x, x_1, x_2 – are expressed in 100' stations.

TO FIND STA. OF P.V.I. WHEN ELEVATIONS OF P_1 AND P_2 ARE KNOWN.

Formula

$x = \frac{\text{elev. } P_1 - \text{elev. } P_2}{A}$

Example

Given:

Elev. $P_1 = 154.50$; Elev. $P_2 = 150.00$; $A = 5.00$

Required:

x = distance in 100' stations from known point to P.V.I.

Solution:

$x = \frac{154.50 - 150.00}{5.00} = 0.90$ (100' stations)

TO FIND LOW POINT ON SAG CURVE

P.V.C.
L
L/2
x
P.V.T.
Low point
$g_1\%$
d
$g_2\%$
P.V.I.

Formulas

$x = g(\text{lesser gradient})\frac{L}{A}$

$d(\text{at low point}) = \frac{x^2 A}{2L}$

Example

Given:

$g_1\% = -3.00\%$; $g_2 = +2.00\%$; $L = 3.00$; $A = 5.00$

Required:

x and d

Solution:

$x = 2.00 \times \frac{3.00}{5.00} = 1.20$

$d = \frac{1.20^2 \times 5.00}{2 \times 3.00} = 1.20'$

VERTICAL SAG CURVE

Length of vertical sag curve should provide headlight illumination for a safe stopping distance. See page 12-15.

NOTE: High point on summit curve can be found by same method.

FIG. A – SYMMETRICAL VERTICAL CURVES.

Formulas

$e = \frac{l_1 l_2}{2(l_1 + l_2)} A$; $y_1 = e\left(\frac{x_1}{l_1}\right)^2$; $y_2 = e\left(\frac{x_2}{l_2}\right)^2$

Example

Given:

$g_1 = +3.00\%$; $g_2 = -2.00\%$; $L = 4.00$; $l_1 = 1.50$; $l_2 = 2.50$; $x_1 = 0.50$; $x_2 = 1.00$

Required:

e, y_1 and y_2

Solution:

$e = \frac{1.50 \times 2.50}{2(1.50 + 2.50)}(3.00 + 2.00) = 2.34'$

$y_1 = 2.35\left(\frac{0.50}{1.50}\right)^2 = 0.26'$

$y_2 = 2.35\left(\frac{1.00}{2.50}\right)^2 = 0.38'$

P.V.I.
$g_1\%$
$g_2\%$
y_1
y_2
e
e
P.V.C.
P.V.T.
x_1
x_2
l_1
l_2
L

FIG. B – UNSYMMETRICAL VERTICAL CURVES.

Used to fit unusual conditions

HIGHWAYS – CURVES-2

ARC DEFINITION.

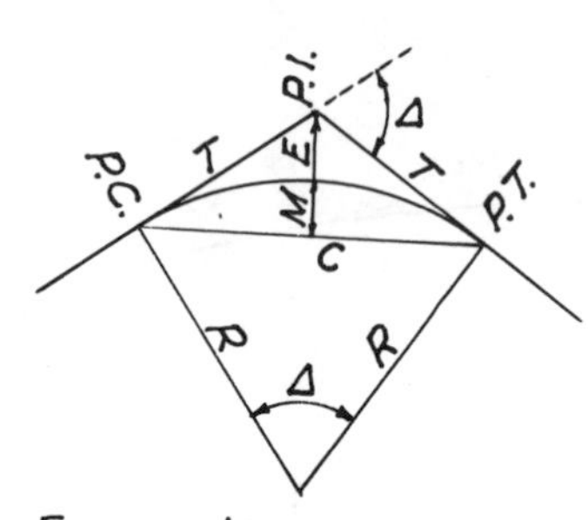
P.C.
T
P.I.
E
M
Δ
T
P.T.
C
R
R
Δ

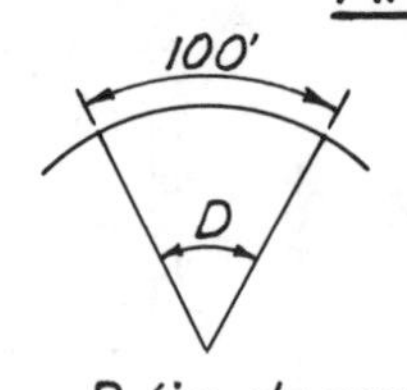
100'
D

D (in degrees) subtends 100' of arc.

$$D = \frac{5729.58}{R}$$

Formulas

$R = \frac{5729.58}{D}$

$T = R \tan\frac{\Delta}{2}$; $T = \frac{\tan 1^\circ \text{ curve for } \Delta}{D}$

$L = \text{Length} = \frac{100\Delta}{D}$

$M = R(1 - \cos\frac{1}{2}\Delta)$

$E = R\left(\frac{1}{\cos\frac{1}{2}\Delta} - 1\right)$; $E = \frac{\text{ext. } 1^\circ \text{ curve for } \Delta}{D}$

$C = 2R\sin\frac{\Delta}{2}$

DEFINITIONS

L = Length of circular curve
P.I. = Point of intersection
P.C. = Point of curvature
P.T. = Point of tangency

Example

Given:

$\Delta = 54°20'$; $D = 7°40'$; P.I. = Sta. 125 + 39.88

Required:

R; T; L and Sta. of P.C. and P.T.

Solution:

$R = \frac{5729.58}{7°40'} = 747.34'$

$T = 747.34\,(\tan 27°10') = 747.34\,(0.513195) = 383.53'$

Also- From Pg. 12-27 (Funct. 1° Curve) by interpolation tan 1° curve for Δ 54° 20' = 2940.41

$\therefore T = \frac{2940.41}{7°40'} = 383.53'$

P.C. = Sta. 125 + 39.88 − 383.53 = Sta. 121 + 56.35

$L = \frac{100\Delta}{D} = \frac{100(54°20')}{7°40'} = 708.70'$

P.T. = Sta. 121 + 56.35 + 708.70 = Sta. 128 + 65.05

DEFLECTIONS

Formulas

Deflection angle = $\frac{D}{2}$ for 100'; $\frac{D}{4}$ for 50', etc.

For "c" feet (in minutes) = 0.3cD

Deflection angle (in minutes) from P.C. to P.T. = 0.3LD

Also- Deflection angle (in degrees) from P.C. to P.T. = $\frac{\Delta}{2}$

Example

Given:

$\Delta = 54°20'$; $D = 7°40'$; L = 708.70; P.C. = Sta. 121+56.35; P.T. = Sta. 128 + 65.05

Required:

Deflection angle from P.C. to Sta. 122+00; Sta. 122+50 and P.T. Sta. 128 + 65.05

Solution:

Sta. 122+00 − P.C. Sta. 121 + 56.35 = 43.65'

∴ Deflection angle to Sta. 122+00 = 0.3 × 43.65 × 7°40' = 100.395' = 1°40.395'

Deflection angle to Sta. 122+50 = 1°40.395' + $\frac{7°40'}{4}$ = 1°40.395' + 1°55' = 3°35.395'

Deflection angle to P.T. Sta. 128 + 65.05 = 0.3 × 708.70 × 7°40' = 27°10'

Also- Deflection angle to P.T. Sta. 128 + 65.05 = $\frac{\Delta}{2} = \frac{54°20'}{2}$ = 27° 10'

EXTERNALS

Example

Given:

$\Delta = 54°20'$; $D = 7°40'$; R = 747.34'

Required:

External "E"

Solution:

$E = 747.34\left(\frac{1}{.8896822} - 1\right) = 92.67'$

Also- From Pg. 12-27 (Funct. 1° curve) by interpolation external 1° curve for Δ 54° 20' = 710.48

$\therefore E = \frac{710.48}{7°40'} = 92.67'$

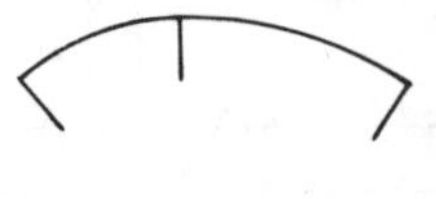

Compound curve
Avoid

Reverse curve
Avoid

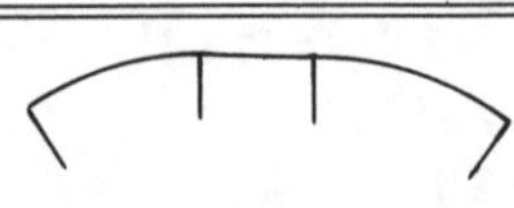

Broken-back curve
Avoid

TANGENT OFFSETS: The approximate offset from the tangent to the curve at any distance from the P.C. = $\frac{\text{distance}^2}{2R}$

* From Geometric Design standards by A.A.S.H.O.

HIGHWAYS – CURVES–3

See Pg. 12-26 for use of table

CENTRAL ANGLE	TANGENT	EXTERNAL	CENTRAL ANGLE	TANGENT	EXTERNAL	CENTRAL ANGLE	TANGENT	EXTERNAL	CENTRAL ANGLE	TANGENT	EXTERNAL
1°	50.00	0.22	31°	1589.0	216.3	61°	3375.0	920.2	91°	5830.5	2444.9
30′	75.00	0.49	30′	1615.9	223.5	30′	3408.8	937.3	30′	5881.7	2481.5
2°	100.01	0.87	32°	1643.0	230.9	62°	3442.7	954.8	92°	5933.2	2518.5
30′	125.02	1.36	30′	1670.0	238.4	30′	3476.8	972.4	30′	5985.3	2556.0
3°	150.04	1.96	33°	1697.2	246.1	63°	3511.1	990.2	93°	6037.8	2594.0
30′	175.06	2.67	30′	1724.4	253.9	30′	3545.6	1008.3	30′	6090.8	2632.6
4°	200.08	3.49	34°	1751.7	261.8	64°	3580.3	1026.6	94°	6144.3	2671.6
30′	225.12	4.42	30′	1779.1	269.9	30′	3615.1	1045.2	30′	6198.3	2711.2
5°	250.16	5.46	35°	1806.6	278.1	65°	3650.2	1063.9	95°	6252.8	2751.3
30′	275.21	6.61	30′	1834.1	286.4	30′	3685.4	1082.9	30′	6307.9	2792.0
6°	300.28	7.86	36°	1861.7	294.9	66°	3720.9	1102.2	96°	6363.4	2833.2
30′	325.35	9.23	30′	1889.4	303.5	30′	3756.5	1121.7	30′	6419.5	2875.0
7°	350.44	10.71	37°	1917.1	312.2	67°	3792.4	1141.4	97°	6476.2	2917.3
30′	375.54	12.29	30′	1945.0	321.1	30′	3828.4	1161.3	30′	6533.4	2960.3
8°	400.65	13.99	38°	1972.9	330.2	68°	3864.7	1181.6	98°	6591.2	3003.8
30′	425.79	15.80	30′	2000.9	339.3	30′	3901.2	1202.0	30′	6649.6	3047.9
9°	450.93	17.72	39°	2029.0	348.6	69°	3937.9	1222.7	99°	6708.6	3092.7
30′	476.10	19.75	30′	2057.2	358.1	30′	3974.8	1243.7	30′	6768.1	3138.1
10°	501.28	21.89	40°	2085.4	367.7	70°	4011.9	1265.0	100°	6828.3	3184.1
30′	526.48	24.14	30′	2113.8	377.5	30′	4049.3	1286.5	30′	6889.2	3230.8
11°	551.70	26.50	41°	2142.2	387.4	71°	4086.9	1308.2	101°	6950.6	3278.1
30′	576.95	28.97	30′	2170.8	397.4	30′	4124.8	1330.3	30′	7012.7	3326.1
12°	602.21	31.56	42°	2199.4	407.6	72°	4162.8	1352.6	102°	7075.5	3374.9
30′	627.50	34.26	30′	2228.1	418.0	30′	4201.2	1375.2	30′	7139.0	3424.3
13°	652.81	37.07	43°	2257.0	428.5	73°	4239.7	1398.0	103°	7203.2	3474.4
30′	678.15	39.99	30′	2285.9	439.2	30′	4278.5	1421.2	30′	7268.0	3525.2
14°	703.51	43.03	44°	2314.9	450.0	74°	4317.6	1444.6	104°	7333.6	3576.8
30′	728.90	46.18	30′	2344.1	461.0	30′	4356.9	1468.4	30′	7399.9	3629.2
15°	754.32	49.44	45°	2373.3	472.1	75°	4396.5	1492.4	105°	7467.0	3682.3
30′	779.77	52.82	30′	2402.6	483.4	30′	4436.4	1516.7	30′	7534.9	3736.2
16°	805.25	56.31	46°	2432.1	494.8	76°	4476.5	1541.4	106°	7603.5	3791.0
30′	830.76	59.91	30′	2461.7	506.4	30′	4516.9	1566.3	30′	7672.9	3846.5
17°	856.30	63.63	47°	2491.3	518.2	77°	4557.6	1591.6	107°	7743.2	3902.9
30′	881.88	67.47	30′	2521.1	530.1	30′	4598.5	1617.1	30′	7814.3	3960.1
18°	907.49	71.42	48°	2551.0	542.2	78°	4639.8	1643.0	108°	7886.2	4018.2
30′	933.13	75.49	30′	2581.0	554.5	30′	4681.3	1669.2	30′	7959.0	4077.2
19°	958.81	79.67	49°	2611.2	566.9	79°	4723.2	1695.8	109°	8032.7	4137.1
30′	984.53	83.97	30′	2641.4	579.5	30′	4765.3	1722.7	30′	8107.3	4197.9
20°	1010.3	88.39	50°	2671.8	592.3	80°	4807.7	1749.9	110°	8182.8	4259.7
30′	1036.1	92.92	30′	2702.3	605.3	30′	4850.5	1777.4	30′	8259.3	4322.4
21°	1061.9	97.58	51°	2732.9	618.4	81°	4893.6	1805.2	111°	8336.7	4386.1
30′	1087.8	102.35	30′	2763.7	631.7	30′	4937.0	1833.6	30′	8415.1	4450.9
22°	1113.7	107.24	52°	2794.5	645.2	82°	4980.7	1862.2	112°	8494.6	4516.6
30′	1139.7	112.25	30′	2825.6	658.8	30′	5024.8	1891.2	30′	8575.0	4583.4
23°	1165.7	117.38	53°	2856.7	672.7	83°	5069.2	1920.5	113°	8656.6	4651.3
30′	1191.8	122.63	30′	2888.0	686.7	30′	5113.9	1950.3	30′	8739.2	4720.3
24°	1217.9	128.00	54°	2919.4	700.9	84°	5159.0	1980.4	114°	8822.9	4790.4
30′	1244.0	133.50	30′	2951.0	715.3	30′	5204.4	2010.8	30′	8907.7	4861.7
25°	1270.2	139.1	55°	2982.7	729.9	85°	5250.3	2041.7	115°	8993.8	4934.1
30′	1296.5	144.8	30′	3014.5	744.6	30′	5296.4	2073.0	30′	9080.9	5007.8
26°	1322.8	150.7	56°	3046.5	759.6	86°	5343.0	2104.7	116°	9169.4	5082.7
30′	1349.2	156.7	30′	3078.7	774.7	30′	5389.9	2136.7	30′	9259.0	5158.8
27°	1375.6	162.8	57°	3110.9	790.1	87°	5437.2	2169.2	117°	9349.9	5236.2
30′	1402.0	169.0	30′	3143.4	805.6	30′	5484.9	2202.2	30′	9442.2	5315.0
28°	1428.6	175.4	58°	3176.0	821.4	88°	5533.1	2235.5	118°	9535.7	5395.1
30′	1455.1	181.9	30′	3208.8	837.3	30′	5581.6	2269.3	30′	9630.7	5476.5
29°	1481.8	188.5	59°	3241.7	853.5	89°	5630.5	2303.5	119°	9727.0	5559.4
30′	1508.5	195.3	30′	3274.8	869.9	30′	5679.9	2338.2	30′	9824.8	5643.8
30°	1535.3	202.1	60°	3308.0	886.4	90°	5729.7	2373.3	120°	9924.0	5729.7
30′	1562.1	209.1	30′	3341.4	903.2	30′	5779.9	2408.9	30′	10025.0	5817.0

HIGHWAYS – CURVES–4

NOTE: The degree of curve is not usually used for the curves involved in street intersections, curbs, road intersections, runway and taxiway fillets, and turnarounds, traffic circles, rotaries, cloverleafs, etc. These curves are defined by the Radius, R, and Central Angle, Δ or θ.

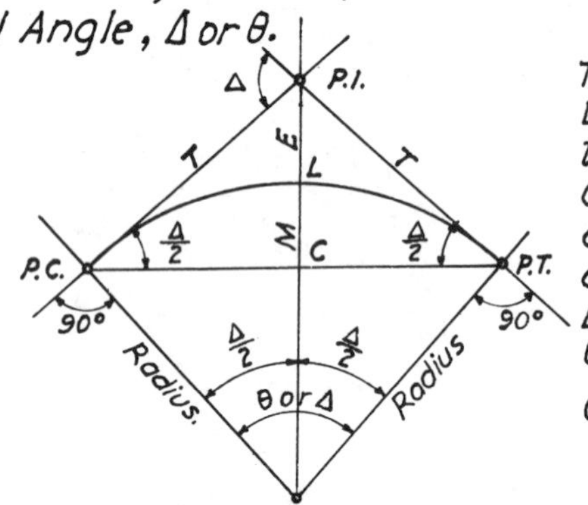

Δ
P.I.
T
E
L
T
P.C.
Δ/2
M
C
Δ/2
P.T.
90°
90°
Radius
Δ/2
Δ/2
Radius
θ or Δ

SHORT RADIUS CURVE.

NOTATION:

- T = Tangent length · P.C. or P.T. to P.I.
- L = Arc length · P.C. to P.T.
- l = Arc length for any sub-chord.
- C = Long chord P.C. to P.T.
- c = Any sub-chord.
- d = Deflection to any point.
- Δ = Central Angle in degrees.
- θ = Central Angle in radians.

One radian $= \frac{360°}{2\pi} = \frac{180°}{\pi} = 57.2958° = 57°17'44.8''$

$\pi = 3.14159$

M = Mid. Ordinate; m for sub-chords.

E = External ; e for sub-chords.

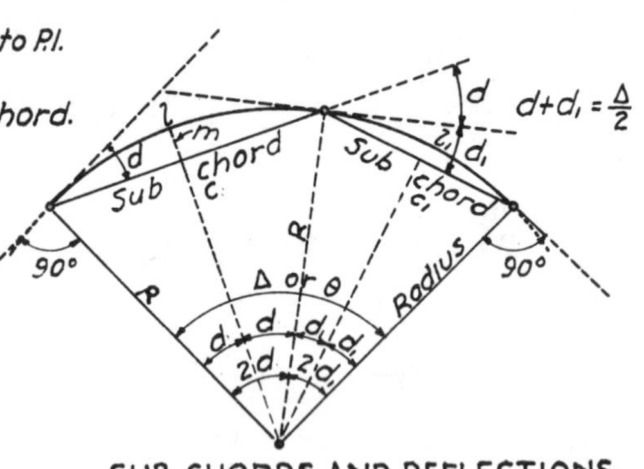

d
d+d₁ = Δ/2
l
m
d
Sub
chord
c
Sub
chord
c₁
l₁
d₁
R
90°
R
Δ or θ
Radius
90°
d
d
d₁
d₁
2d
2d₁

SUB-CHORDS AND DEFLECTIONS.

$R = \frac{L}{\theta} = \frac{L \cdot 180/\Delta}{\pi} = \frac{L}{\Delta} 57.2958 = T \cdot \cot\frac{\Delta}{2} = \frac{C}{2\sin \Delta/2}.$

$\frac{4M^2 + C^2}{8M} = \frac{M^2 + (C/2)^2}{2M}.$

$L = R\theta = \frac{\Delta R \pi}{180} = 0.017453\,\Delta R = \text{Circum.} \cdot \frac{\Delta}{360}.$

$T = R \cdot \tan\frac{\Delta}{2} = E \cdot \cot\frac{\Delta}{4} = \frac{C}{2\cos \Delta/2}.$

$C = 2R \cdot \sin\frac{\Delta}{2} = 2T \cdot \cos\frac{\Delta}{2} = 2\sqrt{M(2R-M)}$

$M = R \cdot \text{vers}\frac{\Delta}{2} = E \cdot \cos\frac{\Delta}{2} = R(1 - \cos\frac{\Delta}{2}).$

$E = R \cdot \text{exsec}\frac{\Delta}{2} = T \cdot \tan\frac{\Delta}{4} = \frac{R}{\cos \Delta/2} - R.$

$\Delta = \frac{180 L}{\pi R} = 57.2958 \frac{L}{R} = \theta \cdot 57.2958.$

$\theta = \frac{L}{R} = \frac{\Delta \pi}{180} = \Delta \cdot 0.017453.$

$\sin\frac{\Delta}{2} = \frac{C}{2R}$; $\cos\frac{\Delta}{2} = \frac{R-M}{R} = \frac{C}{2T}$; $\tan\frac{\Delta}{2} = \frac{T}{R}$

CONCENTRIC CURVES.

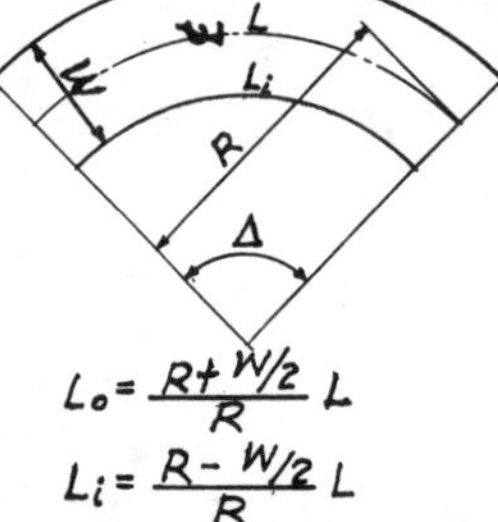

L₀
W
L
Lᵢ
R
Δ

$L_o = \frac{R + W/2}{R} L$

$L_i = \frac{R - W/2}{R} L$

$L_o - L_i = 0.017453\, W \cdot \Delta = W \cdot L/R$

$\text{Area} = L \cdot W$

Sub-chord $= 2R \cdot \sin d = 2(R-M) \cdot \tan d.$

$d(\text{in minutes}) = 1718.873 \frac{l}{R}$. Radius $= \frac{c}{2\sin d}.$

length $= \frac{\pi R d}{90} = 0.034906\, R d$ (d in degrees).

mid-ordinate $= R(1 - \cos d) = 2R \cdot \sin^2\frac{d}{2}$

$\tan d = \frac{1/2\, c}{R - m}$; $\sin d = \frac{1/2\, c}{R}$

Excess of "l" over "c" $= l - c = l - 2R \cdot \sin d.$

Sum of deflection angles, $d_1 + d_2 + \ldots d_n = \frac{\Delta}{2}$

EXAMPLE:

Given: $R = 50'$; $\Delta = 110°\ (\theta = 1.9195)$; $l = 50'$.

Required: $L, l_1, d, d_1, c, \&\ c_1$.

Solution: $L = 50 \times 1.9195 = 95.98'$; $l_1 = 95.98 - 50 = 45.98'$

$d = 1718.873 \times 50/50 = 28°39'$

$d_1 = 1718.873 \times \frac{45.98}{50} = 26°21'$

$c = 2R \sin 28°39' = 47.946'$

$c_1 = 2R \sin 26°21' = 44.385'$

TABLE A – DEFLECTIONS (d) & MIDDLE ORDINATES (m) FOR SUB-CHORDS.*

	CHORD \ RADIUS	10'	12'	15'	18'	20'	25'	30'	35'	40'	45'	50'	60'	70'	80'	90'	100'	120'	150'
DEFLECTION	5'	14°29'	12°01'	9°36'	7°59'	7°11'	5°44'	4°47'	4°06'	3°35'	3°11'	2°52'	2°23'	2°03'	1°47'	1°35'	1°26'	1°12'	0°57'
	10'	30°00'	24°37'	19°28'	16°08'	14°29'	11°32'	9°36'	8°13'	7°11'	6°23'	5°44'	4°47'	4°06'	3°35'	3°11'	2°52'	2°23'	1°55'
	20'		56°26'	41°49'	33°45'	30°00'	23°35'	19°28'	16°36'	14°29'	12°50'	11°32'	9°36'	8°13'	7°11'	6°23'	5°44'	4°47'	3°49'
	25'			56°27'	43°59'	38°41'	30°00'	24°37'	20°55'	18°13'	16°08'	14°29'	12°01'	10°17'	8°59'	7°59'	7°11'	5°59'	4°47'
	50'							56°27'	45°35'	38°41'	33°45'	30°00'	24°37'	20°55'	18°13'	16°08'	14°29'	12°01'	9°36'
M.	10'	1.34	1.09	0.86	0.71	0.64	0.51	0.42	0.36	0.31	0.28'	0.25	0.21	0.18	0.16	0.14	0.13	0.10	0.08
	20'		5.37	3.82	3.03	2.68	2.09	1.72	1.46	1.27	1.12	1.01	0.84	0.72	0.63	0.56	0.50	0.42	0.33

CIRCLE.

$\text{Area} = \pi R^2 = \frac{\pi D^2}{4}$

$\text{Circumference} = 2\pi R = \pi D.$

$R = \frac{\text{Cir.}}{2\pi} = \frac{D}{2} = \sqrt{\frac{\text{Area}}{\pi}}$

$D = 2R = \text{cir.}/\pi$

SECTOR OF CIRCLE.

$\text{Area} = 0.008727 R^2 \Delta = \frac{1}{2} l \cdot R = \pi R^2 \frac{\Delta}{360} = R^2 \cdot \frac{\theta}{2}$

When $\Delta = 90°$: $A = 0.3927 C^2$; $0.7854 R^2$

SEGMENT OF CIRCLE.

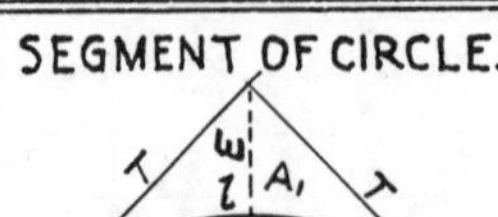

T
W
A₁
T
l
M
A₂
C
R
Δ or θ
A₃
R

$A_1 = R^2(\tan\frac{\Delta}{2} - \frac{\Delta\pi}{360}) = R(T - \frac{l}{2}).$

$A_2 = \frac{lR - c(R-M)}{2} = (\pi R^2 \frac{\Delta}{360}) - [(R\sin\frac{\Delta}{2})(R\cos\frac{\Delta}{2})].$

$A_2 = (\pi R^2 \frac{\Delta}{360}) - \frac{1}{2} c(R-M)$

$A_2 = 2/3\, Mc$ { Correct for parabolic segment, approximate for circular segment.

$A_2 = \frac{1}{2} R^2(\theta - \sin\Delta) = 2/3\, Mc + \frac{M^3}{2c}$

$A_3 = \frac{1}{2} R^2 \sin\Delta = \frac{1}{2} c(R-M) = (R\sin\frac{\Delta}{2})(R\cos\frac{\Delta}{2}).$

When $\Delta = 90°$: $A_1 = 0.2146 R^2 = 1.2594 E^2$

FIG. B – FORMULAS FOR AREAS.

* Adapted from Lefax Society Inc. Phila. Pa.

HIGHWAYS – CURVES-5

FORMULAS:

$T_s = (R_c + p)\tan\frac{\Delta}{2} + k.$

$E_s = (R_c + p)\,\text{exsec}\,\frac{\Delta}{2} + p = \frac{R_c + p}{\cos\frac{\Delta}{2}} - R_c.$

$p = y_c - R_c(1 - \cos\theta_s) = \frac{y_c}{4}$ (approx.).

$k = x_c - R_c \sin\theta_s = \frac{L_s}{2}$ (approx.).

$\theta_s = \frac{L_s D_c}{200}$; $\theta = \left(\frac{L}{L_s}\right)^2 \theta_s$

$\theta = \frac{L^2 D_c}{200 L_s}$

$L_c = \frac{100\,\Delta_c}{D_c}$; $L.C. = \frac{x_c}{\cos\phi_c}$

$\Delta_c = \Delta - \frac{L_s D_c}{100}$

$D = \frac{L'}{L_s} D_c$

$D_c = \frac{200\,\theta_s}{L_s}$

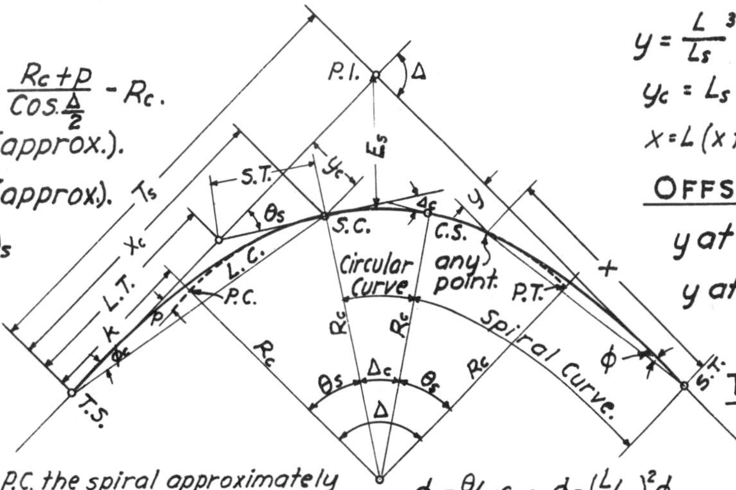

NOTE: At the P.C. the spiral approximately bisects P.

$\phi_c = \theta_s/3 - c$; $\phi = (L/L_s)^2 \phi_c$

OFFSETS TO "x" AND "y":

$y = \left(\frac{L}{L_s}\right)^3 y_c = L\,(y \text{ for } L_s = 1).$

$y_c = L_s\,(y \text{ for } L_s = 1).$

$x = L\,(x \text{ for } L_s = 1)$; $x_c = L_s\,(x \text{ for } L_s = 1).$

OFFSETS TO 1/4 POINTS.

y at 1/4 point $= y_c/4^3$

y at 1/2 point $= y_c/2^3 = P/2$ (approx.)

y at 3/4 point $= y_c/(4/3)^3$

TOTAL LENGTH OF CURVE.

T.S. to S.T. $= 2L_s + 100\frac{\Delta_c}{D_c}$

NOTES: With L_s given or selected from Table "C" below, P, k, x, y, L, T, S.T., and L.C. may be computed for any spiral by multiplying functions for $L_s = 1$ in Table A, page 12-32, by L_s or L in feet. Interpolate for values of θ or θ_s between even degrees. For circular curve layout see page 12-26.

Circular curve may be omitted and curve made transitional throughout in which case S.C. and C.S. coincide at S.C.S., $\theta = \Delta/2$, $\Delta_c = 0$ and T_s & E_s are computed from Table A, page 12-33.

FIG. A - CIRCULAR CURVES WITH SPIRAL TRANSITIONS.

NOTATION

- R_c = Radius of the circular curve.
- P = Offset distance from tangent to the P.C. of the circular curve produced.
- k = Distance from T.S. to P.C. along tangent.
- T_s = Tangent distance.
- E_s = External distance.
- x_c, y_c = Coordinates from T.S. to S.C. & S.T. to C.S.
- θ = Spiral angle at any point on spiral.
- θ_s = Spiral angle at S.C. or C.S.
- L = Length of spiral, T.S. to any point on spiral.
- L_s = Length of spiral, T.S. to S.C. or S.T. to C.S.
- D_c = Degree of circular curve (arc definition).
- D = Degree of curve at any point on spiral.
- x, y = Coordinates from T.S. or S.T. to any point on spiral.
- ϕ_c = Deflection from tangent at T.S. to S.C.
- ϕ = Deflection from tangent at T.S., S.T. or any point on spiral to any other point on spiral.
- L.T., S.T. = Long tangent, short tangent.
- L.C. = Long chord of spiral transition.
- Δ = Intersection and central angle of entire curve.
- Δ_c = Intersection & central angle of circular curve.
- L_c = Length of circular curve, S.C. to C.S.

NOTE: The degree of curvature varies directly as the length, from zero curvature at T.S. to the maximum of D_c at the S.C. The spiral departs from the circular curve at the same rate as from the tangent.

SPIRAL LAYOUT (See page 12-31, also).

Method I: Deflections to even stations by formula $\phi = \theta/3 = \frac{1}{3}\theta_s (L/L_s)^2$. Correct ϕ for c when $\theta > 20°$.

TABLE B: "C" IN FORMULA, $\phi = \theta/3 - c$. (FOR CURVES WITH θ OVER 20°)

θ in degrees.	20	25	30	35	40	45	50
c in minutes.	0.4	0.8	1.4	2.2	3.4	4.8	5.6

Method II: Offsets from tangent. Establish by measuring "x" distances from T.S. and "y" distances from tangent. Compute θ for each point and then compute "x" and "y" coordinates from Table on p. 12-32 or use 1/4 point formulas above.

Method III: Deflection angle from T.S. or S.T. to any point on spiral with coordinates "x" and "y" is the angle whose tangent = y/x.

Method IV: Deflection angles from T.S. to points of 10 equal divisions (10 chord spiral) are: $0.01\phi_c$; $0.04\phi_c$; $0.16\phi_c$; $0.25\phi_c$; $0.36\phi_c$; $0.49\phi_c$; $0.64\phi_c$; $0.81\phi_c$ & ϕ_c.

TABLE C - MINIMUM TRANSITION LENGTHS.

D_c	30 M.P.H. L_s	40 M.P.H. L_s	50 M.P.H. L_s	60 M.P.H. L_s	70 M.P.H. L_s	D_c
1° 30'	150'	150'	150'	150'	150'	1°-30'
2°	"	"	"	"	200'	2°
2° 30'	"	"	"	"	250'	2° 30'
3°	"	"	"	200'	300'	3°
3° 30'	"	"	"	200'	350'	3° 30'
4°	"	"	"	250'	400'	4°
5°	"	"	"	300'		
6°	"	"	200'	350'		
7°	"	"	250'			
8° - 9°	"	"	300'			
10° - 12°	"	200'				
13° - 14°	"	250'				
15° - 23°	"					
24°	200'					

Based on $L_s = \frac{1.6\,V^3}{R_c}$ Where: V = 0.75 Design Speed in M.P.H. Min. L_s = 150 ft.

*Adapted from Transition Curves for Highways by Joseph Barnett, P.R.A.

HIGHWAYS - CURVES-6

INSERTION OF SPIRALS INTO EXISTING ALIGNMENT OF CIRCULAR CURVES.

L_s = Length of spiral - select from Table C-page 12-29.

θ_s = Spiral angle = $\frac{L_s D_c}{200}$, where D_c = Degree of curvature (arc definition).

P = Offset of curve at P.C. to permit spiral introduction - from Table - p.12-32, knowing θ_s.

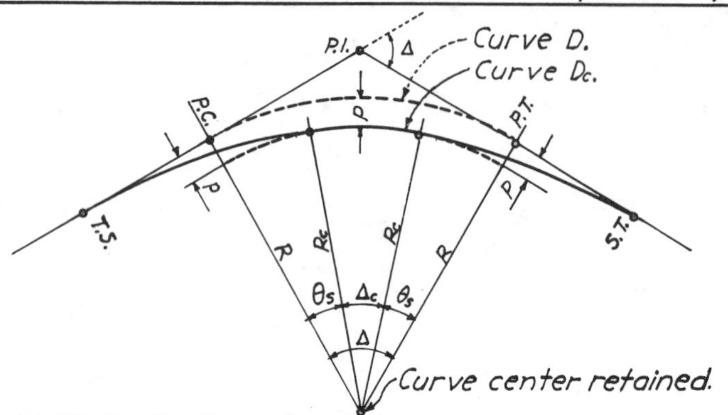

P.I.
Δ
Curve D.
Curve Dc.
P.C.
P.T.
P
T.S.
S.T.
R
Rc
θs
Δc
Δ
Curve center retained.

CASE I - Radius of original circular curve reduced by value of "P" to provide space to insert spiral transition.

1ST Trial: Assume $D_c = D$, find trial "P" as above.

2ND Trial: Compute $D_c = \frac{5727.58}{R_c}$, find correct "P."

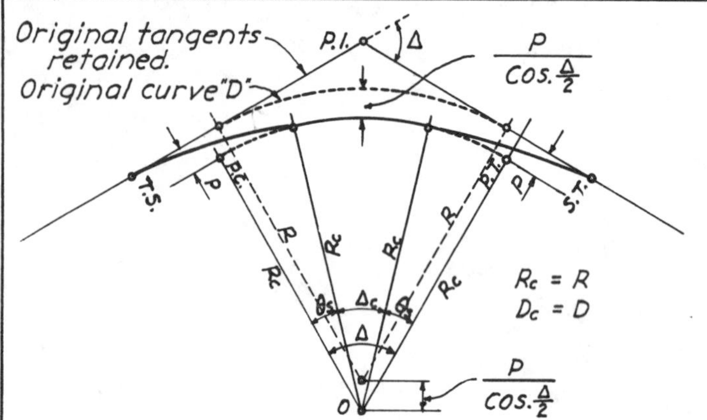

Original tangents retained.
Original curve "D"
P.I.
Δ
P / Cos. Δ/2
T.S.
S.T.
P.C.
P.T.
P
R
Rc
θs
Δc
Δ
Rc = R
Dc = D
O
P / Cos. Δ/2

CASE II - Radius of original curve retained and curve center "O" shifted inward.
Note: Degree of curve retained.

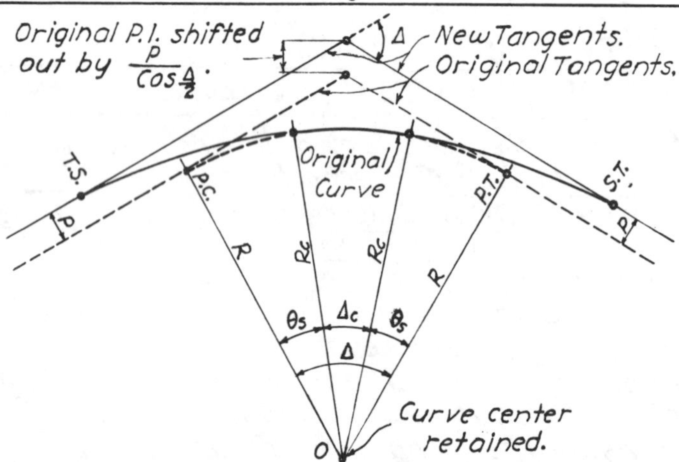

Original P.I. shifted out by P / Cos Δ/2.
Δ
New Tangents.
Original Tangents.
T.S.
S.T.
P.C.
P.T.
Original Curve
P
R
Rc
θs
Δc
Δ
O
Curve center retained.

CASE III - Original circular curve location retained and tangents shifted outward to insert spiral.

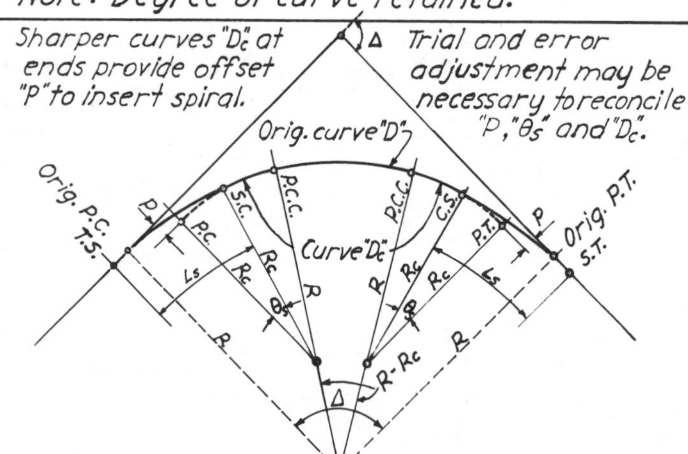

Sharper curves "Dc" at ends provide offset "P" to insert spiral.
Δ
Trial and error adjustment may be necessary to reconcile "P," "θs" and "Dc".
Orig. curve "D"
Orig. P.C. T.S.
Orig. P.T. S.T.
S.C.
P.C.C.
C.S.
P.C.
P.T.
Curve "Dc"
P
Ls
R
Rc
θs
R-Rc
Δ

CASE IV - Original alignment retained as as closely as possible by compounding circular curve at both ends.

θ_a = Equivalent spiral angle. Use in Table page 12-32 to find "P_a".

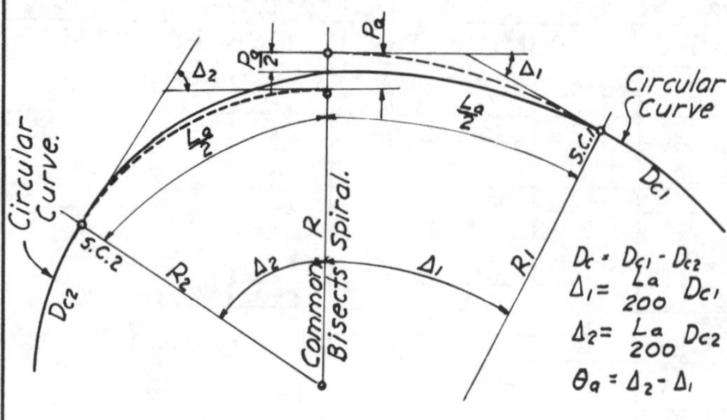

Pa
Pa/2
Δ2
Δ1
Circular Curve
Circular Curve.
La/2
S.C.
S.C.2
Dc1
Dc2
R
R1
R2
Common Spiral. Bisects
Dc = Dc1 - Dc2
Δ1 = La/200 Dc1
Δ2 = La/200 Dc2
θa = Δ2 - Δ1

CASE V - To insert a spiral in a compound curve.

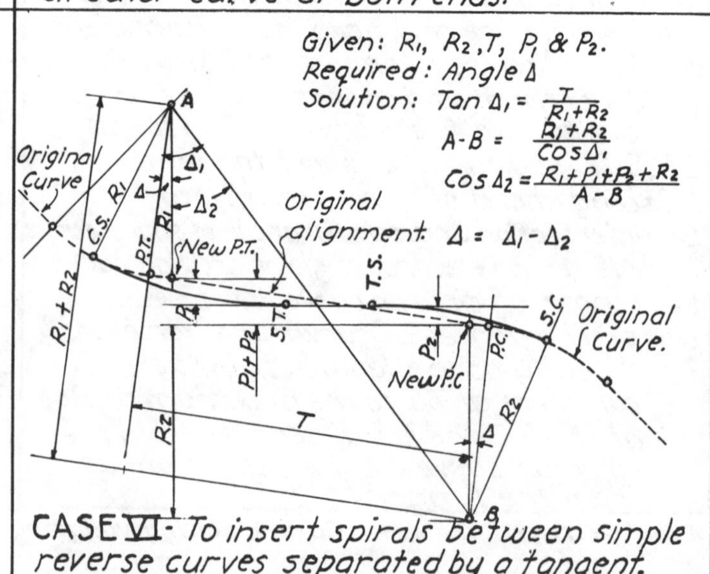

Given: R1, R2, T, P1 & P2.
Required: Angle Δ
Solution: Tan Δ1 = T / (R1+R2)
A-B = (R1+R2) / Cos Δ1
Cos Δ2 = (R1+P1+P2+R2) / (A-B)
Δ = Δ1 - Δ2
A
B
Original Curve
Original alignment
Original Curve.
C.S.
P.T.
New P.T.
S.T.
T.S.
New P.C.
P.C.
S.C.
R1
R2
R1 + R2
P1+P2
P1
P2
T
Δ

CASE VI - To insert spirals between simple reverse curves separated by a tangent.

HIGHWAYS – CURVES–7

PROPERTIES OF SPIRAL.

1. Offsets, y, vary as the cube of L, or length of spiral $\therefore$ y at any point $= \left(\frac{L}{L_s}\right)^3 y_c$. See Fig. A.
2. Spiral angle "θ" varies as L^2 $\therefore$ θ at any point on spiral $= \left(\frac{L}{L_s}\right)^2 \theta_s$.
3. Deflection angle "ϕ" varies as L^2 $\therefore$ $\phi = \left(\frac{L}{L_s}\right)^2 \phi_c$.
   $\phi_c = \frac{1}{3}\theta_s - c$, c being a constant, see Table B - p. 12-29 (May be neglected for ordinary problems).
4. "D", or degree of curve of spiral at any point, varies directly as L $\therefore$ $D = \frac{L}{L_s} D_c$.
5. Spiral bisects "P" very nearly and "k" approximately $= \frac{1}{2} L_s$ $\therefore$ Offset from circular curve or tangent to mid-point of spiral is $\frac{1}{2}$P very nearly.
6. Spiral departs from the circular curve between S.C. and P.C. at the same rate as from the tangent. $\therefore$ Radial offsets from circular curve between S.C. and P.C. to the spiral are the same as perpendicular offsets from the tangent between T.S. and P.C.

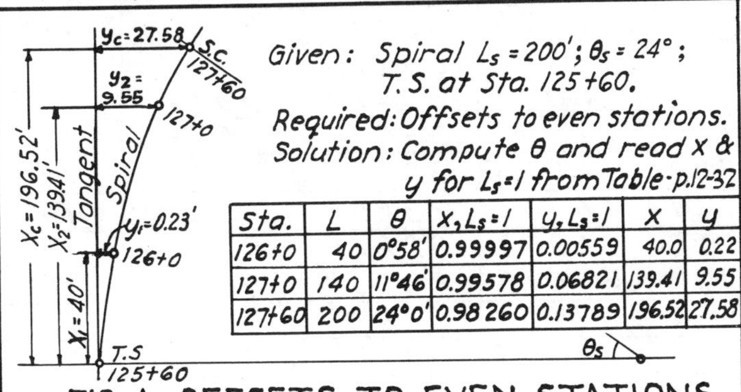

$y_c = 27.58$
S.C.
127+60
$y_2 = 9.55$
127+0
$X_c = 196.52'$
$X_2 = 139.41'$
Tangent
Spiral
$y_1 = 0.23'$
126+0
$X_1 = 40'$
T.S.
125+60
θ_s

Given: Spiral $L_s = 200'$; $\theta_s = 24°$; T.S. at Sta. 125+60.

Required: Offsets to even stations.

Solution: Compute θ and read x & y for $L_s = 1$ from Table - p. 12-32

Sta.	L	θ	$x, L_s=1$	$y, L_s=1$	X	y
126+0	40	0°58'	0.99997	0.00559	40.0	0.22
127+0	140	11°46'	0.99578	0.06821	139.41	9.55
127+60	200	24°0'	0.98260	0.13789	196.52	27.58

FIG. A - OFFSETS TO EVEN STATIONS.

Given: Spiral, $L_s = 200'$; $\theta_s = 24°$.

Required: Offsets to ¼ points.

Solution: From Fig. A, $y_c = 27.58'$

By formula, y at any point $= \left(\frac{L}{L_s}\right)^3 y_c$

At ¼ points, $y = 27.58 \times \frac{1}{64} = 0.43'$.

At ½ points, $y = 27.58 \times \frac{1}{8} = 3.45'$.

S.C. $y = 27.58$

$y = 0.43$

Tangent

Circular Curve

P.C.

$y = 3.45'$

$y = 0.43$

Spiral.

T.S.

FIG. B - OFFSETS TO ¼ POINTS.

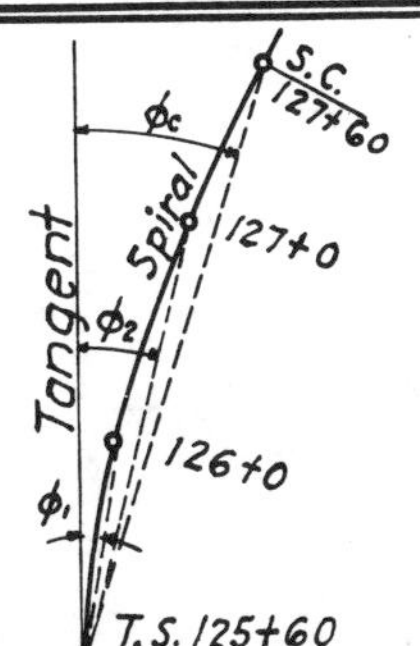

S.C.
127+60
ϕ_c
Spiral
127+0
Tangent
ϕ_2
126+0
ϕ_1
T.S. 125+60

Given: Spiral with $L_s = 200'$ and $\theta_s = 24°$

Required: Deflection angles ϕ_1 to Sta. 126+0; ϕ_2 to Sta. 127+0; ϕ_c to Sta. 127+60.

Solution: By formulas, $\phi_c = \frac{\theta_s}{3} - c$ and $\phi = \frac{1}{3}\left(\frac{L}{L_s}\right)^2 \theta_s - c$.

Sta. 127+60: $\phi_c = 24/3 - 0.8 = 7.9866° = 7° 59.2'$

Sta. 126+00: $\phi_1 = \left(\frac{L}{L_s}\right)^2 \phi_c = \left(\frac{40}{200}\right)^2 \times 7.9866 = 0.3195° = 0° 19'$

Sta. 127+00: $\phi_2 =$ " $\left(\frac{140}{200}\right)^2 \times 7.9866 = 3.9134° = 3° 55'$

Layout: With transit at T.S., foresight along tangent with vernier at 0°. Turn ϕ_1 and measure 40 ft. to Sta. 126+0. Turn ϕ_2 and measure 100 ft. from Sta. 126+0 to Sta. 127+0. Turn ϕ_c and measure 60' from Sta. 127+0 to S.C.

FIG. C - DEFLECTIONS TO EVEN STATIONS.

Given: $\Delta = 90°$; $D_c = 24°$; $L_s = 200'$.
Formulas from pg. 12-29.
Functions of spiral for $L_s = 1$ from p. 12-32.

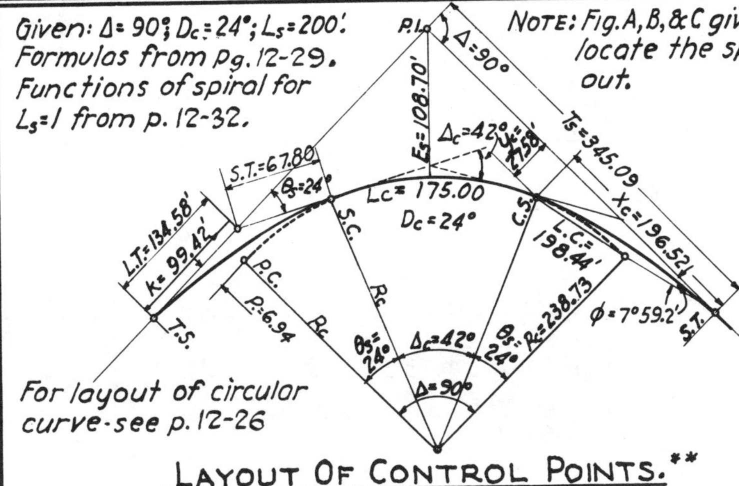

P.I.
$\Delta = 90°$
$E_s = 108.70'$
$\Delta_c = 42°$
$y_c = 27.58'$
$T_s = 345.09$
$x_c = 196.52'$
S.T. = 67.80
$\theta_s = 24°$
$L_c = 175.00$
$D_c = 24°$
S.C.
C.S.
L.C. = 198.44'
L.T. = 134.58'
k = 99.42'
P.C.
P = 6.94
R_c
$R_c = 238.73$
$\phi = 7°59.2'$
S.T.
T.S.
$\theta_s = 24°$
$\Delta_c = 42°$
$\theta_s = 24°$
$\Delta = 90°$

For layout of circular curve - see p. 12-26

NOTE: Fig. A, B, & C give all dimensions usually necessary to plot or locate the spiral. The following example is a curve fully worked out.

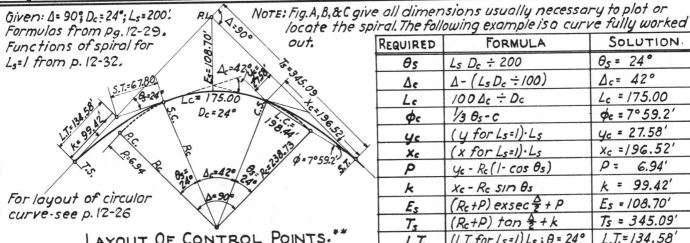

REQUIRED	FORMULA	SOLUTION
θ_s	$L_s D_c \div 200$	$\theta_s = 24°$
Δ_c	$\Delta - (L_s D_c \div 100)$	$\Delta_c = 42°$
L_c	$100 \Delta_c \div D_c$	$L_c = 175.00$
ϕ_c	$\frac{1}{3}\theta_s - c$	$\phi_c = 7°59.2'$
y_c	$(y \text{ for } L_s=1) \cdot L_s$	$y_c = 27.58'$
x_c	$(x \text{ for } L_s=1) \cdot L_s$	$x_c = 196.52'$
P	$y_c - R_c(1 - \cos\theta_s)$	P = 6.94'
k	$x_c - R_c \sin\theta_s$	k = 99.42'
E_s	$(R_c + P)\,\text{exsec}\frac{\Delta}{2} + P$	$E_s = 108.70'$
T_s	$(R_c + P)\tan\frac{\Delta}{2} + k$	$T_s = 345.09'$
L.T.	(L.T. for $L_s=1$) L_s; $\theta = 24°$	L.T. = 134.58'
S.T.	(S.T. for $L_s=1$) L_s; $\theta = 24°$	S.T. = 67.80'
L.C.	(L.C. for $L_s=1$) L_s; $\theta = 24°$	L.C. = 198.44

LAYOUT OF CONTROL POINTS.**

Establish T.S. by measuring "k" from P.O.T. normal to P.C. or by T_s from P.I. Establish S.C. by L.T., θ_s and S.T. or by x_c and y_c from T.S. or by ϕ_c and L.C. from T.S.

FIG. D - COMPUTATIONS FOR SPIRAL TRANSITIONS TO CIRCULAR CURVES.

*Ref. - Transition Curves for Highways by Joseph Barnett, P.R.A.

**Adapted from O'Rourke, General Engineering Handbook, McGraw-Hill.

HIGHWAYS – CURVES-8

Enter Table with value of θ or $θ_s$ and multiply function by L or L_s. *See pg. 12-31 for use of Table.*

θ	p	k	x	y	L.T.	S.T.	L.C.	θ
0°	.00000	.50000	1.00000	.00000	.66667	.33333	1.00000	0°
1°	.00146	.49999	.99997	.00582	.66668	.33334	.99999	1°
2°	.00291	.49998	.99988	.01163	.66671	.33337	.99995	2°
3°	.00435	.49995	.99973	.01745	.66676	.33342	.99988	3°
4°	.00581	.49992	.99951	.02326	.66684	.33349	.99978	4°
5°	.00727	.49987	.99924	.02907	.66693	.33358	.99966	5°
6°	.00872	.49982	.99890	.03488	.66705	.33368	.99951	6°
7°	.01018	.49975	.99851	.04068	.66719	.33381	.99934	7°
8°	.01163	.49967	.99805	.04648	.66735	.33395	.99913	8°
9°	.01308	.49959	.99754	.05227	.66753	.33412	.99890	9°
10°	.01453	.49949	.99696	.05805	.66773	.33430	.99865	10°
11°	.01598	.49939	.99632	.06383	.66796	.33451	.99836	11°
12°	.01743	.49927	.99562	.06959	.66821	.33473	.99805	12°
13°	.01887	.49914	.99486	.07535	.66847	.33498	.99771	13°
14°	.02032	.49901	.99405	.08110	.66877	.33524	.99735	14°
15°	.02176	.49886	.99317	.08684	.66908	.33553	.99696	15°
16°	.02320	.49870	.99223	.09257	.66941	.33583	.99654	16°
17°	.02465	.49854	.99123	.09828	.66977	.33615	.99609	17°
18°	.02608	.49836	.99018	.10398	.67015	.33650	.99562	18°
19°	.02752	.49817	.98906	.10967	.67055	.33687	.99512	19
20°	.02896	.49798	.98788	.11535	.67097	.33725	.99460	20°
21°	.03040	.49777	.98665	.12101	.67142	.33766	.99404	21°
22°	.03183	.49755	.98536	.12665	.67189	.33809	.99346	22°
23°	.03326	.49733	.98401	.13228	.67238	.33854	.99286	23°
24°	.03469	.49709	.98260	.13789	.67290	.33901	.99222	24°
25°	.03611	.49684	.98113	.14348	.67344	.33950	.99157	25°
26°	.03753	.49658	.97960	.14905	.67400	.34001	.99088	26°
27°	.03896	.49632	.97802	.15461	.67459	.34055	.99017	27°
28°	.04037	.49605	.97638	.16014	.67520	.34111	.98943	28°
29°	.04179	.49576	.97469	.16565	.67584	.34169	.98866	29°
30°	.04321	.49546	.97293	.17114	.67650	.34229	.98787	30°
31°	.04462	.49516	.97112	.17661	.67719	.34292	.98705	31°
32°	.04602	.49484	.96926	.18206	.67790	.34356	.98621	32°
33°	.04743	.49452	.96733	.18748	.67863	.34424	.98534	33°
34°	.04883	.49419	.96536	.19288	.67939	.34493	.98444	34°
35°	.05023	.49385	.96332	.19826	.68018	.34565	.98351	35°
36°	.05163	.49349	.96124	.20361	.68100	.34640	.98257	36°
37°	.05301	.49313	.95910	.20893	.68184	.34717	.98159	37°
38°	.05441	.49276	.95690	.21423	.68271	.34796	.98059	38°
39°	.05579	.49238	.95466	.21949	.68360	.34878	.97956	39°
40°	.05718	.49199	.95235	.22473	.68452	.34962	.97851	40°
41°	.05855	.49159	.95000	.22994	.68547	.35049	.97743	41°
42°	.05993	.49118	.94759	.23513	.68645	.35139	.97632	42°
43°	.06130	.49075	.94513	.24028	.68746	.35232	.97519	43°
44°	.06267	.49032	.94262	.24540	.68850	.35327	.97404	44°
45°	.06403	.48990	.94005	.25049	.68957	.35424	.97285	45°
46°	.06538	.48945	.93744	.25555	.69066	.35525	.97165	46°
47°	.06674	.48900	.93477	.26057	.69179	.35629	.97041	47°
48°	.06809	.48852	.93206	.26556	.69295	.35735	.96916	48°
49°	.06944	.48805	.92930	.27052	.69414	.35844	.96787	49°
50°	.07078	.48757	.92649	.27544	.69536	.35957	.96656	50°

* *Adapted from Transition Curves for Highways by Joseph Barnett, P.R.A.*

HIGHWAYS - CURVES-9

TABLE A- TANGENTS AND EXTERNALS FOR $L_s = 1$*

$\Delta°$	T_s	E_s	$\Delta°$	T_s	E_s	$\Delta°$	T_s	E_s
6°	1.00064	0.01747	38°	1.02682	0.11599	70°	1.10214	0.24203
7	1.00087	0.02040	39	1.02832	0.11936	71	1.10561	0.24681
8	1.00114	0.02332	40	1.02987	0.12275	72	1.10917	0.25167
9	1.00144	0.02625	41	1.03146	0.12617	73	1.11281	0.25660
10	1.00178	0.02918	42	1.03310	0.12962	74	1.11654	0.26161
11	1.00216	0.03212	43	1.03479	0.13309	75	1.12036	0.26669
12	1.00257	0.03507	44	1.03653	0.13660	76	1.12427	0.27186
13	1.00302	0.03802	45	1.03831	0.14012	77	1.12828	0.27710
14	1.00350	0.04098	46	1.04015	0.14370	78	1.13240	0.28244
15	1.00402	0.04396	47	1.04204	0.14730	79	1.13661	0.28786
16	1.00458	0.04693	48	1.04399	0.15094	80	1.14092	0.29337
17	1.00518	0.04992	49	1.04598	0.15460	81	1.14535	0.29898
18	1.00581	0.05292	50	1.04804	0.15831	82	1.14988	0.30468
19	1.00648	0.05593	51	1.05014	0.16206	83	1.15453	0.31048
20	1.00719	0.05895	52	1.05230	0.16584	84	1.15930	0.31639
21	1.00794	0.06198	53	1.05452	0.16966	85	1.16418	0.32241
22	1.00873	0.06502	54	1.05680	0.17352	86	1.16919	0.32854
23	1.00955	0.06808	55	1.05913	0.17742	87	1.17433	0.33478
24	1.01042	0.07115	56	1.06153	0.18137	88	1.17960	0.34115
25	1.01132	0.07424	57	1.06399	0.18536	89	1.18500	0.34763
26	1.01226	0.07734	58	1.06651	0.18940	90	1.19054	0.35425
27	1.01324	0.08045	59	1.06909	0.19348	91	1.19623	0.36099
28	1.01427	0.08358	60	1.07174	0.19762	92	1.20207	0.36788
29	1.01533	0.08674	61	1.07446	0.20181	93	1.20806	0.37490
30	1.01644	0.08990	62	1.07724	0.20604	94	1.21421	0.38207
31	1.01758	0.09309	63	1.08010	0.21034	95	1.22052	0.38940
32	1.01877	0.09630	64	1.08302	0.21468	96	1.22700	0.39688
33	1.02000	0.09952	65	1.08602	0.21908	97	1.23366	0.40453
34	1.02128	0.10277	66	1.08909	0.22355	98	1.24050	0.41234
35	1.02260	0.10604	67	1.09223	0.22807	99	1.24753	0.42034
36	1.02396	0.10933	68	1.09546	0.23266	100	1.25475	0.42852
37	1.02537	0.11265	69	1.09876	0.23731			

CASE VII: Given Δ and an external or tangent distance, to determine a curve transitional throughout.

Enter Table A at known Δ and read T_s & E_s values. Then $L_s = E_s / E_s\,value$ and $T_s = L_s \cdot$ Tangent value, or $L_s = T_s / T_s\,value$ and $E_s = L_s \cdot$ External value.

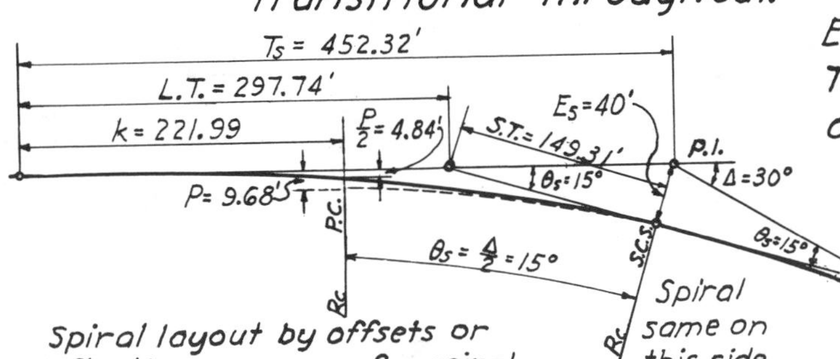

Ts = 452.32'
L.T. = 297.74'
k = 221.99
P/2 = 4.84'
Es=40'
S.T.=149.31'
P.I.
Δ=30°
θs=15°
P= 9.68'
P.C.
θs = Δ/2 = 15°
S.C.S.
θs=15°
Rc
Rc
Spiral same on this side.

Spiral layout by offsets or deflections-same as for spiral transitions to a circular curve.

FIG. B

EXAMPLE

Given: $\Delta = 30°$ and $E_s = 40'$.

Required: L_s, T_s, θ_s, L.T., S.T., D_c, p and k.

Solution: $L_s = 40 \div 0.08990 = 444.9'$ say $445'$

$T_s = 1.01644 \times 445 = 452.32'$. $\theta_s = \frac{\Delta}{2} = 15°$. $D_c = \frac{200\theta_s}{L_s} = 6.74°$

L.T. $= 0.66908 \times 445 = 297.74'$. S.T. $= 0.33553 \times 445 = 149.31'$

$p = 0.02176 \times 445 = 9.68'$. $k = 498.86 \times 445 = 221.99'$.

*Adapted from Transition Curves for Highways by Joseph Barnett, P. R. A.

ROADS & STREETS - TYPICAL CROSS SECTIONS

ROADS

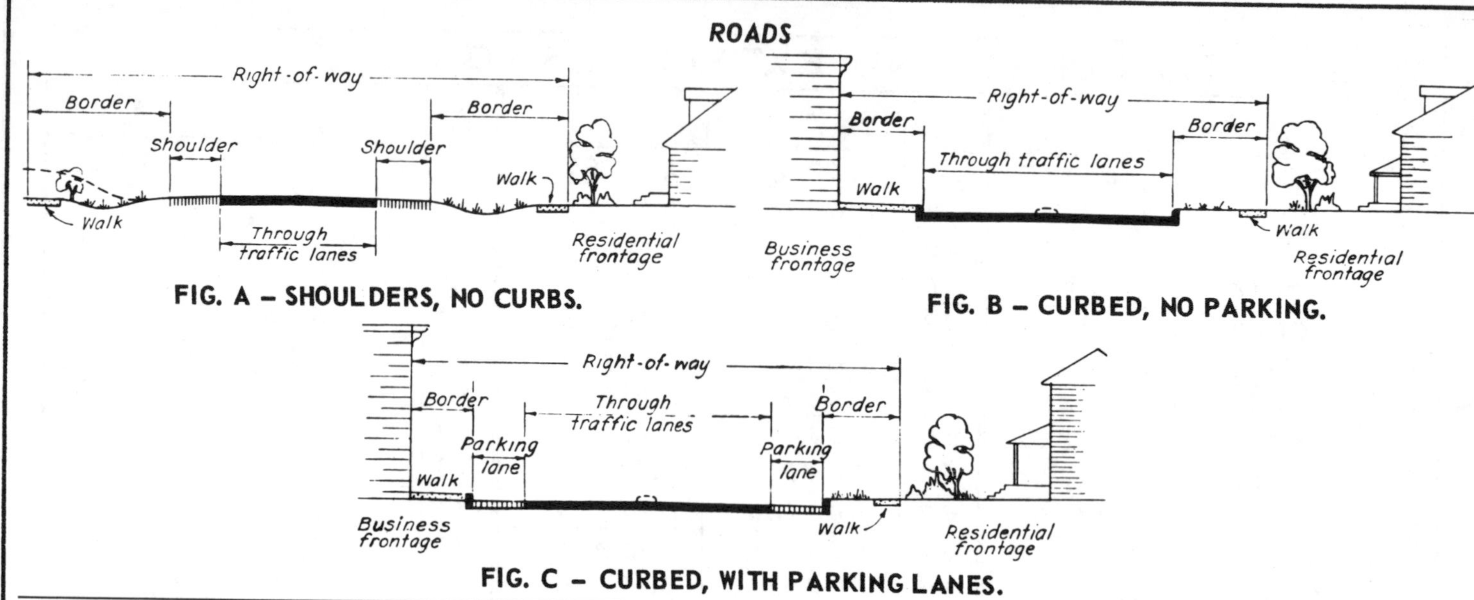

Right-of-way
Border
Shoulder
Walk
Through traffic lanes
Residential frontage
Business frontage
Parking lane

FIG. A – SHOULDERS, NO CURBS.

FIG. B – CURBED, NO PARKING.

FIG. C – CURBED, WITH PARKING LANES.

MAJOR STREETS

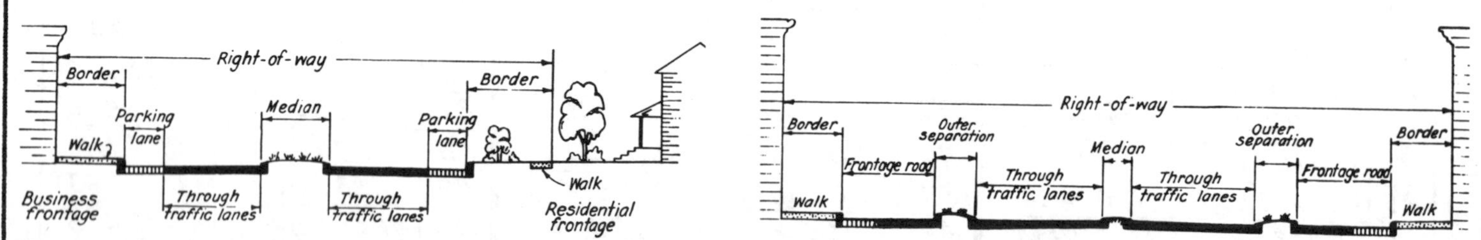

Right-of-way
Border
Parking lane
Median
Walk
Through traffic lanes
Business frontage
Residential frontage
Outer separation
Frontage road

FIG. D – DIVIDED, WITH PARKING LANES.

FIG. E – DIVIDED, WITH SEPARATED PARKING and SERVICE LANES.

TABLE F – MINIMUM RIGHT-OF-WAY WIDTHS FOR MAJOR STREETS

Section	Type of Urban Area	Through-Traffic Lanes*			Width of Other Cross-section Elements, ft.								Right of way width in feet	
		No.	Width, ft.		Median		Curb Parking Lanes†		Additional Elements‡		Border			
			A	B	A	B	A	B	A	B	A	B	A	B
Fig. A, shoulders, no curbs	Residential	2	11	12	0	0	–	–	10	10	12	20	66	84
	Residential	4	11	12	0	14	–	–	10	10	8	12	80	106
Fig. B, curbed, no parking	Business	4	11	12	0	4	–	–	1	2	8	12	62	80
	Residential	4	11	12	0	4	–	–	1	2	12	16	70	88
	Business	6	11	12	0	4	–	–	1	2	8	12	84	104
	Residential	6	11	12	0	4	–	–	1	2	12	16	92	112
Fig. C, curbed, with parking lanes	Business	4	11	12	0	4	10	11	–	–	8	12	80	98
	Residential	4	11	12	0	4	10	10	–	–	12	16	88	104
	Business	6	11	12	0	4	10	11	–	–	8	12	102	122
	Residential	6	11	12	0	4	10	10	–	–	12	16	110	128
Fig. D, divided, with parking lanes	Business	4	11	12	4	14	10	12	–	–	8	12	84	110
	Residential	4	11	12	4	14	10	11	–	–	12	16	92	116
	Business	6	11	12	4	14	10	12	–	–	8	12	106	134
	Residential	6	11	12	4	14	10	11	–	–	12	16	114	140
Fig. E, divided, with separate parking and service lanes	Business or residential	4	11	12	4	14	20	24	4	8	8	12	116	150

A = Acceptable minimum B = Desirable minimum

*Where right-of-way is limited, it may be necessary to use 10-ft. widths for through traffic, but this should not be considered on 2-lane streets.

†For Fig. E, this is combined parking and service-lane width, or frontage road width.

‡For Fig. A, this is width of shoulders; for Fig. B, offset for barrier curb or gutter pan width, and, for Fig. E, width of outer separation.

NOTE: This complete page is adapted from, – *A Policy on Arterial Highways in Urban Areas*, A.S.S.H.O.

ROADS & STREETS – INTERSECTIONS – I

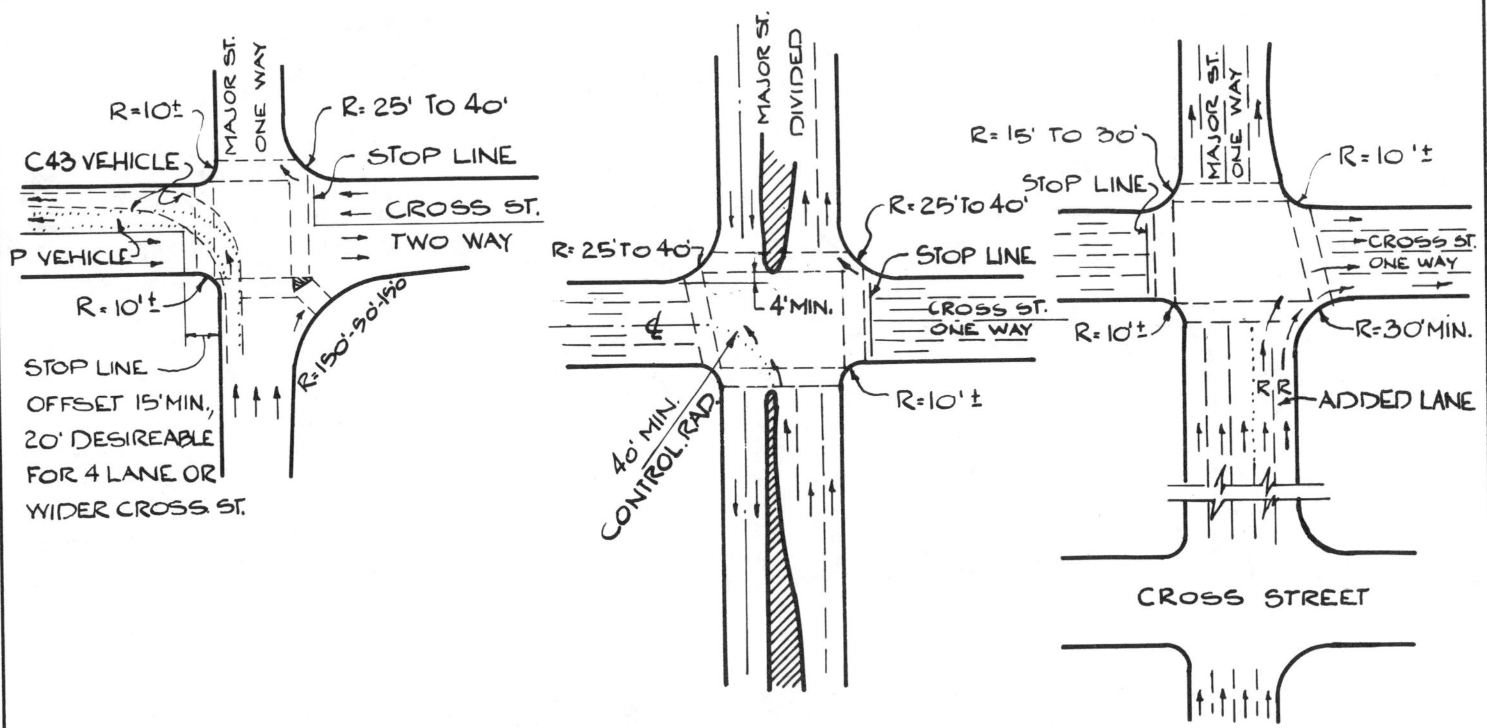
MAJOR ST. ONE WAY
R=10'±
C43 VEHICLE
P VEHICLE
R=10'±
R: 25' TO 40'
STOP LINE
CROSS ST.
TWO WAY
R=150'-50'-150'
STOP LINE OFFSET 15'MIN., 20' DESIREABLE FOR 4 LANE OR WIDER CROSS ST.
MAJOR ST. DIVIDED
R: 25' TO 40'
R: 25' TO 40'
STOP LINE
4'MIN.
CROSS ST. ONE WAY
R=10'±
40' MIN. CONTROL RAD.
MAJOR ST. ONE WAY
R= 15' TO 30'
STOP LINE
R=10'±
R=10'±
CROSS ST. ONE WAY
R=30'MIN.
ADDED LANE
R R
CROSS STREET

MAJOR STREET INTERSECTIONS WITH ONE WAY OPERATION

Δ	Design Vehicle*	d_2 in feet for cases A and B where:									
		R = 15'		R = 20'		R = 25'		R = 30'		R = 40'	
		A	B	A	B	A	B	A	B	A	B
30°	SU	16	14	15	14	15	14	15	14	14	14
	C43	18	16	18	16	18	16	18	16	18	15
	C50	20	17	20	17	20	17	19	16	19	16
60°	SU	25	17	23	17	22	16	20	16	16	15
	C43	27	19	26	19	24	19	23	18	19	17
	C50	32	22	30	21	28	21	26	20	22	19
90°	SU	34	20	30	18	27	17	22	15	15	13
	C43	35	21	31	20	28	19	22	17	17	15
	C50	43	24	38	22	35	21	29	20	21	17
120°	SU	43	23	36	20	30	18	24	16	14†	14†
	C43	41	23	34	21	27	19	20	17†	15†	14†
	C50	50	26	43	24	36	23†	30	21†	19†	17†
150°	SU	50	27	42	23	31	19	23	16	12†	12†
	C43	42	25	33	21	25	18†	17†	15†	11‡	11‡
	C50	52	28	44	25	34	22†	25†	19†	15‡	14‡

NOTE: P design vehicle turns within 11 width where R = 20' or more. a = d_2 or less, except as noted: †a = 25' to 30'; ‡a = 35' to 40'.
No parking lanes on either street.

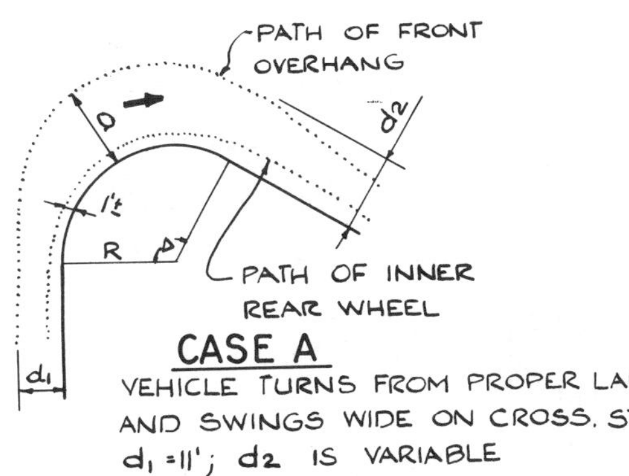
PATH OF FRONT OVERHANG
a
d2
1'±
R
Δ
PATH OF INNER REAR WHEEL
d1

CASE A

VEHICLE TURNS FROM PROPER LANE AND SWINGS WIDE ON CROSS. ST. d_1 = 11'; d_2 IS VARIABLE

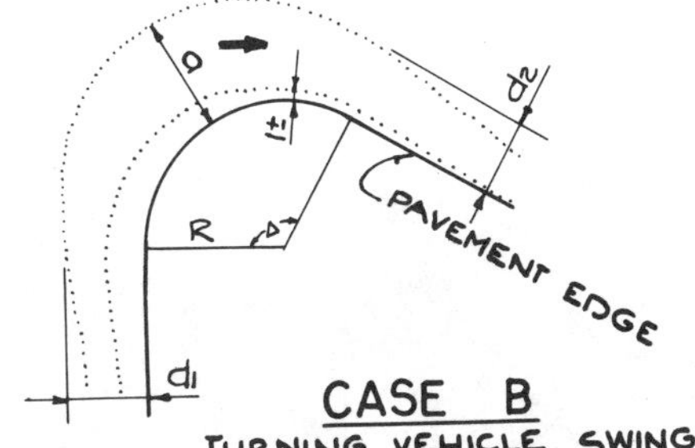
a
d2
1'±
R
Δ
PAVEMENT EDGE
d1

CASE B

TURNING VEHICLE SWINGS EQUALLY WIDE ON BOTH STREETS, $d_1 = d_2$; BOTH VARIABLE

EFFECT OF CURB RADII AND INTERSECTION ANGLES ON TURNING MOVEMENTS

*A.A.S.H.O. designation.

ROADS & STREETS - INTERSECTIONS-2

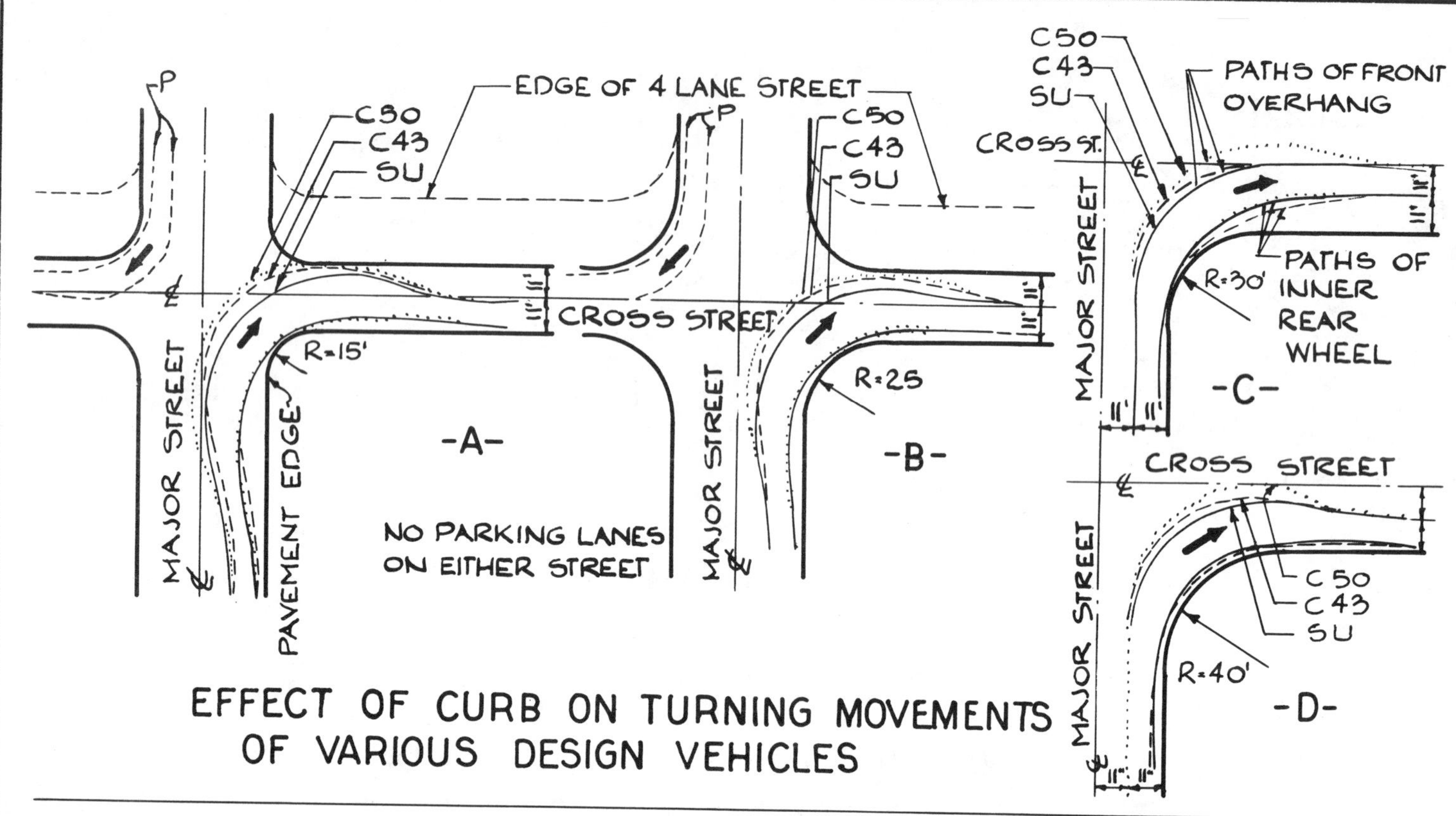
P
EDGE OF 4 LANE STREET
C50
C43
SU
CROSS STREET
R=15'
R=25
MAJOR STREET
PAVEMENT EDGE
-A-
-B-
NO PARKING LANES
ON EITHER STREET
CROSS ST.
PATHS OF FRONT
OVERHANG
PATHS OF
INNER
REAR
WHEEL
R=30'
-C-
R=40'
-D-

EFFECT OF CURB ON TURNING MOVEMENTS OF VARIOUS DESIGN VEHICLES

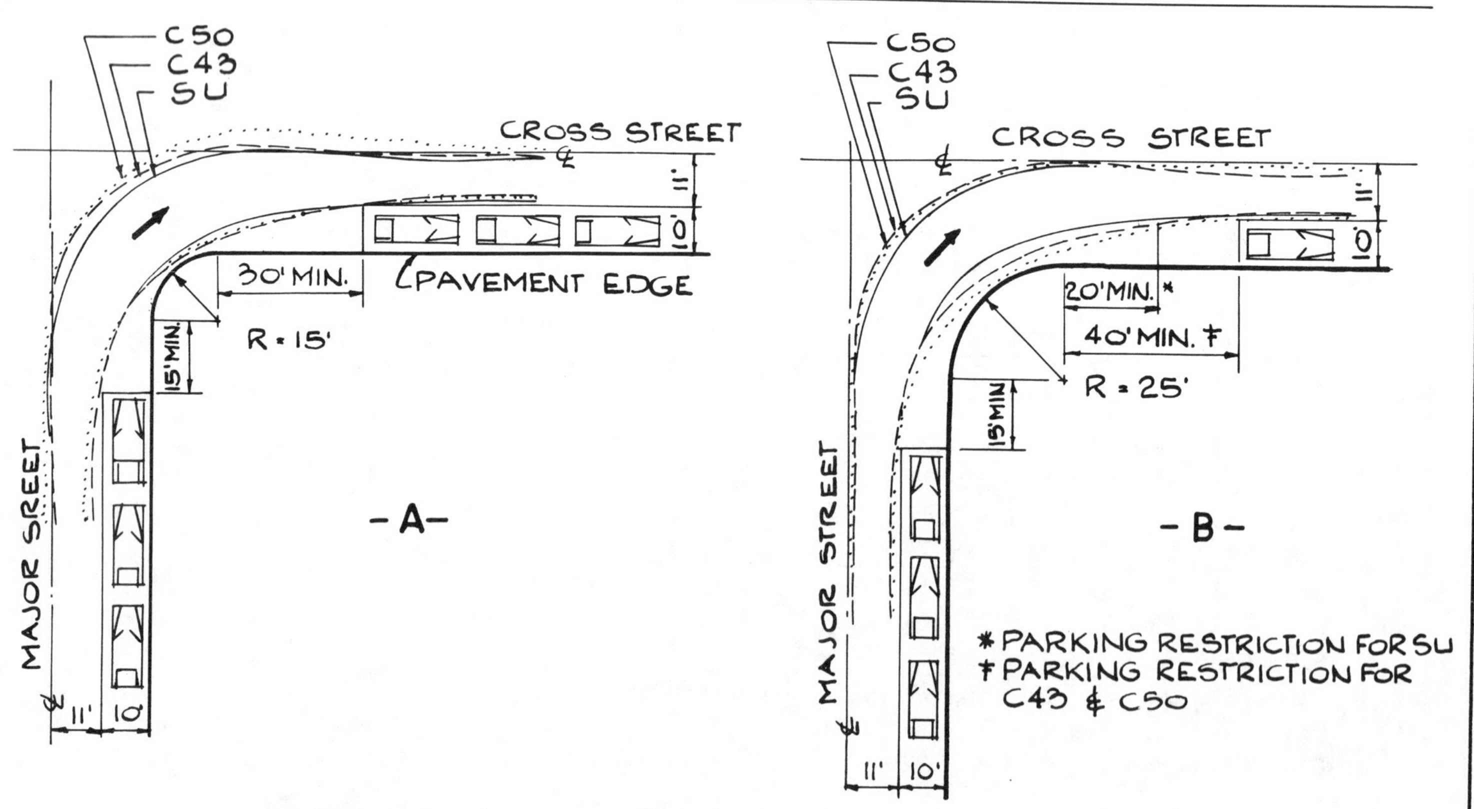
C50
C43
SU
CROSS STREET
30' MIN.
PAVEMENT EDGE
R = 15'
15' MIN.
MAJOR STREET
-A-
20' MIN.*
40' MIN. ‡
R = 25'
-B-
* PARKING RESTRICTION FOR SU
‡ PARKING RESTRICTION FOR C43 & C50

EFFECT OF CURB RADII AND PARKING ON TURNING MOVEMENTS

P= Passenger vehicle
S.U.= Single unit truck or bus.

C-43 = 43 ft. semitrailer combination.
C-50 = 50 ft. semitrailer combination.

ROADS & STREETS - DETAILS - I

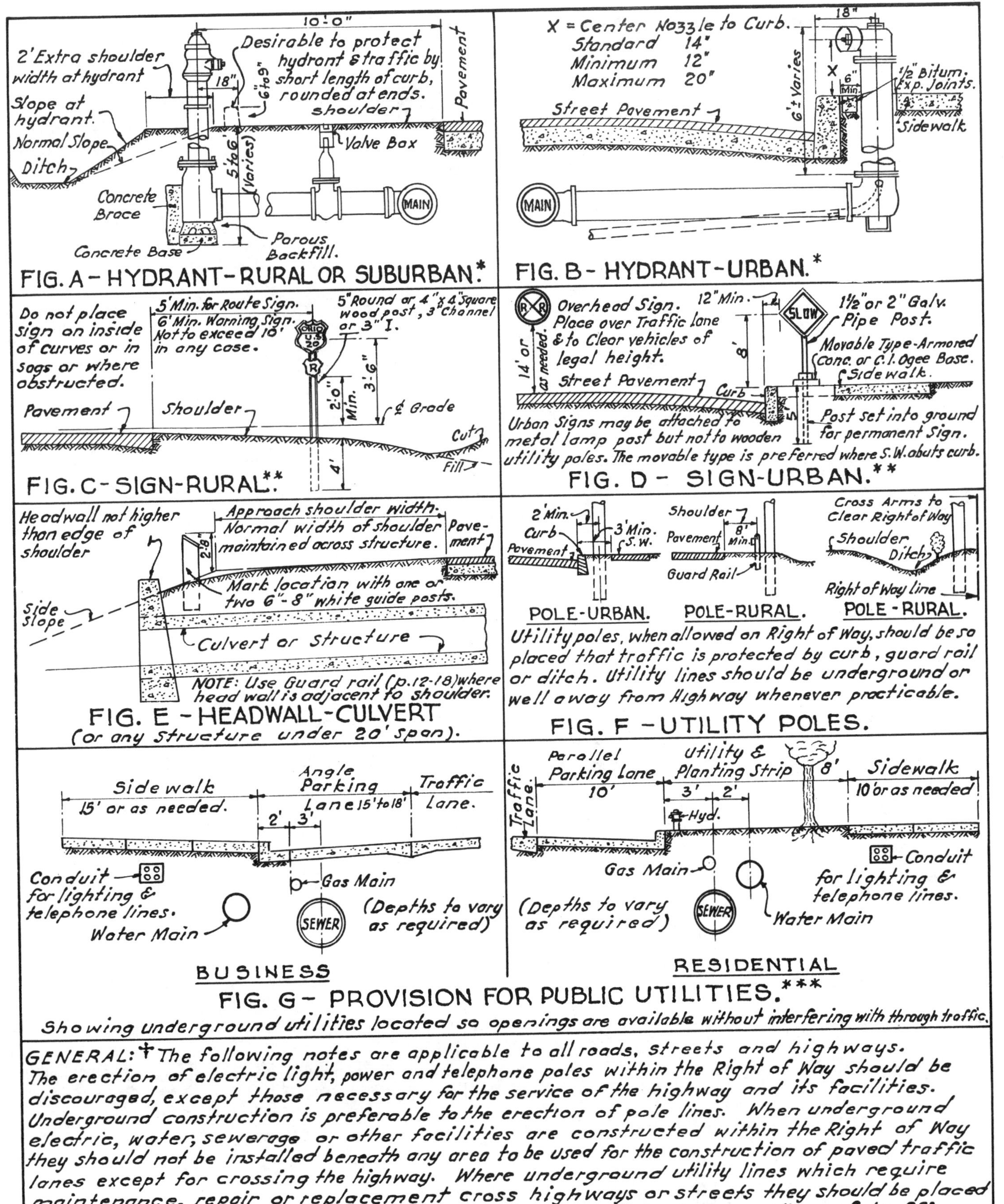

10'-0"
2' Extra shoulder width at hydrant
Desirable to protect hydrant & traffic by short length of curb, rounded at ends.
Pavement
18"
6" to 9"
Slope at hydrant.
Normal Slope
shoulder
Ditch
Valve Box
5' to 6' (Varies)
Concrete Brace
MAIN
Porous Backfill.
Concrete Base
FIG. A - HYDRANT - RURAL OR SUBURBAN.*
X = Center Nozzle to Curb.
Standard 14"
Minimum 12"
Maximum 20"
18"
6' ± Varies
X
6" Min.
1/2" Bitum. Exp. Joints.
Street Pavement
Sidewalk
MAIN
FIG. B - HYDRANT - URBAN.*
Do not place sign on inside of curves or in sags or where obstructed.
5' Min. for Route Sign. 6' Min. Warning Sign. Not to exceed 10' in any case.
5" Round or 4" x 4" Square wood post, 3" Channel or 3" I.
U.S. 20
R
3'-6"
2'-0" Min.
℄ Grade
Pavement
Shoulder
Cut
Fill
4'
FIG. C - SIGN-RURAL.**
R R
Overhead Sign. Place over Traffic lane & to Clear vehicles of legal height.
12" Min.
SLOW
1½" or 2" Galv. Pipe Post.
Movable Type-Armored Conc. or C.I. Ogee Base.
14' or as needed.
8'
Street Pavement
Curb
Sidewalk
Urban Signs may be attached to metal lamp post but not to wooden utility poles. The movable type is preferred where S.W. abuts curb.
Post set into ground for permanent Sign.
FIG. D - SIGN-URBAN.**
Headwall not higher than edge of shoulder
Approach shoulder width. Normal width of shoulder maintained across structure.
Pavement
2'-8"
Mark location with one or two 6"-8" white guide posts.
Side Slope
Culvert or Structure
NOTE: Use Guard rail (p.12-18) where head wall is adjacent to shoulder.
FIG. E - HEADWALL-CULVERT
(or any Structure under 20' span).
2' Min.
Curb
3' Min. S.W.
Pavement
Shoulder
8' Min.
Pavement
Guard Rail
Cross Arms to Clear Right of Way
Shoulder
Ditch
Right of Way line
POLE-URBAN.
POLE-RURAL.
POLE-RURAL.
Utility poles, when allowed on Right of Way, should be so placed that traffic is protected by curb, guard rail or ditch. Utility lines should be underground or well away from Highway whenever practicable.
FIG. F - UTILITY POLES.
Sidewalk 15' or as needed.
Angle Parking Lane 15' to 18'
Traffic Lane.
2'
3'
Conduit for lighting & telephone lines.
Gas Main
Water Main
SEWER
(Depths to vary as required)
BUSINESS
Traffic Lane.
Parallel Parking Lane 10'
Utility & Planting Strip
3'
2'
8'
Sidewalk 10' or as needed
Hyd.
Gas Main
SEWER
Water Main
(Depths to vary as required)
Conduit for lighting & telephone lines.
RESIDENTIAL

FIG. G - PROVISION FOR PUBLIC UTILITIES.***

Showing underground utilities located so openings are available without interfering with through traffic.

GENERAL: † The following notes are applicable to all roads, streets and highways. The erection of electric light, power and telephone poles within the Right of Way should be discouraged, except those necessary for the service of the highway and its facilities. Underground construction is preferable to the erection of pole lines. When underground electric, water, sewerage or other facilities are constructed within the Right of Way they should not be installed beneath any area to be used for the construction of paved traffic lanes except for crossing the highway. Where underground utility lines which require maintenance, repair or replacement cross highways or streets they should be placed in service tunnels under the pavement to insure undisturbed operation of traffic.

* Adapted from Tratman Modern Construction, Eng. News Record. ** Adapted from Manual of Uniform Traffic Control, A. A. S. H. O. *** Adapted from R. E. Toms, P.R.A. Chicago Designs.
† Adapted from Interregional Highways, President of the United States.

ROADS & STREETS – DETAILS – 2

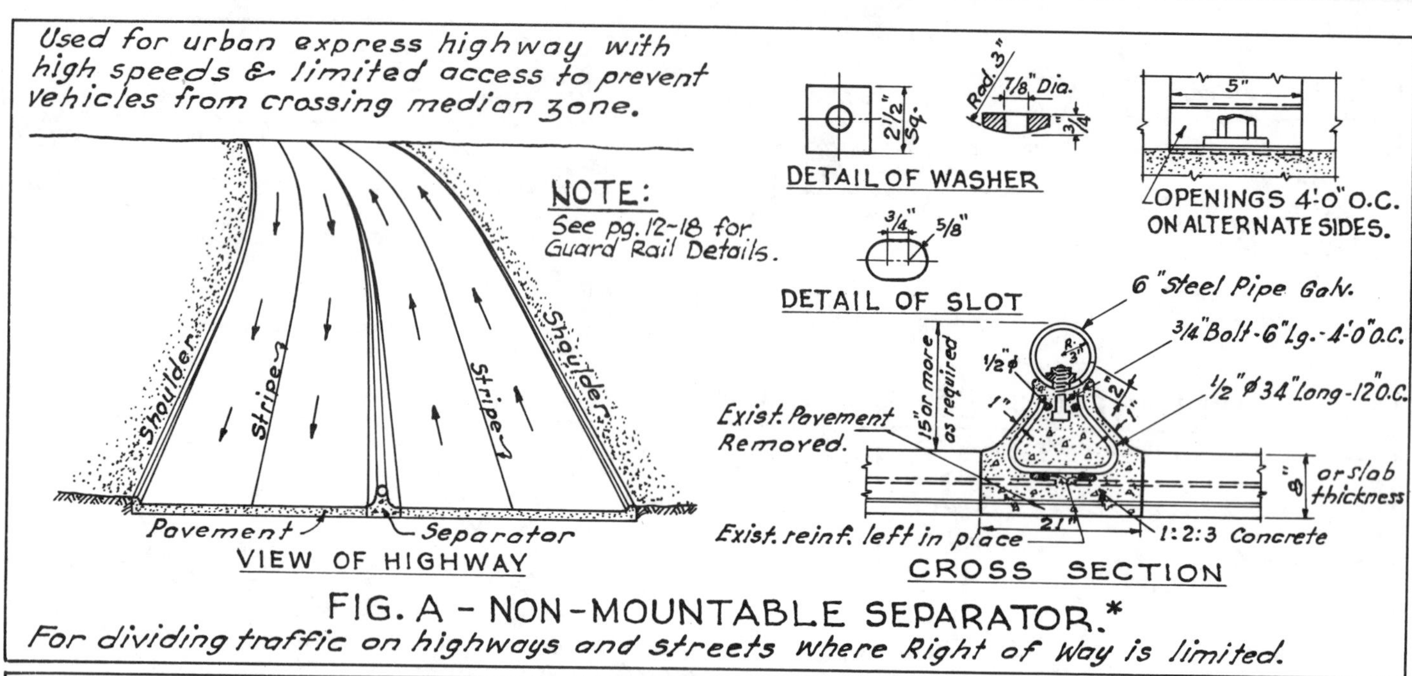

Used for urban express highway with high speeds & limited access to prevent vehicles from crossing median zone.
NOTE:
See pg. 12-18 for Guard Rail Details.
Shoulder
Stripe
Stripe
Shoulder
Pavement
Separator
VIEW OF HIGHWAY
Rad. 3"
7/8" Dia.
2 1/2" Sq.
3/4"
DETAIL OF WASHER
5"
OPENINGS 4'-0" O.C. ON ALTERNATE SIDES.
3/4"
5/8"
DETAIL OF SLOT
6" Steel Pipe Galv.
3/4" Bolt - 6" Lg. - 4'-0" O.C.
1/2" ϕ 34" Long - 12" O.C.
15" or more as required
1/2ϕ
Exist. Pavement Removed.
8" or slab thickness
21"
Exist. reinf. left in place
1:2:3 Concrete
CROSS SECTION

FIG. A – NON-MOUNTABLE SEPARATOR.*
For dividing traffic on highways and streets where Right of Way is limited.

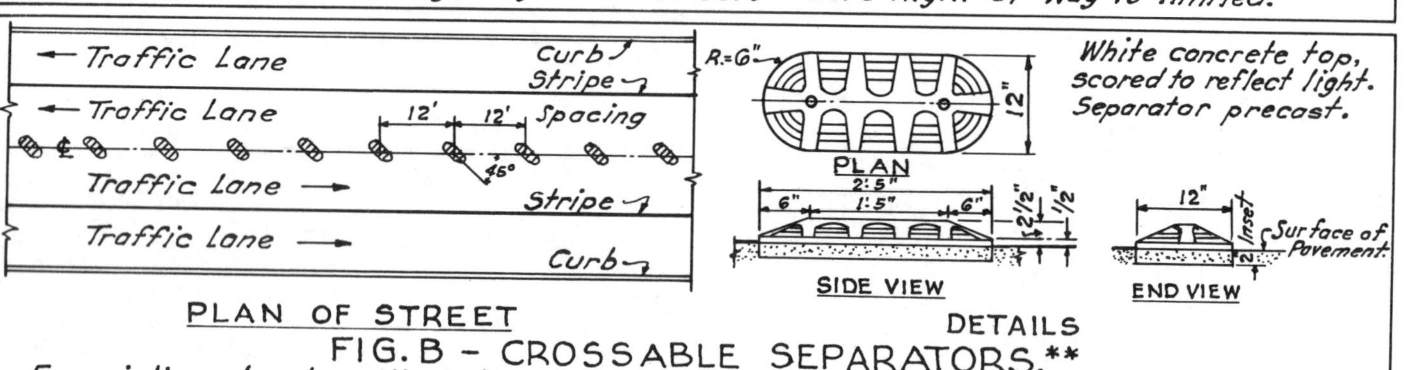

Traffic Lane
Curb
Stripe
Traffic Lane
12'
12'
Spacing
45°
Traffic Lane
Stripe
Traffic Lane
Curb
PLAN OF STREET
R.=6"
12"
PLAN
2' 5"
6"
1' 5"
6"
2 1/2"
1/2"
SIDE VIEW
12"
Inset
Surface of Pavement
2"
END VIEW
White concrete top, scored to reflect light. Separator precast.
DETAILS

FIG. B – CROSSABLE SEPARATORS.**
For existing streets with limited Right of Way where frequent crossing of median zone is necessary. Omit at major intersections.

New Conc. Lane · 1 1/2" Bitum. Top Course · ℄ · New Conc. Median Strip · 4'-0" · 1 1/2" Bitum. Top Course · New Conc. Lane · New Conc. Base · New Curb · Existing Pavement · New Conc. Base · New Curb

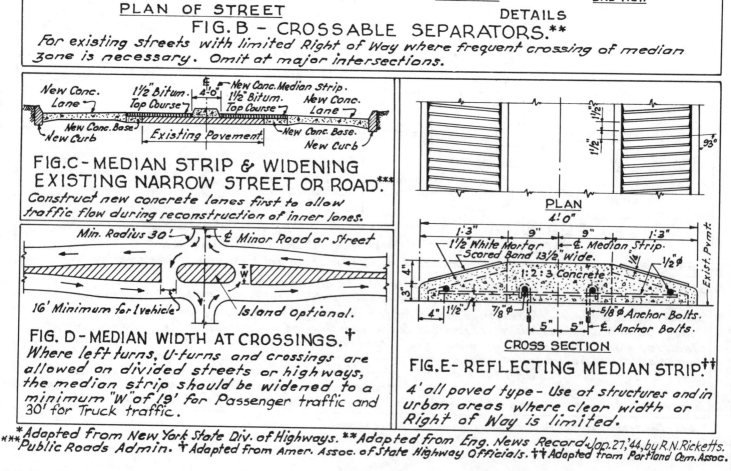

FIG. C – MEDIAN STRIP & WIDENING EXISTING NARROW STREET OR ROAD.***
Construct new concrete lanes first to allow traffic flow during reconstruction of inner lanes.

Min. Radius 30' · ℄ Minor Road or Street · W · 16' Minimum for 1 vehicle · Island optional.

FIG. D – MEDIAN WIDTH AT CROSSINGS.†
Where left turns, U-turns and crossings are allowed on divided streets or highways, the median strip should be widened to a minimum "W" of 19' for Passenger traffic and 30' for Truck traffic.

1 1/2" · 1 1/2" · 1 1/2" · 93° · PLAN

4' 0" · 1' 3" · 9" · 9" · 1' 3" · 1 1/2" White Mortar Scored Band 13 1/2" Wide. · ℄ Median Strip. · 1/4 · 1/2ϕ · 1:2:3 Concrete · Exist. Pvmt. · 4" · 3" · 4" · 1 1/2" · 7/8"ϕ · 5/8"ϕ Anchor Bolts. · 5" · 5" · ℄ Anchor Bolts. · CROSS SECTION

FIG. E – REFLECTING MEDIAN STRIP.††
4' all paved type – Use at structures and in urban areas where clear width or Right of Way is limited.

*Adapted from New York State Div. of Highways. **Adapted from Eng. News Record Jan. 27, '44, by R. N. Ricketts. ***Public Roads Admin. †Adapted from Amer. Assoc. of State Highway Officials. ††Adapted from Portland Cem. Assoc.

ROADS & STREETS – DETAILS – 3

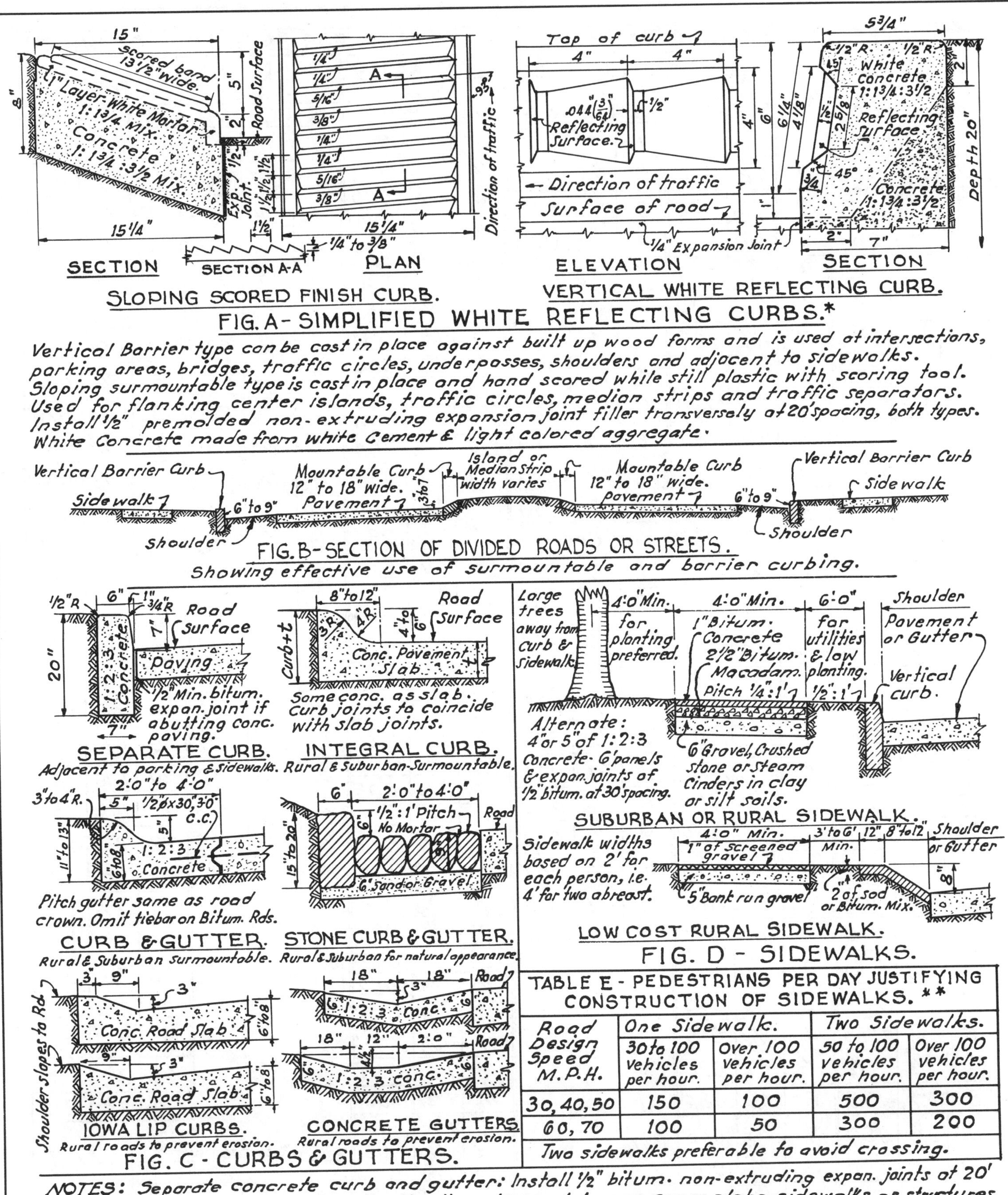

15"
Scored band 13 1/2" wide.
1" Layer-White Mortar 1:1 3/4 Mix.
Concrete 1:1 3/4:3 1/2 Mix.
Road Surface
Exp. Joint.
15 1/4"
SECTION
SECTION A-A
PLAN
1/4" to 3/8"
Direction of traffic
SLOPING SCORED FINISH CURB.
Top of curb
.044"(3/64) Reflecting Surface.
Direction of traffic
Surface of road
1/4" Expansion Joint
ELEVATION
5 3/4"
White concrete 1:1 3/4:3 1/2
Reflecting Surface
Concrete 1:1 3/4:3 1/2
Depth 20"
SECTION
VERTICAL WHITE REFLECTING CURB.
Vertical Barrier Curb
Sidewalk
Shoulder
Mountable Curb 12" to 18" wide.
Pavement
Island or Median Strip width varies
6" to 9"
Road Surface
Paving
1:2:3 Concrete
1/2" Min. bitum. expan. joint if abutting conc. paving.
SEPARATE CURB.
Adjacent to parking & sidewalks.
Conc. Pavement Slab
Same conc. as slab. Curb joints to coincide with slab joints.
INTEGRAL CURB.
Rural & Suburban-Surmountable.
1/2"φx30, 3'0" C.C.
1:2:3 Concrete
Pitch gutter same as road crown. Omit tiebar on Bitum. Rds.
CURB & GUTTER.
Rural & Suburban surmountable.
1/2":1' Pitch No Mortar
6" Sand or Gravel
STONE CURB & GUTTER.
Rural & Suburban for natural appearance.
Shoulder slopes to Rd.
Conc. Road Slab
IOWA LIP CURBS.
Rural roads to prevent erosion.
1:2:3 Conc.
CONCRETE GUTTERS.
Rural roads to prevent erosion.
Large trees away from Curb & Sidewalk
4'0" Min. for planting preferred.
1" Bitum. Concrete 2 1/2" Bitum. Macadam. Pitch 1/4":1'
6'0" for utilities & low planting. 1/2":1'
Shoulder Pavement or Gutter
Vertical curb.
Alternate: 4" or 5" of 1:2:3 Concrete- 6' panels & expan. joints of 1/2" bitum. at 30' spacing.
6" Gravel, Crushed stone or Steam Cinders in clay or silt soils.
SUBURBAN OR RURAL SIDEWALK.
Sidewalk widths based on 2' for each person, i.e. 4' for two abreast.
1" of screened gravel
5" Bank run gravel
2" of sod or Bitum. Mix.
LOW COST RURAL SIDEWALK.

FIG. A – SIMPLIFIED WHITE REFLECTING CURBS.*

Vertical Barrier type can be cast in place against built up wood forms and is used at intersections, parking areas, bridges, traffic circles, underpasses, shoulders and adjacent to sidewalks. Sloping surmountable type is cast in place and hand scored while still plastic with scoring tool. Used for flanking center islands, traffic circles, median strips and traffic separators. Install 1/2" premolded non-extruding expansion joint filler transversely at 20' spacing, both types. White Concrete made from white Cement & light colored aggregate.

FIG. B – SECTION OF DIVIDED ROADS OR STREETS.
Showing effective use of surmountable and barrier curbing.

FIG. C – CURBS & GUTTERS.

FIG. D – SIDEWALKS.

TABLE E – PEDESTRIANS PER DAY JUSTIFYING CONSTRUCTION OF SIDEWALKS.**

Road Design Speed M.P.H.	One Sidewalk.		Two Sidewalks.	
	30 to 100 Vehicles per hour.	Over 100 Vehicles per hour.	50 to 100 Vehicles per hour.	Over 100 Vehicles per hour.
30, 40, 50	150	100	500	300
60, 70	100	50	300	200

Two sidewalks preferable to avoid crossing.

NOTES: Separate concrete curb and gutter: Install 1/2" bitum. non-extruding expan. joints at 20' transverse intervals, and longitudinally adjacent to any conc. slabs, sidewalks or structures. Precast units; length: 4' min. for closures, 5' min. for radials, 10' max. for straights; align with cast-in dowels or keys. Curb or gutter cast integral with or doweled to a pavement must have joints coincide with pavement joints. Vertical curbs are used adjacent to sidewalks. Rounded or sloping curbs may be crossed by vehicles.

* Adapted from Universal Atlas Cement Co. and N.J. State Highway Dept.
** From A.A.S.H.O.

ROADS & STREETS - CROWN DATA

ORDINATES FROM GRADE TANGENT TO SURFACE FOR EACH FOOT OF WIDTH

SURFACE WIDTH OF STREET OR ROAD

20'

1/2" : 1' Crown		
1	1/16"	.004'
2	3/16"	.017'
3	7/16"	.037'
4	13/16"	.067'
5	1 1/4"	.104'
6	1 13/16"	.150'
7	2 7/16"	.204'
8	3 3/16"	.267'
9	4 1/16"	.337'
10	5"	.417'

3/8" : 1' Crown		
1	1/16"	.003'
2	1/8"	.012'
3	5/16"	.028'
4	5/8"	.050'
5	15/16"	.078'
6	1 3/8"	.112'
7	1 13/16"	.153'
8	2 3/8"	.200'
9	3 1/16"	.253'
10	3 3/4"	.312'

1/4" : 1' Crown		
1	0"	.002'
2	1/8"	.008'
3	1/4"	.019'
4	3/8"	.033'
5	5/8"	.052'
6	7/8"	.075'
7	1 1/4"	.102'
8	1 5/8"	.133'
9	2 1/4"	.190'
10	2 1/2"	.208'

22'

3/8" : 1' Crown		
1	1/16"	.003'
2	1/8"	.011'
3	5/16"	.026'
4	9/16"	.045'
5	7/8"	.071'
6	1 1/4"	.102'
7	**1 11/16"**	**.139'**
8	2 3/16"	.182'
9	2 3/4"	.230'
10	3 7/16"	.284'
11	4 1/8"	.344'

1/4" : 1' Crown		
1	0"	.002'
2	1/8"	.008'
3	3/16"	.017'
4	3/8"	.030'
5	9/16"	.047'
6	13/16"	.068'
7	1 1/8"	.093'
8	1 7/16"	.121'
9	1 13/16"	.153'
10	2 1/4"	.189'
11	2 3/4"	.229'

1/8" : 1' Crown		
1	0"	.001'
2	1/16"	.004'
3	1/8"	.009'
4	3/16"	.015'
5	5/16"	.024'
6	7/16"	.034'
7	9/16"	.046'
8	3/4"	.061'
9	15/16"	.077'
10	1 1/8"	.095'
11	1 3/8"	.115'

24'

1/4" : 1' Crown		
1	0"	.002'
2	1/8"	.007'
3	3/16"	.016'
4	5/16"	.028'
5	1/2"	.043'
6	3/4"	.062'
7	1"	.085'
8	1 5/16"	.111'
9	1 11/16"	.140'
10	2 1/16"	.173'
11	2 1/2"	.210'
12	3"	.250'

1/8" : 1' Crown		
1	0"	.001'
2	1/16	.004'
3	1/8	.008'
4	3/16	.014'
5	1/4	.022'
6	3/8	.031'
7	1/2	.043'
8	11/16"	.056'
9	7/8"	.070'
10	1 1/16"	.087'
11	1 1/4"	.105'
12	1 1/2"	.125'

40'

1/4" : 1' Crown		
1	0"	.000'
2	1/16"	.004'
3	1/8"	.009'
4	3/16"	.017'
5	5/16"	.026'
6	7/16"	.038'
7	5/8"	.051'
8	13/16"	.067'
9	1"	.084'
10	1 1/4"	.104'
11	1 1/2"	.126'
12	1 13/16"	.150'
13	2 1/8"	.176'
14	2 7/16"	.204'
15	2 13/16"	.234'
16	3 3/16"	.267'
17	3 5/8"	.301'
18	4 1/16"	.338'
19	4 1/2"	.376'
20	5"	.417'

1/8" : 1' Crown		
1	0"	.001'
2	0"	.002'
3	1/16"	.005'
4	1/8"	.008'
5	3/16"	.013'
6	1/4"	.019'
7	5/16"	.026'
8	3/8"	.033'
9	1/2"	.042'
10	5/8"	.052'
11	3/4"	.063'
12	7/8"	.075'
13	1 1/16"	.088'
14	1 1/4"	.102'
15	1 3/8"	.117'
16	1 5/8"	.133'
17	1 13/16"	.151'
18	2"	.169'
19	2 1/4"	.188'
20	2 1/2"	.208'

44'

1/8" : 1' Crown		
1	0"	.000'
2	0"	.002'
3	1/16"	.004'
4	1/8"	.008'
5	1/8"	.012'
6	3/16"	.017'
7	1/4"	.023'
8	3/8"	.030'
9	7/16"	.038'
10	9/16"	.047'
11	11/16"	.057'
12	13/16"	.068'
13	15/16"	.080'
14	1 1/8"	.093'
15	1 5/16"	.107'
16	1 7/16"	.121'
17	1 5/8"	.137'
18	1 13/16"	.153'
19	2 1/16"	.171'
20	2 1/4"	.189'
21	2 1/2"	.209'
22	2 3/4"	.229'

60'

1/8" : 1' Crown		
1	0"	.000'
2	0"	.001'
3	1/16"	.003'
4	1/16"	.006'
5	1/8"	.009'
6	1/8"	.012'
7	3/16"	.017'
8	1/4"	.022'
9	5/16"	.028'
10	7/16"	.035'
11	1/2"	.042'
12	5/8"	.050'
13	11/16"	.059'
14	13/16"	.068'
15	15/16"	.078'
16	1 1/16"	.089'
17	1 3/16"	.100'
18	1 3/8"	.112'
19	1 1/2"	.125'
20	1 11/16"	.139'
21	1 13/16"	.153'
22	2"	.168'
23	2 3/16"	.184'
24	2 3/8"	.200'
25	2 5/8"	.217'
26	2 13/16"	.235'
27	3 1/16"	.253'
28	3 1/4"	.272'
29	3 1/2"	.292'
30	3 3/4"	.312'

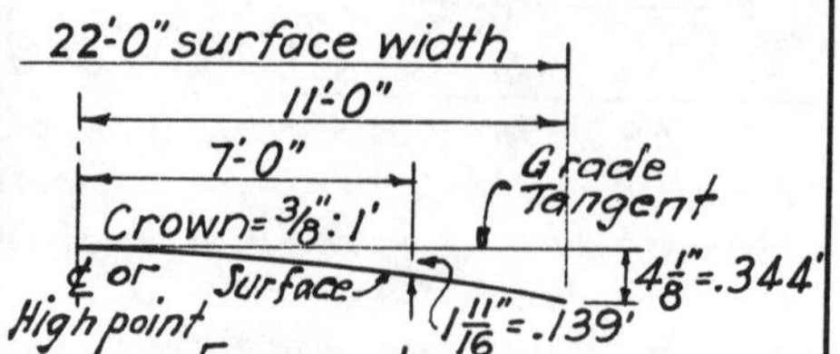

22'-0" surface width
11'-0"
7'-0"
Crown = 3/8" : 1'
Grade Tangent
℄ or High point
Surface
1 11/16" = .139'
4 1/8" = .344'
Example

HIGHWAY & STREET LIGHTING

DESIGN PROCEDURE –

1. Select type of light and lumen output (Table A).
2. Select mounting height (see table B).
3. Determine width of typical roadway section.
4. Select coefficient of utilization (Table C).*
5. For maintenance factor, use 70%.
6. Select average foot candle value desired (Table D).*
7. Calculate spacing between units.
8. Lay out arrangement on typical roadway section.

STREET LIGHTING FORMULA –

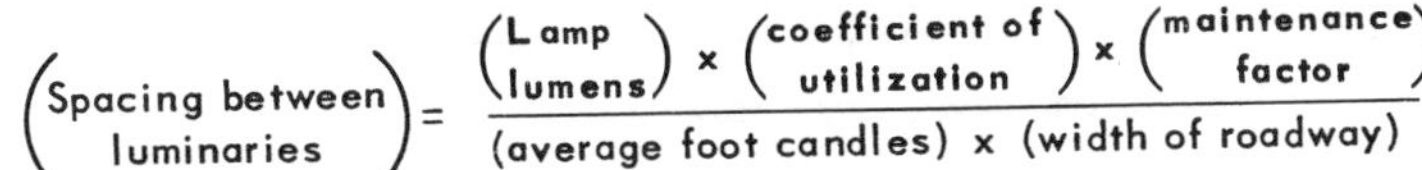

$$\left(\text{Spacing between luminaries}\right) = \frac{\left(\text{Lamp lumens}\right) \times \left(\text{coefficient of utilization}\right) \times \left(\text{maintenance factor}\right)}{(\text{average foot candles}) \times (\text{width of roadway})}$$

TABLE A – LUMEN OUTPUT OF SELECTED UNITS

Mercury (color-corrected)	20,000
Fluorescent	23,000
Incandescent	15,000

TABLE B – MOUNTING HEIGHTS

For highways;	Incandescent and fluorescent 25'-30'	Mercury 30'
For streets:	" and " 15'-30'	" 30'

Coefficient of Utilization (.10, .20, .30, .40, .50) vs. Ratio (Width of Area / Mounting Height) (0, 1, 2, 3, 4, 5)

FIG. C – COEFFICIENT OF UTILIZATION

TABLE D – ILLUMINATION LEVELS

Pedestrian Traffic	Vehicle Traffic Classification (Vehicles per hour in Both Directions)			
	Very Light, 150 or less	Light, 150 - 500	Medium, 500 - 1200	Heavy–Heaviest, 1200 and More
Main business streets (heavy)	0.8	0.8	1.0	1.2
Secondary business streets (medium)	0.6	.6	0.8	1.0
Residential streets and highway (light)	0.2	0.4	0.6	0.8

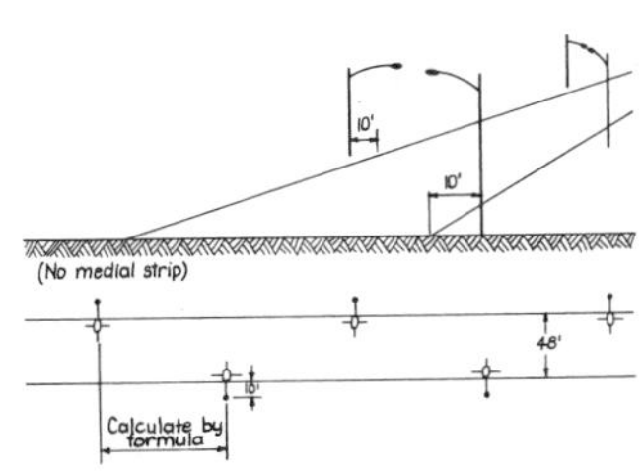
10'
10'
(No medial strip)
48'
10'
Calculate by formula

(1) SINGLE ROADWAY, NO MEDIAN STRIP

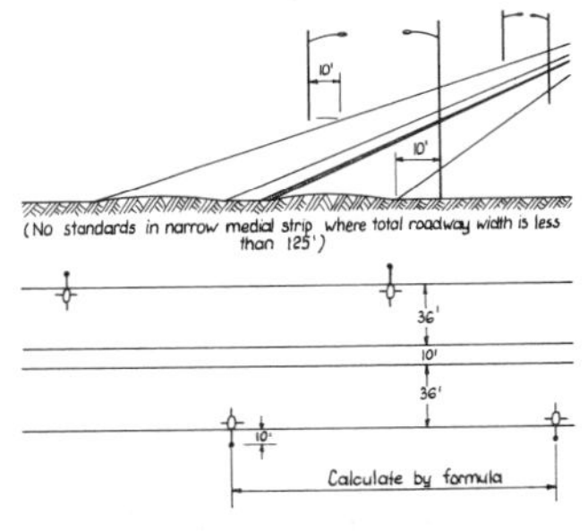
10'
10'
(No standards in narrow medial strip where total roadway width is less than 125')
36'
10'
36'
10'
Calculate by formula

(2) DUAL ROADWAY, NARROW MEDIAN STRIP, OVER-ALL WIDTH LESS THAN 125'

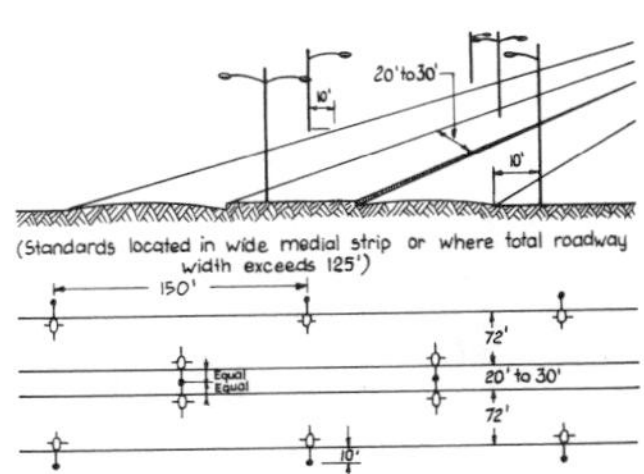
20' to 30'
10'
10'
(Standards located in wide medial strip or where total roadway width exceeds 125')
150'
72'
Equal
Equal
20' to 30'
72'
10'

(3) DUAL ROADWAY, WIDE MEDIAN STRIP, OVER-ALL WIDTH GREATER THAN 125'

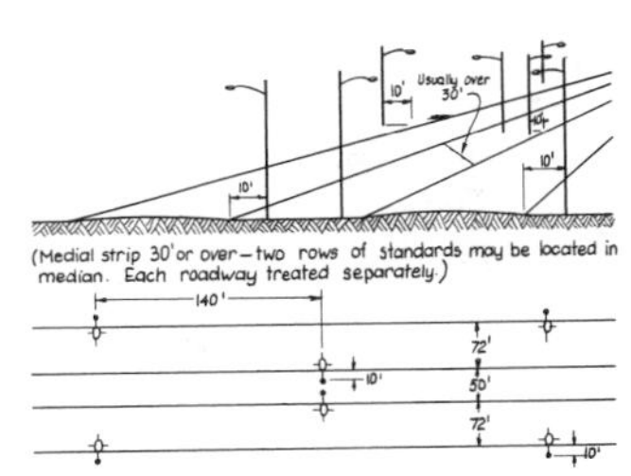
10'
Usually over 30'
10'
10'
(Medial strip 30' or over – two rows of standards may be located in median. Each roadway treated separately.)
140'
72'
10'
30'
72'
10'

(4) DUAL ROADWAY, VERY WIDE MEDIAN STRIP

FIG. E – TYPICAL ARRANGEMENTS FOR VARIOUS HIGHWAYS ARE SHOWN ABOVE

*Adapted from *American Standard Practice for Street and Highway Lighting*, D12.1-1953, Feb. 27, 1953.

PARKING — CURB

PARALLEL CURB PARKING

Automobiles Stall length varies with vehicle size, frequency of turnover, and traffic activity. Standards given are for U. S. automobiles as of 1958.

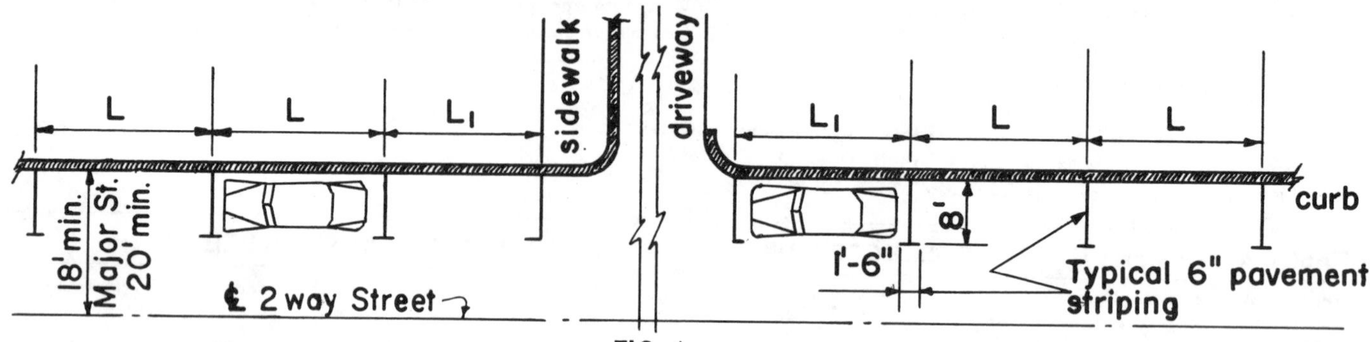
L
L
L1
sidewalk
driveway
L1
L
L
curb
18' min.
Major St. 20' min.
℄ 2 way Street
8'
1'-6"
Typical 6" pavement striping

FIG. A

L = 21' minimum for long-time parking on inactive street.
L = 22'-6" minimum for short-time parking on moderately active street.
L = 24' minimum for short-time parking on active street.
L_1 = L minus 2'. Used where driveway, sidewalk, fire-plug, etc., provide one open end.

TABLE B – TYPICAL VALUES

City	L (57 Design Average)
Baltimore, Md.	22 ft.
Elmira, N. Y.	24 ft.
Kansas City, Mo.	22 ft.
Philadelphia, Pa.	23 ft.
Pittsburgh, Pa.	23 ft.
Los Angeles, Calif.	See Fig. C.

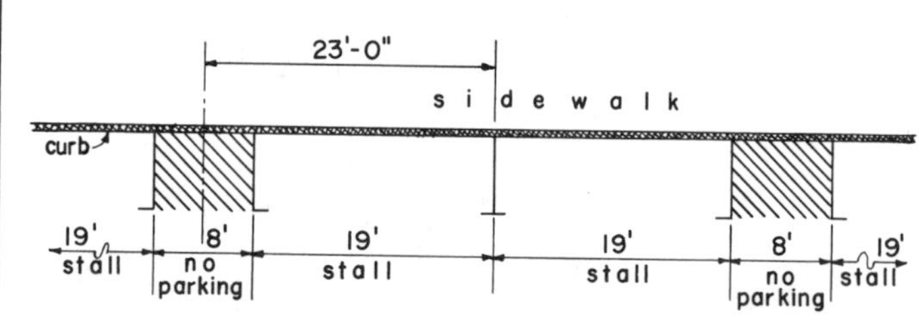
23'-0"
sidewalk
curb
19' stall
8' no parking
19' stall
19' stall
8' no parking
19' stall

Curb space per car = 23'. Some cities, such as Los Angeles, use 20' stalls, increasing curb space to 24'.

FIG. C – PAIR PARKING DETAIL.

TABLE D – MINIMUM CURB BUS-LOADING ZONE LENGTH*

Location	City Buses (35 to 45 Seats)	
	One-Bus Stop	Two-Bus Stop
Far side of intersection	60 ft.	100 ft.
Midblock	105 ft.	145 ft.
Near side of intersection	80 ft.	120 ft.

Trucks Where length restrictions permit metered curb-parking stalls for trucks, the minimum desired unit is two adjacent stalls (made from three auto parking stalls), providing approximately 35 ft. for each stall.

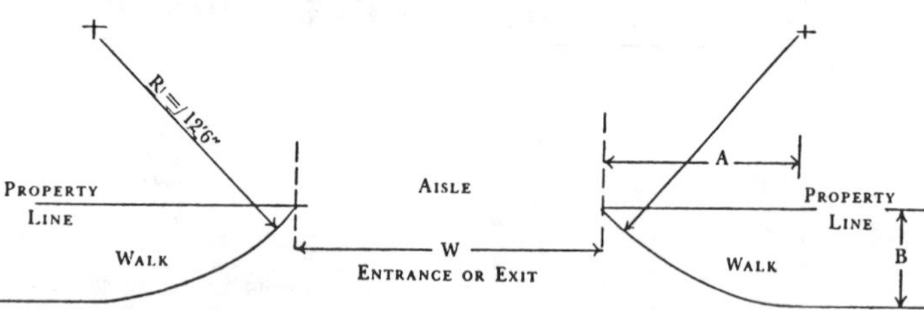
R=12'6"
Property Line
Walk
Aisle
W
Entrance or Exit
A
Property Line
B
Walk

FIG. E

NOTE:
Table F gives curb returns required to permit a car traveling one foot from the curb to make the turn into the parking lot aisle and miss any parked cars by one foot. Exit dimensions permit the reverse of that movement.

TABLE F – EXIT AND ENTRANCE CURB RETURNS FOR FIELDS.†

A	B	Exit W	Entrance W
9'-9"	4'-6"	11'-6"	14'-9"
10'-6"	5'-6"	10'-9"	14'-0"
11'-0"	6'-6"	10'-0"	13'-6"
11'-6"	7'-6"	9'-6"	13'-0"
12'-0"	8'-6"	9'-3"	12'-6"
12'-3"	9'-6"	9'-0"	12'-3"
12'-4"	10'-6"	8'-9"	12'-2"
12'-5"	11'-6"	8'-6"	12'-1"
12'-6"	12'-6"	8'-3"	12'-0"

* Adapted from Recommendations of the American Transit Association.
† Adapted from *Parking*, by the Eno Foundation for Highway Traffic Control.

PARKING — FIELDS

TABLE A – SELECTED STALL AND AISLE DIMENSIONS FOR HEAD-IN SELF-PARKING FIELDS

Remarks	See Fig. B					See Fig. B Unit Parking Depth		
	Col. 1	Col. 2	Col. 3	Col. 4	Col. 5	Col. 6	Col. 7	Col. 8
Notes: Dimensions vary with vehicle size, parking function, and degree of activity. Stall and aisle widths generally have an inverse relationship. Dimensions listed are minimum proven effective for U.S. 1958 autos. (When space is no problem, minimum stall width should be 9'-0").	Angle of Parking (degrees)	Stall Width	Width of stall parallel to aisle	Depth of Stall perpendicular to aisle	Width of Aisle	Parking against Walls or bumper stops	Parking with intermeshing stalls	Parking head-in to curb
General All Purpose Standard, 1958	90	8'- 9"	8'- 9"	18'- 0"	26'- 0"	62'- 0"	–	58'- 2"
All-day and low turnover lots (min: COL. 5-23'; COL. 6-59')	90	8'- 6"	8'- 6"	18'- 0"	26'- 0"	62'- 0"	–	58'- 2"
Large all-day parking lots. (Approx. 170 cars/acre)	90	8'- 6"	8'- 6"	18'- 0"	24'- 0"	60'- 0"	–	–
Inadequate room for opening car doors. Rarely used.	90	8'- 0"	8'- 0"	18'- 0"	32'- 0"	68'- 0"	–	64'- 2"
Typical pay parking field (high turnover)	90	8'- 6"	8'- 6"	18'- 0"	27'- 6"	63'- 6"	–	59'- 8"
Desired dimensions (for 8'-6" stalls)	90	8'- 6"	8'- 6"	18'- 0"	29'- 0"	65'- 0"	–	61'- 2"
Desired dimensions (for 9'-0" stalls)	90	9'- 0"	9'- 0"	18'- 0"	27'- 0"	63'- 0"	–	59'- 2"
Active shopping centers without separate pedestrian walkways and high turnover lots where ample land is available.	90	9'- 0"	9'- 0"	18'- 6"	30'- 0"	67'- 0"	–	63'- 2"
Generally recommended minimum (8'-6" stalls)	60	8'- 6"	9'-10"	19'-10"	18'- 0"	57'- 8"	53'- 5"	52'- 7"
Generally recommended minimum (9'-0" stalls)	60	9'- 0"	10'- 5"	20'- 1"	17'- 0"	57'- 2"	52'- 8"	51'-10"
General all-purpose minimum and in clear span, one-way aisle, self-park garages.	45	8'- 6"	12'- 0"	18'- 9"	12'- 6"	50'- 0"	44'- 0"	44'-10"
Minimum, for special purposes only	45	8'- 0"	11'- 4"	18'- 5"	12'- 0"	48'-10"	43'- 2"	44'- 0"
Minimum for short aisles	45	9'- 0"	12'- 9"	19'- 1"	11'- 4"	49'- 6"	43'- 2"	44'- 0"
Generally recommended minimum (8'-6" stalls)	30	8'- 6"	17'- 0"	16'- 5"	10'- 0"	42'-10"	35'- 9"	40'-11"
Generally recommended minimum (9'-0" stalls)	30	9'- 0"	18'- 0"	16'-10"	9'- 0"	42'- 8"	35'- 6"	40'- 9"
SELECTED STALL AND AISLE DIMENSIONS FOR BACK-IN PARKING. USED PRIMARILY FOR ATTENDANT PARKING.								Parking Back-in to Curb
Attendant parking only	90	8'- 0"	8'- 0"	18'- 0"	22'- 0"	58'- 0"	–	51'- 2"
Minimum for customer self-parking. Aisle and unit depth preferably increased by 1' or 1'-6"	90	8'- 6"	8'- 6"	18'- 0"	21'- 0"	57'- 0"	–	50'- 2"
Used where maximum number of two-way aisles is desired.	90	9'- 0"	9'- 0"	18'- 0"	20'- 0"	56'- 0"	–	49'- 2"

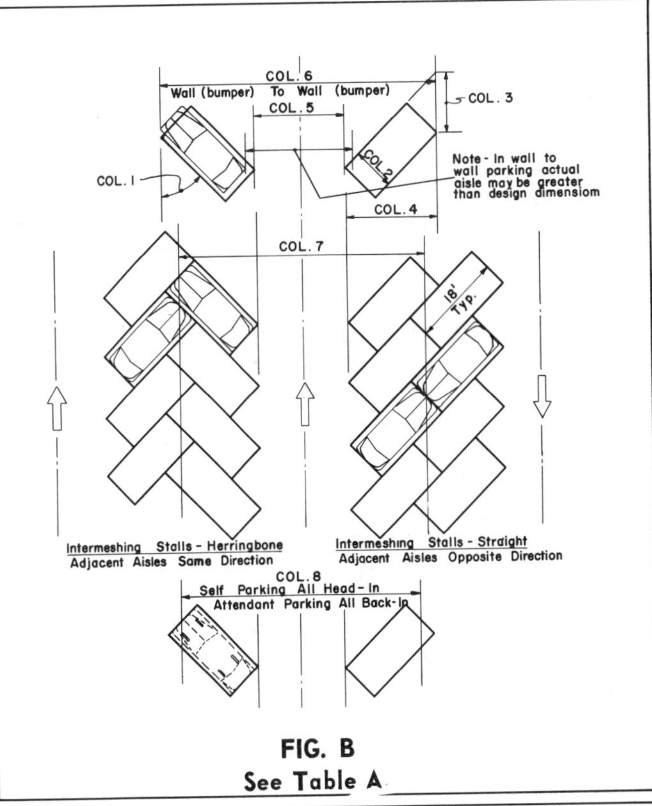

COL. 6
Wall (bumper) To Wall (bumper)
COL. 5
COL. 3
COL. 2
COL. 1
Note - In wall to wall parking actual aisle may be greater than design dimension
COL. 4
COL. 7
18' Typ.
Intermeshing Stalls - Herringbone
Adjacent Aisles Same Direction
Intermeshing Stalls - Straight
Adjacent Aisles Opposite Direction
COL. 8
Self Parking All Head-In
Attendant Parking All Back-In

FIG. B
See Table A

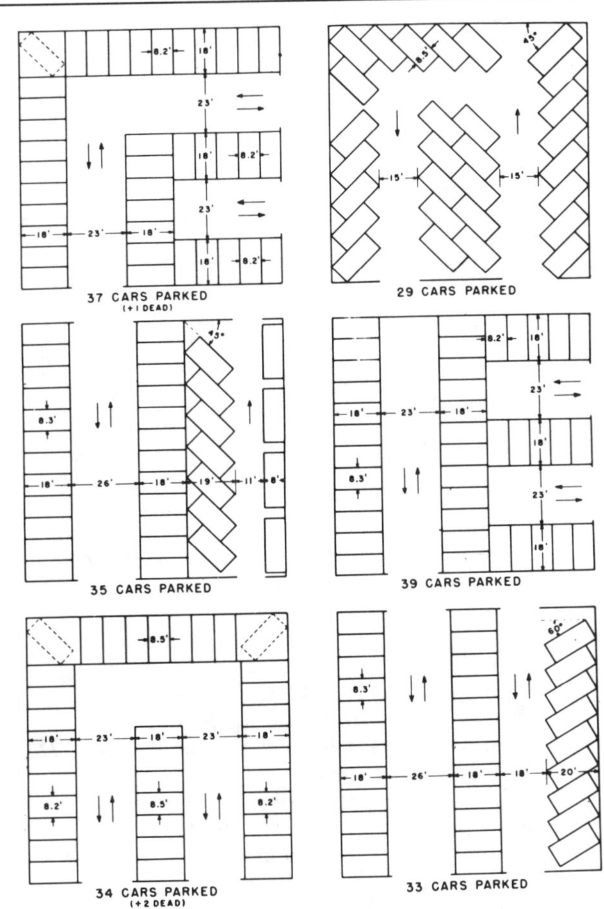

8.2'
18'
23'
37 CARS PARKED (+1 DEAD)
45°
8.5'
15'
29 CARS PARKED
8.3'
26'
19'
11'
35 CARS PARKED
39 CARS PARKED
34 CARS PARKED (+2 DEAD)
60°
20'
33 CARS PARKED

FIG. E – 6 EXAMPLES OF 100' x 100' SELF-PARKING FIELDS

From *Parking* by The Eno Foundation For Highway Traffic Control. 1957.

3" Typ.
Treated wood or metal post
12" Typ.
18" to 27"
24" recommended
Surface
WOOD
Nail
Wall
Old tire (See note)
RUBBER
12 ga. highway guard rail.
12" Typ.
18" to 27"
24" recommended
METAL

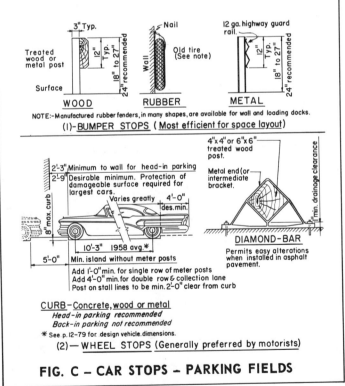

NOTE:-Manufactured rubber fenders, in many shapes, are available for wall and loading docks.

(1)-BUMPER STOPS (Most efficient for space layout)

2'-3" Minimum to wall for head-in parking
2'-9" Desirable minimum. Protection of damageable surface required for largest cars.
Varies greatly
4'-0" des. min.
8" max. curb
10'-3" 1958 avg.*
5'-0" Min. island without meter posts
Add 1'-0" min. for single row of meter posts
Add 4'-0" min. for double row & collection lane
Post on stall lines to be min. 2'-0" clear from curb

4"x4" or 6"x6" treated wood post.
Metal end (or intermediate bracket.
1" min. drainage clearance
DIAMOND-BAR
Permits easy alterations when installed in asphalt pavement.

CURB-Concrete, wood or metal
Head-in parking recommended
Back-in parking not recommended
* See p. 12-79 for design vehicle dimensions.

(2)—WHEEL STOPS (Generally preferred by motorists)

FIG. C – CAR STOPS – PARKING FIELDS

TABLE F – GRADES: PARKING FIELD

Minimum	As required for drainage
Desirable maximum	4 %
Maximum transverse to parking stall . .	6.5%

TABLE G – TYPICAL RANGE OF 90° PARKING FIELD, GROSS AREA PER CAR

Based on commercial lots with greater than 50-car capacity.

Parking Type	Gross, sq.-ft./car
Attendant, back-in	200
Attendant, head-in	225
Customer, minimum	250
Customer, roomy	300

TABLE H – AREA LIGHTING: PARKING FIELD

Desirable minimum: ½ to 2 ft.-candles, increase up to 5 ft.-candles for non-attended meter fields.

PARKING — GARAGES–1

All data on this page are adapted from *Traffic Design of Parking Garages*, 1957, by E.R. Ricker for the Eno Foundation.

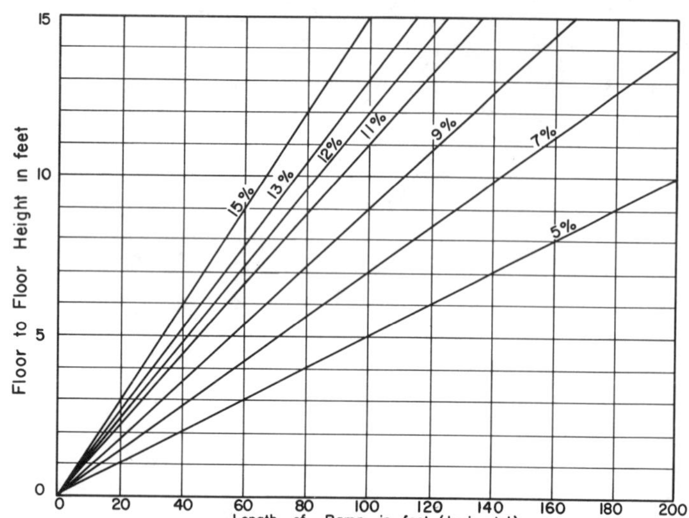
Floor to Floor Height in feet
Length of Ramp in feet (horizontal)
15%
13%
12%
11%
9%
7%
5%

FIG. A – RELATIONSHIP BETWEEN FLOOR-TO-FLOOR HEIGHT, GRADE, AND LENGTH OF RAMP

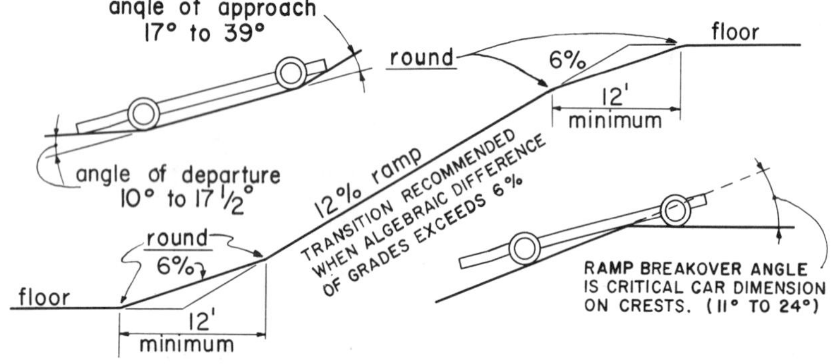
angle of approach
17° to 39°
round
6%
floor
12'
minimum
angle of departure
10° to 17 1/2°
12% ramp
TRANSITION RECOMMENDED WHEN ALGEBRAIC DIFFERENCE OF GRADES EXCEEDS 6%
round
6%
floor
12'
minimum
RAMP BREAKOVER ANGLE IS CRITICAL CAR DIMENSION ON CRESTS. (11° TO 24°)

FIG. B – METHOD OF BLENDING RAMP AND FLOOR GRADE

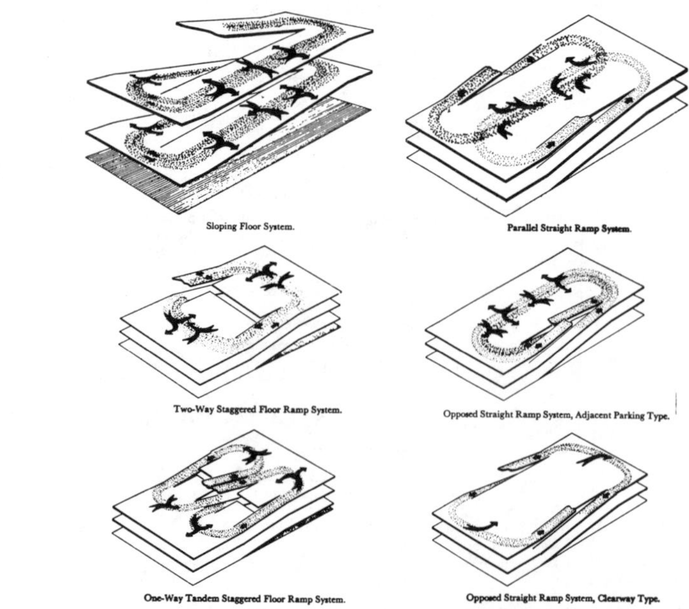
Sloping Floor System.
Parallel Straight Ramp System.
Two-Way Staggered Floor Ramp System.
Opposed Straight Ramp System, Adjacent Parking Type.
One-Way Tandem Staggered Floor Ramp System.
Opposed Straight Ramp System, Clearway Type.

FIG. C – EXAMPLES OF STRAIGHT RAMPED DESIGN

RAMP SLOPE Generally the slope should not exceed 12 or 15%. The maximum practical slope depends largely on the psychological effect on drivers, space available, and floor-to-floor height.

BLENDING OF RAMP AT FLOOR LEVEL The profile design at the top and botton of ramps must be based on the underclearance of current-model automobiles, with ample allowance for future or unusual models. The vehicle dimensions involved are the ramp breakover angle, the angle of Approach, and the angle of departure, as shown in Fig. B.

WIDTH OF RAMPS The usual cross section of one-way straight ramps is a lane 9 to 10 ft. wide, with a low curb 1 ft. wide on each side, giving an over-all inside dimension of 11 to 12 ft. Since the tread of standard automobiles is about 5 ft., and the over-all width less than 8 ft., these dimensions provide adequate room for straight portion of ramps. The end sections, on which cars are completing or starting a turn, must be flared to greater widths. If up-and-down lanes are adjacent, they should be separated by a medial island at least 1 ft. wide. This island, as well as the side curbs, should be not over 6 in. high, to allow clearance of bumpers. Curb faces should be sloped 1 in. from the vertical and rounded off to avoid damage to tires.

GENERAL (NON-MECHANICAL) GARAGE CRITERIA

Preferred construction for flexibility of use:
Clear span for 90° parking (60'-62') 2-way traffic
Clear span for 45° parking (50' ±) 1-way traffic
Where columns are req'd. desirable min. spacing:

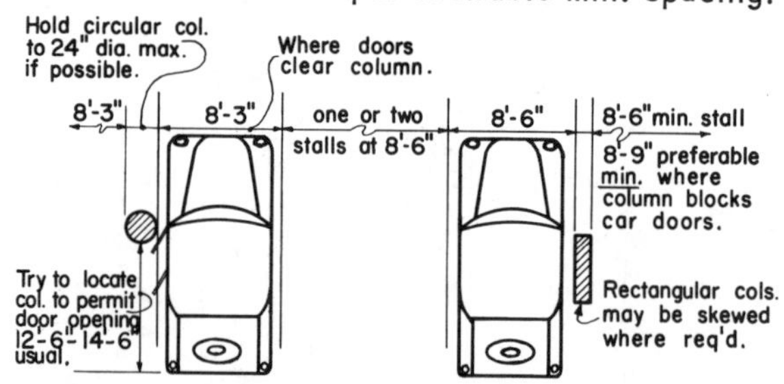
Hold circular col. to 24" dia. max. if possible.
Where doors clear column.
8'-3"
8'-3"
one or two stalls at 8'-6"
8'-6"
8'-6" min. stall
8'-9" preferable min. where column blocks car doors.
Try to locate col. to permit door opening 12'-6"-14'-6" usual.
Rectangular cols. may be skewed where req'd.

Desirable maximum grade of floor in sloping-floor garage–6% parallel to short axis of stall.
Desirable minimum clear ceiling height– 7'-6".
Lighting:
Stall area – 4-ft. candles minimum.
Aisle area – 8-ft. candles minimum.
Entry and exit – 40-ft. candles minimum.
Floor loading per building codes:
Typical live loading for passenger cars, 75 p.s.i. or 2000 concentrated.
50 p.s.i. live load permited by some codes where type of construction is code-specified.
In long-span concrete construction, some codes permit use of A.C.I. Code, 1956, Appendix "Ultimate Design Theory" and see pages 2-76 – 2-81.

Especially suitable where entrances and exits are on separate streets

PARKING — GARAGES-2

All data on this page adapted from *Traffic Design on Parking Garages*, 1957, by Eno Foundation, except as noted.

Very sharp and unrelieved turning will have a dizzying effect on the driver. To reduce this effect, ramp systems are sometimes laid out with a somewhat elliptical travel path. The structural limitation on the movements of cars traveling on a curved path is determined by the radius of the outermost point (usually the front bumper) when the car is turning on a minimum radius. The radius of the inside curb must be less than the minimum radius of the inside rear wheel, but not much smaller, or drivers will try to enter the ramp at too sharp an angle and thus get the car crossways of the lane. Clearance must be provided for the car with the largest outside radius, and all other cars can then travel in the lane provided.

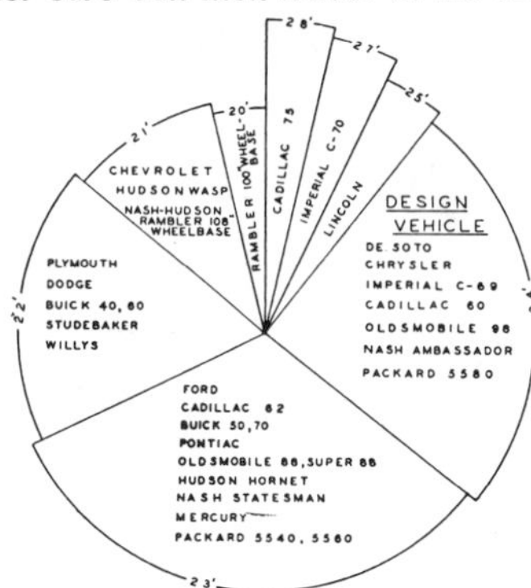
DESIGN VEHICLE
CHEVROLET
HUDSON WASP
NASH-HUDSON RAMBLER 108" WHEELBASE
RAMBLER 100" WHEELBASE
CADILLAC 75
IMPERIAL C-70
LINCOLN
PLYMOUTH
DODGE
BUICK 40, 60
STUDEBAKER
WILLYS
DE SOTO
CHRYSLER
IMPERIAL C-69
CADILLAC 60
OLDSMOBILE 98
NASH AMBASSADOR
PACKARD 5580
FORD
CADILLAC 62
BUICK 50, 70
PONTIAC
OLDSMOBILE 88, SUPER 88
HUDSON HORNET
NASH STATESMAN
MERCURY
PACKARD 5540, 5560

Typical Design Vehicle Showing Vehicle Dimensions Necessary for Computing Turning Movements

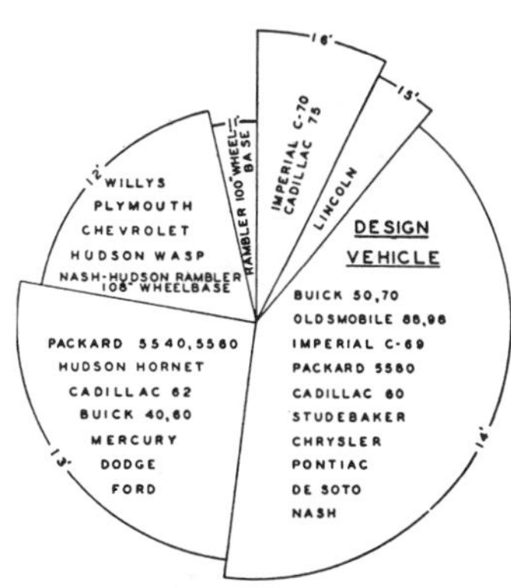
WILLYS
PLYMOUTH
CHEVROLET
HUDSON WASP
NASH-HUDSON RAMBLER 108" WHEELBASE
RAMBLER 100" WHEELBASE
IMPERIAL C-70
CADILLAC 75
LINCOLN
DESIGN VEHICLE
BUICK 50, 70
OLDSMOBILE 88, 98
IMPERIAL C-69
PACKARD 5580
CADILLAC 60
STUDEBAKER
CHRYSLER
PONTIAC
DE SOTO
NASH
PACKARD 5540, 5560
HUDSON HORNET
CADILLAC 62
BUICK 40, 60
MERCURY
DODGE
FORD

Minimum Turning Radius of Extreme Outside Point, Standard 1955-56 Model Automobiles

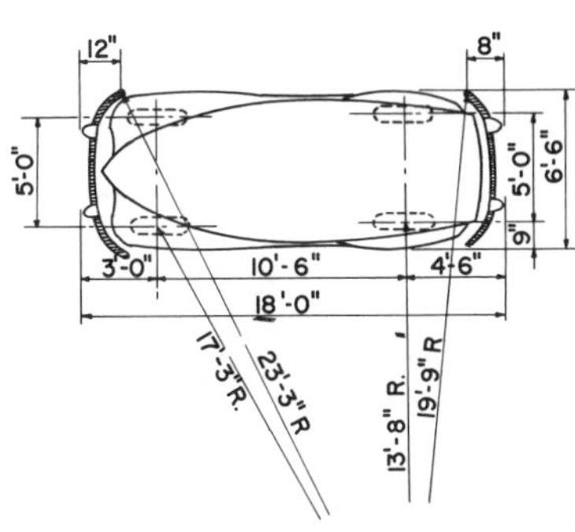
12"
8"
5'-0"
5'-0"
6'-6"
9"
3'-0"
10'-6"
4'-6"
18'-0"
17'-3"R.
23'-3"R
13'-8" R.
19'-9" R

Minimum Turning Radius of Inside Rear Wheel, Standard 1955-56 Model Automobiles

Superelevation on Ramp Turns:

$$e + f = \frac{0.067 V^2}{R}$$

Where e = superelevation in feet per foot of horizontal width
V = speed in miles per hour
R = radius of curve in feet
f = transverse coefficient of friction. (Highest values in attendant garages = 0.5)

The e value of 0.1 foot per foot of width at point of sharpest turning is recommended superelevation with lesser values adjacent to straight sections or storage floors. Care must be taken to avoid unduly distorted floor surfaces.

Curved Ramp Criteria

Desired maximum slope – 10%. Maximum slope – 15%.
Minimum radius measured at face of outer curve of inside lane – 30 feet.
Width between curbs of curved ramps:
Inside lane – minimum 12 feet.
Outside lane – minimum 10 feet, 6 inches.
Minimum curb width – 1 foot.
Minimum median island width – 1 foot, 6 inches.

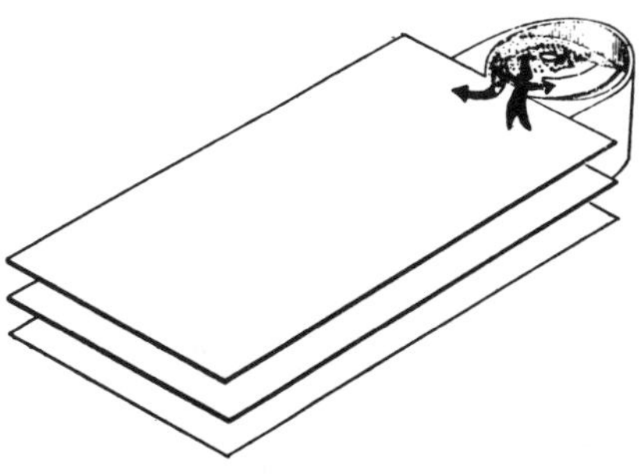

Parallel Circular Ramp
(Common ramp system, separated by median.)

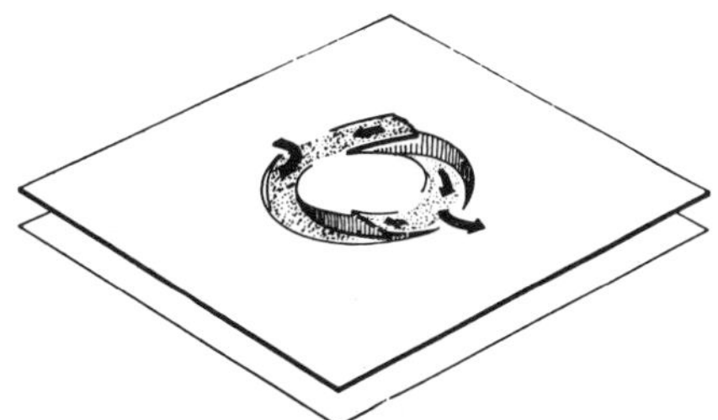

Opposed Circular Ramp
(Independent ramp, opposed slopes.)

2 Examples of Circular Ramp Systems.

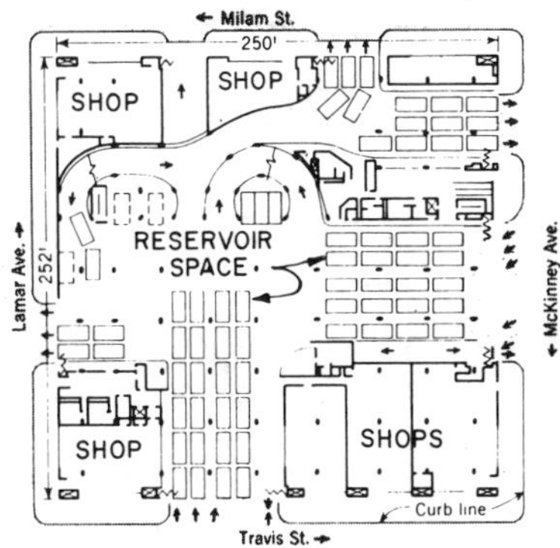
Milam St.
250'
SHOP
SHOP
RESERVOIR SPACE
Lamar Ave.
252'
McKinney Ave.
SHOP
SHOPS
Curb line
Travis St.

GROUND FLOOR PLAN
Note auto reservoir space and street-level shops.

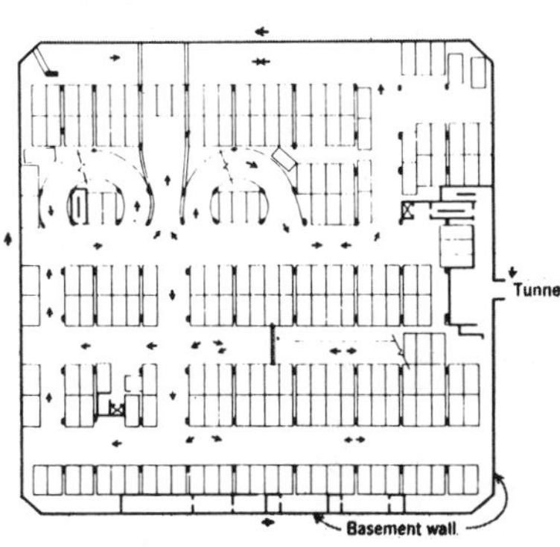
Tunnel
Basement wall

TYPICAL FLOOR PLAN
Note use of odd spaces adjacent to separated ramps.

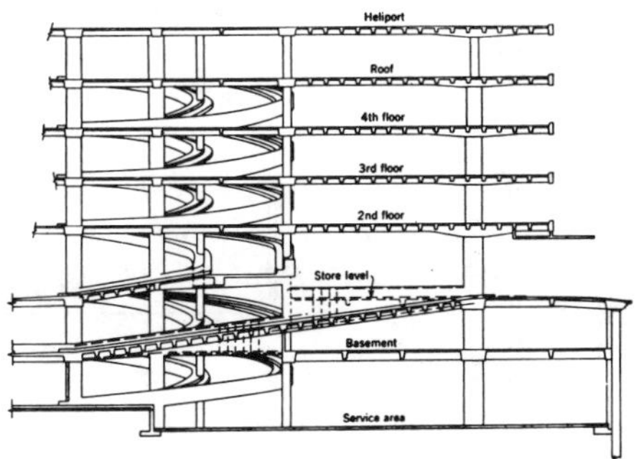
Heliport
Roof
4th floor
3rd floor
2nd floor
Store level
Basement
Service area

PART SECTION
Note use of perimeter cantilever above exterior grade and consideration of roof heliport.

EXAMPLE OF LARGE CURVED RAMP GARAGE

Illustrations courtesy of H. E. Bovay Jr., Consulting Engineers, and *Civil Engineering*, February 1957.

PARKING — GARAGES-3

Entrances and Exits must be planned to work efficiently with the shape and location of site, traffic conditions, and type of mechanical system employed. Ground-floor reservoir area is usually highly desirable. For some mechanical systems, the entrances and exits are preferably designed so that vehicles can enter the elevator on one side and exit from the other. In addition to the elevator-type mechanical garages listed, there are some that operate with different mechanical systems. Among the many types that have been developed are those which use techniques known as "ferris-wheel", "Automatic record changer", and "two-level hydraulic-lift."

Mechanical Type	Typical Examples	Representative Commercial Name	Loading Mechanism	Elevator Movements	Car Positions in Relation to Elevator	Remarks
A. Traveling elevator (see Figs. A and B)	Modified automatic	Bowser	Attendant drives car onto elevator, rides up with it, drives car into stall.	Combined horizontal and vertical movement produces diagonal resultant. Electronic panelboards speed traveling elevators to selected destination.	Right angles	May double park if longtime parking conditions warrant it (see Fig. A).
		Pigeon Hole	Hydraulic dolly on elevator lifts car onto elevator; attendant rides up with car and dolly, operated by attendant, deposits car in stall.		Either right angles or parallel	Fig. B shows parallel position.
	Fully automatic	Speed-Park	Mechanical "fingers" lift car on and off elevator.		Parallel	No parking attendant required on elevator.
B. Fixed elevator	Non-automatic	Commercial Freight Elevator	Attendant drives car on and off elevator.	Vertical movements only.	Any direction	Floor levels are connected by an elevator instead of ramps.
	Fully Automatic (see Fig. C)	Park-O-Mat (Similar to Fig. A)	Dolly with "monkey-wrench"-type bumper clamp rolls car on and off elevator.		Right angles	No parking attendant required on elevator. Adaptable to narrow parcels.
		Squared Circle	Dolly lifts car on and off elevator.		Either right angles or parallel	On each floor, car platforms move on 2 parallel rows of tracks. Each platform can move forward, backward, or sideways under its own power. No attendant required.

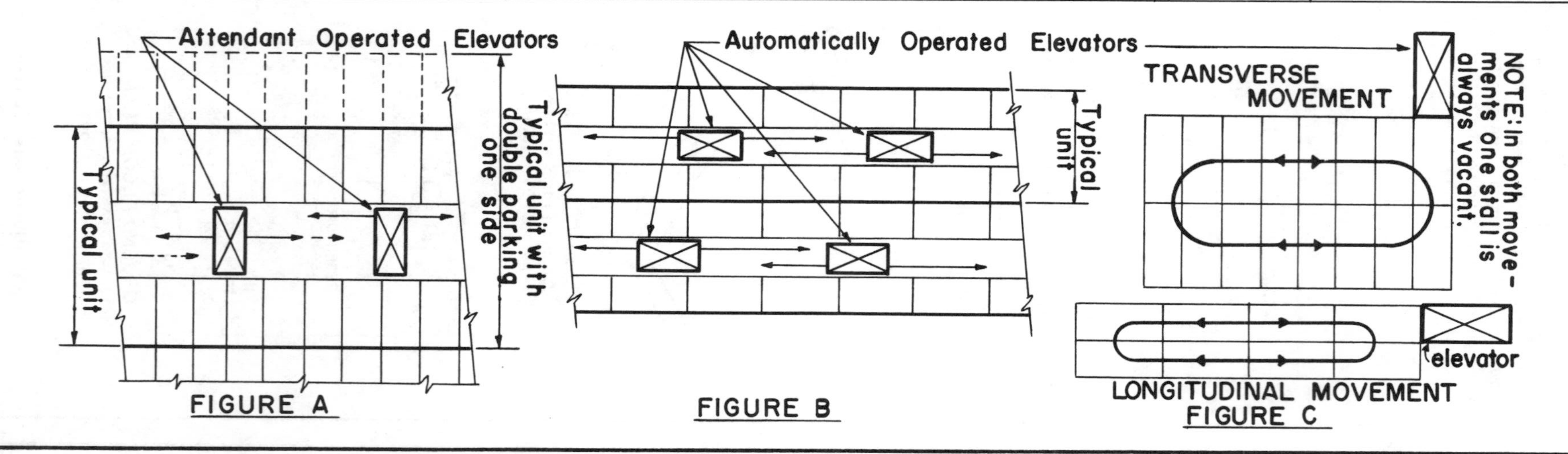
Attendant Operated Elevators
Typical unit with double parking one side
Typical unit
Automatically Operated Elevators
Typical unit
TRANSVERSE MOVEMENT
NOTE: In both movements one stall is always vacant.
elevator
LONGITUDINAL MOVEMENT

FIGURE A

FIGURE B

FIGURE C

PARKING – TRUCKS AND BUSES

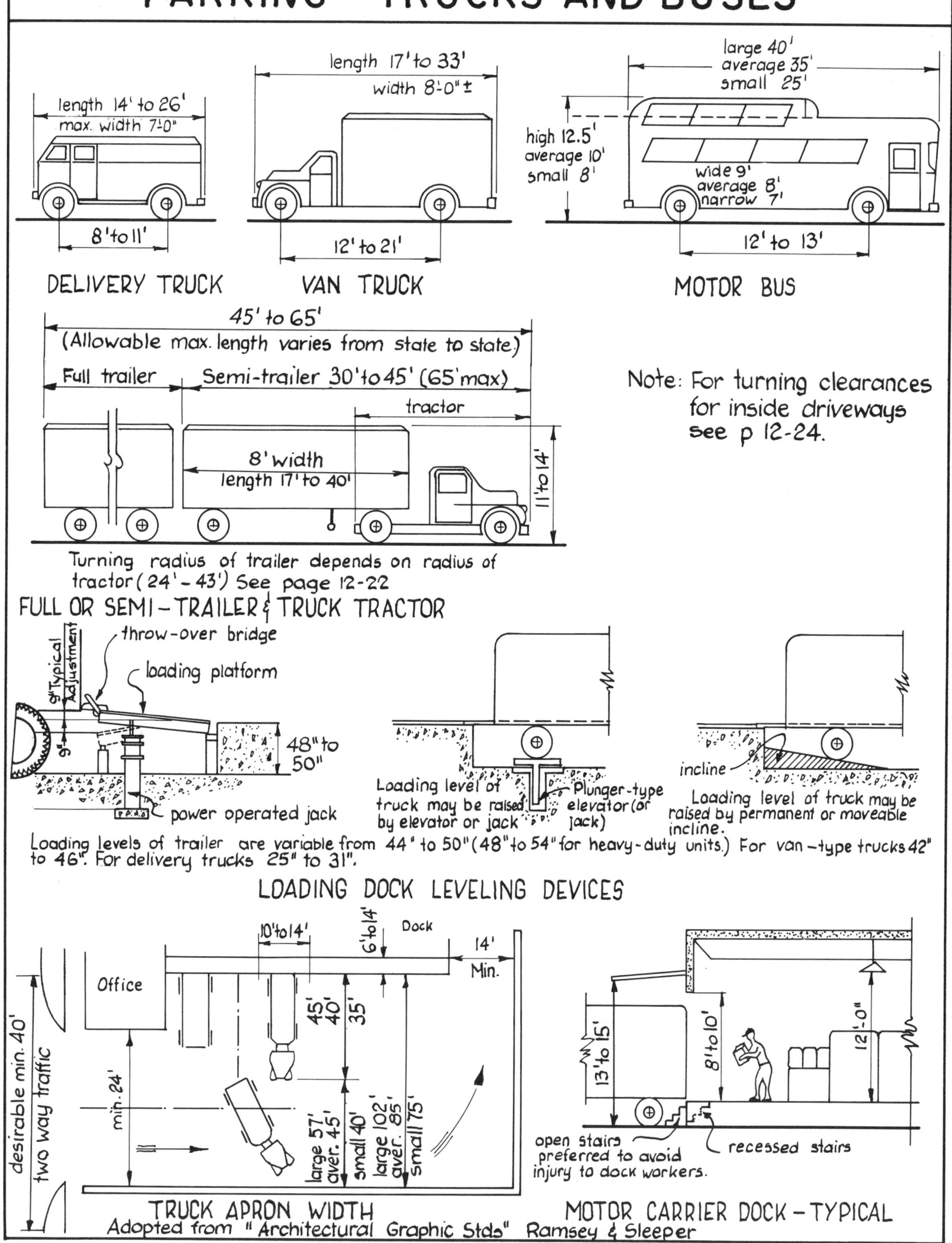
length 14' to 26'
max. width 7'-0"
8' to 11'
DELIVERY TRUCK
length 17' to 33'
width 8'-0" ±
12' to 21'
VAN TRUCK
large 40'
average 35'
small 25'
high 12.5'
average 10'
small 8'
wide 9'
average 8'
narrow 7'
12' to 13'
MOTOR BUS
45' to 65'
(Allowable max. length varies from state to state)
Full trailer
Semi-trailer 30' to 45' (65' max)
tractor
8' width
length 17' to 40'
11' to 14'
Turning radius of trailer depends on radius of tractor (24' - 43') See page 12-22
FULL OR SEMI-TRAILER & TRUCK TRACTOR
Note: For turning clearances for inside driveways see p 12-24.
9" Typical Adjustment
throw-over bridge
loading platform
9"
48" to 50"
power operated jack
Loading level of truck may be raised by elevator or jack
Plunger-type elevator (or jack)
incline
Loading level of truck may be raised by permanent or moveable incline.
Loading levels of trailer are variable from 44" to 50" (48" to 54" for heavy-duty units.) For van-type trucks 42" to 46". For delivery trucks 25" to 31".
LOADING DOCK LEVELING DEVICES
10' to 14'
6' to 14'
Dock
14' Min.
Office
45'
40'
35'
min. 24'
desirable min. 40'
two way traffic
large 57'
aver. 45'
small 40'
large 102'
aver. 85'
small 75'
TRUCK APRON WIDTH
13' to 15'
8' to 10'
12'-0"
open stairs preferred to avoid injury to dock workers.
recessed stairs
MOTOR CARRIER DOCK – TYPICAL

Adopted from "Architectural Graphic Stds" Ramsey & Sleeper

BRIDGES – LIVE LOADS – I

RAILROADS

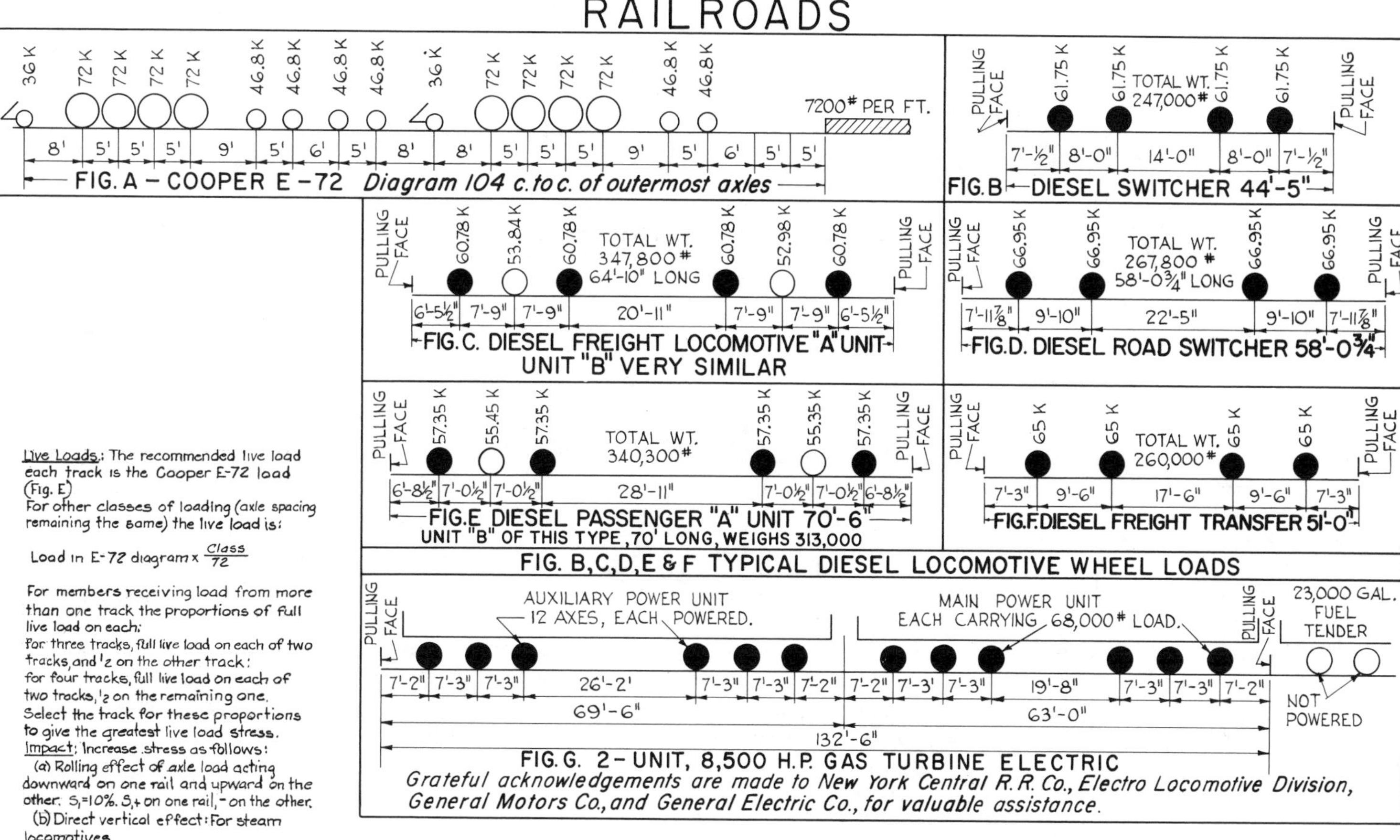
36 K 72 K 72 K 72 K 72 K 46.8 K 46.8 K 46.8 K 46.8 K 36 K 72 K 72 K 72 K 72 K 46.8 K 46.8 K
8' 5' 5' 5' 9' 5' 6' 5' 8' 8' 5' 5' 5' 9' 5' 6' 5' 5'
7200# PER FT.
FIG. A – COOPER E-72 Diagram 104 c. to c. of outermost axles
PULLING FACE
61.75 K 61.75 K 61.75 K 61.75 K
TOTAL WT. 247,000#
7'-½" 8'-0" 14'-0" 8'-0" 7'-½"
FIG. B – DIESEL SWITCHER 44'-5"
60.78 K 53.84 K 60.78 K 60.78 K 52.98 K 60.78 K
TOTAL WT. 347,800# 64'-10" LONG
6'-5½" 7'-9" 7'-9" 20'-11" 7'-9" 7'-9" 6'-5½"
FIG. C. DIESEL FREIGHT LOCOMOTIVE "A" UNIT – UNIT "B" VERY SIMILAR
66.95 K 66.95 K 66.95 K 66.95 K
TOTAL WT. 267,800# 58'-0¾" LONG
7'-11⅞" 9'-10" 22'-5" 9'-10" 7'-11⅞"
FIG. D. DIESEL ROAD SWITCHER 58'-0¾"
57.35 K 55.45 K 57.35 K 57.35 K 55.35 K 57.35 K
TOTAL WT. 340,300#
6'-8½" 7'-0½" 7'-0½" 28'-11" 7'-0½" 7'-0½" 6'-8½"
FIG. E DIESEL PASSENGER "A" UNIT 70'-6"
UNIT "B" OF THIS TYPE, 70' LONG, WEIGHS 313,000
65 K 65 K 65 K 65 K
TOTAL WT. 260,000#
7'-3" 9'-6" 17'-6" 9'-6" 7'-3"
FIG. F DIESEL FREIGHT TRANSFER 51'-0"
FIG. B, C, D, E & F TYPICAL DIESEL LOCOMOTIVE WHEEL LOADS
AUXILIARY POWER UNIT – 12 AXES, EACH POWERED.
MAIN POWER UNIT EACH CARRYING 68,000# LOAD.
23,000 GAL. FUEL TENDER
NOT POWERED
7'-2" 7'-3" 7'-3" 26'-2' 7'-3" 7'-3" 7'-2" 7'-2" 7'-3' 7'-3" 19'-8" 7'-3" 7'-3" 7'-2"
69'-6" 63'-0"
132'-6"
FIG. G. 2-UNIT, 8,500 H.P. GAS TURBINE ELECTRIC

Grateful acknowledgements are made to New York Central R. R. Co., Electro Locomotive Division, General Motors Co., and General Electric Co., for valuable assistance.

<u>Live Loads</u>: The recommended live load each track is the Cooper E-72 load (Fig. E)

For other classes of loading (axle spacing remaining the same) the live load is:

Load in E-72 diagram × $\frac{Class}{72}$

For members receiving load from more than one track the proportions of full live load on each:
for three tracks, full live load on each of two tracks, and ½ on the other track:
for four tracks, full live load on each of two tracks, ½ on the remaining one.
Select the track for these proportions to give the greatest live load stress.

<u>Impact</u>: Increase stress as follows:

(a) Rolling effect of axle load acting downward on one rail and upward on the other. $S_1 = 10\%$. S, + on one rail, − on the other.

(b) Direct vertical effect: For steam locomotives.

(1) For beam spans, stringers, girders, floor beams, truss spans, floor beam hangers (1.a). For truss spans $= \frac{4000}{L+25} + 15$. Where L = less than 100' $= (60 - \frac{L^2}{500}) \times$ axle load ÷ 100. Where L = 100' or more $= (\frac{1800}{L-40} + 10) \times$ axle load ÷ 100.

(2) For Diesel electric locomotives, Tenders alone, etc. $= (40 - \frac{3L^2}{1600}) \times$ axle load ÷ 100 for L < 80'; for L = 80' or more $(\frac{600}{L-30} + 16) \times$ axle load ÷ 100

L = length, in feet, center to center of supports, for stringers, longitudinal girders and trusses (main members), or = length in feet of the larger adjacent longitudinal beam, girder or truss for impact in floor beams, floorbeam hangers, subdiagnols of trusses, transverse girders supports for longitudinal and transverse girders and viaduct columns.

Members receiving load from > one track:
Two Tracks. L = Less than 175', = Full impact on two tracks.
L = 175' to 225', = Full impact on one track + (450 − 2L)% of other.
L = over 225', = Full impact on one track + none on other.
> Two Tracks – Full impact on two tracks

<u>LONGITUDINAL FORCES</u>: Add to live load on one track only at 6' above rail 15% of live load without impact for bracking, or 25% of load on driving without impact for traction, whichever is larger. Decrease 50% where abutments will take part of load.

<u>LATERAL FORCES</u>: Wind Pressure: In any horizontal direction.

(a) Wind application on moving load – 300 lbs. per l.f. on one track at 8' above rail. Wind application on structure – 30 lbs. per sq. ft. as follows: girder spans – 1½ × vertical projection of span; truss spans – vertical projection of span + any portion of leeward trusses not shielded by floor system; viaduct towers and bents – Vertical projections of all columns and tower bracing.

(b) Total wind force on girder and truss spans: not less than 200 lbs. per l.f. on loaded flange or chord (flange or chord supporting floor beams) or 150 lbs per l.f. on unloaded flange or chord.

(c) Total wind force on unloaded bridge not less than 50 lbs per sq. ft. on surfaces as in (a) above.

(d) Design members according to the greater of the following combined stresses: Wind stress from (a) or (b) + stresses from D.L., L.L., impact and centrifugal force; or wind stress from (c) + D.L. stress.

<u>NOSING OF LOCOMOTIVES</u>: A lateral force to be added. 20,000 lbs. at top of rail at any point in either lateral direction (vertical effects of this force to be disregarded).

<u>CENTRIFUGAL FORCE ON CURVES</u>: Apply horizontally 6' above top of rail along a line ⊥ to ℄ of track at each axle. Percentage of axle load to apply = 1.755(E+3), where E = Superelev. in inches.

<u>STABILITY OF SPANS & TOWER</u>: Use a live load of 1,200 lbs. per l.f. on one track without impact. On multiple-track bridges, place this live load on leeward track.

<u>BRACING BETWEEN COMPRESSION MEMBERS</u>: Design for transverse shear in any panel equal to 2½% of total axial stress in both members in that panel + shear from lateral forces.

<u>REVERSAL OF STRESS</u>: Design members subject to reversal of stresses for each stress + 50% of smaller stress. Connection to be designed for sum of maximum stresses.

<u>COMBINED STRESSES</u>: Members subject to both axial & bending stresses, including bending due to floor beam deflection, shall be designed so that combined fiber stresses will not exceed allowable axial stress. In members continuous over panel points, only ¾ of bending stress cumputed as for simple beams shall be added to axial stress.
Members subject to a combination of dead, live, impact loads and centrifugal force with other lateral forces and with longitudinal forces or with bending due to such forces, allow 25% increase in unit stresses; any member shall have sufficient section to resist stresses due to dead, live impact loads + centrifugal force without increase in allowable stresses.

<u>SECONDARY STRESSES</u>: If exceeding 4,000 lbs. per sq. in. for tension members or 3,000 lbs. per sq. in. for compression members to be treated as primary stress. Secondary stresses due to truss distortion usually need not be consideration in any member the width of which, measured parallel to the plane of distortion, is less than 1/10th of its length.

<u>FOR SIDINGS & ROADS WHERE LOW SPEEDS ONLY WILL BE USED:</u>**
Use impact 50% and omit all other rules except the laws of mechanics

*Adopted from A.R.E.A. specifications, 1950 ** By author.

BRIDGES—LIVE LOADS-2

HIGHWAYS

H20-S16 — Express Highways with Commercial Traffic, Thruways, Heavy duty. Commercial streets. All structures on Interregional Highway System.
H20 — Parkways, and similar to above in many states.
H15-S12 & H15 — Used by many states for highways listed above.
H10 — Light traffic only.

H-LOADING | H-S-LOADING

$\overline{W}$ = total weight of loaded truck.

Use loading, truck or lane, that produces maximum stress.

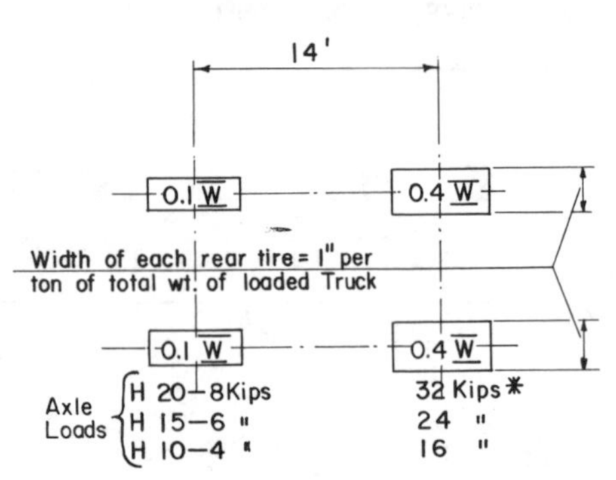

14'
0.1 W
0.4 W
Width of each rear tire = 1" per ton of total wt. of loaded Truck
Axle Loads
H 20—8 Kips
H 15—6 "
H 10—4 "
32 Kips *
24 "
16 "

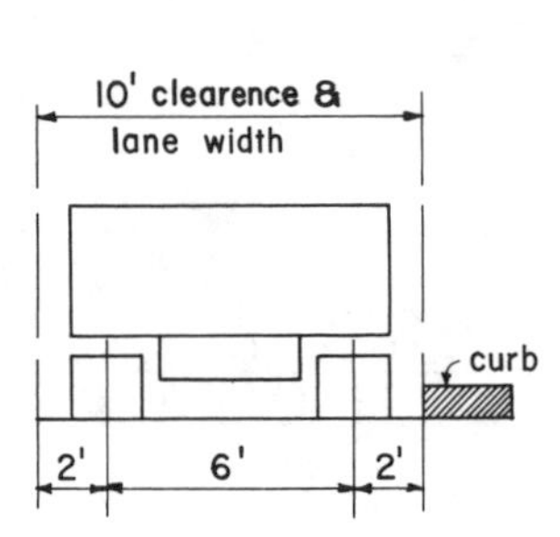

10' clearence & lane width
curb
2'
6'
2'

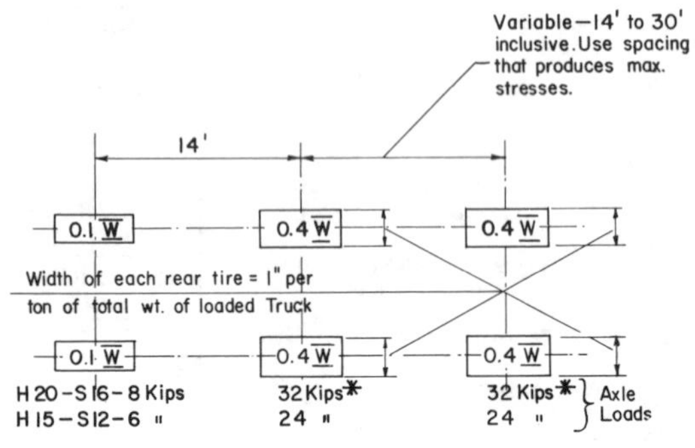

Variable—14' to 30' inclusive. Use spacing that produces max. stresses.
14'
0.1 W
0.4 W
0.4 W
Width of each rear tire = 1" per ton of total wt. of loaded Truck
H 20-S16-8 Kips
H 15-S12-6 "
32 Kips *
24 "
32 Kips *
24 "
Axle Loads
W = Combined weight on the first two axles which is the same as for the corresponding H truck.

For H & HS loadings number after H represents gross weight in tons of truck or tractor. For HS loadings number after S represents gross weight in tons of rear axle load of semitrailer.

* In the design of floors (concrete slabs, steel grid floors, and timber floors) for H-20 or H-20-S-16 loading, one axle load of 24,000 pounds or two axle loads of 16,000 pounds each, spaced 4 feet apart may be used, whichever produces the greater stress, instead of the 32,000 lb. axle shown.

** For slab design the center line of wheel shall be assumed to be 1 ft. from face of curb.

H LANE AND HS LANE LOADINGS

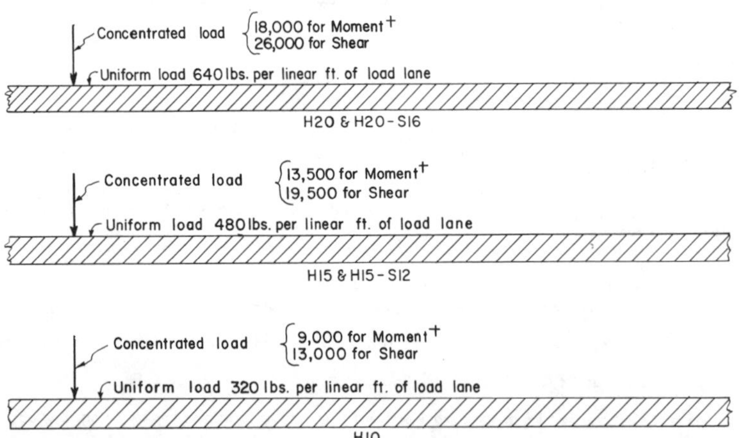

Concentrated load
18,000 for Moment †
26,000 for Shear
Uniform load 640 lbs. per linear ft. of load lane
H20 & H20-S16
Concentrated load
13,500 for Moment †
19,500 for Shear
Uniform load 480 lbs. per linear ft. of load lane
H15 & H15-S12
Concentrated load
9,000 for Moment †
13,000 for Shear
Uniform load 320 lbs. per linear ft. of load lane
H10

† For the loading of continuous spans involving lane loading add another concentrated load of equal weight in one other span in the series in such position as to produce max. negative moment. For maximum positive moment, only one concentrated load shall be used per lane, combined with as many spans loaded uniformly as required to produce max. moment.

REDUCTION IN LOAD INTENSITY.

Where maximum stresses are produced by loading any number of lanes simultaneously, use following percentages of resultant live load stresses:

1 or 2 Lanes	100%	
3 Lanes	90%	
4 Lanes	75%	85% } suggested by S. Hardesty:
5 Lanes or more	75%	75% }

Loads adopted from A.A.S.H.O. specs. for Highway Bridges. 1957

BRIDGES–LIVE LOADS–3

HIGHWAYS*

OVERLOAD

Check design, except flooring, with truck loading increased 100% in one lane only. Resulting dead, live, and impact stresses must not exceed 150% of design stresses. This overload provision does not apply to H20 and H20-S16 loadings.

IMPACT

Increase stress by $S^1 = S \dfrac{50}{L + 125}$

L = span length† of member in feet.
S = stress without impact.
$S + S^1$ = stress with impact.

Maximum Stress Increase Due to Impact

For culverts with cover 0' to 1'-0"	30%
For culverts with cover 1'-1" to 2'-0"	20%
For culverts with cover 2'-1" to 2'-11"	10%

	Where applied	*Where not applied*	
Superstructure including:	Column, towers, rigid-frames above ground surface, concrete or steel piles above ground connected rigidly at top	Abutments and retaining walls Foundation pressure Timber structures Sidewalk loadings	Culverts and substructures with greater than 3' covers

LONGITUDINAL FORCES

5% of lane load with concentrated load for moment applied A[1] above roadway. Also provide for longitudinal force due to friction at expansion bearings.

LATERAL FORCES

Wind: Velocity of 100 miles per hour.

Loads and forces may be reduced or increased in ratio of $\dfrac{V^2}{(100)^2}$ for group II loading.

(a) Superstructure.
- Group II Loading:
  - Trusses and arches: 75 p.s.f. on projected area.
  - Girders and beams: 50 p.s.f. on projected area.
  - Minimum total loads:
    - 300 lb. per lin. ft. – Loaded chord.
    - 150 lb. per lin. ft. – Unloaded chord. Truss span.
    - 300 lb. per lin. ft. – Girder span.
- Group III Loading:
  - Group II loading + moving live load of 100 lb. per lin. ft. applied 6' above deck.

(b) Substructure.
1. Transverse and longitudinal forces transmitted by superstructure:

Skew Angle of Wind, degrees	Group II Loading				Group III Loading	
	Trusses, p.s.f. Area		Girders, p.s.f. Area		70% Group II Loading plus Wind on Moving Live Load (Applied 6' above Deck), lb./lin. ft.	
	Lateral	Longitudinal	Lateral	Longitudinal	Lateral	Longitudinal
0	75	0	50	0	100	0
15	70	12	44	6	88	12
30	65	28	41	12	82	24
45	47	41	33	16	66	32
60	25	50	17	19	34	38
	Applied simultaneously.		Applied simultaneously.		Applied simultaneously.	

In lieu of above, approximate loading for girder and slab bridges with maximum span of 125 ft.

Wind Load on Structure, p.s.f.		Wind Load on Live Load, lb./lin. ft.	
Transverse	Longitudinal	Transverse	Longitudinal
50	12	100	40
Applied simultaneously.		Applied simultaneously.	

* Adapted from A.A.S.H.O., *Standard Specifications for Highway Bridges*, 1957.

† For roadway floors use the design span length.
For transverse members use the distance center to center of supports.
For computing truck-load moments use the span length, except for cantilever arms.
Use the length from moment center to the farthermost axle.
For shear due to truck loads use the length of the loaded portion of span from the point under consideration to the far reaction, except for cantilever arms use 30%.
For continuous spans use length of span under consideration for positive moment, and use the average of two adjacent loaded spans for negative moment.

BRIDGES – LIVE LOADS – 4

HIGHWAYS*

Transverse and longitudinal forces applied directly:
Group II loading 40 p.s.f.
Group III loading 70% of group II loading.
For skewed wind directions, resolve into components perpendicular to end and front, according to functions of skew angle, and apply simultaneously with loads from superstructure.

(c) Overturning forces.
Effect of forces tending to overturn structure plus upward force applied at windward quarter-point of width:
Group II loading 20 p.s.f. of deck and sidewalk area.
Group III loading 6 p.s.f. of deck and sidewalk area.

UPLIFT
Check uplift at any support caused by any combination of loading, increased by 100% of live-plus-impact load.

CENTRIFUGAL FORCE
From trolleys on curved track – apply 10% of trolley live load as lateral force 4' above rail.

SIDEWALK AND SAFETY CURB OVER 2' WIDTH
Slabs, stringers, and immediate supports – 85 p.s.f.
Girders, trusses, arches, etc. – 26' to 100' = 60 p.s.f.

Girders, trusses, arches, etc. over 100' = $(30 + \frac{3000}{L})(\frac{55-W}{50})$ p.s.f.; maximum 60 p.s.f. where L = loaded length of sidewalk in feet, and W = width of sidewalk in feet.

THERMAL FORCES
Provide for stresses or movements resulting from temperature variations, dependent upon local conditions.

	Moderate Climate	Cold Climate
Metal Structures	0°F. to 120°F.	–30°F. to 120°F.
Concrete structures	Rise: 30°F. Fall: 40°F.	Rise: 35°F.. Fall: 45°F.

PRESSURE ON PIERS
Ice: 400 p.s.i. Thickness and height dependent on local conditions (50 kips per sq. ft. on width of pier commonly used†).
Flowing water = KV^2 p.s.f. where V = velocity in feet per second and K = 1⅓ for square ends, ⅔ for circular ends, and ½ for angle ends $\lessgtr 30°$.

EARTH PRESSURE
See p. 7-31. No structure to be designed with less than 30 p.c.f. fluid pressure. When highway traffic can come within a horizontal distance from the top of the structure equal to one-half of its height, the pressure shall have added to it a live load surcharge pressure equal to not less than 2 ft. of earth.
No live load surcharge considered where approach slab is supported by bridge.

SNOW AND ICE LOADS
No allowance to be made.

RAILINGS AND CURBS
See p. 13-06.

* Adapted from A.A.S.H.O., *Standard Specifications for Highway Bridges*, 1957.
† By Author.

BRIDGES – LIVE LOAD DISTRIBUTION – I

NOTATION

E = Width of slab in feet over which wheel load is distributed.
P_1 = Load on one wheel of single axle.
P_2 = Load on one wheel of tandem axle.
P' = Concentrated lane load per lane.
N = Maximum number of lanes of traffic.
W = Width of roadway – curb to curb.
Q = Uniform lane load per linear ft. of lane.

S = Effective span length in feet.
FOR SIMPLE SPANS: C to C of supports, but not over clear span + slab thickness.
FOR CONTINUOUS SPANS – MONOLITHIC: use clear span. OVER STRINGERS OR FLOOR BEAMS: Use clear span + ½ flange width (steel) or ½ beam width (timber). For negative moment use average of two adjacent loaded spans.

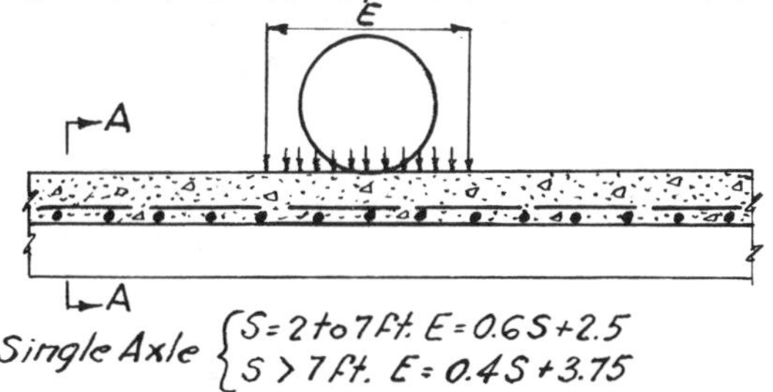

E
A
A

Single Axle: $S = 2$ to 7 ft. $E = 0.6S + 2.5$; $S > 7$ ft. $E = 0.4S + 3.75$
Tandem Axle: $S = 2$ to 7 ft. $E = 0.36S + 2.58$; $S > 7$ ft. $E = 0.063S + 4.65$

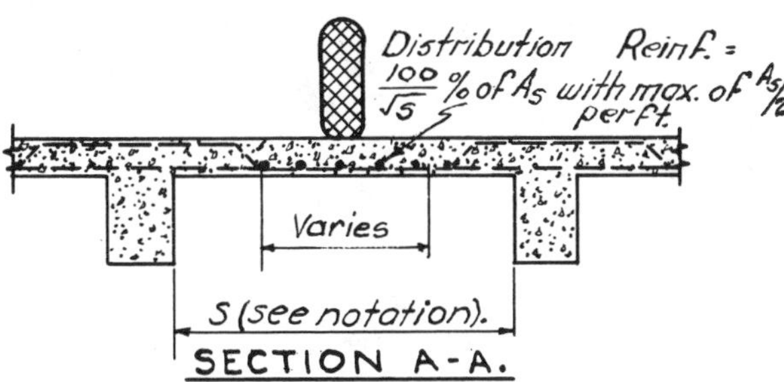

Distribution Reinf. = $\frac{100}{\sqrt{5}}$ % of A_S with max. of $A_S/2$ per ft.
Varies
S (see notation).

SECTION A-A.

FIG. A – MAIN REINFORCEMENT PERPENDICULAR TO TRAFFIC.

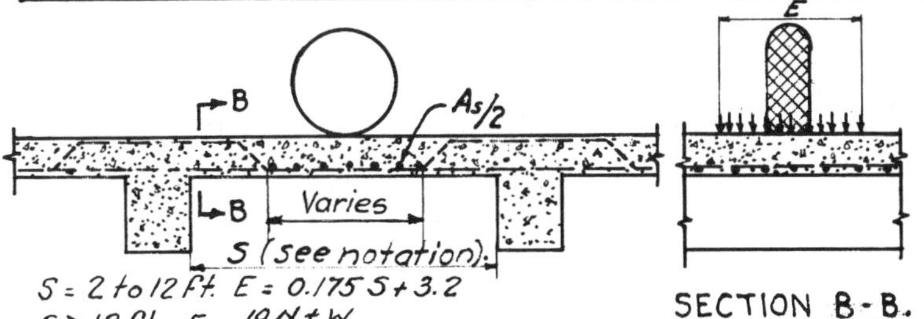

B
$A_S/2$
B
Varies
S (see notation).
E

$S = 2$ to 12 ft. $E = 0.175S + 3.2$
$S > 12$ ft. $E = \frac{10N + W}{4N}$
Load per ft. of slab $= \frac{P}{E}$, placed on span or spans to cause max. + or – moments

SECTION B-B.

P' = Concentrated Load.
Q = Uniform Load.

$S > 12$ ft.

Q distributed $= \frac{NQ}{0.5W + 5N}$ per sq. ft. of slab.

P' distributed $= \frac{NP'}{0.5W + 5N}$ per ft. width of slab.

TRUCK LOADING. LANE LOADING.

For H-20 and H20-S16 loads use single 24,000-lb. axle for $S < 18$ ft.; use two 16,000-lb. axles for $S =$ or > 18 ft.

FIG. B – MAIN REINFORCEMENT PARALLEL TO TRAFFIC.

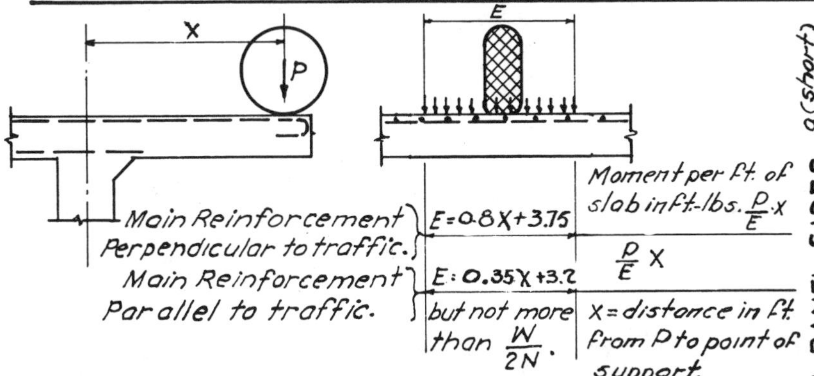

X
P
E

		Moment per ft. of slab in ft-lbs.
Main Reinforcement Perpendicular to traffic.	$E = 0.8X + 3.75$	$\frac{P}{E} X$
Main Reinforcement Parallel to traffic.	$E = 0.35X + 3.2$ but not more than $\frac{W}{2N}$.	$\frac{P}{E} X$

X = distance in ft. from P to point of support.

FIG. C – CANTILEVER SLABS.

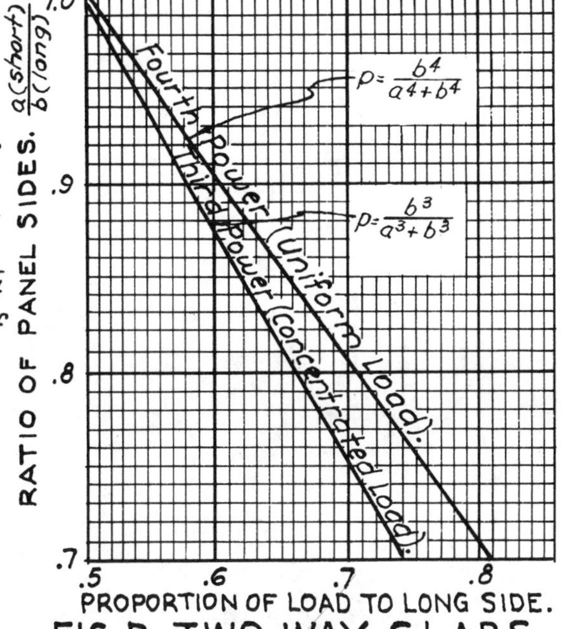

RATIO OF PANEL SIDES. $\frac{a \text{ (short)}}{b \text{ (long)}}$
1.0
.9
.8
.7
.5
.6
.7
.8
PROPORTION OF LOAD TO LONG SIDE.
$P = \frac{b^4}{a^4 + b^4}$
$P = \frac{b^3}{a^3 + b^3}$
Fourth Power (Uniform Load).
Third Power (Concentrated Load).

FIG. D – TWO-WAY SLABS.

Where ratio is less than 0.67 – design as one-way slab. Distribute loads as above.

NOTES

1. Slabs designed for moment with loads distributed as above shall be considered adequate in bond and shear.
2. Provide edge beams where main reinforcement is parallel to traffic; curb section may be used. Live Load $M = 0.10\,PS$ for simple span; reduce 20% for continuous spans.
3. In designing slabs ℄ of wheel load shall be assumed to be 1′ from curb face.

* In accord with A.A.S.H.O. Specs. 1957

BRIDGES-LIVE LOAD DISTRIBUTION-2

MOMENTS ON INTERIOR LONGITUDINAL FLOOR BEAMS & STRINGERS DUE TO WHEEL LOADS.

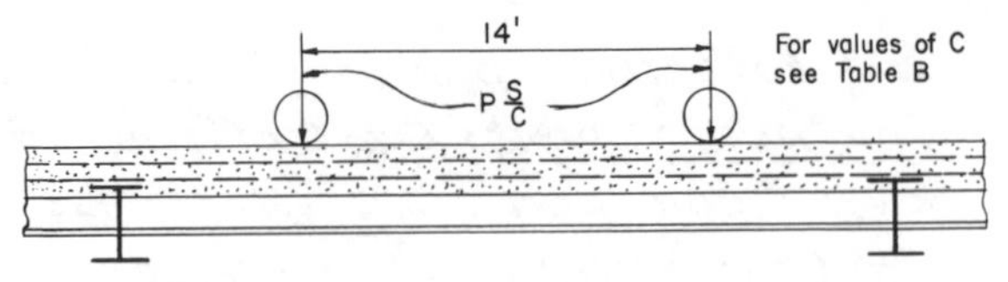

14'
For values of C see Table B
$P\frac{S}{C}$
LONGITUDINAL SECTION

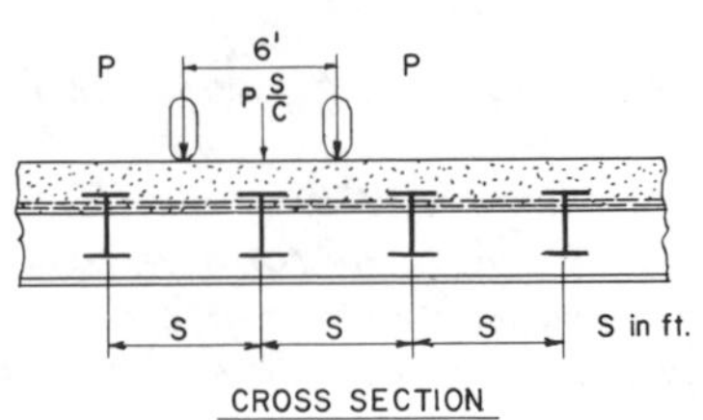

P
6'
P
$P\frac{S}{C}$
S
S
S
S in ft.
CROSS SECTION

FIG. A

TABLE B - VALUES OF C.

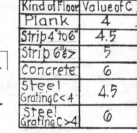

KIND OF FLOOR	STRINGERS TRAFFIC LANES: ONE	STRINGERS TRAFFIC LANES: TWO OR MORE	KIND OF FLOOR	STRINGERS TRAFFIC LANES: ONE	STRINGERS TRAFFIC LANES: TWO OR MORE
PLANK:	4.0	3.75	CONCRETE: On Steel I beam	7.0 (If s > 10' see Note 1)	5.5 (If s > 14' see Note 1)
STRIP: 4" to 6"; Blocks on 4" Planks, Laminated, 5"	4.5	4.0	CONCRETE: On Concrete and Timber	6.0 (if s > 6') see Note 1	5.0 (if s > 10'-6") See Note 1
STRIP: 6" & > See Note 2 below	5.0 (If s > 5' see Note 1)	4.25 (If s > 6'-6" see Note 1)	STEEL GRATING: d < 4"	4.5	4.0
			STEEL GRATING: d ≥ 4"	6.0 (If s > 6' see Note 1)	5.0 (If s > 10'-6" see Note 1)

MOMENTS ON EXTERIOR LONGITUDINAL FLOOR BEAMS AND STRINGERS DUE TO WHEEL LOADS.

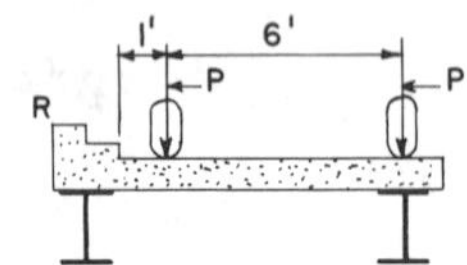

1'
6'
P
P
R

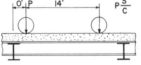

(1) R = Reaction of P or P's, assuming slab acts as a simple beam between stringers.

(2) For four or more steel stringers, R shall not be less than

$\frac{S}{5.5}$ for $S \leq 6'$.

$\frac{S}{4.0 + 0.25S}$ for $S > 6'$ & $S < 14'$.

See Note 1 for $S \geq 14'$.

FIG. C

SHEAR & END REACTIONS FOR LONGITUDINAL & TRANSVERSE BEAMS DUE TO WHEEL LOADS.

0' P 14' $P\frac{S}{C}$

Max. Shear = P + reaction of $P\frac{S}{C}$.

LONGITUDINAL BEAM

NOTE: See Table B for values of C. See Fig. E for values of D.

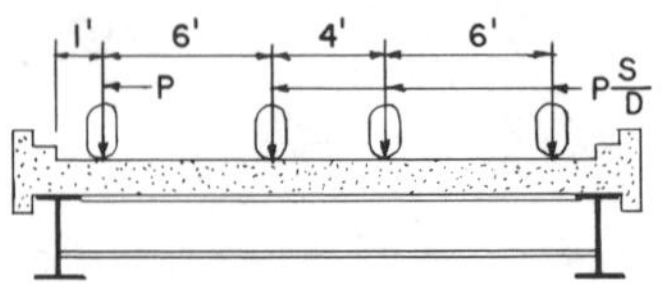

1'
6'
4'
6'
P
$P\frac{S}{D}$

Max. Shear = Reaction of P + reactions of $P\frac{S}{D}$

TRANSVERSE BEAMS NO STRINGERS

FIG. D

MOMENTS ON TRANSVERSE FLOOR BEAMS - NO STRINGERS.

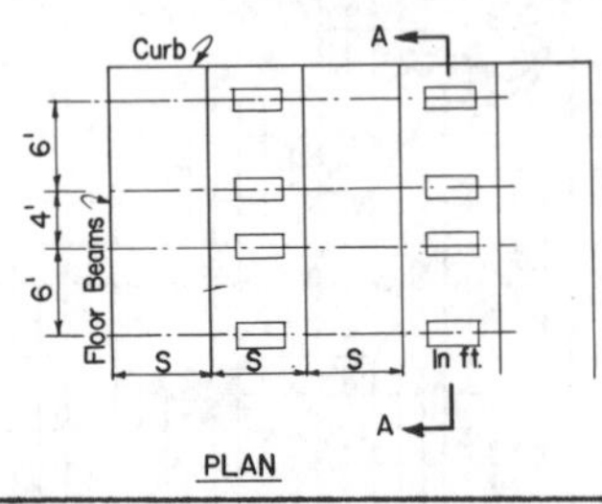

Curb
A
6'
4'
6'
Floor Beams
S
S
S
In ft.
A
PLAN

For Values of D use Table B above - for ONE TRAFFIC LANE. If S greater than D, see Note 1 below.

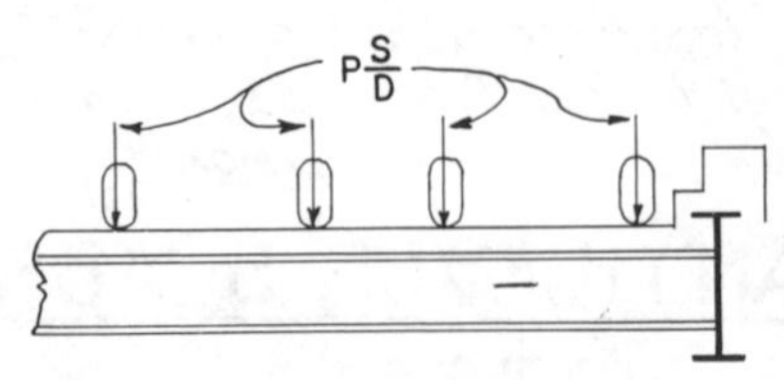

$P\frac{S}{D}$
SECTION A-A

FIG. E

Kind of Floor	Value of C
Plank	4
Strip 4" to 6"	4.5
Strip 6" & >	5
Concrete	6
Steel Grating C < 4	4.5
Steel Grating C > 4	6

Note: 1. In this case the load on each beam or stringer shall be the reaction of the wheel loads, assuming the flooring to act as a simple beam.

2. Strip flooring = planking on edge placed at right angles to ℄ of roadway, spiked together 18" o.c. and at ends; spike length = 2½ plank thickness minimum; toe nails or clips to alternate stringers. Spline, tongue and groove, and doweled flooring shall have the same distribution as strip flooring of equivalent thickness.

3. The combined load capacity of the beams or stringers in a panel shall not be less than the total live and dead load in the panel.

* In accord. with A.A.S.H.O. Specifications, 1957.

BRIDGES – LOAD DISTRIBUTION – 3

TIMBER FLOORING

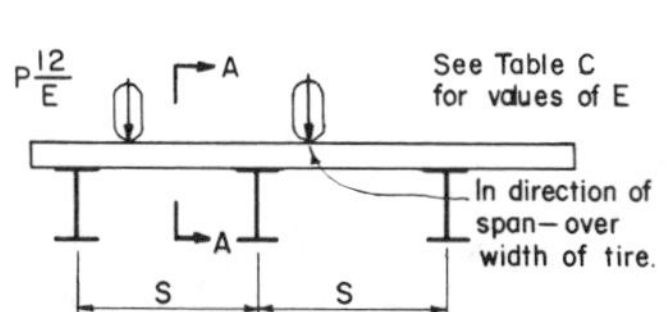

$P\frac{12}{E}$
A
See Table C for values of E
In direction of span – over width of tire.
A
S
S

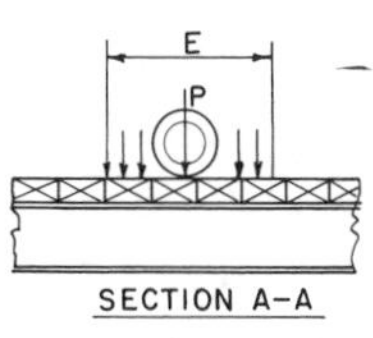

E
P
SECTION A–A

For S – see Note 1. For width of tire, see Note 2.

FIG. A – TRANSVERSE FLOORING

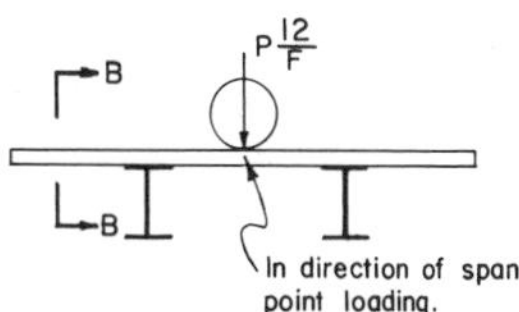

$P\frac{12}{F}$
B
B
In direction of span point loading.

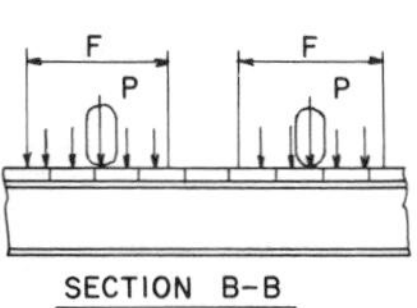

F
F
P
P
SECTION B–B

FIG. B – LONGITUDINAL FLOORING

TABLE C VALUES OF E AND F

Type of Floor	E	F
Plank Free edges.	Width of plank in inches.	Width of plank in inches.
Plank Laminated.	15"	1" × gross weight of truck in tons + floor thickness in inches.
Plank Tongue and groove: Spline doweled not < 5½" thick.	Four times floor thickness in inches.	1" × gross weight of truck in tons + 2 × floor thickness in inches.

Notes: 1. Span length = clear distance between stringers or beams + ½ width of stringer or beam, but not over clear span + floor thickness.
2. Width of tire 10" for H10 loading, 15" for H15 loading, 20" for H20 loading.

DISTRIBUTION THROUGH EARTH FILLS

Case I: <2' fill – same as slabs
Case II: 2' to 8' fill.

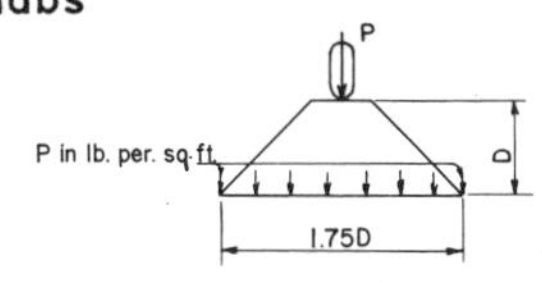

P
P in lb. per sq.ft.
D
1.75D

$$P \text{ in p.s.f.} = \frac{P}{(1.75D)^2}$$

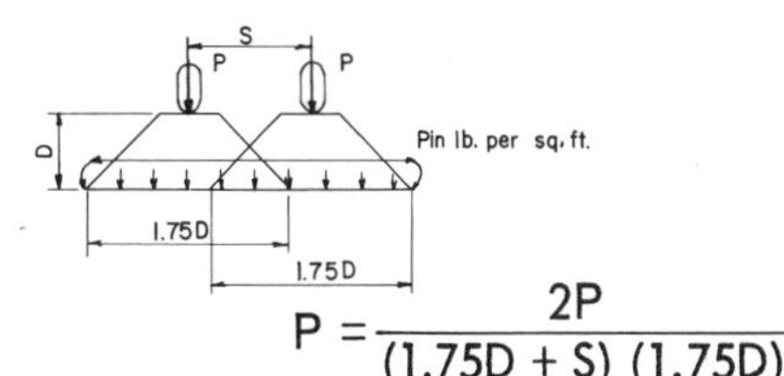

S
P
P
D
Pin lb. per sq. ft.
1.75D
1.75D

$$P = \frac{2P}{(1.75D + S)(1.75D)}$$

Case III: For single spans > 8' fill – neglect if depth > span length.
For multiple spans – neglect if depth > distance between faces of end supports or abutments.

STEEL GRID FLOORS

Case I: Filled with concrete – same as slabs, p. 13-05.
Case II: Open – distribute normal to main bars over width of (1¼" per ton of axle load) + 2 × main bar spacing in inches.

BOX GIRDERS

Distribute to webs in manner similar to floor beams and stringers as is applicable to design of T-beams.

COMPOSITE WOOD-CONCRETE SLABS

Distribute wheel loads over transverse width: 5 ft. for moment. 4 ft. for shear.

ELECTRIC RAILWAY LOADS†

Distribute wheel loads longitudinally 3 ft.; transversely (on ties and ballast construction) 10' per axle load.

CURB AND RAILING LOADS

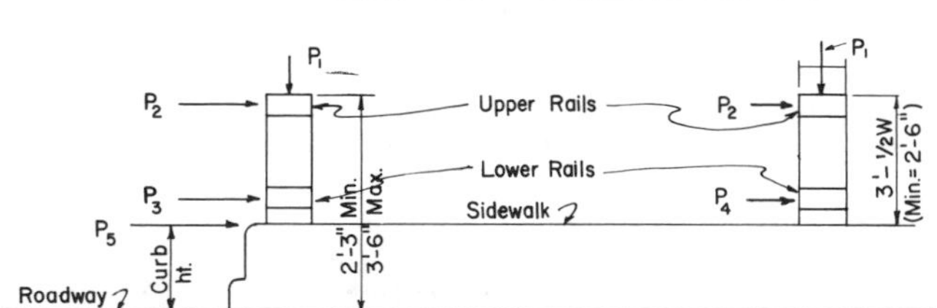

P_1
P_2
Upper Rails
P_2
P_1
P_3
Lower Rails
Sidewalk
P_4
P_5
2'-3" Min.
3'-6" Max.
3'-½W (Min.=2'-6")
Curb ht.
Roadway

Note: See pp. 12-18, 12-55 for additional curb and railing dimensions and criteria.

P_1 = 100 lb. per lin. ft. at top.
P_2 = 150 lb. per lin. ft. at top.
P_3 = Curb height > 9": 300 lb./lin. ft. on lower rail.
Curb height < 9": 300 lb./lin. ft. on lower rail plus 40 lb./lin. ft. for each inch less than 9, with maximum added increment = 200 lb./lin. ft.
If no lower rail, 300 lb./lin. ft. on web members @ 1'-9" above roadway. With curb height > 10", reduce by 15 lb./lin. ft. for each inch less than 10, with minimum of 150 lb./lin. ft.
P_4 = same as P_3, except omit where trusses, girders, arches, or curb railings protect sidewalk railings. Railings without webs and with single rails to be designed for forces P_1 and P_3.
P_5 = 500 lb. per lin. ft. of curb at curb height up to 10" maximum.

*In accord with A.A.S.H.O. Specifications, 1957. † Recommended by Author.

BRIDGES-UNIT STRESSES-I

NOTES: 1. All stresses adapted from A.A.S.H.O. (1957) and A.R.E.A. (1956) specifications.
2. Stresses followed by an asterisk (*) apply to highways; stresses followed by a dagger (†) apply to railroads.
3. For wood stresses use pp. 1-03, 1-04, and 1-05.

STRESS APPLICATION – HIGHWAY BRIDGES

Stress	Group								
	I	II	III	IV	V	VI	VII	VIII	IX
Dead load	X	X	X	X	X	X	X	X	X
Live load	X		X	X		X		X	
Live-load impact	X		X	X		X		X	
Earth pressure	X	X	X	X	X	X	X	X	X
Bouyancy	X	X	X	X	X	X	X	X	X
Wind load on structure		X	X[a]		X	X[a]			X
Wind load on live load 100 lb./lin. ft.			X			X			
Longitudinal force from live load			X			X			
Longitudinal force due to friction			X			X			
Rib shortening				X	X	X			
Shrinkage				X	X	X			
Temperature				X	X	X			
Earthquake							X		
Stream-flow pressure	X	X	X	X	X	X	X	X	X
Ice								X	X
Percentage of unit stress	100	125	125	125	140	140	$133\frac{1}{3}$	140	150

[a] 30% wind load on structure.

STEEL – stress in p.s.i. except as noted

Structural carbon steel (A.S.T.M. – A7-56T)
Rivet steel (A.S.T.M. – A141-55)
Pins and rollers (A.S.T.M. – A235-55 Class C1)*

Axial tension, structural steel, net section 18,000
Tension in extreme fibers of rolled shapes, girders, and built sections subject to bending 18,000
Tension in bolts at root of thread* 13,500
Axial compression, gross section: stiffeners of plate girders 18,000
Columns centrally loaded and with values of L/r not greater than 140: Riveted ends .... $15{,}000 - \frac{1}{4} L^2/r^2$
Pin ends .... $15{,}000 - \frac{1}{3} L^2/r^2$

L = length of member, in inches. r = least radius of gyration of member, in inches.

For compression members with values of L/r greater than 140, and for compression members of known eccentricity, use secant formulas in A.A.S.H.O. or A.R.E.A. Bridge Specifications.

Compression in extreme fibers of rolled shapes, girders and built sections, subject to bending, gross section:

(a) When compression flange is supported laterally its full length by embedment in concrete or by other means:* 18,000

(b) § When compression flange is partially supported or is unsupported [for values of L/b not greater than 30*] ... $18{,}000 - 5(L^2/b^2)$

L = length in inches of unsupported flange between lateral connections, knee braces or other points of support. For continuous beams and girders, L may be taken as the distance from interior support to point of dead load contraflexure if this distance is less than that designated above.

b = flange width in inches.

*§ Continuous or cantilever beams or girders may be proportioned for negative moment at interior supports for an allowable unit stress, 20% higher than permitted by above formula, but in no case exceeding allowable unit stress for compression flange supported its full length. If cover plates are used, the allowable stress at point of cut-off shall be as determined by the formula.

Allowable compression is splice material, gross section* 18,000
Stress in extreme fibers of pins 27,000
Shear in plate girder webs, gross section 11,000
Diagonal tension in webs of girders and rolled beams at sections where maximum shear and bending occur simultaneously 18,000
Shear in power-driven rivets and pins 13,500
Shear in turned bolts and (hand-driven rivets†) (ribbed bolts*) 11,000
Bearing on power-driven rivets, milled stiffeners and other steel parts in contact. (Rivets driven by pneumatically or electrically operated hammers are considered power-driven.) 27,000
Bearing on pins (not subject to rotation*) 24,000
Bearing on pins subject to rotation (not due to deflection) 12,000
Bearing between rockers and rocker pins 12,000
Bearing on turned bolts and (hand-driven rivets†) (ribbed bolts*) 20,000
Bearing on expansion rollers and rockers, pounds per linear inch:

Diameters up to 25 in. $\frac{p - 13{,}000}{20{,}000}\ 600d$

Diameters from 25 to 125 in. $\frac{p - 13{,}000}{20{,}000}\ 3{,}000\sqrt{d}$

d = diameter of roller or rocker, in inches.
p = yield point in tension of steel in the roller or in the base, whichever is the lesser.

In proportioning rivets the nominal diameter shall be used. The effective bearing area of a pin, a bolt, or a rivet shall be its diameter multiplied by the thickness of the metal on which it bears. In metal less than ⅜ in. thick, countersunk rivets shall not be assumed to carry stress; in metal ⅜ in. thick and over, one-half the depth of countersink shall be omitted in calculating bearing area.* For countersunk rivets, deduct one-half depth of countersink.†

BRIDGES-UNIT STRESSES-2

High-Strength Rivets* (A.S.T.M. – A195-52)
Shear 20,000
Bearing 40,000
Wrought Iron* (A.S.T.M. – A207-39, for shapes and bars; A.S.T.M. – A42-55, for plates; A.S.T.M. – A72-52, for pipe)
Tension 14,000
Bending on extreme fiber 14,000
Cast Steel For cast steel the allowable unit stresses in compressions and bearing shall be the same as those for structural steel. Other allowable unit stresses shall be three-fourths of those for structural steel.
Cast Iron* (A.S.T.M. – A48-48-Class 30).
Bending on extreme fiber 3,000
Shear 3,000
Direct compression (short columns) 12,000
Malleable Castings* (A.S.T.M. – A47-52)
Tension 18,000
Bending on extreme fiber 18,000
Modulus of elasticity 25,000,000
Bronze or Copper Alloy*
Bearing on bronze or copper-alloy bearing and expansion plates 2,000
Bearing on Masonry
Bearing on granite masonry 800
Bearing on sandstone and limestone masonry 400
Bearing on Concrete*
Bridge seats, under hinged rockers and bolsters (not subject to high edge loading by a deflecting beam, girder, or truss) 1,000
Bridge seats, under bearing plates or non-hinged shoes (subject to high edge loading by the direct bearing, upon plate or shoe, of a deflecting beam or girder), average 700
(The above bridge-seat unit stresses will apply only where the edge of bridge seat projects at least 3 in. (average) beyond edge of shoe or plate. Otherwise, the unit stresses permitted will be 75% of the above amounts.)
Bearing on concrete† .... $0.25\ f_c^1$
Timber Cross Ties†
Extreme fiber stress in bending:
Yellow pine, dense structural grade 1,500
Douglas fir, close-grain structural grade 1,400
White oak 1,200
White pine, Norway pine, and spruce 800

Slenderness ratio† $= \dfrac{\text{length in inches}}{\text{least radius of gyration}}$; not over 100 for main compression members, 120 for wind and sway bracing in compression, 140 for single lacing, 200 for double lacing, 200 for tension members other than eyebars.

CONCRETE FOR HIGHWAY BRIDGES

The unit stresses are based upon the use of concrete having an ultimate compressive strength at 28 days of 3,000 p.s.i. For concrete having a lesser strength, the unit stresses shall be proportionately reduced.

If concrete having a higher strength than 3,000 lb. at 28 days can be produced, the stresses below, except bond on piling, may be modified as follows:

The allowable stresses may be increased proportionately with the extreme upper limit which may be used in computing working stresses at 4,500 p.s.i.

When stresses other than those specified in the table are used, the allowable stresses used and the ultimate strength of concrete upon which they are based shall be noted on the plans.

The dead-load stresses set forth may be increased 40% for concrete culverts when the depth of cover is greater than 8 ft. and exceeds the span length.

A_g = overall or gross cross-sectional area of a spirally reinforced or tied pier, pedestal, or column, sq. in.

A_c = cross-sectional area of core of spirally reinforced columns measured to the outside diameter of the spiral, sq. in.

A_s = cross-sectional area of longitudinal steel.

$C = \dfrac{fa}{0.40 f'_c}$, a factor used in the design of members subjected to combined axial and bending stresses.

d = least lateral dimension of column, in.

e = eccentricity of resultant load on a column, measured from a gravity axis.

$fa = \dfrac{0.225 f'_c + f'_s P}{1 + (n-1)P}$

f'_c = crushing strength of 6" × 12" concrete test cylinders at age of 28 days, p.s.i.

f_e = maximum allowable compressive stress in members subjected to a combined axial and bending stress, p.s.i.

f_s = nominal working stress in longitudinal reinforcing steel, p.s.i.

f'_s = yield stress of spiral reinforcement (for steel grades not having a definite yield point, the stress causing a 0.2% plastic set), p.s.i.

$K = \dfrac{t^2}{2r^2}$, a factor used in the design of members subjected to combined axial and bending stresses.

L = unsupported length of column, in.

BRIDGES–UNIT STRESSES–3

$n = \dfrac{\text{modulus of elasticity of steel}}{\text{modulus of elasticity of concrete}}$

$f_c' = 2000 - 2400$ $n = 15$
$f_c' = 2500 - 2900$ $n = 12$
$f_c' = 3000 - 3900$ $n = 10$
$f_c' = 4000 - 4900$ $n = 8$
$f_c' = 5000$ – or more $n = 6$

P_e = a load eccentrically applied.
P_p = total load on pier or pedestal, lb.
P_s = total load on spirally reinforced column, lb.
P_{sl} = total load on spirally reinforced long column, lb.
P_t = total load on tied column, lb.
P_{tl} = total load on tied long column, lb.
P = ratio of longitudinal steel area to gross column area.
P' = ratio area of spiral reinforcement to core area.
r = least radius of gyration of section (transformed section).
t = over-all depth of column in the direction of eccentricity or bending.

Coefficients:
Thermal = 0.000006 Shrinkage = 0.0002

Piers and Pedestals (unreinforced): L/d not over 3. Total load: $P_p = 0.25\ Agf_c'$

Spirally Reinforced Columns
Longitudinal reinforcement: Ratio A_s to Ag not less than 0.01, nor more than 0.08 (minimum 6 # 5 Bars)
Spiral reinforcement: $P' = 0.45\,[Ag/A_c - 1]\ f_c'/f_s'$
For column core less than 18" minimum = 1/4" minimum.
For column core more than 18" minimum = 3/8" minimum.
Allowable load: For short column, L/d not over 10 $P_s = 0.225\ f_c'\ Ag + A_s F_s$.
For long column, L/d not over 20 $P_{sl} = P_s\ (1.3 - 0.03\,L/d)$.

Tied Columns
Longitudinal reinforcement: Ratio A_s to Ag not less than 0.01, nor more than 0.04 (minimum 4 # 5).
Hoops and lateral ties: Minimum #2 at 12". (Provide auxiliary ties for intermediate longitudinal bars, spaced more than 2 ft from any tied bar.)
Allowable load: For short column, L/d not over 10 $P_t = 0.8\ (0.225\ f_c'\ Ag + A_s F_s)$.
For long column, L/d not over 20 $P_{tl} = P_t\,(1.3 - 0.003\,L/d)$.

Combined Axial and Bending Stress
Ratio e/t not greater than 0.5 in either plane.

$$f_c = Pe/Ag \left[\frac{1 \pm Ke/t}{1 + (n-1)P}\right]$$

(Column may be designed for equivalent axial load P_s or P_t as given by $P = Pe\,(1 + CKe/t)$.

When bending exists in both planes, Ke/t is the sum of the Ke/t values in both planes.

Maximum Allowable Compressive Stress

$$f_e = f_a \left[\frac{t + Ke}{t + CKe}\right], \text{ where } f_a = \frac{0.225 f_c' + f_s P}{1 + (n-1)\,P}$$

(for spiral columns, and 0.8 that amount for tied columns)

Compression in Extreme Fiber
Extreme fiber stress in flexure .... $f_c = 0.04\ f_c'$

Tension (Extreme Fiber)
In reinforced-concrete members .... None
In plain concrete (primarily footings) .... $f_c = 0.03\ f_c'$

Shearing Stresses
Beams without web reinforcement:
Longitudinal bars not anchored or plain concrete footings .... $f_c = 0.02\ f_c'$ (75 p.s.i. maximum)
Longitudinal bars anchored .... $f_c = 0.03\ f_c'$ (90 p.s.i. maximum)
Beams with web reinforcement .... $V = 0.075\ f_c'\ bjd$
Horizontal shear in shear keys between slab and stem of T-Beams and box girders .... $f_c = 0.15\ f_c'$

Bond on piles (in seals): Timber, Steel or Concrete piles, 10 p.s.i. (providing the pile has the resistance to the pull thereby induced.)

Steel Reinforcement
Tension in flexural members (Intermediate Grade) 20,000
(Structural Grade) 18,000
Tension in web reinforcement (Intermediate Grade) 20,000
(Structural Grade) 18,000
Compression (Intermediate Grade) 16,000
(Structural Grade) 13,200

Bond, Deformed Bars
Straight or hooked ends, exclusive on top bars:
1. In beams, slabs, one-way footings (Intermediate Grade) .... $0.10\ f_c'$ (maximum 350)
(Structural Grade) .... $0.10\ f_c'$ (maximum 350)
2. In two-way footings (Intermediate Grade) .... $0.8\ f_c'$ (maximum 280)
(Structural Grade) .... $0.8\ f_c'$ (maximum 280)

Top Bars
Bars near top of beams and girders having more than 12 inches of concrete under bars (Intermediate Grade) .... $0.06\ f_c'$ (maximum 210)
(Structural Grade) .... $0.06\ f_c'$ (maximum 210)

BRIDGES–LOADING TABLES AND DETAILS

TABLE A–MAXIMUM MOMENTS, SHEARS & REACTIONS–SIMPLE SPANS, ONE LANE (10')*

SPAN IN FT.	H 15		H15-S12		H 20		H20-S16	
	MOMENT	END SHEAR & END REACT. (a)	MOMENT	END SHEAR & END REACT. (a)	MOMENT	END SHEAR & END REACT. (a)	MOMENT	END SHEAR & END REACT. (a)
1	6.0 (b)	24.0 (b)	6.0 (b)	24.0 (b)	8.0 (b)	32.0 (b)	8.0 (b)	32.0 (b)
2	12.0 (b)	24.0 (b)	12.0 (b)	24.0 (b)	16.0 (b)	32.0 (b)	16.0 (b)	32.0 (b)
3	18.0 (b)	24.0 (b)	18.0 (b)	24.0 (b)	24.0 (b)	32.0 (b)	24.0 (b)	32.0 (b)
4	24.0 (b)	24.0 (b)	24.0 (b)	24.0 (b)	32.0 (b)	32.0 (b)	32.0 (b)	32.0 (b)
5	30.0 (b)	24.0 (b)	30.0 (b)	24.0 (b)	40.0 (b)	32.0 (b)	40.0 (b)	32.0 (b)
6	36.0 (b)	24.0 (b)	36.0 (b)	24.0 (b)	48.0 (b)	32.0 (b)	48.0 (b)	32.0 (b)
7	42.0 (b)	24.0 (b)	42.0 (b)	24.0 (b)	56.0 (b)	32.0 (b)	56.0 (b)	32.0 (b)
8	48.0 (b)	24.0 (b)	48.0 (b)	24.0 (b)	64.0 (b)	32.0 (b)	64.0 (b)	32.0 (b)
9	54.0 (b)	24.0 (b)	54.0 (b)	24.0 (b)	72.0 (b)	32.0 (b)	72.0 (b)	32.0 (b)
10	60.0 (b)	24.0 (b)	60.0 (b)	24.0 (b)	80.0 (b)	32.0 (b)	80.0 (b)	32.0 (b)
11	66.0 (b)	24.0 (b)	66.0 (b)	24.0 (b)	88.0 (b)	32.0 (b)	88.0 (b)	32.0 (b)
12	72.0 (b)	24.0 (b)	72.0 (b)	24.0 (b)	96.0 (b)	32.0 (b)	96.0 (b)	32.0 (b)
13	78.0 (b)	24.0 (b)	78.0 (b)	24.0 (b)	104.0 (b)	32.0 (b)	104.0 (b)	32.0 (b)
14	84.0 (b)	24.0 (b)	84.0 (b)	24.0 (b)	112.0 (b)	32.0 (b)	112.0 (b)	32.0 (b)
15	90.0 (b)	24.4 (b)	90.0 (b)	25.6 (b)	120.0 (b)	32.5 (b)	120.0 (b)	34.1 (b)
16	96.0 (b)	24.8 (b)	96.0 (b)	27.0 (b)	128.0 (b)	33.0 (b)	128.0 (b)	36.0 (b)
17	102.0 (b)	25.1 (b)	102.0 (b)	28.2 (b)	136.0 (b)	33.4 (b)	136.0 (b)	37.7 (b)
18	108.0 (b)	25.3 (b)	108.0 (b)	29.3 (b)	144.0 (b)	33.8 (b)	144.0 (b)	39.1 (b)
19	114.0 (b)	25.6 (b)	114.0 (b)	30.3 (b)	152.0 (b)	34.1 (b)	152.0 (b)	40.4 (b)
20	120.0 (b)	25.8 (b)	120.0 (b)	31.2 (b)	160.0 (b)	34.4 (b)	160.0 (b)	41.6 (b)
21	126.0 (b)	26.0 (b)	126.0 (b)	32.0 (b)	168.0 (b)	34.7 (b)	168.0 (b)	42.7 (b)
22	132.0 (b)	26.2 (b)	132.0 (b)	32.7 (b)	176.0 (b)	34.9 (b)	176.0 (b)	43.6 (b)
23	138.0 (b)	26.3 (b)	138.0 (b)	33.4 (b)	184.0 (b)	35.1 (b)	184.0 (b)	44.5 (b)
24	144.0 (b)	26.5 (b)	144.5 (b)	34.0 (b)	192.0 (b)	35.3 (b)	192.7 (b)	45.3 (b)
25	150.0 (b)	26.6 (b)	155.5 (b)	34.6 (b)	200.0 (b)	35.5 (b)	207.4 (b)	46.1 (b)
26	156.0 (b)	26.8 (b)	166.6 (b)	35.1 (b)	208.0 (b)	35.7 (b)	222.2 (b)	46.8 (b)
27	162.7 (b)	26.9 (b)	177.8 (b)	35.6 (b)	216.9 (b)	35.9 (b)	237.0 (b)	47.4 (b)
28	170.1 (b)	27.0 (b)	189.0 (b)	36.0 (b)	226.8 (b)	36.0 (b)	252.0 (b)	48.0 (b)
29	177.5 (b)	27.1 (b)	200.3 (b)	36.6 (b)	236.7 (b)	36.1 (b)	267.0 (b)	48.8 (b)
30	185.0 (b)	27.2 (b)	211.6 (b)	37.2 (b)	246.6 (b)	36.3 (b)	282.1 (b)	49.6 (b)
31	192.4 (b)	27.3 (b)	223.0 (b)	37.7 (b)	256.5 (b)	36.4 (b)	297.3 (b)	50.3 (b)
32	199.8 (b)	27.4 (b)	234.4 (b)	38.3 (b)	266.5 (b)	36.5 (b)	312.5 (b)	51.0 (b)
33	207.3 (b)	27.5 (b)	245.8 (b)	38.7 (b)	276.4 (b)	36.6 (b)	327.8 (b)	51.6 (b)
34	214.7 (b)	27.7	257.7 (b)	39.2 (b)	286.3 (b)	36.9	343.5 (b)	52.2 (b)
35	222.2 (b)	27.9	270.9 (b)	39.6 (b)	296.2 (b)	37.2	361.2 (b)	52.8 (b)
36	229.6 (b)	28.1	284.2 (b)	40.0 (b)	306.2 (b)	37.5	378.9 (b)	53.3 (b)
37	237.1 (b)	28.4	297.5 (b)	40.4 (b)	316.1 (b)	37.8	396.6 (b)	53.8 (b)
38	244.5 (b)	28.6	310.7 (b)	40.7 (b)	326.1 (b)	38.2	414.3 (b)	54.3 (b)
39	252.0 (b)	28.9	324.0 (b)	41.1 (b)	336.0 (b)	38.5	432.1 (b)	54.8 (b)
40	259.5 (b)	29.1	337.4 (b)	41.4 (b)	346.0 (b)	38.8	449.8 (b)	55.2 (b)

SPAN IN FT.	H 15		H15-S12		H 20		H20-S16	
	MOMENT	END SHEAR & END REACT. (a)	MOMENT	END SHEAR & END REACT. (a)	MOMENT	END SHEAR & END REACT. (a)	MOMENT	END SHEAR & END REACT. (a)
42	274.4 (b)	29.6	364.0 (b)	42.0 (b)	365.9 (b)	39.4	485.8 (b)	56.0 (b)
44	289.3 (b)	30.1	390.7 (b)	42.5 (b)	385.8 (b)	40.1	520.9 (b)	56.7 (b)
46	304.3 (b)	30.5	417.4 (b)	43.0 (b)	405.7 (b)	40.7	556.5 (b)	57.3 (b)
48	319.2 (b)	31.0	444.1 (b)	43.5 (b)	425.6 (b)	41.4	592.1 (b)	58.0 (b)
50	334.2 (b)	31.5	470.9 (b)	43.9 (b)	445.6 (b)	42.0	627.9 (b)	58.5 (b)
52	349.1 (b)	32.0	497.7 (b)	44.3 (b)	465.5 (b)	42.6	663.6 (b)	59.1 (b)
54	364.1 (b)	32.5	524.5 (b)	44.7 (b)	485.5 (b)	43.3	699.3 (b)	59.6 (b)
56	379.1 (b)	32.9	551.3 (b)	45.0 (b)	505.4 (b)	43.9	735.1 (b)	60.0 (b)
58	397.6	33.4	578.1 (b)	45.3 (b)	530.1	44.6	770.8 (b)	60.4 (b)
60	418.5	33.9	604.9 (b)	45.6 (b)	558.0	45.2	806.5 (b)	60.8 (b)
62	439.9	34.4	631.8 (b)	45.9 (b)	586.5	45.8	842.4 (b)	61.2 (b)
64	461.8	34.9	658.6 (b)	46.1 (b)	615.7	46.5	878.1 (b)	61.5 (b)
66	484.1	35.3	685.5 (b)	46.4 (b)	645.5	47.1	914.0 (b)	61.9 (b)
68	506.9	35.8	712.3 (b)	46.6 (b)	675.9	47.8	949.7 (b)	62.1 (b)
70	530.3	36.3	739.2 (b)	46.8 (b)	707.0	48.4	985.6 (b)	62.4 (b)
75	590.6	37.5	806.3 (b)	47.3 (b)	787.5	50.0	1,075.1 (b)	63.1 (b)
80	654.0	38.7	873.7 (b)	47.7 (b)	872.0	51.6	1,164.9 (b)	63.6 (b)
85	720.4	39.9	941.0 (b)	48.1 (b)	960.5	53.2	1,254.7 (b)	64.1 (b)
90	789.8	41.1	1,008.3 (b)	48.4 (b)	1,053.0	54.8	1,344.4 (b)	64.5 (b)
95	862.1	42.3	1,074.9 (b)	48.7 (b)	1,149.5	56.4	1,433.2 (b)	64.9 (b)
100	937.5	43.5	1,143.0 (b)	49.0 (b)	1,250.0	58.0	1,524.0 (b)	65.3 (b)
110	1,097.3	45.9	1,277.7 (b)	49.4 (b)	1,463.0	61.2	1,703.6 (b)	65.9 (b)
120	1,269.0	48.3	1,412.5 (b)	49.8 (b)	1,692.0	64.4	1,883.3 (b)	66.4 (b)
130	1,452.8	50.7	1,547.3 (b)	50.7	1,937.0	67.6	2,063.1 (b)	67.6
140	1,648.5	53.1	1,682.1 (b)	53.1	2,198.0	70.8	2,242.8 (b)	70.8
150	1,856.3	55.5	1,856.3	55.5	2,475.0	74.0	2,475.1	74.0
160	2,076.0	57.9	2,076.0	57.9	2,768.0	77.2	2,768.0	77.2
170	2,307.8	60.3	2,307.8	60.3	3,077.0	80.4	3,077.1	80.4
180	2,551.5	62.7	2,551.5	62.7	3,402.0	83.6	3,402.0	83.6
190	2,807.3	65.1	2,807.3	65.1	3,743.0	86.8	3,743.1	86.8
200	3,075.0	67.5	3,075.0	67.5	4,100.0	90.0	4,100.0	90.0
220	3,646.5	72.3	3,646.5	72.3	4,862.0	96.4	4,862.0	96.4
240	4,266.0	77.1	4,266.0	77.1	5,688.0	102.8	5,688.0	102.8
260	4,933.5	81.9	4,933.5	81.9	6,578.0	109.2	6,578.0	109.2
280	5,649.0	86.7	5,649.0	86.7	7,532.0	115.6	7,532.0	115.6
300	6,412.5	91.5	6,412.5	91.5	8,550.0	122.0	8,550.0	122.0

Notes:- Moments given in foot kips; End shears given in kips = 1000 lbs.
Reduce for multiple lane loading. Add for impact.
(a) Concentrated load at support, shear loads used.
(b) Value determined by truck loading (one H-S truck for H-S loadings)

USE OF TABLE

Given H-20 Loading, Clear stringer span 50', 2 Stringers per lane (5 stringer spacing). Required - Max. Stringer Mom. and End Reaction. Solution - End Shear = 42 k, Mom. = 445.6 ft. Kips. Stringer shear = $\frac{42}{2}$ = 21 k; Stringer Mom. = $\frac{445.6}{2}$ = 222.8 ft. kips. For Concrete Slabs Reinforced Parallel to Traffic, divide values in Table by 10 to give shear and moment per ft. of width.

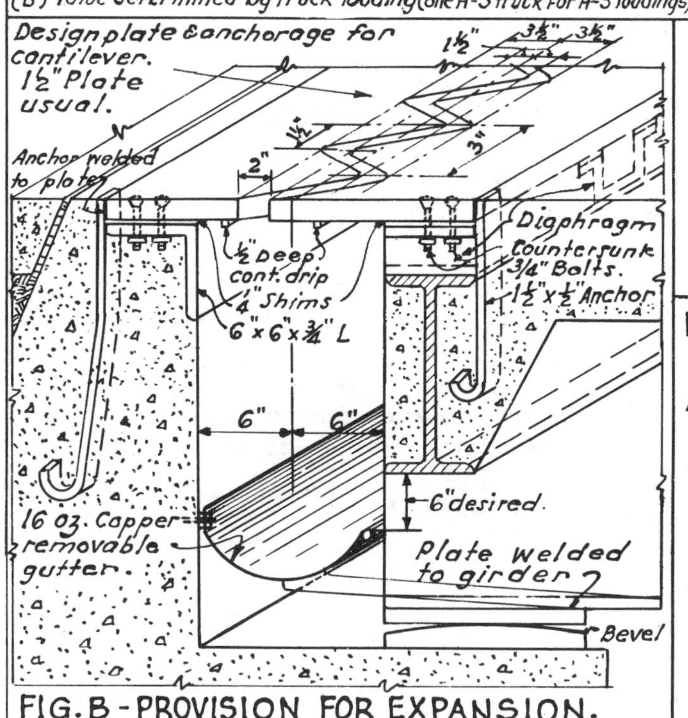

Design plate & anchorage for cantilever. 1½" Plate usual.
Anchor welded to plate
1½"
3½"
3½"
2"
3"
½" Deep cont. drip
¼" Shims
6"×6"×¾" L
Diaphragm
Countersunk ¾" Bolts.
1½"×½" Anchor
6"
6"
6" desired.
16 oz. Copper removable gutter.
Plate welded to girder
Bevel

FIG. B - PROVISION FOR EXPANSION.
Max. span with no provision = 20 feet.
Slip joint - Spans up to 70', Spans over 70' use rollers or rockers.

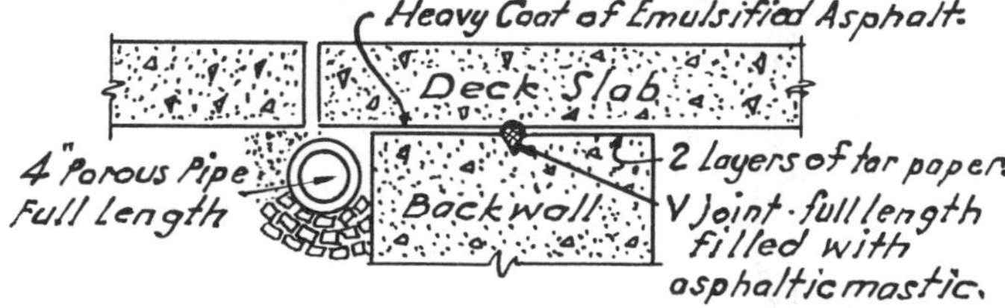

Heavy Coat of Emulsified Asphalt.
Deck Slab
4" Porous Pipe Full Length
Backwall
2 Layers of tar paper.
V Joint - full length filled with asphaltic mastic.

FIG. C - DECK SLAB AT ABUTMENT.
(After detail of Pennsylvania R.R.)

FLUE BLAST PROTECTION OVER RAILROAD.*

For metal or concrete < 20 feet above the tracks Use ¾" C.I. ℄, ⅜" W.I. or alloy ℄, ½" Plain Asbestos Board, or $\frac{7}{16}$" Corrugated Asbestos-Board.
Bolts shall be not less than ⅝" φ.
All fastenings galvanized.

SUPERELEVATION:*

Not > 0.10 foot per foot of roadway width.

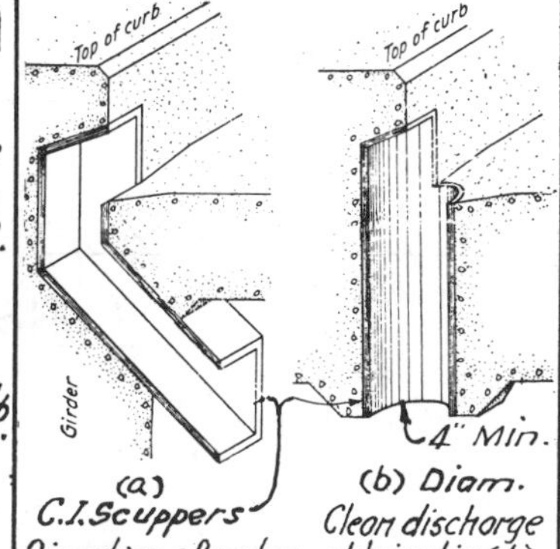

Top of curb
Top of curb
Girder
4" Min.
(a)
C.I. Scuppers
(b) Diam.

Clean discharge obtained in (b)
Direction of water discharge controlled by extending scupper by use of type (a).

FIG. D - SCUPPERS.**

*Adapted from A.A.S.H.O. Specs., 1949 ** Adapted from Portland Cement Assoc. - Conc. Bridge Details, 1949.

BRIDGES — RIGID FRAMES DESIGN DATA

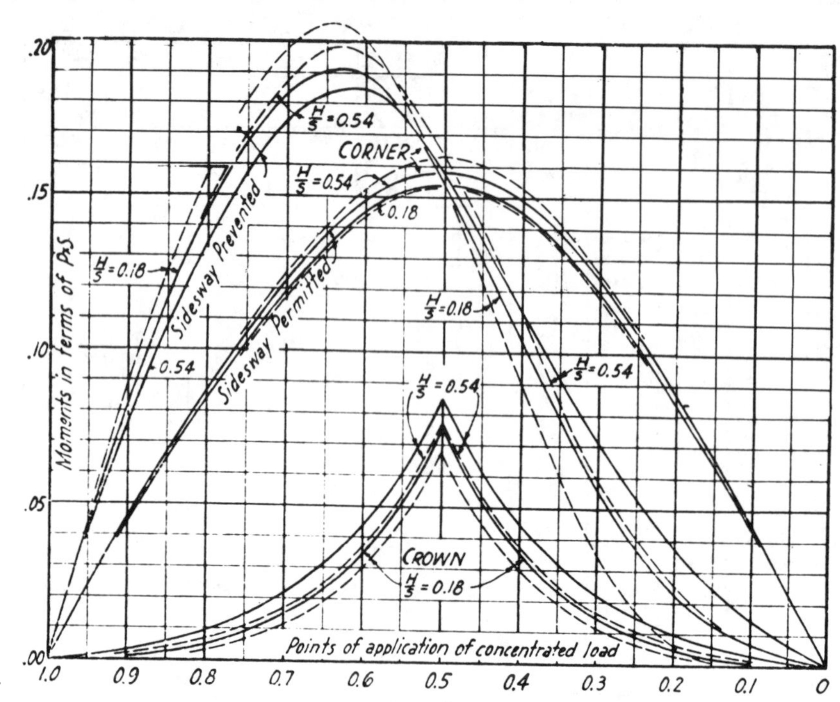

.20
.15
.10
.05
.00
Moments in terms of P×S
H/S=0.54
CORNER
H/S=0.54
0.18
H/S=0.18
Sidesway Prevented
Sidesway Permitted
0.54
H/S=0.18
H/S=0.54
H/S=0.54
CROWN
H/S=0.18
Points of application of concentrated load
1.0 0.9 0.8 0.7 0.6 0.5 0.4 0.3 0.2 0.1 0

Dotted lines for Crown = S/40 & R = 0.03321 S
Solid line for Crown = S/35 & R = 0.02343 S

FIG. A CURVE FOR OBTAINING CROWN AND CORNER MOMENTS

TABLE B - LIMITATIONS FOR FIG. A

Rise of deck axis = R; Crown; Parabola; Corner; Height H; Bott.; Hinge; Span S

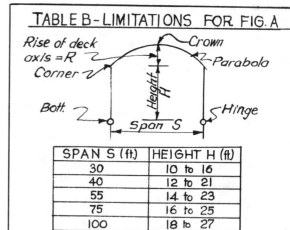

SPAN S (ft.)	HEIGHT H (ft.)
30	10 to 16
40	12 to 21
55	14 to 23
75	16 to 25
100	18 to 27

CROWN	CORNER	BOTTOM	RISE	LONGITUD. CAMBER OF DECK SURFACES
S/40	S/14	S/30+½	.0332×S	S/100
S/35	S/14	S/30+½	.02343×S	S/500

TABLE C - AREA UNDER CURVES FOR MOMENT DUE TO UNIFORM LOAD FOR FIG. A

H/S	CROWN MOMENTS SOLID LINES	CROWN MOMENTS DOTTED LINES	CORNER MOMENTS SOLID LINES	CORNER MOMENTS DOTTED LINES
0.18	0.018 WS²	0.015 WS²	0.095 WS²	0.094 WS²
0.54	0.023 WS²	0.019 WS²	0.098 WS²	0.100 WS²

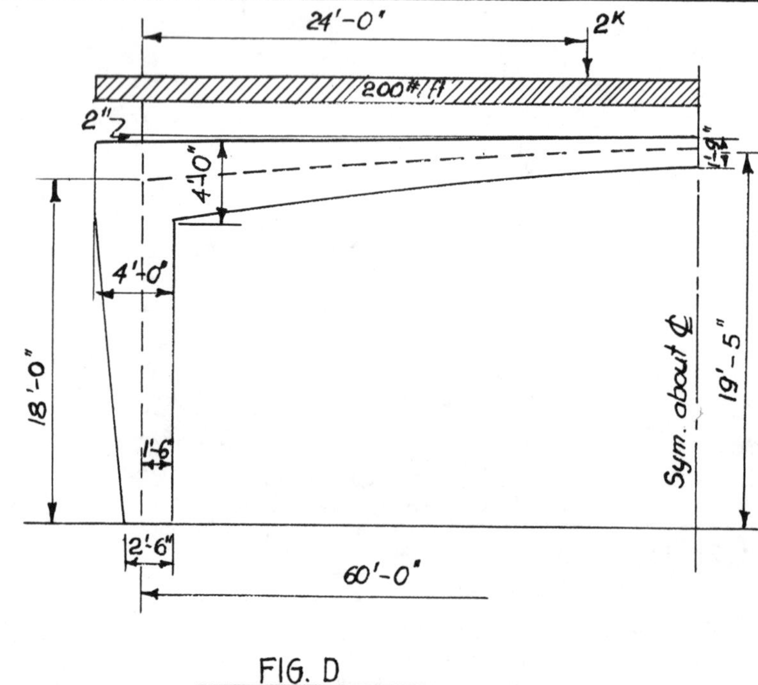

24'-0"
2K
200#/ft
2"
4'-0"
1'-9"
4'-0"
18'-0"
1'-6"
Sym. about ℄
19'-5"
2'-6"
60'-0"

FIG. D

Given: Frame with dimensions and loading as shown.

To Find: Crown and Corner Moments.

Determine:

$\frac{H}{S} = \frac{18}{60} = 0.3$ ∴ Interpolate between curves & tables for $\frac{H}{S} = 0.18$ and 0.54

$\frac{R}{S} = \frac{19.42 - 18.0}{60} = 0.0236$ ∴ Solid lines of Fig. A will be used.

For Crown Moments.

For concentrated loads - enter Fig. A at 0.6 and obtain coefficient 0.038 PS.
$0.038 \times 2 \times 60 = +4.56$ ft.K

For uniform loads - Interpolate in Table C and obtain coefficient $0.02\,WS^2$
$0.02 \times 60^2 = +14.4$ ft.K

For Corner Moments.

For concentrated load - enter Fig. A at 0.6 and obtain coefficient 0.187 PS.
$0.187 \times 2 \times 60 = -22.44$ ft.K

For uniform loads - interpolate in Table C and obtain coefficient $0.096\,WS^2$
$0.096 \times 0.2 \times 60^2 = -69.12$ ft.K

CONSTRUCTION OF AN INFLUENCE DIAGRAM

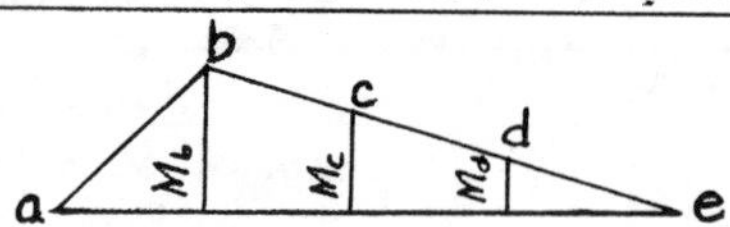

b
c
d
a
e
Mb
Mc
Md

Given: Member in which a moving load P located at points b, c, and d, successively produces moments M_b, M_c, M_d at point b.

Plot M_b, M_c, M_d at point b, c, & d respectively then connecting line abcde is the influence diagram for point b.

BRIDGES—DATA FROM EXISTING CONCRETE RIGID FRAMES

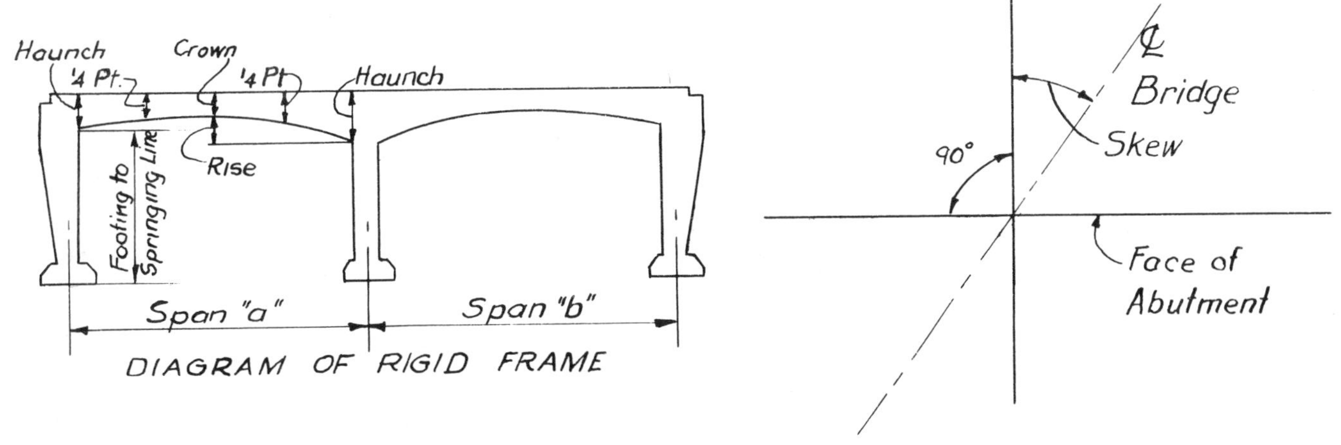

Haunch
Crown
1/4 Pt.
1/4 Pt.
Haunch
Rise
Footing to Springing Line
Span "a"
Span "b"
DIAGRAM OF RIGID FRAME
℄ Bridge
Skew
90°
Face of Abutment

FEATURES OF RIGID FRAME ARCH

			SPAN LENGTH		SKEW	FOOTING TO SPRIN. LINE	THICKNESS				RISE	REINFORCEMENT		LOADING
			ON SQ.	ON SKEW			CROWN	1/4 PT.	INSIDE ABUT.	HAUNCH		INTRADOS AT CROWN	EXTRADOS AT HAUNCH	
SINGLE SPAN	SKEW < 20°		32'-6"	–	0°	12.22'	2'-1"	2'-1"	2'-1"	2'-3"	0	1"φ-12"	1"φ-12"	–
			48'-6"	–	0°	14.0'	1'-3"	1'-11 1/2"	4'-1 1/2"	4'-2"	4'-7"	1"□-12" 1 1/8"□-12"	1"□-12" 1"φ-12"	H20
			50'-4"	–	0°	17.05'	1'-6"	2'-2 1/2"	4'-3"	4'-3 1/2"	2.93'	1"□-6"	1"□-6"	–
			56'-0"	56'-11 1/4"	7°-05'	15'-4"	1'-6"	1'-11"	4'-1"	4'-2"	4'-0 7/8"	7/8"φ-6"	1 1/8"φ-6"	H20
			62'-3 1/2"	–	0°	14'-1"	1'-6"	1'-11"	4'-0"	4'-2"	3'-6"	7/8"φ-5"	1 1/8"□5"	H20
	SKEW > 20°		32'-3"	37'-2 7/8"	30°00'	11.92"	2'-1"	2'-1"	2'-1"	2'-3"	0	1"□-12"	1"□12"	–
			43'-1 5/8"	47'-11"	25°49'-04"	20'-1"	1'-5"	2'-0 1/2"	3'-7 1/4"	4'-2 3/4"	3.55'	1"□-6"	1 1/8"□6"	H20
			50'-4"	61'-5 3/4"	35°-35'	16.77"	1'-11"	2'-3 1/2"	6'-7 1/4"	4'-8 1/2"	3.87'	1 1/8"□-6"	1 1/8"□-6"	–
			56'-0"	66'-6 1/2"	31°-55'	14'-5"	1'-7"	2'-5"	4'-11	4'-11"	4'-0 3/4"	3/4"φ-10" 7/8"φ-10"	7/8"φ-5"	H20
			62'-8"	72'-5"	30°06'-10"	14'-0"	2'-0"	2'-9"	5'-1"	5'-0"	5'-6"	1"□-12" 1"φ-12"	1 1/4"□ 6"	H20
DOUBLE SPAN		a	36.4'	36'-5"	2°-04'	18'-6" 16'-4"	1'-0"	1'-6 1/2" 2'9 1/2"	3'-5"	3'-9 1/2" 6'-9 1/2"	3'-8" 5'-5"	7/8"φ-6"	1"φ-6"	H20
		b	52.13'	52'-2"	2°-04'	17'-1" 14'-9"	1'-3"	2'-3" 2'-3"	3'-8 1/2"	4'-1" 6'-1"	5'-5"	1"φ-12" 7/8"φ-12"	1"□-12" 1"φ-12	H20
		a	38'-10"	39'-5 1/4"	9°-52'	16'-7 1/2"	1'-4 1/4"	2'-1 1/4" 2'-0 1/4"	4'-0"	4'-0"	2'-6"	1"□-12" 1"φ-12"	1"□-6"	H20
		b	43'-10"	44'-6 1/4"	9°-52'	16'-2 1/2"	1'-5"	2'-1 1/2" 2'-2"	4'-0"	4'-0"	2'-6"	1"□-12" 1"φ-12"	1"□-6"	H20
		a	41'-1 3/8"	52'-10 7/16"	38°-57'	15'-8" 16'-0"	1'-8"	1'-11 1/2" 2'-7"	4'-2" 3'-2 9/16"	3'-9" 5'-1"	2'-10 15/16"	1 1/8"□-10" 1"φ-10"	1 1/8□-10" 1"□-10"	H20
		b	41'-1 3/8"	52'-10 7/16	38°-57'	16'-0"	1'-8"	1'-11 1/2" 2'-7"	4'-2" 3'-2 9/16"	3'-9" 5'-1"	2'-10 15/16"	1 1/8"□-10" 1"φ-10"	1 1/8"□10" 1"□ 10"	H20
		a	45'-11"	55'-2"	33°-40'	16'-1" 15'-1" 14'-1"	1'-6 1/4"	2'-5 1/2" 2'-2 3/4"	4'-0"	4'-0"	2'-8 1/2"	1"□-6"	1 1/4"□-12" 1"□-12"	H20
		b	49'-2"	59'-0 7/8"	33°-40'	16'-7"	1'-7"	2'-2 3/4" 2'-3 1/4"	4'-0"	4'-0"	5'-5 1/2"	1"□-6"	1 1/4"□-6"	H20
		a	47'-3"	47'-3 1/2"	2°-14'	13'-6" 14'-6"	1'-5"	2'-3 1/2" 2'-2 1/2"	4'-0"	4'-0"	3'-0 3/4"	7/8"φ-12" 1"φ-12"	1"□-6"	H20
		b	53'-6"	53'-6 1/2"	2°-14'	14'-11" 16'-11"	1'-7 1/2"	2'-3 1/4" 2'-7 3/4"	4'-1 1/2"	4'-11 1/4"	3'-7 1/4"	7/8"φ-6"	1 1/4"□-12" 1"□-12"	H20
		a	50'-3 1/2"	50'-10"	8°-40'	17'-5" 17'-0"	1'-5 3/4"	2'-5" 2'-10"	4'-3 1/8"	4'-5 7/8" 5'-2"	3.62 3.24	1"□-12" 1"φ-12"	1"□-6"	H20
		b	55'-3 1/2"	55'-11"	8°-40'	17'-1" 17'-5"	1'-8 1/4"	3'-0" 2'-8"	4'-3 1/8"	4'-3 1/8" 5'-0 7/8"	3.23 2.92	1 1/8"□12" 1"□-12"	1 1/8"□-12" 1"□-6"	H20

BRIDGES - SIMPLE SPAN

STANDARD BEAM BRIDGES-SIMPLE SPAN

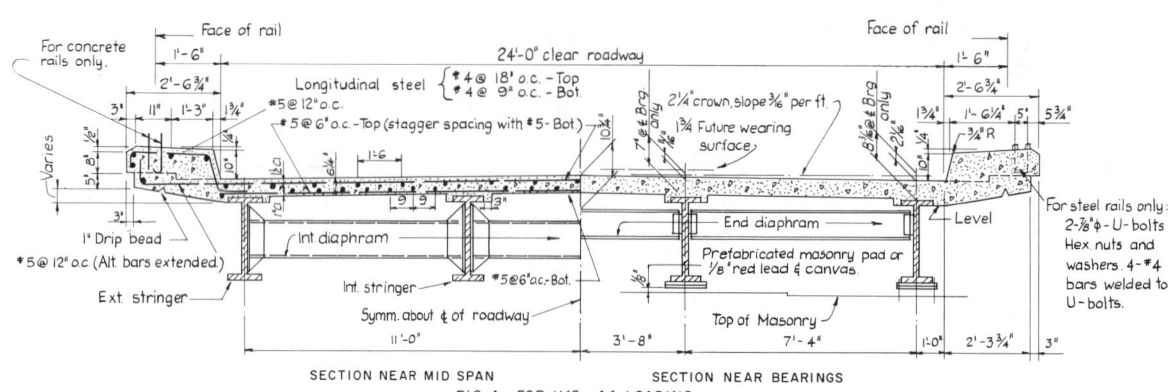

Face of rail
For concrete rails only.
1'-6"
24'-0" clear roadway
2'-6¾"
Longitudinal steel {#4 @ 18" o.c. - Top / #4 @ 9" o.c. - Bot.}
#5 @ 12" o.c.
#5 @ 6" o.c. - Top (stagger spacing with #5 - Bot.)
2¼" crown, slope 3/16" per ft.
1¾ Future wearing surface
Varies
1" Drip bead
#5 @ 12" o.c. (Alt. bars extended)
Ext. stringer
Int. diaphram
Int. stringer
#5 @ 6" o.c. - Bot.
Symm. about ℄ of roadway
End diaphram
Prefabricated masonry pad or 1/8" red lead & canvas.
Top of Masonry
Level
For steel rails only: 2-7/8"φ - U - bolts Hex. nuts and washers. 4-#4 bars welded to U-bolts.
11'-0"
3'-8"
7'-4"
1'-0"
2'-3¾"
3"

SECTION NEAR MID SPAN — SECTION NEAR BEARINGS

FIG. A—FOR H15—44 LOADING

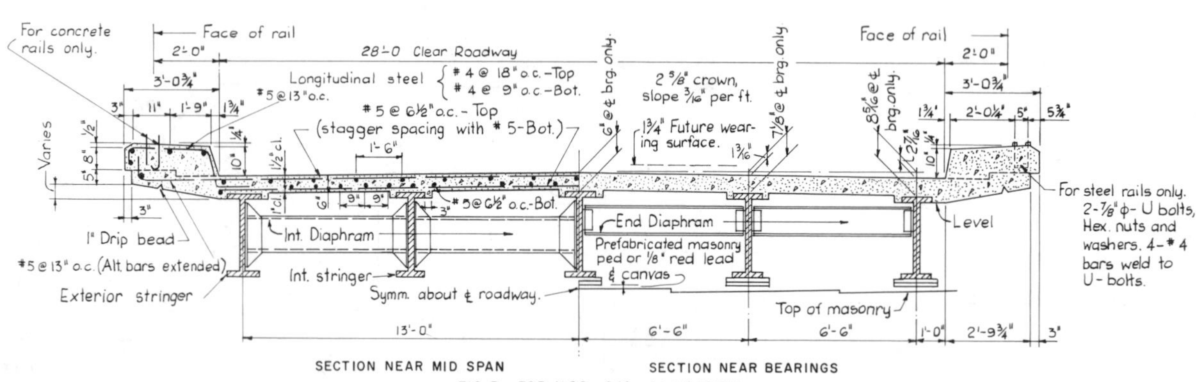

For concrete rails only.
Face of rail
2'-0"
28'-0 Clear Roadway
3'-0¾"
Longitudinal steel {#4 @ 18" o.c. - Top / #4 @ 9" o.c. - Bot.}
#5 @ 13" o.c.
#5 @ 6½" o.c. - Top (stagger spacing with #5 - Bot.)
2 5/8" crown, slope 3/16" per ft.
1¾" Future wearing surface.
Varies
1" Drip bead
#5 @ 13" o.c. (Alt. bars extended)
Exterior stringer
Int. Diaphram
Int. stringer
#5 @ 6½" o.c. - Bot.
Symm. about ℄ roadway.
End Diaphram
Prefabricated masonry ped or 1/8" red lead & canvas
Top of masonry
Level
For steel rails only. 2-7/8"φ - U bolts, Hex. nuts and washers. 4-#4 bars weld to U-bolts.
13'-0"
6'-6"
6'-6"
1'-0"
2'-9¾"
3"

SECTION NEAR MID SPAN — SECTION NEAR BEARINGS

FIG. B—FOR H20—S16—44 LOADING

TABLE - C — STRINGERS

SPAN (FEET)	H 15-44 LOADING		H 20-S16-44 LOADING	
	EXTERIOR	INTERIOR	EXTERIOR	INTERIOR
20	18 WF 50	18 WF 60	21 WF 62	21 WF 62
25	21 WF 62	21 WF 68	24 WF 76	24 WF 76
30	24 WF 76	24 WF 76	27 WF 94	27 WF 94
35	24 WF 84	24 WF 94	30 WF 108	30 WF 108
40	27 WF 94	27 WF 102	30 WF 116	30 WF 124
45	30 WF 108	30 WF 116	33 WF 130	33 WF 141
50	33 WF 130	33 WF 130	36 WF 150	36 WF 160
60	36 WF 150	36 WF 160	36 WF 194	36 WF 230
70	33 WF 220	33 WF 220	36 WF 245	36 WF 245
80	36 WF 260	36 WF 260	—	—

TABLE - D FOR H15 - 44 LOADING

MINIMUM BEAM REACTIONS (KIPS)

SPANS	20 FT.		25 FT		30 FT.		35 FT.		40 FT.		45 FT.		50 FT.		60 FT.		70 FT.		80 FT.	
	EXT	INT	EXT	INT	EXT	INT	EXT	INT	EXT	INT	EXT	INT	EXT	INT	EXT	INT	EXT	INT	EXT	INT
DEAD LOAD	10.70	8.10	13.34	10.20	16.13	12.31	18.90	14.55	21.70	16.66	25.02	19.37	28.37	22.00	34.70	27.20	43.37	34.41	51.55	41.00
LIVE LOAD	11.15	20.97	11.49	21.59	11.75	22.00	13.02	22.29	13.59	22.51	14.15	22.68	14.70	23.20	15.83	24.85	16.95	26.61	18.07	28.38
IMPACT	3.34	6.29	3.45	6.48	3.53	6.60	3.91	6.68	4.08	6.75	4.16	6.67	4.21	6.63	4.28	6.82	4.35	6.82	4.41	6.92
TOTAL	25.19	35.36	28.28	38.27	31.41	40.91	35.83	43.52	39.37	45.92	43.33	48.72	47.28	51.83	54.81	58.87	64.67	67.74	74.03	76.30
⊙ TOTAL ONE ABUT.	89.2		100.3		111.3		122.7		134.9		149.4		163.7		191.6		228.0		262.5	

⊙ Without impact.

TABLE - E FOR H20-S16-44 LOADING

MAXIMUM BEAM REACTIONS (KIPS)

SPANS	20 FT.		25 FT.		30 FT.		35 FT.		40 FT.		45 FT		50 FT		60 FT		70 FT.	
	EXT.	INT.	EXT.	INT.	EXT	INT	EXT.	INT	EXT.	INT.	EXT	INT	EXT	INT	EXT	INT	EXT	INT.
DEAD LOAD	11.54	7.46	14.45	9.54	17.60	10.53	20.70	13.65	23.66	15.84	27.00	18.22	30.55	20.70	38.05	27.96	47.30	32.71
LIVE LOAD	17.60	29.62	19.50	32.53	20.98	34.83	22.31	36.90	23.34	38.46	24.10	39.69	24.72	40.65	25.72	42.10	26.38	43.15
IMPACT	5.28	8.89	5.85	9.76	6.30	10.45	6.70	11.07	7.02	11.54	7.09	11.67	7.07	11.61	6.95	11.37	6.77	11.07
TOTAL	34.42	45.97	39.80	51.83	44.88	55.81	49.71	61.62	54.02	65.84	58.19	69.58	62.34	72.96	70.72	81.43	80.45	86.93
⊙ TOTAL ONE ABUT.	128.7		149.7		166.0		188.0		205.2		222.7		240.2		281.6		317.5	

⊙ Without impact

BRIDGES—CONTINUOUS SPAN

STANDARD BEAM BRIDGES — CONTINUOUS SPAN.

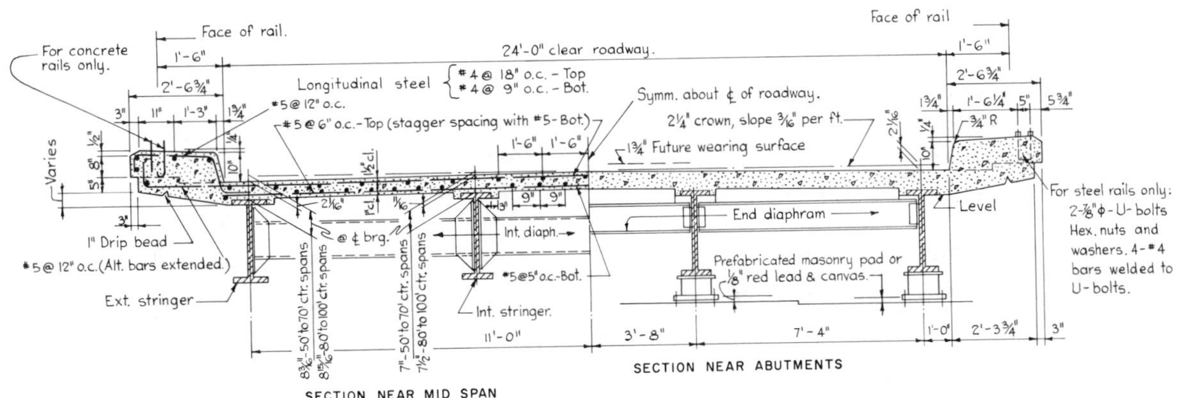

For concrete rails only.
Face of rail.
1'-6"
2'-6 3/4"
24'-0" clear roadway.
Face of rail
Longitudinal steel { #4 @ 18" o.c. - Top; #4 @ 9" o.c. - Bot.
#5 @ 12" o.c.
#5 @ 6" o.c. - Top (stagger spacing with #5-Bot.)
Symm. about ℄ of roadway.
2 1/4" crown, slope 3/16" per ft.
1 3/4" Future wearing surface
Varies
1" Drip bead
#5 @ 12" o.c. (Alt. bars extended.)
Ext. stringer
@ ℄ brg.
Int. diaph.
#5 @ 5" o.c. - Bot.
Int. stringer.
End diaphram
Prefabricated masonry pad or 1/8" red lead & canvas.
Level
For steel rails only: 2-7/8" φ-U-bolts Hex. nuts and washers. 4-#4 bars welded to U-bolts.
11'-0"
3'-8"
7'-4"
1'-0"
2'-3 3/4"
3"

SECTION NEAR MID SPAN

SECTION NEAR ABUTMENTS

FIG. A — FOR H15 — 44 LOADING

SECTION NEAR MID SPAN

SECTION NEAR ABUTMENTS

FIG. B — FOR H20 — S16 — 44 LOADING

TABLE C STRINGERS

SPANS (FEET)	H15-44 LOADING				H20-S16-44 LOADING			
	EXTERIOR		INTERIOR		EXTERIOR		INTERIOR	
	BEAM	COVER PL *	BEAM	COVER PL *	BEAM	COVER PL *	BEAM	COVER PL *
40-50-40	24 WF 76	8 × 1/2 × 11'-0	24 WF 84	8 × 1/2 × 11'-0	27 WF 94	8 × 3/8 × 9'-6"	27 WF 102	8 × 3/8 × 9'-6"
48-60-48	27 WF 94	8 × 1/2 × 13'-0	27 WF 94	8 × 5/8 × 13'-0	30 WF 108	8 × 3/8 × 10'-0	30 WF 124	8 × 3/8 × 10'-0
56-70-56	30 WF 108	8 × 3/4 × 15'-0	30 WF 108	9 × 7/8 × 17'-0	33 WF 130	8 × 1/2 × 11'-6"	33 WF 141	8 × 3/8 × 10'-6"
64-80-64	30 WF 124	9 × 1 1/8 × 21'-6"	30 WF 132	9 × 1 1/8 × 21'-6"	36 WF 150	8 × 5/8 × 12'-0	36 WF 160	8 × 3/8 × 10'-6"
72-90-72	33 WF 141	10 × 1 1/8 × 24'-0	33 WF 152	10 × 1 1/8 × 24'-0	36 WF 170	10 × 3/4 × 16'-0	36 WF 194	8 × 1/2 × 12'-0
80-100-80	36 WF 160	10 × 7/8 × 24'-6" 10 × 1 1/2 × 11'-0	36 WF 170	10 × 3/4 × 24'-6" 10 × 1 3/8 × 11'-0	36 WF 230	10 × 5/8 × 14'-0	36 WF 230	10 × 1/2 × 12'-0"

*All Cover PLs are for top & bottom flanges, and are centered about ℄ of each of two interior supports.

TABLE D MAXIMUM REACTIONS (IN KIPS)

SPANS		H15-44 LOADING				H20-S16-44 LOADING			
		ABUTMENT		PIER		ABUTMENT		PIER	
		EXTERIOR	INTERIOR	EXTERIOR	INTERIOR	EXTERIOR	INTERIOR	EXTERIOR	INTERIOR
40-50-40	D.L.	15.6	12.6	51.9	41.3	16.8	11.5	54.2	37.1
	L.L.	13.2	20.7	18.4	28.9	22.0	36.4	27.0	44.1
	IMP.	4.0	6.2	5.4	8.5	6.6	10.9	7.9	13.0
	TOTAL	32.8	39.5	75.7	78.7	45.4	58.8	89.1	94.2
48-60-48	D.L.	19.2	15.3	63.7	50.8	20.2	14.2	66.1	46.5
	L.L.	14.0	22.0	20.4	32.7	23.3	38.8	27.9	45.6
	IMP.	4.1	6.3	5.7	9.1	6.7	11.2	7.7	12.3
	TOTAL	37.3	43.6	89.8	92.6	50.2	64.2	101.7	104.4
56-70-56	D.L.	22.7	18.1	76.3	61.0	24.1	16.9	79.1	55.7
	L.L.	14.8	23.2	23.3	36.7	24.3	39.9	29.2	47.4
	IMP.	4.1	6.4	6.2	9.8	6.7	11.0	7.8	12.6
	TOTAL	41.6	47.7	105.8	107.5	55.1	67.8	116.1	115.7
64-80-64	D.L.	26.7	21.4	91.1	72.9	28.1	20.0	92.4	65.8
	L.L.	15.6	24.5	25.8	40.4	25.0	41.4	31.0	47.9
	IMP.	4.1	6.5	6.5	10.3	6.6	10.9	7.9	12.2
	TOTAL	46.4	52.4	123.4	123.6	59.7	72.3	131.3	125.9
72-90-72	D.L.	30.5	24.6	104.3	84.3	32.7	23.6	107.5	77.8
	L.L.	16.4	25.8	28.2	44.4	25.6	41.9	34.0	51.8
	IMP.	4.2	6.5	6.9	10.8	6.5	10.6	8.2	12.6
	TOTAL	51.1	56.9	139.4	139.5	64.8	76.1	149.7	142.2
80-100-80	D.L.	35.2	28.5	120.4	97.7	37.8	27.4	125.6	90.9
	L.L.	17.2	27.0	30.6	48.1	26.1	42.7	36.7	56.2
	IMP.	4.2	6.6	7.1	11.2	6.5	10.4	8.5	13.1
	TOTAL	56.6	62.1	158.1	157.0	70.4	80.5	170.8	160.2

BRIDGES—CONCRETE SLAB SIMPLE SPAN

REINF. CONCRETE BRIDGE (PARTICIPATING CURB) SIMPLE SPAN

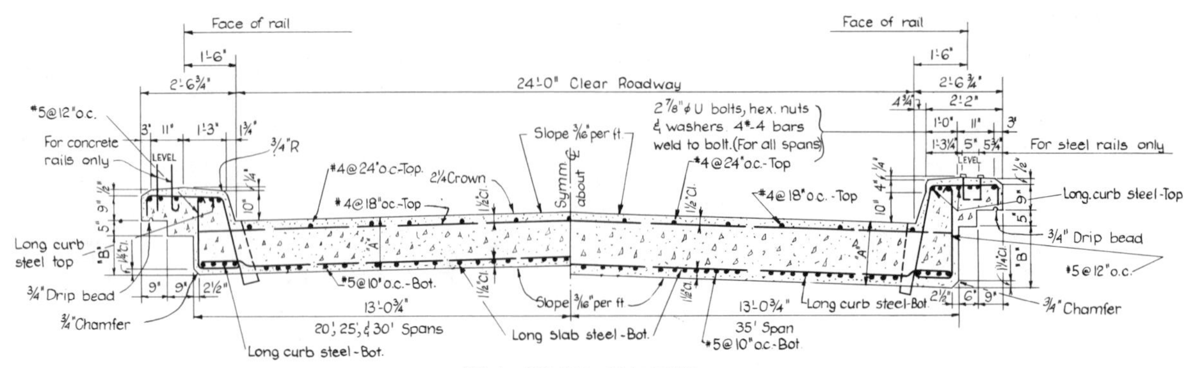
Face of rail
1'-6"
2'-6¾"
24'-0" Clear Roadway
#5@12"o.c.
For concrete rails only
¾"R
Long curb steel top
¾" Drip bead
¾" Chamfer
#4@24"o.c.-Top
#4@18"o.c.-Top
2¼ Crown
Slope 3/16" per ft
Symm about ℄
2⅞"⌀ U bolts, hex. nuts & washers 4-#4 bars weld to bolt. (For all spans)
#4@24"o.c.-Top
#4@18"o.c.-Top
For steel rails only
Long curb steel-Top
¾" Drip bead
#5@12"o.c.
¾" Chamfer
#5@10"o.c.-Bot.
13'-0¾"
20', 25' & 30' Spans
Long curb steel-Bot.
Long slab steel-Bot.
Long curb steel-Bot.
35' Span
#5@10"o.c.-Bot

FIG. A—FOR H15-44 LOADING

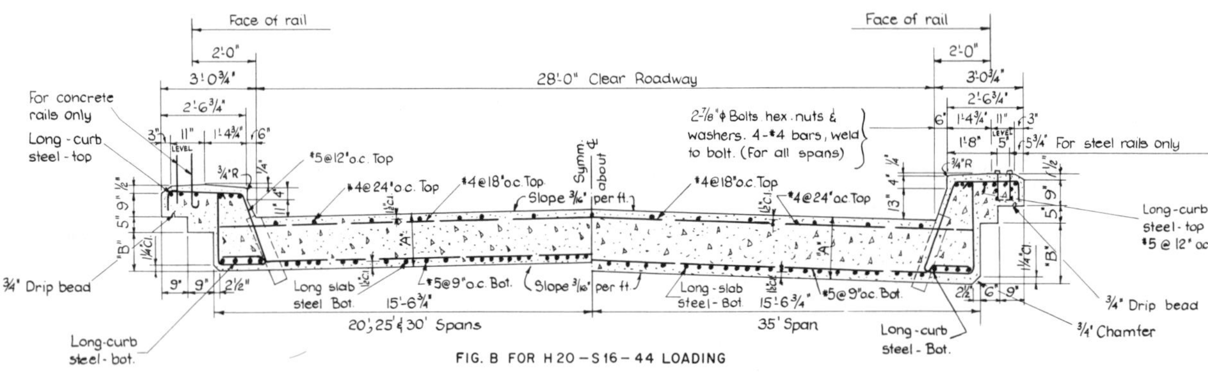
Face of rail
2'-0"
3'-0¾"
28'-0" Clear Roadway
For concrete rails only
Long-curb steel-top
2'-6¾"
#5@12"o.c. Top
#4@24"o.c. Top
#4@18"o.c. Top
Slope 3/16" per ft.
Symm about ℄
2-⅞"⌀ Bolts hex. nuts & washers. 4-#4 bars, weld to bolt. (For all spans)
#4@18"o.c. Top
#4@24"o.c. Top
For steel rails only
Long-curb steel-top
#5 @ 12" oc
¾" Drip bead
¾" Chamfer
¾" Drip bead
Long slab steel Bot.
15'-6¾"
#5@9"o.c. Bot.
Long-slab steel-Bot
15'-6¾"
#5@9"o.c. Bot.
20', 25' & 30' Spans
35' Span
Long-curb steel-bot.
Long-curb steel-Bot.

FIG. B FOR H20-S16-44 LOADING

TABLE C—SLAB & CURB DIMENSIONS & REINFORCING STEEL

SPAN (FEET)	H15-44 LOADING					H20-S16-44 LOADING				
	DIMENSIONS		Long. slab steel-bot.	Long. curb steel-top	Long. curb steel-bot.	DIMENSIONS		Long. slab steel-bot.	Long. curb steel-top	Long. curb steel-bot.
	A"	B"				A"	B"			
20	10½"	6½"	#8@8½"o.c	#8@8"o.c.	#8@3"o.c.	10½"	11½"	#7@6½"o.c.	#4@8"o.c.	#7@3"o.c.
25	12½"	8½"	#8@6"o.c	#8@5"o.c.	#8@3¼"o.c.	12½"	13½"	#8@7"o.c.	#8@5"o.c.	#8@4"o.c.
30	14½"	10½"	#9@7"o.c	#8@4"o.c.	#9@3½"o.c.	14½"	15½"	#9@7"o.c.	#9@5"o.c.	#9@3½"o.c.
35	17½"	17½"	#9@6"o.c.	#9@3¼"o.c.	#9@6"o.c.	18½"	21½"	#9@6"o.c.	#9@3½"o.c.	#9@3¾"o.c.

TABLE D MAXIMUM REACTIONS

	H15-44 LOADING					H20-S16-44 LOADING				
SPAN	D.L	L.L	IMP	TOTAL	PIER*	D.L	L.L	IMP	TOTAL	PIER*
20	49.2	51.9	15.6	116.7	150.3	59.3	85.0	25.5	169.8	208.8
25	69.0	53.5	16.0	138.5	191.5	82.8	93.5	28.0	204.3	266.5
30	92.1	54.6	16.4	163.1	241.1	110.4	100.9	30.3	241.6	328.9
35	126.5	56.5	17.0	200.0	315.0	156.9	107.2	32.2	296.3	427.1

* *Two equal spans, without impact.*

BRIDGES—COMPOSITE BEAM

STANDARD COMPOSITE BEAM BRIDGES — SIMPLE SPAN

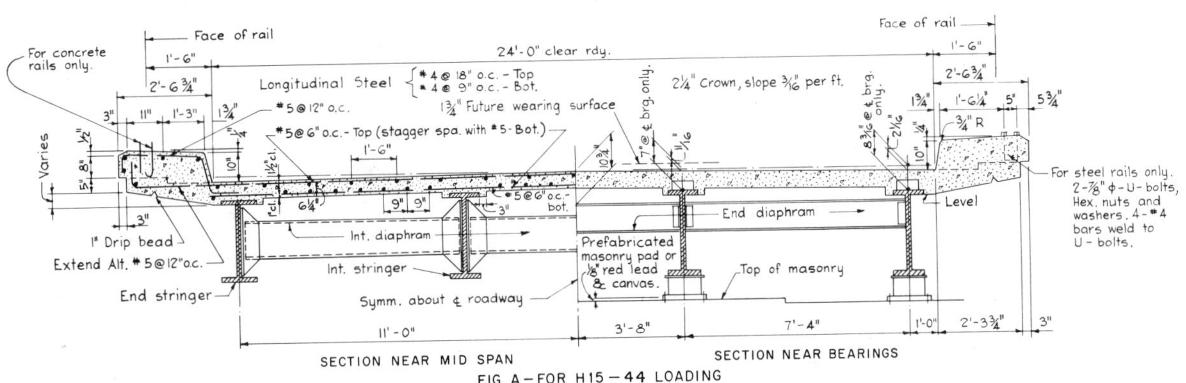

For concrete rails only.
Face of rail
24'-0" clear rdy.
Longitudinal Steel #4 @ 18" o.c. - Top, #4 @ 9" o.c. - Bot.
#5 @ 12" o.c.
1 3/4" Future wearing surface
#5 @ 6" o.c. - Top (stagger spa. with #5 - Bot.)
2 1/4" Crown, slope 3/16" per ft.
Varies
1" Drip bead
Extend Alt. #5 @ 12" o.c.
End stringer
Int. diaphram
Int. stringer
Symm. about ℄ roadway
Prefabricated masonry pad or 1/8" red lead & canvas.
Top of masonry
End diaphram
Level
For steel rails only. 2-7/8" φ - U - bolts, Hex. nuts and washers. 4-#4 bars weld to U - bolts.

SECTION NEAR MID SPAN SECTION NEAR BEARINGS

FIG. A—FOR H15—44 LOADING

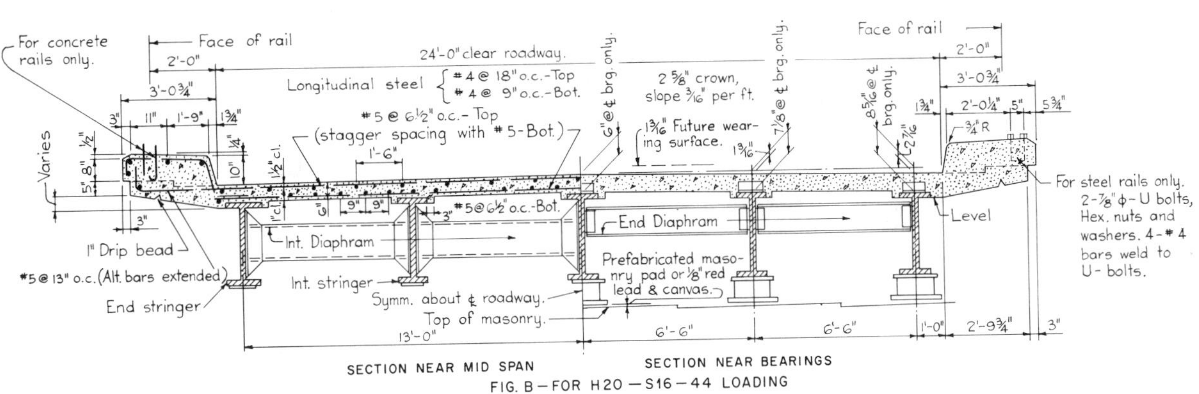

For concrete rails only.
Face of rail
24'-0" clear roadway.
Longitudinal steel #4 @ 18" o.c. - Top, #4 @ 9" o.c. - Bot.
#5 @ 6 1/2" o.c. - Top (stagger spacing with #5 - Bot.)
2 5/8" crown, slope 3/16" per ft.
1 3/16" Future wearing surface.
Varies
1" Drip bead
#5 @ 13" o.c. (Alt. bars extended)
End stringer
Int. Diaphram
Int. stringer
Symm. about ℄ roadway.
Top of masonry.
End Diaphram
Prefabricated masonry pad or 1/8" red lead & canvas.
Level
For steel rails only. 2-7/8" φ - U bolts, Hex. nuts and washers. 4-#4 bars weld to U - bolts.

SECTION NEAR MID SPAN SECTION NEAR BEARINGS

FIG. B—FOR H20—S16—44 LOADING

TABLE-C FOR H15-44 LOADING

MAXIMUM BEAM REACTIONS (KIPS)

SPANS	50 FT.		60 FT.		70 FT.		80 FT.		90 FT.		100 FT.	
	EXT.	INT.	EXT.	INT.	EXT.	INT.	EXT.	INT.	EXT.	INT.	EXT.	INT.
DEAD LOAD	28.20	21.13	35.10	26.04	41.75	31.90	49.63	38.00	60.30	45.20	69.95	52.60
LIVE LOAD	14.71	23.10	15.83	24.85	16.95	26.61	18.07	28.38	19.19	30.13	20.31	31.90
IMPACT	4.20	6.60	4.27	6.71	4.34	6.82	4.41	6.90	4.46	7.00	4.52	7.09
TOTAL	47.11	50.83	55.20	57.60	63.04	65.33	72.11	73.28	83.95	82.33	94.78	91.59
⊙ TOTAL ONE ABUT.	162.2		190.1		219.90		252.		293.2		332.1	

⊙ Without Impact

TABLE-D FOR H20-S16-44 LOADING

MAXIMUM BEAM REACTIONS (KIPS)

SPANS	50 FT.		60 FT.		70 FT.		80 FT.		90 FT.		100 FT.	
	EXT.	INT.	EXT.	INT.	EXT.	INT.	EXT.	INT.	EXT.	INT.	EXT.	INT.
DEAD LOAD	29.84	19.86	36.72	24.68	44.88	30.12	54.81	35.77	65.25	44.30	76.00	53.20
LIVE LOAD	24.73	40.66	25.71	42.12	26.40	43.16	26.90	43.93	27.28	44.54	27.61	45.04
IMPACT	7.06	11.61	6.94	11.37	6.77	11.06	6.56	10.80	6.34	10.36	6.14	10.02
TOTAL	61.63	72.13	69.37	78.17	78.05	84.34	88.27	90.50	98.87	99.20	109.75	108.26
⊙ TOTAL ONE ABUT.	236.3		269.0		304.9		344.1		392.4		442.2	

⊙ Without Impact

TABLE-E STRINGERS

SPAN (FEET)	H15-44 LOADING				H20-S16-44 LOADING			
	EXTERIOR		INTERIOR		EXTERIOR		INTERIOR	
	BEAM	COVER ℄	BEAM	COVER ℄	BEAM	COVER ℄	BEAM	COVER ℄
50	27 WF 102	TOP: NONE BOT: 8 x 1/2 x 27'-0	27 WF 102	TOP: NONE BOT: 8 x 3/8 x 27'-0	30 WF 108	TOP: NONE BOT: 9 x 1/2 x 29'-0	30 WF 108	TOP: NONE BOT: 9 x 1/2 x 29'-0
60	30 WF 108	TOP: NONE BOT: 9 x 3/4 x 38'-0	30 WF 108	TOP: NONE BOT: 9 x 5/8 x 36'-0	33 WF 130	TOP: NONE BOT: 10 x 5/8 x 36'-0	33 WF 130	TOP: NONE BOT:
70	33 WF 130	TOP: 8 x 1/2 x 43'-0 BOT: 10 x 3/4 x 44'-0	33 WF 130	TOP: NONE BOT: 10 x 3/4 x 42'-0	36 WF 170	TOP: NONE BOT: 10 x 3/4 x 40'-0	36 WF 160	TOP: NONE BOT: 10 x 3/4 x 52'-0
80	36 WF 160	TOP: 10 x 1/2 x 50'-0 BOT: 10 x 1 x 51'-0	36 WF 160	TOP: NONE BOT: 10 x 7/8 x 49'-0	36 WF 230	TOP: NONE BOT: 14 x 1/2 x 41'-0	36 WF 194	TOP: NONE BOT: 10 x 1 x 45'-0
90	36 WF 230	TOP: 15 x 3/8 x 50'-0 BOT: 15 x 5/8 x 52'-0	36 WF 194	TOP: NONE BOT: 10 x 1 1/4 x 56'-0	36 WF 280	TOP: NONE BOT: 15 x 1 1/4 x 52'-0	36 WF 230	TOP: NONE BOT: 14 x 1 x 58'-0
100	36 WF 260	TOP: 15 x 1/2 x 63'-0 BOT: 15 x 1/4 x 67'-0	36 WF 230	TOP: 15 x 1/2 x 65'-0 BOT: 15 x 1/8 x 69'-0	36 WF 300	TOP: 15 x 1/2 x 64'-0 BOT: 15 x 1 1/4 x 66'-0	36 WF 300	TOP: NONE BOT: 15 x 1 x 58'-0"

* All cover ℄ are centered about ℄ of span

BRIDGES — BEARING DETAILS

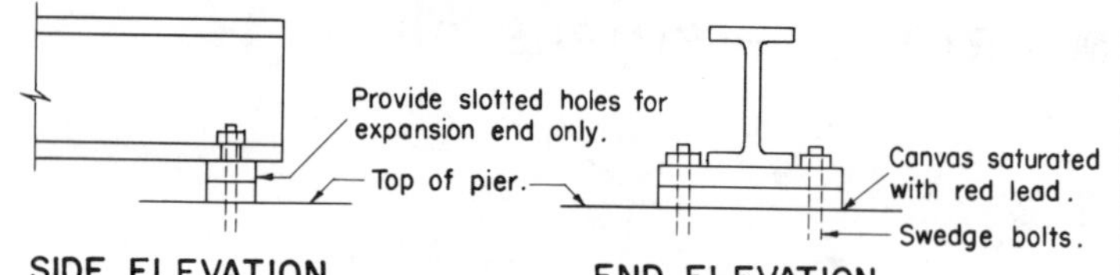

Provide slotted holes for expansion end only.
Top of pier.
Canvas saturated with red lead.
Swedge bolts.
SIDE ELEVATION
END ELEVATION

TYPE I - SIMPLE

FIG. A- FOR STEEL SPANS LESS THAN 50'

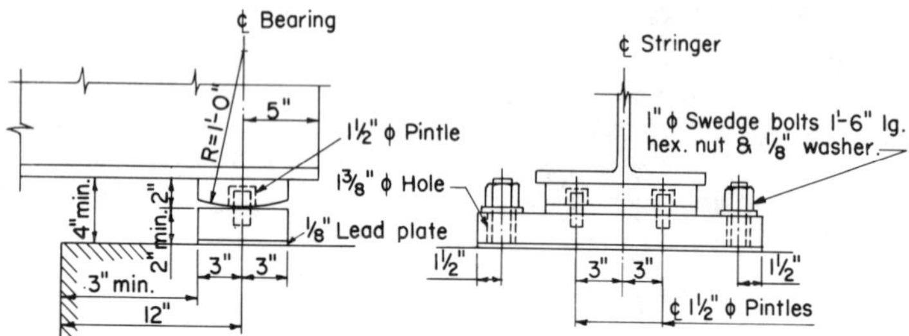

℄ Bearing
℄ Stringer
1½" ϕ Pintle
1⅜" ϕ Hole
⅛" Lead plate
1" ϕ Swedge bolts 1'-6" lg. hex. nut & ⅛" washer.
℄ 1½" ϕ Pintles
SIDE ELEVATION
END ELEVATION

TYPE IIa - FIXED

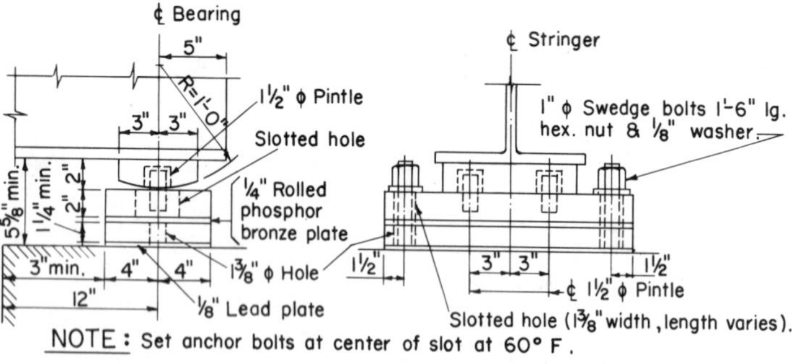

℄ Bearing
℄ Stringer
1½" ϕ Pintle
Slotted hole
¼" Rolled phosphor bronze plate
1⅜" ϕ Hole
⅛" Lead plate
1" ϕ Swedge bolts 1'-6" lg. hex. nut & ⅛" washer.
℄ 1½" ϕ Pintle
Slotted hole (1⅜" width, length varies).
NOTE: Set anchor bolts at center of slot at 60° F.
SIDE ELEVATION
END ELEVATION

TYPE IIb - EXPANSION

FIG. B- FOR STEEL SPANS LESS THAN 75'

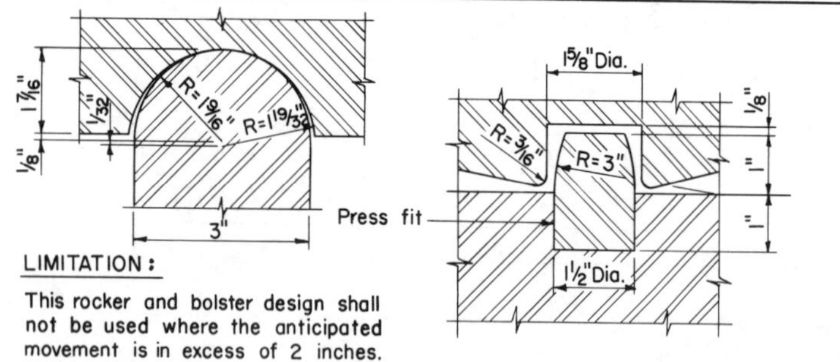

Press fit
1⅝" Dia.
1½" Dia.

LIMITATION: This rocker and bolster design shall not be used where the anticipated movement is in excess of 2 inches.

FIG. C TOP BEARING DETAIL

FIG. D PINTLE DETAIL

The bridge bearings carry the load from the stringers, girders or trusses to the pier or abutment. Their shape affects the direction and magnitude of the forces transmitted to the pier. Each girder is provided with a fixed bearing at one end to take horizontal forces of traction and braking, and a movable bearing at the other to permit expansion.

The simplest type of bearing is the steel plate, Type I (Fig. A). The fixed end is anchored to the concrete. The expansion end permits one steel plate to slide over the other. Sometimes the lower plate is faced with a phosphor bronze wearing surface to prevent erosion and reduce the coefficient of friction. The surfaces of the plate should always be level even if the girder is not. It is recommended that this type be limited to spans of 50 ft. (A.A.S.H.O.).

Type II (Fig. B) is an outgrowth of Type I. The use of a curved surface directly under the girder permits it to deflect without producing an eccentricity on the bearing surface. For spans of 75 ft. or more the expansion bearing should have a rocker as shown in Type III (Fig. E).

Type III b (Fig. E) is the rocker type of expansion bearing. It allows for the deflection of the beam and also allows expansion with a reduced horizontal force. It is used in conjunction with the fixed bearing of Type III a (Fig. E)

In general, bearing details should be so designed that they are easy to inspect, easy to keep clean, and drain water to prevent corrosion.

NOTE: All connections to be welded.

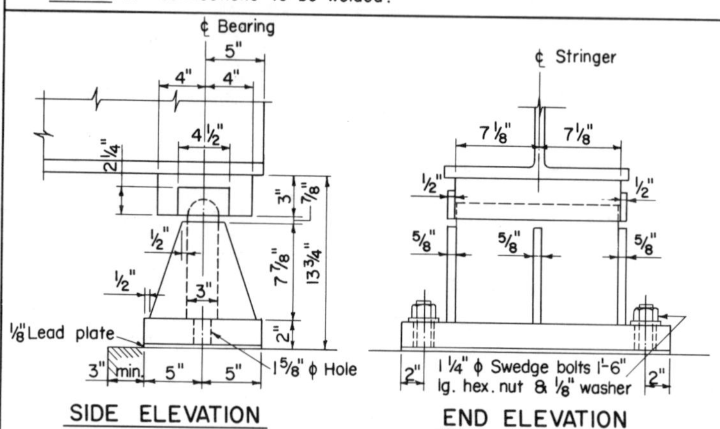

℄ Bearing
℄ Stringer
⅛" Lead plate
1⅝" ϕ Hole
1¼" ϕ Swedge bolts 1'-6" lg. hex. nut & ⅛" washer
SIDE ELEVATION
END ELEVATION

TYPE IIIa - FIXED

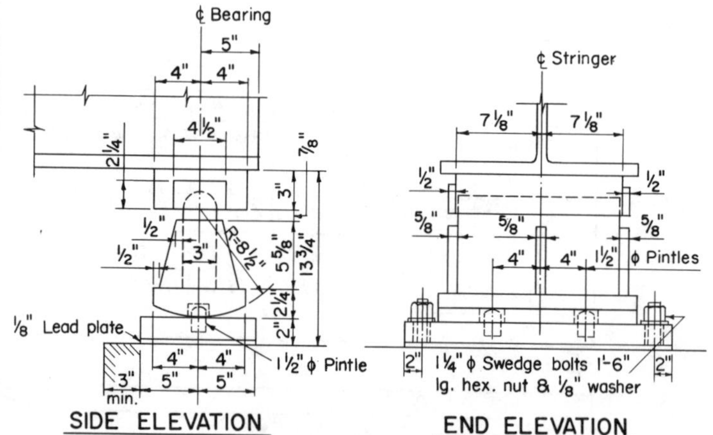

℄ Bearing
℄ Stringer
ϕ Pintles
⅛" Lead plate
1½" ϕ Pintle
1¼" ϕ Swedge bolts 1'-6" lg. hex. nut & ⅛" washer
SIDE ELEVATION
END ELEVATION

TYPE IIIb - EXPANSION

FIG. E- FOR STEEL SPANS LESS THAN 150'

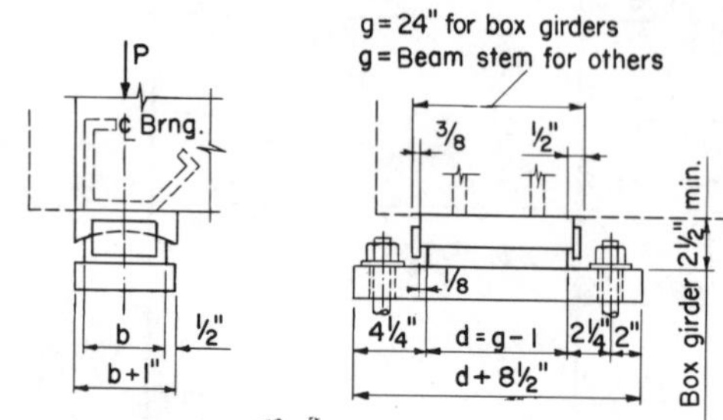

g = 24" for box girders
g = Beam stem for others
P
℄ Brng.
d = g - 1
d + 8½"
Box girder 2½" min.

FIG. F- DETAIL OF BEARINGS FOR CONCRETE* SPANS

* Also may be used with steel.

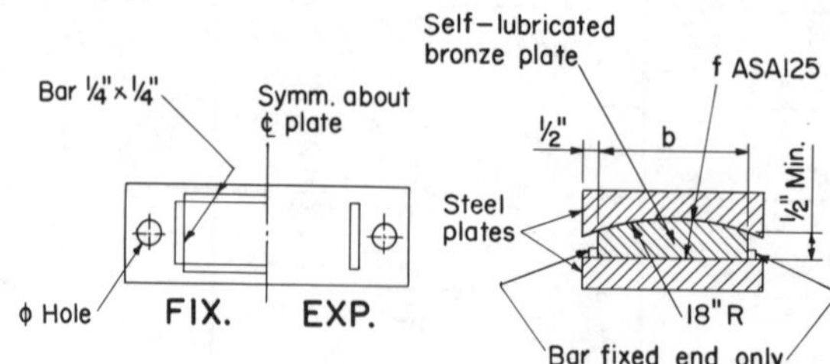

Bar ¼" x ¼"
Symm. about ℄ plate
ϕ Hole
FIX.
EXP.
Self-lubricated bronze plate
f ASA125
Steel plates
18" R
Bar fixed end only
TYPICAL MASONRY PLATE
TYPICAL SECTION

BRONZE PLATE NOTES.

Fixed End: Lubricate curved face only.

Expansion End: Lubricate both faces.

P (Kips) = Reaction (D+L+I).

d (Inches) = Length of bronze plate.

b (Inches) = Width of bronze plate.

$b \geq \frac{P}{3d}$ for box girder.

$b \geq \frac{P}{d}$ for others.

FIG. G- DETAIL OF SELF-LUBRICATED BEARINGS

BRIDGES—EXPANSION JOINT DETAILS

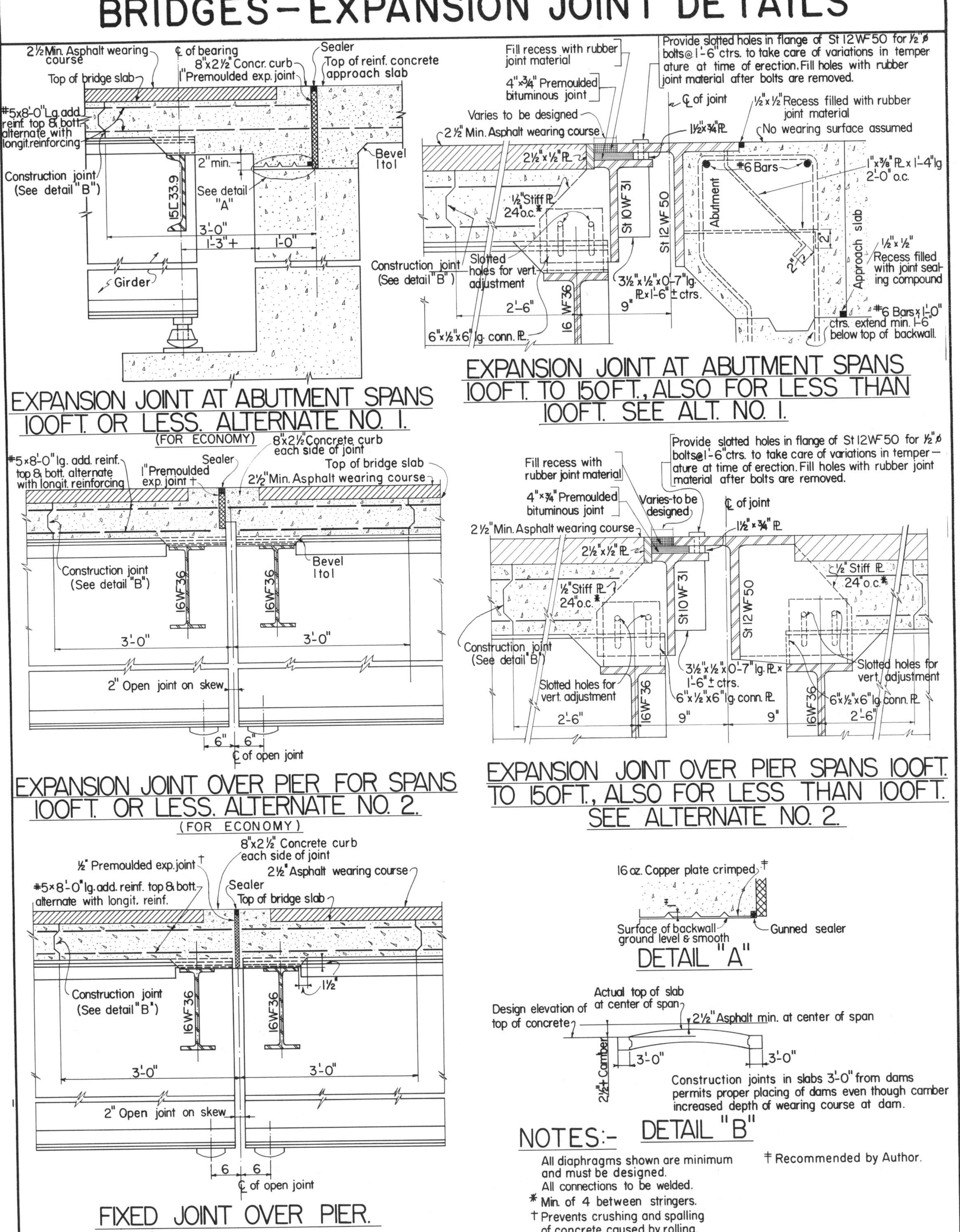

EXPANSION JOINT AT ABUTMENT SPANS 100FT. OR LESS. ALTERNATE NO. 1.
(FOR ECONOMY)
EXPANSION JOINT AT ABUTMENT SPANS 100FT. TO 150FT., ALSO FOR LESS THAN 100FT. SEE ALT. NO. 1.
EXPANSION JOINT OVER PIER FOR SPANS 100FT. OR LESS. ALTERNATE NO. 2.
(FOR ECONOMY)
EXPANSION JOINT OVER PIER SPANS 100FT. TO 150FT., ALSO FOR LESS THAN 100FT. SEE ALTERNATE NO. 2.
FIXED JOINT OVER PIER.
DETAIL "A"
16 oz. Copper plate crimped ‡
Surface of backwall ground level & smooth
Gunned sealer
DETAIL "B"
Construction joints in slabs 3'-0" from dams permits proper placing of dams even though camber increased depth of wearing course at dam.

NOTES:-

All diaphragms shown are minimum and must be designed.
All connections to be welded.
* Min. of 4 between stringers.
† Prevents crushing and spalling of concrete caused by rolling.

‡ Recommended by Author.

RAILROADS — GENERAL

TABLE A – SPUR AND SIDING DESIGN DATA

	RECOMMENDED MAXIMUM DEGREE OF CURVE
HORIZONTAL CURVES See pp. 12-26 to 12-33, 14-06 and 14-07 for curve data functions of 1-degree curve, etc.	*For Steam Locomotives* Recommended general practice up to 14° or 16° Road Engines 18° Switch engines 23° *For Diesel-Electric Locomotives* Switching engines including cars 100' - 150' radius; 60° to 39° Road switching engines 1750 hp. 150' - 200' radius; 39° to 23° "Lead" unit road 2400 hp. 274' raidus 21° *For Cars* Freight cars (normal) Maximum 30°, (special) 50° Passenger cars (normal) maximum 14°, (special) 50°
REVERSE CURVES	Provide a tangent distance between curve, preferably exceeding 100'.
SUPERELEVATION	Superelevation requirements on Table A, p. 14-05.
GRADES	Maximum for road engines : use 1½%. For Diesel-electric engines, use 2%. For unavoidable grades greater than these, consult the using railroad. Both steamers and Diesels, properly geared (mechanically, or electrically) for the service can and do operate on much steeper grades. On heavy trains, about 3% grades being the maximum for main-line service. Maximum 4% for siding (but undesirable).
VERTICAL CURVES	50' minimum length. Use 200' or preferably.
TURNOUTS Details of turnouts on Table A, p. 14-03.	If avoidable, do not locate turnouts on superelevated curves. Use #10 (minimum) turnouts in any main track. Turnouts in ladder tracks #8 (minimum). Turnouts in yards, or from spurs or sidings used by a road engine to be generally #8's, by a switch engine #6 (minimum) only if conditions require. Long cars often uncouple or jump track on #5's. #10 and #11 turnouts are being used on many classification yards now being built.
OVERHEAD AND SIDE CLEARANCES	For diagrams of clearances, see Fig. A p. 14-02. Not less than 16'-0" is necessary to clear tops of highest cars and locomotives. A chart "Legal Requirements – Clearances," revised 12-1-57, published by American Railway Engineering Association, shows clearance laws, rules, or regulations of the states of the United States, including the District of Columbia and Canada.
TIE SPACING	Use 21" if road engines are to be used. 2'-0" maximum.
TIE PLATES	Use on curves and on creosoted ties and on all ties on heavy-service track.
TRACK GAGE	4'-8½" on tangents and curves up to 8°. Add ⅛" per 2° over 8° up to maximum of 4'-9¼".

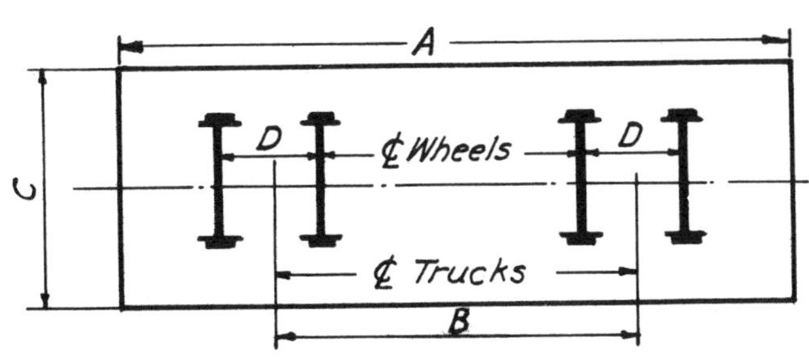
A
D
D
℄ Wheels
C
℄ Trucks
B

DIAGRAMMATIC CAR PLAN

See table for dimensions.

SOME TYPICAL CAR DIMENSIONS (As per Diagram)*

Types of Cars	A	B	C	D	Extreme Height
1. PC – Passenger Coach	82'-5"	59'-6"	10'-0"	9'-0"	13'-6"
2. M & B – Mail & baggage	71'-2"	48'-7"	10'-0¼"	9'-0"	13'-6"
3. MC – Milk	50'-1½"	34'-7"	10'-1½"	8'-0"	13'-2"
4. ASB – All steel box	51'-9½"	40'-9½"	10'-7¼"	5'-6"	15'-1"
5. ASB – All steel box	42'-3¾"	31'-2¾"	9'-11⅝"	5'-6"	15'-5¼"
6. A – Steel auto box	41'-9½"	30'-9½"	10'-7¼"	5'-6"	5'-0⅞"
7. 70-H – 70-Ton hopper	40'-5"	30'-5"	10'-3¼"	5'-8"	11'-4¼"
8. CH – Covered hopper	35'-1⅝"	25'-1⅝"	10'-5"	5'-8"	12'-11 5/16"
9. Ta – 12,500-gal. tank (40'-0⅞" over tank heads – inside diameter of tanks)	43'-2"	32'-2½"	10'-2"	5'-6"	14'-1½"
10. HSG – High side gondola	68'-1"	32'-2½"	9'-1½"	5'-8"	7'-7"
11. ASF – All steel flat	54'-3"	43'-3"	10'-4"	5'-8"	5'-1¼"
12. ASF – All steel flat	42'-10½"	31'-10½"	10'-1¾"	5'-6"	5'-7⅜"

FIG. B. CAR DIMENSIONS

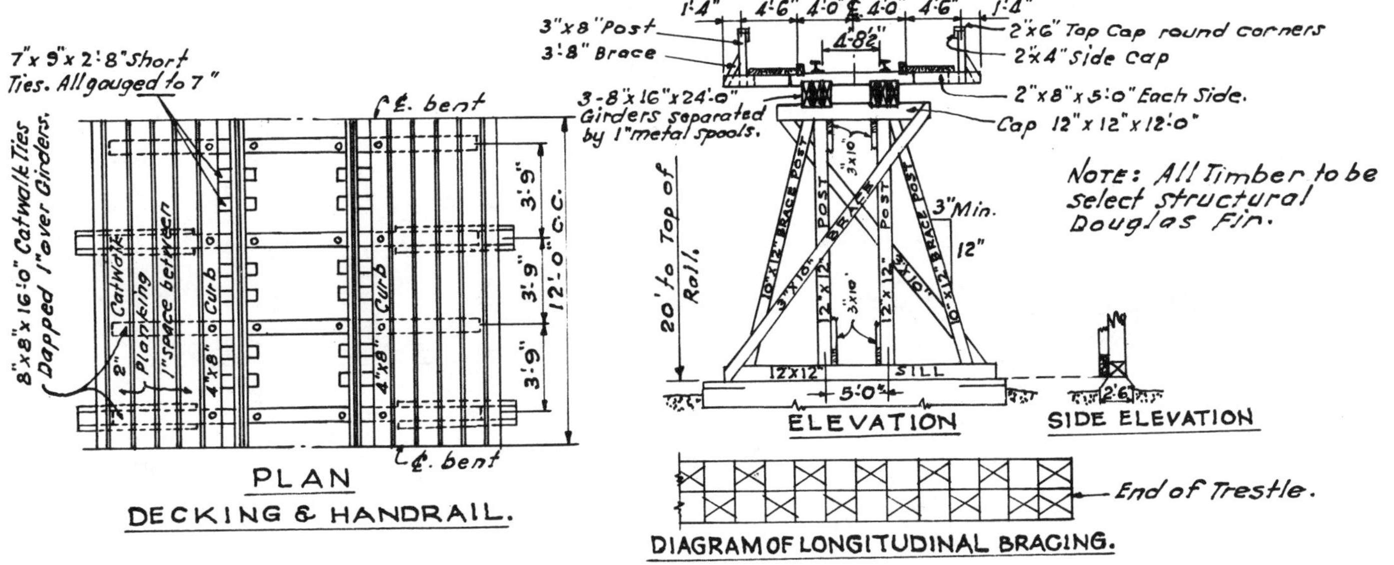
7"x 9"x 2'-8" Short Ties. All gouged to 7"
℄ bent
8"x8"x16'-0" Catwalk Ties Dapped 1" over Girders.
Catwalk
Planking
1" space between
2"
4"x8"
Curb
4"x8"
Curb
3'-9"
3'-9"
3'-9"
12'-0" C.C.
℄ bent
PLAN
DECKING & HANDRAIL.
1'-4"
4'-6"
4'-0"
4'-0"
4'-6"
1'-4"
4'-8½"
3"x8" Post
3'-8" Brace
2"x6" Top Cap round corners
2"x4" Side Cap
3-8"x16"x24'-0" Girders separated by 1" metal spools.
2"x8"x5'-0" Each Side.
Cap 12"x12"x12'-0"
3"x10"
10"x12" BRACE POST
3"x10"
12"x12" POST
BRACE
12"x12" POST
3"x10"
10"x12" BRACE POST
3"x10"
3" Min.
12"
20' to Top of Rail.
12x12
SILL
5'-0"
2'-6"
ELEVATION
SIDE ELEVATION
NOTE: All Timber to be Select Structural Douglas Fir.
End of Trestle.
DIAGRAM OF LONGITUDINAL BRACING.

FIG. C. COAL TRESTLE DETAILS

* From New York Central Rail Road.

RAILROADS – CLEARANCES

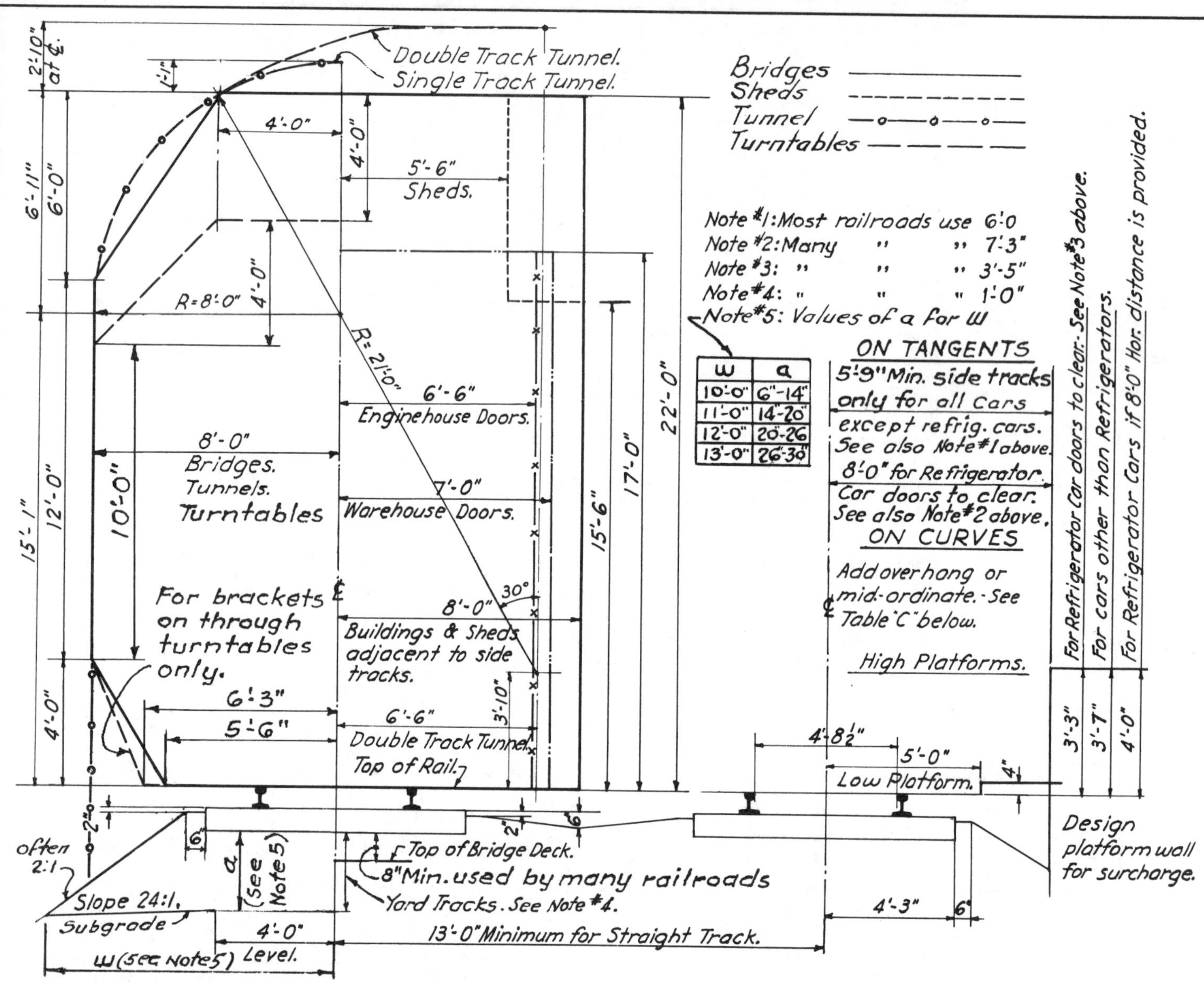

Double Track Tunnel.
Single Track Tunnel.
2'-0" at ℄
1'-1"
4'-0"
4'-0"
5'-6" Sheds.
6'-11"
6'-0"
R=8'-0"
4'-0"
R=21'-0"
6'-6" Enginehouse Doors.
8'-0" Bridges. Tunnels. Turntables
10'-0"
12'-0"
15'-1"
7'-0" Warehouse Doors.
℄
30°
8'-0" Buildings & Sheds adjacent to side tracks.
For brackets on through turntables only.
6'-3"
5'-6"
6'-6" Double Track Tunnel.
3'-10"
Top of Rail.
4'-0"
15'-6"
17'-0"
22'-0"
2'-0"
6"
2"
6"
often 2:1
Slope 24:1.
Subgrade
a (See Note 5)
Top of Bridge Deck.
8" Min. used by many railroads
Yard Tracks. See Note #4.
4'-0" Level.
w (see Note 5)
13'-0" Minimum for Straight Track.

Bridges ————
Sheds - - - - - -
Tunnel —o—o—o—
Turntables — — — —

Note #1: Most railroads use 6'-0
Note #2: Many " " 7'-3"
Note #3: " " " 3'-5"
Note #4: " " " 1'-0"
Note #5: Values of a for w

w	a
10'-0"	6"-14"
11'-0"	14"-20"
12'-0"	20"-26"
13'-0"	26"-30"

ON TANGENTS
5'-9" Min. side tracks only for all cars except refrig. cars. See also Note #1 above.
8'-0" for Refrigerator Car doors to clear. See also Note #2 above.
ON CURVES
Add overhang or mid-ordinate. - See Table "C" below.
℄
High Platforms.
For Refrigerator Car doors to clear. - See Note #3 above.
For cars other than Refrigerators.
For Refrigerator Cars if 8'-0" Hor. distance is provided.
3'-3"
3'-7"
4'-0"
4'-8½"
5'-0"
Low Platform.
4"
Design platform wall for surcharge.
4'-3"
6"

CLEARANCE & BALLAST SECTION. SIDE UNLOADING PLATFORMS.

FIG. A. TANGENT CLEARANCES.*

Note: Allow for curves as indicated in Table C

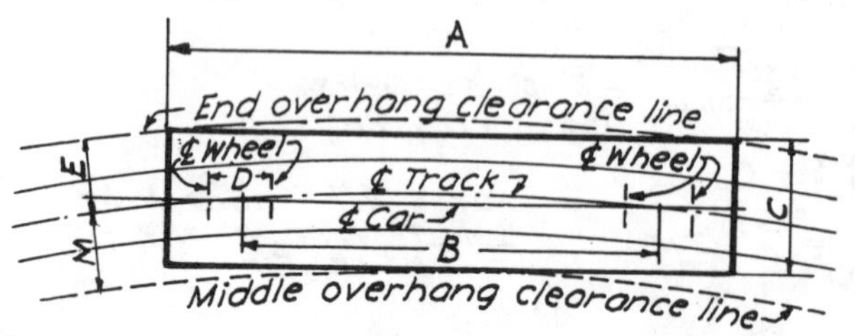

A
End overhang clearance line
℄ Wheel
D
℄ Track
℄ Wheel
℄ Car
B
E
M
C
Middle overhang clearance line

FIG. B - END & MIDDLE OVERHANG.

$$M = R - \sqrt{R^2 - \left[\frac{D^2}{4} + \frac{B^2}{4}\right]}$$

$$E = \sqrt{\left(R - M + \frac{C}{2}\right)^2 + \frac{A^2}{4}} - R$$

Use 14' high car for tilt clearances.

TABLE C – MIDDLE AND END ADDITIONAL "OVERHANGS" ON CURVES

Cars		Curves 2°	5°	9°	13°	18°	23°	30°
1 PC – Passenger coach	M	0.158	0.395	0.711	1.029	1.426	1.824	2.385
	E	5.137	5.441	5.612	5.876	6.201	6.633	6.952
2 M & B – Mail & baggage	M	0.107	0.266	0.480	0.693	0.960	1.228	1.605
	E	5.124	5.295	5.517	5.737	6.008	6.274	6.635
3 MC – Milk	M	0.053	0.133	0.238	0.344	0.477	0.609	0.795
	E	5.119	5.207	5.314	5.422	5.556	5.785	5.869
4 ASB – All steel box	M	0.074	0.185	0.333	0.481	0.666	0.852	1.112
	E	5.345	5.408	5.491	5.573	5.672	5.769	5.900
5 ASB – All steel box	M	0.044	0.110	0.198	0.286	0.396	0.506	0.660
	E	5.018	5.068	5.135	5.200	5.281	5.359	5.466
6 SAB – Steel auto box	M	0.043	0.107	0.192	0.278	0.384	0.491	0.641
	E	5.335	5.385	5.450	5.514	5.593	5.669	5.774
7 70-H – 70 Ton hopper	M	0.042	0.105	0.188	0.272	0.377	0.481	0.628
	E	5.165	5.208	5.266	5.322	5.390	5.458	5.549
8 CH – Covered hopper	M	0.029	0.073	0.131	0.188	0.261	0.333	0.435
	E	5.233	5.269	5.318	5.367	5.425	5.482	5.560
9 Ta – Tank car	M	0.047	0.117	0.210	0.303	0.420	0.536	0.701
	E	5.117	5.168	5.236	5.303	5.383	5.464	5.571
10 HSG – High side gondola	M	0.140	0.350	0.628	0.911	1.262	1.614	2.110
	E	4.624	4.716	4.838	4.953	5.097	5.257	5.424
11 ASF – All steel flat	M	0.083	0.208	0.374	0.540	0.748	0.957	1.249
	E	5.212	5.278	5.366	5.352	5.556	5.657	5.796
12 ASF – All steel flat	M	0.046	0.115	0.206	0.297	0.412	0.526	0.686
	E	5.107	5.157	5.225	5.291	5.371	5.451	5.559

*All data in accord with American Railway Engineers Association (A.R.E.A.) recommendation.

RAILROADS – TURNOUTS & CROSSOVERS

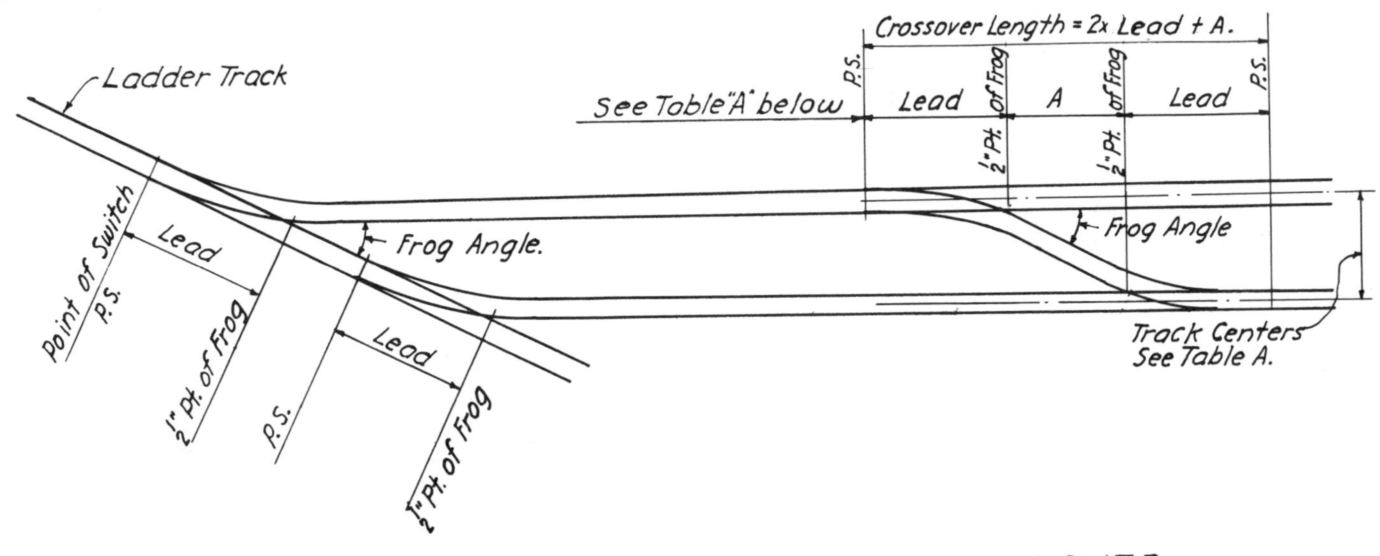
Ladder Track
Point of Switch P.S.
Lead
½" Pt. of Frog
Frog Angle.
P.S.
Lead
½" Pt. of Frog
Crossover Length = 2x Lead + A.
See Table "A" below
P.S.
Lead
½" Pt. of Frog
A
½" Pt. of Frog
Lead
P.S.
Frog Angle
Track Centers See Table A.

TURNOUT

CROSSOVER

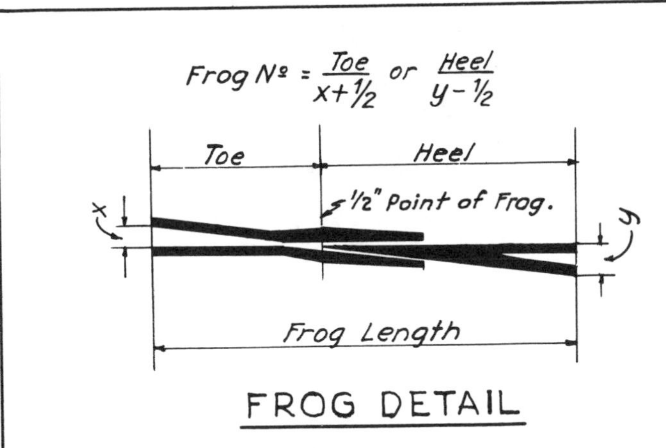
Frog № = Toe / (x+½) or Heel / (y−½)
Toe
Heel
½" Point of Frog.
x
y
Frog Length

FROG DETAIL

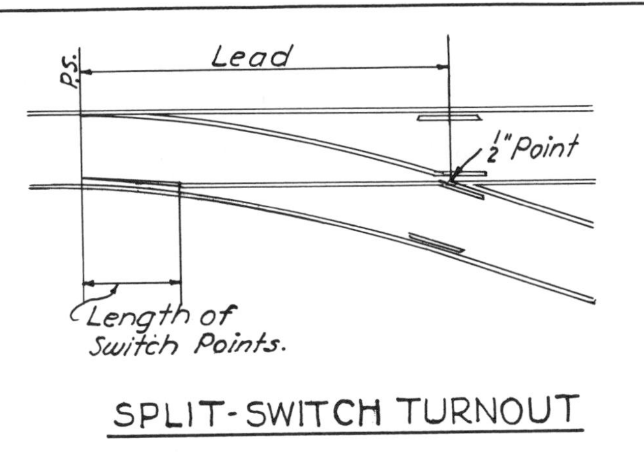
P.S.
Lead
½" Point
Length of Switch Points.

SPLIT-SWITCH TURNOUT

TABLE A – FROGS, SWITCHES AND CROSSOVERS

Straight Switch Turnout		Frog		Curved Switch Turnout		Crossovers Distances A Between ℄ of Parallel Tracks		Q†
Lead	Switch Point	No.	Angle	Switch Point	Lead	13'-0"	14'0"	
42'-6½"	11'-0"	5	11° 25' 16"	13'-0"	46'-6½"	16'-10 13/16"	21'-10¼"	0'-5 15/16"
47'-6"	11'-0"	6	9° 31' 39"	13'-0"	49'-9"	20'-5½"	26'-5"	0'-7⅛"
62'-1"	16'-6"	7	8° 10' 16"	13'-0"	54'-8½"	24'-0⅜"	30'-11 5/16"	0'-8⅜"
68'-0"	16'-6"	8	7° 09' 10"	13'-0"	58'-11½"	27'-7⅛"	35'-6¾"	0'-9 9/16"
72'-3½"	16'-6"	9	6° 21' 35"	19'-6"	74'-1¼"	31'-1⅝"	40'-1 5/16"	0'-10¾"
78'-9"	16'-6"	10	5° 43' 29"	19'-6"	78'-11"	34'-8½"	44'-7 15/16"	0'-11 15/16"
91'-10¼"	22'-0"	11	5° 12' 18"	19'-6"	83'-6"	38'-2½"	49'-2¼"	1'-1 3/16"
96'-8"	22'-0"	12	4° 46' 19"	19'-6"	87'-3½"	41'-8¾"	53'-8½"	1'-2⅜"
107'-0¾"	22'-0"	14	4° 05' 27"	26'-6"	108'-7½"	48'-9¼"	62'-9 1/16"	1'-4 13/16"
126'-4½"	30'-0"	15	3° 49' 06"	26'-6"	113'-5"	52'-3 7/16"	67'-2⅝"	1'-6"
131'-4"	30'-0"	16	3° 34' 47"	26'-6"	118'-5"	55'-9⅜"	71'-11 3/16"	1'-7 3/16"
140'-11½"	30'-0"	18	3° 10' 56"	39'-0"	147'-0½"	62'-9⅞"	80'-9¾"	1'-9 9/16"
151'-11½"	30'-0"	20	2° 51' 21"	39'-0"	156'-0½"	69'-10"	89'-9⅞"	2'-0"
		20		39'-0"	156'-0½"	69'-10"	89'-9⅞"	2'-0"
		20		45'-0"	160'-9 1/10"	69'-9 15/16"	89'-9 3/16"	2'-0"
		20		30'-0"	153'-5½"	–	89'-9 3/16"	2'-0"
		30		45'-0"	211'-8¾"	–	–	–

† Figures in column Q give the amounts to add to tabular figures for every tenth of a foot increase in distance between tracks.

RAILROADS — GRADE CROSSINGS

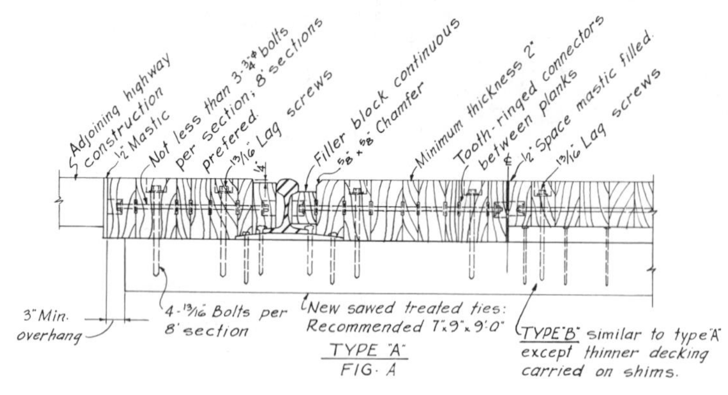

Adjoining highway construction
1/2" Mastic
Not less than 3-3/4" bolts per section; 8' sections prefered
13/16 Lag screws
Filler block continuous 5/8" x 5/8" Chamfer
Minimum thickness 2"
Tooth-ringed connectors between planks
1/2" Space mastic filled
13/16 Lag screws
3" Min. overhang
4-13/16 Bolts per 8' section
New sawed treated ties: Recommended 7"x9"x9'0"
TYPE "B" similar to type "A" except thinner decking carried on shims.

TYPE "A"
FIG. A

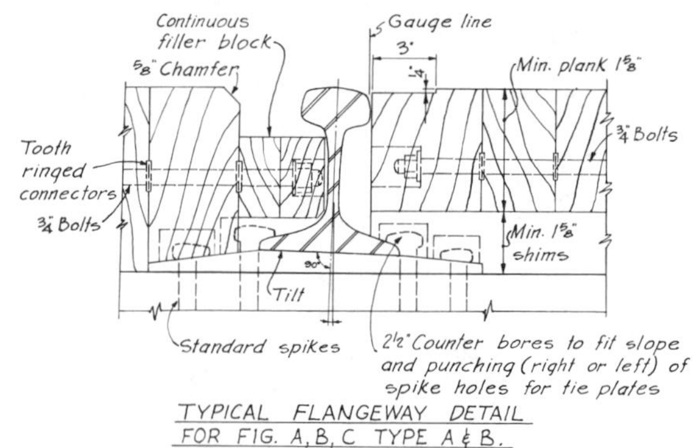

Continuous filler block
5/8" Chamfer
Gauge line
3"
Min. plank 1 5/8"
Tooth ringed connectors
3/4" Bolts
3/4" Bolts
Min. 1 5/8" shims
Tilt
Standard spikes
2 1/2" Counter bores to fit slope and punching (right or left) of spike holes for tie plates

TYPICAL FLANGEWAY DETAIL
FOR FIG. A, B, C TYPE A & B.

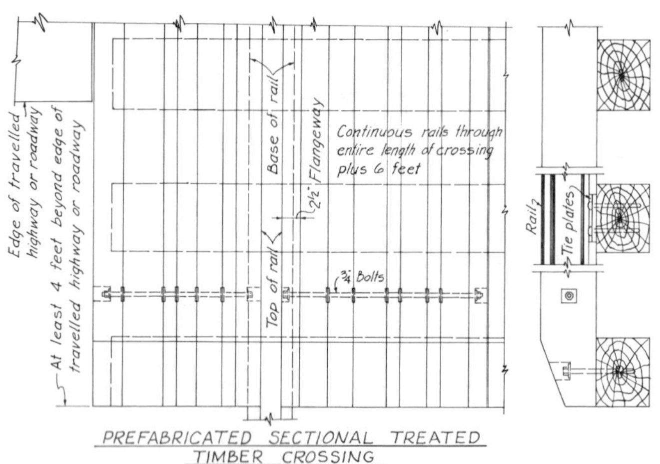

Edge of travelled highway or roadway
At least 4 feet beyond edge of travelled highway or roadway
Base of rail
Top of rail
2 1/2" Flangeway
Continuous rails through entire length of crossing plus 6 feet
3/4" Bolts
Rail
Tie plates

PREFABRICATED SECTIONAL TREATED TIMBER CROSSING

TABLE GRADE CROSSING CLASSIFICATION

CLASS OF HIGHWAY TRAFFIC	R.R. SPEED	BITUMINOUS	WOOD PLANK	PREFAB-RICATED TIMBER	PRECAST CONCRETE	MONOLITHIC CONCRETE
NUMBER		1	2	3	4	5
IMPROVED HEAVY	HS			✓	✓	
	LS	✓		✓	✓	✓
IMPROVED MEDIUM	HS	✓		✓	✓	
	LS	✓		✓	✓	✓
IMPROVED LIGHT	HS		✓	✓	✓	
	LS		✓	✓	✓	✓
UNIMPROVED LIGHT	HS	✓	✓			
	LS	✓	✓			

Numbers 2, 3 & 4 each comprise Types A & B, and Nos. 1 & 5 are single. All are drawn from AREA's specifications and charted to illustrate AREA's recommendations for their use on different type crossings. H.S. = High speed; L.S. = Low speeds.

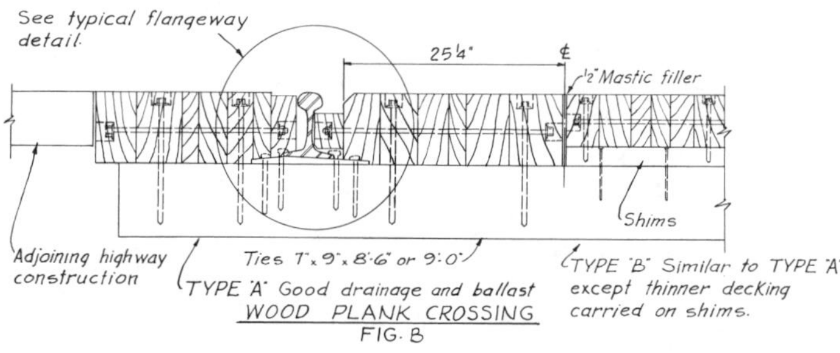

See typical flangeway detail
25 1/4"
℄
1/2" Mastic filler
Shims
Adjoining highway construction
Ties 7"x9"x8'6" or 9'0"
TYPE "B" Similar to TYPE "A" except thinner decking carried on shims.

TYPE "A" Good drainage and ballast
WOOD PLANK CROSSING
FIG. B

All surfaces with seal coat of sand, fine crushed stone and bitumen.

Wearing surface: 1 cu ft 5/8 to 1" stone to 1 gal. bitumen evenly rolled to top of rail

Adjoining highway

Paint

Rock asphalt 2 1/4" rough rolled evenly to top of rail

Paint

2 1/2"

New ballast to top of ties

Ties 7"x9"

Ballast

Bituminous mixture: 1 cu. ft clean graded stone 3/4"-3" to 1/2"-2/3 gal bitumen.

BITUMINOUS CROSSING
FIG. C

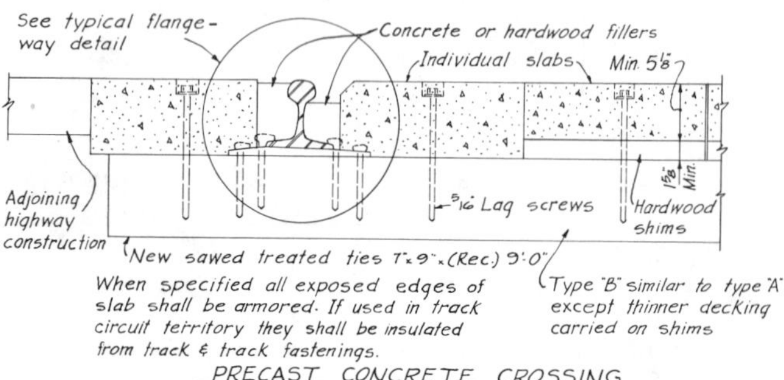

See typical flange-way detail
Concrete or hardwood fillers
Individual slabs
Min. 5 5/8"
1 5/8" Min.
Adjoining highway construction
5/16 Lag screws
Hardwood shims
New sawed treated ties 7"x9"x(Rec.) 9'0"
Type "B" similar to type "A" except thinner decking carried on shims

When specified all exposed edges of slab shall be armored. If used in track circuit territory they shall be insulated from track & track fastenings.

PRECAST CONCRETE CROSSING

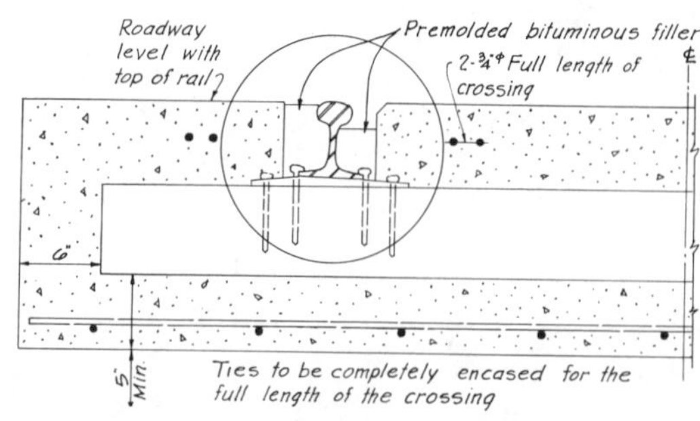

Roadway level with top of rail
Premolded bituminous filler
2-3/4" Full length of crossing
℄
6"
5" Min.
Ties to be completely encased for the full length of the crossing

Track to be supported on brick, stone, etc.
Track to be carefully levelled.

MONOLITHIC CONCRETE CROSSING

LOADING AND DESIGN BASIS FOR PRECAST CONCRETE CROSSING

AA of SHO. H-15 Loading for truck train with max. axle load @ 24,000#, or with H-20 similarly loaded @ 32,000#. Units designed for concentrated wheel loads of 1/2 of said axle loads, placed for max. stresses in the direction of traffic equal to width of slab with max. of 17" - no distribution perpendicular to traffic; with 50% impact in both moment and shear at stresses not greater than 2/3 of the elastic limit for reinf., and 1/2 of the ultimate strength of concrete @ 28 days. Slabs supported on 3 or more ties shall be designed as above, with one intermediate tie not in bearing. Covering of reinf shall be not less than 3/4". Clearances shall be provieded for tieplates and spike heads.

RAILROADS — SUPERELEVATION & ACCESSORIES

TABLE A - SUPERELEVATION OF OUTER RAIL IN INCHES: *(Elevation of Outer rail above Inner rail.) Standard Guage* ✱
NOTE: *For uniform speed traffic use figures as shown. For mixed speed traffic, superelevation is 3" less than table shows; no superelevation where table shows less than 3".*

Deg. of curve	Velocity in miles per hour											
	15	20	25	30	35	40	45	50	55	60	65	70
1	1/8	1/4	3/8	5/8	7/8	1-1/8	1-3/8	1-3/4	2-1/8	2-1/2	2-7/8	3-3/8
2	1/4	1/2	7/8	1-1/4	1-3/4	2-1/4	2-3/4	3-1/2	4-1/8	5	5-7/8	6-3/4
3	1/2	7/8	1-1/4	1-7/8	2-1/2	3-1/4	4-1/4	5-1/8	6-1/4	7-1/2	8-3/4	10-1/8
4	5/8	1-1/8	1-3/4	2-1/2	3-3/8	4-3/8	5-5/8	6-7/8	8-3/8	10	.....	.....
5	3/4	1-3/8	2-1/8	3-1/8	4-1/4	5-1/2	7	8-5/8	.....	.....	.....	.....
6	7/8	1-5/8	2-5/8	3-3/4	5-1/8	6-5/8	8-3/8	.....	.....	.....	.....	.....
7	1-1/8	1-7/8	3	4-3/8	5-7/8	7-3/4	.....	.....	.....	.....	.....	.....
8	1-1/4	2-1/4	3-1/2	5	6-3/4	8-7/8	.....	.....	.....	.....	.....	.....
9	1-3/8	2-1/2	3-7/8	5-5/8	7-5/8	.....	.....	.....	.....	.....	.....	.....
10	1-1/2	2-3/4	4-1/4	6-1/4	8-1/2	.....	.....	.....	.....	.....	.....	.....
11	1-3/4	3	4-3/4	6-7/8								
12	1-7/8	3-1/4	5-1/8	7-1/2	.....	.....	.....	.....	.....	.....	.....	.....
13	2	3-5/8	5-5/8	8-1/8	.....	.....	.....	.....	.....	.....	.....	.....
14	2-1/8	3-7/8	6	8-3/4	.....	.....	.....	.....	.....	.....	.....	.....
15	2-3/8	4-1/8	6-1/2	9-1/4	.....	.....	.....	.....	.....	.....	.....	.....

For mixed speed traffic, subtract 3"; no superelevation where less than 3" shown.

FIG. B - "HAYES" WHEEL STOP. ✱✱

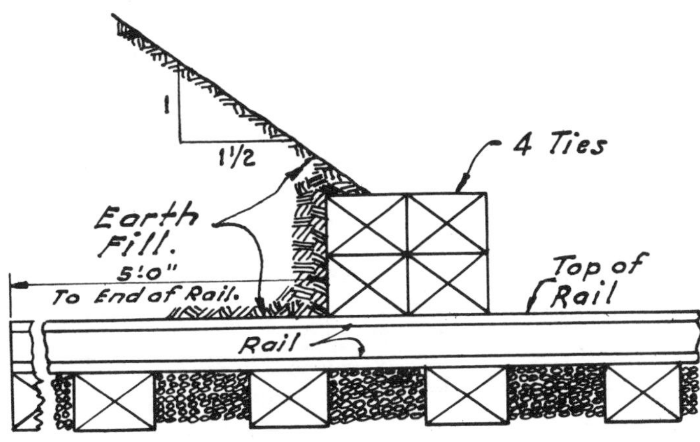
1
1 1/2
4 Ties
Earth Fill.
5'0"
To End of Rail.
Top of Rail
Rail

FIG. C - DETAIL OF TYPICAL CAR BUMPER.

FIG. D - "HAYES" SLIDING TYPE DERAIL, WITH OPERATING STAND ✱✱

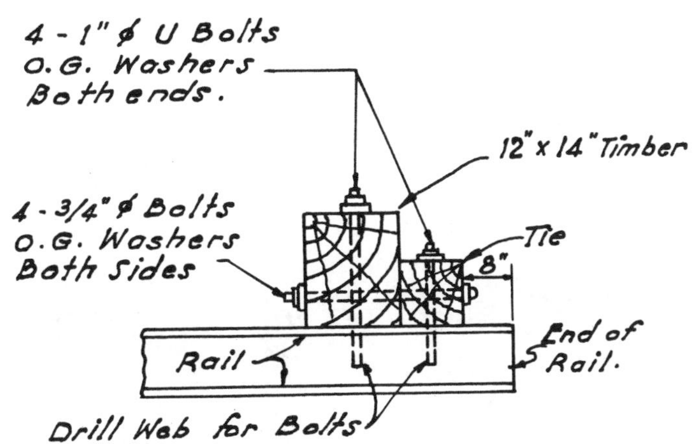
4 - 1" ø U Bolts
O.G. Washers
Both ends.
12" x 14" Timber
4 - 3/4" ø Bolts
O.G. Washers
Both Sides
Tie
8"
Rail
End of Rail.
Drill Web for Bolts

FIG. E - TIMBER WHEEL STOP.

FIG. F - "DURABLE" BUMPER POST ✱✱✱

✱ *Adapted from Amer. Civil Engineers' Handbook by Merriman & Wiggin.*
✱✱ *Adapted from Hayes Track Appliance Co., Richmond, Indiana.*
✱✱✱ *Adapted from, A. B. Letterman Co., Chicago, Ill.*

RAILROADS — CIRCULAR CURVES

CHORD DEFINITION (R. R. CURVE).

D (in degrees) Subtends 100' Chord.

$D = 100\ \Delta / L$

$D = \dfrac{\text{Tan. } 1^\circ \text{ curve}}{T}$ (approx).

$D = \dfrac{\text{Ext. } 1^\circ \text{ curve}}{E}$ (approx.).

Tan Offset $= \dfrac{\text{Chord}^2}{2R}$ = Chord · sin def.

$= \left(\dfrac{\text{Chord}^2}{100}\right)$ Tan Offset Table A.

Chord Offset = 2 Tan Defl. for 100' Chord = 100 sin D°

Tan Def = ½ D $\dfrac{\text{Chord}}{100}$; For "c" feet = 0.3D×c = Def. for 1' in Table A × c.

Chord Def. = 2 Tan Def. = D for 100' Chord.

FORMULAS:

$R = \dfrac{50}{\sin D/2}$; $R = T \cdot \cot \dfrac{\Delta}{2}$; $R = \dfrac{E}{\text{exsec } \Delta/2}$

$T = R \cdot \tan \dfrac{\Delta}{2}$; $T = \dfrac{50 \tan \Delta/2}{\sin \Delta/2}$; $T = \dfrac{\tan 1^\circ \text{ curve}}{D}$ + Corr.*

$L = 100 \dfrac{\Delta}{D}$; $\Delta = \dfrac{DL}{100}$; $M = R(1 - \cos \dfrac{\Delta}{2})$; $M = R \text{ vers } \dfrac{\Delta}{2}$

$E = T \cdot \tan \dfrac{\Delta}{4}$; $E = \dfrac{R}{\cos \Delta/2} - R$; $E = R \cdot \text{exsec } \dfrac{\Delta}{2}$.

$C = 2R \cdot \sin \dfrac{\Delta}{2}$; $E = \dfrac{\text{ext. } 1^\circ \text{ Curve}}{D}$ + Correction.*

$\sin D/2 = \dfrac{50}{R}$; $\sin D/2 = \dfrac{50 \tan \Delta/2}{T}$

Example: Given: Δ = 54° 20'; D = 7°40'
P.I. Sta. 125+39.88.

Required: R, T, L, P.C. & P.T.

Solution: R = 50 ÷ Sin 3°50' = 747.89.

T = 747.89 (tan 27°10') = 383.81.

L = 100 Δ ÷ D = 100 (54° 20') ÷ 7°40' = 708.70.

P.C. = P.I. Sta. 125+39.88 − 383.81 = Sta. 121+56.07

P.T. = Sta. 121+56.07 + 708.70 = Sta. 128+64.77.

* See page 14-07.

TABLE A - RADII, DEFLECTIONS, OFFSETS, ORDINATES, CHORDS & ARCS - 100' CHORD.*

D	RADIUS	DEF.' FOR 1 FOOT	TAN OFF-SET.	MID ORD.	FOR SUB-CHORDS ADD				ACTUAL ARC PER 100' STA.	LONG CHORDS				D
					10'	20'	25'	50'		2 STA.	3 STA.	4 STA.	5 STA.	
30'	11,459.2	0.15	0.436	0.109					100.000	200.000	299.99	399.98	499.96	30'
1°	5,729.65	0.30	0.873	0.218					100.001	199.99	299.97	399.92	499.85	1°
30'	3,819.83	0.45	1.309	0.327					100.003	199.98	299.93	399.83	499.66	30'
2°	2,864.93	0.60	1.745	0.436					100.005	199.97	299.88	399.70	499.39	2°
30'	2,292.01	0.75	2.181	0.545					100.008	199.95	299.81	399.52	499.05	30'
3°	1,910.08	0.90	2.618	0.654					100.011	199.93	299.73	399.32	498.63	3°
30'	1,637.28	1.05	3.054	0.764					100.015	199.91	299.63	399.07	498.14	30'
4°	1,432.69	1.20	3.490	0.872				0.01	100.020	199.88	299.51	398.78	497.57	4°
30'	1,273.57	1.35	3.926	0.982				0.01	100.026	199.85	299.38	398.46	496.92	30'
5°	1,146.28	1.50	4.362	1.091			0.01	0.01	100.032	199.81	299.24	398.10	496.20	5°
30'	1,042.14	1.65	4.798	1.200			0.01	0.01	100.038	199.77	299.08	397.70	495.41	30'
6°	955.37	1.80	5.234	1.309		0.01	0.01	0.02	100.046	199.73	298.90	397.26	494.53	6°
30'	881.95	1.95	5.669	1.418		0.01	0.01	0.02	100.054	199.68	298.71	396.79	493.59	30'
7°	819.02	2.10	6.105	1.528		0.01	0.01	0.02	100.062	199.63	298.51	396.28	492.57	7°
8°	716.78	2.40	6.976	1.746	0.01	0.02	0.02	0.03	100.081	199.51	298.05	395.14	490.31	8°
10°	573.69	3.00	8.716	2.183	0.01	0.02	0.03	0.05	100.127	199.24	296.96	392.42	484.90	10°
12°	478.34	3.60	10.45	2.620	0.02	0.04	0.04	0.07	100.183	198.90	295.63	389.12	478.34	12°
14°	410.28	4.20	12.18	3.058	0.02	0.05	0.06	0.09	100.249	198.51	294.06	385.23	470.65	14°
16°	359.27	4.80	13.92	3.496	0.03	0.06	0.08	0.12	100.326	198.05	292.25	380.76	461.86	16°
18°	319.62	5.40	15.64	3.935	0.04	0.08	0.10	0.15	100.412	197.54	290.21	375.74	452.02	18°
20°	287.94	6.00	17.37	4.374	0.05	0.10	0.12	0.19	100.510	196.96	287.94	370.17	441.15	20°
22°	262.04	6.60	19.08	4.814	0.06	0.12	0.14	0.23	100.617	196.33	285.44	364.06	429.31	22°
24°	240.49	7.20	20.79	5.255	0.07	0.14	0.17	0.28	100.735	195.63	282.71	357.43	416.54	24°
30°	193.18	9.00	25.88	6.583	0.11	0.22	0.29	0.43	101.152	193.19	273.21	334.61	373.21	30°

USE OF TABLES A & B

GIVEN	REQUIRED.	SOLUTION.
D = 2°30'	Deflection for 35 ft.	= 0.75 × 35 = 26.25 = 26' 15"
D = 4°	Tan Offset for 125 ft.	$= 3.49 (\frac{125}{100})^2$ = 5.45 ft.
D = 10°	Mid Ord. for 30 ft. chord	$= 0.0001 \times 30^2 \times 2.183$ = 0.196 ft.
D = 14°	Length of nominal 20 ft. sub chord.	= 20 + 0.05 = 20.05 ft.
D = 20°	Actual length of arc for L = 600 ft. (6 Sta.)	= 100.51 × 6 = 603.06 ft.
D = 3°	Long chord for 3 Sta.	= From Table A = 299.73 ft.
Δ = 27°05'11"	Δ in decimals of °	From Table B = 27 + 0.0833 + 11 × 0.000278 = 27.086°

TABLE B - MINUTES IN DECIMALS OF A DEGREE. SECONDS IN DECIMALS OF A MINUTE.*

1	0.0167	11	0.1833	21	0.3500	31	0.5167	41	0.6833	51	0.8500
2	0.0333	12	0.2000	22	0.3667	32	0.5333	42	0.7000	52	0.8667
3	0.0500	13	0.2167	23	0.3833	33	0.5500	43	0.7167	53	0.8833
4	0.0667	14	0.2333	24	0.4000	34	0.5667	44	0.7333	54	0.9000
5	0.0833	15	0.2500	25	0.4167	35	0.5833	45	0.7500	55	0.9167
6	0.1000	16	0.2667	26	0.4333	36	0.6000	46	0.7667	56	0.9333
7	0.1167	17	0.2833	27	0.4500	37	0.6167	47	0.7833	57	0.9500
8	0.1333	18	0.3000	28	0.4667	38	0.6333	48	0.8000	58	0.9667
9	0.1500	19	0.3167	29	0.4833	39	0.6500	49	0.8167	59	0.9833
10	0.1667	20	0.3333	30	0.5000	40	0.6667	50	0.8333	60	1.0000

Proportional Part for 1" = 0.000278 of 1°

*Adapted from Railroad Curve Tables by Eugene Dietzgen Co., Ref. Field Engineering by Searles & Ives

RAILROADS — CORRECTIONS FOR TANGENTS AND EXTERNALS

For railroad and highway curves laid out by the chord definition these corrections are to be added to the values found, using tables on page 12-27 in order to obtain the corrected tangents and external distances.

For Tangents Add *

Central Angle	Degree of Curve													
	5°	10°	15°	20°	25°	30°	35°	40°	45°	50°	55°	60°	65°	70°
10°	.03	.06	.09	.13	.16	.19	.22	.25	.28	.31	.34	.38	.42	.46
15°	.04	.10	.14	.19	.24	.29	.34	.39	.45	.51	.53	.58	.63	.68
20°	.06	.13	.19	.26	.32	.39	.45	.51	.58	.65	.72	.79	.84	.90
25°	.08	.16	.24	.33	.40	.49	.58	.67	.75	.83	.90	.99	1.06	1.14
30°	.10	.19	.29	.39	.49	.59	.69	.79	.89	.99	1.09	1.20	1.29	1.39
35°	.11	.22	.34	.47	.58	.69	.70	.81	.92	1.04	1.29	1.42	1.54	1.66
40°	.13	.26	.40	.53	.67	.80	.93	1.06	1.20	1.34	1.49	1.64	1.79	1.94
45°	.15	.30	.44	.60	.76	.91	1.06	1.21	1.37	1.52	1.70	1.87	2.04	2.21
50°	.17	.34	.51	.68	.85	1.02	1.19	1.36	1.54	1.72	1.91	2.10	2.29	2.48
55°	.19	.38	.57	.76	.95	1.14	1.32	1.52	1.72	1.92	2.14	2.35	2.56	2.77
60°	.21	.42	.63	.84	1.05	1.27	1.49	1.71	1.94	2.17	2.38	2.60	2.83	3.07
65°	.23	.46	.69	.93	1.16	1.40	1.64	1.88	2.13	2.38	2.63	2.88	3.13	3.39
70°	.25	.51	.76	1.02	1.28	1.54	1.80	2.06	2.33	2.60	2.88	3.16	3.44	3.72
75°	.27	.56	.83	1.12	1.40	1.69	1.98	2.27	2.57	2.87	3.16	3.47	3.78	4.09
80°	.30	.61	.91	1.22	1.53	1.84	2.15	2.46	2.78	3.10	3.44	3.78	4.12	4.46
85°	.33	.66	1.00	1.33	1.68	2.02	2.36	2.70	3.05	3.40	3.77	4.14	4.55	4.89
90°	.36	.72	1.09	1.45	1.83	2.20	2.57	2.94	3.32	3.70	4.10	4.50	4.91	5.32
95°	.39	.79	1.19	1.55	2.00	2.40	2.80	3.20	3.61	4.02	4.40	4.98	5.38	5.83
100°	.43	.86	1.30	1.74	2.18	2.62	3.06	3.50	3.95	4.40	4.88	5.37	5.85	6.34
110°	.51	1.03	1.56	2.08	2.61	3.14	3.67	4.21	4.76	5.31	5.86	6.43	7.01	7.60
120°	.62	1.25	1.93	2.52	3.16	3.81	4.45	5.11	5.77	6.44	7.12	7.80	8.50	9.22

For Externals Add *

Central Angle.	Degree of Curve													
	5°	10°	15°	20°	25°	30°	35°	40°	45°	50°	55°	60°	65°	70°
10°	.001	.003	.004	.006	.007	.008	.009	.011	.012	.014	.015	.017	.018	.020
15°	.003	.007	.010	.014	.018	.023	.027	.029	.032	.035	.039	.043	.047	.051
20°	.006	.011	.017	.022	.028	.034	.038	.045	.051	.057	.063	.070	.076	.083
25°	.009	.018	.027	.036	.046	.056	.065	.074	.083	.093	.106	.120	.127	.135
30°	.013	.025	.038	.051	.065	.078	.090	.103	.116	.129	.149	.170	.179	.188
35°	.018	.035	.054	.072	.086	.109	.131	.153	.175	.197	.213	.230	.247	.264
40°	.023	.046	.070	.093	.117	.141	.172	.203	.234	.265	.277	.290	.315	.341
45°	.030	.060	.093	.119	.153	.184	.216	.254	.289	.325	.351	.378	.411	.445
50°	.037	.075	.116	.151	.189	.227	.266	.305	.345	.384	.425	.467	.508	.550
55°	.046	.093	.142	.188	.236	.283	.332	.381	.420	.479	.530	.582	.641	.700
60°	.056	.112	.168	.225	.283	.340	.398	.457	.516	.575	.636	.697	.774	.851
65°	.067	.135	.204	.273	.343	.412	.483	.554	.625	.697	.711	.845	.922	1.01
70°	.080	.159	.240	.321	.403	.485	.568	.652	.735	.819	.906	.994	1.08	1.17
75°	.095	.182	.286	.383	.480	.578	.678	.777	.877	.977	1.07	1.18	1.29	1.39
80°	.110	.220	.332	.445	.558	.671	.787	.903	1.02	1.13	1.25	1.38	1.50	1.62
85°	.128	.259	.391	.524	.657	.790	.926	1.06	1.20	1.34	1.47	1.62	1.76	1.91
90°	.149	.299	.450	.603	.756	.910	1.07	1.22	1.38	1.54	1.70	1.87	2.03	2.20
95°	.174	.350	.522	.706	.985	1.06	1.25	1.43	1.62	1.80	1.99	2.18	2.38	2.58
100°	.200	.401	.604	.809	1.01	1.22	1.43	1.64	1.85	2.06	2.28	2.50	2.73	2.96
110°	.268	.536	.806	1.08	1.35	1.63	1.91	2.20	2.48	2.76	3.05	3.35	3.66	3.96
120°	.360	.721	1.08	1.45	1.82	2.19	2.57	2.95	3.33	3.72	4.11	4.50	4.91	5.32

* Adapted from: Dietzgen's Railroad Curve Tables by Eugene Dietzgen Co.

NOTE:

AS OF JANUARY 1, 1959 THE CIVIL AERONAUTICS ADMINISTRATION (C.A.A.) WAS INCORPORATED IN THE FEDERAL AVIATION AGENCY (F.A.A.), ALL DESIGN DATA OF THE C.A.A. IS NOW UNDER REVIEW BY THE F.A.A.

AIRPORTS - GENERAL* - 1

FACTORS IN PLANNING

EXISTING AIRPORT FACILITIES	Examine and evaluate existing facilities in area, and determine to what extent they can be economically used for present and future requirements of community or area.
	COMMUNITY SIZE† AND AIRPORT TYPES
SMALL – UP TO 25,000 POPULATION	1. *LOCAL* Airports to serve on local service routes providing service in the Short Haul category normally not exceeding 500 miles.‡
MEDIUM – 25,000 TO 250,000 POPULATION	2. *TRUNK* Airports to serve on airline trunk routes and engage in intermediate-length hauls normally not exceeding 1000 miles.‡
LARGE – OVER 250,000 POPULATION	3. *CONTINENTAL*§ Airports serving long non-stop flights, exclusive of coast-to-coast, normally entirely within the confines of the continental United States. These airports serve non-stop flights up to 2000 miles.‡
GLOBAL CENTERS LONDON, PARIS, NEW YORK, SAN FRANCISCO, WASHINGTON, D.C., ETC.	4. *INTERCONTINENTAL*§ Airports to serve the longest range non-stop flights in the Transcontiental, Transoceanic, and Intercontinental categories.
TYPES OF SERVICES	(a) *SCHEDULED* Passengers, express, slow cargo, airmail, and aerial taxi service. (b) *NON-SCHEDULED* Private and business flying, air training schools, sales and service of planes and parts, aerial taxi, crop dusting, fire protection and conservation, tourist and sport travel, aerial photography, industrial freight, helicopters, air patrol by C.A.A. and others, glider activities, and military. Separation of scheduled, nonscheduled, and military flying is desirable when it is economically feasible and when volume of operations is high.
PLANE MOVEMENTS PER HOUR (p.m.h.)	The peak number of plane movements anticipated, present, and future should be estimated in order to determine the maximum number of runways required for use at one time. Landings require 2 to 3 times the time needed for take-offs. Visual (VFR) operation capacity – one runway 45 to 60 p.m.h. Instrument (IFR) operation capacity – one runway 20 to 40 p.m.h. With the addition of a parallel runway or other runways which will permit noninterfering operation, the estimated capacity can be approximately doubled and can attain a closer balance between volume of landings and take-offs. See p. 15-10. To achieve maximum capacity instrument take-offs, parallel runways should be separated by 3000 to 5000 ft. However, the type of aircraft operation, technical equipment available, and topographical characteristics will govern the extent of separation. Technical advances will bring instrument operation rates closer to visual operation rates.

* Adapted from C.A.A., *Airport Design*, January 1949, as modified by T.S.O. N6b, October 3, 1958.

† An isolated community may best be served by one airport; several communities not too far apart may jointly support a larger airport with increased services. A community that is a trading center for a large area may require a larger airport than would be indicated by its population. Studies on retail sales will reveal this factor. Other service requirements and factors given above will necessitate additional modifications.

‡ Determined by economics of aircraft operation and passenger traffic.

§ Commercial jet transport operations.

AIRPORTS - GENERAL* - 2

FACTORS IN SITE SELECTION	
PROXIMITY TO OTHER AIRPORTS	Avoid interference with traffic patterns of other airports. Radii of airport patterns are as follows: Local, 1 mile to 2 miles; Trunk, 3 miles; Continental and larger, 4 miles. Instrument operation requires greater separation. Study traffic patterns and approach procedures in the area.
FUTURE EXPANSION	Sufficient land available to buy on option to allow for economical development of master plan. See pp. 15-20 to 15-22.
APPROACH CONDITIONS	Safety and dependability of operations are seriously affected by terrain or structures in aerial approaches. Thus, for airports requiring instrument weather operations, approach conditions become the most important single factor. See pp. 15-06 and 15-07.
OBSTRUCTIONS	Practicability of removal of high buildings, stacks, power lines etc. should be determined. Legal zoning steps should be taken to protect approach zones against future erection of obstructions. See approach standards, pp. 15-06 and 15-07.
VISIBILITY	Avoid known areas of frequent fogs and smoke, often found along bodies of water near industrial areas. Avoid lee side from prevailing winds.
WINDS	Sites with prevailing winds from one or two directions may require runways in only one or two directions, thus reducing cost and taxiing time. See p. 15-14.
GROUND TRANSPORTATION	Site should be served by adequate and rapid ground transportation for passenger, freight, employees, and visitors. Generally an express highway is required for the larger airports. Speed, allowable traffic volume, cost, and distance from city are factors.
TOPOGRAPHY AND DRAINAGE	Balanced costs of grading and drainage should be attained. Flat land may require excessive drainage. Hilly land requires excessive grading. Investigate floods at valey sites; destructive winds at elevated or exposed sites. Look for sites slightly above level of flat country or on long easy slopes in hilly country. Avoid sites with high ground-water level for which the necessary subdrainage will be costly.
SOILS	See p. 9-19 for classification. E1 to E5 soils desirable. E5 to E12 usually require substantial base and subbase courses and may require expensive surface drainage systems. E13 is not suitable for subgrade.
CONSTRUCTION MATERIALS	Short hauls to deposits of E1 to E5 soils or stone quarries is important. Availability of bituminous plants, concrete plants, etc. may be important.
ECONOMICS	Cost of land, construction, utilities, and operation against the revenue that the airport will bring, should be studied.
RUNWAY LENGTHS	Increase basic length due to elevation above sea level and effective runway gradient. See p. 15-05.
UTILITIES	Study cost of providing sewer, water, gas, electricity, gasoline, telephone, telegraph etc.

*Adapted from C.A.A. "*Airport Design*", January 1949.

AIRPORTS - GENERAL* - 3

FACTORS IN PREPARING A MASTER PLAN	
PRELIMINARY SURVEY	Prepare property, location, topography, existing airway, obstruction maps. Secure data for Wind Rose (see p. 15-14) and aerial photographs (including "stereopairs"). Soil survey data should include identification of surface and subsoil, soil profile, ground-water table, and location of available construction material deposits. See Soils, Section 9.
DIRECTION AND NUMBER OF LANDING STRIPS	Runways should be oriented so that aircraft may be landed at least 95% of the time with cross-wind components not exceeding 15 miles per hour. Construct Wind Rose, and determine coverage. See p. 15-14. Approach standards (p. 15-06) must be met and an economical grading plan obtained. in determining final orientation. At least one runway should be oriented for ultimate instrument operation (see p. 15-07). Runway and taxiway patterns should be planned for efficient ground movement of aircraft. If traffic forecasts indicate a potential volume in excess of 100,000 operations annually, plans should provide for parallel runways or other configuration which will increase acceptance rate and airport capacity. See p. 15-10.
AIRPORT SIZE AND STANDARDS	See p. 15-04 C.A.A. requirements for runway widths, lengths, grades, etc.
GROUND FACILITIES	Arrangement of ground facilities including Administration Building, hangars, cargo buildings, control tower, fire and crash and maintenance buildings, loading ramps, access and service roads, and auto-parking areas must be planned for maximum efficiency of movement of passengers, aircraft, service vehicles, and employees. Plan must also provide for future expansion of ground facilities and revenue areas. To insure maximum airport capacity, a complete taxiway system must be planned, including holding aprons and bypass taxiways at runway ends and high-speed exits for fast clearance of landing aircraft. See p. 15-11. Administration building and parking should be designed for passengers, employees, and visitors. See pp. 15-23, 15-24, and 15-25.
LIGHTING	See pp. 15-26, 15-27, 15-28, and 15-29.
DESIGN OF PAVEMENTS	For aprons, taxiways, warm-up pads, ends of runways, and all other pavements subject to heavy shearing stresses resulting from short-radius brake turns and vibration of engine run-up, thicker pavements should be provided. If economically feasible, Portland cement concrete pavement for these areas is recommended. See Pavements – Airports, Section 10.
DRAINAGE DESIGN	Each site requires a special drainage solution. Amount of run-off, finished grades, topography, soil, and ground-water level are factors. Surface drainage systems should always be provided to collect and discharge storm run-off, which would otherwise be hazardous to aircraft operations, and would saturate the soil, causing possible frost damage and base instability. Subdrainage systems may be required to lower existing ground-water level to such depth that base stability and frost protection will result. Surface and subdrainage systems should never be combined. See Drainage – Airports, Section 18.
COMPLETED PLANS TO INCLUDE	*GRADING* Plans, profiles, and sections with existing and finished grades for both present and ultimate development. Location of available construction ma erial and disposal area. *DRAINAGE* Plans, profiles, sections, and details for all piping, ditches, subdrainage, and structures for both present and ultimate development. *RUNWAYS, TAXIWAYS, APRONS, PARKING AREAS, ACCESS ROADS* Plans showing location and dimensions, intersection details, jointing details, cross sections, and profiles. All future paved areas to be indicated. *LIGHTING* Size, type, location, and details of lights, cables, transformers, underground ducts, control points, wiring, obstruction lights outside airport, and instrument operation equipment. *UTILITIES* Sewers, water, telephone, power complete for airport, including connections to outside source or disposal. *BUILDINGS* Location of all present and future buildings, including off site proposed development. Plans in detail for present work. *APPROACH PROTECTION* Sufficient to enact local zoning ordinance and/or acquire avigation easements. *STAGE CONSTRUCTION* Show the times when construction of each item or part of master plan should be started and when completed, in order to satisfy estimated future air-traffic requirements. Construction should not interfere with operation of airport.

* Adapted from C.A.A., *Airport Design*, January, 1949.

AIRPORTS - C.A.A. STANDARDS *

TABLE A

	C.A.A. Airfield Disignat ion†	Local	Trunk	Continental	Intercontinental
	Length of runways‡	4,200 ft.	6,000 ft.	7,500 ft.	10,500 lb.
	Pavement loading (equivalent single-wheel load)§	30,000 lb.	60,000 lb.	75,000 lb.	100,000 lb.
Min. Width	Landing strip‖	400 ft.	500 ft.	500 ft.	500 ft.
	Runways	100 ft.	150 ft.	150 ft.	150 ft.
	Taxiways	50 ft.	75 ft.	75 ft.	75 ft.
Max. Grade	Effective¶	1%	1%	1%	1%
	Longitudinal**	1.5%	1.5%	1.5%	1.5%
	Transverse††	1.5%	1.5%	1.5%	1.5%
Minimum Clearances Between Centerline Of‡‡	Parallel runways (IFR)	3,000 ft.	3,000 ft.	3,000 ft.	3,000 ft.
	Parallel runways (VFR)	500 ft.	500 ft.	700 ft.	700 ft.
	Runway and taxiway	200 ft.	300 ft.	350 ft.	450 ft.
	Parallel taxiways	150 ft.	250 ft.	275 ft.	325 ft.
	Taxiway and apron edge	130 ft.	220 ft.	240 ft.	280 ft.
	Taxiway and obstruction	100 ft.	125 ft.	150 ft.	200 ft.
	Runway and building line (IFR)	750 ft.	750 ft.	750 ft.	750 ft.
	Runway and building line (VFR)	300 ft.	425 ft.	500 ft.	650 ft.

† Per C.A.A. T.S.O. N6b, October 3, 1958 – See p. 15-01 for definitions.

‡ Maximum category length, computed for sea-level elevation and zero per cent effective runway gradient, includes runway length adjustment for standard 59°F. temperature + 41°F., and is based on the "balanced runway concept."

§ The equivalent single-wheel loadings are maximum for each category. See Pavements–Airports, Section 10, for design data.

‖ The landing strip is the graded area on which the runway is centered and which extends at least 100 ft. beyond each end of the runway.

¶ The effective runway gradient equals the maximum difference in runway center line elevation divided by the total runway length.

** When necessary, longitudinal taxiway grades may be as high as 3%.

†† Percentages shown are for pavement. To improve run-off, the slopes on unpaved areas may be increased to 2%, and to 5% for a distance of 10 ft. from pavement edge.

‡‡ The clearance dimensions indicated are under review by the F.A.A. (formerly C.A.A.) and may therefore be revised in the near future.

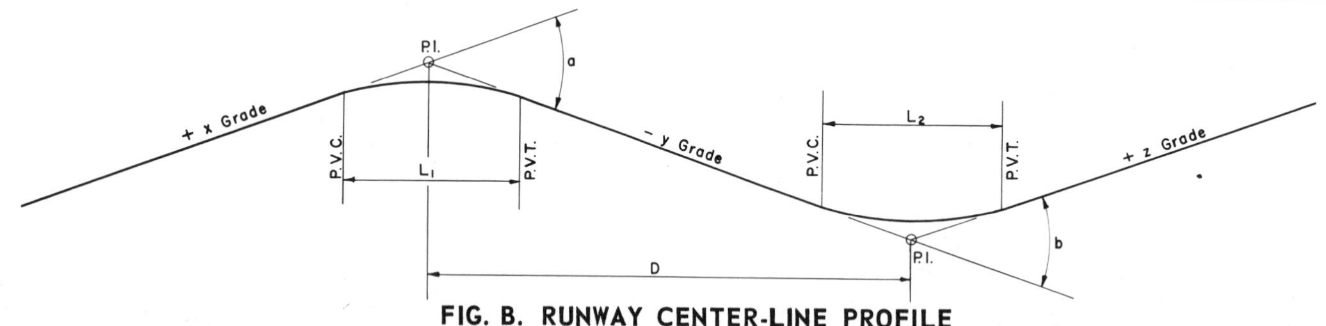
P.I.
a
+ x Grade
P.V.C.
L_1
P.V.T.
− y Grade
L_2
P.V.C.
P.V.T.
+ z Grade
b
D
P.I.

FIG. B. RUNWAY CENTER-LINE PROFILE

1. Vertical curves required when grade changes (*a* or *b*) exceed 0.40%.
2. Maximum grade change not to exceed 1.5%.
3. Length of vertical curve (L_1 or L_2) = 500 ft. for each 1% grade change. (on taxiways use 100 ft. for each 1% grade change)
4. Distance (*D*) between points of intersection of successive vertical curves shall be not less than 50,000 ($a + b$).
5. As a *minimum* requirement, the runway profile should be so designed that there will be an unobstructed line of sight from any point 5 ft. above the runway to any other point 5 ft. above the runway within a distance of at least 500 ft. plus one-half the *ultimate* runway length.

EXAMPLE:

Given: $x = +1.0\%$, $y = -0.5\%$, and $z = +0.4\%$.

Required: L_1, L_2 and D.

Solution: *a* (algebraic difference between x and y) = 1.5%; b = 0.9%.

$L_1 = 500 \times 1.5 = 750'$, $L_2 = 500 \times 0.9 = 450'$.

$D = 50{,}000\ (0.015 + 0.009) = 1{,}200'$.

Increase D if necessary to meet minimum requirements of note 5 above.

* Adapted from C.A.A., *AIRPORT DESIGN*, January 1949, as modified by T.S.O. N-6b, October 3, 1958.

AIRPORTS - CAA RUNWAY LENGTHS*

GENERAL: THE BASIC CATEGORY LENGTH FROM TABLE A ON P. 15-04 MUST BE CORRECTED FOR FIELD ELEVATION ABOVE SEA LEVEL AND EFFECTIVE RUNWAY GRADIENT AS FOLLOWS:

1. A CORRECTION AT THE RATE OF 7% FOR EACH 1,000 FEET OF ELEVATION ABOVE SEA LEVEL MUST BE APPLIED.
2. A FURTHER CORRECTION OF 20% FOR EACH 1% OF EFFECTIVE RUNWAY GRADIENT IS REQUIRED.

EXAMPLE USING FIG. A BELOW GIVEN: FIELD ELEVATION 3,570 FEET ABOVE SEA LEVEL
EFFECTIVE RUNWAY GRADIENT = 1% (SEE NOTE 5, P. 15-04)
BASIC RUNWAY LENGTH OF 5,000 FEET

REQUIRED: CORRECTED RUNWAY LENGTH

SOLUTION: ENTER NOMOGRAPH AT FIELD ELEVATION 3,570, CONTINUE TO EFFECTIVE RUNWAY GRADIENT CURVE OF 1%, READ DOWN TO CORRECTION FACTOR OF 1.5, CONTINUE TO 5,000 BASIC RUNWAY LENGTH LINE AND THEN RIGHT TO CORRECTED RUNWAY LENGTH = 7,500 FEET. i.e. [(3.570 x 0.07) + 1] 1.2 = 1.5; 1.5 x 5000 = 7,500'.

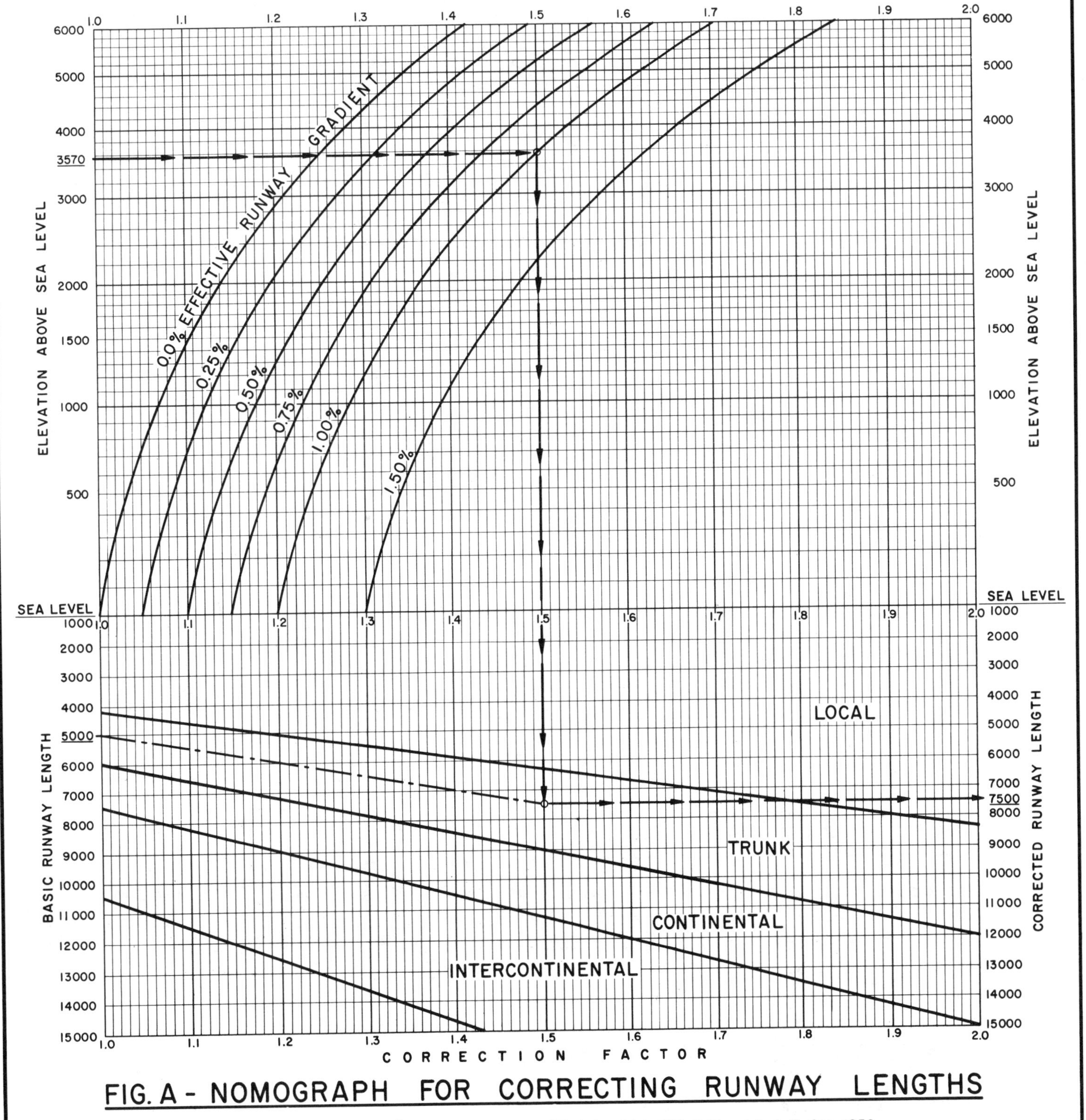

ELEVATION ABOVE SEA LEVEL
6000
5000
4000
3570
3000
2000
1500
1000
500
SEA LEVEL
0.0% EFFECTIVE RUNWAY GRADIENT
0.25%
0.50%
0.75%
1.00%
1.50%
1.0
1.1
1.2
1.3
1.4
1.5
1.6
1.7
1.8
1.9
2.0
BASIC RUNWAY LENGTH
CORRECTED RUNWAY LENGTH
1000
2000
3000
4000
5000
6000
7000
7500
8000
9000
10000
11000
12000
13000
14000
15000
LOCAL
TRUNK
CONTINENTAL
INTERCONTINENTAL
CORRECTION FACTOR

FIG. A - NOMOGRAPH FOR CORRECTING RUNWAY LENGTHS

*Adapted from CAA "AIRPORT DESIGN" Jan. 1949 as modified by CAA TSO N-6b dated 3, Oct. 1958.

AIRPORTS - C.A.A. APPROACH STANDARDS*-1

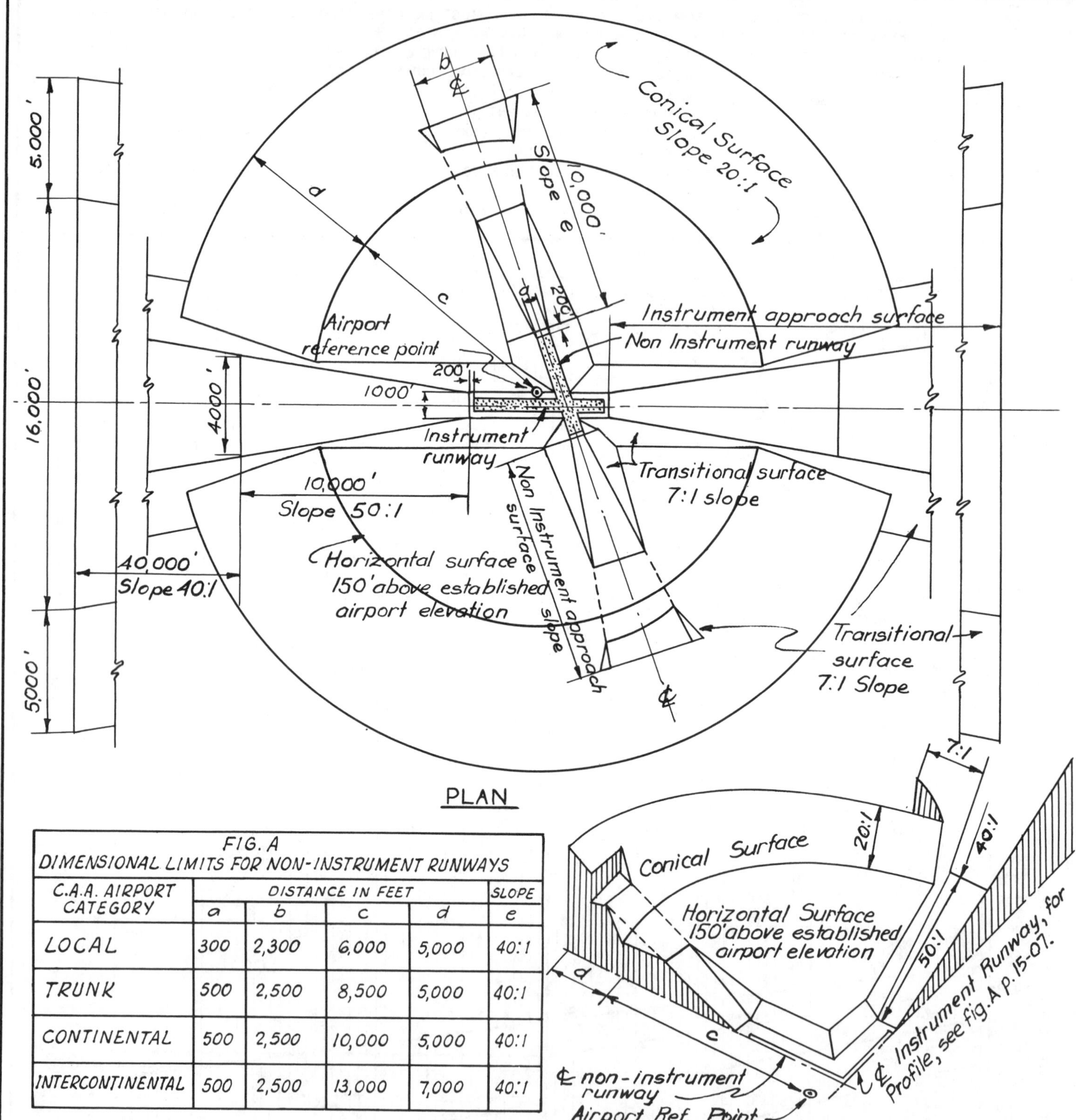
Conical Surface
Slope 20:1
Slope e
10,000'
5,000'
16,000'
5000'
4000'
Airport
reference point
200'
1000'
Instrument approach surface
Non Instrument runway
Instrument
runway
Transitional surface
7:1 slope
10,000'
Slope 50:1
Non Instrument approach surface slope
40,000'
Slope 40:1
Horizontal surface
150' above established
airport elevation
Transitional
surface
7:1 Slope
PLAN
7:1
20:1
40:1
Conical Surface
Horizontal Surface
150' above established
airport elevation
50:1
℄ Instrument Runway, for
Profile, see fig. A p. 15-07.
℄ non-instrument
runway
Airport Ref. Point

FIG. A
DIMENSIONAL LIMITS FOR NON-INSTRUMENT RUNWAYS

C.A.A. AIRPORT CATEGORY	DISTANCE IN FEET				SLOPE
	a	b	c	d	e
LOCAL	300	2,300	6,000	5,000	40:1
TRUNK	500	2,500	8,500	5,000	40:1
CONTINENTAL	500	2,500	10,000	5,000	40:1
INTERCONTINENTAL	500	2,500	13,000	7,000	40:1

1. The approach area standards for instrument runways shall apply to all runways which may be intended for instrument operations and to both ends of such runways.
2. The airport reference point is selected and marked as approximate center of the landing area. The highest point on the usable landing area is the established elevation of the airport.
3. Objects projecting above the landing area, imaginary surfaces shown, or exceeding any of the limiting heights above the ground are obstructions to air navigation. In case of a conflict of limiting heights, the lowest limiting height shall govern.

* These approach stds (pp. 15-06 & 15-07) are under review by the F.A.A. (formely C.A.A.) and subject to revision.

AIRPORTS - C.A.A. APPROACH STANDARDS*- 2

LIMITING HEIGHTS ABOVE GROUND

Objects exceeding the limiting heights above ground, as described below, shall be considered obstructions unless found not to be objectionable after special aeronautical study.

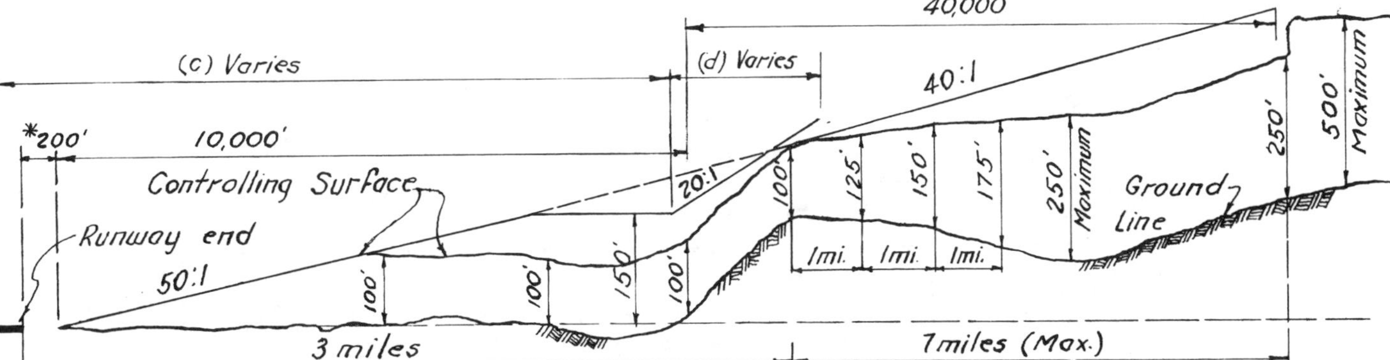

Established elev. of Airport
Reference Point
40,000'
(c) Varies
(d) Varies
40:1
*200'
10,000'
Controlling Surface
20:1
Runway end
50:1
100'
100'
150'
100'
100'
125'
150'
175'
250' Maximum
250'
500' Maximum
Ground Line
1mi.
1mi.
1mi.
3 miles
7 miles (Max.)
Objects in instrument approach areas are obstructions if they extend more than 100' above approach end of runway or ground whichever is higher.
Object height may be increased 25' for each additional mile.

FIG. A - INSTRUMENT APPROACH ZONE PROFILE

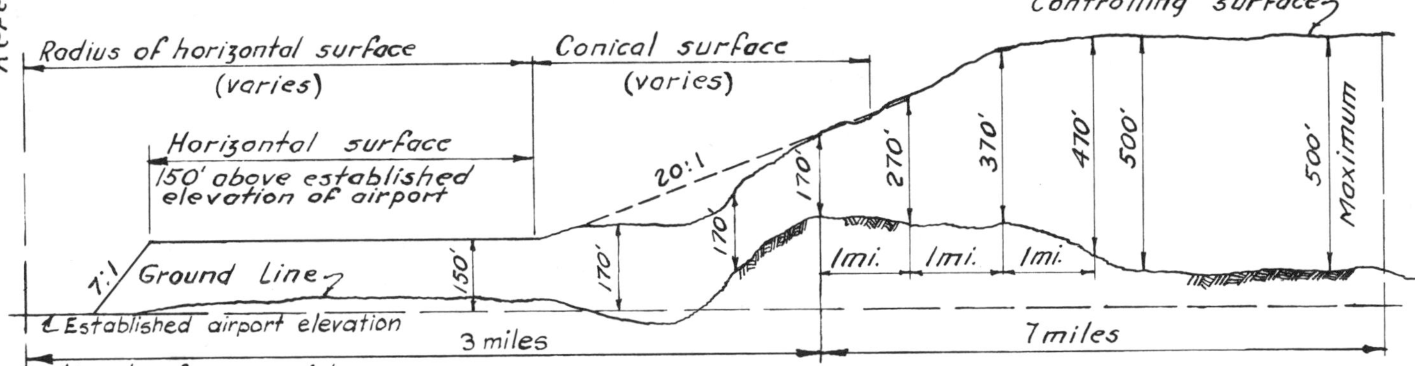

Controlling surface
Radius of horizontal surface (varies)
Conical surface (varies)
Horizontal surface
150' above established elevation of airport
20:1
7:1
Ground Line
150'
170'
170'
170
270'
370'
470'
500'
500' Maximum
1mi.
1mi.
1mi.
Established airport elevation
3 miles
7 miles
Airport reference point.

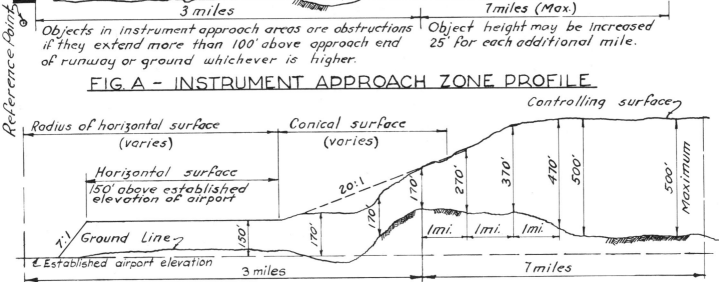

FIG. B TURNING ZONE PROFILE

NOTES:

1 Objects in instrument approach areas that increase the final approach minimum flight altitude which is normally established from the highest point within 5 statute miles of the center line of the final approach course of a radio facility used for final let down to an airport. The above limitation extends for a distance of 10 statute miles along the final approach course outward from the radio facility.

2 Objects in a civil airway or an air traffic control area that extend more than 170 feet above the ground or more than an elevation that would increase an established minimum flight altitude whichever is higher. The minimum flight altitude is normally established from the highest point within 5 statute miles of the center line of the civil airway or air traffic control area affected.

3 Objects that are located within certain established military coastal corridors in which low level flight is required for Department of Defense and Coast Guard air operations conducted from air stations located within 20 statute miles of the Atlantic, Pacific, or Gulf Coast. These corridors will be 10 statute miles in width extending from coastal air stations to sea coast.

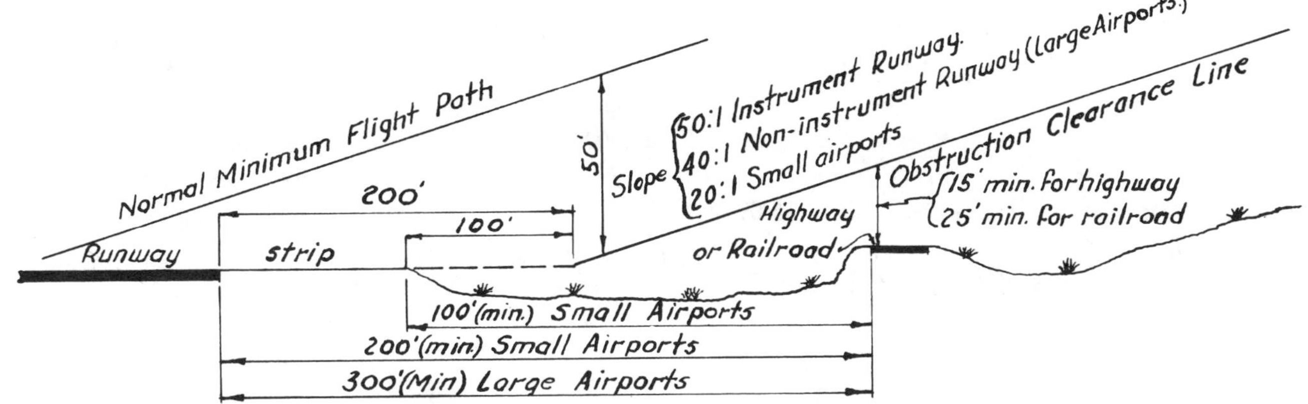

Normal Minimum Flight Path
50'
Slope
50:1 Instrument Runway.
40:1 Non-instrument Runway (Large Airports)
20:1 Small airports
Obstruction Clearance Line
15' min. for highway
25' min. for railroad
200'
100'
Runway
strip
Highway or Railroad
100'(min.) Small Airports
200'(min.) Small Airports
300'(Min) Large Airports

FIG. C HIGHWAY AND RAILROAD CLEARANCES

*Adopted from C.A.A., Figs. A & B adapted from C.A.A. Dwg. No. 814, 4-26-50. Fig. C from "Airport Design", CAA. Jan 1949

See footnote pg. 15-06.

AIRPORTS - NAVIGATIONAL AIDS *

OPTIMUM OPERATIONAL WEATHER MINIMA

Type Navaid	Components	Location	Operating Minima†	
			Ceiling, ft.	Visibility, miles
VHF Omni‡ VOR or TVOR	Range only	Within 5 miles of airport, but not aligned with runway center line.	500	1
	Range only	On the airport.	500	1
	Range and radio marker	Range on the airport. Marker aligned with runway center line and 4 to 5 miles from airport.	400	1
I.L.S.§	VHF localizer, UHF glide slope, Middle marker, Outer marker	On the airport. On the airport. 3500' ± from runway end. 4 to 5 miles from runway end.	300	¾
	Same as above with approach light lane	At approximate end of runway, 1500' to 3000' in length.	200	½

*Contributed by Wm. E. Cullinan, Jr.

†The optimum minima are based on ideal conditions. Substandard aerial approaches due to terrain or structures will necessitate an increase in minimum ceiling, visibility, or both. See pp. 15-06 and 15-07.

‡VHF Omni = VOR = very high frequency omnidirectional radio range station. Frequency range 108 to 118 Mc.

TVOR = terminal VOR on the airfield and used as an approach aid only. The term TVOR is now obsolete since all F.A.A. (Formerly C.A.A.) VOR's are now being installed to "en-route" standards or with such expansion capability.

NOTE: The Air Coordinating Committee has proposed the following uniform terminology for navigational aids:

H, M or L-VOR = VOR as described above.

H, M or L-TACAN = tactical air navigational aid operating in the 960- to 1215-Mc band spectrum and incorporating both azimuth and distance-measuring equipment (DME).

H, M or L-VORTAC = a combined VOR and TACAN on the same site incorporating both VOR azimuth and TACAN azimuth and distance-measuring equipment.

NOTE: The prefix designators H, M, or L indicate the intended normal interference free service volume of the various station types as follows:

Class of VOR, VORTAC, or TACAN	Normal Altitude Service	Normal Interference, Free Distance Service
H	75,000 ft. msl	180 statute miles
M	30,000 ft. msl	90 statute miles
L	15,000 ft. msl	45 statute miles

§I.L.S. = instrument landing system. See p. 15-09.

UHF = ultrahigh frequency from 200 to 400 Mc used by F.A.A. (Formerly C.A.A.) for UHF glide slope component of I.L.S. (328.6 to 335.4 Mc) and for military communications (225 to 400 Mc).

AIRPORTS - NAVIGATIONAL AID SITING

INSTRUMENT RUNWAYS

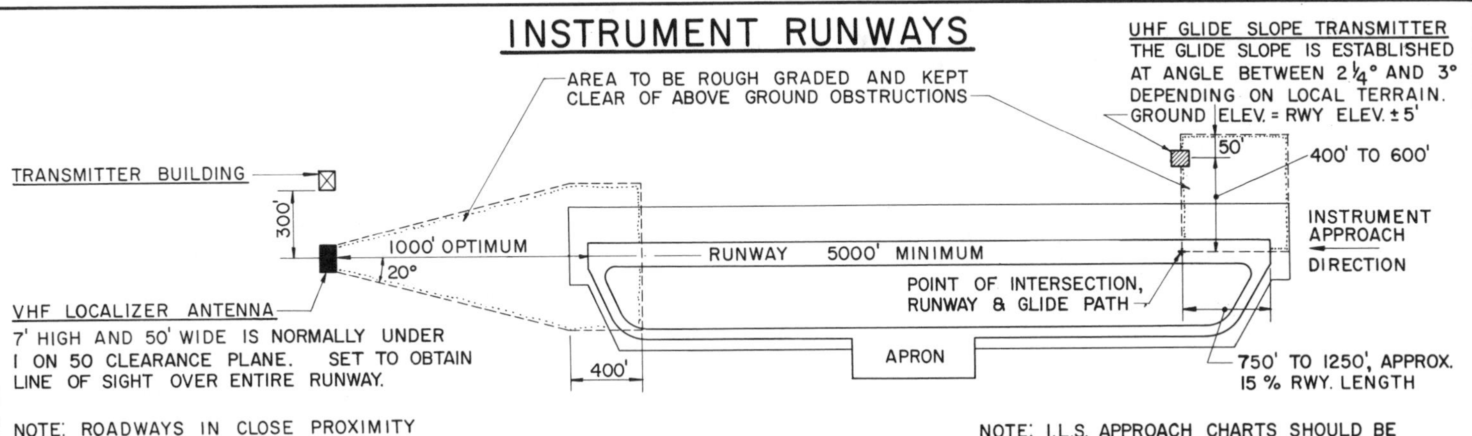
AREA TO BE ROUGH GRADED AND KEPT CLEAR OF ABOVE GROUND OBSTRUCTIONS
UHF GLIDE SLOPE TRANSMITTER
THE GLIDE SLOPE IS ESTABLISHED AT ANGLE BETWEEN 2 1/4° AND 3° DEPENDING ON LOCAL TERRAIN.
GROUND ELEV. = RWY ELEV. ± 5'
50'
400' TO 600'
TRANSMITTER BUILDING
300'
1000' OPTIMUM
20°
RUNWAY 5000' MINIMUM
INSTRUMENT APPROACH DIRECTION
POINT OF INTERSECTION, RUNWAY & GLIDE PATH
VHF LOCALIZER ANTENNA
7' HIGH AND 50' WIDE IS NORMALLY UNDER 1 ON 50 CLEARANCE PLANE. SET TO OBTAIN LINE OF SIGHT OVER ENTIRE RUNWAY.
400'
APRON
750' TO 1250', APPROX. 15 % RWY. LENGTH

NOTE: ROADWAYS IN CLOSE PROXIMITY TO LOCALIZER ANTENNA MAY ADVERSELY AFFECT FRONT OR BACK COURSE OF BEAM.

NOTE: I.L.S. APPROACH CHARTS SHOULD BE CONSULTED TO OBTAIN VARIATIONS OF INDIVIDUAL SYSTEMS

FIG. A - LOCATION AND CLEARANCES FOR INSTRUMENT LANDING SYSTEM LOCALIZER & GLIDE PATH *

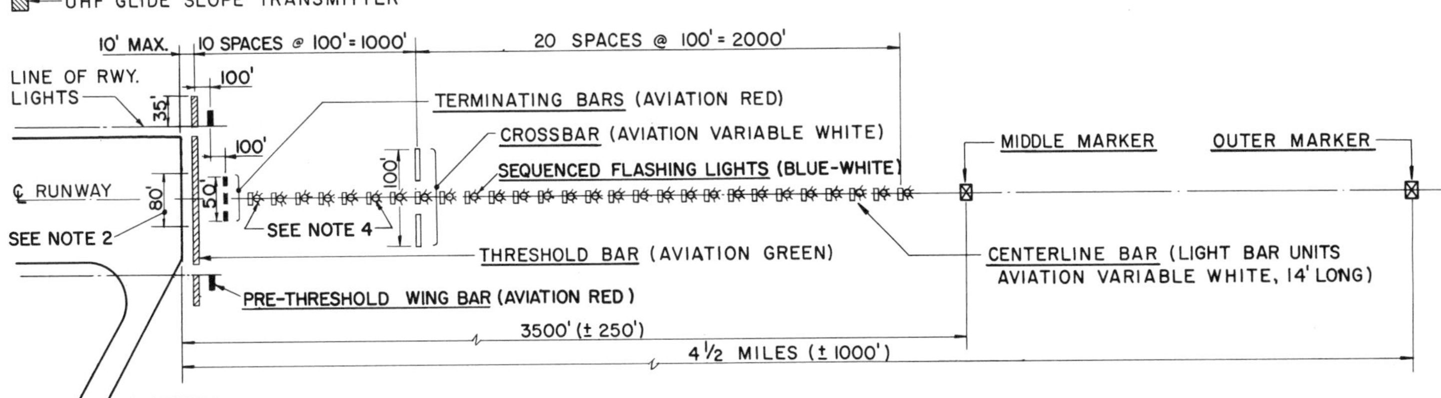
UHF GLIDE SLOPE TRANSMITTER
10' MAX.
10 SPACES @ 100' = 1000'
20 SPACES @ 100' = 2000'
LINE OF RWY. LIGHTS
100'
35'
TERMINATING BARS (AVIATION RED)
CROSSBAR (AVIATION VARIABLE WHITE)
100'
SEQUENCED FLASHING LIGHTS (BLUE-WHITE)
MIDDLE MARKER
OUTER MARKER
℄ RUNWAY
80'
50'
100'
SEE NOTE 4
SEE NOTE 2
THRESHOLD BAR (AVIATION GREEN)
CENTERLINE BAR (LIGHT BAR UNITS AVIATION VARIABLE WHITE, 14' LONG)
PRE-THRESHOLD WING BAR (AVIATION RED)
3500' (± 250')
4 1/2 MILES (± 1000')

NOTES:

1. LOW FREQUENCY RADIO BEACONS AT MIDDLE AND OUTER MARKERS PENDING AVAILABILITY OF DISTANCE MEASURING EQUIPMENT (DME).
2. IF ELEVATED LIGHTS ARE USED FOR THRESHOLD LIGHTING, LEAVE 80-FOOT GAP TO BE LIGHTED WITH FLUSH OR SEMI-FLUSH LIGHTS NOT OVER ONE INCH ABOVE RUNWAY SURFACE.
3. ALL LIGHTS OF THE SYSTEM WITHIN A MINIMUM DISTANCE OF 500 FEET FROM THE RUNWAY THRESHOLD SHALL BE MOUNTED ON FRANGIBLE FITTINGS OR SHALL BE OPERATIONALLY SUITABLE FLUSH OR SEMI-FLUSH UNITS.
4. SEQUENCED FLASHING LIGHTS IN THIS SECTION TO BE AUTOMATICALLY ENERGIZED SIMULTANEOUSLY WITH REMAINDER OF SYSTEM ONLY AT, AND ABOVE, A PRE-DETERMINED MINIMUM, INTENSITY SETTING.
5. WHERE POSSIBLE, ALL LIGHTS SHALL BE IN A HORIZONTAL PLANE AT THRESHOLD ELEVATION. PERMISSABLE LONGITUDINAL SLOPE OF CENTERLINE BARS FROM POINT 200 FEET OFF RUNWAY END: 1% MAX. DOWN, 2% MAX. UP.

FIG. B - APPROACH LIGHT LANE AND RADIO MARKERS †

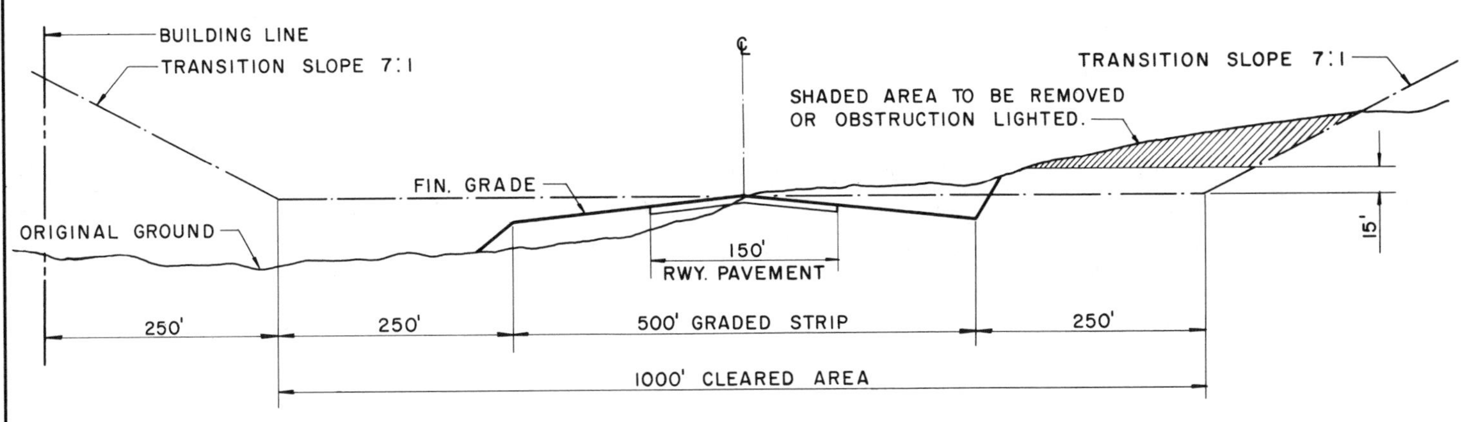
BUILDING LINE
TRANSITION SLOPE 7:1
℄
TRANSITION SLOPE 7:1
SHADED AREA TO BE REMOVED OR OBSTRUCTION LIGHTED.
FIN. GRADE
ORIGINAL GROUND
15'
150'
RWY. PAVEMENT
250'
250'
500' GRADED STRIP
250'
1000' CLEARED AREA

FIG. C - CROSS SECTION INSTRUMENT LANDING STRIP

* ADAPTED FROM DWG No. 1-B-3042 3-23-50 CAA REGION ONE

† ADAPTED FROM NATIONAL STD. FOR APPROACH LIGHTING AGA-NSIa. 4/24/58

AIRPORTS - AIRPORT CAPACITY*-1

THE CAPACITY OF AN AIRPORT IS THE ABILITY OF ITS RUNWAY PATTERN TO ACCOMMODATE A BALANCE OF TAKE-OFF AND LANDING OPERATIONS DURING AN EXTENDED PERIOD. THE CRITICAL, AND THUS THE RATING PERIOD, OCCURS WHEN CEILING AND VISIBILITY CONDITIONS ARE BELOW VFR (VISUAL FLIGHT RULE) WEATHER MINIMA OF 1000 FT. AND 3 MILES. ASSUMING THE AIRPORT HAS AIR TRAFFIC CONTROL, ILS, NAVAIDS AND RADAR, THE AVERAGE IFR (INSTRUMENT FLIGHT RULE) CAPACITIES FOR COMMON RUNWAY CONFIGURATIONS ARE TABULATED BELOW. RATES WILL VARY WITH TYPES AND MIXTURES OF TYPES OF AIRCRAFT AND THE RATIO BETWEEN ARRIVALS AND DEPARTURES AT DIFFERENT PERIODS. FOR VFR CAPACITIES SEE PAGE 15-01.

CONFIGURATION	IFR CAPACITY
INSTRUMENT RUNWAY; TERMINAL AREA; SINGLE RUNWAY PATTERN	30 OPERATIONS /HR
700' MIN.; TERMINAL AREA; PARALLEL RUNWAYS VFR SEPARATION	30 OPERATIONS /HR
LANDING RUNWAY; 3000' MIN.; TAKE-OFF RUNWAY; TERMINAL AREA; PARALLEL INSTRUMENT RUNWAYS IFR SEPARATION	45 OPERATIONS /HR
TAKE-OFF RUNWAY; 1200+; LANDING RUNWAY; TERMINAL AREA; TANDEM-PARALLEL INSTRUMENT RUNWAYS	55 OPERATIONS /HR
TAKE-OFF RUNWAY; TERMINAL AREA; 3000' MIN.; LANDING RUNWAY; OPEN PARALLEL INSTRUMENT RUNWAYS	65 OPERATIONS /HR

* CONDENSED FROM PROCEEDINGS ASCE DEC. 1957 "AIRPORT CONFIGURATION" BY WM. E CULLINAN JR.

AIRPORTS - AIRPORT CAPACITY*-2

BY REDUCING RUNWAY OCCUPANCY TIME, INSTALLATION OF TAXIWAYS CONTRIBUTE TO OPTIMUM AIRPORT CAPACITY. FOR TRANSPORT AIRCRAFT AIRPORTS, HIGH SPEED TAXIWAYS ARE DESIRABLE AND EFFECTIVE WHEN PROPERLY LOCATED, PARTICULARLY FOR JET AIRCRAFT.

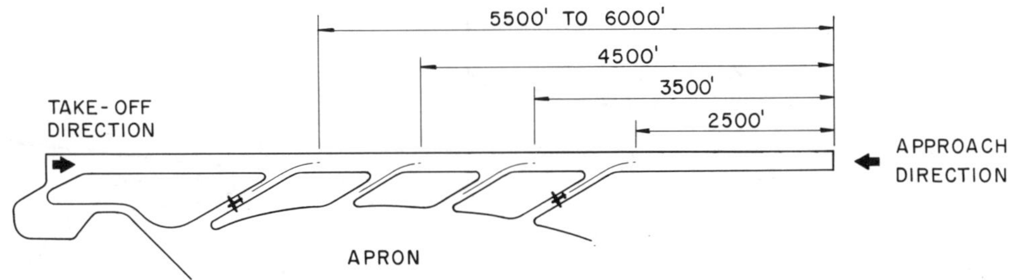

5500' TO 6000'
4500'
3500'
2500'
TAKE-OFF DIRECTION
APPROACH DIRECTION
APRON

UNI-DIRECTIONAL RUNWAY

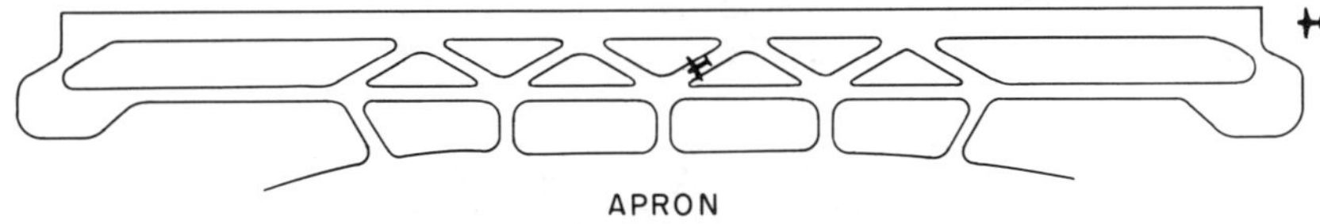

APRON

BI-DIRECTIONAL RUNWAY

LOCATION OF HIGH SPEED TAXIWAYS

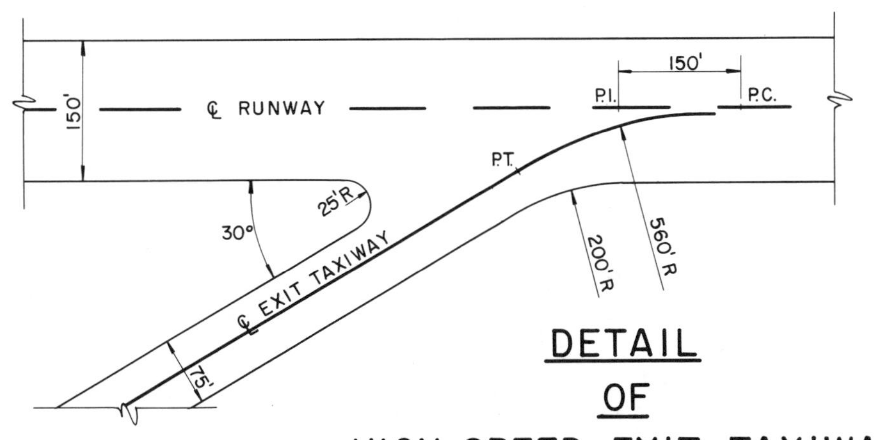

150'
150'
℄ RUNWAY
P.I.
P.C.
P.T.
25' R.
30°
℄ EXIT TAXIWAY
200' R
560' R
75'

DETAIL OF HIGH SPEED EXIT TAXIWAY

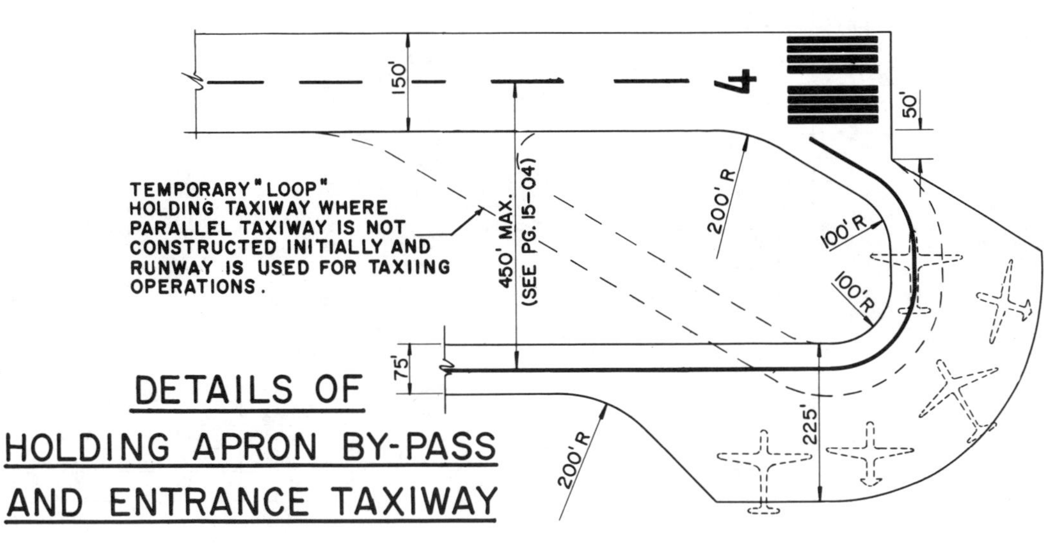

150'
4
50'
TEMPORARY "LOOP" HOLDING TAXIWAY WHERE PARALLEL TAXIWAY IS NOT CONSTRUCTED INITIALLY AND RUNWAY IS USED FOR TAXIING OPERATIONS.
450' MAX. (SEE PG. 15-04)
200' R
100' R
100' R
75'
225'
200' R

DETAILS OF HOLDING APRON BY-PASS AND ENTRANCE TAXIWAY

NOTE:
THIS LAYOUT PERMITS A "ROLLING START" AND ALSO PROVIDES SPACE FOR JET AIRCRAFT, REQUIRING NO WARM-UP, TO BYPASS PISTON AIRCRAFT, WHICH DO.

* FROM PROCEEDINGS ASCE DEC. 1957 "AIRPORT CONFIGURATION" BY WM. E. CULLINAN JR.

AIRPORTS - AIRCRAFT - 1

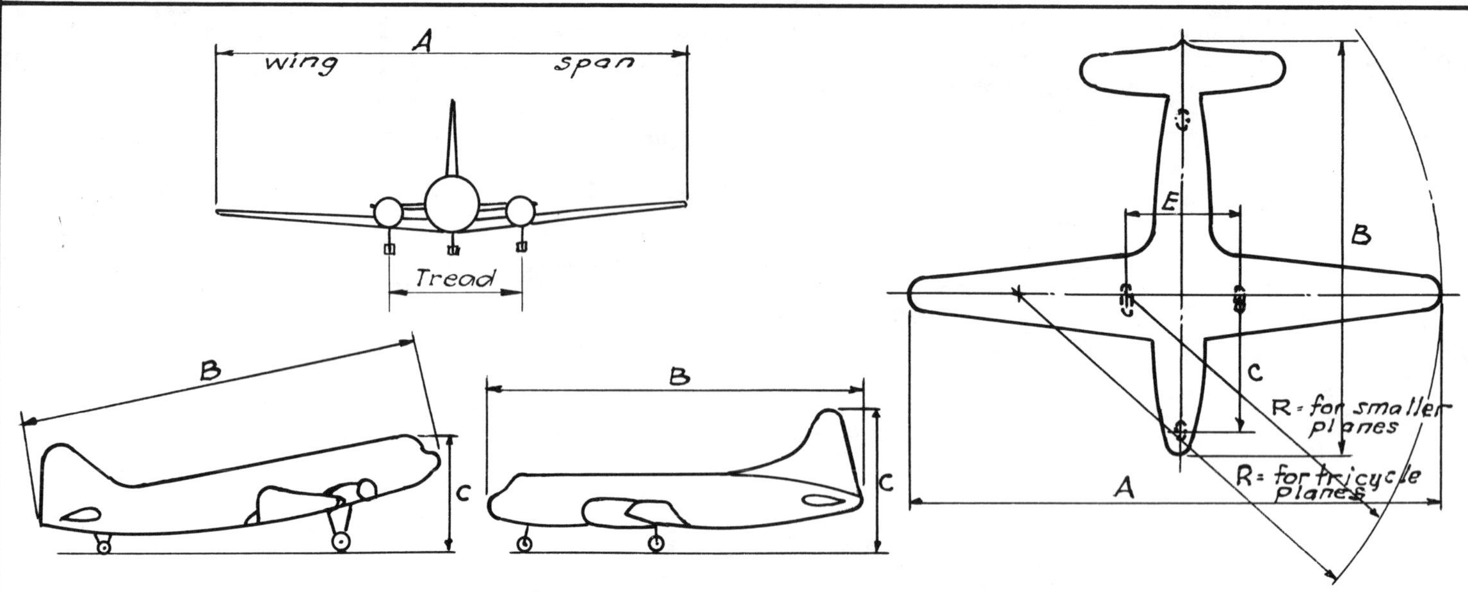

A
wing span
Tread
B
C
B
C
E
B
C
R - for smaller planes
R = for tricycle planes
A

AIRCRAFT DATA*

	Company and Model Designation	A, Wing Span (ft.-in.)	B, Length (ft.-in.)	C, Height (ft.-in.)	R, Turning Radius (ft.-in.)	Max. T.O. Wt. (1000 lb.)	No. of Engines	Landing-Gear Configuration†	
								Type	Tread (ft.-in.)
GENERAL AVIATION AND EXECUTIVE AIRCRAFT	Aeronca Sedan 15 AC	37-6	25-3	7-0	22-3	2.05	1	I	7-0
	Beech G-35 Bonanza	32-10	25-2	6-7	21-3	2.78	1	II	9-7
	Beech D-50 Twin	45-4	31-7	11-4	32-1	6.3	2	II	12-9
	Cessna 180	36-0	26-2	7-6	22-0	2.65	1	I	7-7
	Cessna 195	36-2	27-4	8-0	23-2	3.35	1	I	7-11
	Cessna 310	35-10	27-0	10-6	24-7	4.6	2	II	12-0
	Luscombe 8F Silvaire	35-0	20-0	5-10	22-0	1.4	1	I	6-4
	Piper PA-22 Tri-Pacer	29-4	20-9	7-5	30-3	1.95	1	II	7-0
	Piper PA-23 Apache	37-2	27-5	9-6	46-7	3.5	2	II	11-4
	Ryan Navion	33-5	27-8	8-8	21-0	2.85	1	II	8-8
	Aero 680 Super	44-0	35-5	14-9	35-8	7.0	2	II	13-0
	Beech Super 18	49-8	35-3	10-5	31-4	9.3	2	I	12-11
	Douglas DC-3	95-0	64-6	16-11	56-9	26.2	2	I	18-6
	Curtiss-Wright C-46	108-0	76-4	22-0	67-0	48-0	2	I	25-11
TRANSPORT AIRCRAFT (Conventional)	Boeing 377	141-3	110-4	38-4	84-11	145.8	4	IV	28-6
	Convair 340	105-4	79-2	28-2	65-2	47.0	2	IV	25-0
	Convair 440	105-4	79-2	28-2	65-2	49.1	2	IV	25-0
	Douglas DC-6B Over Water	117-6	106-8	28-8	81-1	107.0	4	III	24-8
	Douglas DC-7C	127-6	113-4	31-10	81-1	143.0	4	III	34-8
	Lockheed 749A	123-0	95-2	23-0	75-6	107.0	4	IV	28-0
	Lockheed 1649A	150-0	116-2	23-5	94-3	156.0	4	IV	38-5
	Martin 404	93-3	74-7	28-8	59-2	43.65	2	III	25-0
TRANSPORT AIRCRAFT (Turbo-Prop)	Bristol Britannia 310	142-3	124-3	36-8	106-3	175.0	4	VI	31-0
	Fairchild Fokker F-27	95-0	73-0	27-6	64-0	34.5	2	III	23-8
	Lockheed Electra	99-0	104-7	32-11	65-1	113.0	4	IV	31-2
	Vickers 840 Viscount	94-0	85-7	26-9	71-2	69.0	4	IV	23-10
	Vickers Vanguard	118-0	122-4	34-11	74-0	135.0	4	IV	30-0
TRANSPORT AIRCRAFT (Pure-Jet)	Boeing 707-320	142-5	152-11	38-11	109-10	280 to 295	4	VI	22-1
	Convair 880	120-0	124-2	36-0	83-3	178.5	4	VI	18-11
	De Havilland Comet IV	115-0	111-6	28-6	86-6	152.2	4	VI	28-2
	Douglas DC-8, Over Water	139-9	148-10	42-4	89-6	287.5	4	VI	20-10
	Douglas DC-9	112-9	118-8	35-4	68-6	138.0	4	VI	17-8
	Fairchild M-185	57-6	56-4	19-9	47-11	28.0	4	II	11-9
	Sud-Est Caravelle 210	112-7	104-10	28-7	90-3	94.8	2	VI	17-2
U.S.A.F. AIRCRAFT	Boeing KC97G Stratofreighter	141-3	110-4	38-3	85-0	175.0	4 Conv.	IV	28-6
	Boeing KC135 Stratotanker	130-10	136-3	38-5	–	300.0±	4 Jet	VI	22-1
	Boeing B47E Stratojet	116-0	107-5½	28-0	–	230.0	6 Jet	VII	–
	Boeing B52 Stratofortress	185-0	156-0	48-0	–	462.1	8 Jet	VIII	8-3
	Convair B36J Heavy Bomber	230-0	162-1	46-9	–	410.0	6 Conv.	VI	46-0
	Convair XB-58 Hustler	55-0	95-0	31-0	–	–	4 Jet	VI	–
	Douglas KC132 Tanker	–	–	–	–	485.6	4 Turbo	IX	21-0
	Douglas C124C Globemaster	174-2	130-0	48-4	–	194.5	4 Conv.	IV	34-2
	Douglas C133A Cargo	179-8	152-8	48-3	–	282.0	4 Turbo	VI	17-4
	Lockheed C130A Hercules	132-7	95-2	38-1	–	124.0	4 Turbo	V	14-3
	Republic F84G Fighter Bomber	36-4	38-5	12-10	–	23.4	1 Jet	II	16-7

* Data on civil aircraft furnished by Wm. E. Cullinan, Jr.

† See Landing Gear Configuration Diagrams on facing page.

AIRPORTS — AIRCRAFT-2

TYPICAL LANDING GEAR CONFIGURATIONS

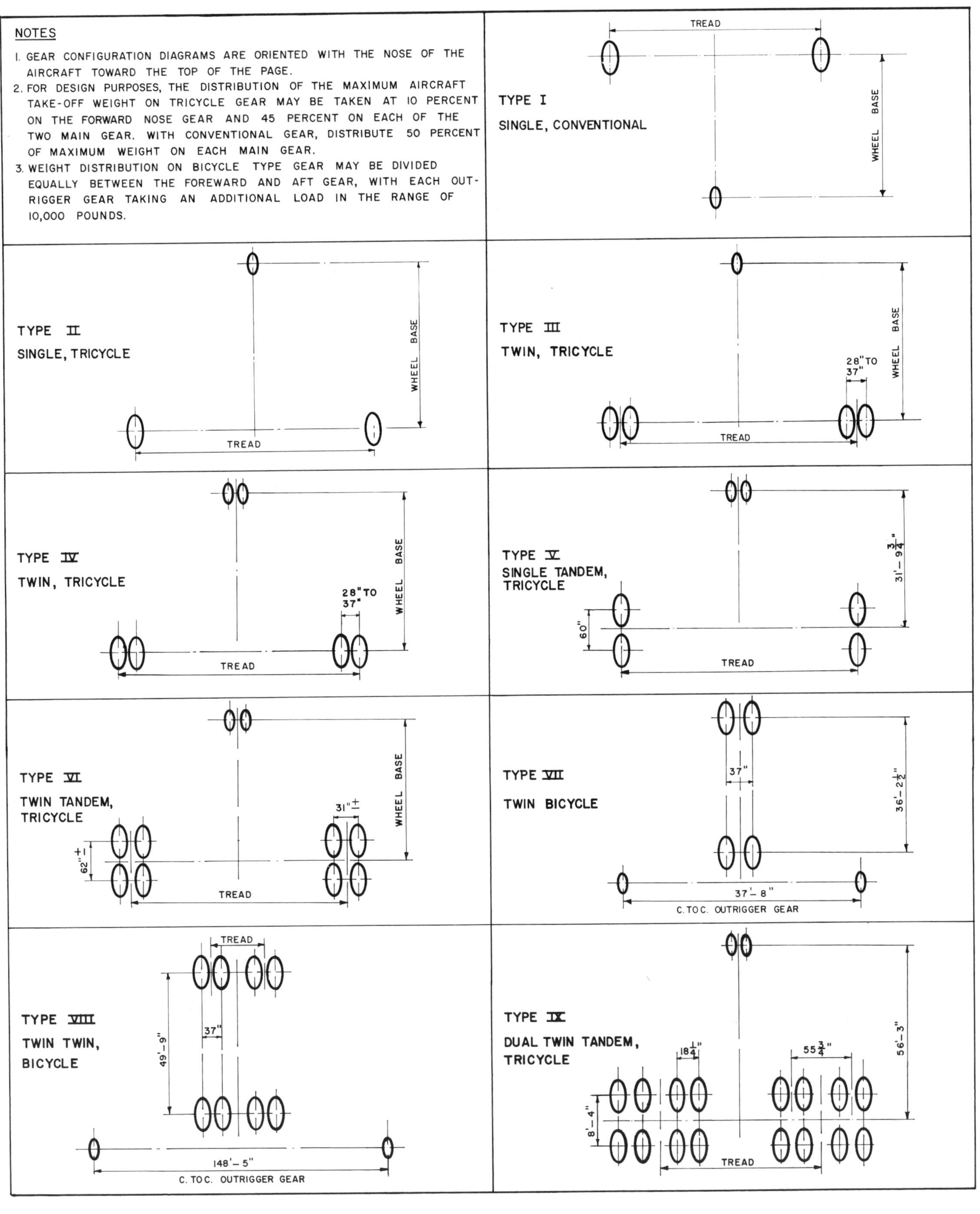
NOTES
1. GEAR CONFIGURATION DIAGRAMS ARE ORIENTED WITH THE NOSE OF THE AIRCRAFT TOWARD THE TOP OF THE PAGE.
2. FOR DESIGN PURPOSES, THE DISTRIBUTION OF THE MAXIMUM AIRCRAFT TAKE-OFF WEIGHT ON TRICYCLE GEAR MAY BE TAKEN AT 10 PERCENT ON THE FORWARD NOSE GEAR AND 45 PERCENT ON EACH OF THE TWO MAIN GEAR. WITH CONVENTIONAL GEAR, DISTRIBUTE 50 PERCENT OF MAXIMUM WEIGHT ON EACH MAIN GEAR.
3. WEIGHT DISTRIBUTION ON BICYCLE TYPE GEAR MAY BE DIVIDED EQUALLY BETWEEN THE FOREWARD AND AFT GEAR, WITH EACH OUTRIGGER GEAR TAKING AN ADDITIONAL LOAD IN THE RANGE OF 10,000 POUNDS.
TYPE I
SINGLE, CONVENTIONAL
TREAD
WHEEL BASE
TYPE II
SINGLE, TRICYCLE
WHEEL BASE
TREAD
TYPE III
TWIN, TRICYCLE
28" TO 37"
WHEEL BASE
TREAD
TYPE IV
TWIN, TRICYCLE
28" TO 37"
WHEEL BASE
TREAD
TYPE V
SINGLE TANDEM, TRICYCLE
31'-9 3/4"
60"
TREAD
TYPE VI
TWIN TANDEM, TRICYCLE
31"±
WHEEL BASE
62"±1
TREAD
TYPE VII
TWIN BICYCLE
37"
36'-2 1/2"
37'-8"
C. TO C. OUTRIGGER GEAR
TYPE VIII
TWIN TWIN, BICYCLE
TREAD
37"
49'-9"
148'-5"
C. TO C. OUTRIGGER GEAR
TYPE IX
DUAL TWIN TANDEM, TRICYCLE
18 1/4"
55 3/4"
56'-3"
8'-4"
TREAD

AIRPORTS - WIND ROSE *

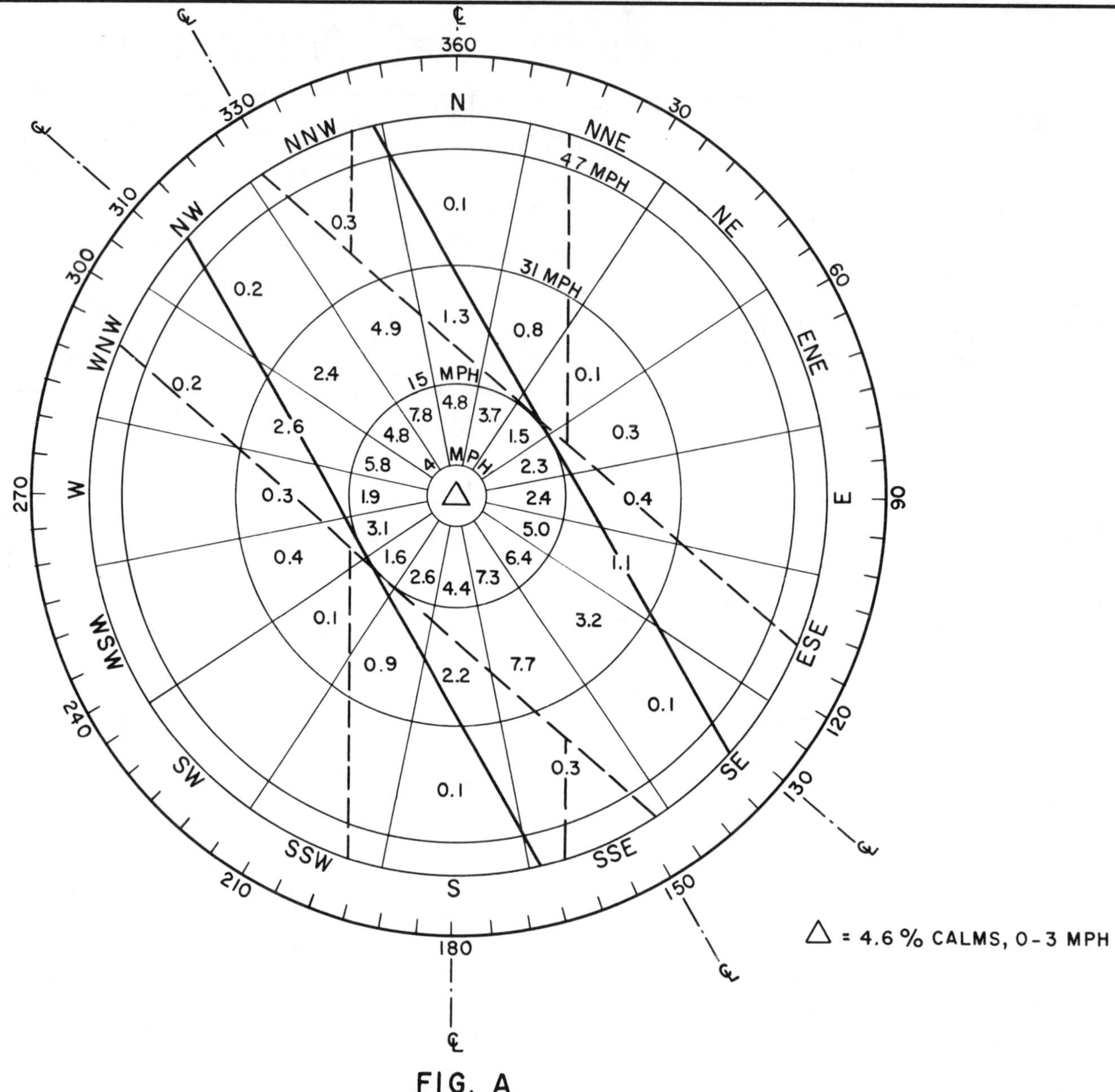

360
330
30
310
60
300
270
90
240
120
210
130
150
180
N
NNE
NE
ENE
E
ESE
SE
SSE
S
SSW
SW
WSW
W
WNW
NW
NNW
47 MPH
31 MPH
15 MPH
4 MPH
△ = 4.6 % CALMS, 0-3 MPH

FIG. A.

EXAMPLE: ***REQUIRED:*** Desirable runway directions.

Solution:

1. Obtain "C.A.A.-type" Wind-Rose data tabulation from U.S. Weather Bureau Records Center, Grove Arcade Building, Ashville, N.C., covering a 4-year period or more.
2. The percentages of time that winds of the indicated direction and velocity can be expected to occur should then be indicated on a wind-rose base as shown in Fig. A.
3. Draw three parallel lines on a transparent strip: the middle line to represent the runway center line, and the distance between it and each outer line to represent the 15 m.p.h. allowable cross-wind component plotted to the same scale as the wind velocity scale on the wind-rose base.
4. Center the transparent strip on the wind rose, and rotate until the sum of the percentages between the two outside lines becomes a maximum. Assume percentages to be distributed equally over segment areas.
   (a) Figure A shows that a single runway (solid lines) oriented 150° – 330° will provide 95% wind coverage. This is the most desirable runway orientation where no additional runways are contemplated.
   (b) Where two runways are to be constructed, the first should be oriented 130° – 310° (dotted lines, Fig. A), thus providing 92% wind coverage, which will be increased to 99% with the addition of a second runway oriented 180° – 360° (dotted lines, Fig. A).
5. Correct selected true bearing to magnetic, and adjust as necessary to meet approach requirements and site topographical conditions.

*Adapted from C.A.A. "Airport Design", Jan. 1949

AIRPORTS - AIR FORCE REGULATIONS* - 1

TABULATION OF FIXED CRITERIA FOR AIRFIELDS

Item Description	Dimensional Criteria	Notes
RUNWAYS		
Runway width	Heavy bomber 300 ft.; Medium Bomber 200 ft.; Fighter 150 ft.	
Width of shoulders	200 ft.	See Note 1
Longitudinal grades of runways and shoulders	Maximum – 1.0%	See Note 2
Transverse grade of runway	Minimum 0.5%; Maximum 1.0%	See Note 2
Transverse grade of shoulders	Minimum 2.0%; Maximum 3.0%	
Width of runway lateral safety	Varies from 400 to 475 ft.	See Note 3
Grades in any direction within lateral safety zones	Maximum – 10%	See Note 3
Clearance from runway center line to fixed and/or movable obstacles without provision of intervening taxiway	Minimum – 750 ft.	See Note 4
Clearance to fixed obstacles from runway center line with intervening existing or possible future taxiway	Minimum of 1000 ft. – Minimum of 1100 ft. to be provided for heavy-bomber airfields	See Note 4
Clearance between center line of parallel runways without intervening taxiway	1000 ft.	
Clearance between parallel runways with provision for intervening taxiway	2000 ft.	
Runway sight distance	Minimum – 5000 ft.	See Note 5
CLEAR ZONES		
Length of clear zone measured on projection of runway center line from end of runway	1000 ft.	See Note 6
Width of clear zone measured at right angles to the extended runway center line	1500 ft.	See Note 6
Longitudinal grade within overrun area	Maximum – 1.5%	See Note 7
Transverse grades within overrun area	Minimum 2.0%; Maximum 3.0%	See Note 7
Grades outside overrun areas (within 750 ft. of extended runway center line) in direction of surface drainage before channelization	Minimum 2.0%; Maximum 10%	See Note 7
TAXIWAYS		
Taxiway width	Minimum 50 ft.; 75 ft. at heavy-bomber airfields	
Width of shoulders	50 ft.	See Note 8
Longitudinal grades of taxiway and shoulders	Maximum – 3.0%	See Note 9
Rate of Longitudinal grade change, per 100 ft.	Maximum – 1.0%	See Note 10
Transverse grade for taxiway pavement	Required – 1.5%	
Transverse grade of taxiway shoulders	Minimum 2.0%; Maximum 3.0%	
Clearance from far edge of taxiway to fixed or movable obstacles	Minimum 250 ft.; Minimum of 350 ft. to be provided for heavy-bomber airfields.	
Grade (in any direction) in taxiway lateral safety zone: i.e., between shoulder and obstacle clearance line	Minimum 2.0%; Maximum 5.0%	See Note 11
Taxiway sight distance	Minimum – 1000 ft.	See Note 12
APRONS		
Pavement grades in any direction	Minimum 0.5%; Maximum 1.5%	See Note 13
Width of apron shoulders, where applicable	25 ft.	See Note 14
Transverse grade of apron shoulders	Minimum 2.0%; Maximum 3.0%	See Note 15
Clearance to fixed or movable obstacles from any edge	Minimum – 125 ft.	See Note 16
APPROACH ZONES		
Glide angle ratio	1 on 50	See Note 17
Length of approach zone	25,000 ft.	
Width at outer end of clear zone	1,500 ft.	
Width at 10,000 ft. from outer end of clear zone	4,000 ft.	
Width at 25,000 ft. from outer end of clear zone	4,000 ft.	

* Extracted from AFR 86-5 and 86-5A

NOTES:

1. Shoulders (non-paved) will be provided adjacent to the longitudinal edges of runways and taxiways and around parking aprons and hardstands. They will not be considered as normal aircraft or vehicular traffic areas, and are intended only for the purpose of minimizing the probability of serious damage to aircraft using these areas accidentally or in cases of emergency. Shoulders will be surfaced with soils selected for stability in wet weather, and thoroughly compacted. Dust and erosion control will be provided by vegetative cover, anchored mulch, coarse-graded aggregate, or liquid palliatives other than asphalt or tars.

2. Grading requirements are dictated by the operational limitations of the aircraft, the need for adequate surface drainage, and the necessity for exercising economy measures in the development of an airfield site. Consistent with these factors, and in consequence of the fact that runway lengths are computed on the basis of generally level pavements, longitudinal sloping of runways will be held to the minimum possible. Maximum slope of runway will not exceed 1%. The maximum rate of longitudinal grade change per 100 ft. of runway will not exceed 0.167%, except for edges of runways and shoulders at runway intersections. This rate of change is produced by vertical curves having 600-ft. length for each per cent of algebraic difference between the two grades. For more than one change in runway grade, the distance between two successive points of intersection will not be less than 1000 ft. The maximum rate of longitudinal grade change per 100 ft. for edges of runways and shoulders at runway intersections will not exceed 0.4%. The transverse-runway grade requirements are not mandatory at runway intersections. Certain exceptions are also made where runway and shoulder gutter construction is involved.

3. Width of these zones depends on runway width involved. Lateral safety zones are the areas located between the runway shoulders and the clearance lines limiting the placement of bulding construction and other obstacles with respect to the runway center line. These areas will be rough-graded to the extent necessary to reduce damage to aircraft, in the event of erratic performance.

4. Fixed obstacles include buildings, trees, rocks, terrain irregularities, and any other feature constituting a possible hazard to moving aircraft. Movable obstacles include moving and parked aircraft, vehicles, railroad cars, and so forth. The prescribed clearances apply with equal force to aprons, roads, highways, and railroad tracks. The following items are permitted to be constructed 500 ft. or more from the runway center line; drainage headwalls, drainage ditches with side slopes of any gradient.

5. Any two points 10 ft. above pavement must be mutually visible for the distance indicated.

6. Clear zones are the areas immediately adjacent to ends of a runway which have been cleared of all above-ground obstructions, and graded to prevent damage to aircraft undershooting or over-running the runway.

7. Overrun areas are portions of the clear zones lying on both sides of the projected center line of the runway having a length equal to that of the clear zone and a width equal to the sum of the widths of the runway and its shoulders. Grading in, and construction of, overrun areas, will be the same as specified for the runway ders.

8. Same treatment as for runway shoulders except paved shoulder at heavy-bomber installations.

9. May be increased to 5% for a length not in excess of 400 ft. Three (3) per cent maximum will not be exceeded on connecting taxiway within 600 ft. of runway entrance. Minimum grade preferred.

10. For more than one change in taxiway grade, the distance between two successive points of intersection will not be less than 500 ft.; changes in longitudinal grades will be accomplished by vertical curves.

11. This area will be rough graded to reduce damage to aircraft as much as possible in the event of erratic aircraft performance.

12. Taxiway grades will be such as not to preclude clear visibility for 1000 ft. between any two points 10 ft. above taxiway pavement.

13. Pavement gradients exceeding the minimum specified are intended for those areas where design of expansive pavements to accommodate unusual runoff dictates such a requirement. Economic factors imposed by difficult terrain features may also require the use of steeper gradients. Arbitrary utilization of gradients in excess of actual or reasonable requirements is not intended. For example, in designing so-call "saw-tooth" surface drainage patterns, extreme care must be exercised to prevent the utilization of steep grades or rapid grade changes at relatively short intervals. Such surface irregularities aggravate normal flexing of aircraft wings while taxiing and may result in damage to low underslung appurtenances on the aircraft.

14. Applies also to hardstands. Same treatment as for runway and taxiway shoulders, with 50-ft. paved shoulder at heavy-bomber installations.

15. Longitudinal grades of shoulders will be governed by grades of apron pavement.

16. For definition of fixed or movable obstacles see Note 4. This clearance distance, on the front edge of the apron, applies so long as the 750-ft. clearance to runway center line is not violated. This clearance distance is not required for hangar access pavement for which the aircraft wing overhang distance is considered adeqate.

17. Glide angle begins at outer end of clear zone and at the same elevation as that for the pavement at the end of runway center line.

AIRPORTS - RUNWAY

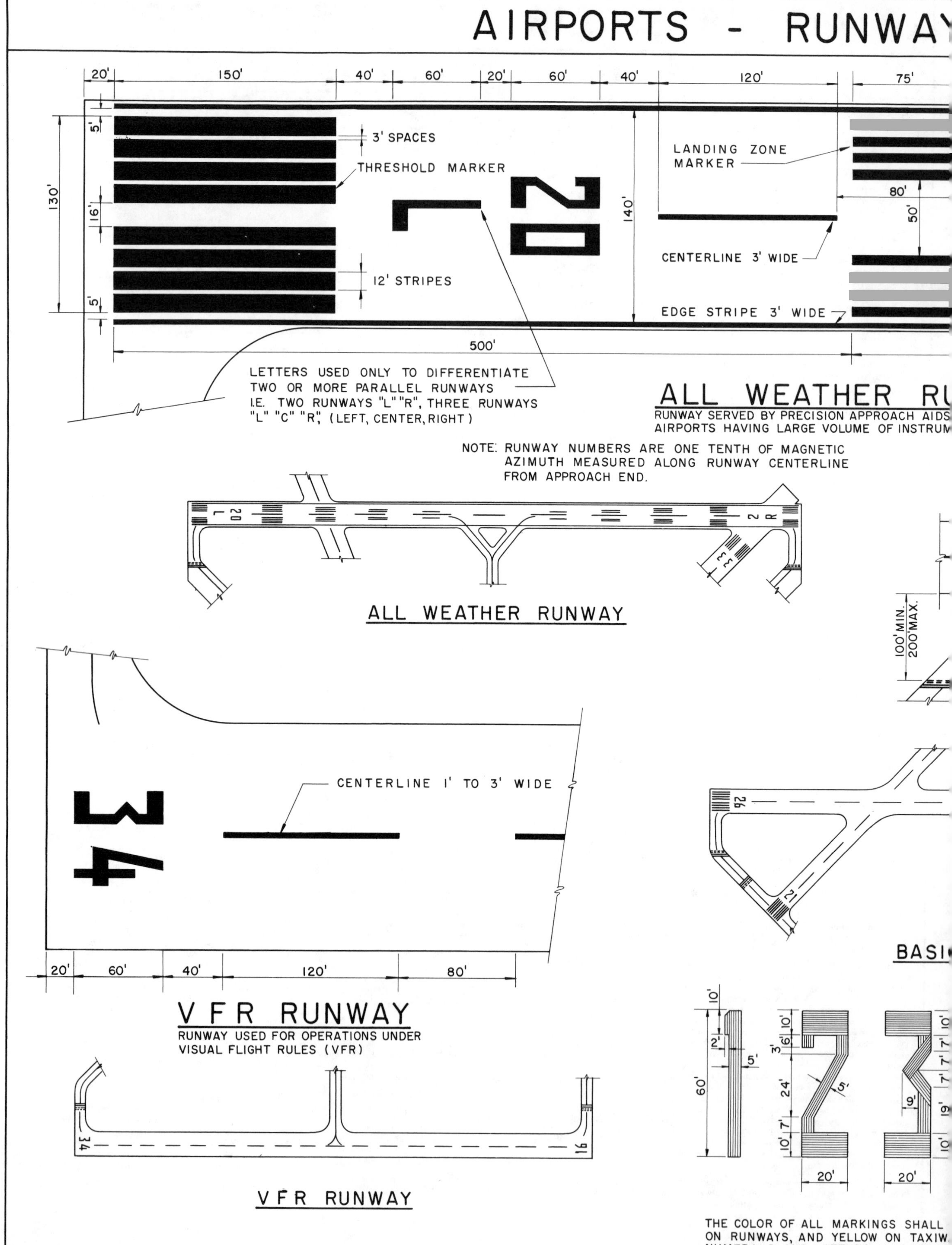

20'
150'
40'
60'
20'
60'
40'
120'
75'
130'
5'
16'
5'
3' SPACES
THRESHOLD MARKER
12' STRIPES
140'
LANDING ZONE MARKER
80'
50'
CENTERLINE 3' WIDE
EDGE STRIPE 3' WIDE
500'
LETTERS USED ONLY TO DIFFERENTIATE TWO OR MORE PARALLEL RUNWAYS I.E. TWO RUNWAYS "L" "R", THREE RUNWAYS "L" "C" "R", (LEFT, CENTER, RIGHT)
ALL WEATHER RU
RUNWAY SERVED BY PRECISION APPROACH AIDS
AIRPORTS HAVING LARGE VOLUME OF INSTRUM
NOTE: RUNWAY NUMBERS ARE ONE TENTH OF MAGNETIC AZIMUTH MEASURED ALONG RUNWAY CENTERLINE FROM APPROACH END.
ALL WEATHER RUNWAY
100' MIN.
200' MAX.
CENTERLINE 1' TO 3' WIDE
20'
60'
40'
120'
80'
VFR RUNWAY
RUNWAY USED FOR OPERATIONS UNDER VISUAL FLIGHT RULES (VFR)
VFR RUNWAY
BASI
10'
60'
2'
5'
3'
6'
24'
7'
9'
19'
20'
THE COLOR OF ALL MARKINGS SHALL
ON RUNWAYS, AND YELLOW ON TAXIW
NUMERALS AND LETTERS MAY BE PA
OR WITH 6" STRIPES, 6" APART.

*ADAPTED FROM CAA TSO-N10a APRIL 27, 1953

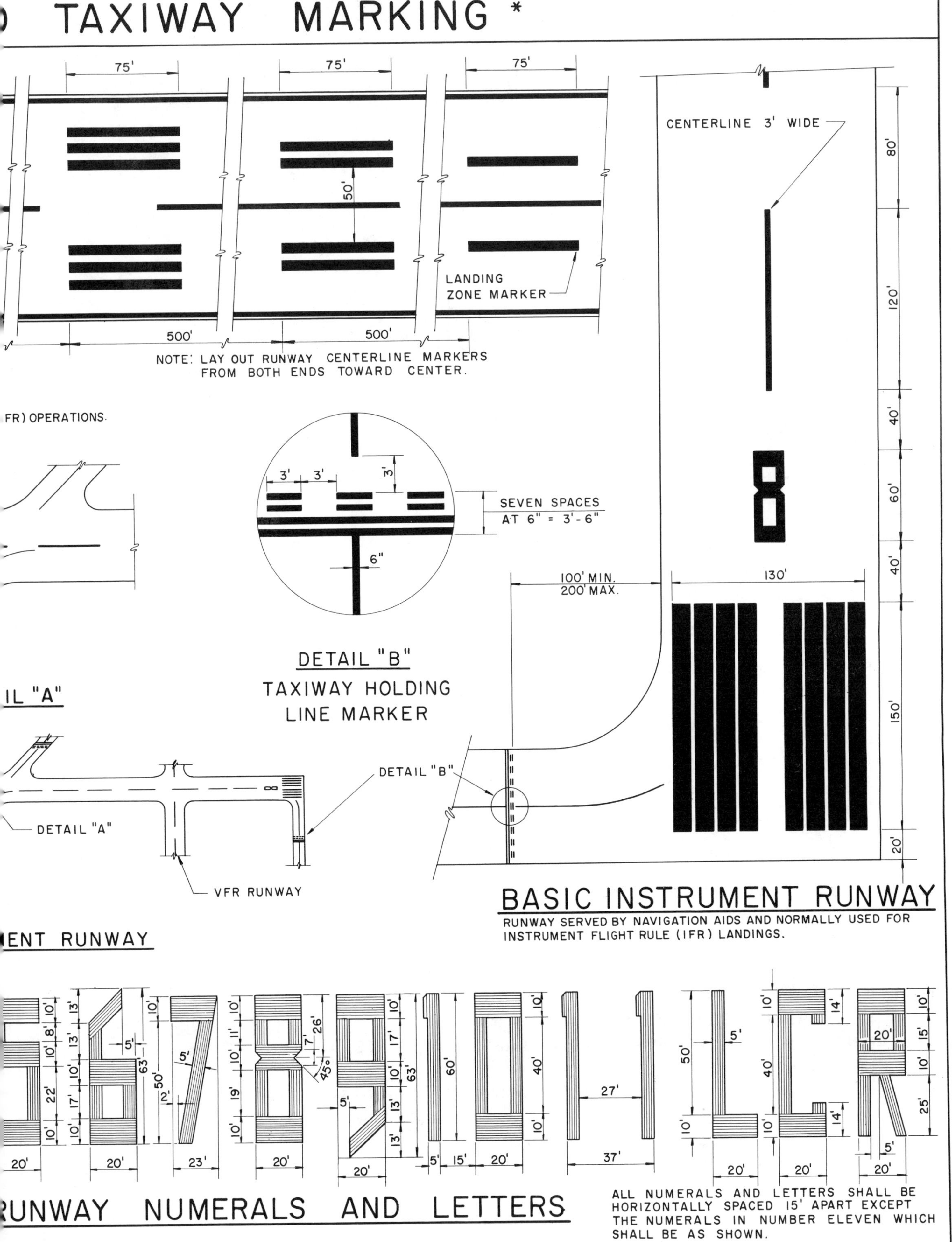
TAXIWAY MARKING *
75'
75'
75'
50'
LANDING ZONE MARKER
500'
500'
NOTE: LAY OUT RUNWAY CENTERLINE MARKERS FROM BOTH ENDS TOWARD CENTER.
FR) OPERATIONS.
CENTERLINE 3' WIDE
80'
120'
40'
60'
40'
150'
20'
130'
100' MIN.
200' MAX.
3'
3'
3'
SEVEN SPACES AT 6" = 3'-6"
6"
DETAIL "B"
TAXIWAY HOLDING LINE MARKER
IL "A"
DETAIL "B"
DETAIL "A"
VFR RUNWAY
BASIC INSTRUMENT RUNWAY
RUNWAY SERVED BY NAVIGATION AIDS AND NORMALLY USED FOR INSTRUMENT FLIGHT RULE (IFR) LANDINGS.
ENT RUNWAY
RUNWAY NUMERALS AND LETTERS
ALL NUMERALS AND LETTERS SHALL BE HORIZONTALLY SPACED 15' APART EXCEPT THE NUMERALS IN NUMBER ELEVEN WHICH SHALL BE AS SHOWN.

AIRPORTS - SMALL AIRFIELDS

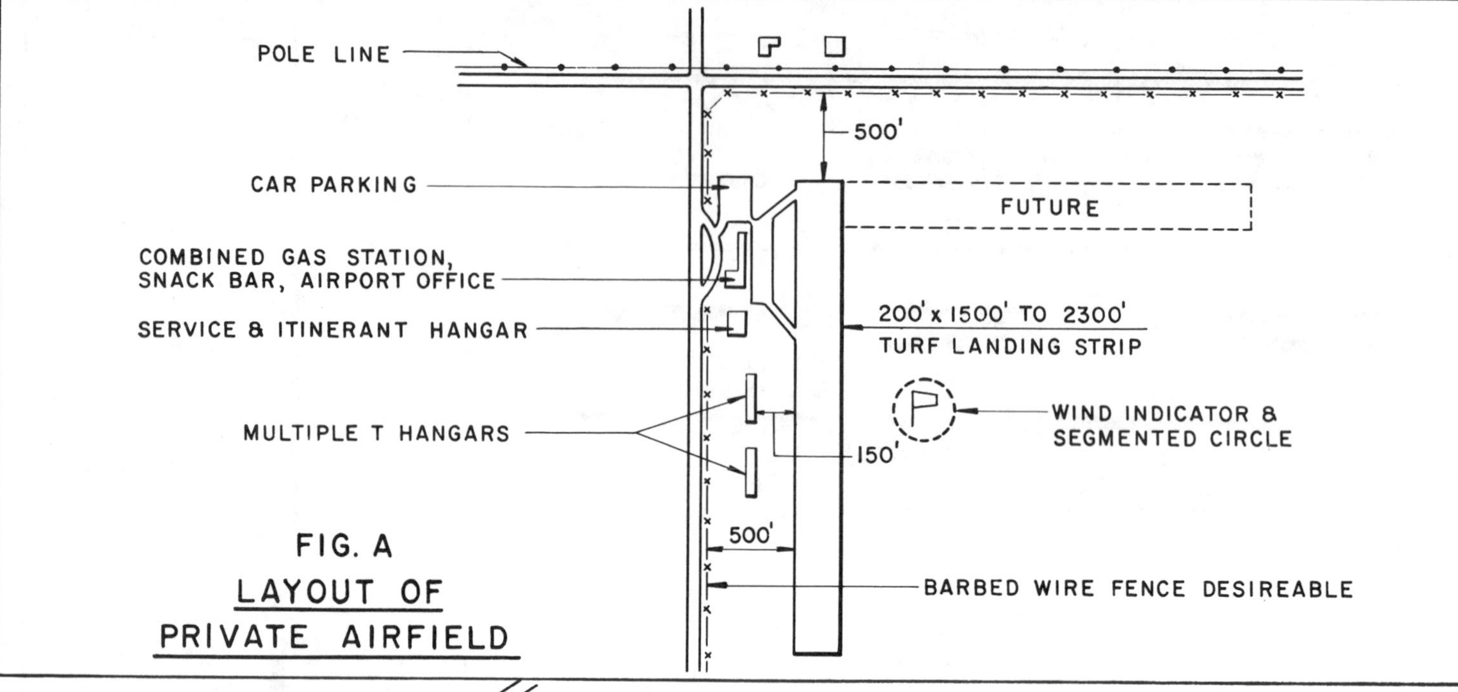
POLE LINE
500'
CAR PARKING
FUTURE
COMBINED GAS STATION, SNACK BAR, AIRPORT OFFICE
SERVICE & ITINERANT HANGAR
200' x 1500' TO 2300'
TURF LANDING STRIP
WIND INDICATOR & SEGMENTED CIRCLE
MULTIPLE T HANGARS
150'
500'
BARBED WIRE FENCE DESIREABLE

FIG. A
LAYOUT OF PRIVATE AIRFIELD

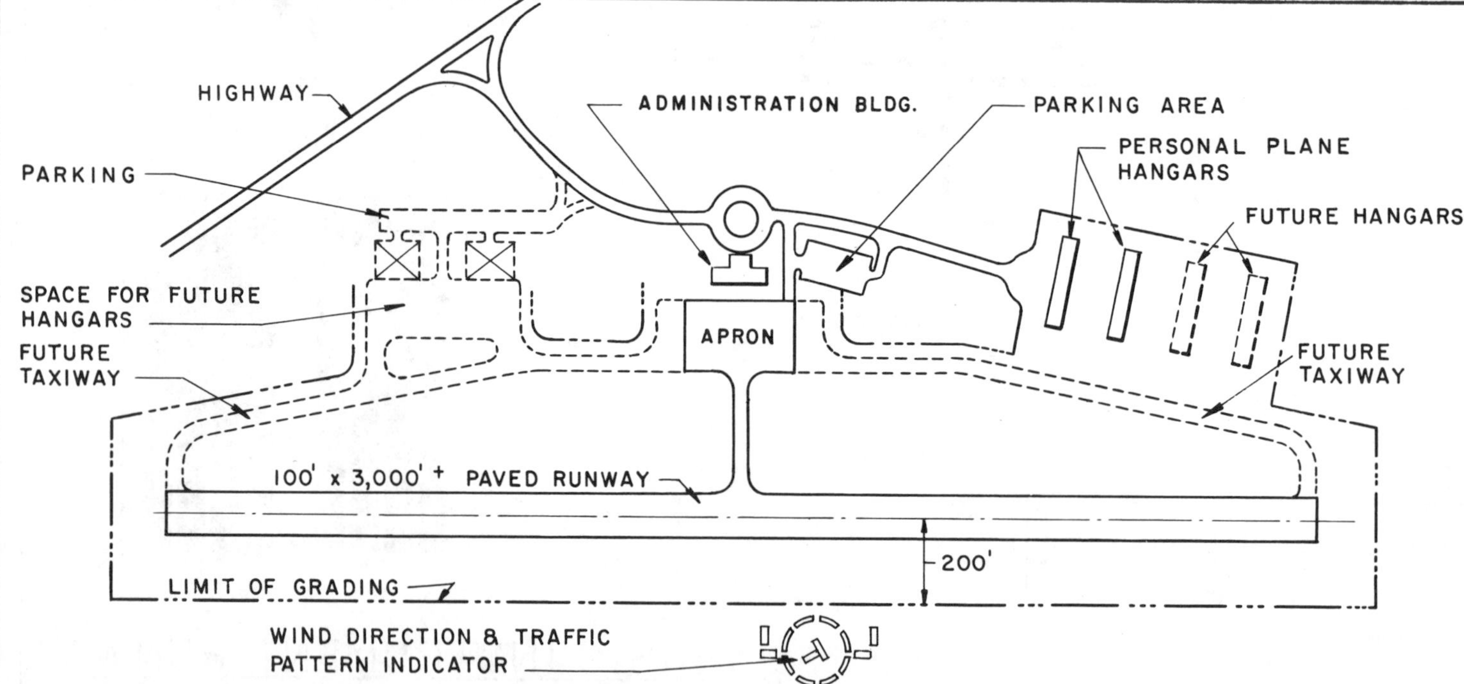
HIGHWAY
ADMINISTRATION BLDG.
PARKING AREA
PERSONAL PLANE HANGARS
PARKING
FUTURE HANGARS
SPACE FOR FUTURE HANGARS
APRON
FUTURE TAXIWAY
FUTURE TAXIWAY
100' x 3,000' + PAVED RUNWAY
200'
LIMIT OF GRADING
WIND DIRECTION & TRAFFIC PATTERN INDICATOR

FIG. B AIRFIELD LAYOUT - LOCAL TYPE

NOTES:

1. SEE LOCAL TYPE IN C.A.A. STANDARDS, P.15-04 AND DISCUSSION ON P. 15-01.
2. DO NOT OVERBUILD, CONTRACT WHAT IS NOW NEEDED AND PROVIDE FOR FUTURE EXPANSION.
3. GRADING AND DRAINAGE IS SIMILAR TO LARGER AIRPORTS AND SHOULD BE KEPT TO A MINIMUM YET ADEQUATE DESIGN.
4. PROVIDE ZONING PROTECTION TO MEET APPROACH STANDARDS, SEE P.15-06
5. SHOW NAME OF AIRPORT ON GROUND SIGN OR HANGAR ROOF.
6. ORIENT RUNWAY PARALLEL TO PREVAILING WINDS.
7. FOR CLEARANCES OFF ENDS OF RUNWAYS, SEE APPROACH STANDARDS, P. 15-07.

AIRPORTS - TRUNK PLAN*

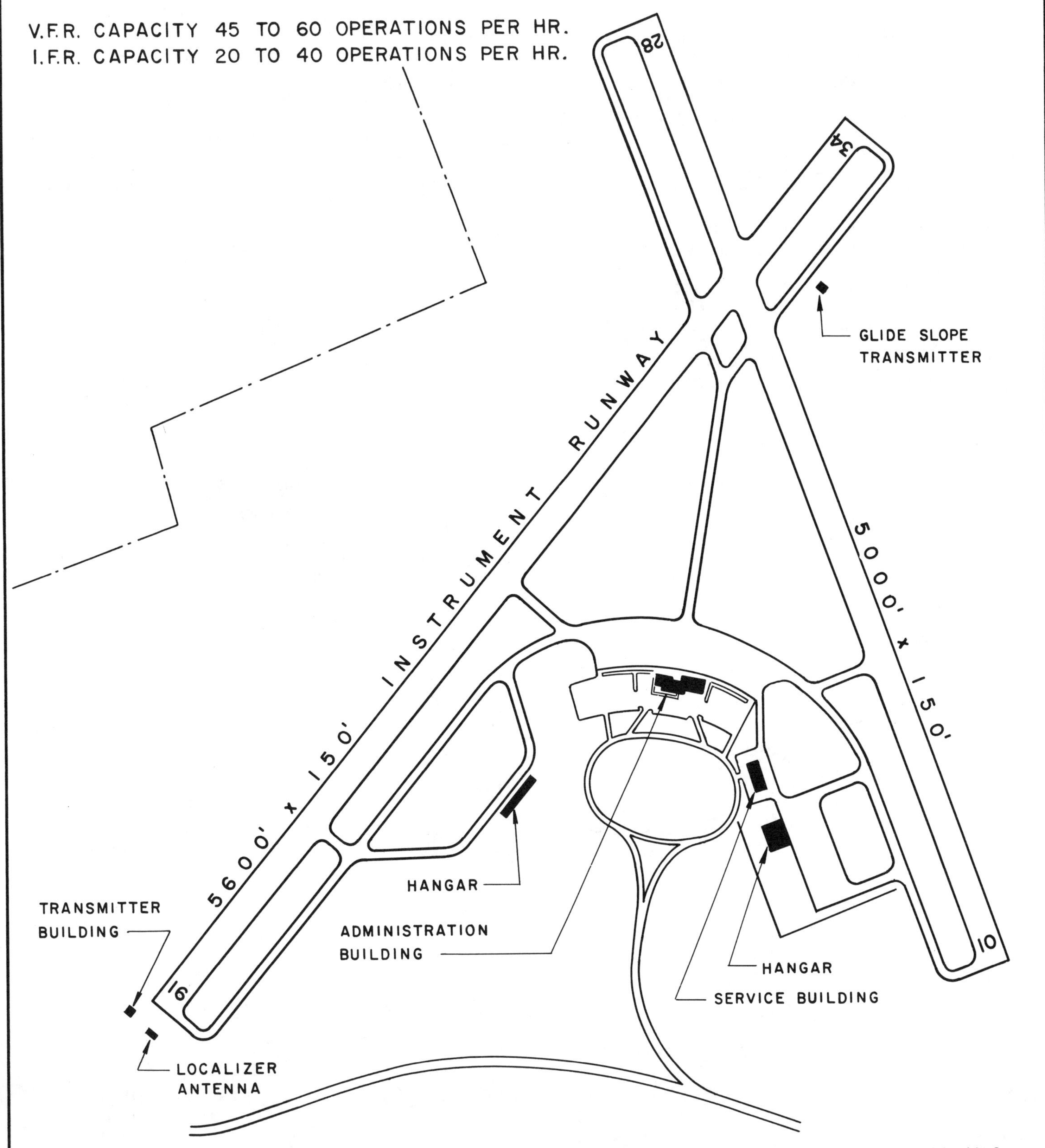
V.F.R. CAPACITY 45 TO 60 OPERATIONS PER HR.
I.F.R. CAPACITY 20 TO 40 OPERATIONS PER HR.
28
34
GLIDE SLOPE TRANSMITTER
5600' x 150' INSTRUMENT RUNWAY
5000' x 150'
HANGAR
TRANSMITTER BUILDING
ADMINISTRATION BUILDING
16
10
HANGAR
SERVICE BUILDING
LOCALIZER ANTENNA

THIS AIRPORT IS TYPICAL OF TRUNK TYPE AIRPORTS, DESIGNED FOR PLANES HAVING EQUIVALENT SINGLE WHEEL LOADS NOT OVER 60,000 LBS.

TRUNK TYPE AIRPORTS USUALLY HAVE ONE PRIMARY INSTRUMENT RUNWAY AND AN ALTERNATE RUNWAY TO BE USED WHEN THE WIND IS NOT IN THE CORRECT DIRECTION FOR THE MAIN RUNWAY.

* AFTER BROOME COUNTY N.Y. (BINGHAMTON) DESIGN BY SEELYE, STEVENSON & VALUE, ENGINEERS.

AIRPORTS - INTERCONTINENTAL PLAN

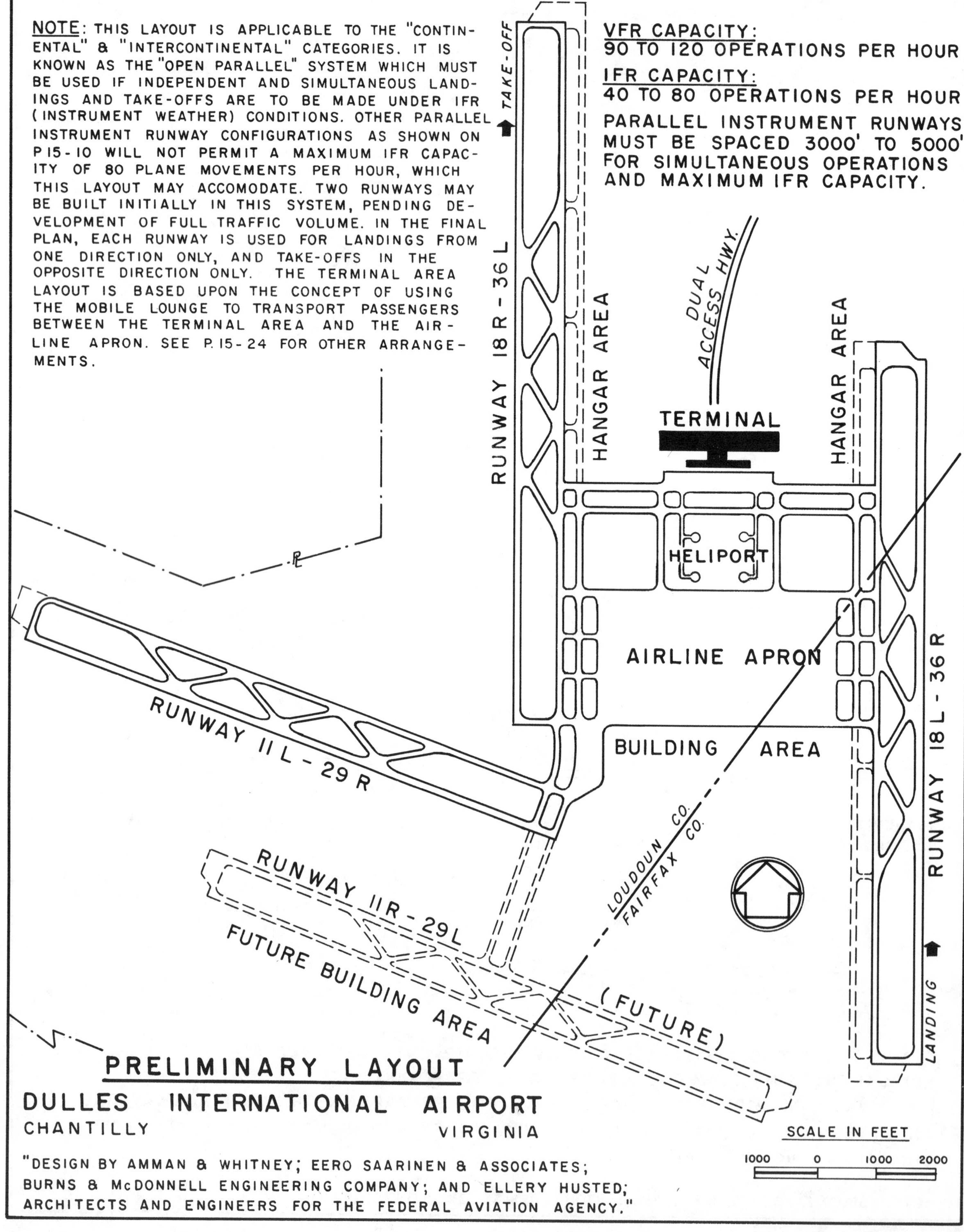
NOTE: THIS LAYOUT IS APPLICABLE TO THE "CONTINENTAL" & "INTERCONTINENTAL" CATEGORIES. IT IS KNOWN AS THE "OPEN PARALLEL" SYSTEM WHICH MUST BE USED IF INDEPENDENT AND SIMULTANEOUS LANDINGS AND TAKE-OFFS ARE TO BE MADE UNDER IFR (INSTRUMENT WEATHER) CONDITIONS. OTHER PARALLEL INSTRUMENT RUNWAY CONFIGURATIONS AS SHOWN ON P 15-10 WILL NOT PERMIT A MAXIMUM IFR CAPACITY OF 80 PLANE MOVEMENTS PER HOUR, WHICH THIS LAYOUT MAY ACCOMODATE. TWO RUNWAYS MAY BE BUILT INITIALLY IN THIS SYSTEM, PENDING DEVELOPMENT OF FULL TRAFFIC VOLUME. IN THE FINAL PLAN, EACH RUNWAY IS USED FOR LANDINGS FROM ONE DIRECTION ONLY, AND TAKE-OFFS IN THE OPPOSITE DIRECTION ONLY. THE TERMINAL AREA LAYOUT IS BASED UPON THE CONCEPT OF USING THE MOBILE LOUNGE TO TRANSPORT PASSENGERS BETWEEN THE TERMINAL AREA AND THE AIRLINE APRON. SEE P. 15-24 FOR OTHER ARRANGEMENTS.
VFR CAPACITY:
90 TO 120 OPERATIONS PER HOUR
IFR CAPACITY:
40 TO 80 OPERATIONS PER HOUR
PARALLEL INSTRUMENT RUNWAYS MUST BE SPACED 3000' TO 5000' FOR SIMULTANEOUS OPERATIONS AND MAXIMUM IFR CAPACITY.
TAKE-OFF
RUNWAY 18R - 36L
HANGAR AREA
DUAL ACCESS HWY.
TERMINAL
HANGAR AREA
HELIPORT
AIRLINE APRON
BUILDING AREA
RUNWAY 18L - 36R
LANDING
RUNWAY 11L - 29R
RUNWAY 11R - 29L
(FUTURE)
FUTURE BUILDING AREA
LOUDOUN CO.
FAIRFAX CO.
PRELIMINARY LAYOUT
DULLES INTERNATIONAL AIRPORT
CHANTILLY
VIRGINIA
SCALE IN FEET
1000 0 1000 2000
"DESIGN BY AMMAN & WHITNEY; EERO SAARINEN & ASSOCIATES; BURNS & McDONNELL ENGINEERING COMPANY; AND ELLERY HUSTED; ARCHITECTS AND ENGINEERS FOR THE FEDERAL AVIATION AGENCY."

AIRPORTS - TERMINAL BUILDINGS - 1

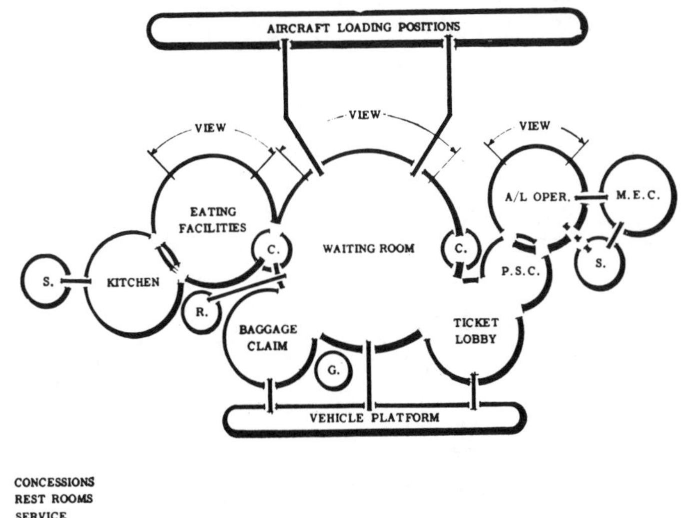
AIRCRAFT LOADING POSITIONS
VIEW
VIEW
VIEW
EATING FACILITIES
WAITING ROOM
A/L OPER.
M.E.C.
S.
KITCHEN
C.
C.
P.S.C.
S.
R.
BAGGAGE CLAIM
TICKET LOBBY
G.
VEHICLE PLATFORM
LEGEND
C. = CONCESSIONS
R. = REST ROOMS
S. = SERVICE
G. = GROUND TRANSPORTATION DISPATCH
M.E.C. = MAIL, EXPRESS, CARGO
P.S.C. = PASSENGER SERVICE COUNTER
A/L OPER. = AIRLINE OPERATIONS
= DIRECT
= SECONDARY

FIG. A – TERMINAL BUILDING SPACE RELATIONSHIP†

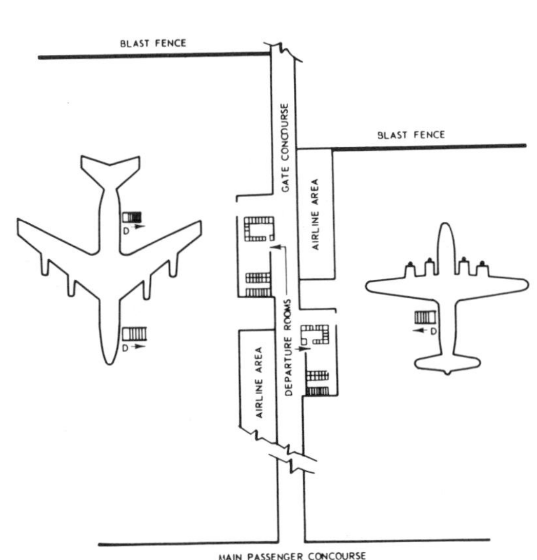
BLAST FENCE
BLAST FENCE
GATE CONCOURSE
AIRLINE AREA
DEPARTURE ROOMS
AIRLINE AREA
D
D
D
MAIN PASSENGER CONCOURSE

FIG. C – TYPICAL SINGLE-LEVEL CONCOURSE*

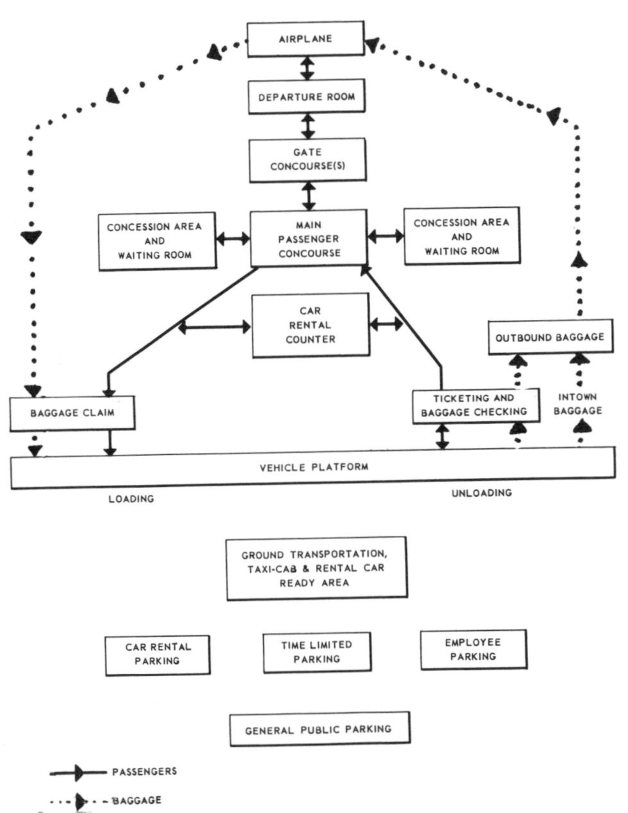
AIRPLANE
DEPARTURE ROOM
GATE CONCOURSE(S)
CONCESSION AREA AND WAITING ROOM
MAIN PASSENGER CONCOURSE
CONCESSION AREA AND WAITING ROOM
CAR RENTAL COUNTER
OUTBOUND BAGGAGE
BAGGAGE CLAIM
TICKETING AND BAGGAGE CHECKING
INTOWN BAGGAGE
VEHICLE PLATFORM
LOADING
UNLOADING
GROUND TRANSPORTATION, TAXI-CAB & RENTAL CAR READY AREA
CAR RENTAL PARKING
TIME LIMITED PARKING
EMPLOYEE PARKING
GENERAL PUBLIC PARKING
PASSENGERS
BAGGAGE

FIG. B – SCHEMATIC ARRANGEMENT OF FACILITIES, PASSENGER, AND BAGGAGE FLOW*

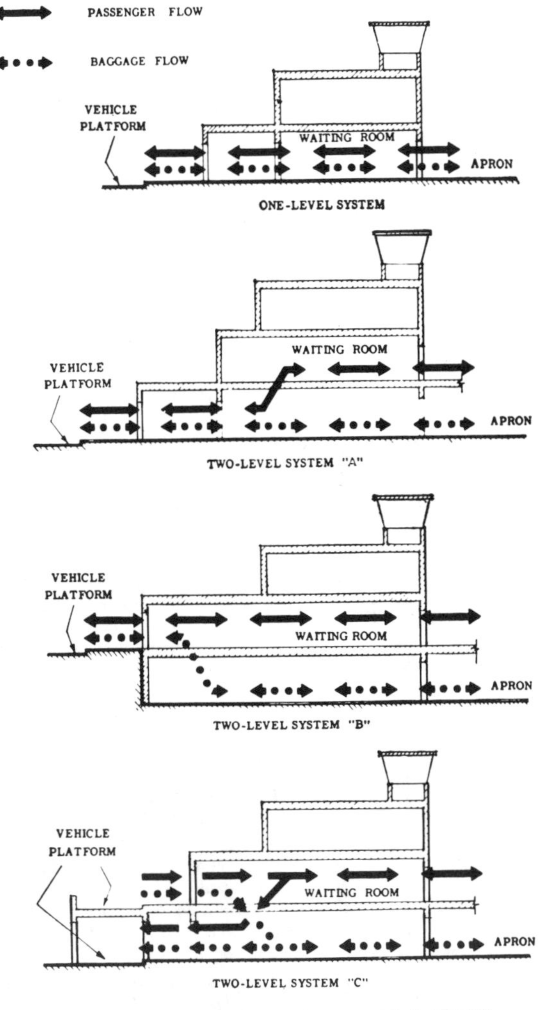
PASSENGER FLOW
BAGGAGE FLOW
VEHICLE PLATFORM
WAITING ROOM
APRON
ONE-LEVEL SYSTEM
VEHICLE PLATFORM
WAITING ROOM
APRON
TWO-LEVEL SYSTEM "A"
VEHICLE PLATFORM
WAITING ROOM
APRON
TWO-LEVEL SYSTEM "B"
VEHICLE PLATFORM
WAITING ROOM
APRON
TWO-LEVEL SYSTEM "C"

FIG. D – ONE- AND TWO-LEVEL BUILDING OPERATIONAL SYSTEM†

* From Air Transport Association of America.
† From *Airport Terminal Buildings*, Civil Aeronautics Administration.

AIRPORTS – TERMINAL BUILDINGS – 2

TYPICAL VARIATIONS OF TWO-LEVEL CONCOURSE*

UPPER (OR PASSENGER) LEVEL SHOWN

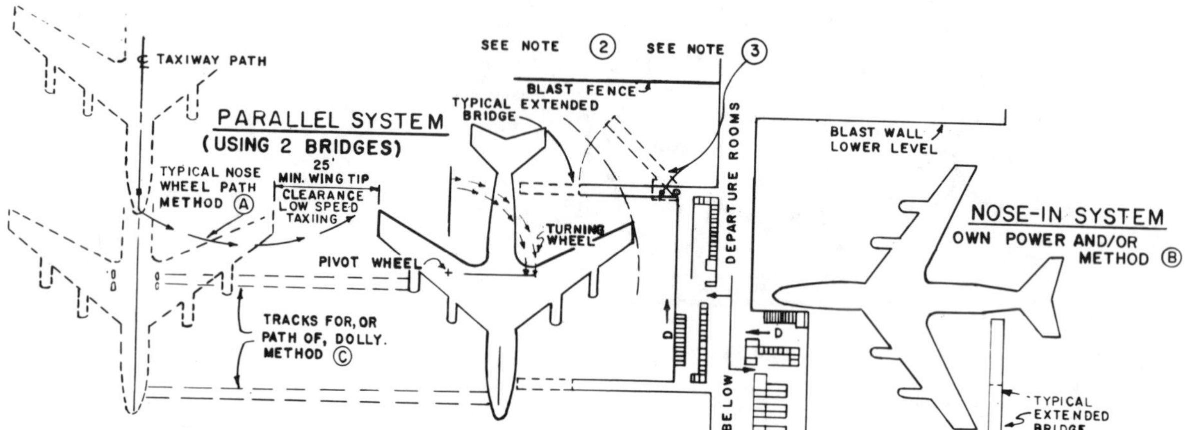

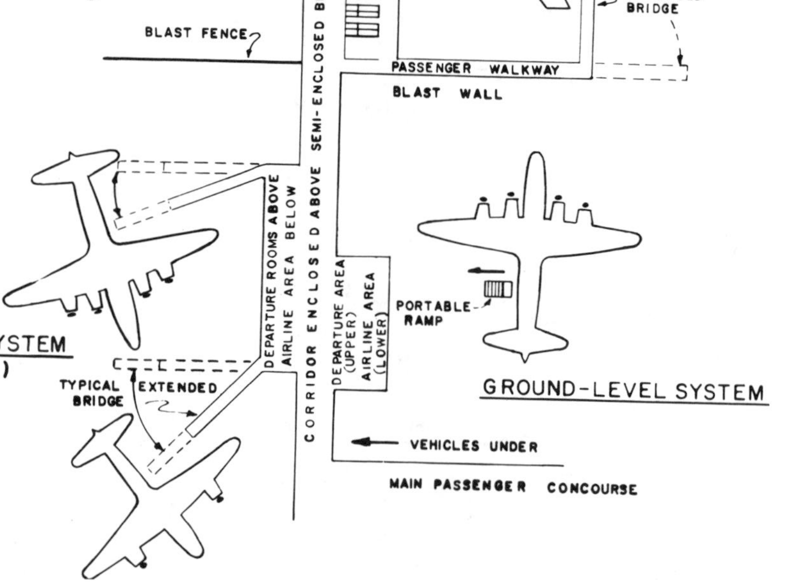

℄ TAXIWAY PATH

SEE NOTE ② SEE NOTE ③

BLAST FENCE

TYPICAL EXTENDED BRIDGE

PARALLEL SYSTEM (USING 2 BRIDGES)

TYPICAL NOSE WHEEL PATH METHOD Ⓐ

25' MIN. WING TIP CLEARANCE LOW SPEED TAXIING

TURNING WHEEL

PIVOT WHEEL

TRACKS FOR, OR PATH OF, DOLLY. METHOD Ⓒ

DEPARTURE ROOMS

D

BLAST WALL LOWER LEVEL

NOSE-IN SYSTEM OWN POWER AND/OR METHOD Ⓑ

TYPICAL EXTENDED BRIDGE

BLAST FENCE

PASSENGER WALKWAY

BLAST WALL

CORRIDOR ENCLOSED ABOVE SEMI-ENCLOSED BELOW

DEPARTURE ROOMS ABOVE AIRLINE AREA BELOW

DEPARTURE AREA (UPPER) AIRLINE AREA (LOWER)

PORTABLE RAMP

PARALLEL SYSTEM (USING 1 BRIDGE)

TYPICAL EXTENDED BRIDGE

GROUND-LEVEL SYSTEM

VEHICLES UNDER MAIN PASSENGER CONCOURSE

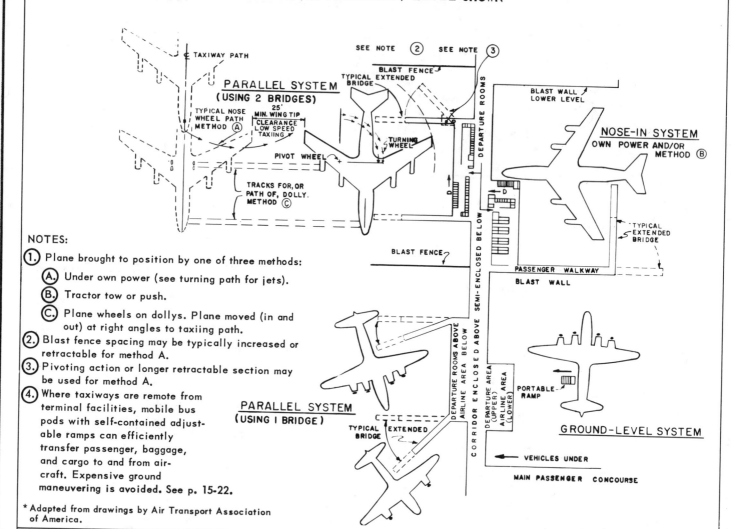

NOTES:

1. Plane brought to position by one of three methods:
   - A. Under own power (see turning path for jets).
   - B. Tractor tow or push.
   - C. Plane wheels on dollys. Plane moved (in and out) at right angles to taxiing path.
2. Blast fence spacing may be typically increased or retractable for method A.
3. Pivoting action or longer retractable section may be used for method A.
4. Where taxiways are remote from terminal facilities, mobile bus pods with self-contained adjustable ramps can efficiently transfer passenger, baggage, and cargo to and from aircraft. Expensive ground maneuvering is avoided. See p. 15-22.

* Adapted from drawings by Air Transport Association of America.

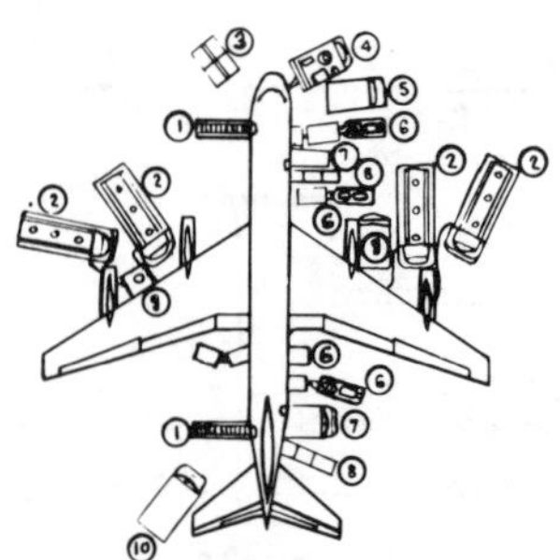

PORTABLE SERVICE FACILITIES AT GATE POSITION FOR HIGH SPEED TURNABOUT OF JET TRANSPORTS†

1. Passenger loading stairs
2. Fuel tank trucks
3. Electrical ground power cart
4. Pneumatic ground power cart
5. Air-conditioning truck
6. Luggage carts and tugs
7. Food service trucks
8. Lavatory service carts
9. Engine injection, water, and miscellaneous fluids
10. Cabin cleaners supplies and equipment

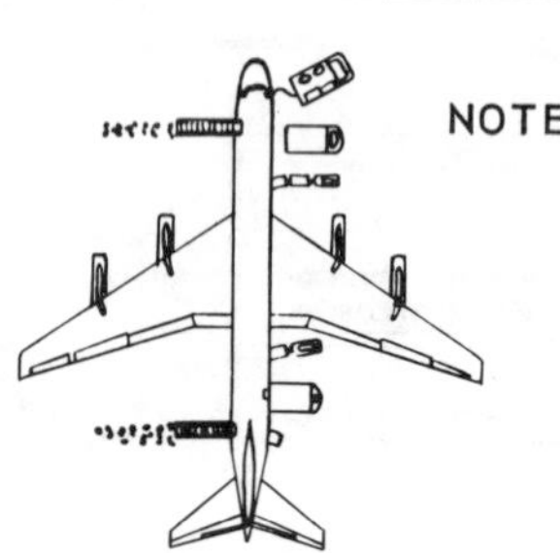

FIXED INSTALLATIONS FOR FUELING HYDRANTS, ELECTRIC POWER†

NOTE: The reduction in mobile service equipment achieved by fixed apron utility installations will substantially reduce apron congestion as well as the incidence of aircraft damage caused by mobile equipment, while maintaining or improving the turnabout time of large jet aircraft.

†Adapted from *Proceedings A.S.C.E.*, December, 1957, Paper 1040 by J. B. Edwards, assistant chief engineer, Douglas Aircraft.

TABLE A – TYPICAL FLOOR AREAS (SMALL TO MEDIUM-SIZE TERMINAL BUILDING) FOR 200 DESIGN PEAK HOUR (D.P.H.) PASSENGERS AND MINIMUM OF THREE ACTIVE CARRIER LOADING GATES

		Area Required, sq. ft.
Air Carrier Space (20' Module)		
Lobby charged to carrier	15' × 20'	300
Counter and aisle	7' × 20'	140
Reservations, communication, load dispatch, and baggage aisle	30' × 20'	600
Sq. ft. for one carrier		1,040
		× 2
		2,080
(Assume minimum of two carriers for 150 to 300 D.P.H. passengers with ready expansion capability for one or two additional carriers.)		
Air Freight and Cargo	50' × 20'	1,000
Public Space		
Waiting Room: 200 D.P.H. passengers × 7.5 sq. ft./pass.		1,500
Toilets (1 men and 1 women)		550
Corridors and vestibules		550
Typical lobby concessions		
Public phone booths: 200 D.P.H. ÷ 50 pass./booth = 4 × 10 sq. ft./booth =		40
Newsstand, gifts, candy		160
Car rental and taxi-limousine		200
		3,000
Restaurant concession: 200 D.P.H. × 7.5 sq. ft./pass.		1,500
Airport Management		750
Building Services (including boiler room, air conditioning, transformer room, storage, and corridors)		1,500
Aviation Office Rental: (Including Base Operators, Cargo Airlines and Business Flying Activities)		750
Typical Minimum Total Terminal Area Required (Rough approximation exclusive of airlines departure rooms, post office, customs and concessions in excess of those listed above)		
200 D.P.H. passengers × 50 sq. ft./pass.	10,000 sq. ft.	

For Weather Bureau and Control Tower Space, see p. 15-25.

AIRPORTS-CONTROL TOWER & TYP. AIRLINE BAY

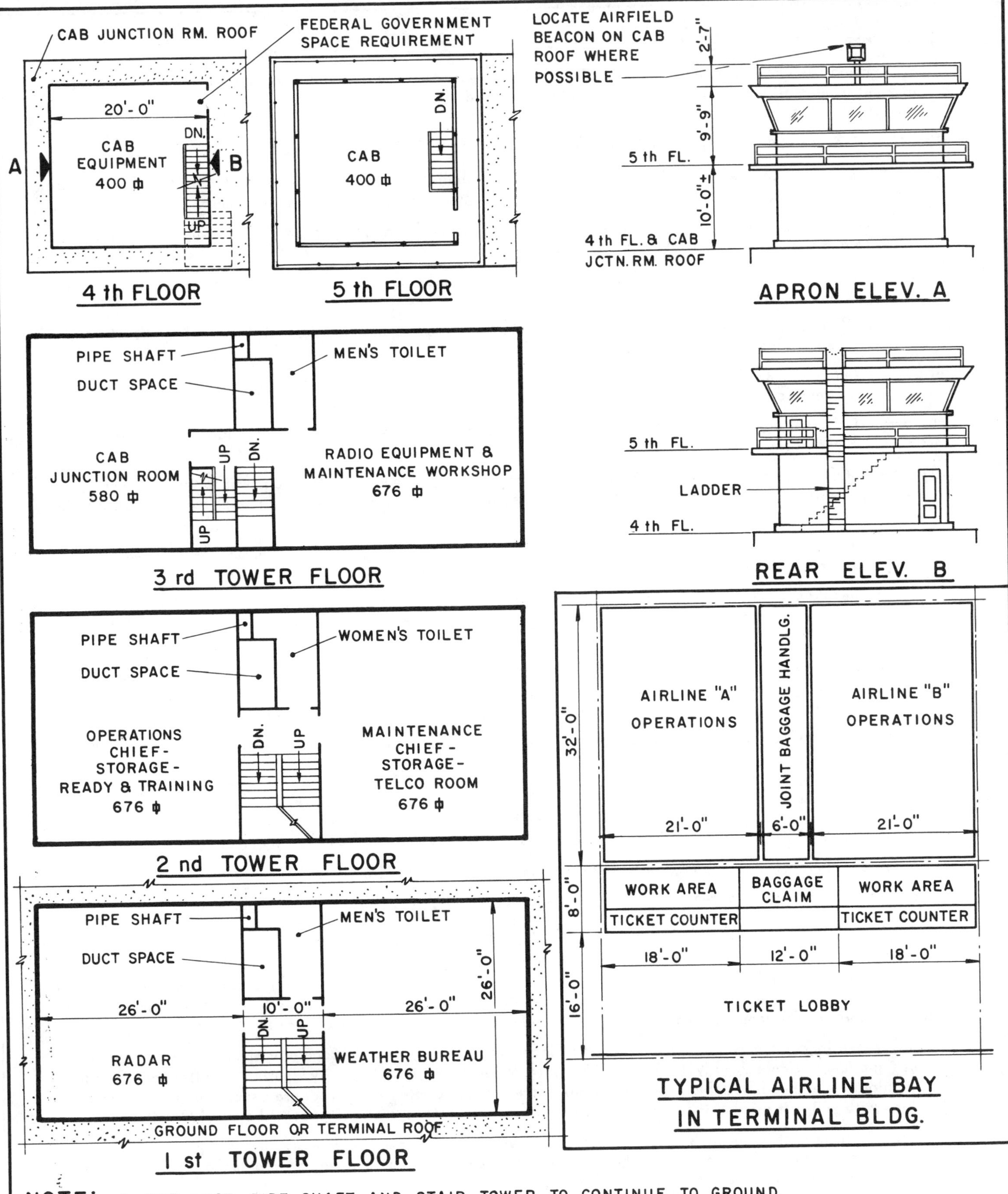
CAB JUNCTION RM. ROOF
FEDERAL GOVERNMENT SPACE REQUIREMENT
20'-0"
CAB EQUIPMENT 400 ф
DN.
UP
A
B
CAB 400 ф
DN.
4 th FLOOR
5 th FLOOR
LOCATE AIRFIELD BEACON ON CAB ROOF WHERE POSSIBLE
2'-7"
9'-9"
10'-0"±
5 th FL.
4 th FL. & CAB JCTN. RM. ROOF
APRON ELEV. A
PIPE SHAFT
DUCT SPACE
MEN'S TOILET
CAB JUNCTION ROOM 580 ф
UP
DN.
RADIO EQUIPMENT & MAINTENANCE WORKSHOP 676 ф
3 rd TOWER FLOOR
5 th FL.
LADDER
4 th FL.
REAR ELEV. B
PIPE SHAFT
DUCT SPACE
WOMEN'S TOILET
OPERATIONS CHIEF-STORAGE-READY & TRAINING 676 ф
DN.
UP
MAINTENANCE CHIEF-STORAGE-TELCO ROOM 676 ф
2 nd TOWER FLOOR
PIPE SHAFT
DUCT SPACE
MEN'S TOILET
26'-0"
10'-0"
26'-0"
26'-0"
DN.
UP
RADAR 676 ф
WEATHER BUREAU 676 ф
GROUND FLOOR OR TERMINAL ROOF
1 st TOWER FLOOR
32'-0"
AIRLINE "A" OPERATIONS
JOINT BAGGAGE HANDLG.
AIRLINE "B" OPERATIONS
21'-0"
6'-0"
21'-0"
8'-0"
WORK AREA
BAGGAGE CLAIM
WORK AREA
TICKET COUNTER
TICKET COUNTER
18'-0"
12'-0"
18'-0"
16'-0"
TICKET LOBBY
TYPICAL AIRLINE BAY IN TERMINAL BLDG.

NOTE:

1. THE DUCT, PIPE SHAFT AND STAIR TOWER TO CONTINUE TO GROUND.
2. GROUND LEVEL SPACE FOR STANDBY POWER USUALLY REQUIRED.
3. CONTROL TOWER SHOWN IS ULTIMATE FOR USE WITH MEDIUM SIZED TERMINALS.
4. DESIRABLE MINIMUM CAB EYE LEVEL OF CONTROL TOWER ABOVE HIGHEST POINT OF RUNWAY AND MAIN APRON AND TAXIWAYS IS 45 FEET. CHECK SIGHT LINE CLEARANCES.

AIRPORTS-LIGHTING- I

1. *STANDARDS*

Airport lighting-system design shall comply with F.A.A. (formerly C.A.A.) requirements, and all items of material and equipment used shall meet F.A.A. standards and be approved by them.

2. *COMPONENTS OF AN AIRPORT LIGHTING SYSTEM*

(a) *Runway Lighting*, consisting of lateral marker lights along both sides at approximately 200-ft. intervals, and green threshold marker lights across each end to indicate the limits of the usable landing pavement.

(b) *Taxiway Lighting*, consisting of blue marker lights along both sides of each taxiway to indicate ground traffic routes to and from runways for take-off and landing.

(c) *Approach Lighting*, consisting of marker lights beyond the thresholds to provide visual aid to supplement instrument landing system equipment.

(d) *Rotating Beacon*, to indicate location of airport. Where the airport is in close proximity to another airport, a code beacon is used to identify the airport quickly and positively.

(e) *Lighted Indicators*, to indicate wind direction or landing direction. These may be wind cones, wind tees, or tetrahedrons, as required to suit airport classification and operational requirements.

(f) *Obstruction Lighting*, to indicate hazards to normal aerial navigation such as buildings, electric-power lines, tanks, towers, smokestacks, etc.

(g) *Power-Supply System*, consisting of provisions to extend the local source of electric power to the switchgear, transformers, regulators, contactors, and other accessories required to energize the airfield lighting system. Power supply is usually 2300 volt, 60 cycle.

(h) *Control System*, consisting of control panels, control cable, circuiting relays, and other accessories required to energize, de-energize, select, and otherwise control the various airfield lighting-system components.

(i) *Transformer Vault*, to house the necessary power-supply system equipment, as well as the control-system relay equipment. The vault is usually a separate building or part of a larger building and is located at least 400 feet from the control tower to minimize radio interference.

(j) *Cable and Duct System*, consisting of wires and cables between the vault and field lighting units, and control cables between the vault and control tower. All wiring and cables are usually installed direct-buried in unpaved areas, in concrete-encased ducts under paved areas and roads, and in rigid steel conduit within buildings.

(k) *Special Visual Aids*, consisting of floodlights etc. where the local airport conditions require additional visual aids not listed above.

3. *DESIGN DATA*

(a) For typical runway and taxiway lighting arrangements and design criteria, see pp. 15-28 and 15-29.

(b) For typical approach lighting arrangement and design criteria, see p. 15-09, Fig. B.

(c) For determining the size of regulators for runway and taxiway lighting, see p. 15-27, Figs. A and B.

(d) For determining obstructions to be lighted see pp. 15-06 and 15-07.

(e) Runway lighting systems are usually designed to be supplied through an automatically regulated constant-current 6.6-ampere series circuit. Since the full-load open-circuit voltage is very high, isolating transformers are provided at each light so that individual lamp burnouts do not affect the basic circuit which is looped through the primary of the transformer. The lamp is connected on the secondary side of the transformer, and thus isolated.

(f) Runway lighting-system cables for series circuit loops may have to carry the regulator open-circuit voltages at times and are therefore Usually No. 8 A.W.G. 3000-volt (5000-volt for high-intensity lighting circuits) cables with polychloroprene sheath suitable for direct burial in the ground. Cables for control are usually multiconductor cables with 600-volt insulation, since control-circuit voltage is always single-phase 60-cycle 120 volts.

(g) Airport lighting-system components are primarily controlled from the control tower. A secondary point of control is also provided for emergency operation and is usually located in the regulator vault. In order to effect the transferring of control points, transfer relays are required.

(h) In areas where lightning storms are prevalent and the isokeraunic incidence is 60 or more, it is advisable to provide a counterpoise network of no. 4 A.W.G. bare copper wire and associated ground rods or its equivalent throughout the system for all series circuits. Additional provisions such as the use of shielded cable are dependent upon a study of the local conditions at the airport.

AIRPORTS – LIGHTING – 2

(Fig. A)

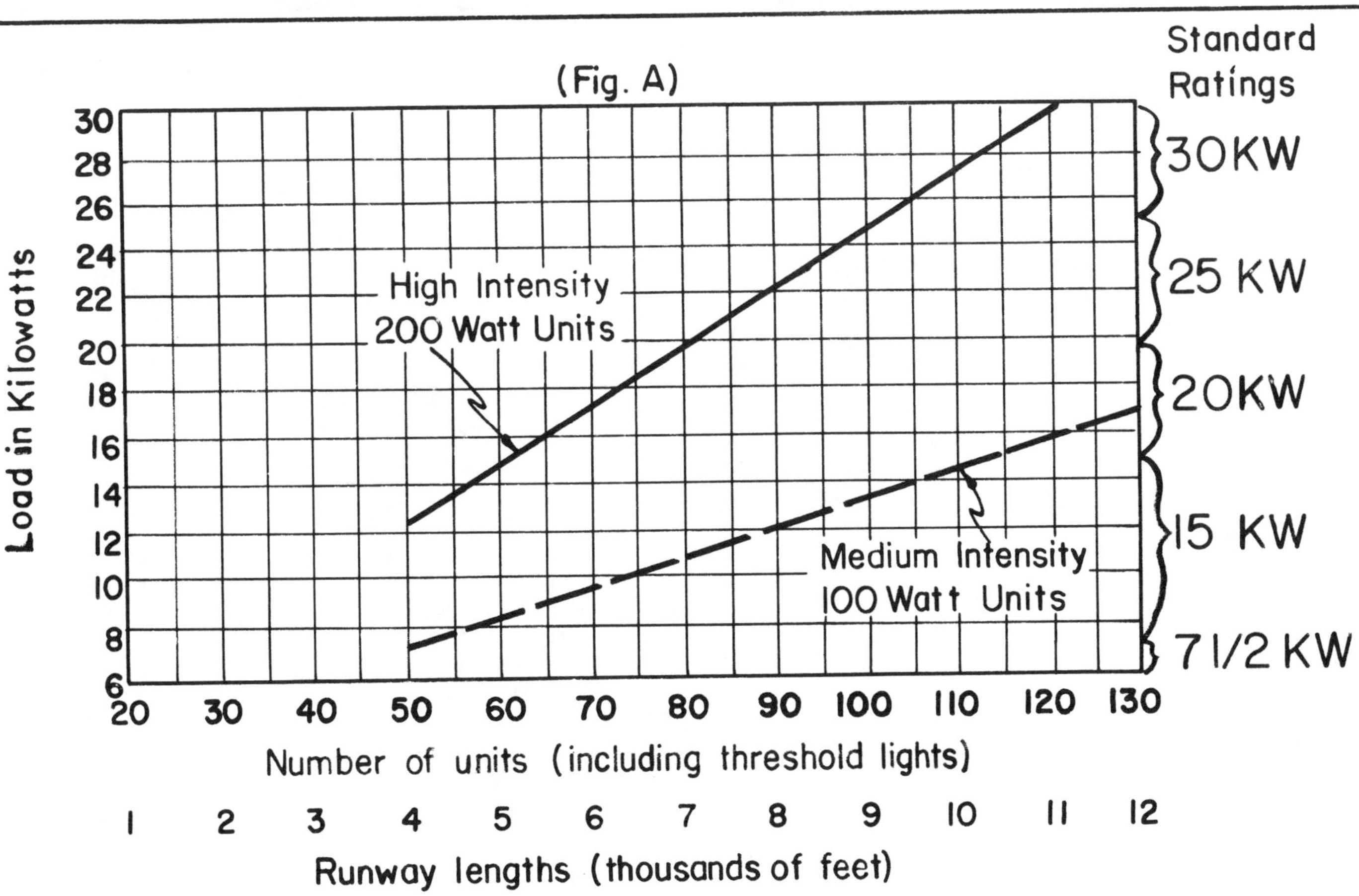

Standard Ratings

30KW

25 KW

20KW

15 KW

7 1/2 KW

Load in Kilowatts

30 28 26 24 22 20 18 16 14 12 10 8 6

High Intensity 200 Watt Units

Medium Intensity 100 Watt Units

20 30 40 50 60 70 80 90 100 110 120 130

Number of units (including threshold lights)

1 2 3 4 5 6 7 8 9 10 11 12

Runway lengths (thousands of feet)

RUNWAY REGULATOR SELECTION

(SERIES CIRCUITS)

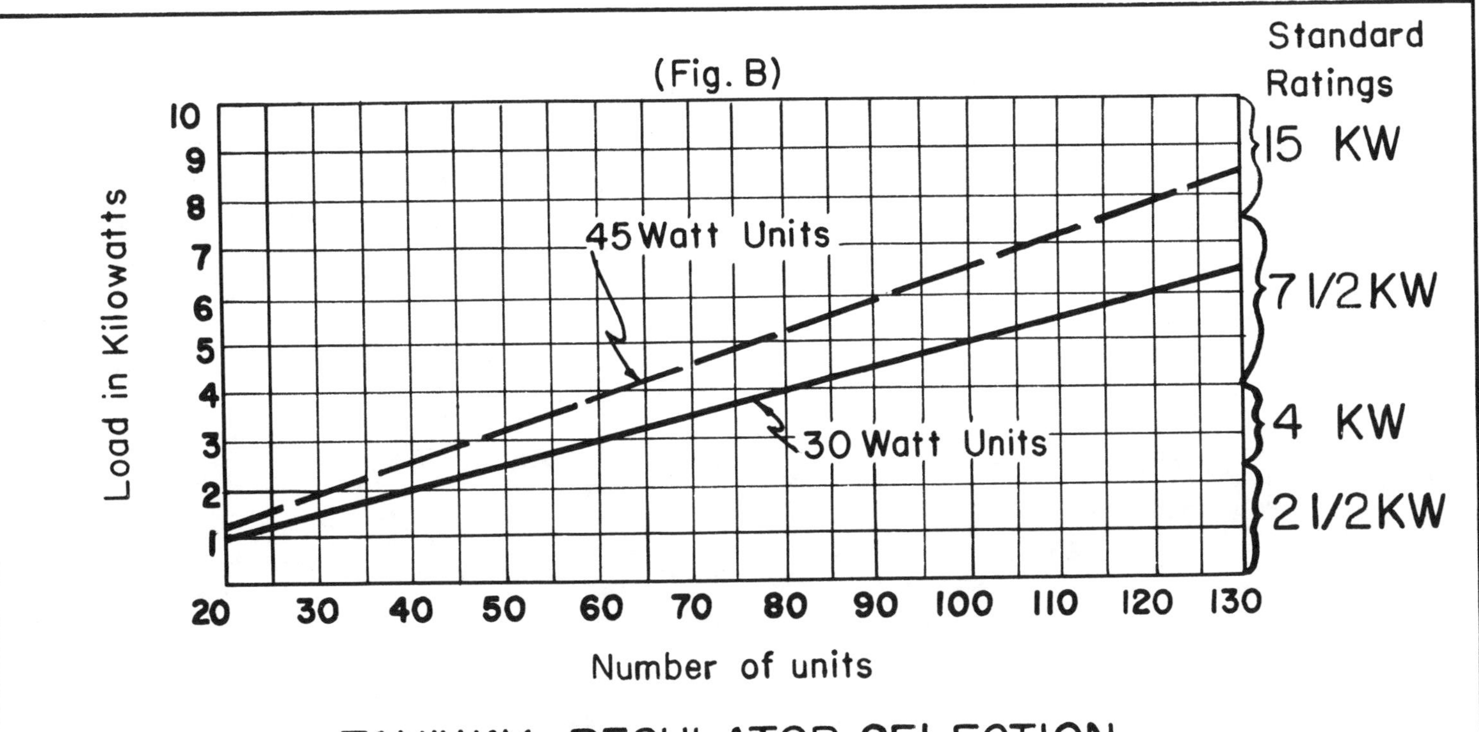
Standard Ratings
(Fig. B)
15 KW
7 1/2 KW
4 KW
2 1/2 KW
Load in Kilowatts
10 9 8 7 6 5 4 3 2 1
45 Watt Units
30 Watt Units
20 30 40 50 60 70 80 90 100 110 120 130
Number of units

TAXIWAY REGULATOR SELECTION

(SERIES CIRCUITS)

Data taken from C.A.A. Airfield Lighting Design Criteria.

AIRPORT

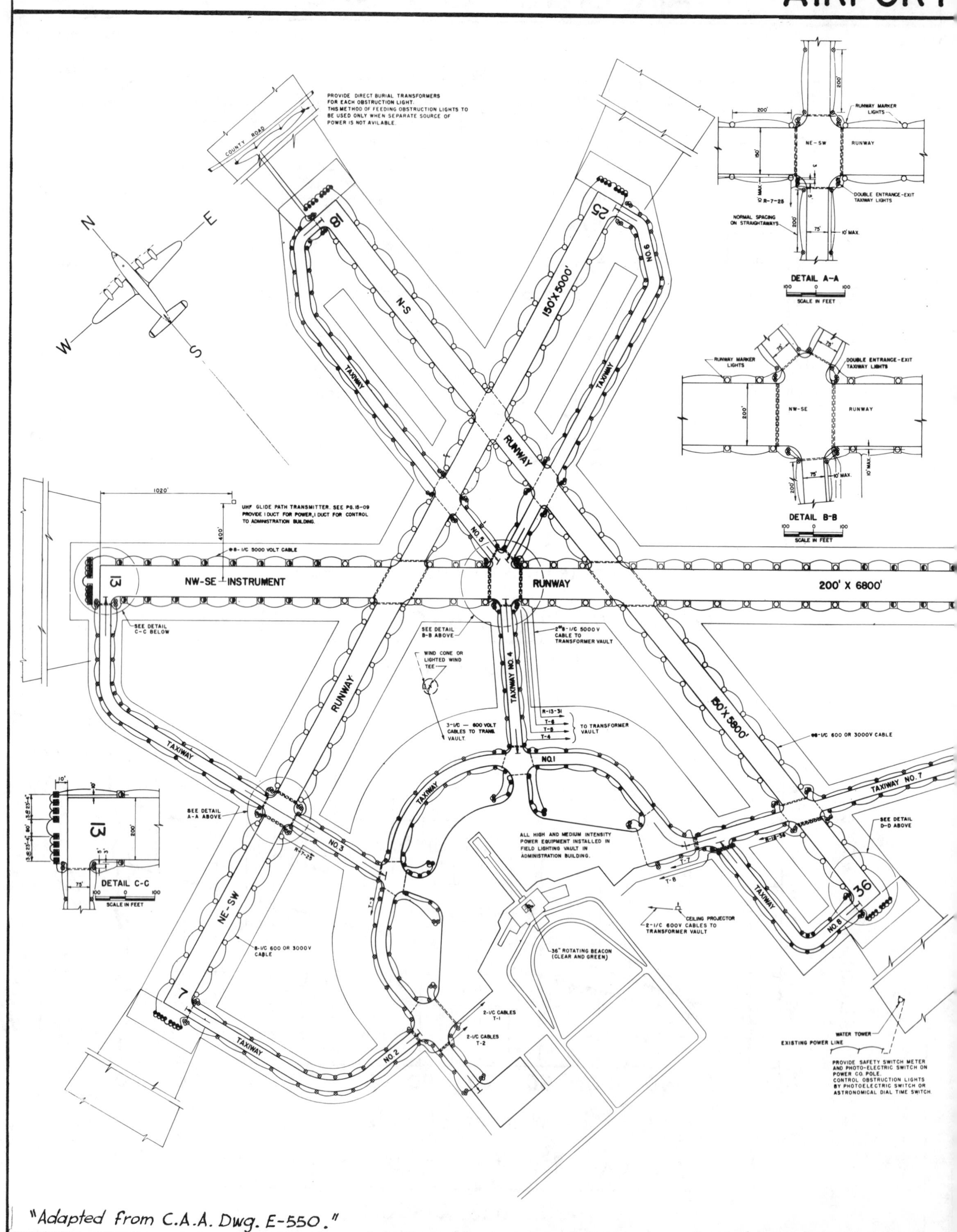

PROVIDE DIRECT BURIAL TRANSFORMERS FOR EACH OBSTRUCTION LIGHT. THIS METHOD OF FEEDING OBSTRUCTION LIGHTS TO BE USED ONLY WHEN SEPARATE SOURCE OF POWER IS NOT AVILABLE.
COUNTY ROAD
N
E
W
S
18
N-S
TAXIWAY
25
NO. 6
150'X 5000'
RUNWAY
TAXIWAY
RUNWAY MARKER LIGHTS
NE-SW
RUNWAY
DOUBLE ENTRANCE-EXIT TAXIWAY LIGHTS
NORMAL SPACING ON STRAIGHTAWAYS
DETAIL A-A
SCALE IN FEET
NW-SE
DETAIL B-B
1020'
UHF GLIDE PATH TRANSMITTER. SEE PG. 15-09 PROVIDE 1 DUCT FOR POWER, 1 DUCT FOR CONTROL TO ADMINISTRATION BUILDING.
8-1/C 5000 VOLT CABLE
NO. 5
13
NW-SE INSTRUMENT
RUNWAY
200' X 6800'
SEE DETAIL C-C BELOW
SEE DETAIL B-B ABOVE
2-8-1/C 5000V CABLE TO TRANSFORMER VAULT
WIND CONE OR LIGHTED WIND TEE
TAXIWAY NO. 4
RUNWAY
150'X 5800'
R-13-31
T-6
T-5
T-4
TO TRANSFORMER VAULT
3-1/C — 600 VOLT CABLES TO TRANS. VAULT.
8-1/C 600 OR 3000V CABLE
TAXIWAY
NO. 1
TAXIWAY NO. 7
SEE DETAIL A-A ABOVE
SEE DETAIL D-D ABOVE
NO. 3
ALL HIGH AND MEDIUM INTENSITY POWER EQUIPMENT INSTALLED IN FIELD LIGHTING VAULT IN ADMINISTRATION BUILDING.
DETAIL C-C
SCALE IN FEET
NE-SW
8-1/C 600 OR 3000V CABLE
T-7
T-8
36
TAXIWAY
NO. 8
CEILING PROJECTOR
2-1/C 600V CABLES TO TRANSFORMER VAULT
36" ROTATING BEACON (CLEAR AND GREEN)
7
2-1/C CABLES T-1
2-1/C CABLES T-2
TAXIWAY
NO. 2
WATER TOWER
EXISTING POWER LINE
PROVIDE SAFETY SWITCH METER AND PHOTO-ELECTRIC SWITCH ON POWER CO. POLE. CONTROL OBSTRUCTION LIGHTS BY PHOTOELECTRIC SWITCH OR ASTRONOMICAL DIAL TIME SWITCH.

"Adapted from C.A.A. Dwg. E-550."

HTING — 3

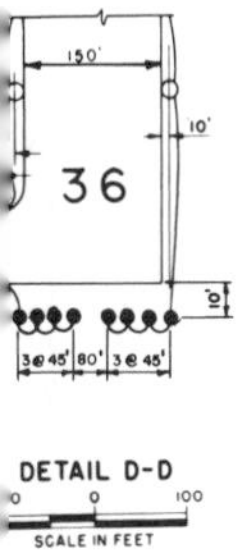

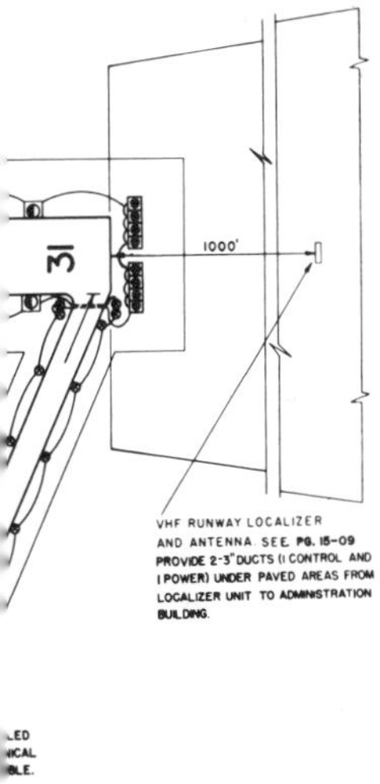

DETAIL D-D

SCALE IN FEET

VHF RUNWAY LOCALIZER AND ANTENNA. SEE PG. 15-09 PROVIDE 2-3" DUCTS (1 CONTROL AND 1 POWER) UNDER PAVED AREAS FROM LOCALIZER UNIT TO ADMINISTRATION BUILDING.

MINIMUM LIGHTING

LIGHTING COMPONENTS:	AIRPORT CATEGORY			
	PRIVATE	LOCAL	TRUNK	CONTINENTAL & INTERCONTINENTAL
BEACON — Small	√	√		
Large			√	√
WIND DIRECTION INDICATOR —				
Illum. Cone	√			
Tee		√		
Tetrahedron (Controllable)			√	√
RUNWAY LIGHTS —				
Medium Intensity		√	√	√
High Intensity			√	√
Bi-Directional, Fixed Focus			(a)	(a)
Bi-Directional, movable beam				(b)
Uni-Directional, Fixed Focus				(c)
TAXIWAY LIGHTS			√	√

NOTES: Approach lights are furnished by C.A.A. as a supplement to the Instrument Landing System. See p. 15-09
(a) For instrument runways on average airport. (b) & (c) For instrument runways on busy terminal airport with operating minimums below 400 ft ceiling & 1 mile visibility.

GENERAL — For lighting requirements at different class airports, see table above except for airports used for emergency landings of large planes or as auxiliary or alternate terminal airports.

ROTARY BEACON — Provide at airports equipped with runway marker lights. It is more desirable than an airways beacon and a code beacon.

THRESHOLD LIGHTS — Space generally as required o.c. at end of runway. For paved runways, lights shall not be more than 10 feet from the end of the runway. Space lights symmetrically in two groups, one on each side of clearance gap, using six units where the rows of runway or strip lights are spaced laterally less than 120 ft. apart and, 8 units for distances equal or over 120 feet. All lights across end of strip shall be on a line at right angles to strip center line. Locate threshold lights as shown on plan at left.

CONTACT & TAXIWAY LIGHTS — Locate as shown on plan at left.

WIND TEE OR CONE — Use a wind tee, 12'-0" tilting type wind-cone or tetrahedron.

ISOLATED OBSTRUCTIONS — Red always used for obstruction lights. If lights are not readily accessible for periodic inspections an electric signalling device, either aural or visual to indicate failure, should be installed.

Obstruction less than 100 feet high: Narrow tower, poles, and trees. Install at least two 100 watt lamps enclosed in aviation red fresnel or prismatic light globes, on top of structure. The lights should burn simultaneously and be so situated as to insure unobstructed visibility.

Obstruction less than 150 feet high: Narrow tower, mast, flagpole, etc. Install at top of structure at least 2-100 watt lamps enclosed in aviation red fresnel or prismatic obstruction light globes. The top lights should burn simutaneously and be so situated as to insure unobstructed visibility. On levels at about two thirds and one third of over all height there should be installed at least two 100 watt lamps. Each light should be placed on diagonally or diametrically opposite positions of the structure.

Obstruction less than 450 feet high: There should be installed at the top of the structure a 300 mm electric code beacon with two 500 watt lamps and aviation red color filters. At levels approximately two thirds and one third of overall height there should be installed at least 2-100 watt lamps on diametrically or diagonally opposite positions.

Obstructions over 450 feet high, Smokestacks, Water Towers, Gas Holders: Consult C.A.A.

AIRPORT LIGHTING SYMBOLS:

- Rotating Beacon
- Elevated Medium Inten. R'way marker
- Elevated, High Inten. R'way marker
- Elevated, High Int. Thres. Marker, green
- Elevated. High Intensity Runway marker light. 180° Aviation Yellow-180° Clear
- Elevated Med. Int. Thres. Light-green
- Elevated Med. Int. Taxiwy marker, blue.
- Obstruction light, Single Unit, Red
- Obstruction light-Duplex Unit, Red
- Obstruction light on pole, Red.
- Wind Cone, Illuminated
- Wind Tee, Illuminated
- Ceiling Projector
- Underground Cable
- Underground ducts (2-3" ducts concrete encased)
- Segmented Circle Airport Marker

HELIPORTS – GENERAL

HELIPORTS

HELIPORT DESIGN is a broad term covering all requirements for the ground operations of air carriers which require very small ground space relative to conventional aircraft. Since the helicopter is the first of the verticraft to come into general use, current (1959) heliport design is based on the current capability of these aircraft, and the design criteria must therefore be considered transitional. The following definitions are applicable to this study:

Verticraft – Commercial scheduled air carriers, including helicopters, convertiplanes, and similar craft, which can operate from a 500-ft. or smaller runway. (Certain verticraft may require longer runways for most economical operation; most types will use smaller runways.)

Heliport – "Airfield" to service simultaneously two or more enplaning and/or deplaning helicopters, convertiplanes, etc., and equipped with support navigation, maintenance, and passenger handling facilities (bus-terminal concept).

Helistop – Landing space or platform, primarily for helicopter-type operation, where passengers may be picked up or deposited (bus-stop concept).

With the advent of even faster commercial transports, the time of travel from outlying airports to downtown areas becomes less tolerable. An analysis of downtown to downtown time is an important element in the decision to install heliports. Figures B and C on p. 15-31 show a theoretical time analysis. Using these charts, if the distance between cities in Fig. A on p. 15-31 were 300 miles, it would take less travel time with a 200-m.p.h. convertiplane than with a 350-m.p.h. fixed-wing aircraft.

ROOFTOP DESIGN CRITERIA

Downtown helistops, or modified heliports, are frequently being master-planned as future rooftop additions to parking garages, bus terminals, and similar large and suitable structures. Building-height zoning restrictions would be required for a minimum of 500-ft. in all directions and as required to maintain a desirable minimum of 8 horizontal to 1 vertical glide slope clearance to edge of landing area in principal landing directions. Landing gear – wheels, skids, pontoons, etc. – used for rooftops would be designed to deform under excessive impact to protect passengers and, indirectly, the rooftop. Dimensions and loading are far from standardized, but current tentative design criteria are as shown in Table B below.

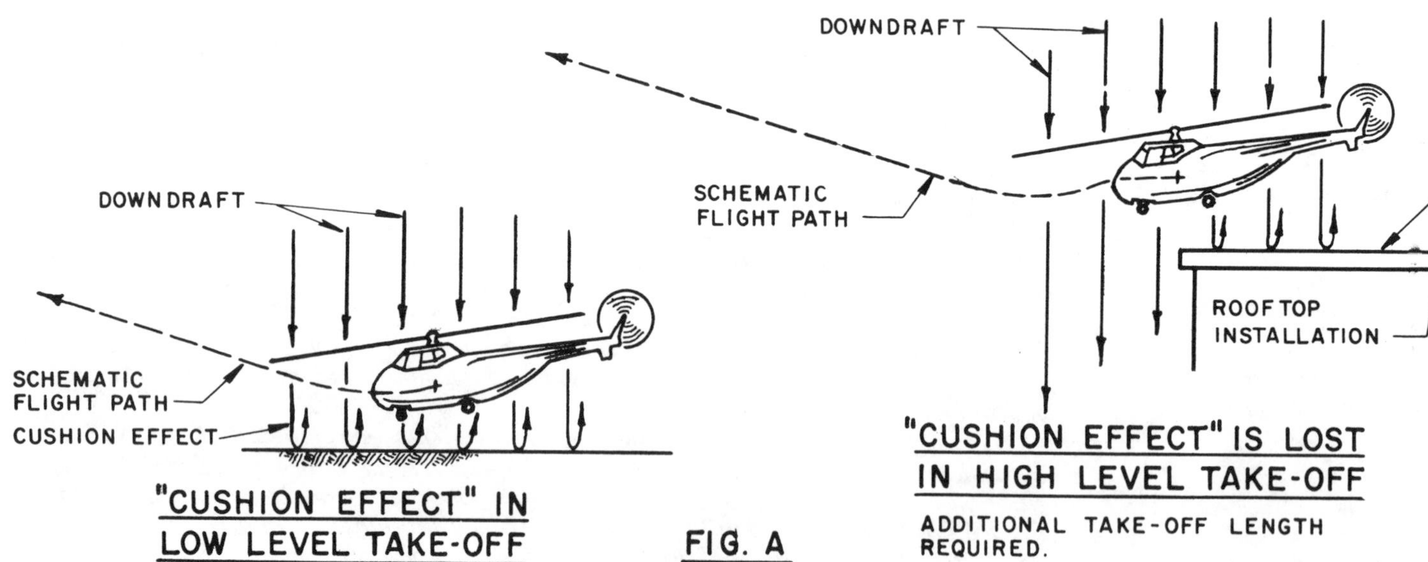
DOWNDRAFT
SCHEMATIC FLIGHT PATH
CUSHION EFFECT
"CUSHION EFFECT" IN LOW LEVEL TAKE-OFF
DOWNDRAFT
SCHEMATIC FLIGHT PATH
ROOF TOP INSTALLATION
"CUSHION EFFECT" IS LOST IN HIGH LEVEL TAKE-OFF
ADDITIONAL TAKE-OFF LENGTH REQUIRED.

FIG. A

TABLE B

Future Verticraft Type	Estimated Maximum Gross Weight Used on Rooftops, lb.	Maximum Impact Concentrated Load, in lb. on 1 S. F. (70% gross wt.)*	Desirable Minimum Impact Landing Surface per Runway, ft.
Personal (1 - 5 passengers)	2,000 - 7,000	1,400 - 4,900	50 × 75
Taxi (5 - 10 passengers)	10,000	7,000	50 × 100
Small transport (10 - 20 passengers)	15,000	10,500	75 × 150
Medium transport (20 - 40 passengers)	30,000	21,000	75 × 200
Large transport (40 - 80 passengers)	60,000	Maximum 25,000 (special landing gear required)	100 × 300

*Impact loading based on conventional landing wheels unless noted.

HELIPORTS - TIME SAVING ANALYSIS

THE TIME SAVING AND THE REVENUE PASSENGER POTENTIAL MUST BE STUDIED. THE CAPACITY OF A TWO-POSITION HELIPORT CAN BE ROUGHLY ESTIMATED AT 30 VERTICRAFT MOVEMENTS PER HOUR. AT 40 PASSENGERS PER MOVEMENT, THIS WOULD BE 15 x 40 = 600 ARRIVING PASSENGERS PER HOUR. WHERE A NUMBER OF HELIPORTS AND HELISTOPS SERVE ONE COMMUNITY, THE SITING MUST CONSIDER THE OVERALL FLIGHT PATTERNS TO AVOID CONGESTION OR CONFLICT.

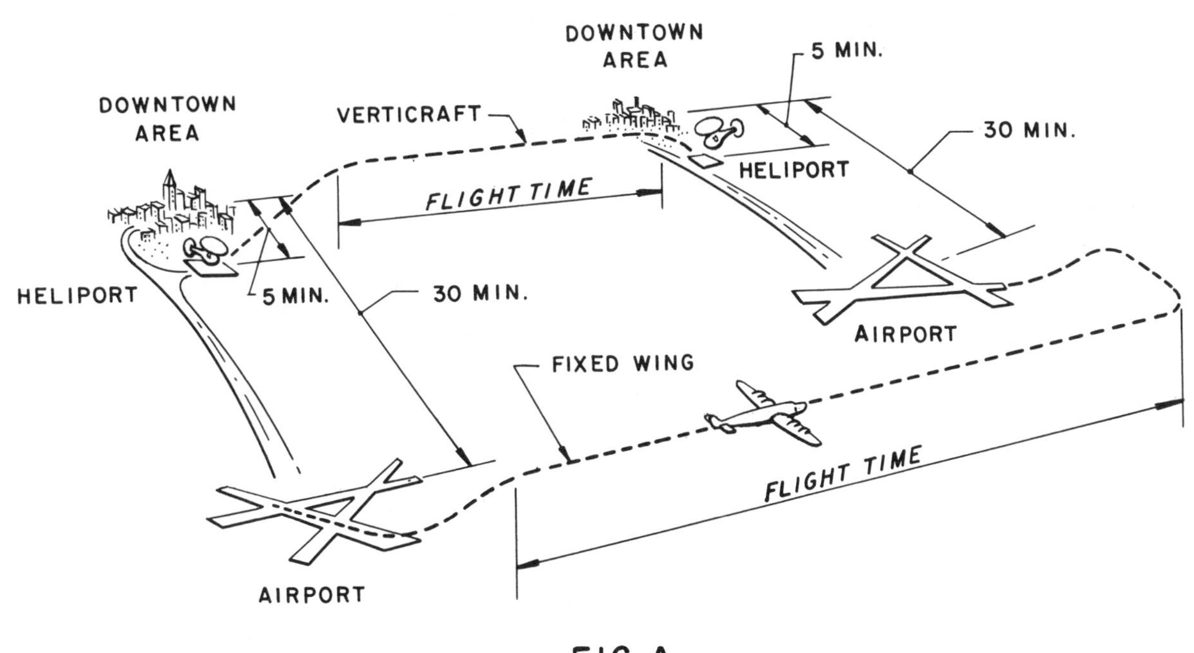

DOWNTOWN AREA
5 MIN.
DOWNTOWN AREA
VERTICRAFT
30 MIN.
HELIPORT
FLIGHT TIME
HELIPORT
5 MIN.
30 MIN.
AIRPORT
FIXED WING
FLIGHT TIME
AIRPORT

FIG. A

FIG. B

DATA BASED ON ACTUAL CONDITIONS

MINUTES REQUIRED	FIXED WING	VERTICRAFT, HELICOPTER & CONVERTIPLANE
TRANSIT TO AND FROM AIRPORT	60'	10'
TAKE OFF CLIMB APPROACH LANDING	8'	4'
* TOTAL MANEUVER & GROUND TRANSIT TIME	68'	14'

* SEE FIG. A

† CONVERTIPLANES COMBINE THE CHARACTERISTICS OF BOTH HELICOPTERS AND FIXED WING AIRCRAFT.

FIG. C

DOWNTOWN TO DOWNTOWN SPEED COMPARISON

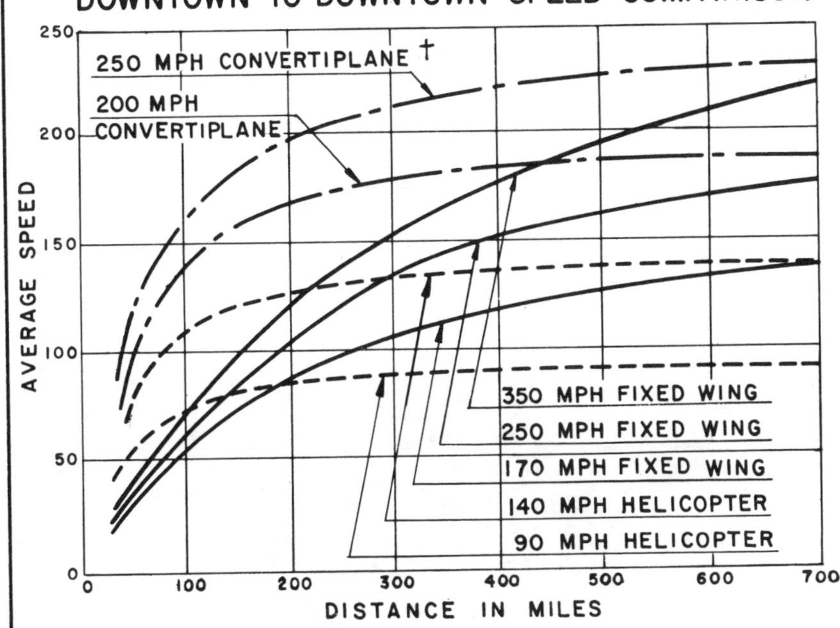

250 MPH CONVERTIPLANE †
200 MPH CONVERTIPLANE
350 MPH FIXED WING
250 MPH FIXED WING
170 MPH FIXED WING
140 MPH HELICOPTER
90 MPH HELICOPTER
AVERAGE SPEED
0
50
100
150
200
250
0
100
200
300
400
500
600
700
DISTANCE IN MILES

$$\text{AVG. SPEED} = \frac{\text{TRAVEL DISTANCE}}{\text{FLIGHT TIME} + \text{TOTAL MANEUVER} + \text{GROUND TRANSIT TIME}}$$

FIG. C – ADAPTED FROM "THE CONVERTIPLANE", OCT. 1954 THE AIR COORDINATING COMMITTEE, WASH. D.C.

HELIPORTS – LAYOUT-1

WITHIN LIMITS OF CONVENTIONAL AIRPORTS

APPROACH ZONES

Length	2000'
Width (at runway end)	250'
Width (at outer end)	500'
Glide-angle ratio (from runway end)	20:1

NOTE: All glide-angle ratios are given thus: Horiz. : Vert.

LANDING AND TAKE-OFF AREAS

RUNWAY

Length	450'
Width	40'
Longitudinal slope	1% max.
Transverse slope	1.5% req.
Lateral clear zone	125'
Minimum distance between center line of parallel runways	300'
Slope in lateral clear zone (between shoulder and lateral clearance line in any direction)	2% min. to 5% max.
Minimum sight distance for two points mutually visible, 5 ft. above pavement.	450'
Shoulder width	30'
Shoulder, longitudinal slope	1% max.
Shoulder, transverse slope	2% min. to 3% max.

APRON

Pavement fillets	R = 25' min.
Slope in any direction	0.5% min., 2% max.
Shoulder width	30'
Shoulder transverse slope	2% min., 3% max.
Lateral clear zone (edge of apron to obstacle)	75'

OVERRUN

Length	100' min.
Width	100'
Longitudinal slope	1% max.
Transverse slope	2% min., 3% max.

TAXIWAY

Width	30'
Longitudinal slope	2% max.
Transverse slope	1.5% req.
Lateral clearance zone (from taxiway-edge to obstacle)	75'
Slope in lateral clearance zone (between taxiway shoulder and lateral clear zone in any direction)	2% min. to 5% max.

NOTE: Runway and taxiway shoulders to have double bituminous surface treatment for dust control.

SELF-CONTAINED HELIPORTS

(Downtown Heliports, Bus-Terminal, or Bus-Stop Concept)

APPROACH ZONES

1. Glide-angle ratio 8:1 to 20:1 in direction of wind, 5:1 in other direction.
2. Additional take-off length required for rooftop installations due to loss of "cushion effect" of air in low-level take-off. See Fig. A, p. 15-30.
3. Attention should be given to downward blast from rotors, especially to downtown areas.
4. Where "straight-in approach" is impossible, 600' to 700' minimum turning radius should be provided.

LANDING AND TAKE-OFF AREAS

100' x 100' minimum. Ground operation facilities additional. 200' x 400' for metropolitan area heliports, capable of absorbing future increase in operation. Ground operation facilities additional.

In designing surface-drainage patterns, change of grades at relatively short intervals cause excessive flexing of aircraft rotor blades during taxiing, and may result in damage to the aircraft.

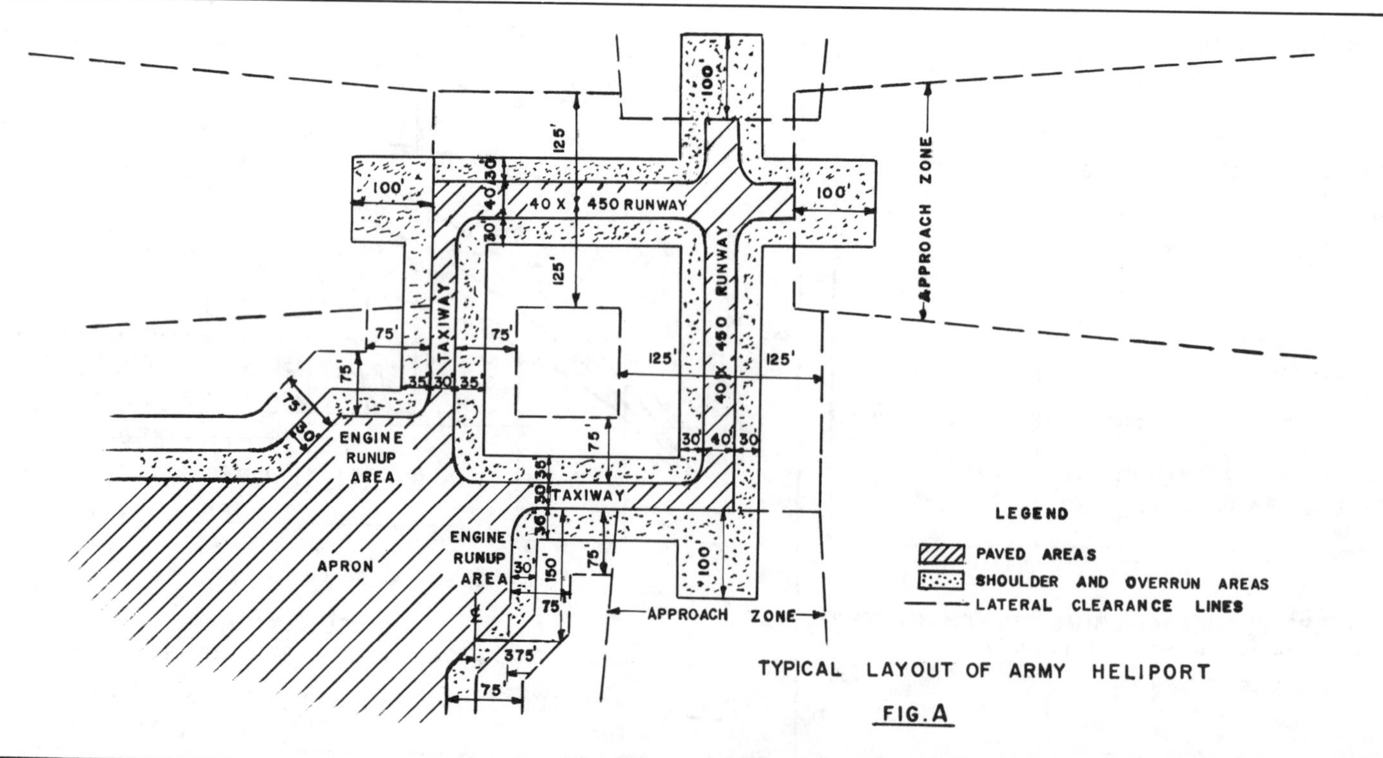

100'
125'
40 X 450 RUNWAY
100'
APPROACH ZONE
RUNWAY
40 X 450
TAXIWAY
75'
125'
125'
75'
35 30 35
30 40 30
ENGINE RUNUP AREA
TAXIWAY
APRON
ENGINE RUNUP AREA
150'
75'
100
375'
APPROACH ZONE
LEGEND
PAVED AREAS
SHOULDER AND OVERRUN AREAS
LATERAL CLEARANCE LINES

TYPICAL LAYOUT OF ARMY HELIPORT

FIG. A

HELIPORTS - LAYOUT-2

LAYOUT

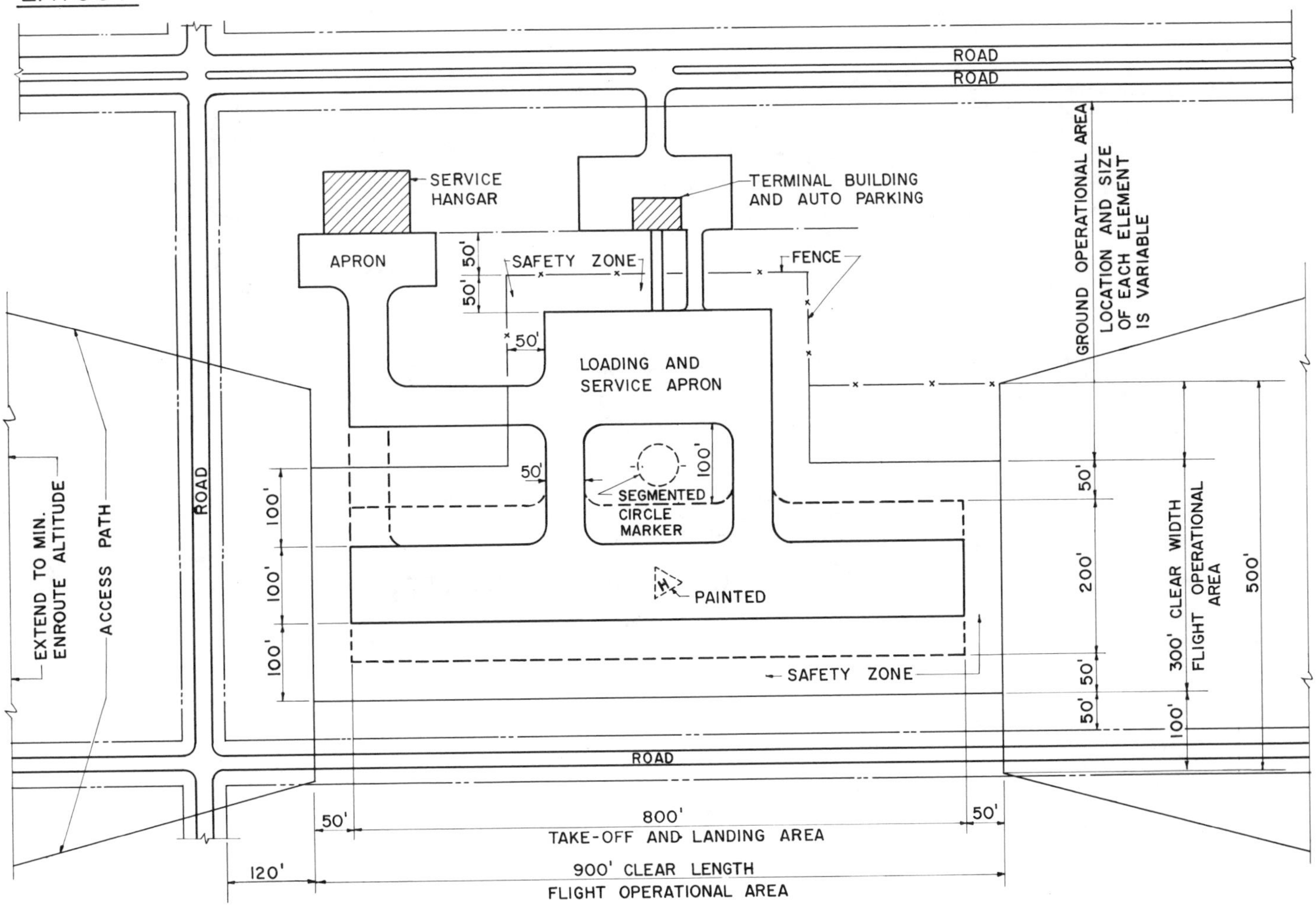

ROAD
ROAD
SERVICE HANGAR
TERMINAL BUILDING AND AUTO PARKING
APRON
SAFETY ZONE
FENCE
50' 50' 50'
LOADING AND SERVICE APRON
GROUND OPERATIONAL AREA
LOCATION AND SIZE OF EACH ELEMENT IS VARIABLE
SEGMENTED CIRCLE MARKER
100'
H PAINTED
ROAD
100' 100' 100'
EXTEND TO MIN. ENROUTE ALTITUDE
ACCESS PATH
50' 200' 50' 50'
300' CLEAR WIDTH FLIGHT OPERATIONAL AREA
100' 500'
SAFETY ZONE
ROAD
50' 800' 50'
TAKE-OFF AND LANDING AREA
120' 900' CLEAR LENGTH
FLIGHT OPERATIONAL AREA

FIG. A - COMMERCIAL HELIPORT

APPROACHES

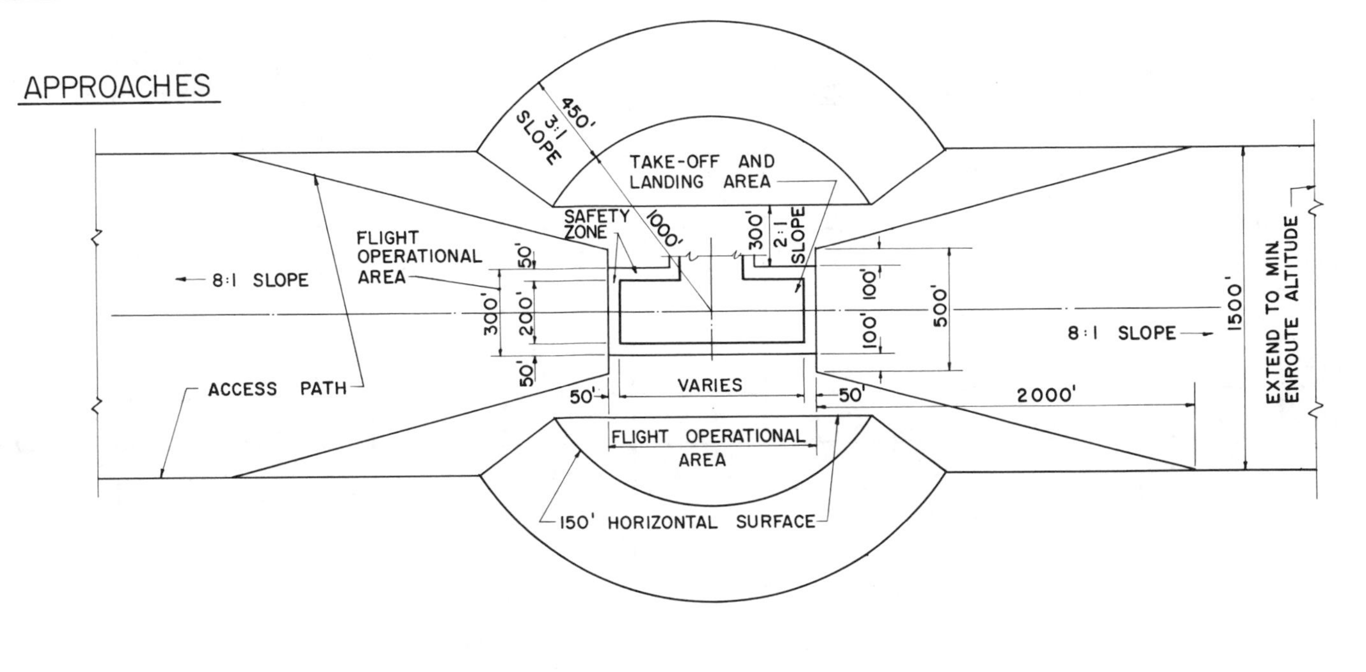

450'
3:1 SLOPE
TAKE-OFF AND LANDING AREA
SAFETY ZONE
1000'
300'
2:1 SLOPE
FLIGHT OPERATIONAL AREA
← 8:1 SLOPE
50'
300'
200'
50'
100'
100'
500'
8:1 SLOPE →
1500'
EXTEND TO MIN. ENROUTE ALTITUDE
ACCESS PATH
VARIES
50'
50'
2000'
FLIGHT OPERATIONAL AREA
150' HORIZONTAL SURFACE

FIG. B - APPROACH CLEARANCES FOR HELIPORTS

HELIPORTS - AIRCRAFT

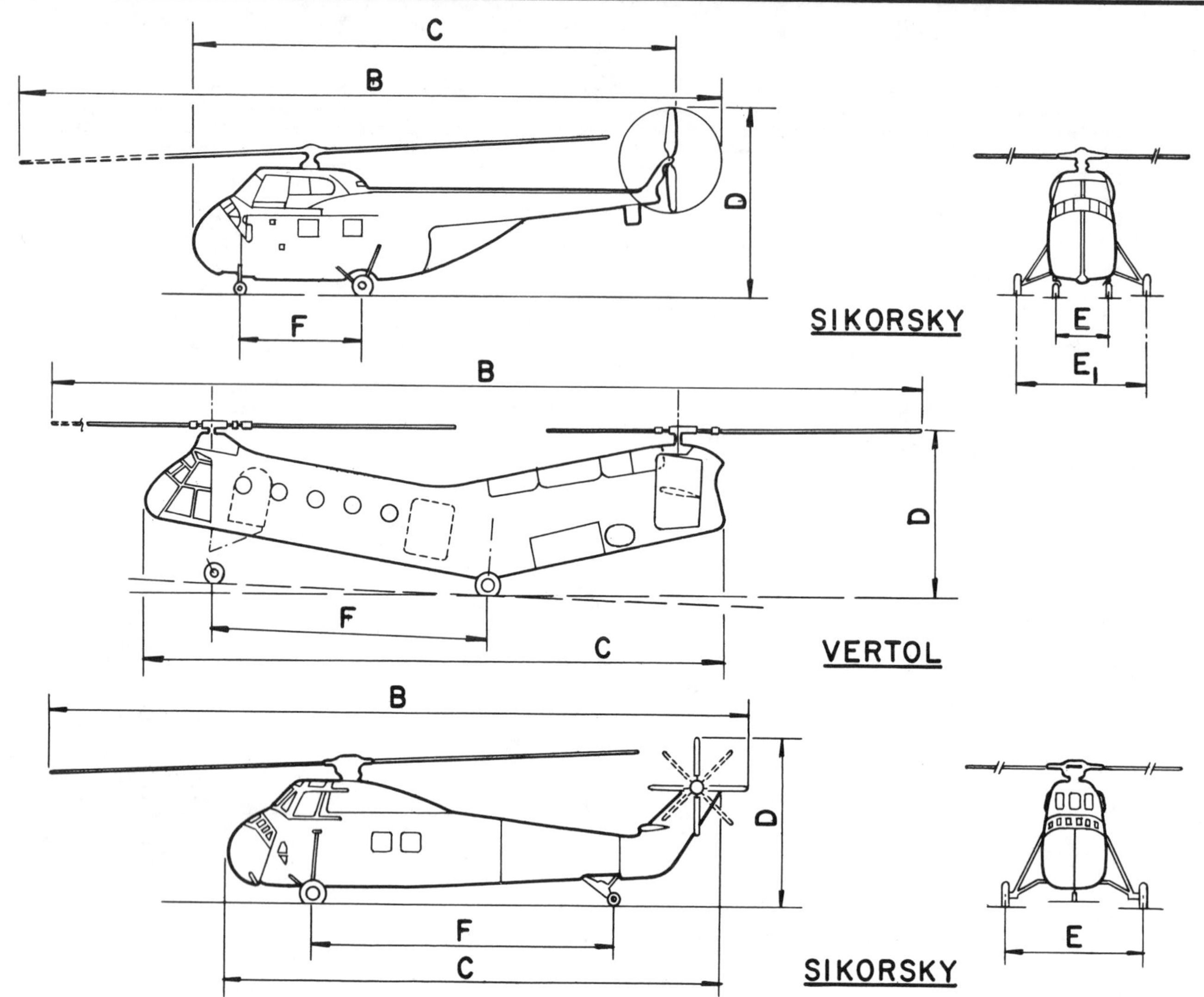

C
B
D
F
SIKORSKY
E
E1
B
D
F
C
VERTOL
B
D
F
C
E
SIKORSKY

HELICOPTER DATA										
COMPANY	MODEL DESIGNATION	A ROTOR DIAM.	B LENGTH O.A.	C LENGTH FUSELAGE	D HEIGHT	E TREAD FORWARD	E_1 TREAD AFT	F WHEEL BASE	MAX. GR. WT. (1000 LBS.)	NO. OF ENGS.
DOMAN	LZ5-2	48'-0"	62'-11"	38'-0"	16'-1"	7'-6"		7'-9"	5.2	1
KAMAN	K-600	47'-0"	* 47'-0"	25'-0"	15'-7"	5'-1"	7'-6"	7'-6"	5.9	1
OMEGA	SB-12	39'-0"	47'-5"	38'-6"	13'-0"	3'-9"	11'-9"	10'-0"	4.35	2
SIKORSKY	S-55A	53'-0"	62'-4"	42'-4"	15'-3"	4'-8"	11'-0"	10'-6"	7.5	1
BELL	47-J	37'-2"	43'-5"	32'-5"	9'-4"	7'-6"		9'-7" SKIDS	2.5	1
CESSNA	CH-1	35'-0"	42'-8"	32'-1"	11'-8"	8'-4"		SKIDS	3.0	1
HILLER	12-C, D	35'-0"	40'-6"	29'-5"	9'-9"	7'-8"		SKIDS	2.5	1
SIKORSKY	S-56	72'-0"	82'-10"	64'-11"	21'-6"	19'-9"		36'-11"	CLASSIFIED	2
SIKORSKY	S-58	56'-0"	65'-10"	47'-2"	15'-10"	12'-0"		28'-3"	12.7	1
POSSIBLE FUTURE HELICOPTER		100'-0"	120'-0"	80'-0"	25'-0"	25'-0"		50'-0"		
VERTOL	42	44'-0"	86'-4"	52'-7"	14'-6"	13'-4"		24'-8"	15.0	1
POSSIBLE FUTURE HELICOPTER		80'-0"	150'-0"	80'-0"	25'-0"	25'-0"		40'-0"	60.0	

* NO TAIL ROTOR

DAMS—FORCES ACTING ON

These data represent average practice and should only be used to supplement scientific analysis based on tests and on models.

PRESSURE DUE TO ICE: *Commonly assumed at 5,000# to 20,000#(Maine) per linear foot.*

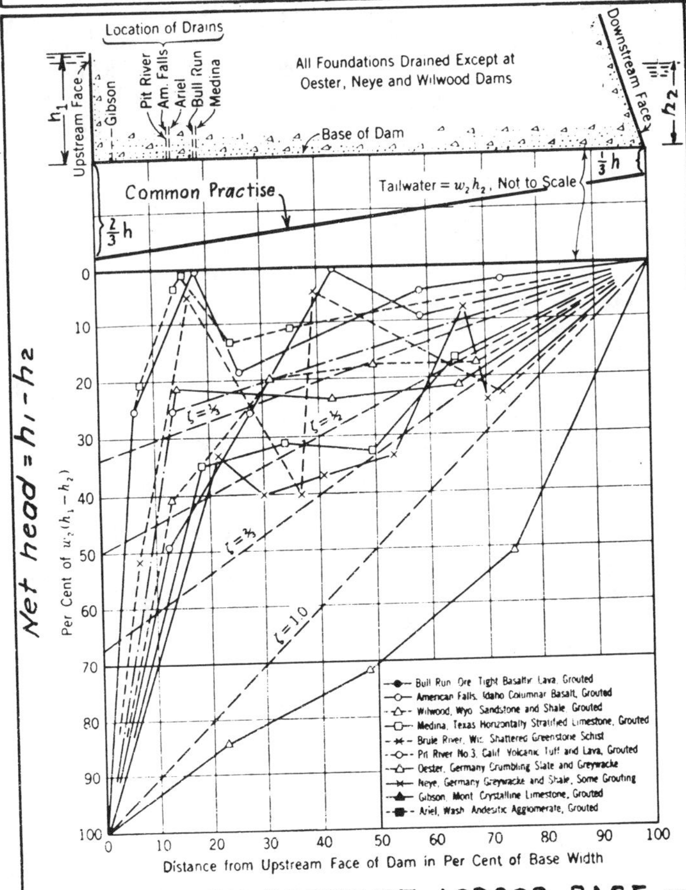

Location of Drains
All Foundations Drained Except at Oester, Neye and Wilwood Dams
Upstream Face
Downstream Face
Base of Dam
Common Practise
Tailwater = $w_2 h_2$, Not to Scale
Net head = $h_1 - h_2$
Per Cent of $w_2(h_1 - h_2)$
Distance from Upstream Face of Dam in Per Cent of Base Width

FIG. A - UPLIFT PRESSURE ACROSS BASE OF MASONRY DAM ON ROCK FOUNDATION.*

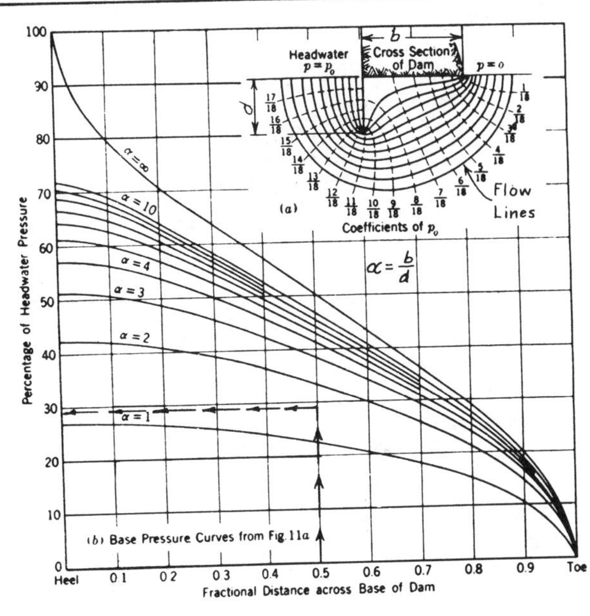

Headwater
Cross Section of Dam
Flow Lines
Coefficients of p_0
$\alpha = \frac{b}{d}$
Percentage of Headwater Pressure
(b) Base Pressure Curves from Fig. 11a
Fractional Distance across Base of Dam
Heel
Toe

EXAMPLE:
Given: Dam with: 10' base; 6' cut off; and 12' net head.
Required: To find uplift pressure at mid-point of base.
Solution: $\frac{b}{d} = \frac{10}{6} = 1.67$; from chart • 29%
$0.29 \times 12'$ head = $3.5'$ head.

FIG. B - UPLIFT PRESSURE ACROSS BASE OF MASONRY DAM ON EARTH FOUNDATION.**

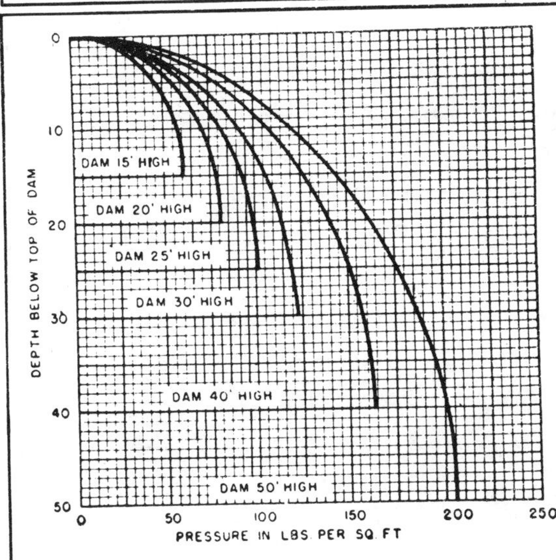

DAM 15' HIGH
DAM 20' HIGH
DAM 25' HIGH
DAM 30' HIGH
DAM 40' HIGH
DAM 50' HIGH
DEPTH BELOW TOP OF DAM
PRESSURE IN LBS PER SQ. FT.

EARTHQUAKE PRESSURES
APPROXIMATE PRESSURES FROM AN EARTHQUAKE OF AMPLITUDE = 0.1G. AND PERIOD = 1 SEC.

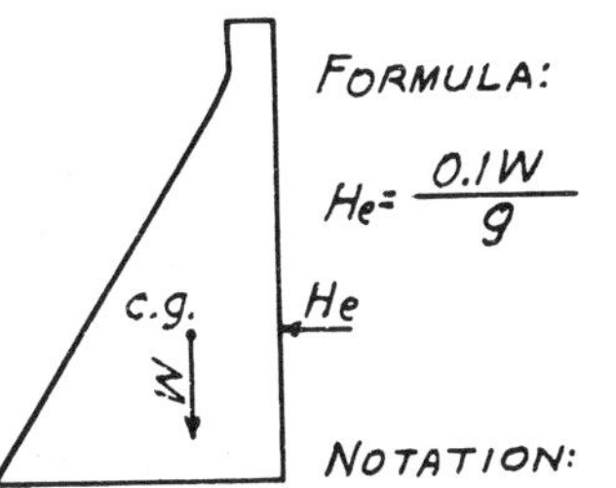

c.g.
W
He

FORMULA:

$$H_e = \frac{0.1W}{g}$$

NOTATION:
He = Longitudinal overturning force applied at c.g.
W = Weight of dam.
g = 32.2.
NOTE: Avoid location across a fault line.
Other factors are: Water pressure due to shock, relations between period of vibration of the foundation and of the structure. As these approach each other, there is danger of resonance.

FIG. C - FORCES CAUSED BY EARTHQUAKES.***

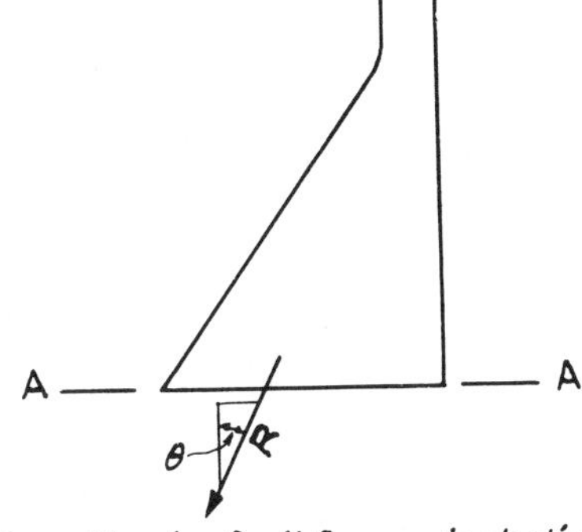

A
A
θ
R

R = Resultant of all forces including uplift acting above plane A-A.
Keep "R" within middle third of base in all cases.
Keep $\tan\theta < 0.75$
Maximum inclined pressure 60,000 lb. per sq. ft. or less depending on nature of rock.

FIG. D - RESISTANCE TO SLIDING - FOR ROCK.

Ivan E. Hoak, Civil Eng. 1932. **L. F. Harza, Trans. A.S.C.E. 1935 * From Low Dams by National Resources Comm.*

DAMS—RESULTANT PRESSURE—PRACTICAL SECTION

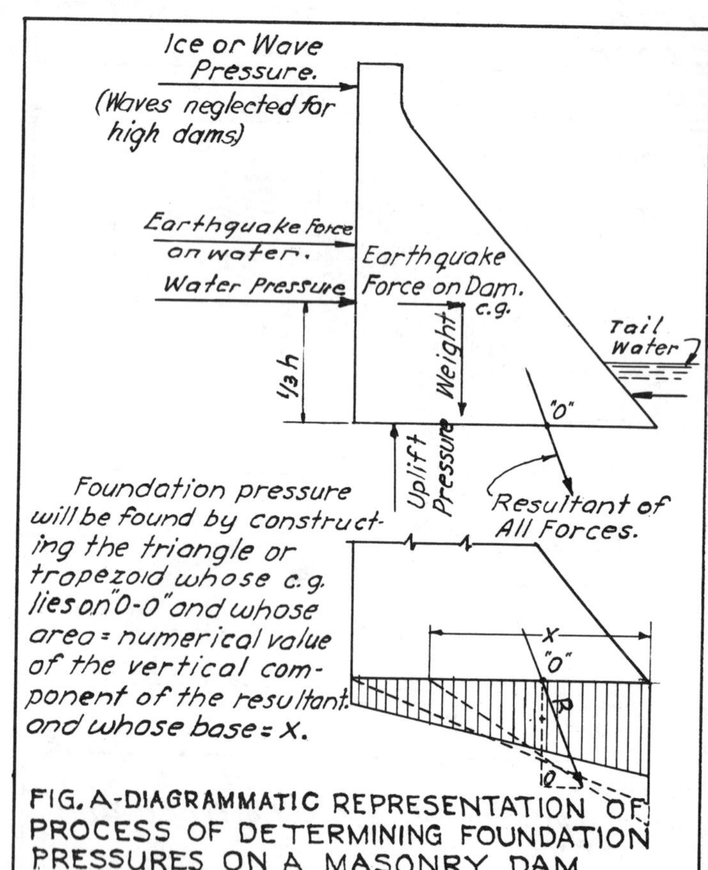
Ice or Wave Pressure.
(Waves neglected for high dams)
Earthquake Force on water.
Water Pressure
Earthquake Force on Dam.
c.g.
1/3 h
Weight
Tail Water
"0"
Uplift Pressure
Resultant of All Forces.
Foundation pressure will be found by constructing the triangle or trapezoid whose c.g. lies on "0-0" and whose area = numerical value of the vertical component of the resultant and whose base = X.
X
"0"
R
0

FIG. A-DIAGRAMMATIC REPRESENTATION OF PROCESS OF DETERMINING FOUNDATION PRESSURES ON A MASONRY DAM.

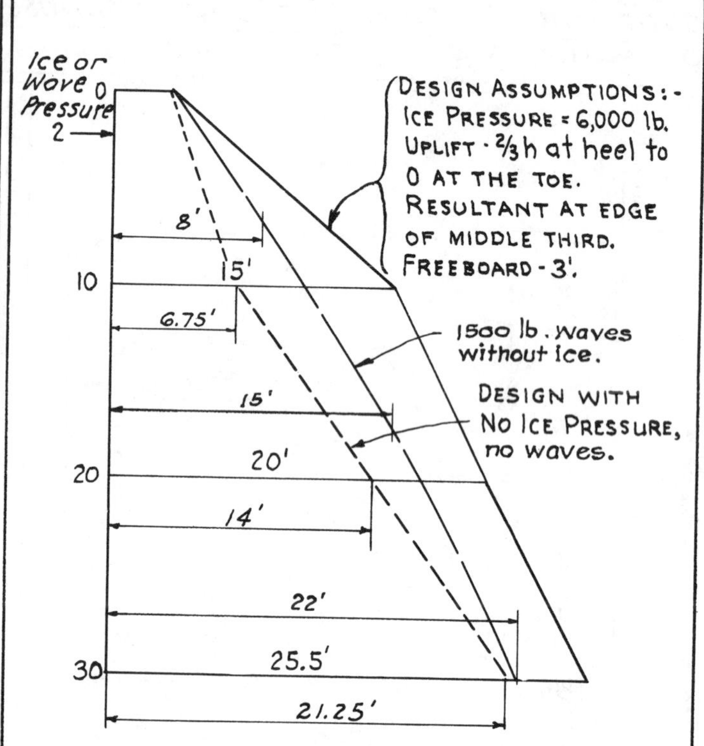
Ice or Wave Pressure
0
2
10
20
30
DESIGN ASSUMPTIONS:-
ICE PRESSURE = 6,000 lb.
UPLIFT - 2/3 h at heel to 0 AT THE TOE.
RESULTANT AT EDGE OF MIDDLE THIRD.
FREEBOARD - 3'.
8'
15'
6.75'
1500 lb. Waves without Ice.
DESIGN WITH No ICE PRESSURE, no waves.
15'
20'
14'
22'
25.5'
21.25'

FIG. B- PRACTICAL SECTION FOR CONCRETE DAMS TO 30' HIGH.

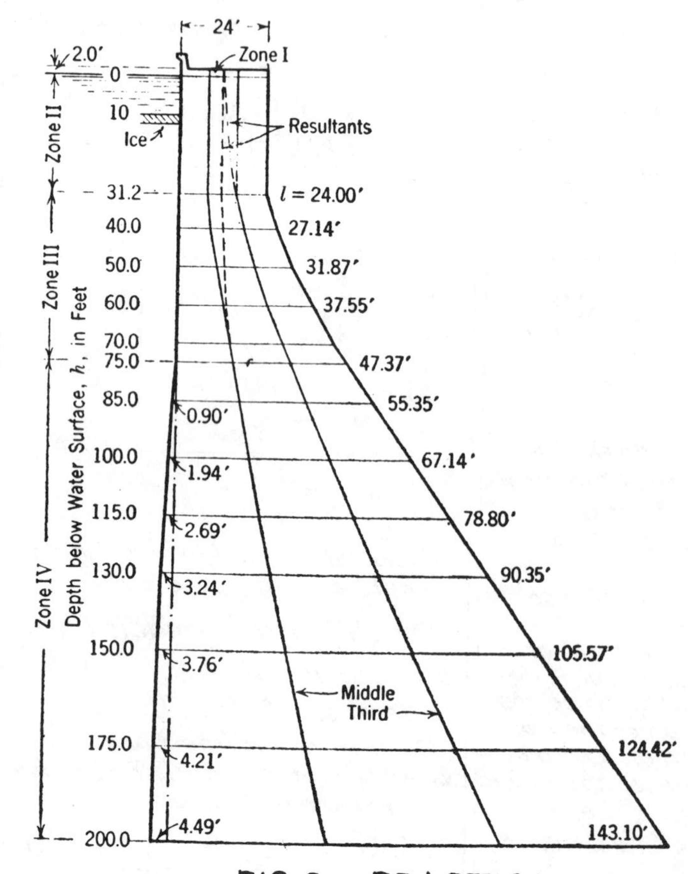
24'
2.0'
Zone I
0
10
Ice
Resultants
Zone II
Zone III
Zone IV
Depth below Water Surface, h, in Feet
31.2
40.0
50.0
60.0
70.0
75.0
85.0
100.0
115.0
130.0
150.0
175.0
200.0
l = 24.00'
27.14'
31.87'
37.55'
47.37'
55.35'
67.14'
78.80'
90.35'
105.57'
124.42'
143.10'
0.90'
1.94'
2.69'
3.24'
3.76'
4.21'
4.49'
Middle Third

DESIGN ASSUMPTIONS:

Maximum Inclined Compressive Stress is 60,000 lb./sq. ft.

Maximum Friction Factor = 0.75

Maximum Uplift Pressure = 0.5 height of water.

Uplift Area - Across full base width.

Wind Velocity = 80 miles per hour.

Ice Pressure = 9,000 lb./lin. ft.

Fetch = 4 miles.

FIG. C- PRACTICAL CONCRETE DAM SECTION.*

*Adapted from Engineering for Dams by Creager, Justin & Hinds.

DAMS-EARTH DAM SECTIONS

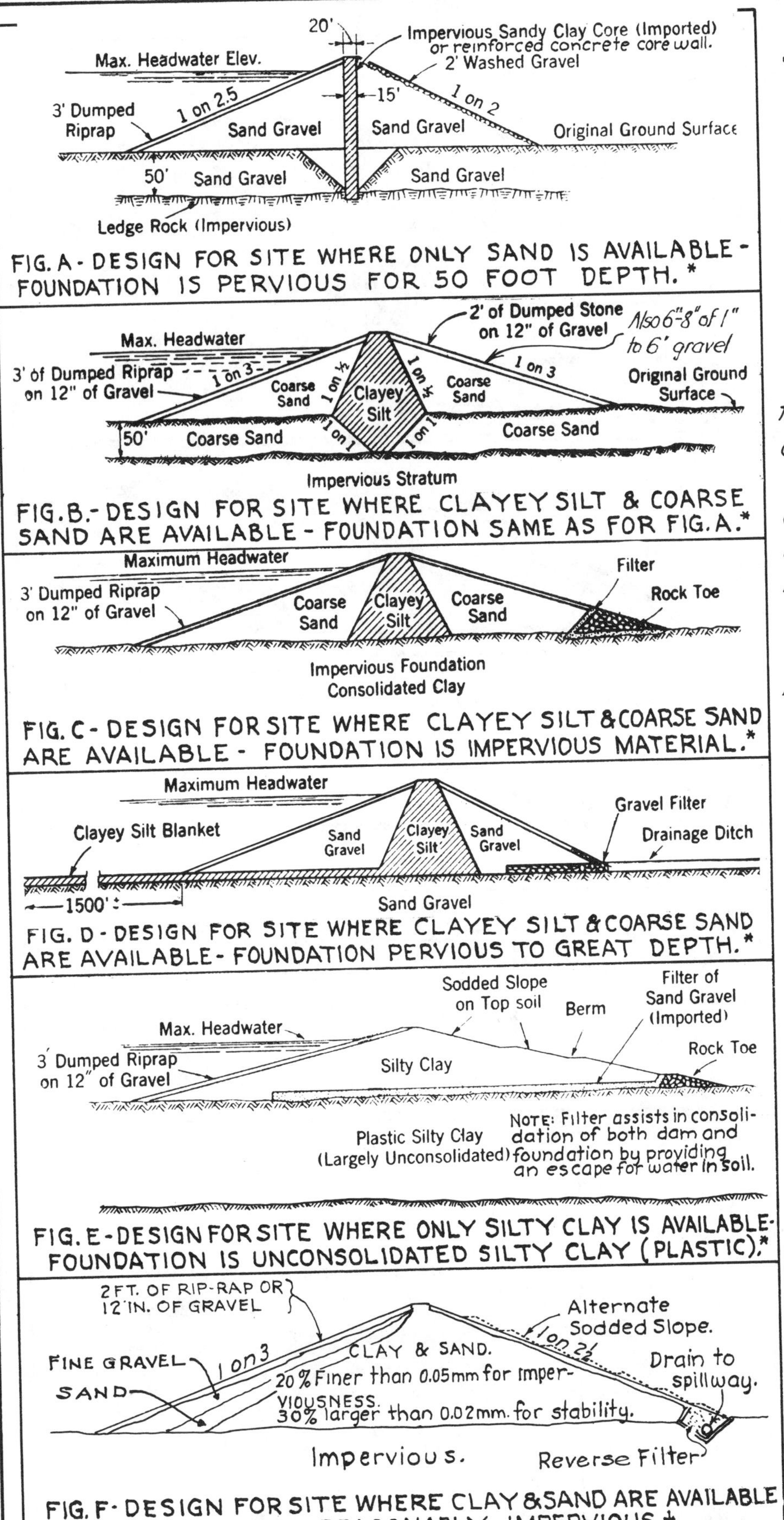

20'
Impervious Sandy Clay Core (Imported)
or reinforced concrete core wall.
2' Washed Gravel
Max. Headwater Elev.
3' Dumped Riprap
1 on 2.5
15'
1 on 2
Sand Gravel
Sand Gravel
Original Ground Surface
50'
Sand Gravel
Sand Gravel
Ledge Rock (Impervious)
2' of Dumped Stone
on 12" of Gravel
Also 6"-8" of 1" to 6' gravel
Max. Headwater
3' of Dumped Riprap
on 12" of Gravel
1 on 3
Coarse Sand
1 on ½
Clayey Silt
1 on ½
Coarse Sand
1 on 3
Original Ground Surface
50'
Coarse Sand
1 on 1
1 on 1
Coarse Sand
Impervious Stratum
Maximum Headwater
3' Dumped Riprap
on 12" of Gravel
Coarse Sand
Clayey Silt
Coarse Sand
Filter
Rock Toe
Impervious Foundation
Consolidated Clay
Maximum Headwater
Clayey Silt Blanket
Sand Gravel
Clayey Silt
Sand Gravel
Gravel Filter
Drainage Ditch
1500'
Sand Gravel
Sodded Slope
on Top soil
Berm
Filter of
Sand Gravel
(Imported)
Max. Headwater
3' Dumped Riprap
on 12" of Gravel
Silty Clay
Rock Toe
NOTE: Filter assists in consolidation of both dam and foundation by providing an escape for water in soil.
Plastic Silty Clay
(Largely Unconsolidated)
2 FT. OF RIP-RAP OR
12 IN. OF GRAVEL
Alternate
Sodded Slope.
1 on 2½
FINE GRAVEL
1 on 3
CLAY & SAND.
Drain to
spillway.
SAND
20% Finer than 0.05mm for imperviousness.
30% larger than 0.02mm. for stability.
Impervious.
Reverse Filter

FIG. A - DESIGN FOR SITE WHERE ONLY SAND IS AVAILABLE - FOUNDATION IS PERVIOUS FOR 50 FOOT DEPTH.*

FIG. B. - DESIGN FOR SITE WHERE CLAYEY SILT & COARSE SAND ARE AVAILABLE - FOUNDATION SAME AS FOR FIG. A.*

FIG. C - DESIGN FOR SITE WHERE CLAYEY SILT & COARSE SAND ARE AVAILABLE - FOUNDATION IS IMPERVIOUS MATERIAL.*

FIG. D - DESIGN FOR SITE WHERE CLAYEY SILT & COARSE SAND ARE AVAILABLE - FOUNDATION PERVIOUS TO GREAT DEPTH.*

FIG. E - DESIGN FOR SITE WHERE ONLY SILTY CLAY IS AVAILABLE - FOUNDATION IS UNCONSOLIDATED SILTY CLAY (PLASTIC).*

FIG. F - DESIGN FOR SITE WHERE CLAY & SAND ARE AVAILABLE FOUNDATION REASONABLY IMPERVIOUS.†

See Pg. 9-16, 9-31 to 9-34 & 16-05 for additional data.

REMARKS ON SECTIONS.

Compaction should be made to maximum density at optimum moisture.

Impervious sections and coarse sections should be within the limits given in Fig. A, Pg. 16-05

Shear across dam sections should be checked by scientific investigation, but the sections shown should give adequate safety factor for the materials noted.

Earth dams require separate overflow structures, as water overtopping dam will destroy it.

Foundations should never be on peat or vegetable loam.

Compact in 6" layers with power equipment and sheepsfoot rollers.

Sheepsfoot rollers should be used to reduce moisture and aid compaction.

Rip-rap with fine gravel & sand on upstream surface to protect against wave action and provide a reverse filter in case of sudden drawdown.

Coarse sand - 85% size >0.8mm.

Check soft foundation assumptions by Elastic Theory, Pg. 9-31.

*Adapted from Engineering for Dams by Justin, Hinds & Creager., † by Author.

DAMS—TYPICAL SECTIONS

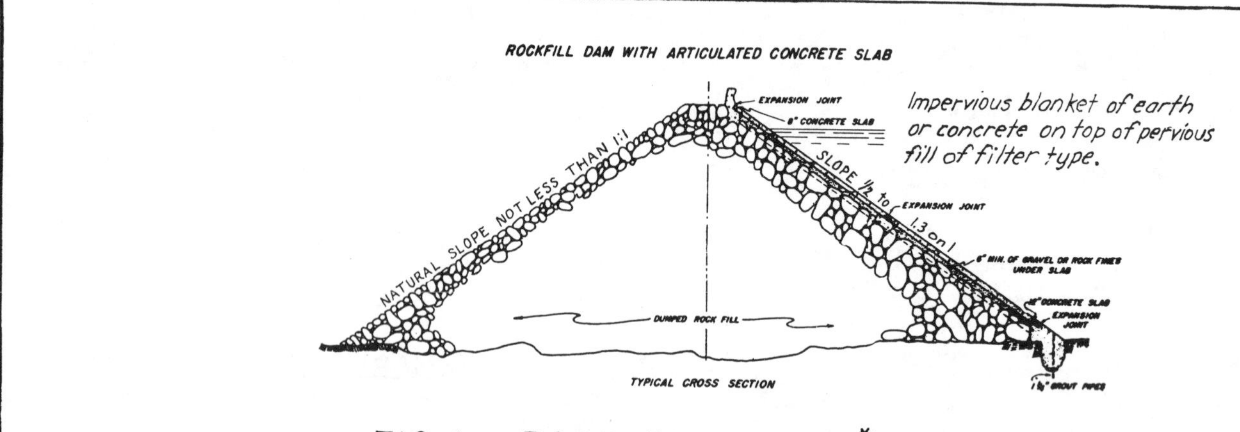
ROCKFILL DAM WITH ARTICULATED CONCRETE SLAB
EXPANSION JOINT
8" CONCRETE SLAB
Impervious blanket of earth or concrete on top of pervious fill of filter type.
NATURAL SLOPE NOT LESS THAN 1:1
SLOPE 1/2 to 1.3 on 1
EXPANSION JOINT
6" MIN. OF GRAVEL OR ROCK FINES UNDER SLAB
12" CONCRETE SLAB
EXPANSION JOINT
DUMPED ROCK FILL
TYPICAL CROSS SECTION

FIG. A - ROCK FILL DAMS.*

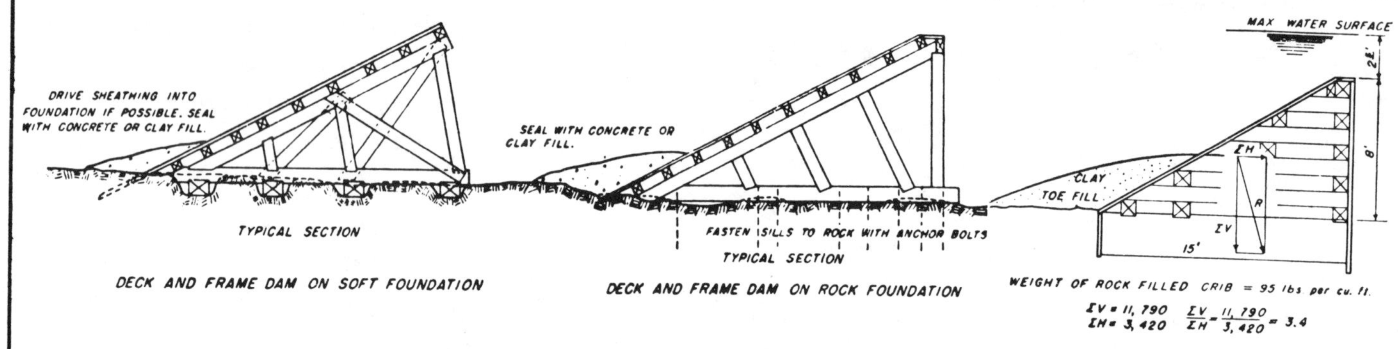
DRIVE SHEATHING INTO FOUNDATION IF POSSIBLE. SEAL WITH CONCRETE OR CLAY FILL.
TYPICAL SECTION
DECK AND FRAME DAM ON SOFT FOUNDATION
SEAL WITH CONCRETE OR CLAY FILL.
FASTEN SILLS TO ROCK WITH ANCHOR BOLTS
TYPICAL SECTION
DECK AND FRAME DAM ON ROCK FOUNDATION
MAX WATER SURFACE
CLAY TOE FILL
15'
WEIGHT OF ROCK FILLED CRIB = 95 lbs per cu. ft.
ΣV = 11,790 ΣH = 3,420 ΣV/ΣH = 11,790/3,420 = 3.4
CRIB AND DECK DAM

FIG. B - LOG CRIB DAMS.*

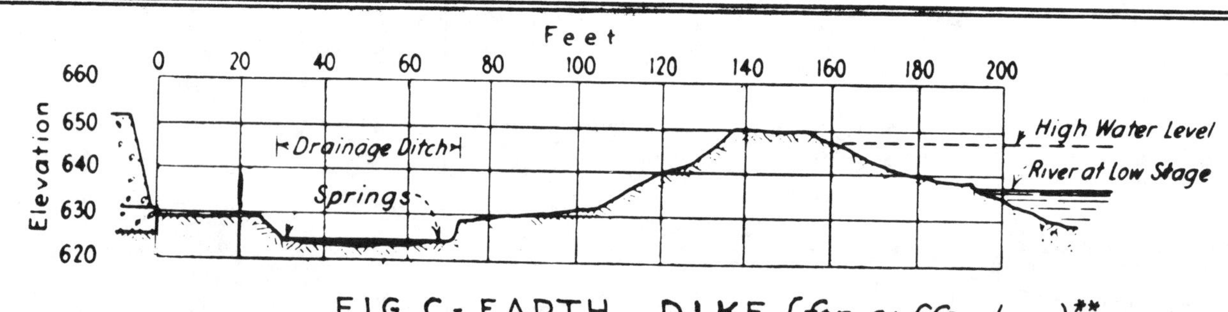
Feet
0 20 40 60 80 100 120 140 160 180 200
Elevation
660 650 640 630 620
Drainage Ditch
Springs
High Water Level
River at Low Stage

FIG. C - EARTH DIKE (for cofferdam).**

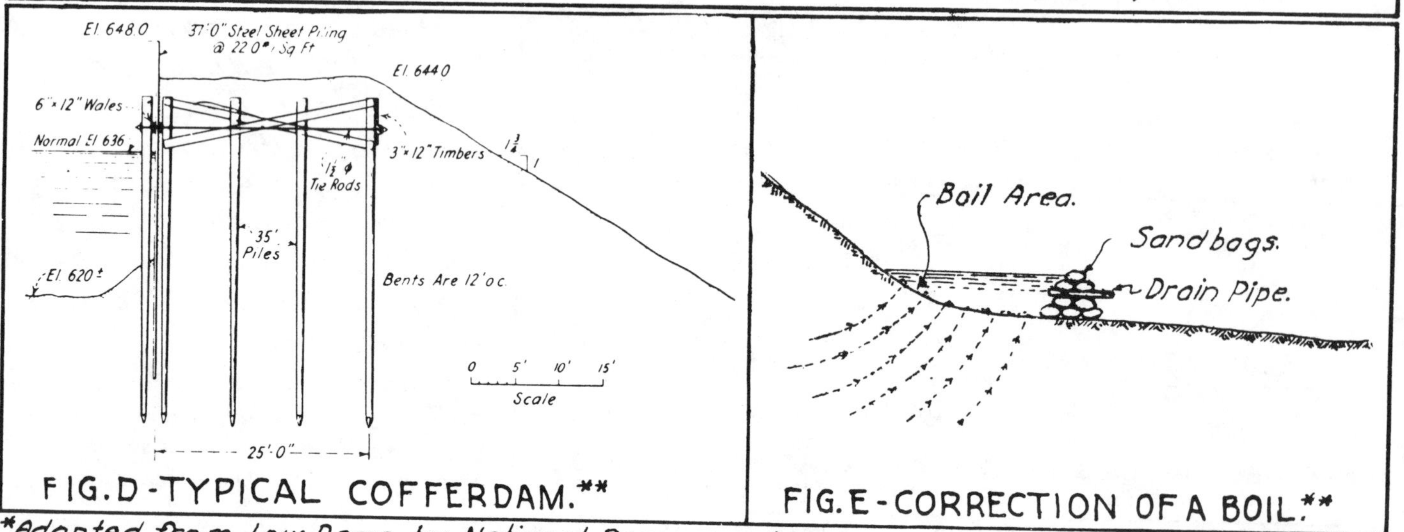
El. 648.0
37'0" Steel Sheet Piling @ 22.0# / Sq Ft
El. 644.0
6" x 12" Wales
Normal El. 636
3" x 12" Timbers
1¼" φ Tie Rods
35' Piles
El. 620 ±
Bents Are 12' o.c.
0 5' 10' 15'
Scale
25'-0"
Boil Area.
Sandbags.
Drain Pipe.

FIG. D - TYPICAL COFFERDAM.**

FIG. E - CORRECTION OF A BOIL.**

*Adopted from Low Dams by National Resources Comm.
**Adopted from White & Prentis, Cofferdams, by permission of Columbia University Press.

DAMS–MISCELLANEOUS CRITERIA

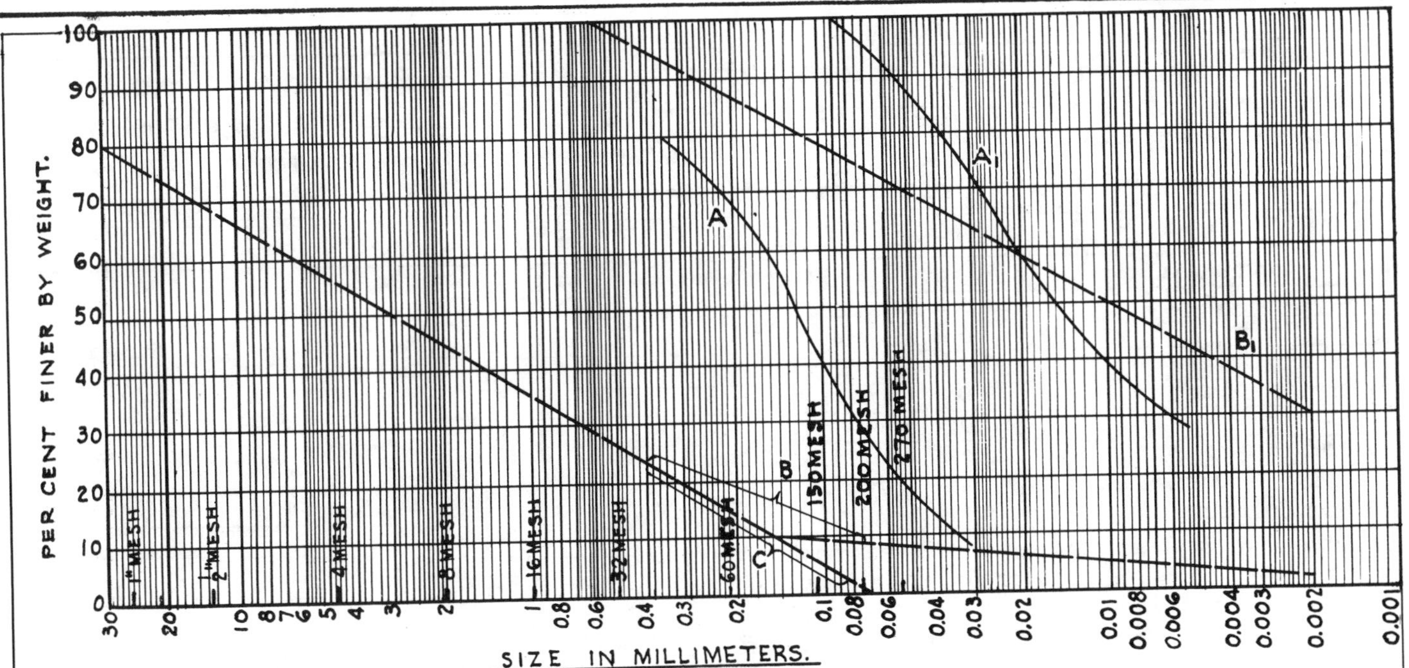

PER CENT FINER BY WEIGHT.
100
90
80
70
60
50
40
30
20
10
0
1" MESH
1/2" MESH
4 MESH
8 MESH
16 MESH
32 MESH
60 MESH
150 MESH
200 MESH
270 MESH
A
A1
B
B1
C
30
20
10
8
7
6
5
4
3
2
1
0.8
0.6
0.4
0.3
0.2
0.1
0.08
0.06
0.04
0.03
0.02
0.01
0.008
0.006
0.004
0.003
0.002
0.001
SIZE IN MILLIMETERS.

A. Possible upper limit of satisfactory core material.

A1. Probable upper limit of safe core material (Urquhart).

B – B1. Proposed limits for ungraded materials suitable for impervious section for rolled fill dams (Lee).

C – B1. Approximate limit of materials for impervious sections of rolled fill dams (Justin).

FIG. A. CHART FOR SELECTION OF MATERIALS.* See also p. 9-16.

TABLE B – SETTLEMENT EXPECTANCY†

Material	Pressure, tons/sq. ft. – No Measurable Settlement	Pressure, tons/sq. ft. – Detrimental Settlement
Stiff clay	2.0	4.4
Silt	1.5	1.6
Sand–clay	2.4	1.6
Fine sand	2.0	1.8

Note inconsistency.

FOUNDATION PRESSURES FOR EARTH DAMS.
For bearing capacities of soils and rocks, see p. 9-30. Soft formations should be checked for shear as shown below.

Rigid Structures: Provide for settlement, shrinkage, and expansion by use of joints and keyways.

Flexible Structures: Allow time for foundation settlement. Allow for consolidation in completing fill.

Basis for Computing Settlement:

$$\text{Settlement} = \frac{e_1 - e_2}{1 + e_1} \times h$$

where e_1 = void ratio before consolidation.
e_2 = void ratio after consolidation.
h = thickness of layer in feet

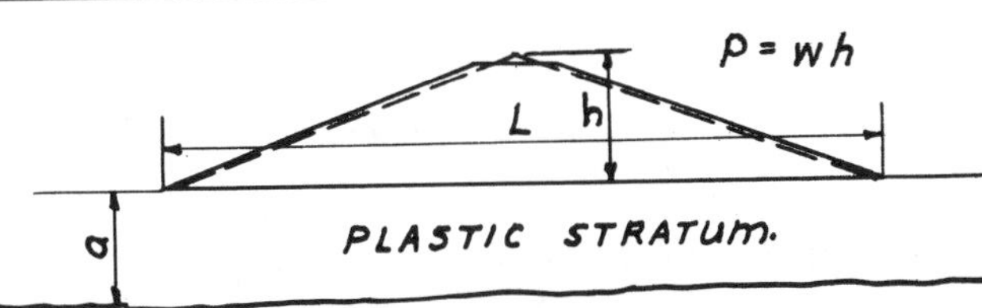

P = wh
L
h
a
PLASTIC STRATUM.

$$S = \frac{P \times a}{L}$$ (Applies only to plastic soils.)‡

Assumes equivalent Δ cross section.
S = maximum unit shear, p.s.i.
P = maximum unit pressure, p.s.i.
For very approximate limits for S, see Table D.

FIG. C. FOUNDATIONS OF PLASTIC MATERIAL

TABLE D – SOIL SHEAR VALUES

Soil	Cohesion per sq. ft.	Angle of Internal Friction ϕ,°
Clay§ – Liquid	100	0
Very soft	200	2
Soft	400	4
Fairly stiff	1,000	6
Very stiff	2,000	12
Sand§ – Wet	0	10
Dry or unmoved	0	34
Silt•	0	20±
Cemented sand and gravel – Wet	500	34
Dry	1,000	34

Shear strength = cohesion + W tan ϕ. Apply a safety factor of 2. Actual determination from soil analysis is advisable.

*From *Low Dams* by National Resources Comm.
†From *Transactions, A.S.C.E.*
‡From *The Application of Theories of Elasticity and Plasticity to Foundation Problems* by Leo Jurgenson, *Journal of the Boston Society of Civil Engineers.*
§Adapted from Hogentogler, *Engineering Properties of Soils,* McGraw Hill. Based on quick shear test. •By author.

DAMS—MISCELLANEOUS REQUIREMENTS-1

SEEPAGE DATA: *The flow of water from the upstream side of dam to the downstream side, through the embankment of earth dams and along the line of creep, must be kept within safe limits in respect to volume and velocity to prevent piping & boiling.*
Creep = Path of water along contact surface between structure and foundation.

TABLE A - RECOMMENDED CREEP RATIOS,* R (After Lane).

MATERIAL	R	MATERIAL	R
Very Fine Sand & Silt.	8.5	Boulders with some Cobbles and Gravel.	2.5
Fine Sand	7.0	Soft Clay	3.0
Medium Sand	6.0	Medium Clay	2.0
Coarse Sand	5.0	Hard Clay	1.8
Fine Gravel	4.0	Very Hard Clay or Hardpan.	1.6
Medium Gravel	3.5		
Coarse Gravel including some Cobbles.	3.0		

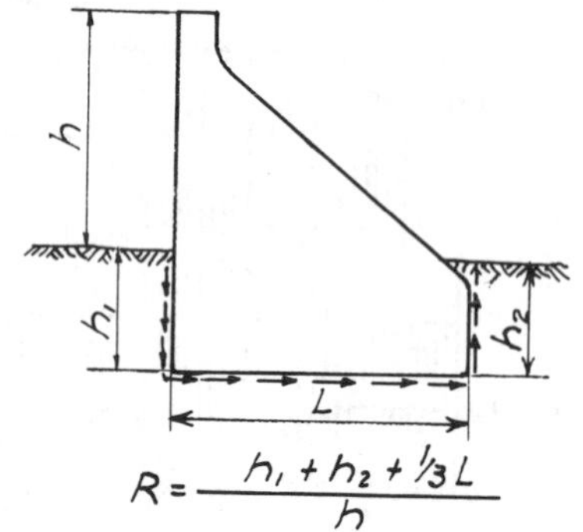
h
h_1
h_2
L

$$R = \frac{h_1 + h_2 + \frac{1}{3}L}{h}$$

R = Creep Ratio.
(A check against piping.)

NOTE: If "R" for a critical section is less than the tabular value, additional flow resistance should be inserted in design section.
Limitations: Does not take into account –
1. Importance of shape - see Fig. C & D p. 16-08
2. Line of creep may not be most direct path of seepage.

TABLE B - SEEPAGE LOSSES FROM COFFERDAMS.**

RATIO OF AREAS a/A	0.1	0.2	0.3	0.4	0.5	0.6	0.7	0.8	0.9
ϕ	0.49	0.62	0.74	0.86	1.00	1.16	1.35	1.62	2.06

MATERIAL	K
Clean Gravel	0.2662 - 0.02778
Clean Sand	$5787 \times 10^{-7} - 5787 \times 10^{-9}$
Fine Sand or Silt	$2778 \times 10^{-9} - 2778 \times 10^{-12}$
Clay.	Less than - 2778×10^{-12}

SCHOKLITSCH FORMULA.

$$Q = 100 K \phi \frac{h}{2}$$

Where: Q = Quantity of flow in cu.ft./sec.
K = Coefficient of permeability in cu.ft./sec. - See Table B, but a soil analysis is indicated.
ϕ = Value in Table B above.

h
A
a

Alternate Formula:†

$$Q_1 = C L h$$

Where:-
Q_1 = Flow in gal. per min.
C = 1.25 x effective (or 10%) grain size of sand. See p. 3-07 for visual determination of grain size
h = Head loss in feet.
L = Length of dam in feet.

NOTE: On dams over 15' high, seepage losses through embankments should be determined by experiments on scale models.

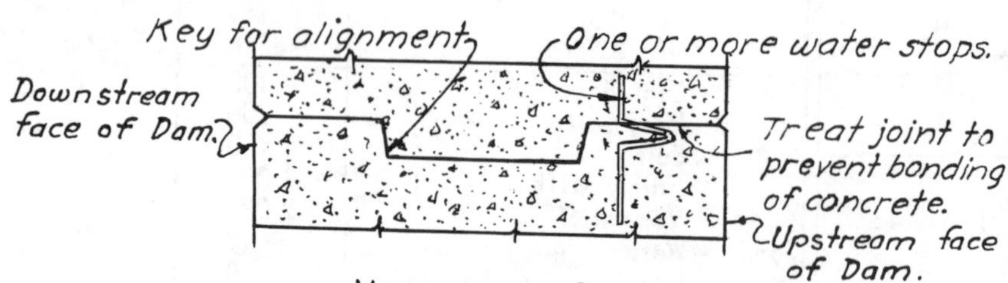
Key for alignment
One or more water stops.
Downstream face of Dam
Treat joint to prevent bonding of concrete.
Upstream face of Dam.

HORIZONTAL SECTION.

Allow lapse of time between concreting adjoining sections to reduce shrinkage. See also p. 16-10

FIG. C - CONSTRUCTION JOINT ESSENTIALS.

FLOTATION GRADIENT.†
Hydraulic Gradient at which boils or piping occur.
Flotation Gradient = F_c
$F_c = \frac{H}{L} = (1-P)(S-1)$
F_c = 1 approximately.
P = Proportion of voids.
S = Specific gravity of sand grains.
H = Hydraulic head.
L = Thickness of bed.

*Adapted from Transactions of A.S.C.E. - 1935, Pg. 1235 by E.W. Lane. ** From Low Dams by National Resources Comm. † Adapted from White & Prentis, Cofferdams, by permission of Columbia University Press.

DAMS – MISCELLANEOUS REQUIREMENTS – 2

STEVENSON FORMULA:

For "F" greater than 20 miles:

$$h = 0.17\sqrt{VF}$$

For "F" less than 20 miles:

$$h = 0.17\sqrt{VF} + 2.5 - F^{\frac{1}{4}}$$

Max. height against vert. surface.
Crest.
$1\frac{1}{3}h$
h
P
$\frac{1}{3}h$
Still Water Level.
$\frac{1}{3}h$
Trough.

Where: h = Height of wave in feet.
V = Wind velocity in miles per hour.
F = Unobstructed length of lake in statute miles (fetch).

Wave Pressure, $P = 125h^2$ in pounds per linear foot.

Maximum height of waves: On a vertical surface = 1.33h.
On an inclined surface = 1.5h.

Minimum Freeboard on lakes less than 2 miles long = 2 feet.

FIG. A - HEIGHT OF FREEBOARD WAVE ACTION.*

NOTE: Waves exert scouring action as they travel up and down slopes. Embankment should be protected by paving, rip-rap, etc.; provision for drainage, in case of rapid drawdown of reservoir, should be made.

ZUIDER ZEE FORMULA:

$$S = \frac{V^2F}{1400d} \times \cos A$$

Where: S = Rise of water level above normal in feet.
V = Wind velocity in miles per hour.
F = Unobstructed length of lake in miles (fetch).
d = Average depth of water in feet.
A = Angle between fetch and direction of wind.

FIG. B - RISE IN WATER LEVEL DUE TO WIND.

Rule for filter material is based on 15% size.

15% size means - 15% by weight of particles are smaller than size given and 85% are larger.

To determine required size - step up grain size by 9.

Layers of material should be 12" min. thickness.

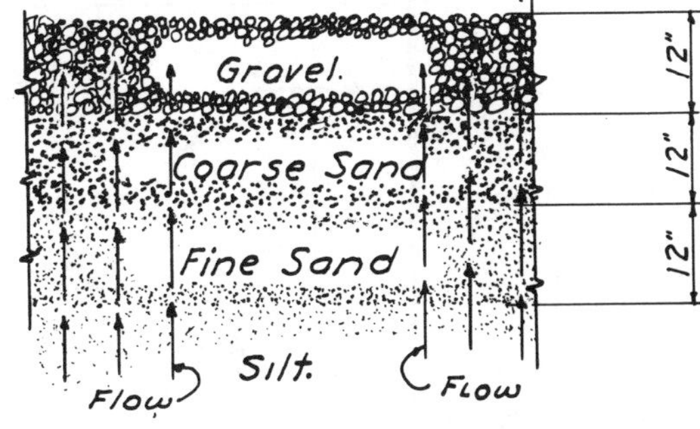

Gravel.
Coarse Sand
Fine Sand
Silt.
Flow
Flow
12"
12"
12"

Example:

Given Silt whose grain size	= 0.01 mm.	Based on 15% size.
First layer - fine sand	= 0.09 mm.	
Second layer - coarse sand	= 0.81 mm.	
Third layer - gravel	= 7.3 mm.	

See also ratio of 15% to 85% sizes, see Pg. 9-37.

FIG. C - SELECTION OF MATERIAL FOR REVERSE FILTER.

*Adapted from Transactions A.S.C.E., 1935, Pg. 984 by D.A. Molitor.

DAMS–WHAT THE FLOW NET SHOWS–1

"While nature is a fickle jade, she is far kinder to those who study her laws." Lazarus White and Edmund Astley Prentis.

Quick material occurs where particles are floated by upward hydraulic force. This occurs in sand approximately where:

$$\frac{\text{Loss of Head}}{\text{Horizontal Distance}} = 1$$

(See page 16-06 for Flotation Gradient).

Quick material tends to lose its supporting power.

Bulls liver is fine sand which retains water and is quaky - 0.005 to 0.01 mm., requires only a small volume of water to boil. Coarse sands require a large volume of water to boil.

Flow nets are obtained by models and/or by analysis. This preparation requires special skill.

The reverse filter allows added stability under greater head of water.

Used to assist in excavation below ground water level.

Use with caution.

Gravel or Broken Stone

Discharge to Open Drain

Sand

Upward Flow

FIG. A · REVERSE FILTER.*

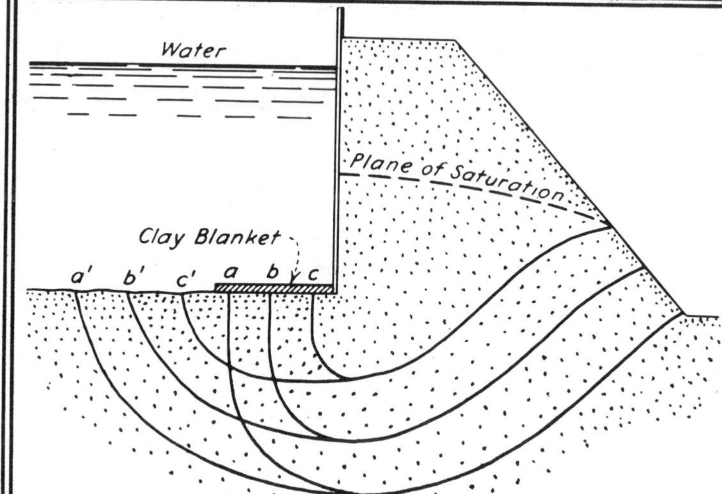
Water
Plane of Saturation
Clay Blanket
a'
b'
c'
a
b
c

FIG. B - ILLUSTRATING SMALL EFFECT OF CLAY BLANKET IN A COFFERDAM.*

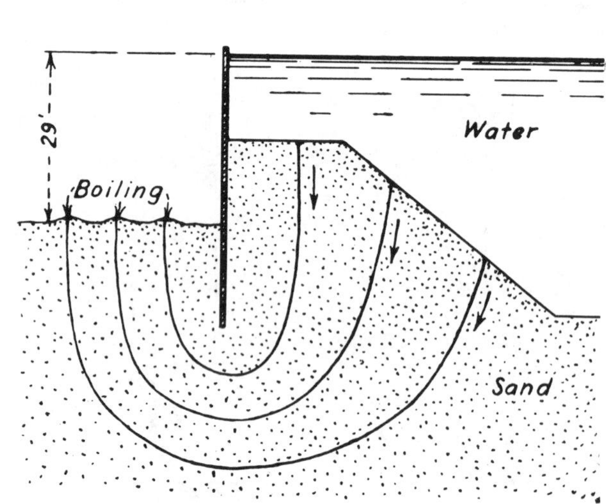
29'
Boiling
Water
Sand

FIG. C - ILLUSTRATING INCORRECT PLACING OF BERM IN A COFFERDAM.*

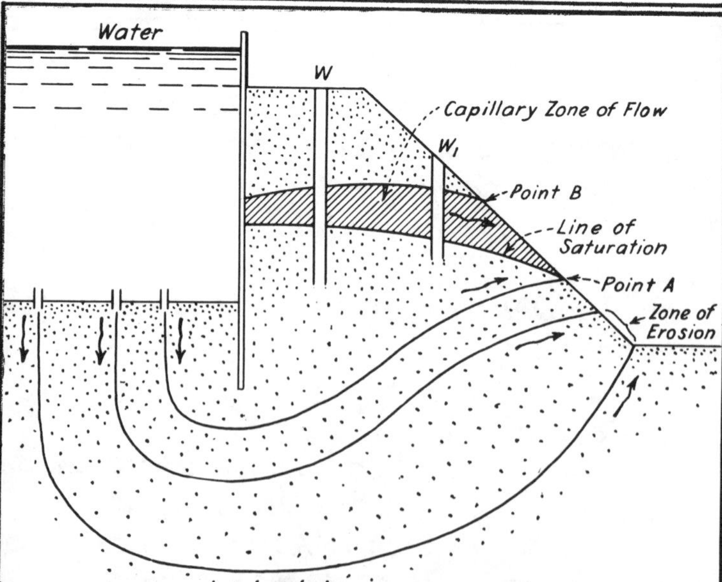
Water
W
W_1
Capillary Zone of Flow
Point B
Line of Saturation
Point A
Zone of Erosion

FIG. D - ILLUSTRATING CORRECT PLACING OF BERM IN A COFFERDAM.*

DAMS–WHAT THE FLOW NET SHOWS–2

NOTE: Full lines represent flow lines after addition of special feature; dotted lines before.

The lines of percolation tend to slough off slope because of saturation.

Remedy: A reverse filter, or rock toe with a drain.

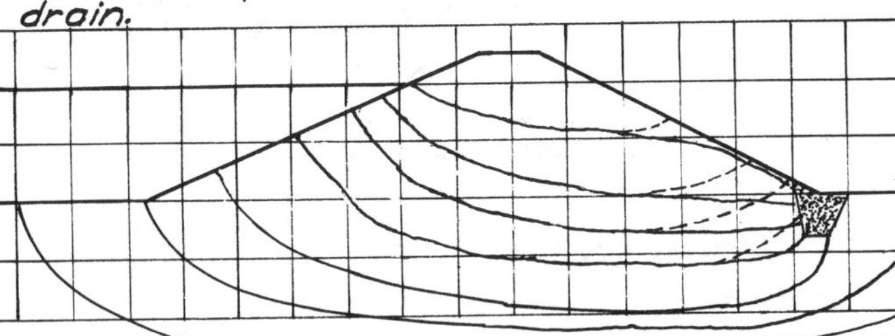

FIG. A- EFFECT OF FILTER AT TOE OF SLOPE.*

Additional fill placed ontop of down-stream slope does not remedy a saturated toe.

FIG. B - INCORRECT REMEDY EFFECT OF ADDED MATERIAL.*

Note improvement in flow lines at toe.

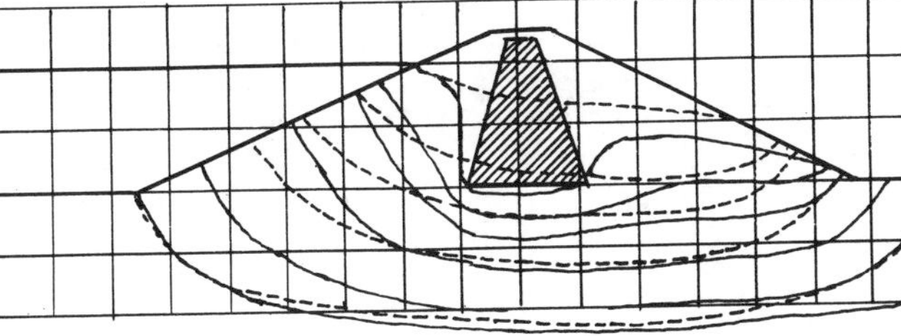

FIG. C- EFFECT OF IMPERVIOUS CORE ON PATH OF PERCOLATION.*

Increased flow concentration at toe is offset by reduced flow pressure.

FIG. D- EFFECT OF IMPERVIOUS BLANKET ON UPSTREAM SLOPE.*

Increased flow concentration at toe is offset by reduced flow pressure.

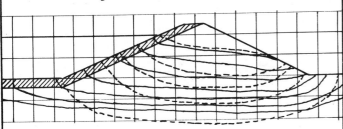

FIG. E- EFFECT OF IMPERVIOUS BLANKET ON UPSTREAM SLOPE & BOTTOM.*

Permeable embankment acts to support the core wall, and as a drain for water percolating through the core.

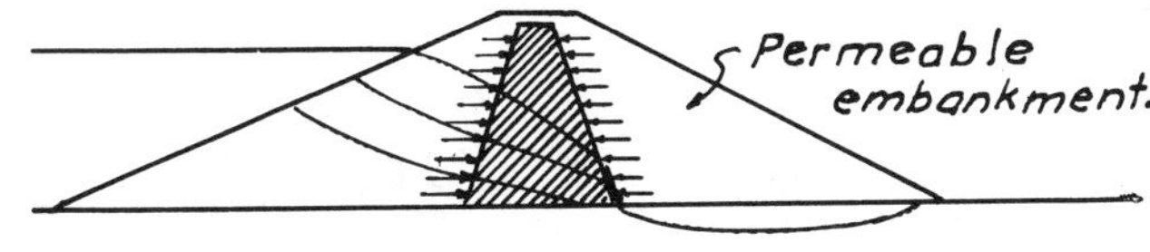
Permeable embankment.

FIG. G- SHOWING HOW CORE WALL IS SUPPORTED BY ADJACENT SLOPES.

Where foundation is pervious, a reverse filter and drain below ground level to carry off waten stabilizes embankment by lowering the water table.

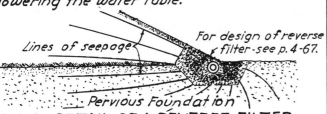

Lines of seepage

For design of reverse filter-see p. 4-67.

Pervious Foundation

FIG. H- DETAIL OF A REVERSE FILTER AT TOE OF SLOPE.

A slight amount of settlement causes pressure on sheet-piling breaking it or shearing the masonry.

The effect of the cut off is to lengthen the path of seepage.

FIG. J - DANGER OF SHEET-PILE CUT OFF UNDER HEEL OF DAM.

**Adapted from Low Dams by National Resources Comm.*

DAMS-WATERSTOPS, DRAINS & DETAILS

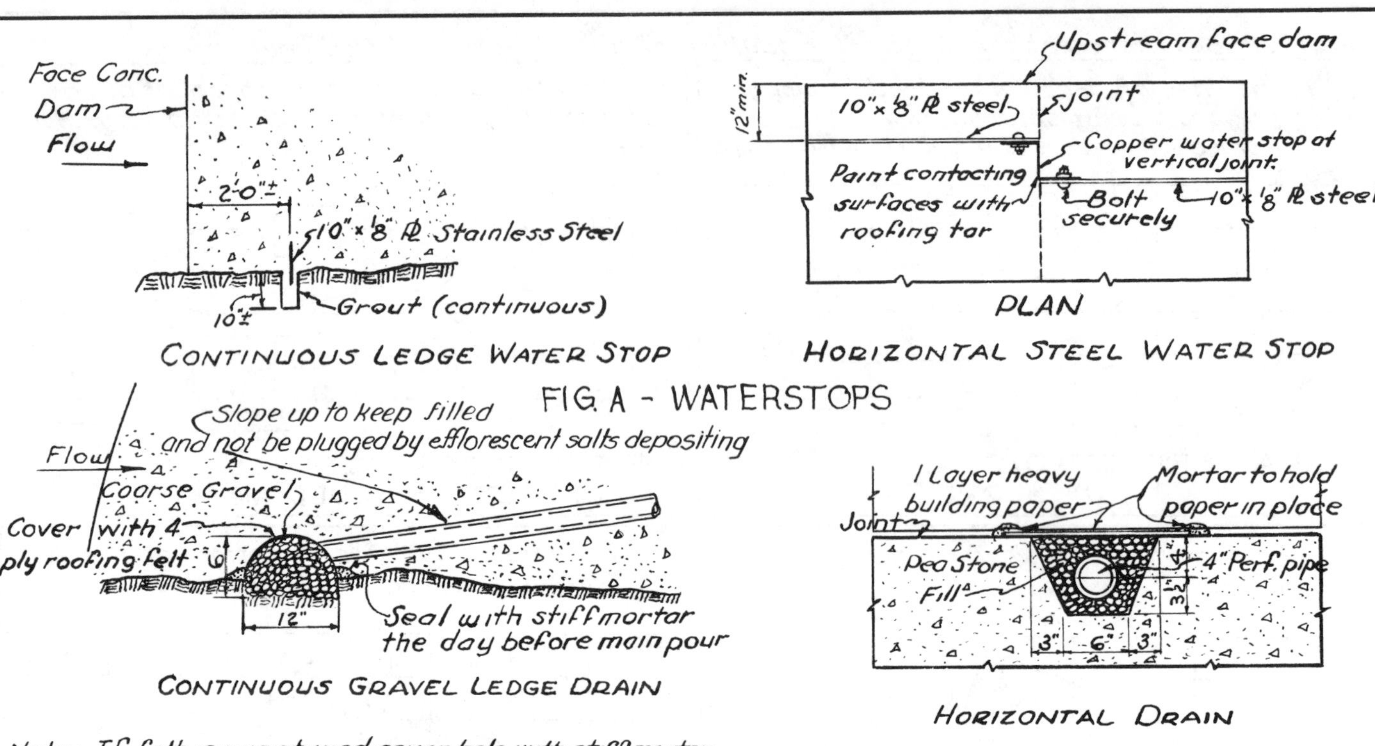

Face Conc.
Dam
Flow
2'-0"±
10"×⅛" Ⅱ Stainless Steel
10±
Grout (continuous)
CONTINUOUS LEDGE WATER STOP
Upstream face dam
12" min.
10"×⅛" Ⅱ steel
Joint
Copper water stop at vertical joint.
Paint contacting surfaces with roofing tar
Bolt securely
10"×⅛" Ⅱ steel
PLAN
HORIZONTAL STEEL WATER STOP
FIG. A - WATERSTOPS
Slope up to keep filled and not be plugged by efflorescent salts depositing
Flow
Coarse Gravel
Cover with 4-ply roofing felt
6
12"
Seal with stiff mortar the day before main pour
CONTINUOUS GRAVEL LEDGE DRAIN
1 Layer heavy building paper
Mortar to hold paper in place
Joint
Pea Stone Fill
4" Perf. pipe
3"
6"
3"
HORIZONTAL DRAIN

Note: If felt is punctured, cover hole with stiff mortar. Cover all joints so there is no leakage from wet concrete to gravel drain.

FIG. B - DRAINS

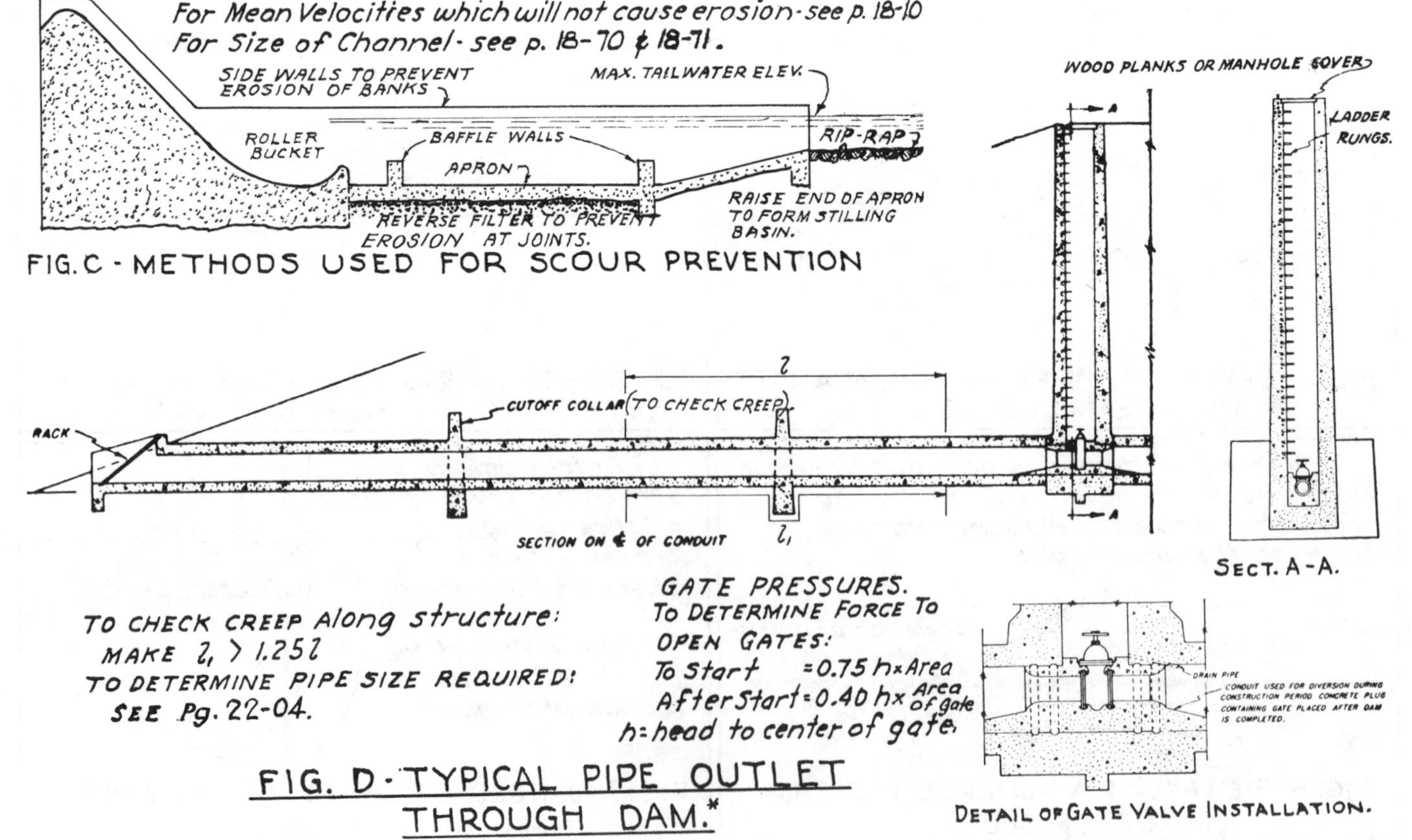

For Mean Velocities which will not cause erosion - see p. 18-10
For Size of Channel - see p. 18-70 & 18-71.
SIDE WALLS TO PREVENT EROSION OF BANKS
MAX. TAILWATER ELEV.
ROLLER BUCKET
BAFFLE WALLS
APRON
RIP-RAP
REVERSE FILTER TO PREVENT EROSION AT JOINTS.
RAISE END OF APRON TO FORM STILLING BASIN.
FIG. C - METHODS USED FOR SCOUR PREVENTION
WOOD PLANKS OR MANHOLE COVER
LADDER RUNGS.
RACK
CUTOFF COLLAR (TO CHECK CREEP)
l
l₁
SECTION ON ℄ OF CONDUIT
SECT. A-A.
DRAIN PIPE
CONDUIT USED FOR DIVERSION DURING CONSTRUCTION PERIOD CONCRETE PLUG CONTAINING GATE PLACED AFTER DAM IS COMPLETED.
DETAIL OF GATE VALVE INSTALLATION.

TO CHECK CREEP Along structure:
MAKE $l_1 > 1.25l$
TO DETERMINE PIPE SIZE REQUIRED:
SEE Pg. 22-04.

GATE PRESSURES.
To DETERMINE FORCE To OPEN GATES:
To Start $= 0.75\, h \times \text{Area}$
After Start $= 0.40\, h \times \text{Area of gate}$
h = head to center of gate.

FIG. D - TYPICAL PIPE OUTLET THROUGH DAM.*

*From Low Dams by National Resources Comm.

DAMS – SPILLWAY DESIGN – I

SPILLWAY DESIGN

Spillways must have sufficient capacity that dam will never be over topped.

Compute storm peak run off. (1) See pages 18-13 to 18-14 ; (2) check with high water history of river ; (3) in addition make a hydrographic study.

Use a liberal factor of safety.

See fig. A & D below to determine shape of Standard crests. Fig. B shows other crests for low heads.

See page 16-12 for discharge formulas and examples.

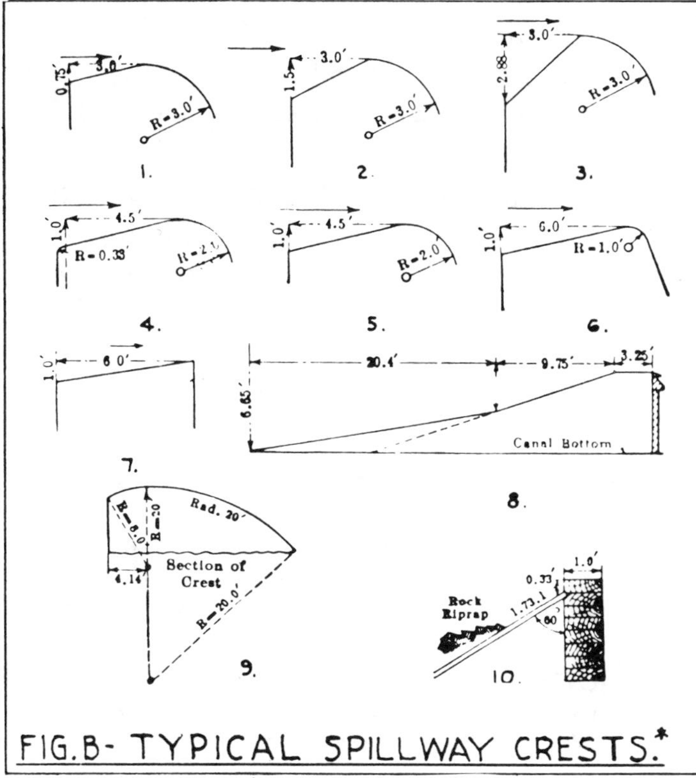

0.75′
3.0′
R=3.0′
1.
1.5′
3.0′
R=3.0′
2.
2.88′
3.0′
R=3.0′
3.
1.0′
4.5′
R=0.33′
R=2.1
4.
1.0′
4.5′
R=2.0
5.
1.0′
6.0′
R=1.0′
6.
1.0′
6.0′
7.
20.4′
9.75′
3.25′
6.65′
Canal Bottom
8.
Rad. 20′
R=6.0
R=20
Section of Crest
4.14
R=20.0′
9.
Rock Riprap
0.33′
1.0′
1.73.1
60
10.

FIG. B – TYPICAL SPILLWAY CRESTS.*

Nº IN FIG. B	HEAD IN FEET.									
	0.5	1.0	1.5	2.0	2.5	3.0	3.5	4.0	4.5	5.0
1	3.32	3.44	3.46	3.42	3.41	3.46	3.50			
2	3.22	3.48	3.61	3.67	3.70	3.72				
3	3.15	3.45	3.64	3.75	3.82	3.87	3.88			
4	3.23	3.34	3.43	3.52	3.59	3.64				
5	3.18	3.30	3.37	3.42	3.46	3.49	3.52	3.54		
6	3.28	3.50	3.54	3.52	3.36	3.31	3.30			
7	3.53	3.54	3.55	3.50	3.35	3.27	3.25	3.25		
8	3.13	3.14	3.10	3.14	3.20	3.26	3.31	3.37		
9	3.09	3.11	3.33							
10		3.80								

TABLE C - DISCHARGE COEFFICIENTS FOR FIG B.*

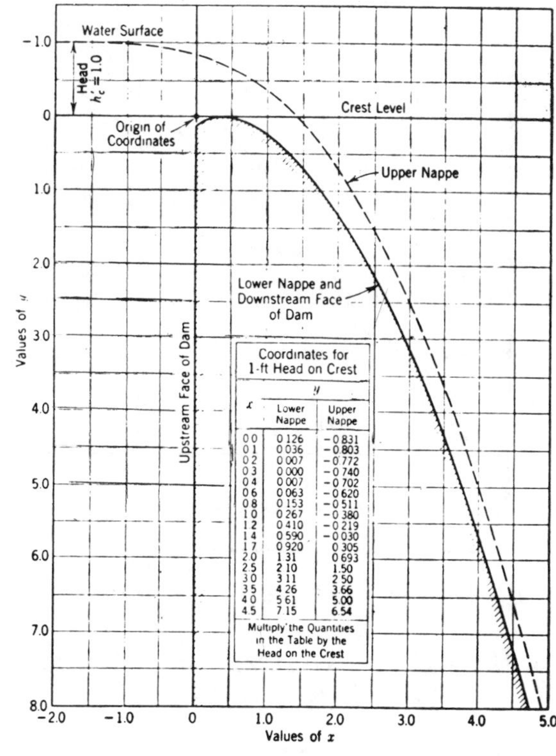

Water Surface
Head $h_c' = 1.0$
Crest Level
Origin of Coordinates
Upper Nappe
Lower Nappe and Downstream Face of Dam
Upstream Face of Dam
Values of y
Values of x
Coordinates for 1-ft Head on Crest
x | y Lower Nappe | y Upper Nappe
0.0 | 0.126 | −0.831
0.1 | 0.036 | −0.803
0.2 | 0.007 | −0.772
0.3 | 0.000 | −0.740
0.4 | 0.007 | −0.702
0.6 | 0.063 | −0.620
0.8 | 0.153 | −0.511
1.0 | 0.267 | −0.380
1.2 | 0.410 | −0.219
1.4 | 0.590 | −0.030
1.7 | 0.920 | 0.305
2.0 | 1.31 | 0.693
2.5 | 2.10 | 1.50
3.0 | 3.11 | 2.50
3.5 | 4.26 | 3.66
4.0 | 5.61 | 5.00
4.5 | 7.15 | 6.54
Multiply the Quantities in the Table by the Head on the Crest

FIG. A. STANDARD CREST, UPSTREAM FACE VERTICAL †

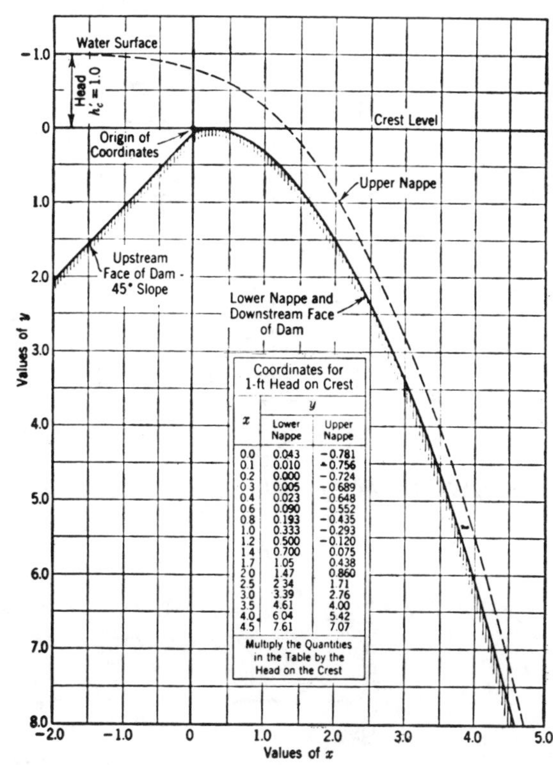

Water Surface
Head $h_c' = 1.0$
Crest Level
Origin of Coordinates
Upper Nappe
Upstream Face of Dam - 45° Slope
Lower Nappe and Downstream Face of Dam
Values of y
Values of x
Coordinates for 1-ft Head on Crest
x | y Lower Nappe | y Upper Nappe
0.0 | 0.043 | −0.781
0.1 | 0.010 | −0.756
0.2 | 0.000 | −0.724
0.3 | 0.005 | −0.689
0.4 | 0.023 | −0.648
0.6 | 0.090 | −0.552
0.8 | 0.193 | −0.435
1.0 | 0.333 | −0.293
1.2 | 0.500 | −0.120
1.4 | 0.700 | 0.075
1.7 | 1.05 | 0.438
2.0 | 1.47 | 0.860
2.5 | 2.34 | 1.71
3.0 | 3.39 | 2.76
3.5 | 4.61 | 4.00
4.0 | 6.04 | 5.42
4.5 | 7.61 | 7.07
Multiply the Quantities in the Table by the Head on the Crest

FIG. D – STANDARD CREST UPSTREAM FACE INCLINED 45° †

† From "Creager, Justin and Hinds" Engineering for Dams" John Wiley & Sons, Inc., 1945

* From King, "Handbook of Hydraulics", Mc Graw-Hill.

DAMS-SPILLWAY DESIGN-2

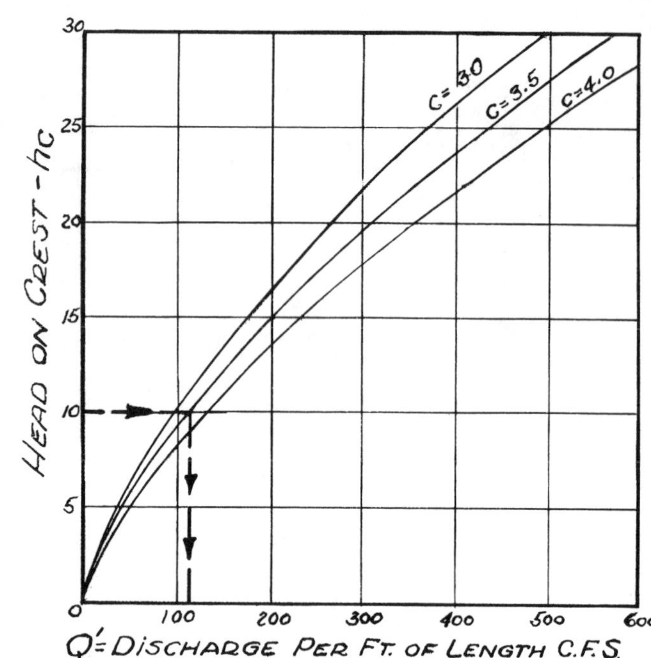

HEAD ON CREST - hc
C=3.0
C=3.5
C=4.0
30
25
20
15
10
5
0
100
200
300
400
500
600
Q'= DISCHARGE PER FT. OF LENGTH C.F.S.

FIG. A - DISCHARGE PER FT. OF LENGTH-NO END CONTRACTION

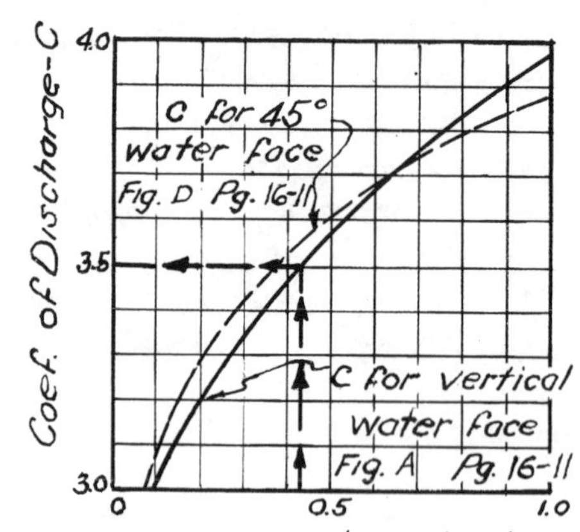

Coef. of Discharge-C
4.0
3.5
3.0
C for 45° water face
Fig. D Pg. 16-11
C for vertical water face
Fig. A Pg. 16-11
0
0.5
1.0
Head Ratio hc/h'c = Head on crest / Design Head

FIG. B-COEFFICIENT OF DISCHARGE FOR STANDARD CREST*
(NO VELOCITY OF APPROACH)

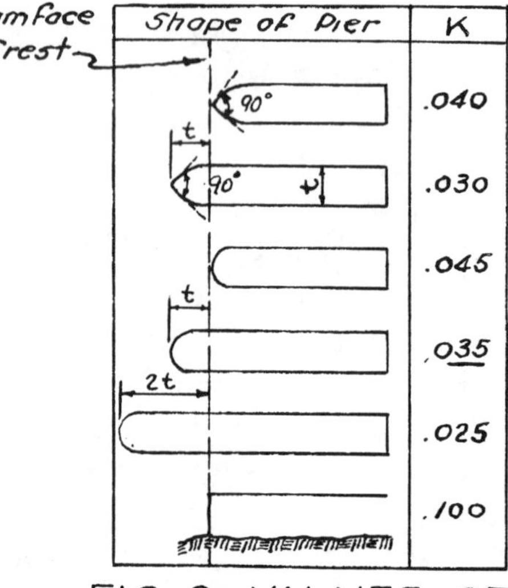

Upstream face of Crest
Shape of Pier
K
90°
.040
t
90°
.030
.045
t
.035
2t
.025
.100

FIG. C - VALUES OF K*

Values of K for all gates open. Multiply K by 2.5 for one gate open.

DISCHARGE FORMULAS

$Q' = Ch_c^{3/2}$ see Fig. A at left

$Q = Q'L$ with no end contraction

$Q = Q'(L - nkh_c)$ with end contraction

Q = Discharge in cubic ft. per sec.
C = Constant - See Fig. B above and Table A page 16-11
h_c = Head on crest in ft.
L = Effective length of crest in ft.
K = Contraction coefficient, Fig. C.
n = Number of end contractions
h'_c = Design head - the head used in determining the shape of crest (Max. flood water)

NOTE:
Velocity of approach neglected in discharge formulas and examples.

EXAMPLE 1 Given: $h_c = 10'$; $h'_c = 23'$; no end contraction; $L = 100'$; vertical face.
Required: Discharge in c.f.s.
Solution: $\frac{h_c}{h'_c} = .043$ Using fig. B, C = 3.5. From fig. A, Q' = 110. Q = 110×100 = 11,000 c.f.s.

EXAMPLE 2 - Given: Same as above except distance between abutments = 200' and plan and shape of piers thus:

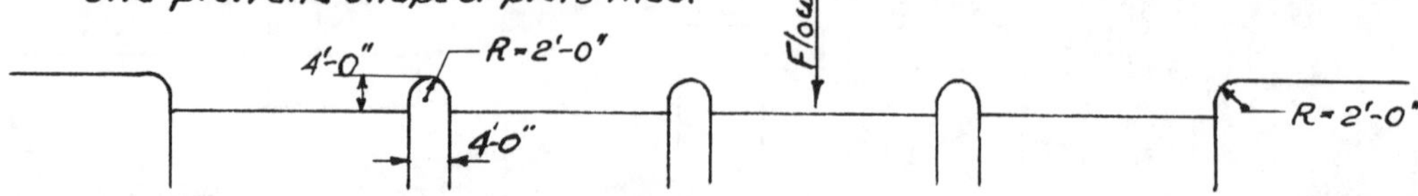

4'-0"
R=2'-0"
4'-0"
Flow
R=2'-0"

Required: Discharge in c.f.s.
Solution: $\frac{h_c}{h'_c} = 0.43$, C = 3.5 and Q' = 110 as in example 1 ∴ $Q = 110(L - nkh_c)$
From sketch - n = 8 and from fig. C k = .035 and L = (200 - 4×3) = 188
∴ Q = 110(188 - 8×.035×10) = 20,372 c.f.s.

*From "Creager, Justin and Hind's" Engineering for Dams" John Wiley & Sons, Inc., 1945

DAMS – STOP LOGS

STOP LOGS – Simplest form of adjustable crest control. Initial cost less than gates. Used as crest control for outlets of log & trash chutes and minor dams. Stop log slots always provided to allow unwatering upstream side of gates when gates become jammed or access is required to upstream side of gate when gate is in lowered position. If gate spans are equal, one set of stop logs may be sufficient.

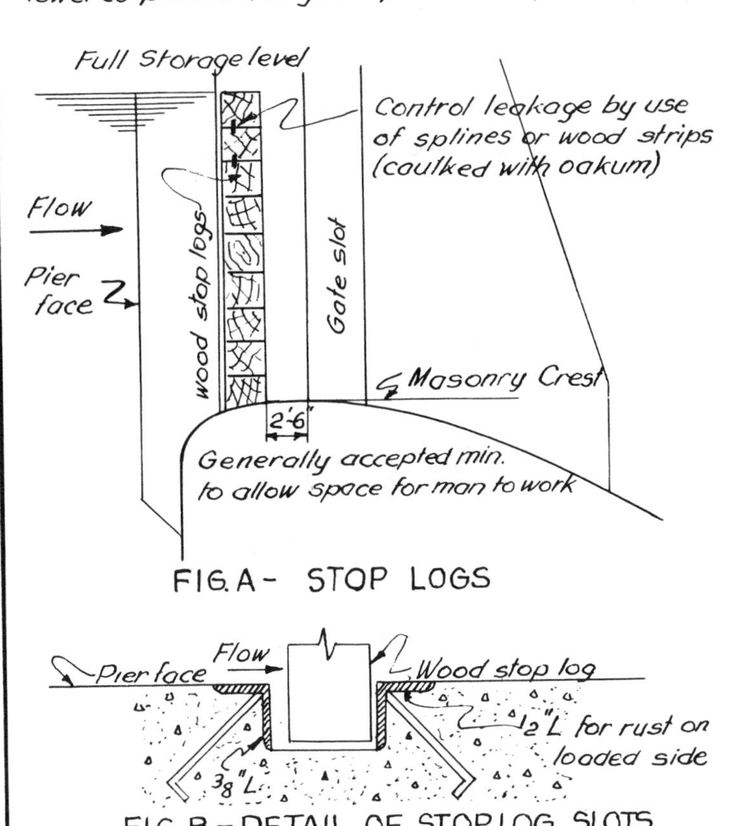

Full Storage level
Control leakage by use of splines or wood strips (caulked with oakum)
Flow
Pier face
wood stop logs
Gate slot
Masonry Crest
2'-6"
Generally accepted min. to allow space for man to work
Pier face
Flow
Wood stop log
½" L for rust on loaded side
⅜" L

FIG. A – STOP LOGS

FIG. B – DETAIL OF STOP LOG SLOTS

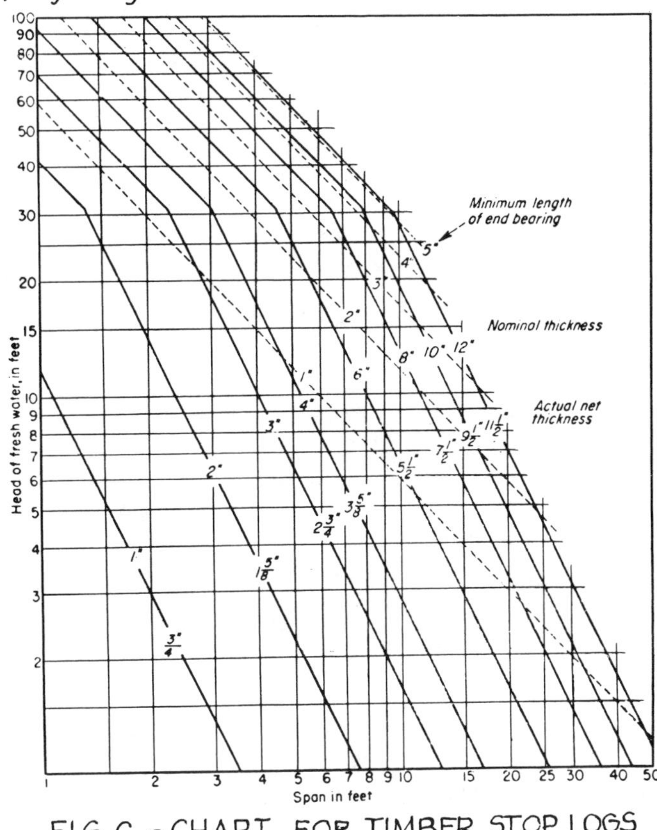

Head of fresh water, in feet
Span in feet
Minimum length of end bearing
Nominal thickness
Actual net thickness

FIG. C – CHART FOR TIMBER STOP LOGS

Chart gives horizontal thickness and length of end bearing for stop log timbers on simple beam spans* Based on stresses of 1000 psi. in bending, 1000 psi. in shear and 150 p.s.i. in bearing. Below a head of 31 ft., bending governs above shear.

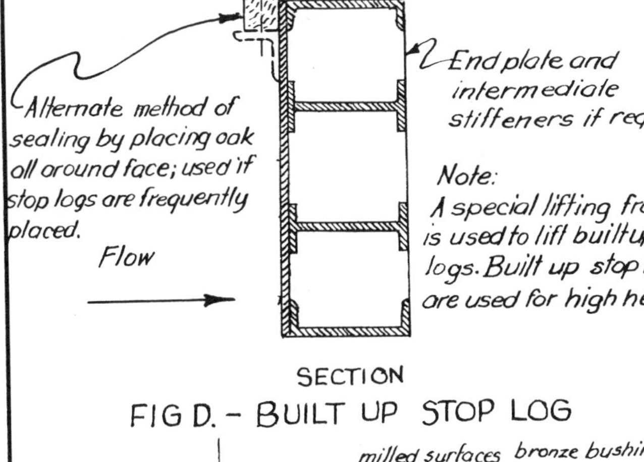

Wood Seal
Lifting lug
End plate and intermediate stiffeners if required
Alternate method of sealing by placing oak all around face; used if stop logs are frequently placed.
Flow
Note:
A special lifting frame is used to lift built up stop logs. Built up stop logs are used for high heads.
SECTION

FIG D. – BUILT UP STOP LOG

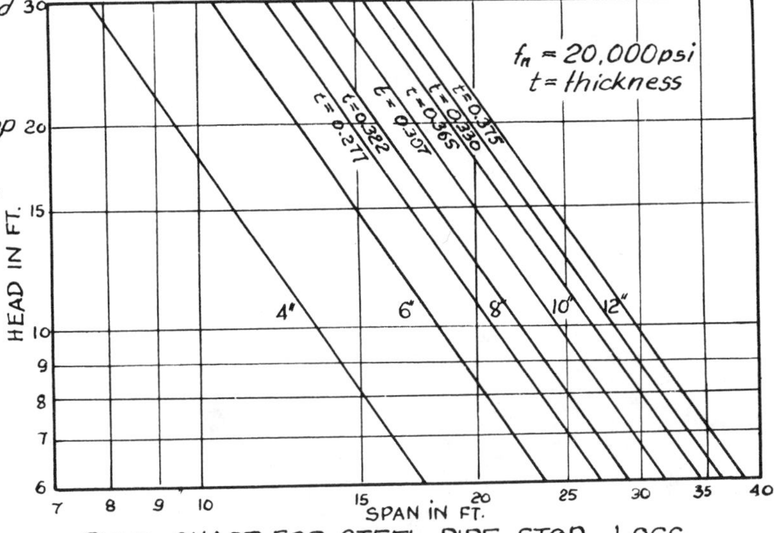

f_n = 20,000 psi
t = thickness
HEAD IN FT.
SPAN IN FT.
t = 0.375
t = 0.330
t = 0.365
t = 0.307
t = 0.322
t = 0.277
4"
6"
8"
10"
12"

FIG. E – CHART FOR STEEL PIPE STOP LOGS

Notes:
Pipe stop logs fitted with square ends can always drop because they spin down, whereas wood stop logs must be jacked down. Pipe stop logs are used for longer spans and to decrease maintenance costs. Wood stop logs are apt to dry rot when not in use.

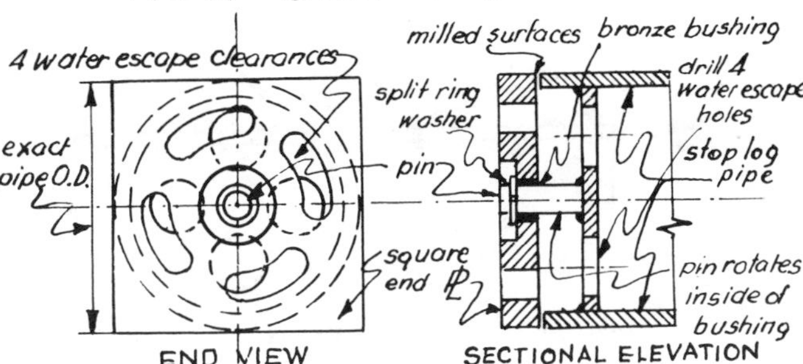

4 water escape clearances
exact pipe O.D.
pin
square end ℙ
END VIEW
milled surfaces
bronze bushing
split ring washer
drill 4 water escape holes
stop log pipe
pin rotates inside of bushing
SECTIONAL ELEVATION

FIG. F END SPIDER FOR PIPE STOP LOGS

* From Field and Office, C.J. Posey; Engineering News Record, Nov. 27, 1947

DAMS — FLASHBOARDS

FLASHBOARDS—Have the lowest initial cost of crest control devices. If men are available to operate permanent flashboards, the permanent type are more economical than temporary flashboards which are lost at each high water.

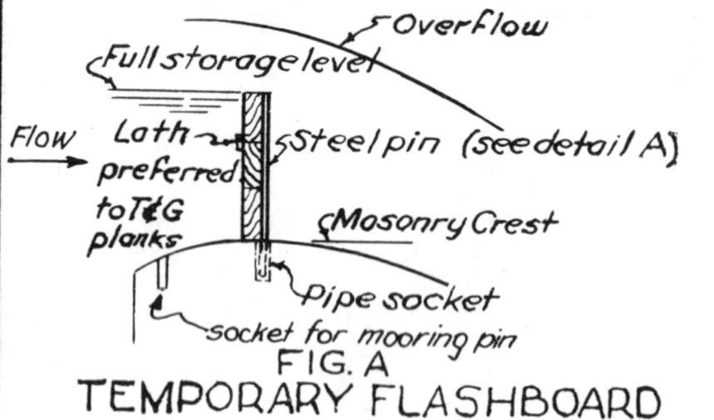

Overflow
Full storage level
Flow
Lath preferred to T&G planks
Steel pin (see detail A)
Masonry Crest
Pipe socket
socket for mooring pin

FIG. A
TEMPORARY FLASHBOARD

Suspension cable used for long spans in place of truss bridge.

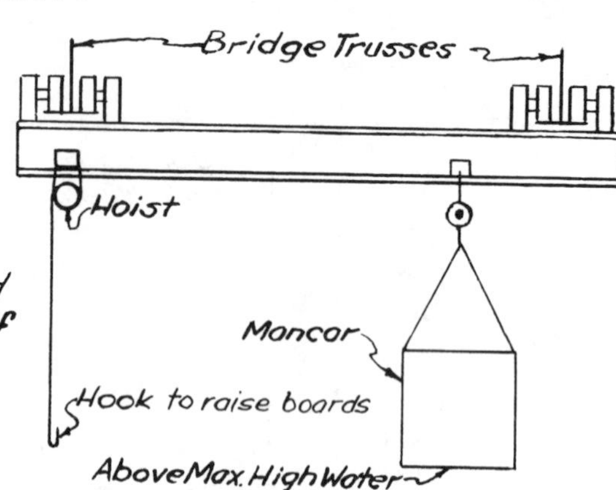

Bridge Trusses
Hoist
Mancor
Hook to raise boards
Above Max. High Water

steel pin
Wood strip wired to pin
Flow
Nail
plank

DETAIL A-
PLAN PIN CONNECTION

Notes: For 3'-0" pin spacing,
1½" Planks to 4'-0" head
2" Planks 4'-0" to 7'-0" head

El. 107'
Full storage level
Skin Pl. (3/8" steel min.)
Wood strut-Compute to break at ultimate stress if not removable
Flow
El. 100.5
Seal
Masonry crest
Flashboard anchor (See Detail B)
Flashboard in lowered position
Strut seat L
Slot for Flasboard hook

FIG C-PERMANENT FLASHBOARD

TABLE B- HEIGHT OF WATER IN FEET ABOVE TOP OF FLASHBOARDS AT WHICH PINS WILL FAIL

SOLID STEEL PIN SIZE	HEIGHT OF FLASHBOARDS					
	1'-6"	2'-0"	2'-6"	3'-0"	3'-6"	4'-0"
3/4"φ	.5					
1"φ	1.9	.7				
1¼"φ	4.0	1.9	.8			
1½"φ	7.4	3.8	2.0	1.0		
1¾"φ		6.4	3.7	2.2	1.1	
2"φ			6.0	3.7	2.2	1.3
2¼"φ				5.7	3.7	2.4
2½"φ					5.5	3.8
2¾"φ						5.5

Note: Pins to fail at stress of 60,000 lbs. per sq. in.
Pins spaced at 3'-0" o.c.

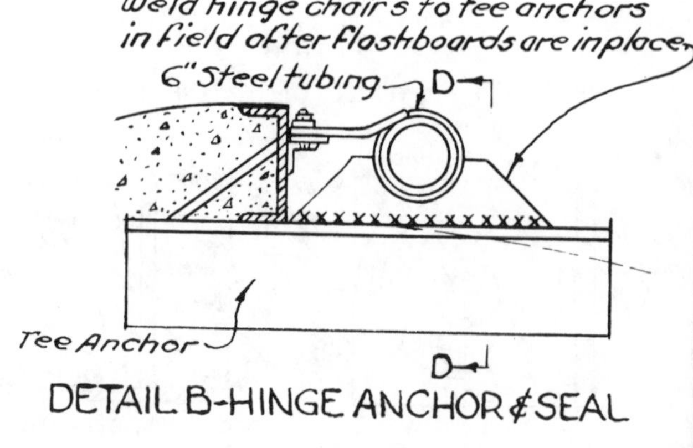

Weld hinge chairs to tee anchors in field after flashboards are in place
6" Steel tubing
D
Tee Anchor
D

DETAIL B-HINGE ANCHOR & SEAL

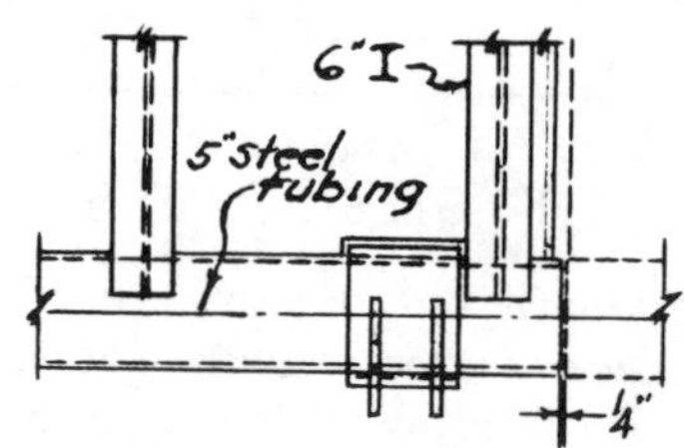

6" I
5" steel tubing
1/4"

SECTION D-D SHOWING FLASHBOARD

DAMS – STANCHION AND BASCULE GATE

STANCHION TYPE FLASHBOARDS - Require piers with walkway spans; boards are held by vertical steel beams or stanchions pivoted near the top so as to swing when raised or tripped, thus clearing the channel and allowing the boards to go downriver. Stanchions usually 5'-0"± on centers. Economical when used in conjunction with gates to provide additional spillway capacity required once in 5 to 15 years.

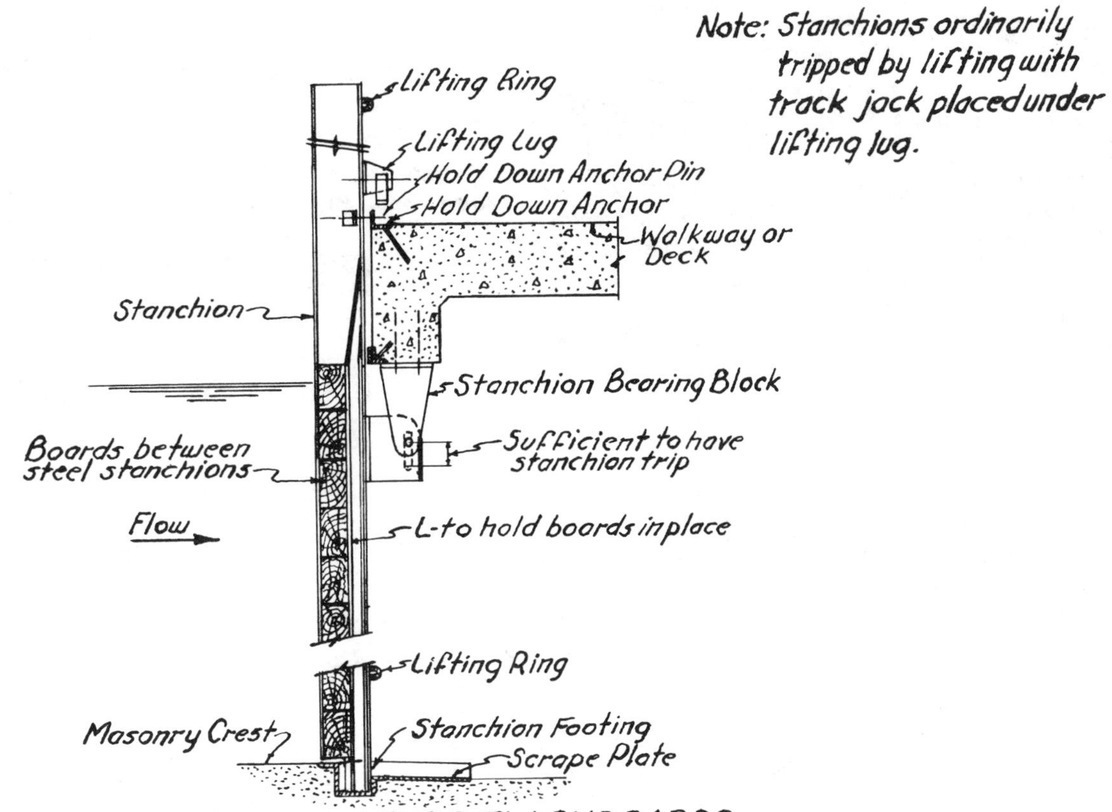
Note: Stanchions ordinarily tripped by lifting with track jack placed under lifting lug.
Lifting Ring
Lifting Lug
Hold Down Anchor Pin
Hold Down Anchor
Walkway or Deck
Stanchion
Stanchion Bearing Block
Sufficient to have stanchion trip
Boards between steel stanchions
Flow
L-to hold boards in place
Lifting Ring
Masonry Crest
Stanchion Footing
Scrape Plate

FIG. A - STANCHION TYPE FLASHBOARDS

BASCULE TYPE CREST GATES - Give best control of crest but initial cost is high. Greater speed in lowering and raising than permanent flashboards and provide unobstructed crest for passing surface debris, also breaking and passing ice in springtime. Generally hydraulically operated by automatic controls and float.

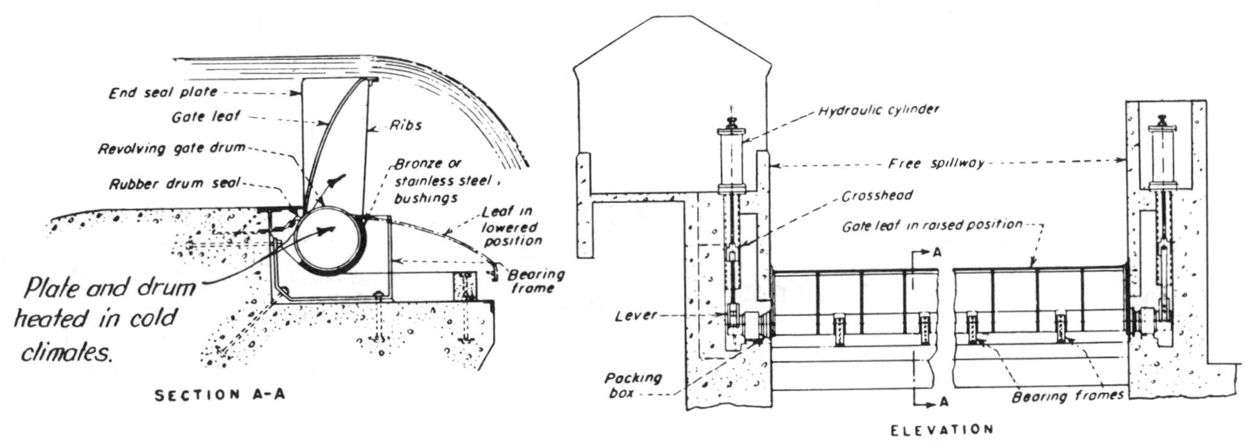
End seal plate
Gate leaf
Ribs
Revolving gate drum
Rubber drum seal
Bronze or stainless steel, bushings
Leaf in lowered position
Bearing frame
Plate and drum heated in cold climates.
SECTION A-A
Hydraulic cylinder
Free spillway
Crosshead
Gate leaf in raised position
A
Lever
Packing box
A
Bearing frames
ELEVATION

FIG. B - BASCULE TYPE CREST GATE*

*Adopted from S. Morgan Smith Co., York, Pa.

DAMS—TAINTER GATES

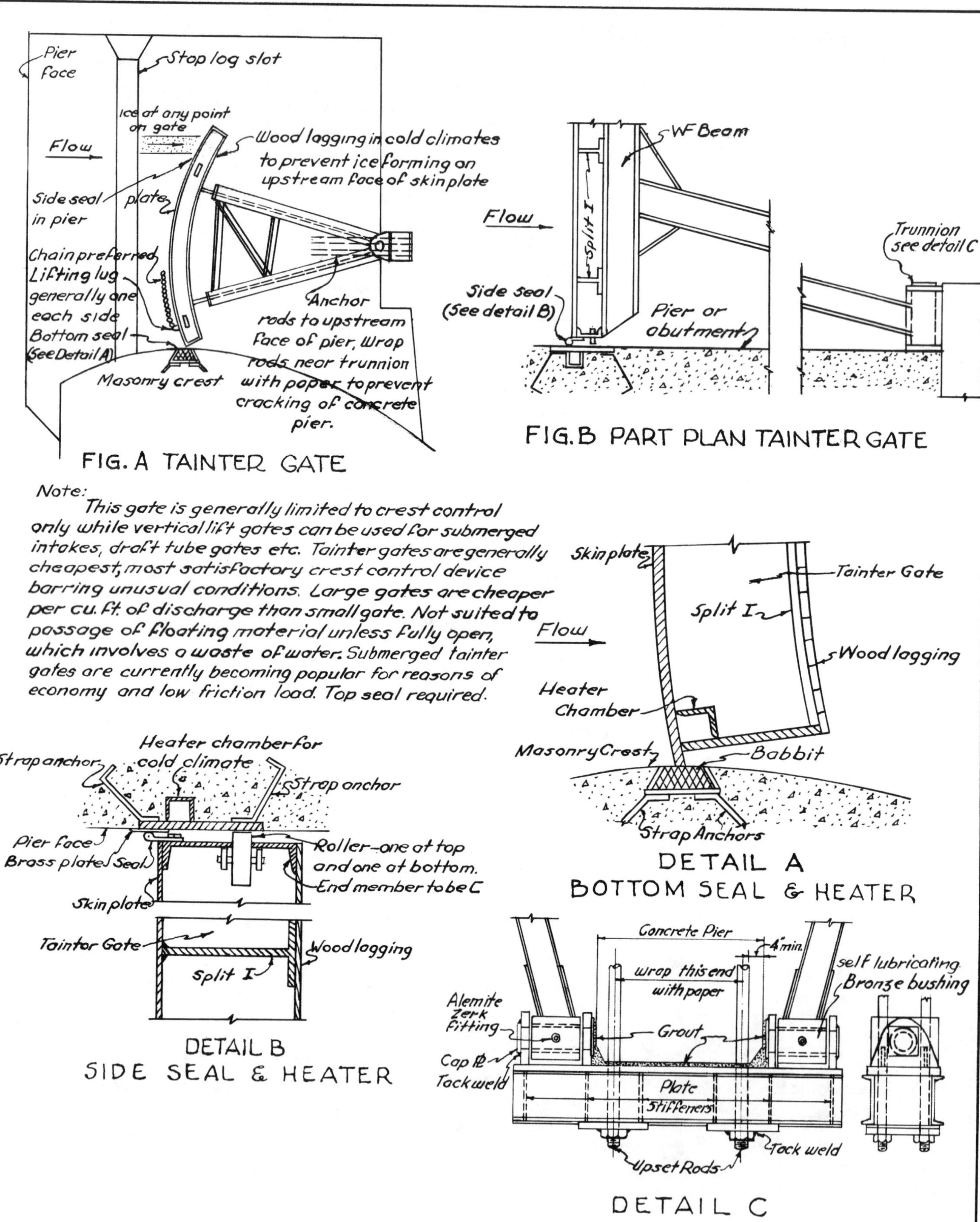
Pier face
Stop log slot
Ice at any point on gate
Flow
Wood lagging in cold climates to prevent ice forming on upstream face of skin plate
Side seal in pier
plate
Chain preferred
Lifting lug generally one each side
Bottom seal (See Detail A)
Masonry crest
Anchor rods to upstream face of pier, Wrap rods near trunnion with paper to prevent cracking of concrete pier.
FIG. A TAINTER GATE
WF Beam
Flow
Split I
Trunnion see detail C
Side seal (See detail B)
Pier or abutment
FIG. B PART PLAN TAINTER GATE
Note:
This gate is generally limited to crest control only while vertical lift gates can be used for submerged intakes, draft tube gates etc. Tainter gates are generally cheapest, most satisfactory crest control device barring unusual conditions. Large gates are cheaper per cu. ft. of discharge than small gate. Not suited to passage of floating material unless fully open, which involves a waste of water. Submerged tainter gates are currently becoming popular for reasons of economy and low friction load. Top seal required.
Skin plate
Tainter Gate
Split I
Flow
Wood lagging
Heater Chamber
Masonry Crest
Babbit
Strap Anchors
DETAIL A
BOTTOM SEAL & HEATER
Heater chamber for cold climate
Strap anchor
Strap anchor
Pier face
Brass plate
Seal
Roller—one at top and one at bottom.
End member to be C
Skin plate
Taintor Gate
Wood lagging
Split I
DETAIL B
SIDE SEAL & HEATER
Concrete Pier
4" min.
wrap this end with paper
self lubricating Bronze bushing
Alemite Zerk Fitting
Grout
Cap ℔
Tack weld
Plate
Stiffeners
Tack weld
Upset Rods
DETAIL C
TRUNNION

DAMS—VERTICAL LIFT GATES

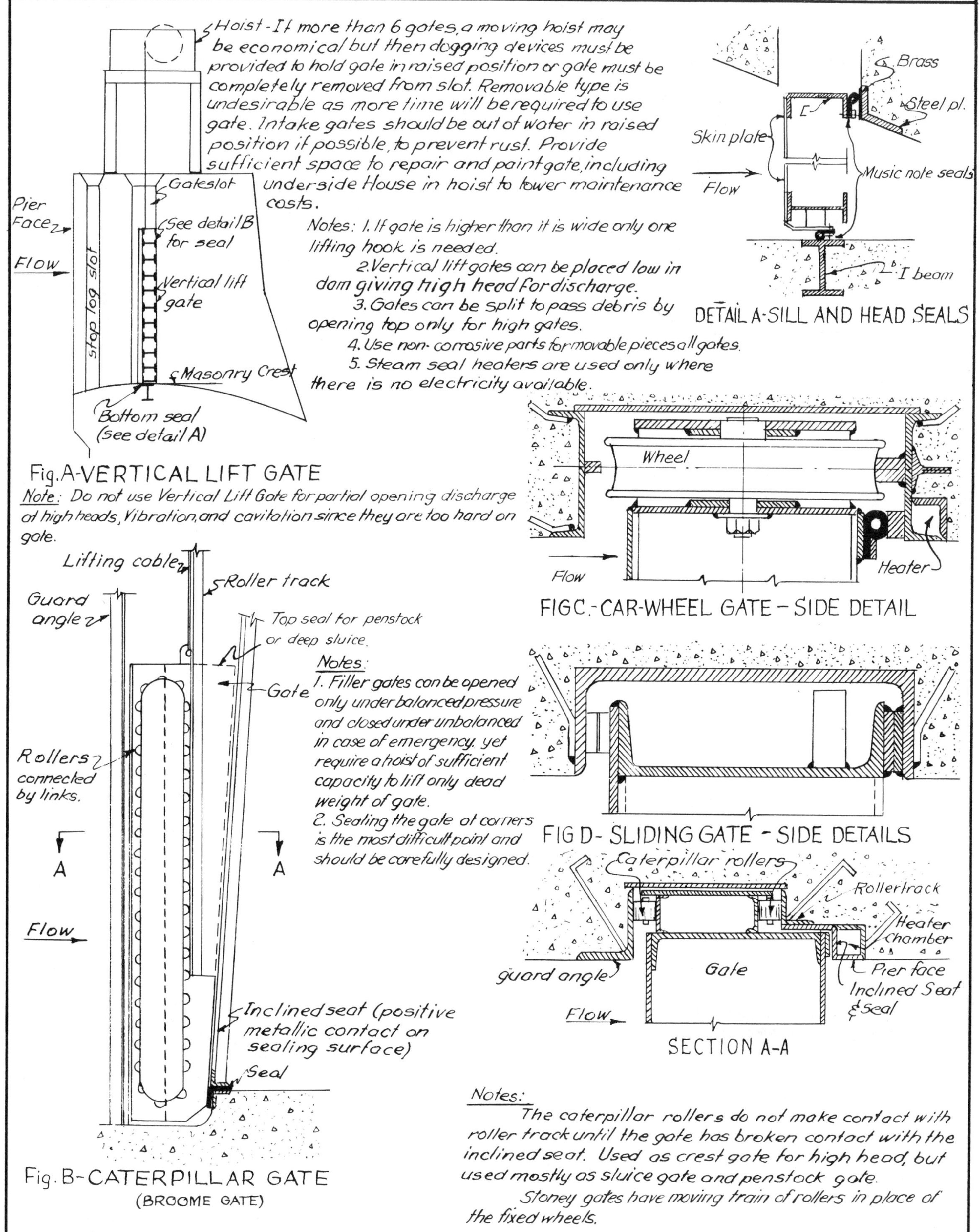
Hoist - If more than 6 gates, a moving hoist may be economical but then dogging devices must be provided to hold gate in raised position or gate must be completely removed from slot. Removable type is undesirable as more time will be required to use gate. Intake gates should be out of water in raised position if possible, to prevent rust. Provide sufficient space to repair and paint gate, including underside. House in hoist to lower maintenance costs.
Gateslot
Pier Face
See detail B for seal
Flow
stop log slot
Vertical lift gate
Masonry Crest
Bottom seal (see detail A)
Notes: 1. If gate is higher than it is wide only one lifting hook is needed.
2. Vertical lift gates can be placed low in dam giving high head for discharge.
3. Gates can be split to pass debris by opening top only for high gates.
4. Use non-corrosive parts for movable pieces all gates.
5. Steam seal heaters are used only where there is no electricity available.
Brass
Steel pl.
Skin plate
Music note seals
Flow
I beam
DETAIL A-SILL AND HEAD SEALS
Wheel
Flow
Heater
Lifting cable
Roller track
Guard angle
Top seal for penstock or deep sluice.
Gate
Notes:
1. Filler gates can be opened only under balanced pressure and closed under unbalanced in case of emergency, yet require a hoist of sufficient capacity to lift only dead weight of gate.
2. Sealing the gate at corners is the most difficult point and should be carefully designed.
Rollers connected by links.
A
A
Flow
Inclined seat (positive metallic contact on sealing surface)
Seal
Caterpillar rollers
Rollertrack
Heater Chamber
Pier face
Inclined Seat & Seal
guard angle
Gate
Flow
SECTION A-A

Fig. A-VERTICAL LIFT GATE

Note: Do not use Vertical Lift Gate for partial opening discharge at high heads, Vibration, and cavitation since they are too hard on gate.

FIG C.-CAR-WHEEL GATE—SIDE DETAIL

FIG D- SLIDING GATE - SIDE DETAILS

Fig. B-CATERPILLAR GATE
(BROOME GATE)

Notes:
The caterpillar rollers do not make contact with roller track until the gate has broken contact with the inclined seat. Used as crest gate for high head, but used mostly as sluice gate and penstock gate.
Stoney gates have moving train of rollers in place of the fixed wheels.

DAMS—MISCELLANEOUS GATES

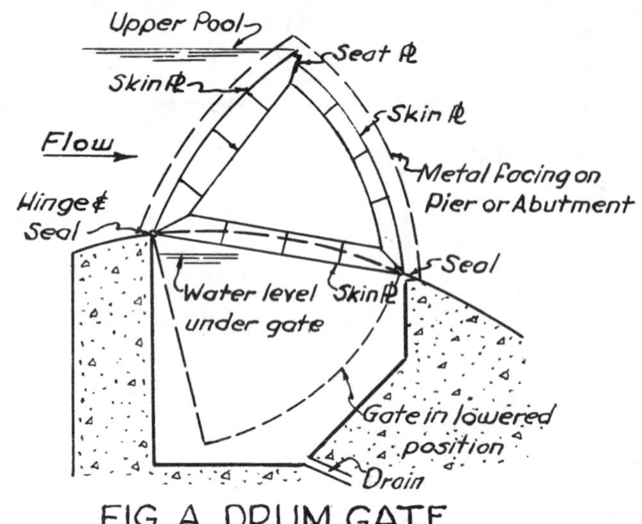
Upper Pool
Seat Pl
Skin Pl
Skin Pl
Flow
Metal facing on Pier or Abutment
Hinge & Seal
Seal
Water level under gate
Skin Pl
Gate in lowered position
Drain

FIG. A DRUM GATE
(UPSTREAM HINGE)

The upstream hinge simplifies the sealing problem.

Note: Not adapted to low dam because of deep excavation. Overhead bridge eliminated and pier height reduced to a minimum. Well adapted to surface regulation and passing debris.

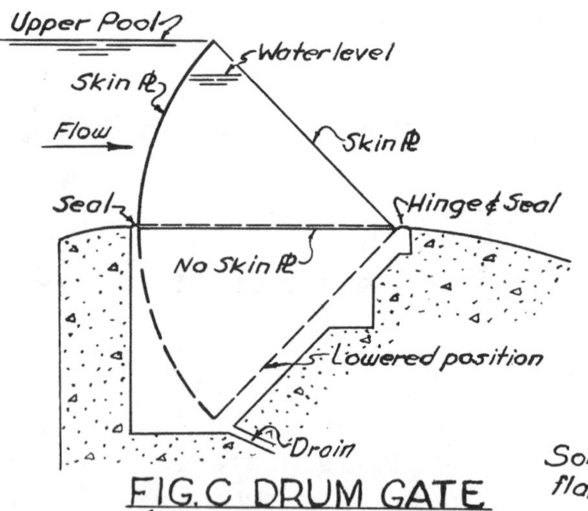
Upper Pool
Water level
Skin Pl
Flow
Skin Pl
Seal
Hinge & Seal
No Skin Pl
Lowered position
Drain

FIG. C DRUM GATE
(DOWNSTREAM HINGE)

Note:
The Drum and Tilting gates are not adapted to regions of severe winters because ice pressure will lower gates.

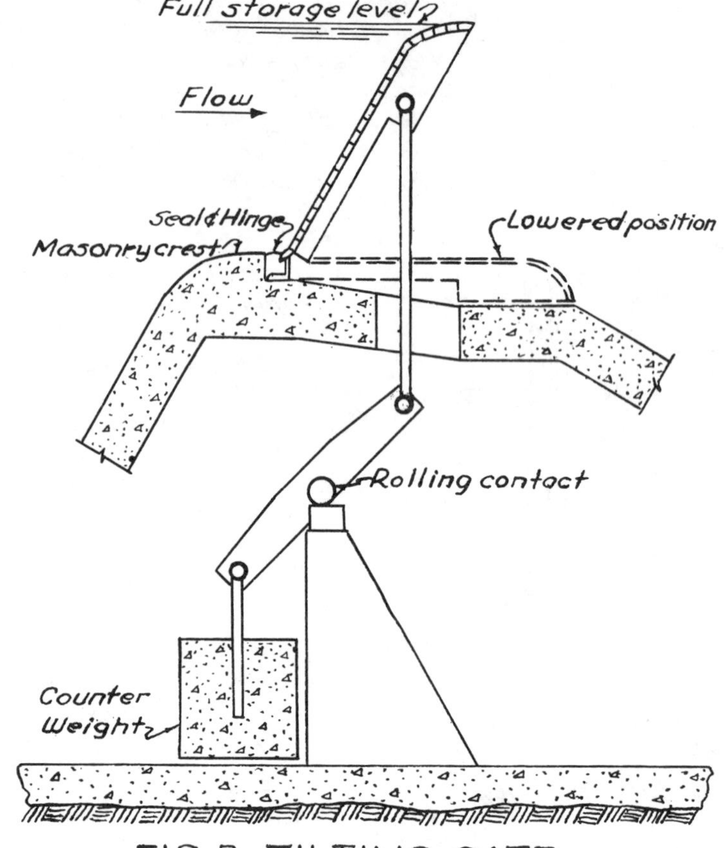
Full storage level
Flow
Seal & Hinge
Masonry crest
Lowered position
Rolling contact
Counter Weight

FIG. B TILTING GATE

Note:
The counterweight may be on the pier above high water level. Used for passing heavy drift. Height of storage level governs position of gate. Facilities for stop logs or needles should be provided above gate.

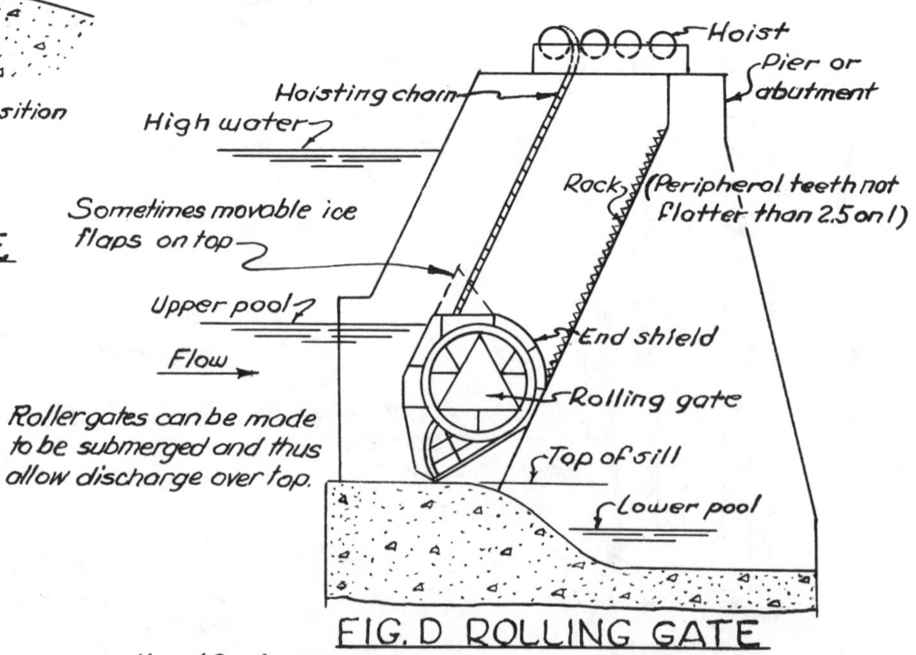
Hoist
Pier or abutment
Hoisting chain
High water
Rack (Peripheral teeth not flatter than 2.5 on 1)
Sometimes movable ice flaps on top
Upper pool
Flow
End shield
Rolling gate
Roller gates can be made to be submerged and thus allow discharge over top.
Top of sill
Lower pool

FIG. D ROLLING GATE

Used for long spans, Raising force usually applied at one end and force needed to raise the other end is transmitted in torsion through the cylinder. Excellent for breaking ice. Long spans allow passing large cakes and wide debris.

DAMS – WEIGHTS OF GATES

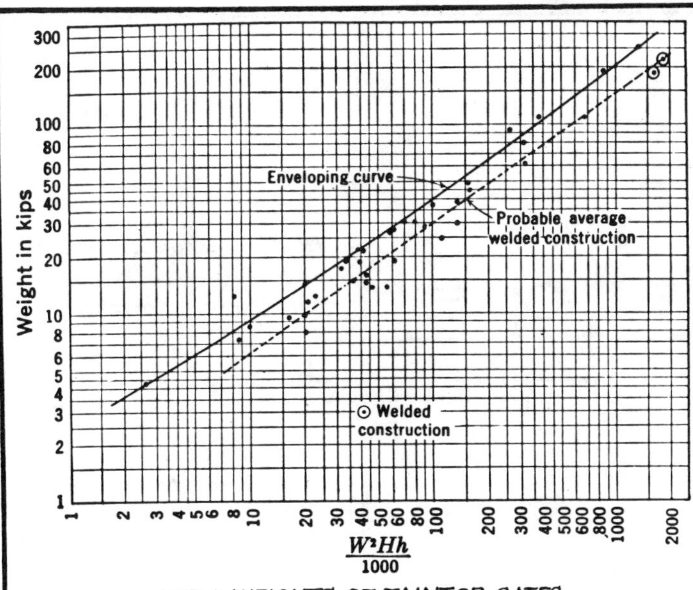

Weight in kips
Enveloping curve
Probable average welded construction
⊙ Welded construction
$\frac{W^2Hh}{1000}$

FIG. A WEIGHTS OF TAINTOR GATES.

Weight includes all moving & stationary parts and complete control

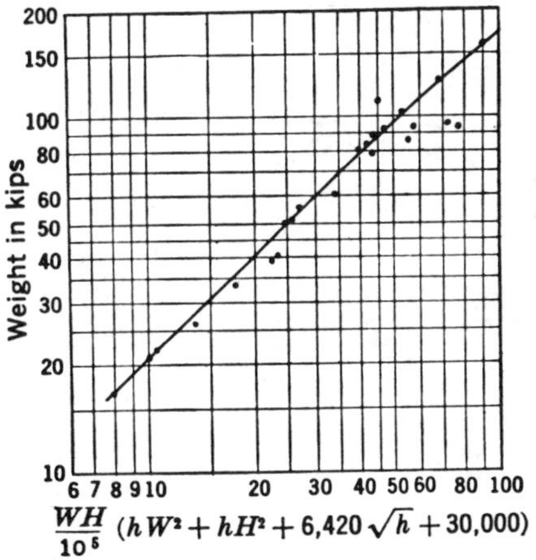

Weight in kips
$\frac{WH}{10^5}(hW^2 + hH^2 + 6{,}420\sqrt{h} + 30{,}000)$

FIG. B WEIGHTS OF SLIDING LIFT GATES WITH HYDRAULIC HOISTS

Weight includes all moving parts with cast iron lining. No hanger, piping or valves included.

SYMBOLS:

W = Width of gate in ft. Clear opening.
H = Height of gate in ft. Clear opening
h = head of water on ℄ of gate in ft.
All weights are in kips. (1000 lbs)

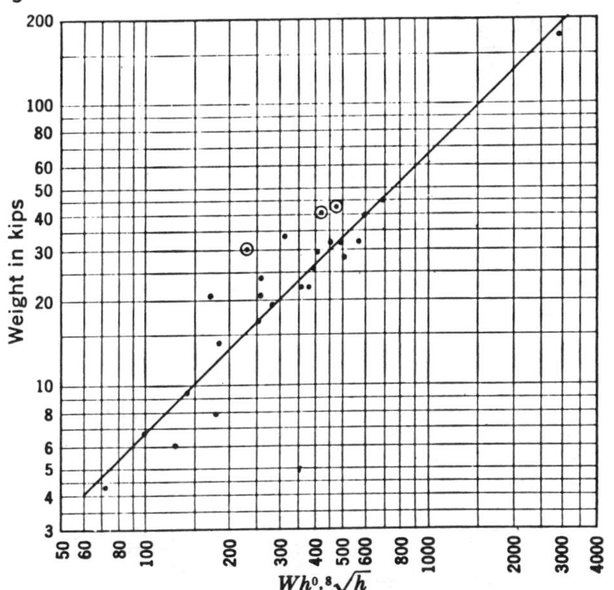

Weight in kips
$Wh^{0.8}\sqrt{h}$

FIG. C WEIGHTS OF CATERPILLAR GATES

Weights include imbedded parts but not conduit lining, hoists and cables. Encircled points have cast frames, others str. steel frames.

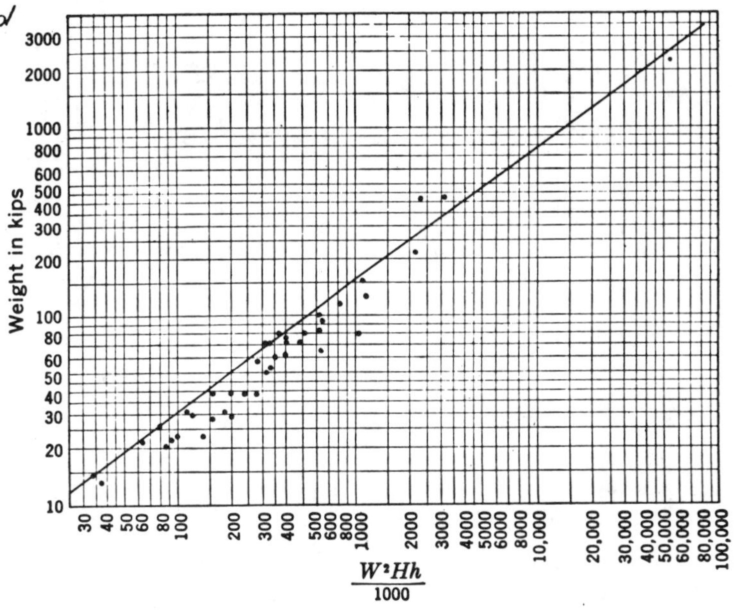

FIG. D - WEIGHTS OF WHEELED AND STONEY GATES

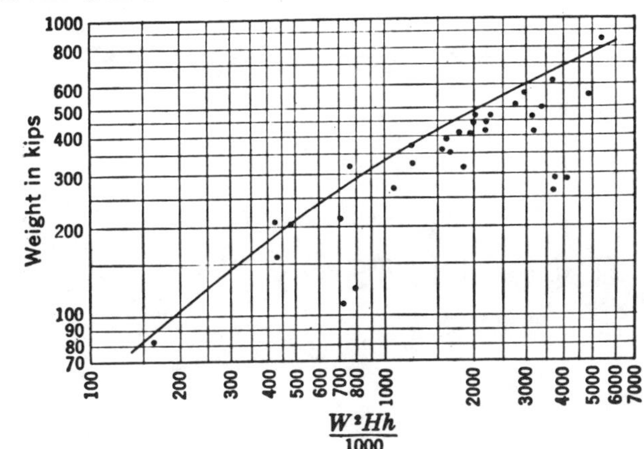

Weight in kips
$\frac{W^2Hh}{1000}$

FIG. E WEIGHTS OF ROLLING GATES

Weights include imbedded metal, exclude weights of hoist & chain

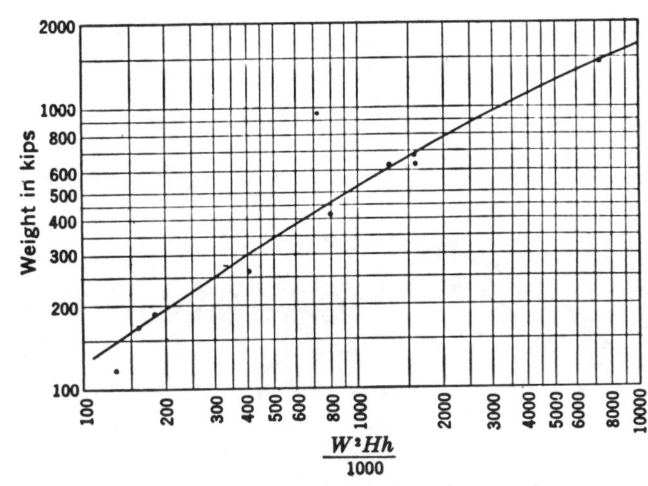

Weight in kips
$\frac{W^2Hh}{1000}$

FIG. F WEIGHTS OF DRUM GATES

Weights include all moving and stationary parts & complete control

* *Data for figures obtained from H.G. Gerdes, P.L. Heslop, E.B. Miller, F.L. Boissonoult, W.P. Creager, and J.D. Justin, the last two authors of "Hydro-Electric Handbook." John Wiley.*

DAMS—SECTION AT POWERHOUSE

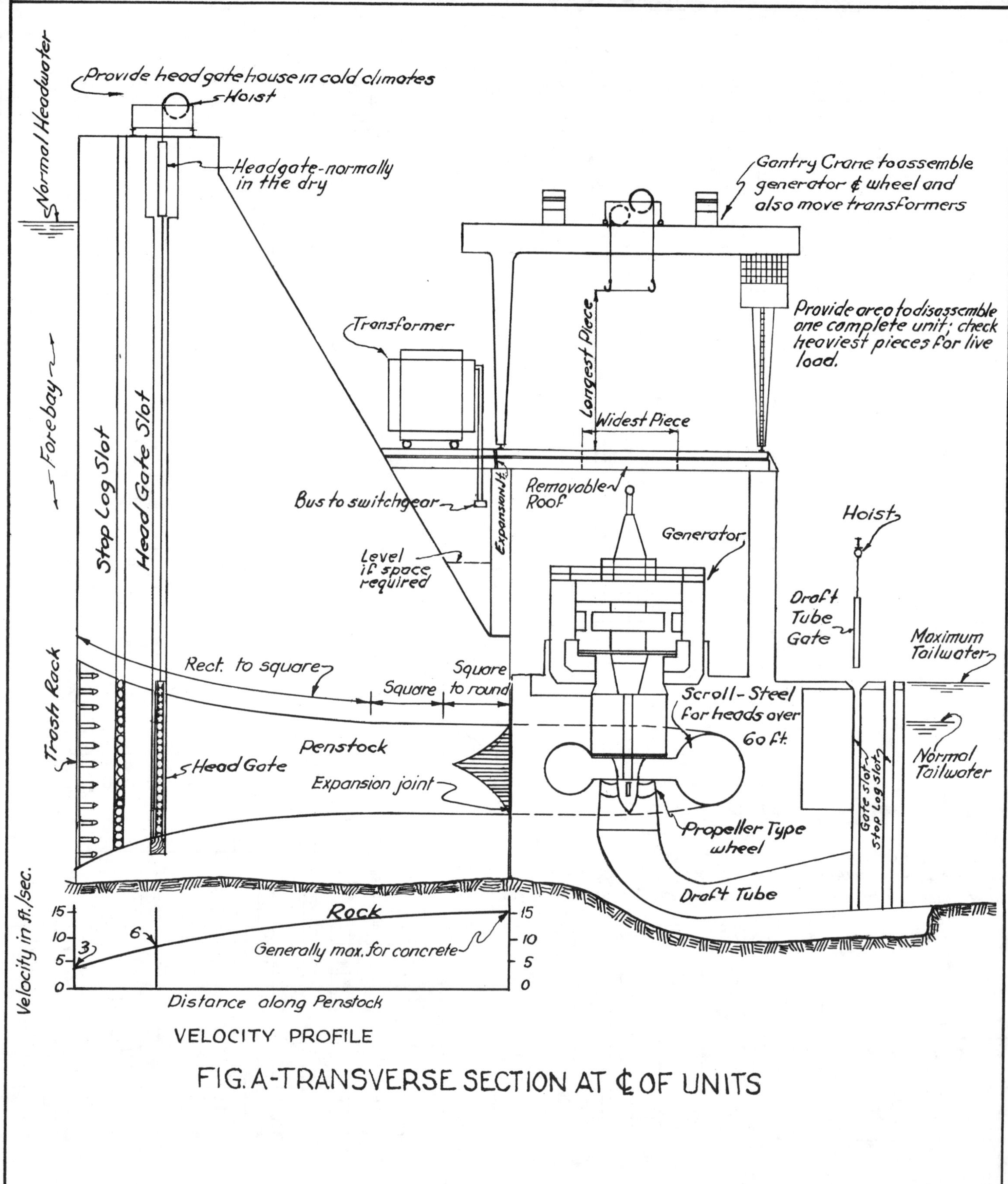

Provide head gate house in cold climates
Hoist
Normal Headwater
Headgate-normally in the dry
Gantry Crane to assemble generator ¢ wheel and also move transformers
Provide area to disassemble one complete unit; check heaviest pieces for live load.
Transformer
Longest Piece
Widest Piece
Forebay
Stop Log Slot
Head Gate Slot
Bus to switchgear
Expansion Jt.
Removable Roof
Generator
Hoist
Level if space required
Draft Tube Gate
Maximum Tailwater
Trash Rack
Rect. to square
Square
Square to round
Scroll-Steel for heads over 60 ft.
Penstock
Head Gate
Expansion joint
Normal Tailwater
Gate slot
Stop Log slot
Propeller Type wheel
Draft Tube
Rock
Velocity in ft./sec.
15
10
5
0
3
6
Generally max. for concrete
Distance along Penstock
VELOCITY PROFILE

FIG. A-TRANSVERSE SECTION AT ¢ OF UNITS

DAMS–POWER CHART

$$K.W.^* = \frac{C.F.S. \times H \times 0.87 \times 0.95 \times 62.5 \times 0.746}{550}$$

H.P. = 1.341 KW

$$K.W.H.^* = \frac{C.F. \times H \times 0.87 \times 0.95 \times 62.5 \times 0.746}{550 \times 3600}$$

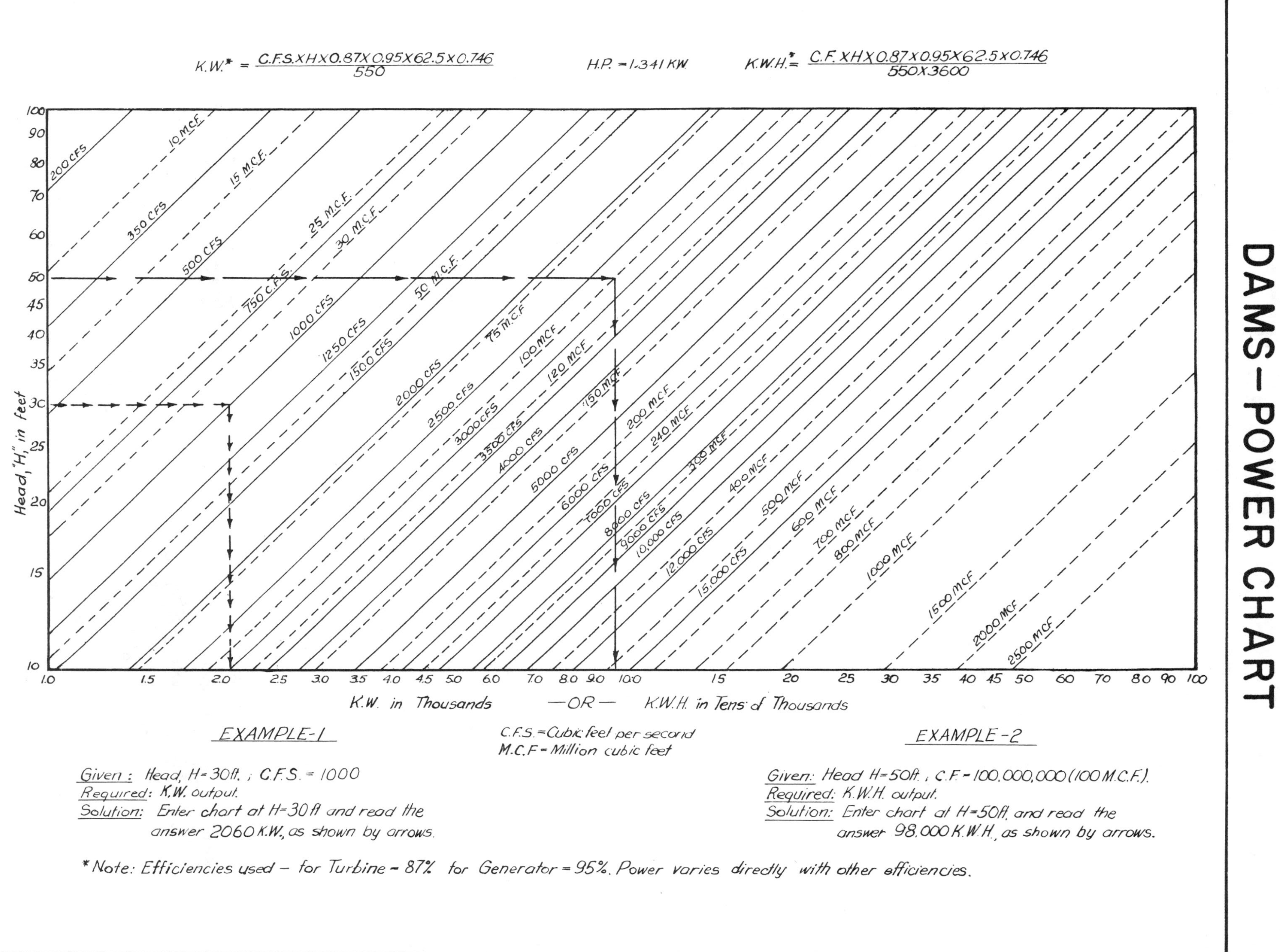

Head, "H", in feet
100
90
80
70
60
50
45
40
35
30
25
20
15
10
1.0
1.5
2.0
2.5
3.0
3.5
4.0
4.5
5.0
6.0
7.0
8.0
9.0
10.0
15
20
25
30
35
40
45
50
60
70
80
90
100
K.W. in Thousands —OR— K.W.H. in Tens of Thousands
200 CFS
10 M.C.F.
350 CFS
15 M.C.F.
500 CFS
25 M.C.F.
30 M.C.F.
750 C.F.S.
1000 CFS
50 M.C.F.
1250 CFS
1500 CFS
75 M.C.F
2000 CFS
100 MCF
120 MCF
2500 CFS
3000 CFS
150 MCF
3500 CFS
4000 CFS
200 MCF
5000 CFS
240 MCF
6000 CFS
300 MCF
7000 CFS
8000 CFS
9000 CFS
10,000 CFS
400 MCF
12,000 CFS
500 MCF
15,000 CFS
600 MCF
700 MCF
800 MCF
1000 MCF
1500 MCF
2000 MCF
2500 MCF

C.F.S. = Cubic feet per second
M.C.F = Million cubic feet

EXAMPLE-1

Given: Head, H = 30ft. ; C.F.S. = 1000
Required: K.W. output.
Solution: Enter chart at H = 30 ft and read the answer 2060 K.W., as shown by arrows.

EXAMPLE-2

Given: Head H = 50ft. ; C.F. = 100,000,000 (100 M.C.F.).
Required: K.W.H. output.
Solution: Enter chart at H = 50ft. and read the answer 98,000 K.W.H., as shown by arrows.

*Note: Efficiencies used – for Turbine = 87% for Generator = 95%. Power varies directly with other efficiencies.

DAMS—LOG SLUICES AND INTAKE RACK

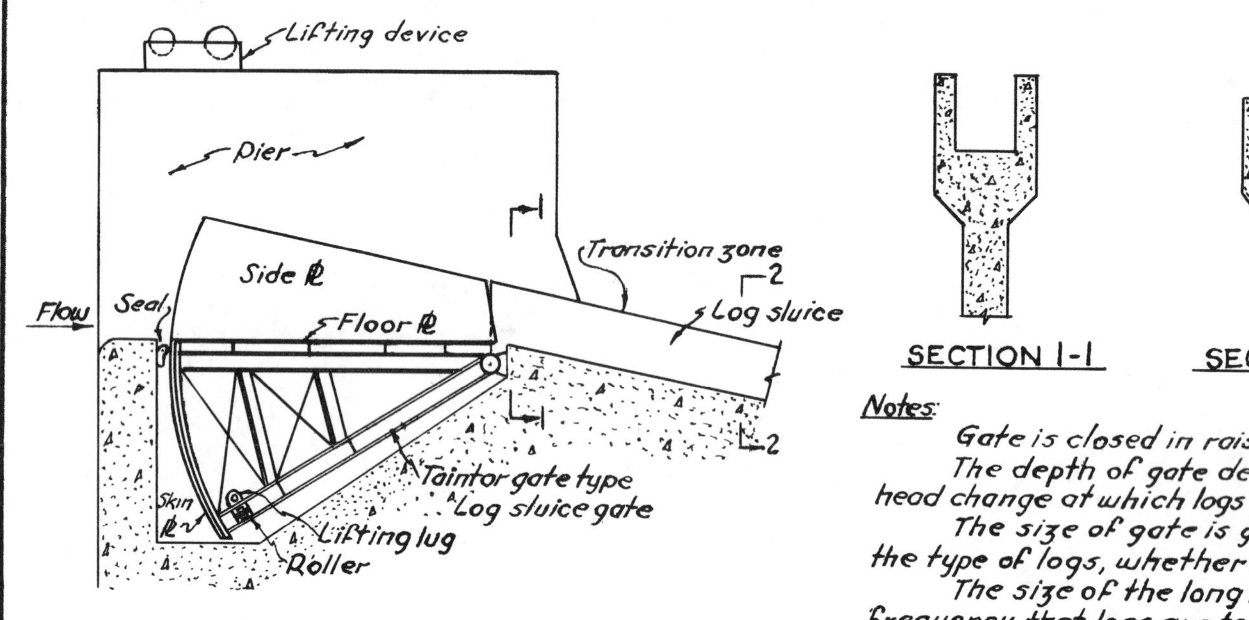
Lifting device
Pier
Side ℞
Transition zone
Flow
Seal
Floor ℞
Log sluice
Skin ℞
Taintor gate type
Log sluice gate
Lifting lug
Roller
SECTION 1-1 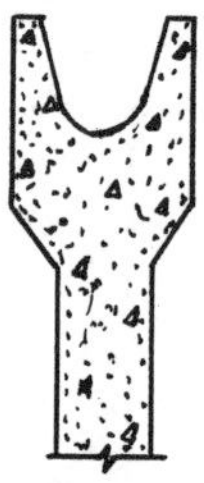
SECTION 2-2

FIG. A-LOG SLUICE-GATE TYPE

Notes:

Gate is closed in raised position.

The depth of gate determines the head change at which logs can be sluiced.

The size of gate is governed by the type of logs, whether long logs or pulp.

The size of the long logs and the frequency that logs are to be sluiced also control the size of the log sluice.

Booms are used to guide logs into log sluice from pond and to keep logs from washing over the spillway.

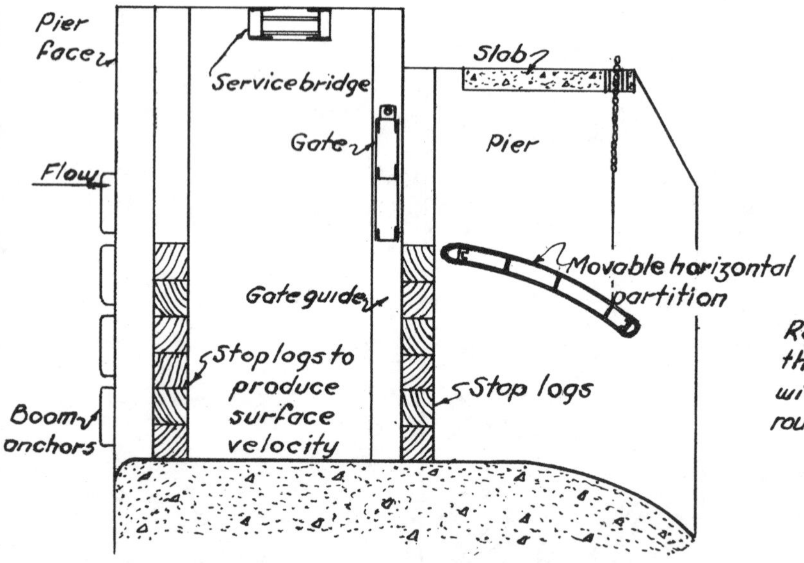

Pier face
Service bridge
Slab
Gate
Pier
Flow
Movable horizontal partition
Gate guide
Stop logs to produce surface velocity
Stop logs
Boom anchors

FIG. B-LOG SLUICE-STOPLOG TYPE

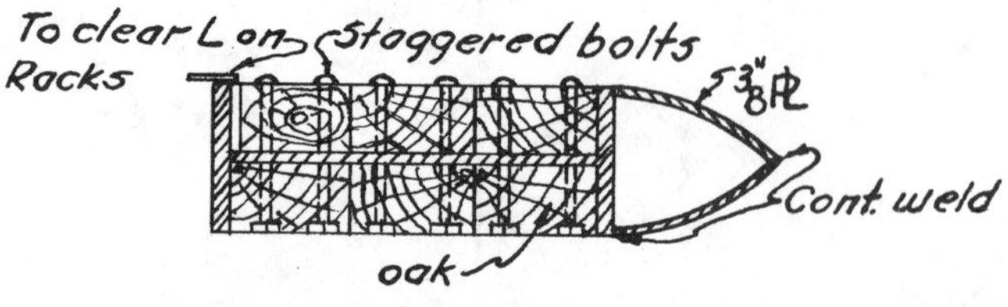

To clear L on Racks
Staggered bolts
3/8" ℞
Cont. weld
oak

DETAIL OF RACK BEAM

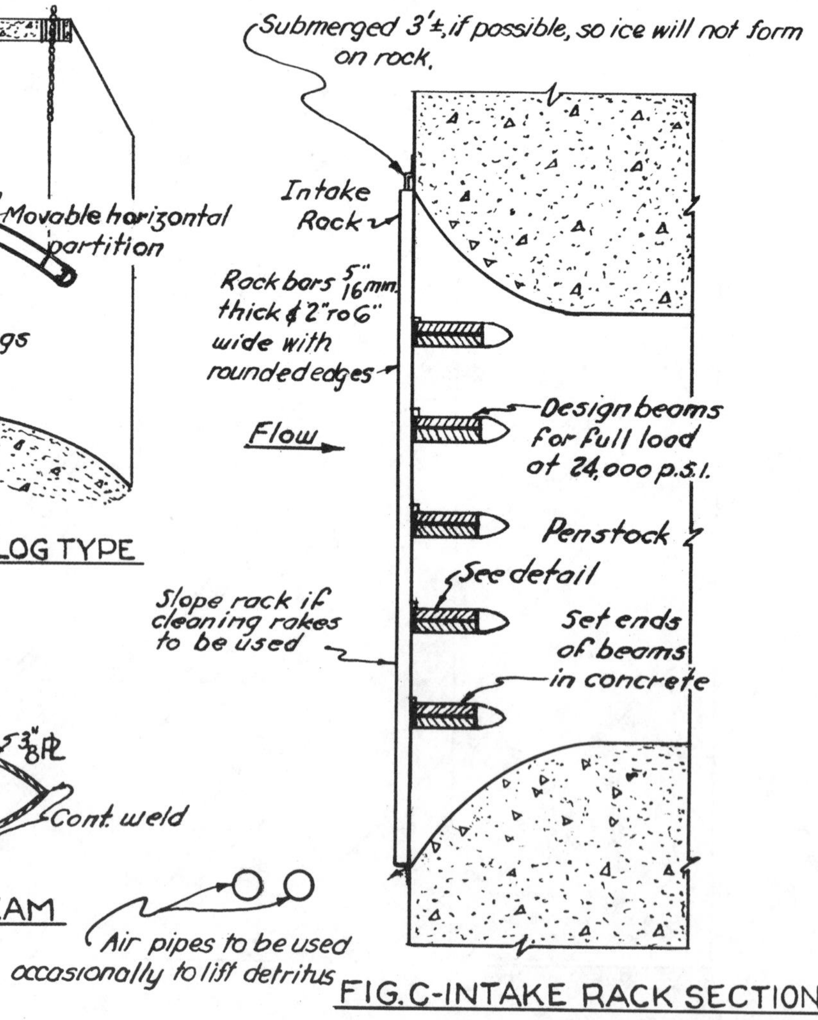

Submerged 3'±, if possible, so ice will not form on rock.
Intake Rack
Rack bars 5/16" min. thick & 2" to 6" wide with rounded edges
Flow
Design beams for full load at 24,000 p.s.i.
Penstock
See detail
Slope rack if cleaning rakes to be used
Set ends of beams in concrete
Air pipes to be used occasionally to lift detritus

FIG. C-INTAKE RACK SECTION

DAMS – FISHWAYS

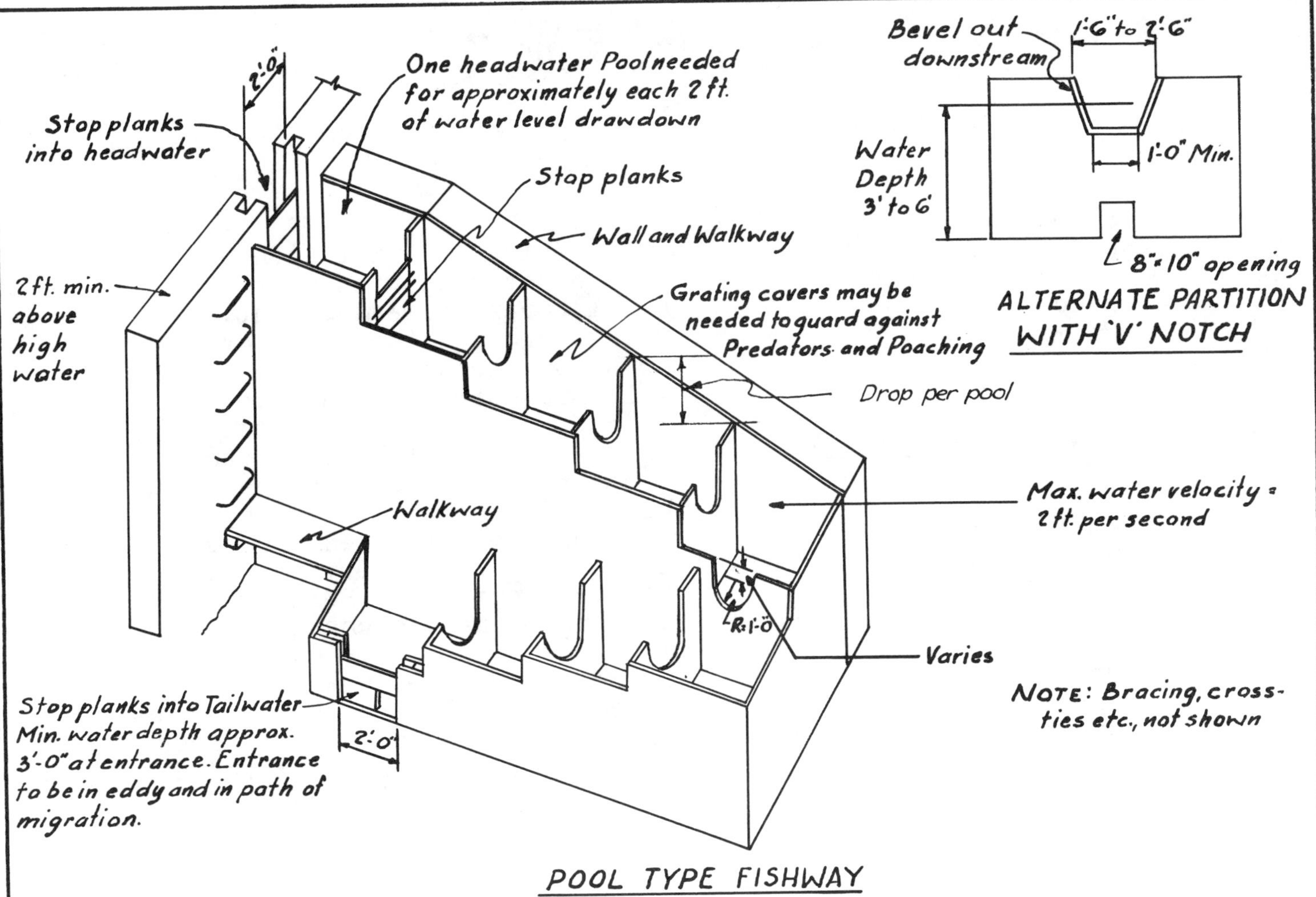
Bevel out downstream
1'-6" to 2'-6"
Water Depth 3' to 6'
1'-0" Min.
8"×10" opening
ALTERNATE PARTITION WITH 'V' NOTCH
2'-0"
Stop planks into headwater
One headwater Pool needed for approximately each 2 ft. of water level drawdown
Stop planks
Wall and Walkway
2 ft. min. above high water
Grating covers may be needed to guard against Predators and Poaching
Drop per pool
Walkway
Max. water velocity = 2 ft. per second
R. 1'-0"
Varies
Stop planks into Tailwater
Min. water depth approx. 3'-0" at entrance. Entrance to be in eddy and in path of migration.
2'-0"
NOTE: Bracing, cross-ties etc., not shown

POOL TYPE FISHWAY

POOL DIMENSIONS	
SALMON ETC.	SMALL FISH
Drop - 12" to 16"	8"
Length - 8'-0"	6'-0"
Width - 6'-0" to 8'-0"	4'-0" to 6'-0"
Depth - 6'-0" min.	4'-0" min.

NOTE: Above for ordinary runs. Special studies required for enormous runs of herring, salmon, etc.

NOTES: Concrete partitions are more expensive than wooden partitions, but require very little maintenance. Precast partitions are cheaper than when cast in place.
Construction may be of concrete, steel plate or timber. Round and smooth all notches and passages.
Provide additional entrance and exit pools if required. Extend access walkways for operation and repair.
Provide rest pool for each 15 ft. to 20 ft. headrise.

By Eudell G. Whitten, Civil Engineer
Ref. Transactions Amer. Fisheries Societies, Dr. George A. Rounsefell

CORROSION OF METALS-GALVANIC ACTION *

GALVANIC SERIES**

Corroded end (anodic)

Magnesium
Aluminum
Duralumin

Zinc
Cadmium

Iron
Chromium-iron
Chromium-nickel-iron

Soft solder
Tin
Lead

Nickel
Brasses
Bronzes
Nickel-copper alloys
Copper

Chromium-iron
Chromium-nickel-iron

Silver solder

Silver
Gold
Platinum

Protected end (cathodic)

This series is built upon actual experience with corrosion and laboratory measurement. Metals grouped together have no strong tendency to produce galvanic corrosion on each other; connecting two metals distant on the list from each other tends to corrode the one higher in the list. Voltage figures are not given because these vary with every new corrosive condition. Relative positions of metals change in many cases but it is unusual for changes to occur across the spaces left blank. The chromium-irons and chromium-nickel-irons change position as indicated depending on oxidizing conditions, acidity, and chloride in solution. The series as it stands is correct for many common dilute water solutions such as sea water, weak acids and alkalies.

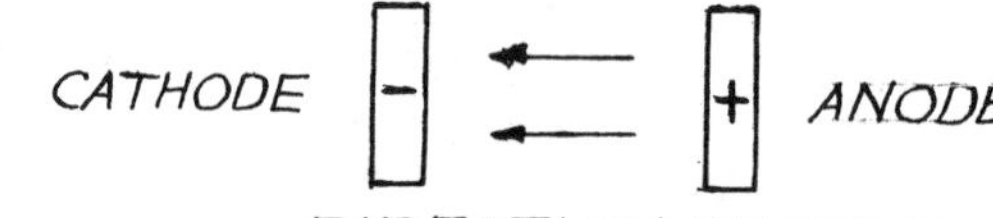
CATHODE
−
+
ANODE

DIRECTION OF FLOW

RED - LIGHTS:

1. Two metals with strong tendency for galvanic action should be used in juxtaposition with caution particularly where moisture containing impurities may tend to complete a galvanic cycle.
2. Use metals that are grouped.
3. Minimize the contact surface between two metals.
4. Separate the contacting surfaces using either: (a) Rubber or Neoprene washers, (b) Felt (c) Bituminous paint.
5. Where Aluminum is in juxtaposition with cement or alkalies it should be separated by felt or bituminous paint.

Note that Cathodic or Anodic is a relative condition depending on the position of the terminals in the above table.

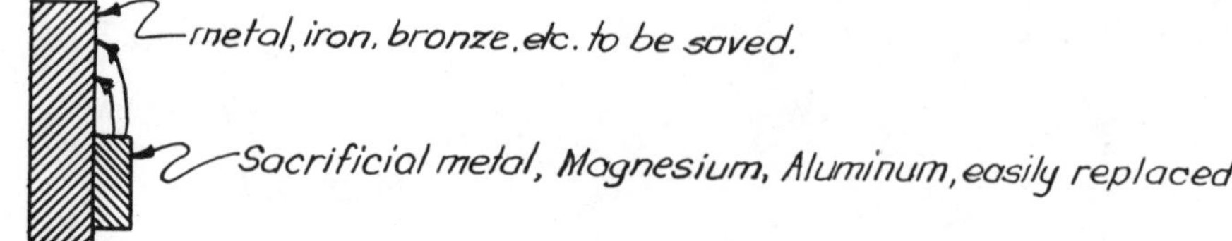
metal, iron, bronze, etc. to be saved.
Sacrificial metal, Magnesium, Aluminum, easily replaced.

PROTECTION FOR METAL IN STRUCTURES IN SALT WATER.

*See also p. 21-18

**Adopted from 'Corrosion Resistance of Metals & Alloys' by McKay & Worthington, Reinhold Pub. Corp.

CATHODIC PROTECTION–I

SURVEYS REQUIRED

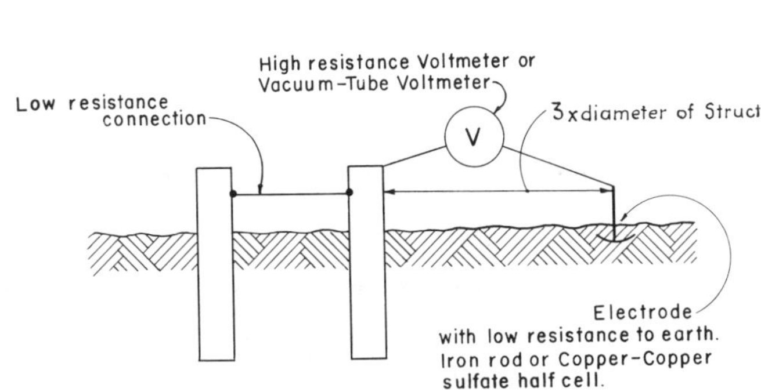
High resistance Voltmeter or
Vacuum-Tube Voltmeter
Low resistance connection
V
3 x diameter of Struct.
Electrode with low resistance to earth. Iron rod or Copper-Copper sulfate half cell.

With copper—copper sulfate half-cell, cathodic protection is required if potential of structure is less negative than – 0.85 volt. If iron rod is used, the critical potential is – 0.3 volt.

FIG. A. MEASURE DIFFERENCE OF POTENTIAL BETWEEN STRUCTURE AND EARTH TO DETERMINE IF CATHODIC PROTECTION IS NECESSARY

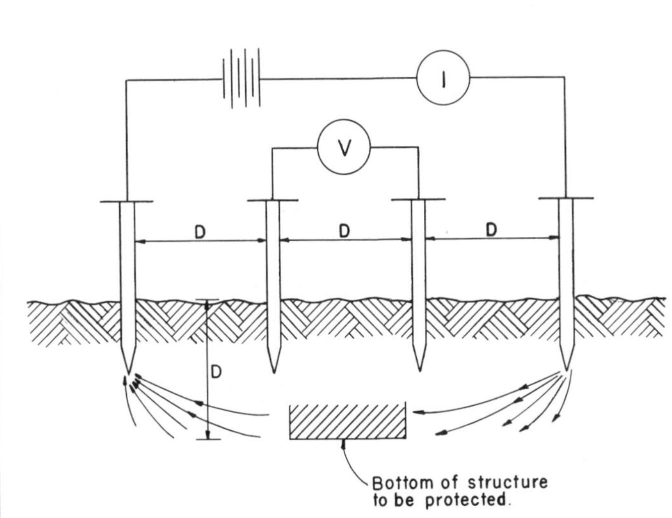
I
V
D
D
D
D
Bottom of structure to be protected.

Resistivity, $\rho = 191\,D \times V / I$

V = voltmeter reading, volts
I = ammeter reading, amperes
D = electrode spacing, feet

Depth of measurement D should extend to bottom of protected structure.

FIG. B. DETERMINE RESISTIVITY OF SOIL IN OHM-CENTIMETERS

CATHODIC PROTECTION

Electrolytic Corrosion of Metals may occur under the following conditions:

A. By local cell action when the bare metal is surrounded by an electrolyte such as sea water or a solution of mineral salts as found in certain soils.
B. By cell action between two dissimilar unprotected metals in the galvanic series (see p.16-50), when surrounded by an electrolyte.
C. By the action of stray earth currents such as may occur in the vicinity of an electric railway.
D. Anaerobic bacteria, differential electrolyte, oxygen deficiency cells.

In each of the above-mentioned cases, the metal subject to corrosion behaves as an anode; that is, a source of positive ions. Positively charged metallic ions flow from the metal into the electrolyte, resulting in pitting and corrosion of the surface of the anode.

Metals are cathodically protected from electrolytic corrosion by being forced to act as a cathode rather than a anode. This is accomplished either by the use of sacrificial anodes closer to the anodic or corroding end of the galvanic series than the metal to be protected, Fig. A p. 16-53, or by the use of an external source of direct current as shown in Fig. A, p. 16-52.

Any metallic structure in contact with an electrolyte can be cathodically protected. In practice, water tanks and tanks containing electrolytic solutions, buried pipelines, steel ships, steel piles, underground cable with metallic sheathing, and many other kinds of structures have been protected from corrosion in this manner.

RED LIGHT

Engineers attempting to apply these data should be warned of the danger of causing hazardous conditions to be established. It is also possible that considerably greater corrosion problems may be experienced on adjacent foreign structures than initially existed on the protected structures due to interference. Bonding, elimination of sparking hazards for ship fueling, insulated couplings, and gas formation are some examples of the items that the experienced corrosion engineer must take into consideration in designing cathodic protection systems and which the inexperienced reader may very easily overlook.

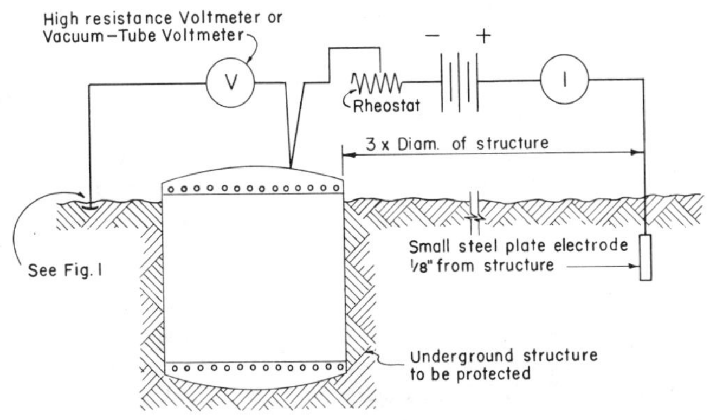
High resistance Voltmeter or
Vacuum-Tube Voltmeter
V
Rheostat
– +
I
3 x Diam. of structure
See Fig. I
Small steel plate electrode 1/8" from structure
Underground structure to be protected

Vary rheostat until potential of structure with respect to earth is more negative than – 0.85 volt or – 0.3 volt, depending on type of electrode. (See note on Fig. A.) Read ammeter.

FIG. C. DETERMINATION OF IMPRESSED CURRENT REQUIRED

CATHODIC PROTECTION-2

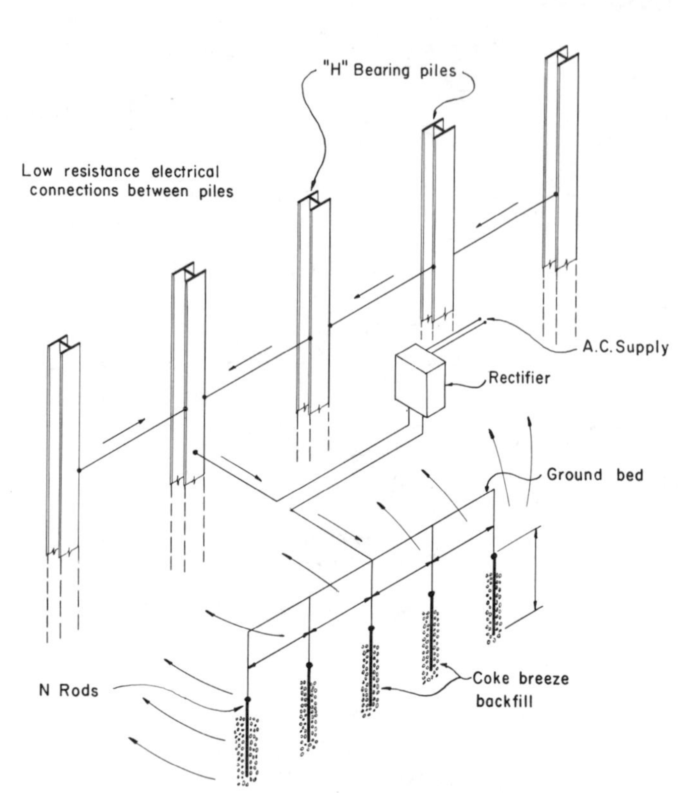

"H" Bearing piles
Low resistance electrical connections between piles
A.C. Supply
Rectifier
Ground bed
N Rods
Coke breeze backfill
Graphite anode for long life, but scrap steel or other metals may be used.

FIG. A. PROTECTION OF STEEL H-PILES BY IMPRESSED CURRENT METHOD

SELECTION OF EXTERNAL POWER SOURCE

1. Determine current I required for protection as in Fig. C, p. 16-51.
2. Determine anode resistance: $r = \frac{0.0342}{L}\rho$

r = resistance of one anode to ground in ohms; 0.0342 is an empirical factor for 3/4" diameter rods.

For multiple rods total $R = \frac{r}{KN}$. (For K, see Table B.)

3. Determine required voltage:

$$\text{Voltage} = IR \text{ volts}$$

Voltage should be kept within economic limits, usually not higher than 15 volts. Vary S and N if voltage requirement is too high.

TABLE B

Distance between anodes, S, ft.	Value for K
20	0.6
40	0.77
60	0.87

TABLE C – CHARACTERISTICS OF MAGNESIUM ANODES

I Anode Mk.	II Description	III Rating	IV Anode Current	V Current MA/LB	VI Anode Wt., Lbs.	VII Anode life, Yrs.
A	16" x 8" diameter	3 amp.-yr.	140-180 ma.	3.1	51	18.4
B	5 3/4" x 23" tapered D	2 amp.-yr.	140-180 ma.	4.7	34	12.1
C	4" x 20" tapered D	1 amp.-yr.	140-180 ma.	9.4	17	6.1
D	1.315" D cored rod	510 amp.-hr./ft.	10-15 ma./ft.	12.0	1.04/ft.	4.7
E	1.05" cored rod	330 amp.-hr./ft.	10-15 ma./ft.	18.0	0.68/ft.	3.2
F	0.84" D cored rib rod	210 amp.-hr./ft.	10-15 ma./ft.	28	0.44/ft.	2.0
G	3/8" x 3/4" cored rib rod	100 amp.-hr./ft.	10-15 ma./ft.	57	0.22/ft.	1.0

Backfill is assumed in cases of anodes A, B, and C if in soil.

Use A, B, or C type anodes in soils or waters with resistivities from 30 to 5,000 ohm-cm. (See Fig. B, p. 16-51.)

For soil resistivities above 5,000 ohm-cm., use type D, E, F, or G anodes. Type A anodes are for use in salt-water environment, or in soils with resistivites below 200 ohm.-cm.

NOTE: The above table is based on soil resistivity of 1000 ohm.-cm. with anodes directly coupled to remote structure. (10 ft. or more from anodes). Where anodes are bolted to structure, they must be insulated from it.

FIG. D. PROTECTION OF PILES IN SEA WATER BY SACRIFICIAL ANODES

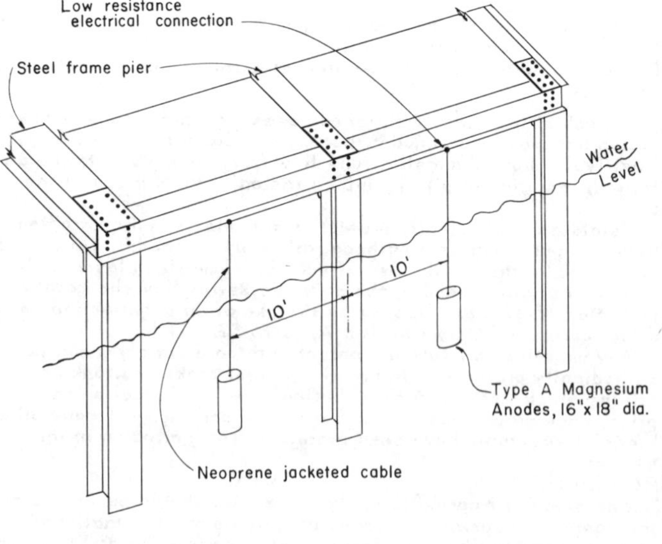

Low resistance electrical connection
Steel frame pier
Water Level
10'
10'
Type A Magnesium Anodes, 16"x 18" dia.
Neoprene jacketed cable

PROBLEM: ***SELECTION OF GALVANIC ANODES***

Assume: Piers on steel H-piles as shown in Fig. D. Piles are in salt water. Surface area of piles to be protected is 500 sq. ft.

Given: In salt water, current required for protection of steel is 10 to 15 ma./sq.ft. during the first 3 months, after which the current can be reduced to 3 ma./sq.ft. because of reduction of corrosion potential of steel. Resistivity of sea water is 30 ohm-cm.

SELECTION OF ANODES See Table C. In 30 ohm-cm. sea-water, use type A anodes. These anodes will have a current output of $\frac{1000}{30} \times 150$ or 4.850 amp. (Table C, col. IV). Use two anodes to supply a total initial current of 9.7 amps., which exceeds initial requirement of $500 \times 15 = 7500$ ma. or 7.5 amp. Anode life will be $\frac{30}{1000} \times 18.4$ yrs. (Table C, col. VII) or approximately 1/2 yr. for initial protection and 1 yr. + for replacement anodes.

CATHODIC PROTECTION-3

FIG. A. CATHODIC PROTECTION OF UNDERGROUND PIPES BY SACRIFICIAL ANODES*

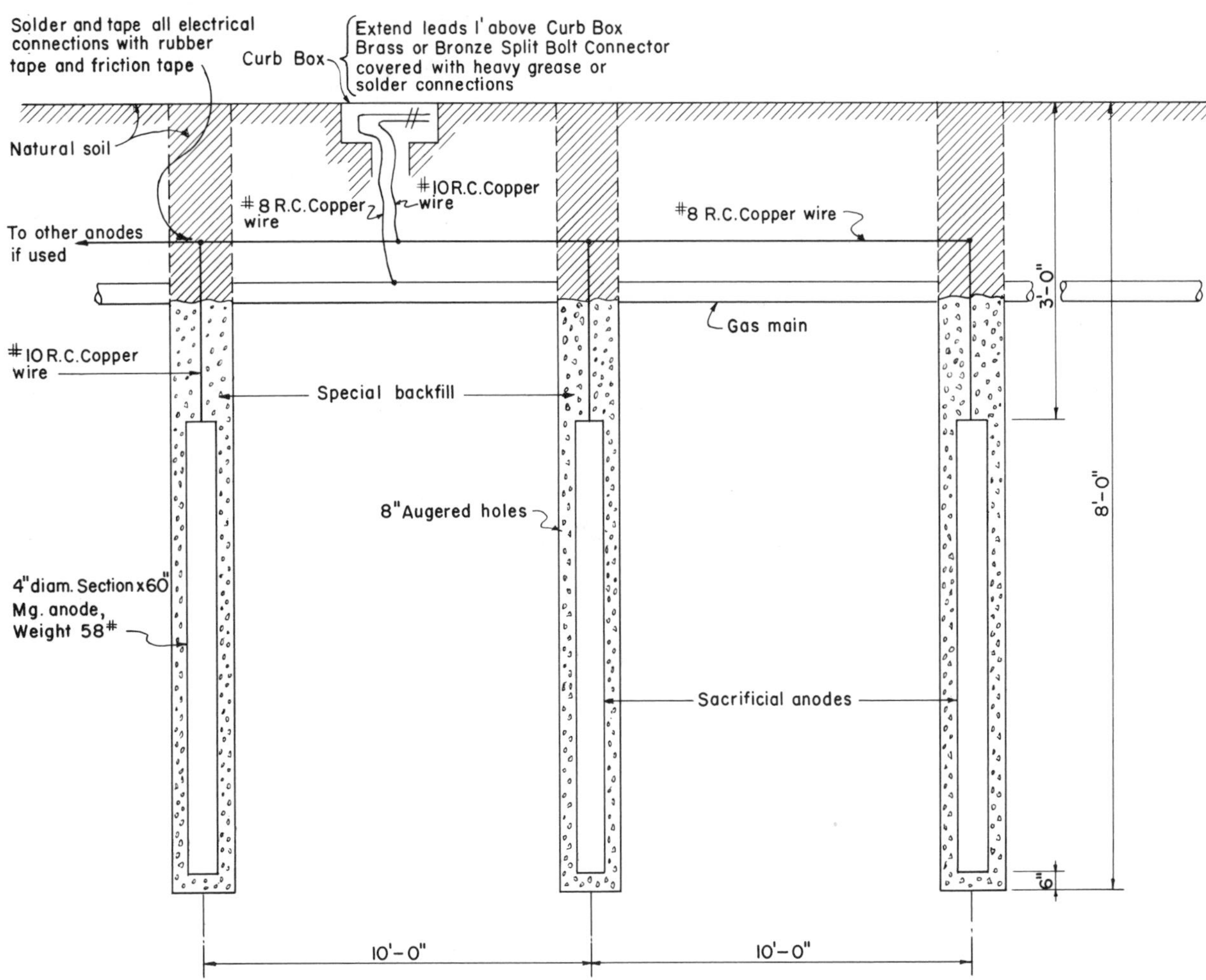

Solder and tape all electrical connections with rubber tape and friction tape
Curb Box
Extend leads 1' above Curb Box
Brass or Bronze Split Bolt Connector covered with heavy grease or solder connections
Natural soil
#8 R.C. Copper wire
#10 R.C. Copper wire
#8 R.C. Copper wire
To other anodes if used
3'-0"
Gas main
#10 R.C. Copper wire
Special backfill
8'-0"
8" Augered holes
4" diam. Section x 60" Mg. anode, Weight 58#
Sacrificial anodes
6"
10'-0"
10'-0"

Backfill consists of a 50/50 mixture of dehydrated gypsum (plaster of Paris) and clay mud. First, pour in backfill mixture to about 2 ft. depth. Then lower anode in center of hole. Then fill hole with backfill until it covers anode by about 18".

Backfill should:

1. Provide firm, low resistance contact to soil.
2. Prevent formation of hydrogen gas around anodes, which will increase their resistance. Addition of a small amount of lime to gypsum will often retard hydrogen formation.

Space magnesium anodes not less than 10' apart, except where limited by space requirements or where it is desired to reduce current output per anode as in the case of low-resistance soils. Where bare pipe is to be protected, anodes should be 10' or more from pipe.

Approximate amount of backfill per anode:

Dry mix	clay	57 lb. (6.5 gal.)
	gypsum	57 lb. (6.5 gal.)
Mud mix	water	11 gal.
	gypsum + clay	10 gal.
	water + gypsum + clay	13 gal.

SUPPLEMENTARY PROTECTION

To reduce the amount of protection required, it is common practice, in pipeline work, to use a wrapped pipe and bitumastic coating. It is then only necessary to protect against pin holes or other small voids in the covering.

Many public utilities and others are beginning to use metallic underground cables having synthetic rubber jackets, such as Neoprene, over the metal which, because of its freedom from voids, requires no cathodic protection at all. These are used in place of the lead-jacketed cables which are subject to anodic action.

*From Henry M. Fanett, and Hugo W. Wahlquist, "Practical use of Galvanic Anodes; *Cathodic Protection, a Symposium*, p. 138, Electrochemical Society and National Association of Corrosion Engineers, 1949.

WATERFRONT STRUCTURES–DYNAMIC FORCES

DYNAMIC FORCES ON STRUCTURES DUE TO BREAKING WAVES – SIMPLIFIED METHOD*

EXAMPLE

Observations Required

1. H – maximum wave height, feet.
2. t_1, t_2, t_3 – range of time for two successive crests to pass a given point during periods of maximum waves – seconds.
3. Obtain depths from hydragraphic charts.

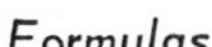

Formulas

$d_b = 1.3H$ in feet

$H_b = 1$ to $2.5H$ in feet

$F = \frac{KE}{V^2}$ in lb. per lin. ft.

$K = 1.5 \times 2g = 96.6$

Given: 9-ft. waves passing at intervals of 7 to 11 seconds.

Procedure

1. Compute breaking depth of wave $1.3 \times 9 = 11.7$ ft. Waves will break on structure located in 11.7 ft. of water.
2. With values of t and d_b, find length of breaking waves, L, on Fig. C.
3. Using values of t and d_b, find velocity of breaking waves, V, on Fig. B.
4. Using values of L and H, find wave energy, E, from Fig. D.
5. Using previous values, find dynamic wave force, F, lb. per lin. ft. of width of structure.

GIVEN		FIND				
		(1)	(2)	(3)	(4)	(5)
H, ft.	t, sec.	d_b = 1.3H ft.	L = ft, Fig. B.	V = f.p.s., FIG. C.	E = ft./lb. FIG. D.	F = $\frac{KE}{V^2}$ = lb./lin.ft.
9	7	11.7	130	18.3	84,000	24,200
9	9	11.7	170	18.9	93,500	25,400
9	11	11.7	210	18.9	105,000	28,200

WAVE CREST
STILL WATER LEVEL
L = WAVE LENGTH
d = DEPTH
H = WAVE HEIGHT
OCEAN BOTTOM
F
H_b
d_b

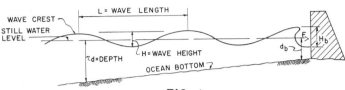

FIG. A

Nomenclature

d_b = breaker depth

H_b = breaker height

F = dynamic wave force on structure

L = wave length

E = wave energy per foot of crest, ft.-lb./ft.

V = velocity of wave, f.p.s.

Wave Forces

1. Breaking on sgructure:
   (a) Dynamic – approaches initial force of wave.
   (b) Hydrostatic – Height of wave.
2. Broken waves:
   (a) Dynamic – Dissipated force of broken wave.
   (b) Hydrostatic – Height of wave.
3. Unbroken wave:
   (a) Hydrostatic – Standing wave.

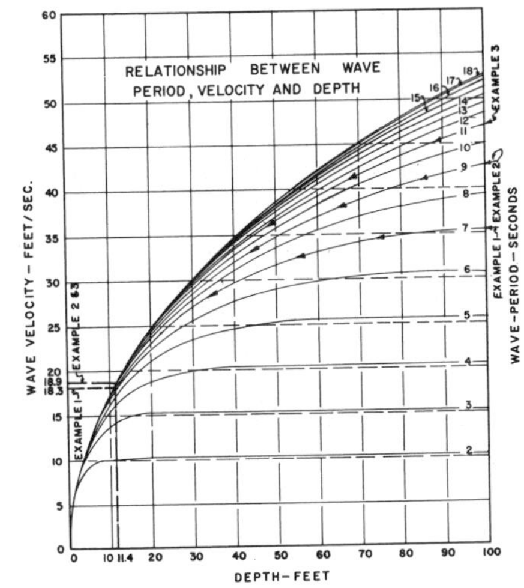

RELATIONSHIP BETWEEN WAVE PERIOD, VELOCITY AND DEPTH
WAVE VELOCITY – FEET/SEC.
DEPTH – FEET
WAVE-PERIOD – SECONDS
EXAMPLE 1
EXAMPLE 2
EXAMPLE 3
EXAMPLE 2 & 3

FIG. B

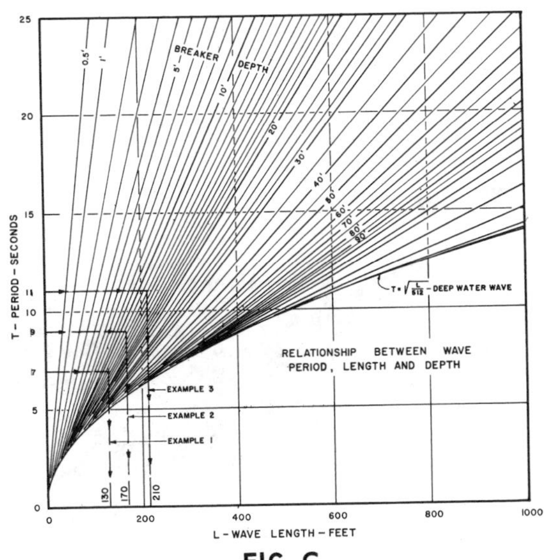

BREAKER DEPTH
$T = \sqrt{\frac{L}{5.12}}$ – DEEP WATER WAVE
RELATIONSHIP BETWEEN WAVE PERIOD, LENGTH AND DEPTH
EXAMPLE 3
EXAMPLE 2
EXAMPLE 1
T – PERIOD – SECONDS
L – WAVE LENGTH – FEET

FIG. C

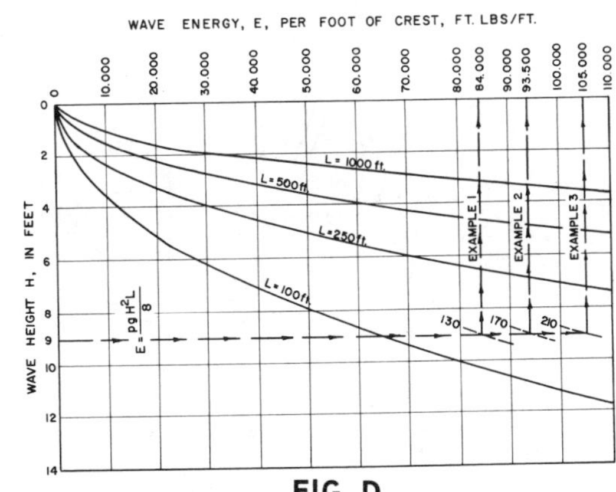

WAVE ENERGY, E, PER FOOT OF CREST, FT. LBS/FT.
L = 1000 ft.
L = 500 ft.
L = 250 ft.
L = 100 ft.
$E = \frac{\rho g H^2 L}{8}$
EXAMPLE 1
EXAMPLE 2
EXAMPLE 3
WAVE HEIGHT H, IN FEET

FIG. D

*By the author. For more exact methods of computing wave forces, see Technical Report No. 4, Beach Erosion Board, Office of the Chief of Engineers, Dept. of the Army.

WATERFRONT STRUCTURES—SEAWALLS

Seawalls, bulkheads, and revetments perform similar functions but are generally distinguished as follows:

SEAWALLS – most massive – primary function to resist waves, secondary to retain fill. Selection of type depends on foundation condition, exposure to wave action, availability of materials, and cost.

1. *Curved Face Wall* (Fig. A) – used under moderately severe wave action where the water level is over the structures base or the beach is narrow, permitting the full wave to hit the wall. Used also when there are poor foundation conditions.
2. *Vertical Face Wall* – concrete and/or stone (not shown) – used under moderate wave action where overtopping is no problem. Used under good foundation conditions and requires extensive toe protection.
3. *Stepped Face Wall* (Fig. B) – used under moderate wave action where waves break before they reach wall. Steps dissipate wave force and prevent scour. Used where there is good foundation conditions.
4. *Composite Wall* (Fig. C) – used under severe wave action where there is a narrow beach and wide tidal range. The curved face prevents overtopping when waves break against the wall, and the steps dissipate energy and prevent scour when waves break near toe. Used when there is deep sand for foundation.
5. *Stone; Rubble Mound* (Fig. D) – used under moderate wave action and where foundation conditions are poor and settlement is anticipated.

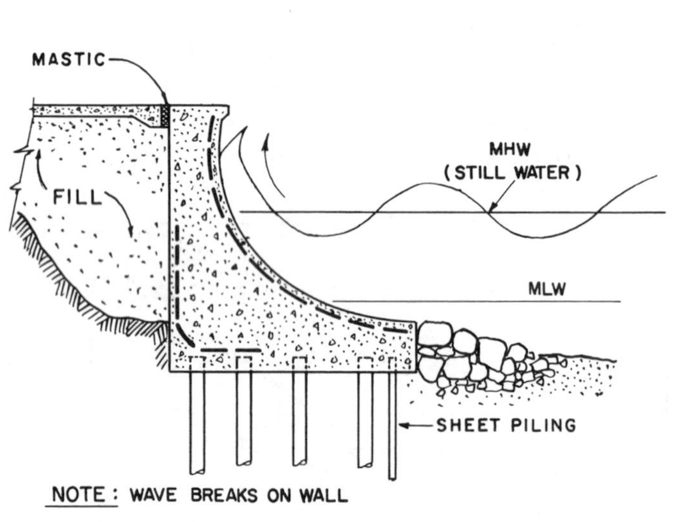
MASTIC
FILL
MHW
(STILL WATER)
MLW
SHEET PILING
NOTE: WAVE BREAKS ON WALL

FIG. A – CURVED FACE WALL

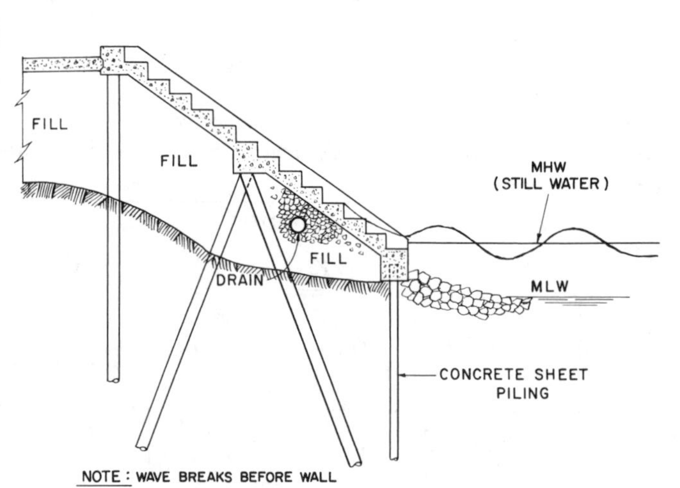
FILL
FILL
MHW
(STILL WATER)
FILL
DRAIN
MLW
CONCRETE SHEET
PILING
NOTE: WAVE BREAKS BEFORE WALL

FIG. B – STEPPED FACE WALL

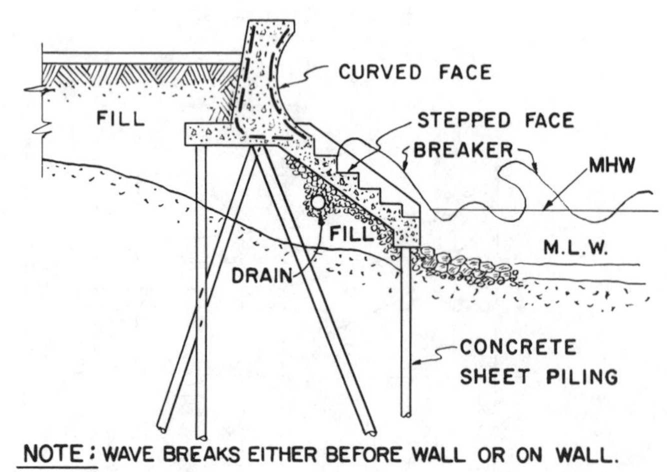
CURVED FACE
FILL
STEPPED FACE
BREAKER
MHW
FILL
M.L.W.
DRAIN
CONCRETE
SHEET PILING
NOTE: WAVE BREAKS EITHER BEFORE WALL OR ON WALL.

FIG. C – COMPOSITE WALL

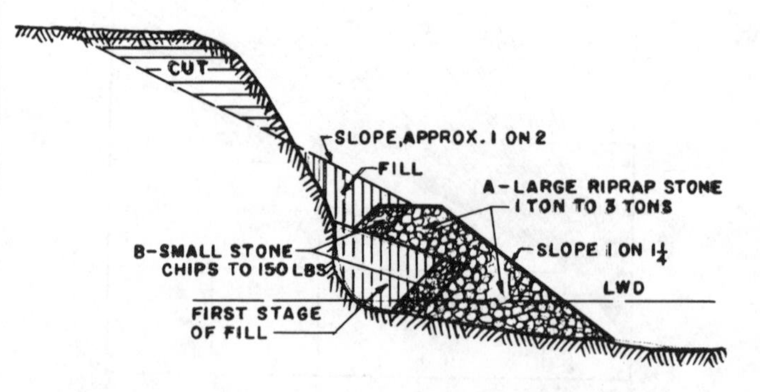
CUT
SLOPE, APPROX. 1 ON 2
FILL
A-LARGE RIPRAP STONE
1 TON TO 3 TONS
B-SMALL STONE
CHIPS TO 150 LBS
SLOPE 1 ON 1¼
LWD
FIRST STAGE
OF FILL

FIG. D – STONE; RUBBLE MOUND

WATERFRONT STRUCTURES – BREAKWATERS

BULKHEADS – lighter construction – primary function to retain fills and secondary to resist wave action.

Selection of types depends on foundation conditions, availability of material, desired life, and cost. The following are generally interchangeable; however, the steel and concrete are used for higher fills.

1. *Steel Sheet Pile* (Fig. A).
2. *Timber Sheet Pile* (not shown).
3. *Concrete Slab and King Pile* (Fig. B).
4. *Timber Piling Cribs Filled with Stone* (not shown).

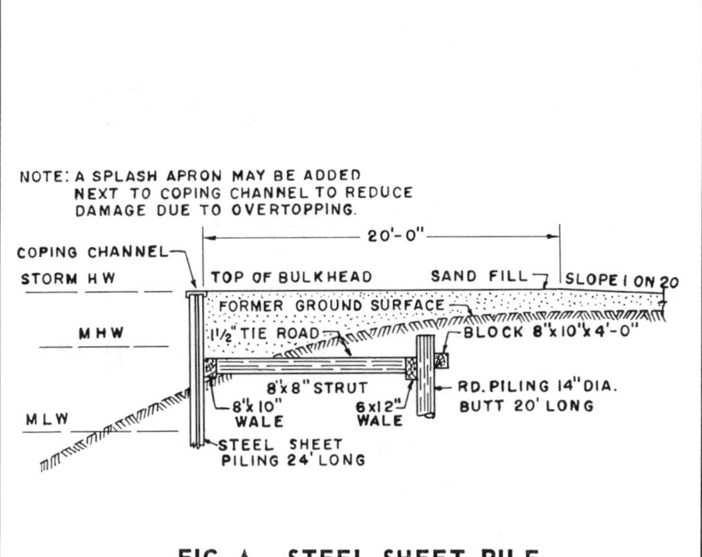
NOTE: A SPLASH APRON MAY BE ADDED NEXT TO COPING CHANNEL TO REDUCE DAMAGE DUE TO OVERTOPPING.
20'-0"
COPING CHANNEL
STORM H W
TOP OF BULKHEAD
SAND FILL
SLOPE 1 ON 20
FORMER GROUND SURFACE
M H W
1½" TIE ROAD
BLOCK 8"x10"x4'-0"
8"x8" STRUT
8"x10" WALE
6x12" WALE
RD. PILING 14" DIA. BUTT 20' LONG
M L W
STEEL SHEET PILING 24' LONG

FIG. A – STEEL SHEET PILE

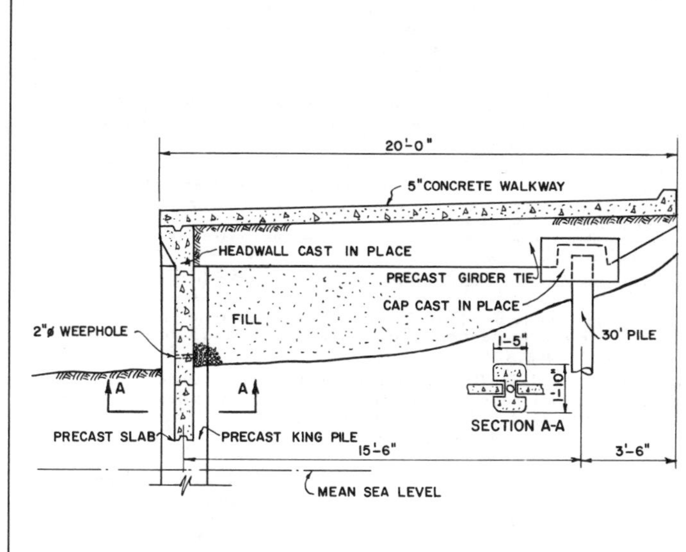
20'-0"
5" CONCRETE WALKWAY
HEADWALL CAST IN PLACE
PRECAST GIRDER TIE
CAP CAST IN PLACE
FILL
2"ø WEEPHOLE
30' PILE
1'-5"
1'-10"
A
A
SECTION A-A
PRECAST SLAB
PRECAST KING PILE
15'-6"
3'-6"
MEAN SEA LEVEL

FIG. B – CONCRETE SLAB AND KING PILE

REVETMENTS – lightest construction – primary function to protect shore or beach against erosion from small wave and tide or current action.

The following types are generally interchangeable and are selected on the basis of availability of material and cost.

1. *Stone, Placed* (Fig. C).
2. *Stone, Dumped* – less severe conditions (not shown).
3. *Portland Cement Concrete* (Fig. D).
4. *Asphalt Cement Concrete* – less severe conditions (not shown).

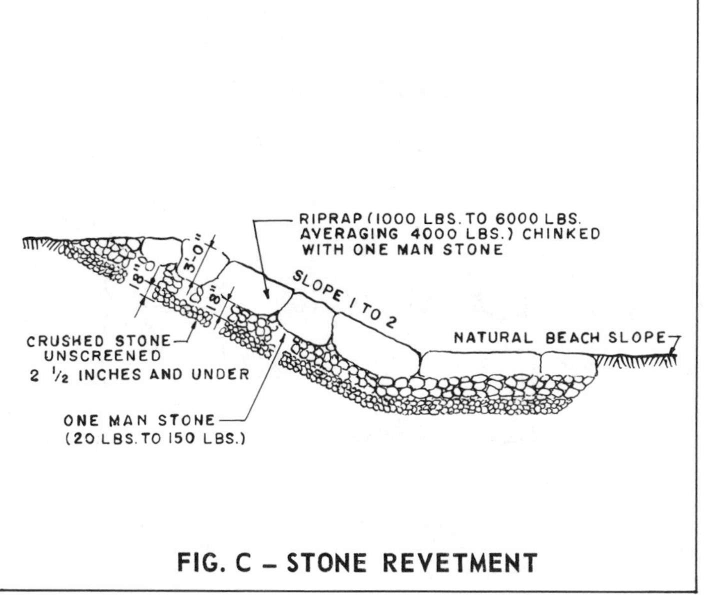
RIPRAP (1000 LBS. TO 6000 LBS. AVERAGING 4000 LBS.) CHINKED WITH ONE MAN STONE
3'-0"
18"
18"
SLOPE 1 TO 2
NATURAL BEACH SLOPE
CRUSHED STONE UNSCREENED 2 ½ INCHES AND UNDER
ONE MAN STONE (20 LBS. TO 150 LBS.)

FIG. C – STONE REVETMENT

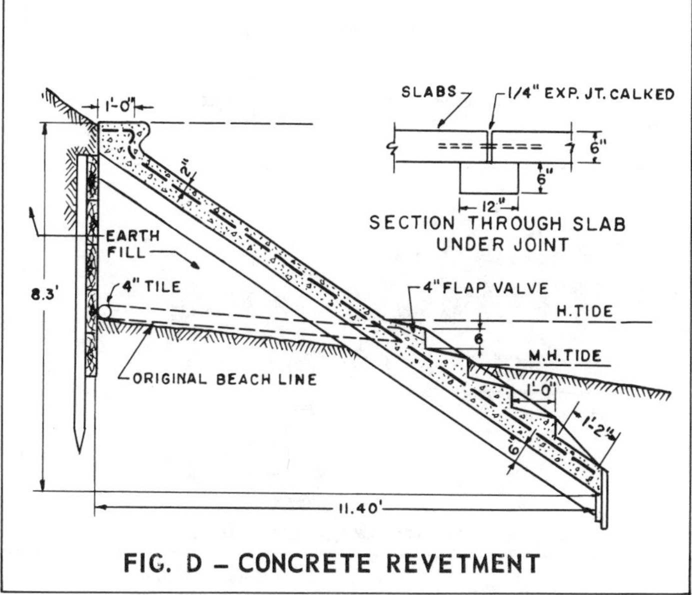
1'-0"
SLABS
1/4" EXP. JT. CALKED
6"
6"
12"
SECTION THROUGH SLAB UNDER JOINT
12"
EARTH FILL
8.3'
4" TILE
4" FLAP VALVE
H. TIDE
M.H. TIDE
ORIGINAL BEACH LINE
1'-0"
1'-2"
11.40'

FIG. D – CONCRETE REVETMENT

WATERFRONT STRUCTURES— GROINS

OFFSHORE BREAKWATERS are structures whose primary purpose is to break or dissipate wave forces and thus protect a harbor, anchorage, or basin. A secondary function is to stabilize a shore line by protecting the toe of a beach or by intercepting littoral transport. This may be accomplished by use of a submerged breakwater.

SUBMERGED BREAKWATERS are structures, the tops of which are below mean low water and whose effect is to cause waves to break at specified location.

JETTIES are structures extending into a body of water to direct and confine the stream or tidal flow at the mouth of a river or entrance to a bay to aid in deepening and stabilizing the channel. Jetties at the entrance to a bay or river also protect the channel from cross currents and storm waves and from littoral drift.

SELECTION of any of the following frequently used types is dependent upon severity of wave force to be resisted, foundation conditions, availability of material, desired life, first cost, and maintenance costs.

1. *Rubble-Mound Type*
   - (a) *Stone Rubble Mound* (fig. A) — adaptable with various side slopes to severe wave action, to any depth of water, and to poor foundation conditions. It is not subject to complete failure since damage and settlement is easily repaired.
   - (b) *Rubble Mound with Earth and Stone* — Similar to the rubble mound with the substitution of a satisfactory clay or sandy material, if readily available at the site, for the base and core. This type would be chosen for economy of material.
   - (c) *Manufactured Stone Rubble Mound* (Fig. A) — Similar to the rubble mound with the substitution of manufactured shapes (tetrahedron or tetrapod) constructed of concrete. Used if natural stone is not economically available.

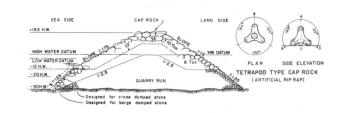

FIG. A – RUBBLE MOUND†

2. *Vertical or Wall Type*
   - (a) *Masonry Wall* — stone or precast concrete blocks (Fig. B) — adaptable to depths of 65 ft. to moderate wave action where the depth of water should be 1.3 to 1.8 times design wave height, and requires a firm foundation. Can be used for berthing ships on the landward face.
   - (b) Concrete *Caisson* (Fig. C) — adaptable to depths up to 35 ft., require a firm foundation and sometimes are set on piles. They are precast on land and floated to site and may or may not have a permanent concrete bottom. They are filled with sand and/or stone for stability.
   - (c) *Timber Crib* (Fig. D) — are either floored or unfloored compartments constructed on land and floated to the site and sunk on a prepared foundation. They are suitable to depths up to 50 ft., do not require as firm a foundation as concrete caissons, and should be limited to fresh water or salt water not subjected to marine borers.
   - (d) *Sheet Piling*
     - (1) *Straight-Wall Type* (Fig. E) — adaptable to depths up to 25 ft., consist of timber, concrete, or steel sheeting, and are used where waves are not severe. Not suitable on rock bottom.
     - (2) *Cellular Steel Type* (Fig. F) — adaptable to depths up to 40 ft. and can be used on all types of foundations.

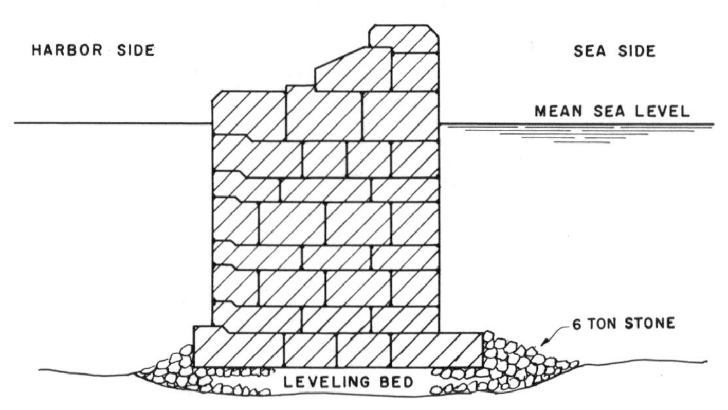
HARBOR SIDE
SEA SIDE
MEAN SEA LEVEL
6 TON STONE
LEVELING BED

FIG. B – MASONRY WALL†

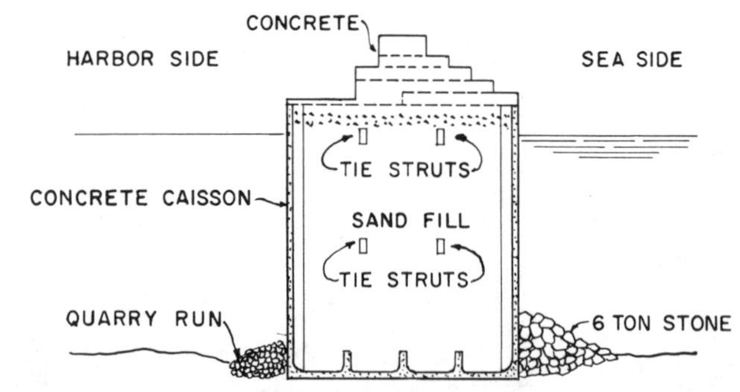
CONCRETE
HARBOR SIDE
SEA SIDE
TIE STRUTS
CONCRETE CAISSON
SAND FILL
TIE STRUTS
QUARRY RUN
6 TON STONE

FIG. C – CAISSON†

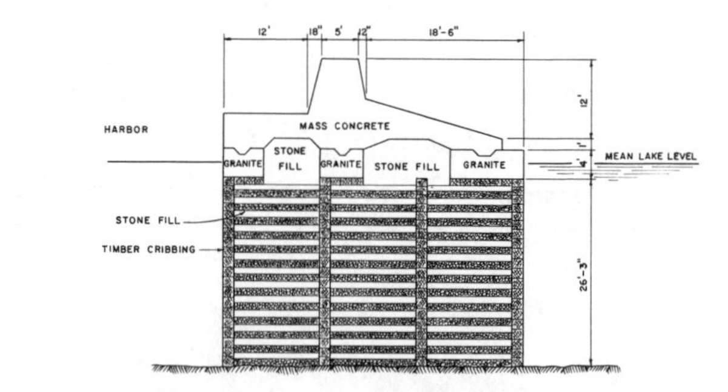
HARBOR
MASS CONCRETE
STONE FILL
GRANITE
STONE FILL
GRANITE
MEAN LAKE LEVEL
STONE FILL
TIMBER CRIBBING

FIG. D – TIMBER CRIB*

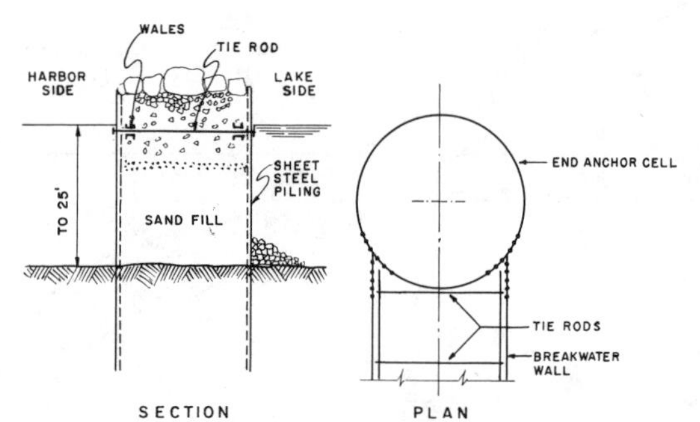
WALES
TIE ROD
HARBOR SIDE
LAKE SIDE
SHEET STEEL PILING
TO 25'
SAND FILL
END ANCHOR CELL
TIE RODS
BREAKWATER WALL
SECTION
PLAN

FIG. E – STRAIGHT-WALL SHEET PILE†

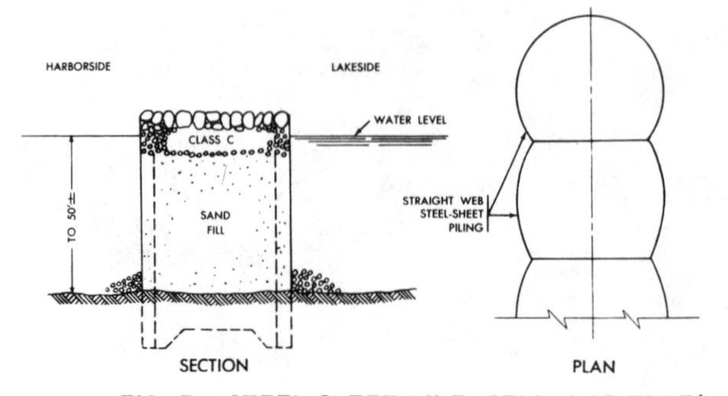
HARBORSIDE
LAKESIDE
WATER LEVEL
CLASS C
SAND FILL
STRAIGHT WEB STEEL-SHEET PILING
SECTION
PLAN

FIG. F – STEEL SHEET PILE, CELLULAR TYPE*

*Adapted from Technical Report No. 4, Beach Erosion Board, Office of the Chief of Engineers, Dept. of the Army.
†Adapted from Waterfront and Harbor Facilities, U.S. Navy, Bureau of Yards and Docks.

WATERFRONT STRUCTURES—JETTIES *

3. *Composite Type* (Figure A) is a combination of wall and mound types and is adaptable to deep-water sites or to locations with large tidal variations. The mound is used either as the foundation for the wall, or as the main structure surmounted by a wall superstructure.

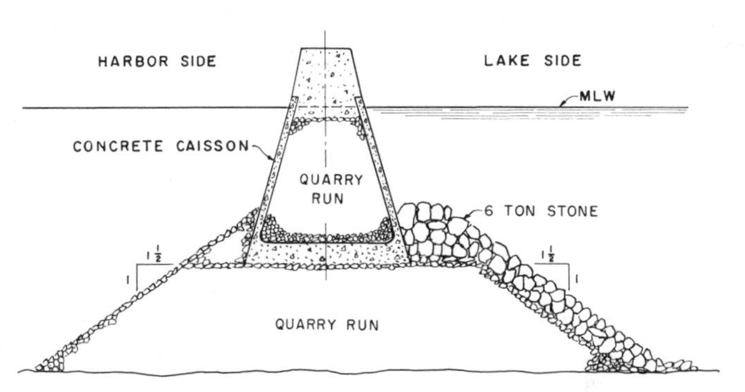
HARBOR SIDE
LAKE SIDE
MLW
CONCRETE CAISSON
QUARRY RUN
6 TON STONE
QUARRY RUN

FIG. A – COMPOSITE TYPE

JETTY ALINEMENT

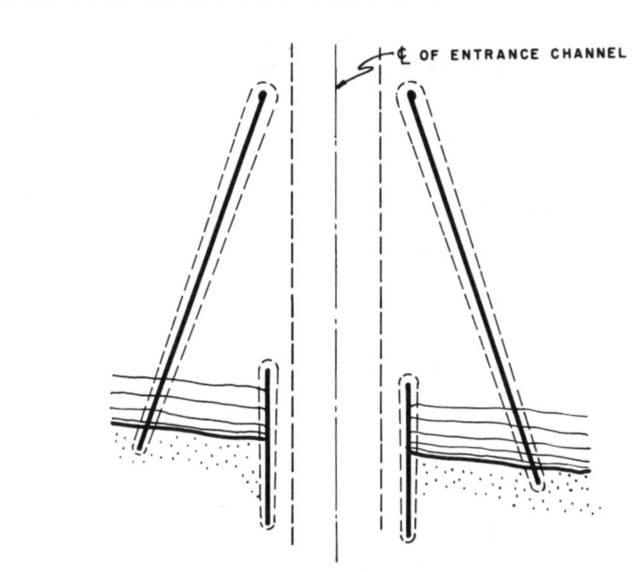
℄ OF ENTRANCE CHANNEL

FIG. B – STRAIGHT

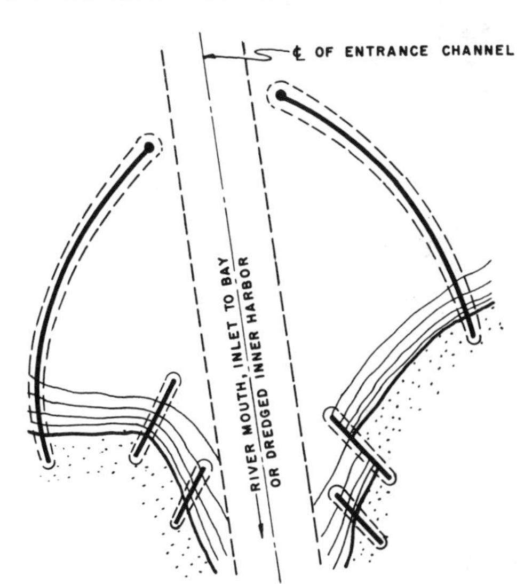
℄ OF ENTRANCE CHANNEL
RIVER MOUTH, INLET TO BAY OR DREDGED INNER HARBOR

FIG. C – CURVED

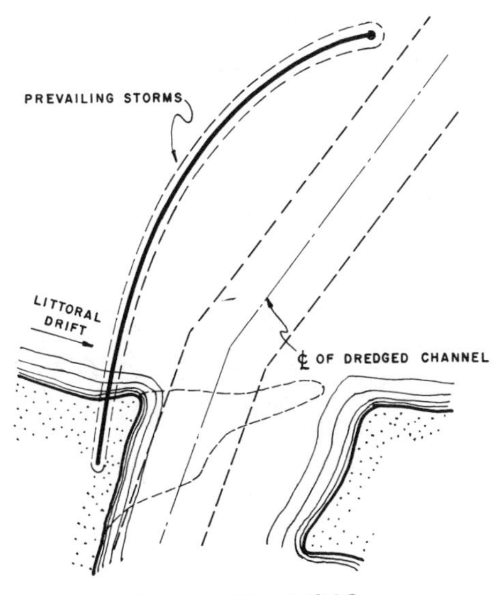
PREVAILING STORMS
LITTORAL DRIFT
℄ OF DREDGED CHANNEL

FIG. D – TRAINING

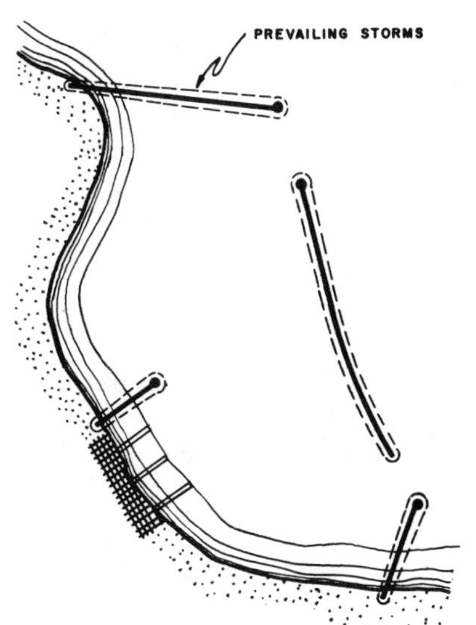
PREVAILING STORMS

FIG. E – HARBOR

* Adapted from Waterfront and Harbor Facilities, U.S. Navy, Bureau of Yards & Docks.

WATERFRONT STRUCTURES—BULKHEADS

Groins are structures built out from shore, generally perpendicular to the current, to trap littoral drift, retard erosion, or prevent scour at base of seawalls. Groins are generally classified as impermeable, semipermeable, and permeable.

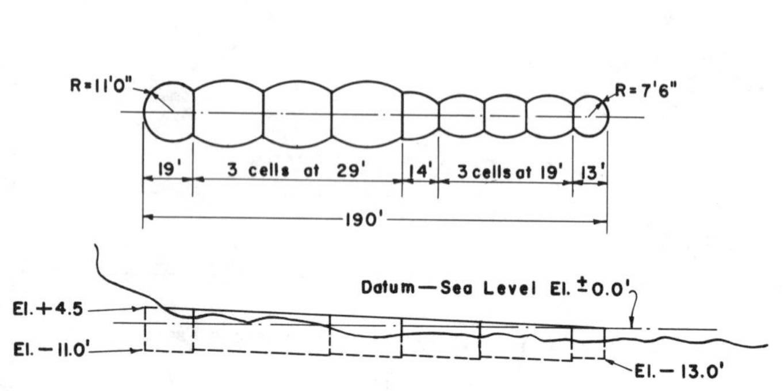
R=11'0"
R=7'6"
19'
3 cells at 29'
14'
3 cells at 19'
13'
190'
Datum—Sea Level El. ±0.0'
El. +4.5
El. −11.0'
El. −13.0'

FIG. A – TYPICAL CELLULAR STEEL GROIN*
(IMPERMEABLE)

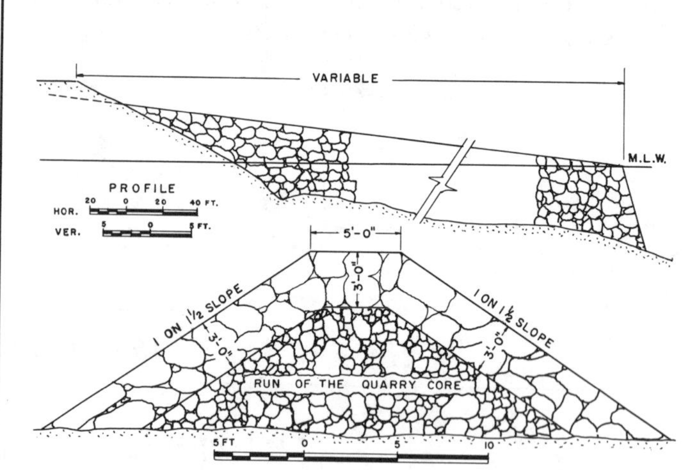
VARIABLE
M.L.W.
PROFILE
HOR.
VER.
5'-0"
3'-0"
1 ON 1½ SLOPE
1 ON 1½ SLOPE
RUN OF THE QUARRY CORE

FIG. B – TYPICAL STONE GROIN*
(SEMIPERMEABLE)

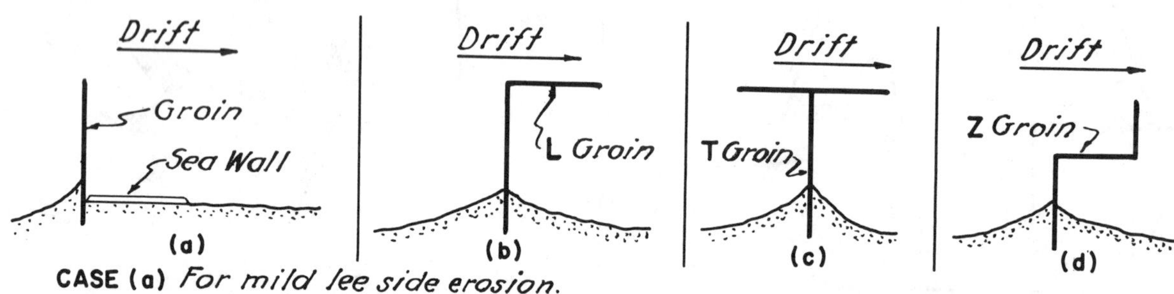
Drift
Groin
Sea Wall
(a)
Drift
L Groin
(b)
Drift
T Groin
(c)
Drift
Z Groin
(d)

CASE (a) *For mild lee side erosion.*
CASES (b)(c)(d) *Employ scouring effect of waves diffracted by outstanding leg of groin to winds on material against groin.*

FIG. C – EXPEDIENTS TO PREVENT FLANKING OF GROIN

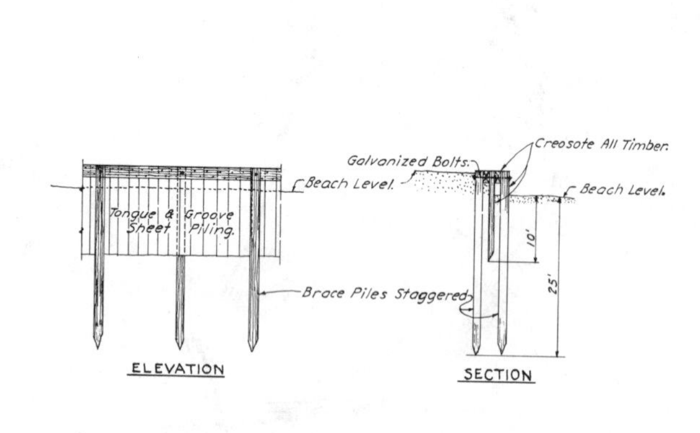
Galvanized Bolts.
Creosote All Timber.
Beach Level.
Beach Level.
Tongue & Groove Sheet Piling.
Brace Piles Staggered
10'
25'
ELEVATION
SECTION

FIG. D – TYPICAL TIMBER GROIN
(IMPERMEABLE)

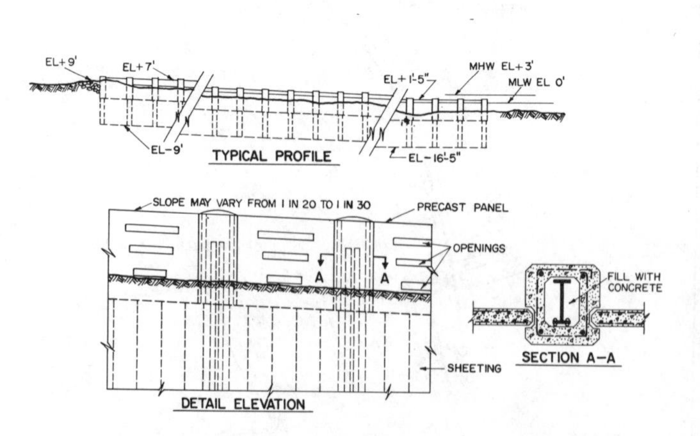
EL+9'
EL+7'
EL+1'-5"
MHW EL+3'
MLW EL 0'
EL-9'
TYPICAL PROFILE
EL-16'-5"
SLOPE MAY VARY FROM 1 IN 20 TO 1 IN 30
PRECAST PANEL
OPENINGS
FILL WITH CONCRETE
SECTION A–A
SHEETING
DETAIL ELEVATION

FIG. E – TYPICAL CONCRETE GROIN
(PERMEABLE)

*Adapted from Technical Report No. 4, Beach Erosion Board, Office of the Chief of Engineers, Dept. of the Army.

WATERFRONT STRUCTURES – WHARVES

Wharves are structures built for the berthing of vessels.

DYNAMIC FORCES AGAINST WHARVES

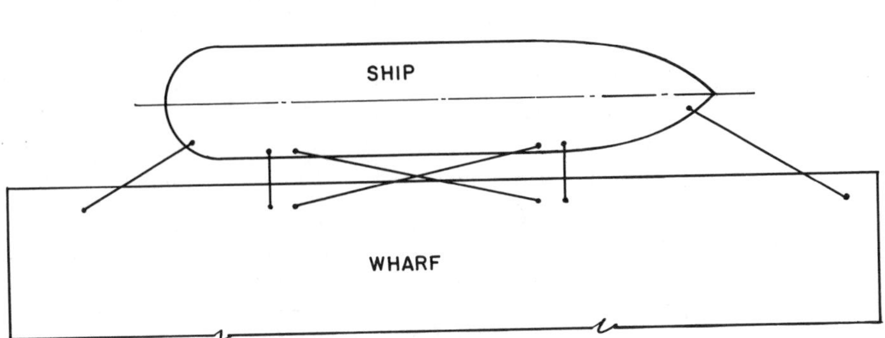
SHIP
WHARF

Procedure:

1. *WAVES* Forces on wharves and vessels subject to waves should be analyzed with respect to wave type, height, length, and direction.

2. *WINDS* Where practical, vessels should not be berthed broadside to prevailing winds.
   Wind pressure = Pw = p.s.f. = $0.00512V^2$
   Where V = wind velocity, m.p.h.
   RED LIGHT Design for minimum hurricane conditions = 75 m.p.h.

3. *CURRENTS* Likewise, where practical, vessels should not be berthed broadside to direction of current. Current pressure
   Current pressure = Pc = p.s.f.
   = $0.75V^2$ against round surfaces
   = $1.50V^2$ against flat surfaces
   where V = velocity of current, f.p.s. use 150 p.s.f. minimum for streams subject to freshets.

4. *SHOCK PRESSURES* A thrust equal to 1% of maximum weight of vessel to be berthed should be provided for, based on a permissible deflection of 5 in. of fender system or wharve and a glancing blow equal to 1/10 of kinetic energy produced by velocity of 1 m.p.h.

5. *ICE* Load from floating ice accumulations must be analyzed as required.

EXAMPLE

Given: Cargo ship class CIB. (See page 17-11.)

Displacement = 12,500 tons
Deadweight = 9000 tons
Length = 400 ft.
Beam = 60 ft.
Draft = 28 ft.
Average height above mean high water = 40 ft.
Maximum wind velocity = 100 m.p.h.
Maximum current = 5 m.p.h.

Solution:

1. Total wind force broadside
   $= 0.00512(100)^2 \times 400 \times 40$
   = 820,000 lb.
2. Total wind force parallel
   $= 0.00512(100)^2 \times 60 \times 40$
   = 123,000 lb.
3. Current force broadside
   $= 1.50(7.35)^2 \times 400 \times 28$
   = 880,000 lb.
   (5 m.p.h. = 7.35 f.p.s.)
4. Current force parallel
   $= 0.75(7.35)^2 \times 60 \times 28$
   = 69,000 lb.
5. Shock pressure
   maximum weight
   $= 12{,}500 \times 2240$
   = 27,000,000 lb.
   1% = 270,000 lb.

WATERFRONT STRUCTURES – BULKHEADS

Bulkheads are wharves of solid construction, generally built parallel to shore. A bulkhead line is a harbor line established by the Corps of Engineers onto which a solid or solid-filled structure may be built.

STATIC FORCES AGAINST BULKHEADS

$$P = \tfrac{1}{2} wh^2 \tan^2(45° - \tfrac{\phi}{2}).$$

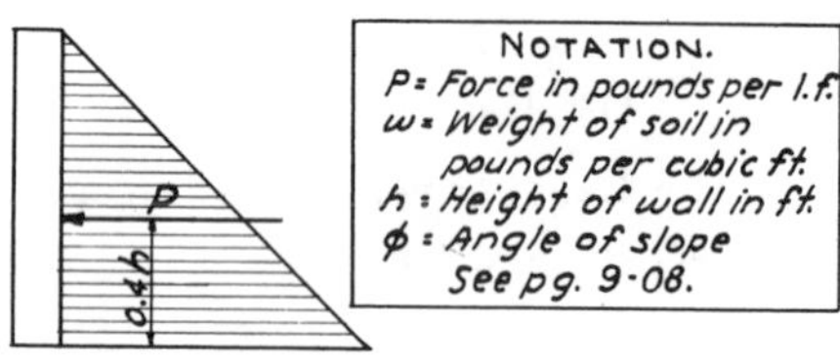
P
0.4h
NOTATION.
P = Force in pounds per l.f.
w = Weight of soil in pounds per cubic ft.
h = Height of wall in ft.
φ = Angle of slope See pg. 9-08.

CASE I – EARTH

$$P = (\tfrac{1}{2} wh^2 + vh) \tan^2(45° - \tfrac{\phi}{2}).$$

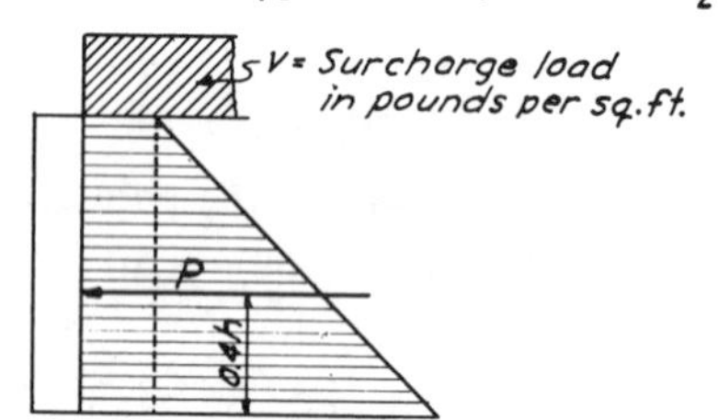
V = Surcharge load in pounds per sq.ft.
P
0.4h

CASE II – EARTH AND SURCHARGE

$$P = \tfrac{1}{2} h^2 [w_1 + w_2 \tan^2(45° - \tfrac{\phi}{2})]$$
$$= \tfrac{1}{2} h^2 W$$

For values of W see Table B.

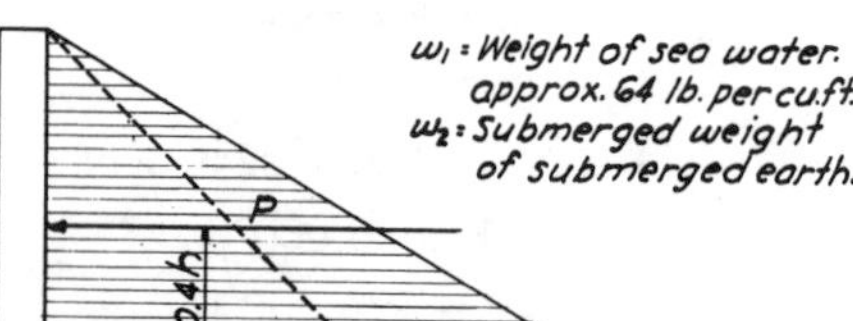
w₁ = Weight of sea water. approx. 64 lb. per cu.ft.
w₂ = Submerged weight of submerged earth.
P
0.4h

CASE III – SUBMERGED EARTH

$$P = vh \tan^2(45° - \tfrac{\phi_1}{2}) + \tfrac{1}{2} h^2 W.$$

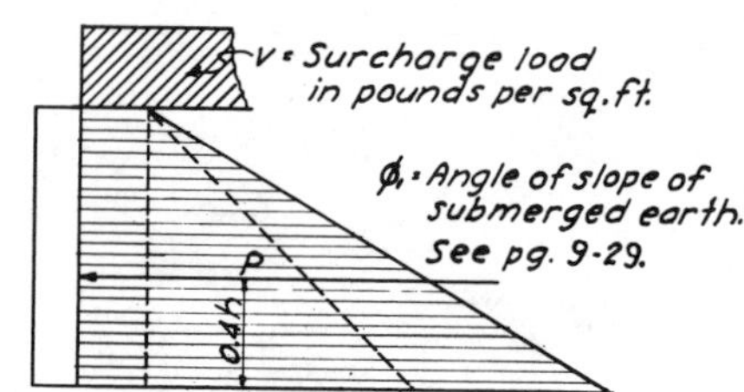
V = Surcharge load in pounds per sq.ft.
φ₁ = Angle of slope of submerged earth. See pg. 9-29.
P
0.4h

CASE IV – SUBMERGED EARTH AND SURCHARGE

TABLE A – WEIGHT OF SOILS SUBMERGED*

Material	Pounds per cubic foot		
	Maximum	Minimum	Average
Gravel and marl		42.0	62.9
Gravel and sand	73	42.0	62.4
Sand	66	42.0	58.3
Gravel, sand and clay	80.9	51.2	70.0
Stiff clay	64.8	38.4	47.8
Stiff clay and gravel	70.3	44.8	52.6

Weights are for soils submerged in sea water.

TABLE B – VALUES OF W*

COMBINED WEIGHT W OF SEA WATER AND EARTH, p.c.f. (EQUIVALENT FLUID PRESSURE)

Slope of Repose of Earth	Weight W_2 of submerged earth, p.c.f. (See Table A.)							
	40	44	48	52	56	60	64	68
1/2: 1	66.2	66.4	66.7	66.9	67.1	67.3	67.6	67.8
3/4: 1	68.4	68.9	69.3	69.8	70.2	70.7	71.1	71.6
1: 1	70.9	71.6	72.2	72.9	73.6	74.3	75.0	75.7
1 1/4: 1	73.2	74.2	75.1	76.0	76.9	77.9	78.8	79.7
1 1/2: 1	75.4	76.6	77.7	78.9	80.0	81.2	82.3	83.5
1 3/4: 1	77.5	78.8	80.2	81.5	82.9	84.2	85.6	86.9
2 : 1	79.3	80.8	82.3	83.9	85.4	86.9	88.4	90.0
2 1/2: 1	82.3	84.2	86.0	87.8	89.7	91.5	93.3	95.2
3 : 1	84.8	86.9	88.9	91.0	93.1	95.2	97.2	99.3
3 1/2: 1	86.8	89.0	91.3	93.6	95.9	98.2	100	102
4 : 1	88.4	90.8	93.3	95.7	98.1	101	103	105
5 : 1	90.9	93.6	96.3	99.0	102	104	107	109
6 : 1	104	108	112	116	120	124	128	132

TABLE C – VALUES OF TAN² (45° – φ/2)

Slope	Angle ϕ	$\tan^2(45° - \phi/2)$
1: on 4	14° – 10'	0.607
1: on 3 1/2	16° – 0'	0.567
1: on 3	18° – 30'	0.518
1: on 2 1/2	21° – 50'	0.458
1: on 2	26° – 50'	0.383
1: on 1 1/2	33° – 40'	0.287
1: on 1 1/3	36° – 50'	0.250
1: on 1	45° – 0'	0.172

*Adopted from *American Civil Engineers Handbook* by Merriman and Wiggin.

WATERFRONT STRUCTURES—PIERS

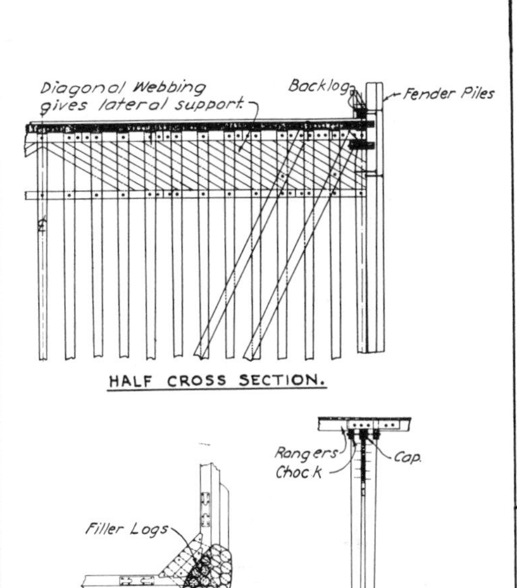
Diagonal Webbing gives lateral support.
Backlog
Fender Piles
HALF CROSS SECTION.
Rangers
Chock
Cap.
Filler Logs
PLAN.
CORNER DETAIL.
SECTION.

FIG. A – ARMY PIER AT PIERMONT, N.Y.

Heavy duty type similar to New York Dock Dept. practice. These piers have an approximate horizontal thrust resistance of 5 tons per lineal foot.

NOTE: Exposed to heavy ice flow of Tappan Zee (Hudson River).

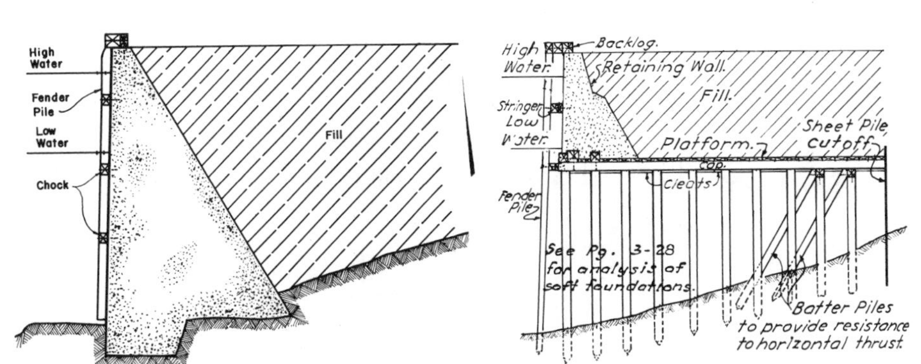
High Water
Fender Pile
Low Water
Chock
Fill
Backlog.
High Water.
Retaining Wall.
Fill.
Stringer
Low Water.
Platform.
Sheet Pile cutoff.
Cap.
Cleats.
Fender Pile
See Pg. 3-28 for analysis of soft foundations.
Batter Piles to provide resistance to horizontal thrust.

FIG. B – RETAINING WALL **FIG. C – RELIEVING PLATFORM**

NOTE: Backfill in these cases should be made from outside toward shore to reduce mud-wave action.

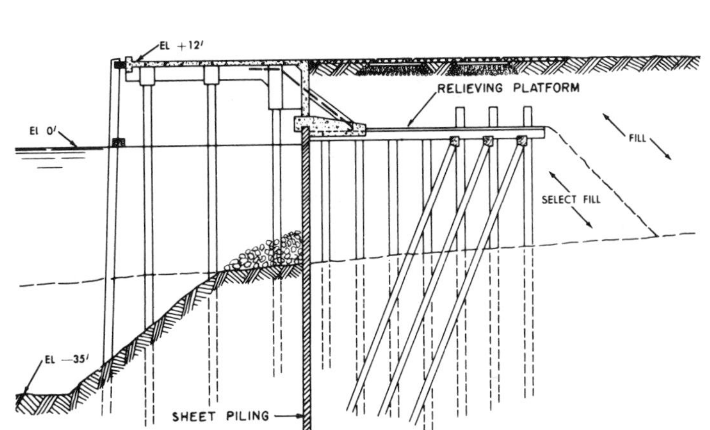
EL +12′
RELIEVING PLATFORM
EL 0′
FILL
SELECT FILL
EL —35′
SHEET PILING

FIG. D – OPEN QUAYWALL WITH RELIEVING PLATFORM*

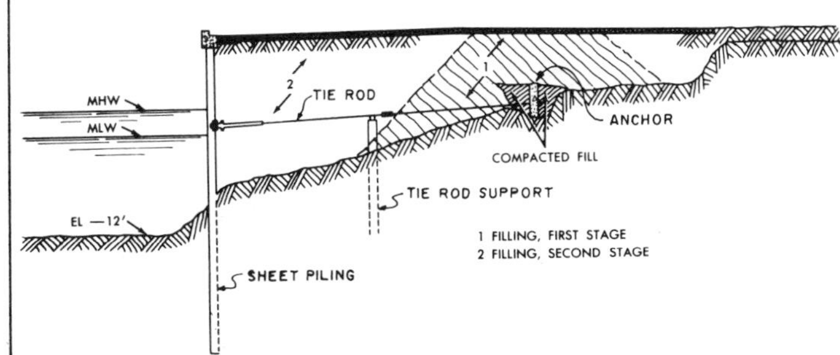
MHW
MLW
2
1
TIE ROD
ANCHOR
COMPACTED FILL
TIE ROD SUPPORT
EL —12′
1 FILLING, FIRST STAGE
2 FILLING, SECOND STAGE
SHEET PILING

FIG. E – OPEN QUAYWALL WITH SHEET-PILING BULKHEADS*

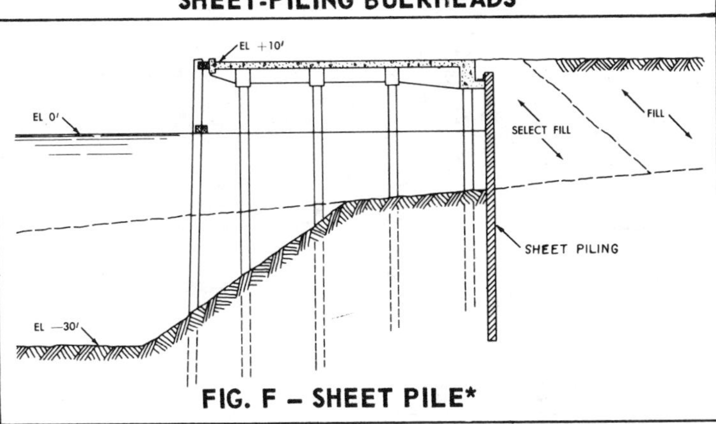
EL +10′
EL 0′
SELECT FILL
FILL
SHEET PILING
EL —30′

FIG. F – SHEET PILE*

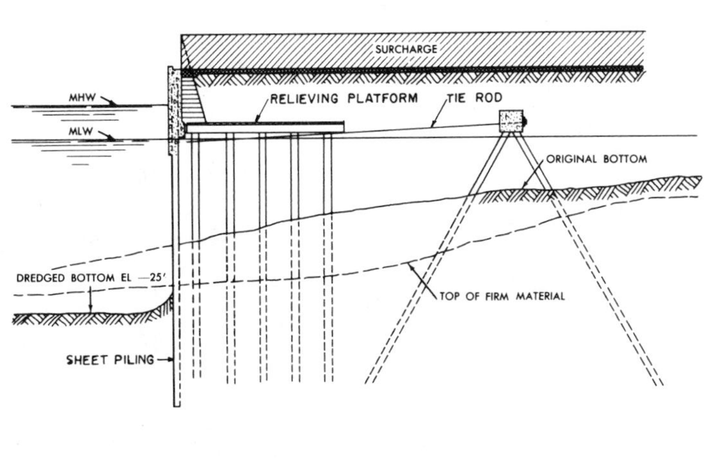
SURCHARGE
MHW
MLW
RELIEVING PLATFORM
TIE ROD
ORIGINAL BOTTOM
DREDGED BOTTOM EL —25′
TOP OF FIRM MATERIAL
SHEET PILING

FIG. G – SHEET PILE WITH RELIEVING PLATFORM*

*Adapted from Waterfront and Harbor Facilities, U. S. Navy, Bureau of Yards and Docks.

WATERFRONT STRUCTURES–PIERS & DETAILS

Piers are wharves of open construction, generally built perpendicular to shore. The pierhead line is a harbor line established by the Corps of Engineers onto which an open-type structure may be built.

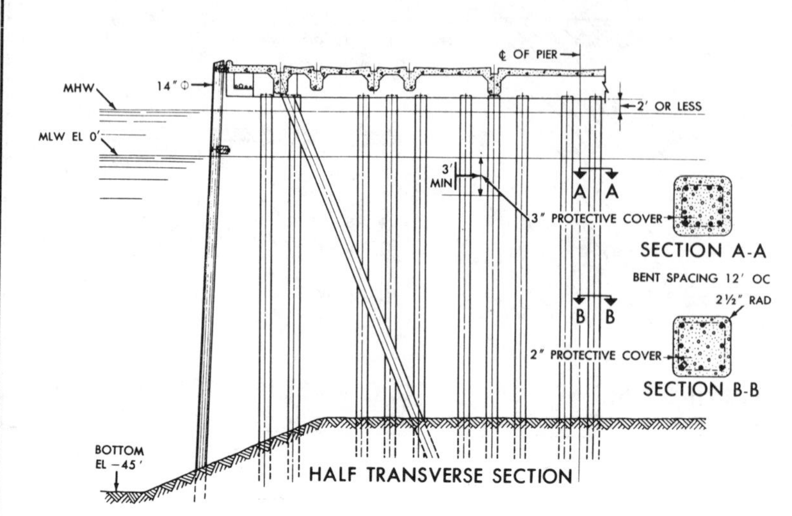

℄ OF PIER
MHW
14" Φ
MLW EL 0'
2' OR LESS
3' MIN
A A
3" PROTECTIVE COVER
SECTION A-A
BENT SPACING 12' OC
2½" RAD
B B
2" PROTECTIVE COVER
SECTION B-B
BOTTOM EL −45'
HALF TRANSVERSE SECTION

FIG. A – PRECAST PILES*

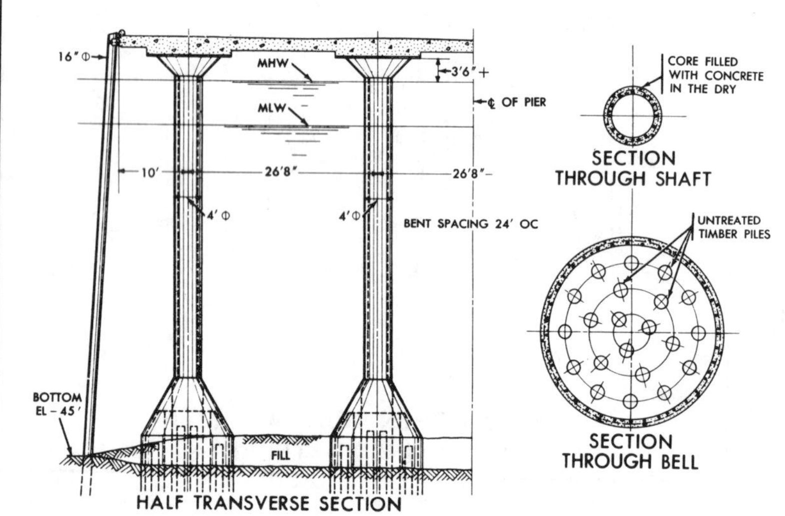

16" Φ
MHW
3'6" +
MLW
℄ OF PIER
10'
26'8"
26'8"
4' Φ
4' Φ
BENT SPACING 24' OC
CORE FILLED WITH CONCRETE IN THE DRY
SECTION THROUGH SHAFT
UNTREATED TIMBER PILES
SECTION THROUGH BELL
BOTTOM EL −45'
FILL
HALF TRANSVERSE SECTION

FIG. B – PRECAST CYLINDERS AND BELLS*

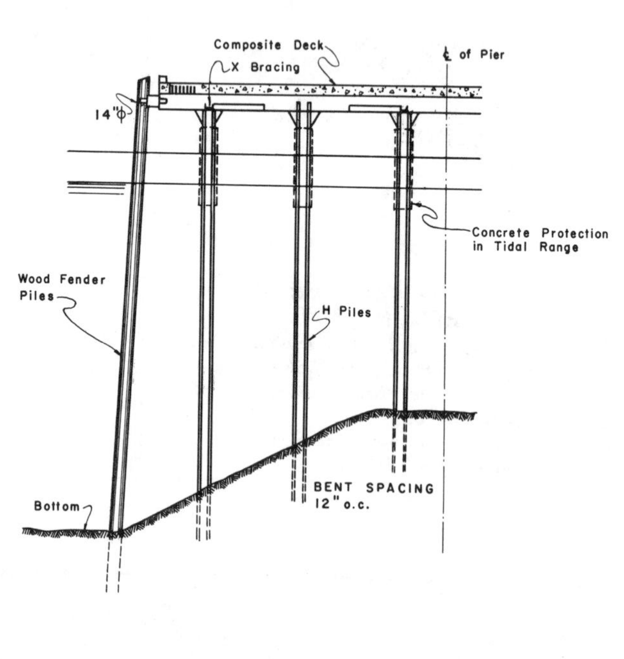

Composite Deck
X Bracing
℄ of Pier
14"Φ
Concrete Protection in Tidal Range
Wood Fender Piles
H Piles
BENT SPACING 12" o.c.
Bottom

FIG. C – STEEL FRAME*

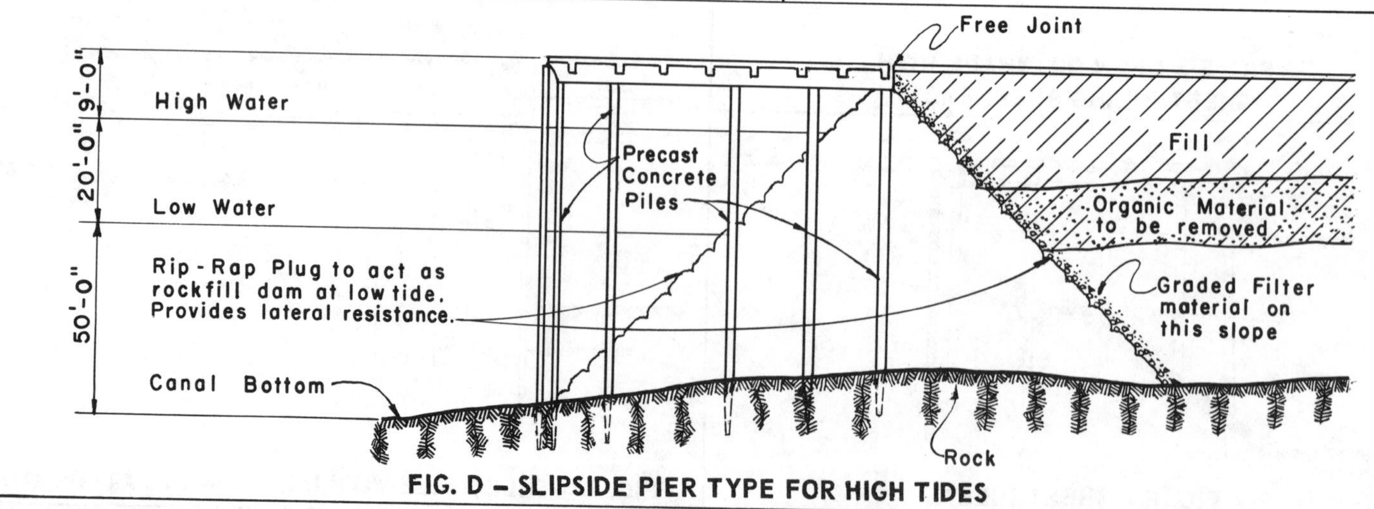

Free Joint
9'-0"
20'-0"
50'-0"
High Water
Low Water
Precast Concrete Piles
Fill
Organic Material to be removed
Rip-Rap Plug to act as rockfill dam at low tide. Provides lateral resistance.
Graded Filter material on this slope
Canal Bottom
Rock

FIG. D – SLIPSIDE PIER TYPE FOR HIGH TIDES

*Adapted from Waterfront and Harbor Facilities, U. S. Navy, Bureau of Yards and Docks.

WATERFRONT STRUCTURES—CARGO PIERS

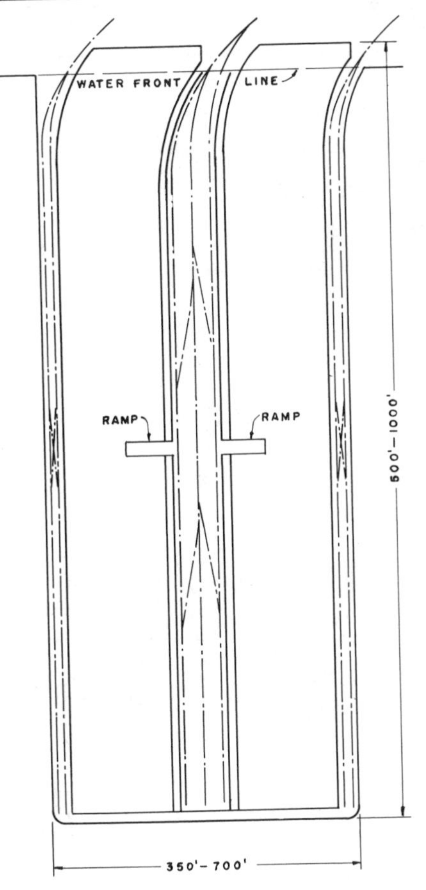
WATER FRONT
LINE
RAMP
RAMP
500'–1000'
350'–700'

FIG. A – TYPICAL LAYOUT FOR CARGO HANDLING PIERS USING SHIPS' GEAR

TABLE B – TYPICAL VESSEL CLASSIFICATION

U. S. Maritime Commission Class	Dead Weight, lb.	Length, ft.	Beam, ft.	Draft, ft.
CIB	9,000	418	60	28
CIM	5,000	339	50	21
C2	8,514	459	63	26
C3	12,929	492	70	29
C4	13,200	520	71	33
EC2	10,800	442	57	28
VC2	10,820	455	62	29
T2	16,500	523	65	32

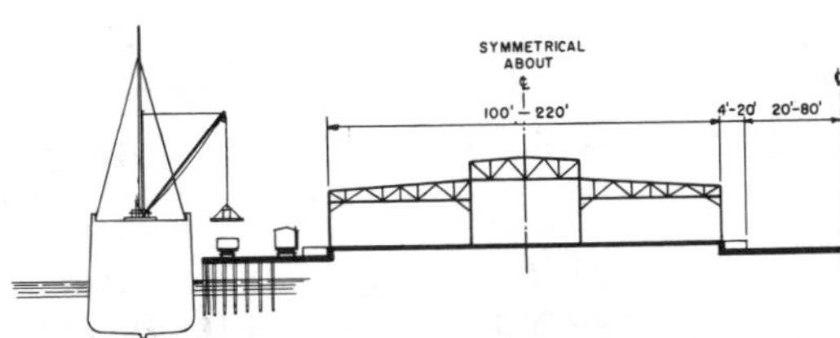
SYMMETRICAL ABOUT ℄
100' – 220'
4'-20'
20'-80'

FIG. C – TYPICAL CROSS SECTION

FREEBOARD

Distance above mean high water. 5 to 6 ft., varying with exposure to waves.

TABLE D – DEPTHS OF WATER

Piers – first-class passenger, 40 ft. at low water.
Piers – first-class freight, 35 ft. at low water.
River steamers – freight, 10 to 25 ft.
Subject to special study to fit nature of traffic expected.

TABLE E – LIVE LOADS

Location	Live Load in pounds per square foot	
	Lower Deck	Upper Deck
New York Dock Dept. Piers	500	350
New York Dock Dept. Bulkhead Walls	1000	
Havana Steamship Docks	250	400
Halifax Concrete Pile Pier No. 2	1000	500
Philadelphia Municipal Wharves	600	300 & 400
San Francisco State Piers	500	
Baltimore Municipal Wharves	600	
Tampico Oil Dock	800	

WATERFRONT STRUCTURES – PIERS & DETAILS

TYPICAL TIMBER PIER

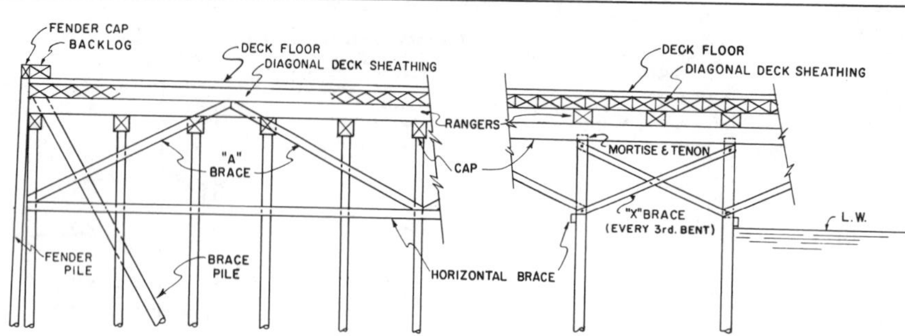
FENDER CAP
BACKLOG
DECK FLOOR
DIAGONAL DECK SHEATHING
"A" BRACE
RANGERS
CAP
FENDER PILE
BRACE PILE
HORIZONTAL BRACE
MORTISE & TENON
"X" BRACE (EVERY 3rd. BENT)
L.W.

CROSS SECTION

SECTION A-A

FIG. A – TYPICAL PILE BENT

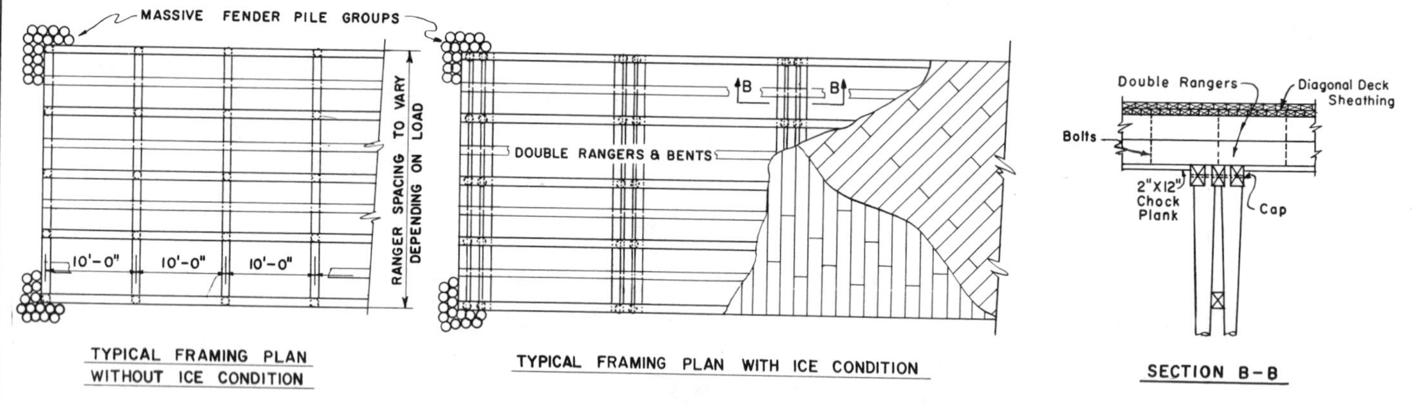
MASSIVE FENDER PILE GROUPS
10'–0"
10'–0"
10'–0"
RANGER SPACING TO VARY DEPENDING ON LOAD
DOUBLE RANGERS & BENTS
B
B
Double Rangers
Diagonal Deck Sheathing
Bolts
2"X12" Chock Plank
Cap

TYPICAL FRAMING PLAN WITHOUT ICE CONDITION

TYPICAL FRAMING PLAN WITH ICE CONDITION

SECTION B–B

FIG. B – TYPICAL BENT SPACING

TABLE C

STRENGTH OF MANILA ROPE		
Hawser, in.	Diameter, in.	Strength, lb.
3	1	7,000
6	2	25,000
8	2 5/8	45,000
9	3	56,000
10	3 1/4	72,000

STANDARD FOUNDRY CAST STEEL FITTINGS	
Fitting	Allowable Load at 30° with Horizontal, lb.
30" cleat	35,000
42" cleat	50,000
Mooring bitts	60,000 to 100,000
Bollards	70,000 to 450,000

NOTE: Anchorage strength of fittings should be designed to hold strength of mooring lines.

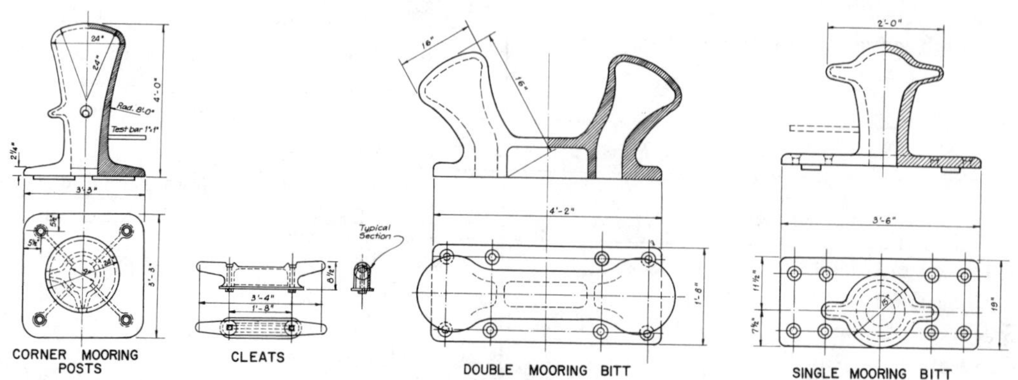
CORNER MOORING POSTS
CLEATS
Typical Section
DOUBLE MOORING BITT
SINGLE MOORING BITT
4'-0"
3'-3"
4'-2"
3'-6"
2'-0"
16"

FIG. D – DOCK FITTINGS*

*Adapted from *American Civil Engineers Handbook* by Merriman and Wiggin.

WATERFRONT STRUCTURES-SMALL CRAFT PIER

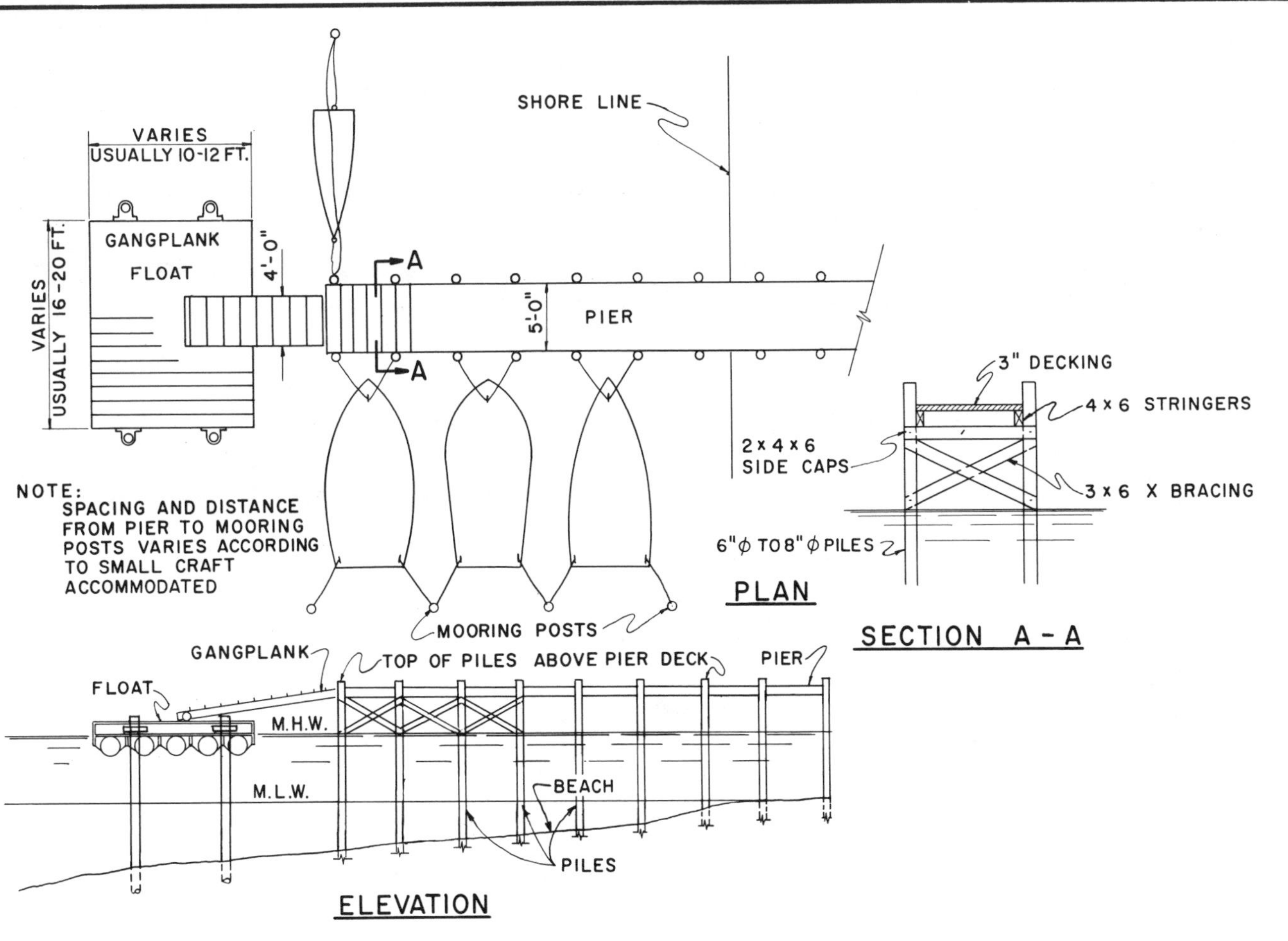
SHORE LINE
VARIES
USUALLY 10-12 FT.
VARIES
USUALLY 16-20 FT.
GANGPLANK
FLOAT
4'-0"
A
A
5'-0"
PIER
3" DECKING
4 x 6 STRINGERS
2 x 4 x 6
SIDE CAPS
3 x 6 X BRACING
6"ϕ TO 8"ϕ PILES
NOTE:
SPACING AND DISTANCE FROM PIER TO MOORING POSTS VARIES ACCORDING TO SMALL CRAFT ACCOMMODATED
PLAN
MOORING POSTS
SECTION A-A
GANGPLANK
TOP OF PILES ABOVE PIER DECK
PIER
FLOAT
M.H.W.
M.L.W.
BEACH
PILES
ELEVATION

MARINE SERVICE STATION

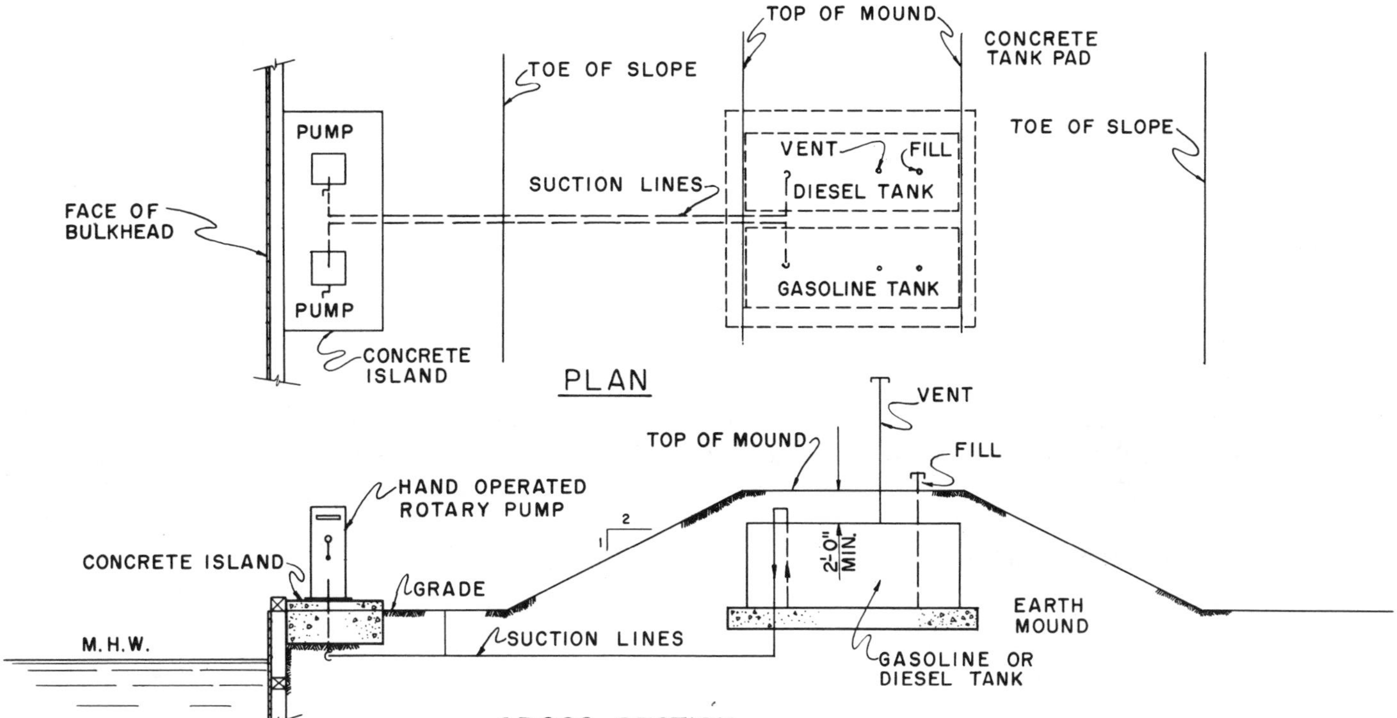

TOP OF MOUND
CONCRETE TANK PAD
TOE OF SLOPE
TOE OF SLOPE
PUMP
VENT
FILL
DIESEL TANK
SUCTION LINES
FACE OF BULKHEAD
GASOLINE TANK
PUMP
CONCRETE ISLAND

PLAN

VENT
TOP OF MOUND
FILL
HAND OPERATED ROTARY PUMP
2
1
2'-0" MIN.
CONCRETE ISLAND
GRADE
EARTH MOUND
M.H.W.
SUCTION LINES
GASOLINE OR DIESEL TANK

CROSS SECTION

WATERFRONT STRUCTURES–PROTECTION & PILES

DOCKS AND PIERS – PROTECTION AND PILES

MARINE BORERS – TEREDO, LIMNORIA, ETC.

Pressure Creosoting Pressure treatment with creosote or creosote coal-tar solutions provides the best means for protecting piles and timber of tide-water structures against marine borers.

Life of Creosoted Piles The life of creosoted piles varies with the latitude, ranging from about 20 years in Gulf and southern coastal waters to 40 to 50 years on the North Atlantic Seaboard. In Pacific coastal waters, life varies from 20 to 25 years.

Heavy Sewage pollution protects piles.

Earth Protection Marine borers cannot exist below the mud line. Earth cover of untreated piles cannot protect against attacks.

RED LIGHT Erosive currents and scour may uncover untreated wood piles and exposes them to attack.

*CORROSION**

The life of a steel pile may be estimated by the formula

$$Y = W/CL$$

Where Y = life of metal in years
W = weight of steel per linear foot
L = perimeter of exposed steel in feet
C = value given in Table A

TABLE A – VALUES OF C

Foul sea air	0.1944	Pure air, or clear river water	0.0125
Clean sea air	0.0970	Air of manufacturing districts and sea air	0.1252
Foul river water	0.1133		

NOTE: Cast iron has considerably more resistance than structural steel. For concrete and steel, waters subject to industrial waste should be examined for corrosive action and adequate protection should be provided.

EROSION†

Concrete structures subject to abrasion by ice or by freezing and thawing should have protection, such as sheathing or riprap.

Wood structures should be protected from ships etc. by dolphins, fender piles, or sheathing.

DECAY†

Wood piles and timbers are subject to decay above low-water level. Pressure-creosote all air-exposed piles and timbers.

SCOUR

Obstructions such as practically solid piers tend to cause river bottom scour and should be made the subject of special study.

Riprap may be used to minimize scour.

PILE FOUNDATIONS‡

"Conditions of foundations or bottom of harbor should be thoroughly known or studied in order to determine length of piles needed, permissible load for each pile or group of piles, and the character of material to be dredged to obtain required depth of water."

FIREPROOFING†

Wood piles should be fireproofed from mean low-water level to concrete deck.

* Adopted from *Wharves and Piers*, by Carleton Greene, McGraw-Hill.
† R. H. Mann, District Engineer, American Wood Preservers Institute.
‡ By Colonel Marcel Garsaud.

DRAINAGE — RUNOFF — I

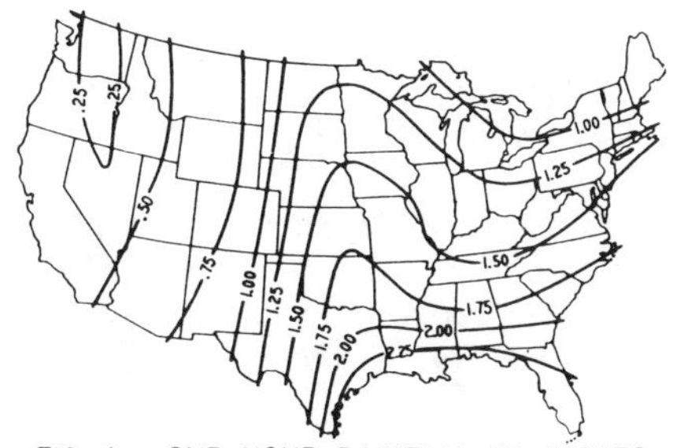

FIG. A. – ONE-HOUR RAINFALL, IN INCHES, TO BE EXPECTED ONCE IN 2 YEARS.

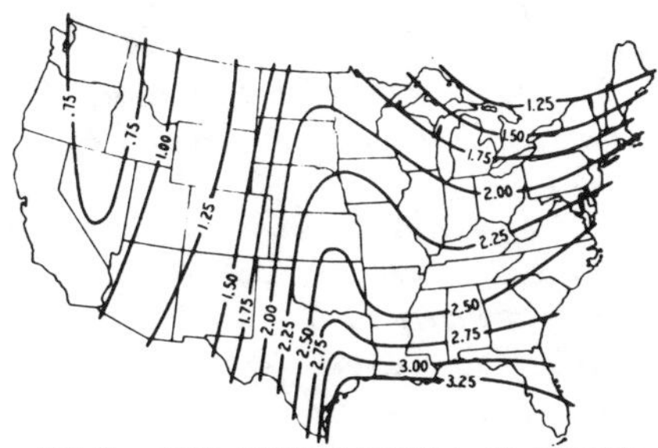

FIG. B. – ONE-HOUR RAINFALL, IN INCHES, TO BE EXPECTED ONCE IN 10 YEARS.

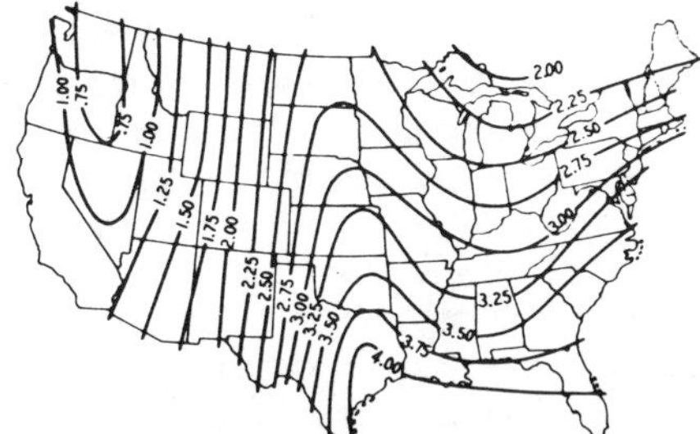

FIG. C. – ONE-HOUR RAINFALL, IN INCHES, TO BE EXPECTED ONCE IN 50 YEARS.

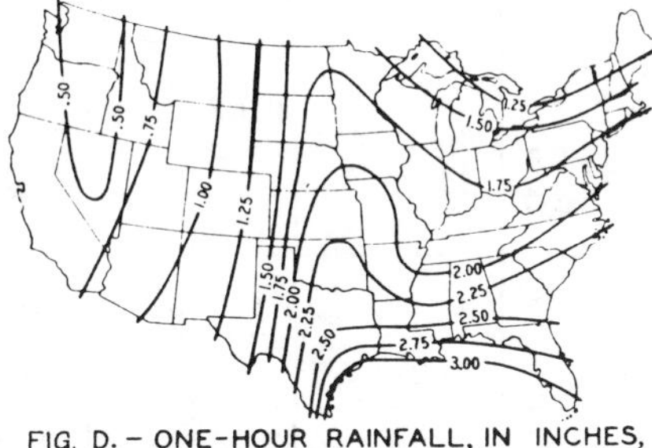

FIG. D. – ONE-HOUR RAINFALL, IN INCHES, TO BE EXPECTED ONCE IN 5 YEARS.

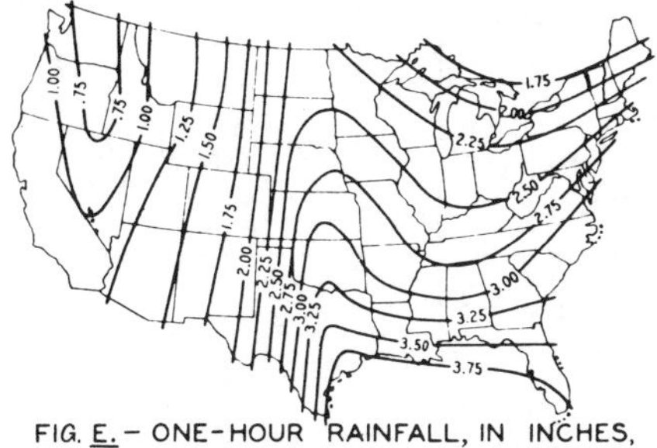

FIG. E. – ONE-HOUR RAINFALL, IN INCHES, TO BE EXPECTED ONCE IN 25 YEARS.

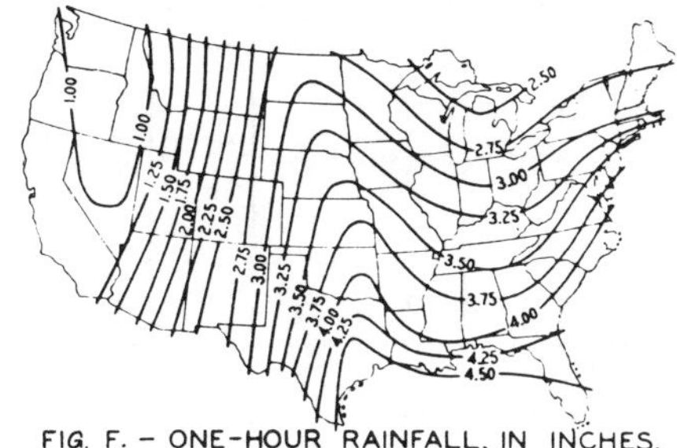

FIG. F. – ONE-HOUR RAINFALL, IN INCHES, TO BE EXPECTED ONCE IN 100 YEARS.

COMPUTATION OF i IN RATIONAL FORMULA.

EXAMPLE: Assume expectancy period = 5 years, see Fig. D, assume locality, find 1 hour intensity = 1.75 in. per hour.

FIG. G-INTENSITY EXPECTATION FOR ONE-HOUR RAINFALL.*

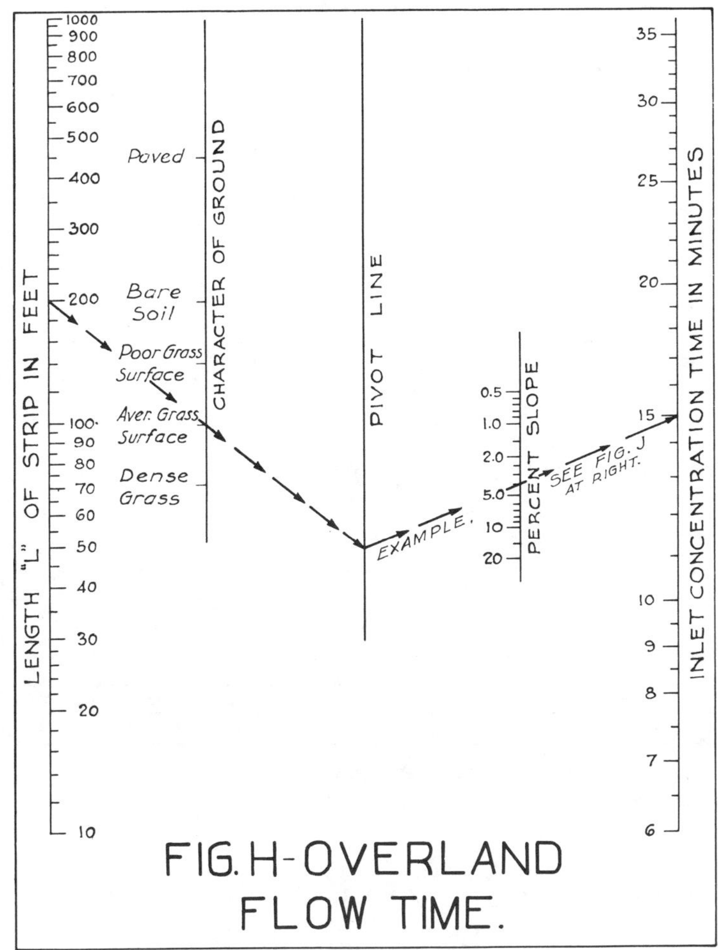

LENGTH "L" OF STRIP IN FEET
CHARACTER OF GROUND
Paved
Bare Soil
Poor Grass Surface
Aver. Grass Surface
Dense Grass
PIVOT LINE
EXAMPLE
PERCENT SLOPE
SEE FIG. J AT RIGHT.
INLET CONCENTRATION TIME IN MINUTES

FIG. H-OVERLAND FLOW TIME.

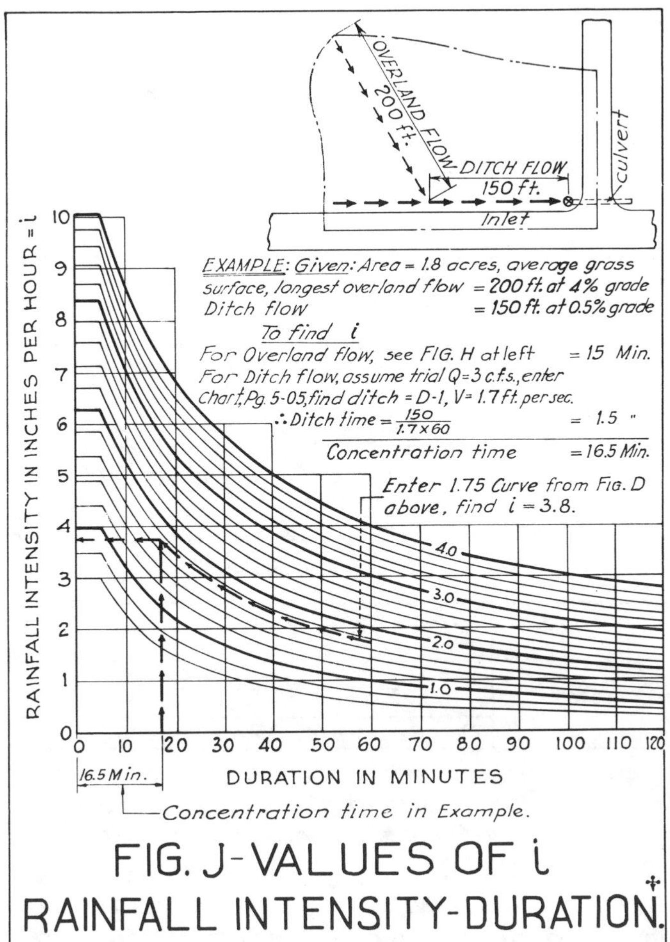

OVERLAND FLOW 200 ft.
DITCH FLOW 150 ft.
culvert
Inlet
EXAMPLE: Given: Area = 1.8 acres, average grass surface, longest overland flow = 200 ft. at 4% grade
Ditch flow = 150 ft. at 0.5% grade
To find i
For Overland flow, see FIG. H at left = 15 Min.
For Ditch flow, assume trial Q=3 c.f.s., enter Chart, Pg. 5-05, find ditch = D-1, V = 1.7 ft. per sec.
∴ Ditch time = 150 / 1.7×60 = 1.5 "
Concentration time = 16.5 Min.
Enter 1.75 Curve from Fig. D above, find i = 3.8.
RAINFALL INTENSITY IN INCHES PER HOUR = i
DURATION IN MINUTES
16.5 Min.
Concentration time in Example.

FIG. J-VALUES OF i RAINFALL INTENSITY-DURATION.†

*Reproduced from Miscellaneous Publication No. 204, U.S. Dept. of Agriculture, by David L. Yarnell.
† Adapted from Engineering Manual of the War Department, Part XIII, Chap. 1, Dec. '45.

DRAINAGE — RUNOFF — 2

Q=Aci RATIONAL FORMULA (Logical approach).

Q = RUNOFF = Peak discharge of watershed in cubic feet per second (c.f.s.) due to maximum storm assumed. See Figs. A to F, Pg. 18-01 (Usually 10 – 25 years).

A = Area of watershed in acres.

C = Coefficient of runoff, Table B below (Measure of losses due to infiltration, etc.).

i = Intensity of rainfall in inches per hour based on concentration time, See Pg. 18-01.

Concentration time = time required for rain falling at most remote point to reach discharge point. Concentration time may include Overland flow time, Fig. H, Pg. 18-01, and Channel flow time, Pg. 18-05, 18-06, 18-69 and 18-71.

TABLE A-COMPUTATION FORM FOR RATIONAL FORMULA.

LOCATION			A		C	TIME OF FLOW-MIN.			L *	Q c.f.s.	DESIGN					PROFILE				
STREET	FROM	TO	INCREMENT	TOTAL		TO INLET	IN CHANNEL	TIME OF CONC			CHANNEL OR PIPE SIZE	SLOPE ft. per ft.	n	CAPACITY FULL c.f.s.	V ft. per Sec.	LENGTH ft.	FALL ft.	OTHER LOSSES ft. †	INV. ELEV. UPPER END	INV. ELEV. LOWER END
FIRST ST.	A	B	1.8	1.8	.44	16.5	0.3	16.5	3.8	3.0	15"	.008	.015	4.6	3.9	60	0.48	0	82.00	81.52
MAIN RD.	B	C	1.9	3.7	.50		2.5	16.8	3.7	6.8	D-2	.011	.030	12.0	2.8	420	4.62	0	81.52	76.90
" "	C	D	2.0	5.7	.50		1.8	19.3	3.5	10.0	21"	.007	.015	11.1	4.5	480	3.36	2.20	74.70	70.34

*Note that the sequence of design as in example, Fig. J, Pg. 18-01 involves trial assumptions in determining i.

† Fall in manhole.

TABLE B – VALUES OF $C = \frac{RUNOFF}{RAINFALL}$

SURFACES			VALUE PROPOSED MIN.	VALUE PROPOSED MAX.	VALUE BY OTHER AUTHORITY MIN.	VALUE BY OTHER AUTHORITY MAX.	
ROOFS, slag to metal.			0.90	1.00	0.70	0.95	①
PAVEMENTS	Concrete or Asphalt.		0.90	1.00	0.95	1.00	③
	Bituminous Macadam, open and closed type.		0.70	0.90	0.70	0.90	③
	Gravel, from clean and loose to clayey and compact.		0.25	0.70	0.15	0.30	①
R.R. YARDS			0.10	0.30	0.10	0.30	①
EARTH SURFACES	SAND, from uniform grain size, no fines, to well graded, some clay or silt.	Bare	0.15	0.50	0.01	0.55	④
		Light Vegetation	0.10	0.40	0.01	0.55	④
		Dense Vegetation	0.05	0.30	0.01	0.55	④
	LOAM, from sandy or gravelly to clayey.	Bare	0.20	0.60			
		Light Vegetation	0.10	0.45			
		Dense Vegetation	0.05	0.35			
	GRAVEL, from clean gravel and gravel sand mixtures, no silt or clay to high clay or silt content.	Bare	0.25	0.65			
		Light Vegetation	0.15	0.50			
		Dense Vegetation	0.10	0.40			
	CLAY, from coarse sandy or silty to pure colloidal clays.	Bare	0.30	0.75	0.10	0.70	④
		Light Vegetation	0.20	0.60	0.10	0.70	④
		Dense Vegetation	0.15	0.50	0.10	0.70	④
COMPOSITE AREAS	City, business areas.		0.60	0.75	0.60	0.95	⑤
	City, dense residential areas, vary as to soil and vegetation.		0.50	0.65	0.30	0.60	⑤
	Suburban residential areas, " " "		0.35	0.55	0.25	0.40	⑤
	Rural Districts, " " "		0.10	0.25	0.10	0.25	②
	Parks, Golf Courses, etc., " " "		0.10	0.35	0.05	0.25	①

NOTE: Values of "C" for earth surfaces are further varied by degree of saturation, compaction, surface irregularity and slope, by character of subsoil, and by presence of frost or glazed snow or ice.

① Bryant & Kuichling, Report, Back Bay Sewerage District, Boston, 1909.

② Metcalf and Eddy, American Sewerage Practice, 1928. McGraw-Hill.

③ Used by City of Boston, reported by Metcalf & Eddy.

④ Used by City of Detroit, reported by Metcalf & Eddy.

⑤ L.C. Urquhart, Civil Engineering Handbook, 1940. McGraw-Hill.

DRAINAGE — RUNOFF — 3

$Q = Aci\sqrt[5]{S/A}$ Mc. MATH FORMULA (Empirical approach).

Q = Runoff in c.f.s.
A = Area of watershed in acres.
c = Coefficient of runoff, see Table B - Pg. 18-02.
*i = Intensity of rainfall in inches per hour for a 20 min. storm**
S = Average slope of watershed in feet per thousand.

EXAMPLE: Given: Area = 12 acres, suburban residential area, clay-loam, light vegetation. Average slope = 25 feet per 1000. Locality southern Ohio. Required to find Q. C = 0.50 from Table B - Pg. 18-02 To find i, enter Fig. D - Pg. 18-01, find intensity for one hour = 1.75; enter Fig. J. Pg. 18-01 on 1.75 curve and find i = 3.4 for 20 min. duration.

$\therefore Q = 12 \times 0.50 \times 3.4 \sqrt[5]{25/12} = 20.4\sqrt[5]{2.1}$ ($\sqrt[5]{2.1} = 1.16$, see Table A below)

$= 20.4 \times 1.16 = 23.7$ c.f.s.

**Recommended.*

TABLE A - FIFTH ROOTS.

NO.	ROOT	NO.	ROOT	NO.	ROOT	NO.	ROOT	NO.	ROOT	NO.	ROOT
.031	.50	.371	.82	1.22	1.04	4.65	1.36	13.4	1.68	32.0	2.00
.038	.52	.418	.84	1.34	1.06	5.00	1.38	14.2	1.70	36.2	2.05
.046	.54	.470	.86	1.47	1.08	5.38	1.40	15.1	1.72	40.8	2.10
.055	.56	.528	.88	1.61	1.10	5.77	1.42	15.9	1.74	45.9	2.15
.066	.58	.590	.90	1.76	1.12	6.19	1.44	16.9	1.76	51.5	2.20
.078	.60	.624	.91	1.93	1.14	6.63	1.46	17.9	1.78	57.7	2.25
.092	.62	.659	.92	2.10	1.16	7.10	1.48	18.9	1.80	64.4	2.30
.107	.64	.696	.93	2.29	1.18	7.59	1.50	20.0	1.82	71.7	2.35
.125	.66	.734	.94	2.49	1.20	8.11	1.52	21.1	1.84	79.6	2.40
.145	.68	.774	.95	2.70	1.22	8.66	1.54	22.3	1.86	88.3	2.45
.168	.70	.815	.96	2.93	1.24	9.24	1.56	23.5	1.88	97.7	2.50
.193	.72	.859	.97	3.18	1.26	9.85	1.58	24.8	1.90	108.	2.55
.222	.74	.904	.98	3.44	1.28	10.5	1.60	26.1	1.92	119.	2.60
.254	.76	.951	.99	3.71	1.30	11.2	1.62	27.5	1.94	131.	2.65
.289	.78	1.00	1.00	4.01	1.32	11.9	1.64	28.9	1.96	143.	2.70
.328	.80	1.10	1.02	4.32	1.34	12.6	1.66	30.4	1.98	157.	2.75

DETERMINATION OF RUNOFF (Q) FROM SURVEY OF EXISTING CHANNEL.

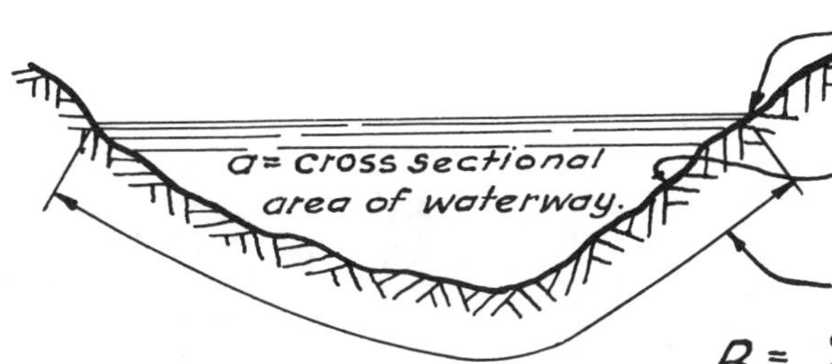
a = cross sectional
area of waterway.

Highwater level from observed marks or records.

n = Coefficient of roughness of existing surface, see Pg. 18-68

p = wetted perimeter, measured

$R = a/p$ = *Hydraulic radius.*

S = Slope of Hydraulic Gradient (water surface) in feet per ft. for reach of channel through cross-section.

FIG. B - EXISTING SECTION AT SITE OF PROPOSED BRIDGE OR CULVERT.

Determine Q by Manning or Kutter Formula, see Pg. 18-69 & Pg. 18-71.

DRAINAGE — RUNOFF — 4

$a = C\sqrt[4]{A^3}$ – TALBOT'S FORMULA – (APPROXIMATE APPROACH)

a = Required section of waterway in square feet.
A = Drainage area in acres; C = Talbot's coefficient.

DRAINAGE AREA		VALUES OF "a"						
		MOUNTAINOUS	HILLY LAND		ROLLING LAND		FLAT LAND	
Acres	Square miles	ASSUME C = 1.00	C = 0.80	C = 0.60	C = 0.50	C = 0.40	C = 0.30	C = 0.20
1	0.0016	1.0	0.8	0.6	0.5	0.4	0.3	0.2
2	0.0031	1.7	1.4	1.0	.8	.7	.5	.3
4	0.0062	2.8	2.2	1.7	1.4	1.1	.8	.6
6	0.0094	3.8	3.0	2.3	1.9	1.5	1.1	.8
8	0.0125	4.8	3.8	2.9	2.4	1.9	1.4	1.0
10	0.016	5.6	4.5	3.4	2.8	2.2	1.7	1.2
15	0.023	7.6	6.1	4.6	3.8	3.0	2.3	1.5
20	0.031	9.5	7.6	5.7	4.7	3.8	2.8	1.9
30	0.047	12.8	10.2	7.7	6.4	5.1	3.8	2.6
40	0.062	15.9	12.7	9.5	8	6.4	4.8	3.2
60	0.094	22	17.6	13	11	8.8	6.6	4.4
ASSUME 80	0.125	27	21.6	16	13	10.8	8.1	5.4
100	0.156	32	25.6	19	16	12.8	9.6	6.4
150	0.234	43	34.4	26	21	17.2	12.9	8.6
200	0.312	53	42.4	32	27	21.2	15.9	10.6
250	0.39	63	50	38	31	25	19	13
300	0.47	72	58	43	36	29	22	14
400	0.62	89	71	53	45	36	27	18
500	0.78	106	85	64	53	42	32	21
600	0.94	121	97	73	61	48	36	24
800	1.25	150	120	90	75	60	45	30
1,000	1.56	178	142	107	89	71	53	36
1,500	2.34	241	193	145	121	96	72	48
2,000	3.12	299	239	179	149	120	90	60
2,500	3.91	354	283	212	177	142	106	71
3,000	4.7	405	324	243	203	162	122	81
4,000	6.2	503	402	302	252	202	151	101
5,000	7.8	595	476	357	297	238	179	119
6,000	9.4	682	546	409	341	273	205	136
8,000	12.5	846	677	508	423	338	254	169
10,000	15.6	1,000	800	600	500	400	300	200
25,000	39.1	1,988	1,590	1,193	994	795	596	398
50,000	78	3,344	2,675	2,006	1,672	1,338	1,003	669
100,000	156	5,623	4,498	3,374	2,812	2,249	1,687	1,125
200,000	312	9,457	7,566	5,674	4,728	3,783	2,837	1,891
400,000	625	15,905	12,724	9,543	7,952	6,362	4,772	3,181

From Manual of Instructions, Surveys and Plans, Division of Highways, California Dept of Public Works, 1931.

Adapted from Urquhart, Civil Engineering Handbook, McGraw-Hill.

DRAINAGE - DITCHES - COMMON SECTIONS - I

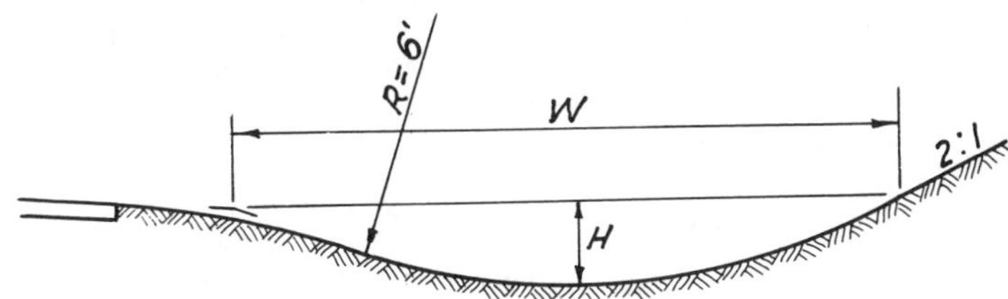
R=6'
W
2:1
H

D-1 SEGMENTAL

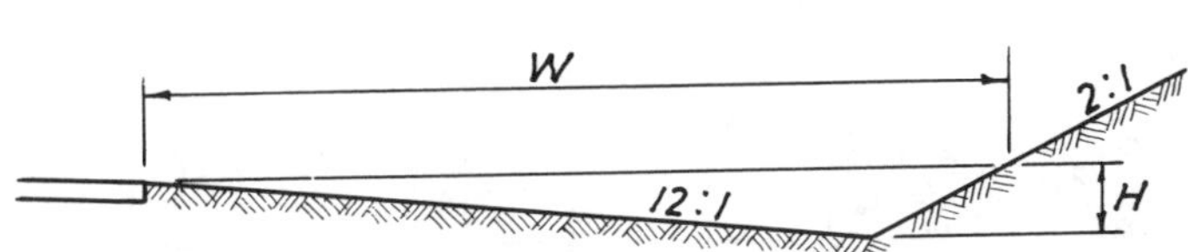
W
2:1
12:1
H

D-1A TRIANGULAR
Unequal side slopes

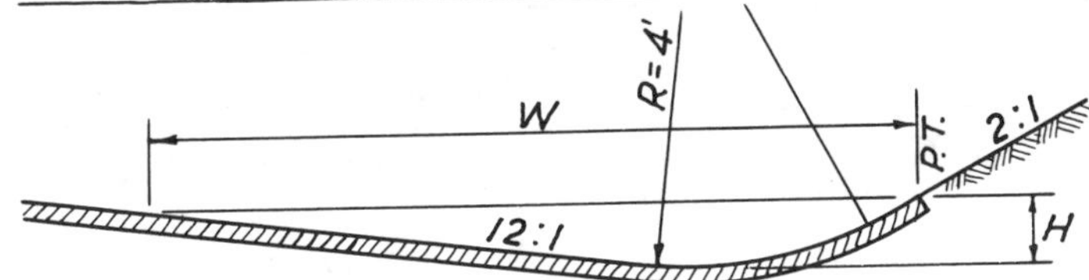
R=4'
W
P.T.
2:1
12:1
H

D-1B BITUMINOUS GUTTER

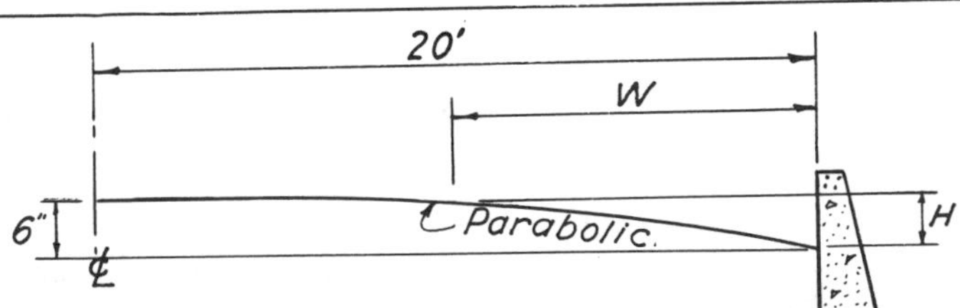
20'
W
6"
Parabolic
H

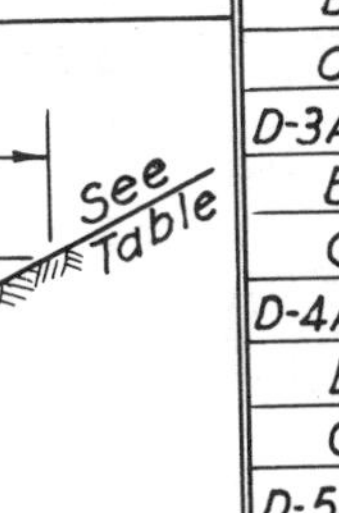

D-1C CURBED CROWNED STREET

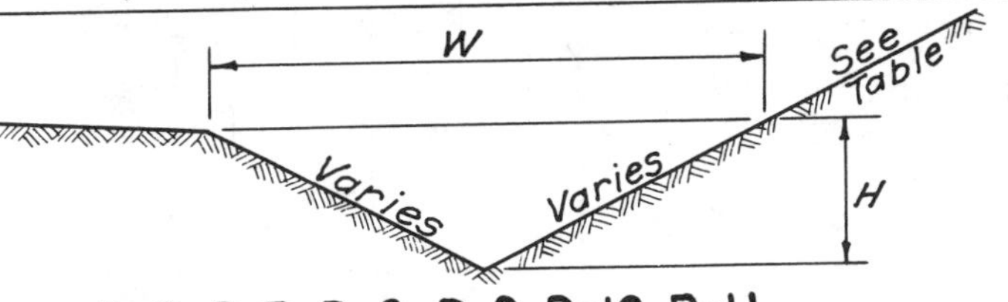

D-2, D-3, D-4, D-5 TRAPEZOIDAL

D-6, D-7, D-8, D-9, D-10, D-11
ISOSCELES TRIANGULAR
D-9, D-10 and D-11 - Airport ditches

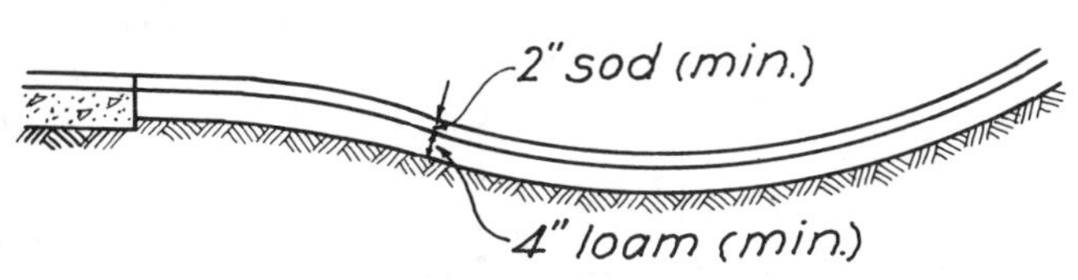
2" sod (min.)
4" loam (min.)

SODDED GUTTER

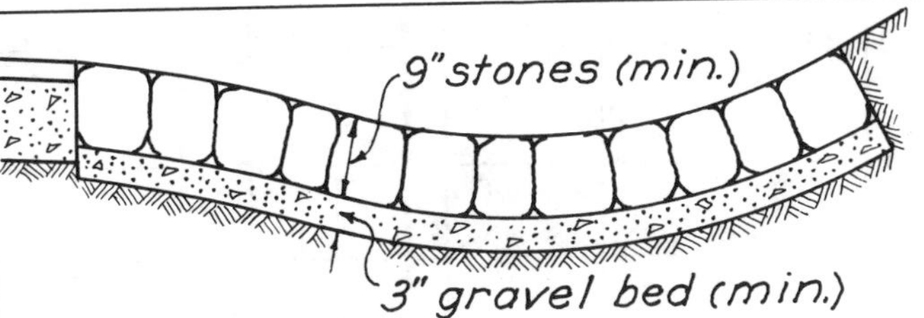
9" stones (min.)
3" gravel bed (min.)

COBBLED GUTTER

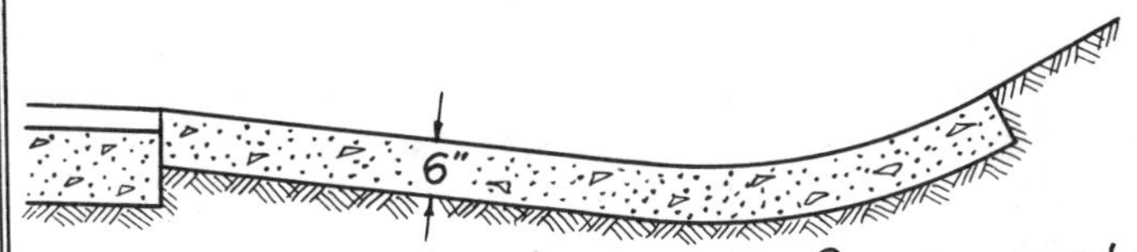
6"

Base required same as for concrete pavement.

CONCRETE GUTTER

TABLE A — PROPERTIES OF DITCHES.

NO.	DIMENSIONS				HYDRAULICS			
	SIDE SLOPES	B	H	W	a	p	R	$R^{2/3}$
D-1	—	—	6½"	5'-0"	1.84	5.16	0.356	0.502
D-1A	12:1 & 2:1	—	6"	7'-0"	1.75	7.14	0.245	0.392
D-1B	12:1 & 2:1	—	5"±	7'-0"	1.64	7.08	0.232	0.377
D-1C	½" to 1'-0"	—	4.5"	10'-0"	1.68	10.38	0.162	0.297
D-2A	1½:1'	2'-0"	1'-0"	5'-0"	3.50	5.61	0.624	0.730
B	2:1	2'-0"	1'-0"	6'-0"	4.00	6.47	0.618	0.726
C	3:1	2'-0"	1'-0"	8'-0"	5.00	8.32	0.601	0.712
D-3A	1½:1	3'-0"	1'-6"	7'-6"	7.88	8.41	0.937	0.958
B	2:1	3'-0"	1'-6"	9'-0"	9.00	9.71	0.927	0.951
C	3:1	3'-0"	1'-6"	12'-0"	11.25	12.49	0.901	0.933
D-4A	1½:1	3'-0"	2'-0"	9'-0"	12.00	10.21	1.175	1.114
B	2:1	3'-0"	2'-0"	11'-0"	14.00	11.94	1.173	1.112
C	3:1	3'-0"	2'-0"	15'-0"	18.00	15.65	1.150	1.097
D-5A	1½:1	4'-0"	3'-0"	13'-0"	25.50	14.82	1.721	1.436
B	2:1	4'-0"	3'-0"	16'-0"	30.00	17.42	1.722	1.437
C	3:1	4'-0"	3'-0	22'-0"	39.00	22.97	1.698	1.423
D-6A	2:1	—	1'-0"	4'-0"	2.00	4.47	0.447	0.584
B	3:1	—	1'-0"	6'-0"	3.00	6.32	0.475	0.609
D-7A	2:1	—	2'-0"	8'-0"	8.00	8.94	0.895	0.929
B	3:1	—	2'-0"	12'-0"	12.00	12.65	0.949	0.965
D-8A	2:1	—	3'-0"	12'-0"	18.00	13.42	1.341	1.216
B	3:1	—	3'-0"	18'-0"	27.00	18.97	1.423	1.265
D-9	7:1	—	1'-0"	14'-0"	7.00	14.14	0.495	0.626
D-10	7:1	—	2'-0"	28'-0"	28.00	28.28	0.990	0.993
D-11	7:1	—	3'-0"	42'-0"	63.00	42.43	1.485	1.302

DRAINAGE – DITCHES – COMMON SECTIONS – 2

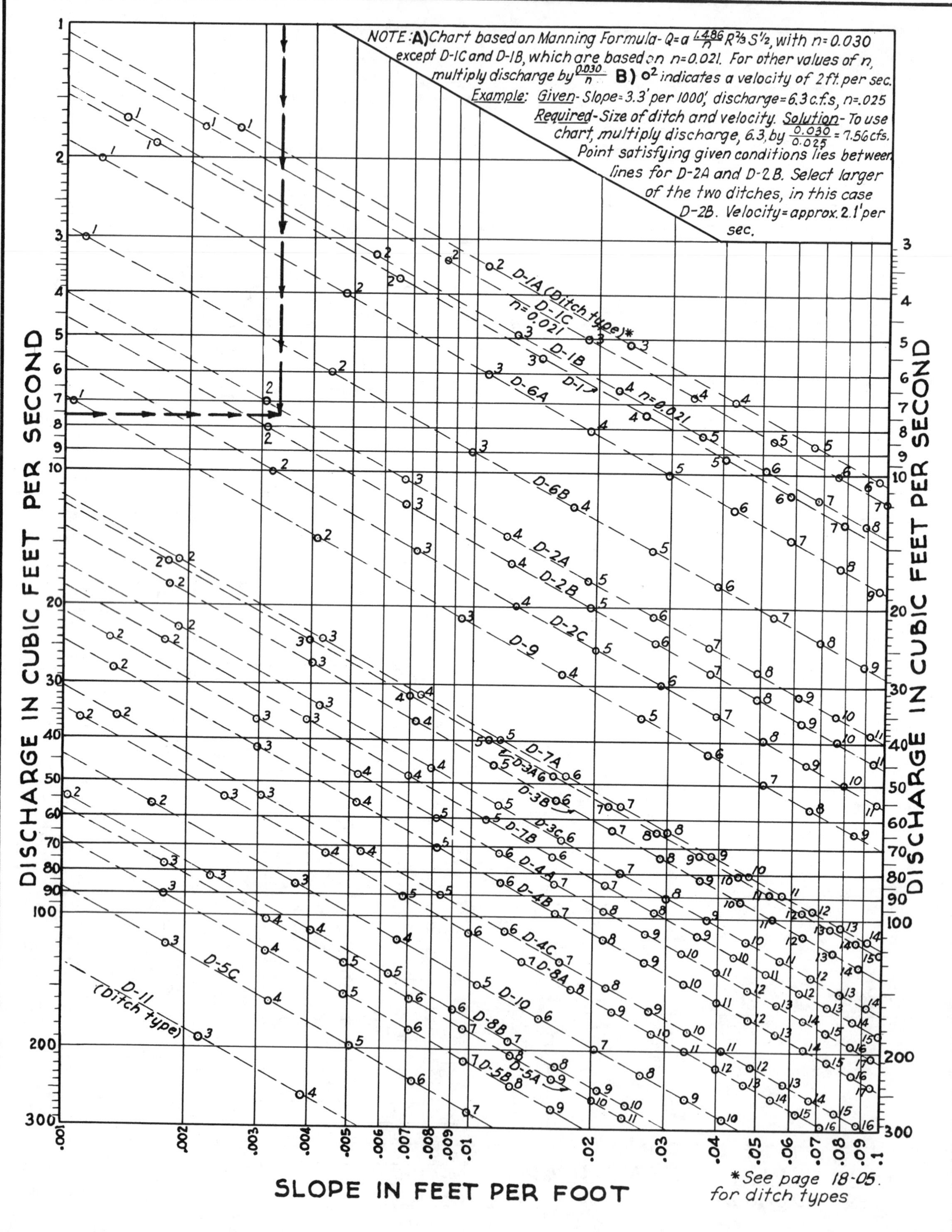
NOTE: A) Chart based on Manning Formula – $Q = a\frac{1.486}{n}R^{2/3}S^{1/2}$, with n = 0.030 except D-1C and D-1B, which are based on n = 0.021. For other values of n, multiply discharge by $\frac{0.030}{n}$. B) o^2 indicates a velocity of 2 ft. per sec.
Example: Given – Slope = 3.3′ per 1000′, discharge = 6.3 c.f.s., n = .025
Required – Size of ditch and velocity. Solution – To use chart, multiply discharge, 6.3, by $\frac{0.030}{0.025}$ = 7.56 cfs. Point satisfying given conditions lies between lines for D-2A and D-2B. Select larger of the two ditches, in this case D-2B. Velocity = approx. 2.1′ per sec.
DISCHARGE IN CUBIC FEET PER SECOND
SLOPE IN FEET PER FOOT
* See page 18-05 for ditch types

DRAINAGE – SUBDRAINAGE§ – I

TABLE A - SUBDRAIN DEPTH AND SPACING * FOR CONTROLLING GROUND WATER LEVEL.

SOIL CLASSES	PERCENTAGE OF EACH CLASS OF SOIL			DEPTH OF BOTTOM OF DRAIN IN FEET	DISTANCE BETWEEN SUBDRAINS IN FEET
	SAND	SILT	CLAY		
Sand	80-100	0-20	0-20	3-4 2-3	150-300 100-150
Sandy Loam	50-80	0-50	0-20	3-4 2-3	100-150 85-100
Loam	30-50	30-50	0-20	3-4 2-3	85-100 75-85
Silt Loam	0-50	50-100	0-20	3-4 2-3	75-85 65-75
Sandy Clay Loam	50-80	0-30	20-30	3-4 2-3	65-75 55-65
Clay Loam	20-50	20-50	20-30	3-4 2-3	55-65 45-55
Silty Clay Loam	0-30	50-80	20-30	3-4 2-3	45-55 40-45
Sandy Clay	50-70	0-20	30-50	3-4 2-3	40-45 35-40
Silty Clay	0-20	50-70	30-50	3-4 2-3	35-40 30-35
Clay	0-50	0-50	30-100	3-4 2-3	30-35 25-30

* From Airport Drainage, by Armco International Corp. Above data to be considered rough approximations only.

TABLE B - SUBSURFACE RUNOFF.**

MEAN ANNUAL† PRECIPITATION (INCHES)	SUBSURFACE RUNOFF	
	INCHES PER DAY PER SQ. FOOT	C.F.S. PER ACRE
30 OR LESS	0.25	0.0105
30-40	0.35	0.0147
40-50	0.50	0.0210
50 OR MORE	0.75	0.0315

** Based on Eng. Manual, U.S. Corps of Engrs.
† See Fig. A, Page 20-04

EXAMPLE:- Given:- Area in central Ohio, Fig. C(e) below, Sandy loam, 800 ft. x 1875 ft., to be subdrained by system of parallel laterals discharging into one submain. Required: Depth and spacing of laterals and discharge from submain = Q. Read in Tab. A, at left, for "Sandy loam" minimum depth = 3 feet, spacing = 100 ft. In Fig. A, Pg. 20-04 find mean annual precipitation = 37.5; enter Tab. B, above, with 37.5 inches mean annual precipitation; read subsurface runoff = 0.0147 c.f.s.

Therefore $Q = \frac{800 \times 1875 \times 0.0147}{43,560} = 0.5$ c.f.s.

For size of Drains see Page 18-66.

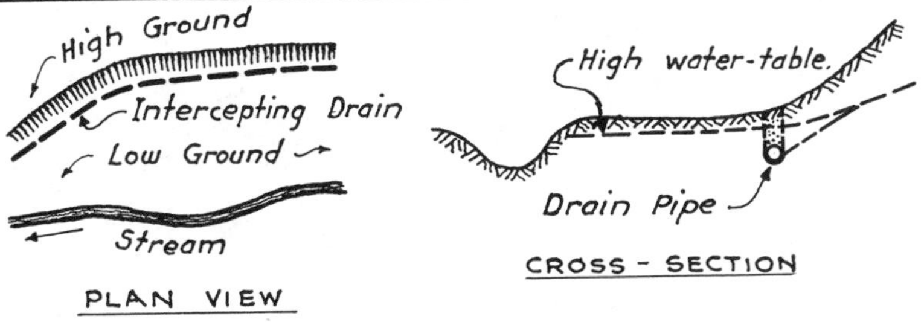

High Ground
Intercepting Drain
Low Ground
Stream
PLAN VIEW
High water-table.
Drain Pipe
CROSS - SECTION
(a) INTERCEPTING SYSTEM

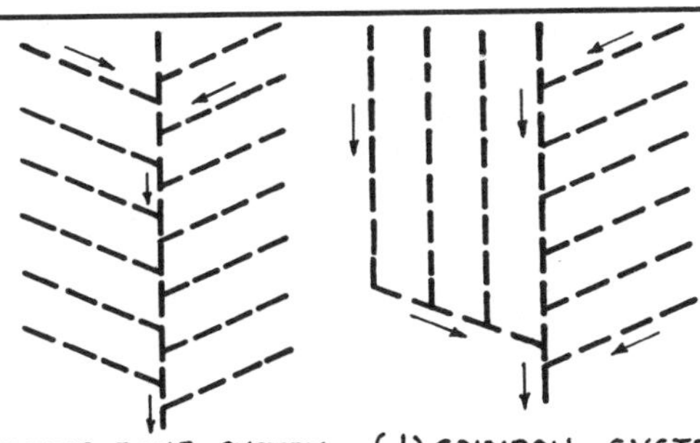

(c) HERRING-BONE SYSTEM

(d) GRIDIRON SYSTEM

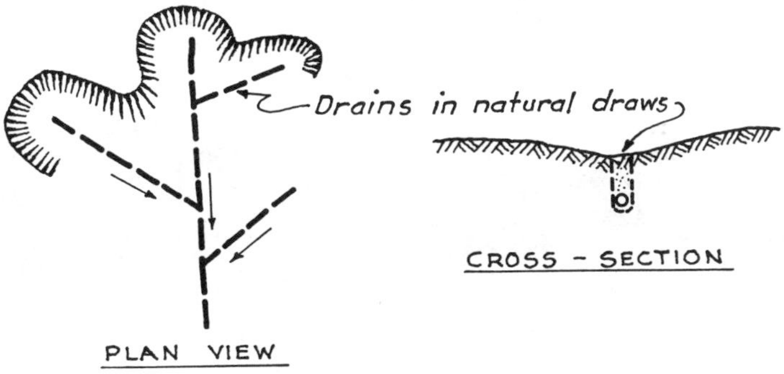

Drains in natural draws
PLAN VIEW
CROSS - SECTION
(b) NATURAL SYSTEM

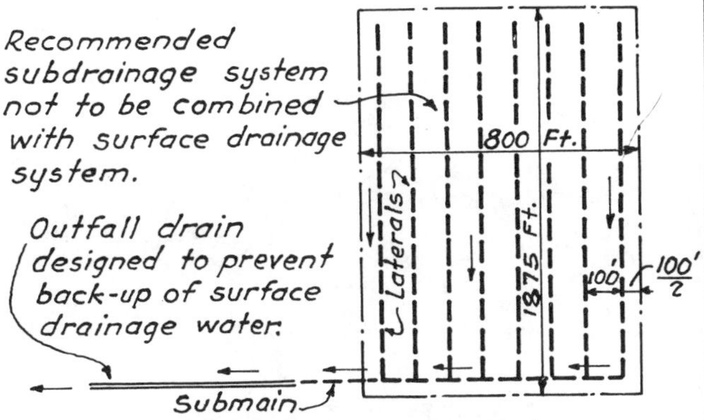

Recommended subdrainage system not to be combined with surface drainage system.
Outfall drain designed to prevent back-up of surface drainage water.
800 Ft.
Laterals
1875 Ft.
100'
100'/2
Submain
(e) PARALLEL SYSTEM

FIG. C - TYPES OF SUBDRAINAGE SYSTEMS.††

†† From Handbook of Culvert and Drainage Practice, by Armco Drainage Products Assoc.
§ For additional data on Subdrainage, see "Soil Mechanics" section.

DRAINAGE — SUBDRAINAGE*—2

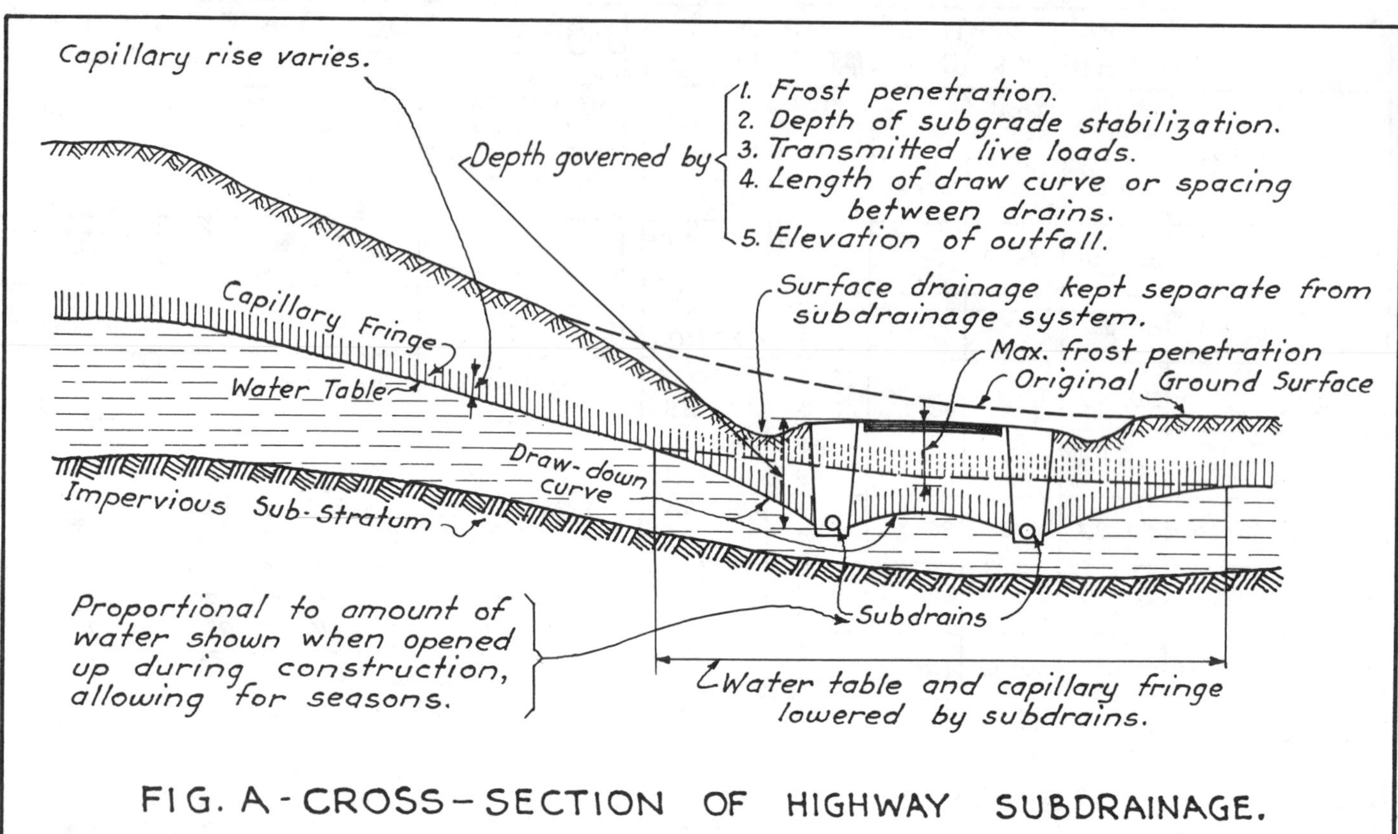
Capillary rise varies.
Depth governed by
1. Frost penetration.
2. Depth of subgrade stabilization.
3. Transmitted live loads.
4. Length of draw curve or spacing between drains.
5. Elevation of outfall.
Surface drainage kept separate from subdrainage system.
Max. frost penetration
Original Ground Surface
Capillary Fringe
Water Table
Draw-down curve
Impervious Sub-Stratum
Proportional to amount of water shown when opened up during construction, allowing for seasons.
Subdrains
Water table and capillary fringe lowered by subdrains.

FIG. A - CROSS-SECTION OF HIGHWAY SUBDRAINAGE.

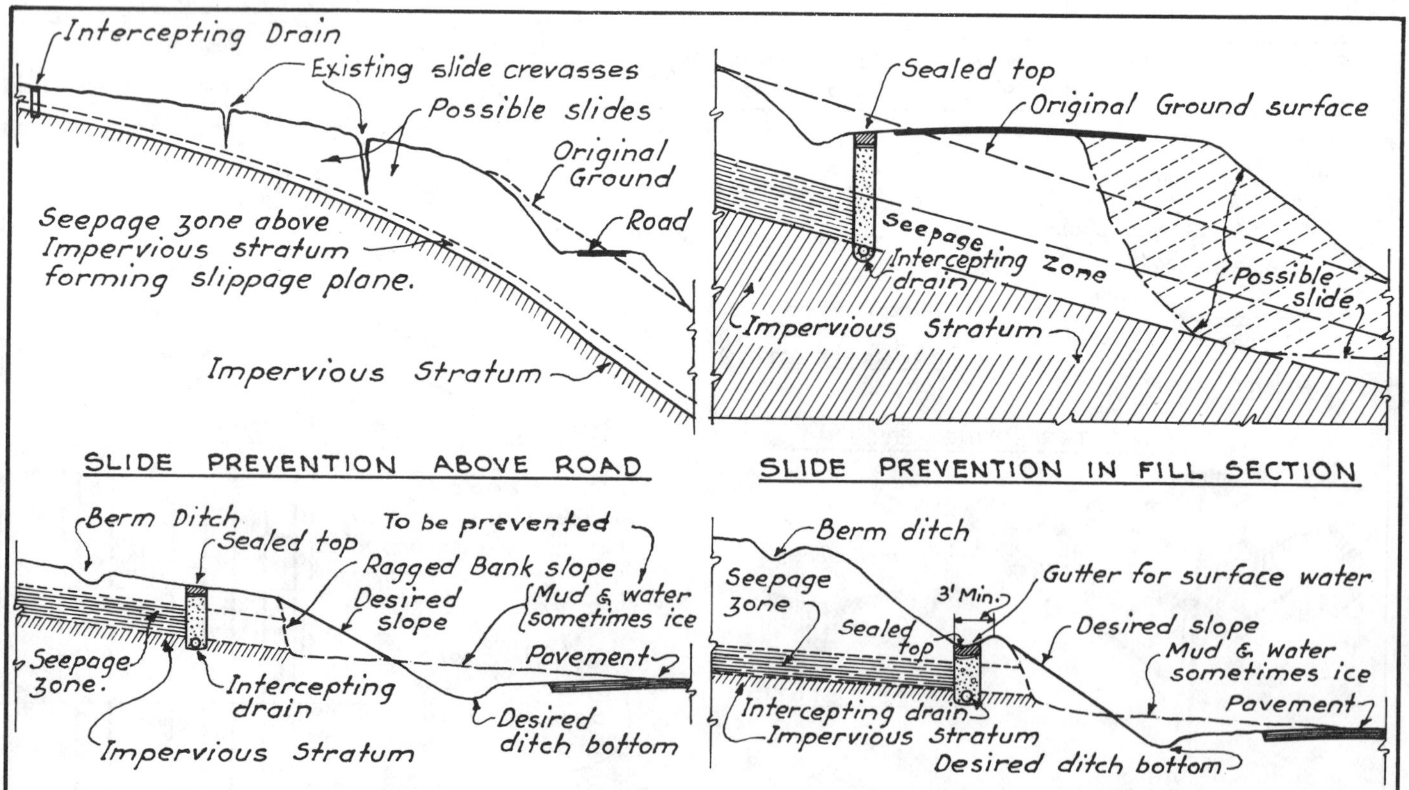

Intercepting Drain
Existing slide crevasses
Possible slides
Original Ground
Road
Seepage zone above Impervious stratum forming slippage plane.
Impervious Stratum

SLIDE PREVENTION ABOVE ROAD

Sealed top
Original Ground surface
Seepage Zone
Intercepting drain
Possible slide
Impervious Stratum

SLIDE PREVENTION IN FILL SECTION

Berm Ditch
Sealed top
To be prevented
Ragged Bank slope
Desired slope
Mud & water sometimes ice
Pavement
Seepage zone.
Intercepting drain
Desired ditch bottom
Impervious Stratum

CUT SLOPE STABILIZATION - LOW BANKS

Berm ditch
Seepage zone
3' Min.
Gutter for surface water
Sealed top
Desired slope
Mud & water sometimes ice
Pavement
Intercepting drain
Impervious Stratum
Desired ditch bottom

CUT SLOPE STABILIZATION - HIGH BANKS

FIG. B - INTERCEPTING SUBDRAINS.†

† Adapted from Handbook of Culvert & Drainage Practice, Armco Drainage Products Assoc.
* For additional data on Subdrainage, see section on "Soils" pg. 9-36

DRAINAGE — HEADWALLS

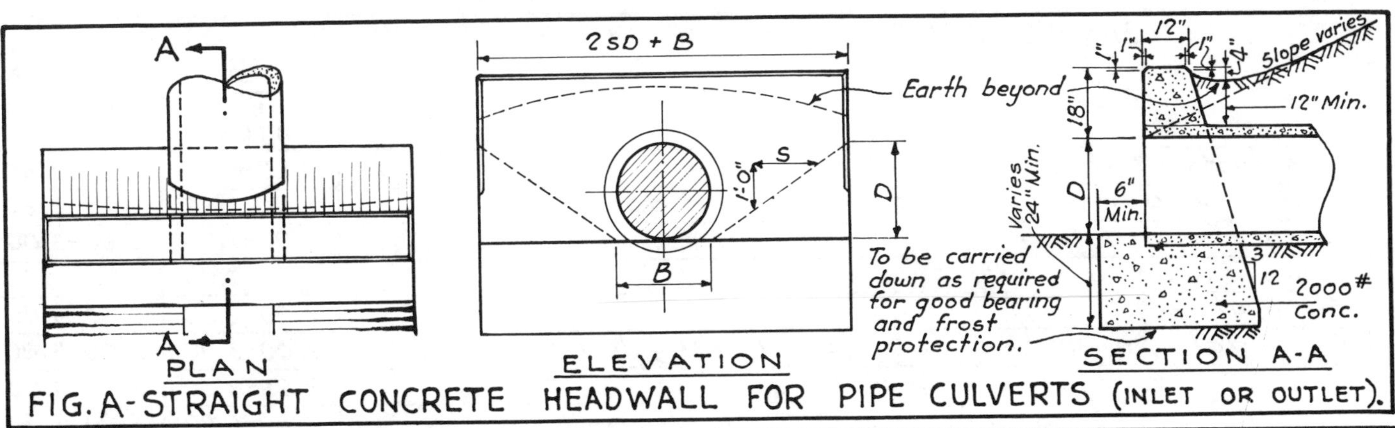

A
A
PLAN
2SD + B
S
1'-0"
D
B
ELEVATION
12"
1"
1"
1"
1"
Slope varies
Earth beyond
12" Min.
18"
D
6"
Min.
Varies 24" Min.
3
12
2000#
Conc.
To be carried down as required for good bearing and frost protection.
SECTION A-A

FIG. A-STRAIGHT CONCRETE HEADWALL FOR PIPE CULVERTS (INLET OR OUTLET).

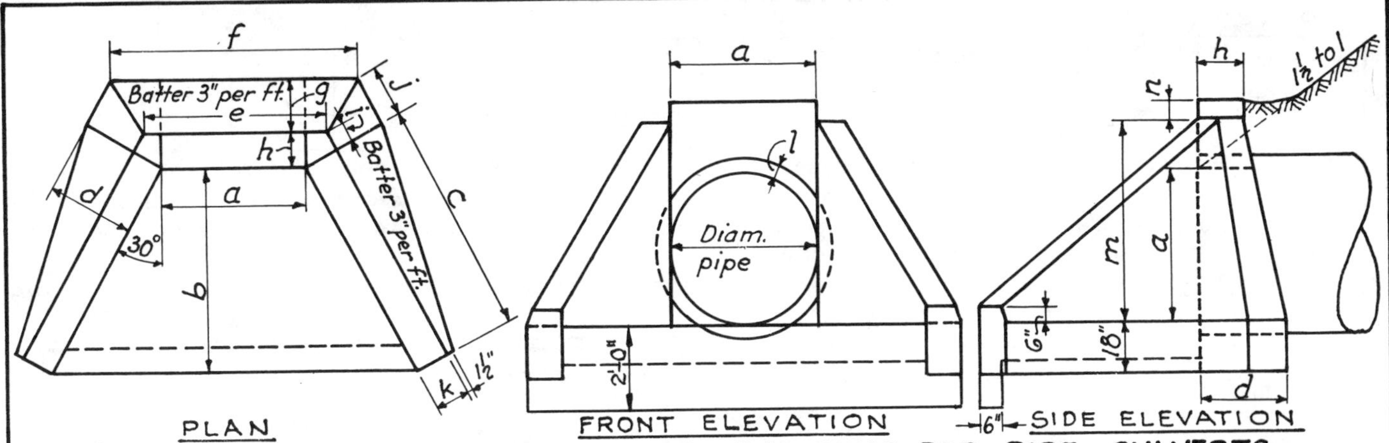

f
Batter 3" per ft.
g
e
h
i
j
Batter 3" per ft.
d
a
30°
b
c
k
1½"
PLAN
a
l
Diam. pipe
2'-0"
FRONT ELEVATION
h
n
1½ to 1
m
a
6"
18"
d
6"
SIDE ELEVATION

FIG. B- CONCRETE HEADWALL WITH WING WALLS FOR PIPE CULVERTS.

TABLE. C - DIMENSIONS FOR HEADWALLS WITH WINGWALLS IN FIG. B

DIAMETER OF PIPE	a	b	c	d	e	f	g	h	i	j	k	l	m	n
42"	42"	5'-0"	5'-9¼"	23½"	4'-5½"	5'-9"	13½"	10"	5¾"	13½"	10"	4½"	4'-6"	3½"
48"	48"	5'-6"	6'-4¼"	25"	4'-11½"	6'-5"	15"	10"	5¾"	14½"	10"	5"	5'-0"	4"
54"	54"	6'-0"	6'-11⅛"	27½"	5'-6¾"	7'-1¾"	16½"	11"	6⅜"	15⅞"	11"	5½"	5'-6"	4½"
60"	60"	6'-6"	7'-6"	30"	6'-2"	7'-10⅝"	18"	12"	7"	17⅜"	12"	6"	6'-0"	5"

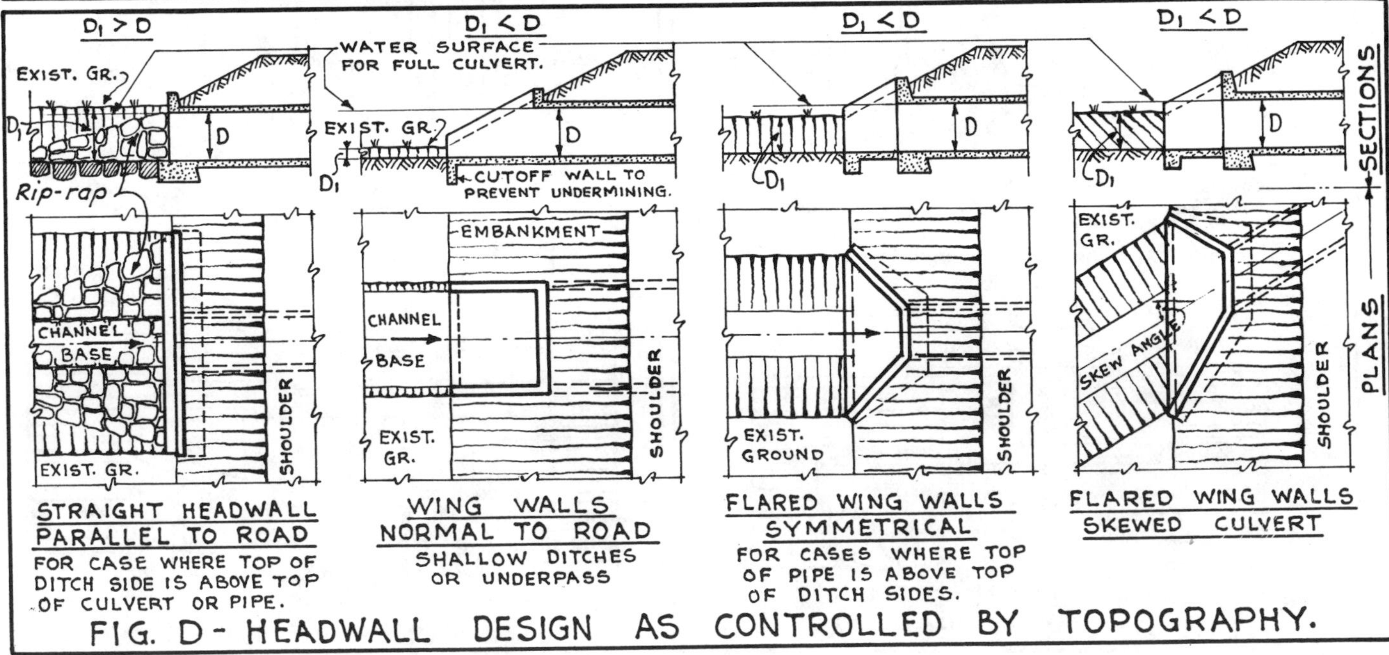

D1 > D
D1 < D
D1 < D
D1 < D
EXIST. GR.
WATER SURFACE FOR FULL CULVERT.
D1
D
EXIST. GR.
D
D1
CUTOFF WALL TO PREVENT UNDERMINING.
D
D1
D
D1
SECTIONS
Rip-rap
CHANNEL BASE
EXIST. GR.
SHOULDER
EMBANKMENT
CHANNEL BASE
EXIST. GR.
SHOULDER
EXIST. GROUND
SHOULDER
EXIST. GR.
SKEW ANGLE
SHOULDER
PLANS
STRAIGHT HEADWALL PARALLEL TO ROAD
FOR CASE WHERE TOP OF DITCH SIDE IS ABOVE TOP OF CULVERT OR PIPE.
WING WALLS NORMAL TO ROAD
SHALLOW DITCHES OR UNDERPASS
FLARED WING WALLS SYMMETRICAL
FOR CASES WHERE TOP OF PIPE IS ABOVE TOP OF DITCH SIDES.
FLARED WING WALLS SKEWED CULVERT

FIG. D- HEADWALL DESIGN AS CONTROLLED BY TOPOGRAPHY.

DRAINAGE – EROSION PROTECTION

TABLE A – MEAN VELOCITIES WHICH WILL NOT ERODE CHANNELS AFTER AGING.

MATERIAL OF CHANNEL BED	VELOCITY IN FEET PER SEC.	
	SHALLOW DITCH	DEEP CANAL *
Fine sand or silt, non-colloidal	0.50-1.50	1.50-2.50
Coarse sand or sandy loam, non-colloidal	1.00-1.50	1.75-2.50
Silty or sand loam " "	1.00-1.75	2.00-3.00
Clayey loam or sandy clay " "	1.50-2.00	2.25-3.50
Fine gravel	2.00-2.50	2.50-5.00
Colloidal clay or non-colloidal gravelly loam	2.00-3.00	3.00-5.00
Colloidal, well-graded gravel	2.25-3.50	4.00-6.00
Pebbles, broken stone, shales or hardpan	2.50-4.00	5.00-6.50
Sodded gutters (See p. 18-05)	3.00-5.00	—
Cobbled gutters, not grouted, or bituminous paving (See p. 18-05)	5.00-7.50	—
Stone masonry	7.50-15.00	—
Solid rock or concrete	15.00-25.00	—

AGING OF CHANNELS increases resistance to erosion as density and stability of channel bed improve due to deposit of silt in interstices and as cohesion increases due to cementation of soil by colloids. New channels may be safely operated at less than maximum design velocities by use of temporary check structures.

VELOCITIES are to be reduced for depths of flow under 6 inches and for water which may transport abrasive materials.

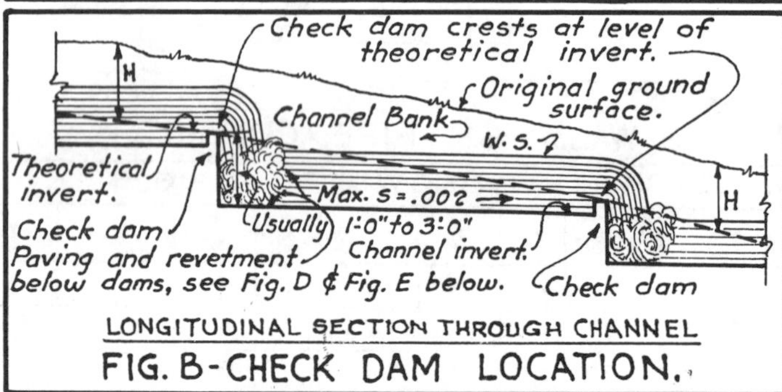

Check dam crests at level of theoretical invert.
Original ground surface.
H
Channel Bank
W.S.
Theoretical invert.
Max. S = .002
H
Check dam
Usually 1'-0" to 3'-0"
Channel invert.
Paving and revetment below dams, see Fig. D & Fig. E below.
Check dam
LONGITUDINAL SECTION THROUGH CHANNEL

FIG. B - CHECK DAM LOCATION.

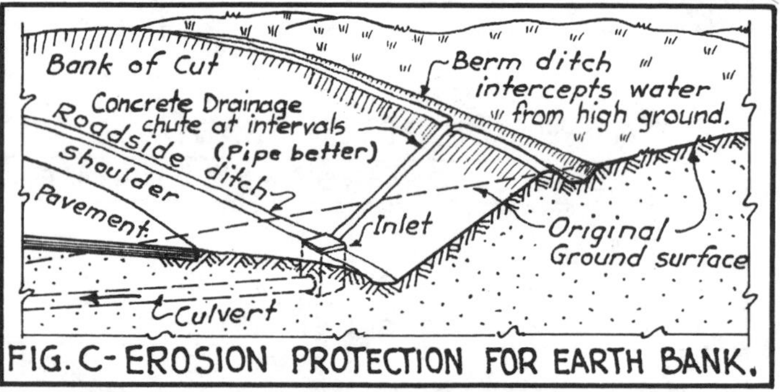

Bank of Cut
Berm ditch intercepts water from high ground.
Concrete Drainage chute at intervals
Roadside ditch
(Pipe better)
Shoulder
Pavement
Inlet
Original Ground surface
Culvert

FIG. C - EROSION PROTECTION FOR EARTH BANK.

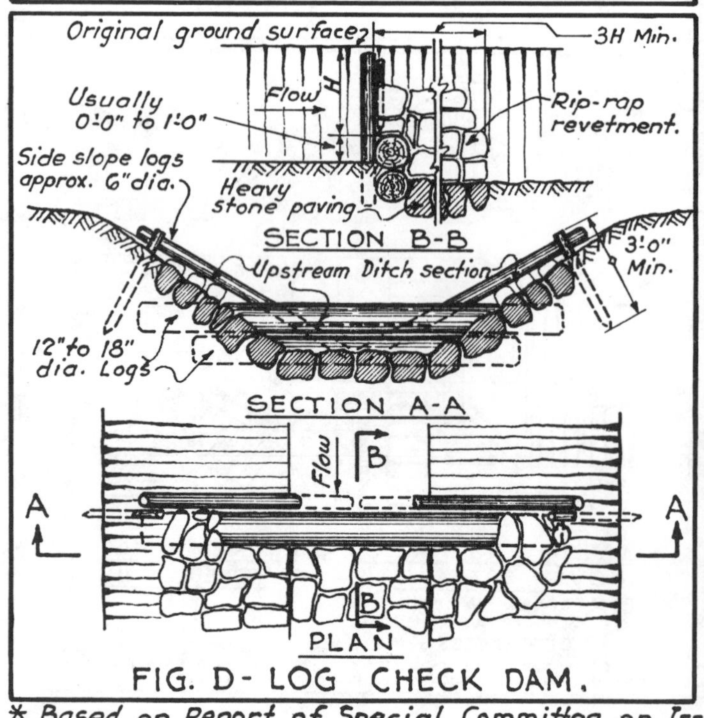

Original ground surface
3H Min.
Usually 0'-0" to 1'-0"
Flow
H
Rip-rap revetment.
Side slope logs approx. 6" dia.
Heavy stone paving
SECTION B-B
3'-0" Min.
Upstream Ditch section
12" to 18" dia. Logs
SECTION A-A
Flow
B
A
A
B
PLAN

FIG. D - LOG CHECK DAM.

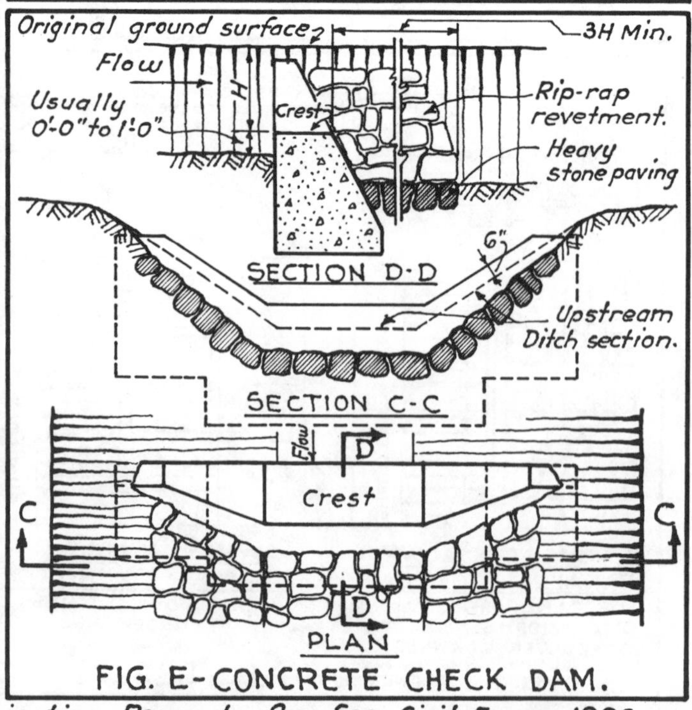

Original ground surface
3H Min.
Flow
Usually 0'-0" to 1'-0"
H
Crest
Rip-rap revetment.
Heavy stone paving
SECTION D-D
6"
Upstream Ditch section.
SECTION C-C
Flow
D
Crest
C
C
D
PLAN

FIG. E - CONCRETE CHECK DAM.

* Based on Report of Special Committee on Irrigation Research, Am. Soc. Civil Engrs. 1926.

DRAINAGE — AIRPORTS — 1 *

GENERAL: EACH SITE REQUIRES A SPECIAL DRAINAGE SOLUTION. AMOUNT OF RUNOFF, FINISHED GRADES, TOPOGRAPHY, SOIL AND GROUND WATER LEVEL ARE FACTORS. SURFACE DRAINAGE SYSTEMS, FIGURES A,B AND C BELOW, SHOULD ALWAYS BE PROVIDED TO REMOVE RAINFALL. WATER REMAINING ON THE SURFACE WOULD BE A HAZARD TO PLANE MOVEMENTS AND WOULD SATURATE THE SOIL, CAUSING POSSIBLE FROST DAMAGE AND BASE INSTABILITY; SUBDRAINAGE SYSTEMS MAY BE REQUIRED TO LOWER THE EXISTING GROUND WATER LEVEL TO SUCH DEPTH THAT BASE STABILITY AND FROST PROTECTION WILL RESULT; SURFACE DRAINAGE SYSTEMS SHOULD NEVER DISCHARGE INTO SUB-DRAINAGE SYSTEM.

DESIGN: SEE PAGES 18-02, 03, 04 FOR METHODS OF COMPUTING RUNOFF AND BASIC DATA REQUIRED FOR COMPUTATIONS.

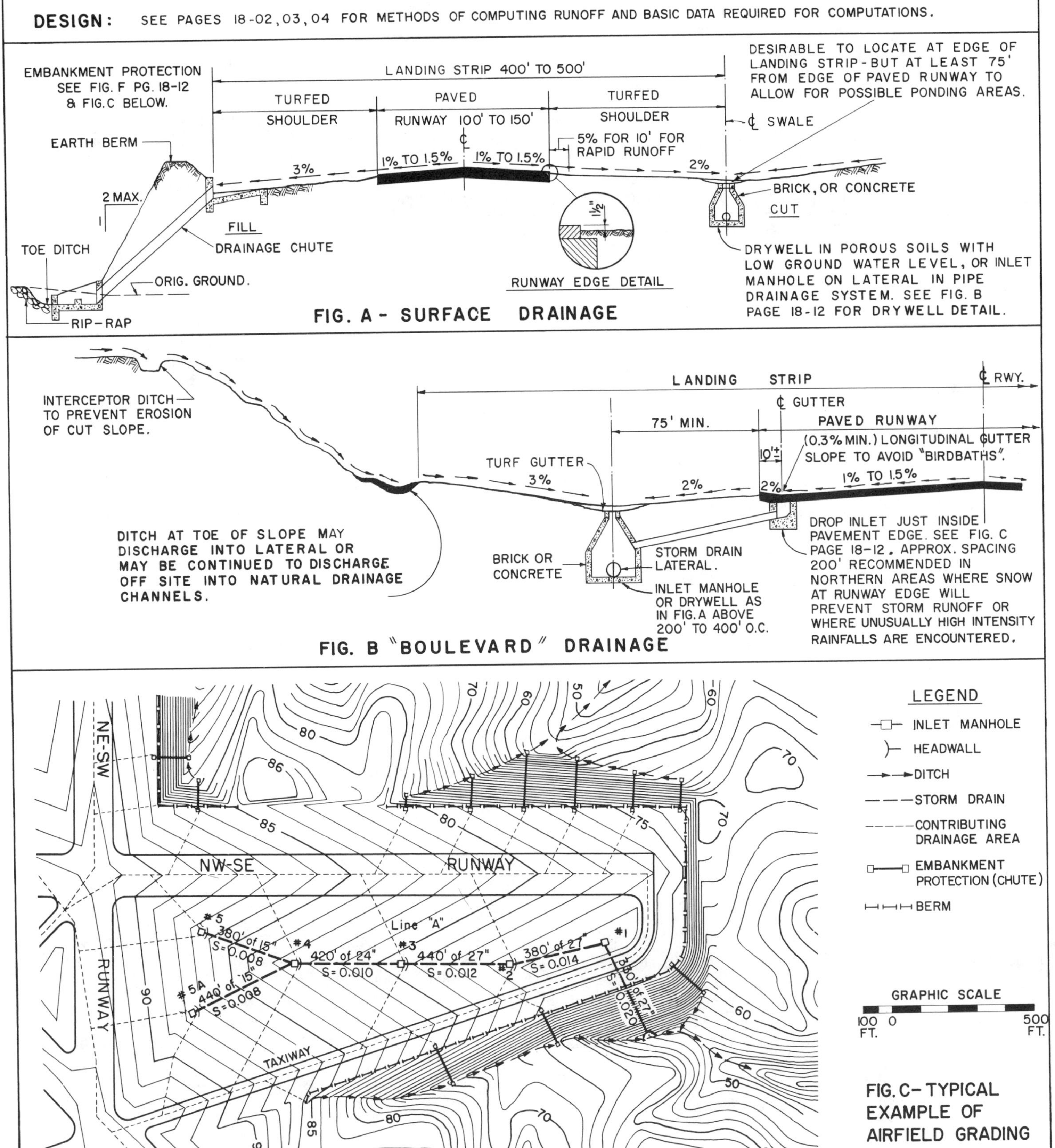

EMBANKMENT PROTECTION SEE FIG. F PG. 18-12 & FIG. C BELOW.
LANDING STRIP 400' TO 500'
TURFED SHOULDER
PAVED RUNWAY 100' TO 150'
TURFED SHOULDER
DESIRABLE TO LOCATE AT EDGE OF LANDING STRIP-BUT AT LEAST 75' FROM EDGE OF PAVED RUNWAY TO ALLOW FOR POSSIBLE PONDING AREAS.
℄ SWALE
EARTH BERM
5% FOR 10' FOR RAPID RUNOFF
3%
1% TO 1.5%
1% TO 1.5%
2%
BRICK, OR CONCRETE
CUT
2 MAX.
1
FILL
1½"
TOE DITCH
DRAINAGE CHUTE
ORIG. GROUND.
RIP-RAP
RUNWAY EDGE DETAIL
DRYWELL IN POROUS SOILS WITH LOW GROUND WATER LEVEL, OR INLET MANHOLE ON LATERAL IN PIPE DRAINAGE SYSTEM. SEE FIG. B PAGE 18-12 FOR DRYWELL DETAIL.

FIG. A - SURFACE DRAINAGE

INTERCEPTOR DITCH TO PREVENT EROSION OF CUT SLOPE.
LANDING STRIP
℄ RWY.
℄ GUTTER
75' MIN.
PAVED RUNWAY
10'±
(0.3% MIN.) LONGITUDINAL GUTTER SLOPE TO AVOID "BIRDBATHS".
TURF GUTTER
3%
2%
2%
1% TO 1.5%
DITCH AT TOE OF SLOPE MAY DISCHARGE INTO LATERAL OR MAY BE CONTINUED TO DISCHARGE OFF SITE INTO NATURAL DRAINAGE CHANNELS.
BRICK OR CONCRETE
STORM DRAIN LATERAL.
INLET MANHOLE OR DRYWELL AS IN FIG. A ABOVE 200' TO 400' O.C.
DROP INLET JUST INSIDE PAVEMENT EDGE. SEE FIG. C PAGE 18-12. APPROX. SPACING 200' RECOMMENDED IN NORTHERN AREAS WHERE SNOW AT RUNWAY EDGE WILL PREVENT STORM RUNOFF OR WHERE UNUSUALLY HIGH INTENSITY RAINFALLS ARE ENCOUNTERED.

FIG. B "BOULEVARD" DRAINAGE

NE-SW
RUNWAY
NW-SE
RUNWAY
TAXIWAY
Line "A"
#5
380' of 15" S=0.008
#4
420' of 24" S=0.010
#3
440' of 27" S=0.012
#2
380' of 27" S=0.014
#1
380' of 27" S=0.020
#5A
440' of 15" S=0.008
80
86
85
80
75
70
60
50
90
LEGEND
INLET MANHOLE
HEADWALL
DITCH
STORM DRAIN
CONTRIBUTING DRAINAGE AREA
EMBANKMENT PROTECTION (CHUTE)
BERM
GRAPHIC SCALE
100 FT.
0
500 FT.

FIG. C– TYPICAL EXAMPLE OF AIRFIELD GRADING & DRAINAGE.

* ADAPTED FROM C.A.A. AIRPORT DRAINAGE NOV. 1956

DRAINAGE — AIRPORTS — 2

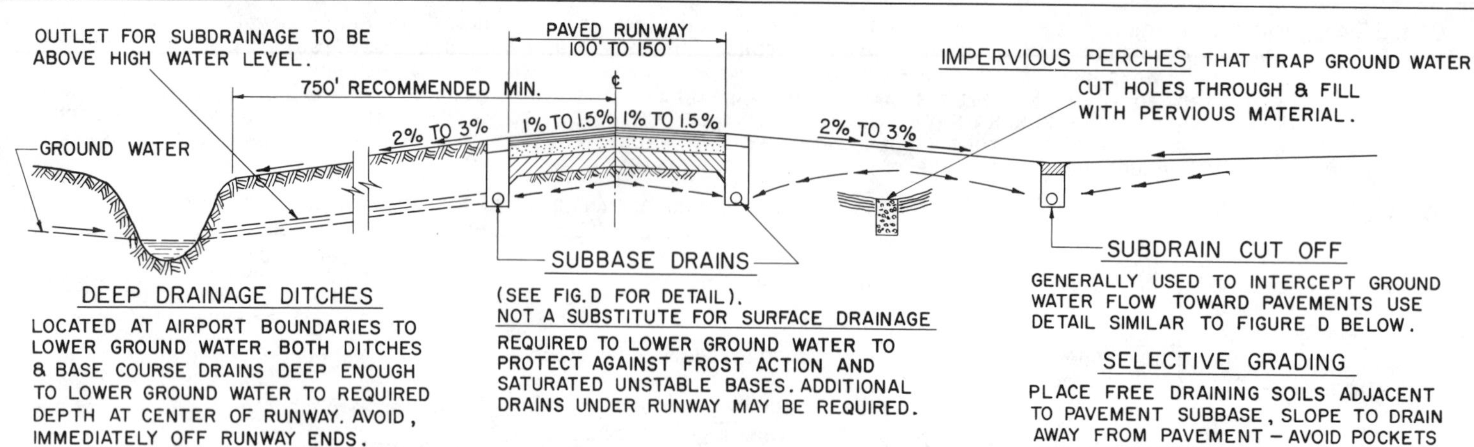
OUTLET FOR SUBDRAINAGE TO BE ABOVE HIGH WATER LEVEL.
PAVED RUNWAY 100' TO 150'
750' RECOMMENDED MIN.
¢
GROUND WATER
2% TO 3%
1% TO 1.5%
1% TO 1.5%
2% TO 3%
IMPERVIOUS PERCHES THAT TRAP GROUND WATER.
CUT HOLES THROUGH & FILL WITH PERVIOUS MATERIAL.
SUBBASE DRAINS
SUBDRAIN CUT OFF

DEEP DRAINAGE DITCHES

LOCATED AT AIRPORT BOUNDARIES TO LOWER GROUND WATER. BOTH DITCHES & BASE COURSE DRAINS DEEP ENOUGH TO LOWER GROUND WATER TO REQUIRED DEPTH AT CENTER OF RUNWAY. AVOID, IMMEDIATELY OFF RUNWAY ENDS.

(SEE FIG. D FOR DETAIL).
NOT A SUBSTITUTE FOR SURFACE DRAINAGE

REQUIRED TO LOWER GROUND WATER TO PROTECT AGAINST FROST ACTION AND SATURATED UNSTABLE BASES. ADDITIONAL DRAINS UNDER RUNWAY MAY BE REQUIRED.

GENERALLY USED TO INTERCEPT GROUND WATER FLOW TOWARD PAVEMENTS USE DETAIL SIMILAR TO FIGURE D BELOW.

SELECTIVE GRADING

PLACE FREE DRAINING SOILS ADJACENT TO PAVEMENT SUBBASE, SLOPE TO DRAIN AWAY FROM PAVEMENT – AVOID POCKETS IN IMPERVIOUS SOILS.

FIG. A – SUBDRAINAGE TO LOWER GROUND WATER.

SEE PAGES 18–07 & 18–08 FOR SUBDRAINAGE DESIGN DATA.

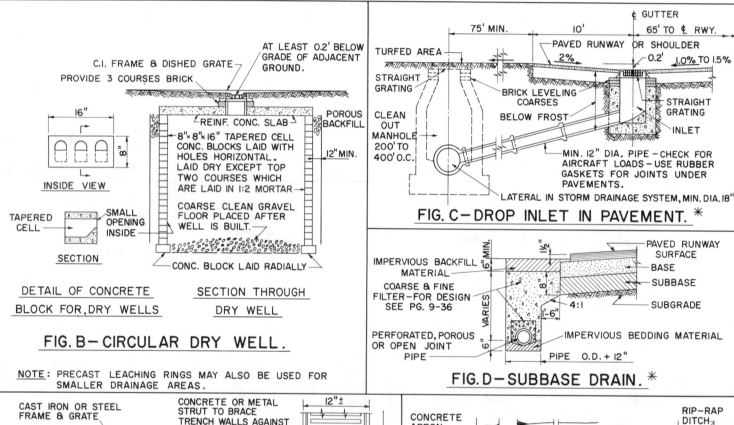

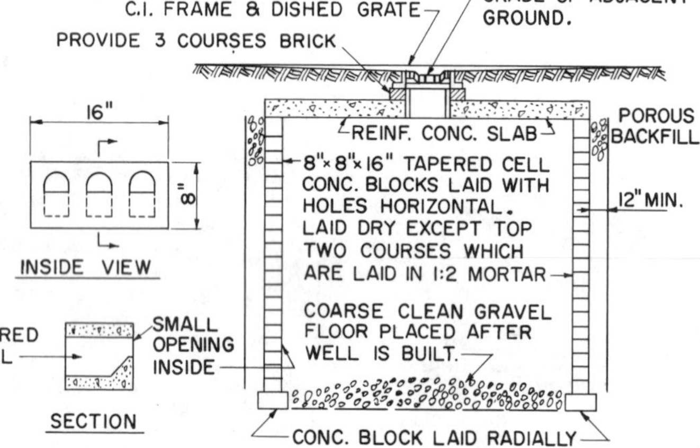
AT LEAST 0.2' BELOW GRADE OF ADJACENT GROUND.
C.I. FRAME & DISHED GRATE
PROVIDE 3 COURSES BRICK
16"
8"
INSIDE VIEW
REINF. CONC. SLAB
POROUS BACKFILL
8"×8"×16" TAPERED CELL CONC. BLOCKS LAID WITH HOLES HORIZONTAL. LAID DRY EXCEPT TOP TWO COURSES WHICH ARE LAID IN 1:2 MORTAR
12" MIN.
COARSE CLEAN GRAVEL FLOOR PLACED AFTER WELL IS BUILT.
TAPERED CELL
SMALL OPENING INSIDE
SECTION
CONC. BLOCK LAID RADIALLY

DETAIL OF CONCRETE BLOCK FOR DRY WELLS

SECTION THROUGH DRY WELL

FIG. B – CIRCULAR DRY WELL.

NOTE: PRECAST LEACHING RINGS MAY ALSO BE USED FOR SMALLER DRAINAGE AREAS.

¢ GUTTER
75' MIN. 10' 65' TO ¢ RWY.
TURFED AREA
PAVED RUNWAY OR SHOULDER
2%
0.2'
1.0% TO 1.5%
STRAIGHT GRATING
BRICK LEVELING COARSES
BELOW FROST
STRAIGHT GRATING
INLET
CLEAN OUT MANHOLE 200' TO 400' O.C.
MIN. 12" DIA. PIPE – CHECK FOR AIRCRAFT LOADS – USE RUBBER GASKETS FOR JOINTS UNDER PAVEMENTS.
LATERAL IN STORM DRAINAGE SYSTEM, MIN. DIA. 18"

FIG. C – DROP INLET IN PAVEMENT. *

IMPERVIOUS BACKFILL MATERIAL
6" MIN.
1½"
PAVED RUNWAY SURFACE
BASE
SUBBASE
SUBGRADE
COARSE & FINE FILTER – FOR DESIGN SEE PG. 9-36
8"
4:1
1'-6"
VARIES
PERFORATED, POROUS OR OPEN JOINT PIPE
6"
IMPERVIOUS BEDDING MATERIAL
PIPE O.D. + 12"

FIG. D – SUBBASE DRAIN. *

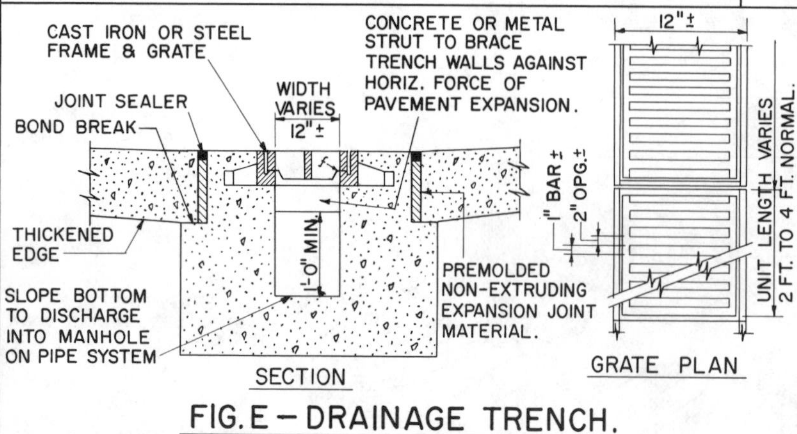
CAST IRON OR STEEL FRAME & GRATE
CONCRETE OR METAL STRUT TO BRACE TRENCH WALLS AGAINST HORIZ. FORCE OF PAVEMENT EXPANSION.
12"±
JOINT SEALER
BOND BREAK
WIDTH VARIES
12"±
THICKENED EDGE
1'-0" MIN.
1" BAR ±
2" OPG. ±
UNIT LENGTH VARIES 2 FT. TO 4 FT. NORMAL.
SLOPE BOTTOM TO DISCHARGE INTO MANHOLE ON PIPE SYSTEM
PREMOLDED NON-EXTRUDING EXPANSION JOINT MATERIAL.
SECTION
GRATE PLAN

FIG. E – DRAINAGE TRENCH.

FOR USE ONLY WHERE WARPING SURFACES WILL NOT GIVE SUFFICIENT SLOPES FOR RUNOFF AND TO ELIMINATE THE "WAFFLE" AND "W" DRAINAGE SECTIONS IN LARGE APRONS WHICH CAUSE UNDESIRABLE FLEXING OF AIRCRAFT WINGS DURING TAXIING OPERATIONS SOMEWHAT MORE EXPENSIVE THAN CONVENTIONAL SYSTEM.

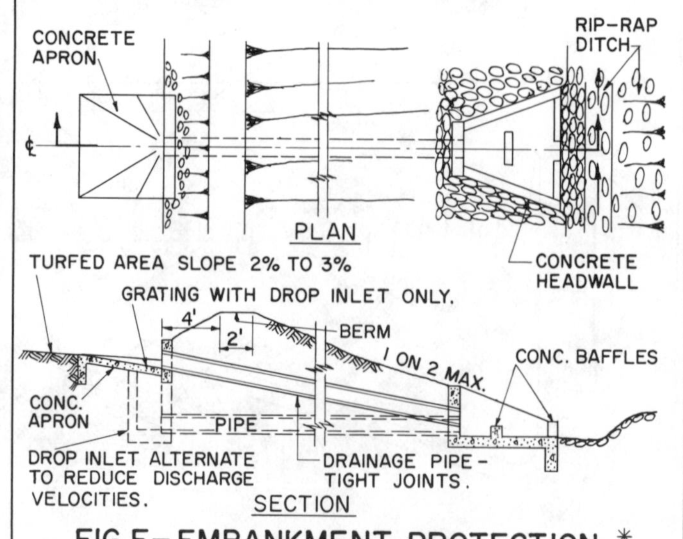
CONCRETE APRON
RIP-RAP DITCH
¢
PLAN
CONCRETE HEADWALL
TURFED AREA SLOPE 2% TO 3%
GRATING WITH DROP INLET ONLY.
4'
2'
BERM
1 ON 2 MAX.
CONC. BAFFLES
CONC. APRON
PIPE
DROP INLET ALTERNATE TO REDUCE DISCHARGE VELOCITIES.
DRAINAGE PIPE – TIGHT JOINTS.
SECTION

FIG. F – EMBANKMENT PROTECTION. *
(DRAINAGE CHUTE)

* ADAPTED FROM C.A.A. AIRPORT DRAINAGE, NOV. 1956.

DRAINAGE–UNIT HYDROGRAPH & FLOOD ROUTING–I

FOR DETERMINING RUNOFF CHARACTERISTICS AND DESIGN OF SPILLWAY

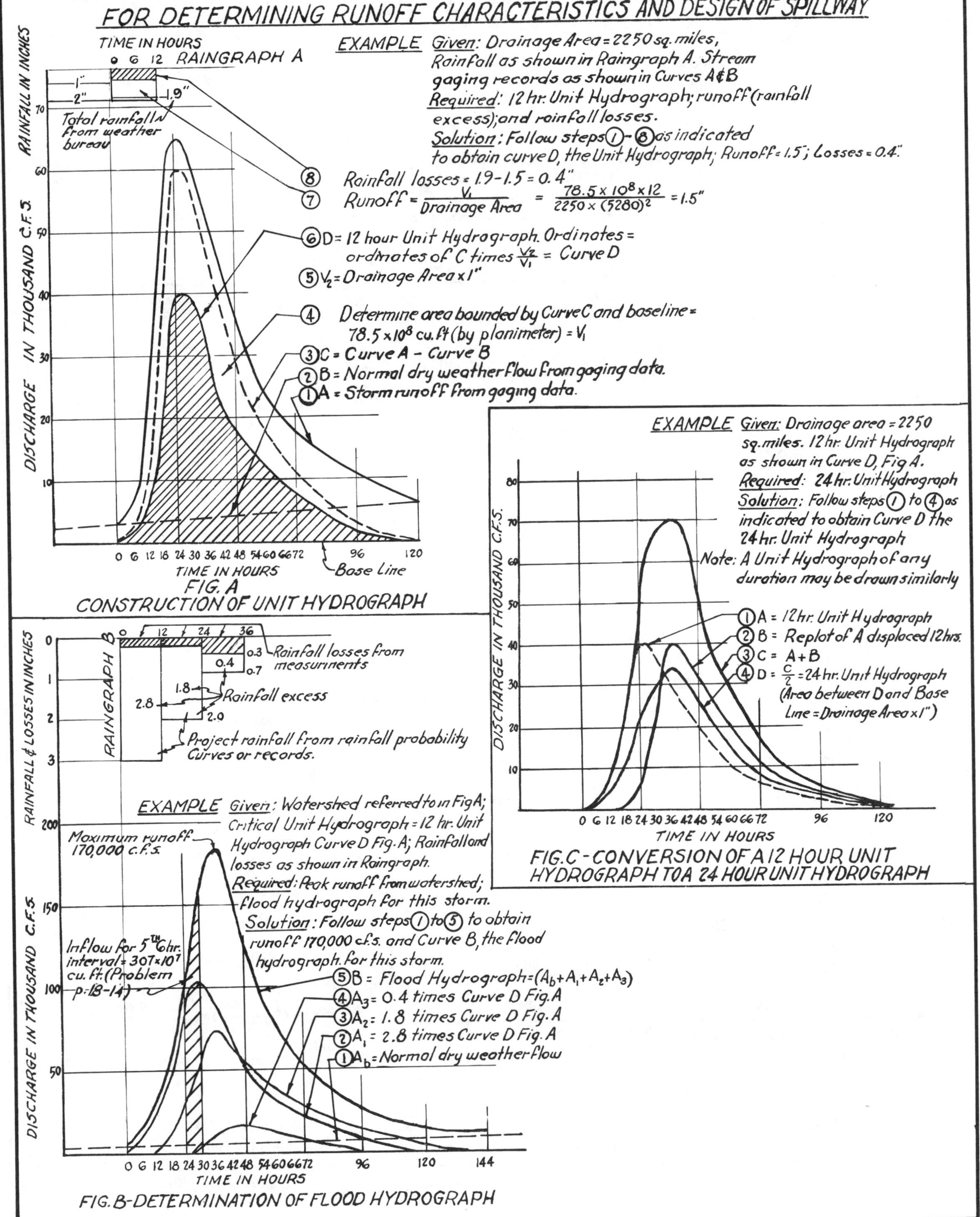

TIME IN HOURS
0 6 12 RAINGRAPH A
RAINFALL IN INCHES
1"
2"
1.9"
Total rainfall from weather bureau
EXAMPLE Given: Drainage Area = 2250 sq. miles, Rainfall as shown in Raingraph A. Stream gaging records as shown in Curves A & B
Required: 12 hr. Unit Hydrograph; runoff (rainfall excess); and rainfall losses.
Solution: Follow steps (1)-(8) as indicated to obtain curve D, the Unit Hydrograph; Runoff = 1.5"; Losses = 0.4"
(8) Rainfall losses = 1.9 - 1.5 = 0.4"
(7) Runoff = V1 / Drainage Area = 78.5 x 10^8 x 12 / 2250 x (5280)^2 = 1.5"
(6) D = 12 hour Unit Hydrograph. Ordinates = ordinates of C times V2/V1 = Curve D
(5) V2 = Drainage Area x 1"
(4) Determine area bounded by Curve C and baseline = 78.5 x 10^8 cu. ft (by planimeter) = V1
(3) C = Curve A - Curve B
(2) B = Normal dry weather flow from gaging data.
(1) A = Storm runoff from gaging data.
DISCHARGE IN THOUSAND C.F.S.
0 6 12 18 24 30 36 42 48 54 60 66 72 96 120
TIME IN HOURS
Base Line
FIG. A
CONSTRUCTION OF UNIT HYDROGRAPH
EXAMPLE Given: Drainage area = 2250 sq. miles. 12 hr. Unit Hydrograph as shown in Curve D, Fig A.
Required: 24 hr. Unit Hydrograph
Solution: Follow steps (1) to (4) as indicated to obtain Curve D the 24 hr. Unit Hydrograph
Note: A Unit Hydrograph of any duration may be drawn similarly
(1) A = 12 hr. Unit Hydrograph
(2) B = Replot of A displaced 12 hrs.
(3) C = A + B
(4) D = C/2 = 24 hr. Unit Hydrograph (Area between D and Base Line = Drainage Area x 1")
DISCHARGE IN THOUSAND C.F.S.
0 6 12 18 24 30 36 42 48 54 60 66 72 96 120
TIME IN HOURS
FIG. C - CONVERSION OF A 12 HOUR UNIT HYDROGRAPH TO A 24 HOUR UNIT HYDROGRAPH
RAINFALL & LOSSES IN INCHES
RAINGRAPH B
0 12 24 36
0.3
0.4
0.7
Rainfall losses from measurements
1.8
2.8
2.0
Rainfall excess
Project rainfall from rainfall probability Curves or records.
EXAMPLE Given: Watershed referred to in Fig A; Critical Unit Hydrograph = 12 hr. Unit Hydrograph Curve D Fig. A; Rainfall and losses as shown in Raingraph.
Required: Peak runoff from watershed; flood hydrograph for this storm.
Solution: Follow steps (1) to (5) to obtain runoff 170,000 c.f.s. and Curve B, the flood hydrograph for this storm.
Maximum runoff 170,000 c.f.s.
Inflow for 5th 6 hr. interval = 307 x 10^7 cu. ft. (Problem p. 18-14)
(5) B = Flood Hydrograph = (Ab + A1 + A2 + A3)
(4) A3 = 0.4 times Curve D Fig. A
(3) A2 = 1.8 times Curve D Fig. A
(2) A1 = 2.8 times Curve D Fig. A
(1) Ab = Normal dry weather flow
DISCHARGE IN THOUSAND C.F.S.
0 6 12 18 24 30 36 42 48 54 60 66 72 96 120 144
TIME IN HOURS
FIG. B-DETERMINATION OF FLOOD HYDROGRAPH

DRAINAGE—UNIT HYDROGRAPH & FLOOD ROUTING-2

EXAMPLE Given 1. Flood Hydrograph as shown on Curve B, Fig. B, p. 18-13
2. Dam spillway 1500′ long with crest at elev. 500 and spillway coefficient = 3.5
3. Area of lake above elev. 500 as shown in col. 3 table A.

Required Maximum discharge over spillway and maximum lake elevation.

Solution 1. Fill in table A
2. Plot Curves of Fig B
3. Plot inflow for 1st 6hr. interval on Fig. B from Fig. B p. 18-13 = 10.8×10^7 cu. ft. and work up successively as shown for 5th 6 hr. interval finding (a) Outflow at end of each interval (b) Height above spillway at end of each interval.

TABLE A - COMPUTATIONS FOR PLOTTING FIG. B

LAKE ELEV. FT.	H HEADOVER SPILLWAY FT.	LAKE AREA SQ. MILES	AVERAGE LAKE AREA SQ MILES	STORAGE PER FT. CU.FT.$\times 10^7$	S STORAGE ABOVE ELEV. 500 CU.FT.$\times 10^7$	Q SPILLWAY DISCH. = $3.5 \times 1500 \times H^{3/2}$ C.F.S.	ΔQ SPILLWAY DISCH. FOR 6 HR. PERIOD CU.FT.$\times 10^7$	$\frac{\Delta Q}{2}$ CU.FT.$\times 10^7$	$S+\frac{\Delta Q}{2}$ CU.FT.$\times 10^7$	$S-\frac{\Delta Q}{2}$ CU.FT.$\times 10^7$
500	0	15								
501	1	16	15 5	43.3	43.3	5,200	11.4	5.7	49.0	37.6
502	2	18	17.0	47.4	90.7	14,700	31.6	15.8	106.5	74.9
503	3	21	19.5	54.4	145.1	27,300	58.9	29.5	174.6	115.6
504	4	26	23.5	65.5	210.6	42,000	90.9	40.5	251.1	170.1

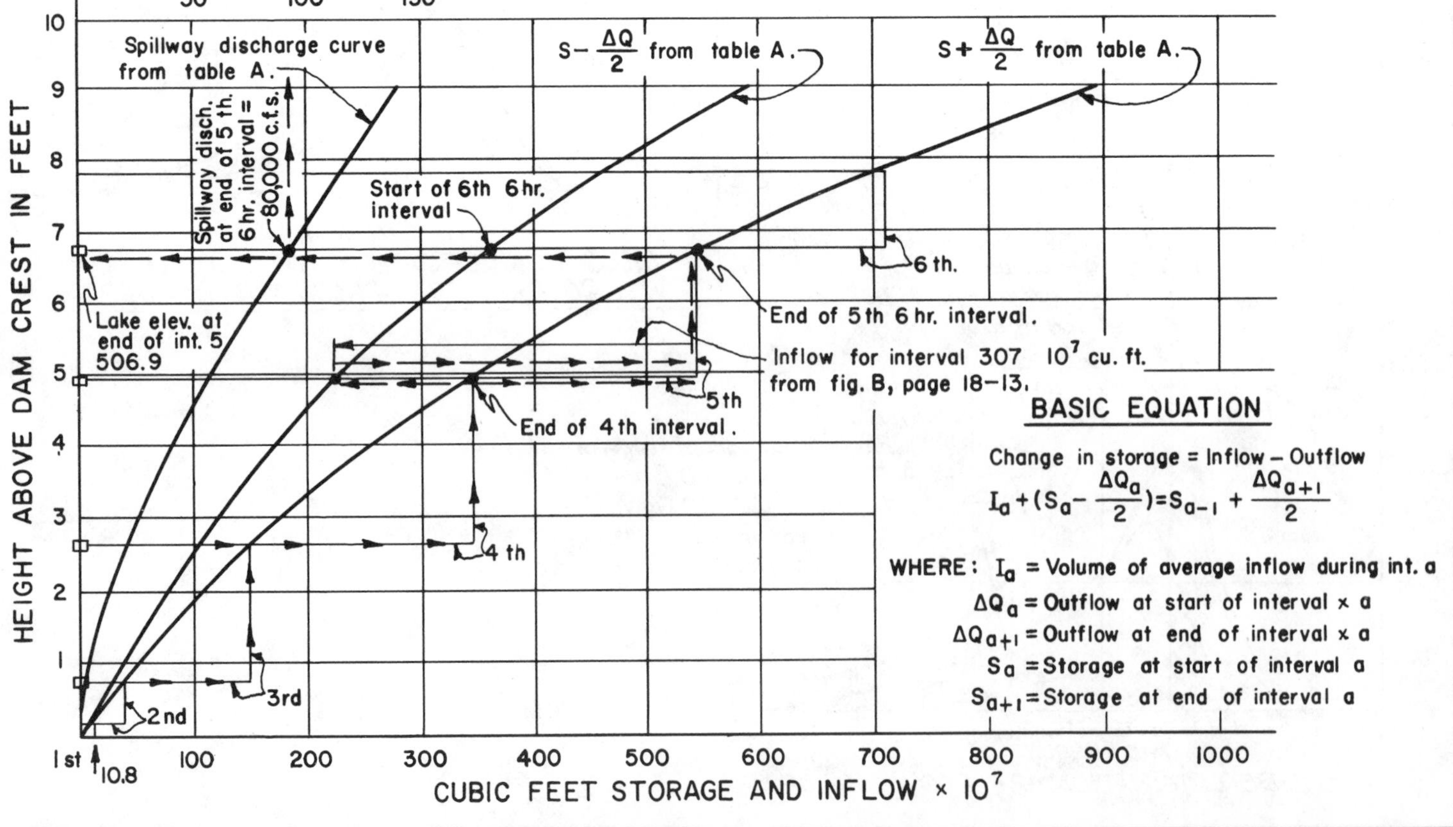

DISCHARGE IN THOUSAND C.F.S.
50
100
150
HEIGHT ABOVE DAM CREST IN FEET
Spillway discharge curve from table A.
S − ΔQ/2 from table A.
S + ΔQ/2 from table A.
Spillway disch. at end of 5 th. 6 hr. interval = 80,000 c.f.s.
Start of 6th 6 hr. interval
6 th.
Lake elev. at end of int. 5 506.9
End of 5th 6 hr. interval.
Inflow for interval 307 10⁷ cu. ft. from fig. B, page 18–13.
5th
End of 4th interval.
4 th
3rd
2nd
1st
10.8
CUBIC FEET STORAGE AND INFLOW × 10⁷
BASIC EQUATION
Change in storage = Inflow − Outflow
$I_a + (S_a - \frac{\Delta Q_a}{2}) = S_{a-1} + \frac{\Delta Q_{a+1}}{2}$
WHERE: I_a = Volume of average inflow during int. a
ΔQ_a = Outflow at start of interval × a
ΔQ_{a+1} = Outflow at end of interval × a
S_a = Storage at start of interval a
S_{a+1} = Storage at end of interval a

DRAINAGE & SEWERAGE — MANHOLES

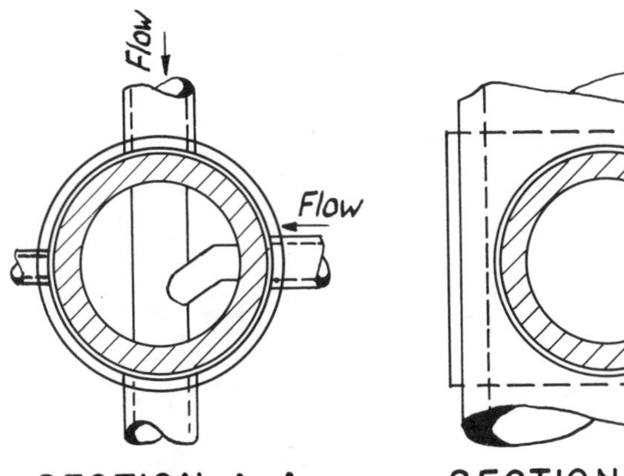
Flow
Flow

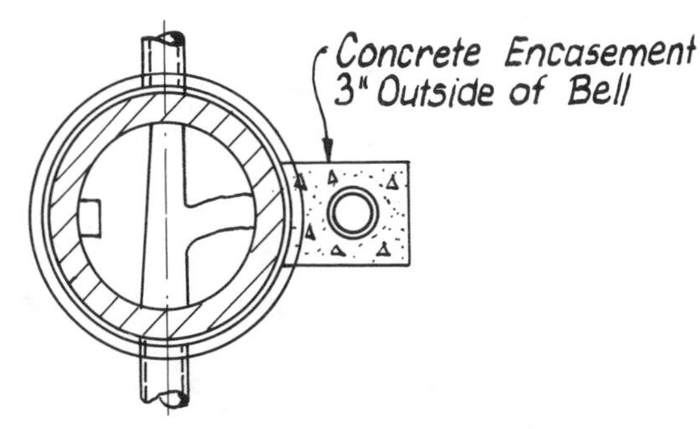
Concrete Encasement
3" Outside of Bell

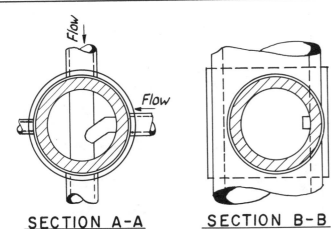

SECTION A-A

SECTION B-B

SECTION C-C

Note: Walls to be 8 inches thick to 13 foot depth, walls to be increased to 12" inch thickness in section below 13 foot depth. Walls constructed preferably of brick. Where use of brick is uneconomical, concrete radial block may be used

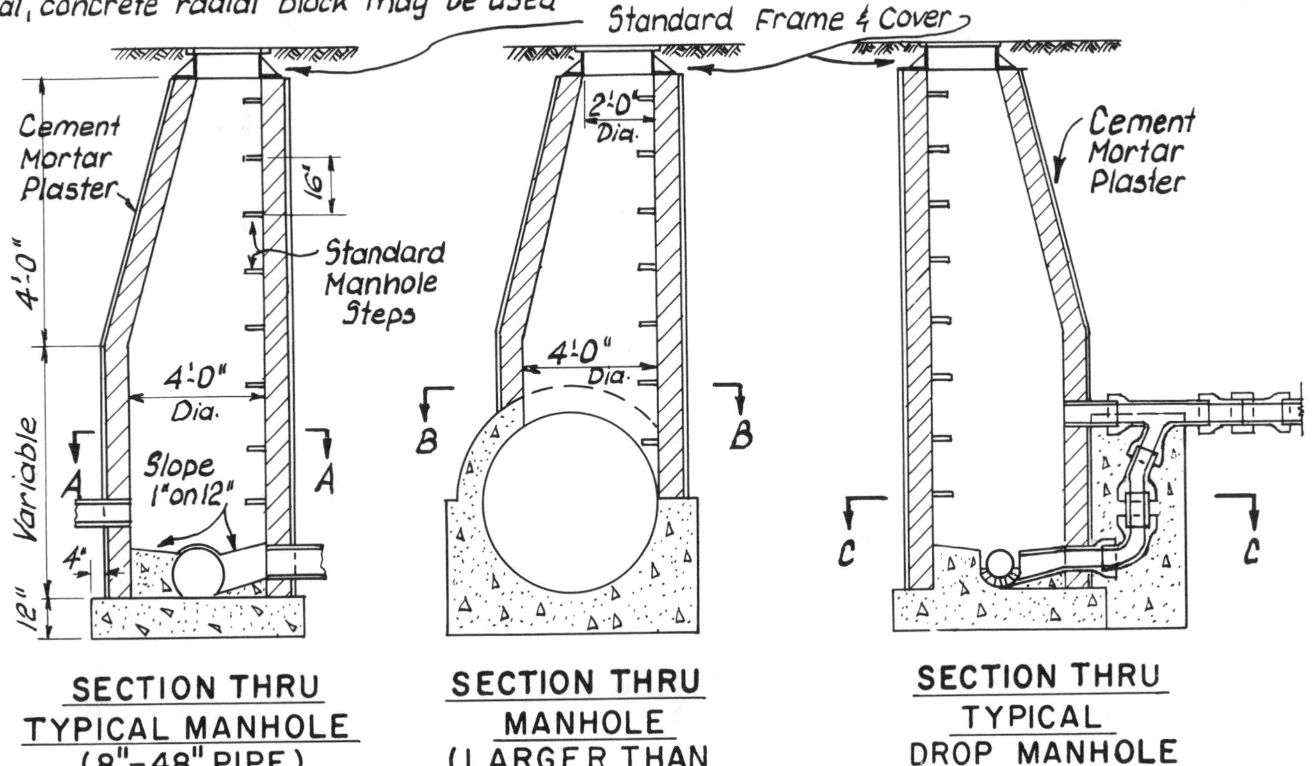
Standard Frame & Cover
Cement Mortar Plaster
4'-0"
16"
Standard Manhole Steps
4'-0" Dia.
Variable
Slope 1" on 12"
4"
12"
A
A
2'-0" Dia.
4'-0" Dia.
B
B
Cement Mortar Plaster
C
C

SECTION THRU TYPICAL MANHOLE (8"–48" PIPE)

SECTION THRU MANHOLE (LARGER THAN 4'-0" DIA. PIPE)

SECTION THRU TYPICAL DROP MANHOLE

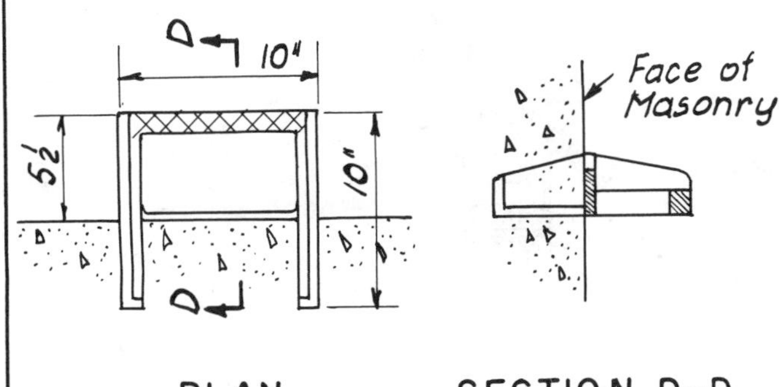
D
10"
5½
10"
D
Face of Masonry

PLAN

SECTION D-D

CAST IRON STEP

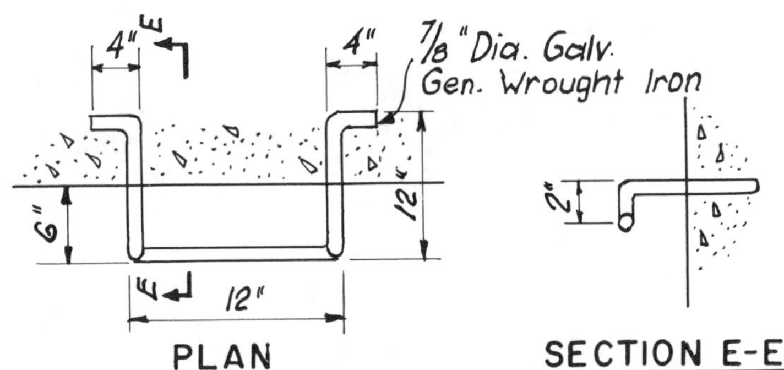
4"
E
4"
⅞" Dia. Galv. Gen. Wrought Iron
6"
12"
E
12"
2"

PLAN

SECTION E-E

WROUGHT IRON STEP

MANHOLE STEPS

DRAINAGE & SEWERAGE — INLETS & CATCH BASINS

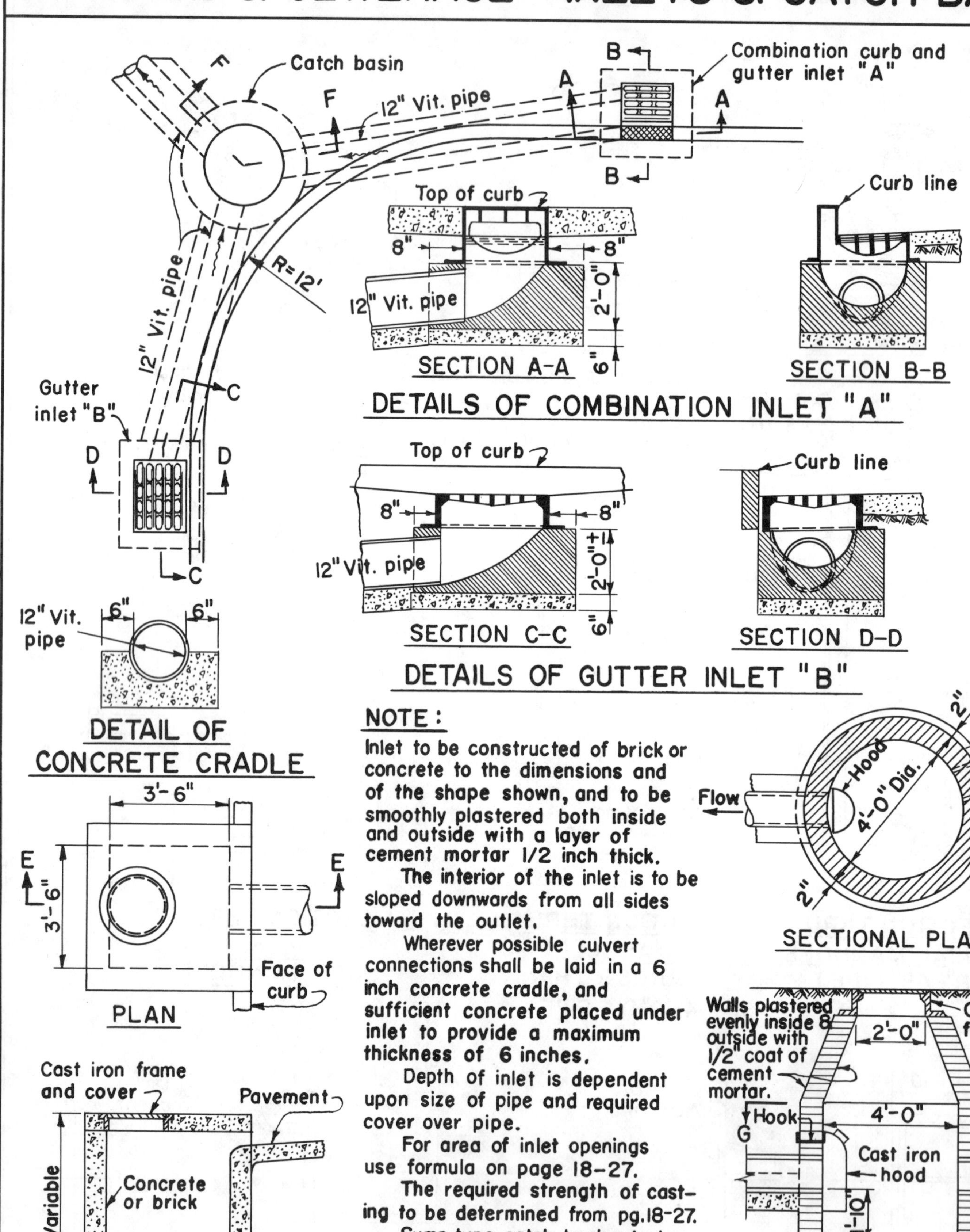

Catch basin
F
F
12" Vit. pipe
A
A
B
B
Combination curb and gutter inlet "A"
R=12'
12" Vit. pipe
Gutter inlet "B"
C
C
D
D
Top of curb
8"
8"
12" Vit. pipe
2'-0"
6"
SECTION A-A
Curb line
SECTION B-B
DETAILS OF COMBINATION INLET "A"
Top of curb
8"
8"
12" Vit. pipe
2'-0"±
6"
SECTION C-C
Curb line
SECTION D-D
DETAILS OF GUTTER INLET "B"
12" Vit. pipe
6"
6"
DETAIL OF CONCRETE CRADLE
3'-6"
3'-6"
E
E
Face of curb
PLAN
Cast iron frame and cover
Pavement
Concrete or brick
Variable
6"
SECTION E-E
DETAIL OF CURB INLET
2"
Flow
Flow
Hood
4'-0" Dia.
2"
SECTIONAL PLAN G-G
Walls plastered evenly inside & outside with 1/2" coat of cement mortar.
2'-0"
Cast iron frame and cover
Brick
Hook
4'-0"
G
G
Cast iron hood
Variable
1'-10"
1'-6"
1'-4"
12"
SECTION F-F
TYPICAL CATCH BASIN

NOTE:

Inlet to be constructed of brick or concrete to the dimensions and of the shape shown, and to be smoothly plastered both inside and outside with a layer of cement mortar 1/2 inch thick.

The interior of the inlet is to be sloped downwards from all sides toward the outlet.

Wherever possible culvert connections shall be laid in a 6 inch concrete cradle, and sufficient concrete placed under inlet to provide a maximum thickness of 6 inches.

Depth of inlet is dependent upon size of pipe and required cover over pipe.

For area of inlet openings use formula on page 18–27.

The required strength of casting to be determined from pg.18–27.

Sump type catch basins to be used only where labor is available for frequent cleaning and where stagnant water is not objectionable. They may be required where flat grades would cause silting up of drainage system.

DRAINAGE & SEWERAGE–MANHOLE, INLET & CATCH BASIN CASTINGS

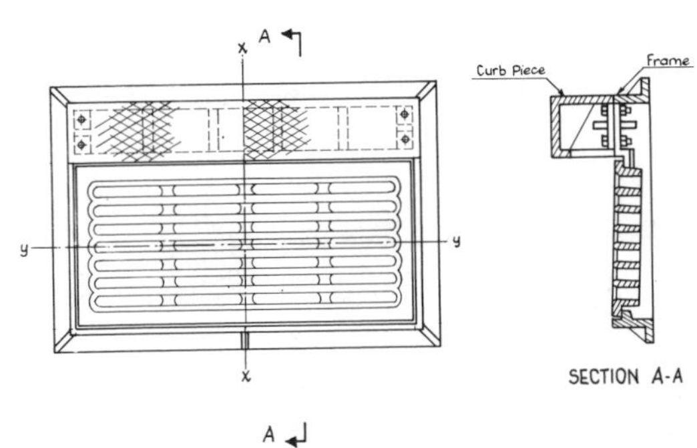
Curb Piece
Frame
SECTION A-A

TYPE 1 – CURB INLET. HIGH-VELOCITY CURB FLOW.

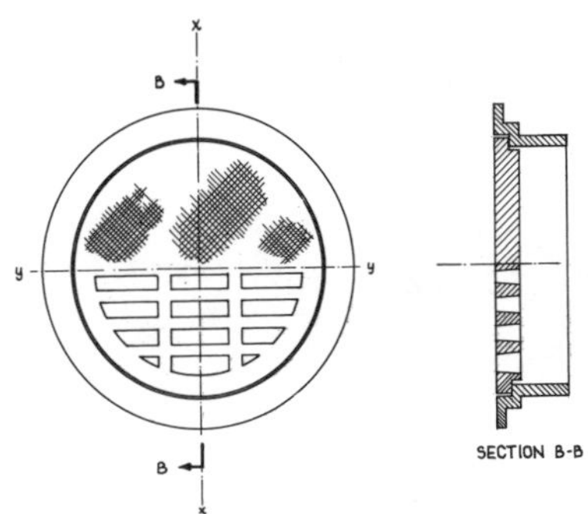
SECTION B-B

TYPE 2 – CIRCULAR PONDING AREA.

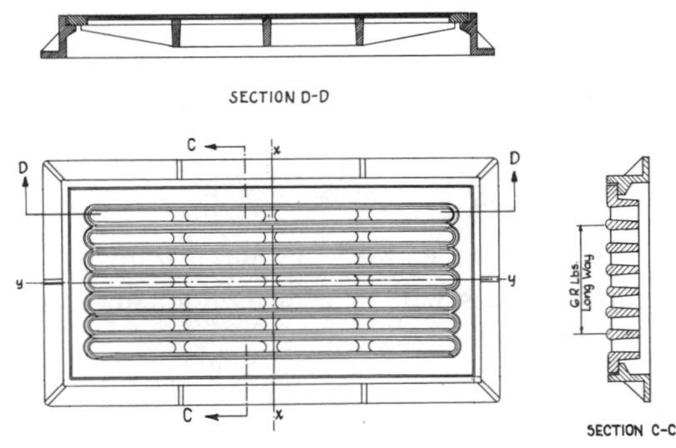
SECTION D-D
G R Lbs Long Way
SECTION C-C

TYPE 3 – RECTANGULAR PONDING. UNEVEN DIRECT FLOW.

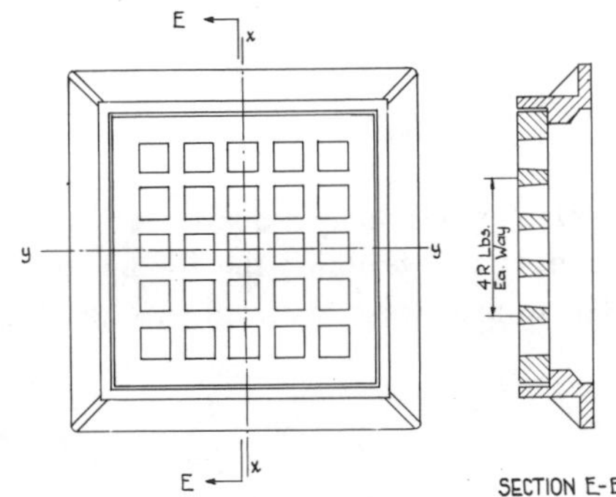
4 R Lbs Ea. Way
SECTION E-E

TYPE 4 – SQUARE PONDING. EVEN DIRECT FLOW.

FIG. C – TYPICAL MANHOLE INLET AND CATCH BASIN DETAILS

TABLE A – COMMONLY USED TYPES

Type	Size of Grating for Types 1 & 3, Frame Opening for Types 2 & 4, inches	Ratio or Number of Ribs Each Way	Area of Grate Opening, sq. in.	Required Section Modulus, cu. in. of Ribs Only, Not Including Edge of Cover: 75-ton Load 37.5 kips per Tire (200 p.s.i.)		H.20 Load 16 kips per Tire (85 p.s.i.)		7 kips per Tire (80 p.s.i.)	
				Sx	Sy	Sx	Sy	Sx	Sy
1	24 × 18¾	5 × 1	145	–	–	15.1	–	8.9	–
	29¾ × 21¾	6 × 2	210	–	–	25.5	–	13.9	–
	35 × 21¾	6 × 2	240	–	–	32.8	–	17.2	–
	47¾ × 21¾	6 × 3	400	–	–	49.5	–	24.4	–
	49½ × 29¾	9 × 3	630	–	–	78	–	39	–
	47¾ × 43½	12 × 3	800	–	–	99	–	49	–
2	21½ Diam.	3½ : 1		22.8	6.3	9.8	2.7	4.9	1.4
	24 Diam.	3½ : 1		26.5	7.4	11.5	3.2	5.6	1.6
	30 Diam.	3½ : 1		35.7	9.8	15.2	4.2	7.3	2.0
	36 Diam.	1 : 1		–	–	17.1	17.1	9.1	9.1
	42 Diam.	1 : 1		–	–	21.4	21.4	11.3	11.3
3	11¾ × 22¾	3 × 1	100	–	–	4.7	2.2	2.3	1.1
	18 × 24	5 × 1	185	–	–	10.2	2.0	5.1	1.0
	21¾ × 35¾	6 × 2	273	–	–	12.2	4.7	5.7	2.4
	26 × 38	7 × 2	366	–	–	15.5	5.1	7.2	2.5
4	26 × 26	1 : 1		20.0	20.0	8.5	8.5	5.0	5.0
	32 × 32	1 : 1		28.0	28.0	11.4	11.4	7.1	7.1
	38 × 38	1 : 1		–	–	16.4	16.4	9.0	9.0
	44 × 44	1 : 1		–	–	21.1	21.1	11.1	11.1
	50 × 50	1 : 1		–	–	25.5	25.5	13.5	13.5

Design bending stress for class 30 or better cast iron = 6000 p.s.i. (For less than class 30, use 3000 p.s.i.).
Required section modulus for net section equals total width of cover minus holes if any. Required section moduli for other loads are approximately proportional. For impact add 50%.

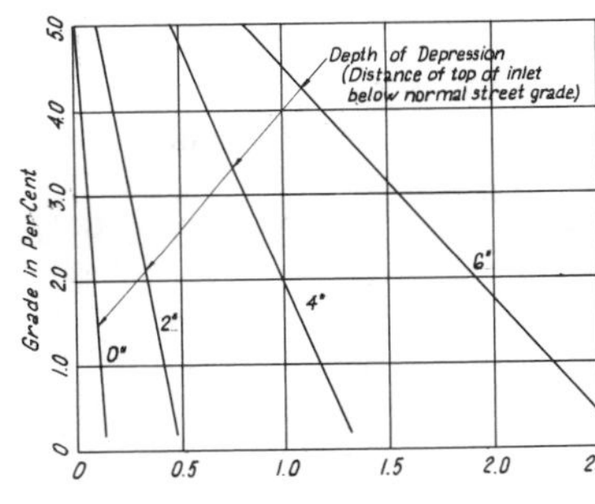
Depth of Depression
(Distance of top of inlet below normal street grade)
0"
2"
4"
6"
Grade in Per Cent
0
1.0
2.0
3.0
4.0
5.0
0
0.5
1.0
1.5
2.0
2.5
Quantity of Water in Cu. Ft. per Sec.

Data for an inlet 4'-6" long and 6" high with water just lapping past. For other size openings the capacities are approximately in proportion.

FIG. B – RELATION BETWEEN STREET GRADES AND GUTTER INLET CAPACITY FOR VARIOUS DEPTHS OF DEPRESSION.

FORMULA FOR DETERMINING THE AREA OF INLET GRATING OPENINGS:

$$Q = cA\sqrt{2gh} \times \text{factor for clogging (Use } 2/3\text{)},$$

where Q = quantity of runoff reaching inlet in cubic feet per second
c = orifice coefficient 0.6 for openings, with square edges. (0.8 for openings with rounded edges)
A = net area in square feet
g = 32.2
h = allowable head on inlet in feet

DRAINAGE & SEWERAGE-LOADING ON PIPES-1

INTRODUCTION

It is suggested by the author that three-edge bearing strength design be used for any corrugated metal pipe, instead of manufacturer's recommended height of cover, when either:

1. Height of fill exceeds 30 feet, or
2. Pipe diameter is 48 inches or larger.

For heights between 30 and 50 feet and/or pipes 48 inches in diameter or greater, on account of scour and corrosion, the author prefers a reinforced-concrete pipe. Above 50 feet we suggest a reinforced-concrete structural culvert.

For all pipes, first-class bedding should be specified.

For rigid pipe of 48 inches diameter and over or with a fill greater than 30 feet, a concrete cradle should be specified. The use of concrete cradles for rigid pipe will depend on the nature of the soil and foundation. In soft soils, concrete cradles should be used for rigid pipe.

Suggestion for corrugated metal pipe on soft soil:

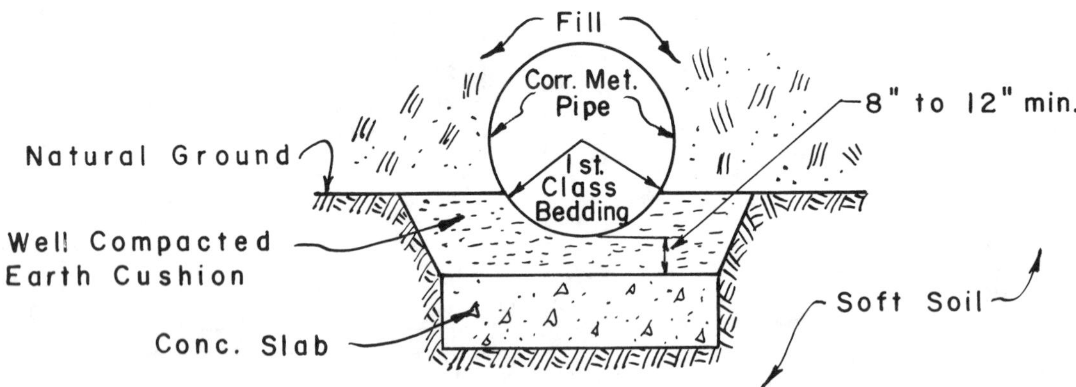
Fill
Corr. Met. Pipe
8" to 12" min.
Natural Ground
1st. Class Bedding
Well Compacted Earth Cushion
Soft Soil
Conc. Slab

For poor soils, it may be necessary to use piles or other forms of support. Care should be taken to avoid relative settlement where the soil adjacent to the pipe settles more than that above the pipe, as this will tend to form a heavy, wedge-shaped load on the pipe.

The above condition and also the similar situation which arises when a rigid pipe is overloaded due to relative settlement of fill, may be allowed for with the use of the imperfect trench. This consists of bringing the fill up to grade and then re-excavating over the pipe, replacing this material with compressible fill to produce a favorable arched loading.

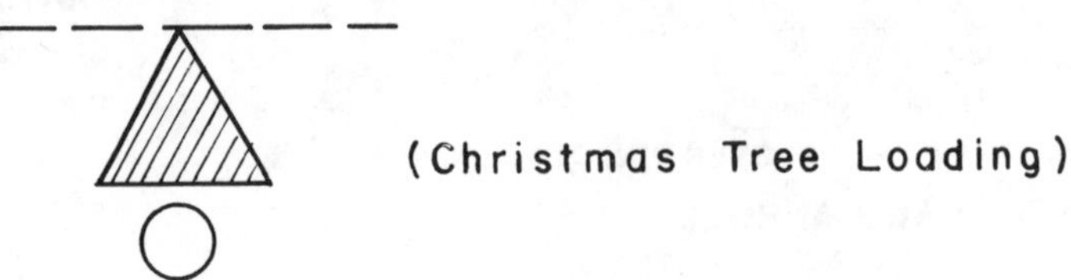
(Christmas Tree Loading)

DRAINAGE & SEWERAGE – LOADING ON RIGID PIPES-2

TABLE A- REQUIRED 3 EDGE BEARING STRENGTH (0.01 INCH CRACK) LBS. PER LIN. FT. OF PIPE FOR VARIOUS DEPTH UNDER H-20 HIGHWAY LOADING.

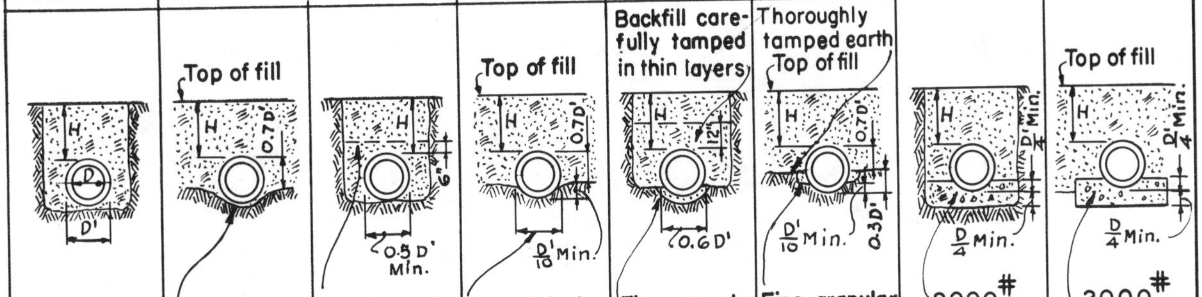

Depth of Cover over pipe (H) (Feet)		BEDDING CONDITIONS							
		IMPERMISSIBLE		ORDINARY		FIRST CLASS		CONCRETE CRADLE	
		Ditch Condition	Projecting Condition	Ditch Condition	Projecting Condition	Ditch Condition	Projecting Condition	Ditch Condition	Projecting Condition
		Backfill untamped.	Top of fill; Not shaped to fit pipe.	0.5 D' Min.; Granular materials, shovel placed & tamped	Top of fill; D'/10 Min.; Accurately shaped to fit pipe.	Backfill carefully tamped in thin layers; 0.6 D'; Fine granular fill.	Thoroughly tamped earth; Top of fill; D'/10 Min.; 0.3D'; Fine granular fill, accurately shaped to fit pipe	D/4 Min.; 2000# Concrete	Top of fill; D/4 Min.; 2000# Concrete
2	DL	275D	275D	180D	180D	150D	150D	115D	115D
	LL	1,735D	1,735D	1,110D	1,250D	945D	1,110D	720D	720D
3	DL	410D	415D	265D	265D	220D	220D	170D	170D
	LL	1,155D	1,155D	740D	840D	625D	745D	480D	480D
4	DL	545D	625D	350D	380D	300D	310D	230D	230D
	LL	850D	775D	545D	615D	460D	545D	350D	350D
5	DL	680D	840D	440D	515D	370D	450D	285D	285D
	LL	660D	595D	420D	480D	360D	430D	275D	275D
6	DL	815D	1,020D	520D	620D	440D	505D	340D	340D
	LL	540D	500D	345D	390D	290D	345D	220D	220D
7	DL	955D	1,280D	610D	790D	515D	650D	400D	425D
	LL	440D	405D	285D	320D	240D	280D	180D	160D
8	DL	1,090D	1,480D	700D	910D	590D	780D	450D	495D
	LL	385D	355D	245D	280D	210D	245D	160D	120D
9	DL	1,230D	1,720D	790D	1,060D	665D	875D	510D	580D
	LL	335D	310D	215D	245D	180D	215D	140D	105D
10	DL	1,370D	1,975D	880D	1,210D	740D	990D	570D	670D
	LL	290D	270D	185D	210D	155D	185D	120D	90D
12	DL	1,640D	2,380D	1,030D	1,480D	885D	1,220D	680D	820D
	LL	230D	210D	145D	165D	120D	145D	95D	70D
14	DL	1,910D	2,840D	1,225D	1,760D	1,030D	1,445D	790D	975D
	LL	180D	170D	115D	130D	100D	115D	75D	60D
16	DL	2,180D	3,320D	1,400D	2,050D	1,175D	1,695D	905D	1,150D
	LL	150D	140D	100D	110D	80D	95D	60D	50D
18	DL	2,450D		1,570D	2,320D	1,320D	1,910D	1,020D	1,300D
	LL	120D		80D	90D	70D	80D	50D	40D
20	DL	2,670D		1,750D	2,580D	1,470D	2,110D	1,130D	1,435D
	LL	100D		80D	60D	60D	70D	45D	35D
25	DL	2,950D		2,160D	3,290D	1,710D	2,710D	1,300D	1,840D
	LL	65D		50D	40D	40D	40D	30D	20D
30	DL	3,160D		2,330D		1,845D	3,270D	1,400D	2,230D
	LL	50D		40D		30D	30D	20D	20D
35	DL			2,450D		1,940D		1,470D	2,600D
	LL			30D		20D		20D	10D
40	DL			2,500D		1,970D		1,515D	2,840D
	LL			30D		20D		10D	10D
45	DL								3,200D
	LL								—

DRAINAGE & SEWERAGE—LOADING ON RIGID PIPES-3

TABLE A—REQUIRED 3 EDGE BEARING STRENGTH (0.01 INCH CRACK) LBS. PER LIN. FT. OF PIPE FOR VARIOUS DEPTH UNDER H-20 HIGHWAY LOADING.

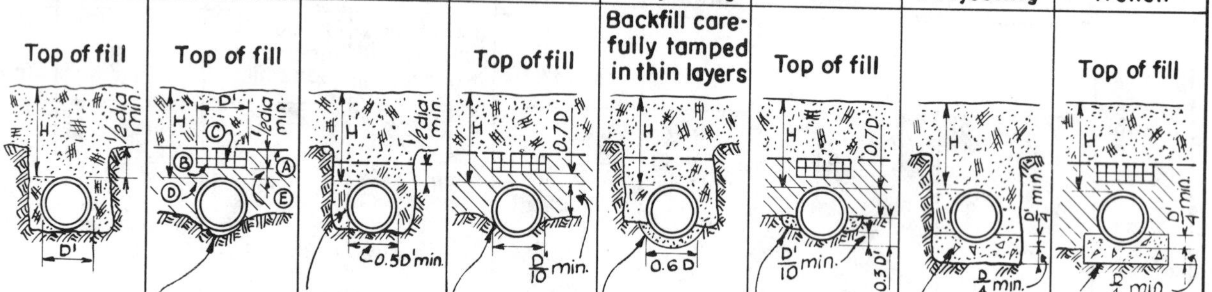

Depth of Cover over pipe (H) (Feet)		BEDDING CONDITIONS							
		IMPERMISSIBLE		ORDINARY		FIRST CLASS		CONCRETE CRADLE	
		Negative Projecting	Imperfect Trench	Negative Projecting	Imperfect Trench	Negative Projecting	Imperfect Trench	Negative Projecting	Imperfect Trench
		Top of fill	Top of fill; Not shaped to fit pipe.	Granular materials, shovel placed & tamped; 0.5D' min.	Top of fill; Accurately shaped to fit pipe; D'/10 min.	Backfill carefully tamped in thin layers; Fine granular fill.; 0.6 D	Top of fill; Fine granular fill.; D'/10 min.	2000# Concrete; D/4 min.	Top of fill; 2000# Concrete; D/4 min.
18	DL	1,780D	1,570D						
	LL	120D	110D						
20	DL	1,930D	1,720D						
	LL	100D	90D						
25	DL	2,450D	2,110D						
	LL	70D	55D						
30	DL	2,970D	2,560D	2,170D	1,490D				
	LL	50D	45D	40D	25D				
35	DL		2,910D	2,530D	1,680D	2,000D	1,330D		
	LL		30D	30D	20D	20D	10D		
40	DL			2,910D	1,910D	2,300D	1,510D		
	LL			20D	10D	10D	—		
45	DL				2,130D	2,510D	1,690D		
50	DL				2,320D	2,830D	1,840D	2,150D	1,260D
60	DL				2,870D	3,310D	2,270D	2,520D	1,560D
70	DL				3,260D		2,580D	2,930D	1,780D
80	DL						2,910D		2,000D
90	DL						3,280D		2,250D

Ⓐ Top of stage construction
Ⓑ Trench dug in compacted fill.
Ⓒ Compressible backfill.
Ⓓ Compacted fill.
Ⓔ 6" min. for pipes from 12" ϕ to 36" ϕ, 12" min. from 36" ϕ up.

DL = dead load
LL = live load
D = inside diameter in feet

EXAMPLE:

A. *For reinforced-concrete pipe;* Assume 36" ϕ reinforced-concrete pipe, 20-ft. depth, ordinary bedding, projection condition with H-20 live load. From Table A, page 18-41 find cracking strength per linear foot. 2,580(D.L.) + 60(L.L.)] × 3 = 7,920 × f.s.
Note: For reinforced-concrete pipe, factor of safety f.s. = 1.0. For other types of pipe f.s. = 1.5.
From page 18-44, select class V reinforced-concrete pipe.

B. For any other pipe: Assume 24" ϕ pipe, 14-ft. depth, concrete cradle bedding and projection condition with H-20 live load. From Table A, page 18-41, find cracking strength per linear foot. [975(D.L.) + 60(L.L.)] × 2 = 2,070 × f.s.
Note: 1.5 is the factor of safety to be used for any kind of pipe except reinforced-concrete pipe. 2,070 × 1.5 = 3,105.
From page 18-45, select vitrified clay pipe, extra strength, or asbestos-cement pipe, extra strength, or non-reinforced concrete pipe, extra strength.
Note: Table A above, all values shown in columns are arched loading (Christmas tree loading). See page 18-41.
From Table A, page 18-43, the values shown in columns headed "Ditch Condition" are columnar or arched loading, and the values in columns headed "Projecting Condition" are wedge-shaped loadings. (See page 18-40.)

DRAINAGE & SEWERAGE—LOADING ON FLEXIBLE PIPES—4

TABLE A—REQUIRED 3 EDGE BEARING STRENGTH @ 5% DEFLECTION LBS. PER LIN. FT. OF PIPE FOR VARIOUS DEPTH UNDER H-20 HIGHWAY LOADING.

Depth of Cover over pipe (H) (Feet)		BEDDING CONDITIONS					
		IMPERMISSIBLE		ORDINARY		FIRST CLASS	
		Ditch Condition	Projecting Condition	Ditch Condition	Projecting Condition	Ditch Condition	Projecting Condition
		Backfill untamped.	Top of fill. Not shaped to fit pipe.	Granular materials, shovel placed & tamped. 0.5 D' min.	Top of fill. Accurately shaped to fit pipe. D'/10 min.	Backfill carefully tamped in thin layers. Fine granular fill. Accurately shaped to fit pipe. 0.6 D'	Thoroughly tamped earth. Top of fill. Fine granular fill. D'/10 min.
2	DL	205 D	205 D	180 D	180 D	175 D	175 D
	LL	1,470 D	1,470 D	1,300 D	1,300 D	1,260 D	1,260 D
3	DL	305 D	305 D	270 D	270 D	260 D	260 D
	LL	975 D	975 D	860 D	860 D	835 D	835 D
4	DL	395 D	395 D	350 D	350 D	340 D	340 D
	LL	715 D	715 D	635 D	635 D	610 D	610 D
5	DL	485 D	485 D	425 D	425 D	415 D	415 D
	LL	555 D	555 D	490 D	490 D	475 D	475 D
6	DL	555 D	555 D	480 D	480 D	475 D	475 D
	LL	450 D	450 D	400 D	400 D	390 D	390 D
7	DL	640 D	640 D	565 D	565 D	550 D	550 D
	LL	370 D	370 D	330 D	330 D	320 D	320 D
8	DL	700 D	700 D	620 D	620 D	600 D	600 D
	LL	320 D	320 D	285 D	285 D	275 D	275 D
9	DL	780 D	800 D	690 D	710 D	670 D	685 D
	LL	280 D	280 D	250 D	250 D	240 D	240 D
10	DL	835 D	885 D	740 D	780 D	715 D	755 D
	LL	245 D	245 D	215 D	215 D	210 D	210 D
12	DL	965 D	1,030 D	855 D	915 D	830 D	885 D
	LL	190 D	190 D	170 D	170 D	160 D	160 D
14	DL	1,055 D	1,200 D	935 D	1,060 D	905 D	1,030 D
	LL	150 D	150 D	135 D	135 D	130 D	130 D
16	DL	1,150 D	1,365 D	1,020 D	1,210 D	985 D	1,170 D
	LL	130 D	130 D	115 D	115 D	110 D	110 D
18	DL	1,250 D	1,515 D	1,110 D	1,345 D	1,070 D	1,300 D
	LL	105 D	105 D	95 D	95 D	90 D	90 D
20	DL	1,300 D	1,700 D	1,150 D	1,500 D	1,115 D	1,460 D
	LL	90 D	90 D	80 D	80 D	75 D	75 D
25	DL			1,290 D	1,860 D	1,245 D	1,795 D
	LL			50 D	50 D	45 D	45 D
30	DL			1,375 D	2,210 D	1,330 D	2,140 D
	LL			30 D	30 D	30 D	30 D

EXAMPLE : Assume 24" φ flexible pipe 14 feet depth, 1st class bedding condition with H-20 live load.
Table A above 3 edge bearing strength @ 5% deflection lbs. per lin. ft.
1,030 (dead load) + 130 (live load) x 2 = 2,320
From page 18-46 select GA # 16

D = Inside diameter in feet, DL = Dead load, LL = Live load.

DRAINAGE & SEWERAGE – PIPE DATA – I

TABLE A DRAINAGE AND SEWERAGE – PIPE DATA

REINFORCED CONCRETE PIPE

INSIDE PIPE DIA. (IN.)	WT. PER LIN. FT (lbs.)	CLASS I A.S.T.M. SPEC. C76-57T SHELL THICKNESS (IN.) WALL "A"	CLASS I WALL "B"	CLASS I *3 EDGE BEARING STRENGTH TO PRODUCE AN 0.01 IN. CRACK #/LIN. FT.	CLASS II A.S.T.M. SPEC. C76-57T SHELL THICKNESS (IN.) WALL "A"	CLASS II WALL "B"	CLASS II *3 EDGE BEARING STRENGTH TO PRODUCE AN 0.01 IN. CRACK #/LIN. FT.	CLASS III A.S.T.M. SPEC. C76-57T SHELL THICKNESS (IN.) WALL "A"	CLASS III WALL "B"	CLASS III 3 EDGE BEARING STRENGTH TO PRODUCE AN 0.01 IN. CRACK #/LIN. FT.	CLASS IV A.S.T.M. SPEC. C76-57T SHELL THICKNESS (IN.) WALL "A"	CLASS IV WALL "B"	CLASS IV WALL "C"	CLASS IV *3 EDGE BEARING STRENGTH TO PRODUCE AN 0.01 IN. CRACK #/LIN. FT.	CLASS V A.S.T.M. SPEC. C76-57T SHELL THICKNESS (IN.) WALL "B"	CLASS V WALL "C"	CLASS V *3 EDGE BEARING STRENGTH TO PRODUCE AN 0.01 IN. CRACK #/LIN. FT.
4		–	–	–	–	–	–	–	–	–	–	–	–	–	–	–	–
6		–	–	–	–	–	–	–	–	–	–	–	–	–	–	–	–
8		–	–	–	–	–	–	–	–	–	–	–	–	–	–	–	–
10		–	–	–	–	–	–	–	–	–	–	–	–	–	–	–	–
12	90	–	–	–	1 3/4*	2*	1,000	1 3/4*	2*	1,350	1 3/4†	2*	–	2,000	2	–	3,000
15	125	–	–	–	1 7/8*	2 1/4*	1,250	1 7/8*	2 1/4*	1,690	1 7/8†	2 1/4*	–	2,500	2 1/4•	–	3,750
18	160	–	–	–	2*	2 1/2*	1,500	2*	2 1/2*	2,025	2†	2 1/2*	–	3,000	2 1/2•	–	4,500
21	205	–	–	–	2 1/4*	2 3/4*	1,750	2 1/4*	2 3/4*	2,360	2 1/4†	2 3/4*	–	3,500	2 3/4•	–	5,250
24	260	–	–	–	2 1/2*	3*	2,000	2 1/2*	3*	2,700	2 1/2†	3*	4*	4,000	3•	4•	6,000
30	370	–	–	–	2 3/4*	3 1/2*	2,500	2 3/4*	3 1/2*	3,375	2 3/4†	3 1/2*	4 1/4*	5,000	3 1/2•	4 1/4•	7,500
36	520	–	–	–	3*	4*	3,000	3*	4*	4,050	–	4*	4 7/8*	6,000	4•	4 7/8•	9,000
42	680	–	–	–	3 1/2*	4 1/2*	3,500	3 1/2*	4 1/2*	4,725	–	4 1/2*	5 1/4*	7,000	4 1/2•	5 1/4•	10,500
48	850	–	–	–	4*	5*	4,000	4*	5*	5,400	–	5*	5 3/4*	8,000	5•	5 3/4•	12,000
54	1,050	–	–	–	4 1/2*	5 1/2*	4,500	4 1/2*	5 1/2*	6,075	–	5 1/2*	6 1/8*	9,000	–	6 1/8•	13,500
60	1,280	5*	6*	4,000	5*	6*	5,000	5*	6*	6,750	–	6†	6 3/4*	10,000	–	6 3/4•	15,000
66	1,480	5 1/2*	6 1/2*	4,400	5 1/2*	6 1/2*	5,500	5 1/2*	6 1/2*	7,425	–	6 1/2†	7 1/4*	11,000	–	7 1/4•	16,500
72	1,835	6*	7*	4,800	6*	7*	6,000	6*	7*	8,100	–	7†	7 3/4†	12,000	–	7 3/4•	18,000
78	2,150	6 1/2	7 1/2	5,200	6 1/2*	7 1/2*	6,500	6 1/2†	7 1/2*	8,775	–	–	8 1/4†	13,000	–	–	–
84	2,300	7*	8*	5,600	7*	8*	7,000	7†	8*	9,450	–	–	8 3/4†	14,000	–	–	–
90	2,600	7 1/2*	8 1/2	6,000	7 1/2*	8 1/2*	7,500	7 1/2†	8 1/2†	10,125	–	–	–	–	–	–	–
96	2,750	8*	9*	6,400	8*	9*	8,000	8†	9†	10,800	–	–	–	–	–	–	–
102	3,050	8 1/2†	9 1/2*	6,800	8 1/2†	9 1/2†	8,500	8 1/2†	9 1/2†	11,475	–	–	–	–	–	–	–
108	3,450	9†	10*	7,200	9†	10†	9,000	9†	10†	12,150	–	–	–	–	–	–	–

* Concrete 4000 p.s.i.

† Concrete 5000 p.s.i.

• Concrete 6000 p.s.i.

Standard laying length: from 12" diameter to 24" diameter, 4' or 6'.
from 24" diameter to 108" diameter 4', 6' or 8'.

Weight per linear foot furnished by American-Marietta Co.

Weights shown are for tongue and groove joints.

DRAINAGE & SEWERAGE – PIPE DATA – 2

TABLE A. PIPE CLASSES & PROPERTIES

	* VITRIFIED CLAY SALT GLAZED						ASBESTOS CEMENT – SEWER PIPE (TRANSITE)									NON.- REINFORCED CONCRETE PIPE				
	STANDARD STRENGTH A.S.T.M. SPEC. C13-54			EXTRA STRENGTH A.S.T.M. SPEC. C200-55T			ʃ STANDARD STRENGTH FED.-SPEC. SS-P-331A REV. 1			ʃ EXTRA STRENGTH FED.-SPEC. SS-P-331A REV. 1			† BUILDING SEWER PIPE			A.S.T.M. SPEC. C-14-57 STD. STRENGTH TABLE I		A.S.T.M. SPEC. C-14-57 EXTRA STRENGTH TABLE II		
INSIDE PIPE DIA. (INCHES)	SHELL THICKNESS (INCHES)	WT. PER LIN. FT. (LBS)	CRUSHING STRENGTH 3-EDGE BEARING LBS PER LIN. FT.	SHELL THICKNESS (INCHES)	WT. PER LIN. FT. (LBS)	CRUSHING STRENGTH 3-EDGE BEARING LBS PER LIN. FT.	SHELL THICKNESS (INCHES)	WT. PER LIN. FT. WITH COUPLING (LBS)	CRUSHING STRENGTH 3-EDGE BEARING LBS. PER LIN. FT.	SHELL THICKNESS (INCHES)	WT. PER LIN. FT. WITH COUPLING (LBS)	CRUSHING STRENGTH 3-EDGE BEARING LBS PER LIN. FT.	SHELL THICKNESS (INCHES)	WT. PER LIN. FT. WITH COUPLING (LBS)	CRUSHING STRENGTH 3-EDGE BEARING LBS PER LIN. FT.	SHELL THICKNESS (INCHES)	CRUSHING STRENGTH 3-EDGE BEARING LBS PER LIN. FT.	SHELL THICKNESS (INCHES)	CRUSHING STRENGTH 3-EDGE BEARING LBS PER LIN. FT.	WT. PER LIN. FT. (LBS)
4	1/2	9	1,000	—	—	—	—	—	—	—	—	—	0.29	4.7	1,740	9/16 ø	1,000	3/4 ø	2,000	
5	—	—	—	—	—	—	—	—	—	—	—	—	0.32	6.3	1,680	—	—	—	—	—
6	5/8	16	1,100	11/16	20	2,000	0.425	9.1	2,600	—	—	—	0.32	7.5	1,420	5/8 ø	1,100	3/4 ø	2,000	25
8	3/4	23	1,300	7/8	30	2,000	0.48	13.2	2,500	—	—	—	—	—	—	3/4 ø	1,300	7/8 ø	2,000	35
10	7/8	36	1,400	1	45	2,000	0.525	18.4	2,200	0.59	19.5	2,800	—	—	—	7/8 ⊙	1,400	1 ⊙	2,000	48
12	1	48	1,500	1 3/16	66	2,250	0.56	24.0	2,200	0.64	26.2	3,000	—	—	—	1 ⊙	1,500	1 3/8 ⊙	2,250	90
14	—	—	—	—	—	—	0.59	29.6	2,200	0.74	34.8	3,400	—	—	—	—	—	—	—	—
15	1 1/4	72	1,750	1 1/2	100	2,750	—	—	—	—	—	—	—	—	—	1 1/4 ⊙	1,750	1 5/8 ⊙	2,750	125
16	—	—	—	—	—	—	0.63	35.9	2,200	0.825	44.2	3,700	—	—	—	—	—	—	—	—
18	1 1/2	107	2,000	1 7/8	149	3,300	0.67	42.4	2,100	0.91	53.5	4,000	—	—	—	1 1/2 ⊙	2,000	2 ⊙	3,300	160
20	—	—	—	—	—	—	0.70	49.0	2,200	0.96	63.5	4,000	—	—	—	—	—	—	—	—
21	1 3/4	145	2,200	2 1/4	205	3,850	—	—	—	—	—	—	—	—	—	1 3/4 ⊙	2,200	2 1/2 ⊙	3,850	205
24	2	192	2,400	2 1/2	280	4,400	0.81	67.6	2,200	1.11	82.4	4,200	—	—	—	2 1/8 ⊙	2,400	3 ⊙	4,000	240
30	2 1/2	308	3,200	3	382	5,000	1.04	105.8	2,800	1.33	130.6	4,600	—	—	—	—	—	—	—	—
36	2 3/4	412	3,900	3 1/2	581	6,000	1.27	154.0	3,300	1.57	184.0	5,000	—	—	—	—	—	—	—	—

ʃ STANDARD LAYING LENGTH – 13 FT.
† LAYING LENGTH – 10 FT.
* STANDARD LAYING LENGTH FROM 4" DIA. TO 12" DIA. – 2 FT., 15" AND OVER - 3 FT.
⊙ STANDARD LAYING LENGTH – 3 FT
ø STANDARD LAYING LENGTH – 2.5 FT.

WEIGHTS PER LIN. FT. FOR NON REINF. CONCRETE PIPE FURNISHED BY AMERICAN – MARIETTA CO.

ALL DATA ON ASBESTOS CEMENT PIPE FURNISHED BY JOHNS – MANVILLE CORP.

PROTECTIVE COATINGS FOR PRECAST CONCRETE PIPE

Where concrete pipe lines are required to convey wastes or liquids containing agents that have disintegrating effects on concrete surfaces or where concrete pipe lines are buried in soils that are similarly injurious, there are two good types of concrete protective-coating materials that are suitable. They have excellent adhesion, good flexibility, and long service life.

1.* High-solids vinyl-copolymer resins, fast-drying hot-spray-applied on clean surfaces. Requires a hot-spray heater, portable compressor, atomizing guns, and a fluid pump of 3.67 to 1 ratio. The material requires heating to about 160°F. Two cross-spray coats having a dry-film thickness of 10 mils is recommended.

2.† Coal-tar-epoxy coating. (*a*) First coat a coal-tar-epoxy primer as a penetrant and sealer on masonry surface applied at the rate of 80 to 100 sq. ft. per gal. (240 to 300 million feet), minimum dry-film thickness 3 mils. (*b*) Then 2 final coats of coal-tar-epoxy with a catalyst thoroughly mixed in, prior to application. (Short pot life.) Applied at a rate of 100 sq. ft. per gal. All coatings may be applied by hand brush, spray, or roller; final dry thickness of 16 mils is recommended.

Coatings should be applied only to exposed areas of precast pipes but not to joint surfaces where adhesion of mortar and other jointing materials would lose bond or adhesion with concrete.

*Courtesy of Corrosion Control Co.
†Courtesy of Pittsburgh Coke & Chemical Co.

DRAINAGE & SEWERAGE - PIPE DATA - 3

TABLE A. PIPE CLASSES & PROPERTIES

CORRUGATED METAL PIPE

INSIDE DIA. OF PIPE	16 GAUGE			14 GAUGE			12 GAUGE			10 GAUGE			8 GAUGE			7 GAUGE			5 GAUGE			3 GAUGE			1 GAUGE		
	⊖ HEIGHT OF COVER (H20) LIVE LOAD	⊗ 3 EDGE BEARING STRENGTH AT 5% DEFLECT	WT. PER LIN. FT.	⊖ HEIGHT OF COVER (H20) LIVE LOAD	⊗ 3 EDGE BEARING STRENGTH AT 5% DEFLECT	WT. PER LIN. FT.	⊖ HEIGHT OF COVER (H20) LIVE LOAD	⊗ 3 EDGE BEARING STRENGTH AT 5% DEFLECT	WT. PER LIN. FT.	⊖ HEIGHT OF COVER (H20) LIVE LOAD	⊗ 3 EDGE BEARING STRENGTH AT 5% DEFLECT	WT. PER LIN. FT.	⊖ HEIGHT OF COVER (H20) LIVE LOAD	⊗ 3 EDGE BEARING STRENGTH AT 5% DEFLECT	WT. PER LIN. FT.	⊖ HEIGHT OF COVER (H20) LIVE LOAD	⊗ 3 EDGE BEARING STRENGTH AT 5% DEFLECT	WT. PER LIN. FT.	⊖ HEIGHT OF COVER (H20) LIVE LOAD	⊗ 3 EDGE BEARING STRENGTH AT 5% DEFLECT	WT. PER LIN. FT.	⊖ HEIGHT OF COVER (H20) LIVE LOAD	⊗ 3 EDGE BEARING STRENGTH AT 5% DEFLECT	WT. PER LIN. FT.	⊖ HEIGHT OF COVER (H20) LIVE LOAD	⊗ 3 EDGE BEARING STRENGTH AT 5% DEFLECT	WT. PER LIN. FT.
(INCHES)	(FT.)	(#/LIN.FT.)	(LBS)	(FT.)	(#/LIN.FT.)	(LBS)	(FT.)	(#/LIN.FT.)	(LBS)	(FT.)	(#/LIN.FT.)	(LBS)	(FT.)	(#/LIN.FT.)	(LBS)	(FT.)	(#/LIN.FT.)	(LBS)	(FT.)	(#/LIN.FT.)	(LBS)	(FT.)	(#/LIN.FT.)	(LBS)	(FT.)	(#/LIN.FT.)	(LBS)
STD. RIVETED																											
8	*50	30,200	11	—	37,700	12	—	—	16	—	—	—	—	—	—												
10	*50	19,300	13	—	24,100	15	—	—	20	—	—	—	—	—	—												
12	*50	13,400	16	—	16,700	18	—	—	23	—	—	—	—	—	—												
15	<-50	8,600	19	51-70	10,700	22	71-100	15,000	29	—	—	—	—	—	—												
18	1-40	5,900	23	41-60	7,400	27	61-100	10,400	34	—	—	41	—	—	—												
21	1-35	4,300	27	36-50	5,400	31	51-80	7,600	39	81-100	9,800	48	—	—	—												
24	1-15	3,300	31	16-45	4,200	36	46-70	5,800	45	71-100	7,500	55	—	—	65												
30				1-30	2,600	43	31-45	3,700	55	46-70	4,800	67	71-100	5,900	78												
36				1-15	1,800	52	16-30	2,600	66	31-45	3,300	80	46-100	4,000	94												
42				—	1,300	61	1-25	1,900	78	26-35	2,400	94	36-100	3,000	110												
48				—	1,000	69	1-20 21-25*	1,400	88	21-25 26-70*	1,800	107	26-35 71-100*	2,300	126												
54							1-15 1-20*	1,100	101	16-20 21-50*	1,500	122	21-30 51-100*	1,800	144												
STRUCTURAL PLATE																											
60							1-25 1-50*	16,200	111	26-50 51-60*	21,000	139	51-70 61-70*	25,800	167	71-80 71-80*	28,900	181	81-100 81-100*	34,000	209	— —	39,200	237	— —	44,500	265
66							1-20 1-45*	13,400	120	21-40 46-55*	17,300	151	41-70 56-70*	21,300	181	71-80 71-80*	23,900	197	81-90 81-90*	28,100	227	91-100 91-100*	32,400	258	— —	36,800	289
72							1-15 1-40*	11,200	130	16-35 41-50*	14,500	163	36-55 51-60*	17,900	196	56-70 61-70*	20,100	213	71-80 71-80*	23,600	246	81-90 81-90*	27,200	280	91-100 91-100*	30,900	313
78							1-15 1-35*	9,600	139	16-30 36-45*	12,400	175	31-50 46-60*	15,200	211	51-60 —	17,100	228	61-70 61-70*	20,100	264	71-80 71-80*	23,200	301	81-90 81-90*	26,300	336
84							6-10 1-35*	8,200	148	1-25 36-45*	10,600	187	26-45 46-55*	13,100	225	46-55 56-60*	14,700	244	56-70 61-70*	17,300	283	71-80 71-80*	20,000	321	81-90 81-90*	22,700	360
96							6-10 1-25*	6,300	175	1-20 26-40*	8,100	220	21-35 41-45*	10,000	265	36-45 46-55*	11,300	287	46-60 56-60*	13,300	332	61-70 61-70*	15,300	377	71-80 71-80*	17,400	422
108							— 1-15*	5,000	194	1-15 16-35*	6,400	244	16-25 36-40*	7,900	294	26-35 41-45*	8,900	319	36-55 46-55*	10,500	369	56-60 56-60*	12,100	420	61-70 61-70*	13,700	469
120							— 6-10*	4,000	213	6-10 1-30*	5,200	268	1-20 31-35*	6,400	323	21-25 36-40*	7,200	351	26-40 41-50*	8,500	406	41-55 51-55*	9,800	461	56-60 56-60*	11,100	516
132							— 6-10*	3,300	240	6-10 1-30*	4,300	301	1-15 31-35*	5,300	363	16-20 36-40*	5,900	393	21-35 41-50*	7,000	455	36-50 51-55*	8,100	517	51-60 56-60*	9,200	578
144							— —	2,800	259	— 1-20*	3,600	326	1-15 21-30*	4,400	392	16-20 31-35*	5,000	425	21-30 36-40*	5,900	492	31-40 41-45*	6,800	560	41-50 46-55*	7,700	625
156							— —	2,400	278	— 1-15*	3,100	350	6-10 16-30*	3,800	421	1-15 —	4,200	457	16-25 31-35*	5,000	528	26-35 36-40*	5,800	601	36-45 41-50*	6,600	672
168							— —	2,000	297	— 1-15*	2,600	373	6-10 16-25*	3,200	450	— 26-30*	3,600	488	1-20 31-35*	4,300	565	21-25 36-40*	5,000	643	26-35 41-45*	5,600	719
180							— —	1,800	324	— 6-10*	2,300	407	— 1-25*	2,800	490	6-10 —	3,200	531	1-15 26-30*	3,700	615	16-25 31-35*	4,300	700	26-30 36-40*	4,900	782

* STRUTTED

NOTE :) WEIGHT OF RIVETED PIPE IS THAT OF : PLAIN GALVANIZED FULL COATED & PAVED.

⊖ WITH LATERAL SUPPORT AS RECOMMENDED BY THE MANUFACTURER. PROJECTION CONDITION NO CONSIDERATION FOR TYPE OF BENDING (USE ONLY FOR <30 FT. OF FILL OR <48"φ)

⊗ LATERAL SUPPORT NEGLECTED-

DRAINAGE & SEWERAGE—BASIS OF DESIGN

STANDARDS FOR DESIGN OF SEWERAGE*

Item	Basis of Design
Design Period	Estimated future tributary population up to 50 years hence.
Average Daily flow	100 gal. per day per capita (includes normal infiltration).
Laterals and submains (flowing full)	400 gal. per day per capita.
Main, trunk and outfall (flowing full)	250 gal. per day per capita.
Industrial plants	Add for sewage and industrial wastes.
Infiltration	Not to exceed 10,000 gal. per day per mile for pipe up to 15 in. diameter.
Minimum size of street sewer	8-in. diameter.
Minimum slope	Sufficient to give mean velocity of 2 f.p.s. when flowing full or half-full, based on "n" of 0.013 in Kutter's or Manning's formula.
Maximum slope	12 to 15 f.p.s. at average flow. Provide protection against erosion and shock for greater velocity.
Manholes	At end of each line; at all changes in grade, size and alignment. Not more than 400 ft. intervals for sewers up to 15 in. dia. Not more than 500 ft. intervals for sewers 18 to 30 in. dia.
Drop-type manhole	Use where entering sewer is 24 in. dia. or more above manhole invert.
Inverted siphons	Minimum pipe size 6-in. diameter. Minimum number of pipes two. Minimum velocity at average flow 3 f.p.s.
Pumping station	Dry-well type preferable. Wet-well type acceptable for installations serving 50 homes or less. At least 2 pumps; one pump permissible for installations serving not more than 50 homes, provided space is allowed for 1 future pump, and provided overflow is permissible. Protect pumps against clogging by installing bar racks with openings not exceeding 2 in. Provide mechanically cleaned bar screen with grinder or comminutor when size of station warrants. Minimum suction and discharge 4-in. diameter Wet-well capacity not to exceed 10 min. detention at average flow. Power supply from 2 independent sources or emergency power should be provided.

SUGGESTED FORM FOR SANITARY SEWER-DESIGN COMPUTATIONS†

SANITARY COMPUTATIONS

Computed by: _ _ _ _ _ _ _ _ "n" = 0.015

Checked by: _ _ _ _ _ _ _ _ _ _ Location ________________ Density of Population = 30/acre

Date: _ _ _ _ _ _ _ _ _ _ _ _ _ Sheet _ _ _ _ _ _ of _ _ _ _ _ _ _

Sewer Location			Tributary Area, acres		Maximum Rate of Sewage Flow			Design				Profile				
Street	From Manhole No.	To Manhole No.	Increacre,	Total	Rate per acre, g.p.d.	Total m.g.d.	Total c.f.s.	Diameter, in.	Slope, ft./ft.	Capacity When Full, c.f.s.	Velocity When Full, f.p.s.	Length, ft.	Fall, ft.	Other Losses, ft.	Invert Elevation, Upper End	Invert Elevation, Lower End
(1)	(2)	(3)	(4)	(5)	(6)	(7)	(8)	(9)	(10)	(11)	(12)	(13)	(14)	(15)	(16)	(17)
Third	10	11	43.3	43.3	12,000	0.52	0.80	10	.0045	1.12	2.2	325	1.46	0.00	95.33	93.87
Third	11	12	14.2	57.5	12,000	0.69	1.07	10	.0082	1.60	2.8	400	3.28	0.08	93.79	90.51
Chestnut	12	13	10.0	67.5	12,000	0.81	1.25	12	.0036	1.75	2.2	350	1.26	0.21	90.30	89.04

* Upper Mississippi River and Great Lakes Boards of Public Health Engineers, *Standards for Sewage Works*, May 1952.
† Adapted from Davis, *Handbook of Applied Hydraulics*, McGraw-Hill.

DRAINAGE & SEWERAGE – PIPE CAPACITIES-1

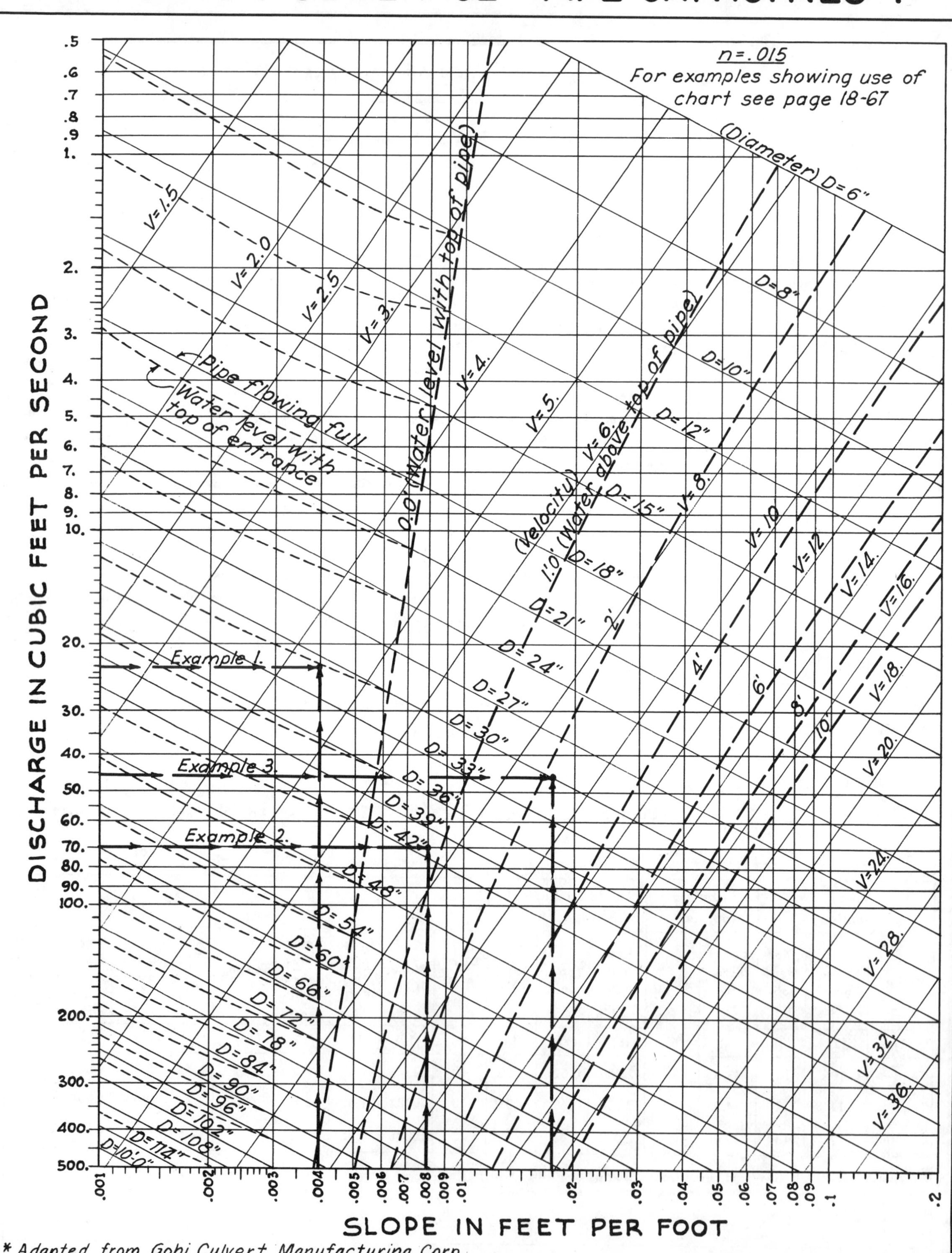

n=.015
For examples showing use of chart see page 18-67
(Diameter) D=6"
D=8"
D=10"
D=12"
D=15"
D=18"
D=21"
D=24"
D=27"
D=30"
D=33"
D=36"
D=39"
D=42"
D=48"
D=54"
D=60"
D=66"
D=72"
D=78"
D=84"
D=90"
D=96"
D=102"
D=108"
D=114"
D=10'0"
V=1.5
V=2.0
V=2.5
V=3.
V=4.
V=5.
(Velocity) V=6.
V=8.
V=10
V=12
V=14.
V=16.
V=18.
V=20.
V=24.
V=28.
V=32.
V=36.
0.0' (Water level with top of pipe)
1.0' (Water above top of pipe)
2'
4'
6'
8'
10'
Pipe flowing full
Water level with top of entrance
Example 1.
Example 2.
Example 3.
DISCHARGE IN CUBIC FEET PER SECOND
SLOPE IN FEET PER FOOT

* Adapted from Gohi Culvert Manufacturing Corp.

DRAINAGE & SEWERAGE – PIPE CAPACITIES-2

EXAMPLES SHOWING USE OF CHART PG. 18-66

Capacities and velocities in chart page 18-66 are for $n = 0.015$. For other values of n, given on page 5-26, multiply charted values by $\frac{0.015}{n}$

Case 1.

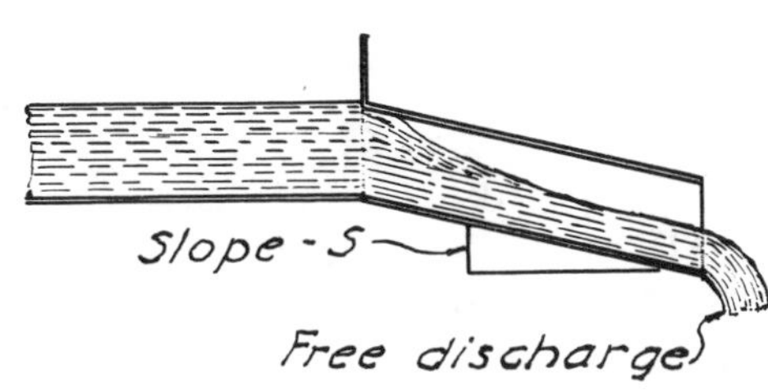

Slope - S
Free discharge

Dash lines to left of 0-0 line give values when water is level with top of pipe at entrance. Velocity of approach and entrance loss neglected.

Example 1. - Given: Q = 23 c.f.s.; S = 0.004 ; n = 0.015.
Required: D and V.
Solution: Enter Chart at 23 c.f.s.; read D = 30" at S = 0.004, and V = 4.4 Ft./Sec.

Case 2.

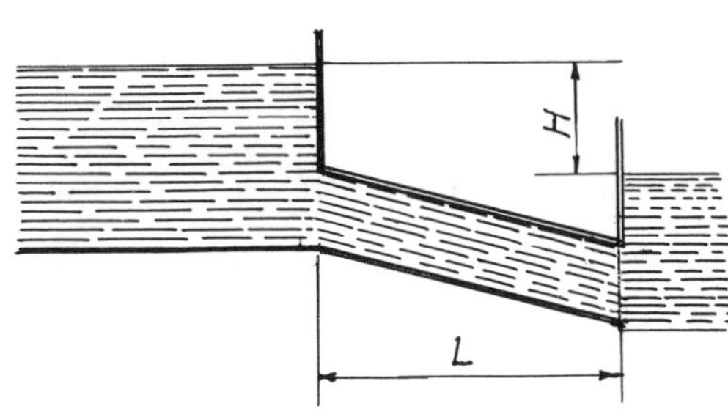

H
L

Solid D lines give values by Manning Formula (see page 18-69) for pipe flowing full. In this case $S = \frac{H}{L}$ = slope of hydraulic gradient. Minor losses neglected.

Example 2. - Given: Q = 70 c.f.s.; H = 4 ft.; L = 500 ft. ∴ $S = \frac{H}{L} = 0.008$
Required: D and V.
Solution: Enter Chart at 70 c.f.s. intersect. S = 0.008. Read D = 42" (nearest adequate size). V = 7.5 Ft./Sec.

Case 3.

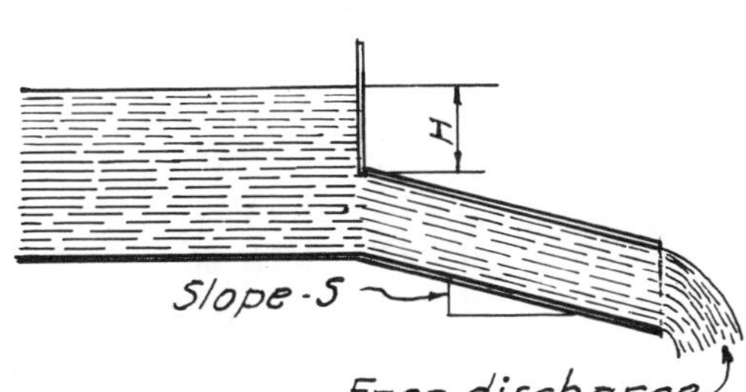

H
Slope - S
Free discharge

Dash lines to right of 0-0 line indicate limits of capacities with inlets submerged to depths shown, from orifice formula $Q = a \times 0.62\sqrt{2gh}$

Example 3. - Given: Q = 46 c.f.s.; S = 0.018.
Required: D with a back up H not more than 3 ft.
Solution: Enter Chart at 46 c.f.s. intersect. S = 0.018 - Read D = 30" (H = 2.3 ft.).

Notation:

- Q = Discharge in cubic feet per second.
- V = Velocity of flow in feet per second.
- S = Slope or hydraulic gradient.
- H = Hydraulic head.
- D = Diameter of pipe.
- L = Length of pipe.
- n = Coefficient of roughness.
- g = Acceleration of gravity = 32.16.

DRAINAGE & SEWERAGE - HYDRAULIC COMPUTATIONS - I

TABLE A-VALUES OF n, TO BE USED WITH KUTTER OR MANNING FORMULAS.†

SURFACE	CONDITION			
	BEST	GOOD	FAIR	BAD
Uncoated cast-iron pipe	0.012	0.013	0.014	0.015
Coated cast-iron pipe	0.011	0.012*	0.013*	
Commercial wrought-iron pipe, black	0.012	0.013	0.014	0.015
Commercial wrought-iron pipe, galvanized	0.013	0.014	0.015	0.017
Smooth brass and glass pipe	0.009	0.010	0.011	0.013
Smooth lockbar and welded OD pipe	0.010	0.011*	0.013*	
Riveted and spiral steel pipe	0.013	0.015*	0.017*	
Vitrified sewer pipe	{0.010, 0.011}	0.013*	0.015	0.017
Common clay drainage tile	0.011	0.012*	0.014*	0.017
Glazed brickwork	0.011	0.012	0.013*	0.015
Brick in cement mortar, brick sewers	0.012	0.013	0.015*	0.017
Neat cement surfaces	0.010	0.011	0.012	0.013
Cement-mortar surfaces	0.011	0.012	0.013*	0.015
Concrete pipe	0.012	0.013	0.015*	0.016
Wood-stave pipe	0.010	0.011	0.012	0.013
Plank flumes:				
Planed	0.010	0.012*	0.013	0.014
Unplaned	0.011	0.013*	0.014	0.015
With battens	0.012	0.015*	0.016	
Concrete-lined channels	0.012	0.014*	0.016*	0.018
Cement-rubble surface	0.017	0.020	0.025	0.030
Dry rubble surface	0.025	0.030	0.033	0.035
Dressed ashlar surface	0.013	0.014	0.015	0.017
Semicircular metal flumes, smooth	0.011	0.012	0.013	0.015
Semicircular metal flumes, corrugated	0.0225	0.025	0.0275	0.030
Canals and ditches:				
Earth, straight and uniform	0.017	0.020	0.0225*	0.025
Rock cuts, smooth and uniform	0.025	0.030	0.033*	0.035
Rock cuts, jagged and irregular	0.035	0.040	0.045	
Winding sluggish canals	0.0225	0.025*	0.0275	0.030
Dredged earth channels	0.025	0.0275*	0.030	0.033
Canals with rough stony beds, weeds on earth banks	0.025	0.030	0.035*	0.040
Earth bottom, rubble sides	0.028	0.030[1]	0.033*	0.035
Natural stream channels:				
1. Clean, straight bank, full stage, no rifts or deep pools	0.025	0.0275	0.030	0.033
2. Same as (1), but some weeds and stones	0.030	0.033	0.035	0.040
3. Winding, some pools and shoals, clean	0.033	0.035	0.040	0.045
4. Same as (3), lower stages, more ineffective slope and sections	0.040	0.045	0.050	0.055
5. Same as (3), some weeds and stones	0.035	0.040	0.045	0.050
6. Same as (4), stony sections	0.045	0.050	0.055	0.060
7. Sluggish river reaches, rather weedy or with very deep pools	0.050	0.060	0.070	0.080
8. Very weedy reaches	0.075	0.100	0.125	0.150

Note: Asbestos-Cement Pipe (Transite) use 0.010.

* *Values commonly used in designing.*

† *Adapted from Handbook of Applied Hydraulics by C.V. Davis, Table compiled by R.E. Horton.*

DRAINAGE & SEWERAGE – HYDRAULIC COMPUTATIONS – 2

MANNING FORMULA (FLOW IN PIPES OR CHANNELS)

$$Q = a \times \frac{1.486}{n} \times R^{2/3} \times S^{1/2}$$

$$V = \frac{1.486}{n} \times R^{2/3} \times S^{1/2}$$

For values of n see Table. A, page 18-68

EXAMPLE: Given: existing waterway illustrated in Fig. A, page 18-73, assuming fairly straight banks with some weeds and stones, $a = 90$ sq. ft., $p = 38$ ft., $s = 0.0012$. Required: Q and V.

SOLUTION: $n = 0.035$ from Table. A, page 18-68, $R = \frac{90}{38} = 2.37$, $R^{2/3} = 1.777$ from Table A below, $S^{1/2} = 0.03464$ from Table B below.

Therefore:- $Q = 90 \times \frac{1.486}{0.035} \times 1.777 \times 0.03464 = 235$ c.f.s.

$V = \frac{235}{90} = 2.6$ feet per second.

TABLE A – VALUES OF $R^{2/3}$.*

Number	.00	.01	.02	.03	.04	.05	.06	.07	.08	.09
.0	.000	.046	.074	.097	.117	.136	.153	.170	.186	.201
.1	.215	.229	.243	.256	.269	.282	.295	.307	.319	.331
.2	.342	.353	.364	.375	.386	.397	.407	.418	.428	.438
.3	.448	.458	.468	.477	.487	.497	.506	.515	.525	.534
.4	.543	.552	.561	.570	.578	.587	.596	.604	.613	.622
.5	.630	.638	.647	.655	.663	.671	.679	.687	.695	.703
.6	.711	.719	.727	.735	.743	.750	.758	.765	.773	.781
.7	.788	.796	.803	.811	.818	.825	.832	.840	.847	.855
.8	.862	.869	.876	.883	.890	.897	.904	.911	.918	.925
.9	.932	.939	.946	.953	.960	.966	.973	.980	.987	.993
1.0	1.000	1.007	1.013	1.020	1.027	1.033	1.040	1.046	1.053	1.059
1.1	1.065	1.072	1.078	1.085	1.091	1.097	1.104	1.110	1.117	1.123
1.2	1.129	1.136	1.142	1.148	1.154	1.160	1.167	1.173	1.179	1.185
1.3	1.191	1.197	1.203	1.209	1.215	1.221	1.227	1.233	1.239	1.245
1.4	1.251	1.257	1.263	1.269	1.275	1.281	1.287	1.293	1.299	1.305
1.5	1.310	1.316	1.322	1.328	1.334	1.339	1.345	1.351	1.357	1.362
1.6	1.368	1.374	1.379	1.385	1.391	1.396	1.402	1.408	1.413	1.419
1.7	1.424	1.430	1.436	1.441	1.447	1.452	1.458	1.463	1.469	1.474
1.8	1.480	1.485	1.491	1.496	1.502	1.507	1.513	1.518	1.523	1.529
1.9	1.534	1.539	1.545	1.550	1.556	1.561	1.566	1.571	1.577	1.582
2.0	1.587	1.593	1.598	1.603	1.608	1.613	1.619	1.624	1.629	1.634
2.1	1.639	1.645	1.650	1.655	1.660	1.665	1.671	1.676	1.681	1.686
2.2	1.691	1.697	1.702	1.707	1.712	1.717	1.722	1.727	1.732	1.737
2.3	1.742	1.747	1.752	1.757	1.762	1.767	1.772	1.777	1.782	1.787
2.4	1.792	1.797	1.802	1.807	1.812	1.817	1.822	1.827	1.832	1.837
2.5	1.842	1.847	1.852	1.857	1.862	1.867	1.871	1.876	1.881	1.886
2.6	1.891	1.896	1.900	1.905	1.910	1.915	1.920	1.925	1.929	1.934
2.7	1.939	1.944	1.949	1.953	1.958	1.963	1.968	1.972	1.977	1.982
2.8	1.987	1.992	1.996	2.001	2.006	2.010	2.015	2.020	2.024	2.029
2.9	2.034	2.038	2.043	2.048	2.052	2.057	2.062	2.066	2.071	2.075
3.0	2.080	2.085	2.089	2.094	2.099	2.103	2.108	2.112	2.117	2.122
3.1	2.126	2.131	2.135	2.140	2.144	2.149	2.153	2.158	2.163	2.167
3.2	2.172	2.176	2.180	2.185	2.190	2.194	2.199	2.203	2.208	2.212
3.3	2.217	2.221	2.226	2.230	2.234	2.239	2.243	2.248	2.252	2.257
3.4	2.261	2.265	2.270	2.274	2.279	2.283	2.288	2.292	2.296	2.301
3.5	2.305	2.310	2.314	2.318	2.323	2.327	2.331	2.336	2.340	2.345
3.6	2.349	2.353	2.358	2.362	2.366	2.371	2.375	2.379	2.384	2.388
3.7	2.392	2.397	2.401	2.405	2.409	2.414	2.418	2.422	2.427	2.431
3.8	2.435	2.439	2.444	2.448	2.452	2.457	2.461	2.465	2.469	2.474
3.9	2.478	2.482	2.486	2.490	2.495	2.499	2.503	2.507	2.511	2.516
4.0	2.520	2.524	2.528	2.532	2.537	2.541	2.545	2.549	2.553	2.558
4.1	2.562	2.566	2.570	2.574	2.579	2.583	2.587	2.591	2.595	2.599
4.2	2.603	2.607	2.611	2.616	2.620	2.624	2.628	2.632	2.636	2.640
4.3	2.644	2.648	2.653	2.657	2.661	2.665	2.669	2.673	2.677	2.681
4.4	2.685	2.689	2.693	2.698	2.702	2.706	2.710	2.714	2.718	2.722
4.5	2.726	2.730	2.734	2.738	2.742	2.746	2.750	2.754	2.758	2.762
4.6	2.766	2.770	2.774	2.778	2.782	2.786	2.790	2.794	2.798	2.802
4.7	2.806	2.810	2.814	2.818	2.822	2.826	2.830	2.834	2.838	2.842
4.8	2.846	2.850	2.854	2.858	2.862	2.865	2.869	2.873	2.877	2.881
4.9	2.885	2.889	2.893	2.897	2.901	2.904	2.908	2.912	2.916	2.920

TABLE B – VALUES OF $S^{1/2}$.*

Number	.---0	.---1	.---2	.---3	.---4	.---5	.---6	.---7	.---8	.---9
.00001	.003162	.003317	.003464	.003606	.003742	.003873	.004000	.004123	.004243	.004359
.00002	.004472	.004583	.004690	.004796	.004899	.005000	.005099	.005196	.005292	.005385
.00003	.005477	.005568	.005657	.005745	.005831	.005916	.006000	.006083	.006164	.006245
.00004	.006325	.006403	.006481	.006557	.006633	.006708	.006782	.006856	.006928	.007000
.00005	.007071	.007141	.007211	.007280	.007348	.007416	.007483	.007550	.007616	.007681
.00006	.007746	.007810	.007874	.007937	.008000	.008062	.008124	.008185	.008246	.008307
.00007	.008367	.008426	.008485	.008544	.008602	.008660	.008718	.008775	.008832	.008888
.00008	.008944	.009000	.009055	.009110	.009165	.009220	.009274	.009327	.009381	.009434
.00009	.009487	.009539	.009592	.009644	.009695	.009747	.009798	.009849	.009899	.009950
.00010	.010000	.010050	.010100	.010149	.010198	.010247	.010296	.010344	.010392	.010440
.0001	.01000	.01049	.01095	.01140	.01183	.01225	.01265	.01304	.01342	.01378
.0002	.01414	.01449	.01483	.01517	.01549	.01581	.01612	.01643	.01673	.01703
.0003	.01732	.01761	.01789	.01817	.01844	.01871	.01897	.01924	.01949	.01975
.0004	.02000	.02025	.02049	.02074	.02098	.02121	.02145	.02168	.02191	.02214
.0005	.02236	.02258	.02280	.02302	.02324	.02345	.02366	.02387	.02408	.02429
.0006	.02449	.02470	.02490	.02510	.02530	.02550	.02569	.02588	.02608	.02627
.0007	.02646	.02665	.02683	.02702	.02720	.02739	.02757	.02775	.02793	.02811
.0008	.02828	.02846	.02864	.02881	.02898	.02915	.02933	.02950	.02966	.02983
.0009	.03000	.03017	.03033	.03050	.03066	.03082	.03098	.03114	.03130	.03146
.0010	.03162	.03178	.03194	.03209	.03225	.03240	.03256	.03271	.03286	.03302
.001	.03162	.03317	.03464	.03606	.03742	.03873	.04000	.04123	.04243	.04359
.002	.04472	.04583	.04690	.04796	.04899	.05000	.05099	.05196	.05292	.05385
.003	.05477	.05568	.05657	.05745	.05831	.05916	.06000	.06083	.06164	.06245
.004	.06325	.06403	.06481	.06557	.06633	.06708	.06782	.06856	.06928	.07000
.005	.07071	.07141	.07211	.07280	.07348	.07416	.07483	.07550	.07616	.07681
.006	.07746	.07810	.07874	.07937	.08000	.08062	.08124	.08185	.08246	.08307
.007	.08367	.08426	.08485	.08544	.08602	.08660	.08718	.08775	.08832	.08888
.008	.08944	.09000	.09055	.09110	.09165	.09220	.09274	.09327	.09381	.09434
.009	.09487	.09539	.09592	.09644	.09695	.09747	.09798	.09849	.09899	.09950
.010	.10000	.10050	.10100	.10149	.10198	.10247	.10296	.10344	.10392	.10440
.01	.1000	.1049	.1095	.1140	.1183	.1225	.1265	.1304	.1342	.1378
.02	.1414	.1449	.1483	.1517	.1549	.1581	.1612	.1643	.1673	.1703
.03	.1732	.1761	.1789	.1817	.1844	.1871	.1897	.1924	.1949	.1975
.04	.2000	.2025	.2049	.2074	.2098	.2121	.2145	.2168	.2191	.2214
.05	.2236	.2258	.2280	.2302	.2324	.2345	.2366	.2387	.2408	.2429
.06	.2449	.2470	.2490	.2510	.2530	.2550	.2569	.2588	.2608	.2627
.07	.2646	.2665	.2683	.2702	.2720	.2739	.2757	.2775	.2793	.2811
.08	.2828	.2846	.2864	.2881	.2898	.2915	.2933	.2950	.2966	.2983
.09	.3000	.3017	.3033	.3050	.3066	.3082	.3098	.3114	.3130	.3146
.10	.3162	.3178	.3194	.3209	.3225	.3240	.3256	.3271	.3286	.3302

* From King, Handbook of Hydraulics, McGraw-Hill.

DRAINAGE & SEWERAGE

SCOBEY CHART FOR THE SOLUTION OF KUTTER'S

KUTTER'S FORMULA

$Q = ac\sqrt{RS}$ in which

$$C = \frac{41.65 + \frac{0.00281}{S} + \frac{1.811}{n}}{1 + \left(41.65 + \frac{0.00281}{S}\right)\frac{n}{\sqrt{R}}}$$

EXAMPLE:

Given: A concrete lined channel,—
a(cross sectional area) = 60 sq. ft.
R(hydraulic radius = $\frac{\text{area}}{\text{wetted perimeter}}$) = 2.6 ft.
n(coefficient of roughness - see Table. A, page
S(longitudinal slope) = 0.00124 ft. per ft.

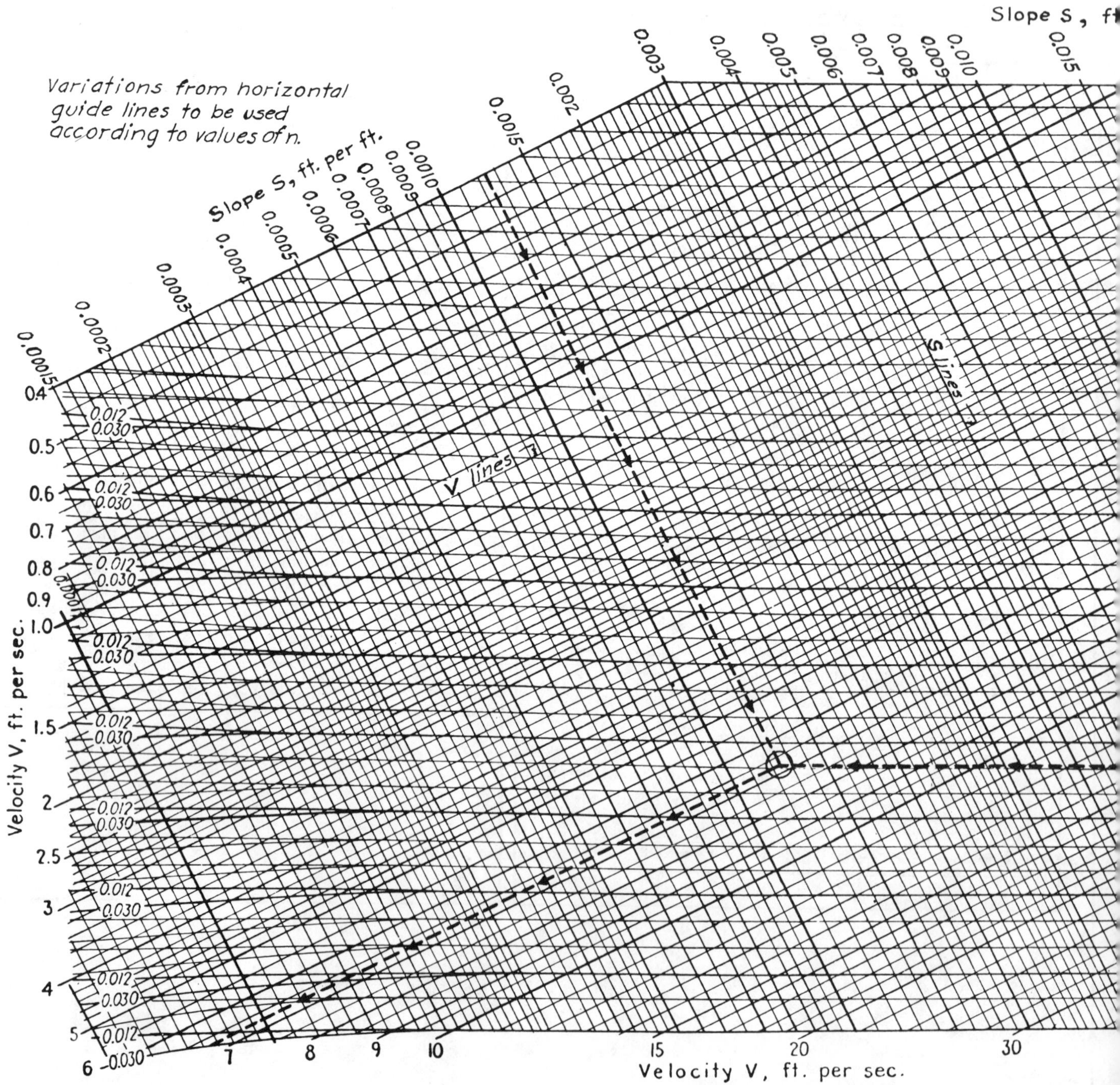

Slope S, ft
0.003
0.004
0.005
0.006
0.007
0.008
0.009
0.010
0.015
0.002
0.0015
Variations from horizontal guide lines to be used according to values of n.
Slope S, ft. per ft.
0.0010
0.0009
0.0008
0.0007
0.0006
0.0005
0.0004
0.0003
0.0002
0.00015
S lines
V lines
0.012
0.030
Velocity V, ft. per sec.
0.4
0.5
0.6
0.7
0.8
0.9
1.0
1.5
2
2.5
3
4
5
6
7
8
9
10
15
20
30
Velocity V, ft. per sec.

For permission to reproduce this diagram the author is indebted to the Division of Irrigation, Soil
printed in Creager and Justin's "Hydroelectric Handbook", 1927.

HYDRAULIC COMPUTATIONS–3

A FOR FLOW IN OPEN CHANNELS.*

REQUIRED: *To find V (velocity in ft. per sec.) and Q (discharge in c.f.s.)*
Enter chart at R = 2.6; proceed parallel with oblique R lines to intersection with n = 0.015; thence horizontally to intersection with S = 0.00124; thence parallel with oblique V line to V scale; read V = 6.75 ft. per sec.
∴ Q = aV = 60 × 6.75 = 405 c.f.s.

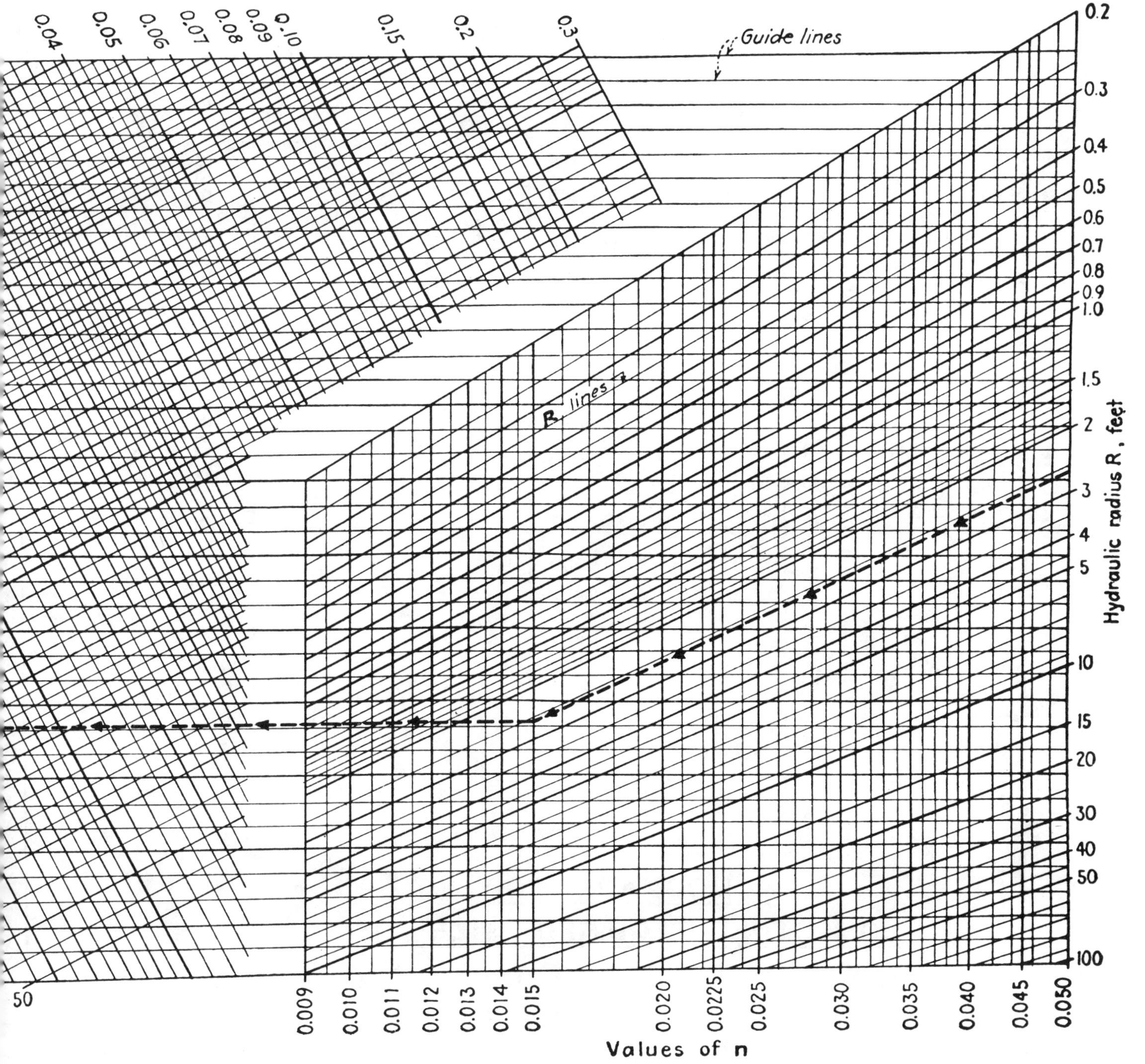

0.04
0.05
0.06
0.07
0.08
0.09
0.10
0.15
0.2
0.3
Guide lines
R lines
0.2
0.3
0.4
0.5
0.6
0.7
0.8
0.9
1.0
1.5
2
3
4
5
10
15
20
30
40
50
100
Hydraulic radius R, feet
50
0.009
0.010
0.011
0.012
0.013
0.014
0.015
0.020
0.0225
0.025
0.030
0.035
0.040
0.045
0.050
Values of n

ion Service, U.S. Department of Agriculture, and to Mr. Scobey. In this form it was first

DRAINAGE & SEWERAGE-HYDRAULIC COMPUTATIONS-4

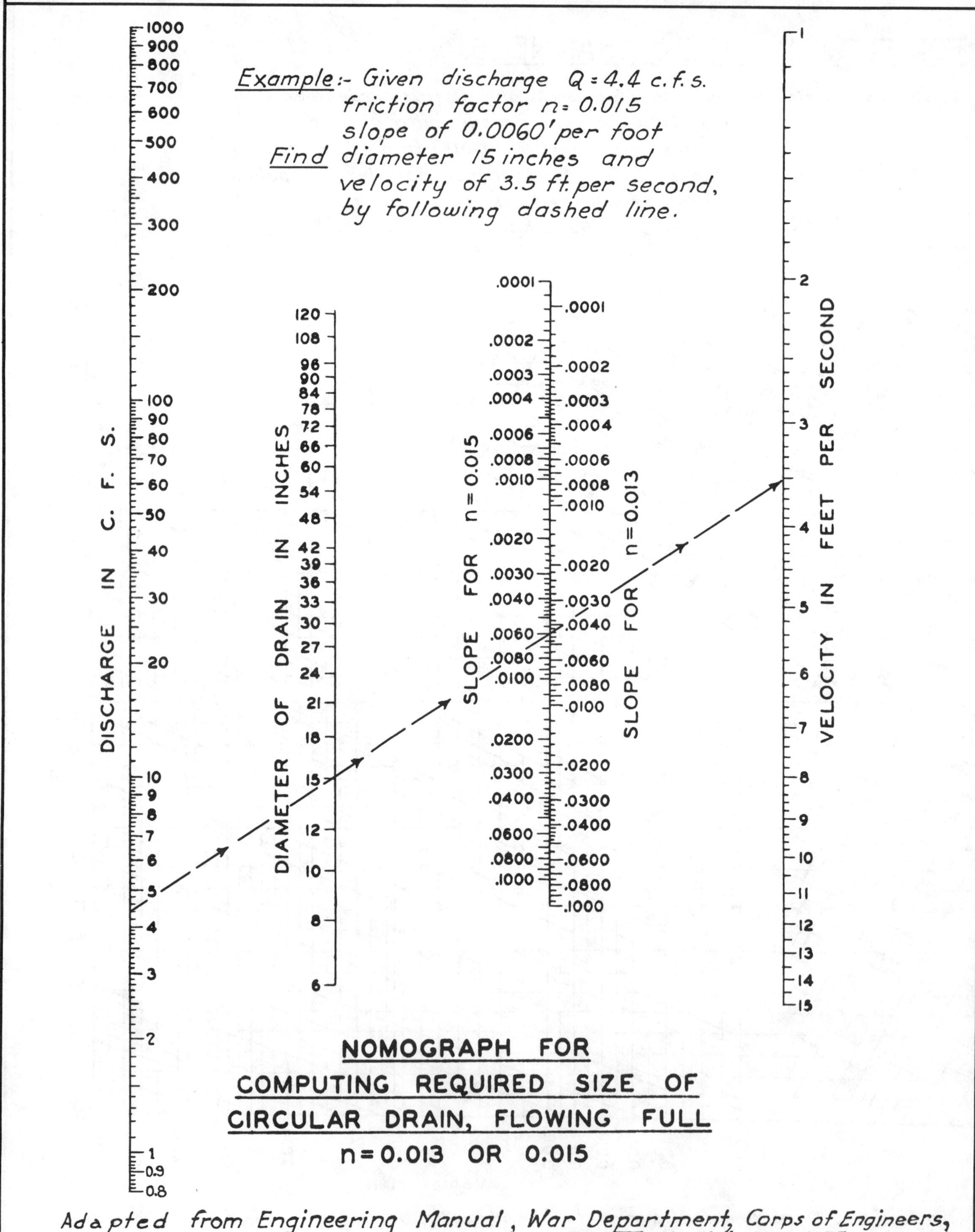
Example:- Given discharge Q = 4.4 c.f.s.
friction factor n = 0.015
slope of 0.0060' per foot
Find diameter 15 inches and
velocity of 3.5 ft. per second,
by following dashed line.
DISCHARGE IN C. F. S.
1000
900
800
700
600
500
400
300
200
100
90
80
70
60
50
40
30
20
10
9
8
7
6
5
4
3
2
1
0.9
0.8
DIAMETER OF DRAIN IN INCHES
120
108
96
90
84
78
72
66
60
54
48
42
39
36
33
30
27
24
21
18
15
12
10
8
6
SLOPE FOR n = 0.015
.0001
.0002
.0003
.0004
.0006
.0008
.0010
.0020
.0030
.0040
.0060
.0080
.0100
.0200
.0300
.0400
.0600
.0800
.1000
SLOPE FOR n = 0.013
.0001
.0002
.0003
.0004
.0006
.0008
.0010
.0020
.0030
.0040
.0060
.0080
.0100
.0200
.0300
.0400
.0600
.0800
.1000
VELOCITY IN FEET PER SECOND
1
2
3
4
5
6
7
8
9
10
11
12
13
14
15

NOMOGRAPH FOR COMPUTING REQUIRED SIZE OF CIRCULAR DRAIN, FLOWING FULL

n = 0.013 OR 0.015

Adapted from Engineering Manual, War Department, Corps of Engineers, Part XIII, Chapt. 1, June 1955.

DRAINAGE & SEWERAGE-HYDRAULIC COMPUTATIONS-5

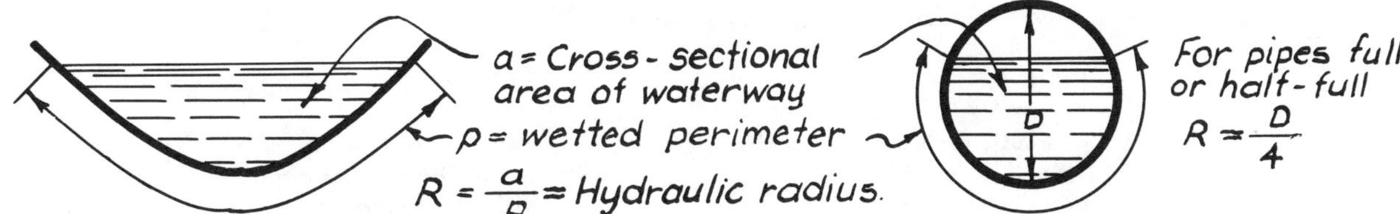
a = Cross-sectional area of waterway
p = wetted perimeter
$R = \frac{a}{p}$ = Hydraulic radius.
D
For pipes full or half-full
$R = \frac{D}{4}$

SECTION OF ANY OPEN CHANNEL

SECTION OF CIRCULAR PIPE

V = Average or mean velocity in feet per second.

Q = a V = Discharge of pipe or channel in cubic feet per second (c.f.s.).

n = Coefficient of roughness of pipe or channel surface, see Table A-Pg.18-68.

S = Slope of Hydraulic Gradient (water surface in open channels or pipes not under pressure, same as slope of channel or pipe invert only when flow is uniform in constant section.

FIG. A-HYDRAULIC ELEMENTS OF CHANNEL SECTIONS.

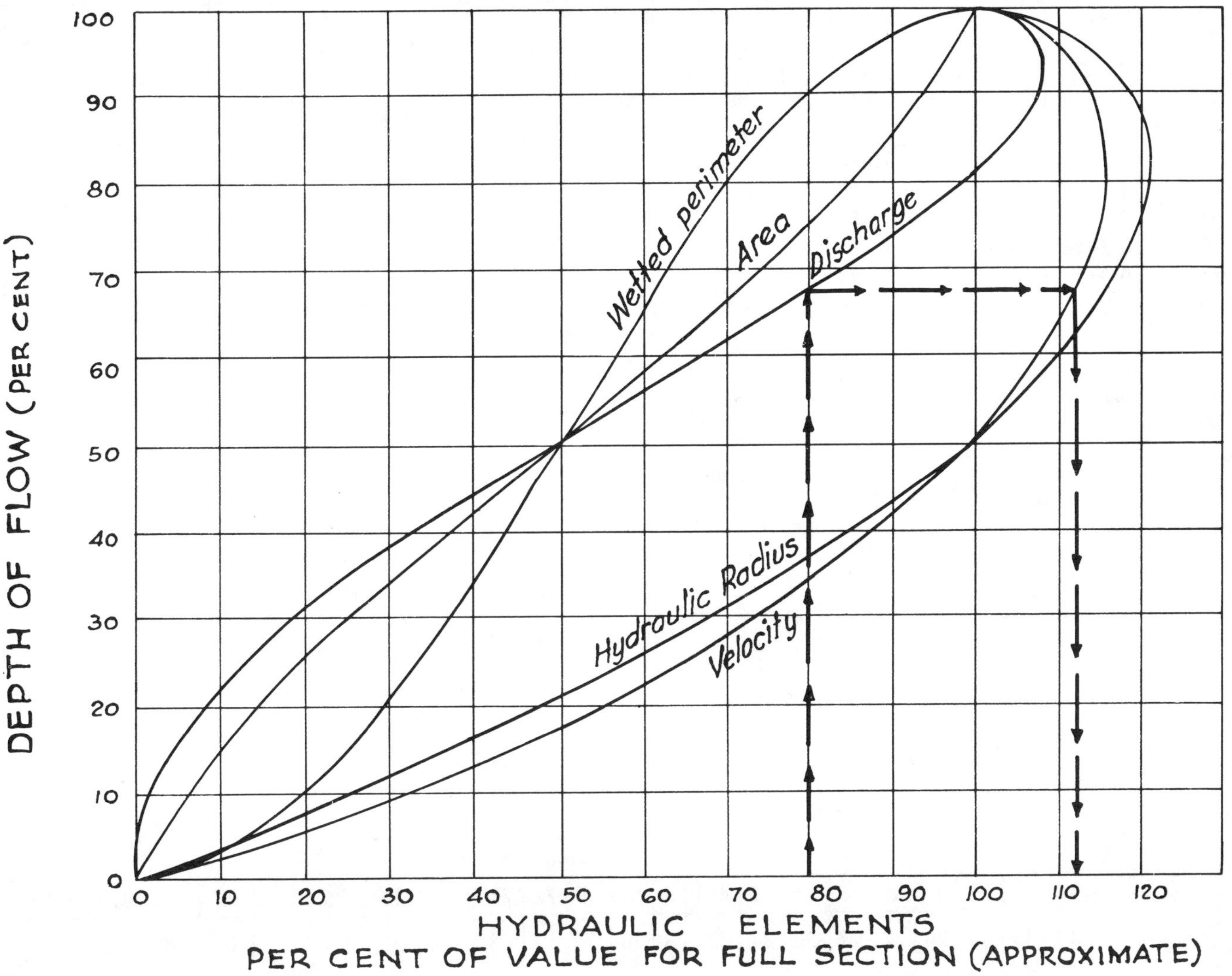

DEPTH OF FLOW (PER CENT): 0, 10, 20, 30, 40, 50, 60, 70, 80, 90, 100

Wetted perimeter

Area

Discharge

Hydraulic Radius

Velocity

HYDRAULIC ELEMENTS: 0, 10, 20, 30, 40, 50, 60, 70, 80, 90, 100, 110, 120

PER CENT OF VALUE FOR FULL SECTION (APPROXIMATE)

EXAMPLE: Given: Discharge = 12 c.f.s. through a pipe which has capacity flowing full of 15 c.f.s. at a velocity of 7.0 ft. per sec. Required to find V for Q = 12 c.f.s. ∴ Percentage of full discharge = $\frac{12}{15}$ = 80%. Enter chart at 80% of value for full section of Hydraulic Elements, find V = 112.5% × 7 = 7.9 ft. per sec.

FIG. B-VALUES OF HYDRAULIC ELEMENTS OF CIRCULAR SECTION FOR VARIOUS DEPTHS OF FLOW.

DRAINAGE & SEWERAGE-HYDRAULIC COMPUTATIONS-6

HYDRAULIC ELEMENTS OF MASSEY FLAT BASE PIPE*

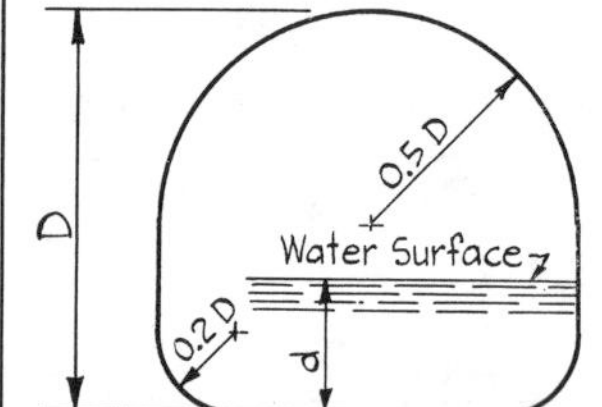
D
0.5 D
Water Surface
0.2 D
d

FULL PIPE
Area = A = $0.876D^2$
Hydraulic radius = R = 0.258D
Velocity = V
Wetted perimeter = P
Discharge = Q

PARTIALLY FILLED PIPE

Small letters are for "d" depth. Diameter of circle with equal carrying capacity = 1.05D

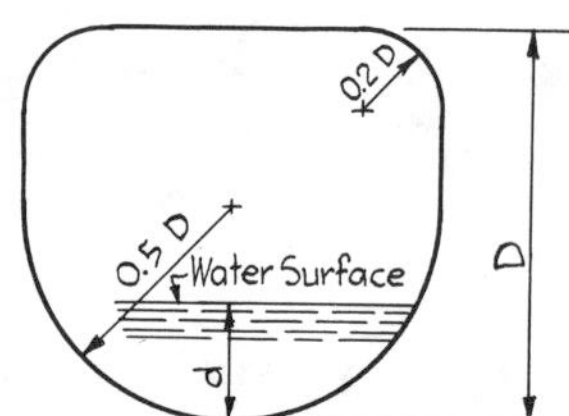
0.2 D
0.5 D
Water Surface
D
d

	TABLE A – FLAT BASE PIPE UPRIGHT					TABLE B – FLAT BASE PIPE INVERTED				
d/D	a/A	p/P	r/R	v/V	q/Q	a/A	p/P	r/R	v/V	q/Q
0	0	0	0	0	0	0	0	0	0	0
0.05	0.043	0.261	0.167	0.301	0.013	0.017	0.132	0.124	0.247	0.004
0.10	0.096	0.300	0.322	0.469	0.045	0.047	0.190	0.248	0.395	0.018
0.15	0.152	0.331	0.457	0.595	0.090	0.084	0.234	0.360	0.506	0.043
0.20	0.209	0.361	0.577	0.694	0.144	0.127	0.272	0.469	0.605	0.077
0.25	0.266	0.391	0.678	0.773	0.205	0.175	0.308	0.566	0.684	0.119
0.30	0.323	0.420	0.767	0.840	0.271	0.226	0.341	0.663	0.760	0.172
0.35	0.380	0.450	0.845	0.894	0.339	0.278	0.372	0.748	0.825	0.230
0.40	0.437	0.479	0.911	0.941	0.411	0.334	0.403	0.829	0.884	0.296
0.45	0.494	0.508	0.973	0.983	0.484	0.390	0.433	0.903	0.936	0.365
0.50	0.551	0.538	1.023	1.017	0.561	0.447	0.462	0.969	0.978	0.437
0.55	0.609	0.567	1.074	1.049	0.639	0.504	0.491	1.023	1.017	0.513
0.60	0.665	0.597	1.116	1.077	0.716	0.562	0.521	1.077	1.052	0.590
0.65	0.721	0.628	1.151	1.099	0.792	0.619	0.550	1.124	1.081	0.670
0.70	0.774	0.659	1.174	1.114	0.862	0.676	0.580	1.163	1.106	0.746
0.75	0.825	0.692	1.194	1.126	0.930	0.733	0.609	1.202	1.131	0.831
0.80	0.873	0.727	1.198	1.128	0.986	0.790	0.639	1.233	1.151	0.907
0.85	0.916	0.766	1.194	1.126	1.031	0.848	0.669	1.267	1.173	0.994
0.90	0.953	0.810	1.178	1.116	1.062	0.904	0.700	1.291	1.185	1.070
0.95	0.983	0.868	1.132	1.086	1.068	0.957	0.739	1.295	1.188	1.135
1.00	1.000	1.000	1.000	1.000	1.000	1.000	1.000	1.000	1.000	1.000

*From Massey Concrete Products Co.

EXAMPLES: Use Manning formula: $V = \frac{1.486}{n} r^{2/3} s^{1/2}$. (See p. 18-69).

Given: D = 80" or 6.67'. s = 0.0016. n = 0.012.
A = 0.876 × 6.67² = 38.90 sq. ft.
R = 0.258 × 6.67 = 1.72.
V = 7.12 f.p.s.
Q = 277 c.f.s.

Find: q when flowing 2 ft. deep = 0.3D

(a) Upright section: r = 0.767 × 1.72 = 1.32.
a = 0.323 × 38.90 = 12.57 sq. ft.
From Table A: v = 0.84 × 7.12 = 5.98 f.p.s.
v = 5.98 f.p.s.
q = av = 5.98 × 12.57 = 74.9 c.f.s.
q = .271 × 277 = 75.0 c.f.s.

(b) Inverted section: r = 0.663 × 1.72 = 1.14
a = 0.226 × 38.90 = 8.85 sq. ft.
From Table B: v = 0.76 × 7.12 = 5.40 f.p.s.
v = 5.40 f.p.s.
q = 8.85 × 5.40 = 47.8 c.f.s.
q = .172 × 277 = 47.6 c.f.s.

TABLE C – ELLIPTICAL REINFORCED CONCRETE PIPE†

Inside Diameter of Round Pipe Equivalent	Dimensions in inches																				
	18	24	27	30	33	36	39	42	48	54	60	66	72	78	84	90	96	102	108	114	120	132	144
Nominal inside height	14	19	22	24	27	29	32	34	38	43	48	53	58	63	68	72	77	82	87	92	97	106	116
Nominal inside width	23	30	34	38	42	45	49	53	60	68	76	83	91	98	106	113	121	128	136	143	151	166	180

† From American-Marietta Co.

For computing capacity of elliptical pipe, use Manning formula for corresponding equivalent round pipe as noted in Table C. For solution of Manning formula, see p. 18-69.

DRAINAGE & SEWERAGE–MINOR LOSSES–INVERTED SIPHONS

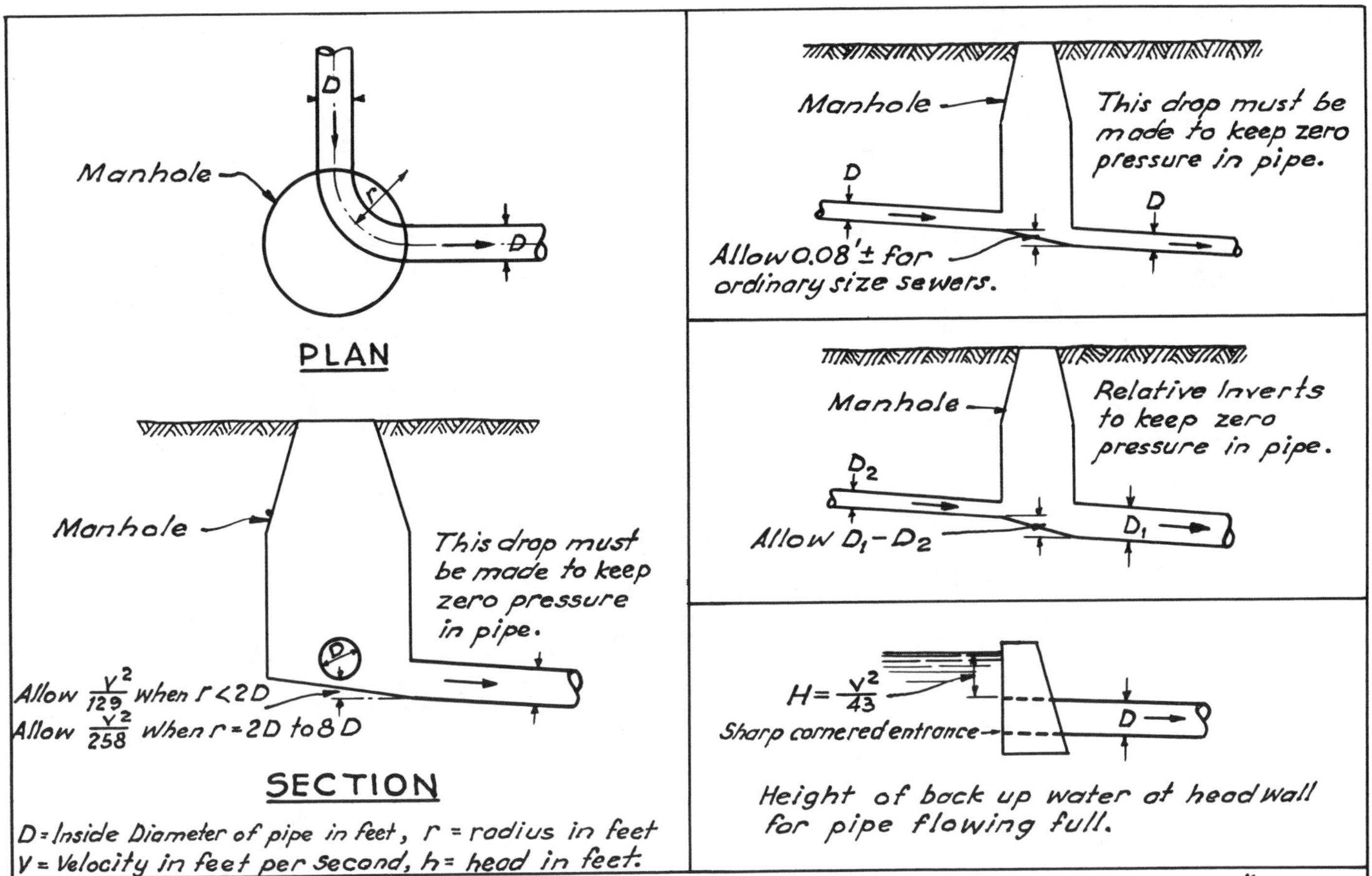

Manhole

D

r

D

PLAN

Manhole

This drop must be made to keep zero pressure in pipe.

D

Allow $\frac{V^2}{129}$ when $r < 2D$

Allow $\frac{V^2}{258}$ when $r = 2D$ to $8D$

SECTION

D = Inside Diameter of pipe in feet, r = radius in feet
V = Velocity in feet per second, h = head in feet.

Manhole

This drop must be made to keep zero pressure in pipe.

D

D

Allow 0.08' ± for ordinary size sewers.

Manhole

Relative Inverts to keep zero pressure in pipe.

D_2

D_1

Allow $D_1 - D_2$

$H = \frac{V^2}{43}$

D

Sharp cornered entrance

Height of back up water at head wall for pipe flowing full.

FIG. A— SPECIAL HEAD LOSS ALLOWANCES IN SEWERS.*

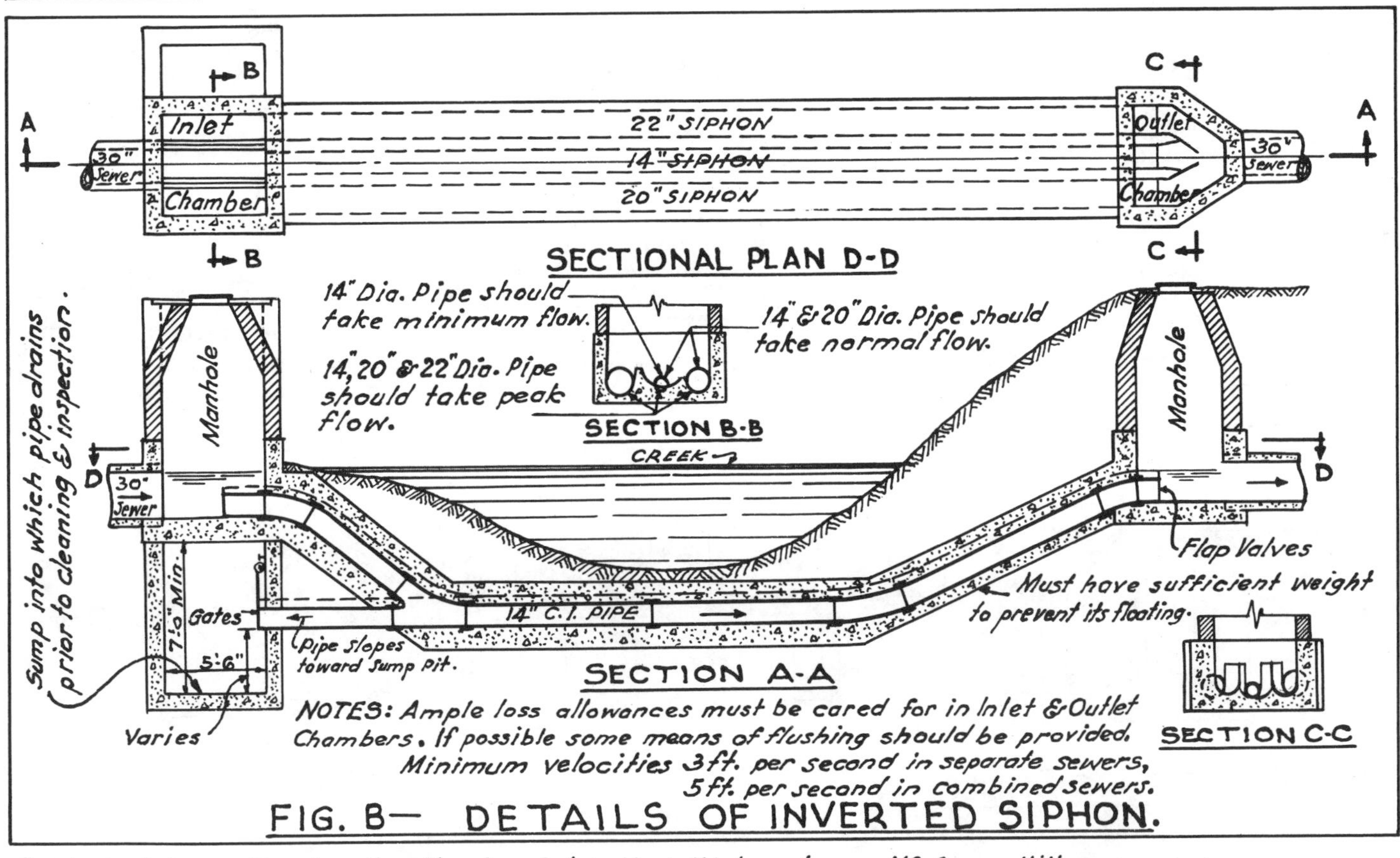

B
B
C
C
A
A
Inlet
Chamber
30" Sewer
22" SIPHON
14" SIPHON
20" SIPHON
Outlet
Chamber
30" Sewer
SECTIONAL PLAN D-D
14" Dia. Pipe should take minimum flow.
14" & 20" Dia. Pipe should take normal flow.
14", 20" & 22" Dia. Pipe should take peak flow.
SECTION B-B
CREEK
Manhole
Manhole
D
D
30" Sewer
Sump into which pipe drains prior to cleaning & inspection.
7'-0" Min.
Gates
5'-6"
Varies
Pipe slopes toward Sump Pit.
14" C.I. PIPE
Flap Valves
Must have sufficient weight to prevent its floating.
SECTION A-A
SECTION C-C

NOTES: Ample loss allowances must be cared for in Inlet & Outlet Chambers. If possible some means of flushing should be provided. Minimum velocities 3 ft. per second in separate sewers, 5 ft. per second in combined sewers.

FIG. B— DETAILS OF INVERTED SIPHON.

*Adapted from Davis, Handbook of Applied Hydraulics, McGraw-Hill.

DRAINAGE & SEWERAGE – PIPE DETAILS – 1

FIG. A – VITRIFIED CLAY PIPE JOINTS & CONNECTIONS.

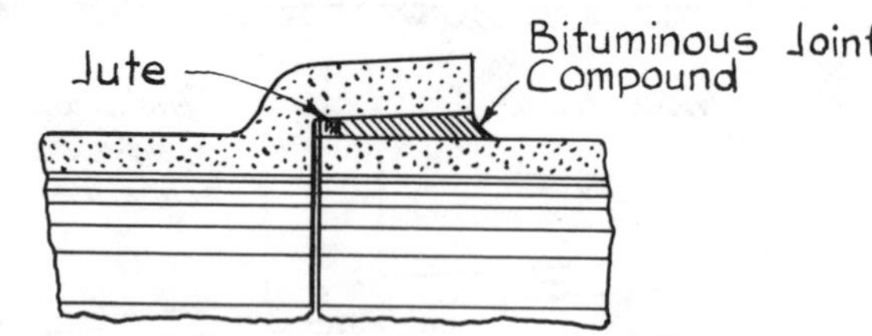
Jute
Bituminous Joint Compound

POORED BITUMINOUS COMPOUND JOINT.

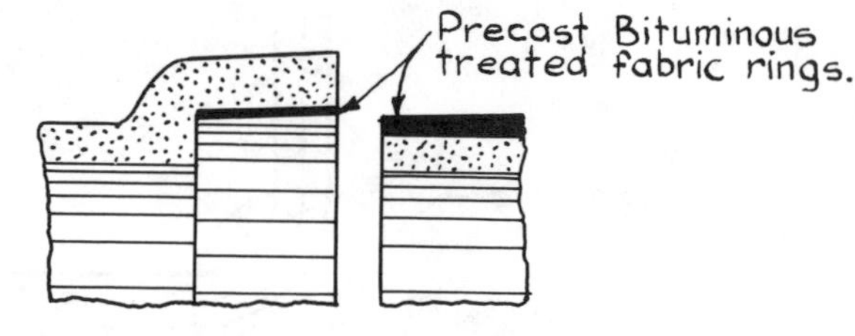
Precast Bituminous treated fabric rings.

SLIP SEAL JOINT APPLIED TO PIPE BEFORE JOINING.

Plug for cleanout if end of line.
A
A
30° or 45° bend
House conn.
Sewer lateral.
PLAN
SECTION A-A

STANDARD SEWER HOUSE CONNECTION.

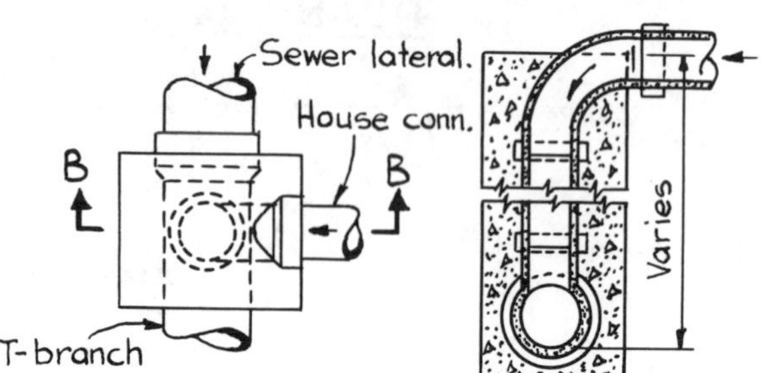
Sewer lateral.
House conn.
B
B
T-branch
Varies
PLAN
SECTION B-B

SEWER HOUSE CONNECTION WITH VERTICAL RISER.

FIG. B – ASBESTOS – CEMENT SEWER PIPE DETAILS.

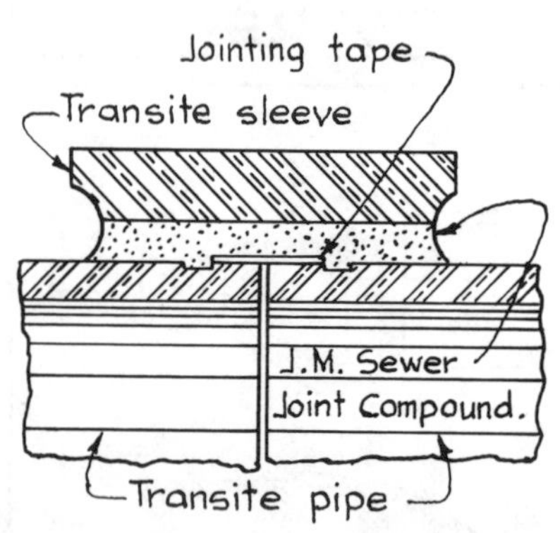
Jointing tape
Transite sleeve
J.M. Sewer Joint Compound.
Transite pipe

FIELD CUT PIPE CONNECTION.*

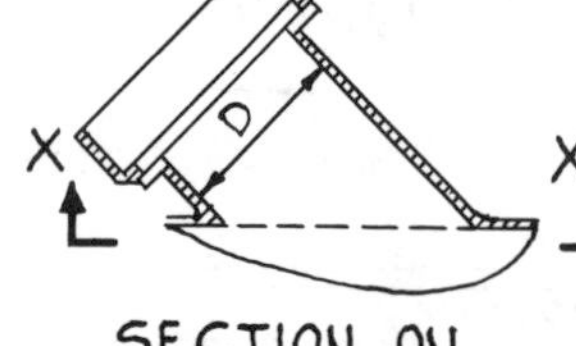
X
X
D

SECTION ON LONG AXIS.

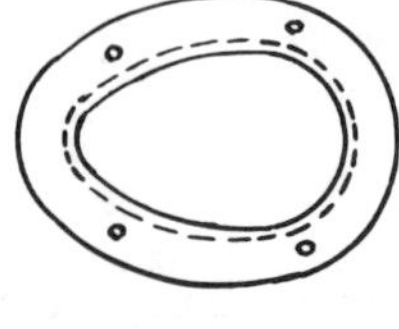

VIEW X-X

NOTE: 4 bolt holes equally spaced, bolts to be ⅜" diameter Monel carriage bolts fitted with lead washers and Monel hex. nuts. 4 supplied with each branch connection.

45-DEGREE TYPE.

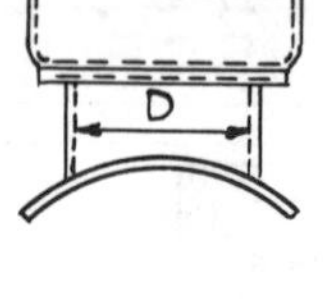
D
Y
Y

SECTION ON LONG AXIS.

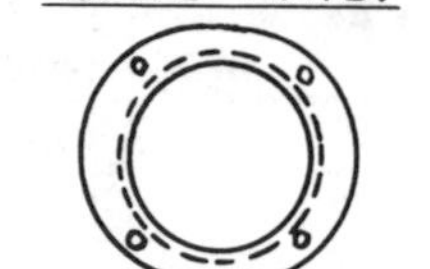

VIEW Y-Y

90-DEGREE TYPE.

CAST IRON CUT-INS FOR HOUSE CONNECTIONS TO TRANSITE PIPE.*

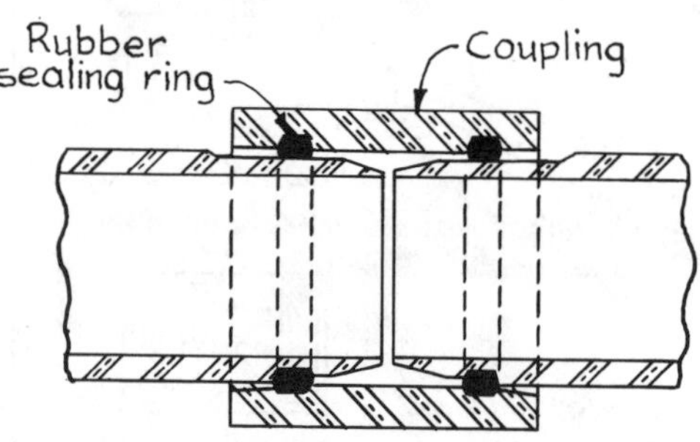
Rubber sealing ring
Coupling

RING TYPE COUPLING CONNECTION.

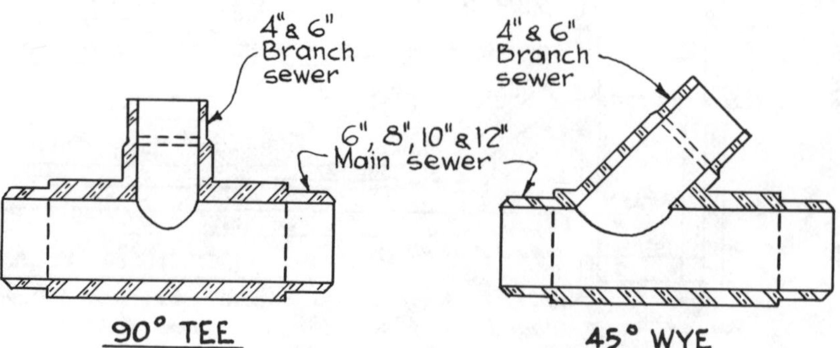
4" & 6" Branch sewer
6", 8", 10" & 12" Main sewer
4" & 6" Branch sewer
90° TEE
45° WYE

SEWER PIPE FITTINGS.†

*From Johns-Manville Corp.
†From Keasbey and Mattison.

DRAINAGE & SEWERAGE – PIPE DETAILS–2

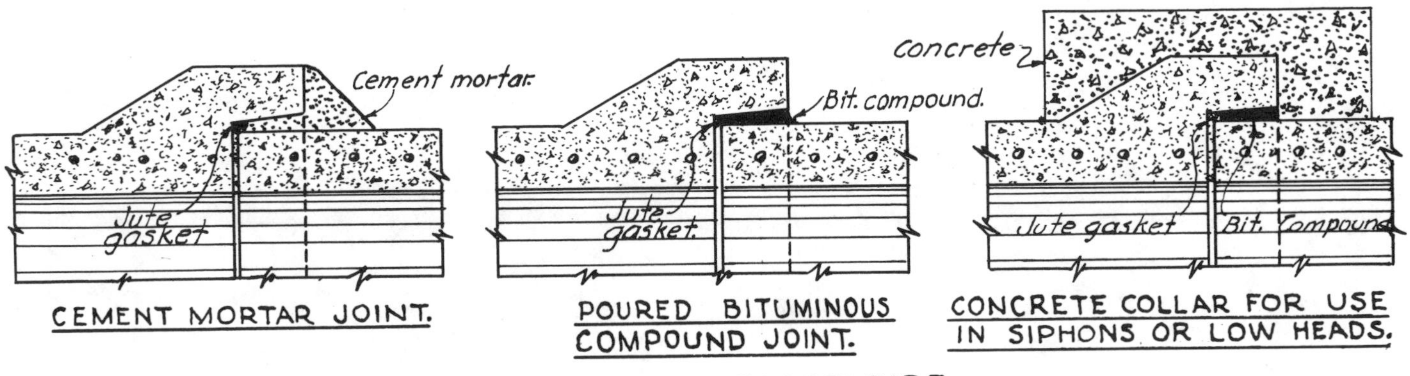
Concrete
Cement mortar.
Bit. compound.
Jute gasket
Jute gasket
Jute gasket
Bit. compound
CEMENT MORTAR JOINT.
POURED BITUMINOUS COMPOUND JOINT.
CONCRETE COLLAR FOR USE IN SIPHONS OR LOW HEADS.

BELL AND SPIGOT PIPE.

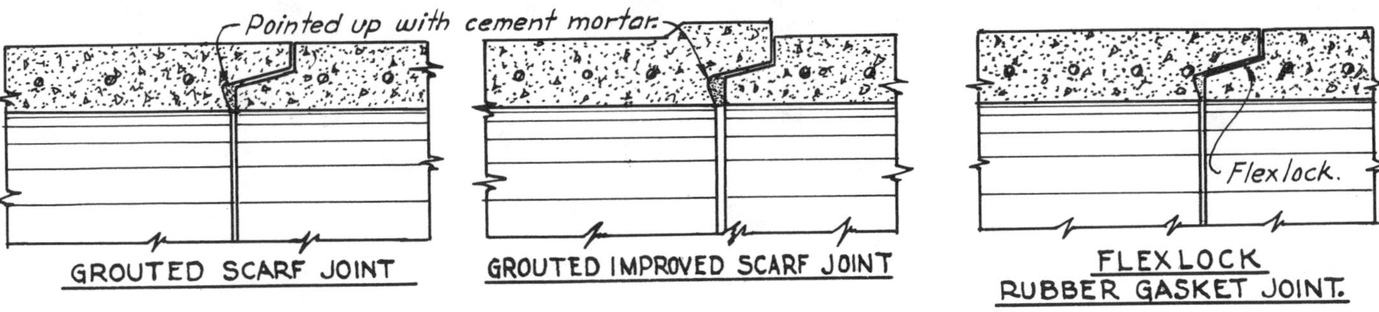
Pointed up with cement mortar.
Flexlock.
GROUTED SCARF JOINT
GROUTED IMPROVED SCARF JOINT
FLEXLOCK RUBBER GASKET JOINT.

TONGUE & GROOVE JOINT PIPE.

FIG. A— CONCRETE PIPE JOINTS.

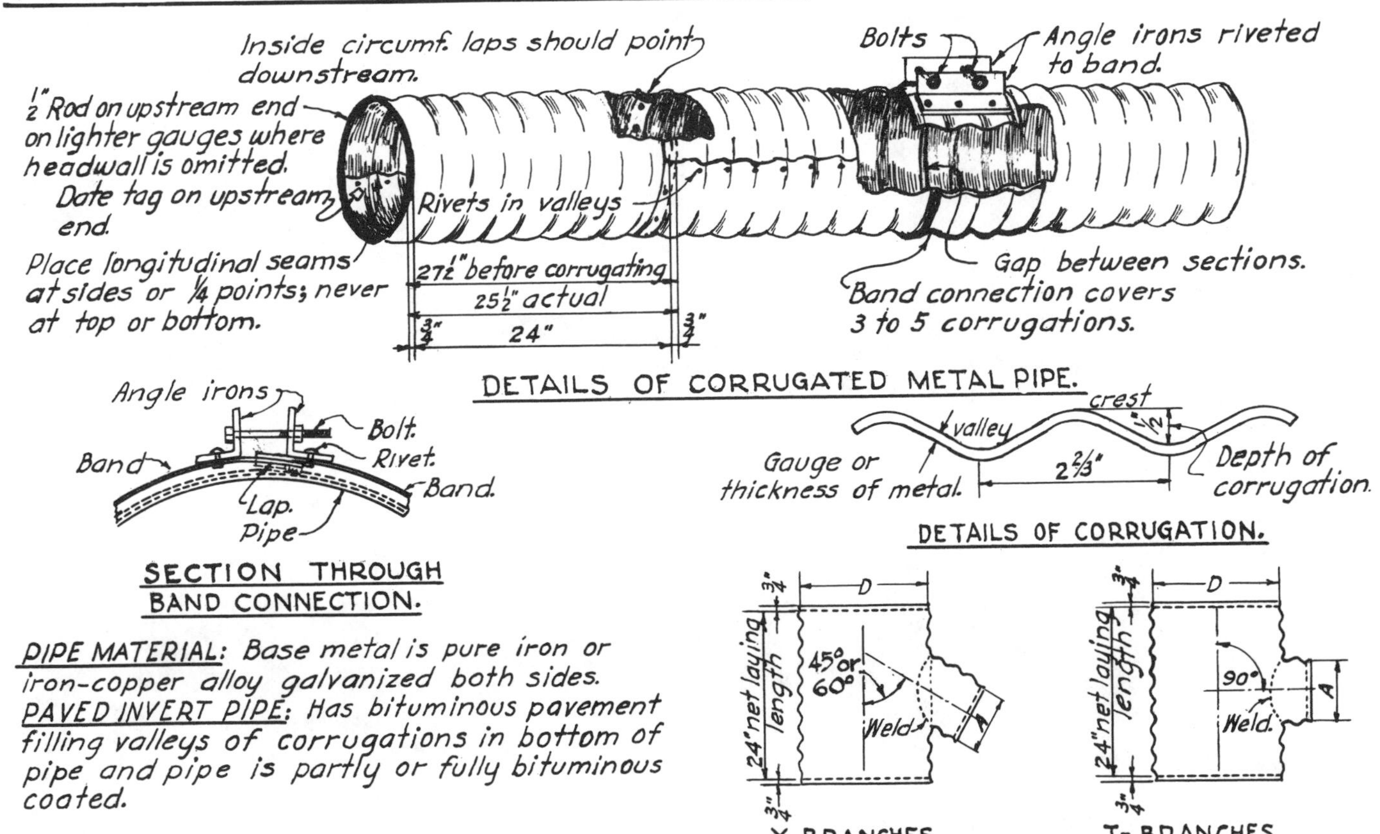
Inside circumf. laps should point downstream.
Bolts
Angle irons riveted to band.
½" Rod on upstream end on lighter gauges where headwall is omitted.
Date tag on upstream end.
Rivets in valleys
Place longitudinal seams at sides or ¼ points; never at top or bottom.
27½" before corrugating
25½" actual
¾"
24"
¾"
Gap between sections.
Band connection covers 3 to 5 corrugations.
DETAILS OF CORRUGATED METAL PIPE.
Angle irons
Bolt.
Rivet.
Band
Band.
Lap.
Pipe
SECTION THROUGH BAND CONNECTION.
crest
valley
½"
Gauge or thickness of metal.
2⅔"
Depth of corrugation.
DETAILS OF CORRUGATION.
D
¾"
24" net laying length
45° or 60°
Weld.
A
Y-BRANCHES
90°
T-BRANCHES.

PIPE MATERIAL: Base metal is pure iron or iron-copper alloy galvanized both sides.
PAVED INVERT PIPE: Has bituminous pavement filling valleys of corrugations in bottom of pipe and pipe is partly or fully bituminous coated.

FIG. B— CORRUGATED METAL PIPE DETAILS.*

* Data from Armco Drainage Products Assoc.

SEWAGE TREATMENT-GENERAL DATA

TABLE A – SEWAGE CHARACTERISTICS*

Suspended Solids (S.S.)	lb. per capita daily
1. Separate sewers	0.17
2. Combined sewers	0.33
5-day biochemical oxygen demand (B.O.D.)	**lbs. per capita daily**
1. Residential areas only, separate sewers	0.12
2. Residential areas, including ordinary commercial establishments, separate sewers	0.17
3. Residential areas, including ordinary commercial establishments, combined sewers	0.25
4. Manufacturing city with no large industrial wastes	0.33
5. Industrial wastes (frequently quite high)	0.33 - 0.50 and over

TABLE B – POPULATION EQUIVALENTS OF SEWAGE†

1. Suspended solids (S.S.) = 0.2 lb./capita/day

2. Five-day biochemical oxygen demand (B.O.D.) = 0.17 lb./capita/day

NOTES: (1) Population equivalents are valuable to determine additional load imposed on sewage treatment plant due to industrial waste.

(2) To obtain population equivalents of industrial or other waste, divide by above figures.

EXAMPLE: *Given* A cannery waste which has 35 gal. of waste per case of no. 2 cans. Waste shows a 5-day B.O.D. of 925 p.p.m. and suspended solids of 225 p.p.m.

Required: Population equivalents of 1 case of no. 2 cans/day.

Solution:

(a) 5-day B.O.D. = $(925)(35)(\frac{62.5}{7.48})(10^{-6})(\frac{1}{0.17})$ = 1.8 people

(b) S. S. = $(225)(35)(\frac{62.5}{7.48})(10^{-6})(\frac{1}{0.2})$ = 0.3 people

TABLE C – APPROXIMATE EFFICIENCIES OF UNITS AND PLANTS‡

Treatment, Operation or Process	% Removal B.O.D.	% Removal Suspended Solids	% Removal B. coli, (coliform)
Fine screening	5-10	2-20	10-20
Chlorination of raw or settled sewage	15-30	–	90-95
Plain sedimentation	25-40	40-70	25-75
Chemical precipitation	50-85	70-90	40-80
High rate trickling filtration preceded and followed by plain sedimentation	65-95	65-92	80-95
Low-rate trickling filtration preceded and followed by plain sedimentation	80-95	70-92	90-95
High-rate activated-sludge treatment preceded and followed by plain sedimentation	65-95	65-95	80-95
Conventional activated-sludge treatment preceded and followed by plain sedimentation	75-95	85-95	90-98
Intermittent sand filtration	90-95	85-95	95-98
Chlorination of biologically treated sewage	–	–	98-99

TABLE D – DILUTION REQUIREMENTS§

To Be Used as Guide Only.

Generally state authorities establish standards for the various waterways under their jurisdiction. Also dilution requirements depend on quality of receiving waters.

	Per 1000 Persons, c.f.s.
Raw sewage	3.5 - 6.0
Settled Sewage	2.0 - 4.0
Chemical treatment effluent	1.5 - 2.5
Trickling-filter effluent	0.5 - 1.0
Activated-sludge effluent	0.3 - 0.5
Sand-filter effluent	0.3 - 0.5

TABLE E – APPROXIMATE RESULTS OBTAINABLE BY PRIMARY AND BY COMPLETE TREATMENT OF AVERAGE DOMESTIC SEWAGE

	Raw Sewage, p.p.m.	After Primary Treatment	After Complete Treatment
B.O.D.	250	140 - 200	15 - 60
Suspended solids	250	100 - 140	15 - 50

* From Davis, *Handbook of Applied Hydraulics*, McGraw-Hill.

† From Glossary of Water and Sewage Control Engineering, joint sponsors: A.S.C.E., A.P.H.A., F.S.W.A., and A.W.W.A.

‡ From Karl Imhoff and Gordon M. Fair, *Sewage Treatment*, Wiley, 1956.

§ From W. Rudolfs, *Principles of Sewage Treatment*, National Lime Association, Washington, D. C.

SEWAGE TREATMENT-FLOW DIAGRAMS-I

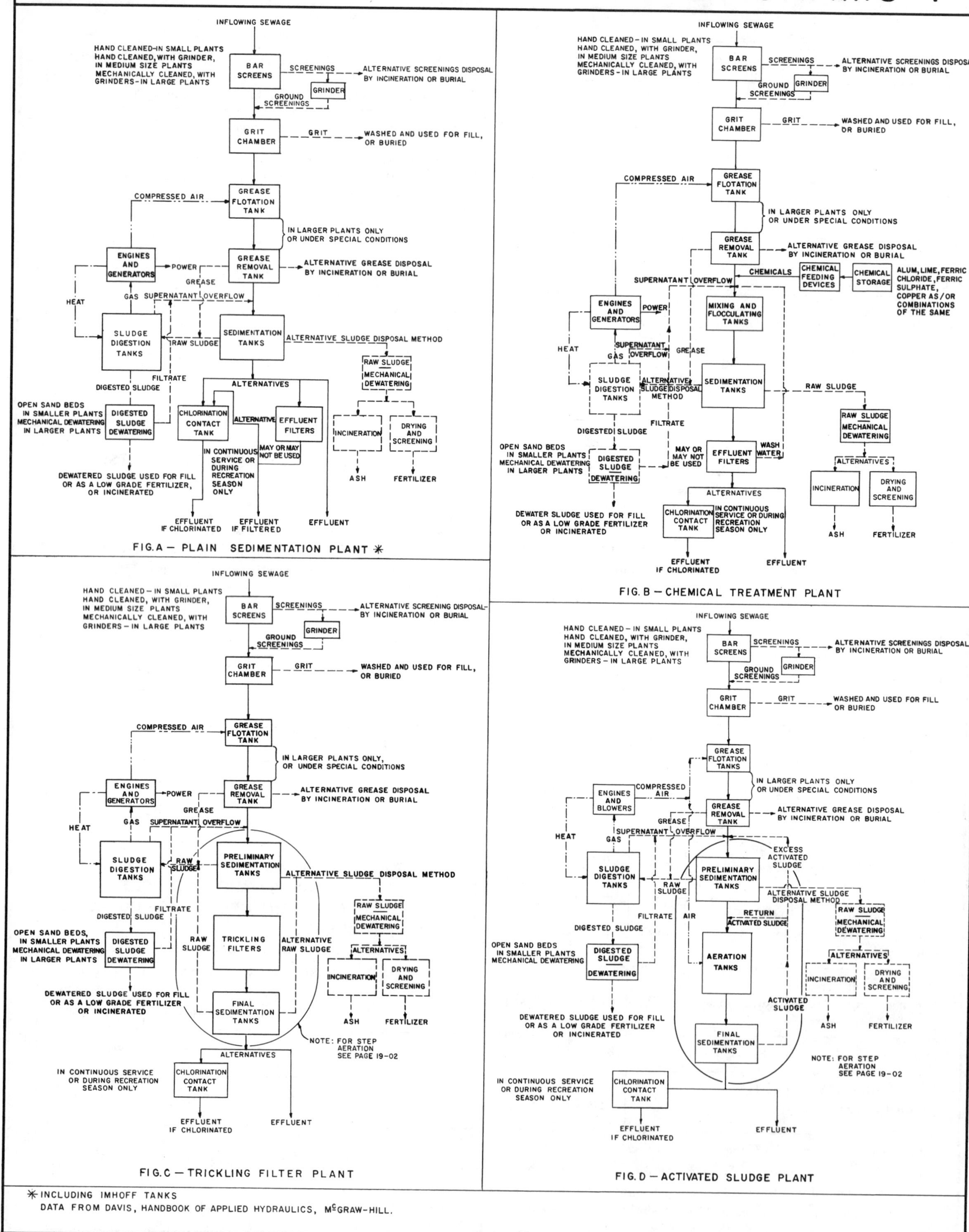
INFLOWING SEWAGE
HAND CLEANED-IN SMALL PLANTS
HAND CLEANED, WITH GRINDER, IN MEDIUM SIZE PLANTS
MECHANICALLY CLEANED, WITH GRINDERS-IN LARGE PLANTS
BAR SCREENS
SCREENINGS
ALTERNATIVE SCREENINGS DISPOSAL BY INCINERATION OR BURIAL
GRINDER
GROUND SCREENINGS
GRIT CHAMBER
GRIT
WASHED AND USED FOR FILL, OR BURIED
COMPRESSED AIR
GREASE FLOTATION TANK
IN LARGER PLANTS ONLY OR UNDER SPECIAL CONDITIONS
ENGINES AND GENERATORS
POWER
GREASE REMOVAL TANK
ALTERNATIVE GREASE DISPOSAL BY INCINERATION OR BURIAL
GREASE
HEAT
GAS
SUPERNATANT OVERFLOW
SLUDGE DIGESTION TANKS
RAW SLUDGE
SEDIMENTATION TANKS
ALTERNATIVE SLUDGE DISPOSAL METHOD
FILTRATE
DIGESTED SLUDGE
ALTERNATIVES
RAW SLUDGE MECHANICAL DEWATERING
OPEN SAND BEDS IN SMALLER PLANTS
MECHANICAL DEWATERING IN LARGER PLANTS
DIGESTED SLUDGE DEWATERING
CHLORINATION CONTACT TANK
ALTERNATIVE
EFFLUENT FILTERS
INCINERATION
DRYING AND SCREENING
IN CONTINUOUS SERVICE OR DURING RECREATION SEASON ONLY
MAY OR MAY NOT BE USED
ASH
FERTILIZER
DEWATERED SLUDGE USED FOR FILL OR AS A LOW GRADE FERTILIZER, OR INCINERATED
EFFLUENT IF CHLORINATED
EFFLUENT IF FILTERED
EFFLUENT
FIG. A — PLAIN SEDIMENTATION PLANT *
CHEMICALS
CHEMICAL FEEDING DEVICES
CHEMICAL STORAGE
ALUM, LIME, FERRIC CHLORIDE, FERRIC SULPHATE, COPPER AS/OR COMBINATIONS OF THE SAME
MIXING AND FLOCCULATING TANKS
ALTERNATIVE SLUDGE DISPOSAL METHOD
WASH WATER
DEWATER SLUDGE USED FOR FILL OR AS A LOW GRADE FERTILIZER OR INCINERATED
FIG. B — CHEMICAL TREATMENT PLANT
PRELIMINARY SEDIMENTATION TANKS
TRICKLING FILTERS
ALTERNATIVE RAW SLUDGE
FINAL SEDIMENTATION TANKS
NOTE: FOR STEP AERATION SEE PAGE 19-02
FIG. C — TRICKLING FILTER PLANT
ENGINES AND BLOWERS
AIR
EXCESS ACTIVATED SLUDGE
RETURN ACTIVATED SLUDGE
AERATION TANKS
ACTIVATED SLUDGE
FIG. D — ACTIVATED SLUDGE PLANT
* INCLUDING IMHOFF TANKS
DATA FROM DAVIS, HANDBOOK OF APPLIED HYDRAULICS, McGRAW-HILL.

SEWAGE TREATMENT-FLOW DIAGRAMS-2

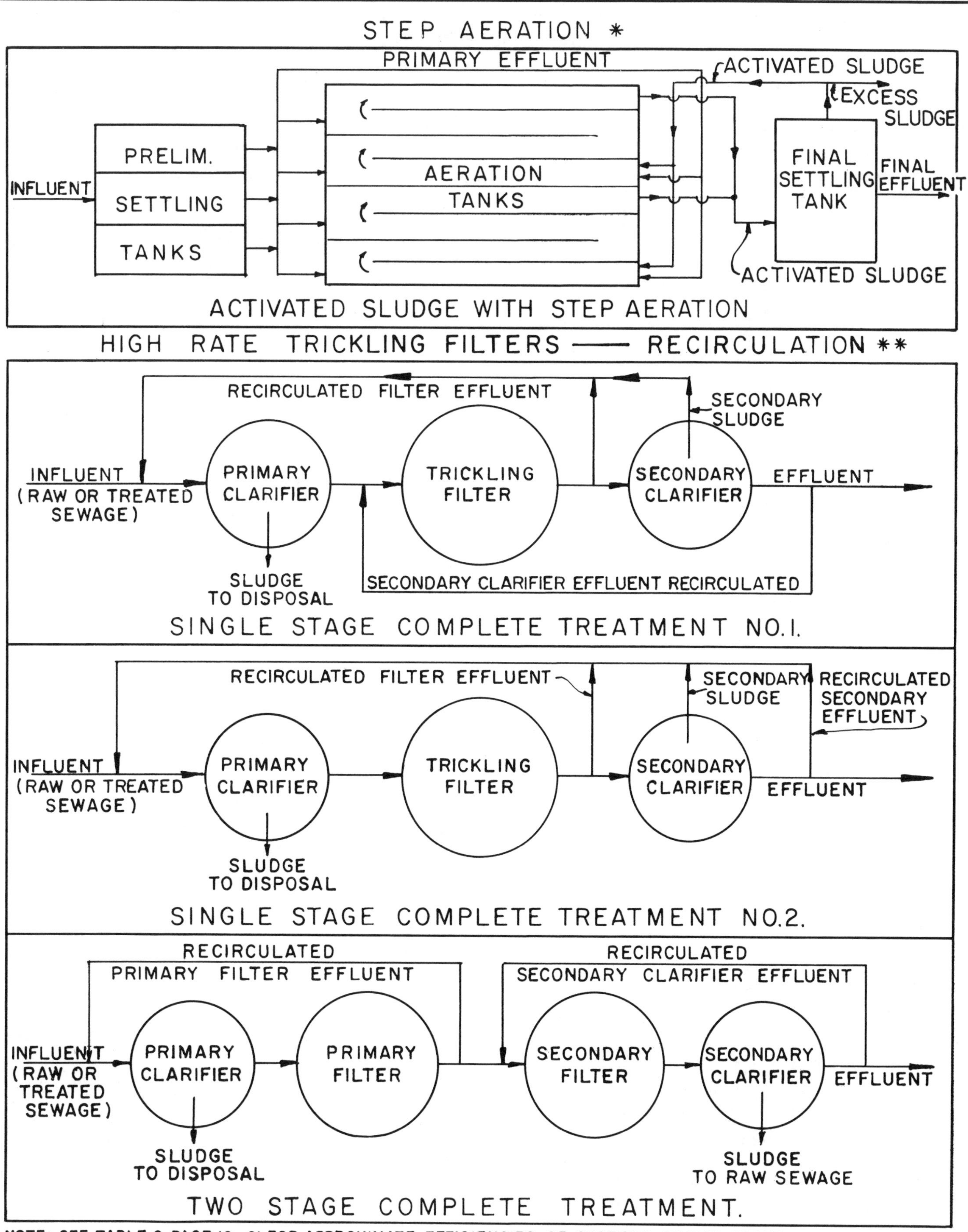
STEP AERATION *
PRIMARY EFFLUENT
ACTIVATED SLUDGE
EXCESS SLUDGE
PRELIM.
SETTLING
TANKS
INFLUENT
AERATION TANKS
FINAL SETTLING TANK
FINAL EFFLUENT
ACTIVATED SLUDGE
ACTIVATED SLUDGE WITH STEP AERATION
HIGH RATE TRICKLING FILTERS —— RECIRCULATION **
RECIRCULATED FILTER EFFLUENT
SECONDARY SLUDGE
INFLUENT (RAW OR TREATED SEWAGE)
PRIMARY CLARIFIER
TRICKLING FILTER
SECONDARY CLARIFIER
EFFLUENT
SLUDGE TO DISPOSAL
SECONDARY CLARIFIER EFFLUENT RECIRCULATED
SINGLE STAGE COMPLETE TREATMENT NO.1.
RECIRCULATED FILTER EFFLUENT
SECONDARY SLUDGE
RECIRCULATED SECONDARY EFFLUENT
INFLUENT (RAW OR TREATED SEWAGE)
PRIMARY CLARIFIER
TRICKLING FILTER
SECONDARY CLARIFIER
EFFLUENT
SLUDGE TO DISPOSAL
SINGLE STAGE COMPLETE TREATMENT NO.2.
RECIRCULATED PRIMARY FILTER EFFLUENT
RECIRCULATED SECONDARY CLARIFIER EFFLUENT
INFLUENT (RAW OR TREATED SEWAGE)
PRIMARY CLARIFIER
PRIMARY FILTER
SECONDARY FILTER
SECONDARY CLARIFIER
EFFLUENT
SLUDGE TO DISPOSAL
SLUDGE TO RAW SEWAGE
TWO STAGE COMPLETE TREATMENT.

NOTE: SEE TABLE C, PAGE 19-01 FOR APPROXIMATE EFFICIENCIES OF SYSTEMS SHOWN ABOVE.

* DATA FROM GOULD, R.H. – TALLMAN'S ISLAND SEWAGE WORKS FROM MUNICIPAL SANITATION.

** DATA FROM DORR OLIVER INC., CATALOGUE.

SEWAGE TREATMENT—SCREENS-1

COARSE SCREEN TYPE

*Clear openings between bars 1" to 2" for manually cleaned screens or as small as 5/8" for mechanically cleaned screens. Total area between bars such as to maintain velocity of 1 f.p.s. average. Maximum velocity = 2.5 f.p.s.

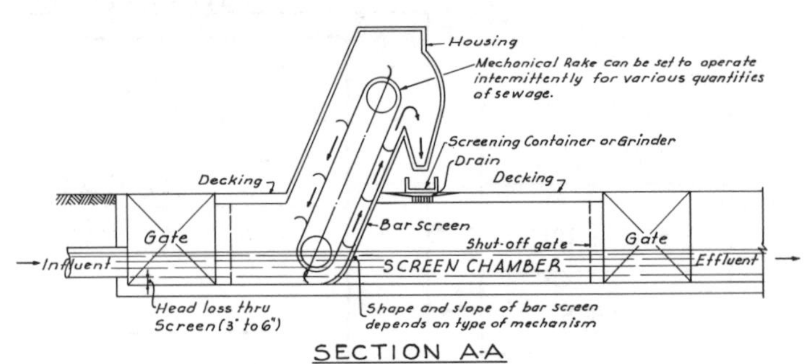
Housing
Mechanical Rake can be set to operate intermittently for various quantities of sewage.
Screening Container or Grinder
Drain
Decking
Decking
Gate
Bar screen
Shut-off gate
Gate
Influent
SCREEN CHAMBER
Effluent
Head loss thru Screen (3" to 6")
Shape and slope of bar screen depends on type of mechanism

SECTION A-A

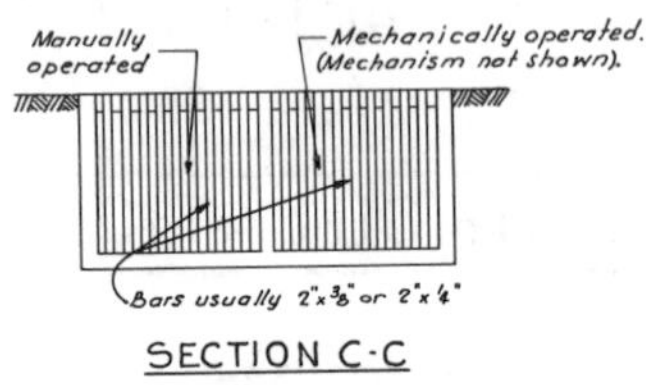
Manually operated
Mechanically operated. (Mechanism not shown).
Bars usually 2"x3/8" or 2"x1/4"

SECTION C-C

QUANTITIES OF SCREENINGS†

0.5 to 6 cu. ft. per million gallons for racks with openings of 2" to 1/2".

5 to 30 cu. ft. per million gallons for fine screens (see below) with openings of 3/4" to 3/32".

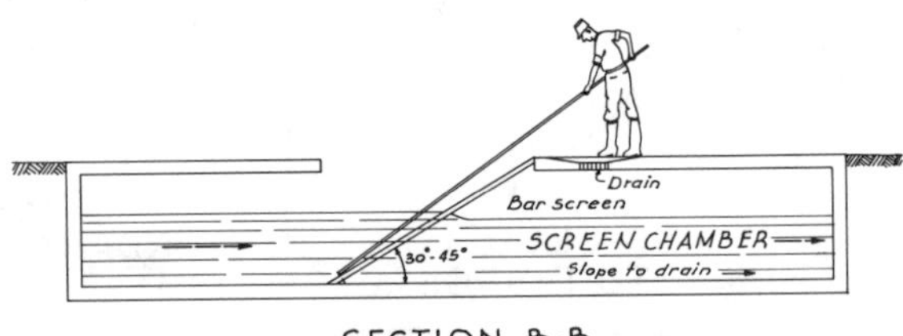
Drain
Bar screen
SCREEN CHAMBER
30°-45°
Slope to drain

SECTION B-B

DISPOSAL OF SCREENINGS

By burial or incineration.

When quantities of screenings are not excessive, shredders and grinders are sometimes used (see below), and ground screenings are replaced in the effluent to pass through rest of treatment units.

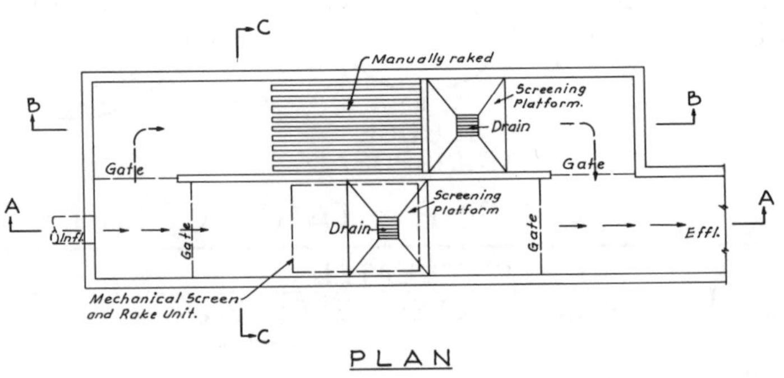
C
Manually raked
Screening Platform.
Drain
B
B
Gate
Gate
A
A
Infl.
Gate
Drain
Screening Platform
Gate
Effl.
Mechanical Screen and Rake Unit.
C

PLAN

FINE SCREEN Use is not recommended.

SHREDDERS AND GRINDERS Manufactured in two types:

(a) For grinding solids already removed from sewage by screenings.

(b) For grinding solids as sewage flows through the device, in some cases replacing the screening process. Head loss for this type = 2" to 15". See p. 19-05 for comminators and barminators.

*Compiled from *Standards for Sewage Works*, 1952. Prepared by Upper Mississippi and Great Lakes Board of Public Health Engineers (Ill., Ind., Iowa, Mich., Minn., Mo., Neb., N.Y., Ohio, Pa., Wis.,).

†Adapted from Metcalf and Eddy, *American Sewerage Practice*, McGraw-Hill.

SEWAGE TREATMENT-SCREENS-2

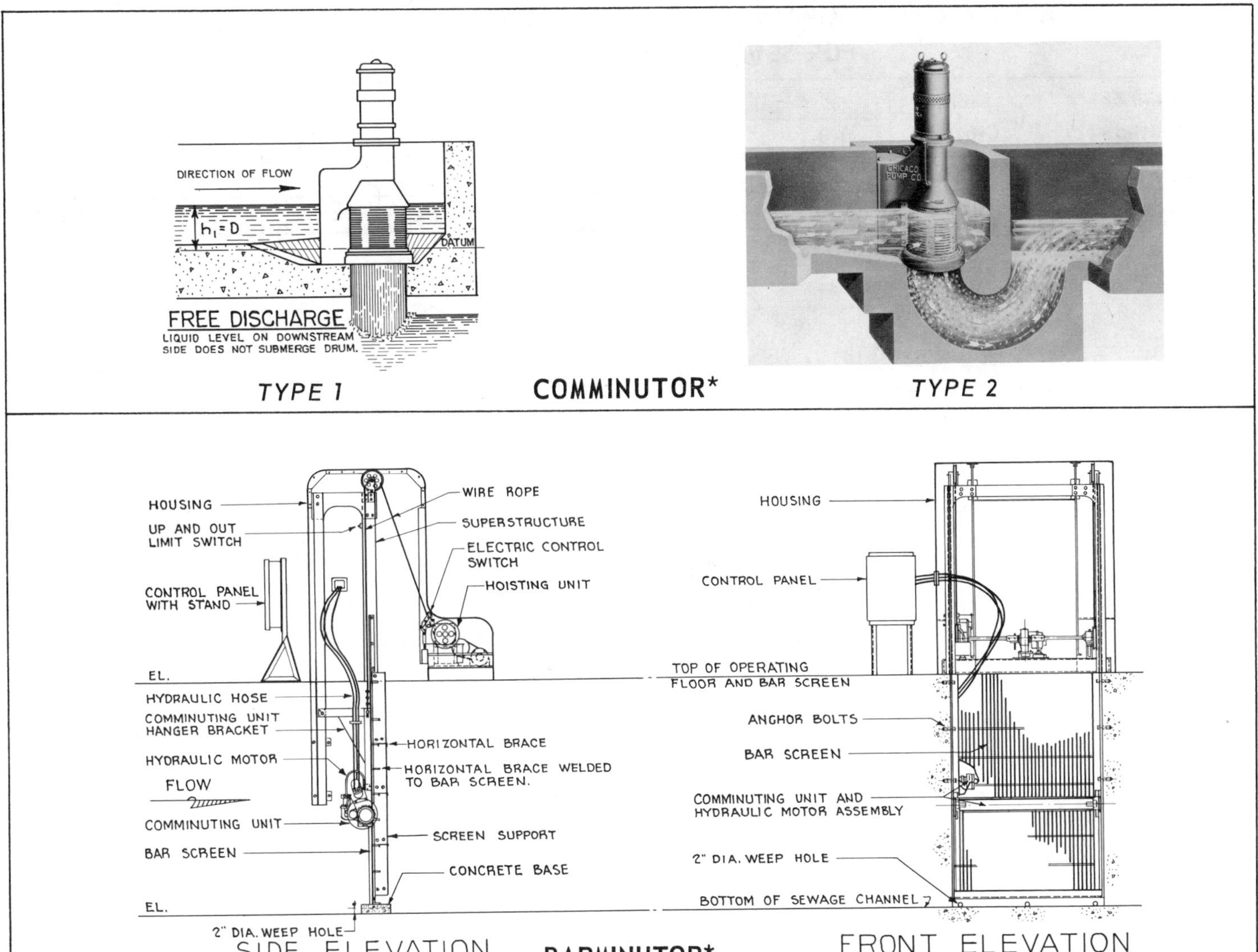
DIRECTION OF FLOW
h_1=D
DATUM
FREE DISCHARGE
LIQUID LEVEL ON DOWNSTREAM SIDE DOES NOT SUBMERGE DRUM.
TYPE 1
COMMINUTOR*
TYPE 2
HOUSING
UP AND OUT LIMIT SWITCH
CONTROL PANEL WITH STAND
WIRE ROPE
SUPERSTRUCTURE
ELECTRIC CONTROL SWITCH
HOISTING UNIT
EL.
HYDRAULIC HOSE
COMMINUTING UNIT HANGER BRACKET
HYDRAULIC MOTOR
FLOW
COMMINUTING UNIT
BAR SCREEN
HORIZONTAL BRACE
HORIZONTAL BRACE WELDED TO BAR SCREEN.
SCREEN SUPPORT
CONCRETE BASE
2" DIA. WEEP HOLE
SIDE ELEVATION
BARMINUTOR*
HOUSING
CONTROL PANEL
TOP OF OPERATING FLOOR AND BAR SCREEN
ANCHOR BOLTS
BAR SCREEN
COMMINUTING UNIT AND HYDRAULIC MOTOR ASSEMBLY
2" DIA. WEEP HOLE
BOTTOM OF SEWAGE CHANNEL
FRONT ELEVATION

Comminuting mechanism provides continuous, automatic screening and cutting of coarse sewage solids into small solids without removing solids from sewage flow.

Size of slots: from ¼" to ¾", depending on type of equipment and manufacturer.

Location of grinder: preceding wet well of pumping station or preceding primary settling tanks. Outdoor locations are generally used except that, for a pumping station, grinder may be installed in wet-well compartment with motor mounted on floor above or out of doors.

Auxiliary screen: Where shredding-screening devices are used, it is general practice to provide an auxiliary hand-cleaned screen arranged in such a manner as to allow automatic diversion of the sewage flow if the mechanical unit should fail.

Comminutor Sizes and Capacities*

No.	Size of Motors	Over-all Capacities, (m.g.d.)	
		Controlled Discharge	Free Discharge
7B	¼	0-0.35	0-0.30
10A	½	0.17-1.1	0.17-0.82
15M	¾	0.4 -2.3	0.4 -1.4
25M	1½	1.0 -6.0	1.0 -3.6
25A	1½	1.0 -11.0	1.0 -6.5
36A	2	1.5 -25.0	1.5 -9.6
54A	Separately designed for each job		

Barminutor Capacities*

Type	Channel Width, ft.	Flow, † m.g.d.
B	1	0.09-1.25†
	1.5	0.24-3.40†
	2	0.34-6.50†
	3	1.70-14.50†
A-2	4	15- 26‡
A	5	29- 48‡
	6	35- 58‡
	7	54- 91‡
	8	62-104‡
	9	87-145‡
	10	97-162‡
	11	128-214‡
	12	140-233‡

* Chicago Pump Co. † Based on 6" liquid-level differential at optimum submergence. ‡ Based on 1.5 to 2.5 f.p.s. upstream velocity.

SEWAGE TREATMENT – GRIT CHAMBERS – I

TABLE A – APPROXIMATE SETTLING VELOCITIES IN INCHES PER MINUTE FOR SEWAGE TEMPERATURE OF 50°F*

Kind of Particle	Specific Gravity	Diameter, mm.						
		1.0	0.5	0.2	0.1	0.05	0.01	0.005
Quartz sand	2.65	330	170	54	16	4	0.2	0.04
Sewage solids	1.01-1.2	1-80	0.2-40	0.01-12	0.01-2	≦0.5	≦0.02	≦0.005

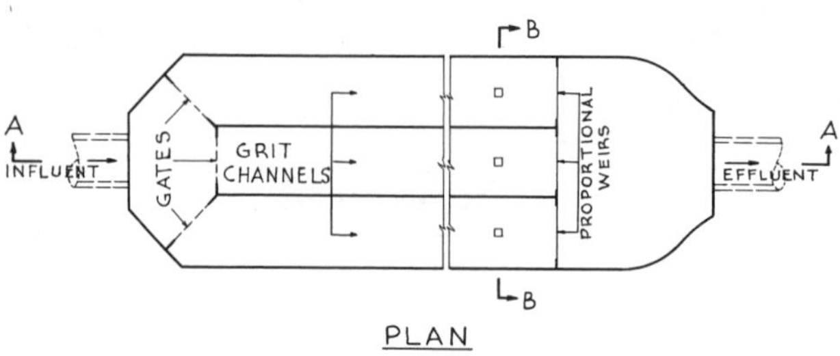
B
A
INFLUENT
GATES
GRIT CHANNELS
PROPORTIONAL WEIRS
EFFLUENT
A
B
PLAN

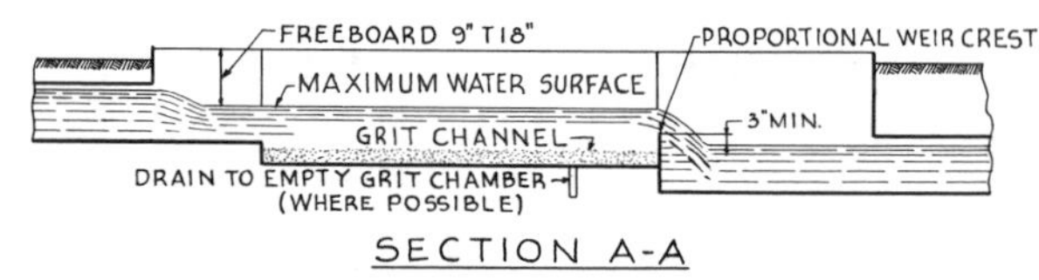
FREEBOARD 9" T 18"
PROPORTIONAL WEIR CREST
MAXIMUM WATER SURFACE
3" MIN.
GRIT CHANNEL
DRAIN TO EMPTY GRIT CHAMBER (WHERE POSSIBLE)
SECTION A-A

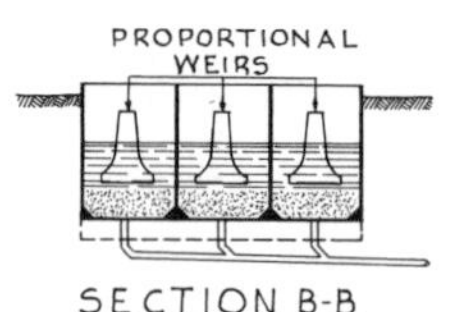
PROPORTIONAL WEIRS
SECTION B-B

GRIT CHAMBER DESIGN

DESIGN

1. *Where Required* At all sewage-treatment plants that receive sewage from combined sewers or from all sanitary sewers where grit is known to be present. Also advocated for plants treating sewage from sanitary sewers where grit may be anticipated.
2. *Location* Location recommended ahead of pumps and comminuting devices. In such cases, coarse bar racks should be placed ahead of mechanically cleaned grit chambers.
3. *Number Required* Duplicate hand-cleaned units or a single mechanically cleaned unit with bypass for plants treating wastes from combined sewers. Mechanically cleaned grit chambers are recommended. Single hand-cleaned channels with bypass are recommended for sewage treatment plants serving sanitary sewerage systems.
4. *Design Factors*

   (a) Velocities: Channels should be designed to provide velocities of not less than ½ f.p.s. and as close as possible to 1 f.p.s.

   (b) Detention Period: 20 seconds to 1 minute.

   (c) Velocity Control: Provision should be made for the regulation of velocity to minimize deposition of organic matter.

   (d) Grit Washing: All chambers not providing positive velocity control should include grit-washing devices for the further separation of organic and inorganic materials.

   (e) Cross Section Required: Equals [rate of sewage flow (c.f.s.) divided by velocity (f.p.s.)] + depth for grit storage and freeboard.

   (f) Overflow rate: Overflow rate (g.p.d. per ft.^{2} of surface) = 900 × settling rate of smallest particle that it is desirable to remove (generally 0.2 mm.).

* From *Sewage Treatment*, by Imhoff and Fair.

SEWAGE TREATMENT – GRIT CHAMBERS – 2

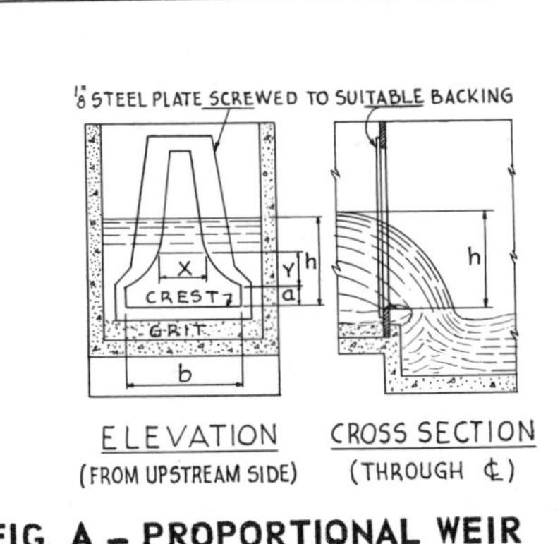

1/8" STEEL PLATE SCREWED TO SUITABLE BACKING
X
Y
h
a
CREST
GRIT
b
h
ELEVATION
(FROM UPSTREAM SIDE)
CROSS SECTION
(THROUGH ℄)

FIG. A – PROPORTIONAL WEIR

FORMULAS FOR DESIGN OF PROPORTIONAL WEIRS*

(1) $Q = 4.97a^{\frac{1}{2}}b\left(h - \frac{a}{3}\right)$

(2) $\frac{x}{b} = 1 - \frac{2}{\pi}\arctan\sqrt{\frac{y}{a}}$

Q = quantity of flow, cu. ft. per sec.

For explanation of other symbols see figures at left.

TABLE B – VALUES OF Y/a AND X/b

Y/a	X/b	Y/a	X/b	Y/a	X/b
0.1	0.805	1.0	0.500	10	0.195
0.2	0.732	2.0	0.392	12	0.179
0.3	0.681	3.0	0.333	14	0.166
0.4	0.641	4.0	0.295	16	0.156
0.5	0.608	5.0	0.268	18	0.147
0.6	0.580	6.0	0.247	20	0.140
0.7	0.556	7.0	0.230	25	0.126
0.8	0.536	8.0	0.216	30	0.115
0.9	0.517	9.0	0.205		

GRIT CHAMBER AND PROPORTIONAL WEIR – EXAMPLE OF DESIGN

Given: A separate sewer system. – Average flow = 4 c.f.s.; Peak flow = 10 c.f.s.; Average accumulation of grit = 6 cu. ft./million gallons of sewage.

Required: Size of channels and proportional weirs.

Solution:

Channel Section

1. Number of units: Grit chamber will consist of three units; two to accommodate peak flow, the third will be a standby unit.
2. Peak Flow in Each Channel = $\frac{10 \text{ c.f.s.}}{2} = 5.0$ c.f.s.
3. Cross-Section Area: Q = 5.0 c.f.s. Velocity = 1 f.p.s. (see note 4 on p. 19-06). Area above crest of weir (exclusive of freeboard) = $\frac{5}{1} = 5.0 \text{ ft.}^2$
4. Settling Velocity of a 0.2-mm. particle of sand from Table A, p. 19-06, is found to be 54"/min.
5. Overflow Rate: From note 4 on p. 19-06,

   Overflow rate = 900 × 54 = 48,600 gal./day/ft.^{2} or 0.07 c.f.s./ft.^{2}
6. Surface Area: Each channel requires $\frac{5 \text{ c.f.s.}}{0.07} = 71.0 \text{ ft.}^2$
7. Depth and Width: Assuming a depth of 2.5 ft. (above crest of weir exclusive of freeboard):

   $$\text{Width} = \frac{5.0 \text{ ft.}^2}{2.5} = 2.0 \text{ ft.}$$
8. Length of Channel:

   $$\text{Length} = \frac{71.0 \text{ ft.}^2 \text{(surface area)}}{2.0 \text{ ft. (width)}} = 35.5 \text{ ft.}$$
9. Volume of Grit Storage: Assume 6" below crest of weir for storage of grit. Storage space in each channel:

   $$\text{Storage space} = \frac{6}{12} \times 35.5 \times 2 = 35.5 \text{ ft.}^3$$

   Storage space required is as follows: 4 c.f.s. = 2.6 million gallons/day. Grit per day = 6 × 2.6 = 15.6 ft.^{3} This means that every second day one channel has to be put out of service to be cleaned.

Proportional Weirs

1. Values of *a* and *b:* Using equation 1 given above, and assuming a value for *a*, solve for *b* (*b* should be of such value as to fit into channel with not less than 4 in. to spare on each side). With Q = 5.0 c.f.s., h = 2.5 ft., and a = 0.12; b is found to be 1.18 ft.
2. Values of X and Y: To find points on curved sides of weir assume values of Y. Then substitute in Table B above; find X/b and solve for X.
3. It is advisable that the designer make a full-scale drawing of the proportional weir to serve as a template.

NOTE: Mechanical means of grit removal are generally provided for large installations. Sometimes combination grit and screening chambers are designed and equipped with mechanism for cleaning and removing grit.

*From E. Soucek, H. E. Howe, and F. T. Mavis, Sutro Weir Investigations, *Engineering News-Record*, Nov. 12, 1936, p. 679.

SEWAGE TREATMENT – GRIT CHAMBERS & PARSHALL FLUME

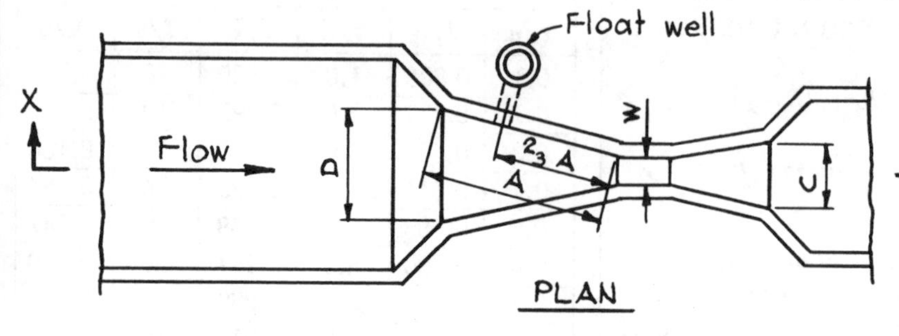

Float well
X
Flow
D
W
$\frac{2}{3}$ A
A
C
X

PLAN

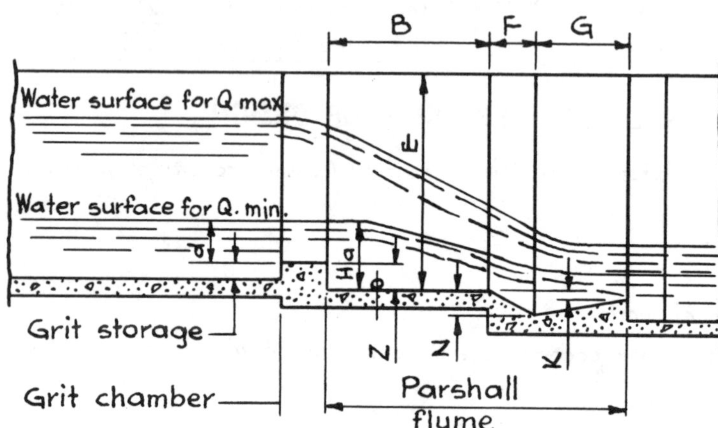

B
F
G
E
Water surface for Q max.
Water surface for Q. min.
d
H_a
Grit storage
N
Z
K
Grit chamber
Parshall flume

SECTION X-X

FLUME DIMENSIONS IN FEET AND INCHES										
W	A	⅔ A	B	C	D	E	F	G	K	N
0-3	1-6 ⅜	1-0 ¼	1-6	0-7	0-10 $\frac{3}{16}$	2-0	0-6	1-0	0-1	0-2 ¼
0-6	2-0 $\frac{7}{16}$	1-4 $\frac{5}{16}$	2-0	1-3 ⅝	1-3 ⅝	2-0	1-0	2-0	0-3	0-4 ½
0-9	2-10 ⅝	1-11 ⅛	2-10	1-3	1-10 ⅝	2-6	1-0	1-6	0-3	0-4 ½
1-0	4-6	3-0	4-4 ⅞	2-0	2-9 ¼	3-0	2-0	3-0	0-3	0-9
1-6	4-9	3-2	4-7 ⅞	2-6	3-4 ⅜	3-0	2-0	3-0	0-3	0-9
2-0	5-0	3-4	4-10 ⅞	3-0	3-11 ½	3-0	2-0	3-0	0-3	0-9
3-0	5-6	3-8	5-4 ¾	4-0	5-1 ⅞	3-0	2-0	3-0	0-3	0-9
4-0	6-0	4-0	5-10 ⅝	5-0	6-4 ¼	3-0	2-0	3-0	0-3	0-9
5-0	6-6	4-4	6-4 ½	6-0	7-6 ⅝	3-0	2-0	3-0	0-3	0-9
6-0	7-0	4-8	6-10 ⅜	7-0	8-9	3-0	2-0	3-0	0-3	0-9
7-0	7-6	5-0	7-4 ¼	8-0	9-11 ⅜	3-0	2-0	3-0	0-3	0-9
8-0	8-0	5-4	7-10 ⅛	9-0	11-1 ¾	3-0	2-0	3-0	0-3	0-9

PROBLEM: To design grit chamber and control flume.

Given: Maximum rate of flow = 7 M.G.D. (10.84 c.f.s.)
Minimum rate of flow = 2 M.G.D. (3.1 c.f.s. and velocities at minimum and maximum rate of flow = 1 f.p.s.

To Find: Dimensions of control flume and grit chamber.

Solution: Assume flume throat width (W) = 6 in. Solve for Z, using formula 3.

$$\frac{3.1}{10.84} = \frac{1.1(3.1/2.05)^{2/3} - Z}{1.1(10.84/2.05)^{2/3} - Z} = \frac{1.45 - Z}{3.34 - Z}$$

Determine depth of flow in grit chamber for various flows:

$$d = 1.1\left(\frac{Q}{4.1W}\right)^{2/3} - Z$$

Compute width of grit chamber from following:

$$b = \frac{Q\ \text{min}}{d\ \text{min}\ V} = \frac{Q\ \text{max}}{d\ \text{max}\ V} = \frac{3.1}{0.76 \times 1} = \frac{10.84}{2.65 \times 1} = 4.08\ \text{ft.}$$

FORMULAS FOR DESIGN OF PARSHALL FLUME AND GRIT CHAMBER

$$Q = 4.1\,WHa^{3/2} \qquad (1)$$

$$d + z = 1.1\,Ha \qquad (2)$$

$$\frac{Q\ \text{min}}{Q\ \text{max}} = \frac{1.1(Q\ \text{min}/4.1W)^{2/3} - Z}{1.1(Q\ \text{max}/4.1W)^{2/3} - Z} \qquad (3)$$

$$V_0 = \frac{2.6(1 - K^{1/3})^{1/2}(K^{1/3} - K)}{(1 - K)^{3/2}}$$

in which

Q = rate of flow, c.f.s.

Q max = maximum rate of flow, c.f.s.

Q min = minimum rate of flow, c.f.s.

V_0 = velocity at largest deviation from V min or V max when V min = V max.

PARSHALL FLUME AS A VELOCITY-CONTROL FLUME FOR GRIT CHAMBERS

ADVANTAGES

1. Simple construction and low cost.
2. Has lowest head loss of commonly used flow meters.
3. Offers unobstructed flow path so that sewage solids will not accumulate.
4. Serves as a means of flow measurement.
5. May be used for proportional control of chlorine feed.

GRIT CHAMBER (VERTICAL SIDES)				
Rate of Flow		Depth of Flow	Cross-Sectional Area	Velocity
M.G.D.	c.f.s.	ft.	ft.^{2}	f.p.s.
2	3.1	0.76	3.10	1.00
3	4.65	1.21	4.94	0.943
4	6.20	1.61	6.67	0.944
5	7.75	1.98	8.08	0.959
6	9.30	2.33	9.50	0.979
7	10.84	2.65	10.84	1.00

Compute velocities at various depths, using $V = \frac{Q}{4.08 \times d}$ and enter results in appropriate columns of table.

SEWAGE TREATMENT-SETTLING TANKS-I

TABLE A-DETENTION PERIOD-BASED ON AVERAGE DAILY FLOW*

Primary Sedimentation	Hours
Plain sedimentation preceding aeration tank in activated sludge	0.5-1.5
Plain sedimentation preceding trickling filters, contact aeration, or sand filters	2.0-2.5
Chemical precipitation or plain sedimentation alone	2.0-4.0
Plain sedimentation preceding trickling filters with recirculation of effluent	5.0-7.5
Final Sedimentation	
Following low-capacity trickling filter	1.0-1.5
Following high-capacity trickling filter	2.0-2.5
Following aeration tank in activated sludge	2.0-2.5

TABLE B-SEDIMENTATION TANK OVERFLOW RATES†

Type of Treatment	Size of Plant	Surface Settling Rates, gal./day/ft.^{2}	
		Tanks	
		Primary	Final
Primary	Up to 1 m.g.d.	Not More than 600	–
Secondary (standard trickling filters)	–	1000	1000
Secondary (high-rate trickling filter)	–	1000	800
Secondary (activated sludge)	Up to 2 m.g.d.	1000	800
Secondary (activated sludge)	Over 2 m.g.d.	1000	1000

TABLE C-SIZE OF TANKS

Rectangular Settling Tanks

The length of rectangular tanks should be from 3 to 5 times the width.

Maximum Length of Rectangular Primary Settling Tank‡

Detention Period, hours	Maximum Length, ft.
1/2	150
3/4	200
1	225
1-1/2	250
2	300

Circular Settling Tanks

Circular settling tanks have been built with a diameter of 200 ft. Most manufacturers make equipment to fit circular tanks varying by 2-ft. intervals up to 30 ft. and by 5-ft. intervals 30 ft. and over.

Depth of Tanks, Rectangular or Circular, ft.	
Minimum	7
Maximum	10 – 15

General Notes:

1. Mechanical equipment for sludge and scum collection is generally used for installations for over 1500 persons.
2. The detention period of hopper-bottomed tanks should be based on the liquid capacity above the hoppers, and for mechanically cleaned tanks should be based on the liquid capacity above a plane passing through the tops of the rotating or moving scrapers.
3. Tanks should have baffled inlets and outlets.
4. Consult manufacturers for sizes of equipment being sold, and select proper width, diameter, and depths for rectangular and circular tanks.

* Compiled from various state Board of Health laws.

† Upper Mississippi and Great Lakes Board of Public Health Engineers (Ill., Ind., Iowa, Mich., Minn., Mo., N.Y., Ohio, Pa., Wisc.).

‡ Data from Link Belt Co.

SEWAGE TREATMENT—SETTLING TANKS—2

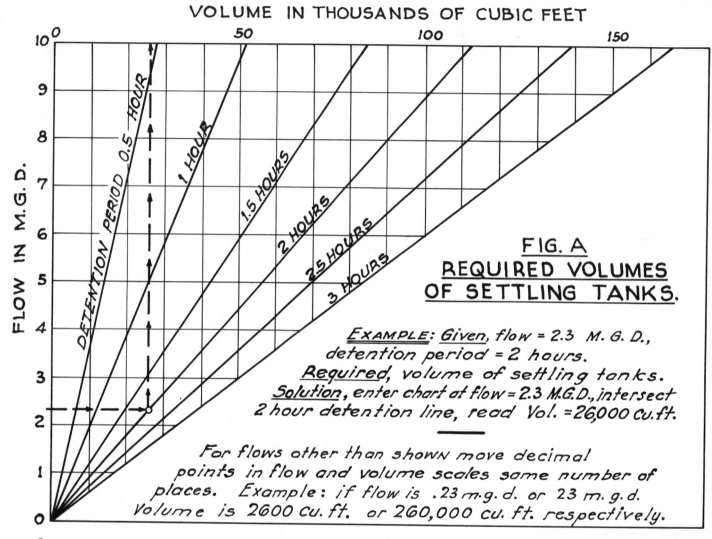

VOLUME IN THOUSANDS OF CUBIC FEET

0 50 100 150

FLOW IN M.G.D.

0 1 2 3 4 5 6 7 8 9 10

DETENTION PERIOD 0.5 HOUR

1 HOUR

1.5 HOURS

2 HOURS

2.5 HOURS

3 HOURS

FIG. A REQUIRED VOLUMES OF SETTLING TANKS.

EXAMPLE: *Given*, flow = 2.3 M.G.D., detention period = 2 hours.
Required, volume of settling tanks.
Solution, enter chart at flow = 2.3 M.G.D., intersect 2 hour detention line, read Vol. = 26,000 cu.ft.

For flows other than shown move decimal points in flow and volume scales same number of places. Example: if flow is .23 m.g.d. or 23 m.g.d. Volume is 2600 cu. ft. or 260,000 cu. ft. respectively.

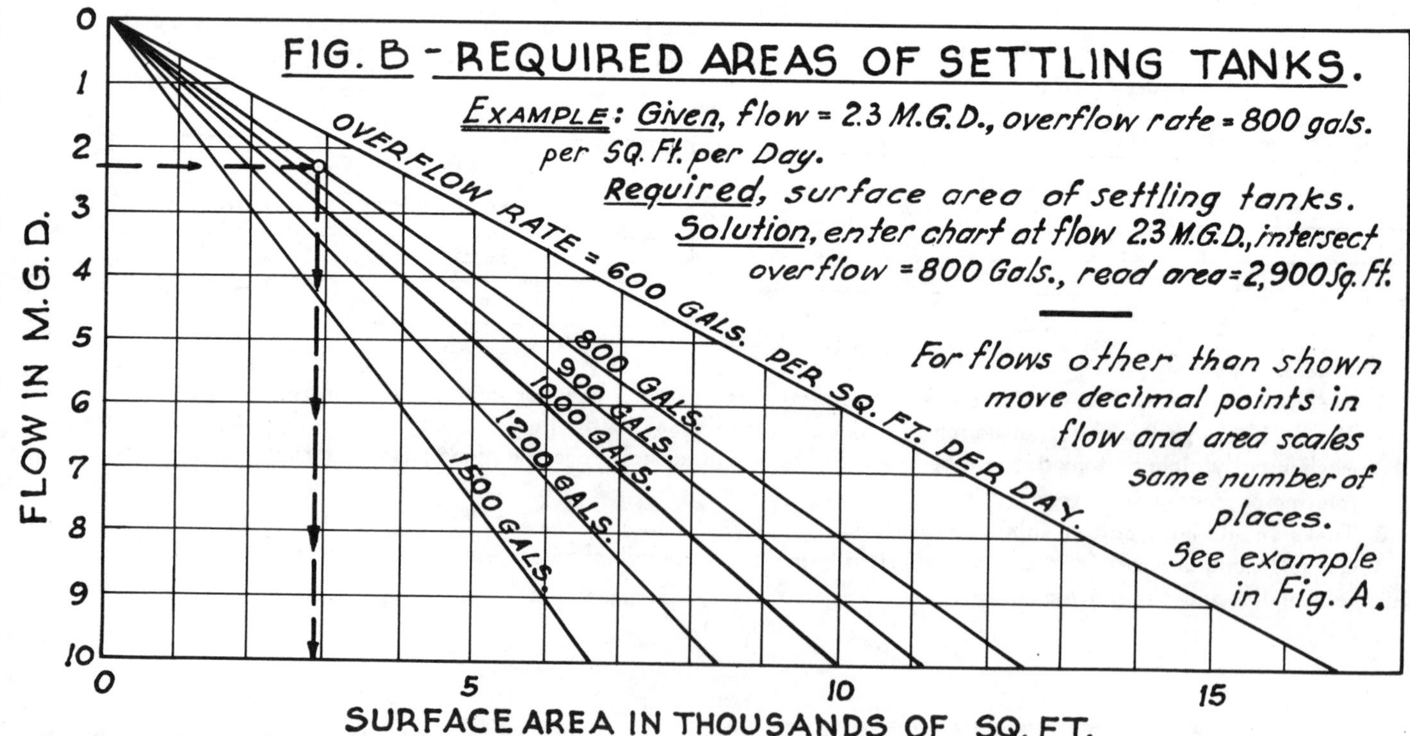
FIG. B - REQUIRED AREAS OF SETTLING TANKS.
EXAMPLE: Given, flow = 2.3 M.G.D., overflow rate = 800 gals. per SQ. Ft. per Day.
Required, surface area of settling tanks.
Solution, enter chart at flow 2.3 M.G.D., intersect overflow = 800 Gals., read area = 2,900 Sq. Ft.
For flows other than shown move decimal points in flow and area scales same number of places. See example in Fig. A.
OVERFLOW RATE = 600 GALS. PER SQ. FT. PER DAY.
800 GALS.
900 GALS.
1000 GALS.
1200 GALS.
1500 GALS.
FLOW IN M.G.D.
0
1
2
3
4
5
6
7
8
9
10
0
5
10
15
SURFACE AREA IN THOUSANDS OF SQ. FT.

NOTE: For recommended detention periods and overflow rates. See Page 19-09.

SEWAGE TREATMENT–IMHOFF TANKS

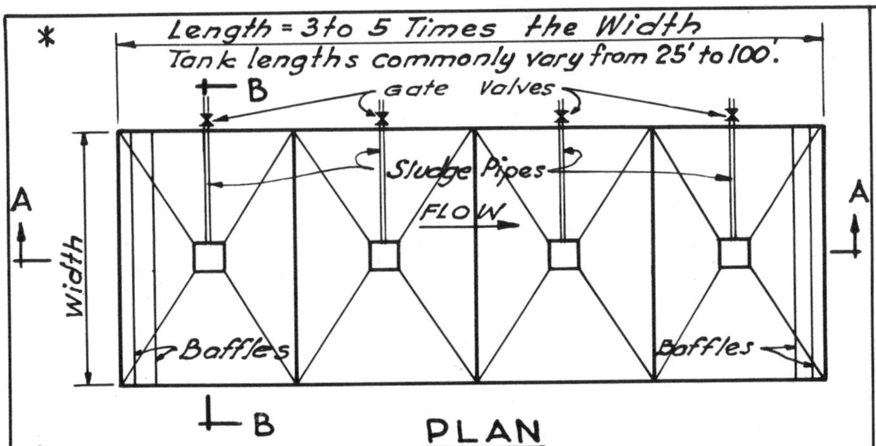

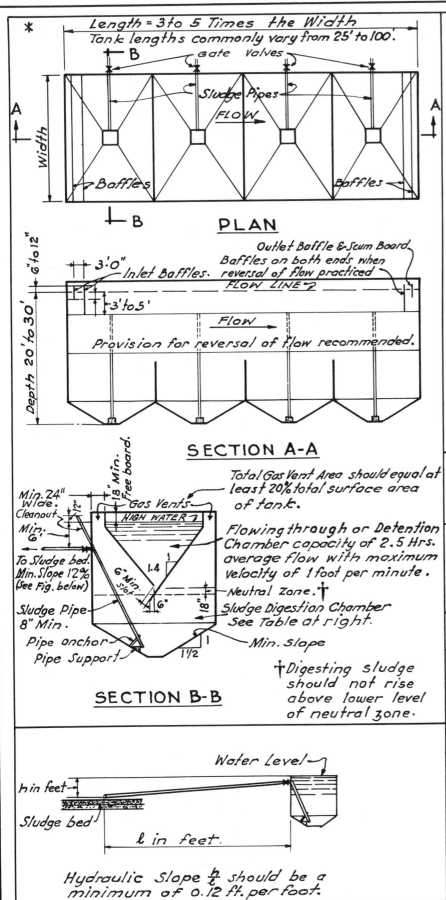

* Length = 3 to 5 Times the Width
Tank lengths commonly vary from 25' to 100'.
B
Gate Valves
Sludge Pipes
FLOW
A
Width
Baffles
Baffles
A
B

PLAN

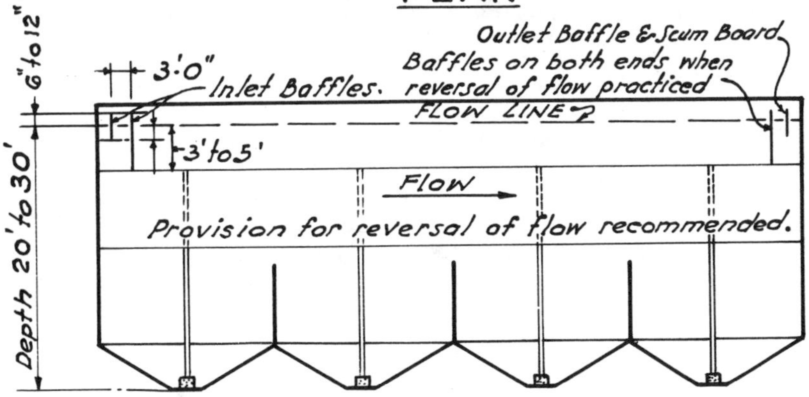

6" to 12"
3'-0"
Inlet Baffles.
Outlet Baffle & Scum Board
Baffles on both ends when reversal of flow practiced
FLOW LINE
3' to 5'
Flow
Provision for reversal of flow recommended.
Depth 20' to 30'

SECTION A-A

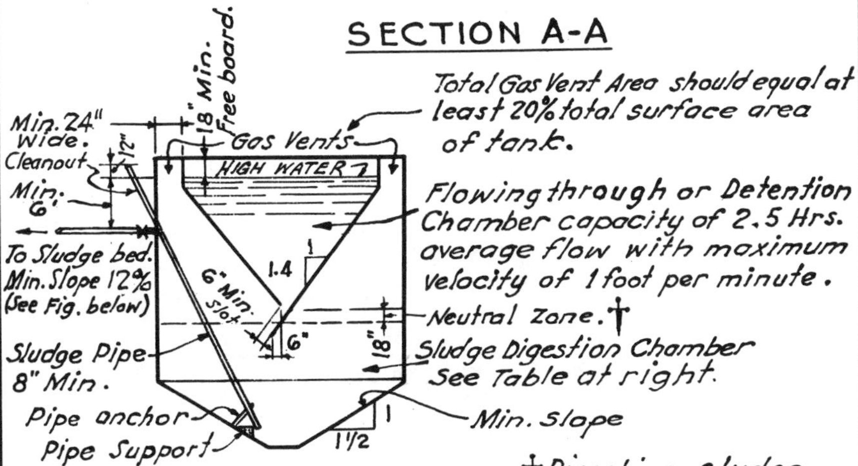

18" Min. Free board.
Min. 24" Wide.
Cleanout
12"
Gas Vents
Total Gas Vent Area should equal at least 20% total surface area of tank.
HIGH WATER
Min. 6'
To Sludge bed. Min. Slope 12% (See Fig. below)
Flowing through or Detention Chamber capacity of 2.5 Hrs. average flow with maximum velocity of 1 foot per minute.
1.4
1
6" Min. Slot
Neutral Zone.†
6"
18"
Sludge Pipe 8" Min.
Sludge Digestion Chamber See Table at right.
Pipe anchor
Pipe Support
1
1½
Min. Slope

SECTION B-B

†Digesting sludge should not rise above lower level of neutral zone.

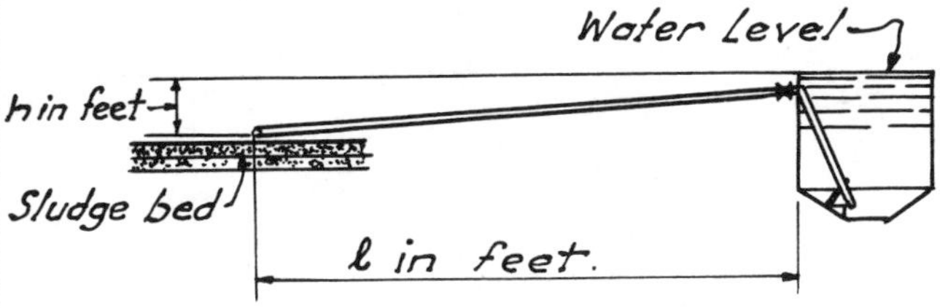

Water Level
h in feet
Sludge bed
ℓ in feet

Hydraulic Slope $\frac{h}{\ell}$ should be a minimum of 0.12 ft. per foot.

USUAL ARRANGEMENT OF IMHOFF TANKS.

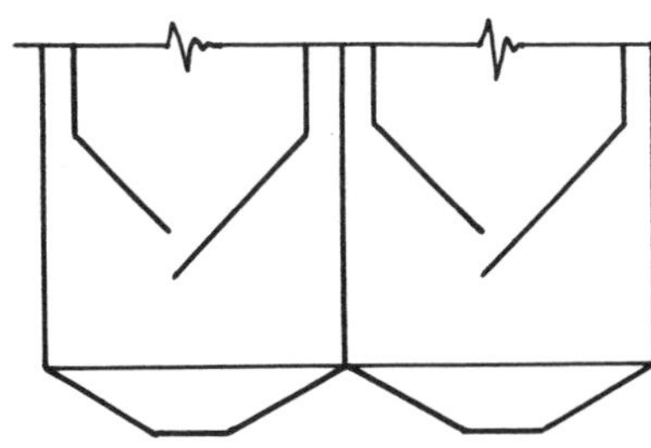

TYPE I

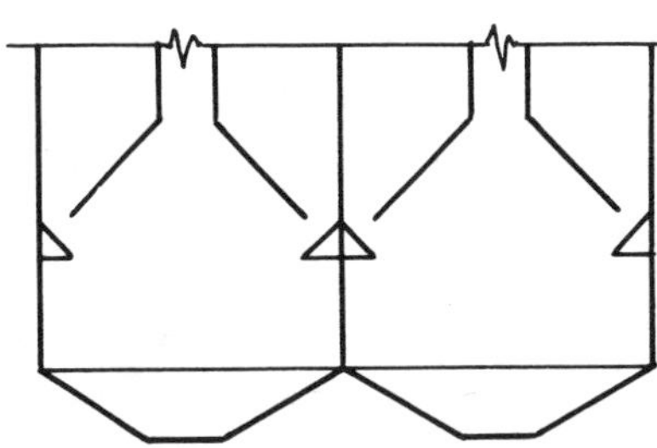

TYPE II

Design data shown on this page apply for both types.

CAPACITIES FOR IMHOFF TANK SLUDGE COMPARTMENTS RECOMMENDED BY IMHOFF**

CU. FT. PER CAPITA.

SIZE OF CITY SERVED AND CHARACTER OF SEWAGE.	NORTHERN STATES WITH LONG WINTERS.	
	SEPARATE SYSTEM	COMBINED SYSTEM WITH GRIT CHAMBER
Small plants with less than 5,000 population.	2.4	3.6
Normal city plants whose sewage contains some trades wastes.	1.2	1.8
City plants whose sewage contains an abnormal amount of sludge forming trades wastes.	1.8	2.7

The recommended capacities for the larger plants are somewhat lower than used in recent American practice.

NOTES: 1., In warmer climates, where digestion is more rapid, smaller volumes may be used.
2., New York State Department of Health requires 3.0 to 4.0 Cu. ft. per capita for separate systems.

* Compiled from Standards for Sewage Works–1952. Prepared by Upper Mississippi and Great Lakes Board of Public Health Engineers (Ill., Ind., Iowa, Mich., Minn., Mo., Neb., N.Y., Ohio, Pa., and Wis.).
** Data from Engl. News, Jan. 16, 1916. Pg. 19, by K. Imhoff.

SEWAGE TREATMENT-CHEMICAL PRECIPITATION

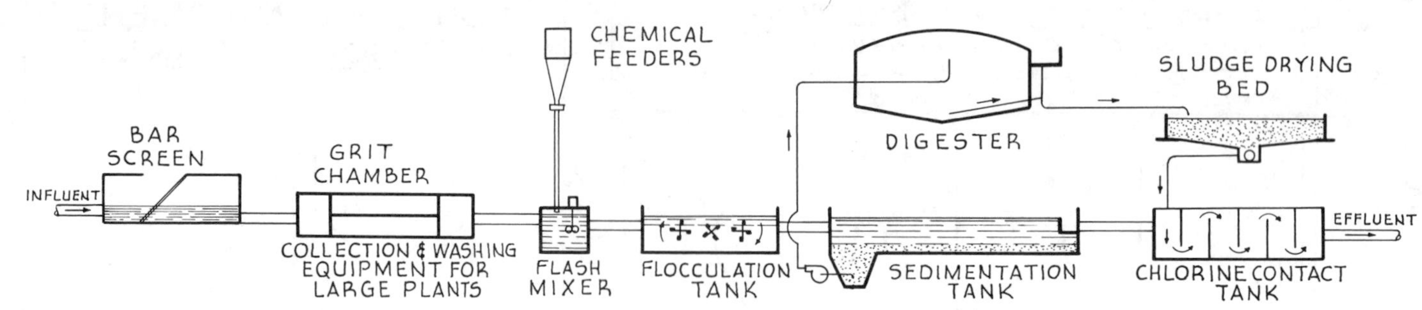

CHEMICAL FEEDERS
DIGESTER
SLUDGE DRYING BED
BAR SCREEN
GRIT CHAMBER
INFLUENT
EFFLUENT
COLLECTION & WASHING EQUIPMENT FOR LARGE PLANTS
FLASH MIXER
FLOCCULATION TANK
SEDIMENTATION TANK
CHLORINE CONTACT TANK

Chemical precipitation process consists of mixing chemicals with sewage to produce a flocculent precipitant, which increases and hastens sedimentation of the suspended sewage solids, settling, sludge digestion, and dewatering. Chemical-precipitation plants are particularly adapted to wide variations in seasonal requirements and conversions of primary plants. Treatment efficiencies lie between those of primary and secondary treatment plants. The principal purpose of chemical treatment is to reduce suspended solids and B.O.D. and to obtain a clarified and colorless effluent. For comparative efficiency, see Table C, p. 19-01. The sewage is analyzed to determine the kind and quantity of chemicals necessary to obtain the required results. After sewage has passed through screens and grit chambers, chemicals are quickly mixed with it. The mixture flows to the flocculation tank, which has a detention period of 20 to 30 min. It then enters the sedimentation tank, where the detention period is 2 to 3 hours. If further bacterial removal is required, the final effluent should be chlorinated. Sludge is digested in a closed tank, and then dewatered on a drying bed or by vacuum filter.

TABLE A-CHARACTERISTICS OF CHEMICAL COAGULANTS*

Chemical	Remarks
Ferric Chloride	Preferable for sludge-conditioning in activated sludge. Most economical for plants requiring more than 7 tons of iron annually, with facilities for handling it. Available as anhydrous solid, lump or as aqueous solution. Can be made locally of chlorine and iron or steel scrap. Floc forms satisfactorily at all temperatures. Suitable for oxidizing H_2S with high *p*H. Highly corrosive and difficult to handle. Solution must be stored and handled in rubber-lined containers and pipes. Useful in odor and corrosion control, forming iron sulfide with H_2S, and not absorbed by organic matter. Cheaper than chlorine for the purpose. Coagulates best with *p*H below 7.0. Optimum at 5.5 Generally considered the best coagulant available. Anticipated percentage removal S.S. = 90 to 95; B.O.D. = 80. Dose at optimum *p*H = 2.0 to 2.5 g. per gal.
Chlorinated Copperas	Good for sludge conditioning in activated sludge. Economical for plants requiring more than 7 tons of iron annually, with facilities for handling it. Poor at *p*H 7.0; good at 5.5 and 9.0 to 9.5 with dose of 2.5 to 5.9 g. per gal. % reduction S.S. = 80 to 90 and B.O.D. = 70 to 80.
Ferrous Sulfate (Copperas)	*p*H greater than 7.7 favors oxidation to ferric hydroxide. Dry feeders not easily used because of caking, but are used. Tendency for cheaper grades to cake during storage. First cost is relatively low. Widely available as waste product from steel mills. Optimum *p*H about 9.0 with dose 2.5 to 5.0 g. per gal. Overdose results in undesirable afterprecipitation. Can be fed in solution form.
Ferric Sulfate (Ferrisul)	More efficacious than copperas or chlorinated copperas when used with lime. *p*H best about 8.0 to 8.5 with dose about 2.5 g. per gal. % reduction S.S. = 80 and B.O.D. = 60. Can be fed dry or as liquid.
Alum	Not yet widely used in sewage treatment. Economical and conveniently handled. Can be used with dry feeder. *p*H range from 6.0 to 8.5; above 7.0 most practical. Dose 5 to 6 g. per gal. with % reduction S.S. = 80 and B.O.D. = 60.
Clay & Other Inert Materials, Bentonite, Sand, Asbestos Fiber, Paper, clay, etc.	Not yet tried on plant scale. Dosages of 100 p.p.m. in lab. have given excellent results. Bentonite flocs readily over wide *p*H range with natural alkalinity. Materials other than bentonite not yet developed practically.
Lime CaO = quick CaOH = hydrated	Can be dry fed. Is non-corrosive. Commonly used for upward adjustment of *p*H. Quicklime must be stored in dry, steel tanks and must be hydrated (slaked) before use. Dry hydrated lime can be stored in any dry place.
Sodium Carbonate (Soda ash)	Non-corrosive. Can be fed dry.
Chlorine	Corrosive and toxic. Can be stored in cast iron, lead, glass, or rubber. Requires special dosing equipment. Useful in odor and concrete-corrosion control and to control flies and ponding on filters.
Aluminum Chloride	Optimum *p*H 5.5 and 9.0; g. per gal. at 9.0 = 5, and at *p*H of 5.5 = 2.5 to 5.0.

TABLE B-AMOUNTS OF CHEMICALS USED AND RESULTS AT VARIOUS PLANTS†

Plant	Sewage Flow, m.g.d.	Chemical	Treatment lb. per m.g.	Suspended Solids		5-day B.O.D.	
				Raw	% Removal	Raw	% Removal
Coney Island, New York	20.1	CoP CaO Cl	191 546 105	154	60	112	39
	22.0	CoP Cl	345 153	169	68.5	108	56
	22.6	F.S. CaO Cl	188 340 101	153	67	98	48
	23.5	CoP Cl	279 135	206	78.7	122	66
New Britain, Conn.	7.8	CoP Cl	281 41.3	173	85.5	156	78.2
New Britain, Conn.	9.1	F.S.	106.3	136	77.9	137	74.4
El Paso, Tex.	6.0	F.C.	72	339	84	287	63.8
Danville, Ill.	4.5	Alum	373	235	81.4	201	61.5
Dearborn, Mich.	2.6	F.C. CaO	305 250	293	92	174	76
Perth Amboy, N.J.	2.5	F.C. CaO	250 650	240	88	257	61
Butler, Pa.	2.1	CoP CaO Cl	494 350 107		73		58.7
Shades Valley, Ala.	1.1	CoP Cl	412	80	86	53	87

CoP = Copperas, CaO = lime, Cl = chlorine, F.S. = ferric sulfate, F.C. = ferric chloride

*Harold E. Babbitt, *Sewerage and Sewage Treatment*, 8th Ed., Wiley, 1958.

†W. Donaldson, Chemical Treatment of Sewage, *Modern Sewage Disposal*, p. 85, Federation of Sewage Works Association.

SEWAGE TREATMENT — INTERMITTENT SAND FILTERS

TABLE A - CAPACITY OF SAND FILTERS.

TYPE OF EFFLUENT	GALLONS PER ACRE PER DAY (MAXIMUM FLOW)
Strong settled sewage (such as Imhoff or Septic tank effluent).	50,000
Settling tank effluent (normal).	125,000
Trickling filter effluent.	500,000
Activated sludge effluent.	500,000

24" Min. Sand — All sand passing 1/4" sieve effective size 0.3 millimeter to 0.6 millimeter. Uniformity coefficient not over 3.5. (As close to 1.0 as possible.)

1/4" to 1/8" Gravel

3/4" to 1/4" Gravel

1 1/2" to 3/4" Gravel

6"

3" Torpedo Sand size 0.8 to 1.2 mm.

1" Gravel

6" Minimum

6" Minimum

Underdrain - open joint Vitrified tile or Farm tile minimum size 4". 1/4" between pipes, cover joints with muslin, burlap or cheesecloth.

TYPICAL SECTION OF SAND BED

SIZE OF DOSING TANK

1. Capacity of dosing tank should be large enough to flood one unit to a depth from 2 to 4 inches.

2. Each filter bed should receive either 1 or 2 doses per day, preferably one.

3. Rate of discharge of siphon at minimum head (See page 19-15) should be at least twice the maximum inflow to the dosing tank from the settling tank.

4. Average rate of dosage should be about 1 cu. ft. per sec. per 5000 sq. ft. area.

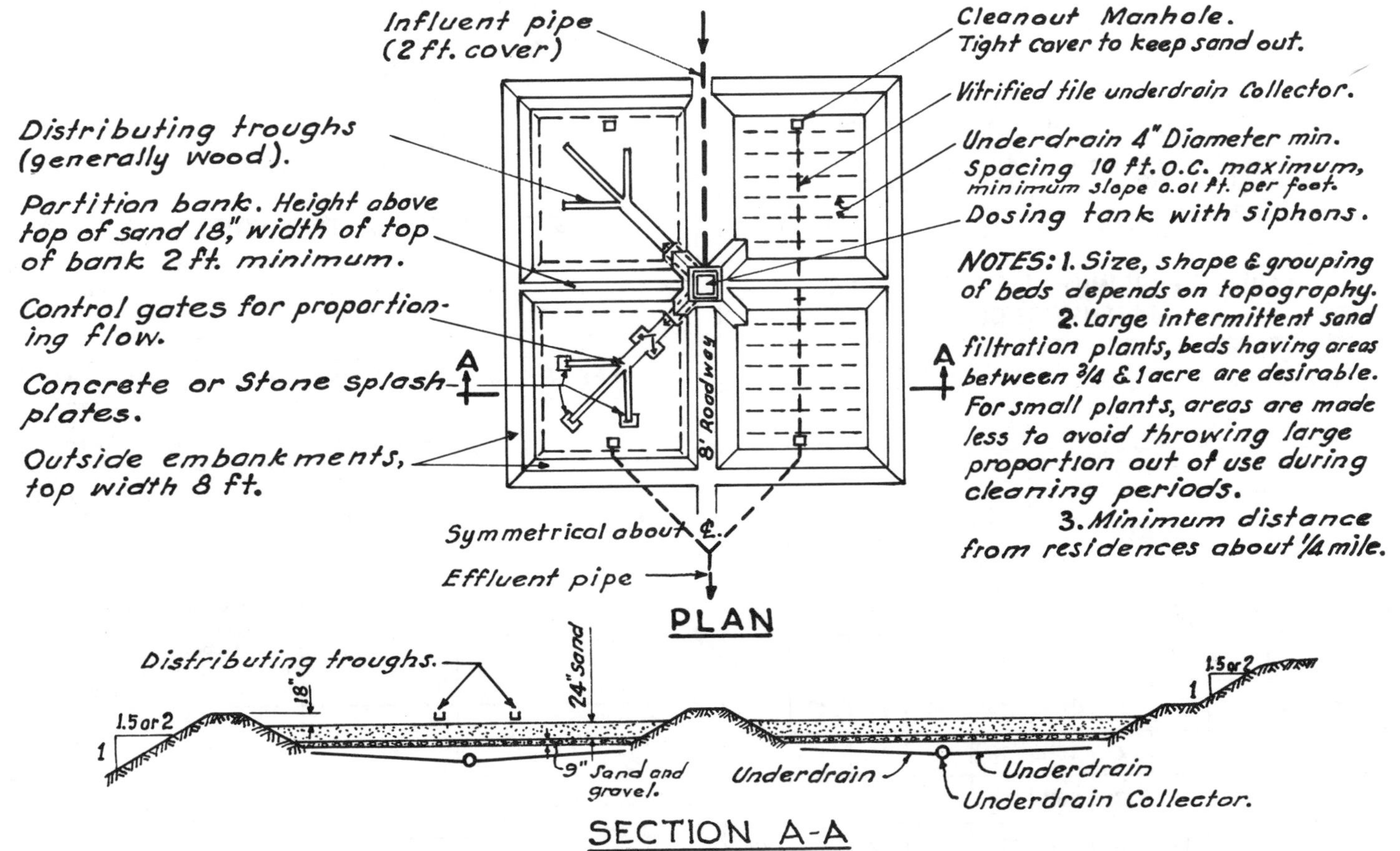

Influent pipe (2 ft. cover)
Distributing troughs (generally wood).
Partition bank. Height above top of sand 18", width of top of bank 2 ft. minimum.
Control gates for proportioning flow.
Concrete or Stone splash-plates.
Outside embankments, top width 8 ft.
Cleanout Manhole. Tight cover to keep sand out.
Vitrified tile underdrain Collector.
Underdrain 4" Diameter min. Spacing 10 ft. O.C. maximum, minimum slope 0.01 ft. per foot.
Dosing tank with siphons.
NOTES: 1. Size, shape & grouping of beds depends on topography.
2. Large intermittent sand filtration plants, beds having areas between 3/4 & 1 acre are desirable. For small plants, areas are made less to avoid throwing large proportion out of use during cleaning periods.
3. Minimum distance from residences about 1/4 mile.
A
A
8' Roadway
Symmetrical about ℄.
Effluent pipe
PLAN
Distributing troughs.
18"
24" Sand
1.5 or 2
1
1.5 or 2
1
9" Sand and gravel.
Underdrain
Underdrain
Underdrain Collector.
SECTION A-A

FIGURE B - TYPICAL LAYOUT OF SAND FILTERS.

Compiled from Standards for Sewage Works of Upper Mississippi River Board of Public Health Engineers and Great Lakes Board of Public Health Engineers.

SEWAGE TREATMENT – SAND FILTER DOSING SIPHONS

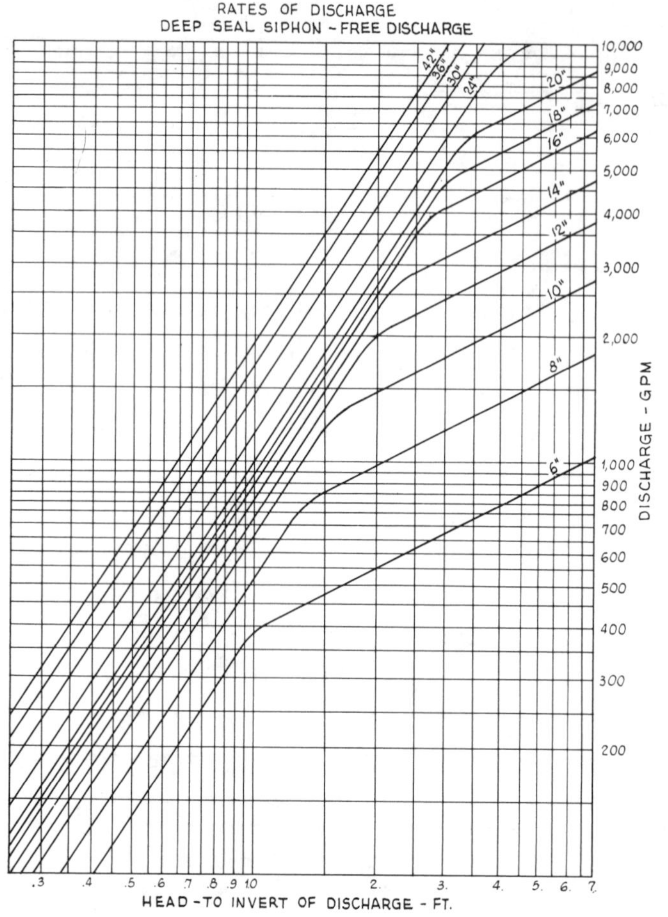
RATES OF DISCHARGE
DEEP SEAL SIPHON – FREE DISCHARGE
DISCHARGE – GPM
HEAD – TO INVERT OF DISCHARGE – FT.

FIG. A – RATES OF DISCHARGE FOR SIPHONS DISCHARGING INTO OPEN TROUGHS

NOTE: Average of rates at minimum and maximum head will give a fair estimate for use in determining time to empty Dosing Tank.

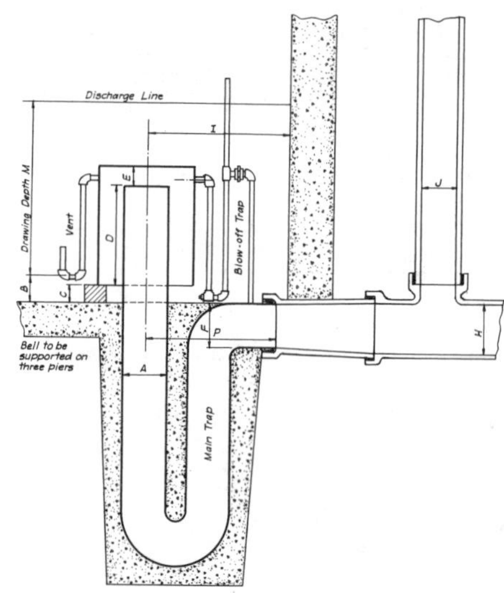
Discharge Line
Drawing Depth M
Vent
Blow-off Trap
Bell to be supported on three piers
Main Trap

FIG. B – SINGLE SIPHON

NOTE: Two siphons in same tank will operate alternately.

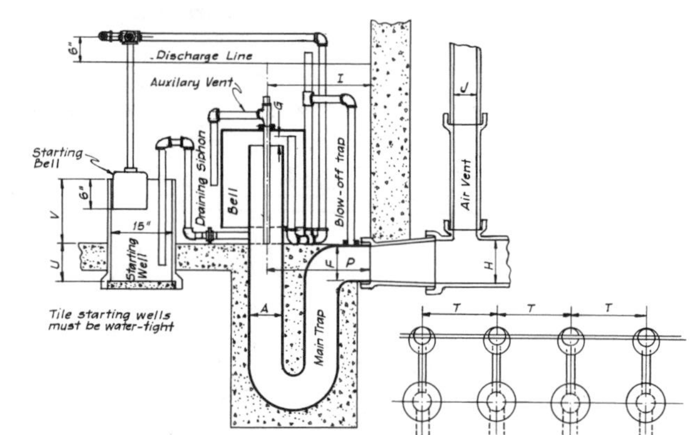
Discharge Line
Auxilary Vent
Starting Bell
Draining Siphon
Bell
Blow-off trap
Air Vent
Starting Well
Main Trap
Tile starting wells must be water-tight

FIG. C – PLURAL ALTERNATING TYPE A

NOTE: Plural alternating type "B" works by means of a starting device on vent lines. Has no starting wells of interconnecting pipes like type "A".

TABLE D – MINIMUM DISCHARGE AND DIMENSIONS OF SIPHONS

A Diam. in.	Discharge in g.p.m. at Min. Head	Dimensions, inches							
		B	C,E	F	H	I,P	J	S	T
6	270	5	3	6½	8	21	6	27	30
8	450	5	3	8½	10	25	6	30	36
10	700	6	4	10½	12	28	8	36	36
12	1200	6	4	12½	14	32	8	36	36
14	1800	7	5	14¾	16	39	10	39	42
16	2750	8	6	16¾	16	44	10	42	48
18		9	6	18¾	18	48	12	42	48
20		10	7	20¾	20	51	12	48	48
24		11	8	24¾	24	57	14	48	60

NOTE: Maximum in flow into dosing tank must be less than siphon discharge at minimum head, or siphon will not close.

TABLE E – RESTRICTIONS ON USE OF SIPHONS

	Single Siphon	Double Alternating Siphon	Plural Alternating Siphon Type "A"	Plural Alternating Siphon Type "B"
Setting		Both siphons accurately set at same elevation	Traps set parallel or at right angles	Can be set to any position
Valves	None	None	Provided to cut any no. of siphons out of operation	Provided to cut any one siphon out of operation when 4 are used
Inter connected piping	No	No	Yes	No
Min. no.	1	2	2	3
Max. no.	1	2	10	4
Min. drawing depth	Depends on size		3'-0"	2'-0"
Max. drawing depth	Depends on size		5'-0"	5'-0"

Data from Pacific Flush Tank, Bulletin 224.

SEWAGE TREATMENT – TRICKLING FILTERS

TABLE A – DESIGN DATA FOR TRICKLING FILTERS.

ALLOWABLE LOADING – B.O.D. IN POUNDS PER ACRE FOOT PER DAY		
B.O.D. OF RAW SEWAGE. p.p.m.	LOW OR STANDARD RATE FILTERS	HIGH RATE * FILTERS. (BASED ON 3' DEPTH BIO-FILTRATION PROCESS) FOR GREATER DEPTH USE LOWER B.O.D. LOADINGS.
Less than 100	200	1300
100 to 150	—	2400
150 to 300	250	3200
300 to 450	—	4000
450 to 600	450	4800

ITEM	STANDARD OR LOW RATE FILTERS	HIGH RATE FILTERS
Depth of Filter Stone.	Min. 5' Max. 7'	3'-0"
Size of stones, uniform throughout.	2" to 3½"	2" to 3½" Primary 1½" to 2½" Secondary
Fixed Nozzles.	Fixed nozzles always used in past.	Never used.
Rotary Distribution.	Present practice favors rotary distr.	Always used.
Dosing Tank.	Always used.	Never used.
Recirculated Flow. (Requires use of Pumps)	Never used.	Always used.

*When using high rate filtration processes (bio-filtration, aero-filtration, etc.) manufacturers should be consulted to obtain charts showing values of loading on filter which they will guarantee to produce the results required. Permissible B.O.D. loading of filter depends on depth of filter, strength of sewage, and the degree of removal desired.

EXAMPLE: Given: 1. Population = 10,000. 2. Residential town including ordinary commercial establishments - separate sanitary sewers.

Required: Volume of rock for standard low rate trickling filter.

Solution: 1. From Tab. A-P. 19-01 - pounds of B.O.D. per capita daily = 0.17.

2. Total B.O.D. requirement of raw sewage = 10,000 × 0.17 = 1700#.

3. From Tab. C. Pg. 19-01 Allow 35% B.O.D. reduction for primary settling tanks, leaving 1700 × 0.65 = 1100# to be applied on filter.

4. From Pg. 18-65 - we find an average flow of 100 Gals. per capita per day.

5. B.O.D. of raw sewage = $\frac{0.17}{100 \times 8.35} \times 1{,}000{,}000 = 204$ p.p.m. From Tab. A above use a value of loading on filter of 250 pounds per acre foot per day.

6. Volume of rock required = $\frac{1100}{250} = 4.4$ acre feet.

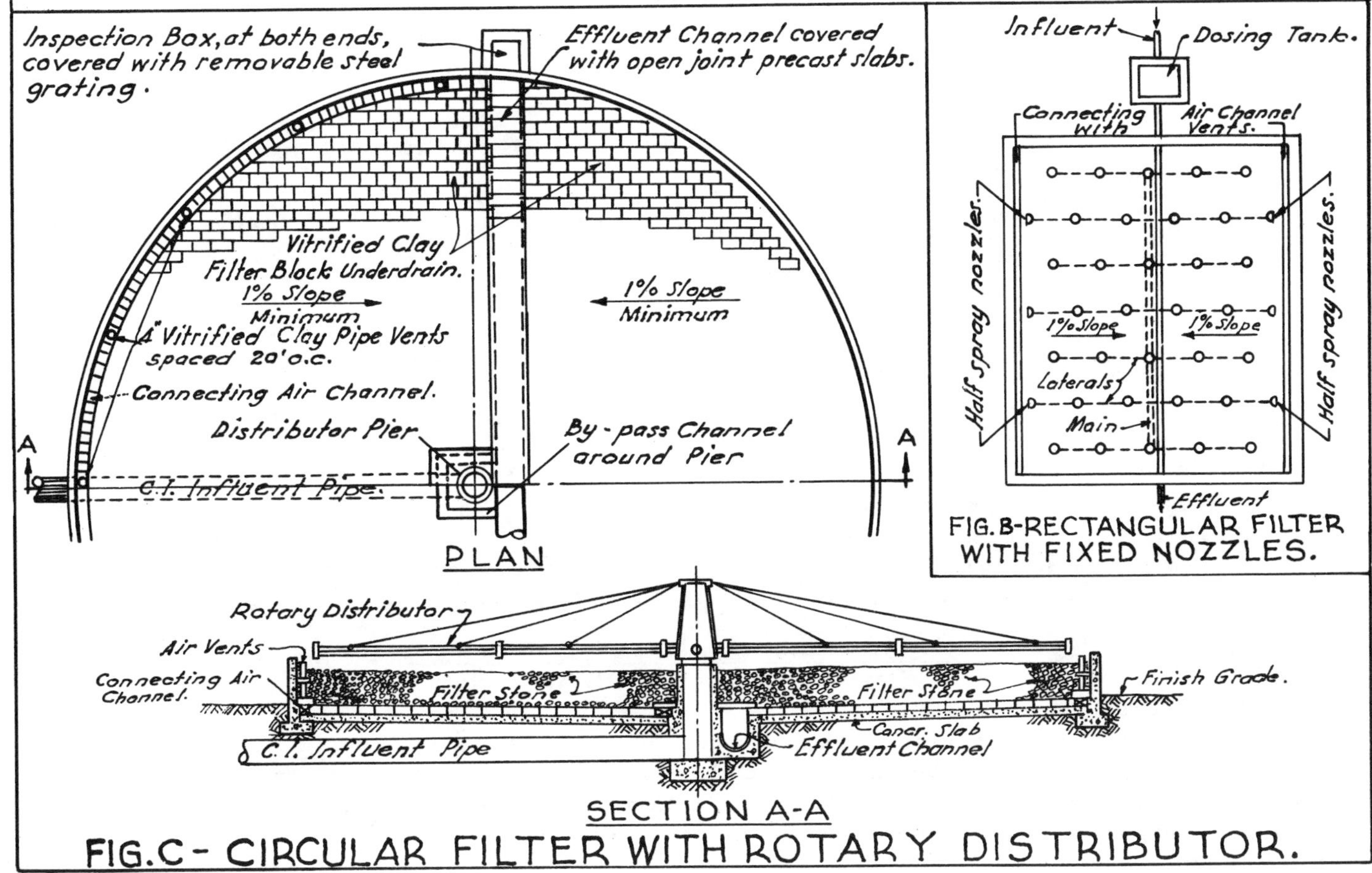

Inspection Box, at both ends, covered with removable steel grating.
Effluent Channel covered with open joint precast slabs.
Vitrified Clay Filter Block Underdrain.
1% Slope Minimum
1% Slope Minimum
4" Vitrified Clay Pipe Vents spaced 20' o.c.
Connecting Air Channel.
Distributor Pier
By-pass Channel around Pier
C.I. Influent Pipe.
A
A
PLAN
Influent
Dosing Tank.
Connecting with
Air Channel Vents.
Half spray nozzles.
Half spray nozzles.
1% Slope
1% Slope
Laterals
Main
Effluent
Rotary Distributor
Air Vents
Connecting Air Channel.
Filter Stone
Filter Stone
Finish Grade.
Concr. Slab
C.I. Influent Pipe
Effluent Channel
SECTION A-A

FIG. B-RECTANGULAR FILTER WITH FIXED NOZZLES.

FIG. C – CIRCULAR FILTER WITH ROTARY DISTRIBUTOR.

SEWAGE TREATMENT–ACTIVATED SLUDGE – I

EXPLANATION OF ACTIVATED SLUDGE PROCESS:
This method is based on the practice of allowing a large number of bacteria to multiply in sludge under conditions that are generally anaerobic (putrefactive). New conditions which are favorable to the development of aerobic bacteria and microscopic organisms are then created by aerating the sludge. When this *activated sludge* is mixed with sewage, and the mixture is aerated, the organic matter in the sewage is changed to harmless compounds by oxidation. After the mixture leaves the aeration tanks, it is discharged into final sedimentation tanks where the sludge settles. A portion of the settled sludge, amounting to 10 to 50% of average sewage flow, is returned to the inlet of the aeration tanks and mixed with the influent sewage. See p. 19-17 for basis of design of different types of systems. See p. 19-18 for air-flow data.

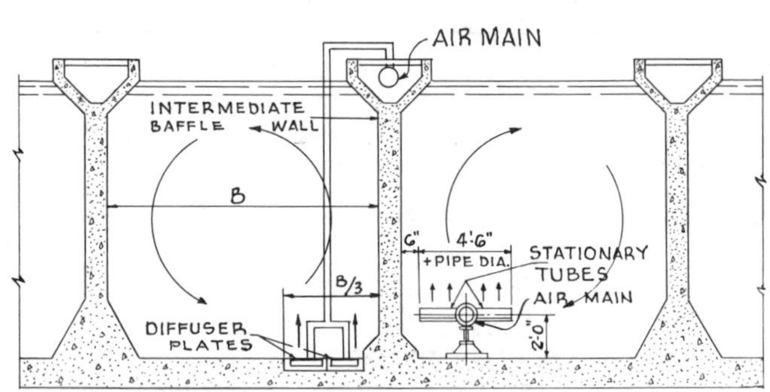
AIR MAIN
INTERMEDIATE BAFFLE WALL
B
B/3
DIFFUSER PLATES
6"
4'-6"
+PIPE DIA.
STATIONARY TUBES
AIR MAIN
2'-0"

FIG. A AERATION TANKS DIFFUSER PLATE OR STATIONARY TUBE TYPE

Diffuser plates or tubes on one side supply air and cause spiral turbulent flow.

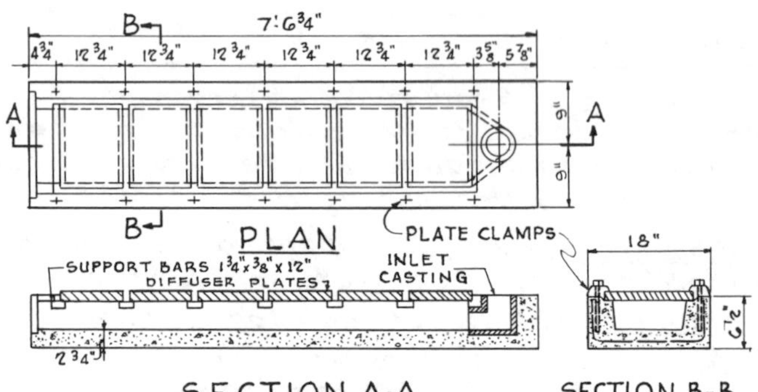
B
7'-6¾"
4¾"
12¾"
3⅝"
5⅞"
A
9"
PLAN
PLATE CLAMPS
18"
SUPPORT BARS 1¾"x⅜"x12"
DIFFUSER PLATES
INLET CASTING
2¾"
6½"
SECTION A-A
SECTION B-B

FIG. B DETAILS OF DIFFUSER PLATE CONTAINER*

Used for spiral flow as shown in Fig. A. These containers may be made of aluminum or reinforced concrete. The plates rest on suitable gaskets and are held in place by clamps.

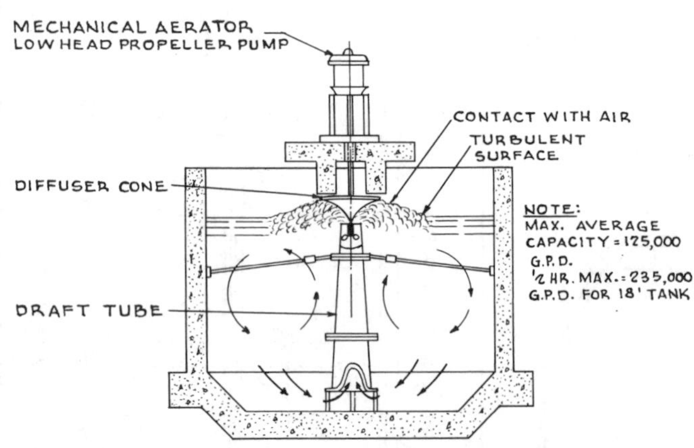
MECHANICAL AERATOR LOW HEAD PROPELLER PUMP
CONTACT WITH AIR
TURBULENT SURFACE
DIFFUSER CONE
DRAFT TUBE
NOTE:
MAX. AVERAGE CAPACITY = 125,000 G.P.D.
½ HR. MAX. = 235,000 G.P.D. FOR 18' TANK

FIG. C AERATION TANK – MECHANICAL†

Low head propeller pump throws sewage in draft tube against diffuser cone, providing aeration of surface of tank by direct contact and surface agitation.

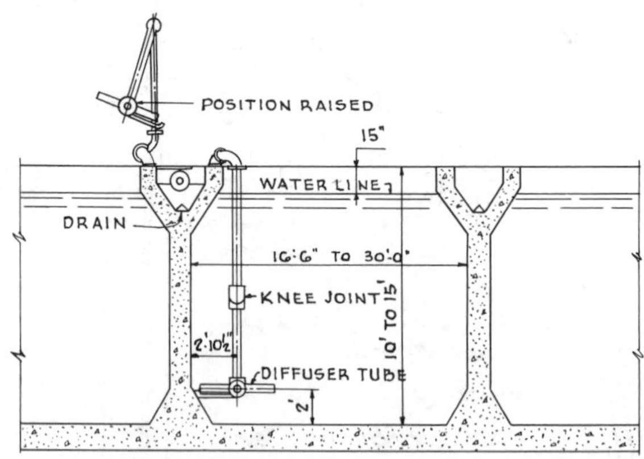
POSITION RAISED
15"
WATER LINE
DRAIN
16'-6" TO 30'-0"
KNEE JOINT
10' TO 15'
2'-10½"
DIFFUSER TUBE
2'

FIG. D AERATION TANK – SWING DIFFUSERS†

Diffuser tubes on one side supply air and cause spiral turbulent flow. Tubes may be examined, respaced, cleaned or replaced without emptying tank.

$$\text{Period of aeration (hours)} = \frac{\text{capacity of tanks (cu. ft.)}}{\text{vol. of sewage flow + vol. of return sludge (cu. ft./hr.)}}$$

Aeration tank capacity based on average flow.

Pumps, pipes, flowmeter, etc., designed for maximum flow.

*The permeability of a diffuser plate is determined for the number of cubic feet of air per minute passing through one sq. ft. of diffuser plate under a pressure equivalent to a 2" column of water when the plate is dry and the temperature is 70° F. Present practice is to use diffuser plates with a permeability from 20 to 50.

†From Chicago Pump Co.

SEWAGE TREATMENT-ACTIVATED SLUDGE -2

TABLE A - BASIS OF DESIGN*

System of Aeration	Economical to use	Sedimentation Tanks: Surface Setting Rates, g.p.d./s.f.		Sedimentation Tanks: Detention Periods, hr.		Aeration Tanks: Diffused Air	Aeration Tanks: Capacity Based on Larger Volume A or B: A Detention			B	Depth, ft.	Activated-Sludge Return	Power Consumption
		Primary	Final	Primary	Final		0.2-0.8 m.g.d., hr.	0.8-1.0 m.g.d., hr.	Over 1.0 m.g.d. hr.				
Mechanically operated systems	For flows up to 1 m.g.d.	Varies from 2000 for 20% B.O.D. removed to 400 for 37% B.O.D. removed	Flow 2.0 m.g.d. or less 800 Over 2.0 m.g.d. 1000	1	2	None	150% of volumes as determined for air-diffusion tanks					10% min. 50% max.	Min. 0.35 kw.-hr./lb. of B.O.D. removed
Air-diffusion spiral-flow system	For flows over 1 m.g.d.	Varies from 2000 for 20% B.O.D. removed to 400 for 37% B.O.D. removed	Flow 2.0 m.g.d. or less 800 Over 2.0 m.g.d. 1000	1	2	1000 cu.ft./lb. of B.O.D. per day	7.5	7.5-6	6	30 cu.ft./lb. of B.O.D. in aerator influent	10 min. 15 max.	10% min. 50% max.	0.35 kw.-hr./lb. of B.O.D. removed

Surface settling rates, diffused air requirements, and aeration tank capacities and depths are minimum design requirements. After plants are in operation, a reduction in air supply and sludge return may produce a satisfactory effluent, Deeper tanks and larger quantities of air for longer periods will produce higher-quality effluents.
Keefer† shows the following normal variations in the United States: air used cu.ft./gal. of sewage 0.35 (weak sewage) to 1.7 (strong sewage); aeration period 3 to 8.4 hr., and sludge return 10.5% to 49.5%.

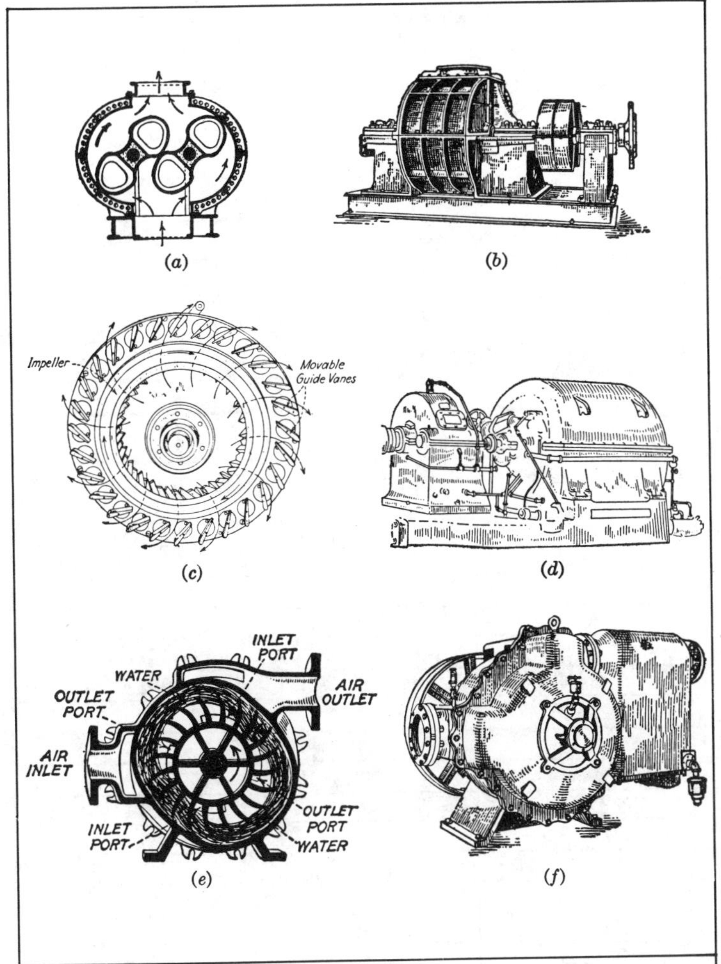

(a)
(b)
Impeller
Movable Guide Vanes
(c)
(d)
INLET PORT
WATER
AIR OUTLET
OUTLET PORT
AIR INLET
OUTLET PORT WATER
INLET PORT
(e)
(f)

FIG. B - TYPES OF AIR COMPRESSORS

AIR FILTRATION

Under normal conditions, air contains large quantities of dust and other impurities.

In order to prevent the clogging of diffuser plates, it is necessary to clean the air thoroughly before it enters the compressors. For this purpose, cloth filters, air washers, or oil cleaners are used. Canton flannel and 10- or 20-oz. duck may be employed for cloth filters.

The allowable rate of filtration through such media is about 4 cu. ft. of air per sq. ft. of cloth per min.

Air washers intercept the dust and other floating particles by means of jets of water sprayed through the air.

In oil cleaners a light oil is applied to a metallic medium. This arrangement, which causes a tortuous passage, removes the dust by retaining on the oily surfaces.

* *Standards for Sewage Works*, 1952, prepared by Upper Mississippi and Great Lakes Boards of Public Health Engineers.

† Keefer, *Sewage Treatment Works*, McGraw-Hill.

SEWAGE TREATMENT-ACTIVATED SLUDGE-3

FLOW OF AIR IN PIPES

$$P = \frac{1.268\, t\, Q^{1.852}}{1{,}000{,}000\, p\, d^{4.973}}$$ FRITZSCHE FORMULA.

P = drop in pressure in pounds per square inch per foot of pipe.
t = absolute temperature in degrees Fahrenheit = recorded temp. in degrees F + 459.6.
Q = cubic feet of free air per minute at 60 degrees Fahrenheit.
p = absolute pressure in pounds per square inch = gage pressure + 14.7.
d = diameter of the pipe in inches.

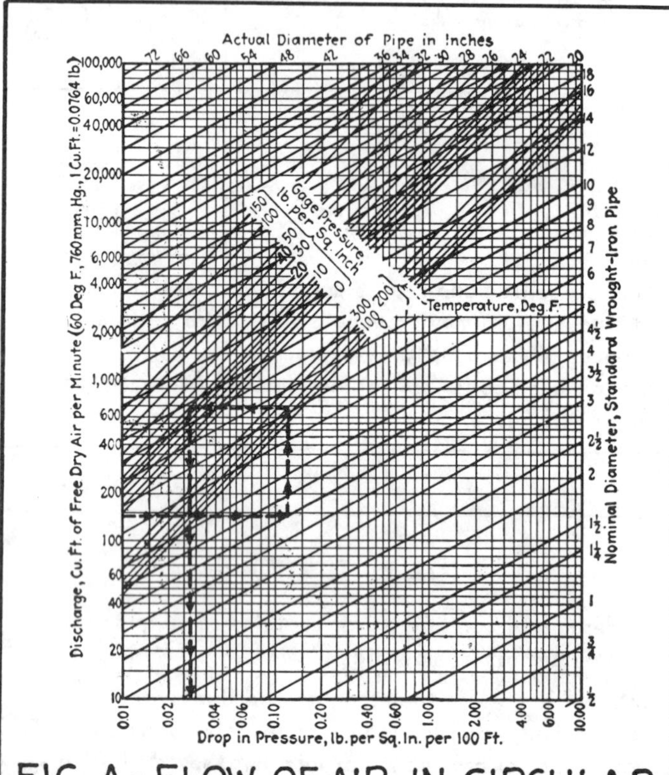

Actual Diameter of Pipe in Inches
Discharge, Cu. Ft. of Free Dry Air per Minute (60 Deg. F., 760 mm. Hg., 1 Cu. Ft. = 0.07641 lb.)
Nominal Diameter, Standard Wrought-Iron Pipe
Gage Pressure lb. per Sq. Inch
Temperature, Deg. F.
Drop in Pressure, lb. per Sq. In. per 100 Ft.

FIG. A - FLOW OF AIR IN CIRCULAR PIPES. MORRILL'S CHART, BASED ON FRITZSCHE FORMULA.*

TABLE B - ECONOMICAL MAXIMUM VELOCITIES OF AIR FLOW IN PIPES.

SIZE OF PIPE INCHES	MAXIMUM VELOCITY FEET PER SEC	SIZE OF PIPE INCHES	MAXIMUM VELOCITY FEET PER SEC.
6	20	16	50
8	28	20	54
10	35	24	57
12	41	30	60
14	46	36	62

TABLE C - POWER REQUIRED FOR COMPRESSING AIR.*

FINAL PRESSURE OF AIR LB. PER SQ. INCH	THEORETICAL WORK TO COMPRESS 1 MIL. CU. FT. OF FREE AIR HP-HR.	FINAL PRESSURE OF AIR LB. PER SQ. INCH	THEORETICAL WORK TO COMPRESS 1 MIL. CU. FT. OF FREE AIR HP-HR.
1	72.3	8	490.2
2	144.0	10	596.5
3	200.8	12	697.1
4	265.2	14	785.4
6	384.7	16	875.9

TABLE D - RESISTANCE TO FLOW OF AIR THROUGH FITTINGS.

DIAMETER OF PIPE IN INCHES	EQUIVALENT LENGTH OF STRAIGHT PIPE IN FEET				
	GATE VALVE	ANGLE VALVE	LONG RADIUS ELBOW	STANDARD ELBOW	SIDE OUTLET TEE
2	1.3	4.8	1.7	3.6	7.1
3	2.1	7.7	2.8	5.7	11.4
4	3.0	10.7	3.9	7.9	15.8
6	4.8	17.4	6.4	12.9	25.6
8	6.7	24.1	8.9	17.9	35.6
10	8.8	31.5	11.5	23.4	46.6
12	10.9	39.3	14.4	29.3	58.6
16	15.4	55.4	20.3	41.3	82.6
20	20.2	72.7	26.6	54.1	108.2
24	25.1	90.4	33.1	67.3	134.6
30	32.8	118.1	43.3	87.9	175.8
36	40.9	147.2	54.0	109.6	219.2
42	49.2	177.1	64.9	131.9	263.8
48	57.7	207.7	76.2	154.6	309.2

*Data from Metcalf & Eddy, American Sewerage Practice, McGraw-Hill.

SEWAGE TREATMENT–SLUDGE DRYING

Sludge is dewatered generally on sand beds (open and glass-covered types) and in vacuum filters. Drying on sand beds is common practice for small and medium-size plants, and mechanical dewatering is used principally in large plants which produce large volumes of sludge.

SIZE OF SLUDGE BEDS

TYPE OF SLUDGE	Area, ft.²/capita	
	Open beds	Covered beds
Primary	1.00*	0.75*
Intermittent-sand filters	1.00	0.75
Standard-rate filter	1.25	1.00
High-rate filter	1.50	1.25
Activated sludge	1.75	1.35
Chemical precipitation	2.00	1.50

6" - 9" sand passing ¼" sieve; effective size 0.3 to 0.5 mm. uniformity coefficient not over 5.0

1" torpedo sand – size 0.8 to 1.2 mm.

¼" to ½" gravel.

½" to 1" gravel.

⅛" to ¼" gravel.

Gravel size ½" to 1".

Underdrain open joint vitrified tile or farm tile. Minimum size 4", Open joints of ¼" covered with muslin, burlap, or cheese cloth.

TYPICAL SECTION OF SLUDGE BED.

Width of sludge beds usually made 15' to 25'.

Underdrain minimum 4" diam. Spacing 8' to 12', on center Minimum slope 0.01' per ft.

Surrounding walls and partition banks may be concrete, wood, and earth embankment.

The recommended height is 18" above top of sand.

NOTE: Size, shape, and grouping of beds, and location of roads (access for equipment to remove sludge) depend on topography.

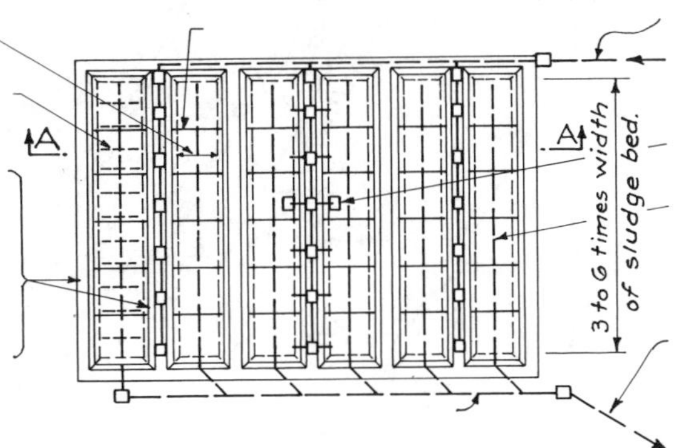
12" high plank
A
A
3 to 6 times width of sludge bed.
Underdrain Collector

PLAN

Minimum size of pipes used to deliver sludge to beds should be 8" cast-iron header, 6" cast-iron distributor.

Concrete or stone splash plates. Underdrain – minimum size 6", slope 0.006' per ft. Underdrainage of sludge beds, if possible, should be returned to primary tanks. The underdrainage should at least be chlorinated before final disposal.

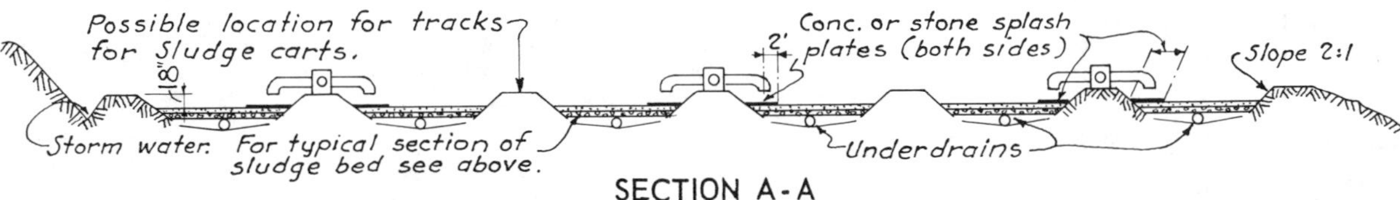
Possible location for tracks for Sludge carts.
18"
2'
Conc. or stone splash plates (both sides)
Slope 2:1
Storm water.
For typical section of sludge bed see above.
Underdrains

SECTION A-A

TYPICAL LAYOUT OF SLUDGE BEDS

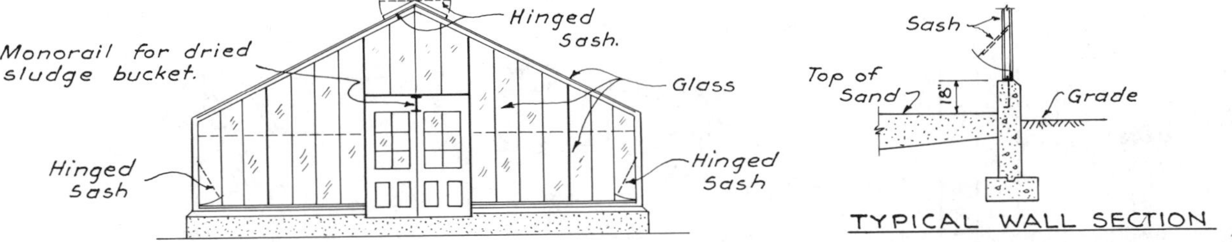
Monorail for dried sludge bucket.
Hinged Sash.
Glass
Hinged Sash
Hinged Sash
Sash
Top of Sand
18"
Grade
TYPICAL WALL SECTION

Sludge bed design to be similar to open beds as indicated above. Framework is of aluminum and steel. Doors are of wood with glass panels. Ventilation is provided by means of hinged sash, manually controlled, or by mechanical operators.

GLASS COVERS FOR SLUDGE BEDS

* Satisfactory for areas between 40° and 45° N. latitude. Increase values by 25% for area north of 45° N. latitude. Decrease values by 25% for area south of 40° N. latitude.

SEWAGE TREATMENT—DIGESTER GAS & HOT WATER LAYOUT **-1

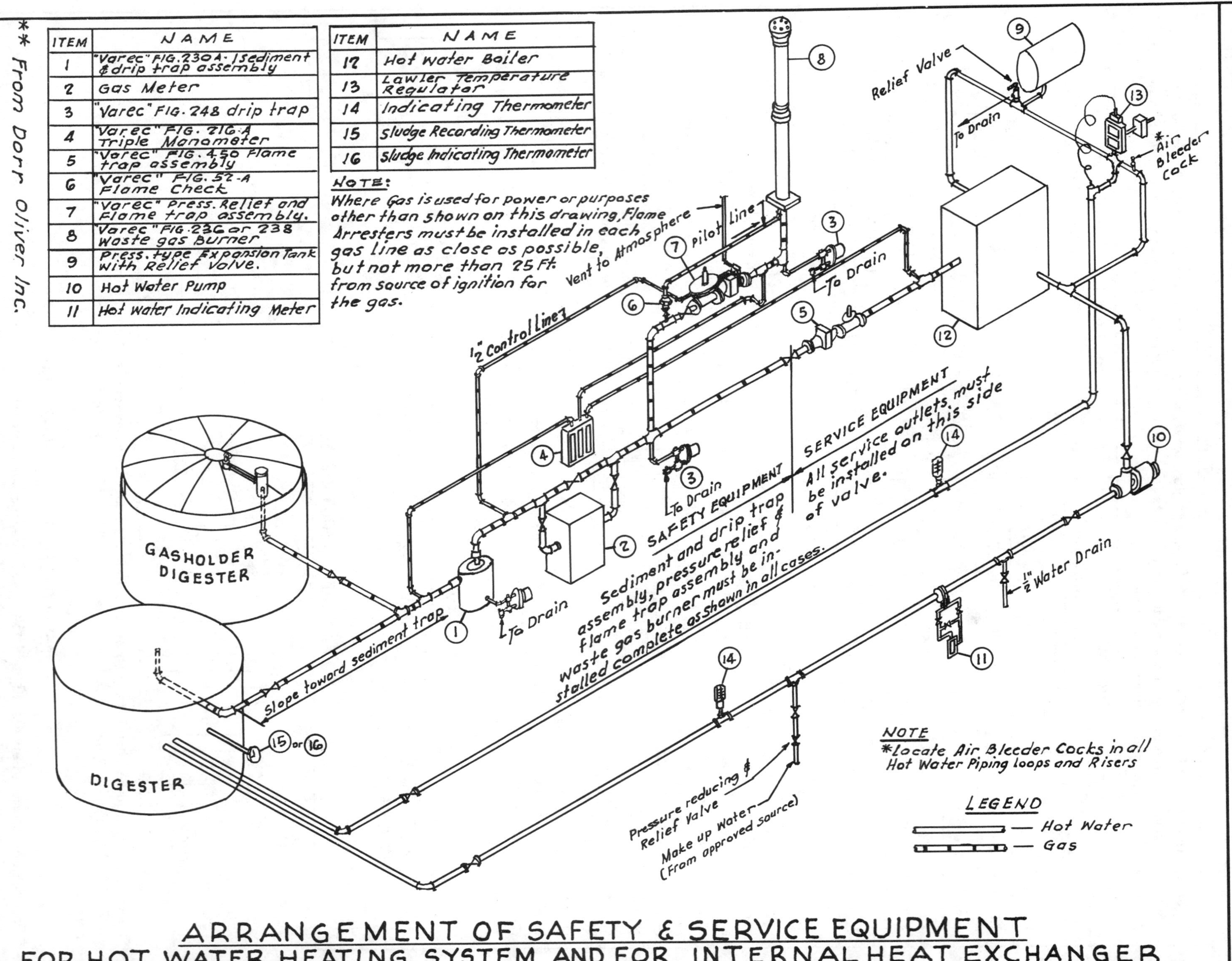

Relief Valve
To Drain
*Air Bleeder Cock
Pilot Line
Vent to Atmosphere
To Drain
½" Control Line
To Drain
SAFETY EQUIPMENT
SERVICE EQUIPMENT
All service outlets must be installed on this side of valve.
Sediment and drip trap assembly, pressure relief & flame trap assembly and waste gas burner must be installed complete as shown in all cases.
To Drain
½" Water Drain
GASHOLDER DIGESTER
Slope toward sediment trap
DIGESTER
15 or 16
Pressure reducing & Relief Valve
Make up Water (From approved source)
NOTE
*Locate Air Bleeder Cocks in all Hot Water Piping loops and Risers
LEGEND
— Hot Water
— Gas
ARRANGEMENT OF SAFETY & SERVICE EQUIPMENT
FOR HOT WATER HEATING SYSTEM AND FOR INTERNAL HEAT EXCHANGER

ITEM	NAME
1	"Varec" FIG. 230A-1 Sediment & drip trap assembly
2	Gas Meter
3	"Varec" FIG. 248 drip trap
4	"Varec" FIG. 216-A Triple Manometer
5	"Varec" FIG. 450 Flame trap assembly
6	"Varec" FIG. 52-A Flame Check
7	"Varec" Press. Relief and Flame trap assembly.
8	"Varec" FIG. 236 or 238 Waste gas Burner
9	Press. type Expansion Tank with Relief Valve.
10	Hot Water Pump
11	Hot water Indicating Meter

ITEM	NAME
12	Hot water Boiler
13	Lawler Temperature Regulator
14	Indicating Thermometer
15	Sludge Recording Thermometer
16	Sludge Indicating Thermometer

NOTE:
Where Gas is used for power or purposes other than shown on this drawing, Flame Arresters must be installed in each gas line as close as possible, but not more than 25 Ft. from source of ignition for the gas.

** From Dorr Oliver Inc.

SEWAGE TREATMENT—DIGESTER GAS & HOT WATER LAYOUT**-2

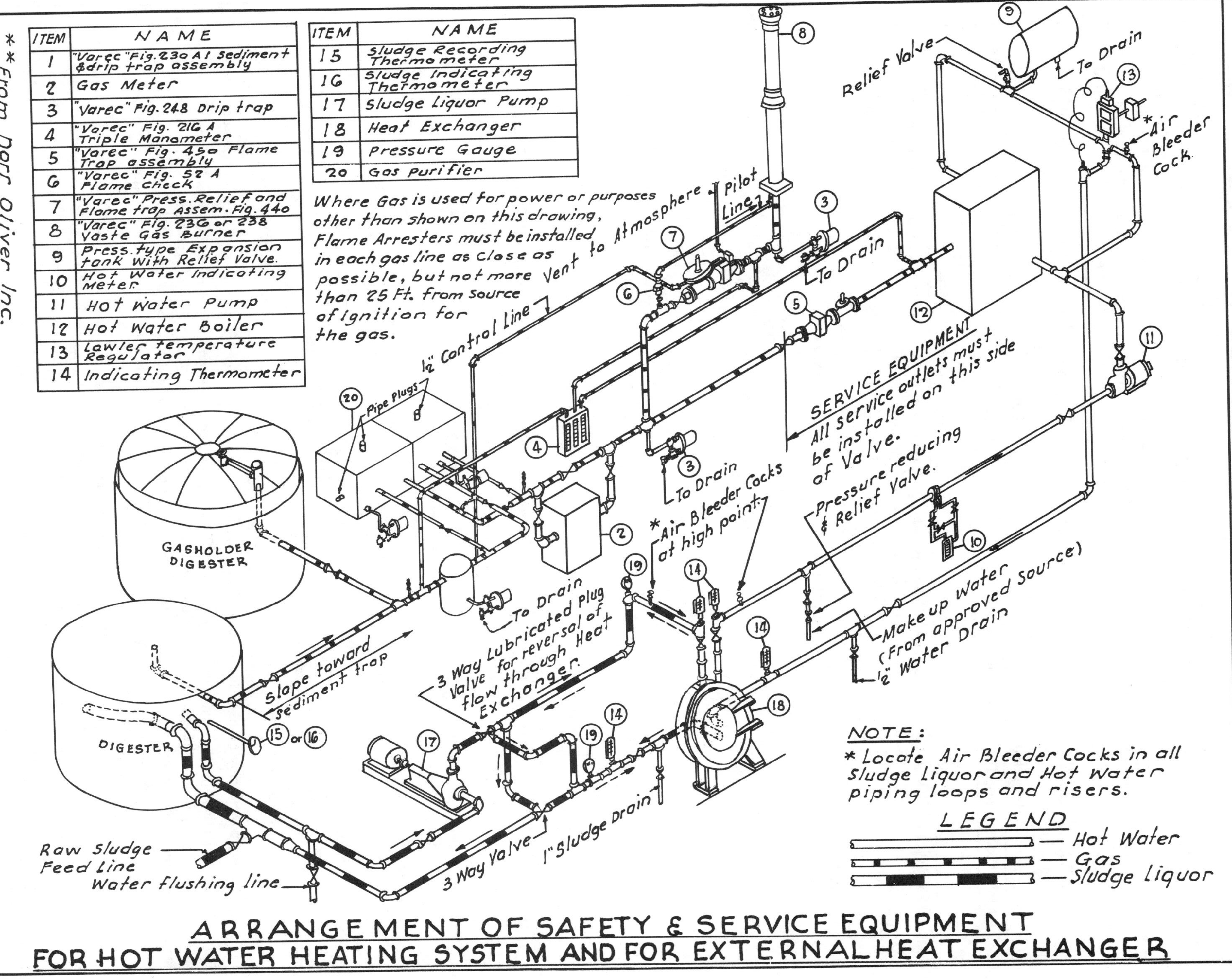

Relief Valve
To Drain
* Air Bleeder Cock
Pilot Line
Vent to Atmosphere
To Drain
½" Control Line
Pipe Plugs
SERVICE EQUIPMENT
All service outlets must be installed on this side of Valve.
Pressure reducing & Relief Valve.
To Drain
* Air Bleeder Cocks at high point.
Make up Water (From approved Source)
½" Water Drain
GASHOLDER DIGESTER
To Drain
Slope toward Sediment trap
3 Way Lubricated Plug Valve for reversal of flow through Heat Exchanger.
DIGESTER
Raw Sludge Feed Line
Water flushing line
3 Way Valve
1" Sludge Drain
NOTE:
* Locate Air Bleeder Cocks in all Sludge Liquor and Hot Water piping loops and risers.
LEGEND
— Hot Water
— Gas
— Sludge Liquor

ITEM	NAME
1	"Varec" Fig. 230 A1 Sediment & drip trap assembly
2	Gas Meter
3	"Varec" Fig. 248 Drip trap
4	"Varec" Fig. 216 A Triple Manometer
5	"Varec" Fig. 450 Flame Trap assembly
6	"Varec" Fig. 52 A Flame Check
7	"Varec" Press. Relief and Flame trap Assem. Fig. 440
8	"Varec" Fig. 236 or 238 Vaste Gas Burner
9	Press. type Expansion tank with Relief Valve.
10	Hot Water Indicating Meter
11	Hot Water Pump
12	Hot Water Boiler
13	Lawler temperature Regulator
14	Indicating Thermometer

ITEM	NAME
15	Sludge Recording Thermometer
16	Sludge Indicating Thermometer
17	Sludge Liquor Pump
18	Heat Exchanger
19	Pressure Gauge
20	Gas Purifier

Where Gas is used for power or purposes other than shown on this drawing, Flame Arresters must be installed in each gas line as close as possible, but not more than 25 Ft. from source of ignition for the gas.

ARRANGEMENT OF SAFETY & SERVICE EQUIPMENT FOR HOT WATER HEATING SYSTEM AND FOR EXTERNAL HEAT EXCHANGER

** From Dorr Oliver Inc.

SEWAGE TREATMENT-SLUDGE DIGESTION

SLUDGE PRODUCTION AND REQUIRED DIGESTION TANK CAPACITIES			
Type of Sewage Treatment Plant	Gallons of Sludge Produced per million gallons of sewage	Digestion Tank Capacity, cubic feet per capita	
		Heated	Unheated
Plain sedimentation	3000	2-3	4-6
Standard (low-capacity) trickling filter	700	3-4	6-8
High-capacity trickling filter	4000-6000	4-5	8-10
Plain sedimentation with chemical precipitation	5100		
Activated sludge	19,400	4-6	8-12

GAS PRODUCTION IN DIGESTION TANKS – The gas which results from the processes of sludge digestion amounts to 8 to 11 cu. ft. per lb. of volatile matter added to the digestion tanks and 0.3 to 1.0 cu. ft. a day per capita served. The gas has a nominal heating value of 500 to 800 B.t.u. per cu. ft. Digested sludge as drawn from tanks commonly has a moisture content of 90 to 95%.

TYPICAL DIGESTION TANKS

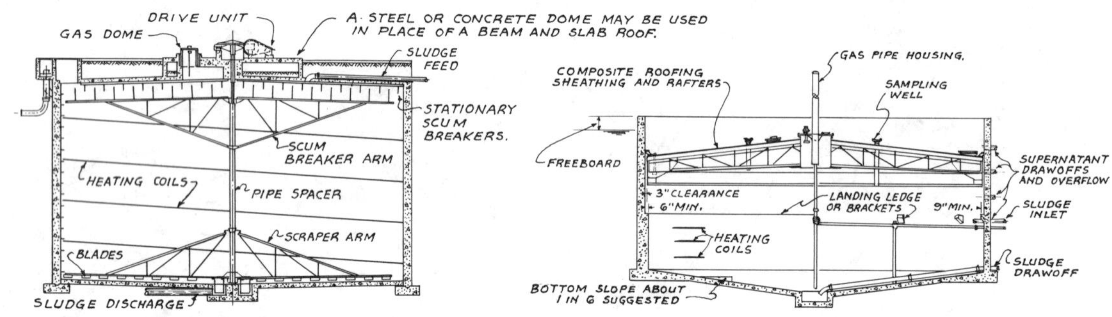
DRIVE UNIT
GAS DOME
A STEEL OR CONCRETE DOME MAY BE USED IN PLACE OF A BEAM AND SLAB ROOF.
SLUDGE FEED
STATIONARY SCUM BREAKERS.
SCUM BREAKER ARM
HEATING COILS
PIPE SPACER
SCRAPER ARM
BLADES
SLUDGE DISCHARGE
GAS PIPE HOUSING.
COMPOSITE ROOFING SHEATHING AND RAFTERS
SAMPLING WELL
FREEBOARD
SUPERNATANT DRAWOFFS AND OVERFLOW
3" CLEARANCE 6" MIN.
LANDING LEDGE OR BRACKETS
9" MIN.
SLUDGE INLET
HEATING COILS
SLUDGE DRAWOFF
BOTTOM SLOPE ABOUT 1 IN 6 SUGGESTED

FIXED-COVER DIGESTER* ***DIGESTER WITH FLOATING COVER†***

The tanks illustrated above are used singly, or in parallel operation when more than one tank is used.

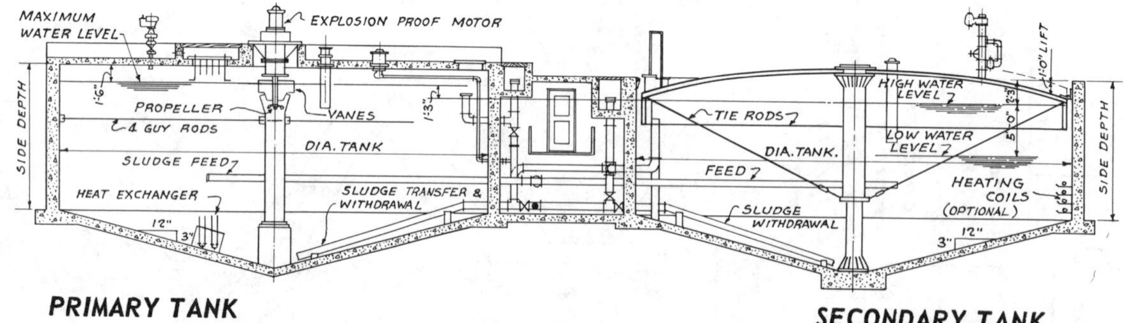
MAXIMUM WATER LEVEL
EXPLOSION PROOF MOTOR
SIDE DEPTH
1'-6"
PROPELLER
VANES
4 GUY RODS
DIA. TANK
SLUDGE FEED
HEAT EXCHANGER
SLUDGE TRANSFER & WITHDRAWAL
12"
3"
1'-0" LIFT
HIGH WATER LEVEL
TIE RODS
LOW WATER LEVEL
DIA. TANK.
FEED
HEATING COILS (OPTIONAL)
SLUDGE WITHDRAWAL

PRIMARY TANK ***SECONDARY TANK***

STAGE DIGESTION*

Floating covers may be used in lieu of covers indicated to provide maximum flexibility and safety of operation.†

Multiple-digestion tanks are recommended. This allows use of one tank if the other is out of service. Raw sludge is pumped into the primary tank where most digestion takes place. When stage digestion is used, sludge is transferred automatically from the primary to the secondary at the rate of feed of the incoming sludge, or when floating covers are incorporated, transfer material can be selected as desired. Supernatant or overflow liquor is normally withdrawn from the secondary digester tank only.

The secondary tank is essentially for gas and sludge storage. The contents of this tank, generally stratified in two layers, are in a quiescent state. The upper layer is a well-settled supernatant liquor which is returned to the incoming raw sewage. The lower layer of thickened and digested sludge is drawn off for drying in a vacuum filter or sludge-drying bed.

*From Dorr-Oliver Inc.
†From Pacific Flush Tank Co.

SEWAGE TREATMENT–SLUDGE CONDITIONING

Methods of treatment to condition sludge in preparation for further drying include digestion, thickening, elutriation, and mixing with chemical or other substances.

DIGESTION has been discussed on p. 19-22.

THICKENING – The purpose of sludge thickening is to reduce the volume of sludge delivered to the digesters, thereby decreasing the capacity of tanks required if no thickening is used. The volume of sludge in the thickener underflow may be decreased as much as 50% compared to normal sedimentation tank sludge.

For primary plants, sludge from the sedimentation tank is continuously pumped to a thickener tank. In secondary plants, sludge from both the primary and secondary sedimentation tanks is continuously pumped to the thickener.. Thickener overflow liquid is returned to the primary sedimentation tanks, and thickened sludge is pumped to the digesters.

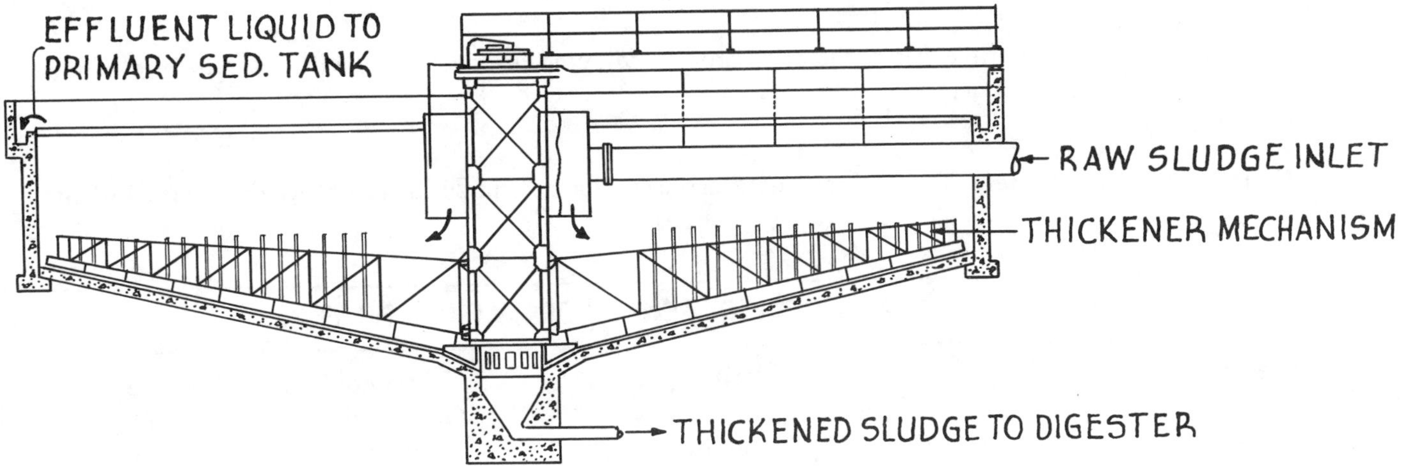
EFFLUENT LIQUID TO PRIMARY SED. TANK
RAW SLUDGE INLET
THICKENER MECHANISM
THICKENED SLUDGE TO DIGESTER

FIG. A – TYPICAL SECTION OF THICKENER*

ELUTRIATION – The purpose of elutriation or washing of sludge is to remove some of the constituents that inhibit coagulation, before dewatering the sludge in vacuum filters. Clean water or treated sewage is used for washing. Advantages of elutriation include a reduction of 50% or more of sludge conditioning chemical such as ferric chloride, eliminating the need for lime, and lowering the ash content of the filter cake. The amount of water used for elutriation is about 2 parts to 1 part of sludge.

Elutriation is performed in a settling tank, and the elutriate or overflow from the tank is returned to the raw sewage.

ELUTRIATION METHODS

1. Single stage, which consists of one-step dilution, sedimentation, and decantation in one tank.
2. Two stage, which consists of two one-step dilutions as above.
3. Countercurrent, which consists of treatment in two settling tanks in series. Washing water is added in the second stage only. Elutriate of the second stage is used as washing water in the first stage.

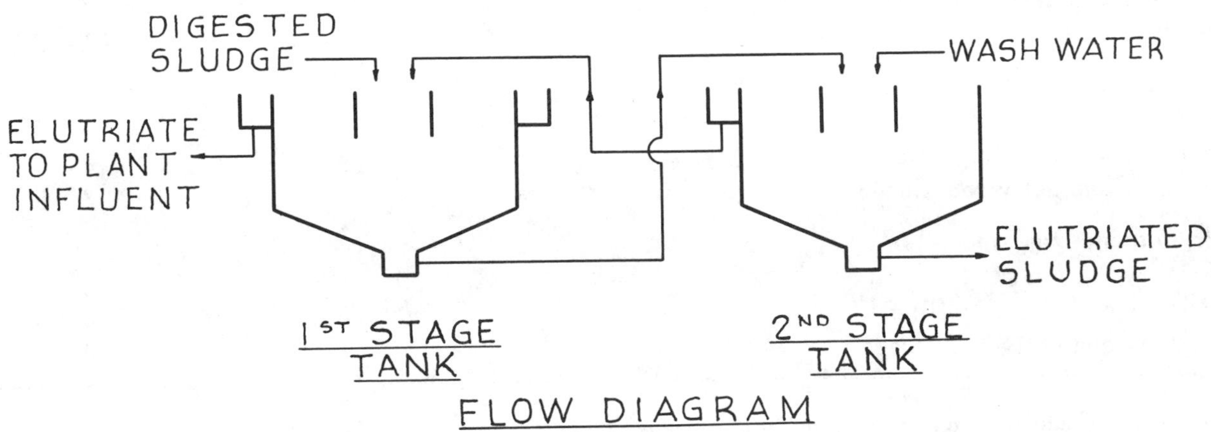
DIGESTED SLUDGE
WASH WATER
ELUTRIATE TO PLANT INFLUENT
ELUTRIATED SLUDGE
1ST STAGE TANK
2ND STAGE TANK
FLOW DIAGRAM

FIG. B – TWO-STAGE COUNTERCURRENT ELUTRIATION

*Dorr-Oliver Incorporated

SEWAGE TREATMENT—SLUDGE DRYING (VACUUM FILTRATION)

Chemicals used for conditioning sludge before vacuum filtration are ferric chloride, ferric sulfate, lime with ferric chloride, chlorinated copperas, and aluminum sulfate. Solutions are prepared in lined tanks and are fed by positive-displacement feeders or pumps into a conditioning tank containing sludge to be filtered.

The moisture content of the filter cake ranges from 60 to 80%, depending on the type of raw sludge and chemicals used.

To determine the size of the filter, the drum area is selected in accordance with the type of sludge and length of filtering period.

Periods are approximately 20 hours per week for large plants, and 30 hours per week for small plants.

APPROXIMATE SPACE REQUIREMENTS FOR VACUUM FILTERS

Filter Size Drum Area, sq. ft.	Filter* Area, sq. ft.	Filter† Appurtenances, sq. ft.	Chemical‡ Storage Space sq. ft.	Total* Area, sq. ft.
100	480	300	320	1100
200	650	530	600	1780
300	760	640	870	2270
400	840	720	1120	2680
600	1100	820	1500	3420

APPROXIMATE FILTER RATES

Type of Sludge	Filter Rate, lb. per hr./sq. ft.
Undigested Sludge	
Primary	8-12
Primary and trickling filter	6- 8
Primary and activated sludge	4- 5
Activated sludge	2.5- 3.5
Digested Sludge	
Primary	5- 8
Primary and trickling filter	4- 5
Primary and activated sludge	4- 5
Elutriated and Digested	
Primary	6- 8
Primary and trickling filter	4- 5
Primary and activated sludge	4- 5

*Areas will be reduced if a package-type filter plant is used.
†Filter appurtenances include sludge, vacuum and filtrate pumps, conditioning and filtrate tanks, and blower.
‡Areas will vary depending on amount and type of chemicals used.

SEWAGE TREATMENT-CHLORINATION-I

TABLE A— CHLORINE DOSAGE BASED ON AVERAGE FLOW.*

TYPE OF EFFLUENT	PARTS PER MILLION P.P.M.	POUNDS PER MILLION GALLONS
Raw Sewage	20-25	166-208
Sedimentation tank effluent	20	166
Trickling filter effluent	15	125
Sand filter effluent	6	50
Activated sludge effluent	8	67

For disinfection, chlorinatior capacity should be adequate to produce a residual of 2.0 p.p.m. in the final effluent. The above values will usually be sufficient for normal domestic sewage.

Pre-chlorination for odor control requires 20 p.p.m. chlorinators should be designed to have a maximum capacity which will satisfy the values listed in table above. 100% standby equipment should be provided.

For small installations hypochlorinators may be used economically.

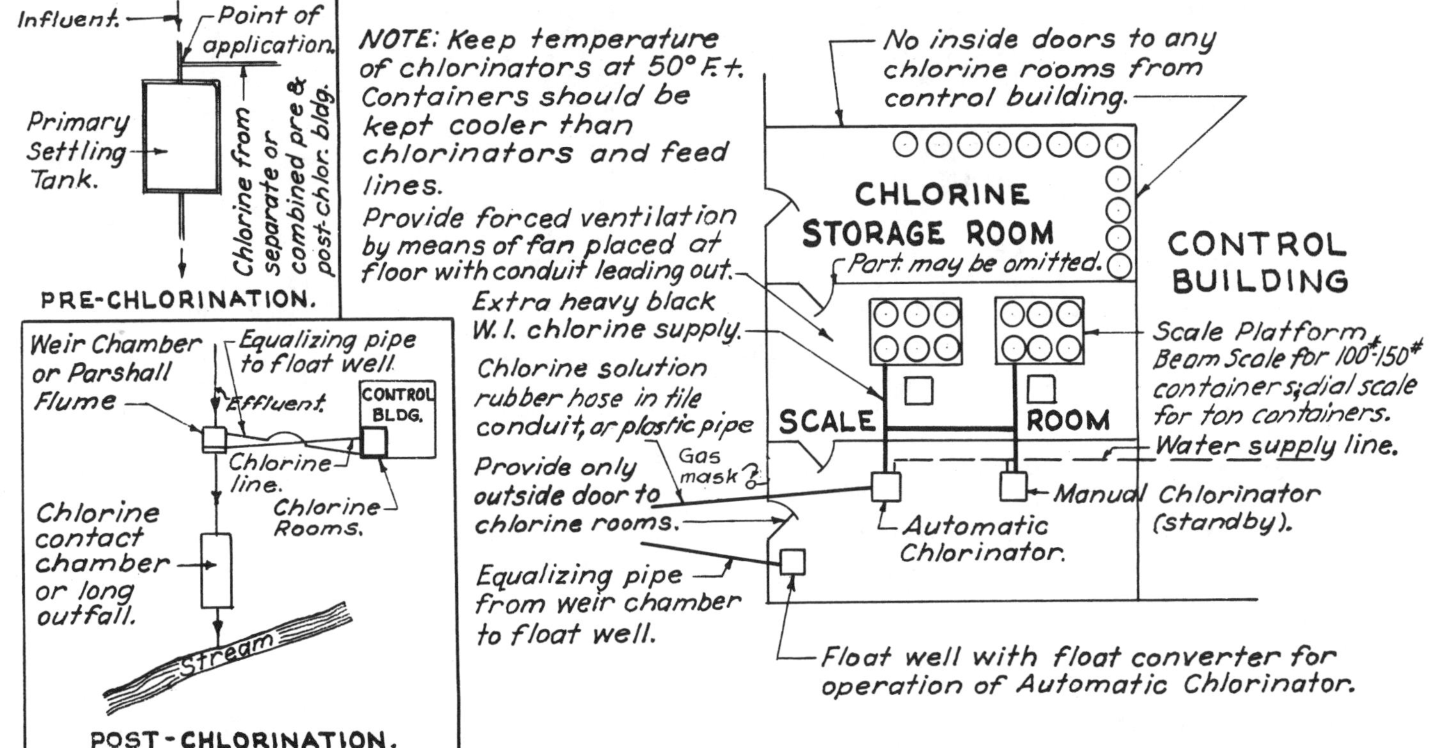

Influent.
Point of application.
Primary Settling Tank.
Chlorine from separate or combined pre & post-chlor. bldg.
PRE-CHLORINATION.
Weir Chamber or Parshall Flume
Equalizing pipe to float well.
Effluent.
CONTROL BLDG.
Chlorine line.
Chlorine Rooms.
Chlorine contact chamber or long outfall.
Stream
POST-CHLORINATION.
NOTE: Keep temperature of chlorinators at 50° F.+. Containers should be kept cooler than chlorinators and feed lines.
Provide forced ventilation by means of fan placed at floor with conduit leading out.
Extra heavy black W.I. chlorine supply.
Chlorine solution rubber hose in tile conduit, or plastic pipe
Provide only outside door to chlorine rooms.
Gas mask
Equalizing pipe from weir chamber to float well.
No inside doors to any chlorine rooms from control building.
CHLORINE STORAGE ROOM
Part may be omitted.
CONTROL BUILDING
SCALE ROOM
Scale Platform Beam Scale for 100#-150# containers; dial scale for ton containers.
Water supply line.
Manual Chlorinator (standby).
Automatic Chlorinator.
Float well with float converter for operation of Automatic Chlorinator.

FIG. B— TYPICAL LAYOUT FOR CHLORINATOR INSTALLATION.

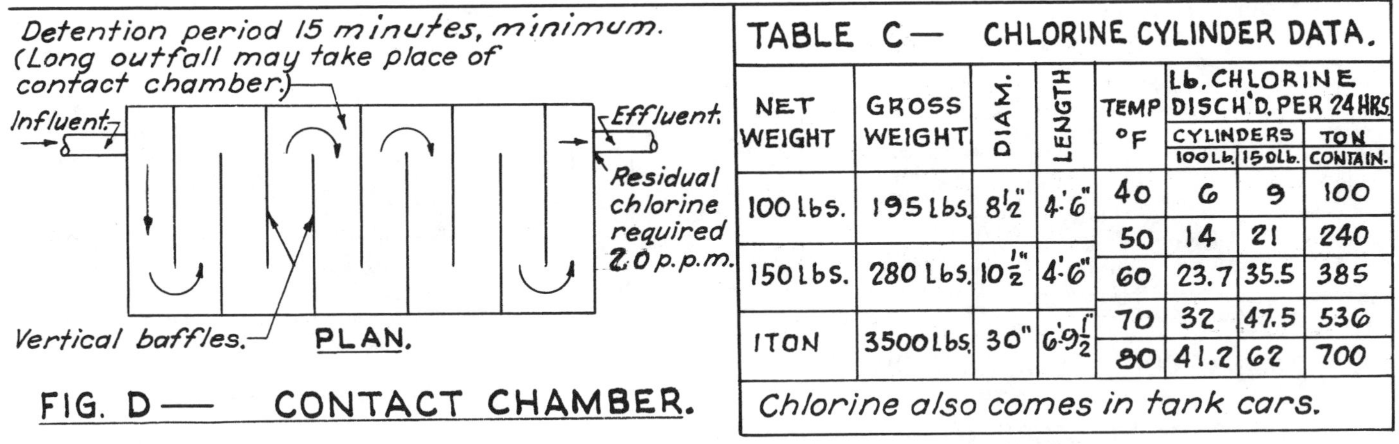

Detention period 15 minutes, minimum. (Long outfall may take place of contact chamber.)
Influent.
Effluent.
Residual chlorine required 2.0 p.p.m.
Vertical baffles.
PLAN.

FIG. D— CONTACT CHAMBER.

TABLE C— CHLORINE CYLINDER DATA.

NET WEIGHT	GROSS WEIGHT	DIAM.	LENGTH	TEMP °F	Lb. CHLORINE DISCH'D. PER 24 HRS. CYLINDERS 100 Lb.	CYLINDERS 150 Lb.	TON CONTAIN.
100 lbs.	195 lbs.	8½"	4'-6"	40	6	9	100
				50	14	21	240
150 lbs.	280 lbs.	10½"	4'-6"	60	23.7	35.5	385
				70	32	47.5	536
1 TON	3500 lbs.	30"	6'-9½"	80	41.2	62	700

Chlorine also comes in tank cars.

* Standards for Sewage Works, May 1952, Upper Mississippi River and Great Lakes Boards of Public Health Engineers.

SEWAGE TREATMENT–CHLORINATION – 2

TABLE A – CHLORINATOR OPERATING WATER REQUIREMENTS*

Chlorine Feed Rate, lb./24 hr.	Injector Data	Back Pressure, p.s.i. 0	10	15	20	25	30	35	40
10	Operating water, p.s.i.	25	45	50	60	70	80	90	100
	Operating water, g.p.m.	0.9	1.1	1.2	1.3	1.4	1.5	1.6	1.6
	Solution discharge, g.p.m.	1.5	1.9	1.8	1.9	2.0	2.1	2.2	2.2
15	Operating water, p.s.i.	25	30	40	50	60	70	80	90
	Operating water, g.p.m.	0.9	1.8	2.0	2.2	2.4	2.6	2.8	3.0
	Solution discharge, g.p.m.	1.5	2.6	2.3	3.0	3.2	3.4	3.6	3.8
20	Operating water, p.s.i.	25	55	55	60	75	85	90	100
	Operating water, g.p.m.	0.9	1.2	2.3	3.5	3.9	4.2	4.4	4.5
	Solution discharge, g.p.m.	1.5	2.1	3.1	4.4	4.8	5.1	5.3	5.4
25	Operating water, p.s.i.	25	60	65	75	80	90	95	100
	Operating water, g.p.m.	0.9	1.3	2.5	2.7	4.0	4.3	4.4	4.6
	Solution discharge, g.p.m.	1.5	1.9	3.3	3.5	4.9	5.2	5.3	5.5
30	Operating water, p.s.i.	25	55	70	85	80	90	100	110
	Operating water, g.p.m.	1.5	2.3	2.6	2.9	4.1	4.3	4.6	4.8
	Solution discharge, g.p.m.	2.4	3.1	3.4	3.7	5.0	5.2	5.5	5.7
40	Operating water, p.s.i.	25	45	60	75	90	100	110	125
	Operating water, g.p.m.	1.6	3.0	3.5	3.9	4.3	4.6	4.8	5.1
	Solution discharge, g.p.m.	2.4	3.9	4.4	4.8	5.2	5.5	5.7	6.0
50	Operating water, p.s.i.	25	50	70	80	95	115	125	110
	Operating water, g.p.m.	1.6	3.3	3.8	4.1	4.4	4.9	5.1	6.2
	Solution discharge, g.p.m.	2.4	4.2	4.7	5.0	5.3	5.8	6.0	7.2
60	Operating water, p.s.i.	25	55	75	90	110	110	120	125
	Operating water, g.p.m.	1.6	3.4	4.0	4.4	4.8	6.2	6.5	6.6
	Solution discharge, g.p.m.	2.4	4.3	4.9	5.3	5.7	7.2	7.5	7.6
75	Operating water, p.s.i.	25	65	85	105	125	145	140	145
	Operating water, g.p.m.	2.3	3.7	4.2	4.7	5.1	5.5	7.0	7.1
	Solution discharge, g.p.m.	3.2	4.6	5.1	5.6	6.0	6.4	8.0	8.1
100	Operating water, p.s.i.	25	60	100	120	150	160	175	145
	Operating water, g.p.m.	3.0	4.5	4.6	5.0	5.6	7.5	7.8	13.3
	Solution discharge, g.p.m.	4.0	5.5	5.5	5.9	6.5	8.5	8.8	14.8
150	Operating water, p.s.i.	25	55	75	90	100	115	130	140
	Operating water, g.p.m.	5.5	8.3	9.6	10.4	11.0	11.8	12.4	13.0
	Solution discharge, g.p.m.	7.0	9.8	11.1	11.9	12.5	12.3	13.9	14.5

Chlorine Feed Rate, lb./24 hr.	Injector Data	Back Pressure, p.s.i. 0	10	15	20	25	30	35	40
200	Operating water, p.s.i.	25	70	90	105	125	145	125	140
	Operating water, p.s.i.	5.4	9.2	10.5	11.4	12.3	13.1	19.1	20.0
	Solution discharge, g.p.m.	6.9	10.7	12.0	12.9	13.8	14.6	20.8	21.7
300	Operating water, p.s.i.	30	105	130	145	135	155	175	190
	Operating water, g.p.m.	9.2	11.0	12.5	13.2	19.8	21.1	22.5	23.5
	Solution discharge, g.p.m.	10.9	12.5	14.0	14.7	20.5	22.8	24.2	24.2
400	Operating water, p.s.i.	25	85	120	105	130	150	160	165
	Operating water, g.p.m.	13.5	15.7	17.8	28.0	31.0	33.2	34.0	35.0
	Solution discharge, g.p.m.	15.5	17.4	19.5	30.0	33.0	35.2	36.0	37.0
500	Operating water, p.s.i.	35	72	95	120	140	160		
	Operating water, g.p.m.	16.0	23.0	26.5	30.0	31.5	34.0		
	Solution discharge, g.p.m.	18.0	26.0	28.5	32.0	33.5	36.0		
750	Operating water, p.s.i.	60	116	150	200	230	255		
	Operating water, g.p.m.	21.0	29.0	33.5	38.5	40.3	42.0		
	Solution discharge, g.p.m.	23.0	31.0	33.5	40.5	42.3	44.0		
1000	Operating water, p.s.i.	25	50	50	60	65	90	100	100
	Operating water, g.p.m.	30	55	60	80	90	90	90	110
	Solution discharge, g.p.m.	30	55	60	80	90	90	90	110
2000	Operating water, p.s.i.	25	40	50	60	80	90	100	100
	Operating water, g.p.m.	60	75	80	90	90	100	100	115
	Solution discharge, g.p.m.	60	75	80	90	90	100	100	115
3000	Operating water, p.s.i.	25	50	75	75	90	100	120	
	Operating water, g.p.m.	80	130	150	190	200	210	240	
	Solution discharge, g.p.m.	80	130	150	190	200	210	240	
4000	Operating water, p.s.i.	25	65	75	85	100	120		
	Operating water, g.p.m.	120	160	190	220	240	260		
	Solution discharge, g.p.m.	120	160	190	220	240	260		
5000	Operating water, p.s.i.	25	75	85	100	120	140		
	Operating water, g.p.m.	140	170	210	230	250	270		
	Solution discharge, g.p.m.	140	170	210	230	250	270		
6000	Operating water, p.s.i.	25	90	100	110	130			
	Operating water, g.p.m.	170	200	240	270	280			
	Solution discharge, g.p.m.	170	200	240	270	280			

NOTES: "Back pressure" is the pressure at point of application plus or minus difference in elevation between point of application and Chlorinator room floor plus friction loss in solution line and diffuser. For initial calculation, assume friction loss of 5 lbs. If possible keep friction loss under 5 lbs. by proper selection of hose or pipe.
Operating water, p.s.i. is water pressure at injector inlet.
Solution discharge, g.p.m. is operating water, g.p.m. plus make-up water.

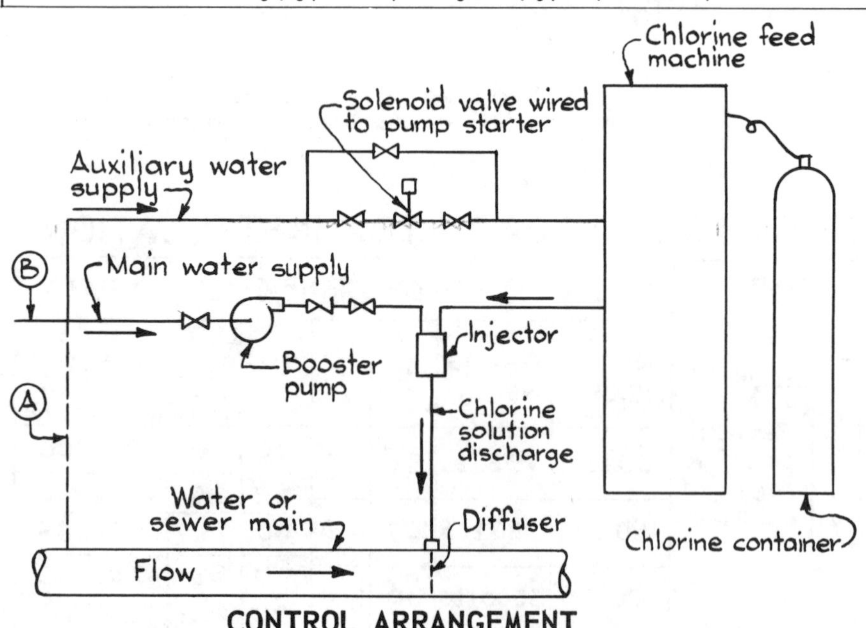

Chlorine feed machine
Solenoid valve wired to pump starter
Auxiliary water supply
Main water supply
B
A
Booster pump
Injector
Chlorine solution discharge
Water or sewer main
Diffuser
Flow
Chlorine container

CONTROL ARRANGEMENT FOR BACK PRESSURE INSTALLATION

A For water system.

B For sewage treatment system.

PROBLEM

Given: 25 p.p.m. (parts per million) of chlorine for raw sewage (see Table A, p. 19-25, and average flow of sewage of 140,000 gal. per day. Back pressure in main of 20 p.s.i.

To Find: Chlorine feed rate per day, operating water pressure, and quantity of water for selecting booster pump.

Solution: Enter Table A above with back pressure of 20 p.s.i. and chlorine feed rate of 30 lb. per 24 hr. Read operating water required in gallons per minute at pounds per square inch, 2.9 g.p.m. at 85 p.s.i.

Total head of pump = 85 p.s.i. − 20 p.s.i. = 65 p.s.i. plus losses due to friction in piping valves, and fittings. Pump should have capacity of, say, 5 g.p.m. at computed head.

* From Wallace & Tiernan, Inc.

SEWAGE TREATMENT-HYPOCHLORINATORS

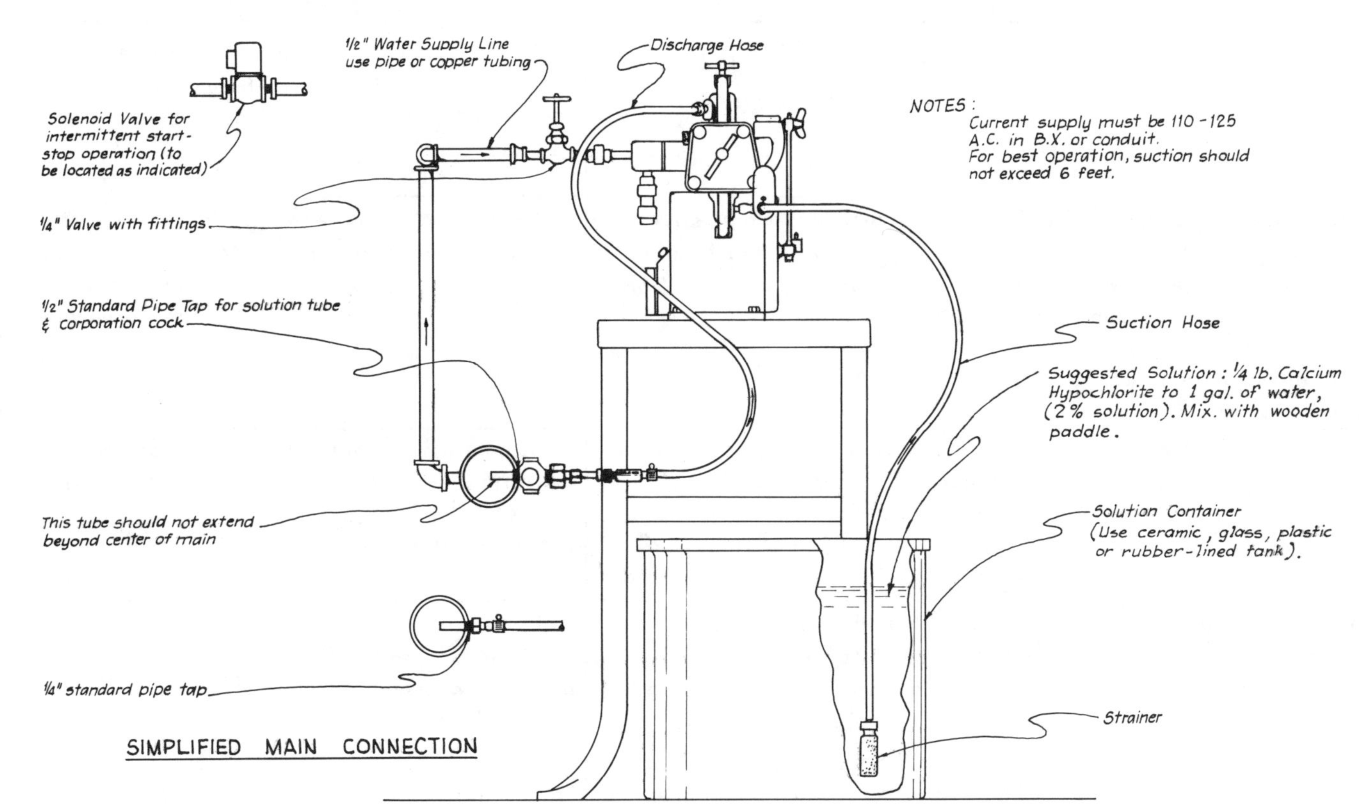

1/2" Water Supply Line use pipe or copper tubing
Discharge Hose
Solenoid Valve for intermittent start-stop operation (to be located as indicated)
NOTES:
Current supply must be 110 - 125 A.C. in B.X. or conduit.
For best operation, suction should not exceed 6 feet.
1/4" Valve with fittings
1/2" Standard Pipe Tap for solution tube & corporation cock
Suction Hose
Suggested Solution: 1/4 lb. Calcium Hypochlorite to 1 gal. of water, (2% solution). Mix. with wooden paddle.
This tube should not extend beyond center of main
Solution Container (Use ceramic, glass, plastic or rubber-lined tank).
1/4" standard pipe tap
Strainer
SIMPLIFIED MAIN CONNECTION

TYPICAL INSTALLATION - ELECTRICALLY OPERATED HYPOCHLORINATOR*

FOR APPLICATION AGAINST PRESSURE 5 TO 125 POUNDS PER SQUARE INCH

TECHNICAL DATA FOR HYPOCHLORINATIONS

Specification	Automatic Series A-416 and A-429	Manual Series A-415	Electrical Types Series A-417		Belt-Driven Series A-537
Capacity solution, gal./24 hr.	60	60	68		68
Pumping range, gal./24 hr.	1-60	5-60	1.9-68		16-68
Milliliters of solution per stroke, maximum	13.2	13.2	15.0		15.0
Allowable main pressure, lb.	10-125	10-125	5-125	Vacuum-5	Vacuum-30
Normal water consumption, g.p.m.	0.5	0.4	0.2	0	0
Drain required	Yes	Yes	Yes	No	No
Current requirements	. .	. .	110-125ν 25 or 60 cycle 1Φ		. .
Power consumption, watts maximum	. .	. .	30		. .
Transformer size	. .	. .	150 VA		. .
Strokes per minute	12 max.	1-12	2 or 12		12 at 720 r.p.m.
Hose size	3/8" I.D.	3/8" I.D.	3/8" I.D.		3/8" I.D.

OPERATION

Essentially, the hypochlorinator is a diaphragm pump constructed of corrosion-resistant materials and so designed that the rate of chlorine application may be easily adjusted to provide the required treatment.

Three types are available. The electric unit utilizes a small integral electric motor drive, and may be controlled manually or may be synchronized with other apparatus for automatic start-and-stop operation. The belt-driven unit, powered through a belt drive by an external motor or engine, may be controlled manually or from the power source controls. The automatic unit is powered by an integral specially designed water motor and is paced by a water meter for fully automatic proportional-flow control.

* From Wallace & Tiernan, Inc.

SEWAGE TREATMENT – CLASSIFICATION OF RECEIVING WATERS

TABLE A*

Use	Standards of Quality of Low-Water Stage	Required Treatment of Sewage Before Discharge	
		Normal	Emergency
(1)	(2)	(3)	(4)
Industrial uses not needing a high-quality water; irrigation of crops not subject to contamination when intended for human consumption; and receipt of wastes without the creation of nuisance.	Absence of nuisance, odors, slick, and unsightly suspended or floating matter; dissolved oxygen present at outfall	Sedimentation except when receiving waters are large in volume	Chlorination for removal of hydrogen sulfide; addition of nitrate to supply oxygen
Fishing; recreational boating; raising of seed oysters; and industrial use after treatment	Absence of slick, odors, and visible floating and suspended solids DO $\geqq$ 3 mg./l and preferably $\geqq$ 5 mg./l; CO_2 < 40 mg/l and preferably < 20 mg/l	Sedimentation, chemical precipitation, or biological treatment depending upon degree of dilution.	Aeration; addition of diluting waters
Bathing, recreation, and shellfish culture	Clear; no visible sewage matter; a coliform indicated number less than 100 to 1,000 per 100 ml depending on length of bathing season; DO near saturation	Sedimentation, chemical precipitation, or biological treatment depending upon degree of dilution; chlorination of effluent	Heavy chlorination
Drinking water and related uses	Chemical standards for substances not removable by common water-treatment methods; clear; DO near saturation; coliform MPN of 50/100 ml when chlorination is the only treatment; up to 20,000/100 ml when treatment is complete	Sedimentation, chemical precipitation or biological treatment depending upon degree of dilution; chlorination of effluent	Heavy chlorination

TABLE B

INTERSTATE COMPACT REQUIREMENTS OF CONNECTICUT, NEW YORK, AND NEW JERSEY FOR TREATMENT OF SEWAGE DISCHARGED INTO THEIR TIDAL WATERS.

	CLASS A Waters expected to be used primarily for recreational purposes, shellfish culture, and development of fish life.	CLASS B All other waters.
1. Floating solids	Full removal	Full removal
2. Suspended solids	60% removal	10% removal or enough to avoid sludge deposits.
3. Coliform bacteria in water samples during bathing season.	Probable number of not more than 1 per c.c. in 50% of 1 c.c. samples.	
4. Dissolved oxygen saturation in vicinity of outfall.	Not less than 50% during any week of the year.	Not less than 30%

*Data from Imhoff and Fair, *Sewage Treatment*.

NOTES: 1. See p. 19-01, Table D, Dilution Requirements.

2. No national standards governing the discharge of sewage and sewage effluents into receiving waters have yet been evolved in the United States. Various State and Interstate Commissions have established standards for waters under their jurisdiction (see Table B).

3. Column 4, Table A, above indicates certain treatment measures to be applied in emergencies, such as shutdown or overloading of treatment works, extreme low water or hot weather, and the emergency use of the receiving water for other than customary purposes.

4. Ordinarily it is permissible to discharge the water-carried wastes from isolated dwellings and small communities without treatment into receiving waters of industrial or fishing uses. Sometimes primary treatment is required.

5. For more analytical methods for determining allowable loading of receiving waters with sewage, see *Sewage Treatment* by Imhoff and Fair.

SEWAGE TREATMENT- PUMP TYPES

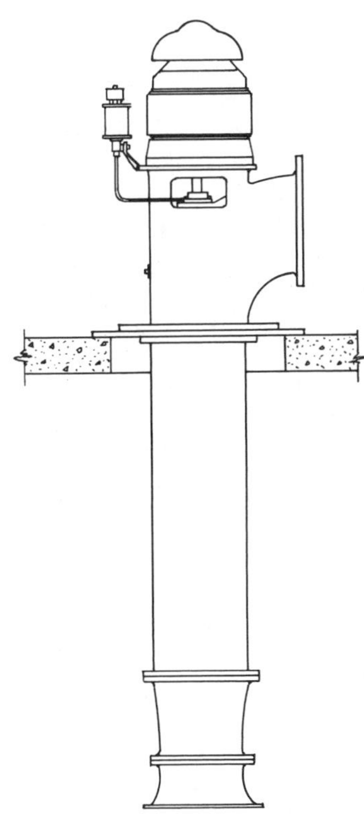

MIXED-FLOW SEWAGE PUMP
(VERTICAL) – ALSO AVAILABLE
IN HORIZONTAL TYPE

MOTOR ABOVE

SLUDGE PUMP
(DIAPHRAGM TYPE WITH BALL VALVES)

CHICAGO PUMP CO.

SLUDGE PUMP
*SCREW-FEED CENTRIFUGAL

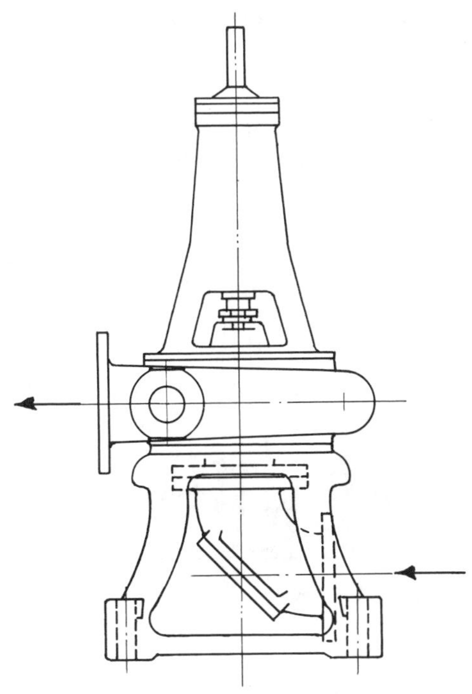

SEWAGE PUMP
VERTICAL CENTRIFUGAL, NON-CLOG
(FOR DRY-PIT INSTALLATION)

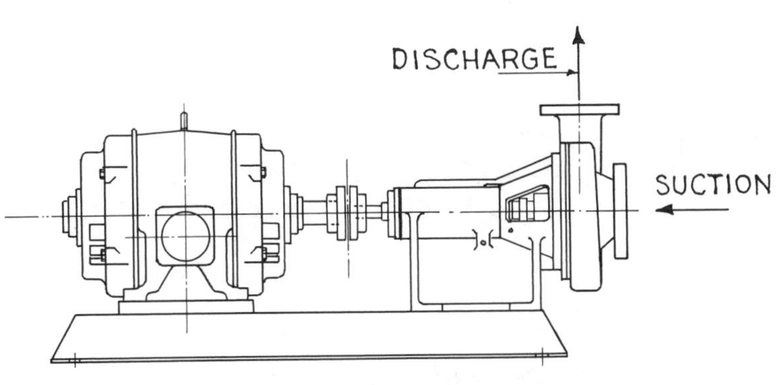
DISCHARGE
SUCTION

SEWAGE PUMP
HORIZONTAL CENTRIFUGAL

*Chicago Pump Co.

SEWAGE TREATMENT-TYPICAL SEWAGE LIFT STATION

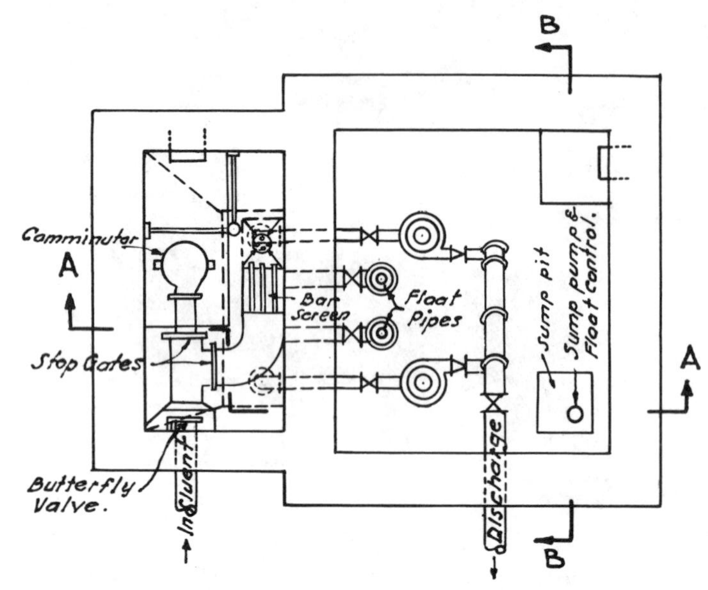
B
Comminutor
A
Bar Screen
Float Pipes
Stop Gates
Sump pit
Sump pump & Float Control.
A
Butterfly Valve.
Influent
Discharge
B

PLAN OF COMMINUTOR & PUMP ROOM.

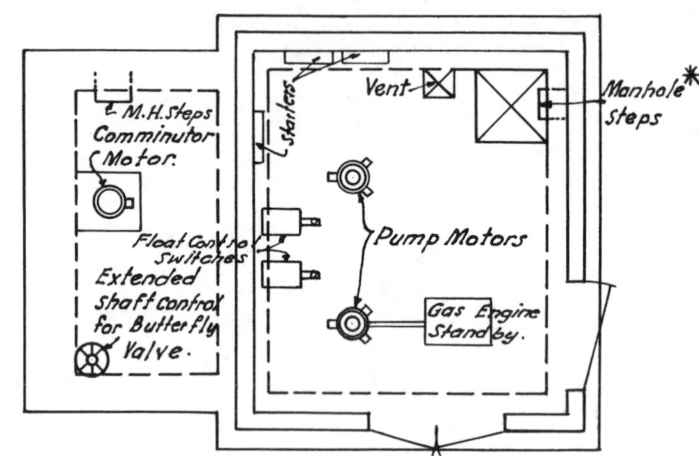
M.H. Steps
Comminutor Motor.
Stairs
Vent
Manhole* Steps
Float Control Switches
Pump Motors
Extended Shaft Control for Butterfly Valve.
Gas Engine Stand by.

PLAN OF CONTROL ROOM.

NOTES FOR DESIGN

For stations with two pumps.

Capacity of pump, each for excess of maximum flow. Size of wet well 10 minutes detention, average daily flow.

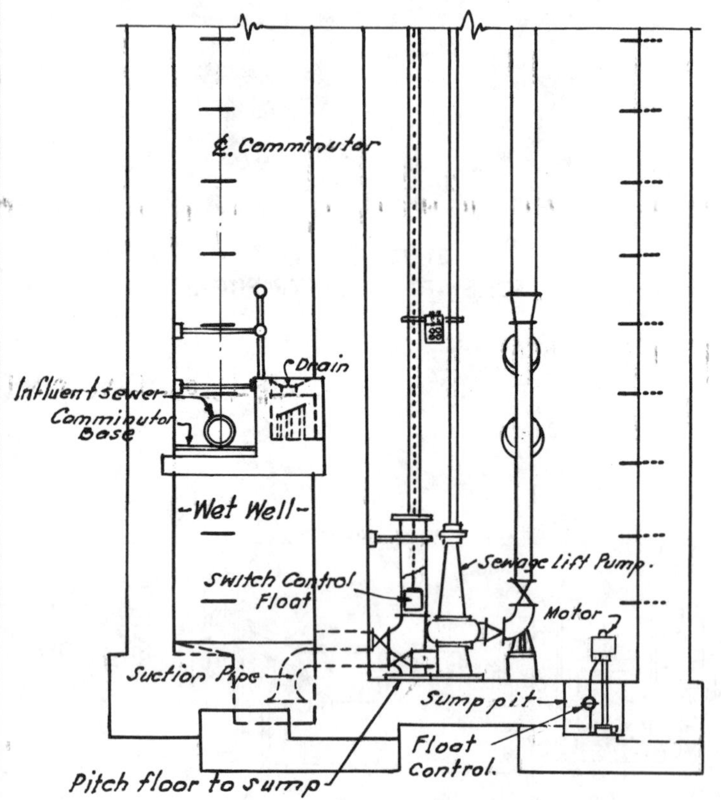
℄ Comminutor
Drain
Influent sewer
Comminutor Base
-Wet Well-
Sewage Lift Pump.
Switch Control Float
Motor
Suction Pipe
Sump pit
Float Control.

Pitch floor to sump

SECTION A-A

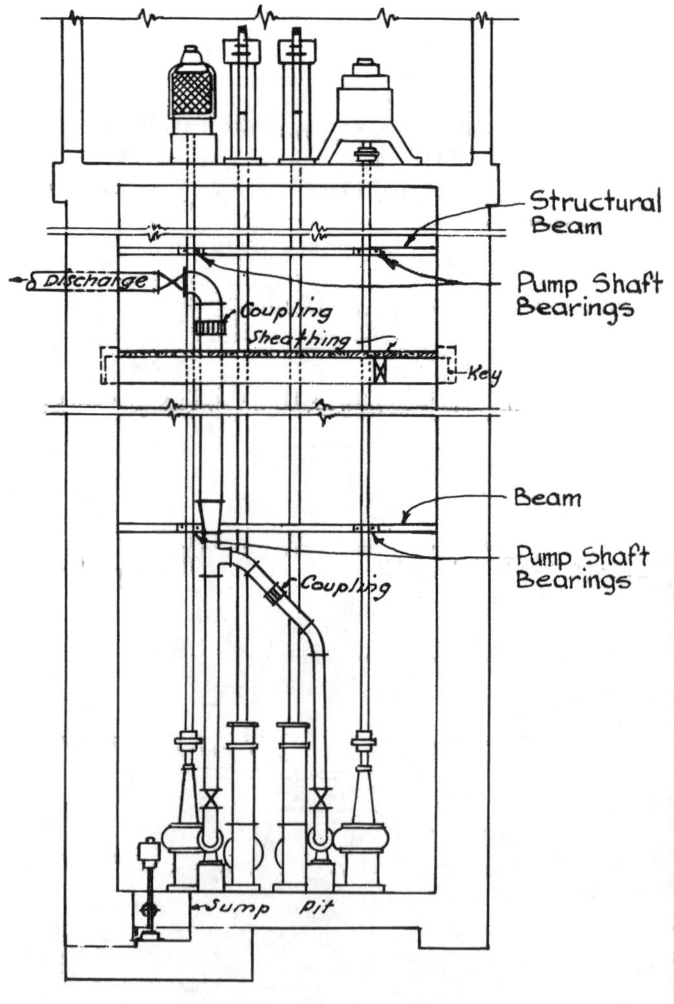
Structural Beam
Discharge
Pump Shaft Bearings
Coupling
Sheathing
Key
Beam
Pump Shaft Bearings
Coupling
Sump pit

SECTION B-B

* Where space is available stairway or ships ladder is preferable.

SEWAGE TREATMENT-PUMPING STATION

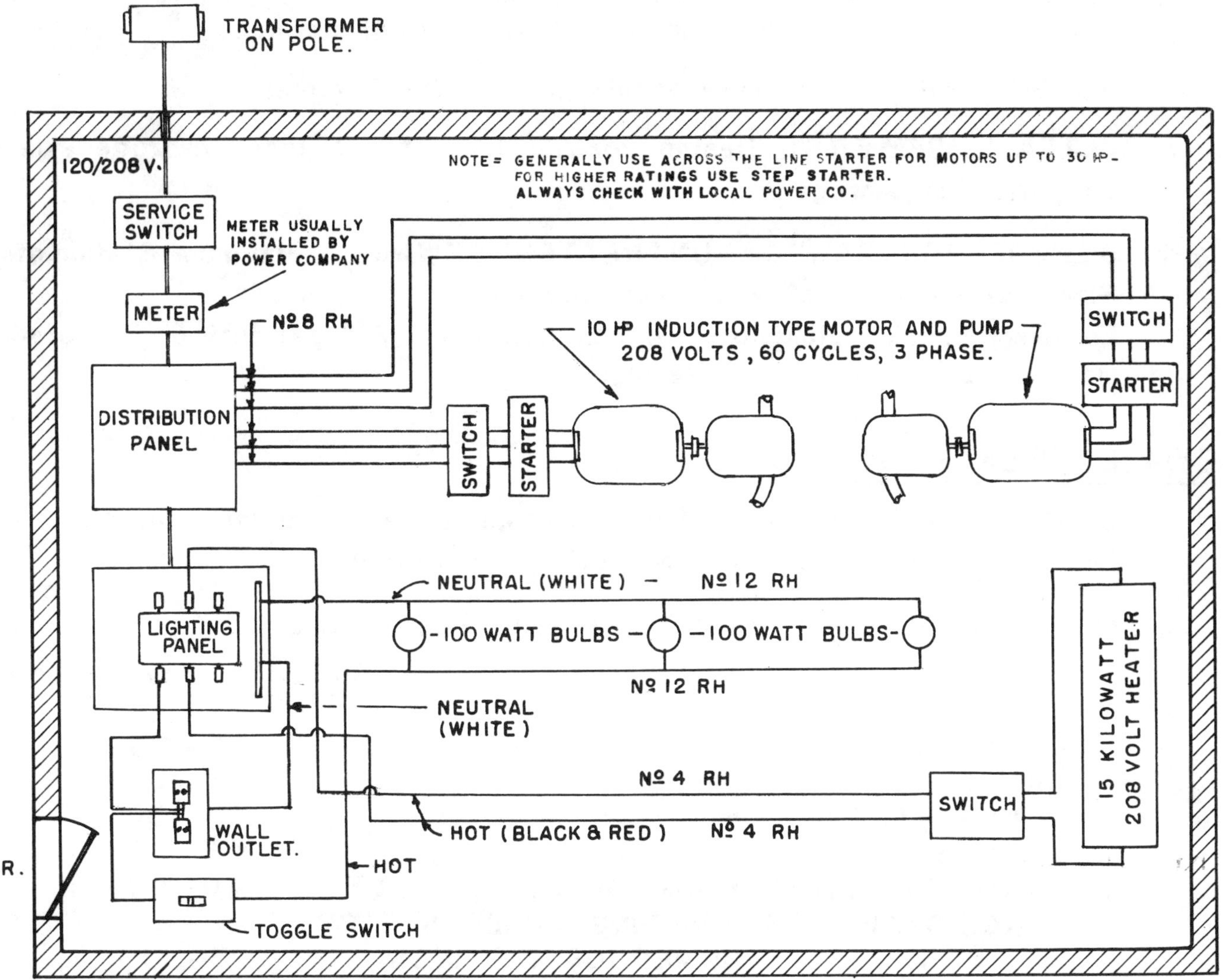
TRANSFORMER ON POLE.
120/208 V.
NOTE = GENERALLY USE ACROSS THE LINE STARTER FOR MOTORS UP TO 30 HP - FOR HIGHER RATINGS USE STEP STARTER. ALWAYS CHECK WITH LOCAL POWER CO.
SERVICE SWITCH
METER USUALLY INSTALLED BY POWER COMPANY
METER
Nº 8 RH
10 HP INDUCTION TYPE MOTOR AND PUMP
208 VOLTS, 60 CYCLES, 3 PHASE.
SWITCH
STARTER
DISTRIBUTION PANEL
SWITCH
STARTER
NEUTRAL (WHITE) - Nº 12 RH
LIGHTING PANEL
-100 WATT BULBS -
-100 WATT BULBS-
Nº 12 RH
15 KILOWATT 208 VOLT HEATER
NEUTRAL (WHITE)
Nº 4 RH
SWITCH
WALL OUTLET.
HOT (BLACK & RED) Nº 4 RH
DOOR.
HOT
TOGGLE SWITCH

TYPICAL ELECTRICAL WIRE-DIAGRAM OF BOOSTER PUMP STATION

SEWAGE TREATMENT—SMALL SYSTEMS-1

DESIGN DATA

SEWERS

1. LATERAL OR MAIN SEWERS: Design when flowing full for not less than 10 times average estimated daily flow of sewage.
2. DESIGN PERIOD: For future population for 10–20 years.
3. OUTFALL SEWERS: Design for not less than 5 times average estimated daily flow of sewage.
4. SEWERS FOR RAW OR UNTREATED SEWAGE: Minimum 6" diameter. Slope at minimum of 1/8" inch per foot or 1 % grade. For very small installations, 4 inch sewer at 1/4" per foot or 2% grade may be used for settled sewage.

SEWAGE TREATMENT

1. CONTRIBUTING POPULATION: Design sewage treatment plant units for the estimated ultimate contributing population.
2. FLOWS SHOULD BE BASED ON THE FOLLOWING:

	Gallons per day per person *
CAMPS, GENERAL	25 - 75
CAMPS, TOURIST	25 - 50
CAMPS, TRAILER	40 - 50
MOTELS	45 - 75
SMALL DWELLINGS, FARM HOUSES, COTTAGES	40 - 60
LARGE DWELLINGS, BOARDING SCHOOLS WITH NUMEROUS FITTINGS	75 - 100
INSTITUTIONS (EXCEPT HOSPITALS)	75 - 125
HOSPITALS	150 - 250
DAY SCHOOLS	15
DAY SCHOOLS WITH SHOWERS	20

FACTORIES - 15-35 GALS. PER PERSON PER SHIFT
DRIVE-IN THEATRES - 5 GALS. PER CAR STALL
MOVIE THEATRES - 5 GALS. PER THEATRE SEAT
GOLF CLUBS - FOR AVERAGE WEEK-END POPUL. { 10 GALS. PER CAP. FOR SHOWERS; 7 GALS. PER CAP. FOR TOILETS, KITCHEN

3. GARBAGE GRINDERS: Add 50% increase in septic tank volume for additional solids. No appreciable increase in flow in sewers.
4. GREASE TRAPS: Provide grease trap in waste line from kitchen for hotels, schools with cafeterias and instutions having large flows from kitchens.

* From New York State Department of Health.

SEWAGE TREATMENT—SMALL SYSTEMS—2

TREATMENT PLANT UNITS

I. SETTLING TANK

PLAIN SETTLING OR SEPTIC TANKS: LIQUID CAPACITY.

(1) For flows up to 500 gals. per day, net volume to be at least 750 gals.

(2) For flows of 500 to 1500 gals. per day, tank volume to be at least 1½ days sewage flow.

(3) For flows larger than 1500 gals. per day, minimum tank liquid volume should equal the following:

$$V = 1125 + 0.75\,Q$$

where V = net volume of tank in gals., and Q = daily flow of sewage in gallons.

(4) For flows above 15000 gals. per day up to 100,000 gals., an Imhoff tank may be more satisfactory for primary treatment than a septic tank.

(5) SEPTIC TANK DESIGN: Plan to be rectangular, length to be 2-4 times width. Minimum depth 4 feet. Tanks less than 8 feet long to have 1 compartment, larger tanks, 2 compartments. Inlet compartment about 75% of total tank capacity.

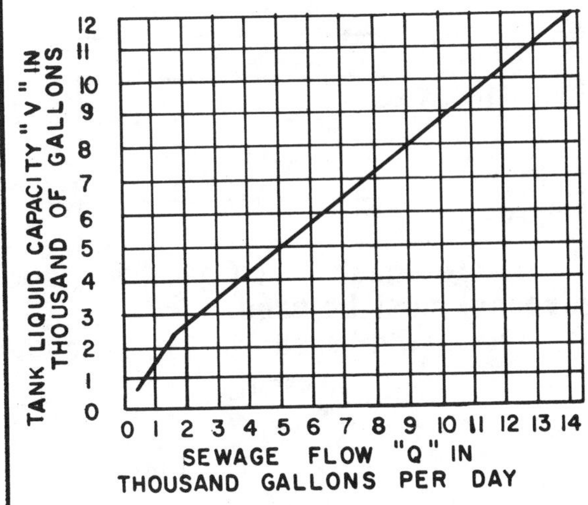

TANK LIQUID CAPACITY "V" IN THOUSAND OF GALLONS
0 1 2 3 4 5 6 7 8 9 10 11 12
0 1 2 3 4 5 6 7 8 9 10 11 12 13 14
SEWAGE FLOW "Q" IN THOUSAND GALLONS PER DAY

SEPTIC TANK CAPACITIES UP TO 14,500 G.P.D. *

* FROM MANUAL OF SEPTIC TANK PRACTICE, U.S. DEPT. OF HEALTH, EDUCATION & WELFARE

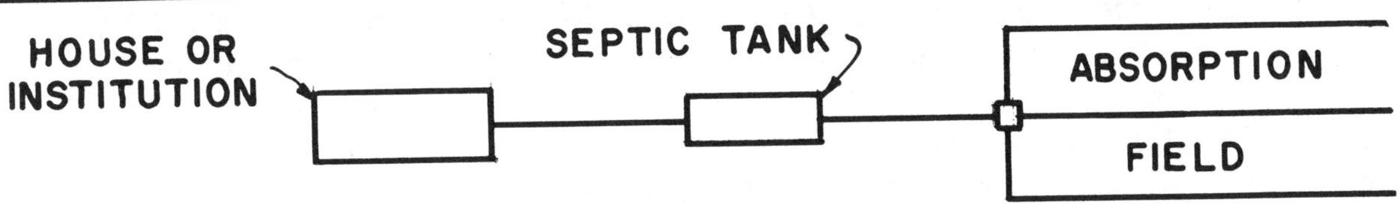

HOUSE OR INSTITUTION
SEPTIC TANK
ABSORPTION
FIELD

A. FLOW DIAGRAM FOR SEPTIC TANK AND SUB-SURFACE DISPOSAL SYSTEM

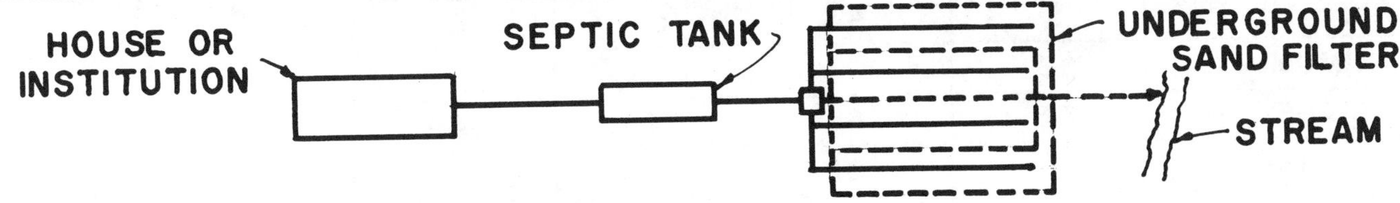

HOUSE OR INSTITUTION
SEPTIC TANK
UNDERGROUND SAND FILTER
STREAM

B. FLOW DIAGRAM OF SEPTIC TANK AND UNDERGROUND SAND FILTER

SEWAGE TREATMENT-SMALL SYSTEMS-3

2. SUB-SURFACE TILE SYSTEM-ALLOWABLE SEWAGE ABSORPTION

TIME FOR WATER TO FALL 1 INCH	ALLOWABLE RATE OF SEWAGE APPLICATION + PER SQ. FT. PER DAY	
	BOTTOM AREA-TRENCHES	PERCOLATION AREA-CESSPOOLS
1 MINUTE OR LESS	4.0 GALS.	5.3 GALS.
2 MINUTES	3.2 "	4.3 "
5 "	2.4 "	3.2 "
10 "	1.7 "	2.3 "
30 "	0.8 "	1.1 "

If rate equals 1 inch per hour, sub-surface disposal is unsuitable, and other means of disposal must be used.

Trench widths should be as follows:

TYPE OF SOIL	TRENCH WIDTH
1-3 MIN. SOIL	18"
4-9 " "	24"
10-60 " "	36"

Min. distance between trench walls — 3 feet.

LATERALS — 4" dia., slope not greater than 0.5%. Use 0.3% if dosing tank is used. Use perforated tile pipe with open joints (1/8"-1/4"). Cover joint with tar paper strip.

3. LEACHING CESSPOOL

May be used in place of, or to supplement tile fields. Preferable where soil below a depth of 2 or 3 feet is more porous than above this depth. Minimum spacing 10 feet. When located in fine sands, surround walls for entire length with layer of graded gravel.

4. SUB-SURFACE SAND FILTER

Use where soil is so tight that sub-surface tile system and leaching cesspools are not practical. A suitable watercourse must be available at the outlet of the filter.

DESIGN RATE: 1.15 gals. per day per sq. ft. of area.

FILTER SAND: Clean, coarse, passing a 1/4-in. mesh screen, an effective size between 0.30 and 0.60, and a uniformity coefficient not greater than 3.5.

5. DOSING TANK AND SIPHON

For filter areas greater than 1800 sq. ft. and for distributor pipes longer than 300 ft., use a dosing tank. Capacity of tank should equal about 75 per cent of volume of distributor pipes.

Use dosing siphon, available in 3, 4, 5 and 6 inch sizes.

For more than 800 ft. of distributor pipe, divide filter in 2 sections and provide alternating siphons to alternate the flow to each filter section.

*SEWAGE SIPHONS - DISCHARGE RATES, GPM.

SIZE, INS.	3"	4"	5"	6"
MAXIMUM	96	227	422	604
AVERAGE	72	165	328	474
MINIMUM	48	102	234	340

*BUL. 125, PACIFIC FLUSH TANK CO. + BUL. No. 1, PART II, N.Y. STATE DEPT. OF HEALTH

SEWAGE TREATMENT – SMALL SYSTEMS – 4

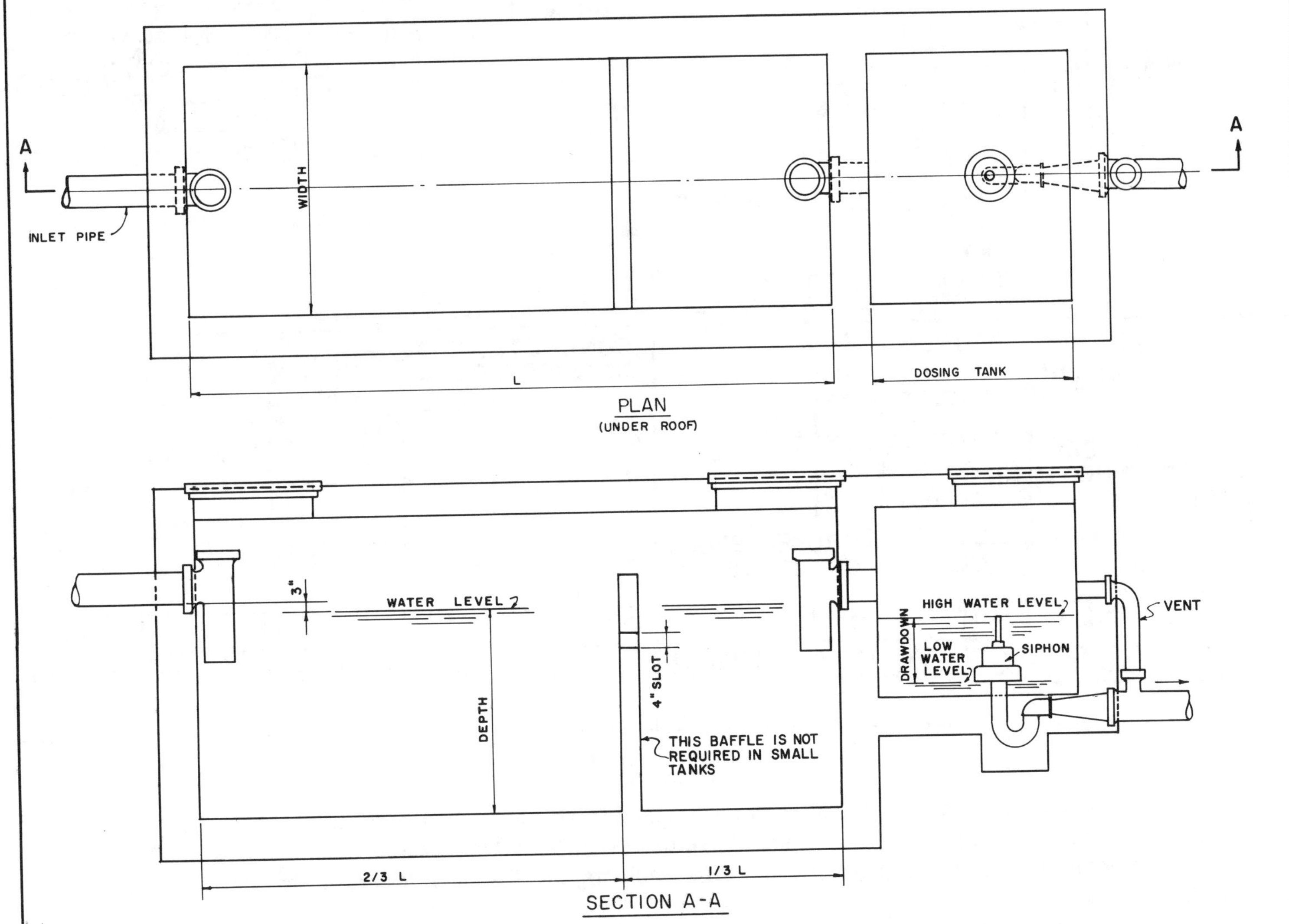

A

A

INLET PIPE

WIDTH

L

DOSING TANK

PLAN

(UNDER ROOF)

3"

WATER LEVEL

DEPTH

4" SLOT

THIS BAFFLE IS NOT REQUIRED IN SMALL TANKS

HIGH WATER LEVEL

DRAWDOWN

LOW WATER LEVEL

SIPHON

VENT

2/3 L

1/3 L

SECTION A-A

DETAILS OF SEPTIC TANK

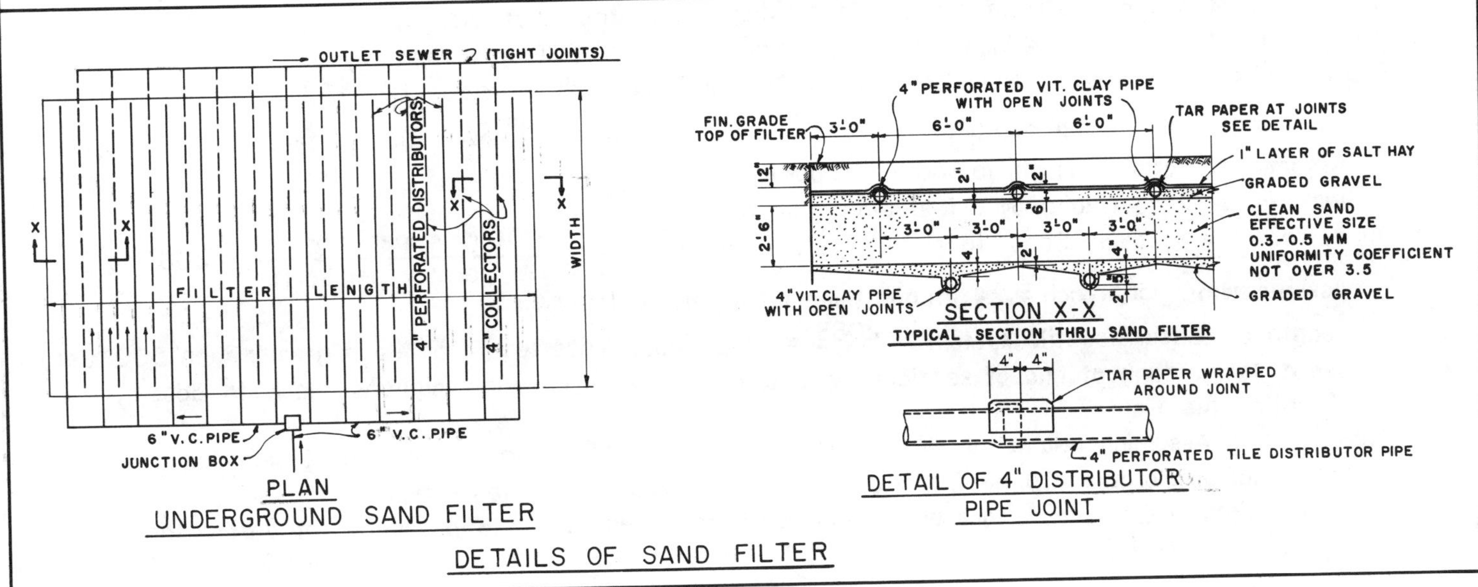
OUTLET SEWER (TIGHT JOINTS)
X
X
X
X
4" PERFORATED DISTRIBUTORS
4" COLLECTORS
FILTER LENGTH
WIDTH
6" V.C. PIPE
6" V.C. PIPE
JUNCTION BOX
PLAN
UNDERGROUND SAND FILTER
4" PERFORATED VIT. CLAY PIPE WITH OPEN JOINTS
FIN. GRADE TOP OF FILTER
3'-0"
6'-0"
6'-0"
TAR PAPER AT JOINTS SEE DETAIL
1" LAYER OF SALT HAY
GRADED GRAVEL
CLEAN SAND EFFECTIVE SIZE 0.3-0.5 MM UNIFORMITY COEFFICIENT NOT OVER 3.5
GRADED GRAVEL
12"
2'-6"
2"
2"
6"
3'-0"
3'-0"
3'-0"
3'-0"
4"
4"
5"
2"
4" VIT. CLAY PIPE WITH OPEN JOINTS
SECTION X-X
TYPICAL SECTION THRU SAND FILTER
4"
4"
TAR PAPER WRAPPED AROUND JOINT
4" PERFORATED TILE DISTRIBUTOR PIPE
DETAIL OF 4" DISTRIBUTOR PIPE JOINT

DETAILS OF SAND FILTER

SEWAGE TREATMENT - SMALL SYSTEMS -5

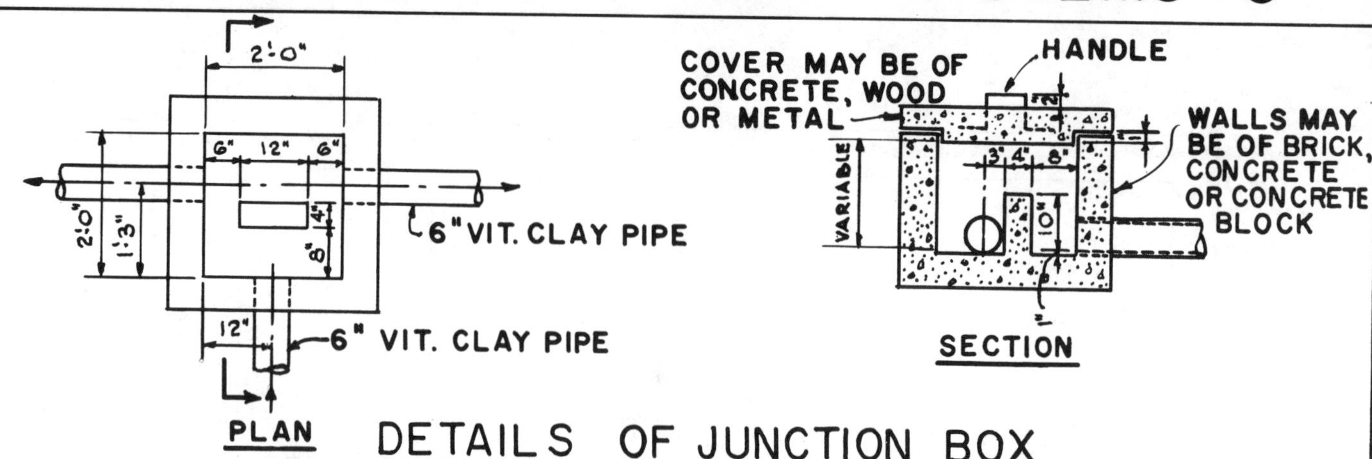
2'-0"
6" 12" 6"
2'-0"
1'-3"
4"
8"
6" VIT. CLAY PIPE
12"
6" VIT. CLAY PIPE
PLAN
COVER MAY BE OF
CONCRETE, WOOD
OR METAL
HANDLE
WALLS MAY
BE OF BRICK,
CONCRETE
OR CONCRETE
BLOCK
VARIABLE
SECTION

DETAILS OF JUNCTION BOX

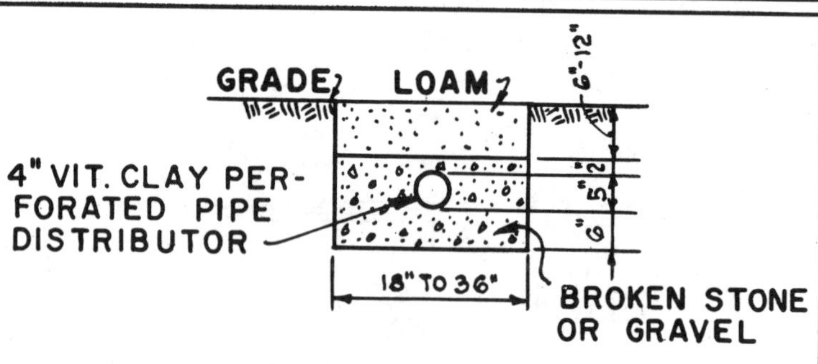
GRADE
LOAM
6"-12"
4" VIT. CLAY PER-
FORATED PIPE
DISTRIBUTOR
18" TO 36"
BROKEN STONE
OR GRAVEL

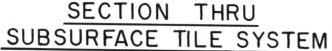

SECTION THRU SUBSURFACE TILE SYSTEM

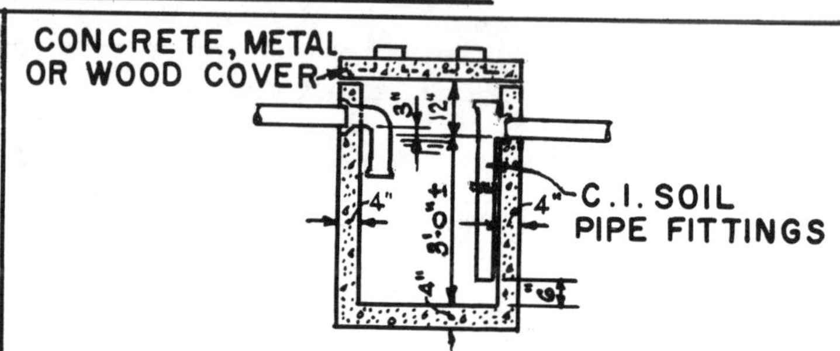
CONCRETE, METAL
OR WOOD COVER
C.I. SOIL
PIPE FITTINGS
3'-0"
12"
4"

SECTION THRU GREASETRAP

(May be square, circular or rectangular in plan.)

PROBLEM

GIVEN: A proposed day school will have a total population of pupils and teachers equal to 250. The school will be equipped with showers. Several test holes have been dug at the site of the sewage treatment facilities and percolation tests indicate the the time for water to fall 1 inch equal to 5 minutes.

TO FIND: The size of a septic tank to serve the school and the area of a subsurface tile system required. As an alternate to a subsurface tile system, a subsurface sand filter will be designed, assuming that a nearby stream is available for disposal of the filter effluent.

SOLUTION: Total daily sewage flow = 250 × 20 gals. per day = 5000 gals. per day.

Septic tank: The volume of a septic tank = 4875 gals. (see curve, page 19-33)

Assuming a tank water depth of 6 ft., tank area will equal $\frac{4875}{7.5 \times 6} = 108$ sq. feet.

Assuming a ratio of length to width of approx. 3, tank will be 18 feet long by 6 feet wide, with depth of 6 feet.

Subsurface tile system: Subsurface tile system required = $\frac{5000 \text{ gals. per day}}{2.4 \text{ gals. per day per sq. ft.}} = 2083$ sq. ft.

Assume width of trench = 24 inches (see table, page 19-34)

Length of subsurface tile system = $\frac{2083}{2} = 1042$ feet. Since length of tile is greater than 500 feet, use a dosing tank at end of septic tank and provide 2 alternating siphons, each to dose $\frac{1}{2}$ of entire tile field.

Dosing tank: Assuming volume of dosing tank to fill tile 75% full and assuming 4 inch dia. tile, Volume of dosing tank = 0.75 × 1042 × 0.087 sq. feet (area of 4 inches) = 68 cu. feet.

With a width of 6 ft. (septic tank width) and a siphon drawdown of 16 in., length of dosing tank = $\frac{68}{6 \times 1.33} = 8.5$ ft.

Subsurface sand filter: Area required = $\frac{5000 \text{ gals. per day}}{1.15 \text{ gals. per day per sq. ft.}} = 4348$ sq. ft.

Assume filter size = 100 ft. long × 44 ft. wide.

SEWAGE TREATMENT-ANALYSES & TESTS-I

CHARACTERISTICS OF SEWAGE

The term "characteristics of sewage" signifies the quantity and character of organic and inorganic matter in domestic and industrial sewages in proportion to the amount of water they contain. A sewage is considered strong or weak according to whether or not the amount of objectionable substances it contains is above or below the average. The strength of domestic sewage depends principally upon the quantity of water used per capita in residences, commercial buildings, and institutions, and also upon the volume of groundwater and storm flow entering the sewers. The variations in the strength of industrial sewages are caused by the difference in the quantity of organic wastes and chemical substances resulting from industrial activities.

TYPICAL CHEMICAL ANALYSES OF SEWAGE, PARTS PER MILLION*

Constituents	Strong	Medium	Weak	Constituents	Strong	Medium	Weak
Solids, Total	1000	500	200	Nitrogen, Total	86	50	25
Volatile	700	350	120	Organic	35	20	10
Fixed	300	150	180	Free ammonia	50	30	15
Suspended, Total	500	300	100	Nitrites (NO_2)	0.10	0.05	0.00
Volatile	400	250	70	Nitrates (NO_3)	0.40	0.20	0.10
Fixed	100	50	30				
Dissolved, Total	500	200	100				
Volatile	300	100	50	Chlorides	175	100	15
Fixed	200	100	50	Alkalinity	200	100	50
				Fats	40	20	0
Biochemical oxygen demand 5-day, 20°C.	300	200	100				
Oxygen consumed	150	75	30				
Dissolved oxygen	0	0	0				

MOST IMPORTANT ANALYSES TO MAKE AT SEWAGE WORKS†

Source of Sample	5-day B.O.D.	Solids Suspended			Nitrogen		Dis-solved Oxygen	Relative Stability	Residual Chlorine	Bacteria	
		Total	Volatile	Settle-able	Total Organic	Nitrate				Total	B. coli, (pre-sumptive test)
Raw sewage	I	I	II	I	II						
Settled sewage	I	I		I	II						
Oxidized effluent	I	I		I	II	I	II	II			
Activated-sludge aeration tank		I[a]		I[a]			I				
Chlorinated effluent									I	II	II
Stream or other diluting water	I						I				III
Sludge-digester overflow	I	I									

[a] The sludge index, which can be calculated from the suspended and settleable solids, is valuable in controlling the operation of activated-sludge aeration tanks.

Key: I. Tests having the greatest value.
II. Tests having less value.
III. Tests that need be made only when conditions warrant.

* Adapted from *Sewerage and Sewage Treatment*, 8th ed., by Harold E. Babbitt and E. Robert Baumann, John Wiley & Sons, 1958.

† From Keefer's, *Sewage Treatment Works*, page 590, McGraw-Hill Book Co.

SEWAGE TREATMENT–ANALYSES & TESTS–2

NOTES: A. For outlined method of tests, see most recent edition of *Standard Methods for the Examination of Water, Sewage, and Industrial Waste*, published by the American Public Health Association.

B. General reasons for tests:

1. sewage or waste analysis may determine type of plant to be used
2. determine plant efficiency.
3. provide basis for plant control.
4. to comply with state laws.
5. provide guide for future design and extensions of existing plants.

(Tests most frequently made are preceded by an asterisk. The numbers in parentheses refer to similar numbers in preceding columns.)

INTERPRETATION OF SEWAGE ANALYSES AND SEWAGE-PLANT OPERATING RECORDS

Test	Units	Sampling Points	Reason for Test	Analysis and Interpretation of Results	Brief Procedure of Test
Screenings (1) Quantity. (2) Moisture content.	(1) Lb. or cu. ft. per million gal. (2) % of solids.	As removed from screens.	(1) a. To determine time periods for intermittent operation of mechanical equipment. b. To determine time periods for collection by trucks. c. To reveal abnormal contributions to system. d. To complete record of solids removal by plants. (2) To determine whether or not screenings have been drained sufficiently to be burned efficiently in incinerator.	(1) Quantity from coarse screens (1/2" – 3" openings) usually 0.5 to 6 cu. ft. per million gal. Quantity from fine screens (3/4" – 3/32" openings) usually 5 to 30 cu. ft. per million gal. Weight per cu. ft. 40 to 60 lb. Variations from above values and the average from the plant in question may indicate large contributions from industrial, commercial, or similar sources. (2) Usual % of moisture between 75 and 90% in screenings by ordinary draining for one day. Dewatering centrifuges and presses usually reduce % moisture to 60 to 65%. To feed incinerator economically, screenings should contain between 60 and 80% moisture. B.t.u. used per lb. of screenings incinerated ranges between 1400 and 3500.	(1) The quantity is measured by weighing a day's collection of the material that has been allowed to drain; or the volume is measured.† (2) The % moisture is determined by weighing and drying a representative sample of about 1 lb. in an oven at 103°C. for 12 to 24 hr. Sample is weighed while still warm, and loss of weight divided by weight of dry solids multiplied by 100 is the % moisture.†
Grit (1) Volume. (2) Size and type of solids being removed. (For sieve test and volatile solids test, see test under item "Sludge.")	Lb. or cu. ft. per million gal.	As removed from grit chambers.	(1) To determine time periods for manual cleaning and setting of time periods for intermittent operation of mechanical grit removers. To complete the records of solids removed by the plant.	(1) For combined sewers, average quantity of grit commonly deposited is 0.5 to 10 cu. ft. per million gal. To allow for maximum storm demands between cleaning operations, 10 to 30 cu. ft. per million gal. grit storage space is generally provided in grit chamber channels. Grit chambers should be cleaned about every 2	(1) Quantity is measured by weighing the amount collected in 1 day after it has been allowed to drain; or the volume is measured.†

†*Standard Methods for the Examination of Water, Sewage, and Industrial Wastes.*

SEWAGE TREATMENT–ANALYSES & TESTS–3

Test	Units	Sampling Points	Reason for Test	Analysis and Interpretation of Results	Brief Procedure of Test
				weeks under normal conditions, and every few days at times of successive storms. Large amounts of grit gathering in screening chamber in sanitary sewage treatment plant may indicate broken lines, presence of illegal storm drain connections, open manholes, etc. Modern tendency is to provide a grit chamber for treatment plants treating sanitary sewage as well as for those treating combined sanitary sewage and storm drainage.	
Scum or skimmings (not a customary test).	Cu. ft. per million gal.	(1) Grease or skimming tanks. (2) Settling tanks. (3) Filters. (4) Aerators. etc-	To determine efficiency of skimming tanks. To determine whether additional amount of air is necessary to increase flotation of grease.	Amount skimmed from tanks varies from 0.1 to 5.0 cu. ft. per million gal. Volume above normal may indicate industrial, garage, etc., waste discharge. Grease causes clogging of pipes and should be removed as early as possible in the treatment process. Grease and scum should not be put in digesters because it interferes with digestion and drying of the sludge.	Can be measured by allowing the collected material to stand in a tank until separation of the water and scum takes place. The water is drawn from the bottom of the tank, and the volume occupied by the scum measured. Care should be taken to ascertain if gases in the scum have escaped, causing the scum to settle as sludge. Sludge settling with the water should be measured as part of the scum.†
*Total suspended solids (includes settleable and non-settleable solids).	Parts per million (p.p.m.).	(1) Raw sewage primary. (2) Settling tank effluent. (3) Filter effluent. (4) Final effluent. (5) Aeration tanks. (6) Supernatant from digesters.	To determine reduction of suspended solids by plant as a whole to satisfy requirements set by state health authorities for water course into which effluent is finally placed. To determine reduction of suspended solids by each unit in the plant. (5) To determine characteristics of activated sludge, or the sludge index, to ascertain the proportion of the material that settles with difficulty.	See Tables A, B, and C, p. 19-01, for average values of raw sewage, and values and efficiency of removal to be expected from the various units and the plant as a whole. Higher amounts of suspended solids in raw sewage may indicate industrial wastes or low per capita flow. Smaller amounts of suspended solids and large flows may indicate infiltration. Lower efficiencies than normally expected may indicate that units should be cleaned and mechanical equipment adjusted more often (tanks and filters). The low efficiencies may also indicate that the recirculation ratio in high-rate trickling filters should be changed, or that the amount of return activated sludge	A Gooch crucible is prepared, ignited, and weighed. A sample of sewage or effluent of 50 to 100 milliliters (ml.) is filtered through, and the crucible is dried for at least 1 hr. in an oven at 103° C. It is then cooled in a desiccator, weighed, and the parts per million (p.p.m.) suspended solids calculated by multiplying the gain in weight in milligrams by 1,000 divided by the ml. of sample taken. (5) Sludge index is the volume in milliliters occupied by one gram of sludge after the aeration liquor has settled for 30 min. A 1-liter sample from the outlet of the aeration tank is settled 30 min. in a 1,000-ml. graduated cylinder, and the volume occupied by the sludge is reported in milliliters.

*Tests most frequently made are preceded by an asterisk.
The numbers in parentheses refer to similar numbers in preceding columns.
†*Standard Methods for the Examination of Water, Sewage, and Industrial Wastes.*

SEWAGE TREATMENT–ANALYSES & TESTS–4

Test	Units	Sampling Points	Reason for Test	Analysis and Interpretation of Results	Brief Procedure of Test
				and air supply in aeration tanks should be increased. (5) Aeration tank samples (activated-sludge process). Limits of suspended solids in mixed liquor should be 1200 to 3000 p.p.m. for diffused aeration plants and 500 to 1200 p.p.m. in mechanical aeration plants. (5) "Sludge index" of 100 is normally expected in diffused air plants. "Sludge index" may be as high as 300 in mechanical aeration plants. "Bulking" occurs with higher "sludge indices." When bulking occurs, sludge rises and flows over effluent weirs.	The dry suspended solids are then determined in parts per million. Computation: (ml. settleable sludge ÷ p.p.m. suspended solids) × 1000.
Total solids (includes screenings, grit, scum, suspended solids, and dissolved solids).	p.p.m.	(1) Raw sewage. (2) Final effluent.	To determine characteristics and condition of the sewage. To reveal industrial wastes.	The following are the average number of parts per million of total solids that different strengths of sewage would contain. Weak 200 p.p.m. Medium 500 p.p.m. Strong 1000 p.p.m. Suspended solids are usually about 1/2 of the total solids. For industrial wastes, some dissolved solids may have to be removed before the effluent is put into waterway.	Total solids are determined by drying at 105° C. a sample of 50 ml. in a silica dish, which previously has been ignited and weighed. The dish containing the dry solids is cooled in a disiccator and weighed. The calculation is the same as that for suspended solids.†
Dissolved solids.	p.p.m.	(1) Raw. (2) Primary settling-tank effluent. (3) Filter effluent. (4) Final effluent.	To reveal breaking down of sewage during treatment and effectiveness of treatment on the portion of organic matter that is in solution.	Indicates age of sewage. The more stale the sewage is, the greater the % of dissolved solids. For domestic sewage, dissolved solids have normally little meaning. Dissolved solids in domestic sewage are usually approximately 2/3 of total solids. For industrial wastes, dissolved solids may have large B.O.D. requirements which must be satisfied.	Dissolved solids are determined by subtracting the total suspended solids from the total solids in parts per million.

The numbers in parentheses refer to similar numbers in preceding columns.
†*Standard Methods for the Examination of Water, Sewage, and Industrial Wastes.*

SEWAGE TREATMENT-ANALYSES & TESTS-5

Test	Units	Sampling Points	Reason for Test	Analysis and Interpretation of Results	Brief Procedure of Test
Volume of settleable solids.	c.c. or ml. per liter.	(1) Raw sewage. (2) Final effluent.	Where primary treatment is used and laboratory facilities are limited, test affords approximate measure of performance.	Imhoff tanks and primary settling tanks should remove about 45 to 60% of suspended solids.	An Imhoff cone is filled to the liter mark with a thoroughly mixed sample. Settled for 0.75 hr.; the material clinging to sides is gently knocked down and settled 0.25 hr. longer. The number of milliliters of settleable solids is then read on the Imhoff cone graduations.‡
*p*H, (hydrogen ion concentration).		(1) Raw sewage (2) Primary-tank effluent. (3) Final effluent. (4) Sludge in settling tank. (5) Sludge in digestion tanks	(1) To reveal staleness and presence of industrial wastes in raw sewage. (1) To determine proper kind of chemicals and dosage in chemical treatment plants for good coagulation and precipitation. (2, 3, 4) To determine *p*H in order to control conditions in the tanks, by adding necessary chemical, so as to promote bacterial activity and prevent sludge bulking. (5) To determine the condition of the sludge in the separate digestion tanks, or in the digestion chamber of an Imhoff tank. (5) To determine dosage for coagulation prior to vacuum filtration.	(1) *p*H values in raw sewage lower than the *p*H of the water supply indicate septic or stale sewage or presence of industrial wastes. *p*H values higher than those of the water supply also indicate industrial wastes. (1) In chemical treatment, the various coagulants give best results for specific ranges of *p*H. Coagulant *p*H range: Aluminum sulfate (alum.) 5.5 to 8.0 Ferrous sulfate 8.5 to 11.0 Ferric sulfate 5.0 to 11.0 Ferric chloride 5.0 to 11.0 (5) For proper digestion of sludge, the *p*H value should be kept above 7.3. Values lower than 7.0 indicate acid condition and should be corrected by addition of lime to digesters.	The hydrogen-ion concentration of sewage and sludge is determined by colorimetric methods (color comparison). Color indicators for *p*H ranges between 5.2 and 9.6 are of greatest use. Sludge samples are diluted with 9 times their volume of distilled water and allowed to settle. The supernatant is decanted and used for the determinations. Electrometric methods are used for more precise determinations.†
Total alkalinity.	p.p.m. in terms of $CaCo_3$	(1) Raw sewage primary. (2) Tank effluent. (3) Final effluent.	To determine efficiency of chemical treatment plants.	High values in raw sewage indicate industrial wastes. Domestic sewage is normally alkaline (approximately 50 to 200 p.p.m).	To a 100-ml. sample of sewage, 2 drops of methyl orange indicator are added, and the sample is titrated to the end point with *N*/50 sulfuric acid. The number of milliliters of acid used is multiplied by 10 to give parts per million total alkalinity as calcium carbonate.‡

The numbers in parentheses refer to similar numbers in preceding columns.
†*Standard Methods for the Examination of Water, Sewage, and Industrial Wastes.*
‡Tests are from National Lime Association's pamphlet on Principles of Sewage Treatment.

SEWAGE TREATMENT-ANALYSES & TESTS-6

Test	Units	Sampling Points	Reason for Test	Analysis and Interpretation of Results	Brief Procedure of Test
Total acidity.	p.p.m. in terms of $CaCo_3$	(1) Raw sewage primary. (2) Tank effluent. (3) Final effluent	To determine efficiency of chemical treatment plants.	High values in raw sewage indicate industrial wastes. Domestic sewage is normally alkaline (approximately 50 to 200 p.p.m).	3 drops of phenolphthalein are added to a 100-ml. sample of sewage, and the sample is titrated with *N*/50 sodium hydroxide until a permanent end point is reached. The calculation is the same as that for alkalinity.‡
Sludge. (1) Moisture and total solids. (2) Ash and volatile. (3) Temperature of sludge in digesters.	(1) Solids and moistures in % of sample of sludge. (2) Ash and volatile matter in % of dry solids.	(1) and (2) Fresh sludge. (1, 2, and 3) In Imhoff tank and separate digestion tanks. (1) Sludge bed before and after drying. (1) Before and after vacuum filtration.	For control of sludge works, drying, vacuum filtration, digestion, or incineration.	(1) Moisture of fresh and digested sludge is usually between 85 and 99%, dependent on the process of treatment. in general, the longer the sludge is digested, and the greater the amount of supernatant removed, the less will be its moisture content. Vacuum filters and centrifuges will produce a sludge cake with a moisture content between 55 and 85%. Sludge must be conditioned with chemicals before vacuum filtration. Doses of lime, 5 to 15% of the dry solids, and doses of ferric chloride, 1 to 5% of the dry solids, are generally effective. When sludge is dewatered to 80% moisture or less, it may be mixed with garbage and rubbish and burned in an incinerator. High moisture content indicates ineffective concentration. (2) High ash reflects grit or industrial wastes. Low reduction of volatiles during digestion indicates low efficiency of digestion. (3) Best temperature for mesophilic (medium temperature) digestion 85 to 95° F. normally should be maintained in digesters. Best temperature for thermophilic (heat-loving) bacteria is 125° F.	(1) Moisture and total solids: Weigh 50 grams of sludge as rapidly as possible in a tared evaporating dish. Evaporate to dryness overnight in an oven at 103° C., cool in a desiccator and reweigh or evaporate on a water bath, dry at 103° C., cool, and weigh to constant weight.‡ (2) Volatile solids and ash: Ignite the residue from the determination of moisture in an electric muffle furnace at 600° C. (dull red heat) for 60 min. Cool in a desiccator, and reweigh.‡

The numbers in parentheses refer to similar numbers in preceding columns.
‡Tests are from National Lime Association's pamphlet on Principles of Sewage Treatment.

SEWAGE TREATMENT-ANALYSES & TESTS-7

Test	Units	Sampling Points	Reason for Test	Analysis and Interpretation of Results	Brief Procedure of Test
Relative stability (putrescibility)	Stability or days required for decolorization.	(1) Filter effluent. (2) Final effluent. (3) Stream above. (4) Stream below.	To determine the % of oxygen available as dissolved nitrite and nitrate oxygen to the total oxygen required to satisfy the B.O.D. To determine adequacy of treatment for discharge into receiving stream. This determination is being replaced by more exact tests, such as dissolved oxygen, nitrite, nitrate, and B.O.D. tests. For small plants with limited facilities, it is useful for indicating satisfactory operation of the biological oxidation processes.		Fill a 150-ml. glass-stoppered bottle with sample, avoiding aeration. Add exactly 0.4 ml. of methylene blue indicator solution below the surface of the liquid. Incubate at 20° C. with a water seal, observing the sample daily until decolorization takes place. Results are reported as follows:‡ Days — % Stability 1 — 21 3 — 50 5 — 68 7 — 80 10 — 90 12 — 94 14 — 96 16 — 97 18 — 98 20 — 99
Oxygen consumed (test now being largely superseded by B.O.D. test).	p.p.m.	(1) Raw sewage. (2) Tank effluent. (3) Final effluent.	To measure plant efficiency in removing organic matter which takes up oxygen from receiving stream. Used in conjunction with B.O.D. test.	As a result of these tests, sewage may be classified as: Weak — 30 Medium — 75 Strong — 150 This classification influences the size of filters.	Place 50 ml. of sewage, or an aliquot diluted to 50 ml. with distilled water, in a round-bottom flask, and add 25 ml. standard dichromate solution. Add 75 ml. conc. H_2SO_4. Attach the flask to the Friedrichs condenser, and reflux the mixture for 2 hrs. Pumice granules or glass beads should be added to the reflux mixture to prevent bumping. Cool, and wash down the condenser with 25 ml. distilled water. Transfer the contents to a 500-ml. Erlenmeyer flask, washing out the reflux flask 4 or 5 times with distilled water. Dilute the mixture to about 350 ml., and titrate the excess dichromate with ferrous

The numbers in parentheses refer to similar numbers in preceding columns.

‡Tests are from National Lime Association's pamphlet on Principles of Sewage Treatment.

SEWAGE TREATMENT–ANALYSES & TESTS –8

Test	Units	Sampling Points	Reason for Test	Analysis and Interpretation of Results	Brief Procedure of Test
					ammonium sulfate, using ferroin indicator. A blank consisting of 50 ml. distilled water instead of the sample, together with the reagents, is refluxed in the same manner. Value of oxygen consumed in p.p.m. is equal to difference of ferrous ammonium sulfate in ml. used for the blank and the sample, times normality of ferrous ammonium sulfate times 8,000 divided by ml. of sample; minus mg/l cl $\times$ 0.23.‡
Dissolved oxygen (D.O.).	p.p.m.	(1) Raw sewage. (2) Filter and aerator effluent. (3) Final effluent. (4) Receiving stream above and below point of discharge of sewage effluent.	To check performance, particularly of aeration tanks and filters.	Presence of 1 p.p.m. or more of dissolved oxygen (D.O.) in raw sewage usually indicates fresh sewage and may indicate infiltration. Stale and septic sewage contains no dissolved oxygen. Solubility of atmospheric oxygen in water in summer is about 5 to 7 p.p.m.; in winter about 10 to 12 p.p.m.	A 300-ml. glass-stoppered bottle is filled with the sample with a minimum of disturbance, and precautions are taken to prevent the entrainment of air bubbles. 0.7 ml. of concentrated sulfuric acid and 1 ml. of potassium permanganate solution are added below the water surface. After the stopper is replaced, the bottle is shaken and allowed to stand for half an hour. The stopper is removed, and 0.5 to 1 ml. of potassium oxalate solution is added to decolorize the permanganate color completely; the stopper is replaced, and the bottle reshaken, and allowed to stand in the dark. When the pink color disappears, the stopper again is removed, and 2 ml. of manganese sulfate and 3 ml. of alkaline potassium iodide solution are added below the surface. The

The numbers in parentheses refer to similar numbers in preceding columns.
‡Tests are from National Lime Association's pamphlet on Principles of Sewage Treatment.

SEWAGE TREATMENT–ANALYSES & TESTS –9

Test	Units	Sampling Points	Reason for Test	Analysis and Interpretation of Results	Brief Procedure of Test
					stopper is immediately replaced, and the bottle thoroughly shaken. The precipitate that forms is allowed to settle, the stopper is removed, and 2 ml. of concentrated sulfuric acid are added. Again the stopper is replaced, and the bottle shaken until the precipitate is completely dissolved. Iodine is liberated in proportion to the amount of dissolved oxygen that was present. A 200-ml. sample is withdrawn from the bottle and titrated with 0.025 *N* sodium thiosulfate solution to the starch end point. The number of milliliters of sodium thiosulfate solution used is equivalent to the p.p.m. dissolved oxygen present in the sample.‡
*Biochemical oxygen demand (B.O.D.).	p.p.m.	(1) Raw sewage. (2) Primary-tank effluent. (3) Filter effluent. (4) Final effluent. (5) Stream above. (6) Stream below. (7) Digester supernatant.	This is one of the basic methods of evaluation of plant efficiency and adequacy of receiving stream for the required dilution of the effluent.	See Tables A, B, and C, p. 19-01, for average values of raw sewage, and values and efficiency of removal to be expected from the various units and the plant as whole. Higher values of B.O.D. in raw sewage may indicate industrial waste or stale or concentrated sewage. For additional comments, see test "Total suspended solids."	The biochemical oxygen demand of a sample of sewage is determined by making 3 dilutions of the sample with buffered distilled water saturated with dissolved oxygen, and placing each dilution in a 300-ml. glass-stoppered bottle which is filled to capacity. A bottle filled with the dilution water alone (blank) is included. The amount of dilution depends on the strength of the sewage or effluent, and is based on the estimated strength range. The bottles are incubated 5 days at 20° C. On removal from the incubator, dissolved oxygen is determined on all samples. The p.p.m. B.O.D. are calculated for each dilution by subtracting the p.p.m. dissolved oxygen in the dilution from those of the dilution water blank, multiplying

*Tests most frequently made are preceded by an asterisk.
The numbers in parentheses refer to similar numbers in preceding columns.
‡Tests are from National Lime Association's pamphlet on Principles of Sewage Treatment.

SEWAGE TREATMENT–ANALYSES & TESTS –10

Test	Units	Sampling Points	Reason for Test	Analysis and Interpretation of Results	Brief Procedure of Test
					by 100, and dividing by the percentage dilution. Results from each dilution are averaged to give the final results.‡
*Chlorine residual.	p.p.m.	Final effluent.	To determine whether sewage is being chlorinated sufficiently.	15 to 30 min. (depending on state health laws) after chlorination, the chlorine residual should be 0.5 p.p.m.	Residual chlorine is generally determined by o-tolidine test. Sometimes the starch–iodine test is used. o-Tolidine forms a yellow color with chlorine in acid solution. In making the determination, 1 ml. of o-tolidine is added to a 10-ml. sample of sewage or effluent which has been warmed to 20° C. Within 5 min., the color formed is compared with a series of standards in which the depth of color is equivalent to known amounts of chlorine.‡
*Presumptive *B. coli* (bacteriological).	Most probable number per millimeter	Final effluent.	To determine whether sewage is being sufficiently aerated and chlorinated.	Depends on size of receiving stream. For additional information, see p. 21-01 for extent of treatment required to change water with various amounts of *B. coli* into drinking water.	See *Standard Methods for Examination of Water and Sewage, and Industrial Wastes.*
Ammonia nitrogen.	p.p.m.	(1) Raw sewage. (2) Filter or aeration tank effluent. (3) Final effluent.	To determine efficiency of filters and aeration tanks in nitrification and stabilization (changing N and NH_3 to NO_2 and NO_3).	The following are the average p.p.m. of ammonia nitrogen in various strengths of raw sewage. Weak 15 p.p.m. Medium 30 p.p.m. Strong 50 p.p.m. High values in raw sewage indicate strong sewage or industrial wastes. Use results obtained for filter effluent and final effluent in conjunction with results of nitrate–nitrite test.	100 ml. of sample are placed in graduated cylinder, and 1 ml. of 10% zinc sulfate solution is added. Mix solution, and add 0.4 to 0.5 ml. of sodium hydroxide solution to obtain a *p*H of 10.5. Mix solution again, and clarify by filtering through filter paper. Dilute 5 ml. or less of the clarified filtrate to the mark in a 50-ml. Nessler tube with Ammonia-free water. Add 1 or 2 drops Rochelle salt solution to prevent cloudy tubes. Add 1 ml. Nessler reagent, and, after 10 min., the color formed is compared with permanent or previously prepared standards of known ammoia–nitrogen content. Calcufrom matched standard the p.p.m. ammonia nitrogen in sample.‡

*Tests most frequently made are preceded by an asterisk.
The numbers in parentheses refer to similar numbers in preceding columns.
‡Tests are from National Lime Association's pamphlet on Principles of Sewage Treatment.

SEWAGE TREATMENT–ANALYSES & TESTS–II

Test	Units	Sampling Points	Reason for Test	Analysis and Interpretation of Results	Brief Procedure of Test
Nitrates (NO_3) and nitrites (NO_2).	p.p.m.	(1) Raw sewage. (2) Filter or aeration tank effluent. (3) Final effluent.	To determine efficiency of filters and aeration tanks in nitrification and stabilization (changing N and NH_3 to NO_2 and NO_3).	Average NO_2 and NO_3 production from N and NH_3 by various processes is as follows: Trickling filters . . . 2 to 13 p.p.m. Sand filters . . . 4 to 12 p.p.m. Activated Sludge . . 0.1 to 6 p.p.m. Sedimentation or Imhoff tanks . . . . . . 0 p.p.m. Chemical treatment . . . . 0 p.p.m.	A 100-ml. sample of effluent is placed in a small casserole, 2 ml. of sodium hydroxide solution are added, and the mixture is slowly boiled down to about 20 ml. The sample is then rinsed into a 50-ml. Nestler tube or graduated test tube. Make up to the mark with ammonia-free water, add 1 strip of aluminum foil, and cover the tube with a stopper equipped with a Bunsen valve. Allow the reduction to take place at 20° C. or above for 6 hr. or overnight. When the reaction has ceased and the liquor is clear, pipette off 50 ml. Aliquots of this sample are Nesslerized and compared with ammonia–nitrogen standards. From the matched standards, the p.p.m. of nitrogen as nitrite–nitrate are calculated. †

Fresh sewage: Contains dissolved oxygen. B.O.D. requirements (p.p.m.) are generally less than total suspended solids (p.p.m.).

Stale sewage: Contains no dissolved oxygen. B.O.D. requirement is generally higher than total suspended solids.

Septic sewage: Undergoing putrefaction in absence of oxygen. B.O.D. requirement is generally equal to or slightly less than total suspended solids.

† Standard Methods for the Examination of water, sewage, and Industrial Wastes.

SEWAGE TREATMENT–LABORATORY FURNITURE*

CLASSIFICATION OF SEWAGE TREATMENT PLANTS FOR SCHEDULING LABORATORY FURNITURE AND EQUIPMENT

Plant Types	Classification per 1000 Population Capacity			
	38 and over	12 to 38	6 to 12	1.5 to 6
Complete treatment with separate sludge digesters	A	A	B	C
Primary tanks and separate digesters, primary treatment only	A	B	C	C
Imhoff tanks with trickling filter or slow sand filter	B	B	C	D
Imhoff tanks only	C	C	D	D

RECOMMENDED LABORATORY FURNITURE FOR SEWAGE TREATMENT PLANTS

For type A plants	Items 1, 2, 3, and 4
For type B and C plants	Items 2, 3, and 5
For type D plants	No furniture

Item 1. One (1) industrial chemistry laboratory table, 72" long, 48" wide, 36" high, equipped with 4 cupboards; not less than 20 drawers; a bottle rack 68" long, 10" wide, 18" high; a lead-lined or stone trough 66" long, 6" wide and 3" to 6" deep; a stone sink 20" long, 12" wide, 12" deep; three (3) straight-way water cocks; one bib water cock; one set of sink fittings, and 2 double gas cocks. The table top should be carbonized. Except as otherwise noted, the table should be oak with natural finish, and should be equivalent to the units designated by catalog numbers of the following companies:

E.H. Sheldon & Co., Muskegon, Mich.	No. 11268
W.W. Kimball & Co., Chicago, Ill.	No. 502
Hamilton Mfg. Co., Two Rivers, Wis.	No. L – 512

Item 2. One (1) balance shelf, 3' long, 2' wide; oak construction except for 1-5/8" thick birch, black carbonized top, equipped with drawer 21" wide, 15" deep, 3-3/4" high. Shelf to be equivalent to the following except as to drawer and size:

E.H. Sheldon & Co.	No. 12520
W.W. Kimball Co.	No. 682
Hamilton Mfg. Co.	No. L – 1174

Item 3. One (1) supply case with natural finish, 48" wide, 15" deep, 80" high, upper section glazed 44" wide, 60" high, with 3 adjustable shelves; cupboard section 44" wide, 12" deep, 13" high, equivalent to the following:

E.H. Sheldon & Co.	No. 41040
W.W. Kimball Co.	No. 9562
Hamilton Mfg. Co.	

Item 4. One (1) lower-section cupboard unit 37-5/8" long, 24" wide, 36" high, containing one double cupboard 34" wide, 28-1/4" high, 20-5/8" deep, with one stone sink 14-1/4" long, 10" wide, 8" deep, with one set of drain fittings, one cold-water pantry cock, and one double electric receptacle for 110-v alternating current. Table top to be scored to drain to sink.

* From *Engineering Manual*, War Dept., Corps of Engineers, June 1950.

SEWAGE TREATMENT-LABORATORY EQUIPMENT-1*

The unit should be of oak construction, natural finish throughout, except table tops which should be 1⅝" birch black carbonized.

Item 5. One (1) chemistry laboratory table, 60" to 96" long 36" high, 30" wide, similar in all respects to item 1 except the table should be constructed for installation against wall with cupboards on one side only.

RECOMMENDED LABORATORY EQUIPMENT FOR SEWAGE TREATMENT PLANTS*

Quantity			Description	Catalog Numbers		
Type of Plant				E.H. Sargent & Co.	Central Scientific Co.	Fisher Scientific Co.
A & B	C	D				
1	1		Analytical balance, chainomatic, notched beam, student model	S–2675	1010	
1	1		Balance weights, set 1gm. to 100gm., gold-plated, no fractional, with case	S–4045		2–215
1	1		Balance, Harvard, metric, 1-kg. beam graduated to 1/10 gm.	S–3215	3470	2–035
1	1		Balance weights, metric, class C 1 to 1000 gm.	S–4285	9140	2–300
1	1		Drying oven, single wall, 140° C. 700w.. 100v.	S–64055	95000A	13–265
1	1		Furnace muffle, electric, Hoskins, type F.D. size no. 202, 110-v.	S–36855	13675A	10–510
1			*p*H meter, electrometric, industrial type, glass electrodes, Beckman model M.	S–30060	21070	
	1	1	Hydrogen-ion apparatus comparator, *p*H, LaMotte or Hellige with bromothymol blue standards	S–41725	21500	
	1		*p*H standards, LaMotte or Hellige cresol red, range K.	S–41735	21550	
	1		Indicator solution, cresol red, LaMotte or Hellige, 100 ml.	S–41745	21562	5–985A
1	1		Mechanical refrigerator, sealed-type mechanism, 6 cy.ft.			
3	3	2	Imhoff cone.	S–84125	29015	15–440
1	1		Incubator, 20°, mechanical cooling Frigidaire type, or Eimer & Amend water bath type			
1	1		Corks XXX assorted sizes, bag of 100	S–23055		7–785
2	2		Funnels: Analytical 75 mm., Bunsen	S–35315	15050	10–325
2	2		Funnels: Short stem, 150 mm.	S–35305	15070	10–320
†1			Funnels: Buechner, porcelain	S–35555	18590	10–355
3	1		Glass tubing, 5 to 10 mm., assorted, 16 oz.	S–40135	14076	11–350
1	1		Glass rod, 6 mm., 16 oz.	S–40075	14050	11–375
2			Glass T-tubes, 5/16" O.D.	S–82725	15650	15–325
2	1		Pipettes, transfer 100 ml.	S–69505	16340	13–650
2	1		Pipettes, transfer 50 ml.	S–69505	16340	13–650
2	1		Pipettes, transfer 25 ml.	S–69505	16340	13–650
2	1		Pipettes, transfer 10 ml.	S–69505	16340	13–650
2	1		Pipettes, transfer 5 ml.	S–69505	16340	13–650
2	2		Pipettes, transfer 1 ml.	S–69505	16340	13–650
6	4		Pipettes, Mohr, 10 ml. graduated to 1/10 ml.	S–69555	16320	13–665
2	1		Rings, iron 4 in.	S–73045	18005	14–050
2	1		Rings, iron 3 in.	S–73045	18005	14–050
4	2		Supports, iron, approx.6" × 9" base, 24" rod	S–78305	19070	14–670
1	1		Spatula, stainless steel, 100 mm.	S–75245	18755	14–365
1	1		Bath, water, 4-hole, for gas heat	S–84355	19840	15–495
72	24		Bottles, B.O.D., glass-stoppered, U.S.	S–83805	29000	
1	1		Desiccator plate, porcelain, 8-hole	S–25185	18605	8–635

* From *Engineering Manual*, War Dept., Corps of Engineers, June 1950

† For activated sludge plants only.

SEWAGE TREATMENT-LABORATORY EQUIPMENT-2

RECOMMENDED LABORATORY EQUIPMENT FOR SEWAGE TREATMENT PLANTS						
Quantity			Description	Catalog Numbers		
Type of Plant				E.H. Sargent & Co.	Central Scientific Co.	Fisher Scientific Co.
A & B	C	D				
1	1		Spoon, horn, 150 mm.	S–75175	18775	14–425
2	1		Thermometers: minus 10° C. to 110° C.	S–80005	19240	14–985
1	1	1	Thermometers: 0° C. to 250° C.	S–80005	19240	14–985
2	1		Thermometers: 10° F. to 220° F.	S–80015	19280	14–990
2	2		Crucible holder, Walter	S–24475	18110	8–285
4	2		Graduated cylinders, double graduated, 1 liter	S–24685	16105	8–555
1	1		Graduated cylinders, double graduated, 500 ml.	S–24685	16105	8–555
1	1		Graduated cylinders, double graduated, 250 ml.	S–24685	16105	8–555
2	2		Graduated cylinders, double graduated, 100 ml.	S–24685	16105	8–555
1	1	1	Low-form cylinders, double graduated, 100 ml.	S–24765	16110	
1	1		Desiccator, Schiebler, with plates, 250 mm.	S–25005	14560	8–595
6			Dishes, evaporating, Coors, porcelain, 150 ml.	S–25505	18575	8–690
2	1		Files, triangular tapered, 4"	S–32225	88325	9–725
12	6		Filter paper, no. 500 per pkg. of 100, 9 cm.	S–32915	13255	9–795
2			Filter paper, no. 500 per pkg. of 100, 12.5 cm.	S–32915	13255	9–795
6	3		Flasks, Erlenmeyer, Pyrex, 500 ml.	S–34155	14905	10–040
6	3		Flasks, 250 ml.	S–34155	14905	
2	1	1	Flasks, filtering, side tubulature, 500 ml.	S–34375	14985	10–175
2	1		Flasks, volumetric, 1 liter	S–34815	16220	10–200
1	1		Flasks, volumetric, 500 ml.	S–34815	16220	10–200
2	1		Flasks, volumetric, 250 ml.	S–34815	16220	10–200
4	2		Flasks, volumetric, 200 ml.	S–34815	16220	10–200
2	1		Flasks, volumetric, 100 ml.	S–34815	16220	10–200
1	1		Flask, volumetric, 200 ml., large mouth for D.O.	S–34995	16295	10–240
1	1		Plate, spot, porcelain, 90 mm. × 110 mm.	S–70025	18600	13–745
2	1		Tongs, crucible, lock joint, 9½"	S–82205	19625	15–185
1	1		Tongs, crucible, lock joint, 14"	S–82215		
2	1		Triangle, Nichrome, 2", round wire	S–82445	19705	15–255
2	1		Wire, gauze, Chromel 4" × 4" 16-mesh	S–85315	19960	15–585
20	10		Rubber tubing, medium wall pure gum, ft., ¼" diam.	S–73505	18270	14–150
10	10		Rubber tubing, medium wall pure gum, ft., ⅜" diam.	S–73505	18270	14–150
10	10		Rubber tubing, vacuum, ¼" diam.	S–73535	18204	14–175
1	1		Filter pump, Richards, size A, ⅜"	S–33575		9–965
6	4		Bottles, dropping, 30 ml.	S–8745	10535	2–980
1	1		Funnel support, 4-place, with clamp	S–78815	19035	14–740
1	1		Brush, test-tube, medium	S–9995	10972	3–590
1	1		Brush, flask, size B	S–9965	10985	3–570
1	1		Mortar and pestle, porcelain, 100 mm.	S–62235	17381	12–960
6	3		Bottles, 5 gal.	S–8435	10310	2–885
1	1		Still water, Stokes, gas-heated	S–27645	12808	9–055
8	6	3	Milk kettles, ivory-enameled, seamless, 4 qt. capacity (for sample storage)			
2			Beakers, Pyrex, 2 liter	S–4675	14265	2–540
2	2		Beakers, Pyrex, 1 liter	S–4675	14265	2–540
2			Beakers, Pyrex, 600 ml.	S–4675	14265	2–540
4	6		Beakers, Pyrex, 400 ml.	S–4675	14265	2–540
4	6		Beakers, Pyrex, 250 ml.	S–4675	14265	2–540
2			Beakers, Pyrex, 50 ml.	S–4675	14265	2–540

* From *Engineering Manual*, War Dept., Corps of Engineers, June 1950.

SEWAGE TREATMENT–LABORATORY EQUIPMENT–3*

RECOMMENDED LABORATORY EQUIPMENT FOR SEWAGE TREATMENT PLANTS

Quantity: Type of Plant A & B	C	D	Description	Catalog Numbers: E.H. Sargent & Co.	Central Scientific Co.	Fisher Scientific Co.
6			Watch glasses, 115 mm., plain, ground, annealed	S–83605	15850	2–610
			Bottles, glass-stoppered, flint glass, machine-made:			
12	6		32 oz.	S–8345	10430	2–915
6	3		16 oz.	S–8345	10430	2–915
12	6		8 oz.	S–8345	10430	2–915
36	36		4 oz.	S–8345	10430	2–915
2	1		Bottle washing, Pyrex, 1000 ml.	S–9365	10710	3–395
2	1		Burettes, Geissler, blue line, 50 ml.	S–10635	15926	
2	1		Burners, Tirril, with stabilizer	S–12265	11026A	3–960
2	1		Clamps, burettes, spring-closing	S–19045	12120	5–770
6	4		Clamps, pinch, flat jaws for 3/8" tubing	S–19045	12186	5–850
3	2		Stoppers, rubber, assorted sizes, solid, 1 lb.	S–73305	18150	14–130
1	1		Cork borer, brass, 5 mm. to 11 mm. approx.	S–23175	12460B	7–845
12	6		Crucibles, Coors, size 1, wide form, 30 ml.	S–23665	18540	7–955
12	6		Crucibles, Gooch, size 3, Coors porcelain, 25 ml.	S–24315	18565	8–195

RECOMMENDED CHEMICALS FOR SEWAGE TREATMENT PLANTS

Quantity: Type of Plant A & B	C	D	Item
1 lb.	1 lb.		Acetic acid, glacial
18 lb.	9 lb.		Sulfuric acid, C.P.
18 lb.			Sulfuric acid, tech.
6 lb.			Hydrochloric acid
100 gm.	100 gm.		Sodium azide†
1 lb.	1 lb.		Sodium bicarbonate
1 lb.			Sodium carbonate, anhyd., C.P.
5 lb.	5 lb.		Sodium hyclorozide pellets, U.S.P.
1 lb.	1 lb.		Sodium thiosulfate, cryst
1 lb.	1 lb.		Potassium iodide, C.P.
1 oz.	1 oz.		Potassium bi-iodate, C.P.
5 lb.	5 lb.		Potassium bichromate, tech.
5 lb.	5 lb.		Manganous sulfate, C.P.
1/4 lb.	1/4 lb.		Potato starch
1 oz.	1 oz.		Methylene blue, U.S.P.
1/2 lb.	1/2 lb.		Asbestos, medium fiber, acid-washed
5 lb.			Ferric chloride, C.P., lump
5 lb.			Copper sulfate, cryst., tech.
1/4 lb.	1/4 lb.		Chloroform, U.S.P.
100 gm.			Orthotolidine
100 ml.			Phenolphthalein indicator
100 ml.			Methyl orange indicator
Obtain in field, ampules			Calcium hypochlorite
5 lb.	5 lb.		Calcium chloride, anhyd.
		1 pt.	Methylene blue solution
1 lb.	1 lb.		Sulfamic acid, C.P.

* From *Engineering Manual*, War Dept., Corps of Engineers, June 1950.

† For secondary-treatment plants only.

SEWAGE TREATMENT–TOOL LIST FOR PLANTS *

MINIMUM REQUIREMENTS FOR CLASS A, B, AND C PLANTS*

Quantity	Description	Catalog No.†
1	Wrenches, pipe: 8"	254 ZAO or equal
1	Wrenches, pipe: 12"	2512 ZAO or equal
2	Wrenches, pipe: 18"	2518 ZAO or equal
1	Wrenches, pipe: 24"	2524 ZAO or equal
1	Wrenches, open, double end, set of 9, 1/4" to 1"	P725 ZAO series or equal
1	Wrenches, socket, hex., set of 10 and handle, 7/16" to 1"	P21 ZAO or equal
1	Wrench, crescent, adjustable, 10"	710 ZA1 or equal
1	Vise, combination jaw and pipe, swivel base	A204½ ZA2 or equal
1	Hammer: blacksmith, 20½ lb.	0272 ZAO or equal
1	Ball peen #2	252 ZAO or equal
1	Claw #1	211 ZAO or equal
1	Pliers, combination, 8"	26 GZA1 or equal
1	Screwdriver, 6"	26 ZAO or equal
2	File, 10" mill bastard	3 ZA3 or equal
1	Cold chisel: 1/2"	200 ZAO or equal
1	1" extra long	205 ZAO or equal
1	Hacksaw	1027 ZA1 or equal
1	Hacksaw blades, doz.	1412 FLZAO or equal
1	Wrecking bar, 24"	95 ZAO or equal
1	Handsaw, 26", 8 point	80 ZA2-08 or equal

* H. Channon Co. Catalog 166, Chicago, Ill.
† From *Engineering Manual*, War Dept., Corps of Engineers, June, 1950.

INDUSTRIAL WASTES–GENERAL

DATA REQUIRED TO MAKE AN INDUSTRIAL WASTE STUDY

a. Complete flow diagram of industry.
b. Volume of waste from each process.
c. Length of time of flow of each process.
d. Complete chemical and biochemical analysis of each waste and combined waste.
e. Analysis of body of water receiving waste.
f. Study of use that is made of body of water receiving waste.

TREATMENT DEPENDING ON WASTE TYPE

INORGANIC → CHEMICAL, MECHANICAL

COMBINED ORGANIC & INORG. → COMBINATION OF TREATMENTS → CHEMICAL, MECHANICAL, BIOLOGICAL

ORGANIC → MECHANICAL, BIOLOGICAL

CHEMICAL

1. Neutralizing with acid or alkali as required.
2. Adding chemicals to form precipitate.
3. Adding chemicals to oxidize.
4. Adding chemicals to reduce.

MECHANICAL

1. Flocculation & Sedimentation
2. Settling
3. Flotation
4. Filtering
5. Skimming
6. Centrifuging
7. Screening
8. Incineration
9. Dilution
10. Drying
11. Lagooning

BIOLOGICAL

1. BIOLOGICAL OXIDATION
   a. Aeration
   b. Trickling filters.
   c. Activated sludge process
   d. Contact aeration process.
   e. Seeding with oxidizing bacteria.
   f. Sand filtration
   g. Chlorination
   h. Lagooning
2. BIOLOGICAL REDUCTION
   a. Anaerobic digestion
   b. Seeding with anaerobic bacteria.
   c. Lagooning
3. DISINFECTION
   a. Chlorination
   b. Ozonization.

RED LIGHTS FOR MUNICIPAL SEWER PLANT PROTECTION

INDUSTRIAL WASTE

Condition	Treatment
Intermittent & Peak Discharge	Holding Tanks
High in suspended matter	Fine Screens
Excess Acidity or Alkalinity	Addition of Lime or Acid
Toxic Substances	Keep out – precipitate
High in carbohydrates	Pretreat-Digest or Aerate
High Grease & Oil Content	Grease Separators

SEWER

QUESTIONS TO BE ANSWERED BY AN INDUSTRIAL WASTE SURVEY

FOR THE INDUSTRY

a. Can any by-products be produced economically from the waste?
b. Can processes be changed to effect any economies in handling waste?
c. Can quantities of waste be reduced; e.g. reusing wash water for many processes?
d. Can any of the wastes be reclaimed and reused?
e. Can some wastes be used to treat other wastes?
f. What other ways are there of remedying the pollutional effect of wastes?
g. Effect of cost of treatment on industrial economy?

FOR THE MUNICIPAL TREATMENT PLANT

a. Will the waste deteriorate the sewerage system?
b. Will the waste respond to existing treatment processes?
c. Is the strength of waste such that it will interfere with the treatment process?
d. Are the various units of the plant of sufficient size to handle the additional load due to the strength, volume, and the solids of the waste?
e. How will the additional load affect the final effluent?

NOTE: The following pages give some data on various industries. These data are approximate and vary from one manufacturing plant to another.

INDUSTRIAL WASTES—DATA AND TREATMENT—I*

PRODUCT	UNIT OF PRODUCTION	WASTE FLOW PER UNIT OF PRODUCTION-GALS.	B.O.D. (5 DAY) PPM	SUSPENDED SOLIDS PPM	TOTAL SOLIDS PPM	POPULATION EQUIVALENT (BOD BASIS)	GENERAL CHARACTERISTIC OF WASTE	POLLUTING EFFECT OF WASTE	METHODS OF TREATMENT	REMARKS
BREWERY:										
(a) Yeast & Water extracted from spent grain.	Barrel (31 Gals.) per day.	180 to 500	1200	650		19	Wastes are composed mainly of liquid pressed from wet grains, liquid from yeasts recovery operations and wash liquid from brew, fermentation, and storage departments.	1. Solids form sludge banks coating river bed, putrefying and robbing river of oxygen. 2. Organic matter depletes oxygen by bacterial action. 3. Bad odors may develop.	1. Fine screens remove suspended solids. 2. Settling tanks remove settleable solids. 3. Digestion of settled sludge in lagoons or digestion tanks. 4. Trickling filters have been effective. 5. Chemical treatment may be good, but chemical costs may be high.	1. Disposal into municipal sewers requires additional capacity but does not appear to affect treatment. 2. Volatile solids are about 460 ppm
(b) Grain sold wet. (Done by many small plants)			800	450		12				
CANNING:							Wastes are composed of wash water used to clean the raw product to be processed and water from washing floors and equipment. Waste contains small particles of the product such as skins, seeds, hulls, juices, etc. Water used for heat sterilization and sealed container cooling.	1. Solids in the waste form sludge banks which coat the river bed and also putrefy robbing the water of oxygen. 2. The dissolved solids remove oxygen from the water through bacterial action. 3. Bad odors may develop.	1. Disposal on land by irrigation. 2. Screening through fine screens (about 40 mesh) 3. Settling in tanks or basins. 4. Sludge digested in tanks or disposed of on dry land. 5. Chemical precipitation & settling (BOD reduction approximately 50%) 6. Pumping to lagoons where sludge is allowed to digest. Volume of storage should be sufficient to provide for season's run. Adding of sodium nitrate has been helpful. 7. Activated sludge, trickling filters and sand beds have been used effectively but short season militates against these processes.	Screenings have been used for feed.
Apples			2700	250						
Apricots		80	1020			4.1				
Asparagus		70	100	30		0.35				
Beets		37	2600	1530		4.8				
Carrots			1110	1830						
Corn-Cream Style		25	620	300		0.75				
Corn-Whole Kernel		25	2000	1250		2.5				
Grapes			720	1650						
Grapefruit Juice	Case of 24 No. 2 Cans per day	5	310	170		0.08				
Grapefruit Sections		56	1850	270		4.8				
Green Beans		35	200	60		0.35				
Lima Beans		250	190	420		2.4				
Peas		25	1700	400		2.1				
Pears and Peaches		65	1340			4.4				
Pork and Beans		35	925	225		1.6				
Pumpkin & Squash		20	5100	1480		5.0				
Sauerkraut		3	6300	630		1.0				
Spinach		160	615			4.9				
Succotash		125	525	250		3.3				
Tomatoes (Whole)		7.5	4000	250		1.5				
Tomato Products		70	1000	2000		3.5				
Cherries	Ton per day	12	1400			0.8				

* Data compiled by J. L. Staunton and H. Hayward

REFERENCES:

Industrial Waste Data—Public Works—Oct. 1949

Proceedings of 1st, 2nd, 3rd, and 4th Industrial Waste Utilization Conferences—Purdue University.

Waste Treatment Guide—Interstate Commission On the Potomac River Basin.

Interim Recommendations for the Disposal of Radioactive Wastes by Off. Commission Users—U.S. Atomic Energy Commission.

Treatment and Disposal of Industrial Waste Waters—B. A. Southgate

Industrial Waste Treatment Practice—E. F. Eldridge.

Sewage and Industrial Wastes—Journal of the Federation of Sewage Works Associations.

INDUSTRIAL WASTES–DATA AND TREATMENT–2

PRODUCT	UNIT OF PRODUCTION	WASTE FLOW PER UNIT OF PRODUCTION (GALS)	BOD–(5 DAY) PPM	SUSPENDED SOLIDS PPM	TOTAL SOLIDS PPM	POPULATION EQUIVALENT (BOD BASIS)	GENERAL CHARACTERISTIC OF WASTE	POLLUTING EFFECT OF WASTE	METHODS OF TREATMENT	REMARKS
COTTON TEXTILES							1. Wastes vary from highly Alkaline to mildly Acid. 2. Wastes from any one plant may continuously change.	1. Solids form sludge banks coating river beds 2. Putrefaction robs river of Oxygen 3. Bad odors may develop 4 Unsightly colors.	1. Recovery of Chemicals 2. Reduction of waste in plant 3. Equalization of Flow and mixing of various wastes 4. Chemical treatment with Iron and Alum Salts requires large amount of chemicals some of which can be recovered. Sludge is voluminous but dries fairly readily. 5. Activated sludge has been successful and effective in removing color when mixed with domestic sewage. 6. Trickling Filters give good treatment but not good color removal. 7. Color may be removed by Chlorine and Lime treatment but is expensive. 8. Color may be removed by acidification followed by neutralization.	1. May be discharged into municipal sewers but (a) Holding Tank having 24 hr. capacity may be necessary (b) pH adjustment may be necessary 2. Proposed disposal standards are: Color <100* pH 6-8 S.S., T.S., BOD Same as for domestic sewage *Some states have more rigorous requirements e.g. Pa. requires color to be 30 ppm or less.
PROCESS:										
Sizing	1000 lbs. of goods per day	60	820			2.4				
Desizing		1100	1750			96				
Kiering		1700	1240			105				
Bleaching		1200	300			18				
Scouring		3400	72			12				
Mercerizing		30,000	55			83				
Composite						20				
DYEING:										
Direct Acid	1000 lbs. of goods per day	6400	220			71				
Direct Basic		18,000	100			90				
Vat		19000	140			130				
Sulphur		5400	1300			347				
Developed		14400	110			120				
Naphthol		4800	250			60				
Aniline Black		15600	55			43				
DAIRY:							1. Wastes result from rinsing washing and spillage in addition to process wastes from Butter, Cheese, Casein, etc.	1. Almost wholly due to oxygen demand of wastes. 2. Fresh wastes may be Acid or Alkaline. 3. On decomposition wastes are acid 4. Black sludge deposits 5. Strong Odors	1. Irrigation for small plants, if sandy soil and constant cultivation are available 2. Odors controlled with Lime or Hypochlorite 3. Anaerobic Digestion followed by Filtration has been successful 4. Trickling Filters with loadings 1.3-2.0 lbs./cu. yd./day may give 80%-90% reduction. 5. Activated Sludge with suitable air volumes and contact periods may give reductions of 90% or greater.	1. Use of Imhoff tanks not satisfactory (Tanks may reduce BOD 30%)
Receiving Station	100 lbs. of milk intake per day	175	500		700	4				
Bottling Plant		250	—		600	6				
Cheese Factory		200	1000	750	3000	10				
Creamery		110	1250	660	1500	7				
Condensing Plant		150	1300	750	1200	10				
Dry Milk Plant		150	480		1200	4				
General Dairy		340	570	540		10				

INDUSTRIAL WASTES–DATA AND TREATMENT–3

PRODUCT	UNIT OF PRODUCTION	WASTE FLOW PER UNIT OF PRODUCTION (GALS.)	BOD (5 DAY) PPM	SUSPENDED SOLIDS (PPM)	TOTAL SOLIDS PPM	POPULATION EQUIVALENT (BOD BASIS)	GENERAL CHARACTERISTIC OF WASTE	POLLUTING EFFECT OF WASTE	METHODS OF TREATMENT	REMARKS
DISTILLERY:	1000 Bushels of grain mashed or equivalent 5000 gallons of 100 proof spirits produced daily.						Fermentation produces highly organic wastes with major portion of solids in solution	1. High BOD major factor. 2. Solids in waste form sludge banks which smother bottom growth and destroy aquatic life.	1. Lagooning 2. Irrigation—allowing 1 acre per 10,000 gallons of slop per day. 3. Digestion and thickening have had some success. 4. Digester effluent has been treated on trickling filters after dilution with 5 parts of filter effluent; rate was 1.5 MGD per acre. 5. Stock feeding 6. Screening. 7. Evaporation of liquid and recovery of solids for stock feeding.	1. Lagooning results in odor and fly breeding. 2. Irrigation requires frequent cultivation. 3. Thin slop may have a pH as low as 3.7 and a chlorine demand of more than 500 ppm.
Grain Distilling		600,000				11,000				
Thin Slop		25,000–35,000	34,000		50,000*	41,600–58,500				
Tailings		1400	740			50				
Evaporator condensate		26,000	1200			1500				
Other Combined wastes						3500				
Molasses Distilling	1000 gallons of 100 proof liquor prod. per day.	8,400 waste & 120,000 cooling water								
Molasses Slop		8,000–10,000	33,000			12,000				
* 90% Volatile Solids.										
GAS	Ton of coal processed per day						Wastes primarily phenols.	1. Phenols cause highly disagreeable tastes and odors in chlorinated waters—0.005 ppm of phenol creates an objectionable taste 2. Less than 1 ppm is toxic to fish. 3. High BOD depletes oxygen.	1. Reuse of some of the liquids may reduce oxygen requirements 30% 2. Phenol removal by extraction or quenching. 3. Phenols may be oxidized by trickling or sand filters or activated sludge, or by chlorine dioxide & ozone or chlorine.	1. Phenols can be recovered but without profit (Medium sized plants might earn operating expenses.) 2. Oxidation processes must have phenol concentrations less than 50 ppm.
Gas		3600	85			15				
Wastes Including cooling water**		3600	85			15				
Wastes without cooling water***		20–35								
** Phenol – 20 ppm. *** Phenol – 1500–2000 ppm										
LAUNDRY	100 lbs. of dry wash per day.	350–500	1000–1500	400–600	1500–2500****	17–37	1. Waste is dirty water from washing machines and is strongly alkaline, turbid, and highly colored. 2. Waste contains soap, soda ash, grease, dirt, dyes, and particles of cloth. 3. Waste has high organic content.	1. Solids settle out forming sludge banks which coat river bed and are highly putrescible. 2. Dissolved organic solids remove oxygen from water by bacterial action. 3. Unsightly scum forms on the surface of the river. 4. Alkalies and soap may affect fish and aquatic life.	1. Chemical precipitation produces sludge volume appr. 10% raw waste. 2. Trickling filters and activated sludge have been successful. 3. Sludge is disposed of on land in lagoons or dried on sand beds.	1. Except in excessive amounts, wastes do not interfere with municipal treatment plants. 2. Slugs of laundry waste upset activated sludge.
**** Volatile Solids 800–1200 ppm Grease: 400–500 ppm.										

INDUSTRIAL WASTES-DATA AND TREATMENT-4

PRODUCT	UNIT OF PRODUCTION	WASTE FLOW PER UNIT OF PRODUCTION (GALS)	BOD (5 DAY) PPM	SUSPENDED SOLIDS-PPM	TOTAL SOLIDS PPM	POPULATION EQUIVALENT (BOD BASIS)	GENERAL CHARACTERISTIC OF WASTE	POLLUTING EFFECT OF WASTE	METHODS OF TREATMENT	REMARKS
MEAT PACKING	Hog Unit per day.						Wastes are generally similar to Domestic Wastes.	1. Oxygen Depletion, Sludge Deposits & Discoloration similar to Domestic Sewage. 2. Danger from Pathogenic Organisms is less than that from Domestic Sewage.	1. Sedimentation (1-3 hrs. detention) may reduce BOD by 35% and suspended solids by 65%. 2. Chemicals give additional removals but may produce a sludge that is difficult to treat. Ferric chloride and alum have reduced BOD by 80% or more, but settled effluent is rarely below 150-200 BOD 3. Activated sludge and trickling filters have been successful. Mixing with carbohydrate wastes prior to treating has been recommended. 4. Superchlorination has been successful.	1. Treatment in Municipal Sewage Disposal Plants is entirely possible provided additional loading is cared for in the design. Regulation to produce a uniform flow is highly desirable.
Packing Plants	Cattle =	550	900			24				
Slaughterhouse	2½ Hog Units each, 1 Hog 1 Calf or 1 Lamb = 1 Hog Unit	150	2200			17				
Stockyards	Acre of Area per day.	2500	65			80				
Poultry Processing	1000 lbs of live weight per day.	2200				300				
METAL PLANTS								1. Oils form films on water coating objects and retarding Reaeration. 2. Cyanides are highly toxic. 3. Metals are toxic, cause tastes, and unsightly odors. 4. Acidity of wastes.	1. Oils removed by skimming and separation. 2. Cyanides may be removed by acid, activated sludge or chlorine treatment. 3. Metals may be removed by formation of insoluble salts and settling. 4. Acidity may be removed by neutralizing.	1. Do not mix Acid & Cyanide Solutions 2. Cyanides and Copper may stop digestion of domestic sludge.
Coolants & Cutting oils					150,000 200,000					
Pickling Wastes H_2SO_4 & $FeSO_4$					5,000 20,000					
Plating Wastes										
a. Cyanides					100					
b. Sulfuric Acid					100					
c. Copper					24					
d. Zinc					20					
e. Chromium					20					
MINING	Acre of Coal mined per day.	60 - 1,200			5,000 - 70,000		Waste is acid drainage from active and abandoned mines containing sulfuric acid and iron salts and washings from coal preparation plants containing silt, sand and fine coal.	1. Acidity of waste. 2. Sludge bank formation	1. Neutralizing Acidity. 2. Settling. 3. Lagooning.	

INDUSTRIAL WASTES—DATA AND TREATMENT-5

PRODUCT	UNIT OF PRODUCTION	WASTE FLOW PER UNIT OF PRODUCTION (GLS)	BOD (5 DAY) PPM	SUSPENDED SOLIDS (PPM)	TOTAL SOLIDS PPM	POPULATION EQUIVALENT (BOD BASIS)	GENERAL CHARACTERISTIC OF WASTE	POLLUTING EFFECT OF WASTE	METHODS OF TREATMENT	REMARKS
OIL REFINERY	Barrel (42 gals.) per. day	700-* 800	10-30	50		0.35-1.2	Waste contains free & emulsified oils & brines.	1. Oils form film on water a. Coating objects. b. Retarding reaeration. 2. Oils & Brines cause tastes & odors.	1. Reuse of cooling water. 2. Maximum separation at refinery 3. Emulsified oils can be broken by acid and lime. 4. Recovery of by products. 5. Corrective measures for mineral acidity, caustic wastes & phenol content. 6. Discharge of brines into underground formations. (This is primarily a problem of oil fields.)	1. If brines contain carbonates & iron they must be removed before discharge into underground formations to prevent clogging. Removal is usually by sedimentation & filtration. 2. Emulsified and gravity separated oils must be kept separate.
* 80%-90% Cooling Water										
PAPER Pulping Mills. Ground wood	Ton of Product per day	5000	645			160	Wastes are caustic and fibrous.	1. High B.O.D. 2. High color. 3. Serious odors from Sulfides (Mercaptans) 4. Digester liquors are the most objectionable wastes produced.	1. Reduction of strength of waste by use of savealls. 2. Chemical treatment may reduce B.O.D. by as much as 65%. 3. Ponding providing 60 days storage has been successful. 4. Sulfides may be removed by aeration. 5. Settling and anaerobic digestion have been successful.	
Soda pulping		85,000	110			460				
Sulfate or kraft pulping		64,000	125			390				
Sulfite pulping		60,000	450			1330				
Paper Mills Paper Misc., no bleach.	Ton of Paper Produced per day	9,000-39,000	155-19	521-70		68-36				
Paper misc, with bleach.		47,000	24			56				
Paperboard		14,000	121			84				
Strawboard		26,000	695-3500	1400-2400		900-4500				
Drinking Waste		83,000	300-900	2670		1250-3670				
PHARMACEUTICALS	Waste varies completely depending on product							High B.O.D.'s	1. Wastes generally respond to biological treatment. 2. Penicillin waste has been reported to be successfully treated by means of 1.4 hrs. aeration followed by high rate trickling filter having a 9-1 recirculation rate and filter loading of 1500 lb. BOD/Acre ft.	1. Dilution with sewage has been helpful especially with respect to Antibiotics.
QUARRY & SAND WASHING							Wastes are composed mainly of settleable solids	1. Sludge bank formation coats river bed robbing river of oxygen.	1. Settling. 2. Lagooning.	
RADIO-ACTIVE WASTES Radio iodine Phosphorus 32 Carbon 14								Radioactivity of waste.	1. Dilution and dispersion. 2. Dilution with media other than stable isotopes of same media. 3. Dilution and confinement: e.g. mixing with concrete and burying. 4. Concentration and confinement: e.g. precipitation and storage until radio activity is reduced to negligible value.	

INDUSTRIAL WASTES-DATA AND TREATMENT-6

PRODUCT	UNIT OF PRODUCTION	WASTE FLOW PER UNIT OF PRODUCTION (GALS)	BOD (5 DAY) PPM	SUSPENDED SOLIDS PPM	TOTAL SOLIDS PPM	POPULATION EQUIVALENT (BOD BASIS)	GENERAL CHARACTERISTIC OF WASTE	POLLUTING EFFECT OF WASTE	METHODS OF TREATMENT	REMARKS
RAYON MFG. *	100 lbs. Produced Per day	2000 -5000	20- 50			2 12		1. Sludge banks rob river of oxygen. 2. High turbidity 3. Bad odors.	1. Chemical treatment has been successful 2. Trickling filters have been successful.	
*See also Cotton Textile - Dyeing.										
SOFT DRINK BOTTLING	100 cases of bottled product per day	1500	150	20	700	11	Wastes are composed of wash waters from plant, equipment, and bottle washing containing dirt, washing compounds, sugars and syrups.	Sugars and syrups remove oxygen from stream through bacterial action	Trickling Filters have been used efficiently. Proper dilution is necessary prior to filtering.	1. Sugar content makes waste difficult to treat 2. Zeolites in water softening cause high bacterial content.
TANNERY	100 lbs. of Raw Hide per day (This will produce approx. 68 lbs. of finished leather)	300- 800 1700	2500- 1200 600	2500- 1000		37 48 50		1. Strong persistent color 2. Oxygen depletion 3. Formation of sludge banks 4. Strongly Alkaline. 5. Formation of Calcium Carbonate scale.	1. Mixing of wastes and settling for periods up to 24 hrs. has removed 40-54% of the BOD and 70-80% of the suspended solids. 2. Intermittent Sand or Cinder Filters are reported efficient in color and BOD removal. 3. Trickling Filters have reduced BOD 65 to 75% and color 65 to 70%.	1. Intermittent discharge of waste accentuates problem of pollution. 2. Chemical precipitation has not been efficient in treatment. 3. Caustic Alkalinity and/or chrome salts may interfere with municipal treatment if the proportion of waste to sewage is high.
WOOL TEXTILES							1. Wastes from any one plant may continuously change.	1. Color 2. Caustic 3. Sludge bank formation. 4. Bad Odors 5. Turbidity	1. Neutralizing washes. 2. Settling. 3. Lagooning Sludge. 4. Color removal by use of Chlorine and lime. 5. Trickling filters have been successful.	
Wool & Yarn Scouring	100 lbs.	126	9300	7300						
Wool Dyeing * Acid Chrome	of Wool		640	100	700					
Vat Dyeing	per day		440	150	5500					
* Including rinse water										

WASTE DISPOSAL-SANITARY FILL

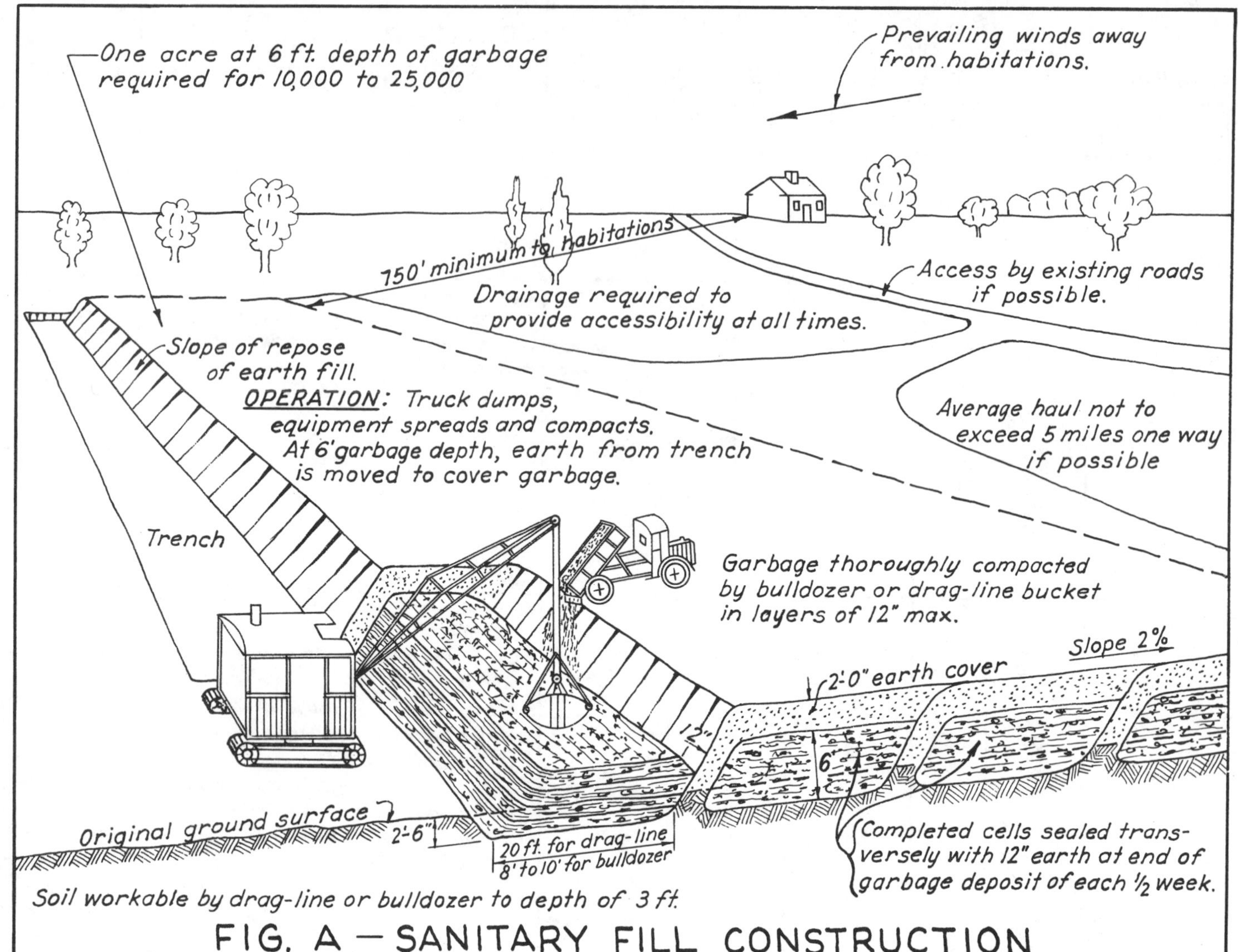

One acre at 6 ft. depth of garbage required for 10,000 to 25,000
Prevailing winds away from habitations.
750' minimum to habitations
Access by existing roads if possible.
Drainage required to provide accessibility at all times.
Slope of repose of earth fill.
OPERATION: Truck dumps, equipment spreads and compacts. At 6' garbage depth, earth from trench is moved to cover garbage.
Average haul not to exceed 5 miles one way if possible
Trench
Garbage thoroughly compacted by bulldozer or drag-line bucket in layers of 12" max.
Slope 2%
2'-0" earth cover
12"
6'
Original ground surface
2'-6"
20 ft. for drag-line
8' to 10' for bulldozer
Completed cells sealed transversely with 12" earth at end of garbage deposit of each ½ week.
Soil workable by drag-line or bulldozer to depth of 3 ft.

FIG. A — SANITARY FILL CONSTRUCTION

TABLE B — EQUIPMENT REQUIRED			
POPULATION	TYPE	BOOM	BUCKET
Under 20,000	Bulldozer	—	1 Cu. Yd.
20,000 to 50,000	Drag-line	30 ft.	½ Cu. Yd.
Over 50,000	" "	40 ft.	¾ Cu. Yd.

TABLE C — QUANTITIES OF REFUSE IN POUNDS PER CAPITA PER YEAR*		
	GARBAGE	RUBBISH
According to economical conditions, locality, climate and method of collection.	100 to 450	30 to 150
Average for large cities.	180	102
NOTE: Peak for short periods = 140% to 200% times average.		

Average unit weights:* Garbage = 40 lbs. per cu. ft.; rubbish = 7.41 lbs. per cu. ft.

* Data based on American Civil Eng. Handbook by Merriman & Wiggin.

WASTE DISPOSAL - INCINERATION *

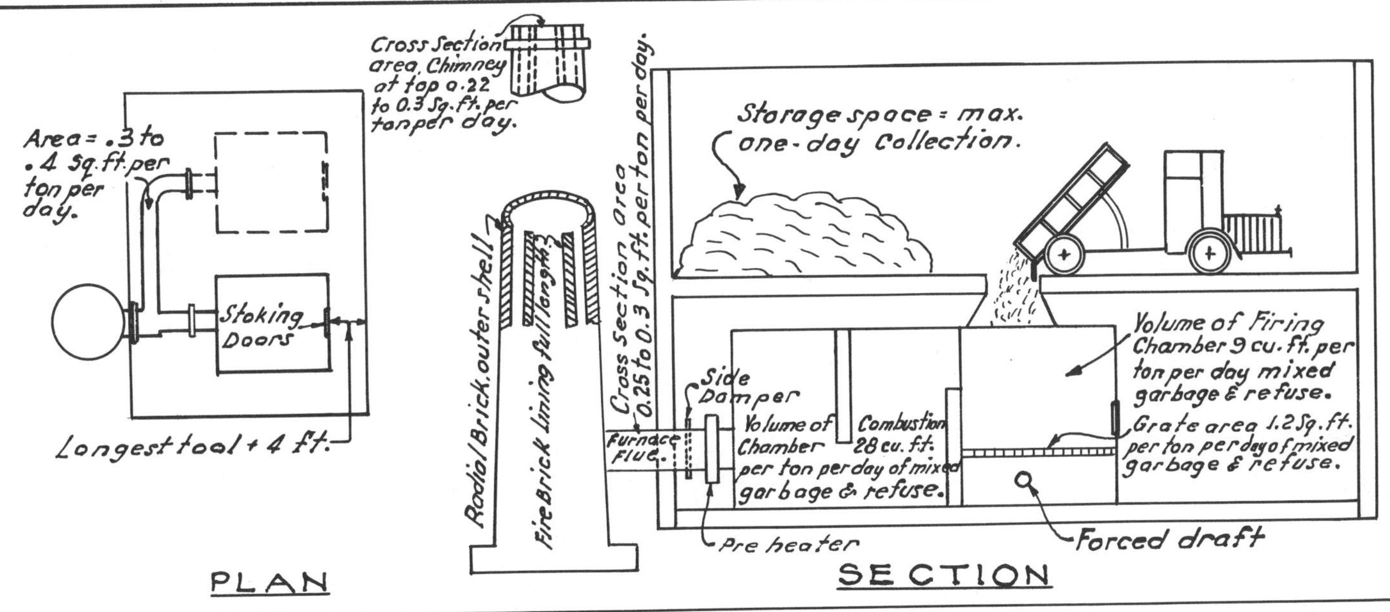

Area = .3 to .4 Sq.ft. per ton per day.
Stoking Doors
Longest tool + 4 ft.
Cross Section area, Chimney at top 0.22 to 0.3 Sq.ft. per ton per day.
Radial Brick, outer shell
Fire Brick lining full length
Cross Section area 0.25 to 0.3 Sq.ft. per ton per day.
Furnace Flue.
Side Damper
Pre heater
Volume of Chamber per ton per day of mixed garbage & refuse.
Combustion 28 cu.ft.
Storage space = max. one-day Collection.
Volume of Firing Chamber 9 cu.ft. per ton per day mixed garbage & refuse.
Grate area 1.2 Sq.ft. per ton per day of mixed garbage & refuse.
Forced draft
PLAN
SECTION

GARBAGE & REFUSE PRODUCTION.

Each 1000 population produces approx. 1 ton of mixed garbage and refuse per day (including stores, restaurants, hotels, etc.); general wide variation depending on efficiency of collection service, etc.

High-Class residential section alone produces 1 lb. per capita per day.
Poor " " " " 3/4 lb. " " " "
Month of February produces 5% of annual total tonnage.
" " August " 11% " " " "

BASIC ELEMENTS OF DESIGN.

Average moisture content of mixed garbage & refuse = 55% by weight.
Average content of combustible material = 45% by weight, of which 5% is average residual ash.
These percentages may vary 50% above or below depending on character of community; a wealthy residential community will produce higher % of combustible material.
Mixed combustible material will produce 7000 to 8000 B.T.U. per pound.
Forced draft required per pound of combustible material = 19 lbs. of air.
Max. H. P. required per ton of combustible material for forced draft = 0.30 H.P.
Optimum temperature of furnaces = 1250° F.
Preliminary drying area or storage hoppers should have capacity to accommodate a full day's (8 hr.) collections.
Clearance in front of stoking doors should be 4 ft. greater than length of longest stoking tool.
Customary chimney ht. = 100 ft. ±; 50' min. Above surrounding bldgs. but not too conspicuous.

EXAMPLE - COMMUNITY OF 50,000.

1. Total waste production @ 2# per capita per day = 2 x 50,000 = 100,000 lbs. daily
2. Average moisture content = 55% x 100,000# = 55,000 lbs. water
3. Average water per hour (8 hr. incinerator operation) = $\frac{55,000}{8}$ = 6,875 lbs. per hour
4. Average combustible content = 40% x 100,000# = 40,000 lbs. combustible
5. Average combustible per hour (8 hour operation) = 5,000 lbs. per hour
6. Average ash content = 5% x 100,000# = 5,000 lbs. ash
7. Average ash per hour (8 hour operation) = 625 lbs. ash
8. Average wt. air req'd. @ 19 lbs. per lb. combustible) = 19 x 5,000# = 95,000 lbs. per hour
9. Cu. ft. of air / min. required = 1/5 x 95,000 lbs. = 19,000 c.f.m.
10. Capacity (approximate) of blower = 19,000 c.f.m. less natural draft
11. Heat requirements (from 0°F. to 1250°F.) per hour.
    a. to raise 6875 lbs. water to 212°F. = 212 x 6,875 = 1,457,000 B.T.U.
    b. to evaporate 6875 lbs. water @ 212°F. = 971 x 6,875 = 6,685,000 "
    c. to raise steam from 212°F. to 1250°F. = .48 x 1,038 x 6,875 = 3,425,000 "
    d. to raise 95,000 lbs. air to 1250°F. = .25 x 95,000 x 1,250 = 29,687,000 "
    e. to raise 625 lbs. ash to 1250°F. = .20 x 625 x 1250 = 156,000 "
    f. to raise 5000 lbs. combustible to 1250°F. = .20 x 5,000 x 1,250 = 1,250,000 "
    Total + 10% radiation losses = 46,926,000 B.T.U.
12. Heat generated per hour
    a. 5,000 lbs. combustible @ 7,500 B.T.U. per lb. = 37,500,000
13. Heat to be supplied by fuel 9,426,000 B.T.U.
14. Gallons oil per hr. @ 151,300 B.T.U. per gal. $\frac{9,426,000}{151,300}$ = 62.5 gals. oil per hr. max.

* By Fred J. Biele, C.E., Sanitary Engineer.

WATER SUPPLY—GENERAL

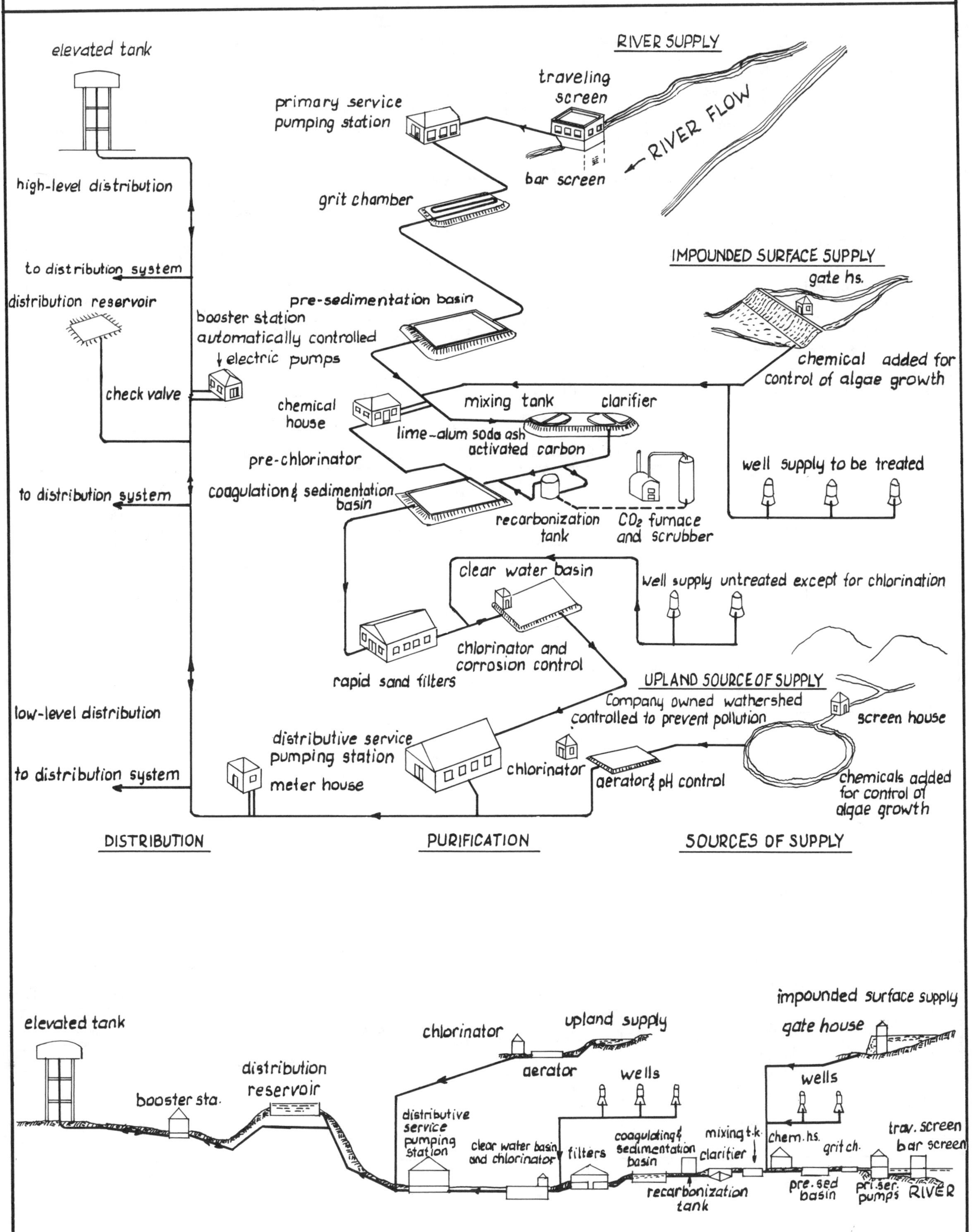

elevated tank
high-level distribution
to distribution system
distribution reservoir
check valve
booster station
automatically controlled
electric pumps
to distribution system
low-level distribution
to distribution system
meter house
RIVER SUPPLY
traveling
screen
primary service
pumping station
RIVER FLOW
bar screen
grit chamber
IMPOUNDED SURFACE SUPPLY
gate hs.
pre-sedimentation basin
chemical added for
control of algae growth
chemical
house
mixing tank
clarifier
lime-alum soda ash
activated carbon
pre-chlorinator
well supply to be treated
coagulation & sedimentation
basin
recarbonization
tank
CO_2 furnace
and scrubber
clear water basin
well supply untreated except for chlorination
rapid sand filters
chlorinator and
corrosion control
UPLAND SOURCE OF SUPPLY
Company owned watershed
controlled to prevent pollution
screen house
distributive service
pumping station
chlorinator
aerator & pH control
chemicals added
for control of
algae growth
DISTRIBUTION
PURIFICATION
SOURCES OF SUPPLY
impounded surface supply
elevated tank
chlorinator
upland supply
gate house
distribution
reservoir
aerator
wells
wells
booster sta.
distributive
service
pumping
station
clear water basin
and chlorinator
filters
coagulating &
sedimentation
basin
mixing t·k.
clarifier
chem. hs.
grit ch.
trav. screen
bar screen
recarbonization
tank
pre. sed
basin
pri. ser.
pumps
RIVER

WATER SUPPLY—DEMANDS—I

1. *FACTORS AFFECTING DEMAND* are population, per capita consumption, pressure, quality, cost, sewer facilities, climate, air conditioning, and use of meters.
2. *INDUSTRIES* or other special large consumers should be carefully considered.
3. *EXTENSIONS OF SYSTEMS* Field survey of pressures and consumption is best guide to design.
4. *FUTURE* Systems and extensions are generally designed for 30 years hence.
5. *CAPACITY* of system should be sufficient to deliver maximum daily flow simultaneously with required fire flow (see p. 20-03). Fire flow generally governs design of distribution system especially for smaller systems.

TABLE A – AVERAGE DESIGN VALUES FOR MUNICIPAL WATER SYSTEMS

City	Population thousands	Average Consumption gallons per day per capita
Residential		100
Residential, Commercial and Industrial	25-50	148
	50-100	133
	Over 100	151

Based on well-regulated and well-operated system, good plumbing, and good metering. Values include public use, waste, and miscellaneous.

TABLE B – FACTORS AFFECTING DEMAND

Factor	Effect in %
Pressure ±	±10
Quality ±	± 5
Cost ±	±20
Absence of sewage facilities	−10
Absence of meters	+20 to 100

Above table extremely approximate, to be used only preliminary to complete analysis and investigation.

TABLE D – COMMERCIAL WATER CONSUMPTION IN THE BOROUGH OF MANHATTAN, N. Y. C.

Type of Building	Gallons per day per 1000 sq. ft.
Hotels	600-1100
Office buildings	100-500
Department stores	100-400
Apartment hotels	200-400
Average	300

TABLE F – AVERAGE DESIGN VALUES FOR INSTITUTIONAL AND PRIVATE SYSTEMS

Description	Gals. per cap. per day
Camps	25-40
Small dwellings, farmhouses, etc.	40-60
Large dwellings, boarding schools, etc.	75-100
Institutions (except hospitals)	75-125
Hospitals	150-250
Day Schools	15
Day Schools with showers	20
Factories	25

TABLE C – FLUCTUATION OF DEMAND

Maximum Demand	% of Average Annual Demand
Monthly	140
Daily	120-240
Hourly: Without fire demand	150-220
Hourly: With Fire demand	200-500

TABLE E – PLUMBING FIXTURES*
(See Fig. C, p. 22-05 for factors of usage.)

Fixture	Excellent Flow g.p.m.	Pressure at Outlets (Faucets Wide Open), p.s.i.
Lavatory faucets, single	4	4
Bathtub faucets, single	6	5
Combination bathtub faucets	8	5
Sink faucets	6	5
Shower heads	6	3
Shower mixing valves	6	30
Water closets, tank type	5	5
Water closets, flush valves	30	25
Garden hose and nozzle	10	30 at hydrant

TABLE G – MISCELLANEOUS REQUIREMENTS†

Domestic Fixtures:	*Gallons*
Filling the ordinary lavatory	1½
Filling the average bathtub	30
Flushing water cabinet closet	6
Shower bath	30
Lawn Fixture:	*Gal. per hr.*
½-inch hose with nozzle	200
¾-inch hose with nozzle	300
Lawn sprinkler	120
Continuous-flowing drinking fountain	90
Farm Animals:	*Gal. per day*
Each cow or horse	10
Each hog or sheep	5
100 chickens	4

*Data from Davis, *Handbook of Applied Hydraulics*, McGraw-Hill.
†Data from Peerless Pump Co.

WATER SUPPLY—DEMANDS—2

FIRE REQUIREMENTS

TABLE A – FIRE FLOW AND HYDRANT SPACING FOR PRINCIPAL MERCANTILE OR BUSINESS DISTRICTS FOR VILLAGES AND CITIES OF VARIOUS TOTAL POPULATIONS*

(See notes below for requirements for other districts.)

Population	Required per Flow, gal. per min.	Hours' Duration	Average Area for Hydrant, sq. ft.
1,000	1,000	4	120,000
1,500	1,250	5	
2,000	1,500	6	
3,000	1,750	7	
4,000	2,000	8	110,000
5,000	2,250	9	
6,000	2,500	10	
10,000	3,000	10	100,000
13,000	3,500	10	
17,000	4,000	10	90,000
22,000	4,500	10	
27,000	5,000	10	85,000
33,000	5,500	10	
40,000	6,000	10	80,000
55,000	7,000	10	70,000
75,000	8,000	10	60,000
95,000	9,000	10	55,000
120,000	10,000	10	48,000
150,000	11,000	10	43,000
200,000	12,000	10	40,000

POPULATIONS OVER 200,000, an additional flow of 2000 to 8000 g.p.m. is required for a second 10-hr. fire.
IN RESIDENTIAL DISTRICTS, the required fire flow depends on the character and congestion of the buildings. Sections where buildings are small and of low height, and with about 1/3 the lots in a block built on, require not less than 500 gallons per minute. With larger or higher buildings, up to 1000 gallons is required, and, where the district is closely built or buildings approach the dimension of hotels or high-value residences, 1500 to 3000 gallons is required, with up to 6000 gallons in densely built sections of 3-story buildings or higher. For residential districts only, required duration may be reduced for required fire flows of 2500 g.p.m., and in no case less than 50% of that given in Table A.

TABLE B – PRESSURE REQUIREMENTS AS RECOMMENDED BY THE NATIONAL BOARD OF FIRE UNDERWRITERS

Condition	Minimum Residual Pressure at Hydrant, p.s.i.
Pumping Engines	10-20
Direct Flow from Hydrants:	
1. High-value districts	75
2. Districts not requiring more than 2500 g.p.m. fire flow, where not more than 10 buildings exceed 3 stories, and in closely built residential sections	60
3. Village mercantile districts where buildings do not exceed 2 stories, and in thinly built residential sections	50

Residual pressure = pressure at hydrant when required flow is being delivered.

NOTE: A.W.W.A. recommends a normal pressure of 60 to 75 p.s.i. in a distribution system at all times.

HYDRANTS

1. Common arrangement of hydrant is one 4½" connection for fire engine pumper, and two 2½" connections for direct hose connection.
2. Connect hydrant to main line with pipe not less than 6 in. diameter, gated and located at street intersection with access from four directions.

TABLE C – METHODS FOR SUPPLYING PRESSURE FOR FIRE STREAMS†

	Method	Recommended for Use
1.	Maintenance of sufficient pressure on the mains at all times for direct hydrant service for hose streams	For villages not provided with full-time fire departments and mobile pumpers. Not usually economical for large communities
2.	Use of emergency fire pumps to boost the pressure in the distribution system during fires	For villages requiring higher pressures for fire fighting than are desirable for normal consumption, and where there are no mobile pumpers.
3.	Use of mobile pumping engines which take suction from the hydrants	For all communities large enough to maintain modern and well-trained fire departments
4.	Use of a separate high-pressure distribution system for fire protection only	In high-value districts of large cities. Supplements main distribution system. Pressures used: 150-300 p.s.i.

* From National Board of Fire Underwriters.

† From Davis, Handbook of Applied Hydraulics, McGraw-Hill.

WATER SUPPLY — RAINFALL & RUNOFF — I

PROCEDURE FOR OBTAINING RUNOFF DATA FOR A DRAINAGE AREA.

1. *Use stream gage data if available. (Best source U.S. Geological Survey.) Supplement by records of stream gaged in same locality*
2. *For general perspective see Fig. C & D, Pg. 20-08.*
3. *For approximations use Fig. A & B below and Table A & B, Pg. 20-05.*
4. *See Pg. 20-06 & 20-07 for reservoir storage.*
5. *For complete analytical analysis see Elements of Hydrology by A. F. Meyer (John Wiley & Sons) and River Discharge by Hoyt & Grover (John Wiley & Sons).*

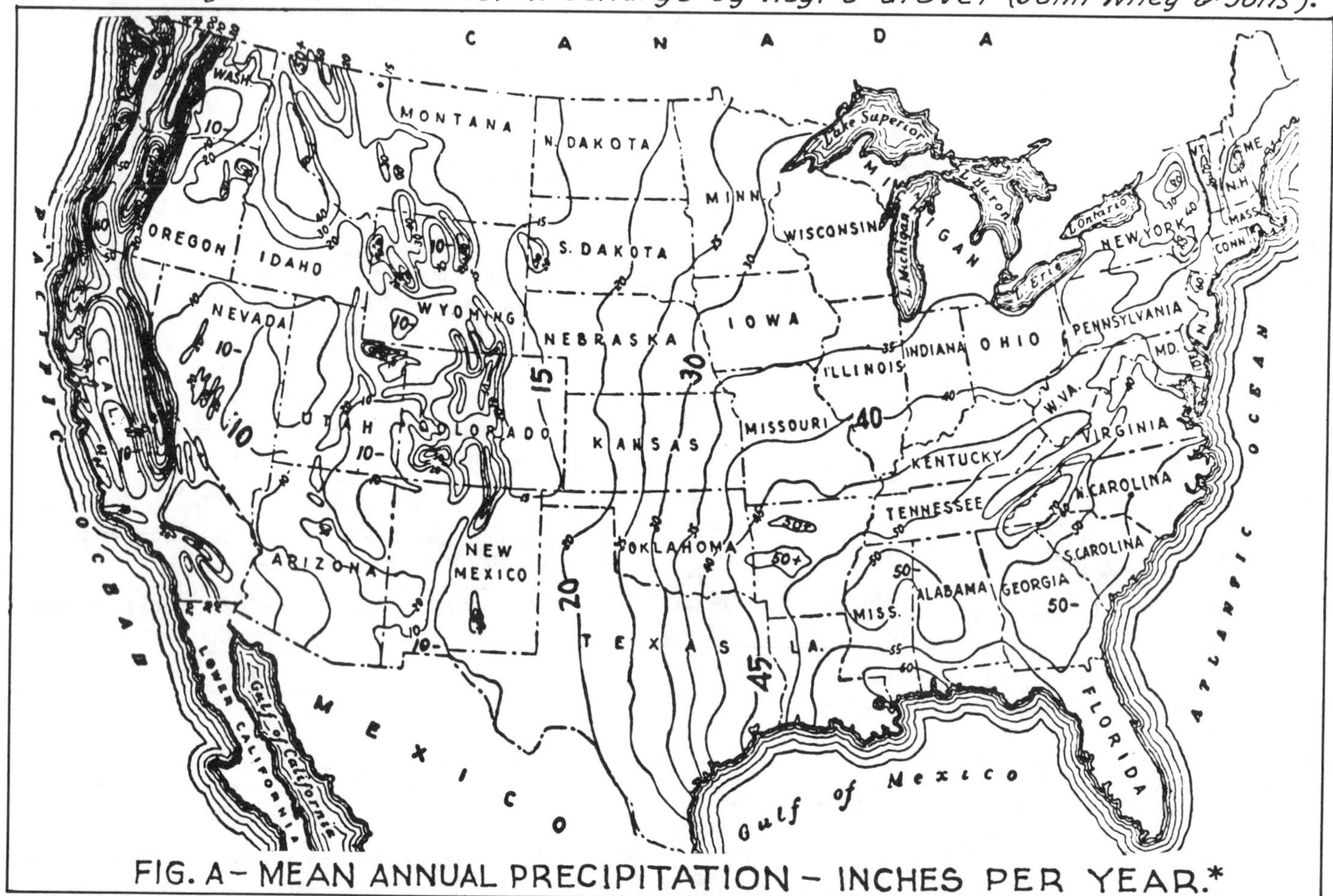
CANADA
MONTANA
N. DAKOTA
S. DAKOTA
MINN
WISCONSIN
MICHIGAN
WASH
OREGON
IDAHO
NEVADA
WYOMING
NEBRASKA
IOWA
ILLINOIS
INDIANA
OHIO
PENNSYLVANIA
NEW YORK
VT
N.H
ME
MASS
CONN
MD.
W.VA.
VIRGINIA
KENTUCKY
TENNESSEE
N.CAROLINA
S.CAROLINA
GEORGIA
ALABAMA
MISS.
LA.
FLORIDA
MISSOURI
KANSAS
OKLAHOMA
TEXAS
NEW MEXICO
ARIZONA
UTAH
COLORADO
CAL
Lake Superior
L. Michigan
L. Huron
Ontario
Erie
LOWER CALIFORNIA
Gulf of California
MEXICO
Gulf of Mexico
ATLANTIC OCEAN
PACIFIC OCEAN
10
15
20
30
40
45
50
50+

FIG. A – MEAN ANNUAL PRECIPITATION – INCHES PER YEAR.*

FIG. B – RAINFALL COEFFICIENTS (AFTER HAZEN).**

MINIMUM RAINFALL.

7 year min. rainfall = Value in Fig. A × (1.00 – Coefficient in Fig. B).
45 year min. rainfall = Value in Fig. A × (1.00 – 2 × Coefficient in Fig. B).

The minimum rainfall for a period of 30 to 40 years is recommended for municipal design.

* *From Handbook of Culvert & Drainage Practice by Armco Mfrs. Assn. (Based on U.S. Weather Bureau records).*
** *From Public Water Supplies by F. E. Turneaure & H. L. Russell.*

WATER SUPPLY—RAINFALL & RUNOFF—2

TABLE A – ANNUAL RUNOFF FROM STREAM DRAINAGE BASINS*

Stream	Locality	Drainage Area, Square Miles	Period of Observation	Average Annual Precipitation, inches	Average Annual Runoff, inches
Kennebec River	Waterville, Me.	4,270	1893-1905	39.6	23.7
Cobbosseecontee River	Gardiner, Me.	240	1891-1905	42.2	17.4
Sudbury River	Massachusetts	75	1875-1932	44.3	21.2
Wachusett River	Massachusetts	109	1897-1932	45.0	23.7
Merrimac River	Lawrence, Mass.	4,461	1880-1934	41.6	20.1
Lake Cochituate	Cochituate, Mass.	18.9	1863-1900	47.1	20.3
Mystic Lake	Boston, Mass.	26.9	1878-95	44.1	20.0
Abbott Run	Rhode Island	27	1908-32	42.9	21.1
Nepaug River	Hartford, Conn.	32	1929-35	42.4	17.3
Pomperaug River	Bennetts Bridge, Conn.	89	1914-16	44.5	19.5
Hudson River	Mechanicville, N. Y.	4,500	1888-1901	44.2	23.3
Croton River	Croton Dam, N. Y.	375	1868-1932	47.7	23.4
Delaware River	Port Jervis, N. Y.	3,070	1902-1930	42.8	25.9
Delaware River	Trenton, N. J.	6,800	1897-1930	44.7	24.9
Neshaminy Creek	Forks, Pa.	139	1884-99	47.6	23.1
Perkiomen Creek	Frederick, Pa.	152	1884-99	48.0	23.6
Tohickon Creek	Point Pleasant, Pa.	102	1888-1911	48.9	26.1
Susquehanna River	Harrisburg, Pa.	28,030	1891-1905	39.4	21.1
Potomac River	Point of Rocks, Md.	9,650	1895-1905	36.8	14.2
James River	Cartersville, Va.	6,240	1899-1934	40.8	15.6
James River	Buchanan, Va.	2,060	1895-1905	41.2	16.9
Roanoke River	Roanoke, Va.	390	1897-1905	42.7	17.7
Shenandoah River	Millville, W. Va.	3,000	1895-1905	38.3	13.6
Ohio River	Wheeling, W. Va.	23,820	1884-1905	41.7	22.7
Miami River	Dayton, Ohio	2,525	25 yrs.	37.1	11.9
Muskingum River	Dresdon, Ohio	5,828	1888-95	39.7	13.1
Tennessee River	Chatanooga, Tenn.	21,400	1881-1934	50.3	24.2
Chattahoochee River	West Point, Ga.	3,550	1896-1934	54.6	22.3
Tombigbee River	Columbus, Miss.	4,440	1901-9	49.2	17.1
Rock River	Rockton, Ill.	6,290	1904-8	33.9	10.0
Wisconsin River	Rhinelander, Wis.	1,110	1909-14	29.6	15.1
Mississippi River	Minneapolis, Minn.	19,500	1897-1913	27.3	5.3
Little Fork River	Little Fork, Minn.	1,720	1909-13	23.9	5.1
Root River	Houston, Minn.	1,560	1908-13	31.4	5.2
Ottertail River	Fergus Falls, Minn.	1,310	1908-13	23.0	2.6
St. Croix River	St. Croix Falls, Minn.	5,930	1902-12	30.0	9.6.
Red River	Grand Forks, N. Dak.	25,500	1882-1934	20.9	1.2
Mississippi River	Keokuk, Iowa	119,000	1878-1934	29.5	7.0
Ralston Creek	Iowa City, Iowa	3	1925-35	33.1	6.8
Drainage Basin A	Wagonwheel Gap, Colo.	.35	1911-26	21.1	6.1
Drainage Basin B	Wagonwheel Gap, Colo.	.31	1911-26	21.0	6.7
Colorado River	Austin, Tex.	37,000	1900-9	26.9	.7
South Fork of Coquille River	Powers, Oreg.	168	1917-26	96.0	63.5
Rogue River	Raygold, Oreg.	2,020	1906-28	43.6	22.6
Pilarcitos and San Andreas Creeks	San Andreas Reservoir, Calif.	.12	1869-1903	45.0	18.0
South Fork of Yuba River	Lake Spaulding, Calif.	123.5	1907-39	60.60	47.4

TABLE B - COMPARISON OF MEAN & MINIMUM ANNUAL RUNOFFS.**

River Locality	Years of Record	Area Drained Sq. Miles.	Mean Annual Flow in Inches Through Sept. 41.	Year of Min. Flow.	Min. Annual Flow in Inches.	Percentage of Average.
Merrimack, N.H.	37	1,507	25.0	1931	16.9	68
Quinebaug, Conn.	24	711	23.8	1930	11.4	48
Sacandaga, N.Y.	30	491	30.7	1934	20.0	65
Genesee, N.Y.	34	1,017	15.9	1934	10.4	65
So. Branch Raritan, N.J.	22	65	23.5	1932	12.0	51
Susquehanna, Pa.	52	24,100	19.3	1931	11.4	59
Potomac, Md.	46	9,650	13.3	1934	6.8	51
James, Va.	42	6,242	15.7	1931	7.8	50
Santee, S.C.	34	14,800	17.0	1927	9.0	53
Chattahoochee, Ga.	44	3,550	21.9	1914	11.0	50
Tombigbee, Miss.	25	4,490	17.5	1904	5.1	29
Mad, Ohio.	26	485	14.1	1934	6.1	43
Cumberland, Ky.	28	4,890	21.0	1931	10.3	49
Tennessee, Tenn.	43	8,930	19.8	1941	10.0	50
Rock, Wis.	28	3,300	7.3	1934	2.4	33
Mississippi at St. Paul.	39	36,800	3.3	1934	0.72	22
Minnesota, Minn.	21	6,300	0.84	1934	0.0095	1
Grand, Mo.	21	2,250	5.7	1938	0.78	14
Neosho, Kan.	20	4,830	5.4	1939	0.92	17
Trinity, Tex.	30	12,840	4.4	1909	0.44	10
Verde, Ariz.	41	6,230	1.7	1934	0.49	29
Humboldt, Nev.	34	5,010	0.87	1934	0.095	11
Sacramento, Calif.	48	9,300	16.3	1924	6.0	37
Kings, Calif.	47	1,694	18.7	1924	4.3	23
Spokane, Wash.	51	4,350	21.2	1930	10.4	49
Salmon, Idaho.	30	13,400	10.3	1931	5.9	57

Notes: Annual flows based on water year ending Sept. 30. Mean annual flow may have little significance in sub-humid, arid, and semi-arid regions. Years of record not consecutive for all stations.

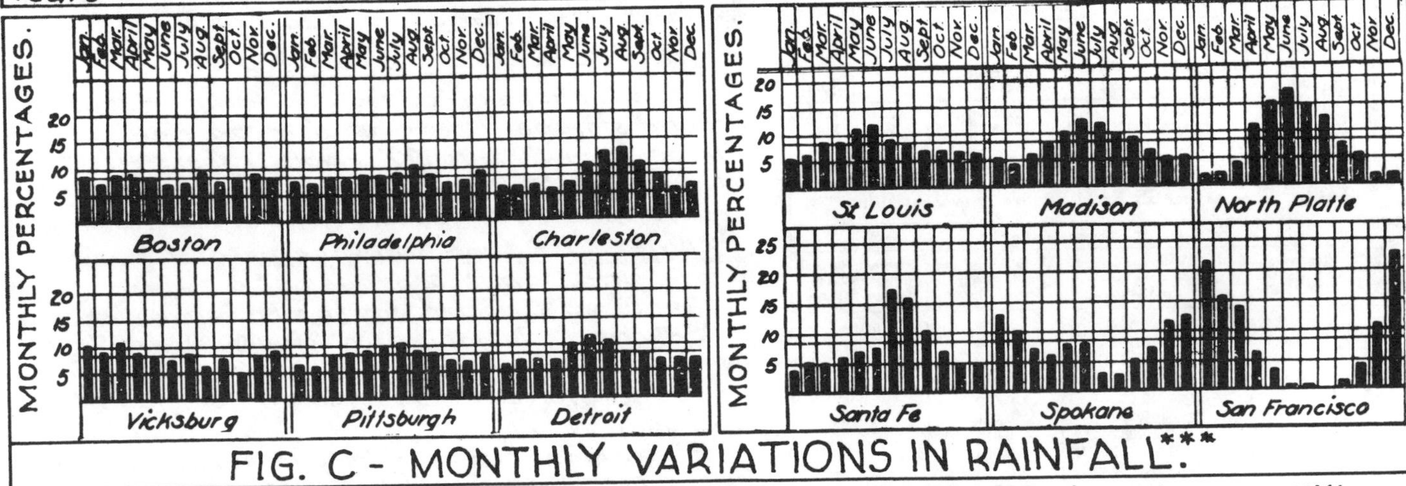
MONTHLY PERCENTAGES.
Jan. Feb. Mar. April May June July Aug. Sept. Oct. Nov. Dec.
Boston
Philadelphia
Charleston
Vicksburg
Pittsburgh
Detroit
St. Louis
Madison
North Platte
Santa Fe
Spokane
San Francisco

FIG. C - MONTHLY VARIATIONS IN RAINFALL.***

* Adapted from Physics of the Earth - IX. Hydrology Published by Mc.Graw-Hill.
** Adapted from water-supply papers published by the Geological Survey, U.S. Dept of Interior.
*** Adapted from Public Water Supplies by F.E. Turneaure & H.L. Russell.

WATER SUPPLY—RESERVOIR STORAGE

PROCEDURE TO FIND REQUIRED VOLUME OF STORAGE FOR A GIVEN DEMAND.

1. Use Fig. A and example below for Maine, Mass., R.I., Conn. and New York States.
2. Use Mass diagram method, Pg. 20-07 if reasonably accurate monthly stream flow records are available for a 20 year or longer period of time.
3. Use "coefficient of variation" method for short record; Table B, C, D & explanation below.

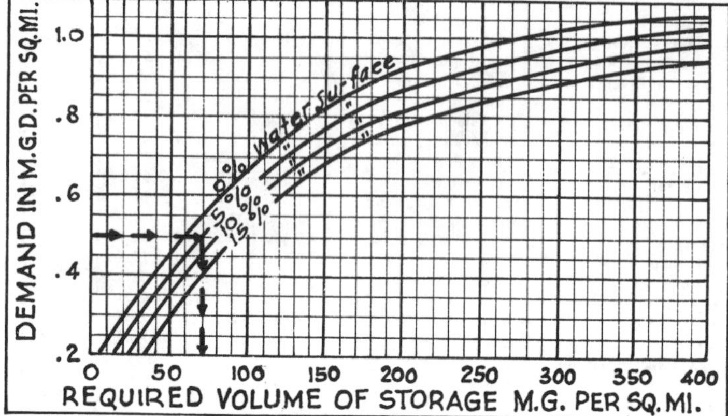

DEMAND IN M.G.D. PER SQ.MI.
1.0
.8
.6
.4
.2
0
50
100
150
200
250
300
350
400
REQUIRED VOLUME OF STORAGE M.G. PER SQ.MI.
0% Water Surface
5%
10%
15%

EXAMPLE: Given: Location in Connecticut, demand of 5,000,000 gals. per day, a drainage area of 10 Sq. miles and 5% water surface (% of water surface = area of all water surfaces when reservoir is full plus 40% area of underdrained swamps plus 30% of drained swamps divided by total drainage area).

Required: Volume of storage.

Solution: $\frac{5{,}000{,}000}{10 \text{ Sq. mi.}} = .5$ m.g.d. per sq. mi.

Enter Fig. A, at left, as indicated and required storage = 70 m.g.d. per sq. mile, 70 × 10 sq. mi. = 700,000,000 gals. Volume of storage required.

FIG. A - YIELD CURVES†. Composite diagram based on Abbott Run (R.I.), Sudbury and Manhan Rivers, Tillotson Brook, Wachusett reservoir (Mass.), Naugatuck River (Conn.), Croton River (N.Y.). Effects of New England drought of 1935 not included.

TABLE B – STORAGE RATIOS*

Reservoir not filling every year. Table usually applies to streams east of Mississippi, Oregon, and Washington.

Per cent of Mean Flow Used	Storage Ratio									Deduction for 30 days' Ground Storage
	c.v. = 0.20	c.v. = 0.22	c.v. = 0.24	c.v. = 0.26	c.v. = 0.28	c.v. = 0.30	c.v. = 0.35	c.v. = 0.40	c.v. = 0.45	
95	1.21	1.33	1.46	1.60	1.74	1.90	2.30	2.70	3.10	0.078
90	0.85	0.92	1.00	1.09	1.20	1.31	1.60	1.88	2.20	0.074
85	0.66	0.71	0.77	0.83	0.91	1.00	1.23	1.47	1.70	0.070
80	0.54	0.57	0.61	0.66	0.71	0.78	0.97	1.19	1.39	0.066
75	0.45	0.47	0.50	0.53	0.57	0.62	0.77	0.95	1.13	0.062
70	0.39	0.40	0.41	0.44	0.47	0.50	0.62	0.76	0.92	0.058
65	0.35	0.35	0.35	0.37	0.39	0.41	0.50	0.61	0.74	0.053
60	0.31	0.31	0.31	0.32	0.33	0.34	0.40	0.49	0.60	0.049
55	0.27	0.27	0.27	0.27	0.28	0.28	0.33	0.39	0.49	0.045
50	0.23	0.23	0.23	0.23	0.23	0.24	0.26	0.32	0.39	0.041

TABLE C – STORAGE RATIOS*

Reservoir not filling every year. Table usually applies to streams west of Mississippi, except Oregon, and Washington.

Per cent of Mean Flow Used	Storage Ratio								
	c.v. = 0.50	c.v. = 0.60	c.v. = 0.70	c.v. = 0.80	c.v. = 0.90	c.v. = 1.00	c.v. = 1.10	c.v. = 1.20	c.v. = 1.50
90	3.00	3.80	4.70	5.60	6.40	–	–	–	–
85	2.30	3.00	3.70	4.50	5.30	6.10	7.00	–	–
80	1.85	2.40	3.10	3.70	4.40	5.10	5.90	6.70	9.30
75	1.55	2.00	2.60	3.15	3.70	4.40	5.00	5.70	8.10
70	1.28	1.70	2.20	2.70	3.20	3.80	4.40	5.00	7.20
65	1.05	1.44	1.85	2.30	2.85	3.40	3.90	4.50	6.50
60	0.89	1.21	1.60	2.00	2.50	3.00	3.50	4.00	6.00
55	0.74	1.02	1.35	1.75	2.20	2.65	3.10	3.60	5.50
50	0.61	0.86	1.15	1.50	1.90	2.35	2.80	3.25	5.00
45	0.51	0.72	0.98	1.30	1.70	2.10	2.50	2.90	4.40
40	0.42	0.61	0.84	1.12	1.45	1.80	2.15	2.50	3.80
35	0.34	0.51	0.72	0.96	1.22	1.50	1.80	2.15	3.30
30	0.27	0.42	0.61	0.80	1.00	1.25	1.50	1.80	2.75

TABLE D – STORAGE RATIOS* See Note 3 at right

Per cent of Mean Flow Used	Storage Ratio			
	Impervious Soils. No Ground Storage	Average Soils. 30 days' Ground Storage	Deep Gravel and Sand. 60 days' Ground Storage	Greatest Natural Storage. 90 days' Ground Storage
50	0.229	0.188	0.147	0.106
45	0.192	0.155	0.118	0.081
40	0.159	0.126	0.093	0.060
35	0.128	0.099	0.070	0.042
30	0.098	0.073	0.049	0.024
25	0.072	0.052	0.031	0.010
20	0.048	0.032	0.015	0
15	0.029	0.017	0.004	0
10	0.014	0.006	0	0

"COEFFICIENT OF VARIATION" METHOD.

1. Compute the "Coefficient of Variation" = C.V.

$$C.V. = \frac{1}{Q_m} \times \sqrt{\frac{(\text{numerical diff. of } Q_m \,\&\, Q_1)^2 + (\text{num. diff. of } Q_m \,\&\, Q_2)^2 + \text{etc.}}{n-1}}$$

in which Q_1, Q_2, Q_3, etc. = total runoff in years 1, 2, 3, etc. (from local data).

n = number of years; $Q_m = \frac{Q_1 + Q_2 + Q_3, \text{etc.}}{n}$

Minimum C.V. to be used = .20.

2. Use Tables B and C as explained below.

EXAMPLE: Given: Location in North Carolina, demand 35 m.g.d., 48 Sq. miles mountainous country (little ground storage), mean annual runoff = 25 in., C.V. = .20.

Required: Total volume of storage.

Solution: Total mean annual runoff = 48 × 25 × 17.39 (a constant to convert in. per sq. mi. to m.g. per year) = 20,868 m.g. per year or 57.1 m.g.d. The % of mean flow used = 35/57.1 = 61%. Use Table B and interpolate to find storage ratio = 0.318. Required storage = 0.318 × 20,868 = 6636 m.g. Add evaporation losses (Use Fig. A - Pg. 20-08 to find total volume of storage required.

3. Use Table D if (1) the % of mean flow used is less than 50%, (2) C.V. from .20 to .25, (3) location is North of Potomac and East of Alleghenies.

GROUND STORAGE.

Table D shows effects of underlying soils on ground storage if per cent of mean flow used varies from 50% to 10%; Table B shows effects from 95% to 50%. These effects (reductions) can be applied to any storage ratio found in Table B or C.

EXAMPLE: Given: Drainage area of deep gravel and sand; storage ratio = .84.

Required: Corrected storage ratio for ground storage.

Solution: In Table D, at left, opposite 45% subtract .118 from .192 = .074, subtract 0.74 from .84 to get storage ratio corrected for ground storage.

It is common practice to disregard the effect of ground water in decreasing the required storage; the presence of ground water then becomes a factor of safety.

*Data from American Civil Engineers' Handbook by Merriman & Wiggin.

† Adapted from Flinn, Weston & Bogert, Waterworks Handbook, Mc Graw-Hill

WATER SUPPLY—MASS DIAGRAM

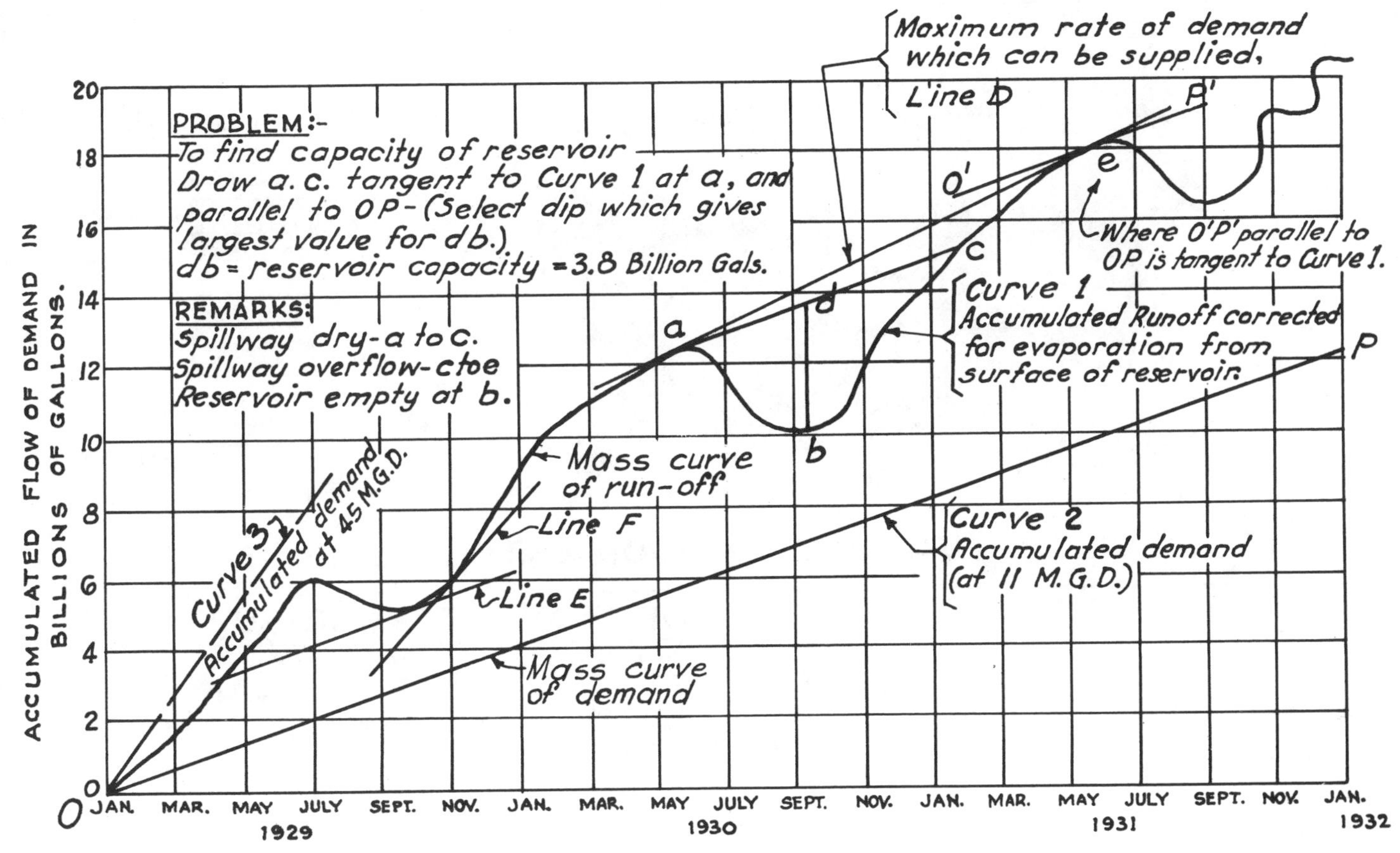

Maximum rate of demand which can be supplied,
Line D
P'
O'
e
Where O'P' parallel to OP is tangent to Curve 1.
PROBLEM:-
To find capacity of reservoir Draw a.c. tangent to Curve 1 at a, and parallel to OP-(Select dip which gives largest value for db.)
db = reservoir capacity = 3.8 Billion Gals.
REMARKS:
Spillway dry-a to c.
Spillway overflow-c to e
Reservoir empty at b.
a
b
c
d
Curve 1
Accumulated Runoff corrected for evaporation from surface of reservoir.
P
Mass curve of run-off
Line F
Line E
Curve 3 Accumulated demand at 45 M.G.D.
Curve 2
Accumulated demand (at 11 M.G.D.)
Mass curve of demand
ACCUMULATED FLOW OF DEMAND IN BILLIONS OF GALLONS.
0
2
4
6
8
10
12
14
16
18
20
O
JAN. MAR. MAY JULY SEPT. NOV. JAN. MAR. MAY JULY SEPT. NOV. JAN. MAR. MAY JULY SEPT. NOV. JAN.
1929
1930
1931
1932

NOTES:-*

1. Impounding reservoir is required when the demand for water is greater than the minimum rate of flow in the stream from which the water is taken. The capacity of the impounding reservoir needed to supply the demand and other data can be determined by the study of mass diagrams of the demand and runoff for the years of minimum flow.
2. Curve 1.- The ordinates represent the total accumulated runoff from the start of observations to the time of the corresponding abscissas, less the sum of the accumulated evaporation from the reservoir and other reservoir losses such as leakage and percolation.
3. Curve 2.- Demand is assumed to be at constant rate.
4. The greatest rate of demand which can be supplied by the stream is represented by the slope of the line, above the Curve 1, which is tangent to it at two or more points but does not intersect Curve 1 at any point. This is shown as Line D on the figure and represents a rate of 15 million gallons per day.
5. If the demand is so great that the reservoir will never fill, a line drawn tangent to and below the runoff curve and parallel to the demand curve will not intersect the runoff curve to the left of the point of tangency. Line E shows that the reservoir will fill for Curve 2. Line F shows that the reservoir will not fill for Curve 3.

*Adapted from Urquhart, Civil Engineering Handbook, Mc Graw-Hill.

WATER SUPPLY – EVAPORATION

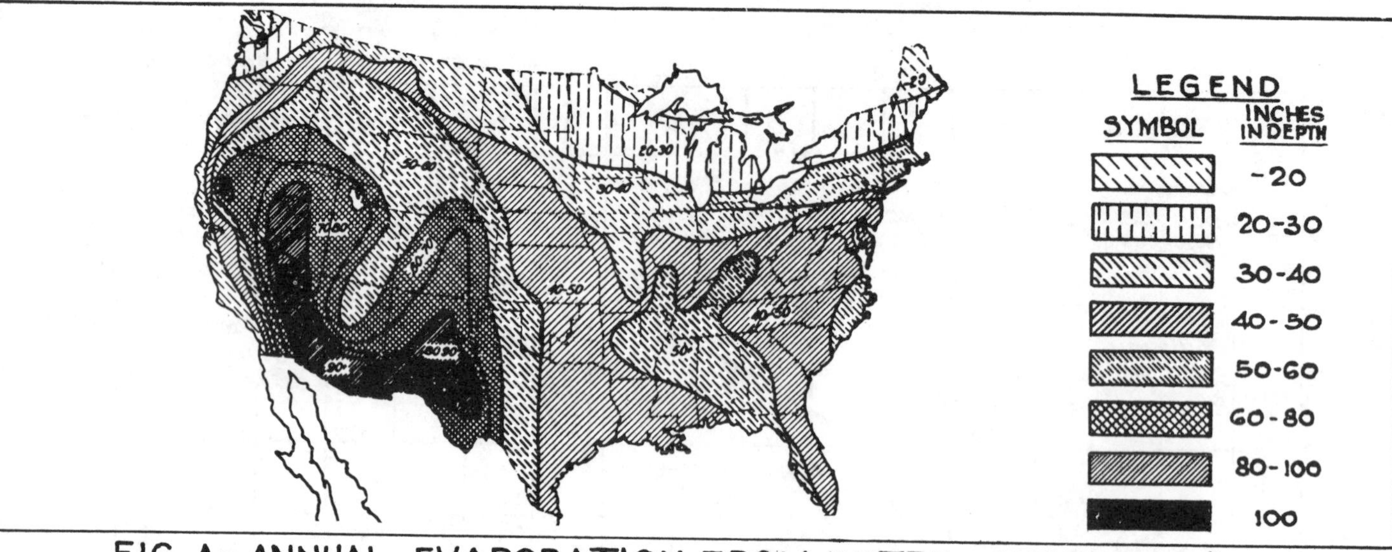
LEGEND
SYMBOL
INCHES IN DEPTH
-20
20-30
30-40
40-50
50-60
60-80
80-100
100

FIG. A- ANNUAL EVAPORATION FROM WATER SURFACE. *

TABLE B-RECORDED EVAPORATION FROM RESERVOIR SURFACES**

Station	Elevation, feet	Years	Temperature of air, degrees Fahrenheit	Wind velocity: Nearby U.S. Weather Bureau Station		Wind velocity: At the pan, miles per hour	Relative humidity, per cent	Reservoir surface evaporation, inches			Percentage range of annual evaporation	Coefficient for pan
				Height, feet	Velocity, miles per hour			April to September	Annual	Maximum per month		
United States:												
Ithaca, N.Y.	800	1918 to 1930	47	100	9.1	1.8	78	17.11	22.54	4.17	90 to 114	0.69
Washington, D.C.	280	1915 to 1917	54	85	6.5	2.3	69	23.52	34.53	4.87	97 to 103	0.69
Chapel Hill, N.C.	500	1921 to 1930	61	...	....	1.1	69	20.03	28.56	4.71	93 to 109	0.69
Madison, Wis.	860	1908 to 1911	46	78	10.0	...	75	12.91	19.82	3.04	97 to 102	0.83
Columbus, Ohio	763	1918 to 1930	52	230	10.7	2.0	74	19.21	26.81	4.94	89 to 108	0.69
Columbus, Mo.	750	1916 to 1927	54	84	8.0	1.5	71	20.28	28.13	5.31	88 to 114	0.69
Grand Forks, N.D.	820	1905 to 1920	38	58	9.8	...	79	21.72	27.07	5.82	90 to 117	0.83
Rapid City, S.D.	3,240	1916 to 1921	46	...	....	2.0	58	25.61	36.43	7.02	92 to 112	0.69
Lincoln, Neb.	1,250	1917 to 1930	51	81	10.0	4.1	69	32.06	42.04	9.92	87 to 118	0.69
Mitchell, Neb.	4,080	1911 to 1929	47	...	....	7.0	..	34.17	41.78	9.28	83 to 118	0.44
Lawrence, Kan.	825	1916 to 1919	55	...	....	4.4	..	31.25	44.80	7.62	....	0.69
Manhattan, Kan.	1,010	1924 to 1929	54	...	....	3.4	..	30.14	42.65	8.06	91 to 112	0.69
Austin, Tex.	475	1916 to 1929	68	148	7.5	2.0	68	28.72	42.47	8.49	77 to 116	0.83
Amarillo, Tex.	3,680	1907 to 1919	56	49	13.0	9.6	56	49.00	66.00	12.63	84 to 109	0.94
Denver, Colo.	5,340	1916	49	...	7.8	4.5	64	38.00	52.15	8.20	....	0.99
Salt Lake, Utah	4,250	1928 to 1930	52	203	6.2	3.6	50	40.67	50.94	10.10	95 to 103	0.69
Yuma, Ariz.	127	1917 to 1929	69	54	5.2	1.3	44	36.29	53.45	9.55	91 to 116	0.69
Independence, Calif.	3,800	1909 to 1911	56	28	6.6	...	45	38.64	55.26	8.00	....	0.83
Salton Sea, Calif.	230	1910	73	...	....	4.1	40	66.35	97.10	13.10	....	0.59
Corvallis, Ore.	235	1922 to 1930	52	...	....	1.5		21.88	30.68	5.92	91 to 106	0.69

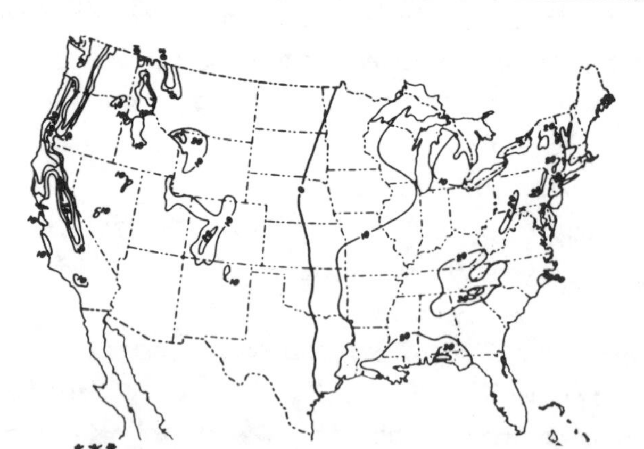

FIG. C-*** Average annual excess of precipitation over the demands of evaporation and transpiration, in Inches. West of the zero line there is generally no annual excess except in mountain areas and in the Pacific North West.

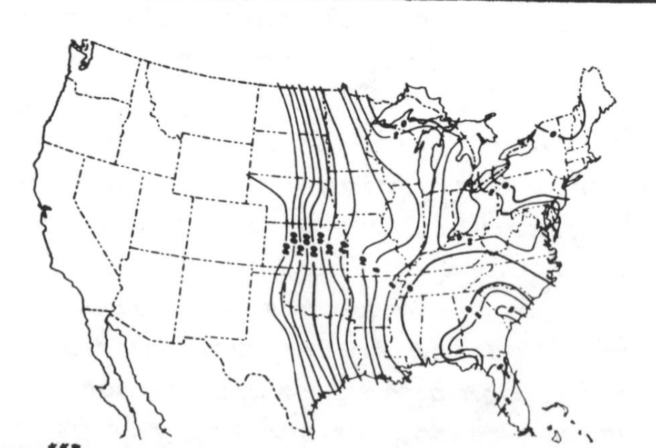

FIG. D***-Percentage of years that annual precipitation has been less than demands for evaporation and transpiration. Throughout the West, except in mountain areas and the Pacific North West, annual demands of evaporation and transpiration, always or nearly always exceed annual precipitation.

* From Public Water Supplies by F.E. Turneaure & H.L. Russell.
** Data from Report-Robert Follansbee-See Trans.- A.S.C.E. Vol. 99 Pg. 708.
*** From Physics of the Earth-IX. Hydrology-Published by Mc Graw-Hill.

WATER SUPPLY—DEEP WELLS

General Notes: 1. Well site should be 200 to 500 feet from possible source of pollution. Locate at elevated point if possible to prevent flooding. Fence in well site.

2. Well building should be fireproof, ventilated and in cold climates insulate thoroughly and provide heat. Elevate above grade to provide drainage away from building.

3. Pumping level is determined by continuous flow test at required well capacity, and the measurement of the resulting drawdown below the static water table.

4. The top of screen should be 50 feet min. below grade to avoid surface pollution unless unusually impervious earth (10 feet of compact clay) occurs at the surface, or well is far removed from possible sources of pollution.

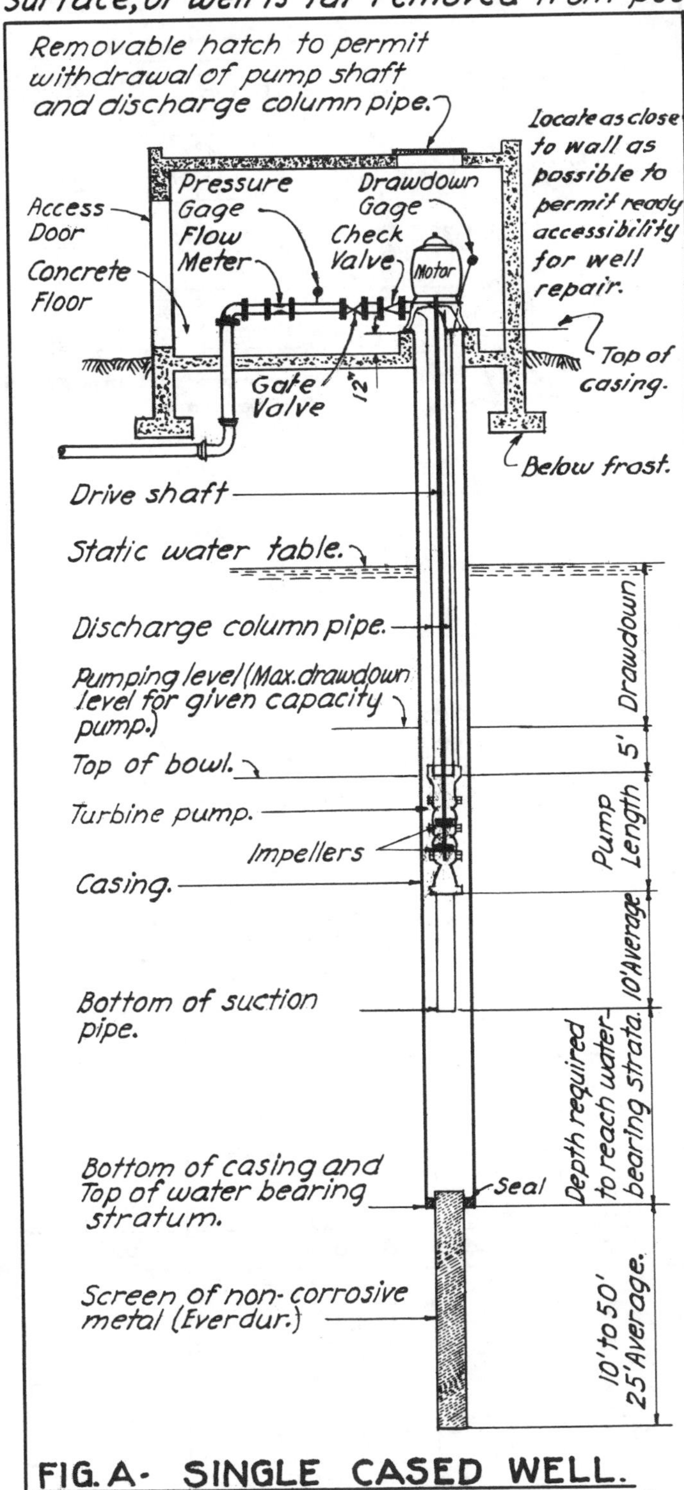
Removable hatch to permit withdrawal of pump shaft and discharge column pipe.
Locate as close to wall as possible to permit ready accessibility for well repair.
Access Door
Pressure Gage
Drawdown Gage
Flow Meter
Check Valve
Motor
Concrete Floor
Top of casing.
Gate Valve
12"
Below frost.
Drive shaft
Static water table.
Discharge column pipe.
Drawdown
Pumping level (Max. drawdown level for given capacity pump.)
5'
Top of bowl.
Turbine pump.
Impellers
Pump Length
Casing.
10' Average
Bottom of suction pipe.
Depth required to reach water-bearing strata.
Bottom of casing and Top of water bearing stratum.
Seal
Screen of non-corrosive metal (Everdur.)
10' to 50' 25' Average.

FIG. A- SINGLE CASED WELL.

For use in ordinary sandy and gravelly soil.

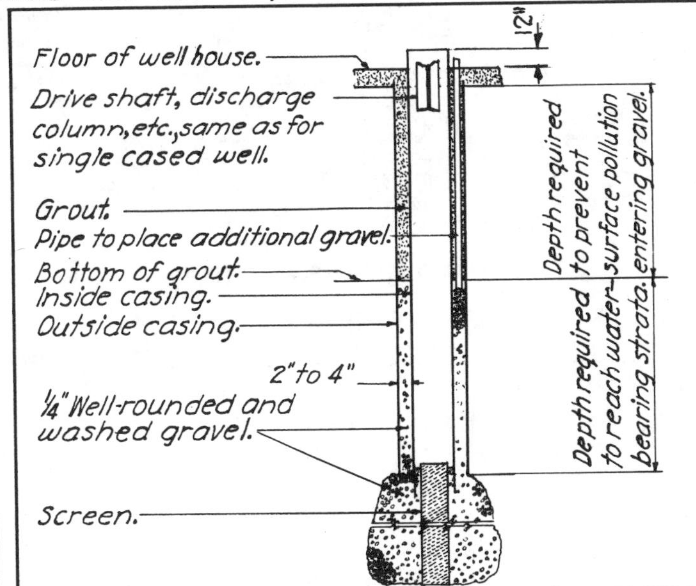
12"
Floor of well house.
Drive shaft, discharge column, etc., same as for single cased well.
Grout.
Pipe to place additional gravel.
Bottom of grout.
Inside casing.
Outside casing.
2" to 4"
¼" Well-rounded and washed gravel.
Screen.
Depth required to prevent surface pollution entering gravel.
Depth required to reach water-bearing strata.

FIG. B-DOUBLE CASED GRAVEL WALL WELL.

For use where water-bearing stratum is of uniform fine sand which is subject to flowing at the velocity the water enters the screen. The gravel provides a much larger area than the screen and the velocity of water at the outer line of gravel may be reduced to such velocity that the sand will not flow.

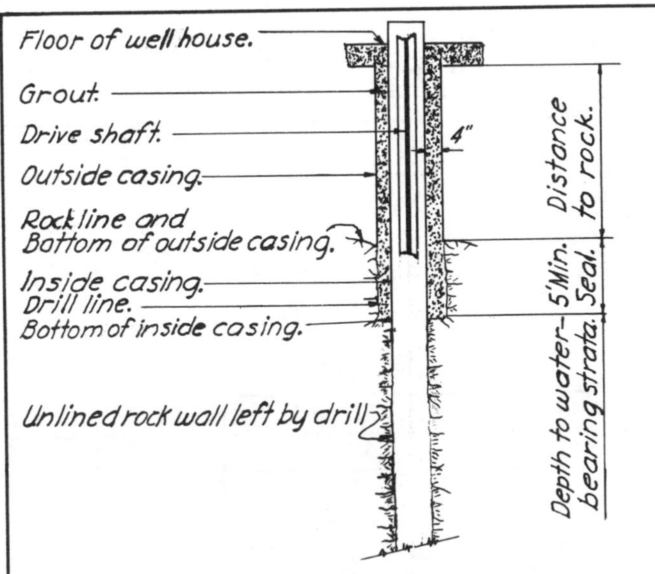
Floor of well house.
Grout.
Drive shaft.
4"
Outside casing.
Rock line and Bottom of outside casing.
Inside casing.
Drill line.
Bottom of inside casing.
Unlined rock wall left by drill.
Distance to rock.
5' Min. Seal.
Depth to water-bearing strata.

Well house, Pumping level, screen and other details are same as for single cased well.

FIG. C- WELL IN ROCK.

WATER SUPPLY—INTAKES, DOMESTIC WELL, WINDMILLS

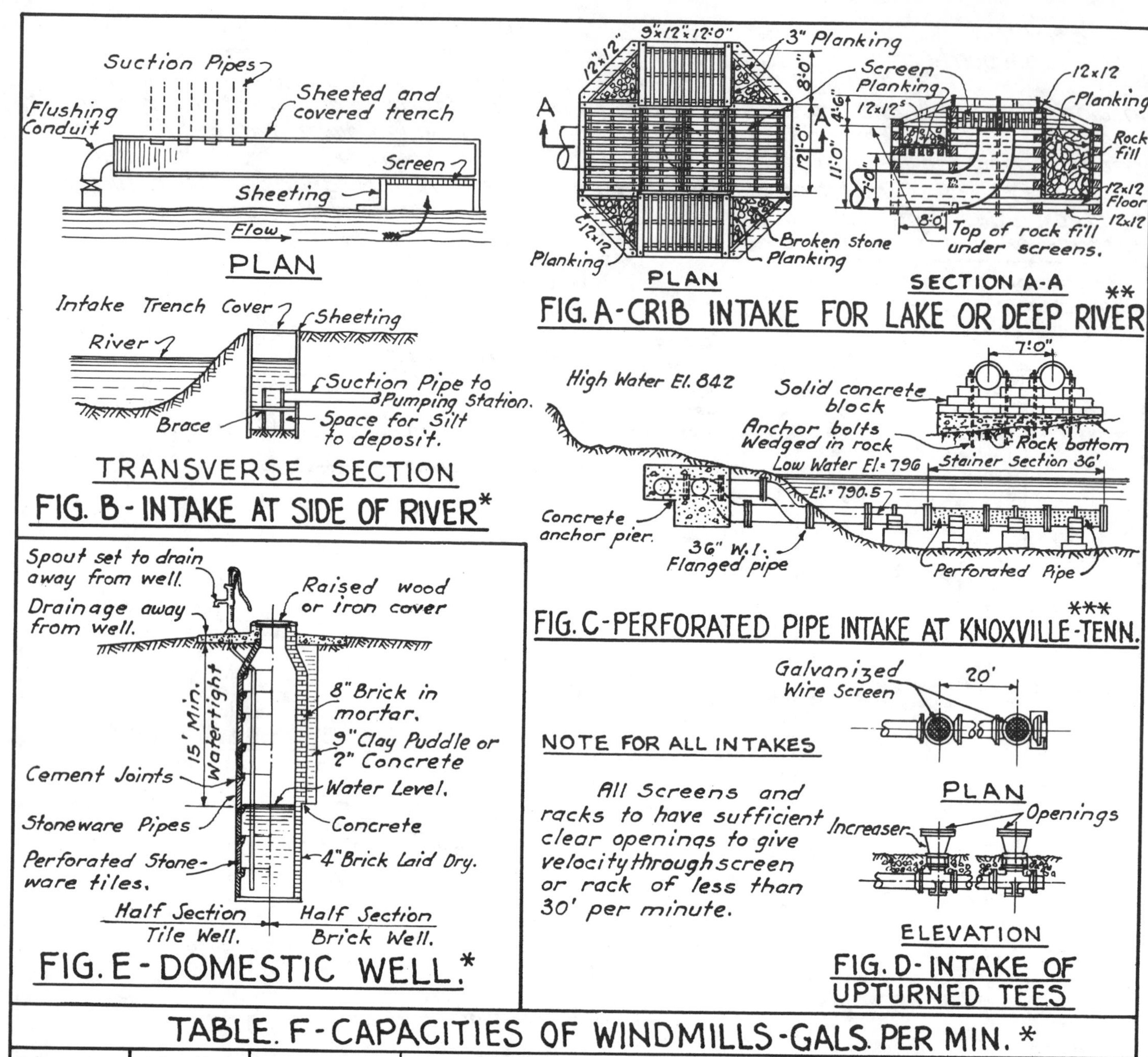

Suction Pipes
Flushing Conduit
Sheeted and covered trench
Screen
Sheeting
Flow
PLAN
Intake Trench Cover
Sheeting
River
Suction Pipe to Pumping Station.
Brace
Space for Silt to deposit.
TRANSVERSE SECTION
FIG. B - INTAKE AT SIDE OF RIVER*
9"x12"x12'0"
12"x12"
3" Planking
8'0"
12'0"
Screen Planking
12x12
Planking
4'6"
11'0"
7'0"
Rock fill
12x12 Floor
12x12
8'0"
Top of rock fill under screens.
Broken Stone
Planking
PLAN
SECTION A-A
FIG. A - CRIB INTAKE FOR LAKE OR DEEP RIVER **
High Water El. 842
7'0"
Solid concrete block
Anchor bolts Wedged in rock
Rock bottom
Low Water El. 796
Stainer Section 36'
El. 790.5
Concrete anchor pier.
36" W.I. Flanged pipe
Perforated Pipe
FIG. C - PERFORATED PIPE INTAKE AT KNOXVILLE-TENN. ***
Spout set to drain away from well.
Drainage away from well.
Raised wood or iron cover
15' Min. Watertight
8" Brick in mortar.
9" Clay Puddle or 2" Concrete
Cement Joints
Water Level.
Stoneware Pipes
Concrete
Perforated Stoneware tiles.
4" Brick Laid Dry.
Half Section Tile Well.
Half Section Brick Well.
FIG. E - DOMESTIC WELL.*
Galvanized Wire Screen
20'
NOTE FOR ALL INTAKES
All Screens and racks to have sufficient clear openings to give velocity through screen or rack of less than 30' per minute.
PLAN
Increaser
Openings
ELEVATION
FIG. D - INTAKE OF UPTURNED TEES

TABLE. F - CAPACITIES OF WINDMILLS - GALS. PER MIN. *

Diameter of blades in feet	Velocity of wind in miles per hour	Revolutions of wheel per minute	ELEVATION - FEET - RAISED 25 ft.	50 ft.	75 ft.	100 ft.	150 ft.	200 ft.	Equivalent actual useful h.p. developed
8½	16	70 to 75	6.19	3.02	------	------	------	-----	0.04
10	16	60 to 65	19.2	9.56	6.64	4.75	-----	-----	0.12
12	16	55 to 60	33.9	17.9	11.8	8.44	5.68	-----	0.21
14	16	50 to 55	45.1	22.6	15.3	11.2	7.81	5.00	0.28
16	16	45 to 50	64.6	31.6	19.5	16.2	9.77	8.08	0.41
18	16	40 to 45	97.7	52.2	32.5	24.4	17.5	12.2	0.61
20	16	35 to 40	125.0	63.7	40.8	31.2	19.3	15.9	0.78
25	16	30 to 35	212.0	107.0	71.6	49.7	37.3	26.7	1.34

*Adapted from Waterworks Handbook by Flinn, Weston & Bogert, McGraw-Hill
**Adapted from Babbitt & Doland, Water Supply Engineering, McGraw-Hill.
***Adapted from Engineering News-Record.

WATER SUPPLY—ELEVATED STORAGE TANKS

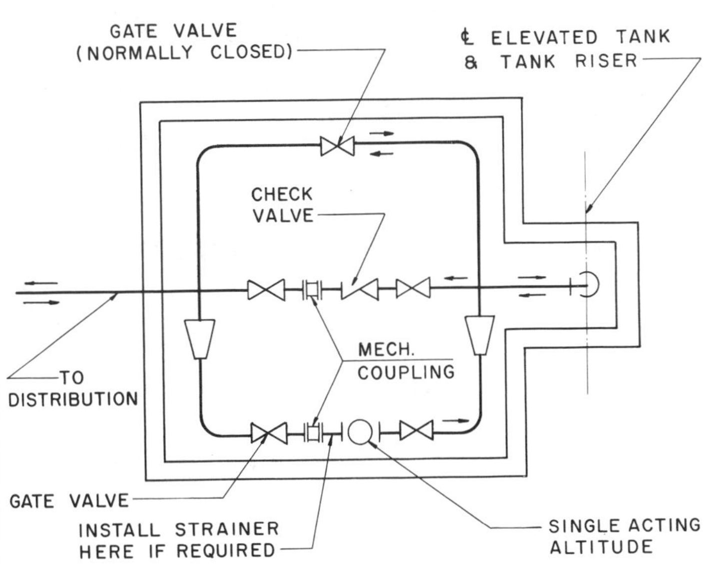
GATE VALVE (NORMALLY CLOSED)
℄ ELEVATED TANK & TANK RISER
CHECK VALVE
MECH. COUPLING
TO DISTRIBUTION
GATE VALVE
INSTALL STRAINER HERE IF REQUIRED
SINGLE ACTING ALTITUDE

FIG. A-TYPICAL VALVE PIT DETAILS

NOTES:
Provisions should be made for heating the valve pit in extremely cold weather. If a heater is not used an additional inside pit cover of 2" plank or its equivalent should be installed at least 4" below the outer cover.
For freezing climates, tank riser pipes are ordinarily 3 ft. to 4 ft. dia. for tanks up to 150,000 gals. and 5ft. to 10 ft. for larger tanks. Heating units are required for this type riser where temperatures reach 5° F. or lower. For general service where water circulates in the tank, heating facilities are not used except in extremely cold climates.
The National Board of Underwriters requires that tanks, located where the water may freeze, shall be adequately heated so that the temperature of the coldest water can be maintained during coldest weather at 42° F.

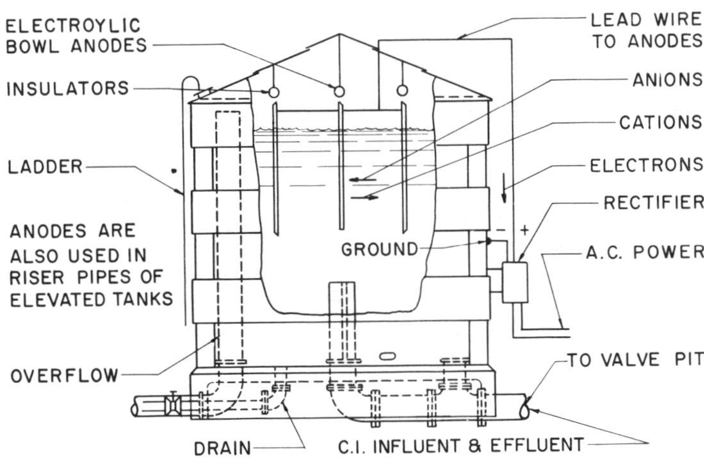
ELECTROYLIC BOWL ANODES
LEAD WIRE TO ANODES
INSULATORS
ANIONS
CATIONS
LADDER
ELECTRONS
RECTIFIER
ANODES ARE ALSO USED IN RISER PIPES OF ELEVATED TANKS
GROUND
A.C. POWER
OVERFLOW
TO VALVE PIT
DRAIN
C.I. INFLUENT & EFFLUENT

FIG. B - TYPICAL STANDPIPE* & CATHODIC PROTECTION+ INSTALLATION

NOTES:
Standpipe may be constructed also of concrete. If concrete is used, cathodic protection is not required.
Useful storage capacity is volume above elevation required to provide adequate pressure in distribution system.
Operation of the cathodic protection system causes the inside wetted surof the steel tank to be coated with a film of hydrogen which prevents oxygen from coming in contact with the steel. The process takes place when low-voltage current is supplied and direct current passes continuously from anodes immersed in the water to the side and bottom tank plates acting as cathodes. Cathodic protection elements immersed in the water inside of tank should be removed during freezing weather.

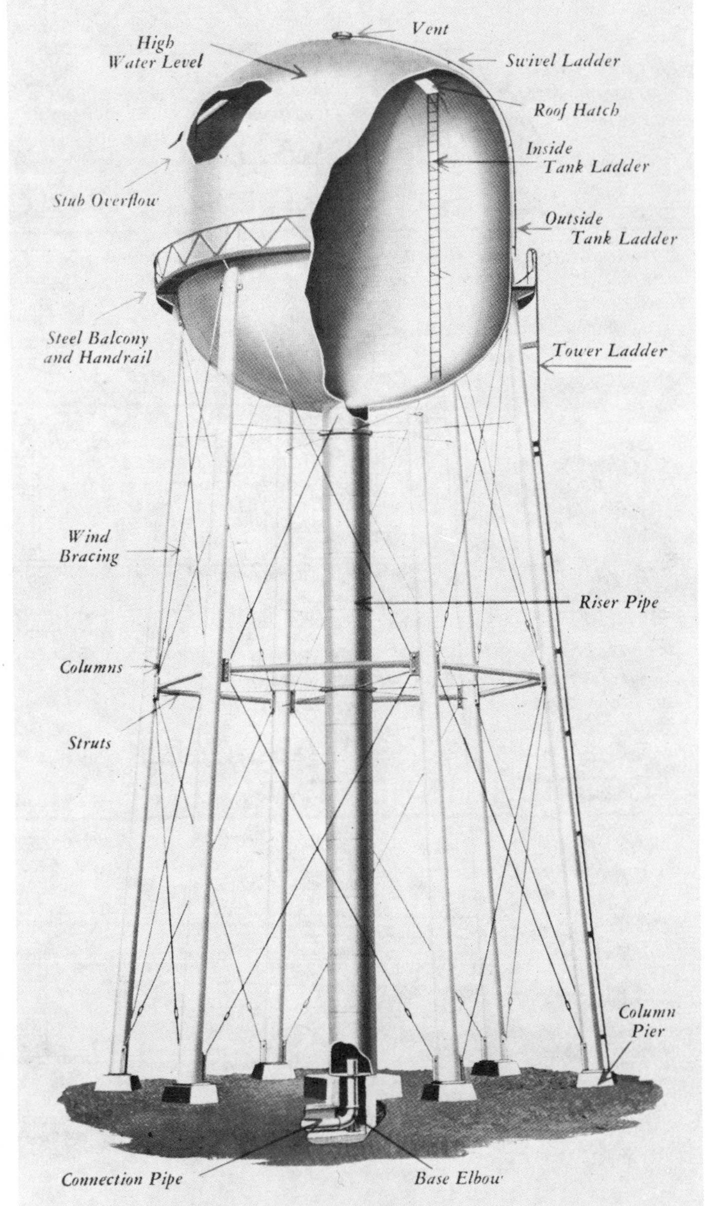
High Water Level
Vent
Swivel Ladder
Roof Hatch
Inside Tank Ladder
Stub Overflow
Outside Tank Ladder
Steel Balcony and Handrail
Tower Ladder
Wind Bracing
Riser Pipe
Columns
Struts
Column Pier
Connection Pipe
Base Elbow

FIG. C - PARTS OF ELEVATED STORAGE TANK

* From Babbit & Doland, Water Supply Engineering, McGraw - Hill.
\+ Adapted from Electro - Rust Proofing Corp., N.J. See also p. 16-50 to 16-53

** From Pittsburgh - Des Moines Steel Co. Bulletin No. 101.

WATER SUPPLY— DISTRIBUTING RESERVOIRS—1

Note:- Inlet & Outlet arranged so that water circulates thru the reservoir.

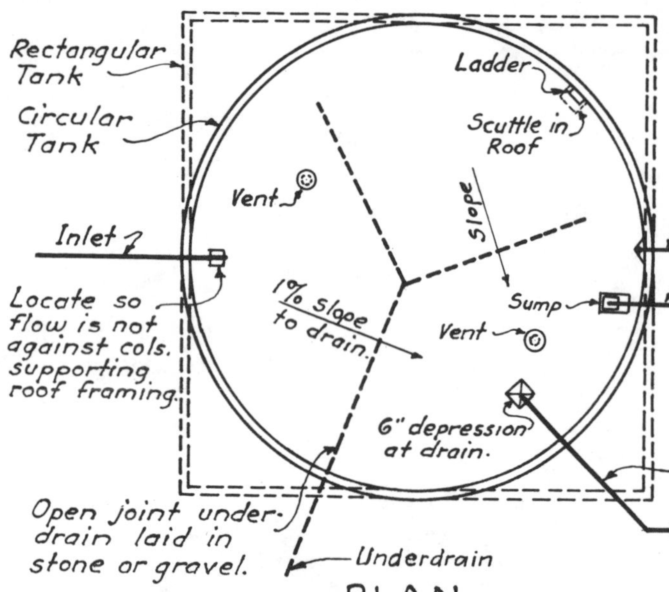

Rectangular Tank
Circular Tank
Ladder
Scuttle in Roof
Vent
Slope
Inlet
Overflow
1% Slope to drain.
Sump
Outlet to distributing system or pumping station.
Vent
Locate so flow is not against cols. supporting roof framing.
6" depression at drain.
Blow off or drain - may be combined with overflow.
Open joint underdrain laid in stone or gravel.
Underdrain
Valve

PLAN

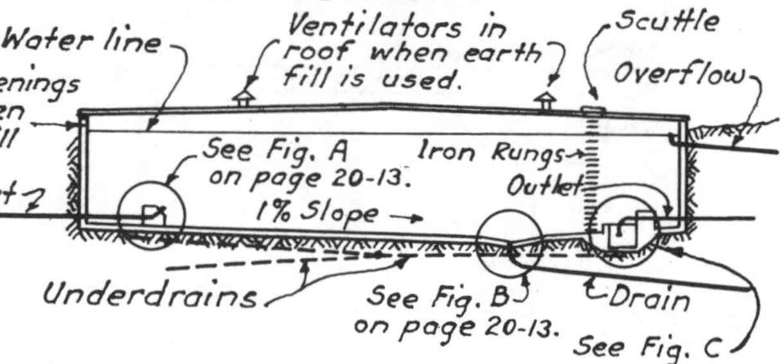

Water line
Ventilators in roof when earth fill is used.
Scuttle
Overflow
Screened openings at sides when no earth fill on roof.
See Fig. A on page 20-13.
Iron Rungs
Outlet
Inlet
1% Slope
Underdrains
See Fig. B on page 20-13.
Drain
See Fig. C on page 20-13.

SECTION SHOWING PIPING

REQUIRED PIPING FOR RESERVOIRS

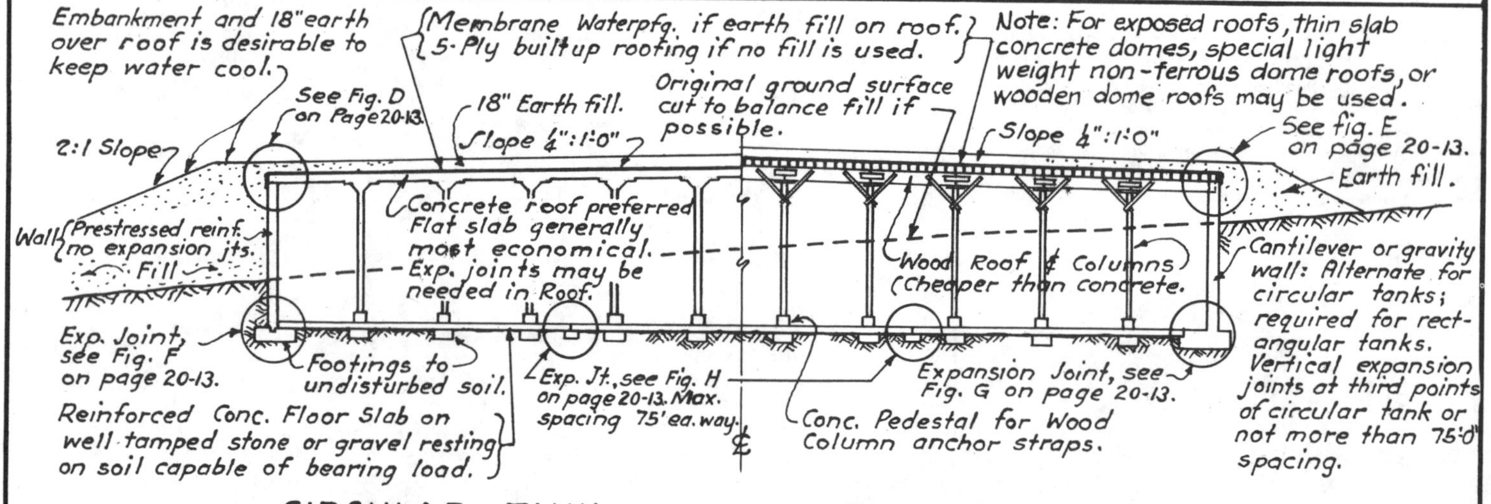

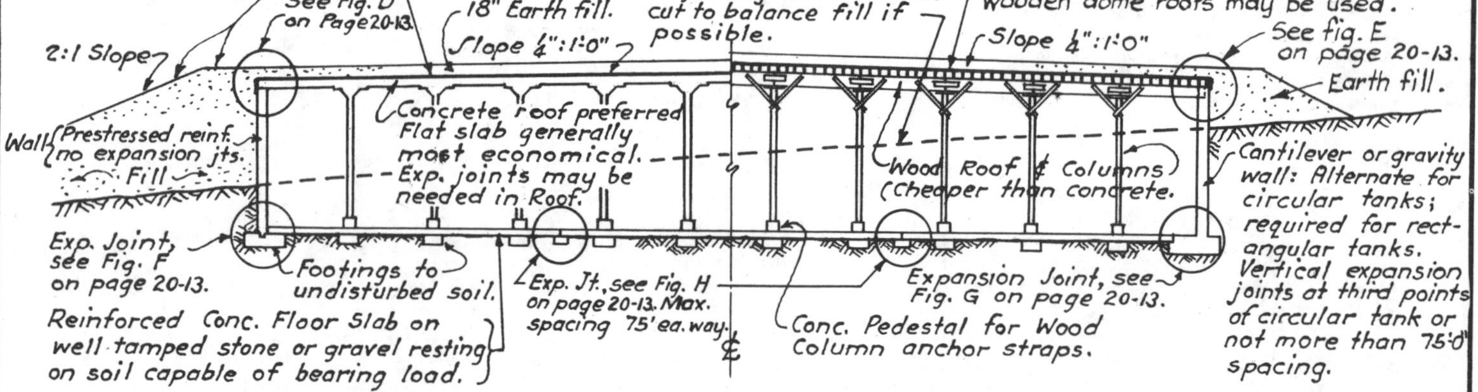
Embankment and 18" earth over roof is desirable to keep water cool.
Membrane Waterpfg. if earth fill on roof.
5-Ply built up roofing if no fill is used.
Note: For exposed roofs, thin slab concrete domes, special light weight non-ferrous dome roofs, or wooden dome roofs may be used.
See Fig. D on Page 20-13.
18" Earth fill.
Original ground surface cut to balance fill if possible.
2:1 Slope
Slope ¼":1'0"
Slope ¼":1'0"
See fig. E on page 20-13.
Earth fill.
Concrete roof preferred Flat slab generally most economical. Exp. joints may be needed in Roof.
Wall
Prestressed reinf. no expansion jts.
Fill
Wood Roof & Columns (Cheaper than concrete.
Cantilever or gravity wall: Alternate for circular tanks; required for rectangular tanks. Vertical expansion joints at third points of circular tank or not more than 75'0" spacing.
Exp. Joint, see Fig. F on page 20-13.
Footings to undisturbed soil.
Exp. Jt., see Fig. H on page 20-13. Max. spacing 75' ea. way.
Expansion Joint, see Fig. G on page 20-13.
Reinforced Conc. Floor Slab on well tamped stone or gravel resting on soil capable of bearing load.
Conc. Pedestal for Wood Column anchor straps.
℄

CIRCULAR TANK RECTANGULAR TANK

SECTION THRU TANK TYPE RESERVOIR - CIRCULAR OR RECTANGULAR

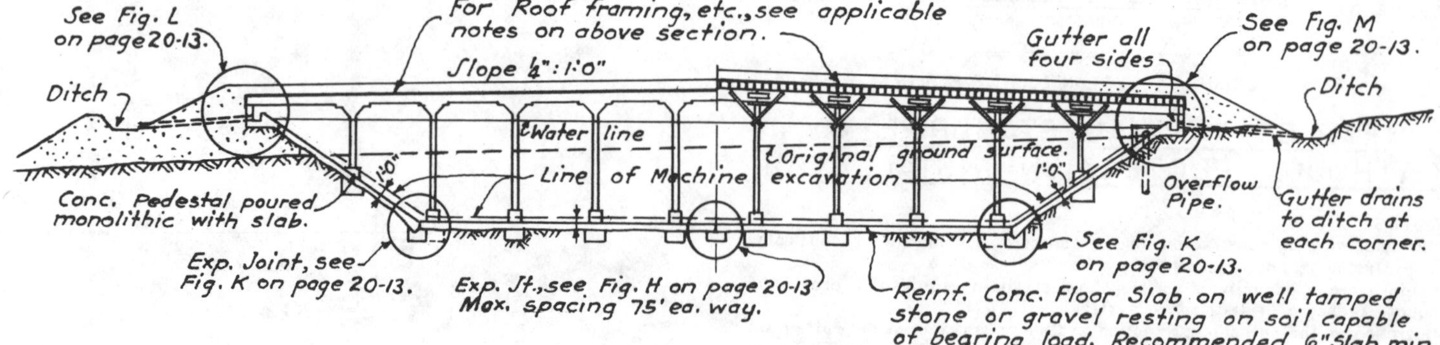
See Fig. L on page 20-13.
For Roof framing, etc., see applicable notes on above section.
Gutter all four sides
See Fig. M on page 20-13.
Slope ¼":1'0"
Ditch
Ditch
Water line
Original ground surface.
Line of Machine excavation
1'0"
1'0"
Overflow Pipe.
Gutter drains to ditch at each corner.
Conc. Pedestal poured monolithic with slab.
Exp. Joint, see Fig. K on page 20-13.
Exp. Jt., see Fig. H on page 20-13 Max. spacing 75' ea. way.
See Fig. K on page 20-13.
Reinf. Conc. Floor Slab on well tamped stone or gravel resting on soil capable of bearing load. Recommended 6" Slab min.

SECTION THRU PAVED CAVITY TYPE RESERVOIR

WATER SUPPLY — DISTRIBUTING RESERVOIRS — 2

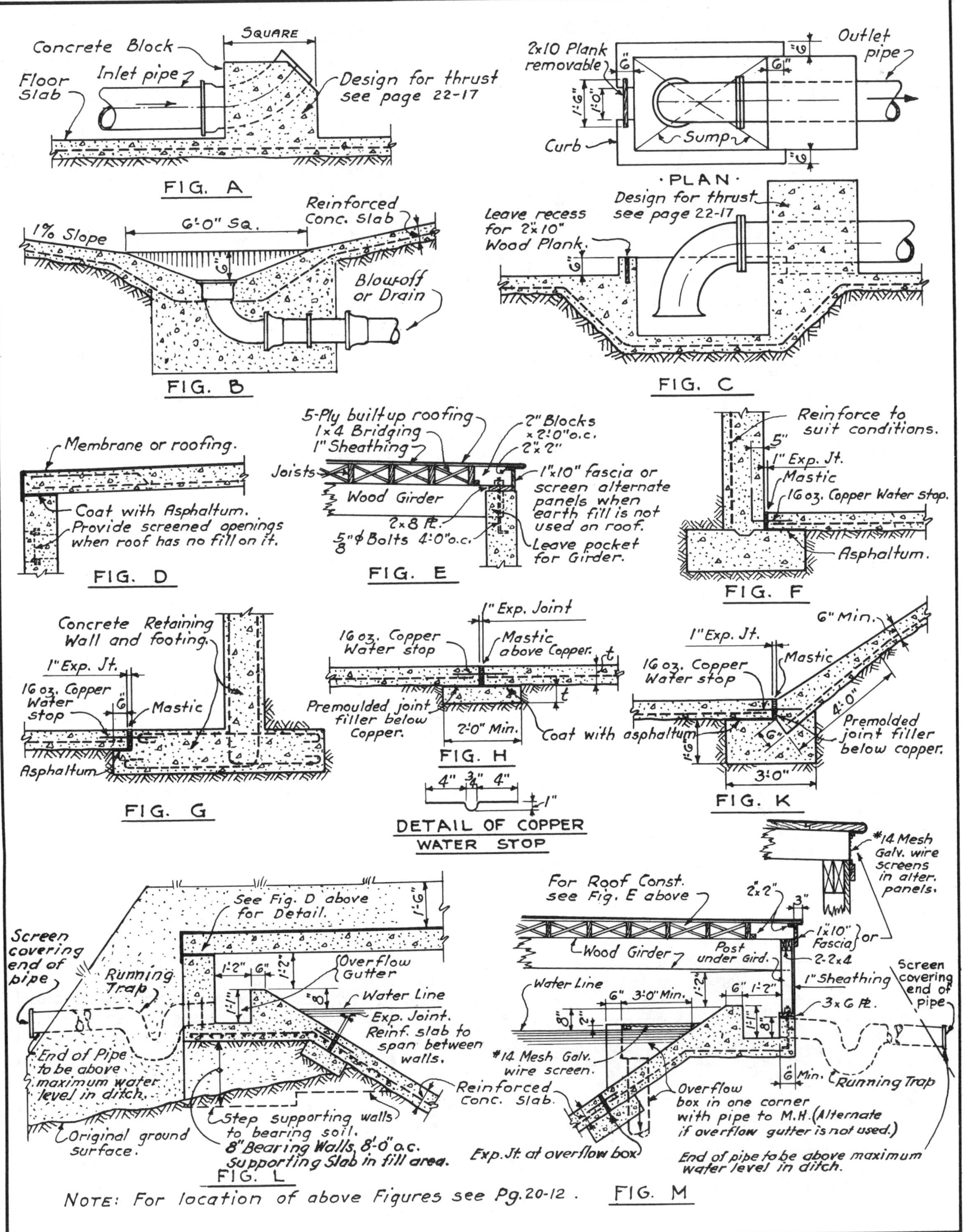

Square
Concrete Block
Floor Slab
Inlet pipe
Design for thrust see page 22-17
FIG. A
2x10 Plank removable
Outlet pipe
Curb
Sump
·PLAN·
Design for thrust see page 22-17
Leave recess for 2x10" Wood Plank.
1% Slope
6'-0" Sq.
Reinforced Conc. Slab
Blowoff or Drain
FIG. B
FIG. C
5-Ply builtup roofing
1x4 Bridging
1" Sheathing
2" Blocks x 2'-0" o.c.
2"x 2"
Joists
Wood Girder
2x8 Pl.
5/8"⌀ Bolts 4'-0" o.c.
1"x10" fascia or screen alternate panels when earth fill is not used on roof.
Leave pocket for Girder.
FIG. E
Membrane or roofing.
Coat with Asphaltum.
Provide screened openings when roof has no fill on it.
FIG. D
Reinforce to suit conditions.
1" Exp. Jt.
Mastic
16 oz. Copper Water stop.
Asphaltum.
FIG. F
Concrete Retaining Wall and footing.
1" Exp. Jt.
16 oz. Copper Water stop
Mastic
Asphaltum
FIG. G
1" Exp. Joint
16 oz. Copper Water stop
Mastic above Copper.
Premoulded joint filler below Copper.
2'-0" Min.
Coat with asphaltum
FIG. H
DETAIL OF COPPER WATER STOP
1" Exp. Jt.
16 oz. Copper Water stop
Mastic
6" Min.
Premolded joint filler below copper.
3'-0"
FIG. K
See Fig. D above for Detail.
Screen covering end of pipe
Running Trap
Overflow Gutter
Water Line
Exp. Joint. Reinf. slab to span between walls.
End of Pipe to be above maximum water level in ditch.
Reinforced Conc. Slab.
Step supporting walls to bearing soil.
Original ground surface.
8" Bearing Walls, 8'-0" o.c. Supporting Slab in fill area.
FIG. L
For Roof Const. see Fig. E above
#14 Mesh Galv. wire Screens in alter. panels.
Wood Girder
Post under Gird.
1"x10" Fascia
or
2-2x4
1" Sheathing
3x6 Pl.
Water Line
3'-0" Min.
Screen covering end of pipe
#14 Mesh Galv. wire screen.
Overflow box in one corner with pipe to M.H. (Alternate if overflow gutter is not used.)
Running Trap
Exp. Jt. at overflow box
End of pipe to be above maximum water level in ditch.
FIG. M

NOTE: For location of above Figures see Pg. 20-12.

WATER SUPPLY – PUMPING

VARIATION IN DEMAND

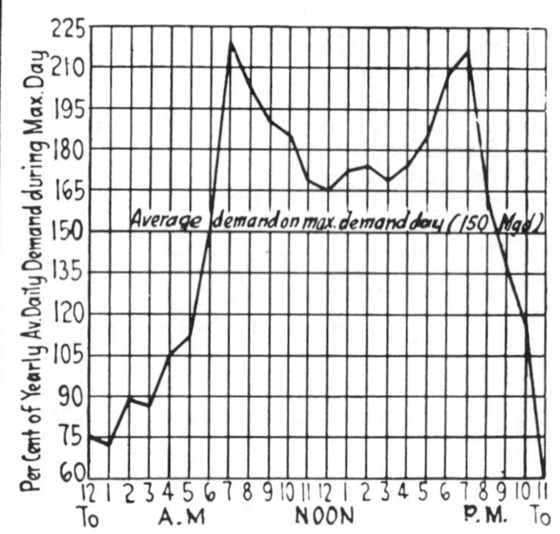
Per Cent of Yearly Av. Daily Demand during Max. Day
Average demand on max. demand day (150 Mgd)
12 To
A.M
NOON
P.M.
11 To

FIG. A–HOURLY IN A MAX. DAY* AT CLEVELAND, OHIO

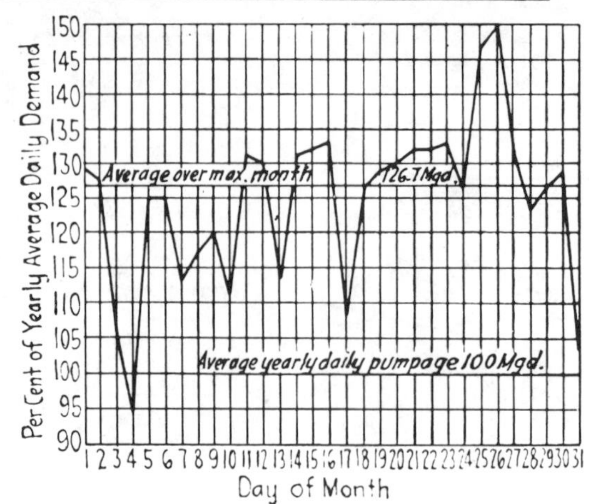
Per Cent of Yearly Average Daily Demand
Average over max. month
126.7 Mgd
Average yearly daily pumpage 100 Mgd.
Day of Month

FIG. B–DAILY IN A MAX. MONTH* AT CLEVELAND, OHIO

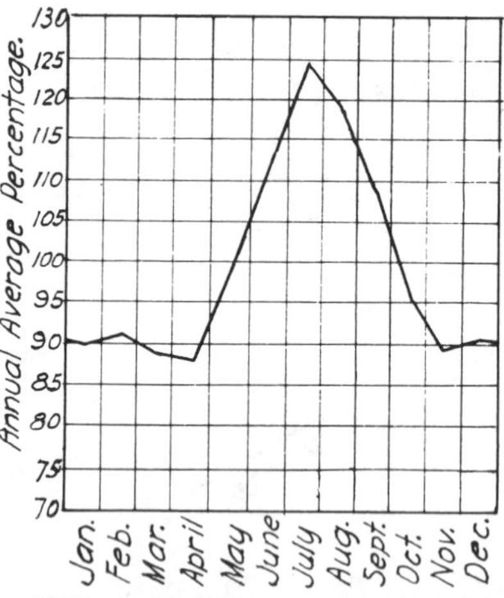
Annual Average Percentage.
Jan.
Feb.
Mar.
April
May
June
July
Aug.
Sept.
Oct.
Nov.
Dec.

FIG. C–ANNUAL VARIATION+ IN MASSACHUSETTS CITIES.

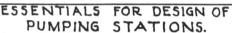

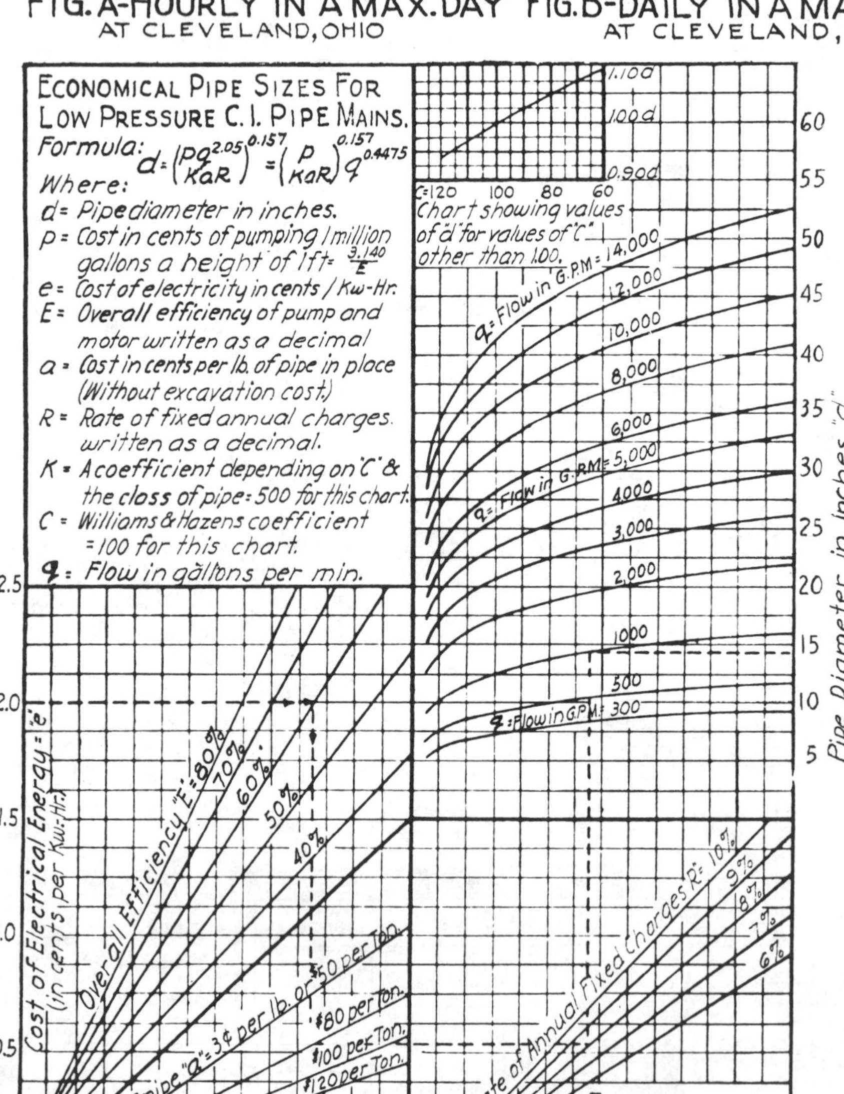
ECONOMICAL PIPE SIZES FOR LOW PRESSURE C.I. PIPE MAINS.
Formula: $d = \left(\frac{pq^{2.05}}{KaR}\right)^{0.157} = \left(\frac{p}{KaR}\right)^{0.157} q^{0.4475}$
Where:
d = Pipe diameter in inches.
p = Cost in cents of pumping 1 million gallons a height of 1 ft = $\frac{3.140}{E}$
e = Cost of electricity in cents / Kw-Hr.
E = Overall efficiency of pump and motor written as a decimal
a = Cost in cents per lb. of pipe in place (Without excavation cost.)
R = Rate of fixed annual charges. written as a decimal.
K = A coefficient depending on "C" & the class of pipe = 500 for this chart.
C = Williams & Hazens coefficient = 100 for this chart.
q = Flow in gallons per min.
Chart showing values of "d" for values of "C" other than 1.00.
q = Flow in G.P.M. = 14,000
q = Flow in G.P.M. = 5,000
q = Flow in GPM = 300
Pipe Diameter in inches "d"
Cost of Electrical Energy = "e" (in cents per Kw-Hr.)
Overall Efficiency "E" = 80%
Cost of pipe "a" = 3¢ per lb. or $60 per Ton.
$80 per Ton.
$100 per Ton.
$120 per Ton.
Rate of Annual Fixed Charges "R" = 10%
EXAMPLE.
Bond Retirement (20 yr) = 5.00%
Interest @ 5% = 2.63%
Depreciation = 0.83%
"R" = 8.46%

FIG. D–CHART FOR DETERMINING ECONOMICAL PIPE SIZES.**

ESSENTIALS FOR DESIGN OF PUMPING STATIONS.

1. Investigate use or construction of storage tanks or reservoirs to allow uniform rates of pumping for long periods of the day.
2. Provide different capacity pumps to obtain maximum efficiencies for variations of demand.
3. Provide in duplicate largest pump and motor unit, or 2 pumps in reserve
4. Provide auxiliary power units (Generally gasoline or Diesel) to meet maximum combined fire and domestic draft.
5. Design for no suction lift if possible.
6. Provide a separate suction intake for each pump if possible. If not, provide tapering header with "Y" branches.
7. Provide eccentric increasers on suction side of pump, if increasers are necessary. Provide exactly horizontal suction pipe, or slope down to intake. Avoid high spots in suction pipe to prevent air entrapment.
8. Provide flexible couplings to facilitate placing and replacing flanged pipe.
9. Provide check valves on discharge side of each pump and gate valves on both sides of each pump.
10. All equipment accessible to an overhead crane or arranged so that it can be removed and replaced without disturbing other pieces of equipment.
11. Provide tie-rods and thrust blocks for pipe.
12. Provide a fireproof, well lighted and ventilated building.

+ From Journal of N.E.W.W.A. Vol. 27, page 85, 1913.
* Data From Flinn, Weston & Bogert, Waterworks Handbook, McGraw-Hill.
**Chart from Cameron Handbook of Hydraulic Data, Courtesy of Geo. B. Gascoigne.

WATER SUPPLY– PUMPING STATION

ESSENTIALS FOR DESIGN OF PUMPING STATIONS – SEE PAGE 6-14

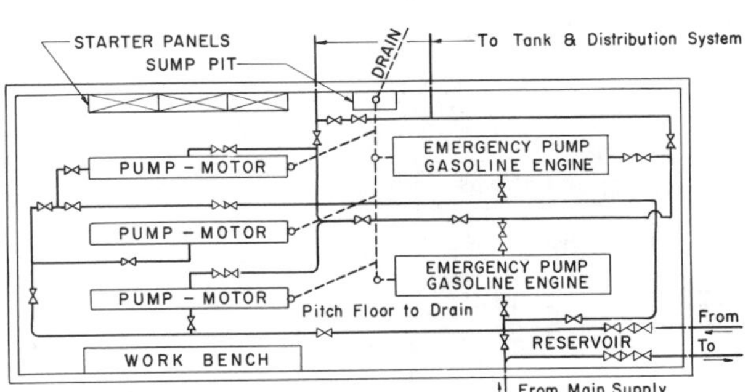
STARTER PANELS
SUMP PIT
DRAIN
To Tank & Distribution System
PUMP – MOTOR
EMERGENCY PUMP GASOLINE ENGINE
PUMP – MOTOR
EMERGENCY PUMP GASOLINE ENGINE
PUMP – MOTOR
Pitch Floor to Drain
From
RESERVOIR
To
WORK BENCH
From Main Supply

FIG. A – TYPICAL LAYOUT WATER PUMPING SYSTEM

DRAIN
SUMP PIT
PRESSURE TANK
AIR COMPRESSOR
NOTE: Air Compressor may be omitted on small jobs and also Pressure Tank may be omitted.
MOTOR PUMP
MOTOR PUMP
EMERGENCY PUMP GASOLINE ENGINE
AIR COMPRESSOR & MOTOR CONTROLS

FIG. B FLOW DIAGRAM – TANK SYSTEM

FIG. A–TYPICAL LAYOUT WATER PUMPING SYSTEM

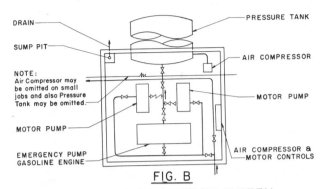

FIG. B–FLOW DIAGRAM – TANK SYSTEM

PNEUMATIC PUMPING SYSTEM

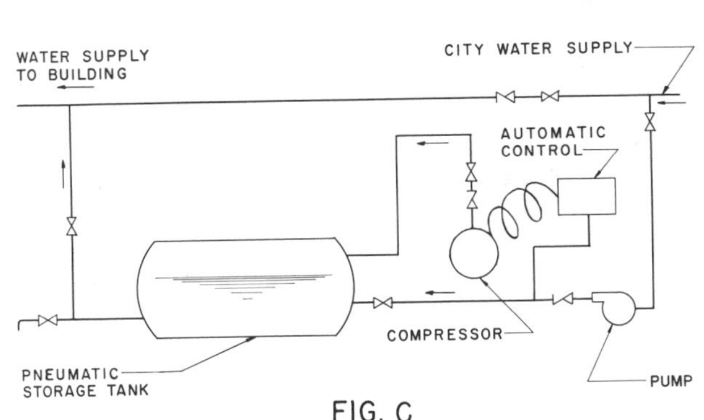
WATER SUPPLY TO BUILDING
CITY WATER SUPPLY
AUTOMATIC CONTROL
COMPRESSOR
PNEUMATIC STORAGE TANK
PUMP

FIG. C FLOW DIAGRAM – TANK SYSTEM

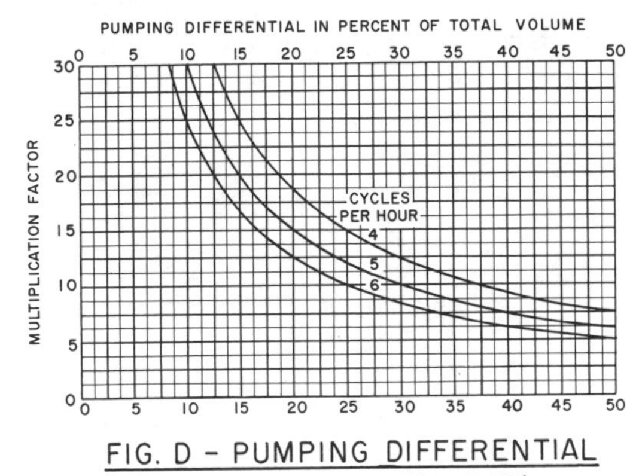
PUMPING DIFFERENTIAL IN PERCENT OF TOTAL VOLUME
0 5 10 15 20 25 30 35 40 45 50
30 25 20 15 10 5 0
MULTIPLICATION FACTOR
CYCLES PER HOUR
4
5
6

FIG. D – PUMPING DIFFERENTIAL (% OF TANK VOLUME)*

FIG. C–FLOW DIAGRAM – TANK SYSTEM

FIG. D–PUMPING DIFFERENTIAL (% OF TANK VOLUME)*

Legend: Gate valve ⋈ Check valve ▷|

PNEUMATIC PUMPING SYSTEM: PROBLEM

Given: Building water requirement of 60 g.p.m., static head from a pressure tank to the highest plumbing fixture of 45 feet, frictional head loss through water piping of 15 feet, and a total dynamic head from the operating water level in a well to the pressure tank of 100 feet. Minimum pressure required at the highest fixture is 10 p.s.i., and operating water system pressure differential is 20 p.s.i.

To Find: Capacity and head for a well pump, and determine capacity of pressure tank.

Solution:

Static head (highest fixture to tank)	= 45 ft.	Differential operating pressure (20 × 2.31)	= 46 ft.
Piping friction loss	= 15 ft.	Tank total working pressure	= 129 ft.
Minimum pressure at fixture 2.31 × 10	= 23 ft.	or 129 × 0.433	= 56 p.s.i.

Therefore tank pressure switch should be selected to operate from 60 p.s.i. cut-off to 40 p.s.i. cut-in.

Well pump should be selected as follows:

Maximum tank working pressure (60 p.s.i. × 2.31)	= 139 ft.
Total dynamic head (tank to water level in well)	= 100 ft.
Total head	= 239 ft.

So select pump with capacity of 60 g.p.m. at 240 ft. total head.

To determine size of pressure tank: Assume that pumping differential is 25% of total tank capacity and that five pumping cycles are desired.

The curve in Fig. D is based on the assumption that the average system demand is equivalent to one-half of the pump capacity. With five cycles per hour and a pumping differential of 25%, a multiplication factor of 12 is found. Total volume of tank is 12 × 60 g.p.m. = 720 gallons. Therefore a tank 120 in. long by 42 in. in diameter may be used.

*From Peerless Pump Division, Food Machinery & Chemical Corp.

WATER SUPPLY—PUMP HORSEPOWER

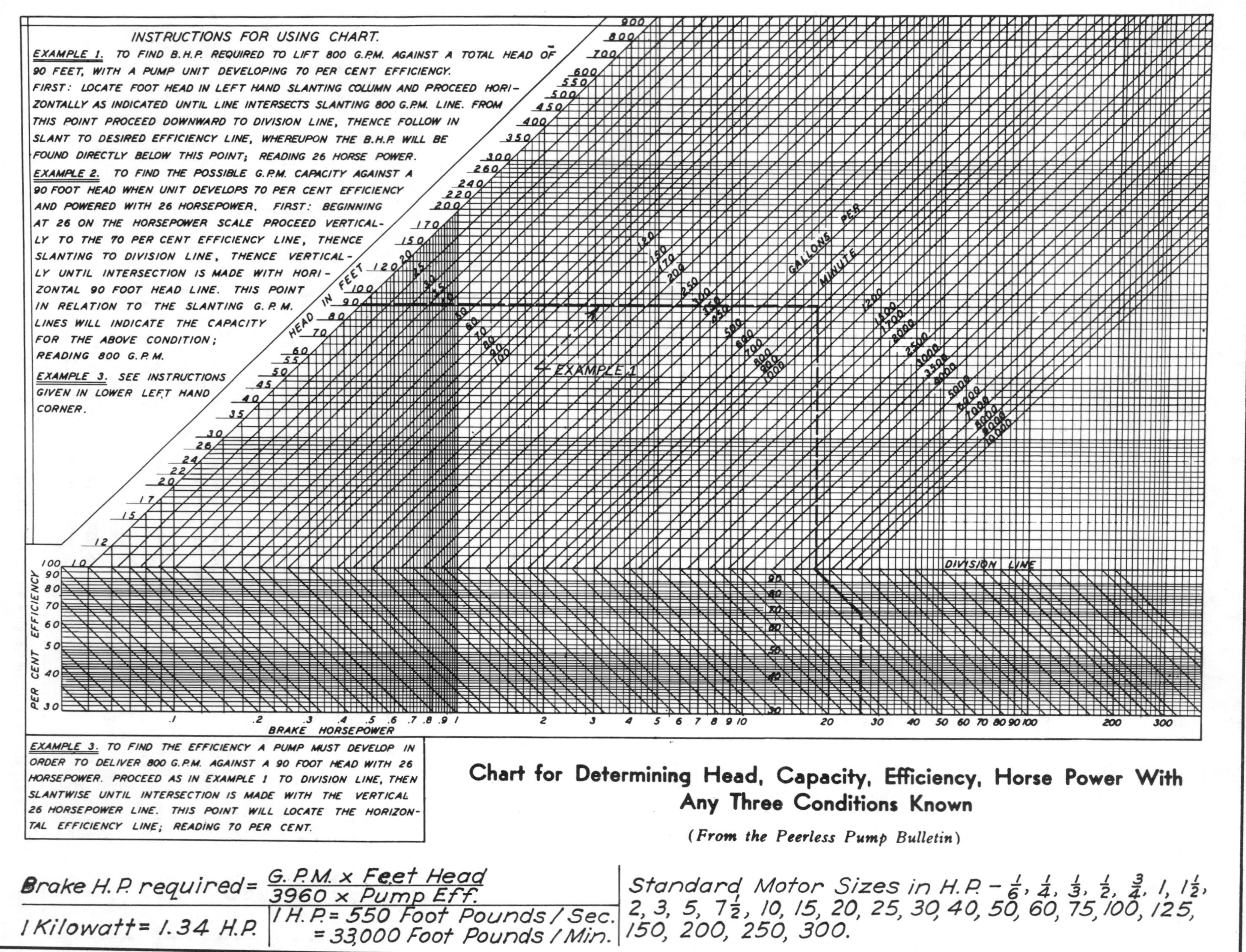
INSTRUCTIONS FOR USING CHART.
EXAMPLE 1. TO FIND B.H.P. REQUIRED TO LIFT 800 G.P.M. AGAINST A TOTAL HEAD OF 90 FEET, WITH A PUMP UNIT DEVELOPING 70 PER CENT EFFICIENCY.
FIRST: LOCATE FOOT HEAD IN LEFT HAND SLANTING COLUMN AND PROCEED HORIZONTALLY AS INDICATED UNTIL LINE INTERSECTS SLANTING 800 G.P.M. LINE. FROM THIS POINT PROCEED DOWNWARD TO DIVISION LINE, THENCE FOLLOW IN SLANT TO DESIRED EFFICIENCY LINE, WHEREUPON THE B.H.P. WILL BE FOUND DIRECTLY BELOW THIS POINT; READING 26 HORSE POWER.
EXAMPLE 2. TO FIND THE POSSIBLE G.P.M. CAPACITY AGAINST A 90 FOOT HEAD WHEN UNIT DEVELOPS 70 PER CENT EFFICIENCY AND POWERED WITH 26 HORSEPOWER. FIRST: BEGINNING AT 26 ON THE HORSEPOWER SCALE PROCEED VERTICALLY TO THE 70 PER CENT EFFICIENCY LINE, THENCE SLANTING TO DIVISION LINE, THENCE VERTICALLY UNTIL INTERSECTION IS MADE WITH HORIZONTAL 90 FOOT HEAD LINE. THIS POINT IN RELATION TO THE SLANTING G.P.M. LINES WILL INDICATE THE CAPACITY FOR THE ABOVE CONDITION; READING 800 G.P.M.
EXAMPLE 3. SEE INSTRUCTIONS GIVEN IN LOWER LEFT HAND CORNER.
HEAD IN FEET
GALLONS PER MINUTE
EXAMPLE 1
DIVISION LINE
PER CENT EFFICIENCY
BRAKE HORSEPOWER
EXAMPLE 3. TO FIND THE EFFICIENCY A PUMP MUST DEVELOP IN ORDER TO DELIVER 800 G.P.M. AGAINST A 90 FOOT HEAD WITH 26 HORSEPOWER. PROCEED AS IN EXAMPLE 1 TO DIVISION LINE, THEN SLANTWISE UNTIL INTERSECTION IS MADE WITH THE VERTICAL 26 HORSEPOWER LINE. THIS POINT WILL LOCATE THE HORIZONTAL EFFICIENCY LINE; READING 70 PER CENT.

Chart for Determining Head, Capacity, Efficiency, Horse Power With Any Three Conditions Known

(From the Peerless Pump Bulletin)

$$\text{Brake H.P. required} = \frac{\text{G.P.M.} \times \text{Feet Head}}{3960 \times \text{Pump Eff.}}$$

1 Kilowatt = 1.34 H.P.

1 H.P. = 550 Foot Pounds / Sec.
= 33,000 Foot Pounds / Min.

Standard Motor Sizes in H.P. – $\frac{1}{6}$, $\frac{1}{4}$, $\frac{1}{3}$, $\frac{1}{2}$, $\frac{3}{4}$, 1, $1\frac{1}{2}$, 2, 3, 5, $7\frac{1}{2}$, 10, 15, 20, 25, 30, 40, 50, 60, 75, 100, 125, 150, 200, 250, 300.

WATER SUPPLY—RECTANGULAR WEIR

NOTE: Weir tables assume no velocity of approach. To correct for velocity of approach add $\frac{V_a^2}{2g}$ to head in feet. (where $g = 32.17$)

Bazins Formula.

$$Q=\left(0.405+\frac{0.00984}{H}\right)\left[1+0.55\frac{H^2}{(P+H)^2}\right]LH\sqrt{2gH}$$

Va = Velocity of approach in feet per sec.

Crest of weir.

Width of weir "L" in feet.

H = Head in feet.

P = Height of weir.

Bottom of channel.

Point where head is measured should be at least 4×H.

This space below overfall (nappe) must have free admission of air. A partial vacuum would increase discharge.

Weir tables can be used only where downstream water surface is below crest of weir.

DISCHARGE VALUES IN CUBIC FEET PER SEC. PER FOOT OF WIDTH OVER RECTANGULAR SHARP-CRESTED SUPPRESSED WEIRS.*

H IN FT.	P=2FT.	P=3FT.	P=4FT.	P=5FT.	P=10FT.	P=20FT.	P=30FT.
0.01	0.01	0.01	0.01	0.01	0.01	0.01	0.01
0.02	0.02	0.02	0.02	0.02	0.02	0.02	0.02
0.04	0.05	0.05	0.05	0.05	0.05	0.05	0.05
0.06	0.07	0.07	0.07	0.07	0.07	0.07	0.07
0.08	0.11	0.11	0.11	0.11	0.11	0.11	0.11
0.10	0.13	0.13	0.13	0.13	0.13	0.13	0.13
0.12	0.17	0.17	0.17	0.17	0.17	0.17	0.17
0.14	0.21	0.21	0.21	0.21	0.21	0.21	0.21
0.16	0.25	0.25	0.25	0.25	0.25	0.25	0.25
0.18	0.29	0.29	0.29	0.29	0.29	0.29	0.29
0.20	0.33	0.33	0.33	0.33	0.33	0.33	0.33
0.22	0.38	0.38	0.38	0.38	0.38	0.38	0.38
0.24	0.43	0.43	0.43	0.43	0.43	0.43	0.43
0.26	0.48	0.48	0.48	0.48	0.48	0.48	0.48
0.28	0.53	0.53	0.53	0.53	0.53	0.53	0.53
0.30	0.58	0.58	0.58	0.58	0.58	0.58	0.58
0.32	0.64	0.64	0.64	0.64	0.64	0.64	0.62
0.34	0.70	0.70	0.70	0.70	0.70	0.69	0.68
0.36	0.76	0.76	0.76	0.76	0.76	0.75	0.74
0.38	0.82	0.82	0.82	0.82	0.81	0.81	0.80
0.40	0.88	0.88	0.88	0.87	0.87	0.87	0.87
0.45	1.06	1.06	1.05	1.05	1.04	1.04	1.03
0.50	1.23	1.22	1.21	1.21	1.21	1.20	1.20
0.55	1.42	1.40	1.39	1.39	1.39	1.39	1.39
0.60	1.62	1.59	1.59	1.58	1.57	1.57	1.57
0.70	2.04	2.01	1.99	1.98	1.97	1.97	1.97
0.80	2.50	2.45	2.43	2.42	2.40	2.40	2.40
0.90	3.00	2.93	2.90	2.88	2.86	2.85	2.85
1.00	3.53	3.44	3.40	3.38	3.34	3.33	3.33
1.10	4.09	3.98	3.92	3.91	3.85	3.84	3.84
1.20	4.68	4.55	4.48	4.47	4.38	4.36	4.36
1.30	5.31	5.15	5.07	5.05	4.94	4.91	4.91
1.40	5.99	5.78	5.68	5.62	5.52	5.49	5.48
1.50	6.68	6.44	6.30	6.23	6.13	6.10	6.09
1.60	7.40	7.12	6.97	6.89	6.74	6.69	6.69
1.70	8.14	7.83	7.66	7.56	7.39	7.33	7.32
1.80	8.93	8.56	8.37	8.25	8.05	7.98	7.96
1.90	9.75	9.32	9.11	8.97	8.74	8.65	8.63
2.00	10.58	10.12	9.87	9.72	9.44	9.34	9.32
2.10	11.45	10.93	10.65	10.48	10.17	10.05	10.02
2.20	12.34	11.77	11.46	11.27	10.91	10.78	10.75
2.30	13.24	12.64	12.29	12.07	11.66	11.52	11.48
2.40	14.20	13.53	13.15	12.91	12.45	12.28	12.24
2.50	15.17	14.45	14.03	13.76	13.26	13.06	13.01
2.60	16.16	15.38	14.92	14.63	14.07	13.85	13.80
2.70	17.18	16.34	15.83	15.52	14.92	14.65	14.60
2.80	18.23	17.32	16.79	16.44	15.76	15.48	15.42
2.90	19.29	18.32	17.77	17.36	16.63	16.33	16.25
3.00	20.39	19.36	18.74	18.33	17.52	17.18	17.10
3.10	21.50	20.40	19.74	19.31	18.42	18.04	17.96
3.20	22.64	21.48	20.77	20.31	19.34	18.93	18.83
3.30	23.81	22.59	21.80	21.33	20.27	19.82	19.73
3.40	24.98	23.70	22.89	22.36	21.24	20.75	20.63
3.50	26.20	24.83	24.00	23.43	22.22	21.69	21.60
3.60	27.41	25.99	25.09	24.49	23.20	22.62	22.48
3.70	28.64	27.17	26.22	25.59	24.20	23.59	23.43
3.80	29.94	28.38	27.38	26.70	25.23	24.56	24.39
4.00	32.54	30.84	29.74	28.99	27.32	26.55	26.35
4.20	35.22	33.39	32.18	31.35	29.48	28.59	28.36
4.40	37.99	36.01	34.70	33.78	31.70	30.66	30.42
4.60	40.83	38.71	37.29	36.29	33.98	32.84	32.53
4.80	43.75	41.49	39.96	38.87	36.33	35.05	34.70
5.00	46.71	44.31	42.67	41.49	38.70	37.28	36.88
5.20	49.81	47.27	45.50	44.23	41.20	39.61	39.17
5.40	52.94	50.23	48.38	47.02	43.71	41.96	41.47
5.60	56.15	53.33	51.34	49.88	46.31	44.38	43.83
5.80	59.42	56.45	54.34	52.79	48.94	46.83	46.22
6.00	62.77	59.65	57.43	55.78	51.64	49.34	48.67

* Data from Hydraulic Tables by Hazen and Williams.

WATER SUPPLY—ORIFICES & WEIRS

DISCHARGE OVER 90° TRIANGULAR V NOTCH SHARP-CRESTED WEIR*

(COMMONLY USED WHERE FLOWS ARE SMALL.)

Thompson Formula

$Q = 2.54 H^{5/2}$

For general notes see drawing of rectangular weir.

PLAN

Q = Discharge in cu.ft. per sec.
H = Head in feet.

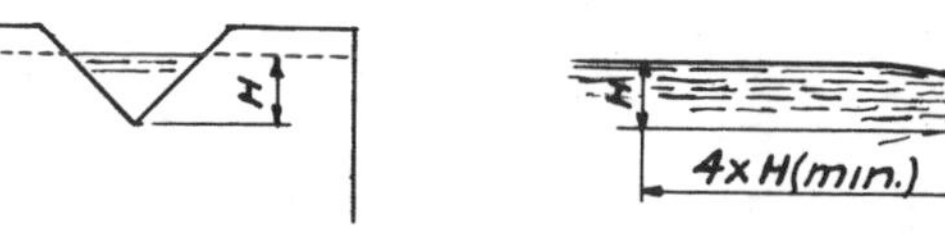

H
4×H(min.)
Q

SECTION A-A   SECTION B-B

H IN FEET	H IN INCHES	Q IN CU.FT. PER SEC.
0.05	5/8	0.0015
0.10	1 3/16	0.0085
0.15	2 3/8	0.022
0.20	2 3/8	0.047
0.30	3 5/8	0.129
0.40	4 13/16	0.262
0.50	6	0.455
0.60	7 3/16	0.714
0.70	8 3/8	1.044
0.80	9 5/8	1.452
0.90	10 13/16	1.943
1.00	12	2.520
1.10	13 3/16	3.189
1.20	14 3/8	3.954
1.30	15 5/8	4.818
1.40	16 13/16	5.785
1.50	18	6.860

DISCHARGE PER FOOT OF WEIR WIDTH OVER CIPOLLETTI WEIR*

(COMMONLY USED WHERE END CONTRACTION CONDITIONS EXIST.)

Cipolletti Formula

$Q = 3.367 L H^{3/2}$

For general notes see drawing of rectangular weir.

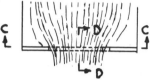

PLAN

Slope 1-Hor. to 4-Vert.

Q = Discharge in cu.ft. per sec.
L = Width of weir in feet.
H = Head in feet.

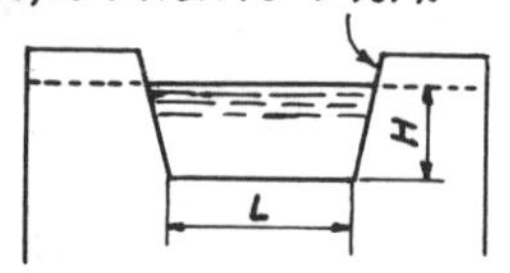

H
L

SECTION C-C   SECTION D-D

4×H(min.)

H IN FEET	Q IN CU.FT./SEC. PER FT. OF WEIR	H IN FEET	Q IN CU.FT./SEC. PER FT. OF WEIR
0.1	0.107	1.6	6.814
0.2	0.301	1.7	7.462
0.3	0.553	1.8	8.130
0.4	0.852	1.9	8.817
0.5	1.190	2.0	9.522
0.6	1.565	2.1	10.245
0.7	1.972	2.2	10.986
0.8	2.409	2.3	11.743
0.9	2.875	2.4	12.517
1.0	3.367	2.5	13.308
1.1	3.884	2.6	14.114
1.2	4.426	2.7	14.936
1.3	4.990	2.8	15.774
1.4	5.577	2.9	16.626
1.5	6.185	3.0	17.494

DISCHARGE THROUGH AN ORIFICE OR TUBE

Diagram	Description	Average discharge coefficient, C_d
	SHORT TUBE OR ORIFICE IN THICK WALL WITH SQUARE-EDGED ENTRY When the stream springs clear from the tube at the upstream corner the flow is the same as for a sharp-edged orifice.	0.61
	RE-ENTRANT TUBE Length about 2½ diameters. Flowing full.	0.73
	RE-ENTRANT TUBE When the length is about one diameter it is called "Borda's Mouthpiece". Stream springs clear of the walls of the tube.	0.52

H = Head in ft. on center of orifice
Q = Discharge in cu.ft./sec.

Orifice Formula.

$Q = C_d A \sqrt{2gH}$

A = Area of orifice in sq.ft.
C_d = Discharge coefficient.
g = 32.17

Diagram	Description	Average discharge coefficient, C_d
	SHARP-EDGED ORIFICE The stream is contracted to about 0.62 of the area of the opening.	0.61
	ORIFICE WITH WELL-ROUNDED ENTRY There is little or no contraction and the stream is about the same size as the opening.	0.98
	SHORT TUBE OR ORIFICE IN THICK WALL WITH SQUARE-EDGED ENTRY When flowing full. When the length of the tube is 2½ diameters it is called a "standard short tube".	0.82

*Adapted from Handbook of Water Control by the California Corrugated Culvert Co.

WATER SUPPLY—PARSHALL FLUME

PARSHALL FLUME*

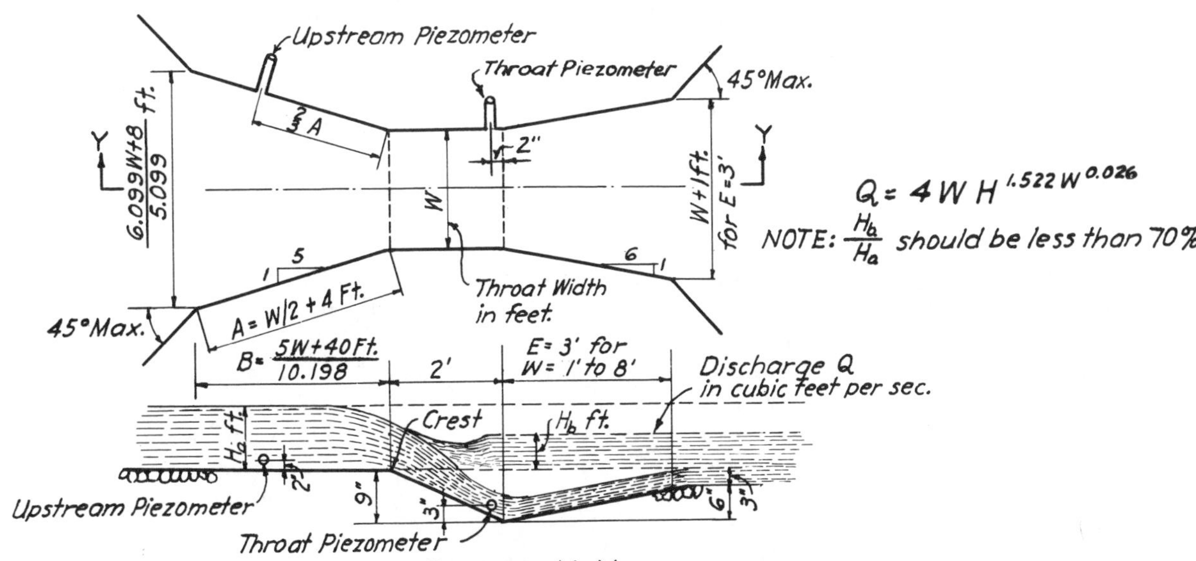
Upstream Piezometer
Throat Piezometer
45°Max.
2/3 A
2"
6.099W+8 / 5.099 ft.
W
W+1 ft. for E=3'
Y
Q = 4 W H^1.522 W^0.026
NOTE: Hb/Ha should be less than 70%
5
1
6
1
A= W/2 + 4 Ft.
Throat Width in feet.
45°Max.
B= 5W+40 Ft. / 10.198
2'
E= 3' for W= 1' to 8'
Discharge Q in cubic feet per sec.
Crest
Hb ft.
Ha ft.
2"
9"
3"
6"
3"
Upstream Piezometer
Throat Piezometer
SECTION Y-Y

TABLE A- DISCHARGE THROUGH PARSHALL FLUMES.**

UPPER HEAD Ha (FT.)	DISCHARGE IN CUBIC FEET PER SECOND — THROAT WIDTH W IN FEET 1	2	3	4	5	6	7	8
0.20	0.35	0.66	0.97	1.26				
0.25	0.49	0.93	1.37	1.80	2.22	2.63		
0.30	0.64	1.24	1.82	2.39	2.96	3.52	4.08	4.62
0.35	0.80	1.57	2.32	3.06	3.78	4.50	5.22	5.93
0.40	0.99	1.93	2.86	3.77	4.68	5.57	6.46	7.34
0.45	1.19	2.32	3.44	4.54	5.63	6.72	7.80	8.87
0.50	1.39	2.73	4.05	5.36	6.66	7.94	9.23	10.5
0.55	1.62	3.17	4.70	6.23	7.74	9.25	10.8	12.2
0.60	1.84	3.62	5.39	7.15	8.89	10.6	12.4	14.1
0.65	2.08	4.11	6.12	8.11	10.1	12.1	14.1	16.0
0.70	2.33	4.60	6.86	9.11	11.4	13.6	15.8	18.0
0.75	2.58	5.12	7.65	10.2	12.7	15.2	17.7	20.1
0.80	2.85	5.66	8.46	11.3	14.0	16.8	19.6	22.4
0.85	3.12	6.22	9.30	12.4	15.5	18.5	21.6	24.6
0.90	3.41	6.80	10.2	13.6	16.9	20.3	23.7	27.0
0.95	3.70	7.39	11.1	14.8	18.4	22.1	25.8	29.5
1.00	4.00	8.00	12.00	16.00	20.00	24.00	28.0	32.0
1.05	4.31	8.63	13.0	17.3	21.6	25.9	30.3	34.6
1.10	4.62	9.27	13.9	18.6	23.3	27.9	32.6	37.3
1.15	4.94	9.94	14.9	19.9	25.0	30.0	35.0	40.1
1.20	5.28	10.6	16.0	21.3	26.7	32.1	37.5	42.9
1.25	5.62	11.3	17.0	22.8	28.5	34.3	40.0	45.8
1.30	5.96	12.0	18.1	24.2	30.3	36.5	42.6	48.6

UPPER HEAD Ha (FT.)	DISCHARGE IN CUBIC FEET PER SECOND — THROAT WIDTH W IN FEET 1	2	3	4	5	6	7	8
1.35	6.32	12.7	19.2	25.7	32.2	38.7	45.3	51.8
1.40	6.68	13.5	20.3	27.2	34.1	41.1	48.0	55.0
1.45	7.04	14.2	21.3	28.8	36.1	43.4	50.8	58.1
1.50	7.41	15.0	22.6	30.3	38.1	45.8	53.6	
1.55	7.80	15.8	23.8	32.0	40.1	48.3	56.5	
1.60	8.18	16.6	25.1	33.6	42.2	50.8	59.4	
1.65	8.57	17.4	26.3	35.3	44.3	53.3		
1.70	8.97	18.2	27.6	37.0	46.4	56.0		
1.75	9.38	19.0	28.8	38.7	48.6	58.6		
1.80	9.79	19.9	30.1	40.5	50.8			
1.85	10.2	20.8	31.5	42.2	53.1			
1.90	10.6	21.6	32.8	44.1	55.4			
1.95	11.1	22.5	34.1	45.9	57.7			
2.00	11.5	23.4	35.5	47.8	60.1			
2.05	11.9	24.3	36.9	49.7				
2.10	12.4	25.3	38.4	51.6				
2.15	12.8	26.2	39.8	53.5				
2.20	13.3	27.2	41.3	55.5				
2.25	13.7	28.1	42.7	57.5				
2.30	14.2	29.1	44.2	59.6				
2.35	14.7	30.1	45.7					
2.40	15.2	31.1	47.3					
2.50	16.1	33.1	50.4					

* Adapted from Schoder & Dawson, Hydraulics, McGraw-Hill.
** Table from Handbook of Water Control by The California Corrugated Culvert Co.

WATER SUPPLY—DRINKING WATER REQUIREMENTS*—I

1. Turbidity (silica scale)† . . . . 10 p.p.m. max.
2. Color (standard platinum cobalt scale). . . . . 20 max.
3. Taste or odor† . . . . None
4. Chromium (hexavalent)† . . . . 0.05 p.p.m. max.
5. Lead†. . . . . 0.1 p.p.m. max.
6. Fluoride†. . . . . 1.5 p.p.m. max.
7. Arsenic† . . . . 0.05 p.p.m. max.
8. Selenium† . . . . 0.05 p.p.m. max.
9. Salts of barium, heavy metal glucosides† . . . . None
10. Copper‡ . . . . 3.0 p.p.m. max.
11. Iron and manganese (together)‡ . . . . 0.3 p.p.m. max.
12. Magnesium‡ . . . . 125 p.p.m. max.
13. Zinc‡ . . . . 15 p.p.m. max.
14. Chloride‡ . . . . 250 p.p.m. max.
15. Sulfate‡ . . . . 250 p.p.m. max.
16. Phenolic compounds (in terms of phenol)‡ . . . . 0.001 p.p.m. max.
17. Total solids‡. . . . . 500-1000 p.p.m. max.
18. Alkalinity (calculates as $CaCO_3$)‡
    (a) Water softened by lime-soda, zeolite, ion-exchange process,. . . or — (a) Total alkalinity produced by the process should not exceed the original hardness by more than 35 p.p.m. (calculated as $CaCO_3$).
    (b) Waters treated chemically . . . . (b_1) The phenolphthalein alkalinity should not be greater than 15 p.p.m. plus 0.4 times total alkalnity.§
    (b_2) Normal carbonate alkalinity should not exceed 120 p.p.m. ||
19. Hardness. . . . . No limit on drinking water. Waters with hardness of over 100 to 150 p.p.m. when used for laundries and boilers usually require softening.

HARDNESS TABLE

p.p.m.	Degree of Hardness	p.p.m.	Degree of Hardness
15	Extremely soft	130	Hard
30	Very soft	170	Very hard
45	Soft	230	Excessively hard
90	Moderately soft	250	Too hard for use
110	Moderately hard		

20. Residual chlorine† . . . . (a) Normally all points in distribution system at least 0.05 to 0.10 p.p.m.
    (b) During water-borne disease outbreaks, minimum of 0.2 to 0.3 p.p.m. in all points of distribution system.
21. *p*H. . . . . Should be as close to 7.0 as economically feasible. Waters below 7.0 (acid waters) cause tuberculation. Waters above 7.0 cause incrustation of pipes.

*Adapted from Standards of the U. S. Public Health Service, 1946.
†Excess of amount indicated shall be grounds for rejection of water.
‡These chemical substances should preferably not occur in excess of the concentrations given where in the judgment of the certifying authority other more suitable supplies are available.
§Requirement under (b_1) limits the permissible *p*H to about 10.6 at 25° C.
||Since the normal alkalinity is a function of the hydrogen-ion concentration and the total alkalinity, this requirement may be met by keeping the total alkalinity within the limits suggested when the water is within the range given. These values apply to water at 25°C.

*p*H Range	Limit for Total Alkalinity, p.p.m. as $CaCO_3$	*p*H Range	Limit for Total Alkalinity, p.p.m. as $CaCO_3$
8.0 to 9.6	400	10.1	210
9.7	340	10.2	190
9.8	300	10.3	180
9.9	260	10.4	170
10.0	230	10.5 to 10.6	160

WATER SUPPLY—DRINKING WATER REQUIREMENTS*—2

10-milliliter Portion

1. Of all portions tested in one month, maximum allowable number of portions showing *B. coli* (+) is 10%.

2. Occasionally 3 or more of the 5 portions of a sample may show *B. coli* (+). This is acceptable except when it occurs in:
   (a) Consecutive samples.
   (b) More than 5% of the standard samples when 20 or more samples have been examined per month.
   (c) More than 1 standard sample when less than 20 samples have been examined per month.

3. When condition 2 exists, daily samples shall be taken and examined until the results from 2 consecutive samples show the water to be satisfactory. With water of unknown quality under condition 2, simultaneous tests should be made on portions of 10 milliliters, 1 milliliter, and 0.1 milliliter.

100-milliliter Portion

1. Of all portions tested in one month maximum allowable number of portions showing *B. coli* (+) is 60%.

2. Occasionally all 5 portions of a sample may show *B. coli* (+). This is acceptable except when it occurs in:
   (a) Consecutive samples.
   (b) More than 20% of the standard sample when 5 or more samples have been examined per month.
   (c) More than 1 standard sample when less than 5 samples have been examined per month.

3. When condition 2 exists, daily samples shall be taken and examined until the results from 2 consecutive samples show the water to be satisfactory. With water of unknown quality under condition 2, simultaneous tests should be made on portions of 100 milliliters, 10 milliliters, and 1 milliliters.

In accordance with the first requirement stated above, namely, that of all the standard 100-milliliter or 10-milliliter portions examined per month in accordance with the specified procedure, not more than 60% or 10%, respectively, shall show the presence of organisms of coliform group. This may be interpreted as implying that the mean density (most probable number M.P.N.) of organisms of the coliform group shall not exceed about 1 per 100 milliliters.

By general convention (+) indicates presence of *B. coli* and (–) indicates absence of *B. coli*.

Standard Portion: 10 milliliters or 100 milliliters
1 standard sample = 5 standard portions

Population Served	**Samples per month, Minimum**
2,500 and under	1
10,000	7
25,000	25
100,000	100
1,000,000	300
2,000,000	390
5,000,000	500

Standard Tests for *Coli aerogenes (B. coli)*

1. Presumptive	For outline procedure of tests, see *Standard*
2. Confirmed	*Method for the Examination of Water and Sewage*, 10th ed., 1955,
3. Completed	published by American Public Health Association.

NOTE: Either confirmed or completed tests are recommended by the Public Health Service for standard methods of testing.

*Adapted from Standards of the U. S. Public Health Service, 1946.

WATER PURIFICATION-EXTENT OF TREATMENT REQUIRED

Group No.	Description	Treatment Required
1	This group limited to underground waters subject to no possibility of contamination, meeting all the requirements of the standards and having a maximum B. coli index of 1.0 per 100 ml. (milliliters) at all times.	Waters requiring no treatment.
2	This group includes underground and surface waters, subject to a low degree of contamination and meeting all the requirements of the standards, except as to coliform bacterial content which should average not more than 50 per 100 ml. in any month.	Waters requiring simple chlorination or its equivalent.
3	This group includes waters requiring filtration treatment for turbidity and color removal, waters of high or variable chlorine demand* and waters polluted by sewage to an extent such as to be inadmissible to Groups 1 and 2, but containing numbers of coliform bacteria averaging not more than 5,000 per 100 ml. in any month and exceeding this number in not more than 20% of the samples examined in any month.	Waters requiring complete rapid sand filtration treatment or its equivalent, together with continuous post-chlorination.
4	This group includes waters meeting the requirements of Group 3 limiting monthly average coliform numbers, but showing numbers exceeding 5,000 per 100 ml. in more than 20% of the samples examined during any month and not exceeding 20,000 per 100 ml. in more than 5% of the samples examined during any month.	Waters requiring auxiliary treatment in addition to complete filtration treatment and post-chlorination.

*Normal demand is usually 0.1 to 1.0 part per million.

Notes: (1) By "auxiliary treatment" is meant presedimentation or prechlorination or their equivalents, either separately or combined. Long-time storage, for periods of 30 days or more, provides a permanent and reliable safeguard, which in many cases offers an effective substitute for presedimentation and prechlorination.

(2) Waters failing to meet the requirements of Groups 1, 2, 3 or 4 are considered as unsuitable for use as a source of water supply, unless they are brought into conformance with these requirements by means of prolonged preliminary storage.

Adapted from Standards of the U. S. Public Health Service, 1946.

WATER PURIFICATION–STORAGE RESERVOIR TREATMENT

TABLE A – APPROXIMATE CONCENTRATIONS OF COPPER SULFATE TO KILL VARIOUS MICRO ORGANISMS*

Organism	Odor (Taste, Color)	Parts per Million Copper Sulfate
Diatomaceae:		
Achnantes		–
Asterionella	Aromatic, geranium, fishy	0.12-0.20
Cyclotella	Faintly aromatic	–
Diatoma	Faintly aromatic	–
Fragilaria		0.25
Melosira		0.20
Meridion	Aromatic	–
Navicula		0.07
Nitzschia		0.50
Synedra	Earthy	0.36-0.50
Stephanodiscus		0.33
Tabellaria	Aromatic, geranium, fishy	0.12-0.50
Chlorophyceae:		
Chara		0.10-0.50
Cladophora		0.50
Closterium		0.17
Coelastrum		0.05-0.33
Conferva		0.25
Desmidium		2.00
Dictyosphaerium	Grassy, nasturtium, fishy	–
Draparnaldia		0.33
Endorina	Faintly fishy	2.00-10.00
Entomophora		0.50
Gloeocystis	Offensive	–
Hydrodictyon	Very offensive	0.10
Microspora		0.40
Nitella flexilis	Objectionable	0.10-0.18
Palmella		2.00
Pandorina	Faintly fishy	2.00-10.00
Protococcus		–
Scenedesmus		1.00
Spirogyra		0.12
Staurastrum	Grassy	1.50
Tetrastrum		–
Ulothrix		0.20
Volvox	Fishy	0.25
Zygnema		0.50
Cyanophyceae:		
Anabaena	Moldy, grassy, vile	0.12-0.48
Aphanizomenon	Moldy, grassy, vile	0.12-0.50
Clathrocystis	Sweet, grassy, vile	0.12-0.25
Coelosphaerium	Sweet, grassy	0.20-0.33
Cylindrospermum	Grassy	0.12
Cyanophyceae:		
Gloeocapsa	Red color	0.24
Microcystis		0.20
Oscillatoria		0.20-0.50
Rivularia	Moldy, grassy	–
Protozoa:		
Bursaria	Irish moss, salt marsh, fishy	–
Ceratium	Fishy, vile (rusty brown color)	0.24-0.33
Chlamydomonas		0.36-1.0
Cryptomonas	Candied violets	0.50
Dinobryon	Aromatic, violets, fishy	0.18
Endamoeba histolytica (cysts)		–
Euglena		0.50
Glenodinium	Fishy	0.50
Mallomonas	Aromatic, violets, fishy	0.50
Peridinium	Fishy, like clamshells	0.50-2.00
Stentor		0.24
Synura	Cucumber, muskmelon, fishy, bitter	0.12-0.25
Uroglenopsis	Fishy, oily, cod-liver oil taste	0.05-0.20
Crustacea:		
Cyclops		–
Daphnia		2.00
Fungi:		
Achyla		–
Beggiatoa	Very offensive, decayed	5.00
Cladothrix (Sphaerotilis dichotomus)		0.20
Crenothrix	Very offensive, decayed, medicinal with chlorine	0.33-0.05
Didymohelix (Gallionella, Spirophyllum)		–
Leptomitis		0.40
Leptothrix		–
Saprolegnia		0.18
Sphaerotilis natans	Very offensive, decayed	0.40
Miscellaneous:		
Chironomus (bloodworm)		–
Chironomus (midges)		–
Craspedacusta (fresh-water jellyfish)		0.3
Nais		–
Potaogeton		0.30-0.80

TABLE B – MAXIMUM CONCENTRATIONS OF COPPER SULFATE SAFE FOR FISH*

Fish	Parts per million	Pounds per million gallons
Trout	0.14	1.2
Carp	0.30	2.5
Suckers	0.30	2.5
Catfish	0.40	3.5
Pickerel	0.40	3.5
Goldfish	0.50	4.0
Perch	0.75	6.0
Sunfish	1.20	10.0
Black bass	2.10	17.0

TABLE C – AMOUNT OF CHLORINE REQUIRED TO DESTROY MICROSCOPIC ORGANISMS

Organisms	Test Concentration of Organisms, Standard Units	Chlorine Dose†		Reduction in Organisms, %	Odors and Tastes Eliminated
		Parts per Million	Pounds per Million Gallons		
Aphanizomenon	1500	0.85	7.1	50	
Cyclotella		1.0	8.3	100	
Melosira	‡	2.0	16.6	100	
Crenothrix		0.54	4.5		
Fungi		0.33	2.7		
Dinobryon	500	0.5	4.2	100	Yes
Uroglenopsis	2000	0.5	4.2	100	Yes
	6000	0.5	4.2	100	No§
Synura	50	0.3	2.5	100	Yes
	100	0.5-0.7	4.2-5.8	100	Yes
	200	>0.7	>5.8	100	Yes
Gnats or blood worms		3.0	24.9	100	

† The amount of chlorine depends somewhat on the chlorine demand of the water.
‡ Present in sufficient numbers to clog filter.
§ Taste not noticeable after 10 to 15 miles flow in aqueduct.

* Adapted from *Manual of Water Quality and Treatment*, by American Water Works Association, 1951.

WATER PURIFICATION – SLOW SAND FILTERS – I

TABLE A - CHARACTERISTICS OF SLOW SAND FILTERS.	
Adaptability.	1. May be economical, if cost of land is low, if raw water load as regards color, bacteria, algae and turbidity is low (see below) and operating costs are important factors (operating costs of slow sand filters generally 1/2 operating costs of rapid sand filters). 2. May be valuable as final process after rapid sand filtration of very polluted waters.
Capacity.	2 m.g.d. per acre for highly polluted waters to 10 m.g.d. per acre for water relatively clear or pre-treated by coagulation and sedimentation.
Efficiency (Bacteria removal).	98 to 99 % if not overloaded.
Flexibility.	Rapid changes in rate not permissible.
Turbidity limitations.	Raw water over 100 p.p.m. can not be handled; over 40 p.p.m. may produce unsatisfactory effluent.
Color limitations.	Raw water over 30 p.p.m. can not be treated satisfactorily.
Bed layout.	Size 1/4 to 1 acre each. Small beds for small plants. Large beds for large plants. Nº. of units sufficient to allow 1 unit in small plants and 2 in large plants to be out of order and still have plant meet maximum demand (less than 4 units not recommended).
Flow Regulators.	Manually operated gates and weirs (no automatic rate of flow controls and measuring devices required).

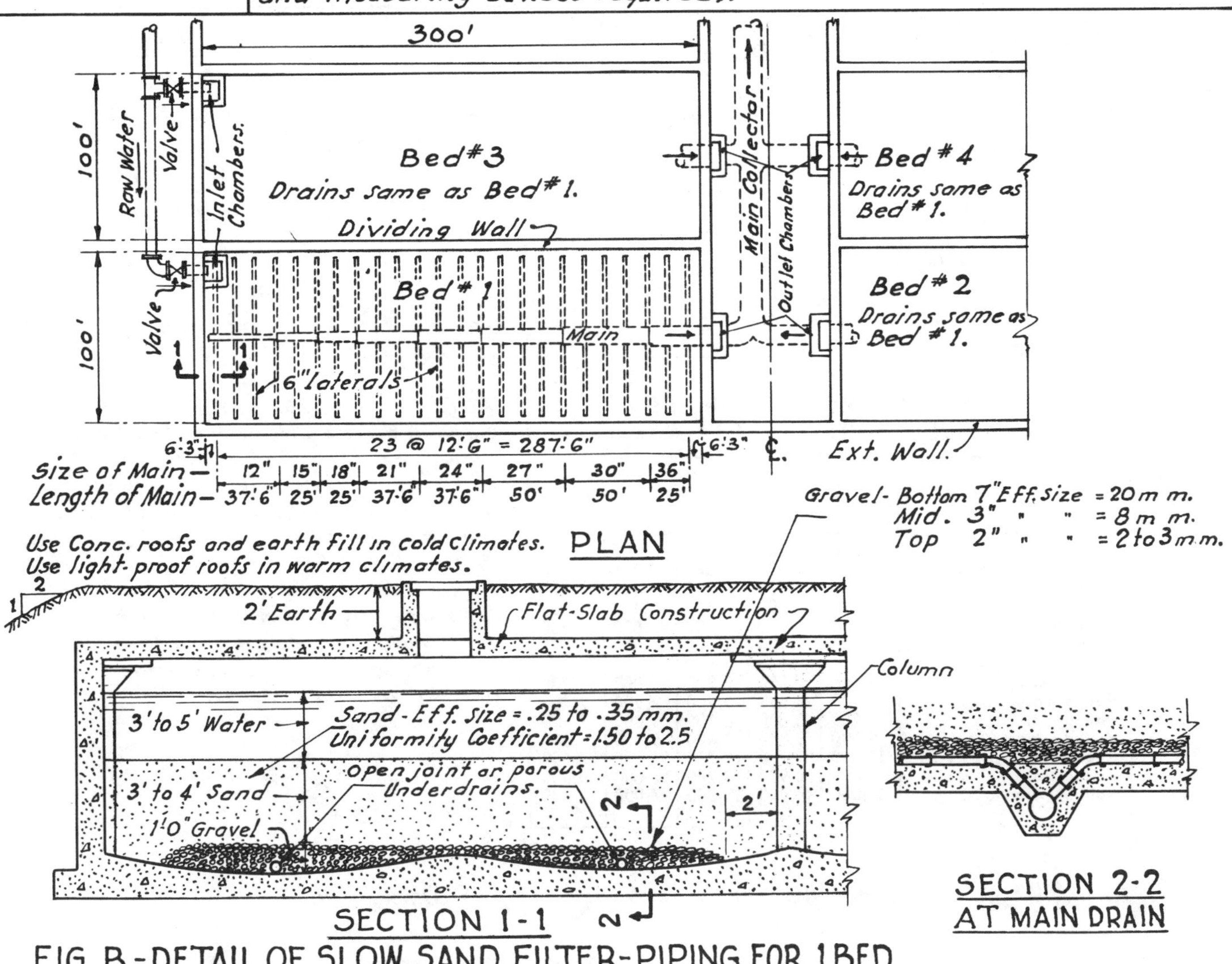

300'
100'
Raw Water
Valve
Inlet Chambers.
Bed #3
Drains same as Bed #1.
Main Collector
Outlet Chambers
Bed #4
Drains same as Bed #1.
Dividing Wall
Bed #1
Bed #2
Drains same as Bed #1.
Main
6" Laterals
Valve
100'
6'3"
23 @ 12'6" = 287'6"
6'3"
℄
Ext. Wall
Size of Main – 12" 15" 18" 21" 24" 27" 30" 36"
Length of Main – 37'6 25' 25' 37'6 37'6 50' 50' 25'
Gravel - Bottom 7" Eff. Size = 20 mm.
Mid. 3" " " = 8 mm.
Top 2" " " = 2 to 3 mm.
Use Conc. roofs and earth fill in cold climates.
Use light-proof roofs in warm climates.
PLAN
2
1
2' Earth
Flat-Slab Construction
Column
3' to 5' Water
Sand - Eff. Size = .25 to .35 mm.
Uniformity Coefficient = 1.50 to 2.5
Open joint or porous Underdrains.
3' to 4' Sand
1'0" Gravel
2'
SECTION 1-1
SECTION 2-2 AT MAIN DRAIN

FIG. B - DETAIL OF SLOW SAND FILTER-PIPING FOR 1 BED.

WATER PURIFICATION – SLOW SAND FILTERS – 2

TABLE A – LOSS OF HEAD IN CLEAN SAND 3 FEET DEEP – TEMPERATURE 50° F.*

Rate of Filtration million gallons per acre per day	Effective Size of Sand, millimeters								
	0.20	0.25	0.30	0.35	0.40	0.45	0.50	0.55	0.60
1	0.10	0.06	0.04	0.03	0.02	0.02	0.02	0.01	0.01
2	0.20	0.13	0.09	0.07	0.05	0.04	0.03	0.03	0.02
3	0.30	0.19	0.13	0.10	0.08	0.06	0.05	0.04	0.03
4	0.40	0.26	0.18	0.13	0.10	0.08	0.06	0.05	0.04
5	0.50	0.32	0.22	0.16	0.12	0.10	0.08	0.07	0.06
6	0.60	0.38	0.27	0.20	0.15	0.12	0.10	0.08	0.07
7	0.70	0.45	0.31	0.23	0.18	0.11	0.11	0.09	0.08
8	0.80	0.51	0.36	0.26	0.20	0.16	0.13	0.11	0.09
9	0.90	0.57	0.40	0.30	0.22	0.18	0.14	0.12	0.10
10	1.00	0.64	0.45	0.33	0.25	0.20	0.16	0.13	0.11
12	1.20	0.77	0.54	0.40	0.30	0.24	0.19	0.16	0.13
14	1.40	0.90	0.63	0.46	0.35	0.28	0.22	0.18	0.15
16	1.60	1.02	0.72	0.53	0.40	0.32	0.26	0.21	0.18
18	1.80	1.15	0.81	0.59	0.45	0.36	0.29	0.24	0.20
20	2.00	1.27	0.89	0.66	0.50	0.40	0.32	0.27	0.22
100	10.02	6.37	4.46	3.28	2.51	1.98	1.61	1.33	1.11
125	12.52	7.96	5.57	4.10	3.13	2.48	2.01	1.67	1.39
150	15.03	9.55	6.69	4.92	3.76	2.97	2.41	1.99	1.67
175	17.54	11.14	7.80	5.74	4.38	3.47	2.81	2.33	1.95
200	20.05	12.74	8.92	6.55	5.01	3.96	3.21	2.65	2.23

NOTES: 1. For loss of head through sand other than the 3' depth shown in Table A, the loss of head varies inversely with the depth.

Example:
Given: Sand depth 40"
Required: Loss of head.
Solution: Loss of head = Value in Table A × $\frac{40}{36}$

2. For temperature other than 50°, loss of head varies as ratio $\frac{60}{°F+10}$.

Example:
Given: Temp. = 70° F.
Required: Loss of head at 70°F.
Solution: Loss of head = Value in Table A × $\frac{60}{70+10}$.

TABLE B – UNDERDRAIN DATA*

1. Rate of filtration, million gallons per acre daily	5	6	8	10	15
2. Average resistance of clean sand, feet	0.150	0.180	0.240	0.300	0.450
3. Total allowable friction and velocity head in underdrainage system, feet	0.037	0.045	0.060	0.075	0.112
4. Approximate ratio of filter area to area of main drain	5100	4700	4200	3800	3200
5. Approximate maximum velocity in main drain (varying somewhat with size), feet per second	0.90	1.00	1.18	1.34	1.68
6. Approximate maximum velocity in laterals (varying somewhat with size), feet per second	0.55	0.61	0.72	0.82	1.04

NOTE: Values on line 3 are 25% of line 2.

UNDERDRAINS

1. Size of underdrains to be such that loss of head in underdrains equals 25% of loss of head through sand. Neglect loss of head through gravel.

2. Underdrains can be laid out using Table C. Check velocities, using maximum velocities indicated in Table B.

TABLE C – MAXIMUM AREAS OF FILTER BEDS DRAINED, SQUARE FEET*

Diameter of Drain, inches	Shape and Kind of Drain	Rate of Filtration, million gallons per acre per day				
		5	6	8	10	15
4	Round lateral	264	245	218	200	168
5	Round lateral	420	390	345	316	266
6	Round lateral	610	570	500	460	390
8	Split lateral	520	490	430	400	320
10	Split lateral	830	770	680	630	530
12	Split lateral	1,200	1,120	1,000	910	770
10	Round main	2,700	2,500	2,200	2,000	1,700
12	Round main	3,900	3,600	3,200	2,900	2,400
15	Round main	6,200	5,800	5,100	4,600	3,900
18	Round main	9,000	8,300	7,400	6,700	5,600
21	Round main	12,300	11,400	10,000	9,100	7,600
24	Round main	16,100	14,900	13,200	12,000	10,000
36	Round main	37,000	34,000	30,000	27,000	22,000

NOTE: Since 1923, few slow sand filters have been constructed in the U.S.A. However, this type of filter is adaptable to warm climates, where covers are not required to protect them from freezing in winter, and slow sand filters also have the following advantages over rapid sand filters: Less loss of head, simpler mechanism, requiring less expert supervision, and higher bacterial efficiency without use of disinfectant.

* From Flinn, Weston & Bogert, Waterworks Handbook, Mc Graw-Hill.

WATER PURIFICATION—RAPID SAND FILTERS—I

TABLE A – CHARACTERISTICS OF RAPID-SAND FILTERS	
Adaptability	Satisfactory means of eliminating turbidity, bacteria, micro-organisms, and, to some extent tastes, odors, and color.
Capacity	Normal rate 2 g.p.m. per sq. ft., except in some cases may be extended to 3 g.p.m. per sq. ft.
Flexibility	Rate of filtration may be varied with demand.
Turbidity Limitations	Effective in treatment of highly polluted waters as well as those subject to extreme variations in turbidity and pollution.
Bed Layout	Minimum of two units should be provided, but three are preferable. Sizes vary from fraction of to multiple of 1 million gallons per day. Maximum size constructed 5 M.G.D. Filters generally constructed rectangular, of concrete.
Underdrain System	Older type consists of a grid of manifolds, headers, and laterals. Newer types consist of false bottom with pyramidal openings (Wheeler), vitrified clay (Leopold) and system of aluminum oxide porous plates (Aloxite) by Carborundum Co.
Wash-Water Troughs	Constructed of steel, cast iron, concrete, Transite, prestressed plastic and Fiberglas, and aluminum. Troughs spaced generally 5 to 7 ft. between weir edges.
Wash Water	Recommended rate of rise = 2 to 3 f.p.m.
Surface Wash	Fixed nozzles or rotating type recommended. Water pressure 45 to 75 p.s.i. at 4 to 8 g.p.m. per sq. ft.

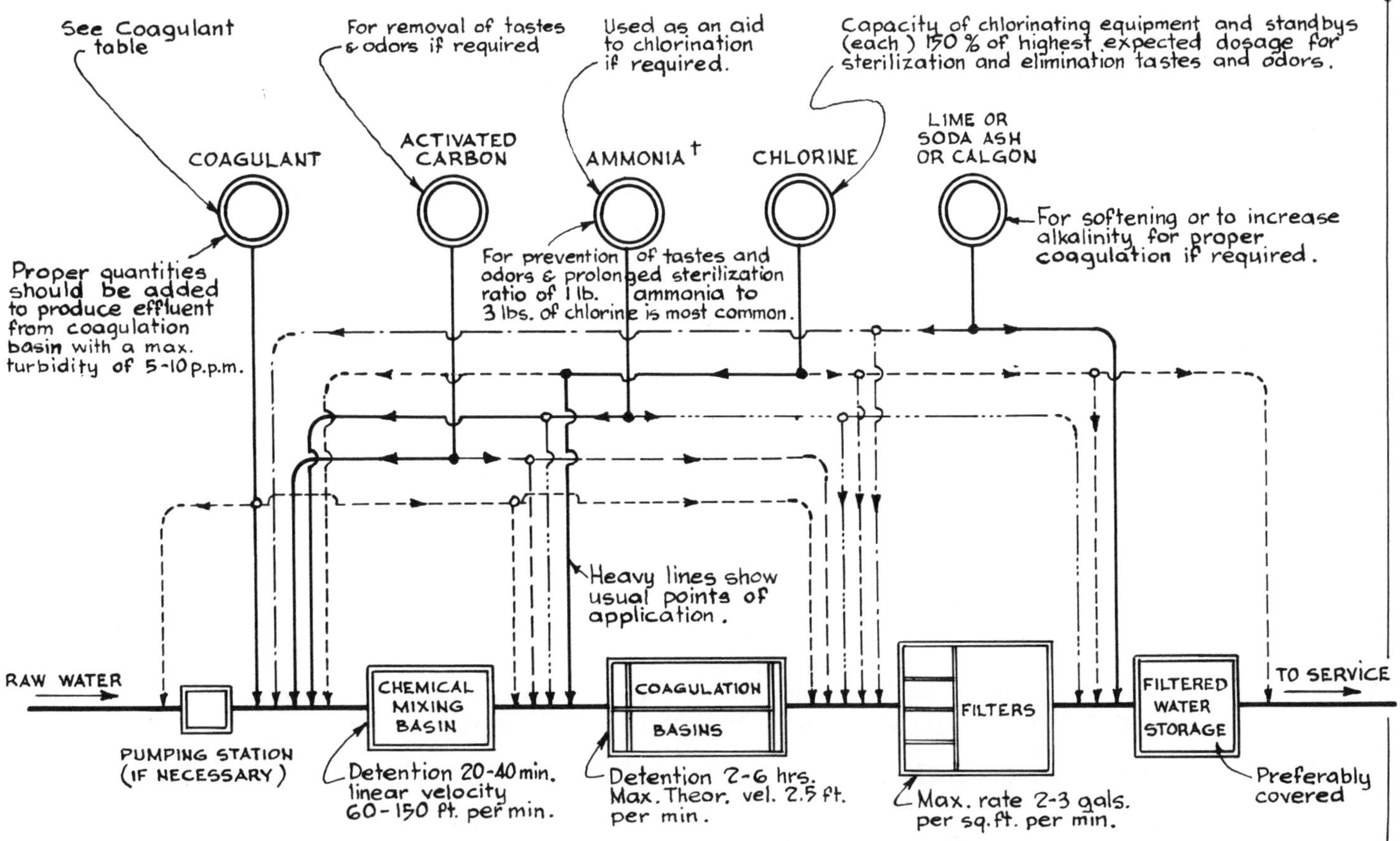

See Coagulant table
For removal of tastes & odors if required
Used as an aid to chlorination if required.
Capacity of chlorinating equipment and standbys (each) 150 % of highest expected dosage for sterilization and elimination tastes and odors.
COAGULANT
ACTIVATED CARBON
AMMONIA †
CHLORINE
LIME OR SODA ASH OR CALGON
For softening or to increase alkalinity for proper coagulation if required.
Proper quantities should be added to produce effluent from coagulation basin with a max. turbidity of 5-10 p.p.m.
For prevention of tastes and odors & prolonged sterilization ratio of 1 lb. ammonia to 3 lbs. of chlorine is most common.
Heavy lines show usual points of application.
RAW WATER
PUMPING STATION (IF NECESSARY)
CHEMICAL MIXING BASIN
COAGULATION BASINS
FILTERS
FILTERED WATER STORAGE
TO SERVICE
Detention 20-40 min. linear velocity 60-150 ft. per min.
Detention 2-6 hrs. Max. Theor. vel. 2.5 ft. per min.
Max. rate 2-3 gals. per sq. ft. per min.
Preferably covered

Plant capacity to equal average daily demand if filtered water storage is large and provides for maximum demand. If filtered water storage is small, plant capacity to be 150 to 225% of average daily demand.

FIG. B – FLOW DIAGRAM FOR MODERN RAPID SAND FILTER PLANT*

† NOTE: Ammonia is used very infrequently.

* Adopted from *Manual of Water Quality and Treatment*, by American Water Works Association, 1941.

WATER PURIFICATION-RAPID SAND FILTERS-2

DIAGRAMMATIC SKETCH SHOWING FILTER OPERATION

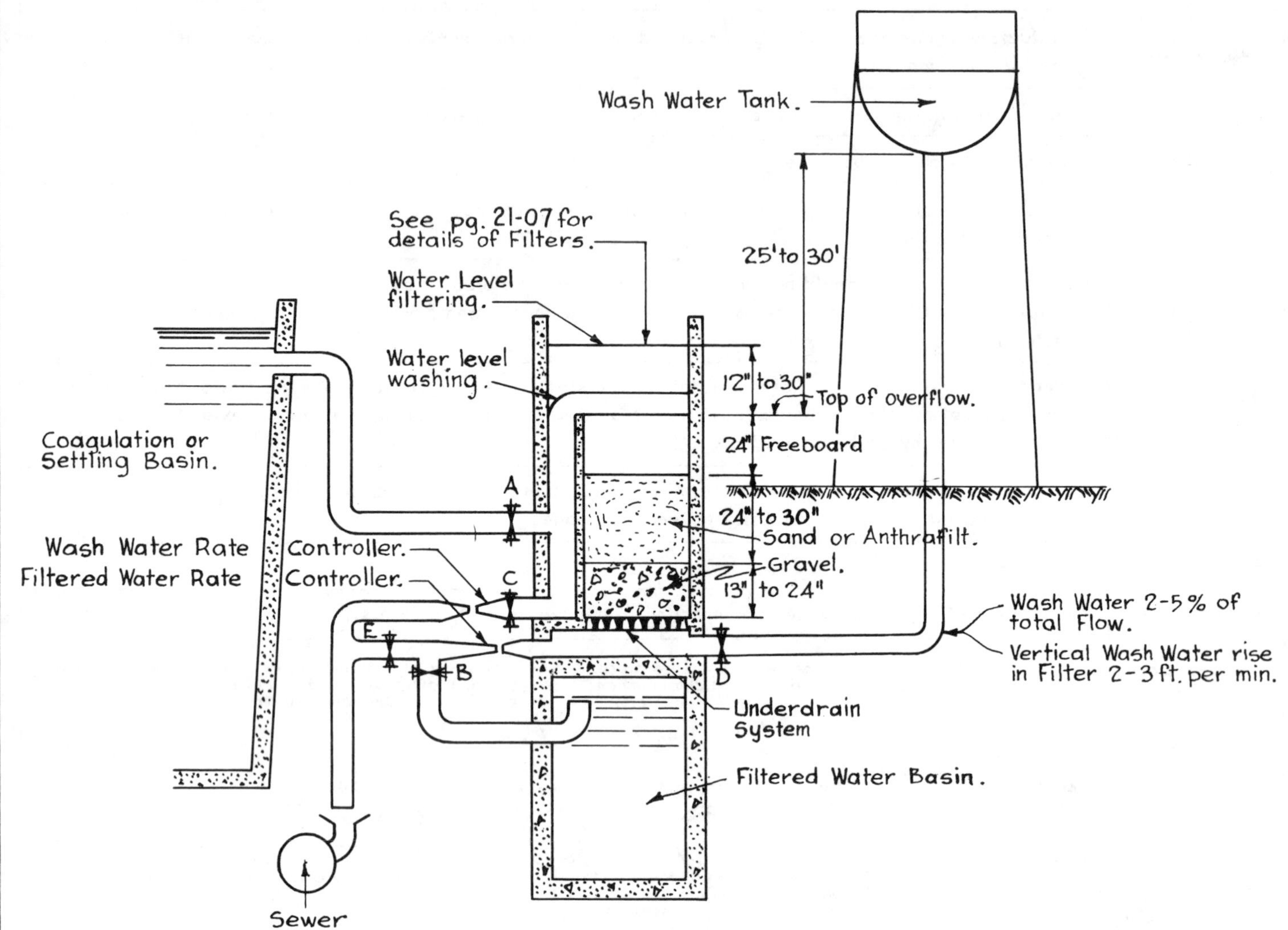
Wash Water Tank.
25' to 30'
See pg. 21-07 for details of Filters.
Water Level filtering.
Water level washing.
12" to 30"
Top of overflow.
24" Freeboard
24" to 30"
Sand or Anthrafilt.
Gravel.
13" to 24"
Coagulation or Settling Basin.
A
C
E
B
D
Wash Water Rate Controller.
Filtered Water Rate Controller.
Wash Water 2-5% of total Flow.
Vertical Wash Water rise in Filter 2-3 ft. per min.
Underdrain System
Filtered Water Basin.
Sewer

How Filter Operates

1. Open valve A. (This allows water from settling basin to flow to filter.)
2. Open valve B. (This allows water to flow through filter to filtered water storage. During filter operation all other valves are closed.)

How Filter Is Washed

1. Close valve A.
2. Close valve B when water in filter filters down to top of overflow.
3. Open valves C and D. (This allows water from wash water tank to flow up through the gravel and sand, loosening up the sand and washing the accumulated dirt from the surface of the sand, out of the filter, and into the sewer.)

How to Filter to Waste

1. Open valves A and E. All other valves closed. Water is sometimes filtered to waste for a few minutes after filter has been washed in order to condition the filter before it is put into service.

Adapted from Water Supply & Treatment by C.P. Hoover. Published by National Lime Assoc. Wash., D.C.

WATER PURIFICATION - RAPID SAND FILTERS-3

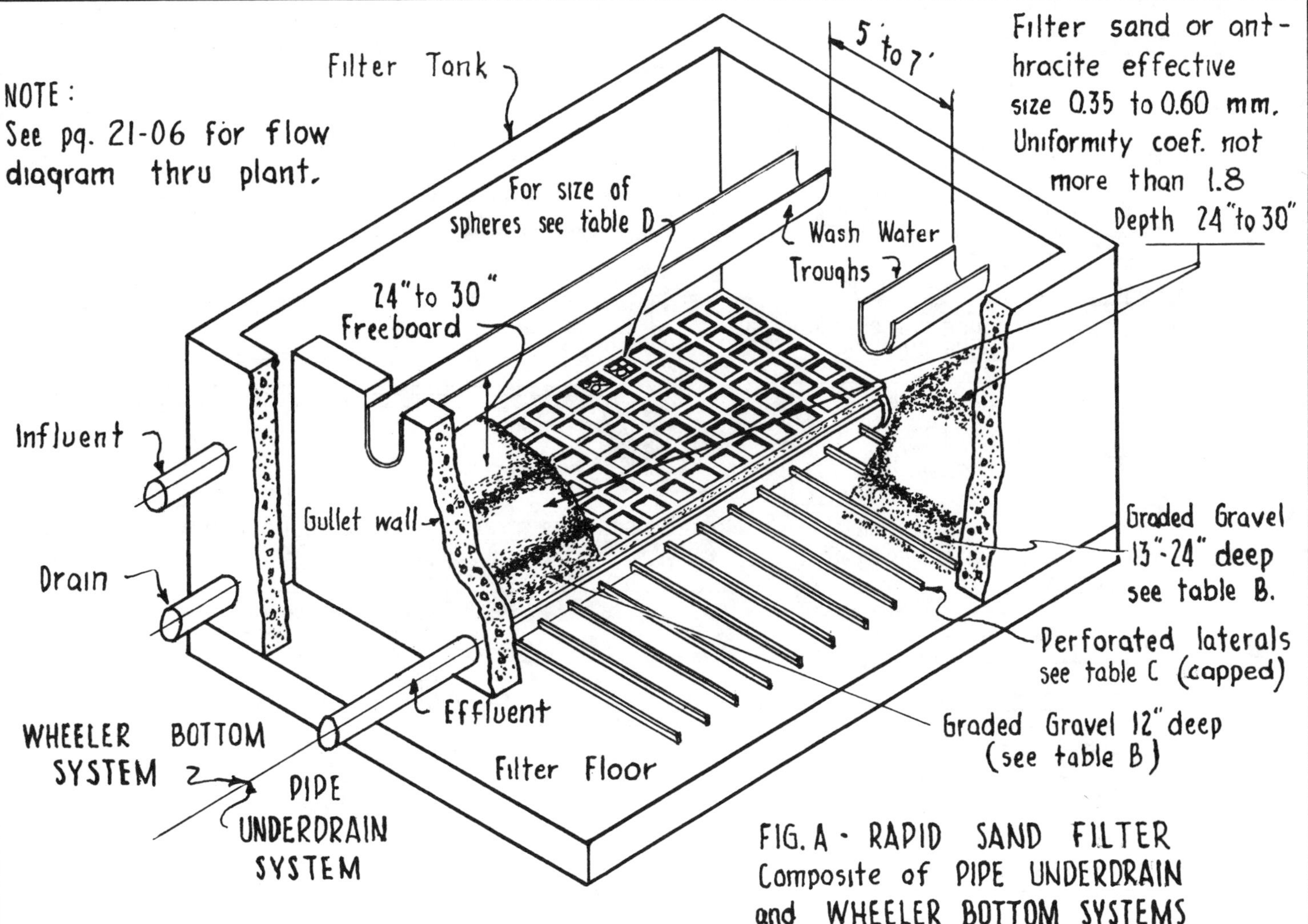

NOTE:
See pg. 21-06 for flow diagram thru plant.
Filter Tank
5' to 7'
Filter sand or anthracite effective size 0.35 to 0.60 mm. Uniformity coef. not more than 1.8
Depth 24" to 30"
For size of spheres see table D
Wash Water Troughs
24" to 30" Freeboard
Influent
Gullet wall
Drain
Graded Gravel 13"-24" deep see table B.
Perforated laterals see table C (capped)
Effluent
Graded Gravel 12" deep (see table B)
WHEELER BOTTOM SYSTEM
PIPE UNDERDRAIN SYSTEM
Filter Floor

FIG. A - RAPID SAND FILTER
Composite of PIPE UNDERDRAIN and WHEELER BOTTOM SYSTEMS

TABLE B - GRADED GRAVEL

FOR PIPE UNDERDRAIN SYSTEM *

LAYER No.	PASSING THRU SCREEN OPENINGS IN INCHES	RETAINED ON SCREEN OPENINGS IN INCHES	DEPTH OF LAYER-IN INCHES
1-(Bottom)	2½	1½	2 to 4
2	1½	¾	3 to 6
3	¾	½	2 to 4
4	½	¼	2 to 4
5	¼	⅛	2 to 4
6 (Top)	(Torpedo) Sand 1.2 to 0.8 m.m.		2

FOR CONCRETE CELL-WHEELER BOTTOM SYSTEM ‡

LAYER No.	PASSING THRU CIRCULAR OPENINGS IN INCHES	RETAINED ON CIRCULAR OPENINGS IN INCHES	DEPTH OF LAYER-IN INCHES
1-Bottom	1	5/8	3
2	5/8	3/8	3
3	3/8	3/16	3
4	3/16	10 Meshes/in #20 Wire.	3

TABLE-C - C.I. UNDERDRAINS **

	¼"	½"
Diameter of perforations.	¼"	½"
Spacing of perforations along laterals.	3"	8"
Max. ratio of total area of perforations to total cross-sectional area of laterals.	0.25	0.5
Minimum total area of perforations per square foot of filter.	0.3a"	0.3a"
Maximum spacing of laterals.	12"	12"
Max. ratio of length of lateral to its diam.	60	60
Rate of washing in Cu.ft. per Sq.Ft. of filter/min.	0.5-3.0	0.5-3.0

TABLE-D PORCELAIN SPHERE SIZES IN WHEELER BOTTOM ‡

NO.	PLACE & SIZE
5	3 in. diam. spheres, one in the bottom of the depression at the apex of the pyramid and four above it.
8	1¼" diam. spheres placed in the spaces between the four large spheres and the sides of the depression.
1	1⅜ in. diameter sphere in the center between the four large spheres.

*From Water Supply and Treatment by C.P. Hoover - Published by National Lime Assoc. Wash. D.C.
** Adapted from An Investigation of Perforated Pipe Filter Underdrains by H.N Jenks - Eng. News Record.
‡ Adapted from Handbook of Builders - Providence Inc. - Providence R.I.

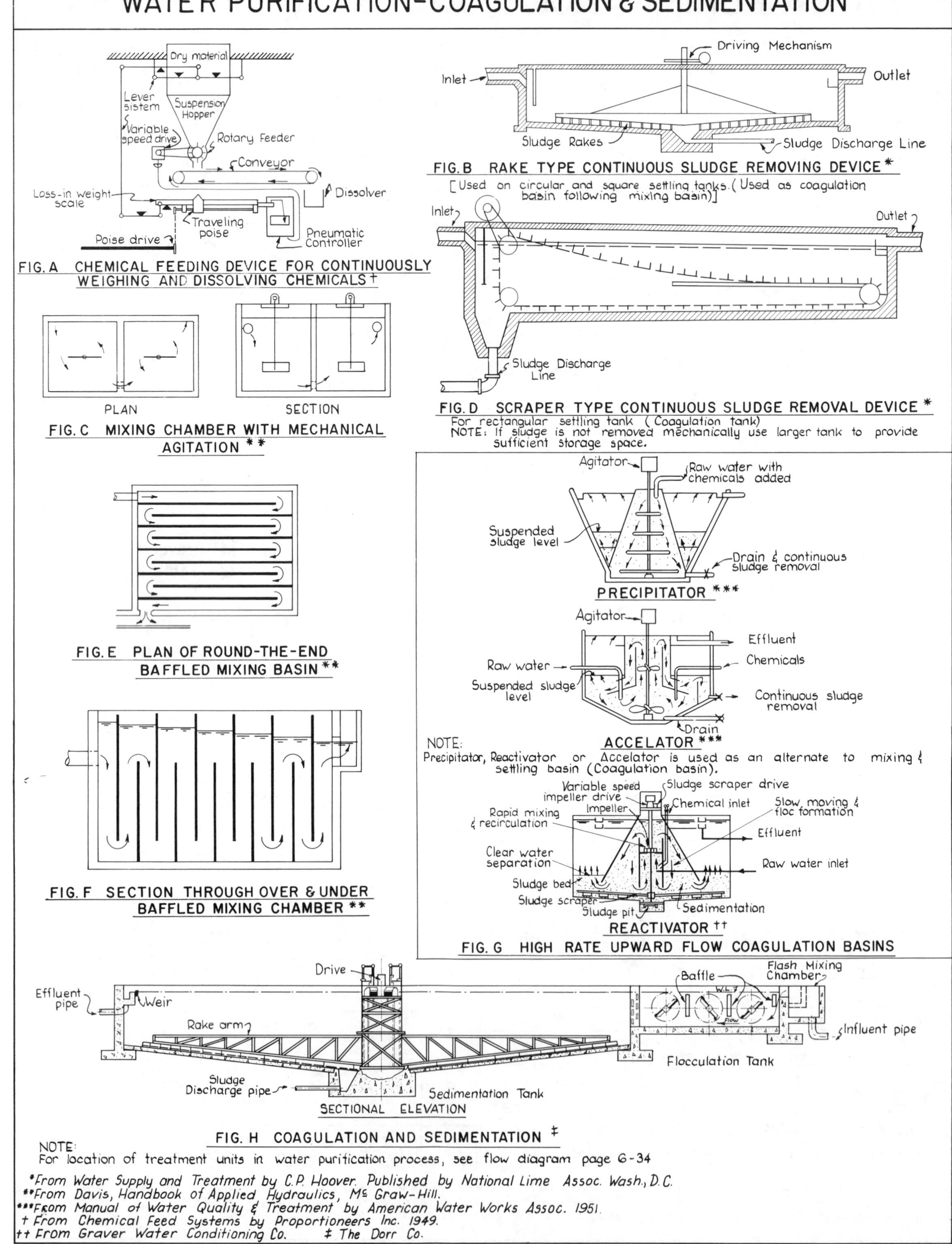
WATER PURIFICATION-COAGULATION & SEDIMENTATION
Dry material
Lever sistem
Suspension Hopper
Variable speed drive
Rotary Feeder
Conveyor
Dissolver
Loss-in weight scale
Traveling poise
Pneumatic Controller
Poise drive
FIG. A CHEMICAL FEEDING DEVICE FOR CONTINUOUSLY WEIGHING AND DISSOLVING CHEMICALS †
Driving Mechanism
Inlet
Outlet
Sludge Rakes
Sludge Discharge Line
FIG. B RAKE TYPE CONTINUOUS SLUDGE REMOVING DEVICE *
[Used on circular and square settling tanks. (Used as coagulation basin following mixing basin)]
PLAN
SECTION
FIG. C MIXING CHAMBER WITH MECHANICAL AGITATION **
Inlet
Outlet
Sludge Discharge Line
FIG. D SCRAPER TYPE CONTINUOUS SLUDGE REMOVAL DEVICE *
For rectangular settling tank (Coagulation tank)
NOTE: If sludge is not removed mechanically use larger tank to provide sufficient storage space.
FIG. E PLAN OF ROUND-THE-END BAFFLED MIXING BASIN **
Agitator
Raw water with chemicals added
Suspended sludge level
Drain & continuous sludge removal
PRECIPITATOR ***
Agitator
Effluent
Chemicals
Raw water
Suspended sludge level
Continuous sludge removal
Drain
ACCELATOR ***
NOTE:
Precipitator, Reactivator or Accelator is used as an alternate to mixing & settling basin (Coagulation basin).
Variable speed impeller drive
Sludge scraper drive
Impeller
Chemical inlet
Slow moving & floc formation
Rapid mixing & recirculation
Effluent
Clear water separation
Raw water inlet
Sludge bed
Sludge scraper
Sludge pit
Sedimentation
REACTIVATOR ††
FIG. F SECTION THROUGH OVER & UNDER BAFFLED MIXING CHAMBER **
FIG. G HIGH RATE UPWARD FLOW COAGULATION BASINS
Drive
Effluent pipe
Weir
Rake arm
Baffle
Flash Mixing Chamber
W.L.
Flow
Influent pipe
Flocculation Tank
Sludge Discharge pipe
Sedimentation Tank
SECTIONAL ELEVATION
FIG. H COAGULATION AND SEDIMENTATION ‡
NOTE:
For location of treatment units in water purification process, see flow diagram page 6-34
*From Water Supply and Treatment by C.P. Hoover. Published by National Lime Assoc. Wash., D.C.
**From Davis, Handbook of Applied Hydraulics, Mc Graw-Hill.
***From Manual of Water Quality & Treatment by American Water Works Assoc. 1951.
† From Chemical Feed Systems by Proportioneers Inc. 1949.
†† From Graver Water Conditioning Co. ‡ The Dorr Co.

WATER PURIFICATION-PRESSURE, DIATOMITE & GRAVITY VACUUM FILTERS

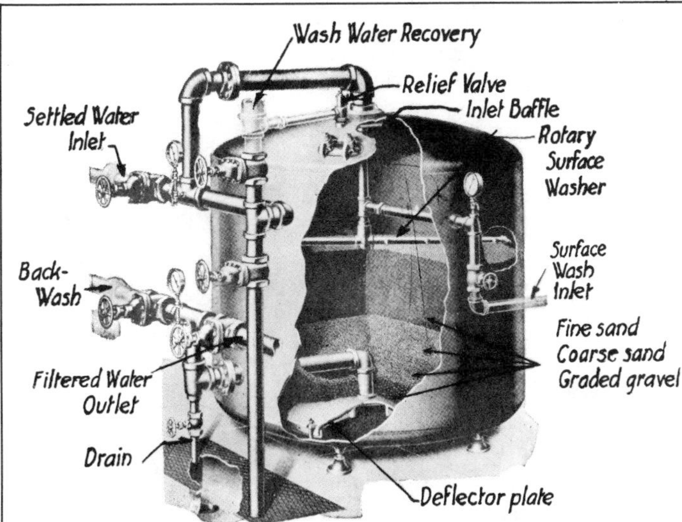
Wash Water Recovery
Relief Valve
Inlet Baffle
Rotary Surface Washer
Settled Water Inlet
Surface Wash Inlet
Back-Wash
Fine sand
Coarse sand
Graded gravel
Filtered Water Outlet
Drain
Deflector plate

FIG. A – TYPE E VERTICAL FILTER WITH ROTARY SURFACE WASHER*

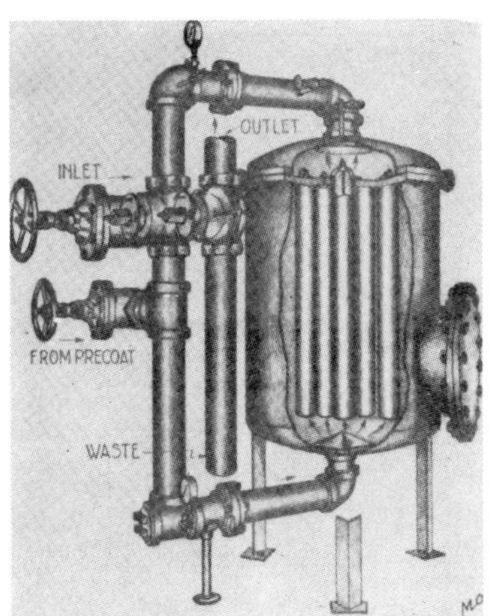
OUTLET
INLET
FROM PRECOAT
WASTE

FIG. C – DIATOMITE FILTER†

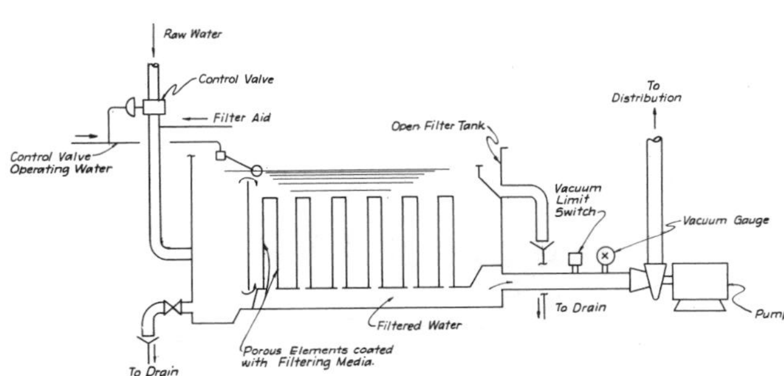
Raw Water
Control Valve
Filter Aid
Control Valve Operating Water
Open Filter Tank
To Distribution
Vacuum Limit Switch
Vacuum Gauge
To Drain
Pump
Filtered Water
Porous Elements coated with Filtering Media.
To Drain

FIG. E – SCHEMATIC ARRANGEMENT OF GRAVITY-VACUUM FILTER‡

* Adapted from the Permutit Co.
† Adapted from the Refinite Corp.
‡ From Proportioneers, Inc.

TABLE B – SMALL FILTERS – SIZES AND CAPACITIES*

Diameter, in.	Area sq. ft.	Capacity, g.p.m.	Backwash g.p.m.	Operating Weight, lb.	Pipe Size, in.
15	1.2	4	12	1300	1
20	2.2	7	20	1900	1
24	3.1	10	30	2400	1¼
30	4.9	15	49	3200	1½
36	7.1	21	71	4500	1½

TABLE D – SIZES AND CAPACITIES TYPE F FILTER*

Diameter, in.	Area, sq. ft.	Capacity, g.p.m. Flow Rates, g.p.m./sq.ft. 2	3	Backwash, g.p.m.	Operating Weight, lb.	Pipe Size, in.
42	9.6	19	29	96	6,100	2
48	12.6	25	38	126	8,000	2½
54	15.9	32	48	160	10,200	2½
60	19.6	39	59	195	12,600	3
66	23.8	48	71	240	15,700	3
72	28.3	57	85	285	18,800	3
78	33.2	66	100	330	22,200	4
84	38.5	77	115	385	26,000	4
90	44.2	88	133	440	30,000	4
96	50.3	101	150	505	34,000	4

NOTES: Graded anthracite may be used in place of sand.
A single multiport valve is often used in place of a nest of filter-control gate valves.
Water required for surface wash is ½ to 1 g.p.m. per sq. ft. of filter area at 50 to 75 p.s.i. pressure.
For large flows it is desirable to use tanks in batteries of 3, to provide required backwash rate, from pump. For example, if 100 g.p.m. filter capacity is required at 3 g.p.m. per sq. ft. rate, use 3 48-in.-diameter units, for which the pump will have a capacity of 100 to 126 g.p.m., the required backwash rate.
Vertical filters are available in sizes of 8 ft. diameter and in lengths up to 28 ft.

TABLE F – SIZES AND CAPACITIES DIATOMITE FILTERS†

Dia. in.	Area, sq. ft.	Capacity, g.p.m. Flow Rates, g.p.m./sq.ft. 3	4	Total No. of Elements	Size of Elements, in.	Backwash, g.p.m.	Tank Hgt., in.	Oper'g. Weight, lb.	Pipe Size, in.
14	11	33	44	7	3×24	30	30	1130	2
22	22	66	88	14	3×24	60	30	1600	2½
24	42.5	128	170	19	3×34	90	36	2150	3
30	69	207	276	31	3×34	150	36	4300	4
42	98	294	392	44	3×34	180	36	4690	4
42	145	435	580	65	3×34	270	36	6950	6

NOTES: Consult other manufacturers for different sizes and capacities.
Gravity-vacuum filters‡ are available in 12 sizes with precoat areas from 80 sq. ft. to 453 sq. ft.
Filtration rates are recommended as follows:
Potable or industrial water supply – 16 g.p.m. per sq. ft. of precoated filter.
Swimming pool recirculated water – 2 to 2.3 g.p.m./sq. ft.
Filter tank underdrain, manifold, baffles, and filter elements are of Fiberglas reinforced plastic.
Filter elements may be cleaned by (1) draining tank and hosing down elements in place or removed, or (2) chemical cleaning with elements in or out of tank.

WATER PURIFICATION-FLOW DIAGRAMS

WATER CONDITIONING FOR INDUSTRIAL USES

COLD PROCESS LIME—SODA SOFTENING

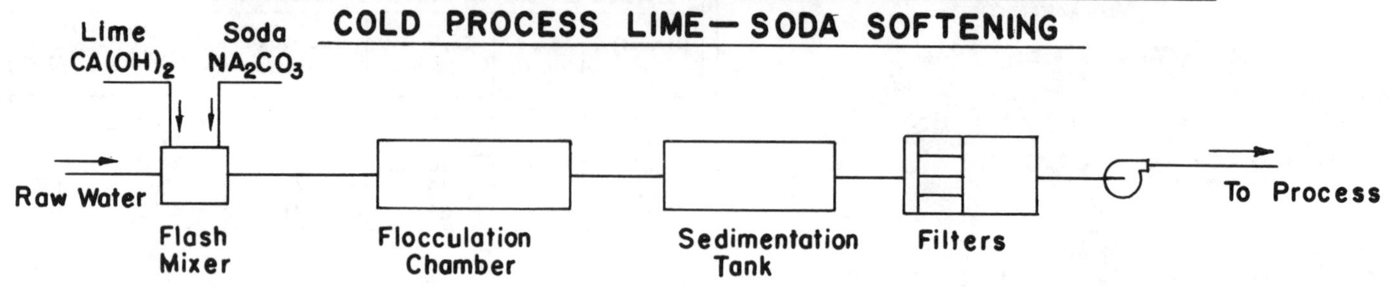
Lime
CA(OH)$_2$
Soda
NA$_2$CO$_3$
Raw Water
Flash Mixer
Flocculation Chamber
Sedimentation Tank
Filters
To Process

HOT PROCESS SOFTENING

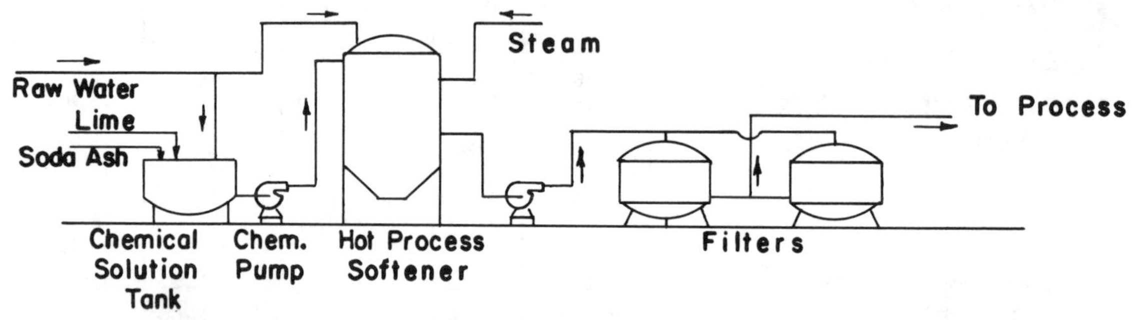
Steam
Raw Water
Lime
Soda Ash
To Process
Chemical Solution Tank
Chem. Pump
Hot Process Softener
Filters

SOFTENING & DEMINERALIZATION BY ION EXCHANGE

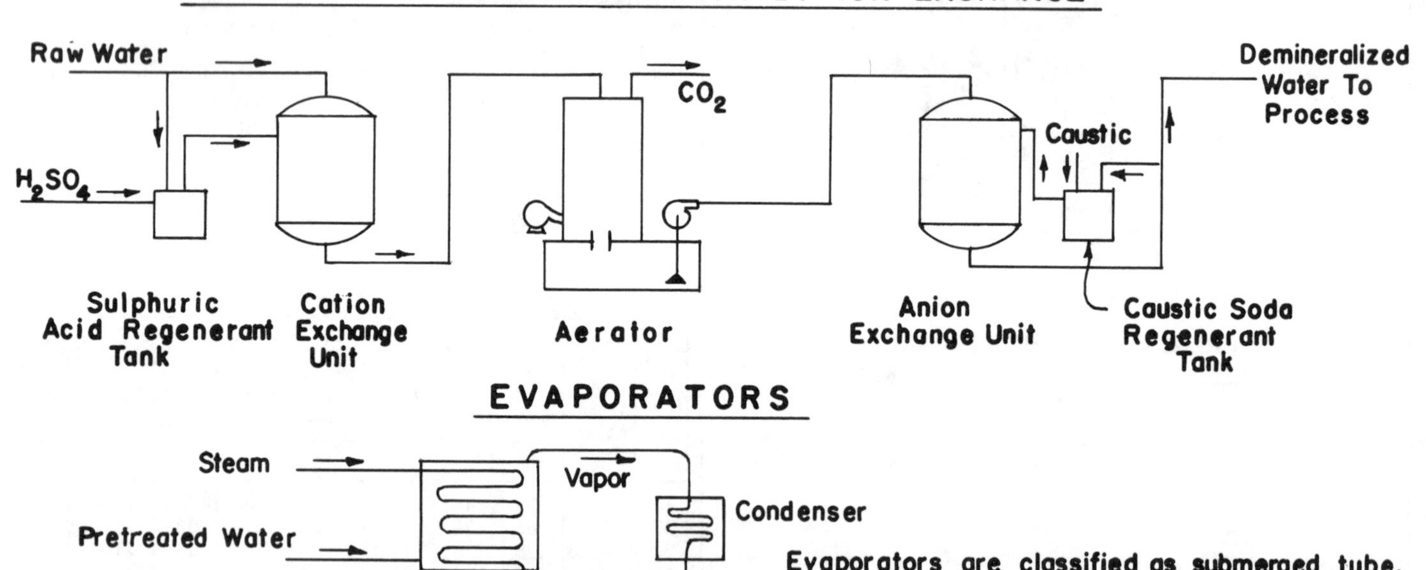
Raw Water
H_2SO_4
CO_2
Demineralized Water To Process
Caustic
Sulphuric Acid Regenerant Tank
Cation Exchange Unit
Aerator
Anion Exchange Unit
Caustic Soda Regenerant Tank

EVAPORATORS

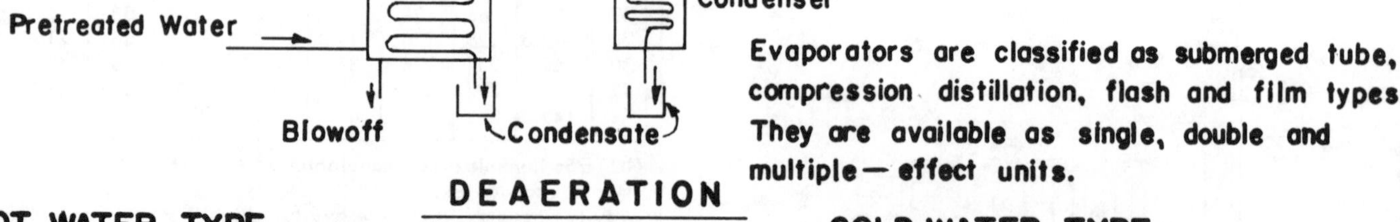
Steam
Vapor
Condenser
Pretreated Water
Blowoff
Condensate

Evaporators are classified as submerged tube, compression distillation, flash and film types. They are available as single, double and multiple—effect units.

DEAERATION

HOT WATER TYPE

COLD WATER TYPE

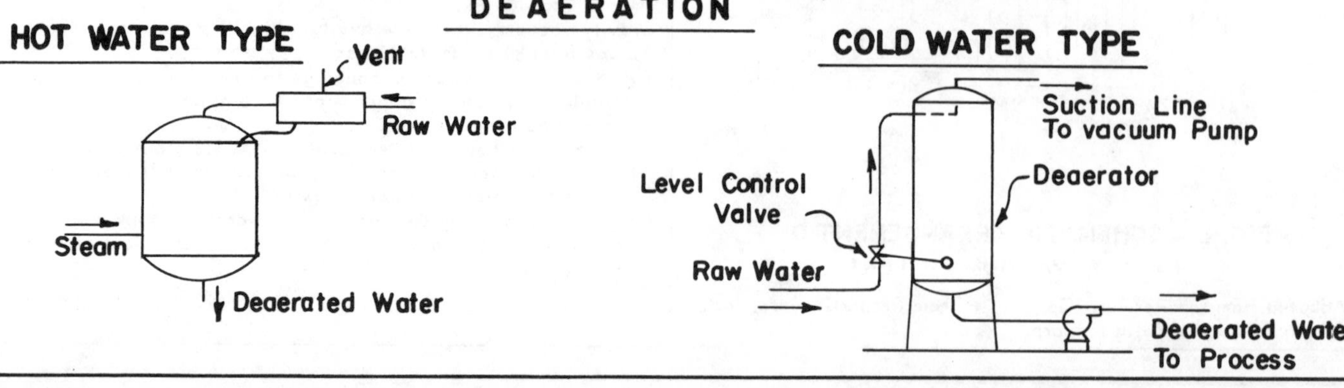
Vent
Raw Water
Steam
Deaerated Water
Suction Line To vacuum Pump
Deaerator
Level Control Valve
Raw Water
Deaerated Water To Process

WATER PURIFICATION-LABORATORY EQUIPMENT-1*

LABORATORY FURNITURE, APPARATUS, AND SUPPLIES
FOR
WATER FILTRATION PLANTS

1. GENERAL - The following schedule of laboratory furniture, apparatus, and supplies has been prepared for the various types of water filtration plants, as follows:

   a. Class A Plants - Filtration plants serving a population of more than 10,000 or treating surface water taken from a polluted source where complete control is required for the operation of the plant. The laboratory equipment provided will permit routine chemical and bacteriological analyses, as well as microscopic examinations.

   b. Class B Plants - Filtration plants taking water from a slightly polluted source or from wells where softening or iron removal is required. The laboratory equipment provided will permit routine chemical control tests, bacteriological tests for coliform organisms, and total plate counts.

   c. Class C Plants - Plants where chlorination only is provided to insure the bacteriological quality of the supply. The laboratory equipment will permit routine residual chlorine, pH, and alkalinity determinations only. The list of laboratory furniture and apparatus is considered to be the minimum required. The equipment listed is complete in itself with all necessary details and accessories.

2. LABORATORY FURNITURE - The following list of laboratory furniture includes those items required in Class A and B plants. It is the intention that this equipment will meet the minimum requirements for laboratory furniture. However, discretion should be exercised in selecting furniture that will fit the space requirements in the laboratory and harmonize with other facilities. It is recommended that the several units be purchased ready-made, as this type is preferable to built-in furniture. The equipment listed is complete in itself with all necessary details and accessories and the items listed are standard with at least three manufacturers. The tops of the table and the benches are of specially treated materials to resist the action of the chemicals. The furniture required is as follows:

   Item 1. One (1) laboratory table of oak, approximately 6'-0" long, 32" wide and 36" high. Table shall be fitted with a maple bottle rack with acid resisting back; rack shall be approximately 60" x 8" x 18". Peg board over sink, 14" x 19½", shall be fitted with 16 pegs. Table top shall be fitted with a lead lined (4 lbs. per sq. ft.) trough, approximately 66" x 4" x 6"/3". All joints shall be burned with pure lead; soldering not permitted. Sink with back at end of table 14" x 18" x 12" mounted on tubular stand shall be of 1¼" selected soapstone. Sink shall have 2" lead "P"

* From Engineering Manual, War Dept., Corps of Engrs., Feb. 1946.

WATER PURIFICATION-LABORATORY EQUIPMENT-2*

trap; for connection see plumbing. Cabinet supporting frame shall be mortised and tenoned, glued and reinforced with bolts. Drawers shall be dovetailed, doors shall be built up and shall have suitable pulls, catches, and hinges. The cabinet shall be equipped with two long drawers at top, four intermediate drawers, and two cupboards. Cupboards shall be fitted with two adjustable removable shelves. Top shall be of Shellstone or approved acid resisting material. Equipment shall have two compression hose bibs over sink for hot and cold water; three straight-way water cocks with hose connection over trough; and $\frac{1}{2}$" pipe conduit with two duplex receptacles with "T" slots, mounted in cast metal conduit fittings. Connections shall be made to floor outlets. All service lines shall be carried to floor with shut-off for each line. Finish for all exposed steel and service piping shall be acid and alkali resisting enamel. Table shall be type #16560 as manufactured by E. H. Sheldon & Co., Muskegon, Mich., or a similar table as manufactured by W. W. Kimball Co., Chicago, Ill., or Hamilton Mfg. Co., Two Rivers, Wis.

Item 2. One (1) balance shelf, 3' long x 2' wide; oak construction, except for 1 5/8" thick birch, black carbonized top; equipped with drawer 21" wide, 15" deep, 3 3/4" high. Shelf to be equivalent to

E. H. Sheldon & Co.	No. 12520
W. W. Kimball Co.	No. 682
Hamilton Mfg. Co.	No. L-1174

Item 3. One (1) supply case, 48" wide, 15" deep, 80" high, upper section glazed 44" wide, 60" high with 3 adjustable shelves; cupboard shelves; cupboard section 44" wide, 12" deep, 13" high, of oak.

E. H. Sheldon & Co.	No. 41040
W. W. Kimball Co.	No. 9562
Hamilton Mfg. Co.	

Item 4. One (1) lower section cupboard unit, 37 5/8" long x 24" wide x 36" high containing 1 double cupboard 34" wide x 28$\frac{1}{4}$" high x 20 5/8" deep, with 1 stone sink 14 1/4" long x 10" wide x 8" deep with 1 set of drain fittings, 1 cold water pantry cock and 1 double electric receptacle for 110-V A.C., table top to be scored to drain to sink.

E. H. Sneldon & Co.	
W. W. Kimball Co.	
Hamilton Mfg. Co.	No. L-804

All furniture shall be of oak construction, natural finish throughout except table tops, which are to be 1 5/8" birch black carbonized.

* From Engineering Manual, War Dept., Corps of Engrs., Feb. 1946.

WATER PURIFICATION—LABORATORY EQUIPMENT-3*

Item 5. In laboratories for Class C plants, the furniture to be provided shall consist of a work table at least 48" long x 36" wide and 30" high, together with a wall cabinet fitted with doors. The cabinet shall be of adequate size to house equipment and chemicals required for a Class C laboratory. These items need not be of special laboratory construction, as kitchen type or built-in-place furniture will suffice. A laboratory sink should be provided convenient to the work table.

3. LABORATORY APPARATUS - The following list of laboratory apparatus includes those items required in Class A, B, and C plants:

Quantity			Description	Catalogue Numbers		
Type of Plant				Central Scientific Co.	E.H. Sargent Co.	Fisher Scientific Co.
A	B	C				
1	1		Balance, analytical, student model, chain type	1020	S–2675	1-905
1	1		Balance cover, rubberized cloth, for above	2020	S–3965	1-990
1	1		Set of balance weights, Class S, 100 mg. to 100 gm.	8171-B	S–4075	2-220
1	1	1	Harvard trip-scale balance	3470	S–3215	2-035
1	1	1	Set of balance weights, 1 to 500 gm.	9140	S–4285	2-300
1	1		37° C electric incubator	46000-A	S–43505-2	11-690-11
1	1		Small refrigerator			
1	1	1	Two-burner hot plate, gas	16685	S–41475-A	4-220
1a	1a	1a	Hot plate, electric, 12" × 18", 3-heat	16650	S–41125	11-500
1	1		Sterilizer oven, gas	48242-A	S–76285-A	14-490-A
1a	1a		Sterilizer oven, electric, 3-heat	48210	S–76265-A	
1	1		Pressure sterilizer, 11" × 24", gas	44120-A	S–76025-A	14-470-A
1a	1a		Pressure sterilizer, electric, 11" × 24"	44120-A	S–76005	14-470-A
1			Electric muffle furnace, Hoskins, Type FD	13675-A	S–36855 Size #202	10-510 Size #202
1	1		Jackson turbidimeter	29105	S–83705	15-380
1	1		Water still, capacity 1 gal. per hr., gas-heated	12760-A	S–27465	9-090
12	12	6	Wide-mouth bottles, glass-stoppered, 30-32 oz.	10450	S–8395	2-910
18	18		Wide-mouth bottles, glass-stoppered, 125 ml.	10450	S–8395	2-910
6			Dropping bottles, 30 ml.	10580	S–8785	3-000
6	6	3	Dropping bottles, 60 ml.	10580	S–8785	3-000
2	2	2	Wash bottles, 1000 ml.	10710	S–9365	3-395
2	2		Pyrex bottles for distilled water, 2½ gal.	10480	S–8475	
4	2	2	Cylinders, double graduated, 100 ml.	16125	S–24695	8-555
2	1		Cylinders, double graduated, 500 ml.	16125	S–24695	8-555
2	1		Cylinders, double graduated, 1000 ml.	16125	S–24695	8-555
1	1		Volumetric flasks, vial mouth and stoppers, 50 ml.	16226	S–34845	10-205
2	1		Volumetric flasks, vial mouth and stoppers, 100 ml.	16226	S–34845	10-205

* From *Engineering Manual*, War Dept., Corps. of Engrs., Feb. 1946.
1a = alternate items.

WATER PURIFICATION–LABORATORY EQUIPMENT–4*

Quantity			Description	Catalogue Numbers		
Type of Plant				Central Scientific Co.	E. H. Sargent Co.	Fisher Scientific Co.
A	B	C				
1	1	1	Thermometer, centigrade, –5° to 205°	19240-D	S–80005	14-985
1	1		Watch glasses, counterpoised (pair), 2½ in. diameter	2250-A	S–3785	2-195
2	2	2	Absorption tube to hold soda lime	755-A	S–28815	9-215
1			Dessicator, 250 mm. diameter	14550	S–25085	8-615
6	4	2	Evaporating dishes, porcelain, 75 mm.	18575	S–25505	8-690
4	2		Evaporating dishes, porcelain, 90 mm.	18575	S–25505	8-690
1	1		Pipette stand	19120	S–78905	14-745
1	1		Burette support, medium	19080	S–78355	14-675
1	1		Burette clamp, steel	12116	S–19105	5-780
1	1		Gas burner, Meeker or Fisher	11105	S–12195	3-900
2	1		Volumetric flasks, vial mouth, and stoppers, 250 ml.	16226	S–34845	10-205
2			Volumetric flasks, vial mouth, and stoppers, 500 ml.	16226	S–34845	10-205
1	1		Volumetric flasks, vial mouth, and stoppers, 1000 ml.	16226	S–34845	10-205
24	24	6	Erlenmeyer flasks, Pyrex, 250 ml.	14905	S–34105	10-040
6	6	4	Erlenmeyer flasks, Pyrex, 500 ml.	14905	S–34105	10-040
2	2		Erlenmeyer flasks, Pyrex, 1000 ml.	14905	S–34105	10-040
4	2		Beakers, Pyrex, 50 ml.	14265	S–4675	2-540
4	2		Beakers, Pyrex, 150 ml.	14265	S–4675	2-540
12	6	4	Beakers, Pyrex, 250 ml.	14265	S–4675	2-540
4	2		Beakers, Pyrex, 400 ml.	14265	S–4675	2-540
6	2		Beakers, Pyrex, 600 ml.	14265	S–4675	2-540
12	4	2	Beaker covers, 3½ in.	15850	S–83605	2-610
1			Filtering flask, 500 ml.	14985	S–34375	10-175
2			Filtering crucibles, Alundum, RA 360, 25 ml.	10065-A	S–24375	8-230
2			Rubber crucible holder	18110	S–24475	8-285
2	2		Funnels, 100 mm. diameter	15070	S–35305	10-320
4	2		Funnels, 65 mm. diameter	15070	S–35305	10-320
2	2	2	Funnels, long stem, 100 mm. diameter	15050	S–35315	10-325
12	12		Nessler tubes, 100 ml. marked at 50 ml.	29060-C	S–21035	7-055
4	4	2	Burettes, graduated in 0.1 ml., 50 ml.	15926-C	S–10635	3-740
2	2		Burettes, graduated in 0.1 ml., 25 ml.	15926-B	S–10635	3-740
4	4	1	Pipettes, volumetric, Exax blue line, 1 ml.	16355	S–69515	13-650
4	4		Pipettes, volumetric, Exax blue line, 5 ml.	16355	S–69515	13-650
4	4		Pipettes, volumetric, Exax blue line, 10 ml.	16355	S–69515	13-650
3	2	1	Pipettes, volumetric, Exax blue line, 25 ml.	16355	S–69515	13-650
3	2	1	Pipettes, volumetric, Exax blue line, 50 ml.	16355	S–69515	13-650

* From *Engineering Manual*, War Dept., Corps of Engrs., Feb. 1946.

WATER PURIFICATION - LABORATORY EQUIPMENT-5*

Quantity			Description	Catalogue Numbers		
Type of Plant				Central Scientific Co.	E.H. Sargent Co.	Fisher Scientific Co.
A	B	C				
2	2	1	Gas burner, Bunsen, Tirrill type	11026-A	S–12295	3-960
2	2	1	Stone jars for waste, 2 gal.	16925-B	S–43945	11-845
4	2		Pinchcocks, Mohr, 2½ in.	12186	S–19495	5-855
2	2	1	Clamps, test tube	12155	S–19555	5-840
1			Spoon, horn	18775-C	S–75175	14-425
1	1		Spatula, stainless steel, 4 in. blade	18755-B	S–75245	14-365
1			Filter pump aspirator		S–33575-A	9-965
2	1	1	File, triangular, 6 in.	88325	S–32235	9-725
1	1	1	Funnel support, hardwood	19035	S–78815	14-740
1	1	1	Iron ring stand; 3 rings, 3 in., 4 in., 5 in.	19072-B	S–78365	14-680
1	1		Nessler tube stand	29070-B	S–21075	7-065
2	2	1	Wire gauze, 4 in.	19970-A	S–85335	15-590
2	2	1	Triangles, 2½ in.	19735-C	S–82415-C	15-280
2	2		Tripod with concentric rings	19775-B	S–82515-B	15-305
2	2	1	Tongs, 9 in.	19600	S–82115	15-200
2	1		Camel's-hair brush, medium	10938-A	S–9725	3-655
2	2	2	Brushes, flask	10985-B	S–9965-B	3-570
2	2	2	Brushes, tube	10970	S–9985	3-575
2	2	2	Brushes, tube	10974	S–10005	3-635
12			Vacuum tubing, ¼ in. bore, feet	18204-C	S–73525	14-165
24	24	12	Rubber tubing, ¼ in. bore, feet	18202-C	S–73515	14-160
12	12	6	Rubber tubing, ⅛ in. bore, feet	18250-A	S–73565	
1	1	1	Corks, bags of assorted sizes, 3 to 16	12404	S–23075	7-785
1	1	1	Cork borer, size 5 to 11 mm., set	12465-B	S–23175	7-845
1	1	1	Rubber stoppers, assorted, lb.	18153-A	S–73305	14-130
4	4		Filter paper, 9 cm., boxes	13250	S–32915	9-795
2	2	1	Filter paper, 12.5 cm., boxes	13250	S–32915	9-795
1	1		Filter paper, quantitative, 9 cm., box	13355	S–32785	9-915
12	6	2	Glass-marking pencils	14015-C	S–65765	13-380
			Hydrogen Ion Comparators			
1	1	1	Hellige comparator, rocket model, 4.0-5-6	21409-C	S–41765-E	
1	1	1	Additional disk, *p*H range 5.2-6.8, chlorphenol Red	21451-E	S–41775-H	
1	1	1	Additional disk, *p*H range 6.0-7.6, Bromthymol Blue	21451-F	S–41775-I	
1	1	1	Additional disk, *p*H range 6.8-8.4, Phenol Red	21451-G	S–41775-K	
1	1	1	Additional disk, *p*H range 8.0-9.6, Thymol Blue	21451-H	S–41775-M	
1	1	1	Chlorine disk, 0.1-1.0 p.p.m.	29175-J		
1a	1a	1a	LaMotte block comparator, *p*H range 3.8-5.4, Bromcreosol Green	21500-E	S–41725-E	
1a	1a	1a	LaMotte block comparator, *p*H range 5.2-6.8, Chlorphenol Red	21500-G	S–41725-G	
1a	1a	1a	LaMotte block comparator, *p*H range 6.0-7.6, Bromthymol Blue	21500-H	S–41725-H	
1a	1a	1a	LaMotte block comparator, *p*H range 7.2-8.8, Cresol Red	21500-K	S–41725-K	
1a	1a	1a	LaMotte block comparator, *p*H range 8.0-9.6, Thymol Blue	21500-L	S–41725-L	
1a	1a	1a	Enslow chlorine comparator, LaMotte 0.05-0.1-0.2-0.5-0.8-1.0 p.p.m.			

* From *Engineering Manual*, War Dept., Corps of Engrs., Feb. 1946.
1a = alternate items

WATER PURIFICATION-LABORATORY EQUIPMENT-6*

Quantity			Description	Catalogue Numbers		
Type of Plant				Central Scientific Co.	E. H. Sargent Co.	Fisher Scientific Co.
A	B	C				
			Bacteriological			
1			Microscope, monocular, Bausch & Lomb	51050-B8	S–52085-H8	12-315-FFS-10
24	24		Pipettes, 1.1 ml. capacity, graduated at 1.0 ml.	24115-B	S-60025	
24	24		Pipettes, 11 ml. capacity, graduated at 10 ml.	24115-H	S-60045	
30	30		Culture dishes with covers, 100 mm. × 15 mm.	44370-D	S–25925-B	8-755
1	1		Culture tubes, large, 7" × ⅞", gross	44500-K	S–79525	
1	1		Vials for large culture tubes, gross	44500-K	S–79525	
1	1		Culture tubes, small, 6" × ¾", gross	44500-J	S–79525	
1	1		Vials for small culture tubes, gross	44500-J	S–79525	
1	1		Thermometer, centigrade, 0° to 110°	19255-A	S–80305	14-985
1			Glass slides, 25 mm. × 75 mm., gross	66310	S–58785	12-550
1			Cover slips, gross	66510-A	S–58715	15-520
1			Lens, reading glass, 4 in.	60410-D	S–44495-C	12-070
1	1		Counter, tallying machine	73320	S–23285	7-905
1	1		Counting apparatus		S–23395	7-925
2	2		Culture dish holder	44400	S–26055	8-775
2	2		Pipette sterilizing boxes, copper, 2½" × 16"	46670-B	S–69815	3-465
2	2		Test-tube support	19200-A	S–79005	14-770
4	4		Test-tube basket	48515-B	S–79835	14-970
1	1		Double boiler, 2 qt.	12970	S–8215	2-750
1			Inoculating needle holder	46220	S–62765	13-090
1			Forceps, cover glass, cornet	66600	S–35225	10-305
1			Inoculating needles, 24 B & S Gauge, Package, Chromel	46210	S–62755	13-095
1			Lens paper, quire	12290	S–44325	11-995
			Microscopical			
1			Sedgewick rafter filter	29032	S–84005	15-405
24			Cloth disks to support filter sand	29034	S–84025	15-415
1			Counting cell	29035	S–84045	15-425
12			Cover glass	29036	S–84055	15-430
1			Eyepiece micrometer disk, whipple	29037	S–84065	15-435
1	1	1	Book *Standard Methods of Water Analysis*, latest edition, A.P.H.A. and A.W.W.A.			
1	1	1	*Laboratory Manual for Chemical and Bacteriological Analysis of Water and Sewage*, Eldridge, Theroux and Mallman, latest edition, McGraw-Hill Book Co.			

*From *Engineering Manual*, War Dept., Corps of Engrs., Feb. 1946

WATER PURIFICATION-LABORATORY EQUIPMENT-7*

STANDARD SOLUTION FOR WATER ANALYSIS ACCORDING TO AMERICAN PUBLIC HEALTH ASSOCIATION STANDARD METHODS OF WATER ANALYSIS, EIGHTH EDITION, 1936

Platinum-cobalt standard, color 500	200 ml.
Standard calcium chloride solution	500 ml.
Standard soap solution	1000 ml.
Standard ferric iron solution	500 ml.
Standard silver nitrate	500 ml.
Soda reagent	1000 ml.
Acid, sulphuric, N/50 solution	1000 ml.
Sodium hydroxide, N/50 solution	500 ml.
Sodium hydroxide, N/44 solution	500 ml.
Potassium thiocyanate	500 ml.
Acid, hydrochloric, dilute, approximately 3 N	500 ml.
Potassium permanganate, approximately N/5	500 ml.
Acid nitric, 6 N Solution	500 ml.
Methyl orange indicator	500 ml.
Phenolphthalein indicator	500 ml.
Erythrosine indicator	500 ml.
Potassium chromate indicator	100 ml.
Orthotolidine indicator	500 ml.
Sodium meta-arsenite 0.5% solution	500 ml.

CULTURE MEDIA
FORMULAS OF "STANDARD METHODS OF WATER ANALYSIS"

Bacto-nutrient agar, dehydrated	1 lb.
Bacto-lactose broth, dehydrated	1 lb.
Levine's-Eosin methylene blue agar, Difco	$\frac{1}{4}$ lb.

BACTERIOLOGICAL STAINING SOLUTIONS
ACCORDING TO "STANDARD METHODS OF WATER ANALYSIS"

Bismark brown	$\frac{1}{4}$ lb.
Carbol fuchsin	$\frac{1}{4}$ lb.
Carbol gentian violet	$\frac{1}{4}$ lb.
Gram's iodine stain	$\frac{1}{4}$ lb.
Methylene blue, Koch's	$\frac{1}{4}$ lb.
Safranin stain	$\frac{1}{4}$ lb.
Potassium iodide	$\frac{1}{2}$ lb.

NOTE: For Class A plants chemicals should be bought in bulk and standard solutions and reagents prepared by the plant chemist. This may apply in part to Class B plants.

† From Engineering Manual, War Dept., Corps of Engrs., Feb. 1946.

WATER PURIFICATION - CORROSION & DEMINERALIZATION CONTROL

TABLE A METHODS USED TO COMBAT THE CORROSIVE ACTION OF WATER

See also "Suggestions for Specifying Paint", DATA BOOK, SPECIFICATIONS AND COSTS, pp. 3-67 and 3-68.

Method	Example
1. Protective coatings.	
a. Galvanizing.	Hot dip galvanizing.
b. Painting.	Red lead, bitumastic, etc.
c. Linings.	Cement, clay, rubber, etc.
d. Plating.	Tin, copper, cadmium, etc.
e. Chemical.	Addition of slight excess of lime, silicate, etc.
2. Cathodic protection.	See pp. 20-11 and 16-50 through 16-53.
3. pH correction.	Addition of lime, caustic soda, etc.
4. Carbon dioxide removal.	Aeration.
5. Dissolved oxygen removal.	
a. Deaeration.	Hot: heating thin layers of water to below boiling. Cold: forming vacuum over thin layers of water.
b. Deactivation.	Heating water and passing over steel scrap.
c. Chemical.	Addition of sodium sulphite.
6. Inhibition.	Addition of polyphosphates (Calgon, etc.).

Definition of demineralization: removal or replacement of mineral content of water.

TABLE B USES OF DEMINERALIZATION

Uses	Process Indicated
1. Water softening.	Ion exchange or lime-soda softening.
2. Removal of tastes and stains.	Ion exchange or aeration followed by settling or filtration.
3. Boiler water treatment.	Ion exchange or lime-soda softening or distillation.
4. Industrial process water treatment.	Ion exchange or lime-soda softening or distillation.
5. Regenerating salt water.	Ion exchange or distillation.

TABLE C LIME-SODA SOFTENING

(The removal of calcium and magnesium by formation of insoluble salts)

Cold Process	Hot Process
1. Addition of lime-soda.	Addition of lime, soda, and phosphate.
2. Mixing.	Mixing.
3. Recarbonation.	Heating by spraying through steam.
4. Settling and/or filtering.	Settling.

Ion exchange process is the replacement of undesirable ions in water by desirable ions from an ion exchanger which has ions of its own to exchange. It is necessary from time to time to reverse the process in order to regenerate the ion exchanger.

TABLE D CLASSES OF ION EXCHANGERS*

1. Cation Exchangers (removes positive ions, e.g., Ca.).
   - A. Inorganic
     1. Natural zeolites (greensand).
     2. Processed greensands.
     3. Synthetic zeolites.
   - B. Organic
     1. Natural (the lignites).
     2. Sulfonated coals (carbonaceous "zeolites").
     3. Synthetic resins.
        - a. Phenolic base.
        - b. Polystyrene.

11. Anion Exchangers (removes negative ions).
   - A. Inorganic
     1. Metallic oxides.
   - B. Organic
     1. Synthetic resins.

Regeneration Methods

Cation exchangers, in general, may be regenerated with common salt (sodium chloride) and be used to remove hardness-forming calcium and magnesium from solutions.

Organic-cation exchangers may also be regenerated with dilute-acid solutions. In the resulting hydrogen condition, they will remove all metalic cations—sodium, potassium, iron, copper, manganese, etc.—from solution.

Anion exchangers, used to remove acid ions—sulfates, chlorides, and nitrates—from solution, may be regenerated with either soda ash (sodium carbonate) caustic soda or ammonia.

(Although both inorganic and organic anion exchangers have been manufactured, only the organics are widely used at present. This is because inorganic materials do not have good operating characteristics for de-ionization.)

*From "Ion Exchange for Water Treatment," E.N.R., June 22, 1950.

WATER PURIFICATION-CHLORINATION

TABLE A – CHLORINE DOSAGES FOR THE TREATMENT OF WATER*

Purpose of Chlorination	Dosage, p.p.m.	Contact Time, min.	Recommended Residual Type, p.p.m.	
Disinfection With combined residual With free residual	1.0-5.0 1.0-5.0	Requirements determined by local health authorities		
Ammonia (NH_3-N) removal	10 × NH_3-N Content	20+	Free	0.1
Taste and odor control	10 × NH_3-N content plus 1-5 p.p.m.	20+	Free	1.0
Hydrogen sulfide (H_2S) Removal	2.22 × S content to free sulfur 8.9 × S content to sulfate	Instantaneous	Free or Combined	0.1
Iron (Fe) removal	0.64 × Fe content	Instantaneous	Combined	0.1
Manganese (Mn) removal	0.65 × Mn content	Variable	Free	0.5
Red water prevention	Maintain a free residual in dead ends	Variable	Free	0.1
Color removal	1.0-10.0	15	Free or Combined	0.1
Algae Control	1.0-10.0	Variable	Free	0.5+
Slime control	1.0-10.0	Residual needed throughout system	Free	0.5+
Control of iron and sulfur bacteria	1.0-10.0		Free	0.1+
Coagulation aid for preparation of: Activated silica (Na_2SiO_3) Chlorinated copperas ($FeSO_4 \cdot 7H_2O$)	 1.4 lb. per gal. of Na_2SiO_3 1 part per 7.8 parts $FeSO_4 \cdot 7H_2O$			

For additional data and application of chlorination see pp. 19-25 and 19-26.

TABLE B – CHLORINATION OF SMALL WATER-SUPPLY SYSTEMS

MAKE-UP OF CHEMICAL SOLUTION FOR CONTAINER†

Water-Supply Pump Capacity, gal. per hr.	Fluid oz. of 5% Sodium Hypochlorite to Be Diluted to 1 gal. with Water — For dosage in p.p.m. ½ p.p.m.	1 p.p.m.	2 p.p.m.	Fluid oz. of 12½% Sodium Hypochlorite to Be Diluted to 1 gal. with Water — For dosage in p.p.m. ½ p.p.m.	1 p.p.m.	2 p.p.m.
60	0.8	1.6	3.2	Use 5% Sodium hypochlorite to make up solution		0.8
80	1.1	2.2	4.3			1.1
100	1.4	2.7	5.4			1.3
120	1.6	3.2	6.5		0.8	1.6
140	1.9	3.8	7.6		0.9	1.9
160	2.2	4.3			1.1	2.2
180	2.4	4.9			1.2	2.4
200	2.7	5.4			1.3	2.7
220	3.0	5.9			1.5	3.0
240	3.2	6.5		0.8	1.6	3.2
260	3.5	7.0		0.9	1.8	3.5
280	3.8	7.6		0.9	1.9	3.8
300	4.0	8.1		1.0	2.0	4.0
320	4.3	Use 12½% sodium hypochlorite to make up solution		1.1	2.2	4.3
340	4.6			1.1	2.3	4.6
360	4.9			1.2	2.4	4.9
380	5.1			1.3	2.6	5.1
400	5.4			1.3	2.7	5.4
420	5.7			1.4	2.8	5.7
440	5.9			1.5	3.0	5.9
460	6.2			1.6	3.1	6.2
480	6.5			1.6	3.2	6.5
500	6.7			1.7	3.4	6.7

Based on 0.0675 fl. oz. 12½%/gal./p.p.m./10 gal. 1 hr. flow
0.270 fl. oz. 5%/gal./p.p.m./10 gal. 1 hr. flow
1 gal. 12½% hypochlorite has 1 lb. available chlorine.

* From Wallace & Tiernan, Inc.
† Proportioners, Inc.

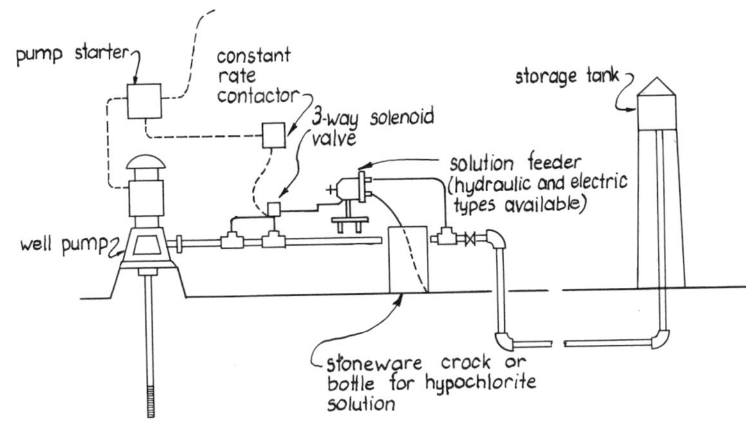
pump starter
constant rate contactor
3-way solenoid valve
storage tank
solution feeder (hydraulic and electric types available)
well pump
stoneware crock or bottle for hypochlorite solution

NOTE: Sodium hypochlorite solution is readily available in glass bottles, 5-gal. carboys, and 30-gal. drums, and is generally known as Javelle water, chlorine bleach, or bleach liquor. The percentage of available chlorine is stated on the container. The materials best suitable for handling hypochlorite solution are rubber, ceramics, glass, and plastics.

Hypochlorite feeders available in electric and hydraulic types. The feeder diaphram is actuated by solenoid impulses controlled by an adjustable contactor. The feeder takes suction from a crock or bottle containing hypochlorite solution and injects the solution into the water main. A pump starter actuates the feeder through a constant rate contactor which causes hypochlorite solution to enter the main while the well pump is in operation.

For very small installations, hypochlorite may be added manually to the tank contents.

EXAMPLE: (See below)

EXAMPLE: Assume that 1 p.p.m. of chlorine is required to provide a residual throughout the water-distribution system of 0.1 p.p.m. (See p. 20-20.)

Given: Well pump capacity of 5 gal. per min. and 2000 gal. of water used per day. (Capacity of pump is stamped on name plate.)

To Find: Amount of 12½% hypochlorite to add to 4 gal. of water to fill a stoneware solution crock or bottle, and amount of hypochlorite used per day.

Solution: 5 gal. per min. = 300 gal. per hr. From Table B, it is found that 2 fluid oz. of 12½% hypochlorite are required for each gallon of water. Therefore for 4 gal. of water, add 4 × 2 or 8 fl. oz. of hypochlorite. Hypochlorite mixes readily with water and requires a slight mixing using a wooden paddle.

Amount of chlorine required per day: $\frac{2000}{1,000,000} \times 8.33 \times 1$ p.p.m. = 0.0167 lb. per day. From Table B, 1 gal. 12½% hypochlorite is equivalent to 1 lb. chlorine. Therefore, amount of hypochlorite required per day = 0.0167 × 1.0 = 0.0167 gal. hypochlorite or 128 oz. × 0.0167 = 2.1 fl. oz. hypochlorite required per day.

WATER DISTRIBUTION–PIPE SYSTEM, FLOW DATA

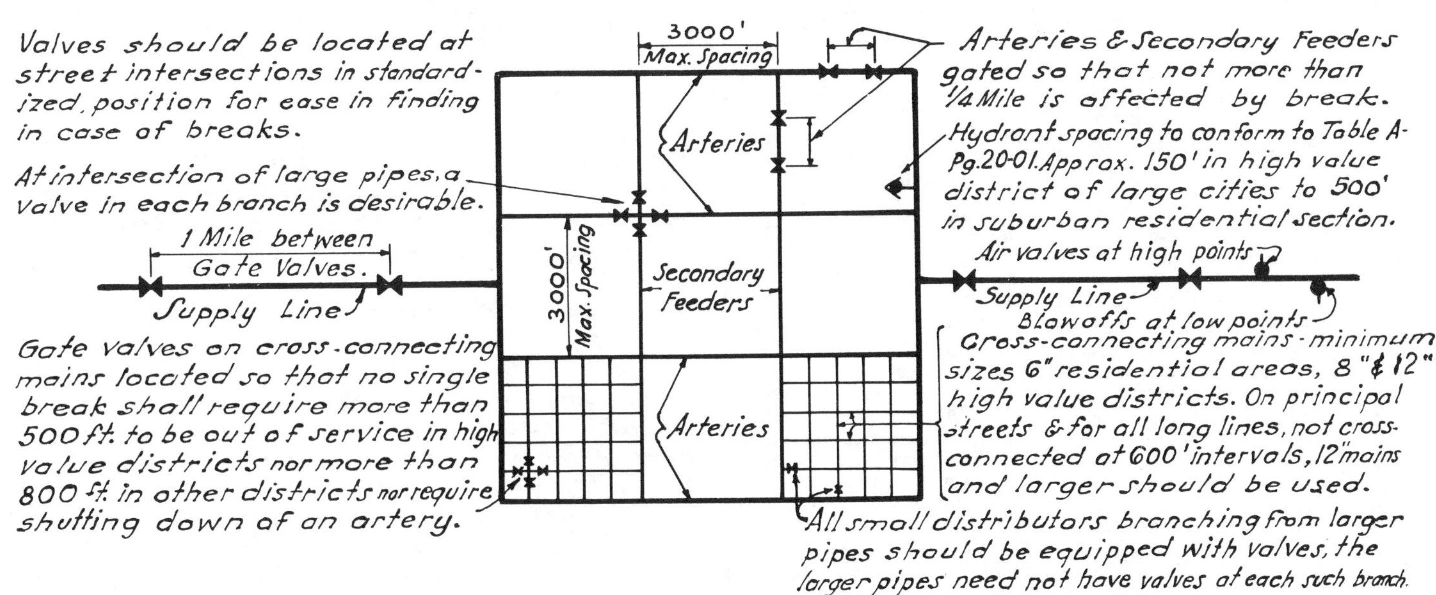
Valves should be located at street intersections in standardized position for ease in finding in case of breaks.
At intersection of large pipes, a valve in each branch is desirable.
1 Mile between Gate Valves.
Supply Line
Gate valves on cross-connecting mains located so that no single break shall require more than 500 ft. to be out of service in high value districts nor more than 800 ft. in other districts nor require shutting down of an artery.
3000' Max. Spacing
Arteries
3000' Max. Spacing
Secondary Feeders
Arteries
Arteries & Secondary Feeders gated so that not more than 1/4 Mile is affected by break.
Hydrant spacing to conform to Table A-Pg.20-01. Approx. 150' in high value district of large cities to 500' in suburban residential section.
Air valves at high points
Supply Line
Blowoffs at low points
Cross-connecting mains - minimum sizes 6" residential areas, 8" & 12" high value districts. On principal streets & for all long lines, not cross-connected at 600' intervals, 12" mains and larger should be used.
All small distributors branching from larger pipes should be equipped with valves, the larger pipes need not have valves at each such branch.

FIG. A - GENERAL ARRANGEMENT OF PIPE SYSTEM.

FIRE FLOW IN PIPES – PRACTICAL APPLICATIONS.

Design of new systems – Select total fire flow from Table A, Pg.20-01. Assume this flow at two or more hydrants at 750 g.p.m. per hydrant in an otherwise closed system. (Allow for domestic use if large.) Compute residual heads by Equivalent Pipe or Hardy Cross method, Pg.22-06 & 22-07; check with requirements of Table B-Pg.20-01. If heads are too low or high, revise size of pipe.

Fire Flow Tests of Existing Systems* – The method of conducting fire flow tests to determine the amount of water available at any given point in a distribution system has been practiced by engineers of the National Board of Fire Underwriters for many years and has been checked by meter delivery and found to be sufficiently accurate for all practical determinations of available fire flow.
The number of hydrants used in a group may vary from 2 to 6, usually 3 or 4 are used, depending on the pressure, size of mains and amount of water required at the point in question. See Table A- Pg.20-01. The flowing pressure at the hydrants is read simultaneously, by Pitot gage. The static pressure at a hydrant at or near the center of the group is read before the other hydrants are opened and the residual pressure read while the hydrants are flowing, the difference being the drop of pressure. Knowing the flowing pressure and the diameter of outlet the flow can be obtained from Table A - Pg.22-02. We now have a certain quantity of water flowing, with a certain drop of pressure, and from this we can obtain the quantity at any other drop in pressure. The most accurate determination is obtained by using the Hazen & Williams formula, or hydraulic slide rule, but for all practical purposes the formula $Q_2 = Q_1\sqrt{h_2/h_1}$ can be used in which h_1 and h_2 are the drop of pressure for quantities Q_1 and Q_2 respectively. In practice it is desired to determine the supply available at 20 pounds, or in some cases at 10 pounds but it is not necessary, or generally desirable, to draw the pressure down to this figure; a drop of pressure at the central hydrant of 10 to 15 pounds is all that is needed and this is generally obtained by flowing not in excess of 1000 gallons from each hydrant, which means a flowing pressure of about 9 pounds from each of two 2½-inch outlets or about 35 pounds from one 2½-inch outlet. The large suction outlet can be used if necessary but it is less desirable because of irregularities in flow. It can be seen from the above that pressure and size of mains, as stated before, must influence the number of hydrants to be opened and quantity of water drawn.
It must be realized that in order to obtain reasonably accurate results at 20 pounds enough water must be drawn to obtain an appreciable drop of pressure, at least 5 pounds. For equipment for tests, see Pg.22-02.

*By R.C. Dennett, Assistant Chief Engr. Natl. Bd of Fire Underwriters.

WATER DISTRIBUTION—HYDRANT & HOSE DISCHARGE

TABLE A - DISCHARGE TABLE FOR HYDRANTS. * †

OUTLET PRESSURE MEASURED BY PITOT GAGE.

FLOWING PRESSURE IN LBS. per SQ. INCH	OUTLET DIAMETER IN INCHES: 2 3/8	2 1/2	2 5/8	2 3/4	2 7/8	3	3 1/8	3 7/8	4	4 3/8	4 1/2	4 5/8
	U.S. GALLONS PER MINUTE.											
1	150	170	180	200	220	240	260	400	430	510	540	580
2	210	240	260	290	310	340	370	570	610	720	770	810
3	260	290	320	350	380	420	450	700	740	890	940	990
4	300	340	370	410	440	480	530	810	860	1030	1090	1150
5	340	380	410	450	500	540	590	900	960	1150	1220	1290
6	370	410	450	500	540	590	640	990	1050	1260	1340	1410
7	400	440	490	540	590	640	690	1070	1140	1360	1440	1520
8	430	480	520	570	630	680	740	1140	1220	1450	1540	1620
9	450	500	550	610	670	730	790	1210	1290	1540	1640	1720
10	480	530	580	640	700	760	830	1280	1360	1630	1730	1820
11	500	560	610	670	730	800	870	1340	1430	1710	1810	1910
12	520	580	640	700	770	840	910	1400	1490	1780	1890	1990
13	550	610	670	730	800	870	950	1450	1550	1850	1960	2070
14	570	630	690	760	830	900	980	1510	1610	1920	2040	2150
15	590	650	720	790	860	940	1020	1560	1660	1990	2110	2220
16	610	670	740	810	890	970	1050	1620	1720	2060	2180	2300
17	620	690	760	840	910	1000	1080	1660	1770	2120	2240	2370
18	640	710	780	860	940	1030	1110	1710	1820	2180	2310	2440
19	660	730	810	890	960	1050	1140	1760	1870	2240	2370	2510
20	680	750	830	910	990	1080	1170	1800	1920	2290	2430	2570
22	710	790	870	950	1040	1130	1230	1890	2020	2400	2550	2700
24	740	820	910	1000	1090	1180	1290	1970	2110	2510	2660	2810
26	770	860	940	1040	1130	1230	1340	2050	2190	2620	2770	2930
28	800	890	980	1070	1170	1280	1390	2130	2280	2720	2880	3040
30	830	920	1010	1110	1210	1320	1430	2210	2350	2820	2980	3150
32	860	950	1050	1150	1260	1370	1480	2280	2430	2910	3080	3250
34	880	980	1080	1180	1290	1410	1530	2350	2510	3000	3170	3350
36	910	1010	1110	1220	1330	1450	1580	2420	2580	3080	3260	3440
38	930	1040	1140	1250	1370	1490	1620	2480	2650	3170	3350	3540
40	960	1060	1170	1290	1400	1530	1660	2550	2720	3250	3440	3630

*Computed with coefficient, C=0.90, to nearest 10 gals. per min.

EQUIPMENT FOR HYDRANT FLOW TESTS.

The equipment necessary consists of a hydrant cap tapped to take a pressure gage, an adequate supply of pitot blades and 50 pound gages, and one 100 or 200 pound gage, depending upon the static pressure on the system. A commercial type pitometer is shown in Fig. C - Pg. 22-20. If the hydrants used have two or more outlets a pressure gage on one outlet while another outlet is flowing will give approximately the same results as the use of a pitot tube. The pitot tube used in determining discharges from hydrant outlets, held at the center of the stream about half the diameter in front of the outlet, has a straight blade about 4 inches long, equipped with a union or threaded connection for the gage; the latter, for most convenient use, is 3 inch graduated in half pounds, from 0 to 50 pounds. **

** By R.C. Dennett, Engr. Natl. Bd. of Fire Underwriters.

TABLE B - FIRE STREAM & HOSE DATA. ††

SIZE OF NOZZLE IN INCHES

Nozzle pressures by Pitot, psi	1				1 1/8				1 1/4				1 3/8				1 1/2			
	Discharge, gpm	Pressure loss 100 ft 2½-in. hose, psi	Vertical reach, ft. of stream	Horizontal reach, ft. of stream	Discharge, gpm	Pressure loss 100 ft 2½-in. hose, psi	Vertical reach, ft	Horizontal reach, ft	Discharge, gpm	Pressure loss 100 ft 2½-in. hose, psi	Vertical reach, ft	Horizontal reach, ft	Discharge, gpm	Pressure loss 100 ft 3-in. hose, psi	Vertical reach, ft	Horizontal reach, ft	Discharge, gpm	Pressure loss 100 ft 3-in. hose, psi	Vertical reach, ft	Horizontal reach, ft.
20	132	4.8	35	37	167	7.3	36	38	206	10.6	36	39	250	5.8	36	40	298	8.1	37	42
25	148	5.8	43	42	187	8.9	44	44	230	13.1	45	46	280	7.2	45	47	333	10.1	46	49
30	162	6.8	51	47	205	10.5	52	50	253	15.5	52	52	307	8.6	53	54	365	11.9	54	56
35	175	7.9	58	51	221	12.1	59	54	273	17.8	59	58	331	9.9	60	59	394	13.7	62	62
40	187	8.9	64	55	237	13.8	65	59	292	20.0	65	62	354	11.2	66	64	422	15.5	69	66
45	198	9.9	69	58	251	15.3	70	63	309	22.2	70	66	376	12.5	72	68	447	17.3	74	71
50	209	10.9	73	61	265	16.8	75	66	326	24.7	75	69	396	13.8	77	72	472	19.1	79	75
55	219	11.9	76	64	277	18.3	79	69	342	27.2	80	72	415	15.1	81	75	494	20.8	83	78
60	229	12.8	79	67	290	19.8	83	72	357	29.6	84	75	434	16.4	85	77	517	22.6	87	80
65	238	13.8	82	70	301	21.3	86	75	372	31.7	87	78	451	17.6	88	79	537	24.3	90	82
70	247	14.8	85	72	313	22.9	88	77	386	33.9	90	80	469	18.8	91	82	558	26.0	92	84
75	256	15.8	87	74	324	24.5	90	79	399	36.1	92	82	485	20.0	93	84	578	27.8	94	86
80	264	16.7	89	76	335	26.1	92	81	413	38.6	94	84	500	21.2	95	86	596	29.5	96	88
85	272	17.7	91	78	345	27.7	94	83	425	40.8	96	87	516	22.5	97	88	614	31.2	98	90
90	280	18.7	92	80	355	29.3	96	85	438	43.1	98	89	531	23.8	99	90	633	32.9	100	91

NOTES:

1. The above values of the discharge and pressure loss in 2½ and 3 inch best quality rubber lined hose are those given by National Board of Fire Underwriters.
2. For inside hand lines 2½ inch hose lines with 1⅛ inch shut off nozzles are generally used.
3. For fighting large fires from the outside, 1½ inch nozzles are usually used with 3 inch hose lines or siamesed 2½ inch lines. Maximum length of hose 400 to 500 feet.
4. For residential areas, 175 g.p.m. fire streams are standardized as satisfactory, and for business districts of ordinary character 250 g.p.m. stream.
5. The ordinary capacity of pumper is 750 g.p.m. which should be supplied by one hydrant.

† From Natl. Bd. of Fire Underwriters. †† From Davis, Handbook of Applied Hydraulics, McGraw-Hill.

WATER DISTRIBUTION—FLOW IN PIPES—I

HAZEN & WILLIAMS FORMULA

$$V = 1.318\,C r^{0.63} S^{0.54}$$

Where V = mean Velocity in feet per second, r = hydraulic radius in feet, S = hydraulic Slope in feet per foot of length, C = Hazen & Williams Coefficient of roughness.

QUANTITY OF FLOW FORMULA

$$Q = AV$$

Where Q = flow in cubic feet per second, A = cross-sectional Area of pipe in square feet, V = mean Velocity in feet per second.

TABLE A – VALUES OF "C" IN HAZEN AND WILLIAMS FORMULA FOR VARIOUS KINDS OF PIPE.

KIND OF PIPE	"C"
CAST IRON	
New well laid	130
4 to 6 years old	120
10 to 12 years old	110
13 to 20 years old	100
26 to 35 years old (4" to 10" sizes)	80
37 to 47 years old (12" to 60" sizes)	80
ASBESTOS-CEMENT	140
CEMENT-LINED PIPE	
Applied by hand	125
Centrifugally applied	140
RIVETED STEEL PIPE	
New (66" to 144" sizes)	110
10 years old (66" to 144" sizes)	100
Over 10 years old	95
2", 2½" & 3" PIPE	
Very smooth & straight Brass, Tin	140
Ordinary straight Brass, Tin, etc.	130
Smooth New Iron	120
Ordinary Iron	100

KIND OF PIPE	"C"
Old Iron	80
Very rough	60
Badly tuberculated	40
WROUGHT IRON (⅛" to 1½" Sizes)	
Very smooth and straight	140
Smooth New Iron	120
Ordinary Iron	100
Old Iron	80
Very rough	60
BRASS PIPE (0.03 to 1.2 Sizes)	130
(May also be used for straight Lead, Tin & drawn Copper Pipes)	
TILE (Good condition)	110
WOOD STAVE OR SMOOTH WOODEN PIPE	120
BRICK	100
CONCRETE	120
FIRE HOSE	
Extremely smooth	143
Rubber lined	125-140
Mill Hose	100-120
Unlined Linen Hose	85-95

TABLE B – RELATIVE CARRYING CAPACITY AND HEAD LOSS FOR VARIOUS VALUES OF "C" FOR USE WITH FLOW CHART, FIG. A-Pg. 6-63.

"C"	"K_1"	"K_2"
40	5.46	0.40
60	2.58	0.60
80	1.51	0.80
90	1.22	0.90
100	1.00	1.00
110	0.84	1.10
120	0.71	1.20
130	0.62	1.30
140	0.54	1.40

Explanation:

1. To determine loss of head with value of "C" other than 100, multiply the loss of head found in Fig. A - Pg. 22-04 by the "K_1" value given in table at left.

2. To determine quantity of flow with value of "C" other than 100, multiply the quantity of flow found in Figure A - Pg. 22-04 by the "K_2" value given in table at left.

WATER DISTRIBUTION—FLOW IN PIPES*—2

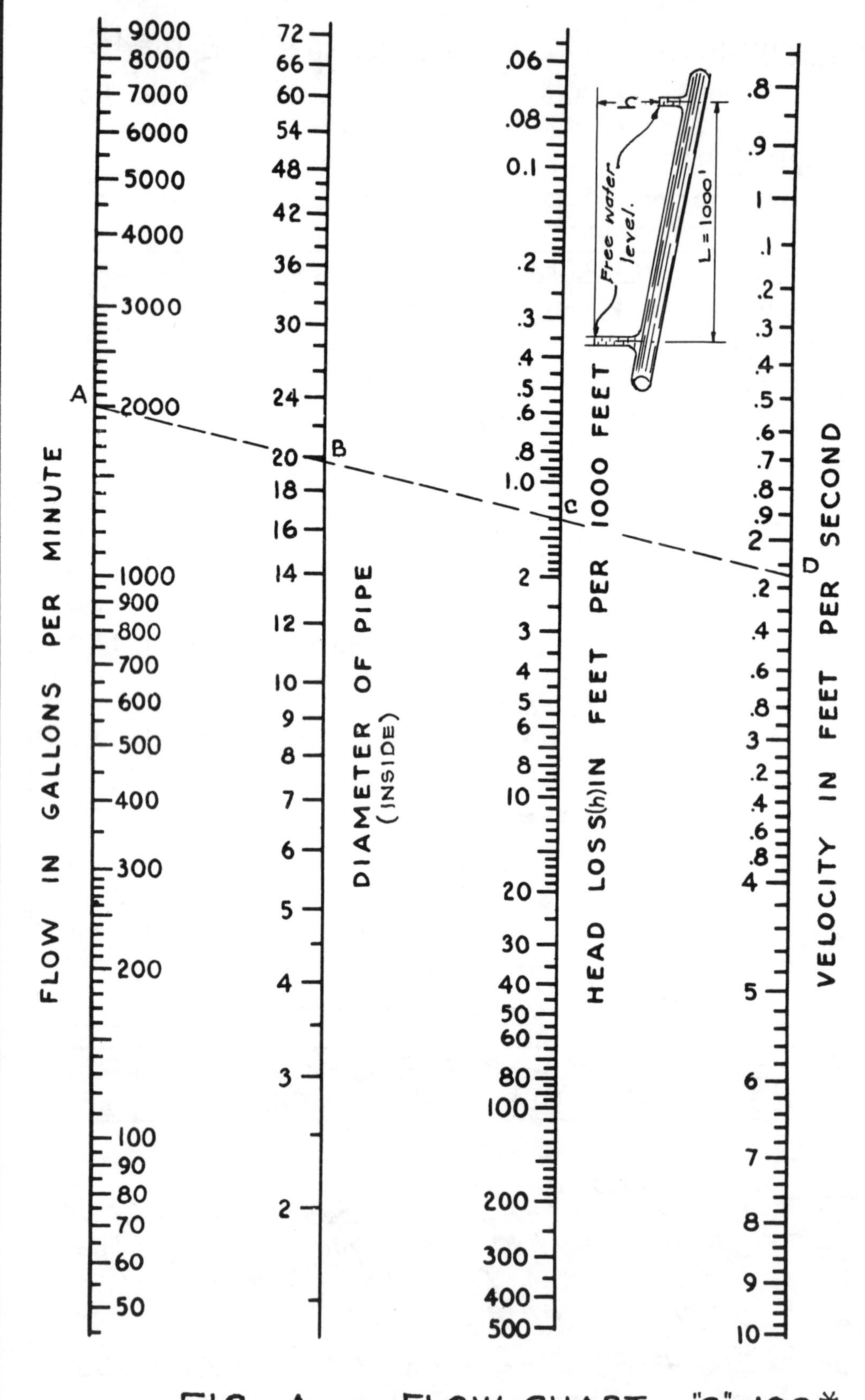
FLOW IN GALLONS PER MINUTE
DIAMETER OF PIPE (INSIDE)
HEAD LOSS(h) IN FEET PER 1000 FEET
VELOCITY IN FEET PER SECOND
A
B
C
D
h
Free water level.
L = 1000'

FIG. A - FLOW CHART "C" = 100*

(BASED ON HAZEN AND WILLIAMS FORMULA)

USE OF FIG. A*

Knowing any two of the following: FLOW, DIAMETER, HEAD LOSS, VELOCITY, the other two values can be found by placing a straight-edge on the factors known and reading desired factors (Known A & B desired C & D.)

TABLE B - CONVERSION OF FLOW.

Gals/Min.	Cu.ft./Sec.	Gallons/Day
10	0.022	14,400
20	0.045	28,800
30	0.067	43,200
40	0.089	57,600
50	0.111	72,000
60	0.134	86,400
70	0.156	100,800
80	0.178	115,200
90	0.201	129,600
100	0.223	144,000
120	0.27	172,800
140	0.31	201,600
160	0.36	230,400
180	0.40	259,200
200	0.45	288,000
220	0.49	316,800
240	0.54	345,600
260	0.58	374,400
280	0.63	403,200
300	0.67	432,000
350	0.78	504,000
400	0.89	576,000
450	1.00	648,000
500	1.11	720,000
550	1.23	792,000
600	1.34	864,000
700	1.56	1,008,000
800	1.78	1,152,000
900	2.01	1,296,000
1000	2.23	1,440,000
1100	2.45	1,584,000
1200	2.67	1,728,000
1300	2.90	1,872,000
1400	3.12	2,016,000
1500	3.34	2,160,000
1600	3.57	2,304,000
1700	3.79	2,448,000
1800	4.01	2,592,000
1900	4.23	2,736,000
2000	4.46	2,880,000
3000	6.68	4,320,000
4000	8.92	5,760,000
5000	11.14	7,200,000
6000	13.36	8,640,000

*See Table B, Page 22-03, for use of Fig. A with values of "C" other than 100.

WATER DISTRIBUTION — FLOW IN PIPES — 3

TABLE A - FLOW OF WATER IN NEW HOUSE SERVICE PIPES.*

Pressure in main, pounds per square inch	Discharge in cubic feet per minute — Nominal internal diameter of pipe (inches) ½	¾	1	1¼	2	3	4
Through 35 feet of service pipe, no back pressure							
30	1.10	3.01	6.13	16.58	33.34	88.16	173.85
40	1.27	3.48	7.08	19.14	38.50	101.80	200.75
50	1.42	3.89	7.92	21.40	43.04	113.82	224.44
60	1.56	4.26	8.67	23.44	47.15	124.68	245.87
75	1.74	4.77	9.70	26.21	52.71	139.39	274.89
100	2.01	5.50	11.20	30.27	60.87	160.96	317.41
130	2.29	6.28	12.77	34.51	69.40	183.52	361.91
Through 100 feet of service pipe, no back pressure							
30	0.66	1.84	3.78	10.40	21.30	58.19	118.13
40	0.77	2.12	4.36	12.01	24.59	67.19	136.41
50	0.86	2.37	4.88	13.43	27.50	75.13	152.51
60	0.94	2.60	5.34	14.71	30.12	82.30	167.06
75	1.05	2.91	5.97	16.45	33.68	92.01	186.78
100	1.22	3.36	6.90	18.99	38.89	106.24	215.68
130	1.39	3.83	7.86	21.66	44.34	121.14	245.91
Through 100 feet of service pipes, and 15 feet vertical rise							
30	0.55	1.52	3.11	8.57	17.55	47.90	96.17
40	0.66	1.81	3.72	10.24	20.95	57.20	116.01
50	0.75	2.06	4.24	11.67	23.87	65.18	132.20
60	0.83	2.29	4.70	12.94	26.48	72.28	146.61
75	0.94	2.59	5.32	14.64	29.96	81.79	165.90
100	1.10	3.02	6.21	17.10	35.00	95.55	193.82
130	1.26	3.48	7.14	19.66	40.23	109.82	222.75
Through 100 feet of service pipe, and 30 feet vertical rise							
30	0.44	1.22	2.50	6.80	14.11	38.63	78.54
40	0.55	1.53	3.15	8.68	17.79	48.68	98.98
50	0.65	1.79	3.69	10.16	20.82	56.98	115.87
60	0.73	2.02	4.15	11.45	23.45	64.22	130.59
75	0.84	2.32	4.77	13.15	26.95	73.76	149.99
100	1.00	2.75	5.65	15.58	31.93	87.38	177.67
130	1.15	3.19	6.55	18.07	37.92	101.33	206.04

TABLE B - EQUIVALENT PIPE SIZES.*

Diam., in.	Diam., in. 3	4	5	6	8	10	12	14	16	18	20	22	24	30	36	48
48	...	...	...	...	...	...	...	...	15.59	11.61	8.92	7 03	5.65	3.24	2.05	1
44	...	...	...	...	...	...	...	17.50	12.54	9.34	7.17	5.66	4.55	2.61	1.65	..
40	...	...	...	...	...	...	20.23	13.47	9.85	7.34	5.64	4.44	3.57	2.05	1.26	..
36	...	...	...	...	...	...	15.58	8.41	7.59	5.65	4.34	3.42	2.74	1.58	1.	..
33	...	...	...	...	34.55	19.78	12.54	8.52	6.11	4.55	3.49	2.75	2.21	1.27	...	..
30	...	...	...	...	27.09	15.54	9.85	6.54	4.80	3.57	2.74	2.16	1.74	1.	...	..
27	...	...	...	42.95	16.61	9.96	7.59	5.16	3.70	2.75	2.11	1.67	1.34	...	...	..
24	...	...	50.50	32.00	15.58	8.92	5.65	3.84	2.75	2.05	1.57	1.24	1.	...	...	..
22	...	70.96	40.65	25.73	12.53	7.17	4.55	3.09	2.16	1.65	1.26	1.	...	...	...	..
20	...	55.96	32.05	20.29	9.88	5.66	3.58	2.43	1.74	1.30	1.	...	...	...	...	..
18	...	42.01	24.63	15.58	7.25	4.34	2.75	1.87	1.34	1.	...	...	...	...	...	..
16	65.77	32.01	18.31	11.60	5.65	3.23	2.05	1.39	1.	...	...	...	...	...	...	..
14	47.14	22.94	13.15	8.32	4.05	2.32	1.47	1.	...	...	...	...	...	...	...	..
12	32.05	15.60	8.93	5.65	2.75	1.57	1.	...	...	...	...	...	...	...	...	..
10	20.31	9.88	5.66	3.58	1.74	1.	...	...	...	...	...	...	...	...	...	..
8	11.63	5.66	3.24	2.05	1.	...	...	...	...	...	...	...	...	...	...	..
6	5.66	2.75	1.58	1.	...	...	...	...	...	...	...	...	...	...	...	..
5	3.58	1.75	1.	...	...	...	...	...	...	...	...	...	...	...	...	..
4	2.05	1.	...	...	...	...	...	...	...	...	...	...	...	...	...	..
3	1.	...	...	...	...	...	...	...	...	...	...	...	...	...	...	..

The figures shown above indicate how many pipes of the sizes printed at the top are equivalent to one pipe of the size in the first column.

EXAMPLE: Given: 30" Pipe.
Find: Equivalent number of 18" Pipes.
Table B. above, gives 3.57, use 4-18" Pipes.

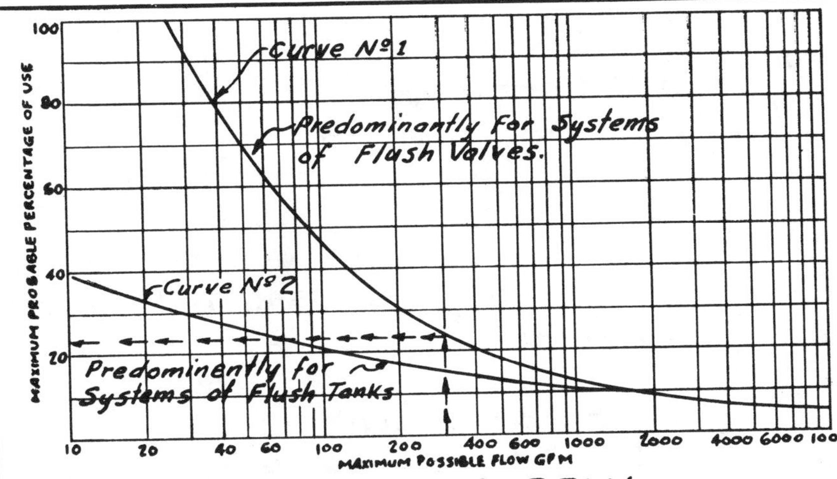

100
80
60
40
20
MAXIMUM PROBABLE PERCENTAGE OF USE
Curve Nº 1
Predominantly for Systems of Flush Valves.
Curve Nº 2
Predominently for Systems of Flush Tanks
10
20
40
60
100
200
400
600
1000
2000
4000
6000
10000
MAXIMUM POSSIBLE FLOW GPM

- BUILDING SUPPLY -

FIG. C - FACTORS OF USAGE.**

EXAMPLE OF USE OF FIG. C

Assume maximum flow 300 g.p.m. (In a specific case this value would be computed using Tables E & G, Pg. 20-02 and number of fixtures involved.)

Using Curve Nº 1 in Fig. C at left (flush valves to be used) the factor of usage will be **24%** and the probable flow will be 300 × **.24** = **72** g.p.m. This **72** g.p.m. is the average flow which should be provided to the bldg. from the outside supply.

NOTE: Chart may not be conservative for plants such as gymnasium and manufacturing plants where total demands occur within certain hours. Factor of usage may be 100%.

* Data from Flinn, Weston & Bogert, Waterworks Handbook, McGraw-Hill.
** Chart from Heating, Ventilating, Air Conditioning Guide, 1943, Chapt. 46.

WATER DISTRIBUTION—EQUIVALENT PIPES

PIPES IN PARALLEL

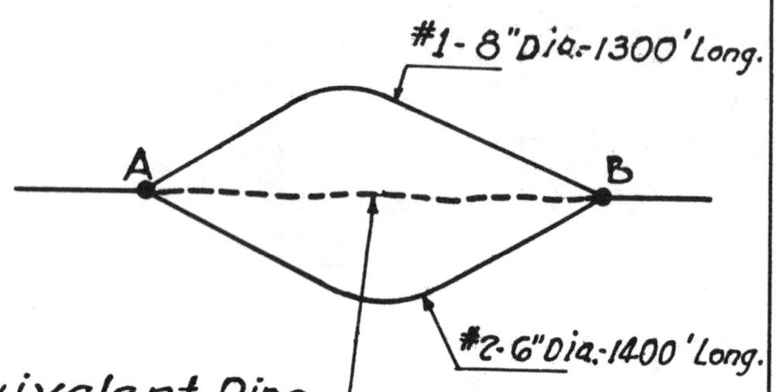
#1- 8"Dia.-1300' Long.
A
B
#2- 6"Dia.-1400' Long.
Equivalent Pipe (Assumed as 1000' in length.)

Example:

Given:

Sizes and lengths of parallel pipes #1 and #2 as shown in above figure.

Required:

Equivalent Pipe (Length = 1000 feet assumed).

Solution:

1. Loss of head through Pipe #1 must always equal loss of head through Pipe #2 between points A & B.
2. Assume any arbitrary head loss (convenient for use in Fig. A, Page 22-04) say 10 feet.
3. Calculate head loss in feet per 1000 feet for pipes #1 & #2.

$$\#1 - \frac{10}{1300} \times 1000 = 7.7'/1000'$$

$$\#2 - \frac{10}{1400} \times 1000 = 7.1'/1000'$$

4. Use Fig. A, Page 22-04 to find flow in gallons per minute.

#1 - 8" Diam.	7.7' loss	495 Gals/Min.
#2 - 6" Diam.	7.1' loss	220 Gals./Min.

Total Q through both pipes = 715 Gals./Min.

5. Using Fig. A, Page 22-04 with head loss = 10'/1000' & Q = 715 Gals/Min. the equivalent pipe size is found to be 8.8" Diameter.

PIPES IN SERIES

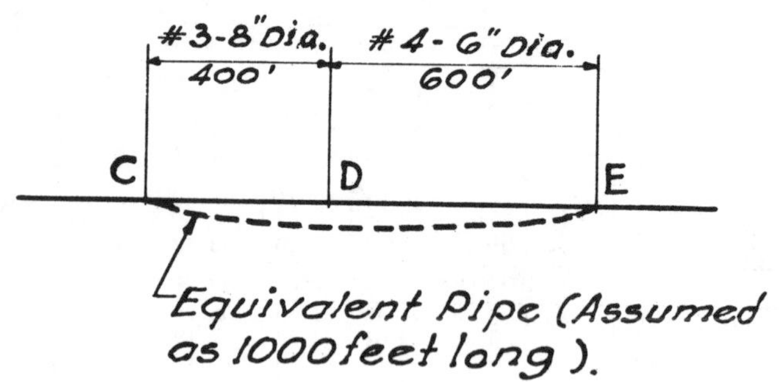
#3-8"Dia.
400'
#4-6"Dia.
600'
C
D
E
Equivalent Pipe (Assumed as 1000 feet long).

Example:

Given:

Sizes and lengths of pipes #3 and #4 as shown in above figure.

Required:

Equivalent Pipe (Length = 1000 feet assumed).

Solution:

1. Quantity of water flowing through pipe #3 and pipe #4 is the same.
2. Assume any arbitrary flow through pipes #3 and #4 (convenient for use with Fig. A, Page 22-04) say 500 Gals./Min.
3. Using Fig. A, Page 22-04 find head loss for pipes #3 and #4.

	Head loss
#3 - 8" Dia. - 400' Long	
500 Gals./Min.	0.4 × 8.0 = 3.2 ft.
#4 - 6" Dia. - 600' Long	
500 Gals./Min.	0.6 × 33.0 = 19.8 "

Total head loss in both pipes = 23.0 ft.

4. Using Fig. A, Page 22-04 with head loss = 23.0'/1000' and Q = 500 Gals./Min. the equivalent pipe size is found to be 6.5" Diameter.

WATER DISTRIBUTION–FLOW IN PIPES–HARDY CROSS†

BASIC EQUATIONS

$$h_f = kQ_m^{1.85}$$

$$C_f = \frac{\text{Difference of } h_f}{1.85 \times \Sigma k Q_m^{0.85}}$$

h_f = loss of head factor.
k = Value in Table B - below. ††
Q = total flow (100 g.p.m. in example below).
Q_m = % of Q in any pipe.
C_f = correction factor to Q_m.

Note: h_f must balance in any loop. This involves a process of trial & error.

EXAMPLE: *Given* system shown in Fig. A, Q = 100 g.p.m. at A & D. *Required*: Loss of head between A & D. *Solution:* Assume Q_m values shown thus (50) in Fig. A. Determine k from Table B. Proceed as in First and Second Trials below.

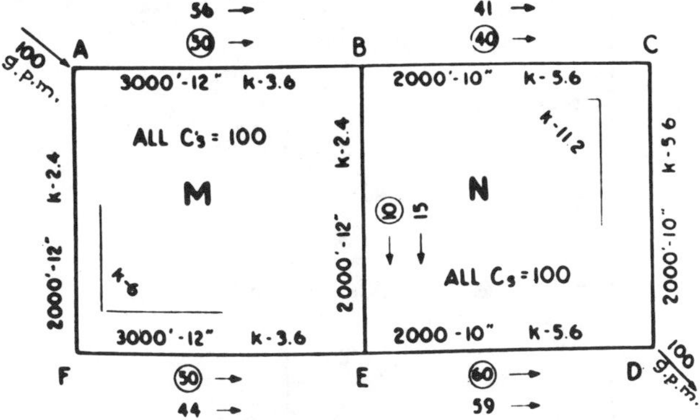
100 g.p.m.
A
B
C
D
E
F
56 →
(50) →
41 →
(40) →
3000'-12" k-3.6
2000'-10" k-5.6
ALL C's = 100
M
N
k-112
k-2.4
2000'-12"
k-2.4
2000'-12"
(10) 15
ALL C's = 100
k-5.6
2000'-10"
3000'-12" k-3.6
2000-10" k-5.6
(50) →
44 →
(60) →
59 →
100 g.p.m.

FIG. A–SYSTEM OF TWO CIRCUITS.

Note: $kQ_m^{1.85} = kQ_m^{0.85} \times Q_m$

First Trial

Circuit M

Loop	$(k \times Q_m^{0.85} = kQ_m^{0.85}) \times Q_m =$	h_f	
AB	(3.6 × 27.8 = 100) × 50 =	5000	
BE	(2.4 × 7.1 = 17) × 10 =	170	5170
AFE	(6.0 × 27.8 = 167) × 50 =	8330	—8330
	284*		3160**

*Adding the $kQ_m^{0.85}$ = 284.
**Subtracting the h_f values = 3160.

$$C_f = \text{Correction} = \frac{3160}{1.85 \times 284} = 6.02$$

First Trial Continued:-

Circuit N

BCD	(11.2 × 23 = 258) × 40 =	10310	—10310
BE	(2.4 × 7.1 = 17) × 10 =	170	
ED	(5.6 × 32.4 = 181) × 60 =	10900	—11070
	456		760

$$C_f = \text{Correction} = \frac{760}{1.85 \times 456} = 0.902$$

In making this correction it should be noted that while circuit M requires a correction of 6, circuit N requires a correction of only 1. This means that the flow from A to B should be increased 6 but at point B this flow should be divided between BCD and BE so that the flow in BCD is increased only 1; therefore, 1 is added to BCD and 5 is added to BE.

Second Trial

Circuit M

Loop	$(k \times Q_m^{0.85} = kQ_m^{0.85}) \times Q_m =$	h_f	
AB	(3.6 × 30.6 = 110) × 56 =	6165	
BE	(2.4 × 10.0 = 24) × 15 =	360	— 6525
AFE	(6.0 × 25.0 = 150) × 44 =	6600	— 6600
	284		75

$$C_f = \text{Correction} = \frac{75}{1.85 \times 284} = 0.143$$

Circuit N

BCD	(11.2 × 23.5 = 263) × 41 =	10800	—10800
BE	(2.4 × 10.0 = 24) × 15 =	360	
ED	(5.6 × 32.0 = 179) × 59 =	10560	—10920
	466		120

$$C_f = \text{Correction} = \frac{120}{1.85 \times 466} = 0.139$$

When h_f balances in a circuit the assumed per cent of flow is correct. Having determined % of flow in piping, determine Q for any pipe by multiplying Q_m × total Q, 100 g.p.m. in example above. With Q and size of pipe known, the total loss of head can be computed using Fig. A, Page 22-04.

TABLE B–VALUES OF K FOR 1000 FT. OF PIPE.††

PIPE DIA.	"C" VALUES (SEE TABLE A PG. 6-61)					
	90	100	110	120	130	140
4	300.0	248.0	208.0	177.0	153.0	133.0
6	41.0	33.7	28.4	24.2	20.9	18.2
8	10.0	8.4	7.0	6.0	5.2	4.5
10	3.4	2.8	2.4	2.0	1.7	1.5
12	1.5	1.2	1.0	.83	.71	.62
14	.66	.55	.46	.39	.34	.30
16	.35	.29	.24	.20	.18	.15
18	.20	.16	.14	.12	.10	.09
20	.12	.10	.08	.07	.06	.05
24	.049	.04	.03	.03	.02	.02
30	.016	.013	.011	.010	.008	.007
36	.0067	.0054	.0046	.0039	.0034	.003

EXAMPLE: *Given:* D = 12", C = 100
Required: k
Solution: k = 1.2

TABLE C–VALUES OF THE 0.85 POWER OF NUMBERS.

No.	0	1	2	3	4	5	6	7	8	9
0	0	1.0	1.8	2.5	3.2	3.9	4.6	5.2	5.9	6.5
10	7.1	7.7	8.3	8.9	9.4	10.0	10.6	11.1	11.6	12.2
20	12.8	13.3	13.8	14.4	14.9	15.4	15.9	16.5	17.0	17.5
30	18.0	18.5	19.0	19.5	20.0	20.5	21.0	21.5	22.0	22.5
40	23.0	23.5	24.0	24.5	25.0	25.4	25.8	26.4	26.8	27.3
50	27.8	28.2	28.7	29.2	29.6	30.1	30.6	31.0	31.4	32.0
60	32.4	33.0	33.3	33.9	34.2	34.7	35.1	35.6	36.0	36.5
70	37.0	37.4	37.9	38.3	38.8	39.1	39.6	40.0	40.5	41.0
80	41.5	42.0	42.4	42.8	43.3	43.7	44.1	44.5	45.0	45.4
90	45.8	46.3	46.7	47.1	47.6	48.0	48.4	48.8	49.2	49.6

EXAMPLE: *Given:* Number 54
Required: $54^{0.85}$
Solution: $54^{0.85} = 29.6$

† Doland System for using Hardy Cross method, from Water Works & Sewage-June '1943 article by R.D. Taylor. †† Based on Hazen-Williams Formula.

WATER DISTRIBUTION–HEAD LOSS IN VALVES, FITTINGS, ETC.

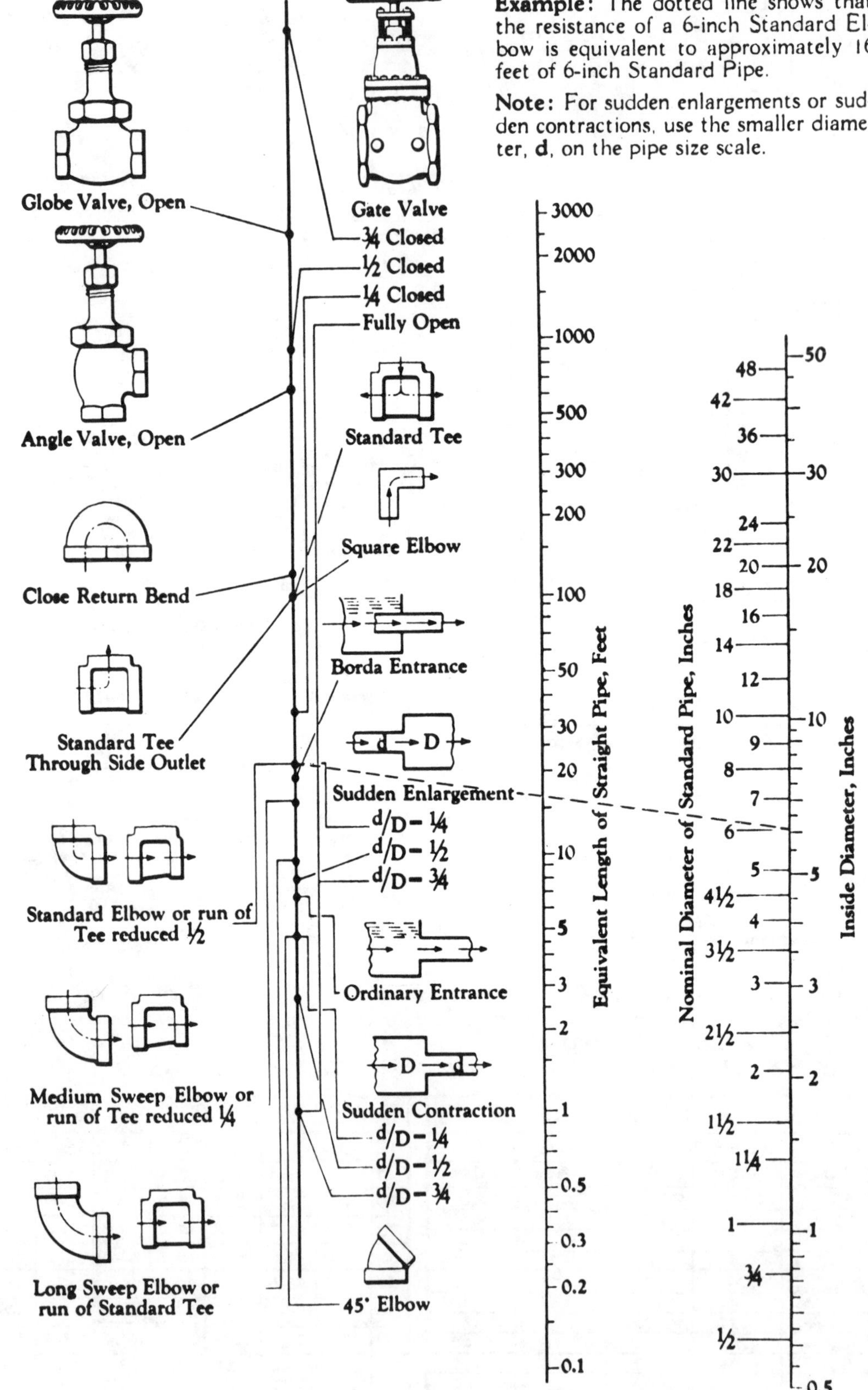
Example: The dotted line shows that the resistance of a 6-inch Standard Elbow is equivalent to approximately 16 feet of 6-inch Standard Pipe.
Note: For sudden enlargements or sudden contractions, use the smaller diameter, d, on the pipe size scale.
Globe Valve, Open
Gate Valve
¾ Closed
½ Closed
¼ Closed
Fully Open
Angle Valve, Open
Standard Tee
Square Elbow
Close Return Bend
Borda Entrance
Standard Tee Through Side Outlet
Sudden Enlargement
d/D – ¼
d/D – ½
d/D – ¾
Standard Elbow or run of Tee reduced ½
Ordinary Entrance
Medium Sweep Elbow or run of Tee reduced ¼
Sudden Contraction
d/D – ¼
d/D – ½
d/D – ¾
Long Sweep Elbow or run of Standard Tee
45° Elbow
Equivalent Length of Straight Pipe, Feet
3000
2000
1000
500
300
200
100
50
30
20
10
5
3
2
1
0.5
0.3
0.2
0.1
Nominal Diameter of Standard Pipe, Inches
48
42
36
30
24
22
20
18
16
14
12
10
9
8
7
6
5
4½
4
3½
3
2½
2
1½
1¼
1
¾
½
Inside Diameter, Inches
50
30
20
10
5
3
2
1
0.5

RESISTANCE OF VALVES AND FITTINGS TO FLOW OF FLUIDS

NOTE: Head loss through check valves varies with types manufactured. Consult manufacturer for correct values.

From Crane Co. Catalog No. 41.

WATER DISTRIBUTION—CAST IRON PIPE—I

TABLE A – AMERICAN STANDARDS ASSOCIATION STANDARD.
STANDARD THICKNESS CLASSES FOR CAST IRON PIT CAST PIPE (Bell and Spigot)*

Size, in.	Class 50, 50 lb. Pressure		Class 100, 100 lb. Pressure		Class 150, 150 lb. Pressure		Class 200, 200 lb. Pressure		Class 250, 250 lb. Pressure		Class 300, 300 lb. Pressure		Class 350, 350 lb. Pressure	
	Thick-ness, in.	Weight per ft., lb.	Thick-ness, in.	Weight per ft., lb.	Thick-ness, in.	Weight per ft., lb.	Thick-ness, in.	Weight per ft., lb.	Thick-ness, in.	Weight per ft., lb.	Thick-ness, in.	Weight per ft., lb.	Thick-ness, in.	Weight per ft., lb.
3	0.37	14.2	0.40	15.0	0.43	15.8	0.46	17.5	0.50	18.8				
4	0.40	19.2	0.43	20.4	0.46	22.5	0.50	24.2	0.54	25.8				
6	0.43	30.0	0.46	31.7	0.50	34.2	0.54	37.5	0.58	40.0	0.63	45.4	0.68	50.0
8	0.46	42.9	0.50	45.8	0.54	49.2	0.58	53.8	0.63	57.5	0.68	61.7	0.73	68.8
10	0.50	57.1	0.54	60.8	0.58	64.6	0.63	72.1	0.68	77.1	0.73	81.7	0.79	92.1
12	0.54	73.3	0.58	77.9	0.63	83.8	0.68	92.1	0.73	97.9	0.79	105.0	0.85	118.3
14	0.54	85.4	0.58	91.3	0.63	100.8	0.68	107.9	0.73	115.0	0.79	123.3	0.85	131.7
16	0.58	105.4	0.63	113.3	0.68	125.0	0.73	132.9	0.79	142.5	0.85	152.1	0.92	162.9
18	0.63	127.9	0.68	136.7	0.73	150.4	0.79	161.3	0.85	172.1	0.92	184.2	0.99	196.7
20	0.66	148.8	0.71	158.8	0.77	170.4	0.83	188.8	0.90	202.5	0.97	216.7	1.05	232.1
24	0.74	198.8	0.80	212.9	0.86	227.1	0.93	252.5	1.00	269.2	1.08	288.3	1.17	309.6
30	0.87	288.3	0.94	311.3	1.02	335.0	1.10	367.1	1.19	402.9	1.29	432.5	1.39	462.1
36	0.97	384.2	1.05	420.8	1.13	449.2	1.22	491.3	1.32	526.7	1.43	578.3	1.54	617.1
42	1.07	497.5	1.16	542.1	1.25	579.2	1.35	637.9	1.46	683.3	1.58	749.2	1.71	802.9
48	1.18	625.8	1.27	668.3	1.37	726.3	1.48	799.6	1.60	856.7	1.73	940.0	1.87	1006.3
54	1.30	777.1	1.40	847.5	1.51	906.3	1.63	997.1	1.76	1006.7	1.90	1168.8	2.05	1248.8
60	1.39	922.5	1.50	1006.3	1.62	1077.1	1.75	1153.8	1.89	1270.9	2.04	1359.6	2.20	1488.3

Water hammer of ordinary intensity allowed for in the above table. Weights based on 12-ft. length.

TABLE B – CENTRIFUGAL CAST IRON PIPE. TYPE II (Bell and Spigot)

Nominal Inside Diameter, in inches	100 lb. Class† or Maximum Working Pressure		150 lb. Class‡ or Maximum Working Pressure		200 lb. Class† or Maximum Working Pressure		250 lb. Class‡ or Maximum Working Pressure	
	Thickness in inches	Approx. Wt. per ft. in lb.	Thickness in inches	Approx. Wt. per ft. in lb.	Thickness in inches	Approx. Wt. per ft. in lb.	Thickness in inches	Approx. Wt. per ft. in lb.
3	100-lb. class weights for class B fittings. 150-, 200-, and 250-lb. classes, weights for class-D fittings.		0.33†	12.6†	0.34	13.0	0.36†	13.8†
4			0.34	16.1	0.36	16.9	0.38	18.1
6			0.37	25.7	0.40	27.1	0.43	28.7
8			0.42	38.6	0.46	41.0	0.50	44.6
10			0.47	52.2	0.52	57.4	0.57	62.3
12			0.50	66.1	0.57	75.0	0.62	81.1
14	0.48	74.6	0.55	86.9	0.62	96.9	0.69	108.0
16	0.52	92.0	0.60	108.1	0.68	121.0	0.75	133.6
18	0.56	111.2	0.65	130.4	0.74	147.2	0.83	164.9
20	0.58	127.9	0.68	152.0	0.78	172.2	0.88	193.6
24	0.64	168.5	0.76	202.9	0.88	231.5	1.00	262.1

Water hammer of ordinary intensity allowed for in the above table. Weights based on 16 ft. length.

TABLE C – WATER HAMMER ALLOWANCE

Diameter of Pipe in inches	Water Hammer, p.s.i.	Diameter of Pipe in inches	Water Hammer, p.s.i.
4" to 10"	120	24	85
12" to 14"	110	30	80
16" to 18"	100	36	75
20"	90	42 to 60	70

NOTE: The above table shows water hammer allowances used by the American Standards Association, 70 East 45th Street, New York, N. Y., in their manual for the computation of the strength and thickness of cast-iron pipe.

Values are for average grids. Each designer of a pipeline should consider whether the conditions in his case may require a more liberal water hammer allowance. For further information, see method including Allieve chart in the *Handbook of Water Control*, by R. Hardesty Manufacturing Co., Denver, Colo.

*Data from *Handbook of Cast Iron Pipe*, by Cast Iron Research Association.

†Data from American Cast Iron Pipe Co.

‡Data from Federal Specifications WW-P-421.

TABLE D – APPROXIMATE QUANTITIES OF MATERIALS USED PER JOINT FOR WATER SERVICE§

Nominal Diameter, in inches	Pounds of Joint Compound 2½" Joint Depth‖	Pounds of Yarn per Joint	Pounds of Lead in Joint 2¼" Deep	Pounds of Lead in Joint 2½" Deep
3		0.18	6.50	7.00
4	2.50	0.21	8.00	8.75
6	3.40	0.31	11.25	12.25
8	4.38	0.44	14.50	15.75
10	5.33	0.53	17.50	19.00
12	6.25	0.61	20.50	22.50
14	7.50	0.81	24.00	26.00
16	10.00	0.94	33.00	35.75
18	10.95	1.00	36.90	40.00
20	12.50	1.25	40.50	44.00
24	15.00	1.50		52.50

§Adapted from U. S. Pipe & Foundry Co.

‖Based on use of Leadite.

NOTE: Weight of lead is based on standard weight = 0.41 lb. per cu. in. This may vary 15% depending on purity.

WATER DISTRIBUTION—CAST IRON PIPE—2

TABLE A – AMERICAN WATER WORKS ASSOCIATION STANDARD BELL AND SPIGOT FITTINGS*

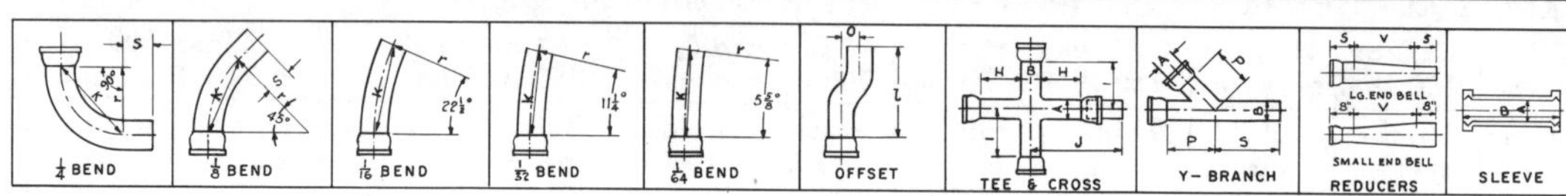

1/4 BEND
1/8 BEND
1/16 BEND
1/32 BEND
1/64 BEND
OFFSET
TEE & CROSS
Y– BRANCH
LG. END BELL
SMALL END BELL
REDUCERS
SLEEVE

Diameter	1/4 Bend			1/8 Bend			1/16 Bend		1/32 Bend		1/64 Bend		Cross and Tee					Y-Branch				Reducers			Sleeves		
	k	r	s	k	r	s	k	r	k	r	k	r	A	B	H	J	I	A	B	P	S	A	V	S	Diameter	A	B
4	22.60	16	8	18.40	24	6	18.70	48	23.52	120			4	4	11	23	11	4	4	10.50	11.50				4	5.80	15
6	22.60	16	8	18.40	24	6	18.70	48	23.52	120			6	6	12	24	12	6	6	13.0	13.0	6	18	8	6	7.90	15
8	22.60	16	10	18.40	24	6	18.70	48	23.52	120			8	8	13	25	13	8	8	16.0	14.0	8	18	8	8	10.10	15
10	22.60	16	12	18.40	24	6	18.70	48	23.52	120			10	10	14	26	14	10	10	18.5	15.5	10	18	8	10	12.20	18
12	22.60	16	12	18.40	24	6	18.70	48	23.52	120			12	12	15	27	15	12	12	21.5	15.5	12	18	8	12	14.30	18
14	25.50	18	12	27.60	36	6	28.10	72	35.28	180			14	14	16	28	16	14	14	24	16	14	20	8	14	16.50	18
16	34.00	24	12	27.60	36	6	28.10	72	35.28	180			16	16	17	29	17	16	16	31	17.5	16	20	8	16	18.90	24
18	34.00	24	12	27.60	36	6	28.10	72	35.28	180			18	18	18	30	18	18	18	34	18	18	20	8	18	21.0	24
20	34.00	24	12	36.70	48	6	37.50	96	47.05	240	47.10	480	20	20	19	31	19	20	20	37	18.75	20	26	8	20	23.1	24
24	42.40	30	12	45.90	60	6	46.80	120	47.05	240	47.10	480	24	24	21	33	21					24	26	8	24	27.4	24
30	50.90	36	12	45.90	60	6	46.80	120	47.05	240	47.10	480	30	30	26	43	26					30 30	26 66	8 8	30	33.8	24
36	67.90	48	12	68.90	90	6	70.20	180	47.05	240	47.10	480	36	36	29	46	29					36 36	32 66	8 8	36	40.2	24

Offset Diameter	O	L
4	6	27
4	12	30
4	18	38
6	6	28
6	12	34
6	18	41
8	6	29
8	12	36
8	18	43
10	6	30
10	12	38
10	18	46
12	6	34
12	12	45
12	18	56
14	6	35
14	12	46
14	18	57
16	6	35
16	12	48
16	18	58

NOTES:

1. All dimensions are in inches.
2. Fittings indicated are class D for 400-ft. head or 173 p.s.i pressure. For 4"-12" pipe, only class-D fittings are furnished. For 14"-24" pipe, class-B and D fittings are furnished. For 30" to 60" pipe, class-A-B, C, and D fittings are furnished.
3. Bell depths
   3" diameter = 3.5"
   4"-24" = 4"
   30"-36" = 4.5"
4. Reducer dimensions are for the larger sizes.
5. Sleeves shown are of the longer laying length.

TABLE B – FITTING DIMENSIONS-AMERICAN STANDARD CLASSES-125 AND 250 LB.†

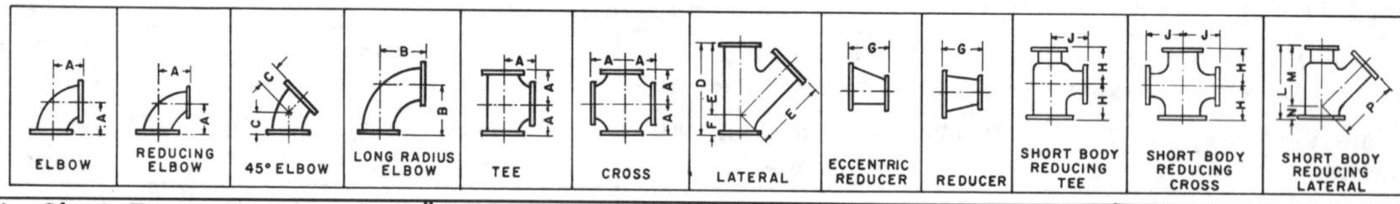

ELBOW
REDUCING ELBOW
45° ELBOW
LONG RADIUS ELBOW
TEE
CROSS
LATERAL
ECCENTRIC REDUCER
REDUCER
SHORT BODY REDUCING TEE
SHORT BODY REDUCING CROSS
SHORT BODY REDUCING LATERAL

125 lb. Class Fittings‡								250 lb. Class Fittings§							
Nominal Pipe Size	A	B	C	D	E	F	G	Nominal Pipe Size	A	B	C	D	E	F	G
3	5½	7¾	3	13	10	3	6	3	6	7¾	3½	14	11	3	6
4	6½	9	4	15	12	3	7	4	7	9	4½	16½	13½	3	7
5	7½	10¼	4½	17	13½	3½	8	5	8	10¼	5	18½	15	3½	8
6	8	11½	5	18	14½	3½	9	6	8½	11½	5½	21½	17½	4	9
8	9	14	5½	22	17½	4½	11	8	10	14	6	25½	20½	5	11
10	11	16½	6½	25½	20½	5	12	10	11½	16½	7	29½	24	5½	12
12	12	19	7½	30	24½	5½	14	12	13	19	8	33½	27½	6	14
14 O.D.	14	21½	7½	33	27	6	16	14 O.D.	15	21½	8½	37½	31	6½	16
16 O.D.	15	24	8	36½	30	6½	18	16 O.D.	16½	24	9½	42	34½	7½	18
18 O.D.	16½	26½	8½	39	32	7	19	18 O.D.	18	26½	10	45½	37½	8	19

NOTES:

1. All dimensions are in inches.
2. Sizes of short-body tees and crosses 125-lb. standard available from 18" to 48" and laterals up to 30" inclusive.
3. Sizes of short-body tees, 250-lb standard available from 18" to 30".
4. For flange diameters, see p. 22-11.
5. 125-lb. fittings reted for liquid and gas service at 175 p.s.i. gage for 3" - 12" diameter and at 150 p.s.i. for 14"-48" diameter 250-lb. fittings rated at 400 p.s.i. gage for 3" - 12" diameter inclusive, and 300 p.s.i. for 14"-48" diameter.

TABLE C – SHORT BODY FITTINGS, 250 P.S.I., WATER PRESSURE AND WATER HAMMER‖

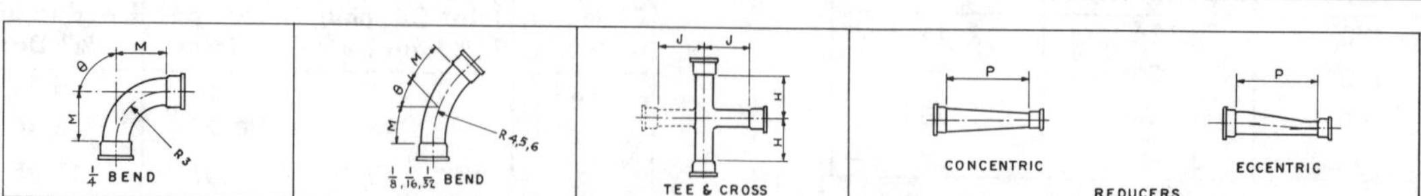

1/4 BEND
1/8, 1/16, 1/32 BEND
TEE & CROSS
CONCENTRIC
ECCENTRIC
REDUCERS

Diameter	1/4 Bend, O = 90°		1/8 Bend, O = 45°		1/16 Bend, O = 22½°		1/32 Bend, O = 11¼°		Tees and Crosses		Reducers
	M	R_3	M	R_4	M	R_5	M	R_6	H	J	P
3	5.50	4.0	3.00	3.62	3.00	7.56	3.00	15.25	5.5	5.5	
4	6.50	4.50	4.00	4.81	4.00	10.06	4.00	20.31	6.5	6.5	7.0
6	8.00	6.00	5.00	7.25	5.00	15.06	5.00	30.50	8.0	8.0	9.0
8	9.00	7.00	5.50	8.44	5.50	17.63	5.50	35.50	9.0	9.0	11.0
10	11.00	9.00	6.50	10.88	6.50	22.62	6.50	45.69	11.0	11.0	12.0
12	12.00	10.00	7.50	13.25	7.50	27.62	7.50	55.81	12.0	12.0	14

NOTE:
All dimensions in inches. Reducer diameter are for the larger end.

* Data from A.W.W.A. Standard Specification C100 – 52T.
† Data from *Handbook of Cast Iron Pipe Research.*
‖ Data from A.S.A. Specification A21.10 – 1952.
‡ A.S.A. Specification B16.1 – 1948.
§ A.S.A. Specification B16b – 1944.

WATER DISTRIBUTION—VALVES & JOINTS

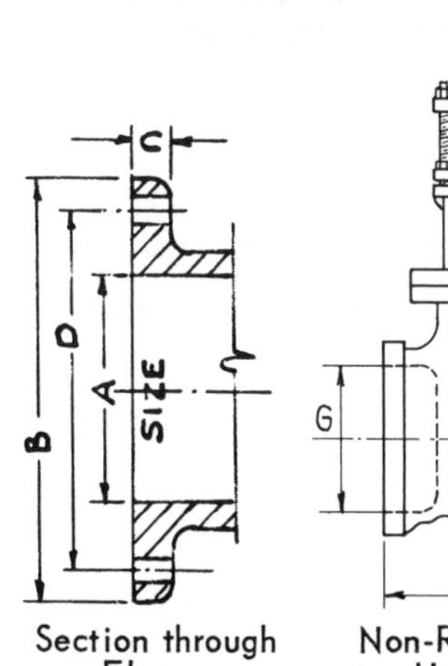
C
D
B
A
SIZE
L
Section through Flange

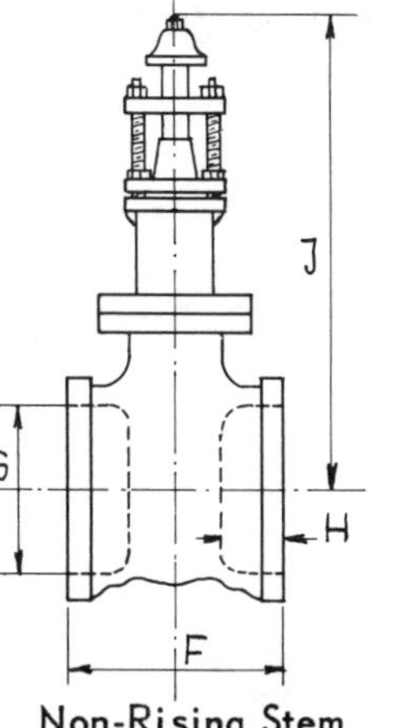
J
G
H
F
Non-Rising Stem Hub Ends

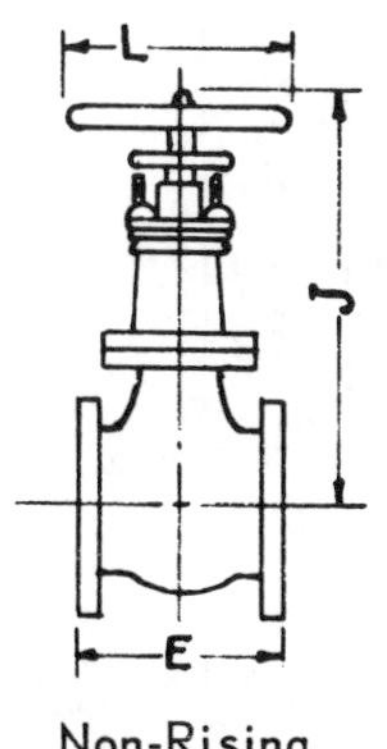
L
J
E
Non-Rising Stem Flanged

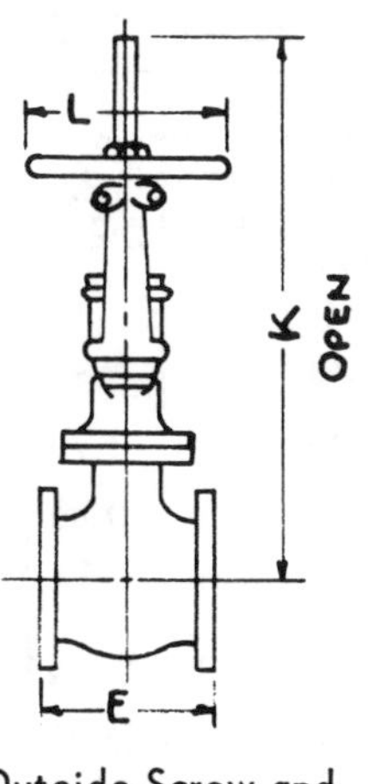
L
K
OPEN
E
Outside Screw and Yoke Flanged

FIG. A. GATE VALVES

FIG. B. CHECK VALVE
150 p.s.i. Working Pressure

TABLE A – CRANE A.W.W.A. STANDARD VALVES														TABLE B – CHAPMAN CHECK TILTING DISK VALVES*					
Size, in.	Flanges A	Flanges B	Flanges C	Flanges D	Bolts No.	Bolts Diam.	E	F	G	H	J	K	L	A	B	C	D	E	Size, in.
3	3	7½	¾	6	4	5/8	8	11	4.66	3.50	16 7/8	21 5/8	9	9½	11½	7½	7½	¾	3
4	4	9	15/16	7½	8	5/8	9	12¼	5.70	4.00	18½	24 5/8	10	11½	14½	9¾	9	15/16	4
6	6	11	1	9½	8	¾	10½	13¼	7.80	4.00	22¼	31¼	12	14	16½	11 7/8	11	1	6
8	8	13½	1 1/8	11¾	8	¾	11½	13¾	10.00	4.00	25¾	41¼	14	19½	20¼	14 7/8	13½	1 1/8	8
10	10	16	1 3/16	14¼	12	7/8	13	14½	12.10	4.00	30½	51½	16	24½	25¼	17¾	16	1 3/16	10
12	12	19	1¼	17	12	7/8	14	15	14.20	4.00	34	58 5/8	18	27½	26½	19½	19	1¼	12
14	14	21	1 3/8	18¾	12	1	15¾	17¼	16.45	4.00	40	67	20	31	29 3/8	22 3/8	21	1 3/8	14
16	16	23½	1 7/16	21¼	16	1	17	18	18.80	4.00	45½	75 5/8	22	30	33 5/8	25 7/8	23½	1 7/16	16
18	18	25	1 9/16	22¾	16	1 1/8	18½	19	20.92	4.00	48 7/8	83½	24	33	37¼	29	25	1 9/16	18

NOTE: Gate valves also available with ends for mechanical joint pipe.

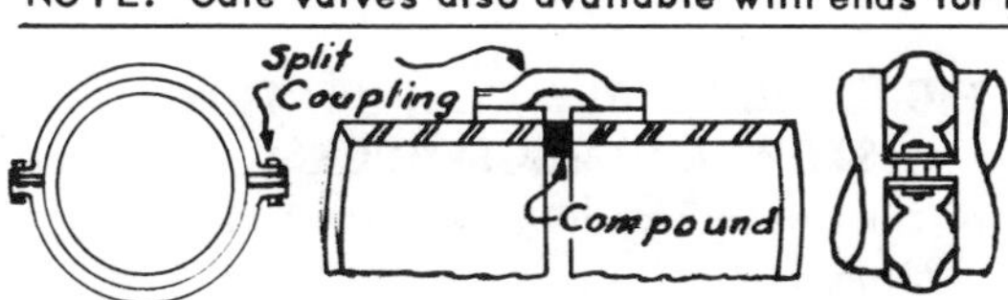
Split Coupling
Compound
FIG. C. JOINT FOR PLAIN-END PIPE

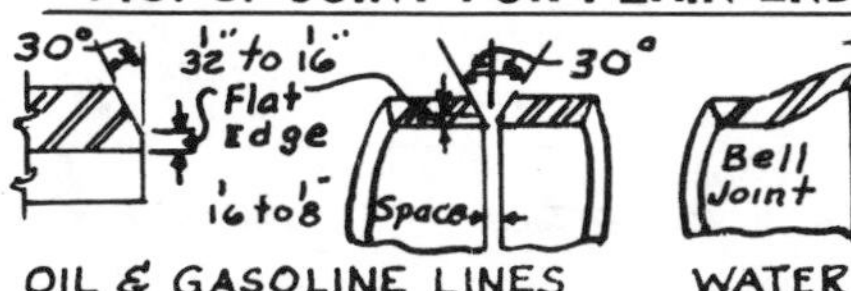
30°
30°
Flat Edge
Space
OIL & GASOLINE LINES

2"
Bell Joint
Spigot Joint
WATER LINES

FIG. D. WELDED JOINT FOR STEEL PIPE

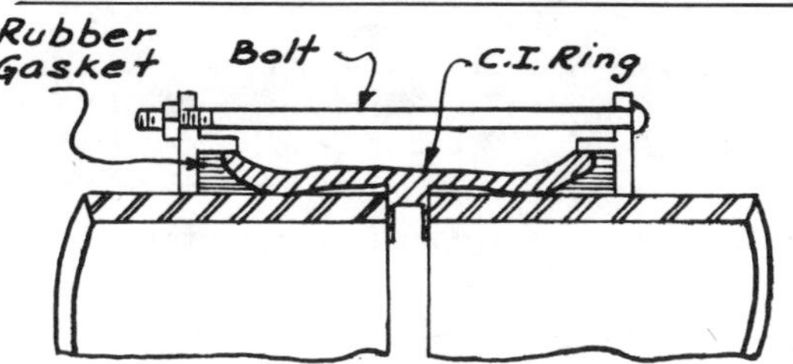
Rubber Gasket
Bolt
C.I. Ring
FIG. E. DRESSER COUPLING
Use on plain-end pipe for flexible joints and expansion joints.

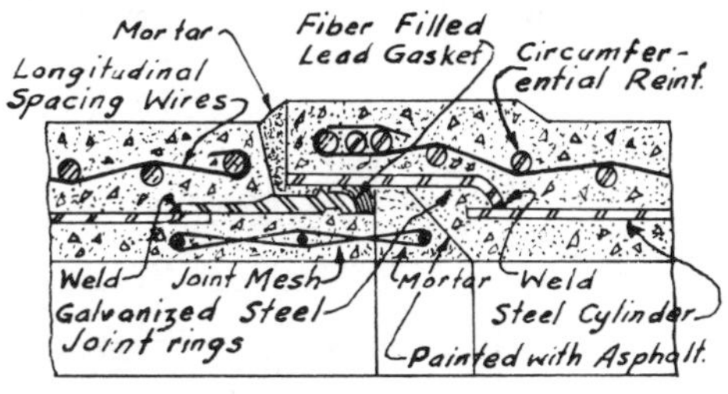
Mortar
Fiber Filled Lead Gasket
Circumferential Reinf.
Longitudinal Spacing Wires
Weld
Joint Mesh
Galvanized Steel Joint rings
Mortar
Weld
Steel Cylinder
Painted with Asphalt
LEAD JOINT

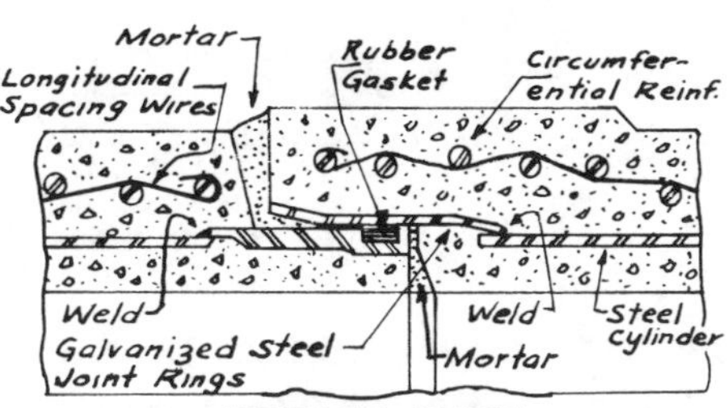
Mortar
Rubber Gasket
Circumferential Reinf.
Longitudinal Spacing Wires
Weld
Weld
Steel cylinder
Galvanized steel Joint Rings
Mortar
RUBBER JOINT

FIG. F. JOINT DETAILS OF CONCRETE PRESSURE PIPE (LOCK JOINT)†

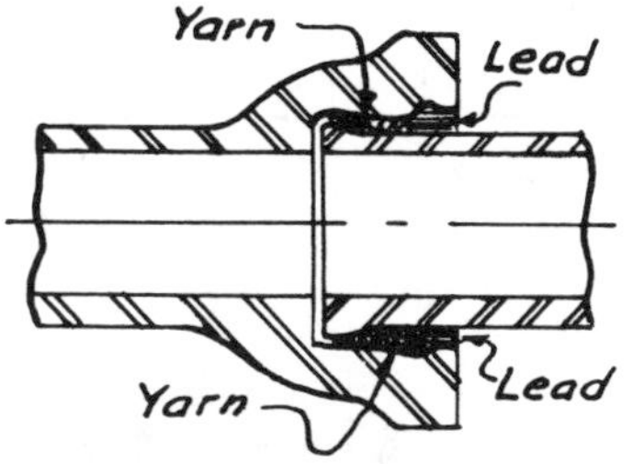
Yarn
Lead
Yarn
Lead
BELL AND SPIGOT JOINT

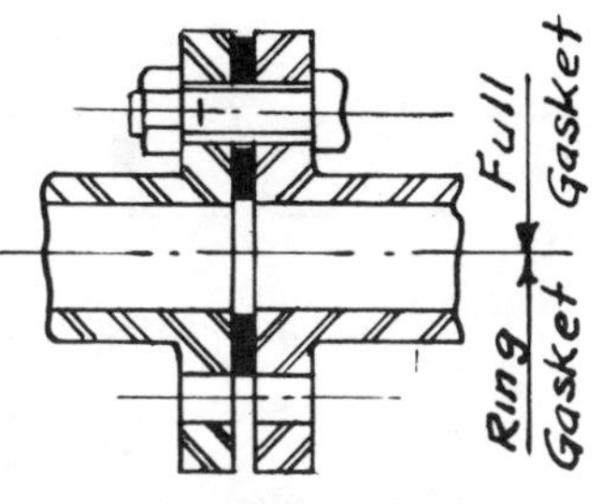
Full Gasket
Ring Gasket
FLANGED JOINT

FIG. G. JOINTS FOR CAST-IRON PIPE

*Manufacturer supplies valve in sizes to 54" inclusive.

† Adapted from Lock Joint Co.

WATER DISTRIBUTION—ASBESTOS—CEMENT PIPE

TABLE A – ASBESTOS-CEMENT PIPE

Nominal Pipe Size, in.	Class 100			Class 150			Class 200		
	Inside Diam., in.	Shell Thickness, in.	Wt. per ft., lb.	Inside Diam., in.	Shell Thickness, in.	Wt. per ft., lb.	Inside Diam., in.	Shell Thickness, in.	Wt. per ft., lb.
4	3.95	0.445	6.3	3.95	0.510	7.6	3.95	0.660	9.3
6	5.85	0.530	10.6	5.85	0.610	13.0	5.70	0.780	15.4
8	7.85	0.590	15.8	7.85	0.710	19.9	7.60	0.920	23.9
10	9.85	0.660	21.8	10.00	0.910	32.0	9.63	1.095	37.2
12	11.70	0.750	29.7	12.00	1.040	43.8	11.56	1.260	51.7
14	13.59	0.820	38.9	14.00	1.190	58.5	13.59	1.470	69.0
16	15.50	0.905	48.8	16.00	1.310	73.0	15.50	1.670	89.2

Class of pipe is same as allowable working pressure in pounds per square inch. Pipe furnished in standard nominal lengths of 13 ft. joints are made using a "ring-tite" coupling made up of an asbestos-cement sleeve and two rubber rings.

TABLE C – LARGEST SIZE CORPORATION STOP RECOMMENDED, INCHES

Tapped Pipe Diameter, in.	Tapped Directly into Pipe Wall (All Classes)	Tapped through Ring-Tite Coupling		Tapped with Service Clamp (All Pipe Classes)	
		Class 150	Class 200	Single Strap	Double Strap
4	¾	¾	¾	–	1
6	¾	¾	1	–	1½
8	1	1	1	2	2
10	1	1	1	2	2
12	1	1	1	2	2
14	1	1	1	2	2
16	1	1	1	2	2

FIG. B. POURED JOINT FOR CAST-IRON FITTING ON TRANSITE PIPE

Cast Iron Fitting
Dry Yarn driven in evenly with yarning iron.
¼" minimum joint space.
Transite pipe.
Lead, Leadite, Hydro-tite or Mineralead joint.
2¼" Min.
3½" to 4½ according to pipe diameter.

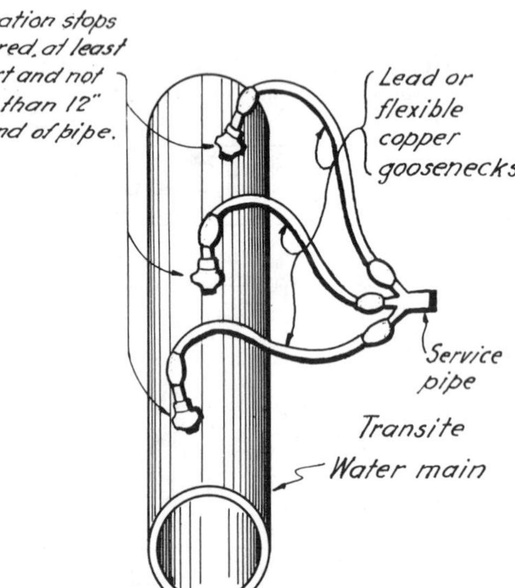

Corporation stops staggered, at least 12" apart and not closer than 12" from end of pipe.
Lead or flexible copper goosenecks
Service pipe
Transite Water main

Recommended for large service connections

FIG. D. MULTIPLE CORPORATION STOPS AND GOOSENECKS
Recommended for large service connections.

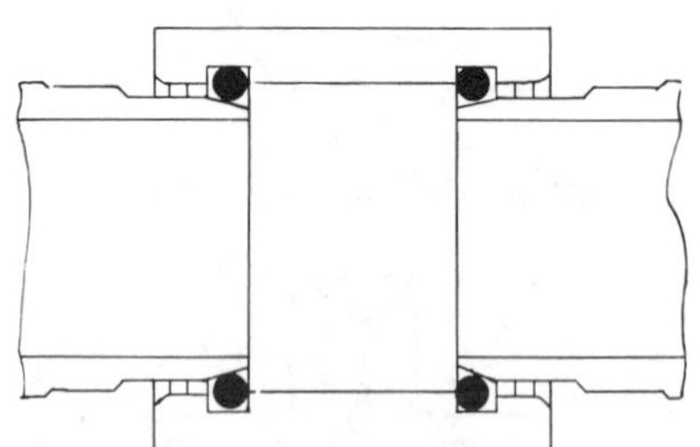

STARTING POSITION

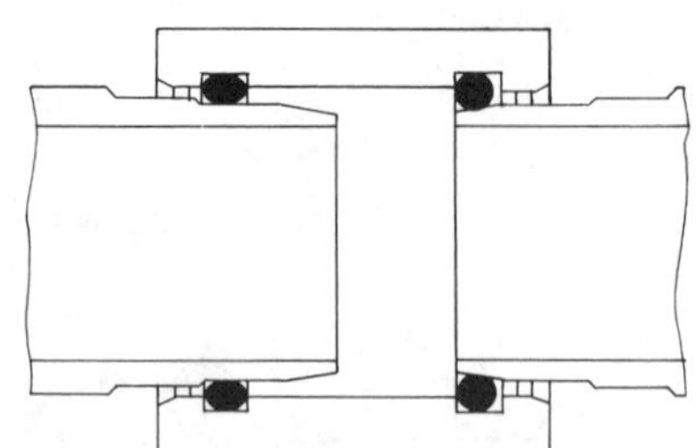

ONE END IN PLACE

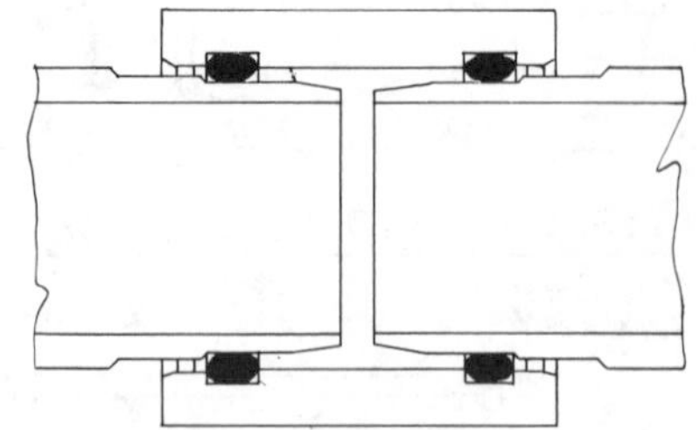

JOINT COMPLETED

FIG. E. RING-TITE COUPLING FOR TRANSITE PRESSURE PIPE-FITTINGS
Standard cast iron with bells at all joints to Transite pipe recommended. Cast-iron bell-end fittings are available for hot-poured and ring-tite joints. Some sizes of Transite pipe require oversize bell fittings to provide minimum ¼" joint space. Standard steel, brass, or copper fittings also used.

All data from Johns-Manville corp.

WATER DISTRIBUTION—STEEL PIPE—I

TABLE A – STEEL PIPE DESIGN

Physical Data of Steel Mill Pipe for Underground Water Service in Accordance with A.W.W.A. Standard Specification C202-49.

Nominal Size, in.	Outside Diameter, in.	Wall Thickness, in.	Weight per lin. ft., lb.	Working Pressure (Safety Factor of 4.0)		Approximate Ultimate Bursting Pressure		Test Pressure		
				40,000 p.s.i. Fiber Stress	50,000 p.s.i. Fiber Stress	40,000 p.s.i. Fiber Stress	50,000 p.s.i. Fiber Stress	Butt Weld Plain or Threaded Ends	Lap Weld, Grade A, Seamless or Electric Welded: Plain Ends	Lap Weld, Grade A, Seamless or Electric Welded: Threaded Ends
¾ *	1.050	0.113	1.13	2150		8610		700	700	700
1 *	1.315	0.133	1.68	2020		8090		700	700	700
1½*	1.900	0.145	2.72	1530		6100		800	1000	1000
2 *	2.375	0.154	3.65	1300	1620	5190	6480	800	1000	1000
2½*	2.875	0.203	5.79	1410	1770	5650	7060	800	1000	1000
3	3.500	0.125	4.51	710	890	2860	3570	800	1300	
		0.156	5.58	890	1110	3560	4460	800	1600	
		0.188	6.63	1070	1340	4300	5370	800	1900	
		0.216	7.58	1230	1540	4940	6170	800	2200	1000
4	4.500	0.125	5.84		690		2780		1000	
		0.156	7.25		870		3470		1200	
		0.188	8.64		1040		4180		1500	
		0.219	10.00		1220		4870		1700	
6	6.625	0.188	12.89		710		2840		1000	
		0.219	14.97		830		3300		1200	
		0.250	17.02		940		3780		1400	
		0.280	18.97		1060		4220		1500	1200
8	8.625	0.188	16.90		550		2180		800	
		0.219	19.64		640		2540		900	
		0.250	22.36		730		2900		1000	
		0.277	24.70		800		3210		1200	1200
10	10.750	0.188	21.15		440		1750		650	
		0.219	24.60		510		2040		750	
		0.250	28.04		580		2330		850	
		0.279	31.20		650		2600		1000	1000
12	12.750	0.188	25.15		370		1480		500	
		0.219	29.28		430		1720		600	
		0.250	33.38		490		1960		700	
		0.281	37.45		550		2200		800	
14	14.000	0.188	27.65		340		1340		450	
		0.219	32.19		390		1560		550	
		0.250	36.71		450		1780		650	
		0.281	41.21		500		2010		700	
16	16.000	0.188	31.66		290		1180		400	
		0.219	36.86		340		1370		500	
		0.250	42.05		390		1560		550	
		0.281	47.22		440		1760		650	
18	18.000	0.188	35.67		260		1040		350	
		0.219	41.54		300		1220		400	
		0.250	47.39		350		1390		500	
		0.281	53.22		390		1560		550	
20	20.000	0.188	39.67		240		940		350	
		0.219	46.21		270		1100		400	
		0.250	52.73		310		1250		450	
		0.281	59.23		350		1400		500	
24	24.000	0.188	47.68		200		780		300	
		0.219	55.56		230		910		300	
		0.250	63.41		260		1040		350	
		0.312	79.06		330		1300		450	
28	28.000	0.188	55.69		170		670		250	
		0.219	64.90		200		780		250	
		0.250	74.09		220		890		300	
		0.281	83.26		250		1000		350	

*These sizes are considered as service pipe. The table does not include all sizes and thicknesses. Weights per foot are unlined and uncoated pipe.

WATER DISTRIBUTION— STEEL PIPE—2

TABLE A – STEEL PIPE, A.S.T.M. SPECIFICATION A-53, WEIGHTS AND DIMENSIONS

Size, Nominal in.	Outside Diameter, in.	Wall Thickness, in.	Weight per Lin. ft., lb.	Number of Threads per inch	Test Pressures, p.s.i. Butt Welded	Test Pressures, p.s.i. Lap Welded or Grade A	Test Pressures, p.s.i. Grade B
STANDARD PIPE – STANDARD-WEIGHT PIPE							
½	0.84	0.109	0.85	14	700	700	700
¾	1.05	0.113	1.13	14	700	700	700
1	1.315	0.133	1.68	11½	700	700	700
1¼	1.660	0.140	2.28	11½	800	1000	1100
1½	1.900	0.145	2.73	11½	800	1000	1100
2	2.375	0.154	3.68	11½	800	1000	1100
2½	2.875	0.203	5.82	8	800	1000	1100
3	3.500	0.216	7.62	8	800	1000	1100
3½	4.000	0.226	9.20	8	1200	1200	1300
4	4.500	0.237	10.89	8	1200	1200	1300
5	5.563	0.258	14.81	8	–	1200	1300
6	6.625	0.280	19.18	8	–	1200	1300
STANDARD PIPE – EXTRA-STRONG PIPE							
½	0.840	0.147	1.09	14	850	850	850
¾	1.050	0.154	1.47	14	850	850	850
1	1.315	0.179	2.17	11½	850	850	850
1¼	1.660	0.191	3.00	11½	1100	1500	1600
1½	1.900	0.200	3.63	11½	1100	1500	1600
2	2.375	0.218	5.02	11½	1100	1500	1600
2½	2.875	0.276	7.66	8	1100	1500	1600
3	3.500	0.300	10.25	8	1100	1500	1600
3½	4.000	0.318	12.51	8	1700	1700	1800
4	4.500	0.337	14.98	8	1700	1700	1800
5	5.563	0.375	20.78	8	–	1700	1800
6	6.625	0.432	28.57	8	–	1700	1800
8	8.625	0.500	43.39	8	–	1700	2400
10	10.750	0.500	54.74	8	–	1600	1900
12	12.750	0.500	65.42	8	–	1600	1900
STANDARD PIPE – DOUBLE-EXTRA-STRONG PIPE							
½	0.840	0.294	1.71	14	1000	1000	1000
¾	1.050	0.308	2.44	14	1000	1000	1000
1	1.315	0.358	3.66	11½	1000	1000	1000
1¼	1.660	0.382	5.31	11½	1200	1800	1900
1½	1.900	0.400	6.41	11½	1200	1800	1900
2	2.375	0.436	9.03	11½	1200	1800	1900
2½	2.875	0.552	13.70	8	1200	1800	1900
3	3.500	0.600	18.58	8	–	1800	1900
4	4.500	0.674	27.54	8	–	2000	2100
5	5.563	0.750	38.55	8	–	2000	2100
6	6.625	0.864	53.16	8	–	2000	2100
8	8.625	0.875	72.42	8	–	2800	2800

WATER DISTRIBUTION — STEEL PIPE — 3

TABLE A - MAXIMUM HEIGHT FOR BACKFILLING IN FEET FOR WELDED STEEL PIPE.

(EITHER TRENCH OR EMBANKMENT CONDITION.)

DIAMETER	14 GAGE.	12 GAGE	10 GAGE	7 GAGE	3 GAGE
4"	138	298			
5"	92	198			
6"	66	142			
7"	50	107			
8"	39	84	148	293	
9"	31	68	119	236	
10"	26	56	98	195	
11"	22	47	82	163	
12"	19	40	70	139	265
14"		30	53	105	198
16"		24	42	82	156
18"		19	33	66	126
20"			28	55	104
22			23	46	87
24"			20	39	74
30"				26	49
36"				19	35
42"					27
48"					21
54"					17
60"					14
66"					12
72"					10

Maximum allowable deflection = 2% of diameter. Table based on bottom and side support obtained by hand tamping.

TABLE B - SAFE SPANS BETWEEN SUPPORTS IN FEET FOR WELDED STEEL PIPE ABOVE GROUND.

DIAMETER	16 GA.	14 GA.	12 GA.	10 GA.	7 GA.	3 GA.
4"	15.5	17.5	18.5	21	22	23.5
5"	16.5	17.5	19.5	22	23.5	25
6"	16.5	18	20.5	22.5	24	26.5
7"	17.5	18.5	21	23	25	27.5
8"	17.5	18.5	22	23.5	26.5	28.5
9"	17.5	19	22.5	24	27	29.5
10"	18	19.5	22.5	24.5	27.5	30
12"	18	19.5	23	25.5	28	31
14"	18.5	20.5	23.5	26.5	29	32
16"	18.5	20.5	23.5	26.5	29.5	33
18"		21	24	27	30	33.5
20"		21	24	27.5	30.5	34
22"			24.5	27.5	30.5	34.5
24"			24.5	28	31	35
30"				28	32	35.5
36"					32.5	36.
42"					33	37
48"						37

Based on strength of pipe as beam and resistance to crushing at supports. For spans below heavy line, reinforcement at supports is recommended to prevent crushing.

DESIGN OF WELDED STEEL PIPE

Internal Pressure Formula:- $t = \frac{P \times d}{2 f_s}$

t = Thickness of pipe in inches.
P = Internal water pressure in p.s.i
d = Outside diameter of pipe in inches.
f_s = Safe working stress in material in p.s.i.
NOTE: Allowance to be made in P for possible excess stress caused by water hammer. See Table C - page 22-09.

EXPANSION IN WELDED STEEL PIPES.

Temperature Stress Formula: $S = T \times E \times n$.

S = Temperature Stress in p.s.i.
T = Range in temperature in degrees F.
E = Modulus of elasticity = 30,000,000.
n = Coefficient of expansion = 0.0000065.

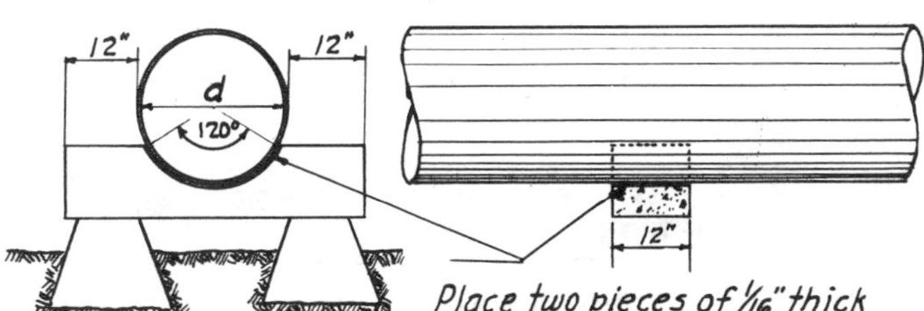

12"
12"
d
120°
12"

Place two pieces of 1/16" thick sheet packing - One fastened to pipe, the other to saddle. Place thin layer of Graphite Grease between.

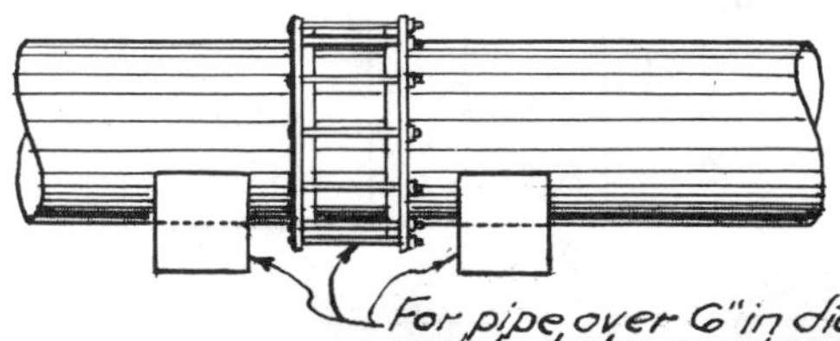

For pipe over 6" in diam. couplings not to take vertical load; Space saddles to support and hold pipe without aid of couplings.

FIG. C- SADDLE SUPPORT FOR WELDED STEEL PIPE.

EXPANSION JOINTS such as Dayton or Dresser couplings (see Fig. E p. 22-11) with long middle rings to be provided for pipes above ground where S > safe working stress of material. Seldom required where continuous flow of water is maintained or where pipe line is very sinuous. Temperature stress can be neglected in buried pipe lines.

THRUST AT BENDS in pipe due to internal pressure - see Table p. 22-17.

Data from Handbook of Water Control, by the Armco Drainage & Metal Products, Inc.

WATER DISTRIBUTION—PIPE CURVATURE

TABLE A – MAXIMUM DEFLECTION OF CAST-IRON BELL AND SPIGOT PIPE*

Nominal Pipe Diameter, in.	Joint Opening, in.	Maximum Deflection in inches with Pipe Length of:			Approximate Radius of Curve in feet Produced by Succession of Joints with Pipe Lengths of:		
		12 ft.	16 ft.	18 ft.	12 ft.	16 ft.	18 ft.
2	0.41	23.6	31.5	35.4	73	98	110
3	0.43	14.8	19.7	22.2	112	149	168
4	0.41	11.1	14.8	16.7	156	208	234
6	0.58	11.1	14.8	16.7	156	208	234
8	0.65	9.7	12.9	14.6	178	238	268
10	0.75	9.3	12.4	14.0	186	248	279
12	0.75	7.9	10.5	11.9	218	292	327
14	0.75	6.7	8.9	10.1	258	345	387
16	0.75	5.9	7.9	8.8	293	390	440
18	0.75	5.3	7.1	8.0	326	434	489
20	0.75	4.8	6.4	7.2	360	480	540
24	0.75	4.0	5.3	6.0	432	577	648
30	0.75	3.3	4.4	5.0	524	699	786
36	0.75	2.8	3.7	4.2	617	824	926

TABLE B – MAXIMUM DEFLECTION OF CAST-IRON MECHANICAL-JOINT PIPE

Pipe Size, in.	Deflection in inches			Approximate Radius in feet of Curve Produced by Succession of Joints		
	12 ft. Length	16 ft. Length	18 ft. Length	12 ft. Length	16 ft. Length	18fft. Length
3	21	28	31	85	110	125
4	21	28	31	85	110	125
6	18	24	27	100	130	145
8	13	18	20	130	170	195
10	13	18	20	130	170	195
12	13	18	20	130	170	195
14	9	12	13½	190	250	285
16	9	12	13½	190	250	285
18	7½	10	11	230	300	340
20	7½	10	11	230	300	340
24	6	8	9	300	400	450
30	6	8	9	300	400	450
36	5	7	8	330	440	500

TABLE C – MAXIMUM RECOMMENDED DEFLECTION OF ASBESTOS-CEMENT PIPE WITH RING-TITE COUPLINGS†

	During Assembly (Belled Pipe Only)			After Assembly		
Pipe Size, in.	Through 12	14	16	Through 12	14	16
Deflection, degrees	2½	0	0	5	3	3

TABLE D – MAXIMUM RECOMMENDED LAYING DEFLECTION FOR DRESSER-STYLE 38 COUPLINGS

Pipe Size	Deflection, degrees			
Up to 2" I.D. inclusive	6			
From 2" I.D. to 14" O.D. inclusive	4			
With middle ring length	5"	7"	8"	10"
Above 14" O.D. to 20" O.D. incl.	2½"	4"	4"	4"
Above 20" O.D. to 30" O.D. incl.	2"	4"	4"	4"
Above 30" O.D. to 37" O.D. incl.	1½"	3"	3½"	3½"
Above 37" O.D. to 42" O.D. incl.		2½"	3½"	3½"

TABLE E – OFFSETS FOR ASBESTOS-CEMENT PIPE ON CURVES†

Angle of Deflection at Coupling, degrees	Offset per Pipe Length, inches		
	1 ft.‡	6½ ft.	13 ft.
1	0.2	1.3	2.7
2	0.4	2.7	5.4
3	0.6	4.0	8.1
4	0.8	5.4	10.9
5	1.0	6.8	13.6

TABLE F – DEFLECTION FROM CENTER LINE FOR VICTAULIC COUPLINGS – STEEL AND WROUGHT-IRON PIPE

Pipe size, inches	4	6	8	10	12	14 O.D.	16 O.D.	18 O.D.	20 O.D.	24 O.D.	28 I.D.	30 I.D.
Deflection, degrees per coupling	3°-11'	2°-10'	1°-40'	1°-20'	1°-7'	1°-2'	0°-54'	0°-48'	0°-43'	0°-36'	0°-30'	0°-28'

*Adopted from A.W.W.A. Specification C600–54T.

Limiting Factors: Joint opening not to exceed 0.75 in. and caulking space at face of bell to be not less than 0.25 in. wide.

†From Johns-Manville. ‡Approximate – to tenths of an inch.

Century Asbestos – Cement Pipe with Simplex Couplings: Allowable deflection in one coupling up to 5° in any direction.

Ball and Socket Joint Cast-Iron Pipe: Approximate maximum deflection at one joint: 15°.

WATER DISTRIBUTION — REACTIONS AT BENDS

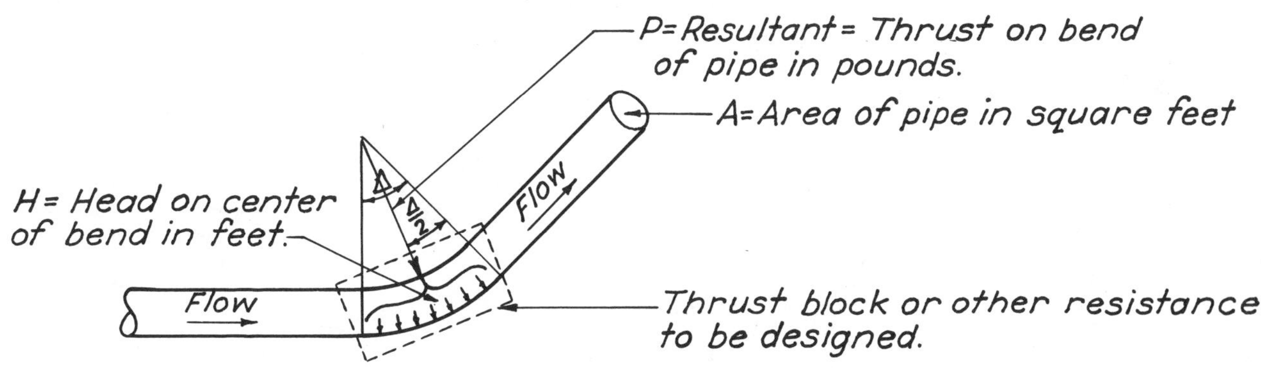
P = Resultant = Thrust on bend of pipe in pounds.
A = Area of pipe in square feet
H = Head on center of bend in feet.
Δ
Δ/2
Flow
Flow
Thrust block or other resistance to be designed.

$$P = 2 \times 62.5\, HA \sin \tfrac{1}{2} \Delta$$

Values given in table = thrust on pipe in pounds.

INSIDE DIAM. OF PIPE IN INCHES	90° BEND			45° BEND			22½° BEND			11¼° BEND		
	H=100'	H=200'	H=300'	H=100'	H=200'	H=300'	H=100'	H=200'	H=300'	H=100'	H=200'	H=300'
2	193	386	579	104	208	312	53	106	159	27	54	81
3	430	860	1,290	235	470	705	120	240	360	61	122	183
4	770	1,540	2,310	418	836	1,250	213	426	639	109	218	327
6	1,730	3,460	5,190	940	1,880	2,820	480	960	1,440	246	492	738
8	3,080	6,160	9,240	1,670	3,340	5,010	850	1,700	2,550	437	874	1,310
10	4,820	9,640	14,500	2,610	5,220	7,830	1,330	2,660	3,990	670	1,340	2,010
12	6,940	13,900	20,800	3,760	7,520	11,300	1,910	3,820	5,730	963	1,920	2,890
14	9,450	18,900	28,400	5,120	10,200	15,400	2,610	5,220	7,830	1,320	2,640	3,960
16	12,300	24,700	37,000	6,680	13,400	20,000	3,400	6,800	10,200	1,720	3,440	5,160
18	15,600	31,200	46,800	8,450	16,900	25,400	4,300	8,600	12,900	2,170	4,340	6,510
20	19,300	38,600	57,900	10,400	20,800	31,200	5,330	10,700	16,000	2,680	5,360	8,040
24	27,800	55,600	83,400	15,000	30,000	45,000	7,650	15,300	23,000	3,850	7,700	11,600
30	43,400	86,800	130,000	23,500	47,000	71,000	12,000	24,000	36,000	6,030	12,100	18,100
36	62,500	125,000	188,000	33,800	67,600	101,000	17,200	34,400	51,600	8,650	17,300	26,000
42	85,000	170,000	255,000	46,000	92,000	138,000	23,500	47,000	70,500	11,800	23,600	35,400
48	111,000	222,000	333,000	60,200	120,000	181,000	30,600	61,200	91,800	15,400	30,800	46,200
54	140,000	280,000	420,000	76,000	152,000	228,000	38,800	77,600	116,000	19,600	39,200	58,800
60	173,000	346,000	519,000	94,000	188,000	282,000	48,000	96,000	144,000	24,100	48,200	72,300
66	210,000	420,000	630,000	114,000	228,000	342,000	58,000	116,000	174,000	29,300	58,600	88,000
72	250,000	500,000	750,000	135,000	270,000	405,000	67,000	134,000	201,000	34,700	69,400	104,000

WATER DISTRIBUTION—VALVES—1

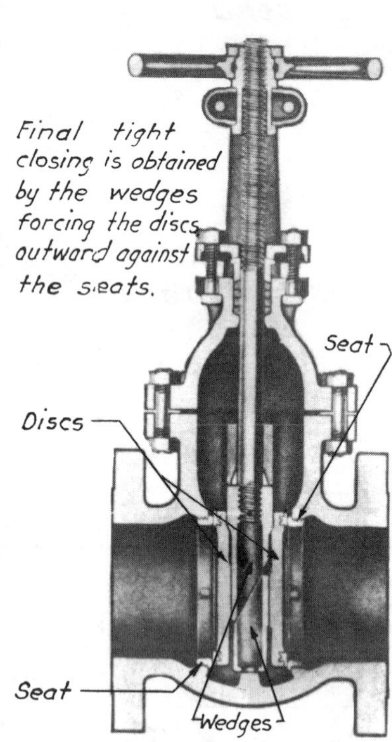
Final tight closing is obtained by the wedges forcing the discs outward against the seats.
Seat
Discs
Seat
Wedges

GATE VALVE*

LOCATION Any place in pipe.

USE Most common type. Use from 2" diameter and larger.

TYPES Double-disk parallel seat (illustrated) and solid-wedge type used in water works practice. Solid-wedge (most durable) advantages are relatively low cost and low head loss when valve is open.

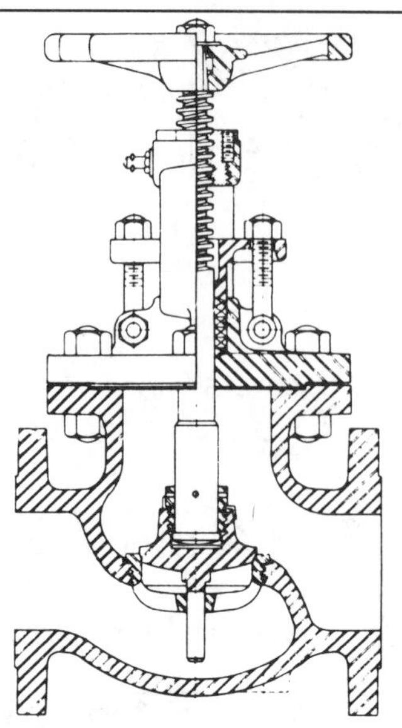

GLOBE VALVE†

LOCATION Any place in pipe.
USE To control rate of flow of fluids, where operation of valve is frequent, generally used under 4" diameter, and limited to plumbing lines.
NOTES If inlet and outlet ends are placed at right angles to each other, valve is called an angle valve. Parts are easily replaced.

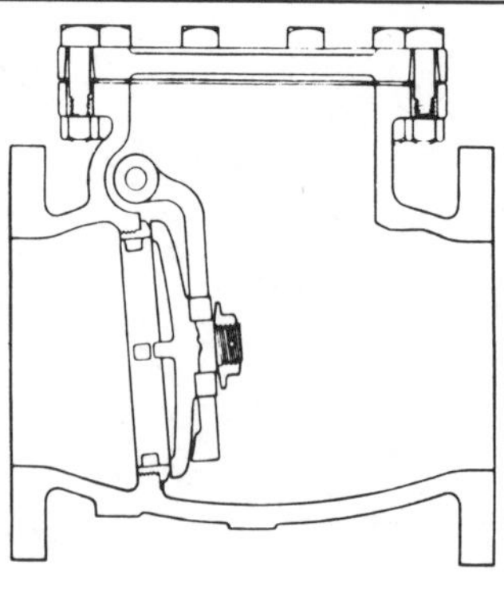

LOCATION Any place in horizontal pipe.

USE To permit fluid to flow in only one direction.

NOTES Swing check valve has, comparatively the least obstruction to flow of all check valves. Valve should preferably be set horizontally.

OTHER TYPE Backwater valve.

SWING CHECK VALVE‡

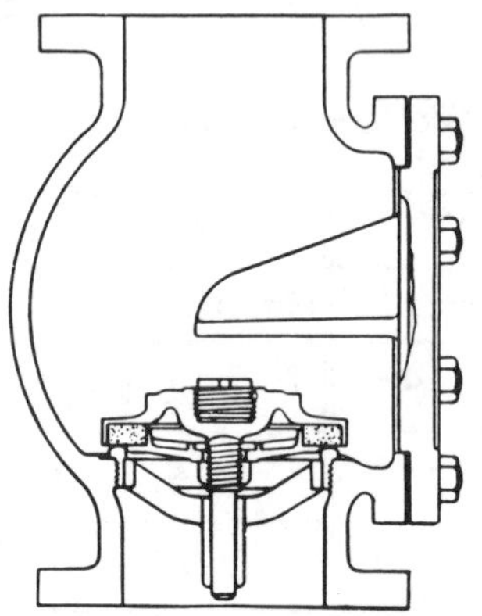

LOCATION Any place in pipe.
USE Lift check valve usually has most positive closure.
TYPES Vertical (illus.) (inlet end on bottom and outlet end on top). Horizontal (inlet and outlet ends set in horizontal plane). Angle (inlet end on bottom and outlet end on side). Foot valve (vertical lift check with screen).
NOTE Never use these valves at pressures and temp. exceeding their maximum ratings.

LIFT CHECK VALVE‡

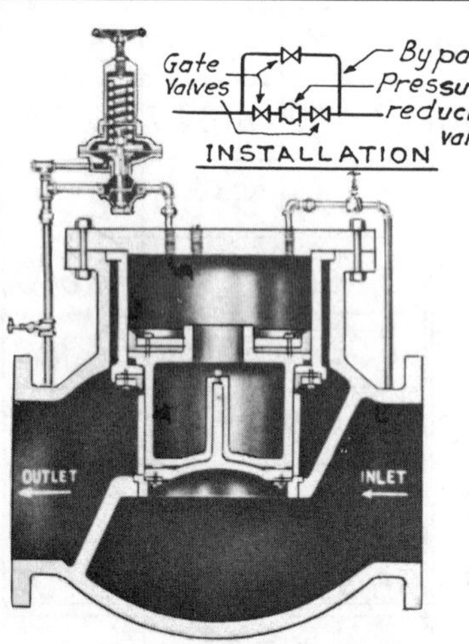
Gate Valves
By pass
Pressure reducing valve.
INSTALLATION
OUTLET
INLET

WATER PRESSURE REDUCING VALVE ||

LOCATION Any place in pipe.
USE To maintain constant downstream pressure against higher inlet head.

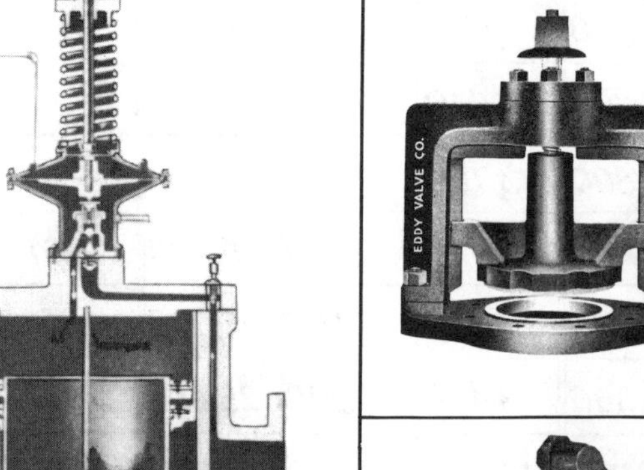
OUTLET
INLET
EDDY VALVE CO.
WATERFORD, N.Y.

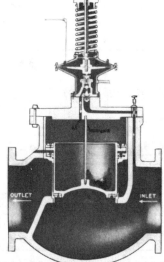

ALTITUDE CONTROL VALVE||

LOCATION In inlet pipes to distribution reservoir elevated tanks and standpipes.

USE To stop flow when water in tank reaches a preset level and let water in again when level goes down.

TYPES Single-acting (illustrated) allows flow only into reservoir. See installation, p. 20-11, Fig. A. Double-acting allows flow both ways installed same as pressure reducing valve.

LOCATION Bottom of sumps, settling basins, channelways, etc.

USE To drain or blow off water and mud.

PLUG DRAIN VALVE§

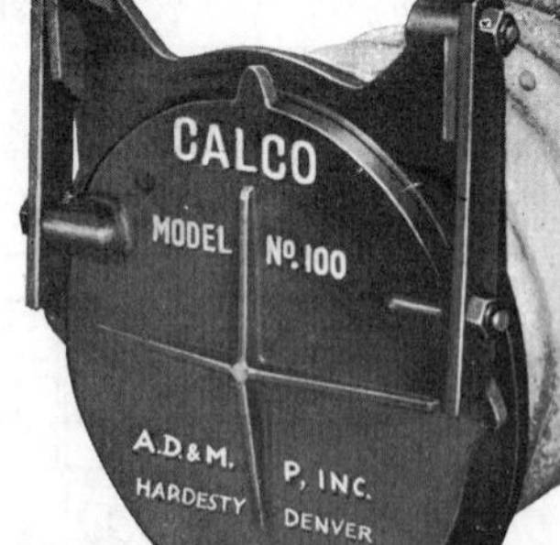
CALCO
MODEL
No. 100
A.D.&M.
HARDESTY
P. INC.
DENVER

AUTOMATIC FLAP GATE¶

LOCATION End of pipelines.

USE See swing check valve, above; used as tide gate.

*From Crane Co.
§From Eddy Valve Co.
† From Ludlow Valve Manufacturing Co.
|| From Golden-Anderson Valve Specialty Co.
‡ From Jenkins Bros.
¶From Hardesty Division, Armco Drainage & Metal Products Inc.

WATER DISTRIBUTION—VALVES—2

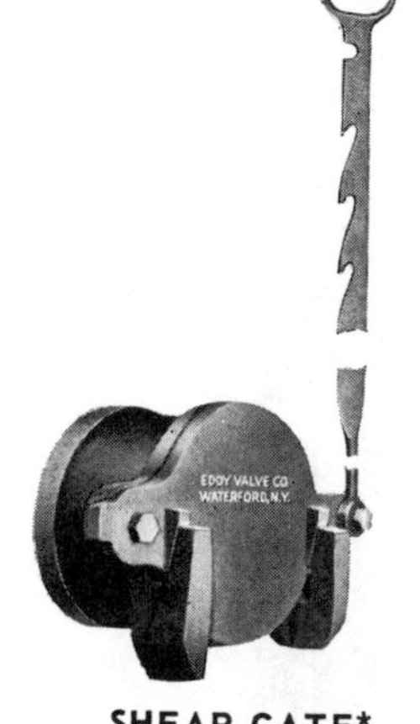
EDDY VALVE CO.
WATERFORD, N.Y.

SHEAR GATE*

LOCATION At ends of pipeline.

USE To control discharge of low pressures.

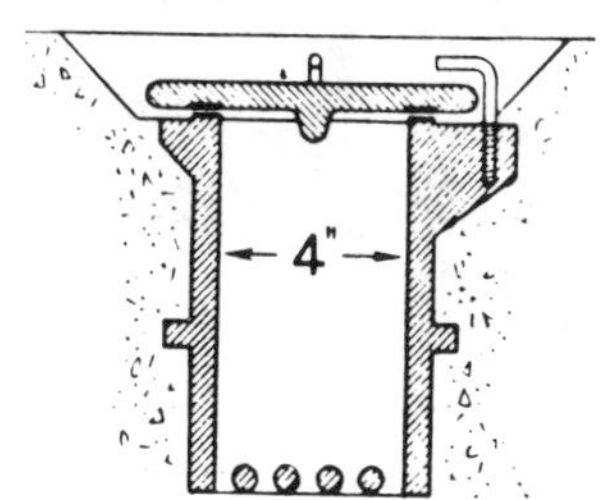
4"

PRESSURE RELIEF VALVE §

LOCATION In bottom of concrete tanks.

USE To prevent tank from floating when it is empty and there is excessive water pressure under and around tank.

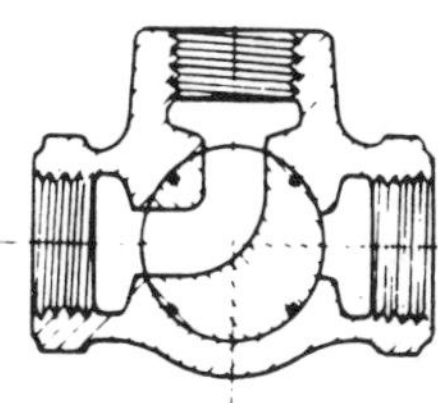

PLUG VALVE ||

LOCATION Any place in pipe.

USE To control flow of fluid. Useful in filter and softener control piping.

NOTES This valve comes in a variety of combinations including straightway; 3-way, 2-port (illus.); 3-way, 3-port; and 4-way, 2-port.

SLIDE OR SLUICE GATES †

LOCATION At ends of pipe or channel.

USE To control discharge at relatively low pressures.

NOTE Gate is frequently rectangular in shape.

BUTTERFLY VALVE ¶

LOCATION Any place in pipe.

USE To regulate rate of flow of fluid at low pressures and as shut off valve, and may be used with a flow meter in place of a rate-of-flow controller in a filter plant.

FLOAT VALVE ‡

LOCATION On supply lines to tanks or reservoirs.

USE To control flow of fluid into tank or reservoir.

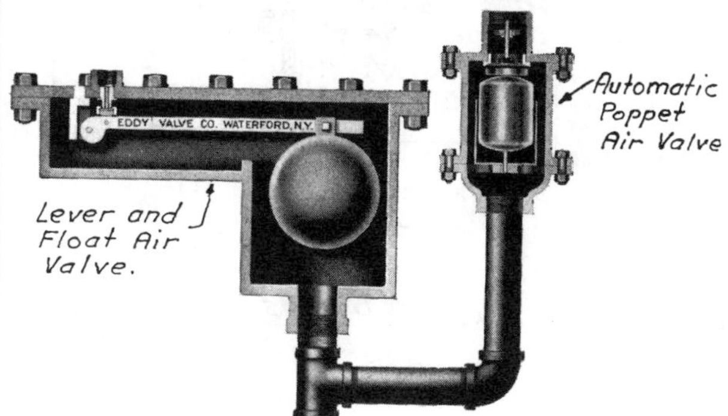
EDDY VALVE CO. WATERFORD, N.Y.
Automatic Poppet Air Valve
Lever and Float Air Valve.

AUTOMATIC FLOAT AND POPPET AIR VALVE*

LOCATION At summits of pipelines.

USE Poppet – To allow air to escape when pipe is being filled.

USE Float – To free pipe of air collecting under pressure when air is carried in water. To allow air to enter steel pipes rapidly to prevent formation of a vacuum causing a collapsing pressure in the event of a break in the steel pipe.

NOTE Generally diameter of valve pipe is from 1/8 to 1/12 of diameter of main pipe.

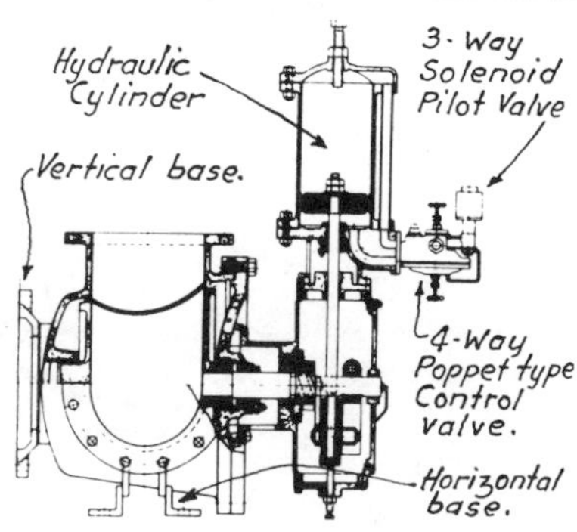
Hydraulic Cylinder
3-Way Solenoid Pilot Valve
Vertical base.
4-Way Poppet type Control Valve.
Horizontal base.

LOCATION At pump.

USE The solenoid control is tied in electrically with pump motor starter to open valve when pump is started and close when pump starter is tripped out. In addition, valve will close on any power failure.

ROTO CHECK VALVE**

*From Eddy Valve Co. † From Hardesty Division, Armco Drainage & Metal Products, Inc. ‡ From Kieley & Mueller, Inc.
§From James B. Clow & Sons. ||From American Car & Foundry Co. ¶From R-S Products Corp.
**From Morgan Smith Co.

WATER DISTRIBUTION—HYDRANT, SERVICE VALVE, PITOMETER

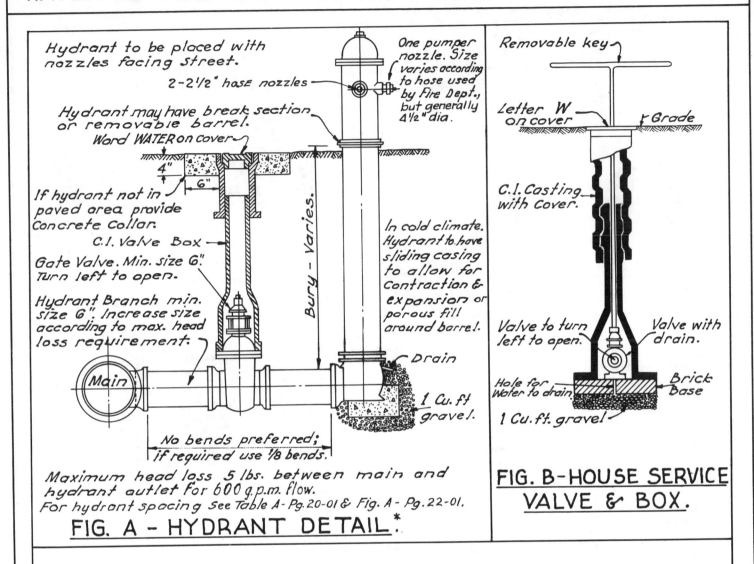

Hydrant to be placed with nozzles facing street.

2-2½" hose nozzles

One pumper nozzle. Size varies according to hose used by Fire Dept., but generally 4½" dia.

Hydrant may have break section or removable barrel.

Word WATER on cover

4"

6"

If hydrant not in paved area provide Concrete Collar.

C.I. Valve Box

Gate Valve. Min. size 6". Turn left to open.

Hydrant Branch min. size 6". Increase size according to max. head loss requirement.

Bury - Varies.

In cold climate. Hydrant to have sliding casing to allow for Contraction & expansion or porous fill around barrel.

Drain

Main

1 Cu. ft gravel.

No bends preferred; if required use 1/8 bends.

Maximum head loss 5 lbs. between main and hydrant outlet for 600 g.p.m. flow.

For hydrant spacing See Table A - Pg. 20-01 & Fig. A - Pg. 22-01.

FIG. A - HYDRANT DETAIL.*

Removable key

Letter W on cover

Grade

C.I. Casting with Cover.

Valve to turn left to open.

Valve with drain.

Hole for Water to drain

Brick Base

1 Cu. ft. gravel

FIG. B-HOUSE SERVICE VALVE & BOX.

FIG. C-COMBINATION HAND AND CLAMP PITOMETER FOR HYDRANT FLOW TESTS.**

* See Hydrant Manufacturers' Catalogs.

** From Northrop & Company, Inc.

WATER DISTRIBUTION—METERS—I

SERVICE METERS (HOUSE CONNECTIONS)

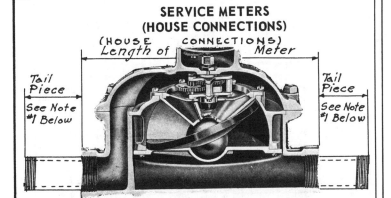

FIG. A DISPLACEMENT-TYPE METER

TABLE C – COLD-WATER METER*
DISPLACEMENT TYPE A.W.W.A. SPECIFICATIONS†

Meter Size, In.	Safe Maximum Operating Capacity, g.p.m.	Maximum Loss of Head, p.s.i.	Minimum Test Flow, g.p.m.	Normal Test Flow, g.p.m.	Meter Length, In.		Recommended Number of Years between Tests, years
					Screw Ends	Flange Ends	
(1)	(2)	(3)	(4)	(5)	(6)	(7)	(8)
5/8	20	15	1/4	1 to 20	7½	–	10
3/4	30	15	1/2	2 to 30	9	–	8
1	50	15	3/4	3 to 50	10¾	–	6
1½	100	20	1½	5 to 100	12⅝	13	4
2	160	20	2	8 to 160	15¼	17	4
3	300	20	4	16 to 300	–	24	3
4	500	20	7	28 to 500	–	29	2
6	1000	20	12	48 to 1000	–	36½	1

NOTES:

1. Coupling tail piece lengths = 5/8" size = 2-3/8", 3/4" size = 2/1/2", 1" size = 2-5/8" long.
2. Accuracy of registration should be ±1.5% when tested within the limits listed in column 5 (Normal Test Flow) with water having a temperature less than 80° F. Tested at Minimum Test Flow (column 4), meter shall not record less than 95% of the actual flow through it.
3. The capacities of the meters given in column 2 are the maximum rates of flow at which water should be passed through meters for short periods of time or the peak loads that should come upon meters only at long intervals. For continuous 24-hour service, meters of the displacement type should not be operated on flows greater than one fifth of the capacity of the meter.

MAIN LINE METERS (VELOCITY TYPE)
TYPES OF METERS COMMONLY USED

Measuring filter plant output to distribution system, and amount of water taken into the primary side of filter plant. Also in sewage plants for measuring sewage and air.

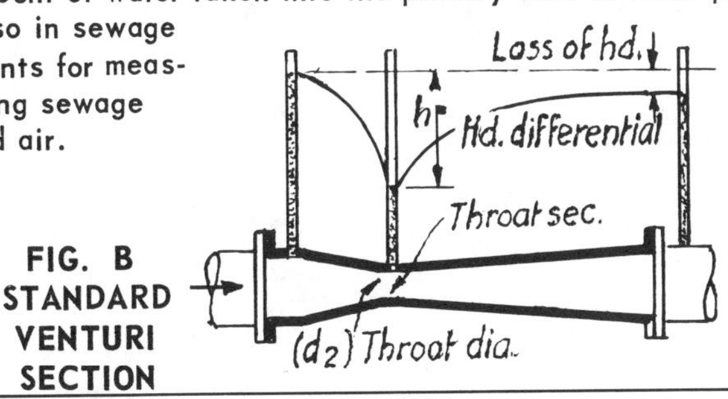
Loss of hd.
h
Hd. differential
Throat sec.
(d_2) Throat dia.

FIG. B STANDARD VENTURI SECTION

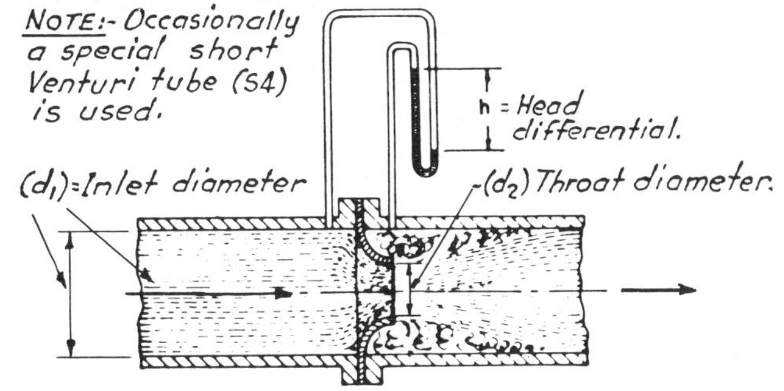
NOTE:- Occasionally a special short Venturi tube (S4) is used.
h = Head differential.
(d_1)=Inlet diameter
(d_2) Throat diameter.

FIG. D FLOW NOZZLE‡

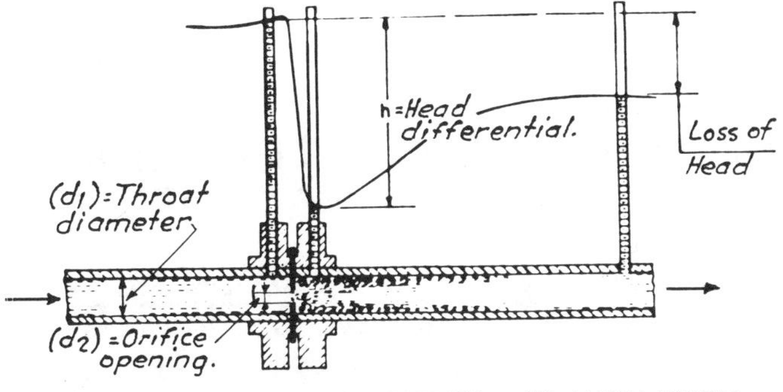
h=Head differential.
Loss of Head
(d_1)=Throat diameter
(d_2) =Orifice opening.

FIG. E FLAT-PLATE ORIFICE – FLANGE TYPE‡

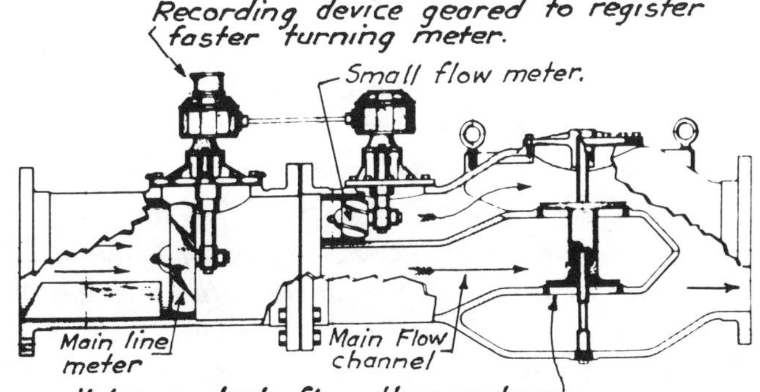
Recording device geared to register faster turning meter.
Small flow meter.
Main line meter
Main Flow channel
Valve controls flow thru meters.

FIG. F COMPOUND METER§

Note for All Meters: It is generally better to select a slightly undersized meter rather than one slightly oversized, as better accuracy will usually then be obtained.

*Adopted from Hersey Manufacturing Co.
†From American Water Works Association Specifications C700-46, May 1946.
§From R.W. Sparling Bulletin 307, 1941.
‡From Water Works and Sewerage, June 1943, "Selection of Main Line Meters" T.C. Thorensen, Builders Providence, Inc., Providence, R.I.

WATER DISTRIBUTION—METERS—2

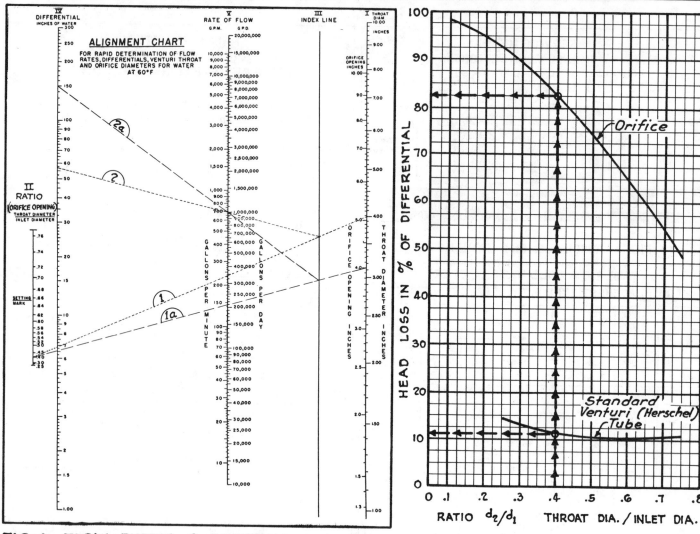

FIG. A- FLOW RATES & DIFFERENTIALS FOR STANDARD VENTURI THROAT & ORIFICE DIAMETERS.*

FIG. B- HEAD LOSS IN STANDARD VENTURI TUBE AND ORIFICE.

EXAMPLE OF THE USE OF FIGURES A & B.

Given- Inlet diameter - d_1 = 10.00 inches.

Throat diameter or orifice opening - d_2 = 4.00 inches.

Rate of flow = 1,000,000 gallons per day (G.P.D.).

Required- 1. Head differential - h betw. inlet and throat when

a - Standard Venturi tube is used.

b - Orifice is used.

2. Loss of head in each case.

Example- Ratio of throat dia. or orifice opening to Inlet dia. d_2/d_1 = 0.4. In Fig. A set straight edge at 4.00 inches in col. I and 0.4 in col. II and mark intersections with col. III (lines 1 and 1a). Set straight edge at intersections on col. III and 1,000,000 G.P.D. col. V (lines 2 and 2a) and read differentials on col. IV of 57 inches for Venturi tube and 150 inches for orifice.

From arrow lines in Fig. B- a. Venturi tube loss = 0.11 x 57 = 6.27 inches.

b. Orifice loss = 0.82 x 150 = 123 inches.

** From Water Works and Sewerage, June 1943, "Selection of Main Line Meters," J.C. Thoresen, Builders Providence, Inc., Prov. R.I.*

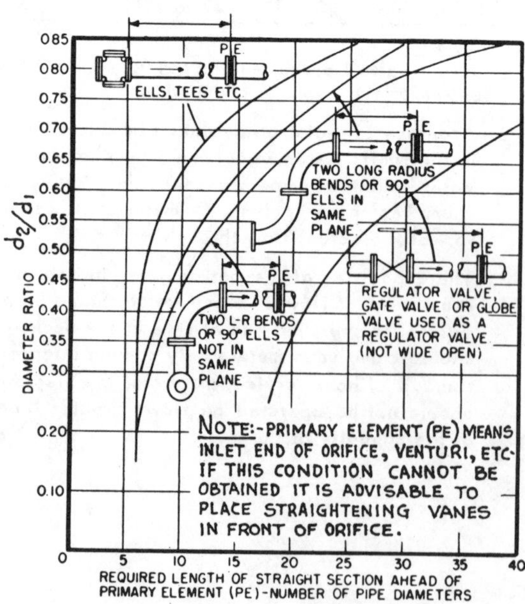

ELLS, TEES ETC
TWO LONG RADIUS BENDS OR 90° ELLS IN SAME PLANE
TWO L-R BENDS OR 90° ELLS NOT IN SAME PLANE
REGULATOR VALVE, GATE VALVE OR GLOBE VALVE USED AS A REGULATOR VALVE (NOT WIDE OPEN)
P.E.
d_2/d_1
DIAMETER RATIO
NOTE:- PRIMARY ELEMENT (PE) MEANS INLET END OF ORIFICE, VENTURI, ETC. IF THIS CONDITION CANNOT BE OBTAINED IT IS ADVISABLE TO PLACE STRAIGHTENING VANES IN FRONT OF ORIFICE.
REQUIRED LENGTH OF STRAIGHT SECTION AHEAD OF PRIMARY ELEMENT (PE) - NUMBER OF PIPE DIAMETERS

METER LOCATION

From Bailey Meter Co. Bulletin 301A.

WATER DISTRIBUTION—METERS—3

TABLE A – CAPACITIES OF VENTURI TUBES*

Size of Tube, inches	Length, feet & inches	Capacity, gal. per day Maximum	Capacity, gal. per day Minimum	Size of Tube, inches	Length, feet & inches	Capacity, gal. per day Maximum	Capacity, gal. per day Minimum
2× $\frac{25}{32}$†	$15\frac{13}{32}$"	53,000	2,600	18×9"	7'-$2\frac{1}{4}$"	7,196,000	360,000
2× $\frac{27}{32}$	$15\frac{13}{32}$"	62,200	3,100	18×11	6'-$2\frac{1}{8}$"	11,220,000	561,000
2×1	$14\frac{1}{16}$"	89,000	4,500	20×8	8'-11"	5,600,000	280,000
4×$1\frac{3}{4}$	2'-4"	270,000	13,500	20×10	7'-$10\frac{7}{8}$"	8,900,000	445,000
4×2	2'-$1\frac{1}{8}$"	355,000	17,700	20×12	6'-$10\frac{11}{16}$	13,200,000	665,000
4×$2\frac{3}{8}$	$23\frac{3}{4}$"	519,000	26,000	20×13	6'-$4\frac{5}{8}$"	16,040,000	802,000
6×$2\frac{1}{2}$	3'-$0\frac{5}{16}$"	546,000	27,300	24×8	11'-$4\frac{5}{16}$"	5,540,000	277,000
6×3	2'-$9\frac{1}{4}$"	800,000	40,000	24×11	9'-$10\frac{1}{8}$"	10,600,000	530,000
6×$3\frac{1}{2}$	2'-$5\frac{7}{16}$"	1,120,000	56,000	24×12	9'-4"	12,760,000	638,000
6×$3\frac{63}{64}$	2'-$3\frac{1}{8}$"	1,520,000	76,000	24×14	8'-$3\frac{7}{8}$"	17,930,000	896,000
8×3	3'-$11\frac{7}{16}$"	782,000	39,000	24×16	7'-$3\frac{11}{16}$"	24,580,000	1,230,000
8×4	3'-$5\frac{3}{8}$"	1,418,000	71,000	30×12	13'-$0\frac{1}{16}$"	12,550,000	627,000
8×5	2'-$11\frac{1}{4}$"	2,336,000	117,000	30×15	11'-$5\frac{3}{4}$"	20,000,000	1,000,000
10×4	4'-$7\frac{1}{2}$"	1,394,000	69,500	30×$16\frac{1}{2}$	10'-$8\frac{5}{8}$"	24,600,000	1,230,000
10×5	4'-$1\frac{3}{8}$"	2,220,000	111,000	30×18	9'-$11\frac{1}{2}$"	30,000,000	1,500,000
10×6	3'-$7\frac{3}{8}$"	3,320,000	166,000	36×15	15'-$1\frac{3}{4}$"	19,700,000	985,000
12×4	5'-$9\frac{11}{16}$"	1,385,000	69,000	36×18	13'-$7\frac{9}{16}$"	28,750,000	1,440,000
12×5	5'-$3\frac{5}{8}$"	2,184,000	109,000	36×20	12'-$7\frac{3}{8}$"	36,000,000	1,800,000
12×6	4'-$9\frac{9}{16}$"	3,198,000	160,000	36×22	11'-$7\frac{3}{16}$"	45,000,000	2,250,000
12×7	4'-$3\frac{3}{8}$"	4,480,000	224,000	42×21	16'-$0\frac{1}{4}$"	39,200,000	1,960,000
12×8	3'-$9\frac{5}{16}$"	6,145,000	307,000	42×22	15'-$6\frac{3}{16}$"	43,300,000	2,165,000
14×6	6'-$0\frac{3}{16}$"	3,150,000	157,500	42×24	14'-6"	52,420,000	2,621,000
14×7	5'-$6\frac{1}{8}$"	4,350,000	217,000	48×20	20'-$2\frac{1}{2}$"	35,000,000	1,750,000
14×8	5'-$0\frac{1}{16}$"	5,820,000	291,000	48×24	18'-2"	51,000,000	2,550,000
14×9	4'-$5\frac{15}{16}$"	7,651,000	380,000	48×28	16'-$1\frac{11}{16}$"	72,000,000	3,600,000
16×6	7'-$2\frac{7}{8}$"	3,127,000	156,500	54×24	21'-10"	50,500,000	2,500,000
16×7	6'-$8\frac{3}{4}$"	4,295,000	214,000	54×30	18'-$9\frac{9}{16}$"	82,000,000	4,100,000
16×8	6'-$2\frac{11}{16}$"	5,685,000	284,000	54×38	15'-$3\frac{7}{8}$"	144,000,000	7,200,000
16×9	5'-$8\frac{5}{8}$"	7,344,000	368,000	60×24	25'-$6\frac{1}{8}$"	50,200,000	2,510,000
16×$9\frac{1}{2}$	5'-$5\frac{9}{16}$"	8,295,000	415,000	60×30	22'-$5\frac{1}{2}$"	80,000,000	4,000,000
18×7	8'-$2\frac{7}{16}$"	4,264,000	213,000	60×36	19'-5"	119,500,000	5,975,000

(See Fig. B on page 22-20.)

NOTES:

1. Table A is a partial list of standard Venturi tubes made by the Simplex Valve & Meter Co.

   These meter sizes and others can be furnished by various manufacturers.

2. The lengths given in the table are for tubes having bell or flanged ends. For spigot ends, approximately 6" to 12" has to be added.

3. The flow range (ratio of maximum capacity to minimum capacity) of Simplex meters is 20:1.

   Other manufacturers make Venturi tubes that have flow ranges of 10:1 and 15:1

† 2" is diameter of main pipe; $\frac{25}{32}$" is diameter of throat.

TABLE B – CAPACITIES OF COMPOUND METERS‡

Size, inches	Normal Range gal. per minute	Over-all Length, inches, Flanged Ends	Over-all Length, inches, Bell & Spigot Ends
6 - 2§	20 - 900	52	$60\frac{1}{2}$
8 - 3	30 - 1200	58	$66\frac{1}{2}$
10 - 3	30 - 1600	68	$76\frac{1}{2}$
12 - 4	50 - 2250	72	$80\frac{1}{2}$
14 - 4	50 - 3000	90	$98\frac{1}{2}$
16 - 4	50 - 3800	102	$110\frac{1}{2}$
18 - 5	75 - 4500	111	$119\frac{1}{2}$
20 - 5	75 - 5500	120	$128\frac{1}{2}$
24 - 6	90 - 8000	138	$146\frac{1}{2}$
30 - 8	100 - 12000	159	$168\frac{1}{2}$
36 - 10	125 - 16000	180	$189\frac{1}{2}$

(See Fig. E on page 22-21.)

NOTES:

1. Table B is a list of the compound meters manufactured by R. W. Sparling. Various other companies supply compound meters having the same and different capacities and sizes.

§ 6" is diameter of main pipe and large measuring unit, and 2" is diameter of small measuring unit.

*Courtesy of Simplex Valve & Meter Co.
‡Courtesy of R. W. Sparling Co.

PETROLEUM PRODUCTS HANDLING-I

TABLE A – PROPERTIES OF COMMON PETROLEUM PRODUCTS AFFECTING STORAGE AND HANDLING

		Col. 1	Col. 2	Col. 3			Col. 4	Col. 5		Col. 6	Col. 7
Designation	Type	Specification	Average Specific Gravity	Kinematic Viscosity			Vapor Pressure	Flash Point, °F	N.B.F.U. Classi-fication	Fire Point	Pour Point, °F
				Temp., °F	C.S.	S.S.U.					
Motor gasoline	72/78 octane, regular	Fed. VV-M-561	0.72	60	0.6-0.9		9, 11, 13	–50	I		
Motor gasoline	80/86 octane, premium	Fed. VV-M-561	0.72	60	0.6-0.9		9, 11, 13	–50	I		
Aviation gasoline	80 octane	Mil.-F-5572	0.72	60	0.6-0.9		5.5-7.0	–50	I		
Aviation gasoline	91/96 octane	Mil.-F-5572	0.72	60	0.6-0.9		5.5-7.0	–50	I		
Aviation gasoline	100/130 octane	Mil.-F-5572	0.72	60	0.6-0.9		5.5-7.0	–50	I		
Aviation gasoline	115/145 octane	Mil.-F-5572	0.72	60	0.6-0.9		5.5-7.0	–50	I		
Jet-engine fuel	JP-1	Mil.-F-5616	0.85	40	10			110	III		
Jet-engine fuel	JP-3	Mil.-F-5624	0.77				5.0-7.0				
Jet-engine fuel	JP-4	Mil.-F-5624	0.82	60	1.1		2-3				
Jet-engine fuel	JP-5	Mil.-F-5624		40	20			140	III		
Kerosene	– –	Fed. VV-K-211	0.80	68	2.69	35		115	III		
Fuel oil	#1	C.S.-12-48	0.82	100		30-40		100	III		0
Fuel oil	#2	C.S.-12-48	0.84	100		40		100	III		5
Fuel oil	#4	C.S.-12-48	0.87	100		45-125		130	III		20
Fuel oil	#5	C.S.-12-48	0.93	100	5.8-26.4			150	III		
Fuel oil	#6	C.S.-12-48	0.97	122	92-638			150	III		
Fuel oil	Navy special	Mil.-F-859	0.99	70		1,400		150	III	200°F	15
				80		925					
				90		600					
Fuel oil	Navy heavy	Mil.-F-859	0.99	70		25,000		150	III	200°F	50
				80		11,000					
				90		7,000					
Diesel oil	1-D	A.S.T.M. D-975-50T	0.80	100	1.4			100	III		20
Diesel oil	2-D	A.S.T.M. D-975-50T	0.80	100	1.8-5.8	32-45		100	III		
Diesel oil	4-D	A.S.T.M. D-975-50T	0.80	100	5.8-26.4	45-125		130	III		
Lubricating oil		S.A.E. – 10	0.91	130		90-120		300-600			
Lubricating oil		S.A.E. – 20	0.91	130		120-185					
Lubricating oil		S.A.E. - 30	0.91	130		185-225					
Lubricating oil		S.A.E. - 40	0.91	130		225-400					
Lubricating oil		S.A.E. - 50	0.91	210		80-105					
Lubricating oil		S.A.E. - 60	0.91	210		105-125					
Lubricating oil		S.A.E. - 70	0.91	210		125-150					

PETROLEUM PRODUCTS HANDLING-2

DEFINITIONS

Col. 1. Fed. indicates Federal Specification; Mil. indicates Military Specification; C.S. indicates Commercial Standard, U.S. Department of Commerce; S.A.E. indicates Society of Automotive Engineers; A.S.T.M. indicates American Society for Testing Materials.

Col. 2. Specific gravity indicates density at 60°F. Water equals 1.0 at this temperature.

Col. 3. Viscosity is a measure of internal friction or resistance to flow. Viscosity varies inversely with temperature and must be determined at temperature ranges at which product is to be handled or pumped. Viscosity in petroleum products is commonly measured by Saybolt viscosimeters and expressed in seconds, as Saybolt Seconds Universal (S.S.U.) or Saybolt Seconds Furol (S.S.F.). Conversion to kinematic viscosity in centistokes (C.S.) is given on the top of Fig. A on page 23-03.

Col. 4. Vapor pressure is a measure of volatility of a liquid, and in petroleum practice is determined by the Reid method at 100° F., in pounds per square inch, and is expressed as Reid vapor-pressure pounds. Vapor pressures of commercial fuels vary with seasons and climate. True vapor pressure may be obtained from Fig. A on page 23-07.

Vapor-pressure considerations are necessary in the study of vapor conservation for the storage of volatile products and for the determination of available net positive suction head in pumping studies of volatile liquids. The net positive suction head (N.P.S.H.) is the total suction head in feet of liquid absolute determined at the suction nozzle and referred to datum, less the vapor pressure of the liquid in feet absolute.

Col. 5. Flash point is the lowest temperature at which product will give sufficient vapor to ignite momentarily when a flame is applied to the vapor and is an index of its fire hazard. National Board of Fire Underwriters (N.B.F.U.) defines flammable liquid as any liquid having a flash point below 200° F. and having a vapor pressure not exceeding 40 p.s.i. abs. at 100° F., and further divides such liquids into three classes:

Class I includes liquids having flash points at/or below 20° F.

Class II includes liquids having flash points above 20° F. and at/or below 70° F.

Class III includes liquids having flash points above 70° F.

Col. 6. Fire point is the lowest temperature at which vapors given off by the product will burn continuously.

Col. 7. Pour point is the lowest point at which the product will just flow and is an index of its ability to be handled or pumped at a particular temperature.

PETROLEUM PRODUCTS HANDLING-3

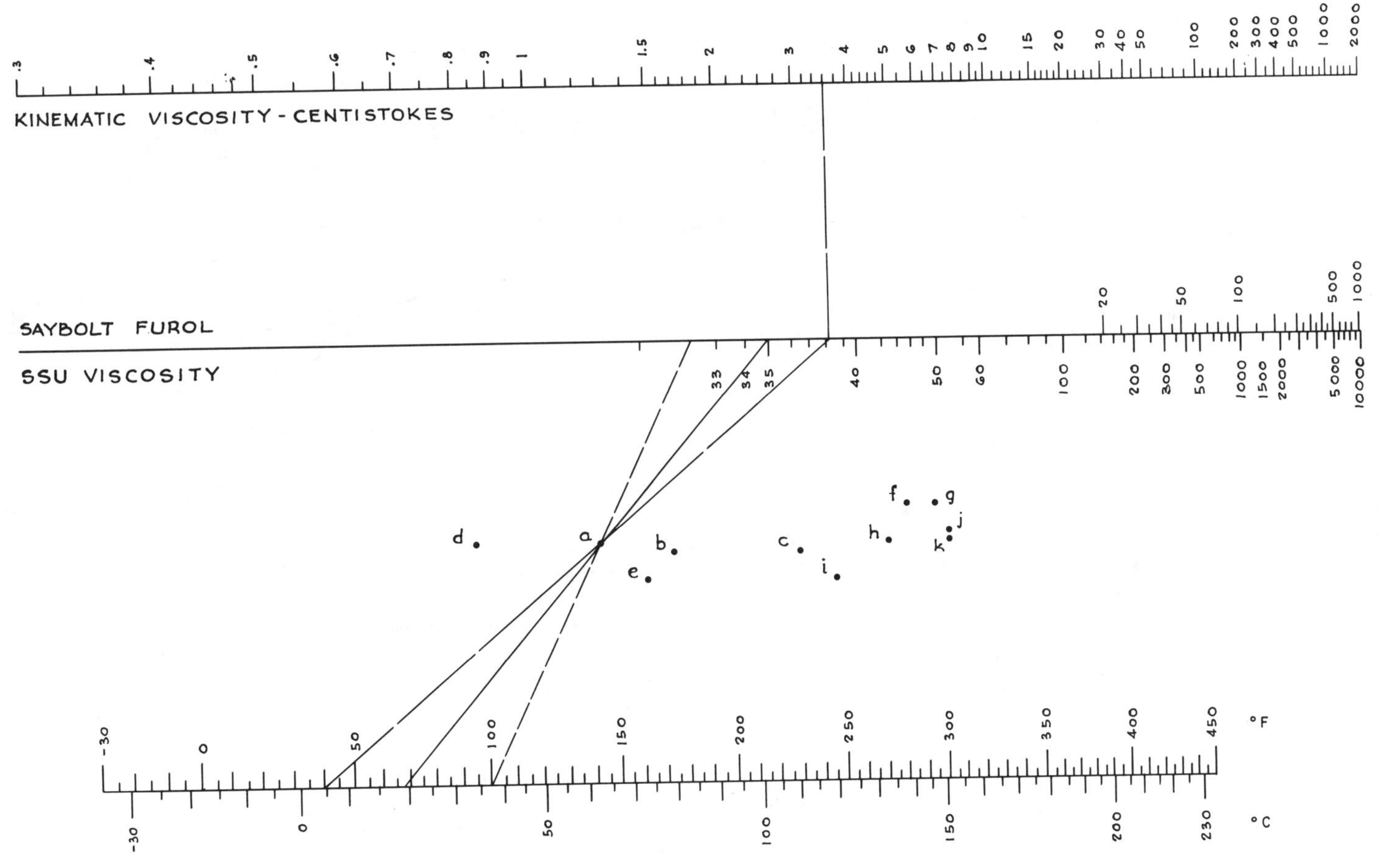
KINEMATIC VISCOSITY - CENTISTOKES
SAYBOLT FUROL
SSU VISCOSITY
°F
°C
a
b
c
d
e
f
g
h
i
j
k

FIG. A – VISCOSITY AND TEMPERATURE CHART

$$\text{Kinematic viscosity} = \frac{\text{absolute viscosity}}{\text{density}}$$

$$1 \text{ centipoise} = \frac{\text{poise}}{100}$$

Unit of kinematic viscosity is the stoke.

$$1 \text{ centistoke} = \frac{\text{stoke}}{100}$$

The unit of absolute viscosity is the poise which is the force in dynes required to move one square centimeter of specific liquid a distance of one centimeter.

If the viscosity of a liquid is given at two temperature points, its viscosity at any temperature can be determined from the above chart:

Given kerosene with viscosity range 35 S.S.U. at 68° F. and 32.6 S.S.U. at 100° F., plot on chart and find pivot point *a*. To find viscosity at 40° F., draw line from 40° F. through *a*, and find S.S.U. viscosity = 38.2. Project at right angles to find 3.6 centistokes.

Pivot points for ordinary viscosity ranges on the following products have been plotted on the chart:

a – Kerosene	*f* – S.A.E. 10
b – #2 fuel oil	*g* – S.A.E. 20
c – #4 fuel oil	*h* – S.A.E. 30
d – J.P. 4	*i* – S.A.E. 40
e – Diesel oil	*j* – S.A.E. 50
	k – S.A.E. 60

PETROLEUM PRODUCTS HANDLING-4

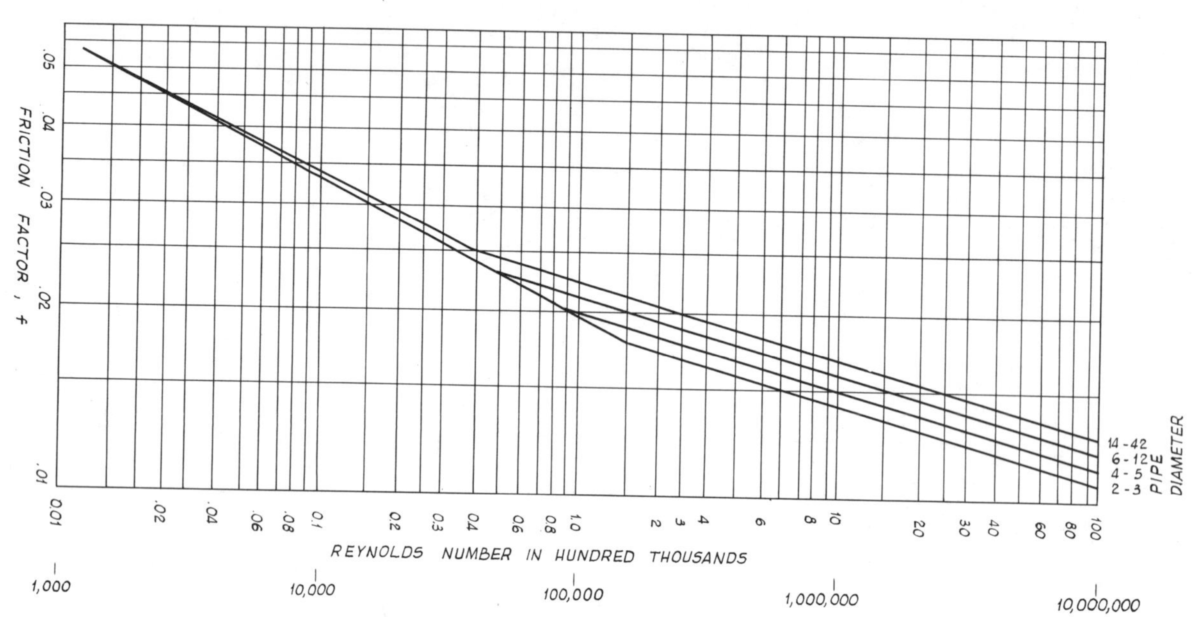

FRICTION FACTOR, f
.05
.04
.03
.02
.01
REYNOLDS NUMBER IN HUNDRED THOUSANDS
0.01 .02 .04 .06 .08 .1 .2 .3 .4 .6 .8 1.0 2 3 4 6 8 10 20 30 40 60 80 100
1,000
10,000
100,000
1,000,000
10,000,000
14-42
6-12
4-5
2-3
PIPE DIAMETER

FIG. A

Reynolds number $= R = \frac{Dv\rho}{\mu}$

ρ = density, p.c.f.

μ = absolute viscosity $= \frac{\text{absolute viscosity in centipoise}}{1488}$

where R is less than 2100, $f = \frac{64}{R}$

Basis for computing TABLE A, p. 23-05

Fanning formula for friction head due to flow of liquids in pipe is $h = \frac{fLv^2}{2GD}$ where:

h = friction loss in feet of product
f = friction factor – see above table
L = equivalent length of pipeline in feet
v = velocity of flow, ft./sec.
g = acceleration of gravity, ft./sec.^{2}
D = inside diameter of pipeline in feet.

Head loss of conventional products in common bulk plant practice is given in Table A, p. 23-05. Velocities should be limited to 7.5 ft./sec. where surge problems might arise. Discharge terminal velocities of 3 to 4 ft./sec. are recommended with highly volatile products to prevent building up of static charges when discharging into large storage tanks and compartments.

PUMPING RATES

Truck filling 3-inch meters 350 g.p.m.
4-inch meters 500-650 g.p.m.

Transfer rates: 1000-2000 g.p.m.

Barge unloading rates: 1000-2000 bbl./hr.

Tanker unloading rates: 8,000-10,000 bbl./hr.

1 gallon per minute (g.p.m.) = 1.428 barrels per hour (bbl./hr)

SUGGESTED TABLE FOR COMPUTING FRICTION LOSSES

Example: Pumping 350 g.p.m. of #2 fuel oil at 20°F*

g.p.m.	Size, in.	Item	No.	PIPING LOSSES			EQUIPMENT LOSSES			TOTAL FRICTION
				Unit Equivalent Length, ft.†	Total Length, ft.	Unit Loss, ft./ft.‡	Unit Loss, p.s.i.§	Viscosity Correction,ǁ	Conversion factor, p.s.i., to ft. of Product	Loss, ft. of Product
350	3	Pipe	–	–	50	× 0.413	–	–	–	20.6
		90° elbow	4	3.1	12.4	–	–	–	Sp.Gr = 0.84	–
		Gate check	1	1.6	1.6	–	–	–	$\frac{2.31}{.84} = 2.75$	–
		Swing check	1	19.8	19.8	–	–	–		–
			–	–	33.8	× 0.413	–	–	–	14.0
		Straight strainer	1	–	–	–	2.2	–	–	–
		Loading arm	1	–	–	–	10.5	–	–	–
							12.7	× 1.3	× 2.75	45.5

*From Table A, p. 23-05, find viscosity = 85 S.S.U.
†From Fig. A, p. 23-08.
‡From Table A, p. 23-05.
§From Table A, p. 23-06.
ǁFrom Fig. B, p. 23-06.

PUMP HORSEPOWER

For nonviscous products $HP = \frac{Q \times H \times \text{sp. gr.}}{E \times 3960}$

where Q = pumping rate in gallons per minute
H = total dynamic head in feet of product (friction + static + velocity head)
E = pump efficiency

In addition, for volatile products the N.P.S.H. available is required to be greater than the minimum N.P.S.H. required, as shown on page 23-07.

For viscous products, pumps are required to be selected as shown on page 23-08.

PETROLEUM PRODUCTS HANDLING-5

TABLE A

Pumping Rate g.p.m.	Pumping Rate bbl./hr.	Pipe Size, in.	Velocity f.p.s.	Velocity Head, ft.	Gasoline JP-4		Kerosene			#2 Fuel Oil			Diesel Oil			#4 Fuel Oil			#5 Fuel Oil	#6 Fuel Oil	
				Temperature, °F	0	60	20	40	60	20	40	60	20	40	60	20	40	60	122	122	160
				S.S.U. Viscosity	33	31.5	42	38.6	35.5	85	65	55	120	80	60	2800	900	600	400	3000	780
					Friction Loss, feet of Product per 1000 feet of Equivalent Steel Pipe Length																
250		3	10.9	1.9	158.0	145.0	178.0	168.0	165.0	227.0	210.0	198.0	246.0	224.0	204.0		345.0		320.0		310.0
		4	6.3	0.6	39.7	37.0	45.0	43.1	42.1	59.1	54.3	50.3	64.9	58.2	52.9	365.0	118.0	78.0	86.0	390.0	103.0
350	500	3	15.2	3.7	296.0	273.0	330.0	313.0	307.0	413.0	385.0	366.0	448.0	407.0	377.0	–	538.0	625.0	580.0	–	680.0
		4	9.8	1.2	75.0	68.5	83.5	80.4	78.0	107.0	99.4	93.9	117.0	106.0	96.9	511.0	130.0	170.0	157.0	547.0	176.0
		6	3.9	0.2	9.8	8.8	11.4	10.9	10.2	15.2	13.9	13.1	16.7	14.9	13.6	99.0	33.0	27.0	23.0	106.0	31.0
510	714	4	10.1	2.5	146.0	134.0	164.0	156.0	151.0	203.0	189.0	181.0	122.0	200.0	185.0	730.0	260.0	320.0	295.0	781.0	350.0
		6	5.6	0.5	18.8	17.1	21.7	20.7	19.7	28.6	26.4	24.8	31.2	28.1	25.8	140.0	49.0	46.0	41.5	151.0	48.0
650	929	4	164.0	4.2	236.0	220.0	261.0	253.0	245.0	324.0	305.0	291.0	354.0	319.0	299.0	–	567.0	513.0	460.0	–	550.0
		6	7.2	0.8	30.5	28.1	34.8	33.4	31.9	45.7	42.0	39.3	49.7	45.1	40.7	184.0	60.0	73.0	65.5	197.0	52.0
700	1,000	6	7.8	0.9	35.0	32.3	39.8	39.0	36.8	52.2	48.0	44.8	56.6	51.5	46.5	198.0	93.0	83.0	76.0	212.0	90.0
		8	4.5	0.3	9.3	8.5	10.7	10.2	9.7	14.0	13.0	12.1	15.3	13.8	12.6	67.0	22.0	22.0	20.3	71.0	18.0
1000	1,428	6	11.1	1.9	68.3	63.8	77.1	73.5	71.1	98.1	89.7	84.8	108.0	96.5	87.5	283.0	172.0	157.0	142.0	302.0	170.0
		8	6.4	0.6	17.9	16.5	20.4	19.5	18.7	26.5	24.3	22.7	28.9	26.1	23.5	95.0	49.0	42.2	32.2	101.0	47.0
1300	1,855	8	8.3	1.1	29.3	27.3	33.1	31.8	30.5	42.6	38.9	36.7	46.4	41.4	37.9	123.0	76.0	67.0	61.0	131.0	74.0
		10	5.3	0.4	9.6	8.9	10.9	10.5	10.1	14.2	13.0	12.2	15.5	14.0	12.6	49.0	26.0	22.6	20.5	53.0	25.0
1400	2,000	8	8.9	1.2	33.5	31.3	37.7	36.4	34.9	47.8	44.1	41.6	52.7	46.9	43.0	133.0	84.0	76.0	69.2	141.0	81.0
		10	5.7	0.5	11.1	10.3	12.5	11.9	11.6	16.2	14.9	14.0	17.7	16.0	14.4	52.0	29.0	25.7	23.3	57.0	28.5
2800	4,000	10	9.7	2.0	40.7	38.0	45.2	43.7	42.2	55.4	51.8	49.6	60.2	54.6	50.8	106.0	98.0	87.6	79.4	114.0	94.0
		12	6.8	1.0	17.0	15.9	18.5	17.8	17.5	23.4	21.9	20.8	25.8	23.0	21.4	51.0	40.0	36.0	33.0	55.0	38.0
5600	8,000	12	16.0	3.9	62.7	58.7	68.8	66.8	64.7	81.8	77.6	74.1	88.5	81.4	76.0	181.0	141.0	127.0	116.0	184.0	139.0
		16	9.7	1.5	17.0	15.8	18.7	18.1	17.6	22.6	21.3	20.3	23.6	22.3	20.9	38.0	41.0	36.0	32.0	39.0	38.0
		20	7.6	0.6	6.3	5.9	7.0	6.4	5.6	8.5	8.0	7.6	9.4	8.4	7.8	15.0	15.2	13.6	12.2	16.0	14.9
7000	10,000	16	12.3	2.4	29.6	27.7	32.5	40.3	30.5	38.4	36.6	34.9	41.6	38.4	35.9	85.0	66.0	60.0	53.2	88.0	63.0
		20	7.8	0.9	9.6	9.0	10.6	10.3	9.9	12.8	12.0	11.5	13.8	12.6	11.8	21.0	22.2	20.4	18.1	22.0	21.3

[Diagonal-slash symbol] Symbol indicates laminar flow

PETROLEUM PRODUCTS HANDLING-6

TABLE A – FRICTION LOSS VALVES, FITTINGS, AND EQUIPMENT – 150-LB. CLASS

Pipe Size, in.	Friction Loss Equivalent, ft. of Straight Pipe										Loss through Equipment, p.s.i.						
	VALVES					T'S		Bends			Maximum Rate of Flow, g.p.m.	Straight Strainers, 80 Mesh	Angle Strainers, 80 Mesh	Back-Pressure Valves	Air Elim-nators	Meters	Loading Arms
	Gate	Swing Check	Angle	Globe	Plug		Side	90°	45°	180°							
3	1.6	19.8	43.0	86.0	5.1	5.2	15.5	3.1	2.0	4.2	350	2.2	3.0	3.4	3.5	5.2	10.5
4	2.1	26.0	57.0	113.0	18.0	6.8	20.3	4.1	2.6	5.5	500-650	7.0	9.0	6.5	3.0	4.0	12.5
6	3.2	39.0	85.0	170.0	22.0	10.2	31.0	6.1	3.9	8.2	1000				1.5	2.0	
8	4.3	52.0	112.0	224.0	55.0	13.4	40.0	8.1	5.2	10.9							
10	5.3	65.0	141.0	281.0	120.0	16.9	51.0	10.2	6.5	13.7							
12	6.4	77.0	168.0	336.0	170.0	20.2	61.0	12.2	7.8	16.3							
16	8.5	104.0	–	–	–	26.9	81.0	16.3	10.4	21.8							
20	12.0	135.0	–	–	–	34.0	120.0	20.0	12.8	26.8							

Friction losses through equipment are generally given for liquids at 31.5 S.S.U. and should be corrected for viscosity at pumping temperature. The following chart gives approximate conversion factors.

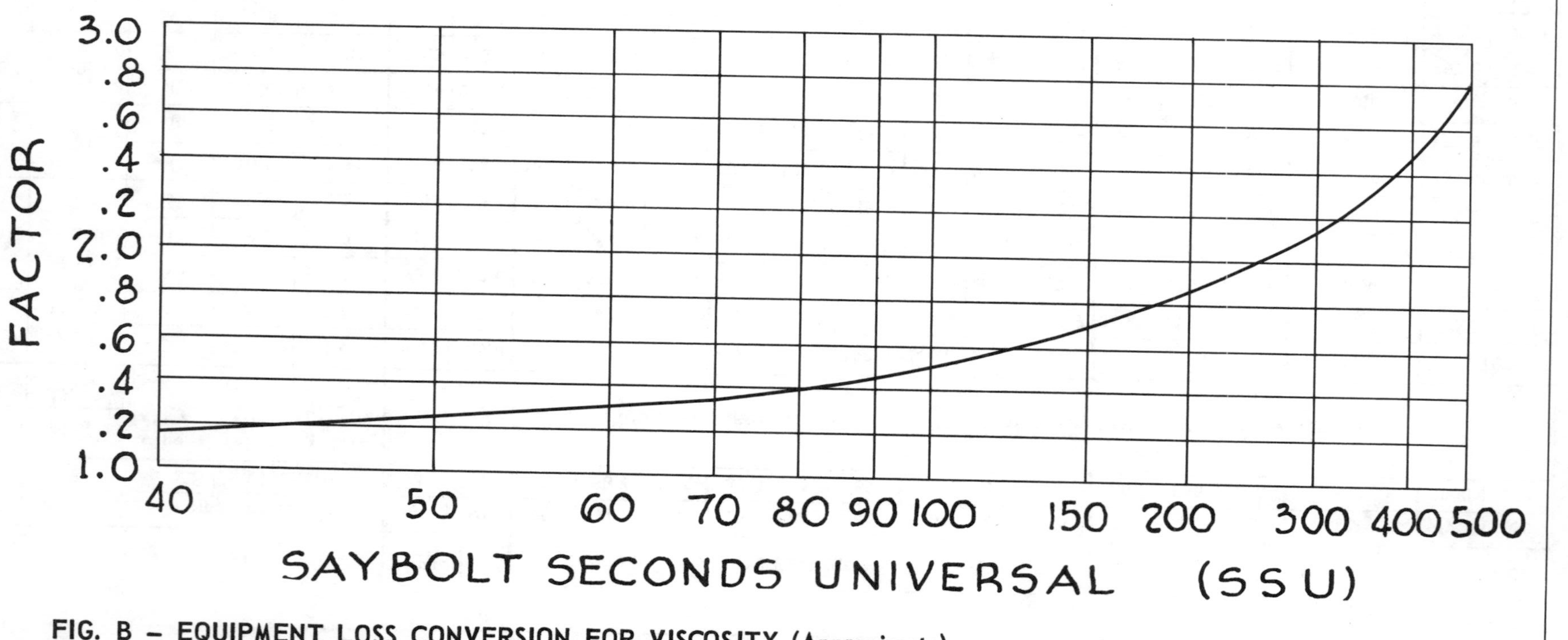
3.0
.8
.6
.4
.2
2.0
.8
.6
.4
.2
1.0
CONVERSION FACTOR
40
50
60
70
80
90
100
150
200
300
400
500
SAYBOLT SECONDS UNIVERSAL (SSU)

FIG. B – EQUIPMENT LOSS CONVERSION FOR VISCOSITY (Approximate)

PETROLEUM PRODUCTS HANDLING-7

REID VAPOR PRESSURE (p.s.i. at 100°F)

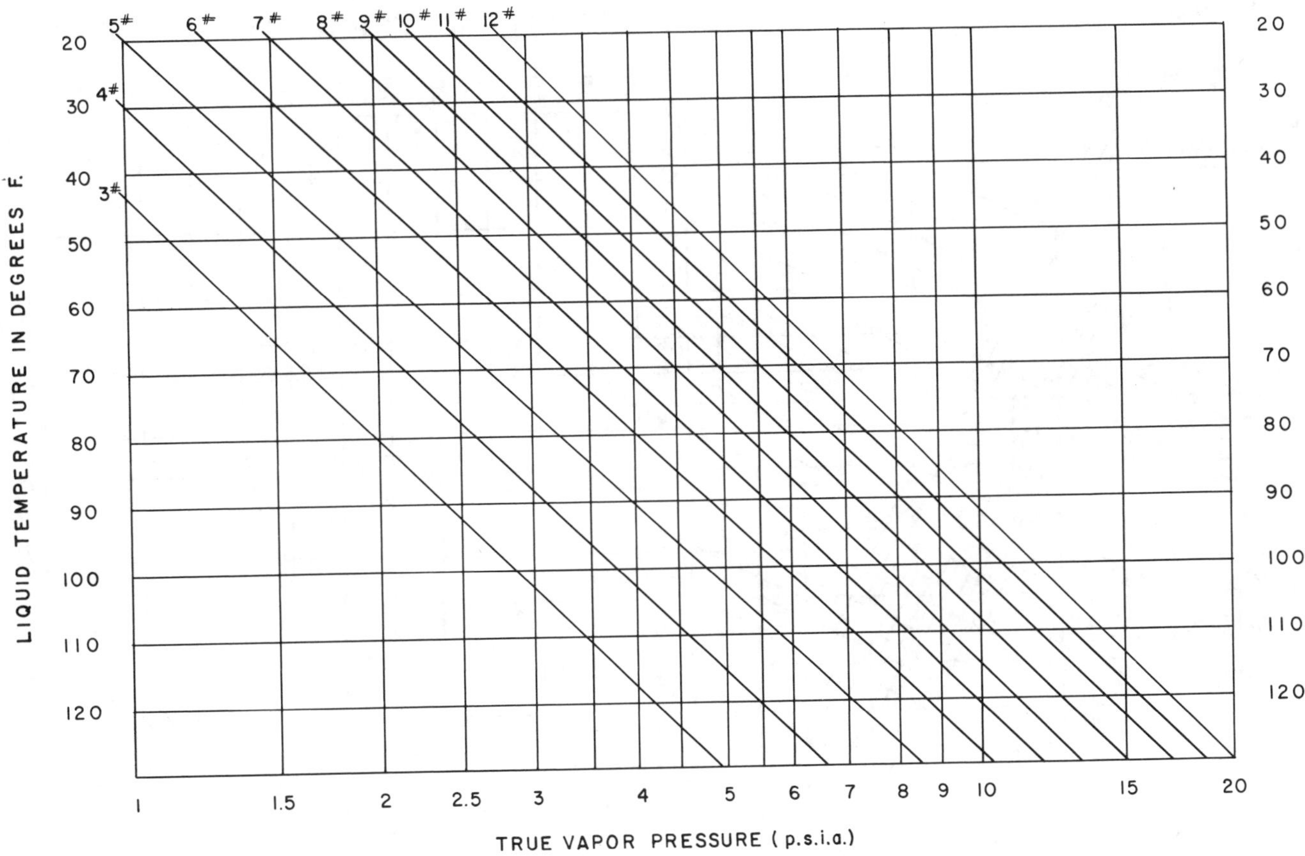
5#
6#
7#
8#
9#
10#
11#
12#
4#
3#
LIQUID TEMPERATURE IN DEGREES F.
20
30
40
50
60
70
80
90
100
110
120
1
1.5
2
2.5
3
4
5
6
7
8
9
10
15
20
TRUE VAPOR PRESSURE (p.s.i.a.)

FIG. A – VAPOR PRESSURES OF MOTOR GASOLINES

Example 1. Pump will pump.
Example 2. Pump will flash (product at eye of impeller will flash to vapor).

A Static suction lift in feet of product.
B Suction friction loss in feet of product.
C Velocity head in feet of product.
D Altitude correction in feet of product = 0.1 p.s.i. for each 200' above sea level.
E Absolute vapor pressure in feet of product.
F Net positive suction head available.
G Net positive suction head required (manufacturers requirement).

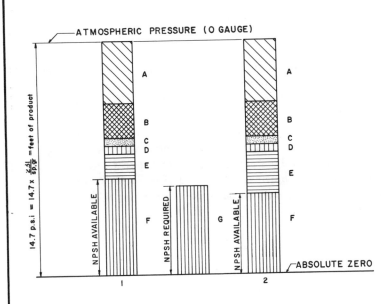

ATMOSPHERIC PRESSURE (O GAUGE)

$14.7 \text{ p.s.i} = 14.7 \times \frac{2.31}{\text{sp. gr}} = \text{feet of product}$

NPSH AVAILABLE — NPSH REQUIRED — NPSH AVAILABLE — ABSOLUTE ZERO

FIG. B

PETROLEUM PRODUCTS HANDLING-8

Centrifugal pump curves are based on water; viscosity = 31.5 S.S.U. Increase in viscosity tends to steepen the slope of head characteristic curve and reduce efficiency. Selection or performance correction of pumps handling viscous liquids may be done with the use of the following chart:

HEAD IN FEET

CORRECTION FACTORS

CAPACITY & EFFICIENCY HEAD

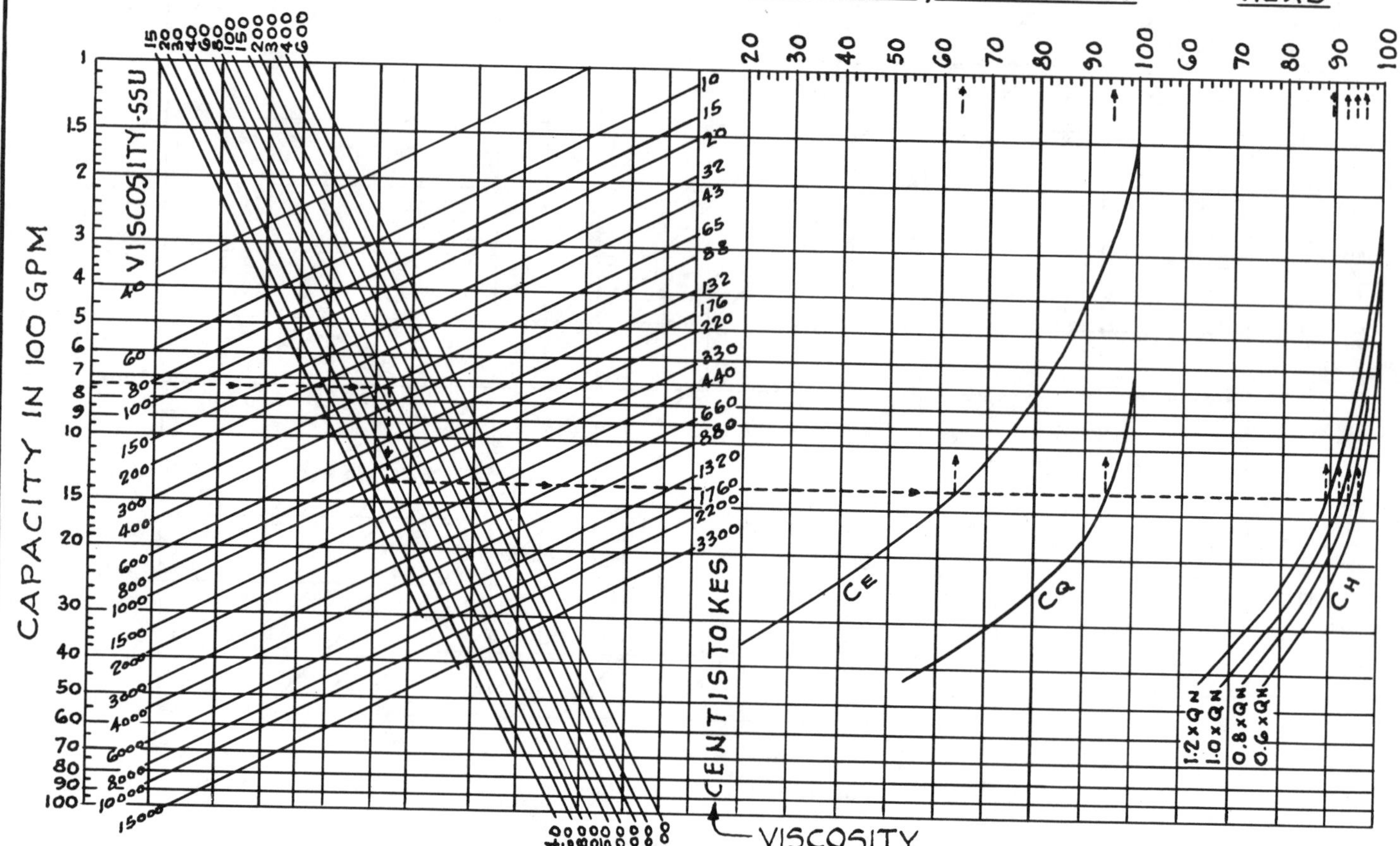

CAPACITY IN 100 GPM
VISCOSITY-SSU
CENTISTOKES
VISCOSITY
PERFORMANCE CORRECTION CHART
C_E
C_Q
C_H
1.2×QN
1.0×QN
0.8×QN
0.6×QN

FIG. A

Curve applicable only to centrifugal pumps of conventional design, open or closed impellers, and adequate N.P.S.H. Not good for axial or mixed-flow pumps or non-uniform liquids. Reprinted from the Tentative Standards of Hydraulic Institute, *Charts for the Determination of Pump Performance When Handling Viscous Liquids*, copyright 1948, by Hydraulic Institute, 122 East 42nd Street, New York 17, N. Y.

PUMP SELECTION

Given: Required capacity = Q g.p.m.
Required total head = H ft. of product.
Viscosity of product in S.S.U. at pumping temperature.
Specific gravity of product.

Enter chart at capacity Q; go horizontally to head H; then vertically to viscosity; then horizontally to find correction factors.

C_Q = capacity correction factor.
C_E = efficiency correction factor
$C_H = (1 \times Q_N)$ = head correction factor

Equivalent water capacity $Q_N = \dfrac{Q}{C_Q}$

Equivalent water head $= H_N = \dfrac{H}{C_H}$

Pump efficiency (viscous) $= E_2$ = pump efficiency (water) $\times C_E$

Required brake horsepower $= \dfrac{QH \times \text{sp. gr.}}{3960 \times E_2}$

PERFORMANCE CORRECTION

Given: Water rated pump curves, specific gravity and viscosity of product at pumping temperature to be used with pump.

Determine water capacity, Q_N, of pump at best operating efficiency. Tabulate values in g.p.m. of 0.6 Q_N, 0.8 Q_N, Q_N, 1.2 Q_N, and corresponding heads and efficiencies from pump curves.

Enter chart at capacity Q_N, go horizontally to head H, then vertically to given viscosity, then horizontally to find C_Q, C_E, and four values of C_H. Apply correction factors to tabulated water capacities, heads, and efficiencies to find corresponding points on pump curve operating with viscous product.

PETROLEUM PRODUCTS HANDLING-9

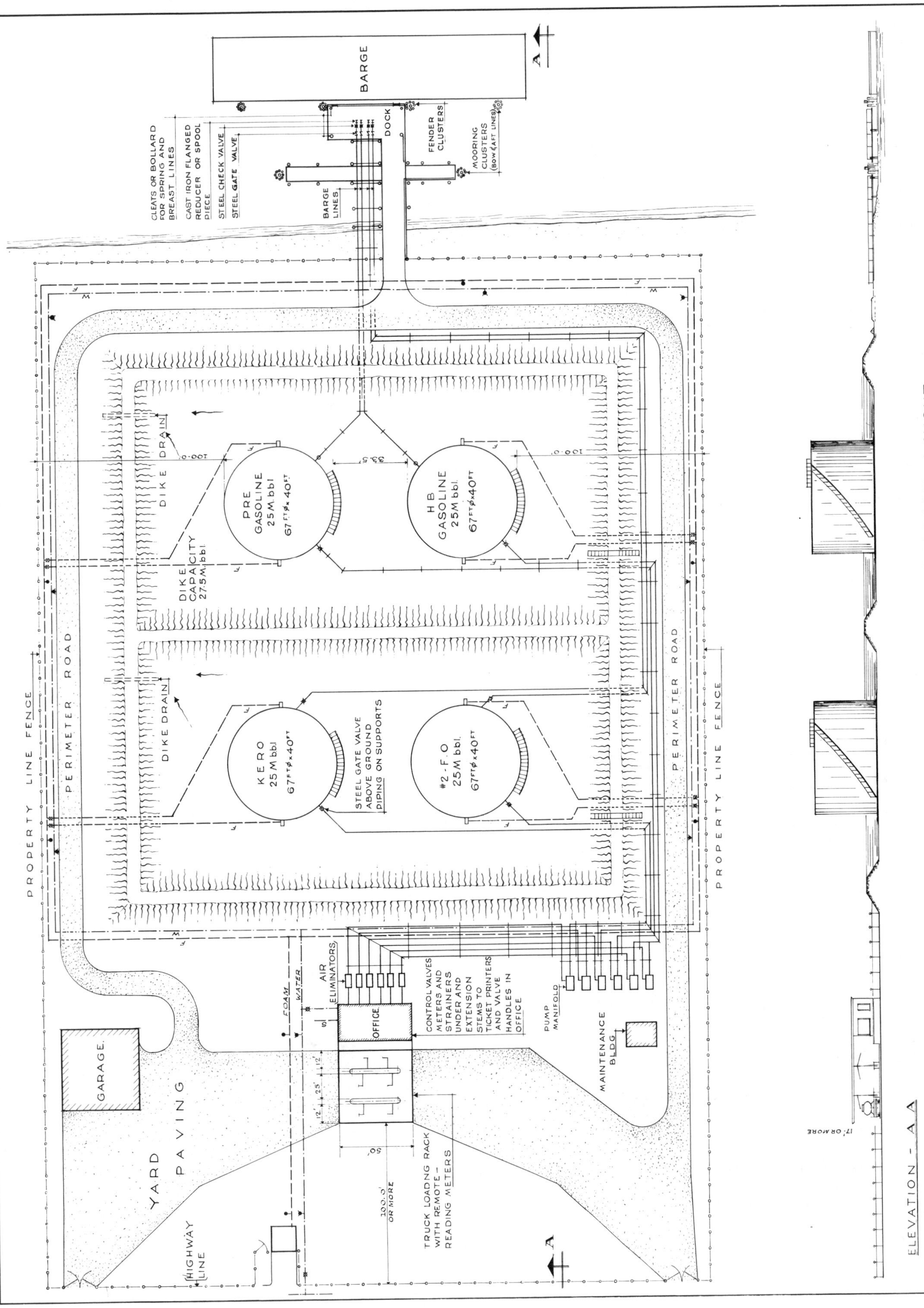

BARGE
DOCK
FENDER CLUSTERS
MOORING CLUSTERS (BOW & AFT LINES)
CLEATS OR BOLLARD FOR SPRING AND BREAST LINES
CAST IRON FLANGED REDUCER OR SPOOL PIECE
STEEL CHECK VALVE
STEEL GATE VALVE
BARGE LINES
DIKE DRAIN
100'-0"
33'-5"
PRE GASOLINE 25M bbl 67 FT ø x 40 FT
H B GASOLINE 25M bbl 67 FT ø x 40 FT
DIKE CAPACITY 27.5M bbl
KERO 25M bbl 67 FT ø x 40 FT
#2-F.O 25M bbl 67 FT ø x 40 FT
STEEL GATE VALVE ABOVE GROUND PIPING ON SUPPORTS
PROPERTY LINE FENCE
PERIMETER ROAD
AIR ELIMINATORS
FOAM
WATER
OFFICE
CONTROL VALVES METERS AND STRAINERS UNDER AND EXTENSION STEMS TO TICKET PRINTERS AND VALVE HANDLES IN OFFICE
PUMP MANIFOLD
MAINTENANCE BLDG
GARAGE
YARD PAVING
12' 25' 12'
50'
100.0' OR MORE
TRUCK LOADING RACK WITH REMOTE-READING METERS
HIGHWAY LINE
A A
17' OR MORE

FIG. A—TYPICAL PETROLEUM STORAGE LAYOUT

ELEVATION - A A

PETROLEUM PRODUCTS HANDLING-10

TABLE A—CAPACITY DIMENSIONS AND WEIGHTS, A.P.I. 12C TANKS

Nominal Capacity, bbl.	Dimensions		Weight in Tons, 3/16" Cone Roof		Ratio, bbl. Storage per ton	
	Diameter, ft.-in.	Height, ft.-in.	1/4" Bottom	5/16" Bottom	1/4" Bottom	5/16" Bottom
500	15-0	16-0	6.85	7.1	73.1	70.5
1,000	21-3	16-0	10.4	10.9	96.0	91
1,500	21-3	24-0	13.55	14.05	111	107
2,000	25-0	24-0	16.4	17.05	122	117
3,000	30-0	24-0	20.65	21.55	145	139
5,000	36-6	32-0	28.8	30.0	173	167
6,000	36-8	32-0	32.4	33.8	185	178
7,000	40-0	32-0	36.65	38.35	191	183
7,500	36-8	40-0	37.55	38.95	200	193
8,500	40-0	40-0	42.25	43.95	201	193
10,000	42-6	40-0	45.85	47.75	218	209
12,000	42-6	48-0	53.1	55.0	226	218
15,000	48-0	48-0	64.1	66.5	234	225
20,000	60-0	40-0	78.2	82.0	256	244
24,000	60-0	48-0	91.8	95.6	261	251
25,000	67-0	40-0	95.2	99.95	262	250
30,000	73-4	40-0	111.65	117.3	269	255
35,500	80-0	40-0	131.7	138.45	270	257
42,500	80-0	48-0	158.15	164.9	269	258
45,000	90-0	40-0	162.05	170.6	277	264
54,000	90-0	48-0	194.9	203.45	277	265
55,000	100-0	40-0	197.05	207.55	279	265
67,000	100-0	48-0	236.6	247.1	283	271
67,000	110-0	40-0	235.2	247.95	285	270
80,000	110-0	48-0	285.0	297.75	280	269
80,000	120-0	40-0	284.5	299.6	281	267
96,000	120-0	48-0	342.0	357.1	280	269
109,500	140-0	40-0	381.0	401.6	287	273
131,500	140-0	48-0	460.0	480.6	286	273
125,000	150-0	40-0	437.0	460.75	285	271
150,000	150-0	48-0	526.0	549.75	285	273
180,000	180-0	40-0	621.5	655.75	290	274
217,000	180-0	48-0	752.5	786.75	289	276
223,000	200-0	40-0	765.0	807.0	291	276
268,000	200-0	48-0	925.0	967.0	290	277

(a) Spiral stairs and platform. (Locate on south side where possible in latitudes subject to ice and snow.)
(b) Roof railing.
(c) Gage hatch.
(d) Plugged opening (for gaging if tank settles out of level).
(e) Roof manhole. (Locate 180° from shell manhole for maximum ventilation when cleaning tank).
(f) Roof vent.
(g) Foam chamber.
(h) Painters eyebolt.
(i) Liquid-level gage.
(j) Swing-line sheave.
(k) Shell manhole — ℄ 30" ϕ manhole 36" off bottom.
(l) Boiler-maker flange for water draw-off valve ℄ 9" from bottom.
(m) Foam delivery riser.
(n) Double-shell suction nozzle for use with swing line. Top of suction nozzle should be below bottom of shell manhole. Swing lines are generally used to allow selective withdrawal when storage is confined to a single tank.
(o) Inlet shell nozzle
(p) Swing-line winch.

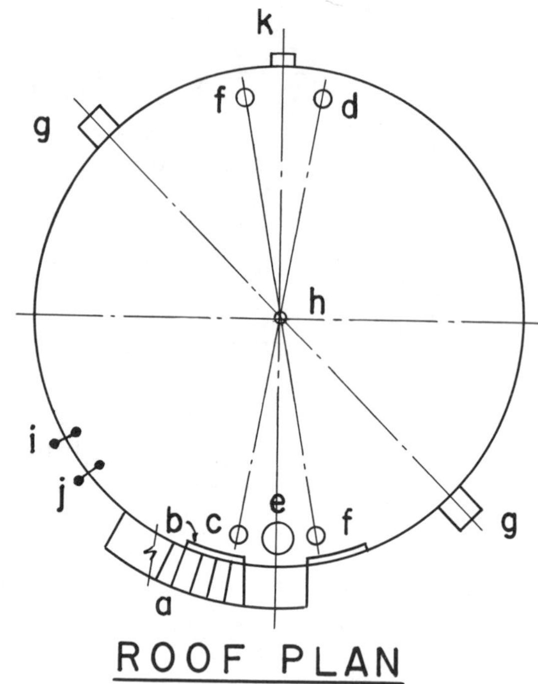
k
f
d
g
h
i
j
e
b
c
f
g
a

ROOF PLAN

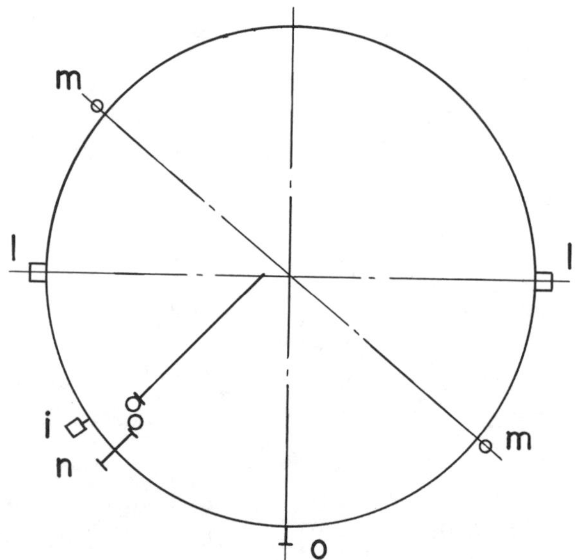
m
l
l
i
n
m
o

SHELL PLAN

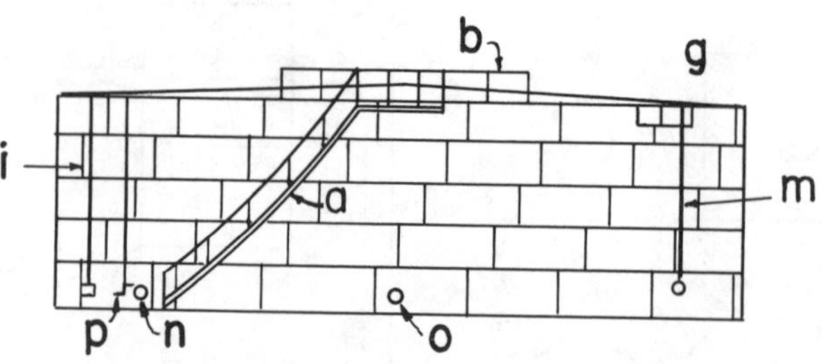
b
g
i
a
m
p
n
o

ELEVATION

FIG. B—TYPICAL TANK FITTING PLANS

PETROLEUM PRODUCTS HANDLING–II

In the absence of more restrictive codes, above-ground tanks for storage of refined petroleum products should be located in accordance with the requirements of N.B.F.U. Pamphlet 30.

TABLE A – LOCATION WITH RESPECT TO PROPERTY LINES

Capacity of Tank, gal.	Class of Flammable Liquid	Distance from Line of Adjoining Property Which May Be Built upon Shall not Be Less than: ft.
0 to 275	III	0
276 to 750	III	5
0 to 750	I, II	10
751 to 12,000	III	10
751 to 12,000	I, II	15
12,001 to 24,000	I, II, III	15
24,001 to 30,000	I, II, III	20
30,001 to 50,000	I, II, III	25
Over 50,000	I, II, III	Greatest dimension of diameter or height of tank, except such distance need not exceed 120 ft. when tank is equipped with approved attached extinguishing system or floating roof, or 1½ times greatest dimension of diameter or height of tank, except such distance need not exceed 175 ft.

A. Spacing between tanks.
   1. Tanks up to 50,000 gal. capacity should have minimum of 3 ft. of space between shells.
   2. Tanks over 50,000 gal. capacity should have a minimum spacing of not less than one-half the diameter of the smaller tank.

B. Tank dikes may be required by local code or topography to contain product within tank-farm area in case of a tank rupture. N.B.F.U. requirements for dike capacity state that net capacity should be equal to that of largest tank plus 10% of aggregate capacity of other tanks served by enclosure. Good practice segregates different classes or products in separate enclosures and limits dike heights to 5 to 7 ft. Dike enclosures should be properly graded and drained. Drains should normally be kept closed and designed to prevent escape of product from enclosure.

Petroleum products are usually stored in all-welded steel cone roof tanks conforming to American Petroleum Institute Specification 12 C (A.P.I. 12 C) unless otherwise required by local code, or consideration of vapor conservation. To conserve vapor and reduce explosive and fire hazards, volatile products are stored in:

1. Cone roof tanks equipped with conservation-type vents.
2. Floating roof tanks (not recommended where severe snow and icing conditions exist).
3. Lifter roof tanks.
4. Flat-bottom tanks equipped with a vapor dome.
5. Cone-roof tanks interconnected to a vapor sphere.
6. Underground tanks.

High-viscosity oils with a pumping range of 120° to 160°F. should be stored in insulated or uninsulated A.P.I. 12C tanks equipped with heating coils or suction heaters.

Open-type vents may be used to provide normal venting capacity of tanks storing products with a flash point of 100° F. and above. Breather-type vents are required on all tanks over 2500 bbl. storing products with a flash point below 100° F.

PETROLEUM PRODUCTS HANDLING-12

Maximum tank working pressure is pressure just below which point roof plates lift.

Working pressure of standard A.P.I. tanks is generally taken at 0.85 oz. per sq. in. the weight of 3/16" roof plates.

Maximum safe vacuum depends on size, design, and use of tank and should be checked with manufacturer. 1.0 oz. per sq. inch vacuum limit is common. Good practice provides for vent pallets set at about 50% of maximum pressure and vacuum, and for 200% of venting requirements divided into two or more units.

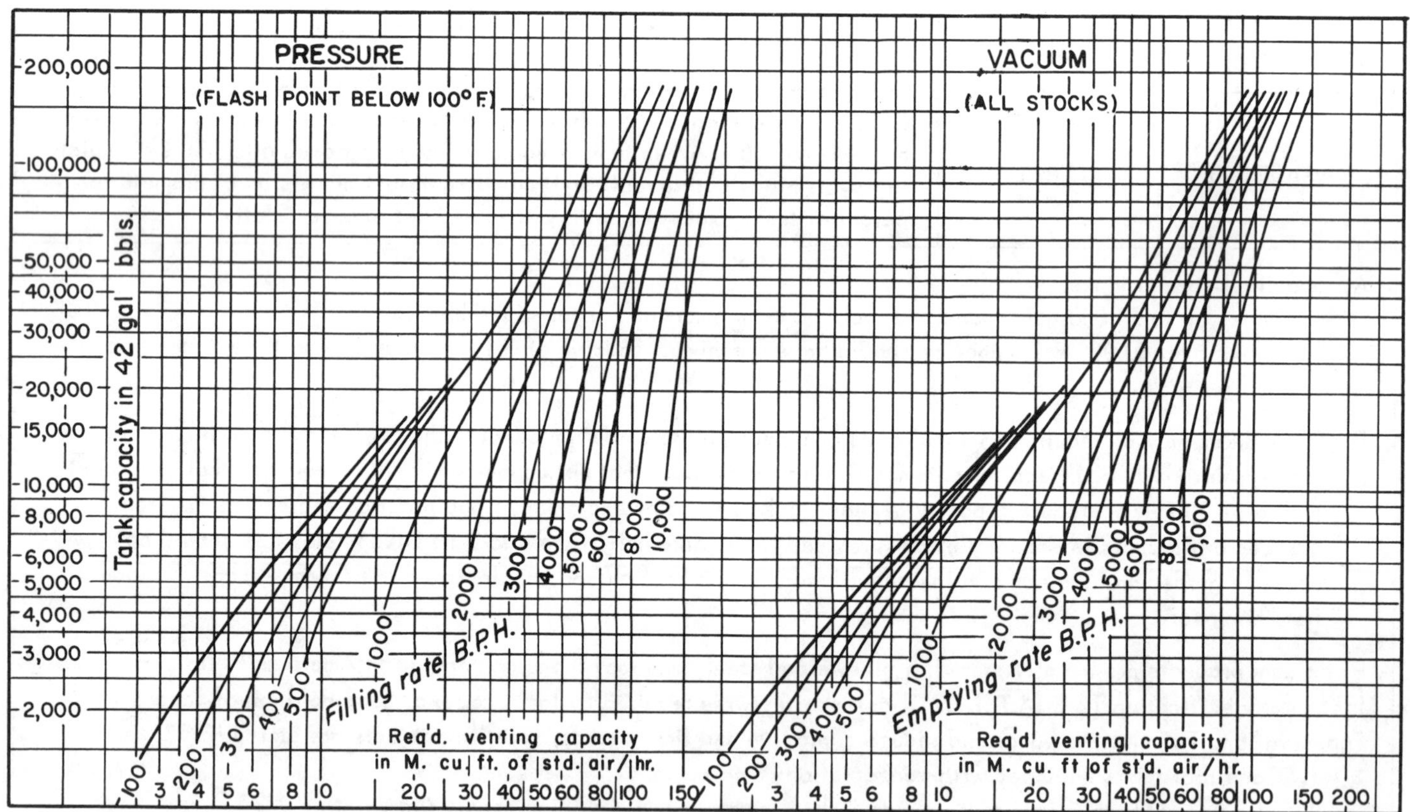
PRESSURE
(FLASH POINT BELOW 100° F.)
VACUUM
(ALL STOCKS)
Tank capacity in 42 gal bbls.
200,000
100,000
50,000
40,000
30,000
20,000
15,000
10,000
8,000
6,000
5,000
4,000
3,000
2,000
Filling rate B.P.H.
100
200
300
400
500
1000
2000
3000
4000
5000
6000
8000
10,000
Emptying rate B.P.H.
Req'd. venting capacity
in M. cu. ft. of std. air/hr.
3 4 5 6 8 10 20 30 40 50 60 80 100 150
3 4 5 6 8 10 20 30 40 50 60 80 100 150 200

FIG. A

EXAMPLE: Given a standard A.P.I. 30 M. bbl. tank, storing gasoline with a maximum filling rate of 4000 bbl. per hr. and an emptying rate of 1500 bbl. per hr.

Find outbreathing requirements of 78 M. cu. ft. of air per hr. and inbreathing requirements of 37 M. cu. ft. of air per hr. Specify two (2) breather valves each with minimum rated capacity 78 M. cu. ft. of air per hr. at 0.85 oz. per sq. in. pressure and 37 cu. ft. of air per hr. at 1.0 oz. sq. in. vacuum with pallet settings at ½ oz. sq. in. pressure and vacuum. Additional emergency venting is not required if tank has A.P.I. Standard 12C weak roof seam construction.

GRANDSTANDS

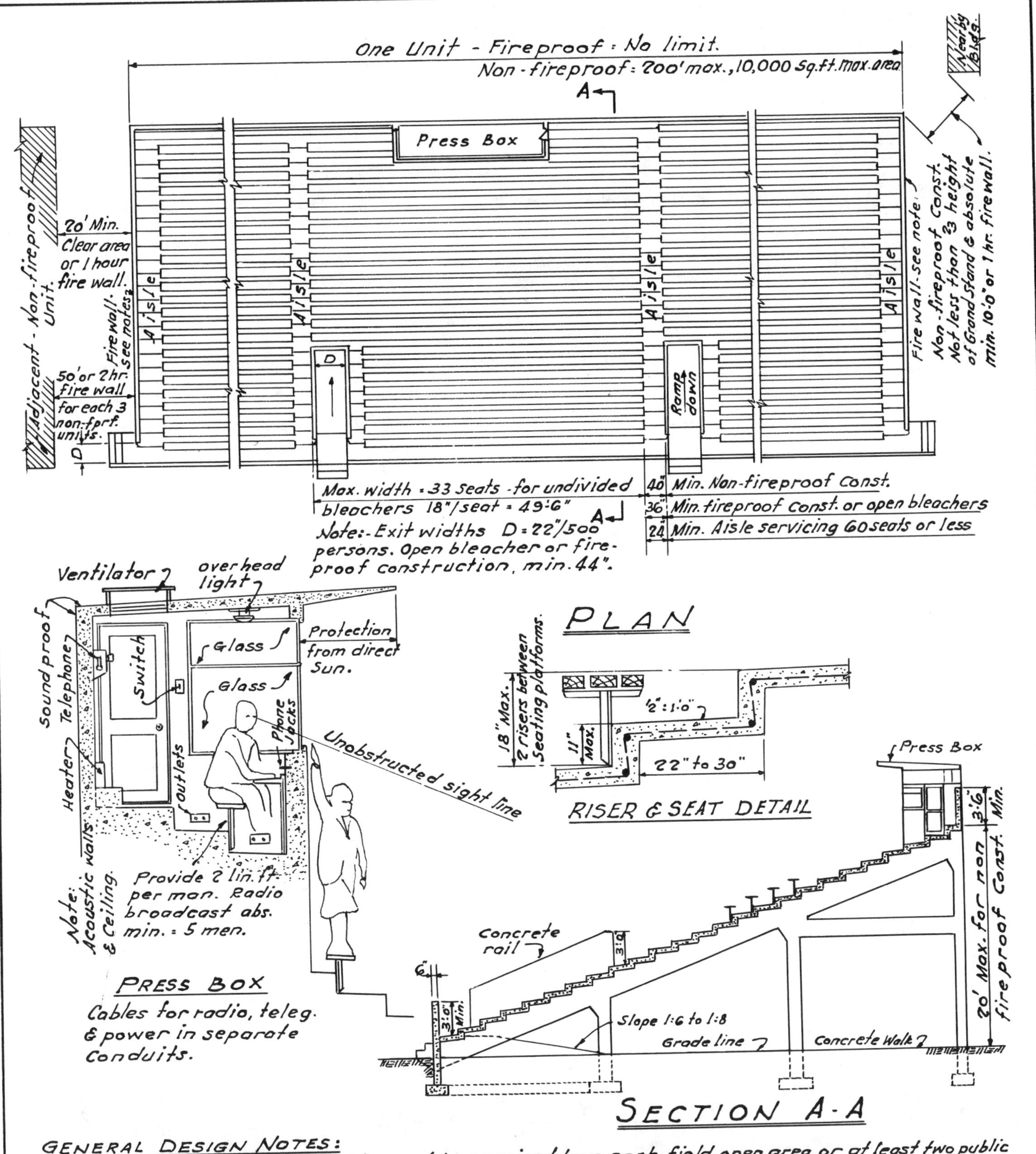
One Unit - Fireproof: No limit.
Non-fireproof: 200' max., 10,000 sq.ft. max. area
A
Press Box
Nearby Bldg.
Adjacent - Non-fireproof Unit.
20' Min. Clear area or 1 hour fire wall.
Fire wall - See notes
50' or 2 hr. fire wall for each 3 non-fprf. units.
D
Aisle
Aisle
Aisle
Aisle
D
Ramp down
Fire wall - See note:
Non-fireproof Const. Not less than 2/3 height of Grand Stand & absolute min. 10'0" or 1 hr. firewall.
Max. width = 33 seats - for undivided bleachers 18"/seat = 49'6"
Note: - Exit widths D = 22"/500 persons. Open bleacher or fireproof construction, min. 44".
40" Min. Non-fireproof Const.
36" Min. fireproof const. or open bleachers
24" Min. Aisle servicing 60 seats or less
PLAN
Ventilator
overhead light
Sound proof Telephone
Switch
Glass
Glass
Protection from direct Sun.
Phone Jacks
outlets
Heater
Unobstructed sight line
Note: Acoustic walls & Ceiling.
Provide 2 lin. ft. per man. Radio broadcast abs. min. = 5 men.
PRESS BOX
Cables for radio, teleg. & power in separate Conduits.
18" Max. 2 risers between Seating platforms.
11" Max.
1/2" : 1'0"
22" to 30"
RISER & SEAT DETAIL
Press Box
3'6" Min.
20' Max. for non fireproof Const.
Concrete rail
3'0"
6"
3'0" Min.
Slope 1:6 to 1:8
Grade line
Concrete Walk
SECTION A-A

GENERAL DESIGN NOTES:

ACCESSIBILITY: Grandstands must be serviced by a park, field, open area or at least two public ways of approach as remote from each other as possible. Aggregate width of Grandstand exits shall not exceed 80% of public approach widths. Lanes from exits to servicing areas shall not be less than 20' wide.

EXITS: For each Unit; provide 2 exits for a capacity up to 1000 persons, 3 exits for a capacity up to 4000, and 4 exits when greater than 4000. Max. path, seat to exit = 150 feet.

CAPACITY: Number of seats plus one person for each 6 sq.ft. of standing room, not including Aisles or exits.

LOADING: See pg. 1-12 *. Based on a publication of the Portland Cement Association.

Based on National Fire Protection Bulletin #102-1949 approved by A.S.A. 1950.

ATHLETIC FIELDS—DRAINAGE & DIMENSIONS

PURPOSE OF DRAINAGE: Surface slopes and drains to take off rainfall. Subdrains (or underdrains) to lower high ground water tables and to minimize frost action. Do not allow rainfall to run into underdrains. A porous fill is generally not subject to frost action, and use of such fill may eliminate need for underdrains. See Pg. 9-35 for further data on frost.

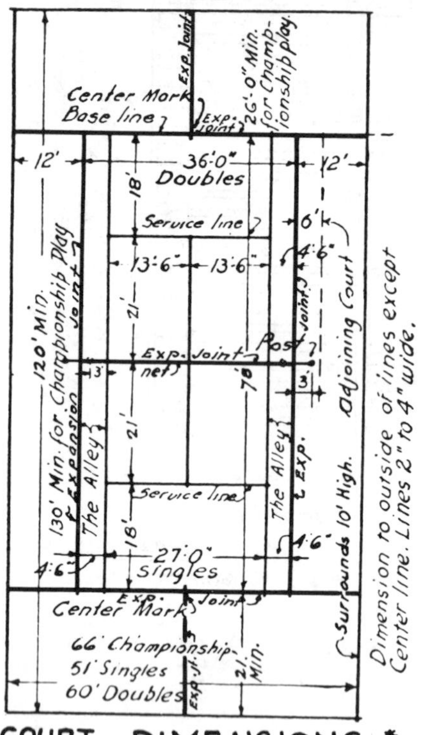

Center Mark
Base line
Exp. Joint
26'-0" Min. for Championship play
12'
36'-0"
Doubles
12'
18'
Service line
6'
4'-6"
13'-6"
13'-6"
21'
Exp. Joint
Post
net
3'
21'
78'
Joint
Adjoining Court
130' Min. for Championship play
120' Min.
Expansion Joint
The Alley
The Alley
Service line
18'
Exp.
27'-0"
Singles
4'-6"
4'-6"
Center Mark
Exp. Joint
Surrounds 10' High.
Dimension to outside of lines except Center line. Lines 2" to 4" wide.
66' Championship
51' Singles
60' Doubles
21' Min.

COURT DIMENSIONS *

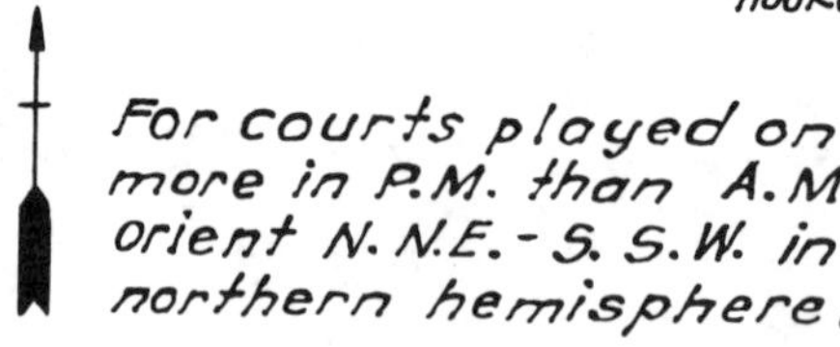

5"
Note: Treat top with Calcium Chloride 400 lbs.± per year.

Clay-Sand mixture. 1 part fine sand, 1 part clay. Materials to be dried, ground and mixed mechanically. Sand to be 85% between 100 & 200 mesh sieve.

1" Bituminous Concrete. See Page 10-10 for mix. or Colprovia type.
Prime - .2 gal. MC1.
6" Gravel, Stone or Cinder Base.

Joints - See Plan.
6"x6" mesh - #5 x #5 wires
4" Conc. Use Type II cem.
6" Gravel, stone or Cinder base.

CLAY COURT — ASPHALT COURT — CONCRETE COURT

TENNIS COURT SECTIONS

Transverse slope - 2" for 60' width - Do not slope longitudinally.

N

For courts played on more in P.M. than A.M. orient N.N.E. - S.S.W. in northern hemisphere.

ORIENTATION

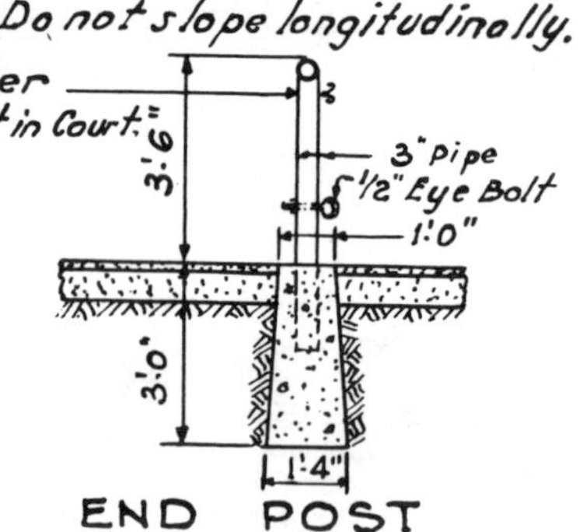

Net 3' at center
Hooked to Cleat in Court.
3'-6"
3" Pipe
1/2" Eye Bolt
1'-0"
3'-0"
1'-4"

END POST

TENNIS COURT DETAILS

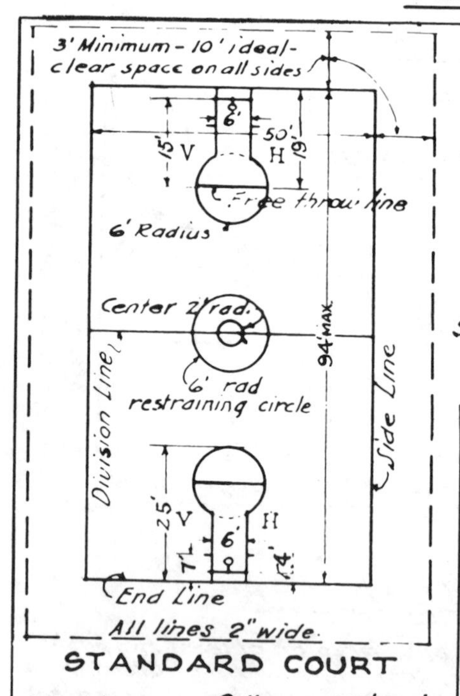

3' Minimum - 10' ideal - clear space on all sides
6'
50'
15'
V
H
19'
Free throw line
6' Radius
Center 2' rad.
94' MAX
6' rad restraining circle
Division Line
Side Line
25'
V
H
6
7'
4'
End Line
All lines 2" wide.

STANDARD COURT

Ideal Court Sizes:	College	94'x50'
	High sch.	84'x50'
	Elem. sch.	60'x35'
Min. Court dimensions		60'x35'
Max. Court dimensions		94'x50'
Junior High School		74'x42'

COURT DIMENSIONS

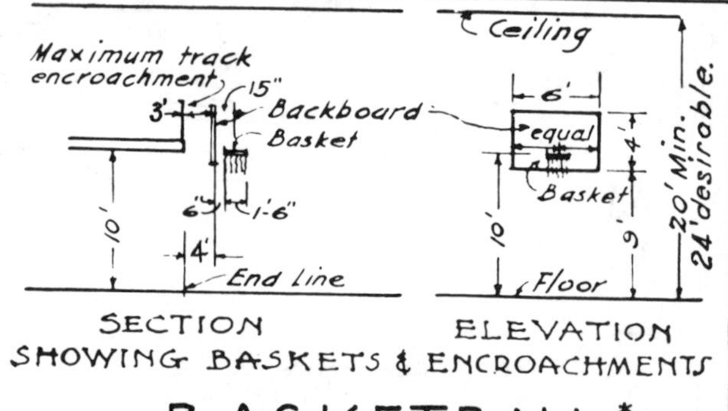

Maximum track encroachment
3'
15"
Backboard
Basket
10'
6"
1'-6"
4'
End Line
Ceiling
6'
equal
4'
Basket
10'
9'
Floor
20' Min. 24' desirable.

SECTION — ELEVATION
SHOWING BASKETS & ENCROACHMENTS

BASKETBALL *

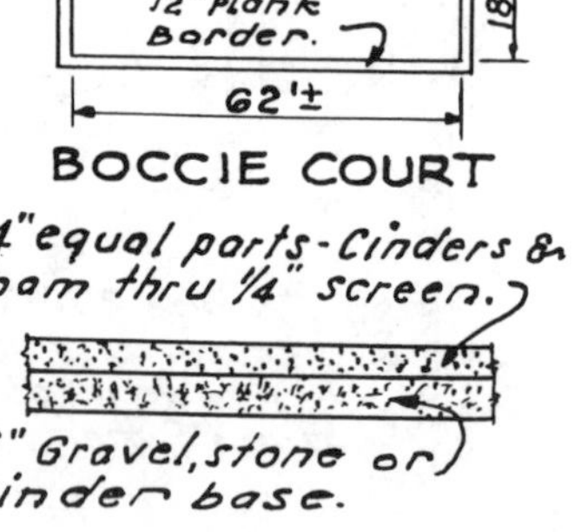

12" Plank Border.
18'
62'±

BOCCIE COURT

4" equal parts - Cinders & loam thru 1/4" screen.

6" Gravel, stone or Cinder base.

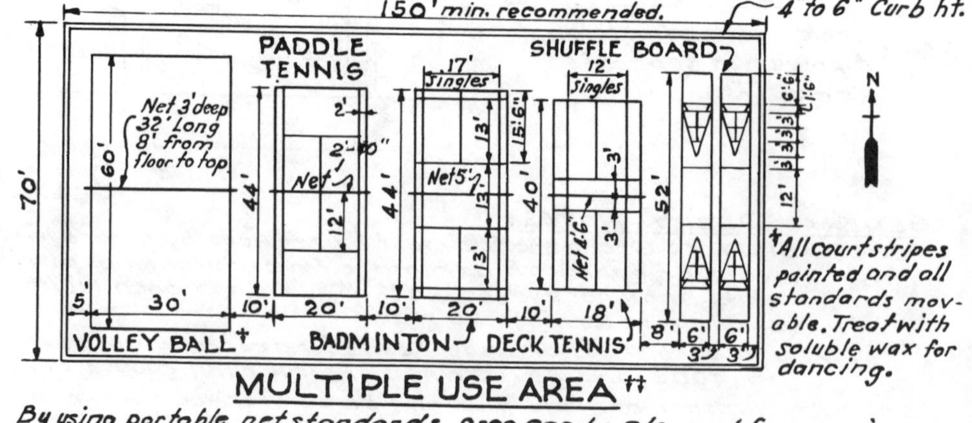

150' min. recommended.
4" to 6" Curb ht.
PADDLE TENNIS
SHUFFLE BOARD
17' Singles
12' Singles
Net 3' deep 32' Long 8' from floor to top
60'
70'
44'
2'
2'-10"
Net
12'
44'
13'
Net 5'
13'
13'
15'-6"
40'
3'
3'
Net 2'-6"
52'
6'-6"
1'-6"
3'
3'
3'
12'
N
5'
30'
10'
20'
10'
20'
10'
18'
8'
6'
6'
3'
3'
VOLLEY BALL †
BADMINTON
DECK TENNIS

† All court stripes painted and all standards movable. Treat with soluble wax for dancing.

MULTIPLE USE AREA ††

By using portable net standards, area can be cleared for various activities such as roller skating, square & Social dancing. Curb allows area to be flooded for ice skating.

* Adapted from Arch. Graphic Stds. by Ramsey & Sleeper. †† Adapted from F. Ellwood Allen.

ATHLETIC FIELDS-DIMENSIONS-1

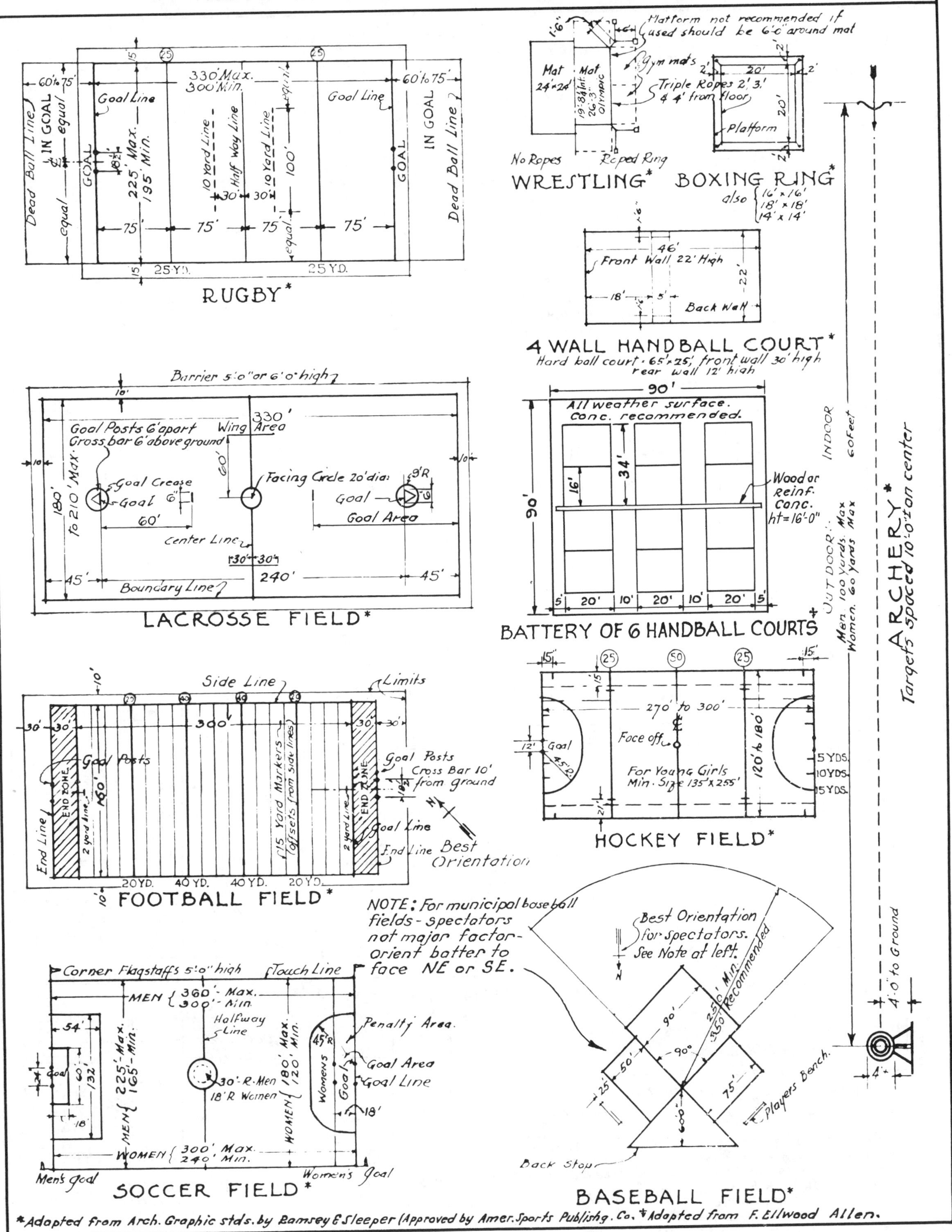

RUGBY*
330' Max. 300' Min.
Goal Line
60' to 75'
IN GOAL
Dead Ball Line
225' Max. 195' Min.
10 Yard Line
Half Way Line
GOAL
25 YD.
WRESTLING*
BOXING RING*
Platform not recommended if used should be 6'-0" around mat
Gym mats
Triple Ropes 2', 3', & 4' from floor
Platform
No Ropes
Roped Ring
also 16'×16' 18'×18' 14'×14'
4 WALL HANDBALL COURT*
Hard ball court - 65'×25', front wall 30' high rear wall 12' high
Front Wall 22' High
Back Wall
LACROSSE FIELD*
Barrier 5'-0" or 6'-0" high
Goal Posts 6' apart
Cross bar 6' above ground
Wing Area
Goal Crease
Goal
Facing Circle 20' dia.
Goal Area
Center Line
Boundary Line
180' to 210' Max.
BATTERY OF 6 HANDBALL COURTS†
All weather surface. Conc. recommended.
Wood or Reinf. Conc. ht=16'-0"
HOCKEY FIELD*
270' to 300'
Face off
For Young Girls Min. Size 135'×255'
120' to 180'
FOOTBALL FIELD*
Side Line
Limits
Goal Posts
END ZONE
2 yard line
15 Yard Markers (offsets from side lines)
Goal Posts Cross Bar 10' from ground
Goal Line
End Line
Best Orientation
ARCHERY*
Targets spaced 10'-0"± on center
INDOOR 60 Feet
OUT DOOR: Men 100 Yards Max Women 60 Yards Max
4'-0" to Ground
NOTE: For municipal baseball fields - spectators not major factor - orient batter to face NE or SE.
SOCCER FIELD*
Corner Flagstaffs 5'-0" high
Touch Line
MEN { 360' - Max. 300' - Min.
Halfway Line
Penalty Area
Goal Area
Goal Line
30' R. Men 18' R. Women
WOMEN { 300' Max. 240' Min.
Men's goal
Women's goal
BASEBALL FIELD*
Best Orientation for spectators. See Note at left.
250' Min. 350' Recommended
Players Bench
Back Stop

*Adapted from Arch. Graphic stds. by Ramsey & Sleeper (Approved by Amer. Sports Publishg. Co.) †Adapted from F. Ellwood Allen.

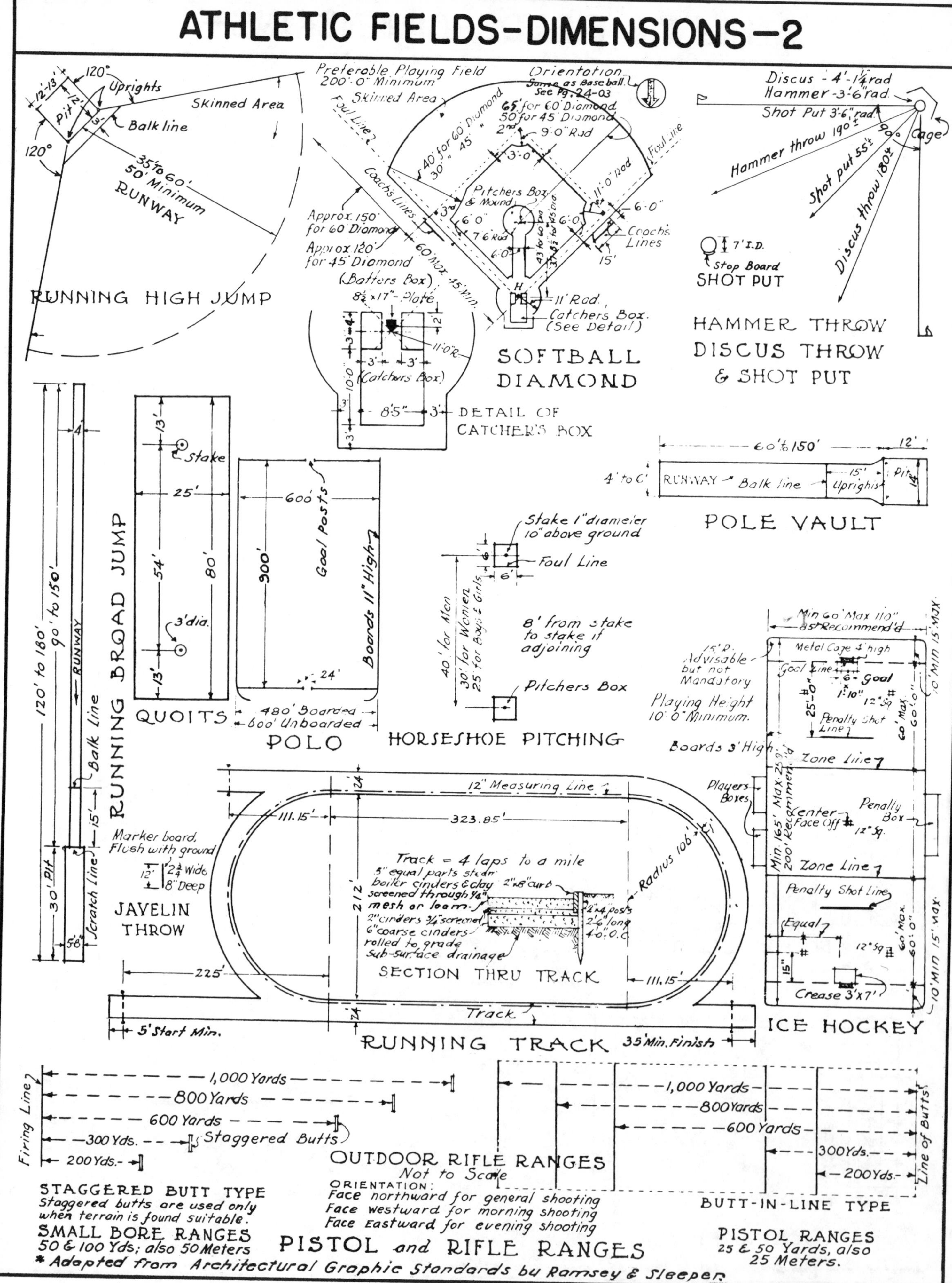
ATHLETIC FIELDS-DIMENSIONS-2
120°
12'-13'
Pit
Uprights
Balk line
Skinned Area
35' to 60'
50' Minimum
RUNWAY
RUNNING HIGH JUMP
Preferable Playing Field
200'-0" Minimum
Skinned Area
Foul Line
Orientation
Same as Baseball
See Pg. 24-03
65' for 60' Diamond
50' for 45' Diamond
9'-0" Rad
40' for 60' Diamond
30' " 45' "
Coach's Lines
Pitchers Box & Mound
Approx. 150' for 60' Diamond
Approx. 120' for 45' Diamond
(Batters Box)
60' Max. 45' Min.
8½"x17" Plate
11' Rad.
Catchers Box.
(See Detail)
(Catchers Box)
DETAIL OF CATCHER'S BOX
SOFTBALL DIAMOND
Discus - 4'-1¼" rad
Hammer - 3'-6" rad.
Shot Put 3'-6" rad.
Cage
Hammer throw 190'±
Shot put 55'±
Discus throw 180'±
7' I.D.
Stop Board
SHOT PUT
HAMMER THROW
DISCUS THROW
& SHOT PUT
120' to 180'
90' to 150'
RUNWAY
Balk Line
Scratch Line
30' Pit
RUNNING BROAD JUMP
Stake
25'
54'
80'
3' dia.
QUOITS
600'
900'
Goal Posts
Boards 11" High
480' Boarded
600' Unboarded
POLO
60' to 150'
12'
4' to 6'
RUNWAY
Balk line
15'
Uprights
Pit
POLE VAULT
Stake 1" diameter 10" above ground
Foul Line
40' for Men
30' for Women
25' for Boys & Girls
8' from stake to stake if adjoining
Pitchers Box
HORSESHOE PITCHING
Min 60' Max 110'
85' Recommend'd
Metal Cage 4' high
Goal Line
6' Goal
1'-10"
12" Sq.
15' R. Advisable but not Mandatory
Playing Height 10'-0" Minimum.
Penalty Shot Line
Boards 3' High
Zone Line
Players Boxes
Min. 165' Max. 250'
200' Recommended
Center Face Off
Penalty Box
Crease 3'x7'
ICE HOCKEY
12" Measuring Line
111.15'
323.85'
Radius 106'
Marker board, Flush with ground
2¾" Wide
8" Deep
JAVELIN THROW
Track = 4 laps to a mile
5" equal parts steam boiler cinders & clay screened through ¼" mesh or loam
2" cinders ¾" screened
6" coarse cinders rolled to grade
Sub-surface drainage
2"x8" curb
2"x4" posts 2'-6" long 4'-0" O.C.
212'
SECTION THRU TRACK
225'
111.15'
Track
5' Start Min.
35' Min. Finish
RUNNING TRACK
Firing Line
1,000 Yards
800 Yards
600 Yards
300 Yds.
200 Yds.
Staggered Butts
OUTDOOR RIFLE RANGES
Not to Scale
ORIENTATION:
Face northward for general shooting
Face westward for morning shooting
Face Eastward for evening shooting
Line of Butts
BUTT-IN-LINE TYPE
STAGGERED BUTT TYPE
Staggered butts are used only when terrain is found suitable.
SMALL BORE RANGES
50 & 100 Yds; also 50 Meters
PISTOL and RIFLE RANGES
PISTOL RANGES
25 & 50 Yards, also 25 Meters.
* Adapted from Architectural Graphic Standards by Ramsey & Sleeper

OUTDOOR SWIMMING POOLS - I

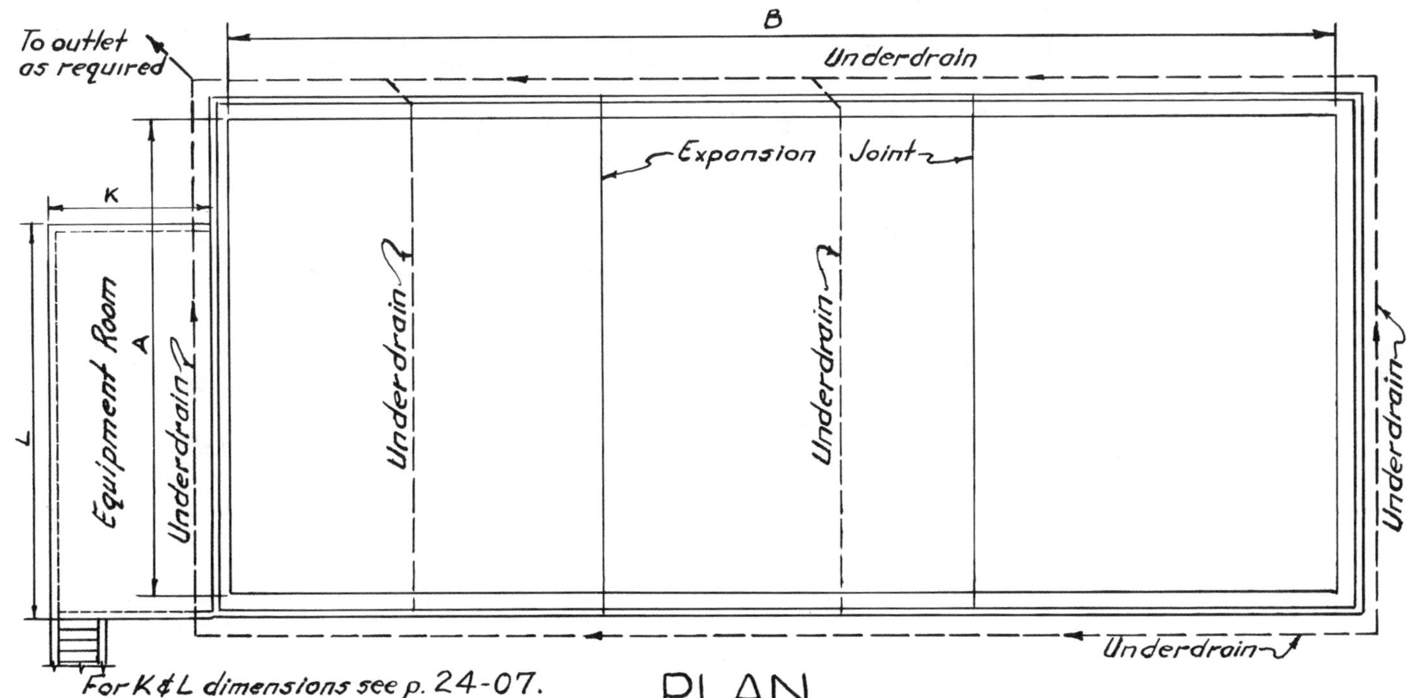

B
To outlet as required
Underdrain
K
Expansion Joint
Equipment Room
A
L
Underdrain
Underdrain
Underdrain
Underdrain
Underdrain

For K&L dimensions see p. 24-07.

PLAN

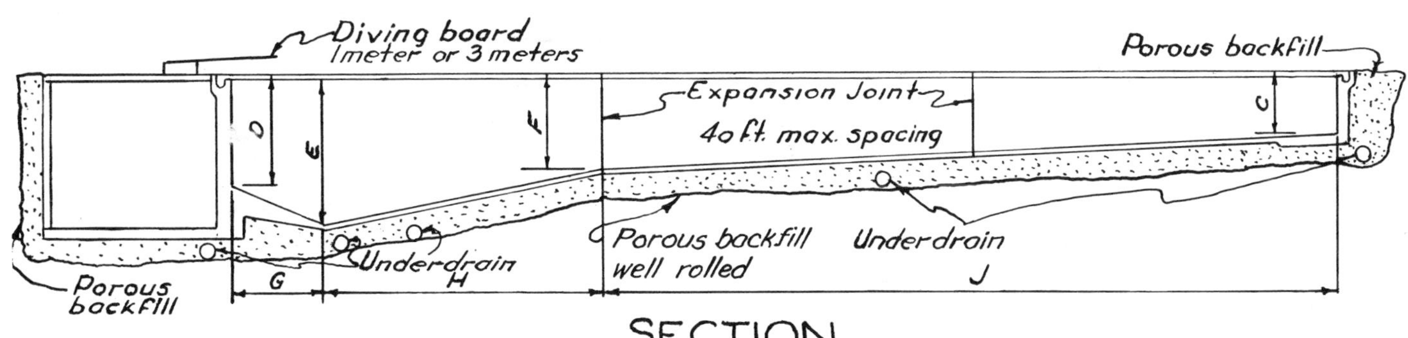

Diving board
1 meter or 3 meters
Porous backfill
Expansion Joint
40 ft. max. spacing
D
E
F
C
Porous backfill well rolled
Underdrain
Underdrain
Porous backfill
G
H
J

SECTION

POOL DESIGN DATA

Pool Size	A	B	C	D	E	F	G	H	J	Volume Gallons
1	25'-0"	60'-0"	3'-3"	6'-0"	8'-0"	5'-0"	5'-0"	15'0"	40'-0"	55,600
2	30'-0"	75'-0"	3'-3"	6'-0"	8'-0"	5'-0"	9'-0"	16'-0"	50'0"	83,800
3 *	35'-0"	82'-6"	3'-3"	8'-0"	10'-0"	5'-0"	9'-0"	19'-6"	54'-0"	116,000
4	50'-0"	100'-0"	3'-3"	8'-0"	10'-0"	5'-0"	15'-0"	20'-0"	65'-0"	208,000
5	60'-0"	150'-0"	3'-3"	8'-0"	10'-0"	5'-0"	15'-0"	25'-0"	110'-0"	348,000
6 *	75'-0"	165'-0"	3'-3"	8'-0"	10'-0"	5'-0"	15'-0"	30'-0"	120'-0"	480,000

* 25 Meter and 50 Meter pools

Max. daily attendance per day = 5% to 10% of population.
Avg. daily attendance per day = 2% to 3% of population
Max. daily attendance at any one time.
Pool area = Max. daily attendance at any one time × 12
Pool and walk area = Max. daily attendance at one time × 20

For ladders, diving boards, slides, etc., consult with m'f'r.
Discontinue overflow gutters about 6" each side of expansion joint so that there is no joint thru gutter.

OUTDOOR SWIMMING POOLS–2

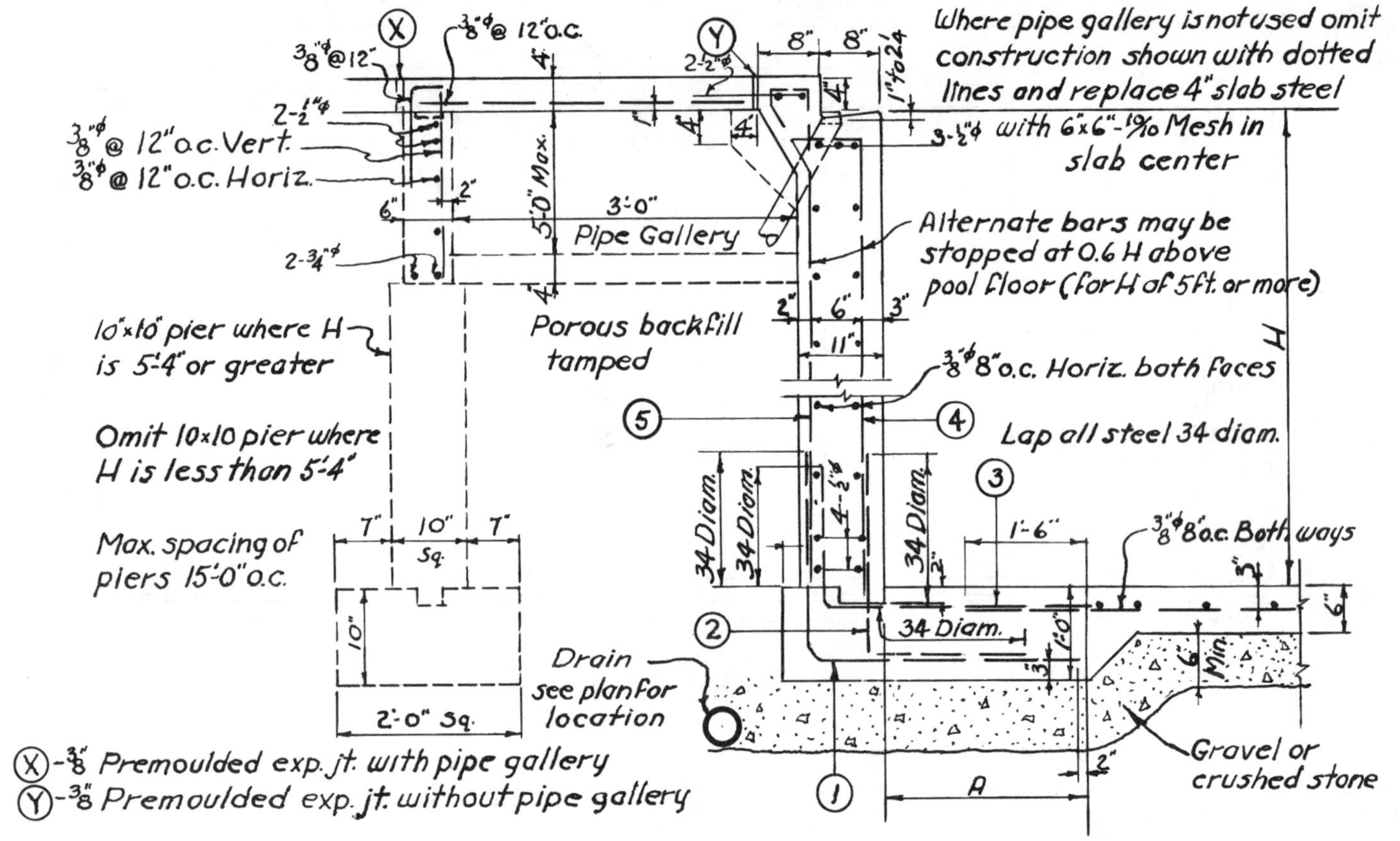

Where pipe gallery is not used omit construction shown with dotted lines and replace 4" slab steel with 6×6"–1% Mesh in slab center

$\frac{3}{8}$"φ @ 12"
$\frac{3}{8}$"φ @ 12" o.c.
2-$\frac{1}{2}$"φ
$\frac{3}{8}$"φ @ 12" o.c. Vert.
$\frac{3}{8}$"φ @ 12" o.c. Horiz.
6"
2"
5'-0" Max.
3'-0"
Pipe Gallery
2-$\frac{3}{4}$"φ
4"
8" 8"
1" to 24"
2-$\frac{1}{2}$"φ
3-$\frac{1}{2}$"φ
Alternate bars may be stopped at 0.6 H above pool floor (for H of 5 ft. or more)
2" 6" 3"
11"
H
10"×10" pier where H is 5'-4" or greater
Porous backfill tamped
$\frac{3}{8}$"φ 8" o.c. Horiz. both faces
Omit 10×10 pier where H is less than 5'-4"
Lap all steel 34 diam.
Max. spacing of piers 15'-0" o.c.
7" 10" Sq. 7"
10"
2'-0" Sq.
34 Diam.
34 Diam.
4-$\frac{1}{2}$"φ
34 Diam.
2"
1'-6"
$\frac{3}{8}$"φ 8" o.c. Both ways
3"
6"
34 Diam.
1'-0"
3"
6" Min.
Drain see plan for location
Gravel or crushed stone
2"
A
(1) (2) (3) (4) (5)

(X)–$\frac{3}{8}$" Premoulded exp. jt. with pipe gallery
(Y)–$\frac{3}{8}$" Premoulded exp. jt. without pipe gallery

TYPICAL WALL SECTION

REINFORCING SCHEDULE

Depth H (FT.)	Base Width A	Reinforcing Bar No.				
		1	2ⓐ	3	4ⓐ	5
LESS THAN 5	2'-2"	$\frac{3}{8}$"φ	$\frac{3}{8}$"φ	$\frac{3}{8}$"φ	$\frac{3}{8}$"φ	$\frac{3}{8}$"φ
5-8	4'-2"	$\frac{1}{2}$"φ	$\frac{5}{8}$"φ	$\frac{5}{8}$"φ	$\frac{5}{8}$"φ	$\frac{1}{2}$"φ
8-9	5'-1"	$\frac{5}{8}$"φ	$\frac{3}{4}$"φ	$\frac{3}{4}$"φ	$\frac{3}{4}$"φ	$\frac{5}{8}$"φ
9-10	6'-1"	$\frac{3}{4}$"φ	$\frac{7}{8}$"φ	$\frac{7}{8}$"φ	$\frac{7}{8}$"φ	$\frac{3}{4}$"φ
10-11	7'-2"	$\frac{7}{8}$"φ	1"φ	1"φ	1"φ	$\frac{7}{8}$"φ
11-12	7'-10"	1"φ	1"□	1"□	1"□	1"φ

Spacing of all vertical reinforcing = 12" o.c.
ⓐ If pipe gallery is used space these bars 8" o.c.

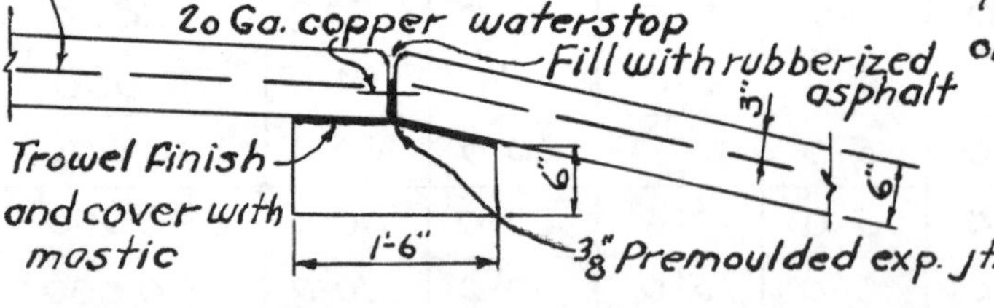

$\frac{3}{8}$"φ 8 o.c. Both ways
20 Ga. copper waterstop
Fill with rubberized asphalt
3"
Trowel finish and cover with mastic
6"
6"
1'-6"
$\frac{3}{8}$" Premoulded exp. jt.

FLOOR JOINT

General Notes:
1 Use stone concrete which shall develop a strength of 3000 p.s.i. @ 28 days.
2 Use air entraining agent to give 3 to 5% entrained air.
3 Max. slump 3".
4. Rubberized asphalt seals shall be as made by U.S. Rubber Co. or approved equal.

20 Ga. copper waterstop 24" wide
$\frac{3}{8}$" Premoulded filler

WALL EXPANSION JOINT

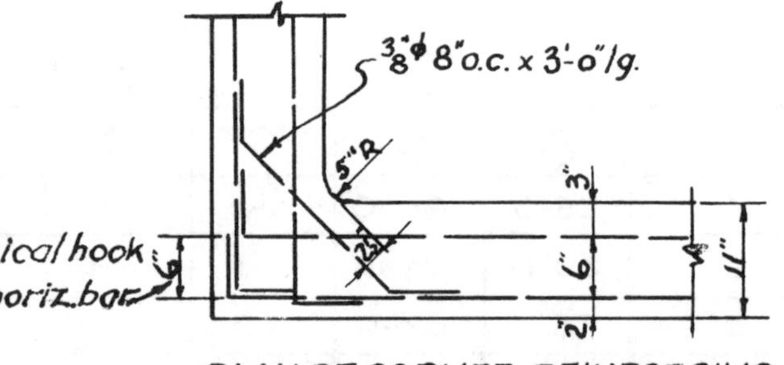

$\frac{3}{8}$"φ 8" o.c. x 3'-0" lg.
5" R
3"
Typical hook on horiz. bar
6"
6"
4"
11"
2"

PLAN OF CORNER REINFORCING

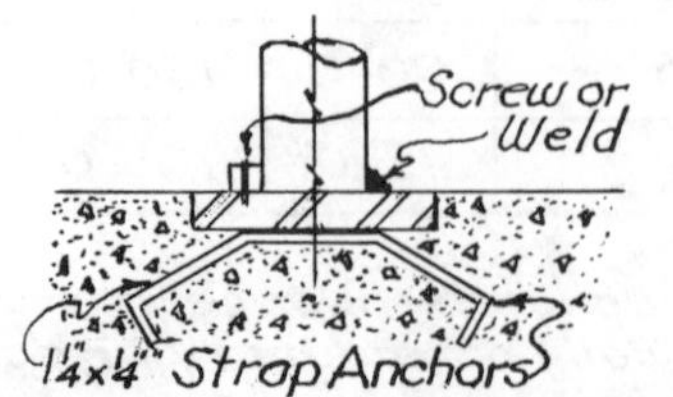

Screw or Weld
1¼×¼" Strap Anchors

INSERT DETAIL FOR DIVING BOARD

6" Drain Tile
3" Pea Gravel
2" Gravel & Sand
1" Sand

UNDERDRAIN DETAIL

OUTDOOR SWIMMING POOLS–3

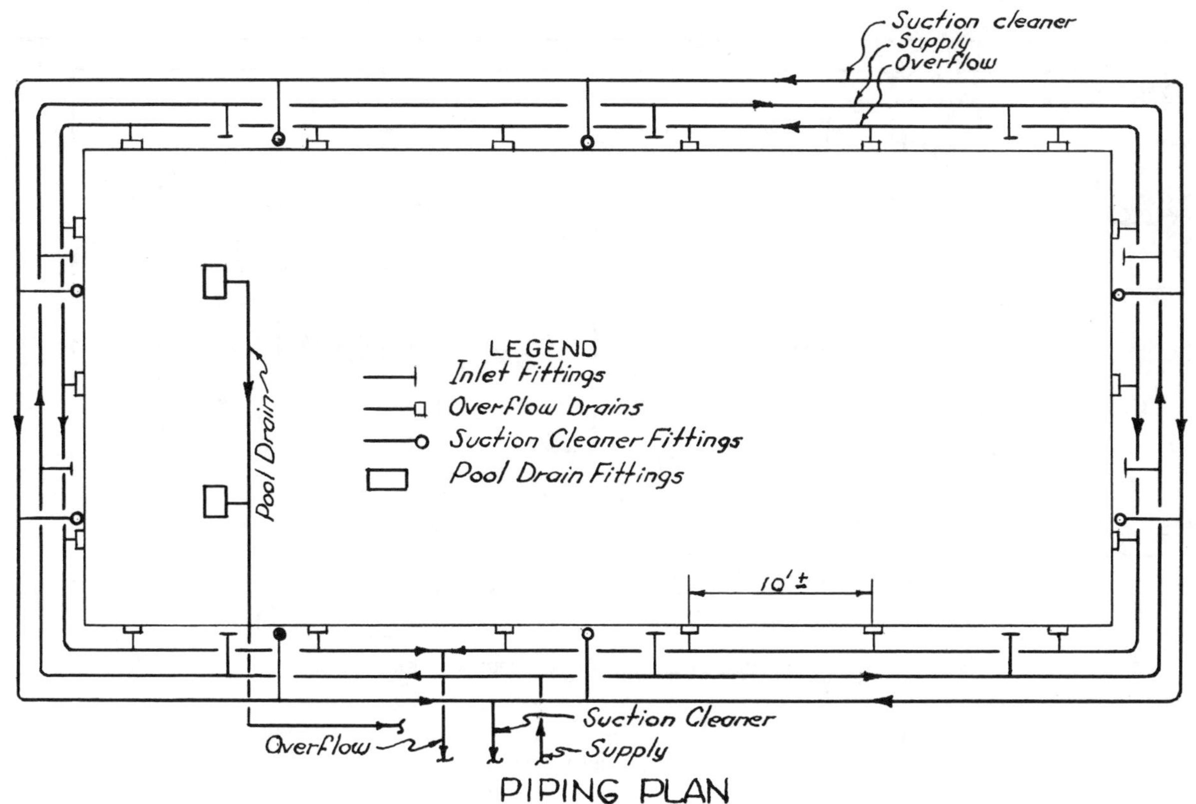
Suction cleaner
Supply
Overflow
LEGEND
Inlet Fittings
Overflow Drains
Suction Cleaner Fittings
Pool Drain Fittings
Pool Drain
10'±
Suction Cleaner
Overflow
Supply

PIPING PLAN

POOL DESIGN DATA

Pool Size	K	L Press Filters	L Diatomite Filters	Pump Rate Gallons per Minute*	No. and size of Pressure Filters	Effective Filter Area, Diatomite Filters	No. and size of Diatomite Filters **	Chlorinator Type	Haircatcher size	Filtered Water Piping	No. of inlets	Pool Drain Pipe Size	No. of Outlets	No. of Overflow Drains	Pool Suction Pipe Size	No. of Pool Suction Fittings
1	12	30	26	115	3-54"ϕ	29	2-22"ϕ×30"	hypo-Chlorinator	6"	2½"ϕ	6	6"	1	17	2"ϕ	4
2	13	33	28	175	3-60"ϕ	44	2-24"ϕ×36"	Gas	6"	3"ϕ	6	6"	2	21	2"ϕ	4
3	15	36	28	242	3-72"ϕ	61	2-24"ϕ×36"	Gas	6"	3"ϕ	8	6"	2	24	2"ϕ	6
4	17	42	30	433	3-96"ϕ	108	2-30"ϕ×36"	Gas	8"	4"ϕ	11	8"	3	30	2"ϕ	8
5	18	55	34	725	4-108"ϕ†	181	2-36"ϕ×36"	Gas	10"	5"ϕ	14	8"	3	42	3"ϕ††	10
6	19	68	42	1,000	5-114"ϕ†	250	2-42"ϕ×36"	Gas	12"	6"ϕ	16	10"	4	48	4"ϕ††	12

* Pump-40' to 60' head for pressure filters and 100' to 150' head for diatomite filters.
† Recommended portable pump. Pool turnover, one every 8 hours.
†† Recommended rapid sand filters and screen chamber.
** Consult manufacturers as size differs.
Effective filter area based on filtration rate of 3 gpm/sq.ft. (Filter rates of 2, 3, & 4 gpm/sq.ft.) have been used.)

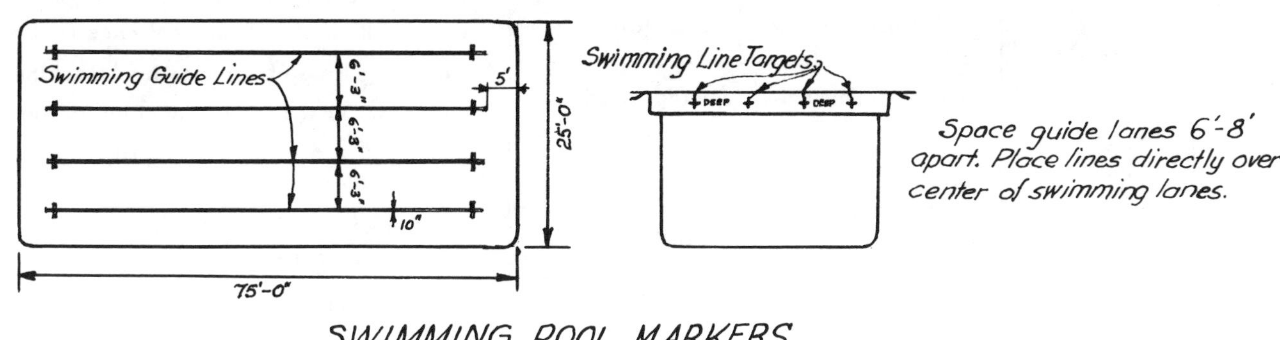
Swimming Guide Lines
6'-3"
6'-8"
6'-3"
5'
10"
25'-0"
75'-0"
Swimming Line Targets
DEEP
DEEP

Space guide lanes 6'-8' apart. Place lines directly over center of swimming lanes.

SWIMMING POOL MARKERS

OUTDOOR SWIMMING POOLS-4

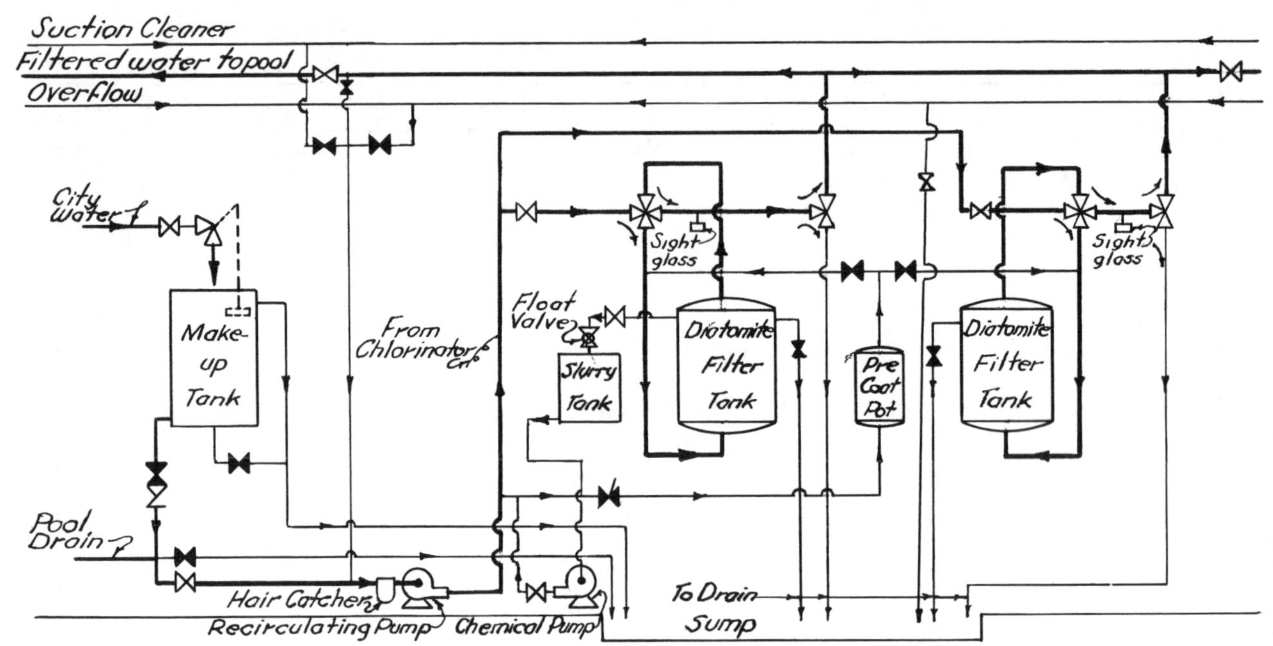
Suction Cleaner
Filtered water to pool
Overflow
City water
Make-up Tank
From Chlorinator
Float Valve
Slurry Tank
Sight glass
Diatomite Filter Tank
Pre Coat Pot
Diatomite Filter Tank
Sight glass
Pool Drain
Hair Catcher
Recirculating Pump
Chemical Pump
To Drain
Sump

FLOW DIAGRAM - DIATOMITE FILTERS
ELEVATION

KEY

- GATE VALVE - NORMALLY OPEN
- GATE VALVE - NORMALLY CLOSED
- QUICK OPENING GATE VALVE
- ANGLE FLOAT VALVE
- FOUR WAY PLUG VALVE
- THREE WAY 2 PORT PLUG VALVE
- CHECK VALVE

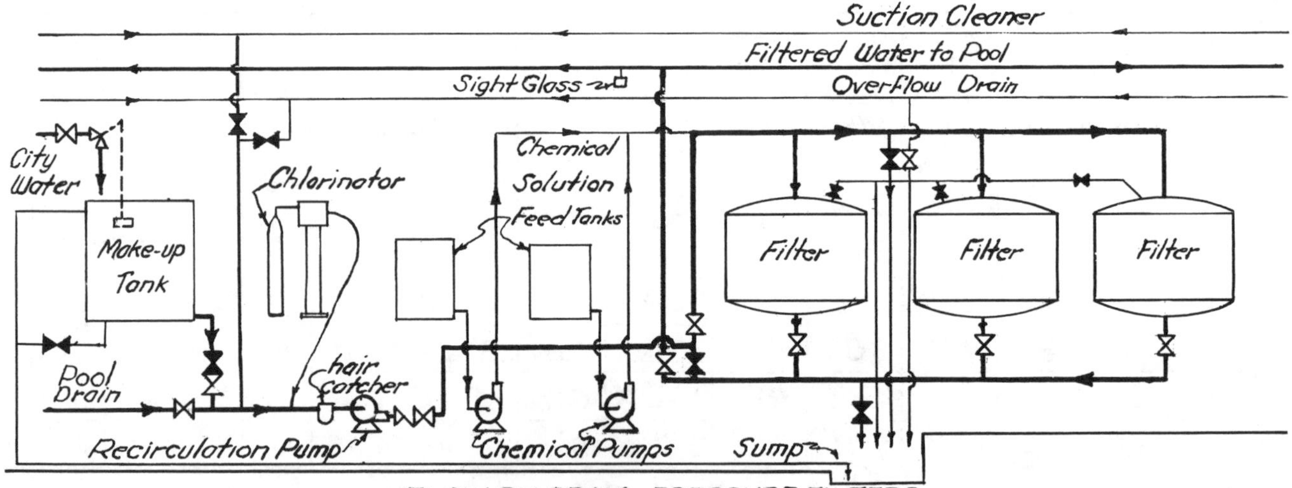
Suction Cleaner
Filtered Water to Pool
Sight Glass
Overflow Drain
City Water
Chlorinator
Chemical Solution Feed Tanks
Make-up Tank
Filter
Filter
Filter
Pool Drain
hair catcher
Recirculation Pump
Chemical Pumps
Sump

FLOW DIAGRAM - PRESSURE FILTERS
ELEVATION

CHLORINE DOSAGES FOR TREATMENT OF SWIMMING POOL WATER

Type of Pool	Chlorine Application Based on Recirculation Rate*	
	Average Minimum	Average Maximum
Indoor	2.0 p.p.m.	5.0 p.p.m.
Outdoor	3.0 p.p.m.	10.0 p.p.m.

*From Wallace & Tiernan, Inc.

PROBLEM

Given: Outdoor pool volume of 75,000 gal. and recirculation of pool content once in 8 hrs.

To Find: Rate of chlorine feed machine

Solution: Maximum dosage chlorine rate = $\frac{75{,}000 \times 3}{1{,}000{,}000} \times 10 \times 8.33$ or 18.7 lb. per day. Machine should have a capacity to feed up to say 20 lb. of chlorine per day.

For additional data on chlorination, see pp. 19-25 to 19-27, 21-19

GENERAL FORMULAS

TRIGONOMETRIC FORMULAS.

Functions of Angle	Opposite	Adjacent	Hyp
sin = Op ÷ Hyp	Hyp × sin		Op ÷ sin
cos = Ad ÷ Hyp		Hyp × cos	Ad ÷ cos
tan = Op ÷ Ad	Ad × tan	Op ÷ tan	
cot = Ad ÷ Op	Ad ÷ cot	Op × cot	
sec = Hy ÷ Ad		Hyp ÷ sec	Ad × sec
cosec = Hy ÷ Op	Hyp ÷ cosec		Op ÷ cosec

Given	To Find	Formula	
ab	A	tan = a ÷ b	cot = b ÷ a
	B	cot = a ÷ b	tan = b ÷ a
ac	A	sin = a ÷ c	cosec = c ÷ a
	B	cos = a ÷ c	sec = c ÷ a
bc	A	sec = c ÷ b	cos = b ÷ c
	B	cosec = c ÷ b	sin = b ÷ c
Aa	b	a cot A	a ÷ tan A
	c	a cosec A	a ÷ sin A
Ab	a	b tan A	b ÷ cot A
	c	b sec A	b ÷ cos A
Ac	a	c sin A	c ÷ cosec A
	b	c cos A	c ÷ sec A

$$S = \frac{a+b+c}{2}$$

Given	To Find	Formula
ABa	b	$a \sin B \div \sin A$
	c	$a \sin(A+B) \div \sin A$
ABb	a	$b \sin A \div \sin B$
	c	$b \sin(A+B) \div \sin B$
ABc	a	$c \sin A \div \sin(A+B)$
	b	$c \sin B \div \sin(A+B)$
ACa	b	$a \sin(A+C) \div \sin A$
	c	$a \sin C \div \sin A$
ACb	a	$b \sin A \div \sin(A+C)$
	c	$b \sin C \div \sin(A+C)$
ACc	a	$c \sin A \div \sin C$
	b	$c \sin(A+C) \div \sin C$
BCa	b	$a \sin B \div \sin(B+C)$
	c	$a \sin C \div \sin(B+C)$
BCb	a	$b \sin(B+C) \div \sin B$
	c	$b \sin C \div \sin B$
BCc	a	$c \sin(B+C) \div \sin C$
	b	$c \sin B \div \sin C$
abc	S	$(a+b+c) \div 2$
abcs	A	$\sin \tfrac{1}{2}A = \sqrt{(s-b)(s-c) \div bc}$
		$\cos \tfrac{1}{2}A = \sqrt{s(s-a) \div bc}$
		$\tan \tfrac{1}{2}A = \sqrt{(s-b)(s-c) \div s(s-a)}$
abcs	B	$\sin \tfrac{1}{2}B = \sqrt{(s-a)(s-c) \div ac}$
		$\cos \tfrac{1}{2}B = \sqrt{s(s-b) \div ac}$
		$\tan \tfrac{1}{2}B = \sqrt{(s-a)(s-c) \div s(s-b)}$
abcs	C	$\sin \tfrac{1}{2}C = \sqrt{(s-a)(s-b) \div ab}$
		$\cos \tfrac{1}{2}C = \sqrt{s(s-c) \div ab}$

Given	To Find	Formula
abcs	C	$\tan \tfrac{1}{2}C = \sqrt{(s-a)(s-b) \div s(s-c)}$
abcs	d	$(b^2+c^2-a^2) \div 2b$
	e	$(a^2+b^2-c^2) \div 2b$
Aab	B	$\sin = b \sin A \div a$
Aac	C	$\sin = c \sin A \div a$
Bab	A	$\sin = a \sin B \div b$
Bbc	C	$\sin = c \sin B \div b$
Cac	A	$\sin = a \sin C \div c$
Cbc	B	$\sin = b \sin C \div c$
Abc	$\tfrac{1}{2}(B+C)$	$90° - \tfrac{1}{2}A$
	$\tfrac{1}{2}(B-C)$	$\tan = [(b-c)\tan(90° - \tfrac{1}{2}A)] \div (b+c)$
	B	$\tfrac{1}{2}(B+C) + \tfrac{1}{2}(B-C)$
	C	$\tfrac{1}{2}(B+C) - \tfrac{1}{2}(B-C)$
	a	$\sqrt{b^2+c^2-2bc \cos A}$
Bac	$\tfrac{1}{2}(A+C)$	$90° - \tfrac{1}{2}B$
	$\tfrac{1}{2}(A-C)$	$\tan = [(a-c)\tan(90° - \tfrac{1}{2}B)] \div (a+c)$
	A	$\tfrac{1}{2}(A+C) + \tfrac{1}{2}(A-C)$
	C	$\tfrac{1}{2}(A+C) - \tfrac{1}{2}(A-C)$
	b	$\sqrt{a^2+c^2-2ac \cos B}$
Cab	$\tfrac{1}{2}(A+B)$	$90° - \tfrac{1}{2}C$
	$\tfrac{1}{2}(A-B)$	$\tan = [(a-b)\tan(90° - \tfrac{1}{2}C)] \div (a+b)$
	A	$\tfrac{1}{2}(A+B) + \tfrac{1}{2}(A-B)$
	B	$\tfrac{1}{2}(A+B) - \tfrac{1}{2}(A-B)$
	c	$\sqrt{a^2+b^2-2ab \cos C}$

Given	To Find	Formula
drB	b	$d \sin^2 B$
	f	$r \sin 2B$
	e	$d \sin B$
drb	Ang B	$\sin B = \sqrt{b \div d}$
	f	$\sqrt{b(d-b)}$
	e	$\sqrt{db}$
dre	Ang B	$\sin B = e \div d$
	b	$e^2 \div d$
	f	$e\sqrt{d^2-e^2} \div d$
bB	r	$\tfrac{1}{2}b \div \sin^2 B$
eB	r	$\tfrac{1}{2}e \div \sin B$
bf	Ang B	$\tan B = b \div f$
	r	$(f^2+b^2) \div 2b$
fe	Ang B	$\sin B = \sqrt{e^2-f^2} \div e$
	r	$\tfrac{1}{2}e^2 \div \sqrt{e^2-f^2}$
be	Ang B	$\sin B = b \div e$
	r	$\tfrac{1}{2}e^2 \div b$
rxy	Ang B	$\cos 2B = (\sqrt{r^2-x^2} - y) \div r$
	b	$r + y - \sqrt{r^2-x^2}$
brx	y	$b + \sqrt{r^2-x^2} - r$
bry	x	$\sqrt{r^2-(r+y-b)^2}$
bxy	r	$[x^2+(b-y)^2] \div (2b-2y)$
r	Circ.	$6.2832 r$
rD	Arc a	$.0174533\ r D°$
	Arc a	$.0002909\ r D'$
	Arc a	$.00000485\ r D''$
r	Area	Circle $= 3.1416 r^2$
d		" $= 0.7854 d^2$
c		" $= 0.0796 c^2$
ar		Sector $= 0.5ar$
arfh		Segment $= 0.5ar - fh$

Given	To Find	Formula
bpw	f	$\sqrt{(b+p)^2+w^2}$
bkv	m	$\sqrt{(b+k)^2+v^2}$
bkp vw	d	$bw(b+k) \div [v(b+p)+w(b+k)]$
	e	$bv(b+p) \div [v(b+p)+w(b+k)]$
bfk pvw	a	$fbv \div [v(b+p)+w(b+k)]$
bkm pvw	c	$bmw \div [v(b+p)+w(b+k)]$
bkpvw	h	$bvw \div [v(b+p)+w(b+k)]$
afw	h	$aw \div f$
cmw	h	$cv \div m$

Given	To Find	Formula
bpw	f	$\sqrt{(b+p)^2+w^2}$
bnw	m	$\sqrt{(b-n)^2+w^2}$
bnp	d	$b(b-n) \div (2b+p-n)$
	e	$b(b+p) \div (2b+p-n)$
bfnp	a	$bf \div (2b+p-n)$
bmnp	c	$bm \div (2b+p-n)$
bnpw	h	$bw \div (2b+p-n)$
afw	h	$aw \div f$
cmw	h	$cw - m$

Given	To Find	Formula
bpw	f	$\sqrt{(b+p)^2+w^2}$
bw	m	$\sqrt{b^2+w^2}$
bp	d	$b^2 \div (2b+p)$
	e	$b(b+p) \div (2b+p)$
bfp	a	$bf \div (2b+p)$
bmp	c	$bm \div (2b+p)$
bpw	h	$bw \div (2b+p)$
afw	h	$aw \div f$
cmw	h	$cw \div m$

Data by American Bridge Co.-from Singleton, Manual of Structural Design, H.M. Ives & Sons.

GENERAL CONVERSION FACTORS-I

TABLES OF CONVERSION FACTORS FOR ENGINEERS

Data are arranged alphabetically.

Unless designated otherwise, the English measures of capacity are those used in the United States, and the units of weight and mass are avoirdupois units.

The word gallon, used in any conversion factor, designates the U. S. gallon. To convert into the Imperial gallon, multiply the U. S. gallon by 0.83267. Likewise, the word ton designates a short ton, 2,000 pounds.

The figures 10^{-1}, 10^{-2}, 10^{-3}, etc. denote 0.1, 0.01, 0.001, etc. respectively.

The figures 10^{1}, 10^{2}, 10^{3}, etc. denote 10, 100, 1000, etc. respectively.

With respect to the properties of water, it freezes at 32°F., and is at its maximum density at 39.2°F. In the conversion factors given below using the properties of water, calculations are based on water at 39.2°F. in vacuo, weighing 62.427 pounds per cubic foot, or 8.345 pounds per U. S. gallon.

"Parts Per Million," designated as P.P.M., is always by weight and is simply a more convenient method of expressing concentration, either dissolved or undissolved material. Usually P.P.M. is used where percentage would be so small as to necessitate several ciphers after the decimal point, as one part per million is equal to 0.0001 per cent.

As used in the Sanitary field, P.P.M. represents the number of pounds of dry solids contained in one million pounds of water, including solids. In this field, one part per million may be expressed as 8.345 pounds of dry solids to one million U. S. gallons of water. In the Metric system, one part per million may be expressed as one gram of dry solids to one million grams of water, or one milligram per liter.

In arriving at parts per million by means of pounds per million gallons or milligrams per liter, it may be mentioned that the density of the solution or suspension has been neglected and if this is appreciably different from unity, the results are slightly in error.

Multiply	*By*	*To Obtain*
Acres	43,560	Square feet
"	4047	Square meters
"	1.562x10^{-3}	Square miles
"	4840	Square yards
Acre-feet	43,560	Cubic feet
" "	325,851	Gallons
" "	1233.49	Cubic meters
Atmospheres	76.0	Cms. of mercury
"	29.92	Inches of mercury
"	33.90	Feet of water
"	10,333	Kgs./sq. meter
"	14.70	Lbs./sq. inch
"	1.058	Tons/sq. ft.
Barrels-oil	42	Gallons-oil
" -cement	376	Pounds-cement
Bags or sacks-cement	94	Pounds-cement
Board-feet	144 sq. in. x 1 in.	Cubic inches
British Thermal Units	0.2520	Kilogram-calories
" " "	777.5	Foot-lbs.
" " "	3.927x10^{-4}	Horse-power-hrs.
" " "	107.5	Kilogram-meters
" " "	2.928x10^{-4}	Kilowatt-hrs.
B.T.U./min	12.96	Foot-lbs./sec.
" / "	0.02356	Horse-power
" / "	0.01757	Kilowatts
" / "	17.57	Watts
Centares (Centiares)	1	Square meters
Centigrams	0.01	Grams
Centiliters	0.01	Liters
Centimeters	0.3937	Inches
"	0.01	Meters
"	10	Millimeters
Centimtrs. of mercury	0.01316	Atmospheres
" " "	0.4461	Feet of water
" " "	136.0	Kgs./sq. meter
" " "	27.85	Lbs./sq. ft.
" " "	0.1934	Lbs./sq. inch
Centimeters/second	1.969	Feet/min.
" / "	0.03281	Feet/sec.
Centimeters/second	0.036	Kilometers/hr.
" / "	0.6	Meters/min.
" / "	0.02237	Miles/hr.
" / "	3.728x10^{-4}	Miles/min.
Chain (Gunters)	66	Feet
Cms./sec./sec	0.03281	Feet/sec./sec.
Cubic centimeters	3.531x10^{-5}	Cubic feet
" "	6.102x10^{-2}	Cubic inches
" "	10^{-6}	Cubic meters
" "	1.308x10^{-6}	Cubic yards
" "	2.642x10^{-4}	Gallons
" "	10^{-3}	Liters
" "	2.113x10^{-3}	Pints (liq.)
" "	1.057x10^{-3}	Quarts (liq.)
Cubic feet	2.832x10^{4}	Cubic cms.
" "	1728	Cubic inches
" "	0.02832	Cubic meters
" "	0.03704	Cubic yards
" "	7.48052	Gallons
" "	28.32	Liters
" "	59.84	Pints (liq.)
" "	29.92	Quarts (liq.)
Cubic feet/minute	472.0	Cubic cms./sec.
" " / "	0.1247	Gallons/sec.
" " / "	0.4720	Liters/sec.
" " / "	62.43	Pounds of water/min.
Cubic feet/second	0.646317	Million gals./day
" " / "	448.831	Gallons/min.
" " / "	13.8 / sq. mi. drainage area	Inches / year
Cubic inches	16.39	Cubic centimeters
" "	5.787x10^{-4}	Cubic feet
" "	1.639x10^{-5}	Cubic meters
" "	2.143x10^{-5}	Cubic yards
" "	4.329x10^{-3}	Gallons
" "	1.639x10^{-2}	Liters
" "	0.03463	Pints (liq.)
" "	0.01732	Quarts (liq.)
Cubic meters	10^{6}	Cubic centimeters
" "	35.31	Cubic feet
" "	61.023	Cubic inches
" "	1.308	Cubic yards
" "	264.2	Gallons
Cubic meters	10^{3}	Liters
" "	2113	Pints (liq.)
" "	1057	Quarts (liq.)
Cubic yards	7.646x10^{5}	Cubic centimeters
" "	27	Cubic feet
" "	46,656	Cubic inches
" "	0.7646	Cubic meters
" "	202.0	Gallons
" "	764.6	Liters
" "	1616	Pints (liq.)
" "	807.9	Quarts (liq.)
Cubic yards/min	0.45	Cubic feet/sec.
" " / "	3.367	Gallons/sec.
" " / "	12.74	Liters/sec.
Decigrams	0.1	Grams
Deciliters	0.1	Liters
Decimeters	0.1	Meters
Degrees (angle)	60	Minutes
" "	0.01745	Radians
" "	3600	Seconds
Degrees/sec	0.01745	Radians/sec.
" / "	0.1667	Revolutions/min.
" / "	0.002778	Revolutions/sec.
Dekagrams	10	Grams
Dekaliters	10	Liters
Dekameters	10	Meters
Drams	27.34375	Grains
"	0.0625	Ounces
"	1.771845	Grams
Fathoms	6	Feet
Feet	30.48	Centimeters
"	12	Inches
"	0.3048	Meters
"	1/3	Yards

Adapted from Tables of Conversion Factors for Engineers by The Dorr Co.

GENERAL CONVERSION FACTORS—2

Multiply	*By*	*To Obtain*
Feet of water	0.02950	Atmospheres
“ “ “	0.8826	Inches of mercury
“ “ “	304.8	Kgs./sq.meter
“ “ “	62.43	Lbs./sq.ft.
“ “ “	0.4335	Lbs./sq.inch
Feet/min.	0.5080	Centimeters/sec.
“ / “	0.01667	Feet/sec.
“ / “	0.01829	Kilometers/hr.
“ / “	0.3048	Meters/min.
“ / “	0.01136	Miles/hr.
Feet/sec.	30.48	Centimeters/sec.
“ / “	1.097	Kilometers/hr.
“ / “	0.5921	Knots
“ / “	18.29	Meters/min.
“ / “	0.6818	Miles/hr.
“ / “	0.01136	Miles/min.
Feet/sec./sec	30.48	Cms./sec./sec.
“ / “ / “	0.3048	Meters/sec./sec.
Foot-pounds	1.286×10^{-3}	British Thermal Units
“ “	5.050×10^{-7}	Horse-power-hrs.
“ “	3.241×10^{-4}	Kilogram-calories
“ “	0.1383	Kilogram-meters
“ “	3.766×10^{-7}	Kilowatt-hrs.
Foot-pounds/min.	1.286×10^{-3}	B. T. Units/min.
“ “ / “	0.01667	Foot-pounds/sec.
“ “ / “	3.030×10^{-5}	Horse-power
“ “ / “	3.241×10^{-4}	Kg.-calories/min.
“ “ / “	2.260×10^{-5}	Kilowatts
Foot-pounds/sec.	7.717×10^{-2}	B. T. Units/min.
“ “ / “	1.818×10^{-3}	Horse-power
“ “ / “	1.945×10^{-2}	Kg.-calories/min
“ “ / “	1.356×10^{-3}	Kilowatts
Gallons	3785	Cubic centimeters
“	0.1337	Cubic feet
“	231	Cubic inches
“	3.785×10^{-3}	Cubic meters
“	4.951×10^{-3}	Cubic yards
“	3.785	Liters
“	8	Pints (liq.)
“	4	Quarts (liq.)
Gallons, Imperial.	1.20095	U.S. gallons
“ U.S.	0.83267	Imperial gallons
Gallons water	8.3453	Pounds of water
Gallons/min.	2.228×10^{-3}	Cubic feet/sec.
“ / “	0.06308	Liters/sec.
“ / “	8.0208	Cu. ft./hr.
“ / “	8.0208	Overflow rate (ft./hr.) / Area (sq. ft.)
Gallons water/min	6.0086	Tons water/24 hrs.
Grains (troy)	1	Grains (avoir.)
“ “	0.06480	Grams
“ “	0.04167	Pennyweights (troy)
“ “	2.0833×10^{-3}	Ounces (troy)
Grains/U.S. gal.	17.118	Parts/million
“ /U.S. gal.	142.86	Lbs./million gal.
“ /Imp. gal.	14.286	Parts/million
Grams	980.7	Dynes
“	15.43	Grains
“	10^{-3}	Kilograms
“	10^{3}	Milligrams
“	0.03527	Ounces
“	0.03215	Ounces (troy)
“	2.205×10^{-3}	Pounds
Grams/cm.	5.600×10^{-3}	Pounds/inch
Grams/cu. cm.	62.43	Pounds/cubic foot
“ / “ “	0.03613	Pounds/cubic inch
Grams/liter	58.417	Grains/gal.
“ / “	8.345	Pounds/1000 gals.
“ / “	0.062427	Pounds/cubic foot
“ / “	1000	Parts/million
Hectares	2.471	Acres
“	1.076×10^{5}	Square feet
Hectograms	100	Grams
Hectoliters	100	Liters
Hectometers	100	Meters
Hectowatts	100	Watts

Multiply	*By*	*To Obtain*
Horse-power	42.44	B.T. Units/min.
“ “	33,000	Foot-lbs./min.
“ “	550	Foot-lbs./sec.
“ “	1.014	Horse-power (metric)
“ “	10.70	Kg.-calories/min.
“ “	0.7457	Kilowatts
“ “	745.7	Watts
Horse-power (boiler)	33,479	B.T.U./hr.
“ “ “	9.803	Kilowatts
Horse-power-hours	2547	British Thermal Units
“ “ “	1.98×10^{6}	Foot-lbs.
“ “ “	641.7	Kilogram-calories
“ “ “	2.737×10^{5}	Kilogram-meters
“ “ “	0.7457	Kilowatt-hours
Inches	2.540	Centimeters
Inches of mercury	0.03342	Atmospheres
“ “ “	1.133	Feet of water
“ “ “	345.3	Kgs./sq. meter
“ “ “	70.73	Lbs./sq. ft.
“ “ “	0.4912	Lbs./sq. inch
Inches of water	0.002458	Atmospheres
“ “ “	0.07355	Inches of mercury
“ “ “	25.40	Kgs./sq. meter
“ “ “	0.5781	Ounces/sq. inch
“ “ “	5.202	Lbs./sq. foot
“ “ “	0.03613	Lbs./sq. inch
“ “ “ / year	sq. mi. / 13.8	Cu. ft. / sec.
Kilograms	980,665	Dynes
“	2.205	Lbs.
“	1.102×10^{-3}	Tons (short)
“	10^{3}	Grams
Kilograms-calories	3.968	British Thermal Units
“ “	3086	Foot-pounds
“ “	1.558×10^{-3}	Horse-power-hours
“ “	1.162×10^{-3}	Kilowatt-hours
Kilogram-calories/min.	51.43	Foot-pounds/sec.
“ “ / “	0.09351	Horse-power
“ “ / “	0.06972	Kilowatts
Kgs./meter	0.6720	Lbs./foot
Kgs./sq. meter	9.678×10^{-5}	Atmospheres
“ / “ “	3.281×10^{-3}	Feet of water
“ / “ “	2.896×10^{-3}	Inches of mercury
“ / “ “	0.2048	Lbs./sq. foot
“ / “ “	1.422×10^{-3}	Lbs./sq. inch
Kgs./sq. millimeter	10^{6}	Kgs./sq. meter
Kiloliters	10^{3}	Liters
Kilometers	10^{5}	Centimeters
“	3281	Feet
“	10^{3}	Meters
“	0.6214	Miles
“	1094	Yards
Kilometers/hr.	27.78	Centimeters/sec.
“ / “	54.68	Feet/min.
“ / “	0.9113	Feet/sec.
“ / “	0.5396	Knots
“ / “	16.67	Meters/min.
“ / “	0.6214	Miles/hr.
Kms./hr./sec.	27.78	Cms./sec./sec.
“ / “ / “	0.9113	Ft./sec./sec.
“ / “ / “	0.2778	Meters/sec./sec.
Kilowatts	56.92	B.T. Units/min.
“	4.425×10^{4}	Foot-lbs./min.
“	737.6	Foot-lbs./sec.
“	1.341	Horse-power
“	14.34	Kg.-calories/min.
“	10^{3}	Watts
Kilowatt-hours	3415	British Thermal Units
“ “	2.655×10^{6}	Foot-lbs.
“ “	1.341	Horse-power-hrs.
“ “	860.5	Kilogram-calories
“ “	3.671×10^{5}	Kilogram-meters
Link (Gunters)	66	Feet
Liters	10^{3}	Cubic centimeters
“	0.03531	Cubic feet
“	61.02	Cubic inches
“	10^{-3}	Cubic meters
“	1.308×10^{-3}	Cubic yards
“	0.2642	Gallons
“	2.113	Pints (liq.)
“	1.057	Quarts (liq.)

Multiply	*By*	*To Obtain*
Liters/min.	5.886×10^{-4}	Cubic ft.-sec.
“ / “	4.403×10^{-3}	Gals.-/sec.
Lumber Width (in.) x Thickness (in.) / 12	Length (ft)	Board Feet
Meters	100	Centimeters
“	3.281	Feet
“	39.37	Inches
“	10^{-3}	Kilometers
“	10^{3}	Millimeters
“	1.094	Yards
Meters/min.	1.667	Centimeters/sec.
“ / “	3.281	Feet/min.
“ / “	0.05468	Feet/sec.
“ / “	0.06	Kilometers/hr.
“ / “	0.03728	Miles/hr.
Meters/sec.	196.8	Feet/min.
“ / “	3.281	Feet/sec.
“ / “	3.6	Kilometers/hr.
“ / “	0.06	Kilometers/min.
“ / “	2.237	Miles/hr.
“ / “	0.03728	Miles/min.
Microns	10^{-6}	Meters
Miles	1.609×10^{5}	Centimeters
“	5280	Feet
“	1.609	Kilometers
“	1760	Yards
Miles/hr.	44.70	Centimeters/sec.
“ / “	88	Feet/min.
“ / “	1.467	Feet/sec.
“ / “	1.609	Kilometers/hr.
“ / “	0.8684	Knots
“ / “	26.82	Meters/min.
Miles/min.	2682	Centimeters/sec.
“ / “	88	Feet/sec.
“ / “	1.609	Kilometers/min.
“ / “	60	Miles/hr.
Milliers	10^{3}	Kilograms
Milligrams	10^{-3}	Grams
Milliliters	10^{-3}	Liters
Millimeters	0.1	Centimeters
“	0.03937	Inches
Milligrams/liter	1	Parts/million
Million gals./day	1.54723	Cubic ft./sec.
Miner's inches	1.5	Cubic ft./min.
Minutes (angle)	2.909×10^{-4}	Radians
Ounces	16	Drams
“	437.5	Grains
“	0.0625	Pounds
“	28.349527	Grams
“	0.9115	Ounces (troy)
“	2.790×10^{-5}	Tons (long)
“	2.835×10^{-5}	Tons (metric)
Ounces, troy	480	Grains
“ “	20	Pennyweights (troy)
“ “	0.08333	Pounds (troy)
“ “	31.103481	Grams
“ “	1.09714	Ounces, avoir.
Ounces (fluid)	1.805	Cubic inches
“ “	0.02957	Liters
Ounces/sq. inch	0.0625	Lbs./sq. inch
Overflow rate (ft./hr.)	0.12468 x area (sq. ft.)	Gals./min.
1 / Overflow rate (ft./hr.)	8.0208	Sq. ft./gal./min.
Parts/million	0.0584	Grains/U.S. gal.
“ / “	0.07016	Grains/Imp. gal.
“ / “	8.345	Lbs./million gal.
Pennyweights (troy)	24	Grains
“ “	1.55517	Grams
“ “	0.05	Ounces (troy)
“ “	4.1667×10^{-3}	Pounds (troy)
Pounds	16	Ounces
“	256	Drams
“	7000	Grains
“	0.0005	Tons (short)
“	453.5924	Grams
“	1.21528	Pounds (troy)
“	14.5833	Ounces (troy)

Adapted from Tables of Conversion Factors for Engineers by The Dorr Co.

GENERAL CONVERSION FACTORS—3

Multiply	*By*	*To Obtain*
Pounds (troy)	5760	Grains
“ “	240	Pennyweights (troy)
“ “	12	Ounces (troy)
“ “	373.24177	Grams
“ “	0.822857	Pounds (avoir.)
“ “	13.1657	Ounces (avoir.)
“ “	3.6735×10^{-4}	Tons (long)
“ “	4.1143×10^{-4}	Tons (short)
“ “	3.7324×10^{-4}	Tons (metric)
Pounds of water	0.01602	Cubic feet
“ “ “	27.68	Cubic inches
“ “ “	0.1198	Gallons
Pounds of water/min.	2.670×10^{-4}	Cubic ft./sec.
Pounds/cubic foot	0.01602	Grams/cubic cm.
“ / “ “	16.02	Kgs./cubic meter
“ / “ “	5.787×10^{-4}	Lbs./cubic inch
Pounds/cubic inch	27.68	Grams/cubic cm.
“ / “ “	2.768×10^{4}	Kgs./cubic meter
“ / “ “	1728	Lbs./cubic foot
Pounds/foot	1.488	Kgs./meter
Pounds/inch	178.6	Grams/cm.
Pounds/sq. foot	0.01602	Feet of water
“ / “ “	4.883	Kgs./sq. meter
“ / “ “	6.945×10^{-3}	Pounds/sq. inch
Pounds/sq. inch	0.06804	Atmospheres
“ / “ “	2.307	Feet of water
“ / “ “	2.036	Inches of mercury
“ / “ “	703.1	Kgs./sq. meter
Quadrants (angle)	90	Degrees
“ “	5400	Minutes
“ “	1.571	Radians
Quarts (dry)	67.20	Cubic inches
Quarts (liq.)	57.75	Cubic inches
Quintal, Argentine	101.28	Pounds
“ Brazil	129.54	Pounds
“ Castile, Peru	101.43	Pounds
“ Chile	101.41	Pounds
“ Mexico	101.47	Pounds
“ Metric	220.46	Pounds
Quires	25	Sheets
Radians	57.30	Degrees
“	3438	Minutes
“	0.637	Quadrants
Radians/sec.	57.30	Degrees/sec.
“ / “	0.1592	Revolutions/sec.
“ / “	9.549	Revolutions/min.
Radians/sec./sec.	573.0	Revolutions/min/min.
“ / “ / “	0.1592	Revolutions/sec./sec.
Reams	500	Sheets
Revolutions	360	Degrees
“	4	Quadrants
“	6.283	Radians
Revolutions/min.	6	Degrees/sec.
“ / “	0.1047	Radians/sec.
“ / “	0.01667	Revolutions/sec.
Revolutions/min./min.	1.745×10^{-3}	Rads./sec./sec.
“ / “ / “	2.778×10^{-4}	Revs./sec./sec.
Revolutions/sec.	360	Degrees/sec.
“ / “	6.283	Radians/sec.
“ / “	60	Revolutions/min.
Revolutions/sec./sec.	6.283	Radians/sec./sec.
“ / “ / “	3600	Revolutions/min/min.
Rods	16.5	Feet
Seconds (angle)	4.848×10^{-6}	Radians
Square centimeters	1.076×10^{-3}	Square feet
“ “	0.1550	Square inches
“ “	10^{-4}	Square meters
“ “	100	Square millimeters
Square Chains (Gunters)	16	Square rods

Multiply	*By*	*To Obtain*
Square feet	2.296×10^{-5}	Acres
“ “	929.0	Square centimeters
“ “	144	Square inches
“ “	0.09290	Square meters
“ “	3.587×10^{-8}	Square miles
“ “	1/9	Square yards
$\frac{1}{\text{Sq. ft./gal./min.}}$	8.0208	Overflow rate (ft./hr.)
Square inches	6.452	Square centimeters
“ “	6.944×10^{-3}	Square feet
“ “	645.2	Square millimeters
Square kilometers	247.1	Acres
“ “	10.76×10^{6}	Square feet
“ “	10^{6}	Square meters
“ “	0.3861	Square miles
“ “	1.196×10^{6}	Square yards
Square meters	2.471×10^{-4}	Acres
“ “	10.76	Square feet
“ “	3.861×10^{-7}	Square miles
“ “	1.196	Square yards
Square miles	640	Acres
“ “	27.88×10^{6}	Square feet
“ “	2.590	Square kilometers
“ “	3.098×10^{6}	Square yards
Square millimeters	0.01	Square centimeters
“ “	1.550×10^{-3}	Square inches
Square yards	2.066×10^{-4}	Acres
“ “	9	Square feet
“ “	0.8361	Square meters
“ “	3.228×10^{-7}	Square miles
Temp. (°C.) + 273	1	Abs. temp. (°C.)
“ “ + 17.78	1.8	Temp. (°F.)
“ (°F.) + 460	1	Abs. temp. (°F.)
“ “ − 32	5/9	Temp. (°C.)
Tons (long)	1016	Kilograms
“ “	2240	Pounds
“ “	1.12000	Tons (short)
Tons (metric)	10^{3}	Kilograms
“ “	2205	Pounds
Tons (short)	2000	Pounds
“ “	32000	Ounces
“ “	907.18486	Kilograms
Tons (short)	2430.56	Pounds (troy)
“ “	0.89287	Tons (long)
“ “	29166.66	Ounces (troy)
“ “	0.90718	Tons (metric)
$\frac{1}{\text{Tons dry solids/24 hrs.}}$	Area (sq. ft.)	Sq.ft./ton/24 hrs.
Tons of water/24 hrs.	83.333	Pounds water/hour
“ “ “ / “	0.16643	Gallons/min.
“ “ “ / “	1.3349	Cu. ft./hr.
Watts	0.05692	B. T. Units/min.
“	44.26	Foot-pounds/min.
“	0.7376	Foot-pounds/sec.
“	1.341×10^{-3}	Horse-power
“	0.01434	Kg.-calories/min.
“	10^{-3}	Kilowatts
Watt-hours	3.415	British Thermal Units
“	2655	Foot-pounds
“	1.341×10^{-3}	Horse-power-hours
“	0.8605	Kilogram-calories
“	367.1	Kilogram-meters
“	10^{-3}	Kilowatt-hours
Yards	91.44	Centimeters
“	3	Feet
“	36	Inches
“	0.9144	Meters

Adapted from Tables of Conversion Factors for Engineers by Dorr Co.

Index

A

B

D

G

M

Q

R

S

T

U

V

W

Y

Z